Look for this in every lesson!

Homework Video Tutor

Prentice Hall Mathematics

Need help with your homework? Prentice Hall has the answer!

Homework Video Tutors—Real teachers, reviewing every lesson!

These narrated, interactive tutorials review the key concepts of every lesson, providing unparalleled homework help. With over 1,000 videos for grades 6–12, we've got the help you need.

See for yourself!

1 Go to PHSchool.com

2 Enter Web Code **bce-0775**

3 Click Go!

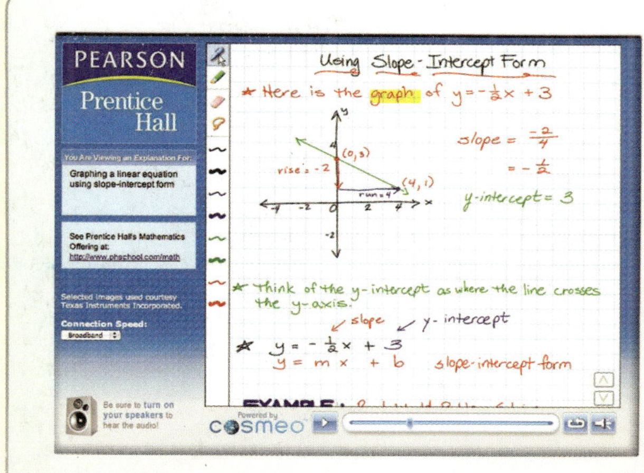

Additional support at PHSchool.com includes

- Vocabulary Quizzes
- Lesson Quizzes
- Data Updates
- Math at Work
- Chapter Tests
- California Standards Tutorials

See page CA29 for a complete list!

Laurie E. Bass
Randall I. Charles
Basia Hall
Art Johnson
Dan Kennedy

Boston, Massachusetts
Upper Saddle River, New Jersey

Series Authors

Dan Kennedy, Ph.D., is a classroom teacher an(he Lupton Distinguished Professor of Mathematics at the Baylor School in Chattooga, Tennessee. A frequent speaker at professional meetings on the su!ct of mathematics education reform, Dr. Kennedy has conducted more th. 50 workshops and institutes for high school teachers. He is coauthor of te)ooks in calculus and precalculus, and from 1990 to 1994 he chaired the Colle Board's AP Calculus Development Committee. He is a 1992 Tandy TechnolocScholar and a 1995 Presidential Award winner.

Randall I. Charles, Ph.D., is Professor Em(tus in the Department of Mathematics and Computer Science at San Jose Staténiversity, San Jose, California. He began his career as a high school mathentics teacher and he was a mathematics supervisor for five years. Dr. Charles has en a member of several NCTM committees and is the former Vice President of tlNational Council of Supervisors of Mathematics. Much of his writing and rearch has been in the area of problem solving. He has authored more than 75athematics textbooks for kindergarten through college.

Basia Hall currently serves as Manager of Instructior Programs for the Houston Independent School District. With 30-years teacng expeience, Ms. Hall has served as a department chair, instructional specialist,istructional supervisor, a school improvement facilitator, and a professional develment (TEXTEAMS) trainer. Ms. Hall has developed curriculum for Algebra 1,eometry, and Algebra 2 and contributed to the development of the Texas Essent Knowedge and Skills. A recipient of the 1992 Presidential Award for Excellence Matematics Teaching, Ms. Hall is also a past president of the Texas Asciatic of Supervisors of Mathematics, and she is a state representative for theationl Council of Supervisors of Mathematics (NCSM).

Acknowledgments appear on pages 886–888, which constitute an extension of this copyright pae.

ISBN 0-13-203122-1
2 3 4 5 8 9 10   11 10 09 08 07

Geometry Authors

Laurie E. Bass is a classroom teacher at the 9–12 division of the Ethical Culture Fieldston School in Riverdale, New York. Ms. Bass has a wide base of teaching experience, ranging from grade 6 through Advanced Placement Calculus. She was the recipient of a 2000 Honorable Mention for the RadioShack National Teacher Awards. She has been a contributing writer of a number of publications, including software-based activities for the Algebra 1 classroom. Among her areas of special interest are cooperative learning for high school students and geometry exploration on the computer. Ms. Bass has been a presenter at a number of local, regional, and national conferences.

Art Johnson, Ed.D., is a professor of mathematics education at Boston University. He is a mathematics educator with 32 years of public school teaching experience, a frequent speaker and workshop leader, and the recipient of a number of awards. Dr. Johnson received the Tandy Prize for Teaching Excellence in 1995, a Presidential Award for Excellence in Mathematics Teaching in 1992, and New Hampshire Teacher of the Year, also in 1992. He was profiled by the Disney Corporation in the American Teacher of the Year Program.

Algebra 1 and Algebra 2 Authors

Allan E. Bellman is a Lecturer/Supervisor in the School of Education at the University of California, Davis. Before coming to Davis, he was a mathematics teacher for 31 years in Montgomery County, Maryland. He has been an instructor for both the Woodrow Wilson National Fellowship Foundation and the T^3 program. Mr. Bellman has a particular expertise in the use of technology in education and speaks frequently on this topic. He was a 1992 Tandy Technology Scholar.

Sadie Chavis Bragg, Ed.D., is Senior Vice President of Academic Affairs at the Borough of Manhattan Community College of the City University of New York. A former professor of mathematics, she is a past president of the American Mathematical Association of Two-Year Colleges (AMATYC), co-director of the AMATYC project to revise the standards for introductory college mathematics before calculus, and an active member of the Benjamin Banneker Association. Dr. Bragg has coauthored more than 50 mathematics textbooks for kindergarten through college.

William G. Handlin, Sr., is a classroom teacher and Department Chairman of Technology Applications at Spring Woods High School in Houston, Texas. Awarded Life Membership in the Texas Congress of Parents and Teachers for his contributions to the well-being of children, Mr. Handlin is also a frequent workshop and seminar leader in professional meetings throughout the world.

California Mathematics Program Advisors

Prentice Hall wishes to thank the following educators for their ongoing advice in the development of this California Edition of Prentice Hall Mathematics. Their valuable insights have helped ensure that this mathematics series meets the needs of California students and their teachers.

Danelle Almaraz
Pre-Algebra Teacher
East Whittier City School District
Whittier, California

Catherine Barker
Algebra Teacher
Los Angeles USD
Long Beach, California

Eric Bitter
Pre-Algebra & Algebra Teacher
Clovis USD
Clovis, California

Joe Brumfield
Consultant
Pasadena, California

Lynn Cevallos
Consultant
Hacienda Heights, California

Manuel E. Chavez
AP Calculus Teacher/Math Coach
Los Angeles USD
Huntington Park, California

Edward Ford
Consultant
Escondido, California

Courtney Glass
Algebra 2 & Pre-Calculus Teacher
Monrovia City USD
Monrovia, California

Bernice Levens
Former Teacher/Consultant
Hacienda Heights, California

Sia Lux
Algebra & Calculus Teacher
Coachella Valley USD
Indio, California

Shawn Neal
Pre-Algebra & Algebra Teacher
Visalia USD
Visalia, California

Christy McAloney
Algebra & Calculus Teacher
Grossmont Union High School District
Spring Valley, California

Brian McElfish
Pre-Algebra & Algebra Teacher
Saddleback Valley USD
Lake Forest, California

Paul Dennis McLaughlin
Algebra Teacher
Walnut Valley USD
Walnut, California

Jason Rose
Algebra Teacher
Temple City USD
Temple City, California

Keith Smith
Pre-Algebra & Algebra Teacher
Grant Joint USD
Sacramento, California

Sarah Thomson
Calculus Teacher & Math Coach
San Bernardino City USD
San Bernardino, California

Contents in Brief

California Mathematics Student Handbook

Tools of Geometry

Student Support

 California Check System

Check Your Readiness 2

Check Skills You'll Need 4, 10, 16, 23, 31, 36, 44, 53, 61

CA Standards Check 4, 5, 6, 10, 11, 12, 17, 19, 23, 24, 25, 32, 33, 36, 37, 38, 39, 44, 45, 46, 47, 54, 55, 62, 63, 64

Checkpoint Quiz 28, 59

Vocabulary

New Vocabulary 4, 10, 16, 23, 31, 36, 44

Vocabulary Tip 10, 17, 18, 24, 26, 31, 34, 36, 53, 61, 62, 63

Vocabulary Builder 29

Vocabulary Review 71

Video Tutor Help 24, 37, 39, 63

Active Math 45, 54, 61

Homework Video Tutor 8, 14, 21, 26, 34, 41, 48, 56, 67

Lesson Quizzes 9, 15, 21, 27, 35, 41, 49, 59, 67

Vocabulary Quiz 71

Chapter Test 74

Geometry at Work 50

California Standards

Geometry

1.0, 3.0, 8.0, 10.0, 16.0

CAHSEE

7NS 1.6, 7NS 1.7, 7NS 2.4, 6SDAP 1.1, 6SDAP 3.3, 7SDAP 1.1, 7AF 1.1, 7AF 1.2, 7AF 1.5, 7AF 2.2, 7AF 3.3, 7AF 4.2, 7MG 1.3, 7MG 2.1, 7MG 2.2, 7MG 2.4, 7MR 1.1, 7MR 1.2, 7MR 2.4, Alg1 3.0, Alg1 6.0, Alg1 7.0, Alg1 9.0, Alg1 10.0

Standards Mastery and Assessment

Page 20

Reasoning and Proof

Student Support

 California Check System

Check Your Readiness 78

Check Skills You'll Need 80, 87, 94, 103, 110

CA Standards Check 80, 81, 82, 87, 88, 89, 94, 95, 96, 104, 105, 111, 112

Checkpoint Quiz 100

New Vocabulary 80, 87, 94, 103, 110

Vocabulary Tip 87, 92, 95, 96

Vocabulary Review 117

GO Online

Active Math 80, 89

Homework Video Tutor 85, 91, 97, 106, 113

Lesson Quizzes 85, 93, 99, 107, 115

Vocabulary Quiz 117

Chapter Test 120

A Point in Time 100

California Standards

Geometry

1.0, 2.0, 3.0, 4.0

CAHSEE

7NS 1.2, 7NS 2.1, 6SDAP 1.1, 7SDAP 1.2, 7AF 3.3, 7MG 2.1, 7MG 2.3, 7MR 1.1, 7MR 2.1, Alg1 4.0, Alg1 8.0, Alg1 15.0

Standards Mastery and Assessment

Page 98

Parallel and Perpendicular Lines

Student Support

 California Check System

 Vocabulary

GO Online

 California Standards

Geometry
1.0, 2.0, 4.0, 7.0, 12.0, 13.0, 16.0

CAHSEE
7NS 1.3, 7NS 1.6, 7NS 1.7, 7NS 2.2, 6SDAP 2.5, 6SDAP 3.3, 7SDAP 1.2, 7AF 2.1, 7AF 3.1, 7AF 3.3, 7AF 3.4, 7MG 2.1, 7MG 3.2, 7MG 3.3, 7MR 1.1, 7MR 2.3, Alg1 7.0

Page 131

Congruent Triangles

Student Support

 California Check System

GO Online

California Standards

Geometry

2.0, 4.0, 5.0, 6.0, 12.0, 16.0

CAHSEE

7NS 2.3, 7SDAP 1.2, 7AF 1.5, 7AF 2.2, 7AF 4.1, 7MG 1.1, 7MG 1.3, Alg1 9.0

Standards Mastery and Assessment

Page 221

Relationships Within Triangles

Student Support

 California Check System

Vocabulary

 Online

 California Standards

Geometry

1.0, 2.0, 4.0, 5.0, 6.0, 17.0, 21.0

CAHSEE

7NS 1.3, 7NS 2.5, 6SDAP 3.3, 6SDAP 3.5, 7SDAP 1.1, 7AF 1.1, 7AF 1.5, 7AF 3.4, 7AF 4.1, 7MG 1.3, 7MG 3.2, 7MG 3.3, 7MR 1.1, Alg1 6.0, Alg1 8.0

Standards Mastery and Assessment

Page 263

Quadrilaterals

Student Support

 California Check System

Check Your Readiness 304

Check Skills You'll Need 306, 312, 321, 329, 336, 343, 348

CA Standards Check 307, 308, 313, 314, 315, 323, 324, 330, 331, 332, 337, 338, 343, 344, 349

Checkpoint Quiz 327, 347

New Vocabulary 306, 312, 336, 348

Vocabulary Tip 307, 312, 329

Vocabulary Review 357

GO Online

Active Math 307, 323, 331

Homework Video Tutor 310, 317, 325, 333, 339, 345, 351

Lesson Quizzes 311, 317, 327, 335, 341, 345, 353

Vocabulary Quiz 357

Chapter Test 360

A Point in Time 347

California Standards

Geometry

1.0, 3.0, 7.0, 12.0, 13.0

CAHSEE

7NS 1.2, 7NS 1.6, 7NS 2.2, 7NS 2.3, 6SDAP 3.1, 7MG 3.4, 7MR 1.2, Alg1 5.0

Page 352

Similarity

Student Support

 California Check System

Check Your Readiness 364

Check Skills You'll Need 366, 373, 382, 391, 398

CA Standards Check 366, 367, 368, 373, 374, 375, 383, 384, 385, 392, 393, 399, 400

Checkpoint Quiz 379, 404

New Vocabulary 366, 373, 382, 391

Vocabulary Tip 366, 367, 392, 394

Vocabulary Builder 389

Vocabulary Review 407

GO Online

Active Math 374, 393

Homework Video Tutor 369, 377, 387, 395, 402

Lesson Quizzes 369, 379, 387, 395, 403

Vocabulary Quiz 407

Chapter Test 410

A Point in Time 371

California Standards

Geometry

2.0, 4.0, 5.0, 7.0

CAHSEE

7NS 1.1, 6SDAP 3.3, 6SDAP 3.5, 7MG 1.2, 7MG 2.2, 7MR 2.1, 7MR 3.3, Alg1 5.0, Alg1 6.0, Alg1 7.0, Alg1 10.0

Standards Mastery and Assessment

Page 366

Right Triangles and Trigonometry

Student Support

 California Check System

GO Online

California Standards

Geometry

14.0, 15.0, 18.0, 19.0, 20.0

CAHSEE

6SDAP 3.3, 7AF 2.2, 7MG 1.1, 7MR 2.1

Standards Mastery and Assessment

Page 433

Transformations

Student Support

California Check System

Vocabulary

Online

California Standards

Geometry

22.0

CAHSEE

7NS 1.1, 6SDAP 3.1, 7SDAP 1.1, 7AF 1.1, 7AF 1.5, 7AF 4.2, 7MG 1.2, 7MG 3.2, 7MG 3.3, 7MR 1.2, 7MR 2.4, Alg1 8.0, Alg1 9.0, Alg1 10.0

Page 475

Area

Student Support

California Check System

Vocabulary

GO Online

California Standards

Geometry

7.0, 8.0, 10.0, 11.0, 19.0, 20.0, 22.0

CAHSEE

7NS 1.7, 7SDAP 1.1, 7AF 1.1, 7AF 1.5

Standards Mastery and Assessment

Page 555

Surface Area and Volume

Student Support

California Check System

Check Your Readiness 596

Check Skills You'll Need 598, 608, 617, 624, 631, 638, 646

CA Standards Check 598, 599, 600, 609, 611, 618, 619, 620, 625, 626, 627, 632, 633, 634, 639, 640, 646, 647, 648

Checkpoint Quiz 623, 644

Vocabulary

New Vocabulary 598, 608, 617, 624, 638, 646

Vocabulary Tip 598, 617, 647

Vocabulary Review 653

Video Tutor Help 626

Active Math 625, 633

Homework Video Tutor 602, 612, 622, 628, 635, 642, 650

Lesson Quizzes 603, 613, 621, 629, 635, 643, 651

Vocabulary Quiz 653

Chapter Test 656

Geometry at Work 637

California Standards

Geometry

8.0, 9.0, 11.0

CAHSEE

7NS 2.1, 7NS 2.4, 6SDAP 1.1, 6SDAP 2.5, 6SDAP 3.3, 7AF 1.2, 7AF 2.2, 7AF 3.1, 7MG 2.3, 7MG 3.2, 7MG 3.3, 7MR 1.2, Alg1 2.0, Alg1 3.0, Alg1 7.0

Standards Mastery and Assessment

Page 646

Circles

Student Support

California Check System

Check Your Readiness 660

Check Skills You'll Need 662, 670, 678, 687, 695, 701

CA Standards Check 663, 664, 665, 670, 671, 673, 679, 680, 681, 688, 689, 690, 695, 696, 697, 701, 702

Checkpoint Quiz 685, 700

New Vocabulary 662, 670, 678, 687, 695, 701

Vocabulary Tip 662, 665, 679, 682, 687, 689, 702

Vocabulary Review 707

GO Online

Active Math 670, 680, 695

Homework Video Tutor 666, 675, 683, 692, 698, 704

Lesson Quizzes 667, 675, 685, 691, 699, 705

Vocabulary Quiz 707

Chapter Test 710

End-of-Course Test 711

Geometry at Work 693

California Standards

Geometry
2.0, 7.0, 12.0, 17.0, 18.0, 21.0

CAHSEE
6SDAP 3.3, 7AF 1.2, 7AF 4.2, 7MG 2.4, Alg1 6.0

Page 702

Learning the California Standards

Your *Prentice Hall Geometry* textbook is designed to help you fully understand and learn the California Mathematics Standards. Here are some features of your textbook that will support your learning throughout the year.

California Check System

Look for the ✓ in each lesson. These questions are opportunities for you and your teacher to make sure you understand the mathematics – so you can be successful with each day's lesson.

GO for Help

Look for this Go for Help arrow throughout every lesson to point you to where you can get help or review important concepts. Be an independent learner – and go for help when you're having trouble.

Get Ready to Learn

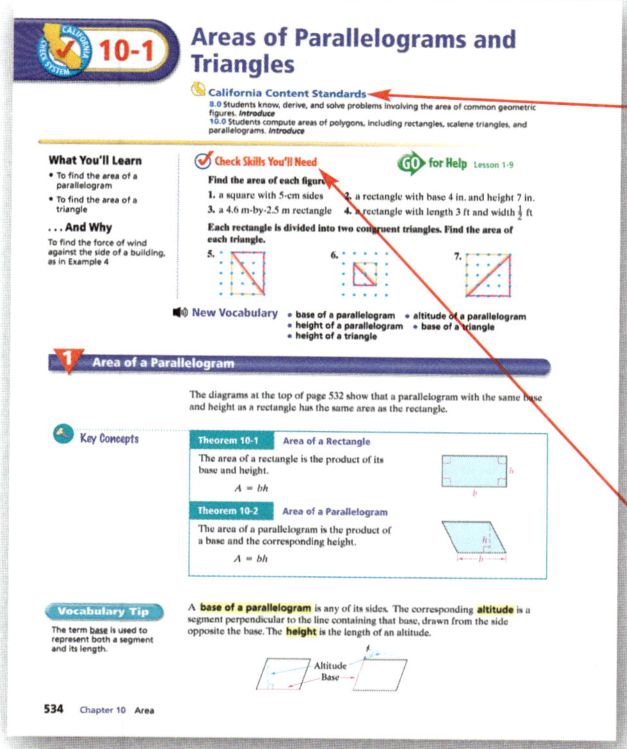

10-1 Areas of Parallelograms and Triangles

California Content Standards

8.0 Students know, derive, and solve problems involving the area of common geometric figures. *Introduce*

10.0 Students compute areas of polygons, including rectangles, scalene triangles, and parallelograms. *Introduce*

What You'll Learn

- To find the area of a parallelogram
- To find the area of a triangle

. . . And Why

To find the force of wind against the side of a building, as in Example 4

Check Skills You'll Need — GO for Help Lesson 1-9

Find the area of each figure.

1. a square with 5-cm sides
2. a rectangle with base 4 in. and height 7 in.
3. a 4.6 m-by-2.5 m rectangle
4. a rectangle with length 3 ft and width $\frac{1}{2}$ ft

Each rectangle is divided into two congruent triangles. Find the area of each triangle.

5. 6. 7.

New Vocabulary
- base of a parallelogram
- altitude of a parallelogram
- height of a parallelogram
- base of a triangle
- height of a triangle

1 Area of a Parallelogram

The diagrams at the top of page 532 show that a parallelogram with the same base and height as a rectangle has the same area as the rectangle.

Key Concepts

Theorem 10-1 Area of a Rectangle

The area of a rectangle is the product of its base and height.

$A = bh$

Theorem 10-2 Area of a Parallelogram

The area of a parallelogram is the product of a base and the corresponding height.

$A = bh$

Vocabulary Tip

The term base is used to represent both a segment and its length.

A **base of a parallelogram** is any of its sides. The corresponding **altitude** is a segment perpendicular to the line containing that base, drawn from the side opposite the base. The **height** is the length of an altitude.

534 Chapter 10 Area

California Content Standards

Your textbook identifies which California standards you'll be covering each day. You'll also know whether you are being introduced to this standard for the first time, developing your understanding of it, or showing your mastery of it.

Check Skills You'll Need

These questions make sure you're ready to start the lesson by reviewing important skills and vocabulary words you'll need. Need help? Go for Help points you to the lesson you can review.

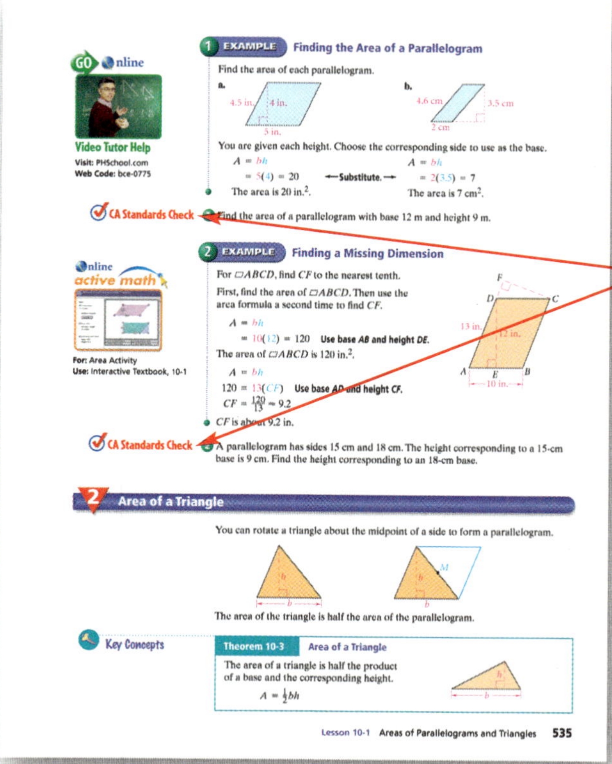

1 EXAMPLE Finding the Area of a Parallelogram

Find the area of each parallelogram.

You are given each height. Choose the corresponding side to use as the base.

$A = bh$ ← Substitute. → $A = bh$

$= 5(4) = 20$ $= 2(3.5) = 7$

The area is 20 in.^{2}. The area is 7 cm^2.

Go Online — Video Tutor Help
Visit: PHSchool.com
Web Code: bce-0775

CA Standards Check 1 Find the area of a parallelogram with base 12 m and height 9 m.

2 EXAMPLE Finding a Missing Dimension

For ▱$ABCD$, find CF to the nearest tenth.

First, find the area of ▱$ABCD$. Then use the area formula a second time to find CF.

$A = bh$

$= 10(12) = 120$ **Use base AB and height DE.**

The area of ▱$ABCD$ is 120 in.^{2}.

$A = bh$

$120 = 13(CF)$ **Use base AD and height CF.**

$CF = \frac{120}{13} \approx 9.2$

CF is about 9.2 in.

Online active math
For: Area Activity
Use: Interactive Textbook, 10-1

CA Standards Check 2 A parallelogram has sides 15 cm and 18 cm. The height corresponding to a 15-cm base is 9 cm. Find the height corresponding to an 18-cm base.

2 Area of a Triangle

You can rotate a triangle about the midpoint of a side to form a parallelogram.

The area of the triangle is half the area of the parallelogram.

Key Concepts

Theorem 10-3 Area of a Triangle

The area of a triangle is half the product of a base and the corresponding height.

$A = \frac{1}{2}bh$

Lesson 10-1 Areas of Parallelograms and Triangles 535

CA Standards Check

These questions, after every example, help you and your teacher make sure you understand the example – and the California Standards being taught.

Standards Mastery in Your Book

Practicing the California Standards

Your *Prentice Hall Geometry* textbook gives you plenty of opportunities to practice, with plenty of homework help along the way.

Standards Practice

Each lesson's Standards Practice gives you different levels of exercises.

- **A** Practice by Example: These exercises give you practice with the examples you just learned. Go for Help points you back to the example for easy reference.
- **B** Apply Your Skills: These exercises give you a chance to practice everything you learned in the day's lesson.
- **C** Challenge: This problem extends your learning and gives you a more challenging problem.

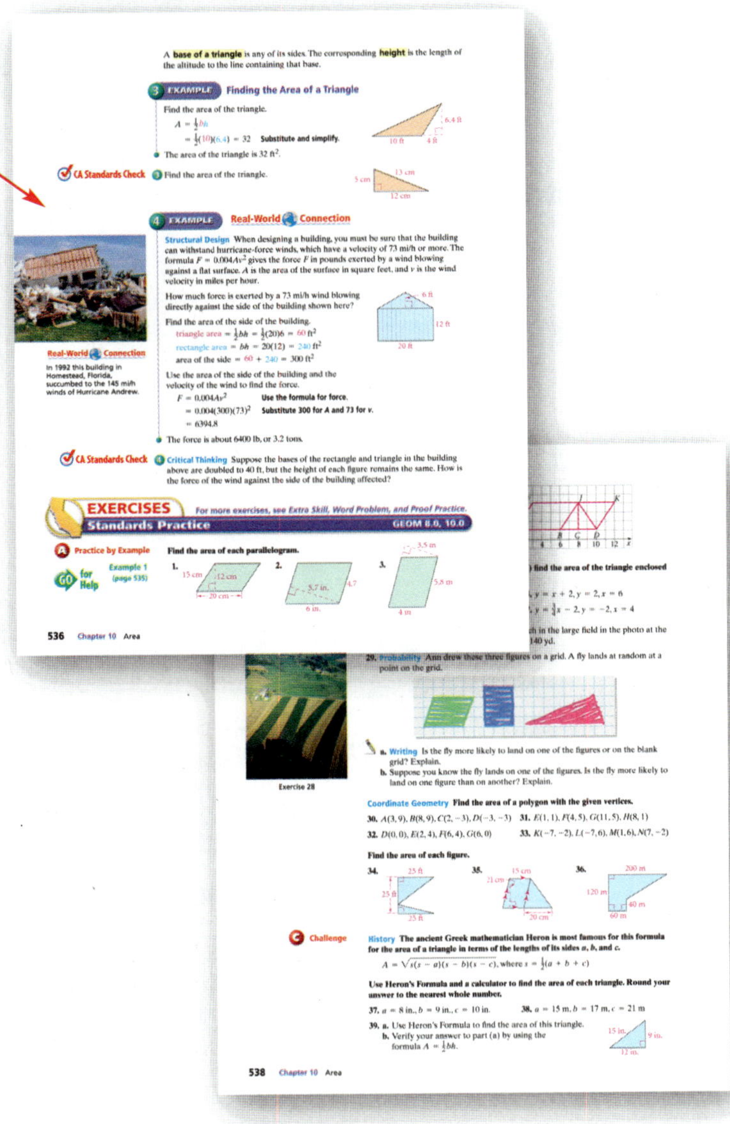

A **base of a triangle** is any of its sides. The corresponding **height** is the length of the altitude to the line containing that base.

3 EXAMPLE Finding the Area of a Triangle

Find the area of the triangle.

$A = \frac{1}{2}bh$

$= \frac{1}{2}(10)(6.4) = 32$ **Substitute and simplify.**

The area of the triangle is 32 ft^2.

CA Standards Check 3 Find the area of the triangle.

4 EXAMPLE Real-World Connection

Structural Design When designing a building, you must be sure that the building can withstand hurricane-force winds, which have a velocity of 73 mi/h or more. The formula $F = 0.004Av^2$ gives the force F in pounds exerted by a wind blowing against a flat surface. A is the area of the surface in square feet, and v is the wind velocity in miles per hour.

How much force is exerted by a 73 mi/h wind blowing directly against the side of the building shown here?

Find the area of the side of the building.

triangle area $= \frac{1}{2}bh = \frac{1}{2}(20)6 = 60$ ft^2

rectangle area $= bh = 20(12) = 240$ ft^2

area of the side $= 60 + 240 = 300$ ft^2

Use the area of the side of the building and the velocity of the wind to find the force.

$F = 0.004Av^2$ **Use the formula for force.**

$= 0.004(300)(73)^2$ **Substitute 300 for A and 73 for v.**

$= 6394.8$

The force is about 6400 lb, or 3.2 tons.

Real-World Connection

In 1992 this building in Homestead, Florida, succumbed to the 145 mi/h winds of Hurricane Andrew.

CA Standards Check 4 Critical Thinking Suppose the bases of the rectangle and triangle in the building above are doubled to 40 ft, but the height of each figure remains the same. How is the force of the wind against the side of the building affected?

EXERCISES For more exercises, see Extra Skill, Word Problem, and Proof Practice.

Standards Practice GEOM 8.0, 10.0

A Practice by Example

GO for Help Example 1 (page 535)

Find the area of each parallelogram.

1. 2. 3.

536 Chapter 10 Area

find the area of the triangle enclosed

$y = x + 2, y = 2, x = 6$

$y = \frac{1}{2}x - 2, y = -2, x = 4$

h in the large field in the photo at the 140 yd.

29. Probability Ann drew these three figures on a grid. A fly lands at random at a point on the grid.

a. Writing Is the fly more likely to land on one of the figures or on the blank grid? Explain.

b. Suppose you know the fly lands on one of the figures. Is the fly more likely to land on one figure than on another? Explain.

Exercise 28

Coordinate Geometry **Find the area of a polygon with the given vertices.**

30. $A(3, 9), B(8, 9), C(2, -3), D(-3, -3)$
31. $E(1, 1), F(4, 5), G(11, 5), H(8, 1)$
32. $D(0, 0), E(2, 4), F(6, 4), G(6, 0)$
33. $K(-7, -2), L(-7, 6), M(1, 6), N(7, -2)$

Find the area of each figure.

34. 35. 36.

C Challenge

History **The ancient Greek mathematician Heron is most famous for this formula for the area of a triangle in terms of the lengths of its sides a, b, and c.**

$A = \sqrt{s(s-a)(s-b)(s-c)}$, where $s = \frac{1}{2}(a + b + c)$

Use Heron's Formula and a calculator to find the area of each triangle. Round your answer to the nearest whole number.

37. $a = 8$ in., $b = 9$ in., $c = 10$ in.
38. $a = 15$ m, $b = 17$ m, $c = 21$ m

39. a. Use Heron's Formula to find the area of this triangle.
b. Verify your answer to part (a) by using the formula $A = \frac{1}{2}bh$.

538 Chapter 10 Area

Homework Video Tutor

What if you need help with your homework? These tutorials show real teachers reviewing each lesson. Your Prentice Hall textbook gives you a unique source for more help – called Homework Video Tutors. Log on to PHSchool.com and enter the Web Code provided – you'll get the homework help you need!

Mastering the California Standards

Your *Prentice Hall Geometry* textbook provides opportunities for you to ensure you've mastered the standards after every lesson and every chapter in your textbook.

Standards Mastery After Every Lesson

Multiple Choice Practice and Mixed Review

The final part of each day's exercises includes practice with multiple choice problems and a chance to review skills you learned earlier in your course. This will help keep your skills sharp and help you receive lots of practice with multiple choice questions. All questions are correlated to the California Standards.

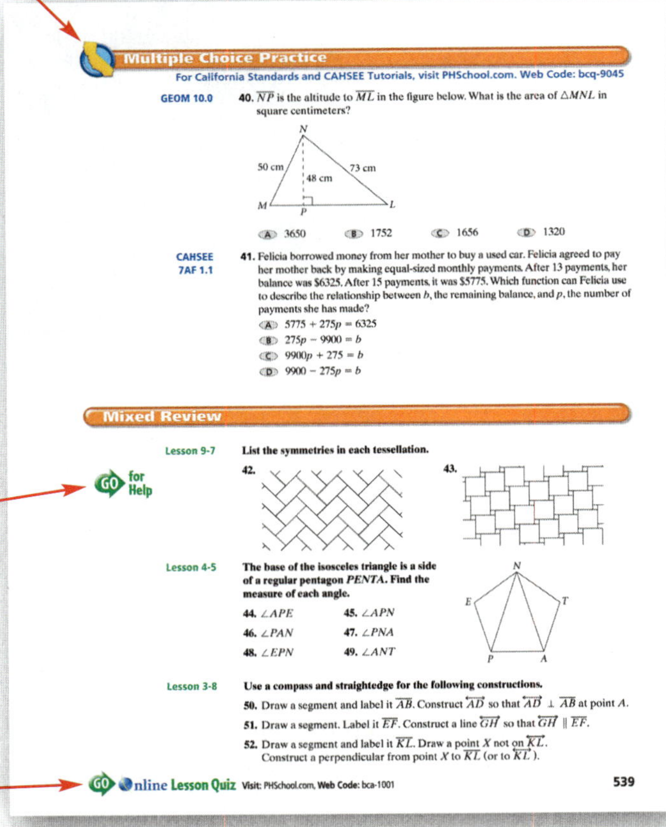

Multiple Choice Practice

For California Standards and CAHSEE Tutorials, visit PHSchool.com. Web Code: bcq-9045

GEOM 10.0

40. $\overline{NP}$ is the altitude to $\overline{ML}$ in the figure below. What is the area of $\triangle MNL$ in square centimeters?

A 3650 B 1752 C 1656 D 1320

CAHSEE 7AF 1.1

41. Felicia borrowed money from her mother to buy a used car. Felicia agreed to pay her mother back by making equal-sized monthly payments. After 13 payments, her balance was \$6325. After 15 payments, it was \$5775. Which function can Felicia use to describe the relationship between b, the remaining balance, and p, the number of payments she has made?

A $5775 + 275p = 6325$
B $275p - 9900 = b$
C $9900p + 275 = b$
D $9900 - 275p = b$

Mixed Review

Lesson 9-7 GO for Help

List the symmetries in each tessellation.

42. **43.**

Lesson 4-5

The base of the isosceles triangle is a side of a regular pentagon *PENTA*. Find the measure of each angle.

44. $\angle APE$ **45.** $\angle APN$
46. $\angle PAN$ **47.** $\angle PNA$
48. $\angle EPN$ **49.** $\angle ANT$

Lesson 3-8

Use a compass and straightedge for the following constructions.

50. Draw a segment and label it $\overline{AB}$. Construct $\overleftrightarrow{AD}$ so that $\overleftrightarrow{AD} \perp \overline{AB}$ at point A.

51. Draw a segment. Label it $\overline{EF}$. Construct a line $\overleftrightarrow{GH}$ so that $\overleftrightarrow{GH} \parallel \overline{EF}$.

52. Draw a segment and label it $\overline{KL}$. Draw a point X not on $\overleftrightarrow{KL}$. Construct a perpendicular from point X to $\overline{KL}$ (or to $\overleftrightarrow{KL}$).

GO Online Lesson Quiz Visit: PHSchool.com, Web Code: bca-1001

539

The Go for Help arrow points to the lesson you can review. You can also go online and take a lesson quiz – another chance to make sure you've mastered the California Standards you learned.

Standards Mastery after Every Chapter

After every chapter, Standards Mastery Cumulative Practice provides an opportunity to demonstrate your understanding of all the California Standards covered so far.

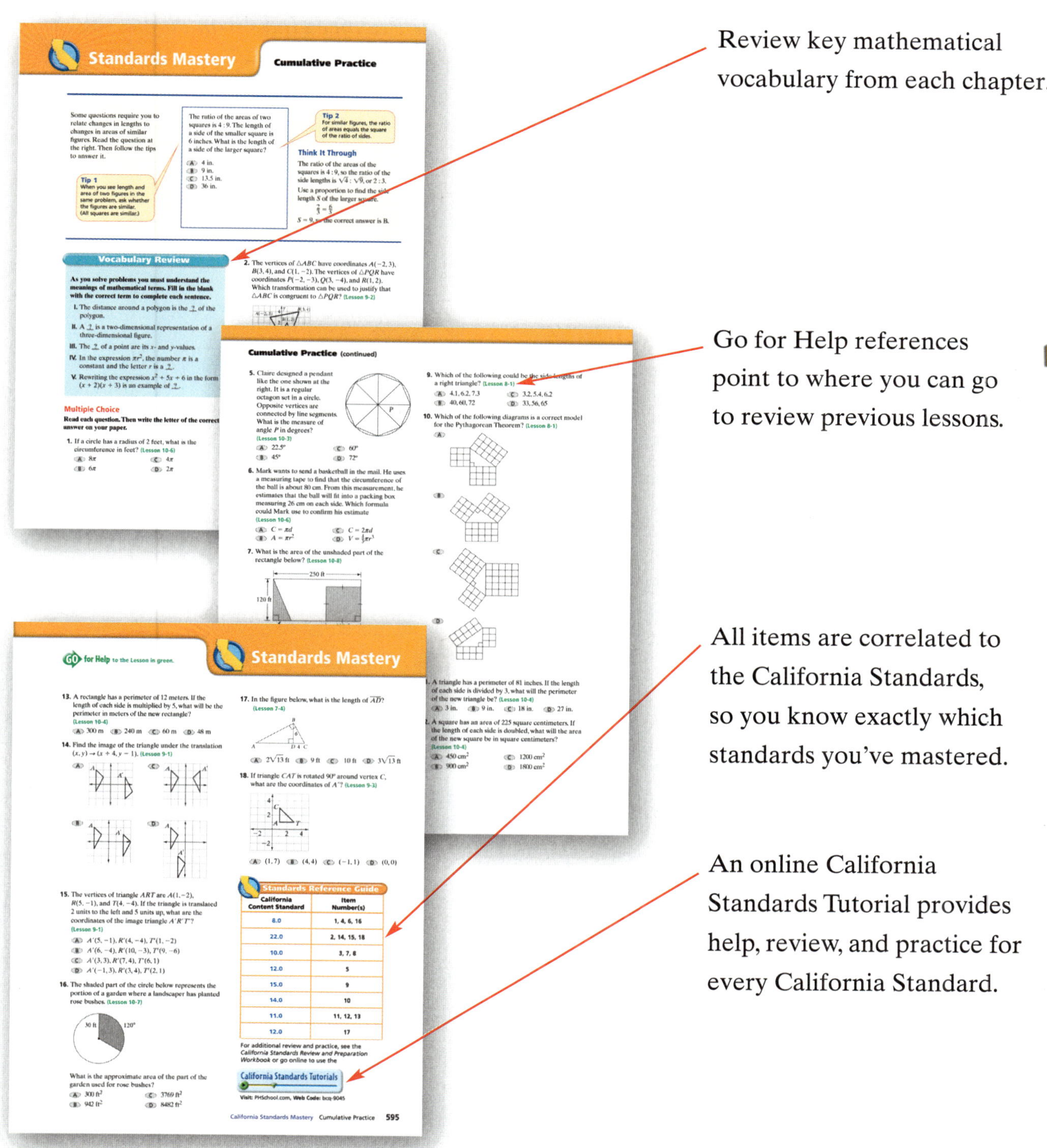

Review key mathematical vocabulary from each chapter.

Go for Help references point to where you can go to review previous lessons.

All items are correlated to the California Standards, so you know exactly which standards you've mastered.

An online California Standards Tutorial provides help, review, and practice for every California Standard.

Standards Mastery in Your Book

Mastering the California Standards

There are other types of features in your *Prentice Hall Geometry* textbook that will help you learn the mathematics and be successful in your class.

These features provide opportunities for you to engage with the mathematics being taught. Each feature states the California Standards being covered.

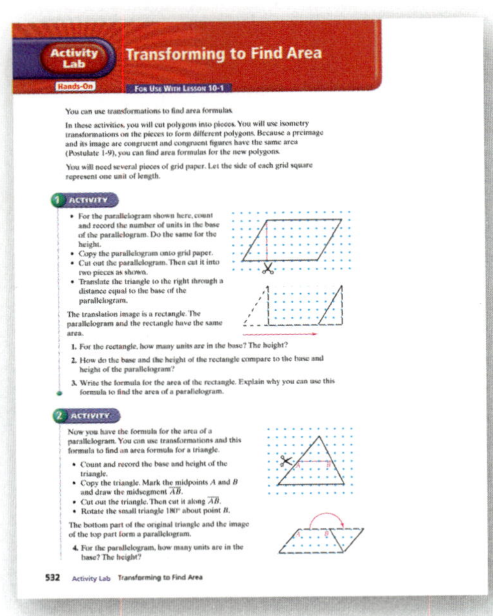

Activity Lab — Transforming to Find Area

Hands-On — For Use With Lesson 10-1

You can use transformations to find area formulas.

In these activities, you will cut polygons into pieces. You will use isometry transformations on the pieces to form different polygons. Because a preimage and its image are congruent and congruent figures have the same area (Postulate 1-9), you can find area formulas for the new polygons.

You will need several pieces of grid paper. Let the side of each grid square represent one unit of length.

1 ACTIVITY

- For the parallelogram shown here, count and record the number of units in the base of the parallelogram. Do the same for the height.
- Copy the parallelogram onto grid paper.
- Cut out the parallelogram. Then cut it into two pieces as shown.
- Translate the triangle to the right through a distance equal to the base of the parallelogram.

The translation image is a rectangle. The parallelogram and the rectangle have the same area.

1. For the rectangle, how many units are in the base? The height?
2. How do the base and the height of the rectangle compare to the base and height of the parallelogram?
3. Write the formula for the area of the rectangle. Explain why you can use this formula to find the area of a parallelogram.

2 ACTIVITY

Now you have the formula for the area of a parallelogram. You can use transformations and this formula to find an area formula for a triangle.

- Count and record the base and height of the triangle.
- Copy the triangle. Mark the midpoints A and B and draw the midsegment $\overline{AB}$.
- Cut out the triangle. Then cut it along $\overline{AB}$.
- Rotate the small triangle 180° about point B.

The bottom part of the original triangle and the image of the top part form a parallelogram.

4. For the parallelogram, how many units are in the base? The height?

532 Activity Lab Transforming to Find Area

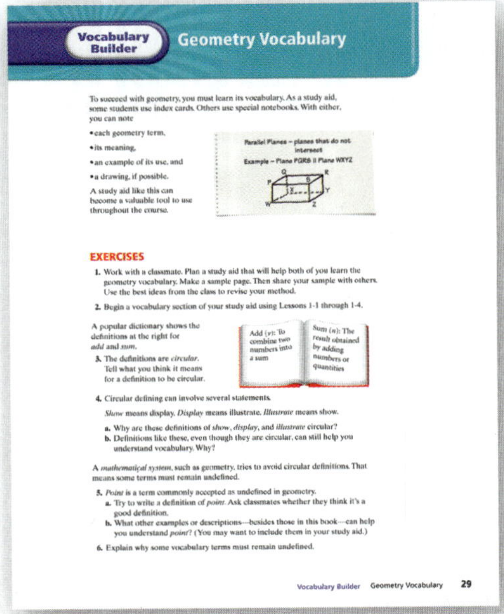

Vocabulary Builder — Geometry Vocabulary

To succeed with geometry, you must learn its vocabulary. As a study aid, some students use index cards. Others use special notebooks. With either, you can note

- each geometry term,
- its meaning,
- an example of its use, and
- a drawing, if possible.

A study aid like this can become a valuable tool to use throughout the course.

EXERCISES

1. Work with a classmate. Plan a study aid that will help both of you learn the geometry vocabulary. Make a sample page. Then share your sample with others. Use the best ideas from the class to revise your method.
2. Begin a vocabulary section of your study aid using Lessons 1-1 through 1-4.

A popular dictionary shows the definitions at the right for *add* and *sum*.

3. The definitions are *circular*. Tell what you think it means for a definition to be circular.
4. Circular defining can involve several statements.

   Show means display. *Display* means illustrate. *Illustrate* means show.

   a. Why are these definitions of *show*, *display*, and *illustrate* circular?
   b. Definitions like these, even though they are circular, can still help you understand vocabulary. Why?

A *mathematical system*, such as geometry, tries to avoid circular definitions. That means some terms must remain undefined.

5. *Point* is a term commonly accepted as undefined in geometry.
   a. Try to write a definition of *point*. Ask classmates whether they think it's a good definition.
   b. What other examples or descriptions—besides those in this book—can help you understand *point*? (You may want to include them in your study aid.)
6. Explain why some vocabulary terms must remain undefined.

Vocabulary Builder Geometry Vocabulary 29

Understanding mathematical vocabulary is an important part of your success in any mathematics class. These features give you even more practice and support in learning these vocabulary words.

GPS Guided Problem Solving

These features give you more support in becoming a good problem solver.

- We provide a great model for talking through the problem-solving process.
- "Think It Through" questions get more in-depth.
- The first two exercises present the steps you should take to solve the problem. They often contain a visual model of the problem.
- The remaining exercises provide independent practice.

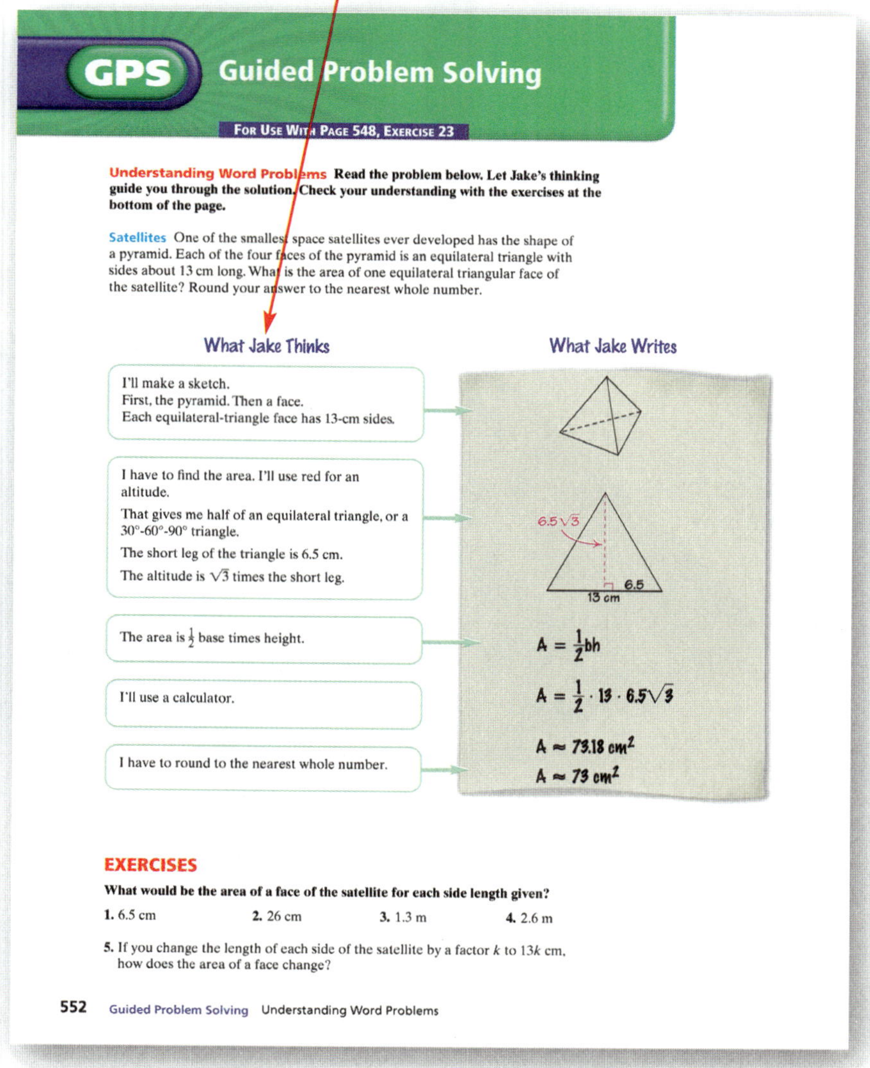

Standards Mastery in Your Book

California
Student Center

Your Prentice Hall Student Center is your one-stop spot for reviewing the math you learned in class and completing your homework.

Interactive Text

Your complete textbook – electronically! Also includes:

- Stepped examples for you to review
- Activities and Videos to review your lesson
- Self-Check tests to make sure you're on the right track.

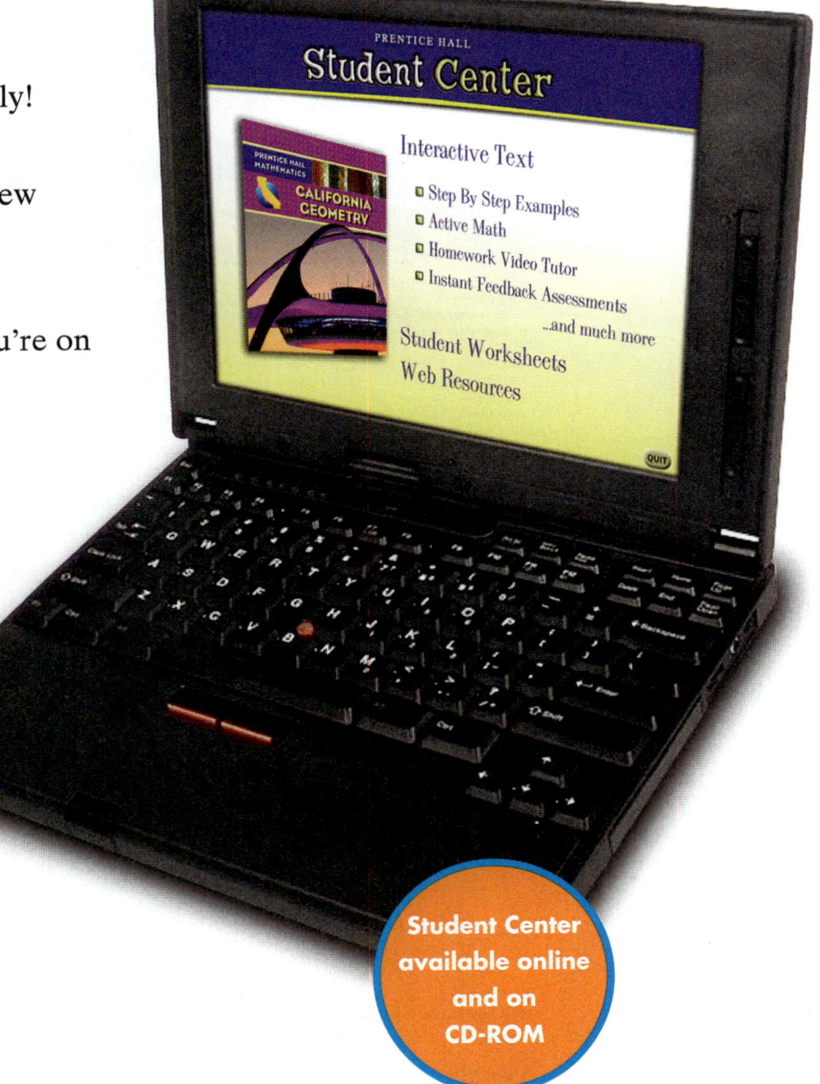

Student Worksheets

All the additional resources your teacher might assign you to extend your learning.

Web Resources

Links out to helpful sites to extend your understanding of math concepts.

Student Center available online and on CD-ROM

Throughout this book you will find links to the Prentice Hall Web site. Use the Web Codes provided with each link to gain direct access to online material. Here's how to *Go Online*:

1. **Go to PHSchool.com**
2. **Enter the Web Code**
3. **Click Go!**

Lesson Web Codes

Lesson Quiz Web Codes: There is an online quiz for every lesson. Access these quizzes with Web Codes bca-0101 through bca-1206 for Lesson 1-1 through Lesson 12-6.

Lesson Quizzes
Web Code format: bca-0204 02 = Chapter 2 04 = Lesson 4

Homework Video Tutor Web Codes: For every lesson, there is additional support online to help students complete their homework. Access the Homework Video Tutors with Web Codes bce-0101 through bce-1206 for Lesson 1-1 through Lesson 12-6.

Homework Video Tutor
Web Code format: bce-0605 06 = Chapter 6 05 = Lesson 5

Chapter Web Codes

Chapter	Vocabulary Quizzes	Chapter Tests	California Standards Tutorial
1	bcj-0151	bca-0152	bcq-9045
2	bcj-0251	bca-0252	bcq-9045
3	bcj-0351	bca-0352	bcq-9045
4	bcj-0451	bca-0452	bcq-9045
5	bcj-0551	bca-0552	bcq-9045
6	bcj-0651	bca-0652	bcq-9045
7	bcj-0751	bca-0752	bcq-9045
8	bcj-0851	bca-0852	bcq-9045
9	bcj-0951	bca-0952	bcq-9045
10	bcj-1051	bca-1052	bcq-9045
11	bcj-1151	bca-1152	bcq-9045
12	bcj-1251	bca-1252	bcq-9045

Additional Web Codes

Video Tutor Help:
Use Web Code bce-0775 to access engaging online instructional videos to help bring math concepts to life.

Data Updates:
Use Web Code bcg-9041 to get up-to-date government data for use in examples and exercises.

Math at Work:
For information about each Math at Work feature, use Web Code bcb-2031.

Your Guide to the Standards

California Mathematics Content Standards

Here is a complete list of the Mathematics Content Standards for Geometry. They are followed by a list of CAHSEE Standards that are included in the book to help you review algebra topics. Use this section as a reference guide as you explore the topics in each chapter.

Geometry

1.0 Students demonstrate understanding by identifying and giving examples of undefined terms, axioms, theorems, and inductive and deductive reasoning.

2.0 Students write geometric proofs, including proofs by contradiction.

3.0 Students construct and judge the validity of a logical argument and give counterexamples to disprove a statement.

4.0 Students prove basic theorems involving congruence and similarity.

5.0 Students prove that triangles are congruent or similar, and they are able to use the concept of corresponding parts of congruent triangles.

6.0 Students know and are able to use the triangle inequality theorem.

7.0 Students prove and use theorems involving the properties of parallel lines cut by a transversal, the properties of quadrilaterals, and the properties of circles.

What You'll Learn

In Chapter 2, you will develop your reasoning skills in problems such as the following.

Identify the hypothesis and conclusion of the conditional statement:

"*If my fans think that I can do everything I say I can do, then they're crazier than I am.*"—Muhammad Ali

Where You'll Learn This

These standards are covered in Chapters 1–7, 10, and 12.

Geometry (continued)

8.0 Students know, derive, and solve problems involving perimeter, circumference, area, volume, lateral area, and surface area of common geometric figures.

9.0 Students compute the volumes and surface areas of prisms, pyramids, cylinders, cones, and spheres; and students commit to memory the formulas for prisms, pyramids, and cylinders.

10.0 Students compute areas of polygons, including rectangles, scalene triangles, equilateral triangles, rhombi, parallelograms, and trapezoids.

11.0 Students determine how changes in dimensions affect the perimeter, area, and volume of common geometric figures and solids.

What You'll Learn

In Chapters 10 and 11, you will explore the volume of solids and the area of geometric figures such as the trapezoid shown below.

Approximate the area of Nevada by finding the area of the trapezoid shown.

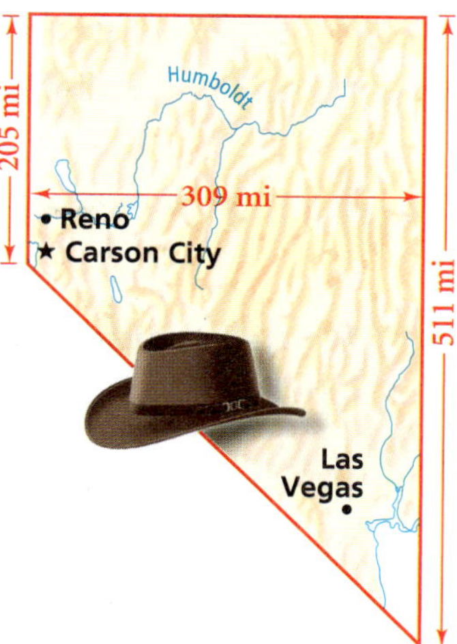

Where You'll Learn This

These standards are addressed in Chapters 1, 10, and 11.

Your Guide to the Standards

Geometry (continued)

12.0 Students find and use measures of sides and of interior and exterior angles of triangles and polygons to classify figures and solve problems.

13.0 Students prove relationships between angles in polygons by using properties of complementary, supplementary, vertical, and exterior angles.

14.0 Students prove the Pythagorean theorem.

15.0 Students use the Pythagorean theorem to determine distance and find missing lengths of sides of right triangles.

16.0 Students perform basic constructions with a straightedge and compass, such as angle bisectors, perpendicular bisectors, and the line parallel to a given line through a point off the line.

17.0 Students prove theorems by using coordinate geometry, including the midpoint of a line segment, the distance formula, and various forms of equations of lines and circles.

What You'll Learn

In Chapter 8, you will use the Pythagorean Theorem to find missing lengths in right triangles.

A painter leans a 15-ft ladder against a house.The base of the ladder is 5 ft from the house. (a) To the nearest tenth of a foot, how high on the house does the ladder reach? (b) The ladder in part (a) reaches too high on the house. By how much should the painter move the ladder's base away from the house to lower the top by 1 ft?

Where You'll Learn This

These standards are covered in Chapters 1, 3, 4, 5, 6, 8, and 12.

Geometry (continued)

18.0 Students know the definitions of the basic trigonometric functions defined by the angles of a right triangle. They also know and are able to use elementary relationships between them. For example, $\tan(x) = \sin(x)/\cos(x)$, $(\sin(x))^2 + (\cos(x))^2 = 1$.

19.0 Students use trigonometric functions to solve for an unknown length of a side of a right triangle, given an angle and a length of a side.

20.0 Students know and are able to use angle and side relationships in problems with special right triangles, such as 30°, 60°, and 90° triangles and 45°, 45°, and 90° triangles.

21.0 Students prove and solve problems regarding relationships among chords, secants, tangents, inscribed angles, and inscribed and circumscribed polygons of circles.

22.0 Students know the effect of rigid motions on figures in the coordinate plane and space, including rotations, translations, and reflections.

What You'll Learn

You can use transformations, which are discussed in Chapter 9, to solve applications such as the trail building problem below.

A hiking club is building a new trail system. It wants to build trails to the Overlook and Balance Rock that will connect at a point on Summit Trail. Working under a tight budget, it wants to minimize the total length of these trails. If the trails cover similar terrain, at what point should they meet on Summit Trail?

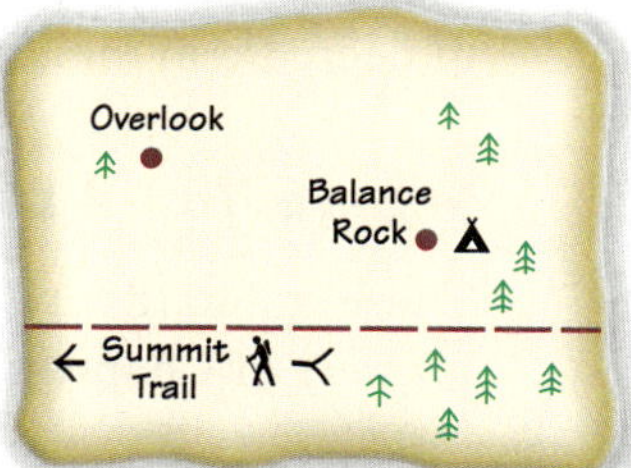

Where You'll Learn This

Standards 18.0 through 22.0 are covered in Chapters 5, 8, 9, 10, and 12.

CAHSEE

7NS 1.0 **Students know the properties of, and compute with, rational numbers expressed in a variety of forms:**

7NS 1.1 Read, write, and compare rational numbers in scientific notation (positive and negative powers of 10) with approximate numbers using scientific notation.

7NS 1.2 Add, subtract, multiply, and divide rational numbers (integers, fractions, and terminating decimals) and take positive rational numbers to whole-number powers.

7NS 1.3 Convert fractions to decimals and percents and use these representations in estimations, computations, and applications.

7NS 1.6 Calculate the percentage of increases and decreases of a quantity.

7NS 1.7 Solve problems that involve discounts, markups, commissions, and profit, and compute simple and compound interest.

What You'll Learn

In Chapter 6, you will solve a percent decrease problem involving population.

In January, a newscaster reported that 462 births were recorded in Bramfor Township last year, which was 63 less than the previous year. What was the ercent of decrease in the total number of births from the previous year?

Where You'll Learn This

These standards are addressed in the *Multiple Choice Practice* or *Standard Mastery* of Chapters 1, 2, 3, 5, 6, 7, 9, and 10.

CAHSEE (continued)

7NS 2.0 **Students use exponents, powers, and roots, and use exponents in working with fractions:**

7NS 2.1 Understand negative whole-number exponents. Multiply and divide expressions involving exponents with a common base.

7NS 2.2 Add and subtract fractions by using factoring to find common denominators.

7NS 2.3 Multiply, divide, and simplify rational numbers by using exponent rules.

7NS 2.4 Use the inverse relationship between raising to a power and extracting the root of a perfect square integer; for an integer that is not square, determine without a calculator the two integers between which its square root lies and explain why.

7NS 2.5 Understand the meaning of the absolute value of a number; interpret the absolute value as the distance of the number from zero on a number line; and determine the absolute value of real numbers.

What You'll Learn

In Chapter 4, you will use exponent rules to simplify rational numbers such as the number below.

$(5^5)^2 =$

Where You'll Learn This

These standards are addressed in the *Multiple Choice Practice* or *Standards Mastery* of Chapters 1, 2, 3, 4, 5, 6, and 11.

CAHSEE (continued)

6SDAP 1.0 **Students compute and analyze statistical measurements for data sets:**

6SDAP 1.1 Compute the mean, median, and mode of data sets.

6SDAP 2.0 **Students use data samples of a population and describe the characteristics and limitations of the samples:**

6SDAP 2.5 Identify claims based on statistical data and, in simple cases, evaluate the validity of the claims.

6SDAP 3.0 **Students determine theoretical and experimental probabilities and use these to make predictions about events:**

6SDAP 3.1 Represent all possible outcomes for compound events in an organized way (e.g., tables, grids, tree diagrams) and express the theoretical probability of each outcome.

6SDAP 3.3 Represent probabilities as ratios, proportions, decimals between 0 and 1, and percentages between 0 and 100 and verify that the probabilities computed are reasonable; know that if P is the probability of an event, $1-P$ is the probability of an event not occurring.

6SDAP 3.5 Understand the difference between independent and dependent events.

7SDAP 1.0 **Student collect, organize, and represent data sets that have one or more variables and identify relationships among variables within a data set by hand and through the use of an electronic spreadsheet software program:**

7SDAP 1.1 Know various forms of display for data sets, use the forms to display a single set of data or to compare two sets of data.

7SDAP 1.2 Represent two numerical variables on a scatterplot and informally describe how the data points are distributed and any apparent relationship that exists between the two variables (e.g., between time spent on homework and grade level).

What You'll Learn

Throughout the book, you will find probabilities and represent them as ratios, percentages, and proportions.

Robin placed 8 red marbles, 10 blue marbles, and 7 white marbles in a bag. He randomly took a marble out. What is the theoretical probability that it will be white?

Where You'll Learn This

These standards are addressed in the *Multiple Choice Practice* or *Standards Mastery* of each chapter in the book.

CAHSEE (continued)

7AF 1.0 **Students express quantitative relationships by using algebraic terminology, expressions, equations, inequalities, and graphs:**

7AF 1.1 Use variables and appropriate operations to write an expression, an equation, an inequality, or a system of equations or inequalities that represents a verbal description (e.g., three less than a number, half as large as area A).

7AF 1.2 Use the correct order of operations to evaluate algebraic expressions such as $3(2x + 5)^2$.

7AF 1.5 Represent quantitative relationships graphically and interpret the meaning of a specific part of a graph in the situation represented by the graph.

7AF 2.0 **Students interpret and evaluate expressions involving integer powers and simple roots:**

7AF 2.1 Interpret positive whole-number powers as repeated multiplication and negative whole-number powers as repeated division or multiplication by the multiplicative inverse. Simplify and evaluate expressions that include exponents.

7AF 2.2 Multiply and divide monomials; extend the process of taking powers and extracting roots to monomials when the latter results in a monomial with an integer exponent.

What You'll Learn

In Chapter 1, you will examine a graph and interpret the meaning of a specific part of the graph.

The graph shows h, the height in feet of a water droplet spraying out from a fountain pipe, versus t, the time in seconds after it leaves the pipe. From the graph, what conclusion can be made about the path of the droplet?

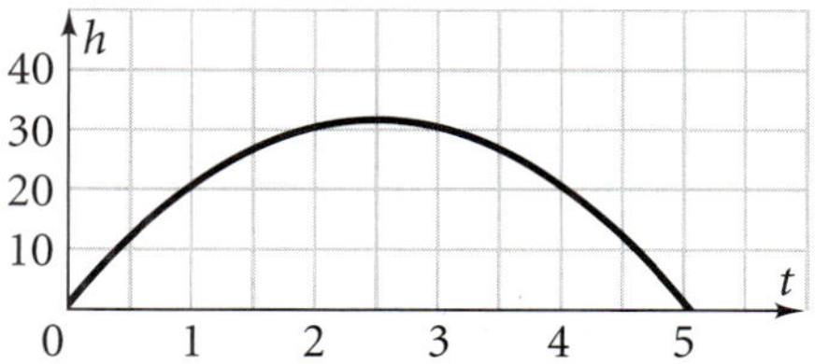

Where You'll Learn This

These standards are addressed in the *Multiple Choice Practice* or *Standards Mastery* of Chapters 1, 3, 4, 5, 8, 9, 10, 11, and 12.

Your Guide to the Standards

CAHSEE (continued)

7AF 3.0 **Students graph and interpret linear and some nonlinear functions:**

7AF 3.1 Graph functions of the form $y = nx^2$ and $y = nx^3$ and use in solving problems.

7AF 3.3 Graph linear functions, noting that the vertical change (change in y-value) per unit of horizontal change (change in x-value) is always the same and know that the ratio ("rise over run") is called the slope of a graph.

7AF 3.4 Plot the values of quantities whose ratios are always the same (e.g., cost to the number of an item, feet to inches, circumference to diameter of a circle). Fit a line to the plot and understand that the slope of the line equals the quantities.

7AF 4.0 **Students solve simple linear equations and inequalities over the rational numbers:**

7AF 4.1 Solve two-step linear equations and inequalities in one variable over the rational numbers, interpret the solution or solutions in the context from which they arose, and verify the reasonableness of the results.

7AF 4.2 Solve multistep problems involving rate, average speed, distance, and time or a direct variation.

What You'll Learn

In Chapter 9, you will solve multistep problems involving rate.

A car consumes 8 gallons of gasoline while traveling 224 miles. At that rate, how many gallons of gasoline will be needed for a 56-mile trip?

Where You'll Learn This

These standards are addressed in the *Multiple Choice Practice* or *Standards Mastery* of Chapters 1–5, 9, 11, and 12.

CAHSEE (continued)

7MG 1.0 **Students choose appropriate units of measure and use ratios to convert within and between measurement systems to solve problems:**

7MG 1.1 Compare weights, capacities, geometric measures, times, and temperatures within and between measurement systems (e.g., miles per hour and feet per second, cubic inches to cubic centimeters).

7MG 1.2 Construct and read drawings and models made to scale.

7MG 1.3 Use measures expressed as rates (e.g., speed, density) and measures expressed as products (e.g., person-days) to solve problems; check the units of the solutions; and use dimensional analysis to check the reasonableness of the answer.

7MG 2.0 **Students compute perimeter, area, and volume of common geometric objects and use the results to find measures of less common objects. They know how perimeter, area and volume are affected by changes of scale:**

7MG 2.1 Use formulas routinely for finding the perimeter and area of basic two-dimensional figures and the surface area and volume of basic three-dimensional figures, including rectangles, parallelograms, trapezoids, squares, triangles, circles, prisms, and cylinders.

7MG 2.2 Estimate and compute the area of more complex or irregular two- and three-dimensional figures by breaking the figures down into more basic geometric objects.

What You'll Learn

Sometimes it is necessary to convert between two measurement systems, as in the following problem:

If Tony is driving at 60 mph, what is his approximate speed in kilometers per hour? (1 mile $\approx$ 1.6 kilometers)

Where You'll Learn This

These standards are addressed in the *Multiple Choice Practice* or *Standards Mastery* of Chapters 1–5, 7, 8, and 9.

CAHSEE (continued)

7MG 2.3 Compute the length of the perimeter, the surface area of the faces, and the volume of a three-dimensional object built from rectangular solids. Understand that when the lengths of all dimensions are multiplied by a scale factor, the surface area is multiplied by the square of the scale factor and volume is multiplied by the cube of the scale factor.

7MG 2.4 Relate the changes in measurement with a change of scale to the units used (e.g., square inches, cubic feet) and to conversions between units (1 square foot = 144 square inches or [1 ft^2] = [144 $in.^2$], 1 cubic inch is approximately 16.38 cubic centimeters or [1 $in.^3$] = [16.38 cm^3]).

7MG 3.0 **Students know the Pythagorean theorem and deepen their understanding of plane and solid geometric shapes by constructing figures that meet given conditions and by identifying attributes of figures:**

7MG 3.2 Understand and use coordinate graphs to plot simple figures, determine lengths and areas related to them, and determine their image under translations and reflections.

7MG 3.3 Know and understand the Pythagorean theorem and its converse and use it to find the length of the missing side of a right triangle and the lengths of other line segments and, in some situations, empirically verify the Pythagorean theorem by direct measurement.

7MG 3.4 Demonstrate an understanding of conditions that indicate two geometrical figures are congruent and what congruence means about the relationships between the sides and angles of the two figures.

What You'll Learn

In Chapter 8, you will use the Pythagorean Theorem to find missing measures in triangles.

A walkway forms the diagonal of a square playground. The walkway is 24 m long. To the nearest tenth of a meter, how long is a side of the playground?

Where You'll Learn This

These standards are addressed in the *Multiple Choice Practice* or *Standards Mastery* of Chapters 1, 2, 3, 5, 6, 9, 11, and 12.

CAHSEE (continued)

7MR 1.0 **Students make decisions about how to approach problems:**

7MR 1.1 Analyze problems by identifying relationships, distinguishing relevant from irrelevant information, identifying missing information, sequencing and prioritizing information, and observing patterns.

7MR 1.2 Formulate and justify mathematical conjectures based on a general description of the mathematical question or problem posed.

7MR 2.0 **Students use strategies, skills, and concepts in finding solutions:**

7MR 2.1 Use estimation to verify the reasonableness of calculated results.

7MR 2.3 Estimate unknown quantities graphically and solve for them by using logical reasoning and arithmetic and algebraic techniques.

7MR 2.4 Make and test conjectures using both inductive and deductive reasoning.

7MR 3.0 **Students determine a solution is complete and move beyond a particular problem by generalizing to other situations:**

7MR 3.3 Develop generalizations of the results obtained and the strategies used and apply them to new problem situations.

What You'll Learn

Thinking about a problem before you begin is a very important part of mathematics. Problems such as the one below will help you develop this skill.

Ken went shopping and spent $14.00 on a book, $6.50 on lunch, and $2.50 on bus fare. On the way home he collected $10 from a friend who owed him money. When he returned home he had $15.00. He guessed that he started with $25.00. How can he verify this?

Where You'll Learn This

These standards are addressed in the *Multiple Choice Practice* or *Standards Mastery* of Chapters 1, 2, 3, 5–9, and 11.

CAHSEE (continued)

Alg1 2.0 Students understand and use such operations as taking the opposite, finding the reciprocal, and taking a root. They understand and use the rules of exponents.

Alg1 3.0 Students solve equations and inequalities involving absolute values.

Alg1 4.0 Students simplify expressions before solving linear equations and inequalities in one variable, such as $3(2x - 5) + 4(x - 2) = 12$.

Alg1 5.0 Students solve multistep problems, including word problems, involving linear equations and linear inequalities in one variable and provide justification for each step.

Alg1 6.0 Students graph a linear equation and compute the x- and y- intercepts (e.g., graph $2x + 6y = 4$).

What You'll Learn

Throughout the book, you will work with linear equations and inequalities as in the following problem.

What is the x-intercept of the line $y = -\frac{3}{4}x - 3$?

Where You'll Learn This

These standards are addressed in the *Multiple Choice Practice* or *Standards Mastery* of Chapters 1, 2, 5, 6, 7, 11, and 12.

CAHSEE (continued)

Alg1 7.0 Students verify that a point lies on a line, given an equation of the line. Students are able to derive linear equations.

Alg1 8.0 Students understand the concepts of parallel lines and how their slopes are related.

Alg1 9.0 Students solve a system of two linear equations in two variables algebraically and are able to interpret the answer graphically. Students are able to solve a system of two linear inequalities in two variables and to sketch the solution sets.

Alg1 10.0 Students add, subtract, multiply, and divide monomials and polynomials. Students solve multistep problems, including word problems, by using these techniques.

Alg1 15.0 Students apply algebraic techniques to solve rate problems, work problems, and percent mixture problems.

What You'll Learn

In Chapter 3, you will learn about parallel and perpendicular lines and the various forms of an equation of a line.

Are lines ℓ_1 and ℓ_2 parallel?

Explain using slope.

Where You'll Learn This

These standards are addressed in the *Multiple Choice Practice* or *Standards Mastery* of Chapters 1–5, 7, 9, and 11.

Entry-Level Assessment

Read each question. Then write the letter of the correct answer on your paper.

1. What is the next figure in the sequence?

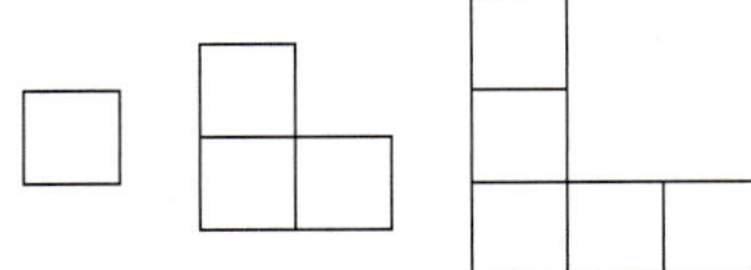

Ⓐ
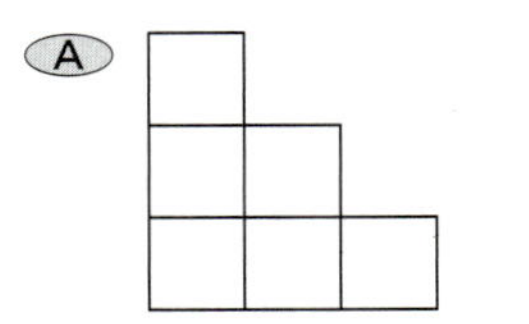

Ⓒ
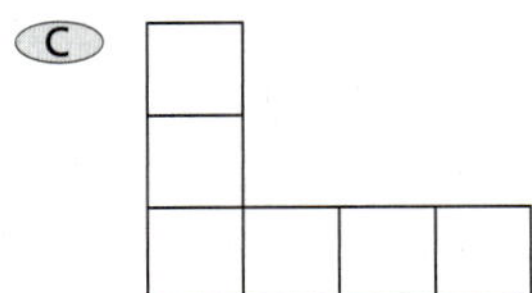

Ⓑ
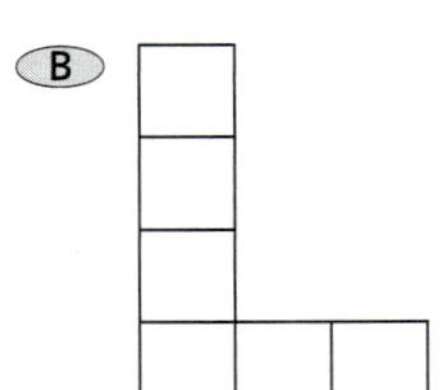

Ⓓ
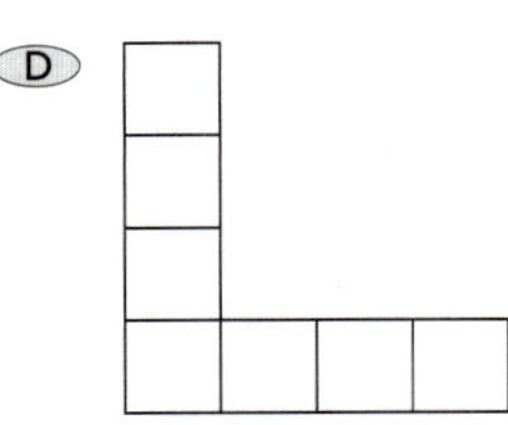

2. What is the distance between points $A(1, 9)$ and $B(4, -2)$?

Ⓐ $\sqrt{58}$ Ⓒ $\sqrt{130}$

Ⓑ 58 Ⓓ 130

3. A farmer leans a 12-ft ladder against a barn. The base of the ladder is 3 ft from the barn. To the nearest foot, how high on the barn does the ladder reach?

Ⓐ 11 ft Ⓒ 13 ft

Ⓑ 12 ft Ⓓ 14 ft

4. You are planning to install a rectangular pool. Its length is 6 feet more than twice its width. You want a walkway 2 feet wide around the pool. Write an expression for the area of the pool and walkway.

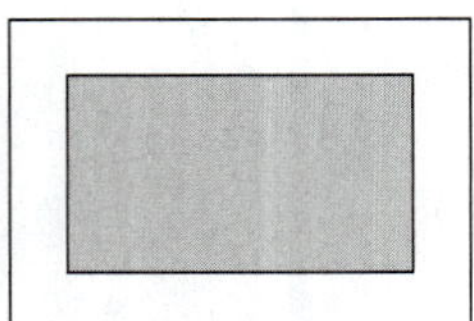

Ⓐ $2x + 14$ Ⓒ $2x^2 + 20x + 40$

Ⓑ $2x^2 + 40$ Ⓓ $2x^2 + 18x + 40$

5. A circular pond has a radius of $2x - 3$. Which expression describes the area of the pond?

Ⓐ $(4x - 6)\pi$

Ⓑ $(4x + 9)\pi$

Ⓒ $(4x^2 + 12x - 9)\pi$

Ⓓ $(4x^2 - 12x + 9)\pi$

6. The graph below shows the distance and time of Roberto's car trip. What does the slope of the line mean?

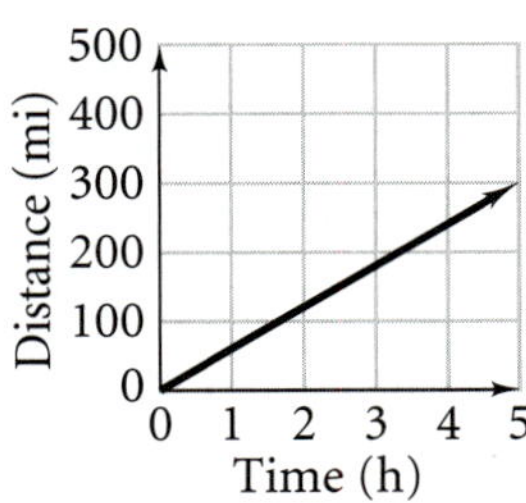

Ⓐ Roberto traveled 0.017 mi/h.

Ⓑ Roberto traveled for 5 h.

Ⓒ Roberto traveled 60 mi/h.

Ⓓ Roberto traveled 300 mi.

7. What is the perimeter of $\triangle ABC$ to the nearest tenth?

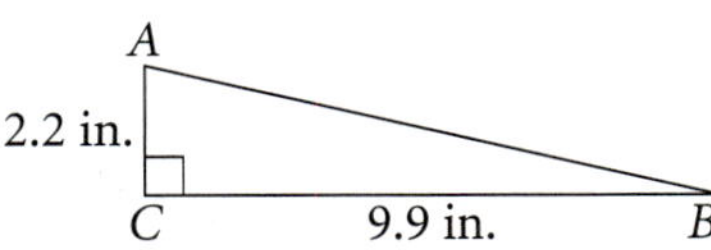

Ⓐ 19.8 in. Ⓒ 9.9 in.

Ⓑ 22.2 in. Ⓓ 20.2 in.

8. In the diagram below, which segment is parallel to $\overline{AB}$?

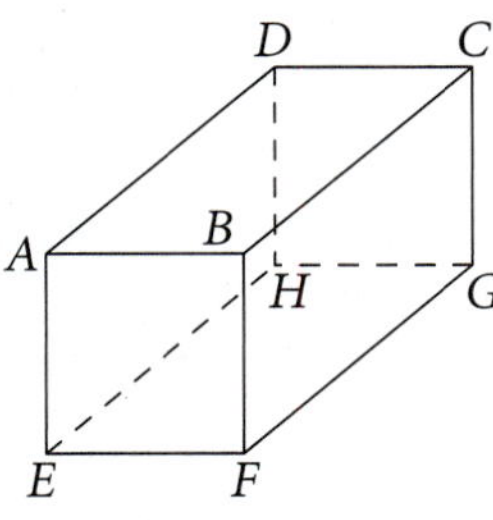

Ⓐ $\overline{BC}$ Ⓒ $\overline{AE}$

Ⓑ $\overline{AD}$ Ⓓ $\overline{EF}$

9. What is the x-coordinate of the midpoint of $\overline{ZU}$?

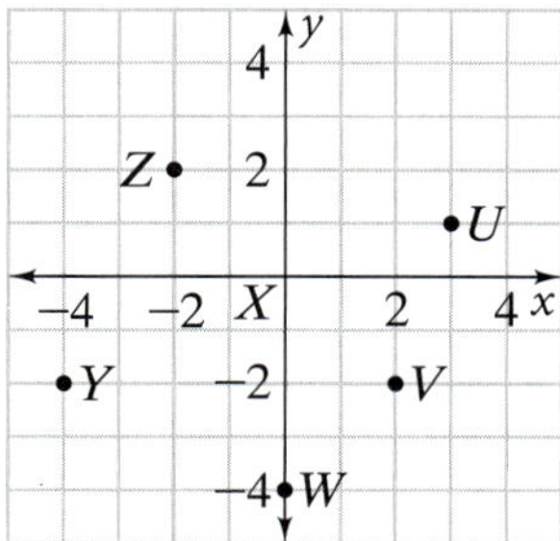

Ⓐ 0
Ⓑ 0.5
Ⓒ 1
Ⓓ 1.5

10. A square has an area of 25 ft^2. What is its area if its perimeter increases by four feet?

Ⓐ 26 ft^2
Ⓑ 29 ft^2
Ⓒ 36 ft^2
Ⓓ 41 ft^2

11. A museum window is formed by adding a semicircle to the top of a rectangle. The base of the rectangle is 2 feet and the height of the rectangle is 5 feet. Find the area of the window. Round your answer to the nearest tenth.

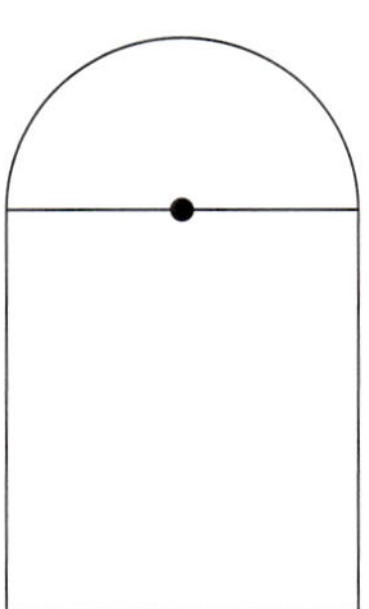

Ⓐ 11.6 square feet
Ⓑ 13.1 square feet
Ⓒ 15.1 square feet
Ⓓ 16.3 square feet

12. Complete the congruence statement, given that quadrilateral $ABCD \cong PQRS$, $\overline{AD} \cong$ __?__

Ⓐ $\overline{PQ}$
Ⓑ $\overline{PS}$
Ⓒ $\overline{PR}$
Ⓓ $\overline{QR}$

13. The ratio of the volume of two spheres is 216 : 125. What is the ratio of their diameters?

Ⓐ 6 : 5
Ⓑ 36 : 25
Ⓒ 91 : 1
Ⓓ 216 : 125

14. Assume that a dart you throw hits the dartboard shown and is equally likely to hit any spot on the board. Find the probability of hitting the shaded region.

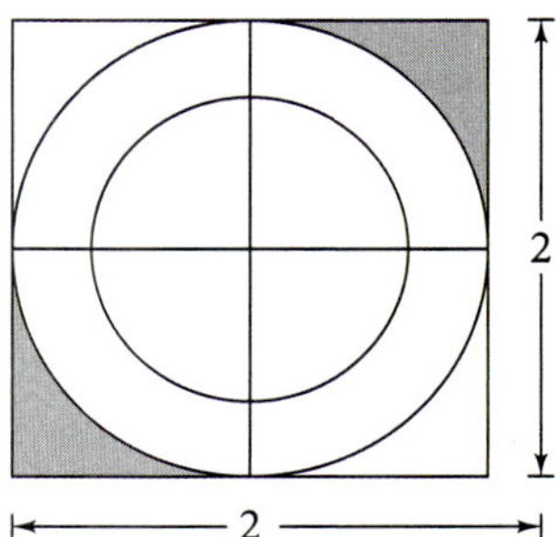

Ⓐ 10.7%
Ⓑ 21.5%
Ⓒ 42.9%
Ⓓ 78.5%

15. N-scale model trains have a scale factor 1 : 160. An N-scale model engine measures 3.75 in. What is the length of the actual engine?

Ⓐ 163.75 in.
Ⓑ 60 in.
Ⓒ $42\frac{2}{3}$ ft
Ⓓ 600 in.

16. Find the glide reflection image of $P(4, 1)$ for the translation $(x, y) \rightarrow (x - 2, y + 2)$ and reflection across $x = 0$.

Ⓐ $P'(-2, 3)$
Ⓑ $P'(-6, 3)$
Ⓒ $P'(2, -3)$
Ⓓ $P'(0, 3)$

17. If the perimeter of the triangle is equal to the perimeter of the square, then the length of a side of the square is __?__.

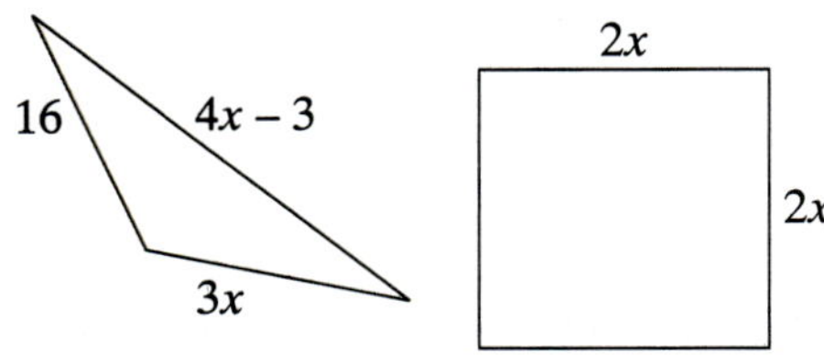

Ⓐ 7
Ⓑ 13
Ⓒ 26
Ⓓ 52

18. The diagram below shows the area that is affected by oil leaking from a tanker. Workers surrounded the oil slick with a retaining material. What is the area covered by the oil slick?

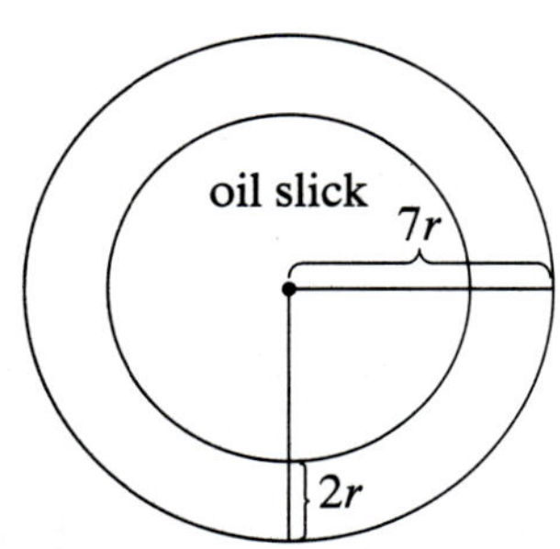

Ⓐ $A = 7r - 2r$
Ⓑ $A = 49r - 4r\pi$
Ⓒ $A = 49r^2 - 4r\pi$
Ⓓ $A = 25\pi r^2$

19. Four circles, each of radius 7 inches, are packed in a square as shown. If you throw a dart and hit the square target, what is the probability that your dart will land inside one of the circles? Express your answer as a percent to the nearest tenth.

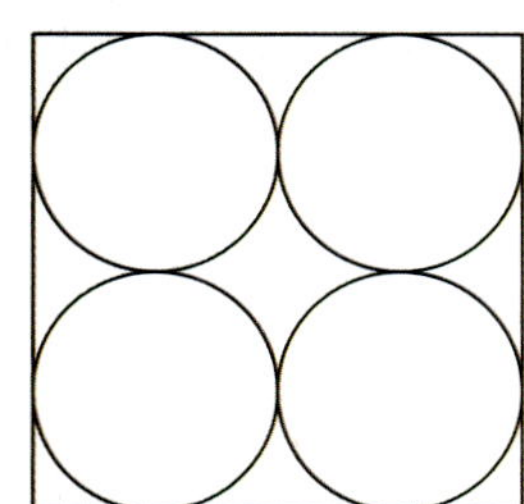

Ⓐ 67.5%
Ⓑ 78.5%
Ⓒ 75.0%
Ⓓ 50.0%

20. The package of a model car states that the ratio of the length of the model to the length of a real car is 1 : 25. What is the ratio of the amount of paint required to cover the model to the amount required to cover the real car?

Ⓐ 1 : 25
Ⓑ 1 : 50
Ⓒ 1 : 625
Ⓓ 1 : 15,625

21. Describe the next term in the sequence in words.

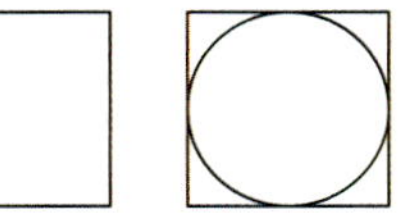
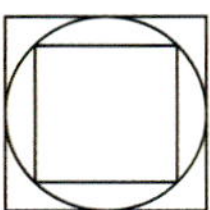

Ⓐ a circle inside a square
Ⓑ a square inside a circle inside a square
Ⓒ a circle inside a square inside a circle inside a square
Ⓓ a square inside a circle inside a square inside a circle

22. Give the equation of the line that is parallel to the line $y = 5x + 2$ and passes through the point $(1, -3)$.

Ⓐ $y = -5x + 2$
Ⓑ $y = 5x + 8$
Ⓒ $y = \frac{1}{5}x - 8$
Ⓓ $y = 5x - 8$

23. Reflect the figure below across the y-axis. What are the new coordinates of Point B?

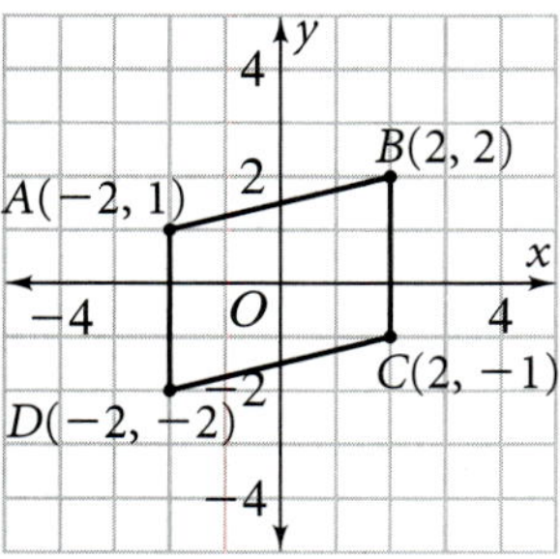

Ⓐ $(2, -2)$
Ⓑ $(-2, -2)$
Ⓒ $(-2, 2)$
Ⓓ $(2, 2)$

24. What is the area of the quarter-circle below if the radius is 5 meters?

Ⓐ $\frac{100\pi}{4}$

Ⓑ $\frac{25\pi}{4}$

Ⓒ $\frac{10\pi}{4}$

Ⓓ $\frac{5\pi}{4}$

25. Write an equation of the line that passes through the point $(1, 3)$ with slope -2.

Ⓐ $y = -2x - 5$

Ⓑ $y = -2x + 7$

Ⓒ $y = -2x + 5$

Ⓓ $y = 2x + 7$

26. What is the equation of the line?

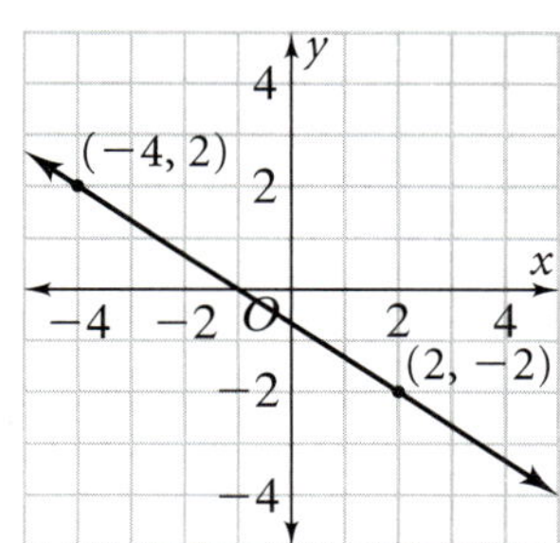

Ⓐ $y = -\frac{2}{3}x - \frac{2}{3}$

Ⓑ $y = -\frac{3}{2}x + 1$

Ⓒ $y = -2x + 2$

Ⓓ $y = \frac{2}{3}x - 1$

27. Use inductive reasoning to find the next term in the pattern.

$\frac{1}{2}, \frac{1}{4}, \frac{1}{16}, \ldots$

Ⓐ $\frac{1}{20}$

Ⓑ $\frac{1}{32}$

Ⓒ $\frac{1}{64}$

Ⓓ $\frac{1}{256}$

28. Use inductive reasoning to describe the pattern.

1, 4, 13, 40, 121

Ⓐ Multiply the previous term by 4.

Ⓑ Multiply the previous term by 3 and add 1.

Ⓒ Add 3 to the previous term.

Ⓓ Multiply the previous term by 2 and add 3.

29. What is the length of the hypotenuse of the triangle below to the nearest tenth?

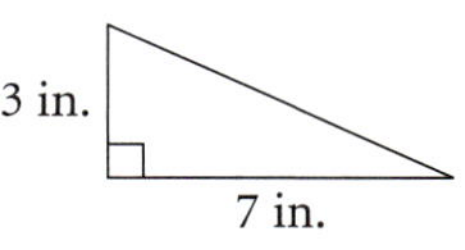

Ⓐ 9.1 in.

Ⓑ 8.6 in.

Ⓒ 8.1 in.

Ⓓ 7.6 in.

30. If $MNPQ \cong ABCD$, which of the following conclusions can you make?

Ⓐ $\angle NMP \cong \angle BCD$

Ⓑ $\angle NPQ \cong \angle BCD$

Ⓒ $\angle PQM \cong \angle BAD$

Ⓓ none of the above

31. A map has a scale of 1 in. : 25 mi. Two cities are 175 mi apart. How far apart are they on the map?

Ⓐ 3 inches

Ⓑ 5 inches

Ⓒ 6 inches

Ⓓ 7 inches

32. $ABCDE \cong VUTSR$. What is $m\angle TUV$?

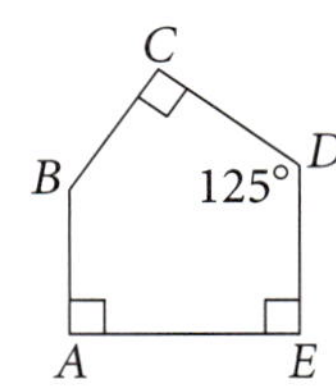

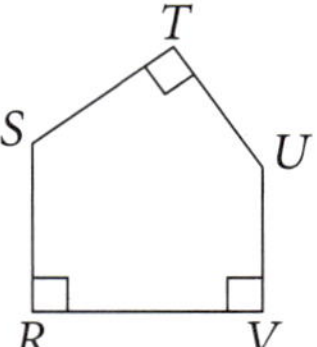

Ⓐ 90

Ⓑ 125

Ⓒ 145

Ⓓ 155

Tools of Geometry

What You've Learned

California Content Standards

NS 1.2 Add, subtract, multiply, and divide rational numbers and take positive rational numbers to whole-number powers. ***Grade 7***

MG 2.1 Use formulas routinely for finding the perimeter and area of basic two-dimensional figures and the surface area and volume of basic three dimensional figures, including rectangles, parallelograms, trapezoids, squares, triangles, circles, prisms, and cylinders. ***Grade 7***

4.0 Students simplify expressions before solving linear equations and inequalities in one variable. ***Algebra 1***

Check Your Readiness

Squaring Numbers (Skills Handbook page 753)

Simplify.

1. 3^2 **2.** 4^2 **3.** 11^2

Simplifying Expressions (Skills Handbook page 754)

Simplify each expression. Use 3.14 for π.

4. $2 \cdot 7.5 + 2 \cdot 11$ **5.** $\pi(5)^2$ **6.** $\sqrt{5^2 + 12^2}$

Evaluating Expressions (Skills Handbook page 754)

Evaluate the following expressions for $a = 4$ and $b = -2$.

7. $\frac{a + b}{2}$ **8.** $\frac{a - 7}{3 - b}$ **9.** $\sqrt{(7 - a)^2 + (2 - b)^2}$

Finding Absolute Value (Skills Handbook page 757)

Simplify each absolute value expression.

10. $|-8|$ **11.** $|2 - 6|$ **12.** $|-5 - (-8)|$

Solving Equations (Skills Handbook page 758)

 **Algebra** Solve each equation.

13. $2x + 7 = 13$ **14.** $5x - 12 = 2x + 6$ **15.** $2(x + 3) - 1 = 7x$

▲ Congruent sculptures in the San Diego Convention Center plaza

What You'll Learn Next

California Content Standards

1.0 Students demonstrate understanding by identifying and giving examples of undefined terms, axioms, and inductive reasoning.

3.0 Students construct and judge the validity of a logical argument and give counterexamples to disprove a statement.

16.0 Students perform basic constructions with a straightedge and compass, such as angle bisectors and perpendicular bisectors.

CAHSEE 7NS 1.6, 7NS 1.7, 7NS 2.4, 6SDAP 1.1, 7AF 1.2, 7AF 1.5, 7AF 2.2, 7AF 3.3, 7AF 4.2, 7MG 1.3, 7MG 2.2, 7MG 2.4, 7MR 1.1, 7MR 2.4, Alg1 3.0, Alg1 7.0, Alg1 10

New Vocabulary

English and Spanish Audio Online

- **acute angle** (p. 37)
- **angle bisector** (p. 46)
- **collinear points** (p. 17)
- **congruent angles** (p. 37)
- **conjecture** (p. 5)
- **construction** (p. 44)
- **coordinate** (p. 31)
- **counterexample** (p. 5)
- **line** (p. 17)
- **midpoint** (p. 32)
- **net** (p. 12)
- **obtuse angle** (p. 37)
- **parallel lines** (p. 24)
- **perpendicular lines** (p. 45)
- **plane** (p. 17)
- **postulate** (p. 18)
- **ray** (p. 23)
- **segment** (p. 23)

1-1 Patterns and Inductive Reasoning

California Content Standards

GEOM 1.0 Students demonstrate understanding by identifying and giving examples of inductive reasoning. ***Introduce***

GEOM 3.0 Students construct and judge the validity of a logical argument and give counterexamples to disprove a statement. ***Introduce***

What You'll Learn

- To use inductive reasoning to make conjectures

. . . And Why

To predict future sales for a skateboard business, as in Example 4

Check Skills You'll Need

GO for Help Skills Handbook page 753

Here is a list of the counting numbers: 1, 2, 3, 4, 5, . . .
Some are even and some are odd.

1. Make a list of the positive even numbers.
2. Make a list of the positive odd numbers.
3. Copy and extend this list to show the first 10 perfect squares.
   $1^2 = 1, 2^2 = 4, 3^2 = 9, 4^2 = 16, \ldots$
4. Which do you think describes the square of any odd number?
   It is odd. It is even.

New Vocabulary • inductive reasoning • conjecture • counterexample

1 Using Inductive Reasoning

Real-World Connection

You can predict growth of the chambered nautilus shell by studying patterns in its cross sections.

Inductive reasoning is reasoning that is based on patterns you observe. If you observe a pattern in a sequence, you can use inductive reasoning to tell what the next terms in the sequence will be.

1 EXAMPLE Finding and Using a Pattern

Find a pattern for each sequence. Use the pattern to show the next two terms in the sequence.

a. 3, 6, 12, 24, . . .

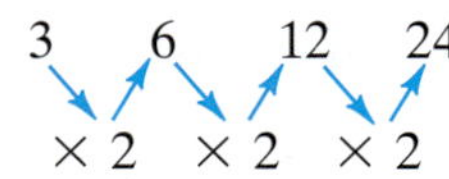

Each term is twice the preceding term. The next two terms are $2 \times 24 = 48$ and $2 \times 48 = 96$.

b.

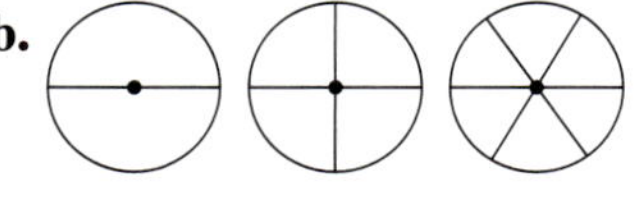

Each circle has one more segment through the center to form equal parts. The next two figures:

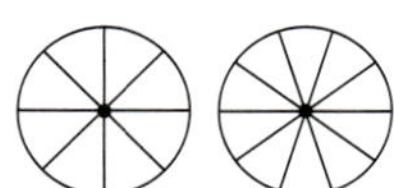

CA Standards Check **1** Write the next two terms in each sequence.

a. 1, 2, 4, 7, 11, 16, 22, . . .

b. Monday, Tuesday, Wednesday, . . .

c.

  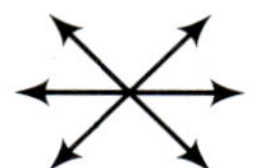

A conclusion you reach using inductive reasoning is called a **conjecture.**

2 EXAMPLE Using Inductive Reasoning

Make a conjecture about the sum of the first 30 odd numbers.

Find the first few sums. Notice that each sum is a perfect square.

$1 = 1 = 1^2$

$1 + 3 = 4 = 2^2$

$1 + 3 + 5 = 9 = 3^2$

$1 + 3 + 5 + 7 = 16 = 4^2$

The perfect squares form a pattern.

Using inductive reasoning, you can conclude that the sum of the first 30 odd numbers is 30^2, or 900.

CA Standards Check 2 Make a conjecture about the sum of the first 35 odd numbers. Use your calculator to verify your conjecture.

Not all conjectures turn out to be true. You can prove that a conjecture is false by finding one counterexample. A **counterexample** to a conjecture is an example for which the conjecture is incorrect.

3 EXAMPLE Finding a Counterexample

Find a counterexample for each conjecture.

a. The square of any number is greater than the original number.
The number 1 is a counterexample because $1^2 \ngtr 1$.

b. You can connect any three points to form a triangle.
If the three points lie on a line, you cannot form a triangle.

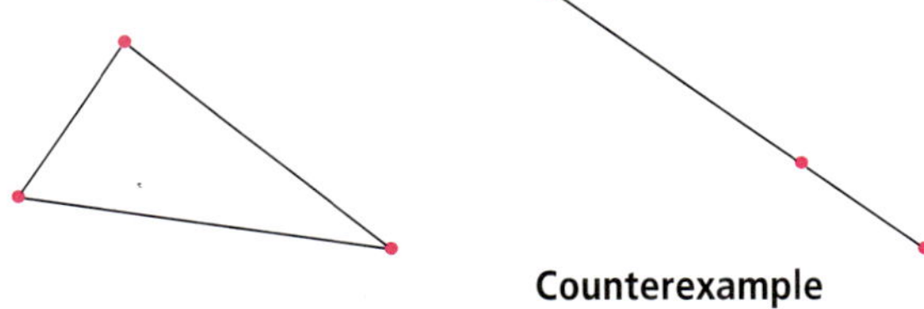

c. Any number and its absolute value are opposites.
The conjecture is true for negative numbers, but not positive numbers.
8 is a counterexample because 8 and $|8|$ are not opposites.

CA Standards Check 3 Alana makes a conjecture about slicing pizza. She says that if you use only straight cuts, the number of pieces will be twice the number of cuts.

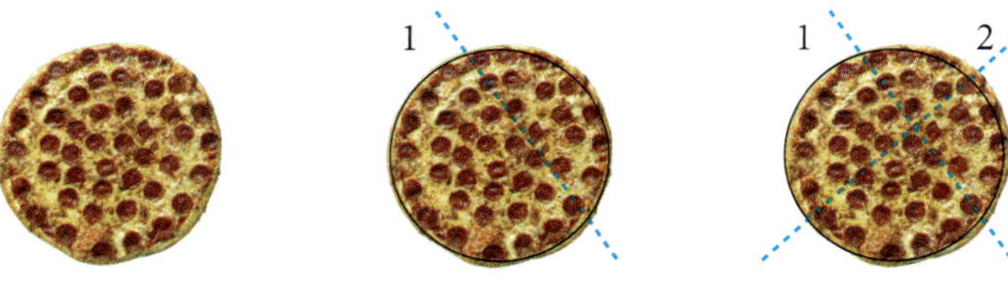

Draw a counterexample that shows you can make 7 pieces using 3 cuts.

4 EXAMPLE Real-World Connection

Business Sales A skateboard shop finds that over a period of five consecutive months, sales of small-wheeled skateboards decreased.

Use inductive reasoning. Make a conjecture about the number of small-wheeled skateboards the shop will sell in June.

The graph shows that sales of small-wheeled skateboards is decreasing by about 3 skateboards each month. By inductive reasoning you might conclude that the shop will sell 42 skateboards in June.

CA Standards Check 4

a. Make a conjecture about the number of small-wheeled skateboards the shop will sell in July.

b. **Critical Thinking** How confident would you be in using the graph to make a conjecture about sales in December? Explain.

EXERCISES

For more exercises, see *Extra Skill, Word Problem, and Proof Practice.*

Standards Practice GEOM 1.0, 3.0

A Practice by Example

Example 1 (page 4)

GO for Help

Find a pattern for each sequence. Use the pattern to show the next two terms.

1. 5, 10, 20, 40, . . .

2. 3, 33, 333, 3333, . . .

3. 1, −1, 2, −2, 3, . . .

4. $1, \frac{1}{2}, \frac{1}{4}, \frac{1}{8}, \ldots$

5. 15, 12, 9, 6, . . .

6. 81, 27, 9, 3, . . .

7. O, T, T, F, F, S, S, E, . . .

8. J, F, M, A, M, . . .

9. 1, 2, 6, 24, 120, . . .

10. 2, 4, 8, 16, 32, . . .

11. $1, \frac{1}{4}, \frac{1}{9}, \frac{1}{16}, \frac{1}{25}, \ldots$

12. $1, \frac{1}{2}, \frac{1}{3}, \frac{1}{4}, \ldots$

13. George, John, Thomas, James, . . .

14. Martha, Abigail, Martha, Dolley, . . .

15. George, Thomas, Abe, Alexander, . . .

16. Aquarius, Pisces, Aries, Taurus, . . .

Draw the next figure in each sequence.

17.   

18. 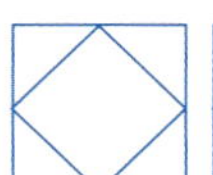 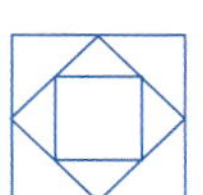

Example 2 (page 5)

Use the table and inductive reasoning. Make a conjecture about each value.

19. the sum of the first 6 positive even numbers

20. the sum of the first 30 positive even numbers

21. the sum of the first 100 positive even numbers

2	= 2 = 1 · 2
2 + 4	= 6 = 2 · 3
2 + 4 + 6	= 12 = 3 · 4
2 + 4 + 6 + 8	= 20 = 4 · 5
2 + 4 + 6 + 8 + 10	= 30 = 5 · 6

22. Use the pattern in Example 2 to make a conjecture about the sum of the first 100 odd numbers.

Predict the next term in each sequence. Use your calculator to verify your answer.

23.
12345679 × 9 = 111111111
12345679 × 18 = 222222222
12345679 × 27 = 333333333
12345679 × 36 = 444444444
12345679 × 45 = ■

24.
1 × 1 = 1
11 × 11 = 121
111 × 111 = 12321
1111 × 1111 = 1234321
11111 × 11111 = ■

Example 3 (page 5)

Find one counterexample to show that each conjecture is false.

25. The sum of two numbers is greater than either number.

26. The product of two positive numbers is greater than either number.

27. The difference of two integers is less than either integer.

28. The quotient of two proper fractions is a proper fraction.

Example 4 (page 6)

29. **Weather** The speed with which a cricket chirps is affected by the temperature. If you hear 20 cricket chirps in 14 seconds, what is the temperature?

Chirps per 14 Seconds

5 chirps	45°F
10 chirps	55°F
15 chirps	65°F

30. **Physical Fitness** Dino works out regularly. When he first started exercising, he could do 10 push-ups. After the first month he could do 14 push-ups. After the second month he could do 19, and after the third month he could do 25. Predict the number of push-ups Dino will be able to do after the fifth month of working out. How confident are you of your prediction? Explain.

B Apply Your Skills

Find a pattern for each sequence. Use the pattern to show the next two terms.

31. 1, 3, 7, 13, 21, . . .

32. 1, 2, 5, 6, 9, . . .

33. 0.1, 0.01, 0.001, . . .

34. 2, 6, 7, 21, 22, 66, 67, . . .

35. 1, 3, 7, 15, 31, . . .

36. $0, \frac{1}{2}, \frac{3}{4}, \frac{7}{8}, \frac{15}{16}, \ldots$

37. M, V, E, M, . . .

38. AL, AK, AZ, AR, . . .

39. H, He, Li, Be, . . .

40. **Writing** Choose two of the sequences in Exercises 31–36 and describe the patterns.

41. Draw two parallel lines on your paper. Locate four points on the paper, each an equal distance from both lines. Describe the figure you get if you continue to locate points, each an equal distance from both lines.

Real-World Connection

Points along the yellow line are equal distances from both sides of the bike trail (Exercise 41).

Draw the next figure in each sequence.

42.

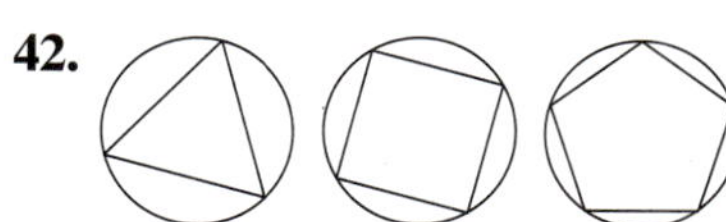

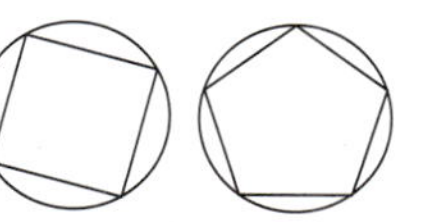

43.

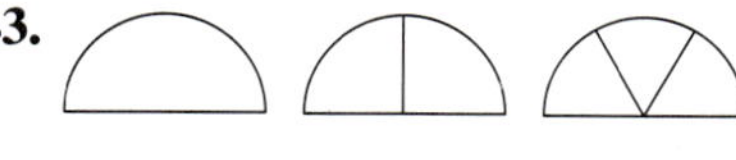

44.

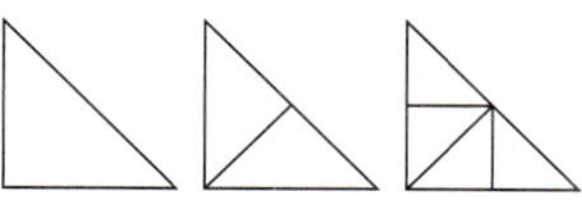

45.

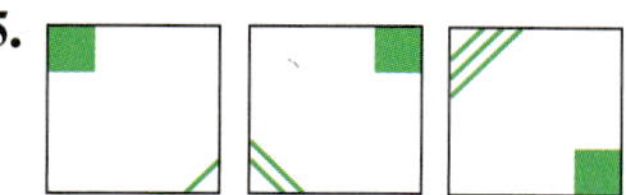

46. **Multiple Choice** Find the perimeter when 100 triangles are put together in the pattern shown. Assume that all triangle sides are 1 cm long.

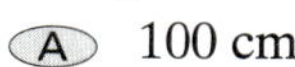

Ⓐ 100 cm Ⓑ 102 cm Ⓒ 202 cm Ⓓ 300 cm

47. **Math in the Media** Read this excerpt from a news article.

> **Top female runners** have been improving about twice as quickly as the fastest men, a new study says. If this pattern continues, women may soon outrun men in competition!
>
> The study is based on world records collected at 10-year intervals, starting in 1905 for men and in the 1920s for women. If the trend continues, the top female and male runners in races ranging from 200 m to 1500 m might attain the same speeds sometime between 2015 and 2055.
>
> Women's marathon records date from 1955 but their rapid fall suggests that the women's record will equal that of men even more quickly.

a. What conclusion was reached in the study?
b. How was inductive reasoning used to reach the conclusion?
c. Explain why the conclusion that women may soon be outrunning men may be incorrect. For which race is the conclusion most suspect? For what reason?

48. **Communications** The table shows the number of commercial radio stations in the United States for a 50-year period.
a. Make a line graph of the data.
b. Use the graph and inductive reasoning to make a conjecture about the number of radio stations in the United States in the year 2010.
c. How confident are you about your conjecture? Explain.

Number of Radio Stations

Year	Stations
1950	2,835
1960	4,224
1970	6,519
1980	7,871
1990	9,379
2000	10,577

SOURCE: Federal Communications Commission

49. **Open-Ended** Write two different number-pattern sequences that begin with the same two numbers.

50. **Error Analysis** For each of the past four years, Paulo has grown 2 in. every year. He is now 16 years old and is 5 ft 10 in. tall. He figures that when he is 22 years old he will be 6 ft 10 in. tall. What would you tell Paulo about his conjecture?

For Exercise 51, you may want to review "Coordinates of a point" in the Glossary.

51. **Coordinate Geometry** You are given x- and y-coordinates for 14 points.
$A(1, 5)$ $B(2, 2)$ $C(2, 8)$ $D(3, 1)$ $E(3, 9)$ $F(6, 0)$ $G(6, 10)$
$H(7, -1)$ $I(7, 11)$ $J(9, 1)$ $K(9, 9)$ $L(10, 2)$ $M(10, 8)$ $N(11, 5)$
a. Graph each point.
b. Most of the points fit a pattern. Which points do not?
c. Describe the figure that fits the pattern.

52. **History** Leonardo of Pisa (about 1175–1258), also known as Fibonacci (fee buh NAH chee), was born in Italy and educated in North Africa. He was one of the first Europeans known to use modern numerals instead of Roman numerals. The special sequence 1, 1, 2, 3, 5, 8, 13, . . . is known as the Fibonacci sequence. Find the next three terms of this sequence.

Visit: PHSchool.com
Web Code: bce-0101

53. **Time Measurement** Leap years have 366 days.
a. The years 1984, 1988, 1992, 1996, and 2000 are consecutive leap years. Look for a pattern in their dates. Then, make a conjecture about leap years.
b. Of the years 2010, 2020, 2100, and 2400, which do you think will be leap years?
c. **Research** Find out whether your conjecture for part (a) and your answer for part (b) are correct. How are leap years determined?

Challenge

54. **History** When he was in the third grade, German mathematician Karl Gauss (1777–1855) took ten seconds to sum the integers from 1 to 100. Now it's your turn. Find a fast way to sum the integers from 1 to 100; from 1 to n. (*Hint:* Use patterns.)

55. a. **Algebra** Write the first six terms of the sequence that starts with 1, and for which the difference between consecutive terms is first 2, and then 3, 4, 5, and 6.

b. Evaluate $\frac{n^2 + n}{2}$ for $n = 1, 2, 3, 4, 5,$ and 6. Compare the sequence you get with your answer for part (a).

c. Examine the diagram at the right and explain how it illustrates a value of $\frac{n^2 + n}{2}$.

d. Draw a similar diagram to represent $\frac{n^2 + n}{2}$ for $n = 5$.

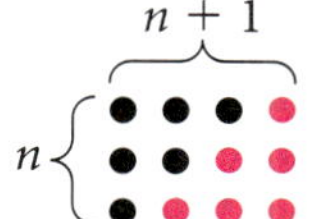

Multiple Choice Practice

For California Standards and CAHSEE Tutorials, visit PHSchool.com. Web Code: bcq-9045

GEOM 1.0

56. What is the next term in the sequence? 5, 13, 20, 26, . . .

Ⓐ 44 Ⓑ 33 Ⓒ 32 Ⓓ 31

CAHSEE 7MR 1.2

57. The numbers in the ones place for the first eight powers of 7 are shown in the table below. For example, in $7^4 = 2401$, 1 is in the ones place.

Number in the Ones Place of Powers of 7

Powers of 7	7^1	7^2	7^3	7^4	7^5	7^6	7^7	7^8
Number in Ones Place	7	9	3	1	7	9	3	1

Which number is in the ones place of 7^{34}?

Ⓐ 1 Ⓑ 3 Ⓒ 7 Ⓓ 9

CAHSEE 6SDAP 1.1

58. At a rummage sale, scarves sold at these prices:

$0.10, $1.25, $1.50, $1.50, $1.75, $2.00, $2.00, $2.00, $2.00, $2.00.

Which measure of central tendency will give the highest price?

Ⓐ Mean Ⓑ Median Ⓒ Mode Ⓓ Range

CAHSEE Alg1 7.0

59. Find the equation that best describes the relationship between x and y in the table at the right.

x	−5	0	15	5
y	−13	7	67	27

Ⓐ $f(x) = \frac{1}{4}x - 7$

Ⓑ $f(x) = 4x + 7$

Ⓒ $f(x) = 7x - 4$

Ⓓ $f(x) = \frac{1}{4}x + 7$

Mixed Review

Skills Handbook

60. Measure the sides DE and EF to the nearest millimeter.

61. Measure each angle of DEF to the nearest degree.

62. Draw a triangle that has sides of length 6 cm and 5 cm with a 90° angle between those two sides.

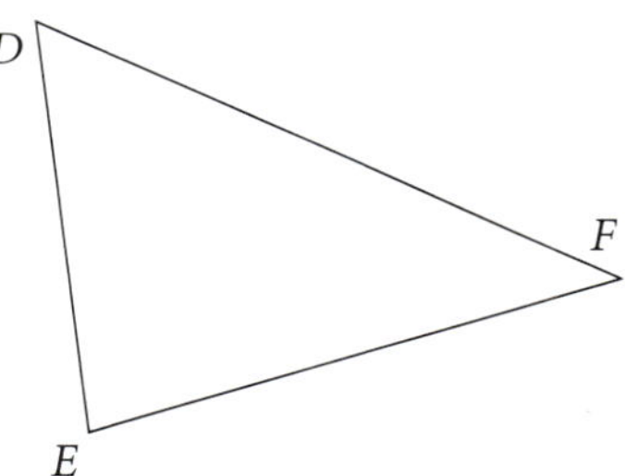

GO Online Lesson Quiz Visit: PHSchool.com, Web Code: bca-0101

1-2 Drawings, Nets, and Other Models

California Content Standards

Prepares for GEOM 8.0 Students know, derive, and solve problems involving perimeter, circumference, area, and volume of common geometric figures.

What You'll Learn

- To make isometric and orthographic drawings
- To draw nets for three-dimensional figures

. . . And Why

To make a foundation drawing, as in Example 3

Check Skills You'll Need

GO for Help Lesson 1-1

Draw the next figure in each sequence.

1.   

2.   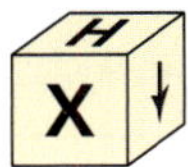

New Vocabulary

- isometric drawing
- orthographic drawing
- foundation drawing
- net

1 Drawing Isometric and Orthographic Views

Vocabulary Tip

In Greek, *isos* means "equal" and *metron* means "measure." In an isometric drawing, all 3-D measurements are scaled equally.

You will study both two-dimensional and three-dimensional figures in geometry. A drawing on a piece of paper is a two-dimensional object. It has length and width. Your textbook is a three-dimensional object. It has length, width, and height. Representing a three-dimensional object on a two-dimensional surface requires special techniques.

You can make an **isometric drawing** on isometric dot paper to show three sides of a figure from a corner view. The simple drawing of a refrigerator at the right is an isometric drawing.

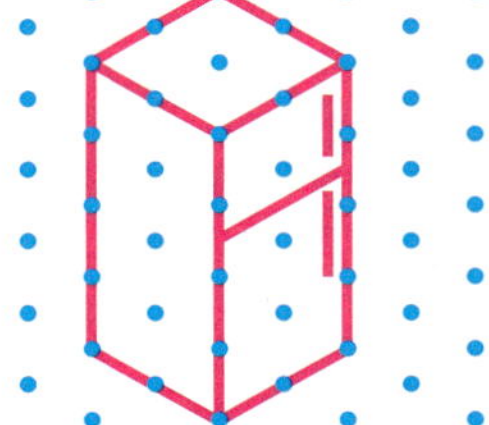

1 EXAMPLE Isometric Drawing

Make an isometric drawing of the cube structure at the left.

Isometric drawing:

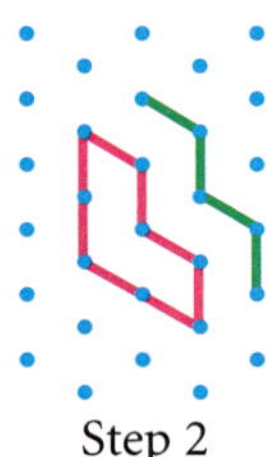

Step 1

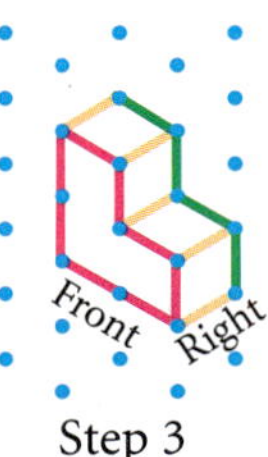

Step 2

Step 3

CA Standards Check

1. On isometric dot paper, make an isometric drawing of the cube structure.

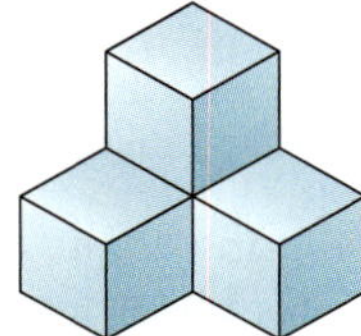

An **orthographic drawing** is another way to show a three-dimensional figure. It shows a top view, front view, and right-side view.

2 EXAMPLE Orthographic Drawing

Make an orthographic drawing from the isometric drawing at the left.

Isometric drawing:

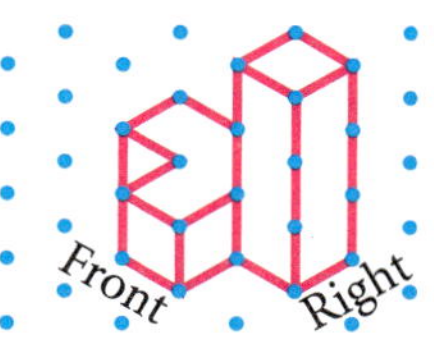

Orthographic drawing:

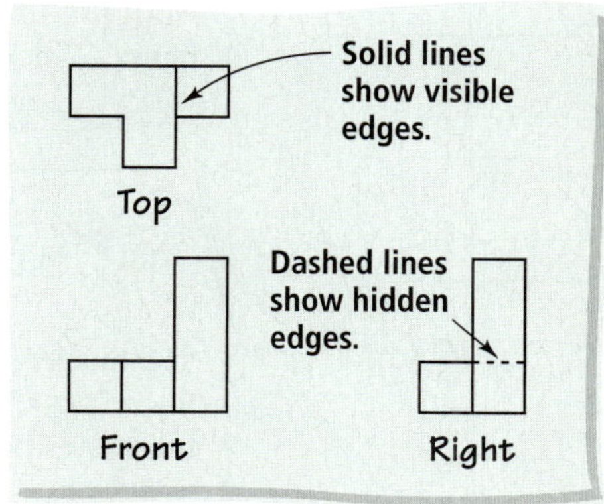

CA Standards Check

2 Make an orthographic drawing from this isometric drawing.

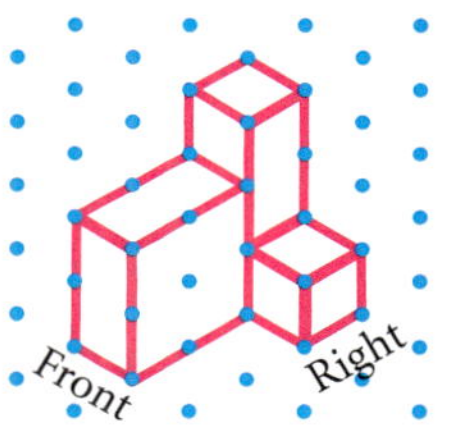

Real-World Connection

The foundation drawing shows four heights in the nine sections of the Sears Tower in Chicago, Illinois.

A **foundation drawing** shows the base of a structure and the height of each part. A foundation drawing of the Sears Tower is shown at the right.

49	89	65
109	109	89
65	89	49

The Sears Tower is made up of nine sections. The numbers tell how many stories tall each section is.

3 EXAMPLE Foundation Drawing

Make a foundation drawing for the isometric drawing at the left.

Isometric drawing:

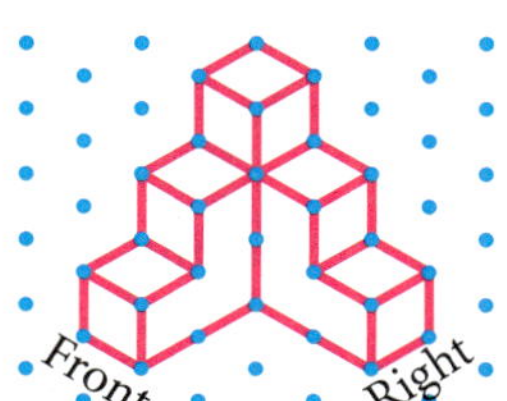

Foundation drawing:

3	2	1
2		
1		

Front / Right

CA Standards Check

3 **a.** How many cubes would you use to make the structure in Example 3?

b. **Critical Thinking** Which drawing did you use to answer part (a), the foundation drawing or the isometric drawing? Explain.

2 Nets for Three-Dimensional Figures

A **net** is a two-dimensional pattern that you can fold to form a three-dimensional figure. A net shows all of the surfaces of a figure in one view.

4 EXAMPLE Identifying Solids From Nets

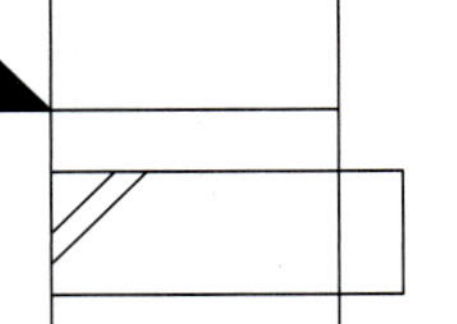

Multiple Choice The net at the left shows all the surfaces of a three-dimensional figure. Which figure can you fold from the net?

Ⓐ 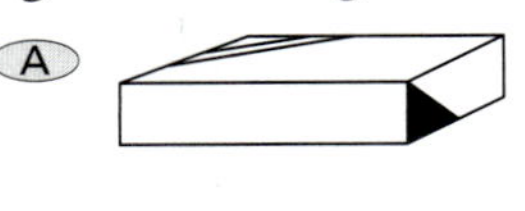

Ⓒ 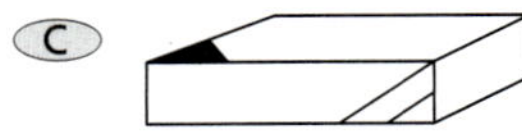

Ⓑ 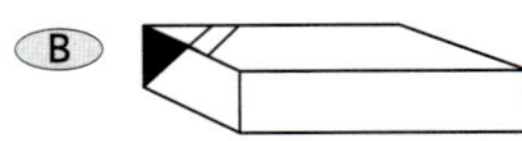

Ⓓ 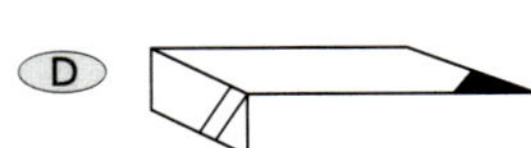

The black triangle is shown on the largest side of the figure instead of the smallest in choices C and D. Those choices cannot be correct. The black triangle will be at the same end as the two diagonal lines when the net is folded into a box. Choice B is correct.

CA Standards Check 4 The net at the right folds into the cube shown beside it. Draw the cube and show which letters will be on its front and top.

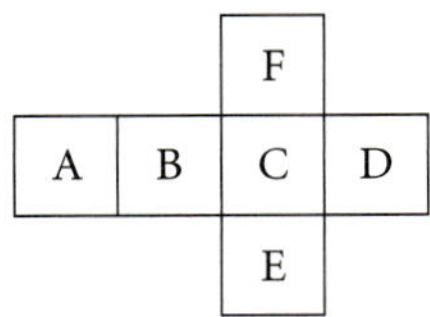

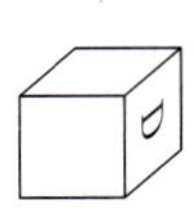

Test-Taking Tip
The number of surfaces on a solid must match the number of regions in its net.

Package designers can use nets to help design containers.

5 EXAMPLE Drawing a Net

Packaging Draw a net for the graham cracker box. Label the net with its dimensions.

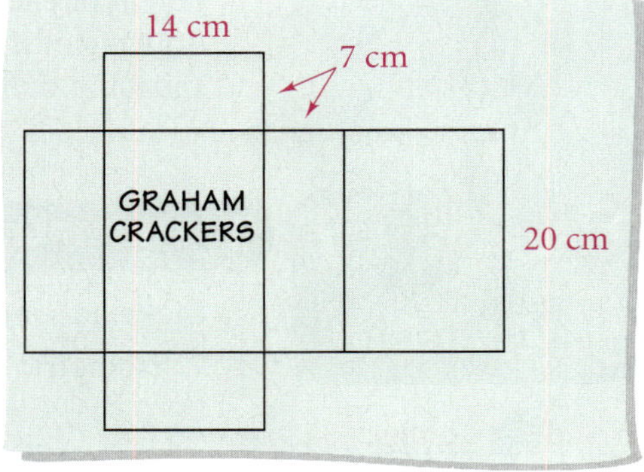

CA Standards Check 5 Draw a net for the solid shown. Label the net with its dimensions.

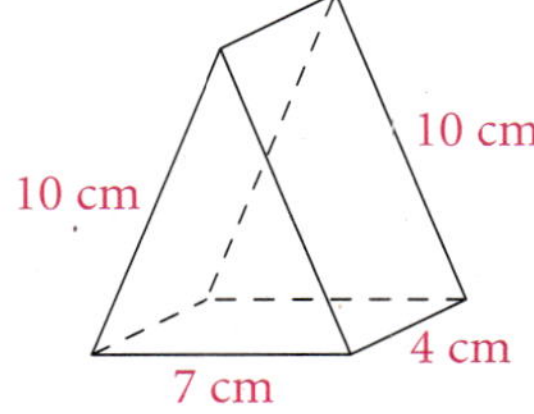

For more exercises, see *Extra Skill, Word Problem, and Proof Practice.*

A Practice by Example

On isometric dot paper, make an isometric drawing of each cube structure.

Example 1 (page 10)

for Help

1. 
2. 
3. 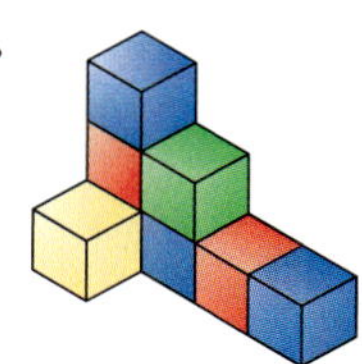

Examples 2, 3 (page 11)

For each figure, make (a) an orthographic drawing, and (b) a foundation drawing.

4. 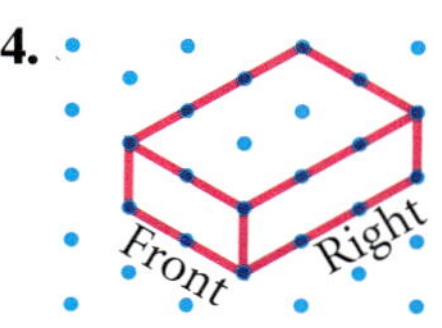

5. 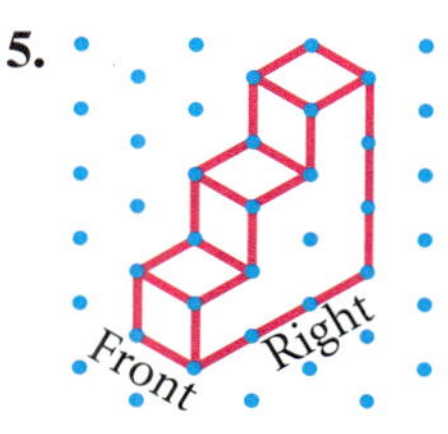

6. 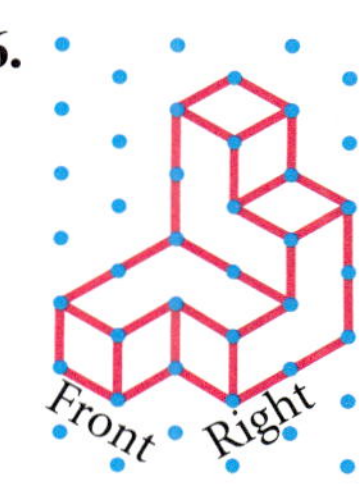

How many cubes would you use to make each of the following?

7. the structure in Exercise 4
8. the structure in Exercise 5
9. the structure in Exercise 6
10. a model of the Sears Tower on page 11

Example 4 (page 12)

Match each three-dimensional figure with its net.

11. 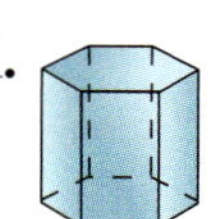
12. 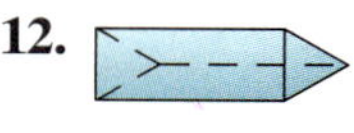
13. 

A. 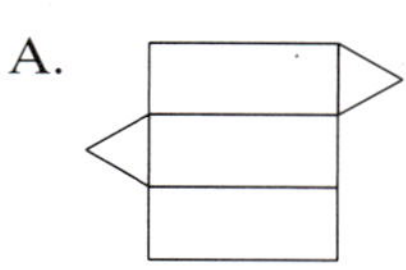

B. 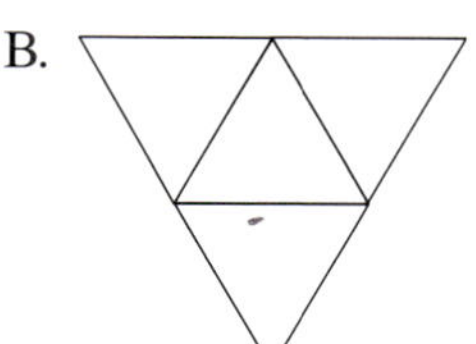

C. 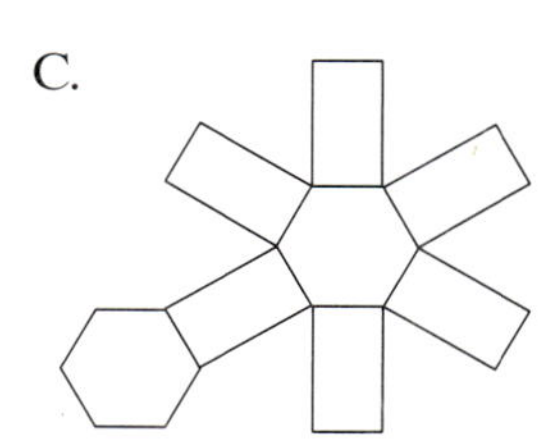

Example 5 (page 12)

Draw a net for each figure. Label the net with its dimensions.

14. 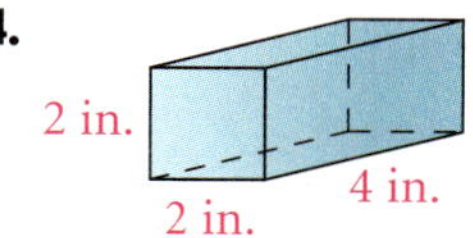

15. 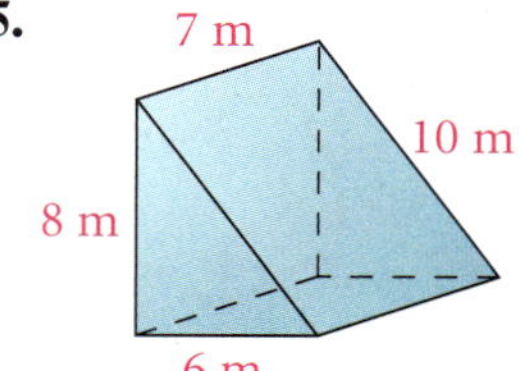

16. 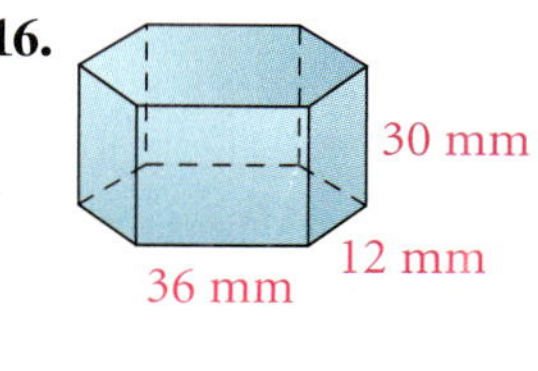

B Apply Your Skills

17. a. **Open-Ended** Make an isometric drawing of a structure that can be built using 8 cubes.
b. Make an orthographic drawing of this structure.
c. Make a foundation drawing for this structure.

For each foundation drawing, make (a) an isometric drawing on dot paper, and (b) an orthographic drawing.

18.

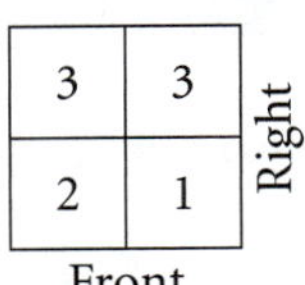

19.

1 2 3
2 1
Right
Front

20.

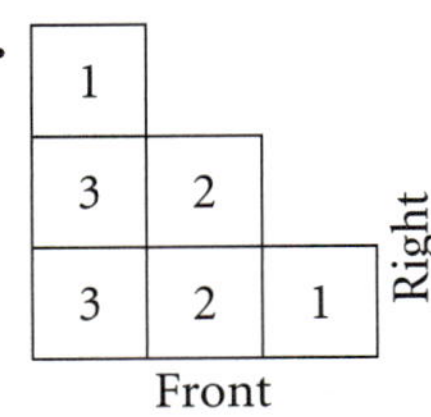

Read the comic strip and complete Exercises 21 and 22.

SHOE by Jeff MacNelly

21. What type of drawing that you've studied in this lesson is a "bird's-eye view"?

22. **Writing** Photographs of the Washington Monument are typically not taken from a bird's-eye view. Describe a situation in which you would want a photo showing a bird's-eye view.

Visualization **Think about how each net can be folded to form a cube. What is the color of the face that will be opposite the red face?**

23.

24.

25.

26.

27. There are eleven different nets for a cube. Four of them are shown above.

a. Draw as many of the other seven as you can. (*Hint:* Two nets are the same if you can rotate or flip one to match the other.)

b. **Writing** If you were going to make 100 cubes for a mobile, which of the eleven nets would you use? Explain why.

Make an orthographic drawing for each isometric drawing.

28.

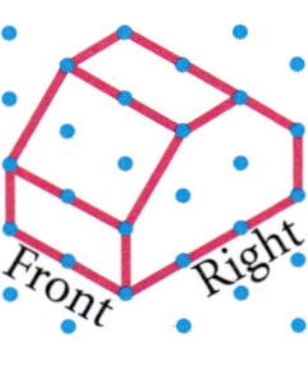

29.

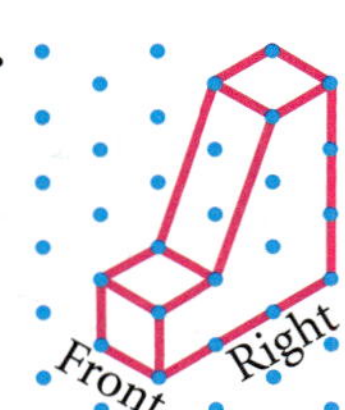

30.

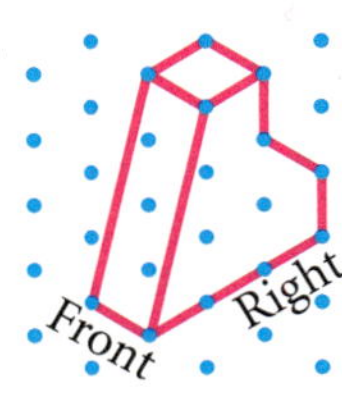

Homework Video Tutor

Visit: PHSchool.com
Web Code: bce-0102

31. Draw a net for a cylinder. (*Hint:* The net needs to show three regions: two circles and a rectangle.)

32. There are eight different nets for a pyramid with a square base. Draw as many of them as you can.

33. **Visualization** Use the orthographic drawing at the right.

a. Make an isometric drawing of the structure.

b. Make an isometric drawing of the structure from part (a) after it has been turned on its base 90° counterclockwise.

c. Make an orthographic drawing of the stucture from part (b).

d. Turn the structure from part (a) 180°. Repeat parts (b) and (c).

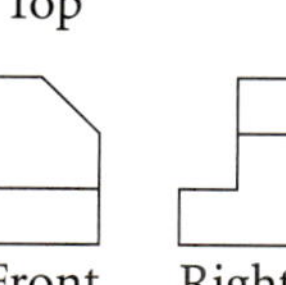

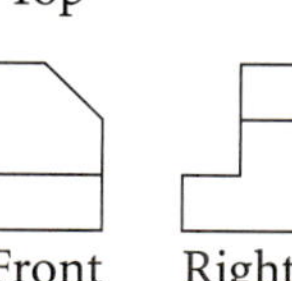

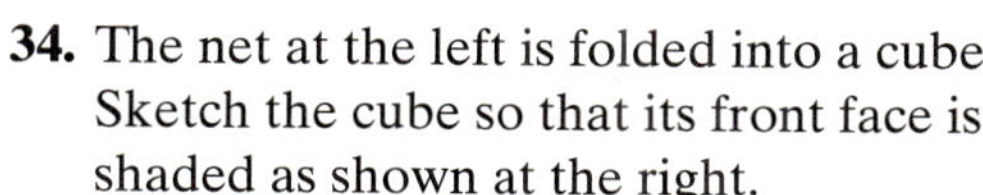

34. The net at the left is folded into a cube. Sketch the cube so that its front face is shaded as shown at the right.

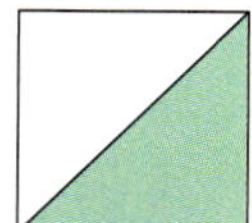

Multiple Choice Practice

For California Standards and CAHSEE Tutorials, visit PHSchool.com. Web Code: bcq-9045

GEOM 3.0

35. Which number is a counterexample to the statement below?
All prime numbers are odd.

Ⓐ 0 Ⓑ 2 Ⓒ 34 Ⓓ 86

CAHSEE Alg1 9.0

36. Monica left home at 9:00 A.M. and drove an average of 45 miles per hour. Her brother Trey left on the same route at 11:00 A.M. and drove an average of 60 miles per hour. The graph below represents this situation. At what time did Trey pass Monica?

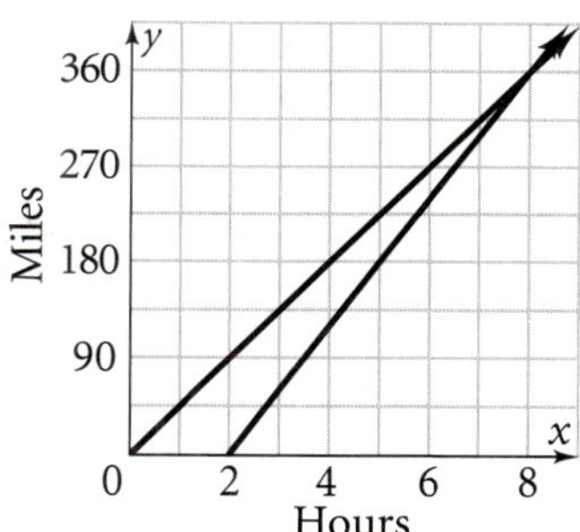

Ⓐ 4:00 P.M. Ⓑ 5:00 P.M. Ⓒ 7:00 P.M. Ⓓ 8:00 P.M.

Mixed Review

Lesson 1-1

37. **Reasoning** What is the last digit of 3^{45}? To answer, make a table, look for a pattern, and use inductive reasoning. Explain the pattern.

Find a pattern for each sequence. Use the pattern to show the next two terms.

38. A, C, E, G, . . .

39. 2, 6, 12, 20, 30, . . .

40. 4, 16, 64, 256, . . .

41. 100, 95, 85, 70, 50, . . .

Skills Handbook

Algebra **Evaluate each expression for the given values.**

42. $a^2 + b^2$ for $a = 3$ and $b = -5$

43. $\frac{1}{2}bh$ for $b = 8$ and $h = 11$

1-3 Points, Lines, and Planes

California Content Standards

GEOM 1.0 Students demonstrate understanding by identifying and giving examples of undefined terms and axioms. ***Introduce, Develop***

What You'll Learn

- To understand basic terms of geometry
- To understand basic postulates of geometry

. . . And Why

To explain why a photographer uses a tripod, as in Exercise 44

Check Skills You'll Need **GO for Help** Skills Handbook page 760

x^2 **Algebra** **Solve each system of equations.**

1. $y = x + 5$
$y = -x + 7$

2. $y = 2x - 4$
$y = 4x - 10$

3. $y = 2x$
$y = -x + 15$

4. Copy the diagram of the four points A, B, C, and D. Draw as many different lines as you can to connect pairs of points.

$A \bullet$ $\bullet B$ $C \bullet$ $\bullet D$

New Vocabulary • point • space • line • collinear points • plane • coplanar • postulate • axiom

1 Basic Terms of Geometry

Hands-On Activity: How Many Lines Can You Draw?

Many constellations are named for animals and mythological figures. It takes some imagination to join the points representing the stars to get a recognizable figure such as Leo the Lion. How many lines can you draw connecting the 10 points in Leo the Lion?

- Make a table and look for a pattern to help you find out.

1. Mark three points on a circle. Now connect the three points with as many (straight) lines as possible. How many lines can you draw?
2. Mark four points on another circle. How many lines can you draw to connect the four points?
3. Repeat this procedure for five points on a circle and then for six points. How many lines can you draw to connect the points?
4. Use inductive reasoning to tell how many lines you can draw to connect the ten points of the constellation Leo the Lion.

In geometry, some words such as *point*, *line*, and *plane* are undefined. It is important however, to have general descriptions of their meanings.

You can think of a **point** as a location. A point has no size. It is represented by a small dot and is named by a capital letter. A geometric figure is a set of points. **Space** is defined as the set of all points.

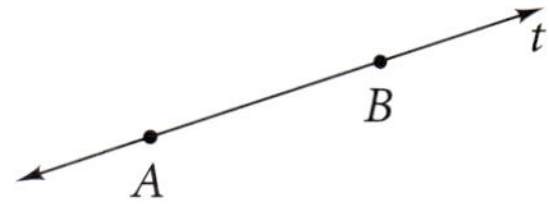

Vocabulary Tip

$\overleftrightarrow{AB}$ ("line AB") and $\overleftrightarrow{BA}$ ("line BA") name the same line.

You can think of a **line** as a series of points that extends in two opposite directions without end. You can name a line by any two points on the line, such as $\overleftrightarrow{AB}$ (read "line AB"). Another way to name a line is with a single lowercase letter, such as line t (see above). Points that lie on the same line are **collinear points.**

1 EXAMPLE Identifying Collinear Points

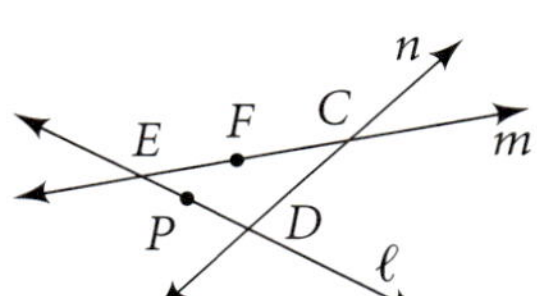

a. Are points E, F, and C collinear? If so, name the line on which they lie.

Points E, F, and C are collinear. They lie on line m.

b. Are points E, F, and D collinear? If so, name the line on which they lie.

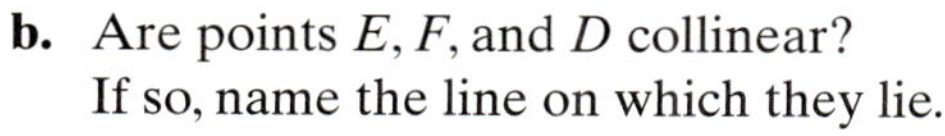

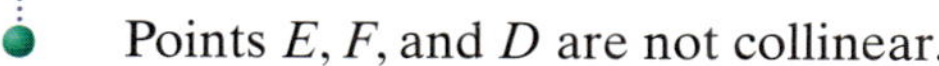

Points E, F, and D are not collinear.

1 **a.** Are points F, P, and C collinear?
b. Name line m in three other ways.
c. **Critical Thinking** Why do you think arrowheads are used when drawing a line or naming a line such as $\overleftrightarrow{EF}$?

A **plane** is a flat surface that has no thickness. A plane contains many lines and extends without end in the directions of all its lines. You can name a plane by either a single capital letter or by at least three of its noncollinear points. Points and lines in the same plane are **coplanar.**

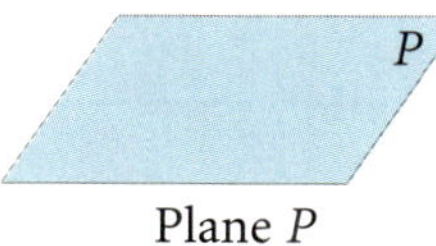

Plane P

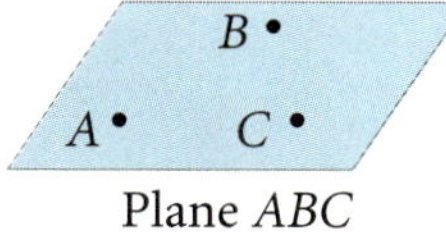

Plane ABC

2 EXAMPLE Naming a Plane

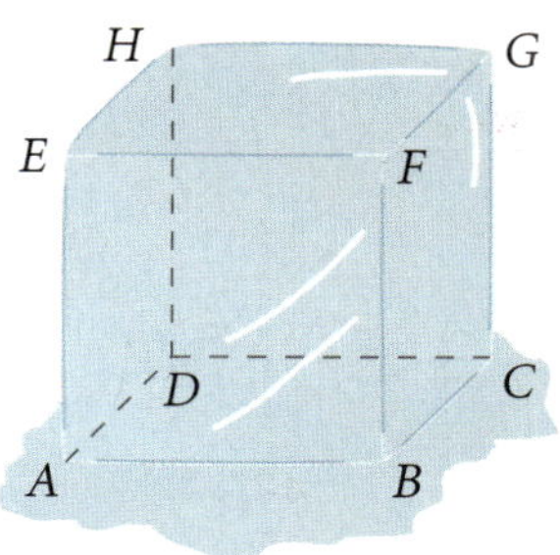

Each surface of the ice cube represents part of a plane. Name the plane represented by the front of the ice cube.

You can name the plane represented by the front of the ice cube using at least three noncollinear points in the plane. Some names are plane AEF, plane AEB, and plane $ABFE$.

2 List three different names for the plane represented by the top of the ice cube.

2 Basic Postulates of Geometry

A **postulate** or **axiom** is an accepted statement of fact.

You have used some of the following geometry postulates in algebra. For example, you used Postulate 1-1 when you graphed an equation such as $y = -2x + 8$. You plotted two points and then drew the line through those two points.

Key Concepts

Postulate 1-1

Through any two points there is exactly one line.

Line t is the only line that passes through points A and B.

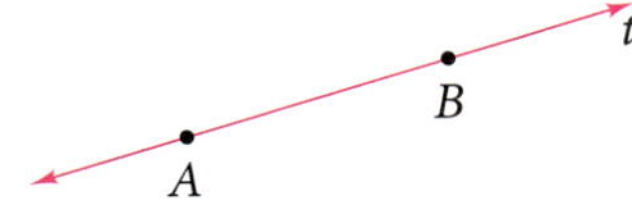

Vocabulary Tip

There is exactly one means "there is one and there is no more than one."

In algebra, one way to solve a system of two equations is to graph the two equations. As the graphs of

$$y = -2x + 8$$

$$y = 3x - 7$$

show, the two lines intersect at a single point, (3, 2). The solution to the system of equations is (3, 2).

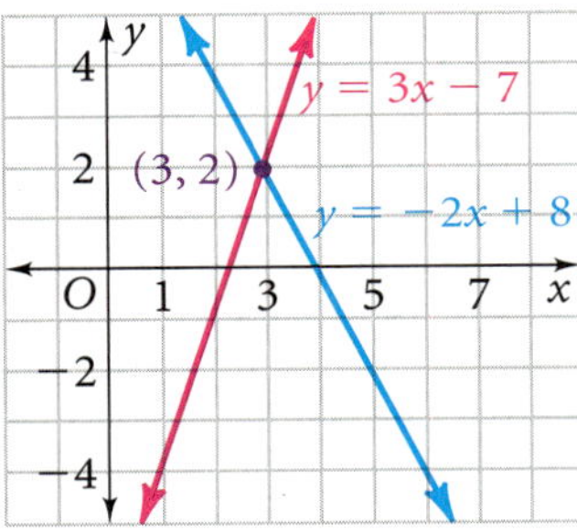

This illustrates Postulate 1-2.

Key Concepts

Postulate 1-2

If two lines intersect, then they intersect in exactly one point.

$\overleftrightarrow{AE}$ and $\overleftrightarrow{BD}$ intersect at C.

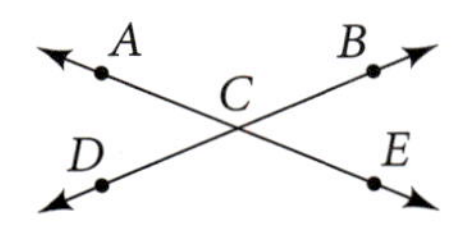

There is a similar postulate about the intersection of planes.

Key Concepts

Postulate 1-3

If two planes intersect, then they intersect in exactly one line.

Plane RST and plane STW intersect in $\overleftrightarrow{ST}$.

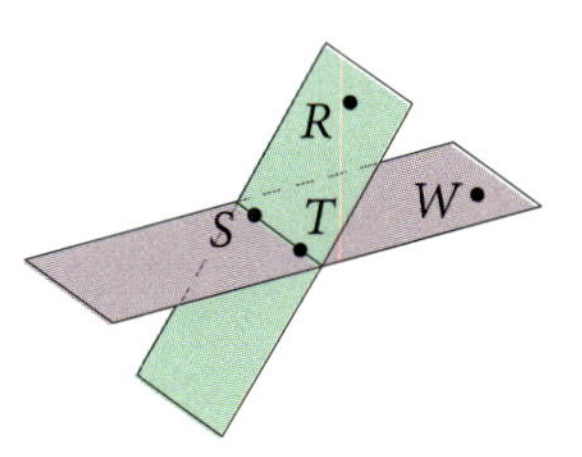

When you know two points in the intersection of two planes, Postulates 1-1 and 1-3 tell you that the line through those points is the line of intersection of the planes.

3 EXAMPLE Finding the Intersection of Two Planes

What is the intersection of plane $HGFE$ and plane $BCGF$?

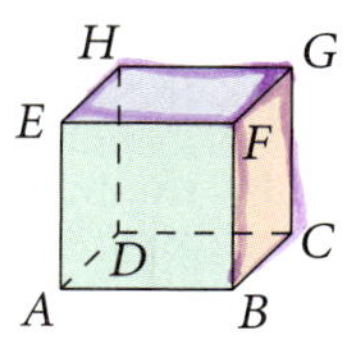

Plane $HGFE$ and plane $BCGF$ intersect in $\overleftrightarrow{GF}$.

3 Name two planes that intersect in $\overleftrightarrow{BF}$.

A three-legged stand will always be stable. As long as the feet of the stand don't lie in one line, the feet of the three legs will lie exactly in one plane.

This illustrates Postulate 1-4.

Key Concepts

Postulate 1-4

Through any three noncollinear points there is exactly one plane.

4 EXAMPLE Using Postulate 1-4

a. Shade the plane that contains A, B, and C.

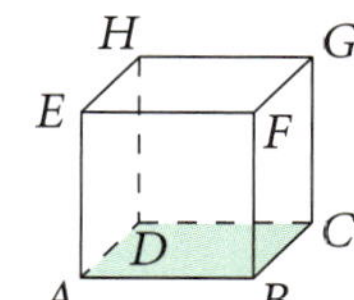

b. Shade the plane that contains E, H, and C.

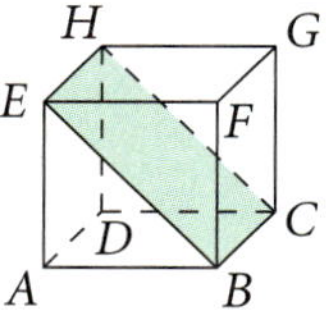

4 **a.** Name another point that is in the same plane as points A, B, and C.

b. Name another point that is coplanar with points E, H, and C.

EXERCISES

For more exercises, see *Extra Skill, Word Problem, and Proof Practice.*

Standards Practice GEOM 1.0

Practice by Example

Example 1 (page 17)

Are the three points collinear? If so, name the line on which they lie.

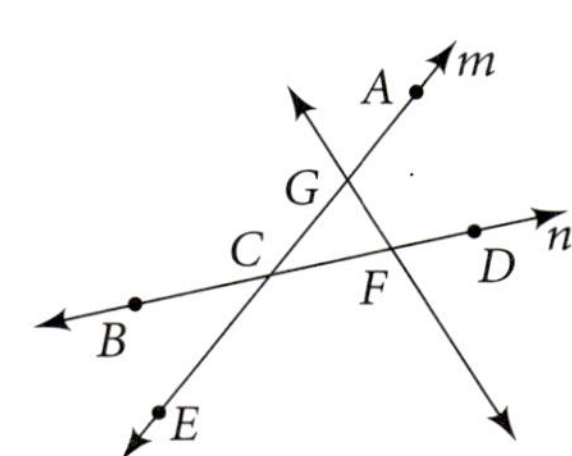

1. A, D, E
2. B, C, D
3. B, C, F
4. A, E, C
5. F, B, D
6. F, A, E
7. G, F, C
8. A, G, C
9. Name line m in three other ways.
10. Name line n in three other ways.

Example 2
(page 17)

Name the plane represented by each surface of the box.

11. the bottom
12. the top
13. the front
14. the back
15. the left side
16. the right side

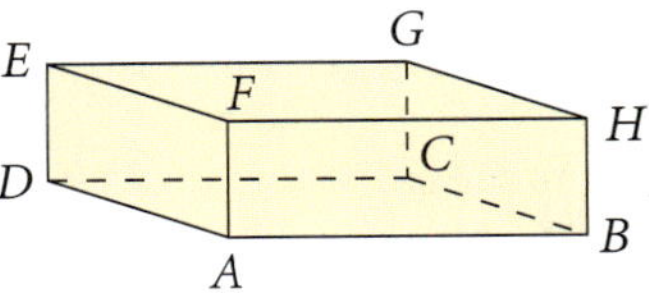

Example 3
(page 19)

Use the figure at the right for Exercises 17–37.
First, name the intersection of each pair of planes.

17. planes *QRS* and *RSW*
18. planes *UXV* and *WVS*
19. planes *XWV* and *UVR*
20. planes *TXW* and *TQU*

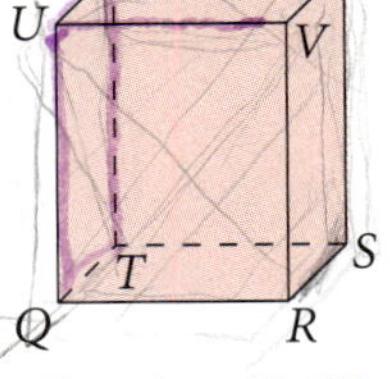

Exercises 17–37

Name two planes that intersect in the given line.

21. $\overleftrightarrow{QU}$
22. $\overleftrightarrow{TS}$
23. $\overleftrightarrow{XT}$
24. $\overleftrightarrow{VW}$

Example 4
(page 19)

Copy the figure. Shade the plane that contains the given points.

25. *R, V, W*
26. *U, V, W*
27. *U, X, S*
28. *T, U, X*
29. *T, V, R*

Name another point in each plane.

30. plane *RVW*
31. plane *UVW*
32. plane *UXS*
33. plane *TUX*
34. plane *TVR*

Is the given point coplanar with the other three points?

35. point *Q* with *V, W, S*
36. point *U* with *T, V, S*
37. point *W* with *X, V, R*

B Apply Your Skills

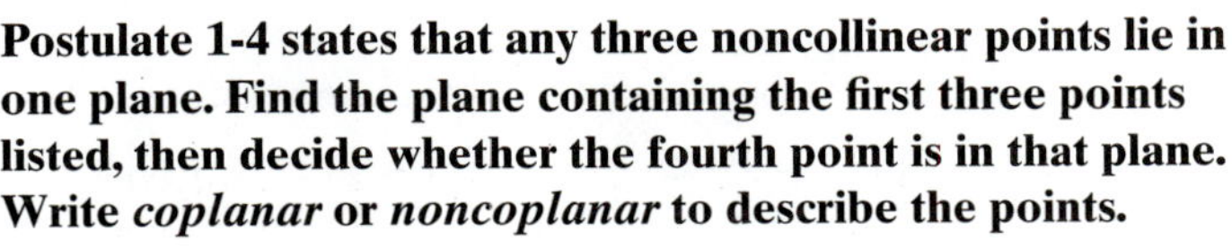

Postulate 1-4 states that any three noncollinear points lie in one plane. Find the plane containing the first three points listed, then decide whether the fourth point is in that plane. Write *coplanar* or *noncoplanar* to describe the points.

38. *Z, S, Y, C*
39. *S, U, V, Y*
40. *X, Y, Z, U*
41. *X, S, V, U*
42. *X, Z, S, V*
43. *S, V, C, Y*

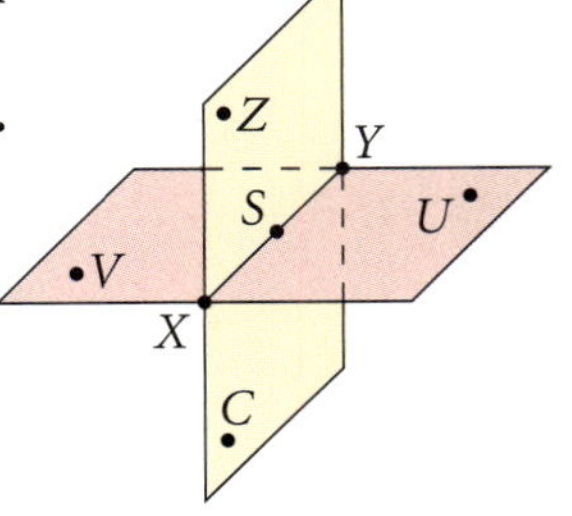

44. **Photography** Photographers and surveyors use a tripod, or three-legged stand, for their instruments. Use one of the postulates to explain why.

45. **Open-Ended** Draw a figure with points *B, C, D, E, F,* and *G* that shows $\overleftrightarrow{CD}$, $\overleftrightarrow{BG}$, and $\overleftrightarrow{EF}$, with one of the points on all three lines.

If possible, draw a figure to fit each description. Otherwise write *not possible.*

46. four points that are collinear
47. two points that are noncollinear
48. three points that are noncollinear
49. three points that are noncoplanar

Coordinate Geometry **Graph the points and state whether they are collinear.**

50. $(3, -3), (2, -3), (-3, 1)$
51. $(2, 2), (-2, -2), (3, 2)$
52. $(2, -2), (-2, -2), (3, -2)$
53. $(-3, 3), (-3, 2), (-3, -1)$

54. **Multiple Choice** Which three points are *not* collinear?
 A. $(0, 0), (0, 2), (0, 4)$
 C. $(0, 0), (0, 2), (3, 0)$
 B. $(0, 0), (3, 0), (5, 0)$
 D. $(2, -2), (2, 2), (2, 3)$

Real-World Connection

The photographer uses a tripod to help assure a clear picture.

Visit: PHSchool.com
Web Code: bce-0103

Use *always, sometimes,* or *never* to make a true statement.

55. Intersecting lines are __?__ coplanar.

56. Two planes __?__ intersect in exactly one point.

57. Three points are __?__ coplanar.

58. A plane containing two points of a line __?__ contains the entire line.

59. Four points are __?__ coplanar.

60. Two lines __?__ meet in more than one point.

61. How many planes contain each line and point?

a. $\overleftrightarrow{EF}$ and point G **b.** $\overleftrightarrow{PH}$ and point E

c. $\overleftrightarrow{FG}$ and point P **d.** $\overleftrightarrow{EP}$ and point G

e. **Make a Conjecture** What do you think is true of a line and a point not on the line?

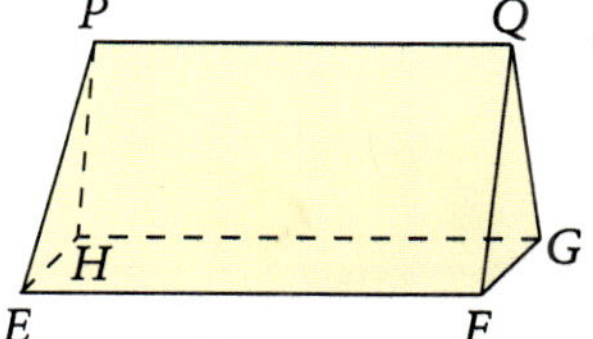

In Exercise 62 and 63, sketch a figure for the given information. Then name the postulate that your figure illustrates.

62. The noncollinear points A, B, and C are all contained in plane N.

63. Planes LNP and MVK intersect in $\overleftrightarrow{NM}$.

64. **Optical Illusions** The diagram (right) is an optical illusion. Which three points are collinear: A, B, and C or A, B, and D? Are you sure? Use a straightedge to check your answer.

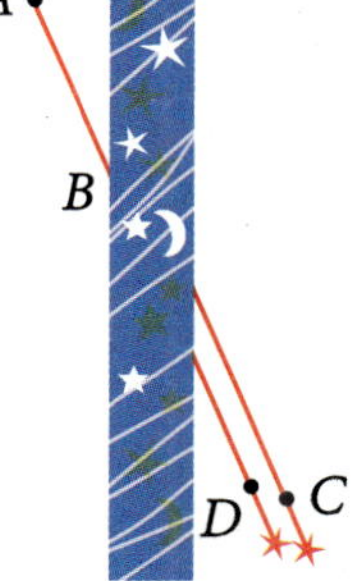

Writing **Use postulates to explain each situation.**

65. A land surveyor can always find a straight line from the point where she stands to any other point she can see.

66. A carpenter knows that a line can represent the intersection of two flat walls.

67. A furniture maker knows that a three-legged table is always steady, but a four-legged table will sometimes wobble.

Coordinate Geometry **Graph the points and state whether they are collinear.**

68. $(1, 1), (4, 4), (-3, -3)$ **69.** $(2, 4), (4, 6), (0, 2)$ **70.** $(0, 0), (-5, 1), (6, -2)$

71. $(0, 0), (8, 10), (4, 6)$ **72.** $(0, 0), (0, 3), (0, -10)$ **73.** $(-2, -6), (1, -2), (4, 1)$

74. How many planes contain the same three collinear points? Explain.

75. **Navigation** Rescue teams use Postulates 1-1 and 1-2 to determine the location of a distress signal. In the diagram, a ship at point A receives a signal from the northeast. A ship at point B receives the same signal from due west. Trace the diagram and find the location of the distress signal. Explain how the two postulates help locate the distress signal.

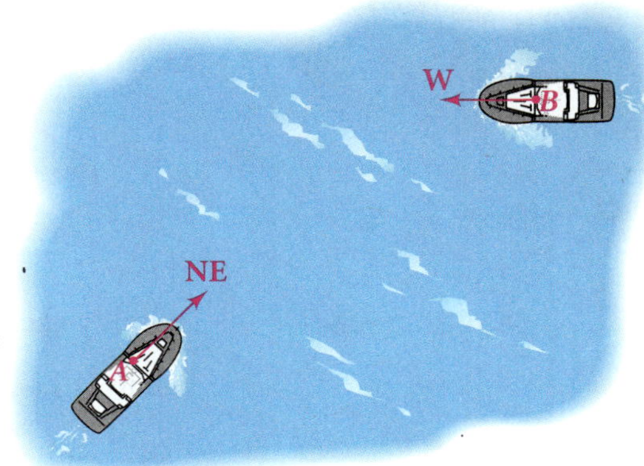

76. a. Open-Ended Suppose two points are in plane P. Explain why it makes sense that the line containing the points would be in the same plane.

b. Suppose two lines intersect. How many planes do you think contain both lines? You may use the diagram and your answer in part (a) to explain your answer.

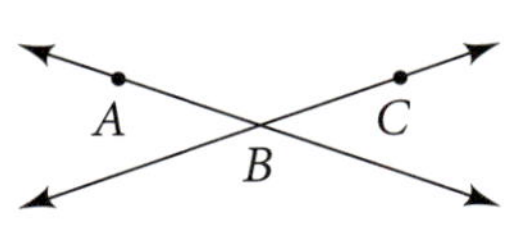

Probability **Points are picked at random from A, B, C, and D, which are arranged as shown. Find the probability that the indicated number of points meet the given condition.**

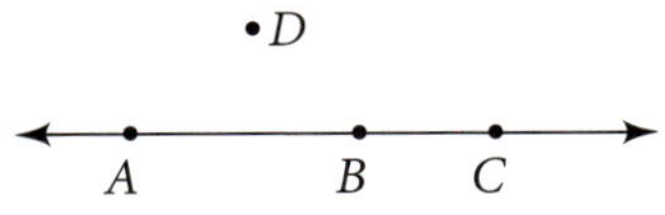

77. 2 points, collinear

78. 3 points, collinear

79. 3 points, coplanar

Multiple Choice Practice

For California Standards and CAHSEE Tutorials, visit PHSchool.com. Web Code: bcq-9045

GEOM 1.0

80. Points A, B, C, D and E are collinear. A is to the right of B, E is to the right of D, and B is to the left of C. Which of the following is NOT a possible arrangement of the points from left to right?

(A) D, B, A, E, C
(B) D, B, A, C, E
(C) B, D, E, C, A
(D) B, A, E, C, D

CAHSEE 6SDAP 3.3

81. The results of each toss of a number cube are shown below. Based on the results, what is the experimental probability of rolling a 5?

1 5 5 6 5 2 4 4 3 5

(A) 15%
(B) 25%
(C) 40%
(D) 50%

CAHSEE 7MG 2.2

82. Rick is painting the walls in a room that is 12 feet long and 10 feet wide, with a ceiling that is 8 feet from the floor. The doorway is 3 feet by 7 feet, and the window is 6 feet by 5 feet. If Rick is NOT painting the doorway or window, what is the approximate area, in square feet, that he is painting?

(A) 301
(B) 322
(C) 331
(D) 352

Mixed Review

Lesson 1-2

Make an orthographic drawing for each figure.

83.

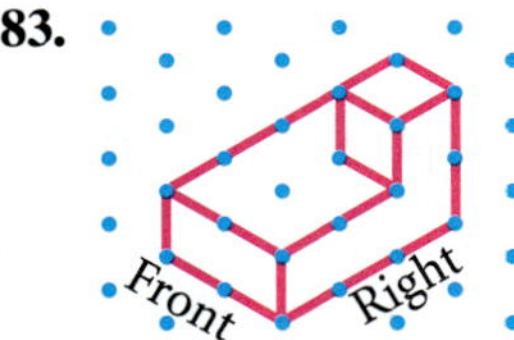

84.

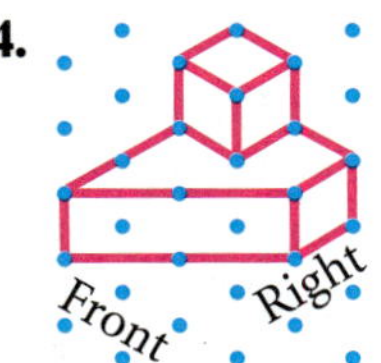

85.

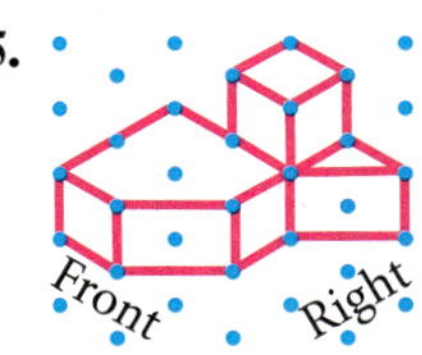

Skills Handbook

Simplify each ratio.

86. 30 to 12

87. $\frac{\pi r^2}{2\pi r}$

88. $\frac{5}{21} : \frac{5}{7}$

Segments, Rays, Parallel Lines and Planes

California Content Standards

GEOM 1.0 Students demonstrate understanding by identifying and giving examples of undefined terms and axioms. ***Master***

What You'll Learn

- To identify segments and rays
- To recognize parallel lines and planes

. . . And Why

To identify compass directions that can be represented by opposite rays, as in Exercise 36

Check Skills You'll Need

GO for Help Lesson 1-3

Judging by appearances, will the lines intersect?

1. 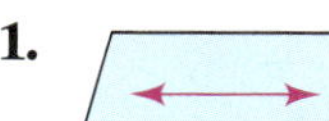

2. 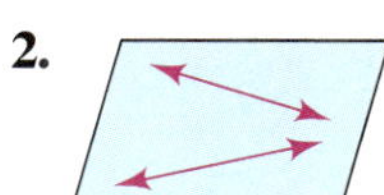

3. 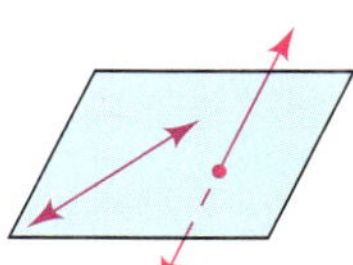

Name the plane represented by each surface of the box.

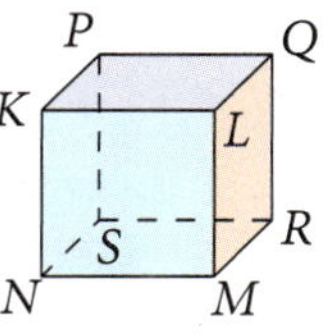

4. the bottom **5.** the top
6. the front **7.** the back
8. the left side **9.** the right side

New Vocabulary • segment • ray • opposite rays • parallel lines • skew lines • parallel planes

1 Identifying Segments and Rays

Real-World Connection

A sunbeam models a ray. The sun is its endpoint.

Many geometric figures, such as squares and angles, are formed by parts of lines called segments or rays. A **segment** is the part of a line consisting of two endpoints and all points between them.

Segment AB

Endpoint A — B Endpoint

$\overline{AB}$

A **ray** is the part of a line consisting of one endpoint and all the points of the line on one side of the endpoint.

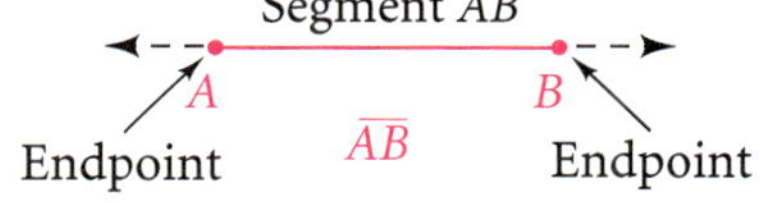

Opposite rays are two collinear rays with the same endpoint. Opposite rays always form a line.

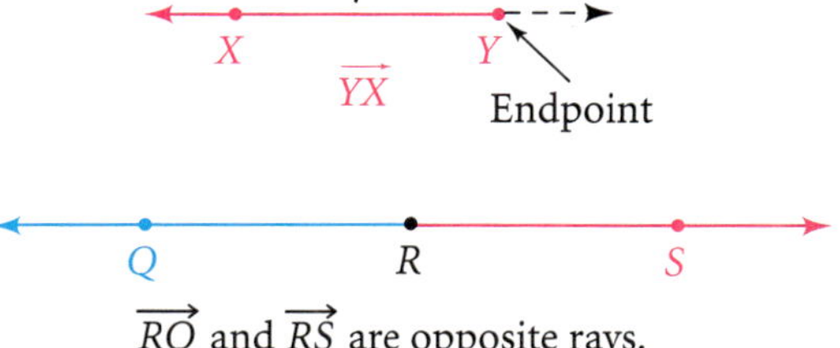

$\overrightarrow{RQ}$ and $\overrightarrow{RS}$ are opposite rays.

1 EXAMPLE Naming Segments and Rays

Name the segments and rays in the figure at the right.

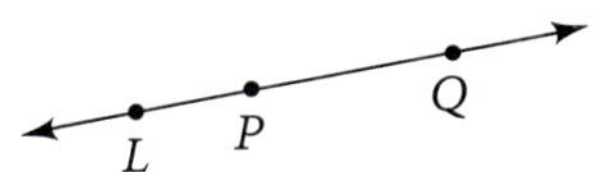

- The three segments are $\overline{LP}$, $\overline{PQ}$, and $\overline{LQ}$.
- The four rays are $\overrightarrow{LP}$ or $\overrightarrow{LQ}$, $\overrightarrow{PQ}$, $\overrightarrow{PL}$, and $\overrightarrow{QP}$ or $\overrightarrow{QL}$.

1 Critical Thinking $\overrightarrow{LP}$ and $\overrightarrow{PL}$ form a line. Are they opposite rays? Explain.

2 Recognizing Parallel Figures

Lines that do not intersect may or may not be coplanar.

Parallel lines are coplanar lines that do not intersect. **Skew lines** are noncoplanar; therefore, they are not parallel and do not intersect.

Vocabulary Tip

You read $\overleftrightarrow{AB} \parallel \overleftrightarrow{EF}$ as "line *AB* is parallel to line *EF*."

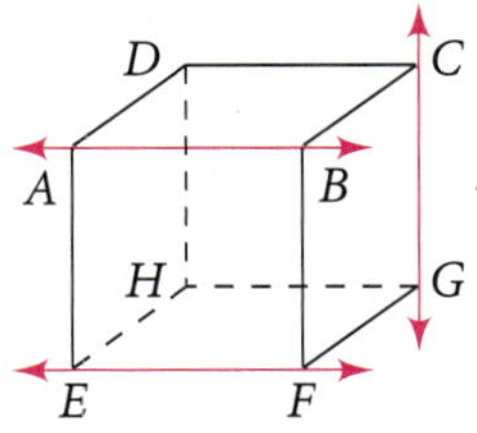

$\overleftrightarrow{AB} \parallel \overleftrightarrow{EF}$
$\overleftrightarrow{AB}$ and $\overleftrightarrow{CG}$ are skew.

Segments or rays are parallel if they lie in parallel lines. They are skew if they lie in skew lines. $\overline{AB}$ and $\overline{CG}$ are skew because $\overleftrightarrow{AB}$ and $\overleftrightarrow{CG}$ are skew.

2 EXAMPLE Identifying Parallel and Skew Segments

Video Tutor Help

Visit: PHSchool.com
Web Code: bce-0775

a. Name all labeled segments that are parallel to $\overline{DC}$.

$\overline{AB}$, $\overline{GH}$, and $\overline{JI}$ are parallel to $\overline{DC}$.

b. Name all labeled segments that are skew to $\overline{DC}$.

$\overline{NJ}$, $\overline{GJ}$, and $\overline{HI}$ are skew to $\overline{DC}$.

CA Standards Check 2 Use the diagram in Example 2.

a. Name all labeled segments that are parallel to $\overline{GJ}$.
b. Name all labeled segments that are skew to $\overline{GJ}$.
c. Name another pair of parallel segments; of skew segments.

Parallel planes are planes that do not intersect. A line and a plane that do not intersect are also parallel.

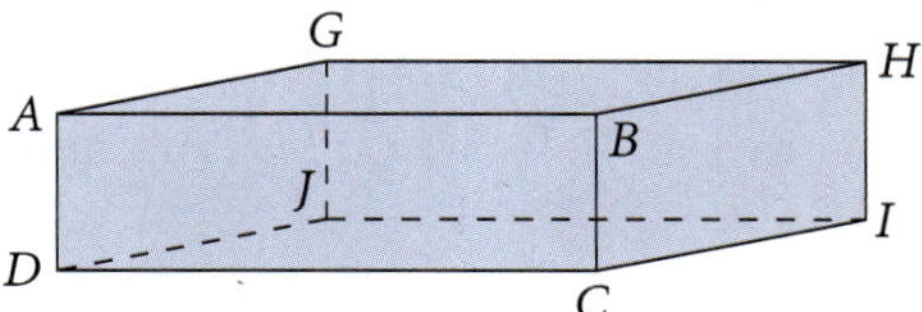

Plane *ABCD* $\parallel$ Plane *GHIJ*.
Plane *ABCD* $\parallel \overleftrightarrow{GH}$.

3 EXAMPLE Identifying Parallel Planes

Use the diagram at the right to name the figures.

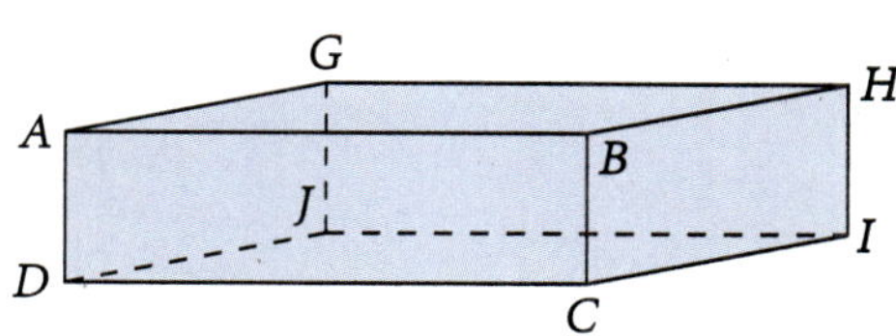

a. two pairs of parallel planes

plane $ABHG \parallel$ plane $DCIJ$
plane $ADJ \parallel$ plane BCI

b. a line that is parallel to plane $GHIJ$

$\overleftrightarrow{AB}$ is parallel to $GHIJ$.

3 Name the figures.

a. three pairs of parallel planes
b. a line that is parallel to plane $QRUV$

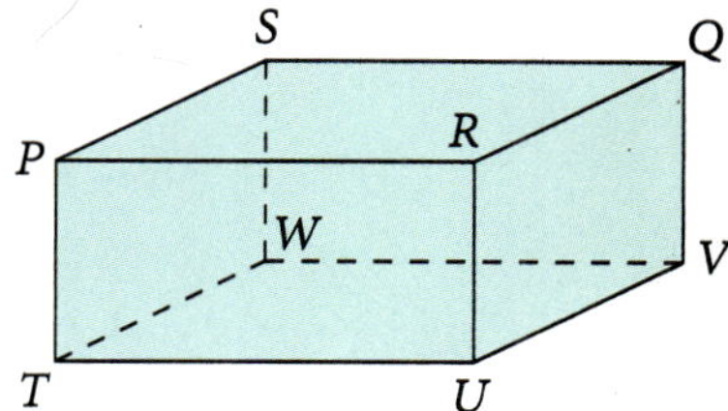

EXERCISES

For more exercises, see *Extra Skill, Word Problem, and Proof Practice.*

Standards Practice GEOM 1.0

Practice by Example

Example 1 (page 23)

Use the figure at the right for Exercises 1–3.

1. Name all the labeled segments.
2. Name all the labeled rays.
3. **a.** Name a pair of opposite rays with T as an endpoint.
   b. Name another pair of opposite rays.

Example 2 (page 24)

Name all segments shown in the diagram that are parallel to the given segment.

4. $\overline{AC}$ 5. $\overline{EF}$ 6. $\overline{AD}$

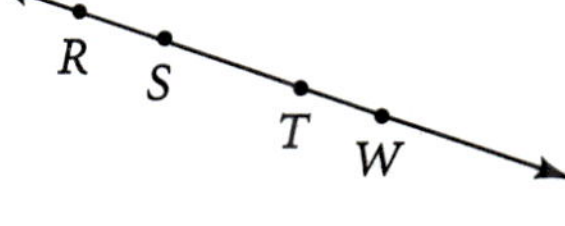

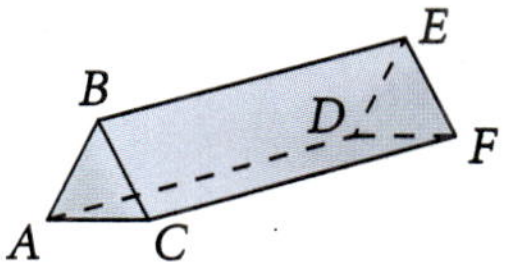

Exercises 4–11

Name all segments shown in the diagram that are skew to the given segment.

7. $\overline{AC}$ 8. $\overline{EF}$ 9. $\overline{AD}$

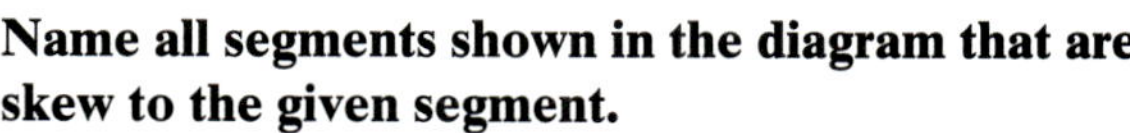

Example 3 (page 25)

Use the diagram above and name a pair of figures to match each description.

10. parallel planes 11. a line and a plane that are parallel

Use the figure at the right to name the following.

12. all lines that are parallel to $\overleftrightarrow{AB}$
13. two lines that are skew to $\overleftrightarrow{EJ}$
14. all lines that are parallel to plane $JFAE$
15. the intersection of plane FAB and plane FAE

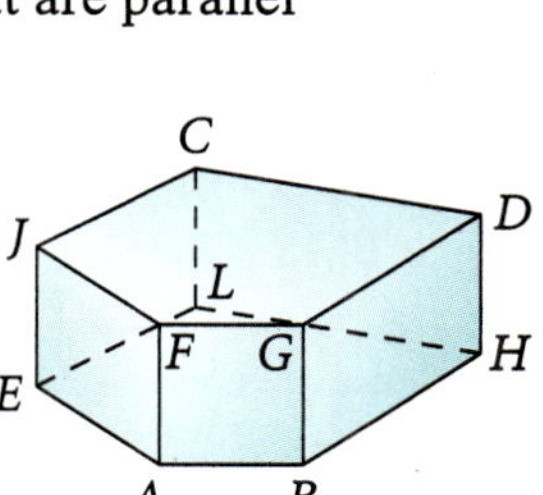

B Apply Your Skills

In Exercises 16–23, describe the statement as true or false. If *false*, explain.

16. $\overleftrightarrow{CB} \parallel \overleftrightarrow{HG}$

17. $\overleftrightarrow{ED} \parallel \overleftrightarrow{HG}$

18. plane $AED \parallel$ plane FGH

19. plane $ABH \parallel$ plane CDF

20. $\overleftrightarrow{AB}$ and $\overleftrightarrow{HG}$ are skew lines.

21. $\overleftrightarrow{AE}$ and $\overleftrightarrow{BC}$ are skew lines.

22. $\overleftrightarrow{CG}$ and $\overleftrightarrow{AI}$ are skew lines.

23. $\overleftrightarrow{CF}$ and $\overleftrightarrow{AJ}$ are skew lines.

24. The following steps show how to draw planes A and B intersecting in $\overleftrightarrow{FG}$.

Step 1

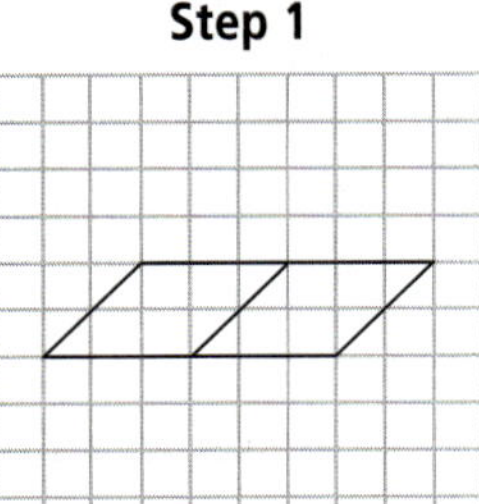

Step 2

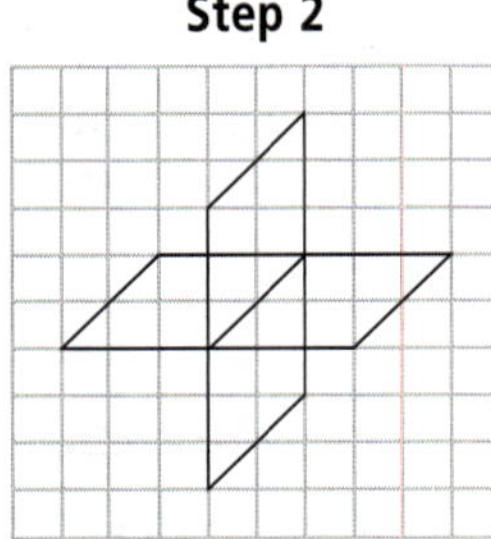

Step 3

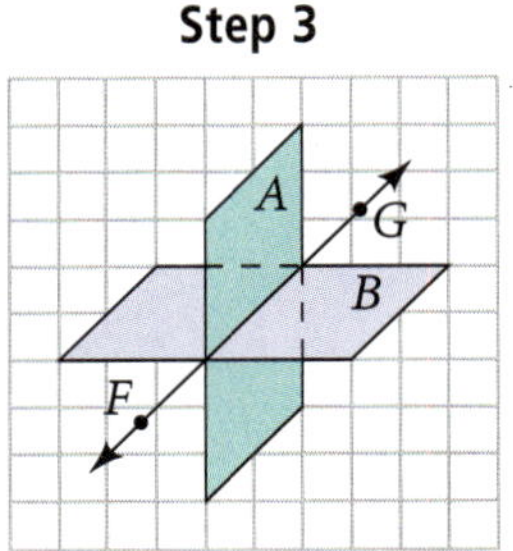

Use similar steps to draw plane DFE and plane DFJ intersecting in $\overleftrightarrow{DF}$.

Vocabulary Tip

Always, sometimes, and never refer to all possible cases, not to intervals of time.

Complete Exercises 25–33 with *always*, *sometimes*, or *never* to make a true statement.

25. Two parallel lines are __?__ coplanar.

26. Two skew lines are __?__ coplanar.

27. Two opposite rays __?__ form a line.

28. $\overleftrightarrow{TQ}$ and $\overleftrightarrow{QT}$ are __?__ the same line.

29. $\overrightarrow{GH}$ and $\overrightarrow{HG}$ are __?__ the same ray.

30. $\overrightarrow{JK}$ and $\overrightarrow{JL}$ are __?__ the same ray.

31. Two planes that do not intersect are __?__ parallel.

32. Two lines that lie in parallel planes are __?__ parallel.

33. Two lines in intersecting planes are __?__ skew.

34. **Multiple Choice** $\overline{FG}$ has endpoints $F(-3, 3)$ and $G(3, 1)$. Which point is also on $\overline{FG}$?

   A $(-6, 4)$ B $(-1, 2)$ C $(0, 2)$ D $(6, 0)$

35. **Coordinate Geometry** $\overrightarrow{AB}$ has endpoint $A(2, 3)$ and contains $B(4, 6)$. Give possible coordinates for point C so that $\overrightarrow{AB}$ and $\overrightarrow{AC}$ are opposite rays. Graph your answer.

36. **Directional Compass** On a directional compass, the directions north and south can be represented by opposite rays.
   a. Name two other compass directions that can be represented by opposite rays.
   b. What other pairs of opposite directions, if any, can you find?

37. **Open-Ended** Summarize the three ways in which two lines may be related. Give examples from the real world that illustrate the relationships.

Visit: PHSchool.com
Web Code: bce-0104

38. **Writing** The term *skew* is a Middle English word meaning "to escape." Explain how this meaning might be appropriate for skew lines.

39. **Critical Thinking** Suppose two parallel planes A and B are each intersected by a third plane C.
    a. Make a conjecture about the intersection of planes A and C and the intersection of planes B and C.
    b. Find examples in your classroom.

Challenge

40. a. Draw a line. Draw points E and F on the line. How many different segments do points E and F determine? Name the segments.
    b. Draw another line. Draw points E, F, and G on the line. How many segments do points E, F, and G determine? Name them.
    c. Continue to draw lines, labeling one more point each time. Make a table showing the number of points and the number of segments determined. Look for and describe a pattern in the data.
    d. Use your pattern to find how many segments are determined if you label 10 points on a line.
    e. If you label n points on a line, how many segments can you name?

Use the figure at the right for Exercises 41 and 42.

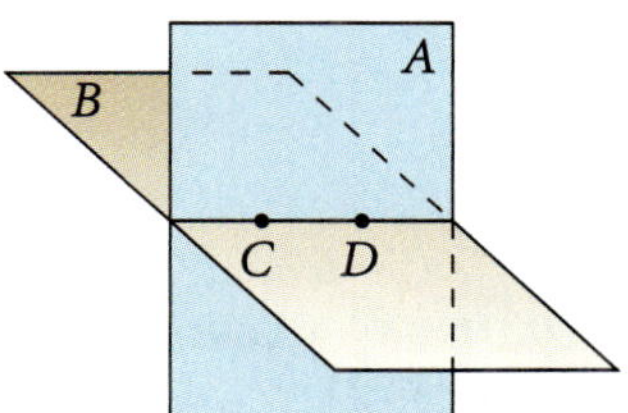

41. Do planes A and B have other lines in common that are parallel to $\overleftrightarrow{CD}$? Explain.

42. **Visualization** Are there planes that intersect planes A and B in lines parallel to $\overleftrightarrow{CD}$? Draw a sketch to support your answer.

The figure at the right is a pyramid.

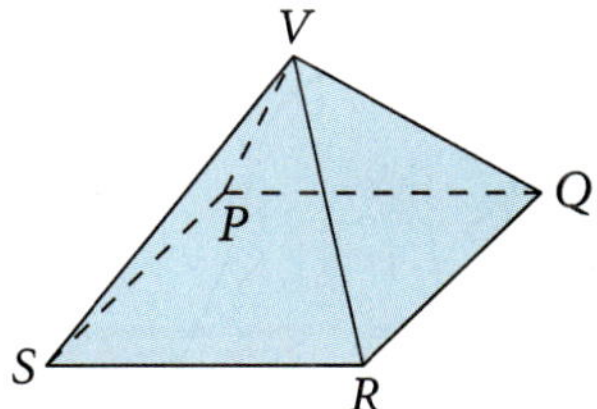

43. Name three lines that intersect at one point.

44. What line could be parallel to $\overleftrightarrow{PS}$?

45. **Visualization** Consider a plane through V that is parallel to plane $PQRS$. Can a line in that plane be parallel to $\overleftrightarrow{SR}$? Can it intersect $\overleftrightarrow{SR}$? Can it be skew to $\overleftrightarrow{SR}$? Explain each answer.

Multiple Choice Practice

For California Standards and CAHSEE Tutorials, visit PHSchool.com. Web Code: bcq-9045

GEOM 1.0

46. Which of the following does NOT extend forever in at least one direction?
    (A) line
    (B) plane
    (C) ray
    (D) segment

CAHSEE 7MR 1.1

47. Amir's high school play was sold out for 3 nights. Tickets cost \$9.00 each. The auditorium for the play has 28 rows with 37 seats in each row. Which of the following is a correct method to find the total amount earned from ticket sales?
    (A) Amir can add 28 and 37, and then multiply by \$9.
    (B) Amir can multiply 3 by 28, then multiply by 37, and then multiply by \$9.
    (C) Amir can add 37 and 3, then multiply by 27 and add \$9.
    (D) Amir can multiply 28 by 37, then multiply 3 by \$9 and add the two products.

CAHSEE 7MG 2.1

48. A rectangle has an area of 84 square inches and a perimeter of 40 inches. What are the dimensions of the rectangle?

(A) 3 in. by 30 in.
(B) 7 in. by 12 in.
(C) 4 in. by 21 in.
(D) 6 in. by 14 in.

Mixed Review

Lesson 1-3

Use the diagram for Exercises 49–56 and name each geometric figure.

49. a line

50. a point

51. the intersection of $\overline{DC}$ and $\overleftrightarrow{CG}$

52. two planes that intersect in $\overleftrightarrow{EF}$

53. the plane represented by the top of the box

54. the plane represented by the front of the box

55. the intersection of planes EFG and DFG

56. another point in plane CGH

Draw the following.

57. $\overleftrightarrow{TR}$

58. $\overline{PQ}$

59. $\overrightarrow{NV}$

Lesson 1-1

Find the next two terms in each sequence.

60. 1, 1.08, 1.16, 1.24, 1.32, . . .

61. −1, −2, −4, −7, −11, −16, . . .

62. AB, BC, CD, DE, EF, . . .

63. A, D, G, J, M, . . .

64. **Reasoning** Raven conjectured: "If you subtract a number from a given number, the result is always less than the given number." Is her conjecture true? Explain.

Checkpoint Quiz 1

Lessons 1-1 through 1-4

Find the next two terms in each sequence.

1. 19, 21.5, 24, 26.5, . . .

2. 3.4, 3.45, 3.456, 3.4567, . . .

3. **Writing** Describe the pattern of each sequence in Exercises 1 and 2.

Use the diagram for Exercises 4–10. In Exercises 4–7, do the points appear to be coplanar? If *yes*, name the plane. If *no*, explain.

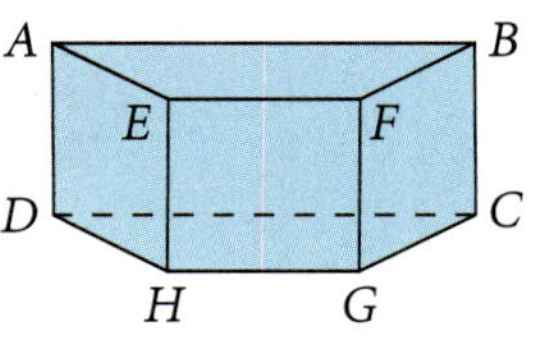

4. Points A, E, F, and B

5. Points D, C, E, and F

6. Points H, G, F, and B

7. Points A, E, B, and C

8. Name all the segments parallel to $\overline{HG}$.

9. Name a pair of skew lines.

10. Draw a net for the figure.

Geometry Vocabulary

To succeed with geometry, you must learn its vocabulary. As a study aid, some students use index cards. Others use special notebooks. With either, you can note

- each geometry term,
- its meaning,
- an example of its use, and
- a drawing, if possible.

A study aid like this can become a valuable tool to use throughout the course.

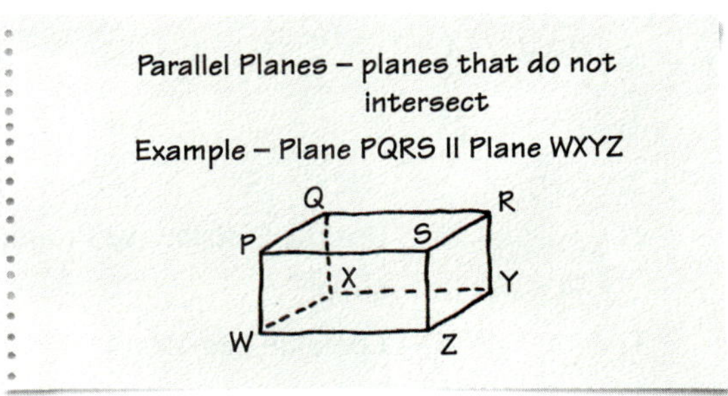

EXERCISES

1. Work with a classmate. Plan a study aid that will help both of you learn the geometry vocabulary. Make a sample page. Then share your sample with others. Use the best ideas from the class to revise your method.
2. Begin a vocabulary section of your study aid using Lessons 1-1 through 1-4.

A popular dictionary shows the definitions at the right for *add* and *sum*.

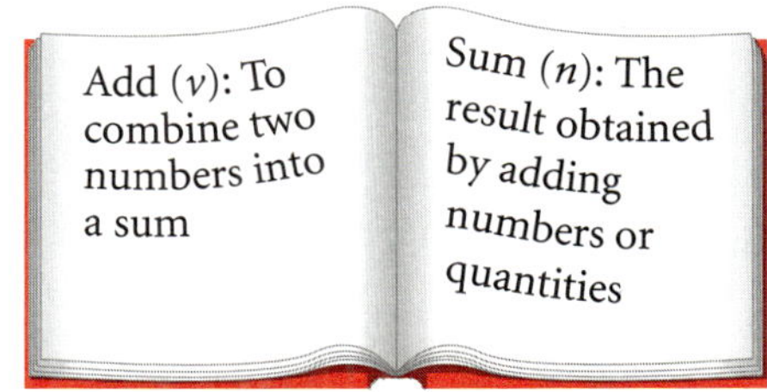

3. The definitions are *circular*. Tell what you think it means for a definition to be circular.
4. Circular defining can involve several statements.

   Show means display. *Display* means illustrate. *Illustrate* means show.

   a. Why are these definitions of *show*, *display*, and *illustrate* circular?
   b. Definitions like these, even though they are circular, can still help you understand vocabulary. Why?

A *mathematical system*, such as geometry, tries to avoid circular definitions. That means some terms must remain undefined.

5. *Point* is a term commonly accepted as undefined in geometry.
   a. Try to write a definition of *point*. Ask classmates whether they think it's a good definition.
   b. What other examples or descriptions—besides those in this book—can help you understand *point*? (You may want to include them in your study aid.)
6. Explain why some vocabulary terms must remain undefined.

Algebra 1 Review

Solving Linear Equations

FOR USE WITH LESSON 1-5

Sometimes you need to combine like terms to solve a linear equation.

1 EXAMPLE

Solve $(5x + 8) - (2x - 9) = 38$.

$(5x + 8) - (2x - 9) = 38$	
$(5x + 8) + (-1)(2x - 9) = 38$	**$-(2x - 9) = (-1)(2x - 9)$**
$(5x + 8) + (-1)2x + (-1)(-9) = 38$	**Use the Distributive Property.**
$5x + 8 - 2x + 9 = 38$	**Simplify.**
$3x + 17 = 38$	**Combine like terms.**
$3x = 21$	**Subtract 17 from each side.**
$x = 7$	**Divide each side by 3.**

To solve an equation with the variable on *both* sides, first convert the equation to one with the variable on *only one* side.

2 EXAMPLE

Solve $4x - 9 = 7x - 15$.

$4x - 9 = 7x - 15$	
$-9 = 3x - 15$	**Subtract 4x from each side.**
$6 = 3x$	**Add 15 to each side.**
$2 = x$	**Divide each side by 3.**

EXERCISES

Solve.

1. $10n + 12 = 14n - 12$

2. $(4w - 28) + (11w + 13) = 180$

3. $(7a + 3) + (-a - 5) = -16$

4. $7y + 44 = 12y + 11$

5. $(7t - 21) + (t + 4) = 15$

6. $8x - 4 - 2x = -10$

7. $(8t + 30) + (-2t - 16) = -22$

8. $6x + 17 = 9x + 2$

9. $(3y - 5) + (5y + 20) = 135$

10. $(11x - 37) + (5x + 59) = 54$

11. $3x - 35 = 9x - 59$

12. $9x - 3 = 8x - 7$

13. $(5w + 24) + (2w + 13) = 156$

14. $(3x + 10) - 5x = 6x - 50$

15. $8y + 12 = 2y - 18$

16. $7t - 8t + 4 = 5t - 2$

17. $13c + 40 = 9c - 20 + c$

18. $(6a - 54) - (5a + 27) = 23$

19. $(2 + 4y) - (y + 9) = 26$

20. $(12c + 35) - (5c - 11) = -2$

Measuring Segments

California Content Standards

Prepares for GEOM 4.0 Students prove basic theorems involving congruence.

What You'll Learn

- To find the lengths of segments

... And Why

To find distance using a highway number line, as in Exercise 23

Check Skills You'll Need  **for Help** Skills Handbook pages 757 and 758

Simplify each absolute value expression.

1. $|-6|$ **2.** $|3.5|$ **3.** $|7 - 10|$

4. $|-4 - 2|$ **5.** $|-2 - (-4)|$ **6.** $|-3 + 12|$

Algebra Solve each equation.

7. $x + 2x - 6 = 6$ **8.** $3x + 9 + 5x = 81$ **9.** $w - 2 = -4 + 7w$

New Vocabulary • coordinate • congruent segments • midpoint

1 Finding Segment Lengths

The distance between points C and D on the ruler is 3 units. You can use the Ruler Postulate to find the distance between points on a number line.

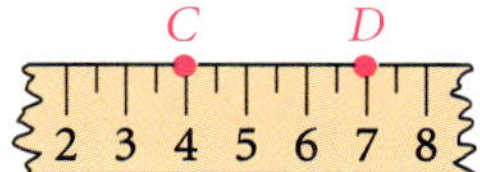

Key Concepts

Postulate 1-5 **Ruler Postulate**

The points of a line can be put into one-to-one correspondence with the real numbers so that the distance between any two points is the absolute value of the difference of the corresponding numbers.

the length of $\overline{AB}$

$$AB = |a - b|$$

coordinate of A **coordinate** of B

(Number line: points A and B with coordinates a and b.)

Vocabulary Tip

The congruence symbol ($\cong$) shows that two figures are equal ($=$) in size and similar ($\sim$) in shape.

Two segments with the same length are **congruent ($\cong$) segments.** In other words, if $AB = CD$, then $\overline{AB} \cong \overline{CD}$. You can use these statements interchangeably.

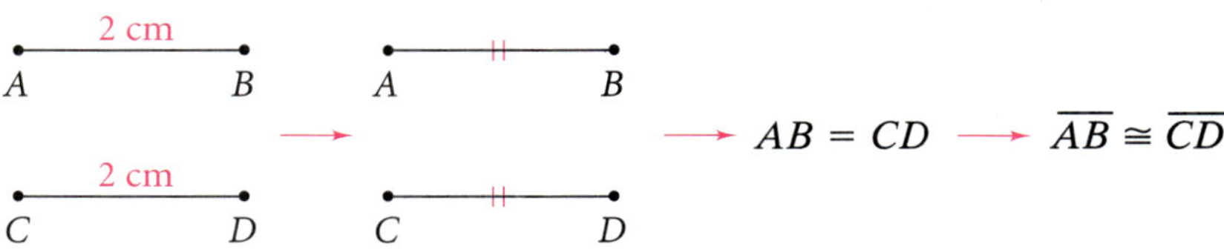

As illustrated above, segments can be marked alike to show they are congruent.

1 EXAMPLE Comparing Segment Lengths

Find AB and BC. Are $\overline{AB}$ and $\overline{BC}$ congruent?

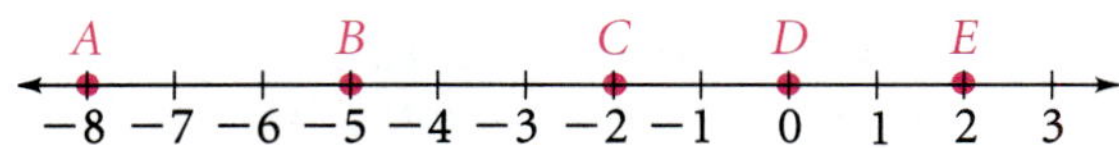

$AB = |-8 - (-5)| = |-3| = 3$

$BC = |-5 - (-2)| = |-3| = 3$

$AB = BC$, so $\overline{AB} \cong \overline{BC}$.

CA Standards Check 1

a. Compare CD and DE. Are the segments congruent?

b. **Critical Thinking** To find AB in Example 1, suppose you subtract −8 from −5. Do you get the same result? Why?

Examine the lengths of $\overline{AB}$ and $\overline{BC}$ in Example 1. Notice that $AB + BC = 6$. Notice that $AC = 6$. This suggests the following postulate.

Key Concepts

Postulate 1-6 **Segment Addition Postulate**

If three points A, B, and C are collinear and B is between A and C, then $AB + BC = AC$.

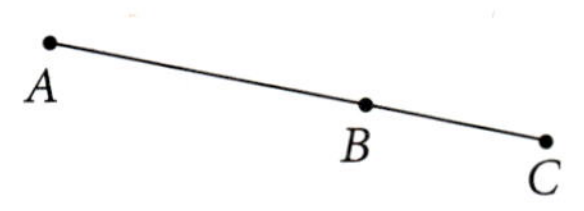

2 EXAMPLE Using the Segment Addition Postulate

Algebra If $DT = 60$, find the value of x. Then find DS and ST.

Segment DT with point S between: $DS = 2x - 8$, $ST = 3x - 12$.

$DS + ST = DT$	**Segment Addition Postulate**
$(2x - 8) + (3x - 12) = 60$	**Substitute.**
$5x - 20 = 60$	**Simplify.**
$5x = 80$	**Add 20 to each side.**
$x = 16$	**Divide each side by 5.**
$DS = 2x - 8 = 2(16) - 8 = 24$	**Substitute 16 for *x*.**
$ST = 3x - 12 = 3(16) - 12 = 36$	

CA Standards Check 2 $EG = 100$. Find the value of x. Then find EF and FG.

Segment EG with point F between: $EF = 4x - 20$, $FG = 2x + 30$.

A **midpoint** of a segment is a point that divides the segment into two congruent segments. A midpoint, or any line, ray, or other segment through a midpoint, is said to *bisect* the segment.

Segment AC with midpoint B: $\overline{AB} \cong \overline{BC}$

3 EXAMPLE Using the Midpoint

Real-World Connection

Gymnasts in the rings event try to keep their arms equidistant from their body for balance.

Algebra C is the midpoint of $\overline{AB}$. Find AC, CB, and AB.

$$\underset{A}{\bullet}\xrightarrow{\quad 2x+1 \quad}\underset{C}{\bullet}\xrightarrow{\quad 3x-4 \quad}\underset{B}{\bullet}$$

$AC = CB$	**Definition of midpoint**
$2x + 1 = 3x - 4$	**Substitute.**
$2x + 5 = 3x$	**Add 4 to each side.**
$5 = x$	**Subtract $2x$ from each side.**
$AC = 2x + 1 = 2(5) + 1 = 11$	**Substitute 5 for x.**
$CB = 3x - 4 = 3(5) - 4 = 11$	

AC and CB are both 11, which is half of 22, the length of $\overline{AB}$.

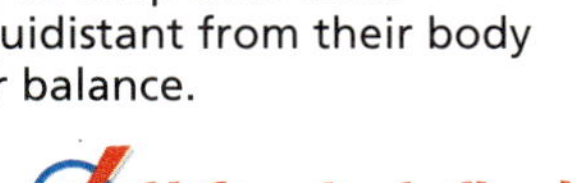

CA Standards Check 3 Z is the midpoint of $\overline{XY}$, and $XY = 27$. Find XZ.

EXERCISES Standards Practice

For more exercises, see *Extra Skill, Word Problem, and Proof Practice.*

A Practice by Example

GO for Help

Example 1 (page 32)

Find the length of each segment. Tell whether the segments are congruent.

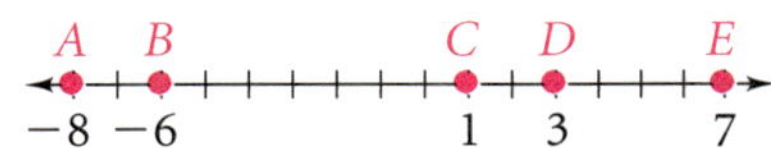

1. $\overline{AC}$ and $\overline{BD}$ **2.** $\overline{BD}$ and $\overline{CE}$

3. $\overline{AD}$ and $\overline{BE}$ **4.** $\overline{BC}$ and $\overline{CE}$

On a number line, the coordinates of X, Y, Z, and W are –7, –3, 1, and 5, respectively. Find the lengths of the two segments and tell whether they are congruent.

5. $\overline{XY}$ and $\overline{ZW}$ **6.** $\overline{ZX}$ and $\overline{WY}$ **7.** $\overline{YZ}$ and $\overline{XW}$

Example 2 (page 32)

Use the figure at the right for Exercises 8–11.

8. If $RS = 15$ and $ST = 9$, then $RT = ■$.

9. If $ST = 15$ and $RT = 40$, then $RS = ■$.

10. a. Algebra If $RS = 3x + 1$, $ST = 2x - 2$, and $RT = 64$, find the value of x.
b. Find RS and ST.

11. a. Algebra If $RS = 8y + 4$, $ST = 4y + 8$, and $RT = 15y - 9$, find the value of y.
b. Find RS, ST, and RT.

Example 3 (page 33)

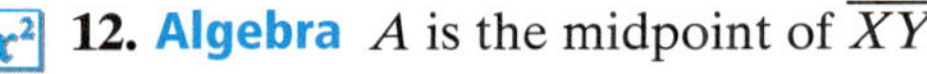

12. Algebra A is the midpoint of $\overline{XY}$.
a Find XA.
b. Find AY and XY.

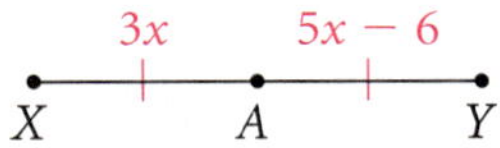

Algebra In Exercises 13–15, use the figure and find PT.

$$\longleftarrow\underset{P}{\bullet}\text{———}\underset{T}{\bullet}\text{———}\underset{Q}{\bullet}\longrightarrow$$

13. $PT = 5x + 3$ and $TQ = 7x - 9$

14. $PT = 4x - 6$ and $TQ = 3x + 4$

15. $PT = 7x - 24$ and $TQ = 6x - 2$

B Apply Your Skills

Use the figure at the right for Exercises 16–19.

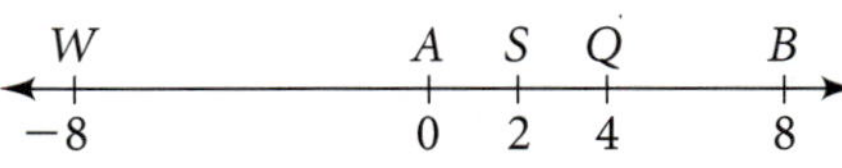

16. Find the midpoint of $\overline{AB}$.

17. What is the coordinate of the midpoint of $\overline{QB}$?

18. What is the coordinate of the midpoint of $\overline{WA}$?

19. What is the coordinate of the midpoint of the segment formed by the two points you found in Exercises 17 and 18?

Suppose the coordinate of A is 0, $AR = 5$, and $AT = 7$. What are the possible coordinates of the midpoint of the given segment?

20. $\overline{AR}$ **21.** $\overline{AT}$ **22.** $\overline{RT}$

Homework Video Tutor

Visit: PHSchool.com
Web Code: bce-0105

23. Mileage Exit numbers on some highways give mileage from one edge of the state. The highway and its exit numbers resemble a number line. You can find distance between exits in the same way that you find distance on a number line.

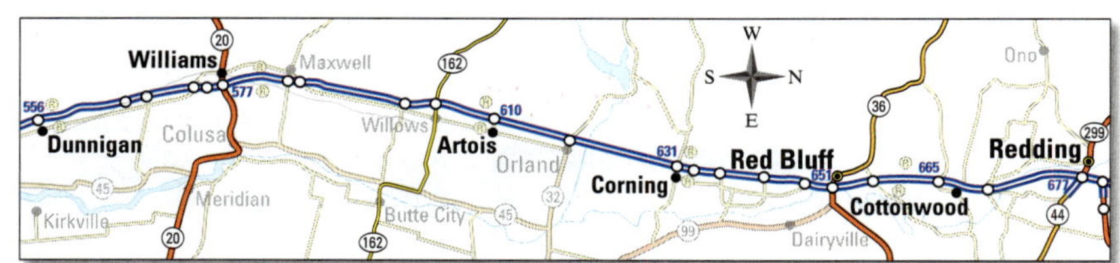

a. Use the map. How many miles is it from Dunnigan to Artois?
b. Which town is 21 miles from Corning?

Visualization **Without using your ruler, sketch a segment with the given length. Then use your ruler to see how well you did.**

24. 3 cm **25.** 3 in. **26.** 6 in. **27.** 10 cm **28.** 65 mm

In Exercises 29–32, describe the statement as *true* or *false*. Explain.

A B C D E
−8 −6 1 3 7

Exercises 29–33

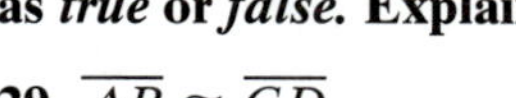

29. $\overline{AB} \cong \overline{CD}$ **30.** $BD < CD$

31. $AC + BD = AD$ **32.** $AC + CD = AD$

33. Suppose $EG = 5$. Find the possible coordinate(s) of point G.

Vocabulary Tip

For Exercise 30, you are to decide whether the distance BD is less than the distance CD.

Algebra **Use the diagram at the right for Exercises 34 and 35.**

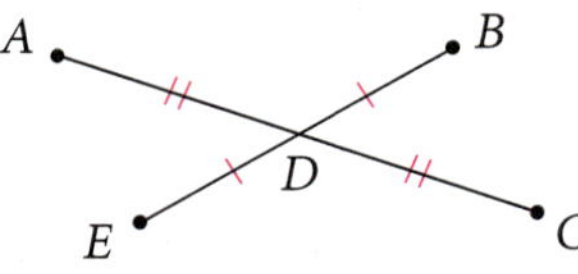

34. If $AD = 12$ and $AC = 4y - 36$, find the value of y. Then find AC and DC.

35. If $ED = x + 4$ and $DB = 3x - 8$, find ED, DB, and EB.

C Challenge

36. C is the midpoint of $\overline{AB}$, D is the midpoint of $\overline{AC}$, E is the midpoint of $\overline{AD}$, F is the midpoint of $\overline{ED}$, G is the midpoint of $\overline{EF}$, and H is the midpoint of $\overline{DB}$. If $DC = 16$, find GH.

37. a. Write an algebraic expression that represents GK.
b. If $GK = 30$, find GH and JK.

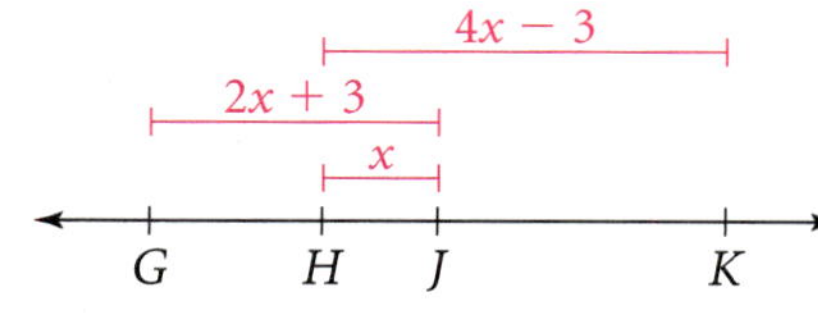

38. Art Project You want to cut pieces of ribbon for an art project. Each piece must be $6\frac{3}{4}$ inches long. For measuring, you have only the old, broken ruler shown at the right.

a. What marks on the ruler would you use to measure the ribbon?

b. Writing You also need 5-in. pieces of string. Describe two ways you can use the broken ruler to measure 5 inches.

Multiple Choice Practice

For California Standards and CAHSEE Tutorials, visit PHSchool.com. Web Code: bcq-9045

GEOM 3.0

39. Which of the following serves as a counterexample to the statement below?
The sum of any two integers is less than or equal to their product.

Ⓐ $-1 + (-1) \leq (-1)(-1)$
Ⓑ $3 + (-1) \leq 3(-1)$
Ⓒ $2 + 2 \leq 2(2)$
Ⓓ $2 + 3 \leq 2(3)$

CAHSEE 7AF 2.2

40. Which expression is equivalent to $\frac{64x^5y^{-4}z^6}{16x^2y^3z^6}$?

Ⓐ $\frac{4x^7y^7}{z^{12}}$ Ⓑ $\frac{4x^3}{y^7}$ Ⓒ $\frac{x^{10}}{4y}$ Ⓓ $\frac{4x^3z^2}{y}$

CAHSEE Alg1 6.0

41. What is the x-intercept of the line $y = -\frac{3}{4}x - 3$?

Ⓐ $(-4, 0)$ Ⓑ $(-3, 0)$ Ⓒ $(0, -3)$ Ⓓ $(0, -4)$

CAHSEE 7MG 1.3

42. Jude ran 500 yards in 2 minutes. How many feet per second is this?

Ⓐ 10 Ⓑ 12.5 Ⓒ 25 Ⓓ 37.5

CAHSEE Alg1 10.0

43. Simplify the expression $(y^2 + 4y - 3) - (2y^2 - 3y + 5)$.

Ⓐ $-y^2 + y - 8$
Ⓑ $-y^2 + 7y + 2$
Ⓒ $y^2 + y - 8$
Ⓓ $-y^2 + 7y - 8$

Mixed Review

Lesson 1-4

Complete each statement with *always*, *sometimes*, or *never* to make a true statement.

44. Skew lines are _?_ coplanar.

45. Skew lines _?_ intersect.

46. Opposite rays _?_ form a line.

47. Parallel planes _?_ intersect.

Lesson 1-3

48. Three points are _?_ coplanar.

49. Two points are _?_ collinear.

50. The intersection of two planes is _?_ a line.

51. Intersecting lines are _?_ parallel.

Lesson 1-1

Find the next two terms in each sequence.

52. 5, 10, 15, 20, . . .

53. 5, 25, 125, 625, . . .

54. 14, 18, 22, 26, . . .

1-6 Measuring Angles

California Content Standards

Prepares for GEOM 13.0 Students prove relationships between angles by using properties of complementary, supplementary, and vertical angles.

What You'll Learn

- To find the measures of angles
- To identify special angle pairs

. . . And Why

To find the measures of angles in flower arrangements, as in Exercise 41

Check Skills You'll Need

GO for Help Skills Handbook page 758

Solve each equation.

1. $50 + a = 130$ **2.** $m - 110 = 20$ **3.** $85 - n = 40$

4. $x + 45 = 180$ **5.** $z - 20 = 90$ **6.** $180 - y = 135$

New Vocabulary

- angle
- acute angle
- right angle
- obtuse angle
- straight angle
- congruent angles
- vertical angles
- complementary angles
- supplementary angles
- adjacent angles

1 Finding Angle Measures

Vocabulary Tip

You may also refer to the angle suggested by the two segments $\overline{BT}$ and $\overline{BQ}$ as $\angle TBQ$.

An **angle** ($\angle$) is formed by two rays with the same endpoint. The rays are the *sides* of the angle. The endpoint is the *vertex* of the angle. The sides of the angle shown here are $\overrightarrow{BT}$ and $\overrightarrow{BQ}$. The vertex is B. You could name this angle $\angle B$, $\angle TBQ$, $\angle QBT$, or $\angle 1$.

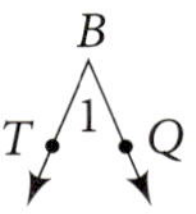

1 EXAMPLE Naming Angles

Name $\angle 1$ in two other ways.

$\angle AEC$ and $\angle CEA$ are other names for $\angle 1$.

CA Standards Check 1 **a.** Name $\angle CED$ two other ways.

b. Critical Thinking Would it be correct to name any of the angles $\angle E$? Explain.

One way to measure an angle is in degrees. To indicate the size or degree measure of an angle, write a lowercase m in front of the angle symbol. The degree measure of angle A is 80. You show this by writing $m\angle A = 80$.

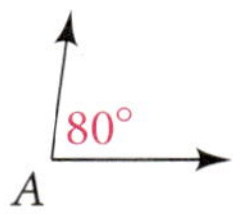

Key Concepts

Postulate 1-7 **Protractor Postulate**

Let $\overrightarrow{OA}$ and $\overrightarrow{OB}$ be opposite rays in a plane. $\overrightarrow{OA}$, $\overrightarrow{OB}$, and all the rays with endpoint O that can be drawn on one side of $\overleftrightarrow{AB}$ can be paired with the real numbers from 0 to 180 so that

a. $\overrightarrow{OA}$ is paired with 0 and $\overrightarrow{OB}$ is paired with 180.

b. If $\overrightarrow{OC}$ is paired with x and $\overrightarrow{OD}$ is paired with y, then $m\angle COD = |x - y|$.

You can classify angles according to their measures.

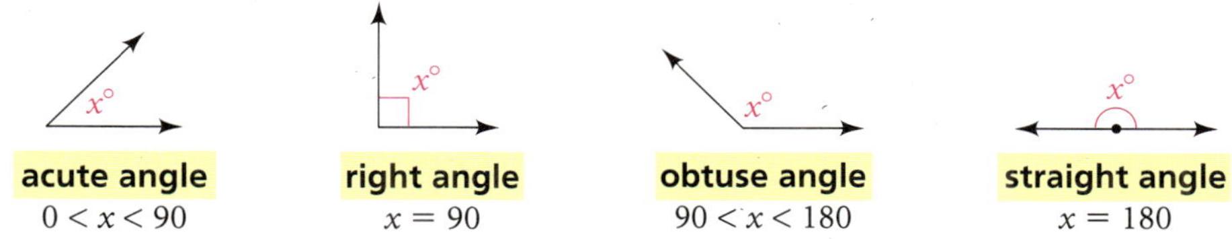

Note the special symbol that's tucked into the corner of the right angle. When you see it, you know that the measure of the angle is 90.

Video Tutor Help

Visit: PHSchool.com

Web Code: bce-0775

2 EXAMPLE Measuring and Classifying Angles

Find the measure of each angle. Classify each as *acute*, *right*, *obtuse*, or *straight*.

a.

120, obtuse

b.

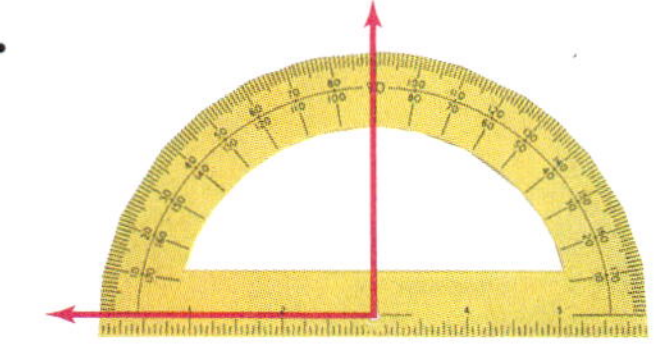

90, right

CA Standards Check

2 Find the measure of each angle. Classify each as *acute*, *right*, *obtuse*, or *straight*.

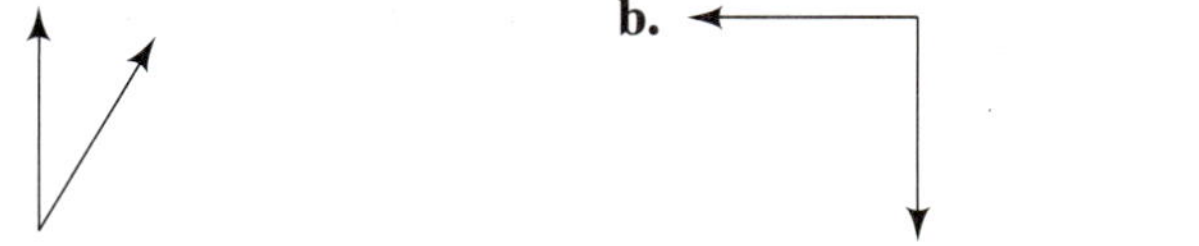

Angles with the same measure are **congruent angles.** In other words, if $m\angle 1 = m\angle 2$, then $\angle 1 \cong \angle 2$. You can use these statements interchangeably.

Angles can be marked alike to show that they are congruent, as in this photograph of the Air Force Thunderbirds precision flying team.

The Angle Addition Postulate is similar to the Segment Addition Postulate.

Postulate 1-8 **Angle Addition Postulate**

If point B is in the interior of $\angle AOC$, then $m\angle AOB + m\angle BOC = m\angle AOC$.

If $\angle AOC$ is a straight angle, then $m\angle AOB + m\angle BOC = 180$.

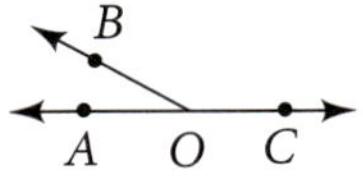

3 EXAMPLE Using the Angle Addition Postulate

What is $m\angle TSW$ if $m\angle RST = 50$ and $m\angle RSW = 125$?

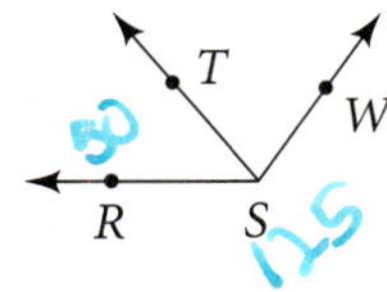

$m\angle RST + m\angle TSW = m\angle RSW$ **Angle Addition Postulate**

$50 + m\angle TSW = 125$ **Substitute.**

$m\angle TSW = 75$ **Subtract 50 from each side.**

3 If $m\angle DEG = 145$, find $m\angle GEF$.

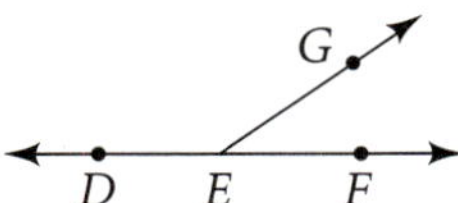

2 Identifying Angle Pairs

Some angle pairs that have special names.

X	Helvetica Condensed
X	Times Roman
X	Eurostile Extended
X	MarkerFelt Thin

In each font, a capital X suggests vertical angles.

vertical angles

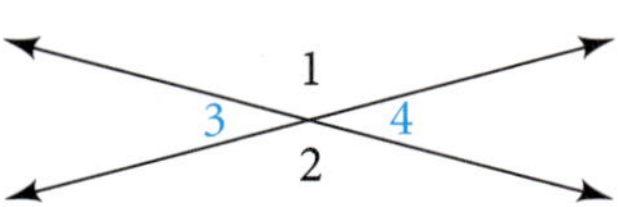

two angles whose sides are opposite rays

adjacent angles

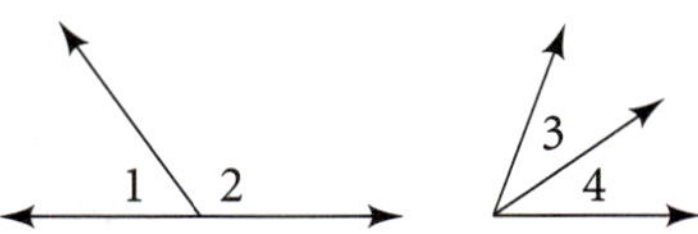

two coplanar angles with a common side, a common vertex, and no common interior points

complementary angles

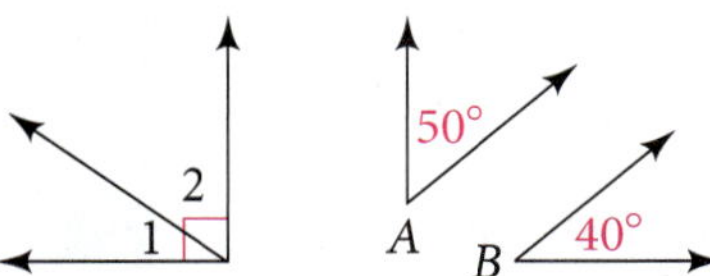

two angles whose measures have sum 90

Each angle is called the *complement* of the other.

supplementary angles

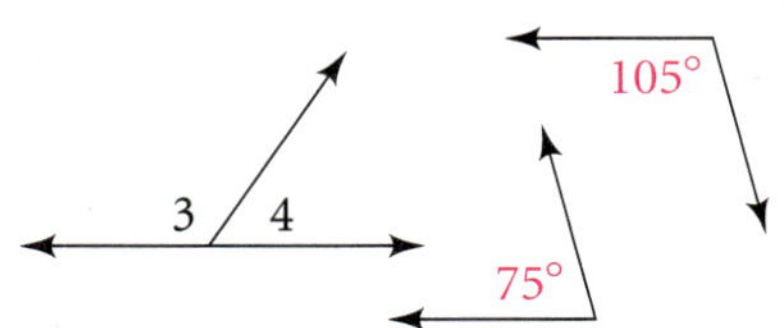

two angles whose measures have sum 180

Each angle is called the *supplement* of the other.

4 EXAMPLE Identifying Angle Pairs

In the diagram identify pairs of numbered angles that are related as follows:

a. complementary
$\angle 2$ and $\angle 3$

b. supplementary
$\angle 4$ and $\angle 5$; $\angle 3$ and $\angle 4$

c. vertical
$\angle 3$ and $\angle 5$

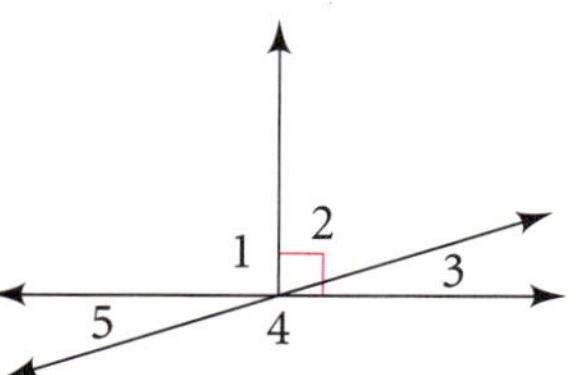

4 **a.** Name two pairs of adjacent angles in the photo below.
b. If $m\angle EFD = 27$, find $m\angle AFD$.

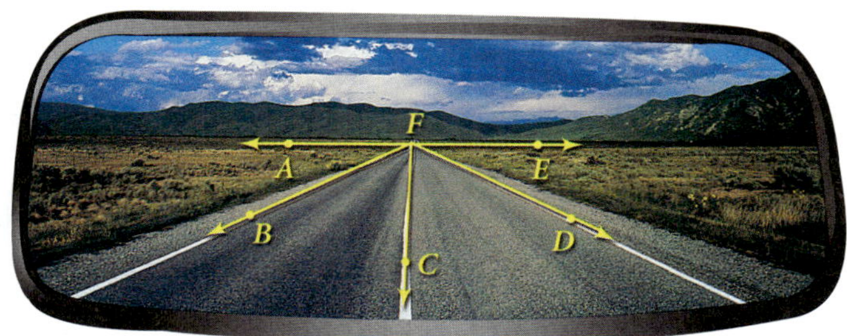

Real-World Connection

Most rear-view mirrors have a viewing angle of about 100°.

Whether you draw a diagram or use a given diagram, you can make some conclusions directly from the diagrams. You *can* conclude that angles are

- adjacent angles
- adjacent supplementary angles
- vertical angles

Unless there are marks that give this information, you *cannot* assume

- angles or segments are congruent
- an angle is a right angle
- lines are parallel or perpendicular

Video Tutor Help
Visit: PHSchool.com
Web Code: bce-0775

5 EXAMPLE Making Conclusions From a Diagram

What can you conclude from the information in the diagram?

- $\angle 1 \cong \angle 2$, by the markings.
- $\angle 2$ and $\angle 3$, for example, are adjacent angles.
- $\angle 4$ and $\angle 5$, for example, are adjacent supplementary angles,
  or $m\angle 4 + m\angle 5 = 180$ by the Angle Addition Postulate.
- $\angle 1$ and $\angle 4$, for example, are vertical angles.

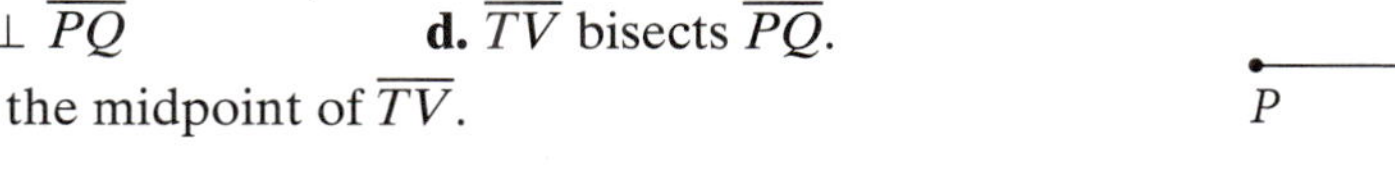

5 Can you make each conclusion from the information in the diagram? Explain.

a. $\overline{TW} \cong \overline{WV}$ **b.** $\overline{PW} \cong \overline{WQ}$
c. $\overline{TV} \perp \overline{PQ}$ **d.** $\overline{TV}$ bisects $\overline{PQ}$.
e. W is the midpoint of $\overline{TV}$.

T
P W Q
V

EXERCISES
Standards Practice

For more exercises, see *Extra Skill, Word Problem, and Proof Practice.*

A Practice by Example

Example 1 (page 36)

GO for Help

Name each angle in three ways.

1.

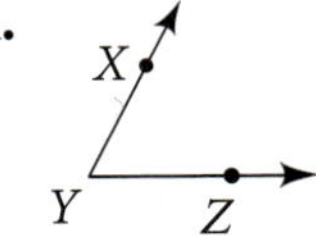

2.

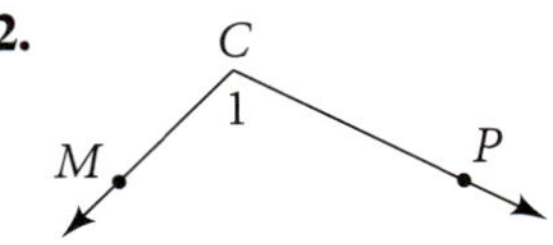

Use the figure at the right. Name the indicated angle in two different ways.

3. $\angle 1$

4. $\angle 2$

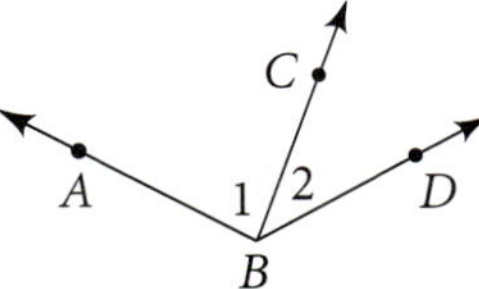

Example 2 (page 37)

Draw and label a figure to fit each description.

5. an obtuse angle, $\angle RST$

6. an acute acute, $\angle BCD$

7. a straight angle, $\angle EFG$

8. a right angle, $\angle GHI$

Use a protractor. Measure and classify each angle.

9.

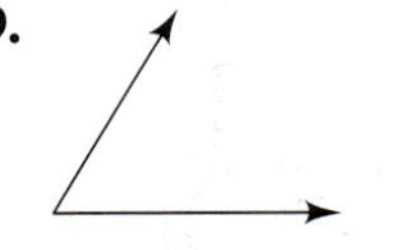
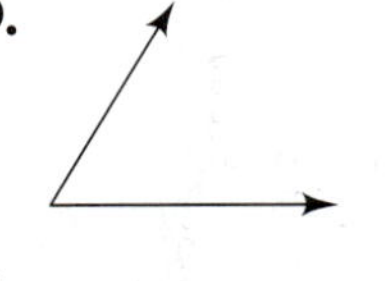

10. the angle formed by the skis

11.

12.

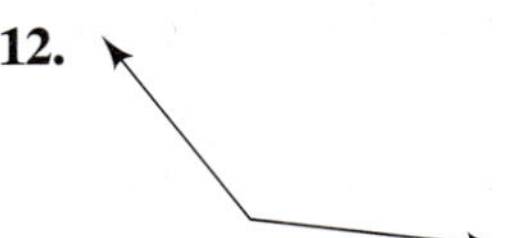

Example 3 (page 38)

13. Find $m\angle CBD$ if $m\angle ABC = 45$ and $m\angle ABD = 79$.

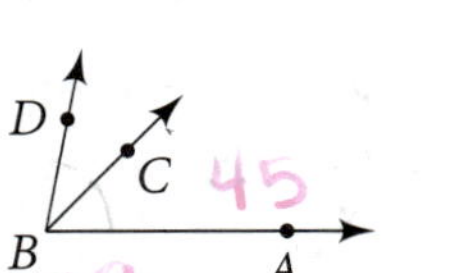

14. Find $m\angle GFJ$ if $m\angle EFG = 110$.

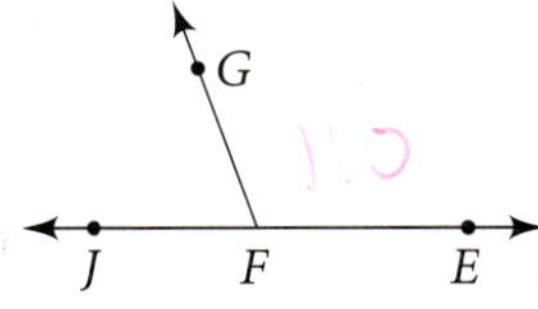

Example 4 (page 39)

Name an angle or angles in the diagram described by each of the following.

15. supplementary to $\angle AOD$

16. adjacent and congruent to $\angle AOE$

17. supplementary to $\angle EOA$

18. complementary to $\angle EOD$

19. a pair of vertical angles

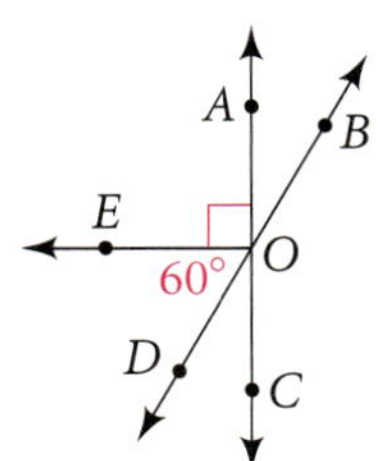

In the diagram above, find the measure of each of the following angles.

20. $\angle EOC$ **21.** $\angle DOC$ **22.** $\angle BOC$ **23.** $\angle AOB$

Example 5 (page 39)

Can you make each conclusion from the information in the diagram? Explain.

24. $\angle J \cong \angle D$
25. $\angle JAC \cong \angle DAC$
26. $\angle JAE$ and $\angle EAF$ are adjacent and supplementary.
27. $m\angle JCA = m\angle DCA$
28. $m\angle JCA + m\angle ACD = 180$
29. $\overline{AJ} \cong \overline{AD}$
30. C is the midpoint of $\overline{JD}$.
31. $\angle EAF$ and $\angle JAD$ are vertical angles.
32. $\overrightarrow{AC}$ bisects $\angle JAD$.

In the diagram, $m\angle ACB = 65$. Find each of the following.

33. $m\angle BCD$
34. $m\angle ECD$

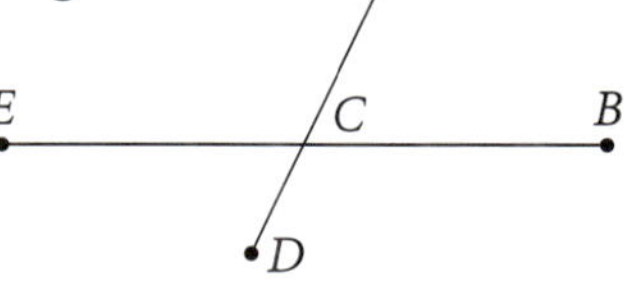

Exercises 33–34

Estimation **Estimate the measure of the angle formed by the hands of a clock at each time.**

35. 6:00
36. 7:00
37. 11:00
38. 4:40
39. 5:20
40. 10:40

41. **Flower Arranging** In Japanese flower arranging, you match a stem that is vertical with 0. You match other stems with numbers from 0 to 90, in both directions from the vertical. What numbers would the flowers shown be paired with on a standard protractor?

Real-World Connection

Japanese flower arranging makes precise use of angles to create a mood.

Algebra **Use the diagram, below right, for Exercises 42–45. Solve for x. Find the angle measures to check your work.**

42. $m\angle AOC = 7x - 2, m\angle AOB = 2x + 8, m\angle BOC = 3x + 14$
43. $m\angle AOB = 4x - 2, m\angle BOC = 5x + 10, m\angle COD = 2x + 14$

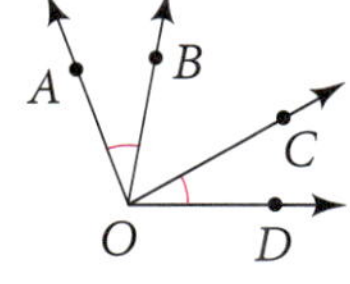

44. $m\angle AOB = 28, m\angle BOC = 3x - 2, m\angle AOD = 6x$
45. $m\angle AOB = 4x + 3, m\angle BOC = 7x, m\angle AOD = 16x - 1$

46. **Multiple Choice** If $m\angle MQV = 90$, which expression can you use to find $m\angle VQP$?
    - (A) $m\angle MQP - 90$
    - (B) $90 - m\angle MQV$
    - (C) $m\angle MQP + 90$
    - (D) $90 + m\angle VQP$

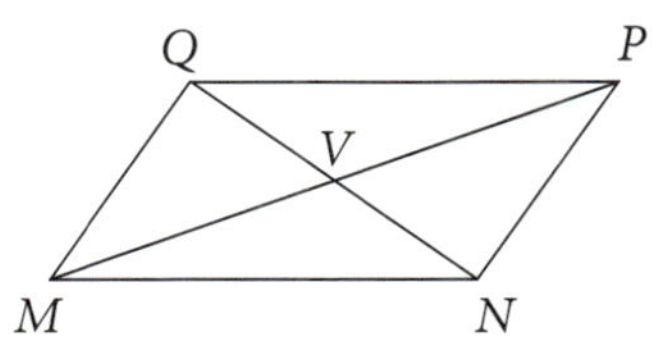

Visit: PHSchool.com
Web Code: bce-0106

47. a. **Algebra** Solve for x if $m\angle RQS = 2x + 4$ and $m\angle TQS = 6x + 20$.
    b. What is $m\angle RQS$? $m\angle TQS$?
    c. Show how you can check your answer.

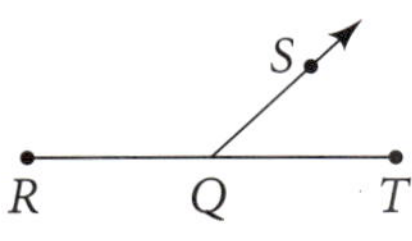

GO Online Lesson Quiz Visit: PHSchool.com, Web Code: bca-0106

C Challenge

48. $\overrightarrow{XC}$ bisects $\angle AXB$, $\overrightarrow{XD}$ bisects $\angle AXC$, $\overrightarrow{XE}$ bisects $\angle AXD$, $\overrightarrow{XF}$ bisects $\angle EXD$, $\overrightarrow{XG}$ bisects $\angle EXF$, and $\overrightarrow{XH}$ bisects $\angle DXB$. If $m\angle DXC = 16$, find $m\angle GXH$.

49. **Technology** Leon constructed an angle. Then he constructed a ray from the vertex of the angle to a point in the interior of the angle. He measured all the angles formed. Then he moved the interior ray. What postulate do the two pictures support?

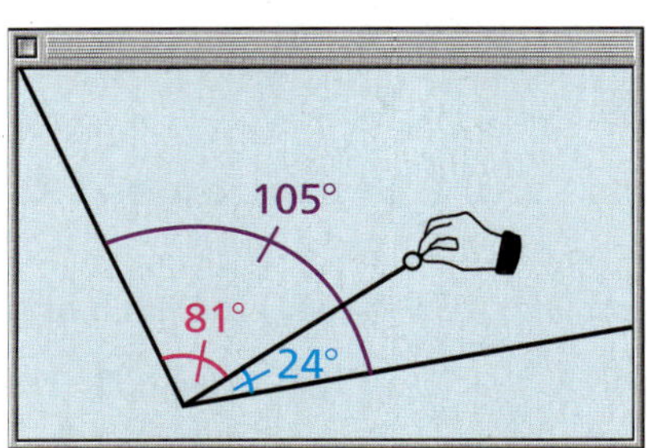

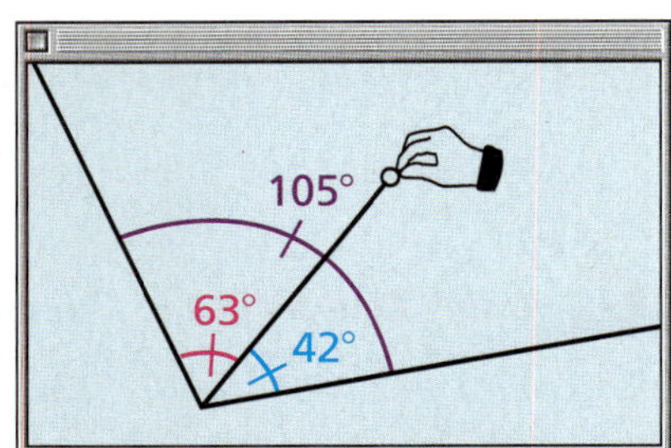

Multiple Choice Practice

For California Standards and CAHSEE Tutorials, visit PHSchool.com. Web Code: bcq-9045

GEOM 1.0

50. If $\angle A$ and $\angle B$ are supplementary angles, what angle relationship between $\angle A$ and $\angle B$ CANNOT be true?

Ⓐ $\angle A$ and $\angle B$ are right angles.
Ⓑ $\angle A$ and $\angle B$ are adjacent angles.
Ⓒ $\angle A$ and $\angle B$ are complementary angles.
Ⓓ $\angle A$ and $\angle B$ are congruent angles.

CAHSEE 7AF 3.3

51. What is the slope of the line shown in the graph at the right?

Ⓐ $-\frac{3}{2}$ Ⓑ $-\frac{2}{3}$ Ⓒ $\frac{2}{3}$ Ⓓ $\frac{3}{2}$

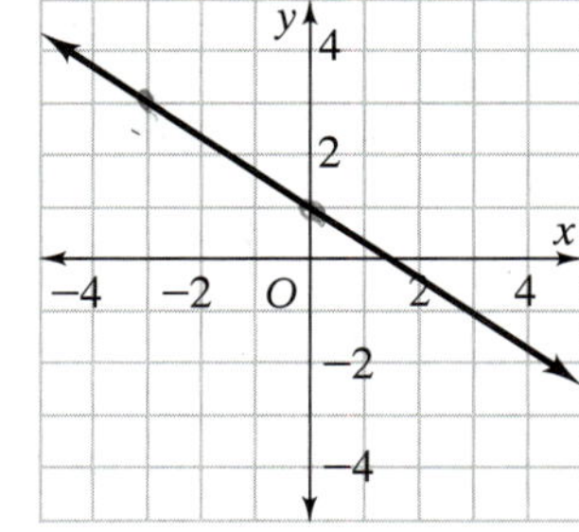

CAHSEE 7MR 1.2

52. Which situation CANNOT be represented by a linear function?

Ⓐ The number of feet in y yards
Ⓑ The number of yards in f feet
Ⓒ The time in hours a car travels 50 miles at r miles per hour
Ⓓ The distance a car travels at 40 miles per hour in h hours

Mixed Review

Lesson 1-5

Use the figure at the right for Exercises 53–54.

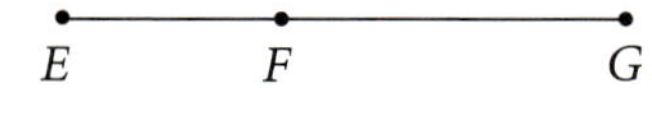

53. If $EG = 75$ and $EF = 28$, what is FG?

54. If $EG = 49$, $EF = 2x + 3$, and $FG = 4x - 2$, find x. Then find EF and FG.

Lesson 1-2

55. **Writing** Explain the difference between an orthographic drawing and an isometric drawing.

Lesson 1-1

Find one counterexample to show that each conjecture is false.

56. The quotient of two integers is not an integer.

57. An even number cannot have 5 as a factor.

Activity Lab Compass Designs

Hands-On

FOR USE WITH LESSON 1-7

Prepares for GEOM 16.0 Students perform basic constructions with a straightedge and compass, such as angle bisectors and perpendicular bisectors.

In Lesson 1-7, you use a compass to construct geometric figures. You can construct figures to show geometric relationships, to suggest new relationships, or simply to make appealing geometric designs like the following.

Step 1: Open your compass to about 2 inches. Make a circle and mark the point at the center of the circle.

Step 2: Keep the opening of your compass fixed. Place the compass point on the circle. With the pencil end, make a small arc to intersect the circle.

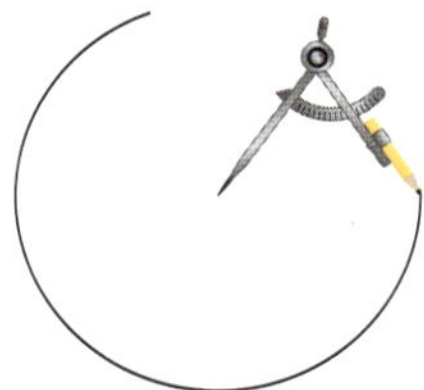

Step 1

Step 3: Place the compass point on the circle at the arc. Mark another arc. Continue around the circle this way to draw four more arcs—six in all.

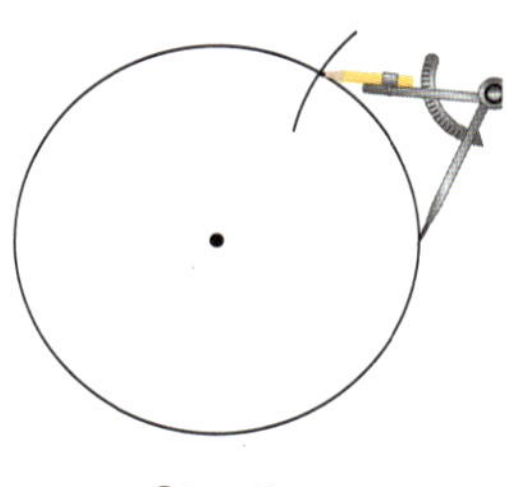

Step 2

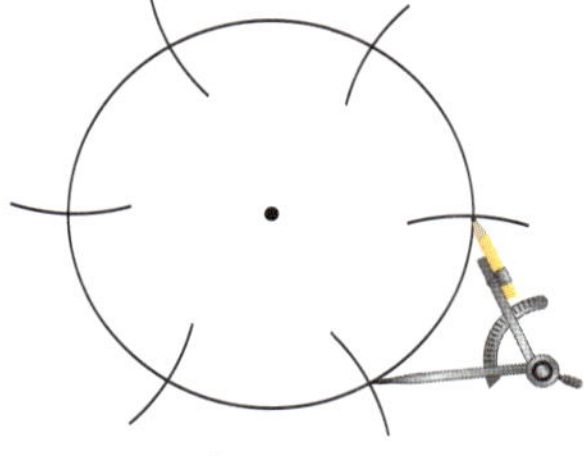

Step 3

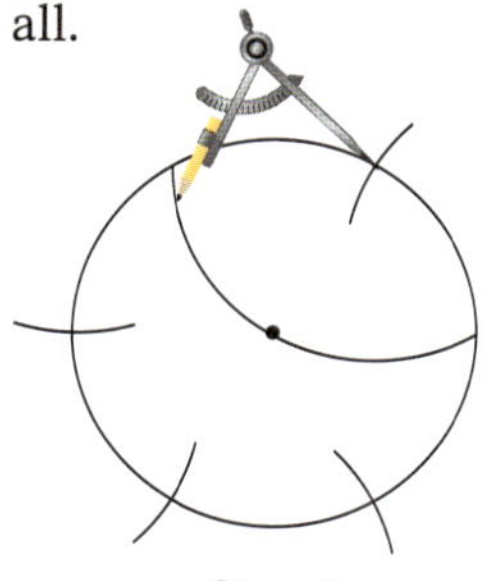

Step 4

Step 4: Place your compass point on an arc you marked on the circle. Place the pencil end at the next arc. Draw a large arc that passes through the circle's center and continues to another point on the circle.

Step 5: Draw six large arcs in this manner, each centered at one of the six points marked on the circle. Some people like to call the final design a daisy. You might like to color your daisy.

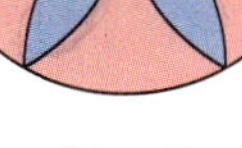

Step 5

EXERCISES

1. In Step 3, did your sixth mark on the circle land precisely on the point where you first placed your compass on the circle?
   a. Survey the class to find how many did.
   b. Because a point has no size, it is very difficult to land on exactly. Try to give a convincing argument that your sixth mark should have landed on your starting point.

2. Extend your design by using one of the six points on the circle as the center for a new circle. Repeat Steps 2–5 with this circle. Repeat several times to make interlocking circles. Color them as you wish.

1-7 Basic Constructions

California Content Standards

GEOM 16.0 Students perform basic constructions with a straightedge and compass, such as angle bisectors and perpendicular bisectors. ***Introduce***

What You'll Learn

- To use a compass and a straightedge to construct congruent segments and congruent angles
- To use a compass and a straightedge to bisect segments and angles

. . . And Why

To construct the bisector of an angle to illustrate angles of incidence and reflection, as in Exercise 18

Check Skills You'll Need

GO for Help Lessons 1-5 and 1-6

In Exercises 1–6, sketch each figure.

1. $\overline{CD}$ **2.** $\overrightarrow{GH}$ **3.** $\overleftrightarrow{AB}$

4. line m **5.** acute $\angle ABC$ **6.** $\overline{XY} \parallel \overline{ST}$

7. $DE = 20$. Point C is the midpoint of $\overline{DE}$. Find CE.

8. Use a protractor to draw a 60° angle.

9. Use a protractor to draw a 120° angle.

New Vocabulary • construction • straightedge • compass • perpendicular lines • perpendicular bisector • angle bisector

1 Constructing Segments and Angles

In a **construction** you use a straightedge and a compass to draw a geometric figure. A **straightedge** is a ruler with no markings on it. A **compass** is a geometric tool used to draw circles and parts of circles called arcs.

Four basic constructions involve constructing congruent segments, congruent angles, and bisectors of segments and angles.

1 EXAMPLE Constructing Congruent Segments

Construct a segment congruent to a given segment.

Given: $\overline{AB}$

Construct: $\overline{CD}$ so that $\overline{CD} \cong \overline{AB}$

A B

Step 1
Draw a ray with endpoint C.

C

Step 2
Open the compass to the length of $\overline{AB}$.

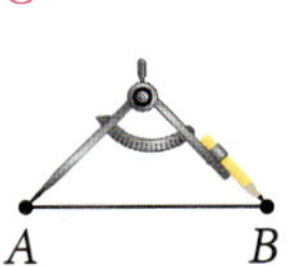

A B

Step 3
With the same compass setting, put the compass point on point C. Draw an arc that intersects the ray. Label the point of intersection D.

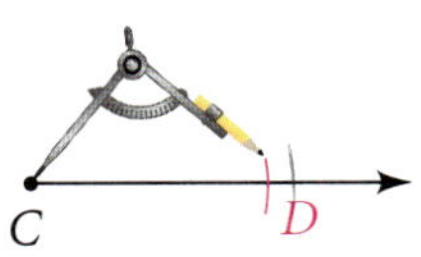

$\overline{CD} \cong \overline{AB}$

CA Standards Check 1 Use a straightedge to draw $\overline{XY}$. Then construct $\overline{RS}$ so that $RS = 2XY$.

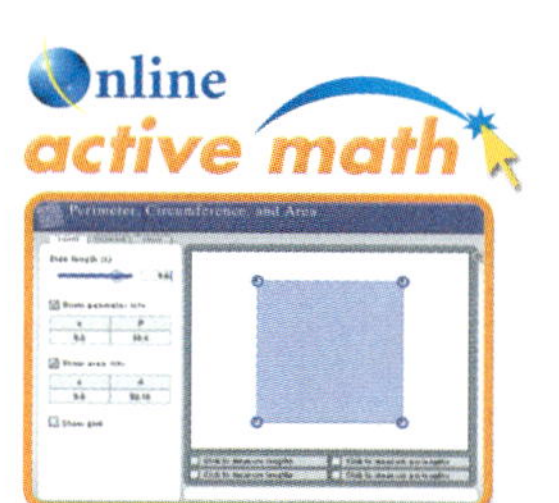

For: Construction Activity
Use: Interactive Textbook, 1-7

2 EXAMPLE Constructing Congruent Angles

Construct an angle congruent to a given angle.

Given: $\angle A$
Construct: $\angle S$ so that $\angle S \cong \angle A$

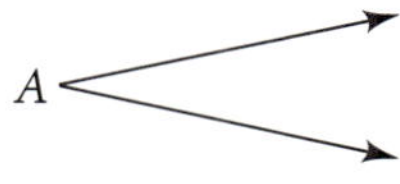

Step 1
Draw a ray with endpoint S.

Step 2
With the compass point on point A, draw an arc that intersects the sides of $\angle A$. Label the points of intersection B and C.

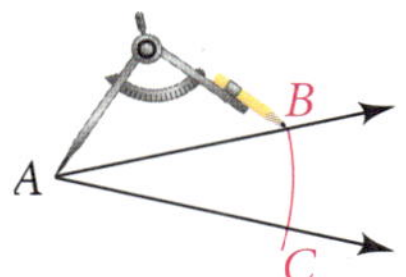

Step 3
With the same compass setting, put the compass point on point S. Draw an arc and label its point of intersection with the ray as R.

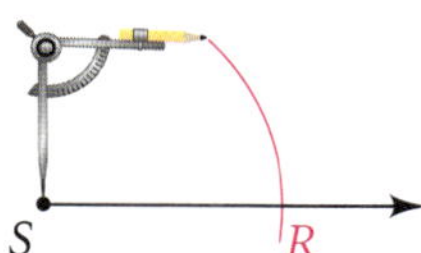

Step 4
Open the compass to the length BC. Keeping the same compass setting, put the compass point on R. Draw an arc to locate point T.

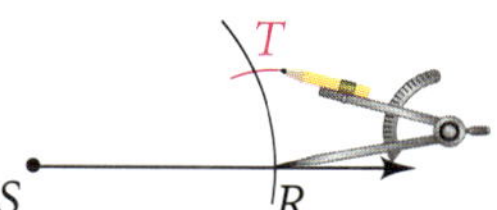

Step 5
Draw $\overrightarrow{ST}$.

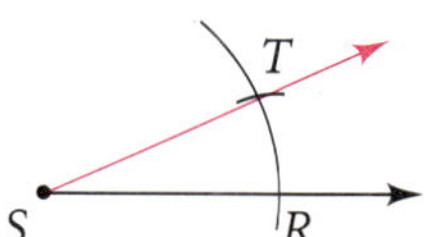

$\angle S \cong \angle A$

CA Standards Check 2 Construct $\angle F$ with $m\angle F = 2m\angle B$.

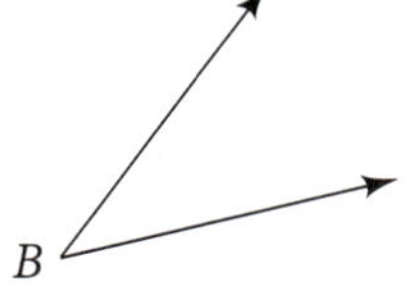

2 Constructing Bisectors

Real-World Connection

Perpendicular hands signal "Time out."

Perpendicular lines are two lines that intersect to form right angles. The symbol $\perp$ means "is perpendicular to." In the diagram at the right, $\overleftrightarrow{AB} \perp \overleftrightarrow{CD}$ and $\overleftrightarrow{CD} \perp \overleftrightarrow{AB}$.

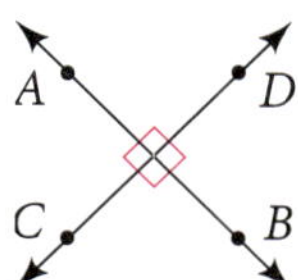

A **perpendicular bisector** of a segment is a line, segment, or ray that is perpendicular to the segment at its midpoint, thereby bisecting the segment into two congruent segments.

As you will learn in Chapter 5, there is just one line that is the perpendicular bisector of a segment in a given plane. Here is a way to construct the perpendicular bisector.

Real-World Connection

Architects use construction tools to work with their designs.

3 EXAMPLE Constructing the Perpendicular Bisector

Construct the perpendicular bisector of a segment.

Given: $\overline{AB}$
Construct: $\overleftrightarrow{XY}$ so that $\overleftrightarrow{XY} \perp \overline{AB}$ at the midpoint M of $\overline{AB}$.

Step 1
Put the compass point on point A and draw a long arc as shown. Be sure the opening is greater than $\frac{1}{2}AB$.

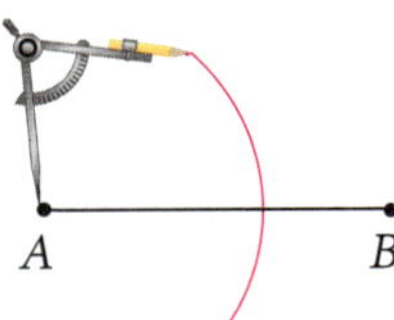

Step 2
With the same compass setting, put the compass point on point B and draw another long arc. Label the points where the two arcs intersect as X and Y.

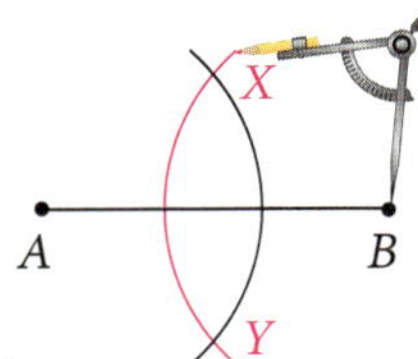

Step 3
Draw $\overleftrightarrow{XY}$. The point of intersection of $\overline{AB}$ and $\overleftrightarrow{XY}$ is M, the midpoint of $\overline{AB}$.

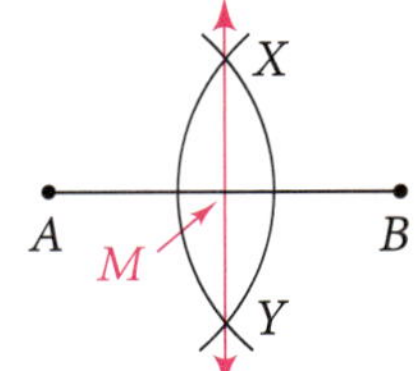

$\overleftrightarrow{XY} \perp \overline{AB}$ at the midpoint of $\overline{AB}$, so $\overleftrightarrow{XY}$ is the perpendicular bisector of $\overline{AB}$.

3 Draw $\overline{ST}$. Construct its perpendicular bisector.

An **angle bisector** is a ray that divides an angle into two congruent coplanar angles. Its endpoint is at the angle vertex. Within the ray, a segment with the same endpoint is also an angle bisector. You may say that the ray or segment *bisects* the angle.

4 EXAMPLE Finding Angle Measures

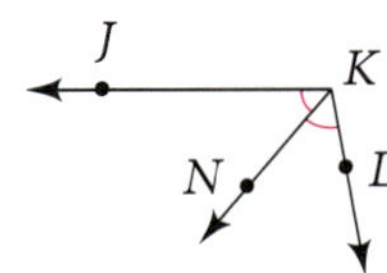

Algebra $\overrightarrow{KN}$ bisects $\angle JKL$ so that $m\angle JKN = 5x - 25$ and $m\angle NKL = 3x + 5$. Solve for x and find $m\angle JKN$.

$m\angle JKN = m\angle NKL$	**Definition of angle bisector**
$5x - 25 = 3x + 5$	**Substitute.**
$5x = 3x + 30$	**Add 25 to each side.**
$2x = 30$	**Subtract 3*x* from each side.**
$x = 15$	**Divide each side by 2.**
$m\angle JKN = 5x - 25 = 5(15) - 25 = 50$	**Substitute 15 for *x*.**

$m\angle JKN = 50$

 Find $m\angle NKL$ and $m\angle JKL$.

5 EXAMPLE Constructing the Angle Bisector

Construct the bisector of an angle.

Given: $\angle A$

Construct: $\overrightarrow{AX}$, the bisector of $\angle A$

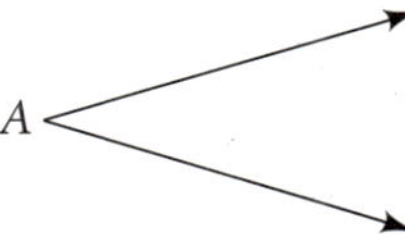

Step 1
Put the compass point on vertex A. Draw an arc that intersects the sides of $\angle A$. Label the points of intersection B and C.

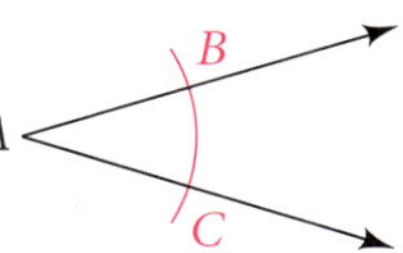

Step 2
Put the compass point on point C and draw an arc. With the same compass setting, draw an arc using point B. Be sure the arcs intersect. Label the point where the two arcs intersect as X.

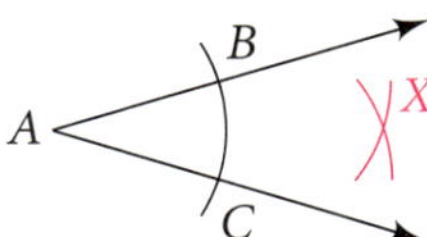

Step 3
Draw $\overrightarrow{AX}$.

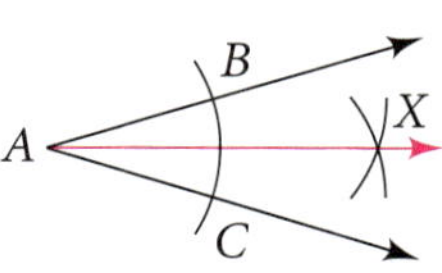

$\overrightarrow{AX}$ is the bisector of $\angle CAB$.

CA Standards Check 5 **a.** Draw obtuse $\angle XYZ$. Then construct its bisector $\overrightarrow{YP}$.
b. Explain how you can use your protractor to check your construction.

EXERCISES Standards Practice

For more exercises, see *Extra Skill, Word Problem, and Proof Practice.*

GEOM 16.0

A Practice by Example

GO for Help

In Exercises 1–8, draw a diagram similar to the given one. Then do the construction. Check your work with a ruler or a protractor.

Example 1 (page 44)

1. Construct $\overline{XY}$ congruent to $\overline{AB}$.

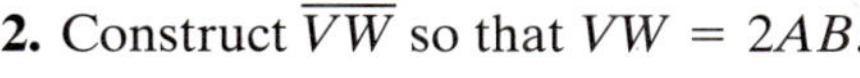

2. Construct $\overline{VW}$ so that $VW = 2AB$.

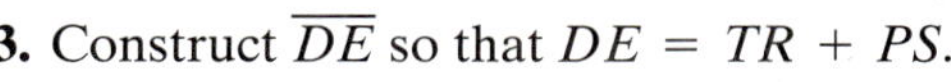

3. Construct $\overline{DE}$ so that $DE = TR + PS$.
4. Construct $\overline{QJ}$ so that $QJ = TR - PS$.

A B

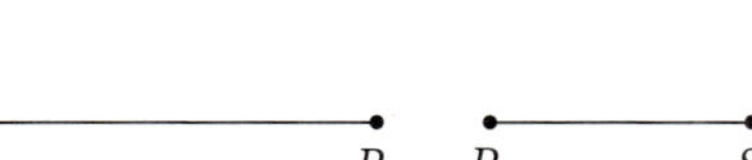

Example 2 (page 45)

5. Construct $\angle D$ so that $\angle D \cong \angle C$.
6. Construct $\angle F$ so that $m\angle F = 2m\angle C$.

C

Example 3 (page 46)

7. Construct the perpendicular bisector of $\overline{AB}$.
8. Construct the perpendicular bisector of $\overline{TR}$.

Example 4 (page 46)

x^2 9. **Algebra** $\overrightarrow{GH}$ bisects $\angle FGI$.
   a. Solve for x and find $m\angle FGH$.
   b. Find $m\angle HGI$.
   c. Find $m\angle FGI$.

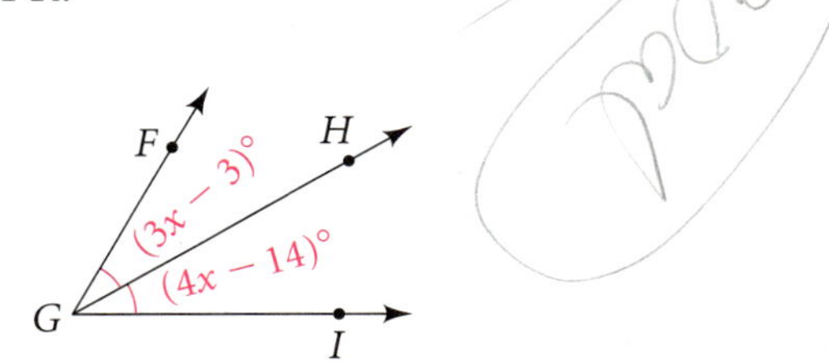

x^2 **Algebra For Exercises 10–12, $\overrightarrow{BX}$ bisects $\angle ABC$. Solve for x and find $m\angle ABC$.**

10. $m\angle ABX = 5x, m\angle XBC = 3x + 10$

11. $m\angle ABC = 4x - 12, m\angle ABX = 24$

12. $m\angle ABX = 4x - 16, m\angle CBX = 2x + 6$

Example 5 (page 47)

13. Draw acute $\angle PQR$. Then construct its bisector.

14. Draw right $\angle TUV$. Then construct its bisector.

B Apply Your Skills

15. Use your protractor and draw $\angle W$ with $m\angle W = 120$. Construct $\angle Z \cong \angle W$. Then construct the bisector of $\angle Z$.

Sketch the figure described. Explain how to construct it. Then do the construction.

16. $\overleftrightarrow{XY} \perp \overleftrightarrow{YZ}$

17. $\overrightarrow{ST}$ bisecting right $\angle PSQ$

18. Optics A beam of light and a mirror can be used to study the behavior of light. Light that strikes the mirror is reflected so that the angle of reflection and the angle of incidence are congruent. In the diagram, $\overline{BC}$ is perpendicular to the mirror, and $\angle ABC$ has a measure of 41°.

Angle of incidence
Angle of reflection
A, B, C, D, E, F

a. Name the angle of reflection and find its measure.

b. Find $m\angle ABD$.

c. Find $m\angle ABE$ and $m\angle DBF$.

19. Use a straightedge and protractor.

a. Draw a mirror and a light beam striking the mirror and reflecting from it.

b. Construct the bisector of the angle formed by the incoming and reflected light beams. Label the angles of incidence and reflection.

PEANUTS® by Charles M. Schulz

20. Open-Ended Snoopy can draw squares with his compass. You can only draw circles. You can, however, construct a square. Explain how to do this. Use sketches if needed. Then do the construction.

21. Answer these questions about a segment in a plane. Explain each answer.

a. How many midpoints does the segment have?

b. How many bisectors does it have? How many lines in the plane are its perpendicular bisectors?

c. How many lines in space are its perpendicular bisectors?

Visit: PHSchool.com
Web Code: bce-0107

For Exercises 22–24, copy $\angle 1$ and $\angle 2$.

1
2

22. Construct $\angle B$ so that $m\angle B = m\angle 1 + m\angle 2$.

23. Construct $\angle C$ so that $m\angle C = m\angle 1 - m\angle 2$.

24. Construct $\angle D$ so that $m\angle D = 2m\angle 2$.

25. **Reasoning** When $\overrightarrow{BX}$ bisects $\angle ABC$, $\angle ABX \cong \angle CBX$. Lani claims there is always a related equation, $m\angle ABX = \frac{1}{2}m\angle ABC$. Denyse claims the related equation is $2m\angle ABX = m\angle ABC$. Which equation is correct? Explain.

26. **Writing** Describe how to construct the midpoint of a segment.

27. Construct a 45° angle.

28. **a.** Draw a large triangle with three acute angles. Construct the bisectors of the three angles. What appears to be true about the three angle bisectors?
   b. Repeat the constructions with a triangle that has one obtuse angle.
   c. **Make a Conjecture** What appears to be true about the three angle bisectors of any triangle?

Use a ruler to draw segments of 2 cm, 4 cm and 5 cm. Then construct each triangle, if possible. If not possible, explain.

29. with 4-cm, 4-cm, and 5-cm sides

30. with 2-cm, 5-cm, and 5-cm sides

31. with 2-cm, 2-cm, and 5-cm sides

32. with 2-cm, 2-cm, and 4-cm sides

33. **a.** Draw a segment, $\overline{XY}$. Construct a triangle with sides congruent to $\overline{XY}$.
   b. Measure the angles of the triangle.
   c. **Writing** Describe how to construct a 60° angle; a 30° angle.

34. **Multiple Choice** Which steps best describe how to construct this pattern?

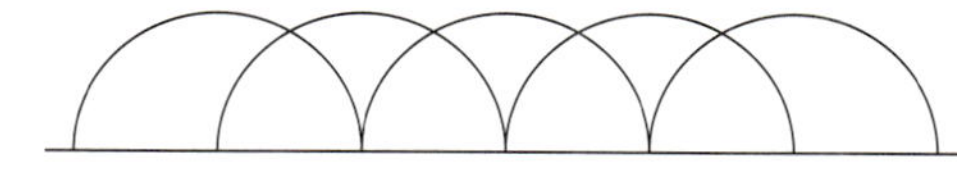

   (A) Use a straightedge to draw the segment and then a compass to draw five half circles.
   (B) Use a straightedge to draw the segment and then a compass to draw six half circles.
   (C) Use a compass to draw five half circles and then a straightedge to join their ends.
   (D) Use a compass to draw six half circles and then a straightedge to join their ends.

Challenge

35. **a.** Use your compass to draw a circle. Locate three points A, B, and C on the circle.
   b. Construct the perpendicular bisectors of $\overline{AB}$ and $\overline{BC}$.
   c. **Critical Thinking** Label the intersection of the two perpendicular bisectors as point O. Make a conjecture about point O.

36. Study the figures. Complete the definition of a line perpendicular to a plane:

   A line is perpendicular to a plane if it is __?__ to every line in the plane that __?__.

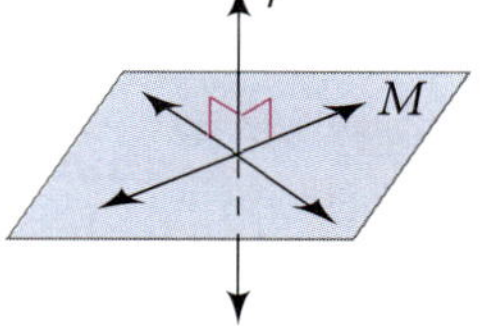

Line $r \perp$ plane M.

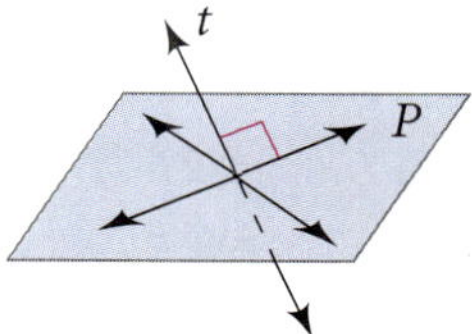

Line t is not $\perp$ plane P.

Multiple Choice Practice

For California Standards and CAHSEE Tutorials, visit PHSchool.com. Web Code: bcq-9045

GEOM 16.0

37. Which construction requires drawing only one arc with a compass?
   (A) constructing congruent segments
   (B) constructing congruent angles
   (C) constructing the perpendicular bisector
   (D) constructing the angle bisector

CAHSEE 7AF 4.2

38. Four years ago Miss Tanya's preschool had 65 students. Now the preschool has 95 students. If it continues to grow at this rate, about how many students will it have three years from now?

(A) 95 (B) 101 (C) 118 (D) 125

CAHSEE Alg1 6.0

39. Find the y-intercept of the graph of $7x + y + 9 = 0$.

(A) -9 (B) -7 (C) 7 (D) 9

CAHSEE Alg1 9.0

40. Solve the system $\begin{cases} 3x - y = -1 \\ 2x + 3y = -30. \end{cases}$

(A) $(8, 3)$ (B) $(-8, -3)$ (C) $(3, 8)$ (D) $(-3, -8)$

Mixed Review

GO for Help

Lesson 1-6

41. Use a protractor to draw a 72° angle.

42. $\angle DEF$ is a straight angle. $m\angle DEG = 80$. Find $m\angle GEF$.

43. $m\angle TUV = 100$ and $m\angle VUW = 80$. Find possible values of $m\angle TUW$.

Lesson 1-5

Use the number line at the right. Find the length of each segment.

44. $\overline{AC}$ **45.** $\overline{AD}$

46. $\overline{CD}$ **47.** $\overline{BC}$

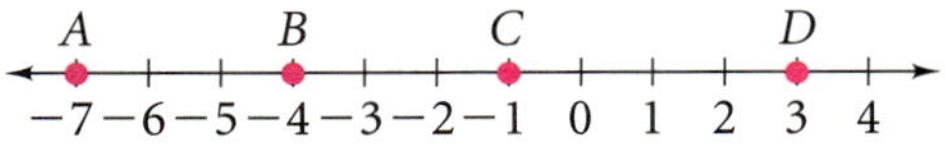

Lesson 1-4

48. Draw $\overleftrightarrow{RS}$.

Use your drawing from Exercise 48. Answer and explain.

49. Are $\overrightarrow{RS}$ and $\overrightarrow{SR}$ opposite rays? **50.** Are $\overline{RS}$ and $\overline{SR}$ the same segment?

Geometry at Work

Cabinetmaker

Cabinetmakers not only make cabinets but all types of wooden furniture. The artistry of cabinetmaking can be seen in the beauty and uniqueness of the finest doors, shelves, and tables. The craft is in knowing which types of wood and tools to use, and how to use them.

The carpenter's square is one of the most useful of the cabinetmaker's tools. It can be applied to a variety of measuring tasks. The figure shows how to use a carpenter's square to bisect $\angle O$.

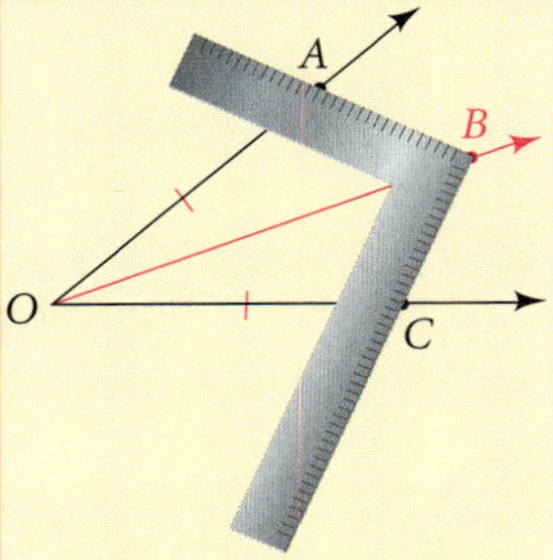

First, mark equal lengths OA and OC on the sides of the angle. Then position the square so that $BA = BC$ to locate point B. Finally, draw $\overrightarrow{OB}$. $\overrightarrow{OB}$ bisects $\angle O$.

Go Online PHSchool.com

For: Information about cabinetmaking
Web Code: bcb-2031

Activity Lab Exploring Constructions

Technology

FOR USE WITH LESSON 1-7

Points, lines, and figures are made in geometry software using Draw tools or Construct tools. A figure made by Draw has no constraints. When the figure is manipulated, it moves or changes size freely. A figure made by Construct is dependent upon an existing object. When you manipulate the existing object, the constructed object moves or resizes accordingly.

In this activity you will explore the difference between Draw and Construct. Before you begin, familiarize yourself with the tools of your software.

Draw and Construct

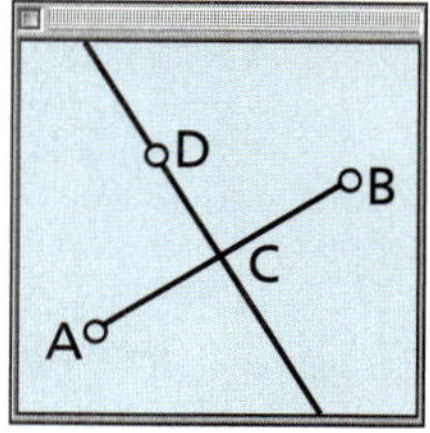

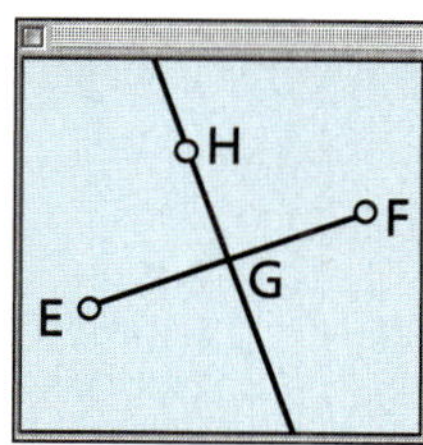

- Draw $\overline{AB}$ and Construct the perpendicular bisector $\overleftrightarrow{DC}$.
- Draw $\overline{EF}$ and Construct G, any point on $\overline{EF}$. Draw $\overleftrightarrow{HG}$. Find EG, GF, and $m\angle HGF$. Try to drag G so that $EG = GF$. Try to drag H so that $m\angle HGF = 90$. Were you able to draw the perpendicular bisector of $\overline{EF}$? Explain.

Investigate

- Drag A and B. Observe AC, CB, and $m\angle DCB$. Is $\overleftrightarrow{DC}$ always the perpendicular bisector of $\overline{AB}$ no matter how you manipulate the figure?
- Drag E and F. Observe EG, GF, and $m\angle HGF$. How is the relationship between $\overline{EF}$ and $\overleftrightarrow{HG}$ different from the relationship between $\overline{AB}$ and $\overleftrightarrow{DC}$?

EXERCISES

1. a. Write a description of the general difference between Draw and Construct.

b. Use your description to explain why the relationship between $\overline{EF}$ and $\overleftrightarrow{HG}$ differs from the relationship between $\overline{AB}$ and $\overleftrightarrow{DC}$.

2. a. Draw $\angle JKL$.

b. Construct its angle bisector, $\overrightarrow{KM}$.

c. Manipulate the figure and observe the different angle measures. Is $\overrightarrow{KM}$ always the angle bisector of $\angle JKL$?

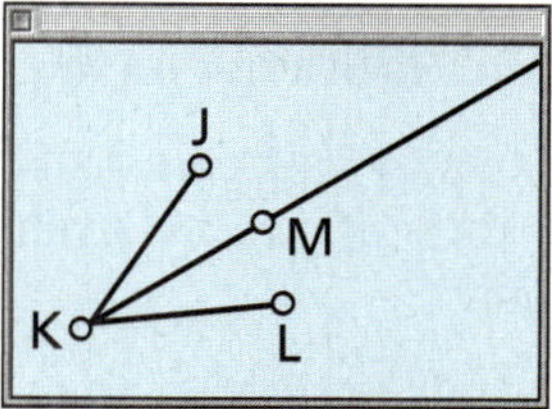

3. a. Draw $\angle NOP$. Draw $\overrightarrow{OQ}$ in the interior of $\angle NOP$. Drag Q until $m\angle NOQ = m\angle QOP$.

b. Manipulate the figure and observe the different angle measures. Is $\overrightarrow{OQ}$ always the angle bisector of $\angle NOP$?

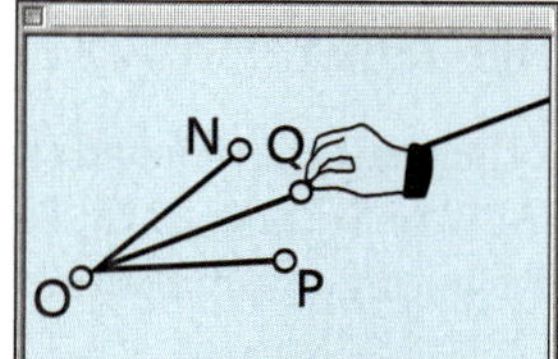

4. For Exercise 2, how can you manipulate the figure on the screen so that it shows a right angle? Justify your answer.

Distance in the Coordinate Plane

Hands-On

For Use With Lesson 1-8

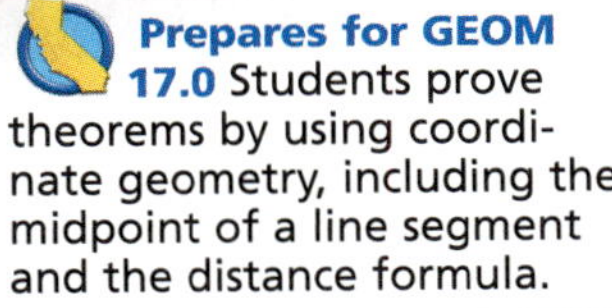

Much of Manhattan is laid out in a rectangular grid, as shown in this map. In general, the streets are parallel running east and west. The avenues are parallel running north and south.

Yvonne's family is at the corner of 44th Street and 7th Avenue. They plan to walk to Madison Square Park at 23rd Street and 5th Avenue. There are several possible routes they can take.

EXERCISES

Use tracing paper to trace the routes on the map. Answer the following questions.

1. Yvonne's father wants to walk east on 44th Street until they reach 5th Avenue. He then plans to walk south on 5th Avenue to Madison Square Park. About how long is his route?
2. Yvonne's mother wants to walk south on 7th Avenue until they reach 23rd Street. She then plans to walk east on 23rd Street to Madison Square Park. About how long is her route?
3. Yvonne notices on the map that Broadway cuts across the grid of streets and leads directly to Madison Square Park. She suggests walking all the way on Broadway. About how long is her route?
4. Whose route is the shortest? Explain.
5. Whose route is the longest? Explain.
6. Give a plausible argument for the following statement: There is no longest route from Times Square to Madison Square Park.

1-8 The Coordinate Plane

California Content Standards

Prepares for GEOM 17.0 Students prove theorems by using coordinate geometry, including the midpoint of a line segment and the distance formula.

What You'll Learn

- To find the distance between two points in the coordinate plane
- To find the coordinates of the midpoint of a segment in the coordinate plane

. . . And Why

To find the distance between two points on a map, as in Example 2

Check Skills You'll Need

for Help Skills Handbook pages 753 and 754

Find the square root of each number. Round to the nearest tenth if necessary.

1. 25 **2.** 17 **3.** 123

Algebra Evaluate each expression for $m = -3$ and $n = 7$.

4. $(m - n)^2$ **5.** $(n - m)^2$ **6.** $m^2 + n^2$

Algebra Evaluate each expression for $a = 6$ and $b = -8$.

7. $(a - b)^2$ **8.** $\sqrt{a^2 + b^2}$ **9.** $\frac{a + b}{2}$

1 Finding Distance on the Coordinate Plane

You can think of a point as a dot, and a line as a series of points. In coordinate geometry you describe a point by an ordered pair (x, y), called the *coordinates of the point.*

Vocabulary Tip

To help review terms shown here, see "Coordinate plane" in the Glossary.

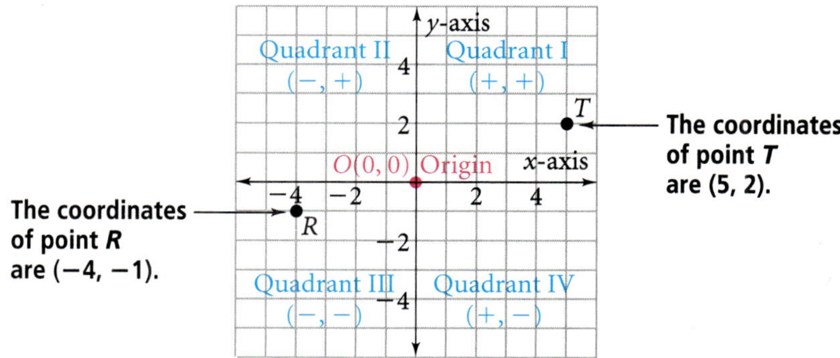

You can use the Ruler Postulate to find the distance between two points if the points are on a horizontal line or a vertical line. To find the distance between two points that are not on a horizontal or vertical line, you can use the Distance Formula.

Key Concepts

Formula **The Distance Formula**

The distance d between two points $A(x_1, y_1)$ and $B(x_2, y_2)$ is

$$d = \sqrt{(x_2 - x_1)^2 + (y_2 - y_1)^2}.$$

You will verify this formula in Chapter 8.

1 EXAMPLE Finding Distance

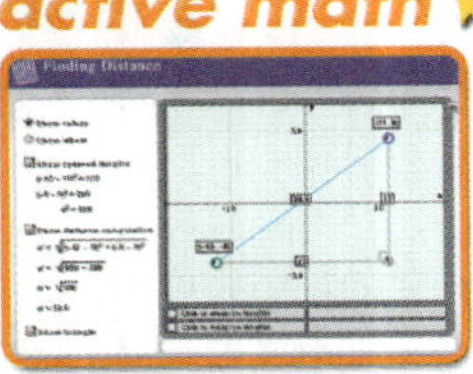

For: Distance Activity
Use: Interactive Textbook, 1-8

Find the distance between $T(5, 2)$ and $R(-4, -1)$ to the nearest tenth.

Let $(5, 2)$ be (x_1, y_1) and $(-4, -1)$ be (x_2, y_2).

$d = \sqrt{(x_2 - x_1)^2 + (y_2 - y_1)^2}$ **Use the Distance Formula.**

$d = \sqrt{(-4 - 5)^2 + (-1 - 2)^2}$ **Substitute.**

$d = \sqrt{(-9)^2 + (-3)^2}$ **Simplify.**

$d = \sqrt{81 + 9} = \sqrt{90}$

90 [√] 9.4868330 **Use a calculator.**

To the nearest tenth, $TR = 9.5$.

1 **a.** $\overline{AB}$ has endpoints $A(1, -3)$ and $B(-4, 4)$. Find AB to the nearest tenth.
b. Critical Thinking In Example 1, suppose you let $(-4, -1)$ be (x_1, y_1) and $(5, 2)$ be (x_2, y_2). Do you get the same result? Why?

2 EXAMPLE Real-World Connection

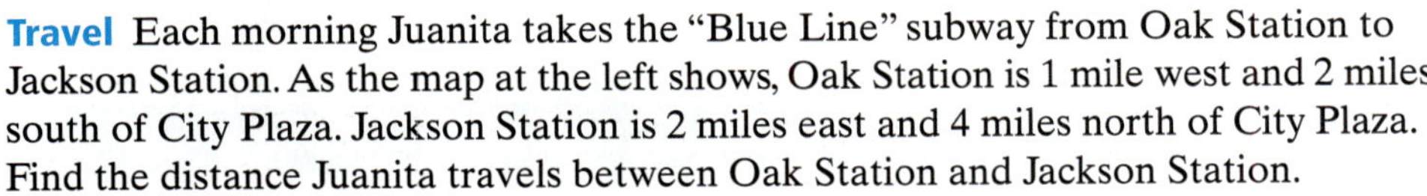

Travel Each morning Juanita takes the "Blue Line" subway from Oak Station to Jackson Station. As the map at the left shows, Oak Station is 1 mile west and 2 miles south of City Plaza. Jackson Station is 2 miles east and 4 miles north of City Plaza. Find the distance Juanita travels between Oak Station and Jackson Station.

Let Oak$(-1, -2)$ be (x_1, y_1) and Jackson $(2, 4)$ be (x_2, y_2).

$d = \sqrt{(x_2 - x_1)^2 + (y_2 - y_1)^2}$ **Use the Distance Formula.**

$d = \sqrt{(2 - (-1))^2 + (4 - (-2))^2}$ **Substitute.**

$d = \sqrt{3^2 + 6^2}$ **Simplify.**

$d = \sqrt{9 + 36} = \sqrt{45}$

45 [√] 6.7082039 **Use a calculator.**

Juanita travels about 6.7 miles between Oak Station and Jackson Station.

2 **a.** Find the distance between Elm Station and Symphony Station.
b. Maple Station is located 6 miles west and 2 miles north of City Plaza. Find the distance between Cedar Station and Maple Station.

2 Finding the Midpoint of a Segment

To find the coordinate of the midpoint of a segment on a number line, find the *average* or *mean* of the coordinates of the endpoints. The coordinate of the midpoint of a segment with endpoints a and b is $\frac{a + b}{2}$.

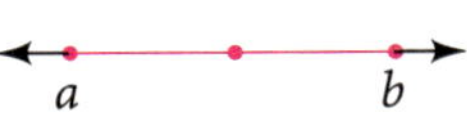

You can extend this process (see next page) to find the coordinates of the midpoint of a segment in the coordinate plane.

Real-World Connection

A well-made seesaw balances (without riders) at its midpoint.

Study the diagram of $\overline{TS}$ with endpoints $T(4, 3)$ and $S(8, 5)$. $\overline{TR}$ is a horizontal segment and $\overline{SR}$ is a vertical segment. The coordinates of R are $(8, 3)$.

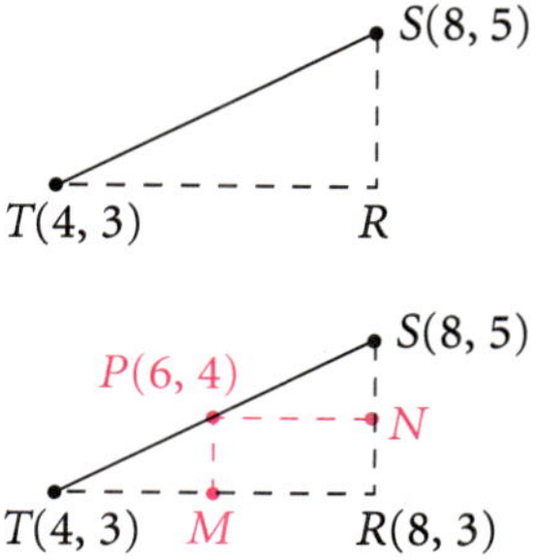

The coordinates of M, the midpoint of $\overline{TR}$, are $(6, 3)$. The coordinates of N, the midpoint of $\overline{SR}$, are $(8, 4)$. A vertical line through M and a horizontal line through N meet at P, the midpoint of $\overline{TS}$.

The coordinates of P are $(6, 4)$.

Thus, you find the coordinates of the midpoint of a segment by averaging the x-coordinates and averaging the y-coordinates of the endpoints.

Key Concepts

Formula **The Midpoint Formula**

The coordinates of the midpoint M of $\overline{AB}$ with endpoints $A(x_1, y_1)$ and $B(x_2, y_2)$ are the following:

$$M = \left(\frac{x_1 + x_2}{2}, \frac{y_1 + y_2}{2}\right)$$

3 EXAMPLE Finding the Midpoint

Algebra $\overline{QS}$ has endpoints $Q(3, 5)$ and $S(7, -9)$. Find the coordinates of its midpoint M.

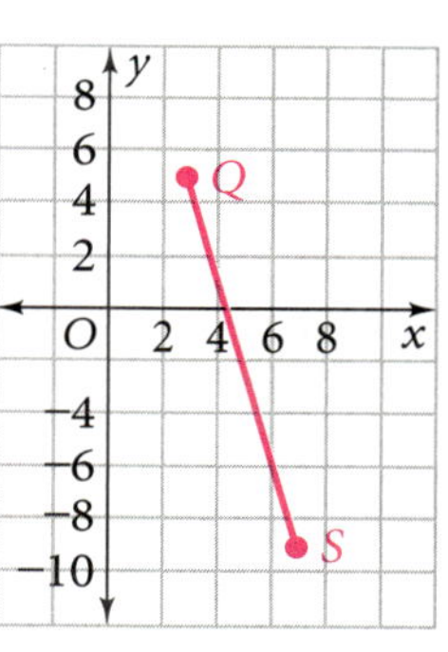

Let $(3, 5)$ be (x_1, y_1) and $(7, -9)$ be (x_2, y_2).

x-coordinate of $M = \dfrac{x_1 + x_2}{2} = \dfrac{3 + 7}{2} = \dfrac{10}{2} = 5$

y-coordinate of $M = \dfrac{y_1 + y_2}{2} = \dfrac{5 + (-9)}{2} = \dfrac{-4}{2} = -2$

The coordinates of midpoint M are $(5, -2)$.

CA Standards Check 3 Find the coordinates of the midpoint of $\overline{XY}$ with endpoints $X(2, -5)$ and $Y(6, 13)$.

4 EXAMPLE Finding an Endpoint

Algebra The midpoint of $\overline{AB}$ is $M(3, 4)$. One endpoint is $A(-3, -2)$. Find the coordinates of the other endpoint B.

Use the Midpoint Formula. Let the coordinates of B be (x_2, y_2).

$3 = \dfrac{-3 + x_2}{2}$	← **Midpoint Formula** →	$4 = \dfrac{-2 + y_2}{2}$
$6 = -3 + x_2$	← **Multiply each side by 2.** →	$8 = -2 + y_2$
$9 = x_2$		$10 = y_2$

The coordinates of B are $(9, 10)$.

CA Standards Check 4 The midpoint of $\overline{XY}$ has coordinates $(4, -6)$. X has coordinates $(2, -3)$. Find the coordinates of Y.

EXERCISES

For more exercises, see *Extra Skill, Word Problem, and Proof Practice.*

Standards Practice

A Practice by Example

Example 1 (page 54)

GO for Help

Find the distance between the points to the nearest tenth.

1. $J(2, -1), K(2, 5)$ **2.** $L(10, 14), M(-8, 14)$ **3.** $N(-1, -11), P(-1, -3)$

4. $A(0, 3), B(0, 12)$ **5.** $C(12, 6), D(-8, 18)$ **6.** $E(6, -2), F(-2, 4)$

7. $Q(12, -12), T(5, 12)$ **8.** $R(0, 5), S(12, 3)$ **9.** $X(-3, -4), Y(5, 5)$

Example 2 (page 54)

Use the map in Example 2 on page 54. Find the distance between the stations.

10. North and South **11.** Oak and Symphony **12.** City Plaza and Cedar

Use the map at the right. Find the distances between the stations to the nearest tenth.

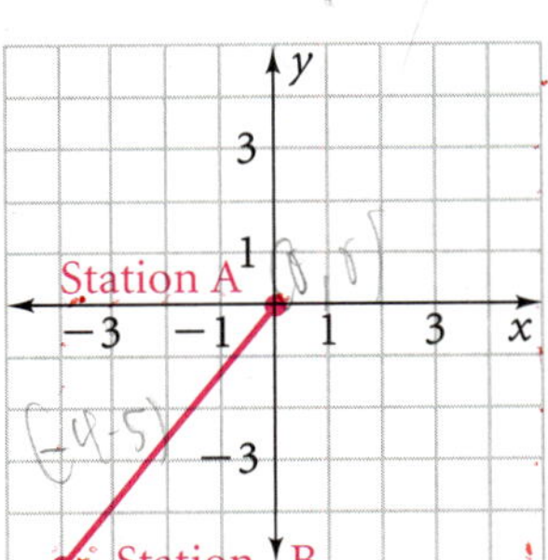

13. Station A and Station B

14. Station B and Station C located at (5, 8)

15. Station B and Station D located at (1, 10)

16. Station E at (2, 12) and Station F at (5, 16)

17. List the stations B, C, D, E, and F in the order of least to greatest distance from Station A.

Example 3 (page 55)

x^2 **Algebra** **Find the coordinates of the midpoint of $\overline{HX}$.**

18. $H(0, 0), X(8, 4)$ **19.** $H(-1, 3), X(7, -1)$

20. $H(13, 8), X(-6, -6)$ **21.** $H(7, 10), X(5, -8)$

22. $H(-6.3, 5.2), X(1.8, -1)$ **23.** $H(5\frac{1}{2}, -4\frac{3}{4}), X(2\frac{1}{4}, -1\frac{1}{4})$

Example 4 (page 55)

x^2 **Algebra** **The coordinates of point T are given. The midpoint of $\overline{ST}$ has coordinates (5, −8). Find the coordinates of point S.**

24. $T(0, 4)$ **25.** $T(5, -15)$ **26.** $T(10, 18)$

27. $T(-2, 8)$ **28.** $T(1, 12)$ **29.** $T(4.5, -2.5)$

An endpoint and a midpoint are given. Find the coordinates of the other endpoint.

30. endpoint (2, 6), midpoint (5, 12) **31.** endpoint (2, 3), midpoint (3, −4)

B Apply Your Skills

Find (a) PQ to the nearest tenth and (b) the coordinates of the midpoint of $\overline{PQ}$.

32. $P(3, 2), Q(6, 6)$ **33.** $P(0, -2), Q(3, 3)$ **34.** $P(-4, -2), Q(1, 3)$

35. $P(-5, 2), Q(0, 4)$ **36.** $P(-3, -1), Q(5, -7)$ **37.** $P(-5, -3), Q(-3, -5)$

38. $P(-4, -5), Q(-1, 1)$ **39.** $P(2, 3), Q(4, -2)$ **40.** $P(4, 2), Q(3, 0)$

41. The midpoint of $\overline{TS}$ is the origin. Point T is located in Quadrant II. What quadrant contains point S?

42. Graph the points $A(2, 1)$, $B(6, -1)$, $C(8, 7)$, and $D(4, 9)$. Draw quadrilateral $ABCD$. Use the Midpoint Formula to find the midpoints of $\overline{AC}$ and $\overline{BD}$. What appears to be true?

Visit: PHSchool.com
Web Code: bce-0108

43. **Multiple Choice** $A(4, -1)$ and $B(-2, 3)$ are points in a coordinate plane. M is the midpoint of $\overline{AB}$. What is the length of $\overline{MB}$ to the nearest tenth of a unit?
Ⓐ 2.5 units Ⓑ 3.6 units Ⓒ 7.2 units Ⓓ 13.0 units

For each graph, find (a) AB to the nearest tenth and (b) the coordinates of the midpoint of $\overline{AB}$.

44. 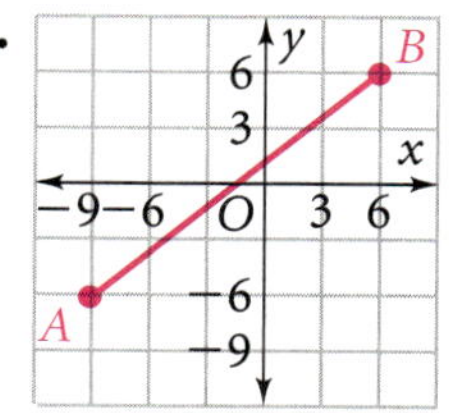

45. 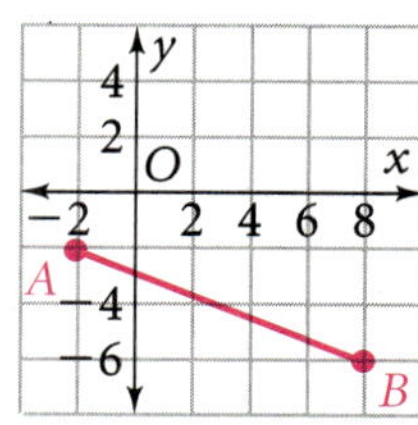

46. 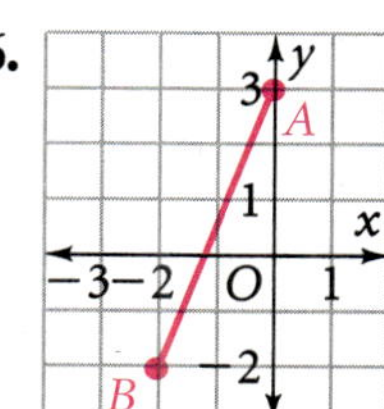

GO for Help

For a guide to solving Exercise 47, see page 60.

47. **Writing** An airplane at $T(80, 20)$ needs to fly to both $U(20, 60)$ and $V(110, 85)$. What is the shortest possible distance for the trip? Explain.

48. **Navigation** A boat at $X(5, -2)$ needs to travel to $Y(-6, 9)$ or $Z(17, -3)$. Which point is closer? What is the distance to the closer point?

Communications The cell phone screen at the right shows coordinates of six cities from a grid placed on North America by a long-distance carrier. The carrier finds distance by the Distance Formula. Each grid unit equals $\sqrt{0.1}$ mile. Find the distance between each pair of cities to the nearest mile.

SOURCE: Peter H. Dana

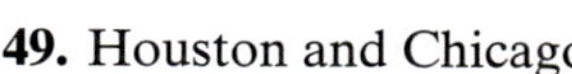

49. Houston and Chicago

50. Denver and New Orleans

51. Boston and San Francisco

52. New Orleans and Houston

Graph $X(-2, 1)$, $Y(2, 3)$, $A(-1, 4)$, $B(0, 2)$, and $C(4, 2)$. For each point described below, give two sets of possible coordinates if they exist. Otherwise, write *exactly one point* and give the coordinates, or *not possible* and explain.

53. point D so that $\overleftrightarrow{AD} \parallel \overleftrightarrow{XY}$

54. E so that $\overleftrightarrow{EC} \parallel \overleftrightarrow{XY}$

55. point F so that $\overleftrightarrow{FB} \perp \overleftrightarrow{XY}$

56. point G so that $\overleftrightarrow{GC} \perp \overleftrightarrow{XY}$

Challenge

57. point H so that $\overleftrightarrow{HX} \parallel \overleftrightarrow{AY}$, and $\overleftrightarrow{HA} \parallel \overleftrightarrow{XY}$

58. point J so that $\overleftrightarrow{JB} \perp \overleftrightarrow{XY}$, and $\overleftrightarrow{JC} \perp \overleftrightarrow{CY}$

59. **Open-Ended** In a coordinate plane, draw any $\overline{AB}$. Draw another segment that is both congruent and parallel to $\overline{AB}$. Label the new segment $\overline{CD}$ in such a way that $ABCD$ is a quadrilateral.
a. Find BC and AD. What do you notice?
b. Write a conjecture that generalizes the result you found in part (a).
c. Find the midpoint of $\overline{AC}$ and the midpoint of $\overline{BD}$. What do you notice?
d. Write a conjecture that generalizes the result you found in part (c).
e. Find the midpoint E of $\overline{AD}$ and the midpoint F of $\overline{BC}$. Find EF and AB. What do you notice?
f. Write a conjecture that generalizes the result you found in part (e).

Exercise 60

Geometry in 3 Dimensions **You can use three coordinates (x, y, z) to locate points in three dimensions. Point P has coordinates $(6, -3.5, 9)$.**

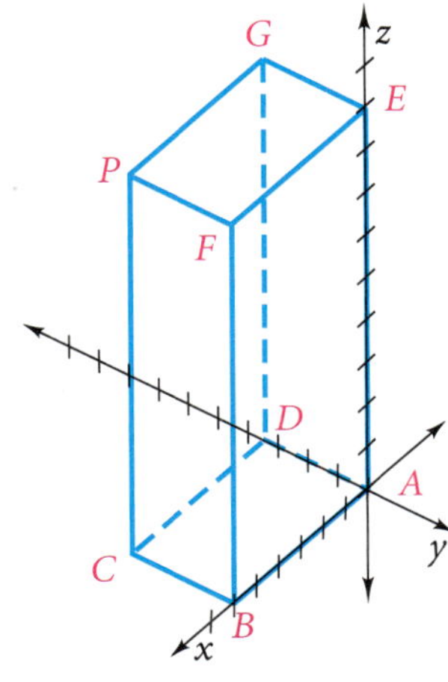

60. Give the coordinates of points $A, B, C, D, E, F,$ and G.

61. Draw three axes like those shown. Then graph $R(4, 5, 9)$.

Distance in 3 Dimensions **In a three-dimensional coordinate system, the distance between two points (x_1, y_1, z_1) and (x_2, y_2, z_2) can be found using this extension of the Distance Formula.**

$$d = \sqrt{(x_2 - x_1)^2 + (y_2 - y_1)^2 + (z_2 - z_1)^2}$$

Find the distance between each pair of points to the nearest tenth.

62. $P(2, 3, 4), B(-2, 4, 9)$

63. $Q(0, 12, 15), Y(-8, 20, 12)$

Multiple Choice Practice

For California Standards and CAHSEE Tutorials, visit PHSchool.com. Web Code: bcq-9045

GEOM 16.0

64. Given: $\angle A$
What is the second step in constructing the angle bisector of $\angle A$?

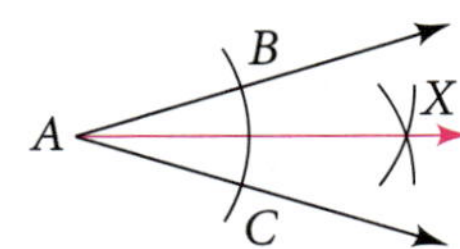

(A) Draw ray $\overline{XY}$.
(B) From points B and C, draw equal arcs that intersect at X.
(C) Draw a line segment connecting points B and C.
(D) From point A, draw an arc that intersects the sides of the angle at points B and C.

CAHSEE Alg1 3.0

65. If y is an integer, what is the solution set for $|y + 5| = 4$?

(A) $\{-1, -9\}$
(B) $\{-1, 9\}$
(C) $\{1, -9\}$
(D) $\{1, 9\}$

CAHSEE 7NS 1.6

66. Mr. Sands received a raise at his job. If he was earning 30,000 dollars per year before the raise, and he now earns 36,000 dollars, what is the percent of increase?

(A) 10% (B) 15% (C) 17% (D) 20%

CAHSEE 7NS 2.4

67. The square root of 263 is between

(A) 15 and 16 (B) 16 and 17 (C) 17 and 18 (D) 18 and 19

CAHSEE 7MG 2.4

68. A gardener needs to cover an area that is 12 feet wide and 18 feet long and 6 inches deep with mulch. If mulch is sold by the cubic yard, how many cubic yards will the gardener need? (1 cubic yard = 27 cubic feet)

(A) 2 (B) 3 (C) 4 (D) 5

Mixed Review

Lesson 1-7 **Use a straightedge and compass.**

GO for Help

69. Draw $\overline{AB}$. Construct $\overline{PQ}$ so that $PQ = 2AB$.
70. Draw $\overline{LK}$. Construct the perpendicular bisector of $\overline{LK}$.
71. Draw an obtuse $\angle B$. Construct $\angle C$ so that $m\angle C = m\angle B$.
72. Draw an acute $\angle RTS$. Construct the bisector of $\angle RTS$.

Lesson 1-6

73. Name $\angle A$ two other ways.

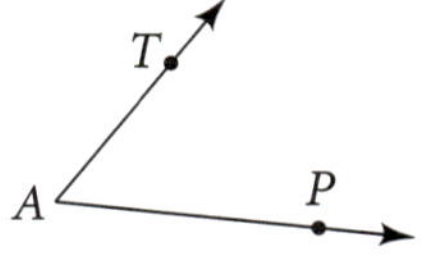

74. $m\angle PQR = 60$. What is $m\angle RQS$?

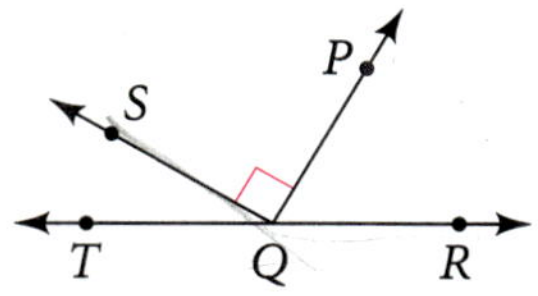

Lesson 1-5

75. **Algebra** The length of $\overline{AC}$ is 45. If $AB = x + 8$ and $BC = 3x - 3$, find the value of x.

A B C

76. Find AB if the coordinate of A is 5 and the coordinate of B is -5.

77. **Algebra** C is the midpoint of $\overline{EF}$. Find EF.

$2x + 10$ $5x - 11$
E C F

Checkpoint Quiz 2 — Lessons 1-5 through 1-8

Use the figure for Exercises 1–3.

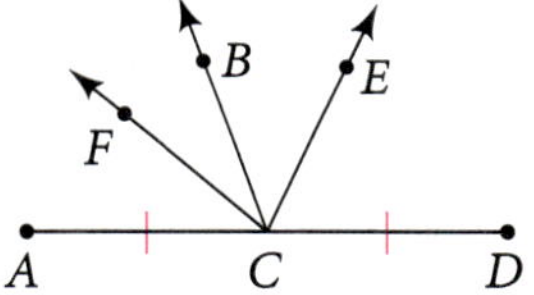

1. If $AC = 4x + 5$ and $DC = 3x + 8$, find AC.
2. If $m\angle BCE = 45$ and $m\angle ECD = 65$, find $m\angle BCD$.
3. If $m\angle FCA = 40$, find $m\angle FCD$.
4. $\overleftrightarrow{AX}$ is the perpendicular bisector of $\overline{QS}$ at M.
   a. What is $m\angle AMS$?
   b. If $QM = 30$, what is QS?
5. $\overrightarrow{PT}$ is the bisector of $\angle APR$. Name two congruent angles.
6. $\overrightarrow{OR}$ is the bisector of right $\angle TOS$. Find $m\angle TOR$.

Use a straightedge to draw three figures like the ones shown at the right. Then do each construction.

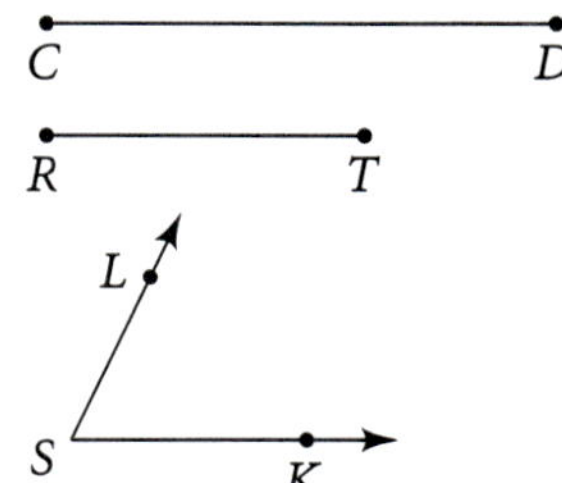

7. Construct $\overline{FG}$ so that $FG = CD + RT$.
8. Construct $\angle HSK$ so that $m\angle HSK = \frac{1}{2}m\angle LSK$.

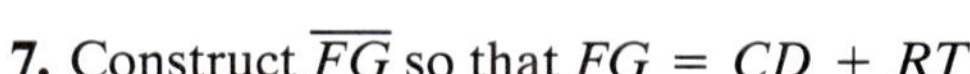

$\overline{AB}$ has endpoints $A(-4, 5)$ and $B(6, -2)$.

9. Find AB to the nearest tenth.
10. Find the coordinates of the midpoint of $\overline{AB}$.

GO Online Lesson Quiz Visit: PHSchool.com, Web Code: bca-0108

GPS Guided Problem Solving

For Use With Page 57, Exercise 47

Understanding Word Problems **Read the problem below. Then let Rosemarie's thinking guide you through the solution. Check your understanding with the exercises at the bottom of the page.**

Writing An airplane at $T(80, 20)$ needs to fly to both $U(20, 60)$ and $V(110, 85)$. What is the shortest possible distance for the trip? Explain.

What Rosemarie Thinks	What Rosemarie Writes
I'll sketch a graph to understand the problem better.	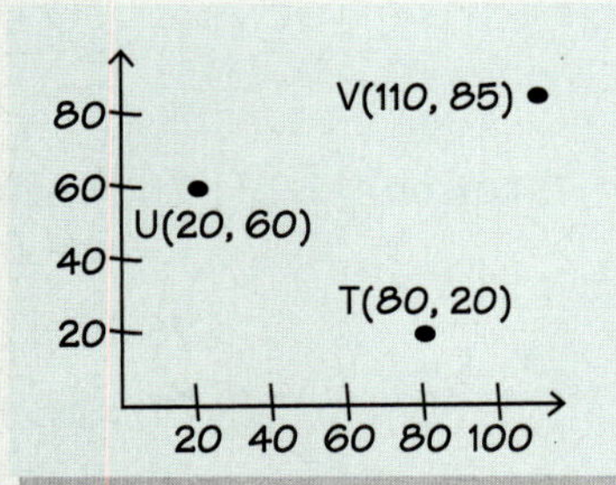
The shortest trip is from T to U to V, or from T to V to U. The total distances for the trips are $TU + UV$ and $TV + VU$. I have to use the Distance Formula.	$TU = \sqrt{(80 - 20)^2 + (20 - 60)^2}$ $TV = \sqrt{(80 - 110)^2 + (20 - 85)^2}$ $UV = VU = \sqrt{(110 - 20)^2 + (85 - 60)^2}$
I can enter the whole expression for $TU + UV$ on my calculator. I press ENTER and get . . .	$TU + UV \approx 165.52$
Then I press 2nd ENTRY. That shows $TU + UV$ as I had entered it. I'll replace the TU values with the TV values. I press ENTER and get . . .	$TV + VU \approx 165.00$
Now I have to explain. Do I use miles or kilometers? I don't know. I'll use "units."	The shortest route for the trip is from T to U to V, or from T to V to U. The T-V-U route is 0.52 unit shorter. The shortest possible distance for the trip is 165 units.

EXERCISES

1. What is the shortest flying distance from $R(65, 72.5)$ to both U and V?
2. An airplane has to fly from T to an airport halfway between U and V. How far must it fly?

Perimeter, Circumference, and Area

California Content Standards

GEOM 8.0 Students know, derive, and solve problems involving perimeter, circumference, and area. ***Introduce***

GEOM 10.0 Students compute areas of polygons, including rectangles. ***Introduce***

What You'll Learn

- To find perimeters of rectangles and squares, and circumferences of circles
- To find areas of rectangles, squares, and circles

. . . And Why

To find the amount of fencing material needed to build a fence, as in Example 1

Check Skills You'll Need

Skills Handbook page 757 and Lesson 1-8

Simplify each absolute value.

1. $|4 - 8|$ **2.** $|10 - (-5)|$ **3.** $|-2 - 6|$

Find the distance between the points to the nearest tenth.

4. $A(2, 3), B(5, 9)$ **5.** $K(-1, -3), L(0, 0)$

6. $W(4, -7), Z(10, -2)$ **7.** $C(-5, 2), D(-7, 6)$

8. $M(-1, -10), P(-12, -3)$ **9.** $Q(-8, -4), R(-3, -10)$

1 Finding Perimeter and Circumference

Hands-On Activity: Finding Perimeter and Area

Draw each figure on centimeter grid paper.

- a rectangle with length 5 cm and width 3 cm
- a rectangle with length 8 cm and height 2 cm
- a rectangle with each side 4 cm

1. To find the perimeter of each rectangle, find the sum of the lengths of the sides. Record the perimeter of each rectangle.

2. To find the area of each rectangle, count the number of square centimeters in its interior. Record the area of each rectangle.

3. Do rectangles with equal perimeters have the same area?

4. Do rectangles with the same area have the same perimeter?

5. Use a piece of string and make a loop. Tie a slip knot. Adjust the loop and fix its total length at 36 cm. Use the loop to approximate different rectangles on your grid paper. Record their lengths, widths, perimeters, and areas. What do you notice?

Vocabulary Tip

You can think of the perimeter of a polygon as the distance around it and the area as the number of square units it encloses.

Online active math

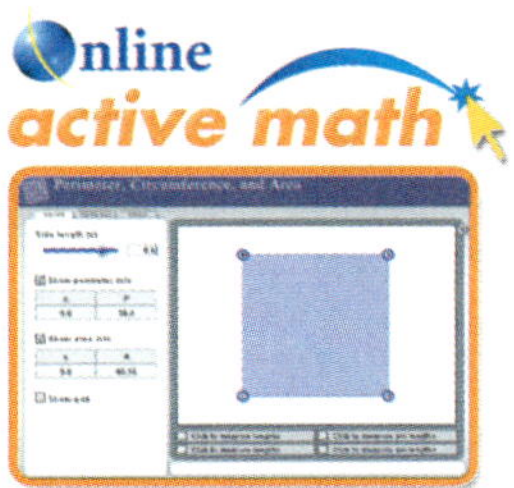

For: Perimeter/Area Activity
Use: Interactive Textbook, 1-9

The perimeter P of a polygon is the sum of the lengths of its sides. The area A of a polygon is the number of square units it encloses. For special figures such as squares, rectangles, and circles, you can use formulas for perimeter (called circumference in circles) and area.

Some formulas for perimeter and area are given in the chart at the top of the next page. You will also find the chart on pages 764 and 765 to be useful at times.

Key Concepts

Summary **Perimeter and Area**

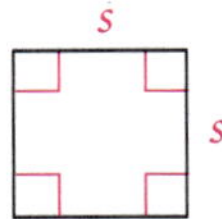

Square with side length s

Perimeter $P = 4s$

Area $A = s^2$

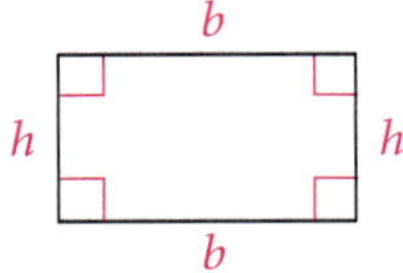

Rectangle with base b and height h

Perimeter $P = 2b + 2h$

Area $A = bh$

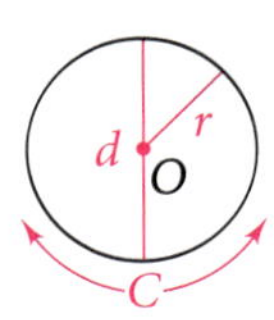

Circle with radius r and diameter d

Circumference $C = \pi d$, or $C = 2\pi r$

Area $= \pi r^2$

The units of measurement for perimeter and circumference include inches, feet, yards, miles, centimeters, meters, and kilometers. When measuring area, use square units such as square inches (in.^{2}), square centimeters (cm^2), square meters (m^2), and square miles (mi^2).

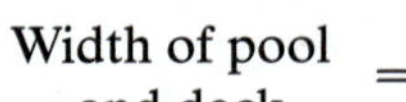

1 EXAMPLE Real-World Connection

Fencing Your pool is 15 ft wide and 20 ft long with a 3-ft wide deck surrounding it. You want to build a fence around the deck. How much fencing will you need?

3 ft
20 ft
15 ft

To find the perimeter of the pool with the deck, first find the width and length of the pool with the deck.

Vocabulary Tip

For a rectangle, "length" and "width" are sometimes used in place of "base" and "height."

Width of pool and deck $= 15 + 3 + 3 = 21$

Length of pool and deck $= 20 + 3 + 3 = 26$

Perimeter of a rectangle $= 2b + 2h$ **Use the formula for the perimeter of a rectangle.**

$P = 2(21) + 2(26)$ **Substitute.**

$P = 42 + 52$ **Simplify.**

$P = 94$

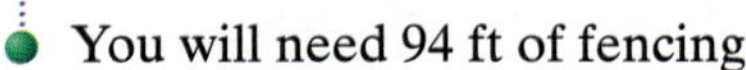

You will need 94 ft of fencing.

CA Standards Check

1 Suppose you want to frame a picture that is 6 in. by 7 in. with a $\frac{1}{2}$-in. wide frame.
a. Find the perimeter of the picture.
b. Find the perimeter of the outside edge of the frame.

Notice that the formulas for a circle involve π. Since the number π is irrational,

$$\pi = 3.1415926\ldots,$$

you cannot write it as a terminating decimal. For an approximate answer, you can use 3.14 or $\frac{22}{7}$ $\left(3.14 \approx \frac{22}{7}\right)$ for π. You can also use the rounded decimal you get by pressing π on your calculator. For an exact answer leave the result in terms of π.

2 EXAMPLE Finding Circumference

Vocabulary Tip

Read $\odot A$ as "circle A."

Find the circumference of $\odot A$ in terms of π. Then find the circumference to the nearest tenth.

$C = \pi d$

$C = 12\pi$ **This is the exact answer.**

12 × π = 37.699112 **Use a calculator.**

$C \approx 37.7$

The circumference of the circle is 12π in., or about 37.7 in.

CA Standards Check 2 **a.** Find the circumference of a circle with a radius of 18 m in terms of π.
b. Find the circumference of a circle with a diameter of 18 m to the nearest tenth.

3 EXAMPLE Finding Perimeter in the Coordinate Plane

Algebra Find the perimeter of $\triangle ABC$.

Find the length of each side. Add the lengths to find the perimeter.

$AB = |5 - (-1)| = 6$

$BC = |6 - (-2)| = 8$ **Use the Ruler Postulate.**

$AC = \sqrt{(5 - (-1))^2 + (6 - (-2))^2}$ **Use the Distance Formula.**

$= \sqrt{6^2 + 8^2} = \sqrt{100} = 10$

$AB + BC + AC = 6 + 8 + 10 = 24$

The perimeter of $\triangle ABC$ is 24 units.

CA Standards Check 3 Graph quadrilateral $KLMN$ with vertices $K(-3, -3)$, $L(1, -3)$, $M(1, 4)$, and $N(-3, 1)$. Find the perimeter of $KLMN$.

2 Finding Area

To find area, you should use the same unit for both dimensions.

Video Tutor Help

Visit: PHSchool.com
Web Code: bce-0775

4 EXAMPLE Finding Area of a Rectangle

You are designing a rectangular banner for the front of the museum. The banner will be 4 ft wide and 7 yd high. How much material do you need?

7 yd = 21 ft **Change yards to feet using 1 yd = 3 ft.**

Area $= bh$ **Use the formula for area of a rectangle.**

$A = 4(21)$ **Substitute 4 for b and 21 for h.**

$A = 84$

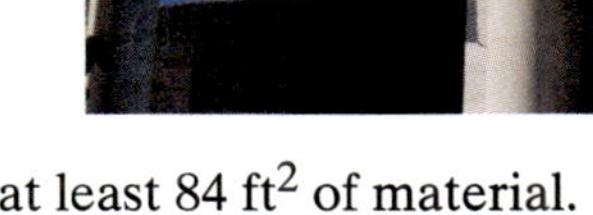

The area of the banner is 84 square feet (ft^2). You need at least 84 ft^2 of material.

CA Standards Check 4 Find the area of the banner in Example 4 by first changing all units to yards. Compare your answer to the one in Example 4. How do they compare?

5 EXAMPLE Finding Area of a Circle

The diameter of a circle is 10 in. Find the area in terms of π.

radius $= \frac{10}{2}$ or 5 $\quad r = \frac{d}{2}$

Area $= \pi r^2$ **Use the formula for area of a circle.**

$A = \pi(5)^2$ **Substitute 5 for *r*.**

$A = 25\pi$

The area of the circle is 25π in.^{2}.

5 The diameter of a circle is 5 ft.

a. Find the area in terms of π.

b. Find the area to the nearest tenth.

The following postulates are useful in finding areas of figures with irregular shapes.

Key Concepts

Postulate 1-9

If two figures are congruent, then their areas are equal.

Postulate 1-10

The area of a region is the sum of the areas of its nonoverlapping parts.

Example 6 applies Postulate 1-10 by summing the areas of the parts of a figure.

6 EXAMPLE Finding Area of an Irregular Shape

Multiple Choice What is the area of the figure at the right?

Ⓐ 12 cm^2 Ⓑ 24 cm^2 Ⓒ 30 cm^2 Ⓓ 36 cm^2

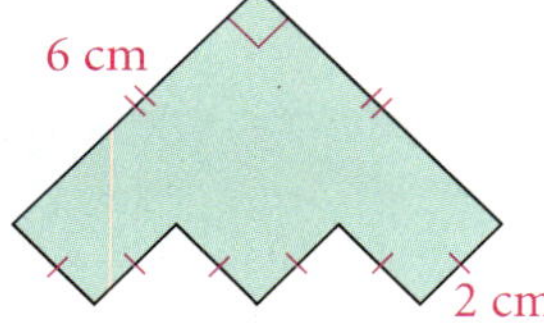

Test-Taking Tip

Marking diagrams on a test can help you understand the problem. If you cannot mark on the test, make a sketch of the diagram on scratch paper.

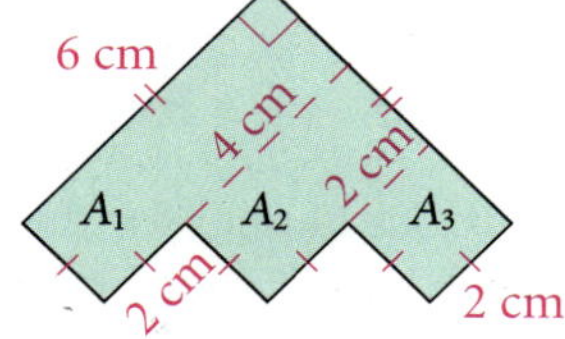

Separate the figure into rectangles.

Area $= bh$ **Use the formula for the area of a rectangle.**

$A_1 = 6 \cdot 2 = 12$ **Find the area of each rectangle.**

$A_2 = 4 \cdot 2 = 8$

$A_3 = 2 \cdot 2 = 4$

Total Area $= 12 + 8 + 4 = 24$ **Add the areas.**

The area of the figure is 24 cm^2. The correct choice is B.

6 Copy the figure in Example 6. Separate it in a different way. Find the area.

EXERCISES

Standards Practice

For more exercises, see *Extra Skill, Word Problem, and Proof Practice.*

GEOM 8.0, 10.0

A Practice by Example

Example 1 (page 62)

GO for Help

Find the perimeter of each figure.

1. 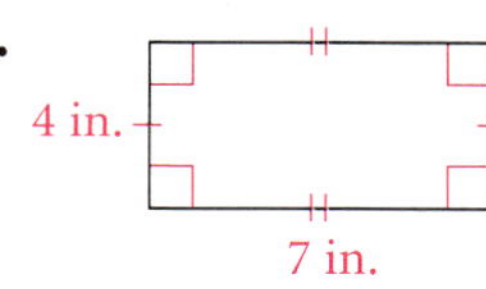

2. 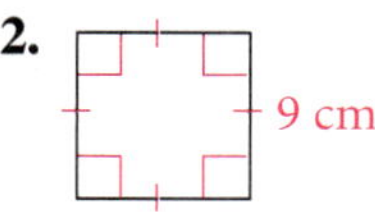

Find the perimeter of each rectangle with the given base and height.

3. 21 in., 7 in.
4. 16 cm, 23 cm
5. 24 m, 36 m

6. **Framing** A rectangular certificate 8 in. by 10 in. will have a frame $1\frac{1}{2}$ in. wide surrounding it. What is the perimeter of the outside edge of the frame?

7. **Fencing** A garden that is 5 ft by 6 ft has a walkway 2 ft wide around it. Find the amount of fencing needed to surround the walkway.

Example 2 (page 63)

Find the circumference of each circle in terms of π.

8. 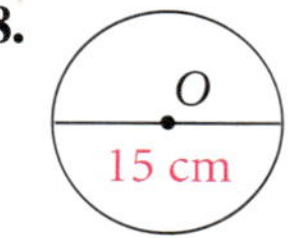

9. 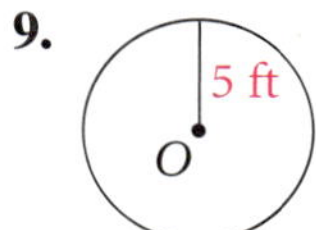

10. 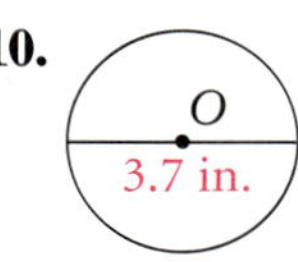

11. 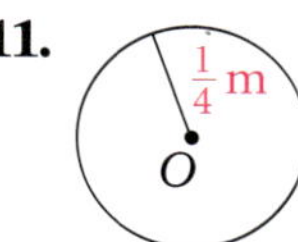

Find the circumference of the circle to the nearest tenth.

12. $r = 9$ in.
13. $d = 7.3$ m
14. $d = \frac{1}{2}$ yd
15. $r = 56$ cm

Example 3 (page 63)

Draw each figure in the coordinate plane. Find the perimeter.

16. $X(0, 2), Y(4, -1), Z(-2, -1)$
17. $A(-4, -1), B(4, 5), C(4, -2)$
18. $L(0, 1), M(3, 5), N(5, 5), P(5, 1)$
19. $S(-5, 3), T(7, -2), U(7, -6), V(-5, -6)$

Example 4 (page 63)

Find the area of each rectangle with the given base and height.

20. 4 ft, 4 in.
21. 30 in., 4 yd
22. 2 ft 3 in., 6 in.
23. 40 cm, 2 m
24. 3 m, 190 cm
25. 240 cm, 5 m

26. Find the area of a section of road pavement that is 20 ft wide and 100 yd long.

Example 5 (page 64)

Find the area of each circle in terms of π.

27. 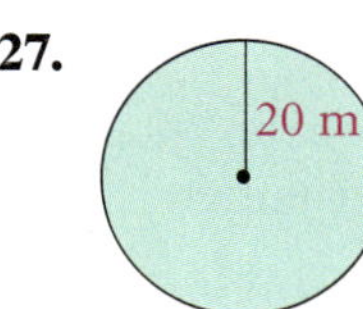

28. 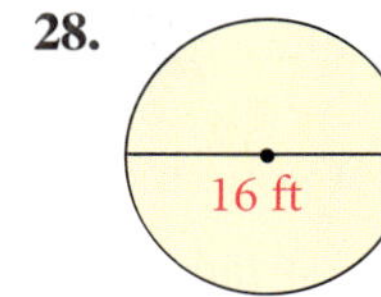

29. 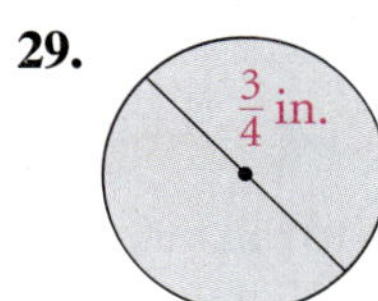

30.

31. 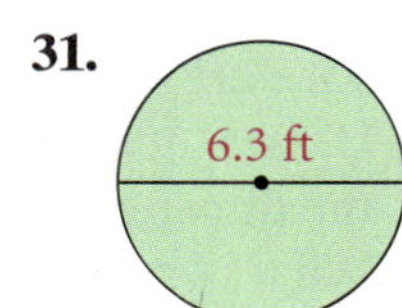

32. 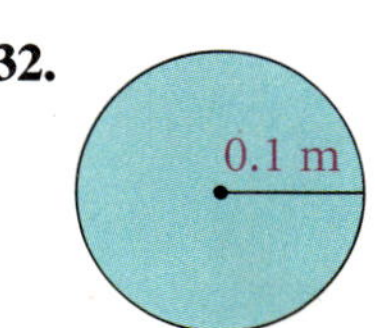

Find the area of each circle to the nearest tenth.

33. $r = 7$ ft **34.** $d = 8.3$ m **35.** $d = 24$ cm **36.** $r = 12$ in.

Example 6 (page 64)

Find the area of the shaded region. All angles are right angles.

37.

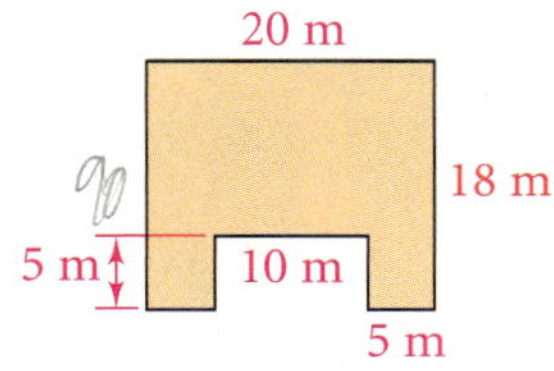

38.

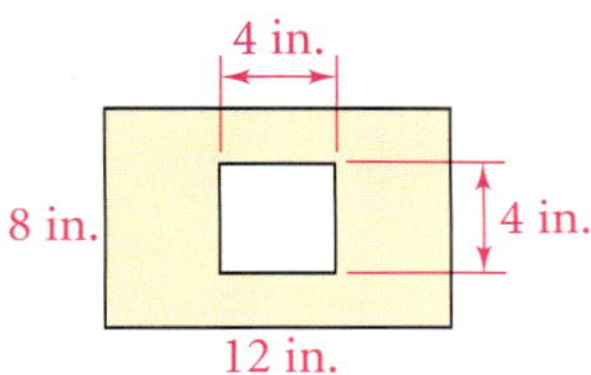

B Apply Your Skills

39. a. What is the area of a square whose sides are 12 in. long?
b. What is the area of a square whose sides are 1 ft long?
c. Reasoning How many square inches are in a square foot? Explain.

40. a. Count squares to find the area of the polygon outlined in blue.
b. Use a formula to find the area of each square outlined in red.
c. How does the sum of your results in part (b) compare to your result in part (a)? Which postulate does this support?

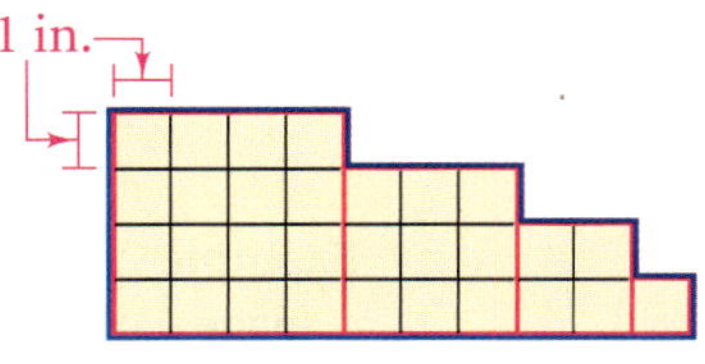

41. Estimation On a postcard from Mexico, Ky sketched the "footprint" of the pyramid known as El Castillo in the ancient Mayan city Chichen Itza. He said he estimated the three different lengths on each side to be 22 m, 6 m, and 11 m. Use those estimates to estimate the area of El Castillo's footprint.

Estimation Estimate the perimeter and area of each object.

42. the front cover of this book **43.** the front cover of your notebook

44. a classroom bulletin board **45.** the top of your desk

46. Writing Choose one exercise from Exercises 42–45 and explain why you chose your unit of length.

47. The area of an 11-cm wide rectangle is 176 cm^2. What is its length?

48. The perimeter of a rectangle is 40 cm and the base is 12 cm. What is its area?

49. A square and a rectangle have equal area. The rectangle is 64 cm by 81 cm. What is the perimeter of the square?

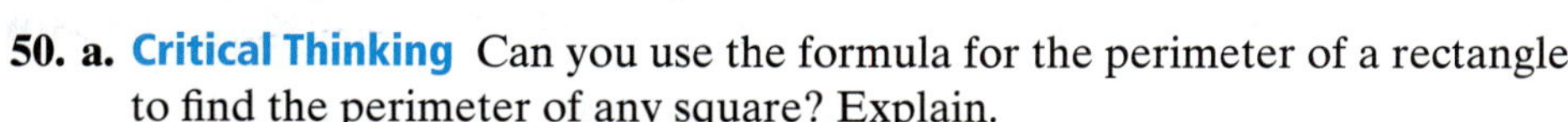

50. a. Critical Thinking Can you use the formula for the perimeter of a rectangle to find the perimeter of any square? Explain.
b. Can you use the formula for the perimeter of a square to find the perimeter of any rectangle? Explain.
c. Use the formula for the perimeter of a square to write a formula for the area of a square in terms of its perimeter.

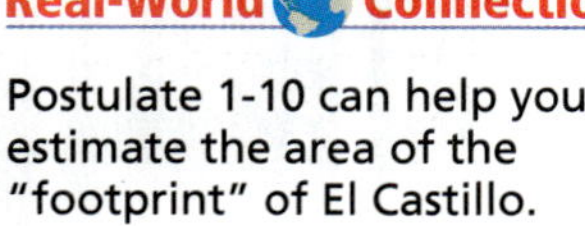

Real-World Connection

Postulate 1-10 can help you estimate the area of the "footprint" of El Castillo.

51. The surface area of a three-dimensional figure is the sum of the areas of all of its surfaces. You can find the surface area by finding the area of a net for the figure.
a. Draw a net for the solid shown. Label the dimensions.
b. What is the area of the net? What is the surface area of the solid?

4 in.
8 in.
6 in.

Real-World Connection

Four 6 in.-by-6 in. tiles will cover 1 ft^2.

52. Tiling The students in the Art Club are tiling a wall that is 8 ft by 16 ft at the entrance to the community center. They are using tiles that are 6 in. by 6 in. to create a multi-colored design. How many tiles do the students need?

Algebra Draw each rectangle in the coordinate plane. Find its perimeter and area.

53. $A(-3, 2), B(-2, 2), C(-2, -2), D(-3, -2)$

54. $A(-2, -6), B(-2, -3), C(3, -3), D(3, -6)$

Coordinate Geometry On graph paper, draw polygon $ABCDEFGH$ with vertices $A(1, 1)$, $B(10, 1)$, $C(10, 8)$, $D(7, 8)$, $E(7, 5)$, $F(4, 5)$, $G(4, 8)$, and $H(1, 8)$.

55. Find the perimeter of the polygon.

56. Divide the polygon into rectangles. Find the area of the polygon.

57. Biology In the Pacific Northwest, a red fox has a circular home range with a radius of about 718 meters. To the nearest thousand square meters, what is the area of the home range of a red fox?

58. Multiple Choice A rectangle has a base of x units. The area is $(4x^2 - 2x)$ square units. What is the height of the rectangle in terms of x?

(A) $(4 - x)$ units
(B) $(4x^3 - 2x^2)$ units
(C) $(x - 2)$ units
(D) $(4x - 2)$ units

Visit: PHSchool.com
Web Code: bce-0109

Home Maintenance To determine how much of each item to buy, tell whether you need to know area or perimeter. Explain your choice.

59. wallpaper for a bedroom

60. weatherstripping for a door

61. fence for a garden

62. paint for a basement floor

63. Coordinate Geometry The endpoints of a diameter of a circle are $A(2, 1)$ and $B(5, 5)$. Find the area of the circle in terms of π.

For: Graphing calculator procedures
Web Code: bcb-2113

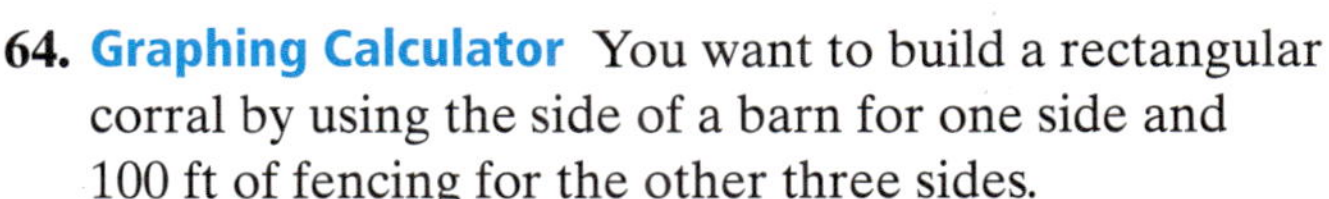

64. Graphing Calculator You want to build a rectangular corral by using the side of a barn for one side and 100 ft of fencing for the other three sides.

a. Make a table on your graphing calculator listing integer values for the base and the corresponding values of the height and area.

b. Make a graph using your table values. Graph the base on the horizontal axis and area on the vertical axis.

c. What are the dimensions of the corral with the greatest area?

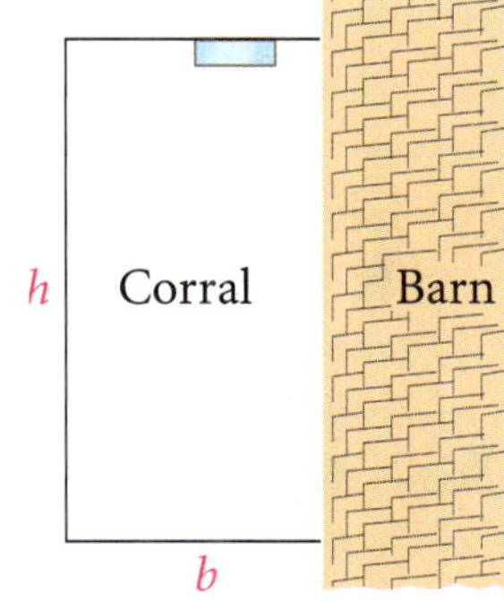

65. How many circles with the given radius are needed for the sum of their areas to equal the area of a circle with the second given radius?

a. 1 in., 3 in. **b.** 2 in., 6 in. **c.** 3 in., 9 in.

d. Make a Conjecture How many circles with a radius of n in. are needed for the sum of their areas to equal the area of a circle with a radius of $3n$ in.?

Algebra Find the area of each figure.

66. a rectangle with side lengths of $\frac{2a}{5b}$ units and $\frac{3b}{8}$ units

67. a square with perimeter $10n$ units

68. a square with side lengths of $(3m - 4n)$ units

Online Lesson Quiz Visit: PHSchool.com, Web Code: bca-0109

69. **Open-Ended** The area of a 5 in.-by-5 in. square is the same as the sum of the areas of a 3 in.-by-3 in. square and a 4 in.-by-4 in. square. Find two or more squares whose total area is the same as the area of an 11 in.-by-11 in. square.

70. **Track** An athletic field is a rectangle, 100 yards by 40 yards, with a semicircle at each of the short sides. A running track 10 yards wide surrounds the field. Find the perimeter of the outside of the running track to the nearest tenth of a yard.

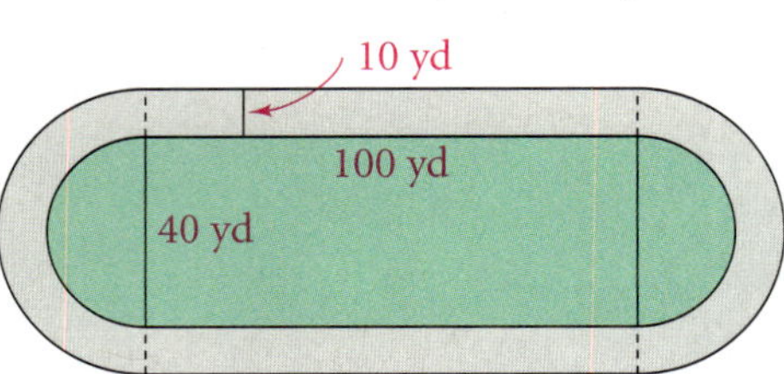

Multiple Choice Practice

For California Standards and CAHSEE Tutorials, visit PHSchool.com. Web Code: bcq-9045

GEOM 8.0

71. A square and a rectangle have equal area. If the rectangle is 36 cm by 25 cm, what is the perimeter of the square?

Ⓐ 30 cm Ⓒ 120 cm
Ⓑ 60 cm Ⓓ 900 cm

CAHSEE 7MG 2.1

72. A net for a small rectangular gift box is shown at the right. Find the surface area of the box.

Ⓐ 468 cm^2 Ⓒ 1026 cm^2
Ⓑ 782 cm^2 Ⓓ 2106 cm^2

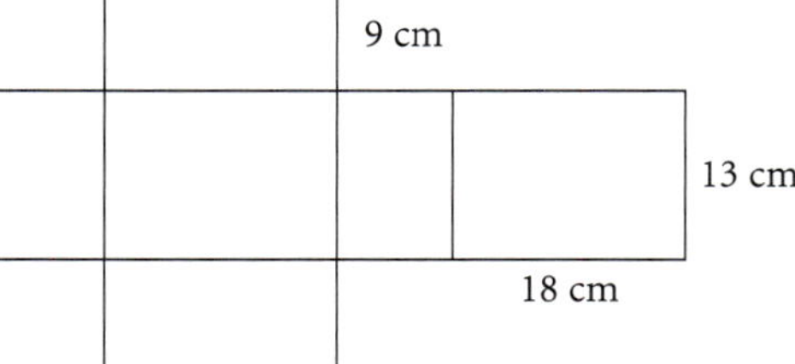

CAHSEE Alg1 6.0

73. In the equation $y = -2x - 15$, what is the value of x when $y = 0$?

Ⓐ -15 Ⓒ 7.5
Ⓑ -7.5 Ⓓ 15

Mixed Review

Lesson 1-8

74. The midpoint of $\overline{CD}$ has coordinates $(5, 6)$. Point C has coordinates $(-5, -1)$. Find the coordinates of point D.

Find (a) AB to the nearest tenth and (b) the coordinates of the midpoint of $\overline{AB}$.

75. $A(4, 1), B(7, 9)$ **76.** $A(0, 3), B(3, 8)$ **77.** $A(9, 2), B(-3, 9)$

78. $A(0, 1), B(-4, 6)$ **79.** $A(4, 10), B(-2, 3)$ **80.** $A(-1, 1), B(-4, -5)$

Lesson 1-7

$\overleftrightarrow{BG}$ is the perpendicular bisector of $\overline{WR}$ at point I.

81. What is $m\angle BIR$? **82.** Name two congruent segments.

83. $\overline{WR}$ has length 124. What is the length of $\overline{IR}$?

Lesson 1-5

For the given coordinates, find PQ.

84. P: 12, Q: -6 **85.** P: 3, Q: 9 **86.** P: -23, Q: 10

Comparing Perimeters and Areas

Technology

FOR USE WITH LESSON 1-9

You can use a graphing calculator or spreadsheet technology to find maximum and minimum values for area and perimeter problems.

You have 32 yards of fencing. You want to make a rectangular pen for the calf you are raising as a 4-H project. What dimensions will give the maximum area? What is the maximum area?

Investigate

Draw some possible rectangular pens and find their areas.

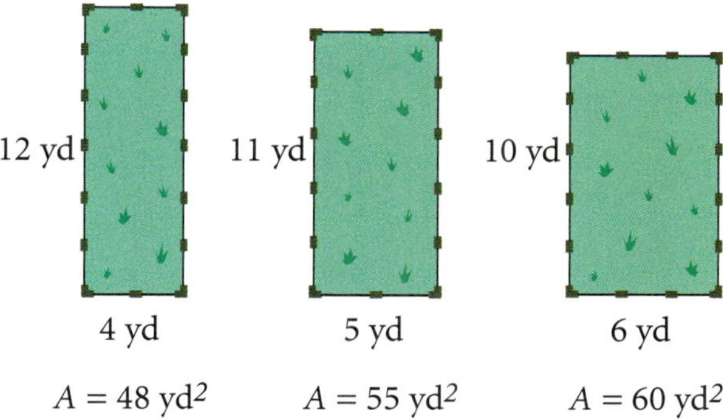

Create a graphing calculator table to find area. Let X represent values for the base. The height then is $16 - \text{X}$, and the area is $\text{X}(16 - \text{X})$. Enter $\text{Y}_1 = 16 - \text{X}$ and $\text{Y}_2 = \text{X}(16 - \text{X})$. Set the table so that X starts at 1 and changes by 1.

X	Y_1	Y_2
4	12	48
5	11	55
6	10	60
7	9	63
8	8	64
9	7	63
10	6	60

X=4

Scroll down the table. Area is maximum when X (or b) is 8. When $b = 8, h = 8$ and $A = 64$.

You can confirm this result by graphing $\text{Y}_1 = \text{X}(16 - \text{X})$. Trace on the graph to find the maximum area.

A square pen with sides of 8 yd will give maximum area for your calf. The maximum area is 64 yd^2.

Y2=X(16-X)

X=8 Y=64

X min = 0 Y min = 0
X max = 18 Y max = 70
X scl = 2 Y scl = 7

EXERCISES

1. **Make a Conjecture** For a fixed perimeter, what rectangular shape will result in a maximum area?
2. Consider that the pen is not restricted to polygon shapes. Determine the area of a circular pen if the circumference is 32 yd. How does this result compare with the maximum square area of 64 yd^2 found in the investigation?
3. You want to make a rectangular garden with an area of 900 ft^2. You want to use a minimum amount of fencing to keep the cost low.
   a. List some possible dimensions for the rectangular garden. Find the perimeter of each rectangle.
   b. Create a graphing calculator table. Use integer values of the base b, and the corresponding values of the height h, to find values for P, the perimeter. What dimensions will give you a garden with the minimum perimeter?

Test-Taking Strategies

Choosing the Process

Not all questions ask you to solve a problem. Some questions ask *how* you would solve a problem. To answer these questions, you must choose a process.

1 EXAMPLE

Tasha is making a circular flower bed in her yard. She has 25 feet of edging to form a border. Tasha wants to find the radius that will allow her to use all of the edging. Which of the following is the best choice for a first step?

A Multiply 25 by 2.
B Multiply 25 by π.
C Divide 25 by π.
D Find the square root of 25.

To use all of the edging, its length must be the circumference of the circular flower bed. There are two formulas involving circumference (C), radius (r), and diameter (d) that Tasha could use.

$$C = 2\pi r \qquad\qquad C = \pi d$$

Solving each:

$$\frac{C}{2\pi} = r \qquad\qquad \frac{C}{\pi} = d$$

For either formula, Tasha can find the radius by first dividing the circumference, 25, by π. So C is the correct answer.

EXERCISES

1. A railroad's dispatch center keeps track of the locations of its trains. The graph at the right shows the distance between a train and the dispatch center as a function of time. Fernando wants to find the train's speed. He chooses two points on the graph, (1, 120) and (2, 200). Which of the following should be his next step?

   A Subtract 2 from 120.
   B Add 200 and 120.
   C Add 1 and 200.
   D Subtract 120 from 200.

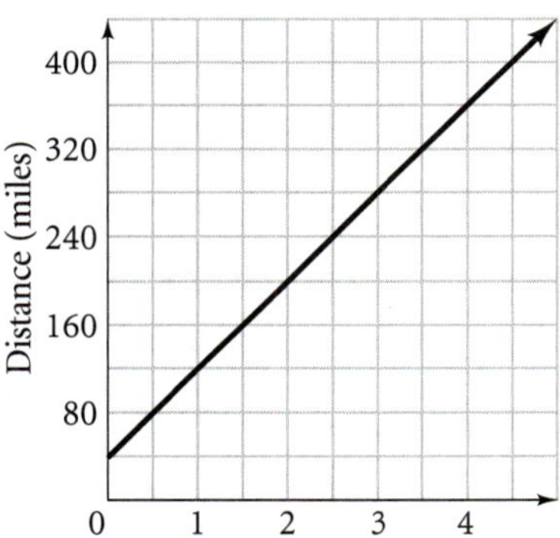

2. The expression $5x + 9$ represents the perimeter of a rectangle. The expression $2x + 1$ represents its length. You need to find an expression for the width. Which of the following is the best choice for a first step?

   A Factor $(5x + 9)$.
   B Subtract $2(2x + 1)$ from $(5x + 9)$.
   C Add $(2x + 1)$ to $(5x + 9)$.
   D Square $(2x + 1)$.

Chapter 1 Review

Vocabulary Review

English and Spanish Audio Online

acute angle (p. 37)
adjacent angles (p. 38)
angle (p. 36)
angle bisector (p. 46)
axiom (p. 18)
collinear points (p. 17)
compass (p. 44)
complementary angles (p. 38)
congruent angles (p. 37)
congruent segments (p. 31)
conjecture (p. 5)
construction (p. 44)
coordinate (p. 31)
coplanar (p. 17)
counterexample (p. 5)
foundation drawing (p. 11)
inductive reasoning (p. 4)
isometric drawing (p. 10)
line (p. 17)
midpoint (p. 32)
net (p. 12)
obtuse angle (p. 37)
opposite rays (p. 23)
orthographic drawing (p. 11)
parallel lines (p. 24)
parallel planes (p. 24)
perpendicular bisector (p. 45)
perpendicular lines (p. 45)
plane (p. 17)
point (p. 17)
postulate (p. 18)
ray (p. 23)
right angle (p. 37)
segment (p. 23)
skew lines (p. 24)
space (p. 17)
straight angle (p. 37)
straightedge (p. 44)
supplementary angles (p. 38)
vertical angles (p. 38)

Choose the correct term to complete each sentence.

1. Figures that are in the same plane are _?_.
2. A(n) _?_ is the part of a line consisting of two endpoints and all points between them.
3. Two segments with the same length are _?_.
4. A(n) _?_ of a segment is a point that divides the segment into two congruent segments.
5. A(n) _?_ is a ray that divides an angle into two congruent angles.
6. A conclusion based upon inductive reasoning is sometimes called a(n) _?_.

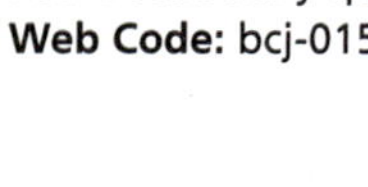

For: Vocabulary quiz
Web Code: bcj-0151

7. A(n) _?_ is an accepted statement of fact.
8. _?_ are coplanar lines that do not intersect.
9. A(n) _?_ is an angle whose measure is between 90 and 180.

Skills and Concepts

Lesson 1-1

- To use inductive reasoning to make conjectures (p. 4)

 GEOM 1.0, 3.0

You use **inductive reasoning** when you make conclusions based on patterns you observe. A **conjecture** describes a conclusion reached using inductive reasoning. A **counterexample** to a conjecture is an example for which the conjecture is incorrect.

Find a pattern for each sequence. Describe the pattern and use it to show the next two terms.

10. 40, 35, 30, 25, . . .
11. 5, –5, 5, –5, . . .
12. 34, 27, 20, 13, 6, . . .
13. 6, 24, 96, 384, . . .
14. 2, 4, 8, 16, 32, . . .
15. 1, 2, 5, 6, 9, . . .
16. Draw the next figure in the sequence.

  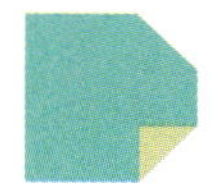

Lesson 1-2

- To make isometric and orthographic drawings (p. 10)
- To draw nets for three-dimensional figures (p. 12)

A **net** is a two-dimensional pattern that you can fold to form a three-dimensional figure. A net shows all surfaces of a figure in one view.

17. The net at the right is for a number cube. What are the three sums of the numbers on opposite surfaces of the cube?

		5	
1	2	3	4
	6		

18. On another number cube, the numbers on each pair of opposite surfaces add to 7. Draw a net for this number cube.

Lessons 1-3 and 1-4

- To understand basic terms of geometry (p. 16)
- To understand basic postulates of geometry (p. 18)
- To identify segments and rays (p. 23)
- To recognize parallel lines and planes (p. 24)

 **GEOM 1.0**

Points that lie on the same line are **collinear points.** Points and lines in the same plane are **coplanar. Segments** and **rays** are parts of lines.

Lines that are coplanar and do not intersect are **parallel lines.** Lines in space that are not parallel and do not intersect are **skew.** Planes that do not intersect are **parallel planes.**

Use the figure at the right for Exercises 19–22.

19. Name two intersecting lines.

20. Name three noncollinear points.

21. Name a pair of parallel planes.

22. Name three lines that intersect at D.

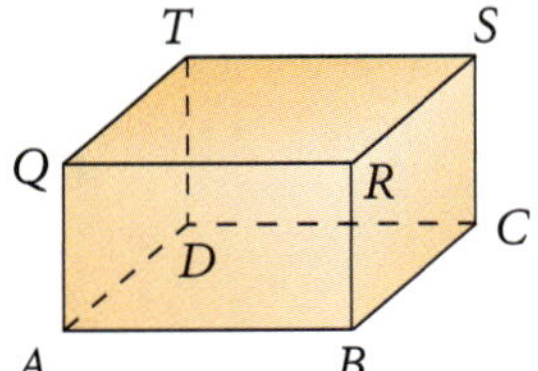

Complete with *always, sometimes,* or *never* to make a true statement.

23. A line and a point are __?__ coplanar.

24. Two segments are __?__ coplanar.

25. Skew lines are __?__ coplanar.

26. Two points are __?__ collinear.

Lessons 1-5 and 1-6

- To find the lengths of segments (p. 31)
- To find the measures of angles (p. 36)
- To identify special angle pairs (p. 38)

Segments with the same length are **congruent segments.** A **midpoint** of a segment divides the segment into two congruent segments.

Two rays with the same endpoint form an **angle.** Angles can be classified as acute, right, obtuse, or straight. Angles with the same measure are **congruent angles.**

27. Find two possible coordinates of Q so that $PQ = 5$.

28. Find the coordinate of the midpoint of $\overline{PH}$.

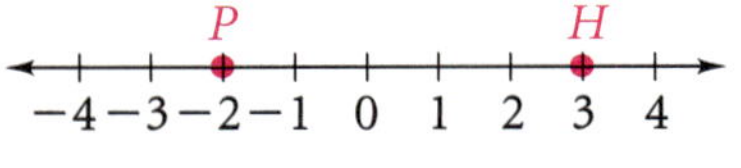

x^2 **Algebra Find the value of each variable.**

29.

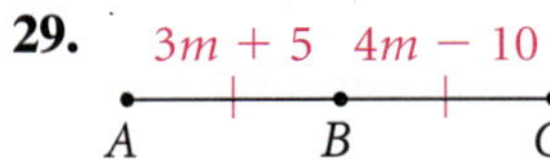

30. $(3x + 31)^\circ$ $(2x - 6)^\circ$

31. Name a pair of each of the following.
a. complementary angles
b. supplementary angles
c. vertical angles

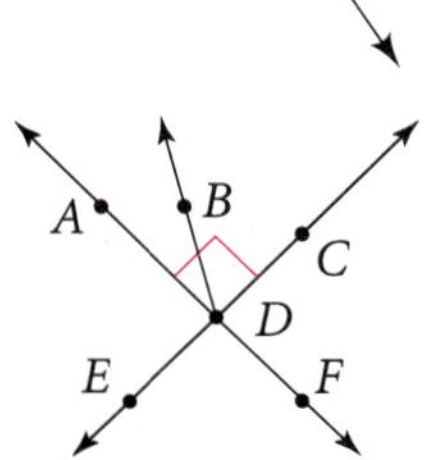

Lesson 1-7

- To use a compass and a straightedge to construct congruent segments and congruent angles (p. 44)
- To use a compass and a straightedge to bisect segments and angles (p. 45)

 **GEOM 16.0**

Construction is the process of making geometric figures using a **compass** and a **straightedge.** Four basic constructions involve constructing congruent segments, congruent angles, and bisectors of segments and angles.

Perpendicular lines intersect at right angles. A **perpendicular bisector** of a segment is perpendicular to the segment at its midpoint, and bisects it into two congruent segments. An **angle bisector** is a ray that divides an angle into two congruent angles.

32. Use a protractor to draw a 64° angle. Then construct an angle congruent to your 64° angle.

33. Use a ruler and draw $\overline{PQ}$.

a. Construct $\overline{AB} \cong \overline{PQ}$.

b. Construct the perpendicular bisector of $\overline{AB}$.

Lesson 1-8

- To find the distance between two points in the coordinate plane (p. 53)
- To find the coordinates of the midpoint of a segment in the coordinate plane (p. 54)

The x-axis and the y-axis intersect at the origin (0, 0) and determine a coordinate plane. You can find the coordinates of the midpoint M of $\overline{AB}$ with endpoints $A(x_1, y_1)$ and $B(x_2, y_2)$ using the **Midpoint Formula.**

$$M = \left(\frac{x_1 + x_2}{2}, \frac{y_1 + y_2}{2}\right)$$

You can find the distance d between points $A(x_1, y_1)$ and $B(x_2, y_2)$ using the **Distance Formula**.

$$d = \sqrt{(x_2 - x_1)^2 + (y_2 - y_1)^2}$$

Find the distance between the points to the nearest tenth.

34. $A(-1, 5), B(0, 4)$ **35.** $C(-1, -1), D(6, 2)$ **36.** $E(-7, 0), F(5, 8)$

$\overline{GH}$ has endpoints $G(-3, 2)$ and $H(3, -2)$.

37. Find the coordinates of the midpoint of $\overline{GH}$.

38. Find GH to the nearest tenth.

Lesson 1-9

- To find perimeters of rectangles and squares, and circumferences of circles (p. 61)
- To find areas of rectangles, squares, and circles (p. 63)

 **GEOM 8.0, 10.0**

The perimeter P of a polygon is the sum of the lengths of its sides. The area A of a polygon is the number of square units it encloses.

Formulas:	Square	Rectangle	Circle
	$P = 4s$	$P = 2b + 2h$	$C = \pi d$ or $C = 2\pi r$
	$A = s^2$	$A = bh$	$A = \pi r^2$

Find the perimeter and the area of each figure.

39.  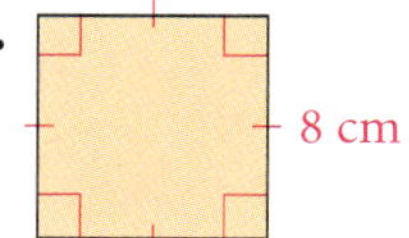

40.  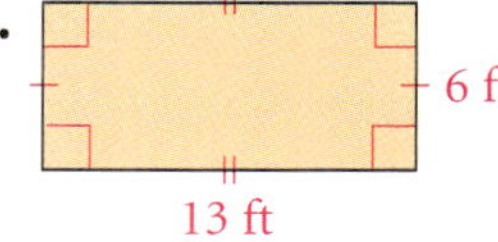

41. 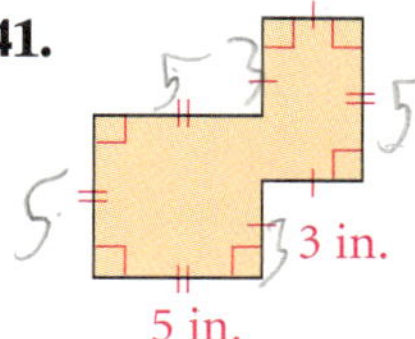

Find the circumference and the area of each circle to the nearest hundredth.

42. $r = 3$ in. **43.** $d = 15$ m **44.** $r = 26$ m

Chapter 1 Test

For: Chapter Test
Web Code: bca-0152

Describe each pattern and find the next two terms of each sequence.

1. $8, -4, 2, -1, \ldots$

2. $0, 2, 4, 6, 8, \ldots$

3. 

4. **Open-Ended** Write two different sequences whose first three terms are 1, 2, 4. Describe each pattern.

5. Draw a net for a cube.

Use the figure for Exercises 6–9.

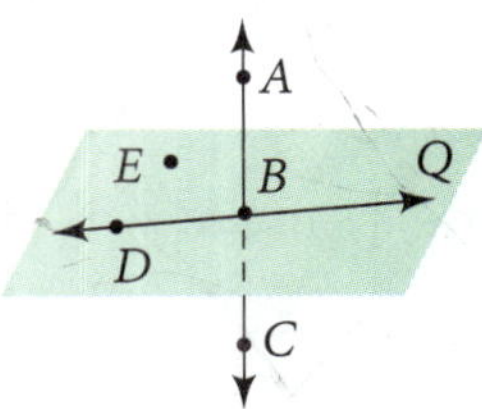

6. Name three collinear points.

7. Name four coplanar points.

8. What is the intersection of $\overleftrightarrow{AC}$ and plane Q?

9. How many planes contain each line and each point?
   a. $\overleftrightarrow{BD}$ and point A
   b. $\overleftrightarrow{AB}$ and point C
   c. $\overleftrightarrow{BE}$ and point C
   d. $\overleftrightarrow{BD}$ and point E

10. **Track** The running track is a rectangle with a half circle on each end. If $\overline{FI}$ and $\overline{GH}$ are diameters, find the area inside the track to the nearest tenth.

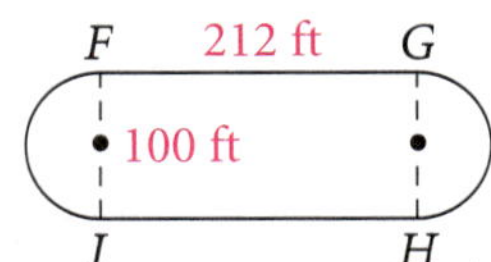

Complete with *always, sometimes,* or *never* to make each statement true.

11. $\overrightarrow{LJ}$ and $\overrightarrow{TJ}$ are _?_ opposite rays.

12. Four points are _?_ coplanar.

13. Skew lines are _?_ coplanar.

14. Two segments that lie in parallel lines are _?_ parallel.

15. The intersection of two planes is _?_ a point.

16. **Algebra** $JK = 48$. Find the value of x.

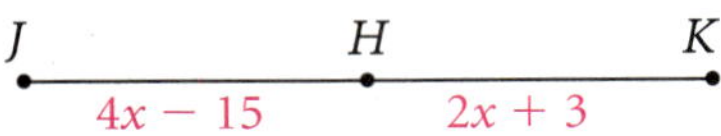

17. **Algebra** $M(x, y)$ is the midpoint of $\overline{CD}$ with endpoints $C(5, 9)$ and $D(17, 29)$.
   a. Find the values of x and y.
   b. Show $MC = MD$.

18. To the nearest tenth, find the perimeter of $\triangle ABC$ with vertices $A(-2, -2)$, $B(0, 5)$, and $C(3, -1)$.

For the given dimensions, find the area of each figure to the nearest hundredth.

19. rectangle
   $b = 4$ m
   $h = 2$ cm

20. square
   $s = 3.5$ in.

21. circle
   $d = 9$ cm

Algebra **Use the figure for Exercises 22–24. In Exercises 22 and 23, find the value of each variable.**

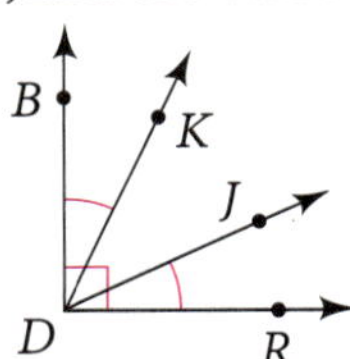

22. $m\angle BDK = 3x + 4, m\angle JDR = 5x - 10$

23. $m\angle BDJ = 7y + 2, m\angle JDR = 2y + 7$

24. Name two complementary angles.

25. **Writing** Why is it useful to have more than one way of naming an angle?

26. Draw an obtuse $\angle ABC$. Use a compass and straightedge to bisect the angle.

Use the figure to complete Exercises 27–30.

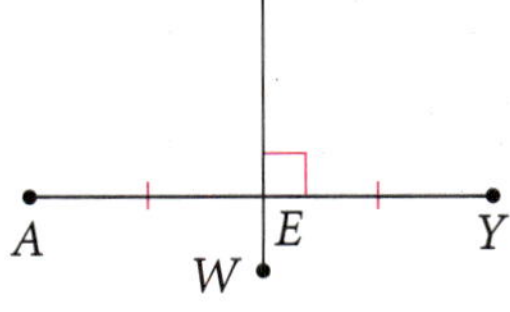

27. $\overline{VW}$ is the _?_ of $\overline{AY}$.

28. If $EY = 3.5$, then $AY =$ _?_.

29. $\frac{1}{2}$ _?_ $= AE$

30. _?_ is the midpoint of _?_.

31. **Carpeting** How many square yards of carpet are needed to carpet a room that is 15 ft long and 20 ft wide?

Standards Mastery

CAHSEE Practice

Some questions ask you to read graphs. Read the question at the right. Then follow the tips to answer it.

Paul drew the circle graph below to show his monthly expenses. If Paul earns $2500 per month, which of the following statements is NOT true?

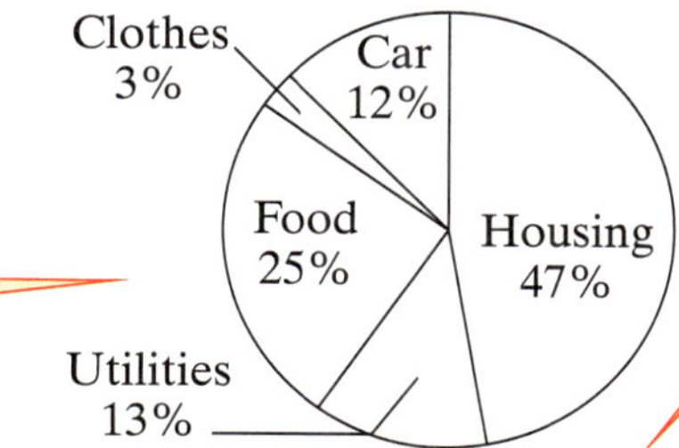

(A) Paul spends over $1000 for housing.

(B) Paul spends about twice as much on food as on utilities.

(C) Paul's car expense is about one third of his monthly income.

(D) Paul spends the least amount of money on clothes.

Tip 1
Study the answer choices. Some require you to compare sectors with each other and some require you to compare to the whole, which is $2500.

Tip 2
Remember that you are trying to find the choice that is NOT true.

Think It Through
For choice C to be true, the sector for car expenses should be about one third of the circle. It is less than one quarter of the circle. So C is the correct answer because it is NOT true.

Vocabulary Review

As you solve problems, you must understand the meanings of mathematical terms. Match each term with its mathematical meaning.

A. natural number	**I.** represented in the coordinate plane by points that lie on a line
B. mean	**II.** a positive integer
C. linear data	**III.** in an (x, y) relationship, the set of values possible for y
D. domain	**IV.** the sum of values divided by the number of values
E. range	**V.** in an (x, y) relationship, the set of values possible for x

Multiple Choice

Read each question. Then write the letter of the correct answer on your paper.

1. Carlos is driving his car at a speed of 25 miles per hour. At this rate, about what distance will Carlos drive in 20 minutes?

(A) 1.2 mi (C) 8.3 mi
(B) 3 mi (D) 500 mi

2. In the equation $d = rt$, the time t varies inversely with rate r for a given distance d. If Marie drives 4.5 hours at 50 miles per hour, find the time in hours to the nearest tenth for the same trip at 40 miles per hour.

(A) 5.1 (C) 5.7
(B) 5.6 (D) 6.1

CAHSEE Practice (continued)

3. The graph shows h, the height in feet of a water droplet spraying out from a fountain pipe, versus t, the time in seconds after it leaves the pipe. From the graph, which conclusion can be made about the path of the droplet?

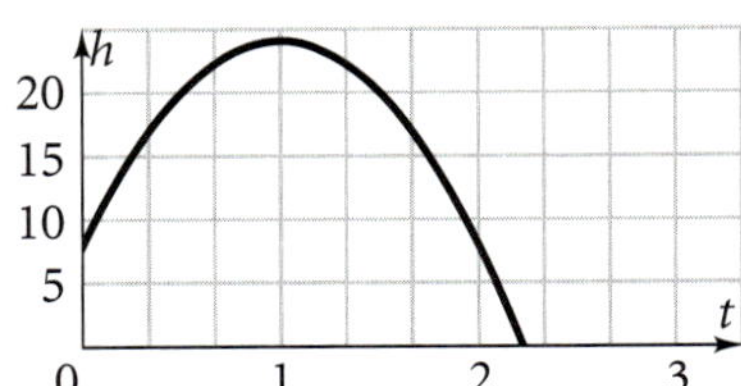

Ⓐ The maximum height was 25 ft.

Ⓑ The pipe is 8 ft off the ground.

Ⓒ The droplet was in the air for less than 2 s.

Ⓓ The droplet began to fall after 2 s.

4. The graph shows h, the height in meters of a tennis ball, versus t, the time in seconds after the ball was struck by a racket. Which conclusion can be made about the flight of the tennis ball?

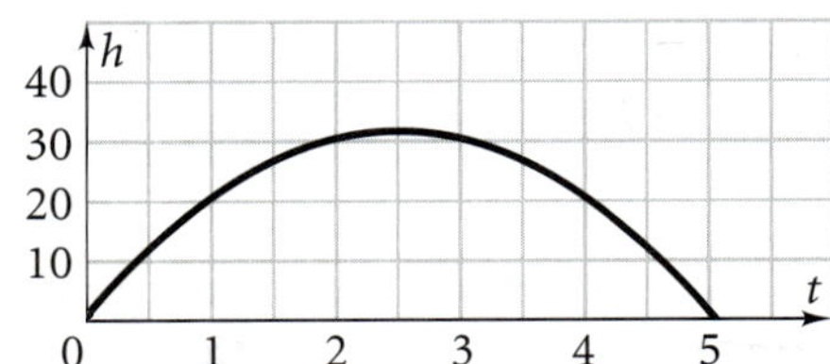

Ⓐ The maximum height was 40 meters.

Ⓑ At 4.5 seconds, the ball was on the ground.

Ⓒ The ball was in flight for 32 seconds.

Ⓓ The ball began to fall after 2.5 seconds.

5. Based on the line graph below, which is a valid prediction for the number of hits Bill's Website receives in the twelfth month?

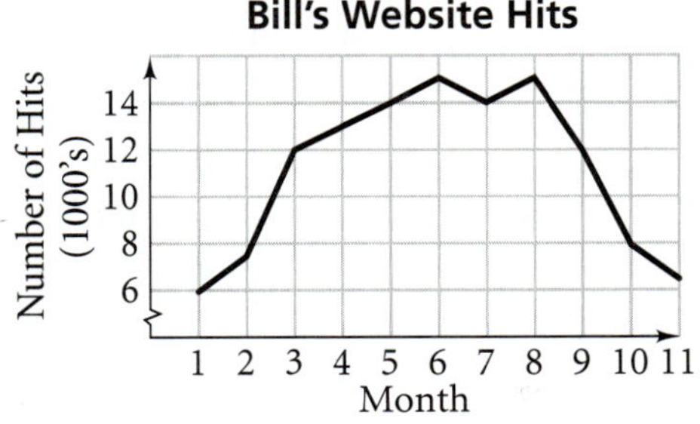

Ⓐ 1200 Ⓒ 6000

Ⓑ 3000 Ⓓ 9000

6. Simplify the expression $5(x + 3) - 2(x + 4)$.

Ⓐ $3x + 23$ Ⓒ $3x + 7$

Ⓑ $7x + 7$ Ⓓ $7x + 23$

7. The graph below shows the number of car sales during the month of August for three different companies.

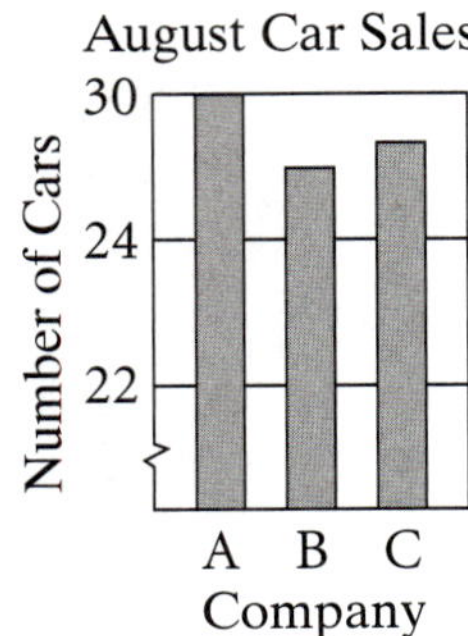

Which statement best explains why the graph may give an incorrect impression to someone reading it?

Ⓐ The title of the graph is misleading.

Ⓑ The intervals for the number of cars are not consistent.

Ⓒ The name of each company is needed.

Ⓓ The bars should not be the same width.

8. Four towns are on the same east-west road. Victory is 8 miles east of Canton. Canton is 2 miles from Milton. Milton is 1 mile from Silverton. What information is needed to find the distance from Victory to Silverton?

Ⓐ The east-west order of only Canton, Milton, and Silverton

Ⓑ The east-west order of only Victory, Canton, and Silverton

Ⓒ The east-west order of only Victory and Silverton

Ⓓ The east-west order of only Canton and Silverton

9. Peaches are \$2.89/lb. Song buys 4.4 pounds of peaches and 6.5 pounds of apples. Sales tax is 4%. Which information is needed to find the total cost of the apples and peaches?

Ⓐ Total number of pounds of apples and peaches

Ⓑ The tax on \$2.89

Ⓒ The product of 6.5 and \$2.89

Ⓓ The price of apples per pound

10. Which table matches the equation below?

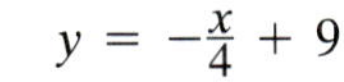

$$y = -\frac{x}{4} + 9$$

Ⓐ

x	−4	0	4	8
y	8	9	10	11

Ⓑ

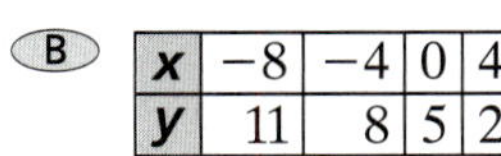

x	−8	−4	0	4
y	11	8	5	2

Ⓒ

x	−8	−4	4	8
y	11	10	9	8

Ⓓ

x	−4	0	4	8
y	10	9	8	7

Standards Mastery

11. The PQR Company experienced a large decrease in sales from month 3 to month 4. Which graph best represents the sales for the PQR Company?

(A)
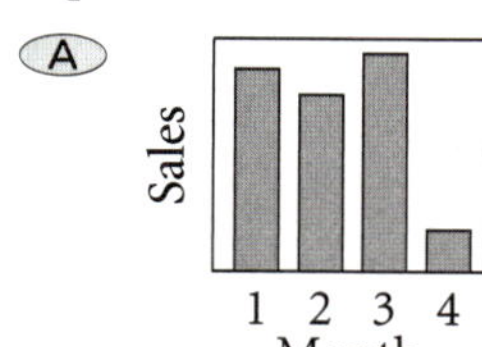

(C)
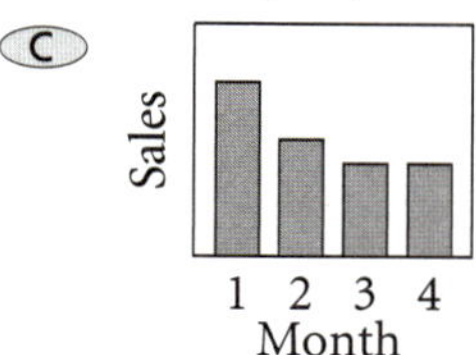

(B)
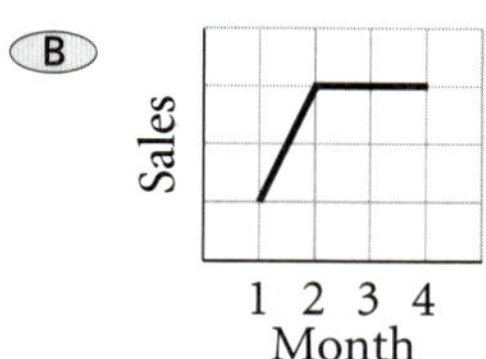

(D)
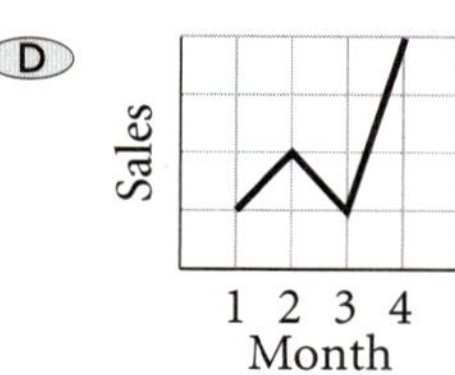

12. Ana purchases music online for \$0.99 per song. Which expression best represents the cost of purchasing s songs?

(A) $0.99s$ (C) $0.99 - s$
(B) $0.99 + s$ (D) $0.99 \div s$

13. The cost of a gallon of milk is \$0.50 more than 5 times the cost of a gallon of water w. Which expression represents the cost of a gallon of milk in terms of w?

(A) $5w + 0.5$ (C) $5(w + 0.5)$
(B) $5w - 0.5$ (D) $5(w - 0.5)$

14. Mr. Beaton received a 4% salary decrease at his job due to poor corporate earnings. If he was earning x dollars per year before, how much is he earning per year now?

(A) $x + 4$ (C) $x + 0.04x$
(B) $x - 0.04$ (D) $x - 0.04x$

15. If the total vacation budget is \$900, which of the following is true?

Vacation Budget

Lodging 33%; Tours 25%; Gifts 17%; Meals 25%

(A) About \$450 is budgeted for lodging.
(B) Less than \$200 is budgeted for tours.
(C) About \$250 is budgeted for meals.
(D) More than \$225 is budgeted for gifts.

16. Which situation is best represented by a circle graph?

(A) A company's annual budget
(B) A company's annual sales growth
(C) A company's employment growth
(D) The production rate of an employee

17. Mrs. Scanella, a car sales person, received a 2% commission on the selling price of a car. If her commission was \$739, what was the selling price of the car?

(A) \$37,689 (C) \$14,780
(B) \$36,950 (D) \$1,478

18. Which expression is equivalent to $4(x + 1)(x - 6) - (8x^2 + x)$?

(A) $-4x^2 - 19x - 24$
(B) $-4x^2 - 4x - 6$
(C) $-4x^2 - x - 24$
(D) $-4x^2 - 21x - 24$

CAHSEE Reference Guide

California Content Standard	Item Number(s)
7AF 4.2	1, 2
7AF 1.5	3, 4, 5
7AF 1.2	6, 18
7MR 1.1	7, 8, 9
7SDAP 1.1	11, 15, 16
7AF 1.1	12, 13, 14
7NS 1.7	17
7MR 2.4	10

For additional review and practice, see the *California Standards Review and Practice Workbook* or go online to use the

Visit: PHSchool.com, **Web Code:** bcq-9045

Reasoning and Proof

What You've Learned

California Content Standards

1.0 Students demonstrate understanding by identifying and giving examples of undefined terms, axioms, theorems, and inductive reasoning.

3.0 Students construct and judge the validity of a logical argument and give counterexamples to disprove a statement.

16.0 Students perform basic constructions with a straightedge and compass, such as angle bisectors and perpendicular bisectors.

for Help to the Lesson in green.

Evaluating Expressions (Skills Handbook page 754)

x^2 **Algebra** Evaluate each expression for the given value of x.

1. $9x - 13$ for $x = 7$
2. $90 - 3x$ for $x = 31$
3. $\frac{1}{2}x + 14$ for $x = 23$

Solving Equations (Algebra 1 Review, page 30)

x^2 **Algebra** Solve each equation.

4. $2x - 17 = 4$
5. $3x + 8 = 53$
6. $(10x + 5) + (6x - 1) = 180$
7. $(x + 21) + (2x + 9) = 90$
8. $3x + 4 = 2x - 1$
9. $3(x + 8) = 12$
10. $2(x + 4) = x + 13$
11. $7x + 5 = 5x + 17$
12. $14x = 2(5x + 14)$
13. $2(3x - 4) + 10 = 5(x + 4)$

Segments and Angles (Lessons 1-5 and 1-6)

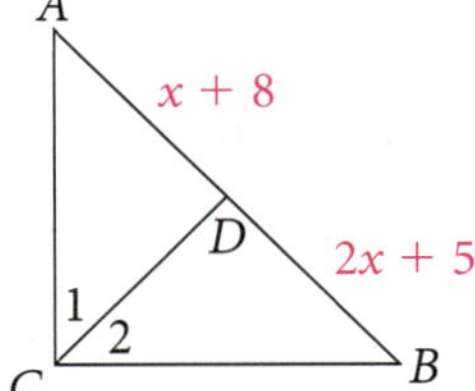

Use the figure at the right.

14. Name $\angle 1$ in two other ways.

15. Name the vertex of $\angle 2$.

16. If D is the midpoint of $\overline{AB}$, find x.

17. If $m\angle ADC$ and $m\angle BDC$ have a sum of 180, name the straight angle.

18. If $\angle 1 \cong \angle 2$, name the bisector of $\angle ACB$.

19. If $m\angle 2 = 45$ and $\angle ACB$ is a right angle, find $m\angle 1$.

20. If $\angle ACB$ is a right angle, $m\angle 1 = 4x$ and $m\angle 2 = 2x + 18$, find $m\angle 1$ and $m\angle 2$.

▲ Cooks "prove" their recipes by testing them.

What You'll Learn Next

California Content Standards

1.0 Students demonstrate understanding by identifying and giving examples of deductive reasoning.

2.0 Students write geometric proofs.

4.0 Students prove basic theorems involving congruence and similarity.

CAHSEE 7NS 1.2, 7NS 2.1, 6SDAP 1.1, 7SDAP 1.2, 7AF 3.3, 7MG 2.1, 7MG 2.3, 7MR 1.1, 7MR 2.1, Alg1 4.0, Alg1 8.0, Alg1 15.0

New Vocabulary

English and Spanish Audio Online

- **biconditional** (p. 87)
- **conclusion** (p. 80)
- **conditional** (p. 80)
- **converse** (p. 81)
- **deductive reasoning** (p. 94)
- **hypothesis** (p. 80)
- **Law of Detachment** (p. 94)
- **Law of Syllogism** (p. 95)
- **paragraph proof** (p. 111)
- **Reflexive Property** (p. 105)
- **Symmetric Property** (p. 105)
- **theorem** (p. 110)
- **Transitive Property** (p. 105)
- **truth value** (p. 81)

2-1 Conditional Statements

California Content Standards

Prepares for GEOM 1.0 Students demonstrate understanding by identifying and giving examples of deductive reasoning.

GEOM 3.0 Students construct and judge the validity of a logical argument and give counterexamples to disprove a statement. ***Develop***

What You'll Learn

- To recognize conditional statements
- To write converses of conditional statements

. . . And Why

To help you read critically, as in Example 7

Check Skills You'll Need GO for Help Skills Handbook pages 757 and 758

x^2 **Algebra** **Solve each equation.**

1. $y + 1 = 0$ **2.** $|x| = 2$ **3.** $17 = 15 + z$

4. $x + 4 = 0$ **5.** $|n| = 0$ **6.** $m - 9 = -10$

New Vocabulary • conditional • hypothesis • conclusion • truth value • converse

1 Conditional Statements

Online active math

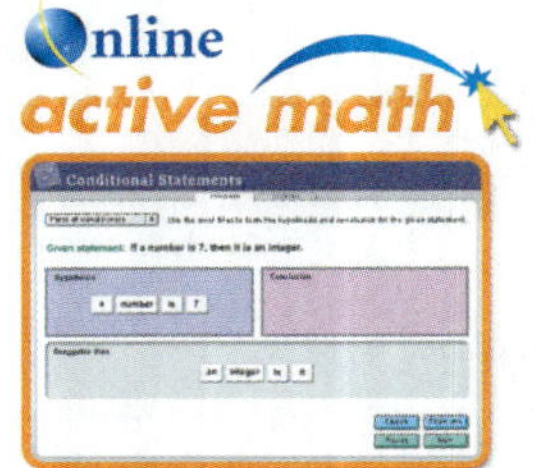

For: Conditionals Activity
Use: Interactive Textbook, 2-1

You have heard *if-then* statements such as this one:

If you are not completely satisfied, then your money will be refunded.

Another name for an *if-then* statement is a **conditional.** Every conditional has two parts. The part following *if* is the **hypothesis,** and the part following *then* is the **conclusion.**

1 EXAMPLE Identifying the Hypothesis and the Conclusion

Identify the hypothesis and the conclusion of this conditional statement:

If the University of Texas won the 2006 Rose Bowl football game, then Texas was college football's 2005 national champion.

Hypothesis: The University of Texas won the 2006 Rose Bowl football game.

Conclusion: Texas was college football's 2005 national champion.

CA Standards Check **1** Identify the hypothesis and the conclusion of this conditional statement:

If $T - 38 = 3$, then $T = 41$.

You can write many sentences as conditionals.

2 EXAMPLE Writing a Conditional

Write each sentence as a conditional.

a. A rectangle has four right angles.
If a figure is a rectangle, then it has four right angles.

b. A tiger is an animal.
If something is a tiger, then it is an animal.

CA Standards Check **2** Write each sentence as a conditional.

a. An integer that ends with 0 is divisible by 5.

b. A square has four congruent sides.

A conditional can have a **truth value** of *true* or *false*. To show that a conditional is true, show that every time the hypothesis is true, the conclusion is also true. To show that a conditional is false, you need to find only one counterexample for which the hypothesis is true and the conclusion is false.

3 EXAMPLE Finding a Counterexample

Show that this conditional is false by finding a counterexample:
If it is February, then there are only 28 days in the month.

To show that this conditional is false, you need to find one counterexample that makes the hypothesis true and the conclusion false.

February in the year 2012 is a counterexample. Because 2012 is a leap year, the month of February has 29 days.

The conditional is false because February 2012 is a counterexample.

CA Standards Check

3 Show that this conditional is false by finding a counterexample:
If the name of a state contains the word *New*, then the state borders an ocean.

You can use a Venn diagram to better understand true conditional statements.

4 EXAMPLE Using a Venn Diagram

Draw a Venn diagram to illustrate this conditional:
If you live in Anaheim, then you live in California.

Residents of California
Residents of Anaheim

The set of things that satisfy the hypothesis lies inside the set of things that satisfy the conclusion.

CA Standards Check

4 Draw a Venn diagram to illustrate this conditional:
If something is a cocker spaniel, then it is a dog.

2 Converses

The **converse** of a conditional switches the hypothesis and the conclusion.

5 EXAMPLE Writing the Converse of a Conditional

Write the converse of the following conditional.

Conditional
If two lines intersect to form right angles, then they are perpendicular.

Converse
If two lines are perpendicular, then they intersect to form right angles.

CA Standards Check

5 Write the converse of the following conditional.
If two lines are not parallel and do not intersect, then they are skew.

In Example 5, both the original conditional and its converse are true. It is possible for a conditional and its converse to have different truth values.

6 EXAMPLE Finding the Truth Value of a Converse

Test-Taking Tip

Read the question carefully. Multiple-choice items usually ask you to find a true or correct answer. Example 6 asks you to find a converse that is *false*.

Multiple Choice Which conditional statement has a false converse?

(A) If two planes are parallel, then they have no points in common.
(B) If a figure is a square, then it has four sides.
(C) If a point has x-coordinate 0, then it lies on the y-axis.
(D) If two angles are congruent, then they have the same measure.

Write the converse of choice B:

If a figure has four sides, then it is a square.

A rectangle is a counterexample, so the converse is false.
Choice B is the correct answer.

CA Standards Check 6 Write the converse of each conditional statement. Determine the truth value of the conditional and its converse. (*Hint*: One of these conditionals is *not* true.)

a. If two lines do not intersect, then they are parallel.
b. If $x = 2$, then $|x| = 2$.

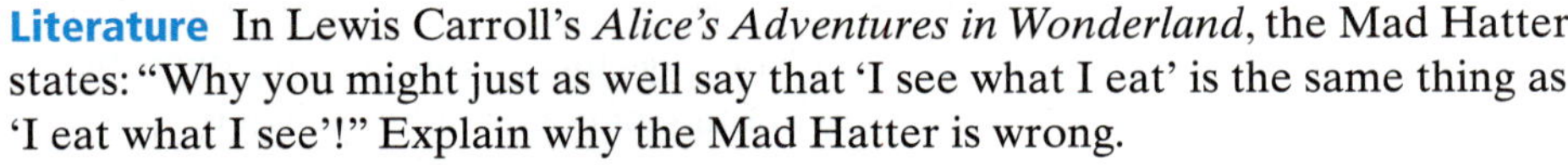

7 EXAMPLE Real-World Connection

This is an illustration by John Tenniel for *Alice's Adventures in Wonderland.*

Literature In Lewis Carroll's *Alice's Adventures in Wonderland*, the Mad Hatter states: "Why you might just as well say that 'I see what I eat' is the same thing as 'I eat what I see'!" Explain why the Mad Hatter is wrong.

The statement "I see what I eat" can be rewritten as a conditional.
"If I eat it, then I see it."

The statement "I eat what I see," can be rewritten as a conditional.
"If I see it, then I eat it."

The two statements are converses of each other. A statement and its converse do not always have the same meaning or the same truth value. The Mad Hatter is wrong to suggest that you can use one just as well as the other.

CA Standards Check 7 In *Alice's Adventures in Wonderland*, the Dormouse states:
"... that 'I breathe when I sleep' is the same thing as 'I sleep when I breathe'!"
Use conditionals to explain why this statement is wrong.

You can use symbolic form to represent a conditional and its converse. In symbolic form, the letter p stands for the hypothesis and the letter q stands for the conclusion.

Key Concepts

Summary Conditional Statements and Converses

Statement	Example	Symbolic Form	You Read It
Conditional	If an angle is a straight angle, then its measure is 180.	$p \rightarrow q$	If p, then q.
Converse	If the measure of an angle is 180, then it is a straight angle.	$q \rightarrow p$	If q, then p.

EXERCISES

For more exercises, see *Extra Skill, Word Problem, and Proof Practice.*

Standards Practice

GEOM 3.0

A Practice by Example

Example 1 (page 80)

GO for Help

1. Identify the hypothesis and the conclusion in the cartoon.

Identify the hypothesis and conclusion of each conditional.

2. If you want to be fit, then you want to get plenty of exercise.

Problem Solving Hint

Some conditionals may omit *then*. You can insert it mentally if you wish.

3. Algebra If $x + 20 = 32$, then $x = 12$.

4. "If you can see the magic in a fairy tale, you can face the future." — Danielle Steel, novelist

5. "If somebody throws a brick at me, I can catch it and throw it back." — Harry S Truman

6. "If you can accept defeat and open your pay envelope without feeling guilty, you're stealing." — George Allen, former NFL coach

7. "If my fans think that I can do everything I say I can do, then they're crazier than I am."— Muhammad Ali

8. ". . . if I could paint that flower in a huge scale, you could not ignore its beauty." — Georgia O'Keeffe, artist

Example 2 (page 80)

Write each sentence as a conditional.

9. Glass objects are fragile.

10. Algebra $3x - 7 = 14$ implies that $3x = 21$.

11. Whole numbers that have 2 as a factor are even.

12. All obtuse angles have measure greater than 90.

13. Good weather makes a picnic enjoyable.

14. Two skew lines do not lie in the same plane.

Example 3 (page 81)

Show that each conditional is false by finding a counterexample.

15. If it is not a weekday, then it is Saturday.

16. Odd integers less than 10 are prime.

17. If you live in a country that borders the United States, then you live in Canada.

18. If you play a sport with a ball and a bat, then you play baseball.

Example 4 (page 81)

Draw a Venn diagram to illustrate each statement.

19. If you live in New England, then you live in the United States.

20. If you play the flute, then you are a musician.

21. If an angle has measure 40, then it is acute.

22. Carrots are vegetables.

Example 5 (page 81)

Write the converse of each conditional statement.

23. If you eat your vegetables, then you grow.

24. If a triangle is a right triangle, then it has a 90° angle.

25. If two segments are congruent, then they have the same length.

26. If you do not work, you do not get paid.

Examples 6 and 7 (page 82)

Write the converse of each conditional statement. Determine the truth values of the original conditional and its converse.

27. If you travel from the United States to Kenya, then you have a passport.

28. **Coordinate Geometry** If a point is in the first quadrant, then its coordinates are positive.

29. **Chemistry** If a substance is water, then its chemical formula is H_2O.

30. **Probability** If the probability that an event will occur is 1, then the event is certain to occur.

31. If you are in Indiana, then you are in Indianapolis.

32. If two angles have measure 90, then the angles are congruent.

B Apply Your Skills

Write a conditional statement that each Venn diagram illustrates.

33.

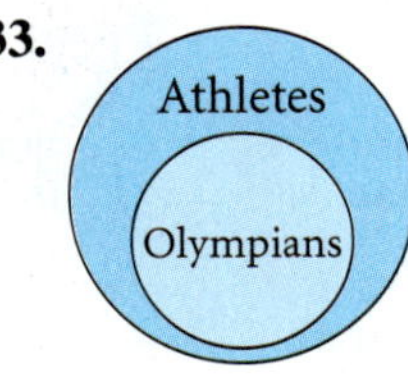

34.

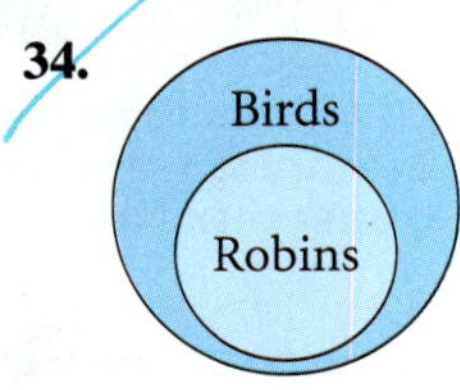

35.

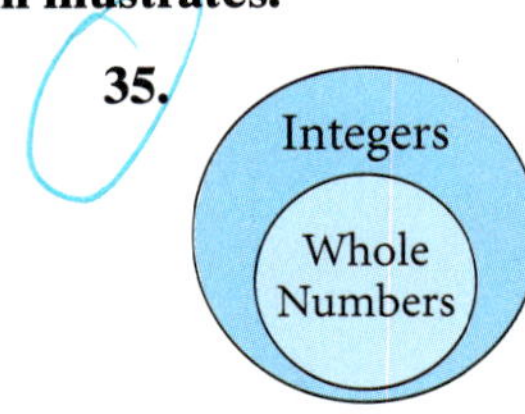

36. **Error Analysis** Ellen claims that both this conditional and its converse are true.

If x is an integer divisible by 3, then x^2 is an integer divisible by 3.

a. Write the converse of the conditional.

b. Only one of the statements is true. Determine which statement is false and provide a counterexample to support your answer.

37. **Multiple Choice** Which conditional and its converse are both true?

(A) If $x = 1$, then $2x = 2$.
(B) If $x = 3$, then $x^2 = 6$.
(C) If $x = 2$, then $x^2 = 4$.
(D) If $x^2 = 4$, then $x = 2$.

Jeanette Rankin was one of nine women among 435 members of Congress at the start of World War II.

Write each statement as a conditional.

38. "We're half the people; we should be half the Congress."— Jeanette Rankin, former U.S. Congresswoman, calling for more women in office

39. "A great work is made out of a combination of obedience and liberty." — Nadia Boulanger, orchestra conductor and musical mentor

40. "A problem well stated is a problem half solved." — Charles F. Kettering, inventor

x^2 **Algebra** **Write the converse of each statement. If the converse is true, write *true*; if not true, provide a counterexample.**

41. If $x - 3 = 15$, then $x = 18$.

42. If y is negative, then $-y$ is positive.

43. If $x = -6$, then $|x| = 6$.

44. If $x < 0$, then $x^2 > 0$.

45. If $x = 2$, then $x^2 = 4$.

46. If $x < 0$, then $x^3 < 0$.

47. **Advertising** Al sees an ad that states, "You want to look good at the beach this summer. Join GoodFit Health Club." Al figures, "I am going to join GoodFit Health Club, so that I will look good at the beach."

a. Write the statement in the ad as a conditional.

b. Write Al's statement as a conditional.

c. **Writing** Explain why the statement in the ad does not have the same meaning as Al's statement.

Reading Math **Let statements *p*, *q*, and *r* be as follows.**

***p*: A figure is square.**

***q*: A figure has four congruent angles.**

***r*: A figure has four congruent sides.**

Write the words for the symbolic statement shown. Determine the truth value of the statement. If it is false, provide a counterexample.

48. $p \rightarrow q$

49. $q \rightarrow p$

50. $r \rightarrow q$

51. $(q \text{ and } r) \rightarrow p$

Advertising **Advertisements often suggest conditional statements. For example, an ad might imply that if you buy a product, you will be popular.**

52. What conditional is implied in the ad at the right?

53. **Open-Ended** Find an ad in which a conditional is used or implied.

Write each postulate as a conditional statement.

54. Two intersecting lines meet in exactly one point.

55. Two intersecting planes meet in exactly one line.

56. Two congruent figures have equal areas.

57. Through any two points there is exactly one line.

58. Through any three noncollinear points there is exactly one plane.

Visit: PHSchool.com
Web Code: bce-0201

Challenge Write a statement beginning with *All*, *Some*, or *No* to match each Venn diagram.

59.

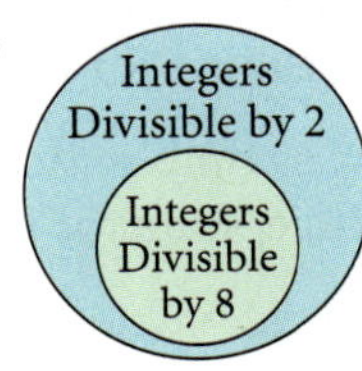

60.

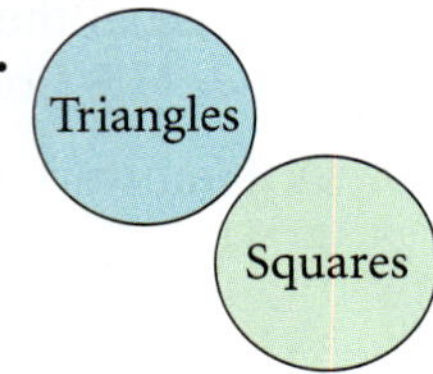

61.

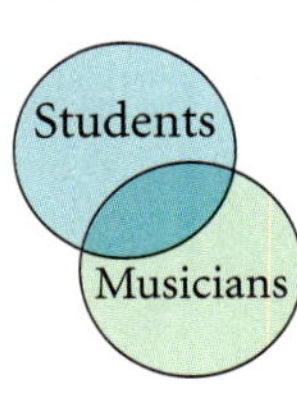

62. **Critical Thinking** You can write many statements that begin with *All* or *No* as conditionals. Give an example of each. (*Hint:* See Exercises 59–61.)

63. Let a represent an integer. Consider the five statements r, s, t, u, and v:
r: a is even. s: a is odd. t: $2a$ is even. u: $2a$ is odd. v: $2a + 1$ is odd.
How many statements of the form $p \rightarrow q$ can you make from these five statements? Decide whether each of the statements is true or false.

Multiple Choice Practice

For California Standards and CAHSEE Tutorials, visit PHSchool.com. Web Code: bcq-9045

GEOM 3.0

64. What value of x is a counterexample to the statement below?
If $x^2 - 8 \geq 0$, then $x \geq 3$.

(A) 4 (B) 2 (C) -2 (D) -3

CAHSEE 7MG 2.1

65. The net for a cylindrical container that will hold a stack of CDs is shown at the right. What is the total surface area to the nearest square centimeter?

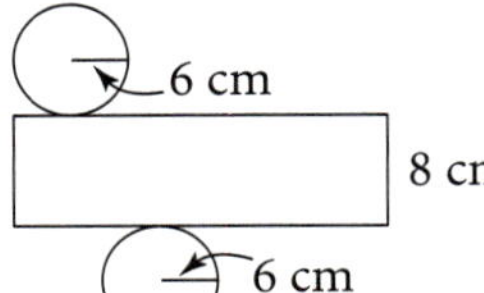

(A) 226 (B) 302 (C) 528 (D) 582

CAHSEE 7NS 1.2

66. $\frac{5}{6} - \left(\frac{3}{4} - \frac{1}{12}\right) =$

(A) $\frac{1}{6}$ (B) $\frac{1}{3}$ (C) $\frac{5}{12}$ (D) $\frac{7}{12}$

CAHSEE 7NS 2.1

67. Which number equals 4^{-3}?

(A) -12 (B) $-\frac{1}{64}$ (C) $\frac{1}{64}$ (D) $\frac{1}{12}$

Mixed Review

Lesson 1-9

Find the perimeter of each rectangle with the given base and height.

68. 6 in., 12 in. **69.** 3.5 cm, 7 cm **70.** $1\frac{3}{4}$ yd, 18 in. **71.** 11 m, 60 cm

72. Find the area of a circle with diameter 10 in. Leave your answer in terms of π.

Lesson 1-8

Find the distance between the points. Round each answer to the nearest tenth.

73. $A(1, 2), B(4, -2)$ **74.** $M(-5, 1), N(0, 5)$ **75.** $R(0, -6), T(2, 3)$

Lesson 1-1

Find the pattern for each sequence. Use the pattern to show the next two terms.

76. $4, 2, 1, \frac{1}{2}, \ldots$ **77.** $5, 2, -1, -4, \ldots$ **78.** N, M, L, K, . . .

2-2 Biconditionals and Definitions

California Content Standards

Prepares for GEOM 1.0 Students demonstrate understanding by identifying and giving examples of deductive reasoning.

What You'll Learn

- To write biconditionals
- To recognize good definitions

. . . And Why

To evaluate definitions of letters used in the American Manual Alphabet, as in Exercises 33–37

Check Skills You'll Need

GO for Help Lesson 2-1

Identify the hypothesis and the conclusion of each conditional statement.

1. If $x > 10$, then $x > 5$.

2. If you live in Milwaukee, then you live in Wisconsin.

Write each statement as a conditional.

3. Squares have four sides.

4. All butterflies have wings.

Write the converse of each statement.

5. If the sun shines, then we go on a picnic.

6. If two lines are skew, then they do not intersect.

7. If $x = -3$, then $x^3 = -27$.

New Vocabulary • biconditional

1 Writing Biconditionals

When a conditional and its converse are true, you can combine them as a true **biconditional.** This is the statement you get by connecting the conditional and its converse with the word *and*. You can write a biconditional more concisely, however, by joining the two parts of each conditional with the phrase *if and only if*.

1 EXAMPLE Writing a Biconditional

Vocabulary Tip

Connect the conditional and its converse with *and*. Then compare with the *if and only if* form.

Consider this true conditional statement. Write its converse. If the converse is also true, combine the statements as a biconditional.

Conditional

If two angles have the same measure, then the angles are congruent.

Converse

If two angles are congruent, then the angles have the same measure.
The converse is also true.

Since both the conditional and its converse are true, you can combine them in a true biconditional by using the phrase *if and only if*.

Biconditional

Two angles have the same measure if and only if the angles are congruent.

CA Standards Check

1 Consider this true conditional statement. Write its converse. If the converse is also true, combine the statements as a biconditional.

Conditional

If three points are collinear, then they lie on the same line.

You can write a biconditional as two conditionals that are converses of each other.

2 EXAMPLE Separating a Biconditional Into Parts

Algebra Write two statements that form this biconditional about whole numbers:
A number is divisible by 3 if and only if the sum of its digits is divisible by 3.

Here are the two statements. They are converses of each other.

If a number is divisible by 3, then the sum of its digits is divisible by 3.

If the sum of a number's digits is divisible by 3, then the number is divisible by 3.

2 Write two statements that form this biconditional about integers greater than 1:
A number is prime if and only if it has only two distinct factors, 1 and itself.

Key Concepts

Summary **Biconditional Statements**

A biconditional combines $p \rightarrow q$ and $q \rightarrow p$ as $p \leftrightarrow q$.

Statement	Example	Symbolic Form	You Read It
Biconditional	An angle is a straight angle if and only if its measure is 180.	$p \leftrightarrow q$	p if and only if q.

2 Recognizing Good Definitions

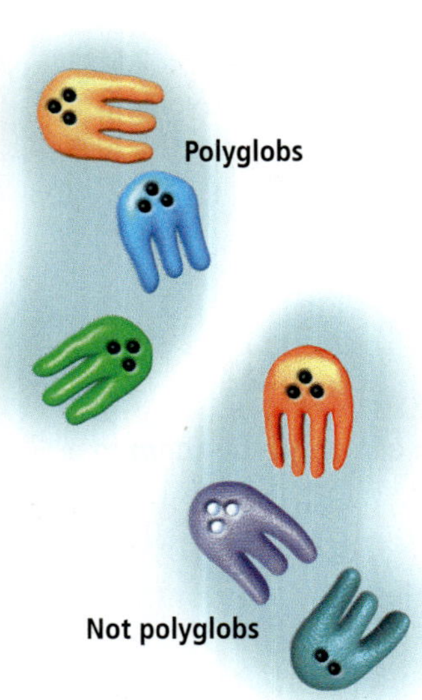

Activity: Writing a Definition

1. Use the examples at the left to identify the figures above that are *polyglobs.*
2. Write a definition of a *polyglob* by describing what a *polyglob* is.

In geometry you start with undefined terms such as point, line, and plane whose meanings you understand intuitively. Then you use those terms to define other terms such as collinear points.

A good definition is a statement that can help you identify or classify an object. A good definition has several important components.

- ✔ A good definition uses clearly understood terms. The terms should be commonly understood or already defined.
- ✔ A good definition is precise. Good definitions avoid words such as *large*, *sort of*, and *almost*.
- ✔ A good definition is reversible. That means that you can write a good definition as a true biconditional.

Real-World Connection

The definitions in a dictionary have to be "good definitions."

3 EXAMPLE Writing a Definition as a Biconditional

Show that this definition of *perpendicular lines* is reversible. Then write it as a true biconditional.

Definition
Perpendicular lines are two lines that intersect to form right angles.

Conditional
If two lines are perpendicular, then they intersect to form right angles.

Converse
If two lines intersect to form right angles, then they are perpendicular.

The two conditionals—converses of each other—are true, so the definition can be written as a true biconditional.

Biconditional
Two lines are perpendicular if and only if they intersect to form right angles.

CA Standards Check 3 Show that this definition of *right angle* is reversible. Then write it as a true biconditional.

Definition
A right angle is an angle whose measure is 90.

One way to show that a statement is *not* a good definition is to find a counterexample.

4 EXAMPLE Real-World Connection

Language Arts Is the given statement a good definition? Explain.

a. An airplane is a vehicle that flies.

The statement is not a good definition because it is not reversible. A helicopter is a counterexample. A helicopter is a vehicle that flies, but a helicopter is not an airplane.

b. A triangle has sharp corners.
The statement is not a good definition because it uses the imprecise word *sharp*, and it is not reversible.

Online active math

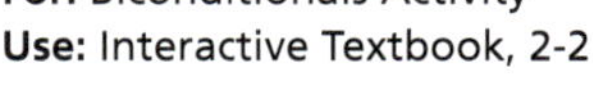

For: Biconditionals Activity
Use: Interactive Textbook, 2-2

CA Standards Check 4 Is the following statement a good definition? Explain.
A square is a figure with four right angles.

EXERCISES

For more exercises, see *Extra Skill, Word Problem, and Proof Practice.*

Standards Practice

A Practice by Example

Example 1 (page 87)

Each conditional statement below is true. Write its converse. If the converse is also true, combine the statements as a biconditional.

1. If two segments have the same length, then they are congruent.

2. **Algebra** If $x = 12$, then $2x - 5 = 19$.
3. If a number is divisible by 20, then it is even.
4. **Algebra** If $x = 3$, then $|x| = 3$.
5. In the United States, if it is July 4th, then it is Independence Day.
6. **Algebra** If $x = -10$, then $x^2 = 100$.

Example 2 (page 88)

Write the two statements that form each biconditional.

7. A line bisects a segment if and only if the line intersects the segment only at its midpoint.
8. An integer is divisible by 100 if and only if its last two digits are zeros.
9. You live in Washington, D. C., if and only if you live in the capital of the United States.
10. Two lines are parallel if and only if they are coplanar and do not intersect.
11. Two angles are congruent if and only if they have the same measure.
12. **Algebra** $x^2 = 144$ if and only if $x = 12$ or $x = -12$.

Example 3 (page 89)

Test each statement below to see if it is reversible. If so, write it as a true biconditional. If not, write *not reversible*.

13. A perpendicular bisector of a segment is a line, segment, or ray that is perpendicular to a segment at its midpoint.
14. Parallel planes are planes that do not intersect.
15. A Tarheel is a person who was born in North Carolina.
16. A rectangle is a four-sided figure with at least one right angle.
17. A midpoint of a segment is a point that divides a segment into two congruent segments.

Example 4 (page 89)

Is each statement below a good definition? If not, explain.

18. A cat is an animal with whiskers.
19. A dog is a good pet.
20. A segment is part of a line.
21. Parallel lines do not intersect.
22. A square is a figure with two pairs of parallel sides.
23. An angle bisector is a ray that divides an angle into two congruent angles.

B Apply Your Skills

24. **Language Arts** Is the following a good definition? Explain.
An obtuse angle is an angle whose measure is greater than 90.
25. **Open-Ended** Choose a definition from a dictionary or from a glossary. Explain what makes the statement a good definition.

26. **Writing** Write a definition of *a line parallel to a plane.*

Visit: PHSchool.com
Web Code: bce-0202

27. Writing Use the figures below to write a good definition of *linear pair*.

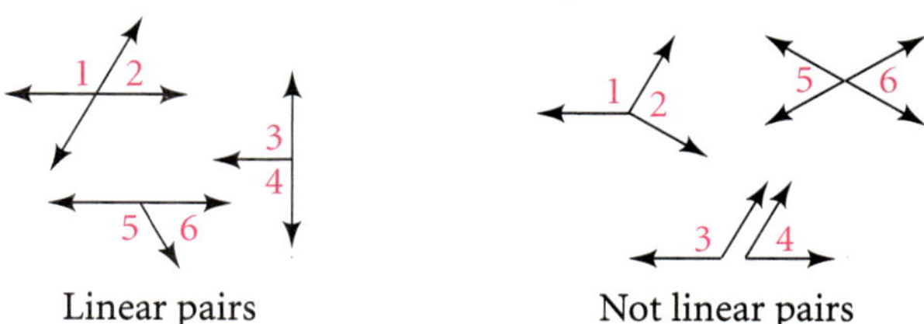

Do angles 1 and 2 form a linear pair? Explain. (*Hint:* See Exercise 27.)

28. **29.** **30.** **31.**

32. Multiple Choice Which conditional and its converse form a true biconditional?

Ⓐ If $x > 0$, then $|x| > 0$.
Ⓑ If $x^3 = 5$, then $x = 125$.
Ⓒ If $x = 3$, then $x^2 = 9$.
Ⓓ If $x = 19$, then $2x - 3 = 35$.

The American Manual Alphabet **For Exercises 33–37, use the chart below. Decide whether the description of each letter is a good definition. If not, provide a counterexample by giving another letter that could fit the definition.**

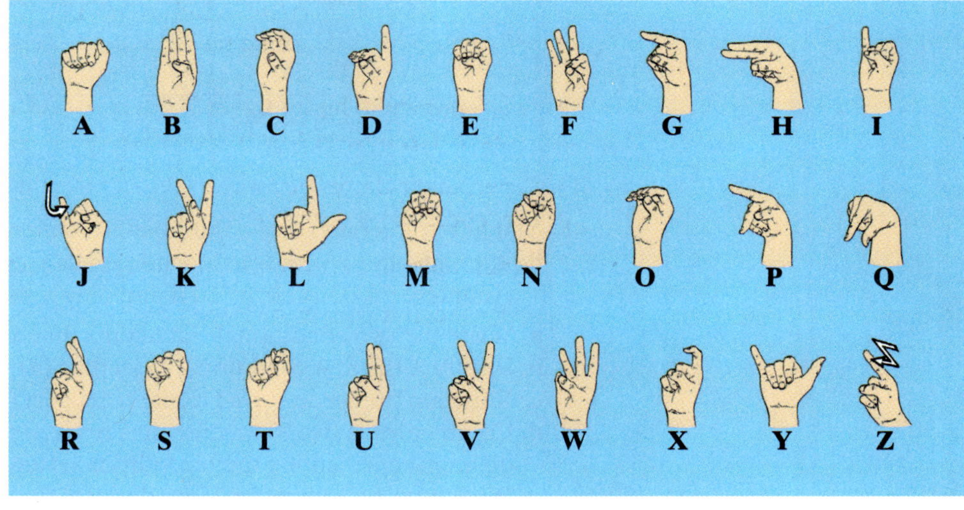

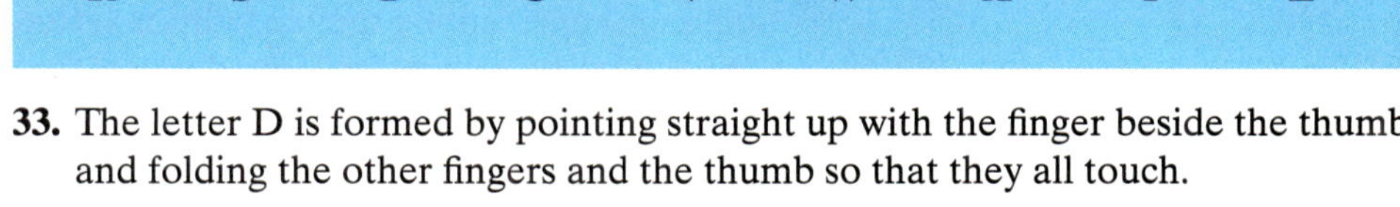

33. The letter D is formed by pointing straight up with the finger beside the thumb and folding the other fingers and the thumb so that they all touch.

34. The letter K is formed by making a V with the two fingers beside the thumb.

35. You have formed the letter Y if and only if the thumb and one finger are extended up and the other fingers are folded into the palm of your hand.

36. You have formed the letter I if and only if the smallest finger is sticking up and the other fingers are folded into the palm of your hand with your thumb folded over them, and your hand is held still.

37. You form the letter B by holding all four fingers tightly together and pointing them straight up while your thumb is folded into the palm of your hand.

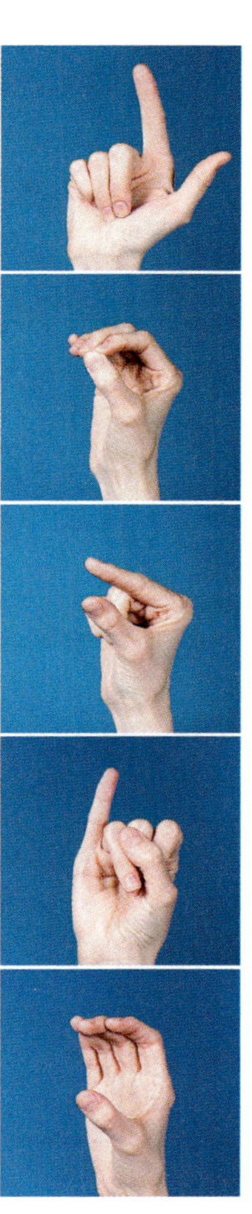

Real-World Connection
The five letters above form a word to think about.

Vocabulary Tip

The expressions vice versa and conversely are synonyms.

Write each statement as a biconditional.

38. Congruent angles are angles with equal measure.

39. When the sum of the digits of an integer is divisible by 9, the integer is divisible by 9 and vice versa.

40. The whole numbers are the nonnegative integers.

Reading Math **Let statements p, q, and r, be as follows.**

p: $\angle A$ and $\angle B$ are right angles. **q: $\angle A$ and $\angle B$ are supplementary angles.**
r: $\angle A$ and $\angle B$ are adjacent angles.

Substitute for p, q, and r, and write each statement the way you would read it.

41. $p \rightarrow q$ **42.** $q \rightarrow p$ **43.** $p \leftrightarrow q$ **44.** $q \leftrightarrow p$ **45.** $p \leftrightarrow r$ **46.** $r \leftrightarrow q$

Challenge

47. Reasoning In a band, Amy, Bob, and Carla are the drummer, guitarist, and keyboard player. Use the clues to find the instrument that each one plays.

Carla and the drummer wear different-colored shirts.
The keyboard player is older than Bob.
Amy, the youngest band member, lives next door to the guitarist.

You can solve this type of logic puzzle by eliminating possibilities. Copy the grid below. Put an X in a box once you eliminate it as a possibility.

Instrument	Amy	Bob	Carla
Drums			
Guitar			
Keyboard			

48. You have illustrated true conditional statements with Venn diagrams. You can do the same thing with true biconditionals. Consider the following statement.
An integer is divisible by 10 if and only if its last digit is 0.

a. Write the two conditional statements that make up this biconditional.
b. Illustrate the first conditional from part (a) with a Venn diagram.
c. Illustrate the second conditional from part (a) with a Venn diagram.
d. Combine your two Venn diagrams from parts (b) and (c) to form a Venn diagram representing the biconditional statement.
e. What must be true of the Venn diagram for any true biconditional statement?
f. Reasoning How does your conclusion in part (e) help to explain why a good definition can be written as a biconditional?

49. Reasoning Alan, Ben, and Cal are seated as shown with their eyes closed. Diane places a hat on each of their heads from a box they know contains 3 red and 2 blue hats. They open their eyes and look forward.

Alan Ben Cal

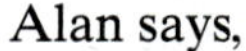

Alan says,
"I cannot deduce what color hat I'm wearing."
Hearing that, Ben says, "I cannot deduce what color I'm wearing, either."
Cal then says, "I know what color I'm wearing!"

How does Cal know the color of his hat?

Multiple Choice Practice

For California Standards and CAHSEE Tutorials, visit PHSchool.com. Web Code: bcq-9045

GEOM 1.0

50. Which statement is NOT true?

- (A) If two lines are parallel, then they lie in one plane and do not intersect.
- (B) Two lines lie in one plane if and only if the lines are parallel.
- (C) If two coplanar lines do not intersect, then the lines are parallel.
- (D) Two lines lie in one plane and do not intersect if and only if the two lines are parallel.

CAHSEE 7SDAP 1.2

51. The graph at the right represents the total number of minutes Tim has been swimming as he completes each successive full lap. Based on this graph, which of the following statements could be an accurate conclusion?

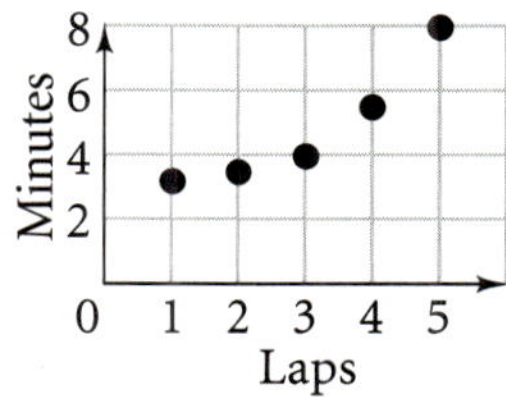

- (A) Tim will take less time to swim the seventh lap than he did to swim the fourth lap.
- (B) Tim will take the same amount of time to swim the seventh lap as he did to swim the fourth lap.
- (C) Tim will take more time to swim the seventh lap than he did to swim the fourth lap.
- (D) Tim will take half as much time to swim the seventh lap as he did to swim the fourth lap.

CAHSEE Alg1 15.0

52. A local hardware store has a 10-gallon can of gray paint left over that is a mixture of 35% black paint and 65% white paint. How many gallons of white paint will be needed to make the gray mixture 20% black?

(A) 5 gallons (B) 7.5 gallons (C) 12 gallons (D) 17.5 gallons

Mixed Review

Lesson 2-1

Write each statement as a conditional.

53. Whole numbers that end in zero are even.

54. When $x = -5, x^2 = 25$.

55. Sunday is a weekend day.

56. All prime numbers greater than 2 are odd.

Lessons 1-5 and 1-6

57. Draw a segment $\overline{XY}$. Construct a bisector of $\overline{XY}$.

58. Draw an acute angle, $\angle 1$. Construct an angle congruent to $\angle 1$.

59. Draw an obtuse angle, $\angle CAD$. Construct the bisector of $\angle CAD$.

Lesson 1-3

Use the figure at the right to name each of the following.

60. two intersecting lines

61. two skew lines

62. two parallel lines

63. two parallel planes

64. three coplanar points

65. two intersecting planes

66. a plane that contains H

67. the intersection of two planes

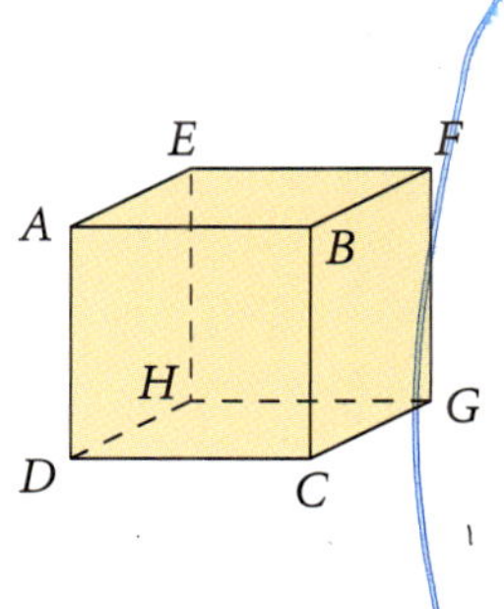

2-3 Deductive Reasoning

California Content Standards

GEOM 1.0 Students demonstrate understanding by identifying and giving examples of deductive reasoning. ***Introduce, Develop***

What You'll Learn

- To use the Law of Detachment
- To use the Law of Syllogism

. . . And Why

To use deductive reasoning to conclude that the Nile River is the longest river in the world, as in Example 5

Check Skills You'll Need

GO for Help Lesson 2-1

Write the converse of each statement.

1. If you don't sleep enough, then your grades suffer.
2. If you want to arrive on time, then you must start early.

Write each statement as a conditional.

3. Leap years have 366 days.
4. Students who do not complete their homework will have lower grades.
5. Two lines that are perpendicular meet to form right angles.
6. Every sixteen-year-old is a teenager.

New Vocabulary
- deductive reasoning
- Law of Detachment
- Law of Syllogism

1 Using the Law of Detachment

Real-World Connection

An auto mechanic uses deductive reasoning as in Example 1.

In Chapter 1 you learned that inductive reasoning is based on observing what has happened and then making a conjecture about what will happen. In this lesson, you will study deductive reasoning.

Deductive reasoning (or logical reasoning) is the process of reasoning logically from given statements to a conclusion. If the given statements are true, deductive reasoning produces a true conclusion.

Many people use deductive reasoning in their jobs. A physician diagnosing a patient's illness uses deductive reasoning. A carpenter uses deductive reasoning to determine what materials are needed at a work site.

1 EXAMPLE Real-World Connection

Auto Maintenance An auto mechanic knows that if a car has a dead battery, the car will not start. A mechanic begins work on a car and finds the battery is dead. What conclusion can she make?

The mechanic can conclude that the car will not start.

CA Standards Check

1 **Critical Thinking** Suppose that a mechanic begins work on a car and finds that the car will not start. Can the mechanic conclude that the car has a dead battery? Explain.

In Example 1 the mechanic is using a law of deductive reasoning called the **Law of Detachment.**

Key Concepts

Vocabulary Tip

You can read $p \rightarrow q$ as "p implies q."

Property	Law of Detachment
If a conditional is true and its hypothesis is true, then its conclusion is true. In symbolic form: If $p \rightarrow q$ is a true statement and p is true, then q is true.	

2 EXAMPLE Using the Law of Detachment

For the given true statements, what can you conclude?

Given: If M is the midpoint of a segment, then it divides the segment into two congruent segments.
M is the midpoint of $\overline{AB}$.

You are given that a conditional and its hypothesis are true. By the Law of Detachment, you can conclude that M divides $\overline{AB}$ into two congruent segments, or $\overline{AM} \cong \overline{MB}$.

CA Standards Check 2 If a baseball player is a pitcher, then that player should not pitch a complete game two days in a row. Vladimir Nuñez is a pitcher. On Monday, he pitches a complete game. What can you conclude?

3 EXAMPLE Real-World Connection

Does the following argument illustrate the Law of Detachment?

Given: If it is snowing, then the temperature is less than or equal to 32°F.
The temperature is 20°F.

You conclude: It must be snowing.

You are given that a conditional and its conclusion are true.
You cannot apply the Law of Detachment and conclude that the hypothesis is true.
You cannot come to any conclusion about whether it is snowing from the information given.

CA Standards Check 3 If possible, use the Law of Detachment to draw a conclusion. If it is not possible to use this law, explain why.

Given: If a road is icy, then driving conditions are hazardous.
Driving conditions are hazardous.

2 Using the Law of Syllogism

Another law of deductive reasoning is the Law of Syllogism. The **Law of Syllogism** allows you to state a conclusion from two true conditional statements when the conclusion of one statement is the hypothesis of the other statement.

Key Concepts

Property	Law of Syllogism
If $p \rightarrow q$ and $q \rightarrow r$ are true statements, then $p \rightarrow r$ is a true statement.	

4 EXAMPLE Using the Law of Syllogism

Vocabulary Tip

2 is a repeated factor of 12 because it appears more than once in the prime factorization of 12.
$12 = 2 \cdot 2 \cdot 3$

Algebra Use the Law of Syllogism to draw a conclusion from the following true statements.

If a number is prime, then it does not have repeated factors.
If a number does not have repeated factors, then it is not a perfect square.

You have two true conditionals where the conclusion of one is the hypothesis of the other. You can use the Law of Syllogism to draw the following conclusion:

If a number is prime, then it is not a perfect square.

CA Standards Check 4 If possible, state a conclusion using the Law of Syllogism. If it is not possible to use this law, explain why.

a. If a number ends in 0, then it is divisible by 10.
If a number is divisible by 10, then it is divisible by 5.

b. If a number ends in 6, then it is divisible by 2.
If a number ends in 4, then it is divisible by 2.

You can use both the Law of Detachment and the Law of Syllogism to draw conclusions.

5 EXAMPLE Real-World Connection

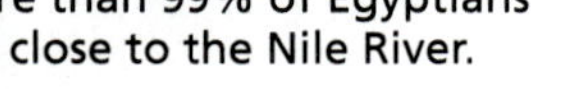

Real-World Connection

More than 99% of Egyptians live close to the Nile River.

Geography Use the Law of Detachment and the Law of Syllogism to draw conclusions from the following true statements.

If a river is more than 4000 mi long, then it is longer than the Amazon.
If a river is longer than the Amazon, then it is the longest river in the world.
The Nile is 4132 mi long.

You can use the first two statements and the Law of Syllogism to conclude:
If a river is more than 4000 mi long, then it is the longest river in the world.

With this, the fact that the Nile is 4132 mi long, and the Law of Detachment, you can also conclude:
The Nile is the longest river in the world.

CA Standards Check 5 Use the Law of Detachment and the Law of Syllogism to draw conclusions.

The Volga River is in Europe.
If a river is less than 2300 mi long, it is not one of the world's ten longest rivers.
If a river is in Europe, then it is less than 2300 mi long.

EXERCISES

For more exercises, see *Extra Skill, Word Problem, and Proof Practice.*

Standards Practice GEOM 1.0

A Practice by Example

Examples 1 and 2
(pages 94, 95)

GO for Help

Use the Law of Detachment to draw a conclusion.

1. If a student gets an A on a final exam, then the student will pass the course.
Felicia gets an A on the music theory final exam.

2. If a student wants to go to college, then the student must study hard.
Rashid wants to go to the University of North Carolina.

3. If two lines are parallel, then they do not intersect.
Line ℓ is parallel to line m.

4. If there is lightning, then it is not safe to be out in the open.
Marla sees lightning from the soccer field.

Example 3 (page 95)

If possible, use the Law of Detachment to draw a conclusion. If not possible, write *not possible*.

5. If a figure is a rectangle, then it has two pairs of parallel sides.
Figure $ABCD$ is a rectangle.

6. **Algebra** If n is a prime number greater than 2, then n^2 is an odd number.
9^2 is an odd number.

7. If three points are on the same line, then they are collinear.
Points X, Y, and Z are on line m.

8. If an angle is obtuse, then it is not acute.
$\angle XYZ$ is not obtuse.

9. If you are a Golden Gopher, you've attended the University of Minnesota.
(See photo.)

Real-World Connection
Hockey wing Nadine Muzerall is a Golden Gopher (Exercise 9).

Example 4 (page 96)

Use the Law of Syllogism to draw a conclusion.

10. **Zoology** If an animal is a red wolf, then its scientific name is *Canis rufus*.
If an animal is named *Canis rufus*, then it is endangered.

11. If two planes intersect, then they intersect in a line.
If two planes are not parallel, then they intersect.

12. If you read a good book, then you enjoy yourself.
If you enjoy yourself, then your time is well spent.

13. If you are studying biology, then you are studying a science.
If you are studying botany, then you are studying biology.

Example 5 (page 96)

Geography **Use the Law of Detachment and the Law of Syllogism to draw conclusions from the following statements.**

14. If a mountain is the highest in Alaska, then it is the highest in the United States.
If an Alaskan mountain is over 20,300 ft high, then it is the highest in Alaska.
Alaska's Mount McKinley is 20,320 ft high.

15. If you live in Lubbock, then you live in Texas. Levon lives in Lubbock.
If you live in Texas, then you live in the 28th state to enter the Union.

B Apply Your Skills

For Exercises 16–21, assume that the following statements are true.

A. If Maria is drinking juice, then it is breakfast time.
B. If it is lunchtime, then Kira is drinking milk and nothing else.
C. If it is mealtime, then Curtis is drinking water and nothing else.
D. If it is breakfast time, then Julio is drinking juice and nothing else.
E. Maria is drinking juice.

Use only the information given above. For each statement, write *must be true*, *may be true*, or *is not true*. Explain your reasoning.

16. Julio is drinking juice.
17. Curtis is drinking water.
18. Kira is drinking milk.
19. Curtis is drinking juice.
20. Maria is drinking water.
21. Julio is drinking milk.

Visit: PHSchool.com
Web Code: bce-0203

Real-World Connection

Average ocean temperature in Key West is about 80°F, a good snorkeling temperature.

For each of the following, write the first statement as a conditional. If possible, use the Law of Detachment to make a conclusion. If not possible, write *not possible*.

22. All national parks are interesting.
Mammoth Cave is a national park.

23. **Weather** The temperature is always above 32°F in Key West, Florida.
The temperature is 62°F.

24. Every high school student likes music.
Ling likes music.

25. All squares are rectangles.
ABCD is a square.

26. **Writing** Give an example of a rule used in your school that could be written as a conditional. Explain how the Law of Detachment is used in applying that rule.

For Exercises 27–31, use the cartoon and deductive reasoning to answer *yes* or *no*. If *no*, explain.

27. Is a person with a red car allowed to park here on Tuesday at 10:00 A. M.?

28. Is a man with a beard allowed to park here on Monday at 10:30 A. M.?

29. Is a woman with a wig allowed to park here on Saturday at 10:00 A. M.?

30. Is a person with a blue car allowed to park here on Tuesday at 9:05 A. M.?

31. Is a person with a convertible with leather seats allowed to park here on Sunday at 6:00 P. M.?

32. **Reasoning** Assume that the following statements are true.

If Anita goes to the concert, Beth will go.
If Beth goes to the concert, Aisha will go.
If Aisha goes to the concert, Ramon will go.

Only two of the four students went to the concert. Who were they?

33. **Critical Thinking** Consider the following given statements and conclusion.

Given: If an animal is a fish, then it has gills.
A turtle does not have gills.
You conclude: A turtle is not a fish.

This argument does not use the Law of Syllogism or the Law of Detachment, but it does use good deductive reasoning.

a. Draw a Venn diagram to illustrate the given information.

b. Use the Venn diagram to help explain why the argument uses good reasoning.

Multiple Choice Practice

For California Standards and CAHSEE Tutorials, visit PHSchool.com. Web Code: bcq-9045

GEOM 1.0

34. **Given:** If Deb gets an A on the test, then she will go to the movies. If Deb goes to the movies, then she will get popcorn. Deb does not get popcorn. What can you conclude?

Ⓐ Deb went to the movies
Ⓑ Deb got a C on the test
Ⓒ Deb got a soda
Ⓓ Deb did not get an A on the test

CAHSEE 7AF 3.3

35. Which line in the graph at the right represents $y = \frac{x}{4}$?

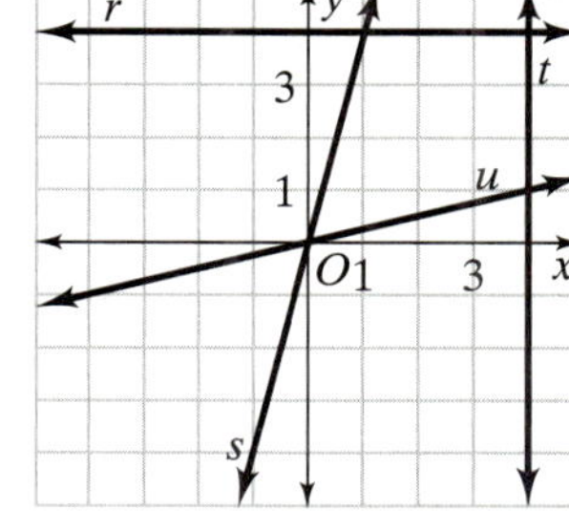

Ⓐ Line r
Ⓑ Line s
Ⓒ Line t
Ⓓ Line u

CAHSEE Alg1 4.0

36. Which of the following is equivalent to $5(x - 4) - 2(x - 3) = 12$?

Ⓐ $5x - 20 - 2x + 6 = 12$
Ⓑ $5x + 20 - 2x - 6 = 12$
Ⓒ $5x - 20 - 2x - 6 = 12$
Ⓓ $5x + 20 - 2x + 6 = 12$

CAHSEE Alg1 8.0

37. What is the slope of a line parallel to the line at the right?

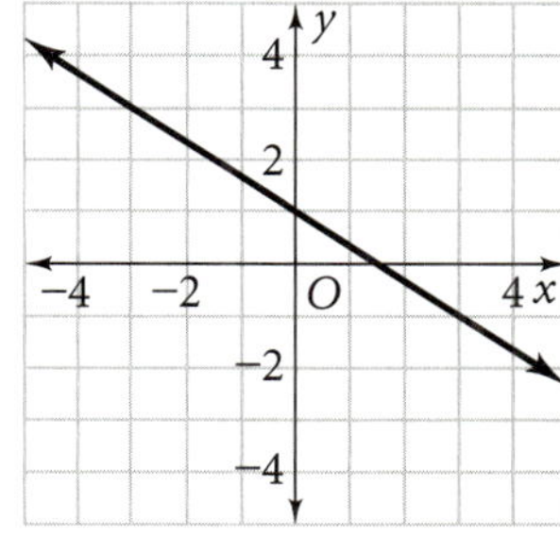

Ⓐ $-\frac{3}{2}$
Ⓑ $-\frac{2}{3}$
Ⓒ $\frac{2}{3}$
Ⓓ $\frac{3}{2}$

Mixed Review

GO for Help

Lesson 2-2

Is each statement a good definition? If not, find a counterexample.

38. An angle is a figure formed by two rays.

39. A ray is an angle bisector if and only if it divides an angle into two congruent angles.

Lesson 2-1

Show that each conditional is false by finding a counterexample.

40. **Geography** If the name of a state contains the word *North*, then the state borders Canada.

41. **Algebra** If you square a fraction, then the result is always greater than the original fraction.

Lesson 1-4

Complete with *always*, *sometimes*, or *never* to make a true statement.

42. Two lines that do not intersect are __?__ parallel.

43. Two lines that intersect are __?__ skew.

44. Two segments that intersect are __?__ coplanar.

GO Online Lesson Quiz Visit: PHSchool.com, Web Code: bca-0203

Checkpoint Quiz 1 — Lessons 2-1 through 2-3

1. Identify the hypothesis and the conclusion of this conditional statement:
   If $x > 5$, then $x^2 > 25$.
2. Write this statement as a conditional: Roses are beautiful flowers.

For Exercises 3 and 4, use this conditional statement:
If an integer ends with 0, then the integer is divisible by 2.

3. Write the converse of the statement.
4. Find a counterexample to show that the converse is *not* true.
5. Write the two conditionals that make up this biconditional:
   An angle is an acute angle if and only if its measure is between 0 and 90.
6. Rewrite this definition as a biconditional:
   Points that lie on the same line are collinear.
7. Find a counterexample to show that the following statement is *not* a good definition:
   A computer is a machine with a keyboard and a memory.

Use the Law of Detachment or the Law of Syllogism to draw a conclusion from each pair of statements. If not possible, write *not possible*.

8. If a student is on the basketball team, then that student has passing grades.
   Theresa is on the basketball team.
9. If a student studies geometry, the student studies mathematics.
   If a student studies mathematics, the student's mind is expanded.
10. If you miss the bus, then you will be late for school.
    You are late for school.

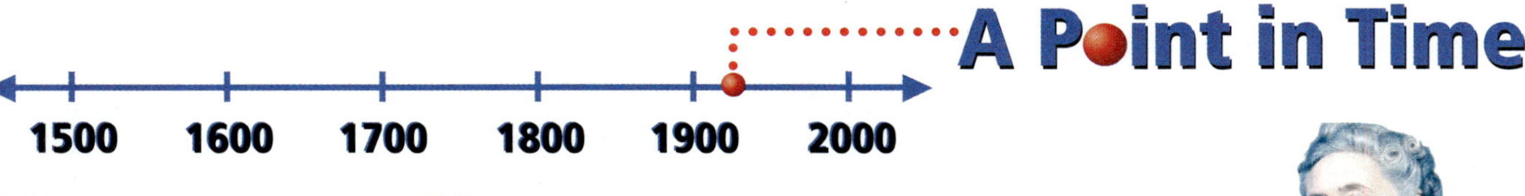

A Point in Time

Hercule Poirot as played by David Suchet, 1989–1997.

Most people are not detectives, but as a young woman, the English writer Agatha Christie (1890–1976) correctly deduced that many people would like to be. In 1920 she published her first book, a detective novel entitled *The Mysterious Affair at Styles* in which she introduced the eccentric and ultra-logical Belgian detective Hercule Poirot. In this and in many subsequent novels, Poirot solves mysteries not with guns or car chases but with logical reasoning.

For: Information about Agatha Christie
Web Code: bcb-2032

Mathematical Systems

FOR USE WITH LESSON 2-3

GEOM 1.0 Students demonstrate understanding by identifying and giving examples of undefined terms, axioms, theorems, and inductive and deductive reasoning. *Develop*

The geometry you are studying is an example of a *mathematical system*.

ACTIVITY

Enter the information at the right into your study aid. (See Vocabulary Builder, page 29.) It tells you about a mathematical system.

You may wish to cross-reference this entry, noting that you learned about:

- undefined terms and definitions on page 29 and in Lesson 2-2,
- postulates in Lessons 1-3 and 1-5, and
- the deductive reasoning framework of geometry in Lesson 2-3.

Mathematical System:

- a vocabulary, and
- a collection of statements in an accepted framework of reasoning.

Example: Euclidean Geometry

Vocabulary	Statements
Undefined terms	Postulates
Defined terms	Theorems

EXERCISES

1. Complete the following:
   a. A mathematical system is made up of a _?_ and a collection of _?_ in an accepted framework of reasoning.
   b. The vocabulary of a mathematical system is made up of _?_ and _?_.
   c. The statements of a mathematical system are _?_ and _?_.

A postulate is a statement that you accept as fact without proof. A *theorem* is a statement that you prove true using deductive reasoning. You base the proof on statements that you already know are true.

2. Explain why some statements must be assumed true.
3. If you haven't already done so, begin a "Statements" section of your study aid for the postulates of Lessons 1-3 and 1-5.

Each statement in a mathematical system is either true or false. A statement that is neither true nor false cannot be in the system. One type of such a statement is a *paradox*.

4. The statement, "This sentence is false" is a well-known paradox. Explain.
5. A statement that contradicts common sense is another type of paradox. The Greek logician Zeno (born circa 490 B.C.) is credited with a paradox that says, in effect, a hare can't catch a tortoise in a race if the tortoise gets a head start. Research Zeno's Paradox. Describe it and explain why it is a paradox.

Activity Lab

Paper-Folding Constructions

Hands-On

FOR USE WITH LESSON 2-4

Prepares for GEOM 16.0 Students perform basic constructions with a straightedge and compass, such as angle bisectors and perpendicular bisectors.

In Chapter 1, you used a compass and straightedge to construct special geometric figures. You can also construct figures by folding paper. Some form of paper that you can see through, such as tracing paper or wax paper, is best.

1 ACTIVITY

- Draw a line segment $\overline{AB}$ on your paper.
- Fold and crease the paper so that B is on top of A.
- Unfold. Label the intersection of $\overline{AB}$ and the crease as C.

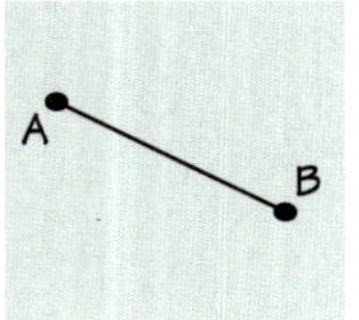

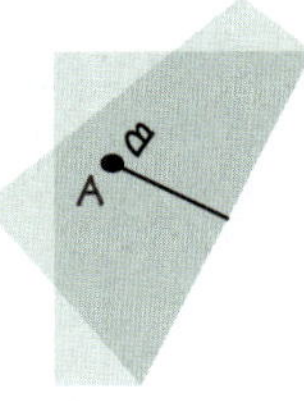

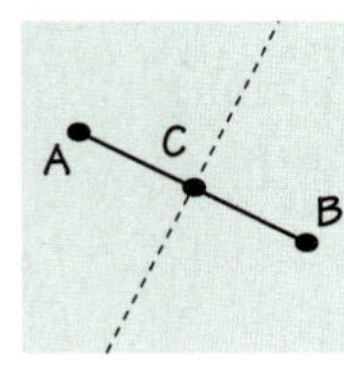

1. Measure lengths AC and CB. What do you notice?
2. Measure the four angles formed by $\overline{AB}$ and the crease. What kind of angles are they?
3. Based on what you found in Exercises 1 and 2, how is the fold line related to $\overline{AB}$?

2 ACTIVITY

- Draw an angle $\angle FGH$ on your paper.
- Fold and crease the paper so that $\overrightarrow{GF}$ lies on top of $\overrightarrow{GH}$.
- Unfold the paper. Label a point J on the fold line and in the interior of $\angle FGH$.

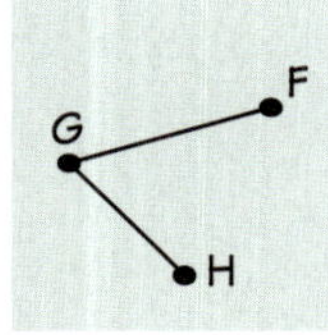

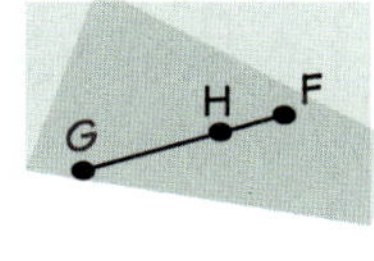

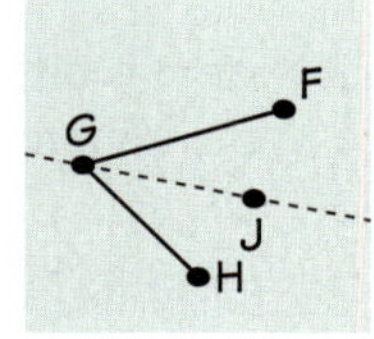

4. Use a protractor. Measure $\angle FGJ$ and $\angle JGH$. What do you notice?
5. Based on what you found in Exercise 4, how is $\overrightarrow{GJ}$ related to $\angle FGH$?

EXERCISES

Fold a line on your paper. Mark a point P on the line and a point Q not on the line. Explain and demonstrate how you can do each of the following.

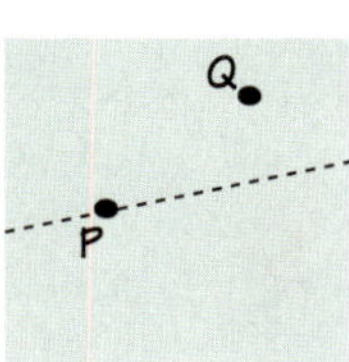

6. Fold a perpendicular to the line at P.
7. Fold a perpendicular to the line through Q.

2-4 Reasoning in Algebra

 **California Content Standards**

GEOM 1.0 Students demonstrate understanding by identifying and giving examples of deductive reasoning. ***Develop***

GEOM 3.0 Students construct and judge the validity of a logical argument. ***Master***

What You'll Learn

- To connect reasoning in algebra and geometry

. . . And Why

To justify steps in a logical argument, as in Example 1

Check Skills You'll Need GO for Help Lesson 1-6

For Exercises 1–5, use the figure at the right.

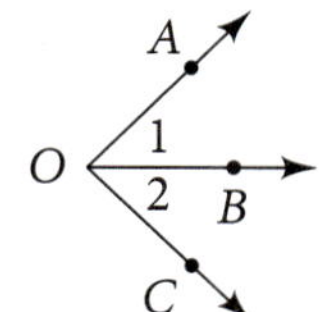

1. Name $\angle 1$ in two other ways.
2. Name the vertex of $\angle 2$.
3. If $\angle 1 \cong \angle 2$, name the bisector of $\angle AOC$.
4. If $m\angle AOC = 90$ and $m\angle 1 = 45$, find $m\angle 2$.
5. If $m\angle AOC = 90$, name two perpendicular rays.

New Vocabulary
- Reflexive Property
- Symmetric Property
- Transitive Property

1 Connecting Reasoning in Algebra and Geometry

In geometry you accept postulates and properties as true. You use deductive reasoning to prove other statements. Some of the properties that you accept as true are the properties of equality from algebra. They are listed below in terms of any numbers a, b, and c.

 **Key Concepts**

Summary — Properties of Equality

Addition Property	If $a = b$, then $a + c = b + c$.
Subtraction Property	If $a = b$, then $a - c = b - c$.
Multiplication Property	If $a = b$, then $a \cdot c = b \cdot c$.
Division Property	If $a = b$ and $c \neq 0$, then $\frac{a}{c} = \frac{b}{c}$.
Reflexive Property	$a = a$
Symmetric Property	If $a = b$, then $b = a$.
Transitive Property	If $a = b$ and $b = c$, then $a = c$.
Substitution Property	If $a = b$, then b can replace a in any expression.

You also assume that other properties from algebra are true.

 **Key Concepts**

Property — The Distributive Property

$a(b + c) = ab + ac$

You use deductive reasoning every time you solve an equation. You can justify each statement that you make with a postulate, a property, or a definition. When you solve problems involving angle measures, you can use the Angle Addition Postulate.

1 EXAMPLE Justifying Steps in Solving an Equation

Algebra Solve for x and justify each step.

Given: $m\angle AOC = 139$

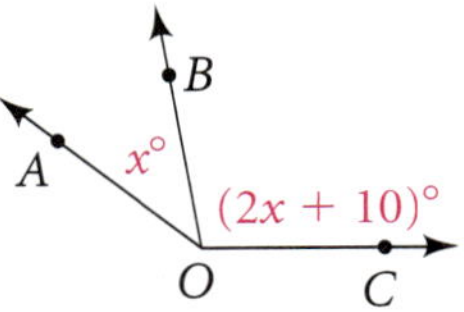

To review the Angle Addition Postulate, go to p. 38.

$m\angle AOB + m\angle BOC = m\angle AOC$	**Angle Addition Postulate**
$x + 2x + 10 = 139$	**Substitution Property**
$3x + 10 = 139$	**Simplify.**
$3x = 129$	**Subtraction Property of Equality**
$x = 43$	**Division Property of Equality**

CA Standards Check

1 Fill in each missing reason.

Given: $\overrightarrow{LM}$ bisects $\angle KLN$.

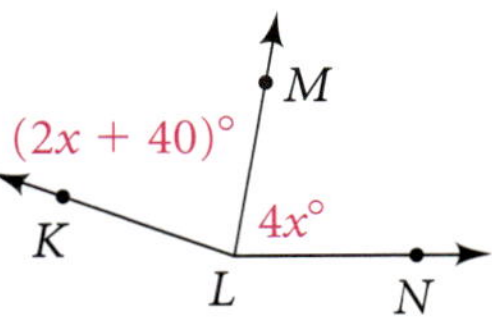

$\overrightarrow{LM}$ bisects $\angle KLN$.	**Given**
$m\angle MLN = m\angle KLM$	**Definition of angle bisector**
$4x = 2x + 40$	?
$2x = 40$	?
$x = 20$	?

You can use the Segment Addition Postulate to justify statements about lengths of segments.

2 EXAMPLE Justifying Steps in Solving an Equation

Algebra Solve for y and justify each step.

Given: $AC = 21$

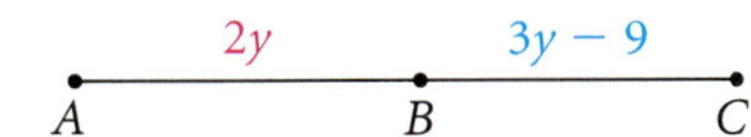

To review the Segment Addition Postulate, go to p. 32.

$AB + BC = AC$	**Segment Addition Postulate**
$2y + (3y - 9) = 21$	**Substitution Property**
$5y - 9 = 21$	**Simplify.**
$5y = 30$	**Addition Property of Equality**
$y = 6$	**Division Property of Equality**

CA Standards Check

2 Find AB and BC by substituting $y = 6$ in the expressions in the diagram above. Check that $AB + BC = 21$.

The Reflexive, Symmetric, and Transitive Properties of Equality have corresponding properties of congruence. You can use properties of congruence to justify statements.

Summary	Properties of Congruence
Reflexive Property	$\overline{AB} \cong \overline{AB}$ $\angle A \cong \angle A$
Symmetric Property	If $\overline{AB} \cong \overline{CD}$, then $\overline{CD} \cong \overline{AB}$. If $\angle A \cong \angle B$, then $\angle B \cong \angle A$.
Transitive Property	If $\overline{AB} \cong \overline{CD}$ and $\overline{CD} \cong \overline{EF}$, then $\overline{AB} \cong \overline{EF}$. If $\angle A \cong \angle B$ and $\angle B \cong \angle C$, then $\angle A \cong \angle C$.

3 EXAMPLE Using Properties of Equality and Congruence

Name the property of equality or congruence that justifies each statement.

a. $\angle K \cong \angle K$
Reflexive Property of Congruence

b. If $2x - 8 = 10$, then $2x = 18$.
Addition Property of Equality

c. If $\overline{RS} \cong \overline{TW}$ and $\overline{TW} \cong \overline{PQ}$, then $\overline{RS} \cong \overline{PQ}$.
Transitive Property of Congruence

d. If $m\angle A = m\angle B$, then $m\angle B = m\angle A$.
Symmetric Property of Equality

3 Name the property of equality or congruence illustrated.
a. $\overline{XY} \cong \overline{XY}$
b. If $m\angle A = 45$ and $45 = m\angle B$, then $m\angle A = m\angle B$.

EXERCISES

For more exercises, see *Extra Skill, Word Problem, and Proof Practice.*

Standards Practice GEOM 1.0, 3.0

Practice by Example

Examples 1 and 2 (page 104)

x^2 Algebra **Fill in the reason that justifies each step.**

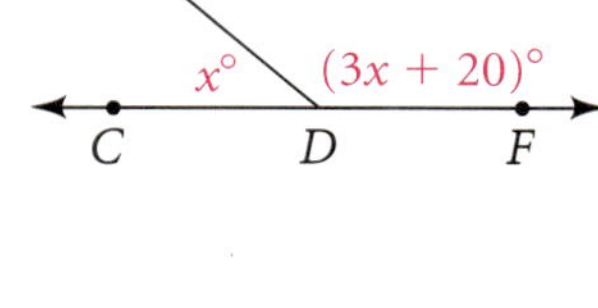

1. Solve for x.

$m\angle CDE + m\angle EDF = 180$	**a.** ?
$x + (3x + 20) = 180$	**b.** ?
$4x + 20 = 180$	**c.** ?
$4x = 160$	**d.** ?
$x = 40$	**e.** ?

2. Solve for n.

3(n + 4) 3n
X Z Y
42

Given: $XY = 42$

$XZ + ZY = XY$	**a.** ?
$3(n + 4) + 3n = 42$	**b.** ?
$3n + 12 + 3n = 42$	**c.** ?
$6n + 12 = 42$	**d.** ?
$6n = 30$	**e.** ?
$n = 5$	**f.** ?

x^2 **Algebra** **Give a reason for each step.**

3. $\frac{1}{2}x - 5 = 10$ Given
$2\left(\frac{1}{2}x - 5\right) = 20$ a. _?_
$x - 10 = 20$ b. _?_
$x = 30$ c. _?_

4. $5(x + 3) = -4$ Given
$5x + 15 = -4$ a. _?_
$5x = -19$ b. _?_
$x = -\frac{19}{5}$ c. _?_

Example 3 (page 105)

Name the property that justifies each statement.

5. $\angle Z \cong \angle Z$

6. $2(3x + 5) = 6x + 10$

7. If $12x = 84$, then $x = 7$.

8. If $\overline{ST} \cong \overline{QR}$, then $\overline{QR} \cong \overline{ST}$.

9. If $m\angle A = 15$, then $3m\angle A = 45$.

10. $XY = XY$

11. If $3x + 14 = 80$, then $3x = 66$.

12. If $KL = MN$, then $MN = KL$.

13. If $2x + y = 5$ and $x = y$, then $2x + x = 5$.

14. If $AB - BC = 12$, then $AB = 12 + BC$.

15. If $\angle 1 \cong \angle 2$ and $\angle 2 \cong \angle 3$, then $\angle 1 \cong \angle 3$.

B Apply Your Skills

Use the given property to complete each statement.

16. Addition Property of Equality
If $2x - 5 = 10$, then $2x =$ _?_.

17. Subtraction Property of Equality
If $5x + 6 = 21$, then _?_ $= 15$.

18. Symmetric Property of Equality
If $AB = YU$, then _?_.

19. Symmetric Property of Congruence
If $\angle H \cong \angle K$, then _?_ $\cong \angle H$.

20. Reflexive Property of Congruence
$\angle PQR \cong$ _?_

21. Distributive Property
$3(x - 1) = 3x -$ _?_

22. Substitution Property
If $LM = 7$ and $EF + LM = NP$, then _?_ $= NP$.

23. Transitive Property of Congruence
If $\angle XYZ \cong \angle AOB$ and $\angle AOB \cong \angle WYT$, then _?_.

Visit: PHSchool.com
Web Code: bce-0204

24. **Multiple Choice** Which expression is equivalent to the left side of this equation? $-4x + 7y + \frac{1}{3}(12x - 3y) = 180$

(A) $8x + 4y$ (B) $6y + 8$ (C) $6y$ (D) $8x$

25. **Writing** Jero claims that the statements $\overline{LR} \cong \overline{RL}$ and $\angle CBA \cong \angle ABC$ are both true by the Reflexive Property of Congruence. Explain why Jero is correct.

26. Use what you know about transitive properties to complete the following:

The Transitive Property of Falling Dominoes:

If domino A causes domino B to fall, and domino B causes domino C to fall, then domino A causes domino _?_ to fall.

27. **Algebra** Fill in the reason that justifies each step.

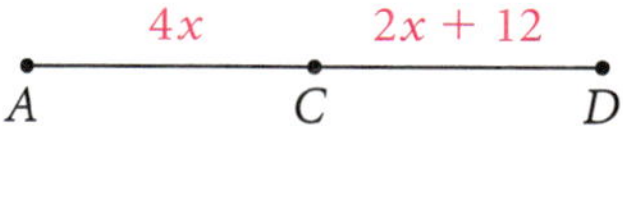

Given: C is the midpoint of $\overline{AD}$.

C is the midpoint of $\overline{AD}$.	**a.** ?
$AC = CD$	**b.** ?
$4x = 2x + 12$	**c.** ?
$2x = 12$	**d.** ?
$x = 6$	**e.** ?

For a guide to solving Exercise 28, see p. 109.

28. **Algebra** In the figure at the right, $KM = 35$.
   a. Solve for x. Justify each step.
   b. Find the length of $\overline{KL}$.

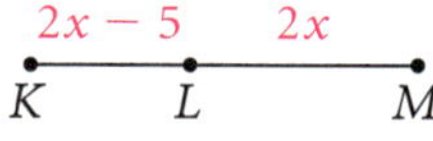

29. **Algebra** In the figure at the right, $m\angle GFI = 128$.
   a. Solve for x. Justify each step.
   b. Find $m\angle EFI$.

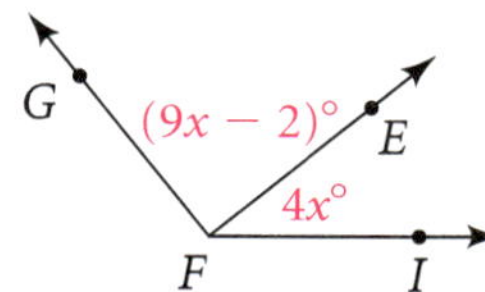

30. **Algebra** Point C is on the crease when you fold $\overrightarrow{BD}$ onto $\overrightarrow{BA}$. Give the reason that justifies each step. (*Hint*: See page 102, Exercises 4 and 5.)

$\overrightarrow{BC}$ bisects $\angle ABD$.	**a.** ?
$m\angle ABC = m\angle CBD$	**b.** ?
$6n + 1 = 4n + 19$	**c.** ?
$2n = 18$	**d.** ?
$n = 9$	**e.** ?

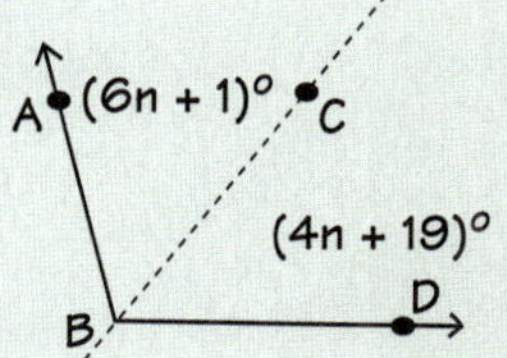

31. **Error Analysis** The steps below "show" that 1 = 2. Find the error.

Given: $a = b$

$a = b$	Given
$ab = b^2$	Multiplication Property of Equality
$ab - a^2 = b^2 - a^2$	Subtraction Property of Equality
$a(b - a) = (b + a)(b - a)$	Distributive Property
$a = b + a$	Division Property of Equality
$a = a + a$	Substitution Property
$a = 2a$	Simplify.
$1 = 2$	Division Property of Equality

Real-World Connection

President Calvin Coolidge, advice columnist Ann Landers, and musician Bill Withers were all born on the Fourth of July. Each one of them "has the same birthday as" either one of the others.

Relationships **The relationships "is equal to" and "is congruent to" are reflexive, symmetric, and transitive. In a later chapter, you will see that this is also true for the relationship "is similar to." Consider the following relationships among people. State whether each relationship is reflexive, symmetric, transitive, or none of these.**

Sample: The relationship "is younger than" is transitive. If Sue is younger than Fred and Fred is younger than Alana, then Sue is younger than Alana. The relationship "is younger than" is not reflexive because Sue is not younger than herself. It is also not symmetric because if Sue is younger than Fred, Fred is not younger than Sue.

32. has the same birthday as
33. is taller than
34. lives in the same state as
35. lives in a different state than
36. is the same height as
37. is a descendant of

Multiple Choice Practice

For California Standards and CAHSEE Tutorials, visit PHSchool.com. Web Code: bcq-9045

GEOM 1.0

38. The Multiplication Property of Equality justifies which statement below?

- (A) If $\frac{3}{4}x = 6$, then $\frac{3x}{4} = 6$.
- (B) If $\frac{3}{4}x + 5 = 6$, then $\frac{3}{4}x = 1$.
- (C) If $\frac{3}{4}x = 6$, then $3x = 24$.
- (D) If $\frac{3}{4}x - 18 = 6$, then $\frac{3}{4}x = 24$.

CAHSEE 6SDAP 1.1

39. The time Max practiced playing his guitar last week is shown in the chart at the right. If Max practiced 10 fewer minutes on each of Monday, Wednesday, and Friday, which measure of central tendency would NOT change?

- (A) Mean
- (B) Median
- (C) Mode
- (D) Range

Max's Guitar Schedule

Day	Time
Monday	80 minutes
Tuesday	20 minutes
Wednesday	90 minutes
Thursday	40 minutes
Friday	30 minutes

CAHSEE 7MG 2.3

40. Jessica has two similar rectangular boxes. The surface area of Box 1 is sixteen times greater than that of Box 2. How many times greater is the volume of Box 1 than Box 2?

(A) 4 (B) 16 (C) 64 (D) 256

CAHSEE 7MR 2.1

41. Miguel wants to buy a new set of headphones for his music player. The headphones are regularly priced at $12.95 and are on sale today for 20% off. If the sales tax is 6%, which is the most reasonable estimate of the total cost of the headphones including tax?

(A) $10.00 (B) $11.00 (C) $11.50 (D) $12.00

Mixed Review

Lesson 2-3

Reasoning **Use logical reasoning to draw a conclusion.**

42. If a student is having difficulty in class, then that student's teacher is concerned. Elena is having difficulty in history class.

43. If a person has a job, then that person is earning money. If a person is earning money, then that person can save money each week.

Lesson 1-6

Use the diagram at the right and find each measure.

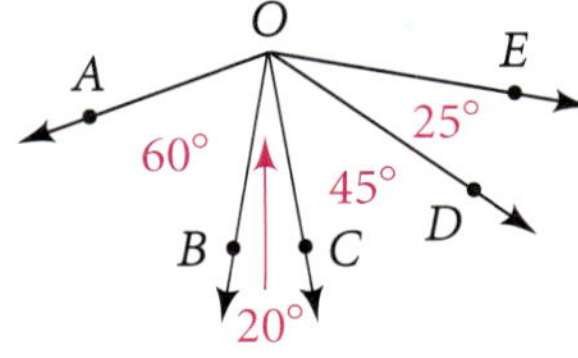

44. $m\angle AOC$

45. $m\angle AOD$

46. $m\angle DOB$

47. $m\angle BOE$

48. In the diagram, name an obtuse angle and a right angle.

Lesson 1-1

Find the next two terms in each sequence.

49. 19, 21.5, 24, 26.5

50. 3.4, 3.45, 3.456, 3.4567

51. −2, 6, −18, 54

52. 8, −4, 2, −1

Guided Problem Solving

FOR USE WITH PAGE 107, EXERCISE 28

Understanding Math Problems **Read the problem below. Then let Larissa's thinking guide you through the solution. Check your understanding with the exercises at the bottom of the page.**

Algebra In the figure at the right, $KM = 35$.

a. Solve for x. Justify each step.

b. Find the length of $\overline{KL}$.

What Larissa Thinks	What Larissa Writes	
What information am I given?	Given: $KM = 35$, $KL = 2x - 5$, $LM = 2x$	
Points K, L, and M are collinear. I can use the Segment Addition Postulate to write an equation.	1. $KL + LM = KM$	1. Segment Addition Postulate
Now I will substitute for KL, LM, and KM. I get an equation that I can solve for x.	2. $(2x - 5) + 2x = 35$	2. Substitute.
I simplify the left side.	3. $4x - 5 = 35$	3. Simplify.
I add 5 to each side.	4. $4x = 40$	4. Addition Property of Equality
I divide each side by 4.	5. $x = 10$	5. Division Property of Equality
Part (b) asks me to find KL. The diagram shows that $KL = 2x - 5$. I know $x = 10$, so I'll substitute to find KL.	$KL = 2x - 5$ $= 2(10) - 5$ $= 20 - 5 = 15$	
Now, I'll write my answer.	The length of $\overline{KL}$ is 15.	

EXERCISES

x^2 **1. Algebra** $\angle 1$ and $\angle 2$ are supplementary; $m\angle 1 = 4y + 15$ and $m\angle 2 = 7y - 11$.

a. Solve for y. Justify each step.

b. Find $m\angle 2$.

2. If $\angle 1$ and $\angle 2$ are complementary and have the (algebraic) measures given in Exercise 1, which angle is larger?

2-5 Proving Angles Congruent

California Content Standards

GEOM 1.0 Students demonstrate understanding by identifying and giving examples of theorems and deductive reasoning. ***Develop***
GEOM 2.0 Students write geometric proofs. ***Introduce***
GEOM 4.0 Students prove basic theorems involving congruence. ***Introduce***

What You'll Learn

- To prove and apply theorems about angles

. . . And Why

To find the measures of angles formed by the legs of a director's chair, as in Exercise 11

Check Skills You'll Need

GO for Help Lesson 1-6

Algebra Find the value of each variable.

1. **2.**

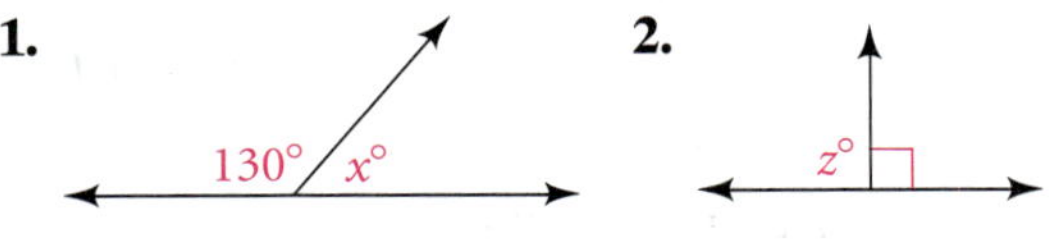

3.

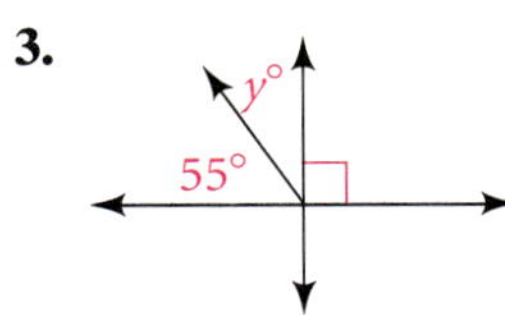

Fill in each blank.

4. Perpendicular lines are two lines that intersect to form _?_.

5. An angle is formed by two rays with the same endpoint. The endpoint is called the _?_ of the angle.

New Vocabulary • theorem • paragraph proof

1 Theorems About Angles

Hands-On Activity: Vertical Angles

- Draw two intersecting lines. Number the angles as shown.
- Fold ∠1 onto ∠2.
- Fold ∠3 onto ∠4.
- Make a conjecture about vertical angles.

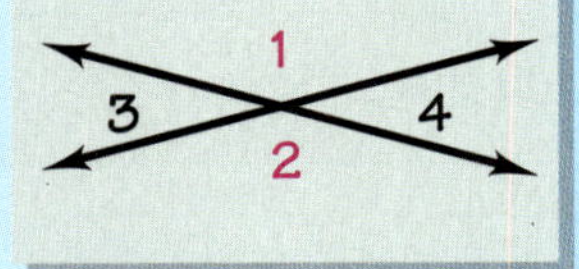

You can use deductive reasoning to show that a conjecture is true. The set of steps you take is called a proof. The statement that you prove true is a **theorem.** The Investigation above leads to a conjecture that becomes the following theorem.

Key Concepts

Theorem 2-1 **Vertical Angles Theorem**

Vertical angles are congruent.

$\angle 1 \cong \angle 2$ and $\angle 3 \cong \angle 4$

In the proof of a theorem, a "Given" list shows you what you know from the hypothesis of the theorem. You prove the conclusion of the theorem. A diagram records the given information visually.

Here is what the start of many proofs will look like.

There are many forms of proofs. A **paragraph proof** is written as sentences in a paragraph. Here is a paragraph proof of Theorem 2-1.

Proof

Given: $\angle 1$ and $\angle 2$ are vertical angles. ← **what you know** →

Prove: $\angle 1 \cong \angle 2$ ← **what you show**

Paragraph Proof: By the Angle Addition Postulate, $m\angle 1 + m\angle 3 = 180$ and $m\angle 2 + m\angle 3 = 180$. By substitution, $m\angle 1 + m\angle 3 = m\angle 2 + m\angle 3$. Subtract $m\angle 3$ from each side. You get $m\angle 1 = m\angle 2$, or $\angle 1 \cong \angle 2$.

You can use the Vertical Angles Theorem to solve for variables and find the measures of angles.

1 EXAMPLE Using the Vertical Angles Theorem

Find the value of x.

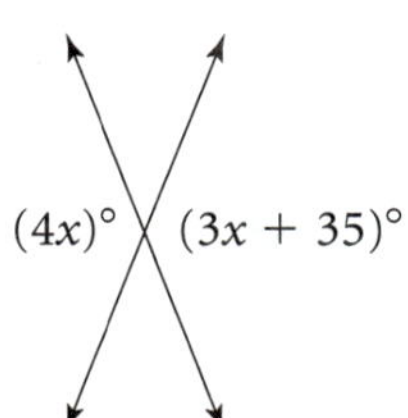

$4x = 3x + 35$ **Vertical angles are congruent.**

$x = 35$ **Subtract $3x$ from each side.**

1 **a.** Find the measures of the labeled pair of vertical angles in the diagram above.
b. Find the measures of the other pair of vertical angles.
c. Check to see that adjacent angles are supplementary.

The Vertical Angles Theorem is actually a special case of the following theorem. A proof of this theorem is shown on the next page. You can write a proof of another form of this theorem in Exercise 27.

Key Concepts

Theorem 2-2 Congruent Supplements Theorem

If two angles are supplements of the same angle (or of congruent angles), then the two angles are congruent.

Proof

2 EXAMPLE Proving Theorem 2-2

Study what you are given, what you are to prove, and the diagram. Write a paragraph proof.

Given: $\angle 1$ and $\angle 2$ are supplementary.
$\angle 3$ and $\angle 2$ are supplementary.

Prove: $\angle 1 \cong \angle 3$

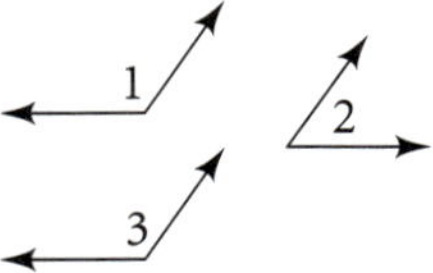

Proof: By the definition of supplementary angles, $m\angle 1 + m\angle 2 = 180$ and $m\angle 3 + m\angle 2 = 180$. By substitution, $m\angle 1 + m\angle 2 = m\angle 3 + m\angle 2$. Subtract $m\angle 2$ from each side. You get $m\angle 1 = m\angle 3$, or $\angle 1 \cong \angle 3$.

2 In the proof above, which Property of Equality allows you to subtract $m\angle 2$ from each side of the equation?

Theorem 2-3 is like the Congruent Supplements Theorem. You can demonstrate its proof in Exercises 7 and 28.

Key Concepts

Theorem 2-3 Congruent Complements Theorem

If two angles are complements of the same angle (or of congruent angles), then the two angles are congruent.

Theorem 2-4

All right angles are congruent.

Theorem 2-5

If two angles are congruent and supplementary, then each is a right angle.

You can complete proofs of Theorems 2-4 and 2-5 in Exercises 14 and 21, respectively.

EXERCISES Standards Practice

For more exercises, see *Extra Skill, Word Problem, and Proof Practice.*

GEOM 1.0, 2.0, 4.0

Example 1 (page 111)

GO for Help

Find the value of each variable.

1.

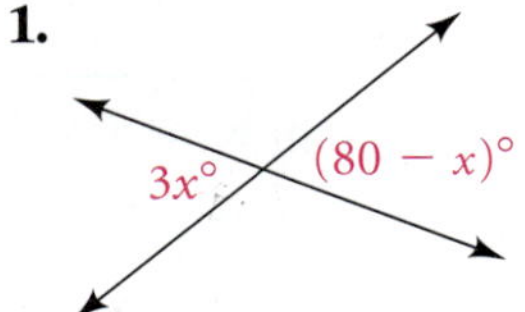

2. $3x°$, $y°$, $75°$

3.

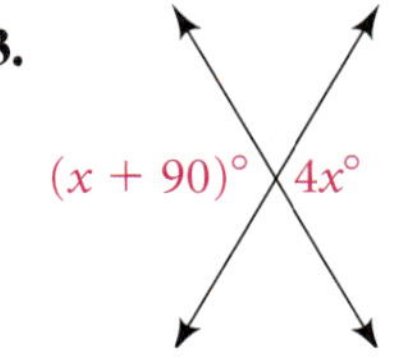

Find the measures of the labeled angles in each exercise.

4. Exercise 1

5. Exercise 2

6. Exercise 3

Example 2 (page 112)

7. **Developing Proof** Complete this proof of one form of Theorem 2-3 by filling in the blanks.

If two angles are complements of the same angle, then the two angles are congruent.

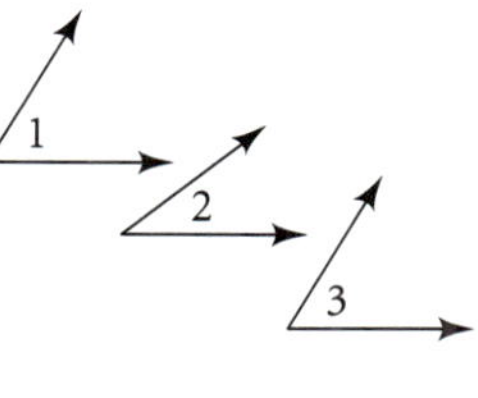

Given: $\angle 1$ and $\angle 2$ are complementary.
$\angle 3$ and $\angle 2$ are complementary.

Prove: $\angle 1 \cong \angle 3$

Proof: By the definition of complementary angles,
$m\angle 1 + m\angle 2 =$ **a.** ___?___ and $m\angle 3 + m\angle 2 =$ **b.** ___?___.
Then $m\angle 1 + m\angle 2 = m\angle 3 + m\angle 2$ by **c.** ___?___.
Subtract $m\angle 2$ from each side. You get $m\angle 1 =$ **d.** ___?___, or $\angle 1 \cong \angle 3$.

8. **Writing** How is a theorem different from a postulate?

9. **Open-Ended** Give an example of vertical angles in your home.

10. **Reasoning** Explain why this statement is true:
If $m\angle 1 + m\angle 2 = 180$ and $m\angle 3 + m\angle 2 = 180$, then $\angle 1 \cong \angle 3$.

11. **Design** The two back legs of the director's chair pictured at the left meet in a 72° angle. Find the measure of each angle formed by the two back legs.

Exercise 11

x^2 **Algebra** **Find the value of each variable and the measure of each labeled angle.**

12.

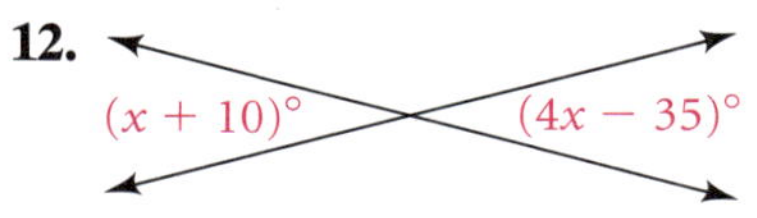

13.

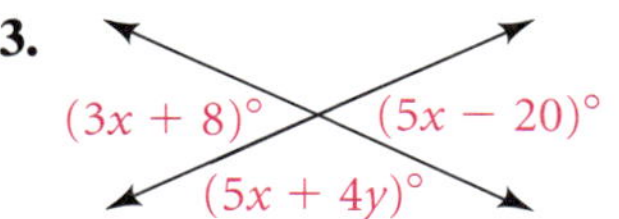

14. **Developing Proof** Complete this proof of Theorem 2-4 by filling in the blanks.

All right angles are congruent.

Given: $\angle X$ and $\angle Y$ are right angles.

Prove: $\angle X \cong \angle Y$

X Y

Proof: By the definition of **a.** ___?___, $m\angle X = 90$ and $m\angle Y = 90$.
By the Substitution Property, $m\angle X =$ **b.** ___?___, or $\angle X \cong \angle Y$.

15. **Multiple Choice** What is the measure of the angle formed by Park St. and Oak St.?

(A) 35° (B) 45° (C) 55° (D) 90°

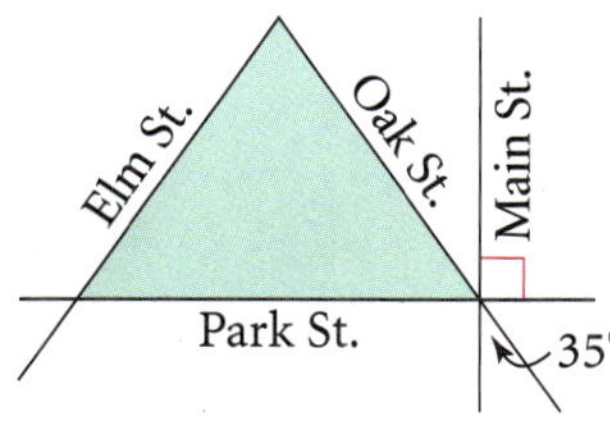

Name two pairs of congruent angles in each figure. Justify your answers.

16.

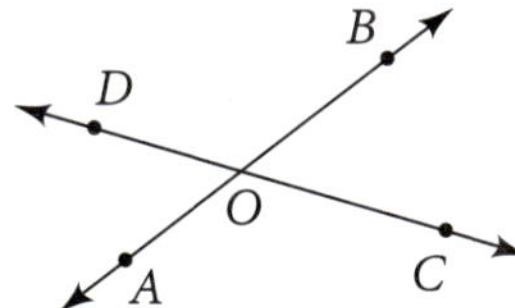

17.

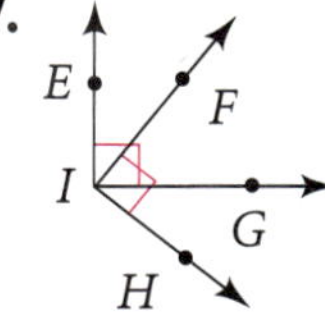

18.

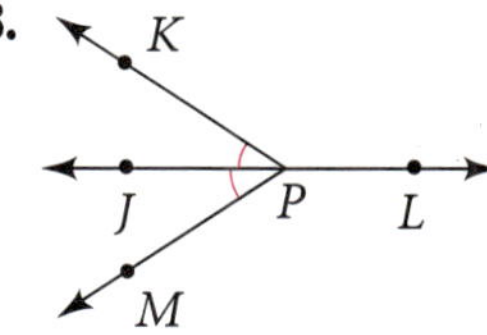

Homework Video Tutor

Visit: PHSchool.com
Web Code: bce-0205

19. **Coordinate Geometry** $\angle DOE$ contains points $D(2, 3)$, $O(0, 0)$, and $E(5, 1)$. Find the coordinates of a point F so that $\overrightarrow{OF}$ is a side of an angle that is adjacent and supplementary to $\angle DOE$.

20. Coordinate Geometry $\angle AOX$ contains points $A(1, 3)$, $O(0, 0)$, and $X(4, 0)$.

a. Find the coordinates of a point B so that $\angle BOA$ and $\angle AOX$ are adjacent complementary angles.

b. Find the coordinates of a point C so that $\overrightarrow{OC}$ is a side of a different angle that is adjacent and complementary to $\angle AOX$.

21. Developing Proof Complete this proof of Theorem 2-5 by filling in the blanks.

If two angles are congruent and supplementary, then each is a right angle.

Given: $\angle W$ and $\angle V$ are congruent and supplementary.

Prove: $\angle W$ and $\angle V$ are right angles.

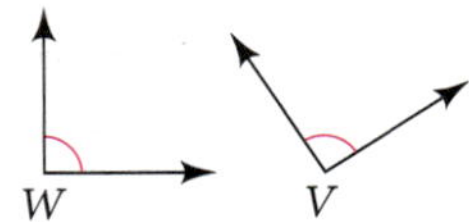

Proof: $\angle W$ and $\angle V$ are congruent, so $m\angle W = m\angle$ **a.** _?_.
$\angle W$ and $\angle V$ are supplementary so $m\angle W + m\angle V =$ **b.** _?_.
Substituting $m\angle W$ for $m\angle V$, you get $m\angle W + m\angle W = 180$, or $2m\angle W = 180$.
By the **c.** _?_ Property of Equality, $m\angle W = 90$.
Since $\angle W \cong \angle V$, $m\angle V = 90$, too. Then both angles are **d.** _?_ angles.

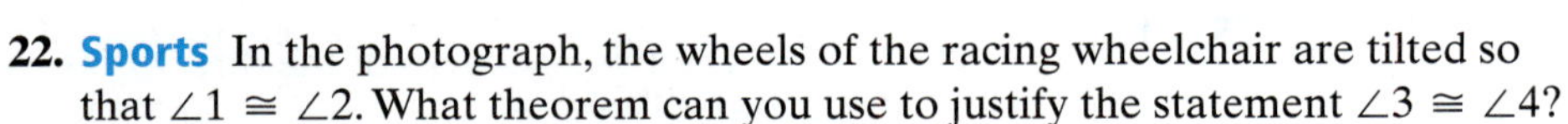

Exercise 22

22. Sports In the photograph, the wheels of the racing wheelchair are tilted so that $\angle 1 \cong \angle 2$. What theorem can you use to justify the statement $\angle 3 \cong \angle 4$?

x^2 **Algebra** **Find the measure of each angle.**

23. $\angle A$ is twice as large as its complement, $\angle B$.

24. $\angle A$ is half as large as its complement, $\angle B$.

25. $\angle A$ is twice as large as its supplement, $\angle B$.

26. $\angle A$ is half as large as twice its supplement, $\angle B$.

Proof **27.** Write a proof for this form of Theorem 2-2.

If two angles are supplements of congruent angles, then the two angles are congruent.

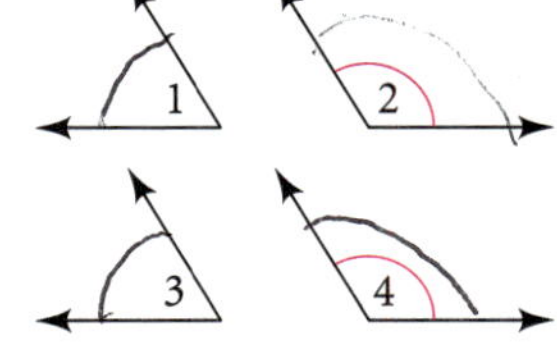

Given: $\angle 1$ and $\angle 2$ are supplementary.
$\angle 3$ and $\angle 4$ are supplementary.
$\angle 2 \cong \angle 4$

Prove: $\angle 1 \cong \angle 3$

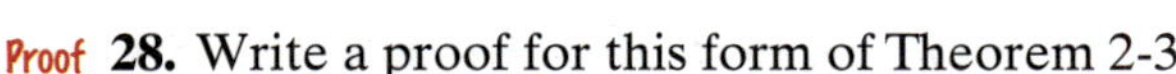

Proof **28.** Write a proof for this form of Theorem 2-3.

If two angles are complements of congruent angles, then the two angles are congruent.

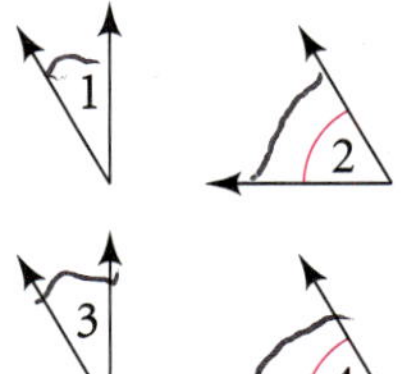

Given: $\angle 1$ and $\angle 2$ are complementary.
$\angle 3$ and $\angle 4$ are complementary.
$\angle 2 \cong \angle 4$

Prove: $\angle 1 \cong \angle 3$

C Challenge

29. Paper Folding After you've done the Activity on page 110, answer these questions.

a. How is the first fold line you make related to angles 3 and 4?

b. How is the second fold line you make related to angles 1 and 2?

c. How are the two fold lines related to each other? Give a convincing argument to support your answer.

Algebra **Find the value of each variable and the measure of each labeled angle.**

30.

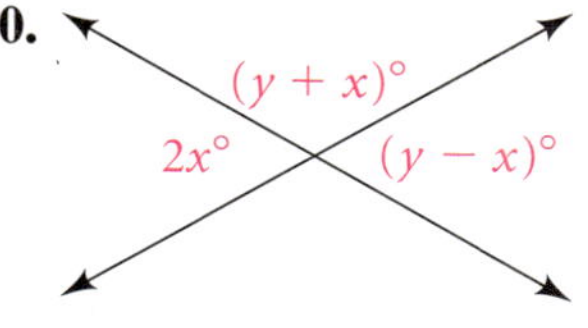

31.

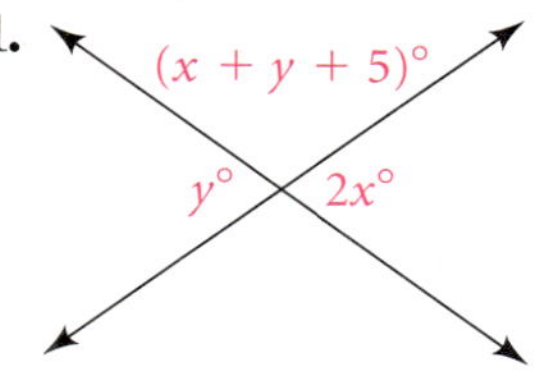

32.

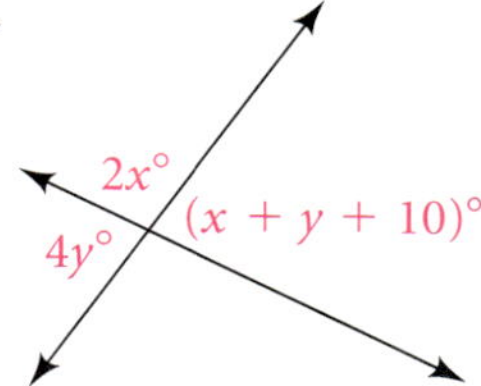

Multiple Choice Practice

For California Standards and CAHSEE Tutorials, visit PHSchool.com. Web Code: bcq-9045

GEOM 1.0

33. $\angle ABE$ and $\angle CBD$ are vertical angles and both are complementary with $\angle FGH$. If $m\angle ABE = (3x - 1)°$, and $m\angle FGH = (4x)°$, what is $m\angle CBD$?

(A) 13° (B) 38° (C) 52° (D) 142°

CAHSEE 7MR 1.1

34. Carmen is painting the four walls and door of a shed. Each wall is 8 feet high and 11 feet wide. She estimates the amount of paint she needs by finding the total area. Paint costs $12.95 per gallon. What additional information will she need to determine the cost of this project?

(A) The color of the paint
(B) The area of the floor
(C) The area that one gallon of paint can cover
(D) The number of pints in one gallon

Mixed Review

Lesson 2-4

Use the given property to complete each statement.

35. Subtraction Property of Equality
If $3x + 7 = 19$, then $3x = \underline{\ ?\ }$.

36. Reflexive Property of Congruence
$\overline{AB} \cong \underline{\ ?\ }$

37. Substitution Property
If $MN = 3$ and $MN + NP = 15$, then $\underline{\ ?\ }$.

Lesson 2-3

Use deductive reasoning to draw a conclusion. If not possible, write *not possible*.

38. If two lines intersect, then they are coplanar.
Lines m and n are coplanar.

39. If two angles are vertical angles, then they are congruent.
$\angle 1$ and $\angle 2$ are vertical angles.

Lesson 2-2

Each conditional statement below is true. Write its converse. If the converse is also true, combine the statements as a biconditional.

40. If $y + 7 = 32$, then $y = 25$.

41. If you live in Australia, then you live south of the equator.

42. If $n > 0$, then $n^2 > 0$.

Test-Taking Strategies

Finding Multiple Correct Answers

In Multiple-Correct-Answer questions, you have to determine whether a number of statements are true or false. As you test each statement, mark it as true or false. You then choose those that are true.

EXAMPLE

The perimeter of a rectangle is 20 cm.
Which of the following could be true?

I. One side of the rectangle is 2 cm.
II. The area of the rectangle is 24 cm^2.
III. The diagonal of the rectangle is 5 cm.

Ⓐ I only Ⓑ II only Ⓒ I and II only Ⓓ II and III only

Test each statement to see if it could be true in a rectangle with perimeter 20 cm.

Statement I is possible in a rectangle with dimensions 2-cm by 8-cm.

Statement II is possible in a rectangle with dimensions 4-cm by 6-cm.

Statement III is not possible. A diagonal is longer than each side. In any rectangle with perimeter 20 cm, at least one side has to be 5 cm or more.

Only statements I and II could be true. The correct choice is C.

EXERCISES

1. In ⊙A, the diameter is 9 cm.
Which of the following are true?

I. The area of the circle is 20.25π cm^2.
II. The radius of the circle is 4.5 cm.
III. The circumference of the circle is 18 cm.

Ⓐ I only Ⓑ II only Ⓒ I and II only Ⓓ II and III only

2. A square and non-square rectangle have the same area.
Which of the following are always true?

I. The perimeter of the square is greater than the perimeter of the rectangle.
II. The perimeter of the square is equal to the perimeter of the rectangle.
III. The diagonal of the square is less than the diagonal of the rectangle.

Ⓐ I only Ⓑ III only Ⓒ I and II only Ⓓ II and III only

3. Two non-adjacent angles are congruent.
Which of the following statements could be true?

I. The angles are vertical angles.
II. The angles are right angles.
III. The angles are supplementary angles.

Ⓐ I only Ⓑ II only Ⓒ I and II only Ⓓ I, II, and III

Chapter 2 Review

Vocabulary Review

English and Spanish Audio Online

biconditional (p. 87)
conclusion (p. 80)
conditional (p. 80)
converse (p. 81)
deductive reasoning (p. 94)
hypothesis (p. 80)
Law of Detachment (p. 94)
Law of Syllogism (p. 95)
paragraph proof (p. 111)
Reflexive Property (p. 105)
Symmetric Property (p. 105)
theorem (p. 110)
Transitive Property (p. 105)
truth value (p. 81)

Choose the correct vocabulary term to complete each sentence.

1. The statement "$\angle A \cong \angle A$" is an example of the _?_ Property of Congruence.
2. In a conditional statement, the part that directly follows *if* is the _?_.
3. "If $\angle A \cong \angle B$ and $\angle B \cong \angle C$, then $\angle A \cong \angle C$" is an example of the _?_ Property of Congruence.
4. When a conditional and its converse are true, they may be written as a single true statement called a _?_.
5. The _?_ of a conditional switches the hypothesis and the conclusion.
6. "If $\angle A \cong \angle B$, then $\angle B \cong \angle A$" is an example of the _?_ Property of Congruence.
7. The part of a conditional statement that follows "then" is the _?_.
8. A conditional has a _?_ of true or false.
9. Reasoning logically from given statements to a conclusion is _?_.
10. A statement that you prove true is a _?_.

Go Online PHSchool.com
For: Vocabulary quiz
Web Code: bcj-0251

Skills and Concepts

Lessons 2-1 and 2-2

- To recognize conditional statements (p. 80)
- To write converses of conditional statements (p. 81)
- To write biconditionals (p. 87)
- To recognize good definitions (p. 88)

GEOM 3.0

An *if-then statement* is a **conditional.** The part following *if* is the **hypothesis.** The part following *then* is the **conclusion.** You find the truth value of a conditional by determining whether it is true or false. The symbolic form of a conditional is $p \rightarrow q$.

The **converse** of a conditional switches the hypothesis and the conclusion. The symbolic form of the converse of $p \rightarrow q$ is $q \rightarrow p$.

When a conditional and its converse are true, you can combine them as a true **biconditional.** To write a biconditional, you join the two parts of each conditional with the phrase *if and only if.* The symbolic form of a biconditional is $p \leftrightarrow q$.

For Exercises 11–13, (a) write the converse and (b) determine the truth value of the conditional and its converse. (c) If both statements are true, write a biconditional.

11. If you are a teenager, then you are younger than 20.
12. If an angle is obtuse, then its measure is greater than 90 and less than 180.
13. If a figure is a square, then it has four sides.

14. Write the following sentence as a conditional: All flowers are beautiful.

A good definition is precise. A good definition uses terms that have been previously defined or are commonly accepted.

15. Rico defines a *book* as something you read. Explain why this is not a good definition.

16. Write this definition as a biconditional:
An *oxymoron* is a phrase that contains contradictory terms.

17. Write this biconditional as two statements, a conditional and its converse:
Two angles are complementary if and only if the sum of their measures is 90.

Lesson 2-3

- To use the Law of Detachment (p. 94)
- To use the Law of Syllogism (p. 95)

 GEOM 1.0

Deductive reasoning is the process of reasoning logically from given statements to a conclusion. If the given statements are true, deductive reasoning produces a true conclusion.

The following are two important laws of deductive reasoning:
Law of Detachment: If $p \rightarrow q$ is a true statement and p is true, then q is true.
Law of Syllogism: If $p \rightarrow q$ and $q \rightarrow r$ are true statements, then $p \rightarrow r$ is true.

Use the Law of Detachment to make a conclusion.

18. If you practice table tennis every day, you will become a better player. Lucy practices table tennis every day.

19. Line ℓ and line m are perpendicular. If two lines are perpendicular, they intersect to form right angles.

20. If two angles are supplementary, then the sum of their measures is 180. $\angle 1$ and $\angle 2$ are supplementary.

Use the Law of Syllogism to make a conclusion.

21. If Kate studies, she will get good grades. If Kate gets good grades, she will graduate.

22. If a, then b. If b, then c.

23. If the weather is wet, the Huskies will not play soccer. If the Huskies do not play soccer, Nathan can stop at the ice cream shop.

Lesson 2-4

- To connect reasoning in algebra and geometry (p. 103)

 GEOM 1.0, 3.0

In algebra, you use deductive reasoning and properties to solve equations. In geometry, each statement in a deductive argument is justified by a property, definition, or postulate. Some of the properties you need are listed below.

Properties of Equality

Addition Property	If $a = b$, then $a + c = b + c$.
Subtraction Property	If $a = b$, then $a - c = b - c$.
Multiplication Property	If $a = b$, then $a \cdot c = b \cdot c$.
Division Property	If $a = b$ and $c \neq 0$, then $\frac{a}{c} \neq \frac{b}{c}$.
Substitution Property	If $a = b$, then b can replace a in any expression.
Distributive Property	$a(b + c) = ab + ac$

Properties of Congruence

Reflexive Property	$\overline{AB} \cong \overline{AB}$ $\angle A \cong \angle A$
Symmetric Property	If $\overline{AB} \cong \overline{CD}$, then $\overline{CD} \cong \overline{AB}$. If $\angle A \cong \angle B$, then $\angle B \cong \angle A$.
Transitive Property	If $\overline{AB} \cong \overline{CD}$ and $\overline{CD} \cong \overline{EF}$, then $\overline{AB} \cong \overline{EF}$. If $\angle A \cong \angle B$ and $\angle B \cong \angle C$, then $\angle A \cong \angle C$.

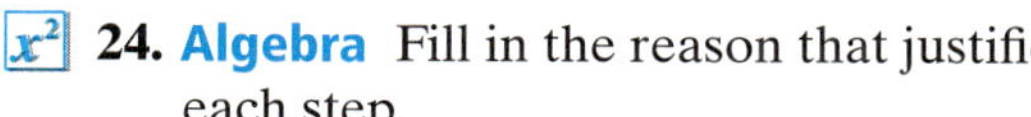

24. Algebra Fill in the reason that justifies each step.

$x + 3$ $2x$ Q R S

Given: $QS = 42$

$QR + RS = QS$	**a.** ?
$x + 3 + 2x = 42$	**b.** ?
$3x + 3 = 42$	**c.** ?
$3x = 39$	**d.** ?
$x = 13$	**e.** ?

Use the given property to complete each statement.

25. Addition Property of Equality
If $x = 5$, then $x + 3 =$? .

26. Division Property of Equality
If $2(AX) = 2(BY)$, then $AX =$? .

27. Reflexive Property of Equality
$m\angle Y =$?

28. Symmetric Property of Equality
If $XY = RS$, then ? .

29. Transitive Property of Equality
If $x = 5$ and $5 = y$, then $x =$? .

30. Distributive Property
$2(4x + 5) = 8x +$?

31. Distributive Property
$3p - 6q = 3($? $)$

32. Reflexive Property of Congruence
$\overline{NM} \cong$?

Lesson 2-5

- To prove and apply theorems about angles (p. 110)

GEOM 1.0, 2.0, 4.0

A statement that you prove true is a **theorem**. A proof written as a paragraph is a **paragraph proof**. You can prove that vertical angles are congruent; that supplements of the same angle are congruent, and that complements of the same angle are congruent.

33. Algebra Find the value of y.

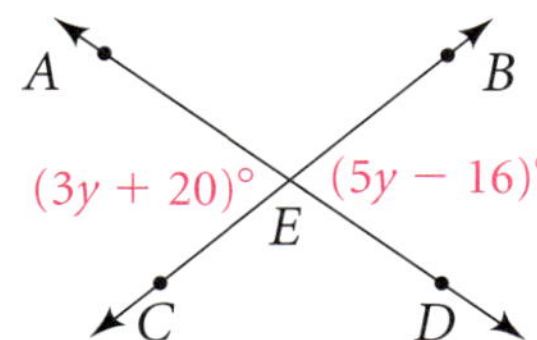

34. For the diagram at the left, find each of the following.

a. $m\angle AEC$
b. $m\angle BED$
c. $m\angle AEB$

35. Complete the following paragraph proof.

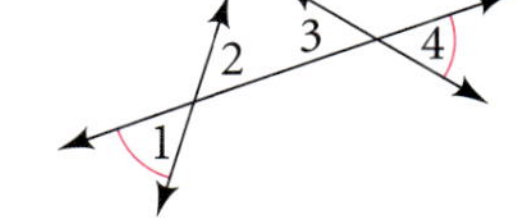

Given: $\angle 1 \cong \angle 4$

Prove: $\angle 2 \cong \angle 3$

Proof: By the Vertical Angles Theorem, $\angle 1 \cong$? and $\angle 4 \cong$? . $\angle 1 \cong \angle 4$ is given, so $\angle 2 \cong \angle 3$ by the ? Property of Congruence.

Chapter 2 Test

For: Chapter test
Web Code: bca-0252

1. Identify the hypothesis and conclusion:

   If $x + 9 = 11$, then $x = 2$.

2. Write this statement as a conditional.

   All babies are cute.

3. Find a counterexample to show that this statement is *not* true.

   If two angles are complementary, then they are not congruent.

For each statement, (a) write the converse and (b) decide whether the converse is true or false.

4. If a figure is a rectangle, then it has two right angles.
5. If two lines intersect, then they lie in the same plane.
6. If it is snowing in South Carolina, then it is not summer.

Writing Explain why each statement is *not* a good definition.

7. A pencil is a writing instrument.
8. Complementary angles are angles that form a right angle.
9. Vertical angles are angles that are congruent.

For Exercises 10–14, name the property that justifies each statement.

10. If $UV = KL$ and $KL = 6$, then $UV = 6$.
11. If $m\angle 1 + m\angle 2 = m\angle 4 + m\angle 2$, then $m\angle 1 = m\angle 4$.
12. $\angle ABC \cong \angle ABC$
13. If $\frac{1}{2}m\angle D = 45$, then $m\angle D = 90$.
14. If $\angle DEF \cong \angle HJK$, then $\angle HJK \cong \angle DEF$.

15. Find the measure of each angle.

   a. $\angle CDM$ b. $\angle KDM$
   c. $\angle JDK$ d. $\angle JDM$
   e. $\angle CDB$ f. $\angle CDK$

16. The measure of an angle is $2z$. What is the measure of its supplement?

17. The measure of an angle is 52 more than the measure of its complement. What is the measure of the angle?

For each diagram, state two pairs of angles that are congruent. Justify your answers.

18. 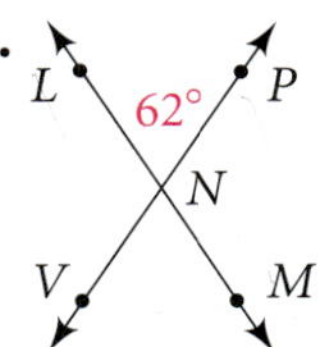

19. 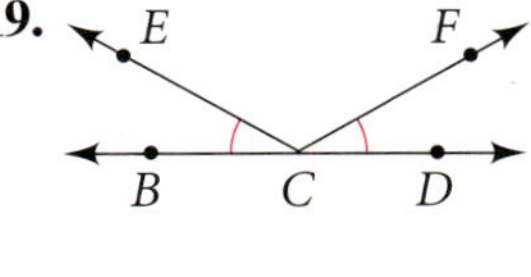

20. Rewrite this biconditional as two conditionals.

   A fish is a bluegill if and only if it is a bluish, freshwater sunfish.

21. Complete this statement.

   If two angles are complements of congruent angles, the angles are __?__.

For Exercises 22–26, use the Law of Detachment and the Law of Syllogism to make any possible conclusion. Write *not possible* if you cannot make any conclusion.

22. People who live in glass houses shouldn't throw stones. Lindsay shouldn't throw stones.
23. James wants to be a chemical engineer. If a student wants to be a chemical engineer, that student must graduate from college.
24. $p \rightarrow q$ and $q \rightarrow r$ are true statements.
25. $p \rightarrow q$ and p are true statements.
26. $p \rightarrow q$ and q are true statements.

27. **Developing Proof** Complete this proof by filling in the blanks.

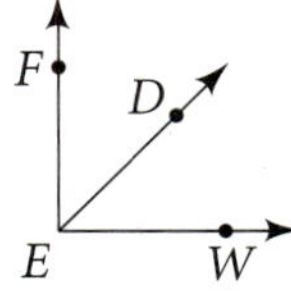

   Given: $\angle FED$ and $\angle DEW$ are complementary.

   Prove: $\angle FEW$ is a right angle.

   Proof: By the definition of complementary angles, $m\angle FED + m\angle DEW =$ **a.** __?__.
   $m\angle FED + m\angle DEW = m\angle FEW$ by the **b.** __?__.
   $90 = m\angle FEW$ by the **c.** __?__ Property of Equality.
   Then, $\angle FEW$ is a right angle, by the **d.** __?__.

Standards Mastery

Cumulative Practice

Some questions ask you to analyze equations. Read the question at the right. Then follow the tips to answer it.

What is the effect on the graph of the equation $y = \frac{1}{2}x^2$ when the coefficient is changed from $\frac{1}{2}$ to 2?

(A) The parabola will be wider.
(B) The parabola will be narrower.
(C) The coordinates of the vertex of the parabola will be $(0, 2)$.
(D) The parabola will be reflected across the x-axis.

Tip 1
Parabolas whose equations have the form $y = ax^2$ all have vertex (0, 0).

Tip 2
The new coefficient is 4 times the original coefficient. The y-values will be 4 times as great.

Think It Through

For a given x-value, the y-value for $y = 2x^2$ is 4 times the y-value for $y = \frac{1}{2}x^2$. The parabola for $y = 2x^2$ rises faster than the parabola for $y = \frac{1}{2}x^2$, so the parabola for $y = 2x^2$ is narrower.

The correct answer is B.

Vocabulary Review

As you solve problems, you must understand the meanings of mathematical terms. Match each term with its mathematical meaning.

A. vertex
B. inverse variation
C. coefficient
D. function
E. equation

I. a mathematical sentence that uses an equal sign
II. a relation that assigns exactly one value in the range to each value in the domain
III. the maximum or minimum point in a parabola
IV. a relation of the form $xy = k$ or $y = \frac{k}{x}$ where $k \neq 0$
V. a constant that multiplies a variable

Multiple Choice

Read each question. Then write the letter of the correct answer on your paper.

1. Which pair of angles must be congruent? (Lesson 2-5)

(A) supplementary angles
(B) complementary angles
(C) adjacent angles
(D) vertical angles

2. Which of the following best defines a postulate? (Lesson 1-3)

(A) A statement accepted without proof
(B) A conclusion reached using inductive reasoning
(C) An example which proves a conjecture false
(D) A statement that you prove true

3.

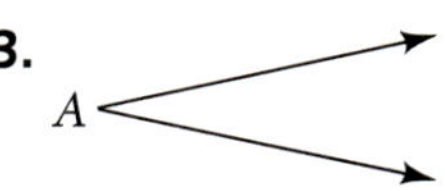

What is the second step in constructing $\angle S$, an angle congruent to $\angle A$? (Lesson 1-7)

(A) S

(B) T S R

(C) S R

(D) A B C

Cumulative Practice (continued)

4. If two angles are adjacent and congruent which of the following statements must be true? **(Lesson 2-4)**

- (A) The angles are right angles.
- (B) The angles are complementary angles.
- (C) The angles are supplementary angles.
- (D) The angles have the same measure.

5.

How many rays are in the next two terms in the sequence? **(Lesson 1-1)**

- (A) 16 and 33 rays
- (B) 17 and 33 rays
- (C) 17 and 34 rays
- (D) 18 and 34 rays

6. Which counterexample shows that the conjecture below is false?

Every square number has exactly three factors. **(Lesson 1-1)**

- (A) The factors of 2 are 1, 2.
- (B) The factors of 4 are 1, 2, 4.
- (C) The factors of 8 are 1, 2, 4, 8.
- (D) The factors of 16 are 1, 2, 4, 8, 16.

7. "If a polygon has exactly five congruent sides, then it is a regular pentagon. If a figure is a regular pentagon, then each interior angle measures 108°." Which of the following could be a conclusion based on the statements above? **(Lesson 2-3)**

- (A) If a polygon is a pentagon, then each interior angle measures 108°.
- (B) If a polygon has an angle that measures 108°, then it is a pentagon.
- (C) If a polygon has an angle that does not measure 108°, then it is not a pentagon.
- (D) If a polygon has exactly five congruent sides, then each interior angle measures 108°.

8. Which of the following statements does not have a counterexample? **(Lesson 1-1)**

- (A) Every month has at least 30 days.
- (B) The product of two fractions is an integer.
- (C) The sum of any two whole numbers is a whole number.
- (D) All paper notes in the U.S. have former presidents on the front.

9. Which Venn diagram represents the statement, "If you live in Sacramento, then you live in California"? **(Lesson 2-1)**

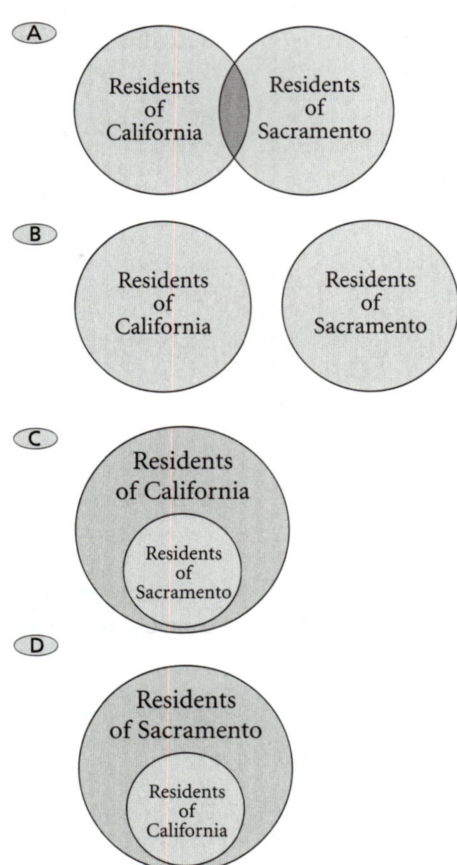

10. A square and rectangle have equal area. If the rectangle measures 32 cm by 18 cm, what is the perimeter of the square? **(Lesson 1-9)**

- (A) 48 cm
- (B) 72 cm
- (C) 96 cm
- (D) 576 cm

11.

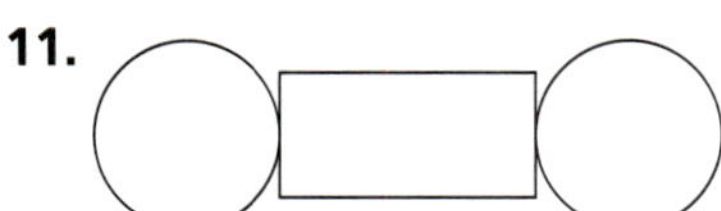

The dumbbell-shaped table above is being used for a banquet. The radius of each of the circular sections is 3 ft and the rectangular portion has an area of 32 ft^2 with the length twice the width. What is the area of the whole table to the nearest tenth? **(Lesson 1-9)**

- (A) 88.5 ft^2
- (B) 124.5 ft^2
- (C) 132.5 ft^2
- (D) 166.5 ft^2

12. An athletic field is a rectangle, 120 yards by 60 yards, with a semicircle at each of the short sides. A running track 15 yards wide surrounds the field. What is the perimeter of the inside edge of the running track to the nearest tenth of a yard? (Lesson 1-9)

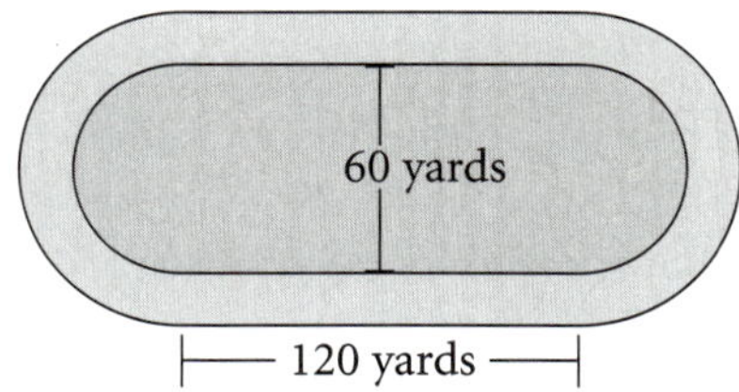

A 334.2 yd
B 360.0 yd
C 428.5 yd
D 617.0 yd

13. What kind of construction is being created below? (Lesson 1-7)

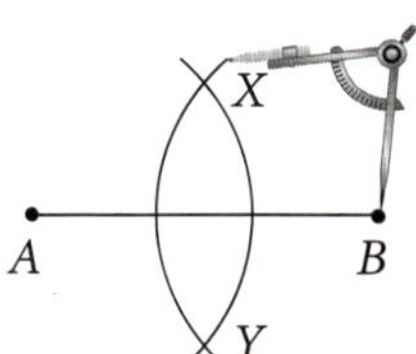

A Angle bisector
B Perpendicular bisector
C Congruent angles
D Congruent segments

14. If $\angle 1 + \angle 2 = 90$ and $\angle 1 + \angle 3 = 90$, which of the following theorems or properties could be used to prove $\angle 2 \cong \angle 3$? (Lesson 2-5)

A Congruent Supplements Theorem
B Congruent Complements Theorem
C Reflexive Property
D Vertical Angles Theorem

15. A rectangle has a base that is 7 cm less than three times its height. If the base is 5 cm, what is the area of the rectangle? (Lesson 1-9)

A 20 cm^2
B 30 cm^2
C 40 cm^2
D 50 cm^2

16. A 3 ft wide walkway is placed all around an animal exhibit at the zoo. If the exhibit is rectangular in shape and has length of 15 ft and width of 8 ft, what is the area of the walkway around the exhibit? (Lesson 1-9)

A 174 ft^2
B 231 ft^2
C 252 ft^2
D 294 ft^2

17. Which of the following conjectures can be disproved with a counterexample? (Lesson 2-1)

A The multiplication of any two even integers is even.
B The addition of any two prime numbers is even.
C Any integer divisible by nine is divisible by three.
D The measure of an obtuse angle must be between 90 and 180 degrees.

Standards Reference Guide

California Content Standard	Item Number(s)
GEOM 1.0	2, 5, 9
GEOM 3.0	4, 6, 7, 8, 17
GEOM 4.0	1, 14
GEOM 8.0	10, 11, 12
GEOM 10.0	15, 16
GEOM 16.0	3, 13

For additional review and practice, see the *California Standards Review and Practice Workbook* or go online to use the

Visit: PHSchool.com, **Web Code:** bcq-9045

Parallel and Perpendicular Lines

What You've Learned

California Content Standards

1.0 Students demonstrate understanding by identifying and giving examples of undefined terms, axioms, theorems, and inductive and deductive reasoning.

16.0 Students perform basic constructions with a straightedge and compass, such as angle bisectors and perpendicular bisectors.

Evaluating Algebraic Expressions (Skills Handbook page 754)

x^2 **Algebra** Evaluate each expression for the given value of n.

1. $\frac{360}{n}; n = 5$ **2.** $(n - 2)180; n = 9$ **3.** $(n - 2)180; n = 17$

Solving Equations (Algebra 1 Review page 30)

x^2 **Algebra** Solve each equation.

4. $3x + 11 = 7x - 5$ **5.** $(2x + 5) + (3x - 10) = 70$ **6.** $(3x + 2) - (2x - 3) = -19$

Writing and Solving an Equation (Skills Handbook page 758)

Write an equation and solve the problem.

7. The sum of the measures of three angles is 180. One measure is twice the size of each of the other two. Find the measure of each angle.

8. The sum of the measures of three angles is 180. One measure is half the size of each of the other two. Find the measure of each angle.

Drawing Parallel and Perpendicular Lines (Lesson 1-4)

9. Draw a picture of a rectangular box and label its eight corners A through H. Name two lines in your picture that appear to be parallel. Name two lines that appear to be perpendicular.

Drawing and Measuring Angles (Lesson 1-6)

Use a straightedge and draw the given type of angle as best you can. Estimate its measure, and then find its measure with a protractor.

10. acute **11.** right **12.** obtuse

▲ Large airports often have parallel runways to maximize the number of planes that can take off or land at a time.

What You'll Learn Next

California Content Standards

2.0 Students write geometric proofs.

4.0 Students prove basic theorems involving congruence.

7.0 Students prove and use theorems involving the properties of parallel lines cut by a transversal.

12.0 Students find and use measures of interior and exterior angles of triangles and polygons to classify figures and solve problems.

CAHSEE 7NS 1.3, 7NS 1.6, 7NS 1.7, 7NS 2.2, 6SDAP 2.5, 6SDAP 3.3, 7SDAP 1.2, 7AF 2.1, 7AF 3.1, 7AF 3.3, 7AF 3.4, 7MG 2.1, 7MG 3.2, 7MG 3.3, 7MR 1.1, 7MR 2.3, Alg1 7.0

New Vocabulary

English and Spanish Audio Online

- **alternate interior angles** (p. 127)
- **corresponding angles** (p. 127)
- **equiangular triangle** (p. 148)
- **equilateral triangle** (p. 148)
- **exterior angle of a polygon** (p. 149)
- **flow proof** (p. 135)
- **isosceles triangle** (p. 148)
- **polygon** (p. 157)
- **regular polygon** (p. 160)
- **remote interior angles** (p. 149)
- **same-side interior angles** (p. 127)
- **scalene triangle** (p. 148)
- **transversal** (p. 127)

Parallel Lines and Related Angles

Technology

FOR USE WITH LESSON 3-1

Prepares for GEOM 7.0 Students prove and use theorems involving the properties of parallel lines cut by a transversal.

Construct

Use geometry software to construct two parallel lines. Check that the lines remain parallel as you manipulate them. Construct a point on each line. Then construct the line through these two points. This line is called a transversal.

Investigate

Measure each of the eight angles formed by the parallel lines and the transversal. Record the measurements. Manipulate the lines. Record the new measurements. What relationships do you notice?

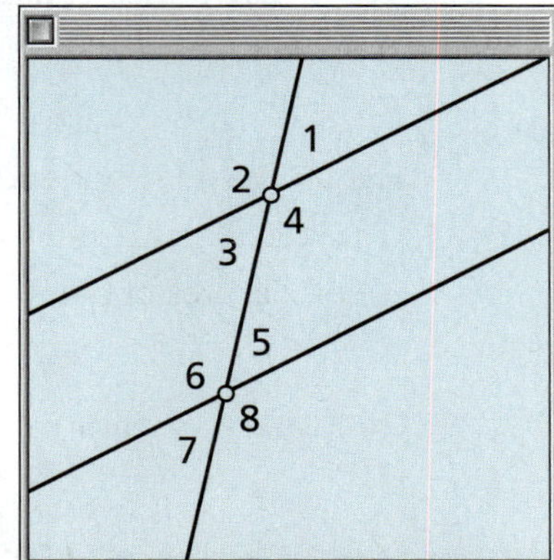

EXERCISES

1. When a transversal intersects parallel lines, what are the relationships among the angles formed? Make as many conjectures as possible.

Extend

2. Use your software to construct three or more parallel lines. Construct a line that intersects all three lines.
 a. What relationships exist among the angles formed?
 b. How many different angle measures are there?

3. Construct two parallel lines and a transversal perpendicular to one of the parallel lines. What angle does it make with the second parallel line?

4. Using geometry software, construct two lines and a transversal, making sure that the two lines are *not* parallel. Locate two angles that are on alternate sides of the transversal and in the interior region between the other two lines. Manipulate the lines so that these angles have the same measure.
 a. Make a conjecture as to the relationship between the two lines.
 b. How is this conjecture different from the conjecture(s) you made in Exercise 1?

5. Again, draw two lines and a transversal, making sure that the two lines are *not* parallel. Locate two angles that are on the same side of the transversal and in the interior region between the two lines. Manipulate the lines so that these angles are supplementary.
 a. Make a conjecture as to the relationship between the two lines.
 b. How is this conjecture different from the conjecture(s) you made in Exercise 1?

6. Construct perpendicular lines a and b. At a point away from the intersection of a and b, construct line c perpendicular to line a. Make a conjecture about lines b and c.

Properties of Parallel Lines

California Content Standards

GEOM 2.0 Students write geometric proofs. ***Develop***
GEOM 4.0 Students prove basic theorems involving congruence. ***Develop***
GEOM 7.0 Students prove and use theorems involving the properties of parallel lines cut by a transversal. ***Introduce***

What You'll Learn

- To identify angles formed by two lines and a transversal
- To prove and use properties of parallel lines

. . . And Why

To describe angles formed by an airport runway that crosses two parallel runways, as in Example 2

Check Skills You'll Need

GO for Help page 30 or Skills Handbook page 758

Algebra Solve each equation.

1. $x + 2x + 3x = 180$
2. $(w + 23) + (4w + 7) = 180$
3. $90 = 2y - 30$
4. $180 - 5y = 135$

Write an equation and solve the problem.

5. The sum of $m\angle 1$ and twice its complement is 146. Find $m\angle 1$.
6. The measures of two supplementary angles are in the ratio 2 : 3. Find their measures.

New Vocabulary
- transversal
- alternate interior angles
- same-side interior angles
- corresponding angles
- two-column proof
- alternate exterior angles
- same-side exterior angles

1 Identifying Angles

A **transversal** is a line that intersects two coplanar lines at two distinct points. The diagram shows the eight angles formed by a transversal t and two lines ℓ and m.

Pairs of the eight angles have special names as suggested by their positions.

$\angle 1$ and $\angle 2$ are **alternate interior angles.**

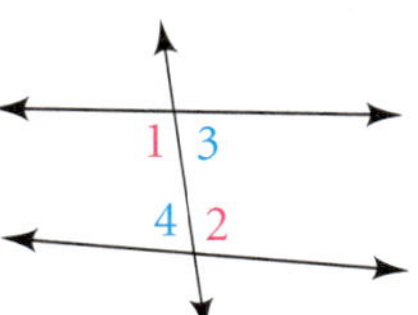

$\angle 1$ and $\angle 4$ are **same-side interior angles.**

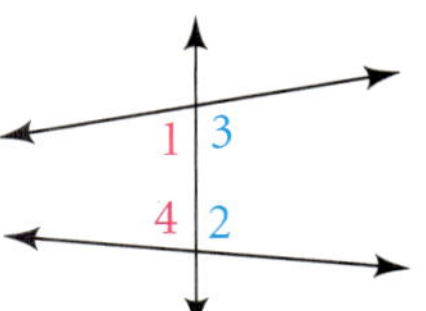

$\angle 1$ and $\angle 7$ are **corresponding angles.**

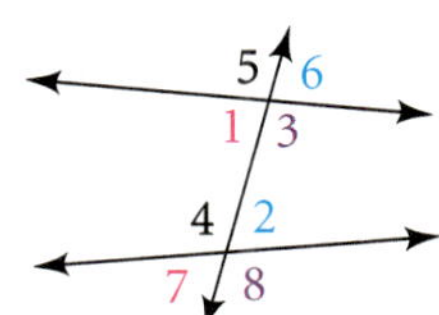

GO Online

Video Tutor Help
Visit: PHSchool.com
Web Code: bce-0775

1 EXAMPLE Identifying Angles

Use the diagrams above. Name another pair of alternate interior angles and another pair of same-side interior angles.

$\angle 3$ and $\angle 4$ are alternate interior angles. $\angle 2$ and $\angle 3$ are same-side interior angles.

1. Name three other pairs of corresponding angles in the diagrams above.

2 EXAMPLE Real-World Connection

Lafayette Regional Airport
Lafayette, Louisiana

Aviation In the diagram of Lafayette Regional Airport, the black segments are runways and the gray areas are taxiways and terminal buildings. Classify ∠1 and ∠2 as alternate interior angles, same-side interior angles, or corresponding angles.

∠1 and ∠2 are corresponding angles.

CA Standards Check **2** Classify ∠2 and ∠3 as alternate interior angles, same-side interior angles, or corresponding angles.

2 Properties of Parallel Lines

In the photograph, the vapor trail of the high-flying aircraft suggests a transversal of the parallel trails of the low-flying aircraft.

Vocabulary Tip

Corresponding objects are related in a special way. Here, corresponding angles are angles that are in similar positions on the same side of a transversal.

The same-size angles that appear to be formed by the vapor trails suggest the postulate and theorems below.

Key Concepts

Postulate 3-1 **Corresponding Angles Postulate**

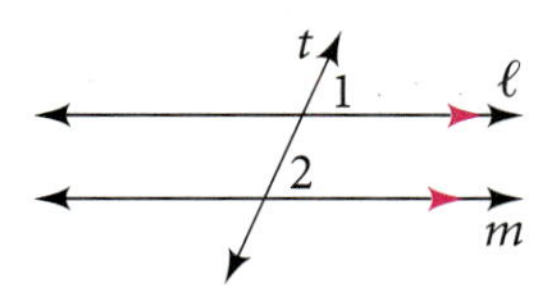

If a transversal intersects two parallel lines, then corresponding angles are congruent.

$\angle 1 \cong \angle 2$

Key Concepts

Theorem 3-1 **Alternate Interior Angles Theorem**

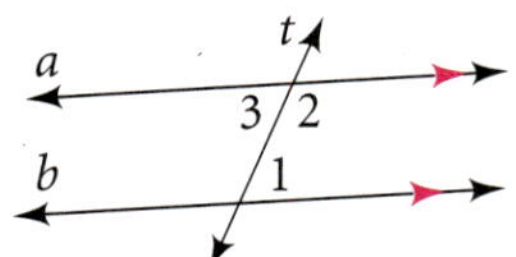

If a transversal intersects two parallel lines, then alternate interior angles are congruent.

$\angle 1 \cong \angle 3$

Theorem 3-2 **Same-Side Interior Angles Theorem**

If a transversal intersects two parallel lines, then same-side interior angles are supplementary.

$m\angle 1 + m\angle 2 = 180$

You can display the steps that prove a theorem in a **two-column proof.**

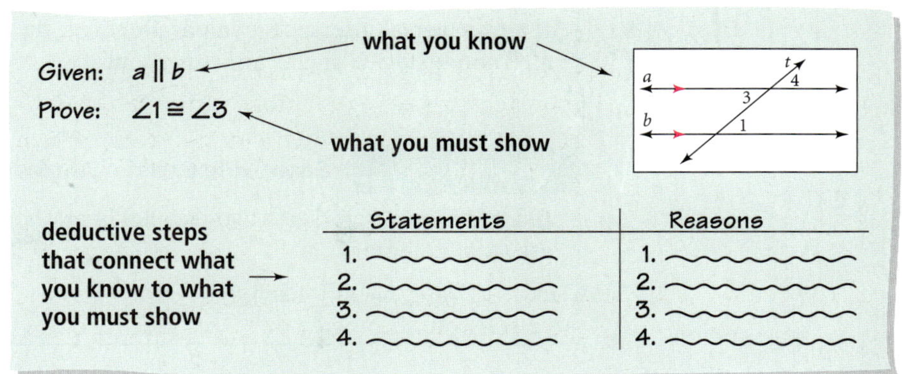

Proof

Proof of Theorem 3-1

If a transversal intersects two parallel lines, then alternate interior angles are congruent.

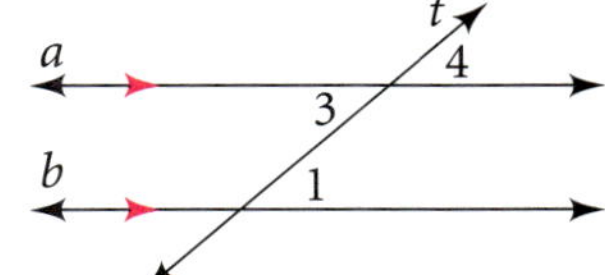

Given: $a \parallel b$

Prove: $\angle 1 \cong \angle 3$

Statements	Reasons
1. $a \parallel b$	**1.** Given
2. $\angle 1 \cong \angle 4$	**2.** If lines are $\parallel$, then corresponding angles are congruent.
3. $\angle 4 \cong \angle 3$	**3.** Vertical angles are congruent.
4. $\angle 1 \cong \angle 3$	**4.** Transitive Property of Congruence

Vocabulary Tip

These symbols indicate lines *a* and *b* are parallel.

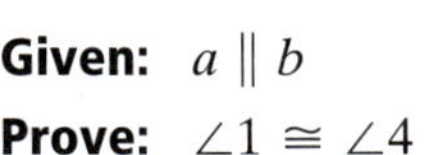

You will prove Theorem 3-2 in Exercise 29.

Proof

3 EXAMPLE Writing a Two-Column Proof

Study what is given, what you are to prove, and the diagram. Then write a two-column proof.

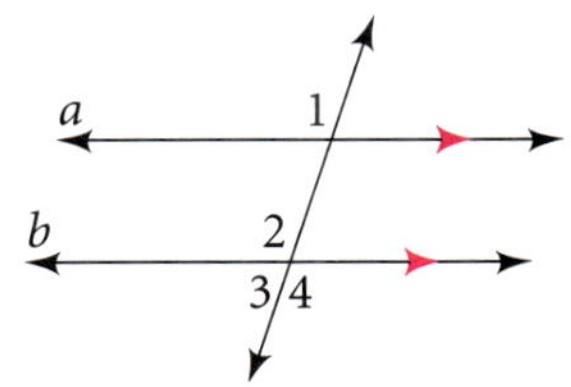

Given: $a \parallel b$

Prove: $\angle 1 \cong \angle 4$

Statements	Reasons
1. $a \parallel b$	**1.** Given
2. $\angle 1 \cong \angle 2$	**2.** If lines are $\parallel$, then corresponding angles are congruent.
3. $\angle 2 \cong \angle 4$	**3.** Vertical angles are congruent.
4. $\angle 1 \cong \angle 4$	**4.** Transitive Property of Congruence

CA Standards Check **3** Using the same given information and diagram from Example 3, prove that $\angle 1$ and $\angle 3$ are supplementary.

In Example 3, $\angle 1$ and $\angle 4$ are **alternate exterior angles.** $\angle 1$ and $\angle 3$ are **same-side exterior angles.** Example 3 and its Quick Check prove Theorems 3-3 and 3-4 as stated on the next page.

Key Concepts

Theorem 3-3 **Alternate Exterior Angles Theorem**

If a transversal intersects two parallel lines, then alternate exterior angles are congruent.

$\angle 1 \cong \angle 3$

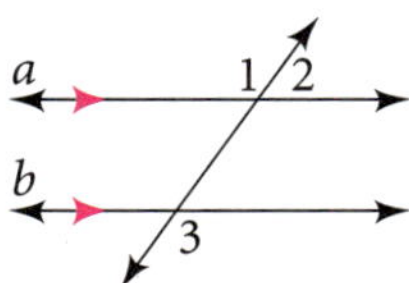

Theorem 3-4 **Same-Side Exterior Angles Theorem**

If a transversal intersects two parallel lines, then same-side exterior angles are supplementary.

$m\angle 2 + m\angle 3 = 180$

When you see two parallel lines and a transversal, and you know the measure of one angle, you can find the measures of all the angles. This is illustrated in Example 4.

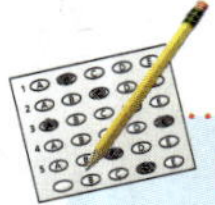

Test-Taking Tip

You can make marks on any diagrams shown on tests. This can help you keep track of known information.

4 EXAMPLE Finding Measures of Angles

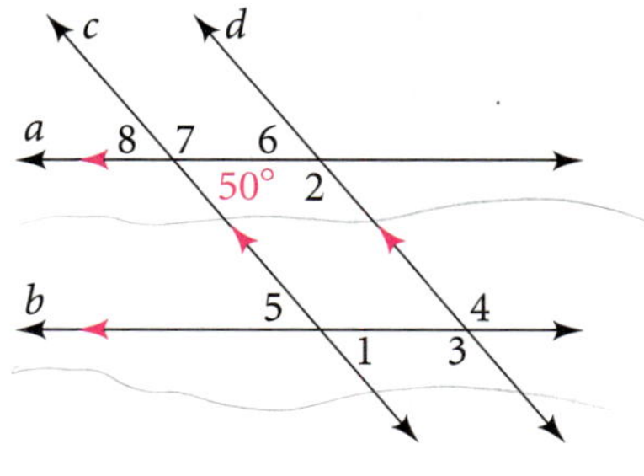

Find $m\angle 1$, and then $m\angle 2$. Which theorem or postulate justifies each answer?

Since $a \parallel b$, $m\angle 1 = 50$ because corresponding angles are congruent (Corresponding Angles Postulate).

Since $c \parallel d$, $m\angle 2 = 130$ because same-side interior angles are supplementary (Same-Side Interior Angles Theorem).

CA Standards Check 4 Find the measure of each angle. Justify each answer.

a. $\angle 3$ **b.** $\angle 4$ **c.** $\angle 5$
d. $\angle 6$ **e.** $\angle 7$ **f.** $\angle 8$

Sometimes you can use algebra to find angle measures.

5 EXAMPLE Using Algebra to Find Angle Measures

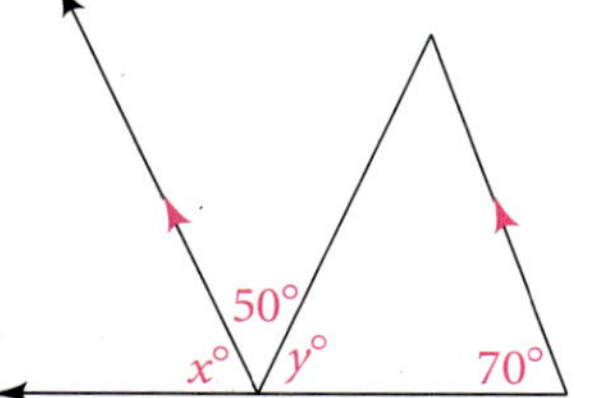

Find the value of y in the diagram at the left.

$x = 70$	**Corresponding angles of parallel lines are ≅.**
$70 + 50 + y = 180$	**Angle Addition Postulate**
$y = 60$	**Subtraction Property of Equality**

CA Standards Check 5 Find the values of x and y. Then find the measures of the angles.

EXERCISES
Standards Practice — GEOM 2.0, 4.0, 7.0

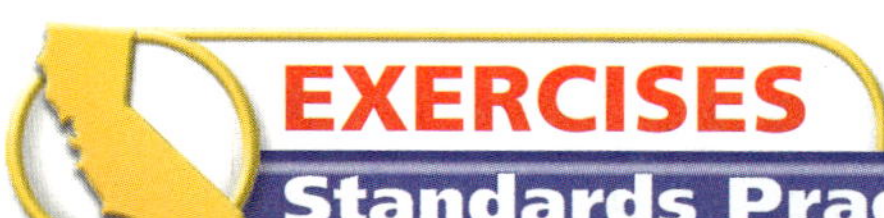

For more exercises, see *Extra Skill, Word Problem, and Proof Practice*.

A Practice by Example

Examples 1, 2 (pages 127, 128)

GO for Help

Name the two lines and the transversal that form each pair of angles. Then classify the pair of angles.

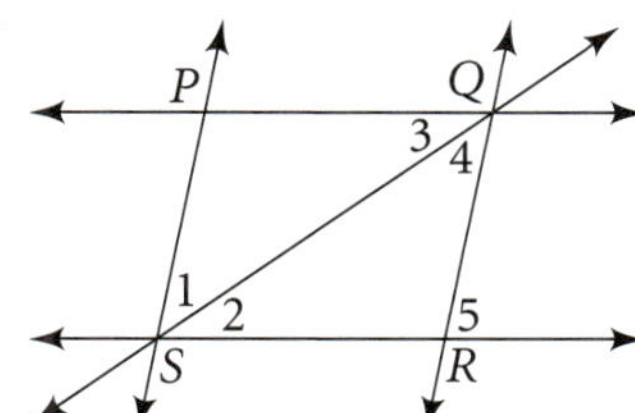

1. ∠2 and ∠3
2. ∠1 and ∠4
3. ∠*SPQ* and ∠*PQR*
4. ∠5 and ∠*PSR*

Classify each pair of angles labeled in the same color as *alternate interior angles*, *same-side interior angles*, or *corresponding angles*.

5.

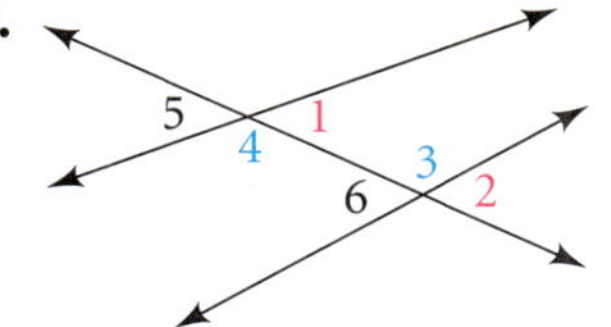

6.

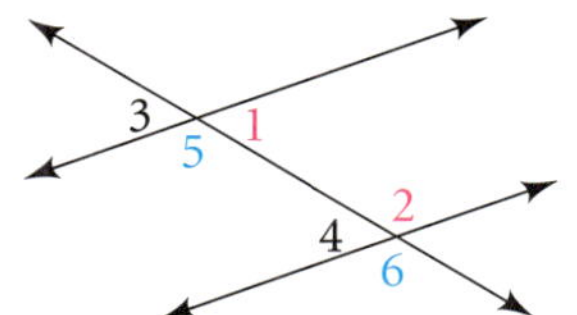

7.

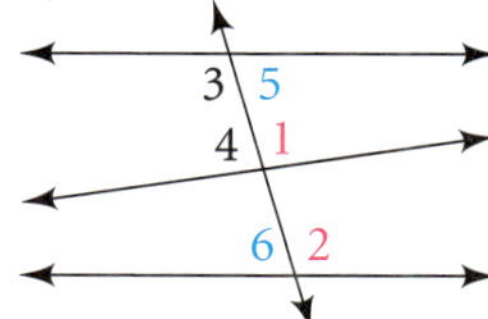

8. The boards securing this barn door suggest two parallel lines and a transversal. Classify ∠1 and ∠2 as alternate interior angles, same-side interior angles, or corresponding angles.

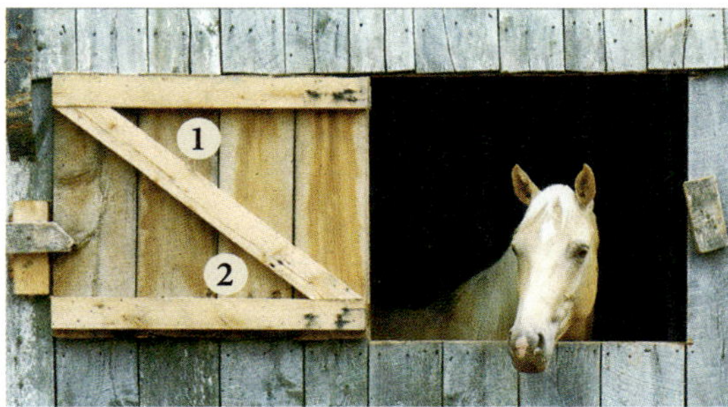

Example 3 (page 129)

9. **Developing Proof** Supply the missing reasons in this two-column proof.

Given: $a \parallel b$, $c \parallel d$

Prove: $\angle 1 \cong \angle 3$

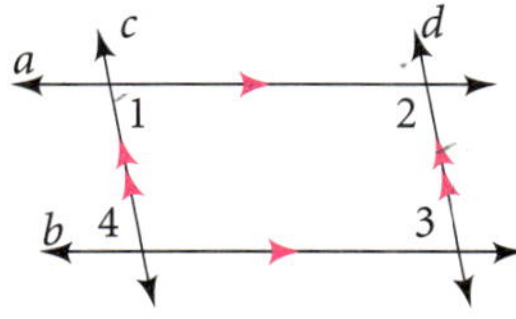

Statements	Reasons
1. $a \parallel b$	1. Given
2. ∠3 and ∠2 are supplementary.	2. ___
3. $c \parallel d$	3. Given
4. ∠1 and ∠2 are supplementary.	4. ___
5. $\angle 1 \cong \angle 3$	5. ___

Proof **10.** Write a two-column proof for Exercise 9 that does not use ∠2.

Example 4 (page 130)

Find *m*∠1, and then *m*∠2. Justify each answer.

11.

12.

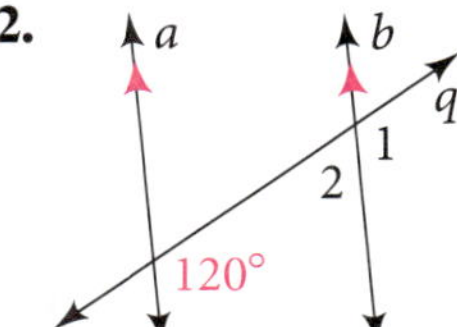

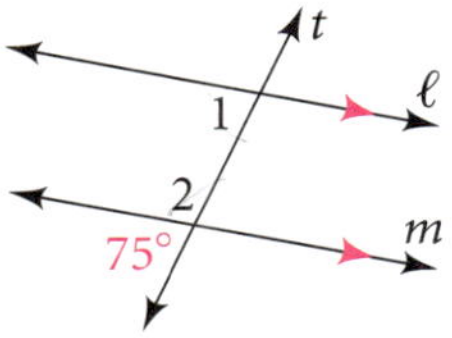

13.

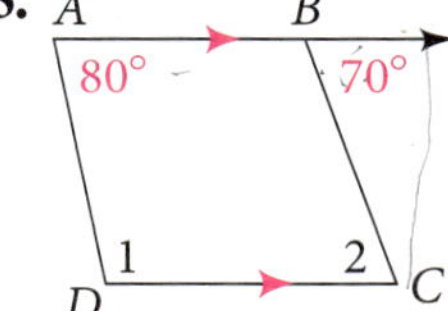

Example 5 (page 130) x^2 **Algebra** **Find the value of x. Then find the measure of each labeled angle.**

14. 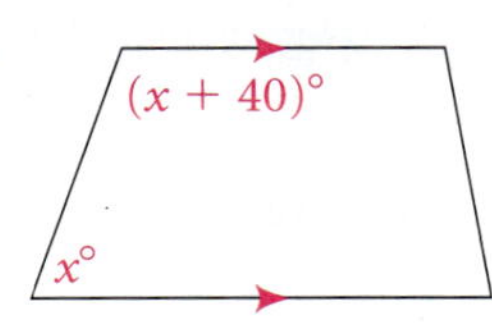

15. 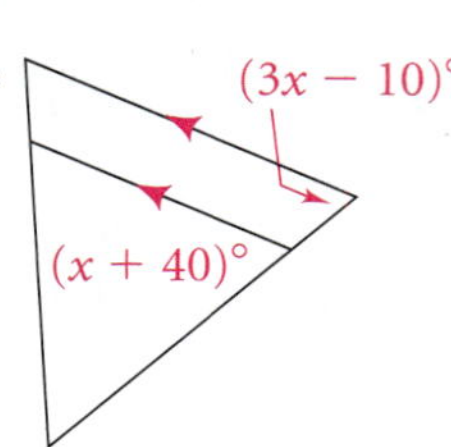

16. 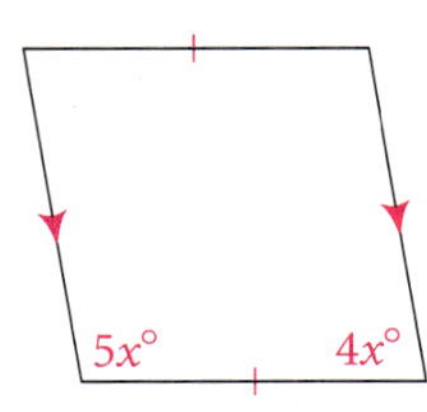

B Apply Your Skills

17. In the figure at the right, $f \parallel g$ and $m \parallel n$. Find the measure of each numbered angle.

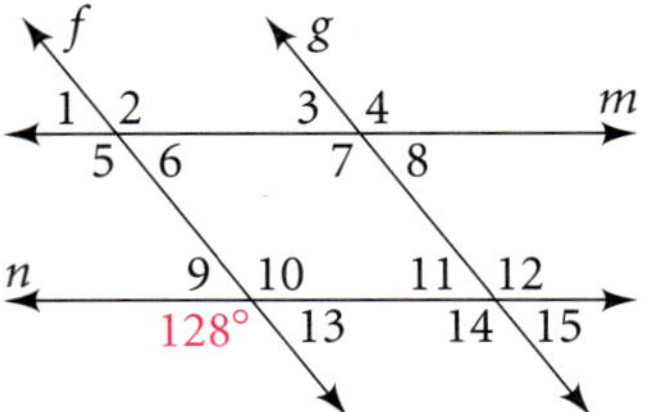

18. Two pairs of parallel segments form the "pound sign" on your telephone keypad. To find the measures of all the angles in the pound sign, how many angles must you measure? Explain.

Two lines and a transversal form how many pairs of the following?

19. alternate interior angles
20. corresponding angles
21. same-side exterior angles
22. vertical angles

x^2 **Algebra** **Find the values of the variables.**

Problem Solving Hint

In Exercise 24, turn your book so the other two parallel lines appear horizontal.

23. 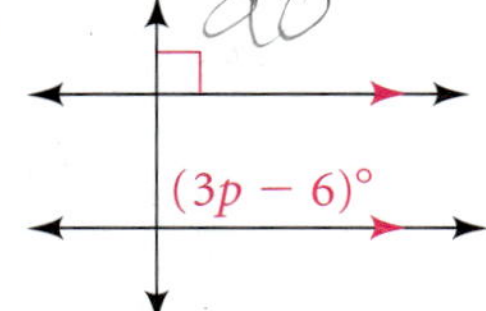

24. 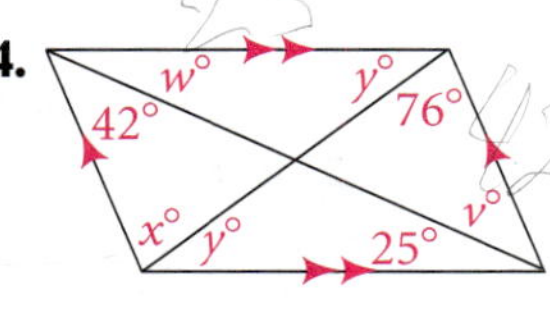

25. 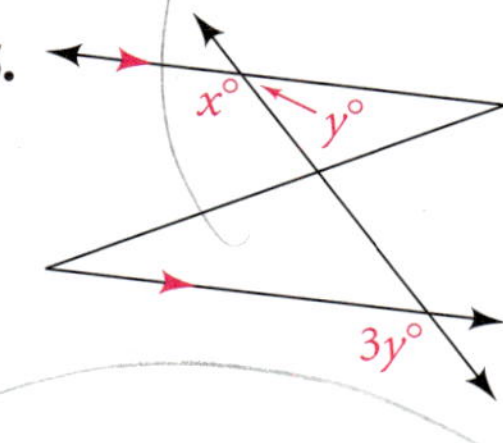

26. **Writing** Look up the meaning of the prefix *trans.* Explain how the meaning of the prefix relates to the word *transversal.*

27. **Open-Ended** The letter Z illustrates alternate interior angles. Find at least two other letters that illustrate the pairs of angles presented in this lesson. Draw the letters, mark the angles, and describe them.

Proof 28. Write a two-column proof.

Given: $a \parallel b$, $\angle 1 \cong \angle 4$

Prove: $\angle 2 \cong \angle 3$

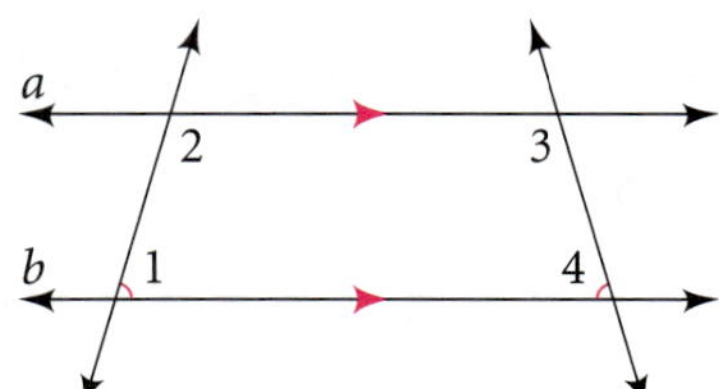

29. Prove Theorem 3-2.

If a transversal intersects two parallel lines, then same-side interior angles are supplementary.

Given: $a \parallel b$

Prove: $\angle 1$ and $\angle 2$ are supplementary.

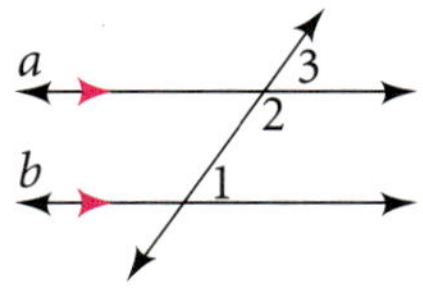

30. **Engineering** Engineers are laying pipe below ground on opposite sides of the street as shown here. To join the pipe, workers on each side of the street work toward the middle.

a. If one team lays pipe at the angle shown, what should the other team use for $m\angle 1$?

b. Are these two angles alternate interior, same-side interior, or corresponding angles?

123°
1

Homework Video Tutor

Visit: PHSchool.com
Web Code: bce-0301

C Challenge

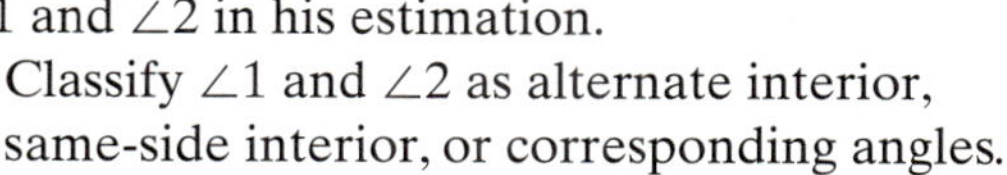

31. History About 220 B.C., Eratosthenes estimated the circumference of Earth. He achieved this remarkable feat by using two locations in Egypt. He assumed that Earth is a sphere and that the sun's rays are parallel. He used the measures of ∠1 and ∠2 in his estimation.

a. Classify ∠1 and ∠2 as alternate interior, same-side interior, or corresponding angles.

b. How did Eratosthenes know that ∠1 ≅ ∠2?

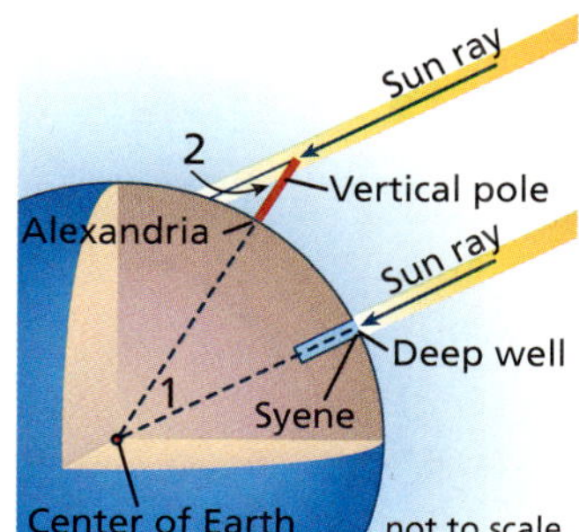

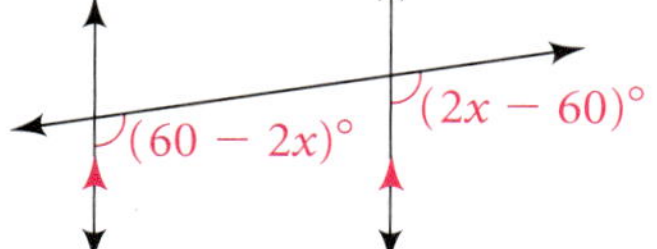

32. Error Analysis The diagram at the left contains contradictory information. What is it? Why is it contradictory?

Line *m* is in plane *A* and line *n* is in plane *B*. Planes *A* and *B* are parallel. Complete each statement with *sometimes, always,* or *never.* Justify each answer.

33. Lines m and n __?__ intersect.

34. Lines m and n are __?__ coplanar.

35. Lines m and n are __?__ parallel.

36. Lines m and n are __?__ skew.

Multiple Choice Practice

For California Standards and CAHSEE Tutorials, visit PHSchool.com. Web Code: bcq-9045

GEOM 7.0

37. What is the value of x in the figure?

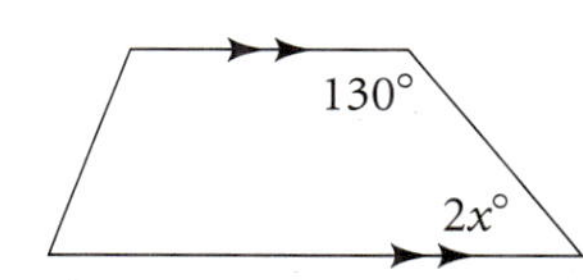

Ⓐ 20
Ⓑ 25
Ⓒ 45
Ⓓ 50

CAHSEE 7NS 1.6

38. A minor league baseball team had a total attendance of 4,000 for Wednesday's game. Tuesday's game had an attendance of 5,000. What is the percent decrease from Tuesday's game to Wednesday's?

Ⓐ 20% Ⓑ 25% Ⓒ 30% Ⓓ 40%

Mixed Review

Lesson 2-5

Find the measure of each angle if $m\angle YDF = 121$ and $\overrightarrow{DR}$ bisects $\angle FDI$.

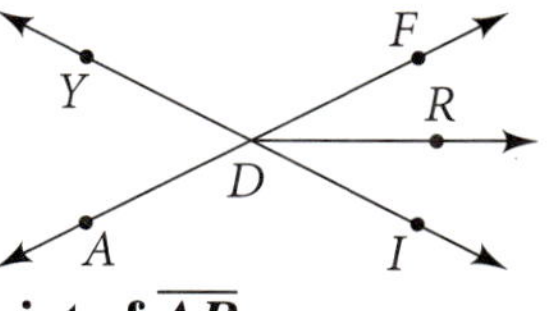

39. $\angle IDA$ **40.** $\angle YDA$ **41.** $\angle RDI$

Lesson 1-8

Coordinate Geometry Find the coordinates of the midpoint of $\overline{AB}$.

42. $A(0, 9), B(1, 5)$ **43.** $A(-3, 8), B(2, -1)$ **44.** $A(10, -1), B(-4, 7)$

Lesson 1-1

Find a pattern for each sequence. Use the pattern to show the next two terms.

45. 4, 8, 12, 16, . . . **46.** 1, −2, 4, −8, . . . **47.** 23, 16, 9, 2, . . .

3-2 Proving Lines Parallel

California Content Standards

GEOM 2.0 Students write geometric proofs. ***Develop***
GEOM 4.0 Students prove basic theorems involving congruence. ***Develop***
GEOM 7.0 Students prove and use theorems involving the properties of parallel lines cut by a transversal. ***Develop***

What You'll Learn

- To use a transversal in proving lines parallel

. . . And Why

To show how to paint parallel lines on a parking lot, as in Example 3

Check Skills You'll Need

for Help page 30 and Lesson 2-1

Algebra **Solve each equation.**

1. $2x + 5 = 27$

2. $8a - 12 = 20$

3. $x - 30 + 4x + 80 = 180$

4. $9x - 7 = 3x + 29$

Write the converse of each conditional statement. Determine the truth value of the converse.

5. If a triangle is a right triangle, then it has a 90° angle.

6. If two angles are vertical angles, then they are congruent.

7. If two angles are same-side interior angles, then they are supplementary.

New Vocabulary • flow proof

1 Using a Transversal

On window blinds like those shown here, you move the tilt bar to let in or shut out the light.

When you move the bar, the slats tilt at the same angle. This keeps them parallel and illustrates the converse of the Corresponding Angles Postulate.

Key Concepts

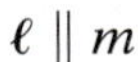

Postulate 3-2 **Converse of the Corresponding Angles Postulate**

If two lines and a transversal form corresponding angles that are congruent, then the two lines are parallel.

$\ell \parallel m$

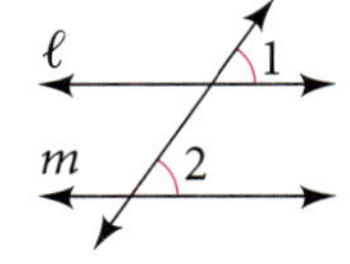

In Lesson 3-1, you proved four theorems based on the Corresponding Angles Postulate. You can also prove theorems that are based on its converse. In fact, Theorems 3-5 and 3-6 happen to be converses of the two theorems from Lesson 3-1, Theorems 3-1 and 3-2.

Key Concepts

Theorem 3-5 **Converse of the Alternate Interior Angles Theorem**

If two lines and a transversal form alternate interior angles that are congruent, then the two lines are parallel.

If $\angle 1 \cong \angle 2$, then $\ell \parallel m$.

Theorem 3-6 **Converse of the Same-Side Interior Angles Theorem**

If two lines and a transversal form same-side interior angles that are supplementary, then the two lines are parallel.

If $\angle 2$ and $\angle 4$ are supplementary, then $\ell \parallel m$.

You have seen two forms of proof—paragraph and two-column. In a third form, called **flow proof,** arrows show the logical connections between the statements. Reasons are written below the statements.

Proof

Proof of Theorem 3-5

If two lines and a transversal form alternate interior angles that are congruent, then the two lines are parallel.

Given: $\angle 1 \cong \angle 2$

Prove: $\ell \parallel m$

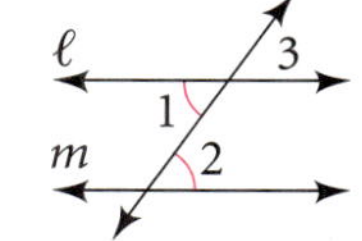

$\angle 1 \cong \angle 2$ (Given) → $\angle 3 \cong \angle 2$

$\angle 1 \cong \angle 3$ (Vertical ∠s are ≅.) → $\angle 3 \cong \angle 2$

$\angle 3 \cong \angle 2$ (Transitive Property of ≅) → $\ell \parallel m$ (Corres. ∠s are ≅.)

You will write a flow proof of Theorem 3-6 in Exercise 40. Theorems 3-6, 3-5, and Postulate 3-2 now provide you with three ways to prove that two lines are parallel.

1 EXAMPLE Using Theorems 3-5 and 3-6

Which lines, if any, must be parallel if $\angle 1 \cong \angle 2$? Justify your answer with a theorem or postulate.

$\overleftrightarrow{DE} \parallel \overleftrightarrow{KC}$ by Theorem 3-5, the Converse of the Alternate Interior Angles Theorem: If alternate interior angles are congruent, then the lines are parallel.

CA Standards Check 1 Which lines, if any, must be parallel if $\angle 3 \cong \angle 4$? Explain.

Theorems 3-3 and 3-4 allow you to make conclusions about alternate exterior angles and same-side exterior angles. Theorems 3-7 and 3-8 on the next page are converses of these two theorems. Like Theorems 3-5 and 3-6, they too can be used to prove that two lines are parallel.

Theorem 3-7 **Converse of the Alternate Exterior Angles Theorem**

If two lines and a transversal form alternate exterior angles that are congruent, then the two lines are parallel.

If $\angle 1 \cong \angle 2$, then $a \parallel b$.

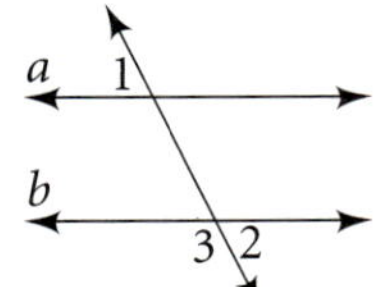

Theorem 3-8 **Converse of the Same-Side Exterior Angles Theorem**

If two lines and a transversal form same-side exterior angles that are supplementary, then the two lines are parallel.

If $\angle 1$ and $\angle 3$ are supplementary, then $a \parallel b$.

Here is a two-column proof of Theorem 3-7. You will prove Theorem 3-8 in Exercise 27.

Proof

Proof of Theorem 3-7

If two lines and a transversal form alternate exterior angles that are congruent, then the two lines are parallel.

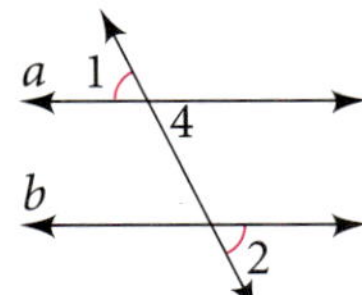

Given: $\angle 1 \cong \angle 2$

Prove: $a \parallel b$

Statements	Reasons
1. $\angle 1 \cong \angle 2$	**1.** Given
2. $\angle 1 \cong \angle 4$	**2.** Vertical angles are congruent.
3. $\angle 2 \cong \angle 4$	**3.** Transitive Property of Congruence
4. $a \parallel b$	**4.** If two lines and a transversal form congruent corresponding angles, then the lines are parallel.

You can use algebra along with Postulates 3-1 and 3-2 and Theorems 3-1 through 3-8 to help you solve problems involving parallel lines.

2 EXAMPLE Using Algebra

Algebra Find the value of x for which $\ell \parallel m$.

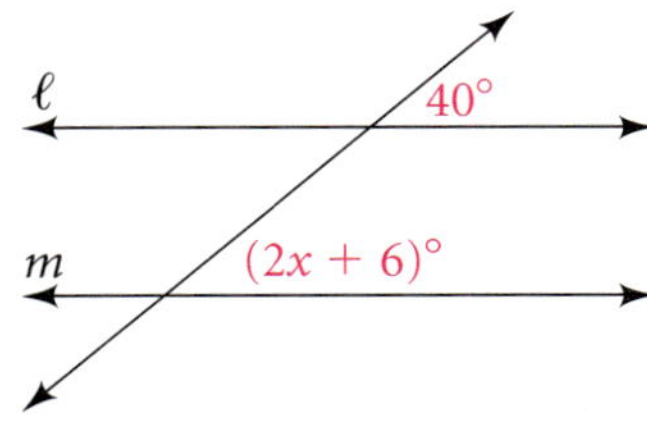

The two angles are corresponding angles.
$\ell \parallel m$ when $2x + 6 = 40$.

$2x + 6 = 40$

$2x = 34$ **Subtract 6 from each side.**

$x = 17$ **Divide each side by 2.**

CA Standards Check **2** Find the value of x for which $a \parallel b$. Explain how you can check your answer.

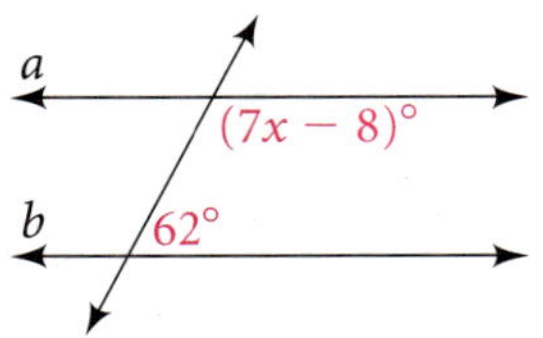

3 EXAMPLE Real-World Connection

Parking Two workers are painting lines for angled parking spaces. The first worker paints a line so that $m\angle 1 = 60^\circ$. The second worker paints a line so that $m\angle 2 = 60^\circ$. Explain why their lines are parallel.

The angles are congruent alternate exterior angles. By the Converse of the Alternate Exterior Angles Theorem, the lines are parallel.

CA Standards Check 3 If the second worker uses $\angle 3$, what should $m\angle 3$ be for parallel lines? Explain.

EXERCISES Standards Practice

For more exercises, see *Extra Skill, Word Problem, and Proof Practice.*

GEOM 2.0, 4.0, 7.0

A Practice by Example

Example 1 (page 135)

Which lines or segments are parallel? Justify your answer.

1.

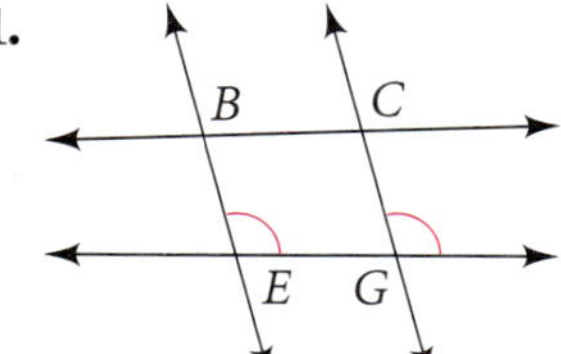

2.

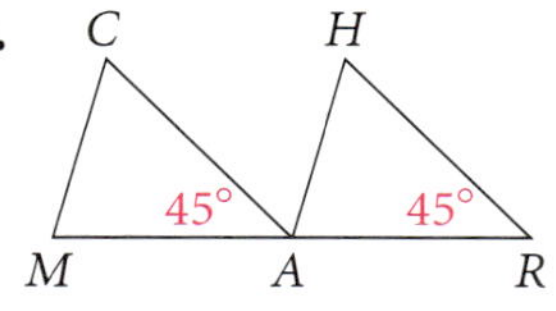

3.

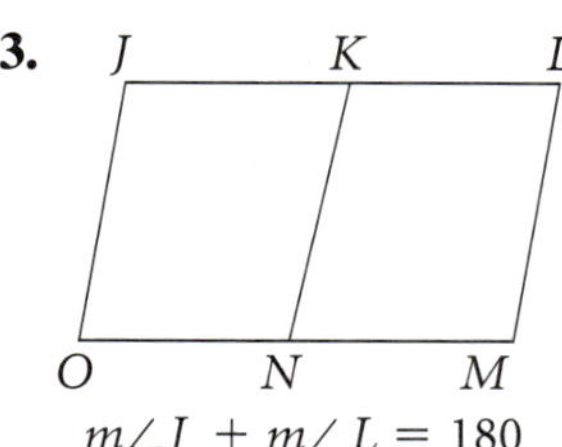

4.

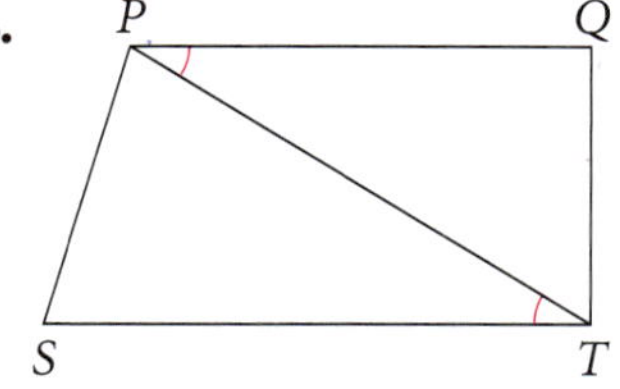

Example 2 (page 136)

x^2 **Algebra** **Find the value of x for which $\ell \parallel m$.**

5.

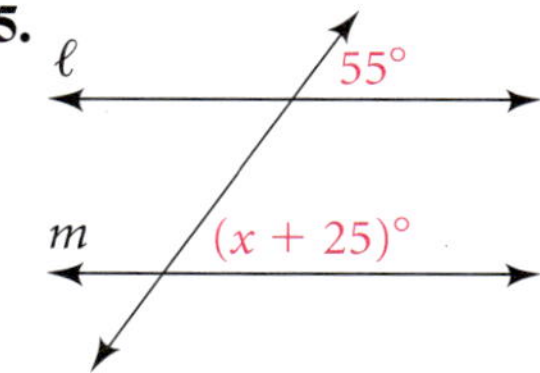

6.

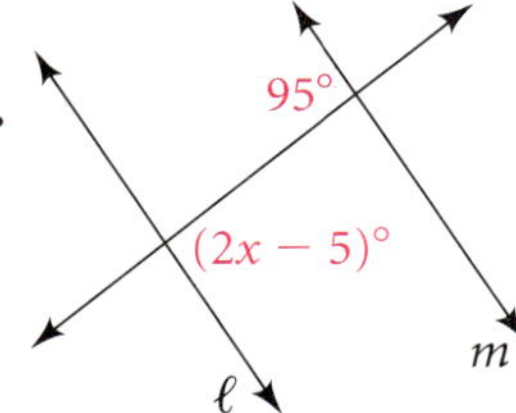

7.

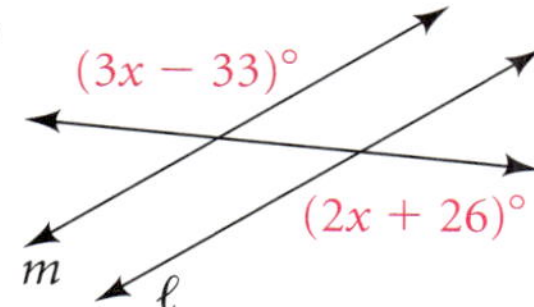

8.

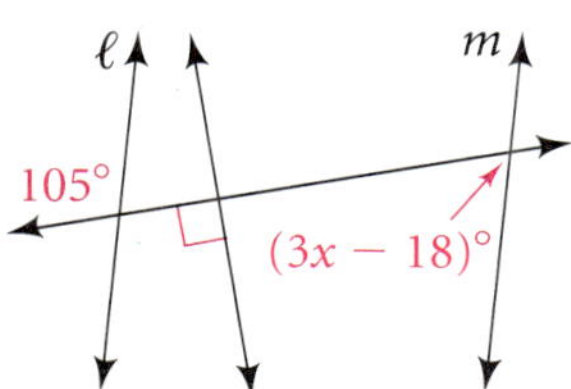

Example 3 (page 137)

9. Drafting An artist uses the drawing tool in the diagram at the right. The artist draws a line, slides the triangle along the flat surface, and draws another line. Explain why the drawn lines must be parallel.

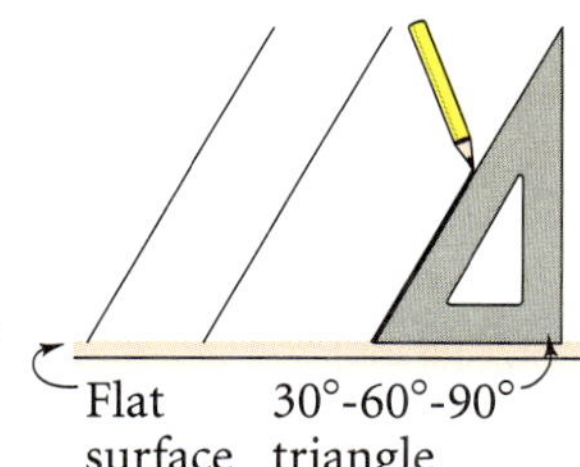

B Apply Your Skills

Developing Proof Using the given information, which lines, if any, can you conclude are parallel? Justify each conclusion with a theorem or postulate.

10. $\angle 2$ is supplementary to $\angle 3$.

11. $\angle 6$ is supplementary to $\angle 7$.

12. $\angle 1$ is supplementary to $\angle 4$.

13. $m\angle 7 = 70, m\angle 9 = 110$

14. $\angle 1 \cong \angle 3$

15. $\angle 9 \cong \angle 12$

16. $\angle 3 \cong \angle 6$

17. $\angle 2 \cong \angle 10$

18. $\angle 1 \cong \angle 8$

19. $\angle 8 \cong \angle 6$

20. $\angle 11 \cong \angle 7$

21. $\angle 5 \cong \angle 10$

22. **Developing Proof** Complete this paragraph proof of Theorem 3-6.

If two lines and a transversal form supplementary same-side interior angles, then the two lines are parallel.

Given: $\angle 1$ and $\angle 2$ are supplementary.

Prove: $\ell \parallel m$

Proof: $\angle 2$ is a supplement of **a.** __?__ and $\angle 3$ is a supplement of **b.** __?__. Since supplements of the same angle are congruent, **c.** __?__ $\cong$ **d.** __?__. Since $\angle 2$ and $\angle 3$ are also corresponding angles, $\ell \parallel m$ by the **e.** __?__ Postulate.

23. **Carpentry** A T-bevel is a tool used by carpenters to draw congruent angles. By loosening the locking lever, the carpenter can adjust the angle. Explain how the carpenter knows that two lines drawn using the T-bevel are parallel.

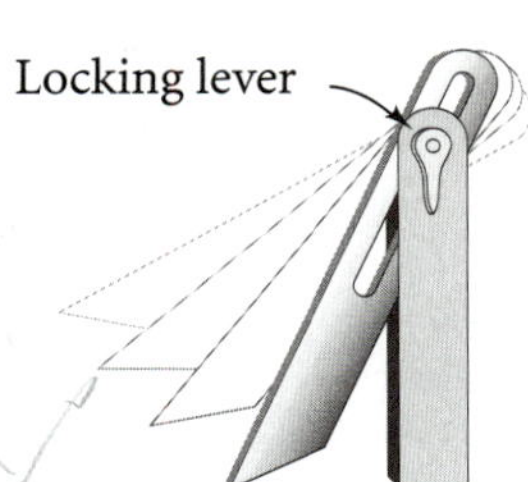

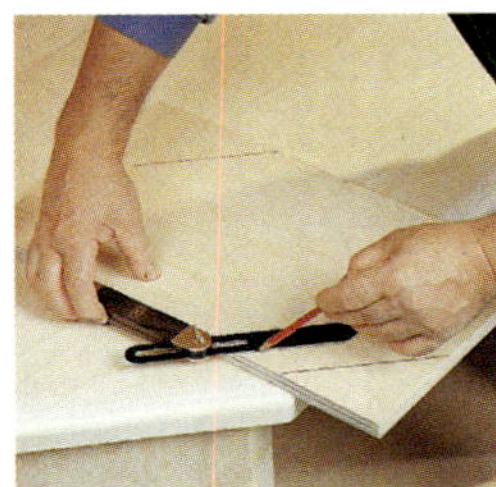

Real-World Connection

A carpenter must draw angles precisely to ensure good fit.

Algebra Find the value of x for which $\ell \parallel m$.

24.

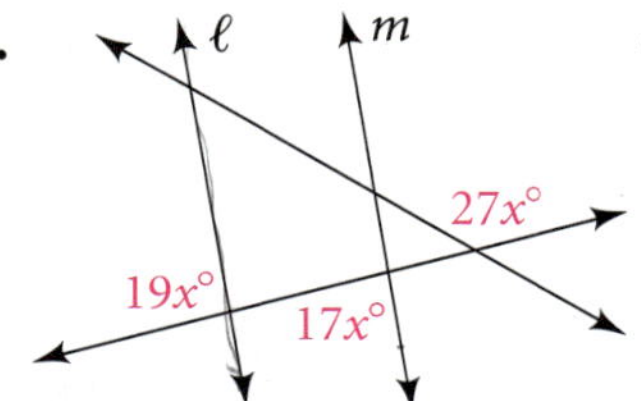

25.

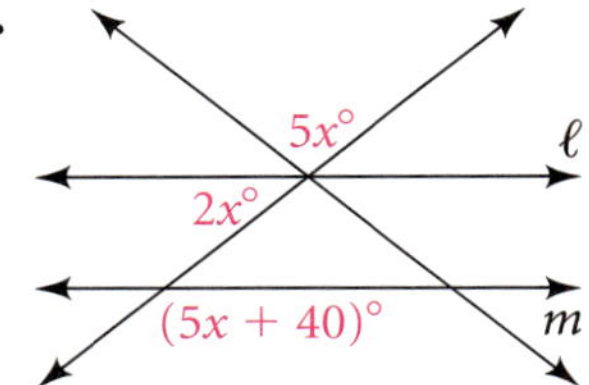

26. **Multiple Choice** Lines f and g are parallel. Which statement can you NOT deduce from this information?

(A) $\angle 1 \cong \angle 8$

(B) $m\angle 2 + m\angle 5 = 180$

(C) $\angle 4 \cong \angle 7$

(D) $m\angle 3 + m\angle 8 = 180$

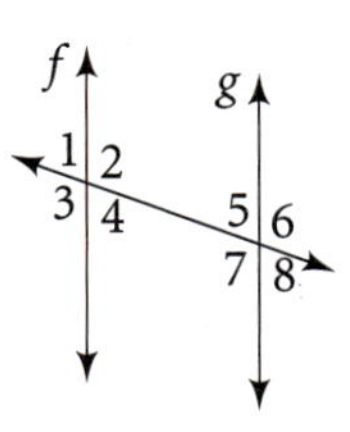

Proof **27.** Prove Theorem 3-8.

If two lines and a transversal form same-side exterior angles that are supplementary, then the lines are parallel.

Given: $m\angle 1 + m\angle 3 = 180$

Prove: $\ell \parallel n$

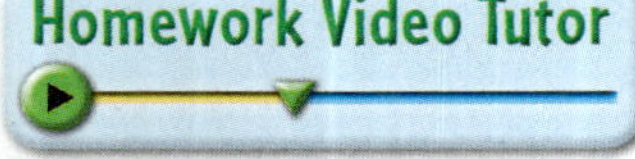

Visit: PHSchool.com
Web Code: bce-0302

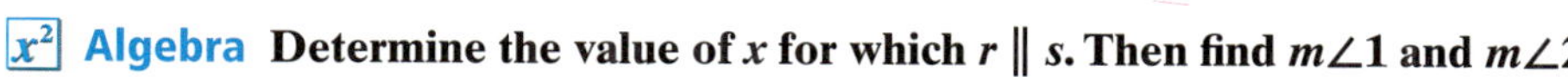

x^2 **Algebra** **Determine the value of x for which $r \parallel s$. Then find $m\angle 1$ and $m\angle 2$.**

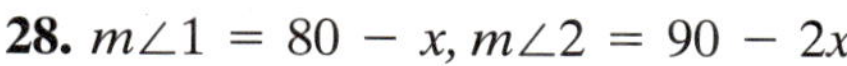

28. $m\angle 1 = 80 - x, m\angle 2 = 90 - 2x$

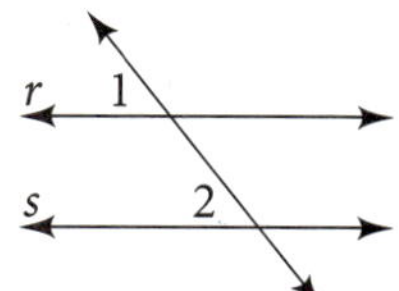

29. $m\angle 1 = 60 - 2x, m\angle 2 = 70 - 4x$

30. $m\angle 1 = 40 - 4x, m\angle 2 = 50 - 8x$

31. $m\angle 1 = 20 - 8x, m\angle 2 = 30 - 16x$

32. **Crew** If the rowing crew at the left strokes in unison, the oars sweep out angles of equal measure. Explain why the oars on each side of the shell stay parallel.

Exercise 32

Open-Ended **In each exercise, information is given about the figure below. State another fact about one of the given angles that will guarantee two lines are parallel. Tell which lines will be parallel and why.**

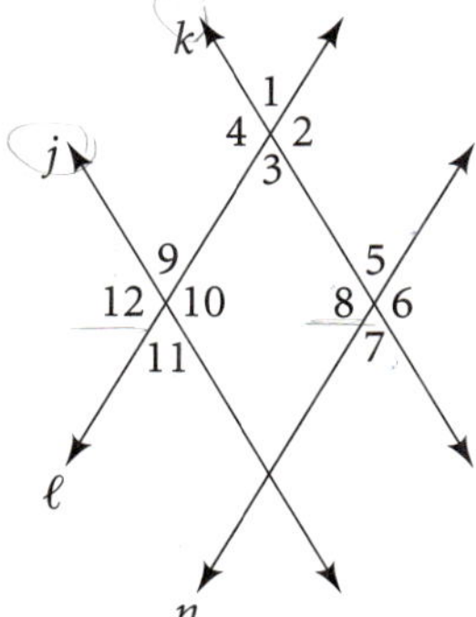

33. $\angle 1 \cong \angle 3$

34. $m\angle 8 = 110, m\angle 9 = 70$

35. $\angle 5 \cong \angle 11$

36. $\angle 11$ and $\angle 12$ are supplementary.

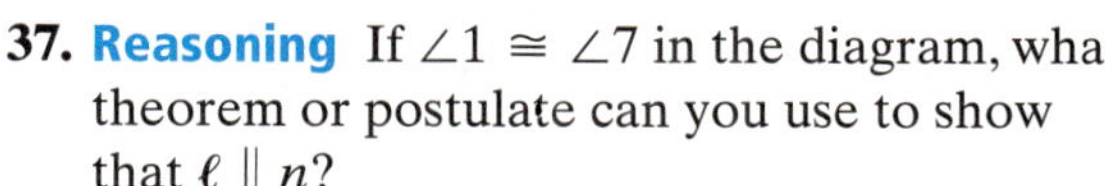

37. **Reasoning** If $\angle 1 \cong \angle 7$ in the diagram, what theorem or postulate can you use to show that $\ell \parallel n$?

Proof **For Exercises 38–39, use the diagram above and write a two-column proof.**

38. Given: $\ell \parallel n, \angle 12 \cong \angle 8$
Prove: $j \parallel k$

39. Given: $j \parallel k, m\angle 8 + m\angle 9 = 180$
Prove: $\ell \parallel n$

Proof **40.** Rewrite this paragraph proof of Theorem 3-6 as a flow proof.

If two lines and a transversal form supplementary same-side interior angles, then the two lines are parallel.

Given: $\angle 1$ and $\angle 2$ are supplementary.

Prove: $\ell \parallel m$

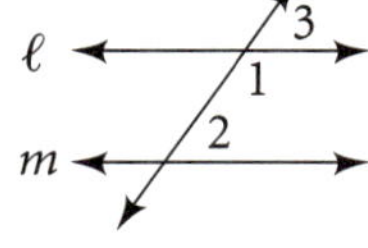

Proof: It is given that $\angle 1$ and $\angle 2$ are supplementary. $\angle 1$ and $\angle 3$ are also supplementary, so $\angle 2 \cong \angle 3$. Since $\angle 2$ and $\angle 3$ are corresponding angles, $\ell \parallel m$.

C **Challenge** **Which sides of quadrilateral *PLAN* must be parallel? Explain.**

41. $m\angle P = 72, m\angle L = 108, m\angle A = 72, m\angle N = 108$

42. $m\angle P = 59, m\angle L = 37, m\angle A = 143, m\angle N = 121$

43. $m\angle P = 67, m\angle L = 120, m\angle A = 73, m\angle N = 100$

44. $m\angle P = 56, m\angle L = 124, m\angle A = 124, m\angle N = 56$

Proof **45.** Prove the following statement is true by following the steps below:

If a transversal intersects two parallel lines, then the bisectors of two corresponding angles are parallel.

a. Draw and label a diagram on paper.
b. State what is given and mark the diagram to keep track of the information.
c. State what you are to prove.
d. Write a plan for proof.
e. Follow your plan and write the proof.

Multiple Choice Practice

For California Standards and CAHSEE Tutorials, visit PHSchool.com. Web Code: bcq-9045

GEOM 7.0

46. Using the diagram at the right, which of the following is a correct conclusion?

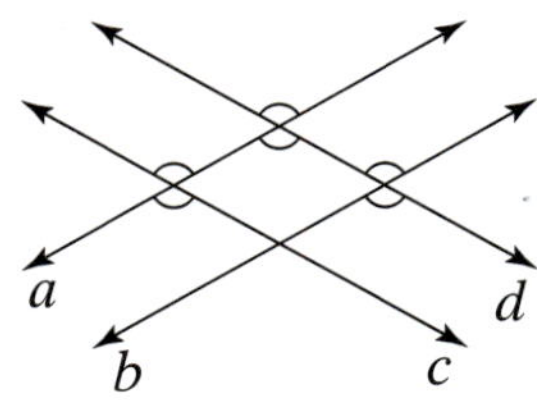

- (A) $a \parallel b$ because alternate interior angles are supplementary.
- (B) $a \parallel b$ because vertical angles are congruent.
- (C) $c \parallel d$ because corresponding angles are congruent.
- (D) $c \parallel d$ because vertical angles are congruent.

CAHSEE 7NS 1.3

47. On his final history exam, Jerry answered 88% of the questions correctly. The exam had 150 questions on it. How many questions did Jerry NOT answer correctly?

- (A) 18
- (B) 62
- (C) 122
- (D) 132

CAHSEE 7MG 2.1

48. A net for a cardboard mail container is shown at the right. What is the total surface area in square inches?

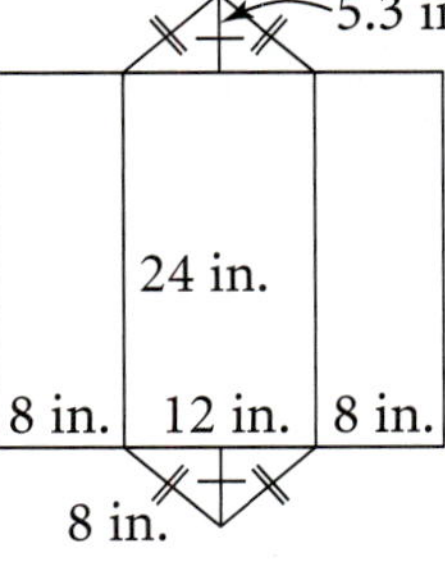

- (A) 63.6
- (B) 640
- (C) 672
- (D) 735.6

Mixed Review

Lesson 3-1

Find $m\angle 1$, and then $m\angle 2$. Justify each answer.

49.

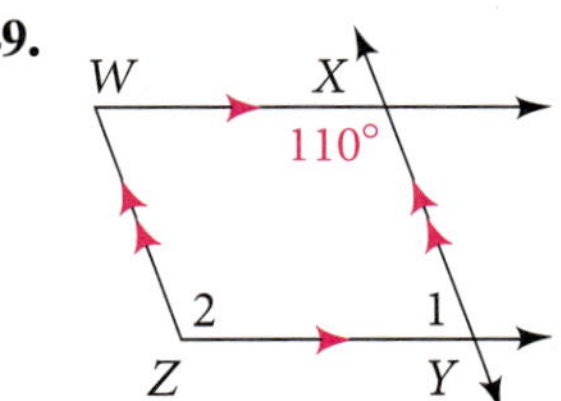

50.

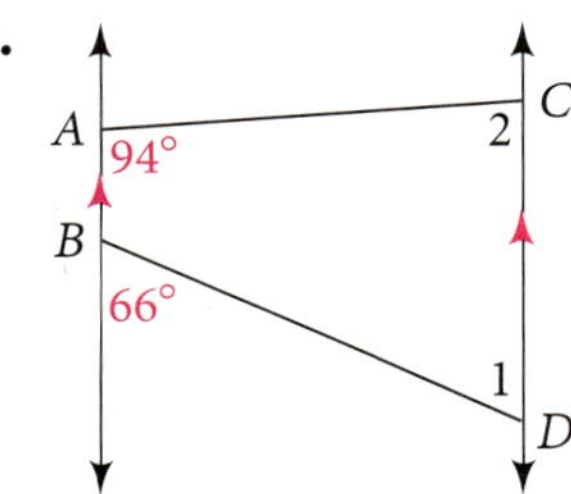

Lesson 2-1

Write the converse of each conditional statement. Determine the truth values of the original conditional and its converse.

51. If you are in Nebraska, you are west of the Mississippi River.

52. If a circle has a diameter of 8 cm, then it has a radius of 4 cm.

53. If a line intersects a pair of parallel lines, then same-side interior angles are supplementary.

54. If you add *ed* to a verb, you form the past tense of a verb.

55. If it is raining, then there are clouds in the sky.

Lesson 1-9

Find the area of each circle. Round to the nearest tenth.

56. $r = 8$ in. **57.** $d = 6$ cm **58.** $d = 9$ ft **59.** $r = 5$ in.

60. $d = 2.8$ m **61.** $r = 1.2$ m **62.** $d = 4.75$ ft **63.** $r = 0.6$ m

3-3 Parallel and Perpendicular Lines

 **California Content Standards**

GEOM 7.0 Students prove and use theorems involving the properties of parallel lines cut by a transversal. ***Master***

What You'll Learn

- To relate parallel and perpendicular lines

. . . And Why

To show how to fit sides of a picture frame together as in Example 1

Check Skills You'll Need

Complete each statement with *always, sometimes,* or *never.*

1. Two lines in the same plane are __?__ parallel.
2. Perpendicular lines __?__ meet at right angles.
3. Two lines in intersecting planes are __?__ perpendicular.
4. Two lines in parallel planes are __?__ perpendicular.

1 Relating Parallel and Perpendicular Lines

The two diagrams suggest ways to draw parallel lines. You can draw them (a) parallel to a given line, or (b) perpendicular to a given line. Theorems 3-9 and 3-10 guarantee that the lines you draw are indeed parallel.

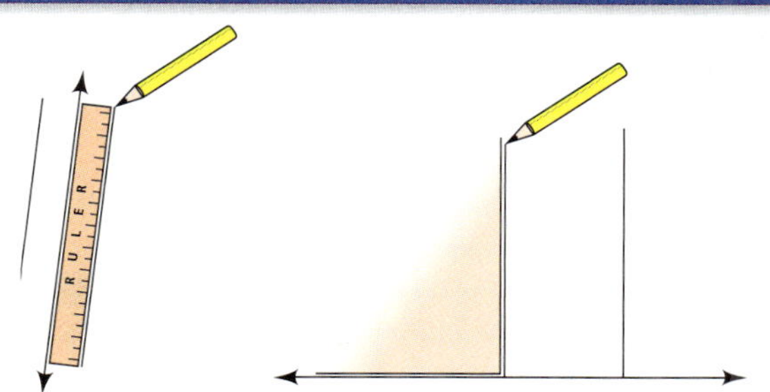

 **Key Concepts**

Theorem 3-9

If two lines are parallel to the same line, then they are parallel to each other.

$a \parallel b$

Theorem 3-10

In a plane, if two lines are perpendicular to the same line, then they are parallel to each other.

$m \parallel n$

Theorem 3-10 includes the phrase *in a plane*. On the other hand, Theorem 3-9 is true for any three such lines, whether they are coplanar (Exercise 3) or noncoplanar.

 Proof

Proof of Theorem 3-10

Study what is given, what you are to prove, and the diagram. Then write a paragraph proof.

Given: $r \perp t, s \perp t$

Prove: $r \parallel s$

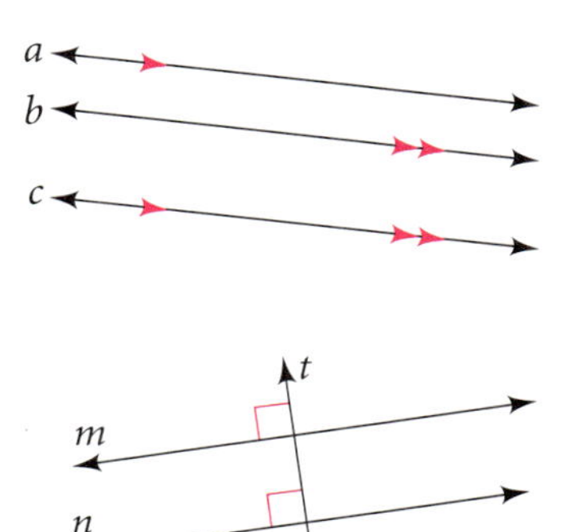

Proof: $\angle 1$ and $\angle 2$ are right angles by the definition of perpendicular, so they are congruent. Since corresponding angles are congruent, $r \parallel s$.

1 EXAMPLE Real-World Connection

Woodworking To make a frame for a painting, a miter box and a backsaw are used to cut the framing at 45° angles. Explain why cutting the framing at this angle ensures that opposite sides of the frame will be parallel.

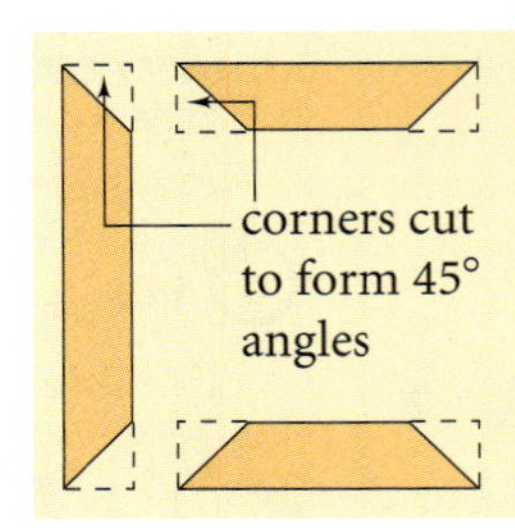

Two adjacent 45° angles form a 90° angle. The opposite sides of the frame are perpendicular to the same side. Thus the opposite sides are parallel because two lines perpendicular to a third line are parallel.

CA Standards Check 1 Can you assemble the framing at the right into a frame with opposite sides parallel? Explain.

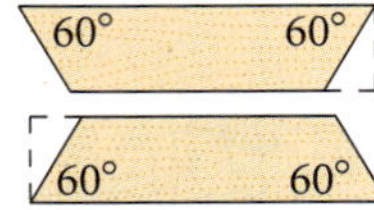

Theorems 3-9 and 3-10 gave conditions by which you can conclude that lines are parallel. Theorem 3-11 provides a way for you to conclude that lines are perpendicular. You will prove Theorem 3-11 in Exercise 11.

Key Concepts

Theorem 3-11

In a plane, if a line is perpendicular to one of two parallel lines, then it is also perpendicular to the other.

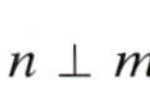

$n \perp m$

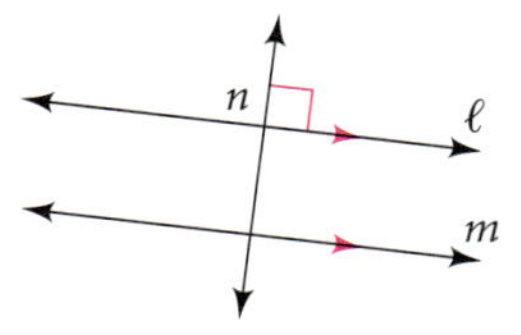

Proof 2 EXAMPLE Using Theorem 3-11

Study what is given, what you are to prove, and the diagram. Then write a paragraph proof.

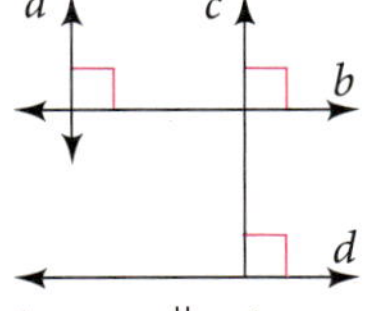

Given: In a plane, $a \perp b$, $b \perp c$, and $c \perp d$.

Prove: $a \perp d$

Proof: Lines a and c are both perpendicular to line b, so $a \parallel c$ because two lines perpendicular to the same line are parallel. It is given that $c \perp d$. Therefore, $a \perp d$ because a line that is perpendicular to one of two parallel lines is perpendicular to the other (Theorem 3-11).

CA Standards Check 2 From what is given in Example 2, can you also conclude $b \parallel d$? Explain.

In the rectangular solid shown here, $\overleftrightarrow{AC}$ and $\overleftrightarrow{BD}$ are parallel. $\overleftrightarrow{EC}$ is perpendicular to $\overleftrightarrow{AC}$, but it is not perpendicular to $\overleftrightarrow{BD}$. This is why Theorem 3-11 states that the lines must be "in a plane."

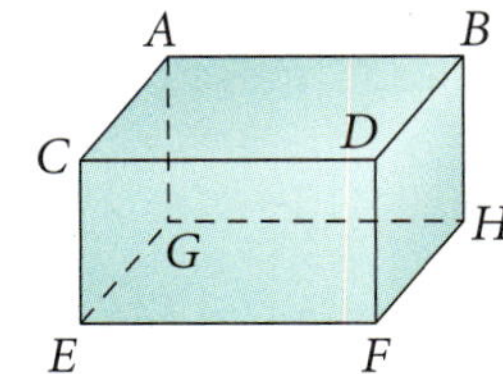

EXERCISES

For more exercises, see *Extra Skill, Word Problem, and Proof Practice.*

Standards Practice

GEOM 7.0

A Practice by Example

Example 1 (page 142)

1. A carpenter is building a trellis for vines to grow on. The completed trellis will have two sets of overlapping diagonal slats of wood.
   a. What must be true of ∠1, ∠2, and ∠3 if slats A, B, and C must be parallel?
   b. The carpenter attaches slat D so that it is perpendicular to slat *A*. Is slat D perpendicular to slats B and C? Justify your answer.

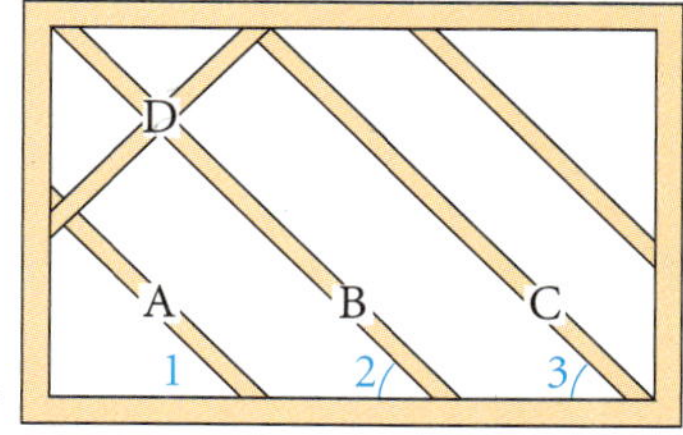

Example 2 (page 142)

Proof 2. Study what is given, what you are to prove, and the diagram. Then write a proof.

Given: In a plane, $a \perp b, b \perp c,$ and $c \parallel d$.

Prove: $a \parallel d$

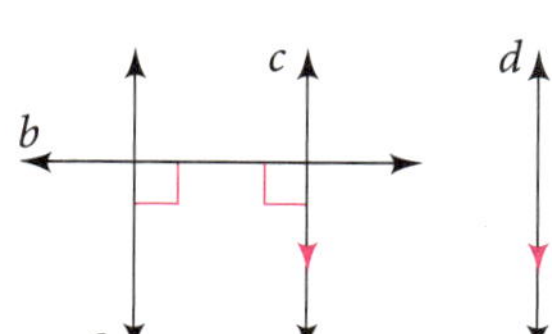

B Apply Your Skills

For a guide to solving Exercise 3, see p. 145.

3. **Developing Proof** Copy and complete this paragraph proof of Theorem 3-9 for three coplanar lines.

If two lines are parallel to the same line, then they are parallel to each other.

Given: $\ell \parallel k$ and $m \parallel k$

Prove: $\ell \parallel m$

Proof: $\ell \parallel k$ means that $\angle 2 \cong \angle 1$ by the **a.** __?__ Postulate. $m \parallel k$ means that **b.** __?__ $\cong$ **c.** __?__ for the same reason. By the Transitive Property of Congruence, $\angle 2 \cong \angle 3$. By the **d.** __?__ Postulate, $\ell \parallel m$.

k
ℓ
m
1
2
3

Each of the following statements describes a ladder. What can you conclude about the rungs, one side, or both sides of each ladder? Explain.

4. The rungs are each perpendicular to one side.
5. The rungs are parallel and the top rung is perpendicular to one side.
6. The sides are parallel. The rungs are perpendicular to one side.
7. The rungs are perpendicular to one side. The other side is perpendicular to the top rung.
8. Each side is perpendicular to the top rung.
9. Each rung is parallel to the top rung.
10. The rungs are perpendicular to one side. The sides are not parallel.

Real-World Connection

This ladder's rungs are perpendicular to each side. Therefore, the rungs are parallel to each other.

Proof 11. Prove Theorem 3-11: In a plane, if a line is perpendicular to one of two parallel lines, then it is also perpendicular to the other.

Given: In a plane, $a \perp b$, and $b \parallel c$.

Prove: $a \perp c$

a
b
c

Proof 12. Prove: If a line is perpendicular to each of two other lines, all in one plane, then the two other lines are parallel.

Given: $t \perp r, t \perp s$

Prove: $r \parallel s$

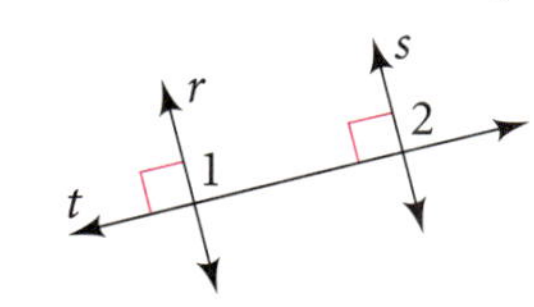

13. **Writing** Theorem 3-10: In a plane, two lines perpendicular to the same line are parallel. Use the rectangular solid at the right to explain why the words *in a plane* are needed.

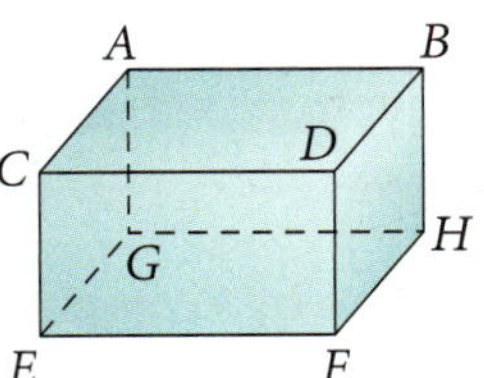

 **Challenge**

a, b, c, and d are distinct lines in the same plane.
Exercises 14-21 show different combinations of relationships between a and b, b and c, and c and d. For each combination of the three relationships, how are a and d related?

14. $a \parallel b, b \parallel c, c \parallel d$
15. $a \parallel b, b \parallel c, c \perp d$
16. $a \parallel b, b \perp c, c \parallel d$
17. $a \perp b, b \parallel c, c \parallel d$
18. $a \parallel b, b \perp c, c \perp d$
19. $a \perp b, b \parallel c, c \perp d$
20. $a \perp b, b \perp c, c \parallel d$
21. $a \perp b, b \perp c, c \perp d$

Homework Video Tutor
Visit: PHSchool.com
Web Code: bce-0303

Critical Thinking **The Reflexive, Symmetric, and Transitive Properties for Congruence ($\cong$) are listed on page 105.**

22. Write reflexive, symmetric, and transitive statements for "is parallel to" ($\parallel$). State whether each statement is true or false and justify your answer.

23. Repeat Exercise 22 for "is perpendicular to" ($\perp$).

Multiple Choice Practice

For California Standards and CAHSEE Tutorials, visit PHSchool.com. Web Code: bcq-9045

GEOM 7.0

24. In a plane, line e is parallel to line f, line f is parallel to line g, and line h is perpendicular to line e. Which of the following MUST be true?

Ⓐ $e \parallel g$ Ⓑ $h \parallel f$ Ⓒ $g \parallel h$ Ⓓ $e \parallel h$

CAHSEE 7NS2.2

25. Which fraction is equivalent to $\frac{3}{8} + \frac{5}{12}$?

Ⓐ $\frac{15}{16}$ Ⓑ $\frac{2}{5}$ Ⓒ $\frac{11}{16}$ Ⓓ $\frac{19}{24}$

Mixed Review

Lesson 3-2 x^2 **Algebra** **Determine the value of x for which $a \parallel b$.**

26.

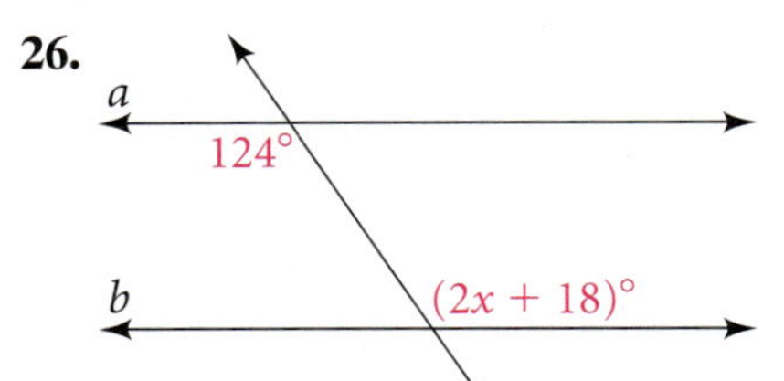

27.

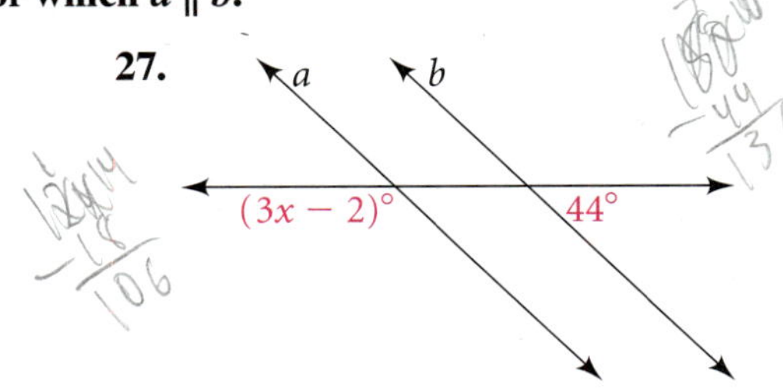

Lesson 2-2 **Each conditional statement below is true. Write its converse. If the converse is also true, combine the statements into a biconditional.**

28. If $x = 7$, then $x^2 = 49$.

29. If two lines in a plane do not meet, then the lines are parallel.

Guided Problem Solving

FOR USE WITH PAGE 143, EXERCISE 3

Understanding Proof Problems **Read the problem below. Then let Collin's thinking guide you through the solution. Check your understanding with the exercise at the bottom of the page.**

Copy and complete the paragraph proof of Theorem 3-9 for three coplanar lines.

If two lines are parallel to the same line, then they are parallel to each other.

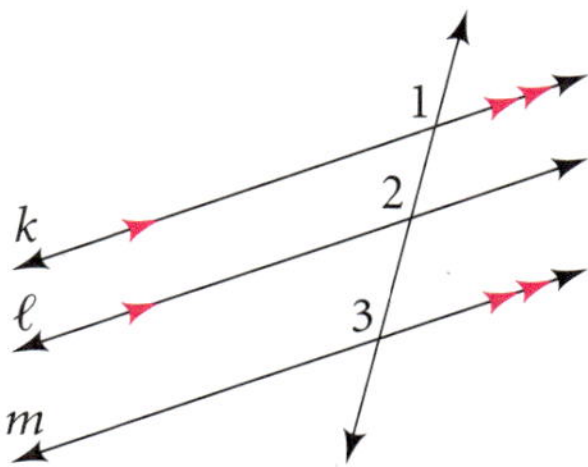

Given: $\ell \parallel k$ and $m \parallel k$

Prove: $\ell \parallel m$

Proof: $\ell \parallel k$ means that $\angle 2 \cong \angle 1$ by the **a.** __?__ Postulate.
$m \parallel k$ means that **b.** __?__ $\cong$ **c.** __?__ for the same reason. By the Transitive Property of Congruence, $\angle 2 \cong \angle 3$. By the **d.** __?__ Postulate, $\ell \parallel m$.

What Collin Thinks	What Collin Writes
Let's see. $\angle 1$ and $\angle 2$ are corresponding angles. If $\ell \parallel k$, then $\angle 1 \cong \angle 2$ by Postulate 3-1 on page 128.	$\ell \parallel k$ means that $\angle 2 \cong \angle 1$ by the a. Corresponding Angles Postulate.
Now I will use the fact that $m \parallel k$. So I'll ignore line ℓ. That leaves me with $\angle 1$ and $\angle 3$ to consider. These also are corresponding angles.	$m \parallel k$ means that b. $\angle 1$ $\cong$ c. $\angle 3$ for the same reason.
Now, $\angle 2 \cong \angle 3$ and they are corresponding angles. This fits the postulate in which the hypothesis is "corresponding angles are congruent." By Postulate 3-2 on page 134, I can conclude "the lines are parallel."	By the Transitive Property of Congruence, $\angle 2 \cong \angle 3$. By the d. Converse of the Corresponding Angles Postulate, $\ell \parallel m$.

EXERCISE

Copy and complete this paragraph proof of Theorem 3-9 for three coplanar lines.

If two lines are parallel to the same line, then they are parallel to each other.

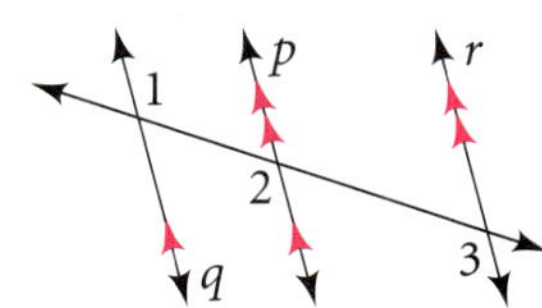

Given: $q \parallel p$ and $p \parallel r$

Prove: $q \parallel r$

Proof: $q \parallel p$ means that $\angle 1 \cong \angle 2$ by the **a.** __?__ Theorem.
$p \parallel r$ means that **b.** __?__ $\cong$ **c.** __?__ by the Corresponding Angles Postulate.
By the Transitive Property of Congruence, $\angle 1 \cong \angle 3$. By the **d.** __?__ Theorem, $q \parallel r$.

Angle Dynamics

Hands-On

FOR USE WITH LESSON 3-4

Prepares for GEOM 12.0 Students find and use measures of sides and of interior and exterior angles of triangles to classify figures and solve problems.

Activity 1 suggests one of the most important ideas of Euclidean geometry.

1 ACTIVITY

- Draw and cut out a large triangle.
- Number the angles and tear them off.
- Place the three angles adjacent to each other to form one angle, as shown at the right.

1. What kind of angle is formed by the three smaller angles? What is its measure?

2. Make a conjecture about the sum of the measures of the angles of a triangle.

- Fold a sheet of paper in half three times. Draw a *scalene triangle* (no two sides congruent) on the folded paper. Carefully cut out the triangle. This will give you eight triangles that are all the same size and shape.
- Number the angles of each triangle 1, 2, and 3. Use the same number on the corresponding angles from one triangle to the next.

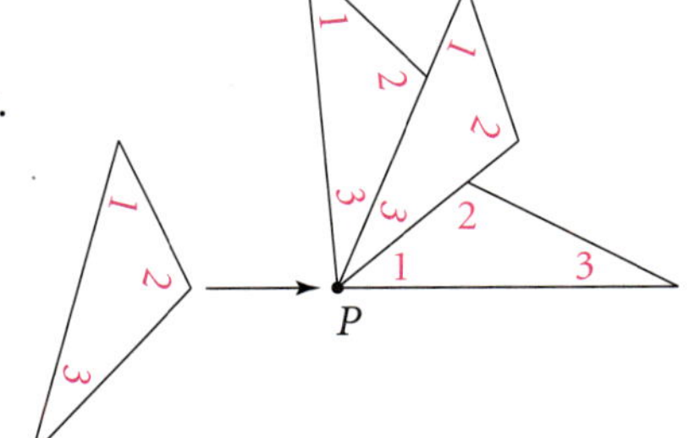

3. Mark a point P in the middle of a blank sheet of paper. How many of your triangles do you think you can fit perfectly about point P without gaps or overlaps?

4. What conjecture are you making about the sum of the measures of the angles about a point?

EXERCISES

5. Use the triangles from Activity 2. Based on your conjectures above, what combination of angles 1, 2, and 3 can you place at a point P to guarantee a perfect fit of triangles about P?

6. There are several ways to arrange the set of six angles 1, 1, 2, 2, 3, 3 to make a perfect fit about P. Make three different listings of the numbers 1, 1, 2, 2, 3, 3 around a point. Arrange your triangles about P to match your lists. Is any arrangement more appealing than the other two? What happens if you turn over some of the triangles?

7. Follow the steps of Activity 1 using a *quadrilateral* (four-sided figure), such as the one shown at the right. Make a conjecture about the sum of the measures of the angles of a quadrilateral.

8. If you fit the angles of a quadrilateral about a point P without gaps or overlaps, what do you think you would discover? Justify your response.

Parallel Lines and the Triangle Angle-Sum Theorem

California Content Standards

GEOM 12.0 Students find and use measures of interior and exterior angles of triangles to classify figures and solve problems. ***Introduce, Develop***

GEOM 13.0 Students prove relationships between angles in polygons by using properties of complementary, supplementary, vertical, and exterior angles. ***Introduce, Develop***

What You'll Learn

- To classify triangles and find the measures of their angles
- To use exterior angles of triangles

. . . And Why

To find the reclining angle of a lounge chair, as in Example 4

Check Skills You'll Need

GO for Help Lesson 1-6

Classify each angle as *acute, right,* or *obtuse.*

1. 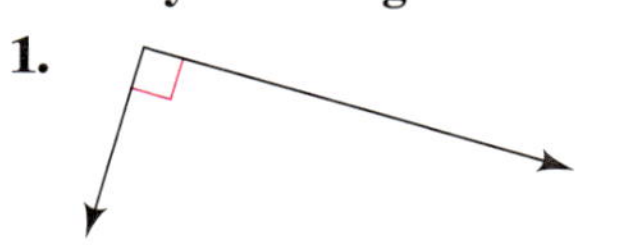

2. 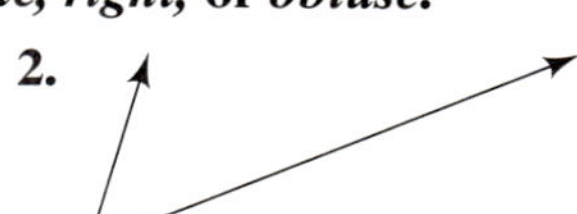

3. 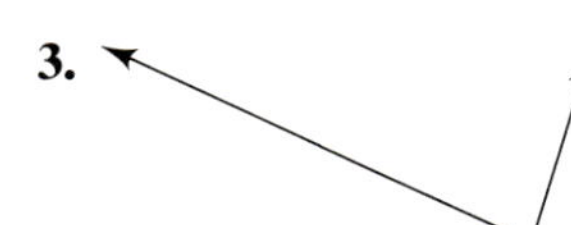

Solve each equation.

4. $30 + 90 + x = 180$

5. $55 + x + 105 = 180$

6. $x + 58 = 90$

7. $32 + x = 90$

New Vocabulary
- acute triangle
- right triangle
- obtuse triangle
- equiangular triangle
- equilateral triangle
- isosceles triangle
- scalene triangle
- exterior angle of a polygon
- remote interior angles

1 Finding Angle Measures in Triangles

The diagrams in Activity 1 on page 146 suggest the Triangle-Angle Sum Theorem.

Key Concepts

Theorem 3-12 **Triangle Angle-Sum Theorem**

The sum of the measures of the angles of a triangle is 180.

$$m\angle A + m\angle B + m\angle C = 180$$

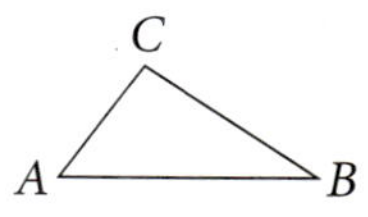

The following proof of Theorem 3-12 relies on the idea that through a point not on a given line you can draw a line parallel to the given line.

Proof

Proof of Theorem 3-12

Given: $\triangle ABC$

Prove: $m\angle A + m\angle B + m\angle 3 = 180$

P C 1 3 2 A B

Proof: By the Protractor Postulate, you can draw $\overrightarrow{CP}$ so that $m\angle 1 = m\angle A$. Then, $\angle 1$ and $\angle A$ are congruent alternate interior angles, so $\overleftrightarrow{CP} \parallel \overline{AB}$. $\angle 2$ and $\angle B$ are also alternate interior angles, so by the Alternate Interior Angles Theorem, $m\angle 2 = m\angle B$. By substitution, $m\angle A + m\angle B + m\angle 3 = m\angle 1 + m\angle 2 + m\angle 3$, which is equal to 180 by the Angle Addition Postulate.

1 EXAMPLE Applying the Triangle Angle-Sum Theorem

Algebra Find the values of x and y.

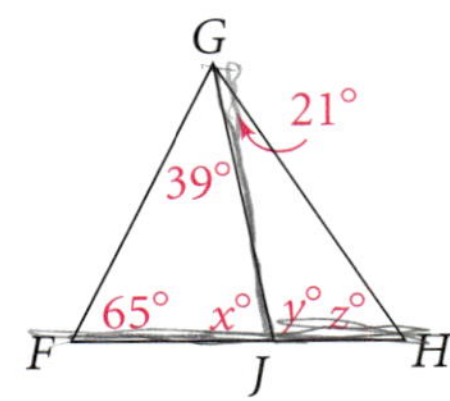

To find the value of x, use $\triangle GFJ$.

$39 + 65 + x = 180$ **Triangle Angle-Sum Theorem**

$104 + x = 180$ **Simplify.**

$x = 76$ **Subtract 104 from each side.**

To find the value of y, look at $\angle FJH$. It is a straight angle.

$m\angle GJF + m\angle GJH = 180$ **Angle Addition Postulate**

$x + y = 180$ **Substitute.**

$76 + y = 180$ **Substitute 76 for x.**

$y = 104$ **Subtract 76 from each side.**

1 Find the value of z in two different ways, each way using the Triangle Angle-Sum Theorem.

GO Online

Video Tutor Help
Visit: PHSchool.com
Web Code: bce-0775

In Chapter 1, you classified an angle by its measure. You can also classify a triangle by its angles and sides.

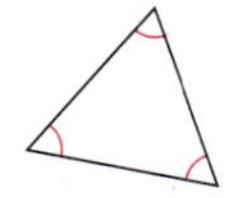

Equiangular
all angles congruent

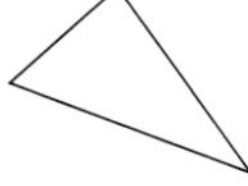

Acute
all angles acute

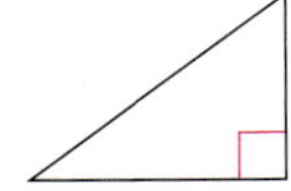

Right
one right angle

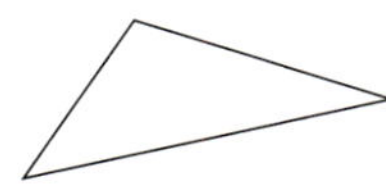

Obtuse
one obtuse angle

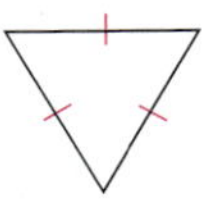

Equilateral
all sides congruent

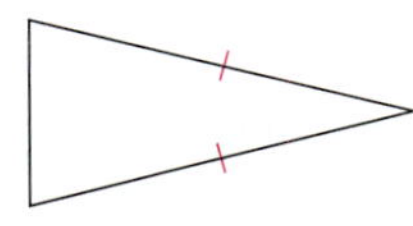

Isosceles
at least two sides congruent

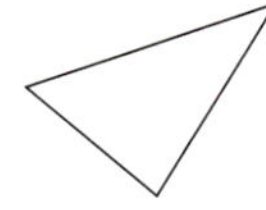

Scalene
no sides congruent

2 EXAMPLE Classifying a Triangle

Classify the triangle by its sides and its angles.

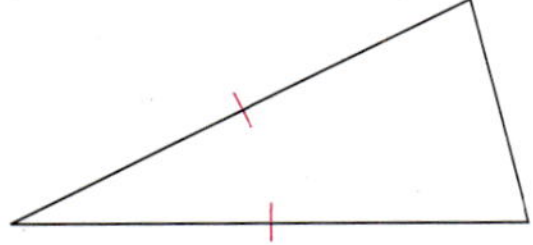

At least two sides are congruent, so the triangle is isosceles. All the angles are acute, so the triangle is acute.

The triangle is an acute isosceles triangle.

2 Draw and mark a triangle to fit each description. If no triangle can be drawn, write *not possible* and explain why.

a. acute scalene **b.** isosceles right **c.** obtuse equiangular

2 Using Exterior Angles of Triangles

Vocabulary Tip

Interior angle of a triangle means the same as "angle of a triangle."

An **exterior angle of a polygon** is an angle formed by a side and an extension of an adjacent side. For each exterior angle of a triangle, the two nonadjacent interior angles are its **remote interior angles.**

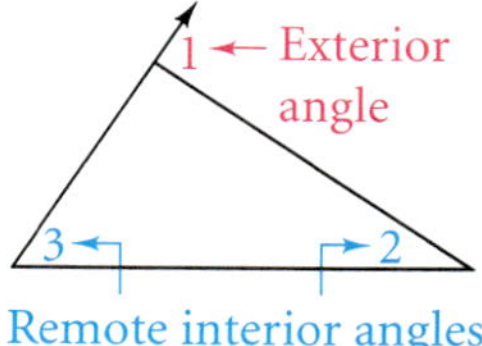

The diagram at the right suggests a relationship between an exterior angle and its two remote interior angles. Theorem 3-13 states this relationship. You will prove this theorem in Exercise 35.

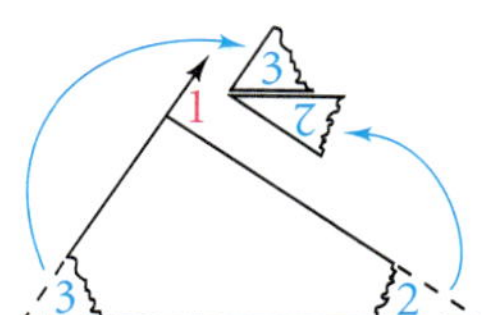

Key Concepts

Theorem 3-13 **Triangle Exterior Angle Theorem**

The measure of each exterior angle of a triangle equals the sum of the measures of its two remote interior angles.

$$m\angle 1 = m\angle 2 + m\angle 3$$

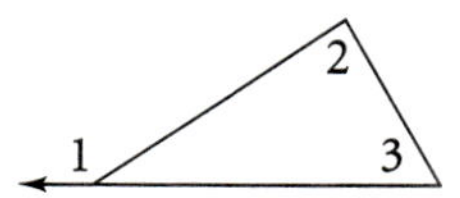

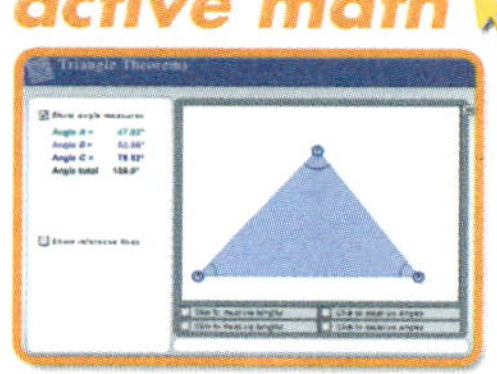

For: Triangle Theorems Activity
Use: Interactive Textbook, 3-4

3 EXAMPLE Using the Exterior Angle Theorem

Algebra Find each missing angle measure.

a.

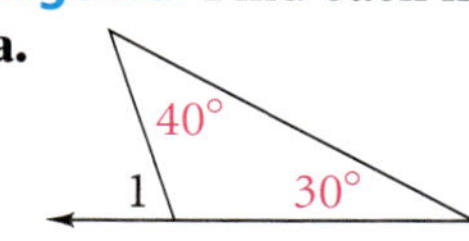

$m\angle 1 = 40 + 30$

$m\angle 1 = 70$

b.

113°
70°
2

$113 = 70 + m\angle 2$

$43 = m\angle 2$

CA Standards Check

3 Two angles of a triangle measure 45. Find the measure of an exterior angle at each vertex.

4 EXAMPLE Real-World Problem Solving

Multiple Choice The lounge chair has different settings that change the angles formed by its parts. Suppose $m\angle 2$ is 32 and $m\angle 3$ is 81. Find $m\angle 1$, the angle formed by the back of the chair and the arm rest.

Ⓐ 67 Ⓒ 113
Ⓑ 81 Ⓓ 180

$m\angle 1 = m\angle 2 + m\angle 3$	**Exterior Angle Theorem**
$m\angle 1 = 32 + 81$	**Substitute.**
$m\angle 1 = 113$	**Simplify.**

The angle formed is a 113° angle.
The correct choice is C.

Problem Solving Hint

To check exterior angle-measure answers, remember that an exterior angle and the adjacent interior angle must be supplementary.

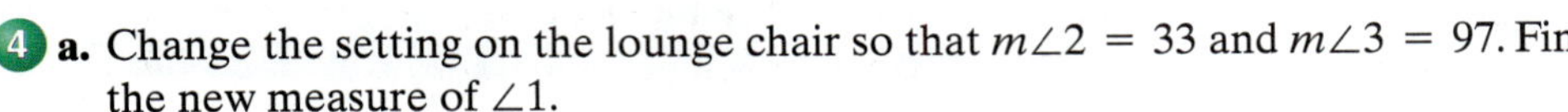

4 a. Change the setting on the lounge chair so that $m\angle 2 = 33$ and $m\angle 3 = 97$. Find the new measure of $\angle 1$.

b. Explain how you can find $m\angle 1$ *without* using the Exterior Angle Theorem.

EXERCISES Standards Practice

For more exercises, see *Extra Skill, Word Problem, and Proof Practice.*

GEOM 12.0, 13.0

A Practice by Example

Example 1 (page 148)

Find $m\angle 1$.

1. 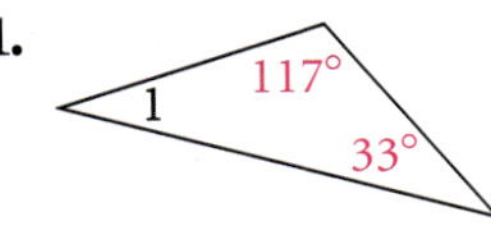

2. 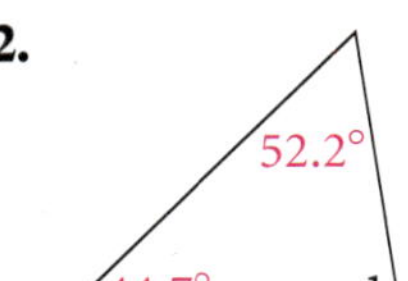

3. 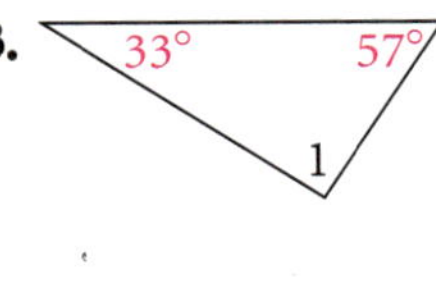

Find the value of each variable.

4. 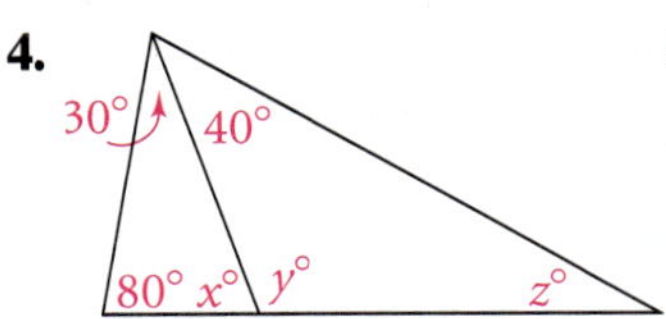

5. 

6. 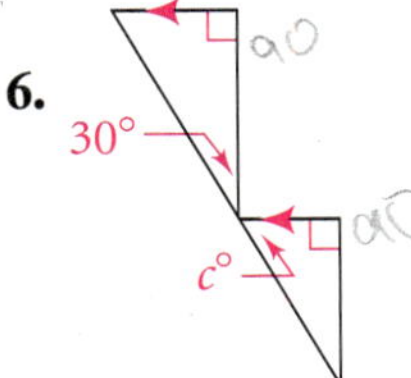

Example 2 (page 148)

Use a protractor and a centimeter ruler to measure the angles and the sides of each triangle. Classify each triangle by its angles and sides.

7. 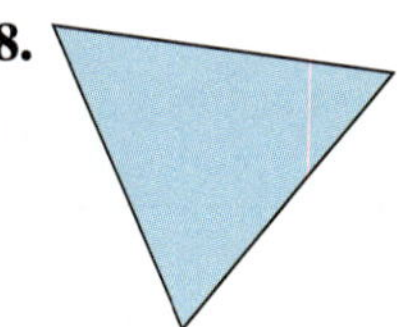

8. 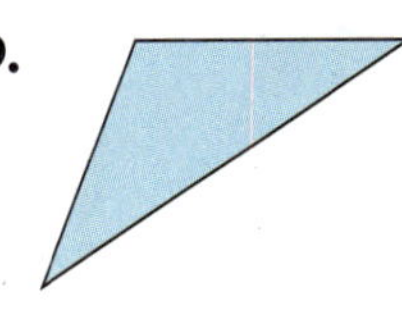

9.

If possible, draw a triangle to fit each description. Mark the triangle to show known information. If no triangle can be drawn, write *not possible* and explain why.

10. acute equilateral
11. equilateral right
12. obtuse scalene
13. obtuse isosceles
14. isosceles right
15. scalene acute

Example 3 (page 149)

16. a. Which of the numbered angles at the right are exterior angles?
 b. Name the remote interior angles for each.
 c. How are exterior angles 6 and 8 related?

17. a. How many exterior angles at the right are at each vertex of the triangle?
 b. How many exterior angles does a triangle have in all?

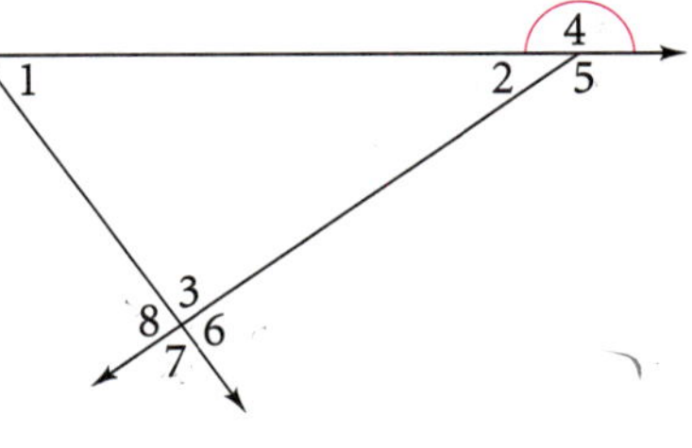

Algebra **Find each missing angle measure.**

18. 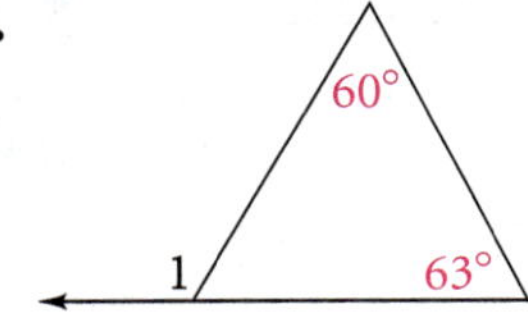

19. 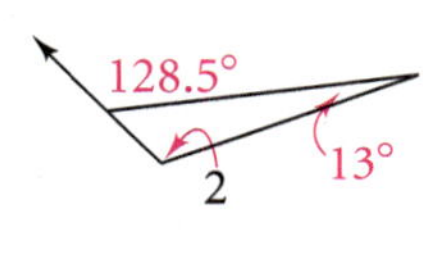

20. 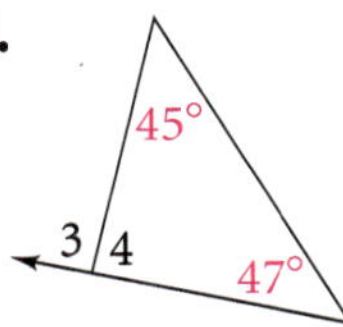

Example 4 (page 149)

21. **Music** The lid of a grand piano is held open by a prop stick whose length can vary, depending upon the effect desired. The longest prop stick makes angles as shown. What are the values of x and y?

22. A short prop stick makes the angles shown below. What are the values of a and b?

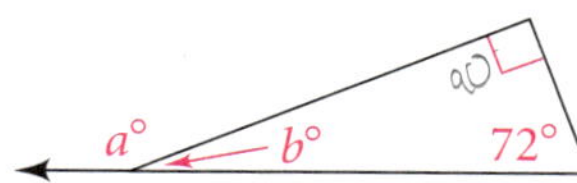

B Apply Your Skills

x^2 **Algebra** **Find the values of the variables and then the measures of the angles. Classify each triangle by its angles. Note that some figures have more than one triangle.**

23. 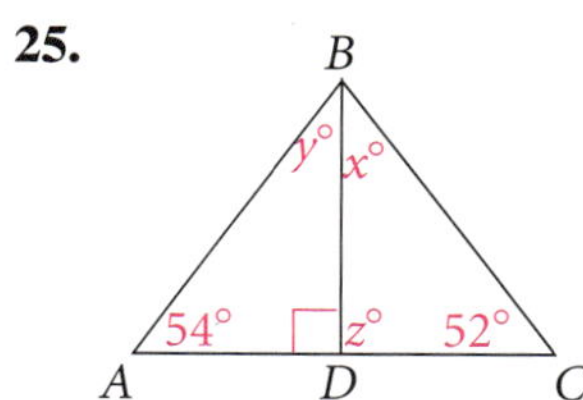

24. 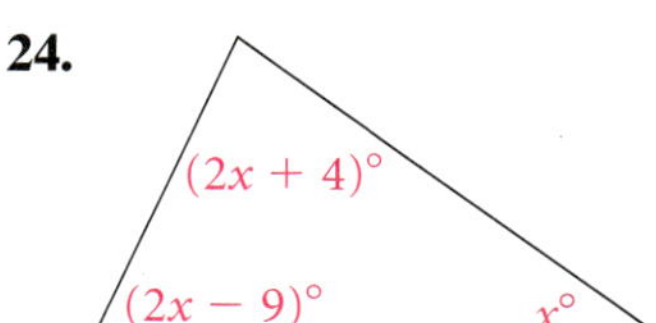

25. 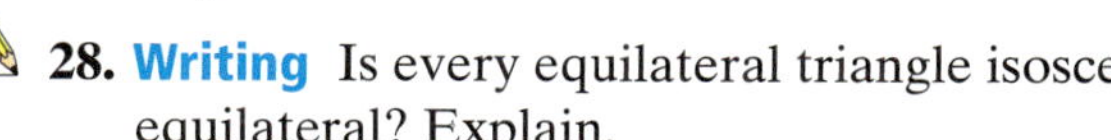

26. 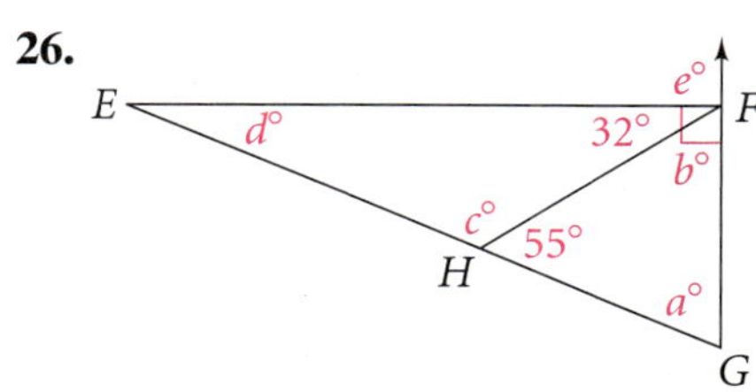

27. **Reasoning** What is the measure of each angle of an equiangular triangle? Explain.

28. **Writing** Is every equilateral triangle isosceles? Is every isosceles triangle equilateral? Explain.

29. **Visualization** The diagram shows a triangle on a 3-by-3 geoboard. How many triangles with different shapes can be made on this geoboard? Classify each triangle by its sides and angles.

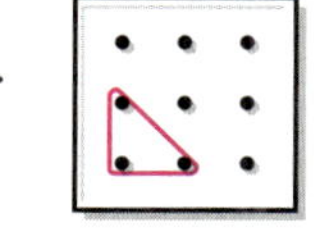

30. **Multiple Choice** The measure of one angle of a triangle is 115. The other two angles are congruent. What is the measure of each?
(A) 32.5 (B) 65 (C) 57.5 (D) 115

Problem Solving Hint

In Exercise 31, use x and $2x$ for the angle measures. In Exercise 32, use $2x$, $3x$, and $4x$.

x^2 31. **Algebra** A right triangle has acute angles whose measures are in the ratio 1 : 2. Find the measures of these angles.

x^2 32. a. **Algebra** The ratio of the angle measures in $\triangle BCR$ is 2 : 3 : 4. Find the angle measures.
b. What type of triangle is $\triangle BCR$?

Visit: PHSchool.com
Web Code: bce-0304

33. Draw any triangle. Label it $\triangle ABC$. Extend both sides of the triangle to form two exterior angles at vertex A. Use the two exterior angles to explain why it does not matter which side of a triangle is extended to form an exterior angle.

Proof **34.** Prove the following theorem.

The acute angles of a right triangle are complementary.

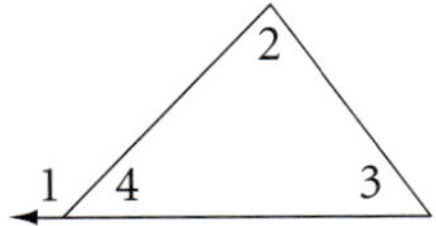

Given: $\triangle ABC$ with right angle C

Prove: $\angle A$ and $\angle B$ are complementary.

Proof **35.** Prove the Triangle Exterior Angle Theorem.

Given: $\angle 1$ is an exterior angle of the triangle.

Prove: $m\angle 1 = m\angle 2 + m\angle 3$

36. Reasoning Two angles of a triangle measure 64 and 48. Find the measure of the largest exterior angle. Explain.

Real-World Connection

Patricia Watson Tsinnie often uses isosceles triangles in her rug designs.

37. Open-Ended Study the design in the Navajo weaving below. Make a design of your own that makes repeated use of isosceles triangles.

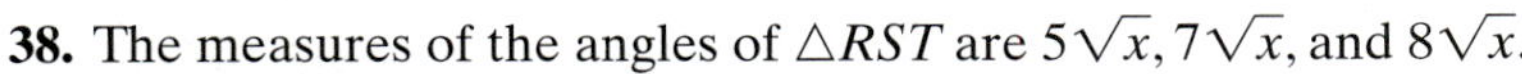
38. The measures of the angles of $\triangle RST$ are $5\sqrt{x}$, $7\sqrt{x}$, and $8\sqrt{x}$.

a. Find the value of x.

b. Give the measure of each angle.

c. What type of triangle is $\triangle RST$?

Challenge

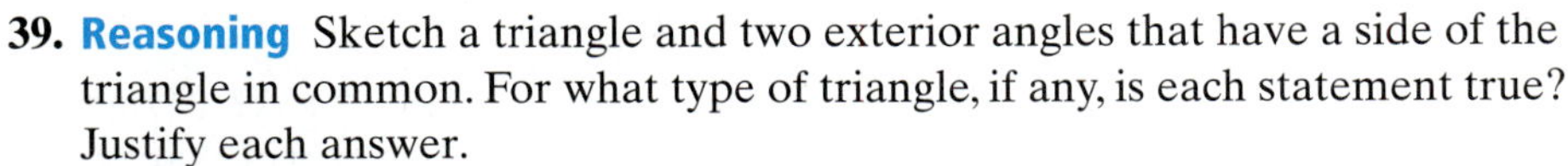
39. Reasoning Sketch a triangle and two exterior angles that have a side of the triangle in common. For what type of triangle, if any, is each statement true? Justify each answer.

a. The bisectors of the two exterior angles are parallel.

b. The bisectors of the two exterior angles are perpendicular.

c. The bisectors of the two exterior angles and the common side of the given triangle form an isosceles triangle.

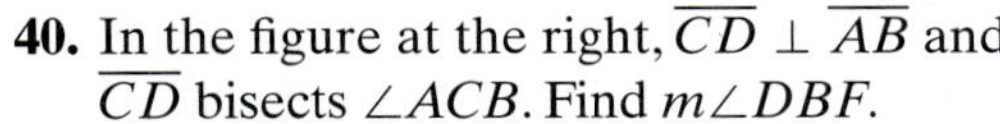
40. In the figure at the right, $\overline{CD} \perp \overline{AB}$ and $\overline{CD}$ bisects $\angle ACB$. Find $m\angle DBF$.

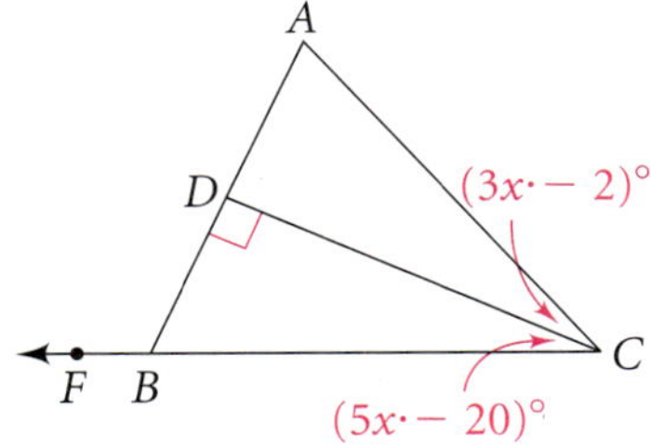

41. What can you conclude about the bisector of an exterior angle of a triangle if the remote interior angles are congruent? Justify your response.

Multiple Choice Practice

For California Standards and CAHSEE Tutorials, visit PHSchool.com. Web Code: bcq-9045

GEOM 12.0

42. In the figure at the right, $m\angle M = 5x$, $m\angle L = 9x$, and $m\angle MKL = 12x - 2$. What is $m\angle JKM$?

Ⓐ 72 Ⓑ 98 Ⓒ 120 Ⓓ 132

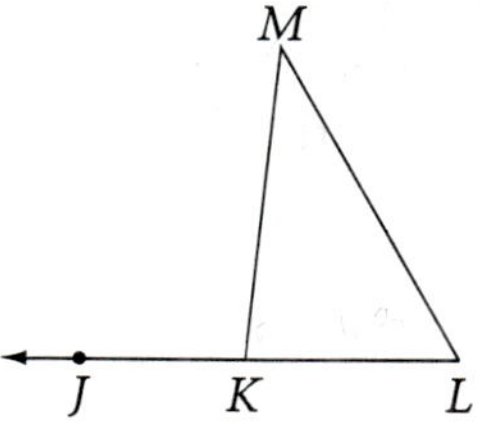

CAHSEE 6SDAP 3.3

43. Robin placed 8 red marbles, 10 blue marbles, and 7 white marbles in a bag. He randomly took a marble out. What is the theoretical probability that it will be white?

Ⓐ 14% Ⓑ 28% Ⓒ 49% Ⓓ 70%

Mixed Review

Lesson 3-3

Use the diagram at the right for Exercises 44–45.

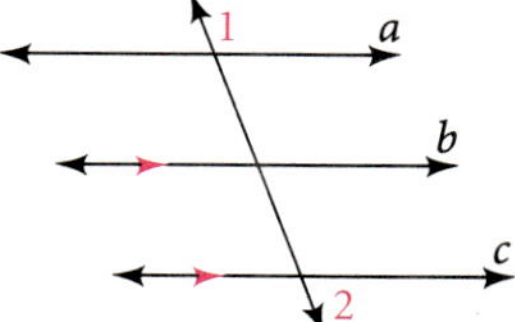

44. If $\angle 1$ and $\angle 2$ are supplementary, what can you conclude about lines a and c? Justify your answer.

45. If $a \parallel c$, what can you conclude about lines a and b? Justify your answer.

Lesson 1-6 x^2

46. **Algebra** In the figure at the right, $m\angle AOB = 3x + 20$, $m\angle BOC = x + 32$, and $m\angle AOC = 80$. Find the value of x.

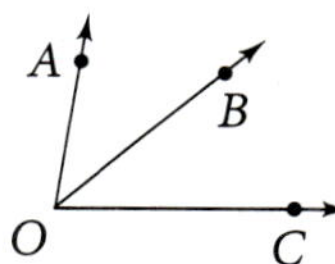

Lesson 1-1

Draw the next figure in each sequence.

47.  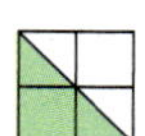 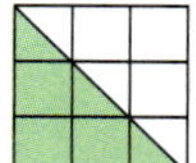

48.  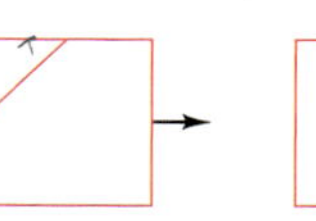 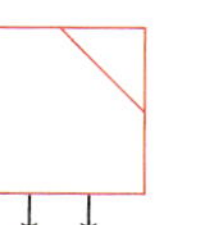 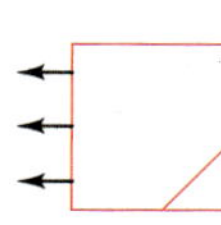

Checkpoint Quiz 1

Lessons 3-1 through 3-4

Use the diagram at the right for Exercises 1–9. State the theorem or postulate that justifies each statement.

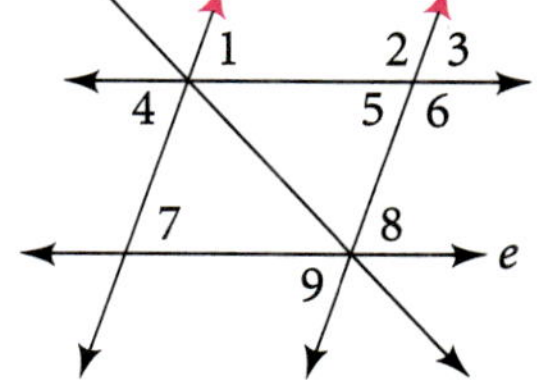

1. $\angle 1 \cong \angle 3$

2. If $\angle 5 \cong \angle 9$, then $d \parallel e$.

3. $m\angle 1 + m\angle 2 = 180$

4. If $\angle 4 \cong \angle 7$, then $d \parallel e$.

5. $\angle 1 \cong \angle 4$

6. $\angle 7 \cong \angle 9$

7. If $\angle 3 \cong \angle 9$, then $d \parallel e$.

8. $\angle 4 \cong \angle 5$

9. If $e \perp b$, then $e \perp c$.

10. Find the measures of the angles of each triangle. Classify each triangle by its angles.

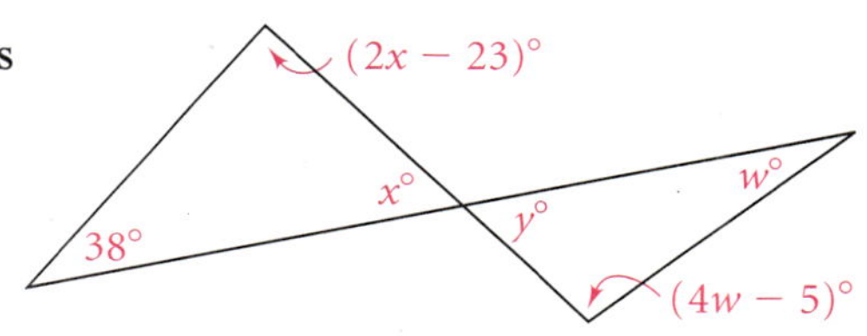

Exploring Spherical Geometry

FOR USE WITH LESSON 3-4

GEOM 1.0 Students demonstrate understanding by identifying and giving examples of undefined terms, axioms, theorems, and inductive and deductive reasoning. ***Master***

Euclidean geometry is the basis for high school geometry courses. Euclidean geometry is the geometry of flat planes, straight lines, and points. In spherical geometry a "plane" is the curved surface of a sphere and a "line" is a great circle. (A *great circle* is the intersection of a sphere and a plane that contains the center of the sphere.)

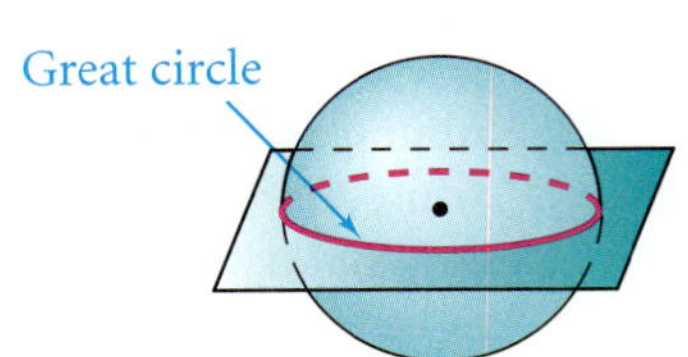

1 ACTIVITY

Lines of latitude and longitude are used to identify positions on Earth. Find these lines on a globe. Which of these lines are great circles?

All lines of longitude are great circles. The equator is the only line of latitude that is a great circle. All other lines of latitude are circles smaller than a great circle.

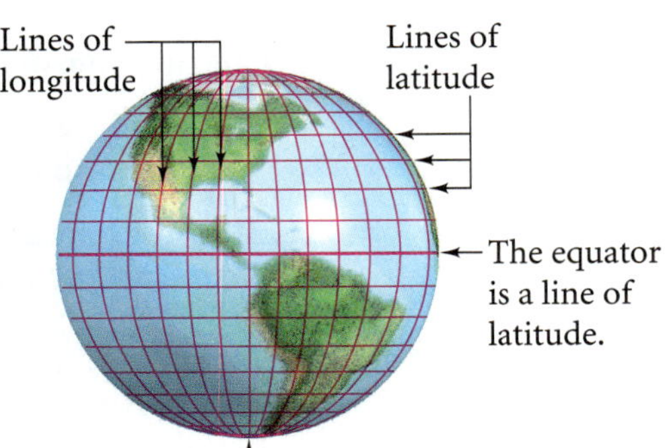

The lines of longitude all pass through the North and South Poles.

In Euclidean geometry,

Through a point not on a line, there is one and only one line parallel to the given line.

This statement is sometimes called Euclid's Parallel Postulate. Since only great circles are lines in spherical geometry, two lines always intersect. In spherical geometry, the Parallel Postulate is quite different:

Through a point not on a line, there is no line parallel to the given line.

2 ACTIVITY

The diagram at the right shows that any two lines on a sphere intersect at *two* points. Locate the points of intersection of lines of longitude on Earth. What is special about these points?

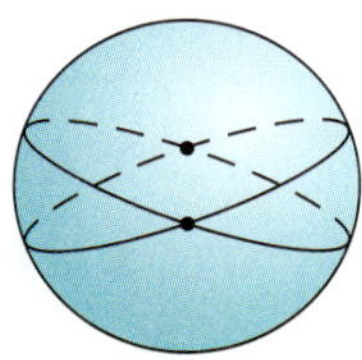

Lines of longitude intersect at the North and South Poles. The poles are on a line that passes through the center of Earth. Thus, the poles lie on Earth's axis and are the endpoints of a diameter of Earth.

One result of Euclid's Parallel Postulate is the Triangle Angle-Sum Theorem of Lesson 3-4. Something quite different happens in spherical geometry as a result of the spherical-geometry Parallel Postulate.

3 ACTIVITY

Hold a string taut between any two points on a sphere. The string forms an arc that is part of a great circle. Make three such arcs to form a triangle on the sphere.

Three such triangles are shown below. Find the sum of the measures of the angles of each of the triangles.

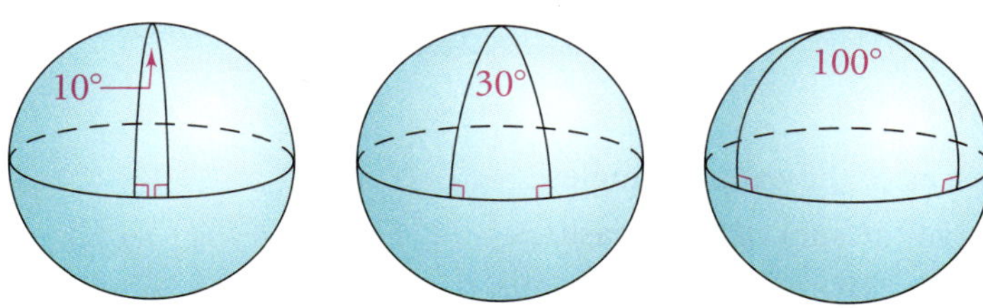

In the first triangle, the sum of the angle measures is 190. In the second triangle, it is 210, and in the third triangle the sum is 280.

EXERCISES

Draw a sketch to illustrate each property of spherical geometry. How does each property compare to what is true in Euclidean geometry?

1. There are pairs of points on a sphere through which more than one line can be drawn.
2. A triangle can have more than one right angle.
3. You can draw two equiangular triangles such that they have different angle measures.

In Exercises 4 and 5, draw a counterexample to show that each of these properties of Euclidean geometry is *not* true in spherical geometry.

4. Two lines that are perpendicular to the same line do not intersect.
5. If two angles of one triangle are congruent to two angles of another triangle, then the third angles are congruent.

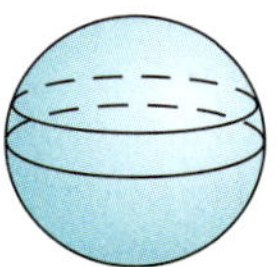

6. The figure at the right appears to show parallel lines on a sphere. Explain why this is not so.
7. Explain why a piece of the top circle in the figure is *not* a line segment. (*Hint:* What must be true of line segments in spherical geometry?)

Each of the following statements is true in Euclidean geometry. Does it seem to be true in spherical geometry? Make figures on a globe, ball, or balloon to support your answer.

8. Vertical angles are congruent.
9. Through a point on a line ℓ there exists one and only one line perpendicular to ℓ.

Exterior Angles of Polygons

Technology

FOR USE WITH LESSON 3-5

GEOM 13.0 Students prove relationships between angles in polygons by using properties of complementary, supplementary, vertical, and exterior angles. *Develop*

Construct

Use geometry software. Construct a polygon similar to the one at the right. Extend each side as shown. To measure the exterior angles you will need to mark a point on each ray.

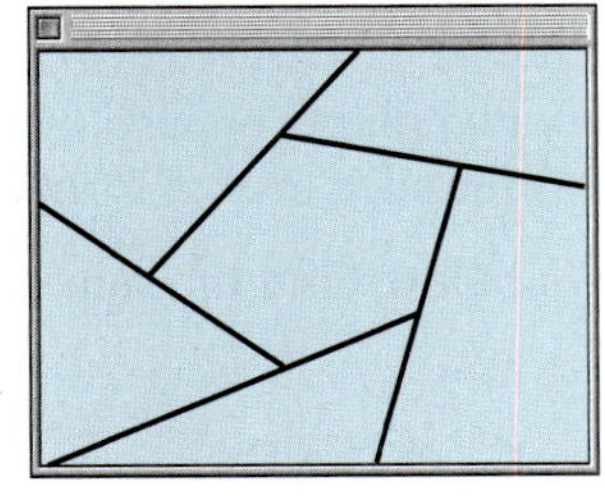

Investigate

- Measure each exterior angle.
- Calculate the sum of the measures of the exterior angles.
- Manipulate the polygon. Observe the sum of the measures of the exterior angles.

EXERCISES

1. Write a conjecture about the sum of the measures of the exterior angles (one at each vertex) of a convex polygon.
2. Test your conjecture with another polygon.

Extend

3. The figures below show a polygon that is decreasing in size until finally it becomes a point. Describe how you could use this to justify your conjecture in Exercise 1.

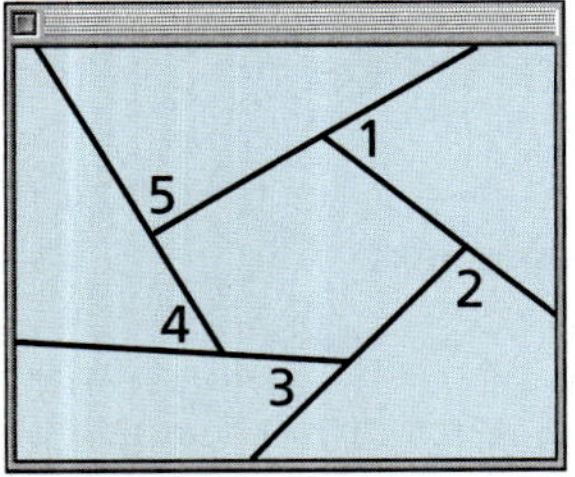

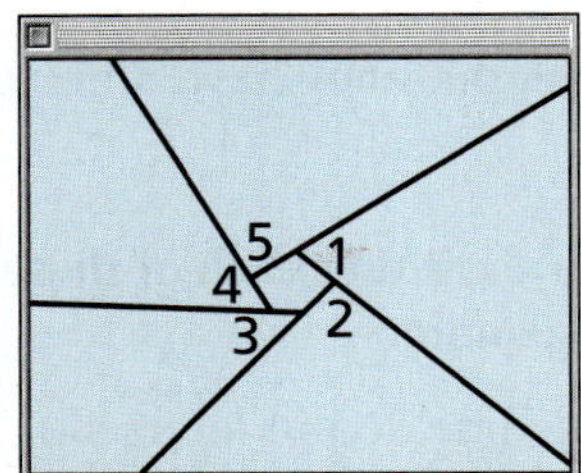

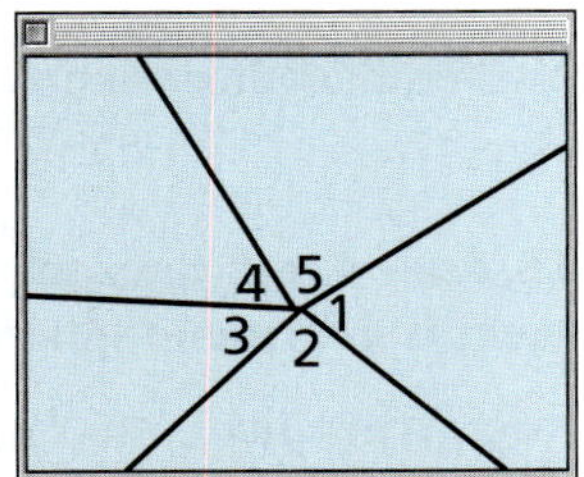

4. The figure at the right shows a square that has been copied several times. Notice that you can use the square to completely cover, or tile, a plane, without gaps or overlaps.
   a. Using geometry software, create several copies of other regular polygons with 3, 5, 6, and 8 sides. Regular polygons have sides of equal length and angles of equal measure.
   b. Which of the polygons you created can tile a plane?
   c. Measure *one* exterior angle of each polygon (including the square).
   d. Write a conjecture about the relationship between the measure of an exterior angle and your ability to tile a plane with a regular polygon.
   e. Test your conjecture with another regular polygon.
5. You can use a square—a rectangle even—to tile a plane. What other four-sided shapes can you use to tile a plane? (*Hint*: See Activity Lab, page 146.)

The Polygon Angle-Sum Theorems

California Content Standards

GEOM 12.0 Students find and use measures of interior and exterior angles of triangles and polygons to classify figures and solve problems. ***Develop***

GEOM 13.0 Students prove relationships between angles in polygons by using properties of exterior angles. ***Develop, Master***

What You'll Learn

- To classify polygons
- To find the sums of the measures of the interior and exterior angles of polygons

. . . And Why

To find the measure of an angle of a triangle used in packaging, as in Example 5

Check Skills You'll Need

GO for Help Lessons 1-6 and 3-4

Find the measure of each angle of quadrilateral *ABCD*.

1.

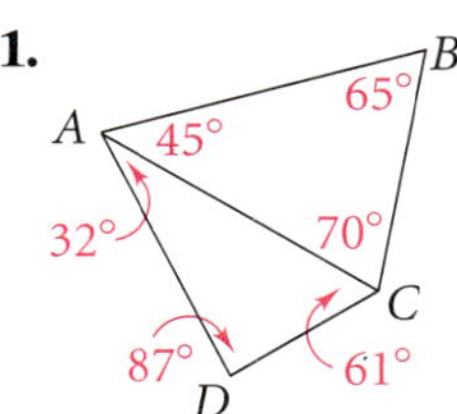

2.

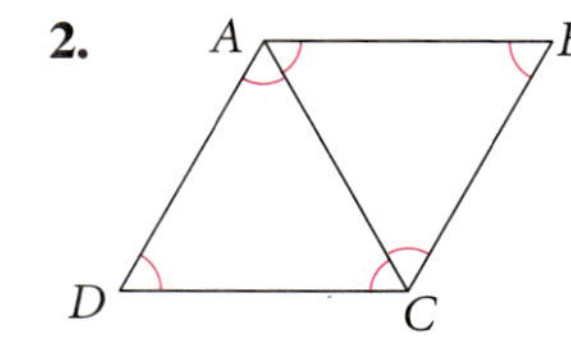

3.

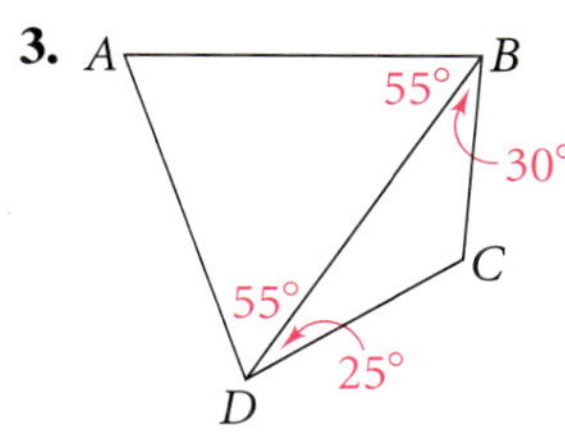

New Vocabulary

- polygon
- convex polygon
- concave polygon
- equilateral polygon
- equiangular polygon
- regular polygon

1 Classifying Polygons

Real-World Connection

Polygons create striking designs on a soccer ball.

A **polygon** is a closed plane figure with at least three sides that are segments. The sides intersect only at their endpoints, and no adjacent sides are collinear.

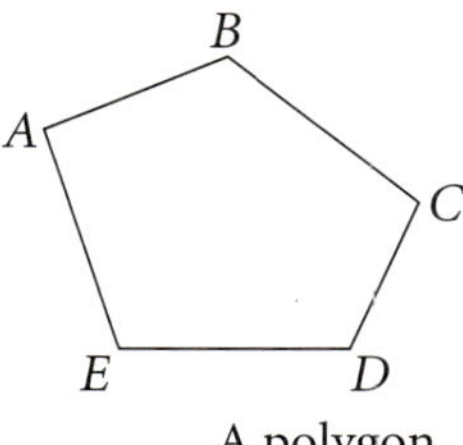

A polygon

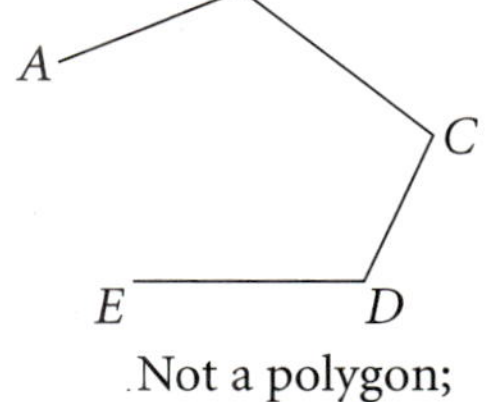

Not a polygon; not a closed figure

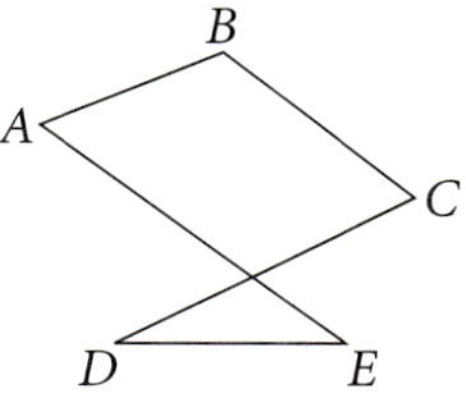

Not a polygon; two sides intersect between endpoints.

To name a polygon, start at any vertex and list the vertices consecutively in a clockwise or counterclockwise direction.

1 EXAMPLE Naming Polygons

Name the polygon. Then identify its vertices, sides, and angles.

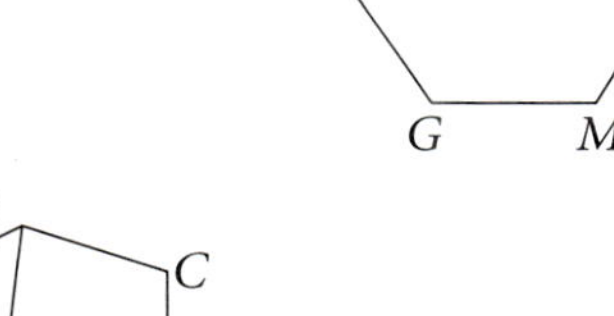

Two names for this polygon are *DHKMGB* and *MKHDBG*.

vertices: D, H, K, M, G, B

sides: $\overline{DH}, \overline{HK}, \overline{KM}, \overline{MG}, \overline{GB}, \overline{BD}$

angles: $\angle D, \angle H, \angle K, \angle M, \angle G, \angle B$

1 Three polygons are pictured at the right. Name each polygon, its sides, and its angles.

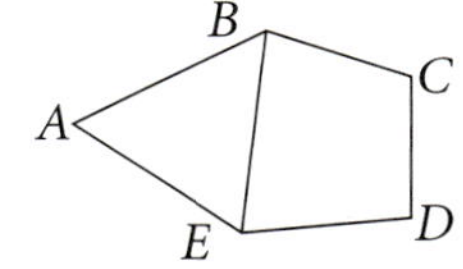

You can classify a polygon by the number of sides it has. The table at the right shows the names of some common polygons.

Sides	Name
3	triangle
4	quadrilateral
5	pentagon
6	hexagon
8	octagon
9	nonagon
10	decagon
12	dodecagon
n	n-gon

Polygons are classified as convex or concave.

Vocabulary Tip

A diagonal of a polygon is a segment that connects two nonconsecutive vertices.

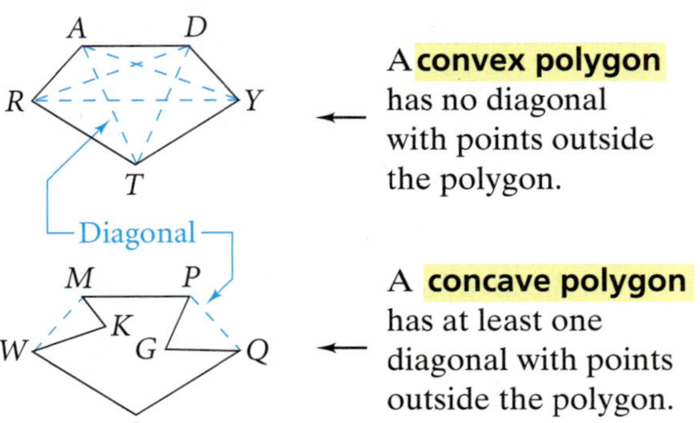

A **convex polygon** has no diagonal with points outside the polygon.

A **concave polygon** has at least one diagonal with points outside the polygon.

In this textbook, a polygon is convex unless stated otherwise.

2 EXAMPLE Real-World Connection

Tilework The tilework in the photo is a combination of different polygons that form a pleasing pattern. Classify the polygon outlined in red by using the table above. Then classify the polygon as convex or concave.

The polygon outlined in red has 6 sides. Therefore, it is a hexagon.

No diagonal of the hexagon contains points outside the hexagon. The hexagon is convex.

CA Standards Check

2 Classify each polygon by its sides. Identify each as convex or concave.

a.

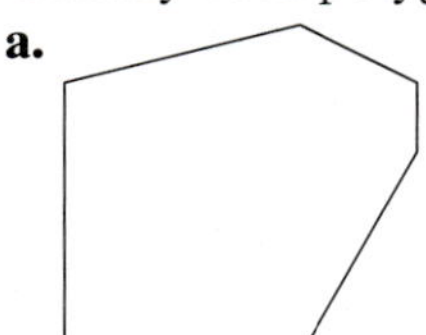

b.

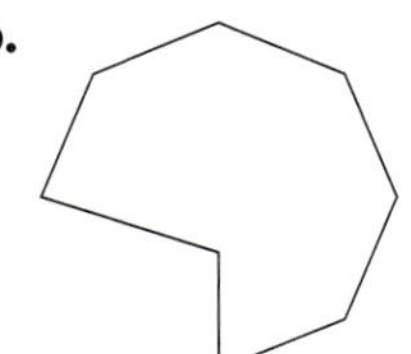

c. the 12-pointed star at the center of the tilework pictured above

Polygon Angle Sums

Activity: The Sum of Polygon Angle Measures

You can use triangles and the Triangle Angle-Sum Theorem to find the sum of the measures of the angles of a polygon. Record your data in a table like the one begun below.

Polygon	Number of Sides	Number of Triangles Formed	Sum of the Interior Angle Measures
▱	4	■	■ • 180 = ■

- Sketch polygons with 4, 5, 6, 7, and 8 sides.
- Divide each polygon into triangles by drawing all diagonals that are possible from one vertex.
- Multiply the number of triangles by 180 to find the sum of the measures of the angles of each polygon.

1. Look for patterns in the table. Describe any that you find.
2. **Inductive Reasoning** Write a rule for the sum of the measures of the angles of an n-gon.

Vocabulary Tip

An n-gon is a polygon with n sides, where n can be 3, 4, 5, 6, . . .

By dividing a polygon with n sides into $n - 2$ triangles, you can show that the sum of the measures of the angles of any polygon is a multiple of 180.

Key Concepts

Theorem 3-14 Polygon Angle-Sum Theorem

The sum of the measures of the angles of an n-gon is $(n - 2)180$.

Video Tutor Help
Visit: PHSchool.com
Web Code: bce-0775

3 EXAMPLE Finding a Polygon Angle Sum

Find the sum of the measures of the angles of a 15-gon.

For a 15-gon, $n = 15$.

$$\begin{aligned} \text{Sum} &= (n - 2)180 && \textbf{Polygon Angle-Sum Theorem} \\ &= (15 - 2)180 && \textbf{Substitute.} \\ &= 13 \cdot 180 && \textbf{Simplify.} \\ &= 2340 \end{aligned}$$

The sum of the measures of the angles of a 15-gon is 2340.

3 **a.** Find the sum of the measures of the angles of a 13-gon.

b. **Critical Thinking** The sum of the measures of the angles of a given polygon is 720. How can you use Sum $= (n - 2)180$ to find the number of sides in the polygon?

You will sometimes use algebra with the Polygon Angle-Sum Theorem to find measures of polygon angles.

4 EXAMPLE Using the Polygon Angle-Sum Theorem

Algebra Find $m\angle Y$ in pentagon $TVYMR$ at the right.
Use the Polygon Angle-Sum Theorem for $n = 5$.

$$m\angle T + m\angle V + m\angle Y + m\angle M + m\angle R = (5 - 2)180$$

$90 + 90 + m\angle Y + 90 + 135 = 540$ **Substitute.**

$m\angle Y + 405 = 540$ **Simplify.**

$m\angle Y = 135$ **Subtract 405 from each side.**

CA Standards Check 4 Pentagon $ABCDE$ has 5 congruent angles. Find the measure of each angle.

You can draw exterior angles at any vertex of a polygon. The figures below show that the sum of the measures of the exterior angles, one at each vertex, is 360. This can be proved as a theorem in a way suggested in Exercise 46.

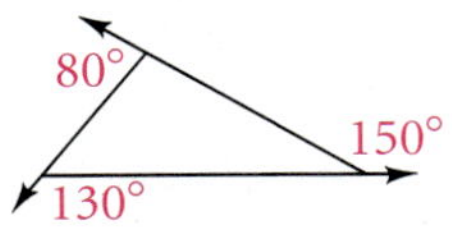

$80 + 150 + 130 = 360$

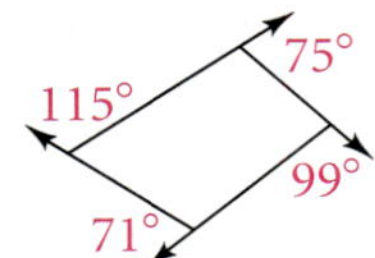

$115 + 75 + 99 + 71 = 360$

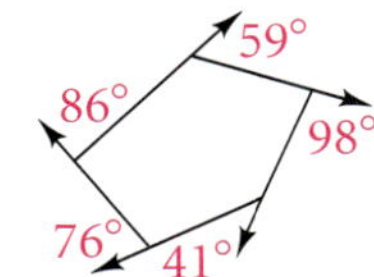

$86 + 59 + 98 + 41 + 76 = 360$

Key Concepts

Theorem 3-15 Polygon Exterior Angle-Sum Theorem

The sum of the measures of the exterior angles of a polygon, one at each vertex, is 360.

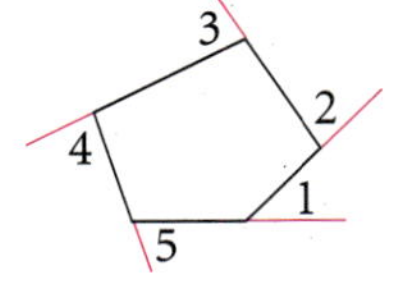

For the pentagon,
$m\angle 1 + m\angle 2 + m\angle 3 + m\angle 4 + m\angle 5 = 360.$

An **equilateral polygon** has all sides congruent. An **equiangular polygon** has all angles congruent. A **regular polygon** is both equilateral and equiangular.

5 EXAMPLE Real-World Connection

Packaging The game board at the right has the shape of a regular hexagon. It is packaged in a rectangular box outlined beneath it. The box uses four right triangles made of foam in its four corners. Find $m\angle 1$ in each foam triangle.

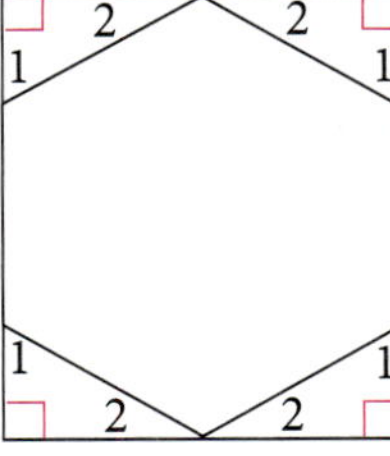

Find the measure of an angle of the hexagon first.

- A regular hexagon has 6 sides and 6 congruent angles. The sum of the measures of the interior angles $= (6 - 2)180$, or 720.
- The measure of one interior angle is $\frac{720}{6}$, or 120.
- The measure of its adjacent exterior angle, $\angle 1$, is $180 - 120$, or 60.

Online active math

For: Regular Polygon Activity
Use: Interactive Textbook, 3-5

CA Standards Check 5 **a.** Find $m\angle 1$ by using the Polygon Exterior Angle-Sum Theorem.
b. Find $m\angle 2$. Is $\angle 2$ an exterior angle? Explain.

For more exercises, see *Extra Skill, Word Problem, and Proof Practice.*

GEOM 12.0, 13.0

A Practice by Example

Example 1 (page 157)

for Help

Is the figure a polygon? If not, tell why.

1. 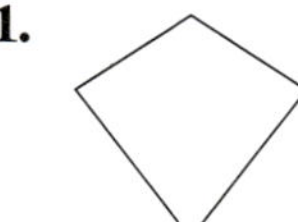

2. 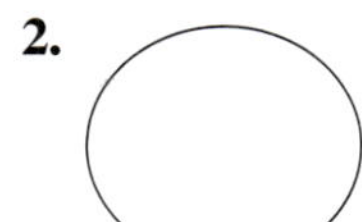

3. 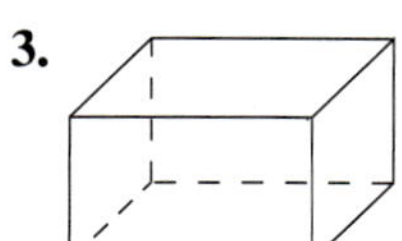

4. 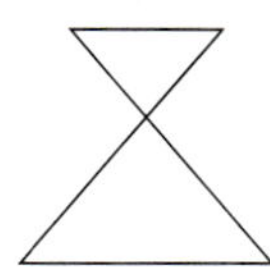

Name each polygon by its vertices. Then identify its sides and angles.

5. 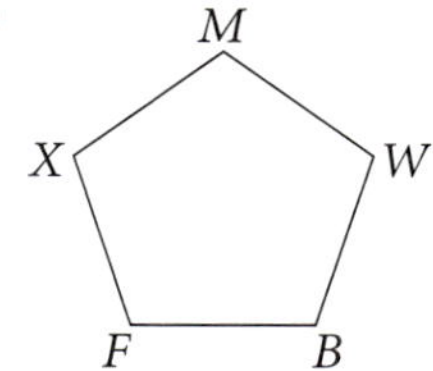

6. 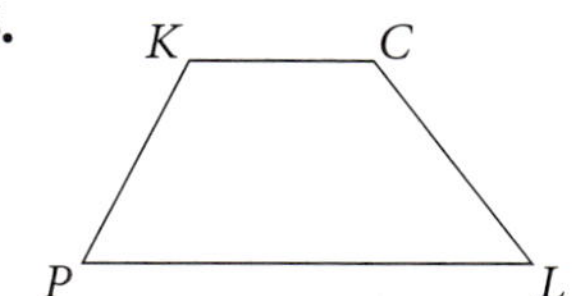

7. 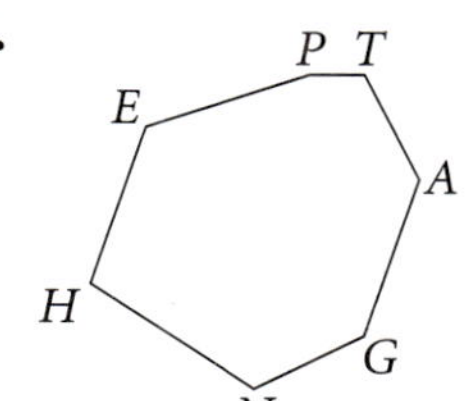

Example 2 (page 158)

Find a polygon in each photograph. Classify the polygon by its number of sides. Tell whether the polygon is convex or concave.

8. 

9. 

10. 

Example 3 (page 159)

Find the sum of the measures of the angles of each polygon.

11. 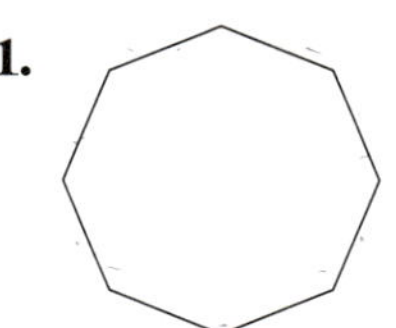

12. dodecagon

13. decagon

14. 20-gon

15. 1002-gon

Example 4 (page 160)

x^2 **Algebra** **Find the missing angle measures.**

16. 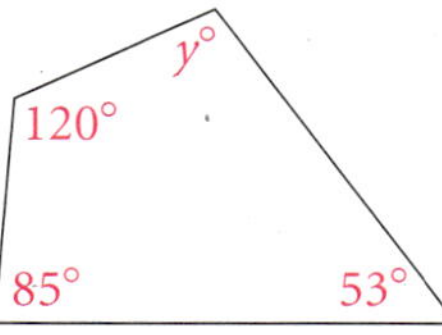

17. 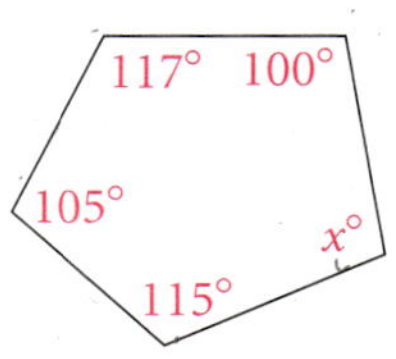

18. 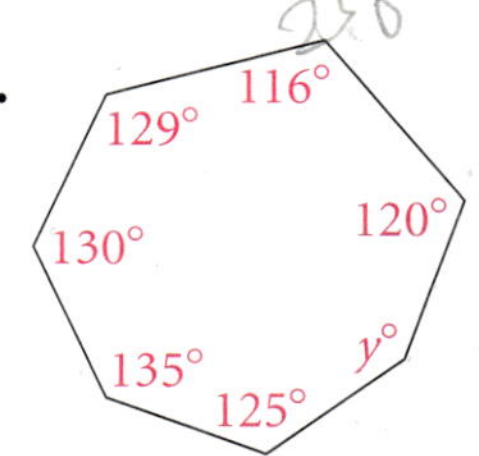

19. 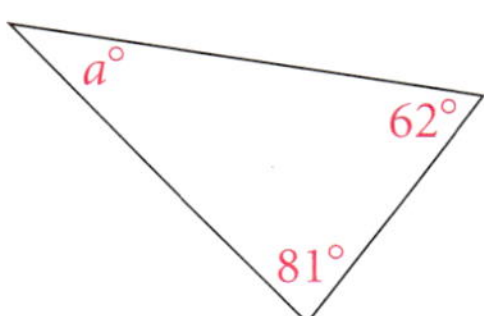

20. 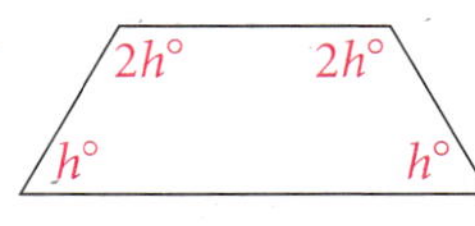

21. 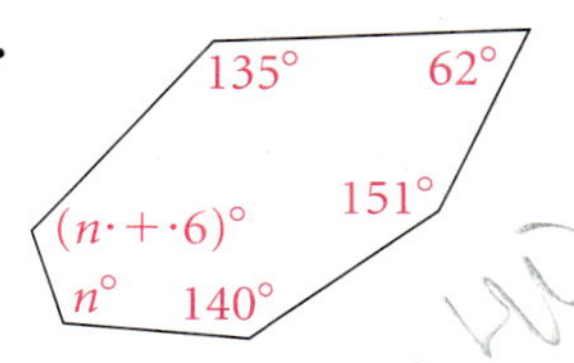

Example 5 (page 160)

Find the measures of an interior angle and an exterior angle of each regular polygon.

22. pentagon

23. dodecagon

24. 18-gon

25. 100-gon

Packaging **The nut container at the right has the shape of a regular octagon. It fits in a square box. A cheese wedge fills each corner of the box.**

26. Find the measure of each angle of a cheese wedge.

27. **Critical Thinking** Show how to rearrange the four pieces of cheese to make a regular polygon. What is the measure of each angle of the polygon?

B Apply Your Skills

Use a protractor. Sketch each type of regular polygon.

Sample: dodecagon

Use the protractor to equally space 12 points around a circle. ($360° \div 12 = 30°$, so mark a point every 30°.) Connect these points to form a regular dodecagon.

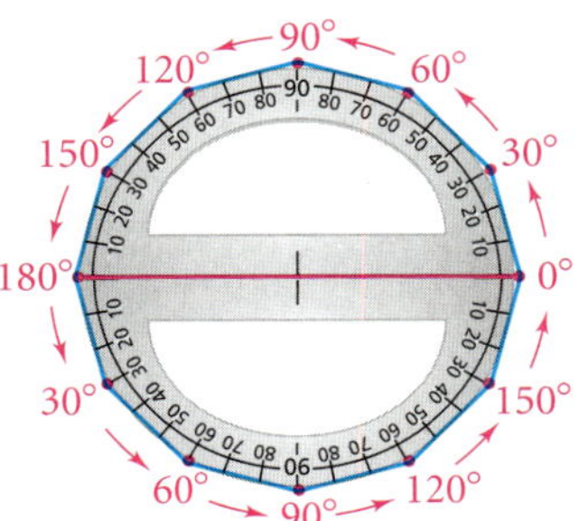

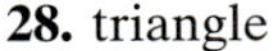

28. triangle **29.** quadrilateral

30. hexagon **31.** octagon

The sum of the measures of the angles of a polygon with *n* sides is given. Find *n*.

32. 180 **33.** 1080 **34.** 1980 **35.** 2880

36. **Multiple Choice** In the figure at right, $\angle B \cong \angle D$ and $m\angle C = 130$. What is $m\angle B$?

Ⓐ 25 Ⓒ 115
Ⓑ 50 Ⓓ 160

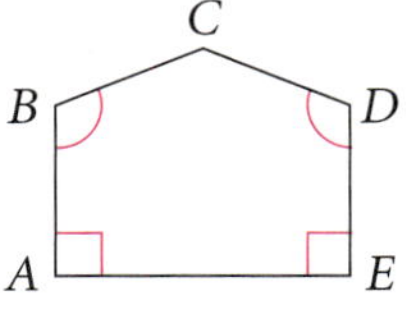

Real-World Connection

At a theater-in-the-round, seats are arranged so the audience surrounds the stage.

37. **Stage Design** The diagram at the right shows platforms constructed for a theater-in-the-round stage. Describe the largest platform by the type of regular polygon it suggests. Find the measure of each numbered angle.

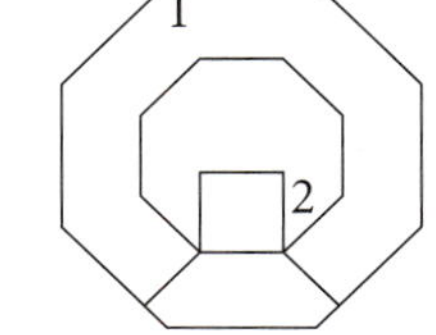

38. **Error Analysis** Miles said that he measured an angle of a regular polygon to be 130°. Explain why this result is impossible.

39. **Critical Thinking** A triangle has two congruent angles and an exterior angle with measure 100. Find two possible sets of measures for the angles of the triangle.

The measure of an exterior angle of a regular polygon is given. Find the measure of an interior angle, and find the number of sides.

40. 72 **41.** 36 **42.** 18 **43.** 30 **44.** x

45. **Probability** Find the probability that the measure of an angle of a regular n-gon is a positive integer if n is an integer and $3 \leq n \leq 12$.

46. **Algebra** A polygon has n sides. An interior angle of the polygon and an adjacent exterior angle form a straight angle.

a. What is the sum of the measures of the n straight angles?
b. What is the sum of the measures of the n interior angles?
c. Using your answers above, what is the sum of the measures of the n exterior angles?
d. What theorem do the steps above lead to?

Homework Video Tutor

Visit: PHSchool.com
Web Code: bce-0305

x^2 **Algebra** **Find each missing angle measure. Then name the polygon.**

47. 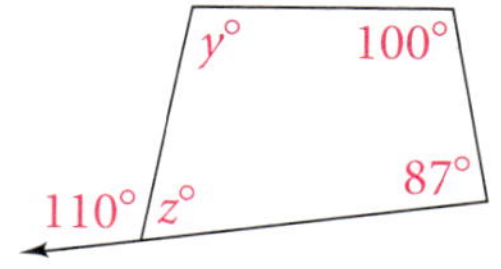

48. 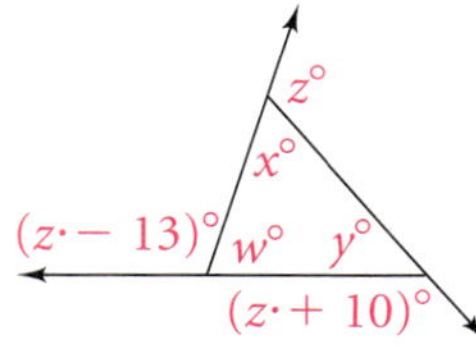

49. 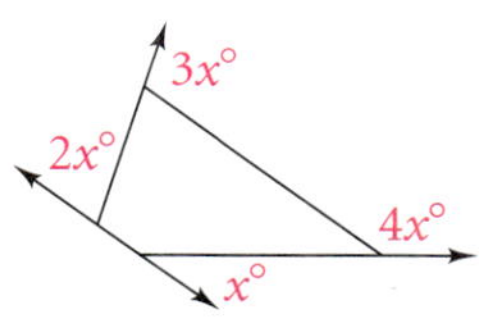

Open-Ended **Sketch each figure described in Exercises 50–53.**

Problem Solving Hint

In Exercises 51-53, sketch a figure to meet the first condition. Then adjust it to meet the second condition.

50. a quadrilateral that is not equiangular

51. an equiangular quadrilateral that is not regular

52. an equilateral polygon that is not equiangular

53. an equiangular polygon that is not equilateral

54. **Critical Thinking** Ellen says she has another way to find the sum of the measures of the angles of a polygon. She picks a point inside the polygon, draws a segment to each vertex, counts the number of triangles, multiplies by 180, and then subtracts 360. Does her method work? Explain.

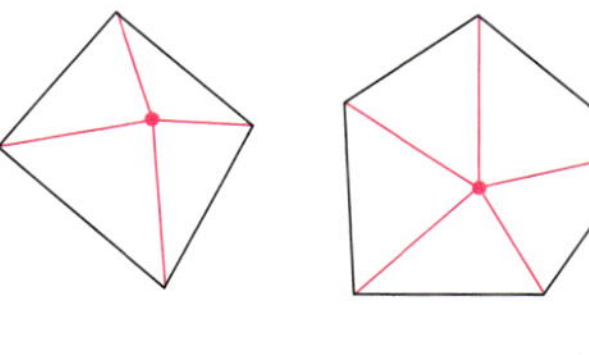

55. **Writing** Tell what you know about the figure at the right.

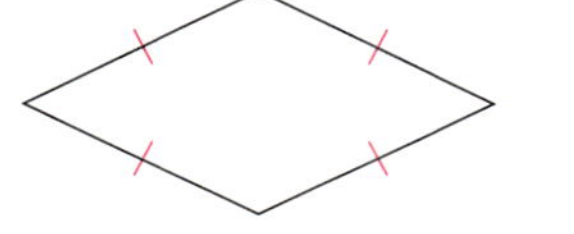

56. The measure of an interior angle of a regular polygon is three times the measure of an exterior angle of the same polygon. What is the name of the polygon?

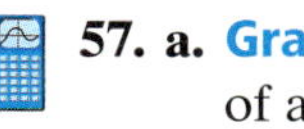

57. a. **Graphing Calculator** Find the measure of an angle of a regular n-gon for $n = 20, 40, 60, 80, \ldots, 200$. Record your results to the nearest tenth as ordered pairs in the form (n, measure of each angle).

b. Plot the ordered pairs using a window like the one shown at the right.

c. **Data Analysis** Based on the graph from part (b), make a statement about the measure of an angle of a regular 1000-gon.

d. Is there a regular n-gon with an angle of 180°? Explain.

Xmin = 0 Ymin = 160
Xmax = 200 Ymax = 184
Xscl = 20 Yscl = 4

To graph the ordered pairs, use STAT and **STAT PLOT** on your graphing calculator.

58. a. Explain why the measure of an angle of a regular n-gon is given by the formulas $\frac{180(n-2)}{n}$ and $180 - \frac{360}{n}$.

b. Use the second formula to explain what happens to the measures in the angles of regular n-gons as n becomes a large number. Explain also what happens to the polygons.

For: Graphing calculator procedures
Web Code: bcb-2113

59. Two rays bisect two consecutive angles of a regular decagon and intersect in the decagon's interior. Find the measure of the acute angles formed by the intersecting rays.

Draw, if possible, the concave quadrilateral described. If not possible, explain.

60. with two pairs of congruent adjacent sides

61. with two pairs of congruent opposite sides

62. with three congruent sides

63. with four congruent sides

Multiple Choice Practice

For California Standards and CAHSEE Tutorials, visit PHSchool.com. Web Code: bcq-9045

GEOM 12.0

64. A company is manufacturing a gear that has the shape of a regular polygon. The measure of each angle of the gear is 162°. How many sides does the polygon have?

Ⓐ 18 Ⓑ 20 Ⓒ 22 Ⓓ 24

GEOM 13.0

65. Exactly three angles of a pentagon are congruent. The other two angles are complementary. What is the measure of one of the three congruent angles?

Ⓐ 150 Ⓑ 135 Ⓒ 120 Ⓓ 108

CAHSEE 6SDAP 2.5

66. The number of shoppers was recorded for each of the first seven days after a new shopping mall opened. Based on the data shown at the right, which of the following is a valid conclusion?

Mall Attendance

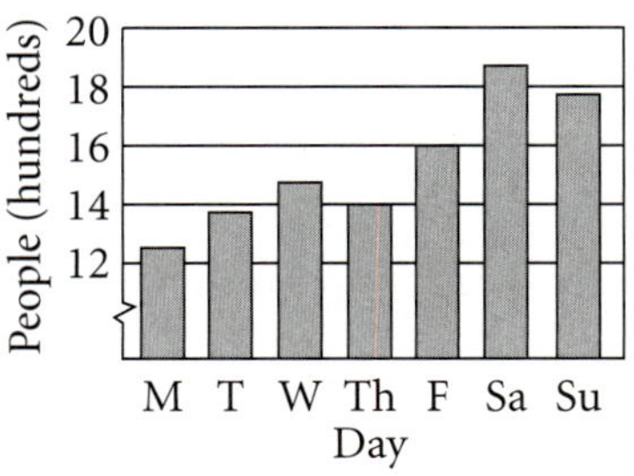

People (hundreds): 12, 14, 16, 18, 20

Day: M T W Th F Sa Su

Ⓐ The attendance increased steadily each of the first seven days.

Ⓑ About twice as many people attended on Sunday as on Monday.

Ⓒ The attendance will likely decrease after the first week.

Ⓓ Attendance was greater for each weekend day than for each weekday.

Mixed Review

Lesson 3-4 x^2 **Find each missing angle measure.**

67.

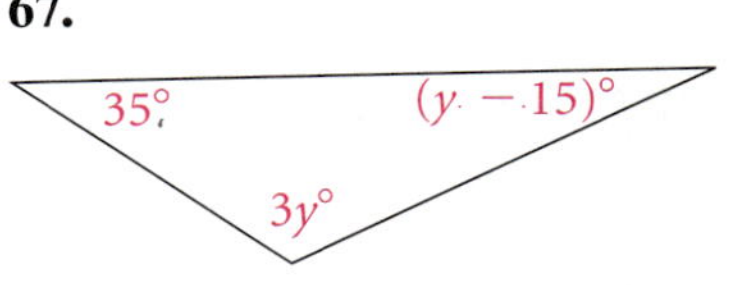

68.

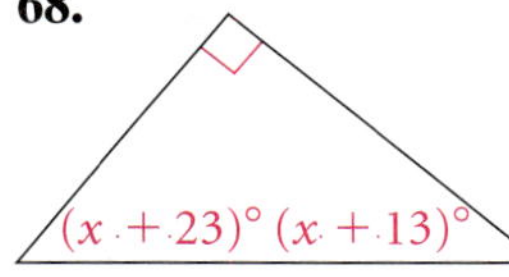

69.

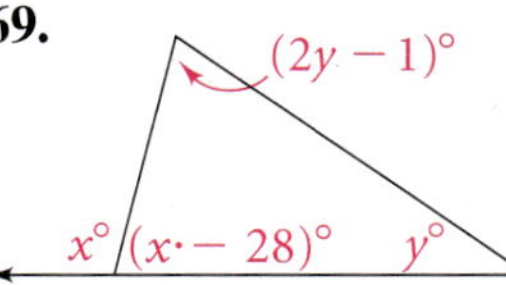

Lesson 2-4 **Name the property that justifies each statement.**

70. $4(2a - 3) = 8a - 12$

71. If $b + c = 7$ and $b = 2$, then $2 + c = 7$.

72. $\overline{RS} \cong \overline{RS}$

73. If $\angle 1 \cong \angle 4$, then $\angle 4 \cong \angle 1$.

74. If $2r = 18$, then $r = 9$.

75. If $AB = BC$ and $BC = 1$, then $AB = 1$.

Lessons 1-4, 1-6 **Identify the following in the diagram.**

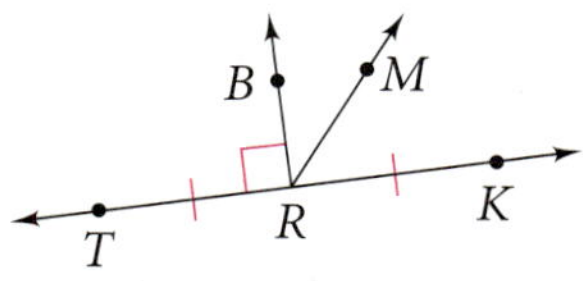

76. a pair of opposite rays

77. two right angles

78. two segments

79. an acute angle

80. an obtuse angle

81. a straight angle

82. a midpoint

Algebra 1 Review

Slope

For Use With Lesson 3-6

Prepares for GEOM 17.0 Students prove theorems by using coordinate geometry, including various forms of equations of lines.

The *slope* of a line is the ratio of the vertical change (rise) to the horizontal change (run) between any two points (x_1, y_1) and (x_2, y_2) of the line.

$$\text{slope} = \frac{\text{vertical change}}{\text{horizontal change}} = \frac{\text{rise}}{\text{run}} = \frac{y_2 - y_1}{x_2 - x_1}$$

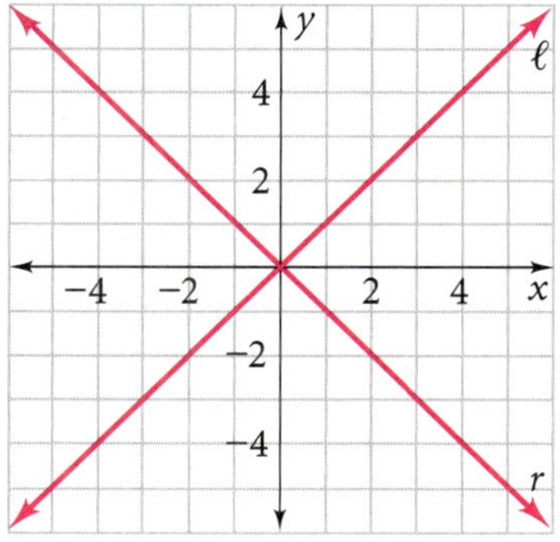

The slope of a line indicates the steepness of the line and whether it rises or falls from left to right. Line ℓ has slope 1, rising from left to right. Line r has slope -1, falling from left to right. Both form 45° angles with the x-axis.

1 EXAMPLE

Find the slope of $\overleftrightarrow{AB}$, which passes through $A(-2, 4)$ and $B(1, -3)$.

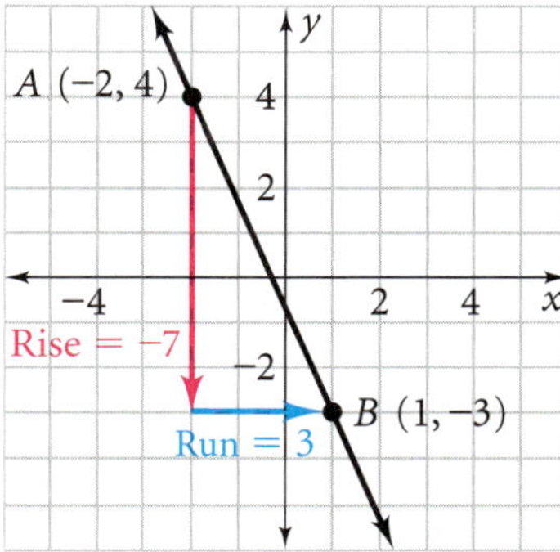

Method 1 Use the formula.

$$\text{slope} = \frac{y_2 - y_1}{x_2 - x_1} = \frac{-3 - 4}{1 - (-2)} = \frac{-7}{3}, \text{ or } -\frac{7}{3}$$

Method 2 Use the graph.

$$\text{slope} = \frac{\text{vertical change (rise)}}{\text{horizontal change (run)}} = \frac{-7}{3}, \text{ or } -\frac{7}{3}$$

2 EXAMPLE

Compare the rise or fall and the steepness of $\overleftrightarrow{AB}$ in Example 1 with that of line ℓ or r at the top of the page.

The slope of $\overleftrightarrow{AB}$ is negative, so $\overleftrightarrow{AB}$ falls from left to right, the same as line r with slope -1. The slope of $\overleftrightarrow{AB}$ has absolute value $\frac{7}{3}$, which is greater than 1, so $\overleftrightarrow{AB}$ is steeper than line r whose slope has absolute value 1.

EXERCISES

Find the slope of $\overleftrightarrow{AB}$. Compare its rise or fall and its steepness with that of line ℓ or r at the top of the page.

1. $A(4, -6), B(7, 2)$

2. $A(7, -6), B(-5, -8)$

3. $A(-3, 7), B(-1, 4)$

4. $A(-2, -5), B(1, -7)$

5. $A(0, 4), B(4, 0)$

6. $A\left(-3\frac{1}{2}, 3\right), B\left(-7, 2\frac{1}{2}\right)$

7. $A(-1.4, -3.7), B(-2.4, 1.3)$

8. $A(3, -2), B(-6, -2)$

9. $A(5, 9), B(5, -6)$

Open-Ended **Graph a line with the given slope m. Compare it with line ℓ or r above.**

10. $m = \frac{1}{3}$

11. $m = -1.7$

12. $m = 37$

13. $m = 0$

14. A horizontal line has slope 0. Why?

15. Slope is undefined for a vertical line. Why?

3-6 Lines in the Coordinate Plane

California Content Standards

Prepares for GEOM 17.0 Students prove theorems by using coordinate geometry, including various forms of equations of lines.

What You'll Learn

- To graph lines given their equations
- To write equations of lines

. . . And Why

To determine whether a wheelchair ramp complies with the law, as in Exercise 52

Check Skills You'll Need GO for Help page 165

Find the slope of the line that contains each pair of points.

1. $A(-2, 2), B(4, -2)$

2. $P(3, 0), X(0, -5)$

3. $R(-3, -4), S(5, -4)$

4. $K(-3, 3), T(-3, 1)$

5. $C(0, 1), D(3, 3)$

6. $E(-1, 4), F(3, -2)$

7. $G(-8, -9), H(-3, -5)$

8. $L(7, -10), M(1, -4)$

New Vocabulary
- slope-intercept form
- standard form of a linear equation
- point-slope form

1 Graphing Lines

Vocabulary Tip

The y-intercept is the y-coordinate of the point where a line crosses the y-axis. The x-intercept is the x-coordinate of the point where a line crosses the x-axis.

In algebra, you learned that the graph of a linear equation is a line. The **slope-intercept form** of a linear equation is $y = mx + b$, where m is the slope of the line and b is the y-intercept. Each line at the right has slope 2, but the lines have y-intercepts of 3, -1, and -4.

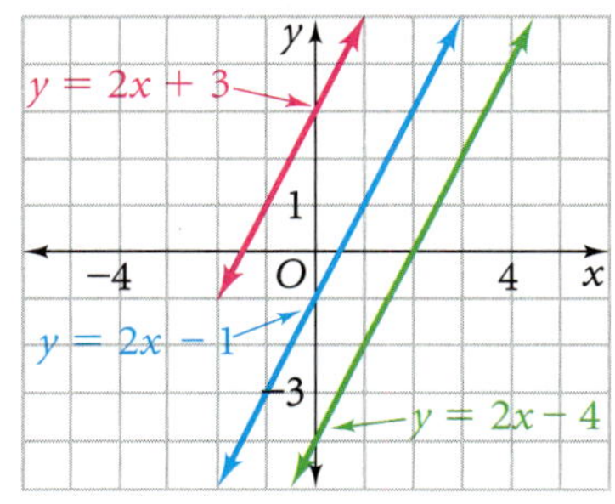

By Postulate 1-1 (two points determine a line), you need only two points to graph a line. The y-intercept gives you one point. You can use the slope to plot another.

1 EXAMPLE Graphing Lines in Slope-Intercept Form

Graph the line $y = \frac{3}{4}x + 2$.

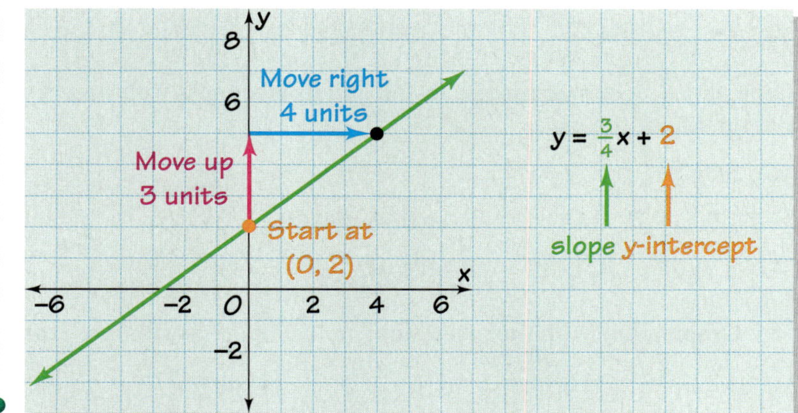

CA Standards Check **1** Graph the line $y = -\frac{1}{2}x - 2$.

Vocabulary Tip

In the standard form, A, B, and C are constants. In Example 2, $6x + 3y = 12$ is in standard form with $A = 6$, $B = 3$, and $C = 12$.

The **standard form of a linear equation** is $Ax + By = C$, where A, B, and C are real numbers and A and B are not both zero. To graph an equation written in standard form, you can readily find two points for the graph by finding the x- and y-intercepts.

2 EXAMPLE Graphing Lines Using Intercepts

Algebra Graph $6x + 3y = 12$.

Step 1 To find the y-intercept, substitute 0 for x; solve for y.

$$6x + 3y = 12$$
$$6(0) + 3y = 12$$
$$3y = 12$$
$$y = 4$$

The y-intercept is 4.
A point on the line is (0, 4).

Step 2 To find the x-intercept, substitute 0 for y; solve for x.

$$6x + 3y = 12$$
$$6x + 3(0) = 12$$
$$6x = 12$$
$$x = 2$$

The x-intercept is 2.
A point on the line is (2, 0).

Step 3 Plot (0, 4) and (2, 0). Draw the line containing the two points.

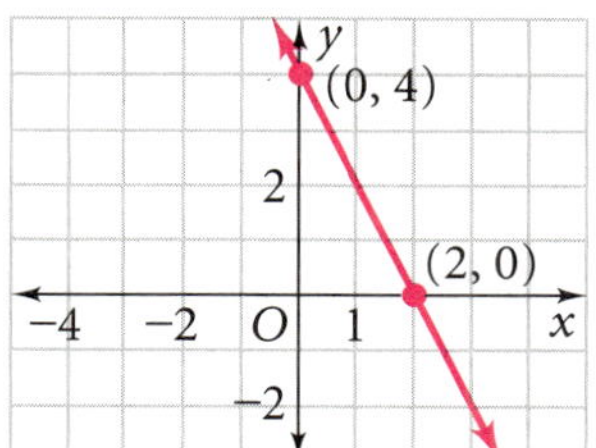

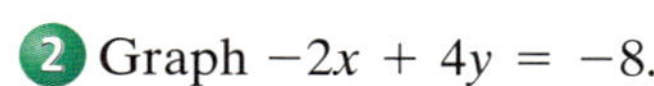
CA Standards Check 2 Graph $-2x + 4y = -8$.

As an alternative, you can graph an equation in standard form by transforming it into slope-intercept form. Knowing the slope and y-intercept beforehand can give you a good mental image of what the graph should look like.

3 EXAMPLE Transforming to Slope-Intercept Form

Algebra Graph $4x - 2y = 9$.

Step 1 Transform the equation to slope-intercept form.

$$4x - 2y = 9$$
$$-2y = -4x + 9$$
$$\frac{-2y}{-2} = \frac{-4x}{-2} + \frac{9}{-2}$$
$$y = 2x - \frac{9}{2}$$

The y-intercept is $-4\frac{1}{2}$ and the slope is 2.

Step 2 Use the y-intercept and the slope to plot two points and draw the line containing them.

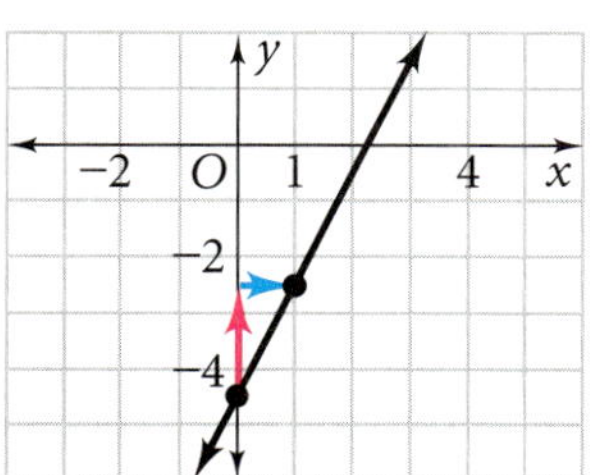

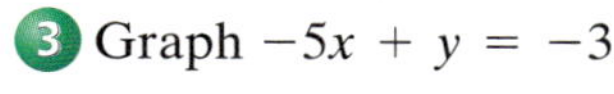
CA Standards Check 3 Graph $-5x + y = -3$.

2 Writing Equations of Lines

A third form for an equation of a line is **point-slope form.** The point-slope form for a nonvertical line through point (x_1, y_1) with slope m is $y - y_1 = m(x - x_1)$.

4 EXAMPLE Using Point-Slope Form

Algebra Write an equation of the line through point $P(-1, 4)$ with slope 3.

$y - y_1 = m(x - x_1)$ **Use point-slope form.**

$y - 4 = 3[x - (-1)]$ **Substitute 3 for m and $(-1, 4)$ for (x_1, y_1).**

$y - 4 = 3(x + 1)$ **Simplify.**

CA Standards Check 4 Write an equation of the line with slope -1 that contains point $P(2, -4)$.

By Postulate 1-1, you need only two points to write an equation of a line.

5 EXAMPLE Writing an Equation of a Line Given Two Points

Algebra Write an equation of the line through $A(-2, 3)$ and $B(1, -1)$.

Step 1 Find the slope.

$m = \frac{y_2 - y_1}{x_2 - x_1}$

$m = \frac{-1 - 3}{1 - (-2)}$ **Substitute $(-2, 3)$ for (x_1, y_1) and $(1, -1)$ for (x_2, y_2).**

$m = -\frac{4}{3}$ **Simplify.**

Step 2 Select one of the points. Write an equation in point-slope form.

$y - y_1 = m(x - x_1)$

$y - 3 = -\frac{4}{3}[x - (-2)]$ **Substitute $(-2, 3)$ for (x_1, y_1) and $-\frac{4}{3}$ for the slope.**

$y - 3 = -\frac{4}{3}(x + 2)$ **Simplify.**

CA Standards Check 5 Write an equation of the line that contains the points $P(5, 0)$ and $Q(7, -3)$.

Recall that the slope of a horizontal line is 0 and the slope of a vertical line is undefined. Thus, horizontal and vertical lines have easily recognized equations.

6 EXAMPLE Equations of Horizontal and Vertical Lines

Write equations for the horizontal line and the vertical line that contain $P(3, 2)$.

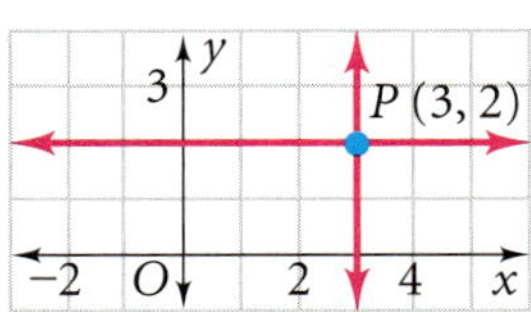

Every point on the horizontal line through $P(3, 2)$ has a y-coordinate of 2. The equation of the line is $y = 2$. It crosses the y-axis at $(0, 2)$.

Every point on the vertical line through $P(3, 2)$ has an x-coordinate of 3. The equation of the line is $x = 3$. It crosses the x-axis at $(3, 0)$.

CA Standards Check 6 Write equations of the horizontal and vertical lines that contain the point $P(5, -1)$.

EXERCISES

For more exercises, see *Extra Skill, Word Problem, and Proof Practice.*

Standards Practice

A Practice by Example

Examples 1, 2 (pages 166, 167)

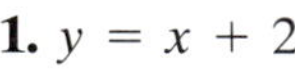

Algebra **Graph each line.**

1. $y = x + 2$
2. $y = 3x + 4$
3. $y = \frac{1}{2}x - 1$
4. $y = -\frac{5}{3}x + 2$

Algebra **Graph each line using intercepts.**

5. $2x + 6y = 12$
6. $3x + y = 15$
7. $5x - 2y = 20$
8. $6x - y = 3$
9. $10x + 5y = 40$
10. $1.2x + 2.4y = 2.4$

Example 3 (page 167)

Algebra **Write each equation in slope-intercept form and graph the line.**

11. $y = 2x + 1$
12. $y - 1 = x$
13. $y + 2x = 4$
14. $8x + 4y = 16$
15. $2x + 6y = 6$
16. $\frac{3}{4}x - \frac{1}{2}y = \frac{1}{8}$

Example 4 (page 168)

Algebra **Write an equation in point-slope form of the line that contains the given point and has the given slope.**

17. $P(2, 3)$, slope 2
18. $X(4, -1)$, slope 3
19. $R(-3, 5)$, slope -1
20. $A(-2, -6)$, slope -4
21. $V(6, 1)$, slope $\frac{1}{2}$
22. $C(0, 4)$, slope 1

Example 5 (page 168)

Write an equation in point-slope form of the line that contains the given points.

23. $D(0, 5), E(5, 8)$
24. $F(6, 2), G(2, 4)$
25. $H(2, 6), K(-1, 3)$
26. $A(-4, 4), B(2, 10)$
27. $L(-1, 0), M(-3, -1)$
28. $P(8, 10), Q(-4, 2)$

Example 6 (page 168)

Write equations for (a) the horizontal line and (b) the vertical line that contain the given point.

29. $A(4, 7)$
30. $Y(3, -2)$
31. $N(0, -1)$
32. $E(6, 4)$

B Apply Your Skills

Graph each line.

33. $x = 3$
34. $y = -2$
35. $x = 9$
36. $y = 4$
37. $y = 6$

Real-World  Connection

NASA's Advanced Communications Technology Satellite has a capacity for 250,000 phone calls.

38. **Telephone Rates** The equation $C = \$.05m + \$4.95$ represents the cost (C) of a long distance telephone call of m minutes.
    - **a.** What is the slope of the line?
    - **b.** What does the slope represent in this situation?
    - **c.** What is the y-intercept (C-intercept)?
    - **d.** What does the y-intercept represent in this situation?
39. **Error Analysis** A classmate claims that having no slope and having a slope of 0 are the same. Is your classmate correct? Explain.
40. **a.** What is the slope of the x-axis? Explain.
    b. Write an equation for the x-axis.
41. **a.** What is the slope of the y-axis? Explain.
    b. Write an equation for the y-axis.

Identify the form of each equation. To graph the line, would you use the given form or change to another form? Explain.

42. $-5x - y = 2$
43. $y = \frac{1}{4}x - \frac{2}{7}$
44. $y + 2 = -(x - 4)$

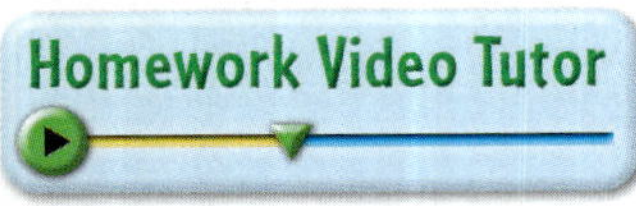

Visit: PHSchool.com
Web Code: bce-0306

Critical Thinking **Graph three different lines having the given property. Describe how the equations of these lines are alike and how they are different.**

45. The lines have slope 2.

46. The lines have y-intercept 2.

47. Graphing Calculator Graphing calculators use slope-intercept form (rather than standard form or point-slope form) to graph lines. Choose either Exercise 45 or Exercise 46 and write three equations for the lines you graphed. Use the Y= window of your graphing calculator to enter your equations. Press GRAPH. Do the graphs on the screen confirm the description you wrote previously?

Graph each pair of lines. Then find their point of intersection.

48. $y = -4, x = 6$ **49.** $x = 0, y = 0$ **50.** $x = -1, y = 3$ **51.** $y = 5, x = 4$

52. Building Access By law, the maximum slope of a ramp in new construction is $\frac{1}{12}$. The plan for the new library shows a 3-ft height from the ground to the main entrance. The distance from the sidewalk to the building is 10 ft. Can you design a ramp for the library that complies with the law? Explain.

Real-World Connection

To visualize a slope of $\frac{1}{12}$, think "one foot over, one inch up."

53. Writing Describe the similarities of and the differences between the graphs of the equations $y = 5x - 2$ and $y = -5x - 2$.

54. Open-Ended Write equations for three different lines that contain the point $(5, 6)$.

55. Critical Thinking The x-intercept of a line is 2 and the y-intercept is 4. Use this information to write an equation for the line.

56. The vertices of a triangle are $A(0, 0)$, $B(2, 5)$, and $C(4, 0)$.
a. Write an equation for the line through A and B.
b. Write an equation for the line through B and C.
c. Compare the slopes and y-intercepts of the two lines.

Do the three points lie on one line? Justify your answer.

57. $A(5, 6), B(3, 2), C(6, 8)$

58. $D(-2, -2), E(4, -4), F(0, 0)$

59. $G(5, -4), H(2, 3), I(-1, 10)$

60. $J(-2, 9), K(1, -1), L(4, -11)$

A line passes through the given points. Write an equation for the line in point-slope form. Then, rewrite the equation in standard form with integer coefficients.

61. $R(-2, 2), S(0, 8)$ **62.** $T(5, 5), W(7, 6)$ **63.** $X(2, 6), Y(5, 8)$

Multiple Choice Practice

For California Standards and CAHSEE Tutorials, visit PHSchool.com. Web Code: bcq-9045

CAHSEE 7MR 2.3

64. Maleo tracked the number of minutes employees spent on personal breaks for several days. He graphed the data in the scatter plot below.

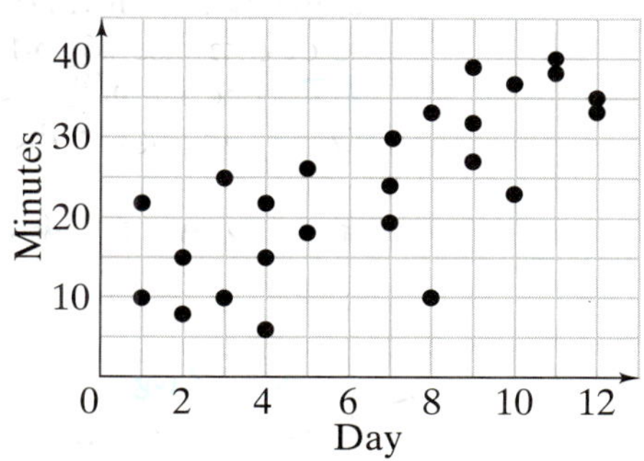

Based on this data, which of the following is the best estimate for the number of minutes the employees spent on personal breaks on Day 6?

(A) 5 (B) 15 (C) 25 (D) 35

CAHSEE 7AF 3.4

65. A math club member pays an initial fee and an additional fee for each monthly dinner meeting attended. The graph at the right represents the total cost for the number of meetings attended. How should the slope of the graph be interpreted?

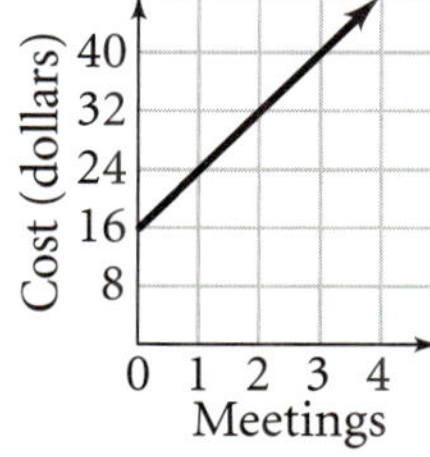

(A) The cost of each dinner meeting is \$8.00.

(B) The initial fee is \$8.00.

(C) The cost of each dinner meeting is \$16.00.

(D) The annual cost is \$48.00.

Mixed Review

Lesson 3-5

Find the sum of the measures of the angles of each polygon.

66. a nonagon **67.** a pentagon **68.** an 11-gon **69.** a 14-gon

Lesson 2-2

Is each statement a good definition? If not, find a counterexample.

70. A quadrilateral is a polygon with four sides.

71. Skew lines are lines that don't intersect.

72. An acute triangle is a triangle with an acute angle.

Lesson 1-7

Algebra **For Exercises 73–76, $\overrightarrow{PQ}$ is the bisector of $\angle MPR$. Solve for a and find the missing angle measure.**

73. $m\angle MPQ = 3a, m\angle QPR = 2a + 5, m\angle MPR = ■$

74. $m\angle MPQ = 7a, m\angle QPR = 4a + 12, m\angle MPR = ■$

75. $m\angle MPQ = 8a - 8, m\angle QPR = 5a - 2, m\angle QPR = ■$

76. $m\angle MPQ = 2a + 9, m\angle QPR = 4a - 3, m\angle MPQ = ■$

Online Lesson Quiz Visit: PHSchool.com, Web Code: bca-0306

Activity Lab

Technology

Solving Linear Equations With Graphs and Tables

FOR USE WITH LESSON 3-6

You enter a linear equation in two variables (a linear function) into a graphing calculator in the form

Y1 = linear expression

The calculator can then give information about Y1 in both graph and table form.

1 ACTIVITY

The windows below use the linear equation $y = 2x - 6$ entered as Y1 = 2X − 6.

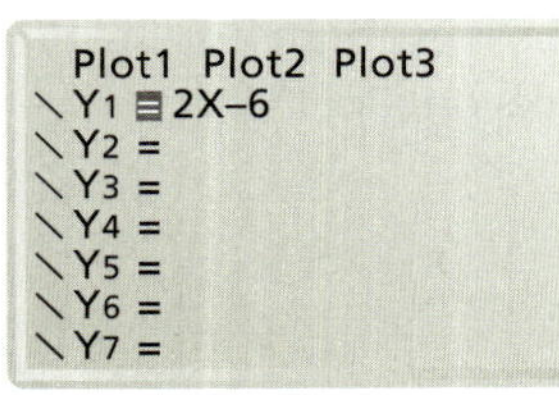

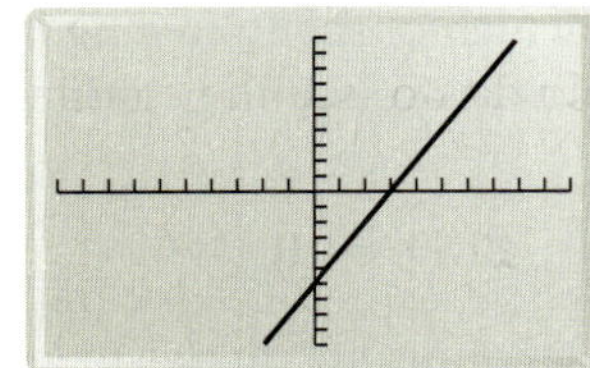

X	Y1
0.00	−6.00
1.00	−4.00
2.00	−2.00
3.00	0.00
4.00	2.00
5.00	4.00
6.00	6.00

X=0

1. How can you find where $2x - 6 = 0$ from the graph? From the table?

2 ACTIVITY

To solve a linear equation from a graph.

let Y1 = linear expression

Then find where Y1 = 0 by using the zero feature in CALC.

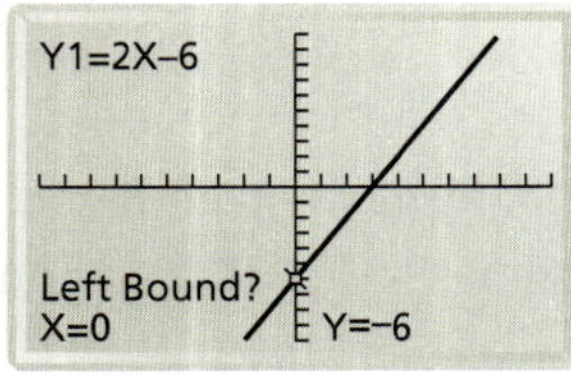

Enter an x-value that is to the left of (<) the solution.

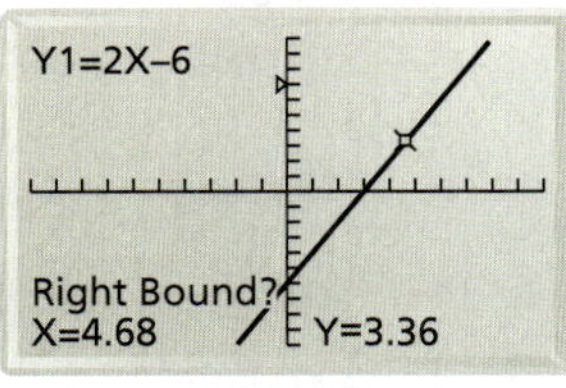

Enter an x-value that is to the right of (>) the solution.

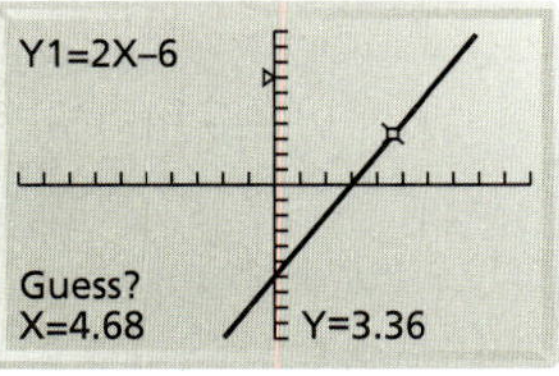

Enter a guess for the solution.

Try it with these linear equations.

2. $2x - 14 = 0$
3. $3x + 1 - (x - 7) = 0$
4. $-x + 10 = -14 + x$
5. $2.5x + 42.5 = 10$

If you haven't already thought of it, here is a terrific shortcut for solving one linear expression = another linear expression.

Let Y1 = one linear expression − (other linear expression),

and find where Y1 = 0. Be sure to enter the expressions carefully into your calculator. Try it with these.

6. $3x - 17 - 5x = -1 - x + 7x$
7. $7 + 4x - (2 + 9x) = x - (6x - 5)$
8. $11 + 3x + 5(7 - x) - 4(3 - (x + 2)) = 3x$

3 ACTIVITY

To help you understand how to solve a linear equation using tables, first think graphically. Look again at the windows in Activity 1.

9. In the graph, to the left of the solution X = 3, are the Y1-values positive or negative? To the right of 3, are the Y1-values positive or negative? Explain.

10. You can say that the Y1-values change _?_ at the solution.

Also, note in the table how Y1 changes sign at the solution X = 3.

Without using your calculator, tell where the Y1-values would change sign in a table for each equation below. State whether the change in Y1 as X increases is from negative to positive or from positive to negative. In each case, tell why. You may assume that the X-values increase by 1 in each step of the table.

11. $y = -x + 4$ **12.** $y = x - 7$ **13.** $y = -2x - 8$ **14.** $y = \frac{1}{2}x + 9$

Now, use a table to solve each equation below. If you do not see the solution, scroll up or down to where you think you'll find Y1 = 0.

15. $0 = 3x - 15$ **16.** $0 = 3x - (x - 4)$ **17.** $0 = 20 + 2x$ **18.** $0 = \frac{1}{3}x - 24$

4 ACTIVITY

Tables become more challenging (and interesting) to use when you see a sign change but you don't see Y1 = 0.

X	Y1
0.00	11.00
1.00	7.00
2.00	3.00
3.00	−1.00
4.00	−5.00
5.00	−9.00
6.00	−13.00

X=0

19. For the table of $Y1 = -4x + 11$ at the right, Y1 changes sign between _?_ and _?_.

20. The solution of $-4x + 11 = 0$ is between _?_ and _?_.

When you have a solution bounded between two X-values that give a sign change, you can perform a *tabular zoom-in* on the solution. If X is between 2 and 3 (above), you can change TblStart to 2 and ΔTbl to 0.1. Study these three screens that capture the solution $x = 2.75$ of the equation $-4x + 11 = 0$.

TABLE SETUP
TbleStart=2
ΔTble=0.1
Indpnt: Auto Ask
Depnd: Auto Ask

X	Y1
2.50	1.00
2.60	.60
2.70	.20
2.80	−.20
2.90	−.60
3.00	−1.00
3.10	−1.40

X=3.1

X	Y1
2.72	.12
2.73	.08
2.74	.04
2.75	0.00
2.76	−.04
2.77	−.08
2.78	−.12

X=2.72

ΔTble=0.01

Use tables to solve each equation.

21. $5x - 14 = x$ **22.** $2(3x + 11) = 2.5$ **23.** $x = 11.75 - x$ **24.** $3x - 100 = 5(0.9 - x)$

25. Use tables to solve $3x - 11 = 0$. Explain what is different about solving this equation from solving those above.

26. Use tables to solve $x\sqrt{2} - \pi = 0$ to as many decimal places as possible on your graphing calculator. Record your answer. Then solve the equation again using a graph.

Slopes of Parallel and Perpendicular Lines

California Content Standards

Prepares for GEOM 17.0 Students prove theorems by using coordinate geometry, including various forms of equations of lines.

What You'll Learn

- To relate slope and parallel lines
- To relate slope and perpendicular lines

. . . And Why

To write an equation that models part of a leaded glass window, as in Example 6

Check Skills You'll Need

page 165 and Lesson 3-6

Find the slope of the line through each pair of points.

1. $F(2, 5), B(-2, 3)$
2. $H(0, -5), D(2, 0)$
3. $E(1, 1), F(2, -4)$

Find the slope of each line.

4. $y = 2x - 5$
5. $x + y = 20$
6. $2x - 3y = 6$
7. $x = y$
8. $y = 7$
9. $y = \frac{2}{3}x + 7$

1 Slope and Parallel Lines

The relationship between slope and parallel lines is summarized below and proved in Lesson 7-3.

Key Concepts

Summary | **Slopes of Parallel Lines**

If two nonvertical lines are parallel, their slopes are equal.

If the slopes of two distinct nonvertical lines are equal, the lines are parallel.

Any two vertical lines are parallel.

Real-World Connection

The ramp and rails are parallel because they have the same slope.

You can test whether nonvertical lines are parallel by comparing their slopes.

1 EXAMPLE Checking for Parallel Lines

Are lines ℓ_1 and ℓ_2 parallel? Explain.

Find and compare the slopes of the lines.

$$\text{slope of } \ell_1 = \frac{5 - (-4)}{1 - (-2)} = \frac{9}{3} = 3$$

$$\text{slope of } \ell_2 = \frac{3 - (-4)}{3 - 1} = \frac{7}{2}$$

Lines ℓ_1 and ℓ_2 are not parallel because their slopes are not equal.

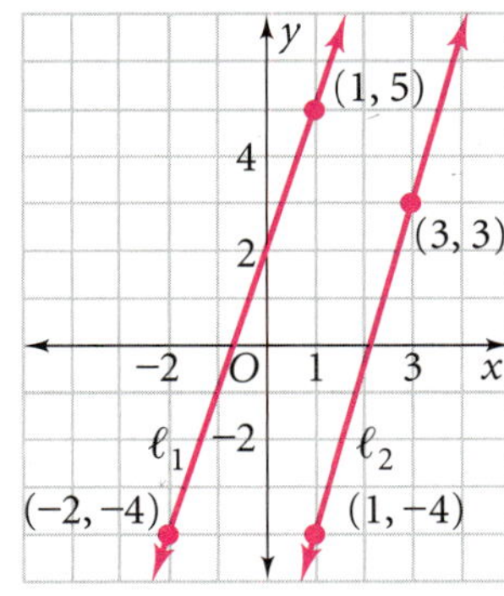

1. Line ℓ_3 contains $A(-4, 2)$ and $B(3, 1)$. Line ℓ_4 contains $C(-4, 0)$ and $D(8, -2)$. Are ℓ_3 and ℓ_4 parallel? Explain.

Slope-intercept form allows you to compare slopes easily in order to decide whether lines are parallel.

2 EXAMPLE Determining Whether Lines Are Parallel

Multiple Choice Which line is parallel to $4y - 12x = 20$?

Ⓐ $y = 3x - 1$

Ⓑ $y = -\frac{1}{3}x - 1$

Ⓒ $4y = -12x + 20$

Ⓓ $y = \frac{1}{3}x + 1$

Write $4y - 12x = 20$ in slope-intercept form.

$$4y - 12x = 20$$

$$4y = 12x + 20 \quad \textbf{Add 12x to each side.}$$

$$y = 3x + 5 \quad \textbf{Divide each side by 4.}$$

The line $4y - 12x = 20$ has slope 3. The line $y = 3x - 1$ is the only answer choice with slope 3. The correct choice is A.

Test-Taking Tip

To compare slopes of two lines, write their equations in slope-intercept form, $y = mx + b$.

CA Standards Check 2 Are the lines parallel? Explain.

a. $y = -\frac{1}{2}x + 5$ and $2x + 4y = 9$ **b.** $y = -\frac{1}{2}x + 5$ and $2x + 4y = 20$

You can write an equation for a line parallel to a given line.

3 EXAMPLE Writing Equations of Parallel Lines

Write an equation for the line parallel to $y = -4x + 3$ that contains $(1, -2)$.

Step 1 Identify the slope of the given line.

$$y = -4x + 3$$

↑ slope

Step 2 Use point-slope form to write an equation for the new line.

$$y - y_1 = m(x - x_1)$$

$$y - (-2) = -4(x - 1) \quad \textbf{Substitute } -4 \textbf{ for } m \textbf{ and } (1, -2) \textbf{ for } (x_1, y_1).$$

$$y + 2 = -4(x - 1) \quad \textbf{Simplify.}$$

CA Standards Check 3 Write an equation for the line parallel to $y = -x + 4$ that contains $(-2, 5)$.

2 Slope and Perpendicular Lines

The relationship between perpendicular lines and their slopes is summarized below. These statements will be proved in Lessons 6-6 and 6-7.

Key Concepts

Summary **Slopes of Perpendicular Lines**

If two nonvertical lines are perpendicular, the product of their slopes is -1.

If the slopes of two lines have a product of -1, the lines are perpendicular.

Any horizontal line and vertical line are perpendicular.

You can test whether lines are perpendicular by first noting whether either line is vertical or horizontal. If not, check their slopes. If the product of the slopes is -1, the lines are perpendicular.

4 EXAMPLE Checking for Perpendicular Lines

Algebra Lines ℓ_1 and ℓ_2 are neither vertical nor horizontal. Are they perpendicular? Explain.

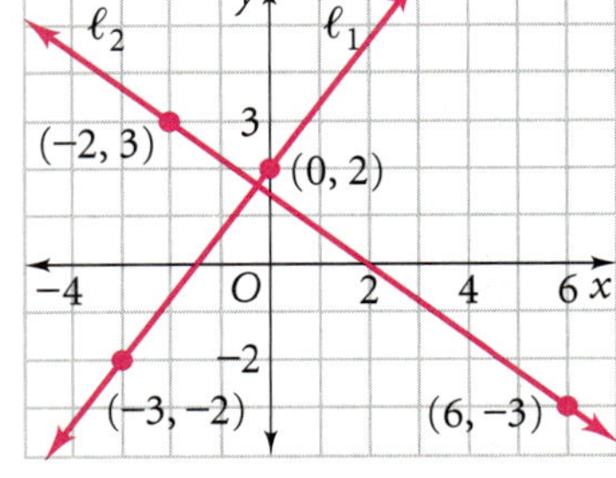

Step 1 Find the slope of each line.

$$m_1 = \text{slope of } \ell_1 = \frac{-2-2}{-3-0} = \frac{-4}{-3} = \frac{4}{3}$$

$$m_2 = \text{slope of } \ell_2 = \frac{3-(-3)}{-2-6} = \frac{6}{-8} = -\frac{3}{4}$$

Step 2 Find the product of the slopes.

$$m_1 \cdot m_2 = \frac{4}{3} \cdot -\frac{3}{4} = -1$$

Lines ℓ_1 and ℓ_2 are perpendicular because the product of their slopes is -1.

CA Standards Check 4 Are ℓ_3 and ℓ_4 perpendicular? Explain.

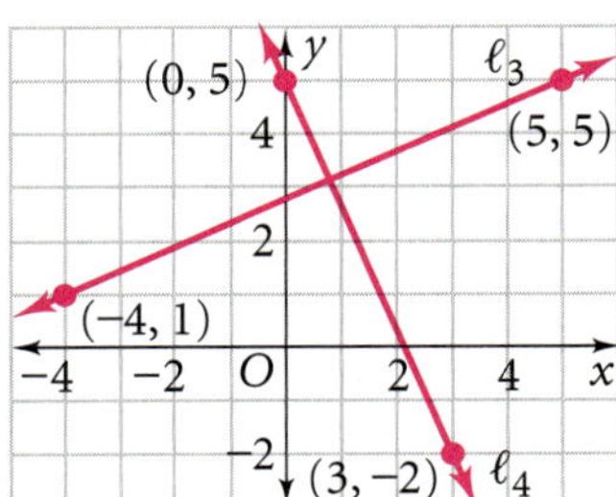

You can write an equation for a line perpendicular to a given line. If the given line is horizontal, write an equation for a vertical line. If the given line is vertical, write an equation for a horizontal line.

5 EXAMPLE Writing Equations for Perpendicular Lines

Write an equation for the line perpendicular to $y = -3x - 5$ that contains $(-3, 7)$.

Step 1 Identify the slope of the given line.

$$y = \underset{\uparrow \text{ slope}}{-3}x - 5$$

Step 2 Find the slope of the line perpendicular to the given line.

Let m be the slope of the perpendicular line.

$-3m = -1$ **The product of the slopes of perpendicular lines is −1.**

$m = \frac{1}{3}$ **Divide each side by −3.**

Step 3 Use point-slope form to write an equation for the new line.

$y - y_1 = m(x - x_1)$

$y - 7 = \frac{1}{3}[x - (-3)]$ **Substitute $\frac{1}{3}$ for m and $(-3, 7)$ for (x_1, y_1).**

$y - 7 = \frac{1}{3}(x + 3)$ **Simplify.**

Vocabulary Tip

Numbers with product -1 are opposite reciprocals. In Example 5, the opposite reciprocal of -3 is $\frac{1}{3}$.

5 Write an equation for the line perpendicular to $5y - x = 10$ that contains $(15, -4)$.

6 EXAMPLE Real-World Connection

The window at the left includes some perpendicular lead strips. The line that contains $\overline{BC}$ has equation $y = -x + 10$. $\overline{AB}$ is perpendicular to $\overline{BC}$. Write an equation for $\overleftrightarrow{AB}$, the line that contains $\overline{AB}$ and point $(-1, 5)$.

The line that contains $\overline{BC}$ has slope -1. Let m be the slope of $\overleftrightarrow{AB}$.

$-1m = -1$ **The product of the slopes is −1.**

$m = 1$

$\overleftrightarrow{AB}$ has slope 1 and can be written in the form $y = 1x + b$, or $y = x + b$.

$y = x + b$ **$\overleftrightarrow{AB}$ has slope 1.**

$5 = -1 + b$ **Substitute 5 for y and −1 for x.**

$6 = b$ **Add 1 to each side.**

The equation for $\overleftrightarrow{AB}$ is $y = x + 6$.

CA Standards Check **6** If the equation for a line containing a lead strip on a different window is $y = -\frac{2}{3}x + 15$, write an equation for the line perpendicular to it that contains $(2, 8)$.

EXERCISES

Standards Practice

For more exercises, see *Extra Skill, Word Problem, and Proof Practice.*

A Practice by Example

Example 1 (page 174)

In Exercises 1–5, are lines ℓ_1 and ℓ_2 parallel? Explain, using slope.

1.

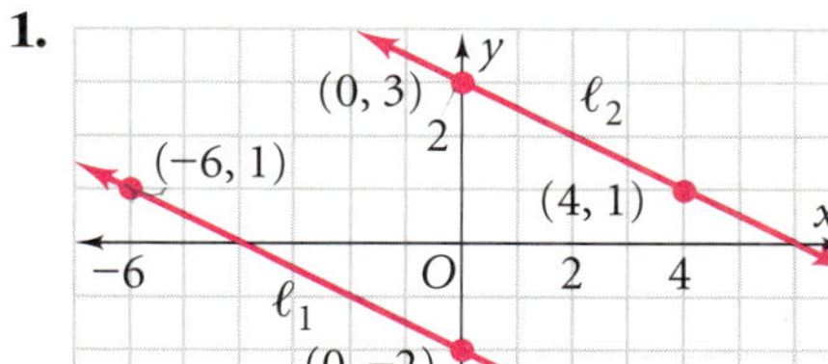

2.

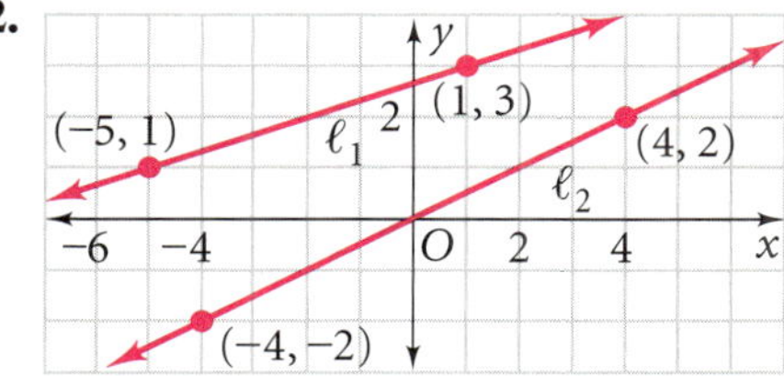

3.

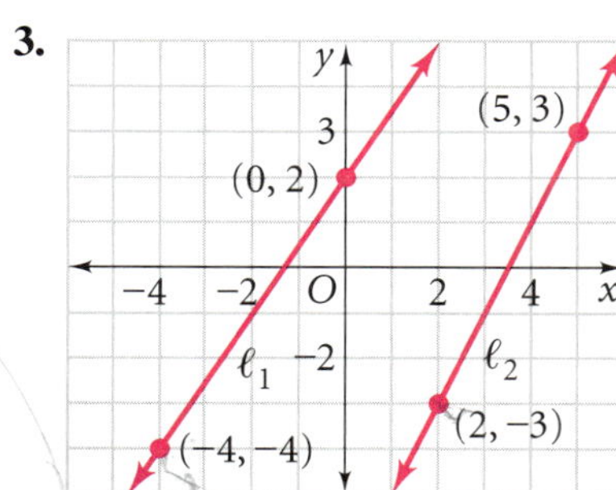

4.

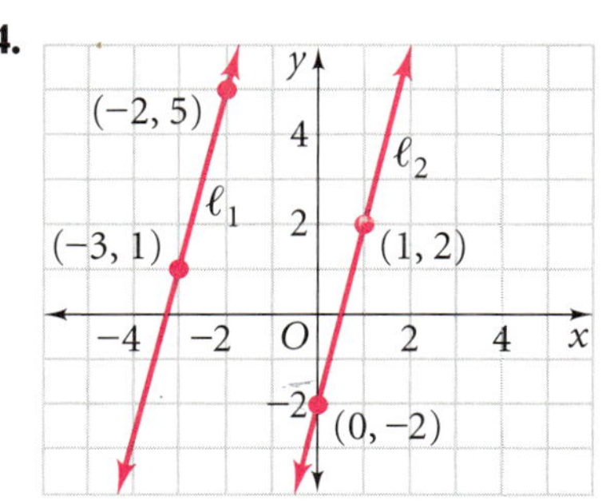

5. Line ℓ_1 contains $A(-3, 6)$ and $B(2, 6)$, and line ℓ_2 contains $C(0, 0)$ and $D(7, 0)$.

Example 2 (page 175)

Algebra **Are the lines parallel? Explain.**

6. $y = 2x + 5$
$y = 2x$

7. $y = \frac{3}{4}x - 10$
$y = \frac{3}{4}x + 2$

8. $y = -x + 6$
$x + y = 20$

9. $y - 7x = 6$
$y + 7x = 8$

10. $3x + 4y = 12$
$6x + 2y = 6$

11. $2x + 5y = -1$
$10y = -4x - 20$

Example 3 (page 175)

Write an equation for the line parallel to $\overleftrightarrow{AB}$ that contains point C.

12. $\overleftrightarrow{AB}$: $y = -2x + 1$, $C(0, 3)$

13. $\overleftrightarrow{AB}$: $y = \frac{1}{3}x$, $C(6, 0)$

14. $\overleftrightarrow{AB}$: $-x + 2y = 4$, $C(-2, 4)$

15. $\overleftrightarrow{AB}$: $3x + 2y = 12$, $C(6, -2)$

Example 4 (page 176)

x^2 **Algebra** **Are lines ℓ_1 and ℓ_2 perpendicular? Explain using slope.**

16.

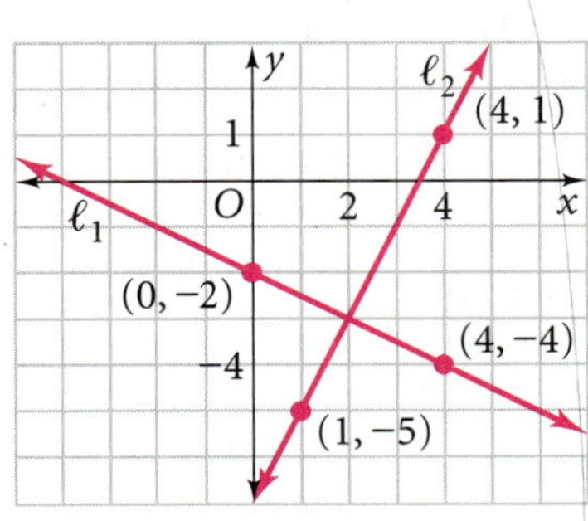

17.

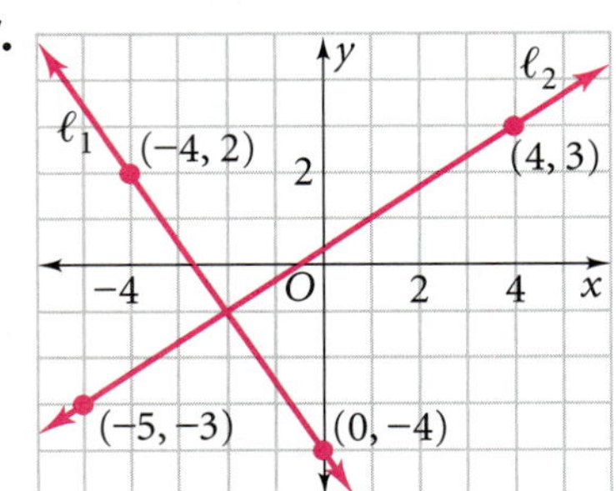

18.

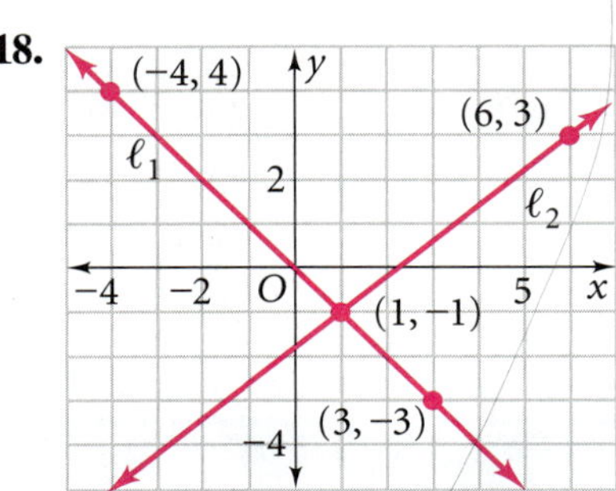

19.

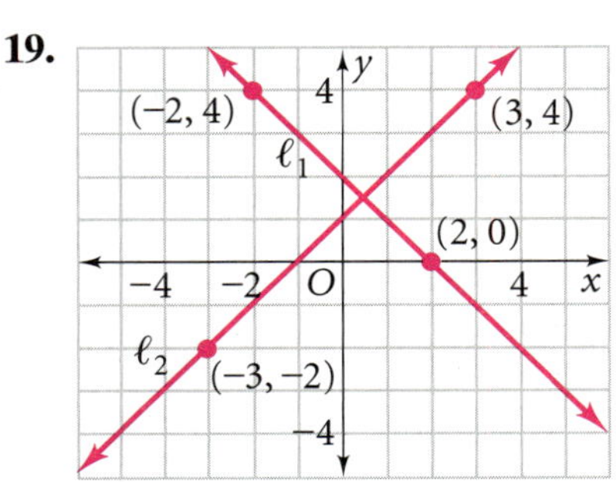

Example 5 (page 176)

Write an equation for the line perpendicular to $\overleftrightarrow{MN}$ that contains point P.

20. $\overleftrightarrow{MN}$: $y = \frac{2}{3}x$, $P(6, 6)$

21. $\overleftrightarrow{MN}$: $y = \frac{1}{2}x - 5$, $P(4, 0)$

22. $\overleftrightarrow{MN}$: $y + 2x = -8$, $P(4, 4)$

23. $\overleftrightarrow{MN}$: $4y + 5x = 20$, $P(0, 0)$

Example 6 (page 177)

24. Highway Construction Highway planners want to construct a road perpendicular to Route 3 at point O. An equation for the Route 3 line is $y = \frac{2}{3}x$. Find an equation for the line for the new road.

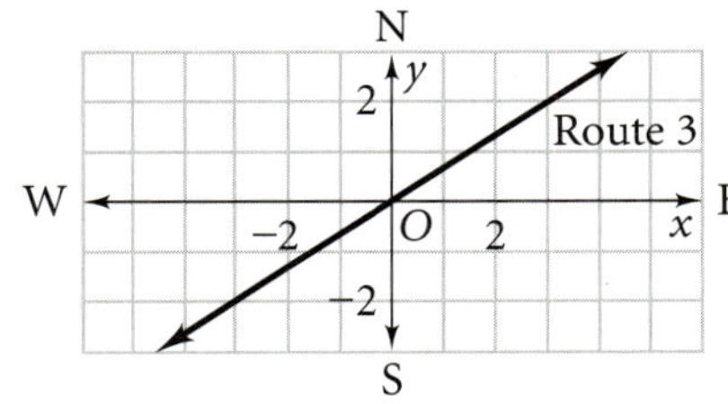

B **Apply Your Skills**

x^2 **Algebra** **Are the lines perpendicular? Explain.**

25. $y = -x - 7$
$y - x = 20$

26. $y = 3$
$x = -2$

27. $2x - 7y = -42$
$4y = -7x - 2$

28. Multiple Choice Which line is perpendicular to $3y + 2x = 12$?

(A) $6x - 4y = 24$

(B) $y + 3x = -2$

(C) $2x + 3y = 6$

(D) $y = -2x + 6$

Use slopes to find whether the opposite sides of quadrilateral $ABCD$ are parallel.

29. $A(0, 2)$, $B(3, 4)$, $C(2, 7)$, $D(-1, 5)$

30. $A(-3, 1)$, $B(1, -2)$, $C(0, -3)$, $D(-4, 0)$

31. $A(1, 1)$, $B(5, 3)$, $C(7, 1)$, $D(3, 0)$

32. $A(1, 0)$, $B(4, 0)$, $C(3, -3)$, $D(-1, -3)$

Homework Video Tutor

Visit: PHSchool.com
Web Code: bce-0307

33. Open-Ended Write equations for two perpendicular lines that have the same y-intercept and do not pass through the origin.

34. Writing Can the y-intercepts of two parallel lines be the same? Explain.

35. Use slope to show that the opposite sides of hexagon $RSTUVW$ at the right are parallel.

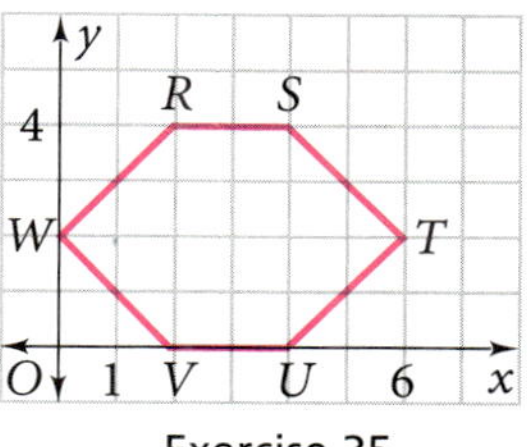

Exercise 35

36. Use slope to determine whether a triangle with vertices $G(3, 2)$, $H(8, 5)$, and $K(0, 10)$ is a right triangle. Explain.

Developing Proof **Use slope to explain why each theorem is true for three lines in the coordinate plane.**

37. Theorem 3-9: If two lines are parallel to the same line, then they are parallel to each other.

38. Theorem 3-10: In a plane, if two lines are perpendicular to the same line, then they are parallel to each other.

Real-World Connection

For a corner kick, the ball is placed within a quarter circle of radius 1 yd.

39. **Soccer** The coordinate system at the right is designed for a soccer field. Each unit represents one yard. Joe is at point $P(35, -20)$. The path of the ball from a corner kick is represented by the equation $y = -\frac{4}{3}x$. To have the best chance for a shot on goal, Joe wants to run toward the ball so that his path meets the path of the ball at a right angle.

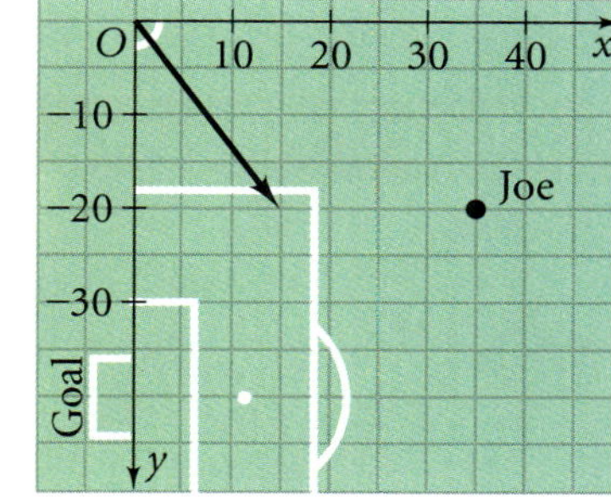

a. Find an equation for the line on which Joe should run.

b. **Critical Thinking** Why is point-slope form the best choice for the equation?

Determine whether $\overleftrightarrow{AB}$ and $\overleftrightarrow{CD}$ are *parallel*, *perpendicular*, or *neither*.

40. $A\left(-1, \frac{1}{2}\right), B(-1, 2), C(3, 7), D(3, -1)$

41. $A(-2, 3), B(-2, 5), C(1, 4), D(2, 4)$

42. $A(2, 4), B(5, 4), C(3, 2), D(0, 8)$

43. $A(-3, 2), B(5, 1), C(2, 7), D(1, -1)$

For: Graphing calculator procedures
Web Code: bcb-2113

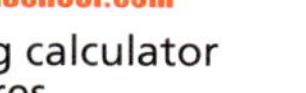

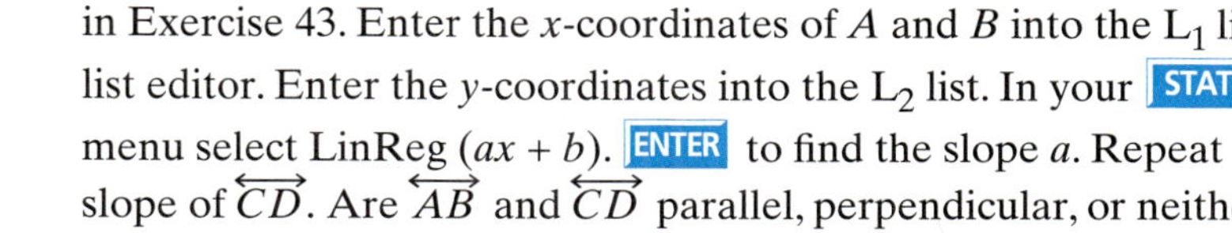

44. **Graphing Calculator** Use your graphing calculator to find the slope of $\overleftrightarrow{AB}$ in Exercise 43. Enter the x-coordinates of A and B into the L_1 list of your list editor. Enter the y-coordinates into the L_2 list. In your STAT CALC menu select LinReg $(ax + b)$. ENTER to find the slope a. Repeat to find the slope of $\overleftrightarrow{CD}$. Are $\overleftrightarrow{AB}$ and $\overleftrightarrow{CD}$ parallel, perpendicular, or neither?

Challenge

45. Show that the diagonals of the figure at the right are congruent.

46. Show that the diagonals of the figure at the right are perpendicular bisectors of each other.

47. a. Graph the points $P(2, 2)$, $Q(7, 4)$, and $R(3, 5)$.

b. Find the coordinates of a point S that, along with points P, Q, and R, will form the vertices of a quadrilateral whose opposite sides are parallel. Graph the quadrilateral.

c. Repeat part (b), finding a different point S and graphing the new quadrilateral.

Exercises 45 and 46

48. A triangle has vertices $L(-5, 6)$, $M(-2, -3)$, and $N(4, 5)$. Write an equation for the line perpendicular to $\overline{LM}$ that contains point N.

Multiple Choice Practice

For California Standards and CAHSEE Tutorials, visit PHSchool.com. Web Code: bcq-9045

GEOM 10.0

49. What is the area in square units of a rectangle with vertices $(-2, 5), (3, 5), (3, -1), (-2, -1)$?

Ⓐ 24 Ⓑ 25 Ⓒ 30 Ⓓ 56

GEOM 3.0

50. Which angles could an obtuse triangle have?

Ⓐ two right angles Ⓒ two obtuse angles
Ⓑ two acute angles Ⓓ two vertical angles

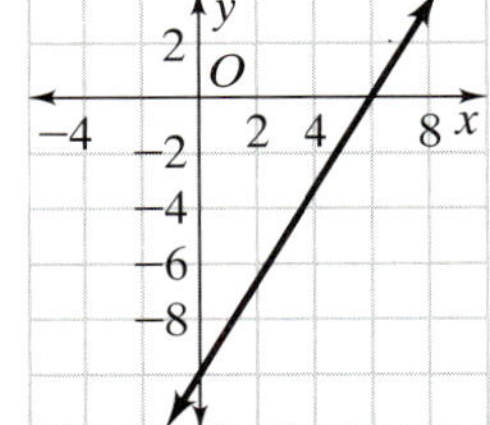

CAHSEE 7AF 3.3

51. What is the slope of the graph at the right?

Ⓐ $-\frac{5}{3}$ Ⓑ $-\frac{3}{5}$ Ⓒ $\frac{3}{5}$ Ⓓ $\frac{5}{3}$

Mixed Review

Lesson 3-6

x^2 **Algebra** **Write an equation for the line containing the given points.**

52. $A(0, 3), B(6, 0)$ **53.** $C(-4, 2), D(-1, 7)$ **54.** $E(3, -2), F(-5, -8)$

Lesson 2-4

Name the property that justifies each statement.

55. $\angle 4 \cong \angle 4$

56. If $m\angle B = 8$, then $2m\angle B = 16$.

57. $-3x + 6 = 3(-x + 2)$

58. If $\overline{RS} \cong \overline{MN}$, then $\overline{MN} \cong \overline{RS}$.

Lesson 2-3

Use the Law of Syllogism to draw a conclusion.

59. If you are in geometry class, then you are in math class. If you are in math class, then you are at school.

60. If you travel to Switzerland, then you travel to Europe. If you travel to Europe, then you have a passport.

Checkpoint Quiz 2

Lessons 3-5 through 3-7

Use the number of sides to name the polygon. Then find the value of each variable.

1.

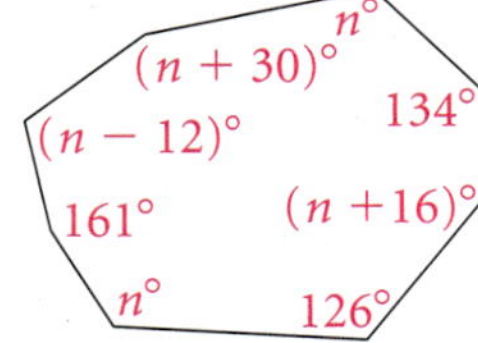

2.

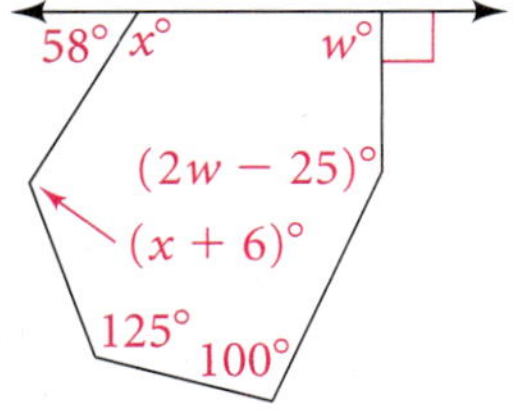

3.

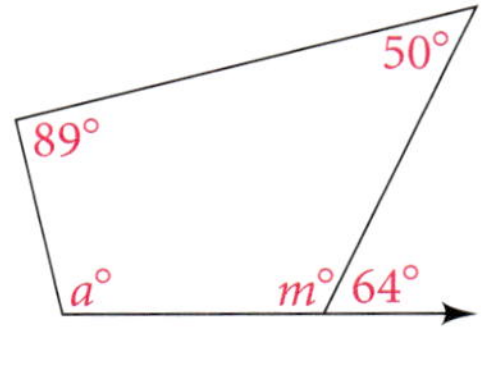

x^2 **Algebra** **Graph each line using intercepts.**

4. $4x + y = -8$ **5.** $-2x + 3y = 12$ **6.** $3x + 5y = 30$

Find the slopes of $\overleftrightarrow{RS}$ and $\overleftrightarrow{TV}$. Then determine whether $\overleftrightarrow{RS}$ and $\overleftrightarrow{TV}$ are parallel, perpendicular, or neither. Explain.

7. $R(-2, 6), S(3, 4), T(3, 5), V(0, 0)$

8. $R(6, -1), S(7, 0), T(3, -4), V(0, -1)$

9. $R(9, 1), S(5, 6), T(3, 8), V(-2, 4)$

10. $R(5, -7), S(-4, -9), T(6, 2), V(-3, 0)$

3-8 Constructing Parallel and Perpendicular Lines

 **California Content Standards**

GEOM 16.0 Students perform basic constructions with a straightedge and compass, such as the line parallel to a given line through a point off the line. ***Master***

What You'll Learn

- To construct parallel lines
- To construct perpendicular lines

. . . And Why

To construct the shortest segment from a point to a line, as in Example 4

Check Skills You'll Need

GO for Help Lesson 1-7

Use a straightedge to draw each figure. Then use a straightedge and compass to construct a figure congruent to it.

1. a segment **2.** an obtuse angle **3.** an acute angle

Use a straightedge to draw each figure. Then use a straightedge and compass to bisect it.

4. a segment **5.** an acute angle **6.** an obtuse angle

1 Constructing Parallel Lines

You can use what you know about parallel lines, transversals, and corresponding angles to construct parallel lines.

1 EXAMPLE Constructing $\ell \parallel m$

Construct the line parallel to a given line and through a given point that is not on the line.

Given: line ℓ and point N not on ℓ

Construct: line m through N with $m \parallel \ell$

Step 1

Label two points H and J on ℓ. Draw $\overleftrightarrow{HN}$.

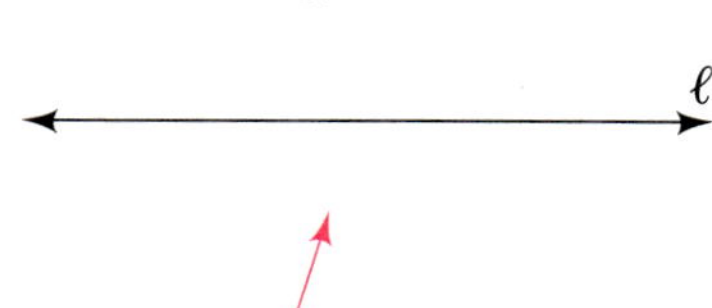

Step 2

Construct $\angle 1$ with vertex at N so that $\angle 1 \cong \angle NHJ$ and the two angles are corresponding angles. Label the line you just constructed m.

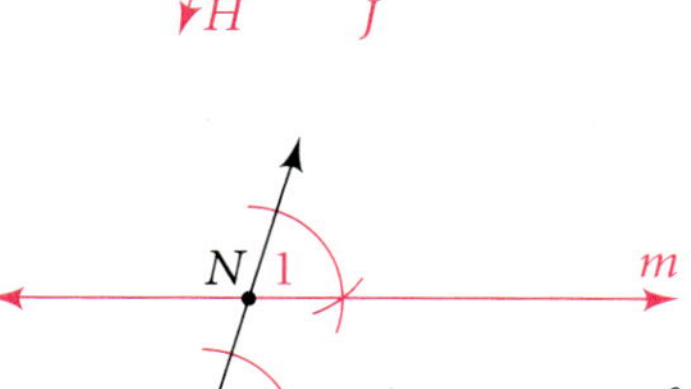

$m \parallel \ell$

Real-World Connection

Architects construct parallel and perpendicular lines when they build models of the buildings they design.

CA Standards Check

1 Critical Thinking Explain why lines ℓ and m must be parallel.

For many constructions, you will find it helpful to first visualize or sketch what the final figure should look like. This will often suggest the construction steps. In Example 2, a sketch is shown at the left of the example.

2 EXAMPLE Constructing a Special Quadrilateral

Construct a quadrilateral with one pair of parallel sides of lengths a and b.

Problem Solving Hint
First, draw a sketch of the figure.

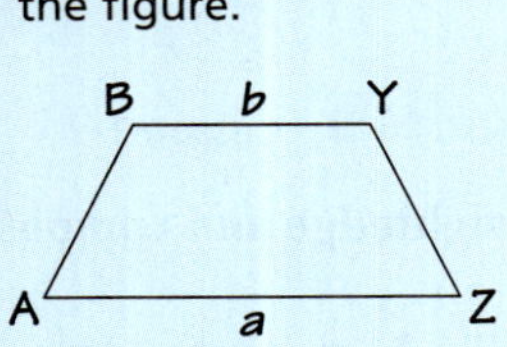

Given: segments of lengths a and b

Construct: quadrilateral $ABYZ$ with $AZ = a, BY = b$, and $\overline{AZ} \parallel \overline{BY}$

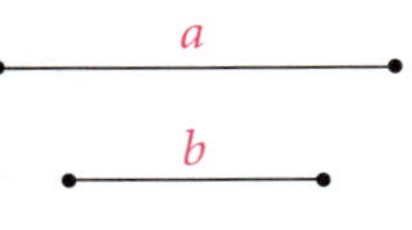

Step 1
Construct $\overline{AZ}$ with length a.

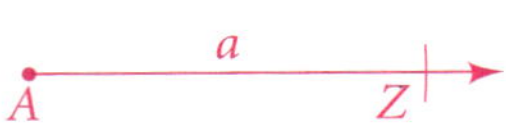

Step 2
Draw a point B not on $\overleftrightarrow{AZ}$. Then draw $\overrightarrow{AB}$.

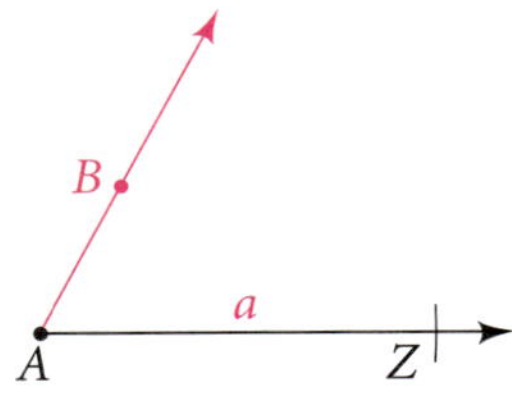

Step 3
Construct a ray parallel to $\overleftrightarrow{AZ}$ through B.

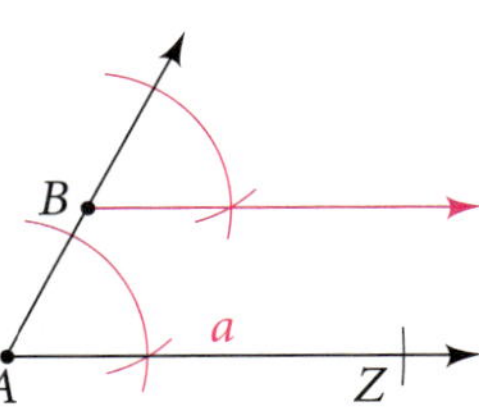

Step 4
Construct Y so that $BY = b$. Then draw $\overline{YZ}$.

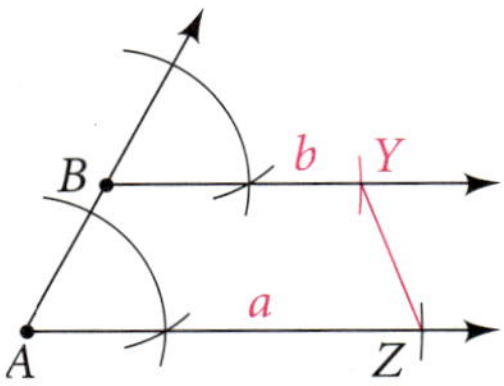

Quadrilateral $ABYZ$ has $AZ = a, BY = b$, and $\overline{AZ} \parallel \overline{BY}$.

2 Draw two segments. Label their lengths c and d. Construct a quadrilateral with one pair of parallel sides of lengths c and $2d$.

2 Constructing Perpendicular Lines

You can construct perpendicular lines using a compass and a straightedge.

3 EXAMPLE Perpendicular at a Point on a Line

Online active math

For: Construction Activity
Use: Interactive Textbook, 3-8

Construct the perpendicular to a given line at a given point on the line.

Given: point P on line ℓ

Construct: $\overleftrightarrow{CP}$ with $\overleftrightarrow{CP} \perp \ell$

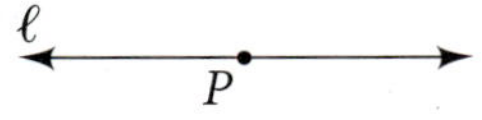

Step 1
Put the compass point on point P. Draw arcs intersecting ℓ in two points. Label the points A and B.

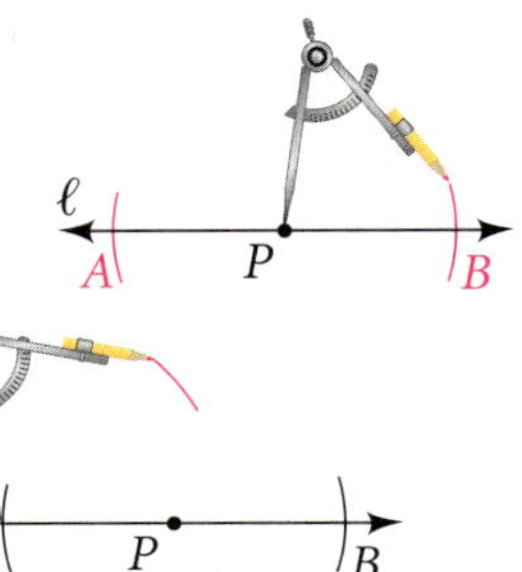

Step 2
Open the compass wider. With the compass tip on A, draw an arc above point P.

Step 3
Without changing the compass setting, place the compass point on point B. Draw an arc that intersects the arc from Step 2. Label the point of intersection C.

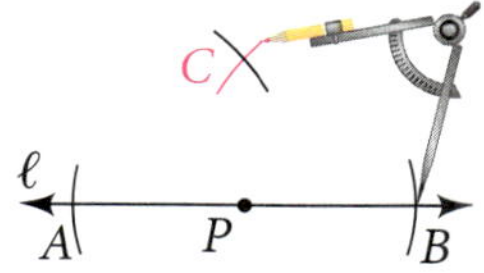

Step 4
Draw $\overleftrightarrow{CP}$.

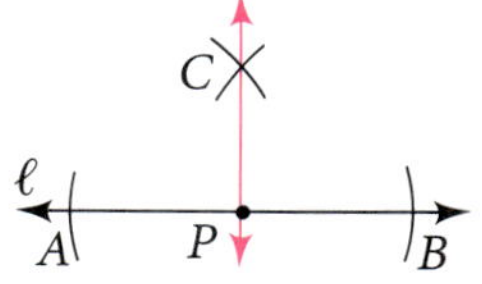

$\overleftrightarrow{CP} \perp \ell$

CA Standards Check 3 Use a straightedge to draw $\overleftrightarrow{EF}$. Construct $\overleftrightarrow{FG}$ so that $\overleftrightarrow{FG} \perp \overleftrightarrow{EF}$ at point F.

You will prove in Chapter 5 that the perpendicular segment is the shortest segment from a point to a line. Here is its construction.

4 EXAMPLE Perpendicular From a Point to a Line

Construct the perpendicular to a given line through a given point not on the line.

Given: line ℓ and point R not on ℓ

Construct: $\overleftrightarrow{RG}$ with $\overleftrightarrow{RG} \perp \ell$

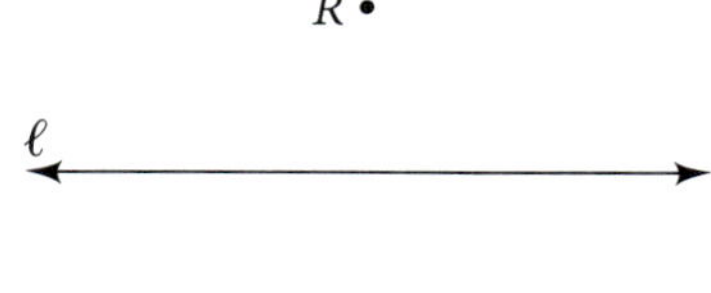

Step 1
Open your compass to a size greater than the distance from R to ℓ. With the compass point on point R, draw an arc that intersects ℓ at two points. Label the points E and F.

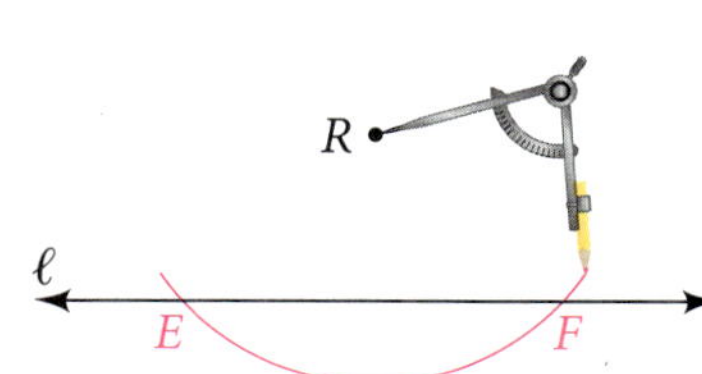

Step 2
Place the compass point on E and make an arc.

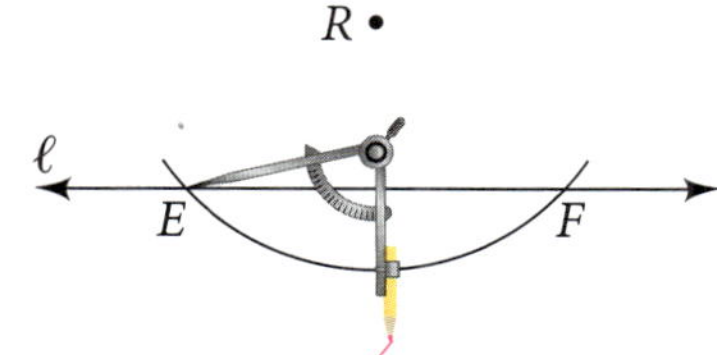

Step 3
Keep the same compass setting. With the compass tip on F, draw an arc that intersects the arc from Step 2. Label the point of intersection G.

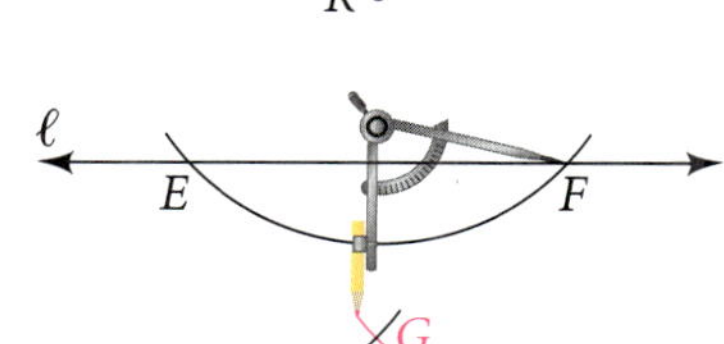

Step 4
Draw $\overleftrightarrow{RG}$.

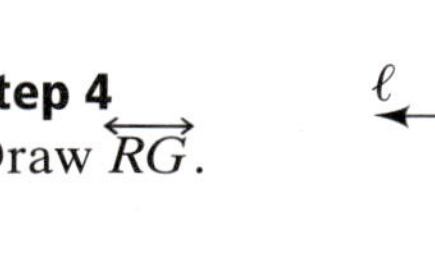

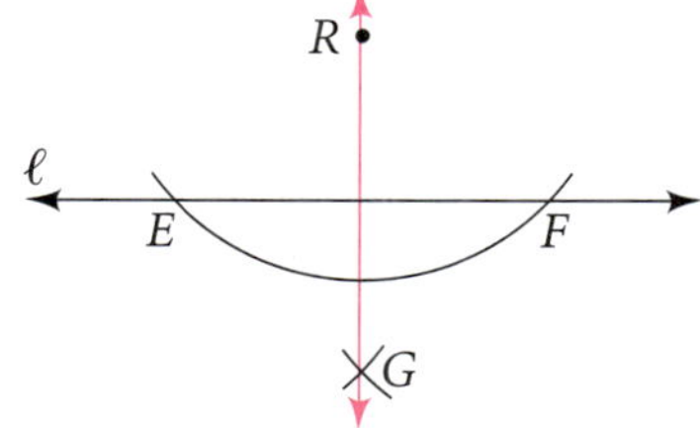

$\overleftrightarrow{RG} \perp \ell$

Real-World Connection

You can draw large circles using a simple, large compass.

CA Standards Check 4 Draw a line $\overleftrightarrow{CX}$ and a point Z not on $\overleftrightarrow{CX}$. Construct $\overleftrightarrow{ZB}$ so that $\overleftrightarrow{ZB} \perp \overleftrightarrow{CX}$.

EXERCISES

For more exercises, see *Extra Skill, Word Problem, and Proof Practice.*

Standards Practice

GEOM 16.0

A Practice by Example

Example 1 (page 181)

GO for Help

In Exercises 1–4, draw a figure like the given one. Then construct the line through point *J* and parallel to $\overleftrightarrow{AB}$.

1.

2.

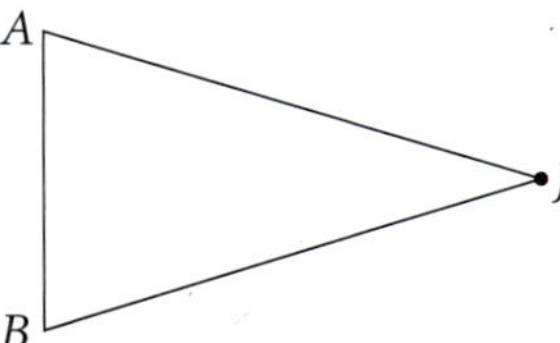

3.

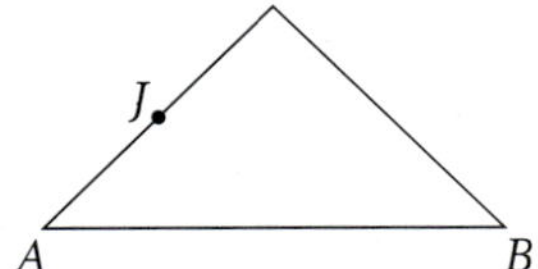

4.

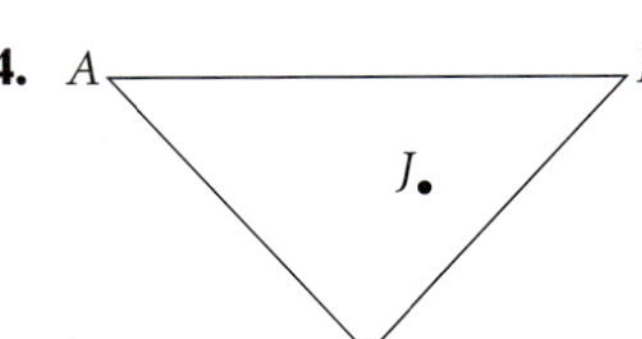

Example 2 (page 182)

For Exercises 5–7, draw two segments. Label their lengths *a* and *b*. Construct a quadrilateral with one pair of parallel sides as described.

5. The sides have lengths a and b.

6. The sides have lengths $2a$ and b.

7. The sides have lengths a and $\frac{1}{2}b$.

Example 3 (pages 182, 183)

In Exercises 8–9, draw a figure like the given one. Then construct the line perpendicular to $\overleftrightarrow{AB}$ at point *P*.

8. A P B

9.

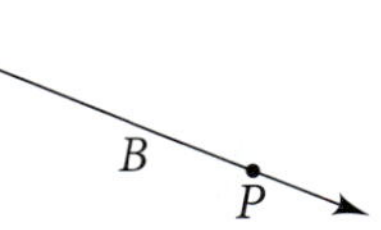

Example 4 (page 183)

In Exercises 10–13, draw a figure like the given one. Then construct the line through point *P* and perpendicular to $\overleftrightarrow{RS}$.

10.

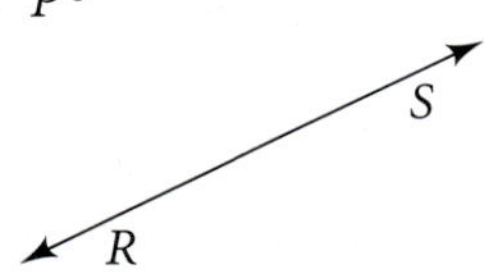

11.

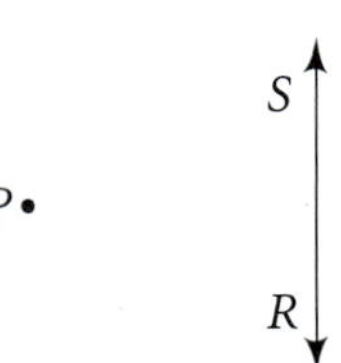

12.

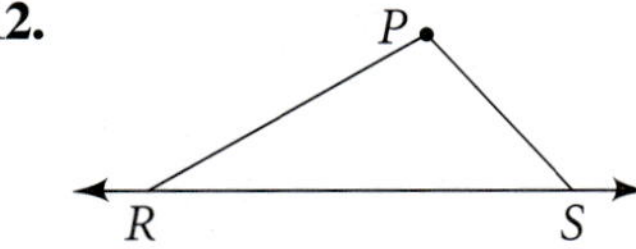

13.

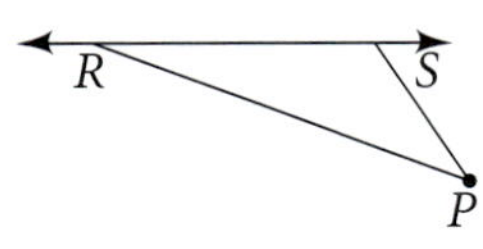

B Apply Your Skills

14. Draw an acute angle. Construct an angle congruent to your angle so that the two angles are alternate interior angles. (*Hint:* Think of the letter Z.)

15. Writing Explain how to use the Converse of the Alternate Interior Angles Theorem to construct a line parallel to a given line through a point not on the line. (*Hint:* See Exercise 14.)

16. Draw obtuse $\triangle ABC$ with obtuse $\angle B$.
    a. Construct line ℓ through point A so that $\ell \parallel \overline{BC}$.
    b. Construct line m through point C so that $m \parallel \overline{AB}$.

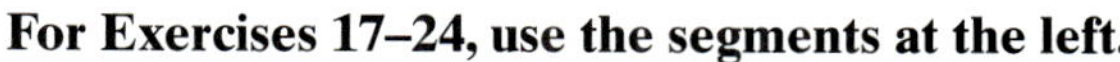

For Exercises 17–24, use the segments at the left.

Exercises 17–24

17. Draw a line m. Construct a segment of length b that is perpendicular to line m.
18. Construct a rectangle with base b and height c.
19. Construct a square with sides of length a.
20. Construct a rectangle with one side length a and a diagonal length b.
21. a. Construct a quadrilateral with a pair of parallel sides of length c.
    b. **Make a Conjecture** What appears to be true about the other pair of sides in the quadrilateral you constructed?
    c. Use a protractor, a ruler, or both to check the conjecture you made in part (b).
22. Construct a right triangle with legs of lengths a and b.
23. Construct a right triangle with legs of lengths b and $\frac{1}{2}b$.
24. a. Construct a triangle with sides of lengths a, b, and c.
    b. Construct the midpoint of each side of the triangle.
    c. Form a new triangle by connecting the midpoints.
    d. **Make a Conjecture** How do the sides of the smaller triangle and the sides of the larger triangle appear to be related?
    e. Use a protractor, a ruler, or both to check the conjecture you made in part (d).

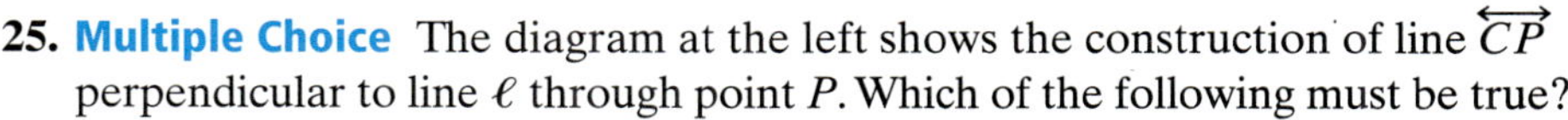

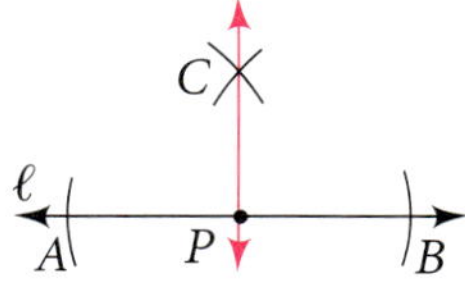

25. **Multiple Choice** The diagram at the left shows the construction of line $\overleftrightarrow{CP}$ perpendicular to line ℓ through point P. Which of the following must be true?

Ⓐ $\overleftrightarrow{CB} \parallel \overleftrightarrow{AB}$ Ⓒ $CP = \frac{1}{2}AB$
Ⓑ $\overleftrightarrow{AC} \perp \overleftrightarrow{CB}$ Ⓓ $\overline{AC} \cong \overline{BC}$

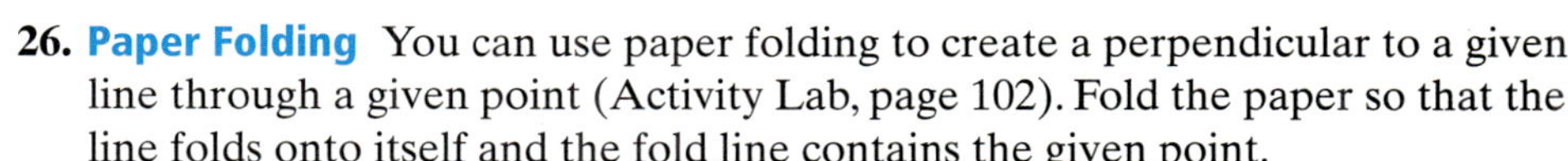

26. **Paper Folding** You can use paper folding to create a perpendicular to a given line through a given point (Activity Lab, page 102). Fold the paper so that the line folds onto itself and the fold line contains the given point.

Visit: PHSchool.com
Web Code: bce-0308

    a. Draw a line m and a point W not on the line. Use paper folding to create the perpendicular to m through W. Label this fold line k.
    b. Next, fold the line perpendicular to k through W. Label this fold line p.
    c. What is true of p and m? Justify your answer.

Challenge

Draw a segment, $\overline{DG}$. Construct a quadrilateral whose diagonals are both congruent to $\overline{DG}$, bisect each other, and meet the additional condition given below. Describe the quadrilateral that you get.

27. The diagonals are not perpendicular. 28. The diagonals are perpendicular.

Construct a rectangle whose side lengths *a* and *b* meet the given condition.

29. $b = 2a$ **30.** $b = \frac{1}{2}a$ **31.** $b = \frac{1}{3}a$ **32.** $b = \frac{2}{3}a$

Construct a triangle whose side lengths *a*, *b*, and *c* meet the given conditions. If such a triangle is not possible, explain.

33. $a = b = c$ **34.** $a = b = 2c$ **35.** $a = 2b = 2c$ **36.** $a = b + c$

Multiple Choice Practice

For California Standards and CAHSEE Tutorials, visit PHSchool.com. Web Code: bcq-9045

GEOM 16.0

37. In the construction shown at the right, the two arcs with centers A and B have the same radius. What must be true of $\overline{PQ}$?

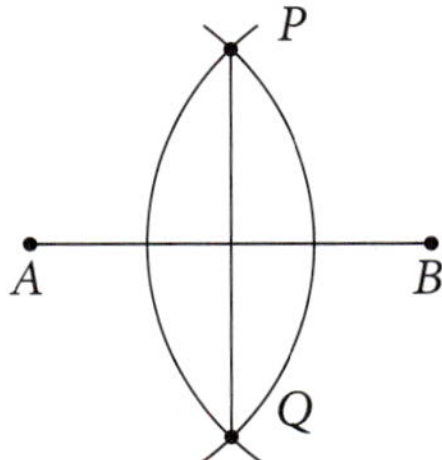

Ⓐ $\overline{PQ}$ bisects $\overline{AB}$.
Ⓑ $\overline{PQ} \parallel \overline{AB}$
Ⓒ $\overline{PQ} \cong \overline{AB}$
Ⓓ $\overline{PQ} \cong \overline{AQ}$

GEOM 16.0

38. Suppose you construct lines ℓ, m, and n so that $\ell \perp m$ and $\ell \parallel n$. Which of the following is true?

Ⓐ $m \parallel n$ Ⓒ $n \perp \ell$
Ⓑ $m \parallel \ell$ Ⓓ $n \perp m$

CAHSEE 7AF 3.1

39. Which equation best describes the graph at the right?

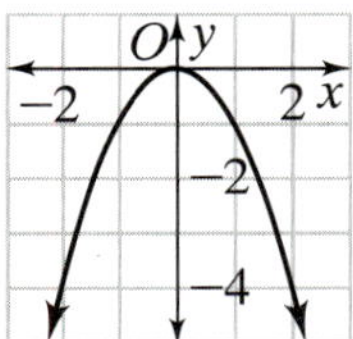

Ⓐ $y = 2x$ Ⓒ $y = x^2$
Ⓑ $y = -2x$ Ⓓ $y = -x^2$

CAHSEE 7MG 3.3

40. The beginner's hill at Powderwonder Ski Resort measures 360 feet horizontally and 81 feet vertically. What is the length of the ski run?

Ⓐ 369 feet Ⓑ 371 feet Ⓒ 441 feet Ⓓ 451 feet

Mixed Review

Lesson 2-1

Are the lines parallel? Explain.

41. $y = -4x - 3$
$y = 4x + 3$

42. $y = \frac{1}{2}x + 1$
$y = -2x - 1$

43. $x + 3y = -6$
$4x + 12y = -6$

Lesson 1-8

Find the distance between the points to the nearest tenth.

44. $W(8, -2)$ and $Z(2, 6)$

45. $W(-4.5, 1.2)$ and $Z(3.5, -2.8)$

Lesson 1-3

Name the intersection of the planes.

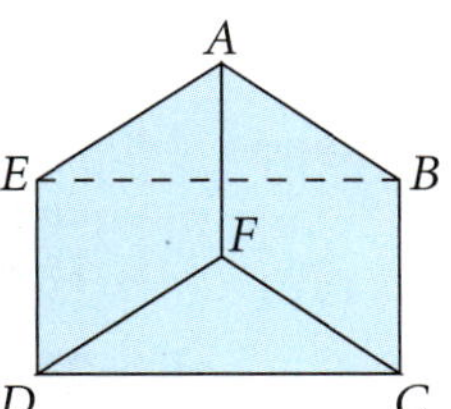

46. plane ABE and plane $EBCD$

47. plane $AFDE$ and plane FCD

Activity Lab: Using Tables and Lists

Technology

FOR USE WITH LESSON 3-8

GEOM 1.0 Students demonstrate understanding by identifying and giving examples of undefined terms, axioms, theorems, and inductive reasoning. ***Master***

Tables and lists on your graphing calculator allow you to study relationships both numerically and graphically. The first example reminds you how to build a table.

1 ACTIVITY

Display a table showing the sums of the measures of the angles of a polygon.

Use the **Y=** screen and write the polygon angle-sum formula in the form $Y_1 = 180(X - 2)$. Use the **TBLSET** feature so that X starts at 3 and changes by 1.

Use the **TABLE** feature to see the table of n-gon angle sums.

X	Y1	
3	180	
4	360	
5	540	
6	720	
7	900	
8	1080	
9	1260	

X=3

The second example suggests a powerful way to use lists. First, press **Y=** **CLEAR**.

2 ACTIVITY

List and plot four ordered pairs for the line $y = 2x - 1$.

On your home screen generate four values for x in list L_1 as follows.

seq(X, X, −2, 4, 2) **STO▶** L_1 **ENTER**.

Enter the corresponding values for y in list L_2 as follows.

$2L_1 - 1$ **STO▶** L_2 **ENTER**.

Access **STAT PLOT**, press 1, and turn "On" Plot 1. Check that your **Xlist** is L_1 and your **Ylist** is L_2. Then **GRAPH** in a standard viewing window as shown at the right.

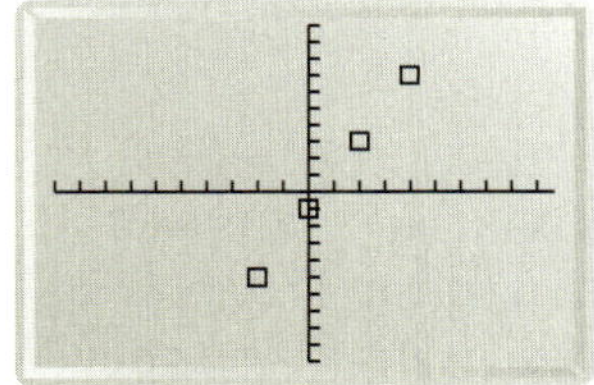

EXERCISES

1. Use lists to plot ordered pairs (x, y) for the relationship $y = 180(x - 2)$. Use the command seq(X, X, 3, 12) to generate L_1 values. Store $180(L_1 - 2)$ into an L_2 list of Y values. Graph the points in an appropriate viewing window.

2. The relationship $y = 180(n - 2)/n$ or $Y_1 = 180(X - 2)/X$ gives the measure of one angle of a regular polygon for each value of n or X.
   a. Display a table of regular-polygon angle measures. Use TblStart = 3.
   b. Scroll down your table. What happens to the angle measures as X gets large?
   c. Make L_1, L_2 lists of (n, y) pairs for $n = 3$ to 20.
   d. Plot the (n, y) points. What happens to the angle measures as n gets large?

3. The formula $y = 360/n$ gives an exterior-angle measure for a regular n-gon. Make L_1, L_2 lists of (n, y) values and plot points. What happens to exterior-angle measures as n gets large?

4. Display $180(n - 2)/n$ values and $360/n$ values side by side in a table. What is the sum of the two values in each row? Show that this is the sum for every value of n.

Test-Taking Strategies

Finding Needed Information

Some questions ask you to identify information you would need to solve a problem. For these questions, first decide how you would solve the problem. Then look at the given information and decide what information you still need.

EXAMPLE

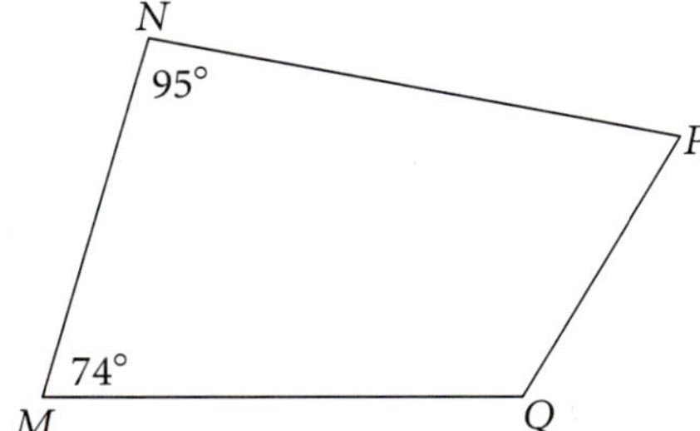

Quadrilateral *MNPQ* is shown at the right. What additional information is needed to find the measure of $\angle P$?

Ⓐ The number of angles in the polygon
Ⓑ The sum of the measures of the angles of the polygon
Ⓒ The measure of $\angle Q$
Ⓓ The complement of 74°

A quadrilateral has four sides and four angles. By the Polygon Angle-Sum Theorem, the sum of the angle measures is $(4 - 2)180 = 360$.

So $m\angle M + m\angle N + m\angle P + m\angle Q = 360$.

$74 + 95 + m\angle P + m\angle Q = 360$ **Substitute.**

$169 + m\angle P + m\angle Q = 360$ **Simplify.**

$m\angle P + m\angle Q = 191$ **Subtract 169 from each side.**

To find $m\angle P$ you need to know $m\angle Q$. The correct answer is C.

EXERCISES

1. Jessica is buying 4 ft-by-8 ft sheets of paneling for the walls of her basement. The walls are 6 feet high. The tiled basement floor is a rectangle 50 feet long and 30 feet wide. What additional information is needed to find the cost of the paneling Jessica needs for the basement?

Ⓐ The volume of the room
Ⓑ The area of the floor
Ⓒ The cost per square foot of the floor tiles
Ⓓ The cost for each piece of paneling

2. Charlie's class needs $300 for a trip to the amusement park. They have $56 in the class fund and hope to raise the rest of the money by having a car wash. What additional information is needed to allow the class to decide how much to charge for each car?

Ⓐ The number of students in his class
Ⓑ The number of cars they expect to wash
Ⓒ The amount of money they need to raise
Ⓓ The number of students washing cars

3. The scale of two similar quadrilaterals is 1 : 2. What additional information do you need to find the perimeter of the larger quadrilateral?

Ⓐ The area of the small quadrilateral
Ⓑ The perimeter of the smaller quadrilateral
Ⓒ Whether the quadrilaterals are squares
Ⓓ Whether the quadrilaterals are rectangles

Chapter 3 Review

Vocabulary Review

English and Spanish Audio Online

acute triangle (p. 148)
alternate exterior angles (p. 129)
alternate interior angles (p. 127)
concave polygon (p. 158)
convex polygon (p. 158)
corresponding angles (p. 127)
equiangular triangle (p. 148)
equiangular polygon (p. 160)
equilateral triangle (p. 148)
equilateral polygon (p. 160)
exterior angle of a polygon (p. 149)
flow proof (p. 135)
isosceles triangle (p. 148)
obtuse triangle (p. 148)
point-slope form (p. 168)
polygon (p. 157)
regular polygon (p. 160)
remote interior angles (p. 149)
right triangle (p. 148)
same-side exterior angles (p. 129)
same-side interior angles (p. 127)
scalene triangle (p. 148)
slope-intercept form (p. 166)
standard form of a linear equation (p. 167)
transversal (p. 127)
two-column proof (p. 129)

For: Vocabulary quiz
Web Code: bcj-0351

Choose the correct vocabulary term to complete each sentence.

1. In a triangle, an angle is right, obtuse, or __?__.
2. A(n) __?__ angle has a measure between 90 and 180.
3. When two coplanar lines are cut by a transversal, two angles that are in similar positions on the same side of the transversal are called __?__.
4. The measure of a(n) __?__ angle of a triangle is equal to the sum of the measures of its two remote interior angles.
5. A polygon is __?__ if no diagonal contains points outside the polygon.
6. A(n) __?__ polygon has all angles congruent.
7. A(n) __?__ polygon is both equiangular and equilateral.
8. The linear equation $y - 3 = 4(x + 5)$ is written in __?__ form.
9. From the __?__ form of a linear equation, you can easily read the value of the slope and the value of the y-intercept.
10. When two coplanar lines are cut by a transversal, the angles between the two lines and on opposite sides of the transversal are called __?__.

Skills and Concepts

Lesson 3-1

- To identify angles formed by two lines and a transversal (p. 127)
- To prove and use properties of parallel lines (p. 128)

 **GEOM 2.0, 4.0, 7.0**

A **transversal** is a line that intersects two coplanar lines at two distinct points.
$\angle 1$ and $\angle 4$ are **corresponding angles.**
$\angle 3$ and $\angle 4$ are **alternate interior angles.**
$\angle 2$ and $\angle 4$ are **same-side interior angles.**

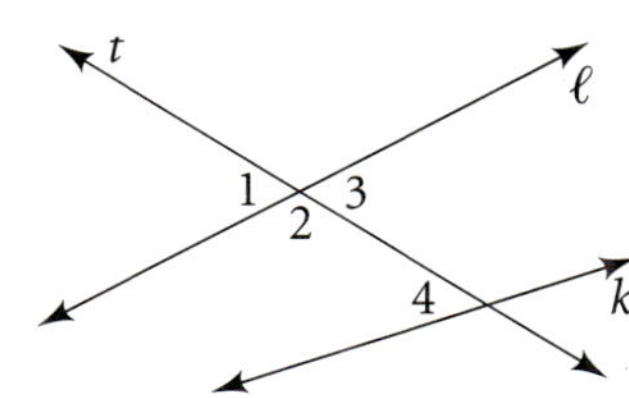

If two parallel lines are cut by a transversal, then
- corresponding angles are congruent.
- alternate interior angles are congruent.
- same-side interior angles are supplementary.

11. Suppose ℓ and k in the diagram above are parallel. If $m\angle 1 = 59$, what are the measures of $\angle 2$, $\angle 3$, and $\angle 4$?

Find $m\angle 1$ and then $m\angle 2$. Justify each answer.

12. 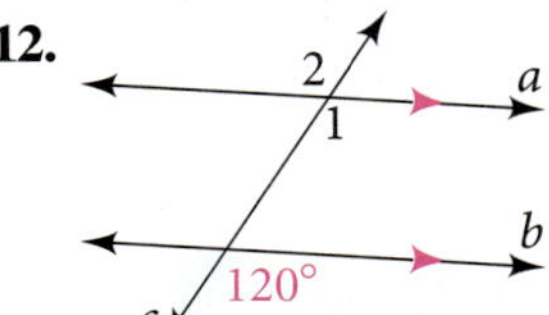

13. 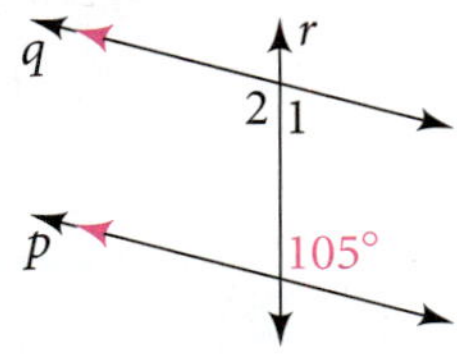

14. 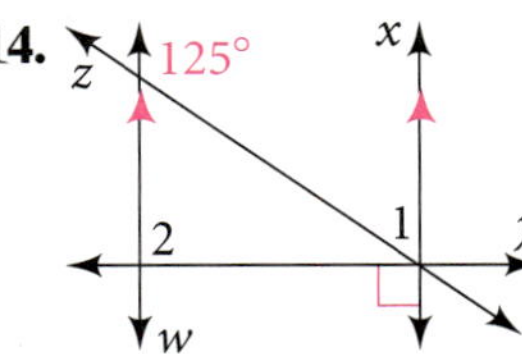

Lessons 3-2 and 3-8

- To use a transversal in proving lines parallel (p. 134)
- To construct parallel lines (p. 181)
- To construct perpendicular lines (p. 182)

 **GEOM 2.0, 4.0, 7.0, 16.0**

Two lines cut by a transversal are parallel if corresponding angles are congruent, alternate interior angles are congruent, or same-side interior angles are supplementary.

You can construct the line parallel to a given line through a given point not on the line. You can also construct the perpendicular to a given line at a given point on the line or through a given point not on the line.

 **Algebra Find the value of x for which $\ell \parallel m$.**

15. 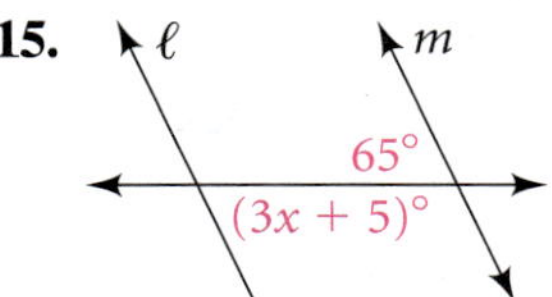

16.

17. 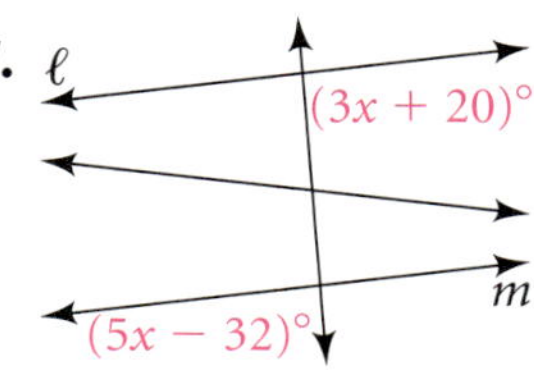

Use the segments at the right for Exercises 18 and 19.

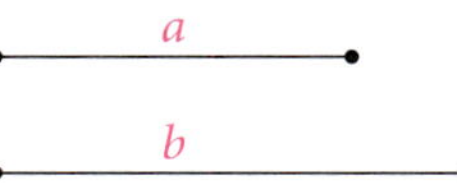

18. Construct a rectangle with side lengths a and b.

19. Construct a quadrilateral with one pair of parallel opposite sides, each side of length $2a$.

Lesson 3-3

- To relate parallel and perpendicular lines (p. 141)

 **GEOM 7.0**

If two lines are perpendicular to the same line, then the two lines are parallel. If a line is perpendicular to one of two parallel lines, then it is perpendicular to both lines. Both of these statements are true only if the lines are coplanar.

Use the diagram at the right for Exercises 20 and 21. State the theorem that justifies each statement.

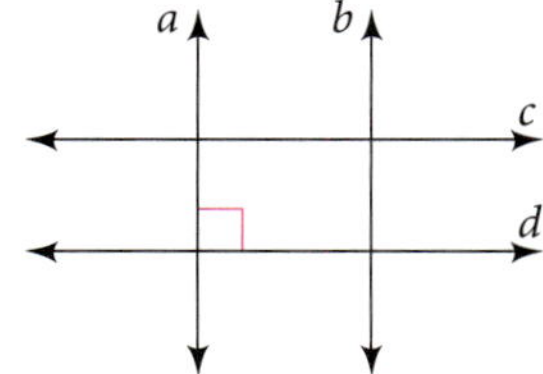

20. If $b \perp c$ and $b \perp d$, then $c \parallel d$.

21. If $c \parallel d$, then $a \perp c$.

Lesson 3-4

- To classify triangles and find the measures of their angles (p. 147)
- To use exterior angles of triangles (p. 149)

 **GEOM 12.0, 13.0**

The sum of the measures of the angles of a triangle is 180. The measure of each **exterior angle** of a triangle equals the sum of the measures of its two **remote interior angles.**

Find the values of the variables. Then classify each triangle by its sides and angles.

22.  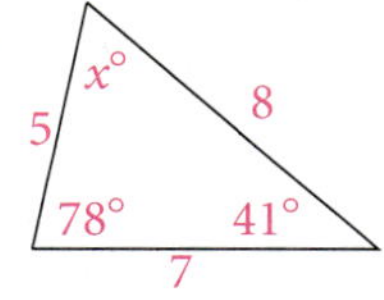

23. 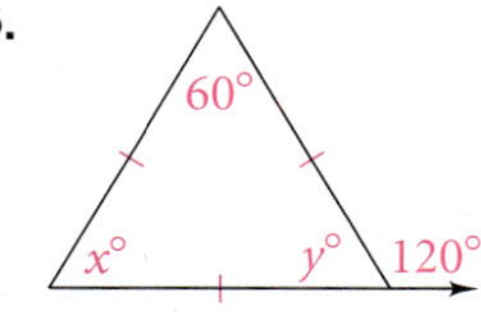

24. 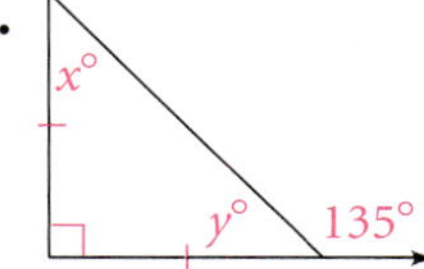

In each of Exercises 25–28, the measures of the three angles of a triangle are given. Find the value of x and then classify the triangle by its angles.

25. $x + 10, x - 20, x + 25$

26. $x, 2x, 3x$

27. $20x + 10, 30x - 2, 7x + 1$

28. $10x - 3, 14x - 20, x + 3$

29. In a right triangle, what is always true about the angles?

Lesson 3-5

- To classify polygons (p. 157)
- To find the sums of the measures of the interior and exterior angles of polygons (p. 159)

 GEOM 12.0, 13.0

A **polygon** is a closed plane figure with at least three sides. To name a polygon, start at any vertex and list the vertices consecutively in a clockwise or counterclockwise direction. A polygon is **convex** if no diagonal contains points outside the polygon. Otherwise, it is **concave.**

An **equilateral polygon** has all sides congruent. An **equiangular polygon** has all angles congruent. A **regular polygon** is equilateral and equiangular.

The sum of the measures of the angles of an n-gon is $(n - 2)180$. The sum of the measures of the exterior angles of an n-gon, one at each vertex, is 360.

Find the measure of an interior angle and an exterior angle of each regular polygon.

30. a hexagon

31. an octagon

32. a decagon

33. a 24-gon

34. What is the sum of the measures of the exterior angles for each polygon in Exercises 30–33?

Lesson 3-6

- To graph lines given their equations (p. 166)
- To write equations of lines (p. 168)

When a linear equation is in **slope-intercept form,** $y = mx + b$, the slope m and the y-intercept b are easily identified. When a linear equation is in **point-slope form,** $(y - y_1) = m(x - x_1)$, point (x_1, y_1) and slope m can easily be identified. The equation $Ax + By = C$, where A and B are not both zero, is in **standard form.** When a linear equation is in standard form, the x- and y-intercepts are readily found.

35. Name the slope and y-intercept of $y = 2x - 1$. Graph the line.

36. Name a point on and the slope of $y - 3 = -2(x + 5)$. Graph the line.

37. Graph $y = -\frac{1}{2}$.

38. Graph $3x - 4y = 12$.

39. Write an equation for the vertical line that contains $A(6, -9)$.

Lesson 3-7

- To relate slope and parallel lines (p. 174)
- To relate slope and perpendicular lines (p. 175)

The slopes of two nonvertical parallel lines are equal. All vertical lines are parallel.

The product of the slopes of two nonvertical perpendicular lines is -1. In a plane, every vertical line is perpendicular to every horizontal line.

Determine whether $\overleftrightarrow{AB}$ and $\overleftrightarrow{CD}$ are *parallel, perpendicular,* or *neither.*

40. $A(-1, -4), B(2, 11), C(1, 1), D(4, 10)$

41. $A(2, 8), B(-1, -2), C(3, 7), D(0, -3)$

42. $A(-3, 3), B(0, 2), C(1, 3), D(-2, -6)$

43. $A(-1, 3), B(4, 8), C(-6, 0), D(2, 8)$

44. Writing For $B(4, 8)$ and $D(2, 8)$, find the slope of $\overleftrightarrow{BD}$. Explain why the slope of any horizontal line is zero.

Chapter 3 Test

For: Chapter test
Web Code: bca-0352

Use a protractor and a centimeter ruler. Classify each triangle by its angles and its sides.

1.

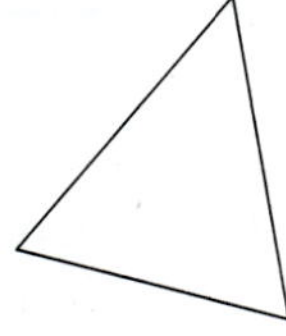

2.

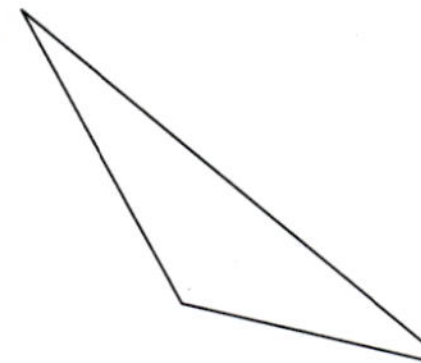

Find $m\angle 1$, then $m\angle 2$. Justify each answer.

3.

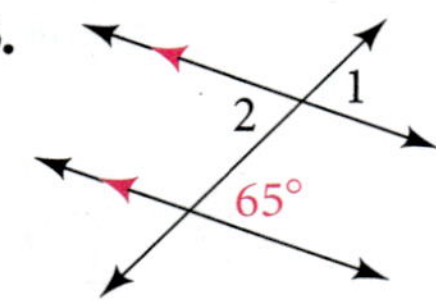

4.

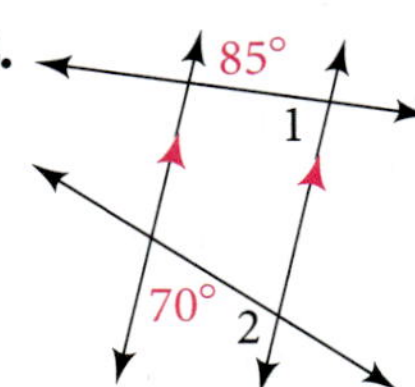

5.

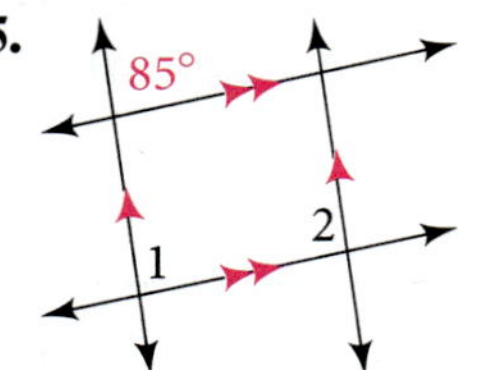

6.

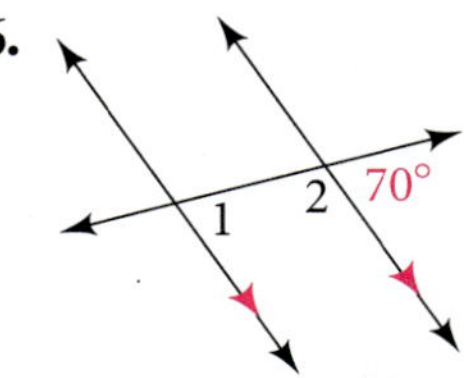

Two lines are parallel and cut by a transversal. Write *yes* or *no* to indicate whether the numbers given could be the measures of a pair of same-side interior angles.

7. 40 and 140

8. 90 and 90

9. 60 and 60

10. 27 and 27

x^2 **Algebra** **Find the value of x for which $\ell \parallel m$.**

11.

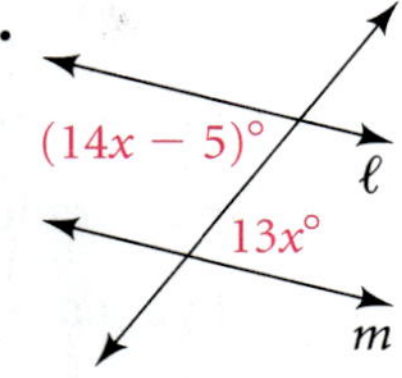

12.

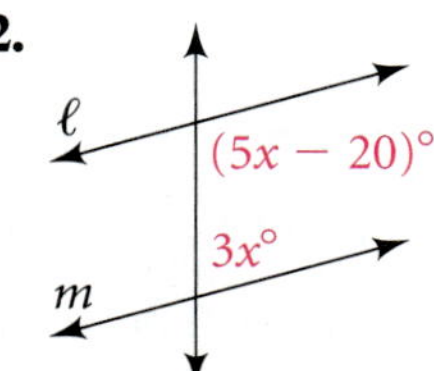

13.

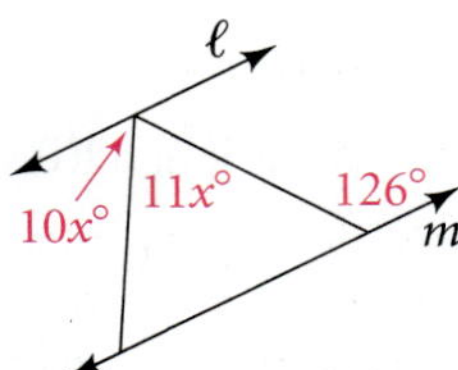

14.

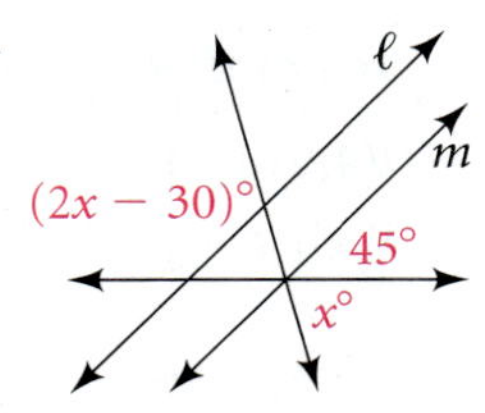

15. Draw a line m and a point T not on the line. Construct the line through T perpendicular to m.

16. Draw an angle, $\angle ABC$. Then construct line m through A so that $m \parallel \overleftrightarrow{BC}$.

17. Open-Ended The letter **F** illustrates a pair of same-side interior angles and a pair of corresponding angles. Find a letter that illustrates alternate interior angles.

18. Open-Ended Describe two corresponding angles formed by lines in your classroom.

19. Supply the reason for each step in the proof.

Given: $\ell \parallel m$ and $\angle 4 \cong \angle 2$

Prove: $n \parallel p$

1. $\ell \parallel m$	**a.** ?
2. $\angle 1 \cong \angle 2$	**b.** ?
3. $\angle 4 \cong \angle 2$	**c.** ?
4. $\angle 1 \cong \angle 4$	**d.** ?
5. $n \parallel p$	**e.** ?

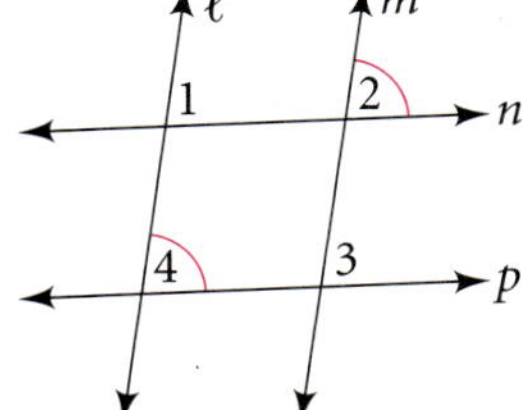

20. Write an equation of the line with slope -5 and containing $A(3, -1)$.

21. Writing Explain how you can determine whether a polygon is concave or convex.

Find the value of each variable.

22.

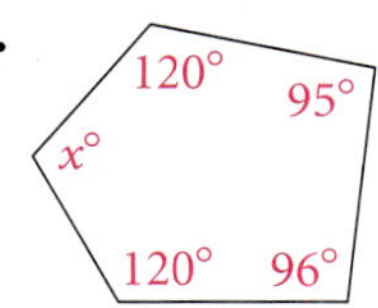

23.

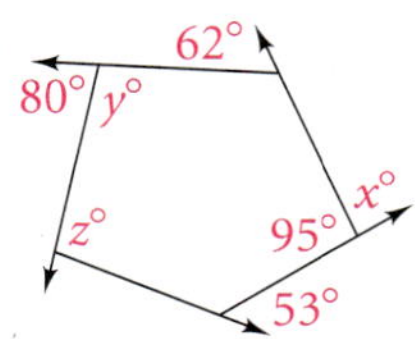

Sketch each pair of lines. Tell whether they are *parallel*, *perpendicular*, or *neither*.

24. $y = 4x + 7$
$y = -\frac{1}{4}x - 3$

25. $y = 3x - 4$
$y = 3x + 1$

26. $y = x + 5$
$y = -5x - 1$

27. $y = -3$
$x = 10$

28. What is the measure of an exterior angle of a regular 12-gon?

Standards Mastery

CAHSEE Practice

Some questions ask you to analyze equations. Read the question at the right. Then follow the tips to answer it.

Tip 2
Find the x-intercept by substituting $y = 0$ into the equation.

The equation of a line is $y = \frac{5}{3}x + 5$. What are x- and y-intercepts of the graph of the line?

(A) The x-intercept is -1, and the y-intercept is 5.

(B) The x-intercept is -3, and the y-intercept is 5.

(C) The x-intercept is 5, and the y-intercept is -1.

(D) The x-intercept is 5, and the y-intercept is -3.

Tip 1
Use the given equation to identify the y-intercept.

Think It Through

The equation $y = \frac{5}{3}x + 5$ is in slope-intercept form, $y = mx + b$, where b is the y-intercept 5.

Substitute 0 for y in the equation

$$0 = \frac{5}{3}x + 5$$

and solve to find that $x = -3$.

The x- and y-intercepts are -3 and 5.

The correct answer is B.

Vocabulary Review

As you solve problems, you must understand the meanings of mathematical terms. Match each term with its mathematical meaning.

A. translation
B. y-intercept
C. x-intercept
D. reflection
E. expression

I. where a line crosses the x-axis
II. the mirror image of a figure
III. all points move the same distance and in the same direction
IV. a mathematical phrase that may include variables, constants, and operation symbols
V. where a line crosses the y-axis

Multiple Choice

Read each question. Then write the letter of the correct answer on your paper.

1. Which expression describes the area of a square that has a side length of $7n^3$?

(A) $14n^6$ (B) $14n^9$ (C) $49n^6$ (D) $49n^9$

2. Which expression is equivalent to $\frac{16a^5b^{-1}c^0}{4ab^6}$?

(A) $\frac{12a^4}{b^5}$ (C) $\frac{4a^4}{b^7}$

(B) $\frac{4a^5c}{b^7}$ (D) $\frac{4a^6bc}{b^7}$

3. The volume of a cube can be found using the formula $V = s^3$. Which expression describes the volume of a cube with a side length of $5xy^2$?

(A) $125x^3y^6$ (C) $125xy^3$

(B) $125x^3y^5$ (D) $125x^3y^5$

4. A line passes through $(3, -4)$ and has a slope of -5. Vic wants to know the y-intercept of the graph of the line. How can he find it?

(A) Substitute -5 for b, -4 for x, and 3 for y in $y = mx + b$. Then solve for m, the y-intercept.

(B) Substitute -5 for b, 3 for x, and -4 for y in $y = mx + b$. Then solve for m, the y-intercept.

(C) Substitute -5 for m, -4 for x, and 3 for y in $y = mx + b$. Then solve for b, the y-intercept.

(D) Substitute -5 for m, 3 for x, and -4 for y in $y = mx + b$. Then solve for b, the y-intercept.

CAHSEE Practice (continued)

5. Which equation describes a line that has a y-intercept of 4 and passes through the point $(6, -1)$?

(A) $y = -\frac{5}{6}x + 4$

(B) $y = -\frac{6}{5}x + 4$

(C) $y = \frac{5}{6}x + 4$

(D) $y = \frac{6}{5}x + 4$

6. Which equation describes the line that passes through point $(4, 6)$ and has a slope of $\frac{1}{2}$?

(A) $y = \frac{1}{2}x + 8$

(B) $y = 2x + 8$

(C) $y = \frac{1}{2}x + 4$

(D) $y = 4x + 4$

7. If $\triangle ABC$, shown below, is reflected across the x-axis what will be the coordinates of its image, $\triangle A'B'C'$?

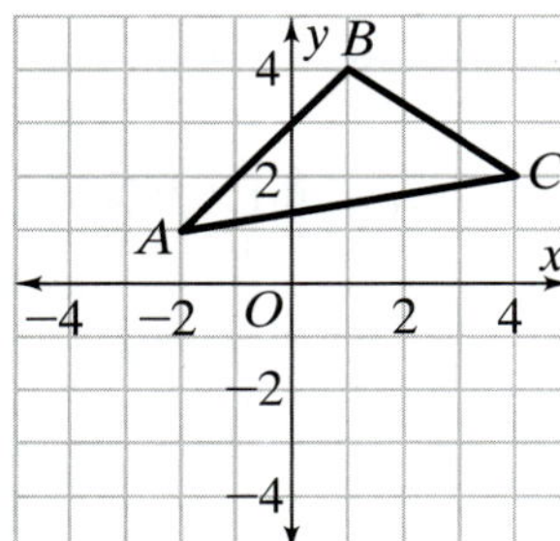

(A) $A'(2, -1), B'(-1, -4)\ C'(-4, -2)$

(B) $A'(2, 1), B'(-1, 4)\ C'(-4, 2)$

(C) $A'(-2, 1), B'(1, -4)\ C'(4, 2)$

(D) $A'(-2, -1), B'(1, -4)\ C'(4, -2)$

8. What is the area, in square units, of $\triangle PQR$?

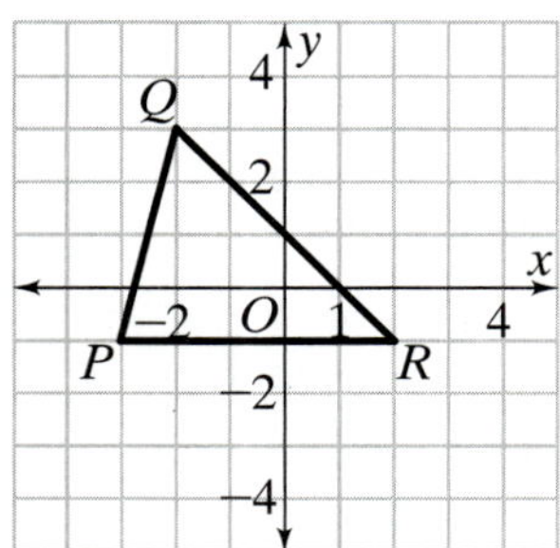

(A) 10

(B) 15

(C) 20

(D) 25

9. Ahkeem puts \$300.00 in a bank account. Each year the account earns 5% in compound interest. How much interest will he have earned after two years?

(A) \$15

(B) \$30

(C) \$30.75

(D) \$31.75

10. Ken went shopping and spent \$14.00 on a book, \$6.50 on lunch, and \$2.50 on bus fare. On the way home he collected \$10 from a friend who owed him money. When he returned home he had \$15.00. He guessed that he started with \$25.00. How can he verify this?

(A) Compute $25 - 14 - 6.5 - 2.5 + 10$ to see if it equals 0.

(B) Compute $25 - 14 - 6.5 - 2.5 + 10$ to see if it equals 15.

(C) Compute $14 + 6.50 + 2.50 - 10$ to see if it equals 25.

(D) Compute $15 - (14 + 6.50 + 2.50) + 10$ to see if it equals 25.

11. What does $y^4 - 2y$ equal when $y = -2$?

(A) -20

(B) -12

(C) 12

(D) 20

12. $x^2y^5 = ?$

(A) $7xy$

(B) $(xy)^7$

(C) $xyyyyxy$

(D) $yyyxxy$

13. A shirt is on sale for 40% off. If the shirt reqularly sells for \$30, what is the sale price of the shirt?

(A) \$10

(B) \$12

(C) \$18

(D) \$24

14. Mrs. Olms surveyed her clarinet students to find how long they practiced each day. The graph shows the relation of each student's age to the number of minutes practiced daily. Based on these results, what type of correlation exists between the age of the student and the number of minutes of daily practice?

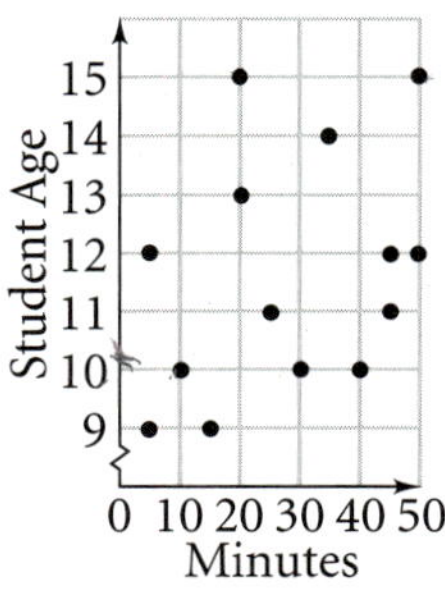

(A) Positive correlation

(B) Negative correlation

(C) Constant correlation

(D) No correlation

Standards Mastery

15. Which graph shows a negative correlation?

Ⓐ 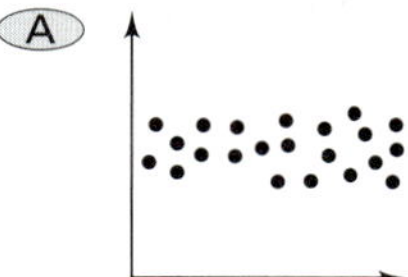

Ⓑ 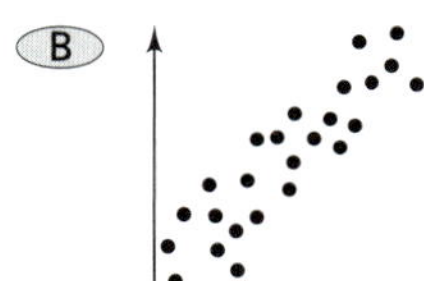

Ⓒ 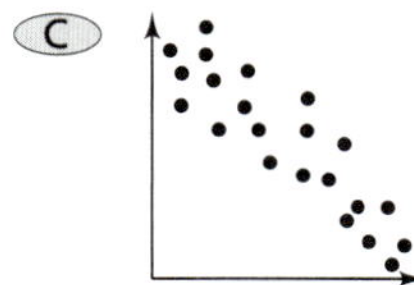

Ⓓ 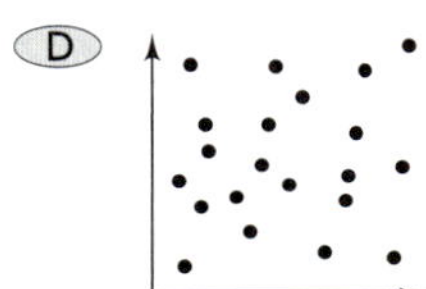

16. Mangos are on sale for 30% off the regular price p. Which expression represents the sale price?

Ⓐ $0.3p$
Ⓑ $-30p$
Ⓒ $p - 0.3$
Ⓓ $p - 0.3p$

17. Owen invested \$15,000 and received \$955 in interest last year. He invested part at 8% interest and the remainder at 4.5%. If x represents the amount Owen invested at 8% and y the amount invested at 4.5%, about how much did he invest at 8%?

Ⓐ \$1200
Ⓑ \$5857
Ⓒ \$8000
Ⓓ \$14,045

18. The perimeter of a rectangular garden is 160 feet. The length is 15 times its width. What is the area of the garden in square feet?

Ⓐ 150
Ⓑ 375
Ⓒ 800
Ⓓ 1600

19. The vertices of $\triangle ABC$ have coordinates $A(-2, 3)$, $B(3, 4)$, and $C(1, -2)$. Which transformation can be used to map point B to the point B' $(3, -4)$?

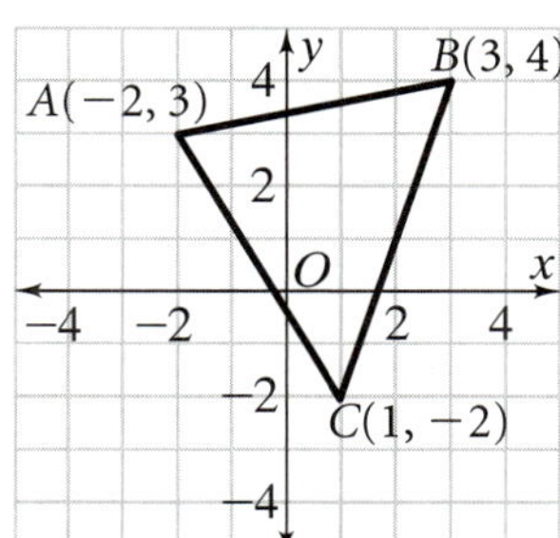

Ⓐ A reflection across the y-axis
Ⓑ A reflection across the x-axis
Ⓒ A rotation 180° clockwise around the origin
Ⓓ A translation 4 units down

CAHSEE Reference Guide

California Content Standard	Item Number(s)
7MG 2.1	1, 3, 18
7AF 2.1	2, 11, 12
Alg1 7.0	4, 5, 6
7MG 3.2	7, 8, 19
7NS 1.7	9, 13, 16, 17
7MR 1.1	10
7SDAP 1.2	14, 15

For additional review and practice, see the *California Standards Review and Practice Workbook* or go online to use the

Visit: PHSchool.com, **Web Code:** bcq-9045

Congruent Triangles

What You've Learned

California Content Standards

1.0 Students demonstrate understanding by identifying and giving examples of undefined terms, axioms, theorems, and inductive and deductive reasoning.

3.0 Students construct and judge the validity of a logical argument and give counterexamples to disprove a statement.

16.0 Students perform basic constructions with a straightedge and compass, such as angle bisectors, perpendicular bisectors, and the line parallel to a given line through a point off the line.

The Distance Formula (Lesson 1-8)

Find the lengths of the sides of $\triangle ABC$.

1. $A(3, 1), B(-1, 1), C(-1, -2)$ **2.** $A(-3, 2), B(-3, -6), C(8, 6)$ **3.** $A(-1, -2), B(6, 1), C(2, 5)$

Proving Angles Congruent (Lesson 2-5)

Draw a conclusion based on the information given.

4. $\angle A$ is supplementary to $\angle B$; $\angle C$ is supplementary to $\angle B$

5. $\angle A$ is supplementary to $\angle B$; $\angle A \cong \angle B$

6. $\angle 1$ is complementary to $\angle 2$

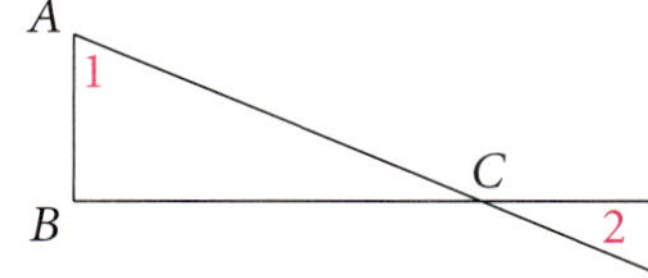

7. $\overrightarrow{FA} \perp \overrightarrow{FC}$; $\overrightarrow{FB} \perp \overrightarrow{FD}$

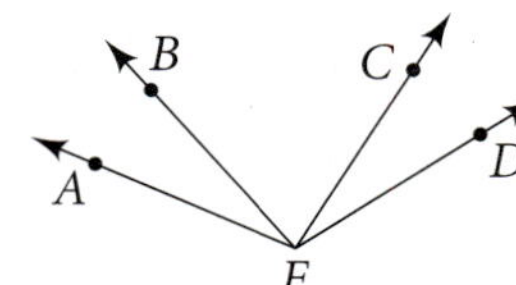

Parallel Lines and the Triangle Angle-Sum Theorem (Lesson 3-3)

What can you conclude from each diagram?

8.

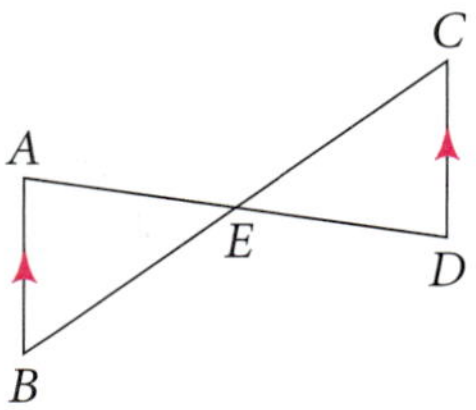

9.

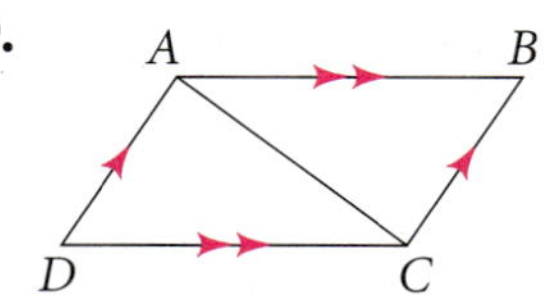

10.

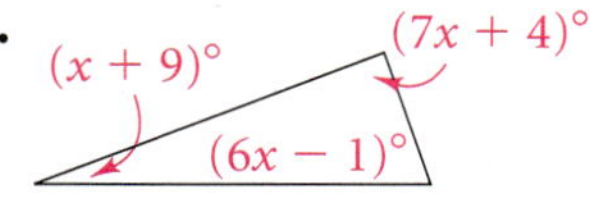

▲ The Hall of Mirrors at France's Palace of Versailles is brightly lit even on a cloudy day, because each of its windows is placed opposite a mirror of the same size and shape.

What You'll Learn Next

California Content Standards

2.0 Students write geometric proofs.

4.0 Students prove basic theorems involving congruence.

5.0 Students prove that triangles are congruent, and are able to use the concept of corresponding parts of congruent triangles.

12.0 Students find and use measures of sides and of interior and exterior angles of triangles and polygons to classify figures and solve problems.

CAHSEE 7NS 2.3, 7SDAP 1.2, 7AF 1.5, 7AF 2.2, 7AF 4.1, 7MG 1.1, 7MG 1.3, Alg1 9.0

New Vocabulary

English and Spanish Audio Online

- **base of an isosceles triangle** (p. 228)
- **base angles of an isosceles triangle** (p. 228)
- **congruent polygons** (p. 198)
- **corollary** (p. 229)
- **CPCTC (corresponding parts of congruent triangles are congruent)** (p. 221)
- **hypotenuse** (p. 235)
- **legs of a right triangle** (p. 235)
- **legs of an isosceles triangle** (p. 228)
- **vertex angle of an isosceles triangle** (p. 228)

Congruent Figures

California Content Standards

GEOM 4.0 Students prove basic theorems involving congruence. ***Develop***
GEOM 5.0 Students prove that triangles are congruent. ***Introduce***
GEOM 12.0 Students find and use measures of sides and of interior angles of triangles to solve problems. ***Develop***

What You'll Learn

- To recognize congruent figures and their corresponding parts

. . . And Why

To use corresponding parts of congruent shapes in the Space Shuttle, as in Example 2

Check Skills You'll Need

for Help Algebra 1 Review, page 30

Algebra **Solve each equation.**

1. $x + 6 = 25$ **2.** $x + 7 + 13 = 33$

3. $5x = 540$ **4.** $x + 10 = 2x$

5. For the triangle at the right, use the Triangle Angle-Sum Theorem to find the value of y.

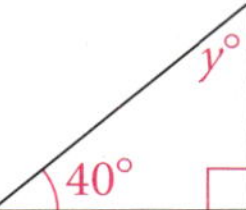

New Vocabulary • congruent polygons

1 Congruent Figures

Congruent figures have the same size and shape. When two figures are congruent, you can move one so that it fits exactly on the other one. Three ways to make such a move—a slide, a flip, and a turn—are shown below. You will learn much more about slides, flips, and turns in Chapter 9.

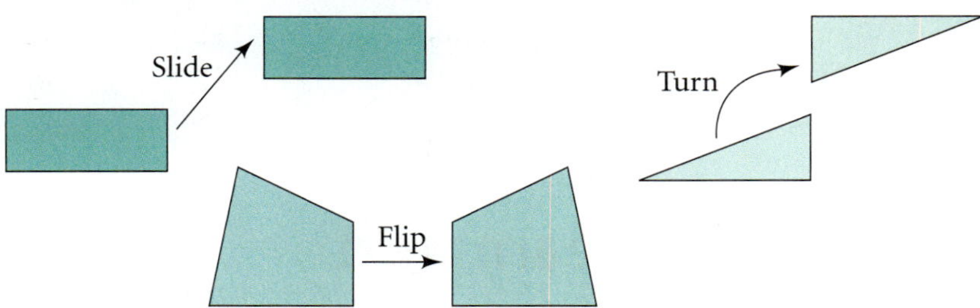

Congruent polygons have congruent corresponding parts—their matching sides and angles. Matching vertices are corresponding vertices. When you name congruent polygons, always list corresponding vertices in the same order.

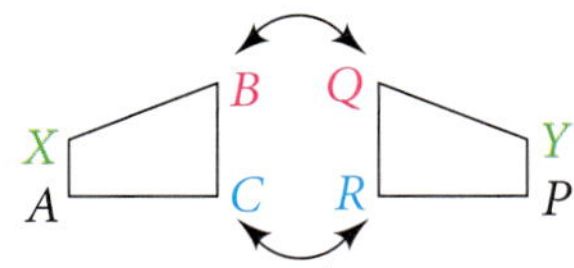

C corresponds to R.
$\angle B$ corresponds to $\angle Q$.
$\overline{AX}$ corresponds to $\overline{PY}$.
$ACBX \cong PRQY$

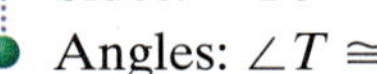

1 EXAMPLE Naming Congruent Parts

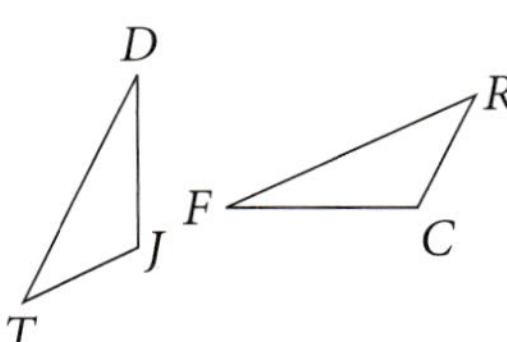

$\triangle TJD \cong \triangle RCF$. List the congruent corresponding parts.

Sides: $\overline{TJ} \cong \overline{RC}$ $\overline{JD} \cong \overline{CF}$ $\overline{DT} \cong \overline{FR}$

Angles: $\angle T \cong \angle R$ $\angle J \cong \angle C$ $\angle D \cong \angle F$

CA Standards Check

1 $\triangle WYS \cong \triangle MKV$. List the congruent corresponding parts. Use three letters to name each angle.

2 EXAMPLE Real-World Connection

Spacecraft The fins of the Space Shuttle suggest congruent pentagons. Find $m\angle B$.

In the congruent pentagons, B corresponds to E, so you know that $\angle B \cong \angle E$. You can find $m\angle B$ by first finding $m\angle E$.

Use the Polygon Angle-Sum Theorem. It tells you that the sum of the measures of the angles of pentagon $SPACE$ is $(5 - 2)180$, or 540.

$m\angle S + m\angle P + m\angle A + m\angle C + m\angle E = 540$	**Polygon Angle-Sum Theorem**
$88 + 90 + 90 + 132 + m\angle E = 540$	**Substitute.**
$400 + m\angle E = 540$	**Simplify.**
$m\angle E = 140$	**Subtract 400 from each side.**

$m\angle B = m\angle E$, so $m\angle B = 140$.

CA Standards Check 2 It is given that $\triangle WYS \cong \triangle MKV$. If $m\angle Y = 35$, what is $m\angle K$? Explain.

Two triangles are congruent when they have three pairs of congruent corresponding sides and three pairs of congruent corresponding angles.

3 EXAMPLE Finding Congruent Triangles

Decide whether the triangles are congruent. Justify your answer.

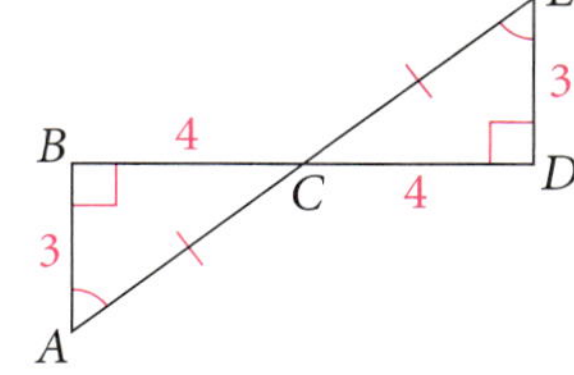

$\overline{AC} \cong \overline{EC}$	**Given**
$\overline{AB} \cong \overline{ED}$	**$AB = 3 = ED$**
$\overline{BC} \cong \overline{DC}$	**$BC = 4 = DC$**
$\angle A \cong \angle E$	**Given**
$\angle B \cong \angle D$	**All right angles are congruent.**
$\angle BCA \cong \angle DCE$	**Vertical angles are congruent.**

$\triangle ABC \cong \triangle EDC$ by the definition of congruent triangles.

CA Standards Check 3 Can you conclude $\triangle JKL \cong \triangle MNL$? Justify your answer.

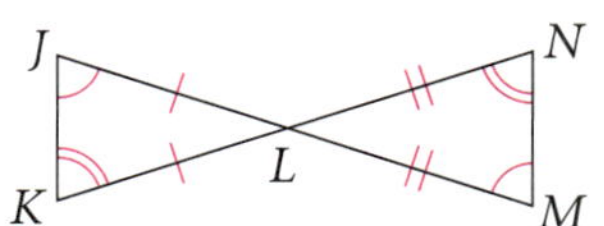

The next theorem follows from the Triangle Angle-Sum Theorem. In Exercise 45, you will show why this theorem is true.

Key Concepts

Theorem 4-1

If two angles of one triangle are congruent to two angles of another triangle, then the third angles are congruent.

$\angle C \cong \angle F$

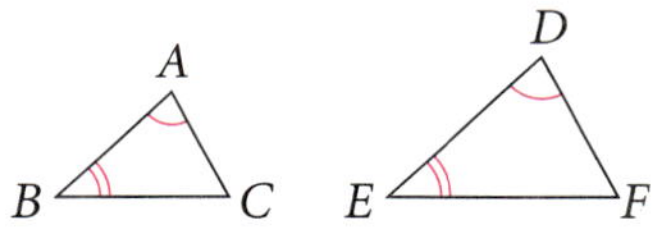

Example 4 shows a typical setup—Given, Prove, and diagram—that requires a proof from you. The form of proof you use is generally a matter of preference.

Proof

4 EXAMPLE Proving Triangles Congruent

Given: $\overline{PQ} \cong \overline{PS}$, $\overline{QR} \cong \overline{SR}$, $\angle Q \cong \angle S$, $\angle QPR \cong \angle SPR$

Prove: $\triangle PQR \cong \triangle PSR$

Statements	Reasons
1. $\overline{PQ} \cong \overline{PS}$, $\overline{QR} \cong \overline{SR}$	1. Given
2. $\overline{PR} \cong \overline{PR}$	2. Reflexive Property of $\cong$
3. $\angle Q \cong \angle S$, $\angle QPR \cong \angle SPR$	3. Given
4. $\angle QRP \cong \angle SRP$	4. Theorem 4-1
5. $\triangle PQR \cong \triangle PSR$	5. Definition of $\cong$ triangles

4 Given: $\angle A \cong \angle D$, $\angle E \cong \angle C$, $\overline{AE} \cong \overline{DC}$, $\overline{EB} \cong \overline{CB}$, $\overline{BA} \cong \overline{BD}$

Prove: $\triangle AEB \cong \triangle DCB$

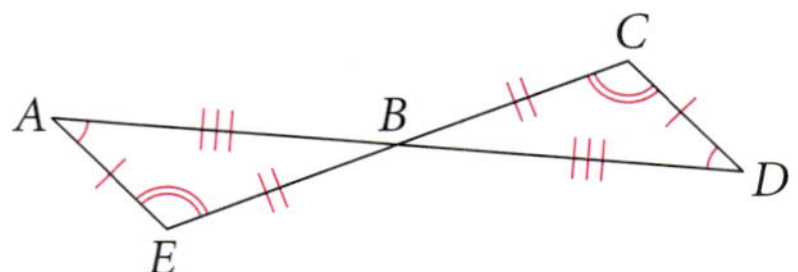

EXERCISES

For more exercises, see *Extra Skill, Word Problem, and Proof Practice.*

Standards Practice

GEOM 4.0, 5.0, 12.0

A Practice by Example

Example 1 (page 198)

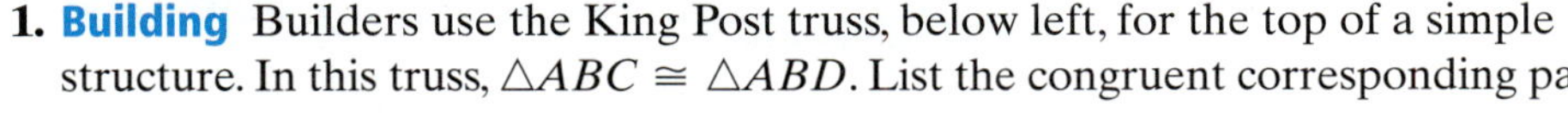

1. Building Builders use the King Post truss, below left, for the top of a simple structure. In this truss, $\triangle ABC \cong \triangle ABD$. List the congruent corresponding parts.

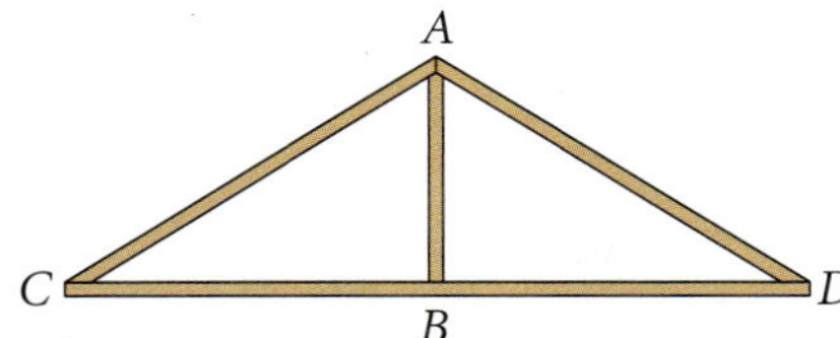

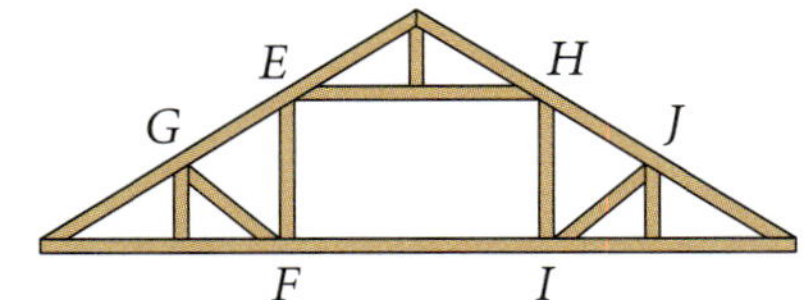

2. The Attic Frame truss, above right, provides open space in the center for storage. In this truss, $\triangle EFG \cong \triangle HIJ$. List the congruent corresponding parts.

Real-World Connection

Exposed beams show the congruent triangles used in Tudor architecture.

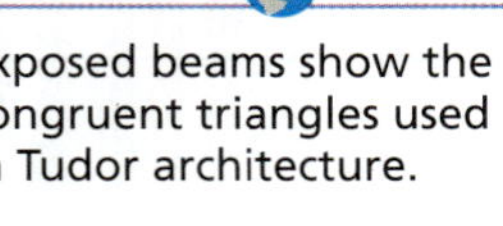

$\triangle LMC \cong \triangle BJK$. Complete the congruence statements.

3. $\overline{LC} \cong$ __?__ **4.** $\overline{KJ} \cong$ __?__

5. $\overline{JB} \cong$ __?__ **6.** $\angle L \cong$ __?__

7. $\angle K \cong$ __?__ **8.** $\angle M \cong$ __?__

9. $\triangle CML \cong$ __?__ **10.** $\triangle KBJ \cong$ __?__

11. $\triangle MLC \cong$ __?__ **12.** $\triangle JKB \cong$ __?__

13. The last piece of the jigsaw puzzle must be put into place. Name the corners that correspond to corners A, B, C, and D.

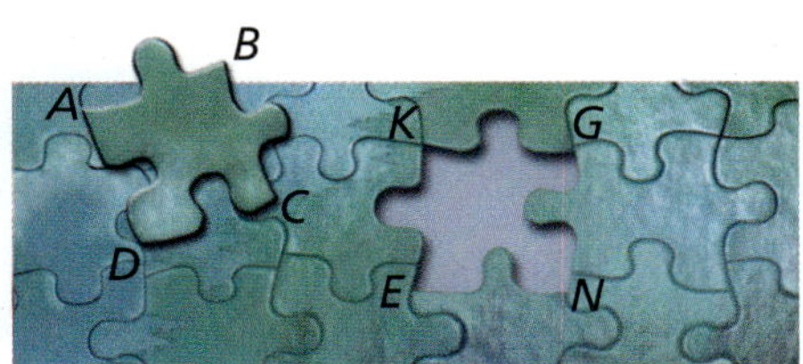

***POLY* $\cong$ *SIDE*. List each of the following.**

14. four pairs of congruent sides

15. four pairs of congruent angles

Example 2 (page 199)

In the two lifeguard chairs, $ABCD \cong FGHI$. Find the measure of the angle or the length of the side.

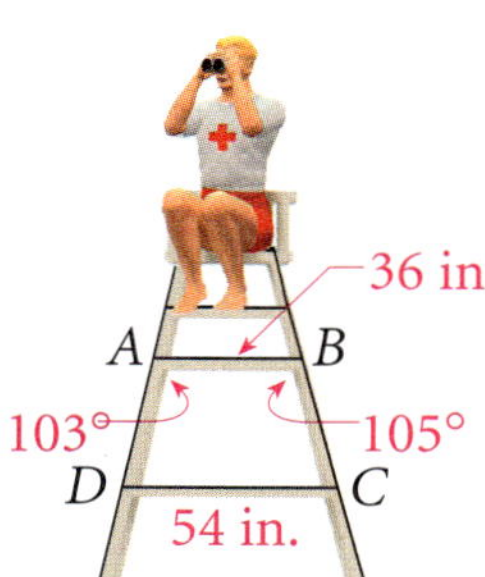

16. $\overline{AD}$
17. $\overline{HI}$
18. $\angle FGH$
19. $\angle ADC$
20. $\overline{FG}$
21. $\overline{BC}$
22. $\angle DCB$
23. $\angle IFG$

Example 3 (page 199)

In Exercises 24–27, can you conclude the figures are congruent? Justify each answer.

24. $\triangle TRK$ and $\triangle TUK$

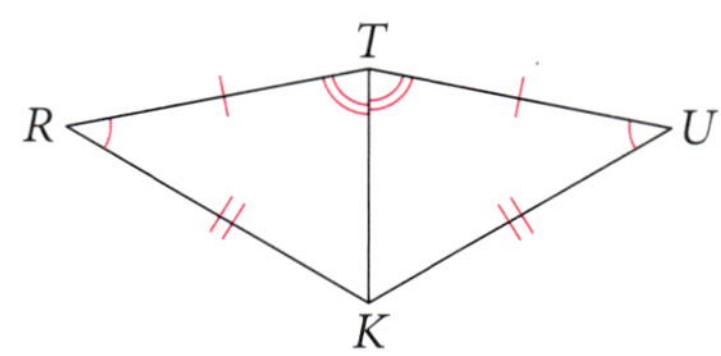

25. $\triangle SPQ$ and $\triangle TUV$

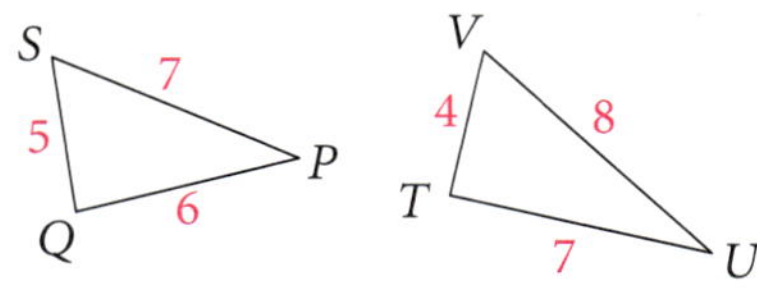

26. $\triangle XYZ$ and $\triangle XYP$

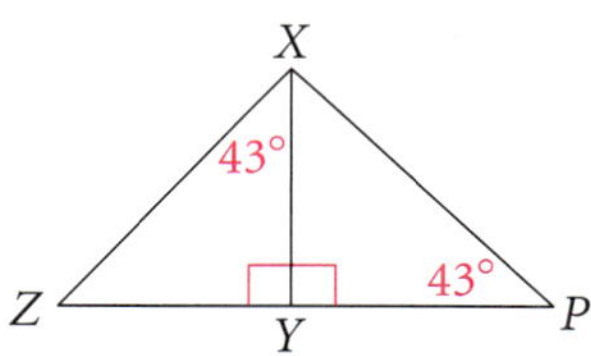

27. $HEJK$ and $GFJK$

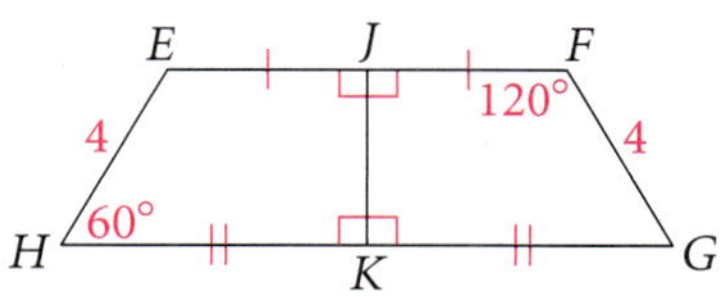

Example 4 (page 200) Proof

28. **Given:** $\overline{AB} \parallel \overline{DC}$, $\angle B \cong \angle D$, $\overline{AB} \cong \overline{DC}$, $\overline{BC} \cong \overline{AD}$

Prove: $\triangle ABC \cong \triangle CDA$

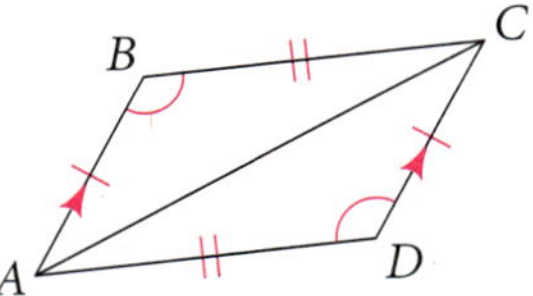

B Apply Your Skills

29. **Multiple Choice** If $\triangle DEF \cong \triangle LMN$, which of the following must be a correct congruence statement?

Ⓐ $\overline{DE} \cong \overline{LN}$
Ⓑ $\angle N \cong \angle F$
Ⓒ $\overline{FE} \cong \overline{NL}$
Ⓓ $\angle M \cong \angle F$

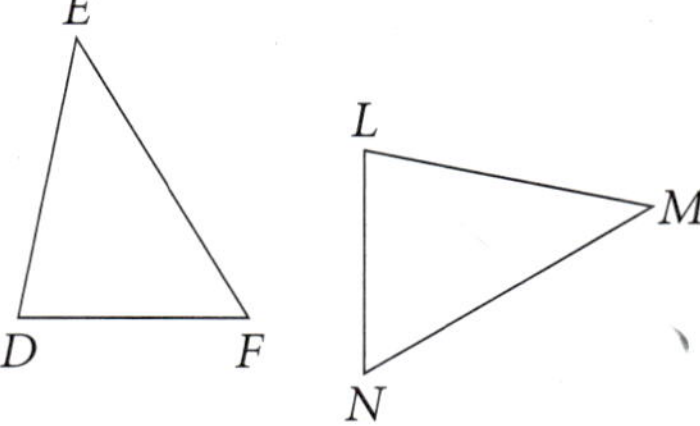

 **Algebra** **Find the values of the variables.**

To review the Triangle Angle-Sum Theorem, go to Lesson 3-4.

30.

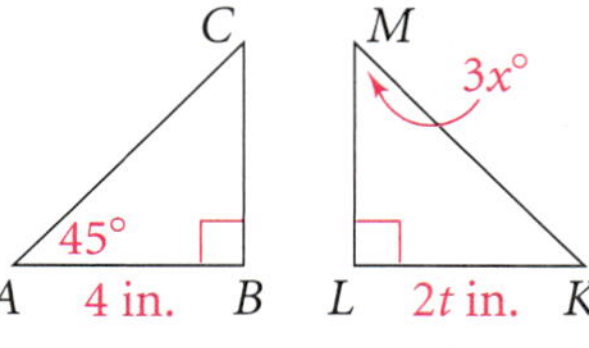

$\triangle ABC \cong \triangle KLM$

31.

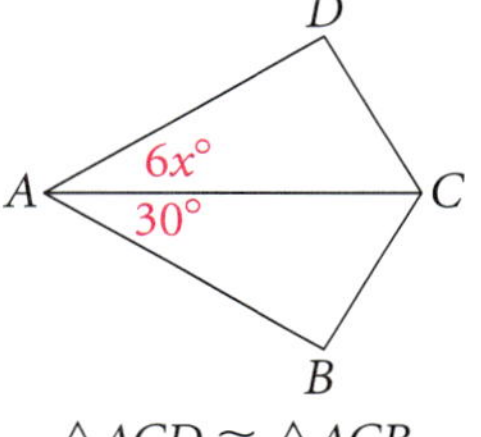

$\triangle ACD \cong \triangle ACB$

Visit: PHSchool.com
Web Code: bce-0401

 **Algebra** **$\triangle ABC \cong \triangle DEF$. Find the measures of the given angles or the lengths of the given sides.**

32. $m\angle A = x + 10, m\angle D = 2x$

33. $m\angle B = 3y, m\angle E = 21$

34. $BC = 3z + 2, EF = z + 6$

35. $AC = 7a + 5, DF = 5a + 9$

36. **Parquet Floor** Explain why it is important that $PACH \cong OLDE$.

37. **Sports Cards** The 225 cards in Tracy's sports card collection are rectangles of three different sizes. Describe how Tracy could quickly sort the cards.

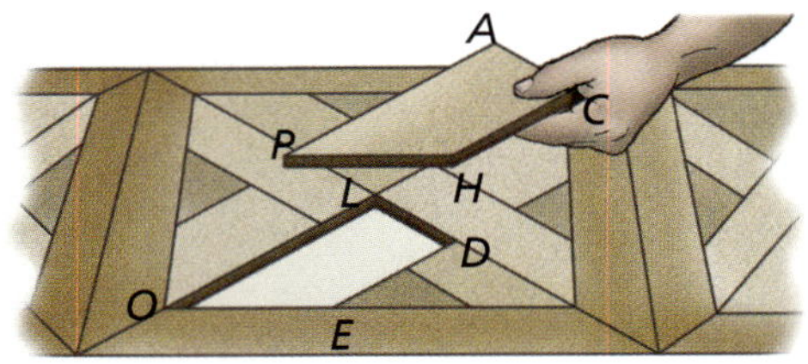

Exercise 36

Write a congruence statement for each pair of triangles.

38.

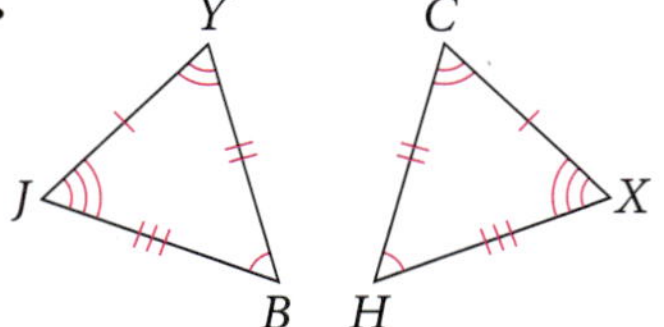

39.

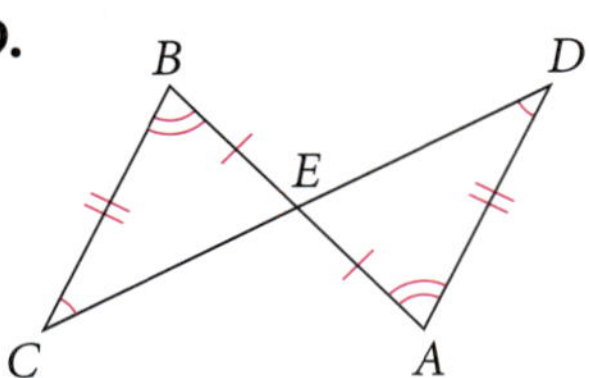

E is the midpoint of $\overline{CD}$.

40.

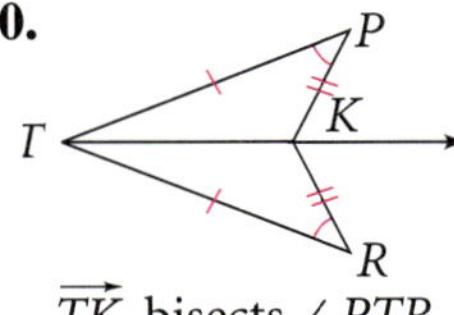

$\overrightarrow{TK}$ bisects $\angle PTR$.

41. Complete in two different ways:

$\triangle JLM \cong$ __?__

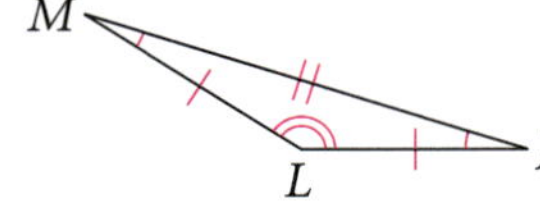

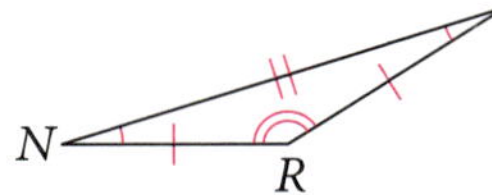

 **42.** **Writing** Die-cast toys are a popular collector's item. Explain why the two die-cast toys that Pearl is studying at the left have congruent shapes.

Exercise 42

43. **Open-Ended** Write a congruence statement for two triangles. List the congruent sides and angles.

Proof **44.** **Given:** $\overline{PR} \parallel \overline{TQ}, \overline{PR} \cong \overline{TQ}$, $\overline{PS} \cong \overline{QS}$, $\overline{PQ}$ bisects $\overline{RT}$.

Prove: $\triangle PRS \cong \triangle QTS$

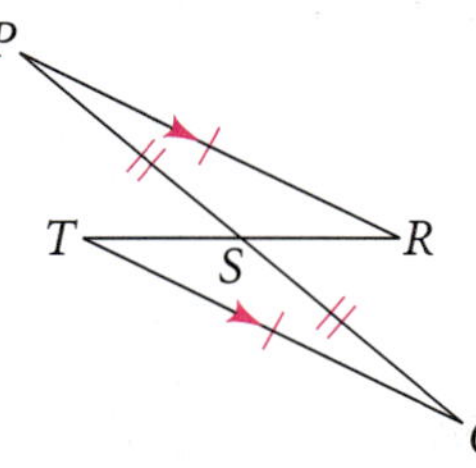

45. Prove Theorem 4-1.

Given: $\angle A \cong \angle D, \angle B \cong \angle E$

Prove: $\angle C \cong \angle F$

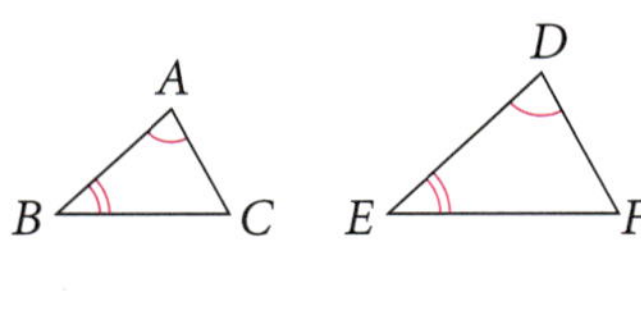

Coordinate Geometry **Vertices of $\triangle GHJ$ are $G(-2, -1), H(-2, 3)$, and $J(1, 3)$.**

46. $\triangle KLM \cong \triangle GHJ$. Find KL, LM, and KM.

47. If L and M have coordinates $L(3, -3)$ and $M(6, -3)$, how many pairs of coordinates are possible for K? Find one such pair.

48. **a.** How many quadrilaterals (convex and concave) with different shapes or sizes can you make on a three-by-three geoboard? One is shown at the right.

b. How many quadrilaterals of each type are there?

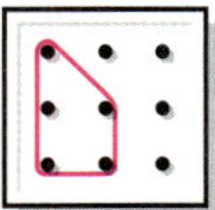

Multiple Choice Practice

For California Standards and CAHSEE Tutorials, visit PHSchool.com. Web Code: bcq-9045

GEOM 12.0

49. What conclusion can be drawn from the figure below?

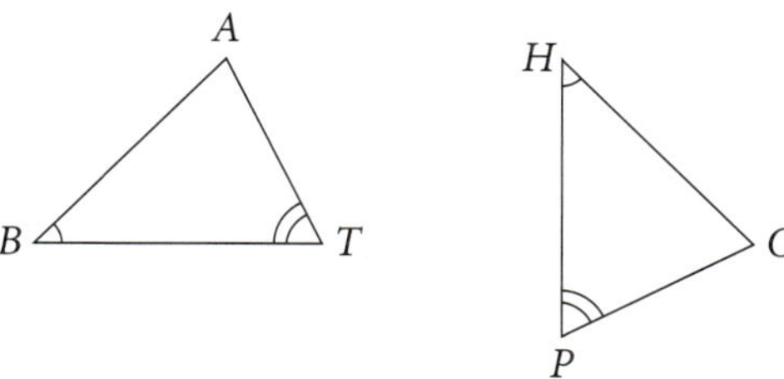

Ⓐ $\overline{BT} \cong \overline{HP}$ Ⓑ $\overline{AT} \parallel \overline{HO}$ Ⓒ $\angle A \cong \angle O$ Ⓓ $m\angle A = 36°$

Mixed Review

Lesson 3-8

Constructions **For Exercises 50 and 51, construct the geometric figure.**

50. a square

51. a rectangle whose length is twice its width

Lesson 3-4

52. Find $m\angle A$ in the figure at the right.

C, 47°, B, 33°, A

Exercise 52

Lesson 2-4

Use the given property to complete each statement.

53. Symmetric Property of Equality

If $PQ = RS$, then __?__.

54. Reflexive Property of Congruence

$\angle 1 \cong$ __?__

55. Addition Property of Equality

If $m\angle A - 4 = 8$, then $m\angle A =$ __?__.

56. Transitive Property of Congruence

If $\overline{AB} \cong \overline{DE}$ and $\overline{DE} \cong \overline{GH}$, then __?__.

Geometry at Work

Die Casting

Two centuries ago, people manufactured articles by hand. Each article produced was slightly different from every other. In 1800, inventor Eli Whitney recognized that he could speed up manufacturing by using congruent parts. Whitney made a die, or mold, for each part of a musket he was producing for the U.S. Army. This allowed workers to rapidly cast the parts and assemble them into standard-sized muskets. It ushered in the era of mass production.

Today, die makers are highly skilled industrial workers who shape dies out of metal, plastic, rubber, and other materials. Machines create and assemble the congruent die-cast parts into standard-sized objects, like the die-cast toy cars at the left. Other workers supply a final inspection and skilled hand finishing.

For: Information about die casting
Web Code: bcb-2031

Building Congruent Triangles

Hands-On

FOR USE WITH LESSON 4-2

GEOM 16.0 Students perform basic constructions with a straightedge and compass. ***Develop***

Are there shortcuts for finding congruent triangles? Explore this with your classmates by building and comparing triangles.

1 ACTIVITY

- Cut three strips of paper with lengths 7 in., 8 in., and 9 in.
- Use the strips to form a triangle on top of a piece of tracing paper. Mark each vertex of the triangle on your tracing paper.
- Remove the strips. Use a straightedge to carefully draw your triangle.

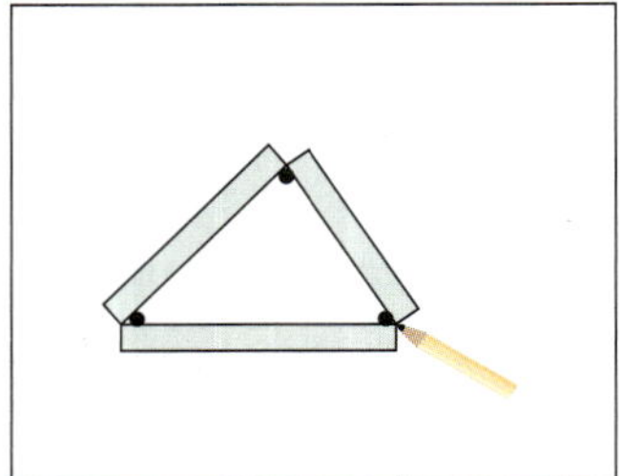

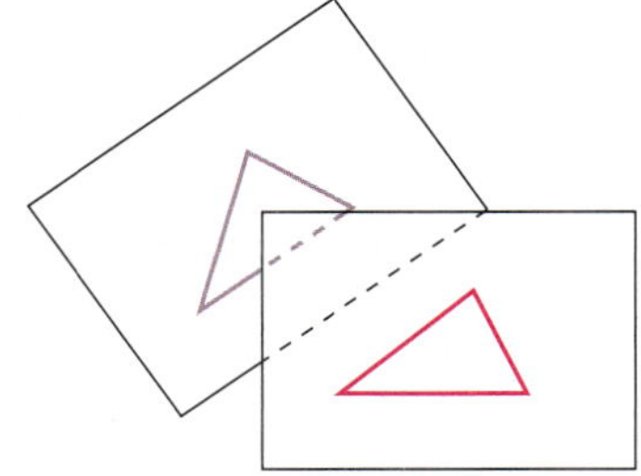

1. Is your triangle congruent to your classmates' triangles? Place your paper on top of a classmate's paper. Try to make your triangle fit exactly on the other triangle.

2. Make a conjecture: What seems to be true about two triangles in which three sides of one are congruent to three sides of the other?

3. As a class, choose three different lengths and repeat the steps above.
a. Are the triangles congruent?
b. Does this support your conjecture from Exercise 2?

2 ACTIVITY

- Use a straightedge to carefully draw a triangle on tracing paper. Label it $\triangle ABC$.
- Use a protractor. Carefully measure $\angle A$ and $\angle B$. Use a ruler to measure the side between them, $\overline{AB}$.
- Write the measurements on an index card and swap cards with a classmate. Draw a triangle using only your classmate's measurements.

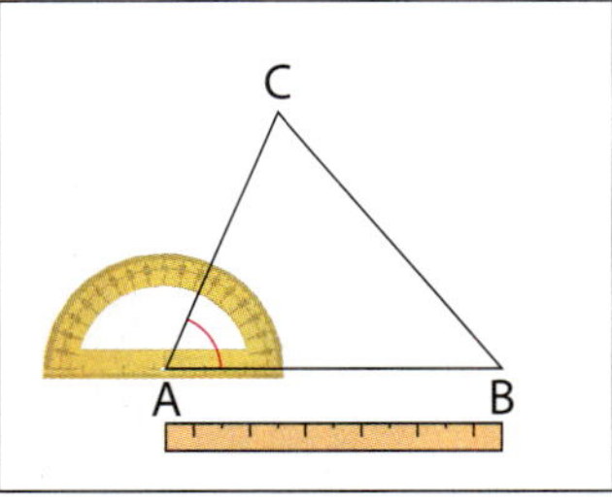

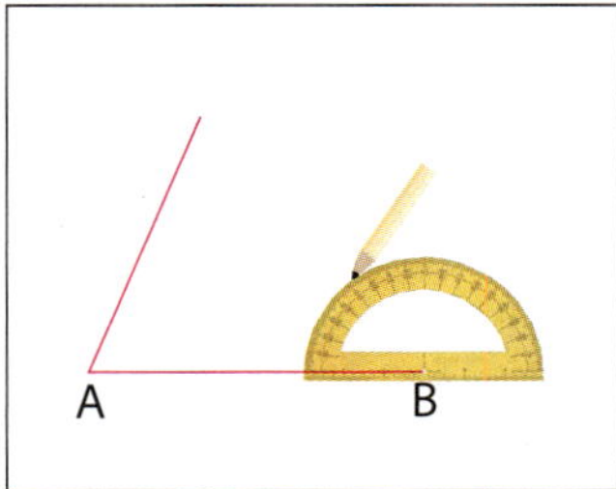

4. Compare your new triangle to your classmate's original $\triangle ABC$. Are the triangles congruent? Is your classmate's new triangle congruent to your original $\triangle ABC$?

5. Make a conjecture: What seems to be true when two triangles have congruent sides between two pairs of congruent angles?

6. Make a conjecture: At most, how many triangle measurements must you know to guarantee that all triangles built with those measurements will be congruent?

Triangle Congruence by SSS and SAS

California Content Standards

GEOM 2.0 Students write geometric proofs. ***Develop***
GEOM 5.0 Students prove that triangles are congruent. ***Develop***

What You'll Learn

- To prove two triangles congruent using the SSS and SAS Postulates

. . . And Why

To prove that two triangles in the framework of a bridge are congruent, as in Example 1

What can you conclude from each diagram?

1.

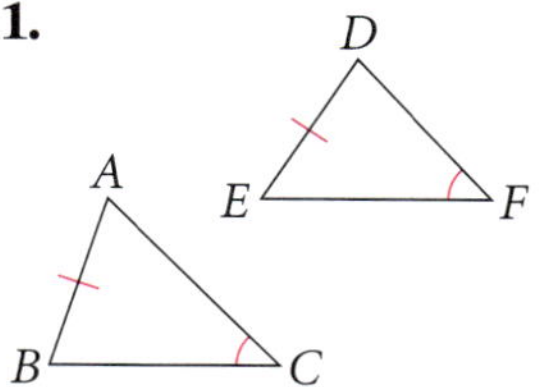

2.

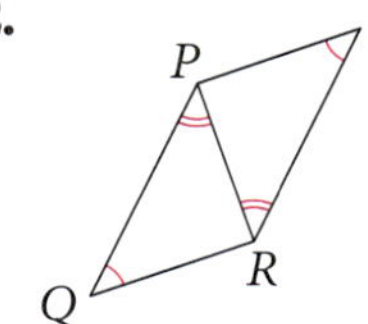

3.

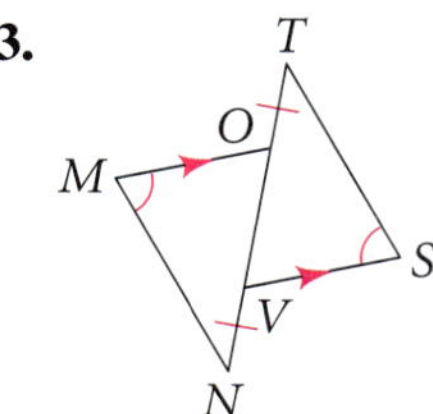

Using the SSS and SAS Postulates

In Lesson 4-1 you learned that if two triangles have three pairs of congruent corresponding angles and three pairs of congruent corresponding sides, then the triangles are congruent.

If you know this,

$\angle A \cong \angle X$	$\overline{AB} \cong \overline{XY}$
$\angle B \cong \angle Y$	$\overline{AC} \cong \overline{XZ}$
$\angle C \cong \angle Z$	$\overline{BC} \cong \overline{YZ}$

then you know this.

$\triangle ABC \cong \triangle XYZ$

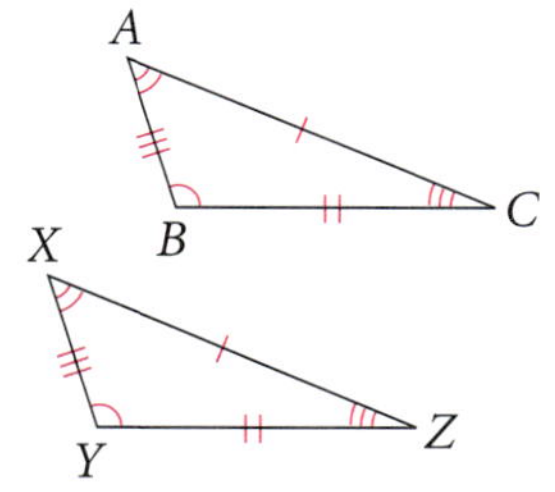

However, you do not need to know that all six corresponding parts are congruent in order to conclude that two triangles are congruent. It is enough to know only that corresponding sides are congruent.

Key Concepts

Postulate 4-1 **Side-Side-Side (SSS) Postulate**

If the three sides of one triangle are congruent to the three sides of another triangle, then the two triangles are congruent.

$\triangle GHF \cong \triangle PQR$

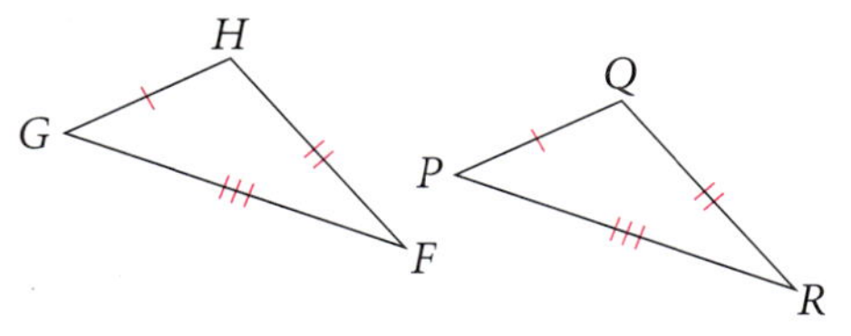

As described on page 101, postulates are the self-evident truths in a mathematical system. Postulate 4-1 is perhaps the most self-evident truth about triangles. It agrees with the notion that triangles are rigid figures, a property demonstrated in bicycle frames and steel bridges.

Proof

1 EXAMPLE Using SSS

Bridge Design The bridge girders are the same size, as marked. Write a flow proof.

Given: $\overline{AB} \cong \overline{CB}, \overline{AD} \cong \overline{CD}$

Prove: $\triangle ABD \cong \triangle CBD$

Plan: To prove the triangles congruent by the SSS Postulate, the three pairs of sides need to be congruent.

Proof:

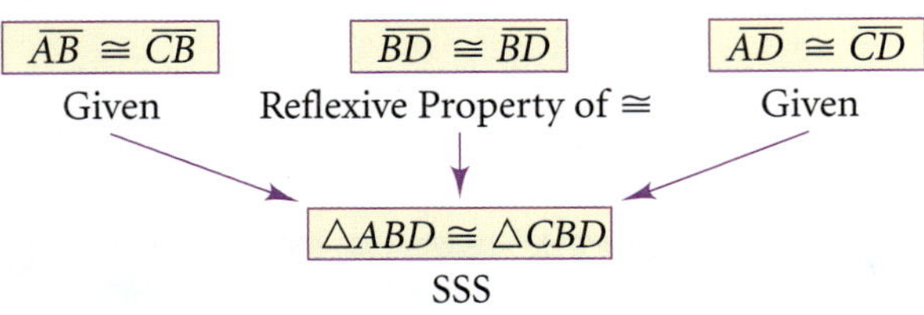

Vocabulary Tip

The Reflexive Property of Congruence tells you that a figure, such as $\overline{BD}$ in this Example, is congruent to itself.

CA Standards Check 1 **Given:** $\overline{HF} \cong \overline{HJ}, \overline{FG} \cong \overline{JK}$, H is the midpoint of $\overline{GK}$.

Prove: $\triangle FGH \cong \triangle JKH$

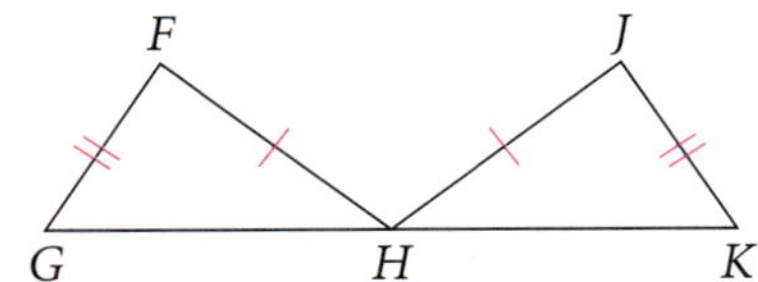

Key Concepts

Postulate 4-2 **Side-Angle-Side (SAS) Postulate**

If two sides and the included angle of one triangle are congruent to two sides and the included angle of another triangle, then the two triangles are congruent.

$\triangle BCA \cong \triangle FDE$

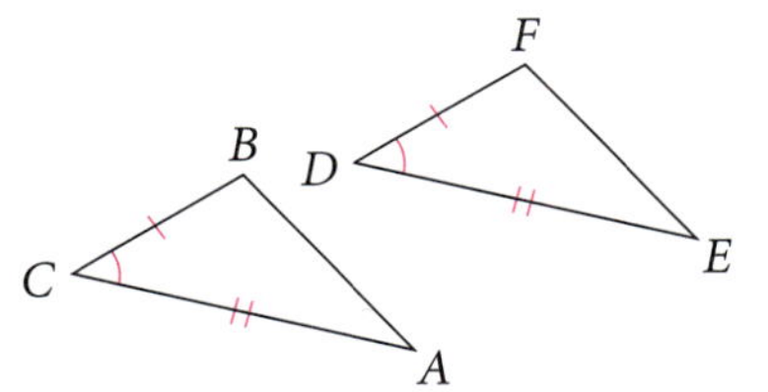

The word *included* is used frequently when referring to the angles and the sides of a triangle.

$\overline{BX}$ is included between $\angle B$ and $\angle X$.

$\angle N$ is included between $\overline{NB}$ and $\overline{NX}$.

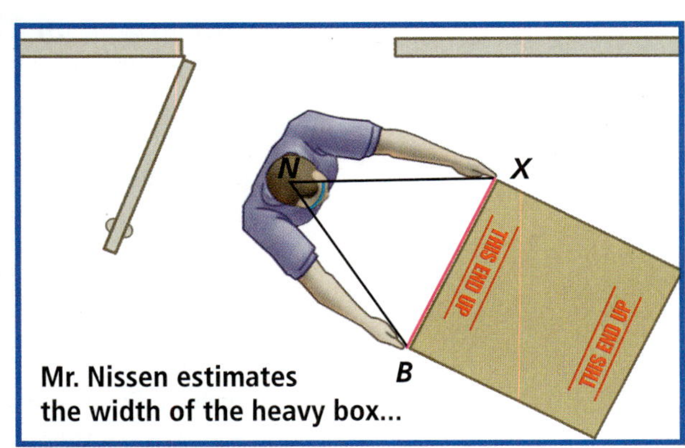

Mr. Nissen kept his arms at a fixed angle as he moved from the box to the doorway. The triangle he used beside the box is congruent to the triangle he used beside the doorway. He knows the two triangles are congruent because two sides and the included angle of one are congruent to two sides and the included angle of the other.

GO Online

Video Tutor Help
Visit: PHSchool.com
Web Code: bce-0775

2 EXAMPLE Using SAS

$\overline{RS} \cong \overline{TK}$. What other information do you need to prove $\triangle RSK \cong \triangle TKS$ by SAS?

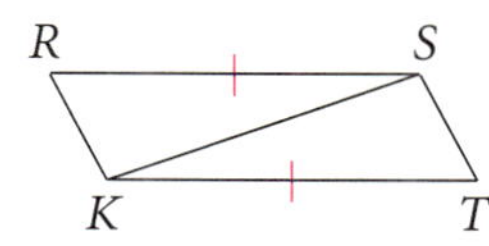

You are given $\overline{RS} \cong \overline{TK}$. Also, $\overline{KS} \cong \overline{KS}$ by the Reflexive Property of Congruence. Therefore, if you know $\angle RSK \cong \angle TKS$, you can prove $\triangle RSK \cong \triangle TKS$ by SAS.

CA Standards Check 2 What other information do you need to prove $\triangle ABC \cong \triangle CDA$ by SAS?

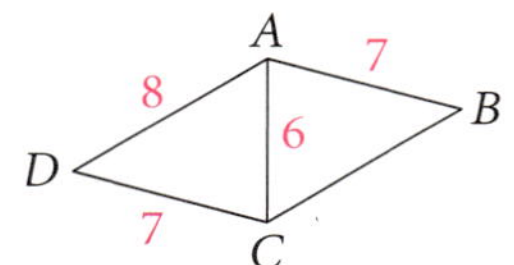

In Lesson 1-7, you learned to construct congruent segments using a compass open to a fixed angle. SAS tells you that the triangles outlined here are congruent, so $\overline{AB} \cong \overline{CD}$.

A B C D

Online active math

For: Congruent Triangles Activity
Use: Interactive Textbook, 4-2

3 EXAMPLE Are the Triangles Congruent?

From the information given, can you prove $\triangle RED \cong \triangle CAT$? Explain.

R E D C A T

Given: $\overline{RE} \cong \overline{CA}$, $\overline{RD} \cong \overline{CT}$, $\angle R \cong \angle T$

No, there is not enough information to prove $\triangle RED \cong \triangle CAT$. $\angle T$ is not included between $\overline{CA}$ and $\overline{CT}$. $\triangle RED$ may or may not be congruent to $\triangle CAT$.

CA Standards Check 3 From the information given, can you prove $\triangle AEB \cong \triangle DBC$? Explain.

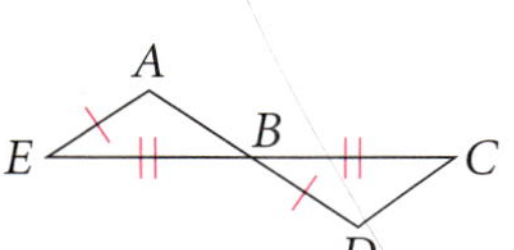

Given: $\overline{EB} \cong \overline{CB}$, $\overline{AE} \cong \overline{DB}$

EXERCISES

For more exercises, see *Extra Skill, Word Problem, and Proof Practice.*

Standards Practice

GEOM 2.0, 5.0

A Practice by Example

Example 1 (page 206)

1. **Developing Proof** Copy and complete the flow proof.

Given: $\overline{JK} \cong \overline{LM}, \overline{JM} \cong \overline{LK}$

Prove: $\triangle JKM \cong \triangle LMK$

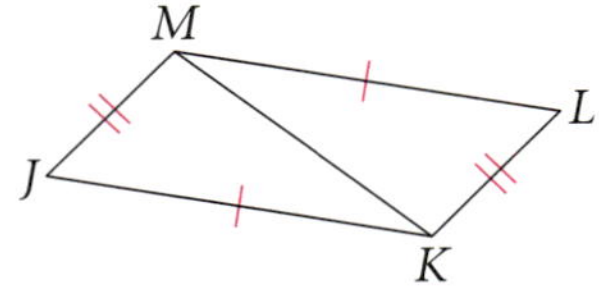

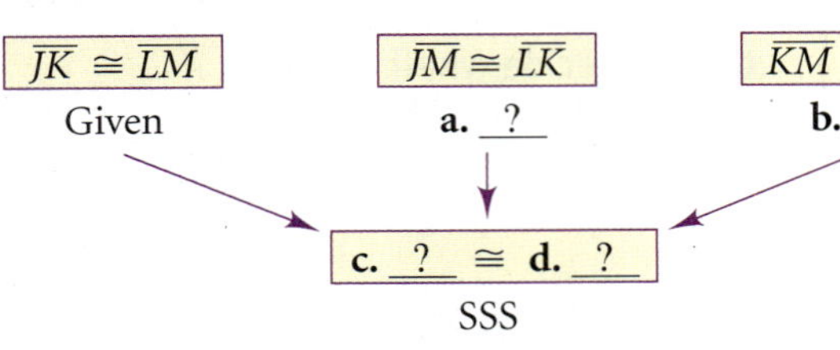

Proof 2. **Given:** $\overline{IE} \cong \overline{GH}, \overline{EF} \cong \overline{HF}$, F is the midpoint of $\overline{GI}$.

Prove: $\triangle EFI \cong \triangle HFG$

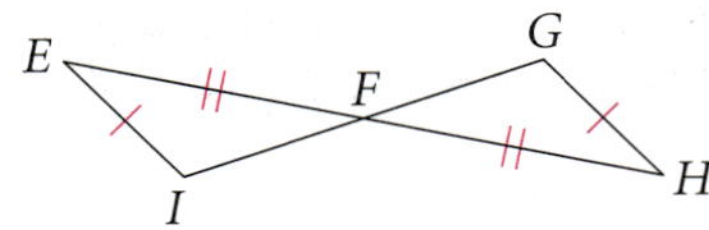

3. **Given:** $\overline{WZ} \cong \overline{ZS} \cong \overline{SD} \cong \overline{DW}$

Prove: $\triangle WZD \cong \triangle SDZ$

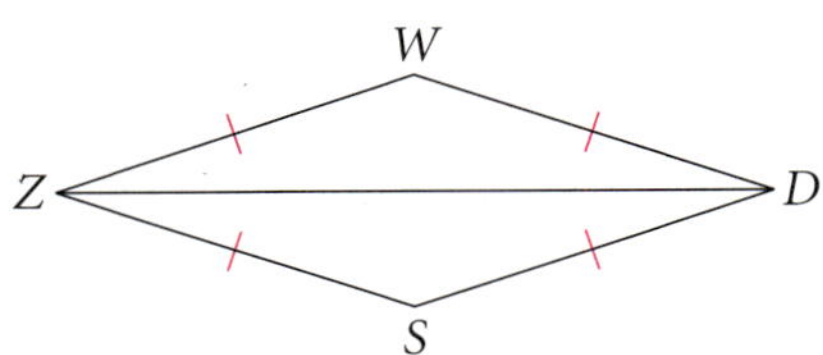

Example 2 (page 207)

Is the information you are given enough for you to prove that the two triangles are congruent? Explain.

4. 

The vertical beam $\overline{OB}$ is perpendicular to the porch roof. P, O, and R are equally spaced.

5. 

The diagonal legs have equal lengths and are joined at their midpoints.

6. 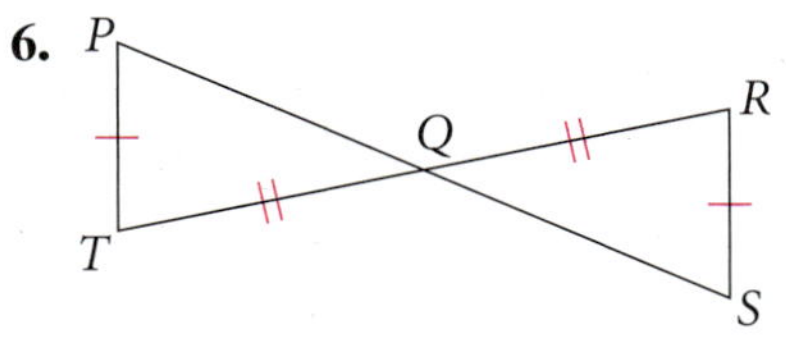

7. A, B, D, C

Example 3 (page 207)

What other information, if any, do you need to prove the two triangles congruent by SSS or SAS?

8. G, L, T, N, M, Q

9. 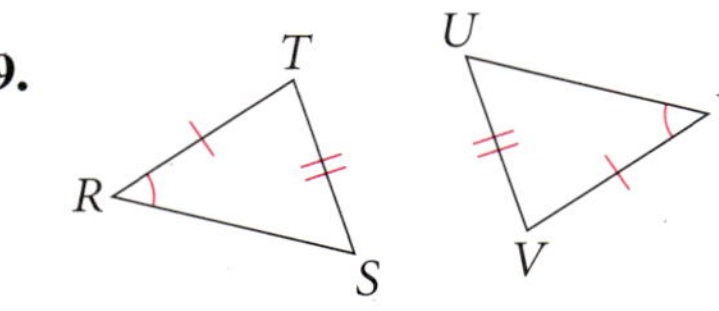

B Apply Your Skills

Copy the triangle. Start at any vertex and label the triangle as $\triangle WVU$.

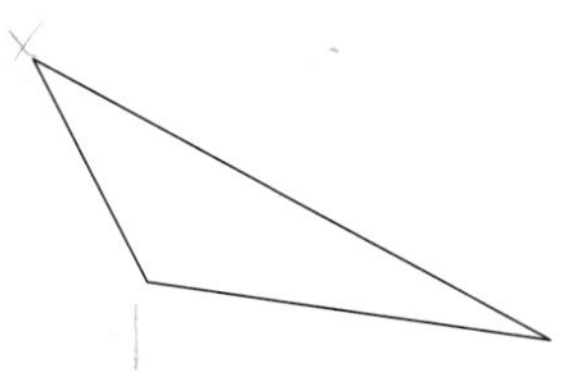

10. What sides include $\angle V$?

11. What angle is included between $\overline{WV}$ and $\overline{WU}$?

12. What angles include $\overline{UV}$?

13. What side is included between $\angle W$ and $\angle U$?

Name the indicated part(s) of $\triangle XYZ$ without drawing $\triangle XYZ$.

14. the angle included between $\overline{XY}$ and $\overline{XZ}$

15. the sides that include $\angle Z$

From the information given in the diagram, can you prove that the two triangles are congruent? Explain.

16.

17.

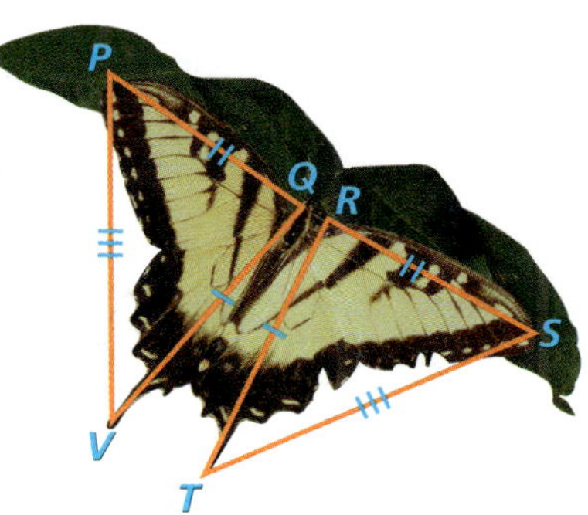

Proof **18. Given:** X is the midpoint of $\overline{AG}$ and of $\overline{NR}$.

Prove: $\triangle ANX \cong \triangle GRX$

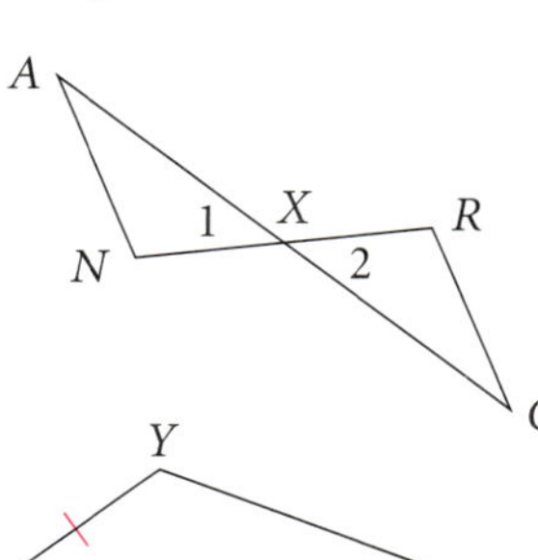

19. Multiple Choice What additional information do you need to prove the two triangles congruent by SAS?

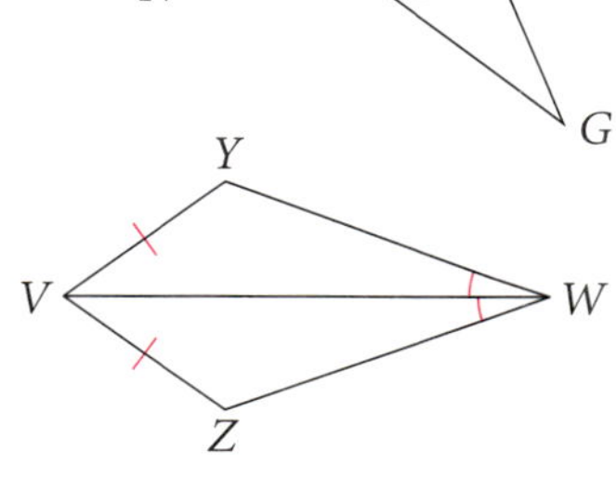

(A) $\overline{YW} \cong \overline{ZW}$

(B) $\angle Y \cong \angle Z$

(C) $\angle WVY \cong \angle ZWV$

(D) no additional information needed

Is there enough information to prove the two triangles congruent? If so, write the congruence statement and name the postulate you would use. If not, write *not possible* and tell what other information you would need.

20.

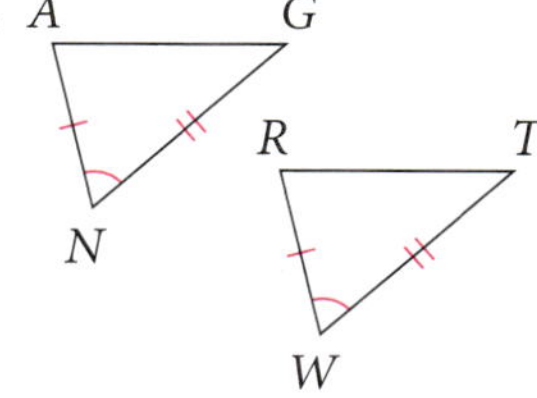

21.

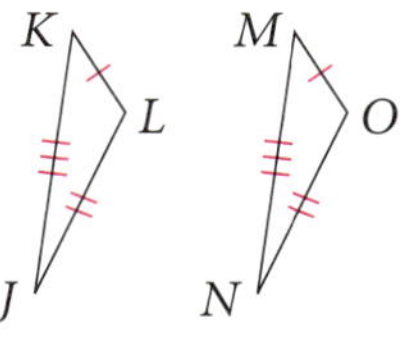

22.

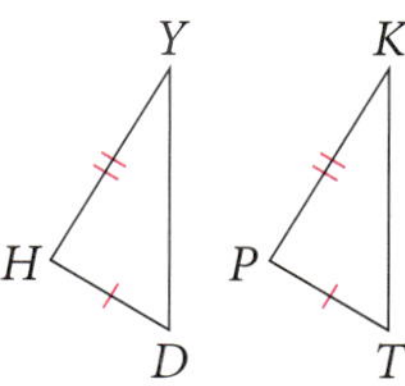

Homework Video Tutor

Visit: PHSchool.com
Web Code: bce-0402

23.

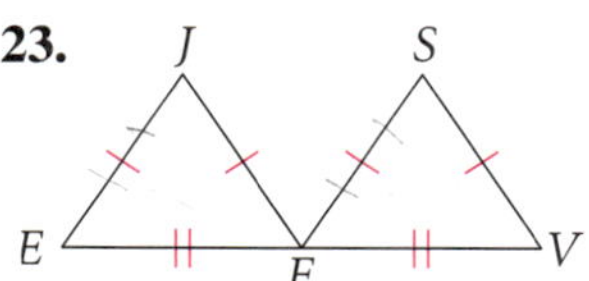

24.

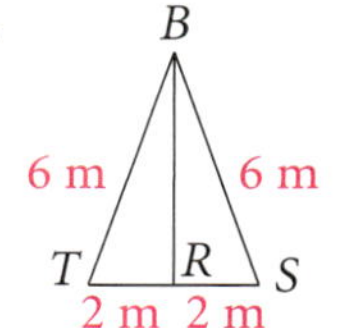

25.

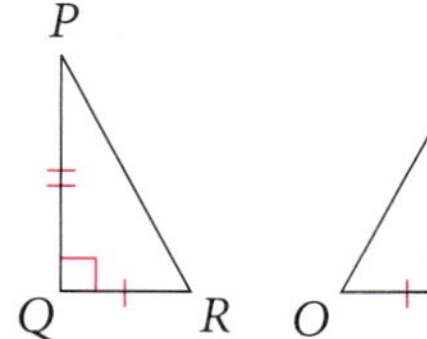

Real-World Connection

Four isosceles triangles cap the Smith Tower in Seattle.

Proof **26. Given:** $\overline{GK}$ bisects $\angle JGM$, $\overline{GJ} \cong \overline{GM}$.

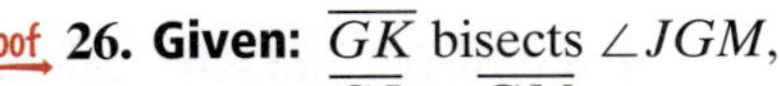

Prove: $\triangle GJK \cong \triangle GMK$

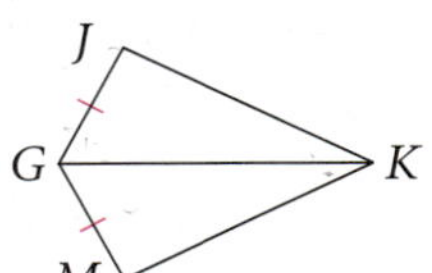

27. Given: $\overline{AE}$ and $\overline{BD}$ bisect each other.

Prove: $\triangle ACB \cong \triangle ECD$

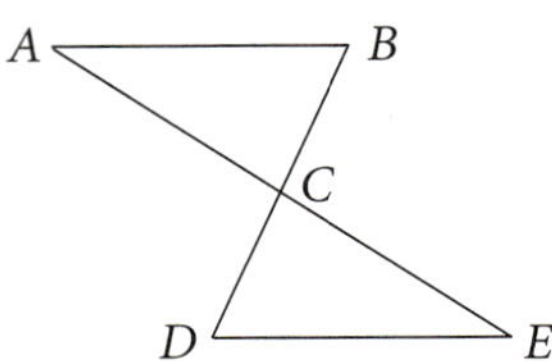

From the information given, can you prove the two triangles congruent? Explain.

28. $\triangle ABC$ and $\triangle DEF$ with $\angle A \cong \angle D, \angle B \cong \angle E, \angle C \cong \angle F$

29. $\triangle GHI$ and $\triangle JKL$ with $\overline{GH} \cong \overline{JK}, \overline{HI} \cong \overline{KL}, \angle I \cong \angle L$

30. $\triangle MNP$ and $\triangle QRS$ with $\overline{MN} \cong \overline{QR}, \angle N \cong \angle R, \overline{NP} \cong \overline{RS}$

Constructions **Use a straightedge to draw $\triangle JKL$. Construct $\triangle MNP \cong \triangle JKL$ using the given postulate.**

31. SSS **32.** SAS

33. a. Open-Ended List three real-life uses of congruent triangles.
b. Writing For each, tell whether you think congruence is necessary and why.

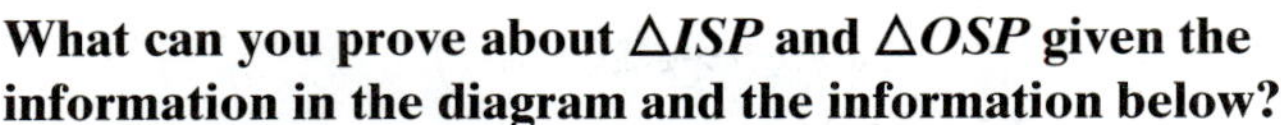

What can you prove about $\triangle ISP$ and $\triangle OSP$ given the information in the diagram and the information below?

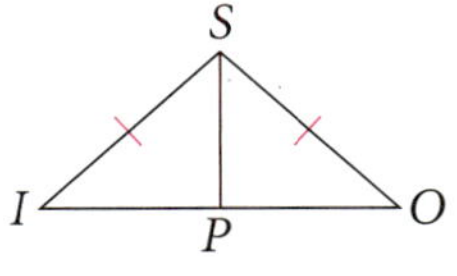

34. $\overline{SP}$ is the bisector of $\angle ISO$.

35. $\overline{SP}$ is a bisector of $\overline{IO}$.

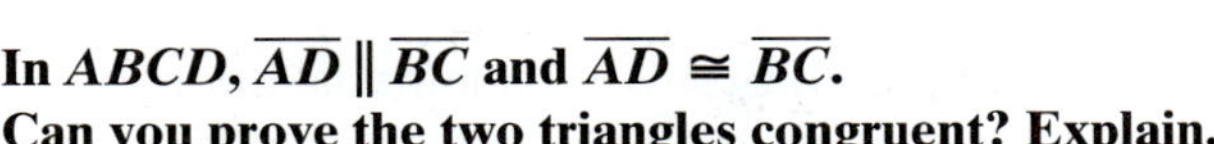

In $ABCD$, $\overline{AD} \parallel \overline{BC}$ and $\overline{AD} \cong \overline{BC}$. Can you prove the two triangles congruent? Explain.

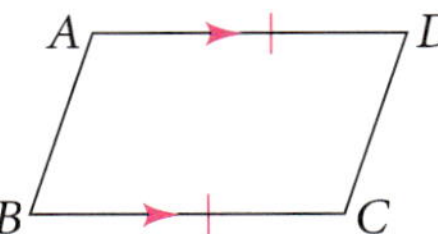

36. $\triangle ADB$ and $\triangle CBD$ **37.** $\triangle ABC$ and $\triangle CDA$

Proof **38. Given:** $\overline{FG} \parallel \overline{KL}, \overline{FG} \cong \overline{KL}$

Prove: $\triangle FGK \cong \triangle KLF$

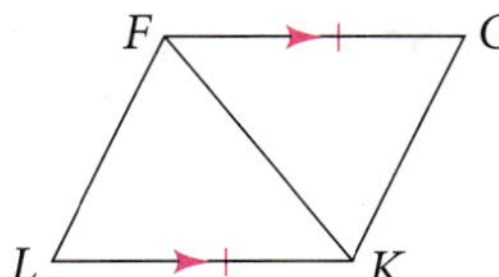

39. Given: $\overline{AB} \perp \overline{CM}, \overline{AB} \perp \overline{DB}$, M is the midpoint of $\overline{AB}, \overline{CM} \cong \overline{DB}$.

Prove: $\triangle AMC \cong \triangle MBD$

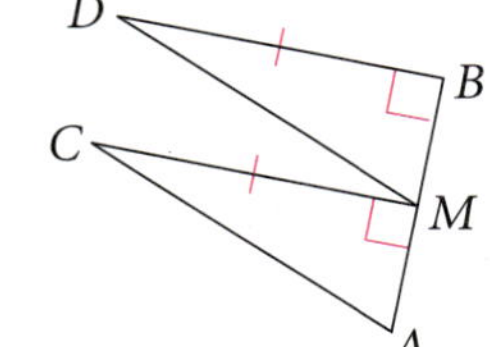
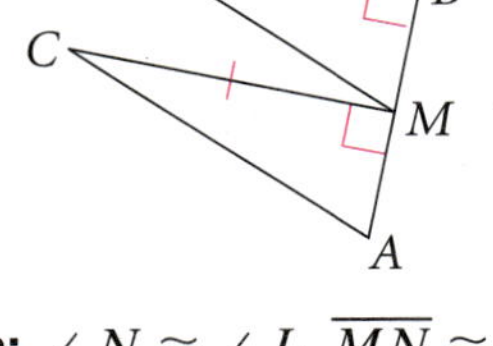

Challenge Proof **40. Given:** $\overline{HK} \cong \overline{LG}, \overline{HF} \cong \overline{LJ}$, $\overline{FG} \cong \overline{JK}$

Prove: $\triangle FGH \cong \triangle JKL$

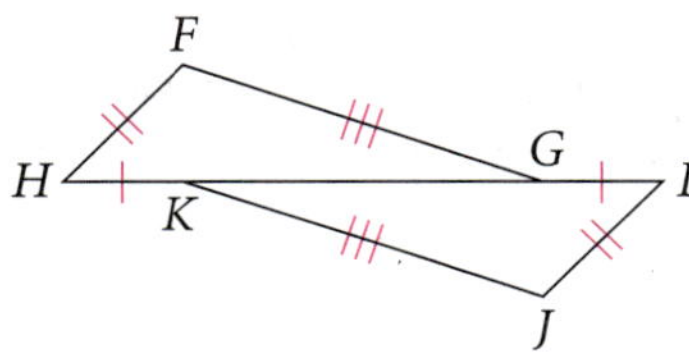

41. Given: $\angle N \cong \angle L, \overline{MN} \cong \overline{OL}$, $\overline{NO} \cong \overline{LM}$

Prove: $\overline{MN} \parallel \overline{LO}$

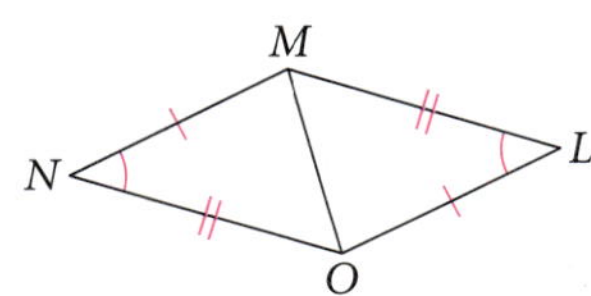

42. Error Analysis A friend conjectures that there should be an AAA Congruence Postulate since there is a SSS Congruence Postulate. Give a counterexample to disprove your friend's conjecture.

43. Critical Thinking Four sides of polygon *ABCD* are congruent, respectively, to the four sides of polygon *EFGH*.

a. Must the two quadrilaterals be congruent? Explain.

b. Is a quadrilateral a rigid figure? Explain.

c. What could you "add" to a quadrilateral to make it a rigid figure? Explain.

Multiple Choice Practice

For California Standards and CAHSEE Tutorials, visit PHSchool.com. Web Code: bcq-9045

GEOM 5.0

44. In triangles *ABC* and *DEF*, $\overline{AB} \cong \overline{DE}$, $\overline{AC} \cong \overline{DF}$. Which additional piece of information is necessary to prove the triangles congruent?

Ⓐ $\angle B \cong \angle E$ Ⓒ $\overline{BA} \cong \overline{DF}$

Ⓑ $\angle A \cong \angle D$ Ⓓ $m\angle B = 25°$ and $m\angle E = 25°$

CAHSEE 7SDAP 1.2

45. Which of the following best describes the correlation between the points in the scatterplot at the right?

Ⓐ Weak positive

Ⓑ Strong negative

Ⓒ No correlation

Ⓓ Weak negative

CAHSEE 7AF 2.2

46. Simplify the expression below.
$(5a^2b^3c)(7a^3bc^4)$

Ⓐ $12a^5b^4c^5$ Ⓒ $35a^5b^4c^5$

Ⓑ $12a^6b^4c^5$ Ⓓ $35a^6b^4c^4$

Mixed Review

Lesson 4-1

***ABCD* ≅ *EFGH*. Name the angle or side that corresponds to the given part.**

47. $\angle A$ **48.** $\overline{EF}$ **49.** $\overline{BC}$ **50.** $\angle G$

Lesson 2-2

51. The following two statements are about lines with defined slopes. Combine them into a single biconditional.

If the product of the slopes of two lines is -1, then the lines are perpendicular.
If two lines are perpendicular, then the product of their slopes is -1.

52. Write the two conditional statements that form this biconditional:

$x = 2$ if and only if $2x = 4$.

Lesson 2-1

Write the converse of the statement. Decide whether the statement and its converse are true or false.

53. If $x = 3$ then $2x = 6$. **54.** If $x = 3$ then $x^2 = 9$.

Vocabulary Builder

High-Use Academic Words

High-use academic words are words that you will see often in textbooks and on tests. These words are not math vocabulary terms, but knowing them will help you to succeed in mathematics.

Words to Learn: Direction Words

Some words tell you what to do in a problem. You need to understand what these words are asking so that you give the correct answer.

Word	Meaning
Explain	State facts and details to make an idea easy to understand.
Justify	Think of this as "Explain why." Give reasons to support a statement, conclusion, or action.

EXERCISES

Use the meanings of the words listed above to help you do the following.

1. *Explain* how to prepare your favorite snack.
2. In the student council meeting, you make a motion to recommend that the school establish a community service requirement.
   a. One reason, you claim, is that this requirement will help students become more responsible adults. What might you say to *justify* this claim.
   b. Determine another way in which you could *justify* your motion.
   c. Write an argument that would *justify* opposition to your motion.

Use the triangles at the right for Exercises 3–5.

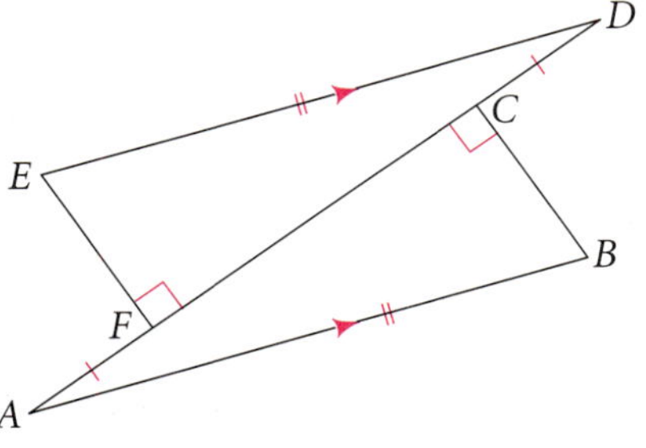

3. Are $\overline{AB}$ and $\overline{ED}$ parallel in the diagram at the right? Explain.
4. Are triangles ABC and DEF congruent? Justify your answer.
5. Can you conclude that $\overline{BC} \cong \overline{EF}$? Justify your answer.
6. **Word Knowledge** Think about the word *conjecture*.
   a. Choose the letter for how well you know the word.
      A. I know its meaning.
      B. I've seen it, but I don't know its meaning.
      C. I don't know it.
   b. **Research** Look up *conjecture* in a dictionary or online. Write its definition.
   c. Write a sentence involving mathematics that uses the word *conjecture*.

Triangle Congruence by ASA and AAS

California Content Standards

GEOM 2.0 Students write geometric proofs. ***Develop***
GEOM 5.0 Students prove that triangles are congruent. ***Develop***

What You'll Learn

- To prove two triangles congruent using the ASA Postulate and the AAS Theorem

. . . And Why

To prove that the two sides of a lacrosse goal are congruent triangles, as in Example 2

Check Skills You'll Need

In $\triangle JHK$, which side is included between the given pair of angles?

1. $\angle J$ and $\angle H$
2. $\angle H$ and $\angle K$

In $\triangle NLM$, which angle is included between the given pair of sides?

3. $\overline{LN}$ and $\overline{LM}$
4. $\overline{NM}$ and $\overline{LN}$

Give a reason to justify each statement.

5. $\overline{PR} \cong \overline{PR}$
6. $\angle A \cong \angle D$

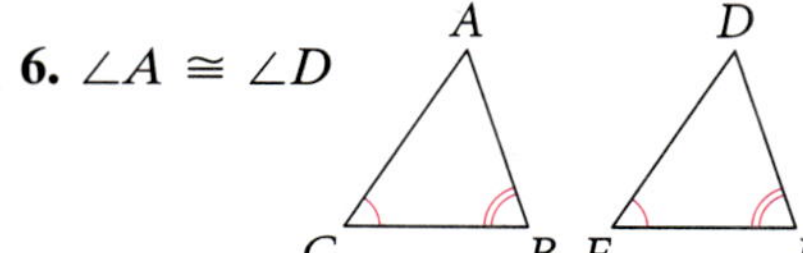

1 Using the ASA Postulate and the AAS Theorem

In Lesson 4-2 you learned that two triangles are congruent if two pairs of sides are congruent and the included angles are congruent (SAS).
The Activity Lab on page 204 suggests that two triangles are also congruent if two pairs of angles are congruent and the included sides are congruent (ASA).

Key Concepts

Postulate 4-3 **Angle-Side-Angle (ASA) Postulate**

If two angles and the included side of one triangle are congruent to two angles and the included side of another triangle, then the two triangles are congruent.

$$\triangle HGB \cong \triangle NKP$$

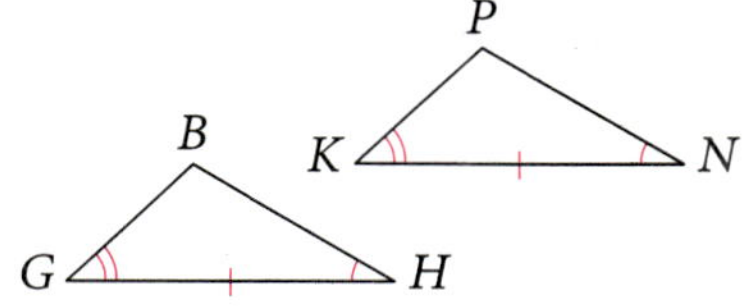

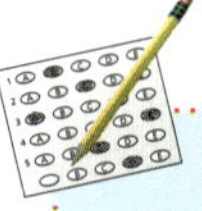

Test-Taking Tip

When there is a correspondence between figures, corresponding vertices should be listed in the same order.

1 EXAMPLE Using ASA

Multiple Choice Which triangle is congruent to $\triangle CAT$ by the ASA Postulate?

Ⓐ $\triangle DOG$ Ⓒ $\triangle GDO$
Ⓑ $\triangle INF$ Ⓓ $\triangle FNI$

$\angle C \cong \angle G$, $\overline{CA} \cong \overline{GD}$, and $\angle A \cong \angle D$.
$\triangle CAT \cong \triangle GDO$ by ASA. Choice C is correct.

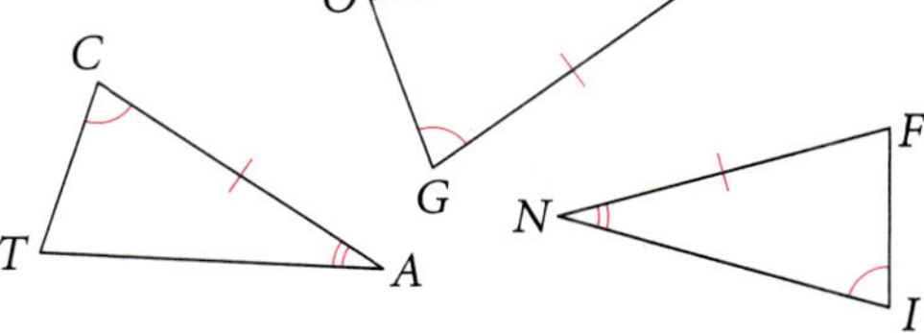

CA Standards Check **1** Can you conclude that $\triangle INF$ is congruent to either of the other two triangles? Explain.

Real-World Connection

Lacrosse is the modern version of the Iroquois game of *baggataway*.

Here is how you can use the ASA Postulate in a proof.

2 EXAMPLE Real-World Connection

Lacrosse Study what you are given and what you are to prove about the lacrosse goal. Then write a proof that uses ASA.

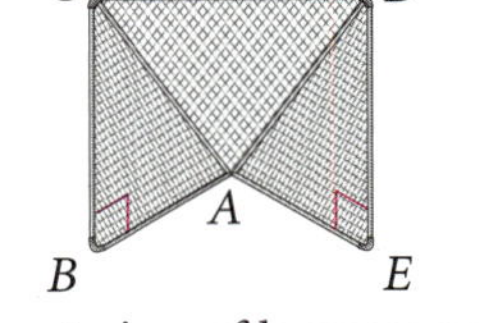

Front view of lacrosse goal

Given: $\angle CAB \cong \angle DAE$, $\overline{AB} \cong \overline{AE}$, $\angle ABC$ and $\angle AED$ are right angles.

Prove: $\triangle ABC \cong \triangle AED$

Proof: $\angle ABC \cong \angle AED$ because all right angles are congruent. You are given that $\overline{AB} \cong \overline{AE}$ and $\angle CAB \cong \angle DAE$. Thus, $\triangle ABC \cong \triangle AED$ by ASA.

2 Write a proof.

Given: $\overline{NM} \cong \overline{NP}$, $\angle M \cong \angle P$

Prove: $\triangle NML \cong \triangle NPO$

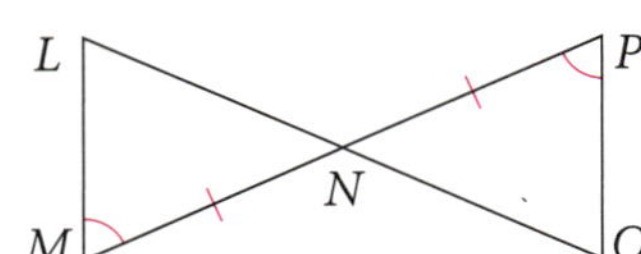

You can use the ASA Postulate to prove the Angle-Angle-Side Congruence Theorem. A flow proof is shown below.

Key Concepts

Theorem 4-2 **Angle-Angle-Side (AAS) Theorem**

If two angles and a nonincluded side of one triangle are congruent to two angles and the corresponding nonincluded side of another triangle, then the triangles are congruent.

$\triangle CDM \cong \triangle XGT$

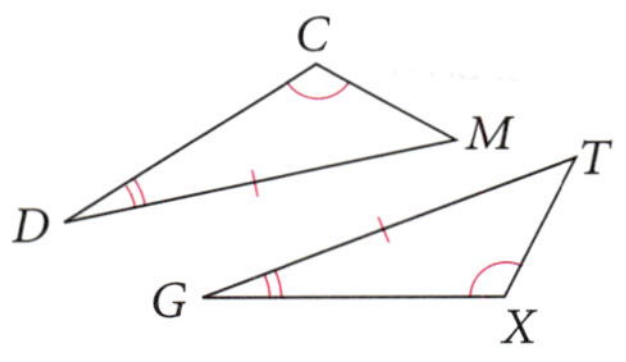

Proof

Proof of the Angle-Angle-Side Theorem

Given: $\angle A \cong \angle X$, $\angle B \cong \angle Y$, $\overline{BC} \cong \overline{YZ}$

Prove: $\triangle ABC \cong \triangle XYZ$

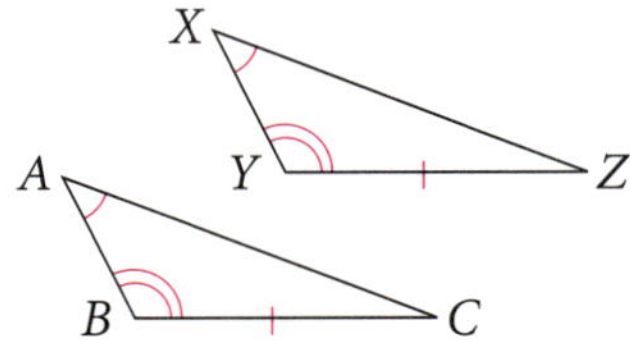

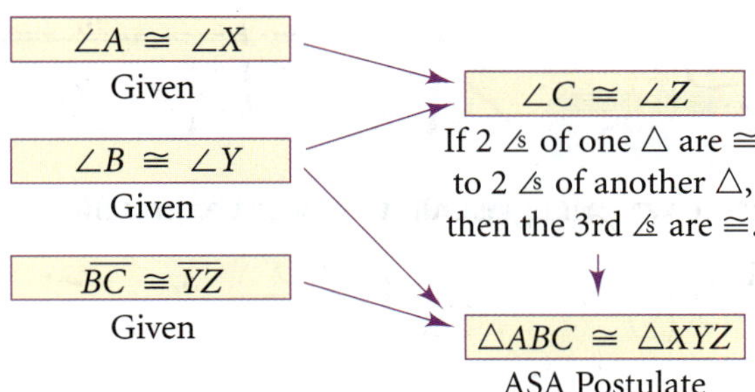

A proof starts in your head with a plan. It may help to jot down your plan. You will find this especially helpful as proofs become more complex. As you become skilled at proof, you will find that a good plan looks a lot like a paragraph proof.

3 EXAMPLE Planning a Proof

Study what you are given and what you are to prove. Then plan a proof that uses AAS.

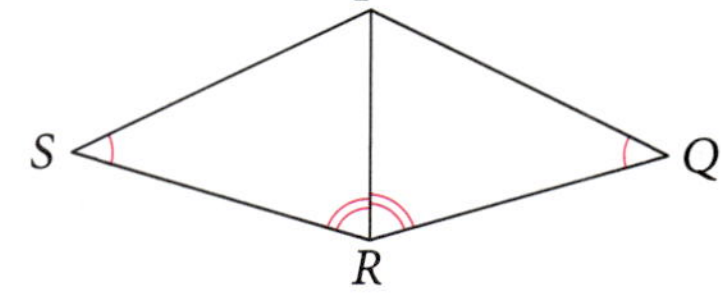

Given: $\angle S \cong \angle Q$, $\overline{RP}$ bisects $\angle SRQ$.

Prove: $\triangle SRP \cong \triangle QRP$

Plan: $\triangle SRP \cong \triangle QRP$ by AAS if $\overline{SP} \cong \overline{QP}$ or $\overline{RP} \cong \overline{RP}$.

The second statement is true by the Reflexive Property of Congruence.

CA Standards Check

3 Use the plan from Example 3 and write a proof.

Proof

4 EXAMPLE Writing a Proof

GO Online

Video Tutor Help

Visit: PHSchool.com
Web Code: bce-0775

Study what you are given and what you are to prove. Then write a proof that uses AAS.

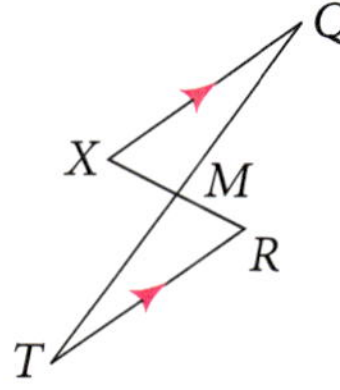

Given: $\overline{XQ} \parallel \overline{TR}$, $\overline{XR}$ bisects $\overline{QT}$.

Prove: $\triangle XMQ \cong \triangle RMT$

Plan: $\overline{XQ} \parallel \overline{TR}$ gives two pairs of congruent alternate interior angles. $\overline{XR}$ bisects $\overline{QT}$ means $\overline{QM} \cong \overline{TM}$. Use AAS.

Statements	Reasons
1. $\overline{XQ} \parallel \overline{TR}$	**1.** Given
2. $\angle Q \cong \angle T$, $\angle X \cong \angle R$	**2.** ?
3. $\overline{XR}$ bisects $\overline{QT}$.	**3.** Given
4. $\overline{QM} \cong \overline{TM}$	**4.** Definition of segment bisector
5. $\triangle XMQ \cong \triangle RMT$	**5.** AAS

CA Standards Check

4 **a.** Supply the reason that justifies Step 2.

b. Critical Thinking Explain how you could prove $\triangle XMQ \cong \triangle RMT$ by ASA.

EXERCISES Standards Practice

For more exercises, see *Extra Skill, Word Problem, and Proof Practice.*

GEOM 2.0, 5.0

A Practice by Example

GO for Help

Example 1 (page 213)

Name two triangles that are congruent by the ASA Postulate.

1.

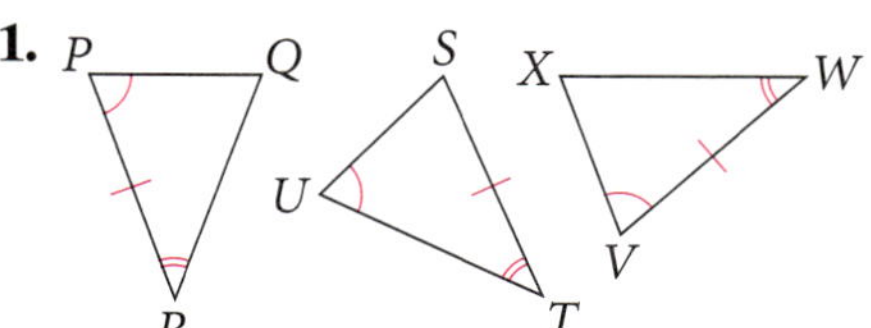

2.

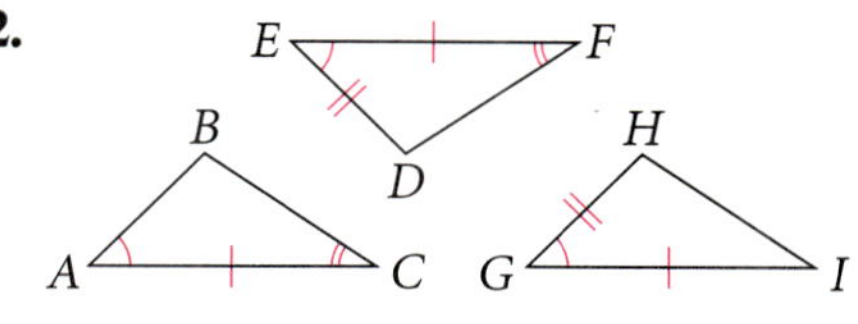

Answer each question without drawing the triangle.

3. Which side is included between $\angle R$ and $\angle S$ in $\triangle RST$?

4. Which angles include $\overline{NO}$ in $\triangle NOM$?

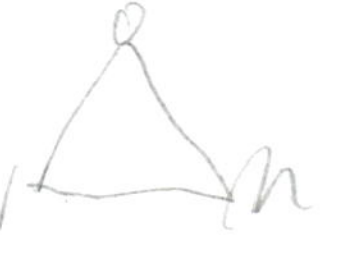

Example 2 (page 214)

5. **Developing Proof** Complete the proof by filling in the blanks.

Given: $\angle LKM \cong \angle JKM$,
$\angle LMK \cong \angle JMK$

Prove: $\triangle LKM \cong \triangle JKM$

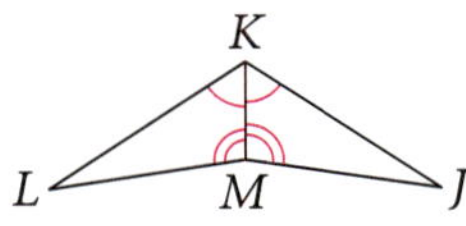

Proof: $\angle LKM \cong \angle JKM$ and $\angle LMK \cong \angle JMK$ are given. $\overline{KM} \cong \overline{KM}$ by the **a.** __?__ Property of Congruence. $\triangle LKM \cong \triangle JKM$ by the **b.** __?__ Postulate.

Proof 6. **Given:** $\angle BAC \cong \angle DAC$
$\overline{AC} \perp \overline{BD}$

Prove: $\triangle ABC \cong \triangle ADC$

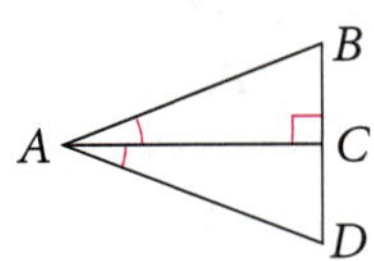

7. **Given:** $\overline{QR} \cong \overline{TS}$, $\overline{QR} \parallel \overline{ST}$

Prove: $\triangle QRT \cong \triangle TSQ$

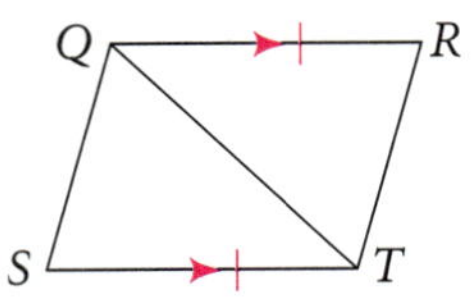

Example 3 (page 215)

8. **Developing Proof** Complete the proof plan by filling in the blanks.

Given: $\angle UWT$ and $\angle UWV$ are right angles,
$\angle T \cong \angle V$.

Prove: $\triangle UWT \cong \triangle UWV$

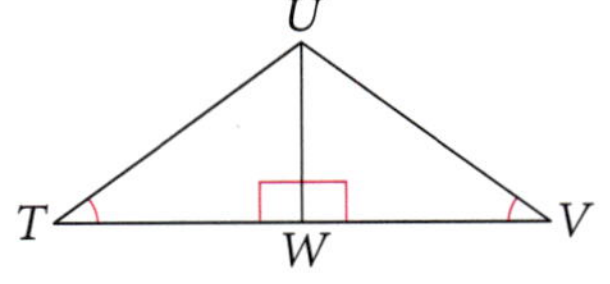

Plan: $\triangle UWT \cong \triangle UWV$ by AAS if $\angle T \cong \angle V$, $\angle UWT \cong$ **a.** __?__, and $\overline{UW} \cong$ **b.** __?__.
$\angle UWT \cong \angle UWV$ because all **c.** __?__ angles are congruent.
$\overline{UW} \cong \overline{UW}$ by the **d.** __?__ Property of Congruence.

Proof 9. Use your plan from Exercise 8 and write a proof.

Example 4 (page 215)

10. **Developing Proof** Complete the proof by filling in the blanks.

Given: $\angle N \cong \angle S$, line ℓ bisects $\overline{TR}$ at Q.

Prove: $\triangle NQT \cong \triangle SQR$

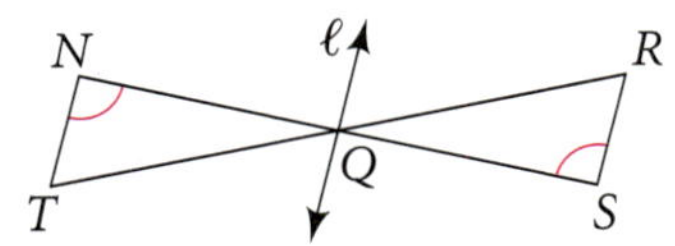

Statements	Reasons
1. $\angle N \cong \angle S$	1. Given
2. $\angle NQT \cong \angle SQR$	2. __?__
3. ℓ bisects $\overline{TR}$ at Q.	3. __?__
4. __?__	4. Definition of bisect
5. $\triangle NQT \cong \triangle SQR$	5. __?__

Proof 11. **Given:** $\angle V \cong \angle Y$,
$\overline{WZ}$ bisects $\angle VWY$.

Prove: $\triangle VWZ \cong \triangle YWZ$

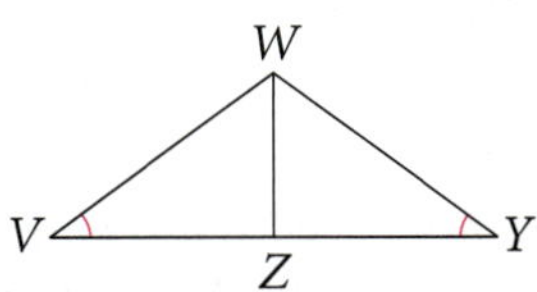

12. **Given:** $\overline{PQ} \perp \overline{QS}$, $\overline{RS} \perp \overline{QS}$,
T is the midpoint of $\overline{PR}$.

Prove: $\triangle PQT \cong \triangle RST$

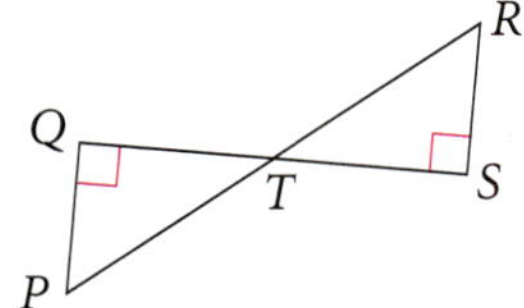

B Apply Your Skills

Real-World Connection

Congruent lapel and collar triangles help you look sharp.

Write a congruence statement for each pair of triangles. Name the postulate or theorem that justifies your statement.

13.

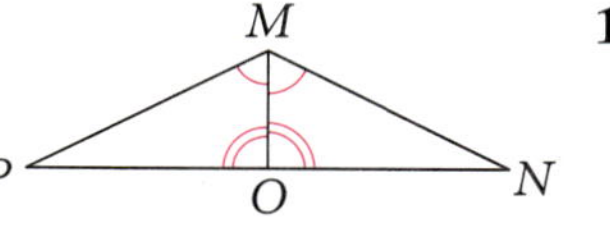

14.

U T R S

15.

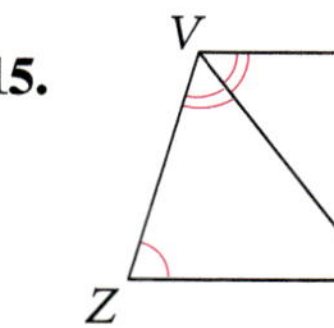

16. Multiple Choice For the triangles shown, $\angle D \cong \angle T$, $\angle E \cong \angle U$, and $\overline{EO} \cong \overline{UX}$. Which of the following statements is true?

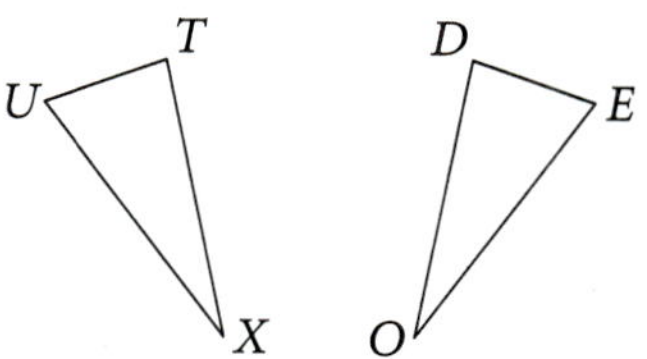

Ⓐ $\triangle TUX \cong \triangle DOE$ by ASA

Ⓑ $\triangle UTX \cong \triangle DEO$ by AAS

Ⓒ $\triangle TXU \cong \triangle ODE$ by ASA

Ⓓ $\triangle TUX \cong \triangle DEO$ by AAS

17. Writing Anita says that you can rewrite any proof that uses the AAS Theorem as a proof that uses the ASA Postulate. Do you agree with Anita? Explain.

18. $\angle E \cong \angle I$ and $\overline{FE} \cong \overline{GI}$. What else must you know to prove $\triangle FDE \cong \triangle GHI$ by AAS? By ASA?

19. Reasoning Can you prove the triangles at the right congruent using ASA or AAS? Justify your answer.

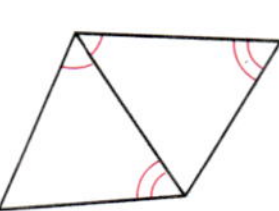

Proof **20. Given:** $\angle N \cong \angle P$, $\overline{MO} \cong \overline{QO}$

Prove: $\triangle MON \cong \triangle QOP$

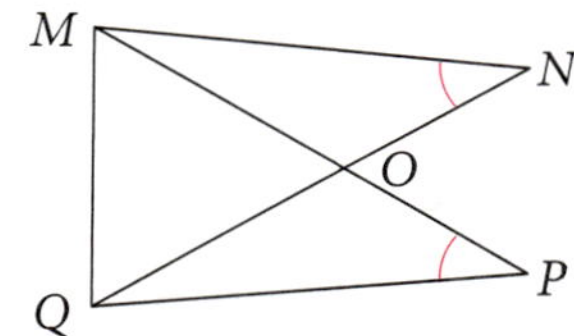

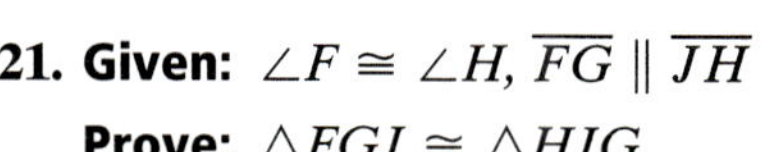

21. Given: $\angle F \cong \angle H$, $\overline{FG} \parallel \overline{JH}$

Prove: $\triangle FGJ \cong \triangle HJG$

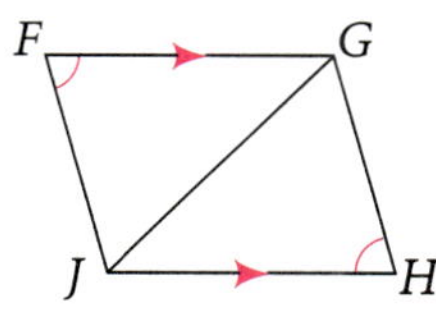

Proof **22. Given:** $\overline{AE} \parallel \overline{BD}$, $\overline{AE} \cong \overline{BD}$, $\angle E \cong \angle D$

Prove: $\triangle AEB \cong \triangle BDC$

E D A B C

23. Given: $\overline{DH}$ bisects $\angle BDF$, $\angle 1 \cong \angle 2$.

Prove: $\triangle BDH \cong \triangle FDH$

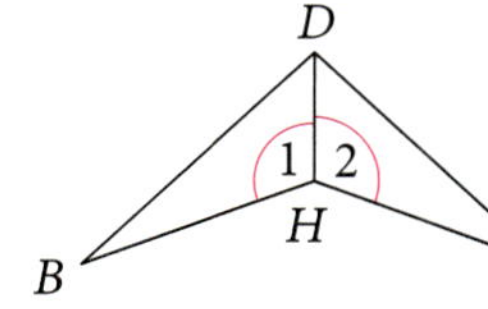

24. Constructions Using a straightedge, draw a triangle. Label it $\triangle JKL$. Construct $\triangle MNP \cong \triangle JKL$ so you know that the triangles are congruent by ASA.

Proof **25. Given:** $\overline{AB} \parallel \overline{DC}$, $\overline{AD} \parallel \overline{BC}$

Prove: $\triangle ABC \cong \triangle CDA$

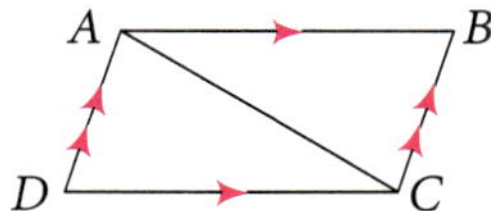

26. Reasoning If possible, draw two noncongruent triangles that have two pairs of congruent angles and one pair of congruent sides. If this is not possible, explain why.

Homework Video Tutor

Visit: PHSchool.com
Web Code: bce-0403

C Challenge

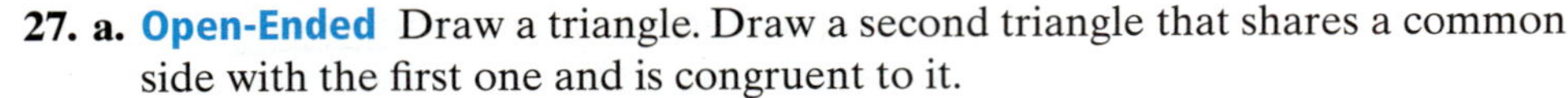

27. a. **Open-Ended** Draw a triangle. Draw a second triangle that shares a common side with the first one and is congruent to it.
    b. Think about how you drew your second triangle. What postulate or theorem did you use to make the second triangle congruent to the first one?

Real-World Connection

The two triangles above are congruent if just one additional condition is met.

Use the figure at the right. Name as many pairs of congruent triangles as you can for the information given.

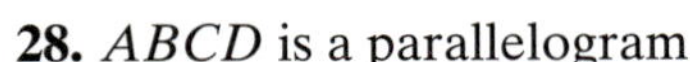

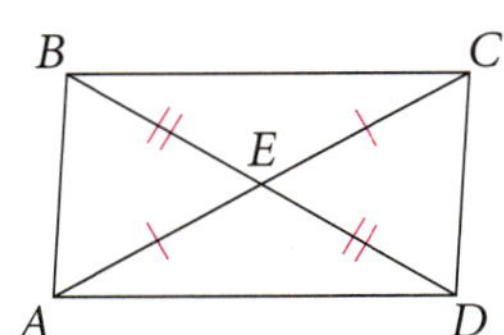

28. $ABCD$ is a parallelogram.

29. $ABCD$ is a rectangle.

30. **Reasoning** $\triangle JKL \cong \triangle MNP$. What additional information about $\overline{KQ}$ and $\overline{NR}$ will allow you to conclude that $\triangle JKQ \cong \triangle MNR$? Explain.

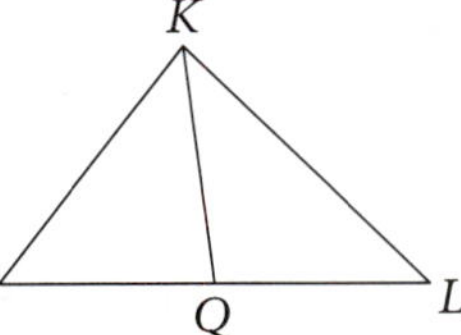

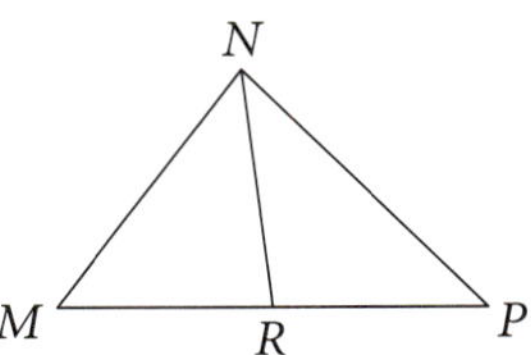

31. **Probability** Here are six congruence statements about the triangles at the right.

    $\angle A \cong \angle X$ $\quad \angle B \cong \angle Y$ $\quad \angle C \cong \angle Z$

    $\overline{AB} \cong \overline{XY}$ $\quad \overline{AC} \cong \overline{XZ}$ $\quad \overline{BC} \cong \overline{YZ}$

    There are 20 ways to choose a group of three statements from these six. What is the probability that three statements chosen at random from the six will guarantee that the triangles are congruent?

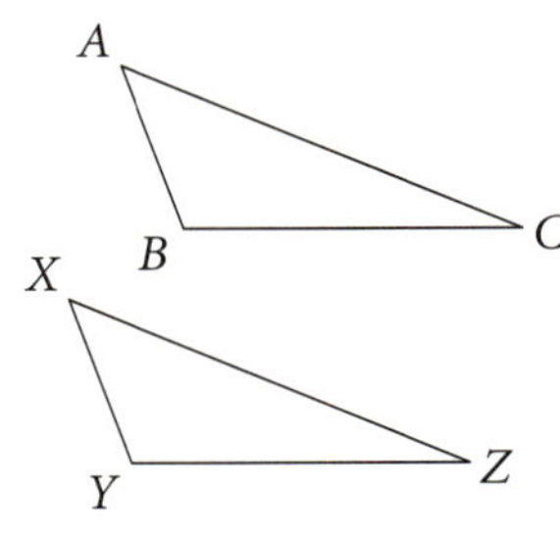

32. $\triangle RST$ at the right has $RS = 5$, $RT = 9$, and $m\angle T = 30$. Show that there is no SSA congruence rule by constructing $\triangle UVW$ with $UV = 5$, $UW = 9$, and $m\angle W = 30$, but with $\triangle UVW \ncong \triangle RST$.

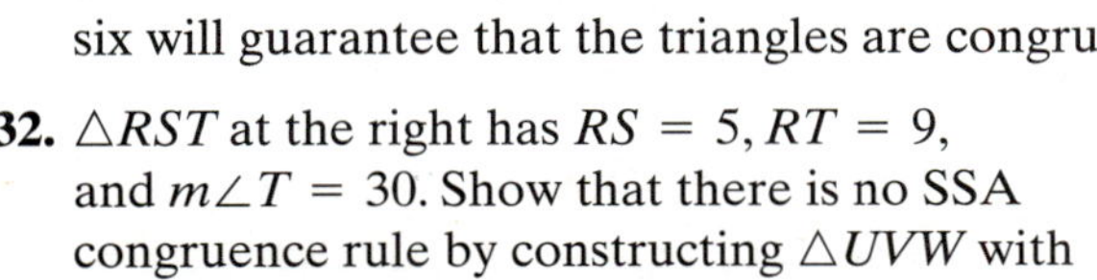

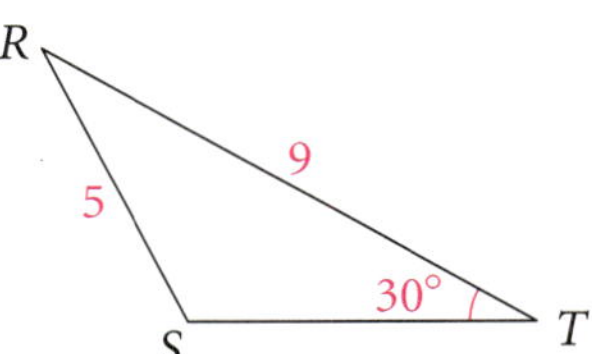

Multiple Choice Practice

For California Standards and CAHSEE Tutorials, visit PHSchool.com. Web Code: bcq-9045

GEOM 2.0

33. The triangles at the right are congruent by which theorem or postulate?

    (A) SAS
    (B) Vertical ∠'s are ≅
    (C) AAS
    (D) CPCTC

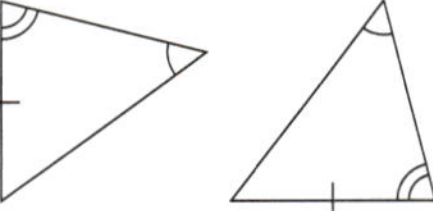

GEOM 5.0

34. In the figure at the right, which additional piece of information would be enough to prove that triangles MNO and QPO are congruent?

    (A) $\angle NOM$ and $\angle POQ$ are acute
    (B) neither triangle is obtuse
    (C) $\overline{MN} \parallel \overline{PQ}$
    (D) $\overline{MN} \cong \overline{QP}$

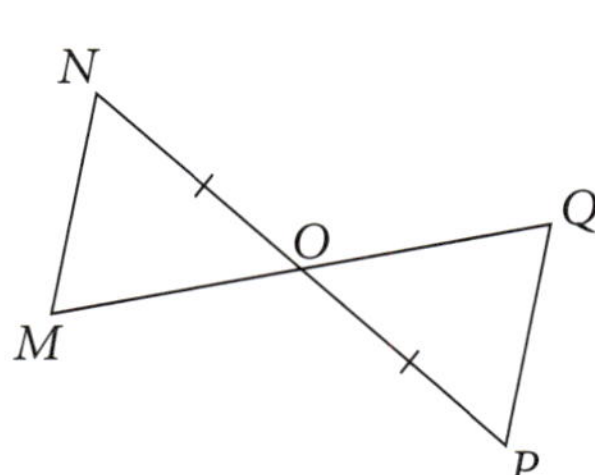

CAHSEE 7MG 1.3

35. Amy weeded 8 square feet of her garden in 24 minutes. At that rate, how many square feet of her garden can she weed in two hours?

    (A) 20 ft^2 (B) 24 ft^2 (C) 32 ft^2 (D) 40 ft^2

CAHSEE 7NS 2.3

36. $(5^5)^2 =$

    (A) 5^3 (B) 5^7 (C) 5^{10} (D) 25^{10}

Mixed Review

GO for Help

Lesson 2-1

In Exercises 37 and 38, decide whether you can use the SSS Postulate or the SAS Postulate to prove the triangles congruent. If so, write the congruence statement and name the postulate. If not, write *not possible.*

37.

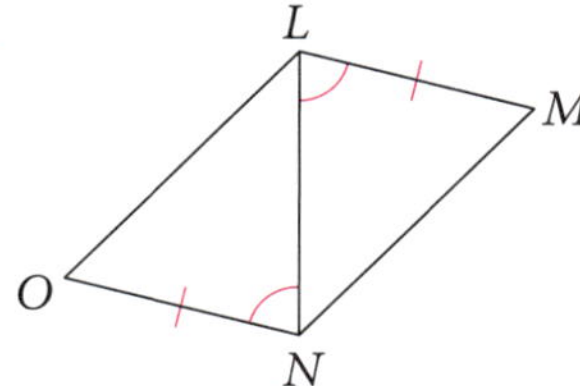

38.

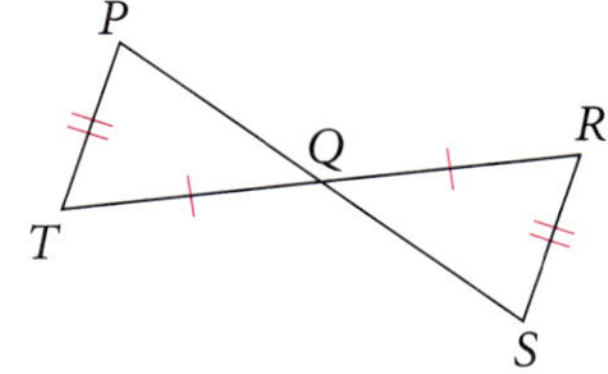

39. For any $\triangle ABC$, which sides are *not* included between $\angle A$ and $\angle B$?

Lesson 3-2

40. State the theorem or postulate that justifies the statement: If $\angle 1 \cong \angle 3$, then $a \parallel b$.

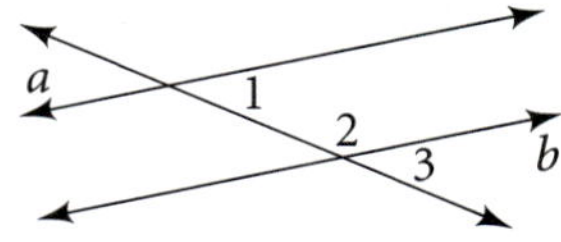

Lesson 1-9

Photography **You want to arrange class-trip photos without overlap to make a 2 ft-by-3 ft poster. You collect 3 in.-by-5 in. and 4 in.-by-6 in. photos. What is the greatest number of each type of photo that you can fit on your poster?**

41. 3 in.-by-5 in.

42. 4 in.-by-6 in.

43. What percent more paper is used for a large photo than a regular photo?

Checkpoint Quiz 1 — Lessons 4-1 through 4-3

1. $\triangle RST \cong \triangle JKL$. List the three pairs of congruent corresponding sides and the three pairs of congruent corresponding angles.

State the postulate or theorem you can use to prove the triangles congruent. If the triangles cannot be proven congruent, write *not possible.*

2.

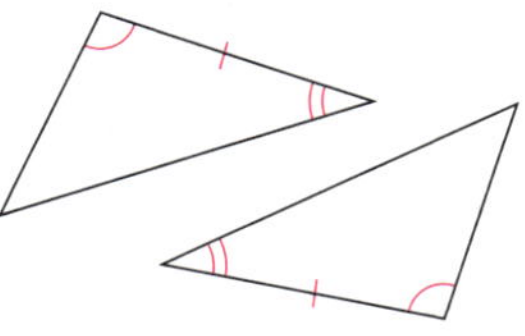

3.

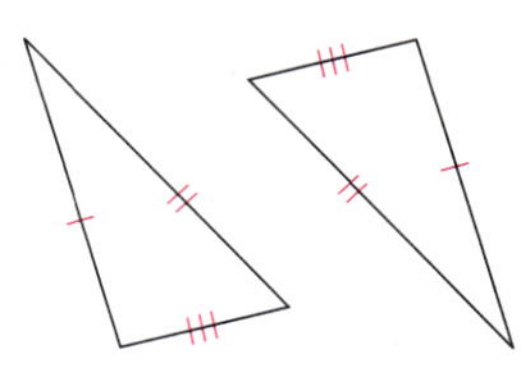

4.

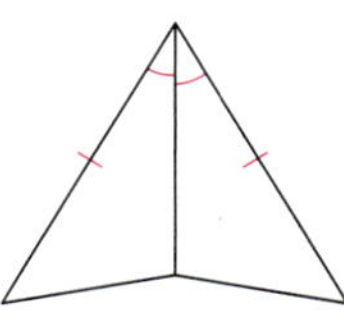

5.

6.

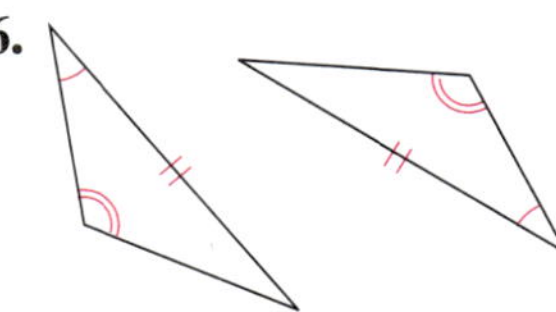

7.

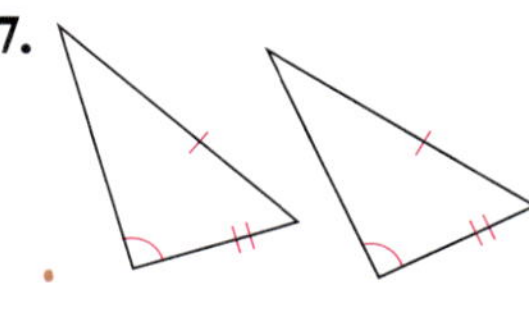

Use the information given in the diagram. Tell why each statement is true.

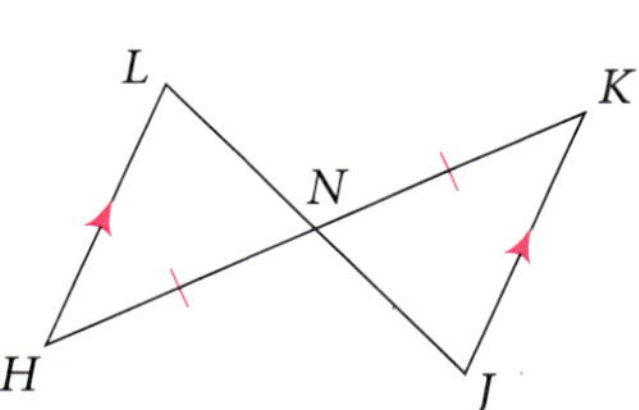

8. $\angle H \cong \angle K$

9. $\angle HNL \cong \angle KNJ$

10. $\triangle HNL \cong \triangle KNJ$

Exploring AAA and SSA

Technology

For Use With Lesson 4-3

Prepares for GEOM 5.0 Students prove that triangles are congruent.

So far, four statements allow you to conclude that two triangles are congruent. You can refer to them as SSS, SAS, ASA, and AAS. It is good mathematics to wonder about the other two possibilities, AAA and SSA.

Construct

Use geometry software to construct $\overrightarrow{AB}$ and $\overrightarrow{AC}$. Construct $\overline{BC}$ to create $\triangle ABC$. Construct a line parallel to $\overline{BC}$ that intersects $\overrightarrow{AB}$ and $\overrightarrow{AC}$ at points D and E to form $\triangle ADE$.

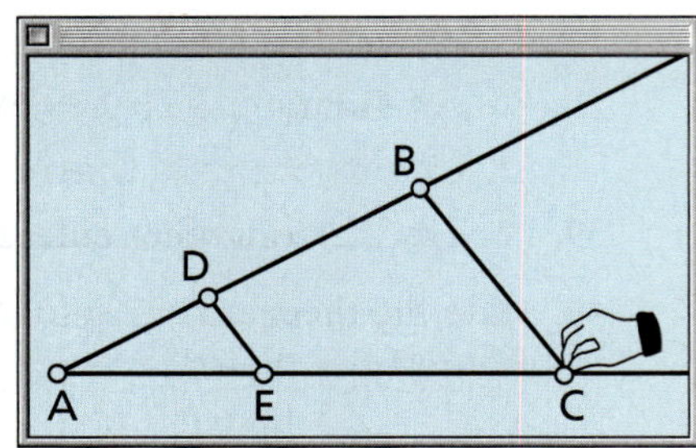

Investigate

Are the three angles of $\triangle ABC$ congruent to the three angles of $\triangle ADE$?
Manipulate the figure to change the positions of $\overline{DE}$ and $\overline{BC}$.
Do the corresponding angles of the triangles remain congruent?
Are the two triangles congruent? Can the two triangles be congruent?

In Exercise 1, you will be asked to make a conjecture about this investigation.

Construct

Construct $\overrightarrow{AB}$. Draw a circle with center C that intersects $\overrightarrow{AB}$ in two points. Construct $\overline{AC}$. Construct a point E on the circle and construct $\overline{CE}$.

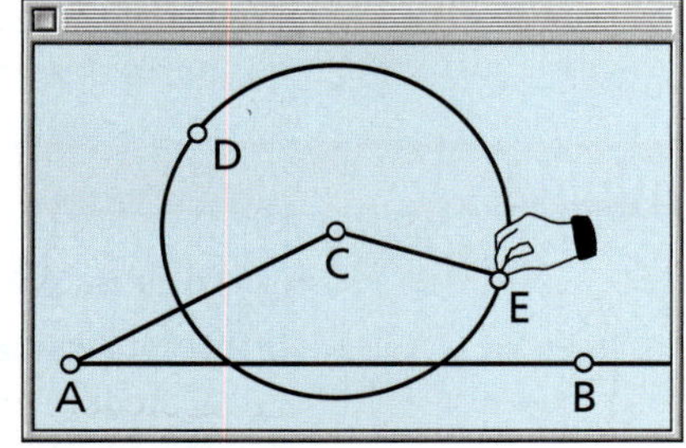

Investigate

Move point E around the circle until E is on $\overrightarrow{AB}$ and forms $\triangle ACE$. Then move E on the circle to the other point on $\overrightarrow{AB}$ to form another $\triangle ACE$.

Compare the measures of $\overline{AC}$, $\overline{CE}$, and $\angle A$ in one triangle with the measures of $\overline{AC}$, $\overline{CE}$, and $\angle A$ in the other triangle. Are two sides and a nonincluded angle of one triangle congruent to two sides and a nonincluded angle of the other triangle? Are the two triangles congruent? Do you get the same results if you change the size of $\angle A$ and the size of the circle?

EXERCISES

1. **Make a Conjecture** Based on your first investigation above, is there an AAA congruence theorem? Explain.

For Exercises 2–4, use what you learned in your second investigation above.

2. **Make a Conjecture** Do you think there is an SSA congruence theorem? Why?
3. Manipulate the figure so that $\angle A$ is obtuse. Decide whether the circle can intersect $\overrightarrow{AB}$ twice to form two triangles. Could there be an SSA congruence theorem if the congruent angles are obtuse? Explain.
4. Suppose you are given $\overline{CE}$, $\overline{AC}$, and $\angle A$. What must be true about CE, AC, and $m\angle A$ so that you can construct exactly one $\triangle ACE$? (*Hint:* Consider cases.)

Using Congruent Triangles: CPCTC

California Content Standards

GEOM 5.0 Students prove that triangles are congruent, and they are able to use the concept of corresponding parts of congruent triangles. ***Develop***
GEOM 6.0 Students know and are able to use the triangle inequality theorem. ***Introduce***

What You'll Learn

- To use triangle congruence and CPCTC to prove that parts of two triangles are congruent

. . . And Why

To measure distance indirectly, as in Example 2

Check Skills You'll Need

GO for Help Lesson 4-1

In the diagram, $\triangle JRC \cong \triangle HVG$.

1. List the congruent corresponding angles.
2. List the congruent corresponding sides.

You are given that $\triangle TIC \cong \triangle LOK$.

3. List the congruent corresponding angles.
4. List the congruent corresponding sides.

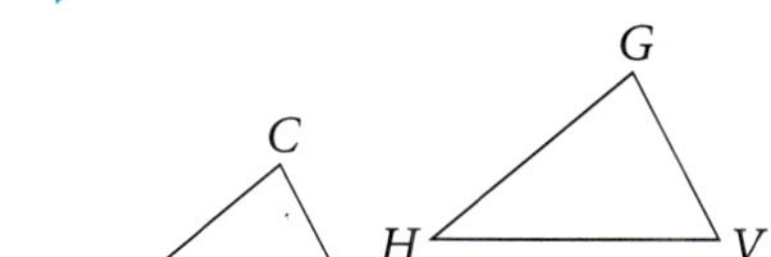

New Vocabulary • CPCTC

1 Proving Parts of Triangles Congruent

With SSS, SAS, ASA, and AAS, you know how to use three parts of triangles to show that the triangles are congruent. Once you have triangles congruent, you can make conclusions about their other parts because, by definition, corresponding parts of congruent triangles are congruent. You can abbreviate this as **CPCTC.**

Real-World Connection

Shapes formed by the ribs, stretchers, and shaft are congruent whether an umbrella is open or closed.

Proof **1 EXAMPLE** **Real-World Connection**

Umbrella Frames In an umbrella frame, the stretchers are congruent and they open to angles of equal measure.

Given: $\overline{SL} \cong \overline{SR}$,
$\angle 1 \cong \angle 2$

Prove that the angles formed by the shaft and the ribs are congruent.

Prove: $\angle 3 \cong \angle 4$

Proof: It is given that $\overline{SL} \cong \overline{SR}$ and $\angle 1 \cong \angle 2$. $\overline{SC} \cong \overline{SC}$ by the Reflexive Property of Congruence. $\triangle LSC \cong \triangle RSC$ by SAS, so $\angle 3 \cong \angle 4$ by CPCTC.

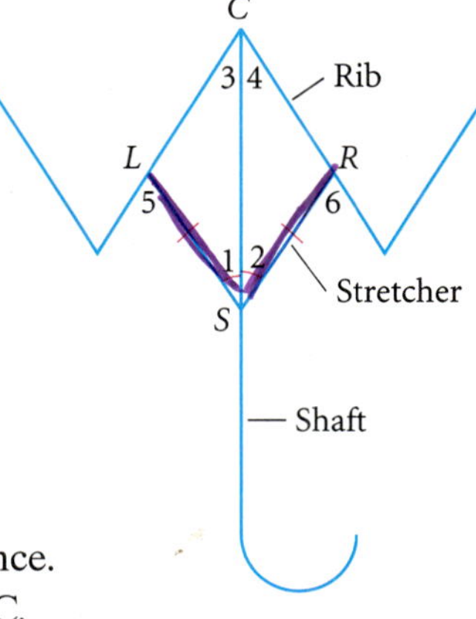

1 **Given:** $\angle Q \cong \angle R$,
$\angle QPS \cong \angle RSP$

Prove: $\overline{SQ} \cong \overline{PR}$

You can use congruent triangles and CPCTC to measure distances, such as the distance across a river, indirectly.

Proof **2 EXAMPLE** **Real-World Connection**

History According to legend, one of Napoleon's officers used congruent triangles to estimate the width of a river. On the riverbank, the officer stood up straight and lowered the visor of his cap until the farthest thing he could see was the edge of the opposite bank. He then turned and noted the spot on his side of the river that was in line with his eye and the tip of his visor.

Given: $\angle DEG$ and $\angle DEF$ are right angles; $\angle EDG \cong \angle EDF$.

The officer then paced off the distance to this spot and declared that distance to be the width of the river! Use congruent triangles to prove that he was correct.

Prove: $\overline{EF} \cong \overline{EG}$

Statements	Reasons
1. $\angle EDG \cong \angle EDF$	**1.** Given
2. $\overline{DE} \cong \overline{DE}$	**2.** Reflexive Property of Congruence
3. $\angle DEG$ and $\angle DEF$ are right angles.	**3.** Given
4. $\angle DEG \cong \angle DEF$	**4.** All right angles are congruent.
5. $\triangle DEF \cong \triangle DEG$	**5.** ASA Postulate
6. $\overline{EF} \cong \overline{EG}$	**6.** CPCTC

CA Standards Check

2 About how wide was the river if the officer stepped off 20 paces and each pace was about $2\frac{1}{2}$ ft long?

EXERCISES

For more exercises, see *Extra Skill, Word Problem, and Proof Practice.*

Standards Practice

GEOM 5.0, 6.0

A Practice by Example

Example 1 (page 221)

1. Developing Proof State why the two triangles are congruent. Give the congruence statement. Then list all the other parts of the triangles that are congruent by CPCTC.

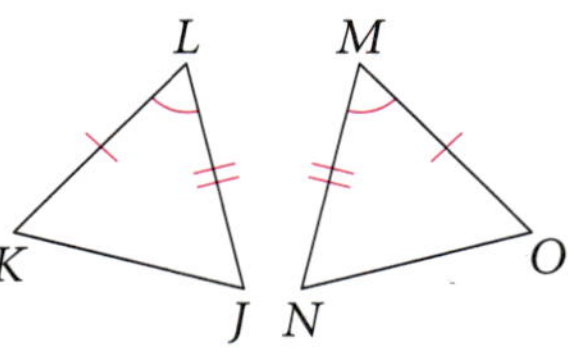

Proof **2. Given:** $\angle ABD \cong \angle CBD$, $\angle BDA \cong \angle BDC$

Prove: $\overline{AB} \cong \overline{CB}$

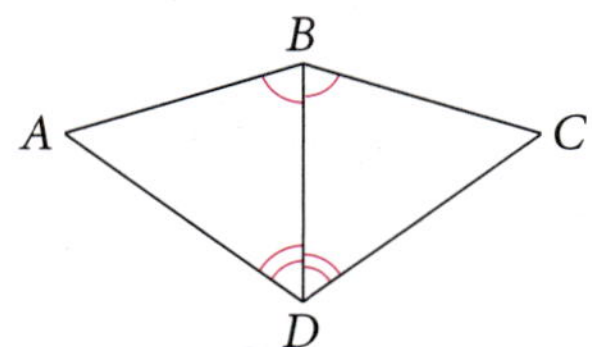

3. Given: $\overline{OM} \cong \overline{ER}$, $\overline{ME} \cong \overline{RO}$

Prove: $\angle M \cong \angle R$

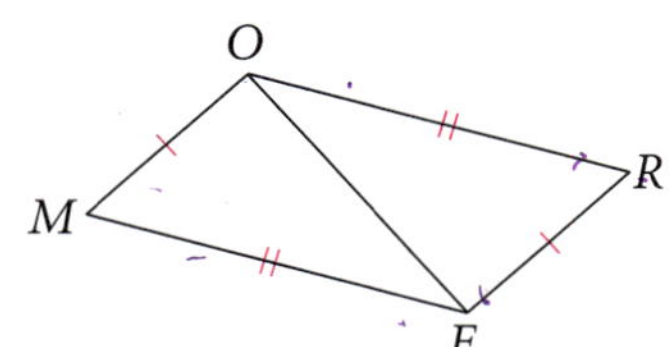

Example 2 (page 222)

4. **Developing Proof** Two cars of the same model have hood braces that are identical, connect to the body of the car in the same place, and fit into the same slot in the hood.

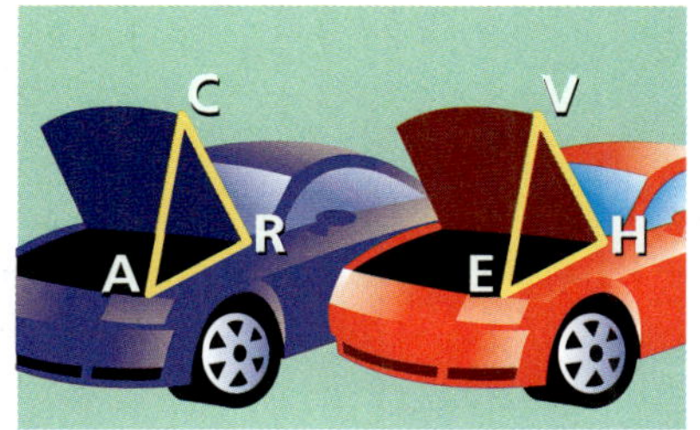

Given: $\overline{CA} \cong \overline{VE}, \overline{AR} \cong \overline{EH}, \overline{RC} \cong \overline{HV}$

Complete the proof that the hood braces hold the hoods open at the same angle.

Prove: $\angle ARC \cong \angle EHV$

Proof: It is given that the three sides of the triangles are congruent, so $\triangle ARC \cong \triangle EHV$ by **a.** __?__. Thus, $\angle ARC \cong \angle EHV$ by **b.** __?__.

B Apply Your Skills

5. **Earth Science** Some distances are best measured indirectly.

Sinkhole Swallows House

The large sinkhole in this photo occurred suddenly in 1981 in Winter Park, Florida, following a severe drought. Increased water consumption lowers the water table. Sinkholes form when caverns in the underlying limestone dry up and collapse.

A geometry class indirectly measured the distance across a sinkhole. The distances they measured are shown in the diagram. Explain how to use their measurements to find the distance across the sinkhole.

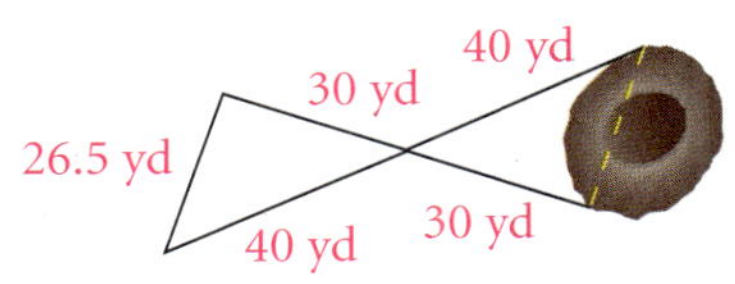

Proof 6. **Given:** $\angle SPT \cong \angle OPT$, $\overline{SP} \cong \overline{OP}$

Prove: $\angle S \cong \angle O$

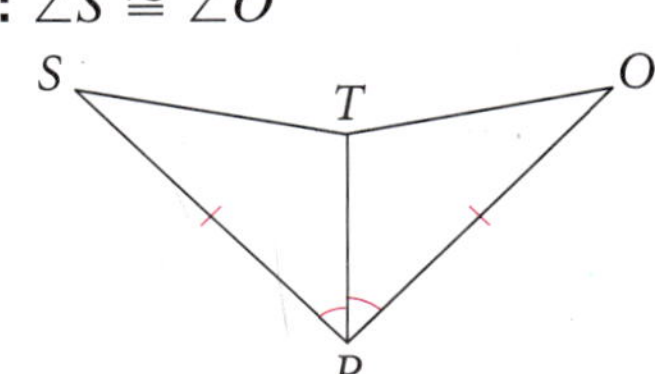

7. **Given:** $\overline{YT} \cong \overline{YP}, \angle C \cong \angle R$, $\angle T \cong \angle P$

Prove: $\overline{CT} \cong \overline{RP}$

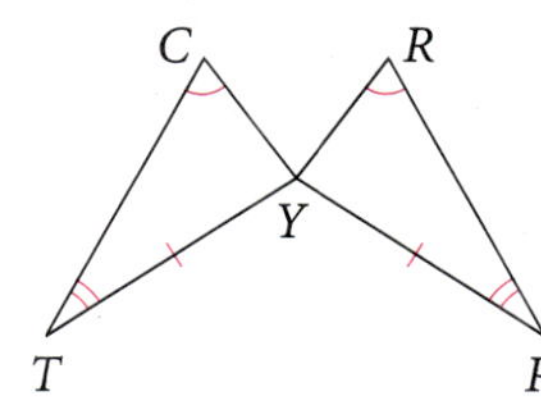

Visit: PHSchool.com
Web Code: bce-0404

Copy and mark the figure to show the given information. Explain how you would use SSS, SAS, ASA, or AAS with CPCTC to prove $\angle P \cong \angle Q$.

K

P L Q

8. **Given:** $\overline{PK} \cong \overline{QK}$, $\overline{KL}$ bisects $\angle PKQ$.

9. **Given:** $\overline{KL}$ is the perpendicular bisector of $\overline{PQ}$.

10. **Given:** $\overline{KL} \perp \overline{PQ}, \overline{KL}$ bisects $\angle PKQ$.

Proof 11. **Given:** $\angle QPS \cong \angle RSP, \angle Q \cong \angle R$

Prove: $\overline{PQ} \cong \overline{SR}$

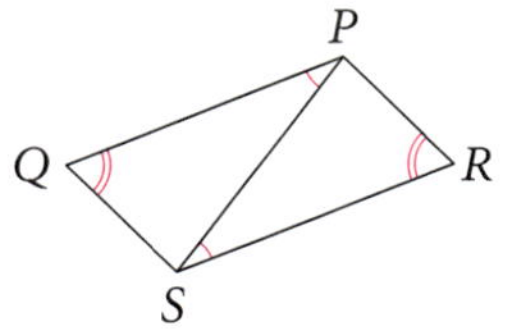

12. **Writing** Karen cut this pattern for the stained glass shown here so that $AB = CB$ and $AD = CD$. Must $\angle A$ be congruent to $\angle C$? Explain.

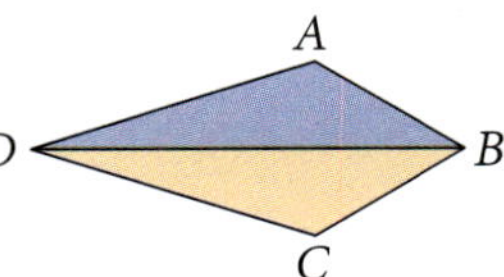

13. **Constructions** The construction of a line perpendicular to line ℓ through point P on ℓ is shown here.
   a. Which lengths or distances are equal by construction?
   b. Explain why you can conclude that $\overleftrightarrow{CP}$ is perpendicular to ℓ. (*Hint:* Do the construction. Then draw $\overline{CA}$ and $\overline{CB}$.)

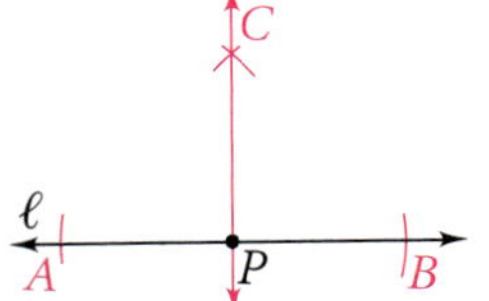

GO for Help

For a guide to solving Exercise 14, see page 226.

14. **Error Analysis** The proof is incorrect. Find the error and tell how you would correct the proof.

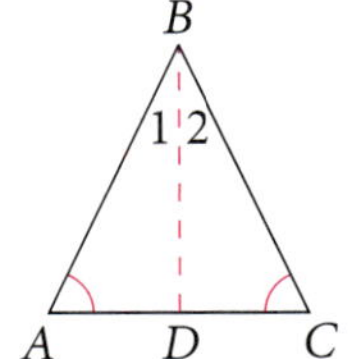

Given: $\angle A \cong \angle C$, $\overline{BD}$ bisects $\angle ABC$.

Prove: $\overline{AB} \cong \overline{CB}$

Statements	Reasons
1. $\angle A \cong \angle C$	1. Given
2. $\overline{BD}$ bisects $\angle ABC$.	2. Given
3. $\angle 1 \cong \angle 2$	3. Definition of bisect
4. $\overline{AD} \cong \overline{CD}$	4. Definition of bisect
5. $\triangle ABD \cong \triangle CBD$	5. AAS Theorem
6. $\overline{AB} \cong \overline{CB}$	6. CPCTC

Proof 15. **Given:** $\overline{BA} \cong \overline{BC}$, $\overline{BD}$ bisects $\angle ABC$.

Prove: $\overline{BD} \perp \overline{AC}$, $\overline{BD}$ bisects $\overline{AC}$.

16. **Given:** $\ell \perp \overline{AB}$, ℓ bisects $\overline{AB}$ at C, P is on ℓ.

Prove: $PA = PB$

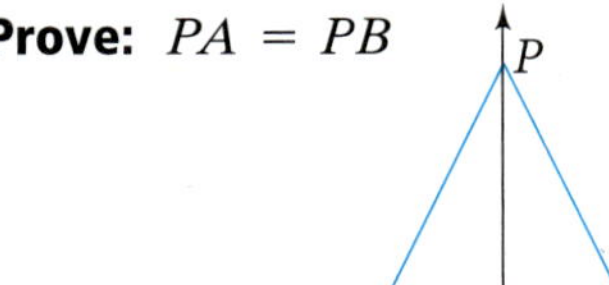

17. **Constructions** In the construction of the bisector of $\angle A$ below, $\overline{AB} \cong \overline{AC}$ because they are radii of the same circle. $\overline{BX} \cong \overline{CX}$ because both arcs had the same compass setting. Tell why you can conclude that $\overrightarrow{AX}$ bisects $\angle BAC$.

Problem Solving Hint

In the third diagram, what two triangles must be congruent, and why?

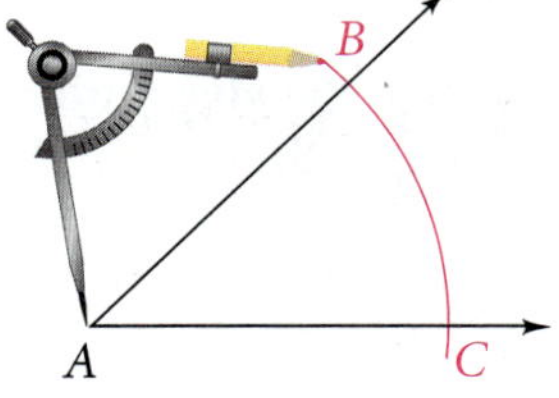

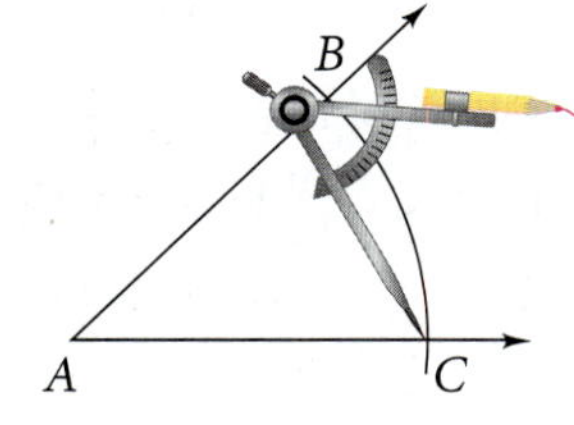

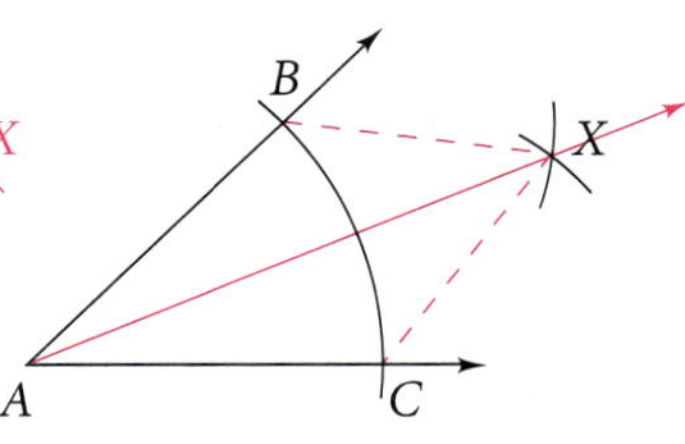

Proof 18. **Given:** $\overline{BE} \perp \overline{AC}$, $\overline{DF} \perp \overline{AC}$, $\overline{BE} \cong \overline{DF}$, $\overline{AF} \cong \overline{EC}$

Prove: $\overline{AB} \cong \overline{DC}$

19. **Given:** $\overline{JK} \parallel \overline{QP}$, $\overline{JK} \cong \overline{QP}$

Prove: $\overline{KQ}$ bisects $\overline{JP}$.

Challenge Proof **20. Given:** $\overline{PR} \parallel \overline{MG}, \overline{MP} \parallel \overline{GR}$

Prove: Each diagonal of $PRGM$ divides $PRGM$ into two congruent triangles.

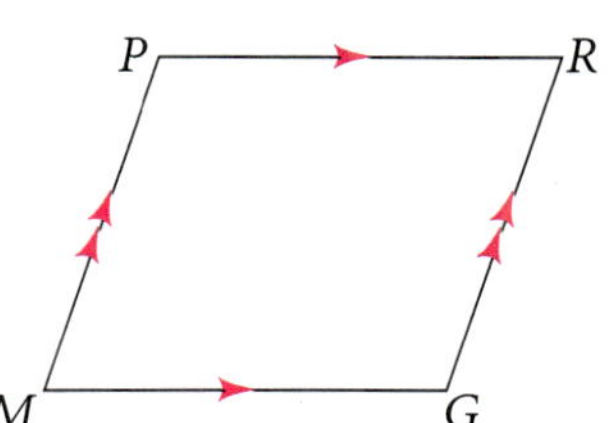

Proof **21. Given:** $\overline{PR} \parallel \overline{MG}, \overline{MP} \parallel \overline{GR}$

Prove: $\overline{PR} \cong \overline{MG}, \overline{MP} \cong \overline{GR}$

(*Hint:* See Exercise 20.)

Multiple Choice Practice

For California Standards and CAHSEE Tutorials, visit PHSchool.com. Web Code: bcq-9045

GEOM 5.0 **22.** If $\triangle SAY \cong \triangle NOW$, which angle is congruent to $\angle S$?

A $\angle Y$
B $\angle W$
C $\angle O$
D $\angle N$

GEOM 5.0 **23.** If $\triangle ABC \cong \triangle XYZ$, by which theorem or postulate is $\overline{CB} \cong \overline{ZY}$?

A CPCTC
B SAS
C Parallel Postulate
D Isosceles Triangle Theorem

CAHSEE Alg1 9.0 **24.** Which method will result in a solution for x in the system of equations below?

$$\begin{cases} 3x - 3y = 5 \\ -x + 6y = 0 \end{cases}$$

A Multiply the first equation by 2 and add the two equations.
B Multiply the first equation by -2 and add the two equations.
C Multiply the second equation by 3 and add the two equations.
D Multiply the second equation by -3 and add the two equations.

Mixed Review

Lesson 4-3 **What postulate or theorem can you use to prove the triangles congruent?**

25.

26.

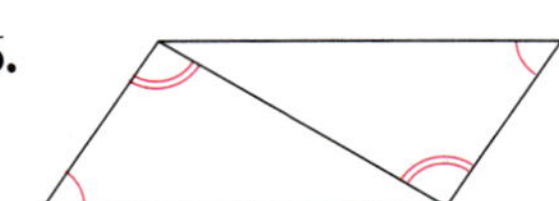

Lesson 2-5 **27.** The measure of an angle is 10 more than the measure of its supplement. Find the measures of both angles.

Lesson 2-3 **If possible, use the Law of Detachment to draw a conclusion. If it is not possible to draw a conclusion, write *not possible.***

28. If two nonvertical lines are parallel, then their slopes are equal. Line m is nonvertical and parallel to line n.

29. If a convex polygon is a quadrilateral, then the sum of its angle measures is 360. Convex polygon $ABCDE$ has five sides.

30. If a quadrilateral is a square, then it has four congruent sides. Quadrilateral $ABCD$ has four congruent sides.

Guided Problem Solving

For Use With Page 224, Exercise 14

Analyzing Errors **Read the problem below. Then let Lucas's thinking guide you through finding the error. Check your understanding with the exercise at the bottom of the page.**

Error Analysis This proof is incorrect. Find the error and tell how you would correct the proof.

Given: $\angle A \cong \angle C$, $\overline{BD}$ bisects $\angle ABC$.

Prove: $\overline{AB} \cong \overline{CB}$

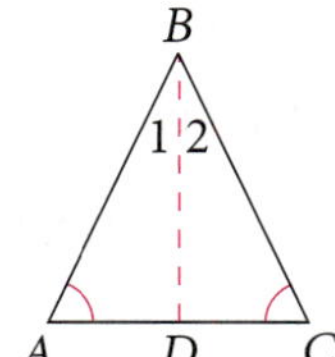

Statements	Reasons
1. $\angle A \cong \angle C$	**1.** Given
2. $\overline{BD}$ bisects $\angle ABC$.	**2.** Given
3. $\angle 1 \cong \angle 2$	**3.** Definition of bisect
4. $\overline{AD} \cong \overline{CD}$	**4.** Definition of bisect
5. $\triangle ABD \cong \triangle CBD$	**5.** AAS Theorem
6. $\overline{AB} \cong \overline{CB}$	**6.** CPCTC

What Lucas Thinks

An error could be anywhere. I'll check some easier parts first.

The first two lines match the given information. The last line matches what is to be proved.

In the last line, $\overline{AB}$ and $\overline{CB}$ are corresponding parts from $\triangle ABD$ and $\triangle CBD$. If the triangles are congruent, then CPCTC is correct.

$\triangle ABD \cong \triangle CBD$ by AAS. I see AAS in lines 1, 3, and 4, and they are in the right order. But I'm still not sure that lines 3 and 4 are correct.

I see the error!

Now I have to correct the proof. I still have two pairs of congruent angles. I know $\overline{BD} \cong \overline{BD}$. I can use this for line 4. AAS still works. I'm done!

What Lucas Writes

The error is in line 4. You cannot say $\overline{AD} \cong \overline{CD}$ by the definition of bisect. $\overline{BD}$ is given to be an angle bisector, not a segment bisector.

Replace line 4 with:

4. $\overline{BD} \cong \overline{BD}$ 4. $\cong$ is reflexive.

EXERCISES

1. In the proof above, let the Given and Statement 2 be "$\overline{BD}$ bisects $\overline{AC}$." Find the error in this proof and tell how you might correct the proof.

2. A proof states "$\triangle RST \cong \triangle VTS$ by SSA." Find the error. Show a counterexample.

Activity Lab Paper-Folding Conjectures

Hands-On

For Use With Lesson 4-5

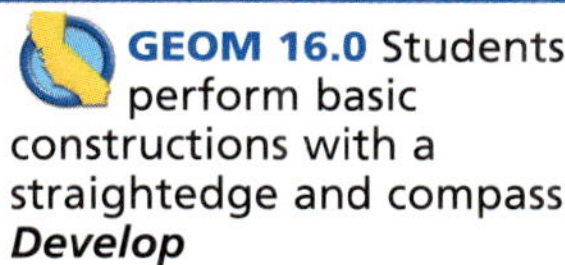

GEOM 16.0 Students perform basic constructions with a straightedge and compass. ***Develop***

In Chapter 3, you learned that isosceles triangles have two congruent sides. Folding one of the sides onto the other will suggest another important property of isosceles triangles.

1 Activity

- Construct an isosceles triangle on tracing paper.

- Name your triangle $\triangle ABC$, with A and B opposite the congruent sides.
- Fold the paper so the two congruent sides fit precisely one on top of the other. Crease the paper. Label the intersection of the fold line and $\overline{AB}$ as point D.

1. What do you notice about $\angle A$ and $\angle B$? Compare your results with others. Make a conjecture about the angles opposite the congruent sides in an isosceles triangle.

2. Study the fold line $\overline{CD}$ and the base $\overline{AB}$.

a. As best you can tell, what type of angle is $\angle CDA$? $\angle CDB$?

b. How do $\overline{AD}$ and $\overline{BD}$ seem to be related?

c. Use your answers to parts (a) and (b) to complete the conjecture: The fold line $\overline{CD}$ is the __?__ of the base $\overline{AB}$ of isosceles $\triangle ABC$.

2 Activity

In Activity 1, you made a conjecture about angles opposite the congruent sides of a triangle. You can also fold paper to study whether the converse is true.

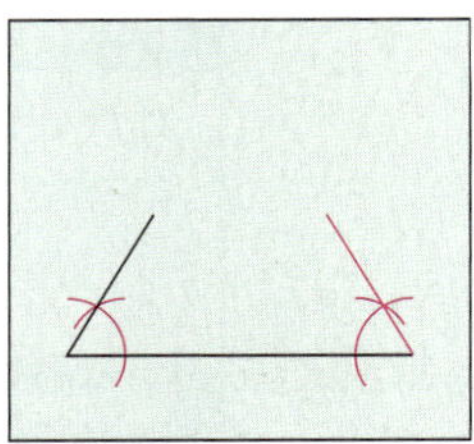

- On tracing paper, draw an acute angle with F as its vertex and one side $\overline{FG}$. Construct $\angle G$ as shown, so that $\angle G \cong \angle F$.
- Fold the paper so the two congruent angles, $\angle F$ and $\angle G$, fit precisely one on top of the other.

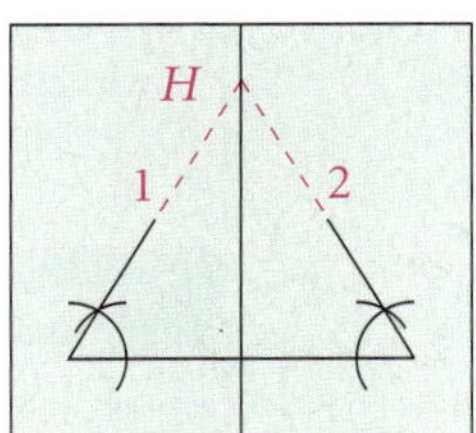

3. Explain why sides 1 and 2 meet at point H on the fold line. Make a conjecture about sides $\overline{FH}$ and $\overline{GH}$ opposite congruent angles in a triangle.

4. Write your conjectures from Steps 1 and 3 as a biconditional.

Isosceles and Equilateral Triangles

California Content Standards

GEOM 4.0 Students prove basic theorems involving congruence. ***Develop***
GEOM 5.0 Students prove that triangles are congruent, and they are able to use the concept of corresponding parts of congruent triangles. ***Develop***
GEOM 12.0 Students find and use measures of sides and of interior and exterior angles of triangles to solve problems. ***Develop***

What You'll Learn

- To use and apply properties of isosceles triangles

. . . And Why

To find the angles of a garden path, as in Example 3

Check Skills You'll Need

GO for Help Lesson 3-4

1. Name the angle opposite $\overline{AB}$.
2. Name the angle opposite $\overline{BC}$.
3. Name the side opposite $\angle A$.
4. Name the side opposite $\angle C$.
5. **Algebra** Find the value of x.

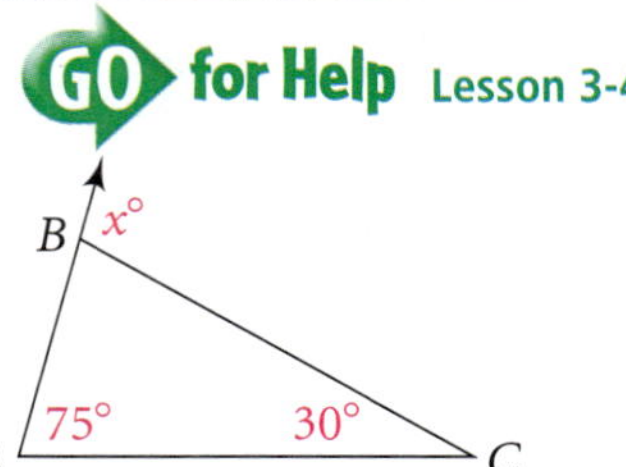

- legs of an isosceles triangle
- base of an isosceles triangle
- vertex angle of an isosceles triangle
- base angles of an isosceles triangle
- corollary

1 The Isosceles Triangle Theorems

Vocabulary Tip

Isosceles is derived from the Greek *isos* for equal and *skelos* for leg.

Isosceles triangles are common in the real world. You can find them in structures such as bridges and buildings. The congruent sides of an isosceles triangle are its **legs.** The third side is the **base.** The two congruent sides form the **vertex angle.** The other two angles are the **base angles.**

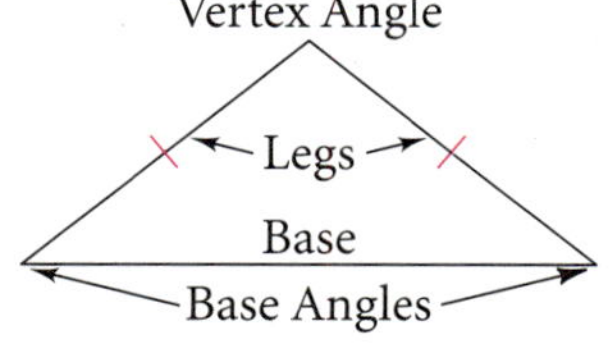

An isosceles triangle has a certain type of *symmetry* about a line through its vertex angle. The theorems below suggest this symmetry, which you will study in greater detail in Lesson 9-4.

Key Concepts

Theorem 4-3 **Isosceles Triangle Theorem**

If two sides of a triangle are congruent, then the angles opposite those sides are congruent.

$\angle A \cong \angle B$

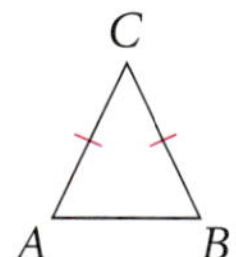

Theorem 4-4 **Converse of Isosceles Triangle Theorem**

If two angles of a triangle are congruent, then the sides opposite the angles are congruent.

$\overline{AC} \cong \overline{BC}$

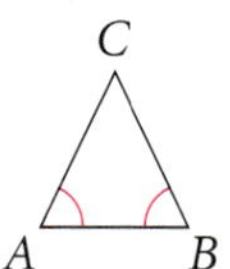

Theorem 4-5

The bisector of the vertex angle of an isosceles triangle is the perpendicular bisector of the base.

$\overline{CD} \perp \overline{AB}$ and $\overline{CD}$ bisects $\overline{AB}$.

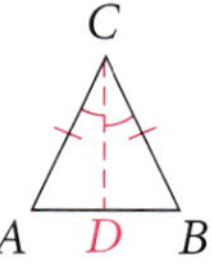

In the following proof of the Isosceles Triangle Theorem, you use a special segment, the bisector of the vertex angle. You will prove Theorems 4-4 and 4-5 in the Exercises.

Proof

Proof of the Isosceles Triangle Theorem

Begin with isosceles $\triangle XYZ$ with $\overline{XY} \cong \overline{XZ}$.
Draw $\overline{XB}$, the bisector of the vertex angle $\angle YXZ$.

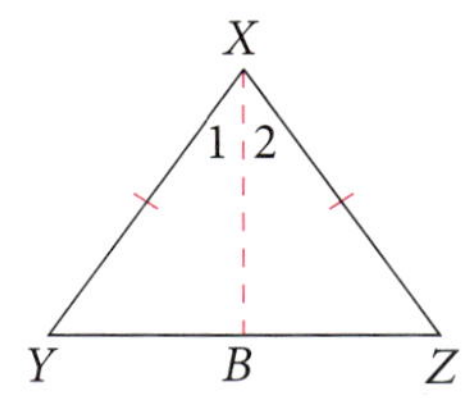

Given: $\overline{XY} \cong \overline{XZ}$, $\overline{XB}$ bisects $\angle YXZ$.

Prove: $\angle Y \cong \angle Z$

Proof: You are given that $\overline{XY} \cong \overline{XZ}$. By the definition of angle bisector, $\angle 1 \cong \angle 2$. By the Reflexive Property of Congruence, $\overline{XB} \cong \overline{XB}$. Therefore, by the SAS Postulate, $\triangle XYB \cong \triangle XZB$, and $\angle Y \cong \angle Z$ by CPCTC.

Real-World Connection

This A-shaped roof has congruent legs and congruent base angles.

1 EXAMPLE Using the Isosceles Triangle Theorems

Explain why each statement is true.

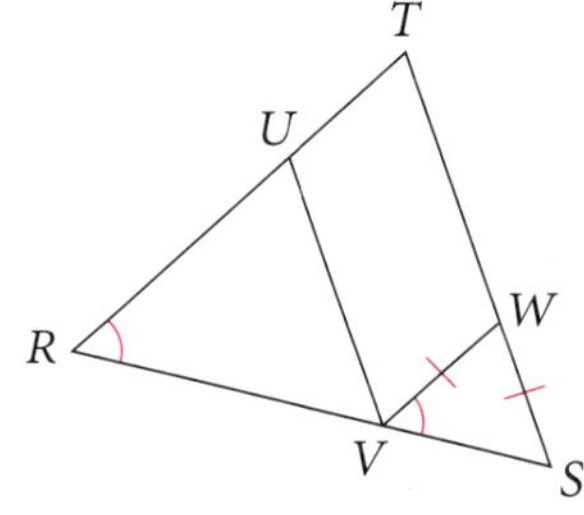

a. $\angle WVS \cong \angle S$

$\overline{WV} \cong \overline{WS}$ so $\angle WVS \cong \angle S$ by the Isosceles Triangle Theorem.

b. $\overline{TR} \cong \overline{TS}$

$\angle R \cong \angle WVS$ and $\angle WVS \cong \angle S$, so $\angle R \cong \angle S$ by the Transitive Property of $\cong$. $\overline{TR} \cong \overline{TS}$ by the Converse of the Isosceles Triangle Theorem.

CA Standards Check 1 Can you deduce that $\triangle RUV$ is isosceles? Explain.

2 EXAMPLE Using Algebra

Multiple Choice Find the value of y.

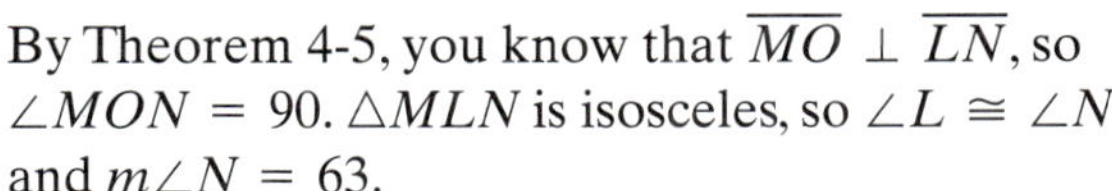

A 17　　C 54
B 27　　D 90

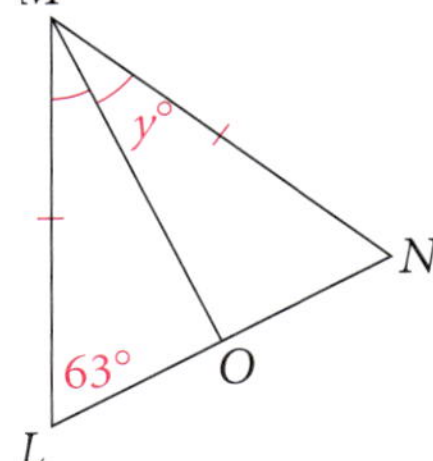

By Theorem 4-5, you know that $\overline{MO} \perp \overline{LN}$, so $\angle MON = 90$. $\triangle MLN$ is isosceles, so $\angle L \cong \angle N$ and $m\angle N = 63$.

$m\angle N + 90 + y = 180$	**Triangle Angle-Sum Theorem**
$63 + 90 + y = 180$	**Substitute for $m\angle N$ and x.**
$y = 27$	**Subtract 153 from each side.**

The correct answer is B.

Test-Taking Tip

Remember that the acute angles of a right triangle are complementary.

CA Standards Check 2 Suppose $m\angle L = 43$. Find the value of y.

A **corollary** is a statement that follows immediately from a theorem. Corollaries to the Isosceles Triangle Theorem and its converse appear on the next page.

Key Concepts

Corollary — Corollary to Theorem 4-3

If a triangle is equilateral, then the triangle is equiangular.

$\angle X \cong \angle Y \cong \angle Z$

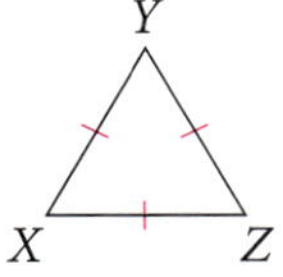

Corollary — Corollary to Theorem 4-4

If a triangle is equiangular, then the triangle is equilateral.

$\overline{XY} \cong \overline{YZ} \cong \overline{ZX}$

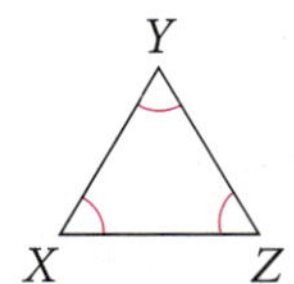

Vocabulary Tip

Equilateral: Congruent sides
Equiangular: Congruent angles

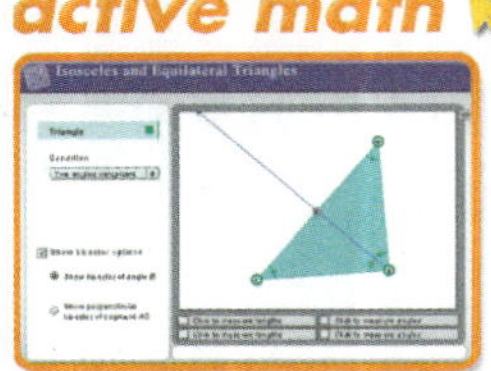

For: Isosceles Triangle Activity
Use: Interactive Textbook, 4-5

3 EXAMPLE Real-World Connection

Landscaping A landscaper uses rectangles and equilateral triangles for the path around the hexagonal garden. Find the value of x.

In a rectangle, an angle measure is 90; in an equilateral triangle, it is 60.

$$x + 90 + 60 + 90 = 360$$
$$x = 120$$

CA Standards Check 3 What is the measure of the angle at each outside corner of the path?

EXERCISES Standards Practice

For more exercises, see *Extra Skill, Word Problem, and Proof Practice.*

GEOM 4.0, 5.0, 12.0

A Practice by Example

Example 1 (page 229)

Complete each statement. Explain why it is true.

1. $\overline{VT} \cong$ __?__
2. $\overline{UT} \cong$ __?__ $\cong \overline{YX}$
3. $\overline{VU} \cong$ __?__
4. $\angle VYU \cong$ __?__

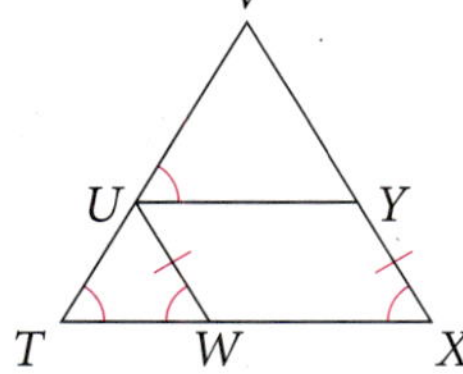

Example 2 (page 229)

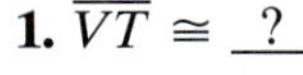

Algebra **Find the values of x and y.**

5.

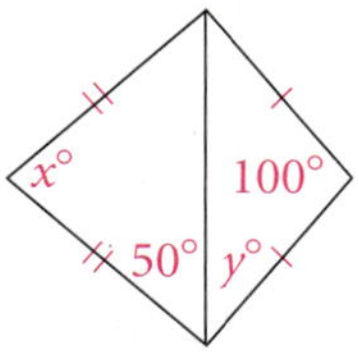

6.

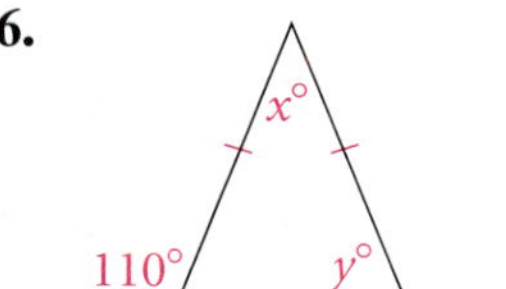

7.

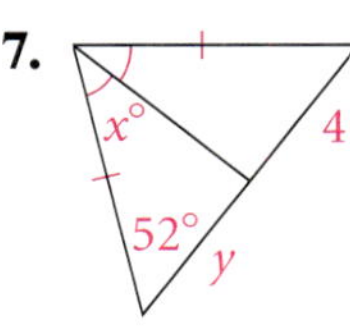

Example 3 (page 230)

8. A square and a regular hexagon are placed so that they have a common side. Find $m\angle SHA$ and $m\angle HAS$.

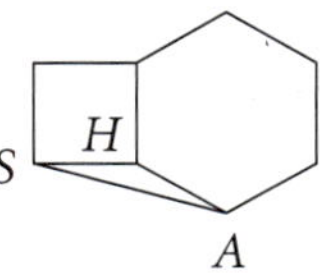

9. Five fences meet at a point to form angles with measures x, $2x$, $3x$, $4x$, and $5x$ around the point. Find the measure of each angle.

Apply Your Skills

Find each value.

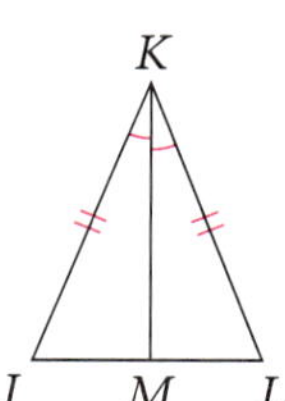

10. If $m\angle L = 58$, then $m\angle LKJ = \blacksquare$.

11. If $JL = 5$, then $ML = \blacksquare$.

12. If $m\angle JKM = 48$, then $m\angle J = \blacksquare$.

13. If $m\angle J = 55$, then $m\angle JKM = \blacksquare$.

Exercise 14

14. Architecture Seventeen spires, pictured at the left, cover the Cadet Chapel at the Air Force Academy in Colorado Springs, Colorado. Each spire is an isosceles triangle with a 40° vertex angle. Find the measure of each base angle.

15. Developing Proof Here is another way to prove the Isosceles Triangle Theorem. Supply the missing parts.

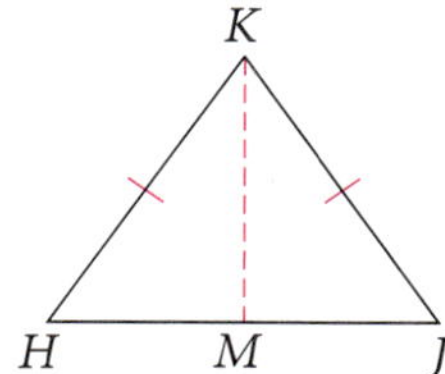

Begin with isosceles $\triangle HKJ$ with $\overline{KH} \cong \overline{KJ}$. Draw **a.** _?_, a bisector of the base $\overline{HJ}$.

Given: $\overline{KH} \cong \overline{KJ}$, **b.** _?_ bisects $\overline{HJ}$.

Prove: $\angle H \cong \angle J$

Statements	Reasons
1. $\overline{KM}$ bisects $\overline{HJ}$.	**c.** _?_
2. $\overline{HM} \cong \overline{JM}$	**d.** _?_
3. $\overline{KH} \cong \overline{KJ}$	**3.** Given
4. $\overline{KM} \cong \overline{KM}$	**e.** _?_
5. $\triangle KHM \cong \triangle KJM$	**f.** _?_
6. $\angle H \cong \angle J$	**g.** _?_

Proof **16.** Supply the missing parts in this statement of the Converse of the Isosceles Triangle Theorem. Then write a proof.

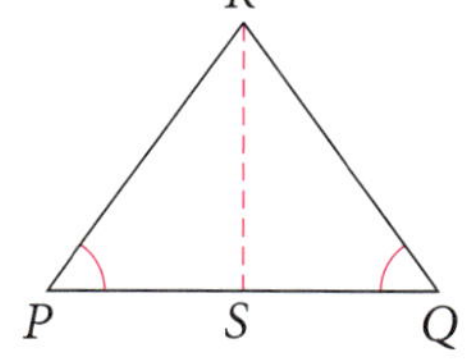

Begin with $\triangle PRQ$ with $\angle P \cong \angle Q$. Draw **a.** _?_, the bisector of $\angle PRQ$.

Given: $\angle P \cong \angle Q$, **b.** _?_ bisects $\angle PRQ$.

Prove: $\overline{PR} \cong \overline{QR}$

17. Graphic Arts A former logo for the National Council of Teachers of Mathematics is shown at the right. Trace the logo onto paper.

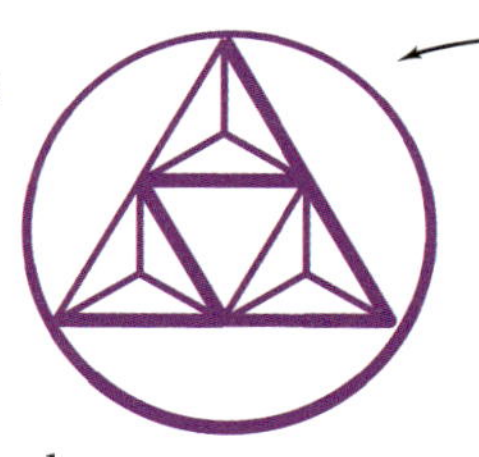

The triangles in the logo have these congruent sides and angles.

a. Highlight an obtuse isosceles triangle in the design. Then find its angle measures.

b. How many different sizes of angles can you find in the logo? What are their measures?

Visit: PHSchool.com
Web Code: bce-0405

18. Writing Explain how each corollary on page 230 follows from its theorem. First, write one explanation and then write the second similar to the first.

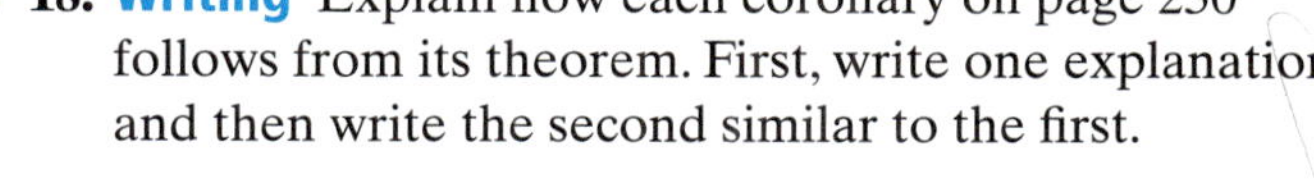

19. Multiple Choice The perimeter of the triangle at the right is 20. Find x.

Ⓐ 3 Ⓑ 5 Ⓒ 6 Ⓓ 7

Algebra **Find the values of x and y.**

20.

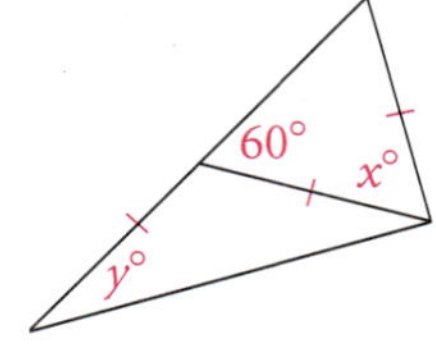

21.

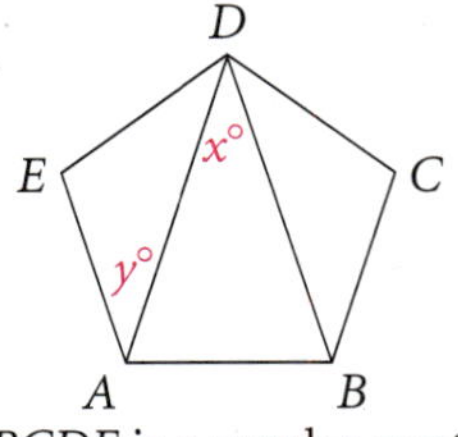

ABCDE is a regular pentagon.

22.

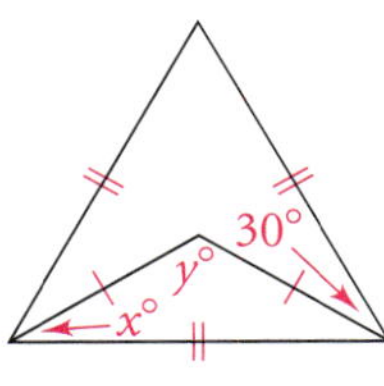

23. Write the Isosceles Triangle Theorem and its converse as a biconditional.

24. **Critical Thinking** An exterior angle of an isosceles triangle has measure 100. Find two possible sets of measures for the angles of the triangle.

25. a. **Communications** In the diagram at the right, what type of triangles are formed by the cables of the same height and the ground?

b. What are the two different base lengths of the triangles?

c. How is the tower related to each of the triangles?

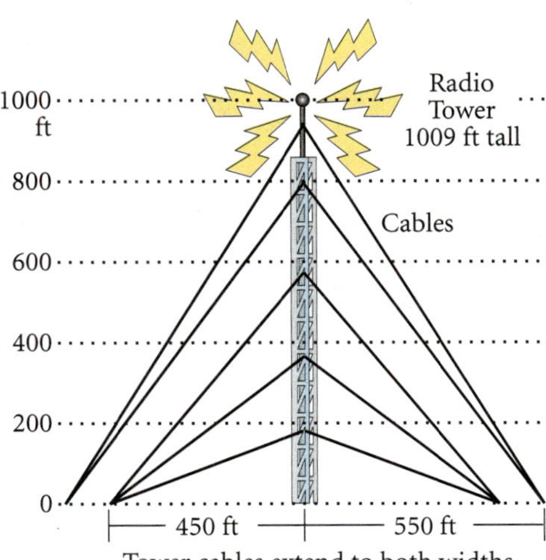

Tower cables extend to both widths.

26. **Critical Thinking** Curtis defines the base of an isosceles triangle as its "bottom side." Is his definition a good one? Explain.

27. **Reasoning** What are the measures of the base angles of an isosceles right triangle? Explain.

Real-World Connection

A radio signal's range is a function of the radio tower's height.

28. Given: $\overline{AE} \cong \overline{DE}$, $\overline{AB} \cong \overline{DC}$

Prove: $\triangle ABE \cong \triangle DCE$

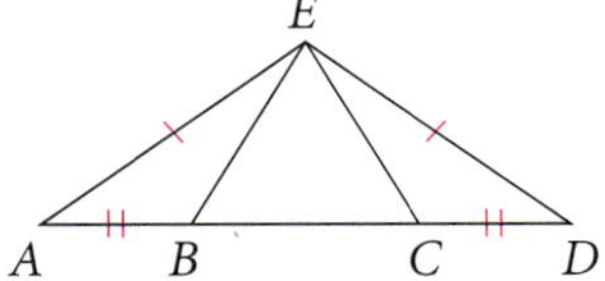

29. Prove Theorem 4-5. Use the diagram next to it on page 228.

Algebra **Find the values of m and n.**

30.

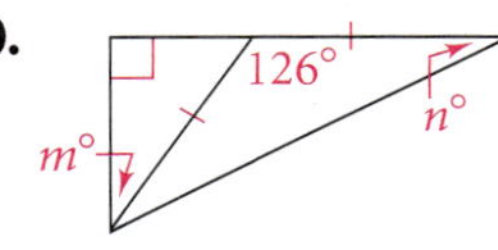

31.

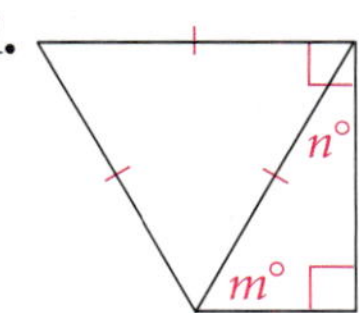

32.

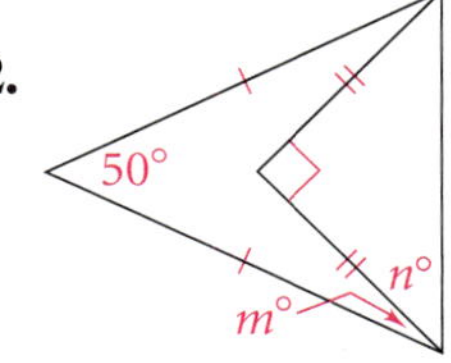

Coordinate Geometry **For each pair of points, there are six points that could be the third vertex of an isosceles right triangle. Find the coordinates of each point.**

33. (4, 0) and (0, 4) **34.** (0, 0) and (5, 5) **35.** (2, 3) and (5, 6)

36. **Algebra** A triangle has angle measures $x + 15$, $3x - 35$, and $4x$.

a. Find the value of x.

b. Find the measure of each angle.

c. What type of triangle is it? Why?

37. State the converse of Theorem 4-5. If the converse is true, write a paragraph proof. If the converse is false, give a counterexample.

Real-World Connection

The circle is a basic shape for many tile designs.

38. **Crafts** The design in Step 3 is used in Hmong crafts and in Islamic and Mexican tiles. To create it, the artist starts by drawing a circle and four equally spaced diameters.

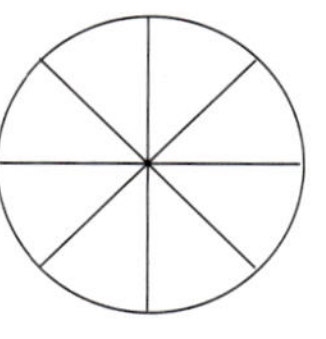

Step 1

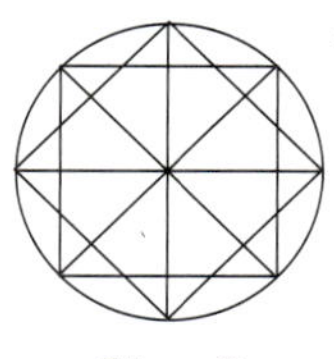

Step 2

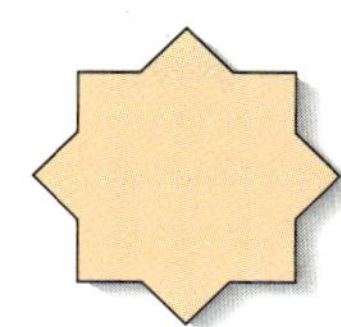

Step 3

a. How many different sizes of isosceles right triangles can you find in Step 2? Trace an example of each onto your paper.

b. How many times does a triangle of each size in part (a) appear in the Step 2 diagram?

Reasoning **What measures are possible for the base angles of each type of triangle? Explain.**

39. an isosceles obtuse triangle

40. an isosceles acute triangle

Multiple Choice Practice

For California Standards and CAHSEE Tutorials, visit PHSchool.com. Web Code: bcq-9045

GEOM 12.0

41. A line divides a square into two equal triangles as in the figure at the right. The triangles formed can be classified as which of the following?

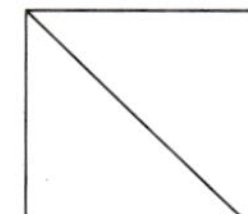

Ⓐ Equilateral
Ⓑ Isosceles
Ⓒ Obtuse
Ⓓ Scalene

GEOM 12.0

42. The following triangle is equilateral. Find the values of x and y.

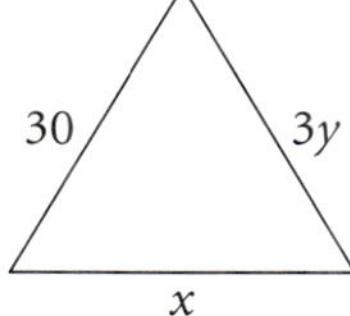

Ⓐ $x = 30$ and $y = 30$
Ⓑ $x = 15$ and $y = 5$
Ⓒ $x = 30$ and $y = 10$
Ⓓ $x = 15$ and $y = 10$

CAHSEE 7MG 1.1

43. Which equation should Margo use to convert y yards to x inches?

Ⓐ $x = 36y$ Ⓑ $y = 36x$ Ⓒ $x = 3y$ Ⓓ $y = 3x$

Mixed Review

Lesson 4-4

44. $m\angle R = 59$, $m\angle T = 93 = m\angle H$, $m\angle V = 28$, and $RT = GH$. What, if anything, can you conclude about RC and GV? Explain.

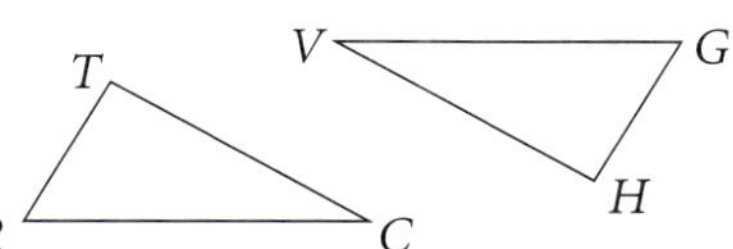

Lessons 4-2, 4-3

Which congruence statement, SSS, SAS, ASA, or AAS, would you use to conclude that the two triangles are congruent?

45.

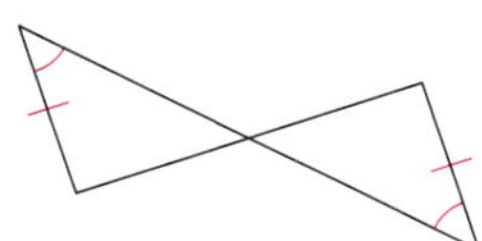

46.

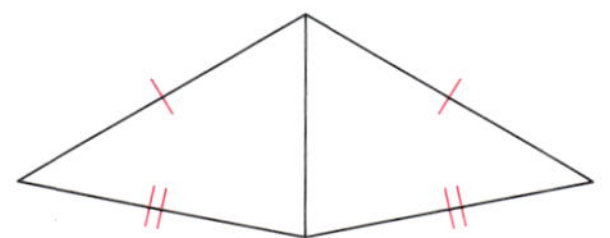

Lesson 3-5

47. How many sides are in a regular polygon whose exterior angles measure 15°

Algebra 1 Review — Systems of Linear Equations

FOR USE WITH LESSON 4-6

You can solve a system of equations in two variables by using substitution to form a one-variable equation.

1 EXAMPLE

x^2 **Algebra** Solve the system: $y = 3x + 5$
$y = x + 1$

$y = x + 1$	**Start with one equation.**
$3x + 5 = x + 1$	**Substitute $3x + 5$ for y.**
$2x = -4$	**Solve for x.**
$x = -2$	

Substitute -2 for x in either equation and solve for y.

$y = x + 1$
$= (-2) + 1 = -1$

Since $x = -2$ and $y = -1$, the solution is $(-2, -1)$. This is the point of intersection of the two lines.

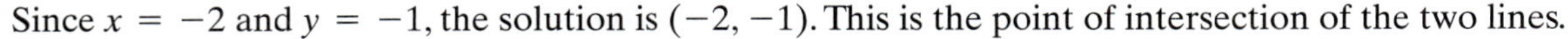

The graph of a linear system with *infinitely many solutions* is one line, and the graph of a linear system with *no solution* is two parallel lines.

2 EXAMPLE

x^2 **Algebra** Solve the system: $x + y = 3$
$4x + 4y = 8$

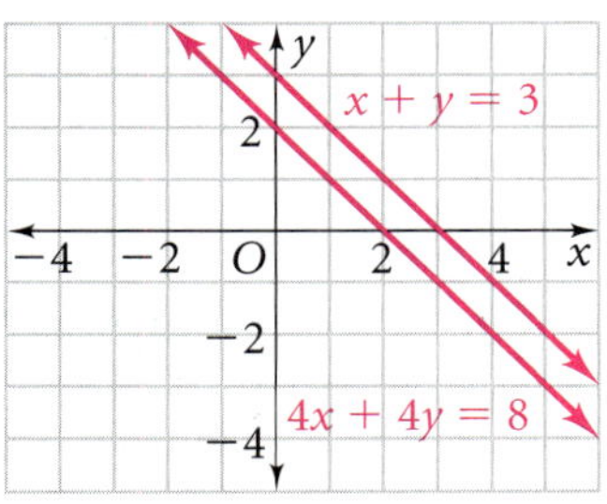

$x + y = 3$	
$x = 3 - y$	**Solve the first equation for x.**
$4(3 - y) + 4y = 8$	**Substitute $3 - y$ for x in the second equation.**
$12 - 4y + 4y = 8$	**Solve for y.**
$12 = 8$	**False!**

Since $12 = 8$ is a false statement, the system has no solution.

EXERCISES

Solve each system of equations.

1. $y = x - 4$
$y = 3x + 2$

2. $2x - y = 8$
$x + 2y = 9$

3. $3x + y = 4$
$-6x - 2y = 12$

4. $2x - 3 = y + 3$
$2x + y = -3$

5. $y = x + 1$
$x = y - 1$

6. $x - y = 4$
$3x - 3y = 6$

7. $y = -x + 2$
$2y = 4 - 2x$

8. $y = 2x + 1$
$y = 3x - 7$

9. $x - y = 2$
$x + y = 1$

10. What is the solution of the system, $x = 4, y = 3$? Interpret the solution using the coordinate plane.

4-6 Congruence in Right Triangles

California Content Standards

GEOM 2.0 Students write geometric proofs. ***Develop***
GEOM 5.0 Students prove that triangles are congruent, and they are able to use the concept of corresponding parts of congruent triangles. ***Master***

What You'll Learn

- To prove triangles congruent using the HL theorem

. . . And Why

To show that one pattern can be used to cut the fabric for the two entrance flaps of a tent, as in Example 1

Check Skills You'll Need

GO for Help Lessons 4-2 and 4-3

Tell whether the abbreviation identifies a congruence statement.

1. SSS **2.** SAS **3.** SSA

4. ASA **5.** AAS **6.** AAA

Can you conclude that the two triangles are congruent? Explain.

7.

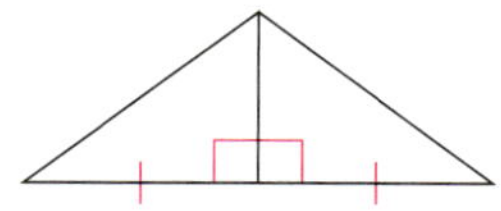

8.

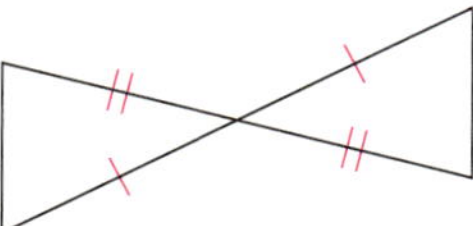

New Vocabulary • hypotenuse • legs of a right triangle

1 The Hypotenuse-Leg Theorem

In a right triangle, the side opposite the right angle is the longest side and is called the **hypotenuse.** The other two sides are called **legs.**

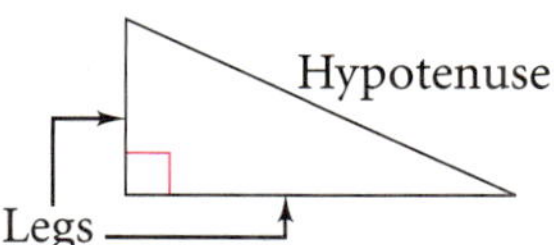

Right triangles provide a special case for which there is an SSA congruence rule. (See Lesson 4-3, Exercise 32.) It occurs when hypotenuses are congruent and one pair of legs are congruent.

Key Concepts

Theorem 4-6 **Hypotenuse-Leg (HL) Theorem**

If the hypotenuse and a leg of one right triangle are congruent to the hypotenuse and a leg of another right triangle, then the triangles are congruent.

Proof

Proof of the HL Theorem

Given: $\triangle PQR$ and $\triangle XYZ$ are right triangles, with right angles Q and Y respectively. $\overline{PR} \cong \overline{XZ}$, and $\overline{PQ} \cong \overline{XY}$.

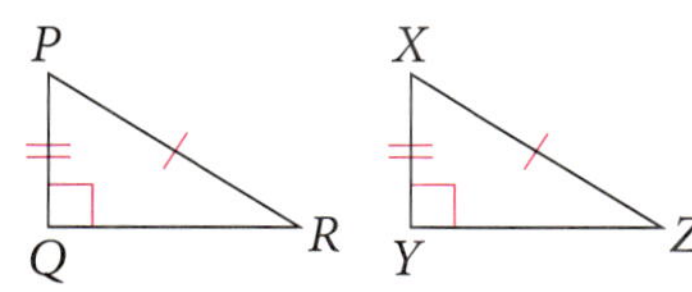

Prove: $\triangle PQR \cong \triangle XYZ$

Proof: On $\triangle XYZ$ at the right, draw $\overrightarrow{ZY}$. Mark point S as shown so that $YS = QR$. Then, $\triangle PQR \cong \triangle XYS$ by SAS. By CPCTC, $\overline{PR} \cong \overline{XS}$. It is given that $\overline{PR} \cong \overline{XZ}$, so $\overline{XS} \cong \overline{XZ}$ by the Transitive Property of Congruence.

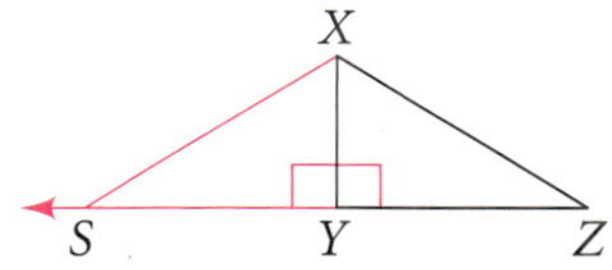

By the Isosceles Triangle Theorem, $\angle S \cong \angle Z$, so $\triangle XYS \cong \triangle XYZ$ by AAS. Therefore, $\triangle PQR \cong \triangle XYZ$ by the Transitive Property of Congruence.

To use the HL Theorem, you must show that three conditions are met.

- There are two right triangles.
- The triangles have congruent hypotenuses.
- There is one pair of congruent legs.

1 EXAMPLE Real-World Connection

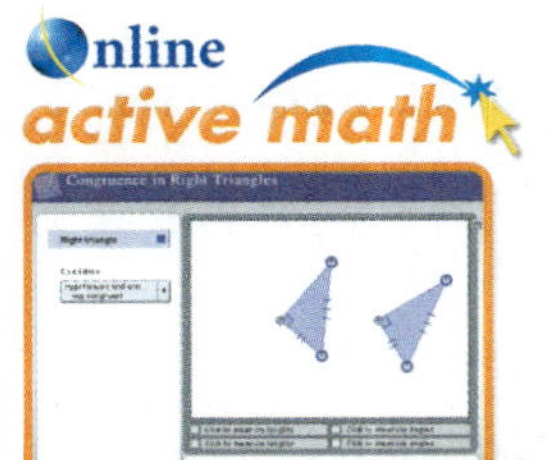

For: Right Triangles Activity
Use: Interactive Textbook, 4-6

Tent Design On the tent, $\angle CPA$ and $\angle MPA$ are right angles and $\overline{CA} \cong \overline{MA}$. Can you use one pattern to cut fabric for both flaps of the tent? Explain.

Check whether the two right triangles meet the three conditions for the HL Theorem.

- You are given that $\angle CPA$ and $\angle MPA$ are right angles. $\triangle CPA$ and $\triangle MPA$ are right triangles.
- The hypotenuses of the triangles are $\overline{CA}$ and $\overline{MA}$. You are given that $\overline{CA} \cong \overline{MA}$.
- $\overline{PA}$ is a leg of both $\triangle CPA$ and $\triangle MPA$. $\overline{PA} \cong \overline{PA}$ by the Reflexive Property of Congruence.

$\triangle CPA \cong \triangle MPA$ by the HL Theorem. The triangles are the same shape and size. You can use one pattern for both flaps.

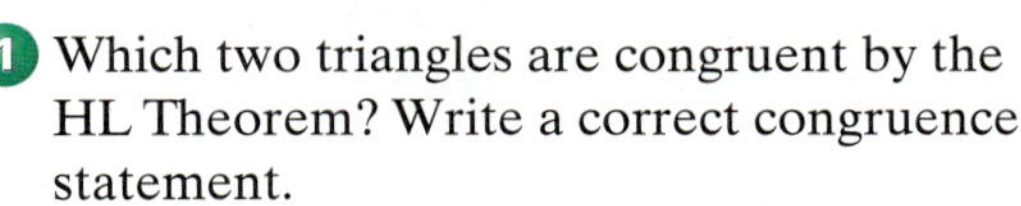

1 Which two triangles are congruent by the HL Theorem? Write a correct congruence statement.

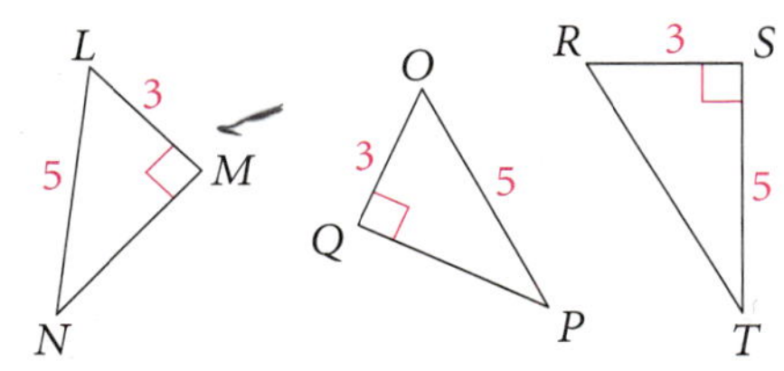

Proof 2 EXAMPLE Using the HL Theorem

Given: $\overline{CD} \cong \overline{EA}$, $\overline{AD}$ is the perpendicular bisector of $\overline{CE}$.

Prove: $\triangle CBD \cong \triangle EBA$

Proof:

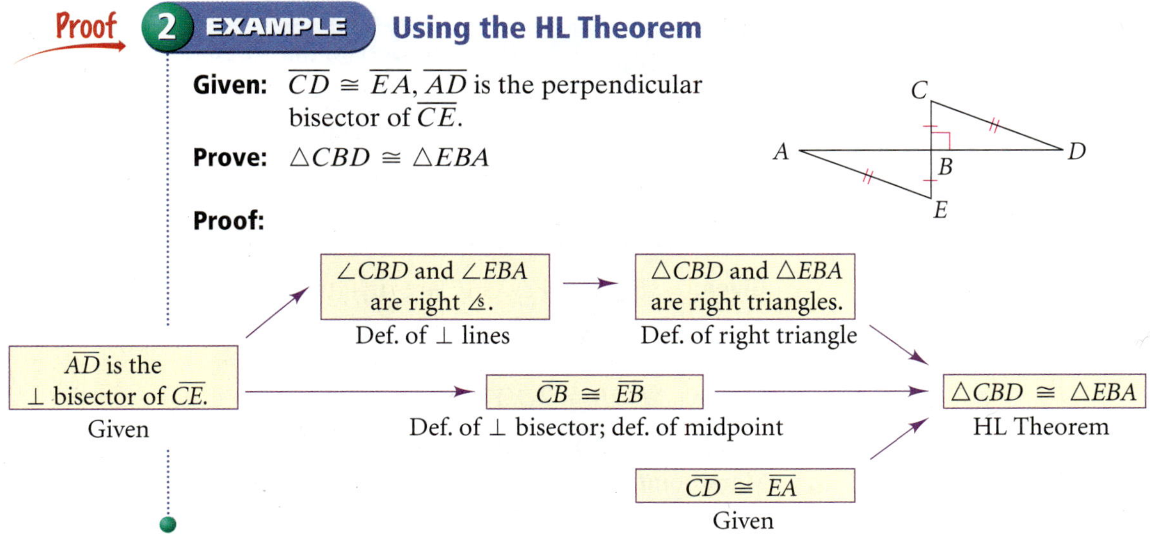

CA Standards Check

2 Prove that the two triangles you named in Quick Check 1 are congruent.

Proof **3 EXAMPLE** **Using the HL Theorem**

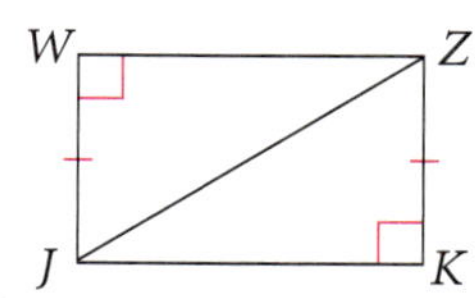

Given: $\overline{WJ} \cong \overline{KZ}$, $\angle W$ and $\angle K$ are right angles.

Prove: $\triangle JWZ \cong \triangle ZKJ$

Statements	Reasons
1. $\angle W$ and $\angle K$ are right angles.	1. Given
2. $\triangle JWZ$ and $\triangle ZKJ$ are right triangles.	2. Definition of right triangle
3. $\overline{JZ} \cong \overline{JZ}$	3. Reflexive Property of Congruence
4. $\overline{WJ} \cong \overline{KZ}$	4. Given
5. $\triangle JWZ \cong \triangle ZKJ$	5. HL Theorem

3 **Given:** $\angle PRS$ and $\angle RPQ$ are right angles, $\overline{SP} \cong \overline{QR}$.

Prove: $\triangle PRS \cong \triangle RPQ$

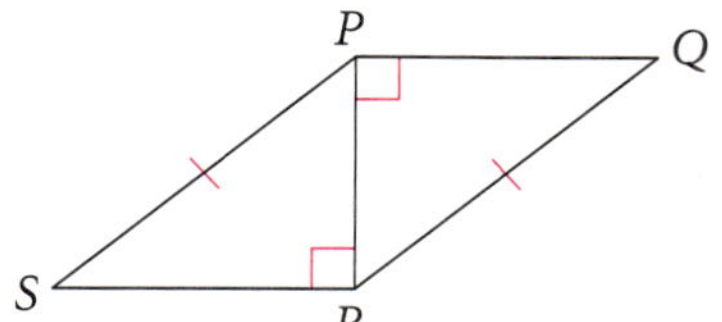

For more exercises, see *Extra Skill, Word Problem, and Proof Practice.*

GEOM 2.0, 5.0

A Practice by Example

for Help

Example 1 (page 236)

Write a short paragraph to explain why the two triangles are congruent.

1.

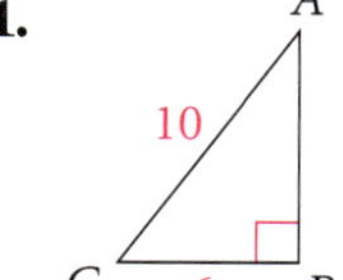

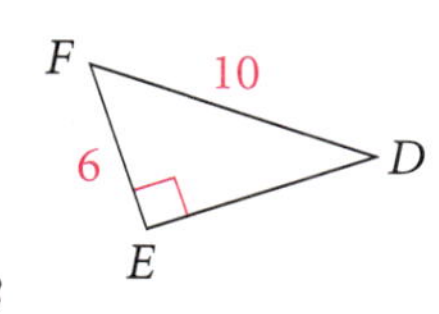

2.

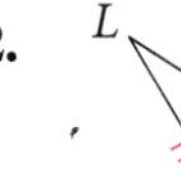

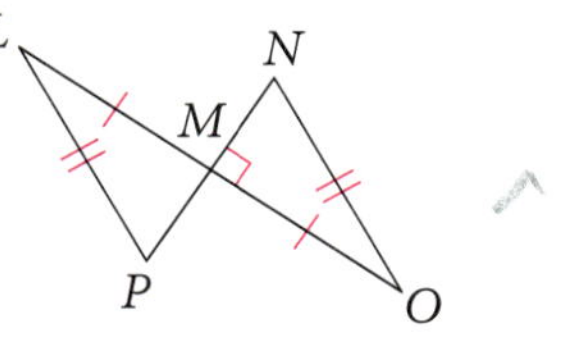

What additional information do you need to prove the triangles congruent by HL?

3. $\triangle BLT$ and $\triangle RKQ$

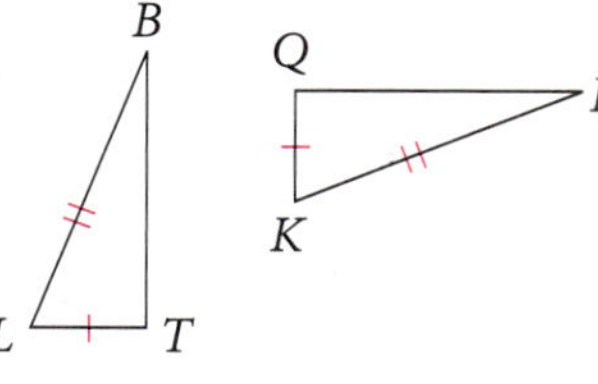

4. $\triangle XRV$ and $\triangle TRV$

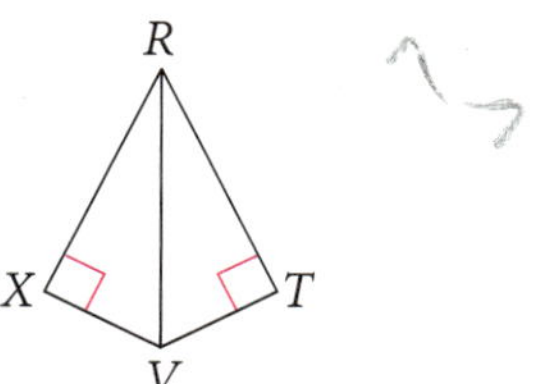

Example 2 (page 236)

5. **Developing Proof** Complete the flow proof.

Given: $\overline{PS} \cong \overline{PT}$, $\angle PRS \cong \angle PRT$

Prove: $\triangle PRS \cong \triangle PRT$

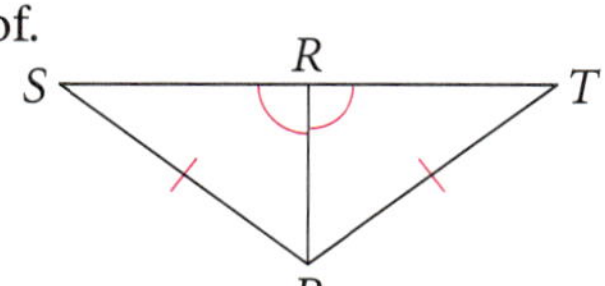

$\angle PRS$ and $\angle PRT$ are $\cong$ and supplementary $\angle$s. → $\angle PRS$ and $\angle PRT$ are right $\angle$s. → $\triangle PRS$ and $\triangle PRT$ are right $\triangle$s.

Given (by diagram) | a. ? | b. ?

$\overline{PS} \cong \overline{PT}$ c. ?

$\overline{PR} \cong \overline{PR}$ d. ?

→ $\triangle PRS \cong \triangle PRT$ e. ?

Proof **6. Given:** $\overline{AD} \cong \overline{CB}$, $\angle D$ and $\angle B$ are right angles.
Prove: $\triangle ADC \cong \triangle CBA$

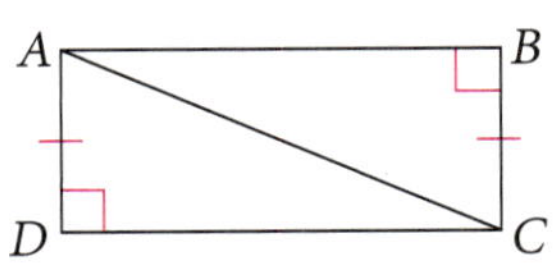

Example 3 (page 237)

7. Developing Proof Complete the two-column proof.

Given: $\overline{JL} \perp \overline{LM}$, $\overline{LJ} \perp \overline{JK}$, $\overline{MJ} \cong \overline{KL}$
Prove: $\triangle JLM \cong \triangle LJK$

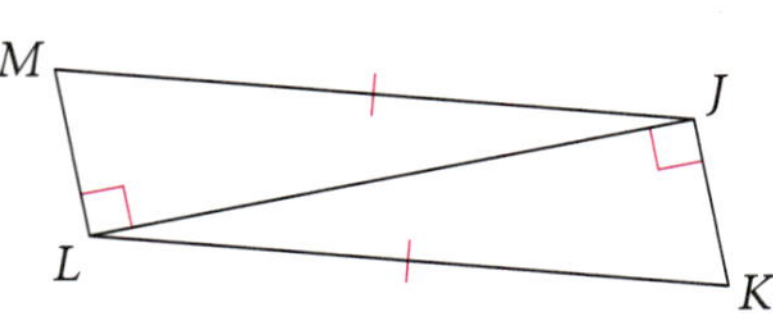

Statements	Reasons
1. $\overline{JL} \perp \overline{LM}$ and $\overline{LJ} \perp \overline{JK}$	a. ?
2. $\angle JLM$ and $\angle LJK$ are right angles.	b. ?
c. ?	3. Definition of a right triangle
4. $\overline{MJ} \cong \overline{KL}$	d. ?
e. ?	5. Reflexive Property of Congruence
6. $\triangle JLM \cong \triangle LJK$	f. ?

Proof **8. Given:** $\overline{HV} \perp \overline{GT}$, $\overline{GH} \cong \overline{TV}$, I is the midpoint of $\overline{HV}$.
Prove: $\triangle IGH \cong \triangle ITV$

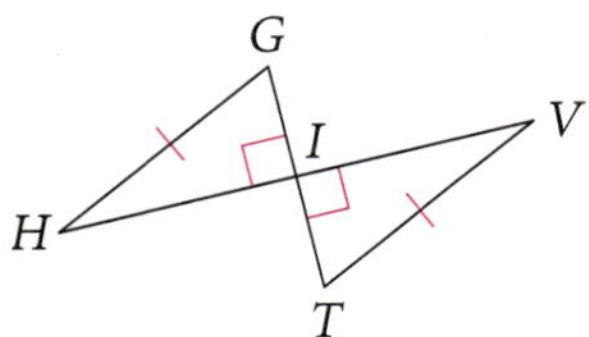

B Apply Your Skills

Real-World Connection

A mechanical clock has gears that transmit motion from a spring to the clock's hands.

9. Antiques To repair an antique clock, a 12-toothed wheel has to be made by cutting right triangles out of a regular polygon that has twelve 4-cm sides. The hypotenuse of each triangle is a side of the regular polygon, and the shorter leg is 1 cm long. Explain why the 12 triangles must be congruent.

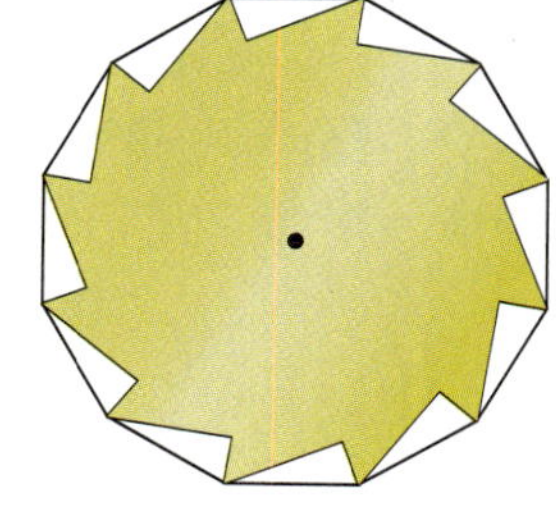

x^2 **Algebra In Exercises 10 and 11, for what values of *x* and *y* are the triangles congruent by HL?**

10.

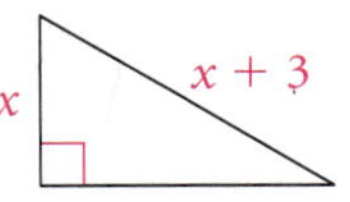

11.

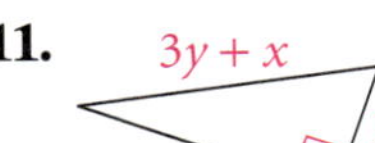

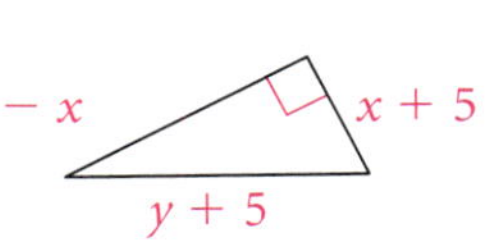

12. Critical Thinking While working for a landscape architect, you are told to lay out a flower bed in the shape of a right triangle with sides of 3 yd and 7 yd. Explain what else you need to know in order to make the flower bed.

Proof **13. Given:** $\overline{RS} \cong \overline{TU}$, $\overline{RS} \perp \overline{ST}$, $\overline{TU} \perp \overline{UV}$, T is the midpoint of $\overline{RV}$.
Prove: $\triangle RST \cong \triangle TUV$

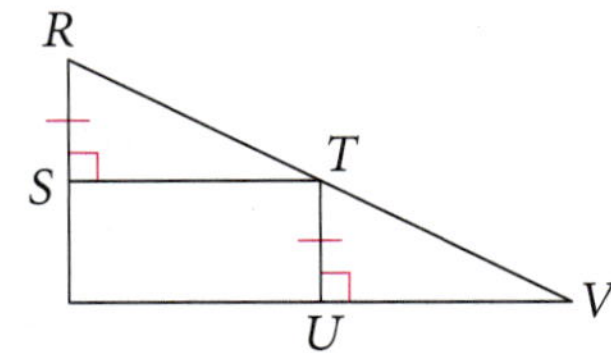

14. Given: $\overline{JM} \cong \overline{WP}$, $\overline{JP} \parallel \overline{MW}$, $\overline{JP} \perp \overline{PM}$
Prove: $\triangle JMP \cong \triangle WPM$

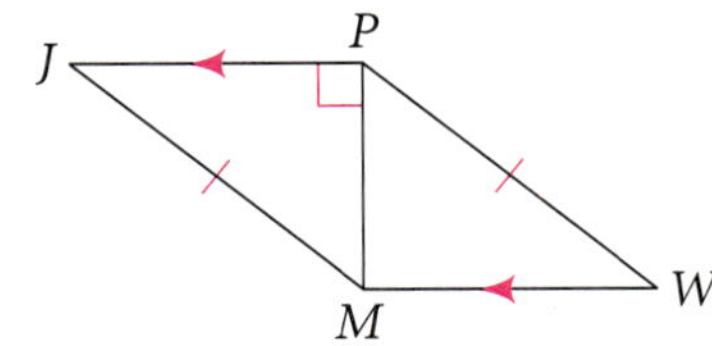

Homework Video Tutor

Visit: PHSchool.com
Web Code: bce-0406

Proof **15.** Study Exercise 5. There is a different set of steps that will prove $\triangle PRS \cong \triangle PRT$. Decide what they are. Then write a proof using these steps.

Constructions **Copy the triangle and construct a triangle congruent to it using the method stated.**

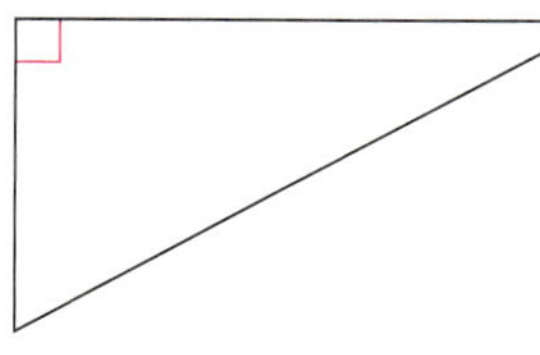

16. by SAS

17. by HL

18. by ASA

19. by SSS

Proof **20. Given:** $\overline{EB} \cong \overline{DB}$, $\angle A$ and $\angle C$ are right angles, and B is the midpoint of $\overline{AC}$.

Prove: $\triangle BEA \cong \triangle BDC$

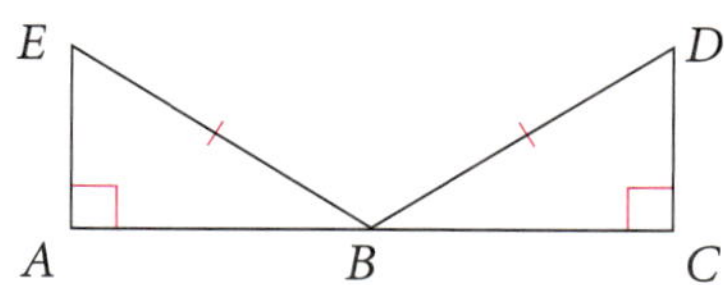

21. Given: $\overline{LO}$ bisects $\angle MLN$, $\overline{OM} \perp \overline{LM}$, and $\overline{ON} \perp \overline{LN}$.

Prove: $\triangle LMO \cong \triangle LNO$

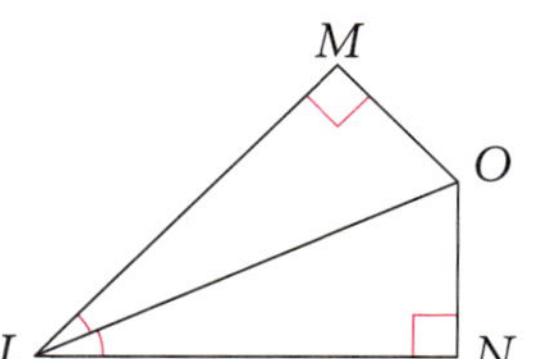

Exercise 22

22. Open-Ended You are the DJ for the school dance. To set up, you have placed one speaker in the corner of the platform. What measurement(s) could you make with a tape measure to make sure that a matching speaker is in the other corner at exactly the same angle? Explain why your method works.

23. a. Coordinate Geometry Use graph paper. Graph the points $E(-1, -1)$, $F(-2, -6)$, $G(-4, -4)$, and $D(-6, -2)$. Connect the points with segments.

b. Find the slope for each of $\overline{DG}$, $\overline{GF}$, and $\overline{GE}$.

c. Use your answer to part (b) to describe $\angle EGD$ and $\angle EGF$.

d. Use the Distance Formula to find DE and FE.

e. Write a paragraph to prove that $\triangle EGD \cong \triangle EGF$.

24. Critical Thinking "A HA!" exclaims Francis. "There is an HA Theorem . . . , something like the HL Theorem!" Explain what Francis is saying and why he is correct or incorrect.

Challenge

Geometry in 3 Dimensions **Use the figure at the right for Exercises 25 and 26.**

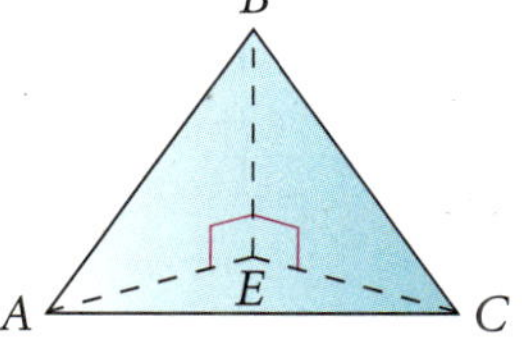

Proof **25. Given:** $\overline{BE} \perp \overline{EA}$, $\overline{BE} \perp \overline{EC}$, $\triangle ABC$ is equilateral.

Prove: $\triangle AEB \cong \triangle CEB$

26. Given: $\triangle AEB \cong \triangle CEB$, $\overline{BE} \perp \overline{EA}$, and $\overline{BE} \perp \overline{EC}$. Can you prove that $\triangle ABC$ is equilateral? Explain.

Multiple Choice Practice

For California Standards and CAHSEE Tutorials, visit PHSchool.com. Web Code: bcq-9045

GEOM 5.0

27. For the following triangles to be congruent by HL, what must be true?

Ⓐ $\overline{ON} = 5$

Ⓑ $\overline{OS} = 10$

Ⓒ $\overline{AS} = 10$

Ⓓ $\overline{AS} = 16$

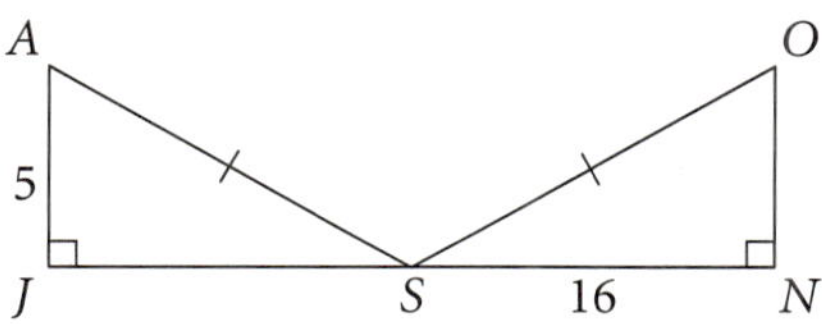

GEOM 5.0

28. Which postulate or theorem justifies the statement $\triangle JLV = \triangle PMK$?

Ⓐ AAS Ⓒ ASA
Ⓑ SAS Ⓓ SSS

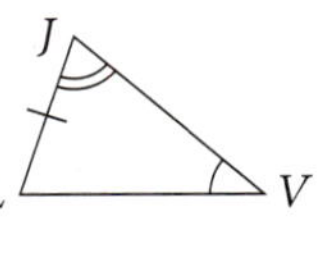

GEOM 4.0

29. Quadrilateral $ABCD \cong$ Quadrilateral $QRST$. Which segment is congruent to TS?

Ⓐ $\overline{AB}$ Ⓒ $\overline{DA}$
Ⓑ $\overline{BC}$ Ⓓ $\overline{DC}$

Mixed Review

Lesson 2-1

For Exercises 30 and 31, what type of triangle must $\triangle XYZ$ be?

30. $\triangle XYZ \cong \triangle ZYX$ **31.** $\triangle XYZ \cong \triangle ZXY$

Lesson 2-1

32. Connect $A(3, 3)$, $B(5, 5)$, $C(9, 1)$, and $D(9, -3)$ in order. Are any sides of the figure parallel? Are any sides perpendicular? Explain.

Lesson 3-1

State the postulate or theorem that justifies each statement.

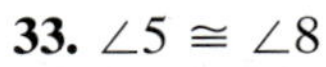

33. $\angle 5 \cong \angle 8$ **34.** $m\angle 4 + m\angle 8 = 180$

35. $\angle 6 \cong \angle 9$ **36.** $\angle 4 \cong \angle 10$

37. $\angle 1 \cong \angle 6$ **38.** $\angle 6$ and $\angle 3$ are supplementary.

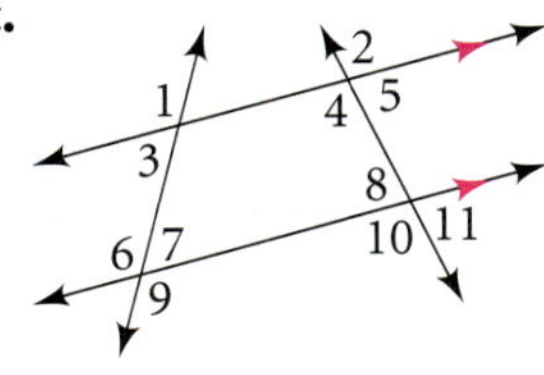

Checkpoint Quiz 1 Lessons 4-4 through 4-6

1. In the diagram at the right, $\triangle PQR \cong \triangle SRQ$ by SAS. What other pairs of sides and angles can you conclude are congruent by CPCTC?

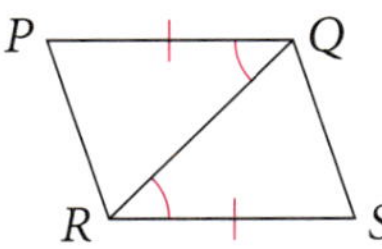
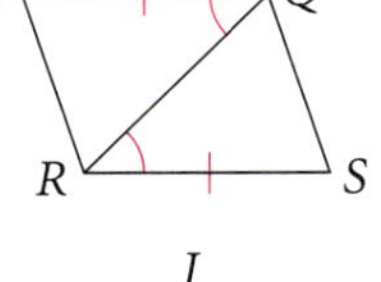

2. Complete the plan for a proof.

Given: Isosceles $\triangle JKL$ with $\overline{JK} \cong \overline{JL}$; $\overline{KM}$ and $\overline{LM}$ are bisectors of the base angles.

Prove: $\triangle KML$ is isosceles.

Plan: Since $\triangle JKL$ is isosceles, $\angle JKL \cong \angle JLK$ by the **a.** __?__ Theorem. Since $\overline{KM}$ and $\overline{LM}$ are angle bisectors, $\angle MKL$ **b.** __?__ $\angle MLK$. Therefore, $\triangle KML$ is isosceles by the **c.** __?__ Theorem.

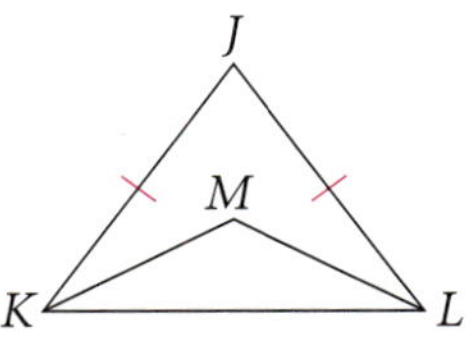

3. Six triangles are pictured in the diagram at the left. Which of the triangles are isosceles? Explain.

E
A B C D
Exercise 3

4. Why are these triangles congruent?

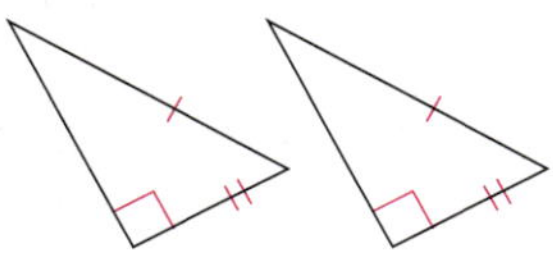

5. Explain why $\overline{GW} \cong \overline{ST}$.

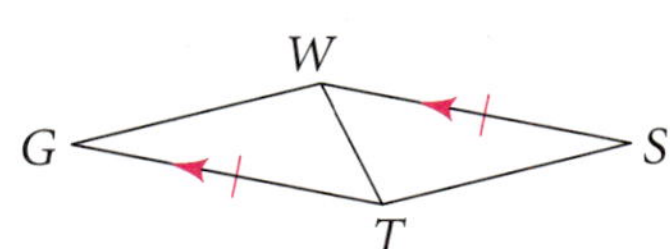

Using Corresponding Parts of Congruent Triangles

California Content Standards

GEOM 5.0 Students prove that triangles are congruent, and they are able to use the concept of corresponding parts of congruent triangles. ***Master***

What You'll Learn

- To identify congruent overlapping triangles
- To prove two triangles congruent by first proving two other triangles congruent

. . . And Why

To identify overlapping triangles in scaffolding, as in Example 1

Check Skills You'll Need

for Help Lessons 1-1 and 4-3

1. How many triangles will the next two figures in this pattern have?

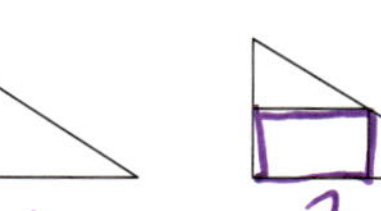

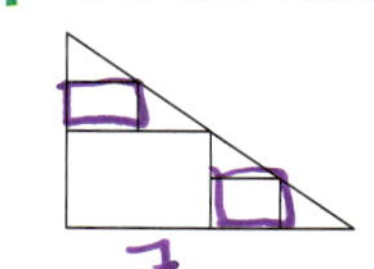

2. Can you conclude that the triangles are congruent? Explain.

a. $\triangle AZK$ and $\triangle DRS$ **b.** $\triangle SDR$ and $\triangle JTN$ **c.** $\triangle ZKA$ and $\triangle NJT$

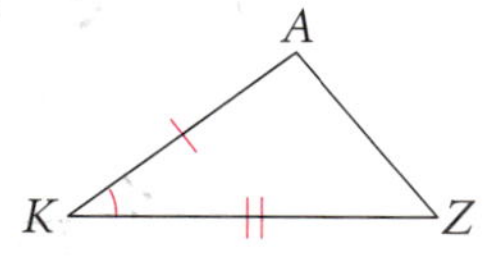

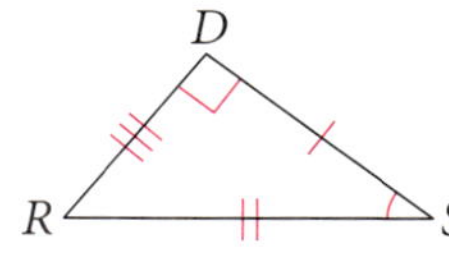

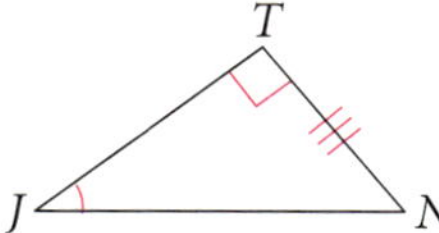

1 Using Overlapping Triangles in Proofs

Vocabulary Tip

Overlapping triangles share part or all of one or more sides.

Some triangle relationships are difficult to see because the triangles overlap. Overlapping triangles may have a common side or angle. You can simplify your work with overlapping triangles by separating and redrawing the triangles.

1 EXAMPLE Identifying Common Parts

Separate and redraw $\triangle DFG$ and $\triangle EHG$. Identify the common angle.

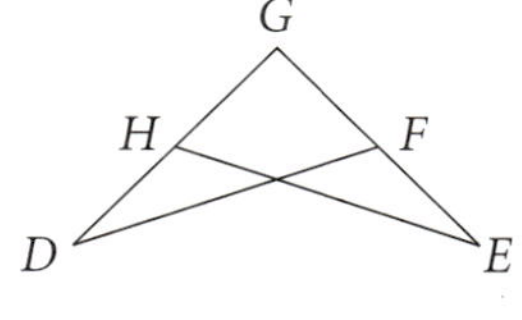

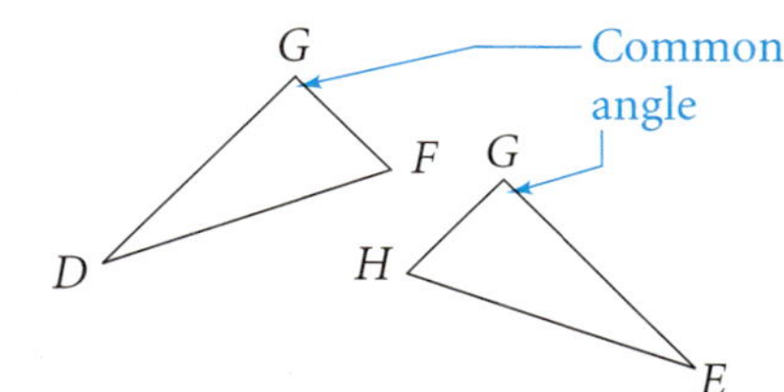

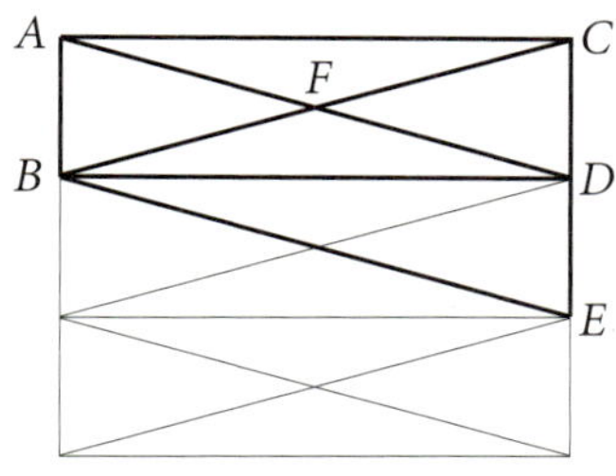

1 Engineering The diagram at the left shows triangles from the scaffolding that workers used when they repaired and cleaned the Statue of Liberty.

a. Name the common side in $\triangle ADC$ and $\triangle BCD$.

b. Name another pair of triangles that share a common side. Name the common side.

In overlapping triangles, a common side or angle is congruent to itself by the Reflexive Property of Congruence.

Proof

2 EXAMPLE Using Common Parts

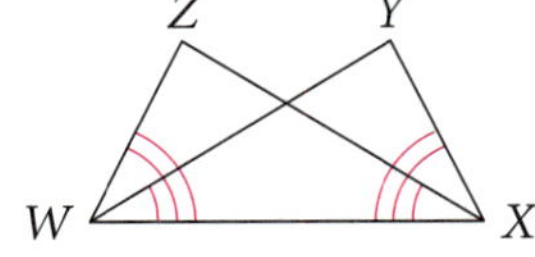

Given: $\angle ZXW \cong \angle YWX$, $\angle ZWX \cong \angle YXW$

Write a plan and then a proof to show that the two "outside" segments are congruent.

Prove: $\overline{ZW} \cong \overline{YX}$

Plan: First, separate the overlapping triangles. $\overline{ZW} \cong \overline{YX}$ by CPCTC if $\triangle ZXW \cong \triangle YWX$. Show this congruence by ASA.

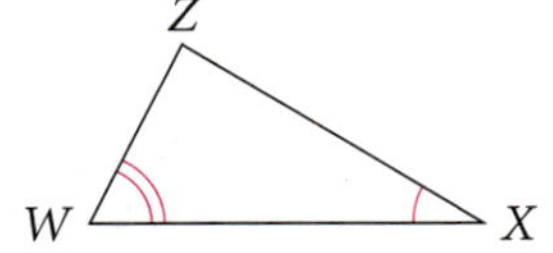

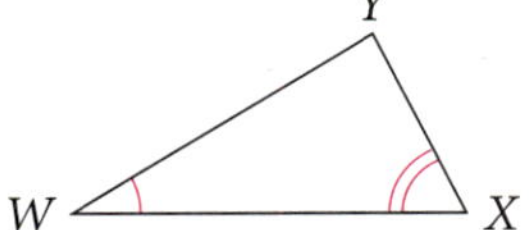

Proof:

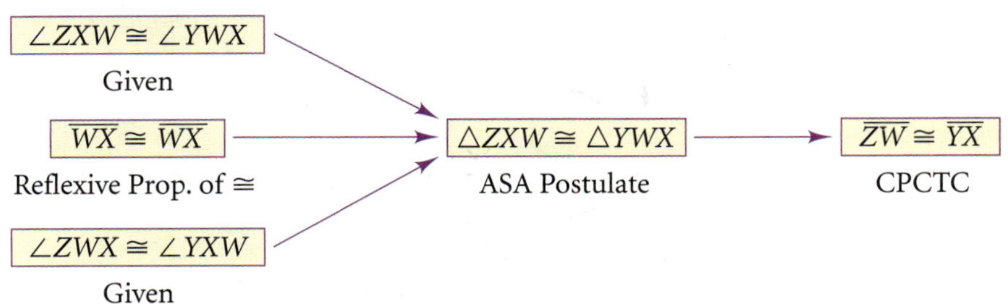

CA Standards Check **2** Write a plan and then a proof.

Given: $\triangle ACD \cong \triangle BDC$

Prove: $\overline{CE} \cong \overline{DE}$

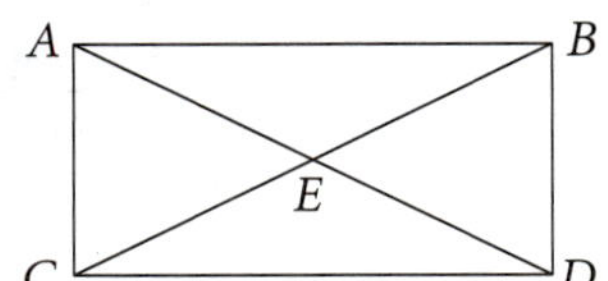

2 Using Two Pairs of Congruent Triangles

Sometimes you can prove one pair of triangles congruent and then use their congruent corresponding parts to prove another pair congruent.

Proof

3 EXAMPLE Using Two Pairs of Triangles

Given: In the quilt, E is the midpoint of $\overline{AC}$ and $\overline{DB}$.

Prove: $\triangle GED \cong \triangle JEB$

Write a plan and then a proof.

Plan: $\triangle GED \cong \triangle JEB$ by ASA if $\angle D \cong \angle B$. These angles are congruent by CPCTC if $\triangle AED \cong \triangle CEB$. These triangles are congruent by SAS.

Proof: E is the midpoint of $\overline{AC}$ and $\overline{DB}$, so $\overline{AE} \cong \overline{CE}$ and $\overline{DE} \cong \overline{BE}$. $\angle AED \cong \angle CEB$ because vertical angles are congruent. Therefore, $\triangle AED \cong \triangle CEB$ by SAS. $\angle D \cong \angle B$ by CPCTC, and $\angle GED \cong \angle JEB$ because they are vertical angles. Therefore, $\triangle GED \cong \triangle JEB$ by ASA.

3 Write a plan and then a proof.

Given: $\overline{PS} \cong \overline{RS}$, $\angle PSQ \cong \angle RSQ$

Prove: $\triangle QPT \cong \triangle QRT$

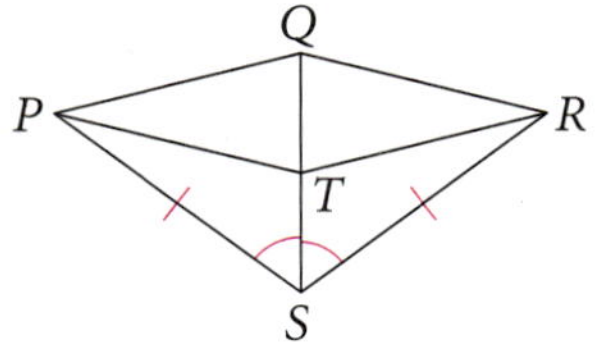

When triangles overlap, you can keep track of information by drawing other diagrams that separate the overlapping triangles.

Proof

Separating Overlapping Triangles

Given: $\overline{CA} \cong \overline{CE}$, $\overline{BA} \cong \overline{DE}$

Write a plan and then a proof to show that two small segments inside the triangle are congruent.

Prove: $\overline{BX} \cong \overline{DX}$

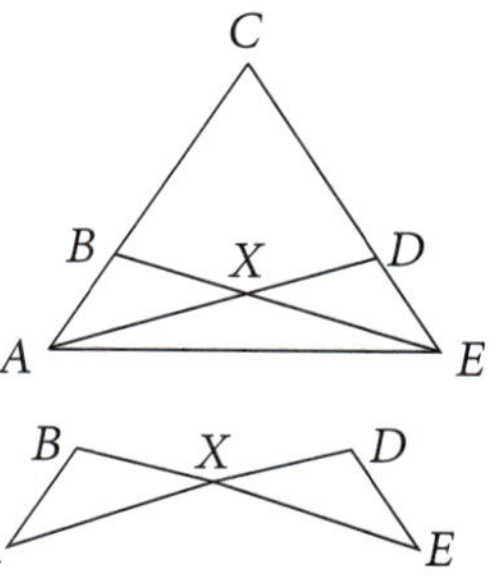

Plan: $\overline{BX} \cong \overline{DX}$ by CPCTC if $\triangle BXA \cong \triangle DXE$. This congruence holds by AAS if $\angle ABX \cong \angle EDX$. These are congruent by CPCTC in $\triangle BAE$ and $\triangle DEA$, which are congruent by SAS.

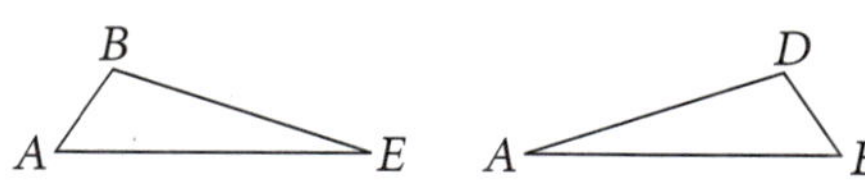

Proof:

Statements	Reasons
1. $\overline{BA} \cong \overline{DE}$	1. Given
2. $\overline{CA} \cong \overline{CE}$	2. Given
3. $\angle CAE \cong \angle CEA$	3. Isosceles Triangle Theorem
4. $\overline{AE} \cong \overline{AE}$	4. Reflexive Property of Congruence
5. $\triangle BAE \cong \triangle DEA$	5. SAS
6. $\angle ABE \cong \angle EDA$	6. CPCTC
7. $\angle BXA \cong \angle DXE$	7. Vertical angles are congruent.
8. $\triangle BXA \cong \triangle DXE$	8. AAS
9. $\overline{BX} \cong \overline{DX}$	9. CPCTC

Real-World Connection

The Japanese paper-folding art of origami involves many overlapping triangles.

CA Standards Check

4 Plan a proof. Separate the overlapping triangles in your plan. Then follow your plan and write a proof.

Given: $\angle CAD \cong \angle EAD$, $\angle C \cong \angle E$

Prove: $\overline{BD} \cong \overline{FD}$

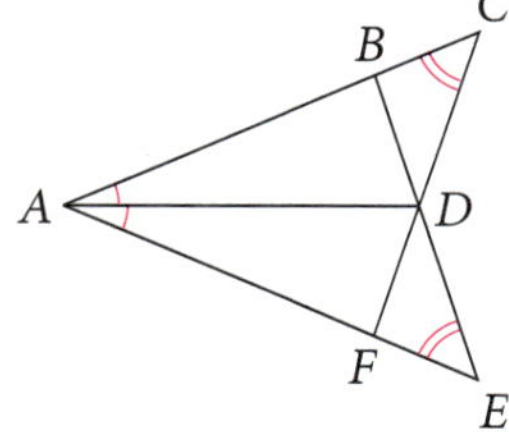

EXERCISES

For more exercises, see *Extra Skill, Word Problem, and Proof Practice.*

Standards Practice GEOM 5.0

A Practice by Example

Example 1 (page 241)

GO for Help

In each diagram, the red and blue triangles are congruent. Identify their common side or angle.

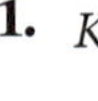

1.

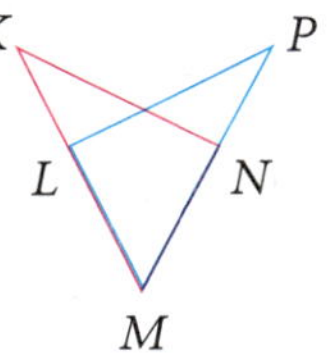

2.

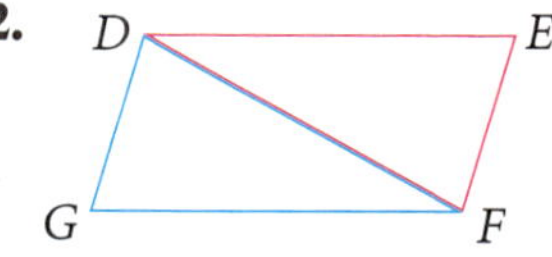

3.

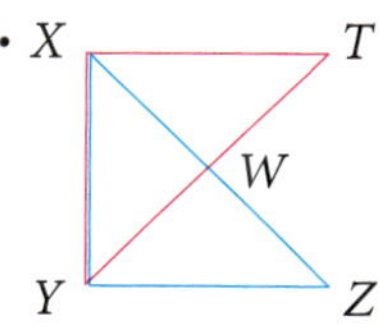

Separate and redraw the indicated triangles. Identify any common angles or sides.

4. $\triangle PQS$ and $\triangle QPR$

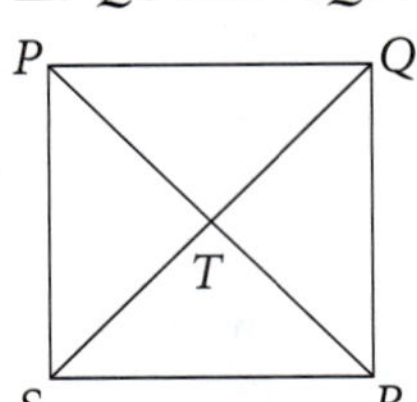

5. $\triangle ACB$ and $\triangle PRB$

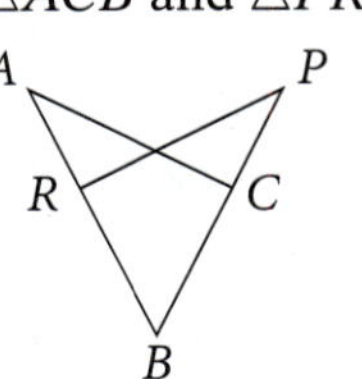

6. $\triangle JKL$ and $\triangle MLK$

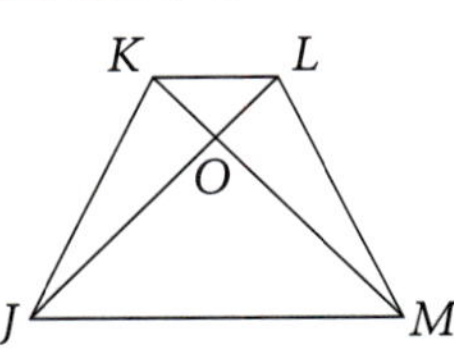

Example 2 (page 242)

7. **Developing Proof** Complete the flow proof.

Given: $\angle T \cong \angle R, \overline{PQ} \cong \overline{PV}$

Prove: $\angle PQT \cong \angle PVR$

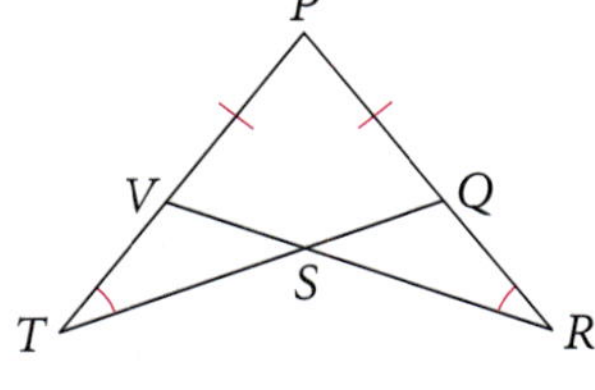

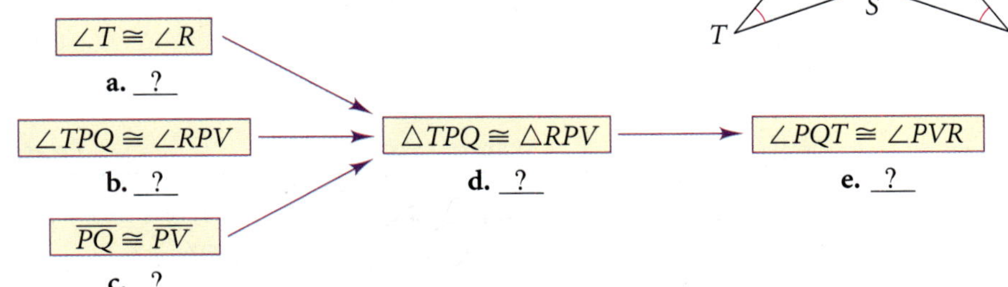

Proof **Write a plan and then a proof.**

8. **Given:** $\overline{RS} \cong \overline{UT}, \overline{RT} \cong \overline{US}$

Prove: $\triangle RST \cong \triangle UTS$

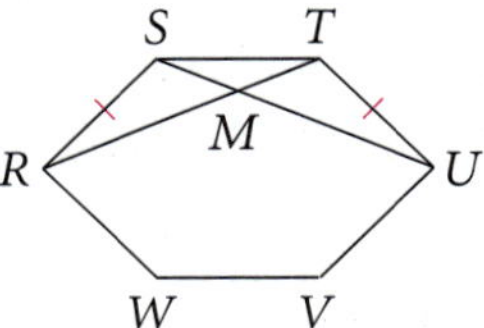

9. **Given:** $\overline{QD} \cong \overline{UA}, \angle QDA \cong \angle UAD$

Prove: $\triangle QDA \cong \triangle UAD$

Examples 3, 4 (pages 242 and 243)

10. **Given:** $\angle 1 \cong \angle 2$, $\angle 3 \cong \angle 4$

Prove: $\triangle QET \cong \triangle QEU$

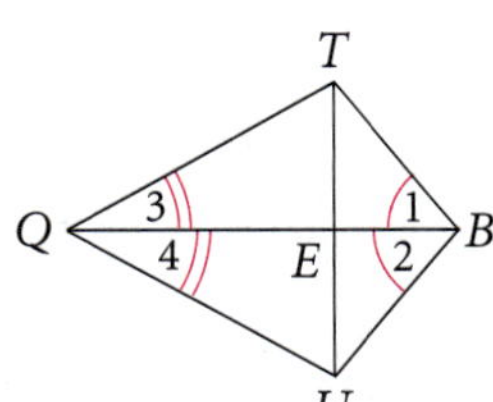

11. **Given:** $\overline{AD} \cong \overline{ED}$,
D is the midpoint of $\overline{BF}$.

Prove: $\triangle ADC \cong \triangle EDG$

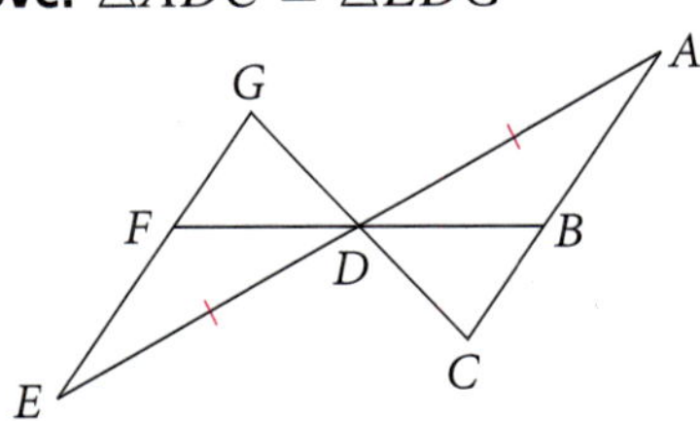

B Apply Your Skills

Open-Ended **Draw the diagram described.**

12. Draw a vertical segment on your paper. On the right side of the segment draw two triangles that share the given segment as a common side.

13. Draw two triangles that have a common angle.

14. Draw two regular pentagons, each with its five diagonals.
 a. In one, shade two triangles that share a common angle.
 b. In the other, shade two triangles that share a common side.

15. Draw two regular hexagons and their diagonals. For these diagrams, do parts (a) and (b) of the preceding exercise.

Homework Video Tutor

Visit: PHSchool.com
Web Code: bce-0407

16. **Multiple Choice** In the diagram, $\angle V \cong \angle S$, $\overline{VU} \cong \overline{ST}$, and $\overline{PS} \cong \overline{QV}$. Which two triangles can you prove congruent by SAS?

Ⓐ $\triangle WVU \cong \triangle RST$
Ⓑ $\triangle PSU \cong \triangle QVT$
Ⓒ $\triangle PWX \cong \triangle QRX$
Ⓓ none of these

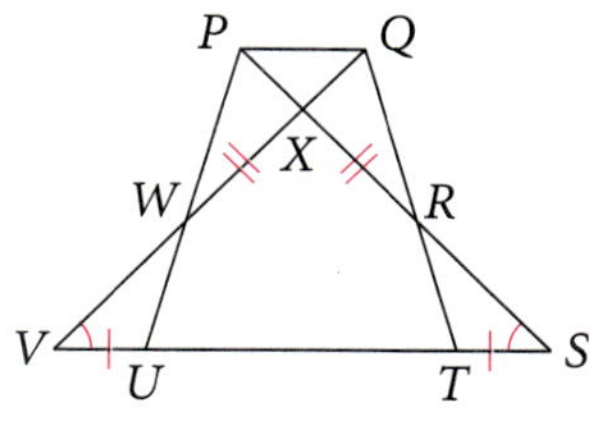

Proof 17. **Given:** $\overline{AC} \cong \overline{EC}$, $\overline{CB} \cong \overline{CD}$

Prove: $\angle A \cong \angle E$

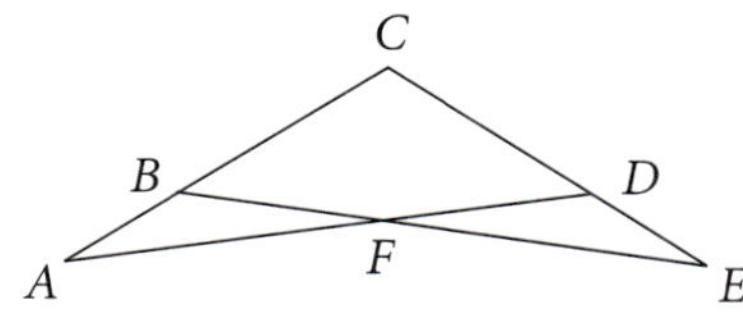

18. **Given:** $\overline{QT} \perp \overline{PR}$, $\overline{QT}$ bisects $\overline{PR}$, $\overline{QT}$ bisects $\angle VQS$.

Prove: $\overline{VQ} \cong \overline{SQ}$

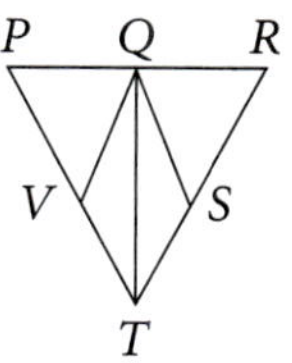

Clothes Design The figure at the right is part of a clothing design pattern. In the figure, $\overline{AB} \parallel \overline{DE} \parallel \overline{FG}$, $\overline{AB} \perp \overline{BC}$, and $\overline{GC} \perp \overline{AC}$. $\triangle DEC$ is isosceles with base $\overline{DC}$, and $m\angle A = 56$.

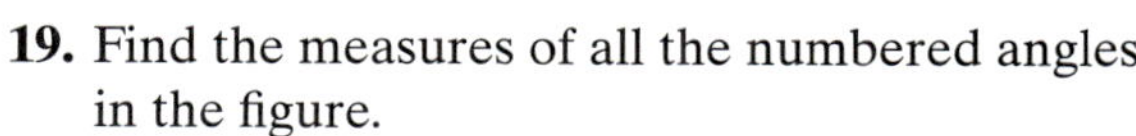
19. Find the measures of all the numbered angles in the figure.

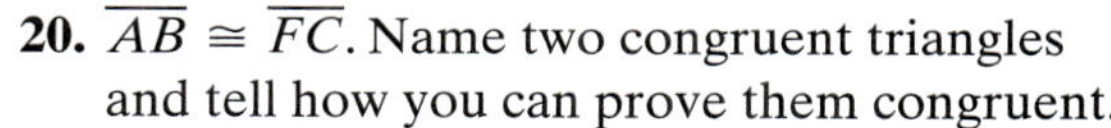
20. $\overline{AB} \cong \overline{FC}$. Name two congruent triangles and tell how you can prove them congruent.

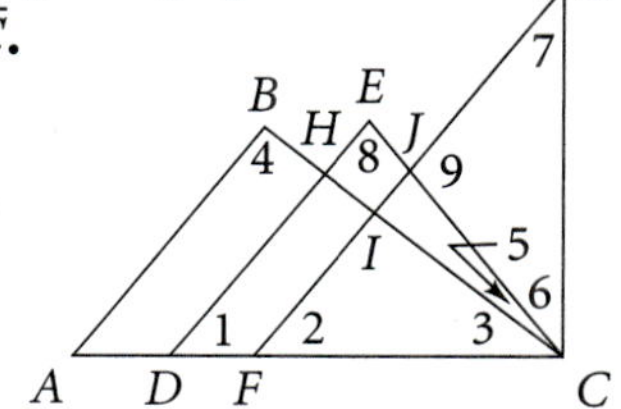

Real-World Connection

A clothing designer must carefully measure angles and segments to create a sewing pattern.

Proof 21. **Given:** $\overline{TE} \cong \overline{RI}$, $\overline{TI} \cong \overline{RE}$, $\angle TDI$ and $\angle ROE$ are right $\angle$s.

Prove: $\overline{TD} \cong \overline{RO}$

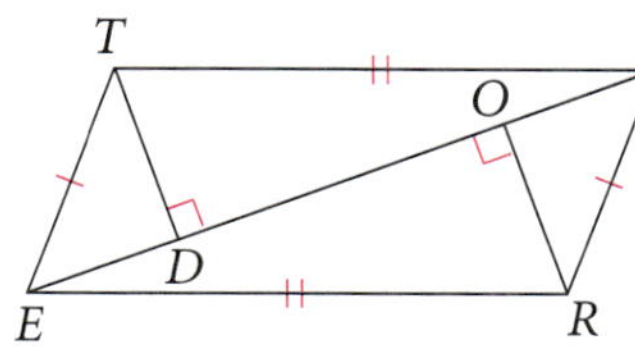

22. **Given:** $\overline{AB} \perp \overline{BC}$, $\overline{DC} \perp \overline{BC}$, $\overline{AC} \cong \overline{DB}$

Prove: $\overline{AE} \cong \overline{DE}$

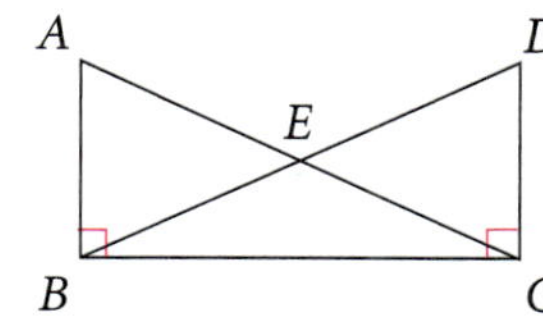

C Challenge

23. **Reasoning** Draw a quadrilateral $ABCD$ with $\overline{AB} \parallel \overline{DC}$ and $\overline{AD} \parallel \overline{BC}$, and its diagonals $\overline{AC}$ and $\overline{DB}$ intersecting at E. Label your diagram to indicate the parallel sides.

a. List all the pairs of congruent segments that you can find in your diagram.
b. **Writing** Explain how you know that the segments you listed are congruent.

Proof **Name a pair of overlapping congruent triangles in each diagram. State whether the triangles are congruent by SSS, SAS, ASA, AAS, or HL. Plan and write a proof.**

24. **Given:** $\overline{AC} \cong \overline{BC}$, $\angle A \cong \angle B$

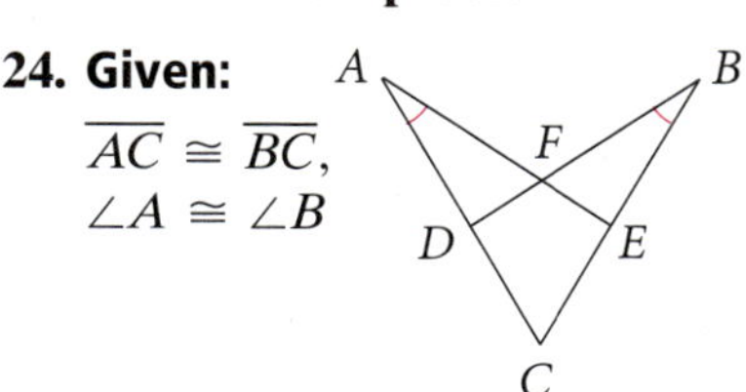

25. **Given:** $\overline{WY} \perp \overline{YX}$, $\overline{ZX} \perp \overline{YX}$, $\overline{WX} \cong \overline{ZY}$

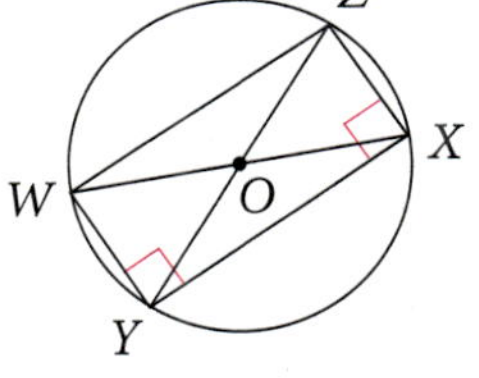

Multiple Choice Practice

For California Standards and CAHSEE Tutorials, visit PHSchool.com. Web Code: bcq-9045

GEOM 5.0

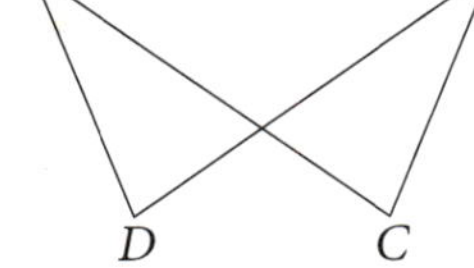

26. In the figure, $\overline{AD} \cong \overline{BC}$ and $\angle BAD \cong \angle ABC$. By which theorem or postulate is $\triangle ABC \cong \triangle BAD$?

(A) AAS (C) SAS

(B) ASA (D) HL

CAHSEE 7AF 4.1

27. Solve for y.

$3y - 14 = 4$

(A) -6 (B) -2 (C) 4 (D) 6

GEOM 2.0

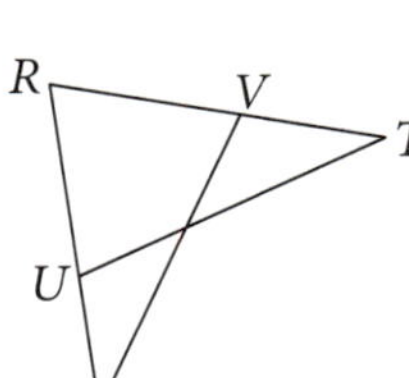

28. In the figure, $\overline{RS} \cong \overline{RT}$ and $\angle S \cong \angle T$. What can be concluded about $\angle RUT$ and $\angle RVS$?

(A) they are right angles by HL

(B) they are congruent by CPCTC

(C) they are right angles by ASA

(D) they are congruent by the vertical angle theorem

CAHSEE 7AF 1.5

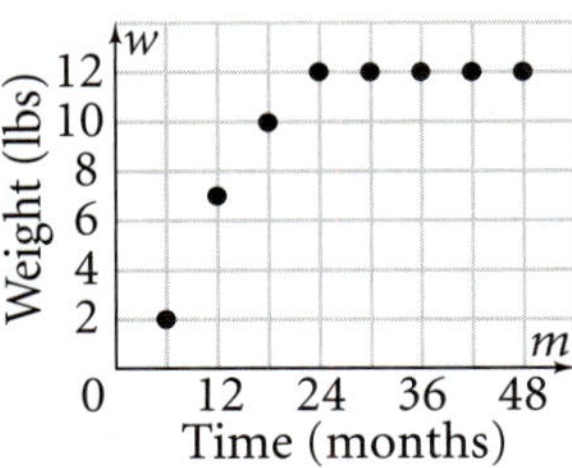

29. The scatter plot at the right shows the weight of a cat at 6-month intervals. Which statement best describes the relationship between the cat's weight w and the number of months m?

(A) As time progressed, the cat remained almost the same weight.

(B) As time progressed, the cat steadily lost weight.

(C) As time progressed, the cat steadily gained weight.

(D) As time progressed, the cat gained weight initially and then maintained its weight.

Mixed Review

Lesson 4-6

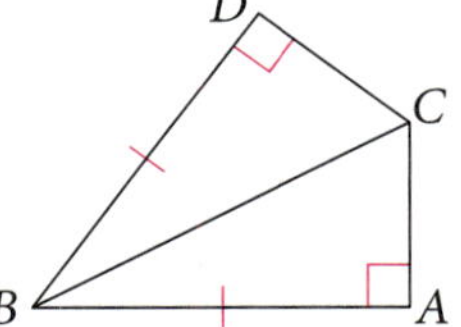

30. Complete the plan for a proof.

Given: $\angle A$ and $\angle D$ are right angles, $\overline{AB} \cong \overline{DB}$.

Prove: $\triangle ABC \cong \triangle DBC$

Plan: $\triangle ABC$ and $\triangle DBC$ are **a.** __?__ triangles with legs that are given to be **b.** __?__. The hypotenuse is congruent to itself by the **c.** __?__ Property of Congruence. $\triangle ABC \cong \triangle DBC$ by the **d.** __?__ Theorem.

Lesson 3-8

Constructions **Draw a line p and a point M not on p. Construct the described line.**

31. line n through M so that $n \perp p$

32. line r through M so that $r \parallel p$

Lesson 3-6

Write an equation in point-slope form of the line that contains the given point and has the given slope.

33. $P(2, -6)$; slope $\frac{1}{2}$

34. $Q(0, 5)$; slope 1

35. $R(-3, 6)$; slope -2

36. $S(0, 0)$; slope $-\frac{1}{3}$

Write an equation in point-slope form of the line that contains the given points.

37. $A(1, 4), B(0, 2)$

38. $E(3, -5), F(6, 0)$

39. $X(-4, -3), Y(2, -8)$

Extension Writing Flow Proofs

For Use With Lesson 4-7

GEOM 2.0 Students write geometric proofs. *Develop*

Proofs can get long and complex. While a two-column proof may appear more organized, a flow proof can show the logic flow better and thus be easier to follow.

To write a flow proof, sketch the logic "paths" of a proof on scratch paper. Then organize your work into a neat, easy-to-follow flow diagram, as in this Example.

EXAMPLE

Given: $\overline{KJ} \cong \overline{LM}$, $\angle KJN \cong \angle LMN$, $\overline{JN} \cong \overline{MN}$

Prove: $\overline{JL} \cong \overline{MK}$

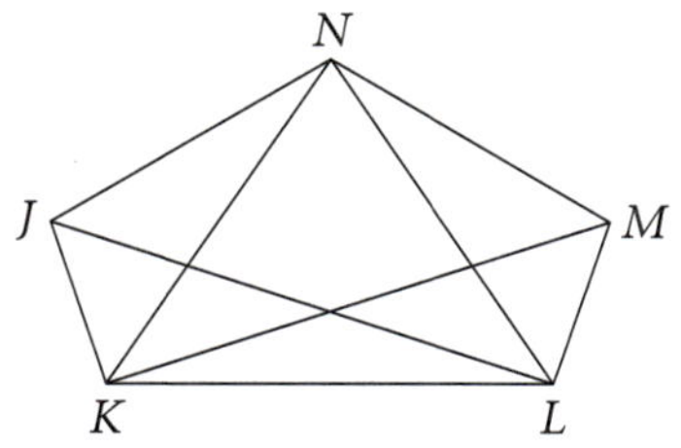

N, J, M, K, L

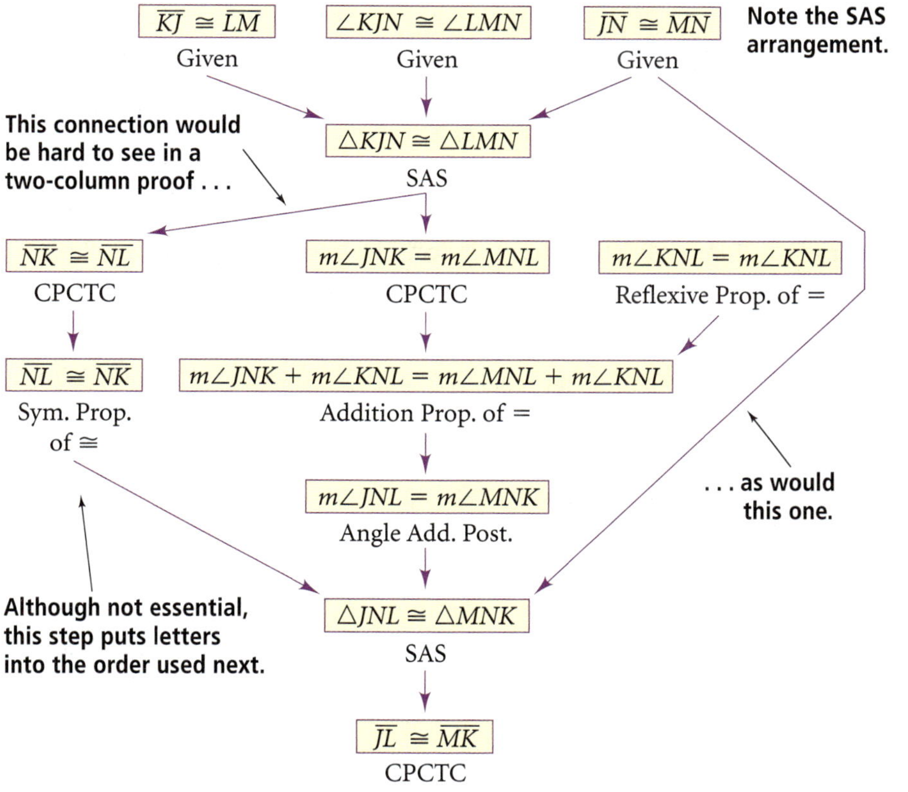

$\overline{KJ} \cong \overline{LM}$ Given | $\angle KJN \cong \angle LMN$ Given | $\overline{JN} \cong \overline{MN}$ Given

Note the SAS arrangement.

$\triangle KJN \cong \triangle LMN$ SAS

This connection would be hard to see in a two-column proof . . .

$\overline{NK} \cong \overline{NL}$ CPCTC | $m\angle JNK = m\angle MNL$ CPCTC | $m\angle KNL = m\angle KNL$ Reflexive Prop. of =

$\overline{NL} \cong \overline{NK}$ Sym. Prop. of $\cong$ | $m\angle JNK + m\angle KNL = m\angle MNL + m\angle KNL$ Addition Prop. of =

. . . as would this one.

$m\angle JNL = m\angle MNK$ Angle Add. Post.

Although not essential, this step puts letters into the order used next.

$\triangle JNL \cong \triangle MNK$ SAS

$\overline{JL} \cong \overline{MK}$ CPCTC

EXERCISES

Write a flow proof. Make the flow of logic as easy to follow as you can.

1. **Given:** $\overline{AB} \cong \overline{DC}$, E is the midpoint of $\overline{AD}$, and $\overline{CE} \cong \overline{BE}$.

   Prove: $\overline{AC} \cong \overline{DB}$

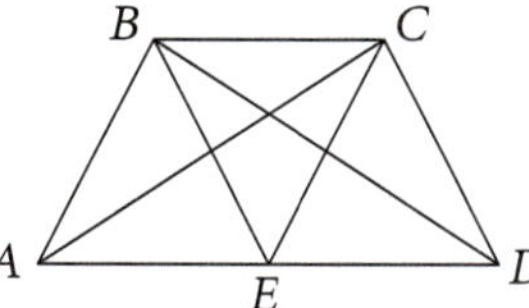

2. **Given:** $\overline{RQ} \cong \overline{RS}$, $\overline{RP} \cong \overline{RT}$, $\overline{QP} \cong \overline{ST}$

   Prove: $\overline{QT} \cong \overline{SP}$

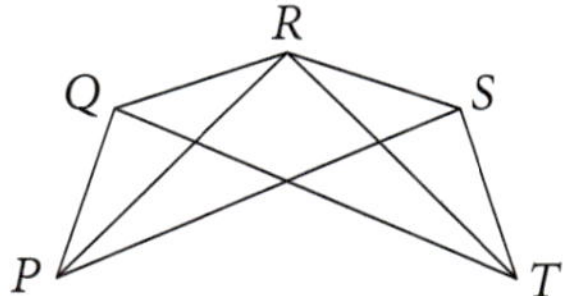

Test-Taking Strategies

Interpreting Data

A test item may ask you to use a data set to determine a relationship between quantities. Often the relationship is linear and may be expressed as a **linear function,** a function that can be represented by a linear equation in two variables. These are the equations you studied in Lesson 3-6.

EXAMPLE

Each row in the table shows the measure b of a base angle and the measure v of the vertex angle of an isosceles triangle. Which function equation describes the relationship between b and v?

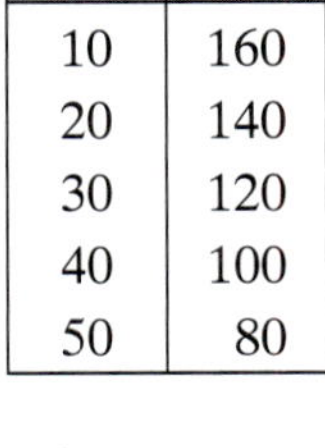

b	v
10	160
20	140
30	120
40	100
50	80

Ⓐ $v = 16b$ Ⓒ $v = 180 - b$

Ⓑ $v = 170 - b$ Ⓓ $v = -2b + 180$

Method 1

The choices suggest a linear relationship. For each change in b of 10, the change in v is -20. The slope must be -2. The correct choice is D.

Method 2

For any isosceles triangle, $b + b + v = 180$ by the Triangle Angle-Sum Theorem. Simplifying, $2b + v = 180$, or $v = -2b + 180$. The correct choice is D.

EXERCISES

1. Jason plotted the weights of several automobile models and their average highway fuel mileage. Based on his data, shown at the right, which of the following is true?

Ⓐ The weight of the car does not affect the fuel mileage.

Ⓑ The more the car weighs, the lower the fuel mileage.

Ⓒ The less the car weighs, the lower the fuel mileage.

Ⓓ The more the car weighs, the greater the fuel mileage.

2. Each row in the table shows the measure a of a base angle of an isosceles triangle. It also shows the measure b of an angle formed by the bisectors of a base angle and the vertex angle. Which equation describes the relationship between a and b?

a	b
10	95
20	100
30	105
40	110
50	115

Ⓐ $2b - a = 180$ Ⓒ $2b + a = 180$

Ⓑ $b + 2a = 115$ Ⓓ $b - 2a = 75$

3. In each row of the table below, what might the values of x and y represent?

Ⓐ The measure x of an angle and the measure y of its supplement.

Ⓑ The measure x of a base angle of an isosceles triangle and the measure y of the vertex angle.

Ⓒ The measure x of a base angle of an isosceles triangle and the measure y of an exterior angle at the vertex.

Ⓓ The measure x of the vertex angle of an isosceles triangle and the sum y of the measures of the two base angles.

x	y
60	120
45	90
30	60
15	30

Chapter 4 Review

Vocabulary Review

English and Spanish Audio Online

base of an isosceles triangle (p. 228)
base angles of an isosceles triangle (p. 228)
congruent polygons (p. 198)
corollary (p. 229)
CPCTC (corresponding parts of congruent triangles are congruent) (p. 221)
hypotenuse (p. 235)
legs of a right triangle (p. 235)
legs of an isosceles triangle (p. 228)
vertex angle of an isosceles triangle (p. 228)

Choose the correct term to complete each sentence.

1. The two congruent sides of an isosceles triangle are the _?_.
2. The two congruent sides of an isosceles triangle form the _?_.
3. If you know that two triangles are congruent, then the corresponding sides and angles of the triangles are congruent because _?_.
4. The side opposite the right angle of a right triangle is the _?_.
5. The angles of an isosceles triangle that are not the vertex angle are called the _?_.
6. A _?_ to a theorem is a statement that follows immediately from the theorem.
7. The _?_ are the two sides of a right triangle that are not the hypotenuse.
8. _?_ have congruent corresponding parts.
9. The side of an isosceles triangle that is not a leg is called the _?_.

For: Vocabulary quiz
Web Code: bcj-0451

Skills and Concepts

Lesson 4-1

- To recognize congruent figures and their corresponding parts (p. 198)

 **GEOM 4.0, 5.0, 12.0**

Congruent polygons have congruent corresponding parts. When you name congruent polygons, always list corresponding vertices in the same order.

Two triangles are congruent when they have three pairs of congruent corresponding sides and three pairs of congruent corresponding angles.

***RSTUV* ≅ *KLMNO*. Complete the congruence statements.**

10. $\overline{TS} \cong$ _?_
11. $\angle N \cong$ _?_
12. $\overline{LM} \cong$ _?_
13. $VUTSR \cong$ _?_

***WXYZ* ≅ *PQRS*. Find the measure of the angle or the length of the side.**

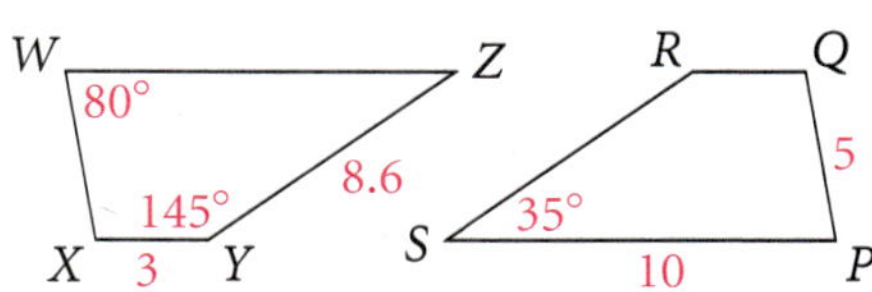

14. $\angle P$
15. $\overline{QR}$
16. $\overline{WX}$
17. $\angle Z$
18. $\angle X$

Lessons 4-2 and 4-3

- To prove two triangles congruent using the SSS and SAS Postulates (p. 205)
- To prove two triangles congruent using the ASA Postulate and the AAS Theorem (p. 213)

 **GEOM 2.0, 5.0**

If three sides of one triangle are congruent to three sides of another triangle, then the two triangles are congruent by the **Side-Side-Side (SSS) Postulate.**

If two sides and the included angle of one triangle are congruent to two sides and the included angle of another triangle, then the two triangles are congruent by the **Side-Angle-Side (SAS) Postulate.**

If two angles and the included side of one triangle are congruent to two angles and the included side of another triangle, then the two triangles are congruent by the **Angle-Side-Angle (ASA) Postulate.**

If two angles and a nonincluded side of one triangle are congruent to two angles and the corresponding nonincluded side of another triangle, then the two triangles are congruent by the **Angle-Angle-Side (AAS) Theorem.**

Which postulate or theorem, if any, could you use to prove the two triangles congruent? If the triangles *cannot* be proven congruent, write *not possible.*

19. 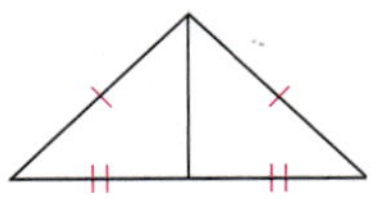

20. 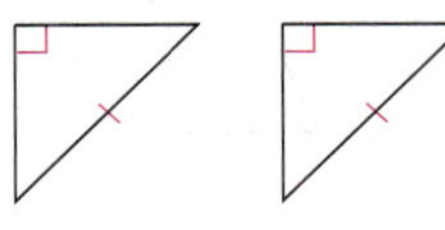

21. 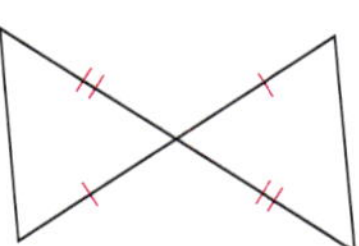

22. 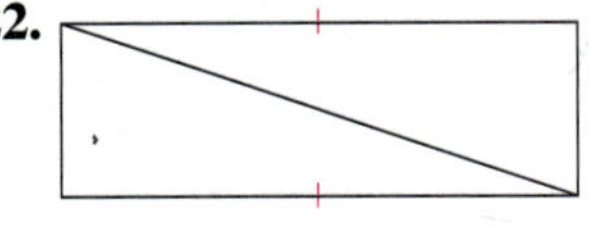

23. 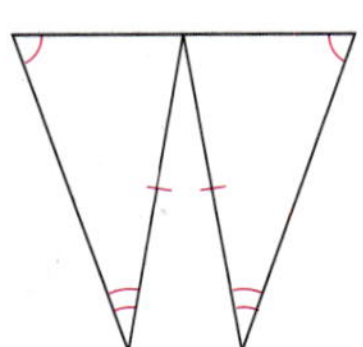

24. 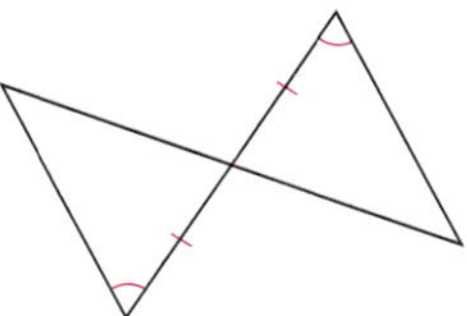

Write a congruence statement for each pair of triangles. Name the postulate or theorem that justifies your statement. If the triangles *cannot* be proven congruent, write *not possible.*

25. 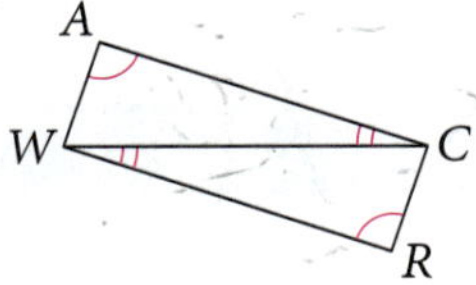

26. 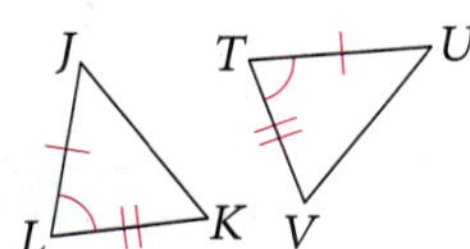

27. 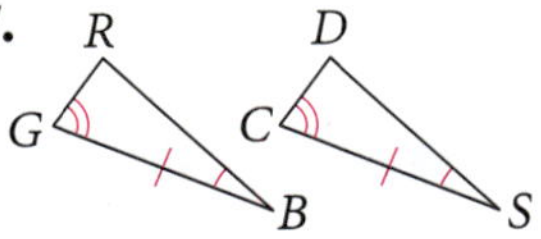

Lesson 4-4

- To use triangle congruence and CPCTC to prove that parts of two triangles are congruent (p. 221)

 **GEOM 5.0, 6.0**

Once you know that triangles are congruent, you can make conclusions about corresponding segments and angles because, by definition, **corresponding parts of congruent triangles are congruent (CPCTC).** You can use congruent triangles in the proofs of many theorems.

Explain how you can use SSS, SAS, ASA, or AAS with CPCTC to prove the statement true.

28. $\overline{TV} \cong \overline{YW}$

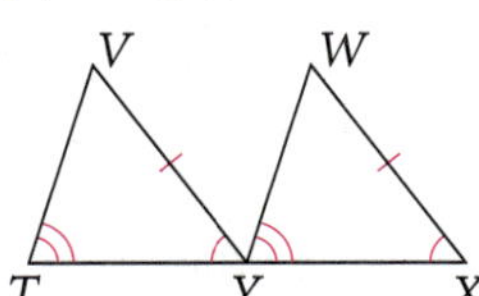

29. $\overline{BE} \cong \overline{DE}$

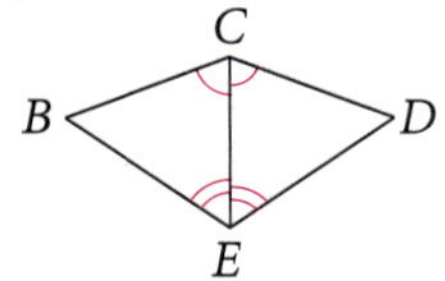

30. $\overline{KN} \cong \overline{ML}$

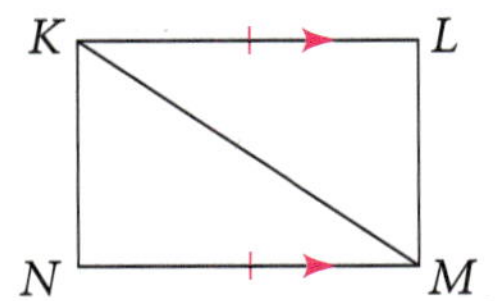

Lessons 4-5 and 4-6

- To use and apply properties of isosceles triangles (p. 228)
- To prove triangles congruent using the HL theorem (p. 235)

If two sides of a triangle are congruent, then the angles opposite those sides are also congruent by the **Isosceles Triangle Theorem.** If two angles of a triangle are congruent, then the sides opposite the angles are congruent by the **Converse of the Isosceles Triangle Theorem.**

The bisector of the vertex angle of an isosceles triangle is the perpendicular bisector of the base.

If the hypotenuse and a leg of one right triangle are congruent to the hypotenuse and a leg of another right triangle, then the triangles are congruent by the **Hypotenuse-Leg (HL) Theorem.**

x^2 Algebra **Find the values of x and y.**

31.

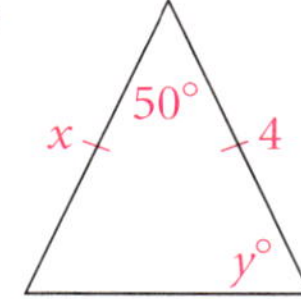

32.

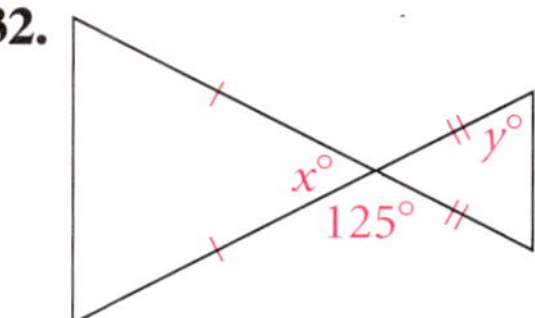

33.

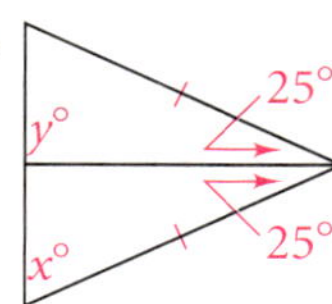

Write a proof for each of the following.

34. Given: $\overline{PS} \perp \overline{SQ}$, $\overline{RQ} \perp \overline{QS}$, $\overline{PQ} \cong \overline{RS}$

Prove: $\triangle PSQ \cong \triangle RQS$

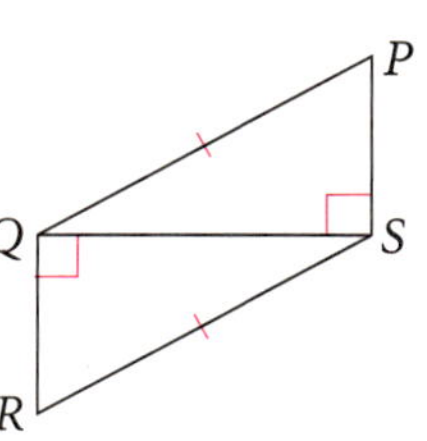

35. Given: $\overline{LN} \perp \overline{KM}$, $\overline{KL} \cong \overline{ML}$

Prove: $\triangle KLN \cong \triangle MLN$

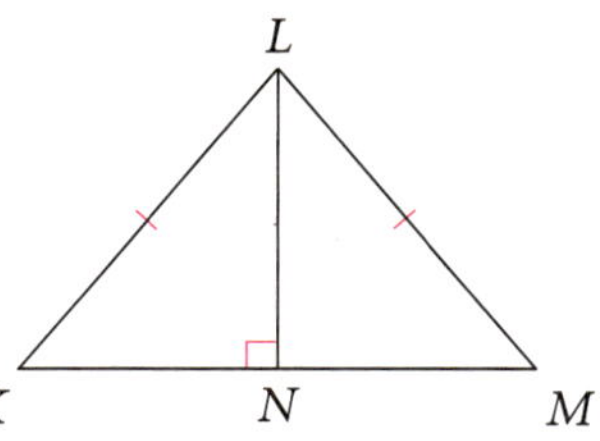

Lesson 4-7

- To identify congruent overlapping triangles (p. 241)
- To prove two triangles congruent by first proving two other triangles congruent (p. 242)

GEOM 5.0

You can prove overlapping triangles congruent. You can also use the common or shared sides and angles of triangles in congruence proofs.

Name a pair of overlapping congruent triangles in each diagram. State whether the triangles are congruent by SSS, SAS, ASA, AAS, or HL.

36.

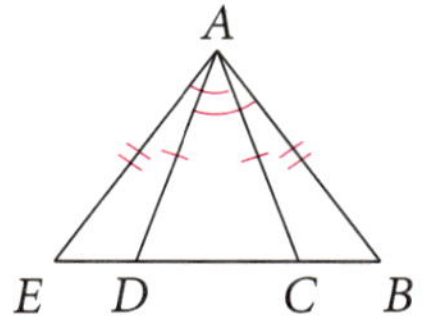

37.

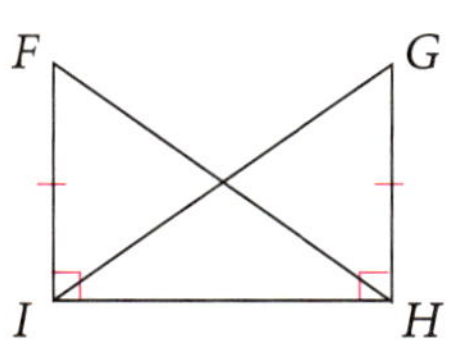

38.

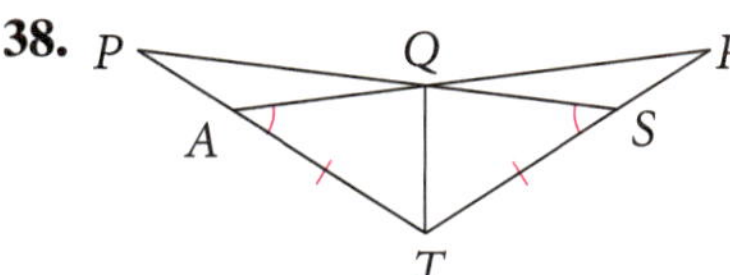

39.

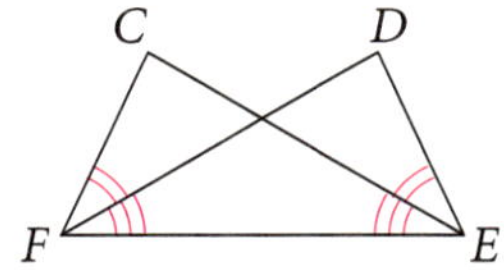

point on the circle

VI. the area of the entire exterior surface of a solid

VII. the distance around a circle

necessary to prove two polygons congruent? (Lesson 4-1)

(A) All the corresponding sides should be congruent.

(B) All the corresponding sides and angles should be congruent.

(C) All the corresponding angles should be congruent.

(D) All sides in each polygon should be congruent.

Chapter 4 Test

For: Chapter Test 4
Web Code: bca-0452

Write a congruence statement for each pair of triangles.

1.

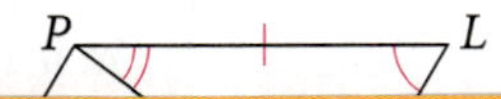

10. Open-Ended Draw a picture to represent $\triangle CEO \cong \triangle HDF$. Name all of the pairs of corresponding congruent parts.

Relationships Within Triangles

What You've Learned

California Content Standards

1.0 Students demonstrate understanding by identifying and giving examples of undefined terms, axioms, theorems, and inductive and deductive reasoning.

4.0 Students prove basic theorems involving congruence.

5.0 Students prove that triangles are congruent, and they are able to use the concept of corresponding parts of congruent triangles.

Check Your Readiness

GO for Help to the Lesson in green.

Inequalities (Previous Course)

Algebra Solve each inequality.

1. $3x + 10 \leq 22$ **2.** $4x - 1 > 2x + 14$ **3.** $30 - 5x \geq x + 24$

Basic Constructions (Lesson 1-7)

Use a compass and straightedge for the following.

4. Construct the perpendicular bisector of a segment.

5. Construct the angle bisector of an angle.

Distance Formula (Lesson 1-8)

Find the distance between each pair of points.

6. $(1, 4), (4, 8)$ **7.** $(-6, 2), (-1, 14)$ **8.** $(-3, -2), (5, -6)$

Midpoint Formula (Lesson 1-8)

Find the midpoint of the segments whose endpoints are given.

9. $(4, 11), (6, 3)$ **10.** $(-8, -3), (2, -4)$ **11.** $(-7, 15), (-2, -10)$

Slope (Algebra 1 Review, page 165)

Find the slope of the line containing each pair of points.

12. $(8, 3), (7, 12)$ **13.** $(3, -2), (0, 6)$ **14.** $(-5, 4), (-2, 4)$

▲ Canals like the Panama Canal are constructed where the distance between major bodies of water is the shortest.

What You'll Learn Next

California Content Standards

2.0 Students write geometric proofs, including proofs by contradiction.

6.0 Students know and are able to use the triangle inequality theorem.

17.0 Students prove theorems by using coordinate geometry, including the midpoint of a line segment, and the distance formula.

CAHSEE 7NS 1.3, 7NS 2.5, 6SDAP 3.3, 6SDAP 3.5, 7SDAP 1.1, 7AF 1.1, 7AF 1.5, 7AF 3.4, 7AF 4.1, 7MG 1.3, 7MG 3.2, 7MG 3.3, 7MR 1.1, Alg1 6.0, Alg1 8.0

New Vocabulary

English and Spanish Audio Online

- **altitude of a triangle** (p. 275)
- **centroid** (p. 274)
- **circumcenter of a triangle** (p. 273)
- **circumscribed about** (p. 273)
- **concurrent** (p. 273)
- **contrapositive** (p. 280)
- **coordinate proof** (p. 260)
- **distance from a point to a line** (p. 266)
- **equivalent statements** (p. 281)
- **incenter of a triangle** (p. 273)
- **indirect proof** (p. 281)
- **indirect reasoning** (p. 281)
- **inscribed in** (p. 273)
- **inverse** (p. 280)
- **median of a triangle** (p. 274)
- **midsegment** (p. 259)
- **negation** (p. 280)
- **orthocenter of a triangle** (p. 275)
- **point of concurrency** (p. 273)

Investigating Midsegments

Technology

FOR USE WITH LESSON 5-1

GEOM 1.0 Students demonstrate understanding by identifying and giving examples of inductive reasoning. ***Master***

Construct

Use geometry software to draw and label $\triangle ABC$. Construct the midpoints D and E of $\overline{AB}$ and $\overline{AC}$. Connect the midpoints with a *midsegment.*

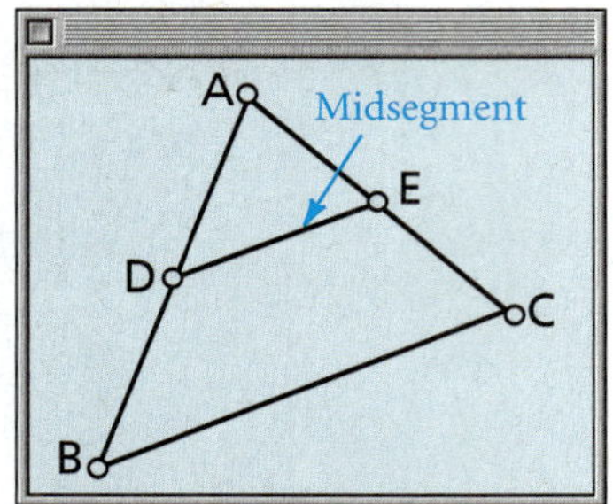

Investigate

- Measure the lengths of $\overline{DE}$ and $\overline{BC}$. Calculate $\frac{DE}{BC}$.
- Measure the slopes of $\overline{DE}$ and $\overline{BC}$.
- Manipulate the triangle and observe the lengths and slopes of $\overline{DE}$ and $\overline{BC}$.

EXERCISES

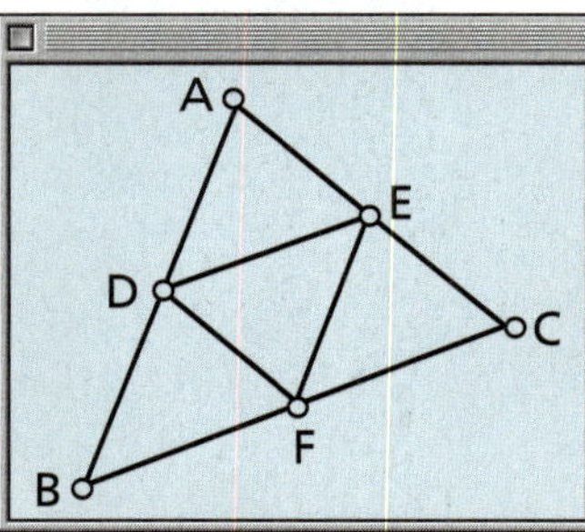

1. Make conjectures about the lengths and slopes of midsegments.

2. Construct the midpoint F of $\overline{BC}$. Then construct the other two midsegments of $\triangle ABC$. Test whether these midsegments support your conjectures in Exercise 1.

3. $\triangle ABC$ and the three midsegments form four small triangles.
a. Measure the sides of the four small triangles and list those that you find are congruent.
b. Use a postulate from Chapter 4 to make a conjecture about the four small triangles.

For the remaining exercises, assume your conjectures in Exercises 1 and 3 are true.

4. What can you say about the areas of the four small triangles in the window above?

5. How does $\triangle ABC$ compare to each small triangle
a. in area?
b. in perimeter?

6. Construct the three midsegments of $\triangle DEF$. Label this triangle $\triangle GHI$. How does $\triangle ABC$ compare to $\triangle GHI$
a. in area?
b. in perimeter?
c. Suppose you construct the midsegment triangle inside $\triangle GHI$. Predict how $\triangle ABC$ would compare to this third midsegment triangle in area and perimeter.

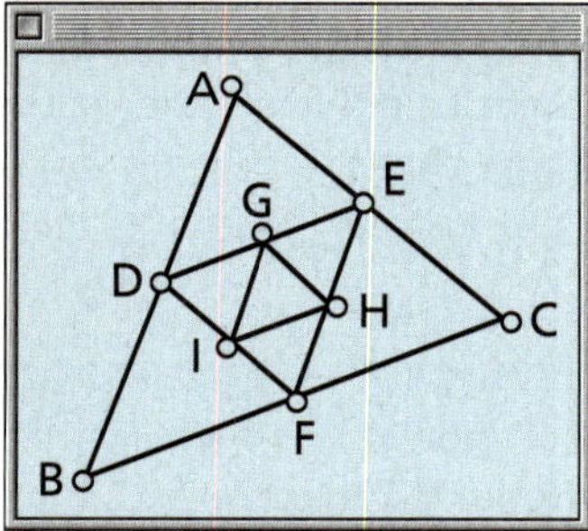

Exercise 6

Extend

7.
- Draw quadrilateral $RSTU$.
- Construct the midpoints of each side of the quadrilateral.
- Join consecutive midpoints to form quadrilateral $YXWV$.
- Manipulate the figure and observe the shape of quadrilateral $YXWV$.

Make a conjecture about the sides of quadrilateral $YXWV$.

Midsegments of Triangles

California Content Standards

GEOM 17.0 Students prove theorems by using coordinate geometry. ***Introduce***

What You'll Learn

- To use properties of midsegments to solve problems

. . . And Why

To use indirect measurement to find the length of a lake, as in Example 3

Check Skills You'll Need

for Help Lesson 1-8 and page 165

Find the coordinates of the midpoint of each segment.

1. $\overline{AB}$ with $A(-2, 3)$ and $B(4, 1)$

2. $\overline{CD}$ with $C(0, 5)$ and $D(3, 6)$

3. $\overline{EF}$ with $E(-4, 6)$ and $F(3, 10)$

4. $\overline{GH}$ with $G(7, 10)$ and $H(-5, -8)$

Find the slope of the line containing each pair of points.

5. $A(-2, 3)$ and $B(3, 1)$

6. $C(0, 5)$ and $D(3, 6)$

7. $E(-4, 6)$ and $F(3, 10)$

8. $G(7, 10)$ and $H(-5, -8)$

New Vocabulary • midsegment • coordinate proof

1 Using Properties of Midsegments

Hands-On Activity: Midsegments of Triangles

Draw, label, and cut out a large scalene triangle. Do the same with other right, acute, and obtuse triangles. Label the vertices *A*, *B*, and *C*.

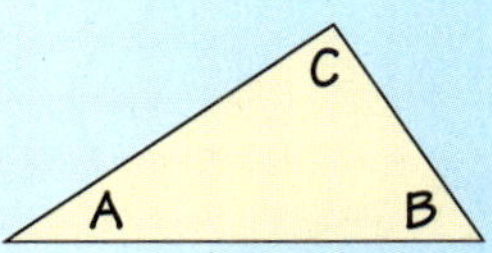

- For each triangle fold *A* onto *C* to find the midpoint of $\overline{AC}$. Do the same for $\overline{BC}$. Label the midpoints *L* and *N*, then draw $\overline{LN}$.

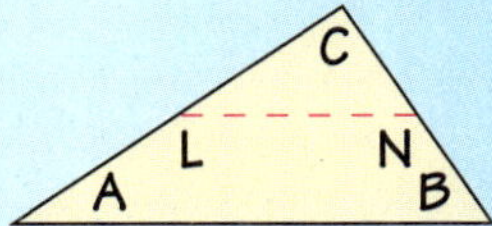

- Fold each triangle on $\overline{LN}$.

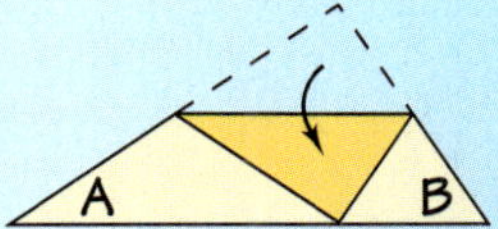

- Fold *A* to *C*. Fold *B* to *C*.

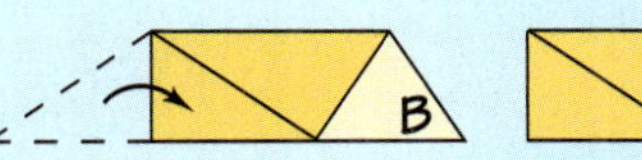

1. How does *LN* compare to *AB*? Explain.

2. Make a conjecture about how the segment joining the midpoints of two sides of a triangle is related to the third side of the triangle.

In $\triangle ABC$ above, $\overline{LN}$ is a triangle midsegment. A **midsegment** of a triangle is a segment connecting the midpoints of two sides.

Key Concepts

Theorem 5-1 **Triangle Midsegment Theorem**

If a segment joins the midpoints of two sides of a triangle, then the segment is parallel to the third side, and is half its length.

One way to prove the Triangle Midsegment Theorem is to use coordinate geometry and algebra. This style of proof is called a **coordinate proof.** You begin the proof by placing a triangle in a convenient spot on the coordinate plane. You then choose variables for the coordinates of the vertices.

Proof

Proof of Theorem 5-1

Given: R is the midpoint of $\overline{OP}$.
S is the midpoint of $\overline{QP}$.

Prove: $\overline{RS} \parallel \overline{OQ}$ and $RS = \frac{1}{2}OQ$

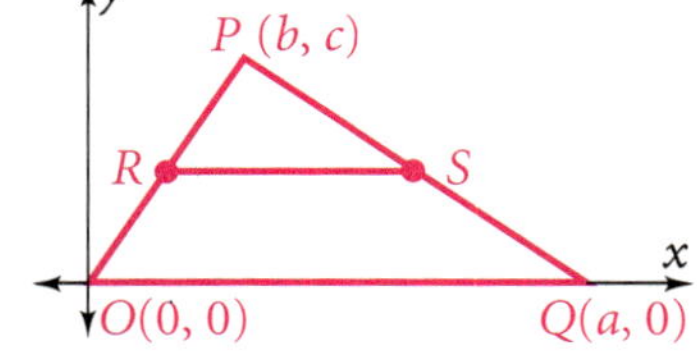

Vocabulary Tip

The Midpoint Formula

$M = \left(\frac{x_1 + x_2}{2}, \frac{y_1 + y_2}{2}\right)$

The Distance Formula

$d = \sqrt{(x_2 - x_1)^2 + (y_2 - y_1)^2}$

- Use the Midpoint Formula to find the coordinates of R and S.

$R: \left(\frac{0 + b}{2}, \frac{0 + c}{2}\right) = \left(\frac{b}{2}, \frac{c}{2}\right)$

$S: \left(\frac{a + b}{2}, \frac{0 + c}{2}\right) = \left(\frac{a + b}{2}, \frac{c}{2}\right)$

- To prove that $\overline{RS}$ and $\overline{OQ}$ are parallel, show that their slopes are equal. Because the y-coordinates of R and S are the same, the slope of $\overline{RS}$ is zero. The same is true for $\overline{OQ}$. Therefore, $\overline{RS} \parallel \overline{OQ}$.

- Use the Distance Formula to find RS and OQ.

$$RS = \sqrt{\left(\frac{a + b}{2} - \frac{b}{2}\right)^2 + \left(\frac{c}{2} - \frac{c}{2}\right)^2}$$

$$= \sqrt{\left(\frac{a}{2} + \frac{b}{2} - \frac{b}{2}\right)^2 + 0^2} = \sqrt{\left(\frac{a}{2}\right)^2} = \frac{a}{2} = \frac{1}{2}a$$

$$OQ = \sqrt{(a - 0)^2 + (0 - 0)^2}$$

$$= \sqrt{a^2 + 0^2} = a$$

Therefore, $RS = \frac{1}{2}OQ$.

1 EXAMPLE Finding Lengths

In $\triangle EFG$, H, J, and K are midpoints. Find HJ, JK, and FG.

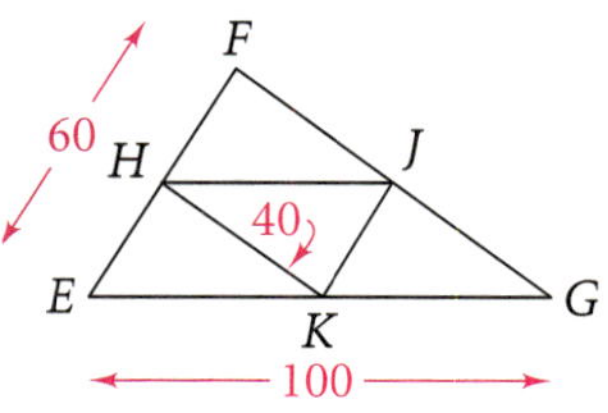

$HJ = \frac{1}{2}EG$ or $\frac{1}{2}(100)$; $HJ = 50$

$JK = \frac{1}{2}EF$ or $\frac{1}{2}(60)$; $JK = 30$

HK or $40 = \frac{1}{2}FG$; $FG = 80$

CA Standards Check

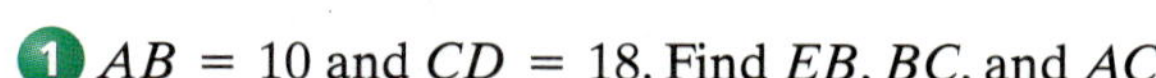
1 $AB = 10$ and $CD = 18$. Find EB, BC, and AC.

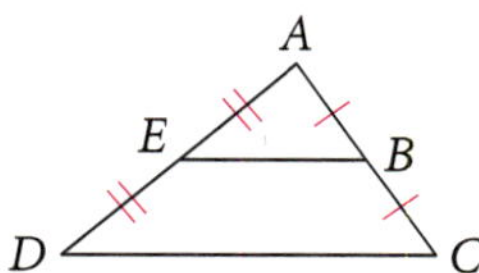

2 EXAMPLE Identifying Parallel Segments

In $\triangle DEF$, A, B, and C are midpoints. Name pairs of parallel segments.

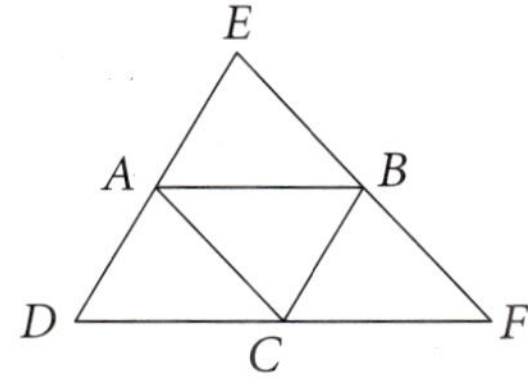

The midsegments are $\overline{AB}$, $\overline{BC}$, and $\overline{CA}$.

By the Triangle Midsegment Theorem, $\overline{AB} \parallel \overline{DF}$, $\overline{BC} \parallel \overline{ED}$, and $\overline{AC} \parallel \overline{EF}$.

CA Standards Check **2** **Critical Thinking** Find $m\angle VUZ$. Justify your answer.

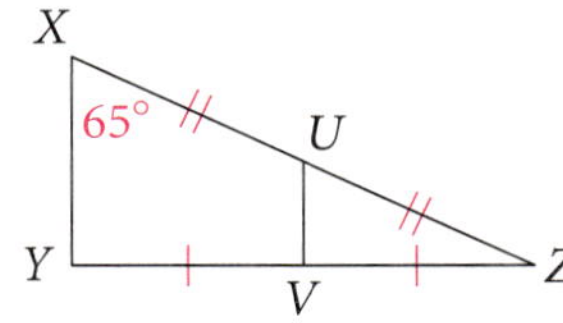

You can use the Triangle Midsegment Theorem to find lengths of segments that might be difficult to measure directly.

3 EXAMPLE Real-World Connection

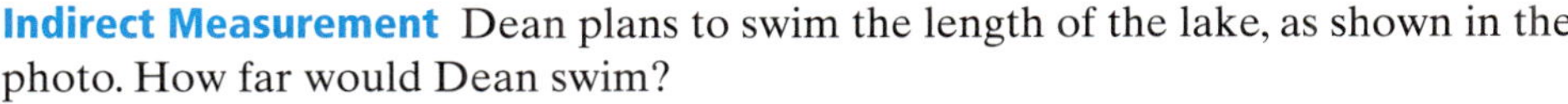

Indirect Measurement Dean plans to swim the length of the lake, as shown in the photo. How far would Dean swim?

Here is what Dean does to find the distance he would swim across the lake.

Step 1: He measures his stride and adjusts it so that it averages about 3 ft.

Step 2: Then he begins at the left edge of the lake (first diagram). He paces 35 strides along the edge of the lake and sets a stake.

Step 3: He paces 35 more strides in the same direction and sets another stake.

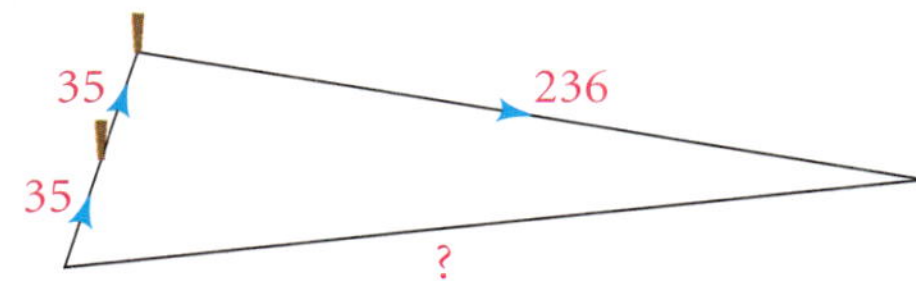

Step 4: He paces to where his swim will end at the other side of the lake, counting 236 strides.

Step 5: Then (second diagram) he paces 118 strides, or half the distance, back towards the second stake.

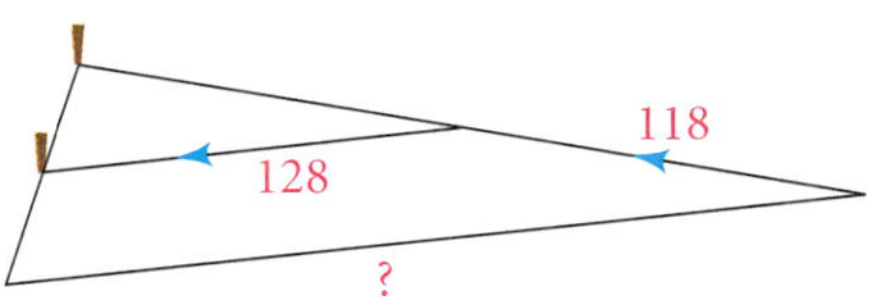

Step 6: He paces to the first stake, counting 128 strides.

Step 7: He converts strides to feet.

$$128 \text{ strides} \times \frac{3 \text{ ft}}{1 \text{ stride}} = 384 \text{ ft}$$

Step 8: He uses Theorem 5-1. The distance across the lake is twice the length of the midsegment.

$$2(384 \text{ ft}) = 768 \text{ ft}$$

Dean would swim approximately 768 ft.

CA Standards Check **3** **a.** $\overline{CD}$ is a new bridge being built over a lake as shown. Find the length of the bridge.

b. How long is the bridge in miles?

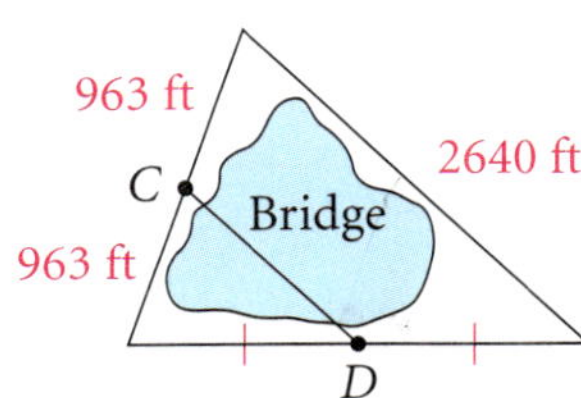

For more exercises, see *Extra Skill, Word Problem, and Proof Practice.*

GEOM 17.0

Example 1 (page 260)

GO for Help

Mental Math **Find the value of x.**

1. 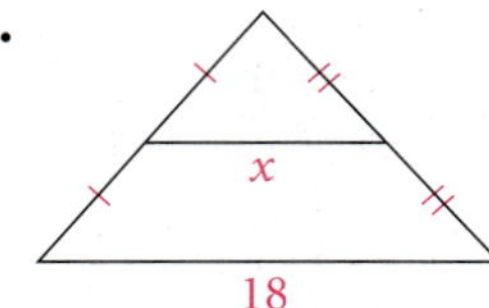

2. 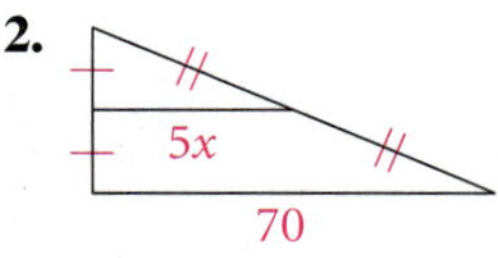

3. 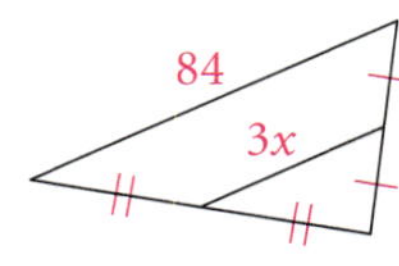

4. 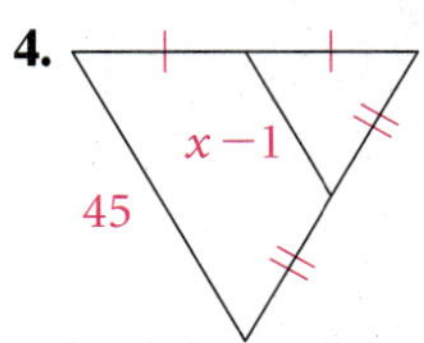

5. 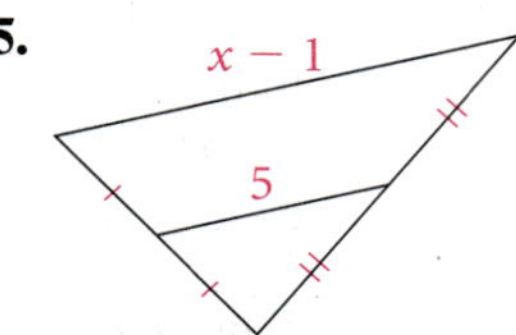

6. 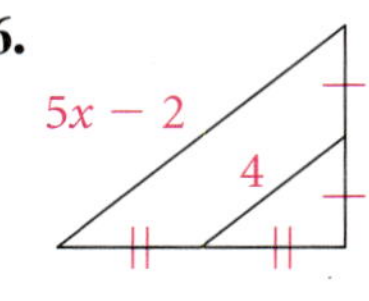

Points E, D, and H are midpoints of $\triangle TUV$. $UV = 80$, $TV = 100$, and $HD = 80$.

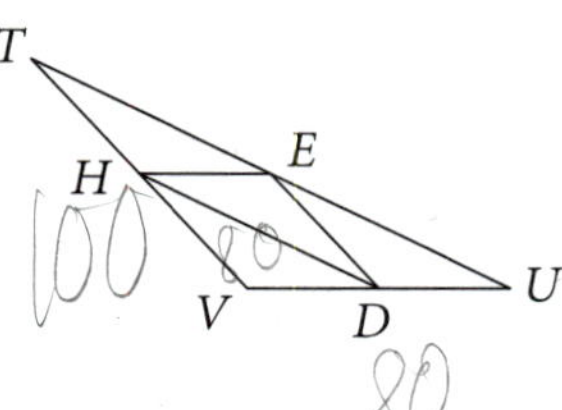

7. Find HE.
8. Find ED.
9. Find TU.
10. Find TE.

Example 2 (page 261)

Identify pairs of parallel segments in each diagram.

11. 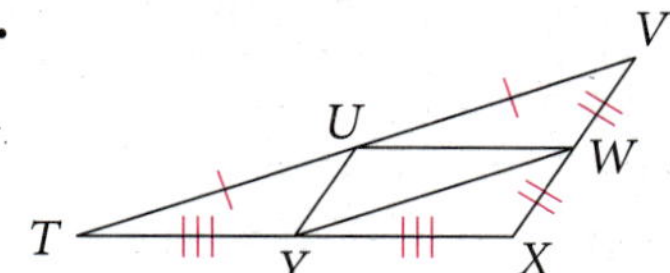

12. 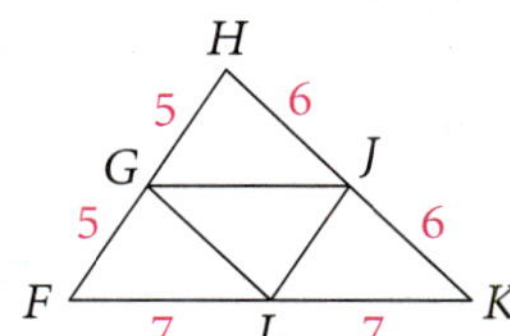

13. a. In the figure at the right, identify pairs of parallel segments.
b. If $m\angle QST = 40$, find $m\angle QPR$.

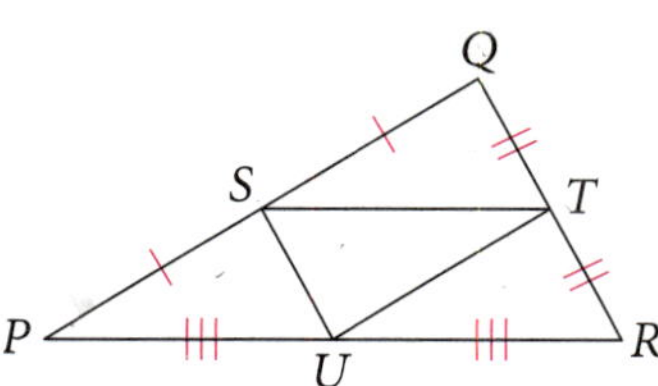

Name the segment that is parallel to the given segment.

14. $\overline{AB}$
15. $\overline{BC}$
16. $\overline{EF}$
17. $\overline{CA}$
18. $\overline{GE}$
19. $\overline{FG}$

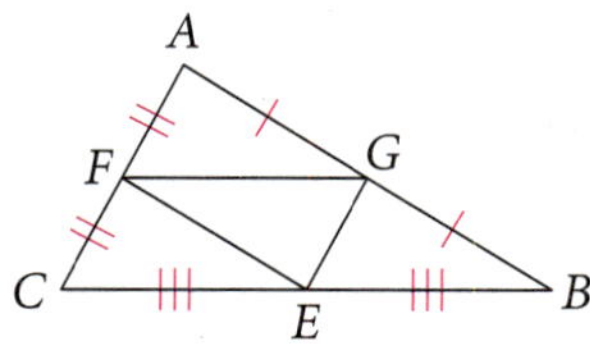

Example 3 (page 261)

20. **Indirect Measurement** Kate wants to paddle her canoe across the lake. To determine how far she must paddle, she paced out a triangle, counting the number of strides, as shown.
a. If Kate's strides average 3.5 ft, what is the length of the longest side of the triangle?
b. What distance must Kate paddle across the lake?

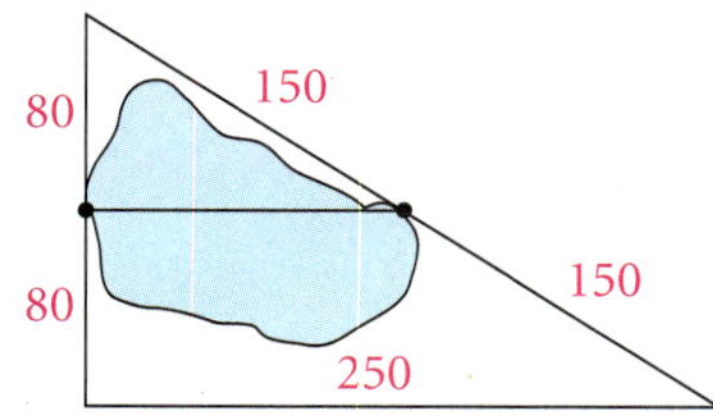

Apply Your Skills

Problem Solving Hint

The highlighted segment is halfway up the face of the Rock and Roll Hall of Fame.

21. a. Architecture The triangular face of the Rock and Roll Hall of Fame in Cleveland, Ohio, is isosceles. The length of the base is 229 ft 6 in. What is the length of the highlighted segment?

b. Writing Explain your reasoning.

X is the midpoint of $\overline{UV}$. Y is the midpoint of $\overline{UW}$.

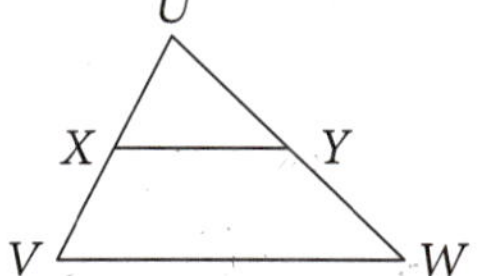

22. If $m\angle UXY = 60$, find $m\angle V$.

23. If $m\angle W = 45$ find $m\angle UYX$.

24. If $XY = 50$, find VW.

25. If $VW = 110$, find XY.

26. Coordinate Geometry The coordinates of the vertices of a triangle are $E(1, 2)$, $F(5, 6)$, and $G(3, -2)$.

a. Find the coordinates of H, the midpoint of $\overline{EG}$, and J, the midpoint of $\overline{FG}$.

b. Verify that $\overline{HJ} \parallel \overline{EF}$.

c. Verify that $HJ = \frac{1}{2}EF$.

$\overline{IJ}$ is a midsegment of $\triangle FGH$. $IJ = 7$, $FH = 10$, and $GH = 13$. Find the perimeter of each triangle.

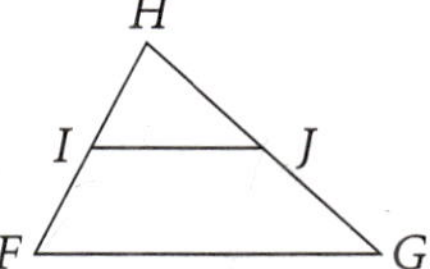

27. $\triangle IJH$

28. $\triangle FGH$

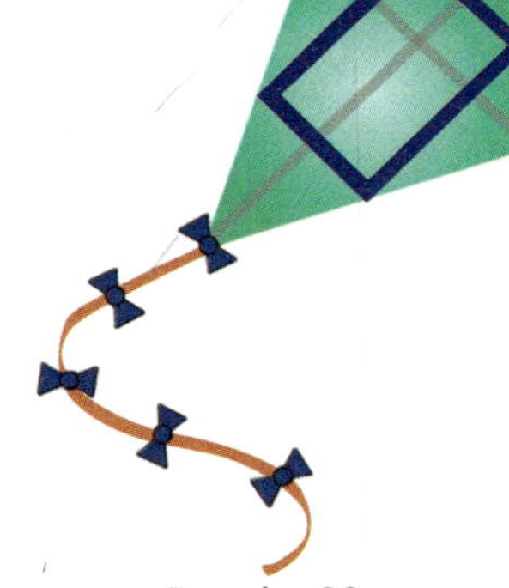

Exercise 29

29. Multiple Choice Marita is designing a kite to look like the one on the left. Its diagonals are to measure 64 cm and 90 cm. She will use ribbon to connect the midpoints of its sides. How much ribbon will Marita need?

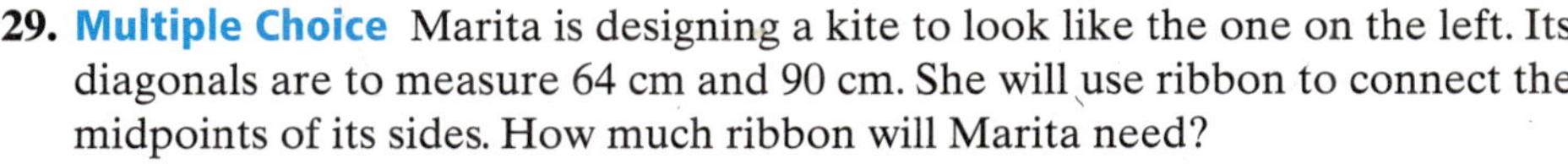

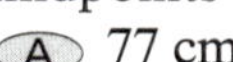

Ⓐ 77 cm Ⓑ 122 cm Ⓒ 154 cm Ⓓ 308 cm

Algebra **Find the value of each variable.**

30.

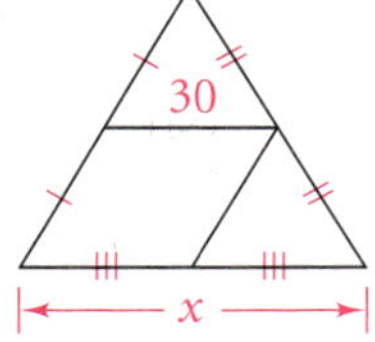

31.

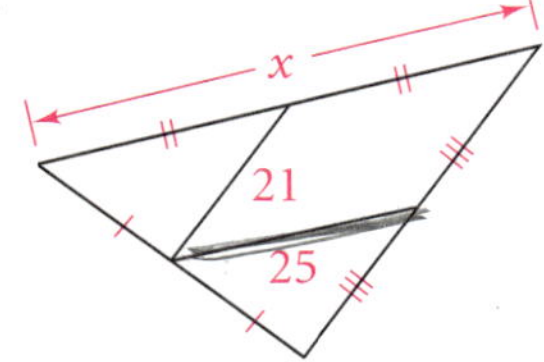

32.

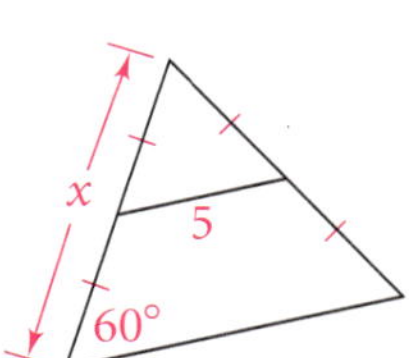

33.

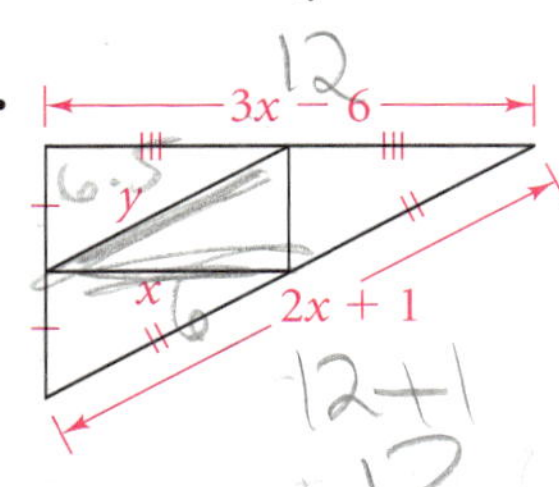

Homework Video Tutor

Visit: PHSchool.com
Web Code: bce-0501

Use the figure at the right for Exercises 34–36.

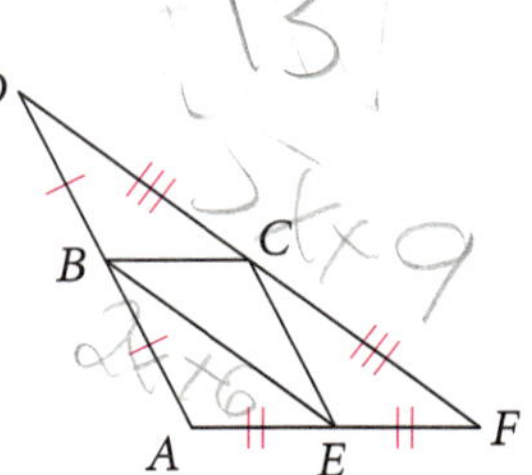

34. If $DF = 24$, $BC = 6$, and $DB = 8$, find the perimeter of $\triangle ADF$.

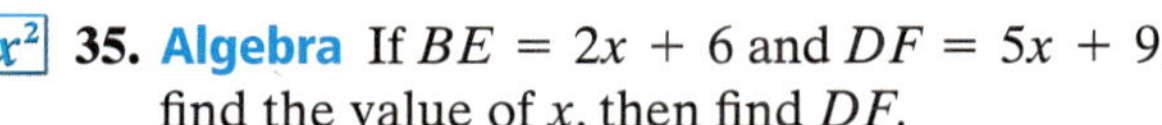

35. Algebra If $BE = 2x + 6$ and $DF = 5x + 9$, find the value of x, then find DF.

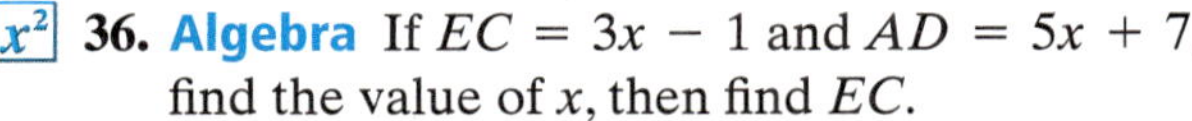

36. Algebra If $EC = 3x - 1$ and $AD = 5x + 7$, find the value of x, then find EC.

37. **Open-Ended** Explain how you could use the Triangle Midsegment Theorem as the basis for this construction. Draw $\overline{CD}$. Draw point A not on $\overline{CD}$. Construct $\overline{AB}$ so that $\overline{AB} \parallel \overline{CD}$ and $AB = \frac{1}{2}CD$.

38. **Coordinate Geometry** In $\triangle GHJ$, $K(2, 3)$ is the midpoint of $\overline{GH}$, $L(4, 1)$ is the midpoint of $\overline{HJ}$, and $M(6, 2)$ is the midpoint of $\overline{GJ}$. Find the coordinates of G, H, and J.

Proof 39. Complete the prove statement and then write a proof.

Given: S, T, and U are midpoints.

Prove: $\triangle YST \cong \triangle TUZ \cong \triangle SVU \cong$ ___?___.

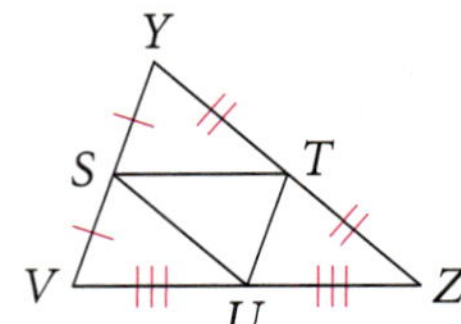

Multiple Choice Practice

For California Standards and CAHSEE Tutorials, visit PHSchool.com. Web Code: bcq-9045

GEOM 17.0

40. In the figure at the right, $A(-2, -4)$, $B(-3, 2)$ and $C(3, 0)$. Find the coordinates of $\triangle DEF$ if D is the midpoint of $\overline{AB}$, E is the midpoint of $\overline{BC}$, and F is the midpoint of $\overline{AC}$.

(A) $D(-2.5, -1)$, $E(0, 1)$, $F(0.5, -2)$
(B) $D(-2, -1)$, $E(0, 0.5)$, $F(0.5, -1)$
(C) $D(-2, -1.5)$, $E(0, 1.5)$, $F(0.8, -2)$
(D) $D(-2.2, -0.8)$, $E(0, 1.5)$, $F(0.5, -2)$

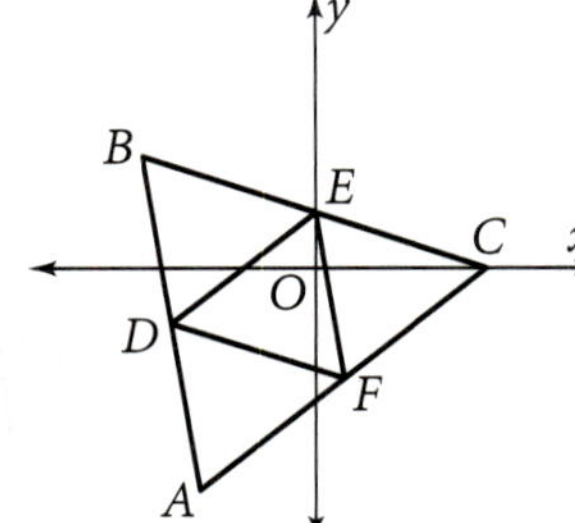

Mixed Review

Lesson 4-7

Name a pair of overlapping congruent triangles in each diagram. State whether the triangles are congruent by SSS, SAS, ASA, AAS, or HL.

41.

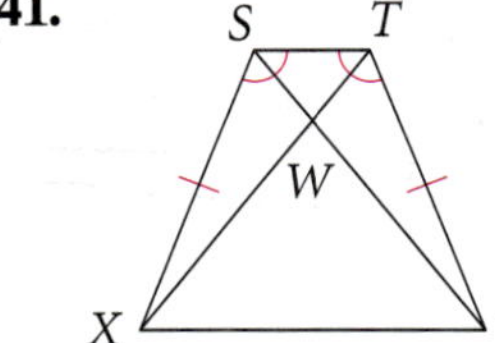

42.

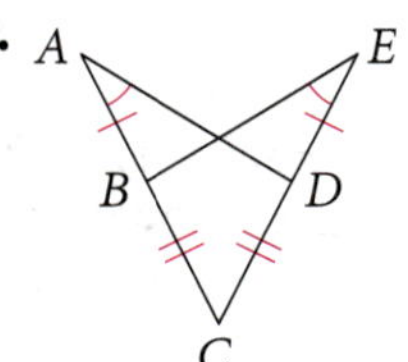

43.

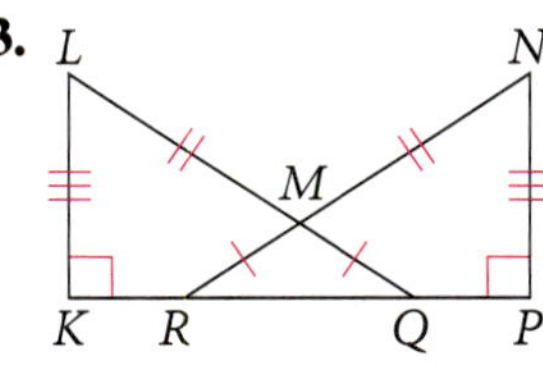

Lesson 3-6 **Algebra** **Graph each line.**

44. $y = x + 2$

45. $y = 3x - 2$

46. $y = -x - 5$

Lesson 3-2 **Algebra** **Determine the value of x for which $\ell \parallel m$.**

47.

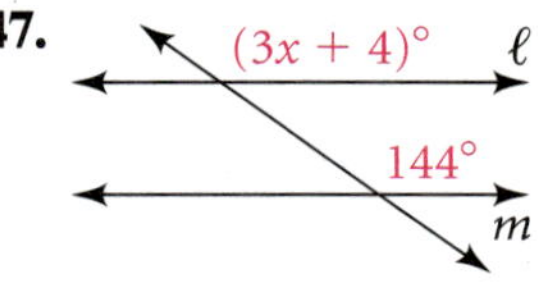

48.

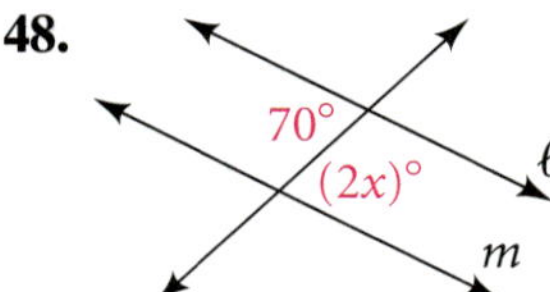

49.

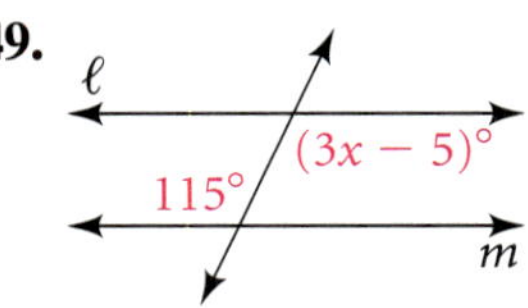

Bisectors in Triangles

California Content Standards

GEOM 2.0 Students write geometric proofs. ***Master***
GEOM 4.0 Students prove basic theorems involving congruence. ***Master***
GEOM 5.0 Students prove that triangles are congruent, and they are able to use the concept of corresponding parts of congruent triangles. ***Master***

What You'll Learn

- To use properties of perpendicular bisectors and angle bisectors

. . . And Why

To locate places equidistant from two given points on a map, as in Example 1

Check Skills You'll Need

Lesson 1-7

Use a compass and a straightedge for the following.

1. Draw a triangle, $\triangle XYZ$. Construct $\triangle STV$ so that $\triangle STV \cong \triangle XYZ$.
2. Draw acute $\angle P$. Construct $\angle Q$ so that $\angle Q \cong \angle P$.
3. Draw $\overline{AB}$. Construct a line $\overleftrightarrow{CD}$ so that $\overleftrightarrow{CD} \perp \overline{AB}$ and $\overleftrightarrow{CD}$ bisects $\overline{AB}$.
4. Draw acute $\angle E$. Construct the bisector of $\angle E$.

$\overrightarrow{TM}$ bisects $\angle STU$ so that $m\angle STM = 5x + 4$ and $m\angle MTU = 6x - 2$.

5. **Algebra** Find the value of x.
6. Find $m\angle STU$.

New Vocabulary • distance from a point to a line

1 Perpendicular Bisectors and Angle Bisectors

Triangles play a key role in relationships involving perpendicular bisectors and angle bisectors.

In the diagram below on the left, $\overleftrightarrow{CD}$ is the perpendicular bisector of $\overline{AB}$. $\overleftrightarrow{CD}$ is perpendicular to $\overline{AB}$ at its midpoint. In the diagram on the right, $\overline{CA}$ and $\overline{CB}$ are drawn to complete the triangles, $\triangle CAD$ and $\triangle CBD$.

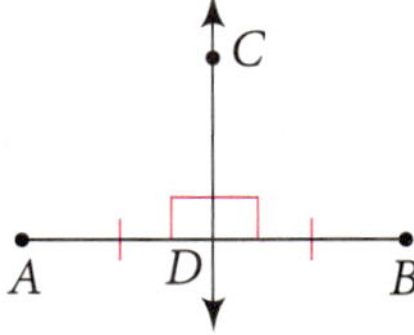

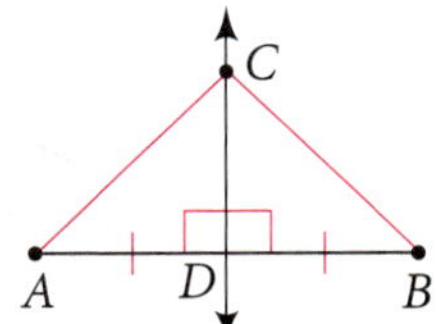

You should recognize from your work in Chapter 4 that $\triangle CAD \cong \triangle CBD$. Thus, you can conclude that $\overline{CA} \cong \overline{CB}$, that $CA = CB$, or simply that C is equidistant from points A and B.

This suggests a proof of Theorem 5-2 below. Its converse is also true and is stated as Theorem 5-3. You will prove these theorems in the exercises.

Key Concepts

Theorem 5-2 — Perpendicular Bisector Theorem

If a point is on the perpendicular bisector of a segment, then it is equidistant from the endpoints of the segment.

Theorem 5-3 — Converse of the Perpendicular Bisector Theorem

If a point is equidistant from the endpoints of a segment, then it is on the perpendicular bisector of the segment.

1 EXAMPLE Real-World Connection

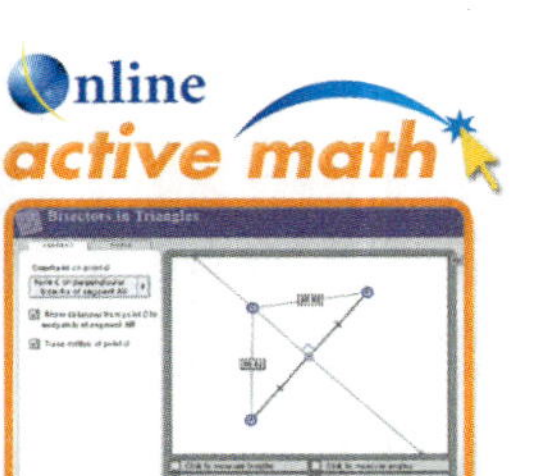

For: Bisector Activity
Use: Interactive Textbook, 5-2

National Landmarks Find the set of points on the map of Washington, D.C. that are equidistant from the Jefferson Memorial and the White House.

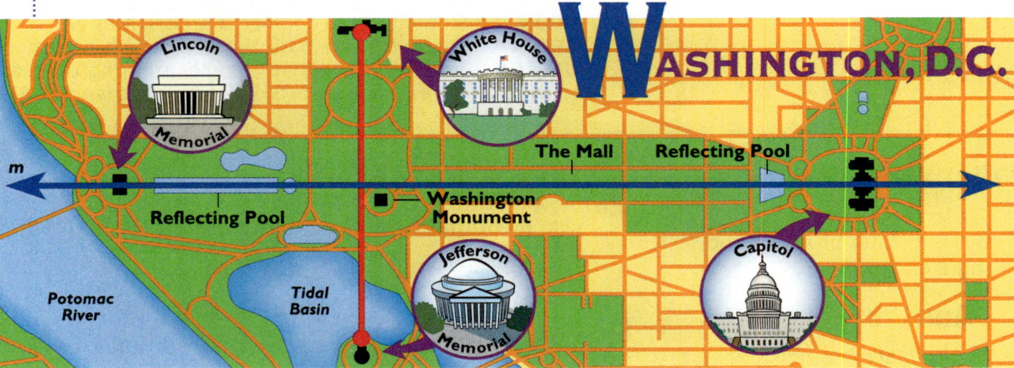

The red segment connects the Jefferson Memorial and the White House. All points on the perpendicular bisector m of this segment are equidistant from the Jefferson Memorial and the White House.

1 Use the information given in the diagram. $\overleftrightarrow{CD}$ is the perpendicular bisector of $\overline{AB}$. Find CA and DB. Explain your reasoning.

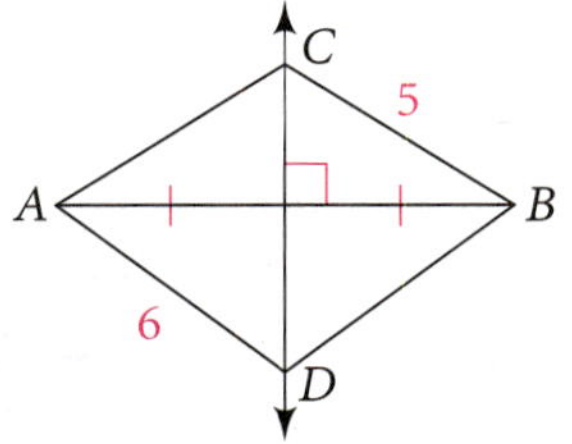

The **distance from a point to a line** is the length of the perpendicular segment from the point to the line. In the diagram, $\overrightarrow{AD}$ is the bisector of $\angle CAB$. If you measure the lengths of the perpendicular segments from D to the two sides of the angle, you will find that the lengths are equal so D is equidistant from the sides.

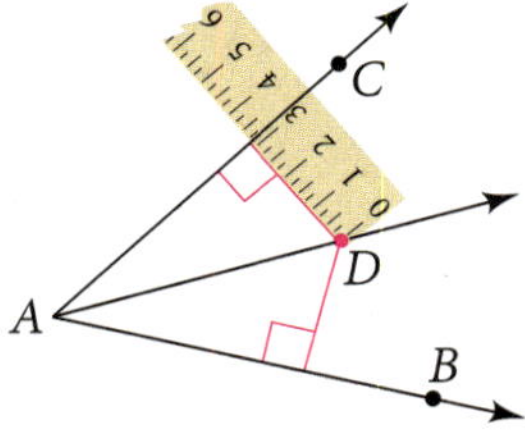

Theorem 5-4 **Angle Bisector Theorem**

If a point is on the bisector of an angle, then the point is equidistant from the sides of the angle.

Theorem 5-5 **Converse of the Angle Bisector Theorem**

If a point in the interior of an angle is equidistant from the sides of the angle, then the point is on the angle bisector.

You will use congruent triangles to prove these theorems in the exercises.

You can combine Theorems 5-4 and 5-5 into a biconditional: A point in the interior of an angle is equidistant from the sides of the angle if and only if it is on the angle bisector.

2 EXAMPLE Using the Angle Bisector Theorem

Test-Taking Tip

Always answer the question asked. In Example 2, choice A is the value of x, but the question asks for FD.

Multiple Choice What is the length of $\overline{FD}$?

(A) 8 (B) 16 (C) 30 (D) 40

From the diagram you can see that F is on the bisector of $\angle ACE$. Therefore, $FB = FD$.

$FB = FD$	
$5x = 2x + 24$	**Substitute.**
$3x = 24$	**Subtract 2*x*.**
$x = 8$	**Divide by 3.**
$FB = 5x = 5(8) = 40$	**Substitute.**
$FD = 40$	**Substitute.**

The correct answer is D.

2 **a.** According to the diagram, how far is K from $\overrightarrow{EH}$? From $\overrightarrow{ED}$?
b. What can you conclude about $\overrightarrow{EK}$?
c. Find the value of x.
d. Find $m\angle DEH$.

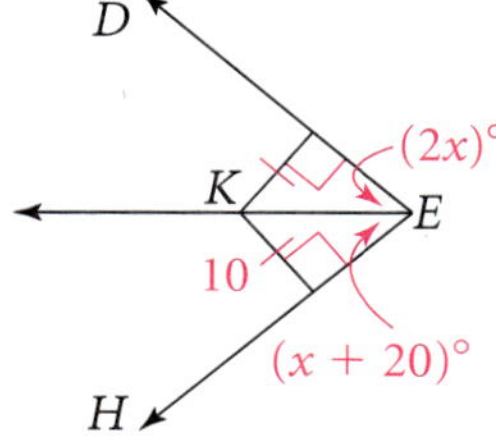

For more exercises, see *Extra Skill, Word Problem, and Proof Practice.*

GEOM 2.0, 4.0, 5.0

Practice by Example

Example 1 (page 266)

Use the figure at the right for Exercises 1–4.

1. From the information given in the figure, how is $\overline{AC}$ related to $\overline{BD}$?

2. Find AB. 3. Find BC. 4. Find ED.

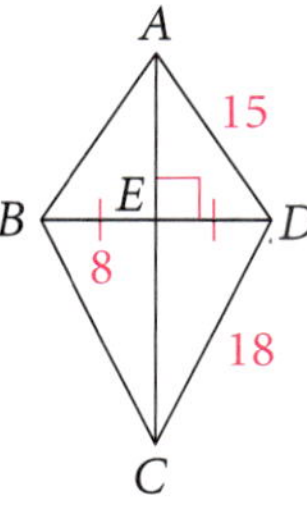

5. On a piece of paper, mark a point H for home and a point S for school. Describe the set of points equidistant from H and S

Example 2 (page 267)

6. **Algebra** Find x, JK, and JM.

7. **Algebra** Find y, ST, and TU.

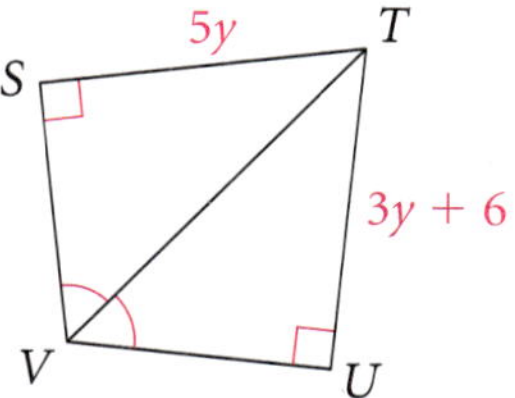

Use the figure at the right for Exercises 8–11.

8. According to the diagram, how far is L from $\overrightarrow{HK}$? From $\overrightarrow{HF}$?

9. How is $\overrightarrow{HL}$ related to $\angle KHF$? Explain.

10. Find the value of y.

11. Find $m\angle KHL$ and $m\angle FHL$.

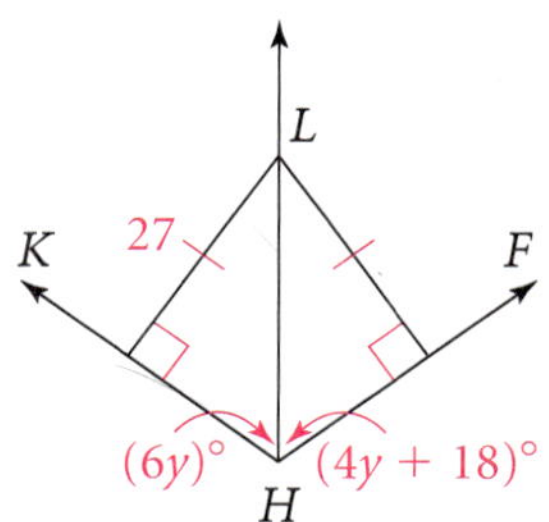

B Apply Your Skills

Algebra **Use the figure, below right, for Exercises 12–16.**

12. Find the value of x.

13. Find TW.

14. Find WZ.

15. What kind of triangle is $\triangle TWZ$? Explain.

16. If R is on the perpendicular bisector of $\overline{TZ}$, then R is __?__ from T and Z, or __?__ = __?__.

17. Write Theorems 5-2 and 5-3 as a single biconditional statement.

$\overleftrightarrow{CD}$ **is the perpendicular bisector of both** $\overline{XY}$ **and** $\overline{ST}$**, and** $CY = 16$**. Find each length.**

18. CT **19.** TY

20. SX **21.** CX

22. MT **23.** ST

24. DY **25.** XY

26. What kind of triangles are $\triangle SCT$ and $\triangle XCY$? Explain.

Homework Video Tutor

Visit: PHSchool.com
Web Code: bce-0502

27. Error Analysis To prove that $\triangle PQR$ is isosceles, a student began by stating that since Q is on the segment perpendicular to $\overline{PR}$, Q is equidistant from the endpoints of $\overline{PR}$. What additional information does the student need in order to make that statement?

Writing **Determine whether point A must be on the bisector of $\angle TXR$. Explain.**

28.

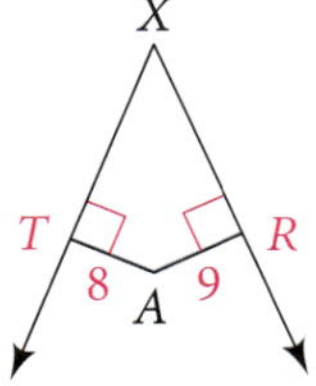

29.

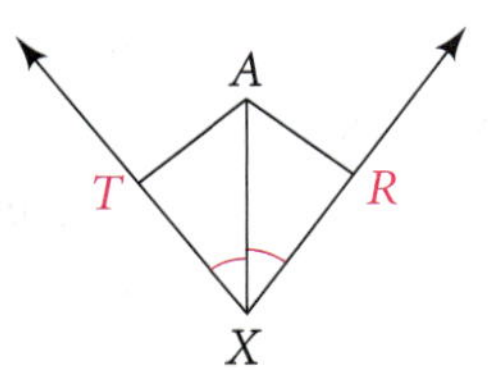

30.

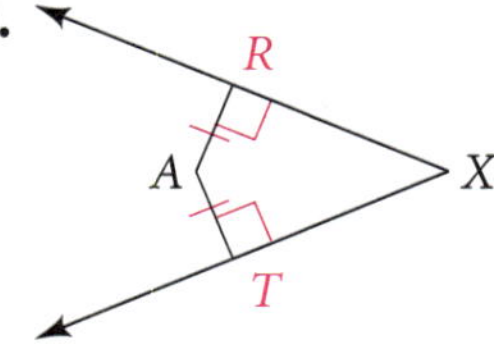

31. Baseball What is the common name for the part of a baseball field that is equidistant from the foul lines and 60 ft 6 in. from home plate?

32. a. Constructions Draw a large triangle, $\triangle CDE$. Construct the angle bisectors of each angle.

b. Make a Conjecture What appears to be true about the angle bisectors?

c. Test your conjecture with another triangle.

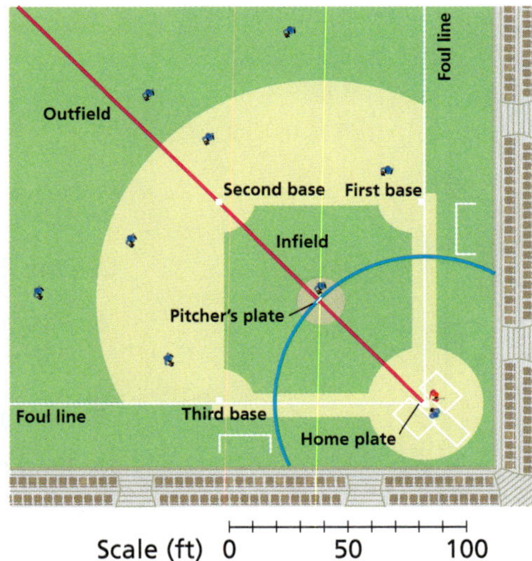

Real-World Connection

On a baseball field, second base is equidistant from the foul lines and 127 feet from home plate.

Problem Solving Hint

In Exercise 33a, your construction may suggest something but be slightly off. If so, test your conjecture *very* carefully in part (c).

Real-World Connection

The picture hangs straight when the hook is on the perpendicular bisector of the picture's top edge.

33. a. Constructions Draw a large acute scalene triangle, $\triangle PQR$. Construct the perpendicular bisectors of each side.

b. Make a Conjecture What appears to be true about the perpendicular bisectors?

c. Test your conjecture with another triangle.

Coordinate Geometry **Find two points on the perpendicular bisector of $\overline{AB}$. Verify your results by showing each point is equidistant from A and B.**

34. $A(0, 0), B(0, 4)$ **35.** $A(0, 2), B(6, 2)$ **36.** $A(3, 3), B(3, -3)$

37. $A(3, 0), B(0, 3)$ **38.** $A(3, 0), B(1, 4)$ **39.** $A(3, 0), B(2, 5)$

Proof **40.** Prove the Perpendicular Bisector Theorem.

Given: $\overleftrightarrow{CD} \perp \overline{AB}, \overleftrightarrow{CD}$ bisects $\overline{AB}$.

Prove: $DA = DB$

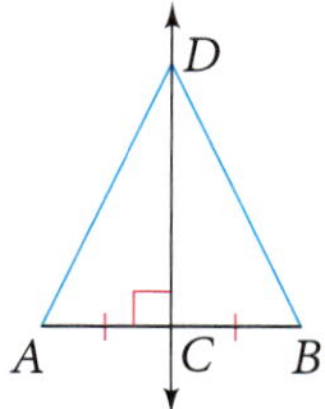

41. Prove the Converse of the Perpendicular Bisector Theorem.

Given: $AP = AQ$ with $\overline{AB} \perp \overline{PQ}$ at B.

Prove: $\overline{AB}$ is the perpendicular bisector of PQ.

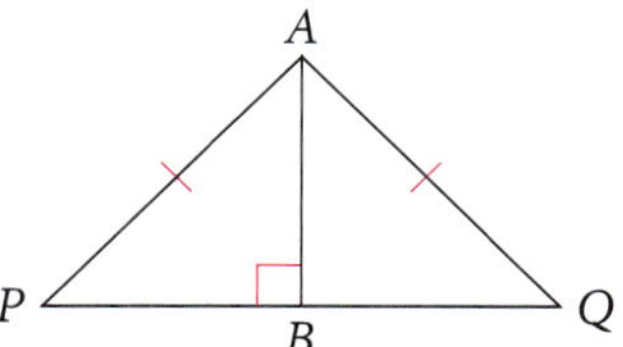

42. Coordinate Geometry You are given points $A(6, 8), O(0, 0)$, and $B(10, 0)$.

a. Write equations of lines ℓ and m such that $\ell \perp \overleftrightarrow{OA}$ at A and $m \perp \overleftrightarrow{OB}$ at B.

b. Find the intersection C of lines ℓ and m.

c. Show that $CA = CB$.

d. Explain why C is on the bisector of $\angle AOB$.

Proof **43.** Prove the Angle Bisector Theorem.

Given: $\overline{PB} \perp \overrightarrow{AB}, \overline{PC} \perp \overrightarrow{AC}$, $\overrightarrow{AP}$ bisects $\angle BAC$.

Prove: $PB = PC$

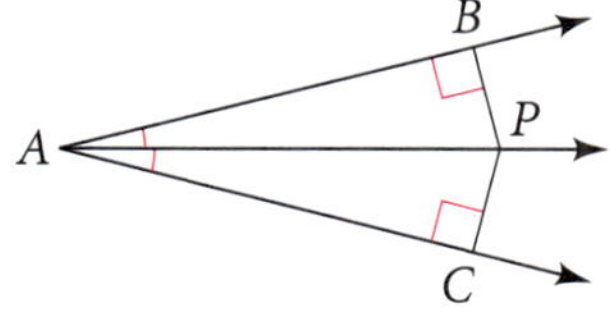

44. Prove the Converse of the Angle Bisector Theorem.

Given: $\overline{SP} \perp \overrightarrow{QP}, \overline{SR} \perp \overrightarrow{QR}$, $SP = SR$

Prove: $\overrightarrow{QS}$ bisects $\angle PQR$.

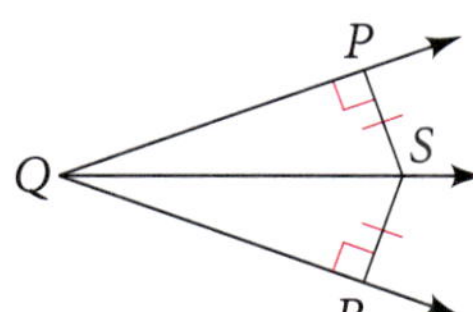

45. Multiple Choice Which equation represents the perpendicular bisector of the segment with endpoints $A(0, 0)$ and B $(6, 0)$?

Ⓐ $y = 3$ Ⓑ $x = 6$ Ⓒ $y = x + 6$ Ⓓ $x = 3$

Coordinate Geometry **Write an equation of the perpendicular bisector of $\overline{AB}$.**

46. $A(0, 0), B(0, 4)$ **47.** $A(1, -1), B(3, 1)$ **48.** $A(-2, 0), B(2, 8)$

49. A, B, and C are three noncollinear points. Describe and sketch a line in plane ABC such that points A, B, and C are equidistant from the line. Justify your response.

50. **Reasoning** M is the intersection of the perpendicular bisectors of two sides of $\triangle ABC$.

a. Explain why M is equidistant from A, B, and C.

b. Line ℓ is perpendicular to plane ABC at M. Explain why a point E on ℓ is equidistant from A, B, and C. (*Hint:* See page 49, Exercise 36. Explain why $\triangle EAM \cong \triangle EBM \cong \triangle ECM$.)

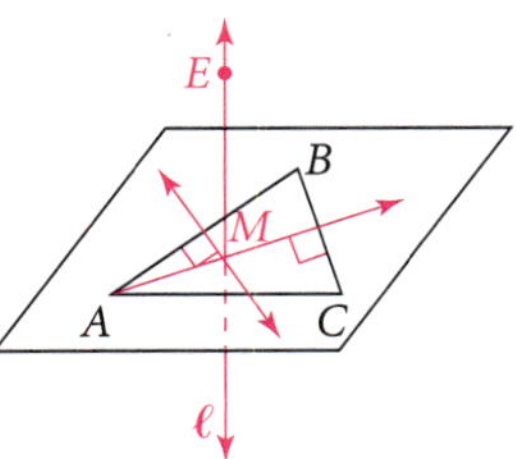

Multiple Choice Practice

For California Standards and CAHSEE Tutorials, visit PHSchool.com. Web Code: bcq-9045

GEOM 5.0

51. Which two triangles are congruent by the AAS postulate?

(A) $\triangle AEB$ and $\triangle EBD$
(B) $\triangle ABE$ and $\triangle BDC$
(C) $\triangle EBD$ and $\triangle CDB$
(D) All three triangles are congruent by AAS

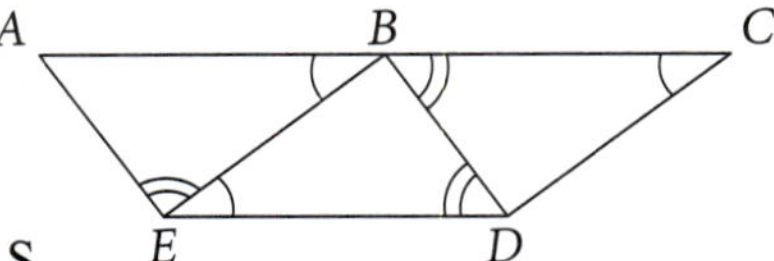

GEOM 12.0

52. If the measure of an exterior angle of a regular polygon is 45°, what is the name for this type of polygon?

(A) Quadrilateral
(B) Pentagon
(C) Hexagon
(D) Octagon

Mixed Review

Lesson 5-1 x^2 **Algebra** **Find the value of x.**

53.

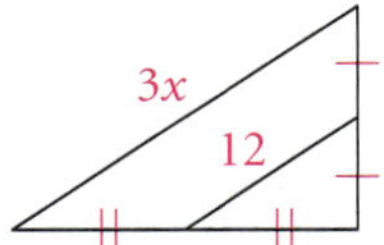

54.

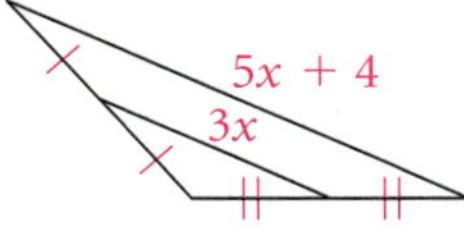

55.

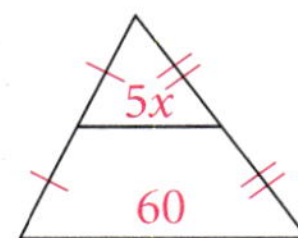

Lesson 2-4 **Name the property that justifies each statement.**

56. $AB = AB$

57. If $2x = 30$, then $x = 15$.

58. If $x = 30 - x$, then $2x = 30$.

59. $3(4x - 1) = 12x - 3$

60. If $m\angle 3 = m\angle 4$ and $m\angle 4 = m\angle 5$, then $m\angle 3 = m\angle 5$.

61. If $\angle 3 \cong \angle 4$ and $\angle 4 \cong \angle 5$, then $\angle 3 \cong \angle 5$.

Lesson 1-8 **Coordinate Geometry** **Find C the midpoint of $\overline{AB}$. Then show that $AC = CB = \frac{1}{2}AB$.**

62. $A(0, 5), B(6, 8)$

63. $A(-2, 8), B(2, -1)$

64. $A(5, 3), B(6, 7)$

Special Segments in Triangles

Technology

FOR USE WITH LESSON 5-3

GEOM 1.0 Students demonstrate understanding by identifying and giving examples of inductive reasoning. ***Master***

Construct

Use geometry software.

- Construct a triangle and the three perpendicular bisectors of its sides.
- Construct a triangle and its three angle bisectors.
- An *altitude* of a triangle is the perpendicular segment from a vertex to the line containing the opposite side. Construct a triangle. Through a vertex of the triangle construct a line that is perpendicular to the line containing the side opposite that vertex. Next construct the altitudes from the other two vertices.
- A *median* of a triangle is the segment joining the midpoint of a side and the opposite vertex. Construct a triangle. Construct the midpoint of one side. Draw the median. Then construct the other two medians.

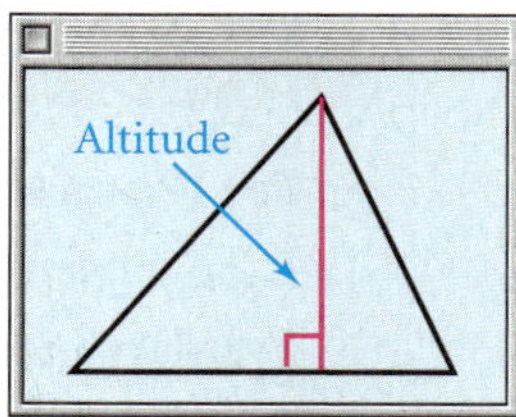

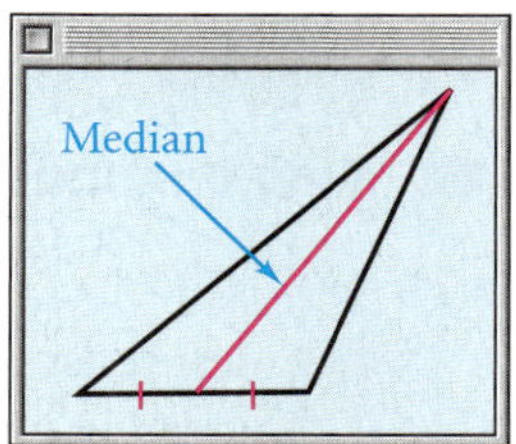

EXERCISES

1. In the constructions above, what property do the perpendicular bisectors, angle bisectors, lines containing altitudes, and medians seem to have?

2. Manipulate the triangles. Does the property still hold as you manipulate the triangles?

3. List your conjectures about the perpendicular bisectors, angle bisectors, lines containing altitudes, and medians of a triangle.

Extend

4. Copy the table. Think about acute triangles, right triangles, and obtuse triangles. Use *inside*, *on*, or *outside* to describe the location of the intersection of the segments or lines for each type of triangle.

	Perpendicular Bisectors	Angle Bisectors	Lines Containing the Altitudes	Medians
Acute Triangle				
Right Triangle				
Obtuse Triangle				

5. What observations, if any, can you make about these special segments for isosceles triangles? Equilateral triangles?

6. Your Exercise 3 conjecture should identify some special points. One of these points is equidistant from the three vertices of its triangle, no matter what shape the triangle has.
a. Use your software. Find which special segments locate this extra-special point.
b. Draw the circle suggested by the properties of this point.

5-3 Concurrent Lines, Medians, and Altitudes

California Content Standards

GEOM 2.0 Students write geometric proofs. ***Master***
GEOM 21.0 Students prove and solve problems regarding relationships among inscribed and circumscribed polygons of circles. ***Introduce***

What You'll Learn

- To identify properties of perpendicular bisectors and angle bisectors
- To identify properties of medians and altitudes of a triangle

. . . And Why

To find a location in a backyard for the largest possible swimming pool, as in Example 2

Check Skills You'll Need

GO for Help Lesson 1-7

For Exercises 1–2, draw a large triangle. Construct each figure.

1. an angle bisector
2. a perpendicular bisector of a side
3. Draw $\overline{GH}$. Construct $\overleftrightarrow{CD} \perp \overline{GH}$ at the midpoint of $\overline{GH}$.
4. Draw $\overleftrightarrow{AB}$ with a point E not on $\overleftrightarrow{AB}$. Construct $\overleftrightarrow{EF} \perp \overleftrightarrow{AB}$.

New Vocabulary
- concurrent
- point of concurrency
- circumcenter of a triangle
- circumscribed about
- incenter of a triangle
- inscribed in
- median of a triangle
- centroid
- altitude of a triangle
- orthocenter of a triangle

1 Properties of Bisectors

Hands-On Activity: Paper Folding Bisectors

- Draw and cut out five different triangles: two acute, two right, and one obtuse.
- Step 1: Use paper folding to create the angle bisectors of each angle of an acute triangle. What do you notice about the angle bisectors?
- Step 2: Repeat Step 1 with a right triangle and an obtuse triangle. Does your discovery from Step 1 still hold true?

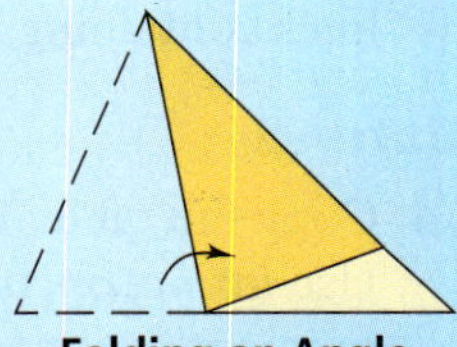

Folding an Angle Bisector

1. Make a conjecture about the bisectors of the angles of a triangle.

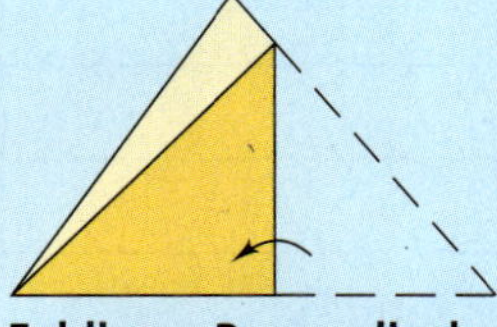

Folding a Perpendicular Bisector

- Step 3: Use paper folding to create the perpendicular bisector of each side of an acute triangle. What do you notice about the perpendicular bisectors?
- Step 4: Repeat Step 3 with a right triangle. What do you notice?

2. Make a conjecture about the perpendicular bisectors of the sides of a triangle.

When three or more lines intersect in one point, they are **concurrent.** The point at which they intersect is the **point of concurrency.** For any triangle, four different sets of lines are concurrent. Theorems 5-6 and 5-7 tell you about two of them.

Key Concepts

Theorem 5-6

The perpendicular bisectors of the sides of a triangle are concurrent at a point equidistant from the vertices.

Theorem 5-7

The bisectors of the angles of a triangle are concurrent at a point equidistant from the sides.

You will prove these theorems in the exercises.

This figure shows $\triangle QRS$ with the perpendicular bisectors of its sides concurrent at C. The point of concurrency of the perpendicular bisectors of a triangle is called the **circumcenter of the triangle.**

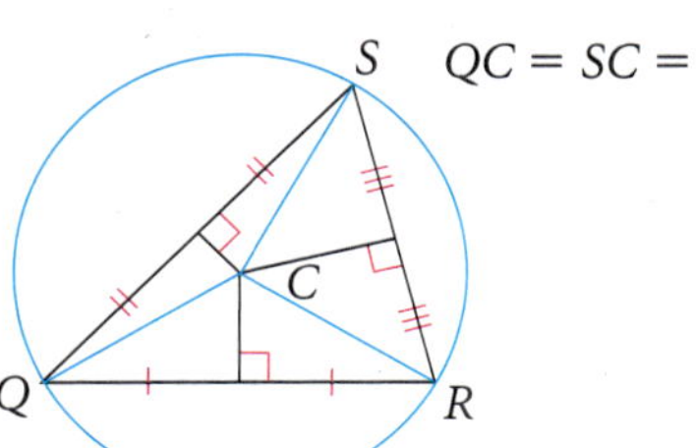

Points Q, R, and S are equidistant from C, the circumcenter. The circle is **circumscribed about** the triangle. Any three non-collinear points can be circumscribed.

Vocabulary Tip

The prefix circum is Latin for "around" or "about."

1 EXAMPLE Finding the Circumcenter

Coordinate Geometry Find the center of the circle that you can circumscribe about $\triangle OPS$.

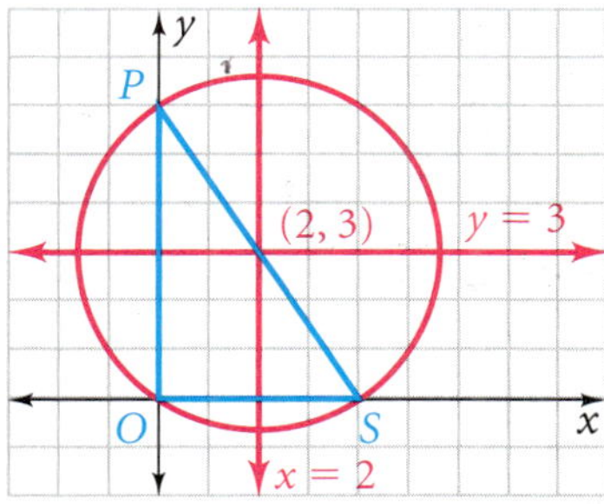

Two perpendicular bisectors of sides of $\triangle OPS$ are $x = 2$ and $y = 3$. These lines intersect at $(2, 3)$. This point is the center of the circle.

CA Standards Check 1 a. Find the center of the circle that you can circumscribe about the triangle with vertices $(0, 0)$, $(-8, 0)$, and $(0, 6)$.

b. **Critical Thinking** In Example 1, explain why it is not necessary to find the third perpendicular bisector.

This figure shows $\triangle UTV$ with the bisectors of its angles concurrent at I. The point of concurrency of the angle bisectors of a triangle is called the **incenter of the triangle.**

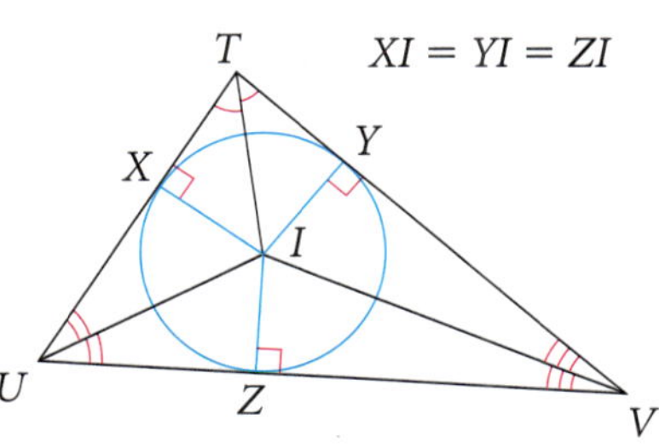

Points X, Y, and Z are equidistant from I, the incenter. The circle is **inscribed in** the triangle.

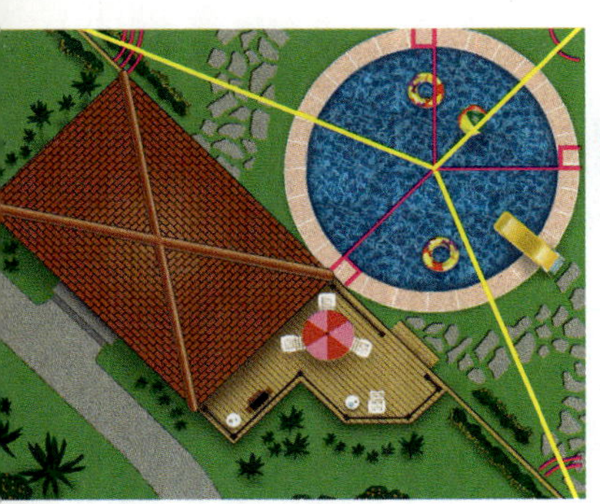

2 EXAMPLE Real-World Connection

Pools The Jacksons want to install the largest possible circular pool in their triangular backyard. Where would the largest possible pool be located?

Locate the center of the pool at the point of concurrency of the angle bisectors. This point is equidistant from the sides of the yard. If you choose any other point as the center of the pool, it will be closer to at least one of the sides of the yard, and the pool will be smaller.

CA Standards Check 2 **a.** The towns of Adamsville, Brooksville, and Cartersville want to build a library that is equidistant from the three towns. Trace the diagram and show where they should build the library.

b. What theorem did you use to find the location?

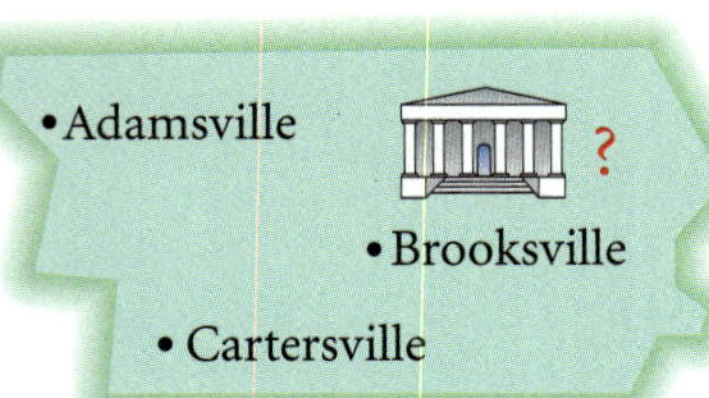

2 Medians and Altitudes

A **median of a triangle** is a segment whose endpoints are a vertex and the midpoint of the opposite side.

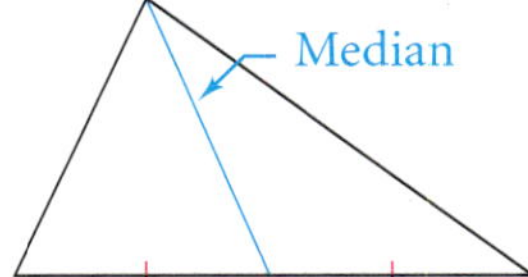

Key Concepts

Theorem 5-8

The medians of a triangle are concurrent at a point that is two thirds the distance from each vertex to the midpoint of the opposite side.

$DC = \frac{2}{3}DJ \quad EC = \frac{2}{3}EG \quad FC = \frac{2}{3}FH$

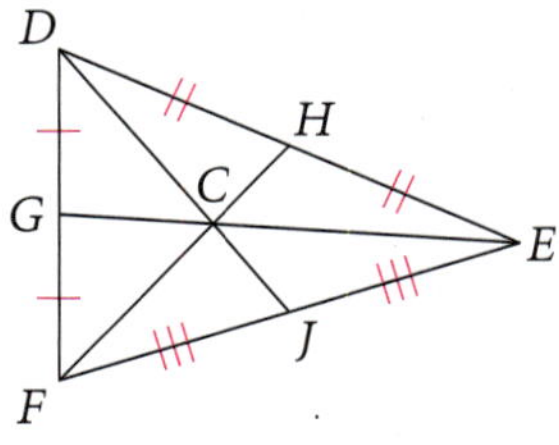

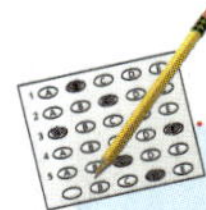

Test-Taking Tip

If you don't remember the meaning of a term, like centroid, the diagram may give a clue.

In a triangle, the point of concurrency of the medians is the **centroid.** The point is also called the center of gravity of a triangle because it is the point where a triangular shape will balance. You will prove Theorem 5-8 in Chapter 6.

3 EXAMPLE Finding Lengths of Medians

In $\triangle ABC$ at the left, D is the centroid and $DE = 6$. Find BE.

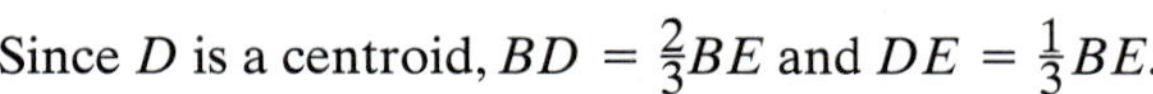

Since D is a centroid, $BD = \frac{2}{3}BE$ and $DE = \frac{1}{3}BE$.

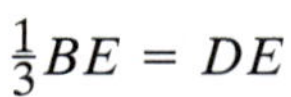

$\frac{1}{3}BE = DE$

$\frac{1}{3}BE = 6$ **Substitute 6 for *DE*.**

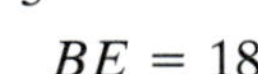

$BE = 18$

CA Standards Check 3

Find BD. Check that $BD + DE = BE$.

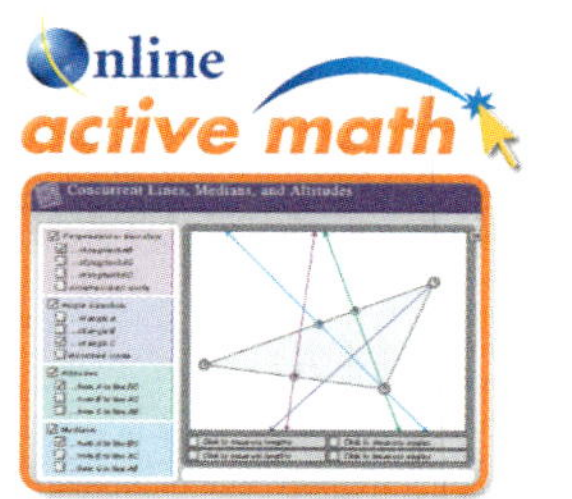

For: Concurrent Lines Activity
Use: Interactive Textbook, 5-3

An **altitude of a triangle** is the perpendicular segment from a vertex to the line containing the opposite side. Unlike angle bisectors and medians, an altitude of a triangle can be a side of a triangle or it may lie outside the triangle.

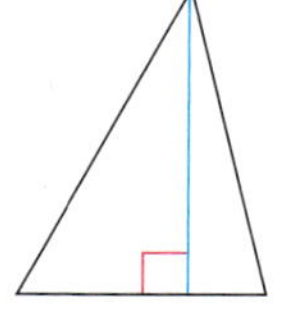
Acute Triangle: Altitude is inside.

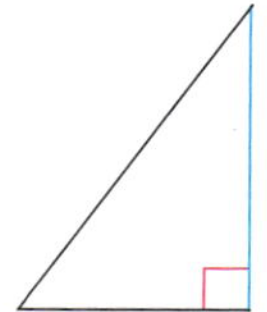
Right Triangle: Altitude is a side.

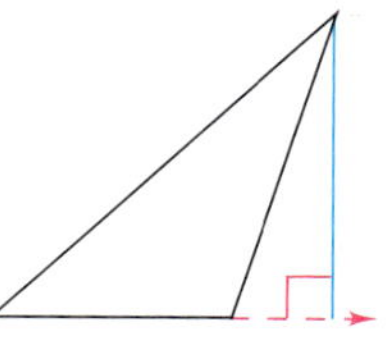
Obtuse Triangle: Altitude is outside.

4 EXAMPLE Identifying Medians and Altitudes

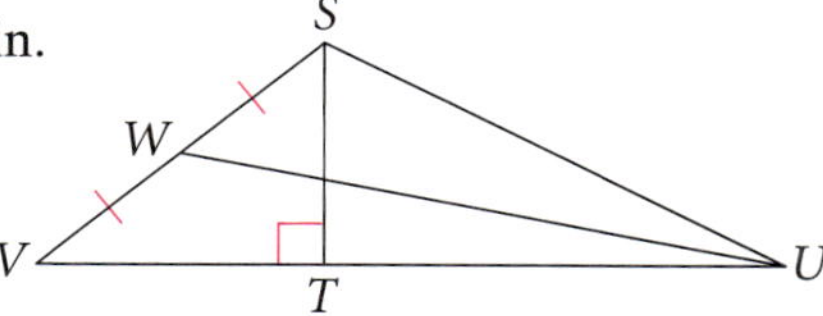

Is $\overline{ST}$ a median, an altitude, or neither? Explain.

$\overline{ST}$ is a segment extending from vertex S to the side opposite S. Also, $\overline{ST} \perp \overline{VU}$. $\overline{ST}$ is an altitude of $\triangle VSU$.

CA Standards Check 4 Is $\overline{UW}$ a median, an altitude, or neither? Explain.

The lines containing the altitudes of a triangle are concurrent at the **orthocenter of the triangle.** A proof of this theorem appears in Chapter 6.

Key Concepts

Theorem 5-9

The lines that contain the altitudes of a triangle are concurrent.

EXERCISES Standards Practice

For more exercises, see *Extra Skill, Word Problem, and Proof Practice.*

GEOM 2.0, 21.0

A Practice by Example

Example 1 (page 273)

Coordinate Geometry **Find the center of the circle that you can circumscribe about each triangle.**

1.

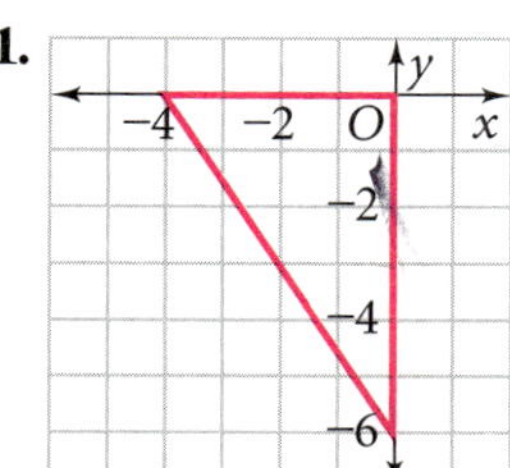

2.

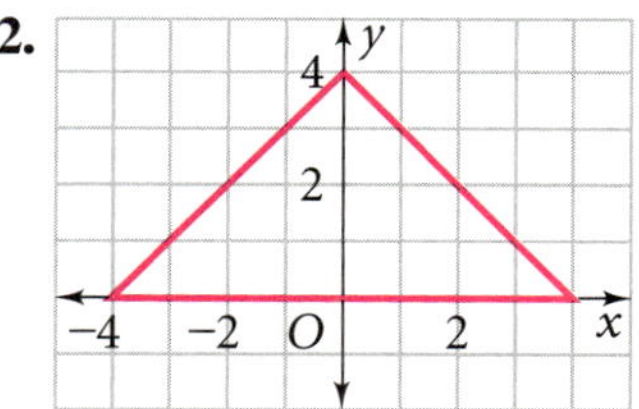

Coordinate Geometry **Find the center of the circle that you can circumscribe about $\triangle ABC$.**

3. $A(0, 0)$	**4.** $A(0, 0)$	**5.** $A(-4, 5)$	**6.** $A(-1, -2)$	**7.** $A(1, 4)$
$B(3, 0)$	$B(4, 0)$	$B(-2, 5)$	$B(-5, -2)$	$B(1, 2)$
$C(3, 2)$	$C(4, -3)$	$C(-2, -2)$	$C(-1, -7)$	$C(6, 2)$

Example 2 (page 274)

Name the point of concurrency of the angle bisectors.

8.

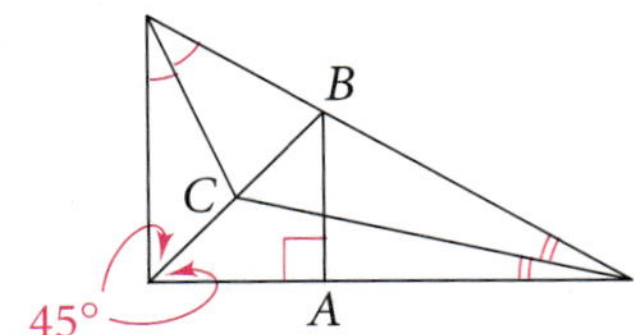

9.

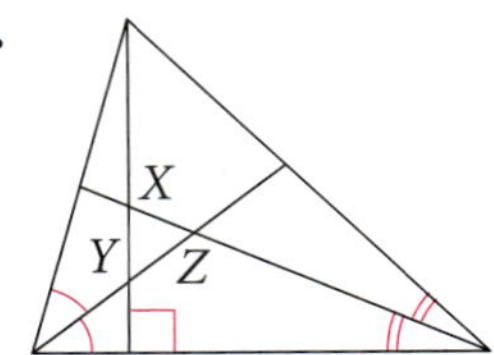

10. **City Planning** Copy the diagram of Altgeld Park. Show where park officials should place a drinking fountain so that it is equidistant from the tennis court, the playground, and the volleyball court.

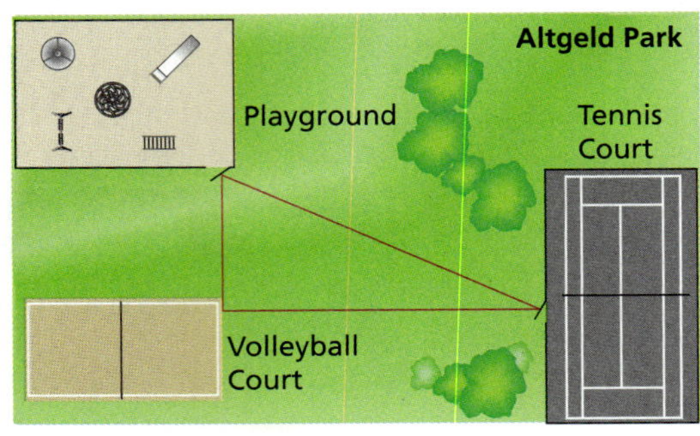

Example 3 (page 274)

In $\triangle TUV$, Y is the centroid.

11. If $YW = 9$, find TY and TW.

12. If $YU = 9$, find ZY and ZU.

13. If $VX = 9$, find VY and YX.

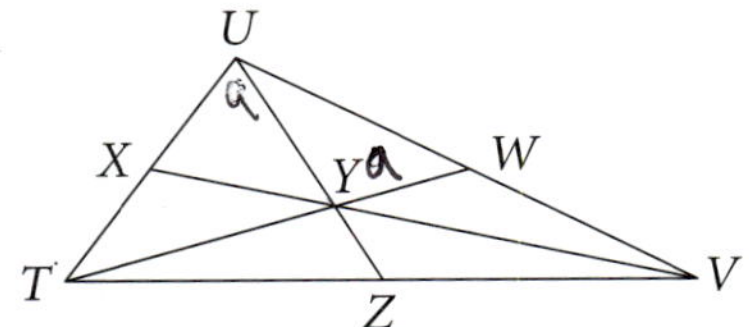

Example 4 (page 275)

Is $\overline{AB}$ a median, an altitude, or neither? Explain.

14.

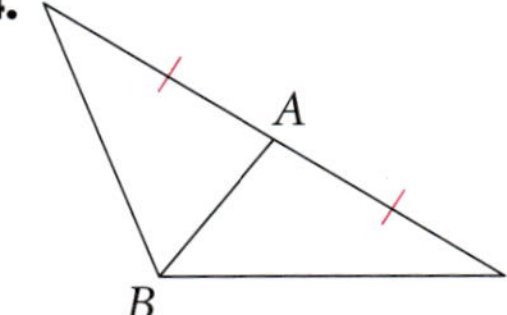

15.

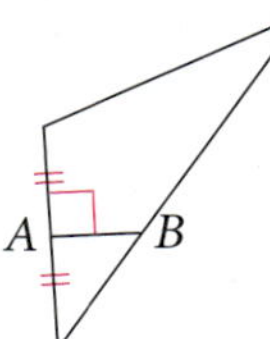

16.

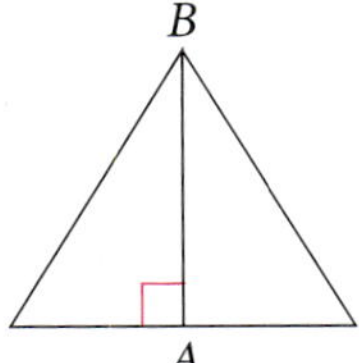

B Apply Your Skills

Constructions **Draw the triangle. Then construct the inscribed circle and the circumscribed circle.**

17. right triangle, $\triangle DEF$

18. obtuse triangle, $\triangle STU$

In Exercises 19–22, name each figure in $\triangle BDF$.

19. an angle bisector

20. a median

21. a perpendicular bisector

22. an altitude

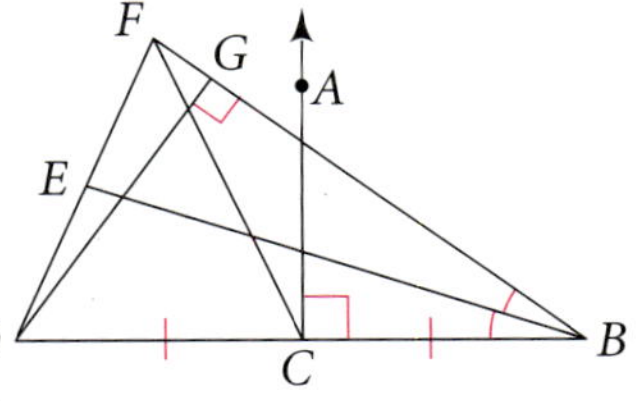

23. **Critical Thinking** A centroid separates a median into two segments. What is the ratio of the lengths of those segments?

24. **Writing** Ivars found a yellowed parchment inside an antique book. It read:

From the spot I buried Olaf's treasure, equal sets of paces did I measure; each of three directions in a line, there to plant a seedling Norway pine. I could not return for failing health; now the hounds of Haiti guard my wealth.—Karl

After searching Caribbean islands for five years, Ivars found one with three tall Norway pines. How might Ivars find where Karl buried Olaf's treasure?

The figures below show how to construct medians and altitudes by paper folding.

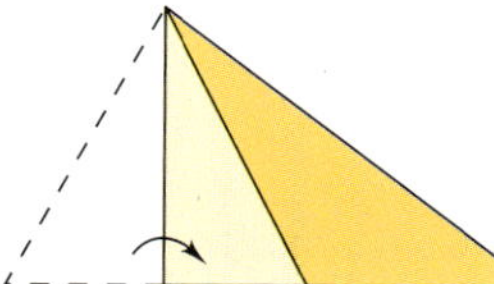

To find an altitude, fold the triangle so that a side overlaps itself and the fold contains the opposite vertex.

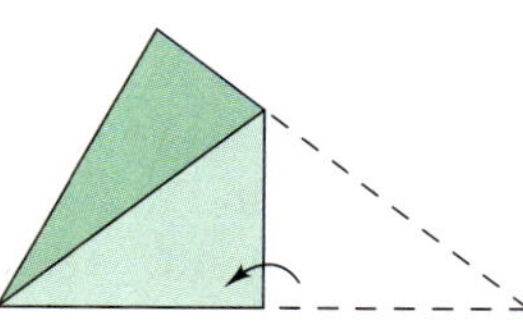

To find a median, fold one vertex to another vertex. This locates the midpoint of a side.

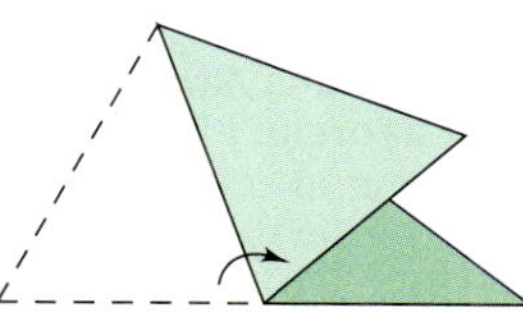

Then fold so that the fold contains the midpoint and the opposite vertex.

Problem Solving Hint

Paper-folding an altitude is the same as paper-folding the perpendicular to a line through a point not on the line.

25. Cut out a large triangle. Paper-fold very carefully to construct the three medians of the triangle and demonstrate Theorem 5-8.

26. Cut out a large acute triangle. Paper-fold very carefully to construct the three altitudes of the triangle and demonstrate Theorem 5-9.

27. **Multiple Choice** C is the centroid of $\triangle DEF$. If $GF = 6x^2 + 9y$, which expression represents CF?

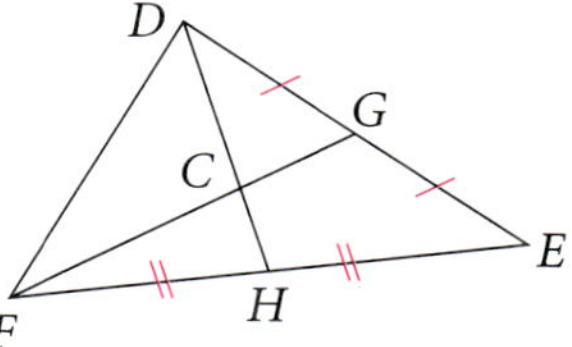

Ⓐ $2x^2 + 9y$　Ⓒ $6x^2 + 9y$

Ⓑ $2x^2 + 3y$　Ⓓ $4x^2 + 6y$

28. Is $\overline{AB}$ a perpendicular bisector, an angle bisector, a median, an altitude, or none of these? Explain.

a.

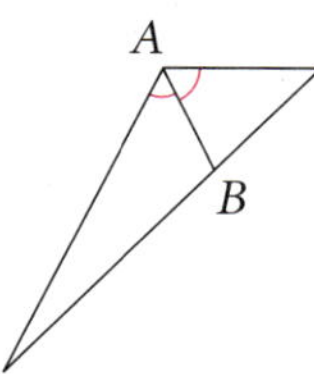

b.

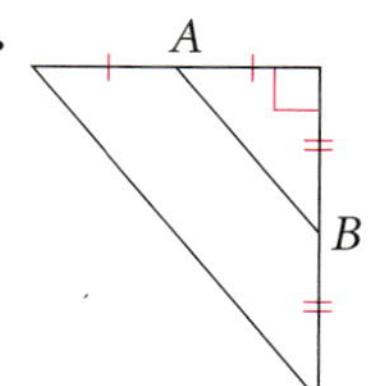

c.

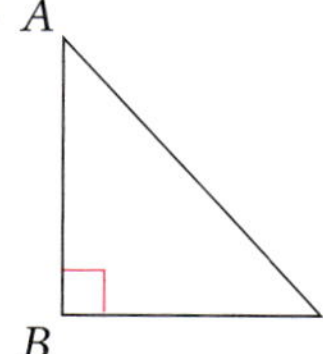

29. **Developing Proof** Complete this proof of Theorem 5-6 by filling in the blanks.

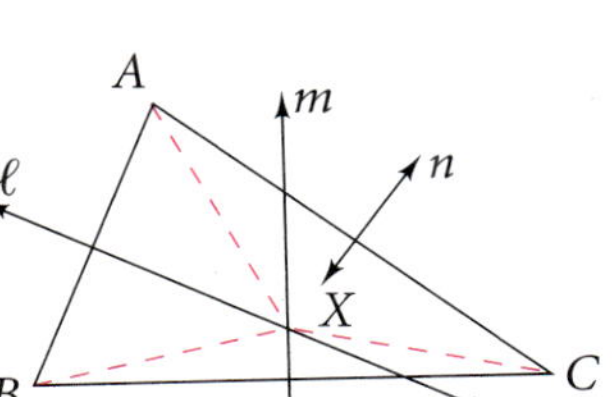

Given: Lines ℓ, m, and n are perpendicular bisectors of the sides of $\triangle ABC$. X is the intersection of lines ℓ and m.

Prove: Line n contains point X, and $XA = XB = XC$.

Proof: Since ℓ is the perpendicular bisector of **a.** __?__, $XA = XB$. Since m is the perpendicular bisector of **b.** __?__, $XB =$ **c.** __?__. Thus $XA = XB = XC$. Since $XA = XC$, X is on line n by the Converse of the **d.** __?__ Theorem.

Proof **30.** Prove Theorem 5-7.

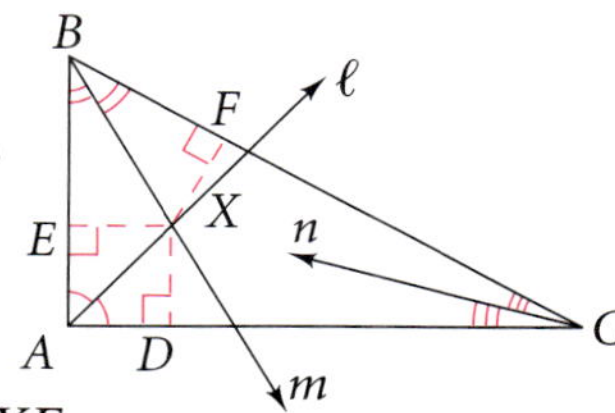

Given: Rays ℓ, m, and n are bisectors of the angles of $\triangle ABC$. X is the intersection of rays ℓ and m and $\overline{XD} \perp \overline{AC}$, $\overline{XE} \perp \overline{AB}$, $\overline{XF} \perp \overline{BC}$.

Prove: Ray n contains point X, and $XD = XE = XF$.

Visit: PHSchool.com
Web Code: bce-0503

31. What kind of triangle has its circumcenter on one of its sides? Explain.

> **Problem Solving Hint**
>
> You can prove Theorem 5-8 for a general $\triangle ABC$ with coordinates $A(0, 0)$, $B(2b, 2d)$, and $C(2c, 0)$ by following the steps for the particular $\triangle ABC$ in Exercise 32.

32. Coordinate Geometry Complete the following steps to locate the centroid.

a. Find the coordinates of midpoints L, M, and N.

b. Find equations of $\overleftrightarrow{AM}$, $\overleftrightarrow{BN}$, and $\overleftrightarrow{CL}$.

c. Find the coordinates of P, the intersection of $\overleftrightarrow{AM}$ and $\overleftrightarrow{BN}$. This is the centroid.

d. Show that point P is on $\overleftrightarrow{CL}$.

e. Use the Distance Formula to show that point P is $\frac{2}{3}$ of the distance from each vertex to the midpoint of the opposite side.

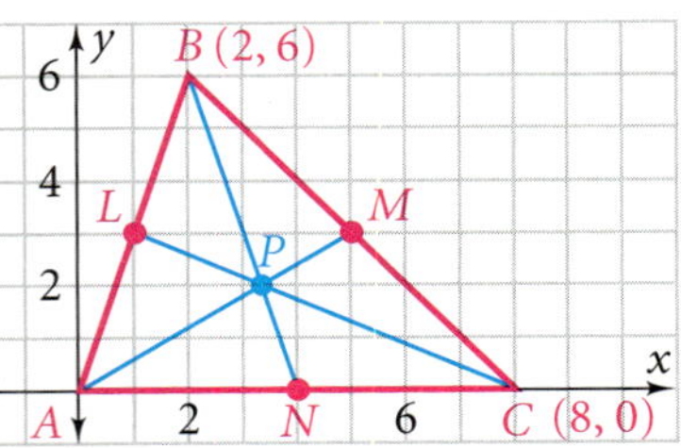

Challenge

For Exercises 33 and 34, points of concurrency have been drawn for two triangles. Match the points with the lines and segments listed in I–IV.

33.

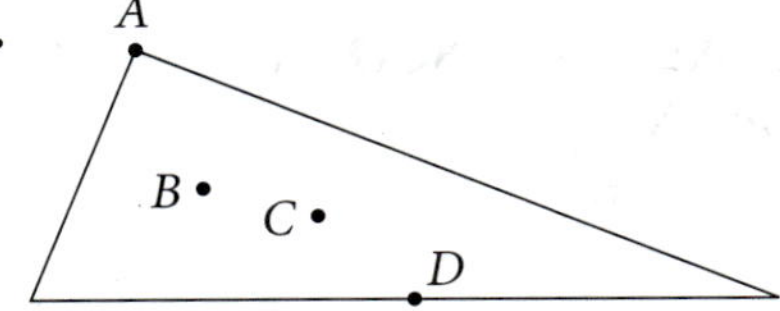

34.

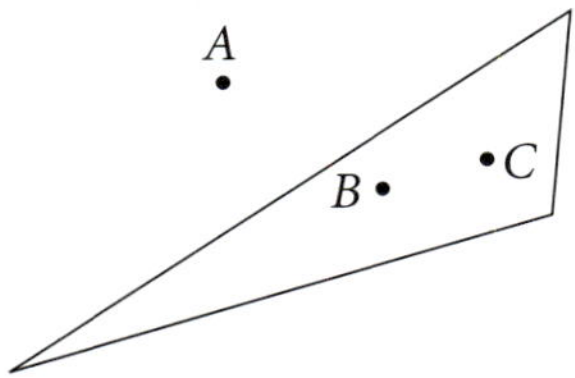

D

I. perpendicular bisectors of sides
II. angle bisectors
III. medians
IV. lines containing altitudes

35. In an isosceles triangle, show that the circumcenter, incenter, centroid, and orthocenter can be four different points but all four must be collinear.

36. History In 1765 Leonhard Euler proved that for any triangle, three of the four points of concurrency are collinear. The line that contains these three points is known as Euler's Line. Use Exercises 33 and 34 to determine which point of concurrency does not necessarily lie on Euler's Line.

Multiple Choice Practice

For California Standards and CAHSEE Tutorials, visit PHSchool.com. Web Code: bcq-9045

GEOM 21.0

37. Find the center of the circle that you can circumscribe about $\triangle ABC$ if $A(-7, 0)$, $B(-3, 8)$, $C(-3, 0)$.

Ⓐ $(-7, -3)$ Ⓑ $(-5, 4)$ Ⓒ $(-4, 3)$ Ⓓ $(-3, 4)$

CAHSEE 7AF 1.5

38. At the circus, a juggler on a tightrope tossed a ball out of reach. The graph shows h, the height in feet of the ball, versus t, the time in seconds after the ball was thrown. From the graph, what conclusion can be made about the flight of the ball?

Ball's Flight

(Graph: Height (feet) h: 10, 20, 30, 40, 50, 60; Time (seconds) t: 0, 1, 2, 3)

Ⓐ The ball was in flight for 4 seconds.

Ⓑ At zero seconds the ball was 50 feet above the ground.

Ⓒ The ball reached its maximum height after 3 seconds.

Ⓓ The ball reached a maximum height of 60 feet.

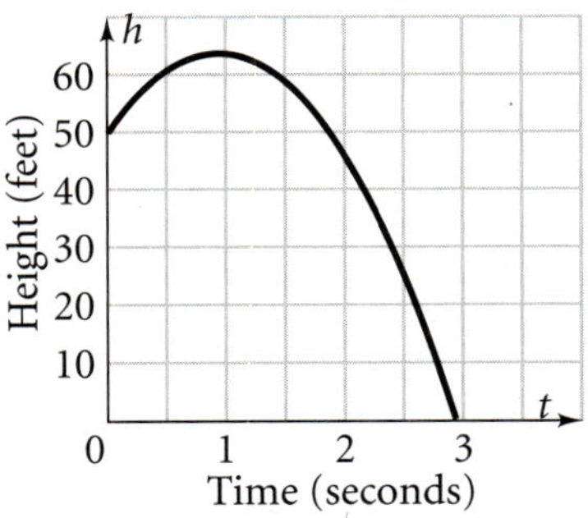

CAHSEE 7NS 2.5

39. What is the absolute value of 6?

Ⓐ -6 Ⓑ $-\frac{1}{6}$ Ⓒ $\frac{1}{6}$ Ⓓ 6

Mixed Review

Lesson 5-2 **Determine whether point B must be on the bisector of $\angle T$. Explain.**

40.

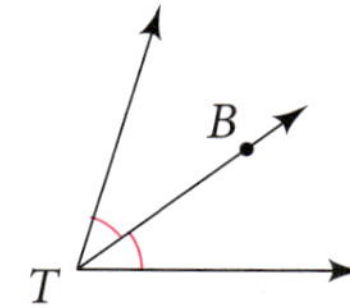

41.

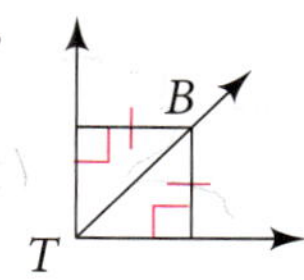

42.

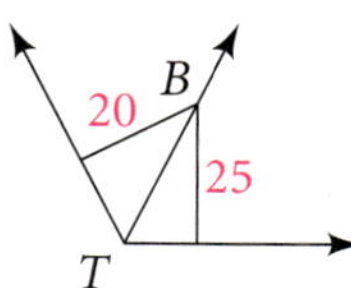

Lesson 3-4 **Classify each $\triangle JKL$ by its angles.**

43. $m\angle J = 37, m\angle K = 53, m\angle L = 90$

44. $m\angle J = 47, m\angle K = 98, m\angle L = 35$

Lesson 1-4 **In the figure at the right, $ABCD$ is a square. Identify each of the following.**

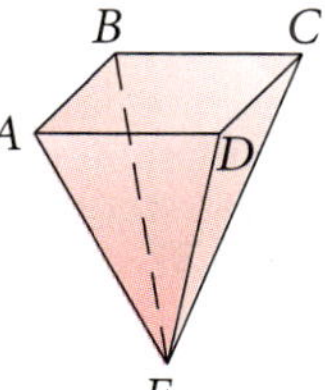

45. a line skew to $\overleftrightarrow{ED}$

46. a line skew to $\overleftrightarrow{EB}$

47. two intersecting planes

48. two parallel segments

49. the intersection of plane ABC and plane BCE

Checkpoint Quiz 1 — Lessons 5-1 through 5-3

Algebra **Find the value of x.**

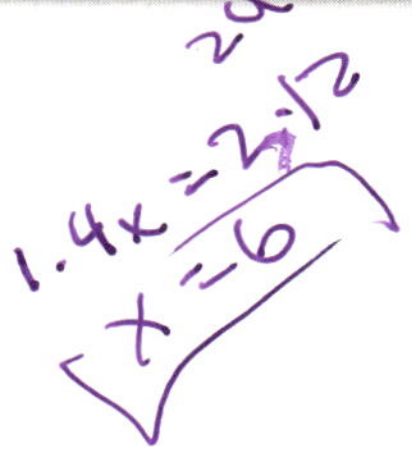

1.

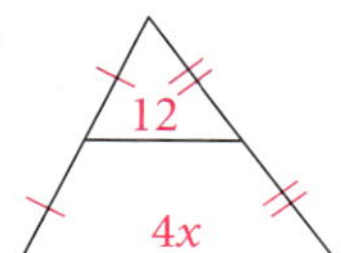

2.

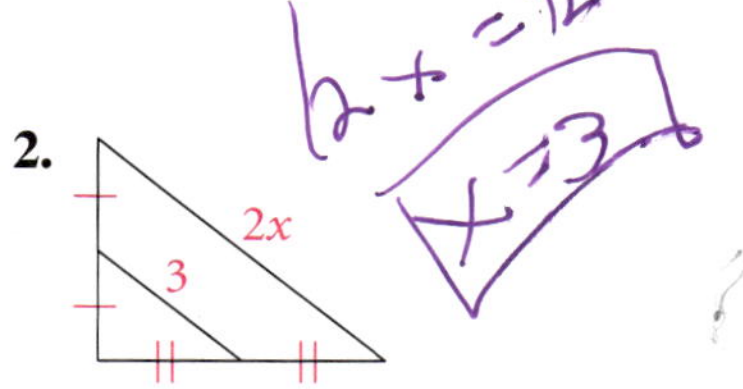

3. a. $\overline{AB}$ is a midsegment of $\triangle XYZ$. $AB = 52$. Find YZ.

b. $AX = 26$ and $BZ = 36$. Find the perimeter of $\triangle XYZ$.

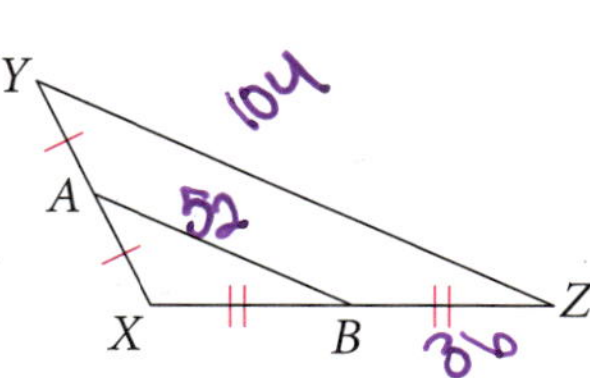

Use the diagram. What can you conclude about each of the following? Explain.

4. $\angle CDB$

5. $\triangle ABD$ and $\triangle CBD$

6. $\overline{AD}$ and $\overline{DC}$

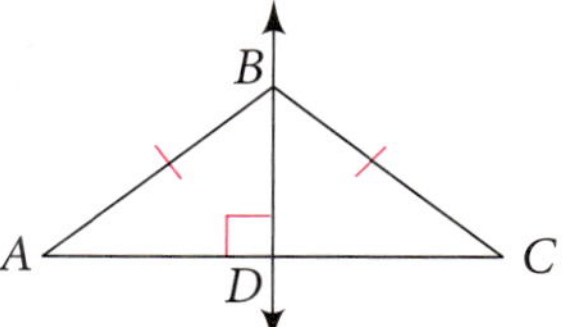

Use the figure at the right.

7. What can you conclude about $\overrightarrow{XY}$? Explain.

8. Find XZ. Justify your response.

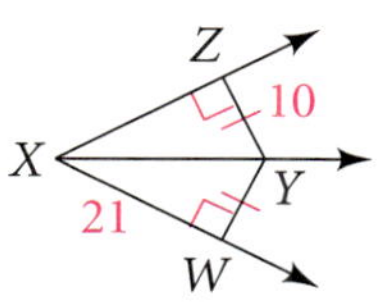

Writing **For a given triangle, describe how you can construct the following.**

9. a median

10. an altitude

Inverses, Contrapositives, and Indirect Reasoning

California Content Standards

GEOM 2.0 Students write geometric proofs, including proofs by contradiction. ***Introduce***

What You'll Learn

- To write the negation of a statement and the inverse and contrapositive of a conditional statement
- To use indirect reasoning

. . . And Why

To describe how an advertisement can be misunderstood, as in Exercise 37

Check Skills You'll Need

GO for Help Lessons 2-1 and 2-2

Write the converse of each statement.

1. If it snows tomorrow, then we will go skiing.
2. If two lines are parallel, then they do not intersect.
3. If $x = -1$, then $x^2 = 1$.

Write two conditional statements that make up each biconditional.

4. A point is on the bisector of an angle if and only if it is equidistant from the sides of the angle.
5. A point is on the perpendicular bisector of a segment if and only if it is equidistant from the endpoints of the segment.
6. You will pass a geometry course if and only if you are successful with your homework.

New Vocabulary • negation • inverse • contrapositive • equivalent statements • indirect reasoning • indirect proof

1 Writing the Negation, Inverse, and Contrapositive

The statement, "Knoxville is the capital of Tennessee," is false. The **negation** of a statement has the opposite truth value. The negation, "Knoxville is not the capital of Tennessee," is true.

1 EXAMPLE Writing the Negation of a Statement

Write the negation of each statement.

a. Statement: $\angle ABC$ is obtuse.

Negation: $\angle ABC$ is not obtuse.

b. Statement: Lines m and n are not perpendicular.

Negation: Lines m and n are perpendicular.

CA Standards Check

1 Write the negation of each statement.

a. $m\angle XYZ > 70$.

b. Today is not Tuesday.

Vocabulary Tip

The prefix contra is Latin for "against."

The **inverse** of a conditional statement negates both the hypothesis and the conclusion. The **contrapositive** of a conditional switches the hypothesis and the conclusion and negates both.

"If you don't stand for something, you'll fall for anything."
—Maya Angelou, poet and author

2 EXAMPLE Writing the Inverse and Contrapositive

Write the inverse and the contrapositive of the conditional statement.

Conditional: If a figure is a square, then it is a rectangle.

Negate both.

Inverse: If a figure is not a square, then it is not a rectangle.

Conditional: If a figure is a square, then it is a rectangle.

Switch and negate both.

Contrapositive: If a figure is not a rectangle, then it is not a square.

CA Standards Check 2 Write (a) the inverse and (b) the contrapositive of Maya Angelou's statement under the photo at the left.

You know that a conditional statement and its converse can have different truth values. A conditional statement and its inverse can also have different truth values. The contrapositive of a conditional statement, however, always has the same truth value as the conditional. A conditional statement and its contrapositive are equivalent. **Equivalent statements** have the same truth value.

Key Concepts

Summary Negation, Inverse, and Contrapositive Statements

Statement	Example	Symbolic Form	You Read It
Conditional	If an angle is a straight angle, then its measure is 180.	$p \rightarrow q$	If p, then q.
Negation (of p)	An angle is not a straight angle.	$\sim p$	Not p.
Inverse	If an angle is not a straight angle, then its measure is not 180.	$\sim p \rightarrow \sim q$	If not p, then not q.
Contrapositive	If an angle's measure is not 180, then it is not a straight angle.	$\sim q \rightarrow \sim p$	If not q, then not p.

2 Using Indirect Reasoning

Suppose your brother tells you, "Susan called a few minutes ago." You think through these three steps.

Step 1 You have two friends named Susan.

Step 2 You know that one of them is at band practice.

Step 3 You conclude that the other Susan must have been the caller.

This type of reasoning is called indirect reasoning. In **indirect reasoning,** all possibilities are considered and then all but one are proved false. The remaining possibility must be true.

A proof involving indirect reasoning is an **indirect proof.** In an indirect proof, a statement and its negation often are the only possibilities.

Summary — Writing an Indirect Proof

Step 1 State as an assumption the opposite (negation) of what you want to prove.

Step 2 Show that this assumption leads to a contradiction.

Step 3 Conclude that the assumption must be false and that what you want to prove must be true.

In the first step of an indirect proof you assume as true the opposite of what you want to prove.

3 EXAMPLE The First Step of an Indirect Proof

Write the first step of an indirect proof.

a. Prove: Quadrilateral $QRWX$ does not have four acute angles.
Assume that quadrilateral $QRWX$ has four acute angles.

b. Prove: An integer n is divisible by 5.
Assume that the integer n is not divisible by 5.

3 You want to prove each statement true. Write the first step of an indirect proof.
a. The shoes cost no more than \$20. **b.** $m\angle A > m\angle B$

Jack, 18, is a good driver. Jack concludes he will get a good insurance rate. His insurance bill is a contradiction.

To do an indirect proof, you have to be able to identify a contradiction.

4 EXAMPLE Identifying Contradictions

Identify the two statements that contradict each other.

I. $\triangle ABC$ is acute. II. $\triangle ABC$ is scalene. III. $\triangle ABC$ is equiangular.

A triangle can be acute and scalene. I and II do not contradict each other.

An equiangular triangle is an acute triangle. I and III do not contradict each other.

An equiangular triangle must be equilateral, so it cannot be scalene. II and III contradict each other.

4 Identify the two statements that contradict each other.

I. $\overline{FG} \parallel \overline{KL}$ II. $\overline{FG} \perp \overline{KL}$ III. $\overline{FG} \cong \overline{KL}$

Proof

5 EXAMPLE Indirect Proof

Read the conditional statement. Think about what is given and what you are to prove. Then give the steps of an indirect proof.

If Jaeleen spends more than \$50 to buy two items at a bicycle shop, then at least one of the items costs more than \$25.

Given: The cost of two items is more than \$50.

Prove: At least one of the items costs more than \$25.

Step 1 Assume as true the opposite of what you want to prove. That is, assume that neither item costs more than \$25.

Step 2 This means that each item costs \$25 or less. This, in turn, means that the two items together cost \$50 or less. This contradicts the given information that the amount spent is more than \$50.

Step 3 Conclude that the assumption is false. One item must cost more than \$25.

CA Standards Check

5 **Critical Thinking** You plan to write an indirect proof showing that $\angle X$ is an obtuse angle. In the first step you assume that $\angle X$ is an acute angle. What have you overlooked?

EXERCISES
Standards Practice

For more exercises, see *Extra Skill, Word Problem, and Proof Practice.*

GEOM 2.0

A Practice by Example

GO for Help

Example 1 (page 280)

Write the negation of each statement.

1. Two angles are congruent.
2. You are not sixteen years old.
3. The angle is not obtuse.
4. The soccer game is on Friday.
5. The figure is a triangle.
6. $m\angle A < 90$

Example 2 (page 281)

Write (a) the inverse and (b) the contrapositive of each conditional statement.

7. If you eat all of your vegetables, then you will grow.
8. If a figure is a square, then all of its angles are right angles.
9. If a figure is a rectangle, then it has four sides.

Example 3 (page 282)

Write the first step of an indirect proof.

10. It is raining outside.
11. $\angle J$ is not a right angle.
12. $\triangle PEN$ is isosceles.
13. At least one angle is obtuse.
14. $\overline{XY} \cong \overline{AB}$
15. $m\angle 2 > 90$

Example 4 (page 282)

Identify the two statements that contradict each other.

16. I. $\triangle PQR$ is equilateral.
    II. $\triangle PQR$ is a right triangle.
    III. $\triangle PQR$ is isosceles.

17. I. In right $\triangle ABC$, $m\angle A = 60$.
    II. In right $\triangle ABC$, $\angle A \cong \angle C$.
    III. In right $\triangle ABC$, $m\angle B = 90$.

18. I. ℓ and m are skew.
    II. ℓ and m do not intersect.
    III. $\ell \parallel m$

19. I. Each of the two items that Val bought costs more than \$10.
    II. Val spent \$34 for the two items.
    III. Neither of the two items that Val bought costs more than \$15.

D $\overline{AB}$ is not congruent to $\overline{MK}$.

Online Lesson Quiz Visit: PHSchool.com, Web Code: bca-0504

Example 5 **20. Developing Proof** Fill in the blanks to prove the following statement.

Using a Variable

You can solve many problems by using a variable to represent an unknown quantity. You use the variable to write an equation or inequality. In the example, Method 1 uses one variable and Method 2 uses two variables.

EXAMPLE

Algebra Points A, B, and C are collinear. BC is 6 less than twice AB, and $AC = 30$. What is the length of $\overline{BC}$?

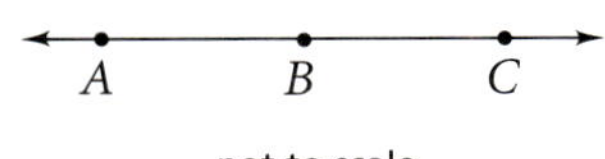

Ⓐ 12 Ⓑ 15 Ⓒ 18 Ⓓ 30

Method 1

Let $AB = x$,
then $BC = 30 - x$. **The length of $\overline{AC}$ is 30.**

BC is 6 less than twice AB,

$$30 - x = 2x - 6$$
$$36 = 3x \quad \textbf{Solve for } x.$$
$$x = 12$$

Therefore $AB = 12$ and $BC = 30 - 12 = 18$. The correct choice is C.

Method 2

Let $AB = x$ and $BC = y$.

Since $AC = AB + BC = 30$, $x + y = 30$.

Since BC is 6 less than twice AB, $y = 2x - 6$.

$$x + (2x - 6) = 30 \quad \textbf{Substitute } 2x - 6 \textbf{ for } y.$$
$$3x = 36 \quad \textbf{Solve for } x.$$
$$x = 12$$

Therefore $AB = 12$ and $BC = 2(12) - 6 = 18$. The correct choice is C.

EXERCISES

1. Points A, B, C, and D are collinear. AB is 15 more than BC, and CD is 8 more than BC. If AD is 5 less than twice AB, what is the length of AB?

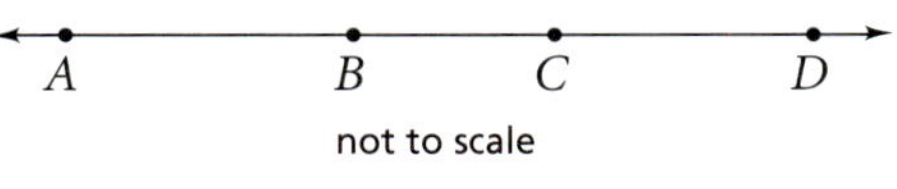

Ⓐ 2 Ⓑ 10 Ⓒ 17 Ⓓ 29

2. Jared wants to draw a triangle with angles that are three consecutive even integers. Which angle measures should he use?

Ⓐ 56, 60, 64 Ⓑ 58, 60, 62 Ⓒ 59, 60, 61 Ⓓ 60, 62, 64

3. What are the coordinates of point P on the x-axis such that the slope of the line through P and $A(1, 4)$ is twice the slope of the line through P and $B(9, 10)$?

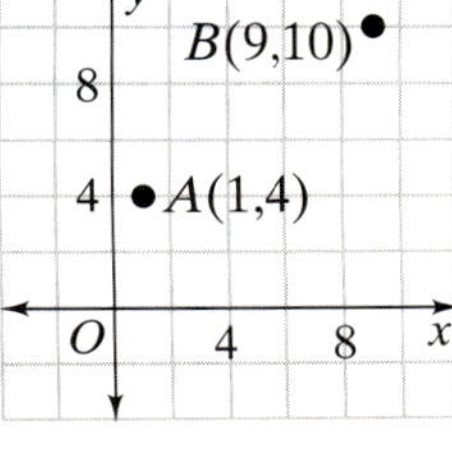

Ⓐ $\left(-\frac{13}{3}, 0\right)$ Ⓒ $(1, 0)$

Ⓑ $(-1, 0)$ Ⓓ $\left(\frac{7}{3}, 0\right)$

Chapter 5 Review

Vocabulary Review

English and Spanish Audio Online

altitude of a triangle (p. 275)
centroid (p. 274)
circumcenter of a triangle (p. 273)
circumscribed about (p. 273)
concurrent (p. 273)
contrapositive (p. 280)
coordinate proof (p. 260)
distance from a point to a line (p. 266)
equivalent statements (p. 281)
incenter of a triangle (p. 273)
indirect proof (p. 281)
indirect reasoning (p. 281)
inscribed in (p. 273)
inverse (p. 280)
median of a triangle (p. 274)
midsegment (p. 259)
negation (p. 280)
orthocenter of a triangle (p. 275)
point of concurrency (p. 273)

Choose the correct vocabulary term to complete each sentence.

1. A *(centroid, median of a triangle)* is a segment whose endpoints are a vertex and the midpoint of the side opposite the vertex.
2. The length of the perpendicular segment from a point to a line is the *(midsegment, distance from the point to the line)*.
3. If T is a point on the perpendicular bisector of $\overline{FG}$, then $TF = TG$ because of the *(Perpendicular Bisector Theorem, Angle Bisector Theorem)*.
4. The *(altitude, median)* of a triangle is a perpendicular segment from a vertex to the line containing the side opposite the vertex.
5. The notation $\sim q \rightarrow \sim p$ is the *(inverse, contrapositive)* of $p \rightarrow q$.
6. To write a(n) *(indirect proof, negation)*, you start by assuming that the opposite of what you want to prove is true.
7. In $\triangle ABC$, $AB + BC > AC$ because of the *(Comparison Property of Inequality, Triangle Inequality Theorem)*.
8. The *(circumcenter, incenter)* of a triangle is the point of concurrency of the angle bisectors of the triangle.
9. The *(Angle Bisector Theorem, Triangle Inequality Theorem)* says that if a point is on the bisector of an angle, then it is equidistant from the sides of the angle.
10. A point where three lines intersect is a *(point of concurrency, incenter)*.

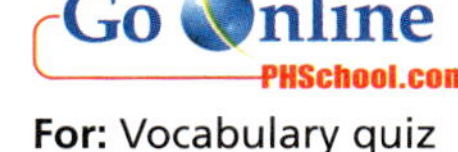

For: Vocabulary quiz
Web Code: bcj-0551

Skills and Concepts

Lessons 5-1 and 5-2

- To use properties of midsegments to solve problems (p. 259)
- To use properties of perpendicular bisectors and angle bisectors (p. 265)

A **midsegment** of a triangle is a segment that connects the midpoints of two sides. A midsegment is parallel to the third side, and is half its length.

In a **coordinate proof,** a figure is drawn on a coordinate plane and formulas are used to prove properties of the figure.

The **distance from a point to a line** is the length of the perpendicular segment from the point to the line. The Perpendicular Bisector Theorem together with its converse states that a point is on the perpendicular bisector of a segment if and only if it is equidistant from the endpoints of the segment. The Angle Bisector Theorem together with its converse states that a point is on the bisector of an angle if and only if it is equidistant from the sides of the angle.

x^2 **Algebra** **Find the value of x.**

11.

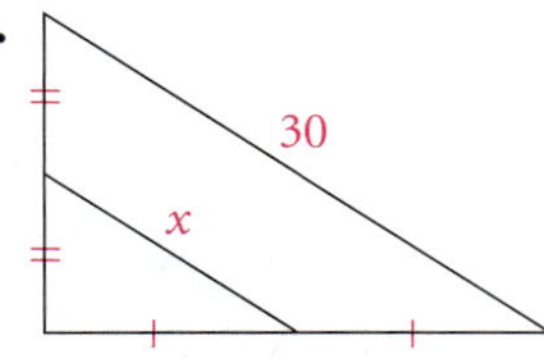

12.

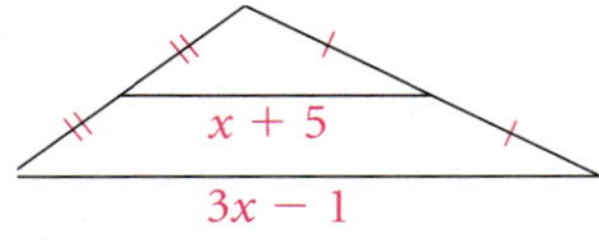

Use the figure to find each segment length or angle measure.

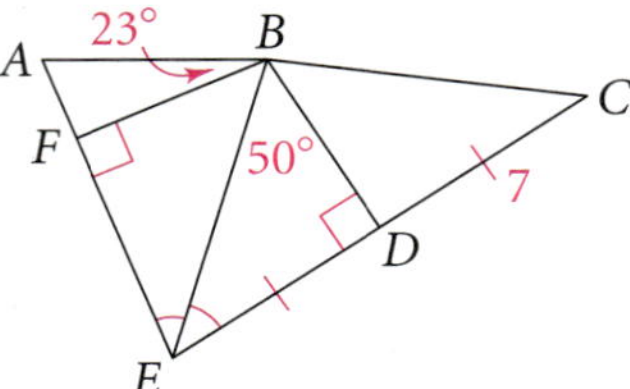

13. $m\angle BEF$ **14.** FE

15. EC **16.** $m\angle CEA$

Lesson 5-3

- To identify properties of perpendicular bisectors and angle bisectors (p. 272)
- To identify properties of medians and altitudes of a triangle (p. 274)

 **GEOM 2.0, 21.0**

When three or more lines intersect in one point, they are **concurrent**.

The **median of a triangle** is a segment whose endpoints are a vertex and the midpoint of the opposite side. The **altitude of a triangle** is a perpendicular segment from a vertex to the line containing the opposite side.

For any given triangle, special segments and lines are concurrent:

- the perpendicular bisectors of the sides at the circumcenter, the center of the circle that can be **circumscribed about** the triangle
- the bisectors of the angles at the incenter, the center of the circle that can be **inscribed in** the triangle
- the medians at the **centroid**
- the lines containing the altitudes at the **orthocenter of the triangle.**

Graph $\triangle ABC$ with vertices $A(2, 3)$, $B(-4, -3)$, and $C(2, -3)$. Find the coordinates of each point of concurrency.

17. circumcenter **18.** centroid **19.** orthocenter

Determine whether $\overline{AB}$ is a perpendicular bisector, an angle bisector, a median, an altitude, or none of these. Explain.

20.

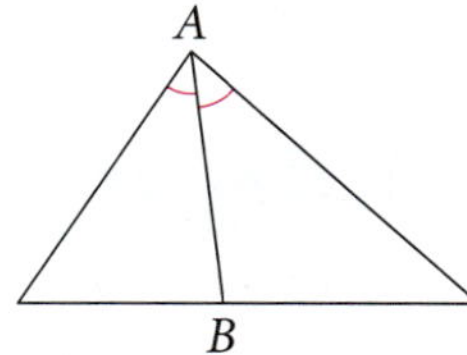

21.

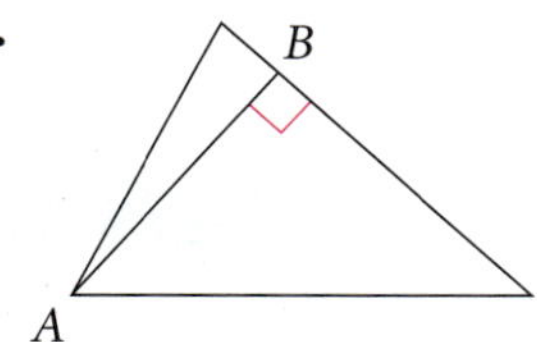

22.

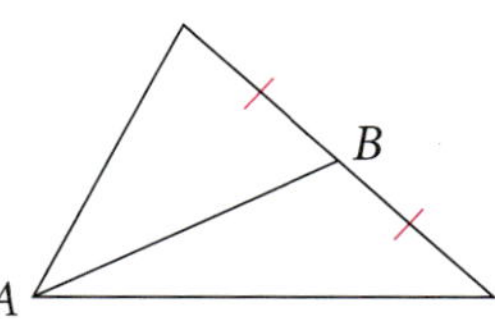

Lesson 5-4

- To write the negation of a statement and the inverse and contrapositive of a conditional statement (p. 280)

The **negation** of a statement has the opposite truth value. The **inverse** of a conditional statement is the negation of both the hypothesis and the conclusion. The **contrapositive** of a conditional statement switches the hypothesis and the conclusion and negates both. Statements that always have the same truth value are **equivalent statements.**

To use **indirect reasoning,** consider all possibilities and then prove all but one false. The remaining possibility must be true.

- To use indirect reasoning (p. 281)

The three steps of an **indirect proof** are:

Step 1 State as an assumption the opposite (negation) of what you want to prove.

Step 2 Show that this assumption leads to a contradiction.

Step 3 Conclude that the assumption must be false and that what you want to prove must be true.

Write the inverse and the contrapositive of each statement.

23. If it is snowing, then it is cold outside.

24. If an angle is obtuse, then its measure is greater than 90 and less than 180.

25. If a figure is a square, then its sides are congruent.

26. If you are in Australia, then you are south of the equator.

Write a convincing argument that uses indirect reasoning.

27. The product of two numbers is even. Show that at least one of the two numbers must be even.

28. Show that a right angle cannot be formed by the intersection of nonperpendicular lines.

29. Show that a triangle can have at most one obtuse angle.

30. Show that an equilateral triangle cannot have an obtuse angle.

Lesson 5-5

- To use inequalities involving angles of triangles (p. 289)
- To use inequalities involving sides of triangles (p. 291)

If two sides of a triangle are not congruent, then the larger angle lies opposite the longer side. The converse is also true. If two angles are not congruent, then the longer side lies opposite the larger angle.

The measure of an exterior angle of a triangle is greater than the measure of each of its remote interior angles. The sum of the lengths of any two sides of a triangle is greater than the length of the third side.

List the angles and sides in order from smallest to largest.

31.

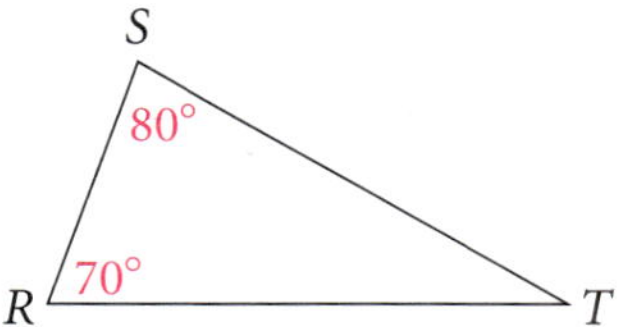

32.

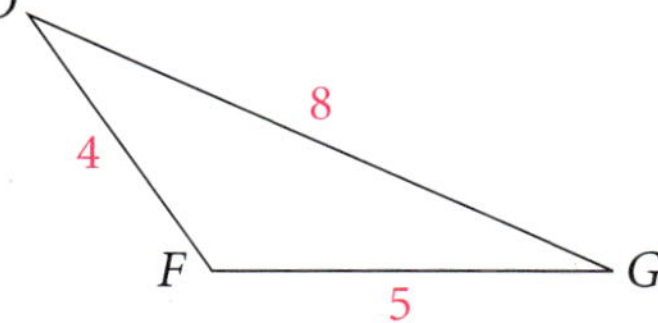

Is it possible for a triangle to have sides with the given lengths? Explain.

33. 5 in., 8 in., 15 in.

34. 10 cm, 12 cm, 20 cm

35. 20 m, 22 m, 24 m

36. 3 ft, 6 ft, 8 ft

37. 1 yd, 1 yd, 3 yd

38. 5 km, 6 km, 7 km

Two side lengths of a triangle are given. Write an inequality to show the range of values, *x*, for the length of the third side.

39. 4 in., 7 in.

40. 8 m, 15 m

41. 2 cm, 8 cm

42. 12 ft, 13 ft

Chapter 5 Test

For: Chapter test
Web Code: bca-0552

Write (a) the inverse and (b) the contrapositive of each statement.

1. If a polygon has eight sides, then it is an octagon.
2. If it is a leap year, then it is an even-numbered year.
3. If it is snowing, then it is not summer.
4. What can you conclude from the diagram? Justify your answer.

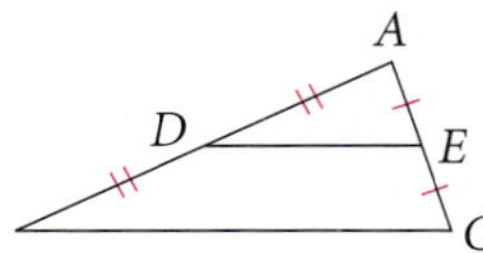

Identify the two statements that contradict each other.

5. I. $\triangle PQR$ is a right triangle.
   II. $\triangle PQR$ is an obtuse triangle.
   III. $\triangle PQR$ is scalene.
6. I. $\angle DAS \cong \angle CAT$
   II. $\angle DAS$ and $\angle CAT$ are vertical.
   III. $\angle DAS$ and $\angle CAT$ are adjacent.

List the angles of $\triangle ABC$ from smallest to largest.

7. $AB = 9, BC = 4, AC = 12$
8. $AB = 10, BC = 11, AC = 9$
9. $AB = 3, BC = 9, AC = 7$
10. **Open-Ended** Write three lengths that cannot be the lengths of the three sides of a triangle. Explain your answer.

List the sides of each triangle in order from shortest to longest.

11.

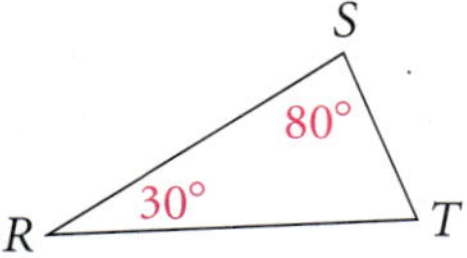

12.

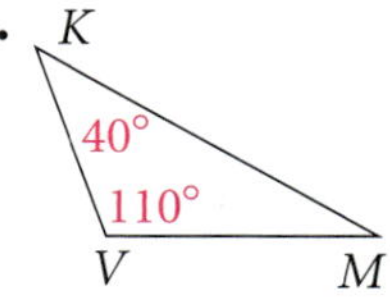

Algebra **Find the value of x in each figure.**

13.

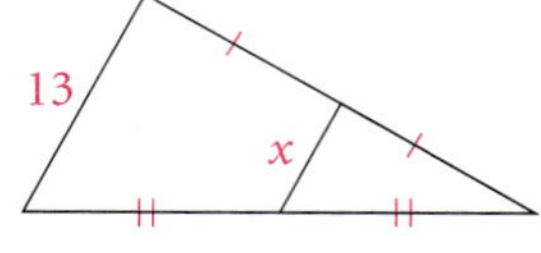

14.

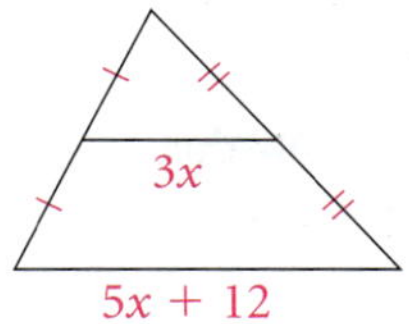

15. **Writing** Use indirect reasoning to explain why the following statement is true: If an isosceles triangle is obtuse, then the obtuse angle is the vertex angle.

Algebra **Find the values of x and KM in each figure.**

16.

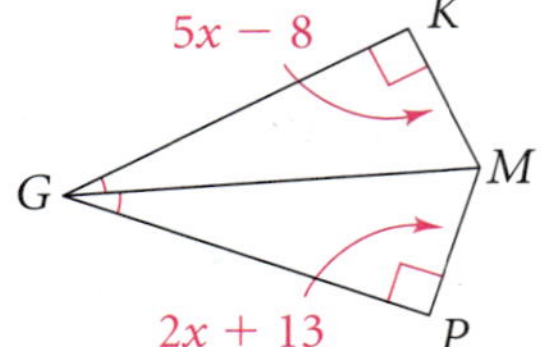

17.

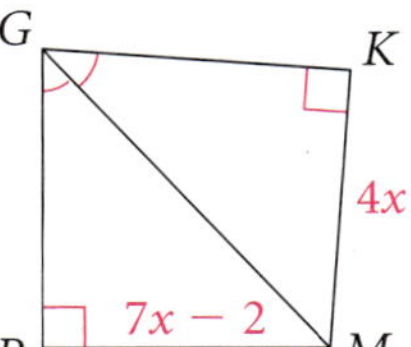

Coordinate Geometry **Find the center of the circle that circumscribes $\triangle ABC$.**

18. $A(0, 0), B(0, -3), C(-5, -3)$
19. $A(0, 5), B(-4, 5), C(-4, -3)$
20. $A(3, -1), B(-2, -1), C(3, -8)$
21. $\triangle ABC$ has vertices $A(2, 5), B(2, -3), C(10, -3)$. What point of concurrency is at $(6, 1)$?
22. **Given:** $\overleftrightarrow{PQ}$ is the perpendicular bisector of $\overline{AB}$. $\overleftrightarrow{QT}$ is the perpendicular bisector of $\overline{AC}$.
    Prove: $QC = QB$

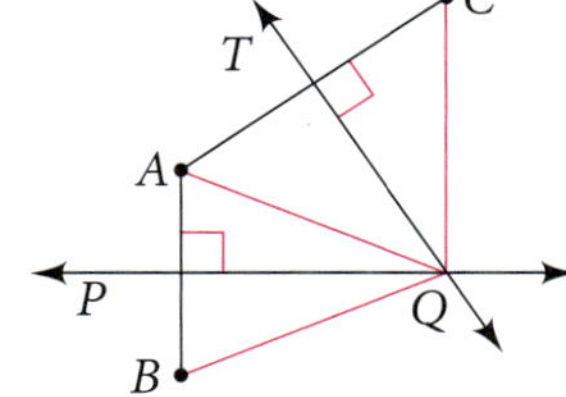

23. What can you conclude about point Y? Explain.

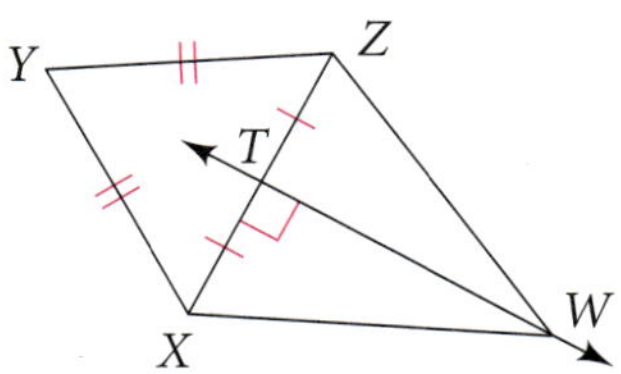

24. In the figure, $WK = KR$. What can you conclude about point A? Explain.

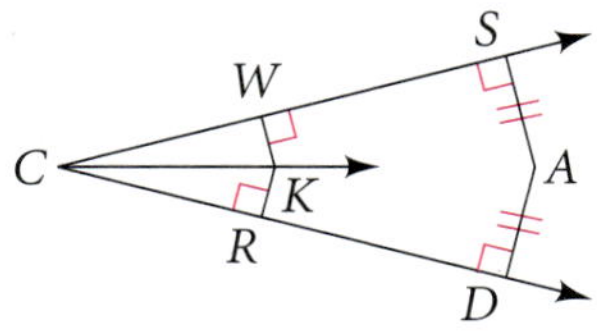

Standards Mastery

CAHSEE Practice

Some questions ask you to choose an equation for a situation. Read the question at the right. Then follow the tips to answer it.

A bus leaves downtown and averages 45 miles per hour. Three hours later, a second bus starts on the same route and averages 65 miles per hour. If t represents the number of hours that the slower bus travels before it is passed by the faster bus, which equation can you use to find t?

(A) $45t = 65(t - 3)$ (C) $45t = 65(t + 3)$
(B) $45(t - 3) = 65t$ (D) $45(3t) = 65$

Tip 1
Recognize this as a distance-rate-time problem. Use $d = rt$.

Tip 2
Recognize the key relationship. When the fast bus overtakes the slow bus, their distances traveled will be equal.

Think It Through

Make a distance-rate-time chart.

	d	r	t
Slower bus	equal	45	t
Faster bus		65	t−3

$$45t = 65(t - 3)$$

The correct choice is A.

Vocabulary Review

As you solve problems, you must understand the meanings of mathematical terms. Choose the correct term to complete each sentence.

I. A (reflection, translation) is a transformation that maps all points the same distance and direction.

II. A transformation that enlarges or shrinks a figure is a (dilation, translation).

III. A (price, rate) is a ratio that compares two quantities that have different units of measure.

IV. In the coordinate plane, a nonvertical line has one point on the y-axis. You can find the (y-intercept, slope) of the line by finding the ratio of vertical change to horizontal change when you move from the point on the y-axis to any other point on the line.

Multiple Choice

Read each question. Then write the letter of the correct answer on your paper.

1. The graph at the right shows the miles a truck traveled over several hours. How many miles per hour did the truck average from the first to the fourth hour?

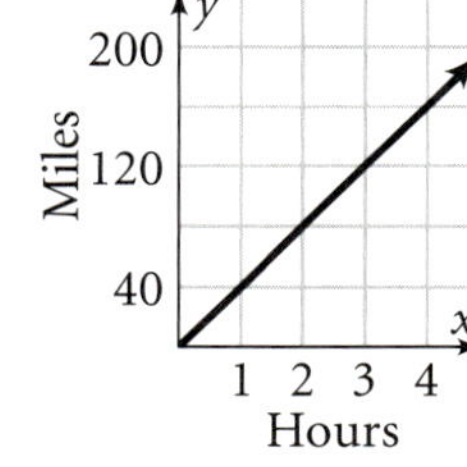

(A) 10 (B) 20 (C) 40 (D) 80

2. The table shows the temperature at several different times last Tuesday afternoon. What was the rate of change of the temperature in degrees per hour?

Time	°F
3 P.M.	45
4 P.M.	48
5 P.M.	51
6 P.M.	54

(A) $\frac{1}{3}$ (B) $-\frac{1}{3}$ 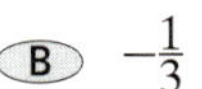 (C) 3 (D) −3

3. Elise places numbered labels on documents at a local business. If she can label about 36 documents a minute, about how many will she label in 25 minutes?

(A) 360 (B) 900 (C) 1260 (D) 2160

4. Sam and Marie sell honey. To determine how much to charge, they compared the following prices per quart from other honeybee farms. The prices were \$2.35, \$2.35, \$3.50, \$1.80, \$2.75, \$3.00, \$2.35, and \$2.80. Which statistical measure gives the highest price?

(A) Mean
(B) Mode
(C) Median
(D) Range

5. What is the slope of a line parallel to the line below?

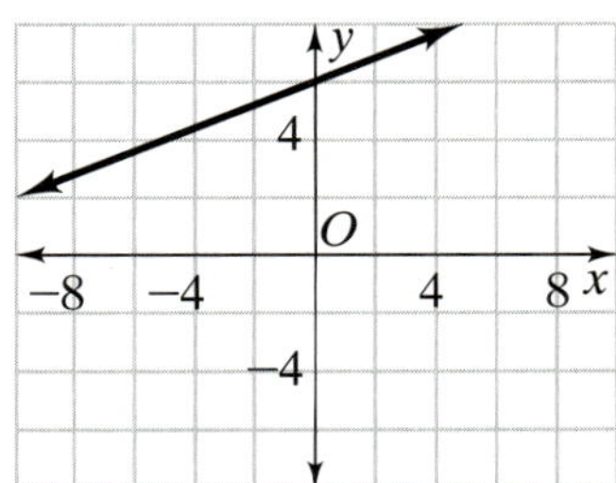

Ⓐ $-\frac{4}{5}$ Ⓒ $\frac{2}{5}$

Ⓑ $-\frac{2}{5}$ Ⓓ $\frac{4}{5}$

6. The algebraic form of a certain linear function is $y = \frac{1}{2}x + 2$. Which of the following represents the same function?

Ⓐ
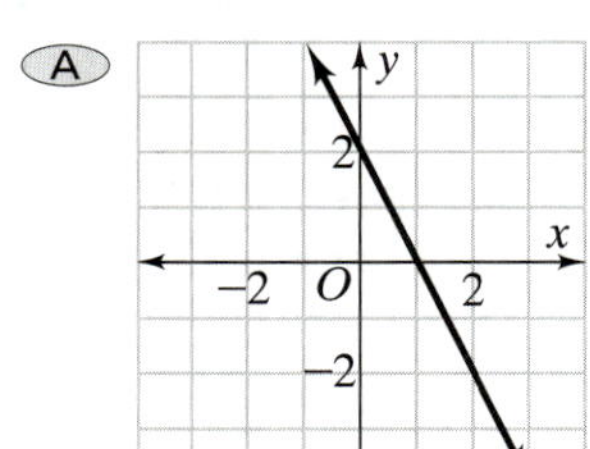

Ⓒ
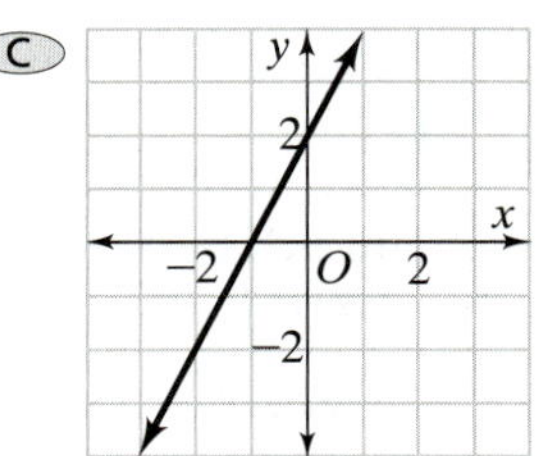

Ⓑ
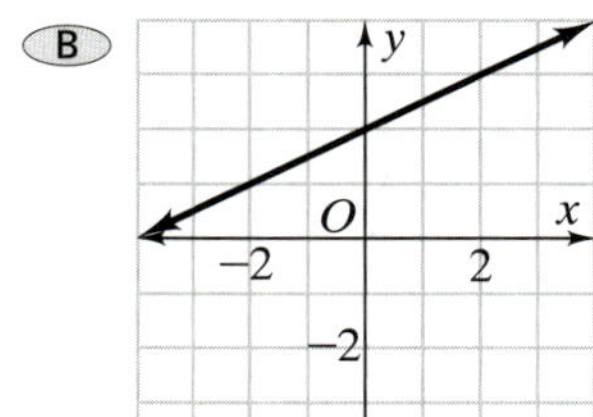

Ⓓ
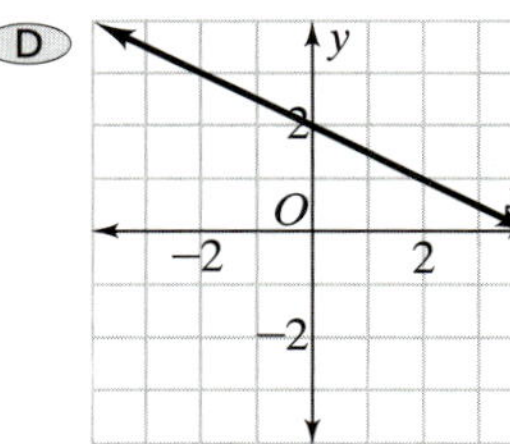

7. Which equation could represent the graph?

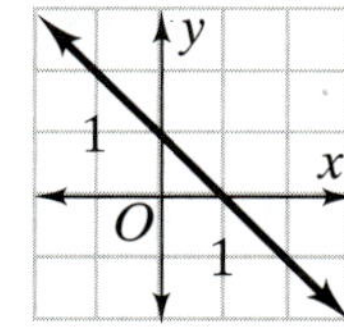

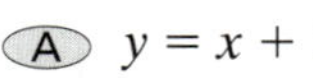

Ⓐ $y = x + 1$

Ⓑ $y = x - 1$

Ⓒ $y = -x + 1$

Ⓓ $y = -x - 1$

8. Which equation has a graph that is parallel to the graph of $y = -2x - 8$?

Ⓐ $y = 2x + 8$ Ⓒ $y = \frac{1}{2}x - 8$

Ⓑ $y = -2x + 8$ Ⓓ $y = \frac{1}{2}x + 8$

9. The table below shows the value in the ones place for different powers of 7. For example, $7^2 = 49$ so the value in the ones place is 9.

Power of 7	Number in the Ones Place
7^1	7
7^2	9
7^3	3
7^4	1
7^5	7
7^6	9

Which digit is in the ones place for 7^{31}?

Ⓐ 1 Ⓑ 3 Ⓒ 7 Ⓓ 9

10. Which statement best describes the pattern of the terms in this sequence?

$$\frac{2}{5}, \frac{6}{25}, \frac{18}{125}, \frac{54}{625}, \ldots$$

Ⓐ Each term is $\frac{2}{5}$ of the previous term.

Ⓑ The numerator is twice a power of 3 and the denominator is a factor of 125.

Ⓒ Each term is $\frac{3}{5}$ of the previous term.

Ⓓ the numerator is a multiple of 3 and the denominator is a power of 5.

11. The table at the right shows the number of minutes Sunee spent watching television each day last week and the number of minutes of commercials she saw. Which equation best describes the relationship between t, the amount of time Sunee watched television, and c, the number of minutes of commercials she saw?

t	c
39	13
42	14
72	24
33	11
63	21
135	45
87	29

Ⓐ $c = t - 26$

Ⓑ $c = 3t$

Ⓒ $t = 3c$

Ⓓ $t = 39c$

12. A bag contains 8 blue marbles, 6 yellow marbles, 4 green marbles, and 9 black marbles. Ginny selects a marble without looking. What is the probability that it is NOT black?

(A) 9% (B) $33\frac{1}{3}\%$ (C) $66\frac{2}{3}\%$ (D) 91%

13. For the data 93, 70, 65, 59, 74, 70, 61, 83, 91, which statement is true?

(A) The range is 43.
(B) Mean, median, and mode are all 70.
(C) The mode and the median are not the same.
(D) The mean is 74.

14. Tina tosses a fair coin 9 times. It lands on heads 4 times and tails 5 times. What is the theoretical probability that it lands on tails on the tenth toss?

(A) 100% (B) 50% (C) 55.6% (D) 25%

15. Each day, Michael goes for a run through the park. The park is shaped like a square. He follows the path shown by the arrows. To the nearest tenth of a meter, how far will Michael run?

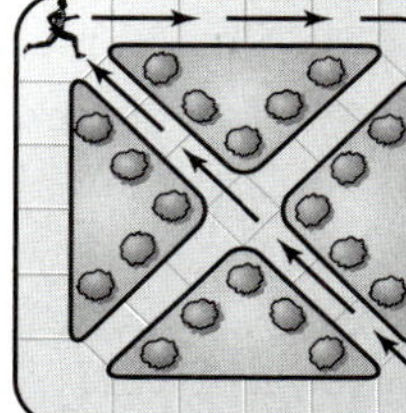

(A) 282.8 (C) 682.8
(B) 600 (D) 800

16. $\triangle TUV$ and $\triangle T'U'V'$ are shown below.

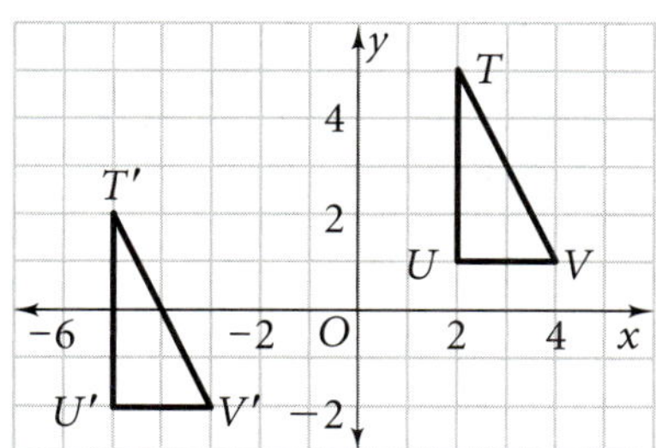

Which of the following best describes the translation from $\triangle TUV$ to $\triangle T'U'V'$?

(A) 3 units down and 5 units to the right
(B) 3 units down and 7 units to the left
(C) 7 units down and 5 units to the right
(D) 7 units down and 7 units to the left

17. What is the area of the circle located on the coordinate grid in terms of π?

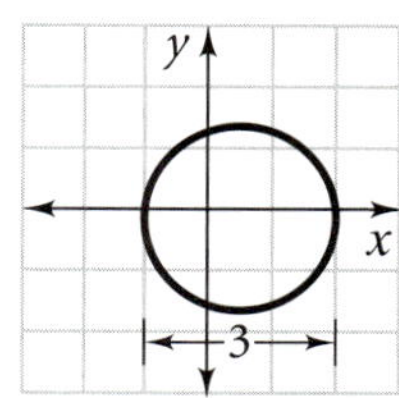

(A) 2.25π
(B) 3π
(C) 6π
(D) 9π

CAHSEE Reference Guide

California Content Standard	Item Number(s)
7AF 1.1	11
7AF 3.4	1
7MG 1.3	2, 3
7SDAP 1.1	4, 13
Alg1 8.0	5, 8
Alg1 6.0	6, 7
7MR 1.1	9, 10
6SDAP 3.3	12, 14
7MG 3.3	15
7MG 3.2	16, 17

For additional review and practice, see the *California Standards Review and Practice Workbook* or go online to use the

Visit: PHSchool.com, **Web Code:** bcq-9045

Quadrilaterals

What You've Learned

California Content Standards

2.0 Students write geometric proofs, including proofs by contradiction.

4.0 Students prove basic theorems involving congruence.

5.0 Students prove that triangles are congruent, and they are able to use the concept of corresponding parts of congruent triangles.

6.0 Students know and are able to use the triangle inequality theorem.

Check Your Readiness

GO for Help to the Lesson in green.

Properties of Parallel Lines (Lesson 3-1)

Algebra Use properties of parallel lines to find the value of x.

1. 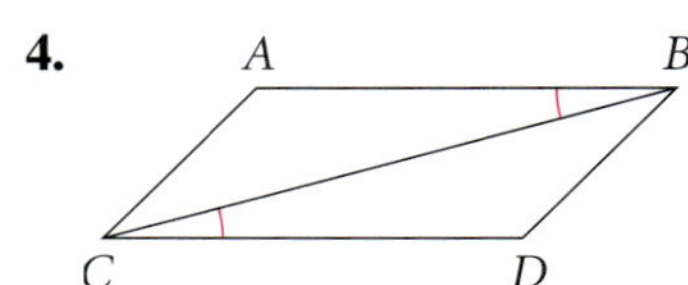

2. 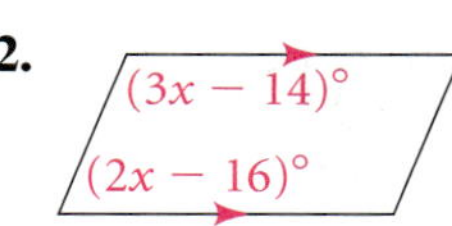

3. 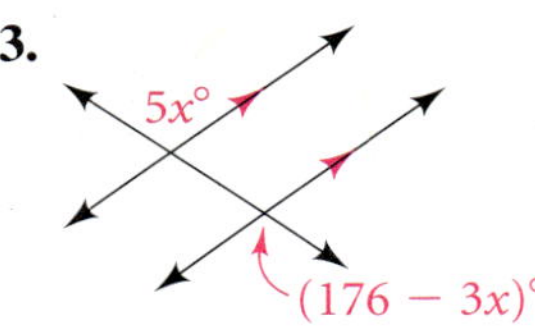

Proving Lines Parallel (Lesson 3-2)

Algebra Determine whether $\overleftrightarrow{AB}$ (or $\overline{AB}$) is parallel to $\overleftrightarrow{CD}$ (or $\overline{CD}$).

4. 

5. 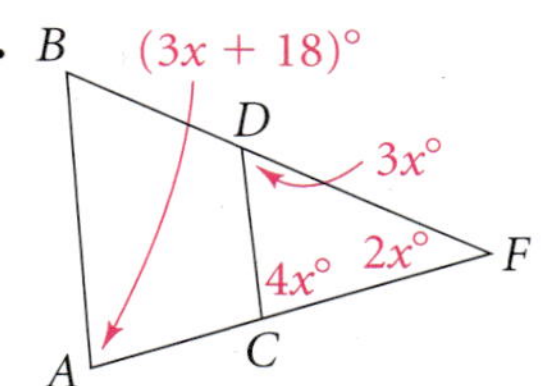

6. 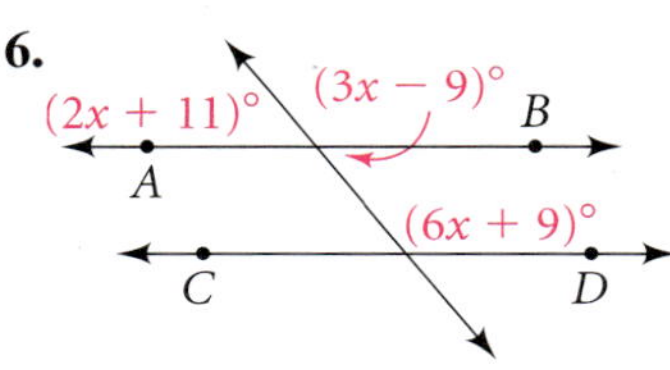

Using Slope to Determine Parallel and Perpendicular Lines (Lesson 3-7)

Algebra Determine whether each pair of lines is parallel, perpendicular, or neither.

7. $y = -2x; y = -2x + 4$

8. $y = -\frac{3}{5}x + 1; y = \frac{5}{3}x - 3$

9. $2x - 3y = 1; 3x - 2y = 8$

Proving Triangles Congruent (Lessons 4-2 and 4-3)

Determine the postulate or theorem that makes each pair of triangles congruent.

10. 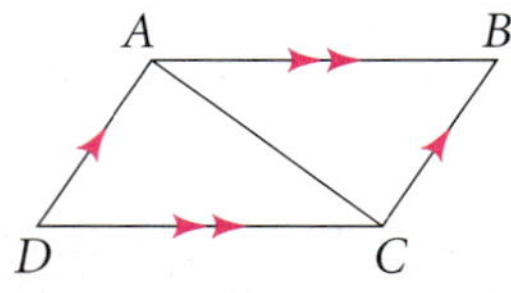

11. 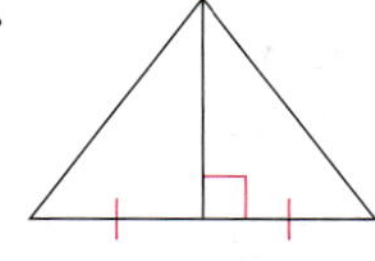

12. 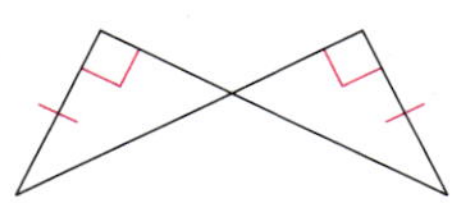

▲ Aerial photographs reveal quadrilateral shapes that are created by farming.

What You'll Learn Next

California Content Standards

7.0 Students prove and use theorems involving the properties of quadrilaterals.

12.0 Students find and use measures of sides and of interior angles of polygons to classify figures and solve problems.

13.0 Students prove relationships between angles in polygons by using properties of complementary and supplementary angles.

17.0 Students prove theorems by using coordinate geometry.

CAHSEE 7NS 1.2, 7NS 1.6, 7NS 2.2, 7NS 2.3, 6SDAP 3.1, 7MG 3.4, 7MR 1.2, Alg1 5.0

New Vocabulary

English and Spanish Audio Online

- **base angles of a trapezoid** (p. 336)
- **consecutive angles** (p. 312)
- **isosceles trapezoid** (p. 306)
- **kite** (p. 306)
- **midsegment of a trapezoid** (p. 348)
- **parallelogram** (p. 306)
- **rectangle** (p. 306)
- **rhombus** (p. 306)
- **square** (p. 306)
- **trapezoid** (p. 306)

6-1 Classifying Quadrilaterals

California Content Standards

GEOM 12.0 Students find and use measures of sides and of interior angles of polygons to classify figures and solve problems. ***Develop***

What You'll Learn

- To define and classify special types of quadrilaterals

. . . And Why

To use the properties of special quadrilaterals with a kite, as in Example 3

Check Skills You'll Need

GO for Help Lesson 1-8 and page 165

Find the distance between the points to the nearest tenth.

1. $M(2, -5), N(-7, 1)$ **2.** $P(-1, -3), Q(-6, -9)$ **3.** $C(-4, 6), D(5, -3)$

Find the slope of the line through each pair of points.

4. $X(0, 6), Y(4, 9)$ **5.** $R(3, 8), S(6, 0)$ **6.** $A(4, 3), B(2, 1)$

New Vocabulary
- parallelogram
- rhombus
- rectangle
- square
- kite
- trapezoid
- isosceles trapezoid

1 Classifying Special Quadrilaterals

Seven important types of quadrilaterals are defined below.

Key Concepts

Definitions **Special Quadrilaterals**

A **parallelogram** is a quadrilateral with both pairs of opposite sides parallel.

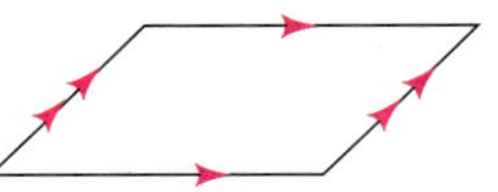

A **rhombus** is a parallelogram with four congruent sides.

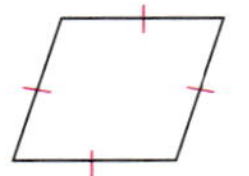

A **rectangle** is a parallelogram with four right angles.

A **square** is a parallelogram with four congruent sides and four right angles.

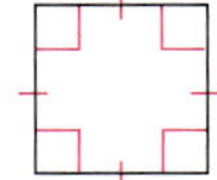

A **kite** is a quadrilateral with two pairs of adjacent sides congruent and no opposite sides congruent.

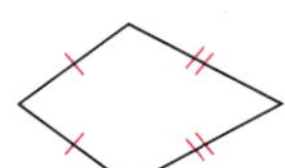

A **trapezoid** is a quadrilateral with exactly one pair of parallel sides. The **isosceles trapezoid** at the right is a trapezoid whose nonparallel opposite sides are congruent.

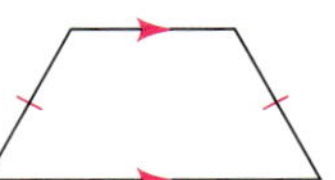

A "kite" is not the only special quadrilateral used to make a kite!

1 EXAMPLE Classifying a Quadrilateral

Judging by appearance, classify $DEFG$ in as many ways as possible.

$DEFG$ is a quadrilateral because it has four sides.

It is a parallelogram because both pairs of opposite sides are parallel.

It is a rectangle because it has four right angles.

CA Standards Check 1

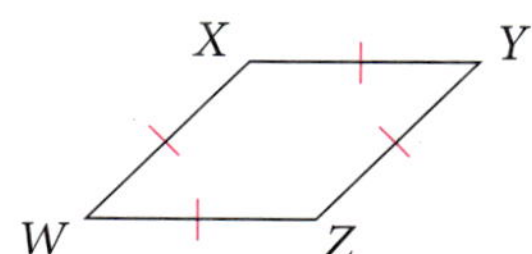

a. Judging by appearance, classify $WXYZ$ at the right in as many ways as possible.

b. Critical Thinking Which name gives the most information about $WXYZ$? Explain.

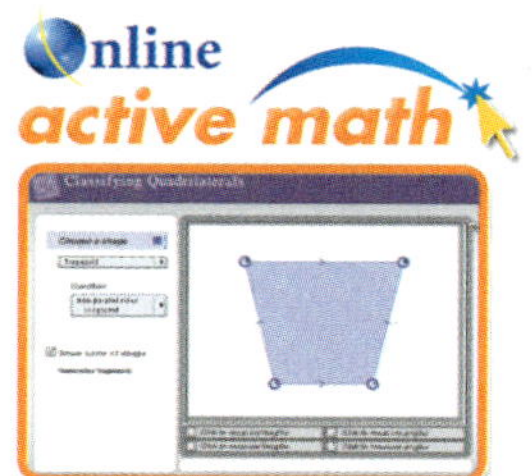

For: Quadrilateral Activity
Use: Interactive Textbook, 6-1

The diagram below shows the relationships among special quadrilaterals.

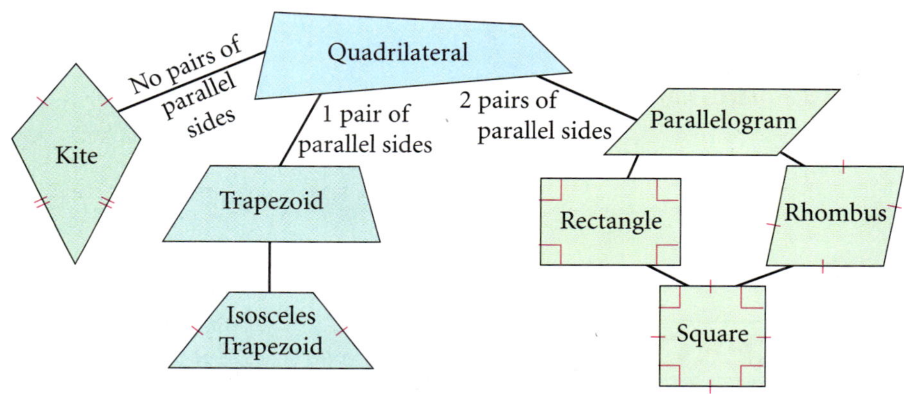

You can use what you know about slope and distance to classify a quadrilateral.

2 EXAMPLE Classifying by Coordinate Methods

Coordinate Geometry Determine the most precise name for quadrilateral $LMNP$.

Step 1 Find the slope of each side.

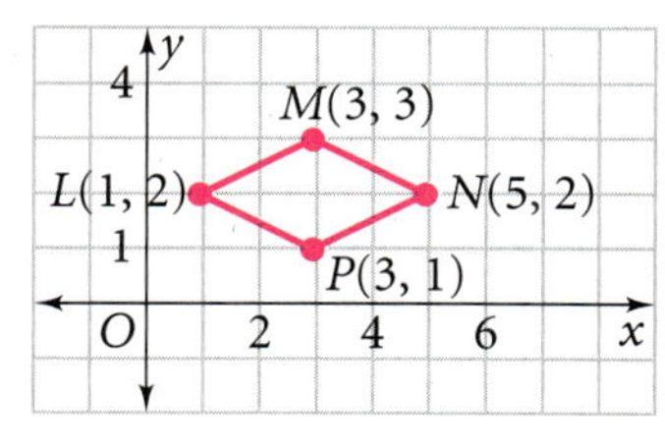

$$\text{slope of } \overline{LM} = \frac{3-2}{3-1} = \frac{1}{2}$$

$$\text{slope of } \overline{NP} = \frac{2-1}{5-3} = \frac{1}{2}$$

$$\text{slope of } \overline{MN} = \frac{3-2}{3-5} = -\frac{1}{2}$$

$$\text{slope of } \overline{LP} = \frac{2-1}{1-3} = -\frac{1}{2}$$

Both pairs of opposite sides are parallel, so $LMNP$ is a parallelogram. No sides are perpendicular, so $LMNP$ is not a rectangle.

Step 2 Use the Distance Formula to see if any pairs of sides are congruent.

$$LM = \sqrt{(3-1)^2 + (3-2)^2} = \sqrt{5} \qquad MN = \sqrt{(3-5)^2 + (3-2)^2} = \sqrt{5}$$

$$NP = \sqrt{(5-3)^2 + (2-1)^2} = \sqrt{5} \qquad LP = \sqrt{(1-3)^2 + (2-1)^2} = \sqrt{5}$$

All sides are congruent, so $LMNP$ is a rhombus.

Although $LMNP$ is a parallelogram, rhombus is the more *precise* name because it gives more information about the quadrilateral.

CA Standards Check 2 Determine the most precise name for quadrilateral $ABCD$ with vertices $A(-3, 3)$, $B(2, 4)$, $C(3, -1)$, and $D(-2, -2)$.

You can use the definitions of special quadrilaterals and algebra to find lengths of sides.

3 EXAMPLE Using the Properties of Special Quadrilaterals

Algebra Find the values of the variables for the kite.

$KB = JB$	**Definition of kite**
$3x - 5 = 2x + 4$	**Substitute.**
$x - 5 = 4$	**Subtract 2*x* from each side.**
$x = 9$	**Add 5 to each side.**
$KT = x + 6 = 15$	**Substitute 9 for *x*.**
$KT = JT$	**Definition of kite**
$15 = 2y + 5$	**Substitute.**
$10 = 2y$	**Subtract 5 from each side.**
$5 = y$	**Divide each side by 2.**

3 Find the values of the variables for the rhombus. Then find the lengths of the sides.

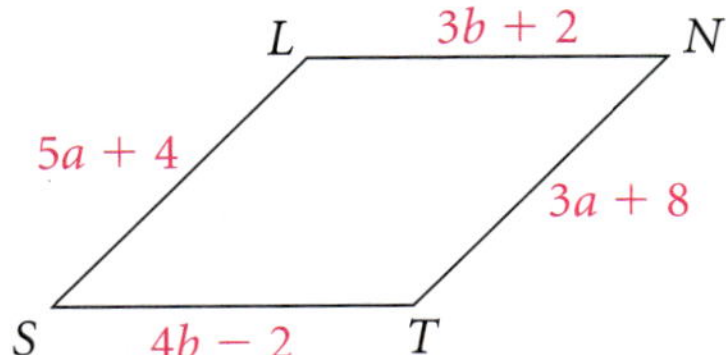

EXERCISES

Standards Practice

For more exercises, see *Extra Skill, Word Problem, and Proof Practice.*

GEOM 12.0

Example 1 (page 307)

These quadrilaterals are made from a toy building set. Judging by appearance, classify each quadrilateral in as many ways as possible.

1. 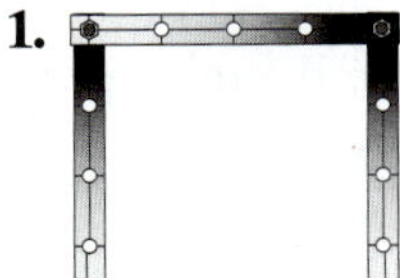

2. 

3. 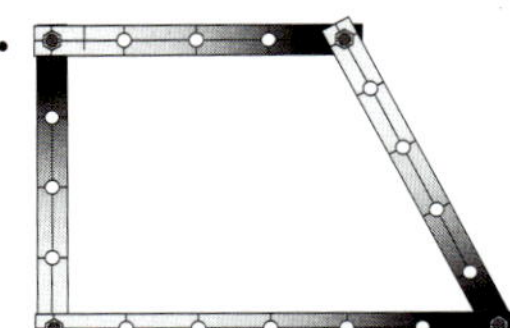

4. 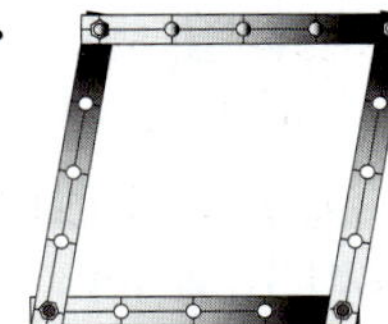

5. 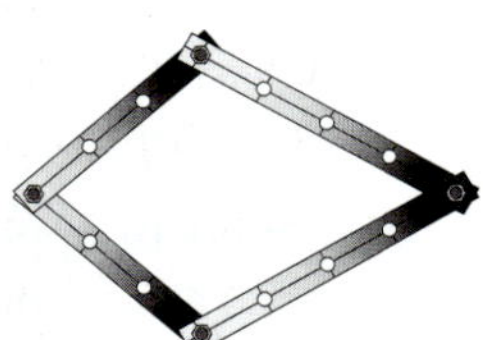

6. 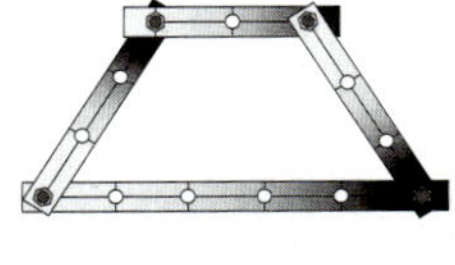

Determine the most precise name for each quadrilateral.

7. 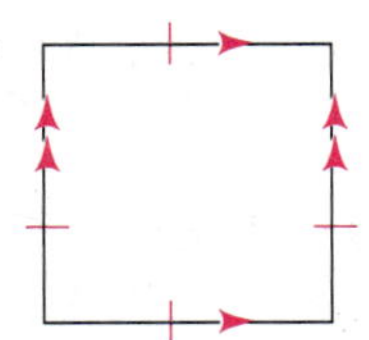

8. 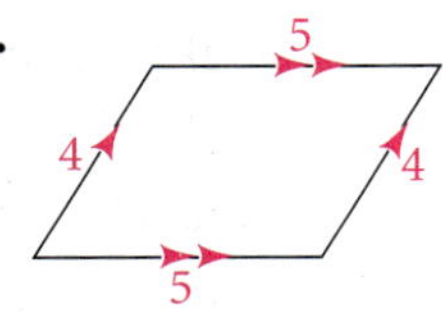

9. 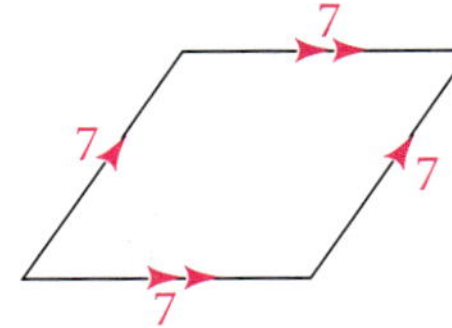

10. 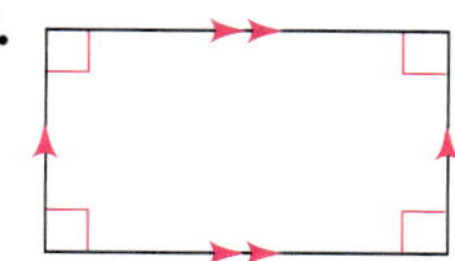

11. 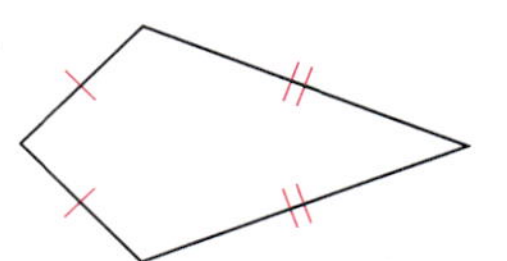

12. 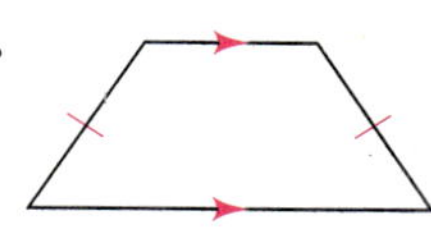

Example 2 (page 307)

Coordinate Geometry **Graph and label each quadrilateral with the given vertices. Then determine the most precise name for each quadrilateral.**

13. $A(3, 5), B(7, 6), C(6, 2), D(2, 1)$

14. $W(-1, 1), X(0, 2), Y(1, 1), Z(0, -2)$

15. $J(2, 1), K(5, 4), L(7, 2), M(2, -3)$

16. $R(-2, -3), S(4, 0), T(3, 2), V(-3, -1)$

17. $N(-6, -4), P(-3, 1), Q(0, 2), R(-3, 5)$

18. $E(-3, 1), F(-7, -3), G(6, -3), H(2, 1)$

Example 3 (page 308)

Algebra **Find the values of the variables. Then find the lengths of the sides.**

19. kite

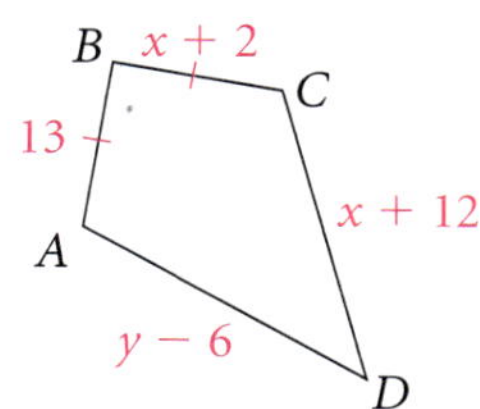

20. kite

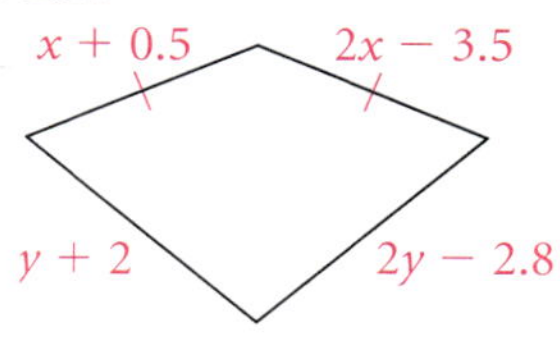

21. kite

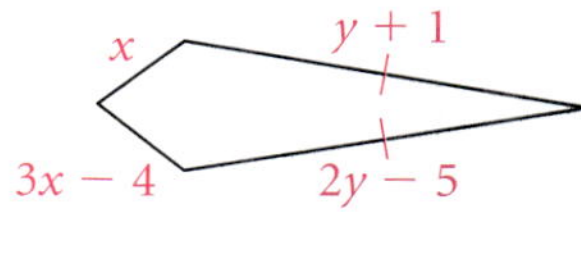

22. isosceles trapezoid

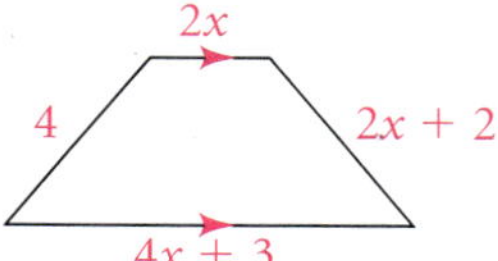

23. rhombus

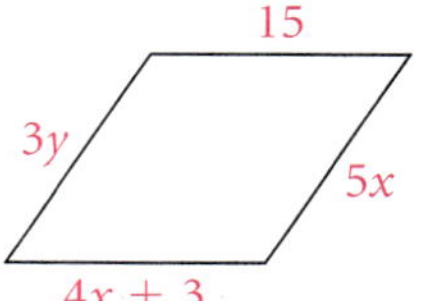

24. square

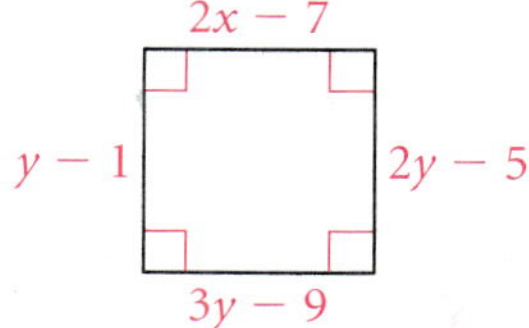

B Apply Your Skills

Algebra **In each figure, find the measures of the angles and the lengths of the sides.**

25. isosceles trapezoid $DEFG$

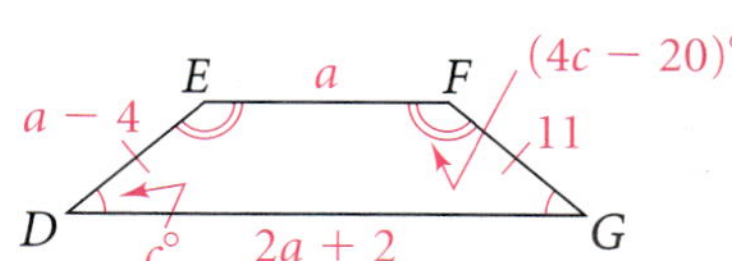

26. rhombus $HKJI$

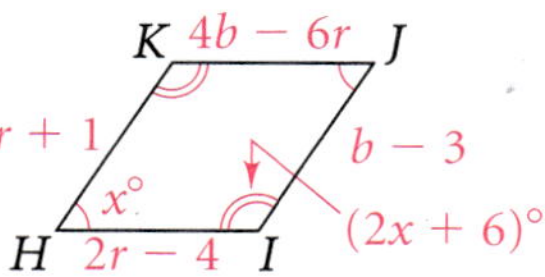

Exercise 27

27. **Art** American artist Charles Demuth created *My Egypt*, the oil painting pictured at the left. It is in an art style called Cubism, in which subjects are made of cubes and other geometric forms. Identify the types of special quadrilaterals you see in the painting.

28. **Multiple Choice** $K(-3, 0)$, $I(0, 2)$, and $T(3, 0)$ are the vertices of a kite. Which point could be the fourth vertex?

Ⓐ $E(0, 2)$ Ⓑ $E(0, 0)$ Ⓒ $E(0, -2)$ Ⓓ $E(0, -10)$

Draw each figure on graph paper. If not possible, explain.

29. a parallelogram that is neither a rectangle nor a rhombus

30. an isosceles trapezoid with vertical and horizontal congruent sides

31. a trapezoid with only one right angle

32. a trapezoid with two right angles

33. a rhombus that is not a square

34. a kite with two right angles

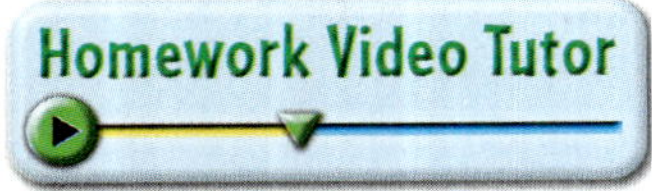

Visit: PHSchool.com
Web Code: bce-0601

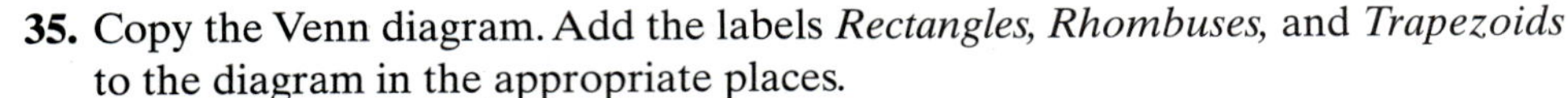

35. Copy the Venn diagram. Add the labels *Rectangles, Rhombuses,* and *Trapezoids* to the diagram in the appropriate places.

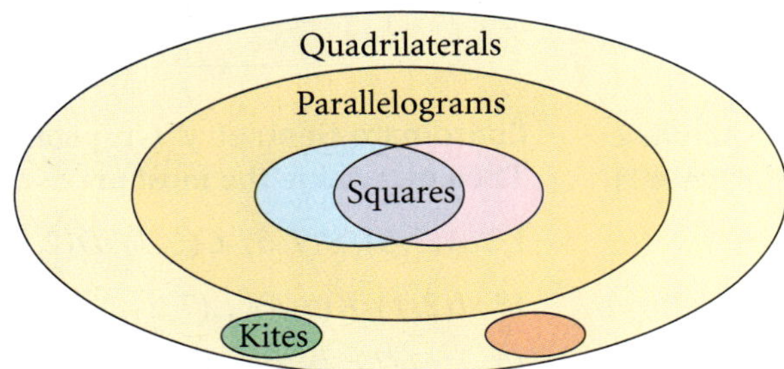

State whether each statement is *true* or *false.* Justify your response. You may find the diagram from Exercise 35 helpful.

36. All squares are rectangles.

37. A trapezoid is a parallelogram.

38. A rhombus can be a kite.

39. Some parallelograms are squares.

40. Every quadrilateral is a parallelogram.

41. All rhombuses are squares.

Real-World Connection

To make this butterfly, a square was folded first on its diagonals.

42. Paper Folding Fold a nonsquare, rectangular piece of paper in half horizontally and then vertically, as shown at the right. Draw and then cut along the line connecting the two opposite corners containing a fold. What quadrilateral do you find when you unfold the paper? Why doesn't it matter what size rectangle you start with?

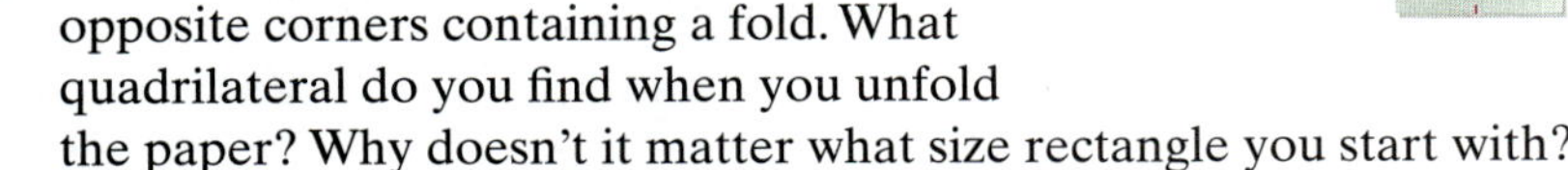

43. Identify a parallelogram, rhombus, rectangle, square, kite, and trapezoid in your classroom. State whether your trapezoid is isosceles.

44. Writing Describe the difference between a rhombus and a kite.

Name each type of special quadrilateral that can meet the given condition. Make sketches to support your answers.

45. exactly one pair of congruent sides

46. two pairs of parallel sides

47. four right angles

48. adjacent sides that are congruent

49. Error Analysis Lauren argues, "A parallelogram has two pairs of parallel sides, so it certainly has one pair of parallel sides. Therefore a parallelogram must also be a trapezoid." What is the error in Lauren's argument?

Name the type(s) of special quadrilateral(s) it appears that you can form by joining the triangles in each pair. Make sketches to support your answers.

Sample two congruent scalene triangles

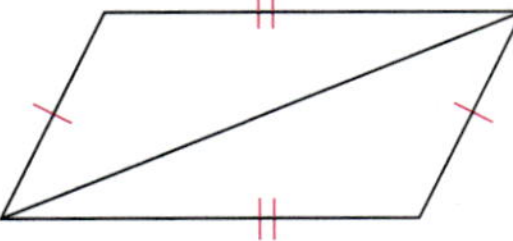

parallelogram

Problem Solving Hint

In Exercises 50–53, if you flip one of the two triangles, you may find different quadrilaterals.

50. two congruent scalene right triangles

51. two congruent equilateral triangles

52. two congruent isosceles right triangles

53. two congruent isosceles acute triangles

Identify a parallelogram, rhombus, rectangle, square, kite and trapezoid at each site. State whether your trapezoid is isosceles.

54. home

55. somewhere other than school and home

Reasoning **A scrap of paper covers part of each quadrilateral. Name all the special quadrilaterals that each could be. Explain each choice.**

56. 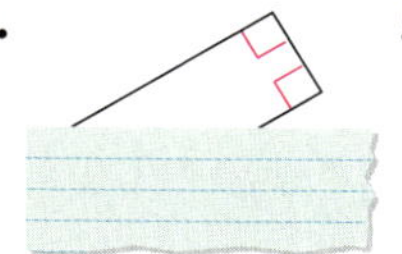

57. 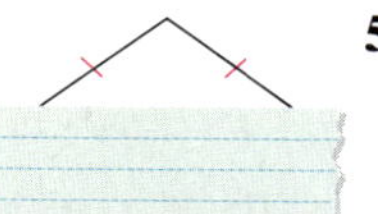

58. 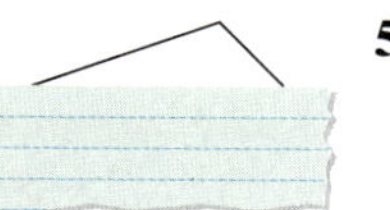

59. 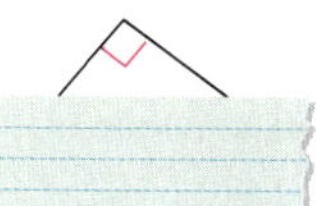

Multiple Choice Practice

For California Standards and CAHSEE Tutorials, visit PHSchool.com. Web Code: bcq-9045

GEOM 12.0

60. What is the most precise name for quadrilateral *EFGH*?

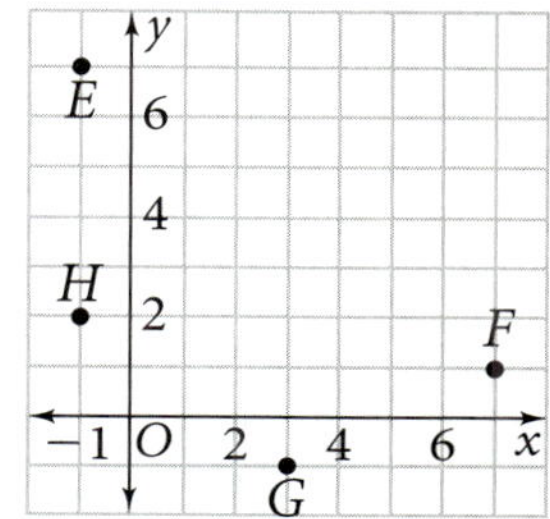

- (A) parallelogram
- (B) kite
- (C) rectangle
- (D) trapezoid

CAHSEE 7MR 1.2

61. The area of a rectangular wall is given by the function $A = bh$. Which statement is true?

- (A) The base of the rectangle depends on the product of the area and height.
- (B) The base of the rectangle depends on the height divided by the area.
- (C) The area of the rectangle depends only on the base.
- (D) The area of the rectangle depends on the product of the base and height.

CAHSEE 7NS 2.2

62. What fraction is equivalent to $\frac{3}{4} - \frac{1}{3}$?

(A) $\frac{4}{7}$ (B) $\frac{5}{12}$ (C) $\frac{2}{1}$ (D) $\frac{1}{4}$

Mixed Review

Lesson 5-5

Can a triangle have sides with the given lengths? Explain.

63. 8 mm, 6 mm, 3 mm

64. 5 ft, 20 ft, 7 ft

65. 3 m, 5 m, 8 m

Lesson 4-1

Quadrilaterals *RSTV* and *NMQP* are congruent. Find the length of the side or the measure of the angle.

66. $\overline{MN}$

67. $\overline{VT}$

68. $\overline{ST}$

69. $\angle S$

70. $\angle V$

71. $\angle R$

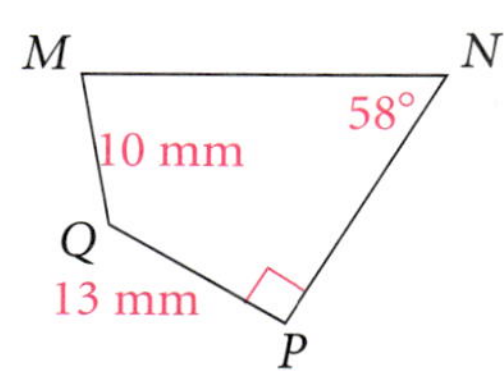

Lesson 3-7

72. Write an equation for the line parallel to $y = -3x - 5$ that contains point $(0, 4)$.

6-2 Properties of Parallelograms

California Content Standards

GEOM 7.0 Students prove and use theorems involving the properties of parallel lines cut by a transversal, and the properties of quadrilaterals. ***Introduce***

GEOM 13.0 Students prove relationships between angles in polygons by using properties of supplementary angles. ***Develop***

What You'll Learn

- To use relationships among sides and among angles of parallelograms
- To use relationships involving diagonals of parallelograms or transversals

. . . And Why

To divide a card into three parts of equal heights without a ruler, as in Example 4

Check Skills You'll Need

GO for Help Lessons 4-1 and 4-3

Use the figure at the right.

1. Name the postulate or theorem that justifies the congruence $\triangle EFG \cong \triangle GHE$.

2. Complete each statement.

a. $\angle FEG \cong$ ___?___ **b.** $\angle EFG \cong$ ___?___

c. $\angle FGE \cong$ ___?___ **d.** $\overline{EF} \cong$ ___?___

e. $\overline{FG} \cong$ ___?___ **f.** $\overline{GE} \cong$ ___?___

3. What other relationship exists between $\overline{FG}$ and $\overline{EH}$?

New Vocabulary • consecutive angles

1 Properties: Sides and Angles

You can use what you know about parallel lines and transversals to prove some theorems about parallelograms.

Key Concepts

Theorem 6-1

Opposite sides of a parallelogram are congruent.

Proof

Proof of Theorem 6-1

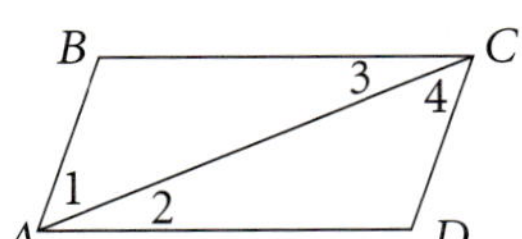

Given: $\square ABCD$

Prove: $\overline{AB} \cong \overline{CD}, \overline{BC} \cong \overline{DA}$

Vocabulary Tip

Read "▱" as "parallelogram." The plural is "▱s."

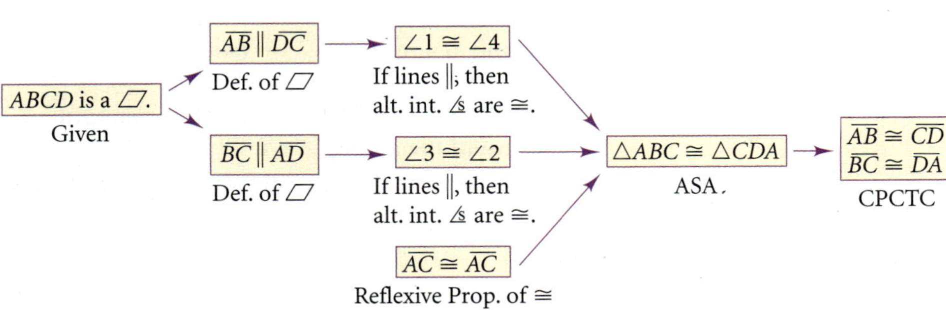

Angles of a polygon that share a side are **consecutive angles.** A parallelogram has opposite sides parallel. Its consecutive angles are same-side interior angles, so they are supplementary. In $\square ABCD$, consecutive angles B and C are supplementary, as are consecutive angles C and D.

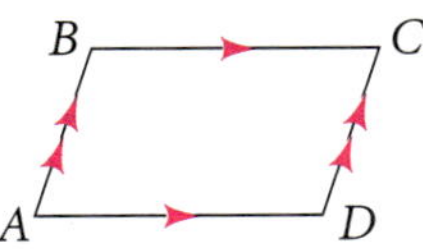

1 EXAMPLE Using Consecutive Angles

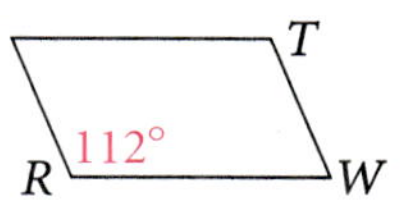

Find $m\angle S$ in ▱$RSTW$.

$\angle R$ and $\angle S$ are consecutive angles of a parallelogram. They are supplementary.

$m\angle R + m\angle S = 180$ **Definition of supplementary angles**

$112 + m\angle S = 180$ **Substitute.**

$m\angle S = 68$ **Subtract 112 from each side.**

CA Standards Check 1 **Critical Thinking** If consecutive angles of a quadrilateral are supplementary, must the quadrilateral be a parallelogram? Explain.

A proof of Theorem 6-2 uses the consecutive angles of a parallelogram and the fact that supplements of the same angle are congruent.

Key Concepts

Theorem 6-2

Opposite angles of a parallelogram are congruent.

Real-World Connection

Opposite angles in the "cat's cradle" parallelogram are congruent.

Plan for Proof of Theorem 6-2

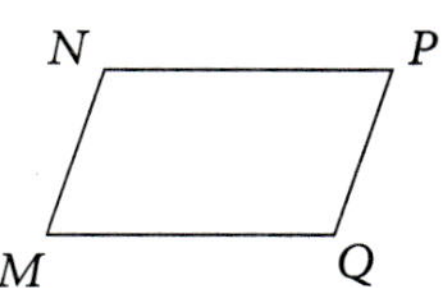

Given: ▱$MNPQ$

Prove: $\angle M \cong \angle P$ and $\angle N \cong \angle Q$

Plan: $\angle M \cong \angle P$ if they are supplements of the same angle, $\angle N$. Each is a supplement of $\angle N$ because same-side interior angles are supplementary. $\angle N \cong \angle Q$ using similar reasoning with $\angle M$.

You can use this plan to write a proof of Theorem 6-2 in Exercise 36. If you choose to write a flow proof, you'll find a guide on page 319.

You can use Theorems 6-1 and 6-2 along with algebra to find unknown values in parallelograms.

2 EXAMPLE Using Algebra

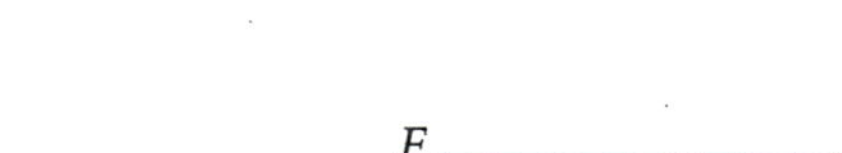

Algebra Find the value of x in ▱$PQRS$. Then find QR and PS.

$3x - 15 = 2x + 3$ **Opposite sides of a ▱ are congruent.**

$x - 15 = 3$ **Subtract 2x from each side.**

$x = 18$ **Add 15 to each side.**

$QR = 3x - 15 = 39$ **Substitute.**

$\overline{PS} \cong \overline{QR}$, so $PS = 39$.

CA Standards Check 2 Find the value of y in ▱$EFGH$. Then find $m\angle E$, $m\angle G$, $m\angle F$, and $m\angle H$.

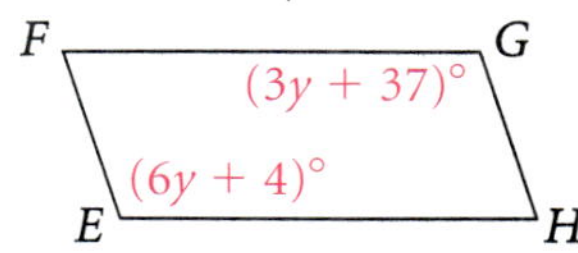

2 Properties: Diagonals and Transversals

The diagonals of parallelograms have a special property.

Key Concepts

> **Theorem 6-3**
>
> The diagonals of a parallelogram bisect each other.

Proof

Real-World Connection

The railing braces form diagonals of parallelograms, and therefore they bisect each other.

Proof of Theorem 6-3

Given: $\square ABCD$

Prove: $\overline{AC}$ and $\overline{BD}$ bisect each other at E.

Statements	Reasons
1. $ABCD$ is a parallelogram.	**1.** Given
2. $\overline{AB} \parallel \overline{DC}$	**2.** Definition of parallelogram
3. $\angle 1 \cong \angle 4$; $\angle 2 \cong \angle 3$	**3.** Parallel lines form $\cong$ alt. int. $\angle$s.
4. $\overline{AB} \cong \overline{DC}$	**4.** Opposite sides of a $\square$ are $\cong$.
5. $\triangle ABE \cong \triangle CDE$	**5.** ASA
6. $\overline{AE} \cong \overline{CE}$; $\overline{BE} \cong \overline{DE}$	**6.** CPCTC
7. $\overline{AC}$ and $\overline{BD}$ bisect each other at E.	**7.** Definition of bisector

You can use Theorem 6-3 to find unknown lengths in parallelograms.

3 EXAMPLE Using Algebra

Solve a system of linear equations to find the values of x and y in $\square ABCD$. Then find AE, EC, BE, and ED.

① $3y - 7 = 2x$
② $y = x + 1$ — **The diagonals of a parallelogram bisect each other.**

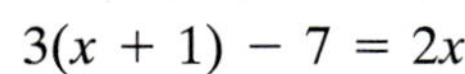

$3(x + 1) - 7 = 2x$ — **Substitute $x + 1$ for y in equation ①.**

$3x + 3 - 7 = 2x$ — **Distribute.**

$3x - 4 = 2x$ — **Simplify.**

$3x = 2x + 4$ — **Add 4 to each side.**

$x = 4$ — **Subtract $2x$ from each side.**

$3y - 7 = 2(4) = 8$
$y = 4 + 1 = 5$ — **Substitute 4 for x in equations ① and ②.**

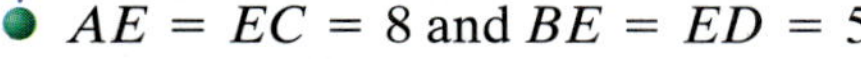

$AE = EC = 8$ and $BE = ED = 5$.

CA Standards Check **3** Find the values of a and b.

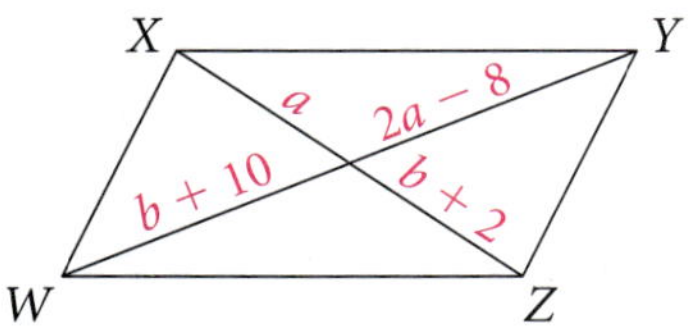

In Exercise 52, you will use Theorem 6-1, opposite sides of a parallelogram are congruent, to prove the following theorem.

Key Concepts

Theorem 6-4

If three (or more) parallel lines cut off congruent segments on one transversal, then they cut off congruent segments on every transversal.

$\overline{BD} \cong \overline{DF}$

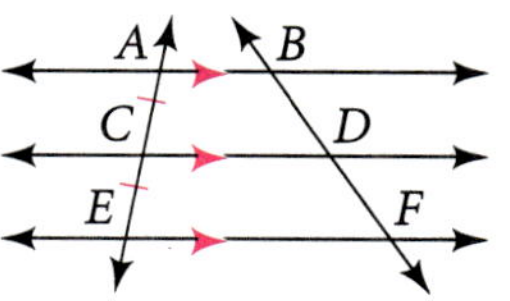

4 EXAMPLE Real-World Connection

Measurement Show how to separate a blank card into three strips that are the same height by using lined paper, a straightedge, and Theorem 6-4.

The lines of the paper are parallel and equally spaced. Place a corner of the top edge of the card on the first line of the paper. Place the corner of the bottom edge on the fourth line. Mark the points where the second and third lines intersect the card. The marks will be equally spaced because the edge of the card is a transversal for the equally spaced parallel lines of the paper. Repeat for the other side of the card. Connect the marks using a straightedge.

CA Standards Check 4 In the figure at the right, $\overleftrightarrow{DH} \parallel \overleftrightarrow{CG} \parallel \overleftrightarrow{BF} \parallel \overleftrightarrow{AE}$, $AB = BC = CD = 2$, and $EF = 2.5$. Find EH.

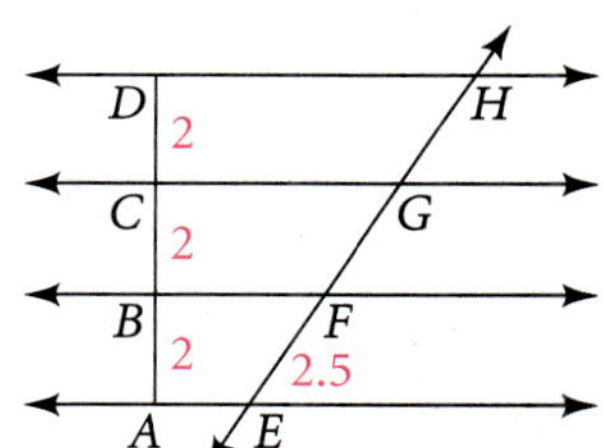

EXERCISES

For more exercises, see *Extra Skill, Word Problem, and Proof Practice.*

Standards Practice

GEOM 7.0, 13.0

A Practice by Example

Algebra Find the value of x in each parallelogram.

Example 1 (page 313)

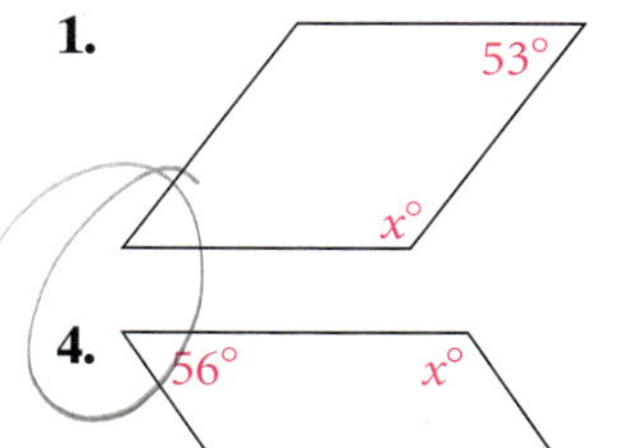

2. 113° x°

5. x° 80°

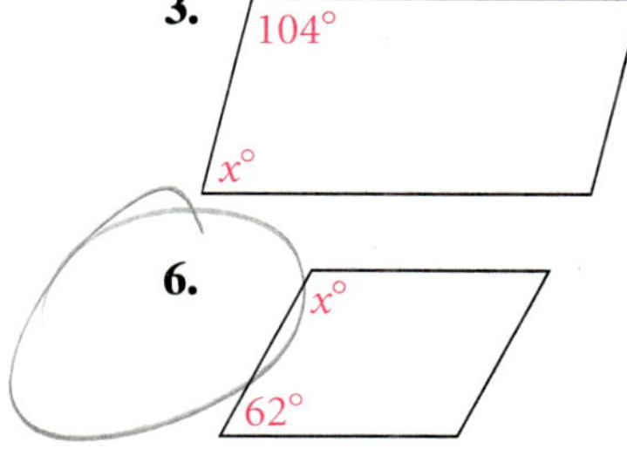

Example 2 (page 313)

Algebra Find the value of x and the length of each side.

7. F, G, M, E; $FG = 7x - 1$, $GM = 2x + 4$, $EM = 6x + 2$, $FE = 10$

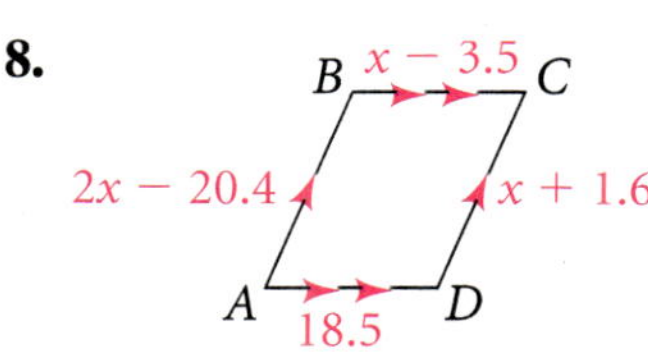

Algebra **Find the value of a.**

9. 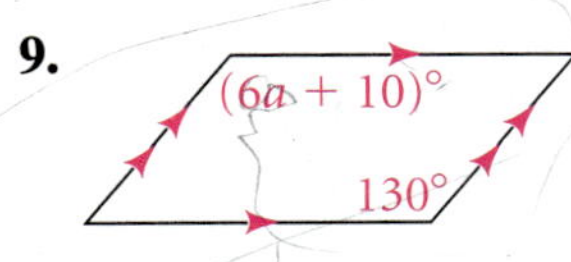

10. 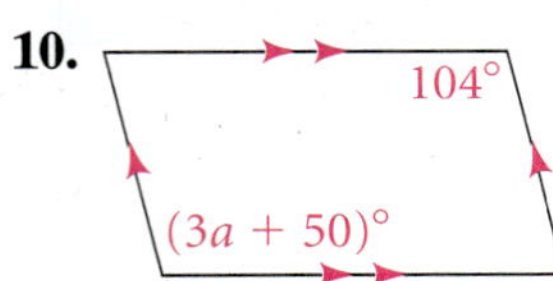

11. 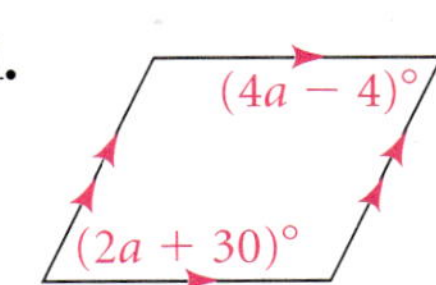

Find the value of a and the measure of each angle in each parallelogram.

12. 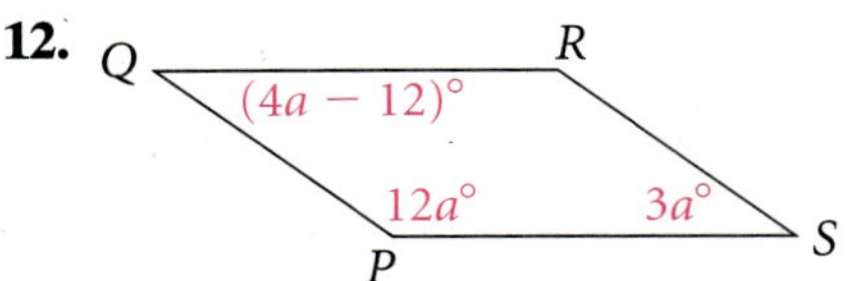

13. 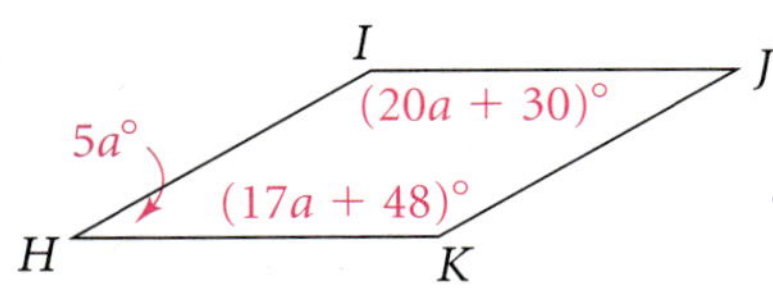

Example 3 (page 314) **Algebra** **Find the values of x and y in $\square PQRS$.**

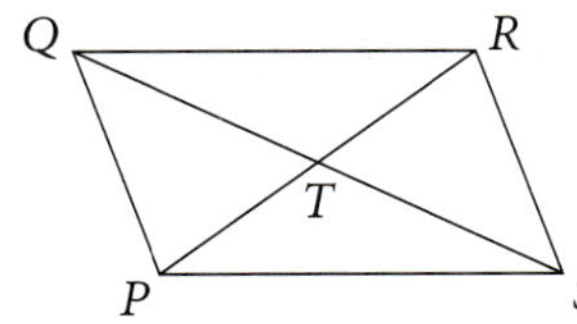

14. $PT = 2x, TR = y + 4, QT = x + 2, TS = y$

15. $PT = x + 2, TR = y, QT = 2x, TS = y + 3$

16. $PT = y, TR = x + 3, QT = 2y, TS = 3x - 1$

17. $PT = 2x, TR = y + 3, QT = 3x, TS = 2y$

18. $PT = 8x, TR = 6y, QT = 2x + 2, TS = 2y$

Example 4 (page 315)

19. Find ED and FD in the figure at the right.

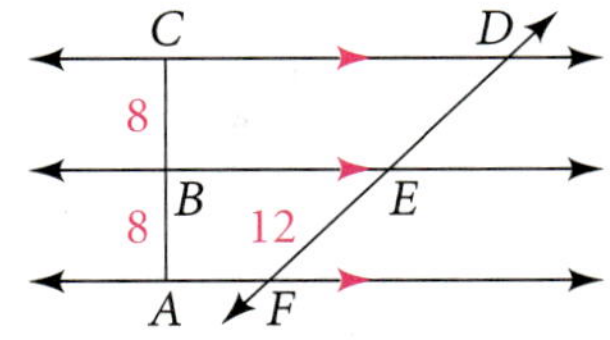

20. **Sewing** Suppose you don't have a ruler. Explain how to space four buttons equally on a shirt if you know where the first and last buttons must be placed and you have a large piece of lined paper.

Exercise 20

In the figure, the horizontal lines are parallel and $PQ = QR = RS$. Find each length.

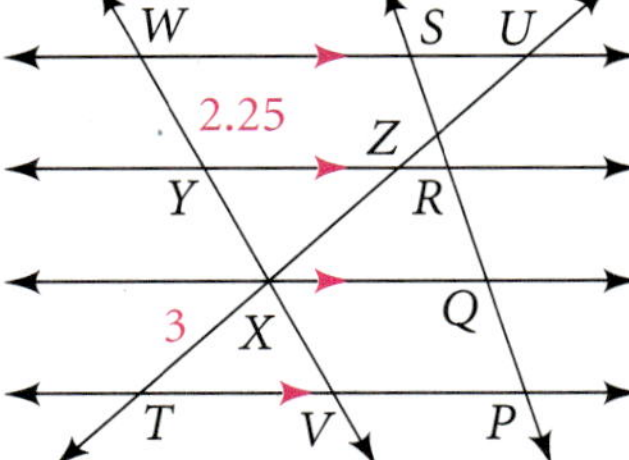

21. ZU	22. XZ
23. XU	24. TZ
25. TU	26. XV
27. YX	28. YV
29. WX	30. WV

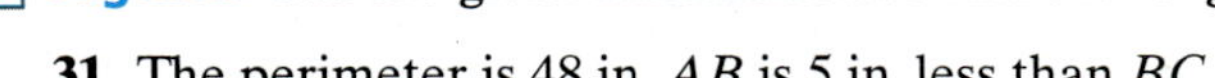

B Apply Your Skills **Algebra** **Use the given information to find the lengths of all four sides of $\square ABCD$.**

31. The perimeter is 48 in. AB is 5 in. less than BC.

32. The perimeter is 92 cm. AD is 7 cm more than twice AB.

33. **Multiple Choice** What is the value of x in the parallelogram?

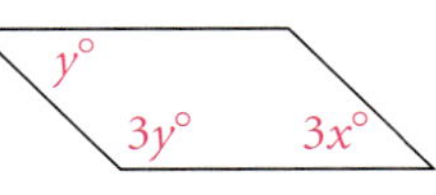

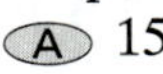

Ⓐ 15 Ⓒ 60
Ⓑ 45 Ⓓ 135

34. **Writing** Explain how to find the measures of the remaining three angles of a parallelogram if you already know the measure of one of the angles.

35. **Developing Proof** Complete this paragraph proof of Theorem 6-1 by filling in the blanks.

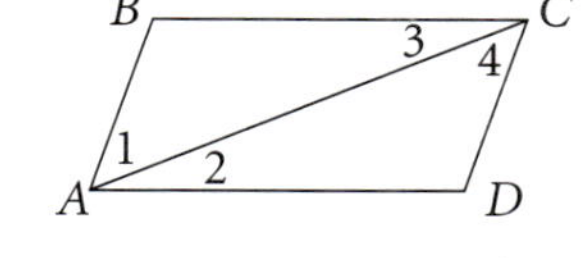

Given: $\square ABCD$

Prove: $\overline{AB} \cong \overline{CD}$ and $\overline{BC} \cong \overline{DA}$

Proof: $ABCD$ is a parallelogram, therefore $\overline{AB} \parallel$ **a.** ? and $\overline{BC} \parallel$ **b.** ? . $\angle 1 \cong \angle 4$ and $\angle 3 \cong \angle 2$, because alternate interior angles are **c.** ? . $\overline{AC} \cong \overline{AC}$ by the **d.** ? Property of Congruence. Therefore $\triangle ABC \cong \triangle CDA$ by **e.** ? . So, $\overline{AB} \cong \overline{CD}$ and $\overline{BC} \cong \overline{DA}$ because **f.** ? .

For a guide to solving Exercise 36, see p. 319.

Proof 36. Write a proof for Theorem 6-2. You may wish to follow the plan on page 313.

Find the measures of the numbered angles for each parallelogram.

37.
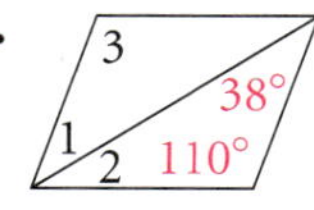

38.
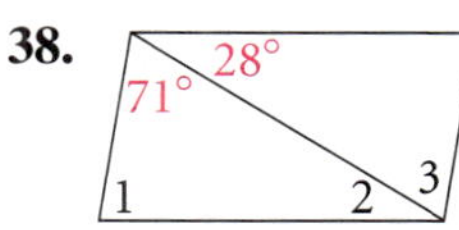

39.
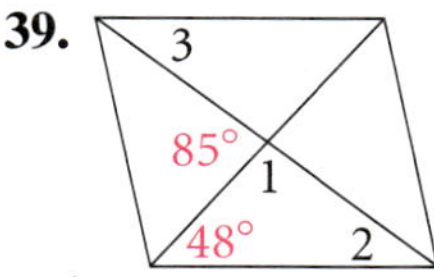

40. **Error Analysis** Brian states that $QV = 10$ cm in the figure at the right. Explain why Brian's statement may not be correct.

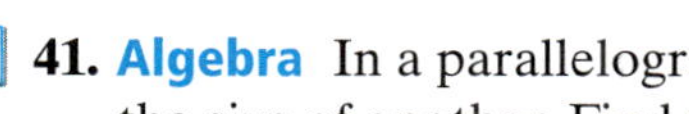

41. **Algebra** In a parallelogram one angle is 9 times the size of another. Find the measures of the angles.

Proof **Write a paragraph proof, flow proof, or two-column proof.**

Problem Solving Hint

For each of Exercises 42–44, sketch the diagram and mark it as you think through a proof.

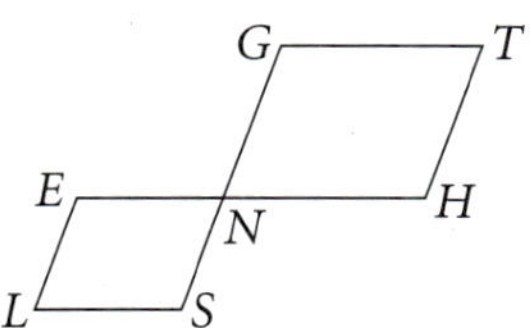

42. **Given:** $\square LENS$ and $\square NGTH$

Prove: $\angle L \cong \angle T$

43. **Given:** $\square LENS$ and $\square NGTH$

Prove: $\overline{LS} \parallel \overline{GT}$

44. **Given:** $\square LENS$ and $\square NGTH$

Prove: $\angle E$ is supplementary to $\angle T$.

Algebra **Find the value(s) of the variable(s) in each parallelogram.**

45.
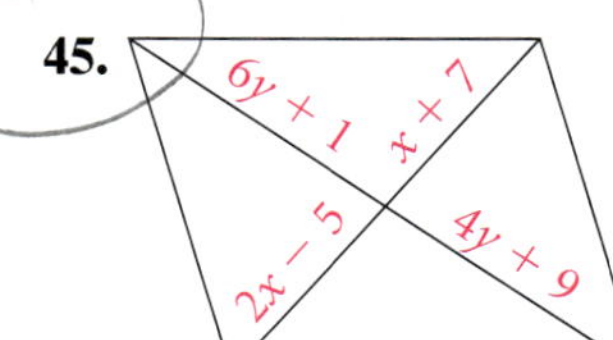

46. $AC = 4x + 10$
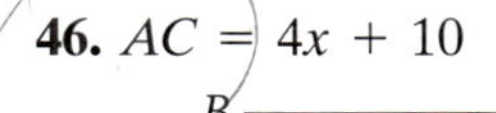

47.
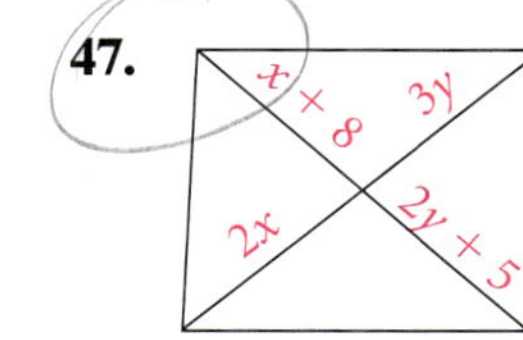

Proof **Write a paragraph proof, flow proof, or two-column proof.**

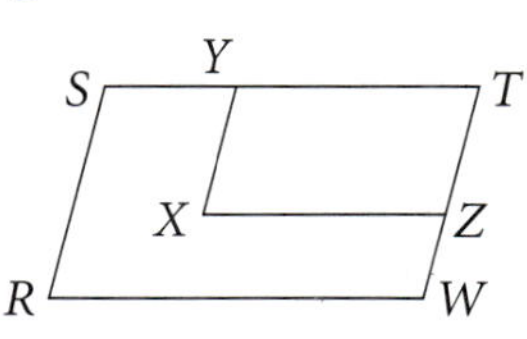

48. **Given:** $\square RSTW$ and $\square XYTZ$

Prove: $\angle R \cong \angle X$

49. **Given:** $\square RSTW$ and $\square XYTZ$

Prove: $\overline{XY} \parallel \overline{RS}$

50. **Given:** $\square ABCD$ and $\overline{AC}$ bisects $\angle DAB$.

Prove: $\overline{AC}$ bisects $\angle DCB$.

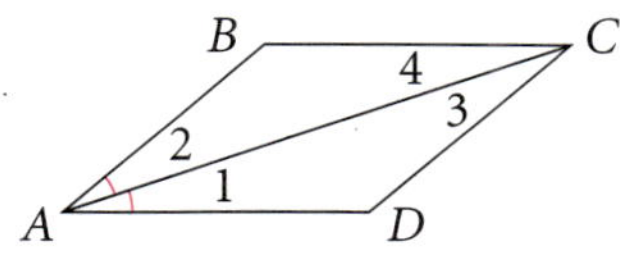

Homework Video Tutor

Visit: PHSchool.com
Web Code: bce-0602

Online Lesson Quiz Visit: PHSchool.com, Web Code: bca-0602

Challenge

51. a. Open-Ended Sketch two parallelograms whose corresponding sides are congruent but whose corresponding angles are not congruent.

b. Critical Thinking Is there an SSSS congruence theorem for parallelograms? Explain.

Proof **52.** Prove Theorem 6-4.

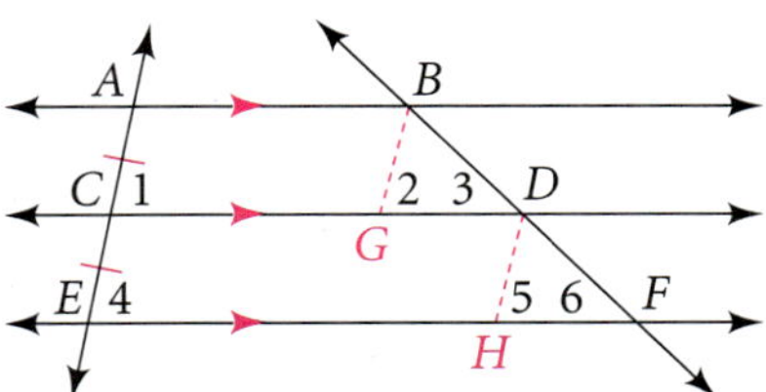

Given: $\overleftrightarrow{AB} \parallel \overleftrightarrow{CD} \parallel \overleftrightarrow{EF}$ and $\overline{AC} \cong \overline{CE}$

Prove: $\overline{BD} \cong \overline{DF}$

(*Hint*: Draw lines through B and D parallel to $\overleftrightarrow{AE}$ and intersecting $\overleftrightarrow{CD}$ at G and $\overleftrightarrow{EF}$ at H.)

Proof **53. a.** Prove that if two sides and the included angle of one parallelogram are congruent to corresponding parts of another parallelogram, then the parallelograms are congruent. (*Hint:* Prove that all the corresponding parts of the parallelograms are congruent.)

b. Is there a theorem similar to SAS for trapezoids? Explain.

Multiple Choice Practice

For California Standards and CAHSEE Tutorials, visit PHSchool.com. Web Code: bcq-9045

GEOM 7.0

54. What values of x and y make the quadrilateral a parallelogram?

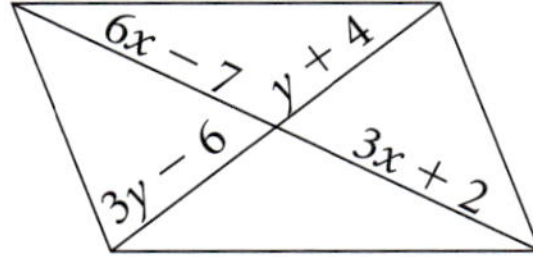

Ⓐ $x = 3, y = 3$ Ⓑ $x = 3, y = 5$ Ⓒ $x = 5, y = 3$ Ⓓ $x = 5, y = 7$

CAHSEE 7NS 2.3

55. $3^5 \div 3^2 =$

Ⓐ 1^3 Ⓑ 1^7 Ⓒ 3^3 Ⓓ 3^7

CAHSEE 7MG 3.4

56. Pentagon $FGHJK$ is congruent to pentagon $PQRST$. Which side is congruent to $\overline{FK}$?

Ⓐ $\overline{PQ}$ Ⓑ $\overline{QR}$ Ⓒ $\overline{ST}$ Ⓓ $\overline{PT}$

Mixed Review

Lesson 6-1

Determine the most precise name for each figure.

57.

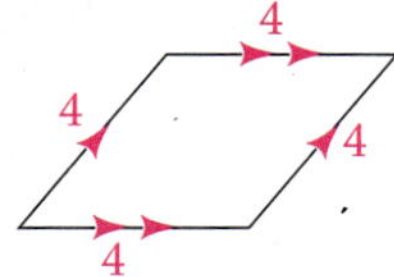

58.

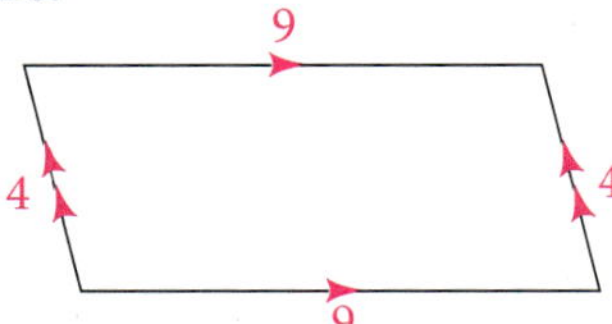

Lesson 4-6

59. What additional information do you need to prove $\triangle ADC \cong \triangle ABC$ by the HL Theorem?

A
D
C
B

Lesson 3-1

In the figure at the right, $\overleftrightarrow{PQ} \parallel \overleftrightarrow{RS}$. Find each measure.

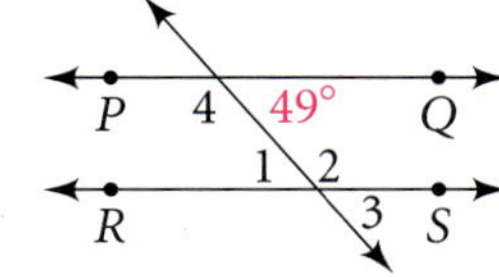

60. $m\angle 1$ **61.** $m\angle 2$ **62.** $m\angle 3$ **63.** $m\angle 4$

Guided Problem Solving

FOR USE WITH PAGE 317, EXERCISE 36

Understanding Proof Problems **Let Ravi's thinking guide you through a flow proof for Theorem 6-2. He follows the plan on page 313, also shown below. Check your understanding with the exercise at the bottom of the page.**

Theorem 6-2 Opposite angles of a parallelogram are congruent.

Given: $\square MNPQ$

Prove: $\angle M \cong \angle P$ and $\angle N \cong \angle Q$

Plan: $\angle M \cong \angle P$ if they are supplements of the same angle, $\angle N$. Each is a supplement of $\angle N$ because same-side interior angles are supplementary. $\angle N \cong \angle Q$ using similar reasoning with $\angle M$.

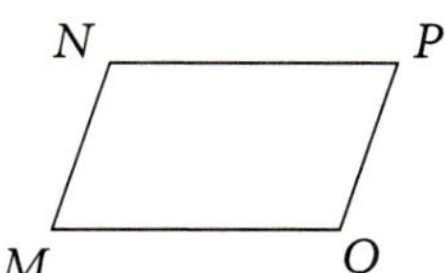

What Ravi Thinks

The second sentence of the plan is

"Each is a supplement of $\angle N$ because same-side interior angles are supplementary."

This tells me where to begin. Opposite sides of a parallelogram are parallel while the other sides act as transversals. The opposite sides and transversals form same-side interior angles.

Same-side interior angles are supplementary and supplements of the same angle are congruent.

This gives me half the proof.

The last sentence of the plan is

"$\angle N \cong \angle Q$ using similar reasoning with $\angle M$."

This tells me I can repeat the first part of the proof, but with different angles.

I have found a clever way to add this second part to my flow proof. It's in red.

What Ravi Writes

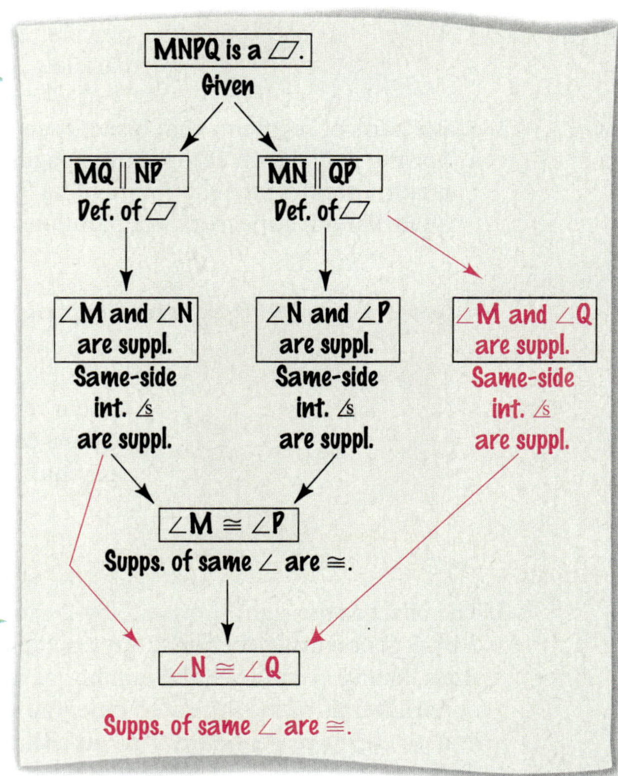

EXERCISE

For Exercise 42 on page 317, write a proof that follows the plan shown below.

Given: $\square LENS$ and $\square NGTH$

Prove: $\angle L \cong \angle T$

Plan: $\angle L \cong \angle T$ if their opposite angles, $\angle ENS$ and $\angle HNG$, are congruent. $\angle ENS \cong \angle HNG$ because they are vertical angles.

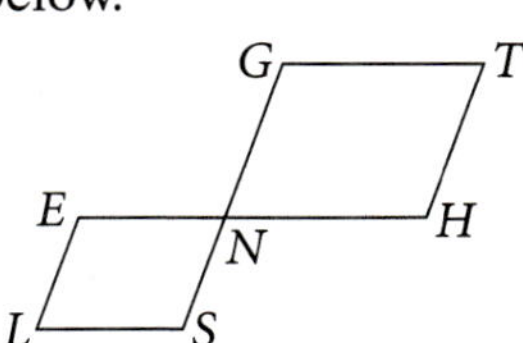

Activity Lab Geo-Models

Hands-On

FOR USE WITH LESSON 6-3

GEOM 1.0 Students demonstrate understanding by identifying and giving examples of inductive reasoning. *Master*

By definition, a quadrilateral is a parallelogram if both pairs of opposite sides are parallel. You can use a geoboard or geopaper to explore other conditions that seem to force a quadrilateral to be a parallelogram.

1. On geopaper or a geoboard, make pairs of segments that are congruent and parallel.

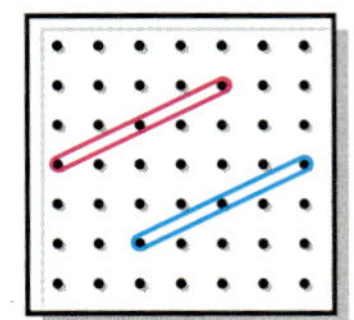

a. The figure at the right shows one pair. Imagine the endpoints connected to form a quadrilateral. Does the quadrilateral appear to be a parallelogram?

b. The photo at the left shows other pairs of congruent parallel segments. A large elastic band wraps around each pair to form a quadrilateral. Does each quadrilateral appear to be a parallelogram?

c. Make a conjecture about a quadrilateral that has one pair of opposite sides congruent and parallel.

2. Make pairs of segments that bisect each other.

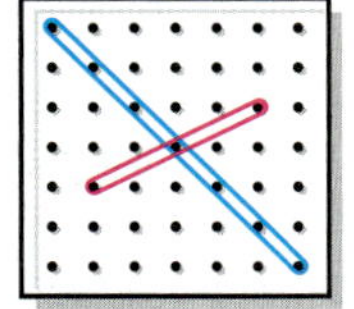

a. For the pair shown at the right, imagine a quadrilateral wrapped around them. Does the quadrilateral appear to be a parallelogram?

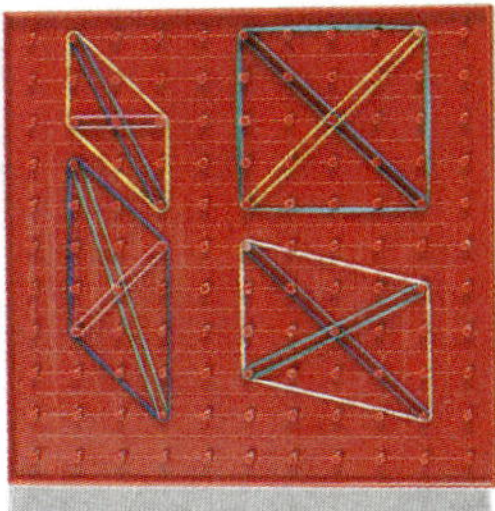

b. The photo at the left shows four more quadrilaterals wrapped around diagonals that bisect each other. Does each quadrilateral appear to be a parallelogram?

c. Make a conjecture about a quadrilateral whose diagonals bisect each other.

3. The chart at the right shows a 2-by-2 geoboard and a 3-by-3 geoboard at the top. Copy and complete the chart, listing in each cell the number of different quadrilaterals of the indicated type that you can find on each size geoboard. Two quadrilaterals are different if they are not congruent. One cell has been completed for you.

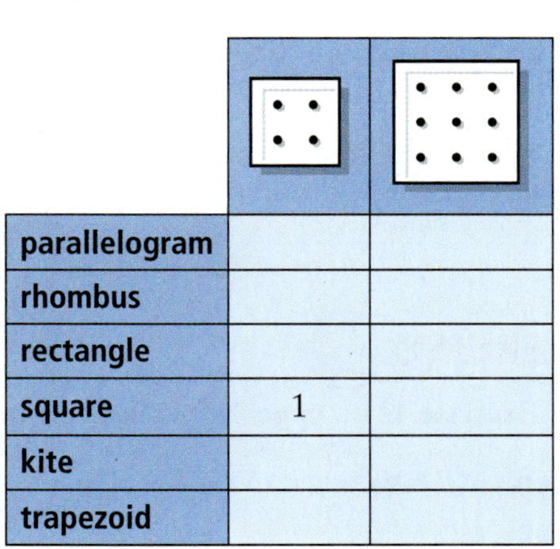

	2-by-2	3-by-3
parallelogram		
rhombus		
rectangle		
square	1	
kite		
trapezoid		

4. How many different squares can you find on a 4-by-4 geoboard?

5. The vertices of a cube form a 2-by-2-by-2 *lattice* in space. How many different squares can you find whose vertices are points of this 2-by-2-by-2 lattice?

6. How many different squares can you find in a 3-by-3-by-3 lattice?

Proving That a Quadrilateral Is a Parallelogram

California Content Standards

GEOM 7.0 Students prove and use theorems involving the properties of quadrilaterals. ***Develop***

GEOM 12.0 Students find and use measures of sides and of interior angles of polygons to classify figures and solve problems. ***Develop***

What You'll Learn

- To determine whether a quadrilateral is a parallelogram

. . . And Why

To use a parallel rule to plot a ship's course, as in Example 3

Check Skills You'll Need

GO for Help Lessons 1-8 and 3-7

Use the figure at the right.

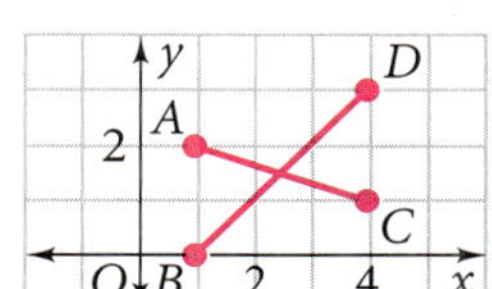

1. Find the coordinates of the midpoints of $\overline{AC}$ and $\overline{BD}$. What is the relationship between $\overline{AC}$ and $\overline{BD}$?
2. Find the slopes of $\overline{BC}$ and $\overline{AD}$. How do they compare?
3. Are $\overline{AB}$ and $\overline{DC}$ parallel? Explain.
4. What type of figure is $ABCD$?

1 Is the Quadrilateral a Parallelogram?

Theorems 6-5 and 6-6 are converses of Theorems 6-1 and 6-2, respectively, from the previous lesson. They provide two ways to conclude that a quadrilateral is a parallelogram.

Key Concepts

Theorem 6-5

If both pairs of opposite sides of a quadrilateral are congruent, then the quadrilateral is a parallelogram.

Real-World  Connection

The frame remains a parallelogram as it is raised and lowered, and the backboard stays vertical.

Proof of Theorem 6-5

Given: $\overline{WX} \cong \overline{ZY}$ and $\overline{XY} \cong \overline{WZ}$

Prove: $WXYZ$ is a parallelogram.

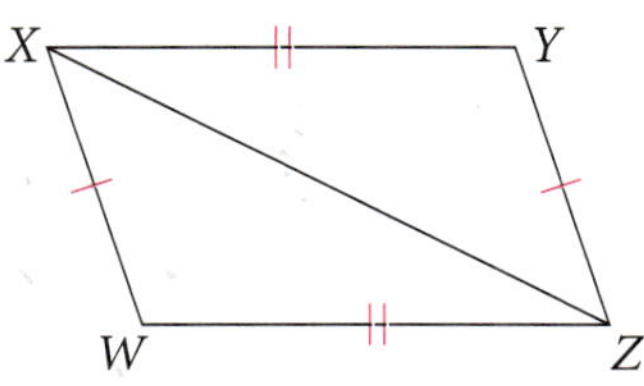

Proof: Draw diagonal $\overline{XZ}$. Since opposite sides of $WXYZ$ are congruent, $\triangle WXZ \cong \triangle YZX$ by SSS. Using CPCTC, $\angle WXZ \cong \angle YZX$, so $\overline{WX} \parallel \overline{ZY}$. Also, $\angle WZX \cong \angle YXZ$, so $\overline{WZ} \parallel \overline{XY}$. $WXYZ$ is a parallelogram by definition.

You will complete a proof of Theorem 6-6 in Exercise 12.

Key Concepts

Theorem 6-6

If both pairs of opposite angles of a quadrilateral are congruent, then the quadrilateral is a parallelogram.

Theorem 6-7 is the converse of Theorem 6-3 of the previous lesson.

Theorem 6-7

If the diagonals of a quadrilateral bisect each other, then the quadrilateral is a parallelogram.

Proof

Proof of Theorem 6-7

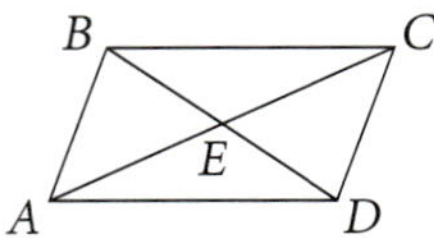

Given: $\overline{AC}$ and $\overline{BD}$ bisect each other at E.

Prove: $ABCD$ is a parallelogram.

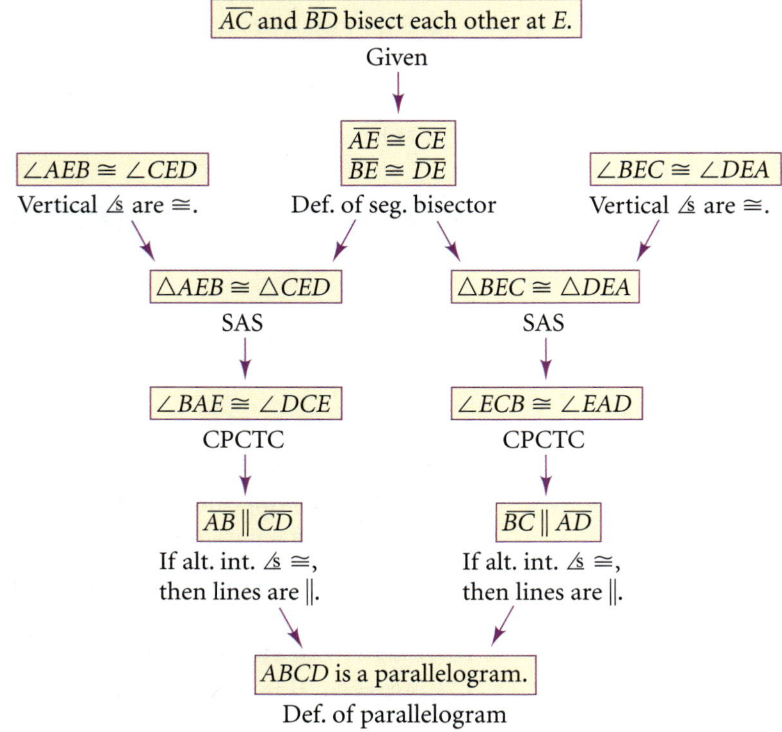

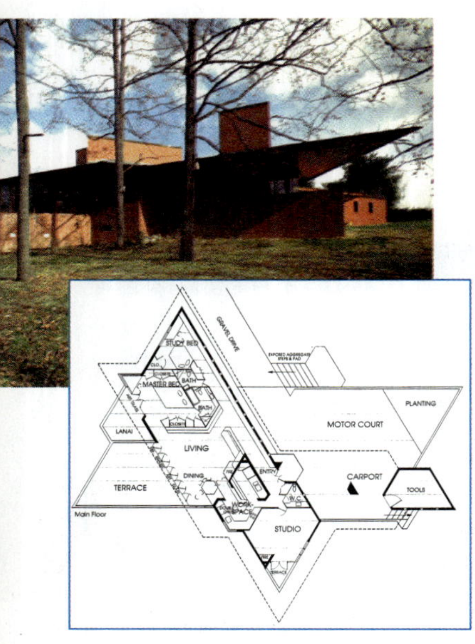

Real-World Connection

Frank Lloyd Wright, a famous architect, used parallelograms in designs of many houses, such as the Kraus House in Kirkwood, Missouri.

Theorem 6-8 suggests that if you keep two objects of the same length parallel, such as cross-country skis, then the quadrilateral determined by their endpoints must be a parallelogram. You will prove Theorem 6-8 in Exercise 13.

Theorem 6-8

If one pair of opposite sides of a quadrilateral is both congruent and parallel, then the quadrilateral is a parallelogram.

You can use algebra and Theorems 6-7 and 6-8 to find values for which quadrilaterals are parallelograms.

1 EXAMPLE Finding Values for Parallelograms

Multiple Choice For what value of x must $MLPN$ be a parallelogram?

(A) 1 (B) 3 (C) 9 (D) 27

Diagonals of a parallelogram bisect each other, so $2y - 7 = y + 2$ and $3x = y$.

$2y - 7 = y + 2$ **Write the equation with one variable.**

$y - 7 = 2$ **Collect the variables on one side.**

$y = 9$ **Solve.**

Substitute 9 for y in the second equation. Find x.

$3x = y$

$3x = 9$ **Substitute 9 for *y*.**

$x = 3$ **Solve.**

For $x = 3$, $MLPN$ is a parallelogram. Answer B is correct.

Test-Taking Tip

Read test questions carefully. In Example 1, the value of y is choice C, but you are asked for the value of x.

CA Standards Check 1 Find the values of a and c for which $PQRS$ must be a parallelogram.

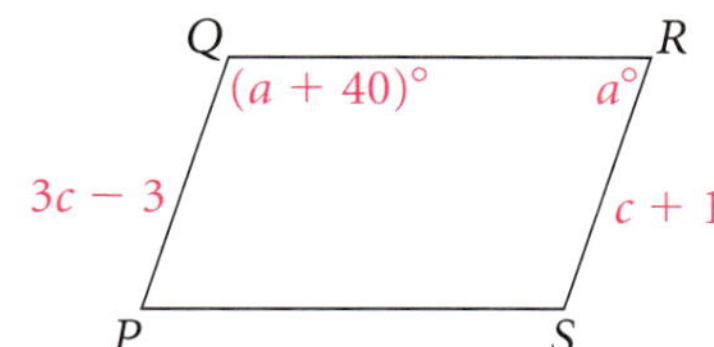

You can conclude that a quadrilateral is a parallelogram if both pairs of opposite sides are parallel. Theorems 6-5 through 6-8 provide four shortcuts to prove that a quadrilateral is a parallelogram.

2 EXAMPLE Is the Quadrilateral a Parallelogram?

Online active math

For: Parallelogram Activity
Use: Interactive Textbook, 6-3

Can you prove the quadrilateral is a parallelogram from what is given? Explain.

a. Given: $\angle ABD \cong \angle CDB$, $\angle BDA \cong \angle DBC$, $\angle A \cong \angle C$

Prove: $ABCD$ is a parallelogram.

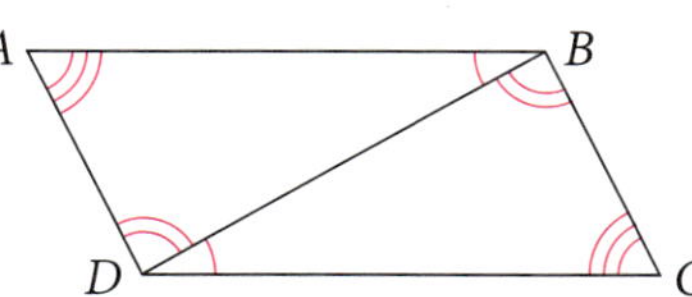

Yes, both pairs of opposite angles are congruent. $ABCD$ is a parallelogram by Theorem 6-6.

b. Given: $\overline{LM} \cong \overline{LO}$, $\overline{NM} \cong \overline{ON}$

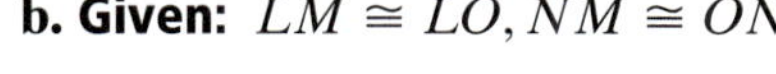

Prove: $LMNO$ is a parallelogram.

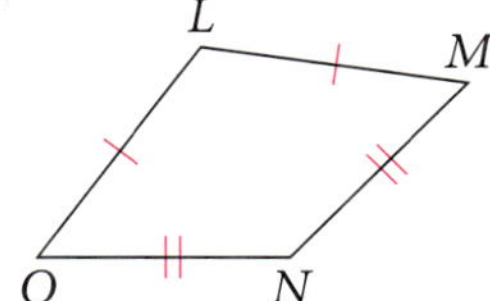

No, the given information is not enough to prove $LMNO$ is a parallelogram.

CA Standards Check 2 Can you prove the quadrilateral is a parallelogram? Explain.

a. Given: $\overline{PQ} \cong \overline{SR}$, $\overline{PQ} \parallel \overline{SR}$

Prove: $PQRS$ is a parallelogram.

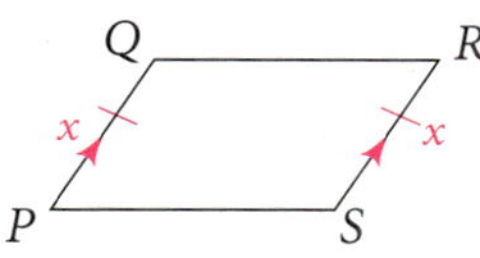

b. Given: $\overline{DH} \cong \overline{GH}$, $\overline{EH} \cong \overline{FH}$

Prove: $DEFG$ is a parallelogram.

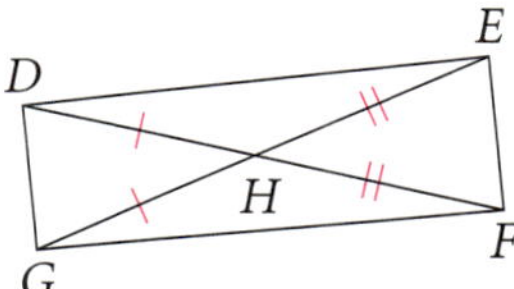

Real-World Connection

This marine navigator is using a parallel rule.

Navigation A parallel rule is a navigation tool that is used to plot ship routes on charts. It is made of two rulers connected with congruent crossbars, such that $AB = DC$ and $AD = BC$. You place one ruler on the line connecting the ship's present position to its destination. Then you move the other ruler onto the chart's compass to find the direction of the route. Explain why this instrument works.

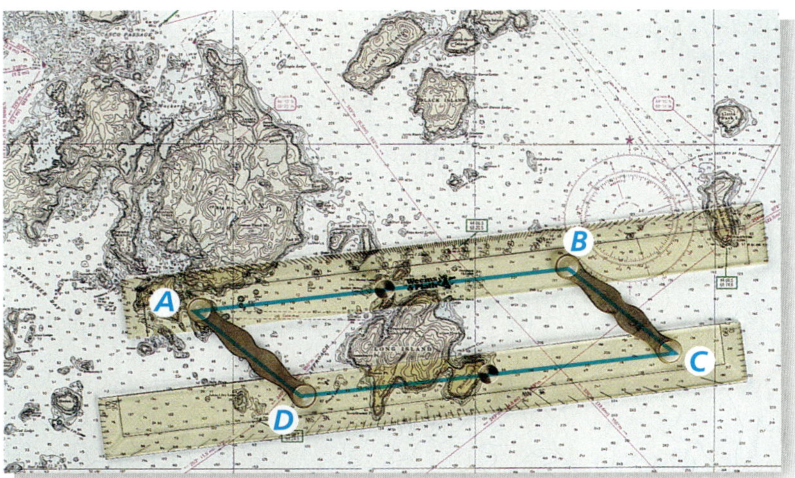

The crossbars and the sections of the rulers are congruent no matter how they are positioned. So, $ABCD$ is always a parallelogram. Since $ABCD$ is a parallelogram, the rulers are parallel. Therefore, the direction the ship should travel is the same as the direction shown on the chart's compass.

CA Standards Check **3 Critical Thinking** Suppose the ruler connecting the ship's position to its destination gets in the way of reading the compass. How can you get the desired reading?

EXERCISES

For more exercises, see *Extra Skill, Word Problem, and Proof Practice.*

Standards Practice GEOM 7.0, 12.0

 **Practice by Example**  **Algebra** **Find the values of x and y for which $ABCD$ must be a parallelogram.**

Example 1 (page 323)

1. 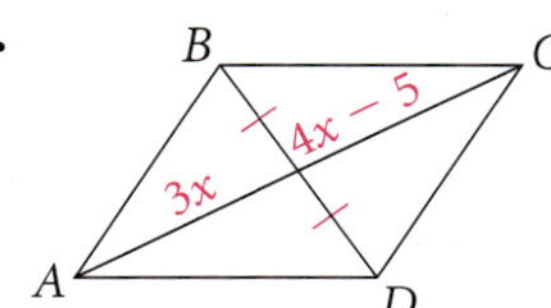

2. 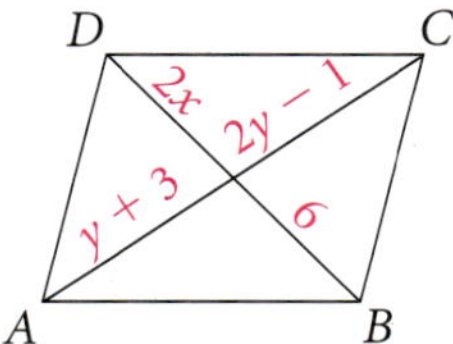

3. 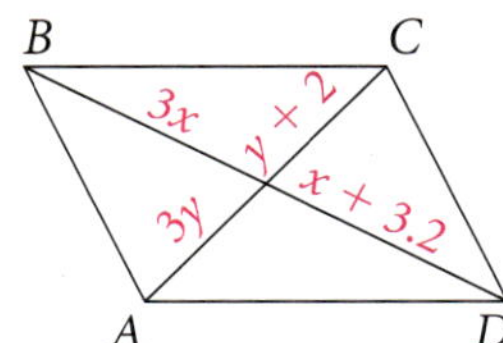

4. 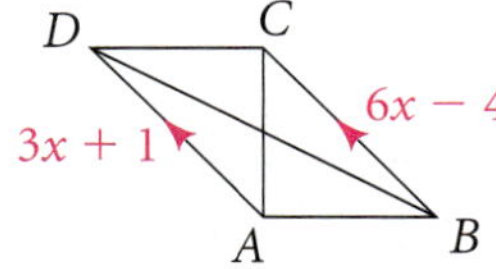

5. 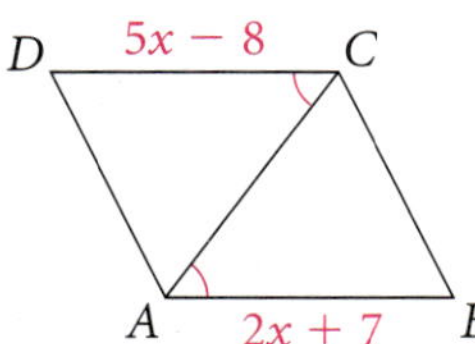

6. 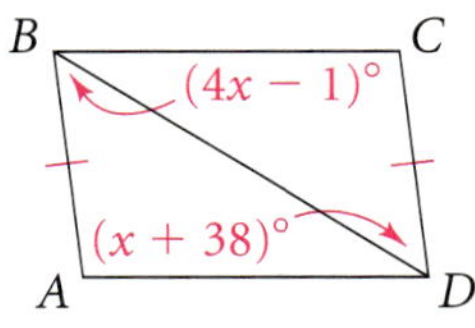

Example 2 (page 323)

Determine whether the quadrilateral must be a parallelogram. Explain.

7. 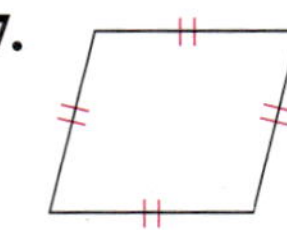

8. 

9. 

Example 3 (page 324)

10. **Fishing** Quadrilaterals are formed on the side of this fishing tackle box by the adjustable shelves and connecting pieces. Explain why the quadrilaterals remain parallelograms no matter what position the shelves are in.

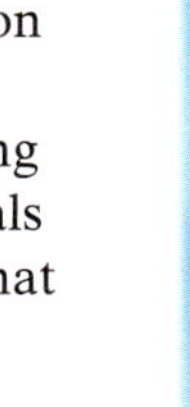

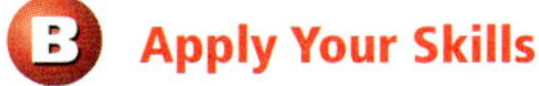

B Apply Your Skills

11. Combine each of Theorems 6-1, 6-2, and 6-3 with Theorems 6-5, 6-6, and 6-7, respectively, into biconditional statements.

12. **Developing Proof** Complete the proof of Theorem 6-6.

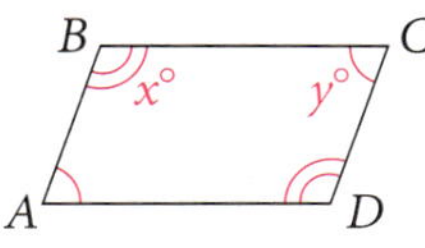

Given: $\angle A \cong \angle C$ and $\angle B \cong \angle D$

Prove: $ABCD$ is a parallelogram.

Statements	Reasons
1. $x + y + x + y = 360$	1. The sum of the measures of the angles of a quadrilateral = 360.
2. $2(x + y) = 360$	a. ?
3. $x + y = 180$	b. ?
4. $\angle A$ and $\angle B$ are supplementary. $\angle A$ and $\angle D$ are supplementary.	4. Definition of supplementary
c. ? $\parallel$?, ? $\parallel$?	d. ?
6. $ABCD$ is a parallelogram.	e. ?

Proof 13. Prove Theorem 6-8.

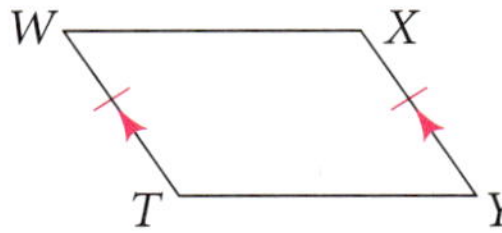

Given: $\overline{TW} \parallel \overline{YX}$ and $\overline{TW} \cong \overline{YX}$

Prove: $TWXY$ is a parallelogram.

(*Hint*: Draw one or both diagonals. Find congruent triangles. Use CPCTC.)

Algebra **Find the values of the variables for which $ABCD$ must be a parallelogram.**

14. 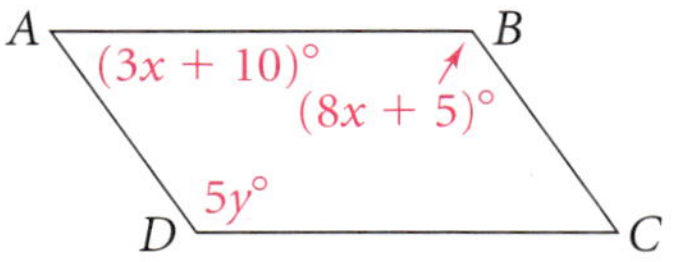 (figure: parallelogram $ABCD$ with angles $(3x + 10)°$, $(8x + 5)°$, $5y°$)

15. 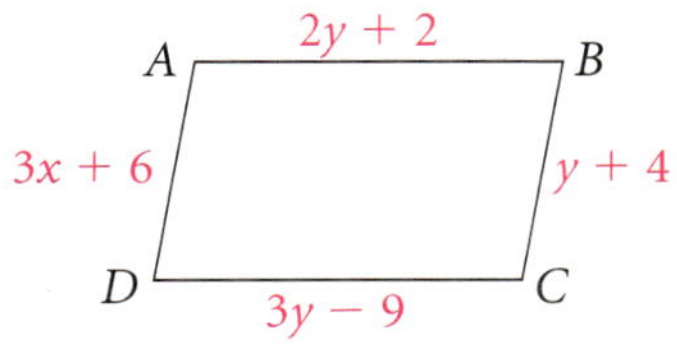 (figure: parallelogram $ABCD$ with sides $AB = 2y + 2$, $BC = y + 4$, $DC = 3y - 9$, $AD = 3x + 6$)

16. 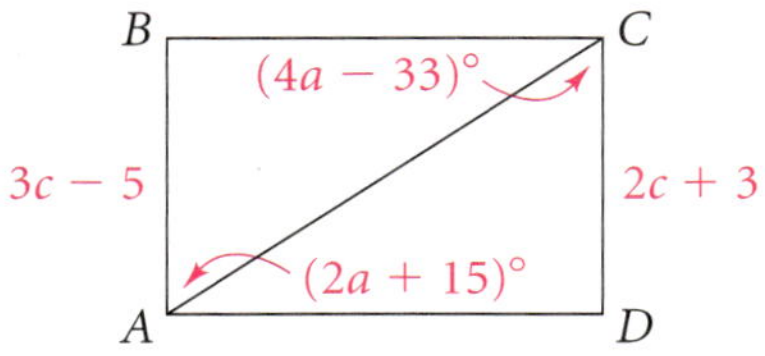 (figure: quadrilateral $ABCD$ with diagonal AC; angles $(4a - 33)°$, $(2a + 15)°$; sides $3c - 5$, $2c + 3$)

17.

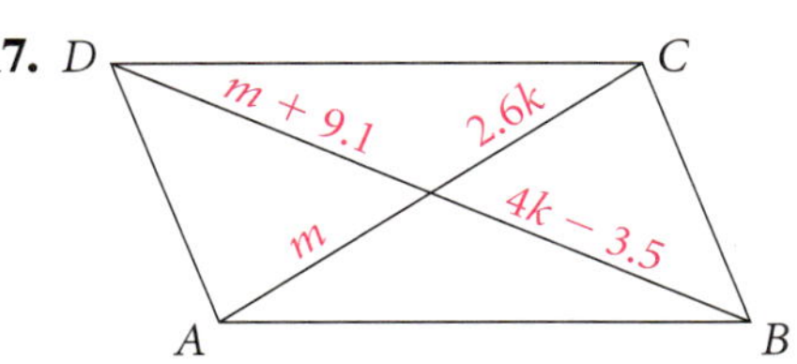

18. **Multiple Choice** From which given information can you conclude that $RSTW$ is a parallelogram?

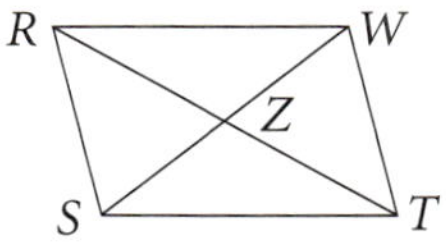

Ⓐ $\overline{RS} \parallel \overline{WT}$, $\overline{RS} \cong \overline{ST}$

Ⓑ $\overline{RS} \parallel \overline{WT}$, $\overline{ST} \cong \overline{RW}$

Ⓒ $\overline{RS} \cong \overline{ST}$, $\overline{RW} \cong \overline{WT}$

Ⓓ $\overline{RZ} \cong \overline{TZ}$, $\overline{SZ} \cong \overline{WZ}$

Homework Video Tutor

Visit: PHSchool.com
Web Code: bce-0603

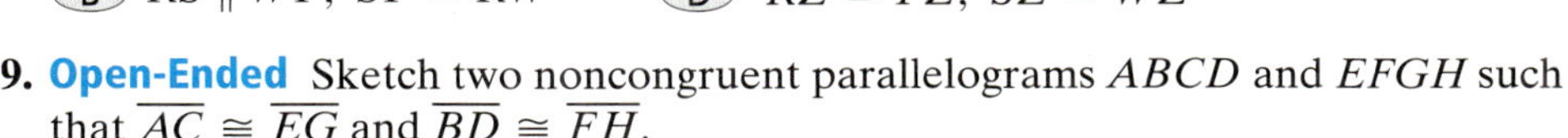

19. **Open-Ended** Sketch two noncongruent parallelograms $ABCD$ and $EFGH$ such that $\overline{AC} \cong \overline{EG}$ and $\overline{BD} \cong \overline{FH}$.

Proof **20. Given:** $\angle JKN \cong \angle LMN$
$\angle LKN \cong \angle JMN$

Prove: $JKLM$ is a parallelogram.

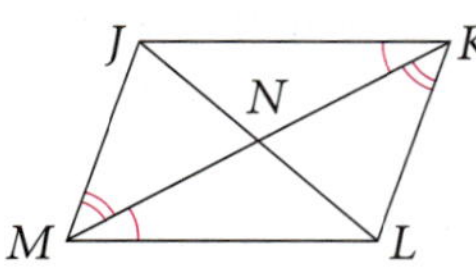

21. Given: $\triangle TRS \cong \triangle RTW$

Prove: $RSTW$ is a parallelogram.

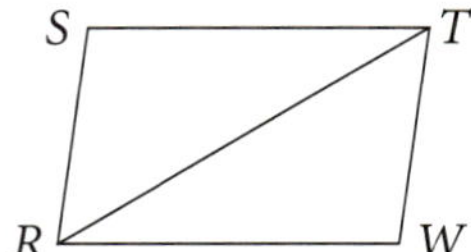

Coordinate Geometry **Given points A, B, and C in the coordinate plane as shown, find the fourth point described below.**

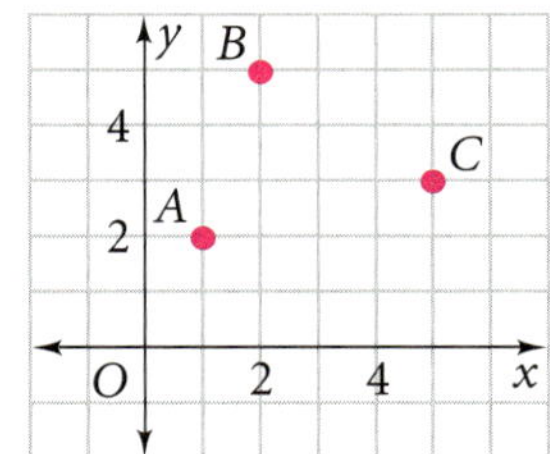

22. point D so that $ABCD$ is a parallelogram

23. point E so that $ABEC$ is a parallelogram

24. point F so that $AFBC$ is a parallelogram

25. Writing Summarize the ways to show that a quadrilateral is a parallelogram.

C Challenge

26. Probability If two opposite angles of a quadrilateral measure 120 and the measures of the other angles are multiples of 10, what is the probability that the quadrilateral is a parallelogram?

Proof **27.** In the figure at the right, point D is constructed by drawing two arcs. One has center C and radius AB. The other has center B and radius AC. Prove that $\overline{AM}$ is a median of $\triangle ABC$.

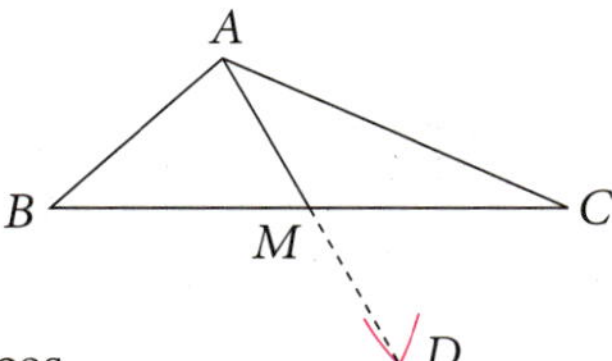

28. Coordinate Geometry The diagonals of quadrilateral $EFGH$ intersect at $D(-1, 4)$. Two vertices of $EFGH$ are $E(2, 7)$ and $F(-3, 5)$. What must be the coordinates of G and H to ensure that $EFGH$ is a parallelogram?

Multiple Choice Practice

For California Standards and CAHSEE Tutorials, visit PHSchool.com. Web Code: bcq-9045

GEOM 7.0

29. What values of x and y make the quadrilateral below a parallelogram?

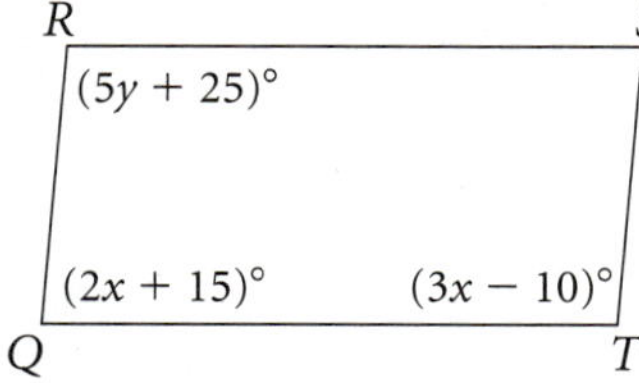

Ⓐ $x = 16, y = 33$
Ⓑ $x = 33, y = 16$
Ⓒ $x = 14, y = 35$
Ⓓ $x = 35, y = 14$

CAHSEE 7NS 1.6

30. In January, a newscaster reported that 462 births were recorded in Bramford Township last year, which was 63 less than the previous year. What was the percent of decrease in the total number of births from the previous year?

Ⓐ 12% Ⓑ 14% Ⓒ 18% Ⓓ 25%

Mixed Review

GO for Help

Lesson 6-2 Algebra **Find the value of each variable in each parallelogram.**

31.

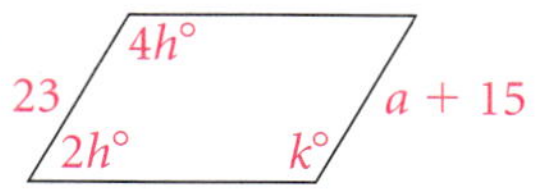

32.

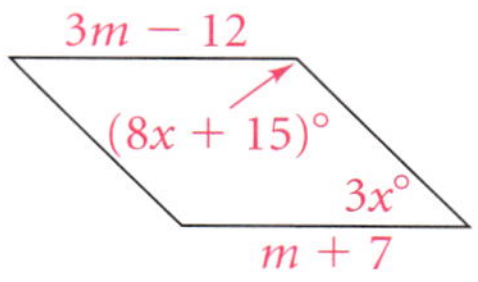

33.

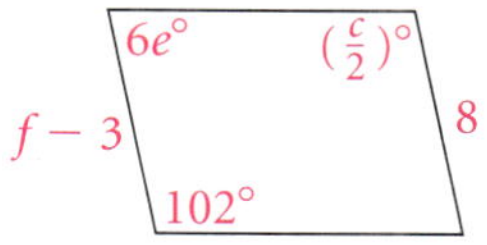

Lesson 4-7

34. Explain how you can use overlapping congruent triangles to prove $\overline{AC} \cong \overline{BD}$.

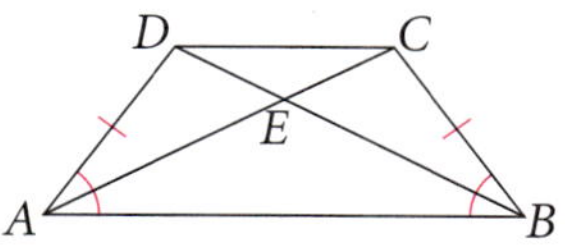

Lesson 2-2 **Write the two conditional statements that make up each biconditional.**

35. The diagonals of a quadrilateral bisect each other if and only if the quadrilateral is a parallelogram.

36. Two lines are parallel if and only if the two lines and a transversal form corresponding angles that are congruent.

37. Two nonvertical lines are perpendicular if and only if the product of their slopes is -1.

Checkpoint Quiz 1 — Lessons 6-1 through 6-3

Find the measures of the numbered angles for each parallelogram.

1. 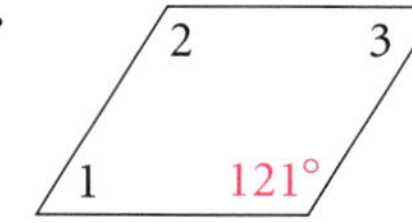

2. 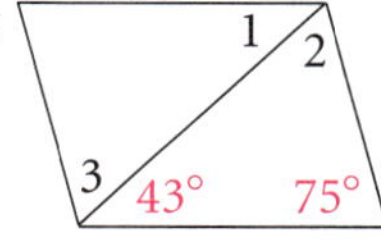

3. 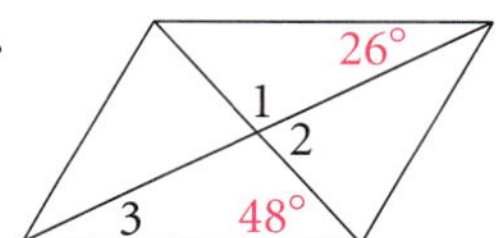

Classify each quadrilateral in as many ways as possible.

4. 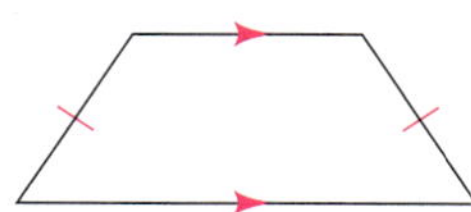

5. 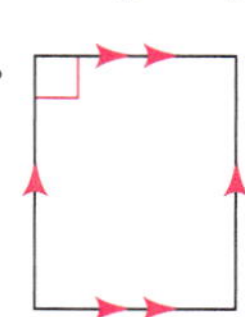

6. 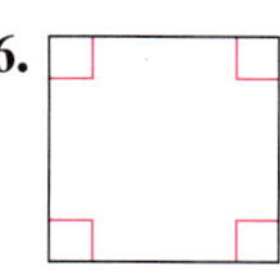

Algebra **Find the values of the variables for which *ABCD* is a parallelogram.**

7.

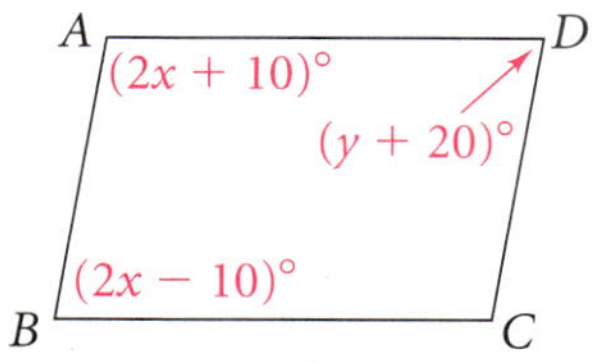

8. 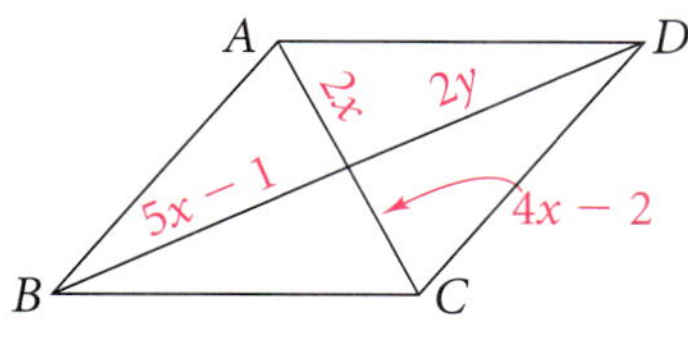

9. In the figure at the right, $\overleftrightarrow{AB} \parallel \overleftrightarrow{CD} \parallel \overleftrightarrow{EF}$. Find AE.

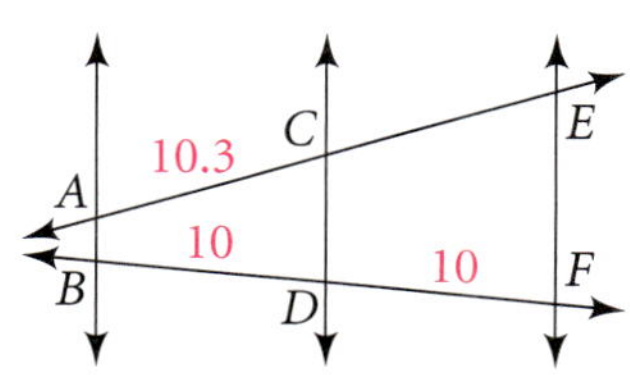

10. What is the most precise name for a quadrilateral with vertices $(3, 5)$, $(-1, 4)$, $(3, -5)$, and $(7, 4)$?

Diagonals of Parallelograms

Technology

FOR USE WITH LESSON 6-4

Construct

Use geometry software to construct a parallelogram.

- Construct segments $\overline{AB}$ and $\overline{BC}$.
- Construct a line through C parallel to $\overline{AB}$ and a line through A parallel to $\overline{BC}$. Label the point where the two lines intersect as D.

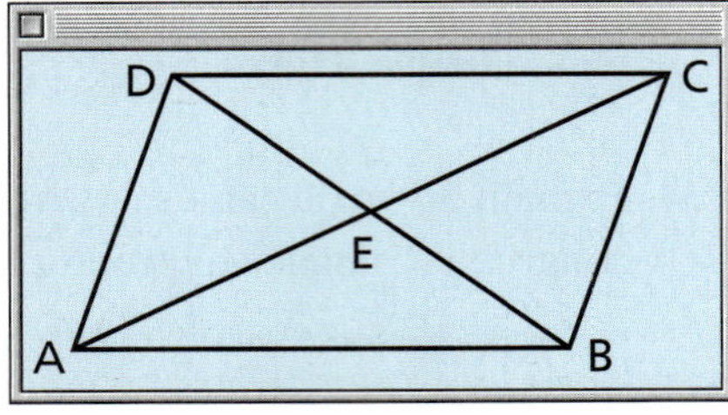

- Hide the lines and construct $\overline{AD}$ and $\overline{CD}$.
- Construct the diagonals of $\square ABCD$ and label their point of intersection E.

Set up the following measurements. Use them where indicated.

M1: $m\angle DAB$, or some other angle of the parallelogram, so that you can tell when the parallelogram is a rectangle

M2: AB and AD, or some other adjacent sides of the parallelogram, so you can tell when it is a rhombus

M3: $m\angle AED$, or some other angle at the intersection of the diagonals, so you can tell when the diagonals are perpendicular

M4: AC and DB, the diagonal lengths, so you can tell when they are equal

M5: the angles at each vertex of the parallelogram, so you can tell when the diagonals bisect the angles of the parallelogram

Investigate

- Manipulate the parallelogram to get a rectangle (M1). Make note of what appear to be any special properties of the diagonals of a rectangle. Manipulate the rectangle to check whether the properties hold (M3–5).
- Manipulate the parallelogram to get a rhombus (M2). Make note of what appear to be any special properties of the diagonals of a rhombus. Manipulate the rhombus to check whether the properties hold (M3–5).

EXERCISES

Make as many conjectures as you can about each of the following.

1. the diagonals of rectangles **2.** the diagonals of rhombuses **3.** the diagonals of squares

Extend

4. Manipulate the diagonals so they are perpendicular (M3). Make a conjecture about the type of parallelogram that is determined by perpendicular diagonals.

5. Manipulate the diagonals so they are congruent (M4). Make a conjecture about the type of parallelogram that is determined by congruent diagonals.

6. Manipulate the diagonals so they bisect the angles of the parallelogram (M5). Make a conjecture about the type of parallelogram that is determined by diagonals that bisect the angles.

7. Construct a trapezoid and then its two diagonals. Manipulate the trapezoid until the diagonals are the same length. Make a conjecture about the type of trapezoid that is determined by congruent diagonals.

Special Parallelograms

California Content Standards

GEOM 7.0 Students prove and use theorems involving the properties of quadrilaterals. ***Develop***
GEOM 12.0 Students find and use measures of sides and of interior angles of polygons to classify figures and solve problems. ***Develop***
GEOM 13.0 Students prove relationships between angles in polygons by using properties of complementary and supplementary angles. ***Develop***

What You'll Learn

- To use properties of diagonals of rhombuses and rectangles
- To determine whether a parallelogram is a rhombus or a rectangle

. . . And Why

To lay out a rectangular patio, as in Example 4

Check Skills You'll Need

GO for Help Lesson 6-2

***PACE* is a parallelogram and *m*∠*PAC* = 109. Complete each of the following.**

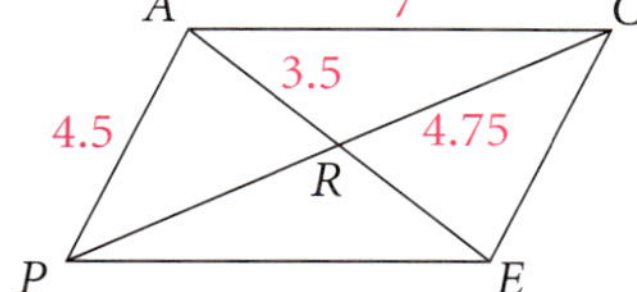

1. $EC = \blacksquare$ **2.** $EP = \blacksquare$
3. $m\angle CEP = \blacksquare$ **4.** $PR = \blacksquare$
5. $RE = \blacksquare$ **6.** $CP = \blacksquare$
7. $m\angle EPA = \blacksquare$ **8.** $m\angle ECA = \blacksquare$

9. Draw a rhombus that is not a square. Draw a rectangle that is not a square. Explain why each is not a square.

1 Diagonals of Rhombuses and Rectangles

If you draw two congruent isosceles triangles with base $\overline{PQ}$, you have drawn a rhombus. Note that $\angle RPQ$, $\angle SPQ$, $\angle RQP$, and $\angle SQP$ are all congruent. This suggests Theorem 6-9 and its proof.

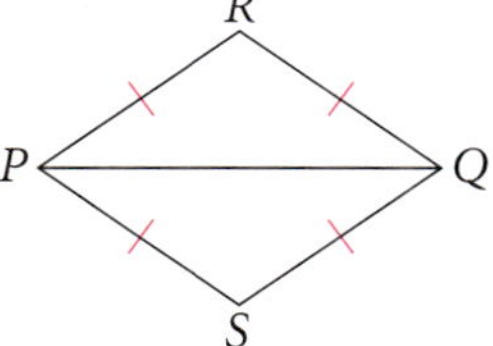

Key Concepts

Theorem 6-9

Each diagonal of a rhombus bisects two angles of the rhombus.

Proof

Proof of Theorem 6-9

Given: rhombus $ABCD$

Prove: $\overline{AC}$ bisects $\angle BAD$ and $\angle BCD$.

Proof: $ABCD$ is a rhombus, so its sides are all congruent. $\overline{AC} \cong \overline{AC}$ by the Reflexive Property of Congruence. Therefore, $\triangle ABC \cong \triangle ADC$ by the SSS Postulate. $\angle 1 \cong \angle 2$ and $\angle 3 \cong \angle 4$ by CPCTC. Therefore, $\overline{AC}$ bisects $\angle BAD$ and $\angle BCD$ by the definition of bisect.

You can show similarly that $\overline{BD}$ bisects $\angle ABC$ and $\angle ADC$.

B C A D 1 2 3 4

Vocabulary Tip

A segment bisects an angle if and only if it divides the angle into two congruent angles.

The diagonals of a rhombus provide an interesting application of the Converse of the Perpendicular Bisector Theorem.

In the rhombus at the right, points B and D are equidistant from A and C. By the Converse of the Perpendicular Bisector Theorem, they are on the perpendicular bisector of $\overline{AC}$. This proves the next theorem. In Exercise 22, you will prove it a second way.

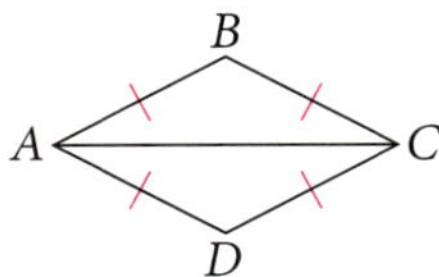

Key Concepts

Theorem 6-10

The diagonals of a rhombus are perpendicular.

$\overline{AC} \perp \overline{BD}$

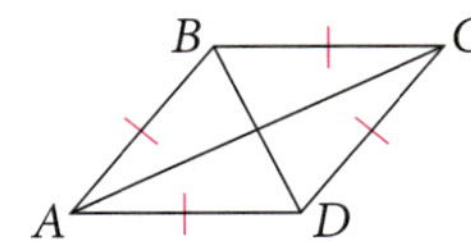

pantograph

Real-World Connection

The diagonals of the rhombus formed by the pantograph stay perpendicular when the pantograph rises or lowers.

You can use Theorems 6-9 and 6-10 to find angle measures in rhombuses.

1 EXAMPLE Finding Angle Measures

$MNPQ$ is a rhombus and $m\angle N = 120$.
Find the measures of the numbered angles.

$m\angle 1 = m\angle 3$	**Isosceles Triangle Theorem**
$m\angle 1 + m\angle 3 + 120 = 180$	**Triangle Angle-Sum Theorem**
$2(m\angle 1) + 120 = 180$	**Substitute.**
$2(m\angle 1) = 60$	**Subtract 120 from each side.**
$m\angle 1 = 30$	**Divide each side by 2.**

Therefore, $m\angle 1 = m\angle 3 = 30$. By Theorem 6-9, $m\angle 1 = m\angle 2$ and $m\angle 3 = m\angle 4$. Therefore, $m\angle 1 = m\angle 2 = m\angle 3 = m\angle 4 = 30$.

N 120° 3 4 P
1 2
M Q

CA Standards Check

1 Find the measures of the numbered angles in the rhombus.

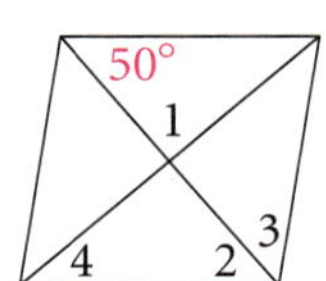

The diagonals of a rectangle, another parallelogram, also have a special property.

Key Concepts

Theorem 6-11

The diagonals of a rectangle are congruent.

Proof

Proof of Theorem 6-11

Given: Rectangle $ABCD$

Prove: $\overline{AC} \cong \overline{BD}$

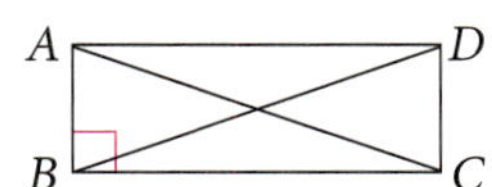

Proof: $ABCD$ is a rectangle, so it is also a parallelogram.
$\overline{AB} \cong \overline{DC}$ because opposite sides of a parallelogram are congruent.
$\overline{BC} \cong \overline{BC}$ by the Reflexive Property of Congruence.
$\angle ABC$ and $\angle DCB$ are right angles by the definition of rectangle.
$\angle ABC \cong \angle DCB$ because all right angles are congruent.
$\triangle ABC \cong \triangle DCB$ by SAS. $\overline{AC} \cong \overline{BD}$ by CPCTC.

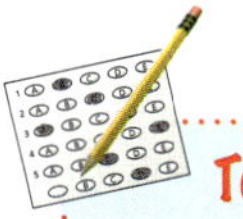

Test-Taking Tip

To check your work, you can often substitute a value you find for an unknown. In Example 2, substitute $\frac{9}{4}$ for y to check that $FD = GE$.

Finding Diagonal Length

Multiple Choice Find the length of $\overline{FD}$ in rectangle $GFED$ if $FD = 2y + 4$ and $GE = 6y - 5$.

Ⓐ $2\frac{1}{2}$ Ⓑ $3\frac{1}{4}$ Ⓒ $4\frac{1}{2}$ Ⓓ $8\frac{1}{2}$

F E G D

$2y + 4 = 6y - 5$	**Diagonals of a rectangle are congruent.**
$9 = 4y$	**Subtract 2y from each side and add 5 to each side.**
$\frac{9}{4} = y$	**Divide each side by 4.**

$FD = 2\left(\frac{9}{4}\right) + 4 = \frac{17}{2}$, or $8\frac{1}{2}$. The correct answer is D.

CA Standards Check 2 Find the length of the diagonals of $GFED$ if $FD = 5y - 9$ and $GE = y + 5$.

2 Is the Parallelogram a Rhombus or a Rectangle?

The following theorems are the converses of Theorems 6-9, 6-10, and 6-11, respectively. You will prove these theorems in the Exercises.

Key Concepts

Theorem 6-12

If one diagonal of a parallelogram bisects two angles of the parallelogram, then the parallelogram is a rhombus.

Theorem 6-13

If the diagonals of a parallelogram are perpendicular, then the parallelogram is a rhombus.

Theorem 6-14

If the diagonals of a parallelogram are congruent, then the parallelogram is a rectangle.

You can use Theorems 6-12, 6-13, and 6-14 to classify parallelograms.

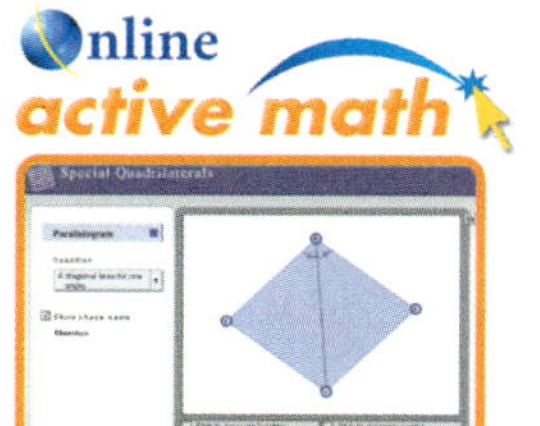

For: Quadrilateral Activity
Use: Interactive Textbook, 6-4

3 EXAMPLE Identifying Special Parallelograms

Can you conclude that the parallelogram is a rhombus or a rectangle? Explain.

a.

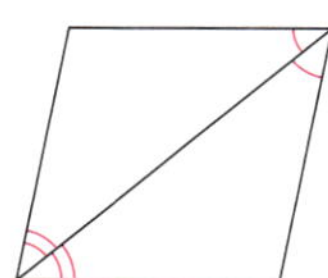

Yes. A diagonal bisects two angles. By Theorem 6-12, this parallelogram is a rhombus.

b.

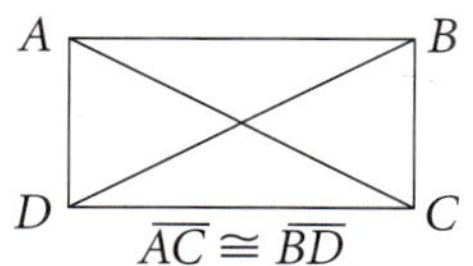

Yes. The diagonals are congruent. By Theorem 6-14, this parallelogram is a rectangle.

CA Standards Check 3 A parallelogram has angles of 30°, 150°, 30°, and 150°. Can you conclude that it is a rhombus or a rectangle? Explain.

You can use properties of diagonals to construct special parallelograms.

4 EXAMPLE Real-World Connection

Community Service Builders use properties of diagonals to "square up" rectangular shapes like building frames and playing-field boundaries.

Suppose you are on the volunteer building team at the right. You are helping to lay out a rectangular patio. Explain how to use properties of diagonals to locate the four corners.

To locate the corners, you can use two theorems:

- Theorem 6-7: If the diagonals of a quadrilateral bisect each other, then the quadrilateral is a parallelogram.
- Theorem 6-14: If the diagonals of a parallelogram are congruent, then the parallelogram is a rectangle.

First, cut two pieces of rope that will be the diagonals of the foundation rectangle. Cut them the same length because of Theorem 6-14. Join them at their midpoints because of Theorem 6-7. Then pull the ropes straight and taut. The ends of the ropes will be the corners of a rectangle.

Real-World Connection

The frame of a house provides a rigid structure for walls.

CA Standards Check 4 Kate thinks that they can adapt this method slightly to stake off a square play area. Is she right? Explain.

EXERCISES

Standards Practice

For more exercises, see *Extra Skill, Word Problem, and Proof Practice.*

GEOM 7.0, 12.0, 13.0

A Practice by Example

Example 1 (page 330)

GO for Help

Find the measures of the numbered angles in each rhombus.

1.

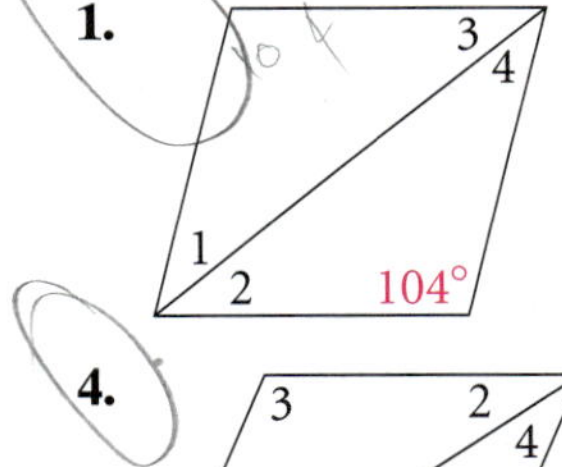

2.

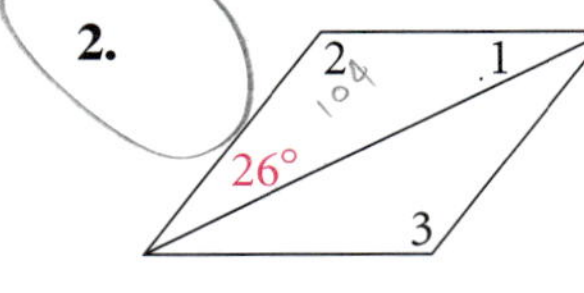

3.

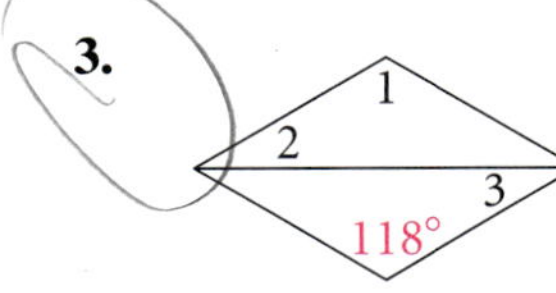

4.

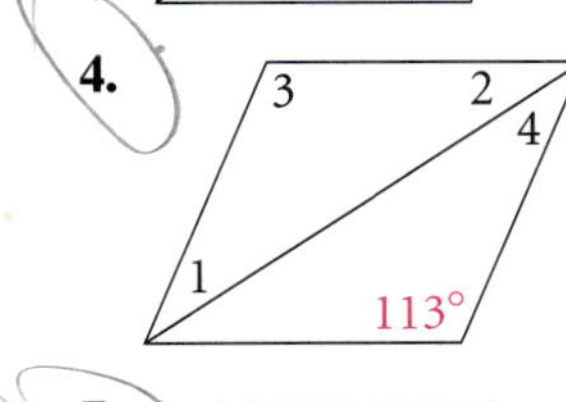

5.

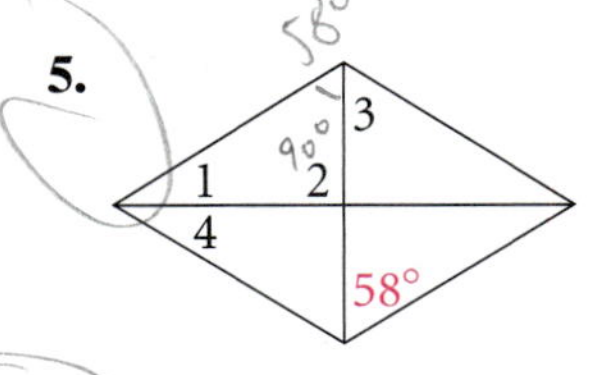

6.

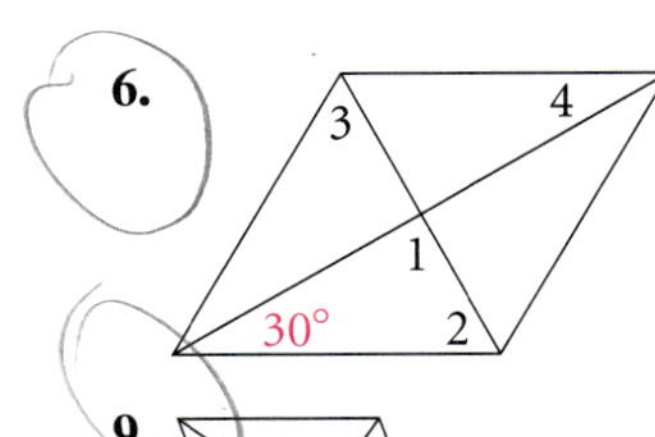

7.

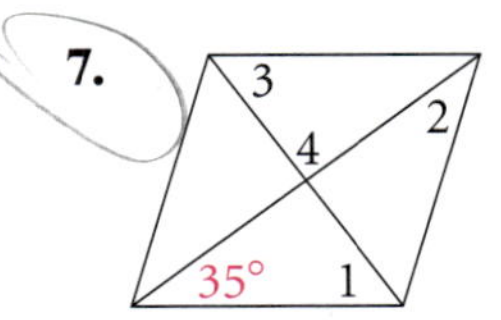

8.

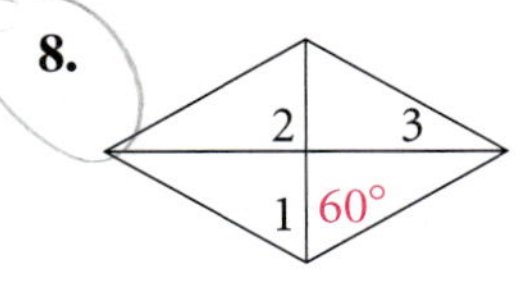

9.

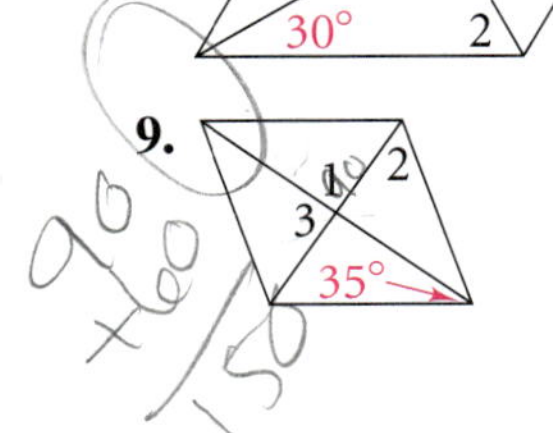

Example 2 (page 331)

Algebra ***LMNP* is a rectangle. Find the value of *x* and the length of each diagonal.**

10. $LN = x$ and $MP = 2x - 4$

11. $LN = 5x - 8$ and $MP = 2x + 1$

12. $LN = 3x + 1$ and $MP = 8x - 4$

13. $LN = 9x - 14$ and $MP = 7x + 4$

14. $LN = 7x - 2$ and $MP = 4x + 3$

15. $LN = 3x + 5$ and $MP = 9x - 10$

Example 3 (page 331)

Is the parallelogram a rhombus or a rectangle? Justify your answer.

16. 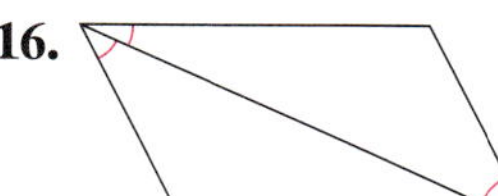

17. 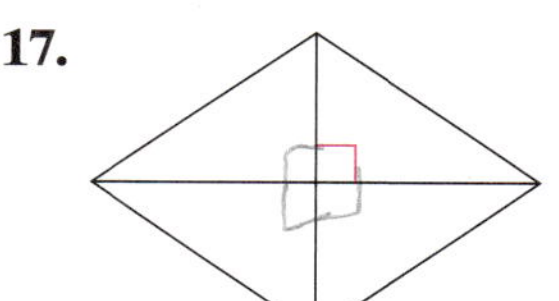

18. 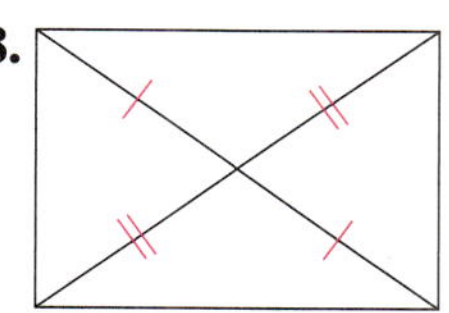

Example 4 (page 332)

19. Hardware You can use a simple device called a turnbuckle to "square up" structures that are parallelograms. For the gate pictured at the right, you tighten or loosen the turnbuckle on the diagonal cable so that the cable stays congruent to the other diagonal. Explain why a frame that normally is rectangular will, when it sags, keep the shape of a parallelogram.

Turnbuckle

20. Carpentry A carpenter is building a bookcase. How can she use a tape measure to check that the bookshelf is rectangular? Justify your answer and name any theorems used.

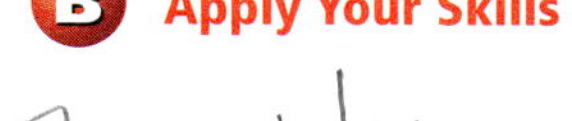

B Apply Your Skills

21. Reasoning Suppose the diagonals of a parallelogram are both perpendicular and congruent. What type of special quadrilateral is it? Explain your reasoning.

22. Developing Proof Complete the flow proof of Theorem 6-10.

Given: $ABCD$ is a rhombus.

Prove: $\overline{AC} \perp \overline{BD}$

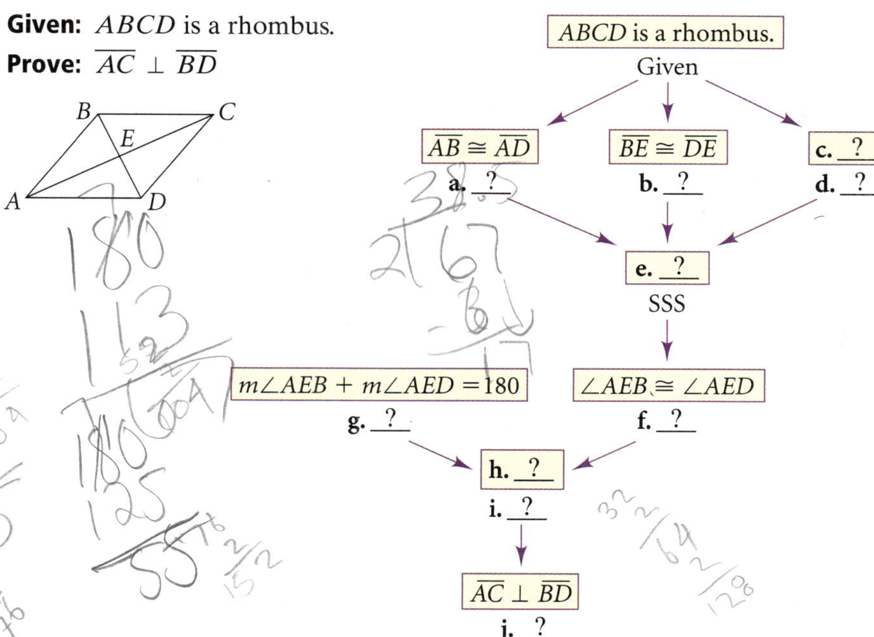

Proof **23.** Prove Theorem 6-13.

Given: $\square ABCD$; $\overline{AC} \perp \overline{BD}$ at E.

Prove: $ABCD$ is a rhombus.

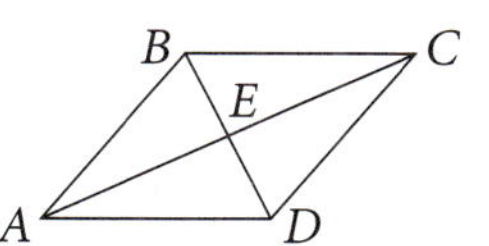

24. Multiple Choice A diagonal of a parallelogram bisects one angle of the parallelogram. What kind of quadrilateral must the figure be?

Ⓐ rhombus Ⓑ rectangle Ⓒ square Ⓓ cannot tell

Homework Video Tutor

Visit: PHSchool.com
Web Code: bce-0604

Using Symbols **Create your own distinctive symbols for a parallelogram, rhombus, rectangle, and square. Then copy the properties in Exercises 25–34. After each property, use your symbols to list the quadrilaterals having that property.**

25. All sides are $\cong$.

26. Opposite sides are $\cong$.

27. Opposite sides are $\parallel$.

28. Opposite $\angle$s are $\cong$.

29. All $\angle$s are right $\angle$s.

30. Consecutive $\angle$s are supplementary.

31. Diagonals bisect each other.

32. Diagonals are $\cong$.

33. Diagonals are $\perp$.

34. Each diagonal bisects opposite $\angle$s.

Which, if any, of the properties in Exercises 25–34 can the following type of quadrilateral have? Draw diagrams to illustrate.

35. a trapezoid

36. a kite

37. a quadrilateral that is not a special quadrilateral

38. **Writing** Summarize the properties of squares that follow from a square being **(a)** a parallelogram, **(b)** a rhombus, and **(c)** a rectangle.

Proof **39.** Prove Theorem 6-12.

Given: $ABCD$ is a parallelogram; $\overline{AC}$ bisects $\angle BAD$ and $\angle BCD$.

Prove: $ABCD$ is a rhombus.

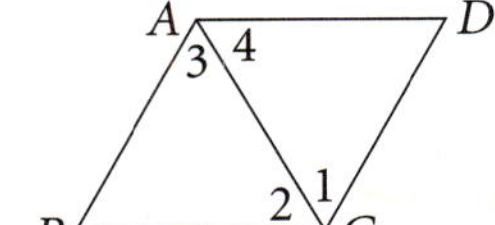

Proof **40.** Prove Theorem 6-14.

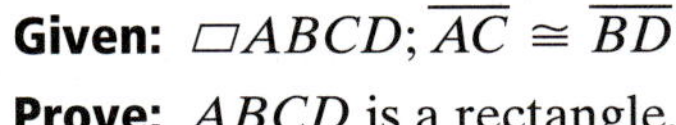

Given: $\square ABCD$; $\overline{AC} \cong \overline{BD}$

Prove: $ABCD$ is a rectangle.

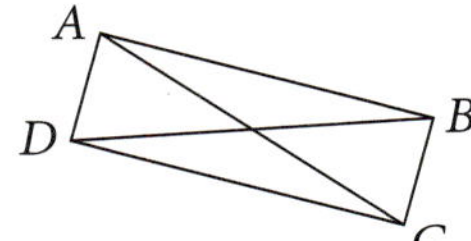

To review what makes a good definition, see Lesson 2-2.

Reasoning **Decide whether each of these is a good definition. Justify your answer.**

41. A rectangle is a quadrilateral with four right angles.

42. A rhombus is a quadrilateral with four congruent sides.

43. A square is a quadrilateral with four right angles and four congruent sides.

Algebra **Find the value(s) of the variable(s) for each parallelogram.**

44. $RZ = 2x + 5$, $SW = 5x - 20$

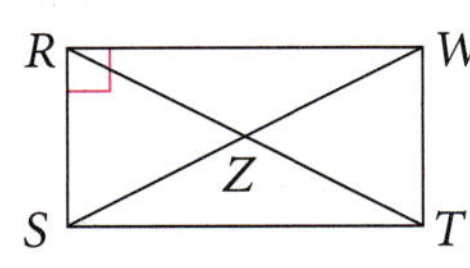

45. $m\angle 1 = 3y - 6$

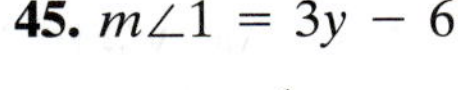

46. $BD = 4x - y + 1$

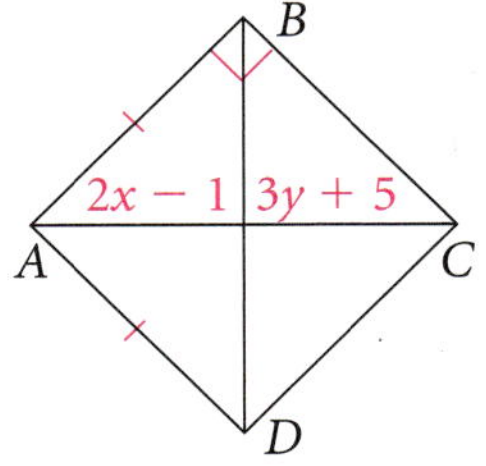

Open-Ended **Given two segments with lengths a and b ($a \neq b$), what special quadrilaterals can you sketch that meet these conditions? Show each sketch.**

47. Both diagonals have length a.

48. The two diagonals have lengths a and b.

49. One diagonal has length a, one side of the quadrilateral has length b.

Problem Solving Hint

In Exercises 54–59, the "given diagonals" are two segments you draw. The constructed figures must have diagonals that match.

Algebra ***ABCD*** **is a rectangle. Find the length of each diagonal.**

50. $AC = 2(x - 3)$ and $BD = x + 5$

51. $AC = 2(5a + 1)$ and $BD = 2(a + 1)$

52. $AC = \frac{3y}{5}$ and $BD = 3y - 4$

53. $AC = \frac{3c}{9}$ and $BD = 4 - c$

Constructions **Explain how to construct each figure, given its diagonals.**

54. parallelogram

55. rectangle

56. rhombus

57. square

58. kite

59. trapezoid

Challenge

Determine whether the quadrilateral can be a parallelogram. If not, write ***impossible.*** **Explain.**

60. The diagonals are congruent, but the quadrilateral has no right angles.

61. Each diagonal is 3 cm long and two opposite sides are 2 cm long.

62. Two opposite angles are right angles, but the quadrilateral is not a rectangle.

Proof **63.** In Theorem 6-12, replace "two angles" with "one angle." Write a paragraph that proves this new statement true or show a counterexample to prove it false.

Multiple Choice Practice

For California Standards and CAHSEE Tutorials, visit PHSchool.com. Web Code: bcq-9045

GEOM 13.0

64. In the rhombus at the right, $m\angle 1$ and $m\angle 3 = 30°$. What is $m\angle 2$?

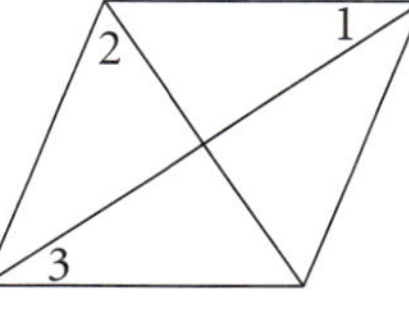

Ⓐ 30°
Ⓑ 60°
Ⓒ 75°
Ⓓ 120°

GEOM 7.0

65. Figure *WXYZ* is a quadrilateral. If the diagonals are congruent and bisect opposite angles, what is the most specific name for this figure?

Ⓐ rectangle
Ⓑ square
Ⓒ rhombus
Ⓓ parallelogram

Mixed Review

Lesson 6-3

Can you conclude that the quadrilateral is a parallelogram? Explain.

66.

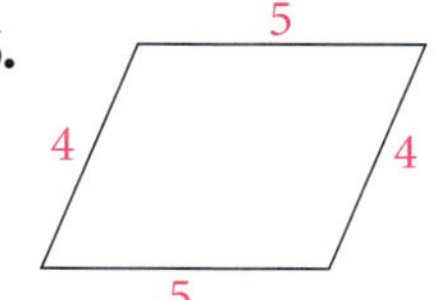

67.

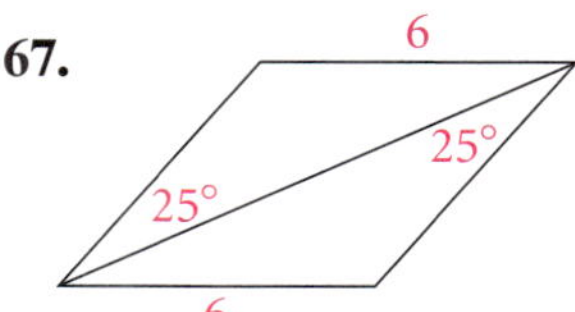

68.

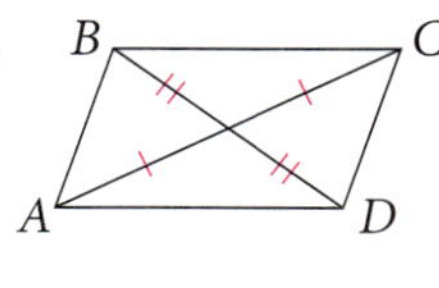

Lesson 5-1

In $\triangle PQR$, points *S*, *T*, and *U* are midpoints. Complete each statement.

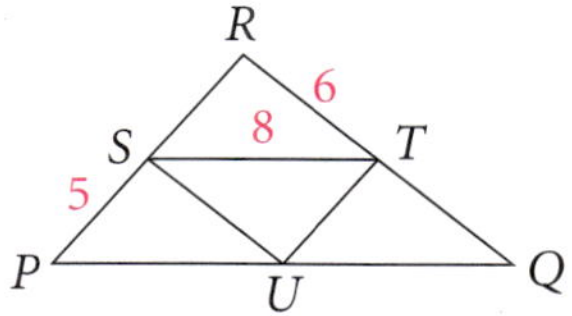

69. $TQ =$ __?__

70. $PQ =$ __?__

71. $TU =$ __?__

72. $\overline{SU} \parallel$ __?__

73. $\overline{TU} \parallel$ __?__

74. $\overline{PQ} \parallel$ __?__

Lesson 3-4

75. **Algebra** Find the value of *c*.

$c°$

$(c + 28)°$

$(c + 65)°$

Online Lesson Quiz Visit: PHSchool.com, Web Code: bca-0604

6-5 Trapezoids and Kites

California Content Standards

GEOM 7.0 Students prove and use theorems involving the properties of quadrilaterals. ***Master***

GEOM 12.0 Students find and use measures of sides and of interior angles of triangles and polygons to solve problems. ***Master***

GEOM 13.0 Students prove relationships between angles in polygons by using properties of complementary and supplementary angles. ***Develop***

What You'll Learn

- To verify and use properties of trapezoids and kites

. . . And Why

To find angle measures of trapezoidal glass panes, as in Example 2

Check Skills You'll Need

GO for Help Lesson 6-1

Algebra **Find the values of the variables. Then find the lengths of the sides.**

1.

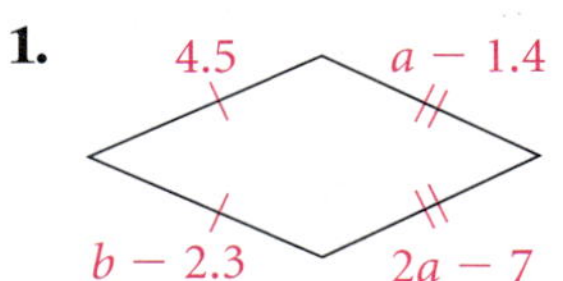

2.

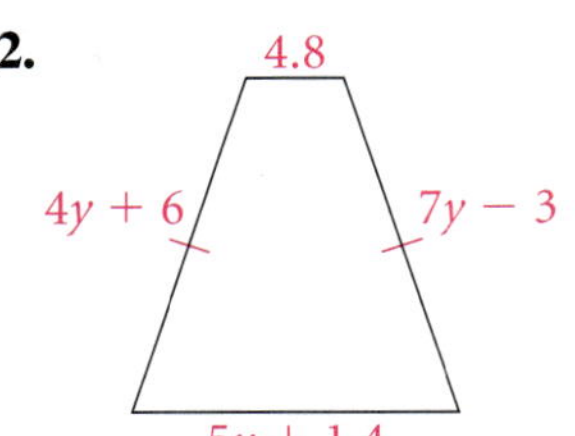

3.

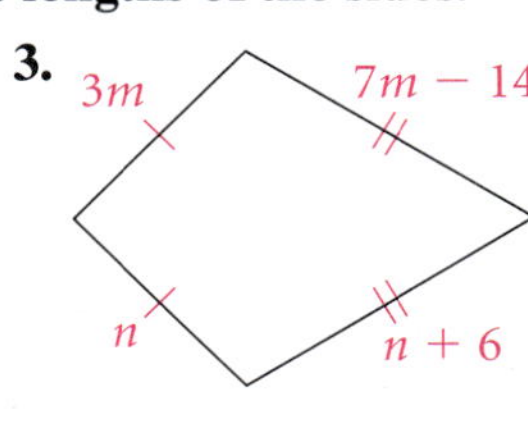

New Vocabulary • base angles of a trapezoid

1 Properties of Trapezoids and Kites

The parallel sides of a trapezoid are its bases. The nonparallel sides are its legs. Two angles that share a base of a trapezoid are **base angles** of the trapezoid.

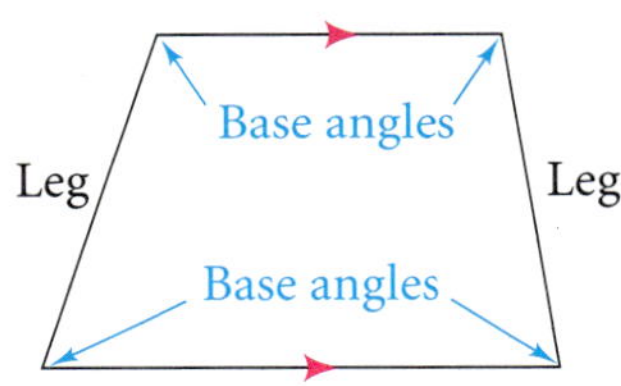

The following theorem is about each pair of base angles. You will be asked to prove it in Exercise 38.

Key Concepts

Theorem 6-15

The base angles of an isosceles trapezoid are congruent.

Real-World Connection

In the isosceles trapezoids at the top of this electric tea kettle, each pair of base angles are congruent.

The bases of a trapezoid are parallel. Therefore the two angles that share a leg are supplementary. This fact and Theorem 6-15 allow you to solve problems involving the angles of a trapezoid.

1 EXAMPLE Finding Angle Measures in Trapezoids

$ABCD$ is an isosceles trapezoid and $m\angle B = 102$. Find $m\angle A$, $m\angle C$, and $m\angle D$.

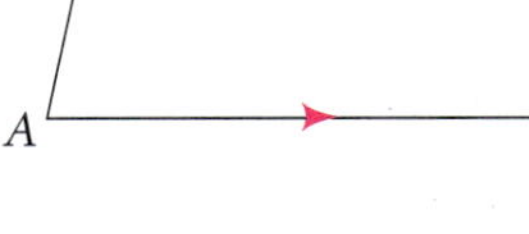

$m\angle A + m\angle B = 180$	**Two angles that share a leg are supplementary.**
$m\angle A + 102 = 180$	**Substitute.**
$m\angle A = 78$	**Subtract 102 from each side.**

By Theorem 6-15, $m\angle C = m\angle B = 102$ and $m\angle D = m\angle A = 78$.

CA Standards Check

1 In the isosceles trapezoid, $m\angle S = 70$. Find $m\angle P$, $m\angle Q$, and $m\angle R$.

2 EXAMPLE Real-World Connection

Architecture The second ring of the ceiling shown at the left is made from congruent isosceles trapezoids that create the illusion of circles. What are the measures of the base angles of these trapezoids?

Each trapezoid is part of an isosceles triangle whose base angles are the acute base angles of the trapezoid. The isosceles triangle has a vertex angle that is half as large as one of the 20 angles at the center of the ceiling.

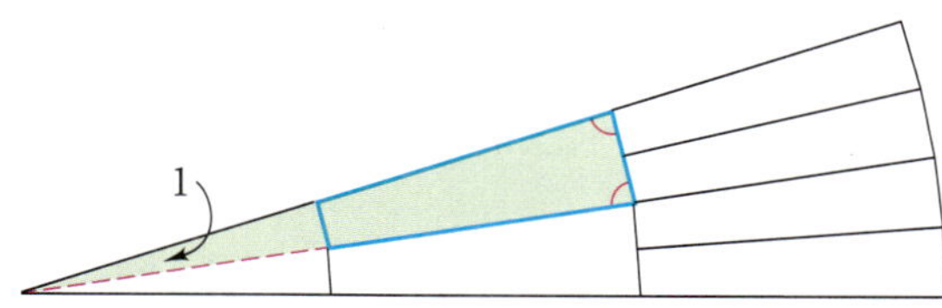

You are looking up at Harbour Centre Tower in Vancouver, Canada.

The measure of each angle at the center of the ceiling is $\frac{360}{20}$ or 18.

The measure of $\angle 1$ is $\frac{18}{2}$, or 9.

The measure of each acute base angle is $\frac{180 - 9}{2}$, or 85.5.

The measure of each obtuse base angle is $180 - 85.5$, or 94.5.

CA Standards Check

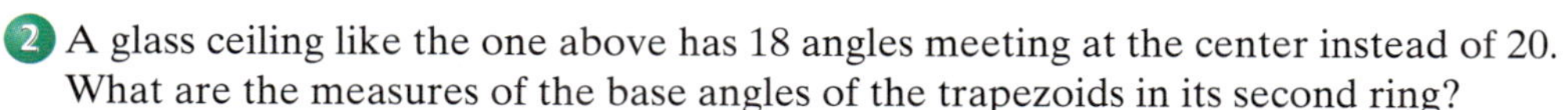

2 A glass ceiling like the one above has 18 angles meeting at the center instead of 20. What are the measures of the base angles of the trapezoids in its second ring?

Like the diagonals of parallelograms, the diagonals of an isosceles trapezoid have a special property.

Key Concepts

Theorem 6-16

The diagonals of an isosceles trapezoid are congruent.

Proof

Proof of Theorem 6-16

Given: Isosceles trapezoid $ABCD$ with $\overline{AB} \cong \overline{DC}$

Prove: $\overline{AC} \cong \overline{DB}$

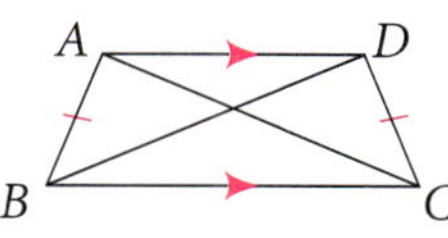

It is given that $\overline{AB} \cong \overline{DC}$. Because the base angles of an isosceles trapezoid are congruent, $\angle ABC \cong \angle DCB$. By the Reflexive Property of Congruence, $\overline{BC} \cong \overline{BC}$. Then, by the SAS Postulate, $\triangle ABC \cong \triangle DCB$. Therefore, $\overline{AC} \cong \overline{DB}$ by CPCTC.

Another special quadrilateral that is not a parallelogram is a kite. The diagonals of a kite, like the diagonals of a rhombus, are perpendicular. A proof of this for a kite (next page) is quite like its proof for a rhombus (at the top of page 330).

Theorem 6-17

The diagonals of a kite are perpendicular.

Proof

Proof of Theorem 6-17

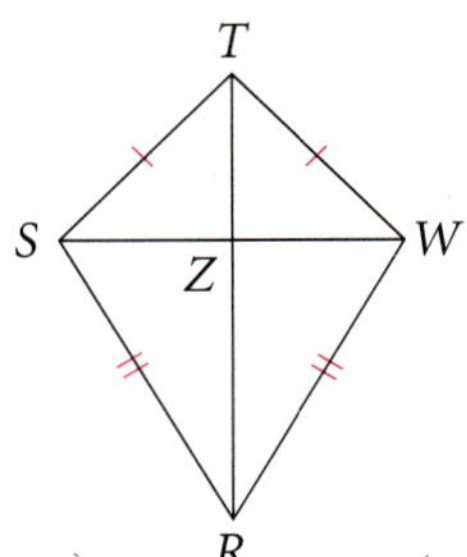

Given: Kite $RSTW$ with $\overline{TS} \cong \overline{TW}$ and $\overline{RS} \cong \overline{RW}$

Prove: $\overline{TR} \perp \overline{SW}$

Both T and R are equidistant from S and W. By the Converse of the Perpendicular Bisector Theorem, T and R lie on the perpendicular bisector of $\overline{SW}$. Since there is exactly one line through any two points (Postulate 1-1), $\overleftrightarrow{TR}$ must be the perpendicular bisector of $\overline{SW}$. Therefore, $\overline{TR} \perp \overline{SW}$.

You can use Theorem 6-17 to find angle measures in kites.

3 EXAMPLE Finding Angle Measures in Kites

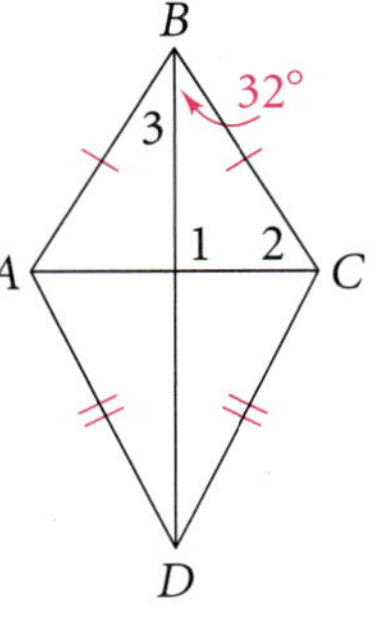

Find $m\angle 1$, $m\angle 2$, and $m\angle 3$ in the kite.

$m\angle 1 = 90$	**Diagonals of a kite are perpendicular.**
$90 + m\angle 2 + 32 = 180$	**Triangle Angle-Sum Theorem**
$122 + m\angle 2 = 180$	**Simplify.**
$m\angle 2 = 58$	**Subtract 122 from each side.**

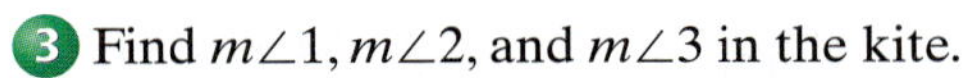

$\triangle ABD \cong \triangle CBD$ by SSS.

By CPCTC, $m\angle 3 = m\angle DBC = 32$.

3 Find $m\angle 1$, $m\angle 2$, and $m\angle 3$ in the kite.

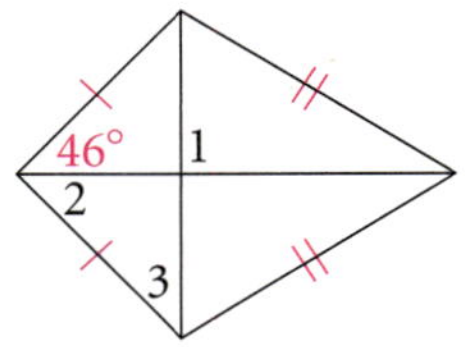

For more exercises, see *Extra Skill, Word Problem, and Proof Practice.*

GEOM 7.0, 12.0, 13.0

Example 1 (page 336)

Each trapezoid is isosceles. Find the measure of each angle.

1.

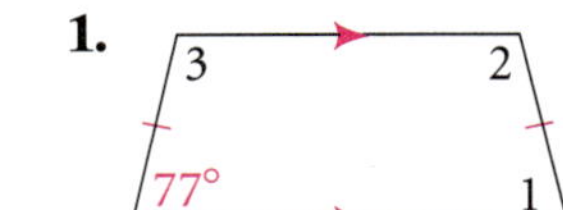

2.

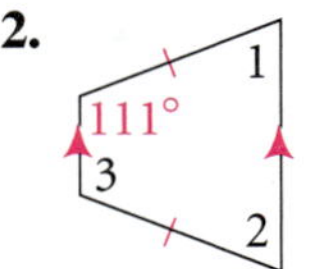

3.

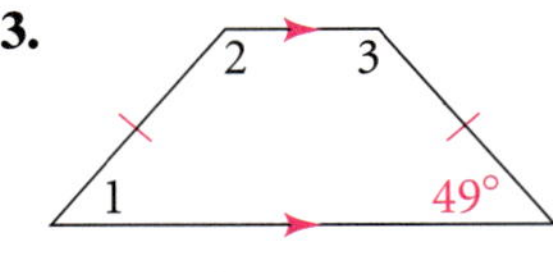

4.

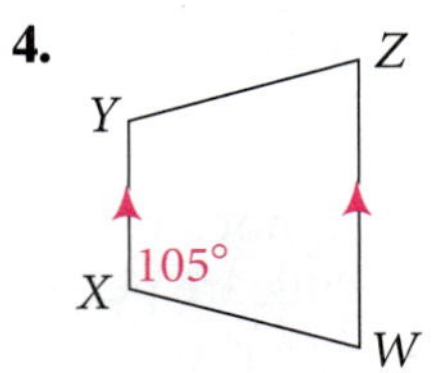

5.

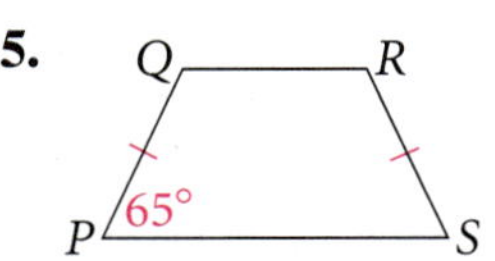

6.

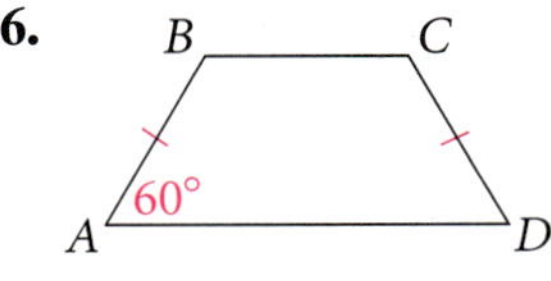

Example 2 (page 337)

7. **Design** Each patio umbrella is made of eight panels that are congruent isosceles triangles with parallel stripes. A sample panel is shown at the right. The vertex angle of the panel measures 42.
   a. Classify the quadrilaterals shown as blue stripes on the panel.
   b. Find the measures of the quadrilaterals' interior angles.

Example 3 (page 338)

Find the measures of the numbered angles in each kite.

8. 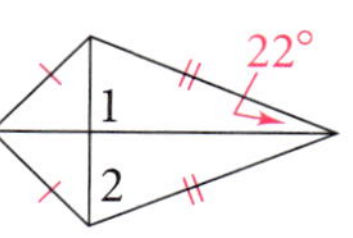

9. 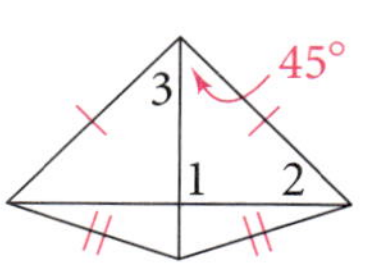

10. 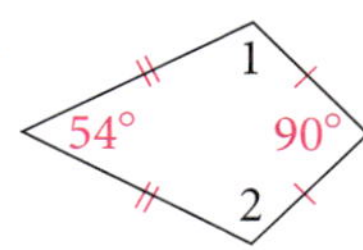

11. 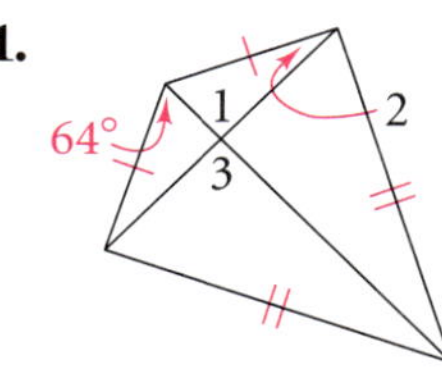

12. 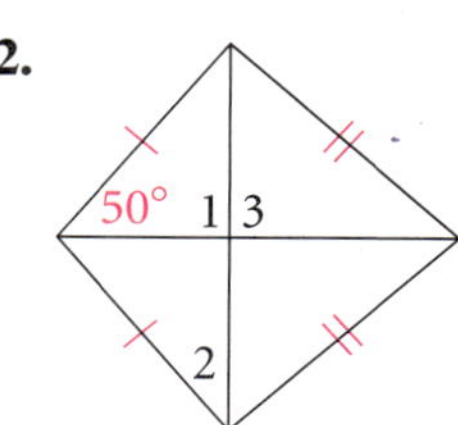

13. 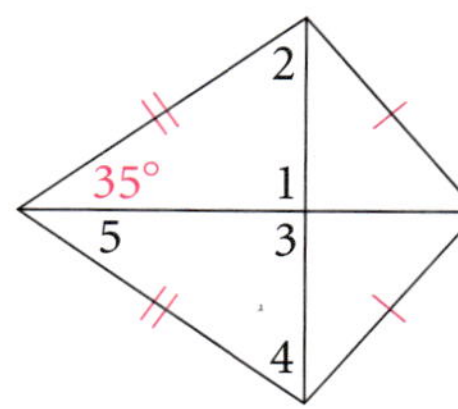

14. 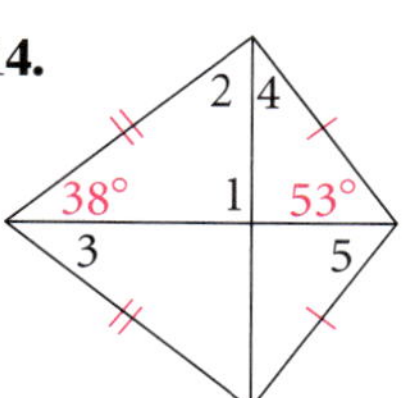

15. 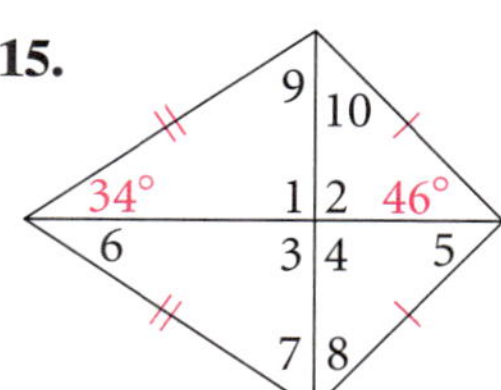

16. 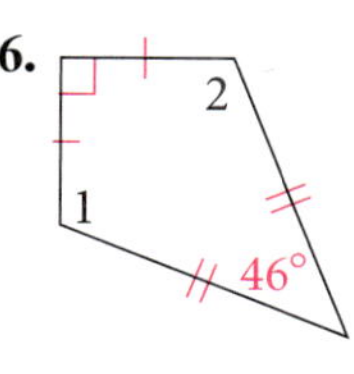

B **Apply Your Skills**

17. **Open-Ended** Sketch two kites that are not congruent, but with the diagonals of one congruent to the diagonals of the other.

18. The perimeter of a kite is 66 cm. The length of one of its sides is 3 cm less than twice the length of another. Find the length of each side of the kite.

19. **Critical Thinking** If $KLMN$ is an isosceles trapezoid, is it possible for $\overline{KM}$ to bisect $\angle LMN$ and $\angle LKN$? Explain.

x^2 **Algebra** **Find the value of the variable in each isosceles trapezoid.**

20. 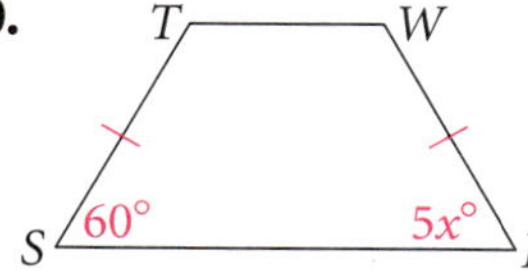

21. 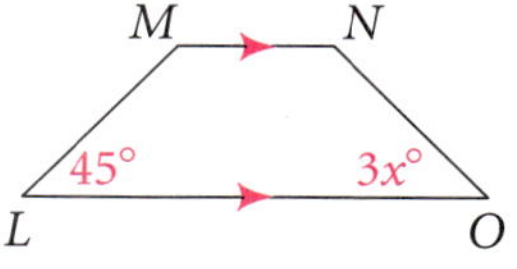

22. 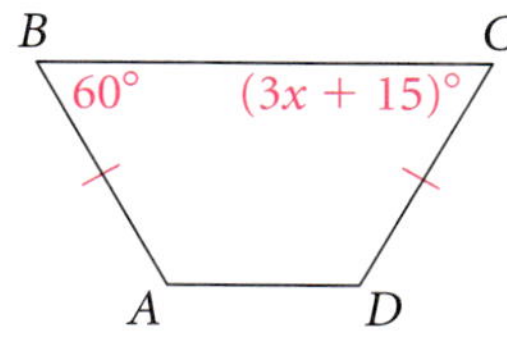

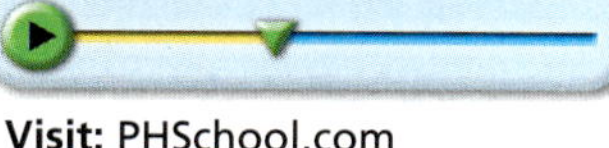

Visit: PHSchool.com
Web Code: bce-0605

23. 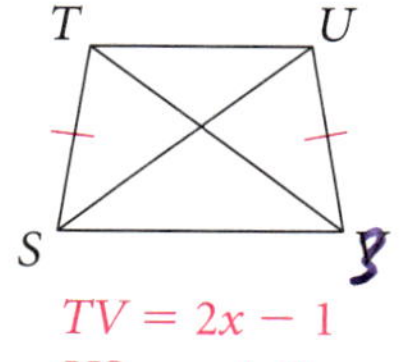

$TV = 2x - 1$
$US = x + 2$

24. 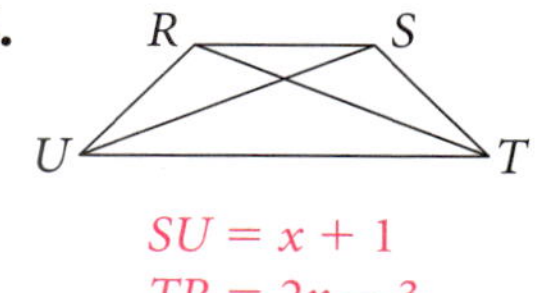

$SU = x + 1$
$TR = 2x - 3$

25. 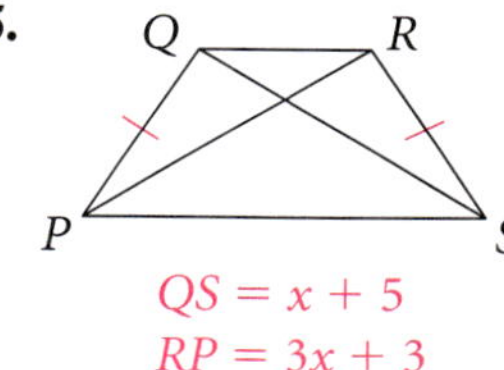

$QS = x + 5$
$RP = 3x + 3$

Algebra **Find the value(s) of the variable(s) in each kite.**

26.

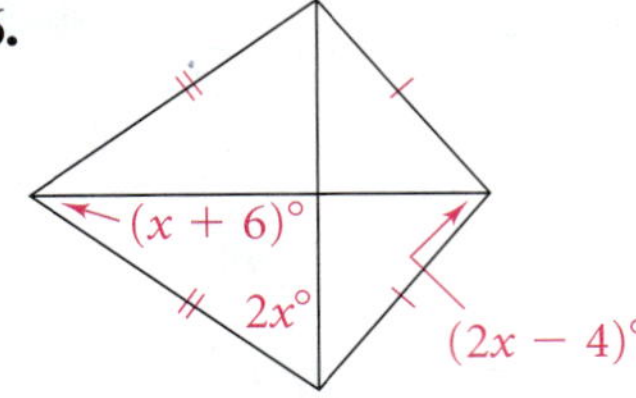

27.

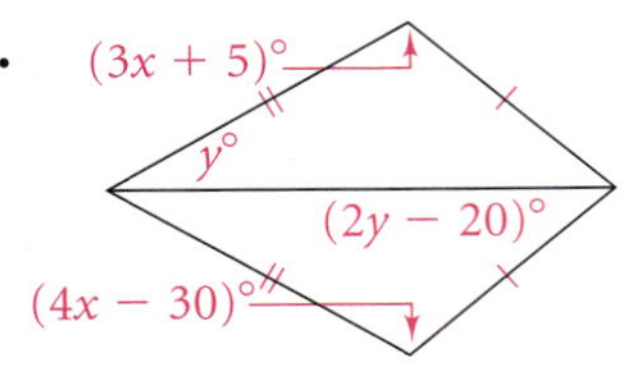

28.

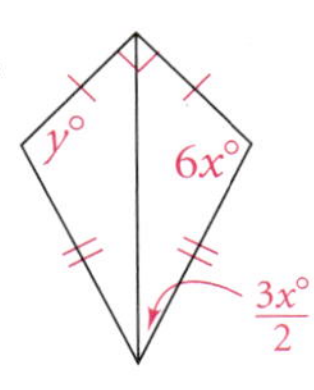

Exercises 29–30

Bridge Design **A quadrilateral is formed by the beams of the bridge at the left.**

29. Classify the quadrilateral. Explain your reasoning.

30. Find the measures of the other interior angles of the quadrilateral.

Critical Thinking **Can two angles of a kite be as follows? Explain.**

31. opposite and acute

32. consecutive and obtuse

33. opposite and supplementary

34. consecutive and supplementary

35. opposite and complementary

36. consecutive and complementary

37. **Writing** A kite is sometimes defined as a quadrilateral with two pairs of consecutive sides congruent. Compare this to the definition you learned in Lesson 6-1. Are parallelograms, trapezoids, rhombuses, rectangles, or squares special kinds of kites according to the changed definition? Explain.

38. **Developing Proof** The plan suggests a proof of Theorem 6-15. Write a proof that follows the plan.

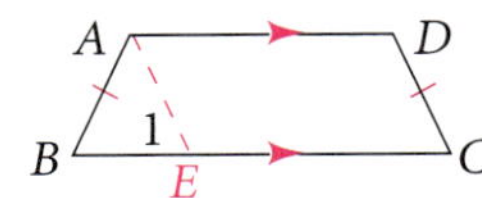

Given: Isosceles trapezoid $ABCD$ with $\overline{AB} \cong \overline{DC}$

Prove: $\angle B \cong \angle C$ and $\angle BAD \cong \angle D$

Plan: Begin by drawing $\overline{AE} \parallel \overline{DC}$ to form parallelogram $AECD$ so that $\overline{AE} \cong \overline{DC} \cong \overline{AB}$. $\angle B \cong \angle C$ because $\angle B \cong \angle 1$ and $\angle 1 \cong \angle C$. Also, $\angle BAD \cong \angle D$ because they are supplements of the congruent angles, $\angle B$ and $\angle C$.

Proof **Write a proof. Use the given figure with additional lines as needed.**

39. Given: Isosceles trapezoid $TRAP$ with $\overline{TR} \cong \overline{PA}$

Prove: $\angle RTA \cong \angle APR$

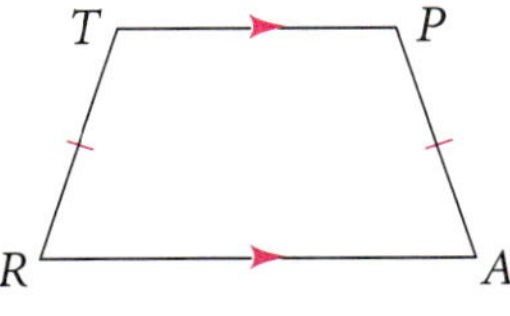

Challenge

40. Given: Isosceles trapezoid $TRAP$ with $\overline{TR} \cong \overline{PA}$; $\overline{BI}$ is the perpendicular bisector of $\overline{RA}$ intersecting $\overline{RA}$ at B and $\overline{TP}$ at I.

Prove: $\overline{BI}$ is the perpendicular bisector of $\overline{TP}$.

For a trapezoid, consider the segment joining the midpoints of the two given segments. How are its length and the lengths of the two parallel sides of the trapezoid related? Justify your answer.

41. the two nonparallel sides

42. the diagonals

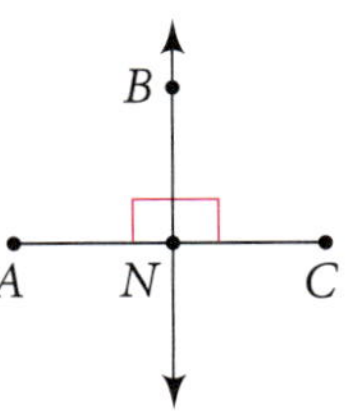

43. $\overleftrightarrow{BN}$ is the perpendicular bisector of $\overline{AC}$ at N. Describe the set of points, D, for which $ABCD$ is a kite.

Proof **44.** Prove that the angles formed by the noncongruent sides of a kite are congruent. (*Hint:* Draw a diagonal of the kite.)

Multiple Choice Practice

For California Standards and CAHSEE Tutorials, visit PHSchool.com. Web Code: bcq-9045

GEOM 7.0

45. Which of the following quadrilaterals does not have a pair of opposite sides congruent?

(A) isosceles trapezoid (B) kite (C) rhombus (D) rectangle

GEOM 12.0

46. In the kite at the right, $m\angle 1 = 24°$ and $m\angle 2 = 44°$. What is $m\angle 3$?

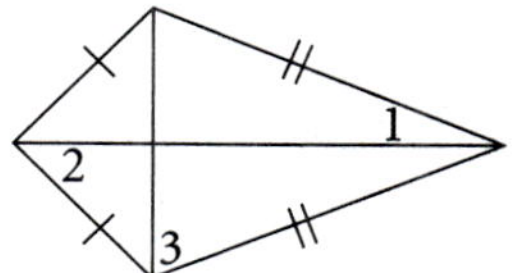

(A) 24°
(B) 44°
(C) 56°
(D) 66°

CAHSEE 6SDAP 3.1

47. A restaurant has simplified its menu for a special promotion. Each diner can choose an entree a vegetable and a dessert. The entree choices are beef, chicken, and fish; the vegetable choices are potatoes, squash, and asparagus; and for dessert the only choice is ice cream. Which of the following tree diagrams represents all of the possible choices for dinner?

(A)

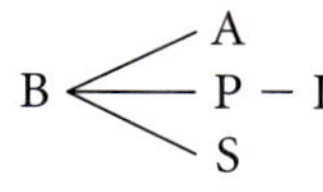

B < A, P – I, S
C < A, P – I, S
F < A, P – I, S

(C)

B – C < A, P, S
B – F < A, P, S
B – B < A, P, S

(B)

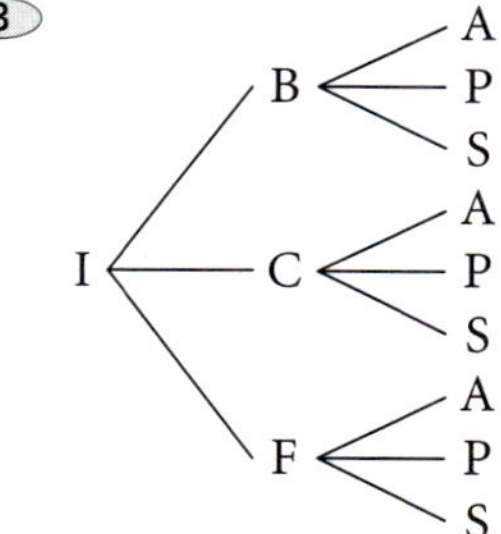

(D)

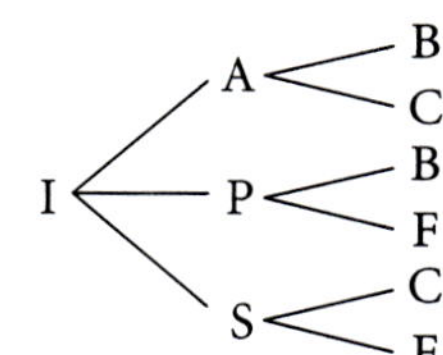

Mixed Review

Lesson 6-4

Find the indicated angle measures for the rhombus.

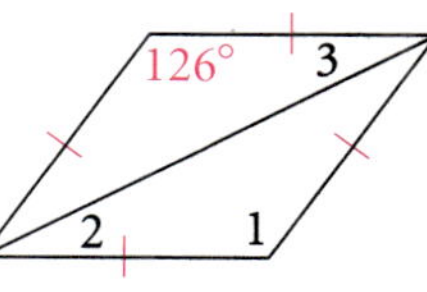

48. $m\angle 1$ **49.** $m\angle 2$ **50.** $m\angle 3$

Lesson 5-2 Algebra **Find the values indicated.**

51. a. a
b. FG
c. GH

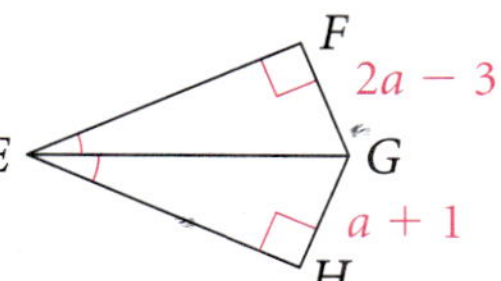

52. a. x
b. CD
c. BC

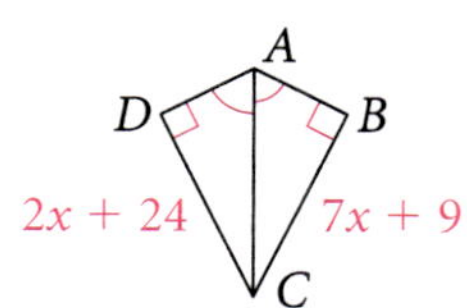

Lesson 4-2

53. State the postulate that justifies the statement $\triangle ABC \cong \triangle DEF$.

B C D A F E

Quadrilaterals in Quadrilaterals

Technology

FOR USE WITH LESSON 6-5

GEOM 3.0 Students construct and judge the validity of a logical argument and give counterexamples to disprove a statement. ***Master***

Construct

- Use geometry software to construct a quadrilateral *ABCD*.
- Construct the midpoint of each side of *ABCD*.
- Construct segments joining the midpoints, in order, to form quadrilateral *EFGH*.

Investigate

- Measure the lengths of the sides of *EFGH* and their slopes.
- Measure the angles of *EFGH*.

What kind of quadrilateral does *EFGH* appear to be?

EXERCISES

1. Manipulate quadrilateral *ABCD*.
  a. Make a conjecture about the quadrilateral whose vertices are the midpoints of the sides of a quadrilateral.
  b. Does your conjecture hold when *ABCD* is concave?
  c. Can you manipulate *ABCD* so that your conjecture doesn't hold?

Extend

2. Draw the diagonals of *ABCD*.
  a. Describe *EFGH* when the diagonals are perpendicular.
  b. Describe *EFGH* when the diagonals are congruent.
  c. Describe *EFGH* when the diagonals are both perpendicular and congruent.

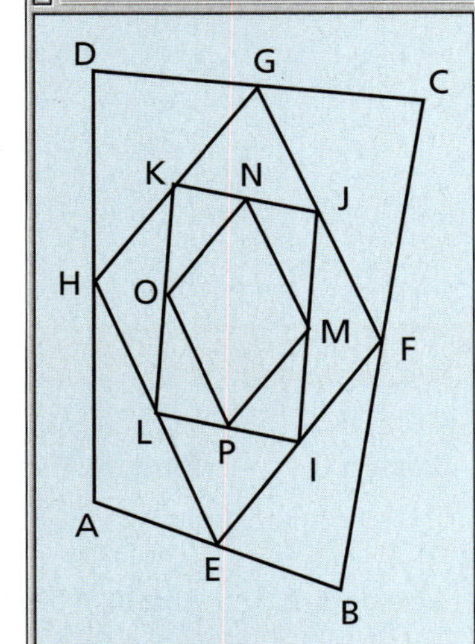

3. Construct the midpoints of *EFGH* and use them to construct quadrilateral *IJKL*. Construct the midpoints of *IJKL* and use them to construct quadrilateral *MNOP*. For *MNOP* and *EFGH*, compare the ratios of the lengths of the sides, perimeters, and areas. How are the sides of *MNOP* and *EFGH* related?

4. Writing During the investigation, you made a conjecture as to the type of quadrilateral *EFGH* appears to be. Prove your conjecture. Include in your proof the Midsegment Theorem, "If a segment joins the midpoints of two sides of a triangle, then the segment is parallel to the third side and half its length."

5. Describe the quadrilateral formed by joining the midpoints, in order, of the sides of each of the following. Justify each response.

a. parallelogram	**b.** rectangle	**c.** rhombus	**d.** square
e. trapezoid	**f.** isosceles trapezoid	**g.** kite	

6-6 Placing Figures in the Coordinate Plane

California Content Standards

Prepares for GEOM 17.0 Students prove theorems by using coordinate geometry.

What You'll Learn

- To name coordinates of special figures by using their properties

. . . And Why

To examine a T-shirt design, as in Example 1

Check Skills You'll Need

GO for Help Lesson 6-1

Draw a quadrilateral with the given vertices. Then determine the most precise name for each quadrilateral.

1. $H(-5, 0), E(-3, 2), A(3, 2), T(5, 0)$

2. $S(0, 0), A(4, 0), N(3, 2), D(-1, 2)$

3. $R(0, 0), A(5, 5), I(8, 4), N(7, 1)$

4. $W(-3, 0), I(0, 3), N(3, 0), D(0, -3)$

1 Naming Coordinates

When working with a figure in the coordinate plane, it generally is good practice to place a vertex at the origin and one side on an axis. You can also use multiples of 2 to avoid fractions when finding midpoints.

1 EXAMPLE Real-World  Connection

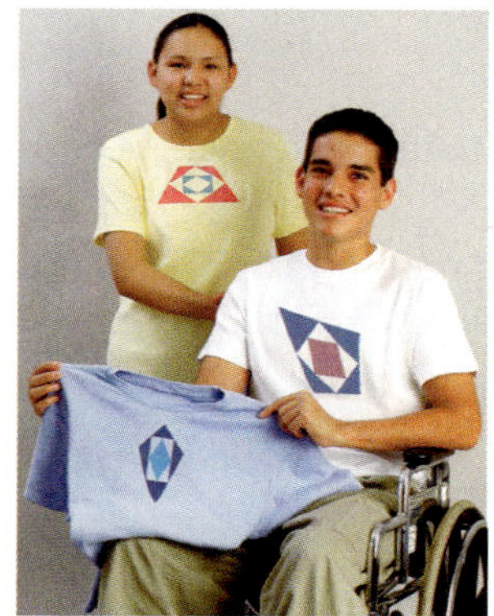

T-Shirt Design An art class creates T-shirt designs by drawing quadrilaterals and connecting their midpoints to form other quadrilaterals. Tiana claims that everyone's inner quadrilateral will be a parallelogram. Is she correct? Explain.

Draw quadrilateral $OACE$ with one vertex at the origin and one side on the x-axis. Since you are finding midpoints, use coordinates that are multiples of 2. Find the coordinates of the midpoints T, W, V, and U.

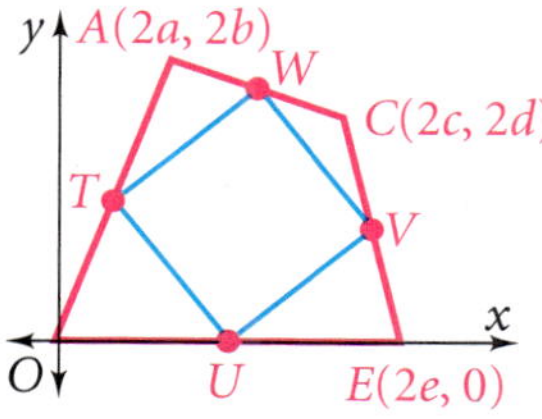

$$T = \text{midpoint of } \overline{OA} = \left(\frac{2a+0}{2}, \frac{2b+0}{2}\right) = (a, b)$$

$$W = \text{midpoint of } \overline{AC} = \left(\frac{2a+2c}{2}, \frac{2b+2d}{2}\right) = (a+c, b+d)$$

$$V = \text{midpoint of } \overline{CE} = \left(\frac{2c+2e}{2}, \frac{2d+0}{2}\right) = (c+e, d)$$

$$U = \text{midpoint of } \overline{OE} = \left(\frac{0+2e}{2}, \frac{0+0}{2}\right) = (e, 0)$$

Find the slopes of the sides of $TWVU$.

$$\text{slope of } \overline{TW} = \frac{b-(b+d)}{a-(a+c)} = \frac{d}{c} \qquad \text{slope of } \overline{VU} = \frac{d-0}{(c+e)-e} = \frac{d}{c}$$

The slopes are equal, so $\overline{TW} \parallel \overline{VU}$. Similarly, $\overline{WV} \parallel \overline{TU}$.

Since both pairs of opposite sides of $TWVU$ are parallel, $TWVU$ is a parallelogram, and Tiana is correct.

CA Standards Check **1** Find the slopes of $\overline{WV}$ and $\overline{TU}$. Verify that $\overline{WV}$ and $\overline{TU}$ are parallel.

Centering figures at the origin can also be helpful when placing figures in the coordinate plane.

2 EXAMPLE Naming Coordinates

Algebra In the diagram, rectangle $KLMN$ is centered at the origin with sides parallel to the axes. Find the missing coordinates.

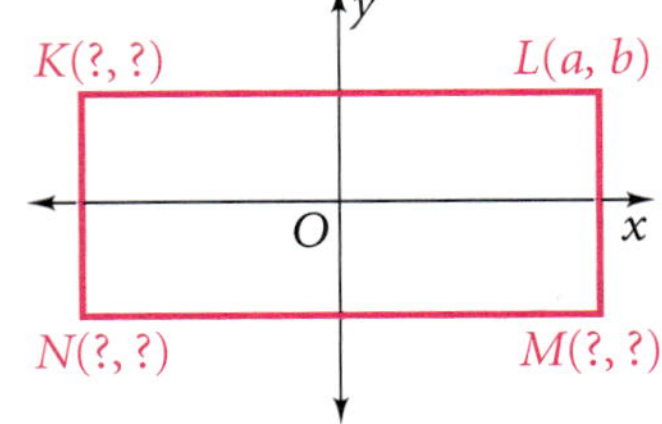

L has coordinates (a, b), so the coordinates of the other vertices are $K(-a, b)$, $M(a, -b)$, and $N(-a, -b)$.

CA Standards Check 2 Use the properties of parallelogram $OPQR$ to find the missing coordinates. Do not use any new variables.

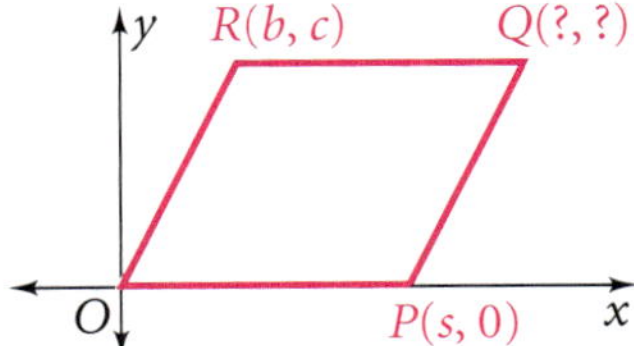

EXERCISES Standards Practice

For more exercises, see *Extra Skill, Word Problem, and Proof Practice.*

A Practice by Example

Example 1 (page 343)

1. Claim: The midpoint of the hypotenuse of a right triangle is equidistant from the vertices of the triangle. Follow Steps 1 and 2 to place a right triangle in the coordinate plane. Complete Steps 3–5 to verify the claim.

Given: Right $\triangle ABC$ with M the midpoint of hypotenuse $\overline{AB}$

Prove: $MA = MB = MC$

Step 1: Draw right $\triangle ABC$ on a coordinate plane. Locate the right angle, $\angle C$, at the origin and leg $\overline{CA}$ on the positive x-axis.

Step 2: You seek a midpoint, so label coordinates using multiples of 2. The coordinates of point A are **a.** ___?___. The coordinates of point B are **b.** ___?___.

Step 3: By the Midpoint Formula, the coordinates of midpoint M are **c.** ___?___.

Step 4: By the Distance Formula, MA = **d.** ___?___, MB = **e.** ___?___, and MC = **f.** ___?___.

Step 5: Conclusion: **g.** ___?___

Example 2 (page 344)

x^2 **Algebra** **Give coordinates for points W and Z without using any new variables.**

2. rectangle

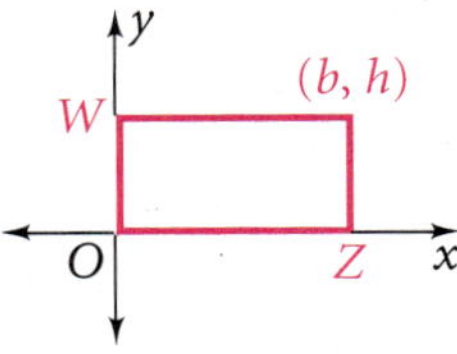

3. square

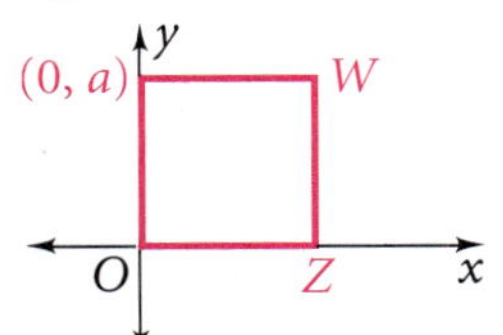

4. square

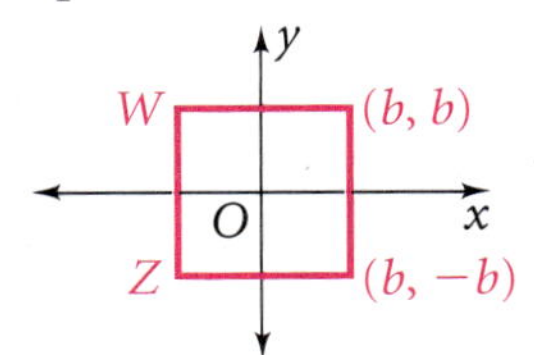

5. parallelogram

y W $(c - a, b)$ O Z $(c, 0)$ x

6. rhombus

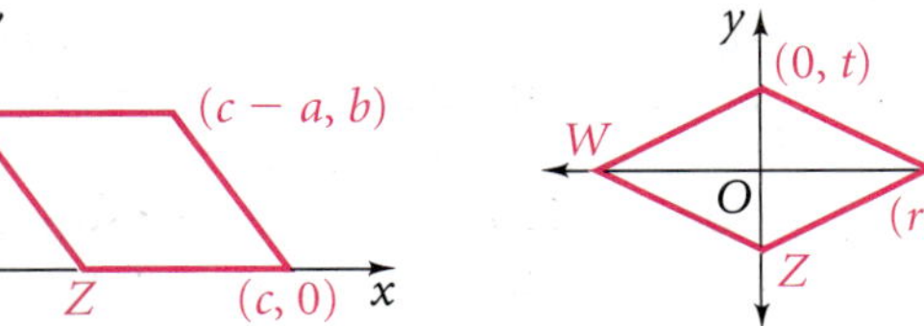

7. isosceles trapezoid

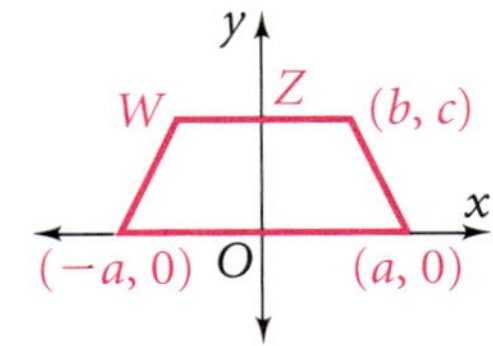

Apply Your Skills 

8. **Writing** Choose values for r and t in Exercise 6. Find the slope and length of each side. State why the figure satisfies the definition of a rhombus.

9. What property of a rhombus makes it convenient to place its diagonals on the x- and y-axes? (See Exercise 6.)

Here are coordinates for eight points in the coordinate plane ($q > p > 0$).

$A(0, 0)$, $B(p, 0)$, $C(q, 0)$, $D(p + q, 0)$, $E(0, q)$, $F(p, q)$, $G(q, q)$, $H(p + q, q)$

Which four points, if any, are the vertices for each type of figure?

10. parallelogram
11. rhombus
12. rectangle
13. square
14. trapezoid
15. isosceles trapezoid

Refer to the diagrams in Exercises 2–7. Use the coordinates given below in place of the ones shown. Then give the coordinates for points W and Z without using any new variables. The new diagram for Exercise 16 is shown here.

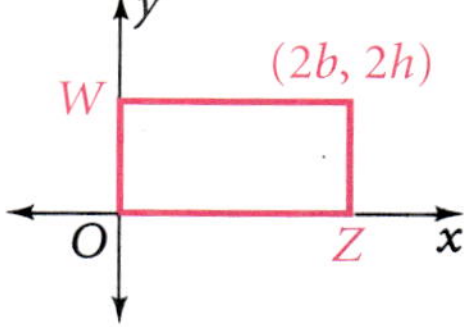

Homework Video Tutor

Visit: PHSchool.com
Web Code: bce-0606

16. Ex. 2, $(2b, 2h)$
17. Ex. 3, $(0, 2a)$
18. Ex. 4, $(2b, 2b)$, $(2b, -2b)$
19. Ex. 5, $(2c, 0)$, $(2c - 2a, b)$
20. Ex. 6, $(2r, 0)$, $(0, 2t)$
21. Ex. 7, $(-2a, 0)$, $(2a, 0)$, $(2b, 2c)$

22. **Multiple Choice** The coordinates of three vertices of a rectangle are $(-2a, 0)$, $(2a, 0)$, and $(2a, 2b)$. What are the coordinates of the midpoint of the diagonal joining one of these points with the fourth vertex?

   (A) $(0, b)$ (B) $(0, 2b)$ (C) $(2a, b)$ (D) $(-2a, 2b)$

Give the coordinates for point P without using any new variables.

23. isosceles trapezoid

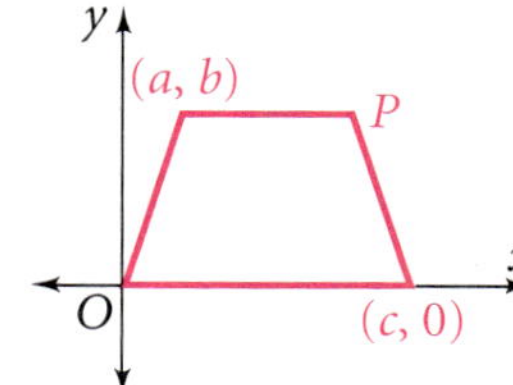

24. trapezoid with a right ∠

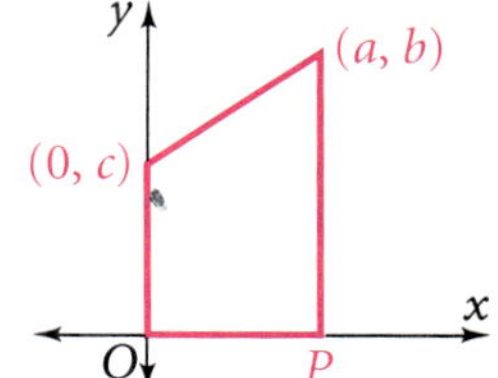

25. kite

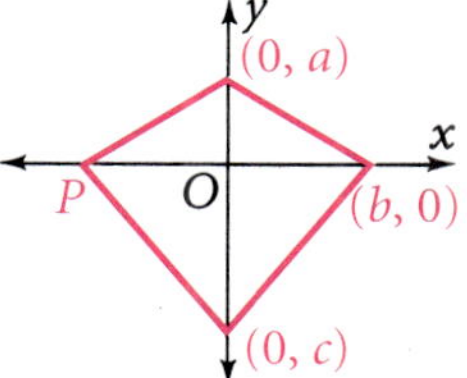

26. a. Draw a square whose diagonals of length $2b$ lie on the x- and y-axes.
   b. Give the coordinates of the vertices of the square.
   c. Compute the length of a side of the square.
   d. Find the slopes of two adjacent sides of the square.
   e. Do the slopes show that the sides are perpendicular? Explain.

27. Make two drawings of an isosceles triangle with base length $2b$ and height $2c$.
   a. In one drawing, place the base on the x-axis with a vertex at the origin.
   b. In the second, place the base on the x-axis with its midpoint at the origin.
   c. Find the lengths of the legs of the triangle as placed in part (a).
   d. Find the lengths of the legs of the triangle as placed in part (b).
   e. How do the results of parts (c) and (d) compare?

Real-World Connection

Marine biologists use coordinate grids to organize data gathering.

28. **Marine Archaeology** Marine archaeologists sometimes use a coordinate system on the ocean floor. They record the coordinates of points where artifacts are found. Assume that each diver searches a square area and can go no farther than b units from the starting point. Draw a model for the region one diver can search. Assign coordinates to the vertices without using any new variables.

Online Lesson Quiz Visit: PHSchool.com, Web Code: bca-0606

29. Coordinate Proof Follow the steps below to prove:

If two nonvertical lines are perpendicular, the product of their slopes is -1.

Step 1: Two nonvertical lines, ℓ_1 and ℓ_2, intersect. Which coordinate point might be the easiest to work with as the point of intersection?

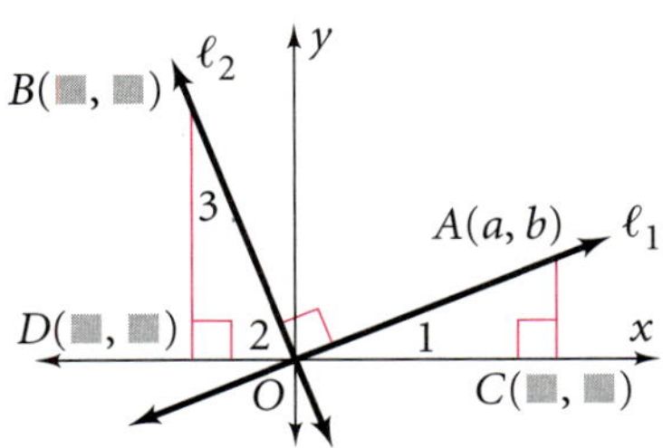

Step 2: To work with the slope of a line, you need two points on the line. Choose one point $A(a, b)$ on ℓ_1. What are the coordinates of C?

Step 3: Notice that $\angle 1$ and $\angle 3$ are both complements of $\angle 2$. Why?

Step 4: This means that the two triangles pictured have congruent angles. Thus, if any pair of sides are congruent, the two triangles are congruent. Congruent triangles are desirable, so what would be a good choice for the coordinates of point D?

Step 5: If you made a choice for D so that $\triangle ACO \cong \triangle ODB$ what must be the coordinates of point B?

Step 6: Now, complete the proof that the product of slopes is -1.

Multiple Choice Practice

For California Standards and CAHSEE Tutorials, visit PHSchool.com. Web Code: bcq-9045

GEOM 17.0

30. Which of the following coordinates of P would make the figure an isosceles trapezoid?

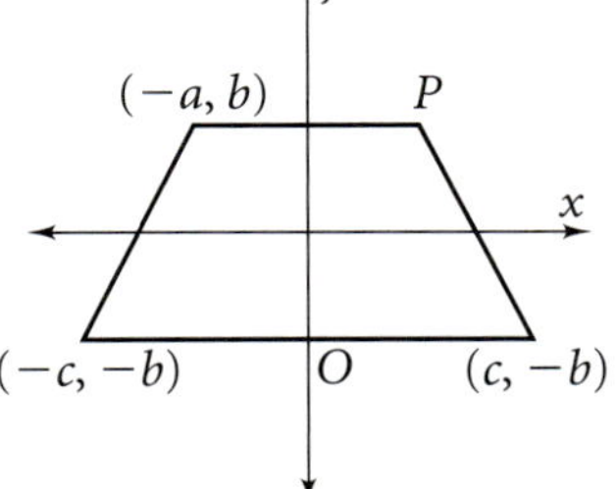

A (a, b)

B $(0, b)$

C (c, b)

D (b, a)

CAHSEE 7NS 1.2

31. $\frac{9}{10} - \left(\frac{1}{2} + \frac{2}{5}\right)$

A -1 B 0 C 1 D 2

Mixed Review

GO for Help

Lesson 6-5

32. Algebra Find the measure of each angle and the value of x in the isosceles trapezoid.

Lesson 5-3

Find the center of the circle that circumscribes $\triangle ABC$.

33. $A(1, 1), B(5, 3), C(5, 1)$

34. $A(-5, 0), B(-1, -8), C(-1, 0)$

Lesson 4-3

35. Given: $\angle ACD \cong \angle ACB$ and $\angle D \cong \angle B$

Prove: $\triangle ADC \cong \triangle ABC$

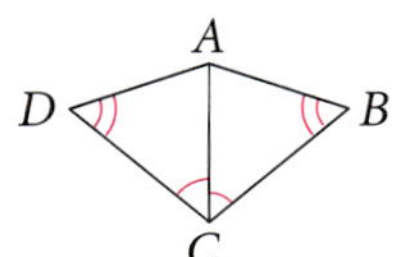

Checkpoint Quiz 2 — Lessons 6-4 through 6-6

Algebra **Find the value(s) of the variable(s).**

1. 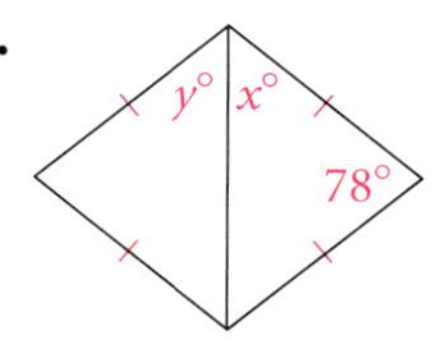

2. 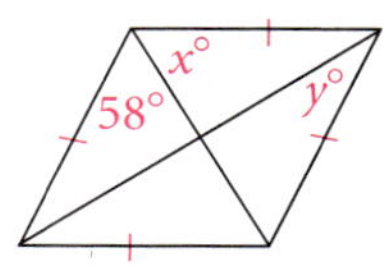

3. 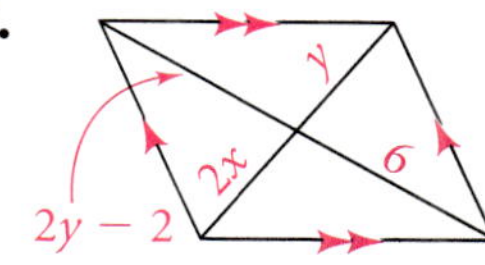

4. 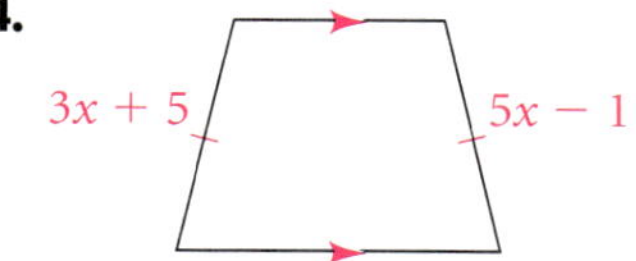

5. 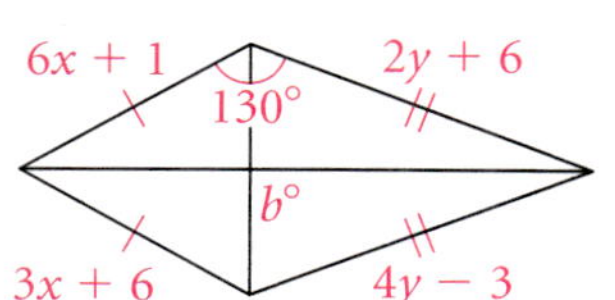

In Exercises 6–8, decide whether the statement is *true* or *false*. If true, explain why. If false, show a counterexample.

6. A quadrilateral with congruent diagonals is an isosceles trapezoid or rectangle.
7. A quadrilateral with congruent and perpendicular diagonals is a square.
8. Each diagonal of a kite bisects two angles of the kite.

Give the coordinates for point *A* without using any new variables.

9. 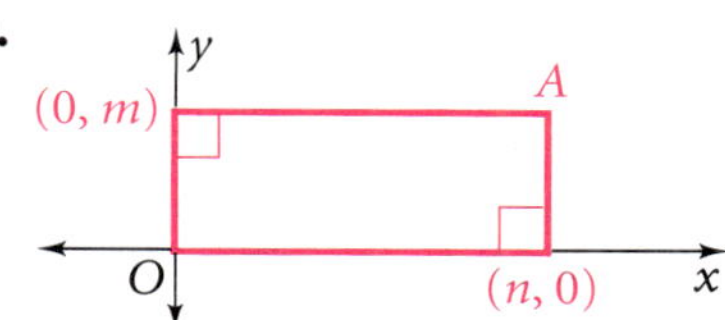

10. 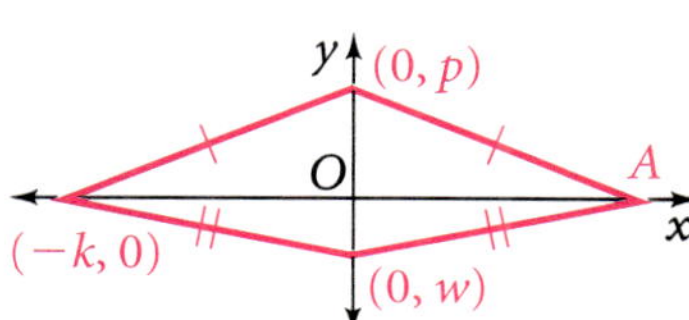

A Point in Time

400 | 200 B.C. | 0 | A.D. 200 | 2000

Many walls in ancient Egypt were decorated with reliefs. The relief in the photo was created in the year 255 B.C. First, the artist sketched the scene on papyrus overlaid with a grid. Next, the wall was marked with a grid the size of the intended sculpture. To draw each line, a tightly stretched string that had been dipped in red ochre was plucked, like a guitar string.

Using the grid squares as guides, the artist transferred the drawing to the wall. Then, a sculptor cut the background away, leaving the scene slightly raised. Finally, an artist painted the scene.

For: Information about Egyptian reliefs
Web Code: bce-2032

Proofs Using Coordinate Geometry

California Content Standards

GEOM 17.0 Students prove theorems by using coordinate geometry. ***Develop, Master***

What You'll Learn

- To prove theorems using figures in the coordinate plane

. . . And Why

To use coordinate geometry to prove that a flag design includes a rhombus, as in Example 2

Check Skills You'll Need

1. Graph the rhombus with vertices $A(2, 2), B(7, 2), C(4, -2)$, and $D(-1, -2)$. Then, connect the midpoints of consecutive sides to form a quadrilateral. What do you notice about the quadrilateral?

x^2 **Algebra Give the coordinates of B without using any new variables.**

2. rectangle

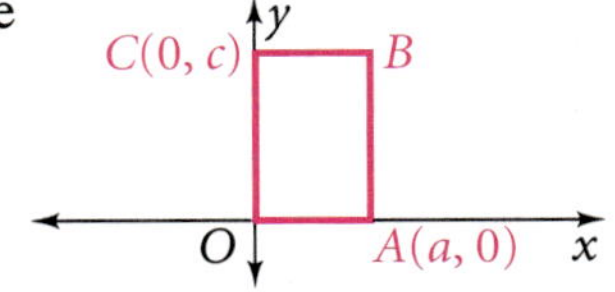

3. isosceles triangle

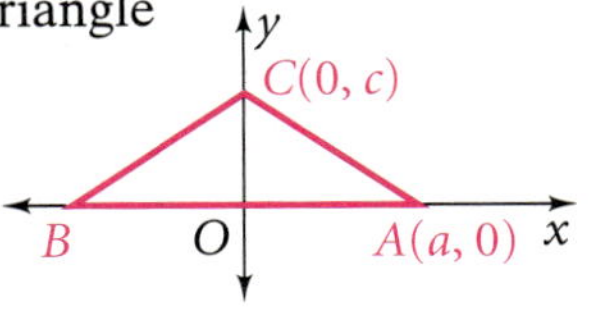

New Vocabulary • midsegment of a trapezoid

1 Building Proofs in the Coordinate Plane

In Lesson 5-1, you learned about midsegments of triangles. A trapezoid also has a midsegment. The **midsegment of a trapezoid** is the segment that joins the midpoints of the nonparallel opposite sides. It has two unique properties.

Key Concepts

Theorem 6-18 **Trapezoid Midsegment Theorem**

(1) The midsegment of a trapezoid is parallel to the bases.
(2) The length of the midsegment of a trapezoid is half the sum of the lengths of the bases.

$$\overline{MN} \parallel \overline{TP}, \overline{MN} \parallel \overline{RA}, \text{ and } MN = \tfrac{1}{2}(TP + RA).$$

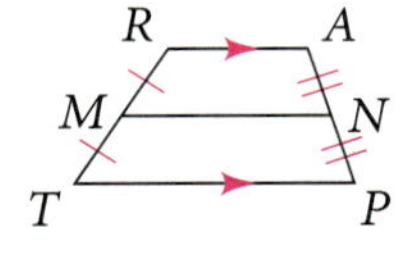

Formulas for slope, midpoint, and distance are used in a proof of Theorem 6-18.

1 EXAMPLE Planning a Coordinate Geometry Proof

Developing Proof Plan a coordinate proof of Theorem 6-18.

Given: $\overline{MN}$ is the midsegment of trapezoid $TRAP$.

Prove: $\overline{MN} \parallel \overline{TP}, \overline{MN} \parallel \overline{RA}$, and $MN = \frac{1}{2}(TP + RA)$.

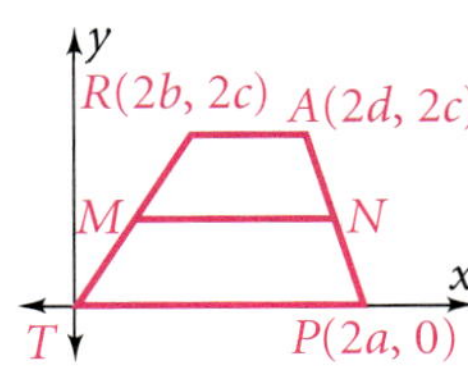

Plan: Place the trapezoid in the coordinate plane with a vertex at the origin and a base along the x-axis. Since midpoints will be involved, use multiples of 2 to name coordinates. To show lines are parallel, check for equal slopes. To compare lengths, use the Distance Formula.

1. Complete the coordinate proof of Theorem 6-18.
   a. Find the coordinates of midpoints M and N. How do the multiples of 2 help?
   b. Find and compare the slopes of $\overline{MN}$, $\overline{TP}$, and $\overline{RA}$.
   c. Find and compare the lengths MN, TP, and RA.
   d. In parts (b) and (c), how does placing a base along the x-axis help?

2 EXAMPLE Real-World Connection

Algebra The rectangular flag at the left is constructed by connecting the midpoints of its sides. Use coordinate geometry to prove that the quadrilateral formed by connecting the midpoints of the sides of a rectangle is a rhombus.

Given: $MNPO$ is a rectangle.
T, W, V, U are midpoints of its sides.

Prove: $TWVU$ is a rhombus.

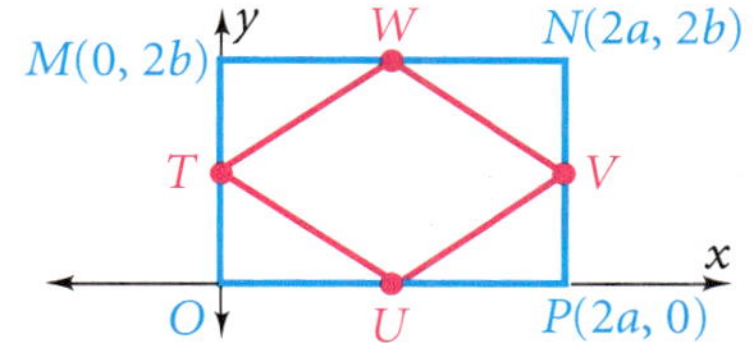

Plan: Place the rectangle in the coordinate plane with two sides along the axes. Use multiples of 2 to name coordinates.
A rhombus is a parallelogram with four congruent sides. From Lesson 6-6, Example 1, you know that $TWVU$ is a parallelogram. To show $\overline{TW} \cong \overline{WV} \cong \overline{VU} \cong \overline{UT}$, use the Distance Formula.

Coordinate Proof: By the Midpoint Formula, the coordinates of the midpoints are $T(0, b)$, $W(a, 2b)$, $V(2a, b)$, and $U(a, 0)$. By the Distance Formula,

$$TW = \sqrt{(a - 0)^2 + (2b - b)^2} = \sqrt{a^2 + b^2}$$

$$WV = \sqrt{(2a - a)^2 + (b - 2b)^2} = \sqrt{a^2 + b^2}$$

$$VU = \sqrt{(a - 2a)^2 + (0 - b)^2} = \sqrt{a^2 + b^2}$$

$$UT = \sqrt{(0 - a)^2 + (b - 0)^2} = \sqrt{a^2 + b^2}$$

$\overline{TW} \cong \overline{WV} \cong \overline{VU} \cong \overline{UT}$, so parallelogram $TWVU$ is a rhombus.

2. **Critical Thinking** Explain why the proof using $M(0, 2b)$, $N(2a, 2b)$, $P(2a, 0)$, and $O(0, 0)$ is easier than a proof using $M(0, b)$, $N(a, b)$, $P(a, 0)$, and $O(0, 0)$.

EXERCISES

For more exercises, see *Extra Skill, Word Problem, and Proof Practice.*

Standards Practice GEOM 17.0

A Practice by Example

Example 1 (page 348)

1. W and Z are the midpoints of $\overline{OR}$ and $\overline{ST}$, respectively. In parts (a)–(c), find the coordinates of W and Z.

a.

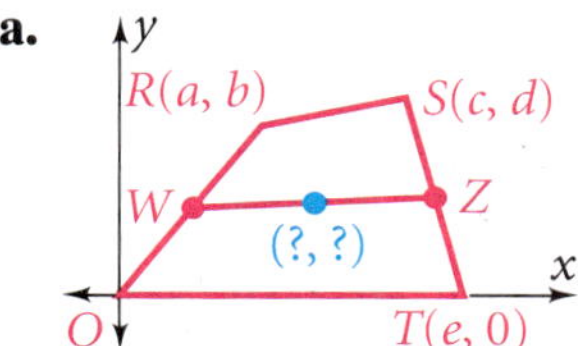

b.

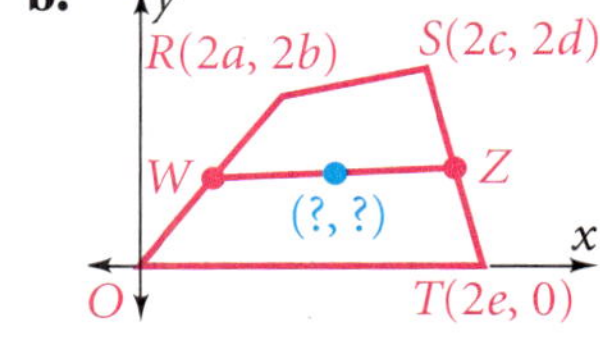

c.

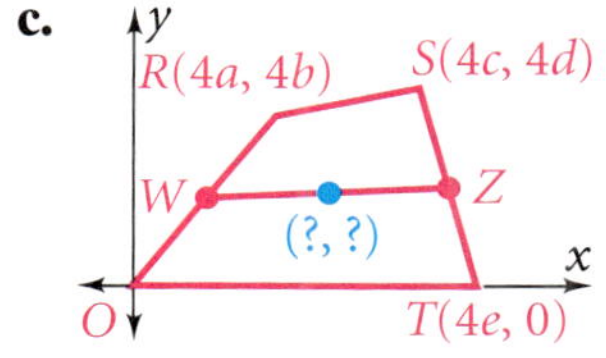

d. You are to plan a coordinate proof involving the midpoint of $\overline{WZ}$. Which of the figures (a)–(c) would you prefer to use? Explain.

Problem Solving Hint

When you read large blocks of math text, cover all but a few lines to help you focus.

Developing Proof **Complete the plan for each coordinate proof.**

2. The diagonals of a parallelogram bisect each other (Theorem 6-3).

Given: Parallelogram $ABCD$

Prove: $\overline{AC}$ bisects $\overline{BD}$, and $\overline{BD}$ bisects $\overline{AC}$.

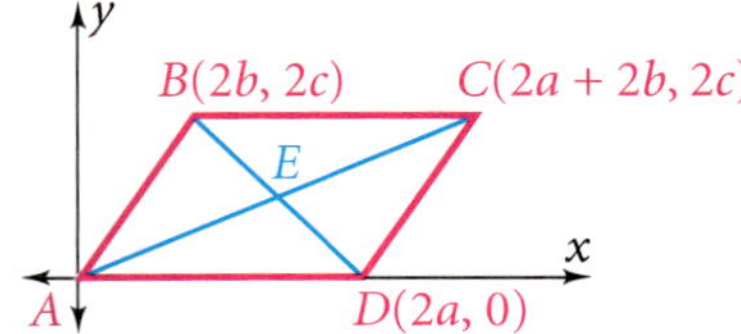

Plan: Place the parallelogram in the coordinate plane with a vertex at the **a.** _?_ and a side along the **b.** _?_. Since midpoints will be involved, use multiples of **c.** _?_ to name coordinates. To show segments bisect each other, show the midpoints have the same **d.** _?_.

3. The diagonals of an isosceles trapezoid are congruent (Theorem 6-16).

Given: Trapezoid $EFGH$ with $\overline{FE} \cong \overline{GH}$

Prove: $\overline{EG} \cong \overline{HF}$

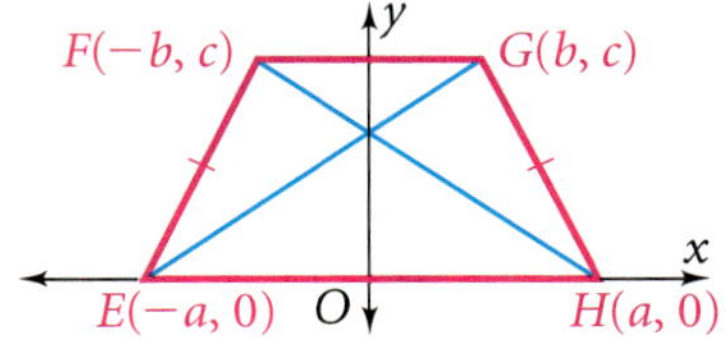

Plan: The trapezoid is isosceles, so place one base on the x-axis so that the **a.** _?_ bisects its bases. To show the diagonals are congruent, use the **b.** _?_ Formula.

4. The median to the hypotenuse of a right triangle is half the hypotenuse.

Given: $\triangle MNO$ is a right triangle with right $\angle MON$. P is the midpoint of $\overline{MN}$.

Prove: $OP = \frac{1}{2}MN$

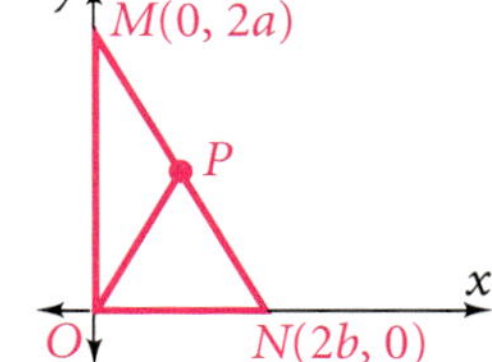

Plan: Place the right triangle in the coordinate plane with the vertex of the **a.** _?_ at the origin and the **b.** _?_ along each axis. Since midpoints will be involved, use **c.** _?_ to name coordinates for points **d.** _?_ and **e.** _?_. Use the **f.** _?_ Formula to find the coordinates of P. To compare lengths, use the **g.** _?_ Formula.

5. The segments joining the midpoints of consecutive sides of an isosceles trapezoid form a rhombus.

Given: Trapezoid $TRAP$ with $\overline{TR} \cong \overline{PA}$; D, E, F, and G are midpoints of the indicated sides.

Prove: $DEFG$ is a rhombus.

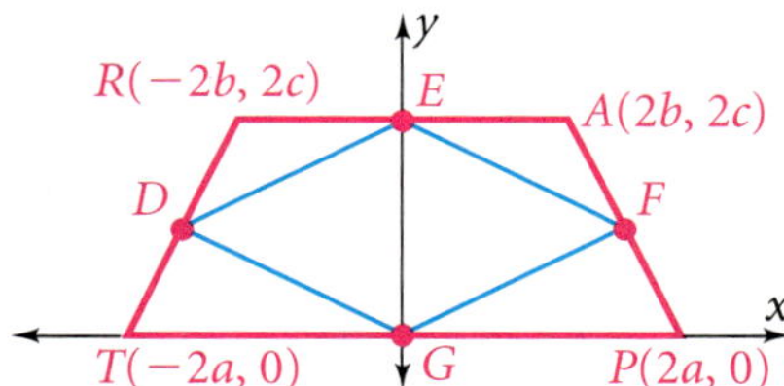

Plan: The trapezoid is **a.** _?_, so place one base on the **b.** _?_ so that the **c.** _?_ bisects its bases. Use multiples of 2 to name coordinates since **d.** _?_ will be involved. A rhombus is a parallelogram with four **e.** _?_. To show opposite sides are parallel, show that their **f.** _?_ are the same. To show sides are congruent, use **g.** _?_.

Example 2 (page 349)

Developing Proof **Follow the plans above to complete the coordinate proofs.**

6. (Exercise 3) The diagonals of an isosceles trapezoid are congruent.

Proof: By the Distance Formula, EG = **a.** _?_ and HF = **b.** _?_. Therefore, $\overline{EG} \cong \overline{HF}$ by the definition of congruence.

7. (Exercise 4) The median from the vertex of the right angle of a right triangle is half as long as the hypotenuse.

Proof: By the Distance Formula, OP = **a.** _?_ and MN = **b.** _?_. Therefore, $OP = \frac{1}{2}MN$.

8. (Exercise 5) The segments joining the midpoints of consecutive sides of an isosceles trapezoid form a rhombus.

Proof: The midpoints have coordinates **a.** $D(\underline{?}, \underline{?})$, $E(\underline{?}, \underline{?})$, $F(\underline{?}, \underline{?})$, and $G(\underline{?}, \underline{?})$. By the Distance Formula, $DE =$ **b.** $\underline{?}$, $EF =$ **c.** $\underline{?}$, $FG =$ **d.** $\underline{?}$, and $GD =$ **e.** $\underline{?}$. The slope of $DE =$ **f.** $\underline{?}$ and the slope of $FG =$ **g.** $\underline{?}$. The slope of $EF =$ **h.** $\underline{?}$ and that of $GD =$ **i.** $\underline{?}$. Thus, $DEFG$ is a parallelogram with congruent **j.** $\underline{?}$, so **k.** $\underline{?}$ is a rhombus by the definition of rhombus.

9. **Developing Proof** Use coordinate geometry to prove that the diagonals of a rectangle bisect each other.

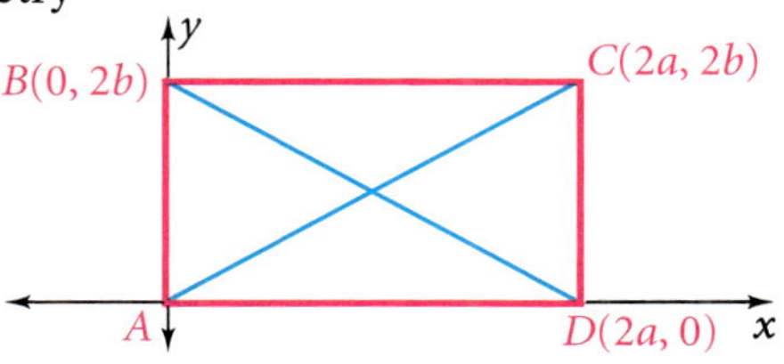

Proof: The midpoint of $\overline{AC}$ is **a.** $\underline{?}$.
The midpoint of $\overline{BD}$ is **b.** $\underline{?}$.
The midpoints are **c.** $\underline{?}$,
so the diagonals bisect each other.

B Apply Your Skills

10. **Open-Ended** Give an example of a statement that you think is easier to prove with a coordinate geometry proof than with a paragraph, flow, or two-column proof. Explain your choice.

Problem Solving Hint
Lines with undefined slope, such as $\overline{KN}$ and $\overline{LM}$, are vertical lines. All vertical lines are parallel.

Proof 11. Prove: The midpoints of the sides of a kite determine a rectangle.

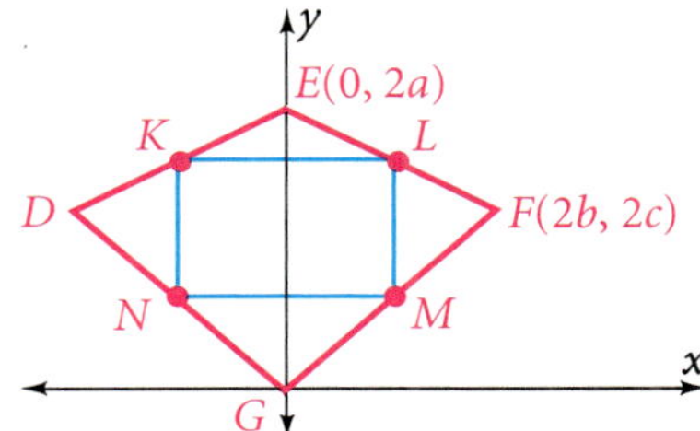

Given: Kite $DEFG$ with $DE = EF$ and $DG = GF$; K, L, M, and N are midpoints of the sides.

Prove: $KLMN$ is a rectangle.

State whether each type of conclusion shown here could be reached using coordinate methods. Give a reason for each answer.

12. $\overline{AB} \cong \overline{CD}$
13. $\overline{AB} \parallel \overline{CD}$
14. $\overline{AB} \perp \overline{CD}$
15. $\overline{AB}$ bisects $\overline{CD}$.
16. $\overline{AB}$ bisects $\angle CAD$.
17. $\angle A \cong \angle B$
18. $\angle A$ is a right angle.
19. $AB + BC = AC$
20. $\triangle ABC$ is isosceles.
21. $\angle A$ and $\angle B$ are supplementary.
22. $\overline{AB}, \overline{CD}$, and $\overline{EF}$ are concurrent.
23. Quadrilateral $ABCD$ is a rhombus.

24. **Multiple Choice** $DEFG$ is a rhombus. What is the slope of its diagonal $\overline{DF}$?

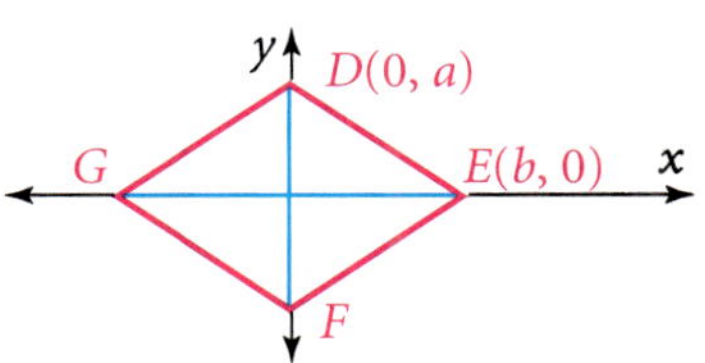

Ⓐ 0
Ⓒ 1
Ⓑ  $\left(\frac{a}{b}\right)$
Ⓓ undefined

Visit: PHSchool.com
Web Code: bce-0607

A and B have coordinates -2 and 10 on a number line. Find the coordinates of the points that separate $\overline{AB}$ into the given number of congruent segments.

25. 4
26. 6
27. 10
28. 50
29. n

The endpoints of $\overline{AB}$ are $A(-3, 5)$ and $B(9, 15)$. Find the coordinates of the points that separate $\overline{AB}$ into the given number of congruent segments.

30. 4 **31.** 6 **32.** 10 **33.** 50 **34.** n

Proof **35.** You learned in Lesson 5-3 (Theorem 5-8) that the centroid of a triangle, the point where the medians meet, is $\frac{2}{3}$ of the distance from each vertex to the midpoint of the opposite side. Complete the following steps to prove this theorem.

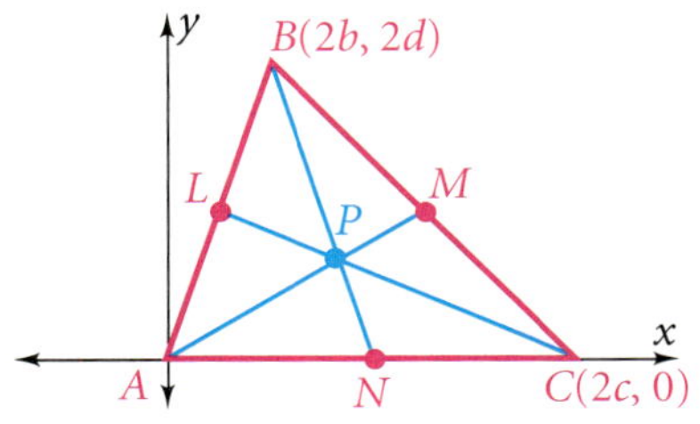

a. Find the coordinates of points L, M, and N, the midpoints of the sides of $\triangle ABC$.

b. Find equations of $\overleftrightarrow{AM}$, $\overleftrightarrow{BN}$, and $\overleftrightarrow{CL}$.

c. Find the coordinates of point P, the intersection of $\overleftrightarrow{AM}$ and $\overleftrightarrow{BN}$.

d. Show that point P is on $\overleftrightarrow{CL}$.

e. Use the Distance Formula to show that point P is $\frac{2}{3}$ of the distance from each vertex to the midpoint of the opposite side.

Problem Solving Hint

To show three lines are concurrent, you can show

1) one point is on all three lines

2) the intersection point of two lines is on the third line, or

3) the intersection of one pair of lines is the same as the intersection of another pair.

Proof **36.** Complete the following steps to prove Theorem 5-9. You are given $\triangle ABC$ with altitudes p, q, and r. Show that p, q, and r intersect in a point (called the orthocenter of the triangle).

a. The slope of $\overline{BC}$ is $\frac{c}{-b}$. What is the slope of line p?

b. Show that the equation of line p is $y = \frac{b}{c}(x - a)$.

c. What is the equation of line q?

d. Show that lines p and q intersect at $\left(0, \frac{-ab}{c}\right)$.

e. The slope of $\overline{AC}$ is $\frac{c}{-a}$. What is the slope of line r?

f. Show that the equation of line r is $y = \frac{a}{c}(x - b)$.

g. Show that lines r and q intersect at $\left(0, \frac{-ab}{c}\right)$.

h. Give the coordinates of the orthocenter of $\triangle ABC$.

The endpoints of $\overline{AB}$ are as given. Find the coordinates of the points that separate $\overline{AB}$ into n congruent segments.

37. A has coordinate a and B has coordinate b on a number line.

38. A has coordinates (a, c) and B has coordinates (b, d) in the coordinate plane.

39. Use the diagram at the right.

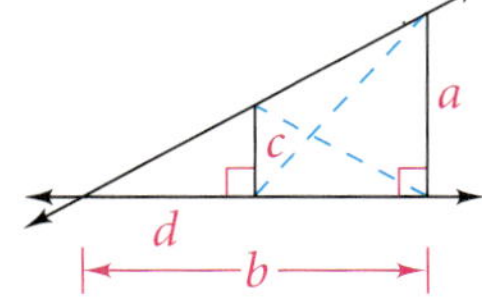

a. Explain using area why $\frac{1}{2}ad = \frac{1}{2}bc$ and hence $ad = bc$. (*Hint:* Area of triangle $= \frac{1}{2} \cdot \text{base} \cdot \text{height}$)

b. Use slope and part (a) to show: If $\frac{a}{b} = \frac{c}{d}$, then $ad = bc$.

Carefully balanced metal shapes hang from wires in the mobiles of Alexander Calder (1898–1976).

40. Physics For a mobile to be in balance, you suspend each part of the mobile at its center of mass. The center of mass, or centroid, is the point around which the weight of an object appears to be evenly distributed. You learned a method for finding the centroid of a triangle in Lesson 5-3. Now use it to help you find the centroid of a quadrilateral. (*Hint:* Where is the centroid of a segment?)

41. Write a coordinate proof of the theorem:

If the slopes of two lines have product -1, the lines are perpendicular.

a. First, argue that neither line can be horizontal or vertical.

b. Then, tell why the lines must intersect. (*Hint:* Use indirect reasoning.)

c. Knowing that they do intersect, place the lines in the coordinate plane, choose a point on ℓ_1, find a related point on ℓ_2, and complete the proof.

Multiple Choice Practice

For California Standards and CAHSEE Tutorials, visit PHSchool.com. Web Code: bcq-9045

GEOM 17.0

42. What coordinates of A make $ABCD$ a rhombus?

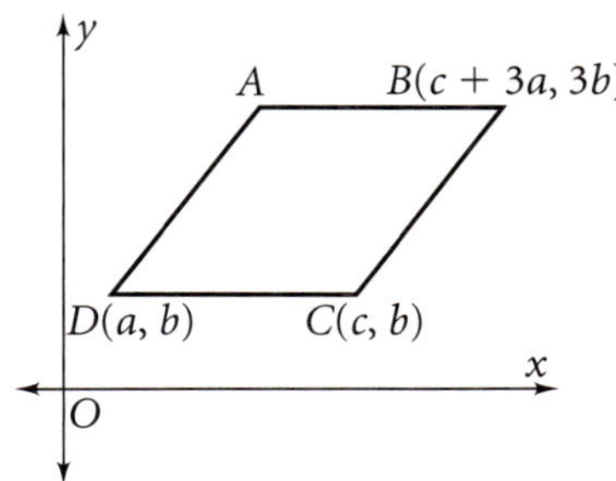

Ⓐ $(a + c, 3b)$

Ⓑ $(c, 3a + 3b)$

Ⓒ $(4a, 3b)$

Ⓓ $(a + c, a + b)$

CAHSEE Alg1 5.0

43. Solve for x.

$$4(2x - 3) + 5 \leq 3x - 2$$

Ⓐ $x \leq -\frac{9}{5}$

Ⓑ $x \leq -1$

Ⓒ $x \leq 1$

Ⓓ $x \leq \frac{9}{5}$

Mixed Review

Lesson 6-6

44. Rectangle $LMNP$ at the right is centered at the origin. Give coordinates for point P without using any new variables.

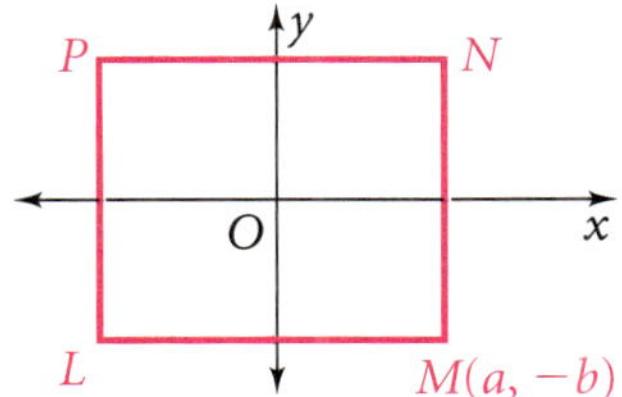

Lesson 5-4

Write (a) the inverse and (b) the contrapositive of each statement.

45. If the sum of the angles of a polygon is not 360°, then the polygon is not a quadrilateral.

46. If $x = 51$, then $2x = 102$.

47. If $a = 5$, then $a^2 = 25$.

48. If $b < -4$, then b is negative.

49. If $c > 0$, then c is positive.

Lesson 4-4

Explain how you can use SSS, SAS, ASA, or AAS with CPCTC to prove each statement true.

50. $\overline{AB} \cong \overline{CB}$

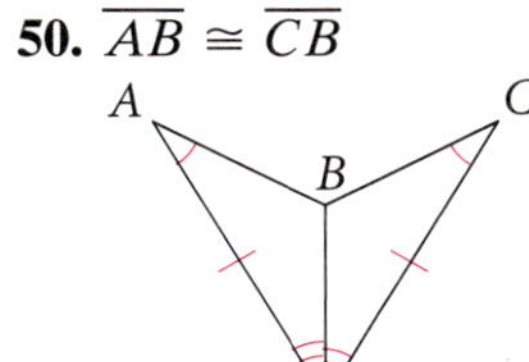

51. $\angle 1 \cong \angle 2$

52. $\angle K \cong \angle M$

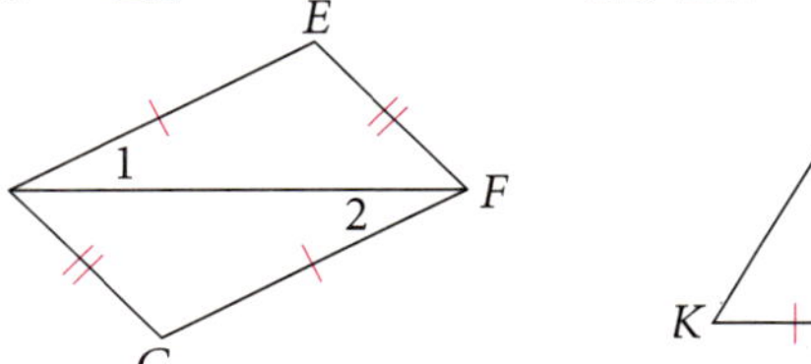

Algebra 1 Review

Quadratics

For Use With Lesson 6-7

The graph of the quadratic function $y = ax^2$ $(a > 1)$ is a vertical stretch of $y = x^2$ by the factor a. Each y-value for $y = ax^2$ is a times the corresponding y-value for $y = x^2$.

For $0 < a < 1$, the function $y = ax^2$ is a vertical shrink of $y = x^2$ by the factor a.

1 EXAMPLE

Graph $y = x^2$, $y = 3x^2$, and $y = \frac{1}{3}x^2$. Compare the graphs.

First, draw $y = x^2$. For $y = 3x^2$, draw each point to be 3 times as high as the corresponding point in $y = x^2$.

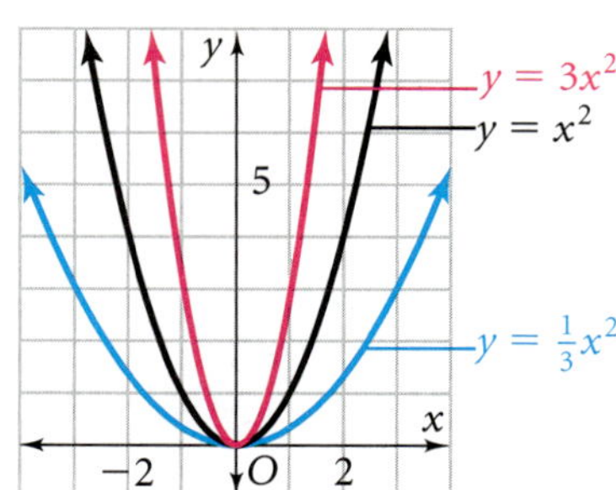

To draw $y = \frac{1}{3}x^2$, draw each point to be one third as high as the corresponding point in $y = x^2$.

The graph of $y = 3x^2$ is a stretch of the graph of $y = x^2$ by the factor 3.

The graph of $y = \frac{1}{3}x^2$ is a shrink of the graph of $y = x^2$ by the factor $\frac{1}{3}$.

The graph of $y = -ax^2$ is a *reflection* (mirror image) of the graph of $y = ax^2$ across the x-axis. Each y-value for $y = -ax^2$ is the opposite of the corresponding y-value for $y = ax^2$.

2 EXAMPLE

Graph $y = 3x^2$ and $y = -3x^2$. Compare the graphs.

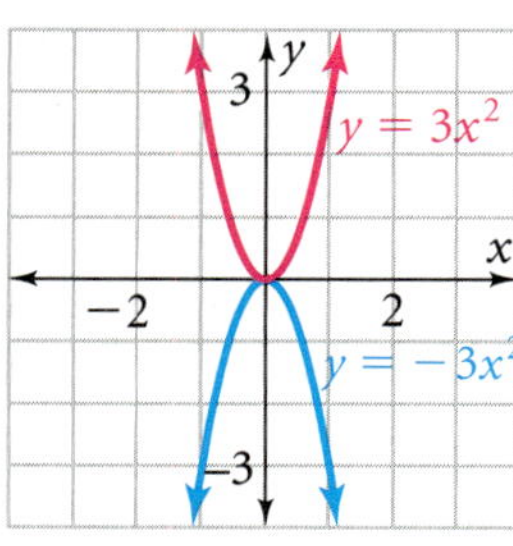

Draw $y = 3x^2$ as in Example 1. For $y = -3x^2$, draw each point so its y-value is the opposite of the corresponding y-value for $y = 3x^2$.

The graph of $y = -3x^2$ is the reflection across the x-axis of the graph of $y = 3x^2$.

EXERCISES

Draw graphs of $y = x^2$ and the given function in the same coordinate plane. Compare the graphs.

1. $y = 2x^2$ **2.** $y = \frac{1}{2}x^2$ **3.** $y = 0.75x^2$ **4.** $y = \frac{3}{2}x^2$

For the given value of a, draw graphs of $y = ax^2$ and $y = -ax^2$ in the same coordinate plane. Compare the graphs.

5. $a = 4$ **6.** $a = \frac{1}{4}$ **7.** $a = \frac{2}{3}$ **8.** $a = \frac{5}{2}$

The graph of $y = x^2 + k$ is a vertical shift, or *translation*, of $y = x^2$ by $|k|$ units. If $k > 0$, the shift is upward. If $k < 0$, the shift is downward.

3 EXAMPLE

Graph $y = x^2$, $y = x^2 + 2$, and $y = x^2 - 3$. Compare the graphs.

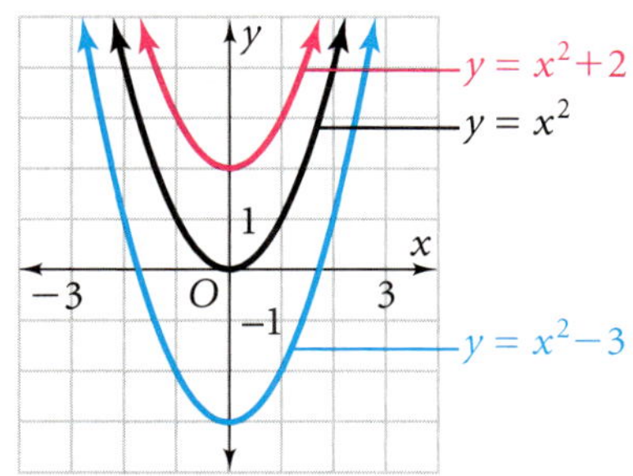

Draw $y = x^2$ as in Example 1. For $y = x^2 + 2$, draw each point to be 2 units higher than the corresponding point in $y = x^2$.

For $y = x^2 - 3$ draw each point to be 3 units lower than the corresponding point in $y = x^2$.

The graph of $y = x^2 + 2$ is 2 units above the graph of $y = x^2$.
The graph of $y = x^2 - 3$ is 3 units below the graph of $y = x^2$.

EXERCISES

Draw graphs of $y = x^2$ and the given function in the same coordinate plane. Compare the graphs.

9. $y = x^2 + 1$ **10.** $y = x^2 - 1$ **11.** $y = x^2 + \frac{5}{2}$ **12.** $y = x^2 - 3.5$

Wherever the graph of the quadratic function $y = ax^2 + bx + c$ crosses the x-axis, y has value 0. This means that each x-intercept is a solution of $ax^2 + bx + c = 0$, a quadratic equation in standard form. The zeros of the quadratic function are the same as the solutions of the quadratic equation. Thus, you can solve quadratic equations by graphing the corresponding function and finding its zeros.

4 EXAMPLE

Solve $x^2 + x - 6 = 0$.

Method 1

$x^2 + x - 6 = 0$

$(x + 3)(x - 2) = 0$ **Factor.**

$x + 3 = 0$ or $x - 2 = 0$ **If $a \cdot b = 0$, then $a = 0$ or $b = 0$.**

$x = -3$ or $x = 2$ **Solve for x.**

Method 2

Graph $y_1 = x^2 + x - 6$. Use the zero feature in the CALC menu to find where $y = 0$.

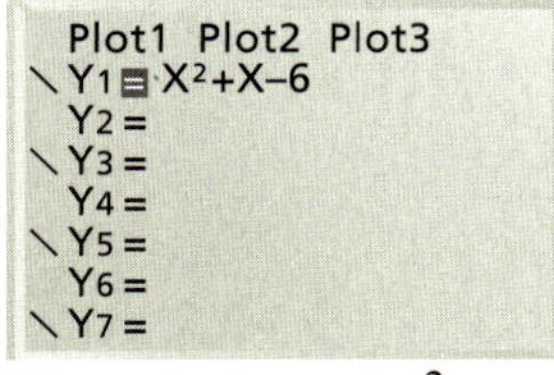

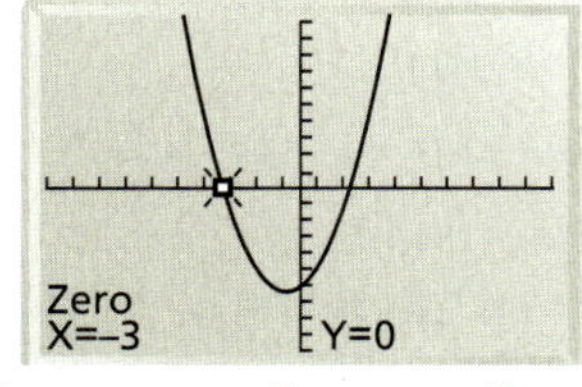

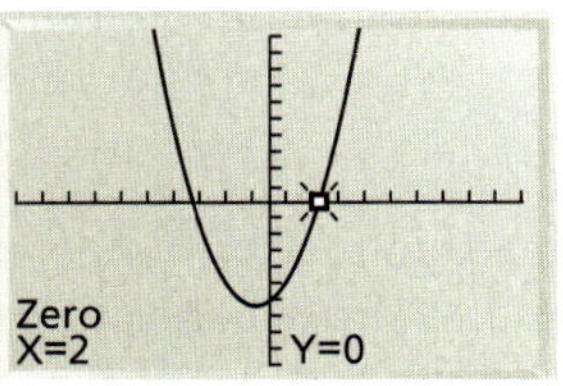

The solutions of $x^2 + x - 6 = 0$ are $x = -3$ or $x = 2$.

EXERCISES

Solve each quadratic equation.

13. $x^2 - 4x = 0$ **14.** $x^2 - x - 6 = 0$ **15.** $x^2 - 25 = 0$ **16.** $x^2 - x - 210 = 0$

17. $x^2 + 4x + 3 = 0$ **18.** $2x^2 + 3x - 20 = 0$ **19.** $2x^2 + 11x - 6 = 0$ **20.** $6x^2 - 13x + 6 = 0$

Drawing a Diagram

Sometimes, if a problem does not have a diagram, you should draw one. A diagram helps you to see the given information and make inferences from it.

1 EXAMPLE

Side $\overline{MN}$ of square $MNPQ$ has endpoints $M(0, 0)$ and $N(5, 2)$. What are the possible locations of point P?

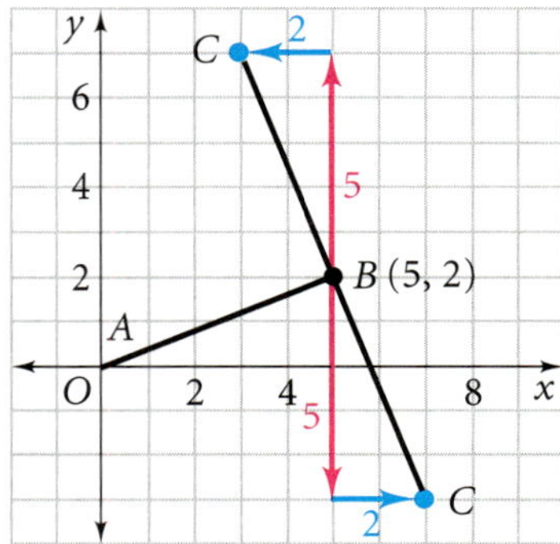

Ⓐ $(-2, 5)$ or $(3, 7)$ Ⓒ $(-2, 5)$ or $(2, -5)$

Ⓑ $(2, -5)$ or $(7, -3)$ Ⓓ $(3, 7)$ or $(7, -3)$

You need to draw a diagram to do this problem. Point P can be in two different locations.

First draw the segment $\overline{MN}$, and then begin the two possible squares that have $\overline{MN}$ as a side. To locate P from N you can either go up 5 units and left 2 units, or you can go down 5 units and right 2 units. The coordinates of point P can be either $(3, 7)$ or $(7, -3)$. The correct choice is D.

2 EXAMPLE

Marie is making a design for a quilt using rhombuses. She connects the midpoints of each rhombus to form a quadrilateral inside. What is the most specific name for the inner quadrilateral?

Ⓐ parallelogram Ⓑ rhombus Ⓒ rectangle Ⓓ square

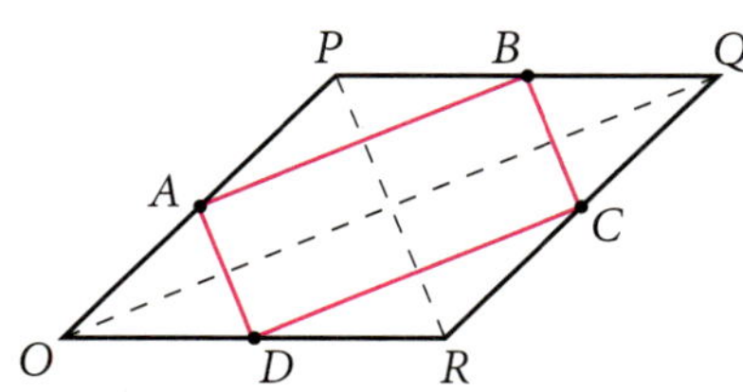

Draw a rhombus and join the midpoints of the sides. The midpoint quadrilateral is a parallelogram since both pairs of sides are parallel to a diagonal of the rhombus. The diagonals of the rhombus are perpendicular, so adjacent sides of the parallelogram are perpendicular to each other.

The quadrilateral formed is a rectangle. The correct choice is C.

EXERCISES

1. Three vertices of parallelogram $PGRM$ are $P(0, 0)$, $G(5, 2)$, and $R(-2, 4)$. What is the location of point M?

   Ⓐ $(3, 6)$ Ⓑ $(-6, 3)$ Ⓒ $(7, -2)$ Ⓓ $(-7, 2)$

2. Quinn needs to reach a window that is 20 feet above the ground. A shrub below the window is 3 feet from the house and 5 feet high. How long is the shortest ladder Quinn can use to reach the window without touching the top of the shrub?

   Ⓐ 20 feet Ⓑ 21 feet Ⓒ 22 feet Ⓓ 23 feet

3. Quadrilateral $S'T'U'V'$ is similar to quadrilateral $STUV$. The vertices of $STUV$ are $S(1, 2)$, $T(2, 5)$, $U(5, 4)$, and $V(4, 1)$. If $S'(3, 3)$, $T'(5, 9)$, and $U'(11, 7)$ are vertices of $S'T'U'V'$, what are the coordinates of V'?

   Ⓐ $(8, 2)$ Ⓑ $(9, 2)$ Ⓒ $(9, 1)$ Ⓓ $(10, 1)$

Chapter 6 Review

Vocabulary Review

English and Spanish Audio Online

base angles of a trapezoid (p. 336)
consecutive angles (p. 312)
isosceles trapezoid (p. 306)
kite (p. 306)
midsegment of a trapezoid (p. 348)
parallelogram (p. 306)
rectangle (p. 306)
rhombus (p. 306)
square (p. 306)
trapezoid (p. 306)

To complete each definition, find the appropriate word in the second column.

1. A(n) _?_ is a parallelogram with four right angles.
2. A(n) _?_ is a quadrilateral with two pairs of adjacent sides congruent and no opposite sides congruent.
3. Angles of a polygon that share a common side are _?_.
4. A(n) _?_ is a quadrilateral with exactly one pair of parallel sides.
5. A(n) _?_ is a parallelogram with four congruent sides.
6. The _?_ of a trapezoid is the segment that joins the midpoints of the nonparallel opposite sides.
7. A(n) _?_ is a quadrilateral with both pairs of opposite sides parallel.
8. A(n) _?_ is a parallelogram with four congruent sides and four right angles.
9. A(n) _?_ is a trapezoid whose nonparallel opposite sides are congruent.
10. Two angles that share a base of a trapezoid are its _?_.

A. parallelogram
B. trapezoid
C. square
D. base angles
E. isosceles trapezoid
F. rectangle
G. consecutive angles
H. kite
I. rhombus
J. midsegment

For: Vocabulary quiz
Web Code: bcj-0651

Skills and Concepts

Lesson 6-1

- To define and classify special types of quadrilaterals (p. 306)

 GEOM 12.0

Special quadrilaterals are defined by their characteristics.

A **parallelogram** is a quadrilateral with both pairs of opposite sides parallel.
A **rhombus** is a parallelogram with four congruent sides.
A **rectangle** is a parallelogram with four right angles.
A **square** is a parallelogram with four congruent sides and four right angles.
A **kite** is a quadrilateral with two pairs of adjacent sides congruent and no opposite sides congruent.
A **trapezoid** is a quadrilateral with exactly one pair of parallel sides.
An **isosceles trapezoid** is a trapezoid whose nonparallel opposite sides are congruent.

Draw and label each quadrilateral with the given vertices. Then determine the most precise name for each quadrilateral.

11. $N(-1, 2), M(3, 4), K(5, 0), L(1, -2)$

12. $P(-4, 2), Q(-1, 3), R(7, 0), S(4, -1)$

Algebra **Find the values of the variables and the lengths of the sides.**

13. isosceles trapezoid $ABCD$

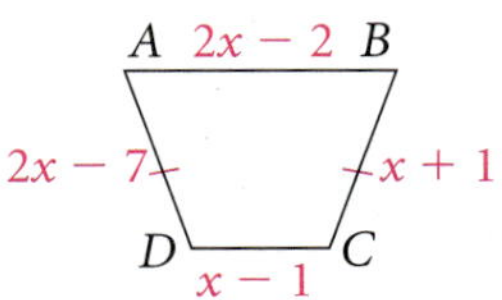

14. kite $KLMN$

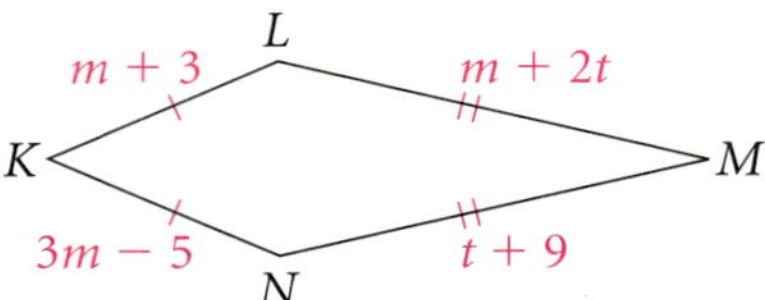

Lessons 6-2 and 6-3

- To use relationships among sides and among angles of parallelograms (p. 312)
- To use relationships involving diagonals of parallelograms or transversals (p. 314)
- To determine whether a quadrilateral is a parallelogram (p. 321)

 GEOM 7.0, 12.0, 13.0

Opposite sides and opposite angles of a parallelogram are congruent. The diagonals of a parallelogram bisect each other.

If three (or more) parallel lines cut off congruent segments on one transversal, then they cut off congruent segments on every transversal.

A quadrilateral is a parallelogram if any one of the following is true.
Both pairs of opposite sides are congruent.
Both pairs of opposite angles are congruent.
The diagonals bisect each other.
One pair of opposite sides is both congruent and parallel.

Find the measures of the numbered angles for each parallelogram.

15.

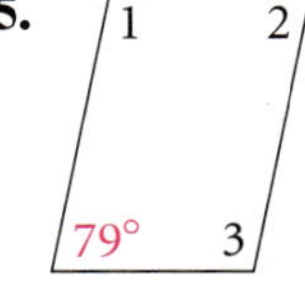

16.

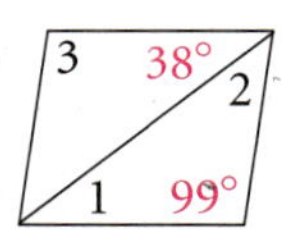

17.

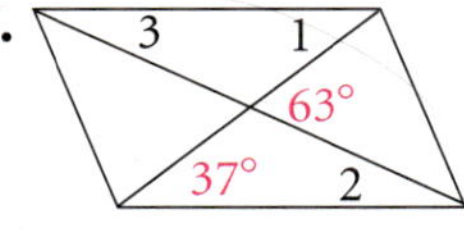

Determine whether the quadrilateral must be a parallelogram.

18.

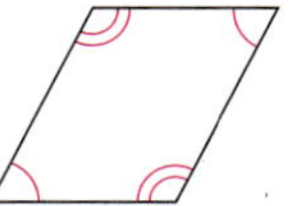

19.

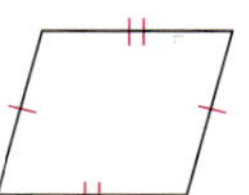

20.

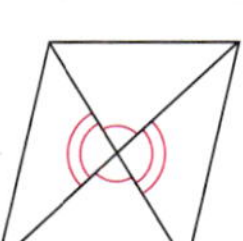

21.

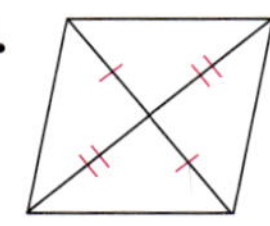

Algebra **Find the values of the variables for which $ABCD$ must be a parallelogram.**

22.

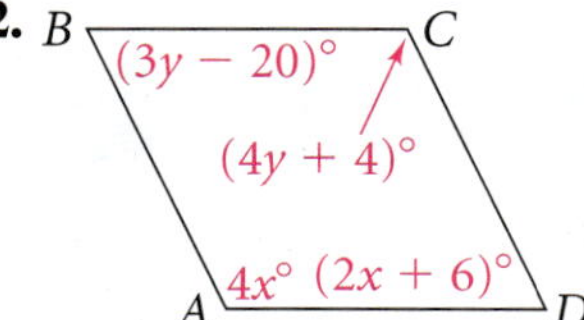

23.

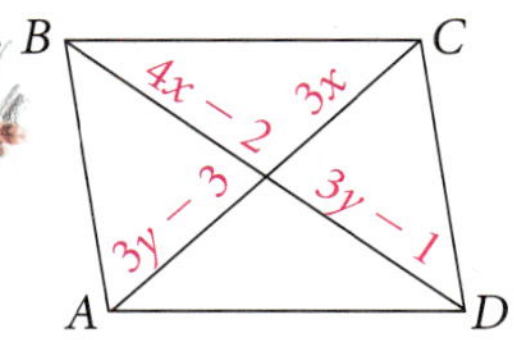

Lessons 6-4 and 6-5

- To use properties of diagonals of rhombuses and rectangles (p. 329)
- To determine whether a parallelogram is a rhombus or a rectangle (p. 331)
- To verify and use properties of trapezoids and kites (p. 336)

 GEOM 7.0, 12.0, 13.0

Each diagonal of a rhombus bisects two angles of the rhombus. The diagonals of a rhombus are perpendicular.

The diagonals of a rectangle are congruent.

If one diagonal of a parallelogram bisects two angles of the parallelogram, then the parallelogram is a rhombus. If the diagonals of a parallelogram are perpendicular, then the parallelogram is a rhombus. If the diagonals of a parallelogram are congruent, then the parallelogram is a rectangle.

The parallel sides of a trapezoid are its bases and the nonparallel sides are its legs. Two angles that share a base of a trapezoid are **base angles** of the trapezoid.

The base angles of an isosceles trapezoid are congruent. The diagonals of an isosceles trapezoid are congruent.

The diagonals of a kite are perpendicular.

Find the measures of the numbered angles in each quadrilateral.

24. 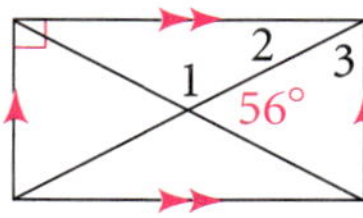

25. 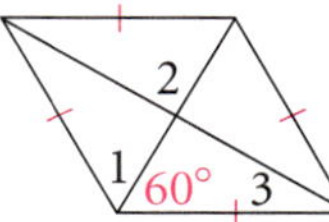

26. 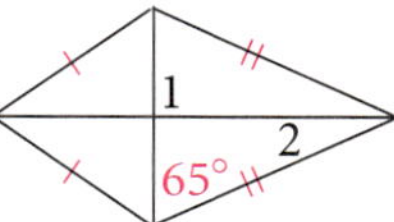

Find *AC* for each quadrilateral.

27. 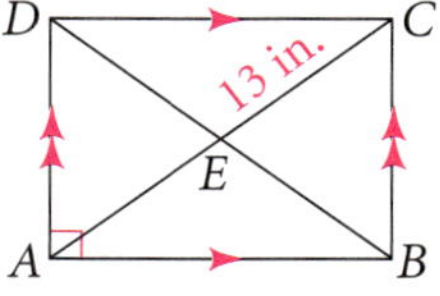

28. 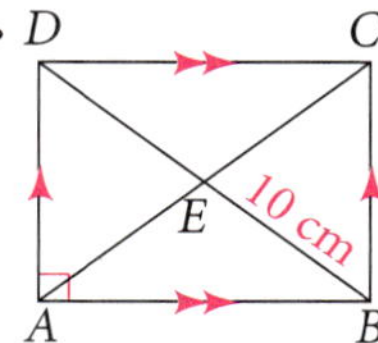

29. 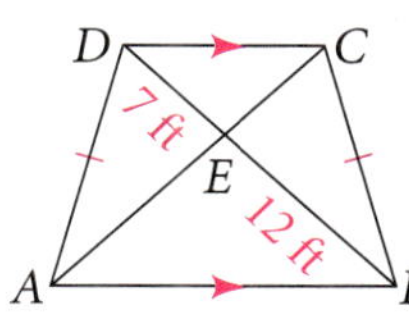

Lessons 6-6 and 6-7

- To name coordinates of special figures by using their properties (p. 343)
- To prove theorems using figures in the coordinate plane (p. 348)

 **GEOM 17.0**

In coordinate proofs, it generally is good practice to center the figure on the origin or place a vertex at the origin and one side of the figure on an axis.

The segment that joins the midpoints of the nonparallel sides of a trapezoid is the **midsegment** of the trapezoid. It is parallel to the bases and half as long as the sum of the lengths of the bases.

The formulas for slope, midpoint, and distance are used in coordinate proofs.

Give the coordinates of point *P* without using any new variable.

30. rectangle

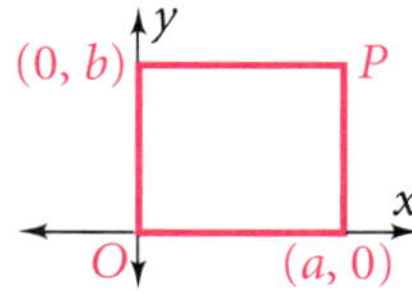

31. square

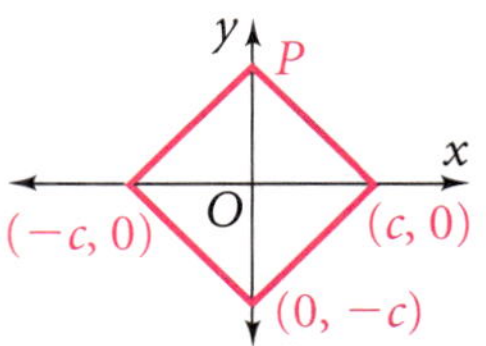

32. parallelogram

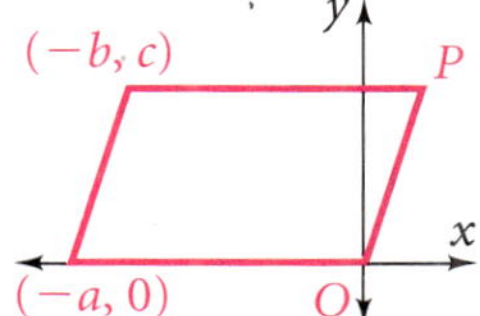

Complete each coordinate proof.

33. The diagonals of a square are perpendicular.

Given: Square $FGHI$ with vertices $I(0, 0)$, $H(a, 0)$, $G(a, a)$, and $F(0, a)$

Prove: $\overline{FH} \perp \overline{GI}$

The slope of $\overline{FH}$ is **a.** _?_. The slope of $\overline{GI}$ is **b.** _?_. $\overline{FH} \perp \overline{GI}$ because **c.** _?_.

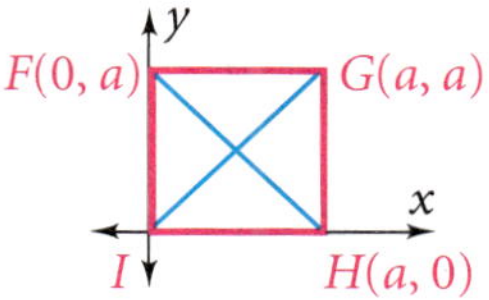

34. The diagonals of a rectangle are congruent.

Given: Rectangle $ABCD$

Prove: $\overline{AC} \cong \overline{BD}$

Vertex A is at the origin with coordinates $(0, 0)$. Name B as $(a, 0)$, C as (**a.** _?_, b), and D as **b.** _?_. AC = **c.** _?_ and BD = **d.** _?_. So, AC = **e.** _?_, and $\overline{AC} \cong \overline{BD}$.

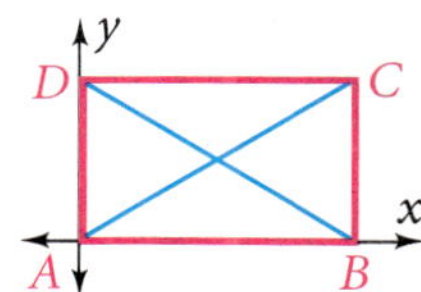

Chapter 6 Test

For: Chapter test
Web Code: bca-0652

Graph each quadrilateral $ABCD$. Then determine the most precise name for it.

1. $A(1, 2), B(11, 2), C(7, 5), D(4, 5)$
2. $A(3, -2), B(5, 4), C(3, 6), D(1, 4)$
3. $A(1, -4), B(1, 1), C(-2, 2), D(-2, -3)$
4. **Open-Ended** Write the coordinates of four points that determine each figure with the given conditions. One vertex is at the origin and one side is 3 units long.
   a. square **b.** parallelogram
   c. rectangle **d.** trapezoid

Find AN for each parallelogram.

5. 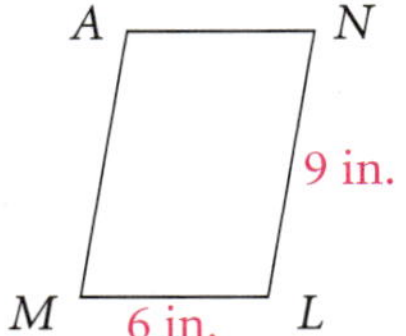

6. 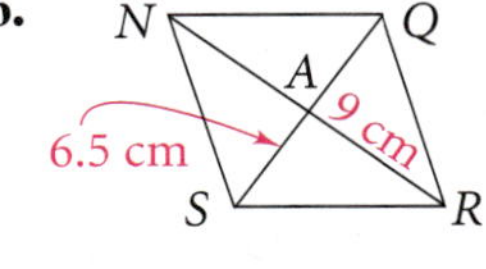

7. Sketch two noncongruent parallelograms $ABCD$ and $EFGH$ such that $\overline{AC} \cong \overline{BD} \cong \overline{EG} \cong \overline{FH}$.

 **Algebra Find the values of the variables for each parallelogram.**

8. 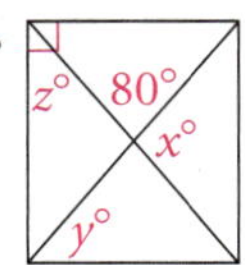

9. 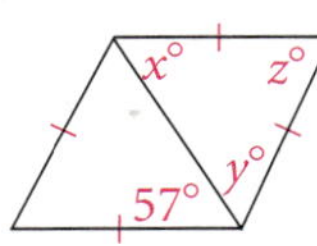

10. 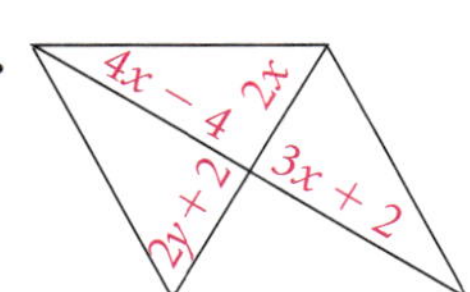

11. 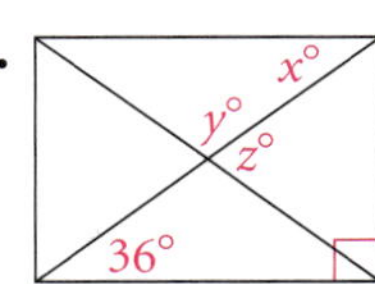

Does the information allow you to prove that $ABCD$ is a parallelogram? Explain.

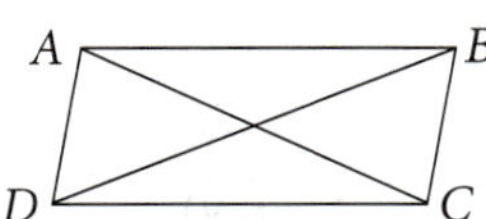

12. $\overline{AC}$ bisects $\overline{BD}$.
13. $\overline{AB} \cong \overline{DC}$; $\overline{AB} \parallel \overline{DC}$
14. $\overline{AB} \cong \overline{DC}$; $\overline{BC} \cong \overline{AD}$
15. $\angle DAB \cong \angle BCD$ and $\angle ABC \cong \angle CDA$

16. **Writing** Explain why a square cannot be a kite.

17. **Algebra** Determine the values of the variables for which $ABCD$ is a parallelogram.

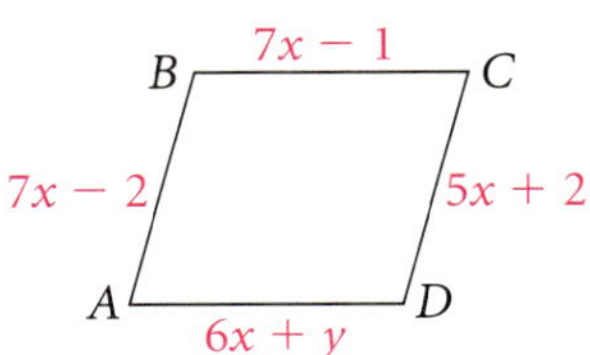

Find the measures of $\angle 1$ and $\angle 2$.

18. 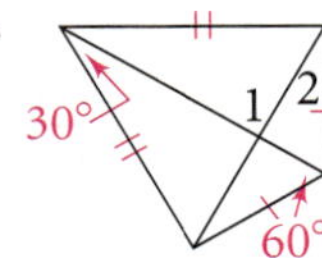

19. 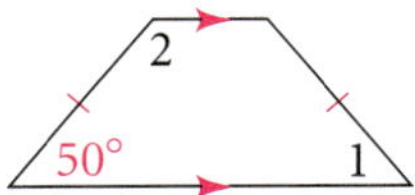

Give the coordinates for points S and T without using any new variables. Then find the midpoint and the slope of $\overline{ST}$.

20. rectangle

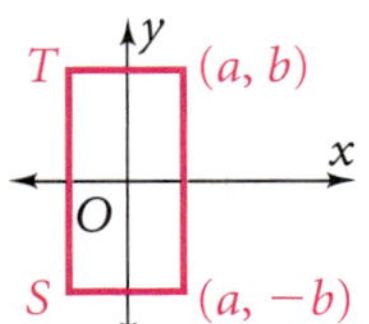

21. parallelogram

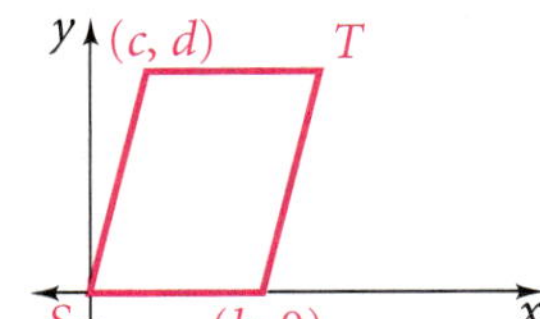

22. You want a garden border halfway between the front of your house and the city sidewalk, which are parallel to each other. You've measured those two edges. Find the length of the garden border.

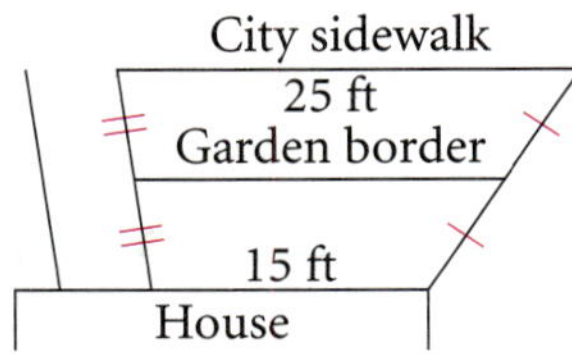

23. Complete this coordinate proof that the diagonals of a square are congruent.

Given: Square $ABCD$ with vertices $A\ (0, 0), B(a, 0), C(a, a)$, and $D(0, a)$

Prove: $\overline{AC} \cong \overline{BD}$

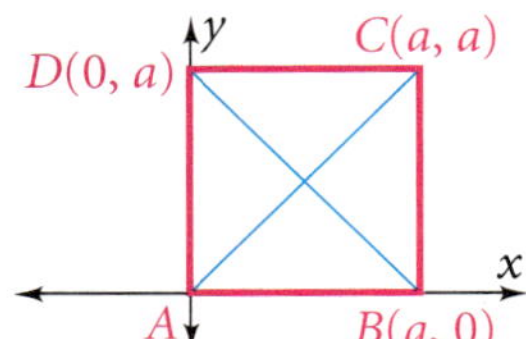

$AC =$ **a.** _?_, and $BD =$ **b.** _?_. So, **c.** _?_ $= BD$ and $\overline{AC} \cong \overline{BD}$.

Standards Mastery

Cumulative Practice

Some questions ask you to match an equation to a graph. Read the question at the right. Then follow the tips to answer it.

Which linear function best describes the graph shown below?

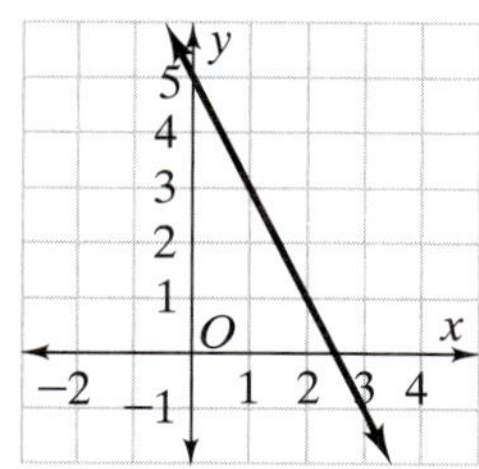

Ⓐ $y = -2x + 5$
Ⓑ $y = 2x + 5$
Ⓒ $y = -5x + 2$
Ⓓ $y = 5x - 2$

Tip 1
To identify the equation of the graph of a line, first find the y-intercept.

Tip 2
Determine whether the slope is positive or negative.

Think It Through

The y-intercept b is 5. This eliminates choices C and D.

The graph slants downward from left to right. The slope is negative. This eliminates choice B (and D again).

The correct choice is A.

Vocabulary Review

As you solve problems, you must understand the meanings of mathematical terms. Choose the correct term to complete each sentence.

I. Two lines that intersect to form right angles are (parallel, perpendicular) lines.

II. The most frequently occurring value in a data set is the (median, mode).

III. In the same plane, two lines that never intersect are (parallel, skew) lines.

IV. A (permutation, probability) is a number describing the chance that an event will occur.

V. A (cone, pyramid) has a polygon as a base and triangles forming its other surfaces.

Multiple Choice

Read each question. Then write the letter of the correct answer on your paper.

1. Which list could represent the lengths of the sides of a triangle? **(Lesson 5-5)**

Ⓐ 1 ft, 2 ft, 4 ft
Ⓑ 3 m, 5 m, 7 m
Ⓒ 4 in., 6 in., 10 in.
Ⓓ 7 cm, 10 cm, 25 cm

2. If $STUV$ is a parallelogram, what are the coordinates of point U? **(Lesson 6-6)**

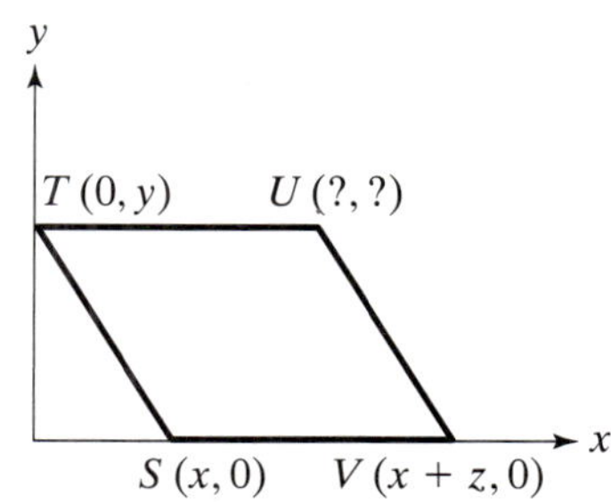

Ⓐ (x, y)
Ⓑ $(x + z, y)$
Ⓒ (y, z)
Ⓓ (z, y)

Cumulative Practice (continued)

3. Given: $\overline{AD} \cong \overline{AE}$, $\overline{DB} \cong \overline{EC}$, $\angle ABC \cong \angle ACB$ Which statement must be proved first by CPCTC to prove $\triangle ABE \cong \triangle ACD$ by SSS? **(Lesson 4-7)**

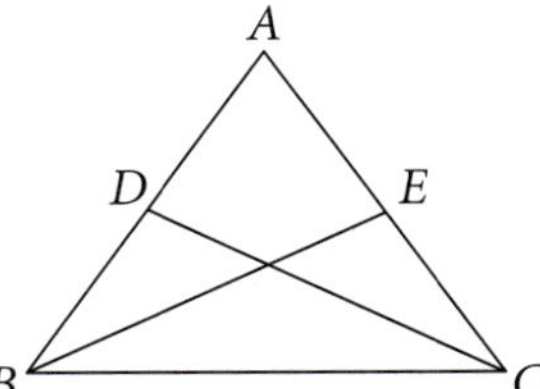

Ⓐ $\angle A \cong \angle A$
Ⓑ $\overline{AB} \cong \overline{BC}$
Ⓒ $\overline{BC} \cong \overline{BC}$
Ⓓ $\overline{DC} \cong \overline{EB}$

4. Which of the following quadrilaterals does not have a pair of adjacent sides congruent? **(Lesson 6-1)**

Ⓐ square
Ⓑ kite
Ⓒ rhombus
Ⓓ rectangle

5. Figure *FGHJ* is a quadrilateral. If at least one pair of opposite angles in the figure is congruent, which statement is false? **(Lesson 6-5)**

Ⓐ Quadrilateral *FGHJ* is a trapezoid.
Ⓑ Quadrilateral *FGHJ* is a rhombus.
Ⓒ Quadrilateral *FGHJ* is a kite.
Ⓓ Quadrilateral *FGHJ* is a parallelogram.

6. Which quadrilateral cannot contain four right angles? **(Lesson 6-1)**

Ⓐ square
Ⓑ rhombus
Ⓒ trapezoid
Ⓓ rectangle

7. What is the measure of $\angle 1$ in the figure below? **(Lesson 3-4)**

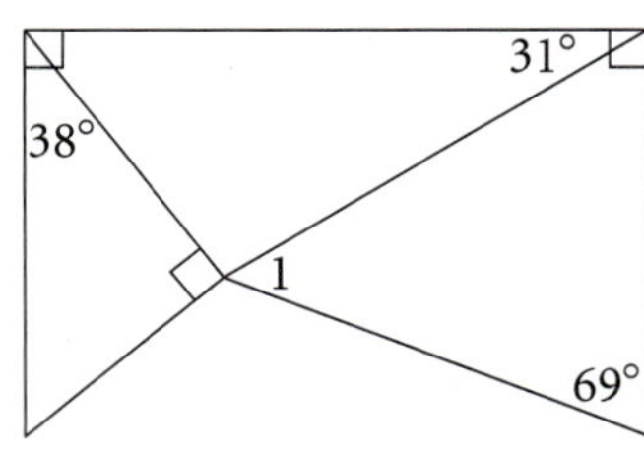

Ⓐ 52°
Ⓑ 73°
Ⓒ 83°
Ⓓ 93°

8. Given $\triangle ABC$ below, what is $m\angle B$? **(Lesson 3-4)**

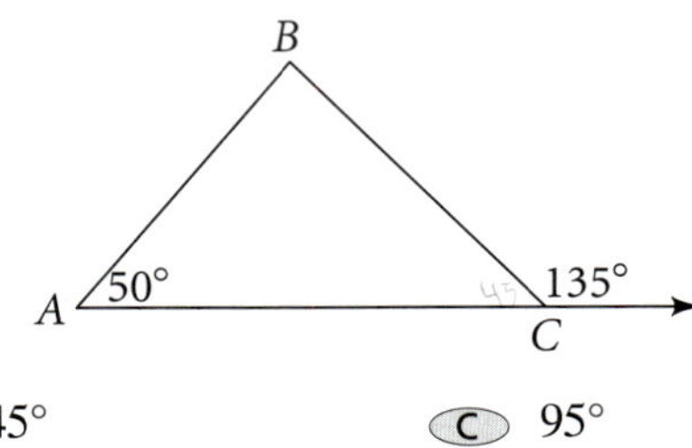

Ⓐ 45°
Ⓑ 85°
Ⓒ 95°
Ⓓ 100°

9. If the measure of each interior angle in a regular polygon is 140°, what type of polygon is it? **(Lesson 3-5)**

Ⓐ hexagon
Ⓑ decagon
Ⓒ octagon
Ⓓ nonagon

10. In triangle *GHJ*, $\overline{GH} \cong \overline{HJ}$. Using the indirect proof method, you attempt to prove that $\angle G$ and $\angle J$ are right angles. What theorem will contradict this claim? **(Lesson 5-4)**

Ⓐ Triangle Angle-Sum Theorem
Ⓑ Side-Angle-Side Theorem
Ⓒ Converse of the Isosceles Triangle Theorem
Ⓓ Angle-Angle-Side Theorem

11. *ABCD* is a rhombus. To prove that the diagonals of a rhombus are perpendicular, which pair of angles below must first be proved congruent by CPCTC? **(Lesson 6-4)**

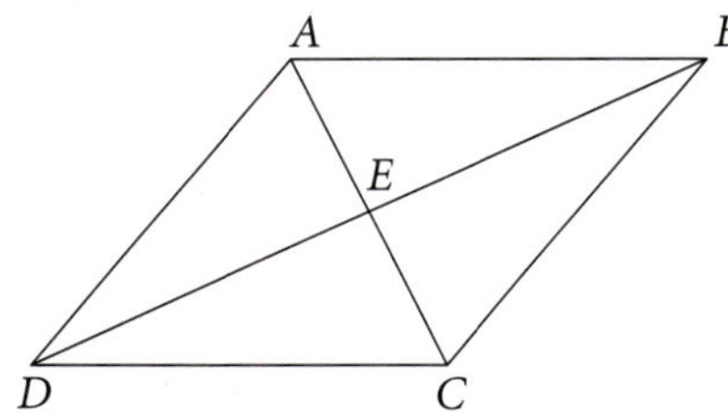

Ⓐ $\angle AEB$ and $\angle DEC$
Ⓑ $\angle AEB$ and $\angle AED$
Ⓒ $\angle BEC$ and $\angle AED$
Ⓓ $\angle DAB$ and $\angle ABC$

12. Which quadrilateral must have congruent diagonals? **(Lesson 6-4)**

Ⓐ kite
Ⓑ rectangle
Ⓒ parallelogram
Ⓓ rhombus

Standards Mastery

13. What are the possible values for n to make ABC a valid triangle? **(Lesson 5-5)**

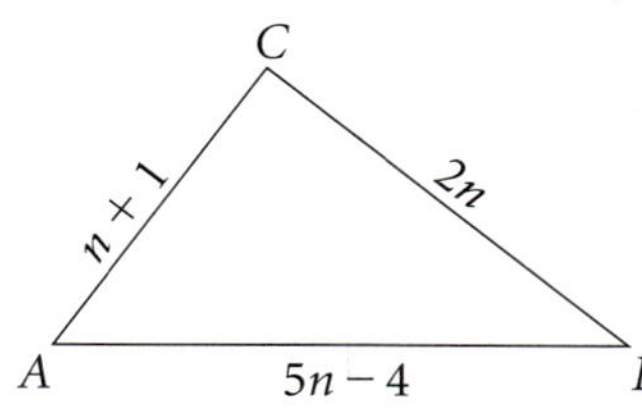

Ⓐ $-\frac{5}{2} < n < \frac{5}{6}$ Ⓒ $\frac{3}{4} < n < \frac{5}{6}$

Ⓑ $\frac{3}{4} < n < \frac{5}{2}$ Ⓓ $\frac{5}{6} < n < \frac{5}{2}$

14. An isosceles triangle has two angles measuring 48 and 84. What is the measure of the third angle? **(Lesson 4-5)**

Ⓐ 84 Ⓑ 51 Ⓒ 49 Ⓓ 48

15. For which value of x are lines g and h parallel? **(Lesson 3-2)**

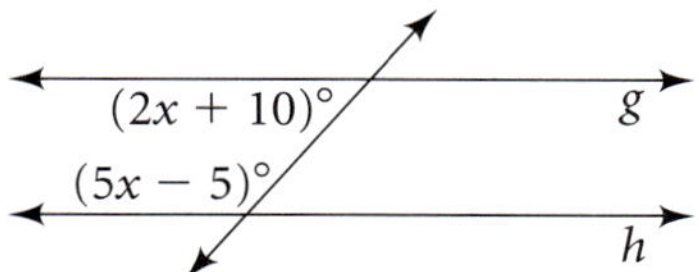

Ⓐ 25 Ⓑ 18 Ⓒ 15 Ⓓ 12

16. Which statement is true for the figure? **(Lesson 5-5)**

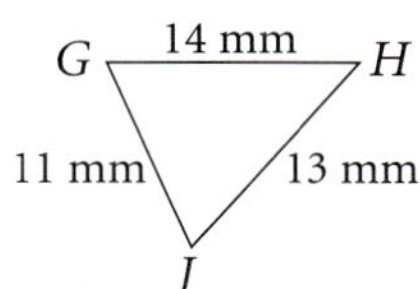

Ⓐ $m\angle J < m\angle G < m\angle H$
Ⓑ $m\angle H < m\angle G < m\angle J$
Ⓒ $m\angle J < m\angle H < m\angle G$
Ⓓ $m\angle H < m\angle J < m\angle G$

17. What is the value of x in the kite below? **(Lesson 6-5)**

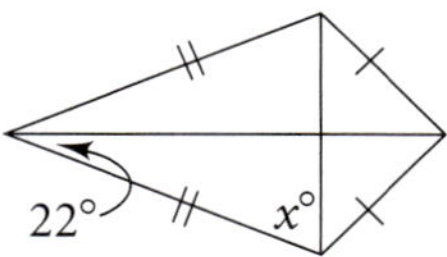

Ⓐ 48 Ⓑ 52 Ⓒ 62 Ⓓ 68

18. What is the value of x in the figure below? **(Lesson 3-5)**

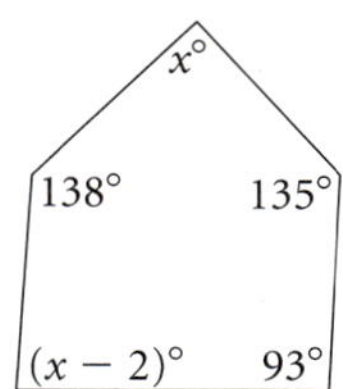

Ⓐ 88 Ⓑ 86 Ⓒ 78 Ⓓ 68

Standards Reference Guide

California Content Standard	Item Number(s)
GEOM 6.0	1, 13
GEOM 17.0	2
GEOM 5.0	3, 11
GEOM 7.0	4–6, 12, 15
GEOM 13.0	7, 17
GEOM 12.0	8, 9, 14, 16, 18
GEOM 2.0	10

For additional review and practice, see the *California Standards Review and Practice Workbook* or go online to use the

California Standards Tutorials

Visit: PHSchool.com, **Web Code:** bcq-9045

Similarity

What You've Learned

California Content Standards

7.0 Students prove and use theorems involving the properties of quadrilaterals.

12.0 Students find and use measures of sides and of interior angles of polygons to classify figures and solve problems.

13.0 Students prove relationships between angles in polygons by using properties of complementary and supplementary angles.

17.0 Students prove theorems by using coordinate geometry.

Check Your Readiness

Simplifying Ratios (Skills Handbook page 756)

Simplify each ratio.

1. 12 : 18 **2.** $\frac{55}{11}$ **3.** $\frac{20a^2}{15a^5}$ **4.** $\frac{3x^2 - 12x}{x^3 - x^2}$

Polygon Angle-Sum Theorems (Lesson 3-4)

Determine the measure of an angle of each regular polygon.

5. pentagon **6.** octagon **7.** decagon **8.** 27-gon

Congruent Figures (Lesson 4-1)

$\triangle PAC \cong \triangle DHL$. Complete the congruence statements.

9. $\overline{PC} \cong$ __?__ **10.** $\angle H \cong$ __?__ **11.** $\angle PCA \cong$ __?__ **12.** $\triangle HDL \cong$ __?__

Triangle Congruence (Lessons 4-2 and 4-3)

State why each pair of triangles are congruent.

13.

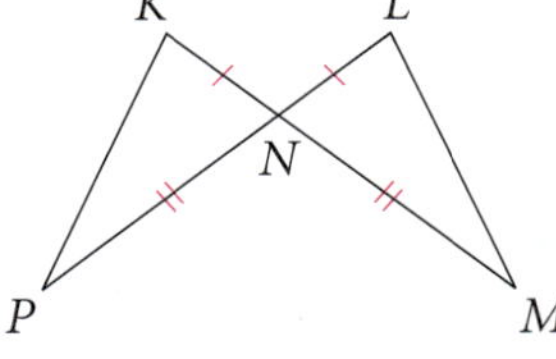

14.

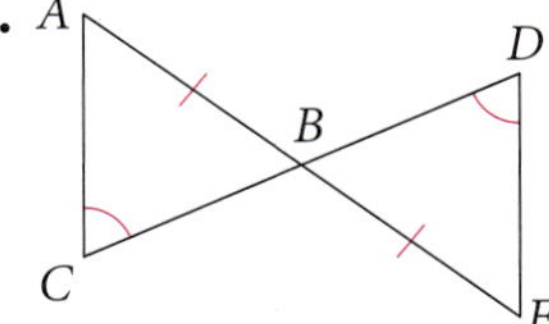

15.

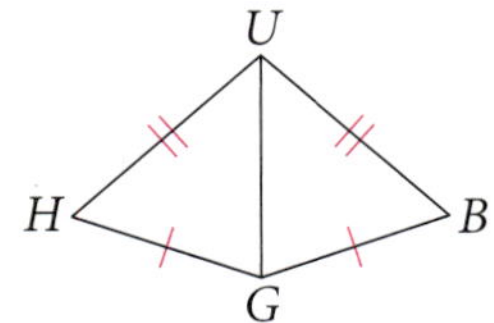

Midsegments of Triangles (Lesson 5-1)

Use the figure at the right for Exercises 16–17.

16. If $BC = 12$, then $BF =$ __?__ and $DE =$ __?__.

17. If $EF = 4.7$, then $AD =$ __?__ and $AC =$ __?__.

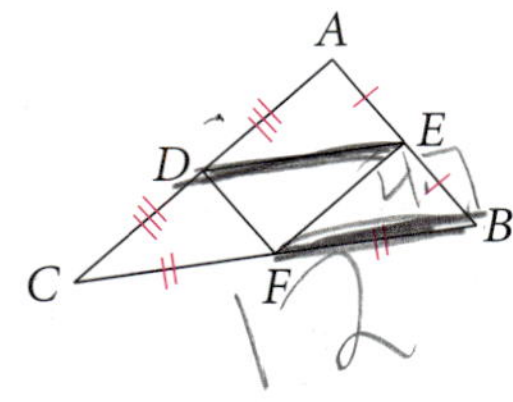

▲ Dew drops come in different sizes, but similar shapes.

What You'll Learn Next

California Content Standards

4.0 Students prove theorems involving similarity.

5.0 Students prove that triangles are similar.

7.0 Students prove and use theorems involving the properties of parallel lines cut by a transversal.

CAHSEE 7NS 1.1, 6SDAP 3.3, 6SDAP 3.5, 7MG 1.2, 7MG 2.2, 7MR 2.1, 7MR 3.3, Alg1 5.0, Alg1 6.0, Alg1 7.0, Alg1 10.0

New Vocabulary

English and Spanish Audio Online

- **Cross-Product Property** (p. 367)
- **extended proportion** (p. 367)
- **geometric mean** (p. 392)
- **golden ratio** (p. 375)
- **golden rectangle** (p. 375)
- **indirect measurement** (p. 384)
- **proportion** (p. 367)
- **ratio** (p. 366)
- **scale** (p. 368)
- **scale drawing** (p. 368)
- **similar** (p. 373)
- **similarity ratio** (p. 373)

7-1 Ratios and Proportions

What You'll Learn

- To write ratios and solve proportions

. . . And Why

To find dimensions from a scale drawing, as in Example 4

Check Skills You'll Need

Simplify each ratio.

1. $\frac{2}{4}$ **2.** $\frac{8}{12}$ **3.** $\frac{6}{8}$ **4.** $\frac{10}{10}$

5. 20 : 30 **6.** 8 to 2 **7.** 2 to 8 **8.** 12 : 9

9. Draw a triangle. Then draw its three midsegments to form a smaller triangle. How do the lengths of the sides of the smaller triangle compare to the lengths of the sides of the larger triangle?

New Vocabulary • ratio • proportion • extended proportion • Cross-Product Property • scale drawing • scale

1 Using Ratios and Proportions

Vocabulary Tip

You can read $a : b$ as the ratio a to b.

A **ratio** is a comparison of two quantities. You can write the ratio of a to b or $a : b$ as the quotient $\frac{a}{b}$ when $b \neq 0$. Unless otherwise stated, the terms and expressions appearing in ratios in this book are assumed to be nonzero.

1 EXAMPLE Real-World Connection

Photography A photo that is 8 in. wide and $5\frac{1}{3}$ in. high is enlarged to a poster that is 2 ft wide and $1\frac{1}{3}$ ft high. What is the ratio of the width of the photo to the width of the poster?

$$\frac{\text{width of photo}}{\text{width of poster}} = \frac{8 \text{ in.}}{2 \text{ ft}} = \frac{8 \text{ in.}}{24 \text{ in.}} = \frac{8}{24} = \frac{1}{3}$$

The ratio of the width of the photo to the width of the poster is 1 : 3 or $\frac{1}{3}$.

1 What is the ratio of the height of the photo to the height of the poster?

Vocabulary Tip

You can read both $\frac{a}{b} = \frac{c}{d}$ and $a : b = c : d$ as <u>a is to b as c is to d</u>.

A **proportion** is a statement that two ratios are equal. You can write a proportion in these forms:

$$\frac{a}{b} = \frac{c}{d} \text{ and } a : b = c : d$$

When three or more ratios are equal, you can write an **extended proportion.** For example, you could write the following:

$$\frac{6}{24} = \frac{4}{16} = \frac{1}{4}$$

Two equations are equivalent when either can be deduced from the other using the Properties of Equality. Several equations are equivalent to a proportion. Some of them are important enough to be called Properties of Proportions.

Key Concepts

Property	Properties of Proportions	
$\frac{a}{b} = \frac{c}{d}$ is equivalent to	(1) $ad = bc$	(2) $\frac{b}{a} = \frac{d}{c}$
	(3) $\frac{a}{c} = \frac{b}{d}$	(4) $\frac{a + b}{b} = \frac{c + d}{d}$

Multiplying both sides of $\frac{a}{b} = \frac{c}{d}$ by bd results in the first property, called the **Cross-Product Property.** You may state this property as "The product of the extremes is equal to the product of the means."

means

$a : b = c : d$

extremes

$\frac{a}{b} = \frac{c}{d}$

$ad = bc$

2 EXAMPLE Properties of Proportions

Algebra If $\frac{x}{y} = \frac{5}{6}$, complete each statement.

a. $6x = \blacksquare$

$6x = 5y$

b. $\frac{y}{x} = \frac{\blacksquare}{\blacksquare}$

$\frac{y}{x} = \frac{6}{5}$

c. $\frac{x}{5} = \frac{\blacksquare}{\blacksquare}$

$\frac{x}{5} = \frac{y}{6}$

d. $\frac{x + y}{y} = \frac{\blacksquare}{\blacksquare}$

$\frac{x + y}{y} = \frac{11}{6}$

CA Standards Check **2** **Critical Thinking** Write two proportions that are equivalent to $\frac{m}{4} = \frac{n}{11}$.

You solve a proportion by finding the value of the variable.

3 EXAMPLE Solving for a Variable

Algebra Solve each proportion.

a. $\frac{x}{5} = \frac{12}{7}$

$7x = 5(12)$ ← **Cross-Product Property**

$7x = 60$

$x = \frac{60}{7}$

b. $\frac{y + 3}{8} = \frac{y}{4}$

Cross-Product Property → $4(y + 3) = 8y$

$4y + 12 = 8y$

$12 = 4y$

$y = 3$

CA Standards Check **3** Solve each proportion.

a. $\frac{5}{z} = \frac{20}{3}$

b. $\frac{18}{n + 6} = \frac{6}{n}$

In a **scale drawing,** the **scale** compares each length in the drawing to the actual length. The lengths used in a scale can be in different units. A scale might be written as 1 in. to 100 mi, 1 in. = 12 ft, or 1 mm : 1 m. You can use proportions to find the actual dimensions represented in a scale drawing.

4 EXAMPLE — Real-World Connection

Multiple Choice Use a ruler to measure the length ℓ of the bedroom in the scale drawing. What is the length of the actual bedroom?

Ⓐ 10 ft Ⓑ 14 ft Ⓒ $18\frac{2}{7}$ ft Ⓓ $24\frac{2}{5}$ ft

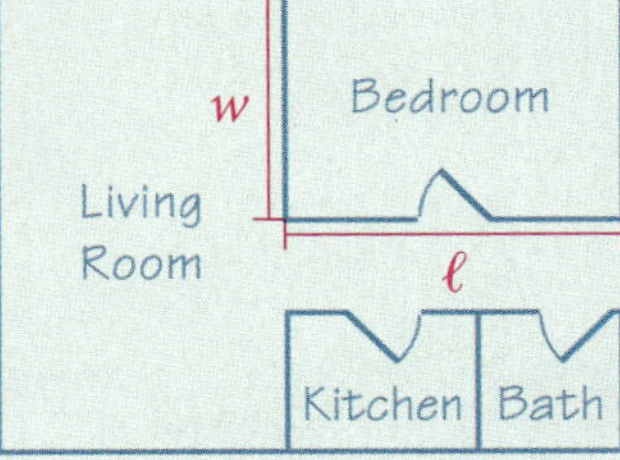

Scale: 1 in. = 16 ft

An inch ruler shows that $\ell = \frac{7}{8}$ in.

$\frac{1}{16} = \frac{\frac{7}{8}}{\ell}$ **drawing length (in.) / actual length (ft)**

$\ell = 16\left(\frac{7}{8}\right)$ **Cross-Product Property**

$\ell = 14$

The actual bedroom is 14 ft long. The correct answer is B.

Test-Taking Tip

If you use a ruler and your answer does not match any answer choice, first check that you measured correctly.

CA Standards Check 4 Find the width of the actual bedroom.

EXERCISES Standards Practice

For more exercises, see *Extra Skill, Word Problem, and Proof Practice.*

A Practice by Example

Example 1 (page 366)

GO for Help

1. The base of the pyramid at the right is a square whose sides measure 0.675 m. The intent was for the sides to measure 675 m. What is the ratio of the length of a base side in the small pyramid to the length of a base side in the intended pyramid?

2. **Models** The Leaning Tower of Pisa in Italy is about 185 ft tall. A model of the Leaning Tower is 6 in. tall. What is the ratio of the height of the model to the height of the real tower?

"We had a little problem with the decimal point."

Example 2 (page 367)

x^2 **Algebra If $\frac{a}{b} = \frac{3}{4}$, complete each statement.**

3. $4a = \blacksquare$
4. $\frac{b}{a} = \frac{\blacksquare}{\blacksquare}$
5. $\frac{a}{3} = \frac{\blacksquare}{\blacksquare}$
6. $\frac{4}{3} = \frac{\blacksquare}{\blacksquare}$
7. $\frac{4}{b} = \frac{\blacksquare}{\blacksquare}$
8. $3b = \blacksquare$
9. $\frac{a+b}{b} = \frac{\blacksquare}{\blacksquare}$
10. $\frac{a}{a+b} = \frac{\blacksquare}{\blacksquare}$
11. $\frac{a+3}{3} = \frac{\blacksquare}{\blacksquare}$

Example 3 (page 367)

x^2 **Algebra** **Solve each proportion.**

12. $\frac{x}{2} = \frac{8}{4}$
13. $\frac{9}{5} = \frac{3}{x}$
14. $\frac{1}{3} = \frac{x}{12}$
15. $\frac{5}{x} = \frac{8}{11}$
16. $\frac{4}{x} = \frac{5}{9}$
17. $\frac{5}{6} = \frac{6}{x}$
18. $\frac{x + 3}{3} = \frac{10 + 4}{4}$
19. $\frac{x + 7}{7} = \frac{15}{5}$
20. $\frac{3}{5} = \frac{6}{x + 3}$

Example 4 (page 368)

21. **Geography** The scale for this map of Louisiana is 1 in. = 40 mi. On the map, the distance from Lake Charles to Baton Rouge is about $3\frac{1}{8}$ in. About how far apart are the two cities?

Map Reading **Measure the map distance. Then find the actual distance.**

22. Morgan City to Rayne
23. Vinton to New Roads
24. Kaplan to Plaquemine

25. **Design** You want to make a scale drawing of your bedroom to help you arrange your furniture. You decide on a scale of 3 in. = 2 ft. Your bedroom is a 12 ft-by-15 ft rectangle. What should be its dimensions in your scale drawing?

B Apply Your Skills

For each rectangle, find the ratio of the longer side to the shorter side.

26. 30 m, 65 m

27. 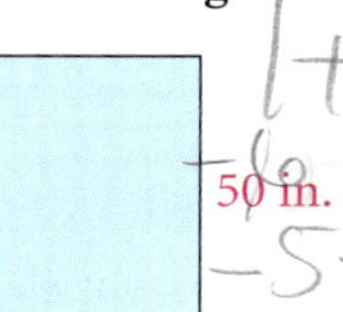
50 in., 40 in.

28. 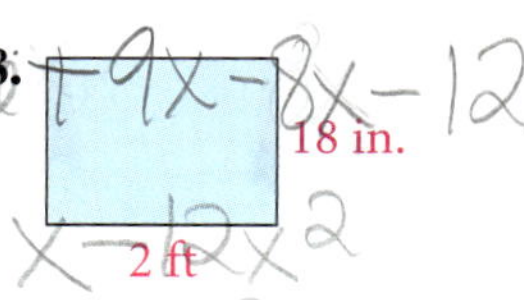
18 in., 2 ft

29. **Multiple Choice** The diameter of a dinner plate is 1 ft. In a dollhouse set, the diameter of a dinner plate is $1\frac{1}{4}$ in. What is the ratio of the diameter of the dollhouse plate to the diameter of the full-size plate?

Ⓐ $\frac{5}{48}$ Ⓑ $\frac{4}{5}$ Ⓒ $\frac{5}{4}$ Ⓓ $\frac{48}{5}$

Complete each statement.

30. If $\frac{x}{7} = \frac{y}{3}$, then $\frac{x}{y} = \blacksquare$.
31. If $4m = 9n$, then $\frac{m}{n} = \blacksquare$.
32. If $\frac{30}{t} = \frac{18}{r}$, then $\frac{t}{r} = \blacksquare$.
33. If $\frac{a + 5}{5} = \frac{b + 2}{2}$, then $\frac{a}{5} = \blacksquare$.

Homework Video Tutor

Visit: PHSchool.com
Web Code: bce-0701

34. **Writing** Use a map in your classroom or a map from a textbook. Explain how to use a ruler and the scale of the map to approximate an actual distance. Give an example. (If you do not have access to another map, use the map above.)

x^2 **Algebra** **Solve each proportion.**

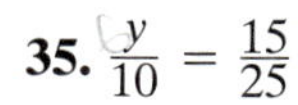

35. $\frac{y}{10} = \frac{15}{25}$
36. $\frac{9}{24} = \frac{12}{n}$
37. $\frac{11}{14} = \frac{b}{21}$
38. $\frac{5}{x - 3} = \frac{10}{x}$
39. $\frac{8}{n + 4} = \frac{4}{n}$
40. $\frac{2b - 1}{5} = \frac{b}{12}$
41. $\frac{x - 3}{3} = \frac{2}{x + 2}$
42. $\frac{3 + 4x}{1 + 5x} = \frac{1}{2 + 3x}$

Exercise 43

43. **Models** The sandwich shop at the left is 40 ft tall. The shop is an enlargement of an actual milk bottle. The scale used in construction is 5 ft = 2 cm. Find the height of the actual milk bottle.

44. **Geography** Students at the University of Minnesota in Minneapolis built a model globe 42 ft in diameter using a scale of 1 : 1,000,000. About how tall is Mount Everest on the model? (Mount Everest is about 29,000 ft tall.)

Complete each extended proportion.

45. $\frac{8}{12} = \frac{6}{■} = \frac{12}{■}$

46. $\frac{■}{15} = \frac{15}{25} = \frac{■}{20}$

47. $\frac{14}{■} = \frac{■}{12} = \frac{35}{20}$

Games **Choose a scale and make a scale drawing of the playing region.**

48. A pool table is 5 ft by 10 ft.

49. A bowling lane is 3.5 ft by 60 ft.

50. A basketball court is 92 ft by 50 ft.

51. A football field is 160 ft by 120 yd.

52. **Error Analysis** One rectangle has length 3 in. and width 4 ft. Another rectangle has length 3 ft and width 4 yd. Elaine claims that the two rectangles are similar because their corresponding angles are congruent and their corresponding sides are in proportion. Explain why Elaine's reasoning is incorrect.

If $\frac{a}{b} = \frac{c}{d}$, complete each statement.

53. $\frac{a + b}{c + d} = \frac{■}{■}$

54. $\frac{a + c}{b + d} = \frac{■}{■}$

55. $\frac{a + 2b}{b} = \frac{■}{■}$

Challenge **Algebra** **Justify the indicated property of proportions.**

56. Property (2): If $\frac{a}{b} = \frac{c}{d}$, then $\frac{b}{a} = \frac{d}{c}$.

57. Property (3): If $\frac{a}{b} = \frac{c}{d}$, then $\frac{a}{c} = \frac{b}{d}$.

58. Property (4): If $\frac{a}{b} = \frac{c}{d}$, then $\frac{a + b}{b} = \frac{c + d}{d}$.

Solve each extended proportion for x and y with $x > 0$ and $y > 0$.

59. $\frac{x}{6} = \frac{x + 10}{18} = \frac{4x}{y}$

60. $\frac{x}{5} = \frac{9}{y} = \frac{y}{25}$

61. $\frac{1}{x} = \frac{4}{x + 9} = \frac{7}{y}$

Multiple Choice Practice

For California Standards and CAHSEE Tutorials, visit PHSchool.com. Web Code: bcq-9045

GEOM 7.0

62. What values of x and y will make quadrilateral $PQRS$ a rhombus?

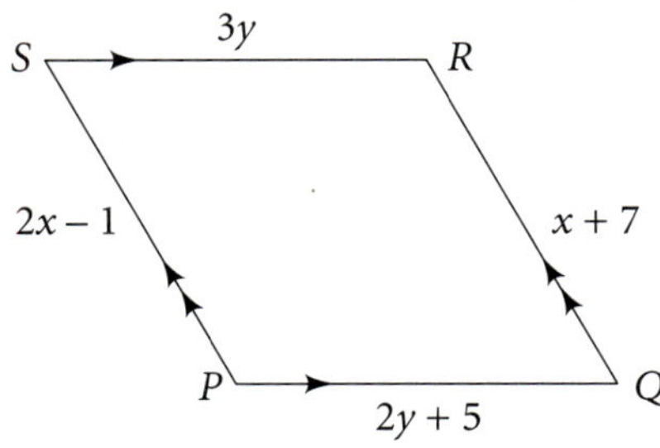

(A) $x = 8, y = 5$
(B) $x = 5, y = 8$
(C) $x = 7, y = 6$
(D) $x = 3, y = 7$

CAHSEE Alg1 7.0

63. Which equation best describes the relationship between x and y in the table shown below?

x	−3	−1	2	5
y	−9	−1	11	23

(A) $y = 4x - 3$
(B) $y = \frac{1}{4}x - 3$
(C) $y = 4x + 3$
(D) $y = \frac{1}{4}x + 3$

Mixed Review

GO for Help

Lesson 6-7

Give the coordinates for point P without using any new variables.

64. square

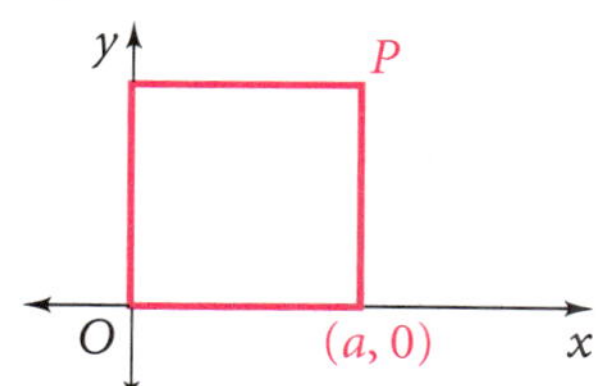

65. parallelogram

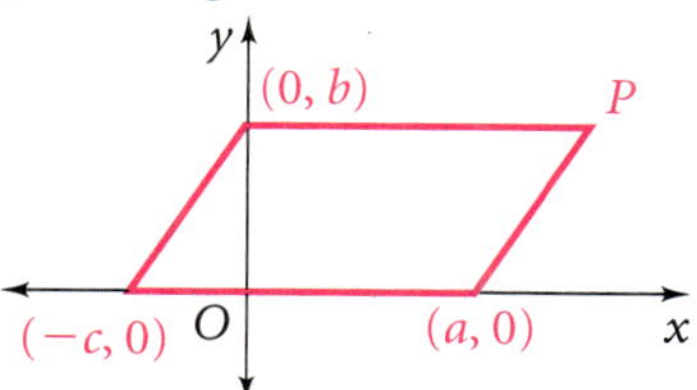

Lesson 6-1

Graph each quadrilateral $ABCD$. Classify $ABCD$ in as many ways as possible.

66. $A(-1, -2), B(3, -2), C(1, 4), D(-3, 4)$

67. $A(2, -1), B(6, 2), C(8, 2), D(10, -1)$

68. $A(-7, 1), B(-5, 3), C(0, -2), D(-2, -4)$

69. $A(1, 1), B(-4, 4), C(1, 7), D(6, 4)$

Lesson 5-4

In each exercise, identify two statements that contradict each other.

70. I. $\triangle PQR$ is isosceles.
II. $\triangle PQR$ is an obtuse triangle.
III. $\triangle PQR$ is scalene.

71. I. $\angle 1 \cong \angle 2$
II. $\angle 1$ and $\angle 2$ are complementary.
III. $m\angle 1 + m\angle 2 = 180$

Write (a) the inverse and (b) the contrapositive of each statement.

72. If an angle is acute, then it has measure between 0 and 90.

73. If two lines are parallel, then they are coplanar.

74. If two angles are complementary, then both angles are acute.

A Point in Time

1500 1600 1700 1800 1900 2000

In 1675, Danish astronomer Ole Römer used proportions to estimate the speed of light. He carefully measured the movements of Jupiter's moons. With Earth at point *B*, a moon emerged from behind Jupiter 16.6 min later than when Earth was at point *A*. He reasoned that it must have taken 16.6 min for the light to travel from point *A* to point *B*. Using proportions, Römer estimated the speed of light to be 150,000 mi/s. This estimate is about 81% of today's accepted value of 186,282 mi/s.

Jupiter
Sun
A
B
Earth
Moon
not drawn to scale

For: Information about the speed of light
Web Code: bce-2032

Algebra 1 Review

Solving Quadratic Equations

FOR USE WITH LESSON 7-2

There are several ways to solve a quadratic equation in standard form:

$ax^2 + bx + c = 0, a \neq 0.$

Page 355 reviewed graphical and algebraic methods. Another method—one that works for all quadratic equations—is the Quadratic Formula:

$$x = \frac{-b \pm \sqrt{b^2 - 4ac}}{2a}$$

1 EXAMPLE

Solve for x: $7x^2 + 6x - 1 = 0$. **The equation is in standard form.**

$a = 7, b = 6, c = -1$

$$x = \frac{-6 \pm \sqrt{6^2 - 4(7)(-1)}}{2(7)}$$ **Substitute in the Quadratic Formula.**

$$x = \frac{-6 \pm \sqrt{36 + 28}}{14}$$ **Simplify.**

$$x = \frac{-6 \pm \sqrt{64}}{14}$$

$$x = \frac{-6 + 8}{14} \text{ or } x = \frac{-6 - 8}{14}$$

$$x = \frac{1}{7} \text{ or } x = -1$$

Sometimes you may need a calculator to approximate the solutions.

2 EXAMPLE

Solve for x: $-3x^2 - 5x + 1 = 0$.

$a = -3, b = -5, c = 1$

$$x = \frac{-(-5) \pm \sqrt{(-5)^2 - 4(-3)(1)}}{2(-3)}$$ **Substitute in the Quadratic Formula.**

$$x = \frac{5 \pm \sqrt{25 + 12}}{-6}$$ **Simplify.**

$$x = \frac{5 + \sqrt{37}}{-6} \text{ or } x = \frac{5 - \sqrt{37}}{-6}$$

$x \approx -1.85$ or $x \approx 0.18$ **Use a calculator and round.**

EXERCISES

Solve each quadratic equation by the method of your choice. Round answers that are not integers to the nearest hundredth.

1. $x^2 + 5x - 14 = 0$

2. $4x^2 - 13x + 3 = 0$

3. $2x^2 + 7x + 3 = 0$

4. $5x^2 + 2x - 2 = 0$

5. $6x^2 + 10x = 5$

6. $1 = 2x^2 - 6x$

7. $x^2 - 6x = 27$

8. $2x^2 - 10x + 11 = 0$

9. $8x^2 - 2x - 3 = 0$

7-2 Similar Polygons

California Content Standards

Prepares for GEOM 4.0 Students prove basic theorems involving similarity.

What You'll Learn

- To identify similar polygons
- To apply similar polygons

... And Why

To find the size of a video image, as in Example 4

Check Skills You'll Need

GO for Help Lessons 4-1 and 7-1

1. $\triangle ABC \cong \triangle HIJ$. Name three pairs of congruent sides.

Algebra **Solve each proportion.**

2. $\frac{3}{4} = \frac{x}{8}$
3. $\frac{2}{x} = \frac{8}{24}$
4. $\frac{x}{9} = \frac{1}{3}$
5. $\frac{10}{25} = \frac{2}{x}$

New Vocabulary • similar • similarity ratio • golden rectangle • golden ratio

1 Similar Polygons

Two figures that have the same shape but not necessarily the same size are similar ($\sim$). Two polygons are **similar** if (1) corresponding angles are congruent and (2) corresponding sides are proportional. The ratio of the lengths of corresponding sides is the **similarity ratio.**

1 EXAMPLE Understanding Similarity

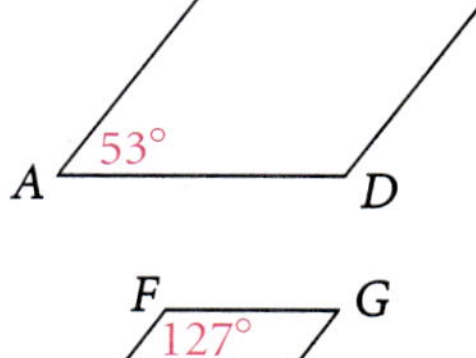

$ABCD \sim EFGH$. Complete each statement.

a. $m\angle E = ■$

$m\angle E = m\angle A = 53$ **Corresponding angles are $\cong$.**

b. $\frac{AB}{EF} = \frac{AD}{■}$

$\frac{AB}{EF} = \frac{AD}{EH}$ **Corresponding sides are proportional.**

CA Standards Check **1** Complete: $m\angle B = ■$ and $\frac{GH}{CD} = \frac{FG}{■}$

2 EXAMPLE Determining Similarity

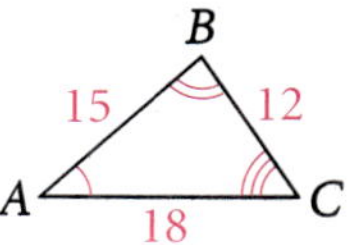

Determine whether the triangles are similar. If they are, write a similarity statement and give the similarity ratio.

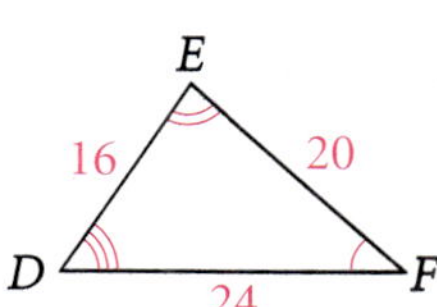

Three pairs of angles are congruent.
Also, corresponding sides are proportional.

$\frac{AC}{FD} = \frac{18}{24} = \frac{3}{4}$ $\quad \frac{AB}{FE} = \frac{15}{20} = \frac{3}{4}$ $\quad \frac{BC}{ED} = \frac{12}{16} = \frac{3}{4}$

$\triangle ABC \sim \triangle FED$ with a similarity ratio of $\frac{3}{4}$ or 3 : 4.

CA Standards Check **2** Sketch $\triangle XYZ$ and $\triangle MNP$ with $\angle X \cong \angle M$, $\angle Y \cong \angle N$, and $\angle Z \cong \angle P$. Also, $XY = 12$, $YZ = 14$, $ZX = 16$, $MN = 18$, $NP = 21$, and $PM = 24$. Can you conclude that the two triangles are similar? Explain.

You can use proportions to find unknown lengths in similar polygons.

3 EXAMPLE Using Similar Figures

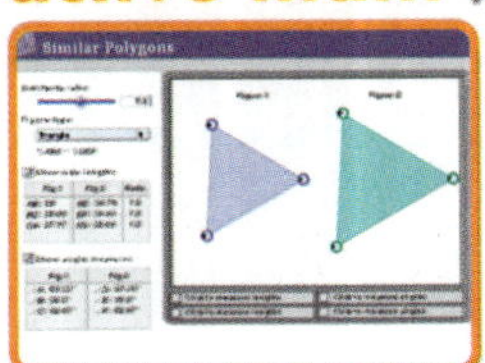

For: Similar Polygon Activity
Use: Interactive Textbook, 7-2

Algebra $LMNO \sim QRST$
Find the value of x.

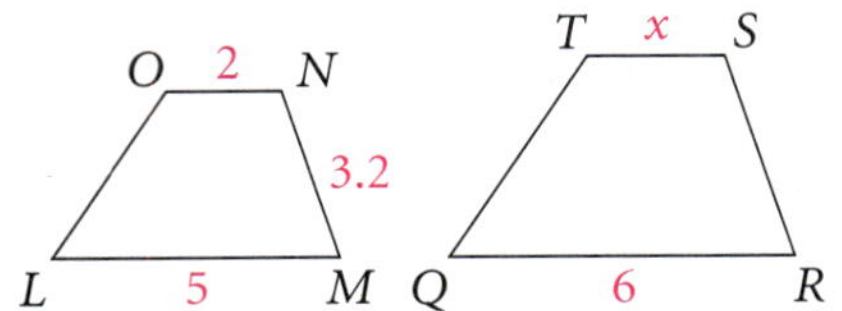

Write a proportion.

$\frac{LM}{QR} = \frac{ON}{TS}$ **Corresponding sides of ~ polygons are proportional.**

$\frac{5}{6} = \frac{2}{x}$ **Substitute.**

$5x = 12$ **Cross-Product Property**

$x = 2.4$ **Solve for x.**

CA Standards Check 3 Use the figures above. Find SR to the nearest tenth.

2 Applying Similar Polygons

You can use similar polygons to find measures when using enlarged or reduced images.

4 EXAMPLE Real-World Connection

Technology If you have a vision problem, a magnification system can help you read. You choose a level of magnification. Then you place an image under the viewer. A similar, magnified image appears on the video screen.

The video screen pictured is 16 in. wide by 12 in. tall. What is the largest complete video image possible for a block of text that is 6 in. wide by 4 in. tall?

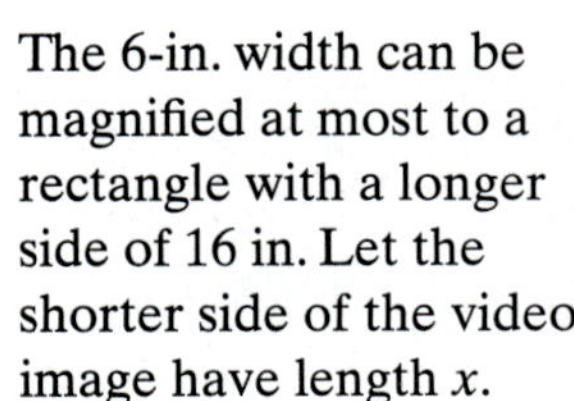

Problem Solving Hint

To calculate image size, estimate which screen dimension, width or height, will fill first as the image expands.

The 6-in. width can be magnified at most to a rectangle with a longer side of 16 in. Let the shorter side of the video image have length x.

$\frac{6}{16} = \frac{4}{x}$ **Corresponding sides of ~ polygons are proportional.**

$6x = 64$ **Cross-Product Property**

$x = 10\frac{2}{3}$ **Solve for x.**

The 4-in. height of the block of text enlarges to $10\frac{2}{3}$ in. This is less than the 12-in. height of the video screen, so the entire block of text fits on the screen.

The largest complete video image possible for the block of text is 16 in. by $10\frac{2}{3}$ in.

CA Standards Check 4 On the video screen in Example 4, what is the largest complete image possible for a photograph that is 3 in. wide by 5 in. tall?

Real-World Connection

The United Nations Secretariat building in New York City suggests a golden rectangle.

A **golden rectangle** is a rectangle that can be divided into a square and a rectangle that is similar to the original rectangle. A pattern of repeated golden rectangles is shown at the right. Each golden rectangle that is formed is copied and divided again. Each golden rectangle is similar to the original rectangle.

In any golden rectangle, the length and width are in the **golden ratio** which is about 1.618 : 1. You will derive this ratio in Exercise 51.

The golden rectangle is considered pleasing to the human eye. It has appeared in architecture and art since ancient times. It has intrigued artists including Leonardo da Vinci (1452–1519). Da Vinci illustrated *The Divine Proportion,* a book about the golden rectangle.

You can apply the golden ratio to real-life design problems.

5 EXAMPLE

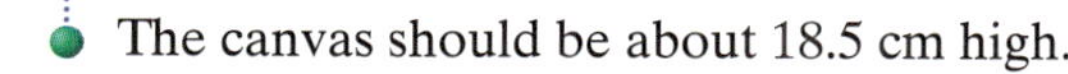

Art An artist plans to paint a picture. He wants the canvas to be a golden rectangle with its longer horizontal sides 30 cm wide. How high should the canvas be?

Let h be the height of the canvas.

$\frac{30}{h} = \frac{1.618}{1}$ **Write a proportion.**

$1.618h = 30$ **Cross-Product Property**

$h = \frac{30}{1.618}$ **Solve for h.**

$h = 18.541409$ **Use a calculator.**

The canvas should be about 18.5 cm high.

5 A golden rectangle has shorter sides of length 20 cm. Find the length of the longer sides.

EXERCISES
Standards Practice

For more exercises, see *Extra Skill, Word Problem, and Proof Practice.*

Example 1 (page 373)

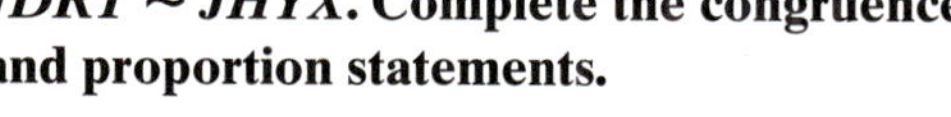

***JDRT* ~ *JHYX*. Complete the congruence and proportion statements.**

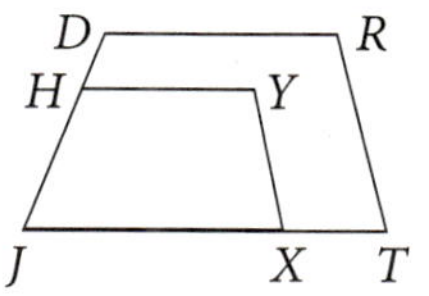

1. $\angle D \cong$? **2.** $\angle Y \cong$? **3.** $\angle T \cong$?

4. $\frac{JD}{JH} = \frac{DR}{■}$ **5.** $\frac{RT}{YX} = \frac{■}{JX}$ **6.** $\frac{■}{DR} = \frac{YX}{RT}$

Example 2 (page 373)

Are the polygons similar? If they are, write a similarity statement and give the similarity ratio. If they are not, explain.

7. Rectangle ABCD: A, B, C, D; BC = 20, DC = 36. Rectangle EFGH: E, F, G, H; FG = 30, HG = 52.

8.

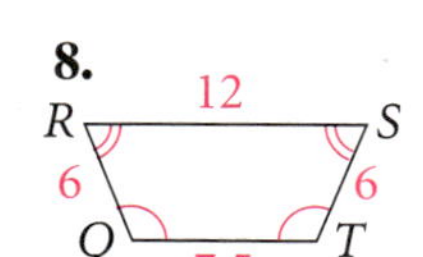

Are the polygons similar? If they are, write a similarity statement and give the similarity ratio. If they are not, explain.

9.

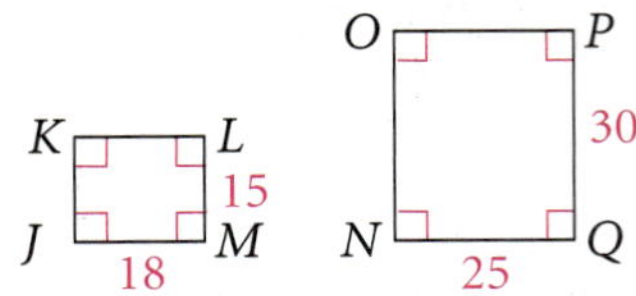

10.

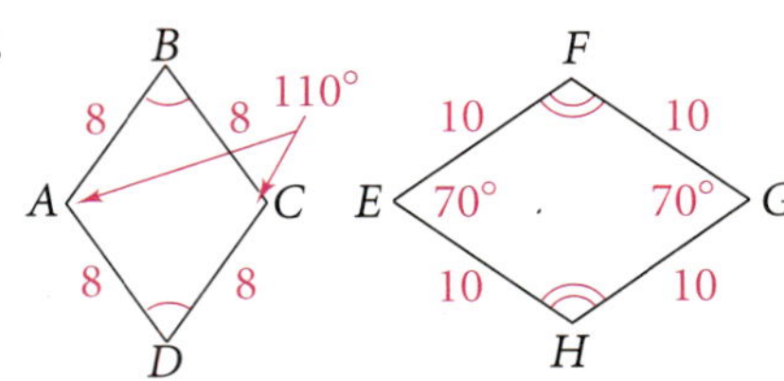

11.

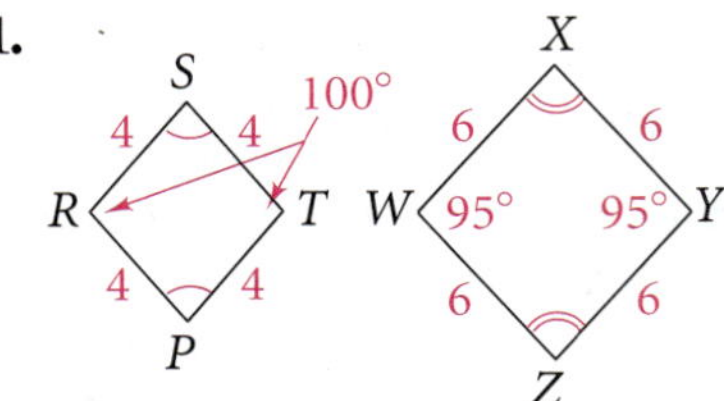

12.

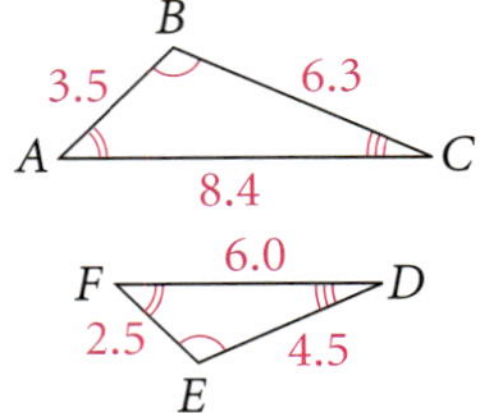

Example 3 (page 374) Algebra **The polygons are similar. Find the value of each variable.**

13.

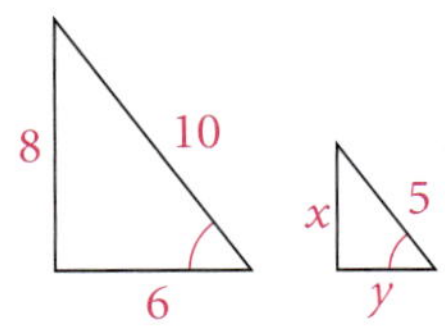

14.

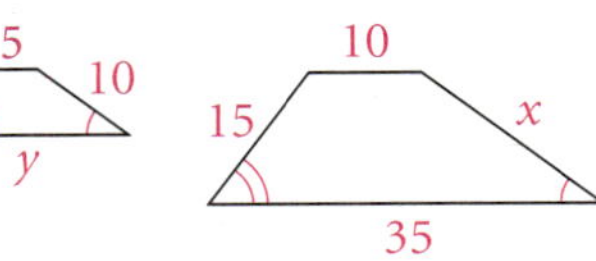

15.

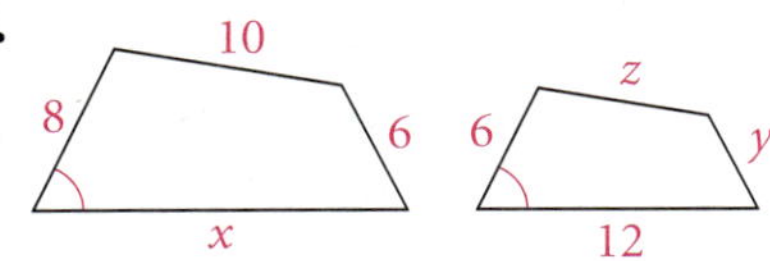

16.

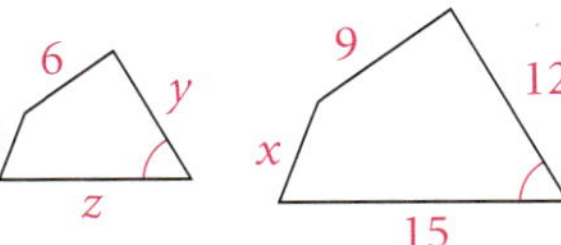

Example 4 (page 374)

17. **Drawing** You want to draw an enlargement of a design that is painted on a 3 in.-by-5 in. card. You plan to draw on an $8\frac{1}{2}$ in.-by-11 in. piece of paper. What are the dimensions of the largest complete enlargement you can draw?

18. A map has dimensions 9 in. by 15 in. You want to reduce the map so that it will fit on a 4 in.-by-6 in. index card. What are the dimensions of the largest possible complete map that you can fit on the index card?

Example 5 (page 375)

19. **Electrical Equipment** A switch plate for a standard wall switch has the shape of a golden rectangle. The longer side of the switch plate is about 114 mm. How long is the shorter side? Round your answer to the nearest millimeter.

20. **Design** You want the banner you are creating from one piece of cloth to be a golden rectangle. The cloth will be cut from a bolt that is 54 in. wide. What are the dimensions of the largest banner that you can make?

B Apply Your Skills

$\triangle DFG \sim \triangle HKM$. Use the diagram to find the following.

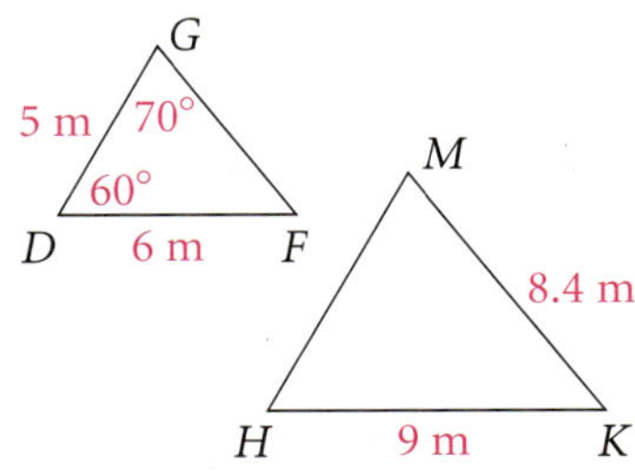

21. the similarity ratio of $\triangle DFG$ to $\triangle HKM$

22. the similarity ratio of $\triangle HKM$ to $\triangle DFG$

23. $m\angle F$ **24.** $m\angle K$ **25.** $m\angle M$

26. $\frac{DF}{HK}$ **27.** HM **28.** GF

29. **Writing** Are two congruent figures similar? Explain.

PEANUTS

By CHARLES SCHULZ

30. a. **Reading Math** What two symbols combine to form the congruence symbol?
b. Explain why the congruence symbol makes sense.

31. **Multiple Choice** An art class is painting a rectangular mural for a community festival. The students planned the mural with a diagram that is 80 in. long and 16 in. high. The mural is 4 ft high. What is its length?

(A) 5 ft 4 in. (B) 10 ft (C) 20 ft (D) 26 ft 8 in.

Exercise 31

Algebra Find the values of the variables.

32.

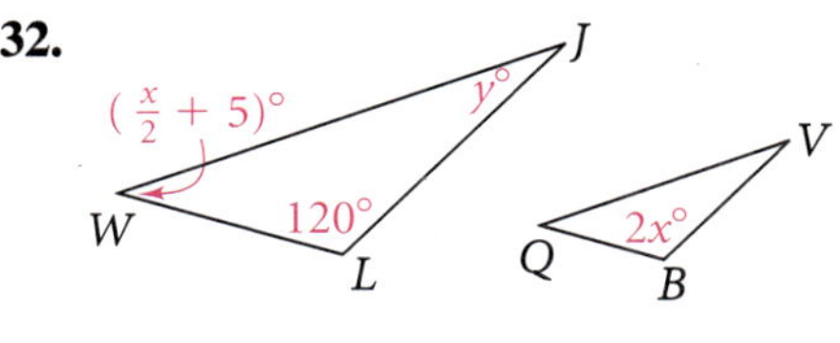

$\triangle WLJ \sim \triangle QBV$

33.

R, S, T, P, Q; x; 5.2 cm; 3.5 cm; 7 cm

$\triangle PRQ \sim \triangle SRT$

The polygons in each exercise are similar. Find the similarity ratio of the first to the second.

34.

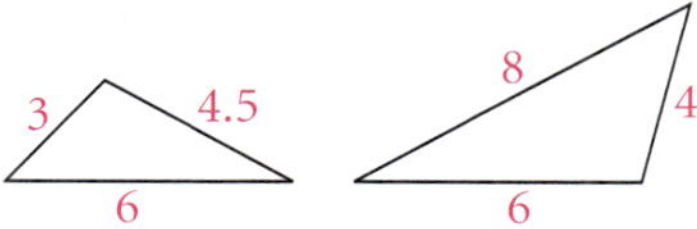

35.

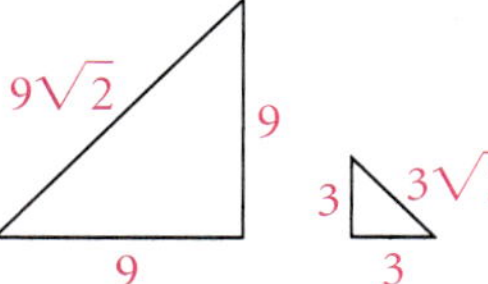

36. Exercise 13
37. Exercise 14
38. Exercise 15
39. Exercise 16

Logo Design A company logo is a rhombus with 4-cm sides and angles of 60° and 120°. Find the angle measures and side lengths if the logo is changed as follows.

40. reduced by 50%
41. reduced to 50%
42. reduced by 20%
43. reduced to 20%
44. reduced by 75%
45. reduced to 75%

Problem Solving Hint

A value reduced *by* n% is reduced *to* $(100 - n)$%, and vice versa.

Find the other side length of the golden rectangle to the nearest tenth of an inch.

46. The shorter side is 10 in.
47. The longer side is 10 in.

48. **Money** From 1861 to 1928, U. S. paper currency measured 7.42 in. by 3.13 in. The dimensions of a current bill are shown here. Are the old and new bills similar rectangles? Explain.

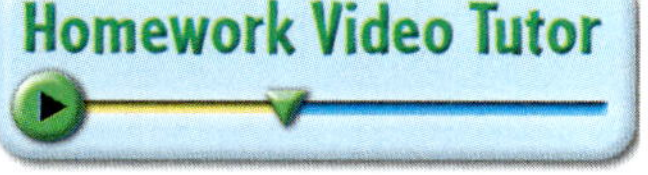

Visit: PHSchool.com
Web Code: bce-0702

49. **Critical Thinking** Are all circles similar? Explain.

50. **Open-Ended** Draw two polygons with sides in the ratio 2 : 1 that are not similar.

Challenge

51. a. **Algebra** You know that the golden ratio is about 1.618 : 1. You can use the definition of a golden rectangle to derive this ratio. Let $ABCD$ at the left be a golden rectangle with width 1. By the definition of golden rectangle, $ABCD \sim BCFE$. Fill in the reasons in the following argument.

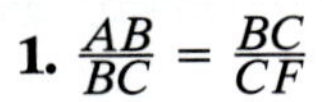

1. $\frac{AB}{BC} = \frac{BC}{CF}$	1. ?
2. $\frac{x}{1} = \frac{1}{x - 1}$	2. ?
3. $x^2 - x = 1$	3. ?
4. $x^2 - x - 1 = 0$	4. ?

b. Using the quadratic formula to solve the equation in part (a), you get $x = \frac{1 \pm \sqrt{5}}{2}$. Explain why $x = \frac{1 - \sqrt{5}}{2}$ does not make sense in this situation.

c. The golden ratio is the ratio $x : 1$, or $\frac{1 + \sqrt{5}}{2} : 1$. Use a calculator to find the value of $x = \frac{1 + \sqrt{5}}{2}$ to the nearest ten-thousandth.

52. a. In the Fibonacci Sequence (see Lesson 1-1), each term after the first two is the sum of the two preceding terms. The first seven terms of the Fibonacci Sequence are 1, 1, 2, 3, 5, 8, and 13. Find the next seven terms.

b. Start with the second term. Here is the ratio of each term to the prior term.

$\frac{1}{1} = 1$ $\quad \frac{2}{1} = 2$ $\quad \frac{3}{2} = 1.5$ $\quad \frac{5}{3} \approx 1.6667$

Find the next nine ratios. Round to the nearest ten thousandth.

c. Compare the ratios that you found in part (b) to the golden ratio.

Multiple Choice Practice

For California Standards and CAHSEE Tutorials, visit PHSchool.com. Web Code: bcq-9045

GEOM 12.0

53. If quadrilateral $ABCD$ has at least one pair of sides congruent and a pair of angles measuring 53°, what type of quadrilateral can $ABCD$ not be?

(A) parallelogram (B) trapezoid (C) kite (D) rectangle

GEOM 2.0

54. Given $\triangle CAT$ and $\triangle OGD$, you have proven that $\angle C \cong \angle D$, $\angle A \cong \angle O$, and $\overline{CA} \cong \overline{DO}$. What can you conclude?

(A) $\angle T \cong \angle D$ by ASA

(B) $\triangle CAT \cong \triangle DOG$ by ASA

(C) $\triangle CAT \cong \triangle OGD$ by ASA

(D) $\overline{CA} \cong \overline{DO}$ by AAS

CAHSEE Alg1 6.0

55. Which linear function best describes the graph at the right?

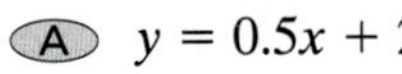

(A) $y = 0.5x + 2$

(B) $y = 0.5x - 2$

(C) $y = 2x + 0.5$

(D) $y = 2x - 0.5$

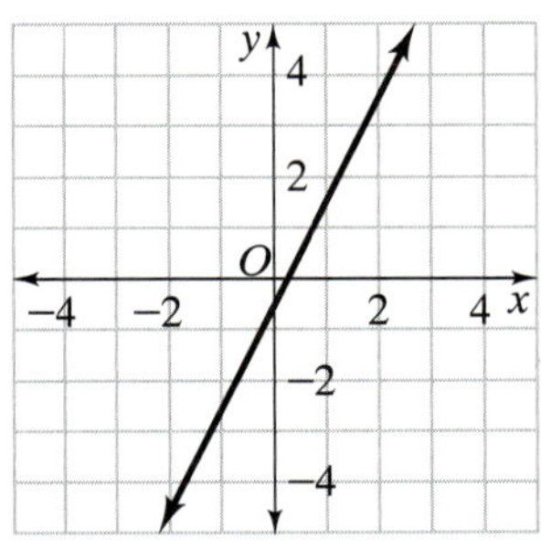

CAHSEE 7NS 1.1

56. $5.8 \times 10^5 =$

(A) 0.000058 (C) 5,800

(B) 58 (D) 580,000

Mixed Review

Lesson 7-1 **If $\frac{x}{7} = \frac{y}{9}$, complete each of the following using properties of proportions.**

57. $9x = ■$ **58.** $\frac{x}{y} = \frac{■}{■}$ **59.** $\frac{x + 7}{7} = \frac{■}{■}$

Lesson 6-3 **Can you conclude that the quadrilateral is a parallelogram? Explain.**

60.

61. 

62. 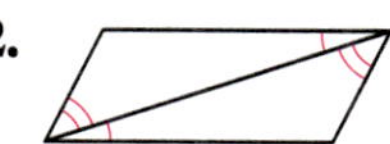

Lesson 4-5 **Use the marked $\triangle CEA$ for Exercises 63–66.**

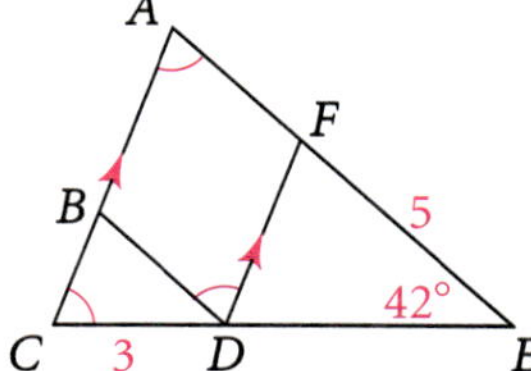

63. Name the isosceles triangles in the figure.

64. $\overline{CD} \cong$ __?__ $\cong$ __?__

Find the value of each of the following.

65. AE **66.** $m\angle A$

Checkpoint Quiz 1 — Lessons 7-1 through 7-2

1. **Models** A table is 4 ft high. A small model of the table is 6 in. high. What is the ratio of the height of the model table to the height of the real table?

2. If $\frac{a}{b} = \frac{9}{10}$, complete this statement: $\frac{a}{9} = ■$

3. Are the two polygons similar? If so, give the similarity ratio of the first polygon to the second. If not, explain.

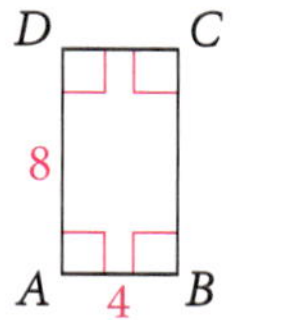

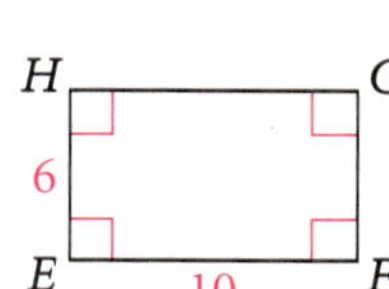

Solve each proportion.

4. $\frac{y}{6} = \frac{18}{54}$ 5. $\frac{5}{7} = \frac{x - 2}{4}$

6. The scale of a scale drawing is 2 in. = 5 ft. A room is 5 in. long on the scale drawing. Find the actual length of the room.

$\triangle ABC \sim \triangle DBF$ at the right. Complete each statement.

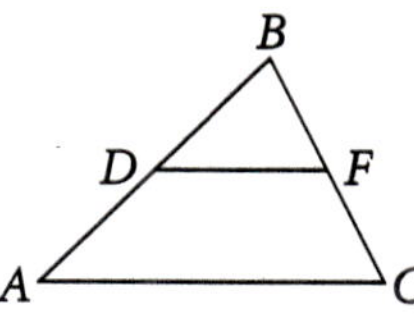

7. $m\angle A = m\angle$ __?__ 8. $\frac{AB}{DB} = \frac{BC}{■}$

9. A postcard is 6 in. by 4 in. A printing shop will enlarge it so that the longer side is any length up to 3 ft. Find the dimensions of the biggest enlargement.

10. **Algebra** The polygons at the right are similar. Find the value of x.

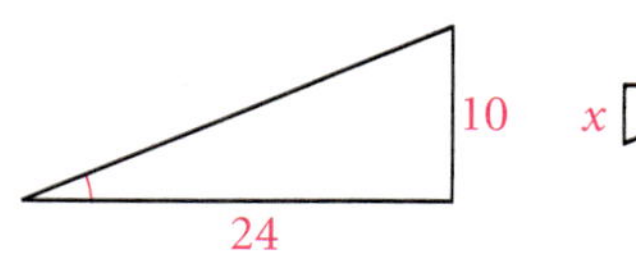

Fractals

FOR USE WITH LESSON 7-2

Fractals are objects that have three important properties:

- You can form them by repeating steps—a process called *iteration*.
- They require infinitely many iterations. In practice, you can continue until the objects become too small to draw. Even then the steps could continue in your mind.
- At each stage, a portion of the object is a reduced copy of the entire object at the previous stage. This property is called *self-similarity*.

1 EXAMPLE

The segment below of length 1 unit is Stage 0 of a fractal tree. Draw Stage 1 and Stage 2 of the tree. For each stage, draw two branches from the top third of each segment.

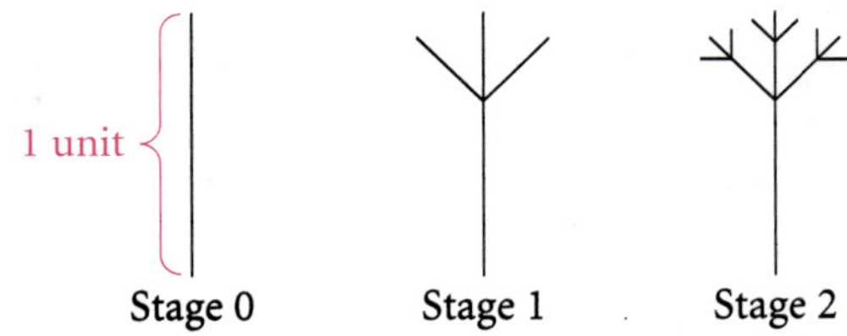

Amazingly, some fractals are used to describe natural formations such as mountain ranges and clouds. In 1904, Swedish mathematician Helge von Koch created the Koch Curve, a fractal that is used to model coastlines.

2 EXAMPLE

The segment at the right of length 1 unit is Stage 0 of a Koch Curve. Draw Stages 1–4 of the curve. For each stage, replace the middle third of each segment with two segments, both equal in length to the middle third.

- For Stage 1, replace the middle third with two segments, both $\frac{1}{3}$ unit long.
- For Stage 2, replace the middle third of each segment with two segments, both $\frac{1}{9}$ unit long.
- Continue with a third and fourth iteration.

Stages 0–4 are shown at the right.

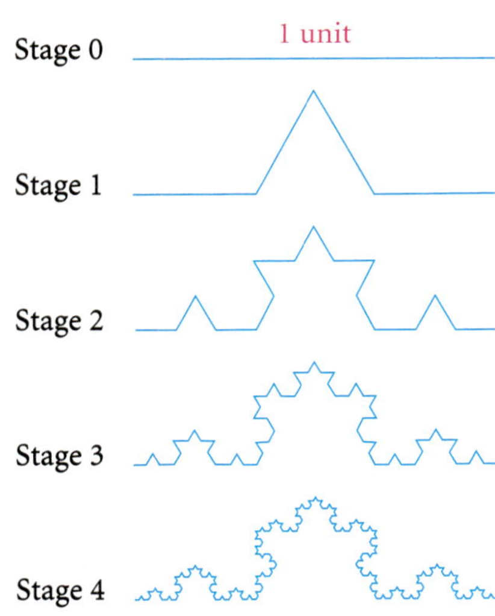

You can construct a Koch Curve on each side of an equilateral triangle and get a Koch Snowflake.

3 EXAMPLE

The equilateral triangle at the right is Stage 0 of a Koch Snowflake. Draw Stage 1.

- Draw an equilateral triangle on the middle third of each side.
- Erase the middle third of each side to get Stage 1 of the snowflake. (Continued in Exercises 5–9.)

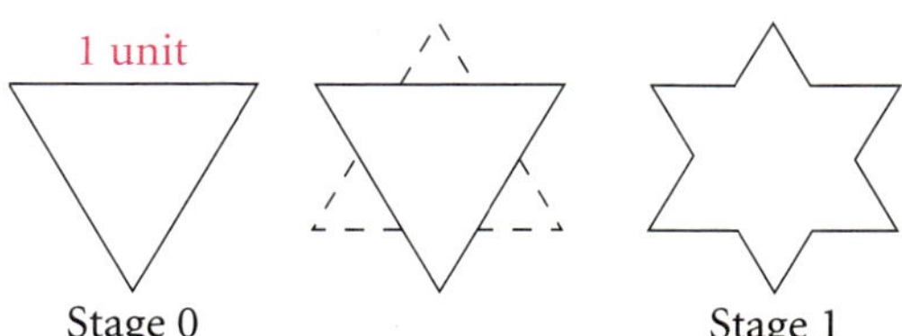

Stage 0 Stage 1

EXERCISES

1. Draw Stage 3 of the fractal tree in Example 1.

Use the Koch Curve in Example 2 for Exercises 2–4.

2. Complete the table to find the length of the Koch Curve at each stage.

Stage	0	1	2
Length	1	■	■

3. Examine the results of Exercise 2 and look for a pattern. Use this pattern to predict the length of the Koch Curve at Stage 3; at Stage 4.
4. Suppose you are able to complete a Koch Curve to Stage n.
   a. Write an expression for the length of the curve.
   b. What happens to the length of the curve as n gets large?

5. Draw Stage 2 of the Koch Snowflake in Example 3.

Stage 3 of the Koch Snowflake is shown at the right. Use it and the earlier stages to answer Exercises 6–8.

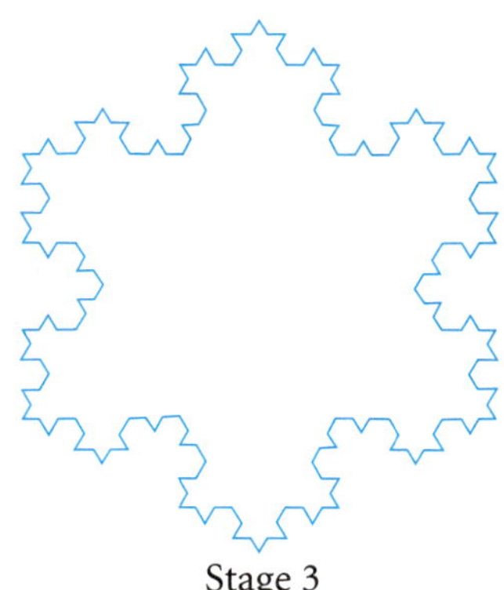
Stage 3

6. At each stage, is the snowflake equilateral?
7. **a.** Complete the table to find the perimeter at each stage.

Stage	Number of Sides	Length of a Side	Perimeter
0	3	1	3
1	■	$\frac{1}{3}$	■
2	48	■	■
3	■	■	■

   b. Predict the perimeter at Stage 4.
   c. Will there be a stage at which the perimeter is greater than 100 units? Explain.

8. Exercises 4 and 7 suggest that there is no bound on the perimeter of the Koch snowflake. Is this true about the area of the Koch snowflake? Explain.

9. To draw the Sierpinski Triangle fractal, start with an equilateral triangle. For each stage, connect the midpoints of all of the triangles "pointed upwards." Stages 0–2 are shown at the right. Draw Stage 3.

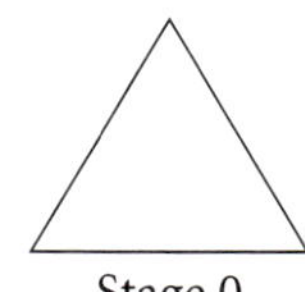
Stage 0

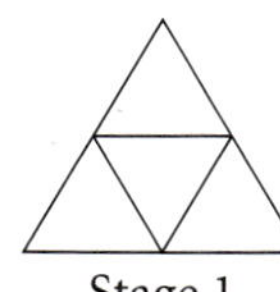
Stage 1

Stage 2

7-3 Proving Triangles Similar

California Content Standards

GEOM 2.0 Students write geometric proofs. ***Master***
GEOM 4.0 Students prove basic theorems involving similarity. ***Introduce***
GEOM 5.0 Students prove that triangles are similar. ***Introduce***

What You'll Learn

- To use AA, SAS, and SSS similarity statements
- To apply AA, SAS, and SSS similarity statements

. . . And Why

To measure height indirectly, as in Example 4

Check Skills You'll Need

GO for Help Lessons 4-2 and 4-3

Name the postulate or theorem you can use to prove the triangles congruent.

1. 2. 3.

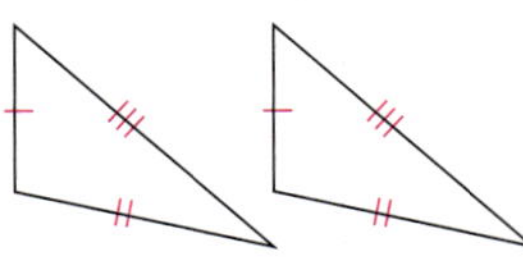

New Vocabulary • indirect measurement

1 The AA Postulate and the SAS and SSS Theorems

Real-World Connection

The gables on the historic House of the Seven Gables in Salem, Massachusetts, suggest similar triangles.

Activity: Triangles with Two Pairs of Congruent Angles

- Draw two triangles of different sizes, each with a 50° angle and a 60° angle.
- Measure the sides of each triangle to the nearest millimeter.
- Find the ratio of the lengths of each pair of corresponding sides.

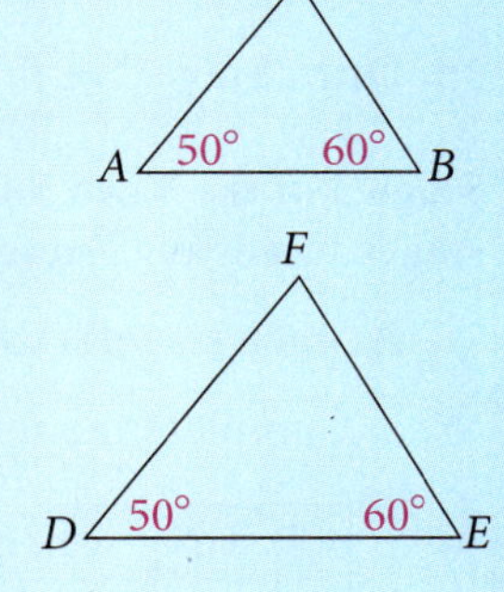

1. What conclusion can you make about the two triangles?
2. Complete this conjecture:

   If two angles of one triangle are congruent to two angles of another triangle, then the triangles are __?__.

In this lesson, you will show triangles are similar without using the definition of similar triangles. The two triangles shown above suggest the following postulate.

Key Concepts

Postulate 7-1 **Angle-Angle Similarity (AA ~) Postulate**

If two angles of one triangle are congruent to two angles of another triangle, then the triangles are similar.

$\triangle TRS \sim \triangle PLM$

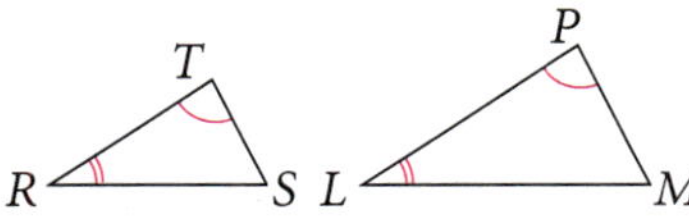

1 EXAMPLE Using the AA ~ Postulate

Explain why the triangles are similar. Write a similarity statement.

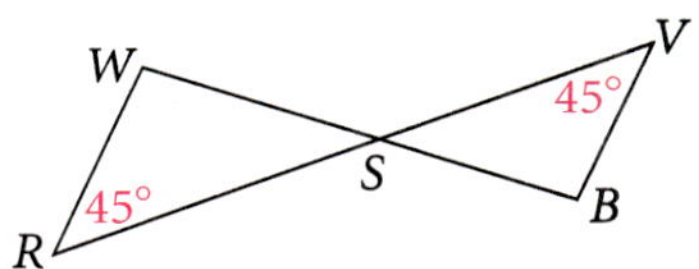

$\angle RSW \cong \angle VSB$ because vertical angles are congruent. $\angle R \cong \angle V$ because their measures are equal. $\triangle RSW \sim \triangle VSB$ by the Angle-Angle Similarity Postulate.

CA Standards Check

1 Critical Thinking In Example 1, you have enough information to write a similarity statement. Do you have enough information to find the similarity ratio? Explain.

The next two theorems follow from the AA Similarity Postulate.

Key Concepts

Theorem 7-1 **Side-Angle-Side Similarity (SAS ~) Theorem**

If an angle of one triangle is congruent to an angle of a second triangle, and the sides including the two angles are proportional, then the triangles are similar.

Proof

Proof of Theorem 7-1

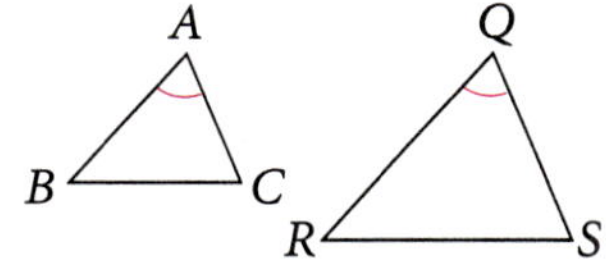

Given: $\angle A \cong \angle Q$, $\frac{AB}{QR} = \frac{AC}{QS}$

Prove: $\triangle ABC \sim \triangle QRS$

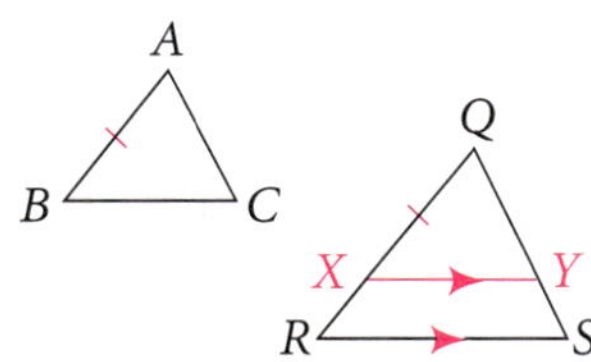

Choose X on $\overline{QR}$ so that $QX = AB$. (See figures at left.) Draw $\overline{XY} \parallel \overline{RS}$. Then $\angle QXY \cong \angle R$. By the AA ~ Postulate, $\triangle QXY \sim \triangle QRS$. Thus $\frac{QX}{QR} = \frac{QY}{QS}$. Combining this proportion, the given proportion, and the fact that $AB = QX$ shows that $\frac{AC}{QS} = \frac{QY}{QS}$. Hence, $AC = QY$.

Then $\triangle ABC \cong \triangle QXY$ by the SAS Congruence Postulate. $\angle B \cong \angle QXY$ by CPCTC, and $\angle B \cong \angle R$ by the Transitive Property. Thus, $\triangle ABC \sim \triangle QRS$ by the AA ~ Postulate.

Key Concepts

Theorem 7-2 **Side-Side-Side Similarity (SSS ~) Theorem**

If the corresponding sides of two triangles are proportional, then the triangles are similar.

Proof

Proof of Theorem 7-2

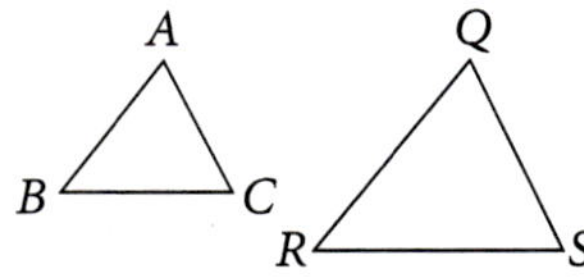

Given: $\frac{AB}{QR} = \frac{BC}{RS} = \frac{AC}{QS}$

Prove: $\triangle ABC \sim \triangle QRS$

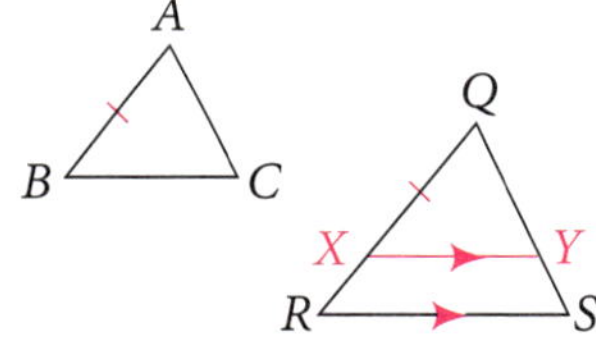

Draw $\overline{XY}$ as in the proof of Theorem 7-1. By the AA ~ Postulate, $\triangle QXY \sim \triangle QRS$. Thus $\frac{QX}{QR} = \frac{XY}{RS} = \frac{QY}{QS}$. Combining this proportion, the given proportion, and the fact that $AB = QX$ shows that $BC = XY$ and $AC = QY$.

Then $\triangle ABC \cong \triangle QXY$ by the SSS Congruence Postulate and, as in the proof above, $\triangle ABC \sim \triangle QRS$ by the AA ~ Postulate.

2 EXAMPLE Using Similarity Theorems

Explain why the triangles must be similar. Write a similarity statement.

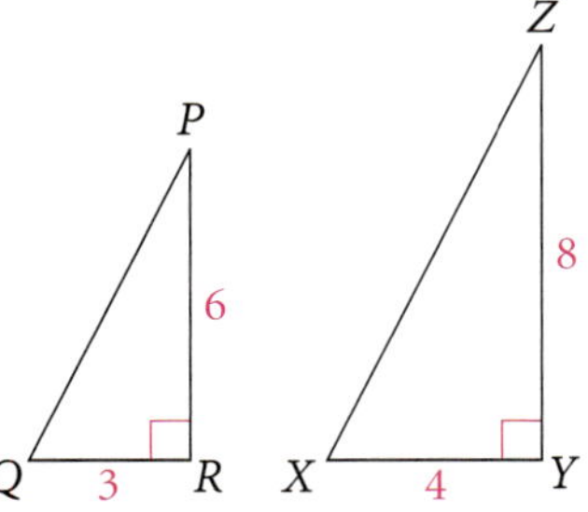

$\angle QRP \cong \angle XYZ$ because they are right angles.

$\frac{QR}{XY} = \frac{3}{4}$ and $\frac{PR}{ZY} = \frac{6}{8} = \frac{3}{4}$.

Therefore, $\triangle QRP \sim \triangle XYZ$ by the SAS $\sim$ Theorem.

2 Explain why the triangles must be similar. Write a similarity statement.

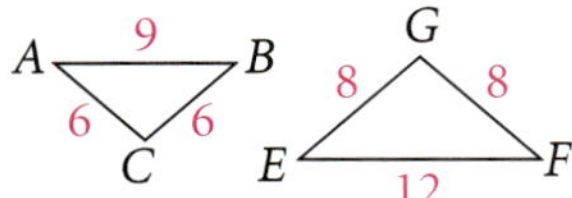

2 Applying AA, SAS, and SSS Similarity

You can apply the AA Similarity Postulate and the SAS and SSS Similarity Theorems to find the lengths of sides in similar triangles.

3 EXAMPLE Finding Lengths in Similar Triangles

Test-Taking Tip

In similarity and congruence problems, try to name the figures to show the correct correspondence between vertices.

Multiple Choice If the triangles are similar, find DE.

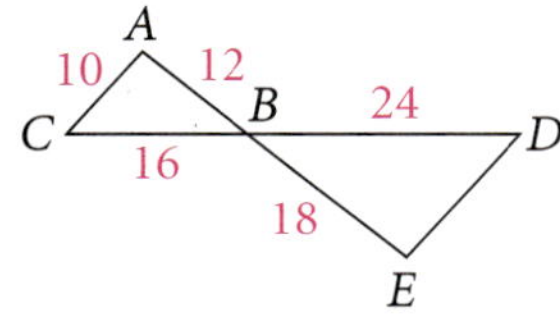

Ⓐ 10 Ⓒ 32

Ⓑ 15 Ⓓ The triangles are not similar.

$\angle ABC \cong \angle EBD$ because vertical angles are congruent.

$\frac{AB}{EB} = \frac{12}{18} = \frac{2}{3}$ and $\frac{CB}{DB} = \frac{16}{24} = \frac{2}{3}$

Therefore, $\triangle ABC \sim \triangle EBD$ by the SAS $\sim$ Theorem.

$\frac{CA}{DE} = \frac{2}{3}$ **Corresponding sides of $\sim$ triangles are proportional.**

$\frac{10}{DE} = \frac{2}{3}$ **Substitute.**

$2DE = 30$ **Cross-Product Property**

$DE = 15$ The correct answer is B.

3 Find the value of x in the figure at the right.

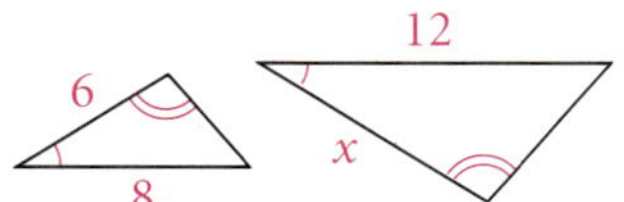

You can use similar triangles and measurements to find distances that are difficult to measure directly. This is called **indirect measurement.**

One method of indirect measurement uses the fact that light reflects off a mirror at the same angle at which it hits the mirror. A second method uses the similar triangles that are formed by certain figures and their shadows.

Both methods are illustrated in Example 4.

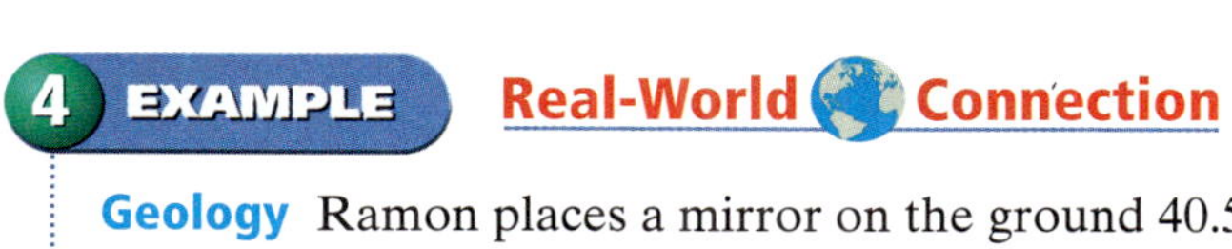

Geology Ramon places a mirror on the ground 40.5 ft from the base of a geyser. He walks backwards until he can see the top of the geyser in the middle of the mirror. At that point, Ramon's eyes are 6 ft above the ground and he is 7 ft from the image in the mirror. Use similar triangles to find the height of the geyser.

$\angle HVT \cong \angle JVS$ follows from the fact that the angle of incidence equals the angle of reflection. See p. 48, Exercise 18.

$\triangle HTV \sim \triangle JSV$	**AA ~ Postulate**
$\frac{HT}{JS} = \frac{TV}{SV}$	**Corresponding sides of ~ triangles are proportional.**
$\frac{6}{x} = \frac{7}{40.5}$	**Substitute.**
$243 = 7x$	**Cross-Product Property**
$34.7 \approx x$	**Solve for *x*.**

The geyser is about 35 ft high.

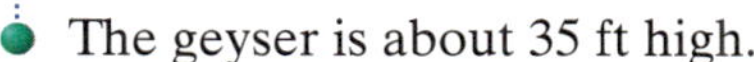

4 In sunlight, a cactus casts a 9-ft shadow. At the same time a person 6 ft tall casts a 4-ft shadow. Use similar triangles to find the height of the cactus.

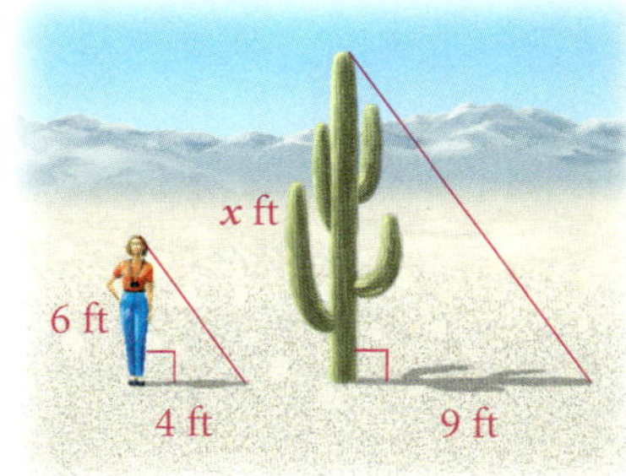

EXERCISES

For more exercises, see *Extra Skill, Word Problem, and Proof Practice.*

Standards Practice

GEOM 2.0, 4.0, 5.0

Examples 1 and 2 (pages 383 and 384)

Can you conclude the triangles are similar? If so, write a similarity statement and name the postulate or theorem you used. If not, explain.

1.

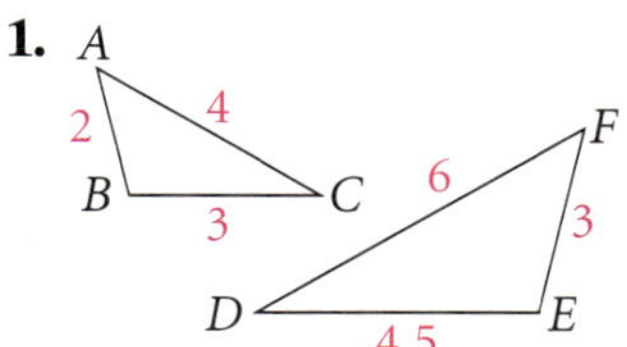

2.

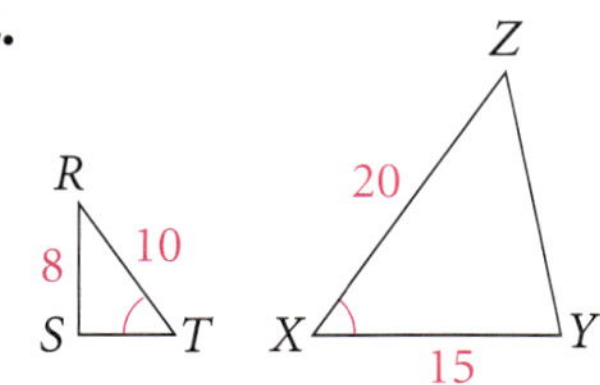

3. If possible, find the similarity ratio for each pair of similar triangles in Exercises 1 and 2. If not possible, explain.

Are the triangles similar? If so, write a similarity statement and name the postulate or theorem you used. If not, explain.

4.

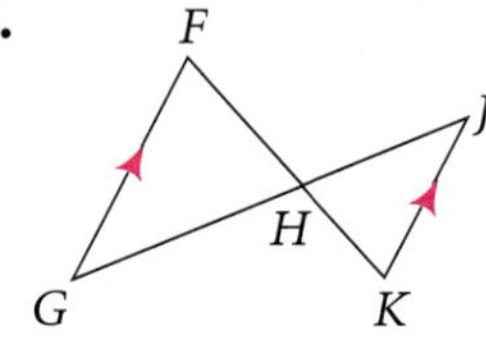

5.

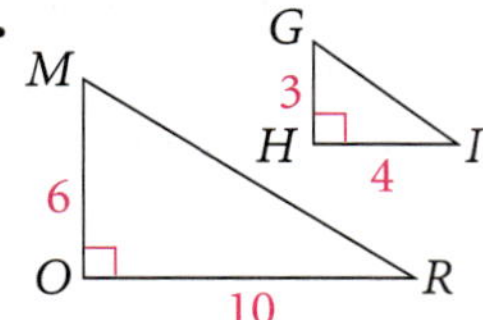

6.

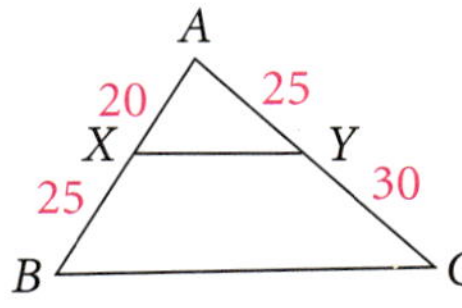

7.

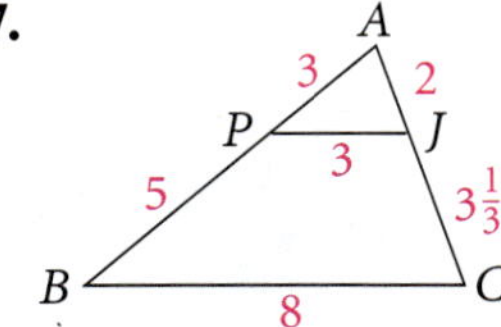

8.

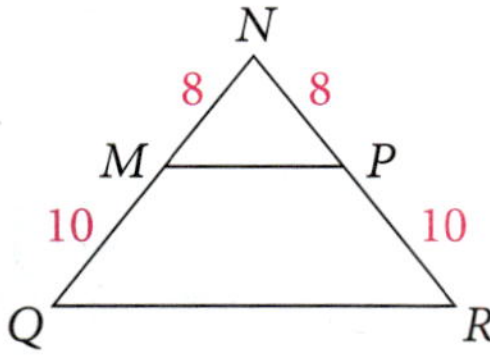

9.

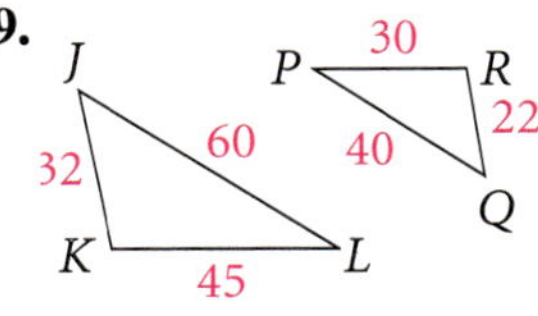

Example 3 (page 384) **Algebra** **Explain why the triangles are similar. Then find the value of *x*.**

10.

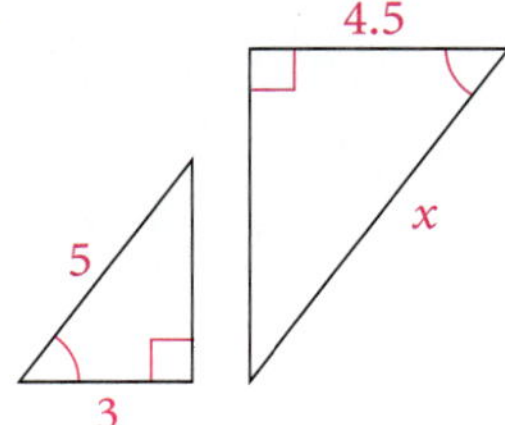

11.

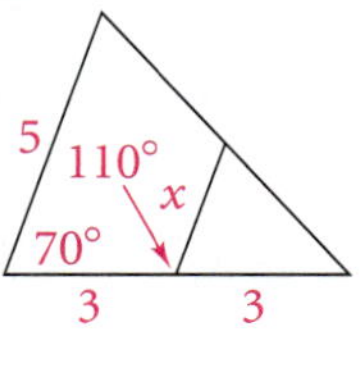

12.

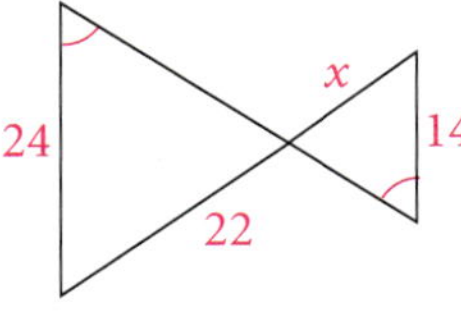

13.

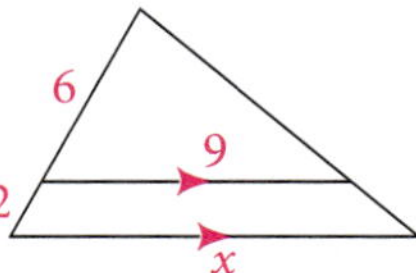

14.

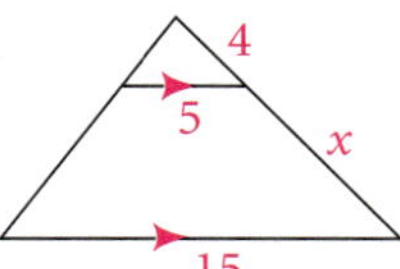

15.

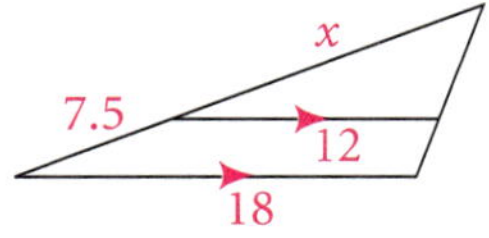

Example 4 (page 385) **Indirect Measurement** **Explain why the triangles are similar. Then find the distance represented by *x*.**

16.

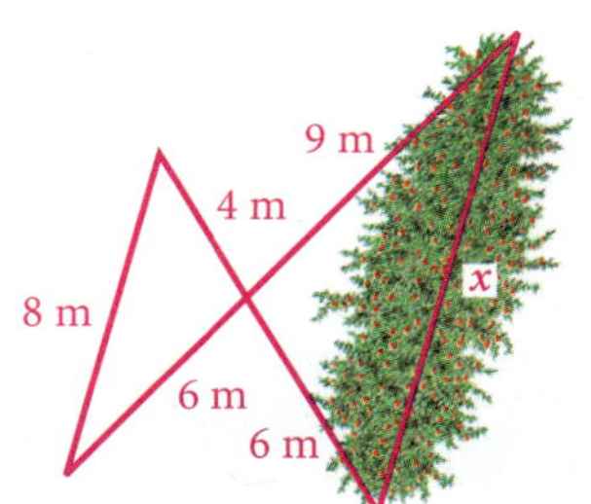

17.

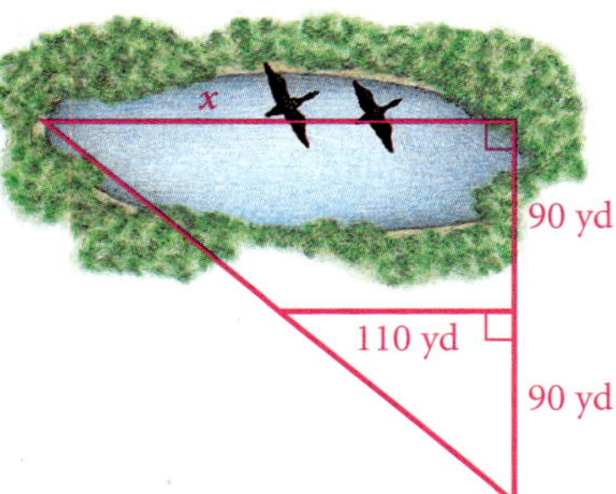

18.

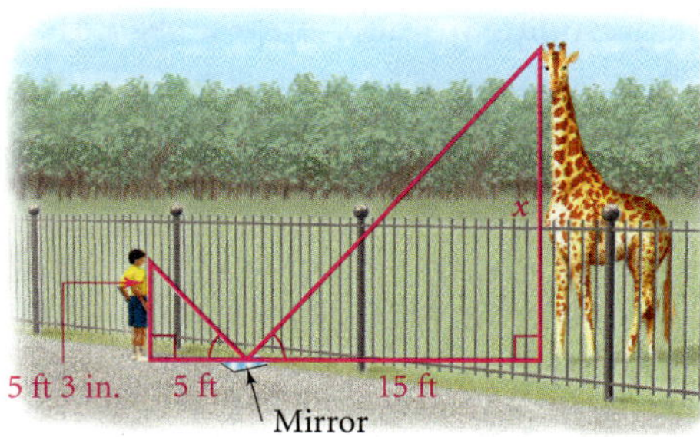

19.

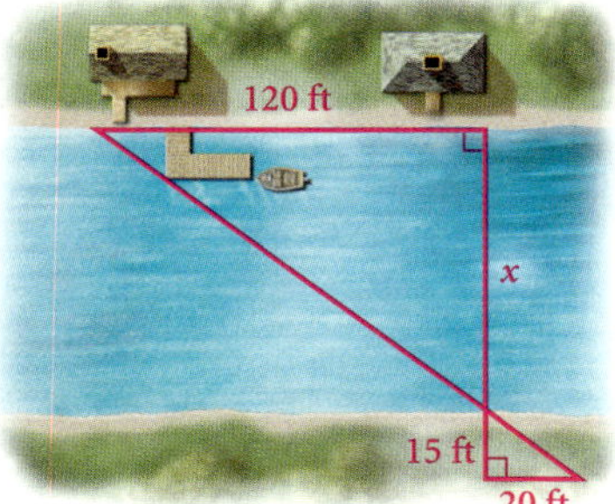

Apply Your Skills

Skyscraper? Nein!

Hannelore Kraus of Frankfurt, Germany, stopped the development of a skyscraper because she wanted her apartment to get its fair amount of sunlight. To halt the construction she used a German law that specifies that every homeowner is entitled to sunlight.

Kraus was offered 1.6 million dollars to drop her lawsuit, but she refused. The skyscraper was scheduled to be 265 m tall and was to be built only 60 m from Kraus's apartment.

Tall Buildings **Use the news article for Exercises 20 and 21.**

20. Writing Explain how Hannelore Kraus could use indirect measurement to estimate the length of the shadow of the building at a particular time of day.

21. Indirect Measurement Suppose Ms. Kraus is 1.75 m tall. When her shadow is 1 m long, about how long would the shadow of the proposed building be?

22. a. Classify $RSTW$.

b. Must any of the triangles shown be similar? Explain.

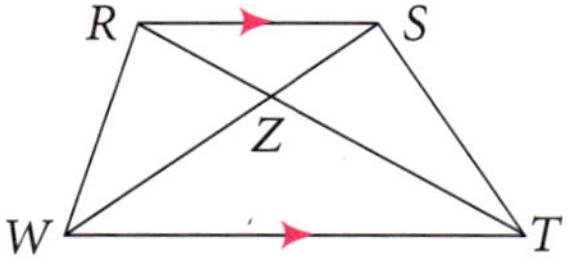

23. a. Critical Thinking Are two isosceles triangles always similar? Explain.

b. Are two isosceles right triangles always similar? Explain.

Can you conclude that the triangles are similar? If so, write a similarity statement and name the postulate or theorem you used. If not, explain.

24.

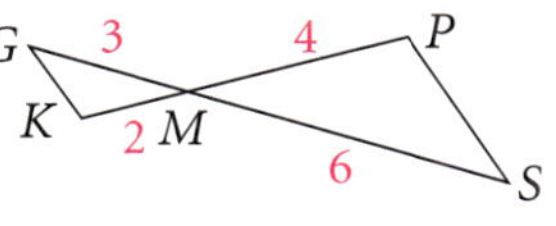

25.

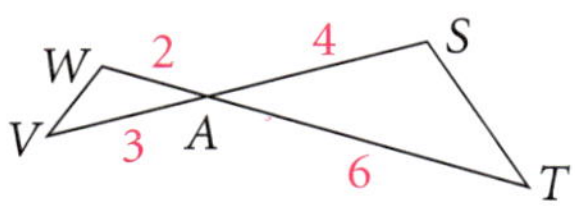

26.

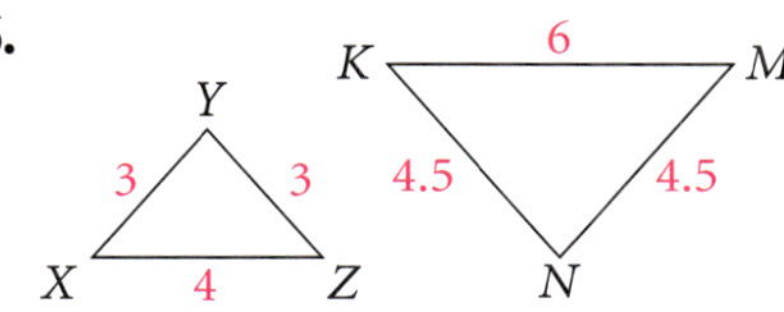

27.

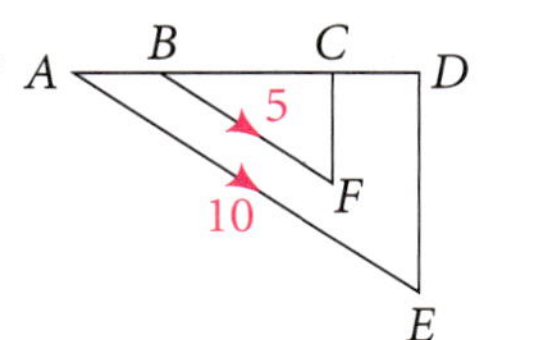

28. Indirect Measurement In sunlight, a vertical yardstick casts a 1-ft shadow at the same time that a nearby tree casts a 15-ft shadow. How tall is the tree?

29. Open-Ended Name something with a height that would be difficult to measure directly. Describe how you could measure it indirectly.

Find the similarity ratio of the larger to the smaller triangle in each exercise.

30. Ex. 10	**31.** Ex. 11	**32.** Ex. 12	**33.** Ex. 13	**34.** Ex. 14
35. Ex. 15	**36.** Ex. 16	**37.** Ex. 17	**38.** Ex. 18	**39.** Ex.19

Homework Video Tutor

Visit: PHSchool.com
Web Code: bce-0703

40. Constructions Draw any $\triangle ABC$. Use a straightedge and a compass to construct $\triangle RST$ so that $\triangle ABC \sim \triangle RST$ with similarity ratio 1 : 3.

Proof **41. Given:** $RT \cdot TQ = MT \cdot TS$

Prove: $\triangle RTM \sim \triangle STQ$

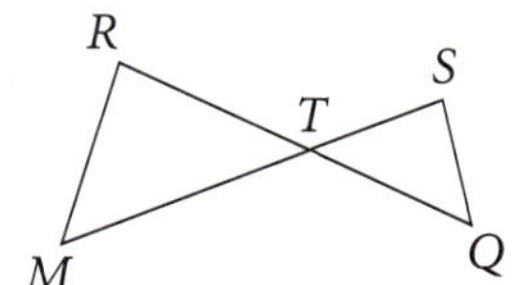

Challenge

Proof **42.** Write a proof of the following: Any two nonvertical parallel lines have equal slopes.

Given: nonvertical lines ℓ_1 and ℓ_2, $\ell_1 \parallel \ell_2$, $\overline{EF}$ and $\overline{BC} \perp$ to the x-axis

Prove: $\frac{BC}{AC} = \frac{EF}{DF}$

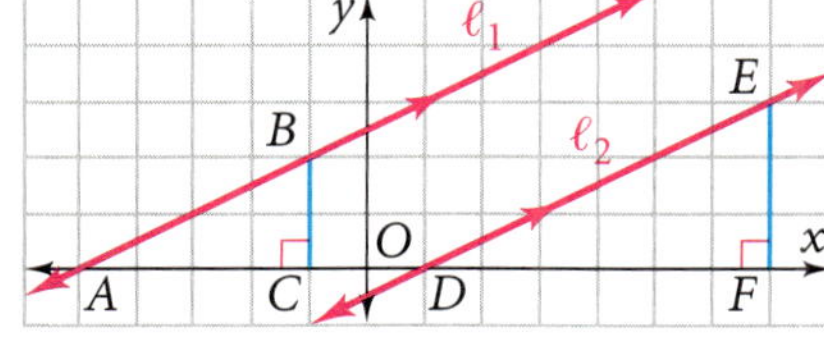

Hint: Use the x-axis as a transversal to show that $\angle BAC \cong \angle EDF$.

Proof **43.** Use the diagram in Exercise 42 on the preceding page.
Prove: Any two nonvertical lines with equal slopes are parallel.

44. There are three similar triangles in the figure at the right. Find them and prove that they are similar.

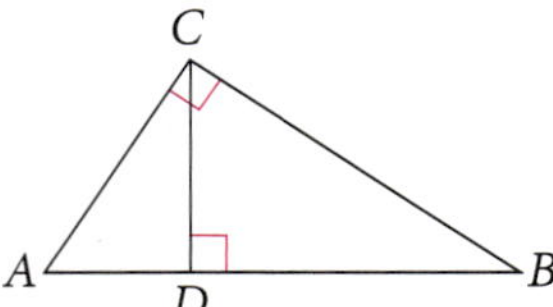

Multiple Choice Practice

For California Standards and CAHSEE Tutorials, visit PHSchool.com. Web Code: bcq-9045

GEOM 4.0

45. Suppose $\triangle VLQ \sim \triangle PSX$, $m\angle V = 48$, and $m\angle L = 80$. Find $m\angle X$.

(A) 48
(B) 52
(C) 100
(D) 128

GEOM 5.0

46. Complete the statement $\triangle ABC \sim$ _______ and identify the reason why the triangles are similar.

(A) $\triangle LGC$; SSS~
(B) $\triangle GLC$; SSS~
(C) $\triangle LGC$; AA~
(D) $\triangle GLC$; AA~

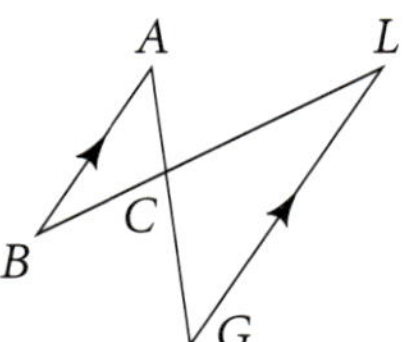

CAHSEE 7MG 1.2

47. A flower border along a stone wall is 30 feet long. Use the landscape architect's scale drawing below to find the actual width.

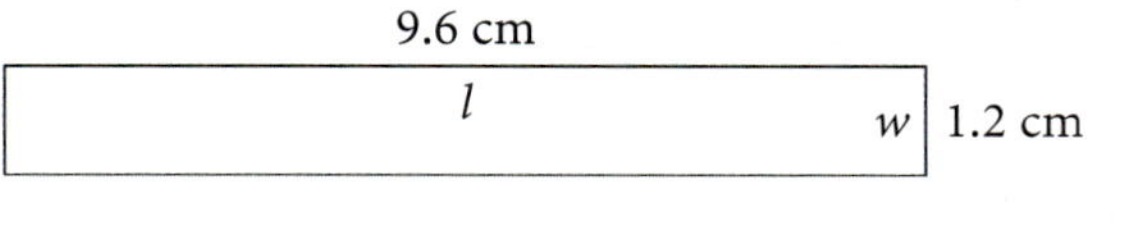

(A) 15 ft
(B) 3.75 ft
(C) 15 cm
(D) 2 cm

Mixed Review

Lesson 7-2

***TRAP* ~ *EZYD*. Complete each statement.**

48. $\angle T \cong$?
49. $\angle D \cong$?
50. $\angle A \cong$?

51. $\frac{AP}{YD} = \frac{RA}{\blacksquare}$
52. $\frac{TR}{RA} = \frac{\blacksquare}{ZY}$
53. $\frac{DE}{PT} = \frac{\blacksquare}{AR}$

Lesson 6-6

Give possible coordinates of points *W* and *Z* without using any new variables.

54.

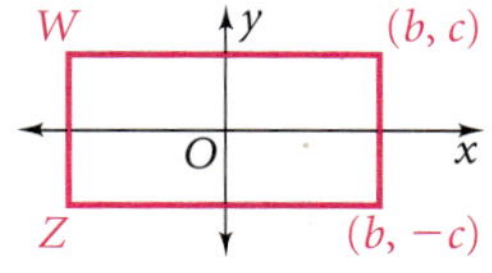

Rectangle

55.

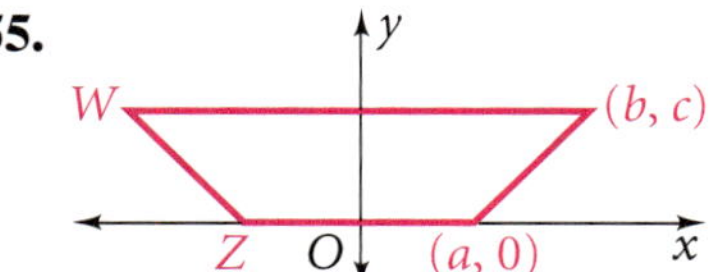

Isosceles trapezoid

Lesson 5-5

56. A triangle has sides with lengths 9 m and 15 m. Write an inequality that shows the range of possible lengths for the third side.

Vocabulary Builder

Abbreviations and Symbols

Many cellular phones and other modern communication devices can send and receive short text messages. Because of their small screen sizes and the need to send information quickly, people have found ways to communicate using new abbreviations and symbols.

1 EXAMPLE

Decode the short text message shown at the right.

In words:

Got to go. Talk to you later.

Abbreviations and symbols have always been part of the written vocabulary of mathematics. As with electronic text messaging, the purposes of "mathematical text messaging" include speed and economy in the recording of information.

2 EXAMPLE

Write this statement using words. Then write it again using symbols from the diagram.

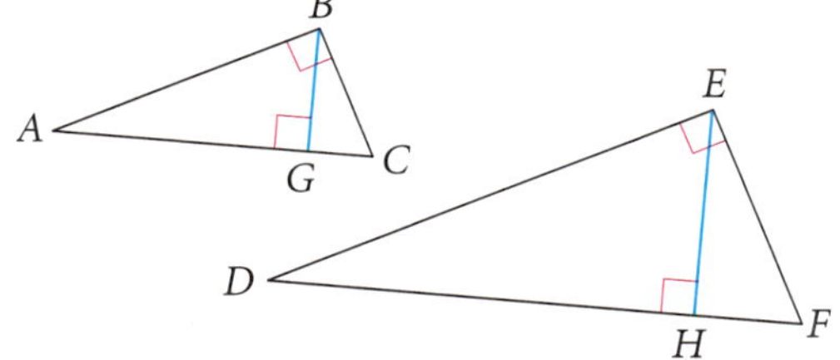

In ~ rt. $\triangle$s, corres. ⊥ seg. are prop. to corres. sides.

In words:
In similar right triangles, corresponding perpendicular segments are proportional to corresponding sides.

In symbols from the diagram:

$\triangle ABC \sim \triangle DEF, \overline{BG} \perp \overline{AC}, \overline{EH} \perp \overline{DF} \rightarrow \frac{BG}{EH} = \frac{AC}{DF}$

EXERCISES

Write each short text message using words.

1. SUP **2.** BCNU **3.** THNQ **4.** ILBL8 **5.** RURDY

Write each math statement using words.

6. Alt. int. $\angle$s formed by 2 $\|$ lines and a transv. are $\cong$.

7. A ▱ is a rhom. $\leftrightarrow$ 1 diag. bis. opp. $\angle$s.

Write each math statement using symbols from the diagram.

8. Parallel lines intercept proportional segments on all transversals.

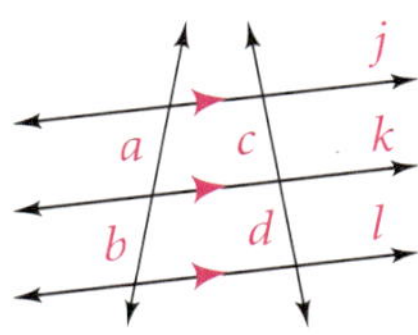

9. The altitude to the hypotenuse of a right triangle divides the triangle into two triangles that are similar to the original triangle and to each other.

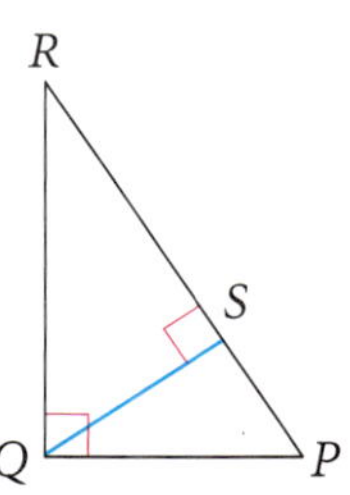

Algebra 1 Review: Simplifying Radicals

FOR USE WITH LESSON 7-4

You can multiply and divide numbers that are under radical signs.

1 EXAMPLE

Simplify the expressions $\sqrt{2} \cdot \sqrt{8}$ and $\sqrt{294} \div \sqrt{3}$.

$\sqrt{2} \cdot \sqrt{8} = \sqrt{2 \cdot 8}$	← **Rewrite using one radical sign.** →	$\sqrt{294} \div \sqrt{3} = \sqrt{\frac{294}{3}}$
$= \sqrt{16}$	← **Simplify the expression under the radical.** →	$= \sqrt{98}$
$= 4$	← **Factor out perfect squares and simplify.** →	$= \sqrt{49 \cdot 2}$
		$= 7\sqrt{2}$

A radical expression is in simplest form when all the following are true.

- The number under the radical sign has no perfect square factors other than 1.
- The number under the radical sign does not contain a fraction.
- A denominator does not contain a radical expression.

2 EXAMPLE

Write $\sqrt{\frac{4}{3}}$ in simplest form.

$\sqrt{\frac{4}{3}} = \frac{\sqrt{4}}{\sqrt{3}}$	**Rewrite the single radical as a quotient.**
$= \frac{2}{\sqrt{3}}$	**Simplify the numerator.**
$= \frac{2}{\sqrt{3}} \cdot \frac{\sqrt{3}}{\sqrt{3}}$	**Multiply by a form of 1 to rationalize the denominator.**
$= \frac{2\sqrt{3}}{3}$	**Simplify.**

Check using a calculator: $\sqrt{\frac{4}{3}} \approx 1.1547005$ and $\frac{2\sqrt{3}}{3} \approx 1.1547005$.

EXERCISES

Simplify each expression.

1. $\sqrt{5} \cdot \sqrt{10}$ **2.** $\sqrt{243}$ **3.** $\sqrt{128} \div \sqrt{2}$ **4.** $\sqrt{\frac{125}{4}}$

5. $\sqrt{6} \cdot \sqrt{8}$ **6.** $\frac{\sqrt{36}}{\sqrt{3}}$ **7.** $\frac{\sqrt{144}}{\sqrt{2}}$ **8.** $\sqrt{3} \cdot \sqrt{12}$

9. $\sqrt{72} \div \sqrt{2}$ **10.** $\sqrt{169}$ **11.** $28 \div \sqrt{8}$ **12.** $\sqrt{300} \div \sqrt{5}$

13. $\sqrt{12} \cdot \sqrt{2}$ **14.** $\frac{\sqrt{24}}{\sqrt{3}}$ **15.** $\sqrt{\frac{75}{3}}$ **16.** $\sqrt{18} \cdot \sqrt{2}$

17. $\sqrt{68}$ **18.** $\sqrt{3} \cdot \sqrt{15}$ **19.** $\frac{\sqrt{20}}{\sqrt{5}}$ **20.** $45 \div \sqrt{3}$

21. $\sqrt{\frac{25}{20}}$ **22.** $\sqrt{\frac{8}{28}}$ **23.** $\frac{\sqrt{6} \cdot \sqrt{3}}{\sqrt{9}}$ **24.** $\frac{\sqrt{3} \cdot \sqrt{15}}{\sqrt{2}}$

25. Find and simplify the difference of the golden ratio (see page 378, Exercise 51) and its multiplicative inverse.

Similarity in Right Triangles

California Content Standards

GEOM 4.0 Students prove basic theorems involving similarity. ***Develop***
GEOM 5.0 Students prove that triangles are similar. ***Develop***

What You'll Learn

- To find and use relationships in similar right triangles

. . . And Why

To find a distance indirectly, as in Example 3

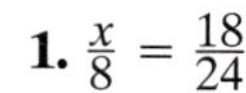

Check Skills You'll Need

GO for Help Lesson 7-1 and page 390

Algebra **Solve each proportion.**

1. $\frac{x}{8} = \frac{18}{24}$ **2.** $\frac{2}{3} = \frac{x}{7}$ **3.** $\frac{15}{4} = \frac{18}{x}$ **4.** $\frac{51}{x} = \frac{17}{13}$

5. $\frac{4}{10} = \frac{x}{5}$ **6.** $\frac{3}{m} = \frac{9}{8}$ **7.** $\frac{w}{2} = \frac{20}{9}$ **8.** $\frac{9}{6} = \frac{27}{a}$

9. Draw a right triangle. Label the triangle $\triangle ABC$ with right angle $\angle C$. Draw the altitude to the hypotenuse. Label the altitude $\overline{CD}$. Name the two smaller right triangles that are formed.

New Vocabulary • geometric mean

1 Using Similarity in Right Triangles

Hands-On Activity: Similarity in Right Triangles

- Draw one diagonal on a rectangular sheet of paper. Cut the paper on the diagonal to make two congruent right triangles.
- In one of the triangles, use paper folding to locate the altitude to the hypotenuse. Cut the triangle along the altitude to make two smaller right triangles.
- Label the angles of the three triangles as shown.
- Compare the angles of the three triangles by placing the angles on top of one another.

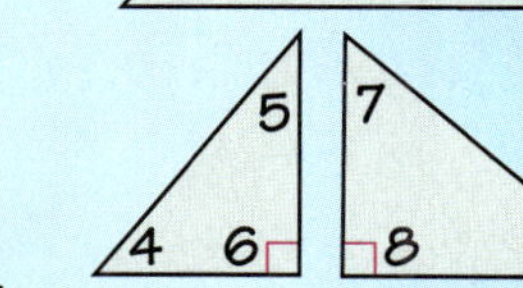

1. Which angles have the same measure as $\angle 1$?

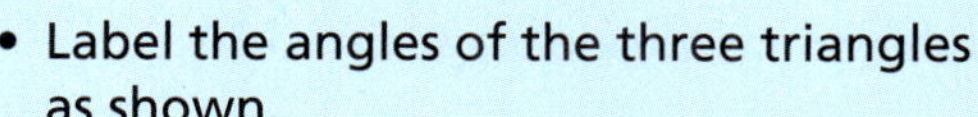

2. Which angles have the same measure as $\angle 2$?

3. Which angles have the same measure as $\angle 3$?

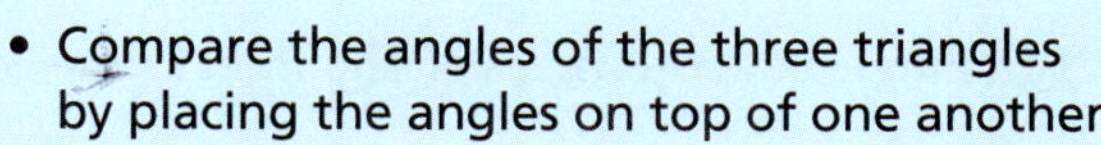

4. Based on your results, what is true about the three triangles?

5. Use the diagram at the right to complete the similarity statement. $\triangle RST \sim \triangle \underline{?} \sim \triangle \underline{?}$

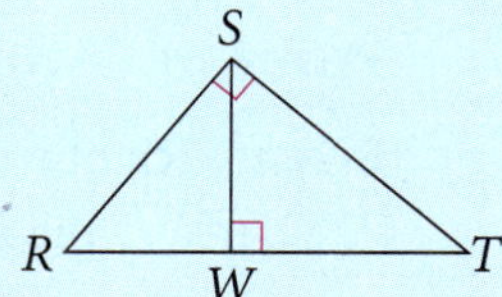

In a right triangle, the altitude to the hypotenuse yields three similar triangles.

Key Concepts

Theorem 7-3

The altitude to the hypotenuse of a right triangle divides the triangle into two triangles that are similar to the original triangle and to each other.

Proof

Proof of Theorem 7-3

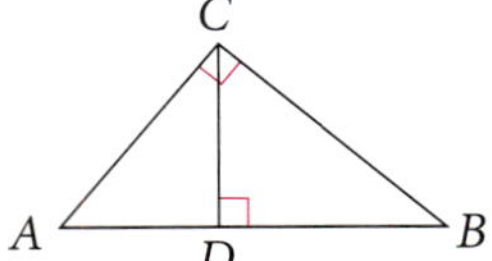

Given: Right triangle, $\triangle ABC$, with $\overline{CD}$ the altitude to the hypotenuse

Prove: $\triangle ABC \sim \triangle ACD \sim \triangle CBD$

Proof: Both smaller triangles are right triangles. Each also shares an angle with $\triangle ABC$. Thus each smaller triangle is similar to $\triangle ABC$ by the AA ~ Postulate. Since both smaller triangles are similar to $\triangle ABC$, their corresponding angles are congruent. Thus they are similar to each other.

Vocabulary Tip

In the proportion $\frac{a}{b} = \frac{c}{d}$, b and c are the means.

Proportions in which the means are equal occur frequently in geometry. For any two positive numbers a and b, the **geometric mean** of a and b is the positive number x such that $\frac{a}{x} = \frac{x}{b}$. Note that $x = \sqrt{ab}$.

1 EXAMPLE Finding the Geometric Mean

Algebra Find the geometric mean of 4 and 18.

$\frac{4}{x} = \frac{x}{18}$	**Write a proportion.**
$x^2 = 72$	**Cross-Product Property**
$x = \sqrt{72}$	**Take the square root.**
$x = 6\sqrt{2}$	**Write in simplest radical form.**

The geometric mean of 4 and 18 is $6\sqrt{2}$.

CA Standards Check

1 Find the geometric mean of 15 and 20.

Two important corollaries of Theorem 7-3 involve a geometric mean.

Key Concepts

Corollary **Corollary 1 to Theorem 7-3**

The length of the altitude to the hypotenuse of a right triangle is the geometric mean of the lengths of the segments of the hypotenuse.

Proof

Proof of Corollary 1

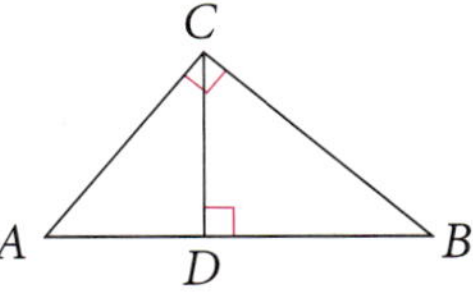

Given: Right triangle, $\triangle ABC$, with $\overline{CD}$ the altitude to the hypotenuse

Prove: $\frac{AD}{CD} = \frac{CD}{DB}$

Proof: By Theorem 7-3, $\triangle ACD \sim \triangle CBD$. Since corresponding sides of similar triangles are proportional, $\frac{AD}{CD} = \frac{CD}{DB}$.

Key Concepts

Corollary | **Corollary 2 to Theorem 7-3**

The altitude to the hypotenuse of a right triangle separates the hypotenuse so that the length of each leg of the triangle is the geometric mean of the length of the adjacent hypotenuse segment and the length of the hypotenuse.

Proof

Proof of Corollary 2

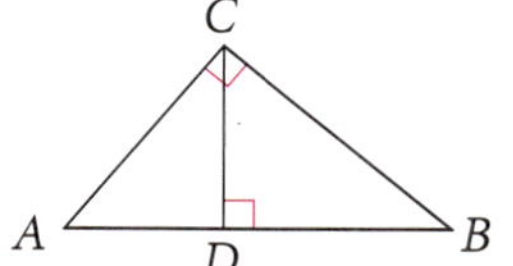

Given: Right triangle, $\triangle ABC$, with $\overline{CD}$ the altitude to the hypotenuse

Prove: $\frac{AD}{AC} = \frac{AC}{AB}$, $\frac{DB}{CB} = \frac{CB}{AB}$

Proof: By Theorem 7-3, $\triangle ACD \sim \triangle ABC$. Their corresponding sides are proportional, so $\frac{AD}{AC} = \frac{AC}{AB}$. Similarly, $\triangle CBD \sim \triangle ABC$ and $\frac{DB}{CB} = \frac{CB}{AB}$.

2 EXAMPLE Applying Corollaries 1 and 2

Online active math

For: Similarity Activity
Use: Interactive Textbook, 7-4

Algebra Solve for x and y.

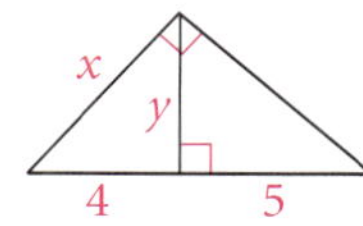

Use Corollary 2 to solve for x:		Use Corollary 1 to solve for y:
$\frac{4}{x} = \frac{x}{4+5}$	← Write a proportion. →	$\frac{4}{y} = \frac{y}{5}$
$x^2 = 36$	← Cross-Product Property →	$y^2 = 20$
$x = 6$	← Take the square root. →	$y = 2\sqrt{5}$

CA Standards Check

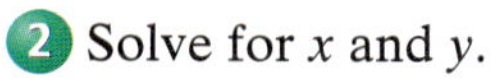

2 Solve for x and y.

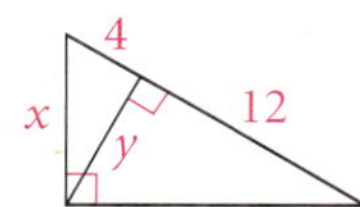

3 EXAMPLE Real-World Connection

Real-World Connection

Canoes can be rented for recreation at some lakes.

Recreation The 300-m path to the information center and the 400-m path to the canoe rental dock meet at a right angle at the parking lot. Marla walks straight from the parking lot to the lake as shown, where a sign tells her that she is 320 m from the dock. How far is Marla from the information center?

$\frac{x}{300} = \frac{300}{x + 320}$	Corollary 2
$x^2 + 320x = 90{,}000$	Cross-Product Property
$x^2 + 320x - 90{,}000 = 0$	Standard Form Quadratic Equation
$x = 180$	Solve. Use a calculator (see p. 372).

Marla is 180 m from the information center.

CA Standards Check

3 How far did Marla walk from the parking lot to the lake?

EXERCISES Standards Practice

For more exercises, see *Extra Skill, Word Problem, and Proof Practice.*

GEOM 4.0, 5.0

Practice by Example

Algebra **Find the geometric mean of each pair of numbers.**

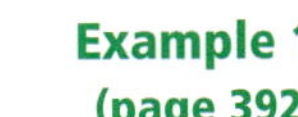

Example 1 (page 392)

GO for Help

1. 4 and 9 **2.** 4 and 10 **3.** 4 and 12 **4.** 3 and 48

5. 7 and 56 **6.** 5 and 125 **7.** 9 and 24 **8.** 7 and 9

Example 2 (page 393)

Algebra **Refer to the figure to complete each proportion.**

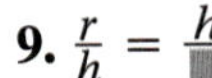

9. $\frac{r}{h} = \frac{h}{\blacksquare}$ **10.** $\frac{c}{a} = \frac{a}{\blacksquare}$ **11.** $\frac{\blacksquare}{b} = \frac{b}{s}$

12. $\frac{r}{\blacksquare} = \frac{\blacksquare}{c}$ **13.** $\frac{r}{h} = \frac{\blacksquare}{s}$ **14.** $\frac{s}{b} = \frac{\blacksquare}{c}$

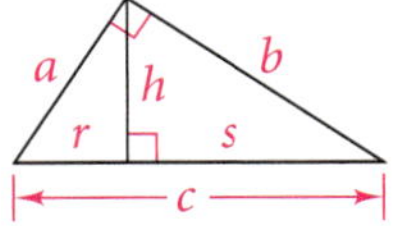

Algebra **Solve for x.**

15.

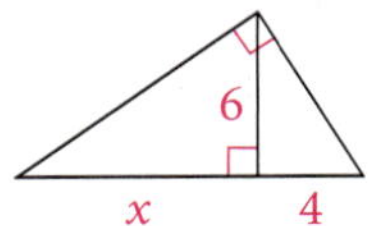

16.

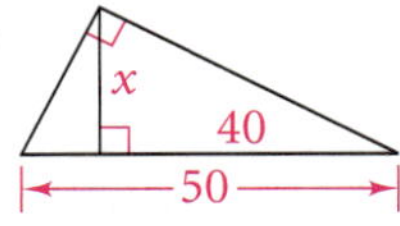

17.

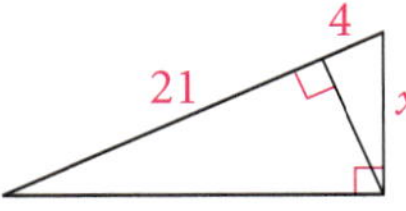

18.

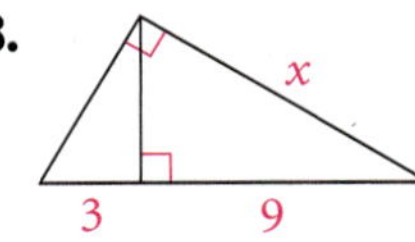

19.

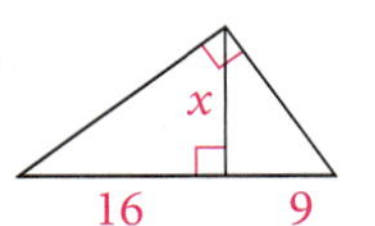

20.

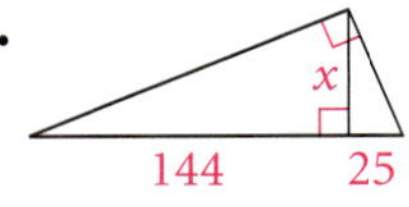

Example 3 (page 393)

21. a. Civil Engineering Study the plan at the right. A service station will be built on the highway, and a road will connect it with Cray. How far from Blare should the service station be located so that the proposed road will be perpendicular to the highway?

b. How long will the new road be?

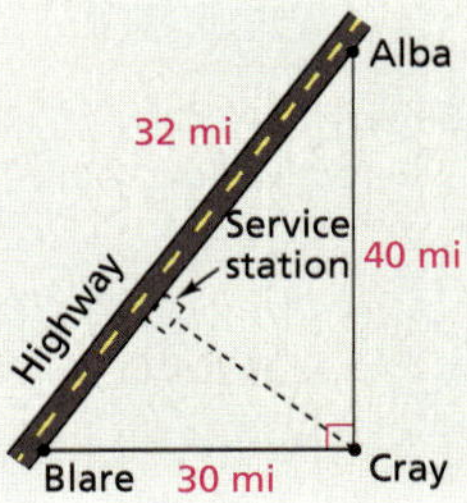

Apply Your Skills

22. Complete:
$\triangle JKL \sim \triangle$ <u>?</u> $\sim \triangle$ <u>?</u>

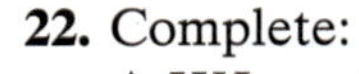

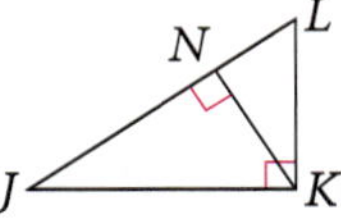

23. a. The altitude to the hypotenuse of a right triangle divides the hypotenuse into segments 2 cm and 8 cm long. Find the length h of the altitude.

b. Drawing Use the value you found for h in part (a), along with the lengths 2 cm and 8 cm, to draw the right triangle accurately.

c. Writing Explain how you drew the triangle in part (b).

24. a. Open-Ended Draw a right triangle so that the altitude from the right angle to the hypotenuse bisects the hypotenuse.

b. How does the length of the altitude compare with the lengths of the segments of the hypotenuse? Explain.

25. Coordinate Geometry $\overline{CD}$ is the altitude to the hypotenuse of right $\triangle ABC$. The coordinates of A, D, and B are (4, 2), (4, 6), and (4, 15), respectively. Find all possible coordinates of point C.

Vocabulary Tip

Respectively in Exercise 25 means you match the lists in the order named: $A(4, 2)$, $D(4, 6)$, $B(4, 15)$.

Algebra **Find the geometric mean of each pair of numbers.**

26. 3 and 16 **27.** 4 and 49 **28.** $\sqrt{8}$ and $\sqrt{2}$ **29.** $\sqrt{28}$ and $\sqrt{7}$

30. $\frac{1}{2}$ and 2 **31.** 5 and 1.25 **32.** 1 and 1000 **33.** 11 and 1331

Algebra **Find the values of the variables.**

34. 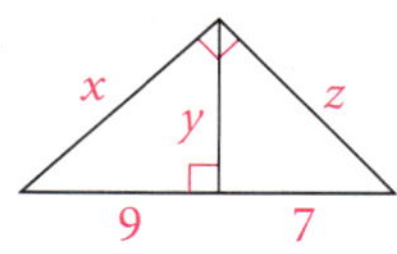

35. 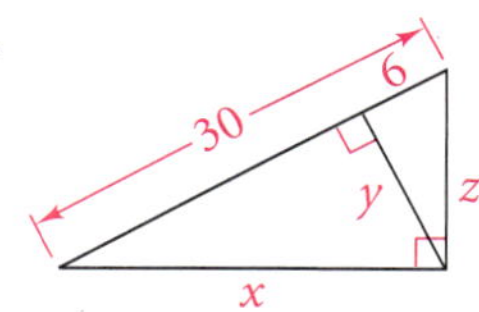

36. 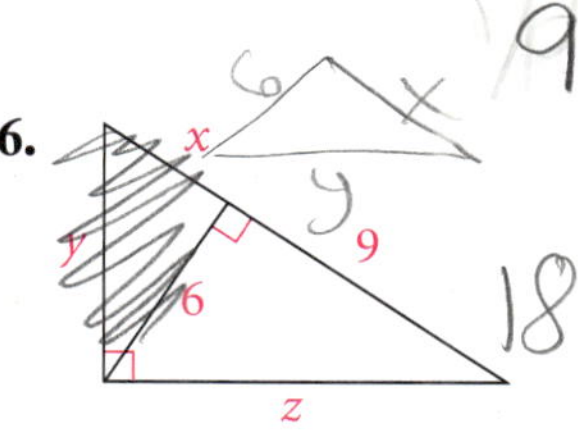

Exercise 38

37. **Algebra** The altitude to the hypotenuse of a right triangle divides the hypotenuse into segments with lengths in the ratio 1 : 2. The length of the altitude is 8. How long is the hypotenuse?

38. **Multiple Choice** To estimate the height of a totem pole, Jorge uses a small square of plastic. He holds the square up to his eyes and walks backward from the pole. He stops when the bottom of the pole lines up with the bottom edge of the square and the top of the pole lines up with the top edge of the square. Jorge's eye level is about 2 m from the ground. He is about 3 m from the pole. Which is the best estimate for the height of the totem pole?

Ⓐ 4.5 m Ⓑ 5 m Ⓒ 6.5 m Ⓓ 9 m

For a right triangle, denote lengths as follows: ℓ_1 and ℓ_2 the legs, h the hypotenuse, a the altitude, and h_1 and h_2 the hypotenuse segments determined by the altitude. For the two given measures, find the other four. Use simplest radical form.

39. $h = 2, h_1 = 1$
40. $h_1 = 4, h_2 = 9$
41. $a = 6, h_1 = 6$
42. $\ell_1 = 2, h_2 = 3$
43. $h = 13, \ell_2 = 12$
44. $\ell_1 = 4, h_1 = 3$
45. $a = 8, h_1 = 16$
46. $h_1 = 3, \ell_2 = 6\sqrt{3}$

Proof 47. **Given:** isosceles right $\triangle ABC$
Prove: $AB = x\sqrt{2}$

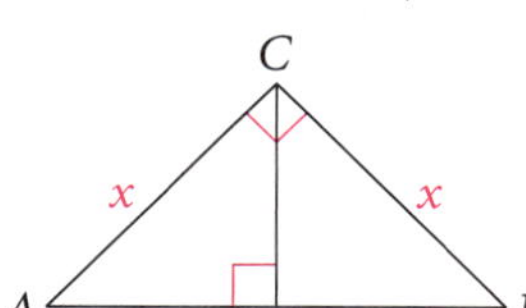

48. **Given:** equilateral $\triangle ABC$
Prove: $h = 10\sqrt{3}$

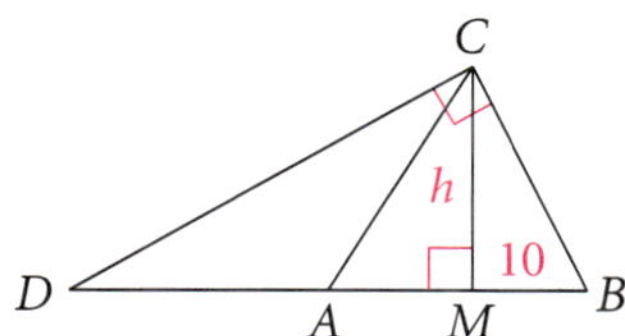

Visit: PHSchool.com
Web Code: bce-0704

Algebra **Find the value of x.**

49. 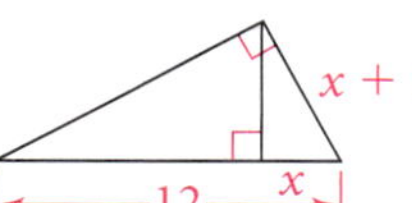

50. 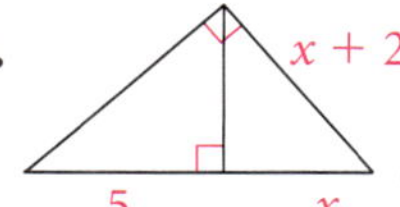

51. 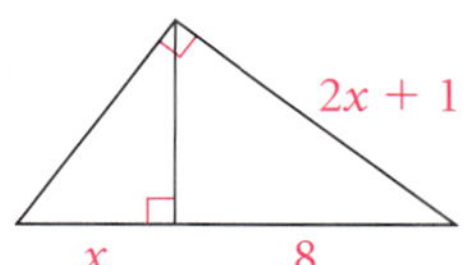

Challenge

52. a. Lauren thinks she has found a new corollary: The product of the lengths of the two legs of a right triangle is equal to the product of the lengths of the hypotenuse and the altitude to the hypotenuse. Draw a figure for this corollary. Write the *Given* information and what you are to *Prove*.
b. **Critical Thinking** Is Lauren's corollary true? Explain.

53. In the diagram $c = q + r$. Also, Corollary 2 to Theorem 7-3 gives you two more equations involving a, b, c, q, and r. The result is a system of three equations in five variables.
a. Reduce the system to one equation in three variables by eliminating q and r.
b. State in words what the one resulting equation tells you.

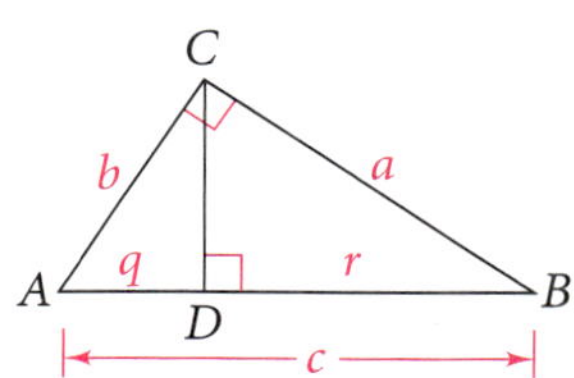

Proof **54. Given:** equilateral $\triangle ABC$

Prove: $h = x\sqrt{3}$

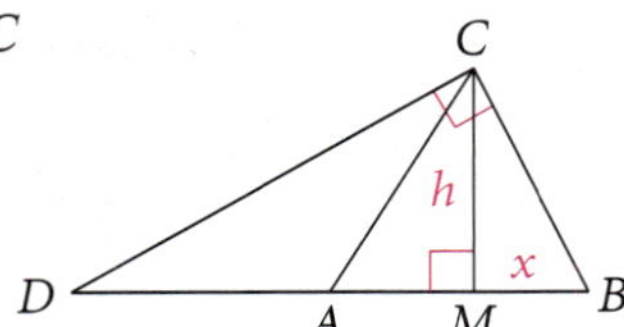

Multiple Choice Practice

For California Standards and CAHSEE Tutorials, visit PHSchool.com. Web Code: bcq-9045

CAHSEE 7MG 2.2

55. In the figure shown at the right, all the corners form right angles. What is the area of the figure in square units?

A 55
B 63
C 66
D 88

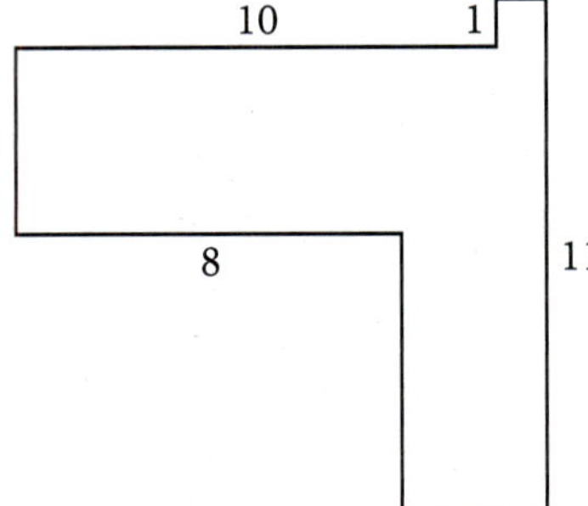

GEOM 4.0

56. The altitude to the hypotenuse of a right triangle divides the triangle into two triangles. Which of the following must be true of the two new triangles?

A They are obtuse triangles.
B They are congruent to each other.
C They are not right triangles.
D They are similar to each other.

Mixed Review

Lesson 7-3

If the triangles are similar, (a) write a similarity statement and (b) name the postulate or theorem you used. If the triangles are not similar, write *not similar*.

57.

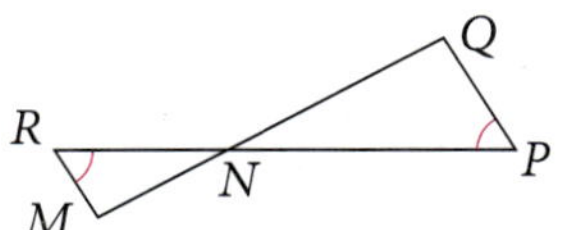

58.

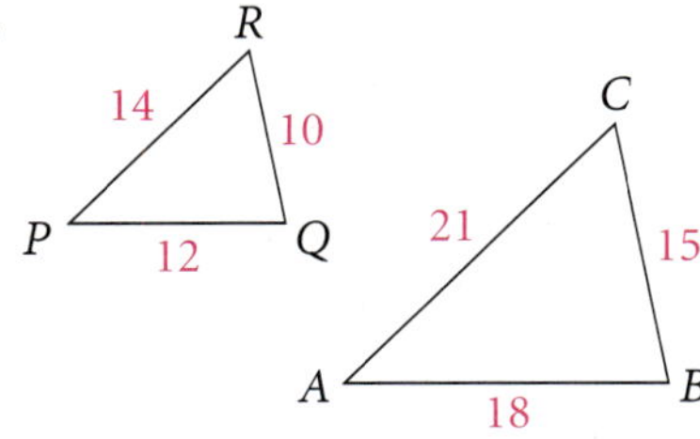

59.

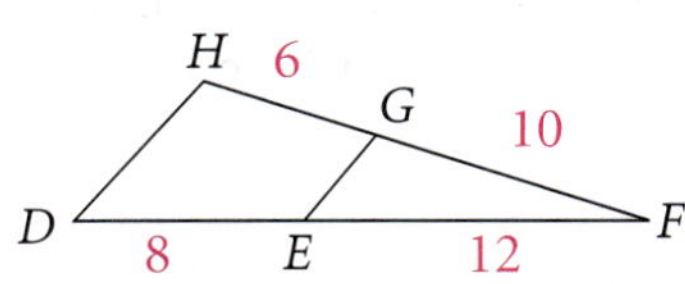

Lesson 6-2 x^2 Algebra **Find the values of x and y in $\square RSTV$.**

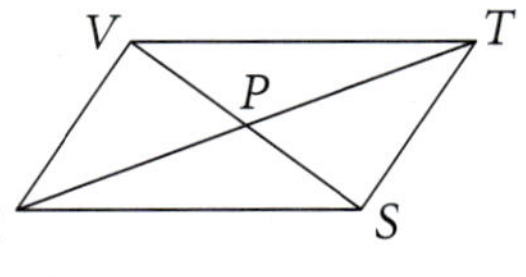

60. $RP = 2x, PT = y + 2, VP = y, PS = x + 3$

61. $RP = 4x, PT = 3y - 3, VP = 2x + 3, PS = y + 6$

62. $RV = 2x + 3, VT = 5x, TS = y + 5, SR = 4y - 1$

Lesson 5-1 **Find the value of x.**

63.

64.

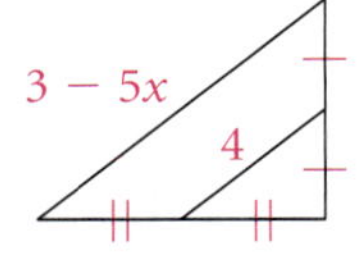

Exploring Proportions in Triangles

Technology

FOR USE WITH LESSON 7-5

Prepares for GEOM 7.0 Students prove and use theorems involving the properties of parallel lines cut by a transversal.

Construct

- Use geometry software. Draw $\triangle ABC$ and construct point D on $\overline{AB}$.
- Construct a line through D parallel to $\overline{AC}$.
- Construct the intersection E of the parallel line with $\overline{BC}$.

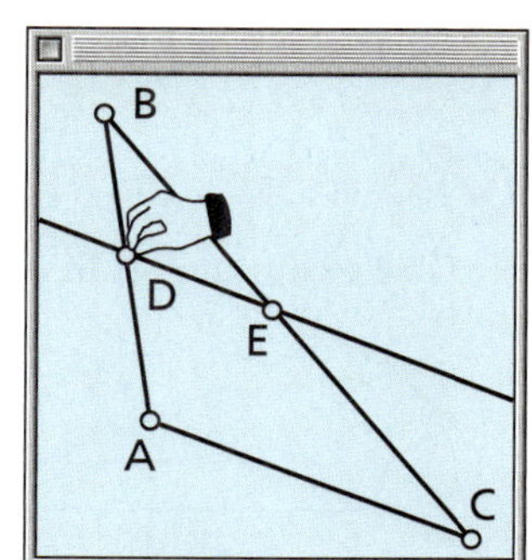

Investigate

- Measure $\overline{BD}$, $\overline{DA}$, $\overline{BE}$, and $\overline{EC}$.
- Calculate the ratios $\frac{BD}{DA}$ and $\frac{BE}{EC}$.
- Manipulate $\triangle ABC$ and observe the ratios $\frac{BD}{DA}$ and $\frac{BE}{EC}$. (Save your observations for Exercise 1.)

Construct

- Use geometry software. Draw $\triangle ABC$. Construct the bisector of $\angle A$. Construct point D, the intersection of the bisector and $\overline{CB}$.

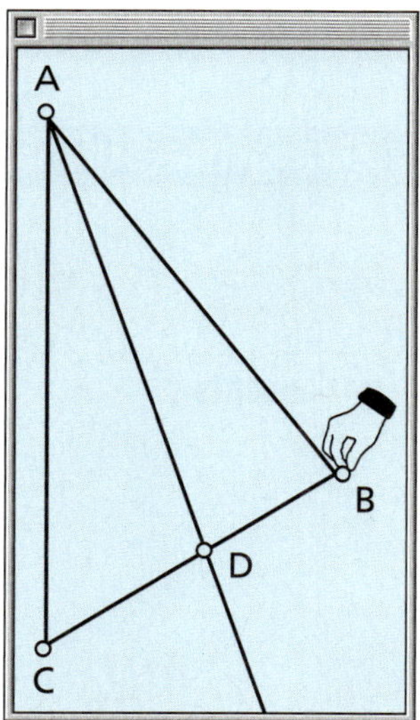

Investigate

- Measure $\overline{AC}$, $\overline{AB}$, $\overline{CD}$, and $\overline{DB}$.
- Calculate the ratios $\frac{AC}{AB}$ and $\frac{CD}{DB}$.
- Manipulate $\triangle ABC$ and observe the ratios $\frac{AC}{AB}$ and $\frac{CD}{DB}$. (Save your observations for Exercise 2.)

EXERCISES

1. Suppose a line parallel to one side of a triangle intersects the other two sides. Make a conjecture about the four segments formed.
2. The bisector of an angle of a triangle divides the opposite side into two segments. Make a conjecture about the two segments and the other two sides of the triangle.

Extend

- Construct $\overleftrightarrow{AB} \parallel \overleftrightarrow{CD}$. Then construct lines $\overleftrightarrow{AC}$ and $\overleftrightarrow{BD}$.
- Construct point E on $\overleftrightarrow{AC}$.
- Construct $\overleftrightarrow{EF} \parallel \overleftrightarrow{AB}$ with F the intersection of $\overleftrightarrow{EF}$ and $\overleftrightarrow{BD}$.
- Measure $\overline{AC}$, $\overline{CE}$, $\overline{BD}$, and $\overline{DF}$.
- Calculate the ratios $\frac{AC}{CE}$ and $\frac{BD}{DF}$.
- Manipulate the locations of A and B. Observe $\frac{AC}{CE}$ and $\frac{BD}{DF}$.

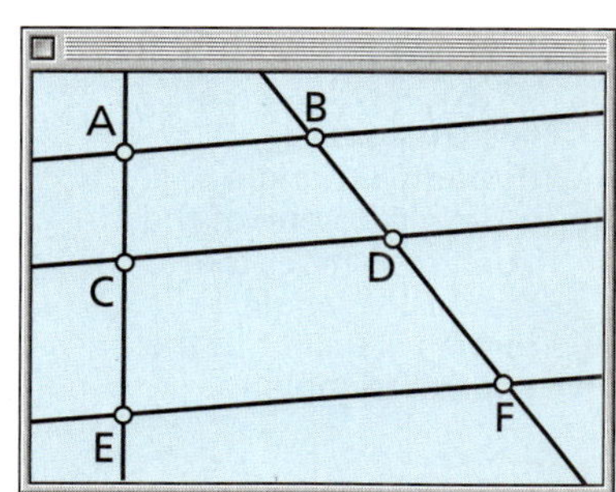

3. Suppose three parallel lines intersect two transversals. Make a conjecture about the segments of the transversals.
4. Suppose that four or more parallel lines intersect two transversals. Make a conjecture about the segments of the transversals.

Proportions in Triangles

California Content Standards

GEOM 4.0 Students prove theorems involving similarity. ***Master***
GEOM 5.0 Students prove that triangles are similar. ***Master***
GEOM 7.0 Students prove and use theorems involving the properties of parallel lines cut by a transversal. ***Master***

What You'll Learn

- To use the Side-Splitter Theorem
- To use the Triangle-Angle-Bisector Theorem

. . . And Why

To design a sail, as in Example 2

Check Skills You'll Need

GO for Help Lesson 7-2

The two triangles in each diagram are similar. Find the value of x in each.

1.
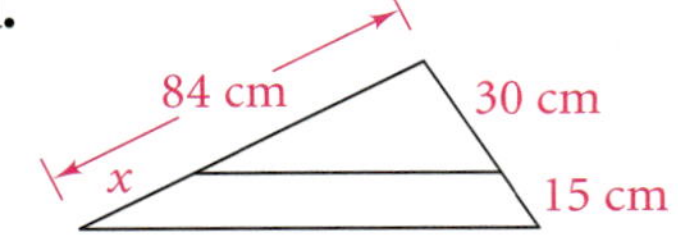

2.
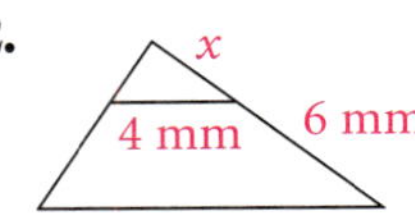

3.
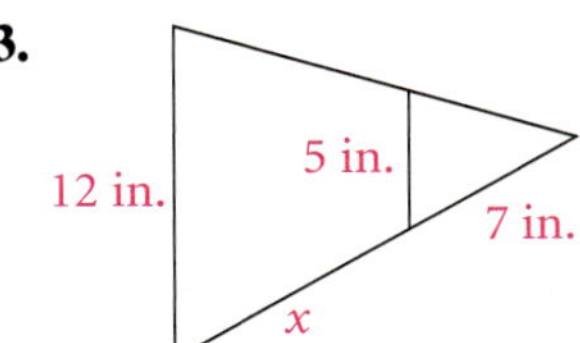

4.
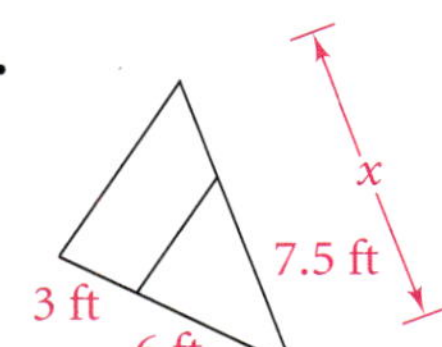

1 Using the Side-Splitter Theorem

You can use similar triangles to prove the following theorem.

Key Concepts

Theorem 7-4 **Side-Splitter Theorem**

If a line is parallel to one side of a triangle and intersects the other two sides, then it divides those sides proportionally.

Proof

Proof of Theorem 7-4

Given: $\triangle QXY$ with $\overleftrightarrow{RS} \parallel \overleftrightarrow{XY}$

Prove: $\frac{XR}{RQ} = \frac{YS}{SQ}$

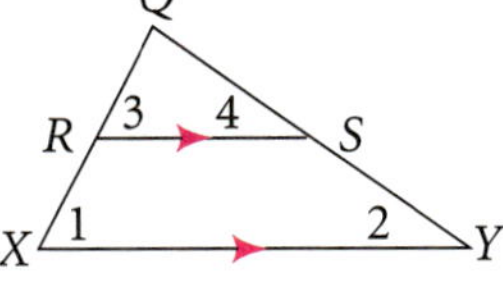

Statements	Reasons
1. $\overleftrightarrow{RS} \parallel \overleftrightarrow{XY}$	1. Given
2. $\angle 1 \cong \angle 3, \angle 2 \cong \angle 4$	2. If lines are $\parallel$, then corr. $\measuredangle$s are $\cong$.
3. $\triangle QXY \sim \triangle QRS$	3. AA $\sim$ Postulate
4. $\frac{XQ}{RQ} = \frac{YQ}{SQ}$	4. Corr. sides of $\sim$ $\triangle$s are proportional.
5. $XQ = XR + RQ, YQ = YS + SQ$	5. Segment Addition Postulate
6. $\frac{XR + RQ}{RQ} = \frac{YS + SQ}{SQ}$	6. Substitute.
7. $\frac{XR}{RQ} = \frac{YS}{SQ}$	7. A Property of Proportions

Problem Solving Hint

Whenever you see a line that passes through a triangle and is parallel to one of the sides, remember that the small triangle formed is similar to the original triangle.

1 EXAMPLE Using the Side-Splitter Theorem

Find the value of x.

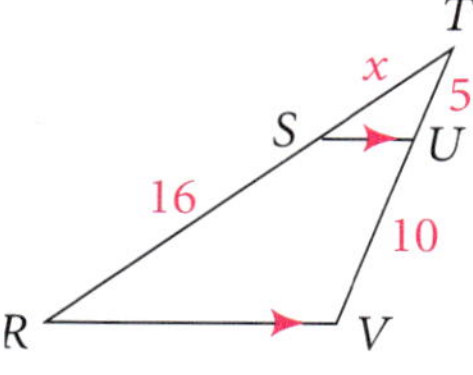

$\frac{TS}{SR} = \frac{TU}{UV}$ **Side-Splitter Theorem**

$\frac{x}{16} = \frac{5}{10}$ **Substitute.**

$x = \frac{5}{10} \cdot 16$ **Solve for *x*.**

$x = 8$

CA Standards Check 1 Use the Side-Splitter Theorem to find the value of x.

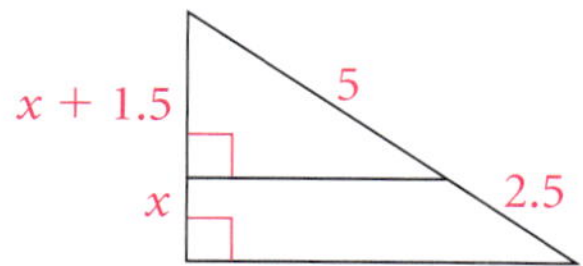

The following corollary to the Side Splitter Theorem says that parallel lines divide all transversals proportionally. You will prove this corollary in Exercise 35.

Key Concepts

Corollary **Corollary to Theorem 7-4**

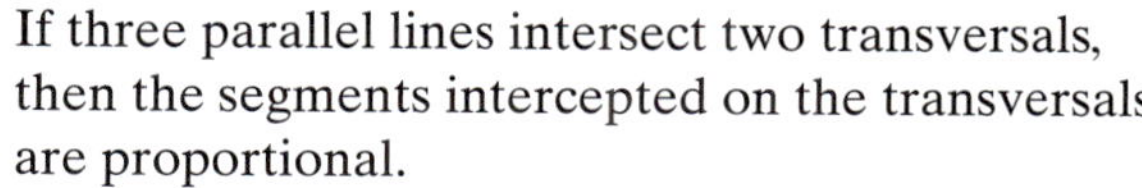

If three parallel lines intersect two transversals, then the segments intercepted on the transversals are proportional.

$$\frac{a}{b} = \frac{c}{d}$$

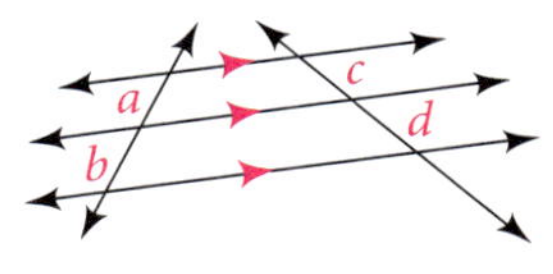

Real-World Connection

Some windsurfing sails are made of a polyester material that is transparent and lightweight.

2 EXAMPLE Real-World Connection

Sail Making Sailmakers sometimes use a computer to create a pattern for a sail. After they cut out the panels of the sail, they sew them together to form the sail.

The seams of the panels in the sail at the right are parallel. Find the lengths x and y.

$\frac{2}{x} = \frac{1.7}{1.7}$ **Side-Splitter Theorem**

$x = 2$

$\frac{3}{2} = \frac{y}{1.7}$ **Corollary to the Side-Splitter Theorem**

$\frac{3}{2}(1.7) = y$ **Solve for *y*.**

$2.55 = y$

Length x is 2 ft and length y is 2.55 ft.

CA Standards Check 2 Solve for x and y.

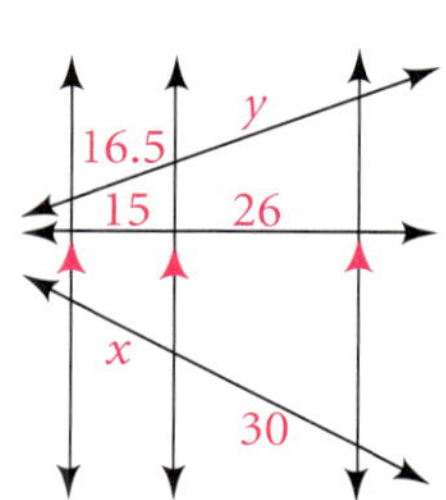

Using the Triangle-Angle-Bisector Theorem

You can use the Side-Splitter Theorem to prove the following relationship.

Key Concepts

Theorem 7-5 **Triangle-Angle-Bisector Theorem**

If a ray bisects an angle of a triangle, then it divides the opposite side into two segments that are proportional to the other two sides of the triangle.

Proof

Proof of Theorem 7-5

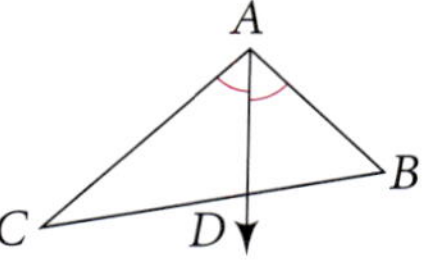

Given: $\triangle ABC$, $\overrightarrow{AD}$ bisects $\angle CAB$.

Prove: $\frac{CD}{DB} = \frac{CA}{BA}$

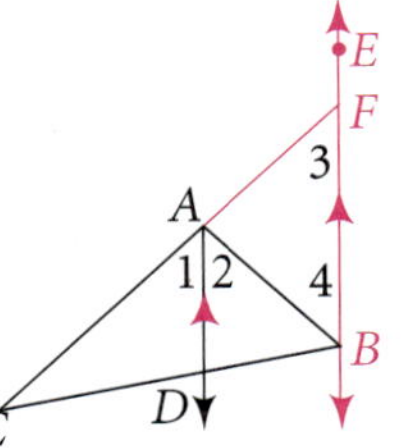

Draw $\overleftrightarrow{BE} \parallel \overline{DA}$. Extend $\overline{CA}$ to meet $\overleftrightarrow{BE}$ at point F.

Proof: By the Side-Splitter Theorem, $\frac{CD}{DB} = \frac{CA}{AF}$. By the Corresponding Angles Postulate, $\angle 3 \cong \angle 1$. Since $\overrightarrow{AD}$ bisects $\angle CAB$, $\angle 1 \cong \angle 2$. By the Alternate Interior Angles Theorem, $\angle 2 \cong \angle 4$. Using the Transitive Property of Congruence, you know that $\angle 3 \cong \angle 4$. By the Converse of the Isosceles Triangle Theorem, $BA = AF$. Substituting BA for AF, $\frac{CD}{DB} = \frac{CA}{BA}$.

Problem Solving Hint

Drawing $\overleftrightarrow{BE} \parallel \overline{DA}$ sets up $\triangle BCF$ for the Side-Splitter Theorem, as well as congruent $\angle$s 1, 2, 3, and 4.

3 EXAMPLE Using the Triangle-Angle-Bisector Theorem

Algebra Find the value of x.

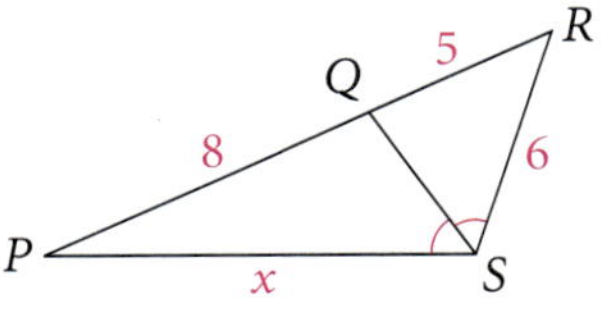

$\frac{PS}{SR} = \frac{PQ}{RQ}$	**Triangle-Angle-Bisector Theorem**
$\frac{x}{6} = \frac{8}{5}$	**Substitute.**
$5x = 48$	**Cross-Product Property**
$x = 9.6$	**Solve for x.**

CA Standards Check **3** Find the value of y.

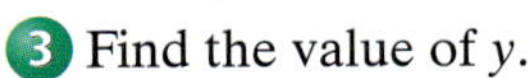

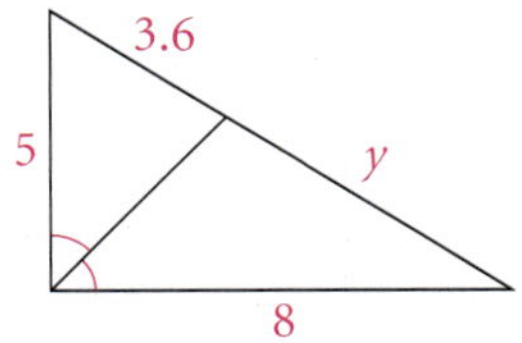

EXERCISES
Standards Practice

For more exercises, see *Extra Skill, Word Problem, and Proof Practice*.

GEOM 4.0, 5.0, 7.0

A Practice by Example x^2 **Algebra** **Solve for x.**

GO for Help **Example 1** (page 399)

1.

2.

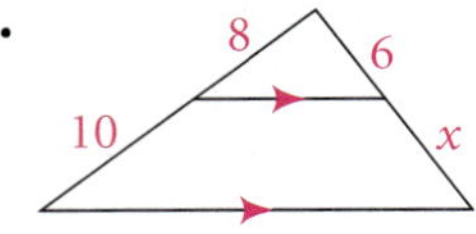

3.

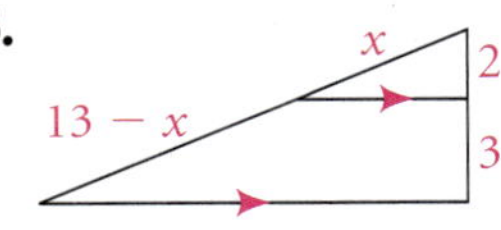

Example 2 (page 399)

Use the figure at the right to complete each proportion.

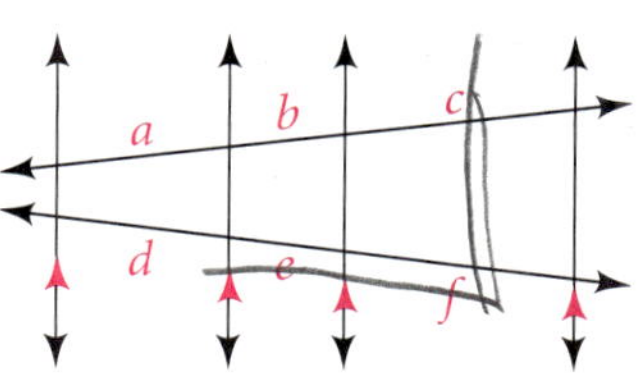

4. $\frac{a}{b} = \frac{\blacksquare}{e}$
5. $\frac{b}{\blacksquare} = \frac{e}{f}$
6. $\frac{f}{e} = \frac{c}{\blacksquare}$
7. $\frac{a}{b + c} = \frac{\blacksquare}{e + f}$

Algebra **Solve for x.**

8. 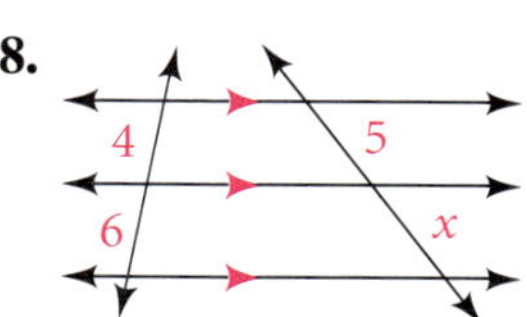

9.

10. 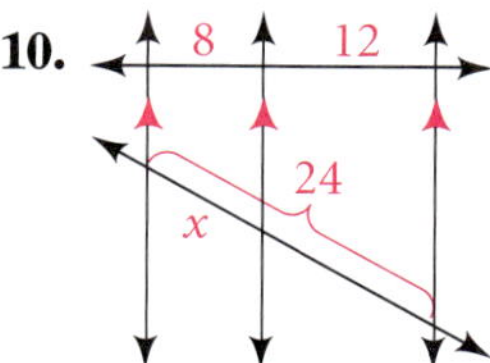

Example 3 (page 400)

Algebra **Solve for x.**

11. 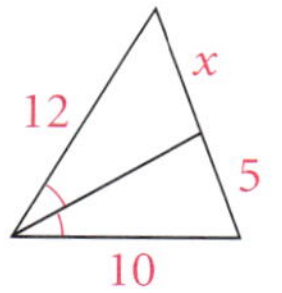

12. 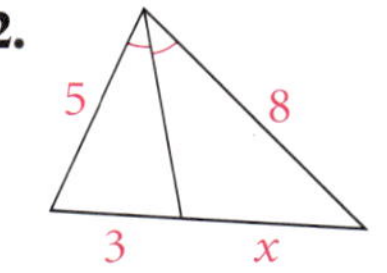

13. 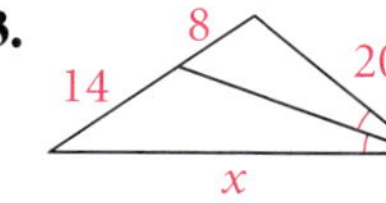

14. 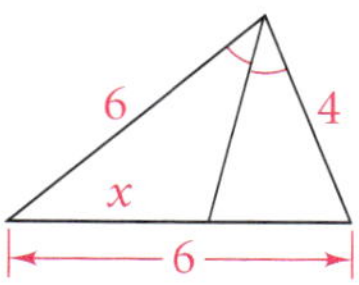

15. 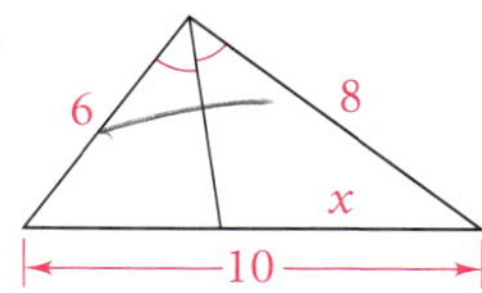

16. 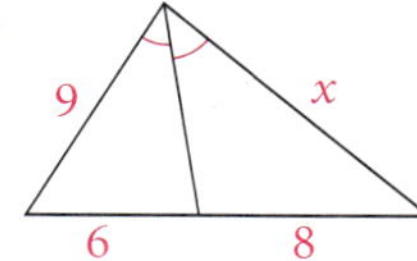

Use the figure at the right to complete each proportion.

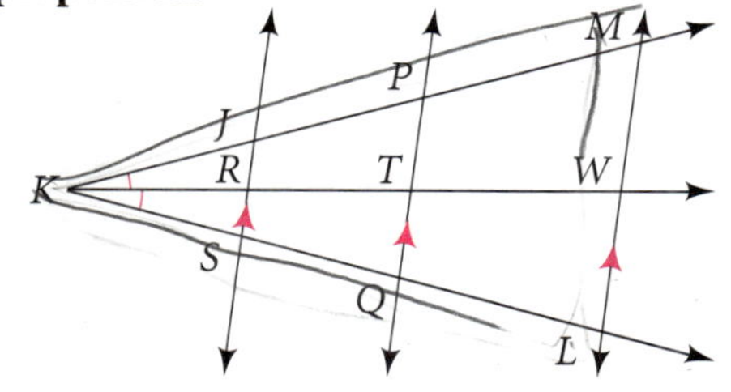

17. $\frac{RS}{\blacksquare} = \frac{JR}{KJ}$
18. $\frac{KJ}{JP} = \frac{KS}{\blacksquare}$
19. $\frac{QL}{PM} = \frac{SQ}{\blacksquare}$
20. $\frac{PT}{\blacksquare} = \frac{TQ}{KQ}$
21. $\frac{KL}{LW} = \frac{\blacksquare}{MW}$
22. $\frac{\blacksquare}{KP} = \frac{LQ}{KQ}$
23. $\frac{\blacksquare}{SQ} = \frac{JK}{KS}$
24. $\frac{KL}{KM} = \frac{\blacksquare}{MW}$

B Apply Your Skills

Real-World Connection

In historic cities such as Washington, D.C., planners must work around and with pre-existing structures.

Urban Design **In Washington, D.C., 17th, 18th, 19th, and 20th Streets are parallel streets that intersect Pennsylvania Avenue and I Street.**

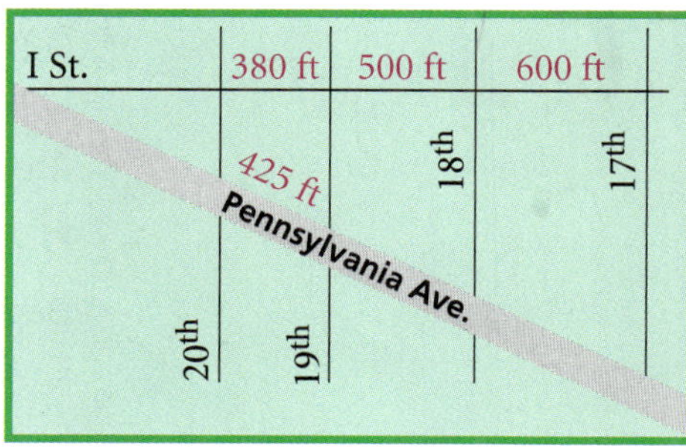

25. How long (to the nearest foot) is Pennsylvania Avenue between 19th Street and 18th Street?

26. How long (to the nearest foot) is Pennsylvania Avenue between 18th Street and 17th Street?

27. The sides of a triangle are 5 cm, 12 cm, and 13 cm long. Find the lengths, to the nearest tenth, of the segments into which the bisector of each angle divides the opposite side.

28. **Open-Ended** In a triangle, the bisector of an angle divides the opposite side into two segments with lengths 6 cm and 9 cm. How long could the other two sides of the triangle be? (*Caution:* Make sure the three sides satisfy the Triangle Inequality Theorem.)

GO for Help

For a guide to solving Exercise 29, see p. 405.

29. Surveying The perimeter of the triangular lot at the right is 50 m. The surveyor's tape bisects an angle. Find the lengths x and y.

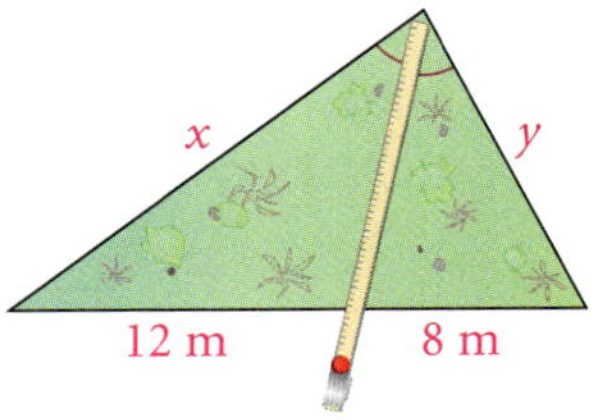

30. Critical Thinking Sharell draws $\triangle ABC$. She finds that the bisector of $\angle C$ bisects the opposite side.

a. Sketch $\triangle ABC$ and the bisector.

b. Writing What type of triangle is $\triangle ABC$? Explain your reasoning.

Algebra **Solve for x.**

31.

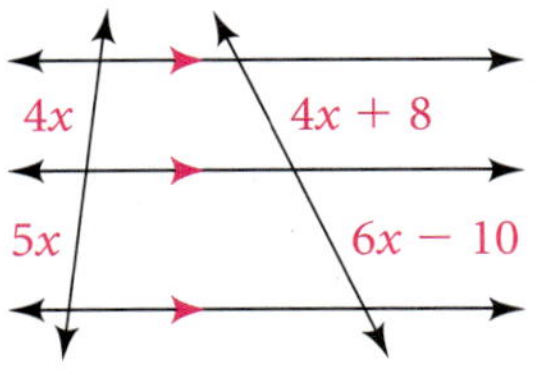

32.

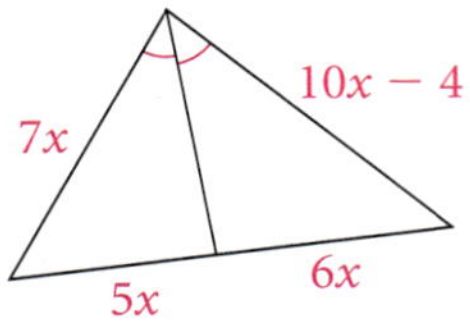

33.

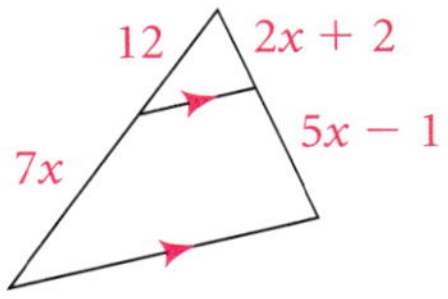

34. Developing Proof Copy and complete this two-column proof of the Converse of the Side-Splitter Theorem: If a line divides two sides of a triangle proportionally, then it is parallel to the third side.

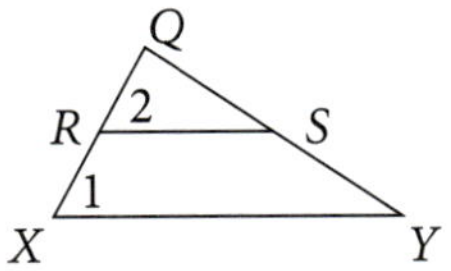

Given: $\frac{XR}{RQ} = \frac{YS}{SQ}$

Prove: $\overline{RS} \parallel \overline{XY}$

Statements	Reasons
1. $\frac{XR}{RQ} = \frac{YS}{SQ}$	a. __?__
2. $\frac{XR + RQ}{RQ} = \frac{YS + SQ}{SQ}$	b. __?__
3. $\frac{XQ}{RQ} = \frac{YQ}{SQ}$	c. __?__
4. $\angle Q \cong \angle Q$	d. __?__
5. $\triangle XQY \sim \triangle RQS$	e. __?__
6. $\angle 1 \cong \angle 2$	f. __?__
7. $\overline{RS} \parallel \overline{XY}$	g. __?__

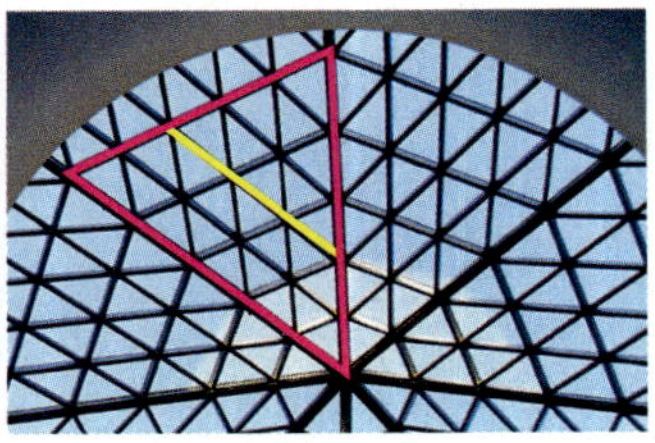

Real-World Connection

In this glass roof, parallel lines divide the sides of triangles proportionally.

Proof **35.** Follow the steps below. Write a proof of the Corollary to the Side-Splitter Theorem found on page 399.

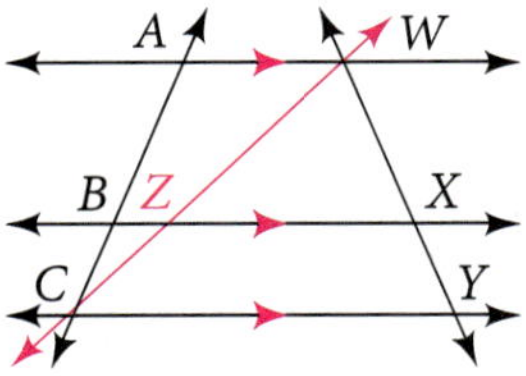

Given: $\overleftrightarrow{AW} \parallel \overleftrightarrow{BX} \parallel \overleftrightarrow{CY}$

Prove: $\frac{AB}{BC} = \frac{WX}{XY}$

Begin by drawing $\overleftrightarrow{WC}$, intersecting $\overline{BX}$ at point Z.

a. Apply the Side-Splitter Theorem to $\triangle ACW$: $\frac{\blacksquare}{\blacksquare} = \frac{WZ}{ZC}$.

b. Apply the Side-Splitter Theorem to $\triangle CWY$: $\frac{WZ}{ZC} = \frac{\blacksquare}{\blacksquare}$.

c. Substitute to prove the corollary.

Determine whether the red segments are parallel. Explain each answer. You can use the theorem proved in Exercise 34.

36.

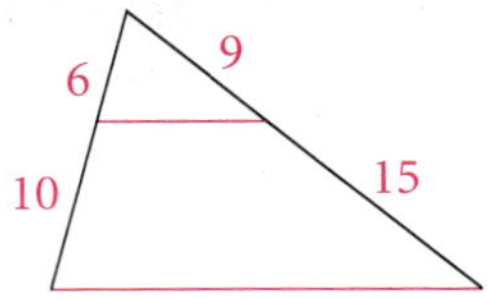

37.

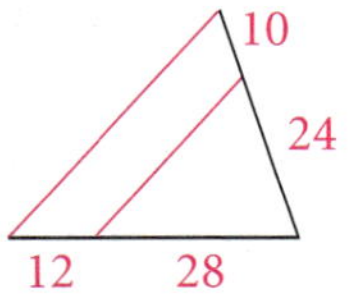

38.

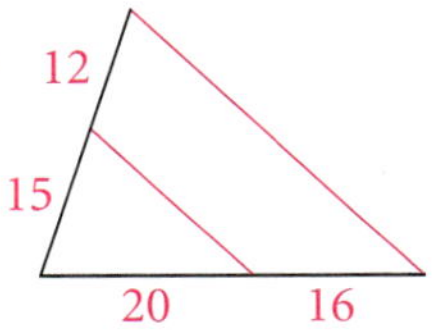

Homework Video Tutor

Visit: PHSchool.com
Web Code: bce-0705

Real-World Connection

You measure an oil spill to find the size of the boom you'll need to contain it.

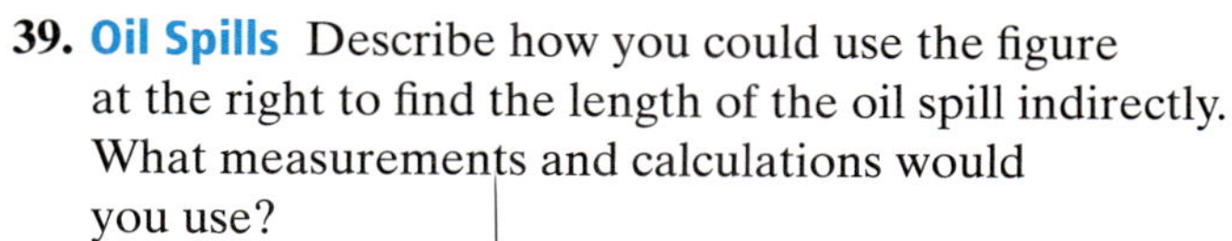

39. Oil Spills Describe how you could use the figure at the right to find the length of the oil spill indirectly. What measurements and calculations would you use?

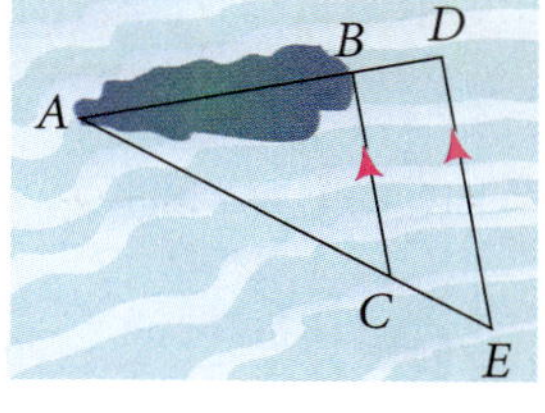

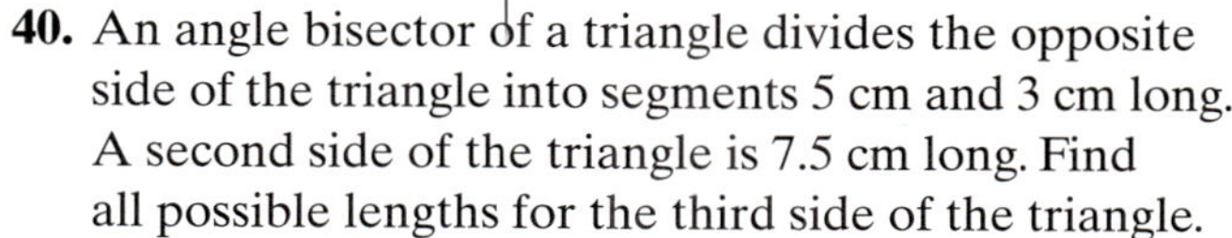

40. An angle bisector of a triangle divides the opposite side of the triangle into segments 5 cm and 3 cm long. A second side of the triangle is 7.5 cm long. Find all possible lengths for the third side of the triangle.

Geometry in 3 Dimensions **In the figure at the right, $\overleftrightarrow{FG} \parallel \overleftrightarrow{AB}$, $\overleftrightarrow{GH} \parallel \overleftrightarrow{BC}$, $AF = 2$, $FE = 4$, and $BG = 3$.**

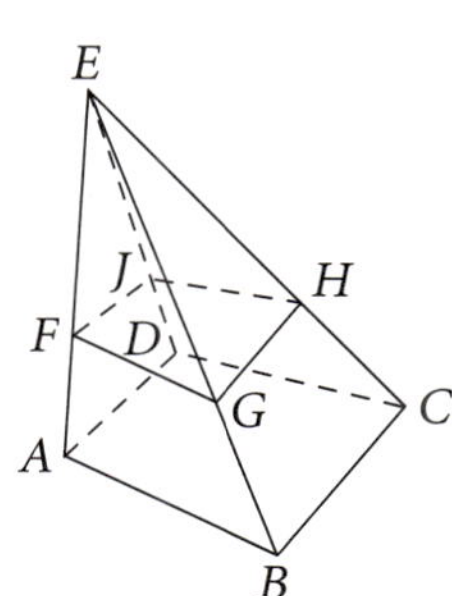

41. Find GE.

42. If $EH = 5$, find HC.

43. If $FG = 3$, find the perimeter of $\triangle ABE$.

Proof **44.** One side of a triangle is k times as long as a second side. The bisector of their angle cuts the third side into two segments. Prove that one of those segments is k times as long as the other.

Challenge Proof **45.** Use the definition in part (a) to prove the statements in parts (b) and (c).

a. Write a definition for a midsegment of a parallelogram.
b. A parallelogram midsegment is parallel to two sides of the parallelogram.
c. A parallelogram midsegment bisects the diagonals of a parallelogram.

46. State the converse of the Triangle-Angle-Bisector Theorem. Give a convincing argument that the converse is true or a counterexample to prove that it is false.

47. In $\triangle ABC$, the bisectors of $\angle A$, $\angle B$, and $\angle C$ cut the opposite sides into lengths a_1 and a_2, b_1 and b_2, and c_1 and c_2, respectively, labeled in this order counterclockwise around $\triangle ABC$. Find the perimeter of $\triangle ABC$ for each of the following.

a. $a_1 = \frac{5}{3}, a_2 = \frac{10}{3}, b_1 = \frac{15}{4}$ **b.** $a_1 = \frac{10}{3}, b_1 = \frac{10}{9}, c_1 = \frac{8}{7}$

Multiple Choice Practice

For California Standards and CAHSEE Tutorials, visit PHSchool.com. Web Code: bcq-9045

GEOM 5.0

48. Complete the statement $\triangle ABC \sim$ ____ , and identify the reason why the triangles are similar.

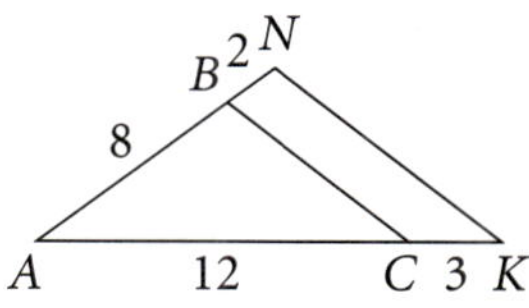

(A) $\triangle AKN$; SSS~   (C) $\triangle ANK$; SAS~
(B) $\triangle AKN$; SAS~   (D) $\triangle ANK$; AA~

GEOM 7.0

49. Find n given the figure at the right.

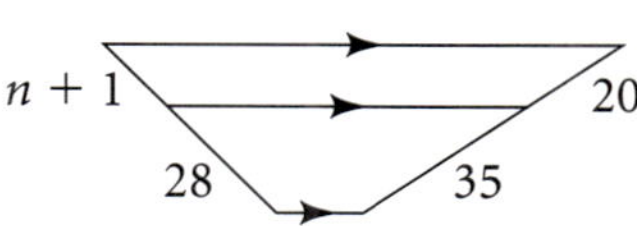

(A) 24   (B) 19   (C) 16   (D) 15

CAHSEE 6SDAP 3.3

50. In a certain town, $\frac{3}{5}$ of the population is female. An advertiser randomly selects 1500 people from that town. About how many of them are male?

(A) 900   (B) 600   (C) 500   (D) 300

Mixed Review

Lesson 7-4 **Refer to the figure to complete each proportion.**

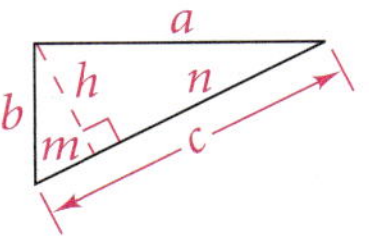

51. $\frac{n}{h} = \frac{h}{\square}$ **52.** $\frac{\square}{b} = \frac{b}{c}$

53. $\frac{n}{a} = \frac{a}{\square}$ **54.** $\frac{m}{h} = \frac{\square}{n}$

Lesson 6-4 **Algebra** ***RSTV* is a rectangle. Find the lengths of the diagonals $\overline{RT}$ and $\overline{SV}$.**

55. $RT = 5x + 8, SV = x + 32$ **56.** $RT = 42 - x, SV = 9x - 8$

57. $RT = 8x - 4, SV = 6x + 9$ **58.** $RT = 3x + 5, SV = 5x + 4$

Lesson 5-3 **Find the center of the circle that you can circumscribe about each $\triangle ABC$.**

59. $A(0, 0)$, $B(6, 0)$, $C(0, -6)$

60. $A(2, 5)$, $B(-2, 5)$, $C(-2, -1)$

61. $A(-2, 0)$, $B(5, 5)$, $C(-2, 5)$

For $\triangle ABC$, name the point of intersection associated with each set of segments.

62. the medians **63.** the angle bisectors **64.** the altitudes

Checkpoint Quiz 2 — Lessons 7-3 through 7-5

Determine whether the triangles are similar. If so, write the similarity statement. Also, write the postulate or theorem that proves they are similar.

1.

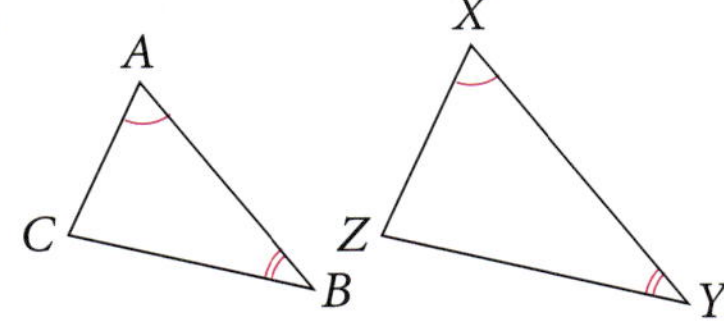

2.

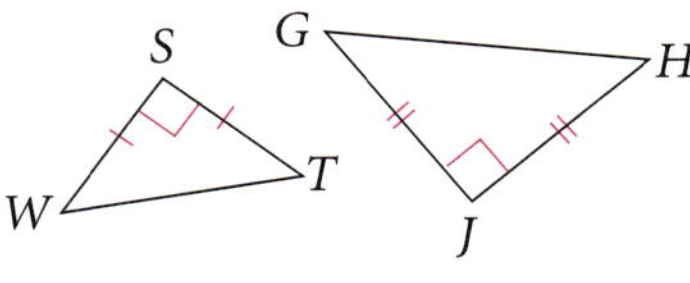

Algebra **The polygons are similar. Find the value of each variable.**

3.

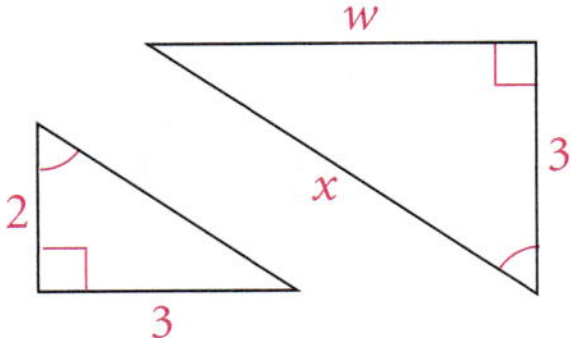

4.

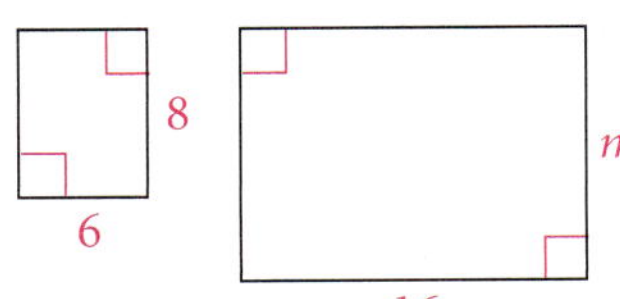

Algebra **Find the value of each variable.**

5.

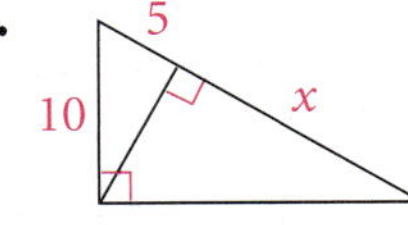

6.

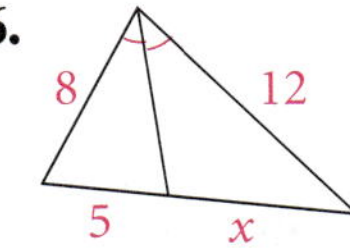

7.

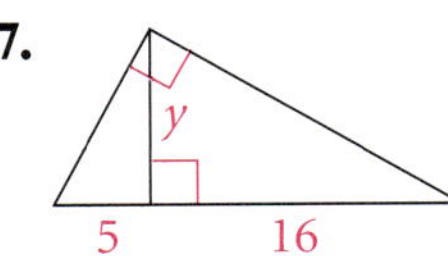

8.

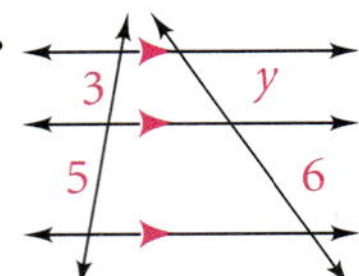

9.

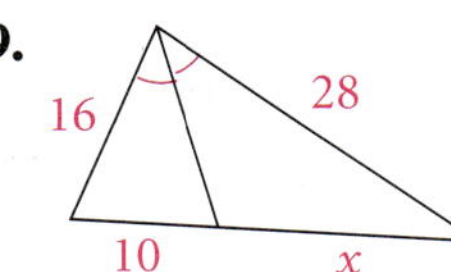

10.

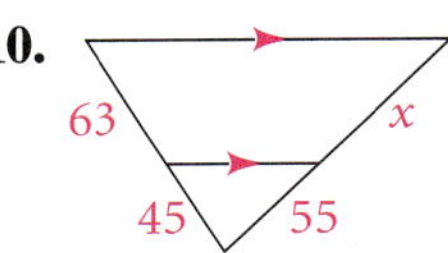

GPS Guided Problem Solving

FOR USE WITH PAGE 402, EXERCISE 29

Understanding Word Problems **Read the problem below. Then let Kyla's thinking guide you through the solution. Check your understanding with the exercises at the bottom of the page.**

Surveying The perimeter of the triangular lot at the right is 50 m. The surveyor's tape bisects an angle. Find the lengths x and y.

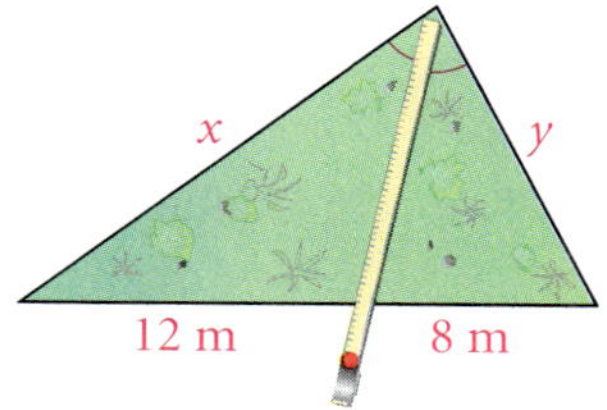

What Kyla Thinks	What Kyla Writes
The perimeter of the triangular lot is 50. The perimeter is the distance around the lot. I can write and simplify an equation.	$x + y + 8 + 12 = 50$ $x + y + 20 = 50$ $x + y = 30$ ✔
The surveyor's tape bisects the top angle. By the Triangle-Angle-Bisector Theorem, the tape divides the bottom side into two lengths that are proportional to the other two sides. I can write and simplify a proportion.	$\frac{12}{8} = \frac{x}{y}$ $12y = 8x$ $y = \frac{2}{3}x$ ✔
I have a system of two equations (✔ and ✔) in two unknowns. I can substitute $\frac{2}{3}x$ for y in the first equation and solve for x.	$x + \frac{2}{3}x = 30$ $\frac{5}{3}x = 30$ $x = 18$
Now I can solve for y.	Since $x + y = 30$, y must be 12.

EXERCISES

Refer to the exercise above. If the surveyor's tape forms segments of the following lengths, what are x and y?

1. 10 m and 10 m

2. 8 m and 12 m

3. 15 m and 5 m

4. 2 m and 18 m

5. Again, refer to the problem Kyla solved. If the cuts measure a and b, and $a + b = 20$, show that $x = \frac{3}{2}a$ and $y = \frac{3}{2}b$.

Test-Taking Strategies: Testing Multiple Choices

Your choices in a multiple-choice question include a correct answer. A strategy is to test each choice in the original problem. You may find mental math to be particularly useful for this.

EXAMPLE

Algebra What is the value of x in the diagram at the right?

Ⓐ 3 Ⓑ 6 Ⓒ 10 Ⓓ 12

The triangles are similar. The proportion $\frac{3}{5} = \frac{x}{x+4}$ is apparent in the diagram. You can test answer choices using mental math.

Let $x = 3$.
$\frac{x}{x+4} = \frac{3}{3+4} = \frac{3}{7} \neq \frac{3}{5}$
A is not the answer.

Let $x = 6$.
$\frac{x}{x+4} = \frac{6}{6+4} = \frac{6}{10} = \frac{3}{5}$
Yes! B is likely the answer.

To help make sure, let $x = 10$.
$\frac{x}{x+4} = \frac{10}{10+4} = \frac{10}{14} = \frac{5}{7} \neq \frac{3}{5}$
C is not the answer.

Finally, let $x = 12$.
$\frac{x}{x+4} = \frac{12}{12+4} = \frac{12}{16} = \frac{3}{4} \neq \frac{3}{5}$
D is not the answer.

You can mark B as your answer.

EXERCISES

1. What number is a solution to $\frac{x}{12} = \frac{x-3}{8}$?

Ⓐ 0 Ⓑ 3 Ⓒ 6 Ⓓ 9

2. What is the value of x?

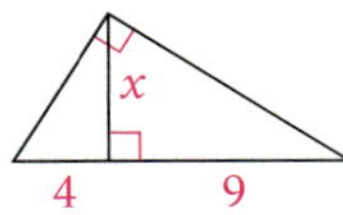

Ⓐ 4 Ⓑ 6 Ⓒ 8 Ⓓ 10

3. Segment AB is shown at the right. Which pair of points are endpoints of a segment that is parallel to and the same length as segment AB?

Ⓐ (4, −1), (12, 3) Ⓒ (4, −2), (12, 2)
Ⓑ (2, −2), (9, 3) Ⓓ (2, 2), (10, −2)

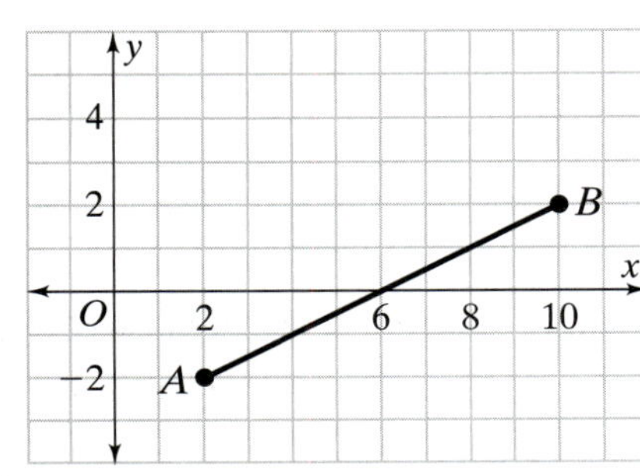

4. What are the roots of the quadratic equation $2x^2 - 6x = 108$?

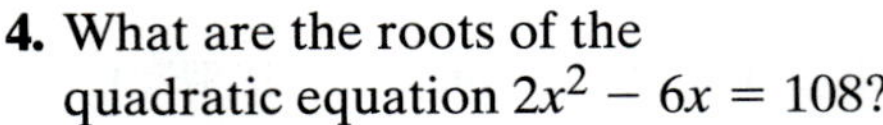

Ⓐ 2, −108 Ⓒ −6, 12
Ⓑ −6, 9 Ⓓ −12, 9

Chapter 7 Review

Vocabulary Review

English and Spanish Audio Online

Cross-Product Property (p. 367)	**golden rectangle** (p. 375)	**scale** (p. 368)
extended proportion (p. 367)	**indirect measurement** (p. 384)	**scale drawing** (p. 368)
geometric mean (p. 392)	**proportion** (p. 367)	**similar** (p. 373)
golden ratio (p. 375)	**ratio** (p. 366)	**similarity ratio** (p. 373)

For: Vocabulary quiz
Web Code: bcj-0751

Choose the correct term to complete each sentence.

1. Two polygons are _?_ if corresponding angles are congruent and corresponding sides are proportional.
2. The _?_ states that the product of the extremes is equal to the product of the means.
3. A _?_ is a rectangle that can be divided into a square and a rectangle that is similar to the original rectangle.
4. The ratio of the lengths of corresponding sides of two similar figures is the _?_.
5. A _?_ is a statement that two ratios are equal.
6. Finding distances using similar triangles is called _?_.
7. The length and width of a golden rectangle are in the _?_.

Skills and Concepts

Lesson 7-1

- To write ratios and solve proportions (p. 366)

A ratio is a comparison of two quantities by division. You can write the ratio of a to b or $a : b$ as the quotient $\frac{a}{b}$ when $b \neq 0$.

A **proportion** is a statement that two ratios are equal. According to the **Properties of Proportions,** $\frac{a}{b} = \frac{c}{d}$ is equivalent to

(1) $ad = bc$ (2) $\frac{b}{a} = \frac{d}{c}$ (3) $\frac{a}{c} = \frac{b}{d}$ (4) $\frac{a+b}{b} = \frac{c+d}{d}$

Property 1, above, illustrates the **Cross-Product Property,** which states that the product of the extremes is equal to the product of the means.

When three or more ratios are equal, you can write an **extended proportion.**

In a **scale drawing,** the **scale** compares each length in the drawing to the actual length being represented.

Dollhouses **Dollhouse furnishings come in different sizes depending on the size of the dollhouse. For each exercise, write a ratio of the size of the dollhouse item to the size of the larger item.**

8. dollhouse sofa: $1\frac{1}{2}$ in. long; real sofa: 6 ft long
9. dollhouse piano: $1\frac{3}{4}$ in. high real piano: 3 ft 6 in. high

If $\frac{p}{q} = \frac{2}{5}$, tell whether each equation must be true. Explain.

10. $2q = 5p$
11. $\frac{5}{2} = \frac{q}{p}$
12. $5q = 2p$
13. $\frac{p}{2} = \frac{q}{5}$

Lessons 7-2 and 7-3

- To identify similar polygons (p. 373)
- To apply similar polygons (p. 374)
- To use AA, SAS, and SSS similarity statements (p. 382)
- To apply AA, SAS, and SSS similarity statements (p. 384)

 **GEOM 2.0, 4.0, 5.0**

Similar polygons have congruent corresponding angles and proportional corresponding sides. The ratio of the lengths of corresponding sides is the **similarity ratio.**

A **golden rectangle** is a rectangle that can be divided into a square and a rectangle that is similar to the original rectangle. In any golden rectangle, the length and the width are in the **golden ratio,** which is $\frac{1 + \sqrt{5}}{2} : 1$, or about $1.618 : 1$.

If two angles of one triangle are congruent to two angles of another triangle, then the triangles are similar by the **Angle-Angle Similarity Postulate** (AA ~).

If an angle of one triangle is congruent to an angle of a second triangle, and the sides including the two angles are proportional, then the triangles are similar by the **Side-Angle-Side Similarity Theorem** (SAS ~).

If the corresponding sides of two triangles are proportional, then the triangles are similar by the **Side-Side-Side Similarity Theorem** (SSS ~).

Methods of **indirect measurement** use similar triangles and measurements to find distances that are difficult to measure directly.

14. If $\triangle MNP \sim \triangle RST$, which angles are congruent? Write an extended proportion to indicate the proportional corresponding sides of the triangles.

15. Art An artist is creating a stained glass window and wants it to be a golden rectangle. To the nearest inch, what should be the length if the width is 24 in.?

The triangles are similar. Find the similarity ratio of the first to the second.

16.

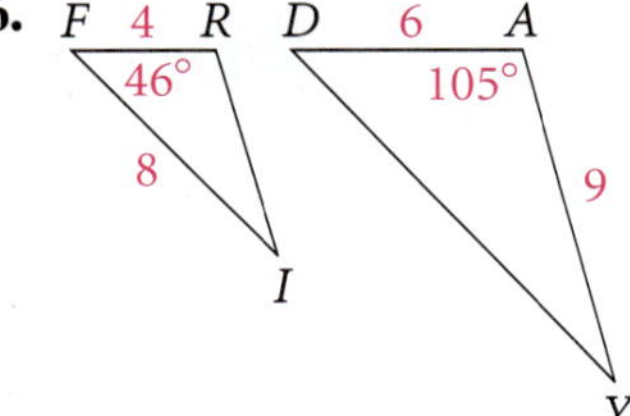

17.

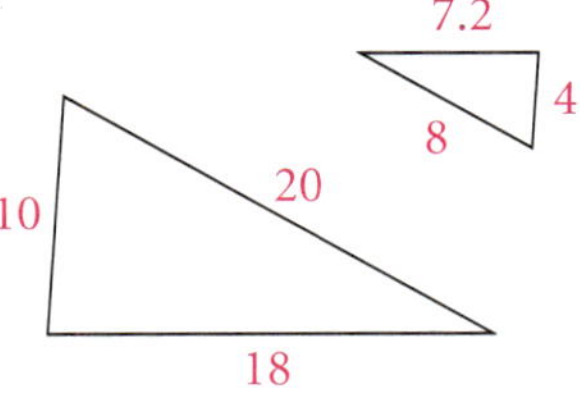

Algebra **The polygons are similar. Find the value of each variable.**

18.

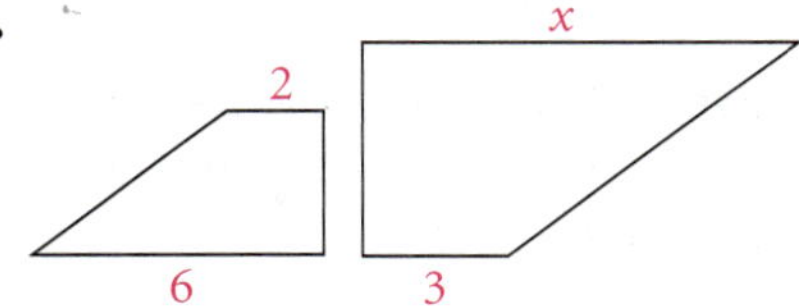

19.

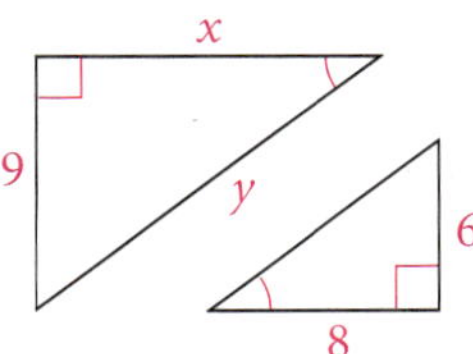

Are the triangles similar? If so, write the similarity statement and name the postulate or theorem you used. If not, explain.

20.

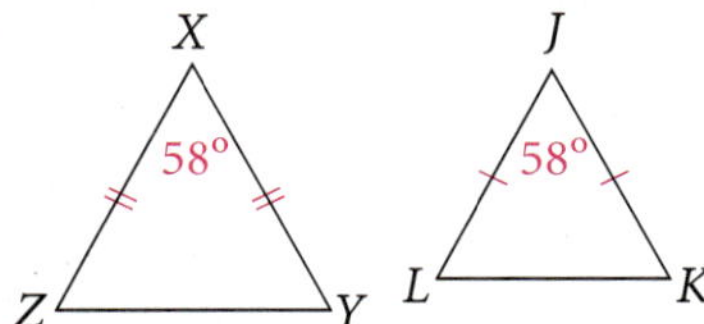

21.

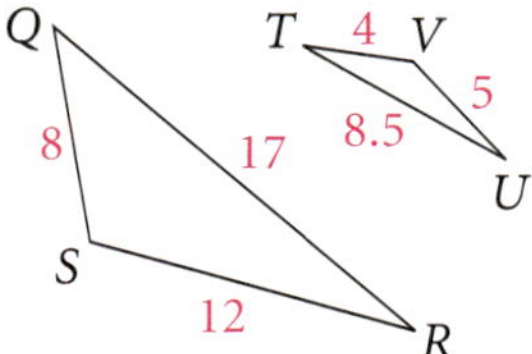

22. Two right triangles have an acute angle with the same measure. Name the theorem or postulate that is the most direct way to prove the triangles similar.

23. **Indirect Measurement** A crate is 1.5 ft high and casts a 2-ft shadow. At the same time, an apple tree casts an 18-ft shadow. How tall is the tree?

Lesson 7-4

- To find and use relationships in similar right triangles (p. 391)

 **GEOM 4.0, 5.0**

The **geometric mean** of two positive numbers a and b is the positive number x such that $\frac{a}{x} = \frac{x}{b}$.

When the altitude is drawn to the hypotenuse of a right triangle:

- the two triangles formed are similar to the original triangle and to each other;
- the length of the altitude is the geometric mean of the lengths of the segments of the hypotenuse; and
- the length of each leg is the geometric mean of the length of the adjacent hypotenuse segment and the length of the hypotenuse.

x^2 **Algebra Find the geometric mean of each pair of numbers.**

24. 4 and 25
25. 3 and 300
26. 5 and 12
27. $\frac{1}{9}$ and 28
28. 0.36 and 4
29. $2\sqrt{3}$ and $\sqrt{12}$

x^2 **Algebra Find the values of the variables. When an answer is not a whole number, leave it in simplest radical form.**

30.
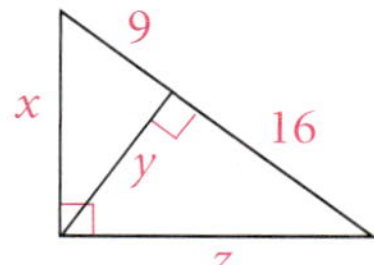

31.
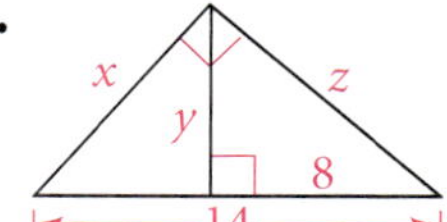

32.
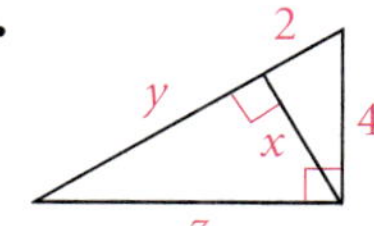

Lesson 7-5

- To use the Side-Splitter Theorem (p. 398)
- To use the Triangle-Angle-Bisector Theorem (p. 400)

 **GEOM 4.0, 5.0, 7.0**

The **Side-Splitter Theorem** states that if a line is parallel to one side of a triangle and intersects the other two sides, then it divides those sides proportionally. If three parallel lines intersect two transversals, then the segments intercepted on the transversals are proportional.

The **Triangle-Angle-Bisector Theorem** states that if a ray bisects an angle of a triangle, then it divides the opposite side into two segments that are proportional to the other two sides of the triangle.

x^2 **Algebra Find the value of x.**

33.
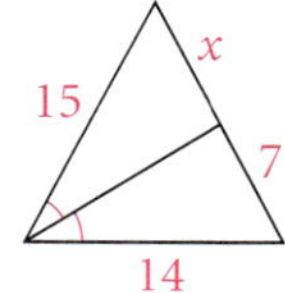

34.
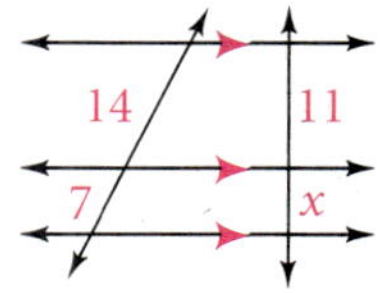

35.
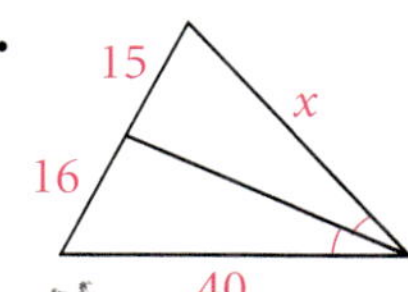

36.
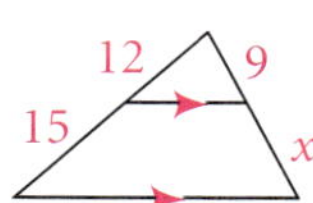

37.
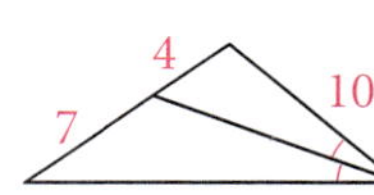

38.
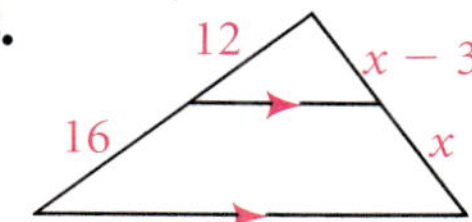

Chapter 7 Test

For: Chapter test
Web Code: bca-0752

Algebra **Solve each proportion.**

1. $\frac{4}{5} = \frac{x}{20}$ **2.** $\frac{6}{x} = \frac{10}{7}$ **3.** $\frac{x}{3} = \frac{8}{12}$

Algebra **The figures in each pair are similar. Find the value of each variable.**

4.

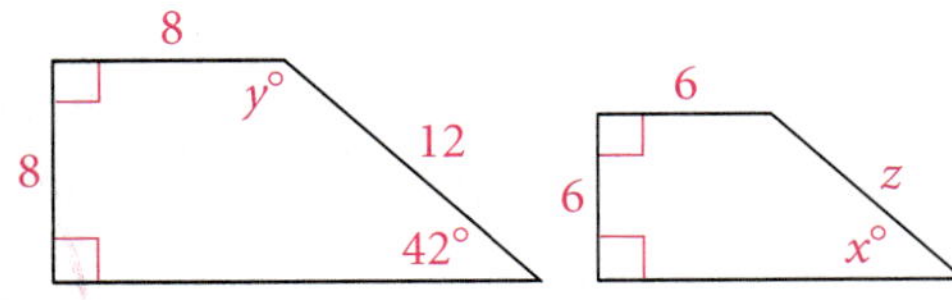

5.

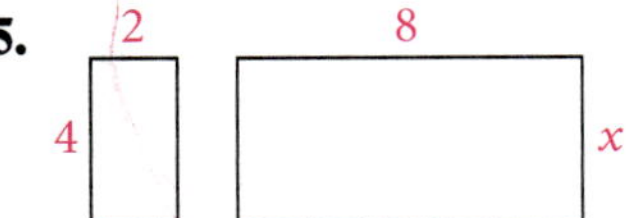

6.

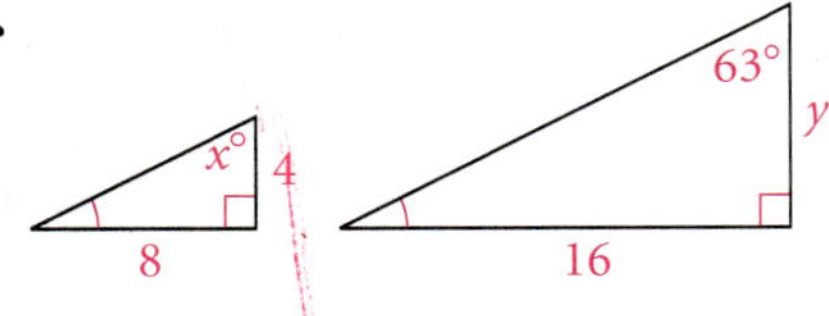

Are the triangles similar? If *yes*, write the similarity statement and name the postulate or theorem you can use to prove they are similar. If *no*, explain.

7.

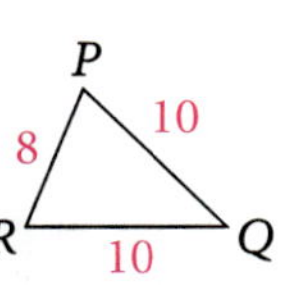

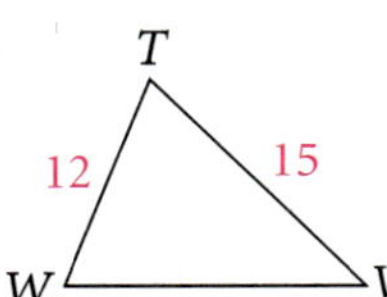

8.

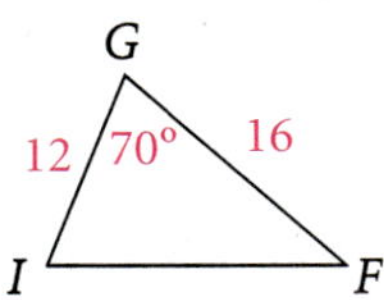

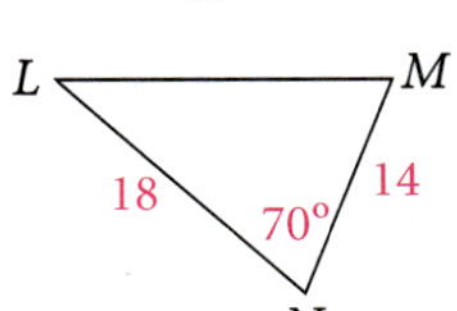

9.

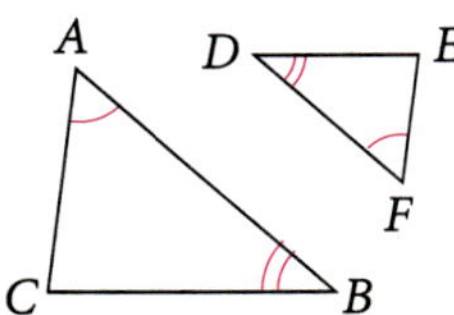

Use the information shown in the diagram and write a proportion for each triangle.

10.

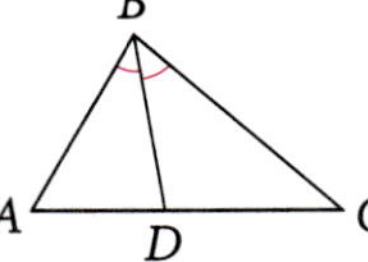

11.

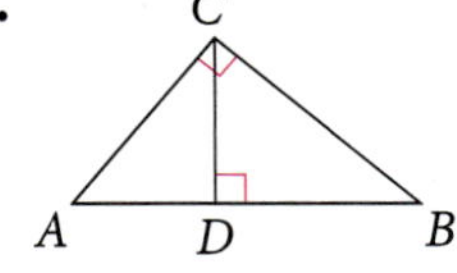

12. Indirect Measurement A meter stick is held perpendicular to the ground. It casts a shadow 1.5 m long. At the same time, a telephone pole casts a shadow that is 9 m long. How tall is the telephone pole?

13. Photography A photographic negative is 3 cm by 2 cm. If a similar print from the negative is 9 cm long on its shorter side, what is the length of its longer side?

Find the geometric mean of each pair of numbers. If the answer is not a whole number, write it in simplest radical form.

14. 10, 15 **15.** 4, 9 **16.** 6, 12

17. Open-Ended Draw an isosceles triangle, $\triangle ABC$. Then draw $\triangle DEF$ so that $\triangle ABC \sim \triangle DEF$. State the similarity ratio of $\triangle ABC$ to $\triangle DEF$.

Algebra **Find the value of *x*.**

18.

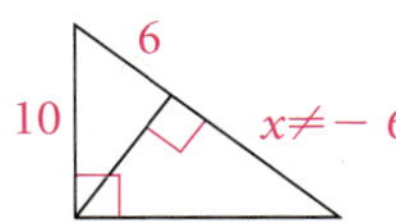

19.

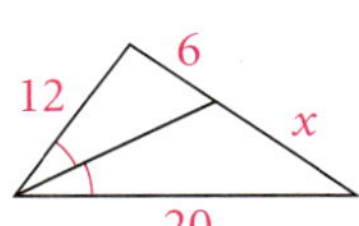

20.

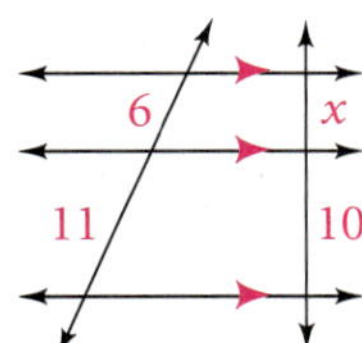

21.

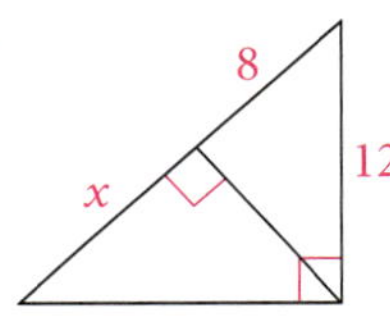

22. Writing Describe an object whose height or length would be difficult to measure directly. Then describe a method for measuring the object that involves using similar triangles.

Explain why the triangles are similar. Then find the value of *x*.

23.

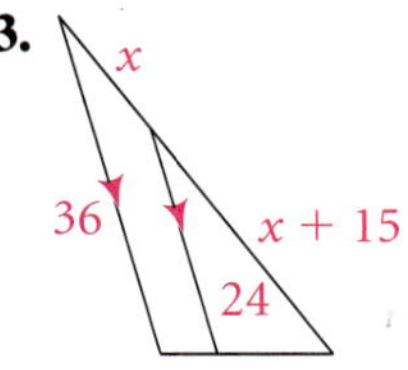

24.

12.5
10
7
x

Standards Mastery

CAHSEE Practice

Some questions ask you to find an expression to represent a situation. Read the question at the right. Then follow the tips to answer it.

In a certain state, the sales tax is 7.5% of the selling price. If c represents the selling price of a scooter, which expression represents how much will have to be paid for the scooter?

(A) $0.075c$
(B) $1.75c$
(C) $c + 0.075$
(D) $c + 0.075c$

Tip 1
Recall the decimal form of a percent:
7.5% = 0.075.

Tip 2
Total cost is the selling price plus the tax.

Think It Through

Total cost
= selling price + sales tax
= c + $0.075c$.
The correct answer is D.

Vocabulary Review

As you solve problems, you must understand the meanings of mathematical terms. Choose the correct term to complete each sentence.

I. A (frequency diagram, Venn diagram) shows relationships between sets of data.

II. All rational numbers and all irrational numbers together are the (real numbers, whole numbers).

III. A (function, relation) is a set of ordered pairs in which two different ordered pairs may have the same first component.

IV. An (assignment, inequality) is a statement that compares two different values.

V. In a (positive correlation, negative correlation) the values of one variable increase as the values of another variable also increase.

Multiple Choice

Read each question. Then write the letter of the correct answer on your paper.

1. What is the x-intercept of the equation graphed at the right?

(A) -5
(B) -2
(C) 5
(D) 2

2. For the linear function shown in the table at the right, what is the value of x when y is zero?

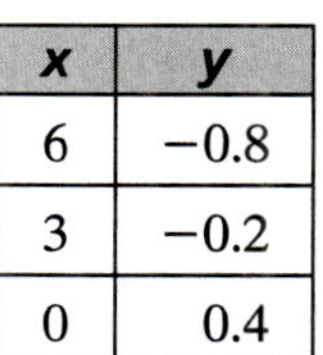

x	y
6	−0.8
3	−0.2
0	0.4

(A) 0.4
(B) 1
(C) 1.5
(D) 2

3. Find the y-intercept of the graph of the equation $9x - 4y = 1$.

(A) $-\frac{9}{4}$
(B) $-\frac{1}{4}$
(C) $\frac{1}{9}$
(D) $\frac{9}{4}$

4. Simplify $3a^7b^4 - 5a^7b^4$.

(A) $-8a^2b^1$
(B) -2
(C) $-2a^7b^4$
(D) $2a^0b^0$

5. Alexandra drew the figure below by drawing a polygon and then connecting the midpoints of each of the sides.

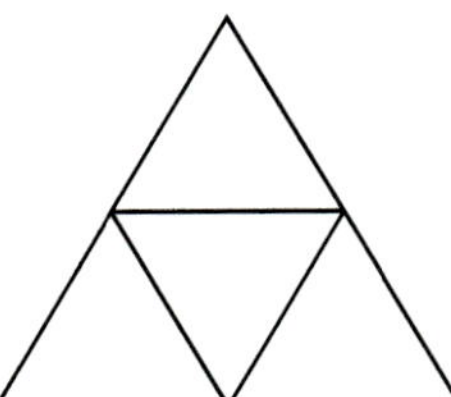

Which of the following figures could she make by using the same method?

Ⓐ 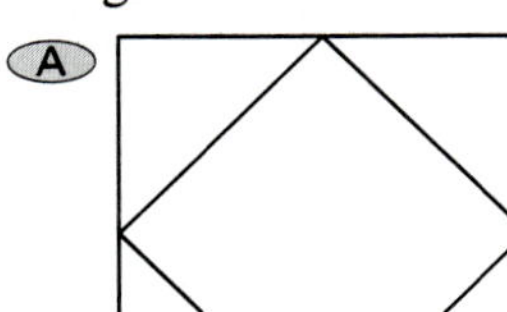

Ⓒ 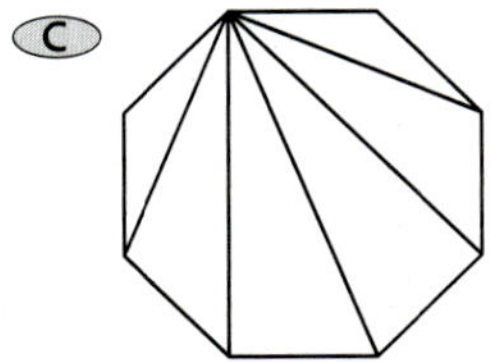

Ⓑ 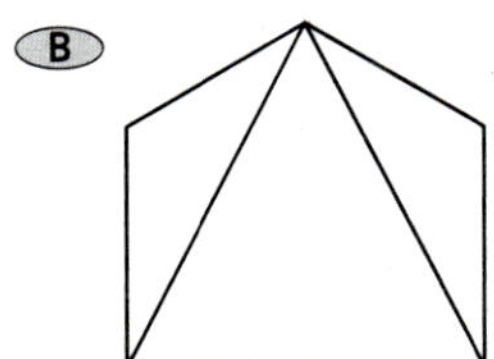

Ⓓ 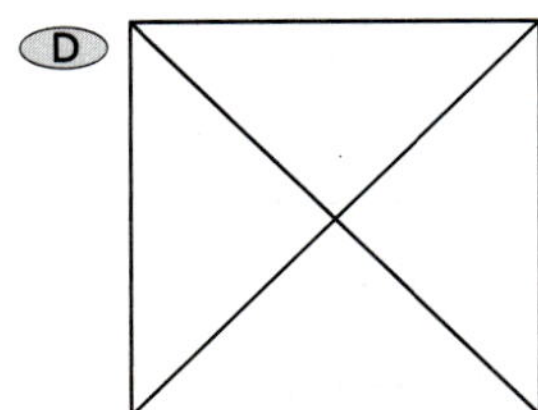

6. Kyle has a red and a green number cube. The sides of each cube are numbered from 1 to 6. If Kyle tosses both cubes, what is the probability that there will be an even number on the red cube and a multiple of 3 on the green cube?

Ⓐ $\frac{1}{8}$ Ⓒ $\frac{1}{5}$
Ⓑ $\frac{1}{6}$ Ⓓ $\frac{1}{4}$

7. For a school play, 6 boys and 6 girls volunteer to be ushers. If two ushers are chosen at random by drawing names from a bowl, what is the probability that both ushers will be girls?

Ⓐ $\frac{5}{24}$ Ⓒ $\frac{1}{4}$
Ⓑ $\frac{5}{22}$ Ⓓ $\frac{1}{2}$

8. Bruno has a bag of 25 marbles: 5 red, 5 white, 5 blue, 5 black, and 5 yellow. Bruno randomly pulls out a marble. Then he pulls out a second one without replacing the first. What is the probability that both marbles are white?

Ⓐ $\frac{4}{125}$ Ⓒ $\frac{1}{25}$
Ⓑ $\frac{1}{30}$ Ⓓ $\frac{1}{20}$

9. Two race cars travel at constant speeds in the same direction around a track. The faster car travels 187 miles per hour and completes 50 laps each time the slower car completes 48.7 laps. Which range is a reasonable estimate of the slower car's speed?

Ⓐ Between 180 mph and 185 mph
Ⓑ Between 175 mph and 180 mph
Ⓒ Between 145 mph and 150 mph
Ⓓ Between 135 mph and 140 mph

10. Nioma is making a patchwork quilt using three different types of fabric. Each type of fabric is cut into rectangles. The first fabric is cut into pieces measuring 4 inches by 6 inches. The second fabric's rectangular dimensions are 115% of those of the first type, and the third fabric's dimensions are 115% of those of the second. Which of the following dimensions are the best estimate for the third type of fabric?

Ⓐ 4.60 in. by 6.90 in.
Ⓑ 5.29 in. by 7.94 in.
Ⓒ 5.67 in. by 6.90 in.
Ⓓ 5.75 in. by 7.25 in.

11. Joanna solved the inequality $3(x - 2) < 5$ using the following steps.

Given: $3(x - 2) < 5$
Step 1: $6x - 6 < 5$
Step 2: $6x < 11$
Step 3: $x < \frac{11}{6}$

What did Joanna do to get from Step 1 to Step 2?

Ⓐ subtracted 6 from both sides
Ⓑ divided both sides by 6
Ⓒ multiplied both sides by 6
Ⓓ added 6 to both sides

12. Solve for x. $4(x + 2) - 3x > 7$

Ⓐ $x > 9$ Ⓒ $x > -1$
Ⓑ $x > 5$ Ⓓ $x > -15$

13. A front view of a barn is shown. Using the scale of 1 inch (in.) = 7 feet (ft), what is the height, in feet, of the barn's doorway?

Ⓐ 11 ft
Ⓑ 14 ft
Ⓒ 21 ft
Ⓓ 28 ft

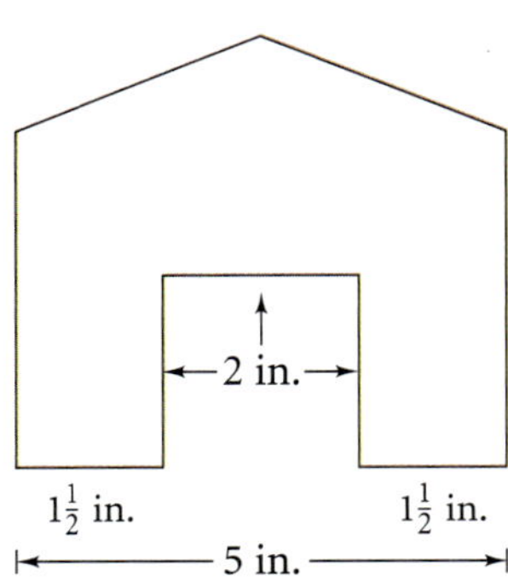

14. A scale drawing of a swimming pool and deck is shown below. Use the scale 1 inch (in.) = 2 meters (m). What is the width of the pool in meters?

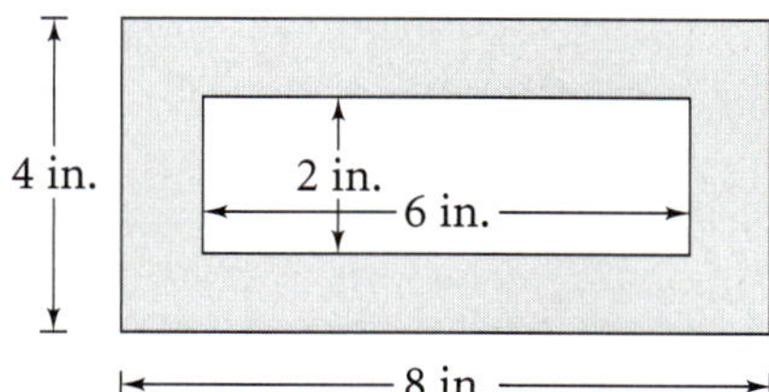

A 4 m
B 8 m
C 12 m
D 16 m

15. Which expression is equivalent to $\frac{3}{5}(5x - 15y) + (8y - 6x)$?

A $11x - 15y$
B $-3x - 21y$
C $11x - 9y$
D $-3x - y$

16. A tire completes 300 revolutions per minute at a certain speed. After 8 minutes how many revolutions has the tire completed?
Which of the following problems can be solved using the same method that was used in the problem above?

A If a farmer picks three bushels of pears a day, how many days will it take to pick 35 bushels?
B An apple weighs three ounces. How many apples will you need to fill a five-pound bag?
C If an electric car averages 40 miles per hour, what distance can it cover in 30 minutes?
D A test contains 30 problems. If the number of problems is evenly distributed over 5 pages, how many problems occur on each page of the test?

17. Mindi tossed a fair coin 16 times. It came up heads 4 times and tails all the other times. To the nearest whole percent, what is the probability that the coin will come up tails the next time she tosses it?

A 100%
B 75%
C 50%
D 25%

18. Which expression best represents the simplification of $(6a^3b^4)(-2a^{-3}b^2)$?

A $\frac{12}{a^6b^6}$
B $-12a^6b^6$
C $-12b^6$
D $\frac{12}{b^6}$

19. The area of a rectangle is $256r^8s^{12}$ square units. If the width of the rectangle is $8r^3s^4$ units, what is the rectangle's length?

A $2048r^{11}s^{16}$ units
B $64r^{11}s^{16}$ units
C $32r^5s^8$ units
D $24r^5s^8$ units

CAHSEE Reference Guide

California Content Standard	Item Number(s)
Alg1 6.0	1–3
Alg1 10.0	4, 15, 18, 19
7MR 3.3	5, 16
6SDAP 3.5	6, 7, 8, 17
7MR 2.1	9, 10
Alg1 5.0	11, 12
7MG 1.2	13, 14

For additional review and practice, see the *California Standards Review and Practice Workbook* or go online to use the

Visit: PHSchool.com, **Web Code:** bcq-9045

Right Triangles and Trigonometry

What You've Learned

California Content Standards

4.0 Students prove theorems involving similarity.

5.0 Students prove that triangles are similar.

7.0 Students prove and use theorems involving the properties of parallel lines cut by a transversal.

Check Your Readiness

for Help to the Lesson in green.

Solving Proportions (Lesson 7-1)

Algebra Solve for x. Round answers to the nearest thousandth.

1. $0.2734 = \frac{x}{17}$ **2.** $0.5858 = \frac{24}{x}$ **3.** $0.8572 = \frac{5271}{x}$ **4.** $0.5 = \frac{x}{3x + 5}$

Proving Triangles Similar (Lesson 7-3)

Name the postulate or theorem that proves each pair of triangles similar.

5. $\overline{CD} \parallel \overline{AB}$

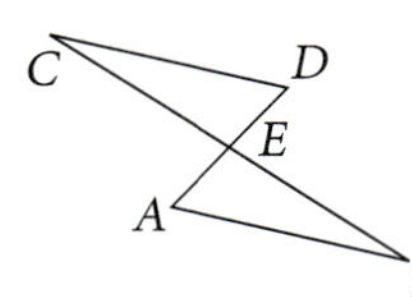

6.

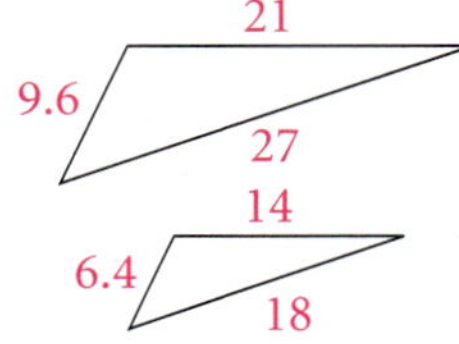

7. $\overline{JK} \perp \overline{ML}$

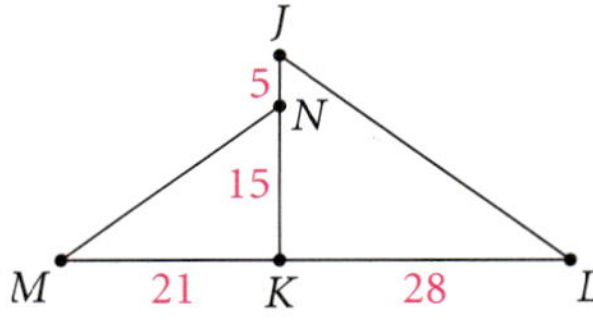

Similarity in Right Triangles (Lesson 7-4)

Algebra Find the unknown quantity in $\triangle ABC$ with right $\angle ACB$ and altitude $\overline{CD}$.

8.

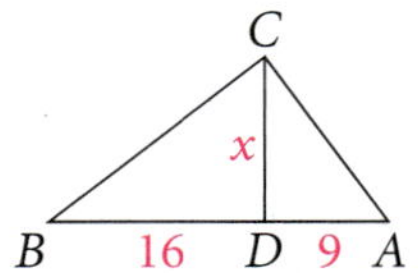

9.

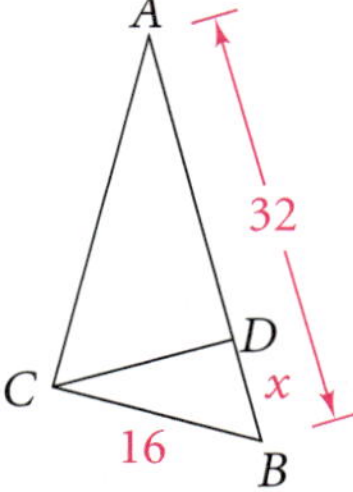

10.

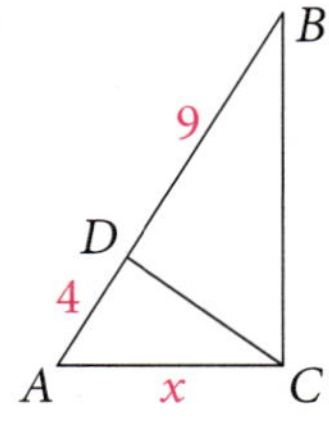

11.

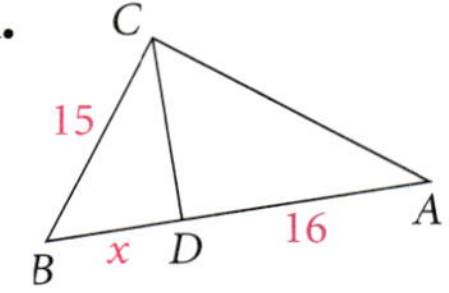

▲ Meteorologists use right triangle trigonometry to determine the height of storm clouds.

What You'll Learn Next

California Content Standards

14.0 Students prove the Pythagorean theorem.

15.0 Students use the Pythagorean theorem to determine distance and find missing lengths of sides of right triangles.

18.0 Students know the definitions of the basic trigonometric functions defined by the angles of a right triangle.

19.0 Students use trigonometric functions to solve for an unknown length of a side of a right triangle, given an angle and a length of a side.

CAHSEE 6SDAP 3.3, 7AF 2.2, 7MG 1.1, 7MR 2.1

New Vocabulary

English and Spanish Audio Online

- **angle of depression** (p. 445)
- **angle of elevation** (p. 445)
- **cosine** (p. 439)
- **identity** (p. 440)
- **initial point** (p. 452)
- **magnitude** (p. 452)
- **Pythagorean triple** (p. 417)
- **resultant** (p. 454)
- **sine** (p. 439)
- **tangent** (p. 432)
- **terminal point** (p. 452)
- **vector** (p. 452)

Activity Lab: The Pythagorean Theorem

Hands-On

FOR USE WITH LESSON 8-1

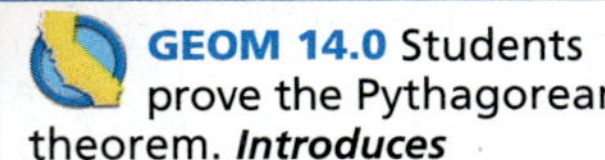

GEOM 14.0 Students prove the Pythagorean theorem. ***Introduces***

The Pythagorean Theorem is introduced in Lesson 8-1. The Activity and exercises below will help you understand why the theorem works.

ACTIVITY

Using graph paper, draw any rectangle. Label the sides a and b. Cut four rectangles with length a and width b from the graph paper. Then cut each rectangle on its diagonal, c, forming eight congruent triangles.

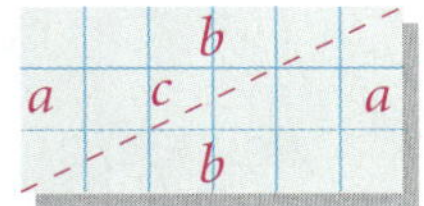

Cut three squares from colored paper, one with sides of length a, one with sides of length b, and one with sides of length c.

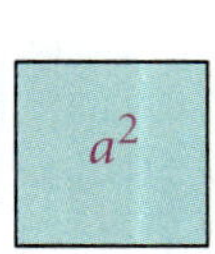

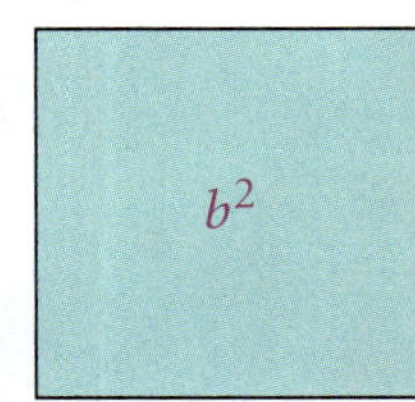

Separate the eleven pieces into groups.

Group 1: four triangles and the two smaller squares

Group 2: four triangles and the largest square

Arrange the pieces of each group to form a square.

1. **a.** How do the areas of the two squares you formed in the last step above compare?
   b. Write an algebraic expression for the area of each of these squares.
   c. What can you conclude about the areas of the three squares you cut from colored paper?
2. Repeat this investigation using a new rectangle with different a and b values. What do you notice?
3. Express your conclusion as an algebraic equation.
4. Use your ruler with any rectangle to find actual measures for a, b, and c. Do these measures confirm that $a^2 + b^2 = c^2$?

EXERCISES

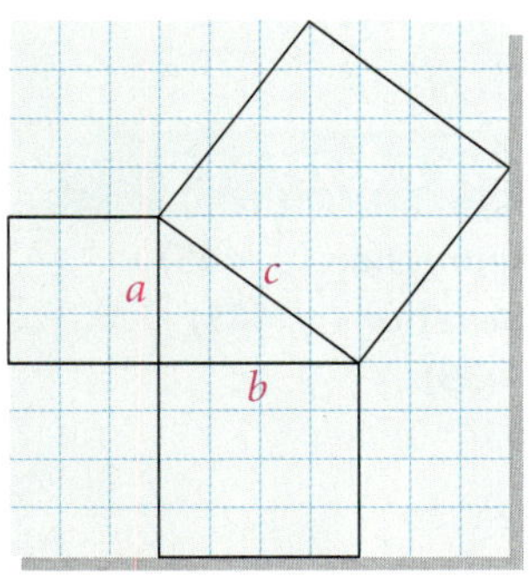

5. Explain how the diagram at the right represents your conclusion in Exercise 3.
6. Does the equation from Exercise 4 work for triangles other than right triangles? Explore and explain.

The Pythagorean Theorem and Its Converse

California Content Standards

GEOM 15.0 Students use the Pythagorean theorem to determine distance and find missing lengths of sides of right triangles. ***Introduce***

What You'll Learn

- To use the Pythagorean Theorem
- To use the Converse of the Pythagorean Theorem

. . . And Why

To find the distance between two docks on a lake, as in Example 3

Check Skills You'll Need

GO for Help Skills Handbook, p. 753

Square the lengths of the sides of each triangle. What do you notice?

1.

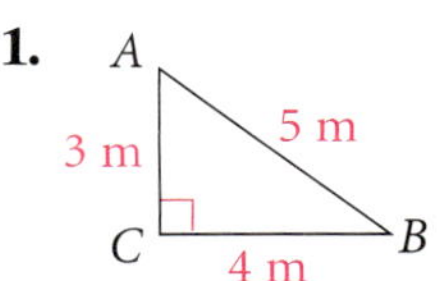

2.

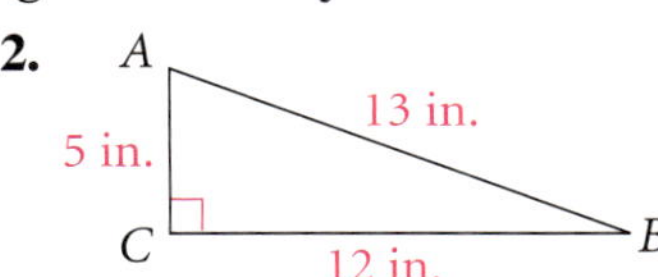

3.

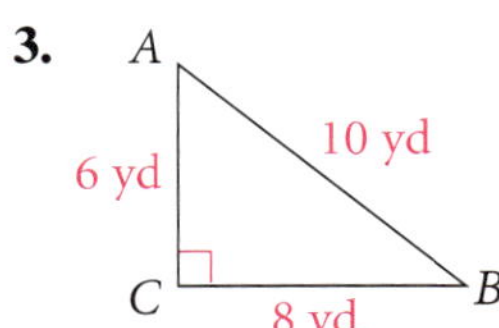

4.

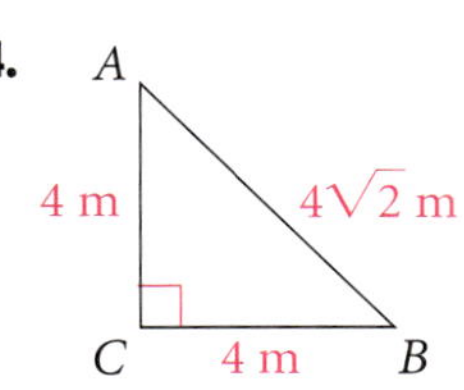

New Vocabulary • Pythagorean triple

1 The Pythagorean Theorem

The well-known right triangle relationship called the Pythagorean Theorem is named for Pythagoras, a Greek mathematician who lived in the sixth century B.C. We now know that the Babylonians, Egyptians, and Chinese were aware of this relationship before its discovery by Pythagoras.

There are many proofs of the Pythagorean Theorem. You will see one proof in Exercise 48 and others later in the book.

Key Concepts

Theorem 8-1 **Pythagorean Theorem**

In a right triangle, the sum of the squares of the lengths of the legs is equal to the square of the length of the hypotenuse.

$$a^2 + b^2 = c^2$$

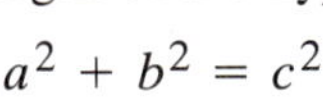

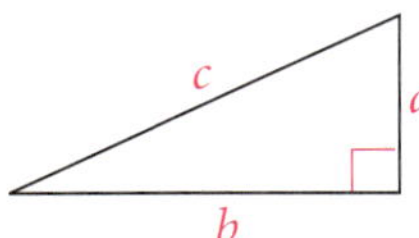

Vocabulary Tip

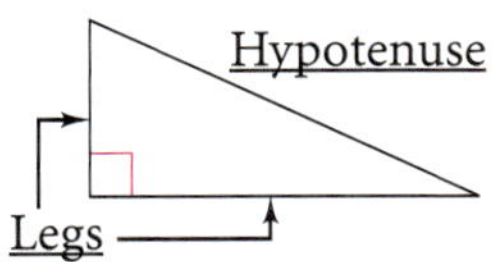

A **Pythagorean triple** is a set of nonzero whole numbers a, b, and c that satisfy the equation $a^2 + b^2 = c^2$. Here are some common Pythagorean triples.

3, 4, 5   5, 12, 13   8, 15, 17   7, 24, 25

If you multiply each number in a Pythagorean triple by the same nonzero whole number, the three numbers that result also form a Pythagorean triple.

1 EXAMPLE Pythagorean Triples

Test-Taking Tip

Memorizing the common Pythagorean triples, like those at the bottom of p. 417, can help you solve problems more quickly.

Find the length of the hypotenuse of $\triangle ABC$. Do the lengths of the sides of $\triangle ABC$ form a Pythagorean triple?

$a^2 + b^2 = c^2$	**Use the Pythagorean Theorem.**
$21^2 + 20^2 = c^2$	**Substitute 21 for *a* and 20 for *b*.**
$441 + 400 = c^2$	**Simplify.**
$841 = c^2$	
$c = 29$	**Take the square root.**

The length of the hypotenuse is 29. The lengths of the sides, 20, 21, and 29, form a Pythagorean triple because they are whole numbers that satisfy $a^2 + b^2 = c^2$.

CA Standards Check **1** A right triangle has a hypotenuse of length 25 and a leg of length 10. Find the length of the other leg. Do the lengths of the sides form a Pythagorean triple?

In some cases, you will write the length of a side in simplest radical form.

2 EXAMPLE Using Simplest Radical Form

Algebra Find the value of x. Leave your answer in simplest radical form (page 390).

$a^2 + b^2 = c^2$	**Pythagorean Theorem**
$8^2 + x^2 = 20^2$	**Substitute.**
$64 + x^2 = 400$	**Simplify.**
$x^2 = 336$	**Subtract 64 from each side.**
$x = \sqrt{336}$	**Take the square root.**
$x = \sqrt{16(21)}$	**Simplify.**
$x = 4\sqrt{21}$	

CA Standards Check **2** The hypotenuse of a right triangle has length 12. One leg has length 6. Find the length of the other leg. Leave your answer in simplest radical form.

3 EXAMPLE Real-World Connection

The Parks Department rents paddle boats at docks near each entrance to the park. To the nearest meter, how far is it to paddle from one dock to the other?

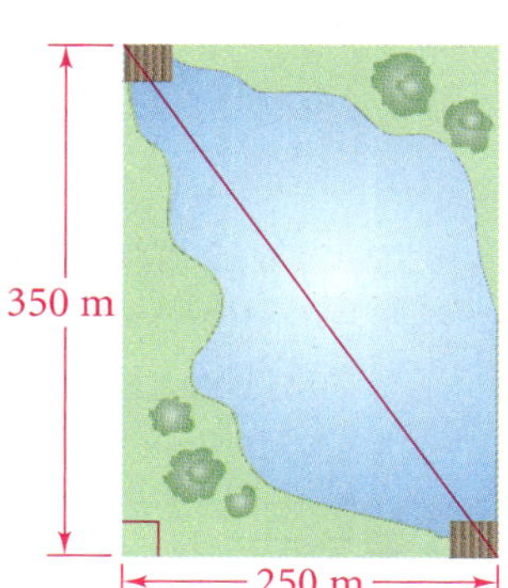

$a^2 + b^2 = c^2$	**Pythagorean Theorem**
$250^2 + 350^2 = c^2$	**Substitute.**
$185{,}000 = c^2$	**Simplify.**
$c = \sqrt{185{,}000}$	**Take the square root.**
$c = 430.11626$	**Use a calculator.**

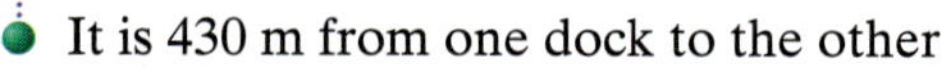

It is 430 m from one dock to the other.

CA Standards Check **3** **Critical Thinking** When you want to know how far you have to paddle a boat, why is an approximate answer more useful than an answer in simplest radical form?

The Converse of the Pythagorean Theorem

You can use the Converse of the Pythagorean Theorem to determine whether a triangle is a right triangle. You will prove Theorem 8-2 in Exercise 58.

Key Concepts

Theorem 8-2 **Converse of the Pythagorean Theorem**

If the square of the length of one side of a triangle is equal to the sum of the squares of the lengths of the other two sides, then the triangle is a right triangle.

Online active math

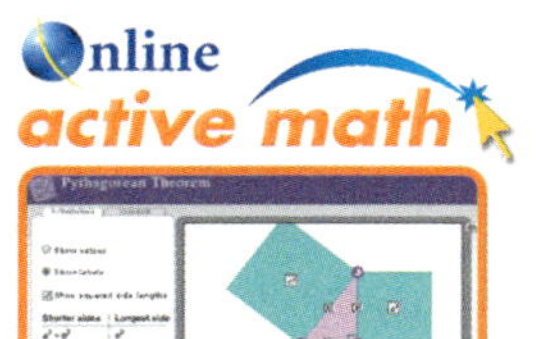

For: Pythagorean Activity
Use: Interactive Textbook, 8-1

4 EXAMPLE Is It a Right Triangle?

Is this triangle a right triangle?

$c^2 \stackrel{?}{=} a^2 + b^2$

$85^2 \stackrel{?}{=} 13^2 + 84^2$ **Substitute the greatest length for c.**

$7225 \stackrel{?}{=} 169 + 7056$ **Simplify.**

$7225 = 7225$ ✓

$c^2 = a^2 + b^2$, so the triangle is a right triangle.

CA Standards Check 4 A triangle has sides of lengths 16, 48, and 50. Is the triangle a right triangle?

You can also use the squares of the lengths of the sides of a triangle to find whether the triangle is acute or obtuse. The following two theorems tell how.

Key Concepts

Theorem 8-3

If the square of the length of the longest side of a triangle is greater than the sum of the squares of the lengths of the other two sides, the triangle is obtuse.

If $c^2 > a^2 + b^2$, the triangle is obtuse.

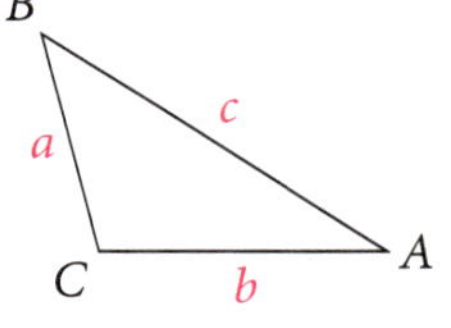

Theorem 8-4

If the square of the length of the longest side of a triangle is less than the sum of the squares of the lengths of the other two sides, the triangle is acute.

If $c^2 < a^2 + b^2$, the triangle is acute.

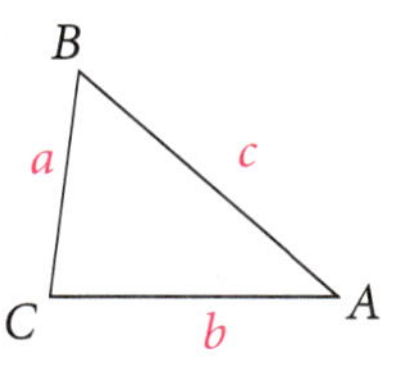

Real-World Connection

The horizontal brace is kept short to ensure an acute angle at the top of the ladder, so the ladder won't collapse.

5 EXAMPLE Classifying Triangles as Acute, Obtuse, or Right

Classify the triangle whose side lengths are 6, 11, and 14 as acute, obtuse, or right.

$14^2 \stackrel{?}{=} 6^2 + 11^2$ **Compare c^2 to $a^2 + b^2$. Substitute the greatest length for c.**

$196 \stackrel{?}{=} 36 + 121$

$196 > 157$

Since $c^2 > a^2 + b^2$, the triangle is obtuse.

CA Standards Check 5 A triangle has sides of lengths 7, 8, and 9. Classify the triangle by its angles.

For more exercises, see *Extra Skill, Word Problem, and Proof Practice.*

GEOM 15.0

Practice by Example

Example 1 (page 418)

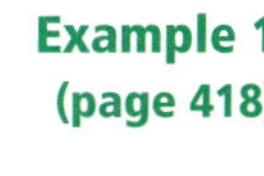

1. 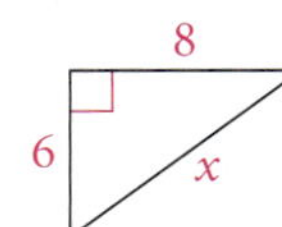

2. 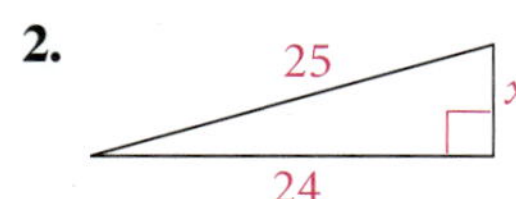

3. 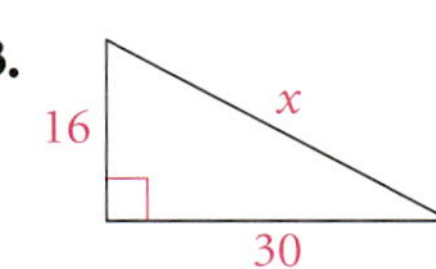

4. 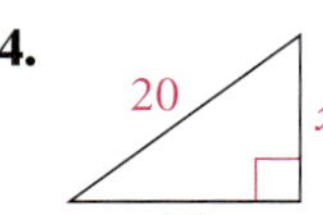

5. 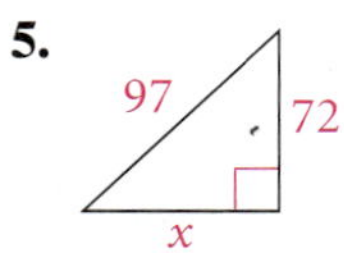

6. 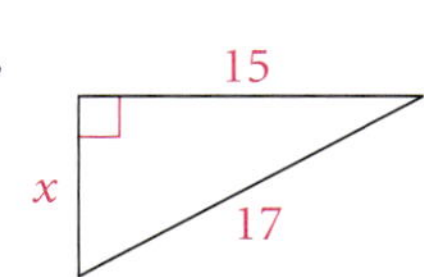

Does each set of numbers form a Pythagorean triple? Explain.

7. 4, 5, 6 **8.** 10, 24, 26 **9.** 15, 20, 25

Example 2 (page 418)

Algebra **Find the value of x. Leave your answer in simplest radical form.**

10. 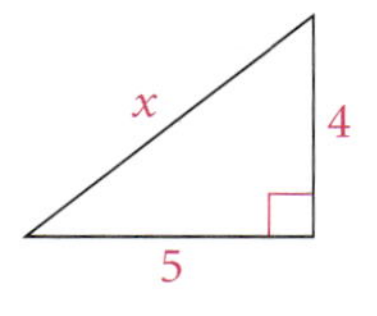

11. 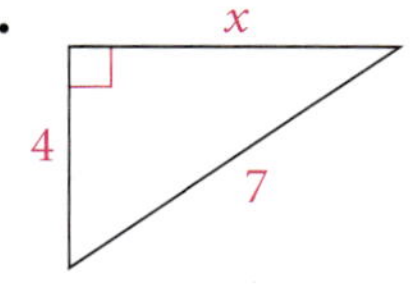

12. 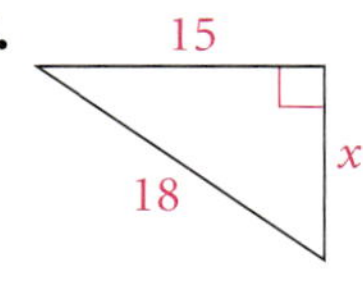

13. 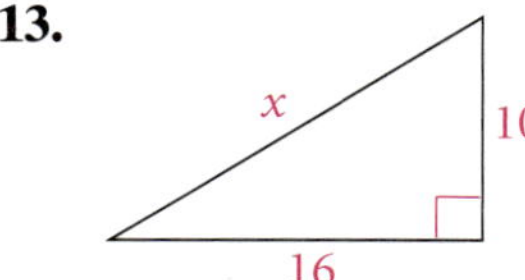

14. 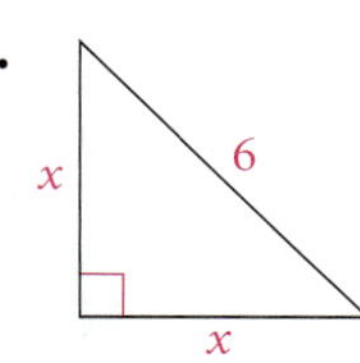

15. 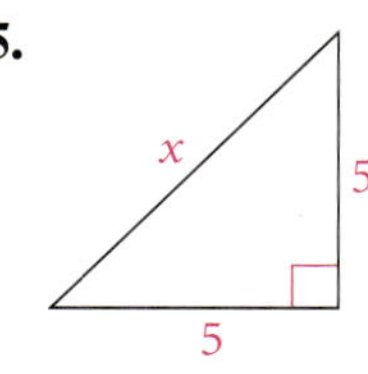

Example 3 (page 418)

16. **Home Maintenance** A painter leans a 15-ft ladder against a house. The base of the ladder is 5 ft from the house.

a. To the nearest tenth of a foot, how high on the house does the ladder reach?

b. The ladder in part (a) reaches too high on the house. By how much should the painter move the ladder's base away from the house to lower the top by 1 ft?

17. A walkway forms the diagonal of a square playground. The walkway is 24 m long. To the nearest tenth of a meter, how long is a side of the playground?

Example 4 (page 419)

Is each triangle a right triangle? Explain.

18. 20, 28, 19

19. 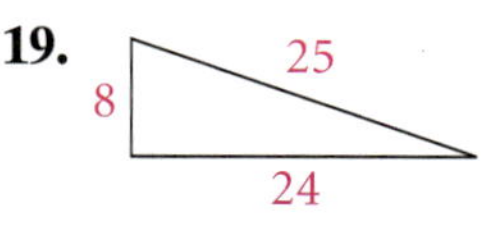

20. 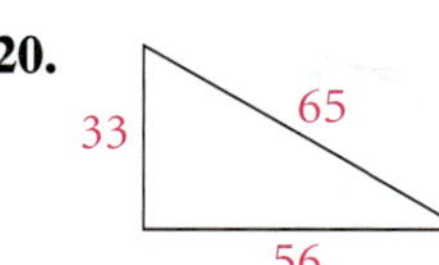

Example 5 (page 419)

The lengths of the sides of a triangle are given. Classify each triangle as acute, right, or obtuse.

21. 4, 5, 6 **22.** 0.3, 0.4, 0.6 **23.** 11, 12, 15

24. $\sqrt{3}, 2, 3$ **25.** 30, 40, 50 **26.** $\sqrt{11}, \sqrt{7}, 4$

B **Apply Your Skills** x^2 **Algebra** **Find the value of x. Leave your answer in simplest radical form.**

27.

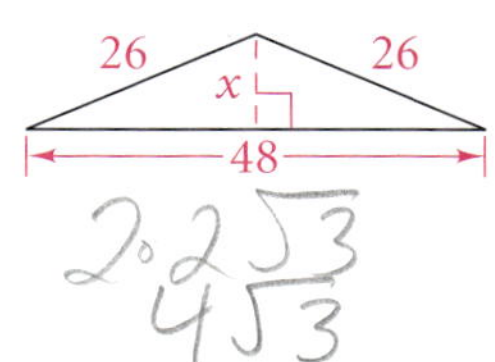

28.

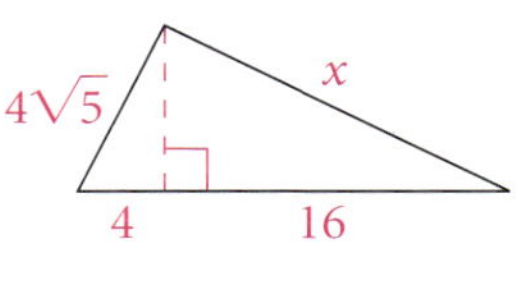

29.

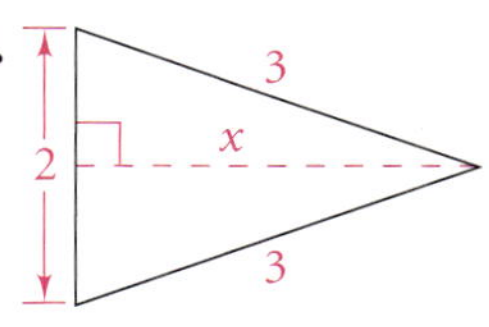

30. **Writing** Each year in an ancient land, a large river overflowed its banks, often destroying boundary markers. The royal surveyors used a rope with knots at 12 equal intervals to help reconstruct boundaries. Explain how a surveyor could use this rope to form a right angle. (*Hint:* Use the Pythagorean triple 3, 4, 5.)

31. **Multiple Choice** Which triangle is *not* a right triangle?

(A)

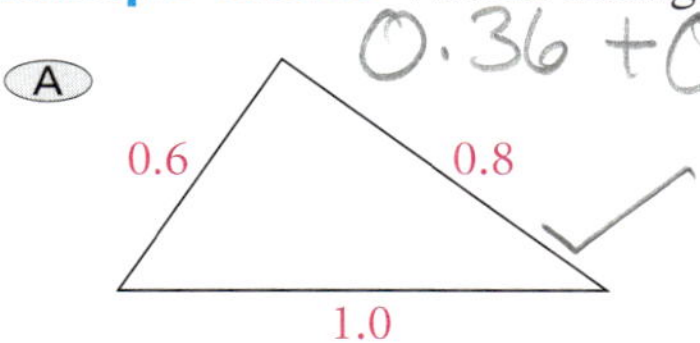

(C)

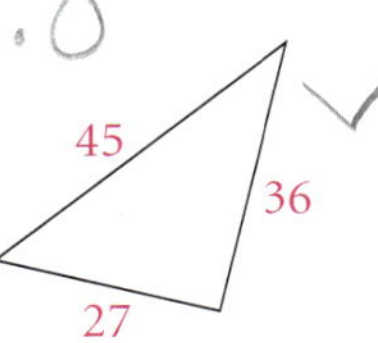

(B)

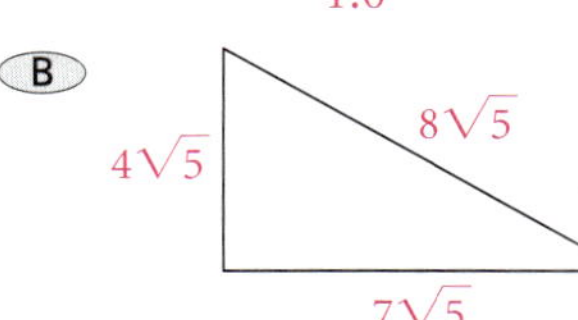

(D)

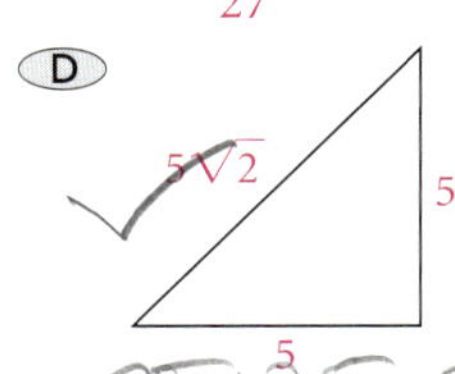

GO for Help

For a guide to solving Exercise 32, see p. 424.

32. **Embroidery** You want to embroider a square design. You have an embroidery hoop with a 6 in. diameter. Find the largest value of x so that the entire square will fit in the hoop. Round to the nearest tenth.

33. In parallelogram $RSTW$, $RS = 7$, $ST = 24$, and $RT = 25$. Is $RSTW$ a rectangle? Explain.

Proof **34.** **Coordinate Geometry** You can use the Pythagorean Theorem to prove the Distance Formula. Let points $P(x_1, y_1)$ and $Q(x_2, y_2)$ be the endpoints of the hypotenuse of a right triangle.

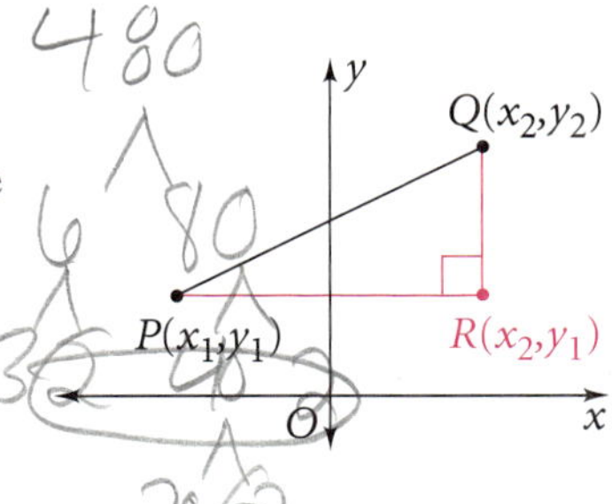

a. Write an algebraic expression to complete each of the following: $PR = \blacksquare$ and $QR = \blacksquare$.

b. By the Pythagorean Theorem, $PQ^2 = PR^2 + QR^2$. Rewrite this statement substituting the algebraic expressions you found for PR and QR in part (a).

c. Complete the proof by taking the square root of each side of the equation that you wrote in part (b).

Homework Video Tutor

Visit: PHSchool.com
Web Code: bce-0801

35. **Constructions** Explain how to construct a segment of length $\sqrt{n}$, where n is any positive integer, and you are given a segment of length 1. (*Hint:* See the diagram.)

Find a third whole number so that the three numbers form a Pythagorean triple.

36. 20, 21 **37.** 14, 48 **38.** 13, 85 **39.** 12, 37

Find integers j and k so that (a) the two given integers and j represent the lengths of the sides of an acute triangle and (b) the two given integers and k represent the lengths of the sides of an obtuse triangle.

40. 4, 5 **41.** 2, 4 **42.** 6, 9 **43.** 5, 10

44. 6, 7 **45.** 9, 12 **46.** 8, 17 **47.** 9, 40

Proof **48.** Prove the Pythagorean Theorem.

Given: $\triangle ABC$ is a right triangle

Prove: $a^2 + b^2 = c^2$

(*Hint*: Begin with proportions suggested by Theorem 7-3 or its corollaries.)

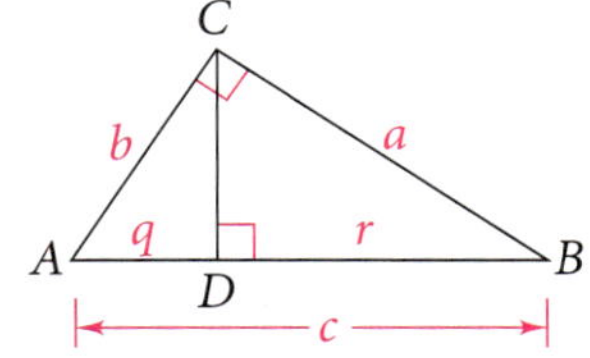

49. Astronomy The Hubble Space Telescope is orbiting Earth 600 km above Earth's surface. Earth's radius is about 6370 km. Use the Pythagorean Theorem to find the distance x from the telescope to Earth's horizon. Round your answer to the nearest ten kilometers.

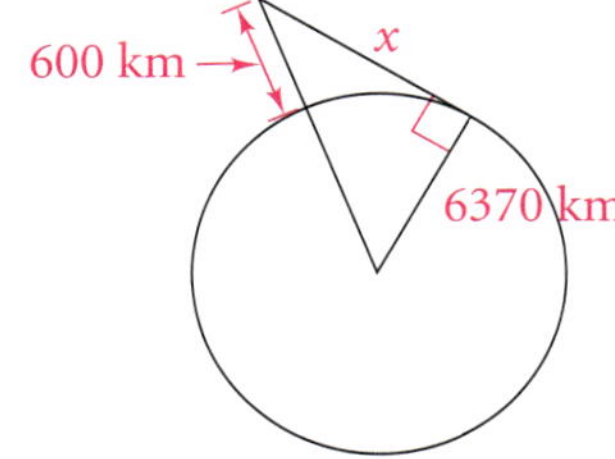

not to scale

Real-World Connection

Research by Edwin Hubble (1889–1953), here guiding a telescope in 1923, led to the Big Bang Theory of the formation of the universe.

The figures below are drawn on centimeter grid paper. Find the perimeter of each shaded figure to the nearest tenth.

50.

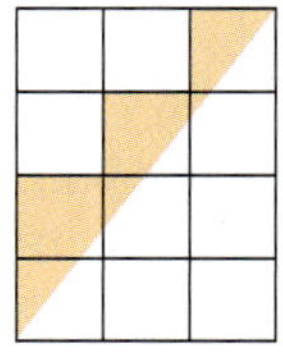

51.

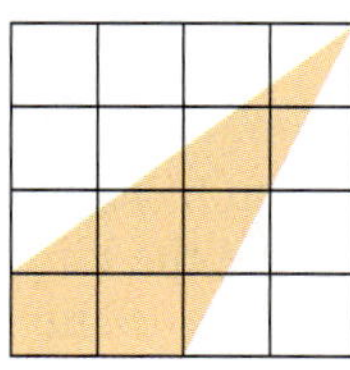

52.

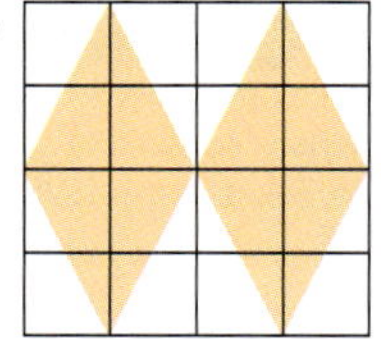

53. a. The ancient Greek philosopher Plato used the expressions $2n$, $n^2 - 1$, and $n^2 + 1$ to produce Pythagorean triples. Choose any integer greater than 1. Substitute for n and evaluate the three expressions.

b. Verify that your answers to part (a) form a Pythagorean triple.

c. Show that, in general, $(2n)^2 + (n^2 - 1)^2 = (n^2 + 1)^2$ for any n.

Challenge

54. Geometry in 3 Dimensions The box at the right is a rectangular solid.

a. Use $\triangle ABC$ to find the length d_1 of the diagonal of the base.

b. Use $\triangle ABD$ to find the length d_2 of the diagonal of the box.

c. You can generalize the steps in parts (a) and (b). Use the facts that $AC^2 + BC^2 = d_1{}^2$ and $d_1{}^2 + BD^2 = d_2{}^2$ to write a one-step formula to find d_2.

d. Use the formula you wrote to find the length of the longest fishing pole you can pack in a box with dimensions 18 in., 24 in., and 16 in.

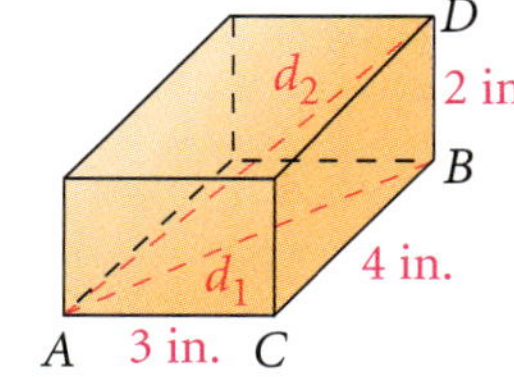

Geometry in 3 Dimensions **Points $P(x_1, y_1, z_1)$ and $Q(x_2, y_2, z_2)$ at the left are points in a three-dimensional coordinate system. Use the following formula to find PQ. Leave your answer in simplest radical form.**

$$d = \sqrt{(x_2 - x_1)^2 + (y_2 - y_1)^2 + (z_2 - z_1)^2}$$

55. $P(0, 0, 0)$, $Q(1, 2, 3)$ **56.** $P(0, 0, 0)$, $Q(-3, 4, -6)$ **57.** $P(-1, 3, 5)$, $Q(2, 1, 7)$

Proof **58.** Use the plan and write a proof of Theorem 8-2, the Converse of the Pythagorean Theorem.

Given: $\triangle ABC$ with sides of length a, b, and c where $a^2 + b^2 = c^2$

Prove: $\triangle ABC$ is a right triangle.

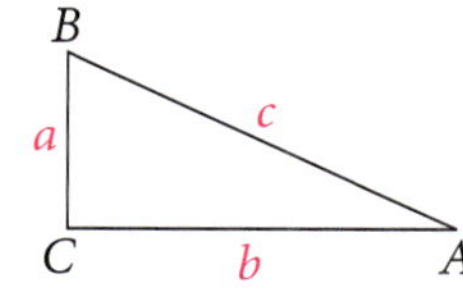

Plan: Draw a right triangle (not $\triangle ABC$) with legs of lengths a and b. Label the hypotenuse x. By the Pythagorean Theorem, $a^2 + b^2 = x^2$. Use substitution to compare the lengths of the sides of your triangle and $\triangle ABC$. Then prove the triangles congruent.

Multiple Choice Practice

For California Standards and CAHSEE Tutorials, visit PHSchool.com. Web Code: bcq-9045

GEOM 15.0

59. A highway detour affects a company's delivery route. The plan showing the old route and the detour is at the right. How many extra miles will the trucks travel once the detour is established?

old shipping route
3.9 miles
5.2 miles

Ⓐ 6.5 Ⓒ 9.1

Ⓑ 2.6 Ⓓ 1.3

GEOM 12.0

60. Determine the value of x in the figure at the right.

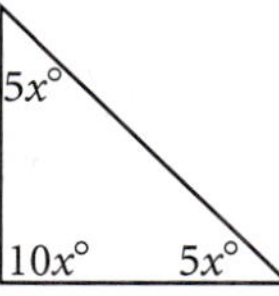

Ⓐ 8 Ⓒ 18

Ⓑ 9 Ⓓ 20

Mixed Review

Lesson 4-7

For the figure at the right, complete the proportion.

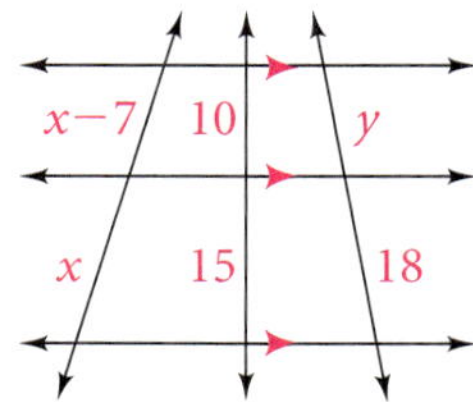

61. $\frac{10}{■} = \frac{y}{18}$ **62.** $\frac{■}{x} = \frac{y}{18}$

63. Find the values of x and y.

Lesson 5-2

In the second figure, $\overrightarrow{PS}$ bisects $\angle RPT$. Solve for each variable. Then find RS.

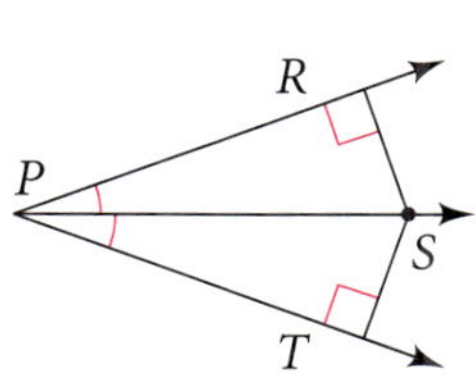

64. $RS = 2x + 19, ST = 7x - 16$; $x =$ ■, $RS =$ ■

65. $RS = 2(7y - 11), ST = 5y + 5$; $y =$ ■, $RS =$ ■

Lesson 4-1

$\triangle PQR \cong \triangle STV$. Solve for each variable.

66. $m\angle P = 4w + 5, m\angle S = 6w - 15$ **67.** $RQ = 10y - 6, VT = 5y + 9$

68. $m\angle T = 2x - 40, m\angle Q = x + 10$ **69.** $PR = 2z + 3, SV = 4z - 11$

GPS Guided Problem Solving

FOR USE WITH PAGE 421, EXERCISE 32

Understanding Word Problems **Read the problem below. Let Sharleen's thinking guide you through the solution. Check your understanding with the exercises at the bottom of the page.**

Embroidery You want to embroider a square design. You have an embroidery hoop with a 6 in. diameter. Find the largest value of x so that the entire square will fit in the hoop. Round to the nearest tenth.

What Sharleen Thinks	What Sharleen Writes
I'll draw the diagram without the embroidery and mark in it the information I know. In particular, I'll show a 6-in. diameter that *also* happens to be a diagonal of the square.	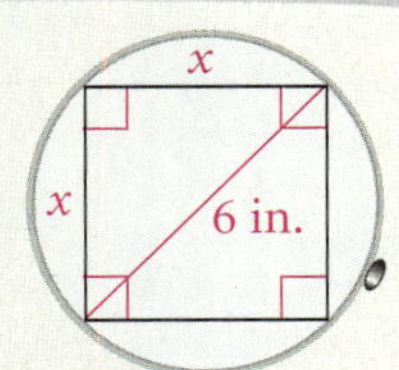
Now I have two right triangles. Each right triangle has two legs of length x in. and a hypotenuse of 6 in. I can apply the Pythagorean Theorem to find x.	Diameter = 6, so length of diagonal = 6 $a^2 + b^2 = c^2$ $a = x$, $b = x$, and $c = 6$ $x^2 + x^2 = 6^2$
Combine like terms. Divide each side by 2. Take the square root of each side. Use a calculator. Then round to the nearest tenth.	$2x^2 = 36$ $x^2 = 18$ $x = \sqrt{18}$ $x \approx 4.2426$
The largest square that you can embroider with this hoop has sides of 4.2 in. (Don't forget to include the unit of measure.)	$x \approx$ **4.2 in.**

EXERCISES

1. In the problem above, what is the value of x in simplest radical form?

Suppose the embroidery hoop had the given diameter. What is the largest value of x possible? Give your answer in simplest radical form.

2. 12 in. **3.** 18 in. **4.** 9 in. **5.** 3 in.

6. Based on your results for Exercises 1–5, make a conjecture about a relationship between a side and a diagonal of a square.

Special Right Triangles

California Content Standards

GEOM 15.0 Students use the Pythagorean theorem to determine distance and find missing lengths of sides of right triangles. ***Develop***

GEOM 20.0 Students know and are able to use angle and side relationships in problems with special right triangles, such as 30°, 60°, and 90° triangles and 45°, 45°, and 90° triangles. ***Introduce, Develop***

What You'll Learn

- To use the properties of 45°-45°-90° triangles
- To use the properties of 30°-60°-90° triangles

. . . And Why

To find the distance from home plate to second base on a softball diamond, as in Example 3

Check Skills You'll Need

Use a protractor to find the measures of the angles of each triangle.

1.

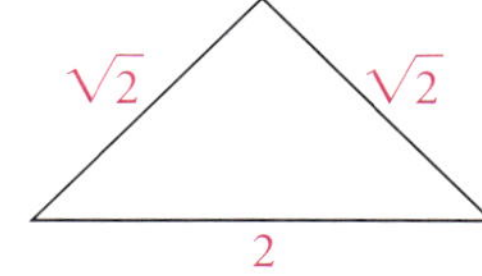

2.

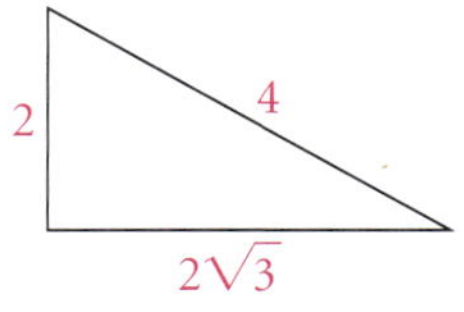

3.

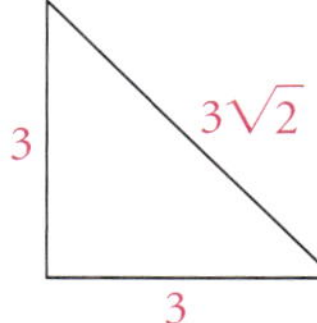

1 Using 45°-45°-90° Triangles

The acute angles of an isosceles right triangle are both 45° angles. Another name for an isosceles right triangle is a 45°-45°-90° triangle. If each leg has length x and the hypotenuse has length y, you can solve for y in terms of x.

$x^2 + x^2 = y^2$ **Use the Pythagorean Theorem.**

$2x^2 = y^2$ **Simplify.**

$x\sqrt{2} = y$ **Take the square root of each side.**

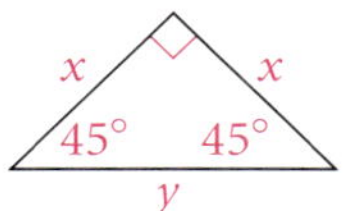

You have just proved the following theorem.

Key Concepts

Theorem 8-5 **45°-45°-90° Triangle Theorem**

In a 45°-45°-90° triangle, both legs are congruent and the length of the hypotenuse is $\sqrt{2}$ times the length of a leg.

$$\text{hypotenuse} = \sqrt{2} \cdot \text{leg}$$

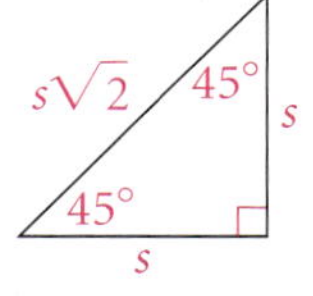

Test-Taking Tip

If you forget the formula for a 45°-45°-90° triangle, use the Pythagorean Theorem. The triangle is isosceles, so the legs have the same length.

1 EXAMPLE Finding the Length of the Hypotenuse

Find the value of each variable.

a.

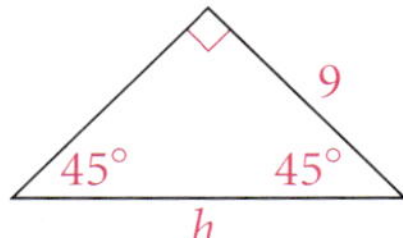

b.

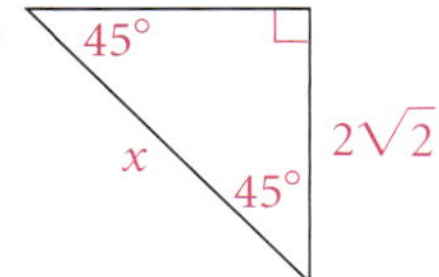

$h = \sqrt{2} \cdot 9$ ← **hypotenuse = $\sqrt{2}$ · leg** → $x = \sqrt{2} \cdot 2\sqrt{2}$

$h = 9\sqrt{2}$ ← **Simplify.** → $x = 4$

1 Find the length of the hypotenuse of a 45°-45°-90° triangle with legs of length $5\sqrt{3}$.

2 EXAMPLE Finding the Length of a Leg

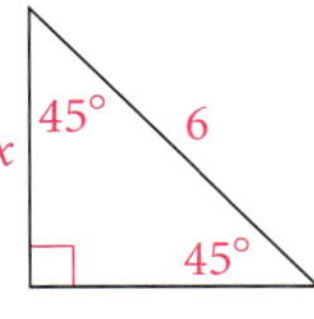

Multiple Choice What is the value of x?

Ⓐ 3 Ⓑ $3\sqrt{2}$ Ⓒ 6 Ⓓ $6\sqrt{2}$

$6 = \sqrt{2} \cdot x$ **hypotenuse = $\sqrt{2}$ · leg**

$x = \frac{6}{\sqrt{2}}$ **Divide each side by $\sqrt{2}$.**

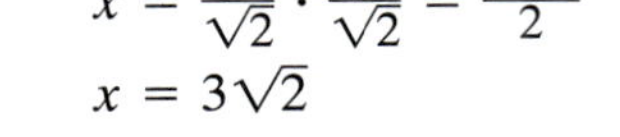

$x = \frac{6}{\sqrt{2}} \cdot \frac{\sqrt{2}}{\sqrt{2}} = \frac{6\sqrt{2}}{2}$ **Multiply by a form of 1.**

$x = 3\sqrt{2}$ **Simplify.**

The correct answer is B.

CA Standards Check 2 Find the length of a leg of a 45°-45°-90° triangle with a hypotenuse of length 10.

When you apply the 45°-45°-90° Triangle Theorem to a real-life example, you can use a calculator to evaluate square roots.

3 EXAMPLE Real-World Connection

Softball A high school softball diamond is a square. The distance from base to base is 60 ft. To the nearest foot, how far does a catcher throw the ball from home plate to second base?

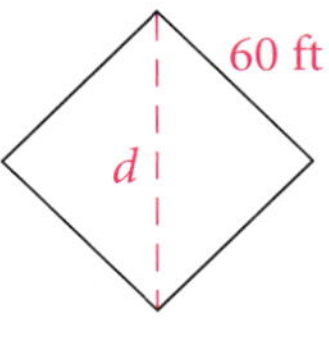

The distance d from home plate to second base is the length of the hypotenuse of a 45°-45°-90° triangle.

$d = 60\sqrt{2}$ **hypotenuse = $\sqrt{2}$ · leg**

$d = 84.852814$ **Use a calculator.**

On a high school softball diamond, the catcher throws the ball about 85 ft from home plate to second base.

CA Standards Check 3 A square garden has sides 100 ft long. You want to build a brick path along a diagonal of the square. How long will the path be? Round your answer to the nearest foot.

2 Using 30°-60°-90° Triangles

Another type of special right triangle is a 30°-60°-90° triangle.

Key Concepts

Theorem 8-6 **30°-60°-90° Triangle Theorem**

In a 30°-60°-90° triangle, the length of the hypotenuse is twice the length of the shorter leg. The length of the longer leg is $\sqrt{3}$ times the length of the shorter leg.

hypotenuse = 2 · shorter leg

longer leg = $\sqrt{3}$ · shorter leg

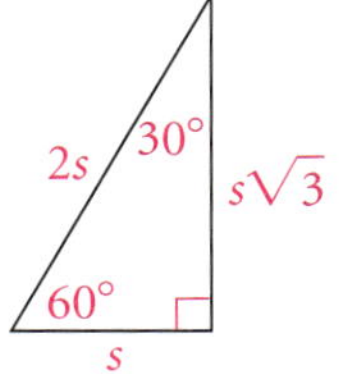

To prove Theorem 8-6, draw a 30°-60°-90° triangle using an equilateral triangle.

Proof

Proof of Theorem 8-6

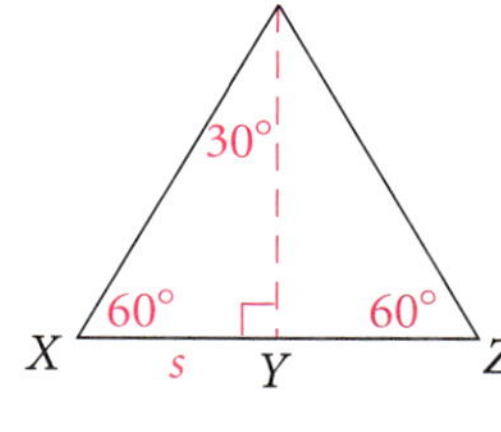

For 30°-60°-90° $\triangle WXY$ in equilateral $\triangle WXZ$, $\overline{WY}$ is the perpendicular bisector of $\overline{XZ}$. Thus, $XY = \frac{1}{2}XZ = \frac{1}{2}XW$, or $XW = 2XY = 2s$. Also,

$XY^2 + YW^2 = XW^2$	**Use the Pythagorean Theorem.**
$s^2 + YW^2 = (2s)^2$	**Substitute *s* for *XY* and 2*s* for *XW*.**
$YW^2 = 4s^2 - s^2$	**Subtract s^2 from each side.**
$YW^2 = 3s^2$	**Simplify.**
$YW = s\sqrt{3}$	**Find the square root of each side.**

The 30°-60°-90° Triangle Theorem, like the 45°-45°-90° Triangle Theorem, lets you find two sides of a right triangle when you know the length of the third side.

4 EXAMPLE Using the Length of One Side

Algebra Find the value of each variable.

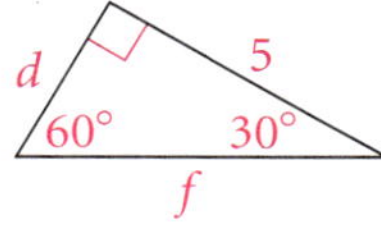

$5 = d\sqrt{3}$	**longer leg = $\sqrt{3}$ · shorter leg**
$d = \frac{5}{\sqrt{3}} \cdot \frac{\sqrt{3}}{\sqrt{3}} = \frac{5\sqrt{3}}{3}$	**Solve for *d*.**
$f = 2d$	**hypotenuse = 2 · shorter leg**
$f = 2 \cdot \frac{5\sqrt{3}}{3} = \frac{10\sqrt{3}}{3}$	**Substitute $\frac{5\sqrt{3}}{3}$ for *d*.**

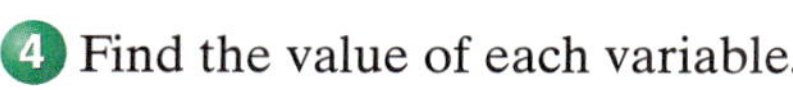

4 Find the value of each variable.

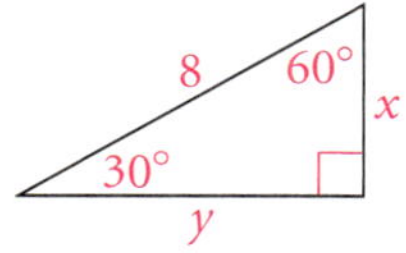

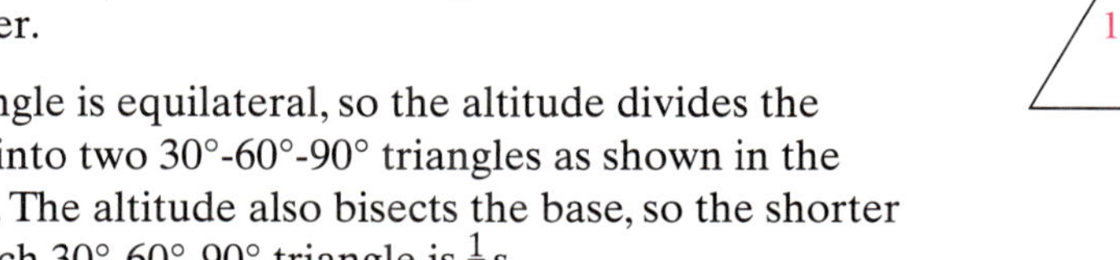

Road Signs The moose warning sign at the left is an equilateral triangle. The height of the sign is 1 m. Find the length *s* of each side of the sign to the nearest tenth of a meter.

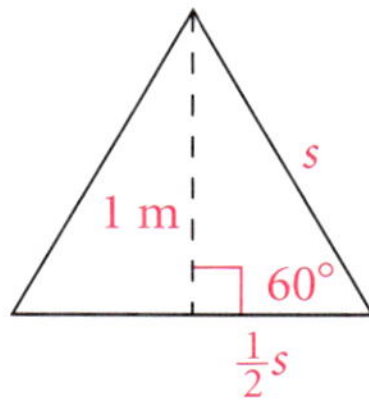

The triangle is equilateral, so the altitude divides the triangle into two 30°-60°-90° triangles as shown in the diagram. The altitude also bisects the base, so the shorter leg of each 30°-60°-90° triangle is $\frac{1}{2}s$.

$1 = \sqrt{3}\left(\frac{1}{2}s\right)$	**longer leg = $\sqrt{3}$ · shorter leg**
$\frac{2}{\sqrt{3}} = s$	**Solve for *s*.**
$s \approx 1.155$	**Simplify. Use a calculator.**

Each side of the sign is about 1.2 m long.

5 If the sides of the sign are 1 m long, what is the height?

EXERCISES

For more exercises, see *Extra Skill, Word Problem, and Proof Practice.*

Standards Practice

GEOM 15.0, 20.0

A Practice by Example

Example 1 (page 425)

Find the value of each variable. If your answer is not an integer, leave it in simplest radical form.

1.

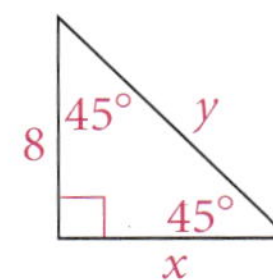

2.

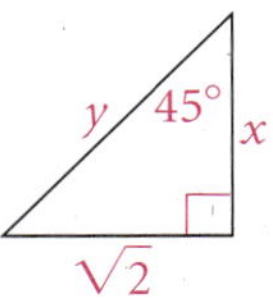

3.

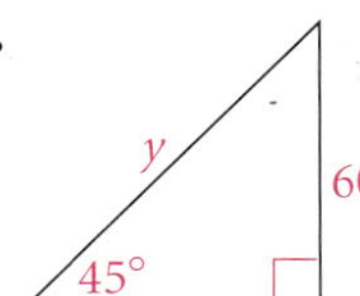

Examples 2 and 3 (page 426)

4. y, $15\sqrt{2}$, 45°, x

5. 45°, 8, 45°, x

6.

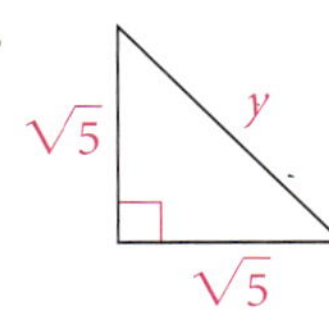

Exercise 7

7. Dinnerware Design You are designing dinnerware. What is the length of a side of the smallest square plate on which a 20-cm chopstick can fit along a diagonal without any overhang? Round your answer to the nearest tenth of a centimeter.

8. Helicopters The four blades of a helicopter meet at right angles and are all the same length. The distance between the tips of two adjacent blades is 36 ft. How long is each blade? Round your answer to the nearest tenth.

Examples 4 and 5 (page 427)

x^2 **Algebra Find the value of each variable. If your answer is not an integer, leave it in simplest radical form.**

9.

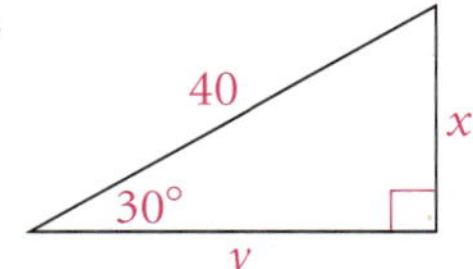

10.

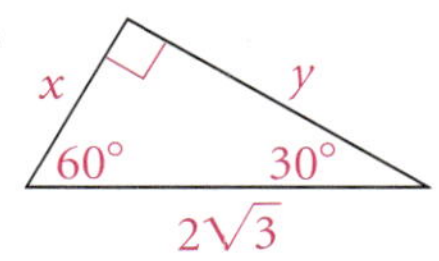

11.

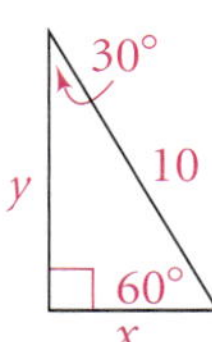

12.

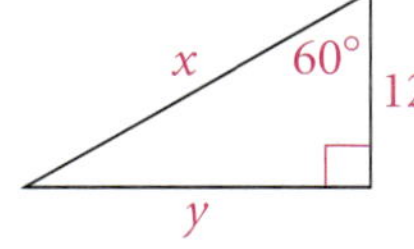

13.

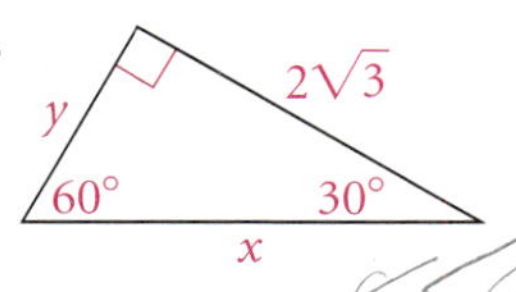

14.

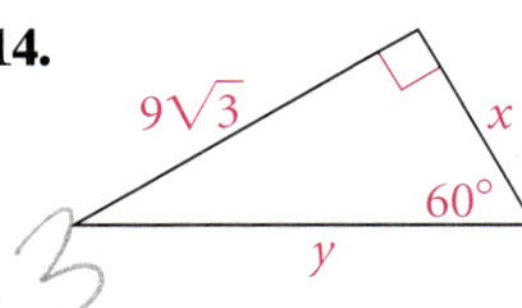

15. Architecture An escalator lifts people to the second floor, 25 ft above the first floor. The escalator rises at a 30° angle. How far does a person travel from the bottom to the top of the escalator?

25 ft, 30°

16. City Planning Jefferson Park sits on one square city block 300 ft on each side. Sidewalks join opposite corners. About how long is each diagonal sidewalk?

B Apply Your Skills

x^2 **Algebra Find the value of each variable. Leave your answer in simplest radical form.**

17. $7\sqrt{2}$, a, b, 45°, 30°, c, d

18.

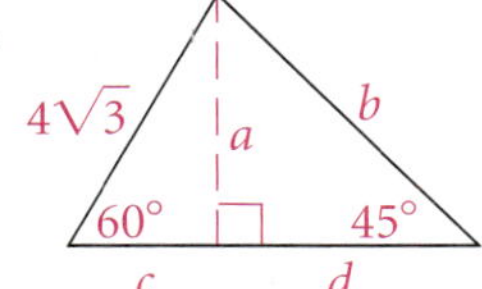

19.

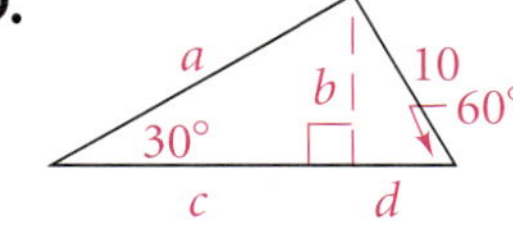

Visit: PHSchool.com
Web Code: bce-0802

20.

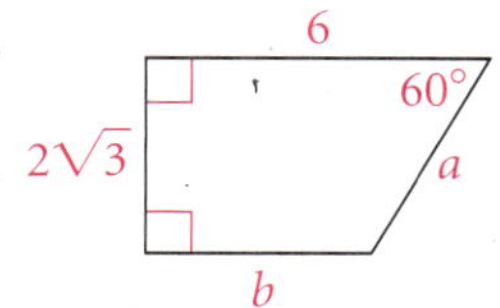

21.

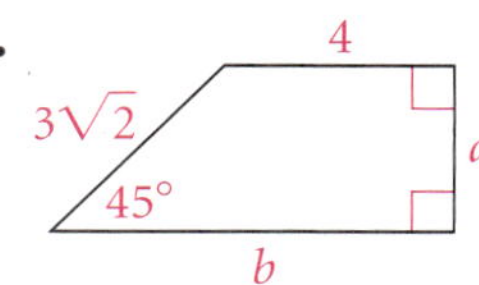

22.

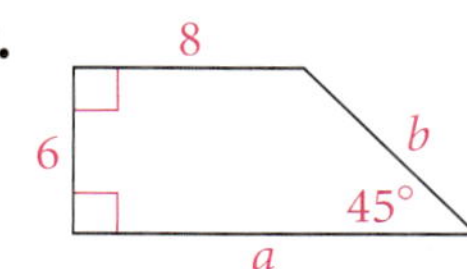

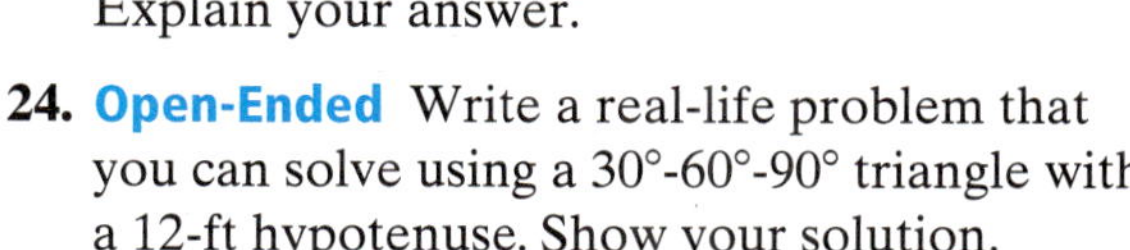

23. Error Analysis Sandra drew the triangle at the right. Rika said that the lengths couldn't be correct. With which student do you agree? Explain your answer.

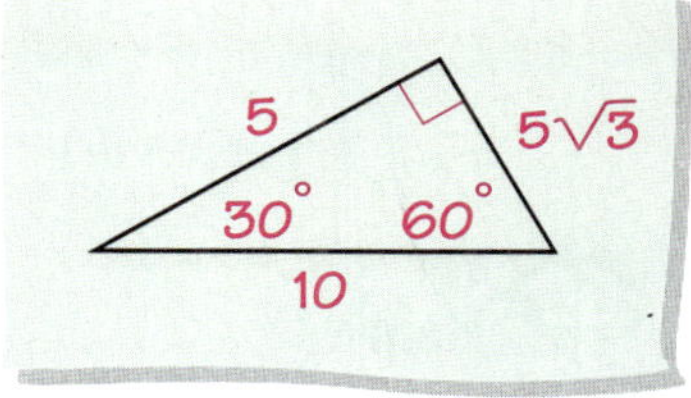

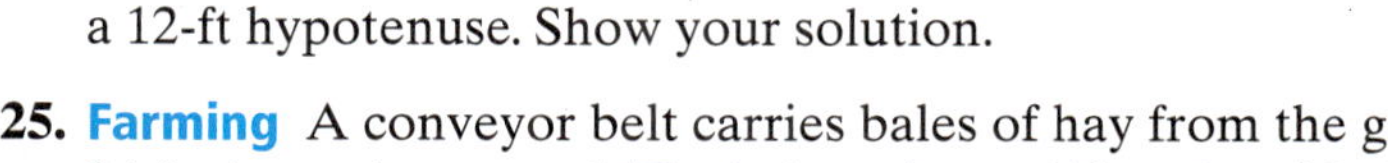

24. Open-Ended Write a real-life problem that you can solve using a 30°-60°-90° triangle with a 12-ft hypotenuse. Show your solution.

25. Farming A conveyor belt carries bales of hay from the ground to the barn loft 24 ft above the ground. The belt makes a 60° angle with the ground.

a. How far does a bale of hay travel from one end of the conveyor belt to the other? Round your answer to the nearest foot.

b. The conveyor belt moves at 100 ft/min. How long does it take for a bale of hay to go from the ground to the barn loft?

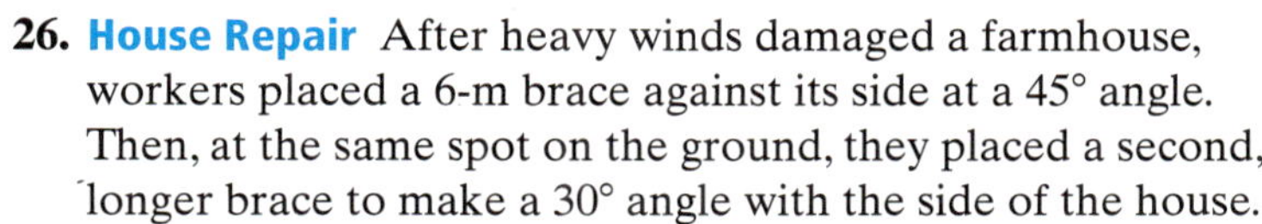

Exercise 25

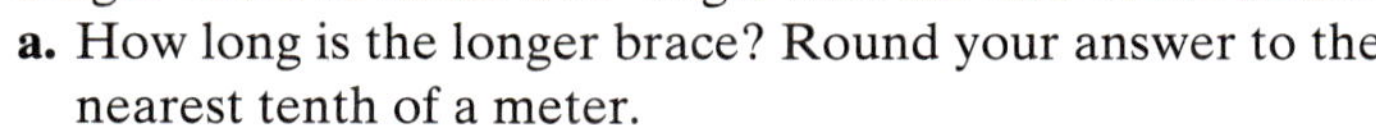

26. House Repair After heavy winds damaged a farmhouse, workers placed a 6-m brace against its side at a 45° angle. Then, at the same spot on the ground, they placed a second, longer brace to make a 30° angle with the side of the house.

a. How long is the longer brace? Round your answer to the nearest tenth of a meter.

b. About how much higher on the house does the longer brace reach than the shorter brace?

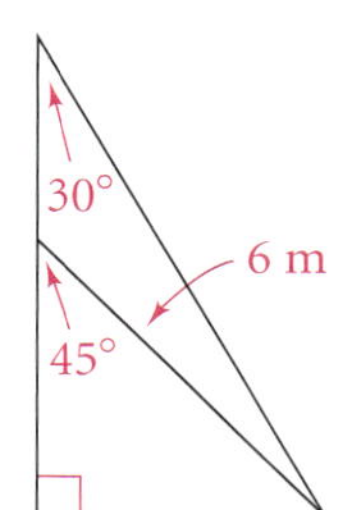

Challenge

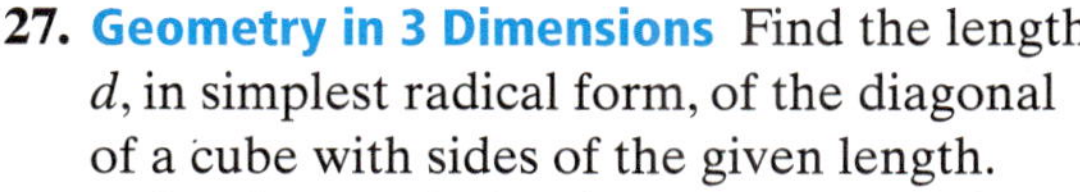

27. Geometry in 3 Dimensions Find the length d, in simplest radical form, of the diagonal of a cube with sides of the given length.

a. 1 unit **b.** 2 units **c.** s units

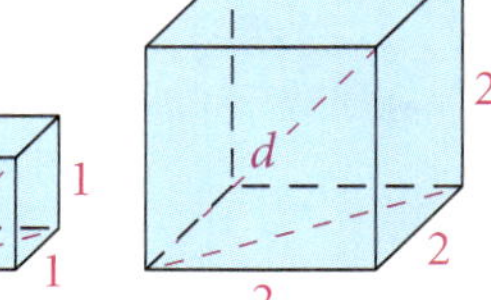
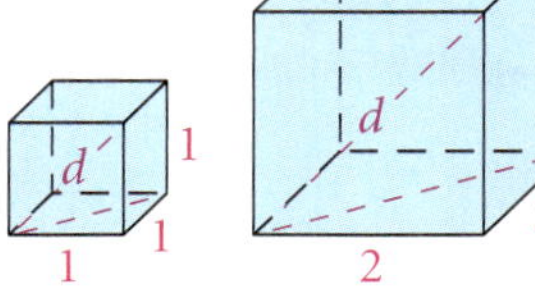

28. Constructions Construct a 30°-60°-90° triangle given a segment that is

a. the shorter leg.

b. the hypotenuse.

c. the longer leg.

Multiple Choice Practice

For California Standards and CAHSEE Tutorials, visit PHSchool.com. Web Code: bcq-9045

GEOM 20.0

29. A ladder rests against a peach tree at a 60° angle as in the picture at the right. If the ladder is 15 feet long, how long is x, the distance from the ladder to the base of the tree?

(A) 7.5 ft

(B) $7.5\sqrt{3}$ ft

(C) 10 ft

(D) 30 ft

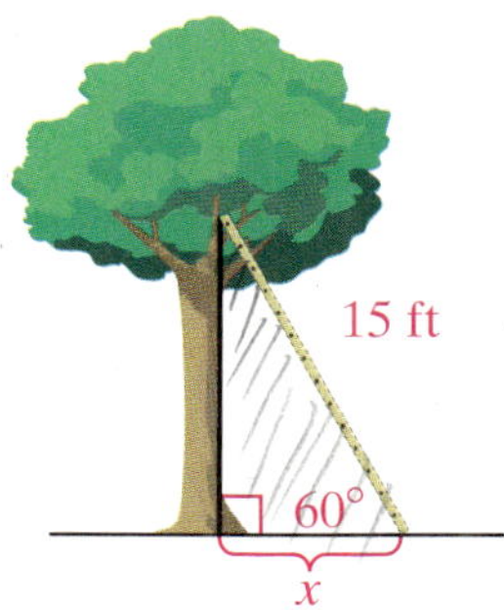

CAHSEE 7AF 2.2

30. Which expression is equivalent to $\frac{8a^3b^4c^0}{4a^5b^{-3}}$?

(A) $\frac{2a^7}{b^8}$ (B) $\frac{b^7}{2a^8}$ (C) $\frac{2b^7c}{a^8}$ (D) $\frac{2b^7}{a^2}$

Mixed Review

Lesson 8-1 **For Exercises 31 and 32, leave your answers in simplest radical form.**

31. A right triangle has a 6-in. hypotenuse and a 5-in. leg. Find the length of the other leg.

32. An isosceles triangle has 20-cm legs and a 16-cm base. Find the length of the altitude to the base.

Lesson 6-4 **Determine whether each quadrilateral must be a parallelogram. If not, provide a counterexample.**

33. The diagonals are congruent and perpendicular to each other.

34. Two opposite angles are right angles and two opposite sides are 5 cm long.

35. One pair of sides is congruent and the other pair of sides is parallel.

Lesson 4-3 **Can you conclude that $\triangle TRY \cong \triangle ANG$ from the given conditions? If so, name the postulate or theorem that justifies your conclusion.**

36. $\angle A \cong \angle T, \angle Y \cong \angle G, \overline{TR} \cong \overline{AN}$

37. $\angle T \cong \angle A, \angle R \cong \angle N, \angle Y \cong \angle G$

38. $\angle R \cong \angle N, \overline{TR} \cong \overline{AN}, \overline{TY} \cong \overline{AG}$

39. $\angle G \cong \angle Y, \angle N \cong \angle R, \overline{RY} \cong \overline{NG}$

Checkpoint Quiz 1 — Lessons 8-1 through 8-2

x^2 Algebra **Find the value of each variable. Leave your answer in simplest radical form.**

1. 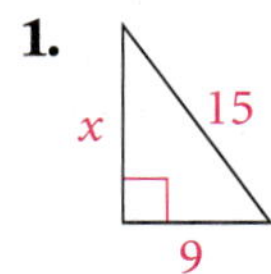

2. 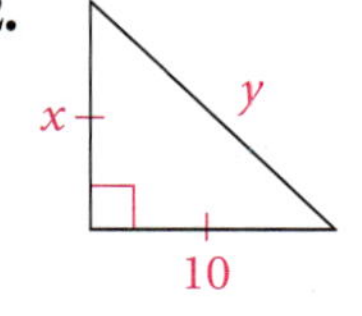

3. 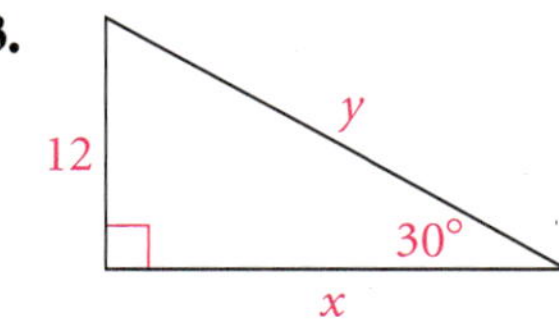

4. $45°$, $9\sqrt{2}$, a

5. 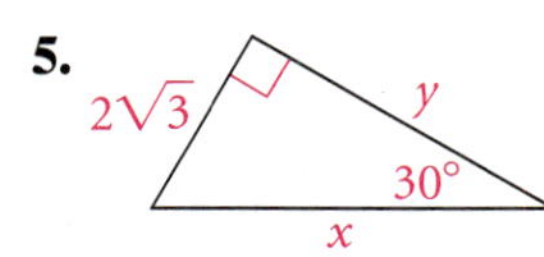

6. 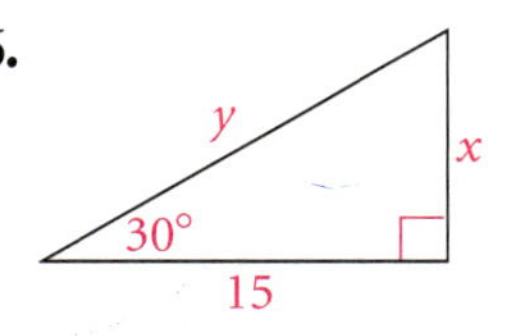

The lengths of the sides of a triangle are given. Classify each triangle as *acute*, *obtuse*, or *right*.

7. 7, 8, 9

8. 15, 36, 39

9. 10, 12, 16

10. A square has a 40-cm diagonal. How long is each side of the square? Round your answer to the nearest tenth of a centimeter.

Proving the Pythagorean Theorem

FOR USE WITH LESSON 8-2

GEOM 14.0 Students prove the Pythagorean theorem. ***Develop, Master***

In Lesson 8-1, you learned about the Pythagorean Theorem. Though the theorem is generally attributed to Pythagoras, a Babylonian tablet from the 20th century B.C. provides the first known written record of the theorem.

The Pythagorean Theorem can be proved using different concepts involving inscribed and circumscribed figures, rigid motions, similarity, and circles.

1 ACTIVITY

Use the properties of quadrilaterals to prove the Pythagorean Theorem.

Draw four congruent right triangles with sides a, b, c, and arrange the triangles to form a square like the one at the right.

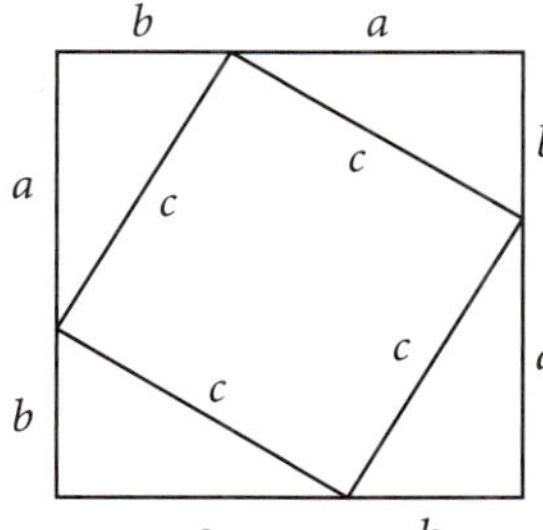

1. Write an expression of the area of the larger square.

2. Write an expression of the area of the four triangles plus the area of the inner square formed by the four triangles.

3. Write an equation using your results from Exercises 1 and 2.

4. Simplify the equation from Exercise 3, combining like terms.

5. Try changing the length of the legs in your right triangles. Does this change the equation that results? Explain.

2 ACTIVITY

Use inscribed triangles and the geometric mean to prove the Pythagorean Theorem.

Copy the figure at the right. $\triangle STV$ is a right triangle that is inscribed in the circle with center P. $\overline{SR}$ is the altitude to the hypotenuse of $\triangle STV$. Label $\overline{SR}$ as a, the base of $\triangle SPR$ as b, and the hypotenuse of $\triangle SPR$ as c.

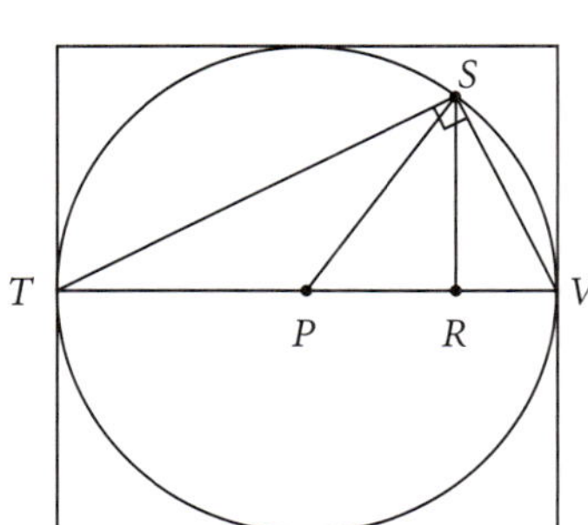

6. a. What is the length of $\overline{TP}$ in terms of a, b, and c?
b. What is the length of $\overline{RV}$ in terms of a, b, and c?

Use Corollary 1 of Theorem 7-3 to set up a geometric mean equation for $\triangle STV$.

7. What is the resulting equation?

8. Simplify your equation from Exercise 7. What is the result?

9. Research and present another Pythagorean Theorem proof.

8-3 The Tangent Ratio

California Content Standards

GEOM 18.0 Students know the definitions of the basic trigonometric functions defined by the angles of a right triangle. ***Introduce***

GEOM 19.0 Students use trigonometric functions to solve for an unknown length of a side of a right triangle, given an angle and a length of a side. ***Introduce***

What You'll Learn

- To use tangent ratios to determine side lengths in triangles

. . . And Why

To use the tangent ratio to estimate distance to a distant object, as in Example 2

Check Skills You'll Need

GO for Help Lessons 7-1 and 8-2

Find the ratios $\frac{BC}{AB}$, $\frac{AC}{AB}$, and $\frac{BC}{AC}$. Round answers to the nearest hundredth.

1.

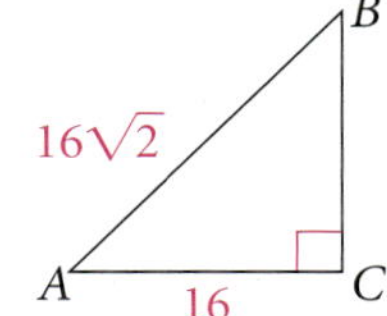

2.

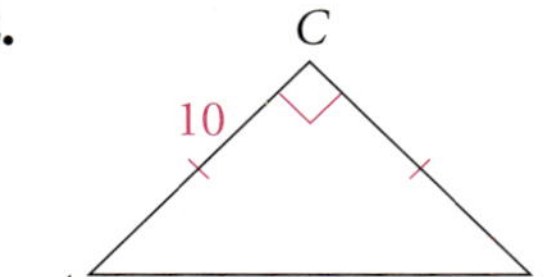

3.

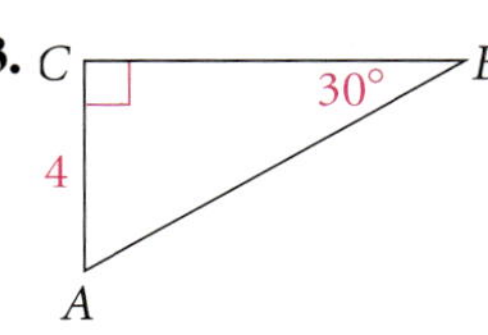

Algebra Solve each proportion.

4. $\frac{x}{3} = \frac{4}{7}$ **5.** $\frac{6}{11} = \frac{x}{9}$ **6.** $\frac{8}{15} = \frac{4}{x}$ **7.** $\frac{5}{x} = \frac{7}{12}$

New Vocabulary • tangent

1 Using Tangents in Triangles

Activity: Tangent Ratios

Work in groups of three or four.

- Have your group select one angle measure from {10°, 20°, . . . , 80°}. Then have each member of your group draw a right triangle, $\triangle ABC$, where $\angle A$ has the selected measure. Make the triangles different sizes.
- Measure the legs of each $\triangle ABC$ to the nearest millimeter.

1. Compute the ratio $\frac{\text{leg opposite } \angle A}{\text{leg adjacent to } \angle A}$ and round to two decimal places.

2. Compare the ratios in your group. Make a conjecture.

Vocabulary Tip

Trigonometry comes from the Greek words *trigonon* and *metria* meaning "triangle measurement."

For each pair of complementary angles, $\angle A$ and $\angle B$, there is a family of similar right triangles. In each such family, the ratio

$$\frac{\text{length of leg opposite } \angle A}{\text{length of leg adjacent to } \angle A}$$

is constant no matter the size of $\triangle ABC$. This trigonometric ratio is called the tangent ratio.

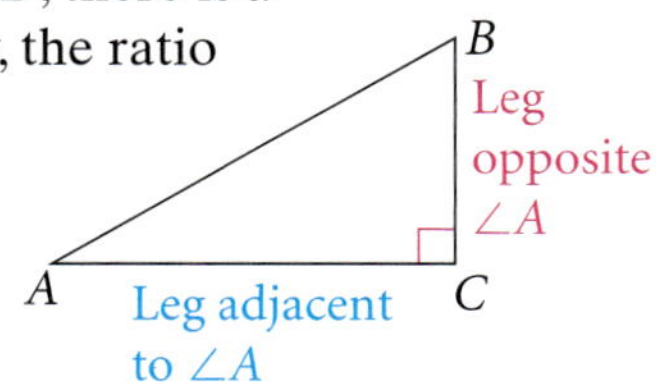

tangent of $\angle A = \frac{\text{length of leg opposite } \angle A}{\text{length of leg adjacent to } \angle A}$

You can abbreviate this equation as $\tan A = \frac{\text{opposite}}{\text{adjacent}}$.

1 EXAMPLE Writing Tangent Ratios

Write the tangent ratios for $\angle T$ and $\angle U$.

$\tan T = \frac{\text{opposite}}{\text{adjacent}} = \frac{UV}{TV} = \frac{3}{4}$

$\tan U = \frac{\text{opposite}}{\text{adjacent}} = \frac{TV}{UV} = \frac{4}{3}$

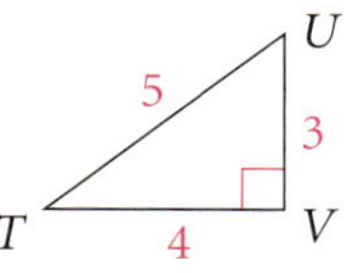

1 **a.** Write the tangent ratios for $\angle K$ and $\angle J$.
b. How is tan K related to tan J?

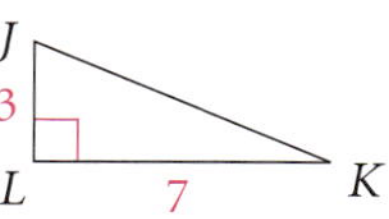

As stated on the facing page, the tangent ratio for an acute angle does not depend on leg lengths of a right triangle. To see why this is so, consider the congruent angles, $\angle T$ and $\angle T'$, in the two right triangles shown here.

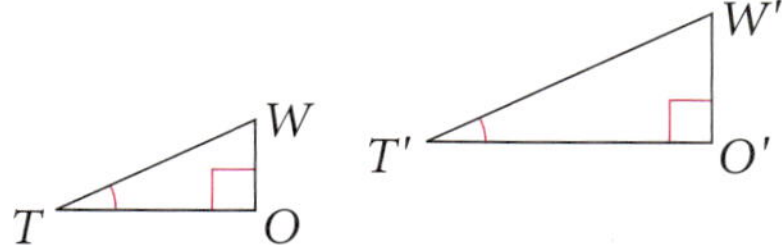

$\triangle TOW \sim T'O'W'$	**AA Similarity Postulate**
$\frac{OW}{TO} = \frac{O'W'}{T'O'}$	**Corresponding sides of ~ triangles are proportional.**
$\tan T = \tan T'$	**Substitute.**

You can use the tangent ratio to measure distances that would be difficult to measure directly.

2 EXAMPLE Real-World Connection

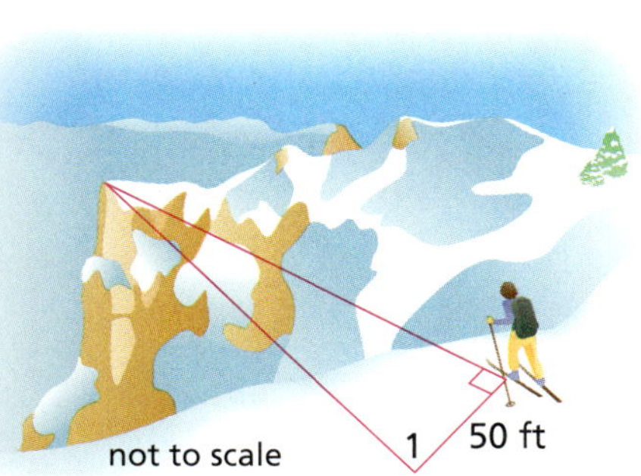

Cross-Country Skiing Your goal in Bryce Canyon National Park is the distant cliff. About how far away is the cliff?

Step 1 Point your compass at a distinctive feature of the cliff and note the reading.

Step 2 Turn 90° and stride 50 ft in a straight path.

Step 3 Turn and point the compass again at the same feature seen in Step 1. Take a reading.

Suppose in Step 3, you find that $m\angle 1 = 86$. The distance you walked in Step 2 was 50 ft. To find the distance to the cliff use the tangent ratio.

$\tan 86^\circ = \frac{x}{50}$	**Use the tangent ratio.**
$x = 50(\tan 86^\circ)$	**Solve for *x*.**
50 TAN 86 ENTER 715.03331	**Use a calculator.**

The cliff is about 715 ft away.

2 Find the value of w to the nearest tenth.

a.

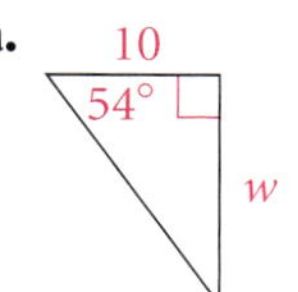

b.

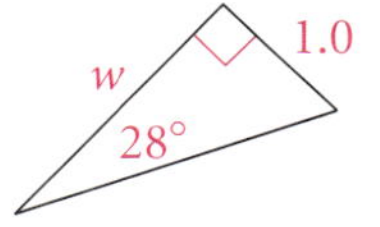

c.

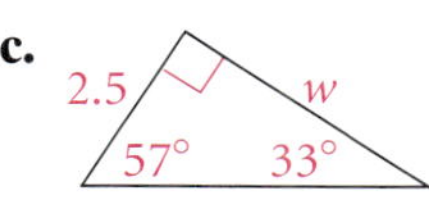

If you know leg lengths for a right triangle, you can find the tangent ratio for each acute angle. Conversely, if you know the tangent ratio for an angle, you can use inverse of tangent, $\tan^{-1}$, to find the measure of the angle.

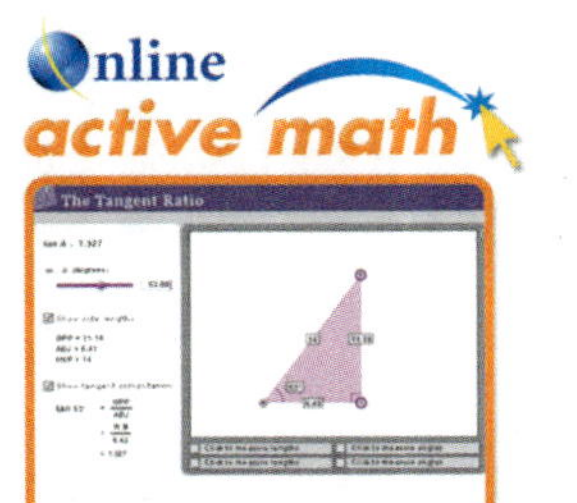

For: Tangent Activity
Use: Interactive Textbook, 8-3

3 EXAMPLE Using the Inverse of Tangent

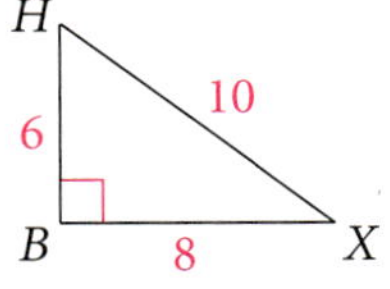

The lengths of the sides of $\triangle BHX$ are given. Find $m\angle X$ to the nearest degree.

$\tan X = \frac{6}{8} = 0.75$ **Find the tangent ratio.**

$m\angle X = \tan^{-1}(0.75)$ **Use the inverse of tangent.**

TAN^{-1} 0.75 ENTER 36.869898 **Use a calculator.**

So $m\angle X \approx 37$.

3 Find $m\angle Y$ to the nearest degree.

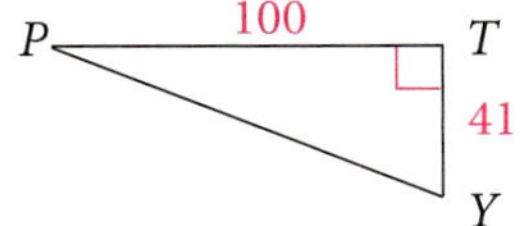

For more exercises, see *Extra Skill, Word Problem, and Proof Practice.*

Standards Practice

GEOM 18.0, 19.0

Practice by Example

Example 1 (page 433)

Write the tangent ratios for $\angle A$ and $\angle B$.

SOH CAH TOA

1. 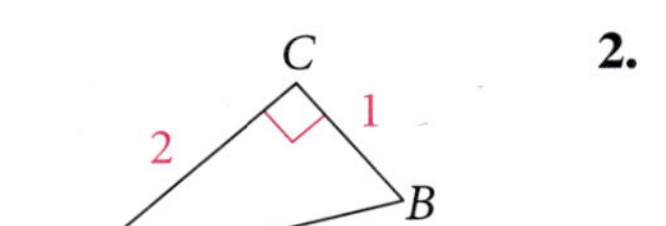

2. 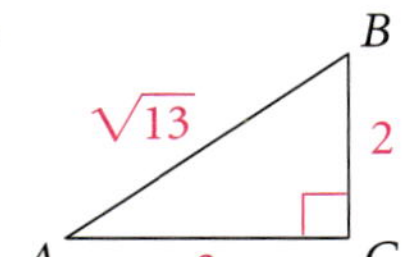

3. 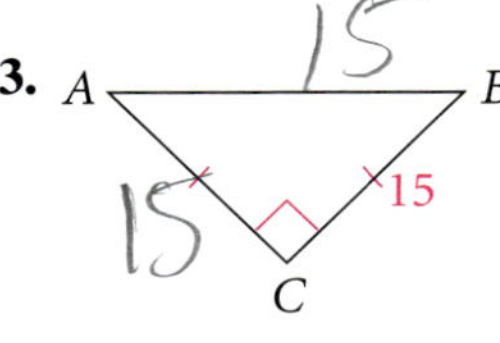

Example 2 (page 433)

Find the value of x to the nearest tenth.

4. 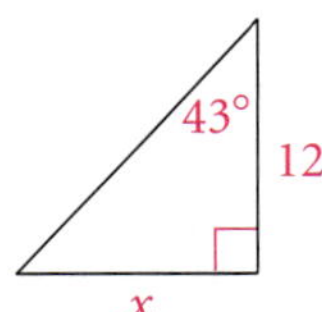

5. 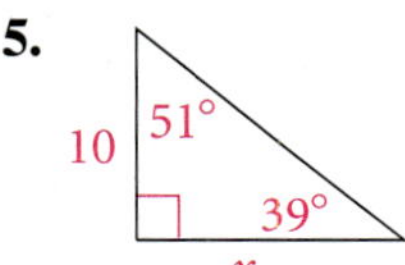

6. 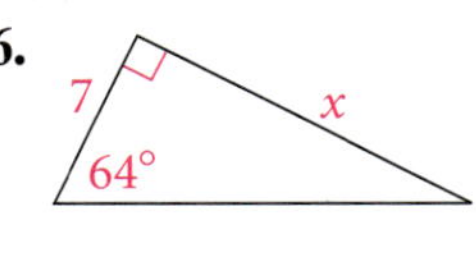

Krogsdale

7. 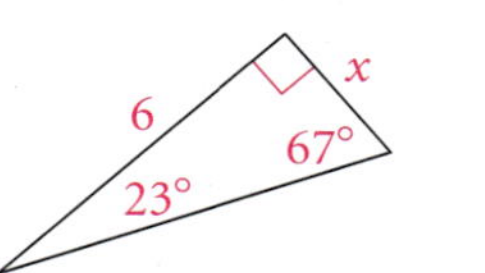

8. 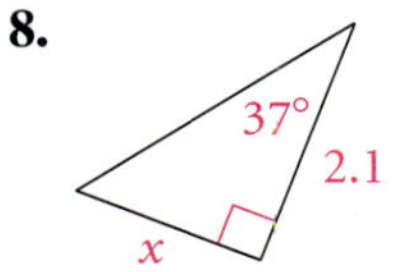

9. 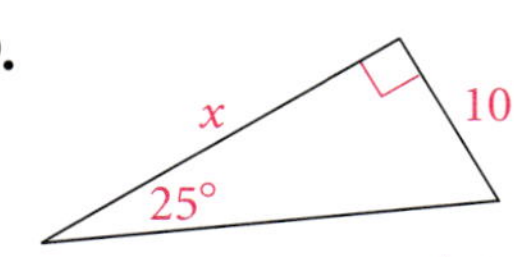

10. **Surveying** To find the distance from the boathouse on shore to the cabin on the island, a surveyor measures from the boathouse to point X as shown. He then finds $m\angle X$ with an instrument called a transit. Use the surveyor's measurements to find the distance from the boathouse to the cabin.

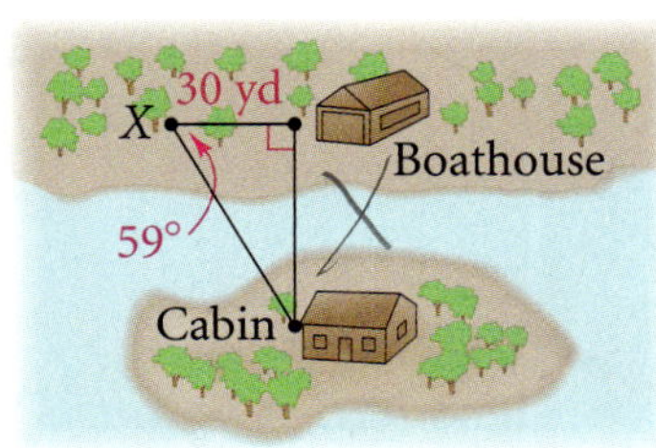

Example 3
(page 434)

SOH
CAH
TOA

Find the value of x to the nearest degree.

11.
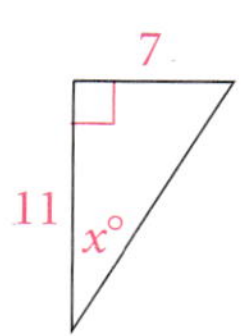

12.
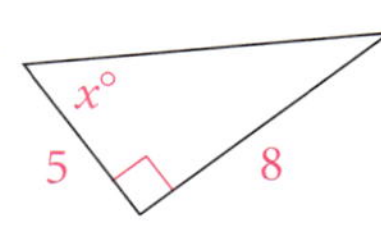

13.
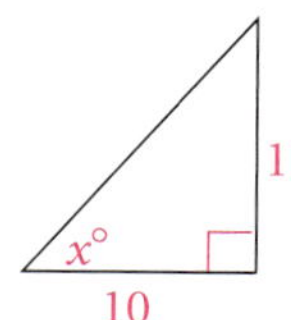

14.
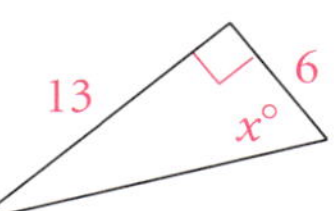

15.
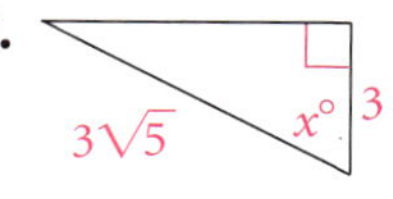

16.
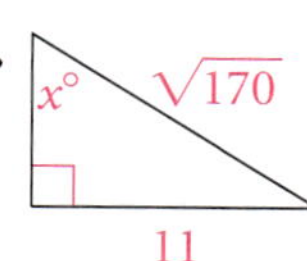

Find each missing value to the nearest tenth.

17. $\tan \blacksquare^\circ = 3.5$ **18.** $\tan 34^\circ = \dfrac{\blacksquare}{20}$ **19.** $\tan 2^\circ = \dfrac{4}{\blacksquare}$ **20.** $\tan \blacksquare^\circ = 90$

21. The lengths of the diagonals of a rhombus are 2 in. and 5 in. Find the measures of the angles of the rhombus to the nearest degree.

22. Pyramids All but two of the pyramids built by the ancient Egyptians have faces inclined at 52° angles. Suppose an archaeologist discovers the ruins of a pyramid. Most of the pyramid has eroded, but she is able to determine that the length of a side of the square base is 82 m. How tall was the pyramid, assuming its faces were inclined at 52°? Round your answer to the nearest meter.

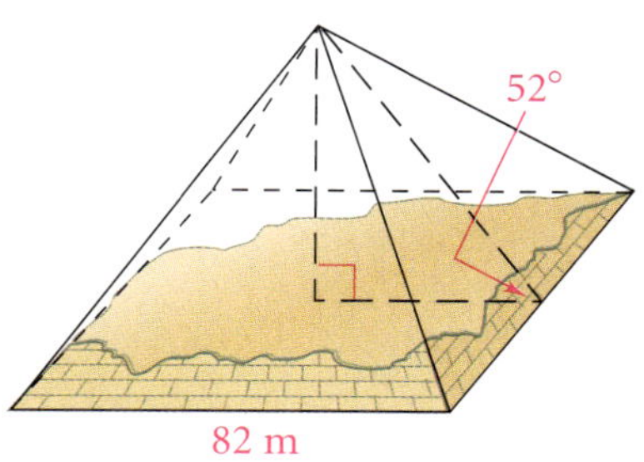

23. Multiple Choice The legs of a right triangle have lengths 5 and 12. What is the tangent of the angle opposite the leg with length 12?

Ⓐ $\frac{5}{13}$ Ⓑ $\frac{12}{13}$ Ⓒ $\frac{13}{12}$ Ⓓ $\frac{12}{5}$

24. Writing Explain why $\tan 60^\circ = \sqrt{3}$. Include a diagram with your explanation.

25. Explain why $\tan^{-1} \frac{\sqrt{2}}{\sqrt{2}} = 45^\circ$.

26. A rectangle is 80 cm long and 20 cm wide. To the nearest degree, find the measures of the angles formed by the diagonals at the center of the rectangle.

Find the value of w, then x. Round lengths of segments to the nearest tenth. Round angle measures to the nearest degree.

27.
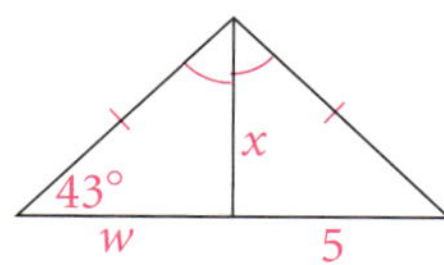

28.
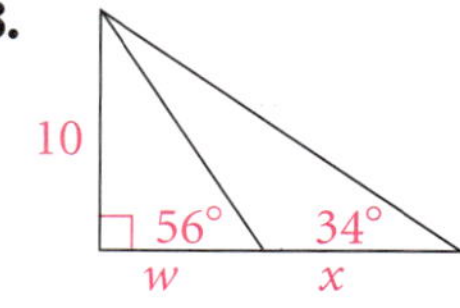

29.
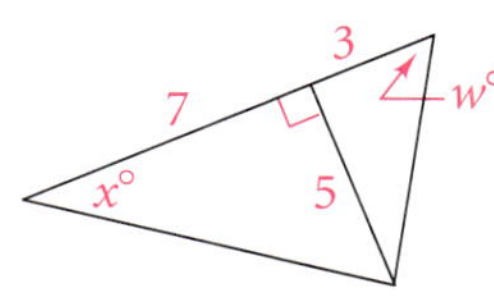

30. a. Coordinate Geometry Complete the table of values at the right. Give table entries to the nearest tenth.

b. Plot the points $(x, \tan x^\circ)$ on the coordinate plane. Connect the points with a smooth curve.

c. What happens to the tangent ratio as the angle measure x approaches 0? Approaches 90?

d. Use your graph to estimate each value.

$\tan \blacksquare^\circ = 7$ $\tan 68^\circ = \blacksquare$ $\tan \blacksquare^\circ = 3.5$

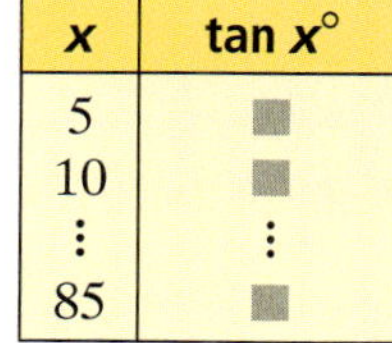

x	$\tan x^\circ$
5	■
10	■
⋮	⋮
85	■

Homework Video Tutor

Visit: PHSchool.com
Web Code: bce-0803

Real-World Connection

The world's steepest railway is the Katoomba Scenic Railway in Australia's Blue Mountains.

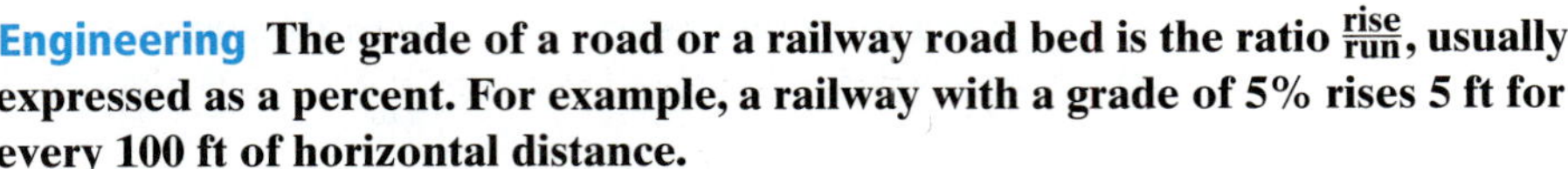

Engineering **The grade of a road or a railway road bed is the ratio $\frac{\text{rise}}{\text{run}}$, usually expressed as a percent. For example, a railway with a grade of 5% rises 5 ft for every 100 ft of horizontal distance.**

31. The Katoomba Railway, pictured at left, has a grade of 122%. What angle does its roadbed make with the horizontal?

32. The Johnstown, Pennsylvania, inclined railway was built as a "lifesaver" after the Johnstown flood of 1889. It has a 987-ft run at a 71% grade. How high does this railway lift its passengers?

33. The Fenelon Place Elevator railway in Dubuque, Iowa, lifts passengers 189 ft to the top of a bluff. It has an 83% grade. How long is this railway?

34. The Duquesne Incline Plane Company's roadway in Pittsburgh, Pennsylvania, climbs Mt. Washington, located above the mouth of the Monongahela River. It reaches a height of 400 ft with a 793-ft incline. What is its grade?

Find the missing value to the nearest tenth.

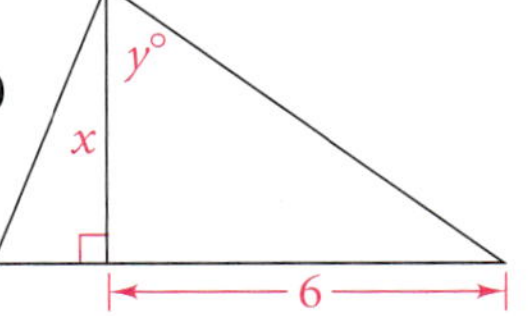

35. $x = 2, y = ■$
36. $x = 2\sqrt{3}, y = ■$
37. $x = 6, y = ■$
38. $x = 6\sqrt{3}, y = ■$
39. $x = ■, y = 15$
40. $x = ■, y = 30$
41. $x = ■, y = 45$
42. $x = ■, y = 60$
43. $x = ■, y = 75$

44. a. **Critical Thinking** Does $\tan A + \tan B = \tan(A + B)$ when $A + B < 90$? Explain.
    b. **Reasoning** Does $\tan A - \tan B = \tan(A - B)$ when $A - B > 0$? Use part (a) and indirect reasoning to explain.

45. **Graphing Calculator** Use the **TABLE** feature of your graphing calculator to study tan X as X gets close to 90. In the **Y=** screen, enter Y1 = tan X.
    a. Use the **TBLSET** feature so that X starts at 80 and changes by 1. Access the **TABLE**. From the table, what is tan X for X = 89?
    b. Perform a "numerical zoom in." Use the **TBLSET** feature, so that X starts with 89 and changes by 0.1. What is tan X for X = 89.9?
    c. Continue to numerically zoom in on values close to 90. What is the greatest value you can get for tan X on your calculator? How close is X to 90?
    d. **Writing** Use right triangles to explain the behavior of tan X found above.

Go Online PHSchool.com

For: Graphing calculator procedures
Web Code: bcb-2113

 **Challenge**

46. **Graphing Calculator** Use the **TABLE** and graphing features of your graphing calculator to study the product tan X · tan (90 − X). In the **Y=** screen, enter Y1 = tan X · tan (90 − X).
    a. Use the **TBLSET** feature so that X starts at 1 and changes by 1. Access the **TABLE**. What do you notice?
    b. Press **GRAPH**. What do you notice?
    c. **Proof** Make a conjecture about tan X · tan (90 − X) based on parts (a) and (b). Write a paragraph proof of your conjecture.

Use the given information and $\tan^{-1}$ to find $m\angle A$ to the nearest whole number.

47. $\tan 2A = 9.5144$
48. $\tan \frac{A}{3} = 0.4663$
49. $(\tan 5A)^2 = 0.3333$
50. $\frac{\tan A}{1 + \tan A} = 0.5437$

Problem Solving Hint

Think of $\tan^{-1}(x)$ as "the angle whose tangent is x."

Simplify each expression.

51. $\tan(\tan^{-1} x)$
52. $\tan^{-1}(\tan X)$

Coordinate Geometry **You can use the slope of a line to find the measure of the acute angle that the line forms with any horizontal line.**

$$\text{slope} = \frac{\text{rise}}{\text{run}} = 3$$

$$\tan A = \frac{\text{opposite}}{\text{adjacent}} = 3$$

$$m\angle A = \tan^{-1}(3) \approx 71.6$$

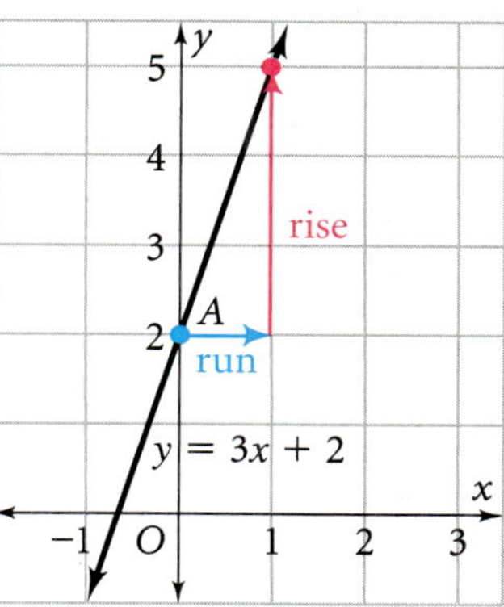

To the nearest tenth, find the measure of the acute angle that the line forms with a horizontal line.

53. $y = \frac{1}{2}x + 6$

54. $y = 6x - 1$

55. $y = 5x - 7$

56. $y = \frac{4}{3}x - 1$

57. $3x - 4y = 8$

58. $-2x + 3y = 6$

Multiple Choice Practice

For California Standards and CAHSEE Tutorials, visit PHSchool.com. Web Code: bcq-9045

GEOM 18.0

59. In $\triangle ABC$, $\cos A = \frac{8}{17}$. Use this information to find $\tan B$ and $\sin B$.

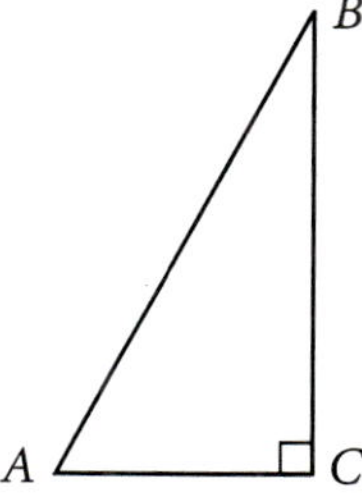

A) $\tan B = \frac{8}{15}$ and $\sin B = \frac{15}{17}$

B) $\tan B = \frac{8}{15}$ and $\sin B = \frac{8}{17}$

C) $\tan B = \frac{15}{8}$ and $\sin B = \frac{17}{8}$

D) $\tan B = \frac{15}{17}$ and $\sin B = \frac{8}{17}$

CAHSEE 7MG 1.1

60. If Tony is driving at 60 mph, what is his approximate speed in kilometers per hour? (1 mile ≈ 1.6 kilometers)

A) 36

B) 37.5

C) 61.6

D) 96

Mixed Review

Lesson 8-2

61. A diagonal of a square is 10 units. Find the length of a side of the square. Leave your answer in simplest radical form.

Lesson 8-1

The lengths of the sides of a triangle are given. Classify each triangle as *acute*, *right*, or *obtuse*.

62. 5, 8, 4

63. 15, 15, 20

64. 0.5, 1.2, 1.3

Lesson 6-7

65. For the kite pictured at the right, give the coordinates of the midpoints of its sides.

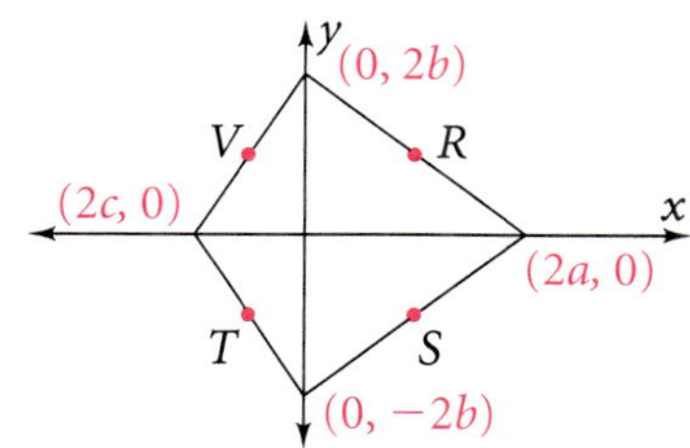

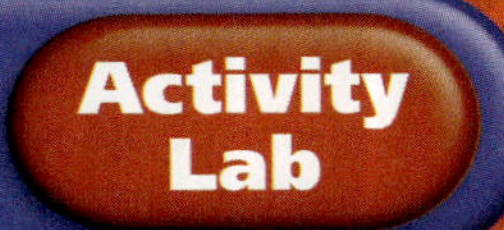

Technology

Exploring Trigonometric Ratios

FOR USE WITH LESSON 8-4

Prepares for GEOM 18.0 Students know the definitions of the basic trigonometric functions defined by the angles of a right triangle.

Construct

Use geometry software to construct $\overrightarrow{AB}$ and $\overrightarrow{AC}$ so that $\angle A$ is acute. Through a point D on $\overrightarrow{AB}$ construct a line perpendicular to $\overrightarrow{AB}$ that intersects $\overrightarrow{AC}$ in point E.

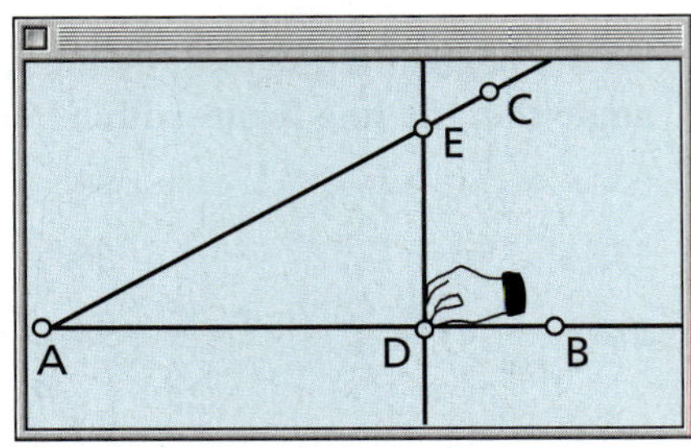

Moving point D enlarges or reduces $\triangle ADE$. Moving point C changes the size of $\angle A$.

Hypotenuse
Leg opposite $\angle A$
Leg adjacent to $\angle A$

EXERCISES

1. • Measure $\angle A$.
   • Find the lengths of the sides of $\triangle ADE$.
   • Calculate the ratio $\frac{\text{leg opposite } \angle A}{\text{hypotenuse}}$, which is $\frac{ED}{AE}$.
   • Move point D to change the size of the right triangle without changing the size of $\angle A$.

   What do you observe about the ratio as the size of $\triangle ADE$ changes?

2. • Move point C to change the size of $\angle A$.
   a. What do you observe about the ratio as the size of $\angle A$ changes?
   b. What value does the ratio approach as $m\angle A$ approaches 0? As $m\angle A$ approaches 90?

3. • Make a table that shows values for $m\angle A$ and the ratio $\frac{\text{leg opposite } \angle A}{\text{hypotenuse}}$. In your table, include 10, 20, 30, . . . , 80 for $m\angle A$.
   • Compare your table with the table of trigonometric ratios on page 769.

   Do your values for $\frac{\text{leg opposite } \angle A}{\text{hypotenuse}}$ match the values in one of the columns of the table? What is the name of this ratio in the table?

Extend

4. Repeat Exercises 1–3 for the ratio $\frac{\text{leg adjacent to } \angle A}{\text{hypotenuse}}$, which is $\frac{AD}{AE}$.

5. Repeat Exercises 1–3 for the ratio $\frac{\text{leg opposite } \angle A}{\text{leg adjacent to } \angle A}$, which is $\frac{ED}{AD}$.

6. • Choose a measure for $\angle A$ and determine the ratio $r = \frac{\text{leg opposite } \angle A}{\text{hypotenuse}}$. Record $m\angle A$ and this ratio.
   • Manipulate the triangle so that $\frac{\text{leg adjacent to } \angle A}{\text{hypotenuse}}$ has the same value r. Record this $m\angle A$ and compare it with your first value of $m\angle A$.
   • Repeat this procedure several times.
   • Look for a pattern in the two measures of $\angle A$ that you found for the different values of r.

   Make a conjecture.

8-4 Sine and Cosine Ratios

California Content Standards

GEOM 18.0 Students know the definitions of the basic trigonometric functions defined by the angles of a right triangle. They are able to use elementary relationships between them. ***Develop***

GEOM 19.0 Students use trigonometric functions to solve for an unknown length of a side of a right triangle, given an angle and a length of a side. ***Develop***

What You'll Learn

- To use sine and cosine to determine side lengths in triangles

. . . And Why

To use the sine ratio to estimate astronomical distances indirectly, as in Example 2

Check Skills You'll Need

GO for Help Lesson 8-3

For each triangle, find (a) the length of the leg opposite $\angle B$ and (b) the length of the leg adjacent to $\angle B$.

1.

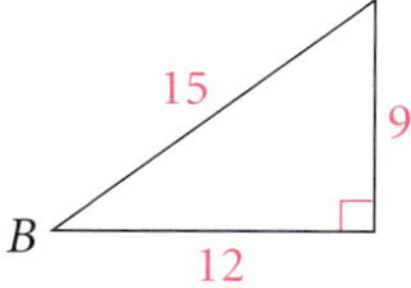

2.

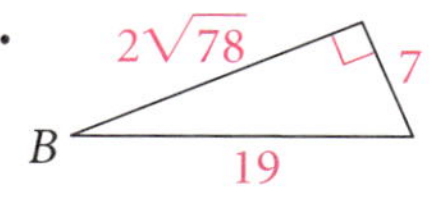

3.

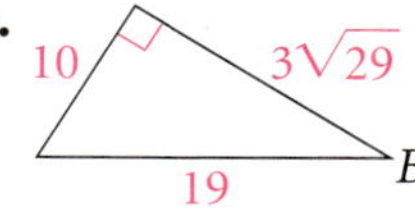

New Vocabulary • sine • cosine • identity

1 Using Sine and Cosine in Triangles

Real-World Connection For an angle of a given size, the sine and cosine ratios are constant, no matter where the angle is located.

The tangent ratio, as you have seen, involves both legs of a right triangle. The sine and cosine ratios involve one leg and the hypotenuse.

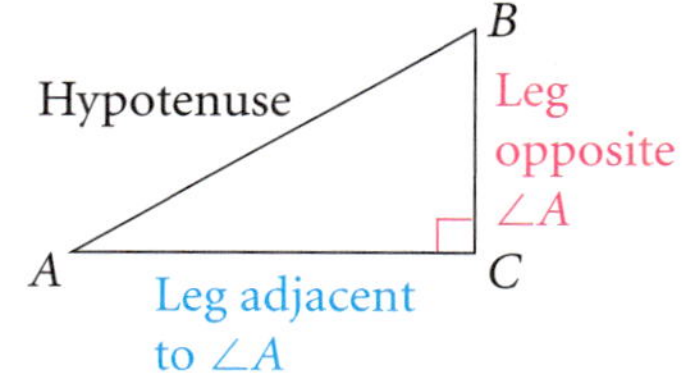

sine of $\angle A = \dfrac{\text{leg opposite } \angle A}{\text{hypotenuse}}$

cosine of $\angle A = \dfrac{\text{leg adjacent to } \angle A}{\text{hypotenuse}}$

These equations can be abbreviated:

$$\sin A = \frac{\text{opposite}}{\text{hypotenuse}} \qquad \cos A = \frac{\text{adjacent}}{\text{hypotenuse}}$$

1 EXAMPLE Writing Sine and Cosine Ratios

Use the triangle to write each ratio.

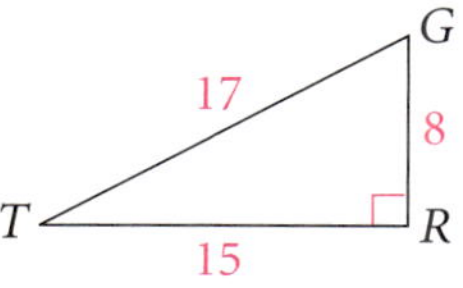

a. $\sin T$ $\quad \sin T = \dfrac{\text{opposite}}{\text{hypotenuse}} = \dfrac{8}{17}$

b. $\cos T$ $\quad \cos T = \dfrac{\text{adjacent}}{\text{hypotenuse}} = \dfrac{15}{17}$

c. $\sin G$ $\quad \sin G = \dfrac{\text{opposite}}{\text{hypotenuse}} = \dfrac{15}{17}$

d. $\cos G$ $\quad \cos G = \dfrac{\text{adjacent}}{\text{hypotenuse}} = \dfrac{8}{17}$

CA Standards Check 1 **a.** Write the sine and cosine ratios for $\angle X$ and $\angle Y$.

b. Critical Thinking When does $\sin X = \cos Y$? Explain.

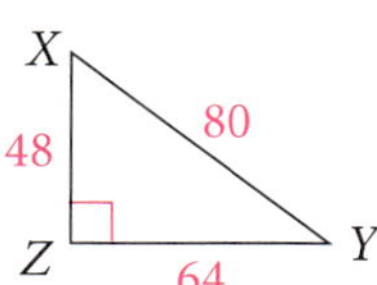

One way to describe the relationship of sine and cosine is to say that $\sin x° = \cos (90 - x)°$ for values of x between 0 and 90. This type of equation is called an **identity** because it is true for all the allowed values of the variable. You will discover other identities in the exercises.

2 EXAMPLE Real-World Connection

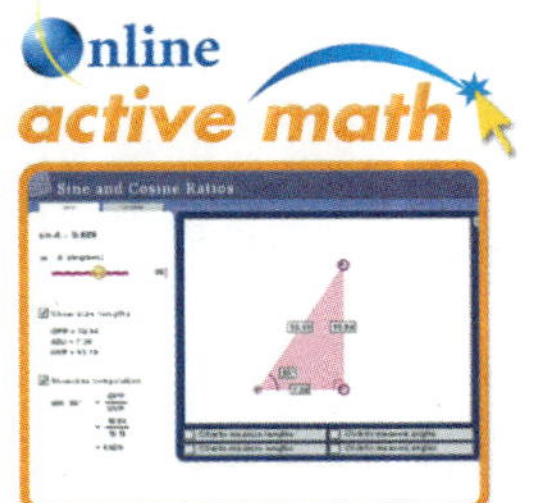

For: Sine and Cosine Activity
Use: Interactive Textbook, 8-4

Astronomy The trigonometric ratios have been known for centuries by peoples in many cultures. The Polish astronomer Nicolaus Copernicus (1473–1543) developed a method for determining the sizes of orbits of planets closer to the sun than Earth. The key to his method was determining when the planets were in the position shown in the diagram, and then measuring the angle to find a.

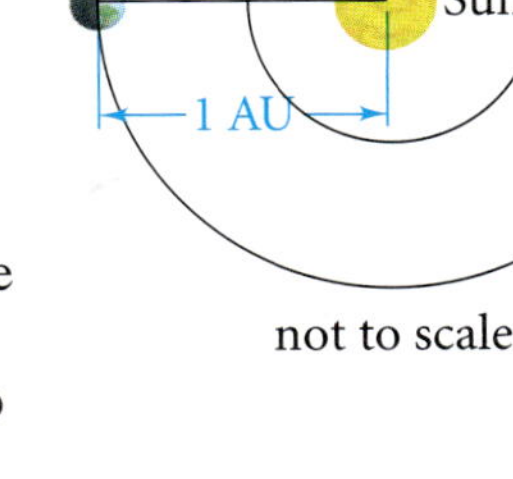

not to scale

If $a = 22.3$ for Mercury, how far is Mercury from the sun in astronomical units (AU)? One astronomical unit is defined as the average distance from Earth to the center of the sun, about 93 million miles.

$\sin 22.3° = \frac{x}{1}$ **Use the sine ratio.**

$x = \sin 22.3°$ **Solve for x.**

SIN 22.3 ENTER .37945616 **Use a calculator.**

Mercury is about 0.38 AU from the sun.

2 **a.** If $a = 46$ for Venus, how far is Venus from the sun in AU?
b. About how many miles from the sun is Venus? Mercury?

When you know the leg and hypotenuse lengths of a right triangle, you can use inverse of sine and inverse of cosine to find the measures of the acute angles.

3 EXAMPLE Using the Inverse of Cosine and Sine

Find $m\angle L$ to the nearest degree.

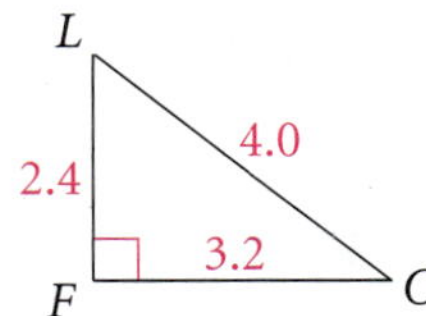

Problem Solving Hint

Think of $\cos^{-1}\left(\frac{2.4}{4.0}\right)$ as "the angle whose cosine is $\frac{2.4}{4.0}$," and $\sin^{-1}\left(\frac{3.2}{4.0}\right)$ as "the angle whose sine is $\frac{3.2}{4.0}$."

Method 1		**Method 2**
$\cos L = \frac{2.4}{4.0}$	← **Find the trigonometric ratio.** →	$\sin L = \frac{3.2}{4.0}$
$m\angle L = \cos^{-1}\left(\frac{2.4}{4.0}\right)$	← **Use the inverse.** →	$m\angle L = \sin^{-1}\left(\frac{3.2}{4.0}\right)$
COS^{-1} 2.4 ÷ 4.0 ENTER 53.130102	← **Use a calculator.** →	SIN^{-1} 3.2 ÷ 4.0 ENTER 53.130102
$m\angle L \approx 53$		$m\angle L \approx 53$

3 Find the value of x. Round your answer to the nearest degree.

a.

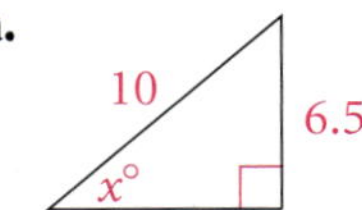

b.

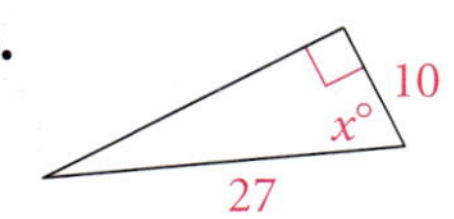

EXERCISES

For more exercises, see *Extra Skill, Word Problem, and Proof Practice.*

Standards Practice

GEOM 18.0, 19.0

A Practice by Example

Example 1 (page 439)

GO for Help

Write the ratios for sin *M* and cos *M*.

1.

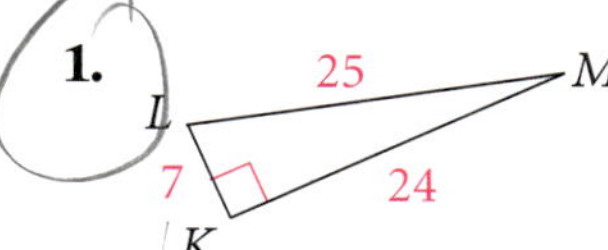

2.

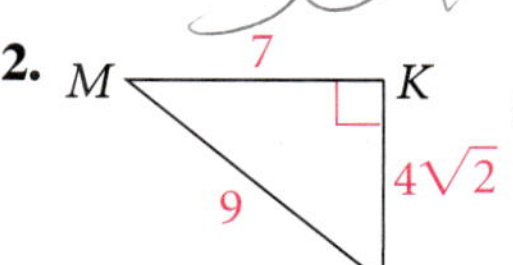

3.

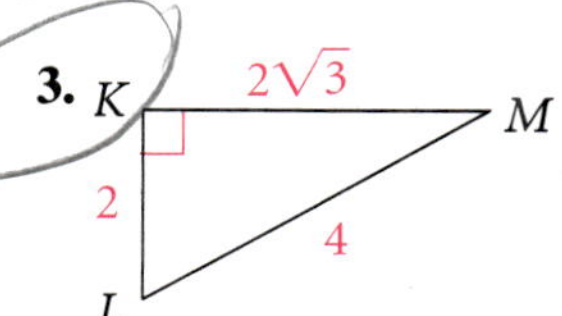

Example 2 (page 440)

Find the value of *x*. Round answers to the nearest tenth.

4. **7.**

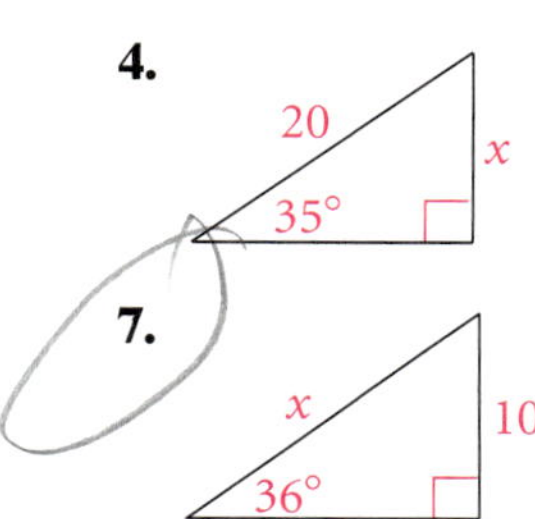

5.

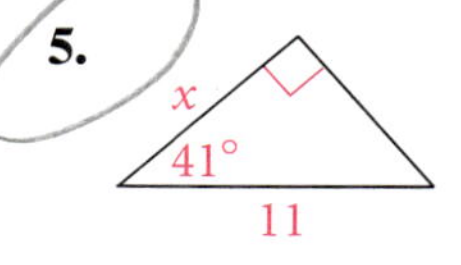

8. 12, *x*, 21°

6. **9.**

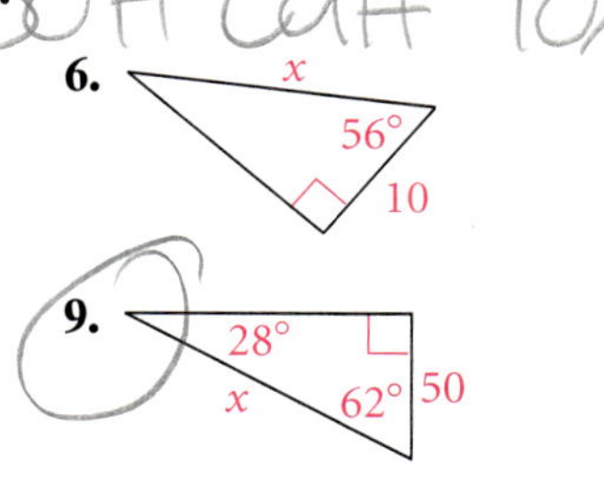

10. Escalators An escalator in the subway system of St. Petersburg, Russia, has a vertical rise of 195 ft 9.5 in., and rises at an angle of 10.4°. How long is the escalator? Round your answer to the nearest foot.

Example 3 (page 440)

Find the value of *x*. Round answers to the nearest degree.

11.

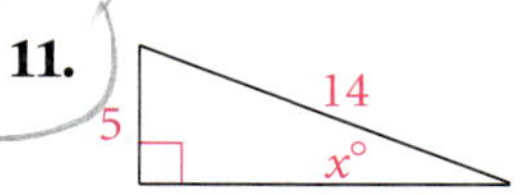

12.

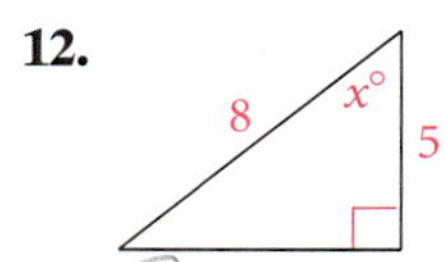

13.

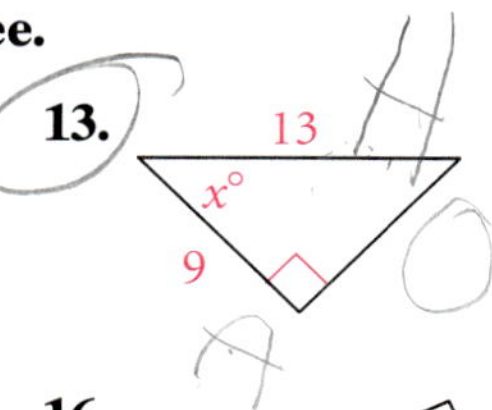

14.

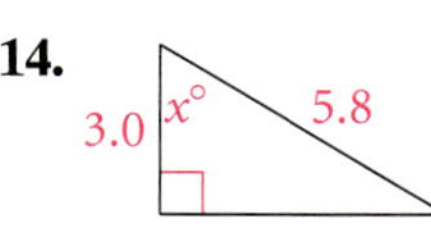

15.

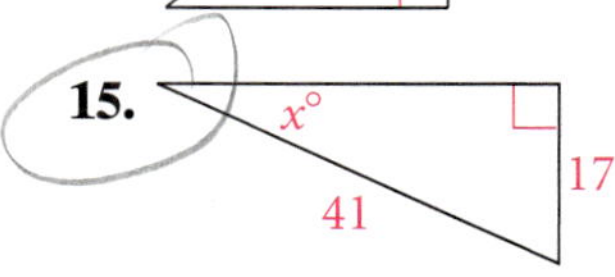

16.

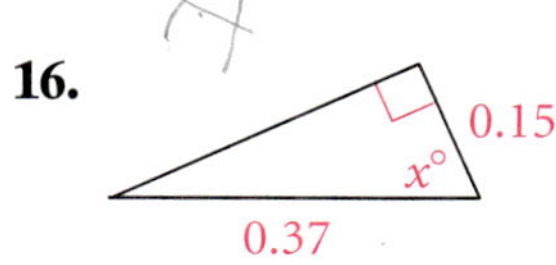

B Apply Your Skills

17. Construction Carlos is planning to build a grain bin with a radius of 15 ft. He reads that the recommended slant of the roof is 25°. He wants the roof to overhang the edge of the bin by 1 ft. What should the length *x* be? Give your answer in feet and inches.

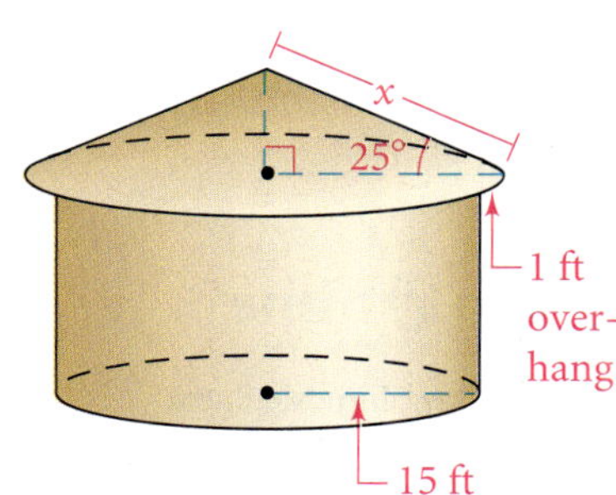

Use what you know about trigonometric ratios (and other identities) to show that each equation is an identity.

18. $\tan X = \frac{\sin X}{\cos X}$ **19.** $\sin X = \cos X \cdot \tan X$ **20.** $\cos X = \frac{\sin X}{\tan X}$

21. Error Analysis A student states that $\sin A > \sin X$ because the lengths of the sides of $\triangle ABC$ are greater than the lengths of the sides of $\triangle XYZ$. Is the student correct? Explain.

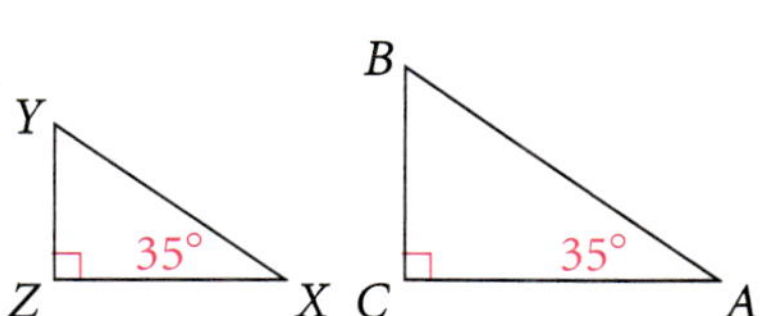

Visit: PHSchool.com
Web Code: bce-0804

Find the values of w and then x. Round lengths to the nearest tenth and angle measures to the nearest degree.

22. 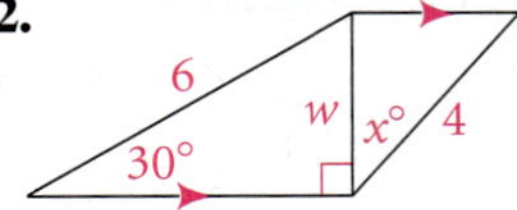

23. 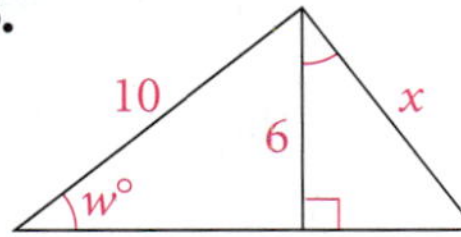

24. 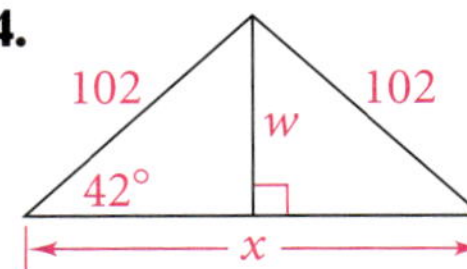

25. a. In $\triangle ABC$, how does $\sin A$ compare to $\cos B$? Is this true for the acute angles of other right triangles?

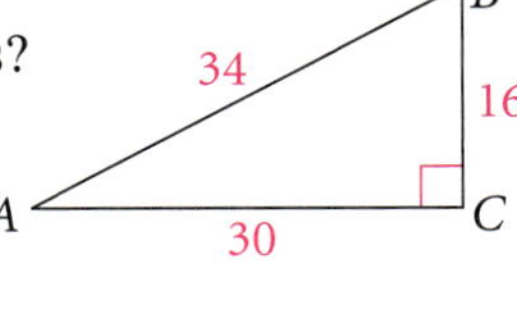

b. Reading Math The word cosine is derived from the words *complement's sine* (see page 694). Which angle in $\triangle ABC$ is the complement of $\angle A$? Of $\angle B$?

c. Explain why the derivation of the word cosine makes sense.

26. Find each ratio.

a. $\sin P$ **b.** $\cos P$

c. $\sin R$ **d.** $\cos R$

e. Make a conjecture about how the sine and cosine of a 45° angle are related.

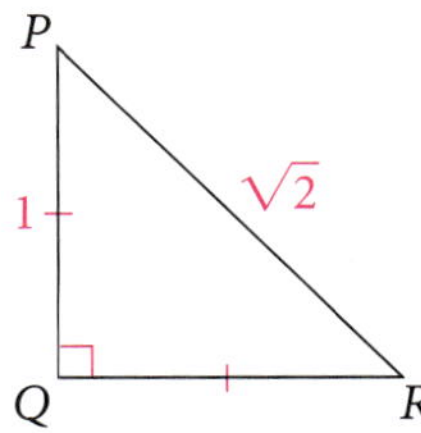

27. Writing Leona said that if she had a diagram that showed the measure of one acute angle and the length of one side of a right triangle, she could find the measure of the other acute angle and the lengths of the other sides. Is she correct? Explain.

28. Find each ratio.

a. $\sin S$ **b.** $\cos S$

c. $\sin T$ **d.** $\cos T$

e. Make a conjecture about how the sine and cosine of a 30° angle are related.

f. Make a conjecture about how the sine and cosine of a 60° angle are related.

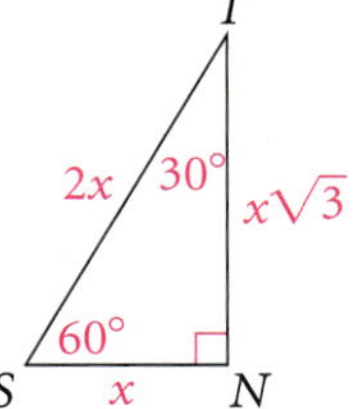

29. For right $\triangle ABC$ with right $\angle C$, prove each of the following.

a. $\sin A < 1$, no matter how large $\angle A$ is.

b. $\cos A < 1$, no matter how small $\angle A$ is.

30. Graphing Calculator Use the **TABLE** feature of your graphing calculator to study sin X as X gets close (but $\neq$) to 90. In the  screen, enter Y1 = sin X.

a. Use the **TBLSET** feature so that X starts at 80 and changes by 1. Access the **TABLE**. From the table, what is sin X for X = 89?

b. Perform a "numerical zoom in." Use the **TBLSET** feature, so that X starts with 89 and changes by 0.1. What is sin X for X = 89.9?

c. Continue to numerically zoom in on values close to 90. What is the greatest value you can get for sin X on your calculator? How close is X to 90? Does your result contradict what you are asked to prove in Exercise 29a?

d. Writing Use right triangles to explain the behavior of sin X found above.

For: Graphing calculator procedures
Web Code: bcb-2113

Challenge

Show that each equation is an identity by showing that each expression on the left simplifies to 1.

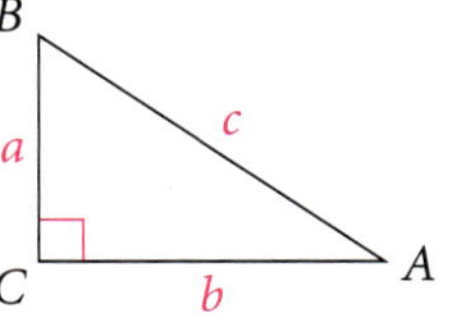

31. $(\sin A)^2 + (\cos A)^2 = 1$ **32.** $(\sin B)^2 + (\cos B)^2 = 1$

33. $\frac{1}{(\cos A)^2} - (\tan A)^2 = 1$ **34.** $\frac{1}{(\sin A)^2} - \frac{1}{(\tan A)^2} = 1$

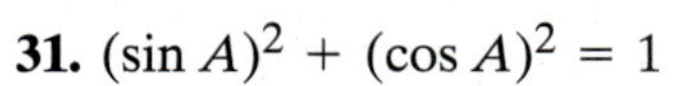

35. Show that $(\tan A)^2 - (\sin A)^2 = (\tan A)^2 (\sin A)^2$ is an identity.

Real-World Connection

Poland honored Copernicus with this 1000-zloty note, last used in 1995.

36. Astronomy Copernicus devised a method different from the one in Example 2 in order to find the sizes of the orbits of planets farther from the sun than Earth. His method involved noting the number of days between the times that a planet was in the positions labeled A and B in the diagram. Using this time and the number of days in each planet's year, he calculated c and d.

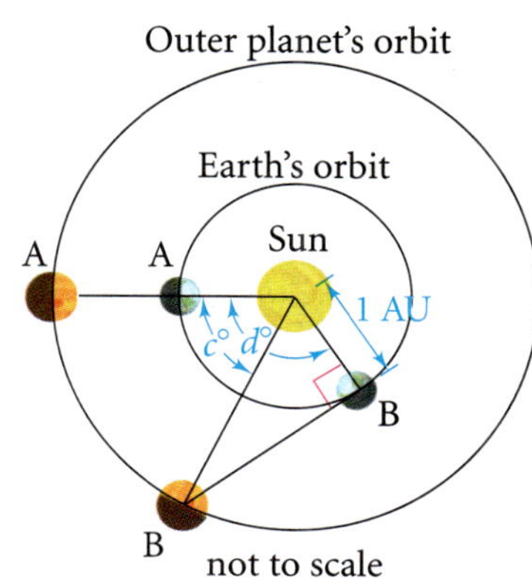

a. For Mars, $c = 55.2$ and $d = 103.8$. How far is Mars from the sun in astronomical units (AU)?

b. For Jupiter, $c = 21.9$ and $d = 100.8$. How far is Jupiter from the sun in astronomical units?

Multiple Choice Practice

For California Standards and CAHSEE Tutorials, visit PHSchool.com. Web Code: bcq-9045

GEOM 19.0

37. In the diagram at the right, $m\angle A = 46°$ and $\overline{AB} = 83$. Which equation could be used to find x in $\triangle ABC$?

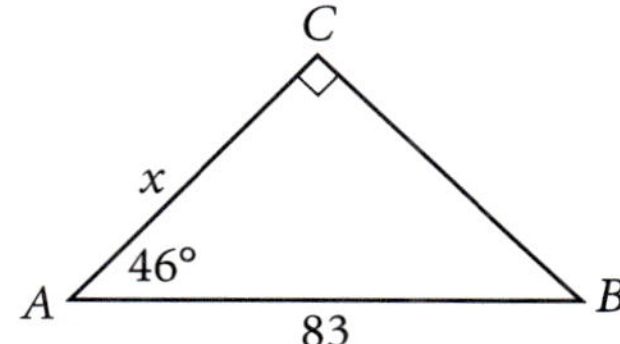

(A) $x = \frac{83}{\cos 46°}$

(B) $x = 83(\cos 46°)$

(C) $x = \frac{\tan 46°}{83}$

(D) $x = 83(\tan 46°)$

CAHSEE 6SDAP 3.3

38. Maya has a spinner and a cube. The spinner has 5 equal sections numbered from 1 to 5. The surfaces of the cube are numbered from 1 to 6. If Maya spins the spinner and tosses the cube, what is the probability that she will get an even number on both the spinner and the cube?

(A) 0.03 (B) 0.2 (C) 0.25 (D) 0.9

Mixed Review

Lesson 8-3

Find the value of x. Round answers to the nearest tenth.

39.

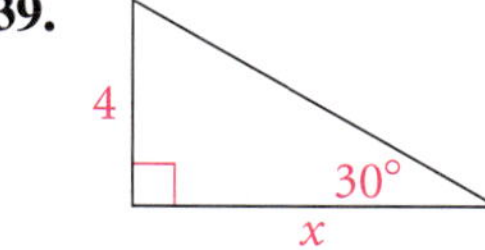

40.

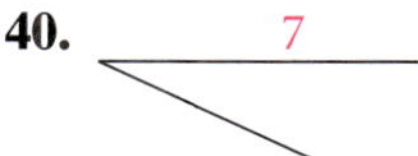

41.

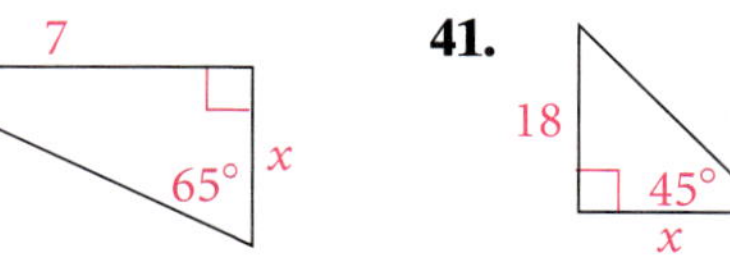

Lesson 7-2

42. The wall of a room is in the shape of a golden rectangle. If the height of the wall is 8 ft, what are the possible lengths of the wall to the nearest tenth?

Lesson 6-2

Find the value of x for each parallelogram.

43.

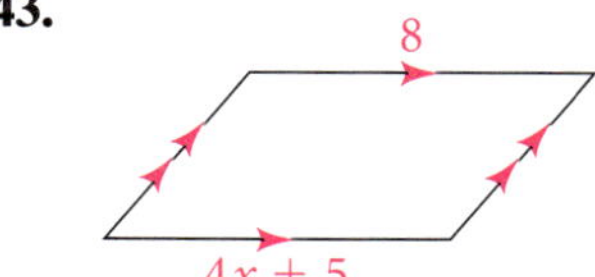

44.

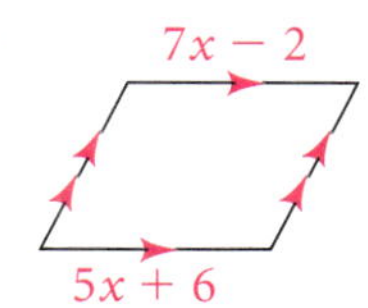

45.

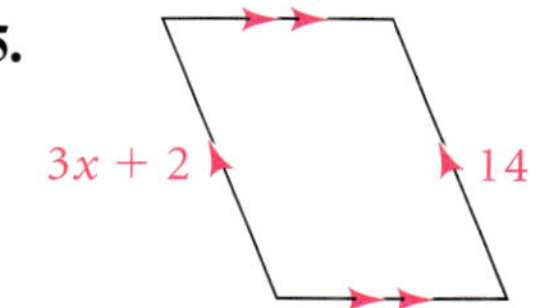

Activity Lab

Hands-On

Measuring From Afar

For Use With Lesson 8-5

In this activity, you will make a tool to find heights. You will use similar right triangles and trigonometric ratios.

ACTIVITY

1. The device shown below is called an *inclinometer.* Make your own inclinometer using a protractor, a piece of string, and a washer.

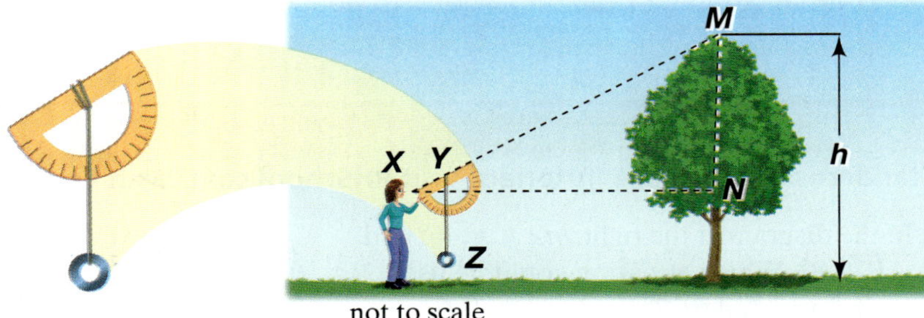

not to scale

2. The string on the inclinometer shows $m\angle XYZ$. Explain how you can find $m\angle X$ if you know $m\angle XYZ$.

3. Think about how you can calculate an approximate height of the tree using the inclinometer and a trigonometric ratio.

a. What other measurement do you need besides the $m\angle X$?

b. What trigonometric ratio do you need?

c. Show how to find the height of the tree.

4. Use an inclinometer to find the height of a tall object at your school.

EXERCISES

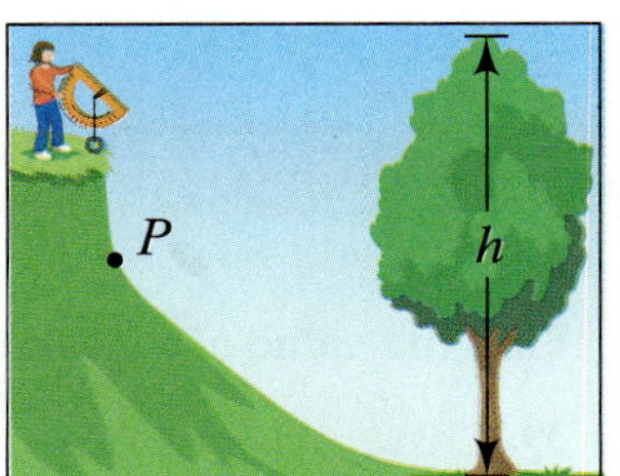

not to scale

5. You are on a steep hillside directly across from the top of a tree. Explain how you could use an inclinometer, a trigonometric ratio, and the distance from the hilltop to the base of the tree to find the approximate height of the tree.

6. Suppose you could climb up only to point P on the hillside. Explain how you could use an inclinometer and trigonometric ratios to find the height of the tree from that spot.

7. Use the diagram at the right. Show that the height of the tree can be found using the formula

$$h = d(\cos b)(\tan a + \tan b).$$

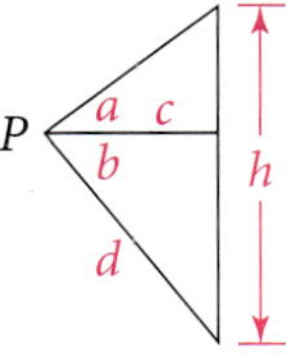

Angles of Elevation and Depression

 **California Content Standards**

GEOM 18.0 Students know the definitions of the basic trigonometric functions defined by the angles of a right triangle. ***Master***

GEOM 19.0 Students use trigonometric functions to solve for an unknown length of a side of a right triangle, given an angle and a length of a side. ***Develop***

What You'll Learn

- To use angles of elevation and depression to solve problems

. . . And Why

To use the angle of elevation to calculate the height of a natural wonder, as in Example 2

Check Skills You'll Need

GO for Help Lesson 6-1

Refer to rectangle *ABCD* to complete the statements.

1. $\angle 1 \cong$ ■ **2.** $\angle 5 \cong$ ■

3. $\angle 3 \cong$ ■ **4.** $m\angle 1 + m\angle 5 =$ ■

5. $m\angle 10 + m\angle 3 =$ ■ **6.** $\angle 10 \cong$ ■

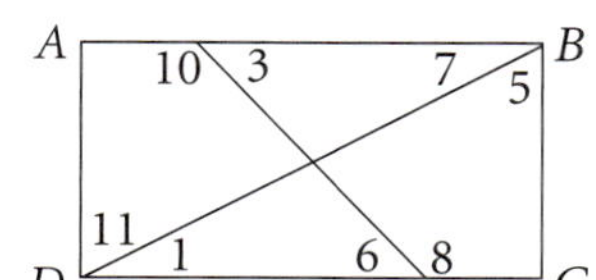

 **New Vocabulary** • angle of elevation • angle of depression

1 Using Angles of Elevation and Depression

Suppose a person on the ground sees a hot-air balloon gondola at a 38° angle above a horizontal line.

This angle is the **angle of elevation.**

At the same time, a person in the hot-air balloon sees the person on the ground at a 38° angle below a horizontal line.

This angle is the **angle of depression.**

Examine the diagram. The angle of elevation is congruent to the angle of depression because they are alternate interior angles.

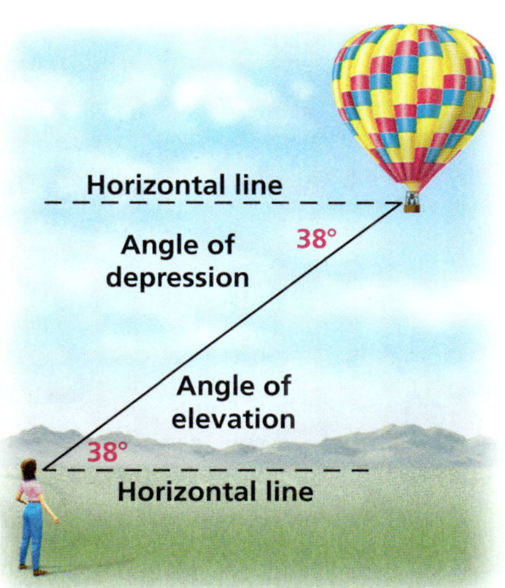

1 EXAMPLE Identifying Angles of Elevation and Depression

Describe each angle as it relates to the situation shown.

a. $\angle 1$ $\angle 1$ is the angle of depression from the peak to the hiker.

b. $\angle 4$ $\angle 4$ is the angle of elevation from the hut to the hiker.

1 Describe each angle as it relates to the situation in Example 1.

a. $\angle 2$ **b.** $\angle 3$

Surveyors use two instruments, the transit and the theodolite, to measure angles of elevation and depression. On both instruments, the surveyor sets the horizon line perpendicular to the direction of gravity. Using gravity to find the horizon line ensures accurate measures even on sloping surfaces.

Horizon line

Pull of gravity

2 EXAMPLE Real-World Connection

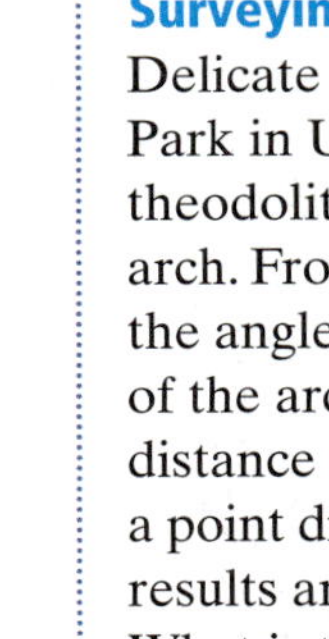

Surveying To find the height of Delicate Arch in Arches National Park in Utah, a surveyor levels a theodolite with the bottom of the arch. From there, she measures the angle of elevation to the top of the arch. She then measures the distance from where she stands to a point directly under the arch. Her results are shown in the diagram. What is the height of the arch?

$\tan 48° = \frac{x}{36}$ **Use the tangent ratio.**

$x = 36(\tan 48°)$ **Solve for *x*.**

36 TAN 48 ENTER 39.982051 **Use a calculator.**

So $x \approx 40$. To find the height of the arch, add the height of the theodolite. Since $40 + 5 = 45$, Delicate Arch is about 45 feet high.

CA Standards Check

2 You sight a rock climber on a cliff at a 32° angle of elevation. The horizontal ground distance to the cliff is 1000 ft. Find the line-of-sight distance to the rock climber.

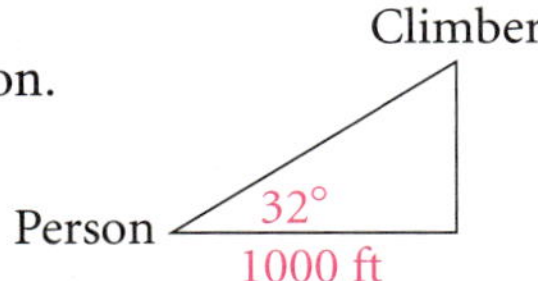

Test-Taking Tip

For problems with angles of elevation or depression, draw a detailed diagram to help you visualize the given information.

3 EXAMPLE Real-World Connection

Multiple Choice To approach runway 17 of the Ponca City Municipal Airport in Oklahoma, the pilot must begin a 3° descent starting from an altitude of 2714 ft. The airport altitude is 1007 ft. How many miles from the runway is the airplane at the start of this approach?

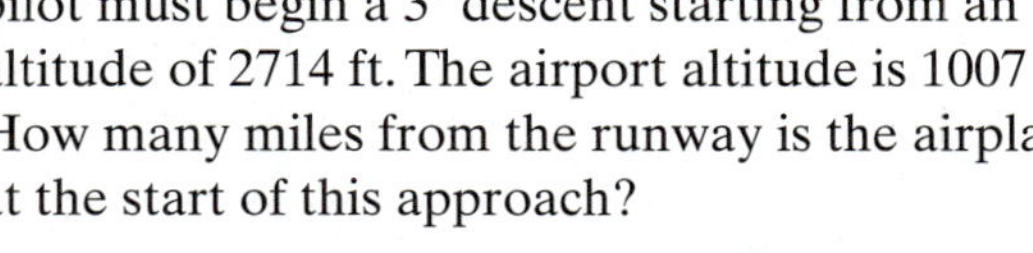

Ⓐ 3.6 mi Ⓑ 5.7 mi Ⓒ 6.2 mi Ⓓ 9.8 mi

The airplane is 2714 − 1007, or 1707 ft above the level of the airport.

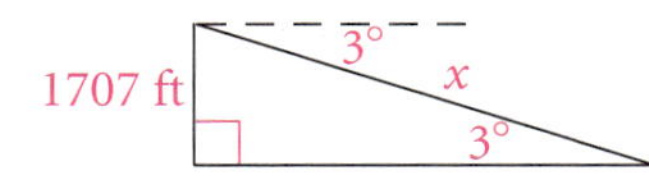

$\sin 3° = \frac{1707}{x}$ **Use the sine ratio.**

$x = \frac{1707}{\sin 3°}$ **Solve for *x*.**

1707 ÷ SIN 3 ENTER 32616.2 **Use a calculator.**

÷ 5280 ENTER 6.1773105 **Divide by 5280 to convert feet to miles.**

The airplane is about 6.2 mi from the runway. The correct answer is C.

CA Standards Check

3 An airplane pilot sights a life raft at a 26° angle of depression. The airplane's altitude is 3 km. What is the airplane's horizontal distance *d* from the raft?

EXERCISES
Standards Practice

For more exercises, see *Extra Skill, Word Problem, and Proof Practice.*

GEOM 18.0, 19.0

Practice by Example

Example 1 (page 445)

GO for Help

Describe each angle as it relates to the situation in the diagram.

1. $\angle 1$ **2.** $\angle 2$ **3.** $\angle 3$ **4.** $\angle 4$ **5.** $\angle 5$ **6.** $\angle 6$ **7.** $\angle 7$ **8.** $\angle 8$

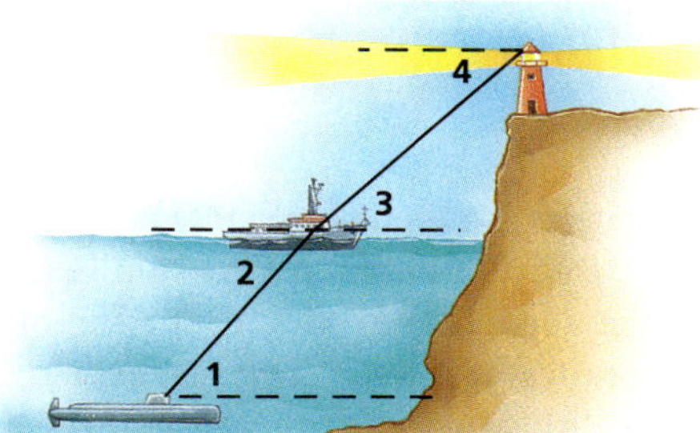

Example 2 (page 446)

Find the value of x. Round the lengths to the nearest tenth of a unit.

9. 100 ft; 20°; x

10.

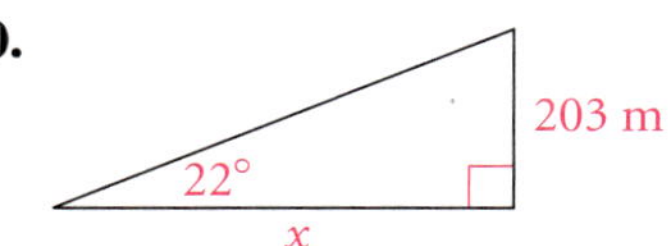

11. Meteorology A meteorologist measures the angle of elevation of a weather balloon as 41°. A radio signal from the balloon indicates that it is 1503 m from his location. To the nearest meter, how high above the ground is the balloon?

Example 3 (page 446)

Find the value of x. Round the lengths to the nearest tenth of a unit.

12. 580 yd; 27°; x

13.

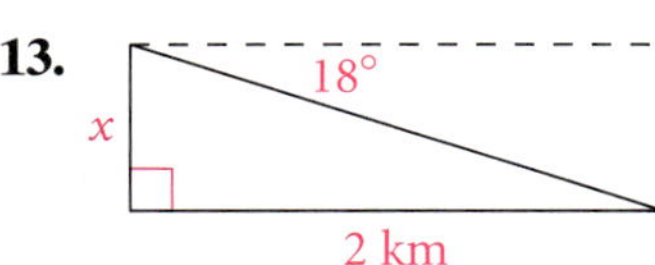

14. Indirect Measurement Miguel looks out from the crown of the Statue of Liberty approximately 250 ft above ground. He sights a ship coming into New York harbor and measures the angle of depression as 18°. Find the distance from the base of the statue to the ship to the nearest foot.

Apply Your Skills

15. Flagpole The world's tallest unsupported flagpole is a 282-ft-tall steel pole in Surrey, British Columbia. The shortest shadow cast by the pole during the year is 137 ft long. To the nearest degree, what is the angle of elevation of the sun when the shortest shadow is cast?

16. Engineering The Americans with Disabilities Act states that wheelchair ramps can have a slope no greater than $\frac{1}{12}$. Find the angle of elevation of a ramp with this slope. Round your answer to the nearest tenth.

17. Construction Two office buildings are 51 m apart. The height of the taller building is 207 m. The angle of depression from the top of the taller building to the top of the shorter building is 15°. Find the height of the shorter building to the nearest meter.

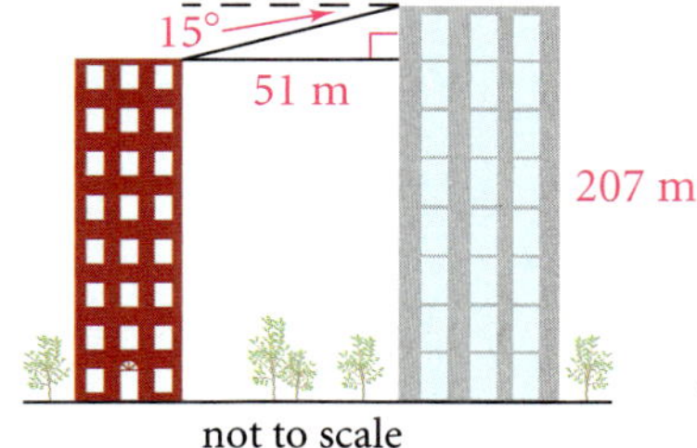

not to scale

Homework Video Tutor

Visit: PHSchool.com
Web Code: bce-0805

For a guide to solving Exercise 18, see p. 451.

18. **Aerial Television** A blimp is providing aerial television views of a football game. The television camera sights the stadium at a 7° angle of depression. The blimp's altitude is 400 m. What is the line-of-sight distance from the TV camera to the stadium, to the nearest hundred meters?

 **Algebra** **The angle of elevation e from A to B and the angle of depression d from B to A are shown below. Find the measure of each angle.**

19. e: $(7x - 5)°$, d: $4(x + 7)°$

20. e: $(3x + 1)°$, d: $2(x + 8)°$

21. e: $(x + 21)°$, d: $3(x + 3)°$

22. e: $5(x - 2)°$, d: $(x + 14)°$

23. **Multiple Choice** An engineer is 980 ft from the base of a fountain at Fountain Hills, Arizona. The angle of elevation to the top of the column of water is 29.7°. The surveyor's angle measuring device is at the same level as the base of the fountain. Find the height of the column of water to the nearest 10 ft.

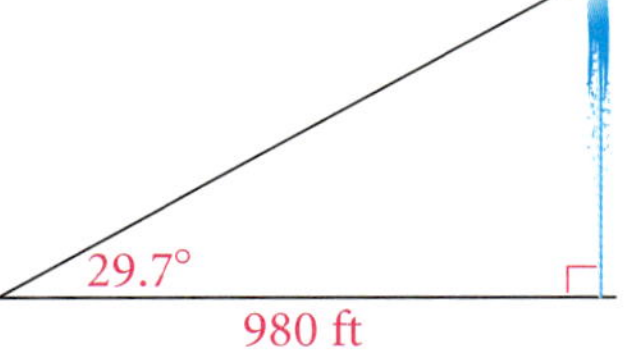

 490 ft 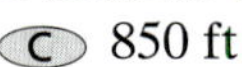 560 ft 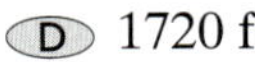 850 ft Ⓓ 1720 ft

24. **Writing** A communications tower is located on a plot of flat land. The tower is supported by several guy wires. Assume that you are able to measure distances along the ground, as well as angles formed by the guy wires and the ground. Explain how you could estimate each of the following measurements.
a. the length of any guy wire
b. how high on the tower each wire is attached

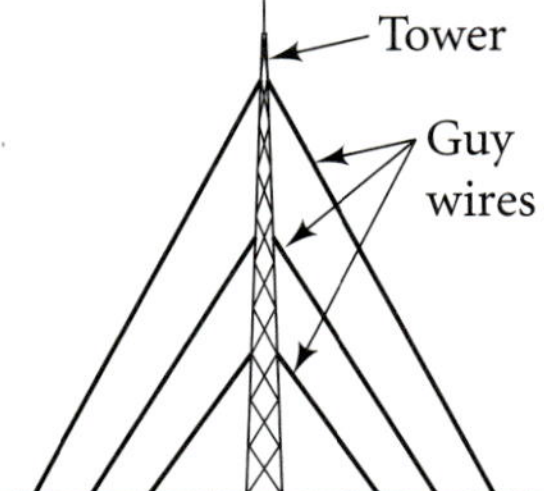

Flying **An airplane at altitude a flies distance d towards you with velocity v. You watch for time t and measure its angles of elevation, $\angle E_1$ and $\angle E_2$, at the start and end of your watch. Find the missing information.**

25. $a = ■$ mi, $v = 5$ mi/min, $t = 1$ min, $m\angle E_1 = 45$, $m\angle E_2 = 90$

26. $a = 2$ mi, $v = ■$ mi/min, $t = 15$ s, $m\angle E_1 = 40$, $m\angle E_2 = 50$

27. $a = 4$ mi, $d = 3$ mi, $v = 6$ mi/min, $t = ■$ min, $m\angle E_1 = 50$, $m\angle E_2 = ■$

28. **Meteorology** One method that meteorologists could use to find the height of a layer of clouds above the ground is to shine a bright spotlight directly up onto the cloud layer and measure the angle of elevation from a known distance away. Find the height of the cloud layer in the diagram to the nearest 10 m.

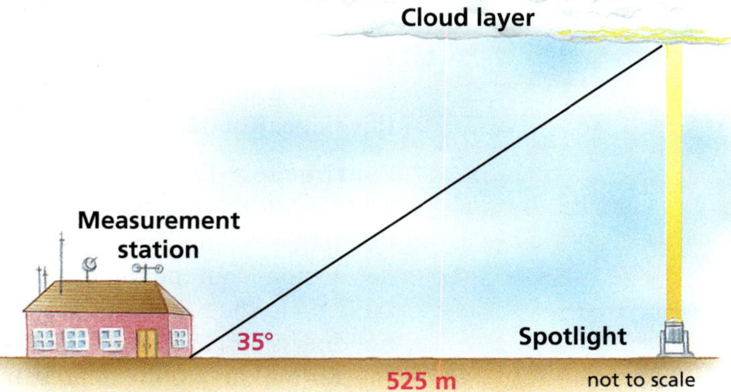

Challenge

29. Firefighting A firefighter on the ground sees fire break through a window near the top of the building. There is voice contact between the ground and firefighters on the roof. The angle of elevation to the windowsill is 28°. The angle of elevation to the top of the building is 42°. The firefighter is 75 ft from the building and her eyes are 5 ft above the ground. What roof-to-windowsill distance can she report to the firefighters on the roof?

28° 42° 75 ft

not to scale

30. Geography For locations in the United States, the relationship between the latitude ℓ and the greatest angle of elevation a of the sun at noon on the first day of summer is $a = 90° - \ell + 23\frac{1}{2}°$. Find the latitude of your town. Then determine the greatest angle of elevation of the sun for your town on the first day of summer.

Multiple Choice Practice

For California Standards and CAHSEE Tutorials, visit PHSchool.com. Web Code: bcq-9045

GEOM 19.0

31. A bird-watcher sees a California Condor fly over a pond at an angle of elevation of 35°. If the distance from the bird-watcher to the pond is 150 feet, approximately how many feet above the pond is the condor flying? You must take into account the height of the bird-watcher, which is 5 feet.

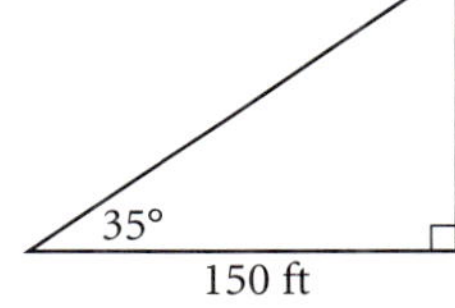

Ⓐ 110 Ⓑ 128 Ⓒ 159 Ⓓ 220

GEOM 20.0

32. In a 45°-45°-90° triangle, the hypotenuse measures $20\sqrt{2}$ feet. What is the length in feet of the two legs?

Ⓐ $40, 20\sqrt{2}$ Ⓑ 40, 40 Ⓒ 20, 20 Ⓓ $\sqrt{2}, 20$

CAHSEE 7MR 2.1

33. Which is the best estimate of 237×189?

Ⓐ 400 Ⓑ 4,000 Ⓒ 40,000 Ⓓ 400,000

CAHSEE 7MR 2.1

34. Anatole plans to buy a sweater that is \$39.99. If the sweater is on sale for 10% off, what is the best estimate of the final cost?

Ⓐ \$35.00 Ⓑ \$36.00 Ⓒ \$37.00 Ⓓ \$38.00

Mixed Review

Lesson 8-4

Find the value of *x*. Round answers to the nearest tenth.

35.

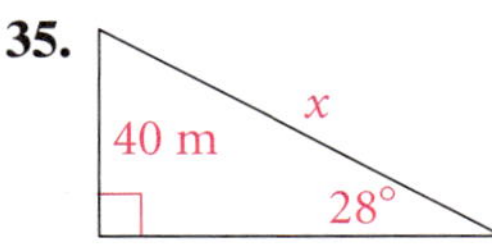

36.

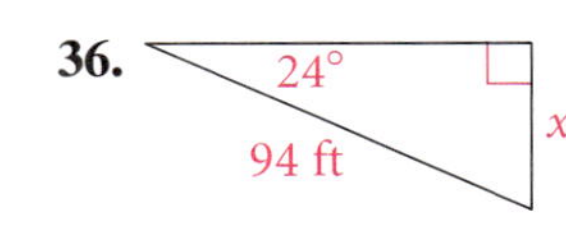

37.

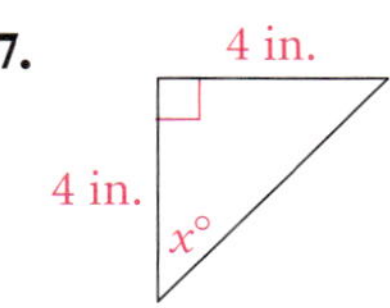

Lesson 6-1 x^2 **Algebra** **Find the value of each variable. Then find the length of each side.**

38.

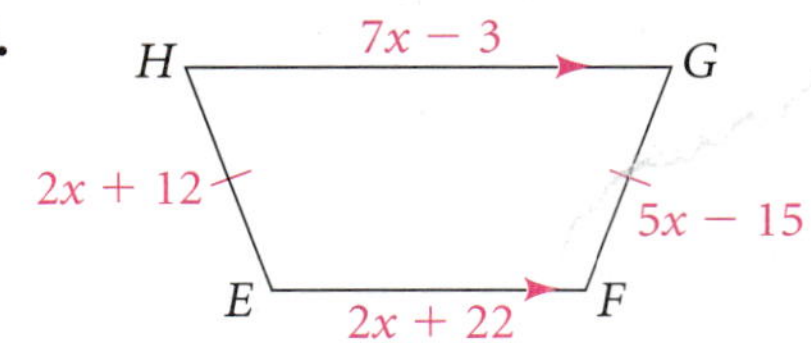

39.

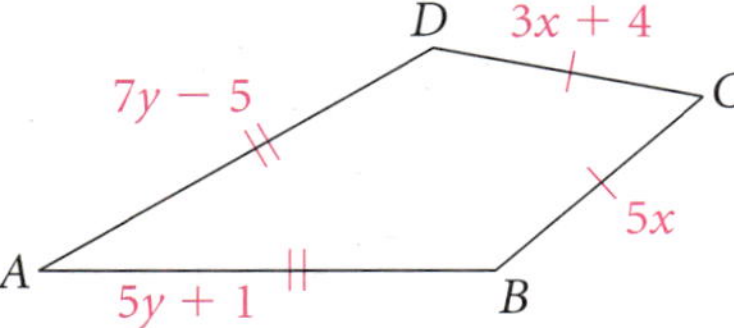

Lesson 4-4 **40. Given:** $\angle QPS \cong \angle RSP$, $\angle Q \cong \angle R$

Prove: $\overline{PQ} \cong \overline{SR}$

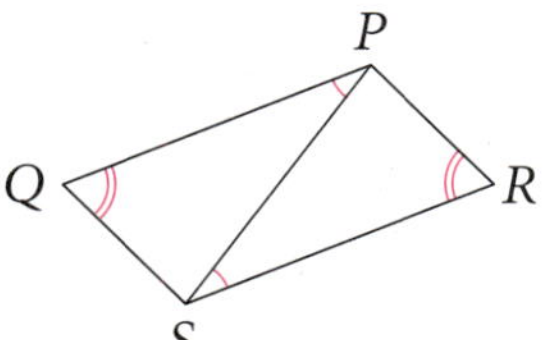

Checkpoint Quiz 2 — Lessons 8-3 through 8-5

Write the tangent, sine, and cosine ratios for $\angle A$ and $\angle B$.

1.

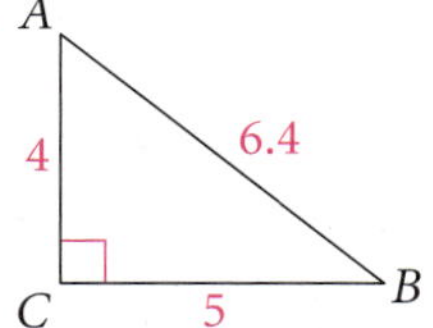

2.

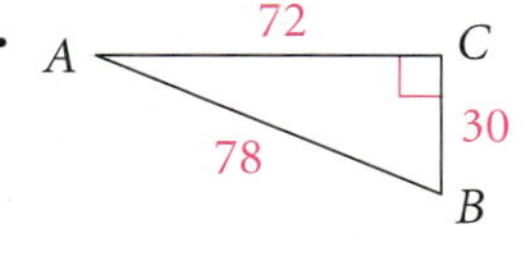

3.

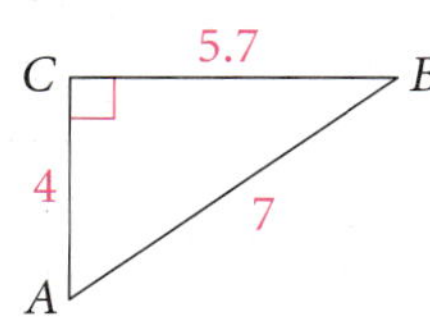

x^2 **Algebra** **Find the value of x. Round each segment length to the nearest tenth and each angle measure to the nearest whole number.**

4.

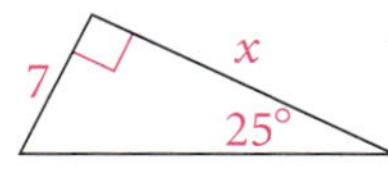

5.

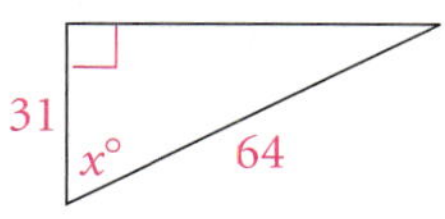

6.

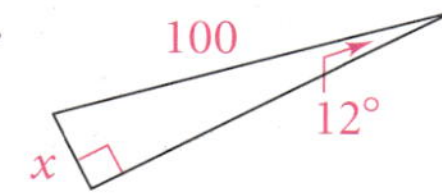

7. Landmarks The Leaning Tower of Pisa, shown at the right, reopened in 2001 after a 10-year project reduced its tilt from vertical by 0.5°. How far from the base of the tower will an object land if it is dropped the 150 ft shown in the photo?

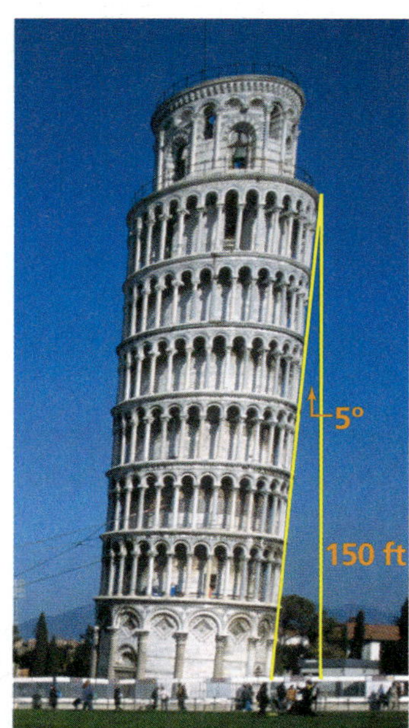

8. Navigation A captain of a sailboat sights the top of a lighthouse at a 17° angle of elevation. A navigation chart shows the height of the lighthouse to be 120 m. How far is the sailboat from the lighthouse?

9. Writing How do you decide which trigonometric ratio to use to solve a problem?

10. Hang Gliding Students in a hang gliding class stand on the top of a cliff 70 m high. They watch a hang glider land on the beach below. The angle of depression to the hang glider is 72°. How far is the hang glider from the base of the cliff?

Guided Problem Solving

For Use With Page 448, Exercise 18

Understanding Word Problems **Read the problem below. Then let Curtis's thinking guide you through the solution. Check your understanding with the exercises at the bottom of the page.**

A blimp is providing aerial television views of a football game. The television camera sights the stadium at a 7° angle of depression. The blimp's altitude is 400 m. What is the line-of-sight distance from the TV camera to the stadium, to the nearest hundred meters?

What Curtis Thinks

To start, I'll draw and label a sketch.
Angle of depression = 7°
Altitude of blimp = 400 m

I'm supposed to find the line-of-sight distance from the TV camera to the stadium. That's the diagonal segment from the blimp's gondola to the 50-yd line. I'll label it x.

To find x, I'll probably need a trig ratio. I've labeled a leg and the hypotenuse of a right triangle 400 m and x. I can use a *sine* or *cosine* ratio if I can find one of the acute angles.

Because alternate interior angles must be congruent, the smaller acute angle is 7°.

Now I'll write an equation.

I'll solve the equation for x.

I'll use a calculator and round.

Now I can state the answer.

What Curtis Writes

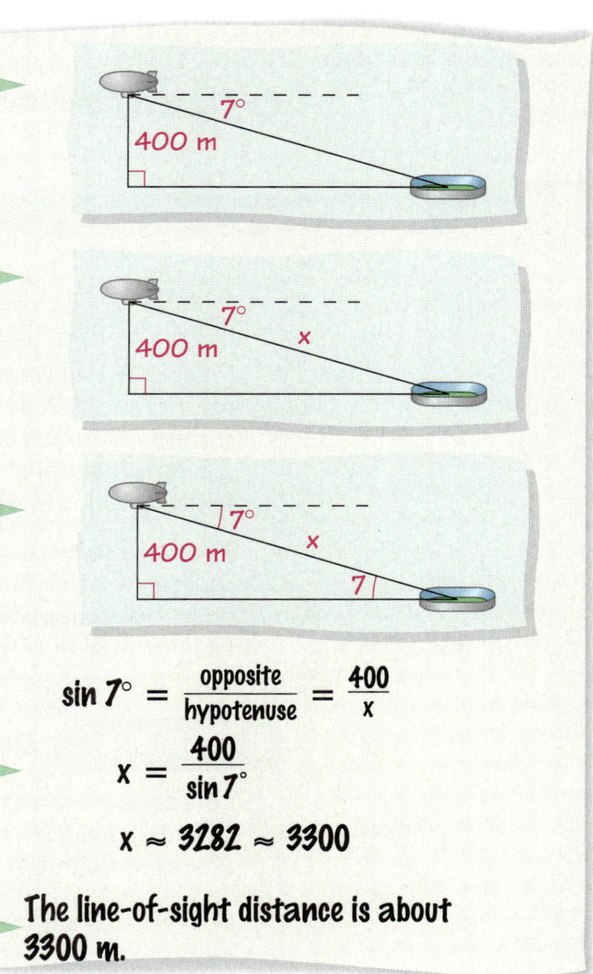

EXERCISES

1. The blimp moves straight up and the angle of depression for the TV camera increases by 1°. How high did the blimp rise?

2. A pedestrian sights the top of a building at an angle of elevation of 75°. She is standing 50 ft from the base of the building. How high above her eye level is the top of the building to the nearest foot?

8-6 Vectors

California Content Standards

GEOM 15.0 Students use the Pythagorean theorem to determine distance and find missing lengths of sides of right triangles. ***Master***

GEOM 19.0 Students use trigonometric functions to solve for an unknown length of a side of a right triangle, given an angle and a length of a side. ***Master***

What You'll Learn

- To describe vectors
- To solve problems that involve vector addition

. . . And Why

To use vectors to describe the distance and direction of an airplane flight, as in Example 3

Lesson 8-1

Algebra **Find the value of x. Leave your answers in simplest radical form.**

1. 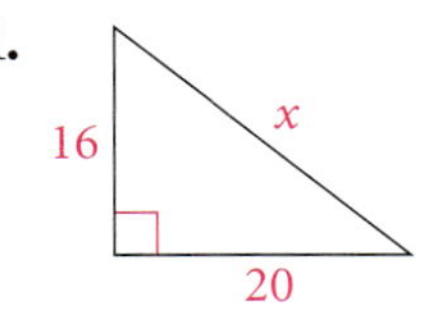

2. 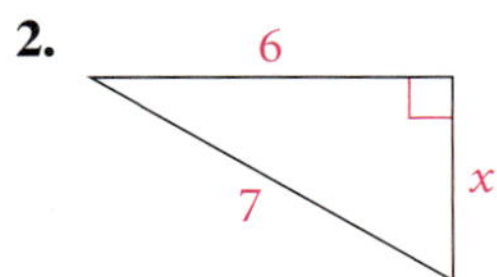

3. 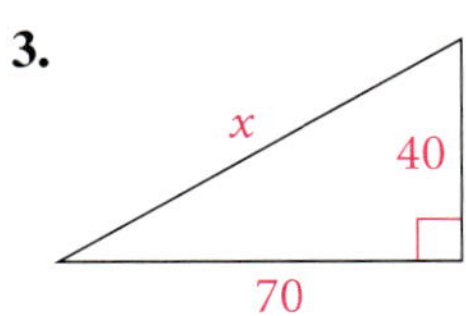

New Vocabulary • vector • magnitude • initial point • terminal point • resultant

1 Describing Vectors

A **vector** is any quantity with magnitude (size) and direction. There are many models for a vector.

You can use an arrow for a vector as shown by the velocity vector $\overrightarrow{KW}$ in the photo. The **magnitude** corresponds to the distance from **initial point** K to the **terminal point** W. The direction corresponds to the direction in which the arrow points.

You can also use an ordered pair $\langle x, y\rangle$ in the coordinate plane for a vector. The magnitude and direction of the vector correspond to the distance and direction of $\langle x, y\rangle$ from the origin.

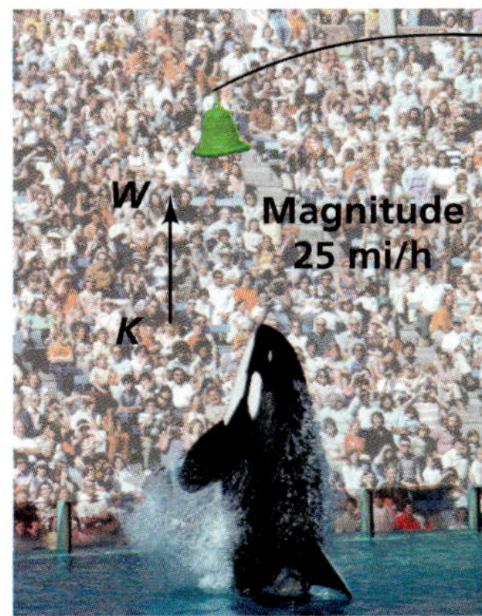

1 EXAMPLE Describing a Vector

Coordinate Geometry Describe $\overrightarrow{OL}$ as an ordered pair. Give the coordinates to the nearest tenth.

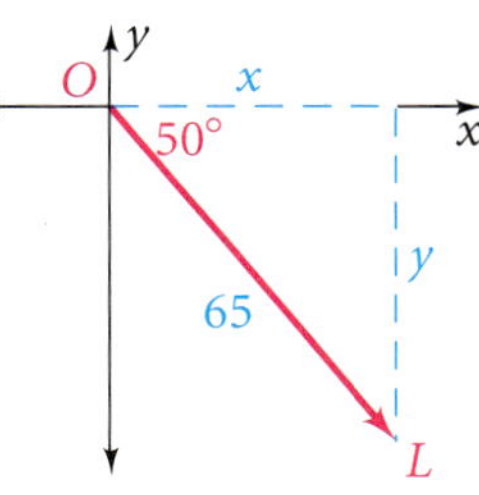

Use the sine and cosine ratios to find the values of x and y.

$\cos 50° = \frac{x}{65}$	$\sin 50° = \frac{y}{65}$	**Use sine and cosine.**
$x = 65(\cos 50°)$	$y = 65(\sin 50°)$	**Solve for the variable.**
≈ 41.78119463	≈ 49.7928888	**Use a calculator.**

L is in the fourth quadrant so the y-coordinate is negative. $\overrightarrow{OL} \approx \langle 41.8, -49.8\rangle$.

CA Standards Check **1** Describe the vector at the right as an ordered pair. Give the coordinates to the nearest tenth.

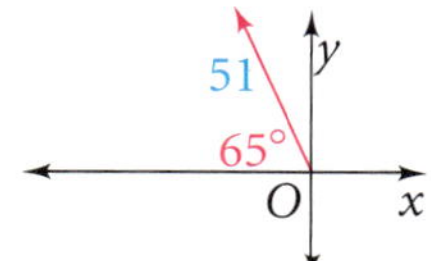

Real-World Connection

A velocity vector for a "bullet train" can have magnitude 275 km/h paired with any direction point on a compass.

In many applications of vectors, you use the compass directions north, south, east, and west to describe the direction of a vector.

2 EXAMPLE Describing a Vector Direction

Use compass directions to describe the direction of each vector.

a.

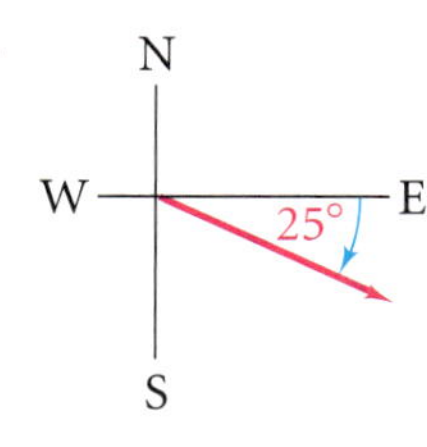

25° south of east

b.

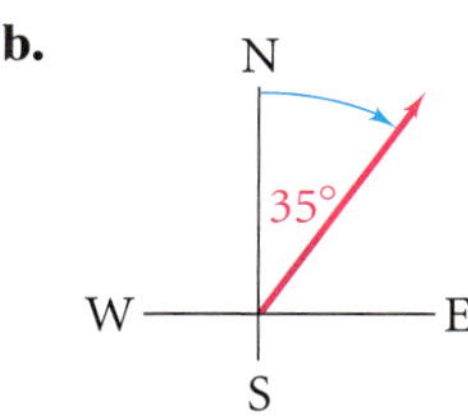

35° east of north

CA Standards Check **2** **a.** Sketch a vector that has the direction 30° west of north.
b. Critical Thinking Give a second description for the direction of this vector.

Example 3 shows how to describe a vector's magnitude and direction when you are given its description as an ordered pair.

3 EXAMPLE Real-World Connection

Aviation An airplane lands 40 km west and 25 km south from where it took off. The result of the trip can be described by the vector $\langle -40, -25 \rangle$. Use distance (for magnitude) and direction to describe this vector a second way.

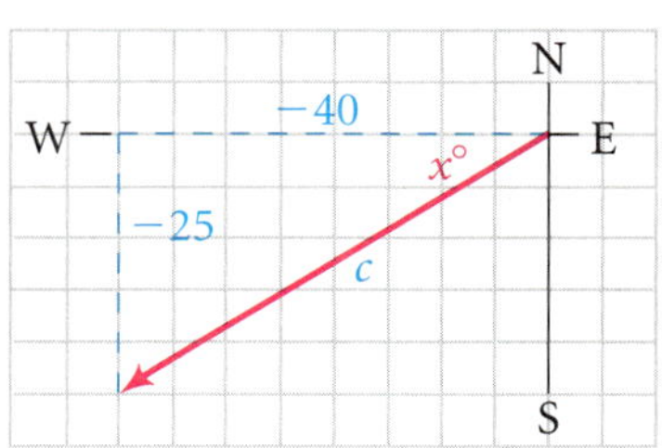

To find the distance, use the Distance Formula:

$d = \sqrt{(-40 - 0)^2 + (-25 - 0)^2}$

$d = \sqrt{1600 + 625}$ **Simplify.**

$d = \sqrt{2225}$

$d \approx 47.169906$ **Use a calculator to find the square root.**

To find the direction of the flight, find the angle of the vector south of west.

$\tan x° = \frac{25}{40}$ **Find the tangent ratio.**

$x = \tan^{-1}\left(\frac{25}{40}\right)$ **Use the inverse of tangent.**

TAN^{-1} 25 ÷ 40 ENTER 32.005383 **Use a calculator.**

The airplane flew about 47 km at 32° south of west.

CA Standards Check **3** A small airplane lands at a point 246 mi east and 76 mi north of the point from which it took off. Describe the magnitude and the direction of its flight vector.

2 Adding Vectors

You can also use a single lowercase letter, such as $\vec{\mathbf{u}}$, to name a vector.

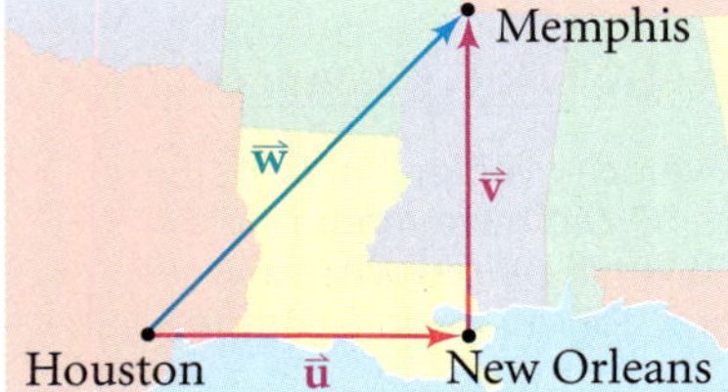

This map shows vectors representing a flight from Houston to Memphis with a stopover in New Orleans. The vector from Houston to Memphis is called the sum, or **resultant,** of the other two vectors. You write this as

$$\vec{\mathbf{w}} = \vec{\mathbf{u}} + \vec{\mathbf{v}}.$$

You can add vectors by adding their coordinates. You can also show the sum geometrically.

Key Concepts

Property	Adding Vectors
For $\vec{\mathbf{a}} = \langle x_1, y_1 \rangle$ and $\vec{\mathbf{c}} = \langle x_2, y_2 \rangle$, $\vec{\mathbf{a}} + \vec{\mathbf{c}} = \langle x_1 + x_2, y_1 + y_2 \rangle$.	

Online active math

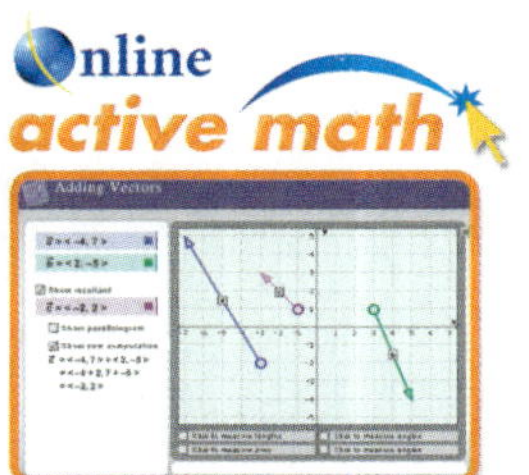

For: Adding Vectors Activity
Use: Interactive Textbook, 8-6

4 EXAMPLE Adding Vectors

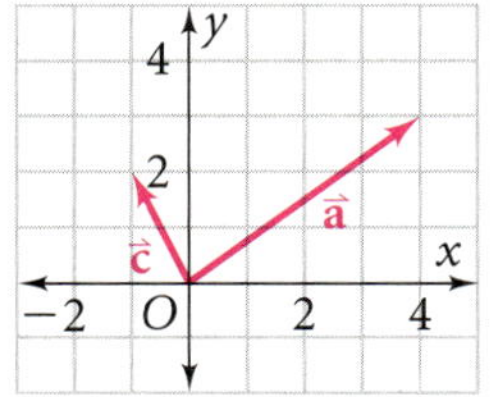

Vectors $\vec{\mathbf{a}}$ $\langle 4, 3 \rangle$ and $\vec{\mathbf{c}}$ $\langle -1, 2 \rangle$ are shown in the diagram. Write the sum of the two vectors as an ordered pair. Then draw $\vec{\mathbf{e}}$, the sum of $\vec{\mathbf{a}}$ and $\vec{\mathbf{c}}$.

$\vec{\mathbf{a}} + \vec{\mathbf{c}} = \langle 4, 3 \rangle + \langle -1, 2 \rangle$

$= \langle 4 + (-1), 3 + 2 \rangle$ **Add the coordinates.**

$= \langle 3, 5 \rangle$ **Simplify.**

$\langle 3, 5 \rangle$ is the resultant.

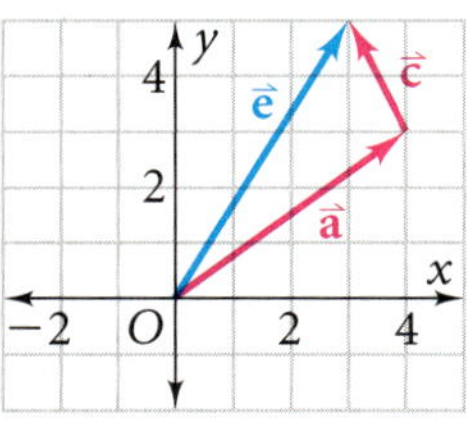

Draw $\vec{\mathbf{a}}$ with its initial point at the origin. Then draw $\vec{\mathbf{c}}$ with its initial point at the terminal point of $\vec{\mathbf{a}}$. Finally, draw the resultant $\vec{\mathbf{e}}$ from the initial point of $\vec{\mathbf{a}}$ to the terminal point of $\vec{\mathbf{c}}$.

CA Standards Check 4 Write the sum of the two vectors $\langle 2, 3 \rangle$ and $\langle -4, -2 \rangle$ as an ordered pair.

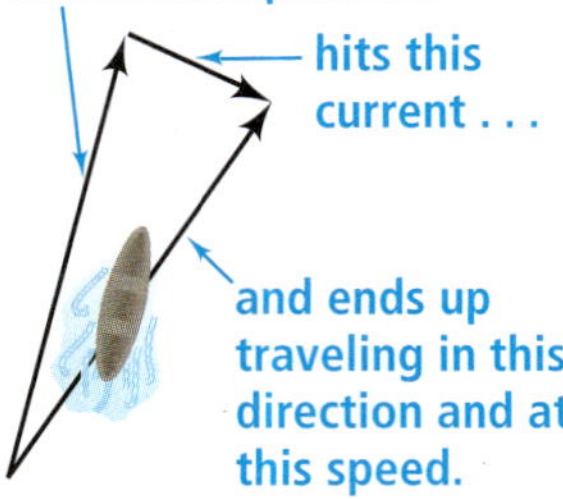

A vector sum can show the result of vectors that occur in sequence, such as in the airplane flight described above.

A vector sum can also show the result of vectors that act at the same time, such as when you row in a direction different from that of the current. See diagram at left.

The velocity of the canoe is the vector sum of the velocities of the paddlers and the stream.

 Real-World Connection

Water currents, wind, and motor capacity must be taken into account when planning ferry schedules.

Navigation A ferry shuttles people from one side of a river to the other. The speed of the ferry in still water is 25 mi/h. The river flows directly south at 7 mi/h. If the ferry heads directly west, what are the ferry's resultant speed and direction?

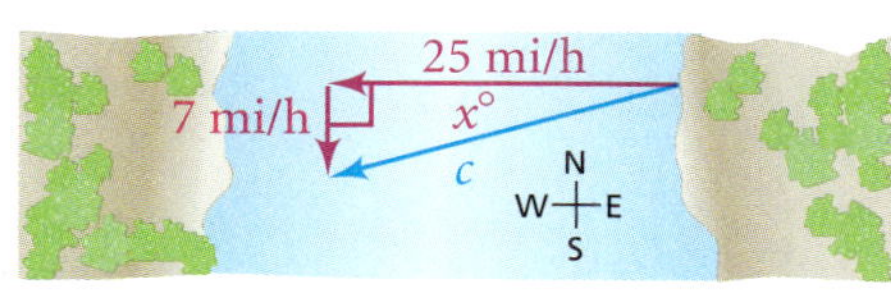

The diagram shows the sum of the two vectors. To find the ferry's resultant speed, use the Pythagorean Theorem.

$c^2 = 25^2 + 7^2$ **The lengths of the legs are 25 and 7.**

$c^2 = 674$ **Simplify.**

$c \approx 25.961510$ **Use a calculator.**

To find the ferry's resultant direction, use trigonometry.

$\tan x° = \frac{7}{25}$ **Use the tangent ratio.**

$x = \tan^{-1}\left(\frac{7}{25}\right)$ **Use the inverse of the tangent.**

$x \approx 15.642246$ **Use a calculator.**

The ferry's speed is about 26 mi/h. Its direction is about 16° south of west.

CA Standards Check **5 Critical Thinking** Use the diagram to find the angle at which the ferry must head upriver in order to travel directly across the river.

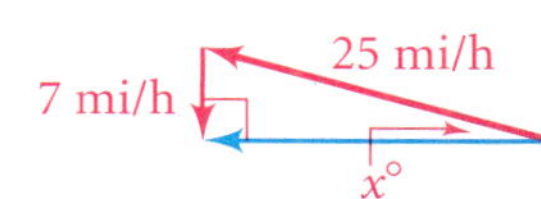

EXERCISES
Standards Practice

For more exercises, see *Extra Skill, Word Problem, and Proof Practice.*

GEOM 15.0, 19.0

A Practice by Example

Example 1 (page 452)

Describe each vector as an ordered pair. Give the coordinates to the nearest tenth.

1.

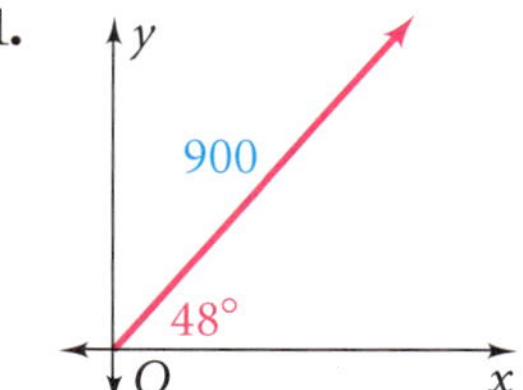

2.

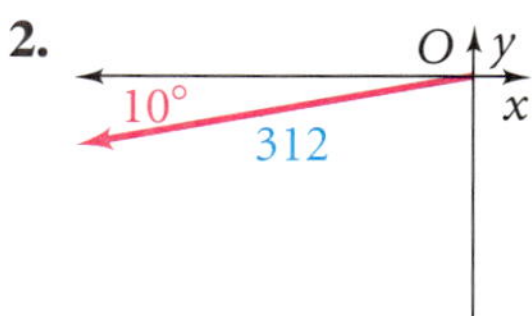

3.

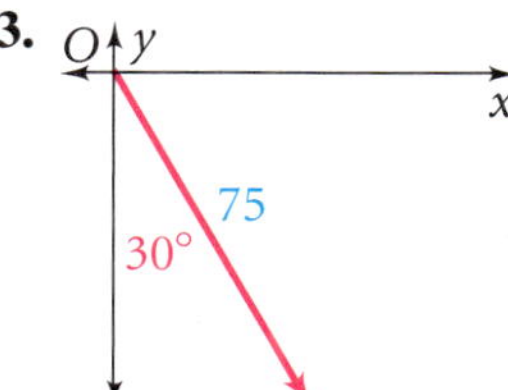

Example 2 (page 453)

Use compass directions to describe the direction of each vector.

4.

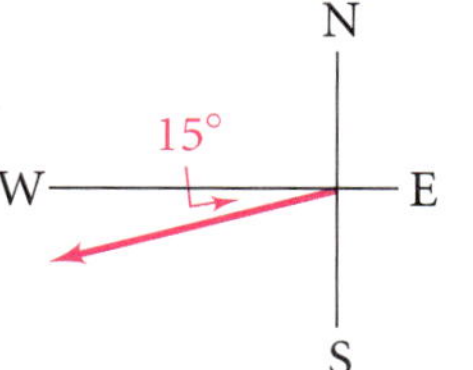

5.

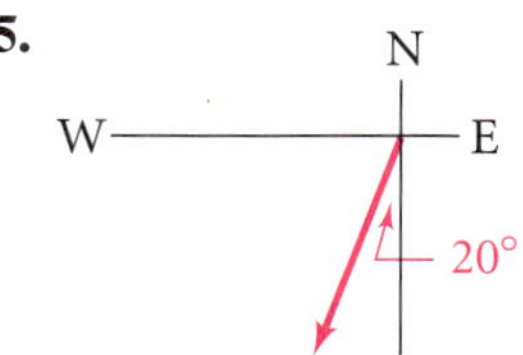

6.

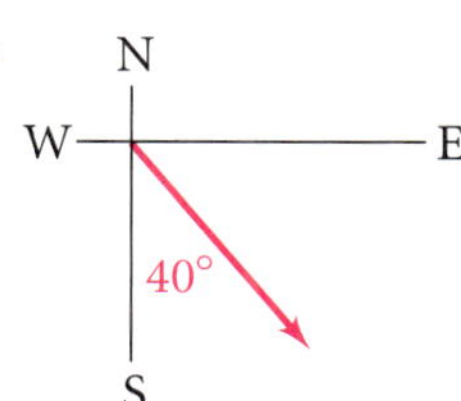

Sketch a vector that has the given direction.

7. 50° south of east
8. 20° north of west
9. 45° northeast
10. 70° west of north
11. 45° southwest
12. 10° east of south

Example 3 (page 453)

13. **History** Homing pigeons have the ability or instinct to find their way home when released hundreds of miles away from home. Homing pigeons carried news of Olympic victories to various cities in ancient Greece. Suppose one such pigeon took off from Athens and landed in Sparta, which is 73 mi west and 64 mi south of Athens. Find the distance and direction of its flight.

Find the magnitude and direction of each vector.

14.

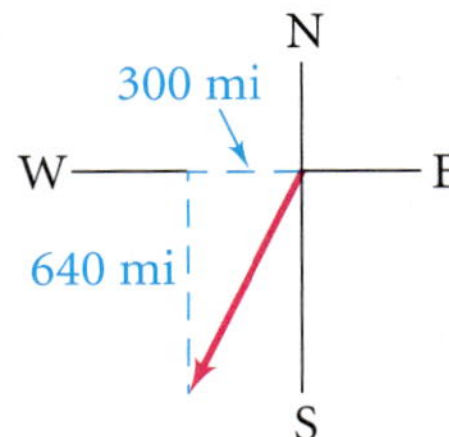

15.

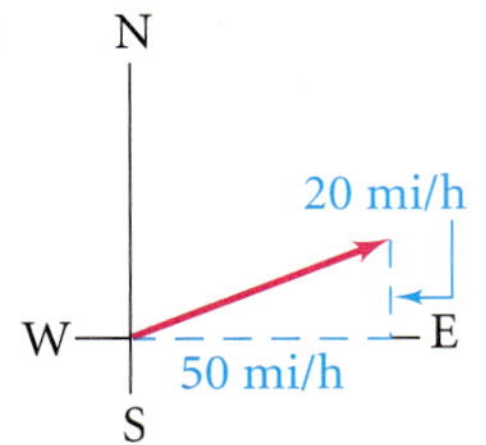

16.

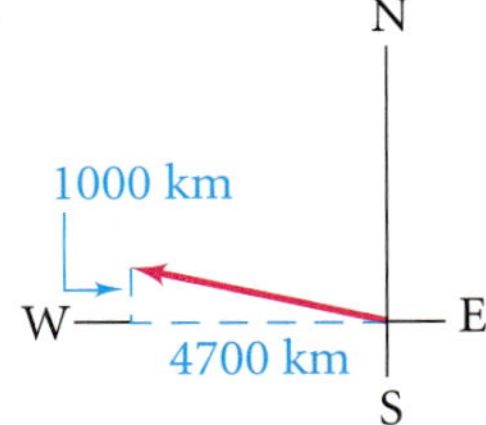

Example 4 (page 454)

In Exercises 17–22, (a) write the resultant as an ordered pair and (b) draw the resultant.

17.

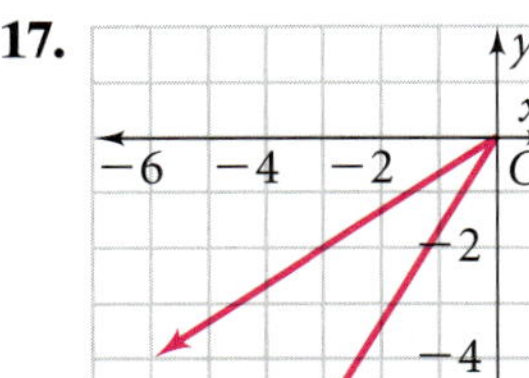

18.

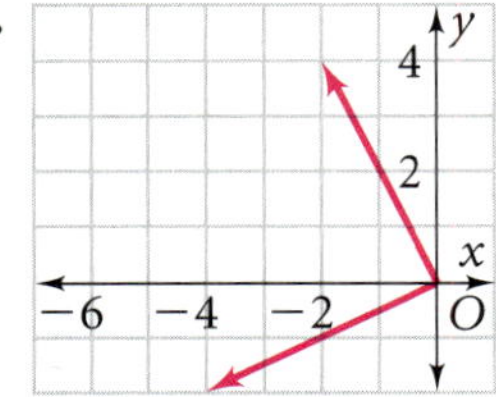

19.

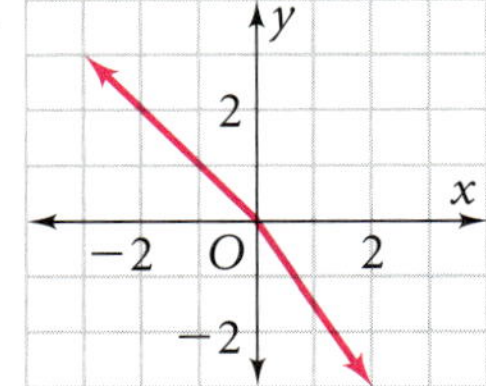

20.

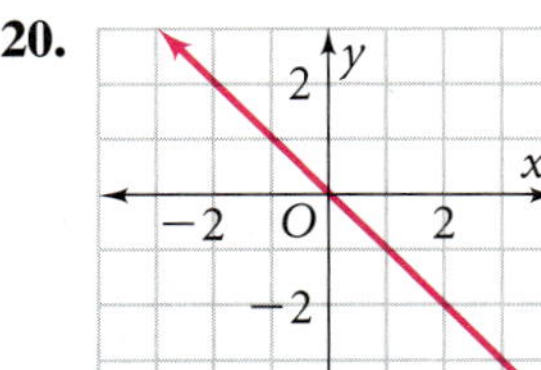

21.

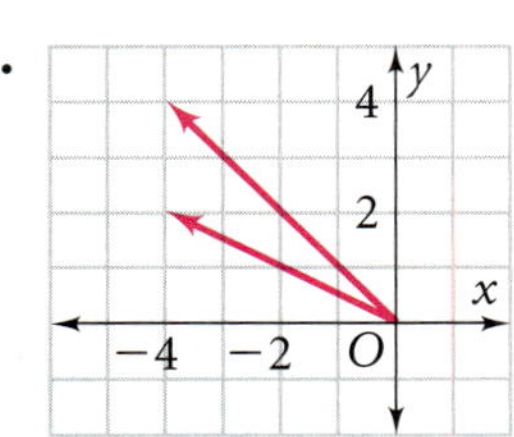

22.

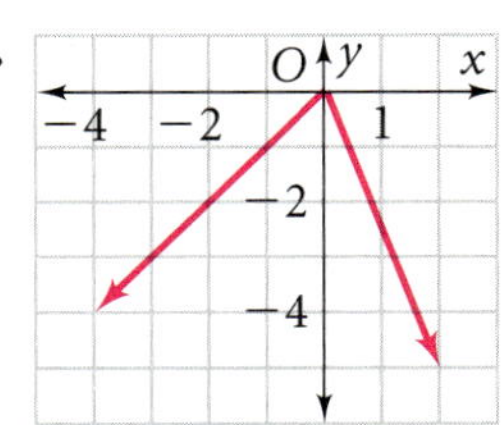

Example 5 (page 455)

Write the sum of the two vectors as an ordered pair.

23. $\langle 2, 1\rangle$ and $\langle -3, 2\rangle$

24. $\langle 0, 0\rangle$ and $\langle 4, -6\rangle$

25. $\langle -1, 1\rangle$ and $\langle -1, 2\rangle$

Navigation **The speed of a powerboat in still water is 35 mi/h. It is traveling on a river that flows directly south at 8 mi/h.**

26. The boat heads directly west across the river. What are the resulting speed and direction of the boat? Round answers to the nearest tenth.

27. At what angle should the boat head upriver in order to travel directly west?

28. **Aviation** A twin-engine airplane has a speed of 300 mi/h in still air. Suppose this airplane heads directly south and encounters a 50 mi/h wind blowing due east. Find the resulting speed and direction of the plane. Round your answers to the nearest unit.

B Apply Your Skills

29. **Critical Thinking** Valerie described the direction of a vector as 35° south of east. Pablo described it as 55° east of south. Could the two be describing the same vector? Explain.

30. **Error Analysis** Ely says that the magnitude of vector $\langle 6, 1\rangle$ is 3 times that of vector $\langle 2, 1\rangle$ since 6 is 3 times 2. Explain why Ely's statement is incorrect.

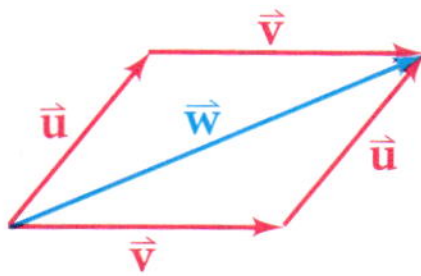

Problem Solving Hint
You can also model vector addition with the *Triangle Rule* as shown in Example 4, and by either triangular half of the diagram above.

31. The diagram at the left shows that you can add vectors in any order. That is, $\vec{u} + \vec{v} = \vec{v} + \vec{u}$. Notice also that the four vectors shown in red form a parallelogram. The resultant $\vec{w}$ is the diagonal of the parallelogram. This representation of vector addition is called *The Parallelogram Rule*.

a. Copy the diagram at the right. Draw a parallelogram that has the given vectors as adjacent sides.

b. Find the magnitude and direction of the resultant.

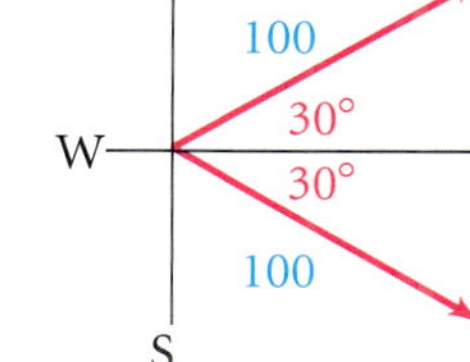

32. Use the diagrams below to write a definition of *equal vectors*.

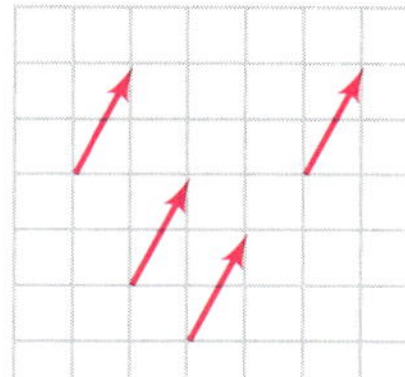

These vectors are equal.

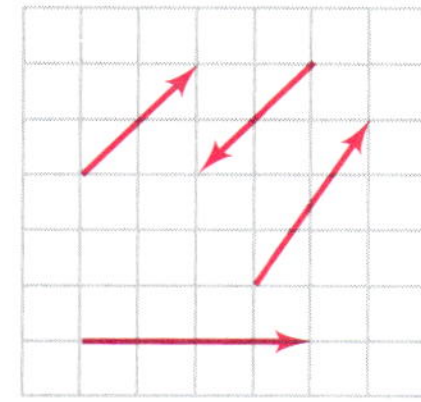

No two of these vectors are equal.

33. Use the diagrams below to write a definition of *parallel vectors*.

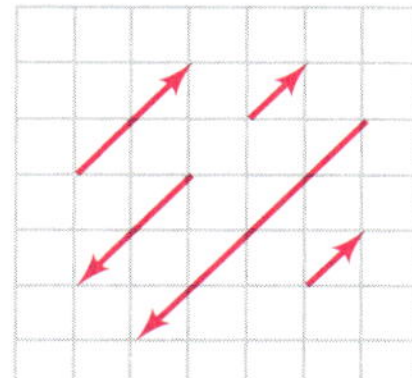

These vectors are parallel.

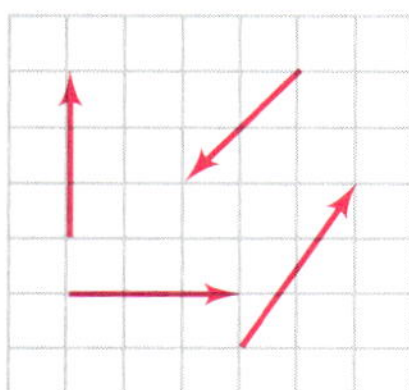

No two of these vectors are parallel.

34. Multiple Choice A Red Cross helicopter takes off and flies 75 km at 20° south of west. There, it drops off some relief supplies. It then flies 130 km at 20° west of north to pick up three medics. What is the helicopter's distance from its point of origin?

Ⓐ 75 mi Ⓑ 130 mi Ⓒ 150 mi Ⓓ 205 mi

35. a. Find the sum of $\vec{a}$ and $\vec{c}$, where $\vec{a} = \langle 45, -60 \rangle$ and $\vec{c} = \langle -45, 60 \rangle$.

b. Writing Based on your answer to part (a), how can you describe $\vec{a}$ and $\vec{c}$?

36. Aviation In still air, the WP-3D (see below) flies at 374 mi/h. Suppose that a WP-3D flies due west and meets a hurricane wind blowing due south at 95 mi/h. What are the resultant speed and direction of the airplane to the nearest unit?

Flying into a Hurricane

When most pilots hear a forecast for gale force winds, they don't think, "Time to fly." Then again, most pilots don't work for the National Oceanic and Atmospheric Administration. NOAA fly their four-engine WP-3D turboprops directly into hurricanes. These aircraft carry eight crew members, up to ten scientists, and a load of data-collection equipment. Some of this equipment is in the WP-3D's long "snout," which also serves as a lightning rod. In a routine flight, the WP-3D is struck by lightning three or four times. Surprisingly, small burn holes are the only damage from these strikes. To help overcome temporary blindness caused by lightning flashes, the pilot sets the cockpit lights at the brightest level.

Visit: PHSchool.com
Web Code: bce-0806

The vector $\langle -5, 5\rangle$ can be written as the *column matrix* $\begin{bmatrix} -5 \\ 5 \end{bmatrix}$. Find the sum of the vectors in column matrix form.

37. $\begin{bmatrix} 2 \\ -4 \end{bmatrix} + \begin{bmatrix} -3 \\ 2 \end{bmatrix}$ **38.** $\begin{bmatrix} 8 \\ -1 \end{bmatrix} + \begin{bmatrix} 3 \\ -4 \end{bmatrix}$ **39.** $\begin{bmatrix} 4 \\ -5 \end{bmatrix} + \begin{bmatrix} -5 \\ 5 \end{bmatrix}$

Problem Solving Hint

In Exercise 40, remember that any vector is equal to one whose initial point is the origin.

40. Aviation An airplane takes off from a runway in the direction 10° east of south. When it reaches 5000 ft, it turns right 45°. It cruises at this altitude for 60 mi. Then it turns left 160°, descends, and lands. Match each vector with the appropriate portion of the flight.

I. N W E S

II.

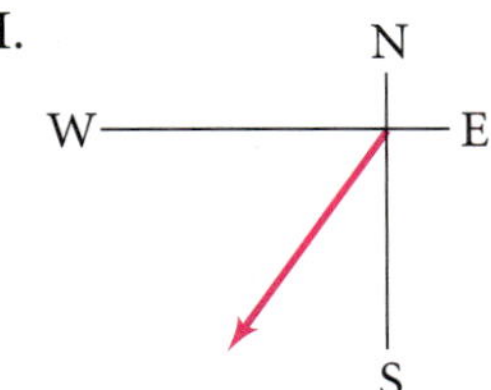

III.

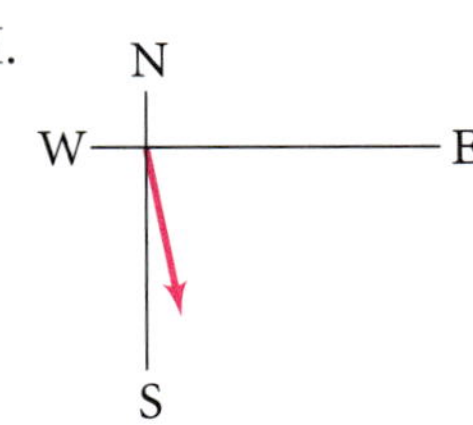

A. The plane takes off. B. The plane cruises. C. The plane lands.

41. Aviation The cruising speed of a Boeing 767 in still air is 530 mi/h. Suppose that a 767 is cruising directly east when it encounters an 80 mi/h wind blowing 40° south of west.

a. Sketch the vectors for the velocities of the airplane and the wind.
b. Express both vectors from part (a) in ordered pair notation.
c. Find the sum of the vectors from part (b).
d. Find the magnitude and direction of the vector from part (c).

Give the sum of $\vec{a}$ and $\vec{b}$. Show $\vec{a}$ and $\vec{b}$ and their sum in the coordinate plane.

42. $\vec{a}\ \langle -5, -2\rangle, \vec{b}\ \langle 2, -5\rangle$ **43.** $\vec{a}\ \langle 5, -2\rangle, \vec{b}\ \langle -5, -2\rangle$ **44.** $\vec{a}\ \langle 5, -5\rangle, \vec{b}\ \langle -2, 2\rangle$

45. Writing How are vectors $\overrightarrow{AB}$ and $\overrightarrow{BA}$ alike? How are they different?

46. Open-Ended Name four other vectors with the same magnitude as $\langle -7, -24\rangle$.

47. Navigation A fishing boat leaves its home port and travels 150 mi directly east. It then changes course and travels 40 mi due north.

a. In what direction should the boat head to return to home port?
b. How long will the return trip take if the boat averages 23 mi/h?

THE FAR SIDE® By GARY LARSON

"Well, lemme think. ... You've stumped me, son. Most folks only wanna know how to go the other way."

Exercise 45

48. Navigation A boat left dock A, traveled north for 10 miles, then 45° east of north for 20 miles, and docked at B.

a. How far north did the boat travel? How far east did it travel?
b. Find the magnitude and direction of the direct-path vector $\overrightarrow{AB}$.

49. Geometry in 3 Dimensions A hot-air balloon traveled 2000 ft north and 900 ft east, while rising 400 ft. This trip can be described with the three-coordinate vector $\langle 2000, 900, 400\rangle$. What is the magnitude of the vector? What is the angle of elevation of the balloon from its starting point?

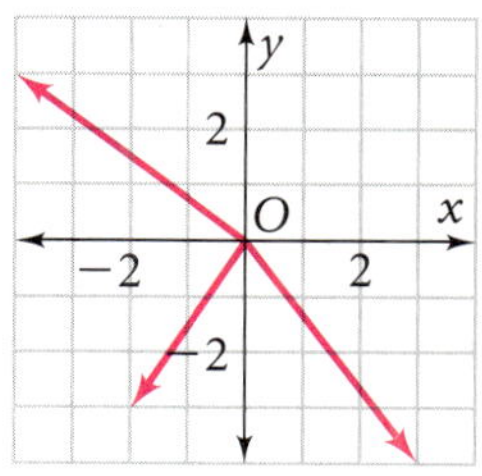

50. a. Probability You choose two of the vectors at the right at random. Find the probability that the magnitude of their resultant vector is greater than that of the third vector.

b. Open-Ended Draw three vectors of your own. Then do part (a) for your vectors.

51. Writing Think of the number zero and its properties. Define a *zero vector* and justify your definition.

52. Aviation A helicopter starts at (0, 0) and makes three parts of a flight represented by the vectors $\langle 10, 10\rangle$, $\langle 5, -4\rangle$, and $\langle -3, 5\rangle$, in that order.

a. If another helicopter starts at (0, 0) and flies the same three parts in a different order, would it end in the same place? Justify your answer.

b. If yet another helicopter flew the three parts of the flight in a different order from the original trip, could the second part of the flight end at the same place as the second part of the original trip? Justify your answer.

Multiple Choice Practice

For California Standards and CAHSEE Tutorials, visit PHSchool.com. Web Code: bcq-9045

GEOM 15.0

53. A hiker is traveling north through the Anza Borrego State Park. When he is 0.8 miles from his destination, he veers off course for 1 mile. Use the diagram at the right to determine how many miles, x, the hiker is from his destination.

N
x
0.8 miles
1 mile
E

Ⓐ 1.8   Ⓒ 0.36

Ⓑ 0.6   Ⓓ 0.2

GEOM 6.0

54. If a triangle has sides of length 6 inches and 10 inches, which of the following could ***not*** be the length of the third leg?

Ⓐ 2 in.   Ⓑ 8 in.   Ⓒ 10 in.   Ⓓ 12 in.

Mixed Review

Lesson 8-5

55. Indirect Measurement A hot-air balloon pilot sights the landing field from a height of 2000 ft. The angle of depression is 24°. To the nearest foot, what is the ground distance from the hot-air balloon to the landing field?

Lesson 8-2

Find the value of each variable.

56.

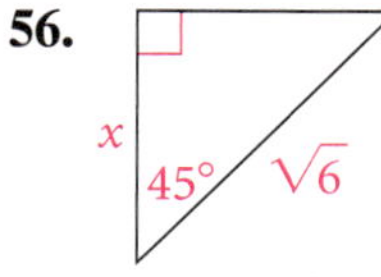

57.

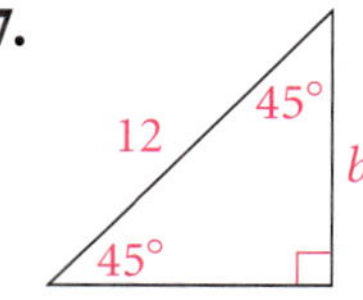

58.

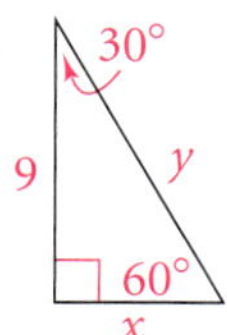

Lesson 6-1

59. Classify the quadrilateral with vertices $A(-1, -5)$, $B(6, -5)$, $C(9, 3)$, and $D(2, 3)$.

Test-Taking Strategies

Eliminating Answers

Before you begin working a problem in earnest, or if you do not know how to do a problem, you usually can eliminate some answer choices. Cross out the answers you eliminate. But do this in the test booklet, not on the answer sheet.

1 EXAMPLE

The length of a diagonal of a square is 12 cm. What is the area of the square?

A) 49 cm^2 B) 72 cm^2 C) 100 cm^2 D) 145 cm^2

The area of a square with side length s is s^2. Since the length of the side of the square is less than the length of the diagonal, you can eliminate answer D. Its area is greater than $12^2 = 144$. Answer A is the area of a square whose side is 7 cm, and this square does not have a diagonal whose length is 12 cm. One of B or C must be the correct answer.

2 EXAMPLE

A kite at the end of a 100-ft string has an angle of elevation of 28°. The end of the string is staked to the ground. Which is the best estimate of the kite's height?

A) 38 ft B) 47 ft C) 49 ft D) 52 ft

Draw a diagram. Since 28° is a little less than 30°, you can estimate the answer by using a 30°-60°-90° triangle. If the angle of elevation were 30°, then the height of the kite would be 50 ft. Since the actual angle is less than 30°, you know the actual height is less than 50 ft. You can eliminate answer D since it is greater than 50 ft. The correct answer must be A, B, or C.

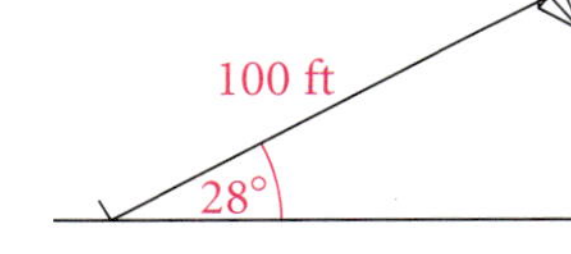

EXERCISES

Identify at least two answers that you can eliminate. Explain your reasoning.

1. Which equation best describes the graph shown at the right?

   A) $y = 0.25x - 5$ C) $y = -0.25x - 5$
   B) $y = -0.25x + 5$ D) $y = 0.25x + 5$

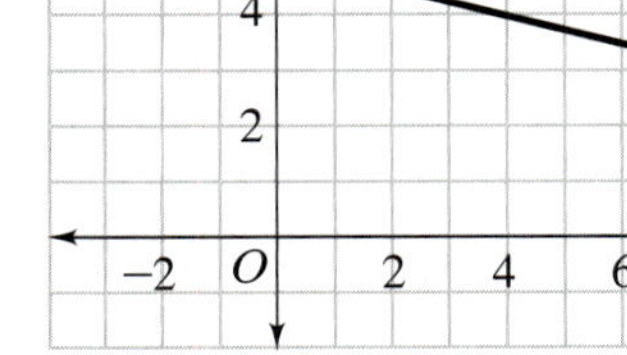

2. The equation $f(x) = x^2 + 9$ describes a function. If the replacement set for the independent variable is $\{-2, 1, 4, 7\}$, which of the following numbers is in the related set for the dependent variable?

   A) −4 B) 3 C) 11 D) 25

3. Clara designed a custom window like the one shown at the right. What is the perimeter of the rhombus to the nearest tenth of a foot?

   A) 8.7 ft C) 14.4 ft
   B) 11.8 ft D) 20.0 ft

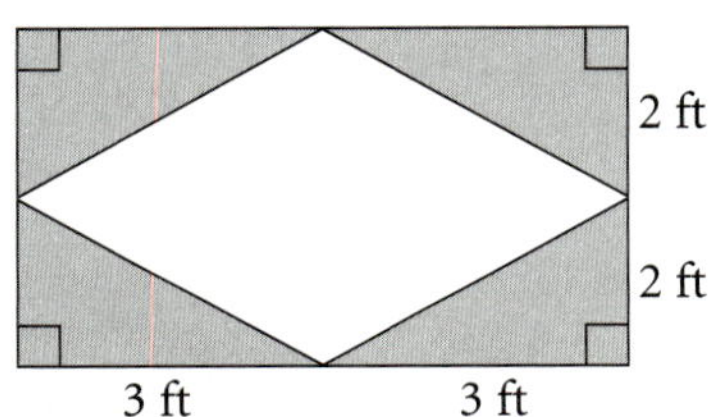

Chapter 8 Review

Vocabulary Review

English and Spanish Audio Online

angle of depression (p. 445)
angle of elevation (p. 445)
cosine (p. 439)
identity (p. 440)
initial point (p. 452)
magnitude (p. 452)
Pythagorean triple (p. 417)
resultant (p. 454)
sine (p. 439)
tangent (p. 432)
terminal point (p. 452)
vector (p. 452)

Go Online PHSchool.com
For: Vocabulary quiz
Web Code: bcj-0851

Choose the correct term to complete each sentence.

1. Any quantity that has magnitude and direction is called a(n) _?_.

2. The _?_ is the angle formed by a horizontal line and the line of sight to an object above that horizontal line.

3. The _?_ of $\angle A$ is $\frac{\text{leg adjacent } \angle A}{\text{hypotenuse}}$.

4. The _?_ of $\angle A$ is $\frac{\text{leg opposite } \angle A}{\text{hypotenuse}}$.

5. An equation that is true for all allowed values of the variable is called a(n) _?_.

6. The sum of two vectors is the _?_.

7. A set of three nonzero whole numbers that satisfy the equation $a^2 + b^2 = c^2$ form a(n) _?_.

Skills and Concepts

Lessons 8-1 and 8-2

- To use the Pythagorean Theorem (p. 417)
- To use the Converse of the Pythagorean Theorem (p. 419)
- To use the properties of 45°-45°-90° triangles (p. 425)
- To use the properties of 30°-60°-90° triangles (p. 426)

 **GEOM 15.0, 20.0**

The **Pythagorean Theorem** states that in a right triangle, the sum of the squares of the lengths of the legs equals the square of the length of the hypotenuse, or $a^2 + b^2 = c^2$.

Hypotenuse
Legs
$a^2 + b^2 = c^2$

Positive integers a, b, and c form a **Pythagorean triple** if $a^2 + b^2 = c^2$.

The **Pythagorean Theorem Converse** states that if the square of the length of one side of a triangle is equal to the sum of the squares of the lengths of the other two sides, then the triangle is a right triangle.

In a 45°-45°-90° triangle, the legs are congruent and the length of the hypotenuse is $\sqrt{2}$ times the length of a leg.

In a 30°-60°-90° triangle, the length of the hypotenuse is twice the length of the shorter leg. The length of the longer leg is $\sqrt{3}$ times the length of the shorter leg.

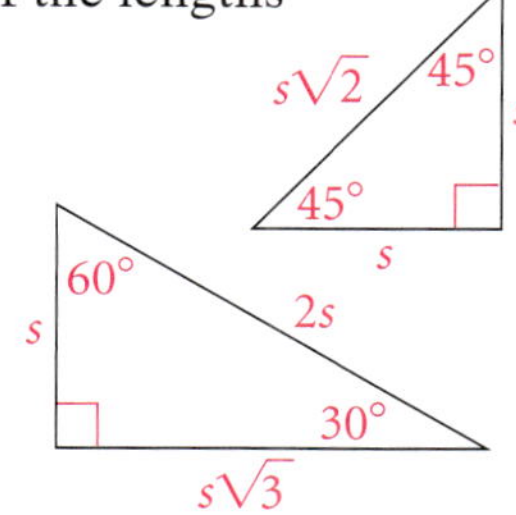

Find the value of each variable. If your answer is not an integer, leave it in simplest radical form.

8. x, 20, 12

9.

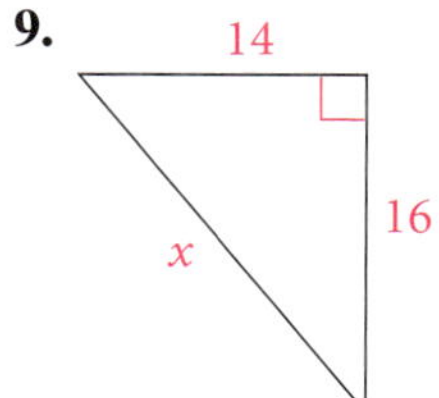

10.

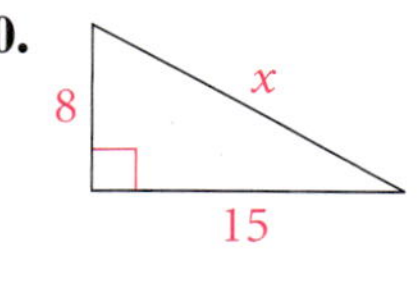

Find the value of each variable. If your answer is not an integer, leave it in simplest radical form.

11.

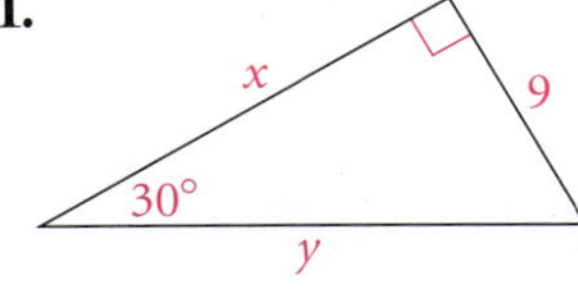

12.

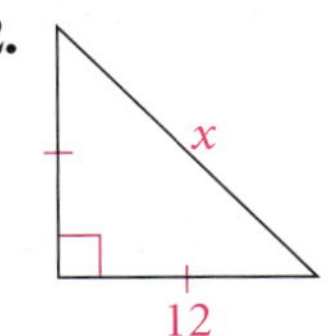

13.

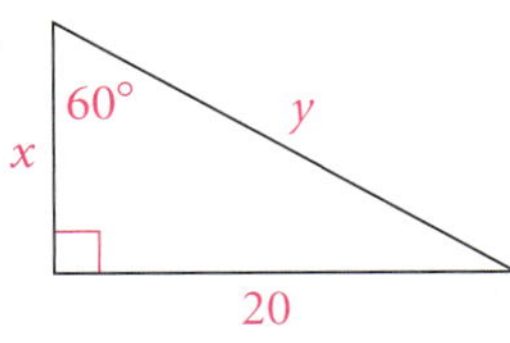

Lessons 8-3 and 8-4

- To use tangent ratios to determine side lengths in triangles (p. 432)
- To use sine and cosine to determine side lengths in triangles (p. 439)

 **GEOM 18.0, 19.0**

In right $\triangle ABC$,

$$\textbf{sine of } \angle A = \sin A = \frac{\text{leg opposite } \angle A}{\text{hypotenuse}}$$

$$\textbf{cosine of } \angle A = \cos A = \frac{\text{leg adjacent to } \angle A}{\text{hypotenuse}}$$

$$\text{and } \textbf{tangent of } \angle A = \tan A = \frac{\text{leg opposite } \angle A}{\text{leg adjacent to } \angle A}$$

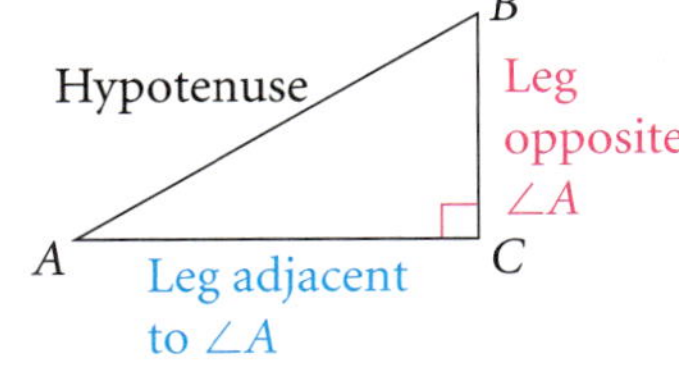

You can use the inverses of sine, cosine, and tangent to find the measures of the acute angles of a right triangle.

A trigonometric **identity** is an equation that is always true for all the allowed values of the variable.

Find each missing value to the nearest whole number.

14. $\tan ■° = 0.9$ **15.** $\sin 17° = \frac{■}{7}$ **16.** $\tan 27° = \frac{1}{■}$ **17.** $\cos ■° = 0.39$

18. $\sin ■° = 0.39$ **19.** $\sin ■° = \frac{2}{3}$ **20.** $\tan 76° = \frac{■}{3}$ **21.** $\cos 83° = \frac{1}{■}$

Express sin *A*, cos *A*, and tan *A* as ratios.

22.

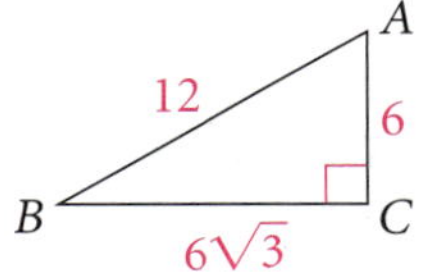

23.

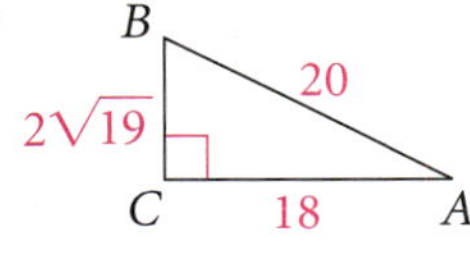

24.

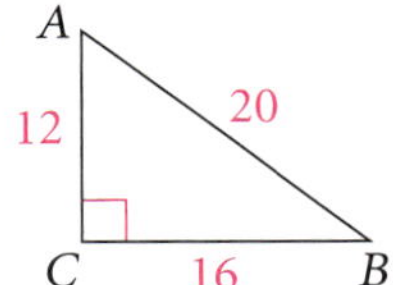

Find the value of *x*. Round lengths of segments to the nearest tenth. Round angle measures to the nearest degree.

25.

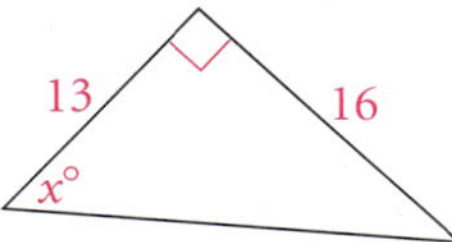

26.

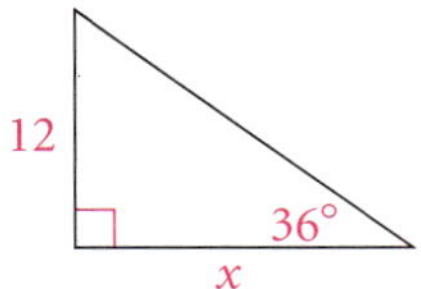

27.

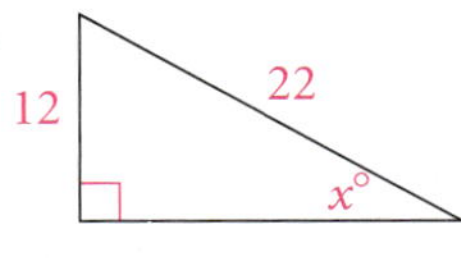

Lesson 8-5

- To use angles of elevation and depression to solve problems (p. 445)

 **GEOM 18.0, 19.0**

A horizontal line and the line of sight to an object above the horizontal line form an **angle of elevation.** A horizontal line and the line of sight to an object below that horizontal line form an **angle of depression.**

Solve each problem.

28. Elevation Two hills are 2 mi apart. The taller hill is 2707 ft high. The angle of depression from the top of the taller hill to the top of the shorter hill is 7°. Find the height of the shorter hill to the nearest foot. (1 mi = 5280 ft)

not to scale

7°

2707 ft

2.0 mi

29. **Surveying** A surveyor is 305 ft from the base of the new courthouse. Her angle measuring device is 5 ft above the ground. The angle of elevation to the top of the courthouse is 42°. Find the height of the courthouse to the nearest foot.

30. **Indirect Measurement** Linda is flying a kite. She lets out 45 yd of string and anchors it to the ground. She determines that the angle of elevation of the kite is 58°. What is the height of the kite from the ground?

Lesson 8-6

- To describe vectors (p. 452)
- To solve problems that involve vector addition (p. 454)

 **GEOM 15.0, 19.0**

A **vector** is any quantity that has magnitude and direction. You can describe a vector by its horizontal and vertical change (ordered pair) or by its size and direction. Direction is often described in relation to north, south, east, and west.

The sum of two vectors is the **resultant.** Vector sums can show the result of vector actions that take place one after the other. Additionally, vector sums can show the result of two vector actions that occur at the same time. You can add vectors by adding their coordinates. You can also show the sum geometrically.

Describe each vector using ordered pair notation.

31. 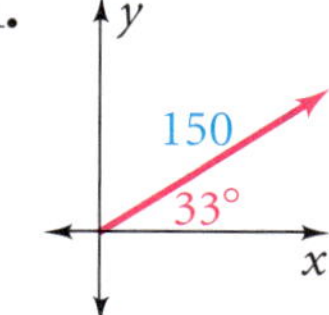

32. 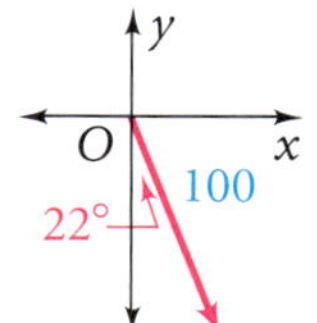

33. 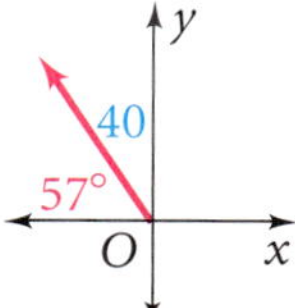

Find the magnitude and direction of each vector.

34. 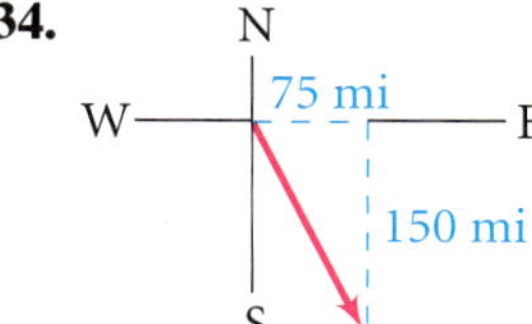

35. 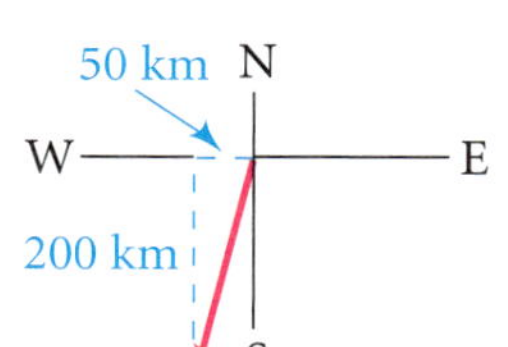

36. 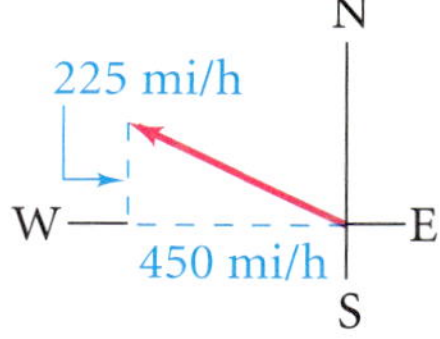

37. **Open-Ended** Write three vectors with the same direction as $\langle 3, 4 \rangle$. Explain how you found your answers.

Sketch a vector that has the given direction.

38. 25° east of north

39. 45° north of east

40. 60° west of south

Find the sum of each pair of vectors. Give your answers in ordered pair notation.

41. 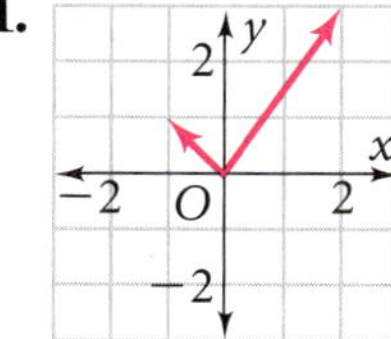

42. 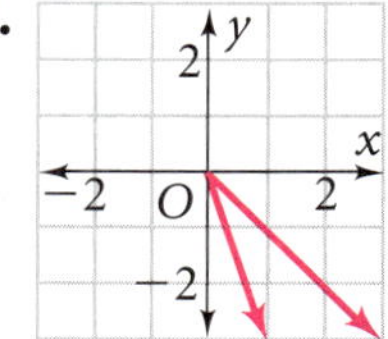

43. 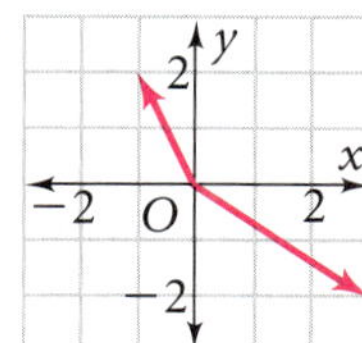

44. **Navigation** A whale-watching tour leaves port and travels 12 mi directly north. The tour then travels 5 mi due east.
   a. In what direction should the boat head to return directly to port?
   b. How long will the return trip take if the boat averages 20 mi/h?

Chapter 8 Test

For: Chapter Test
Web Code: bca-0852

 **Algebra Find the value of each variable. Leave your answer in simplest radical form.**

1. 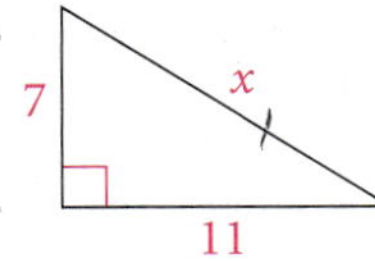

2. 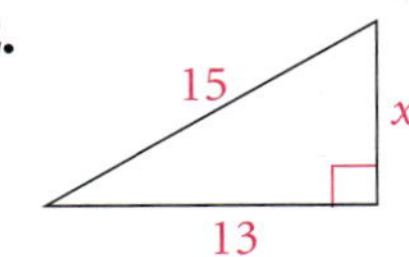

3. 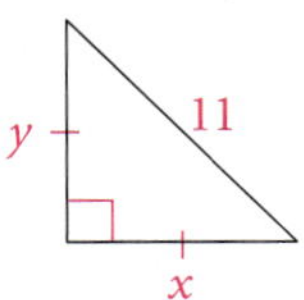

4. 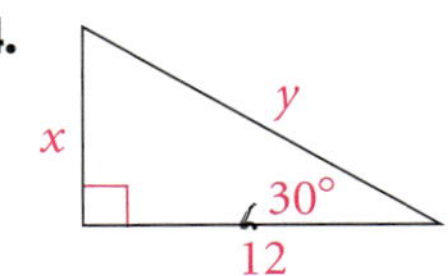

The lengths of three sides of a triangle are given. Describe each triangle as *acute, right,* or *obtuse*.

5. 9 cm, 10 cm, 12 cm
6. 8 m, 15 m, 17 m
7. 5 in., 6 in., 10 in.

Express sin *B*, cos *B*, and tan *B* as ratios.

8. 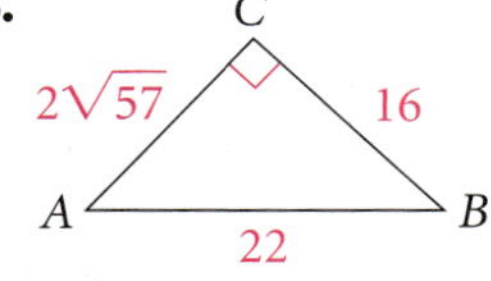

9. 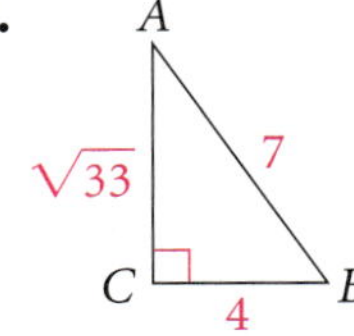

Find each missing value to the nearest tenth.

10. $\tan ■° = 1.11$
11. $\tan 18° = \frac{■}{87}$
12. $\sin 34° = \frac{5}{■}$
13. $\sin ■° = 0.996$
14. $\cos ■° = \frac{12}{15}$
15. $\cos 76° = \frac{■}{24}$

Find the value of *x*. Round lengths to the nearest tenth and angle measures to the nearest degree.

16. 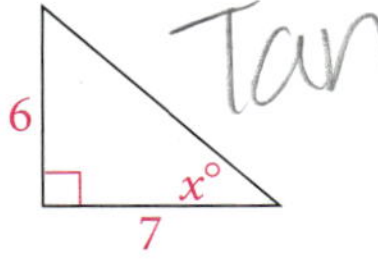

17. 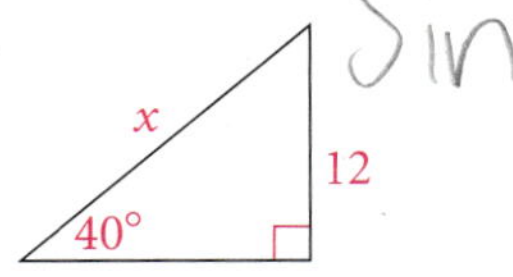

18. 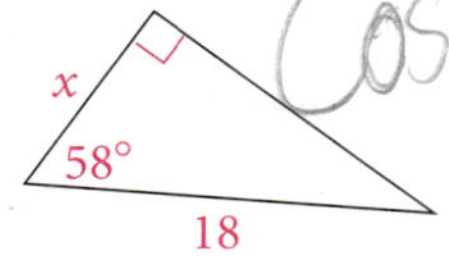

19. 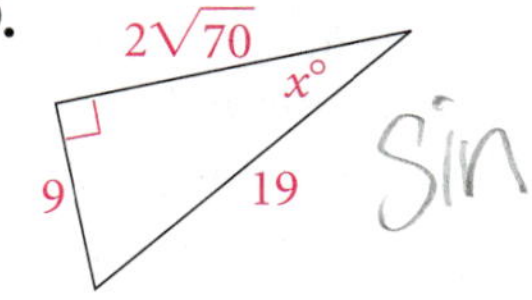

Indirect Measurement Solve each problem.

20. A surveyor measuring the tallest tree in a park is 100 ft from the tree. His angle-measuring device is 5 ft above the ground. The angle of elevation to the top of the tree is 48°. How tall is the tree?

21. Twenty minutes after being launched, a hot-air balloon has risen to an altitude of 300 m. The pilot can still see the starting point on the ground at a 25° angle of depression. How many meters is the balloon from the starting point?

22. A 5-foot-tall woman stands 15 ft from a statue. She must look up at an angle of 60° to see the top of the statue. How tall is the statue?

23. **Writing** Explain why $\sin x° = \cos(90 - x)°$. Include a diagram with your explanation.

Describe each vector using ordered pair notation. Round the coordinates to the nearest unit.

24. 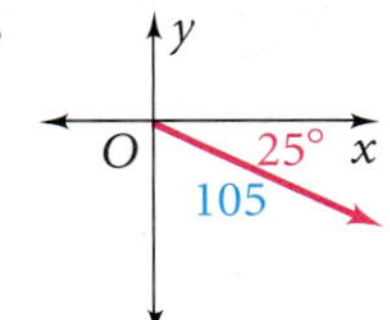

25. 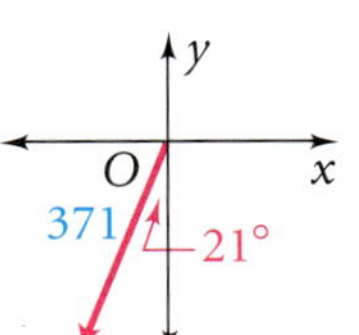

Find the magnitude and direction of each vector.

26. 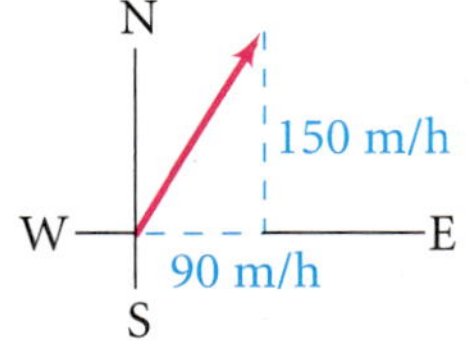

27. 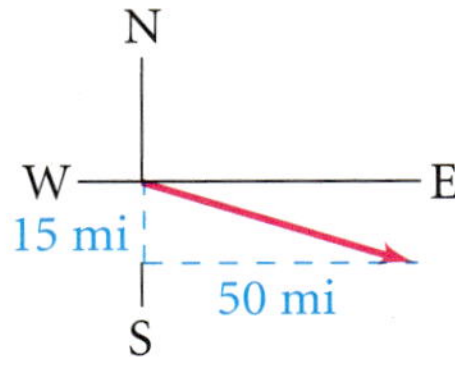

Describe the resultant as an ordered pair.

28. 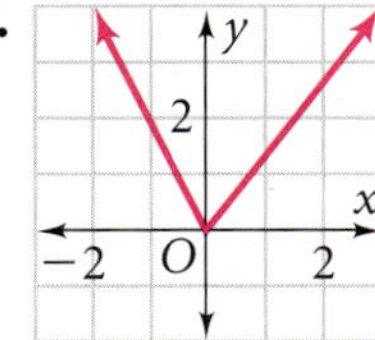

29. 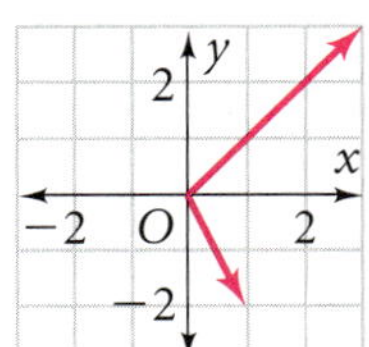

30. A canoe heading 30° west of north is being paddled at a rate of 7 mi/h. The current is pushing the canoe 20° south of west at a rate of 3 mi/h. Find the resulting speed and direction of the canoe.

Standards Mastery

Cumulative Practice

Some questions ask you to analyze formulas. Read the question at the right. Then follow the tips to answer it.

Tip 1
Read the choices. Note that each asks about area *A*, the left side of the formula.

A rhombus is a parallelogram with four sides of equal length s. One formula for calculating the area of a rhombus is $A = \frac{1}{2}d_1d_2$, where d_1 and d_2 are the lengths of the diagonals of the rhombus. Which statement is true?

(A) The area depends on the value of s only.
(B) The area depends on the value of d_1 only.
(C) The area depends on the values of d_1 and d_2 only.
(D) The area depends on the values of s, d_1, and d_2.

Tip 2
The value of *A* on the left side of the formula depends on what can vary on the right side of the formula.

Think It Through

In $A = \frac{1}{2}d_1d_2$, both d_1 and d_2 can vary. A depends on the values of d_1 and d_2 only.

Choice C is the correct answer.

Vocabulary Review

As you solve problems, you must understand the meanings of mathematical terms. Choose the correct term to complete each sentence.

I. A number that can be expressed as a fraction is a (real number, rational number).

II. The (roots, sources) of a quadratic equation are the places where the graph of the related quadratic function crosses the x-axis.

III. The (median, mean) is the middle number in a set of data.

IV. The (circumference, diameter) is the distance across a circle through its center.

V. In a(n) (scale, isometric) drawing of an object, the dimensions of the object are multiplied by a factor.

Multiple Choice

Read each question. Then write the letter of the correct answer on your paper.

1. A roofer leans a 20 ft ladder against a house. The base of the ladder is 7 ft from the house. To the nearest tenth of a foot, how high on the house does the ladder reach? **(Lesson 8-1)**

(A) 13 (B) 20.9 (C) 18.7 (D) 21.2

2. Which of the following facts would be sufficient to prove $\triangle ACE \sim \triangle BCD$? **(Lesson 7-3)**

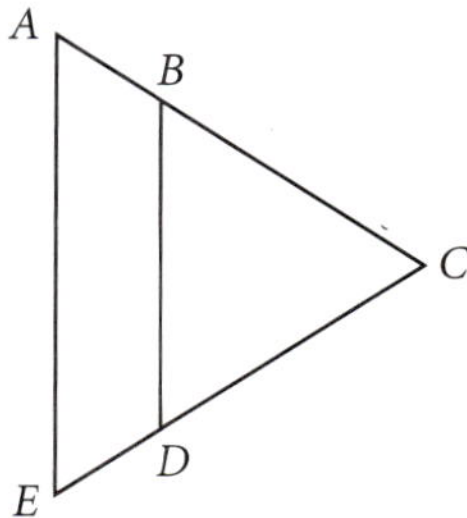

(A) $\triangle BCD$ is a right triangle
(B) $AB \cong ED$
(C) $m\angle A = m\angle E$
(D) $AE \parallel BD$

3. If the measure of an interior angle of a regular polygon is 135°, how many sides does the polygon have? **(Lesson 3-5)**

(A) 5 (B) 6 (C) 7 (D) 8

4. A cross country skier is moving south through Soda Springs. Five miles from his destination, he veers off course for 8 miles. Use the diagram below to determine about how many miles, x, the skier is from his destination. **(Lesson 8-6)**

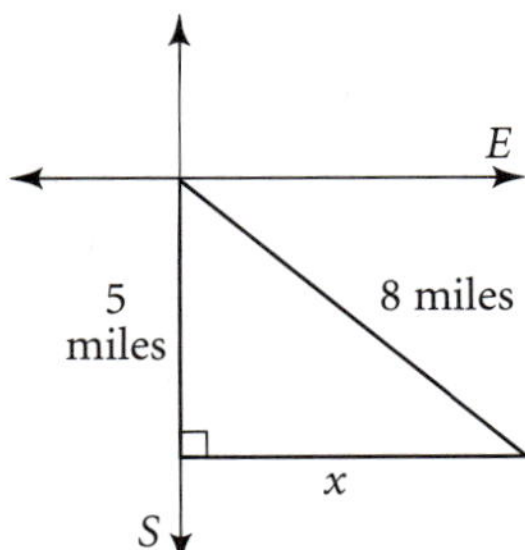

(A) 6.2 (B) 3.0 (C) 39.0 (D) 21.6

5. From the top of a 45-foot-tall building, the angle of depression to the edge of a parking lot is 48°. About how many feet is the base of the building from the edge of the parking lot? **(Lesson 8-5)**

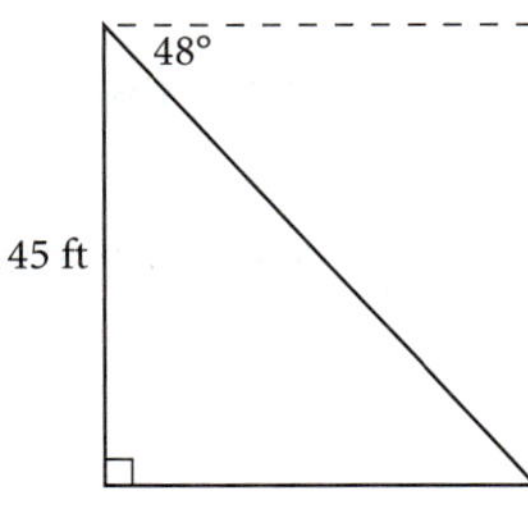

(A) 20.7 (B) 40.5 (C) 50 (D) 60

6. In a 30°-60°-90° triangle, the shortest leg measures 13 inches. How many inches long is the longer leg? **(Lesson 8-2)**

(A) $13\sqrt{3}$ (B) 13 (C) 26 (D) $13\sqrt{2}$

7. If $m\angle 2 = m\angle 3$ in the following diagram, which statement must be true? **(Lesson 3-1)**

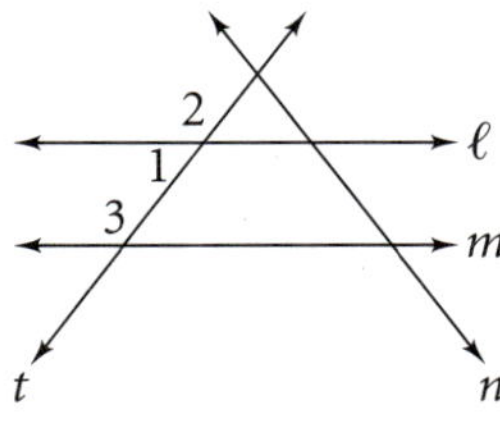

(A) $\ell \parallel m$
(B) $t \perp m$
(C) $t \perp \ell$
(D) $m\angle 1 = m\angle 2$

8. A loading plank onto a ship is 15 feet long and makes an angle of 18° with the dock. Approximately how many feet above the dock is the ship's deck? **(Lesson 8-4)**

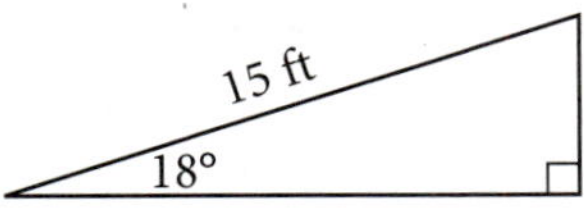

(A) 4.6
(B) 23.4
(C) 48.5
(D) 14.3

9. A bike messenger has just been asked to make an additional stop. Now, instead of biking straight from the law office to the court, she is going to stop at City Hall in between. Using the picture below, determine how many additional miles she will bike. **(Lesson 8-1)**

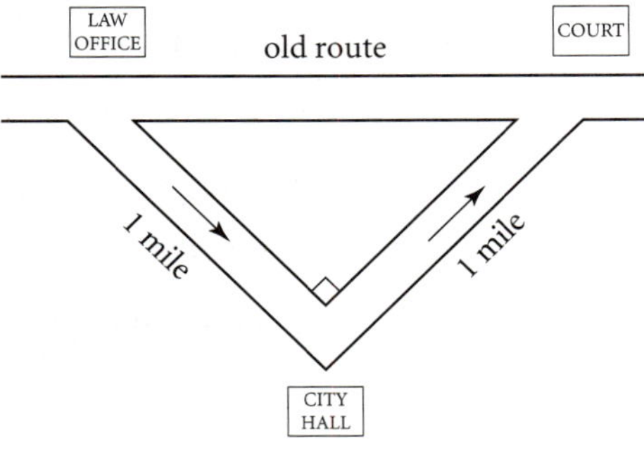

(A) $\sqrt{2} - 2$
(B) $2 - \sqrt{2}$
(C) 2
(D) $2 + \sqrt{2}$

10. Determine the approximate area of the rectangle at the right. **(Lesson 8-3)**

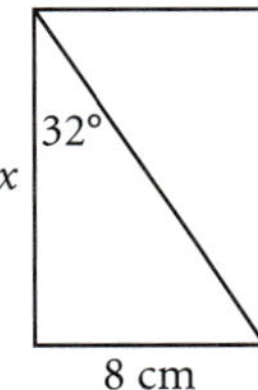

(A) 102 cm^2
(B) 45 cm^2
(C) 63 cm^2
(D) 75 cm^2

11. Determine the value of x in the following figure. **(Lesson 7-5)**

3, 4, 21, x

(A) 34
(B) 22
(C) 28
(D) 12

Standards Mastery

12. In the figure below, the labels refer to the length of the line segment. By which theorem or postulate does $x = 5$? **(Lesson 7-5)**

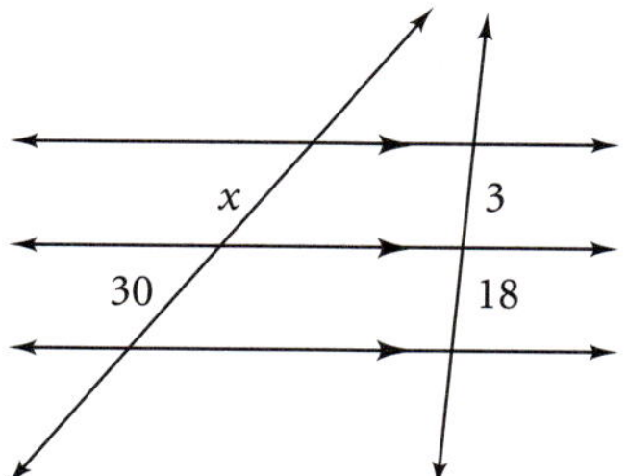

(A) SAS Triangle Congruence
(B) If three parallel lines intersect two transversals, then the segments intercepted on the transversals are proportional.
(C) Opposite sides of a parallelogram are congruent.
(D) If two lines are parallel to the same line, then they are parallel to each other.

13. Determine the value of y in the following isosceles triangle. **(Lesson 8-2)**

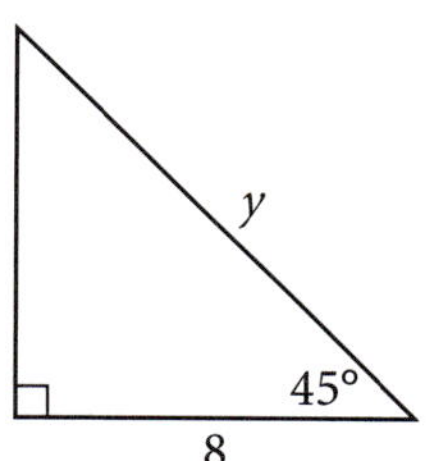

(A) 16 (B) $8\sqrt{2}$ (C) 8 (D) $8\sqrt{3}$

14. Determine which angle is congruent to $\angle DCB$. Note that CD is an altitude of triangle ACB. **(Lesson 7-4)**

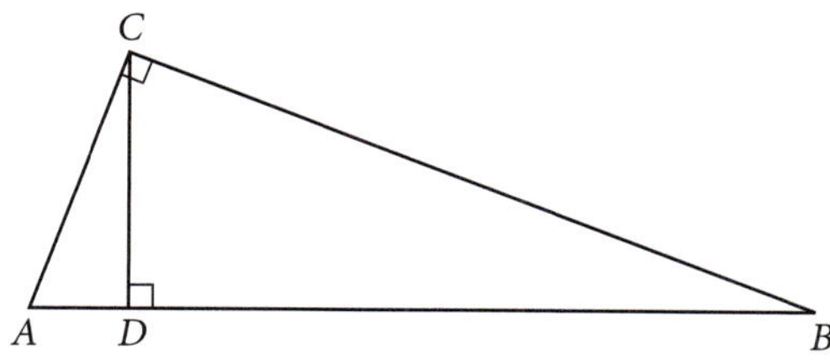

(A) $\angle B$
(B) $\angle CDB$
(C) $\angle A$
(D) $\angle ACD$

15. Find the values of x and of y in the following parallelogram. **(Lesson 6-2)**

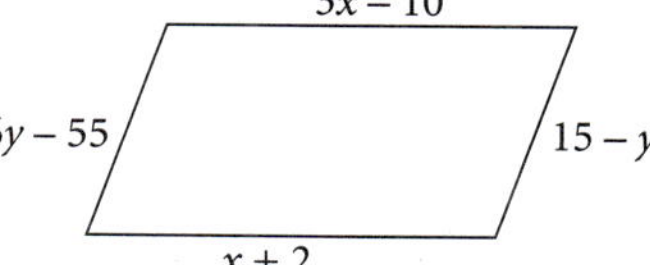

(A) $x = 3, y = 5$
(B) $x = 6, y = 10$
(C) $x = 6, x = 14$
(D) $x = 2, y = 2$

16. In the figure below, $\sin A = 0.34$. What is the approximate value of x? **(Lesson 8-4)**

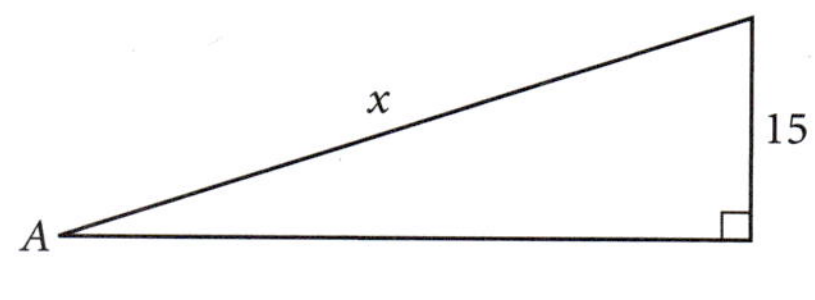

(A) 25 (B) 16 (C) 44 (D) 30

17. Which combination could not be the lengths of the sides of a right triangle? **(Lesson 8-1)**

(A) 3, 4, 5
(B) 8, 15, 17
(C) 6, 8, 12
(D) 7, 24, 25

Standards Reference Guide

California Content Standard	Item Number(s)
GEOM 15.0	1, 4, 9, 17
GEOM 4.0	2, 14
GEOM 12.0	3, 11
GEOM 19.0	5, 8, 10
GEOM 20.0	6, 13
GEOM 7.0	7, 12, 15
GEOM 18.0	16

For additional review and practice, see the *California Standards Review and Practice Workbook* or go online to use the

Visit: PHSchool.com, **Web Code:** bcq-9045

Transformations

What You've Learned

California Content Standards

5.0 Students prove that triangles are congruent or similar, and they are able to use the concept of corresponding parts of congruent triangles.

12.0 Students find and use measures of sides and of angles of triangles and polygons to classify figures and solve problems.

17.0 Students prove theorems by using coordinate geometry, including the midpoint of a line segment, the distance formula, and various forms of equations of lines.

Regular Polygons (Lesson 3-5)

Determine the measure of an angle of the given regular polygon.

1. pentagon **2.** octagon **3.** decagon **4.** 18-gon

Congruent Figures (Lesson 4-1)

The triangles are congruent. Complete the congruence statement, $\triangle ABC \cong$ __?__.

5.

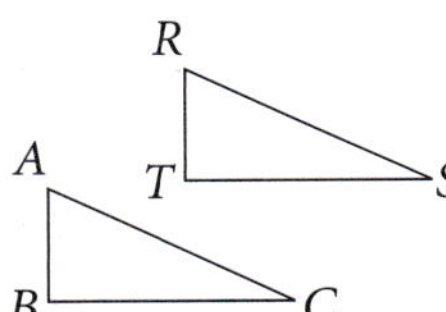

6.

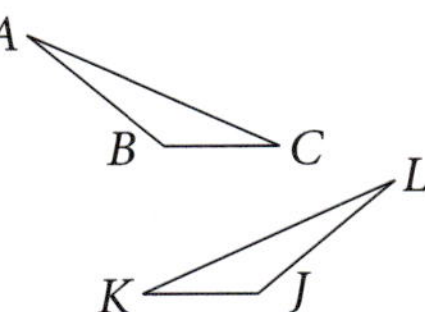

7.

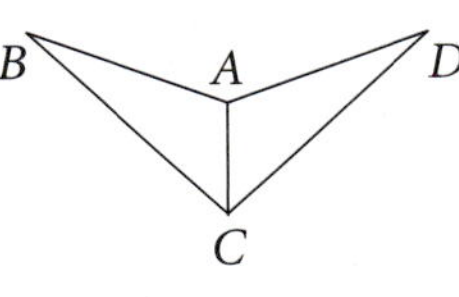

8.

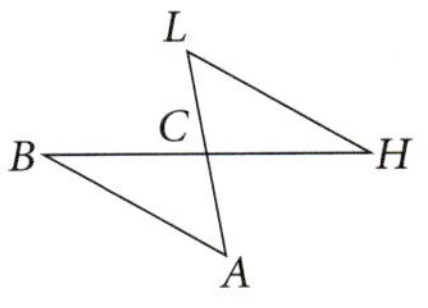

Quadrilaterals (Lessons 6-2, 6-4, and 6-5)

Determine whether a diagonal of the given quadrilateral *always, sometimes,* or *never* produces congruent triangles.

9. rectangle **10.** isosceles trapezoid **11.** kite **12.** parallelogram

Scale Drawing (Lesson 7-1)

The scale of a blueprint is 1 in. = 20 ft.

13. What is the actual length of a wall that measures 2.5 in. in the blueprint?

14. What is the blueprint length of an entrance that is 5 ft wide?

Vectors (Lesson 8-6)

Write the sum of the two vectors as an ordered pair.

15. $\langle 3, 1\rangle$ and $\langle -2, 4\rangle$ **16.** $\langle -1, 2\rangle$ and $\langle -5, 0\rangle$ **17.** $\langle 0, 0\rangle$ and $\langle -6, 3\rangle$

What You'll Learn Next

▲ Computer animated images such as this dragon are created in a 3-dimensional coordinate plane, and then made to move by applying various transformations to the coordinates.

California Content Standards

22.0 Students know the effect of rigid motions on figures in the coordinate plane and space, including rotations, translations, and reflections.

CAHSEE 7NS 1.1, 6SDAP 3.1, 7SDAP 1.1, 7AF 1.1, 7AF 1.5, 7AF 4.2, 7MG 1.2, 7MG 3.2, 7MG 3.3, 7MR 1.2, 7MR 2.4, Alg1 8.0, Alg1 9.0, Alg1 10.0

New Vocabulary

English and Spanish Audio Online

- **center of a regular polygon** (p. 484)
- **composition** (p. 472)
- **dilation** (p. 498)
- **enlargement** (p. 498)
- **glide reflection** (p. 508)
- **glide reflectional symmetry** (p. 516)
- **image** (p. 470)
- **isometry** (p. 470)
- **line symmetry** (p. 492)
- **point symmetry** (p. 493)
- **preimage** (p. 470)
- **reduction** (p. 498)
- **reflection** (p. 478)
- **reflectional symmetry** (p. 492)
- **rigid motion** (p. 470)
- **rotation** (p. 484)
- **rotational symmetry** (p. 493)
- **symmetry** (p. 492)
- **tessellation** (p. 515)
- **tiling** (p. 515)
- **transformation** (p. 470)
- **translation** (p. 471)
- **translational symmetry** (p. 516)

9-1 Translations

California Content Standards

GEOM 22.0 Students know the effect of rigid motions on figures in the coordinate plane and space, including translations. ***Introduce, Develop***

What You'll Learn

- To identify isometries
- To find translation images of figures

. . . And Why

To combine translations, as in Example 5

Check Skills You'll Need

GO for Help Lesson 4-1

$\triangle ABC \cong \triangle EFG$.

Complete the congruence statements.

1. $\overline{AB} \cong$ __?__ **2.** $\overline{EG} \cong$ __?__

3. $\overline{FG} \cong$ __?__ **4.** $\angle C \cong$ __?__

5. $\angle E \cong$ __?__ **6.** $\angle B \cong$ __?__

7. Complete: If $\triangle KTQ \cong \triangle LGR$, then $\overline{TK} \cong$ __?__ and $\angle TQK \cong$ __?__.

New Vocabulary • transformation • preimage • image • isometry • rigid motion • translation • composition

1 Identifying Isometries

Vocabulary Tip

In general, the word transformation can refer to any kind of change in appearance.

A **transformation** of a geometric figure is a change in its position, shape, or size. When you assemble a jigsaw puzzle, you often move the puzzle pieces by flipping them, sliding them, or turning them. Each move is a type of transformation. The photos below illustrate some basic transformations that you will study.

The figure flips.

The figure slides.

The figure turns.

The original figure is the **preimage.** The resulting figure is an **image.** An **isometry** is a transformation in which the preimage and image are congruent. Each transformation above is an isometry. You can also think of an isometry as a **rigid motion,** a motion that preserves distance and angles.

1 EXAMPLE Identifying Isometries

Does the transformation appear to be an isometry? Explain.

No, this transformation involves a change in size. The sides of the preimage square and the sides of its image are not congruent.

Preimage Image

1 Does the transformation appear to be an isometry? Explain.

a.

b.

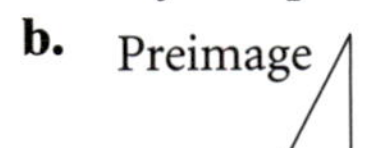

Vocabulary Tip

Read $K \to K'$ as "K maps onto K prime."

A transformation maps a figure onto its image and may be described with arrow ($\to$) notation. Prime ($'$) notation is sometimes used to identify image points. In the diagram at the right, K' is the image of K ($K \to K'$).

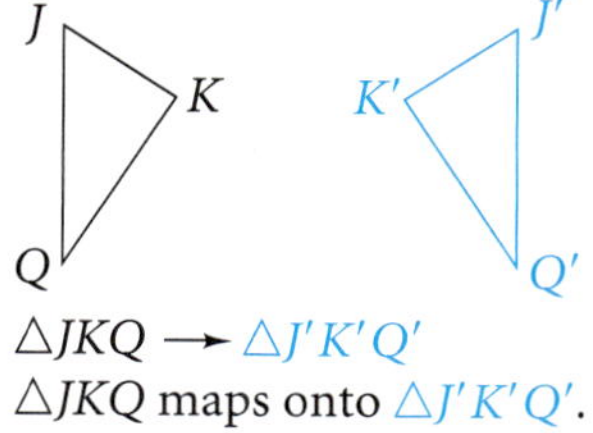

Notice that you list corresponding points of the preimage and image in the same order, as you do for corresponding points of congruent or similar figures.

2 EXAMPLE Naming Images and Corresponding Parts

In the diagram, $E'F'G'H'$ is an image of $EFGH$.

a. Name the images of $\angle F$ and $\angle H$.

$\angle F'$ is the image of $\angle F$.
$\angle H'$ is the image of $\angle H$.

b. List all pairs of corresponding sides.

$\overline{EF}$ and $\overline{E'F'}$; $\overline{FG}$ and $\overline{F'G'}$;
$\overline{EH}$ and $\overline{E'H'}$; $\overline{GH}$ and $\overline{G'H'}$

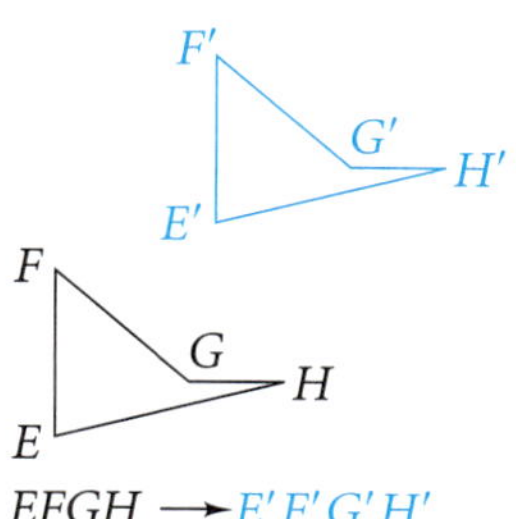

2 In the diagram, $NID \to SUP$.

a. Name the images of $\angle I$ and point D.

b. List all pairs of corresponding sides.

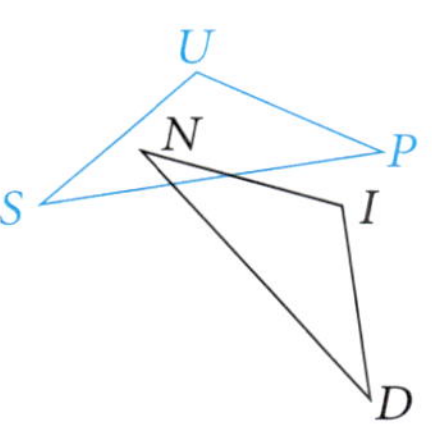

2 Translations

Real-World Connection

It is easier to check parts when each is a translation image of the others.

A **translation** (or *slide*) is an isometry that maps all points of a figure the same distance in the same direction. The diagram at the right shows a translation of the black square by 4 units right and 2 units down. Using variables, you can say that each (x, y) pair in the original figure is mapped in a rigid motion to $(x + 4, y - 2)$.

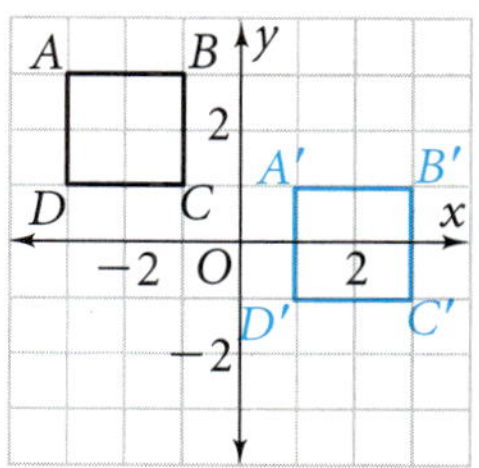

3 EXAMPLE Finding a Translation Image

Find the image of $\triangle XYZ$ under the translation $(x, y) \to (x - 2, y - 5)$.

Use the rule to find each vertex in the translated image.

$X(2, 1)$ translates to $(2 - 2, 1 - 5)$, or $X'(0, -4)$.

$Y(3, 3)$ translates to $(3 - 2, 3 - 5)$, or $Y'(1, -2)$.

$Z(-1, 3)$ translates to $(-1 - 2, 3 - 5)$, or $Z'(-3, -2)$.

The image of $\triangle XYZ$ is $\triangle X'Y'Z'$ with $X'(0, -4)$, $Y'(1, -2)$, and $Z'(-3, -2)$.

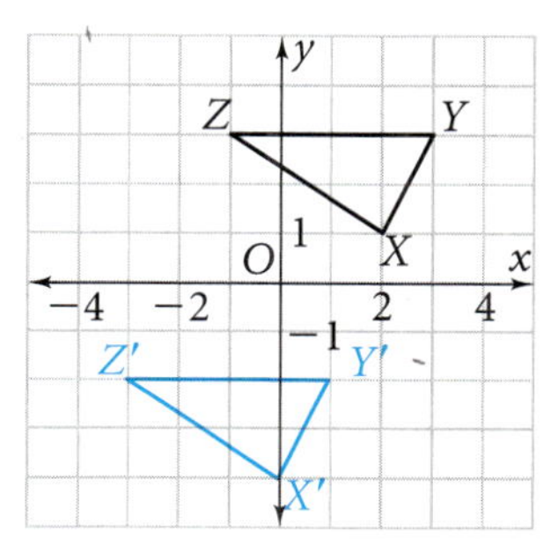

CA Standards Check

3 Find the image of $\triangle XYZ$ for the translation $(x, y) \to (x - 4, y + 1)$.

4 EXAMPLE Writing a Rule to Describe a Translation

Write a rule to describe the translation $PQRS \rightarrow P'Q'R'S'$.

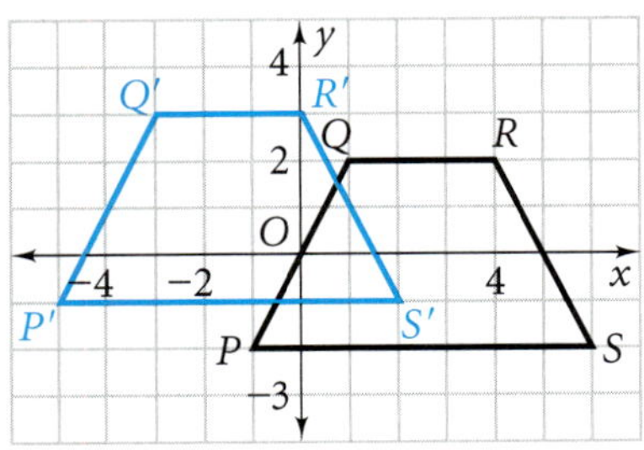

Use $P(-1, -2)$ and its image $P'(-5, -1)$.

Horizontal change: $-5 - (-1) = -4$
Vertical change: $-1 - (-2) = 1$

The rule is $(x, y) \rightarrow (x - 4, y + 1)$.

CA Standards Check **4** The translation image of $\triangle LMN$ is $\triangle L'M'N'$ with $L'(1, -2)$, $M'(3, -4)$, $N'(6, -2)$. Write a rule to describe the translation.

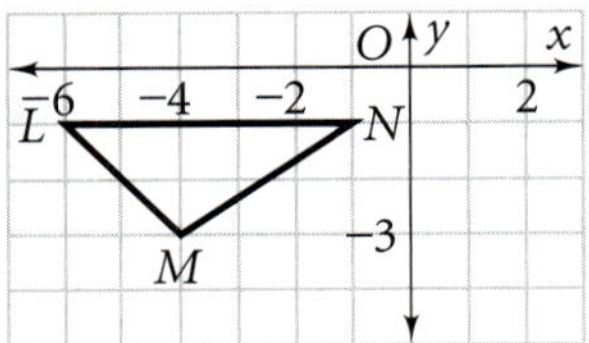

A **composition** of transformations is a combination of two or more transformations. In a composition, each transformation is performed on the image of the preceding transformation.

In a knight's move on a chessboard, the translation indicated in blue is the composition of two translations indicated in red.

In general, the composition of any two translations is a translation.

5 EXAMPLE Real-World Connection

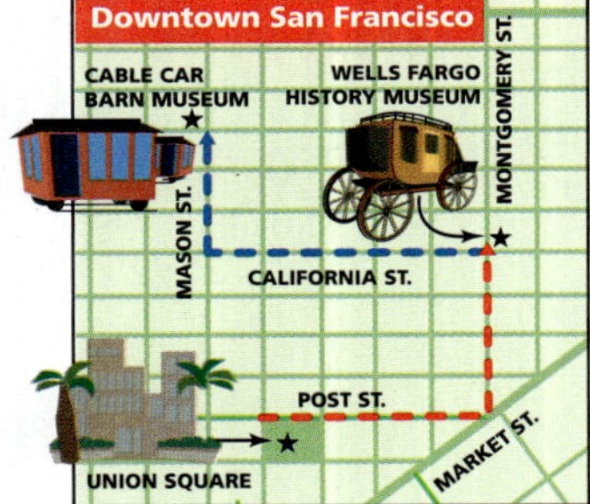

Tourism Yolanda Perez is visiting San Francisco. From her hotel near Union Square, she walks 4 blocks east and 4 blocks north to the Wells Fargo History Museum to see a stagecoach and relics of the Gold Rush. Then she walks 5 blocks west and 3 blocks north to the Cable Car Barn Museum. Now how many blocks is she from her hotel?

Use $(0, 0)$ to represent Yolanda's hotel.

$(x, y) \rightarrow (x + 4, y + 4)$ represents a walk of 4 blocks east and 4 blocks north.
$(x, y) \rightarrow (x - 5, y + 3)$ represents a walk of 5 blocks west and 3 blocks north.

Yolanda's current position is the composition of the two translations. First,

$(0, 0)$ translates to $(0 + 4, 0 + 4)$ or $(4, 4)$. Then,
$(4, 4)$ translates to $(4 - 5, 4 + 3)$ or $(-1, 7)$.

Yolanda is 1 block west and 7 blocks north of her hotel.

5 Yolanda next walks to a restaurant 2 blocks east and 4 blocks south of the Cable Car Barn Museum. Now how many blocks is she from her hotel?

EXERCISES

For more exercises, see *Extra Skill, Word Problem, and Proof Practice.*

Standards Practice

GEOM 22.0

A Practice by Example

GO for Help

Example 1 (page 470)

State whether the transformation appears to be an isometry. Explain.

1.

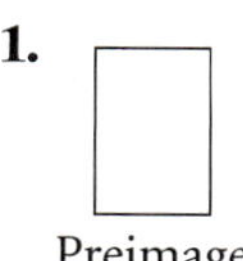

Image

2.

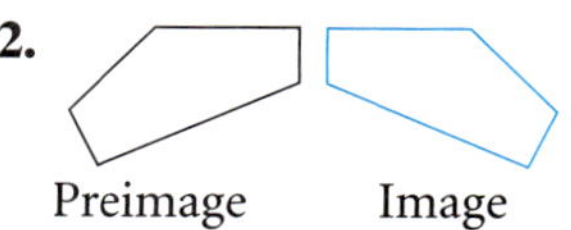

3.

Image

Example 2 (page 471)

In each diagram, the blue figure is an image of the black figure.
(a) Choose an angle or point from the preimage and name its image.
(b) List all pairs of corresponding sides.

4.

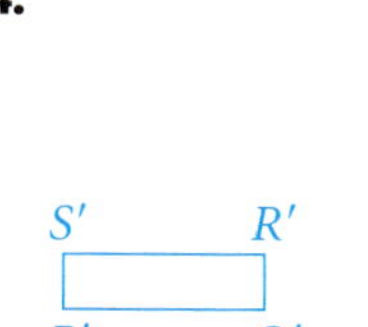

5.

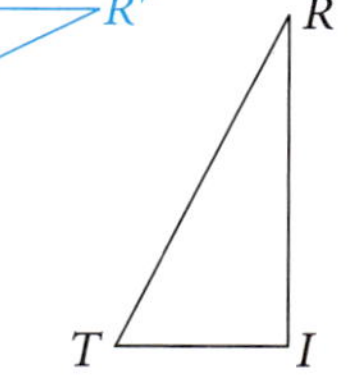

6.

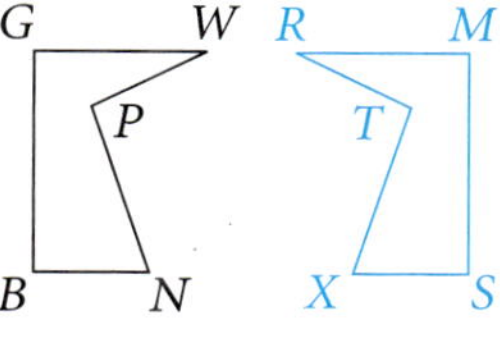

Example 3 (page 471)

Find the image of each figure under the given translation.

7. $(x, y) \rightarrow (x + 3, y + 2)$

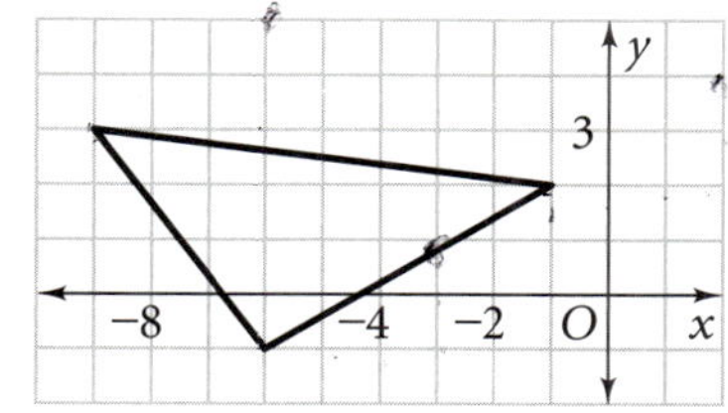

8. $(x, y) \rightarrow (x + 5, y - 1)$

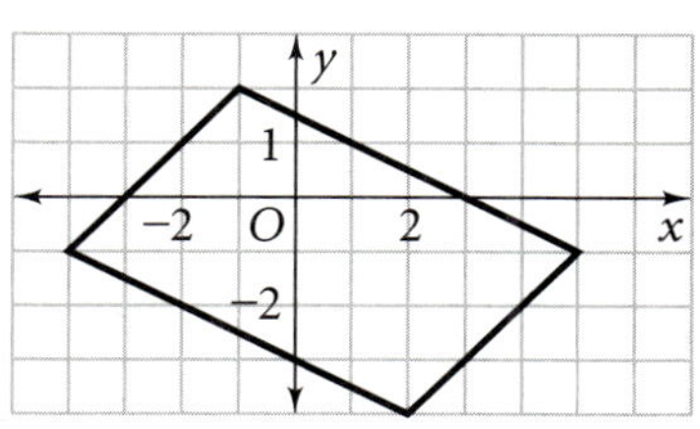

9. $(x, y) \rightarrow (x - 2, y + 5)$

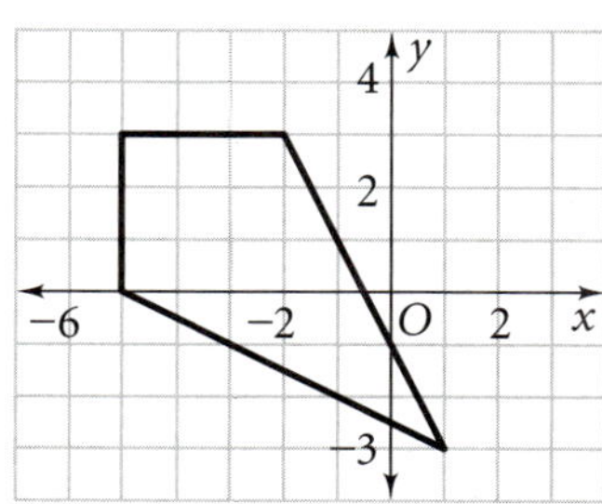

10. $(x, y) \rightarrow (x - 4, y + 3)$

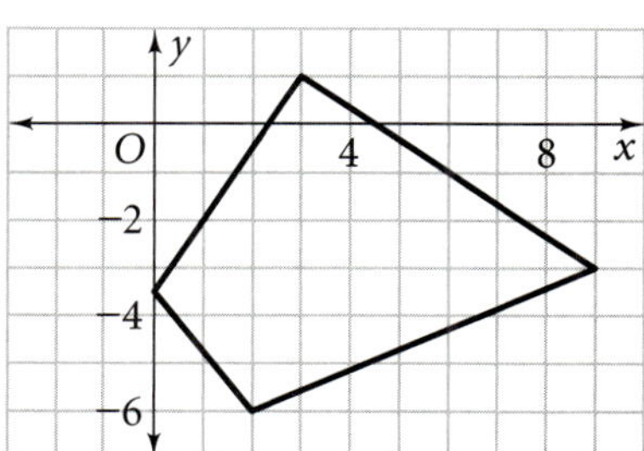

Example 4 (page 472)

In Exercises 11–14, the blue figure is a translation image of the red figure. Write a rule to describe each translation.

11.

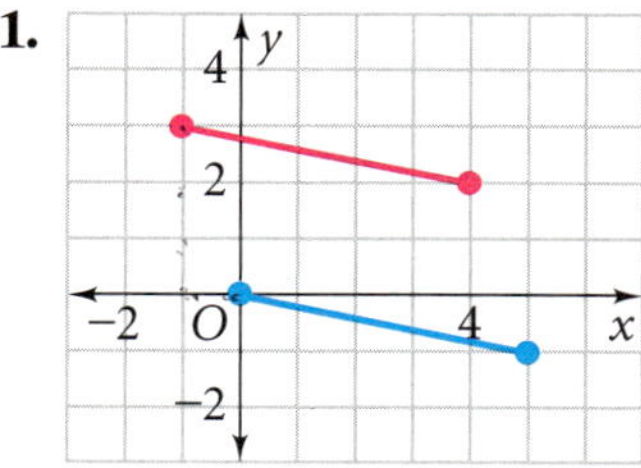

12.

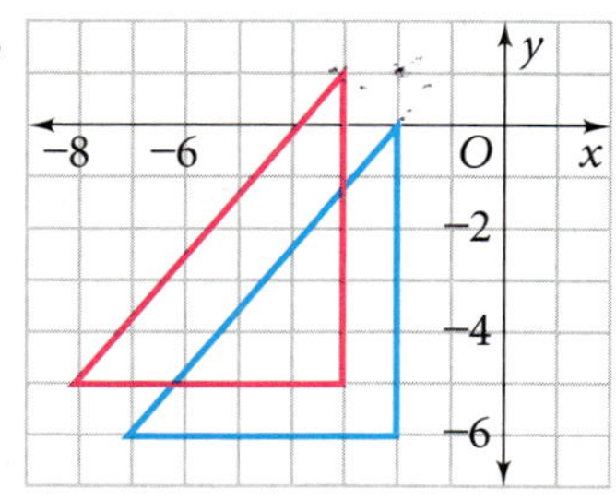

13. 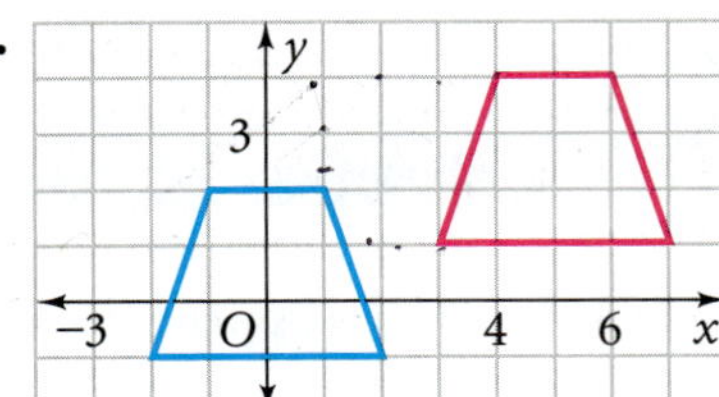

14. 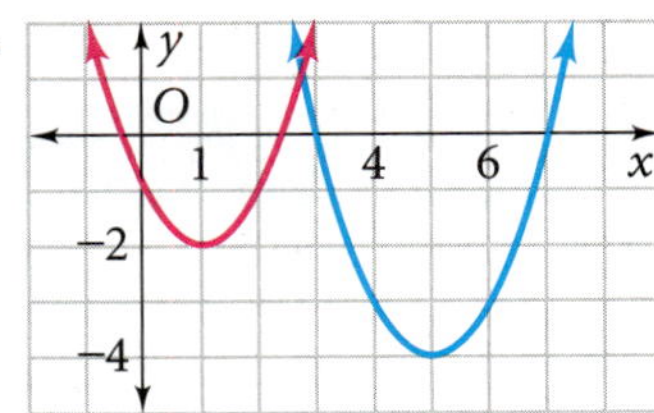

Example 5 (page 472)

15. Emily left Galveston Bay at the east jetty and sailed 4 km north to an oil rig. She then sailed 5 km west to Redfish Island. Finally, she sailed 3 km southwest to Spinnaker Restaurant.
   a. Draw a diagram on graph paper that shows her journey.
   b. Describe where Spinnaker Restaurant is from where Emily started.

16. Nakesha and her parents are visiting colleges. They leave their home in Enid, Oklahoma, and drive to Tulsa, which is 107 mi east and 18 mi south of Enid. From Tulsa, they go to Norman, 83 mi west and 63 mi south of Tulsa. Draw a diagram to show their trip. Then, tell where Norman is in relation to Enid.

B Apply Your Skills

The orange figure is a translation image of the red figure. Write a rule to describe each translation.

17. 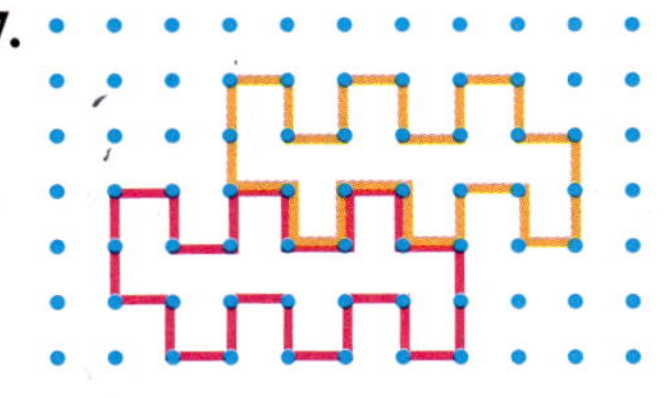

18. 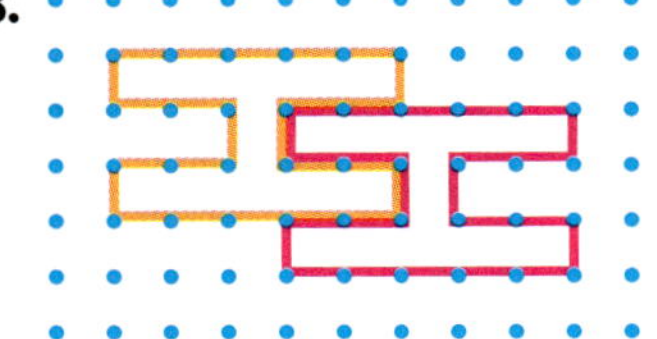

19. **Multiple Choice** For which translation does the image of $\triangle JKL$ have a vertex at the origin?

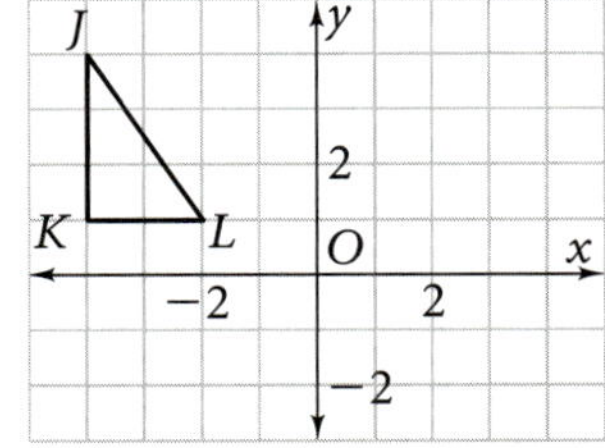

   (A) 2 units to the left and 2 units down
   (B) 4 units to the right and 1 unit up
   (C) 2 units to the right and 3 units down
   (D) 4 units to the right and 4 units down

Graph each polygon and its image for the given translation.

20. $\triangle ACE$ with vertices $A(7, 2)$, $C(-8, 5)$, $E(0, -6)$
translation $(x, y) \rightarrow (x - 9, y + 4)$

21. $\square PLAT$ with vertices $P(-2, 0)$, $L(-1, 1)$, $A(0, 1)$, $T(-1, 0)$
translation $(x, y) \rightarrow (x + 1, y)$

22. $\square NILE$ with vertices $N(2, -5)$, $I(2, 2)$, $L(-3, 4)$, $E(-3, -3)$
translation $(x, y) \rightarrow (x - 3, y - 4)$

23. **Landscaping** The Michelsons want to build a storage shed in their backyard. The diagram at the right shows the original site plan. Local law, however, requires the shed to set back at least 15 feet from property lines.
   a. Describe in words how the Michelsons should move the shed.
   b. Write a translation rule for moving the shed.

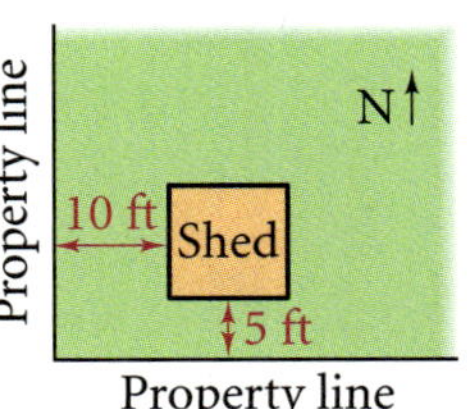

Real-World Connection

A time-lapse photo captures translations of different points of the truck.

24. **Photography** When you snap a photograph, a shutter opens to expose the film to light. The amount of time that the shutter remains open is known as the shutter speed. For the photograph at the left, the photographer used a long shutter speed. It created an image that suggests a translation. Draw a picture of your own that suggests a translation.

25. **Coordinate Geometry** $\triangle MUG$ has coordinates $M(2, -4)$, $U(6, 6)$, and $G(7, 2)$. A translation maps point M to $M'(-3, 6)$. Find the coordinates of U' and G' under this translation.

26. **Coordinate Geometry** $\square ABCD$ has vertices $A(3, 6)$, $B(5, 5)$, $C(4, 2)$, and $D(2, 3)$. The figure is translated so that the image of point C is the origin.
    a. Find the rule that describes the translation.
    b. Graph $\square ABCD$ and its image.

27. **Writing** Is the transformation at the right, $\triangle HYP \rightarrow \triangle H'Y'P'$, a translation? Explain.

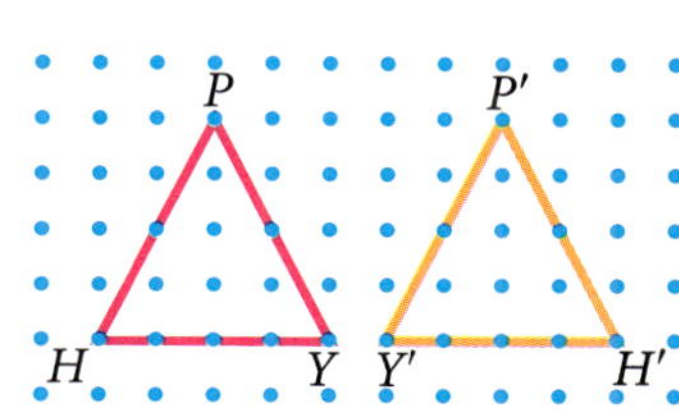

Find a single translation that has the same effect as each composition of translations.

28. $(x, y) \rightarrow (x + 2, y + 5)$ followed by $(x, y) \rightarrow (x - 4, y + 9)$

29. $(x, y) \rightarrow (x + 12, y + 0.5)$ followed by $(x, y) \rightarrow (x + 1, y - 3)$

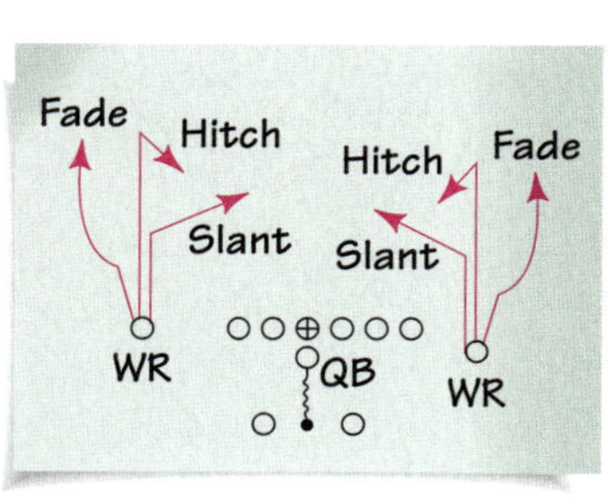

30. **Football** The play chart at the left shows routes that a wide receiver (WR) can choose to run when the team is in the "red zone" (within 20 yards of the goal line). The quarterback (QB) drops back two steps to make the pass to the wide receiver.
    a. Suppose a wide receiver runs a slant. Describe the two translations involved and the composition of those two translations.
    b. Describe the two intended translations of the football during the play, and the composition of the translations.
    c. What is the intended outcome of the two compositions in parts (a) and (b)?

Geometry in 3 Dimensions **Use each figure, graph paper, and the given translation to draw a three-dimensional figure.**

SAMPLE Use the rectangle and $(x, y) \rightarrow (x + 3, y + 1)$ to draw a box.

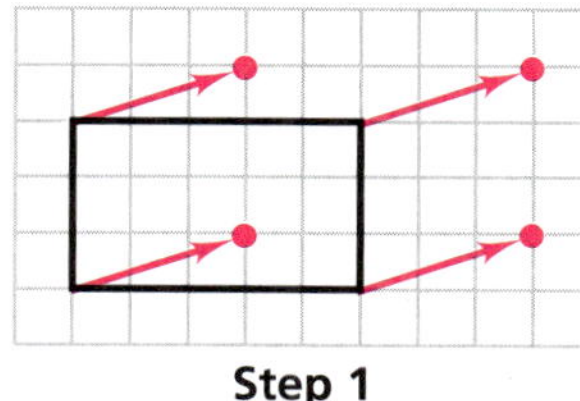

Step 1

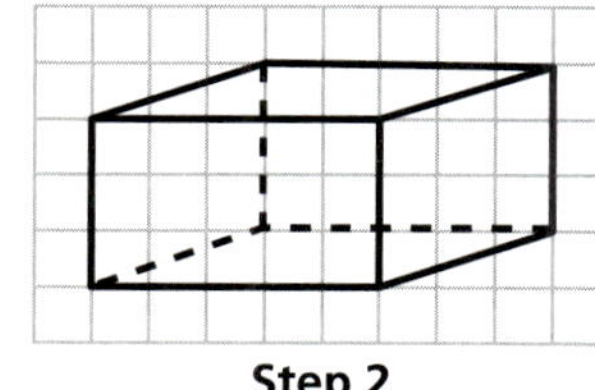

Step 2

31. 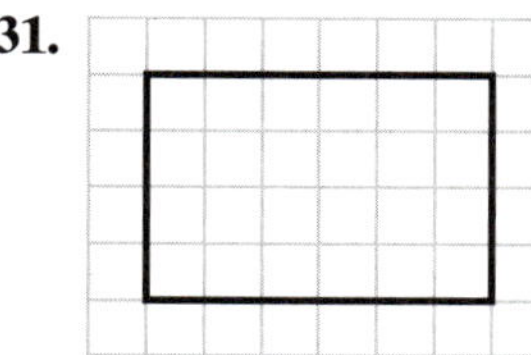

$(x, y) \rightarrow (x + 2, y - 1)$

32. 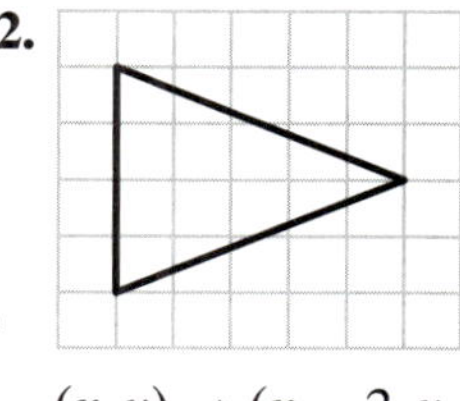

$(x, y) \rightarrow (x - 2, y + 2)$

33. 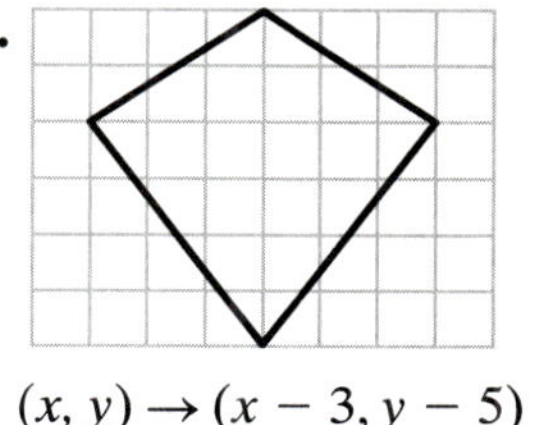

$(x, y) \rightarrow (x - 3, y - 5)$

Homework Video Tutor

Visit: PHSchool.com
Web Code: bce-0901

34. **Open-Ended** You work for a company that specializes in creating unique, artistic designs for business stationery. One of your clients is Totter Toy Co. You have been assigned to create a border design for the top of their stationery. Create a design that involves translations to present to your client.

35. The cube at the right has coordinates $A(0, 2, 8)$, $B(0, 6, 8)$, $C(-4, 6, 8)$, $D(-4, 2, 8)$, $E(-4, 2, 4)$, $F(0, 2, 4)$, $G(0, 6, 4)$, $H(-4, 6, 4)$. A translation maps point A to $A'(3, 2, 4)$. Find the coordinates of B', C', D', E', F' G', and H' under this translation.

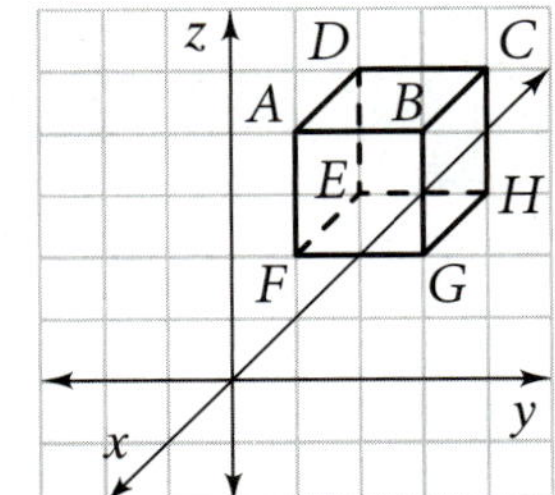

Challenge

36. **Writing** Explain how a parallelogram could be defined in terms of translations.

Multiple Choice Practice

For California Standards and CAHSEE Tutorials, visit PHSchool.com. Web Code: bcq-9045

GEOM 22.0

37. Find the image of triangle CAT under the translation $(x, y) \rightarrow (x + 5, y + 2)$.

Ⓐ $C'(4, 3), A'(3, -1), T'(7, 1)$
Ⓑ $C'(-6, 3), A'(-7, -1), T'(-3, 1)$
Ⓒ $C'(6, 3), A'(7, 5), T'(7, 3)$
Ⓓ $C'(1, 6), A'(0, 2), T'(4, 4)$

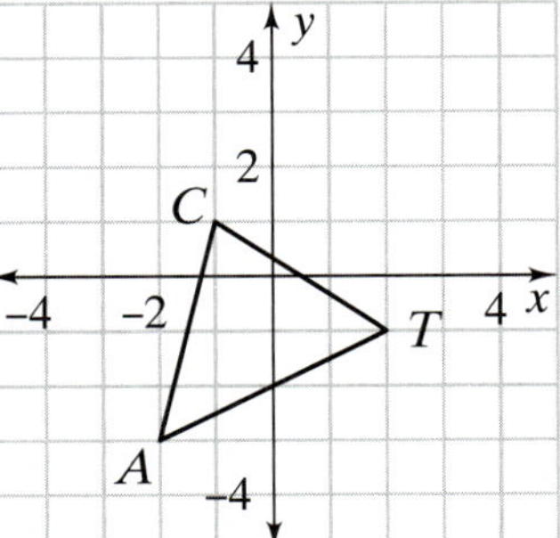

GEOM 20.0

38. If triangle DAN is a 30°-60°-90° triangle and the hypotenuse is equal to 40 cm, what is the length of the longest leg?

Ⓐ 80 cm
Ⓑ $20\sqrt{3}$ cm
Ⓒ $\frac{40\sqrt{3}}{3}$ cm
Ⓓ 20 cm

Mixed Review

Lesson 8-6

39. **Navigation** An airplane lands at a point 100 km east and 420 km south from where it took off. Describe the magnitude and the direction of its flight vector.

Lesson 7-5

Solve for x.

40.

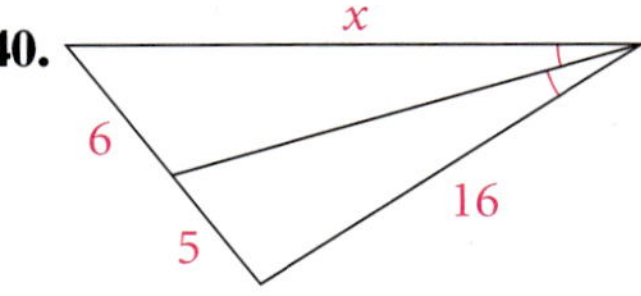

41.

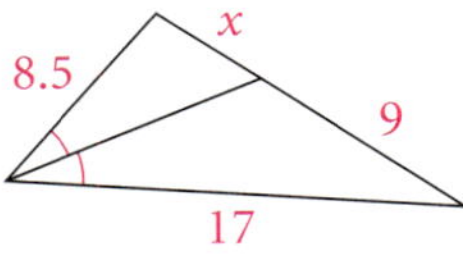

Lesson 4-4

42. **Given:** $\overline{BC} \cong \overline{EF}$, $\overline{BC} \parallel \overline{EF}$
$\overline{AD} \cong \overline{DC} \cong \overline{CF}$

Prove: $\overline{AB} \cong \overline{DE}$

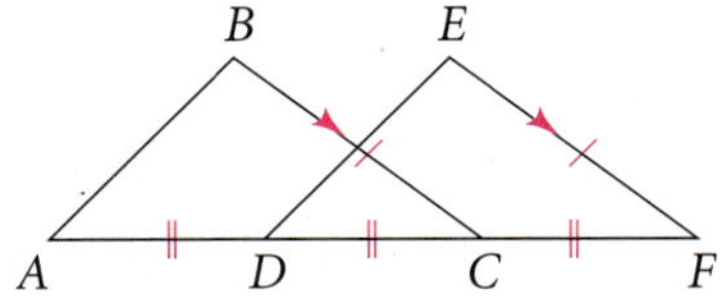

43. **Given:** $\overline{AB} \cong \overline{CB}$, $\overline{BD} \perp \overline{AC}$

Prove: $\angle ABD \cong \angle CBD$

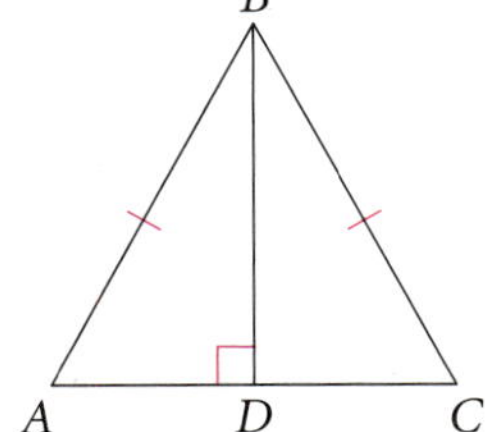

Paper Folding and Reflections

Hands-On

FOR USE WITH LESSON 9-2

Prepares for GEOM 22.0 Students know the effect of rigid motions on figures in the coordinate plane, including reflections.

Activity 1 will help you understand how a figure and its reflection image are related. Then in Activity 2 you will use the properties to construct a reflection image using a compass and a straightedge.

1 ACTIVITY

- Use a page of tracing paper and a straightedge. Using less than half the page, draw a large, scalene triangle. Label its vertices A, B, and C.
- Fold the paper so that your triangle is covered. Trace $\triangle ABC$ carefully using a straightedge.
- Unfold the paper. Label the traced points corresponding to A, B, and C as A', B', and C', respectively. $\triangle A'B'C'$ is a reflection image of $\triangle ABC$. The fold is the reflection line.

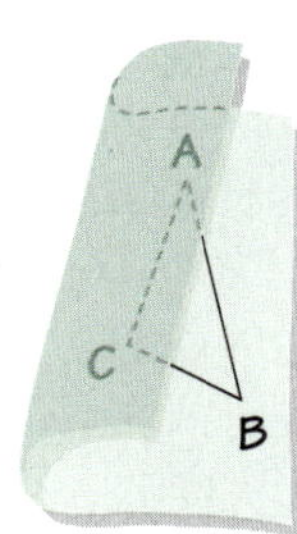

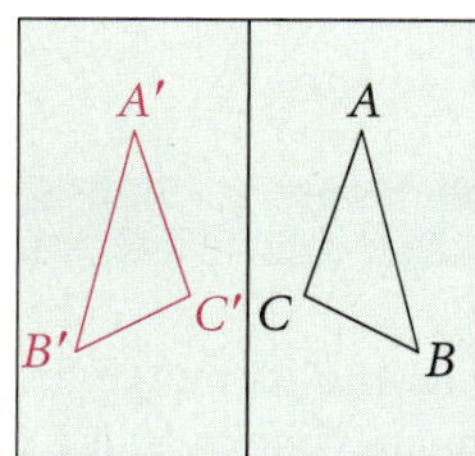

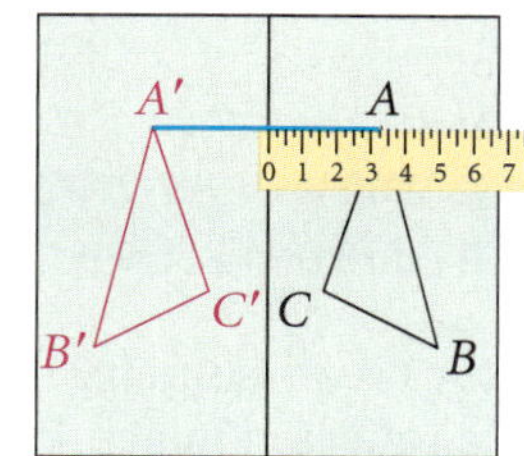

1. Use a ruler to draw $\overline{AA'}$. Measure the perpendicular distances from A to the fold and from A' to the fold. What do you notice?
2. Measure the angles formed by the fold and $\overline{AA'}$.
3. Repeat Exercises 1 and 2 for B and B' and for C and C'.
4. Make a conjecture: How is the reflection line related to the segment joining a point and its reflected image?

2 ACTIVITY

- On graph paper, draw a simple shape or design made of segments (on less than half the page). See the example at the right.
- Draw a reflection line near your figure.
- Use a compass and straightedge to construct a perpendicular to the reflection line through one point of your drawing.

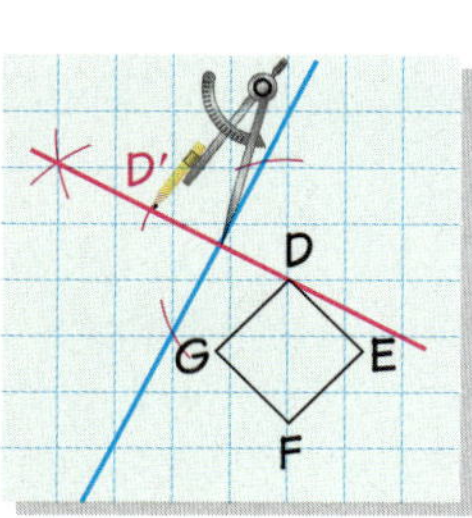

5. Explain how you can use a compass and the perpendicular you drew to find the reflection image of the point you chose.
6. Find the reflection images for several points of your shape. Connect the image points corresponding to the connections in your original figure and complete the image. Check the accuracy of your reflection image by folding the paper along the reflection line and holding the paper to the light.

Reflections

California Content Standards

GEOM 22.0 Students know the effect of rigid motions on figures in the coordinate plane and space, including reflections. ***Introduce, Develop***

What You'll Learn

- To find reflection images of figures

. . . And Why

To minimize distances in construction, as in Example 3

Check Skills You'll Need

for Help Lesson 3-7

Write an equation for the line through point A that is perpendicular to the given line.

1. 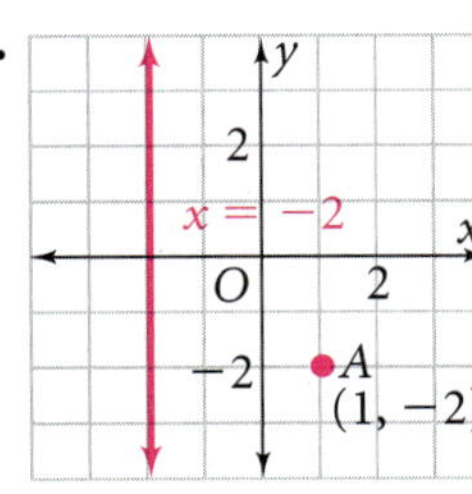

2. 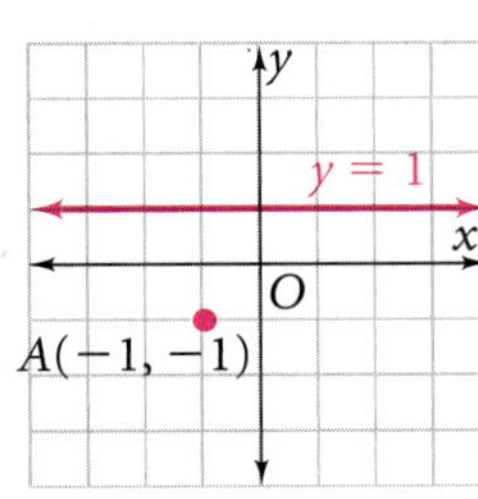

3. 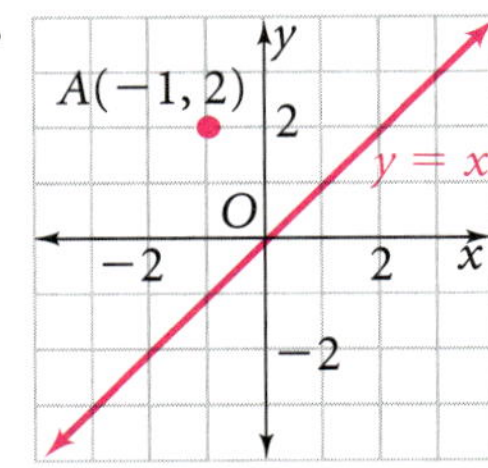

New Vocabulary • reflection

1 Finding Reflection Images

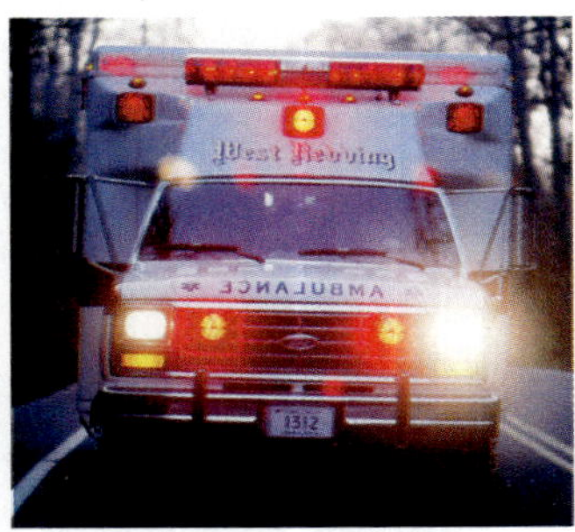

Real-World  **Connection**

In your rear-view mirror you see the reflection "AMBULANCE."

A **reflection** (or *flip*) is an isometry in which a figure and its image have opposite orientations. Thus, a reflected image in a mirror appears "backwards." In the diagram below, $\triangle BUG$ is reflected across a line to produce $\triangle B'U'G'$. Since the reflection is an isometry, $\triangle BUG \cong \triangle B'U'G'$.

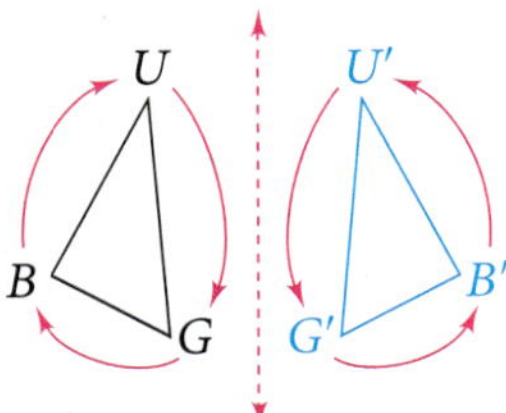

You can use the following two rules to reflect a figure across a line r.

- If a point A is on line r, then the image of A is A itself (that is, $A' = A$).
- If a point B is not on line r, then r is the perpendicular bisector of $\overline{BB'}$.

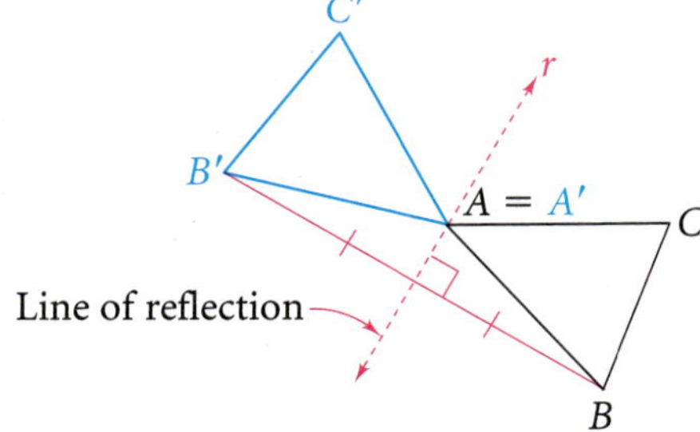

As the diagram shows, a point on the line of reflection is mapped onto itself. Otherwise, a point and its image are equidistant from the line of reflection.

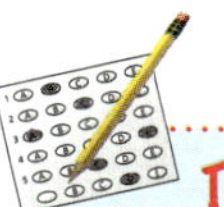

Test-Taking Tip

Remember that equations of the form $y = a$ represent horizontal lines. The form $x = b$ represents vertical lines.

1 EXAMPLE Finding Reflection Images

Multiple Choice If point $P(2, -1)$ is reflected across the line $y = 1$, what are the coordinates of its reflection image?

Ⓐ $(2, 1)$ Ⓑ $(0, -1)$ Ⓒ $(2, 3)$ Ⓓ $(-2, 3)$

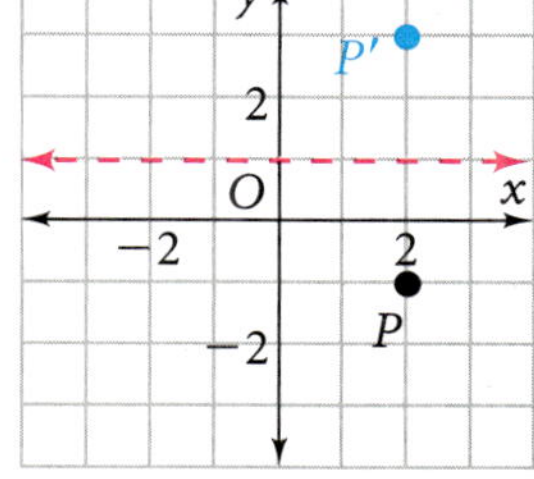

P is 2 units below the reflection line, so its image P' is 2 units above the reflection line. The reflection line is the perpendicular bisector of $\overline{PP'}$ if P' is at $(2, 3)$. The correct answer is C.

1 What are the coordinates of the images of P if the reflection line is $x = 1$?

2 EXAMPLE Drawing Reflection Images

Online active math

For: Reflection Activity
Use: Interactive Textbook, 9-2

Coordinate Geometry Given points $A(-3, 4)$, $B(0, 1)$, and $C(2, 3)$, draw $\triangle ABC$ and its reflection image across each line.

a. the x-axis

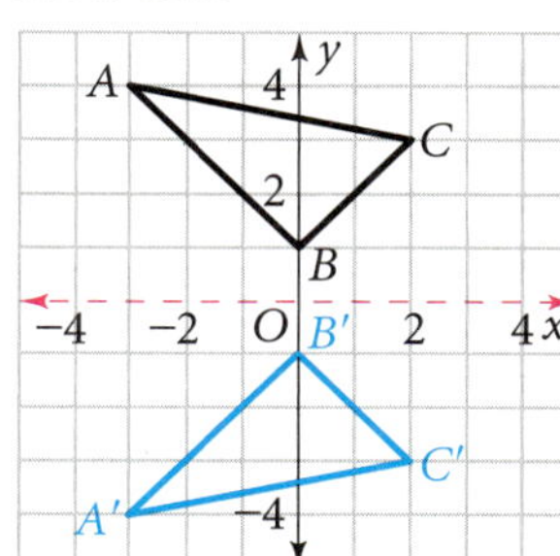

Locate points A', B', and C' such that the line of reflection is the perpendicular bisector of $\overline{AA'}$, $\overline{BB'}$, and $\overline{CC'}$.

b. the y-axis

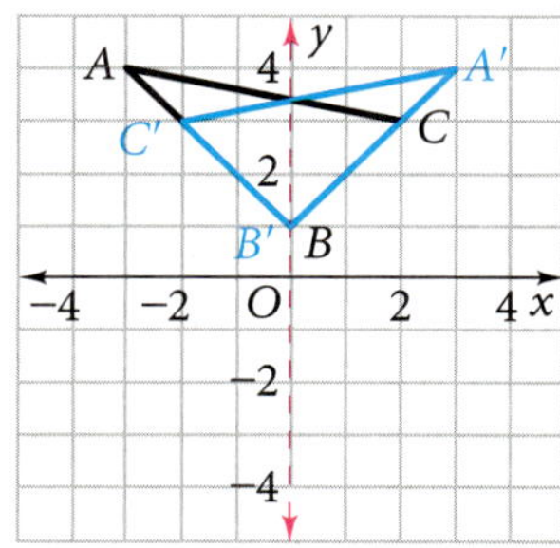

2 Draw $\triangle ABC$ of Example 2. Then draw its reflection image across the line $x = 3$.

You can use the properties of reflections to solve real-world problems.

3 EXAMPLE Real-World Connection

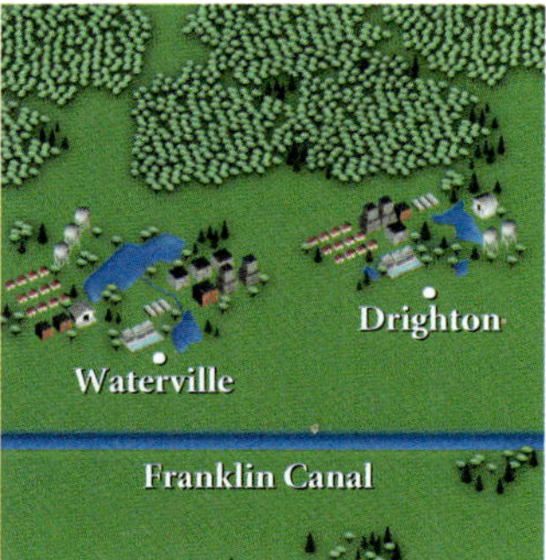

Engineering Town officials in Waterville and Drighton (see aerial view at left) are planning to construct a water pumping station along the Franklin Canal. The station will provide both towns with water. Where along the canal should the officials build the pumping station to minimize the amount of pipe needed?

You need to find the point P on ℓ such that $WP + PD$ is as small as possible. Locate D', the reflection image of D in ℓ. Because a reflection is an isometry, $PD = PD'$, and $WP + PD = WP + PD'$. By the Triangle Inequality Theorem, the sum $WP + PD'$ is smallest when W, P, and D' are collinear. So, the pump should be located at the point P where $\overline{WD'}$ intersects ℓ.

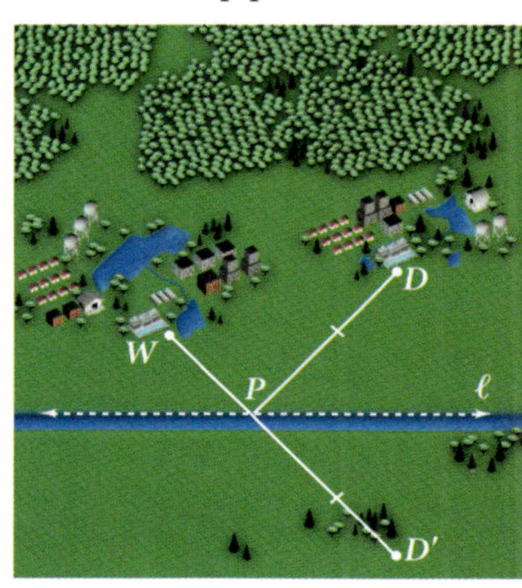

3 **Critical Thinking** Angela began to solve the problem above by reflecting point W in line ℓ. Will her method work? Explain.

EXERCISES

For more exercises, see *Extra Skill, Word Problem, and Proof Practice.*

Standards Practice

GEOM 22.0

A Practice by Example

Example 1 (page 479)

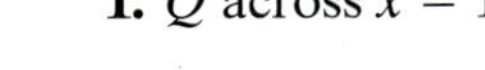

Each point is reflected across the line indicated. Find the coordinates of the images.

1. Q across $x = 1$

2. R across $y = -1$

3. S across the y-axis

4. T across $y = 0.5$

5. U across $x = -3$

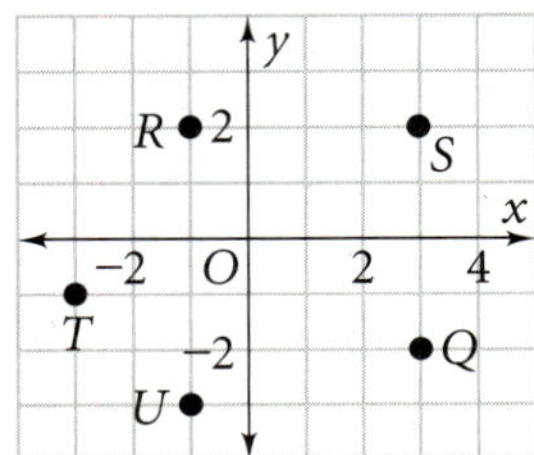

Example 2 (page 479)

Coordinate Geometry **Given points $J(1, 4)$, $A(3, 5)$, and $R(2, 1)$, draw $\triangle JAR$ and its reflection image across each line.**

6. the x-axis **7.** the y-axis **8.** $y = 2$ **9.** $y = 5$

10. $x = -1$ **11.** $x = 2$ **12.** $y = -x$ **13.** $y = x - 3$

Example 3 (page 479)

14. **Trail Building** A hiking club is building a new trail system. It wants to build trails to the Overlook and Balance Rock that will connect at a point on Summit Trail. Working under a tight budget, it wants to minimize the total length of these trails. If the trails cover similar terrain, at what point should they meet on Summit Trail?

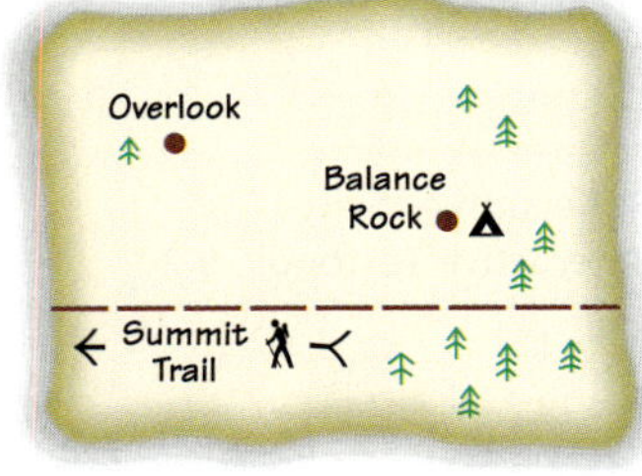

15. **Security** You are installing a security camera. At what point on the mirrored wall should you aim camera C in order to videotape door D?

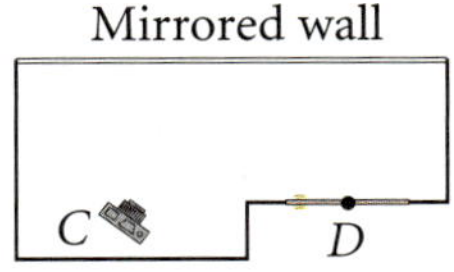

B Apply Your Skills

Copy each pair of figures. Then draw the line of reflection you can use to map one figure onto the other.

16.

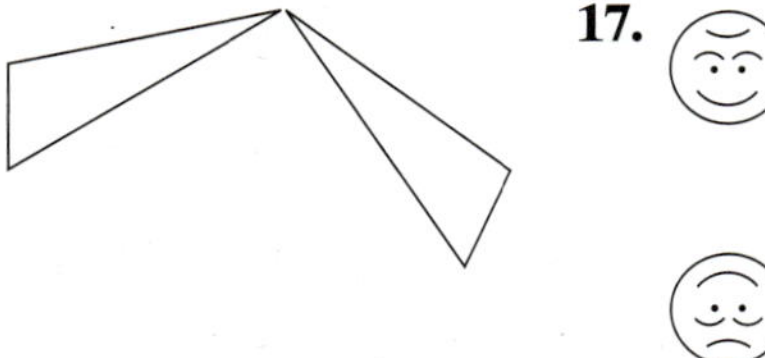

17.

18.

19. **Multiple Choice** If you reflect $\triangle FGH$ across the x-axis, what will be the coordinates of the vertices of the image $\triangle F'G'H'$?

Ⓐ $F'(-2, 1)$, $G'(2, -2)$, $H'(4, 3)$

Ⓑ $F'(2, -1)$, $G'(-2, 2)$, $H'(-4, 3)$

Ⓒ $F'(-1, 2)$, $G'(2, -2)$, $H'(3, -4)$

Ⓓ $F'(1, -2)$, $G'(-2, 2)$, $H'(-3, 4)$

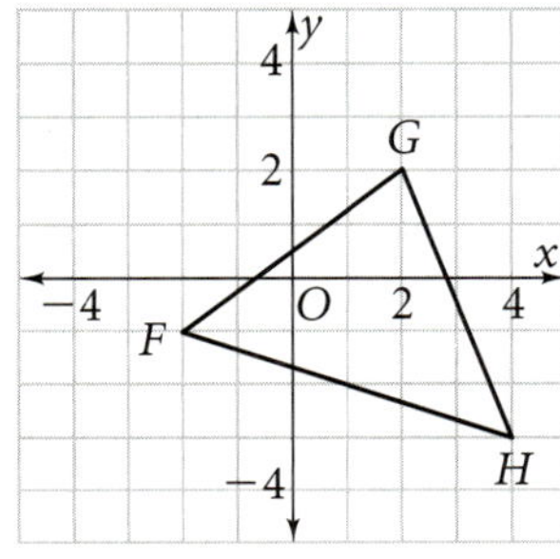

Homework Video Tutor

Visit: PHSchool.com
Web Code: bce-0902

Copy each figure and line ℓ. Then draw each figure's reflection image across line ℓ.

20. 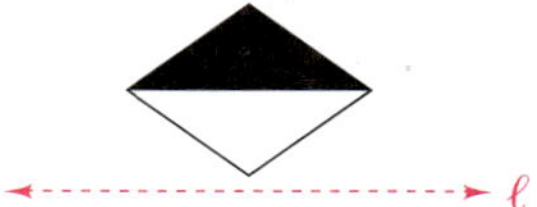

21. 

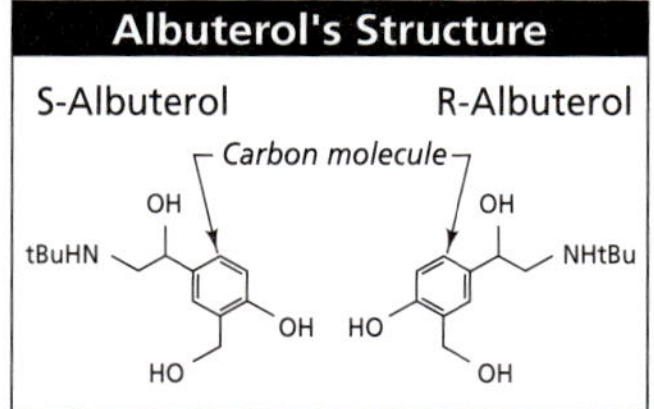

SOURCE: *Wall Street Journal*

The *R-isomer* of the drug albuterol relieves asthma. Its twin has been shown to increase the chances of having future attacks.

22. Pharmaceuticals Most drugs are made of two versions of the same molecule, each a mirror image of the other. One version is known as an *R-isomer* and the other as an *S-isomer*. While one isomer can help with what ails you, the other can create unwanted side effects. Models of two isomers are shown above. For this drug to cure an illness, it needs to fit into the "receptor molecule." Which isomer will cure the illness?

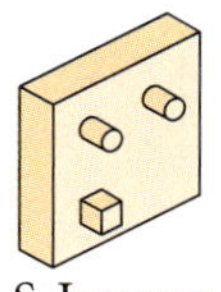
S-Isomer

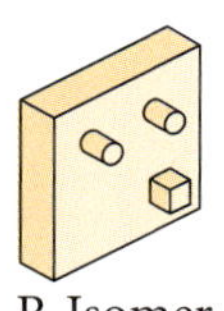
R-Isomer

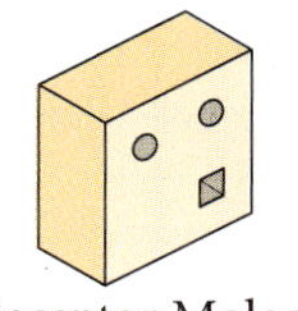
Receptor Molecule

23. Open-Ended Give three examples from everyday life of objects that come in a left-handed version and a right-handed version.

Coordinate Geometry A point is reflected across the given line. How are the coordinates of the point and its image related? Explain.

24. x-axis

25. y-axis

26. the line $y = x$

27. History The work of artist and scientist Leonardo da Vinci (1452–1519) has an unusual characteristic. His handwriting is a mirror image of normal handwriting.

a. Write the mirror image of the sentence, "Leonardo da Vinci was left-handed." Use a mirror to check how well you did.

b. Explain why the fact about da Vinci in part (a) might have made mirror writing seem natural to him.

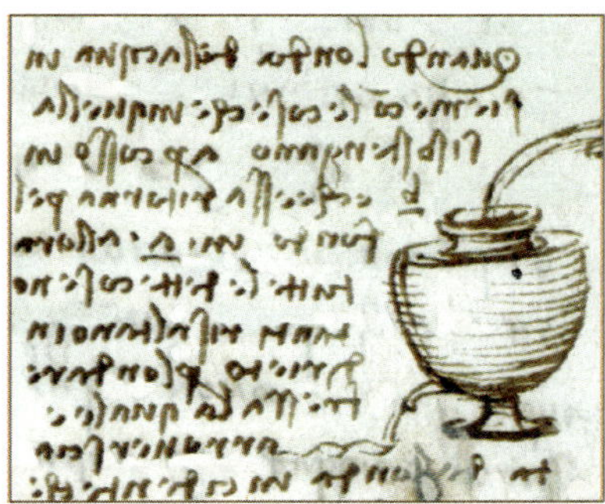

Find the image of $O(0, 0)$ after two reflections, first across ℓ_1 and then across ℓ_2.

28. $\ell_1: y = 3, \ell_2$: x-axis

29. $\ell_1: x = -2, \ell_2$: y-axis

30. ℓ_1: x-axis, ℓ_2: y-axis

31. $\ell_1: x = -2, \ell_2: y = 3$

32. $\ell_1: y = 3, \ell_2: x = -2$

33. $\ell_1: x = -2, \ell_2: y = x$

34. $\ell_1: x = a, \ell_2: y = b$

35. $\ell_1: x = a, \ell_2: y = x$

36. $\ell_1: y = b, \ell_2: y = x$

37. Critical Thinking Given that the transformation $\triangle ABC \rightarrow \triangle A'B'C'$ is an isometry, list everything you know about the two figures.

Challenge

Writing Can you form the given type of quadrilateral by drawing a triangle and then reflecting one or more times? Explain.

38. parallelogram

39. isosceles trapezoid

40. kite

41. rhombus

42. rectangle

43. square

44. Coordinate Geometry Show that $B(b, a)$ is the reflection image of $A(a, b)$ across the line $y = x$. (*Hint:* Show that $y = x$ is the perpendicular bisector of $\overleftrightarrow{AB}$.)

45. Coordinate Geometry Find the line of reflection that maps $A(a, b)$ to $C(c, d)$.

46. Use the diagram at the right. Find the coordinates of the given point across the given line.
  a. A', the reflection image of A across $y = x$
  b. A'', the reflection image of A' across $y = -x$
  c. A''', the reflection image of A'' across $y = x$
  d. A'''', the reflection image of A''' across $y = -x$
  e. How are A and A'''' related?

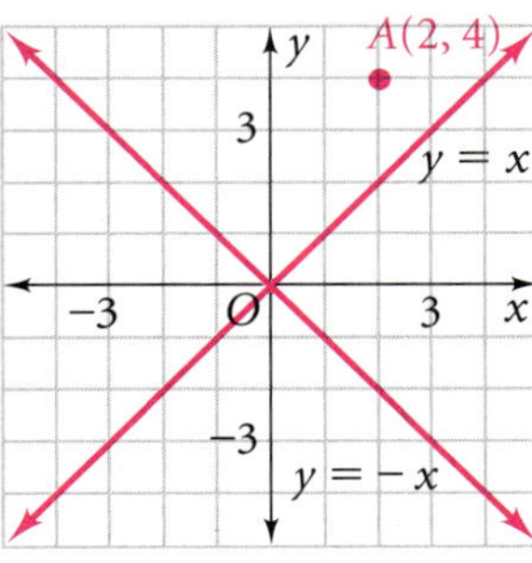

Multiple Choice Practice

For California Standards and CAHSEE Tutorials, visit PHSchool.com. Web Code: bcq-9045

GEOM 22.0

47. If the point $(3, -5)$ is reflected about the x-axis, what are the coordinates of its image?

Ⓐ $(3, 5)$ Ⓑ $(-3, -5)$ Ⓒ $(-3, 5)$ Ⓓ $(5, 3)$

GEOM 15.0

48. To measure the greatest distance across a pond, a hiker paces off 1000 feet south from one end of the pond and then paces off 750 feet east to the opposite end of the pond. What is the length of the pond in feet?

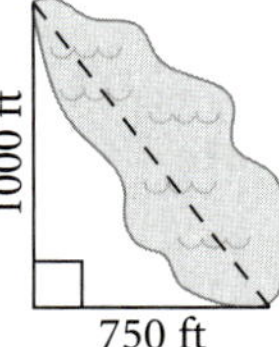

Ⓐ 42 Ⓑ 1000 Ⓒ 1250 Ⓓ 1750

GEOM 10.0

49. Nancy's rectangular garden has a length of 22 feet and a width of 10 feet. If she decreases the length by x feet and increases the width by the same amount, her new garden will have an area of 252 square feet. Find the length of her new garden in feet.

Ⓐ 10 Ⓑ 12 Ⓒ 18 Ⓓ 22

GEOM 5.0

50. Standing in the sun, a 5-foot-tall woman casts a shadow 12 feet long. At the same time, a flagpole casts a shadow 42 feet long. How tall is the flagpole in feet?

Ⓐ 12.5 Ⓑ 17.5 Ⓒ 52.5 Ⓓ 100.8

Mixed Review

Lesson 9-1

In each diagram, the blue figure is the translation image of the red figure. Write a rule to describe each translation.

51.

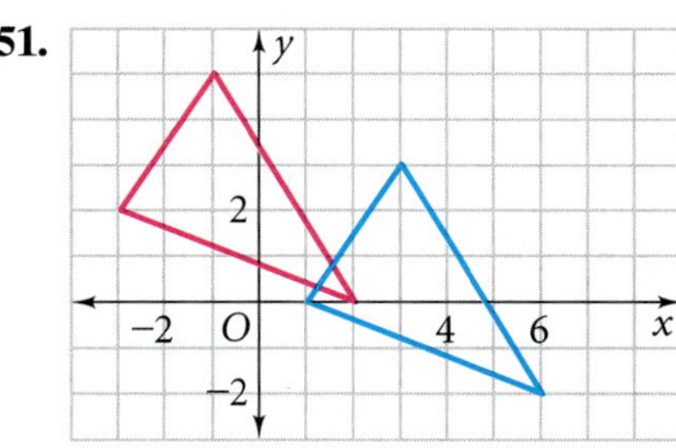

52.

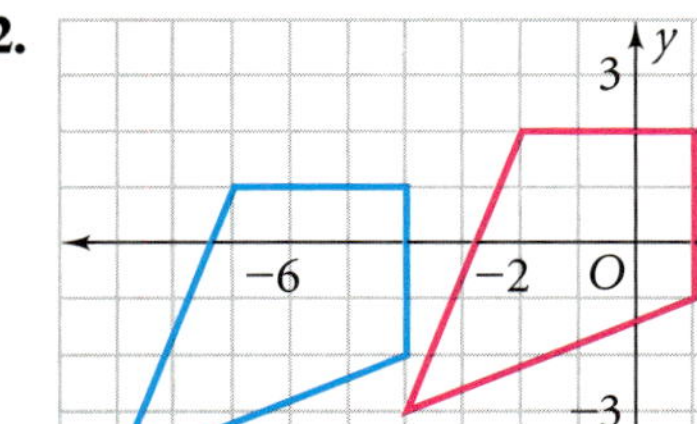

Lesson 7-1

53. **Maps** A map of Alberta, Canada, is drawn to the scale 1 cm = 25 km. On the map, the distance from Calgary to Edmonton is about 11.1 cm.
  a. About how far apart are the two cities?
  b. If 1 km = 0.62 mi, about how many miles apart are the cities?

Lesson 3-6

Write an equation in point-slope form of the line that contains the given points.

54. $M(2, 3), N(-2, 5)$

55. $A(-1, -1), B(0, 4)$

56. $X(1, -3), Y(7, 6)$

9-3 Rotations

California Content Standards

GEOM 22.0 Students know the effect of rigid motions on figures in the coordinate plane and space, including rotations. ***Introduce, Develop***

What You'll Learn

- To draw and identify rotation images of figures

. . . And Why

To identify rotation images in art designs, as in Example 3

Check Skills You'll Need

for Help Skills Handbook, p. 746, Lesson 1-7

Use a protractor to draw an angle with the given measure.

1. 120 **2.** 90 **3.** 72 **4.** 60 **5.** 45 **6.** 36

7. Draw a segment, $\overline{AB}$. Then construct $\overline{A'B'}$ congruent to $\overline{AB}$.

New Vocabulary • rotation • center of a regular polygon

1 Drawing and Identifying Rotation Images

Hands-On Activity: Making Designs With Rotations

You will need at least three pieces of tracing paper for this Activity.

Step 1 Place a piece of paper over the figure at the right. Trace the six points on the circle, the center of the circle, and the triangle.

Step 2 Place the point of your pencil on the center of the circle and then rotate the paper until the six points align again. Trace the triangle in its new location.

Step 3 Repeat Step 2 until there are six triangles on your paper. Compare your drawings to others to be sure that your results look the same.

Step 4 Now make your own triangle pattern. Place a piece of paper over the figure above, trace the six points on the circle and the center of the circle, and then draw your own triangle on the paper.

Step 5 Place the paper from Step 4 on your desk. On a blank piece of paper, repeat Steps 1–3. Color your drawing to make a design.

Turn the above diagram 60° about the center of the circle and each dot on the circle maps to the next dot on the circle. This is an example of a type of transformation, a rigid motion known as a rotation (or *turn*).

Vocabulary Tip

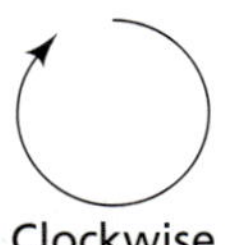

Clockwise

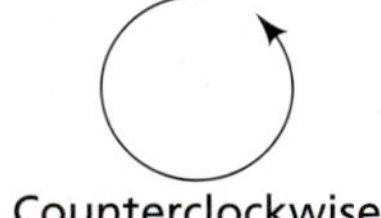

Counterclockwise

To describe a **rotation**, you need to know the center of rotation (a point), the angle of rotation (a positive number of degrees), and whether the rotation is clockwise or counterclockwise. Unless stated otherwise, rotations in this book are counterclockwise.

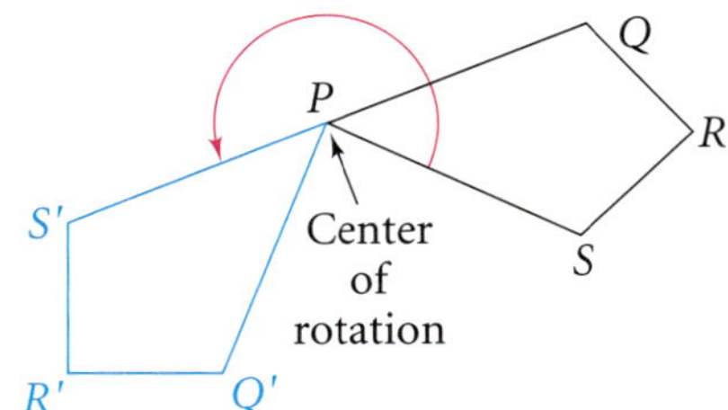

You can use the following two rules to rotate a figure through $x°$ about a point R:

- The image of R is itself (that is, $R' = R$).
- For any point V, $RV' = RV$ and $m\angle VRV' = x$.

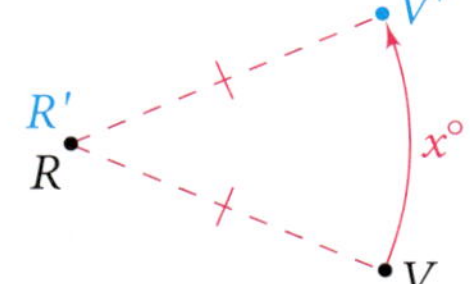

1 EXAMPLE Drawing a Rotation Image

Draw the image of $\triangle LOB$ for a 100° rotation about C.

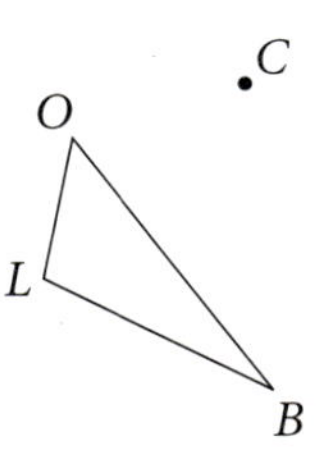

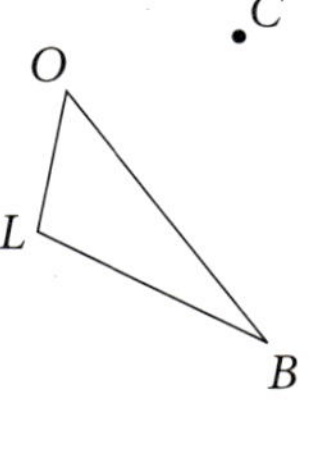

Step 1
Use a protractor to draw a 100° angle with vertex C and side $\overline{CO}$.

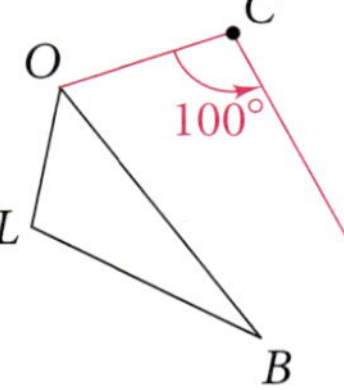

Step 2
Use a compass to construct $\overline{CO'} \cong \overline{CO}$.

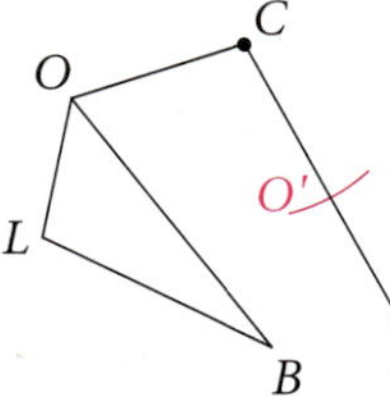

Step 3
Locate B' and L' in a similar manner. Draw $\triangle L'O'B'$.

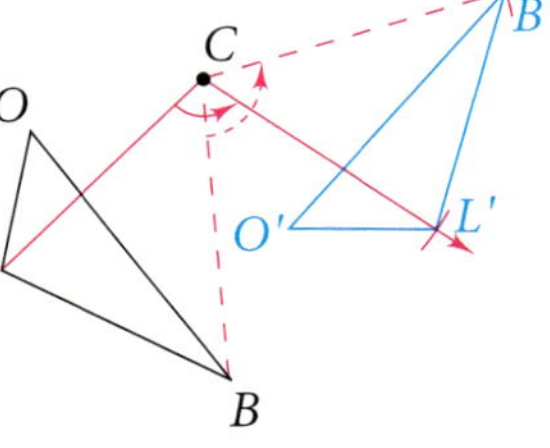

CA Standards Check **1** Draw the image of $\triangle LOB$ from Example 1 for a 50° rotation about B. Label the vertices of the image.

A regular polygon has a **center** that is equidistant from its vertices. Segments that connect the center to the vertices divide the polygon into congruent triangles. You can use this fact to find rotation images of regular polygons.

2 EXAMPLE Identifying a Rotation Image

Regular pentagon $PENTA$ is divided into five congruent triangles.

a. Name the image of E for a 72° rotation about X.

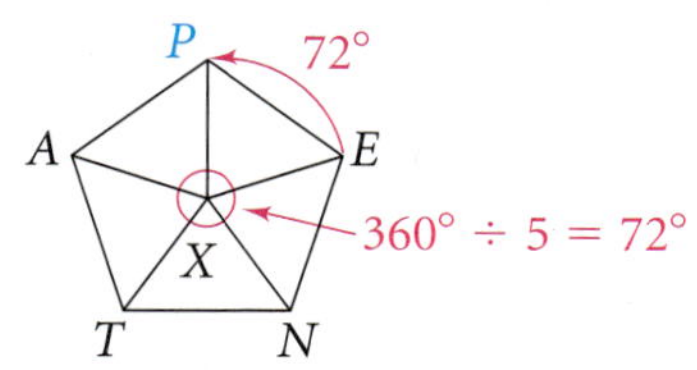

P is the image of E.

b. Name the image of P for a 216° rotation about X.

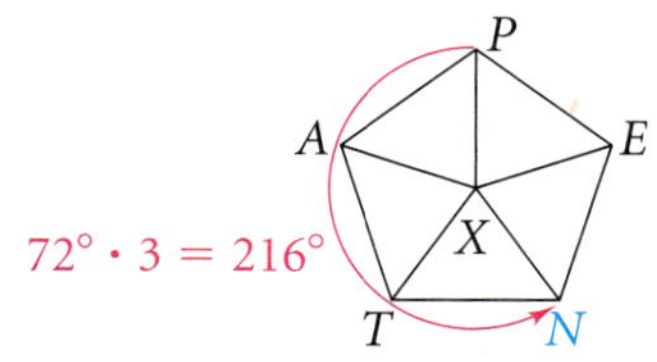

N is the image of P.

CA Standards Check **2** Name the image of T for a 144° rotation about X.

Figures that are rotation images of themselves have a special beauty that you will study more closely in Lesson 9-4.

3 EXAMPLE Real-World Connection

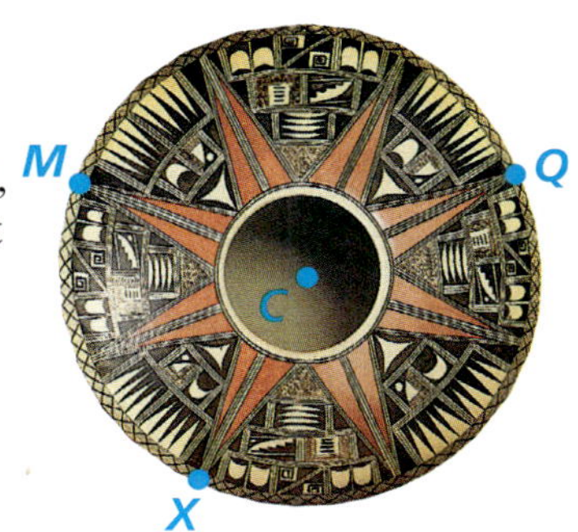

You can find designs with rotation images in some types of Native American art. In the design pictured, how many degrees are in the angle of rotation about C that maps Q to X?

The eight-pointed star in the design divides the circle into eight congruent parts. $360 \div 8 = 45$, so each part has a 45° angle at the center. The angle of rotation that maps Q to X is $5 \cdot 45$, or 225°.

CA Standards Check 3 In the design above, find the angle of rotation about C that maps Q to M.

A composition of rotations about the same point is itself a rotation about that point. To sketch the image, add the angles of rotation to find the total rotation.

4 EXAMPLE Compositions of Rotations

Draw the image of the kite at the left for a composition of a 30° rotation and a 60° rotation, both about point K.

The total rotation is 90°. Draw the kite. Locate image points of the vertices for a 90° rotation. Use the image points to sketch the entire image.

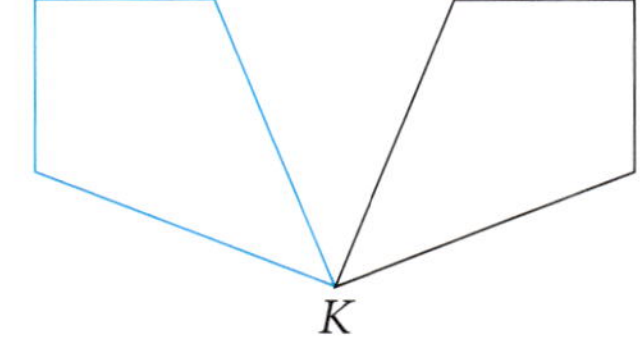

CA Standards Check 4 Draw the kite at the left. Then draw its image for a composition of two 90° rotations about point K.

EXERCISES Standards Practice

For more exercises, see *Extra Skill, Word Problem, and Proof Practice.*

GEOM 22.0

Practice by Example

Example 1 (page 484)

Copy each figure and point P. Draw the image of each figure for the given rotation about P. Label the vertices of the image.

1. 60°

2. 90°

3. 90°

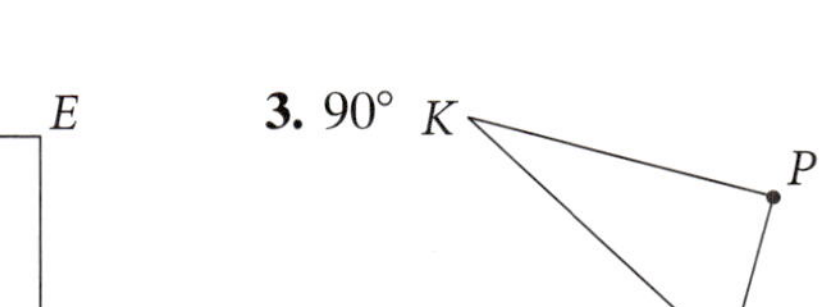

4. 180°

5. 140°

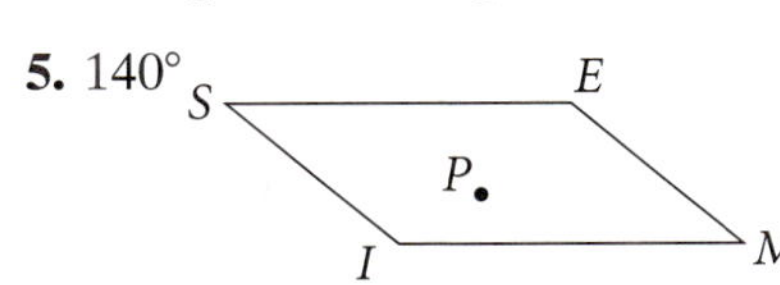

Copy each figure. Then draw the image of $\overline{JK}$ for a 180° rotation about P.

6. J P K

7. K J P

8. J K P

9. K $J \cdot = \cdot P$

Example 2 (page 484)

The large triangle, quadrilateral, and hexagon are regular. Find the image of each point or segment for the given rotation. (*Hint:* Green segments form 30° angles.)

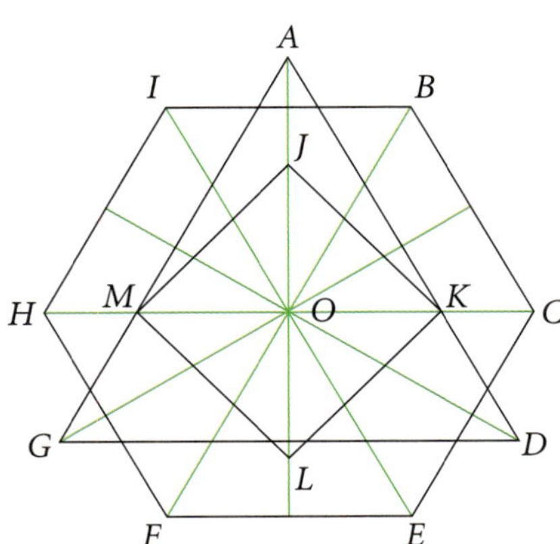

10. 120° rotation of B about O

11. 270° rotation of L about O

12. 60° rotation of E about O

13. 300° rotation of $\overline{IB}$ about O

14. 240° rotation of G about O

15. 180° rotation of $\overline{JK}$ about O

16. 120° rotation of F about H

17. 270° rotation of M about L

Example 3 (page 485)

Native American Art **Find the angle of rotation about C that (a) maps Q to X and (b) maps X to Q.**

18.

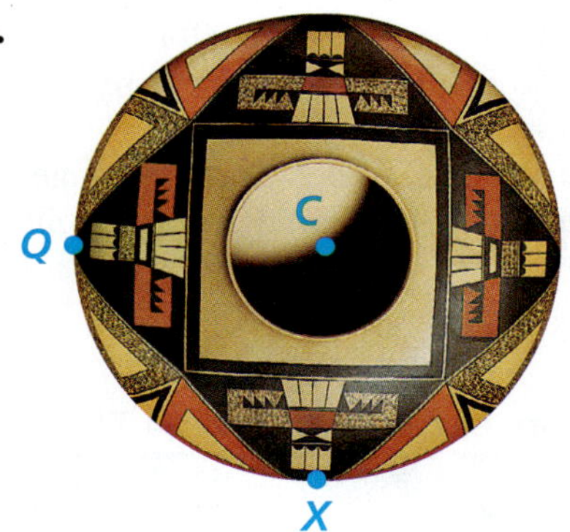

19.

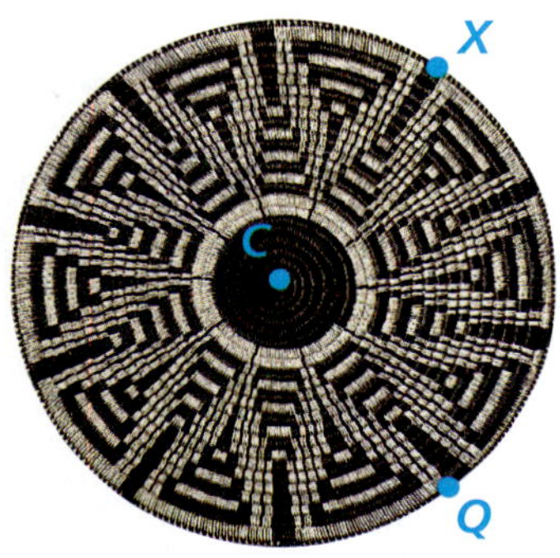

Example 4 (page 485)

For each of Exercises 20–25, copy $\triangle XYZ$. Draw the image of $\triangle XYZ$ for the given composition of rotations about the given point.

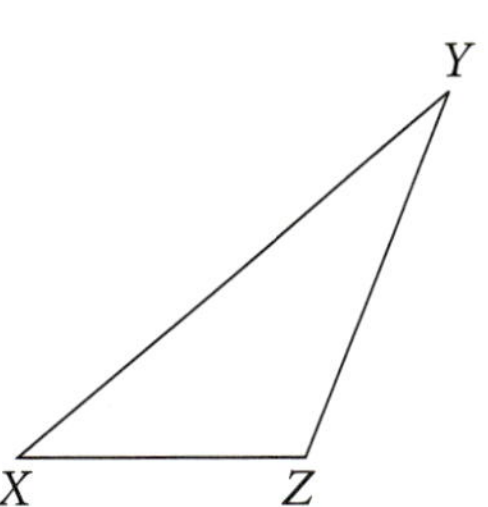

20. 45°, then 45°; X

21. 45°, then 45°; Y

22. 30°, then 30°; Z

23. 20°, then 160°; Z

24. 135°, then 135°; Y

25. 180°, then 180°; X

Apply Your Skills

26. $\overline{M'N'}$ is the rotation image of $\overline{MN}$ about point E. Name all pairs of congruent angles and all pairs of congruent segments in the diagram.

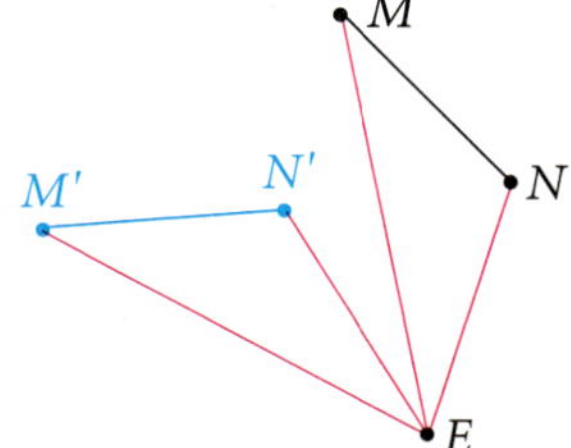

27. **Language Arts** The symbol ə is called a *schwa*. It is used in dictionaries to represent neutral vowel sounds such as *a* in *ago*, *i* in *sanity*, and *u* in *focus*. What transformation maps a ə to a lowercase e?

Visit: PHSchool.com
Web Code: bce-0903

Find the angle of rotation about C that maps the black figure onto the blue figure.

28. 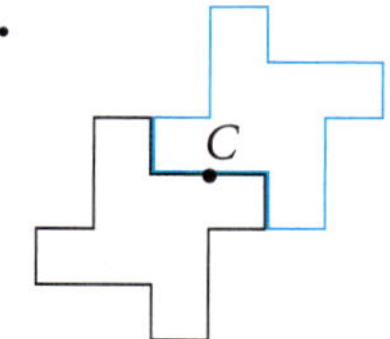

29. 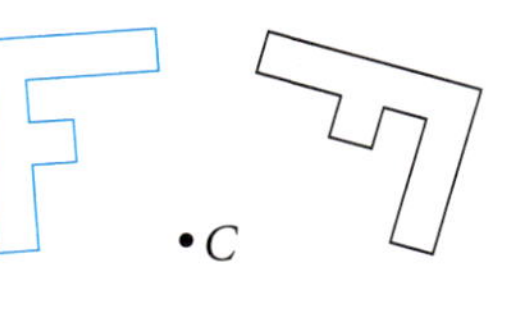

30. 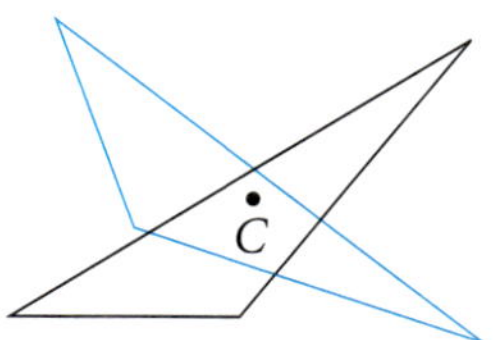

31. a. **Coordinate Geometry** Graph $A(5, 2)$. Then graph B, the image of A for a 90° rotation about the origin O. (*Hint:* Consider the slope of $\overline{OA}$.)
b. Graph C, the image of A for a 180° rotation about O.
c. Graph D, the image of A for a 270° rotation about O.
d. What type of quadrilateral is $ABCD$? Explain.

For a guide to solving Exercise 32, see p. 489.

32. **Reasoning** If you are given a figure and a rotation image of the figure, how can you find the center and angle of rotation?

33. **Writing** Describe compositions of rotations that have the same effect as a 360° rotation about a point X.

Challenge

34. **Coordinate Geometry** Draw $\triangle LMN$ with vertices $L(2, -1)$, $M(6, -2)$, and $N(4, 2)$. Find the coordinates of the vertices after a 90° rotation about the origin and about each of the points L, M, and N.

Multiple Choice Practice

For California Standards and CAHSEE Tutorials, visit PHSchool.com. Web Code: bcq-9045

GEOM 22.0

35. If triangle CAT is rotated 180 degrees about the origin, what are the coordinates of C'?

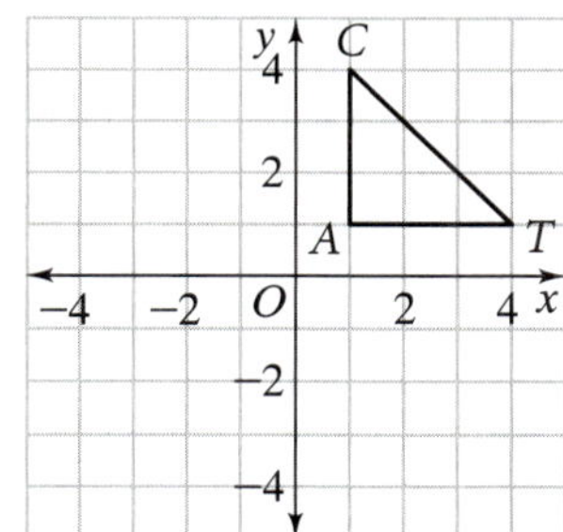

Ⓐ $(1, -4)$
Ⓑ $(-1, -4)$
Ⓒ $(-4, -1)$
Ⓓ $(-1, 4)$

CAHSEE Alg1 9.0

36. Which of the following is the x-value of the solution to the system of equations $\begin{cases} 0.6x - 0.4y = -1 \\ x + 2y = 4.2 \end{cases}$?

Ⓐ -0.2　Ⓑ 0.95　Ⓒ 2.3　Ⓓ -3

CAHSEE 7AF 1.1

37. Rico's cat is at least 2 years older than three times his dog's age. Let c represent the cat's age and d represents the dog's age. Which inequality best represents this situation?

Ⓐ $c \leq 3d + 2$
Ⓑ $c \geq 3d + 2$
Ⓒ $c < 3d + 2$
Ⓓ $c > 3d + 2$

GEOM 22.0

38. Which set of points describes the vertices for a reflection of triangle ABC across the x-axis?

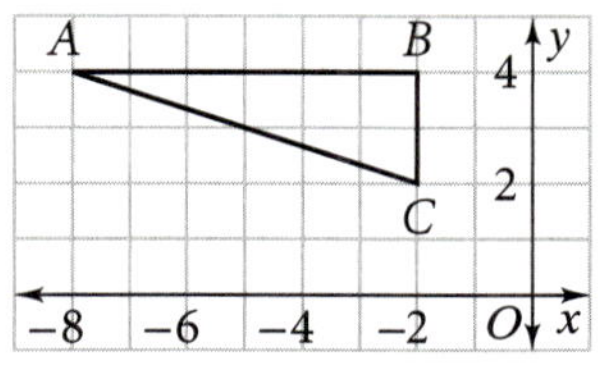

Ⓐ $A'(2, 4), B'(8, 4), C'(8, 2)$
Ⓑ $A'(8, 4), B'(2, 4), C'(2, 2)$
Ⓒ $A'(-8, -4), B'(-2, -4), C'(-2, -6)$
Ⓓ $A'(-8, -4), B'(-2, -4), C'(-2, -2)$

Mixed Review

Lesson 9-2 $\triangle BIG$ **has vertices** $B(-4, 2)$, $I(0, -3)$, **and** $G(1, 0)$. **Draw** $\triangle BIG$ **and then its reflection image across the given line.**

39. the y-axis **40.** the x-axis **41.** $x = 4$

Lesson 8-5 **Find the value of x. Round answers to the nearest tenth.**

42.

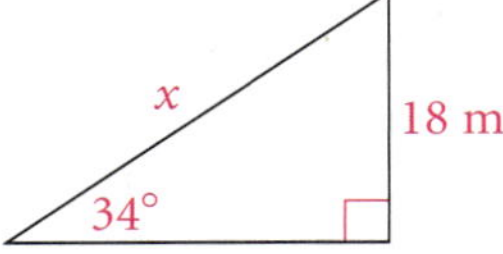

43.

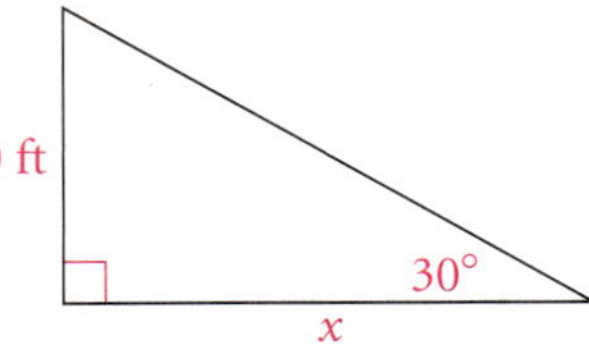

Lesson 3-7 **Are the given lines *parallel, perpendicular,* or *neither*? Explain.**

44. $y - \frac{2}{3}x = 2$
$3x - 4 = 2y$

45. $x = -3$
$x = 3$

46. $-2x + 4y = -8$
$y = -2x + 13$

Checkpoint Quiz 1 — Lessons 9-1 through 9-3

State whether the transformation appears to be an isometry. Explain.

1.

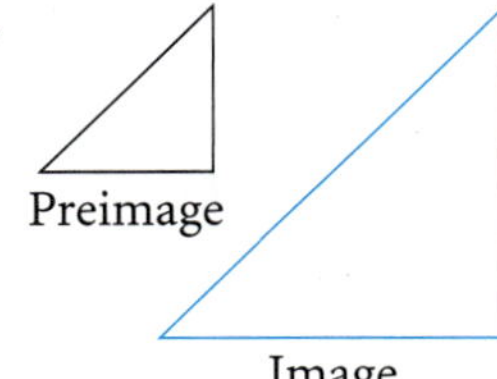

2.

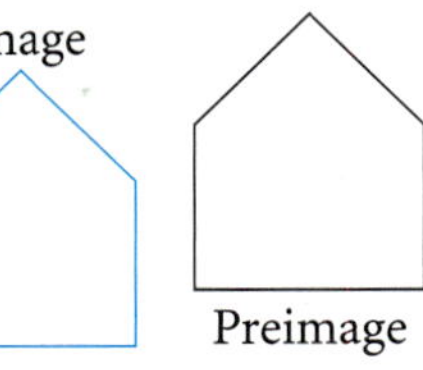

3.

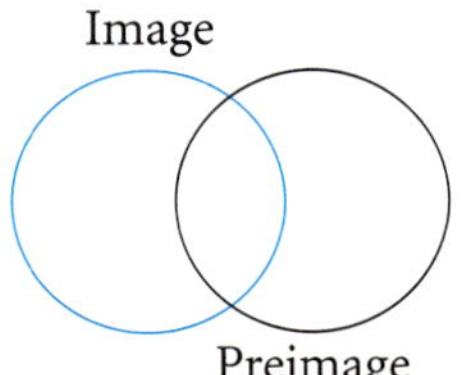

4. Describe in words the translation represented by $(x, y) \to (x - 3, y + 5)$.

5. Write a rule to describe a translation 5 units left and 10 units up.

6. Describe in words the result of the translation $(x, y) \to (x + 7, y - 2)$ followed by the translation $(x, y) \to (x - 3, y + 2)$.

Sketch each figure and point A. Draw the image of each figure for the given angle of rotation about A. Label the vertices of the image.

7. 40°

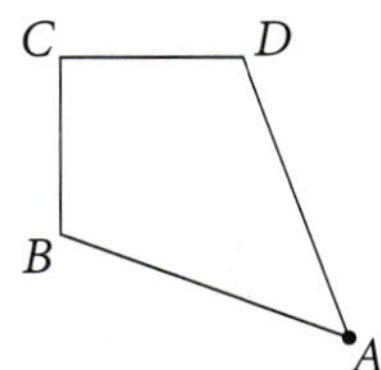

8. 90°

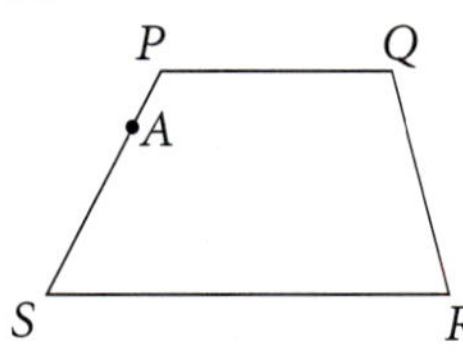

9. 180°

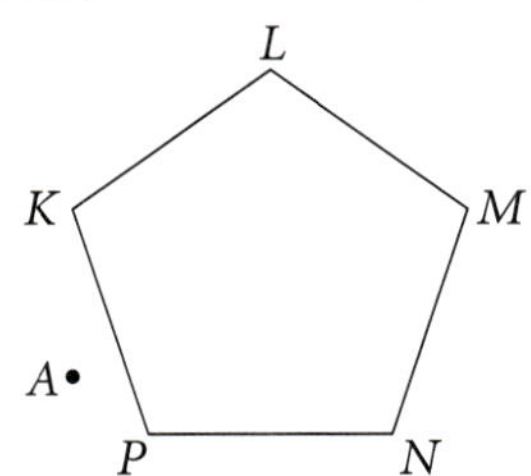

10. $\triangle WXY$ has vertices $W(-4, 1)$, $X(2, -7)$, and $Y(0, -3)$. Find its image for the translation $(x, y) \to (x - 2, y + 5)$.

Guided Problem Solving

For Use With Page 487, Exercise 32

Understanding Math Problems **Read the problem below. Then let Kate's thinking guide you through the solution. Check your understanding with the exercises at the bottom of the page.**

Reasoning If you are given a figure and a rotation image of the figure, how can you find the center and angle of rotation?

What Kate Thinks

I'll use tracing paper and draw two congruent figures that have the same orientation.
I like the letter K, so I think I'll use it.
I'll tilt one K so that it's not a translation image.

I think I see a rotation. It will rotate the left K counterclockwise onto the right K. It will rotate point A to point A' and point B to point B'.

That means that A and A' are on a circle whose center is the center of rotation. The same is true for B and B'. The perpendicular bisectors of both $\overline{AA'}$ and $\overline{BB'}$ go through the center. I'll use blue to construct the perpendicular bisectors.

Point C must be a center of rotation.

I can check this using tracing paper.

To find the angle of rotation, I just have to draw $\angle ACA'$ and measure it.

What Kate Draws

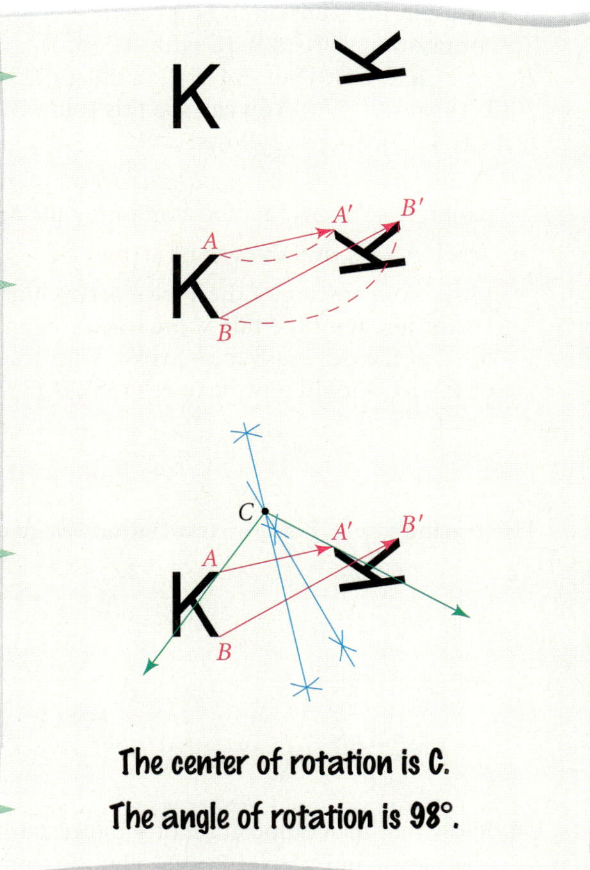

EXERCISES

1. In a newspaper or magazine that you can cut, find two large identical letters that are the same size and type. Cut them out and place them at random on a large piece of paper. Then find the center and angle of rotation that maps one onto the other.

2. **Writing** Can a triangle be both a translation image and a rotation image of another triangle? Explain.

Activity Lab: Tracing-Paper Transformations

Hands-On

FOR USE WITH LESSON 9-3

GEOM 22.0 Students know the effect of rigid motions on figures in the coordinate plane, including rotations, translations, and reflections. *Develop*

In Lesson 9-1, you learned how to describe a translation using variables. In the Activity on page 477, you drew a reflection in a plane by using tracing paper. Now, you will use tracing paper to perform translations and rotations. You will also discover ways to describe certain rotations and reflections using variables.

1 ACTIVITY

You can use the vector arrow shown in the diagram to represent the translation $(x, y) \rightarrow (x + 4, y + 2)$. The translation shifts $\triangle ABC$ with $A(-3, 3)$, $B(-1, 1)$, and $C(1, 4)$ to $\triangle A'B'C'$ with $A'(1, 5)$, $B'(3, 3)$, and $C'(5, 6)$. You can see this translation using tracing paper as follows:

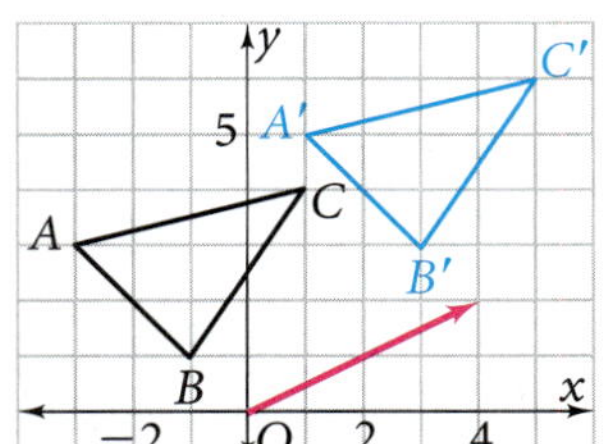

- Draw $\triangle ABC$ and the vector arrow on graph paper. Also, show the line containing the arrow.
- Trace $\triangle ABC$ and the vector arrow.
- Move your tracing of the vector arrow along the vector line until the tail of the tracing is on the head of the original vector arrow. Your tracing of $\triangle ABC$ should now be at $A'(1, 5)$, $B'(3, 3)$, and $C'(5, 6)$.

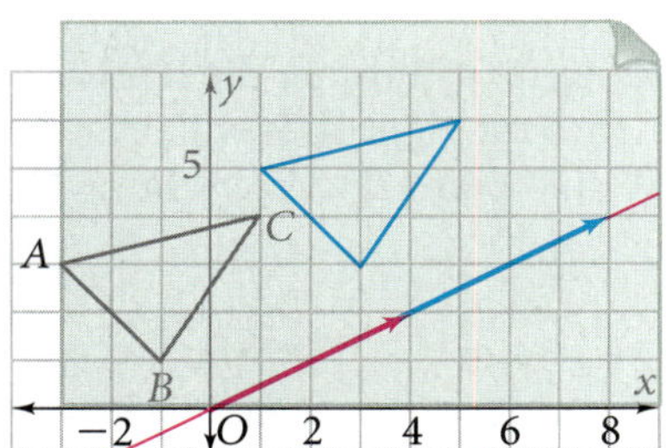

Use tracing paper. Find the translation image of each triangle for the given vector.

1.

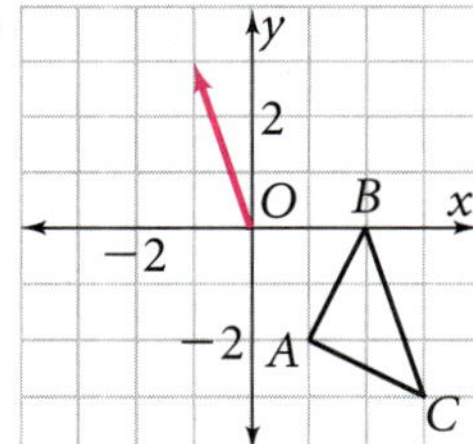

2.

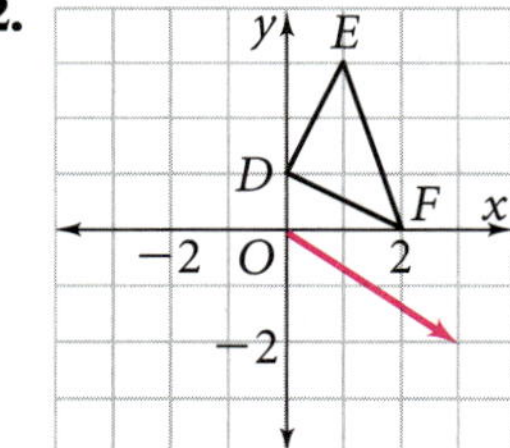

3.

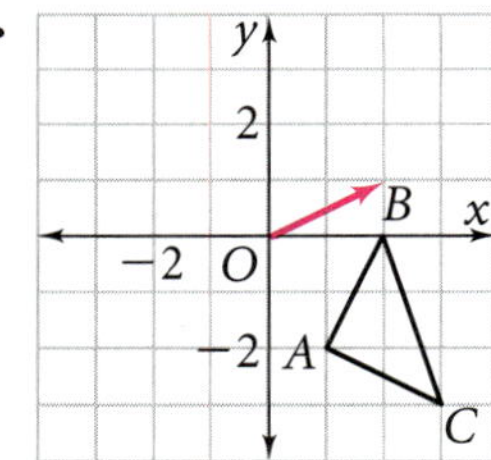

4. Show that the composition of the translation in Exercise 1 followed by the translation in Exercise 2 gives you the translation in Exercise 3.

2 ACTIVITY

To rotate a figure 90° about the origin, trace the figure, one axis, and the origin. Then turn your tracing paper counterclockwise, keeping the origin in place and aligning the traced axis with the other original axis.

5. Use $\triangle ABC$ from Activity 1 with $A(-3, 3)$, $B(-1, 1)$, and $C(1, 4)$. What is the image of $\triangle ABC$ for a 90° turn about the origin?

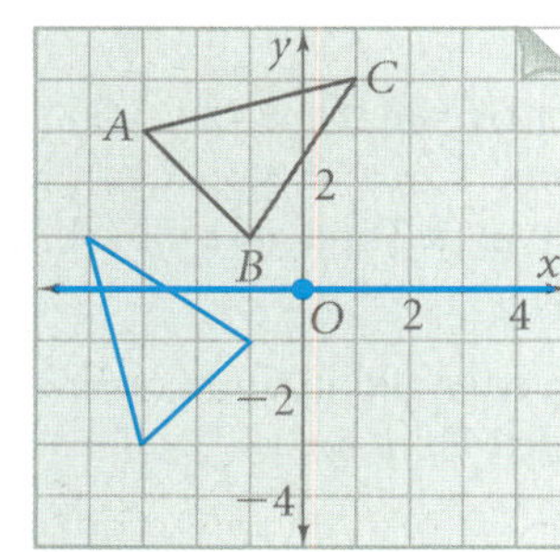

6. Copy and complete the table. Use tracing paper to find point P', the image of point P, for a turn of 90° about the origin.

P	P'
(3, 4)	
(−3, 4)	
(−3, −4)	
(3, −4)	
(3, 0)	
(0, 4)	

7. Study the pattern in your table. Complete this rule for a turn of 90° about the origin.

$(x, y) \rightarrow (\blacksquare, \blacksquare)$

8. Test your rule with tracing paper. For $\triangle TRN$ with $T(0, 2)$, $R(3, 0)$, and $N(4, 5)$, a 90° turn about the origin should result in $\triangle T'R'N'$ with $T'(-2, 0)$, $R'(0, 3)$, and $N'(-5, 4)$. Does it?

9. In parts (a) and (b) below, what should be the result of each composition?
   a. a 90° turn about the origin followed by a 90° turn about the origin
   b. $(x, y) \rightarrow (-y, x)$ followed by $(x, y) \rightarrow (-y, x)$
   c. Use tracing paper. Test your conjectures from parts (a) and (b) on $\triangle ABC$ from Activity 1.

3 ACTIVITY

To reflect a figure across an axis using tracing paper, trace the figure, the axis, and the origin. Then turn over your tracing paper, keeping the origin in place and aligning the traced axis with the original axis.

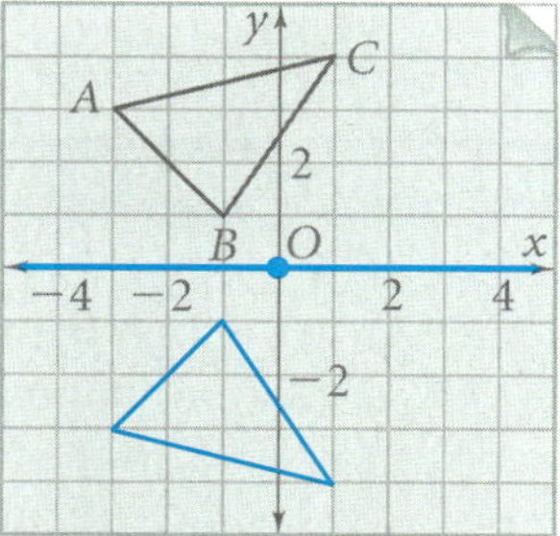

10. What is the reflection image of $\triangle ABC$ across the x-axis? Across the y-axis?

11. Copy and complete the table. Use tracing paper to find points P_x and P_y, the reflection images of point P across the x-axis and y-axis, respectively.

P	P_x	P_y
(3, 4)		
(−3, 4)		
(−3, −4)		
(3, −4)		
(3, 0)		
(0, 4)		

Study the patterns in the table. Complete these rules for reflections across the axes.

12. x-axis: $(x, y) \rightarrow (\blacksquare, \blacksquare)$

13. y-axis: $(x, y) \rightarrow (\blacksquare, \blacksquare)$

14. Test your rules on $\triangle KLM$ with $K(-2, 3)$, $L(3, 1)$, and $M(1, -2)$.
    a. Across the x-axis, $\triangle KLM$ should map to _?_. Does it?
    b. Across the y-axis, $\triangle KLM$ should map to _?_. Does it?

15. In parts (a) and (b) below, what should be the result of each composition?
    a. $(x, y) \rightarrow (x, -y)$ followed by $(x, y) \rightarrow (-x, y)$
    b. a reflection across the x-axis followed by a reflection across the y-axis
    c. Use tracing paper. Test your conjectures from parts (a) and (b) on $\triangle ABC$ from Activity 1.

EXERCISES

16. Use tracing paper to find a rule for a reflection across the line $y = x$. Test your rule on $\triangle ABC$ from Activity 1.

17. Compare the results of Exercises 9 and 15. Make a conjecture about the compositions suggested by each.

18. What other transformation is equivalent to a reflection across the line $y = x$ followed by a reflection across the line $y = -x$? Explain.

9-4 Symmetry

CALIFORNIA CHECK SYSTEM

What You'll Learn

- To identify the type of symmetry in a figure

. . . And Why

To identify types of symmetry in real-life objects, as in Example 3

Check Skills You'll Need

GO for Help Lessons 9-2 and 9-3

The regular octagon at the right is divided into eight congruent triangles. Find the image of the given point or segment for the given rotation or reflection.

1. point A; a 90° rotation about the center
2. point H; a 180° rotation about the center
3. $\overline{AB}$; a reflection in $\overleftrightarrow{AE}$
4. $\overline{GH}$; a reflection in $\overleftrightarrow{AE}$

New Vocabulary • symmetry • reflectional symmetry • line symmetry • rotational symmetry • point symmetry

Identifying Types of Symmetry in Figures

A figure has **symmetry** if there is an isometry that maps the figure onto itself. If the isometry is the reflection of a plane figure, the figure has **reflectional symmetry** or **line symmetry.** One half of the figure is a mirror image of its other half. Fold the figure along the line of symmetry and the halves match exactly.

The image of the Inuit sculpture at the right has reflectional symmetry about a vertical line down the middle of the face.

It is possible for a figure to have more than one line of symmetry.

1 EXAMPLE Identifying Lines of Symmetry

Draw all lines of symmetry for a regular hexagon.

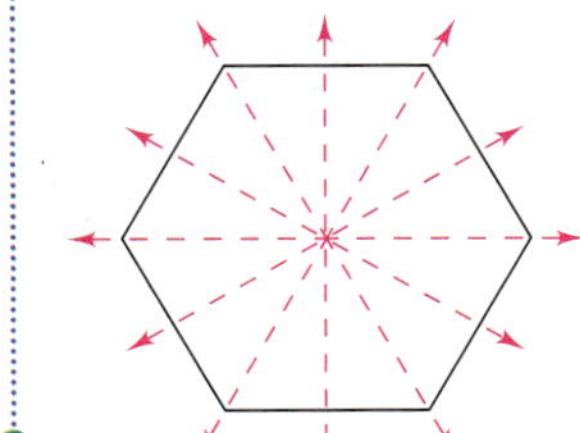

Draw a regular hexagon.

Then draw lines on the hexagon that make mirror-image congruent halves.

There are 6 lines of symmetry.

CA Standards Check 1 Draw a rectangle and all of its lines of symmetry.

A figure that has **rotational symmetry** is its own image for some rotation of 180° or less. A figure that has **point symmetry** has 180° rotational symmetry. A square has 90° and 180° rotational symmetry with the center of rotation at the center of the square. Thus, a square also has point symmetry.

2 EXAMPLE Identifying Rotational Symmetry

Vocabulary Tip

The angle of rotation for rotational symmetry is the smallest angle needed for the figure to rotate onto itself.

Judging from appearance, tell whether each triangle has rotational symmetry. If so, give the angle of rotation.

a.

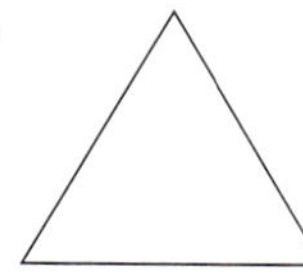

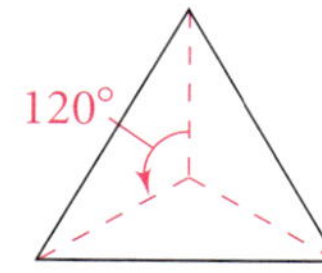

The equilateral triangle has rotational symmetry. The angle of rotation is 120°.

b.

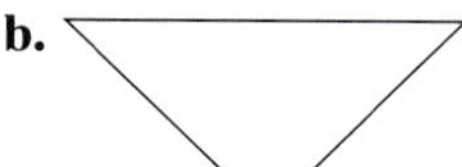

This isosceles triangle does not have rotational symmetry.

CA Standards Check 2 **a.** Judging from appearance, tell whether the figure at the right has rotational symmetry. If so, give the angle of rotation.

b. Does the figure have point symmetry?

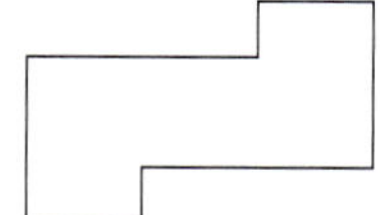

Three-dimensional objects can have various types of symmetry, including rotational symmetry about a line and reflectional symmetry in a plane.

3 EXAMPLE Real-World Connection

Symmetric Design Tell whether each object has rotational symmetry about a line and/or reflectional symmetry in a plane.

a.

The paddle has both rotational and reflectional symmetry.

b.

The cup has reflectional symmetry.

CA Standards Check 3 Tell whether the umbrella has rotational symmetry about a line and/or reflectional symmetry in a plane.

EXERCISES

For more exercises, see *Extra Skill, Word Problem, and Proof Practice.*

Standards Practice

A Practice by Example

Examples 1, 2 (pages 492, 493)

Tell what type(s) of symmetry each figure has. If it has line symmetry, sketch the figure and the line(s) of symmetry. If it has rotational symmetry, state the angle of rotation.

1. 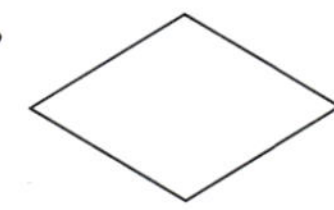
2. 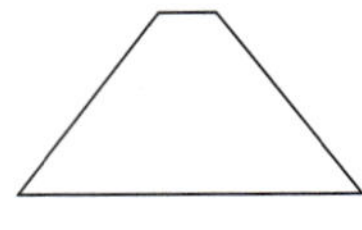
3. 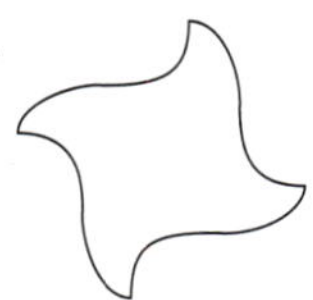
4. 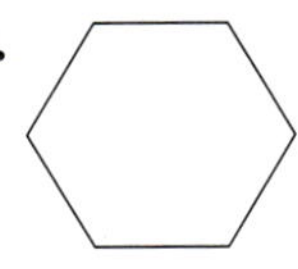
5. 
6. 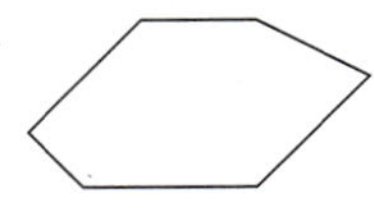
7. 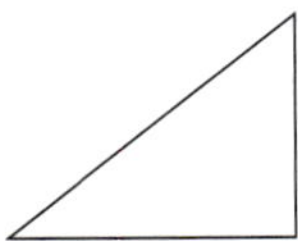
8. 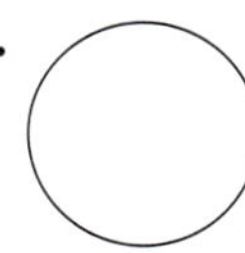
9. 
10. 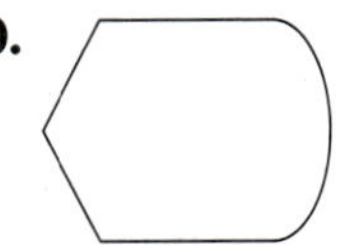
11. 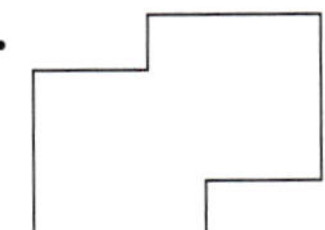
12. 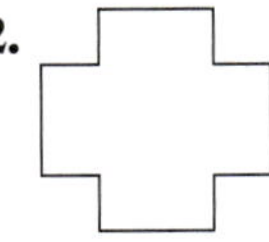

Draw each quadrilateral. Then draw all of its lines of symmetry.

13. rhombus
14. kite
15. square
16. parallelogram

Example 3 (page 493)

Tell whether each three-dimensional object has rotational symmetry about a line and/or reflectional symmetry in a plane.

17. 
18. 

B Apply Your Skills

19. **Open-Ended** The word CHECKBOOK has a horizontal line of symmetry. Find two other words for which this is true.

20. **Open-Ended** Stack the letters of MATH vertically and upright, and you can find a vertical line of symmetry. Find two other words for which this is true.

21. a. **Alphabets** Copy the chart. Use it to classify the letters of the English and Greek alphabets below. You will list some letters in more than one category.
   b. Which alphabet can you say is more symmetrical? Explain.

Type of Symmetry

Language	Horizontal Line	Vertical Line	Point
English			
Greek			

English: ABCDEFGHIJKLMNOPQRSTUVWXYZ

Greek: ΑΒΓΔΕΖΗΘΙΚΛΜΝΞΟΠΡΣΤΥΦΧΨΩ

Visit: PHSchool.com
Web Code: bce-0904

Tell what type(s) of symmetry each image has. For line symmetry, sketch the image and the line(s) of symmetry. For rotational symmetry, state the angle of rotation.

22. 

23. 

24. Open-Ended The equation $\frac{10}{10} - 1 = 0 \div \frac{83}{83}$ is not only true, but also symmetrical (horizontally). Write four other equations or inequalities that are both true and symmetrical.

Logos **Describe the types of symmetry, if any, of each automobile logo.**

25. 

26. 

27. 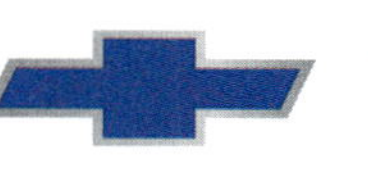

28. 

29. 

30. 

31. 

32. 

Problem Solving Hint

In Exercises 33–35, you can conclude a statement is false by finding a counterexample.

33. Is the line that contains the bisector of an angle a line of symmetry of the angle? Explain.

34. Is the line that contains the bisector of an angle of a triangle a line of symmetry of the triangle? Explain.

35. Is a bisector of a segment a line of symmetry of the segment? Explain.

36. Multiple Choice Which statement is true about the figure at the right?

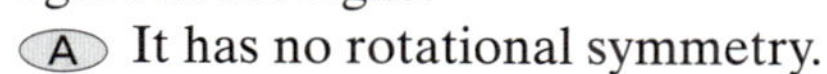

(A) It has no rotational symmetry.

(B) It has no reflectional symmetry.

(C) It has rotational symmetry with an angle of rotation of 45°.

(D) It has reflectional symmetry with six lines of symmetry.

Coordinate Geometry **A figure has a vertex at (3, 4). If the figure has the given type of symmetry, state the coordinates of another vertex of the figure.**

37. line symmetry about the y-axis

38. line symmetry about the x-axis

39. point symmetry about the origin

40. line symmetry about the line $y = x$

Coordinate Geometry **Graph each equation. Describe the symmetry of each graph.**

41. $y = x$

42. $y = x^2$

43. $x = y^2$

44. $x^2 + y^2 = 9$

45. $y = (x + 2)^2$

46. $y = x^3$

47. $y = |x|$

48. $x = |y|$

GO Online Lesson Quiz Visit: PHSchool.com, Web Code: bca-0904

For each three-dimensional figure, draw a net that has rotational symmetry and a net that has 1, 2, or 4 lines of symmetry.

49. 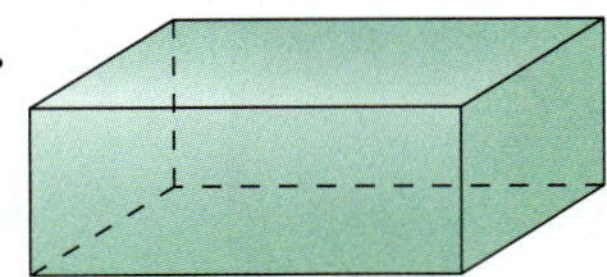

50. 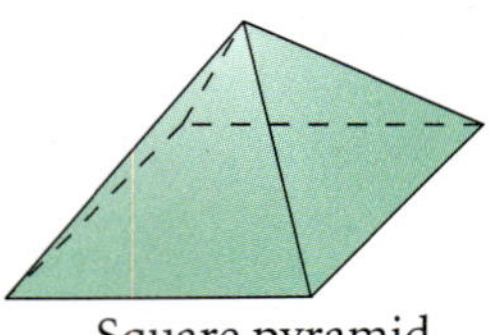

Square pyramid

Multiple Choice Practice

For California Standards and CAHSEE Tutorials, visit PHSchool.com. Web Code: bcq-9045

GEOM 22.0

51. Find the reflection image of triangle XYZ if it is reflected about the line $y = 2$.

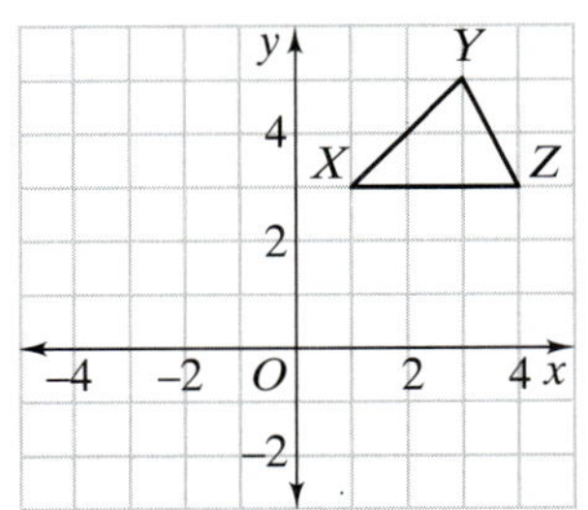

Ⓐ $X(1, 1), Y(3, -1), Z(4, 1)$
Ⓑ $X(3, 3), Y(1, 5), Z(0, 3)$
Ⓒ $X(1, -3), Y(3, -1), Z(4, -3)$
Ⓓ $X(1, -3), Y(3, -5), Z(4, -3)$

CAHSEE 7MR 2.4

52. A veterinarian weighed a newborn puppy on the day after it was born and then each week for the first 4 weeks. The table shows the number of ounces the puppy weighed each time.

Day	1	8	15	22	29
Weight (oz)	8	22	36	50	64

Which equation best describes the relationship between the number of days since birth d and the weight of the puppy w?

Ⓐ $w = 6d + 2$ Ⓑ $d = 6w - 2$ Ⓒ $w = 2d + 6$ Ⓓ $w = \frac{w + 6}{2}$

CAHSEE Alg1 10.0

53. Which equation is equivalent to the function $f(x) = 4x(10 - 3x)$?

Ⓐ $f(x) = 4x^2(10 - 3x)$
Ⓑ $f(40x - 12x) = x$
Ⓒ $y = 40x - 3x$
Ⓓ $y = 4x(10 - 3x)$

Mixed Review

Lesson 9-3

54. Which capital letters of the alphabet are rotation images of themselves? Draw each letter and give an angle of rotation (< 360°).

Lesson 9-1

The blue figure is a transformation image of the black figure. Does the transformation appear to be an isometry? Explain.

55. 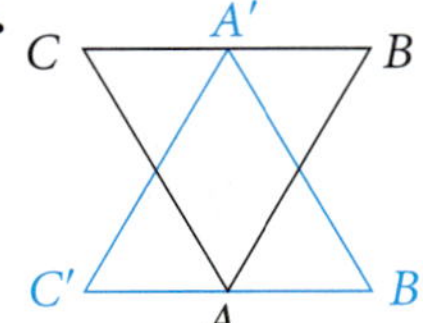

56. 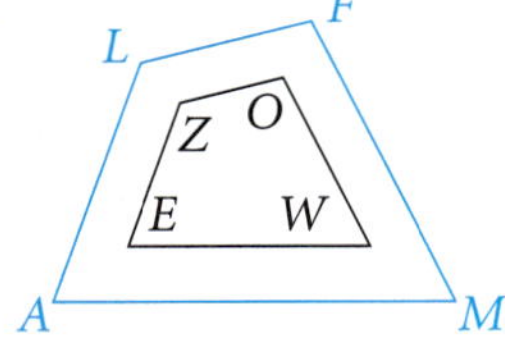

57. 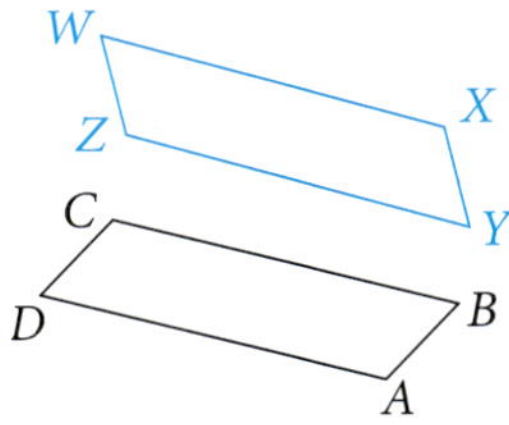

Lesson 6-6

58. Three vertices of an isosceles trapezoid are $(-2, 1)$, $(1, 4)$, and $(4, 4)$. Find possible pairs of coordinates for the fourth vertex.

Vocabulary Builder

Extend and Generalize

To *extend* an idea means to enlarge the scope of the idea. One way is to substitute a new word for a key word. Another is to change the frame of reference. In each case, you then decide whether the new statement is true.

1 EXAMPLE

Is the following extension of the Pythagorean Theorem true?

The *semicircles* on the legs of a right triangle are equal in area to the *semicircle* on the hypotenuse.

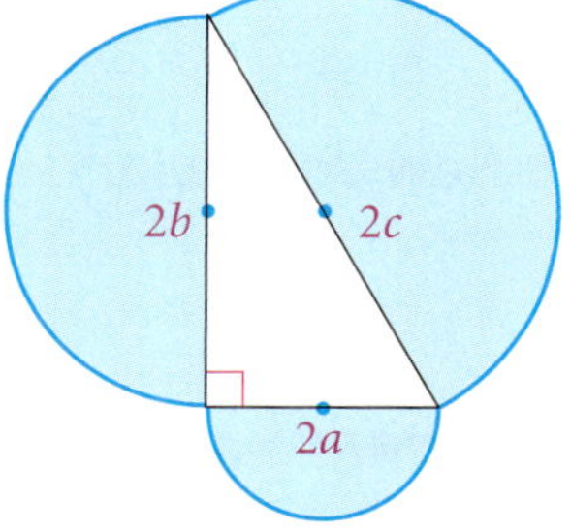

$(2a)^2 + (2b)^2 = (2c)^2$ by the Pythagorean Theorem.

Using the Multiplication Property of Equality, it follows that $\frac{1}{2}\pi a^2 + \frac{1}{2}\pi b^2 = \frac{1}{2}\pi c^2$.

This extension of the Pythagorean Theorem to semicircles is true.

2 EXAMPLE

Extend the definition of *circle* to spherical geometry (see pp. 154–155). Give an unusual result.

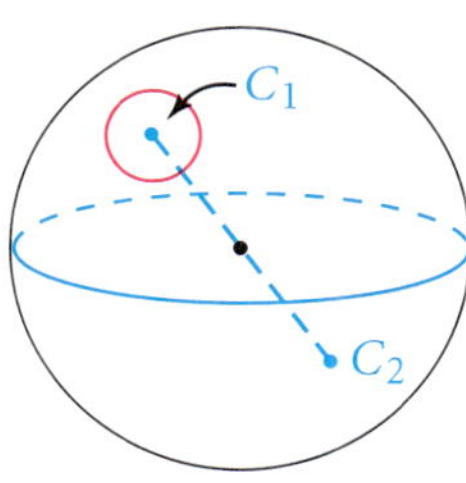

Let a circle on a sphere be the set of all points equidistant from a given point. For this definition, one circle will have two distinct centers.

EXERCISES

1. Suppose you have similar triangles on the three sides of a right triangle. Are the triangles on the legs equal in area to the triangle on the hypotenuse? Explain.
2. Determine whether the Pythagorean Theorem extends as follows: The squares on the two shortest sides of an acute triangle are equal in area to the square on the longest side. Justify your response.
3. Example 2 extends the idea of a circle to spherical geometry and gives an unusual result. Find another unusual fact that follows from the definition in Example 2.

You *generalize* an idea if you can extend it to all possible cases (of a particular type).

4. The midpoint of the segment from a to b on a number line is $\frac{a+b}{2}$.

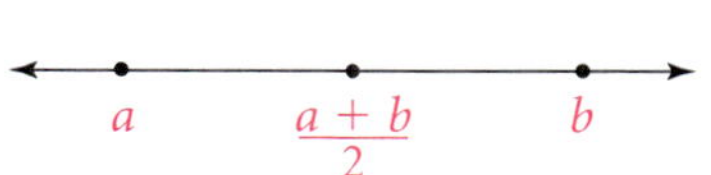

   a. Explain how you can extend this idea to both two and three dimensions.
   b. Explain how you can generalize the Midpoint Formula to be a definition of a midpoint in n dimensions, for any whole number n.
5. Explain how to extend the Midpoint Formula to points of trisection and points of quadrisection. Can you generalize?
6. Explain how to extend and generalize the Triangle Angle-Sum Theorem.

Dilations

What You'll Learn

- To locate dilation images of figures

. . . And Why

To find the dimensions of a car using the dimensions of its model, as in Example 2

Check Skills You'll Need

GO for Help Lesson 7-1

Determine the scale drawing dimensions of a room using a scale of $\frac{1}{4}$ in. = 1 ft.

1. kitchen: 12 ft by 16 ft

2. bedroom: 8 ft by 10 ft

3. laundry room: 6 ft by 9 ft

4. bathroom: 5 ft by 7 ft

New Vocabulary • dilation • enlargement • reduction

1 Locating Dilation Images

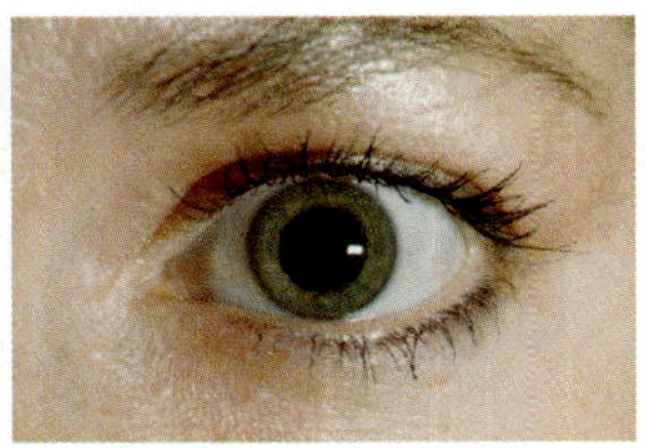

Real-World Connection

Look closely at your pupil in a mirror, and as light fluctuates you can watch your eye dilate.

A **dilation** is a transformation whose preimage and image are similar. Thus, a dilation is a similarity transformation. It is *not*, in general, an isometry.

Every dilation has a center and a scale factor n, $n > 0$. The scale factor describes the size change from the original figure to the image.

To find a dilation with center C and scale factor n, you can use the following two rules.

- The image of C is itself (that is, $C' = C$).
- For any point R, R' is on $\overrightarrow{CR}$ and $CR' = n \cdot CR$.

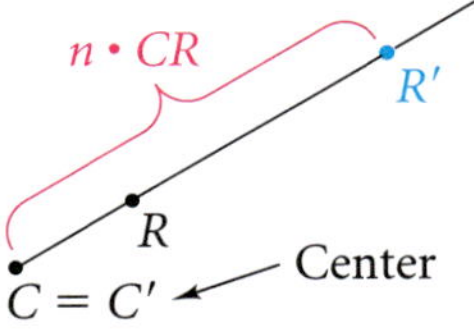

The dilation is an **enlargement** if the scale factor is greater than 1. The dilation is a **reduction** if the scale factor is between 0 and 1.

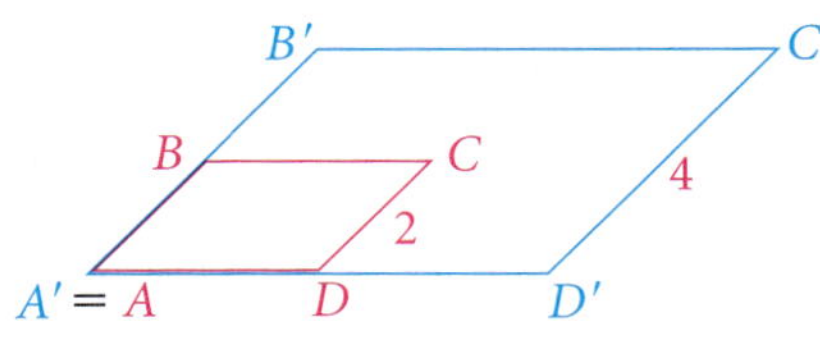

Enlargement
Center A, scale factor 2

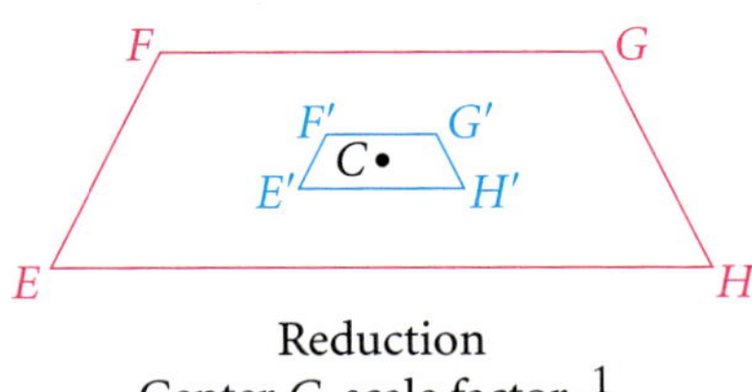

Reduction
Center C, scale factor $\frac{1}{4}$

1 EXAMPLE Finding a Scale Factor

Test-Taking Tip

When finding a scale factor, note first whether the dilation is an enlargement or a reduction.

The blue triangle is a dilation image of the red triangle. Describe the dilation.

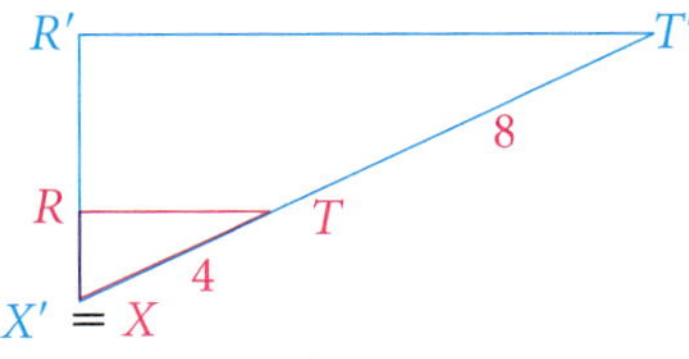

The center is X. The image is larger than the preimage, so the dilation is an enlargement.

$$\frac{X'T'}{XT} = \frac{4 + 8}{4} = 3$$

The dilation has center X and scale factor 3.

CA Standards Check **1** The blue quadrilateral is a dilation image of the red quadrilateral. Describe the dilation.

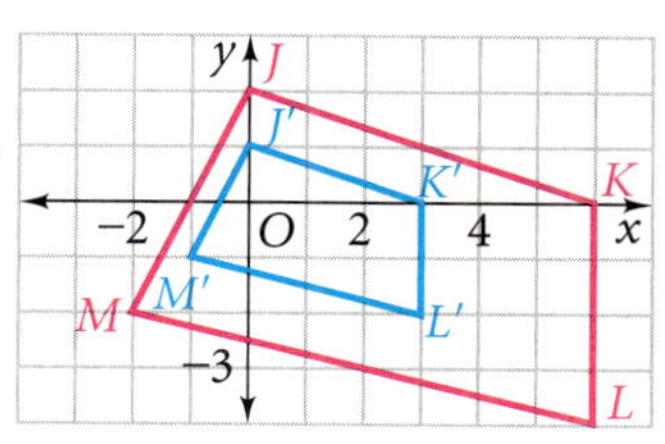

Scale factors help you understand scale models, both large and small.

2 EXAMPLE Real-World Connection

Scale Models The packaging lists a model car's length as 7.6 cm. It also gives the scale as 1 : 63. What is the length of the actual car?

To "enlarge" the model car to the actual car, use the scale factor 63. Multiply 7.6 cm by 63 to get 478.8 cm, or about 4.8 m, for the length of the actual car.

CA Standards Check **2** The height of a tractor-trailer truck is 4.2 m. The scale factor for a model of the truck is $\frac{1}{54}$. Find the height of the model to the nearest centimeter.

Suppose a dilation is centered at the origin. You can find the dilation image of a point by multiplying its coordinates by the scale factor.

Scale factor 4, $(x, y) \longrightarrow (4x, 4y)$

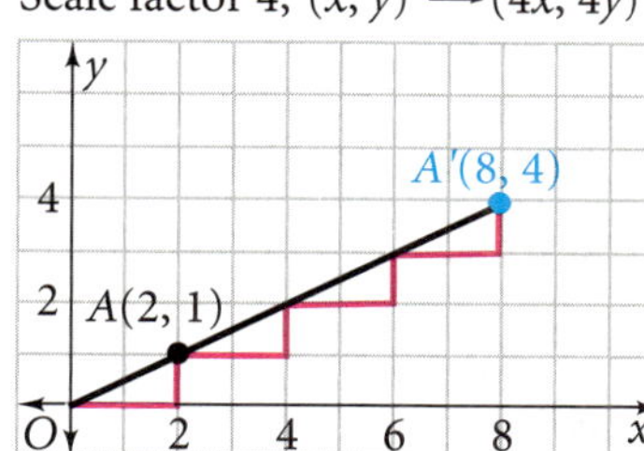

Scale factor $\frac{1}{3}$, $(x, y) \longrightarrow \left(\frac{1}{3}x, \frac{1}{3}y\right)$

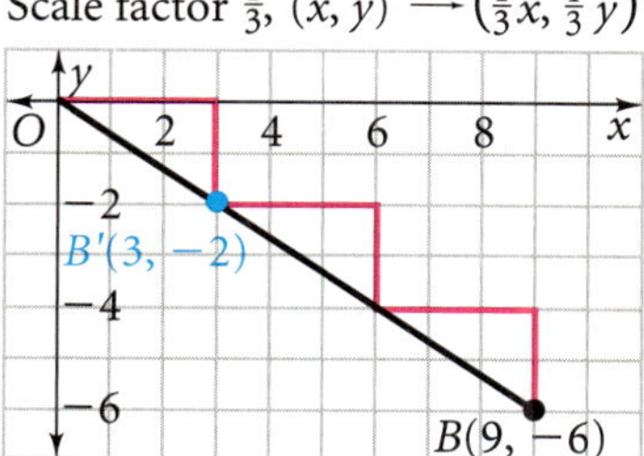

To dilate a triangle from the origin, find the dilation images of its vertices.

3 EXAMPLE Graphing Dilation Images

Multiple Choice $\triangle PZG$ has vertices $P(2, 0)$, $Z\left(-1, \frac{1}{2}\right)$, and $G(1, -2)$. What are the coordinates of the image of P for a dilation with center $(0, 0)$ and scale factor 3?

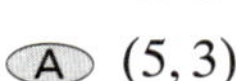

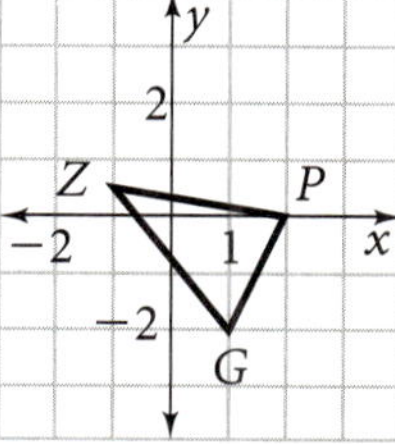

Ⓐ $(5, 3)$ Ⓒ $\left(\frac{2}{3}, 0\right)$

Ⓑ $(6, 0)$ Ⓓ $(3, -6)$

The scale factor is 3, so use the rule $(x, y) \rightarrow (3x, 3y)$.

$P(2, 0) \rightarrow P'(3 \cdot 2, 3 \cdot 0)$, or $P'(6, 0)$. The correct answer is B.

Z' is $\left(3 \cdot (-1), 3 \cdot \frac{1}{2}\right)$ and G' is $(3 \cdot 1, 3 \cdot (-2))$, so the vertices of the enlargement at the left are $P'(6, 0)$, $Z'\left(-3, \frac{3}{2}\right)$, and $G'(3, -6)$.

CA Standards Check **3** Find the image of $\triangle PZG$ for a dilation with center $(0, 0)$ and scale factor $\frac{1}{2}$. Draw the reduction.

For more exercises, see *Extra Skill, Word Problem, and Proof Practice.*

A Practice by Example

Example 1 (page 498)

GO for Help

The blue figure is a dilation image of the red figure. Describe the dilation.

1. 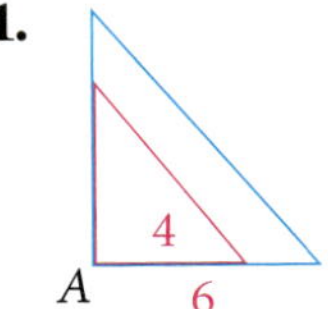

2. 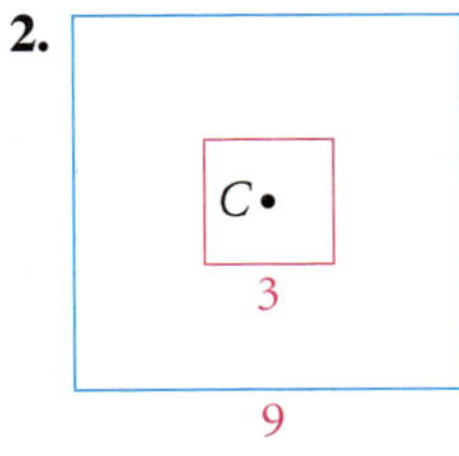

3. 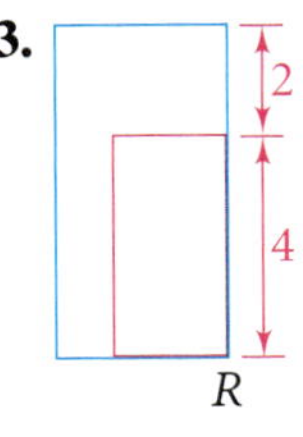

4. 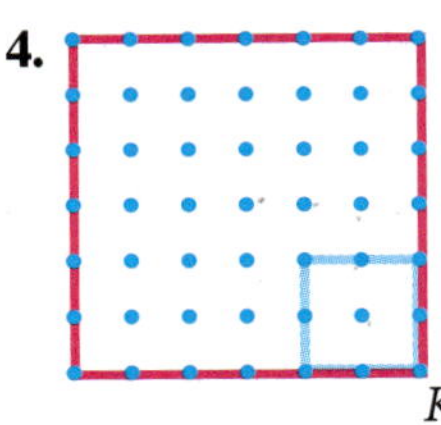

5. 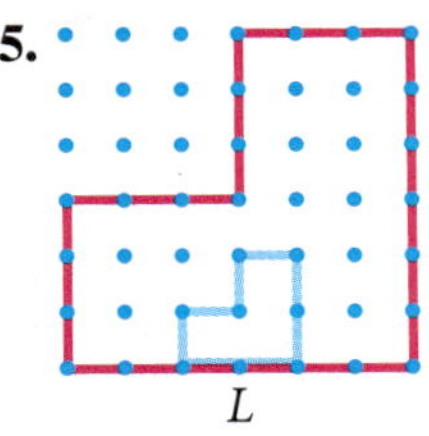

6. 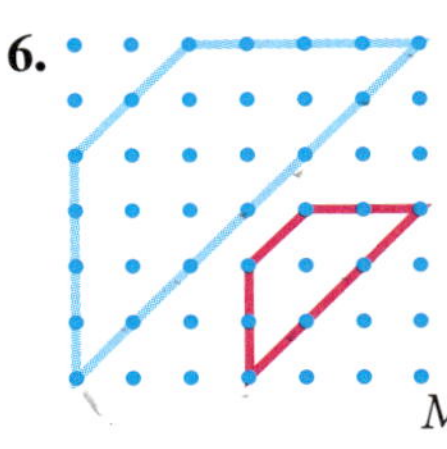

7. 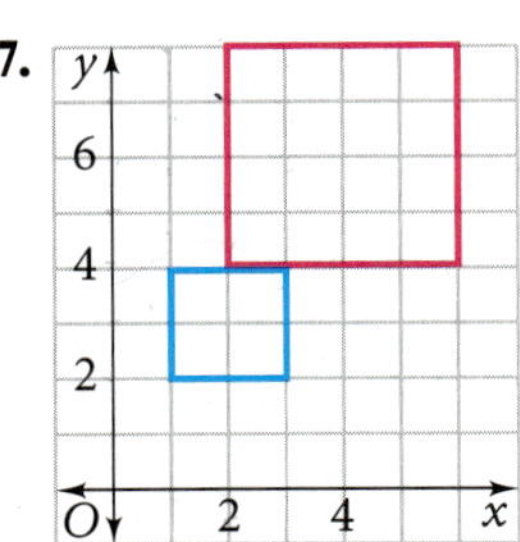

8. 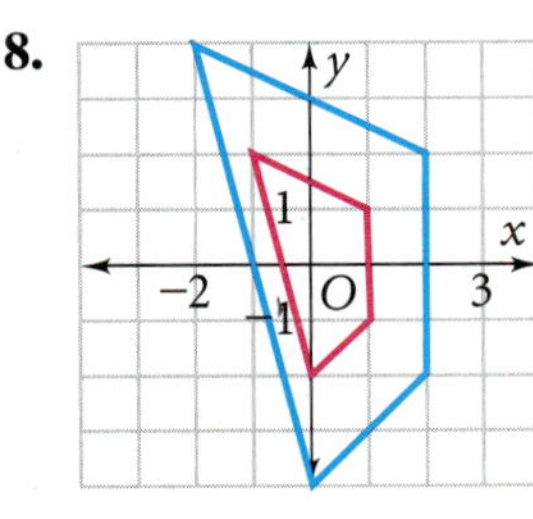

9. 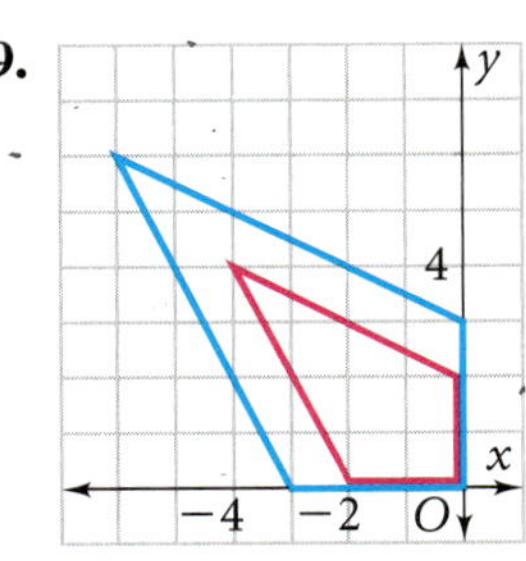

Example 2 (page 499)

Model Railroads **The table shows scales for different types of model railroads. For each model in Exercises 10–12, what would be the actual measurement?**

Model Railroad Scales

Scale Name	Scale Ratio
N	1 : 160
HO	1 : 87.1
S	1 : 64
O	1 : 48
G	1 : 22.5

10. An HO-scale tank car is 1.4 in. high.

11. An S-scale boxcar has length 8 in.

12. A model of an engineer in a G-scale model train layout is 3 in. tall.

13. A diesel engine is 60 feet long. How long is its O-scale model?

14. Actual railroad tracks are 4 ft 8.5 in. apart. How far apart are N-scale tracks?

Example 3 (page 499)

Find the image of $\triangle PQR$ for a dilation with center (0, 0) and the scale factor given. Draw the image.

15. 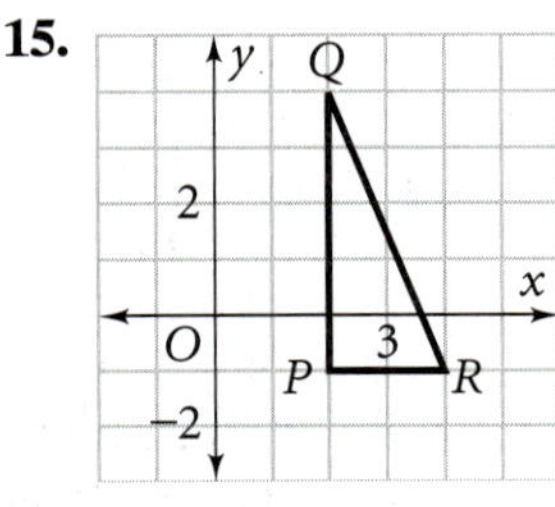

scale factor 3

16. 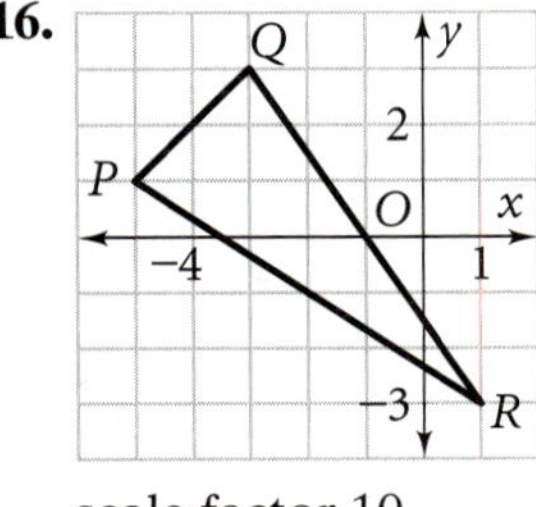

scale factor 10

17. 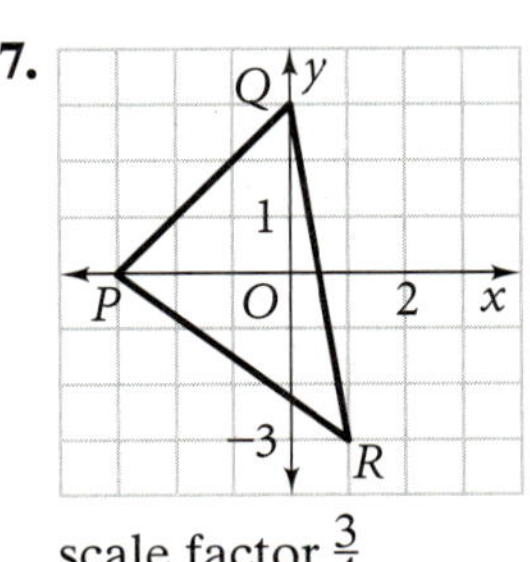

scale factor $\frac{3}{4}$

A dilation has center (0, 0). Find the image of each point for the scale factor given.

18. $D(1, -5); 2$ **19.** $L(-3, 0); 5$ **20.** $A(-6, 2); 1.5$

21. $T(0, 6); 3$ **22.** $M(0, 0); 10$ **23.** $N(-4, -7); 0.1$

24. $F(-3, -2); \frac{1}{3}$ **25.** $B(\frac{5}{4}, -\frac{2}{3}); \frac{1}{10}$ **26.** $Q(6, \frac{\sqrt{3}}{2}); \sqrt{6}$

B Apply Your Skills

Find the image of $QRTW$ for a dilation with center (0, 0) and the scale factor given.

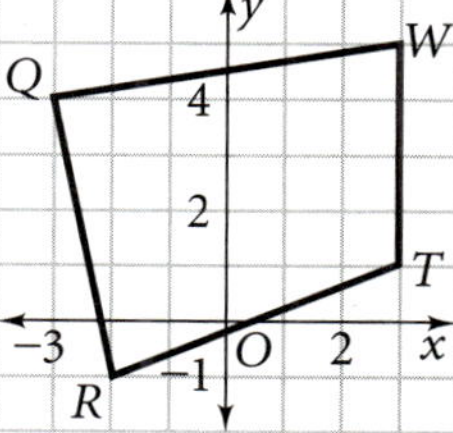

27. 3 **28.** 2 **29.** $\frac{1}{2}$ **30.** $\frac{1}{4}$

31. 0.6 **32.** 0.9 **33.** 10 **34.** 100

35. Writing An equilateral triangle has 4-in. sides. Describe its image for a dilation with scale factor 2.5. Explain.

36. Multiple Choice $\triangle D'E'F'$ is a dilation of $\triangle DEF$ with center (0, 0). Which of the following statements is *not* true?

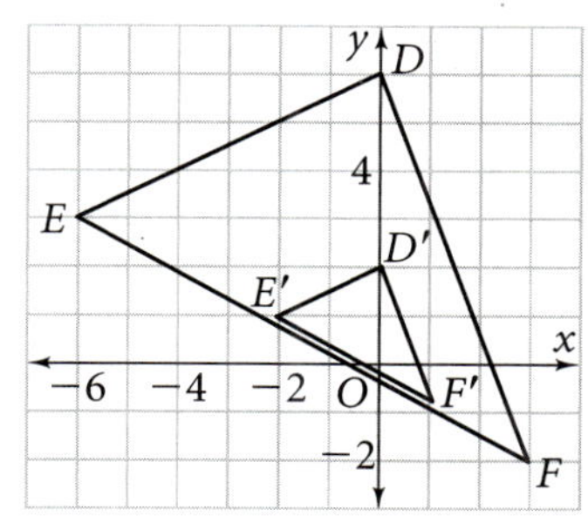

Ⓐ $\triangle DEF$ and $\triangle D'E'F'$ are similar.

Ⓑ $m\angle F'D'E' = \frac{1}{3} \cdot m\angle FDE$

Ⓒ The scale factor of the dilaton is $\frac{1}{3}$.

Ⓓ $\angle EFD \cong \angle E'F'D'$

Coordinate Geometry **Graph $MNPQ$ and its image $M'N'P'Q'$ for a dilation with center (0, 0) and the scale factor given.**

37. $M(-1, -1), N(1, -2), P(1, 2), Q(-1, 3)$; scale factor 2

38. $M(1, 3), N(-3, 3), P(-5, -3), Q(-1, -3)$; scale factor 3

39. $M(0, 0), N(4, 0), P(6, -2), Q(-2, -2)$; scale factor $\frac{1}{2}$

40. $M(2, 6), N(-4, 10), P(-4, -8), Q(-2, -12)$; scale factor $\frac{1}{4}$

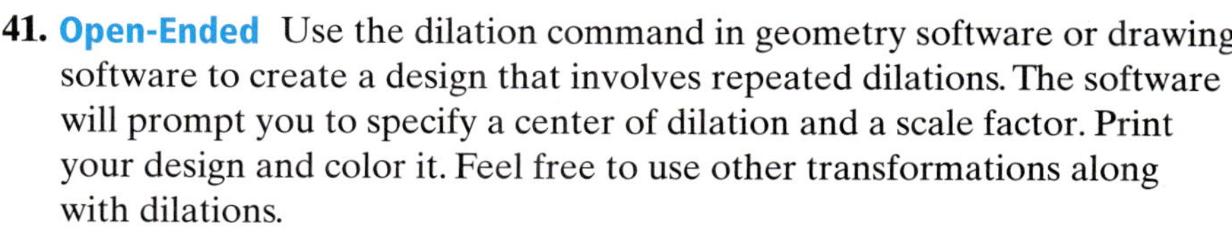

41. Open-Ended Use the dilation command in geometry software or drawing software to create a design that involves repeated dilations. The software will prompt you to specify a center of dilation and a scale factor. Print your design and color it. Feel free to use other transformations along with dilations.

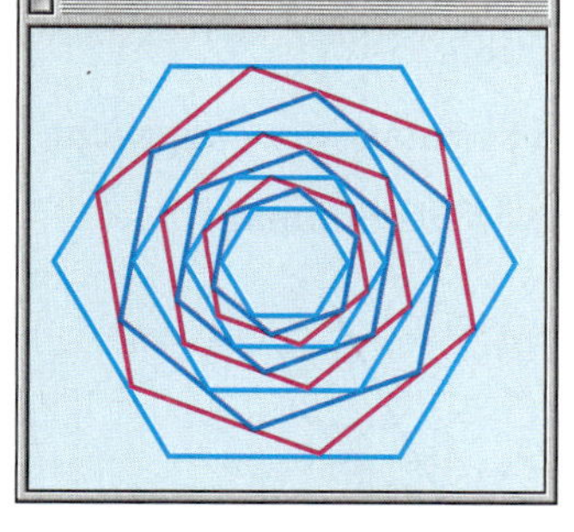

42. Copy Reduction Your copy of your family crest is 4.5 in. wide. You need a reduced copy for the front page of the family newsletter. The copy must fit in a space 1.8 in. wide. What scale factor should you use on the copy machine?

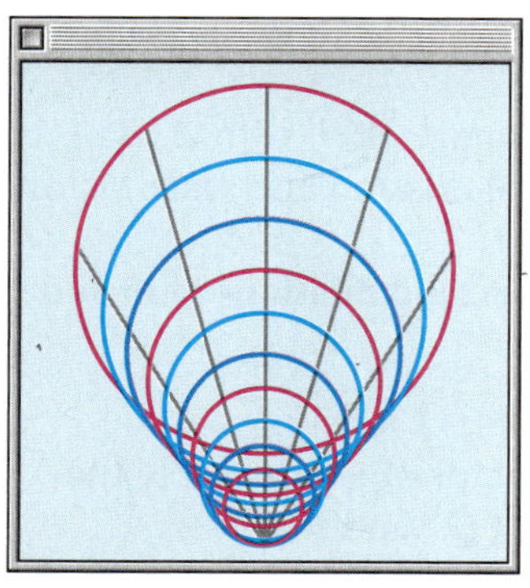

Exercise 41

A dilation maps $\triangle HIJ$ to $\triangle H'I'J'$. Find the missing values.

43.	44.	45.
$HI = 8$ in.	$HI = 7$ cm	$HI =$ ■ ft
$IJ = 5$ in.	$IJ = 7$ cm	$IJ = 30$ ft
$HJ = 6$ in.	$HJ =$ ■ cm	$HJ = 24$ ft
$H'I' = 16$ in.	$H'I' = 5.25$ cm	$H'I' = 8$ ft
$I'J' =$ ■ in.	$I'J' =$ ■ cm	$I'J' =$ ■ ft
$H'J' =$ ■ in.	$H'J' = 9$ cm	$H'J' = 6$ ft

46. Error Analysis Brendan says that when a rectangle with length 6 cm and width 4 cm is dilated by a scale factor of 2, the perimeter and area of the rectangle are doubled. Explain what is incorrect about Brendan's statement.

Visit: PHSchool.com
Web Code: bce-0905

The diagram at the right shows $\triangle LMN$ and its image $\triangle L'M'N'$ for a dilation with center P.

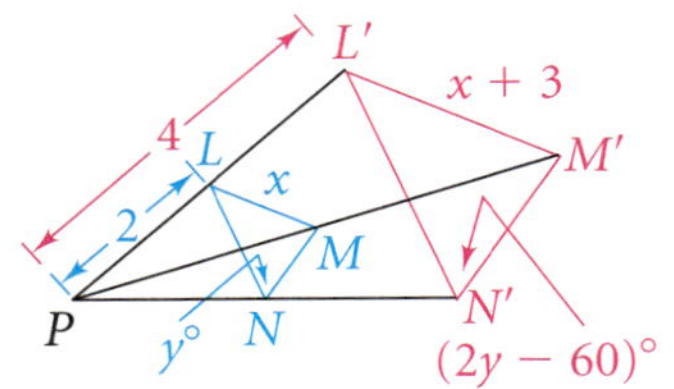

47. **Algebra** Find the values of x and y.

48. Evaluate y and $2y - 60$. Explain why the two values must be equal.

Copy $\triangle TBA$ and point O for each of Exercises 49–52. Draw the dilation image $\triangle T'B'A'$ for the given center and scale factor.

49. center O, scale factor $\frac{1}{2}$

50. center B, scale factor 3

51. center T, scale factor $\frac{1}{3}$

52. center O, scale factor 2

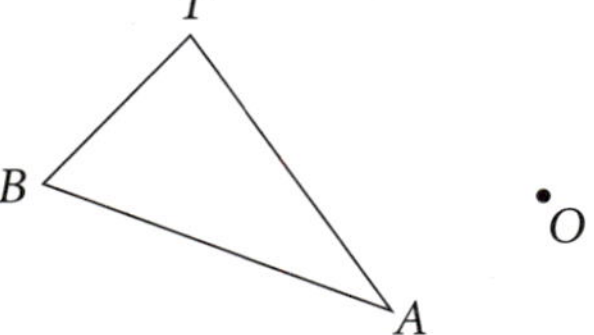

An overhead projection is a dilation only when the mirror in the projector's head is tilted at a 45° angle. Turning the head distorts the images.

53. **Constructions** Copy $\triangle GHI$ and point X onto your paper. Use a compass and straightedge to construct the image of $\triangle GHI$ for a dilation with center X and scale factor 2.

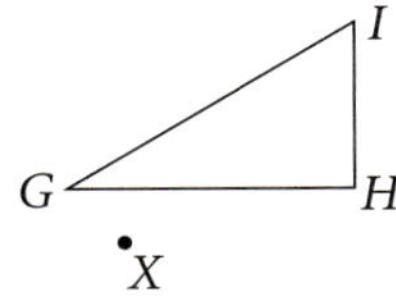

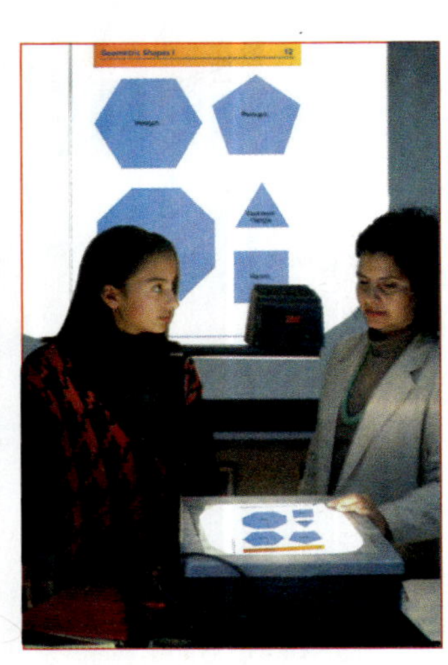

Overhead Projection **An overhead projector can dilate figures on transparencies.**

54. A segment on a transparency is 2 in. long. Its image on the screen is 2 ft long. What is the scale factor of the dilation?

55. The height of a parallelogram on the transparency is 4 cm. The scale factor is 15. What is the height of the parallelogram on the screen?

56. The area of a square on the screen is 3 ft^2. The scale factor is 16. What is the area of the square on the transparency?

Write *true* or *false* for Exercises 57–61. Explain your answers.

57. A dilation is an isometry.

58. A dilation changes orientation.

59. A dilation with a scale factor greater than 1 is a reduction.

60. For a dilation, corresponding angles of the image and preimage are congruent.

61. A dilation image cannot have any points in common with its preimage.

62. A flashlight projects an image of rectangle $ABCD$ on a wall so that each vertex of $ABCD$ is 3 ft away from the corresponding vertex of $A'B'C'D'$. The length of $\overline{AB}$ is 3 in. The length of $\overline{A'B'}$ is 1 ft. How far from each vertex of $ABCD$ is the light?

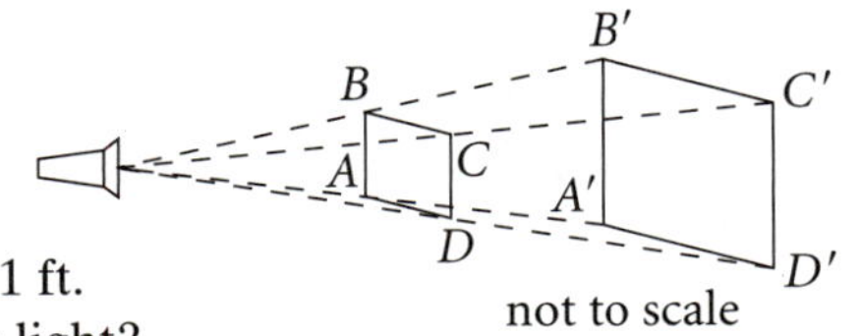

63. **Critical Thinking** You are given $\overline{AB}$ and its dilation image $\overline{A'B'}$ with A, B, A', and B' noncollinear. Explain how to find the center of dilation and scale factor.

Coordinate Geometry **In the coordinate plane you can extend dilations to include scale factors that are negative numbers.**

64. **a.** Graph $\triangle PQR$ with vertices $P(1, 2)$, $Q(3, 4)$, and $R(4, 1)$.
b. For a dilation centered at the origin with a scale factor of -3, multiply the coordinates in part (a) by -3. List the results as P', Q', and R'.
c. Graph $\triangle P'Q'R'$ on the same set of axes.

Problem Solving Hint

For Exercise 65(b), recall the meaning of reflection in a line.

65. a. A dilation with center at the origin and scale factor -1 (see Exercise 64) may be called a *reflection in a point*. For $\triangle PQR$ of Exercise 64, find the image $\triangle P'Q'R'$ for such a dilation.

b. Writing Explain why the dilation described in part (a) may be called a *reflection in a point*. Extend your explanation to a new definition of point symmetry. Compare your new definition with the definition given on page 493.

66. Constructions Draw acute $\triangle ABC$. Construct square $DEFG$ so that $\overline{DG}$ is on $\overline{AC}$, and E and F are on the other two sides of $\triangle ABC$. (*Hint:* First, try the special case with a right angle at A and use a dilation.)

Multiple Choice Practice

For California Standards and CAHSEE Tutorials, visit PHSchool.com. Web Code: bcq-9045

GEOM 22.0

67. The vertices of $\triangle ART$ are $A(1, 1)$, $R(3, 4)$, and $T(5, 1)$. If $\triangle ART$ is translated 6 units down and 4 units to the left, what are the coordinates of the image $\triangle A'R'T'$?

(A) $A'(-6, -9), R'(-4, -6), T'(-2, -9)$
(B) $A'(5, -5), R'(7, -2), T'(9, -5)$
(C) $A'(-6, -4), R'(-4, -1), T'(-2, -4)$
(D) $A'(-3, -5), R'(-1, -2), T'(1, -5)$

GEOM 7.0

68. Given the parallelogram at the right determine the value of x.

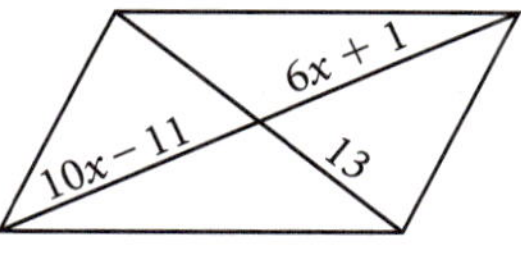

(A) 2 (C) 3
(B) 2.5 (D) 10

CAHSEE 7MG 1.2

69. A miniature replica of the city is displayed inside City Hall on Main Street, a street that is 1.8 miles long. In the replica, Main Street is 12 feet long. The tallest building replica stands 3 inches, or $\frac{1}{4}$ foot, high. How tall is the actual building?

(A) 198 ft (B) 891 ft (C) 1782 ft (D) 3168 ft

Mixed Review

Lesson 9-4

Coordinate Geometry **A figure has a vertex at $(-2, 7)$. If the figure has the given type of symmetry, state the coordinates of another vertex of the figure.**

70. line symmetry about the x-axis
71. line symmetry about the y-axis
72. point symmetry about the origin
73. line symmetry about the line $y = x$

Lesson 9-1

Write a rule for each translation.

74. 0 units to the right, 4 units up
75. 2 units to the left, 1 unit down
76. 3 units to the left, 6 units down
77. 8 units to the right, 10 units up

Lesson 8-5

For the angle of depression shown, find the value of x to the nearest tenth of a unit.

78.

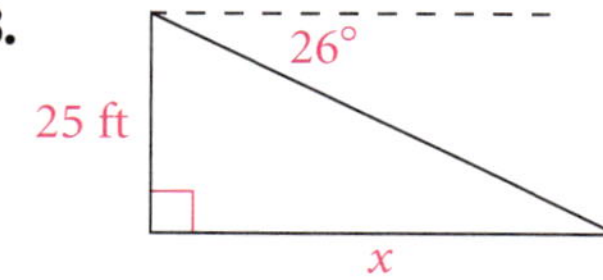

79.

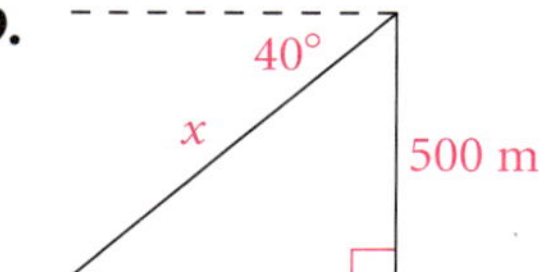

GO Online Lesson Quiz Visit: PHSchool.com, Web Code: bca-0905

Extension

Transformations Using Vectors and Matrices

For Use With Lesson 9-5

In Lesson 8-6, you learned that vectors are quantities that have both direction and magnitude. A translation moves all the points of a figure in the same direction and the same distance, so you can use a vector to describe a translation.

1 EXAMPLE Using Vectors to Find Translation Images

a. Find the image of $D(1, -2)$ under the translation described by the vector $\langle 1, 4\rangle$.

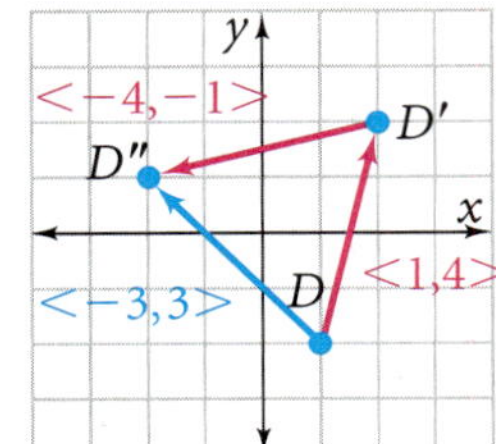

$\langle 1, 4\rangle$ is a translation right 1 unit and up 4 units, the same as $(x, y) \rightarrow (x + 1, y + 4)$.
$D(1, -2)$ translates to $(1 + 1, -2 + 4)$, or $(2, 2)$.

b. Find the image of $D(1, -2)$ under the composition of translations: $\langle 1, 4\rangle$ followed by $\langle -4, -1\rangle$.

Use vector addition: $\langle 1, 4\rangle + \langle -4, -1\rangle = \langle 1 + (-4), 4 + (-1)\rangle = \langle -3, 3\rangle$
$D(1, -2)$ translates to $(1 - 3, -2 + 3)$, or $(-2, 1)$.

You can use a matrix to find the transformation images of several points at once. A *matrix* is a rectangular arrangement of numbers in rows and columns.

The matrix at the right is called a 2×3 (two-by-three) matrix because it has 2 rows and 3 columns. Each entry in a matrix is an *element*. This matrix has 6 elements.

$$\begin{bmatrix} -1 & 4 & 9 \\ 3 & -5 & 7 \end{bmatrix}$$

To find the translation image of a triangle, use a matrix made up of the three vertices and a translation matrix that shows the translation vector three times.

2 EXAMPLE Using Matrices to Find Translation Images

Use matrices to find the image of $\triangle MFH$ under the translation $\langle 4, -5\rangle$.

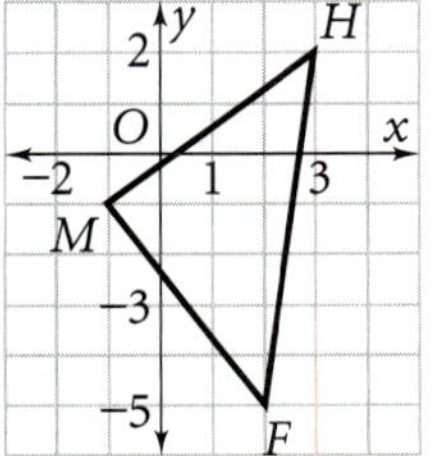

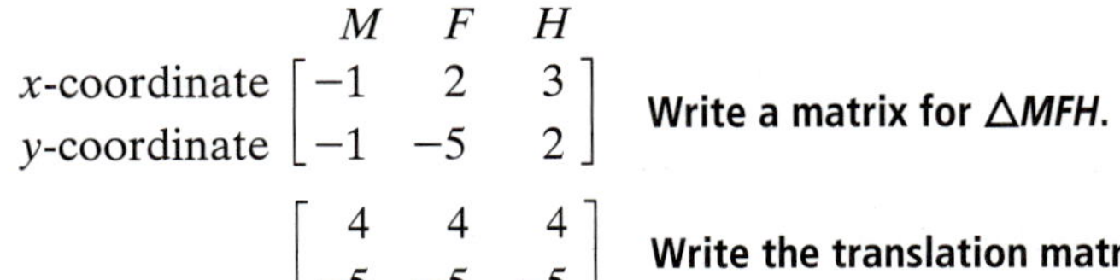

$$\begin{matrix} \text{x-coordinate} \\ \text{y-coordinate} \end{matrix} \overset{\quad M \quad F \quad H}{\begin{bmatrix} -1 & 2 & 3 \\ -1 & -5 & 2 \end{bmatrix}}$$ **Write a matrix for $\triangle MFH$.**

$$\begin{bmatrix} 4 & 4 & 4 \\ -5 & -5 & -5 \end{bmatrix}$$ **Write the translation matrix.**

$$\begin{bmatrix} -1 & 2 & 3 \\ -1 & -5 & 2 \end{bmatrix} + \begin{bmatrix} 4 & 4 & 4 \\ -5 & -5 & -5 \end{bmatrix} = \begin{bmatrix} 3 & 6 & 7 \\ -6 & -10 & -3 \end{bmatrix}$$ **Add the matrices.**

The image of $\triangle MFH$ is $\triangle M'F'H'$ with $M'(3, -6)$, $F'(6, -10)$, and $H'(7, -3)$.

To find the image of a triangle for a dilation from the origin, multiply each element in the vertex matrix for the triangle by the scale factor. Multiplying a matrix by a number in this way is called *scalar multiplication*.

3 EXAMPLE Using Scalar Multiplication

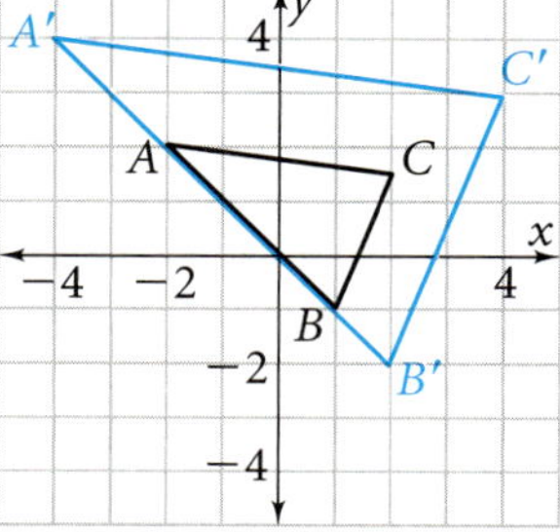

$\triangle ABC$ has vertices $A(-2, 2)$, $B(1, -1)$, and $C\left(2, 1\frac{1}{2}\right)$. Use scalar multiplication to find the image of $\triangle ABC$ for a dilation with center $(0, 0)$ and scale factor 2. Draw the enlargement.

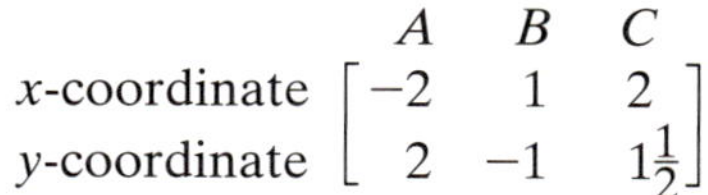

$$\begin{array}{l} \\ x\text{-coordinate} \\ y\text{-coordinate} \end{array} \begin{array}{c} \begin{array}{ccc} A & B & C \end{array} \\ \begin{bmatrix} -2 & 1 & 2 \\ 2 & -1 & 1\frac{1}{2} \end{bmatrix} \end{array}$$

Write a matrix for $\triangle ABC$.

$$2 \cdot \begin{bmatrix} -2 & 1 & 2 \\ 2 & -1 & 1\frac{1}{2} \end{bmatrix} = \begin{bmatrix} -4 & 2 & 4 \\ 4 & -2 & 3 \end{bmatrix}$$

Multiply each element of the matrix by the scale factor, or scalar, 2.

The vertices of the enlargement are $A'(-4, 4)$, $B'(2, -2)$, and $C'(4, 3)$.

EXERCISES

Find the image of each figure under the translation described by the given vector.

1. $\langle 4, -1 \rangle$

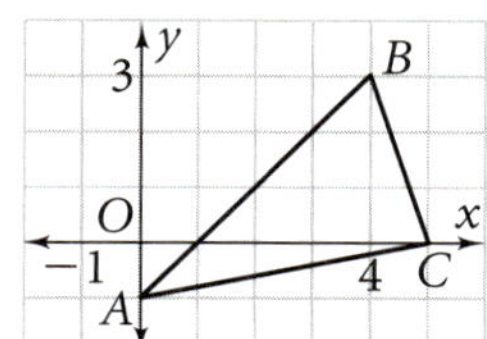

2. $\langle 2, 0 \rangle$

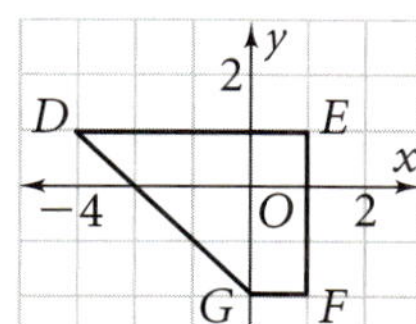

3. $\langle -3, -2 \rangle$

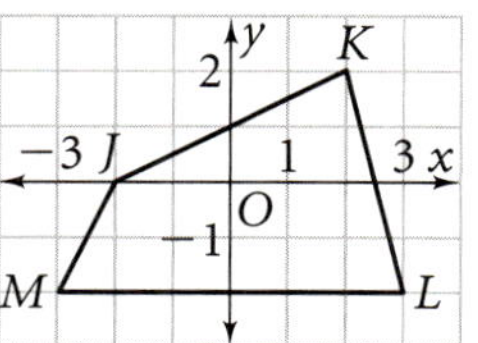

Find a single translation that has the same effect as each composition of translations.

4. $\langle -2, 3 \rangle$ followed by $\langle 1, 5 \rangle$
5. $\langle -6, 0 \rangle$ followed by $\langle -4, 7 \rangle$
6. $\langle 3, -4 \rangle$ followed by $\langle -4, 3 \rangle$
7. $\langle 8, 10 \rangle$ followed by $\langle -7, 6 \rangle$

Use matrices to find the image of each figure under the given translation.

8. $\langle 2, -1 \rangle$

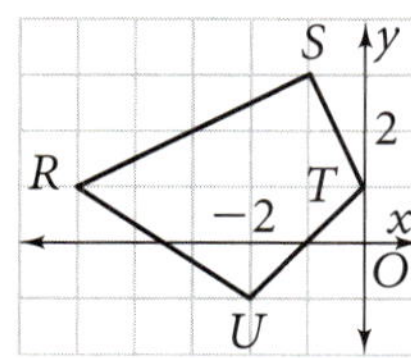

9. $\langle 4, -3 \rangle$

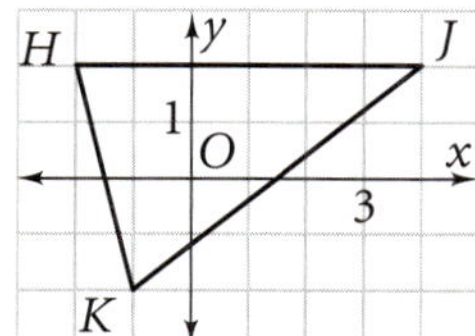

10. $\langle 0, 2 \rangle$

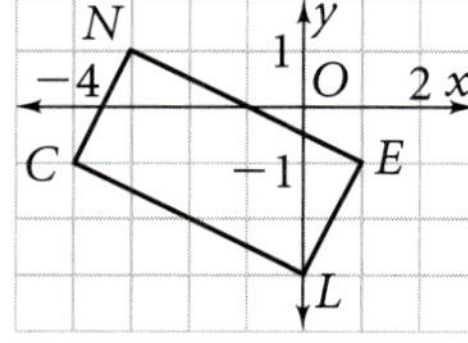

Use a matrix and scalar multiplication. Find the image of each figure for a dilation with center (0, 0) and the scale factor given. Use graph paper. Draw the figure and its image.

11. 3

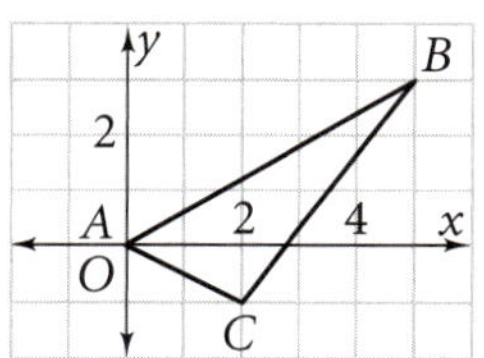

12. $\frac{1}{3}$

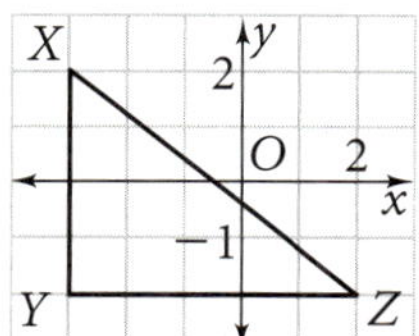

13. $1\frac{1}{2}$

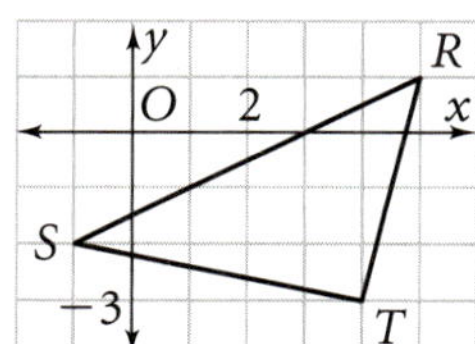

Compositions of Reflections

California Content Standards

GEOM 22.0 Students know the effect of rigid motions on figures in the coordinate plane and space, including rotations, translations, and reflections. ***Master***

What You'll Learn

- To use a composition of reflections
- To identify glide reflections

. . . And Why

To classify isometries, as in Example 5

Check Skills You'll Need

for Help Lessons 9-1 and 9-2

Given points $R(-1, 1)$, $S(-4, 3)$, and $T(-2, 5)$, draw $\triangle RST$ and its reflection image in each line.

1. the y-axis **2.** the x-axis **3.** $y = 1$

Draw $\triangle RST$ described above and its translation image for each translation.

4. $(x, y) \rightarrow (x, y - 3)$ **5.** $(x, y) \rightarrow (x + 4, y)$

6. $(x, y) \rightarrow (x + 2, y - 5)$

7. Copy the figure at the right. Draw the image of the figure for a reflection across $\overleftrightarrow{DG}$.

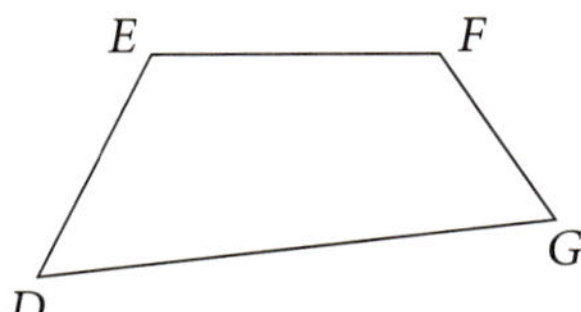

New Vocabulary • glide reflection

1 Compositions of Reflections

If two figures are congruent, there is a transformation that maps one onto the other. If no reflection is involved, then the figures are either translation or rotation images of each other.

1 EXAMPLE Recognizing the Transformation

The two figures are congruent. Is one figure a translation image of the other, a rotation image, or neither? Explain.

The orientations of these congruent figures do not appear to be opposite, so one is a translation image or a rotation image of the other. Clearly, it's not a translation image, so it must be a rotation image.

CA Standards Check 1 The two figures are congruent. Is one figure a translation image of the other, a rotation image, or neither? Explain.

Any translation or rotation can be expressed as the composition of two reflections.

Key Concepts

Theorem 9-1

A translation or rotation is a composition of two reflections.

The examples that illustrate Theorems 9-2 and 9-3 suggest a proof of Theorem 9-1 (how to find two reflections for a given translation or rotation).

Theorems 9-2 and 9-3 together form the converse of Theorem 9-1.

Theorem 9-2

A composition of reflections across two parallel lines is a translation.

Theorem 9-3

A composition of reflections across two intersecting lines is a rotation.

2 EXAMPLE Composition of Reflections Across Parallel Lines

Real-World Connection

Each mirror shows a reverse image. But bend the mirrors like this ⁄‾\ and you get compositions of reflections.

Find the image of R for a reflection across line ℓ followed by a reflection across line m. Describe the resulting translation.

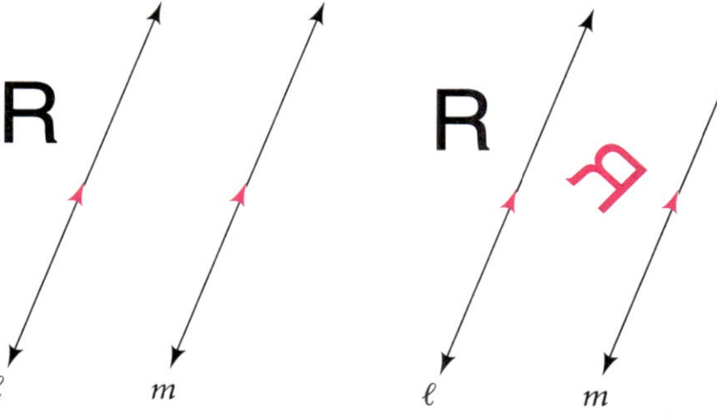

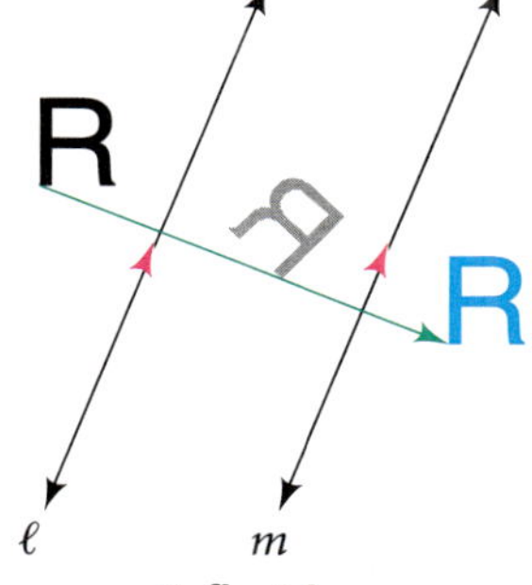

R is translated the distance and direction shown by the green arrow. The arrow is perpendicular to lines ℓ and m with length equal to twice the distance from ℓ to m.

2 Draw lines ℓ and m as shown above. Draw R between ℓ and m. Find the image of R for a reflection across line ℓ and then across line m. Describe the resulting translation.

3 EXAMPLE Composition of Reflections in Intersecting Lines

Lines a and b intersect in point C and form acute $\angle 1$ with measure 35. Find the image of R for a reflection across line a and then a reflection across line b. Describe the resulting rotation.

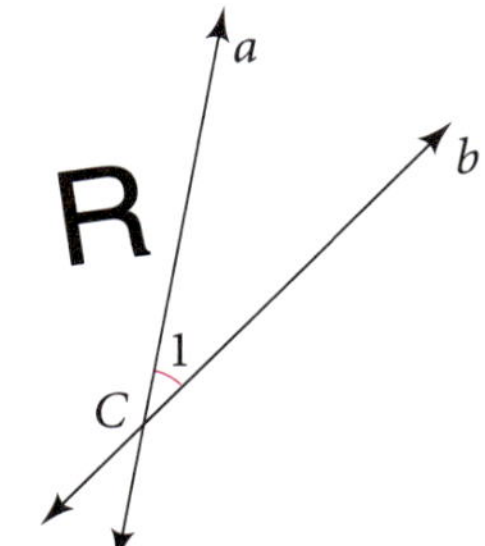

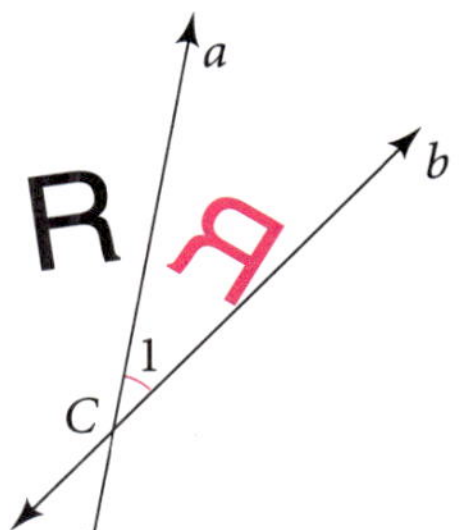

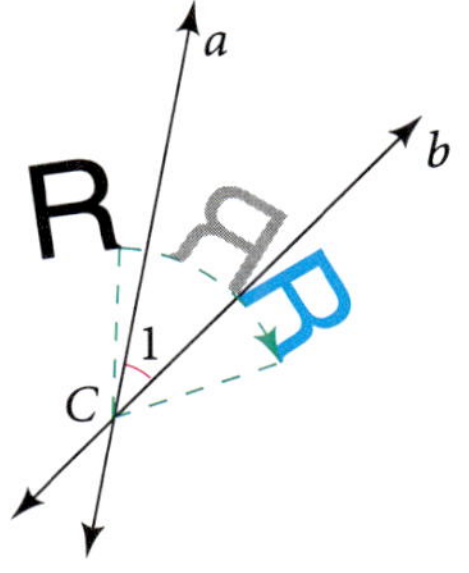

R rotates clockwise through the angle shown by the green arrow. The center of rotation is C and the measure of the angle is twice $m\angle 1$, or 70.

3 Repeat Example 3, but begin with R in a different position.

2 Glide Reflections

Two plane figures A and B can be congruent with opposite orientations. Reflect A and you get a figure A′ that has the same orientation as B. Thus, B is a translation or rotation image of A′. By Theorem 9-1, two reflections map A′ to B. The net result is that three reflections map A to B.

This is summarized in what is sometimes called the Fundamental Theorem of Isometries.

Key Concepts

Theorem 9-4 **Fundamental Theorem of Isometries**

In a plane, one of two congruent figures can be mapped onto the other by a composition of at most three reflections.

If two figures are congruent and have opposite orientations (but are not simply reflections of each other), then there is a slide and a reflection that will map one onto the other. A **glide reflection** is the composition of a glide (translation) and a reflection across a line parallel to the direction of translation.

Real-World Connection

A computer can translate an image and then reflect it, or vice versa. The two rabbit images are glide reflection images of each other.

4 EXAMPLE Finding a Glide Reflection Image

Coordinate Geometry Find the image of $\triangle TEX$ for a glide reflection where the translation is $(x, y) \rightarrow (x, y - 5)$ and the reflection line is $x = 0$.

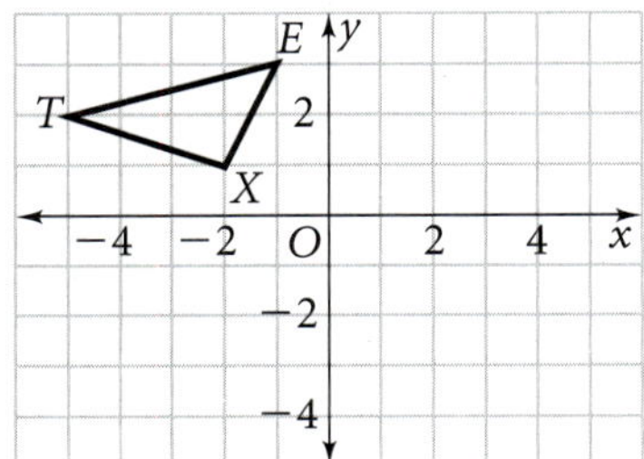

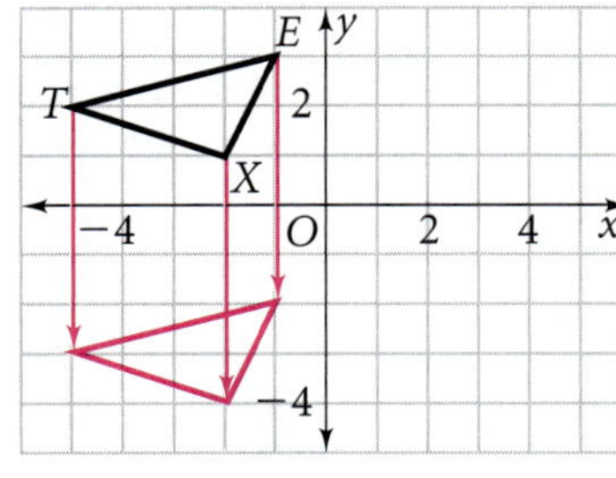

Translate $\triangle TEX$.

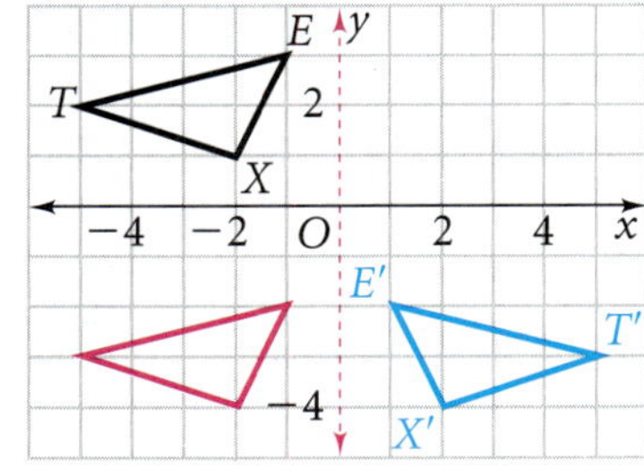

Reflect the image in $x = 0$.

CA Standards Check

4 Use $\triangle TEX$ from Example 4 above.

a. Find the image of $\triangle TEX$ under a glide reflection where the translation is $(x, y) \rightarrow (x + 1, y)$ and the reflection line is $y = -2$.

b. **Critical Thinking** Would the result of part (a) be the same if you reflected $\triangle TEX$ first, and then translated it? Explain.

You can map one of any two congruent figures onto the other by a single reflection, translation, rotation, or glide reflection. Thus, you are able to classify any isometry.

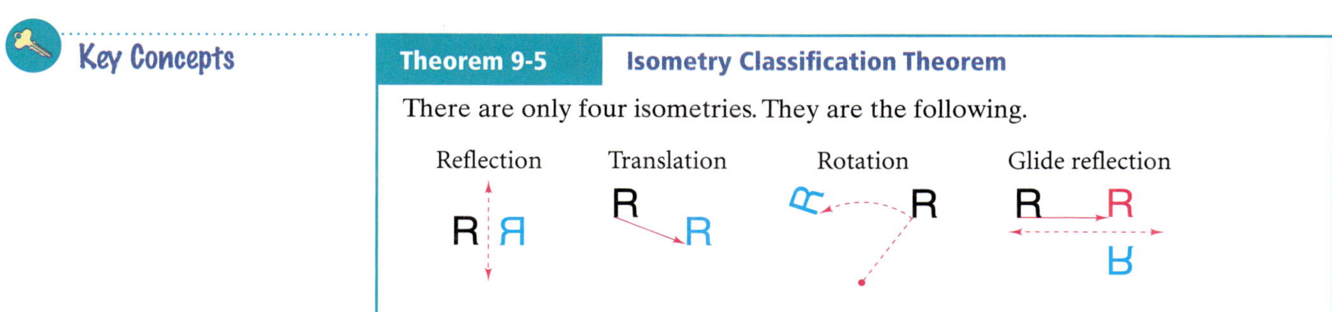

Theorem 9-5 **Isometry Classification Theorem**

There are only four isometries. They are the following.

5 EXAMPLE Classifying Isometries

Each figure is an isometry image of the figure at the left. Tell whether their orientations are the same or opposite. Then classify the isometry.

a. opposite; a reflection

b. 
opposite; a glide reflection

c. 
same; a translation

d. 
same; a rotation

CA Standards Check 5 Classify the isometry.

EXERCISES

For more exercises, see *Extra Skill, Word Problem, and Proof Practice.*

Standards Practice GEOM 22.0

A Practice by Example

GO for Help

Example 1 (page 506)

The two figures in each pair are congruent. Is one figure a translation image of the other, a rotation image, or neither? Explain.

1.

2. 

3. 

Example 2 (page 507)

Find the image of each letter for a reflection across line ℓ and then a reflection across line m. Describe the resulting translation or rotation.

4. 

5. 

6. 

Example 3 (page 507)

7. 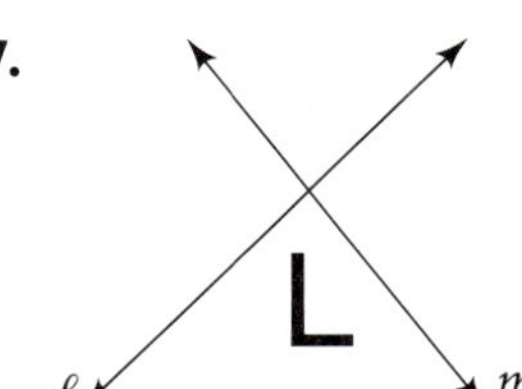

8. 

9. 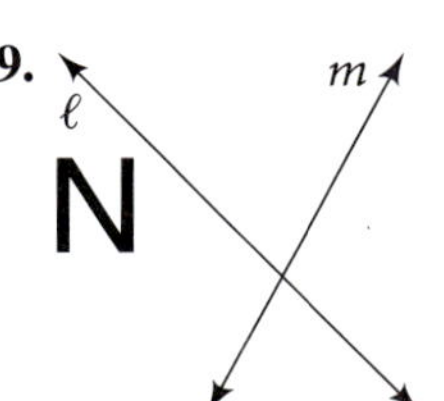

Example 4 (page 508)

Find the glide reflection image of $\triangle PNB$ for the given translation and reflection line.

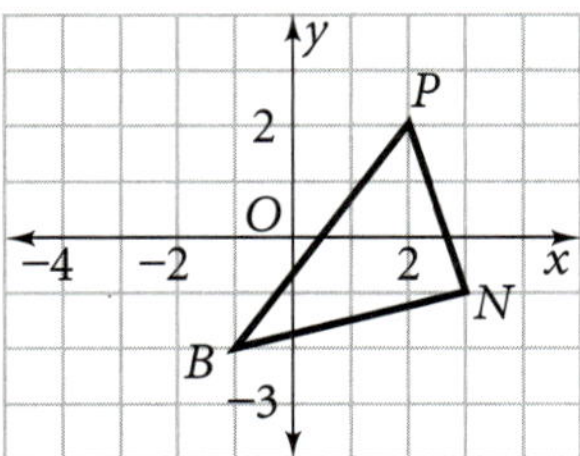

10. $(x, y) \rightarrow (x + 2, y); y = 3$

11. $(x, y) \rightarrow (x, y - 3); x = 0$

12. $(x, y) \rightarrow (x + 2, y + 2); y = x$

13. $(x, y) \rightarrow (x - 1, y + 1); y = -x$

14. $(x, y) \rightarrow (x, y - 1); x = 2$

15. $(x, y) \rightarrow (x - 2, y - 2); y = x$

Example 5 (page 509)

Each figure is an isometry image of the figure at the left. Tell whether their orientations are the same or opposite. Then classify the isometry.

16. 

17. 

18. 

19. 

20.

21.

22.

23.

B Apply Your Skills

The two figures are congruent. Name the isometry that maps one onto the other.

24. 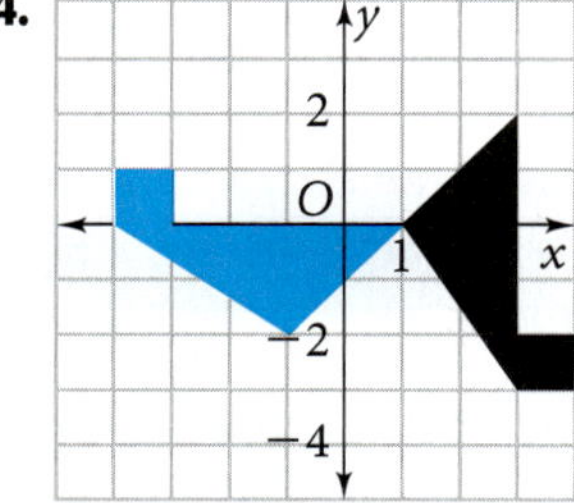

25. 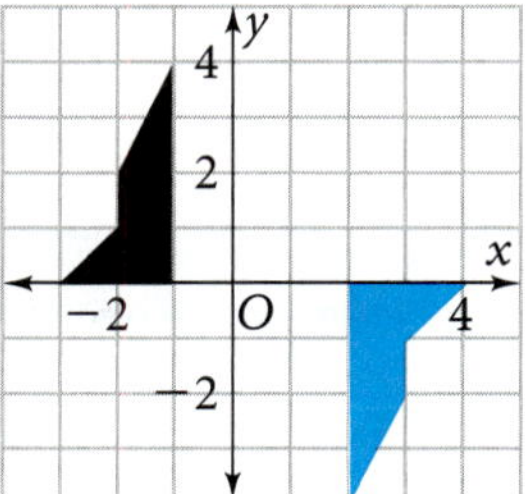

26. Multiple Choice Which transformation maps the black triangle onto the blue triangle?

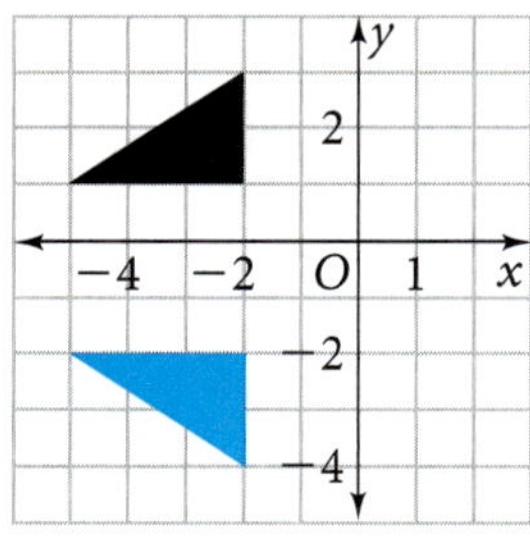

Ⓐ a translation $(x, y) \rightarrow (x, y - 3)$ followed by a reflection across $x = -2$

Ⓑ a rotation of 180° about the origin

Ⓒ a reflection across $y = -\frac{1}{2}$

Ⓓ a reflection across the y-axis followed by a 180° rotation about the origin

27. Writing Reflections and glide reflections are *odd isometries*, while translations and rotations are *even isometries*. Use what you learned in this lesson to explain why these categories make sense.

28. Open-Ended Draw $\triangle ABC$. Then, describe a reflection, a translation, a rotation, and a glide reflection, and draw the image of $\triangle ABC$ for each transformation.

Homework Video Tutor

Visit: PHSchool.com
Web Code: bce-0906

29. For center of rotation P, does an $x°$ rotation followed by a $y°$ rotation give the same image as a $y°$ rotation followed by an $x°$ rotation? Explain.

30. Does an $x°$ rotation about a point P followed by a reflection in a line ℓ give the same image as a reflection in ℓ followed by an $x°$ rotation about P? Explain.

To learn more about kaleidoscopes, see p. 513.

Kaleidoscopes **The vibrant images of a kaleidoscope are produced by compositions of reflections in intersecting mirrors. Determine the angle between the mirrors in each kaleidoscope image.**

31. 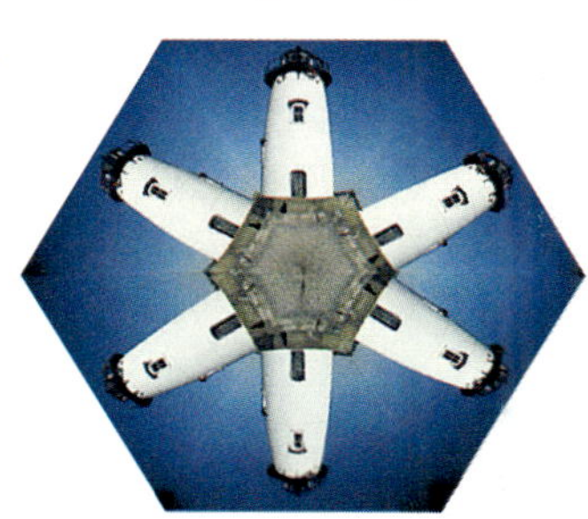

32. 

33. 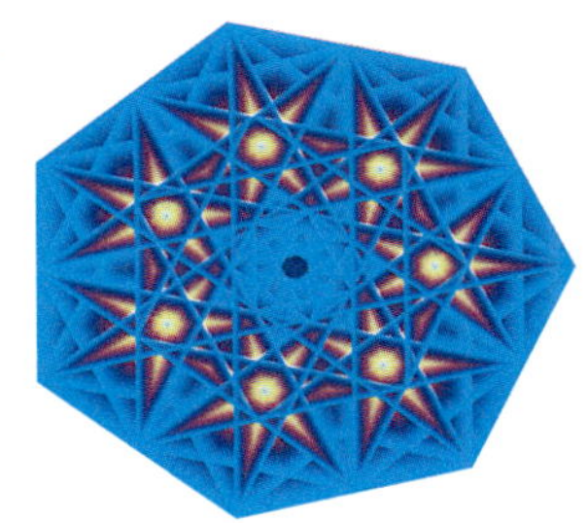

34. 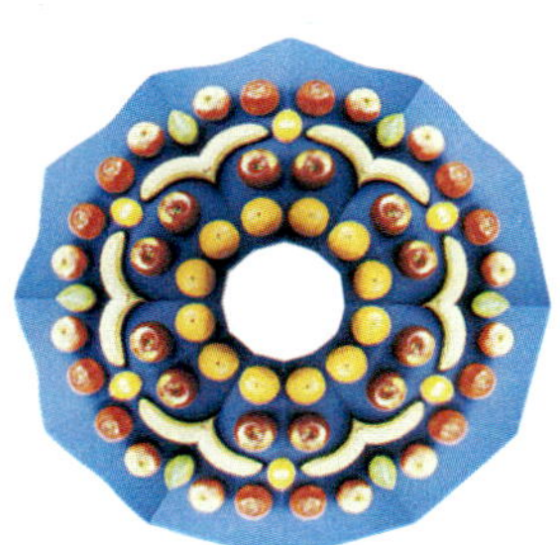

Identify each mapping as a reflection, translation, rotation, or glide reflection. Find the reflection line, translation rule, center and angle of rotation, or glide translation and reflection line.

35. $\triangle ABC \to \triangle EDC$

36. $\triangle EDC \to \triangle PQM$

37. $\triangle MNJ \to \triangle EDC$

38. $\triangle HIF \to \triangle HGF$

39. $\triangle PQM \to \triangle JLM$

40. $\triangle MNP \to \triangle EDC$

41. $\triangle JLM \to \triangle MNJ$

42. $\triangle PQM \to \triangle KJN$ **43.** $\triangle KJN \to \triangle ABC$ **44.** $\triangle HGF \to \triangle KJN$

45. Describe a glide and a reflection that maps the red R to the blue R.

For the given transformation mapping $\overline{XY}$ to $\overline{X'Y'}$, give a convincing argument why $\overline{XY} \cong \overline{X'Y'}$.

46. a reflection **47.** a translation **48.** a rotation

49. The definition states that a glide reflection is the composition of a translation and a reflection. Explain why these can occur in either order.

50. For lines of reflection r and s, does a reflection in r followed by a reflection in s give the same image as a reflection in s followed by a reflection in r? Explain.

$P \to P'(3, -1)$ for the given translation and reflection line. Find the coordinates of P.

51. $(x, y) \to (x - 3, y); y = 2$

52. $(x, y) \to (x, y - 3); y = 2$

53. $(x, y) \to (x - 3, y - 3); y = x$

54. $(x, y) \to (x + 4, y - 4); y = -x$

Multiple Choice Practice

For California Standards and CAHSEE Tutorials, visit PHSchool.com. Web Code: bcq-9045

GEOM 22.0

55. Given triangle MAP, determine the coordinates of the image M' under a glide reflection where the translation is $(x, y) \rightarrow (x, y + 4)$ followed by a reflection about the y-axis.

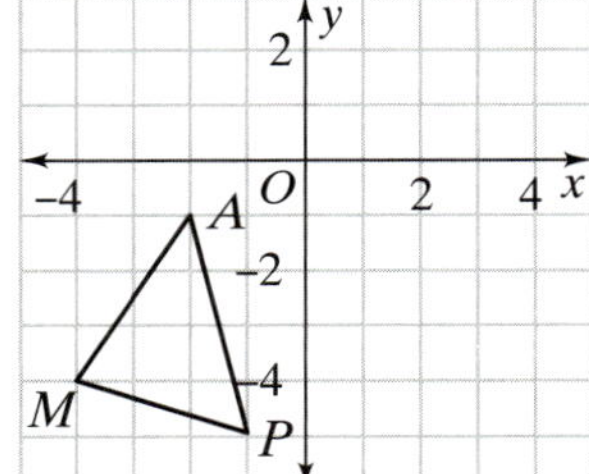

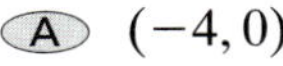

Ⓐ $(-4, 0)$ Ⓒ $(0, -4)$

Ⓑ $(4, 0)$ Ⓓ $(0, 0)$

CAHSEE 7AF 1.5

56. Which function described below is *not* linear?

Ⓐ The sum of x and y is 3.

Ⓑ The difference of x and y is 3.

Ⓒ The product of x and y is 3.

Ⓓ The quotient of x and y is 3.

GEOM 22.0

57. Figure $QRST$ is shown in the coordinate plane. Which transformation creates an image with a vertex at the origin?

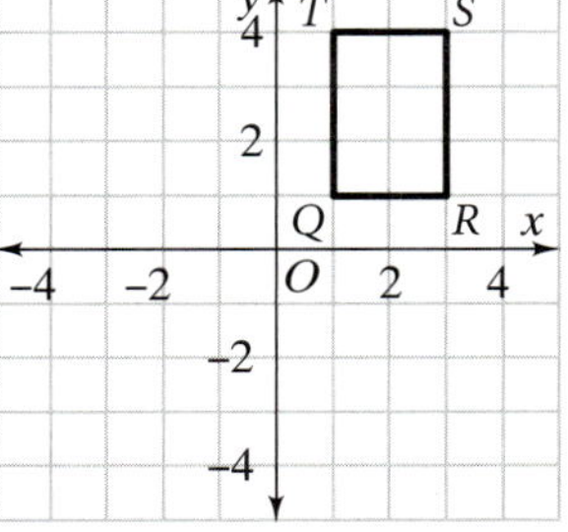

Ⓐ Rotate figure $QRST$ by 90° around Q.

Ⓑ Reflect figure $QRST$ across the line $y = 1$.

Ⓒ Reflect figure $QRST$ across the line $x = 1$.

Ⓓ Translate figure $QRST$ to the left 3 and down 4.

CAHSEE 7MR 1.2

58. Which of the following situations is best represented by a linear function?

Ⓐ The cost of buying t tickets for a show

Ⓑ The speed s of a car over time as it travels in heavy traffic

Ⓒ The population p of the United States each year since 1800

Ⓓ The altitude a of an airplane from takeoff to landing

Mixed Review

Lesson 9-5

Coordinate Geometry **Find the image of $\triangle ABC$ for a dilation with center (0, 0) and the scale factor given.**

59. $A(0, 4), B(0, 0), C(3, 0); 3$

60. $A(4, 2), B(0, 0), C(6, 0); 1.5$

61. $A(2, 3), B(-4, -2), C(5, -3); 2$

62. $A(7, 8), B(5, 4), C(9, 6); 0.5$

Lesson 7-4

Algebra **Find the value of x.**

63.

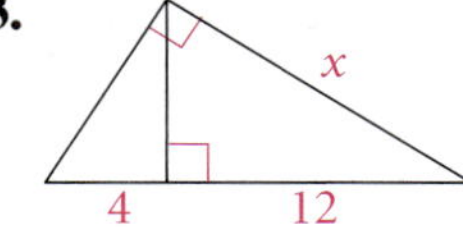

64.

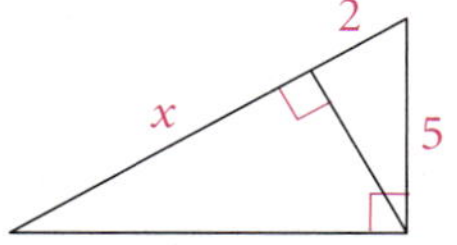

65.

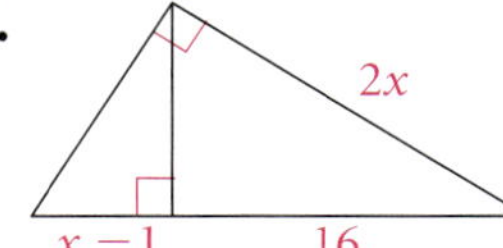

Lesson 5-4

Identify the two statements that contradict each other.

66. I. $\triangle ABC$ is right.

II. $\triangle ABC$ is equiangular.

III. $\triangle ABC$ is isosceles.

67. I. In right $\triangle ABC, m\angle B = 90$.

II. In right $\triangle ABC, m\angle A = 80$.

III. In right $\triangle ABC, m\angle C = 90$.

Kaleidoscopes

Technology

For Use With Lesson 9-6

GEOM 22.0 Students know the effect of rigid motions on figures in the coordinate plane and space, including rotations, translations, and reflections. ***Master***

The mirrors in a kaleidoscope provide compositions of reflections to create a *symmetrical* design. You can create your own kaleidoscope.

Construct

- Use geometry software. Draw a line and construct a point on the line. Rotate the line 60° about the point and repeat to get a third line.
- Construct a polygon in the interior of an angle, as shown. Reflect the polygon in the closest line, then in the next line, and so on until the kaleidoscope is filled and you have a symmetrical design. Then hide the lines of reflection.

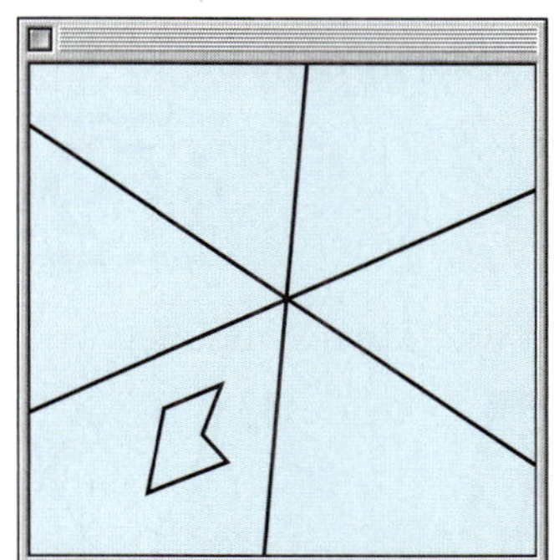

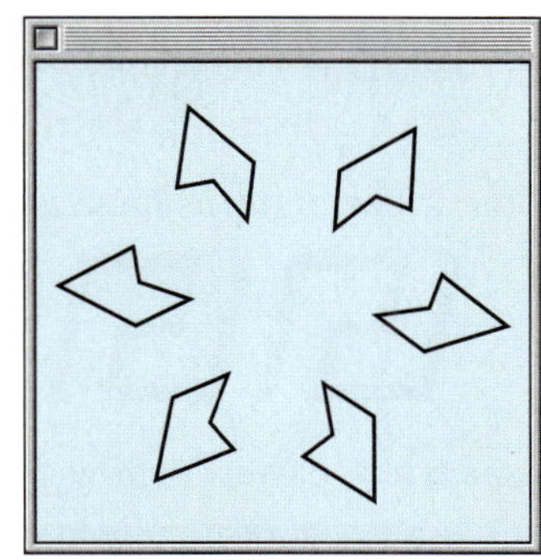

Investigate

- Manipulate the original figure by dragging any of its vertices or selecting and moving it. As you manipulate the figure, what happens to the images? Does the design remain symmetrical? Continue manipulating the original figure until you are satisfied with your design. Print the design and color it.
- Now add other figures beside the original polygon. Reflect these figures to create a more interesting design, as shown at the right. (You may need to temporarily show the lines of reflection.) Print your design and color it.

EXERCISES

1. Create a kaleidoscope with four lines of reflection. Draw a line and construct a point on the line. Rotate the line 45° about the point and repeat two more times to get four lines. Add a figure to the interior of an angle and reflect it as described above.

A *tessellation* is another type of interesting design that can be constructed using geometry software. An example is shown in Exercise 2 and tessellations are explained more fully in Lesson 9-7.

2. Construct a regular hexagon. Translate the hexagon several times to create the tessellation shown. Can you make a similar tessellation with other regular polygons? Can you find two or more polygons that, when placed together repeatedly, form a tessellation?

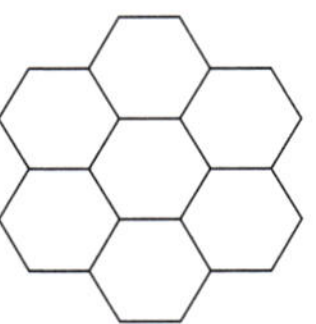

Frieze Patterns

FOR USE WITH LESSON 9-7

GEOM 22.0 Students know the effect of rigid motions on figures in the coordinate plane and space, including rotations, translations, and reflections. *Master*

A frieze pattern is a design that repeats itself along a straight line. Frieze patterns are popular forms of design trim, often appearing along the edges of buildings or rooms.

Every frieze pattern can be mapped onto itself by a translation. Some frieze patterns can be mapped onto themselves by reflections.

A. 

B. 

Some frieze patterns have repeated rotational symmetry.

C. 

D. 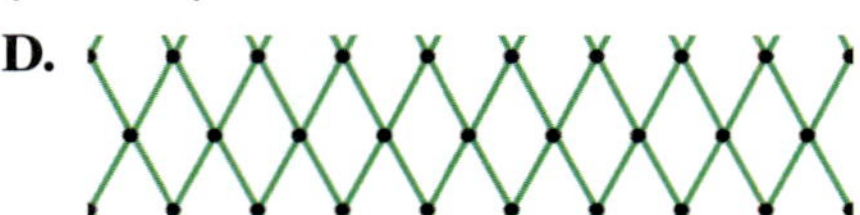

Some frieze patterns show glide reflections.

E. 

F. 

EXERCISES

For Exercises 1–4, refer to the frieze patterns above. You may find tracing paper helpful.

1. For each frieze pattern, find a portion (as small as possible) that you could translate repeatedly to form the entire pattern.
2. Which patterns show reflectional symmetry? Find their reflection lines.
3. Which patterns show rotational symmetry? Find their centers of rotation.
4. Which patterns show glide reflections? Find a glide translation and reflection line.

For each frieze pattern below, describe all the transformations that map the pattern onto itself.

5. 

6. 

7. 

8. 

9. 

10. 

9-7 Tessellations

California Content Standards

GEOM 22.0 Students know the effect of rigid motions on figures in the coordinate plane and space, including rotations, translations, and reflections. ***Master***

What You'll Learn

- To identify transformations in tessellations, and figures that will tessellate
- To identify symmetries in tessellations

. . . And Why

To identify a tessellation in art, as in Example 1

Check Skills You'll Need

GO for Help Lesson 3-5

Classify the polygon with the given number of sides.

1. five **2.** eight **3.** twelve

Find the measure of an angle of each regular polygon.

4. triangle **5.** quadrilateral **6.** hexagon

7. octagon **8.** decagon **9.** 14-gon

New Vocabulary • tessellation • tiling • translational symmetry • glide reflectional symmetry

1 Identifying Transformations in Tessellations

Vocabulary Tip

A figure that creates a tessellation is said to tessellate.

A **tessellation,** or **tiling,** is a repeating pattern of figures that completely covers a plane, without gaps or overlaps. You can create tessellations with translations, rotations, and reflections. You can find tessellations in art (see below), nature (cells in a honeycomb), and everyday life (tiled floors).

1 EXAMPLE Identifying the Transformation in a Tessellation

Art Identify a transformation and the repeating figures in this tessellation.

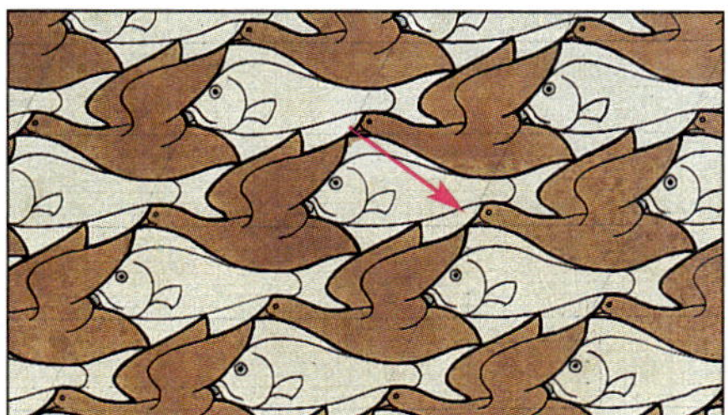

Repeating figures

The arrow shows a translation.

CA Standards Check

1 Identify a transformation and the repeating figures in each tessellation below.

a.

b.

Because the figures in a tessellation do not overlap or leave gaps, the sum of the measures of the angles around any vertex must be 360°. If the angles around a vertex are all congruent, then the measure of each angle must be a factor of 360.

2 EXAMPLE Determining Figures That Will Tessellate

Determine whether a regular 18-gon tessellates a plane.

$a = \frac{180(n - 2)}{n}$ **Use the formula for the measure of an angle of a regular polygon.**

$a = \frac{180(18 - 2)}{18}$ **Substitute 18 for *n*.**

$a = 160$ **Simplify.**

Since 160 is not a factor of 360, the 18-gon will not tessellate.

CA Standards Check 2 Explain why you can tessellate a plane with an equilateral triangle.

A figure does not have to be a regular polygon to tessellate.

Key Concepts

Theorem 9-6

Every triangle tessellates.

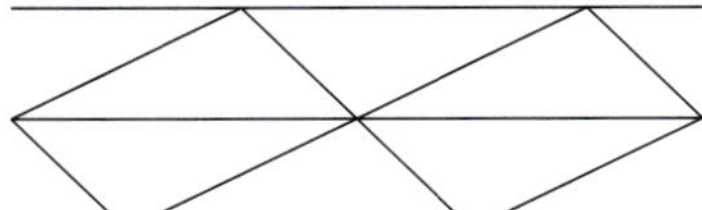

Theorem 9-7

Every quadrilateral tessellates.

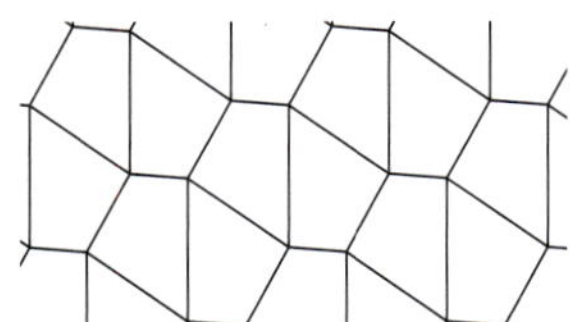

2 Identifying Symmetries in Tessellations

Real-World Connection

Physicists apply the symmetries of tessellations to study subatomic particles. Here they use a particle detector to study quarks.

The tessellation with regular hexagons at the right has reflectional symmetry in each of the blue lines. It has rotational symmetry centered at each of the red points. The tessellation also has translational symmetry and glide reflectional symmetry, as shown below.

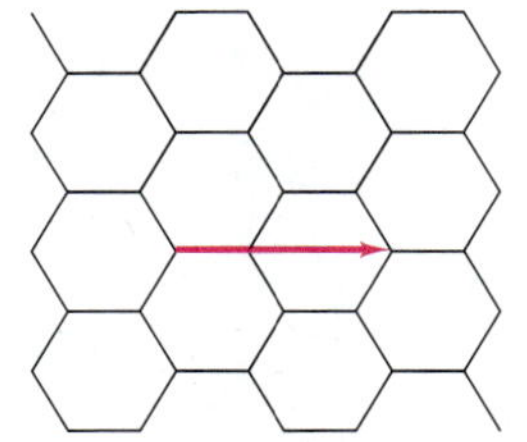

Translational Symmetry

A translation maps the tessellation onto itself.

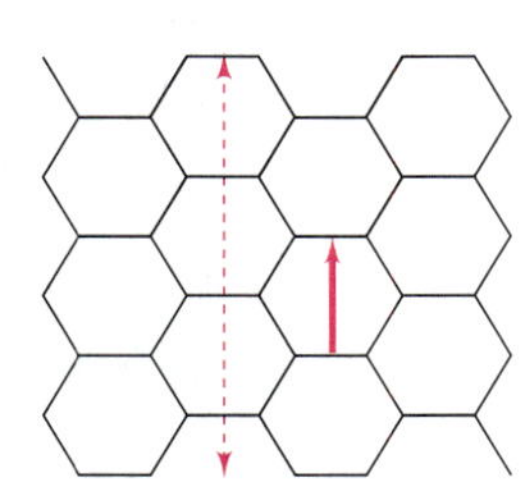

Glide Reflectional Symmetry

A glide reflection maps the tessellation onto itself.

3 EXAMPLE Identifying Symmetries in Tessellations

List the symmetries in the tessellation.

Rotational symmetry centered at each red point
Translational symmetry (blue arrow)

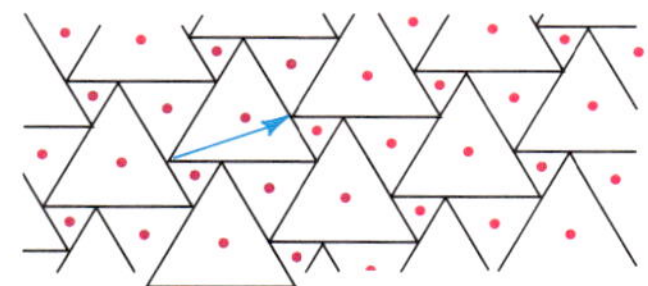

CA Standards Check

3 List the symmetries in the tessellation at the right.

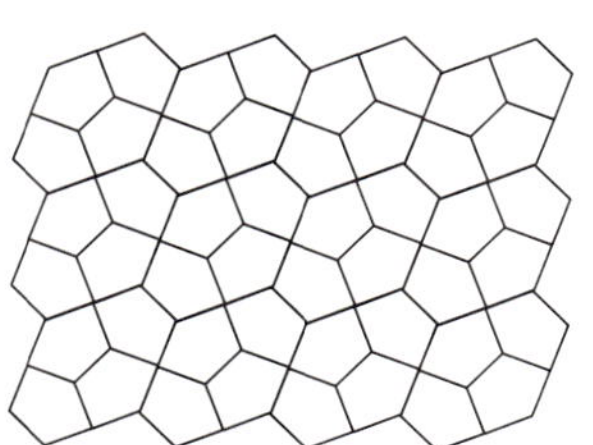

The following Activity shows the steps for making creative tessellations.

Hands-On Activity: Drawing Tessellations

- Draw a 1.5-inch square on a blank piece of paper and cut it out.
- Draw a curve joining two consecutive vertices.

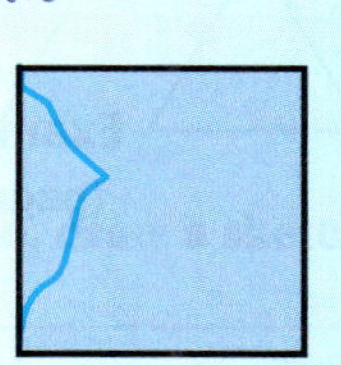

- Cut along the curve you drew and slide the cutout piece to the opposite side of the square. Tape it in place.

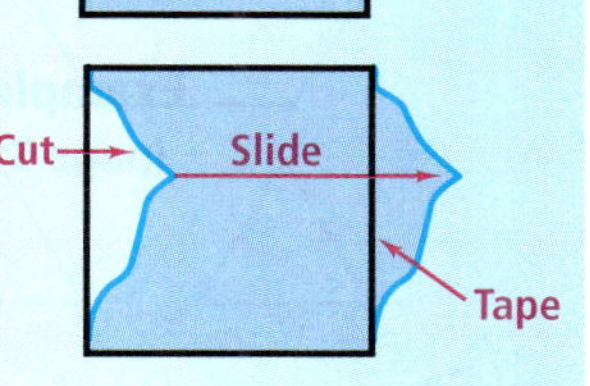

- Repeat this process using the other two opposite sides of the square.
- Rotate the resulting figure. What does your imagination suggest it looks like? Is it a penguin wearing a hat or a knight on horseback? Could it be a dog with floppy ears? Draw the image on your figure.

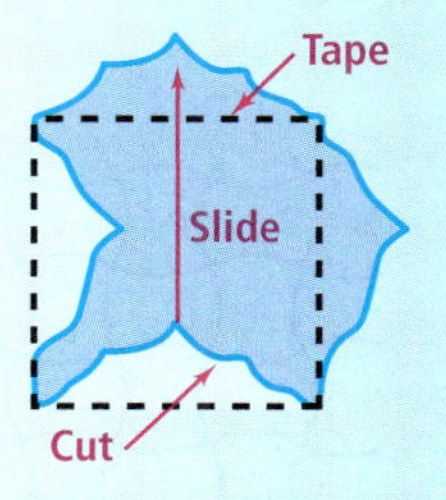

- Draw a tessellation using your figure.

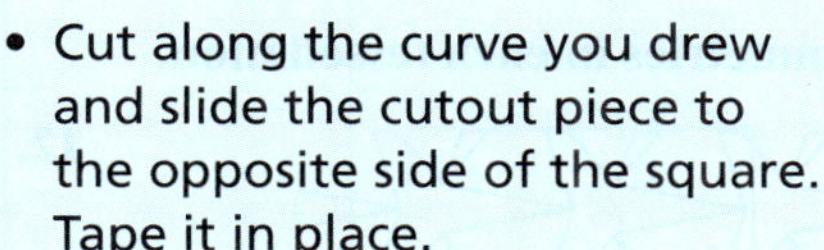

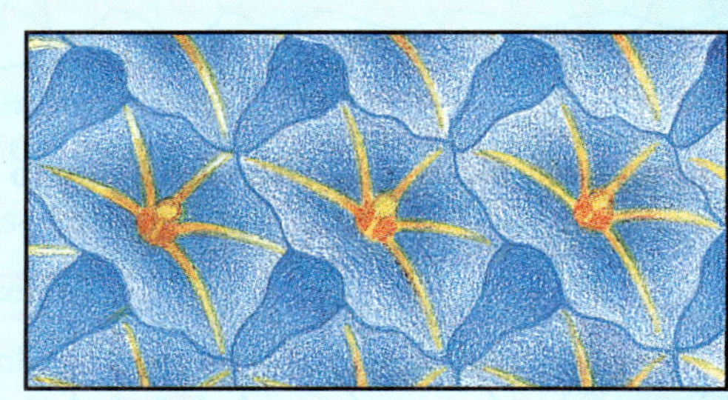

Answering the Question Asked

When answering a question, be sure to answer the question that is asked. Read the question carefully and identify the quantity that you are asked to find. Some answer choices are answers to related questions, so you have to be careful.

1 EXAMPLE

The point $L(a, b)$ in Quadrant I is rotated 90° about the origin and is then reflected across the y-axis. What is the x-coordinate of the image?

Ⓐ a Ⓑ $-b$ Ⓒ b Ⓓ $-a$

The question asks for the x-coordinate of the image. The x-coordinate of the preimage is a. After the 90° rotation about the origin, the x-coordinate is $-b$. After a reflection across the y-axis, the x-coordinate of the image is b.

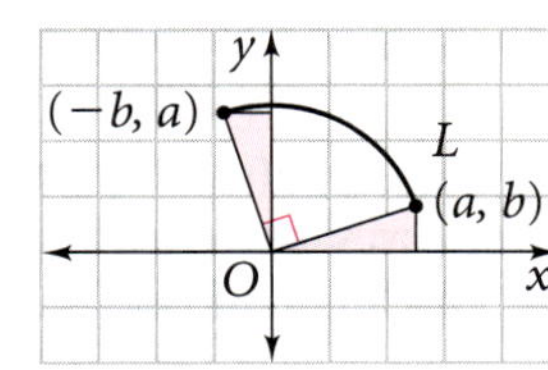

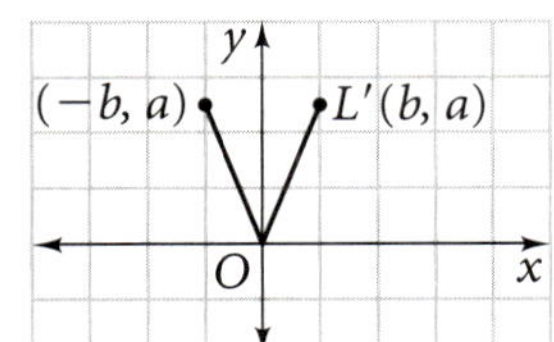

The correct answer is C. Choice A is the y-coordinate of the image, but that is not what is asked for.

2 EXAMPLE

What is the image of $M(4, 3)$ for the translation $(x, y) \rightarrow (x - 2, y + 1)$ followed by a reflection across the line $x = -1$?

Ⓐ $(2, -6)$ Ⓑ $(-4, 4)$ Ⓒ $(0, 4)$ Ⓓ $(-8, 4)$

The translation puts the point at $(2, 4)$. A reflection across $x = -1$ then puts the point at $(-4, 4)$. The correct answer is B.

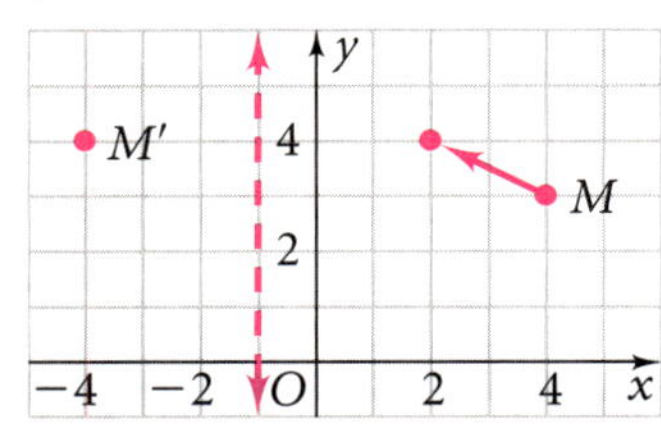

Answers A and C result from the translation followed by a reflection across a line other than $x = -1$. In fact, if you reflect $(2, 4)$ across $x = 1$, which is one unit in the negative direction from $(2, 4)$, you get $(0, 4)$, or choice C, as your answer. If you do the steps in the wrong order, you will get answer D.

EXERCISES

Answer the question. Then ask a related question that would lead to an incorrect choice.

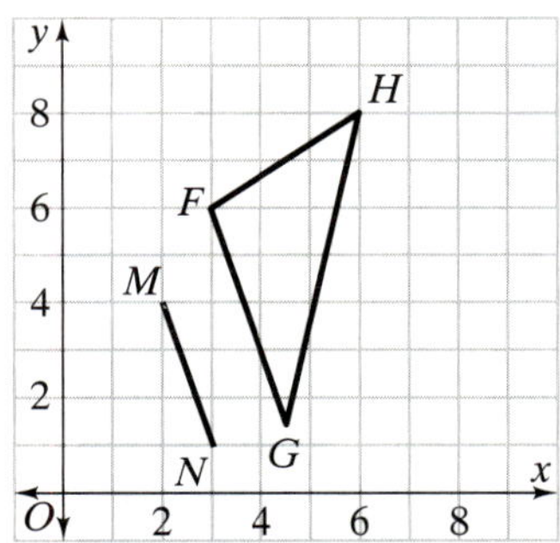

1. For the diagram at the right, the coordinates of point P are such that triangle MNP is similar to triangle FGH. Which of the following could be the y-coordinate of P?

   Ⓐ $\frac{2}{3}$ Ⓒ $5\frac{1}{3}$

   Ⓑ 4 Ⓓ 6

2. The point $T(5, -1)$ is reflected across the y-axis. Its image is point T'. What is the distance between T and T'?

   Ⓐ 1 Ⓒ 10

   Ⓑ 5 Ⓓ $\sqrt{26}$

Chapter 9 Review

Vocabulary Review

English and Spanish Audio Online

center of a regular polygon (p. 484)
composition (p. 472)
dilation (p. 498)
enlargement (p. 498)
glide reflection (p. 508)
glide reflectional symmetry (p. 516)
image (p. 470)
isometry (p. 470)
line symmetry (p. 492)
point symmetry (p. 493)
preimage (p. 470)
reduction (p. 498)
reflection (p. 478)
reflectional symmetry (p. 492)
rigid motion (p. 470)
rotation (p. 484)
rotational symmetry (p. 493)
symmetry (p. 492)
tessellation (p. 515)
tiling (p. 515)
transformation (p. 470)
translation (p. 471)
translational symmetry (p. 516)

To complete each definition, find the appropriate word in the second column.

1. A(n) _?_ is a change in position, shape, or size of a figure.
2. A(n) _?_ is a transformation in which the preimage and its image are congruent.
3. A _?_ is an isometry in which a figure and its image have opposite orientations.
4. A _?_ is an isometry in which all points of a figure move the same distance in the same direction.
5. A(n) _?_ is a translation followed by a reflection in a line parallel to the translation vector.
6. A(n) _?_ is a repeating pattern of figures that completely covers a plane, without gaps or overlaps.
7. A(n) _?_ is a transformation that proportionally reduces or enlarges a figure.

A. dilation
B. glide reflection
C. tessellation
D. isometry
E. reflection
F. transformation
G. translation

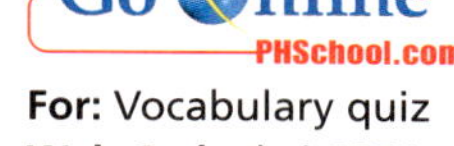

For: Vocabulary quiz
Web Code: bcj-0951

Skills and Concepts

Lessons 9-1 and 9-2

- To identify isometries (p. 470)
- To find translation images of figures (p. 471)
- To find reflection images of figures (p. 478)

 GEOM 22.0

A **transformation** of a geometric figure is a change in its position, shape, or size. An **isometry** is a transformation in which the **preimage** and **image** are congruent. A transformation maps a figure onto its image.

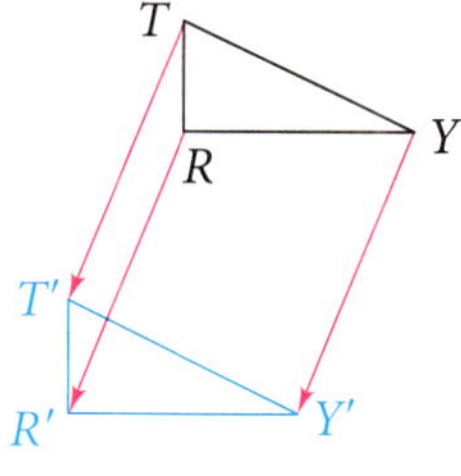

A **translation** is an isometry that maps all points of a figure the same distance in the same direction. A translation is an isometry that does not change orientation.

The second diagram shows a **reflection** of B to B' across line r. A reflection is an isometry in which a figure and its image have opposite orientations.

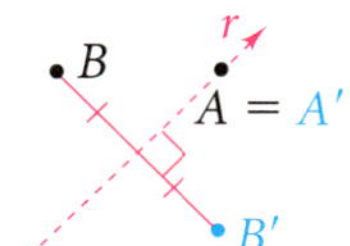

A **composition** of transformations is a combination of two or more transformations. Each transformation is performed on the image of the preceding transformation.

CAHSEE Practice (continued)

3. What is the y-coordinate of the solution to the system of linear equations below?

$$\begin{cases} 4x - 3y = 18 \\ 6x + 3y = 32 \end{cases}$$

Ⓐ 5 Ⓑ $\frac{2}{3}$ Ⓒ -5 Ⓓ $-\frac{2}{3}$

4. In the system of equations below, which expression can Julia substitute for y in the equation $2x - 9y = -2$?

$$\begin{cases} 9x + 3y = 18 \\ 2x - 9y = -2 \end{cases}$$

Ⓐ $18 + 3x$ Ⓒ $6 + 3x$

Ⓑ $18 - 3x$ Ⓓ $6 - 3x$

5. Which graph could be used to find the solution of the system $\begin{cases} y = \frac{1}{2}x - 1 \\ 2x + 2y = 4 \end{cases}$?

Ⓐ

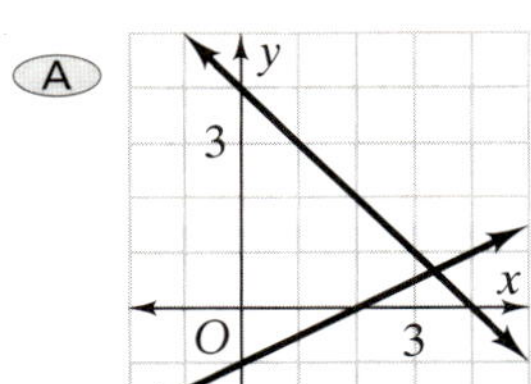

Ⓒ

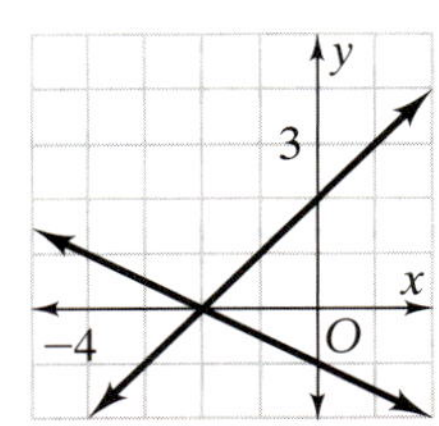

Ⓑ

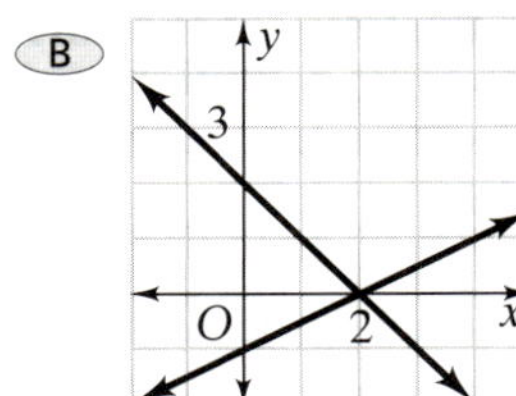

Ⓓ

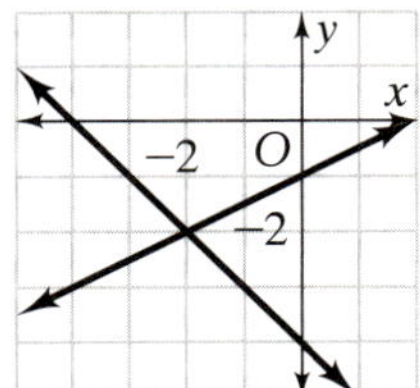

6. The circle graph represents the number of homes for sale offering 1, 2, 3, 4, and 5 bedrooms (BR). Which table best matches the information shown in the graph?

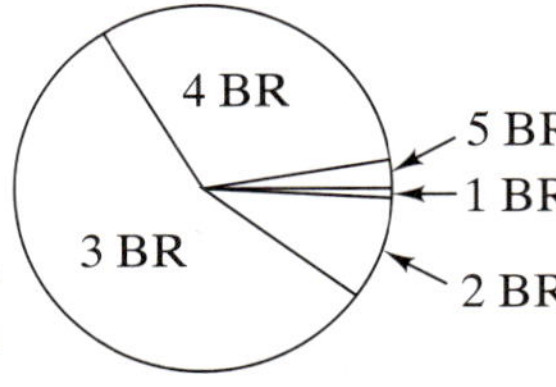

Ⓐ

1 BR	2 BR	3 BR	4 BR	5 BR
25	12	230	108	19

Ⓑ

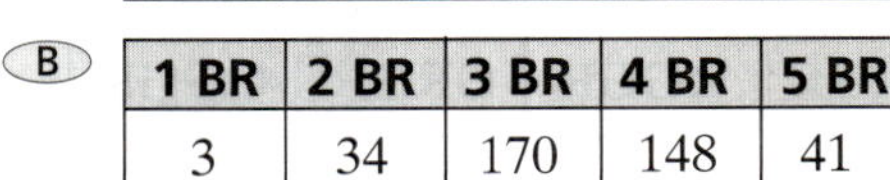

1 BR	2 BR	3 BR	4 BR	5 BR
3	34	170	148	41

Ⓒ

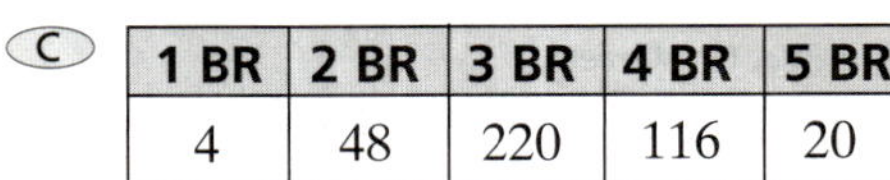

1 BR	2 BR	3 BR	4 BR	5 BR
4	48	220	116	20

Ⓓ

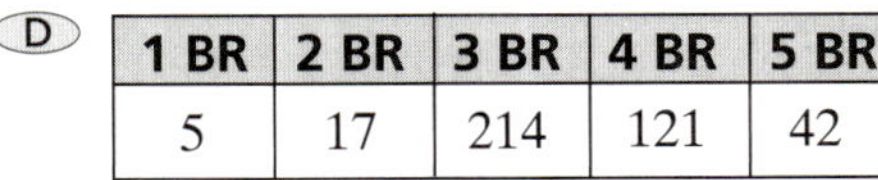

1 BR	2 BR	3 BR	4 BR	5 BR
5	17	214	121	42

7. The stem-and-leaf plot at the right shows the heights in inches of the girls on the swimming team. What is the mode?

7	1 4
6	0 0 3 5 5 5 8 8 9
5	4 6 6 8 8 9
4	8

7 | 1 = 71 in.

Ⓐ 60 Ⓑ 65 Ⓒ 68 Ⓓ 69

8. Abby had to drive to a meeting 210 miles away. She planned to leave at 1 P.M., drive at an average speed of 60 mi/h, and arrive 30 minutes prior to the meeting. She heard a weather report that suggested she should drive at a lower average speed of 50 mi/h. At the slower speed, how many minutes earlier should Abby leave in order to arrive 30 minutes prior to the meeting?

Ⓐ 72 Ⓑ 70 Ⓒ 42 Ⓓ 10

9. A portion of parallelogram $MQAC$ is shown on the grid at the right. Vertex Q is located so that $\overline{MQ}$ is parallel to $\overline{AC}$ and $\overline{MC}$ is parallel to $\overline{QA}$. What are the coordinates of vertex Q?

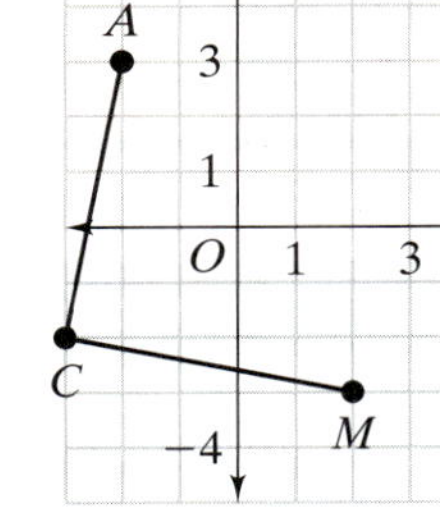

Ⓐ $(2, 1)$ Ⓒ $(3, 2)$

Ⓑ $(3, 1)$ Ⓓ $(4, 2)$

10. Three vertices of parallelogram $ABCD$ are $A(1, 7)$, $B(0, 0)$, and $C(7, -1)$. Where is the fourth vertex, D?

Ⓐ $(6, 6)$ Ⓒ $(7, 7)$

Ⓑ $(6, 8)$ Ⓓ $(8, 6)$

11. Line q is shown below. Which equation could represent a line parallel to line q?

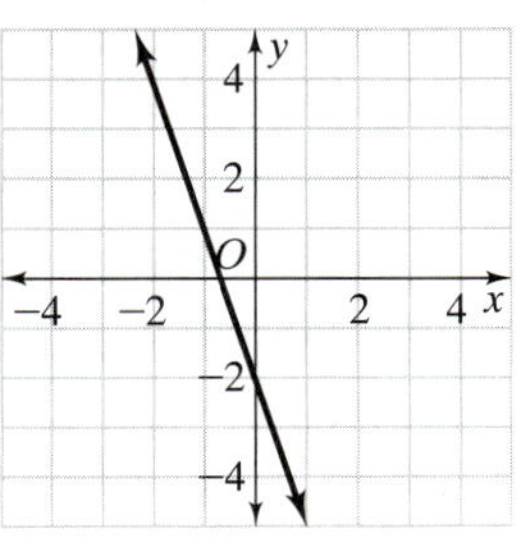

Ⓐ $y = -\frac{1}{3}x + 4$ Ⓒ $y = \frac{1}{3}x + 5$

Ⓑ $y = -3x + 3$ Ⓓ $y = 3x - 1$

12. What is the area of the square, in square units, in the figure below?

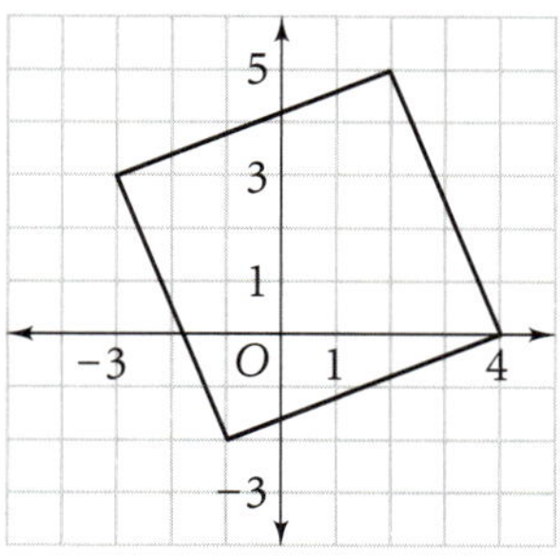

(A) 21 (B) 25 (C) 29 (D) 30

13. Erik typed a report from a handwritten copy. The table shows how many pages he had typed after certain numbers of minutes. At this rate, how many pages had Erik typed after 10 minutes?

Time (minutes)	Number of Pages
4	5
8	10
16	20

(A) 6.5 pages
(B) 8 pages
(C) 12.5 pages
(D) 15 pages

14. At a constant speed, the distance d a bicycle travels is directly proportional to the time t it travels. Ira rode his bicycle 9 miles in 48 minutes. At this rate, in how many minutes would he ride $\frac{2}{3}$ of that distance?

(A) 16 (B) 32 (C) 36 (D) 72

15. A car consumes 8 gallons of gasoline while traveling 224 miles. At that rate, how many gallons of gasoline will be needed for a 56-mile trip?

(A) 32 (B) 16 (C) 2 (D) 0.5

16. Lucia is making a triangular garden in the corner of her fenced rectangular backyard. She has 25 feet of edging to go along the unfenced edge of the garden. One of the fenced sides of the garden is 15 feet long. To the nearest tenth, what is the length in feet of the other fenced side of the garden?

(A) 10 (B) 20 (C) 29.2 (D) 400

17. Mica and Joy are standing at corner A of the rectangular field shown below.

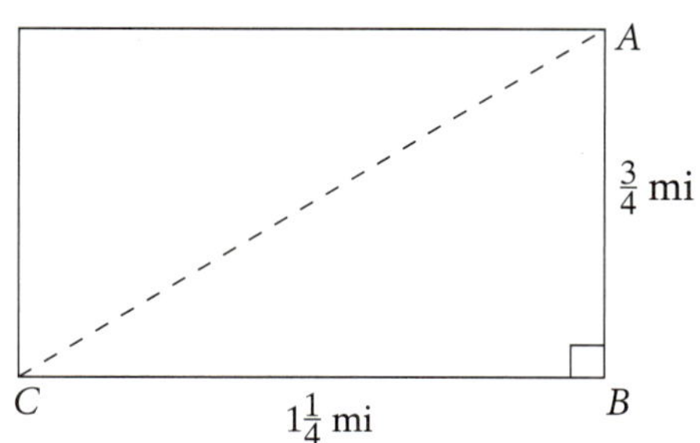

Mica walks diagonally across the field, from corner A to corner C. Joy walks from corner A to corner B, and then to corner C. To the nearest hundredth of a mile, how much farther did Joy walk than Mica?

(A) 0.54 (B) 1 (C) 1.46 (D) 2

CAHSEE Reference Guide

California Content Standard	Item Number(s)
7NS 1.1	1
6SDAP 3.1	2
Alg1 9.0	3, 4, 5
7SDAP 1.1	6, 7
7AF 4.2	8, 13, 14, 15
7MG 3.2	9, 10, 12
Alg1 8.0	11
7MG 3.3	16, 17

For additional review and practice, see the *California Standards Review and Practice Workbook* or go online to use the

Visit: PHSchool.com, **Web Code:** bcq-9045

Area

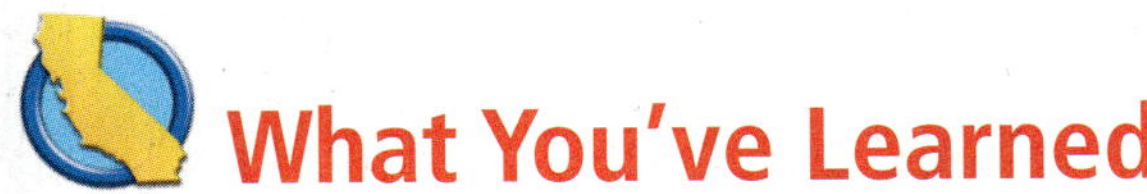

What You've Learned

California Content Standards

4.0 Students prove basic theorems involving similarity.

12.0 Students find and use measures of sides and of interior angles of polygons to classify figures and solve problems.

19.0 Students use trigonometric functions to solve for an unknown length of a side of a right triangle.

Squaring Numbers and Finding Square Roots (Skills Handbook page 753)

Simplify.

1. 3^2 **2.** 8^2 **3.** 12^2 **4.** 15^2

5. $\sqrt{16}$ **6.** $\sqrt{64}$ **7.** $\sqrt{100}$ **8.** $\sqrt{169}$

Solve each quadratic equation. Round to the nearest tenth or whole number.

9. $x^2 = 36$ **10.** $a^2 = 104$ **11.** $x^2 - 48 = 0$ **12.** $b^2 - 65 = 0$

Simplifying Radicals (Skills Handbook page 755)

Simplify. Leave your answer in simplest radical form.

13. $\sqrt{8}$ **14.** $\sqrt{27}$ **15.** $\sqrt{48}$ **16.** $6\sqrt{72}$

Probability (Skills Handbook page 762)

A jar contains 3 red balls and 2 green balls. You draw one ball at random. Determine the probability of selecting a ball of the given color.

17. red **18.** green **19.** blue **20.** red or green

Classifying Quadrilaterals (Lesson 6-1)

Classify each quadrilateral as specifically as possible.

21.

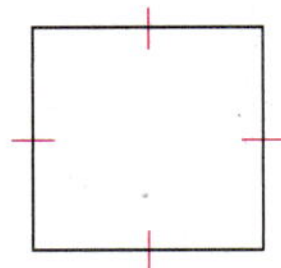

22.

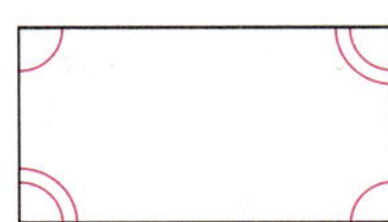

23.

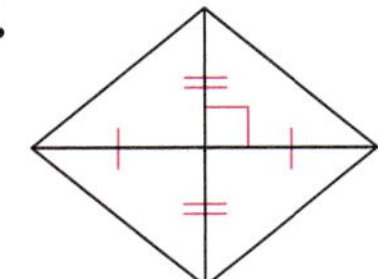

What You'll Learn Next

California Content Standards

7.0 Students prove and use theorems involving the properties of circles.

8.0 Students know, derive, and solve problems involving perimeter, circumference, and area of common geometric figures.

10.0 Students compute areas of polygons, including rectangles, scalene triangles, equilateral triangles, rhombi, parallelograms, and trapezoids.

11.0 Students determine how changes in dimensions affect the perimeter and area of common geometric figures.

CAHSEE 7NS 1.7, 7SDAP 1.1, 7AF 1.1, 7AF 1.5, 7AF 4.1

▲ Solar panels have been developed to maximize the surface area of the panel that is directed toward the sun. When the sun strikes the panel, light is converted into electricity that can power a spacecraft.

New Vocabulary

English and Spanish Audio Online

- **adjacent arcs** (p. 567)
- **apothem** (p. 546)
- **arc length** (p. 569)
- **central angle** (p. 566)
- **circumference** (p. 568)
- **concentric circles** (p. 568)
- **congruent arcs** (p. 569)
- **diameter** (p. 566)
- **geometric probability** (p. 582)
- **major arc** (p. 567)
- **minor arc** (p. 567)
- **radius** (p. 566)
- **radius of a regular polygon** (p. 546)
- **sector of a circle** (p. 576)
- **segment of a circle** (p. 577)
- **semicircle** (p. 567)

Transforming to Find Area

Hands-On

FOR USE WITH LESSON 10-1

GEOM 22.0 Students know the effect of rigid motions on figures in the coordinate plane. ***Master***

You can use transformations to find area formulas.

In these activities, you will cut polygons into pieces. You will use isometry transformations on the pieces to form different polygons. Because a preimage and its image are congruent and congruent figures have the same area (Postulate 1-9), you can find area formulas for the new polygons.

You will need several pieces of grid paper. Let the side of each grid square represent one unit of length.

1 ACTIVITY

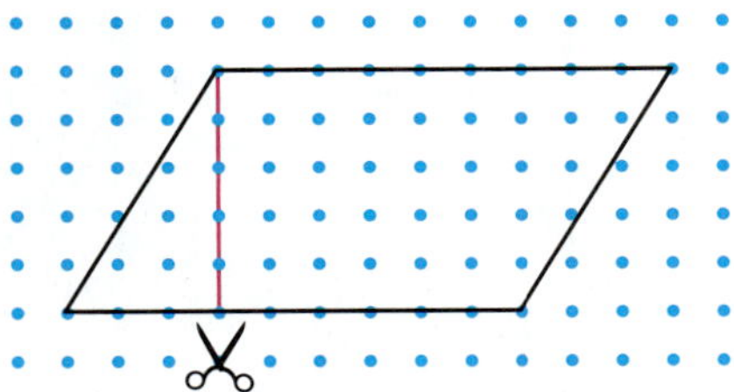

- For the parallelogram shown here, count and record the number of units in the base of the parallelogram. Do the same for the height.
- Copy the parallelogram onto grid paper.
- Cut out the parallelogram. Then cut it into two pieces as shown.
- Translate the triangle to the right through a distance equal to the base of the parallelogram.

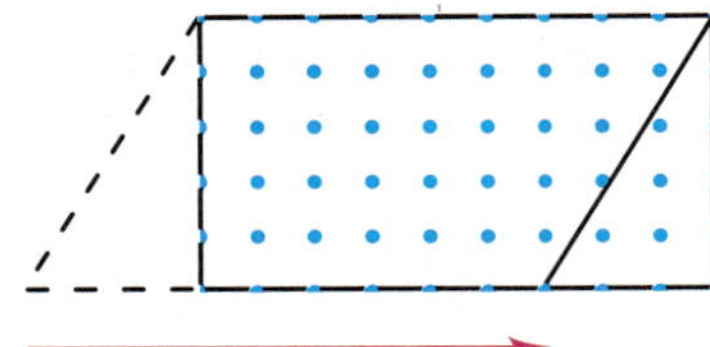

The translation image is a rectangle. The parallelogram and the rectangle have the same area.

1. For the rectangle, how many units are in the base? The height?
2. How do the base and the height of the rectangle compare to the base and height of the parallelogram?
3. Write the formula for the area of the rectangle. Explain why you can use this formula to find the area of a parallelogram.

2 ACTIVITY

Now you have the formula for the area of a parallelogram. You can use transformations and this formula to find an area formula for a triangle.

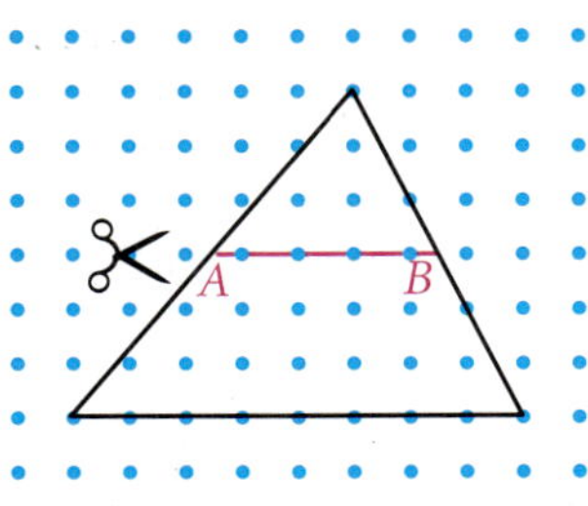

- Count and record the base and height of the triangle.
- Copy the triangle. Mark the midpoints A and B and draw the midsegment $\overline{AB}$.
- Cut out the triangle. Then cut it along $\overline{AB}$.
- Rotate the small triangle 180° about point B.

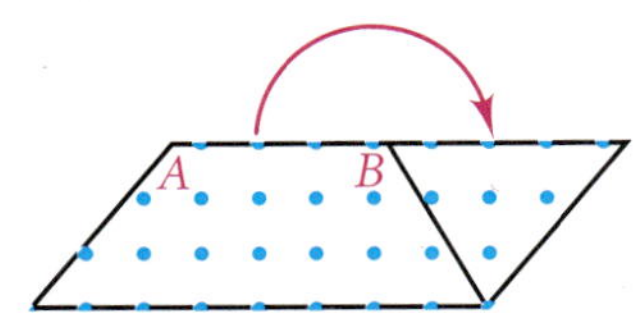

The bottom part of the original triangle and the image of the top part form a parallelogram.

4. For the parallelogram, how many units are in the base? The height?

5. How do the base and height of the parallelogram compare to the base and height of the original triangle? Write an expression for the height of the parallelogram in terms of the height, h, of the triangle.

6. Write your formula for the area of a parallelogram from Exercise 3. To find an area formula for a triangle, substitute into the formula the expression you wrote in Exercise 5 for the height of the parallelogram.

3 ACTIVITY

You can find an area formula for a trapezoid using a transformation similar to the one you used for a triangle.

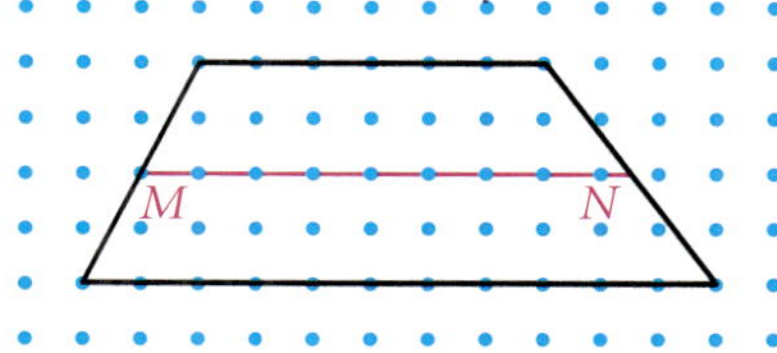

- Draw a trapezoid like the one shown here.
- Count and record its bases and height.
- Find the midpoints of the legs and the midsegment of the trapezoid.
- Cut out the trapezoid. Then cut it along the midsegment.

7. What transformation can you apply to the top piece of the trapezoid to form a parallelogram?

8. **a.** Write an expression for the base of the parallelogram in terms of the two bases, b_1 and b_2, of the trapezoid.
   b. Write an expression for the height of the parallelogram in terms of h, the height of the trapezoid.
   c. To find an area formula for a trapezoid, substitute your answers for parts (a) and (b) into your area formula for a parallelogram.

EXERCISES

9. In Activity 2, is there another rotation of the small triangle that will form a parallelogram? Explain.

10. Make another copy of the Activity 2 triangle. Mark the midpoints A and B. Find a rotation of the entire triangle so that the preimage and image together form a parallelogram. Show how to use it and your formula for the area of a parallelogram to find the formula for the area of a triangle.

11. For Exercise 10, there are in fact three rotations of the entire triangle that you can use. Find all three and describe them.

12. Here is the trapezoid with a different cut. What transformation can you apply to the top piece to form a triangle from the trapezoid? Use your formula for the area of a triangle to find a formula for the area of a trapezoid.

13. Show how you can find an area formula for a kite using a reflection. (*Hint:* Reflect half of the kite across its line of symmetry d_1 by folding the kite along d_1. How is the area of the triangle formed related to the area of the kite?)

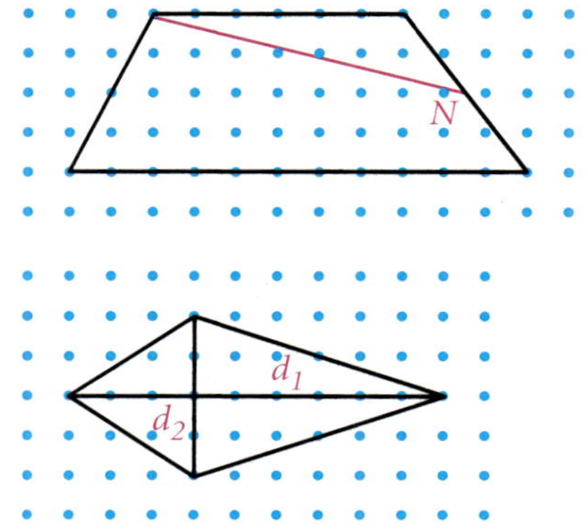

Areas of Parallelograms and Triangles

 **California Content Standards**

8.0 Students know, derive, and solve problems involving the area of common geometric figures. ***Introduce***
10.0 Students compute areas of polygons, including rectangles, scalene triangles, and parallelograms. ***Introduce***

What You'll Learn

- To find the area of a parallelogram
- To find the area of a triangle

. . . And Why

To find the force of wind against the side of a building, as in Example 4

Check Skills You'll Need

GO for Help Lesson 1-9

Find the area of each figure.

1. a square with 5-cm sides
2. a rectangle with base 4 in. and height 7 in.
3. a 4.6 m-by-2.5 m rectangle
4. a rectangle with length 3 ft and width $\frac{1}{2}$ ft

Each rectangle is divided into two congruent triangles. Find the area of each triangle.

5. 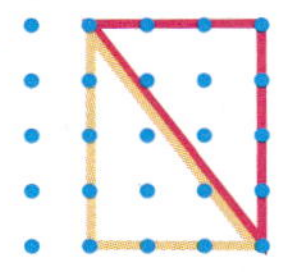
6. 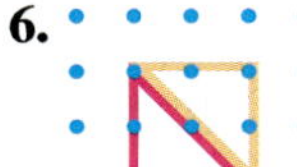
7. 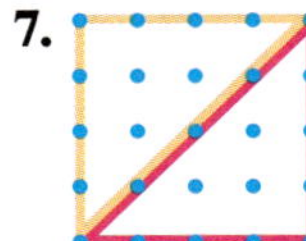

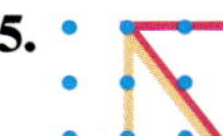

New Vocabulary

- base of a parallelogram
- altitude of a parallelogram
- height of a parallelogram
- base of a triangle
- height of a triangle

1 Area of a Parallelogram

The diagrams at the top of page 532 show that a parallelogram with the same base and height as a rectangle has the same area as the rectangle.

Key Concepts

Theorem 10-1 **Area of a Rectangle**

The area of a rectangle is the product of its base and height.

$$A = bh$$

Theorem 10-2 **Area of a Parallelogram**

The area of a parallelogram is the product of a base and the corresponding height.

$$A = bh$$

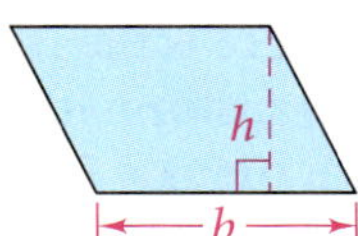

Vocabulary Tip

The term base is used to represent both a segment and its length.

A **base of a parallelogram** is any of its sides. The corresponding **altitude** is a segment perpendicular to the line containing that base, drawn from the side opposite the base. The **height** is the length of an altitude.

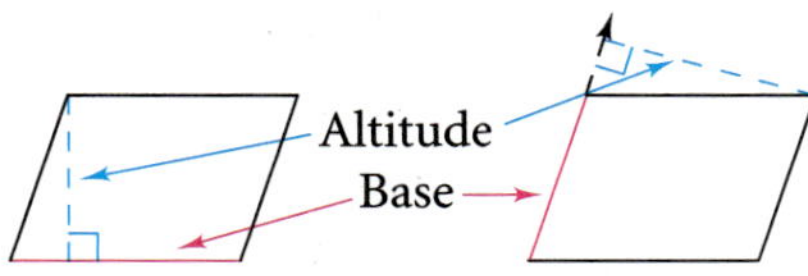

Video Tutor Help

Visit: PHSchool.com
Web Code: bce-0775

1 EXAMPLE Finding the Area of a Parallelogram

Find the area of each parallelogram.

a.

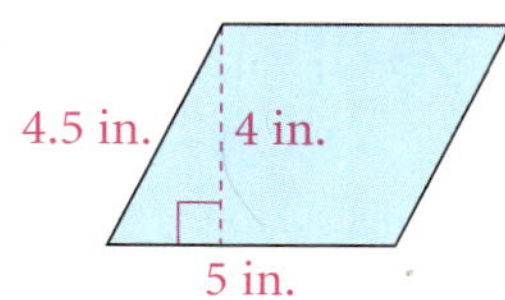

b.

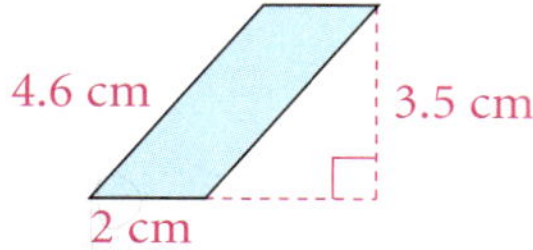

You are given each height. Choose the corresponding side to use as the base.

$A = bh$		$A = bh$
$= 5(4) = 20$	←**Substitute.**→	$= 2(3.5) = 7$
The area is 20 in.^{2}.		The area is 7 cm^2.

CA Standards Check **1** Find the area of a parallelogram with base 12 m and height 9 m.

2 EXAMPLE Finding a Missing Dimension

Online active math

For: Area Activity
Use: Interactive Textbook, 10-1

For ▱$ABCD$, find CF to the nearest tenth.

First, find the area of ▱$ABCD$. Then use the area formula a second time to find CF.

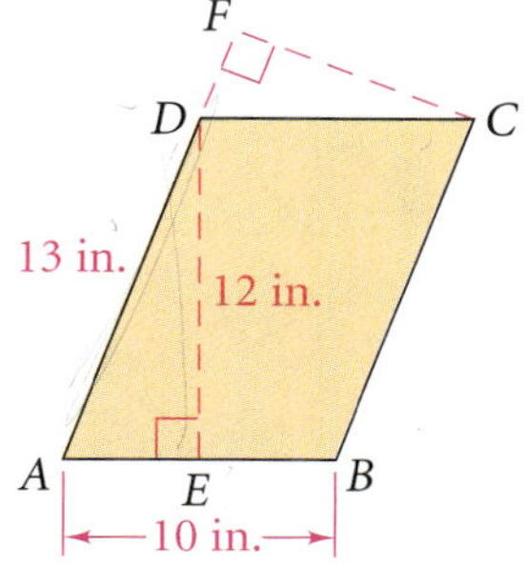

$A = bh$

$= 10(12) = 120$ **Use base AB and height DE.**

The area of ▱$ABCD$ is 120 in.^{2}.

$A = bh$

$120 = 13(CF)$ **Use base AD and height CF.**

$CF = \frac{120}{13} \approx 9.2$

CF is about 9.2 in.

CA Standards Check **2** A parallelogram has sides 15 cm and 18 cm. The height corresponding to a 15-cm base is 9 cm. Find the height corresponding to an 18-cm base.

2 Area of a Triangle

You can rotate a triangle about the midpoint of a side to form a parallelogram.

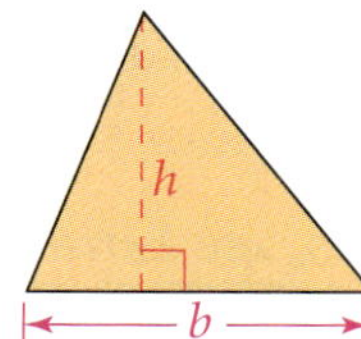

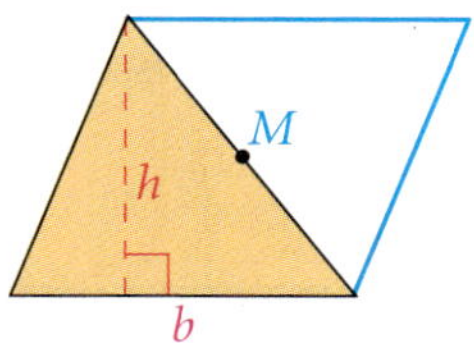

The area of the triangle is half the area of the parallelogram.

Key Concepts

Theorem 10-3 **Area of a Triangle**

The area of a triangle is half the product of a base and the corresponding height.

$$A = \frac{1}{2}bh$$

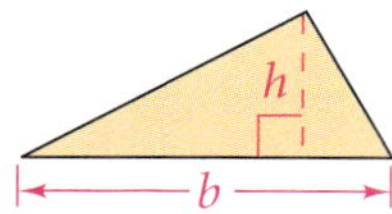

A **base of a triangle** is any of its sides. The corresponding **height** is the length of the altitude to the line containing that base.

3 EXAMPLE Finding the Area of a Triangle

Find the area of the triangle.

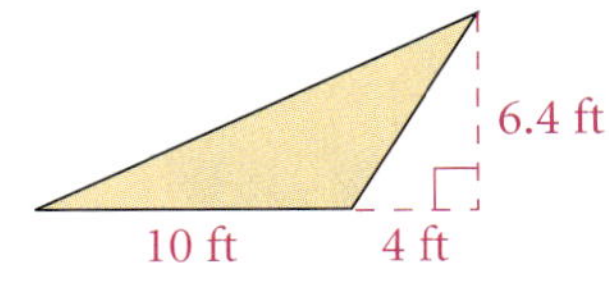

$A = \frac{1}{2}bh$

$= \frac{1}{2}(10)(6.4) = 32$ **Substitute and simplify.**

The area of the triangle is 32 ft^2.

3 Find the area of the triangle.

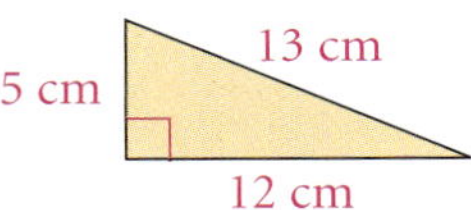

4 EXAMPLE Real-World Connection

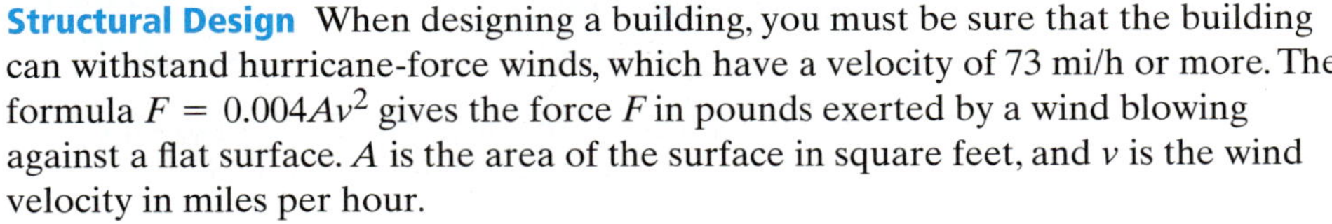

Structural Design When designing a building, you must be sure that the building can withstand hurricane-force winds, which have a velocity of 73 mi/h or more. The formula $F = 0.004Av^2$ gives the force F in pounds exerted by a wind blowing against a flat surface. A is the area of the surface in square feet, and v is the wind velocity in miles per hour.

Real-World Connection

In 1992 this building in Homestead, Florida, succumbed to the 145 mi/h winds of Hurricane Andrew.

How much force is exerted by a 73 mi/h wind blowing directly against the side of the building shown here?

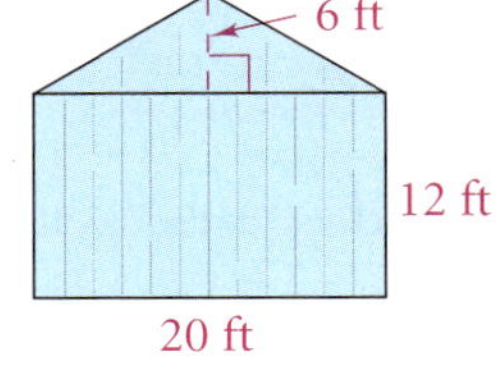

Find the area of the side of the building.

triangle area $= \frac{1}{2}bh = \frac{1}{2}(20)6 = 60$ ft^2

rectangle area $= bh = 20(12) = 240$ ft^2

area of the side $= 60 + 240 = 300$ ft^2

Use the area of the side of the building and the velocity of the wind to find the force.

$F = 0.004Av^2$ **Use the formula for force.**

$= 0.004(300)(73)^2$ **Substitute 300 for *A* and 73 for *v*.**

$= 6394.8$

The force is about 6400 lb, or 3.2 tons.

4 **Critical Thinking** Suppose the bases of the rectangle and triangle in the building above are doubled to 40 ft, but the height of each figure remains the same. How is the force of the wind against the side of the building affected?

EXERCISES

For more exercises, see *Extra Skill, Word Problem, and Proof Practice.*

Standards Practice **GEOM 8.0, 10.0**

A Practice by Example

Example 1 (page 535)

Find the area of each parallelogram.

1. 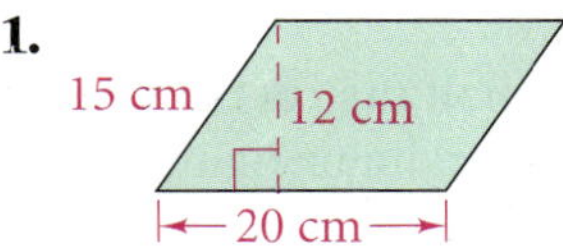

2. 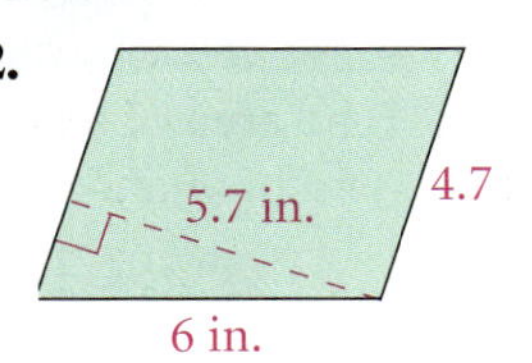

3. 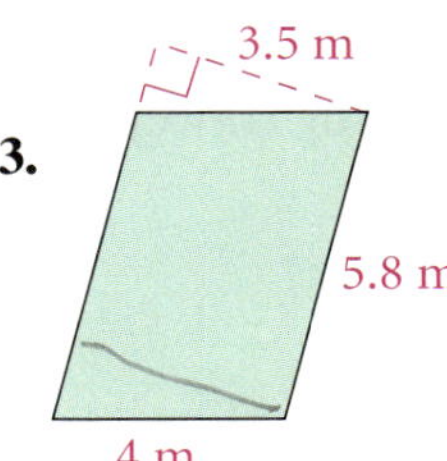

Example 2 (page 535)

Find the value of h for each parallelogram.

4.

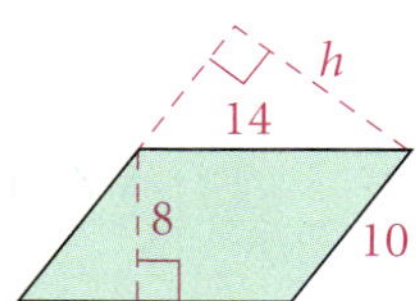

5.

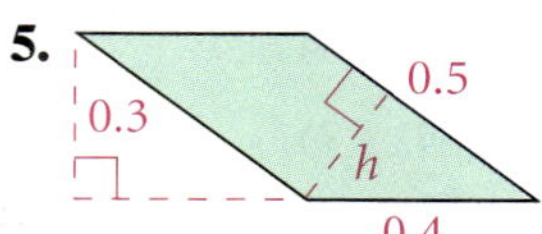

6.

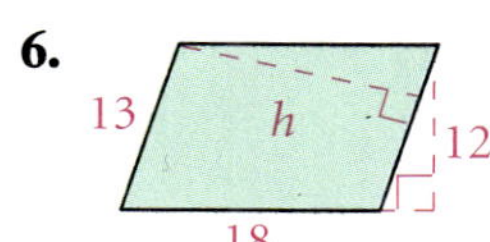

Example 3 (page 536)

Find the area of each triangle.

7.

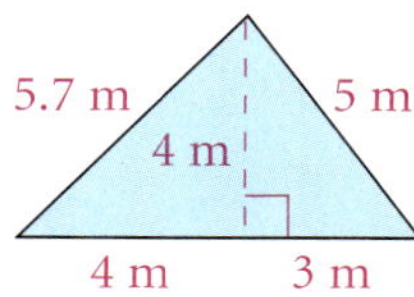

8. 4.5 yd, 6 yd, 7.5 yd

9.

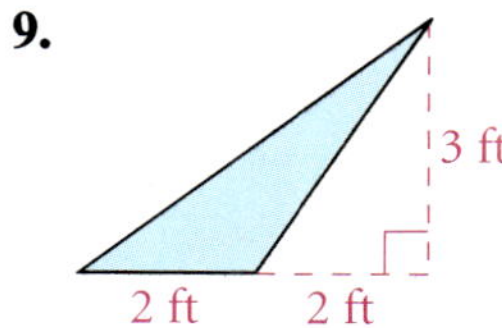

Example 4 (page 536)

10. Landscaping Taisha's Bakery has a plan for a 50 ft-by-31 ft parking lot. The four parking spaces are congruent parallelograms, the driving region is a rectangle, and the two unpaved areas for flowers are congruent triangles.

a. Find the area of the surface to be paved by adding the areas of the driving region and the four parking spaces.

b. Describe another method for finding the area of the surface to be paved.

c. Use your method from part (b) to find the area. Then compare answers from parts (a) and (b) to check your work.

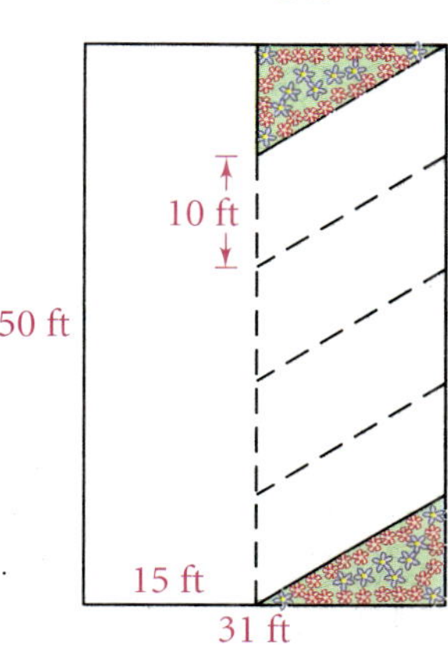

Apply Your Skills

11. The area of a parallelogram is 24 in.^{2} and the height is 6 in. Find the corresponding base.

12. Multiple Choice What is the area of the figure at the right?

Ⓐ 64 cm^2 Ⓒ 96 cm^2

Ⓑ 88 cm^2 Ⓓ 112 cm^2

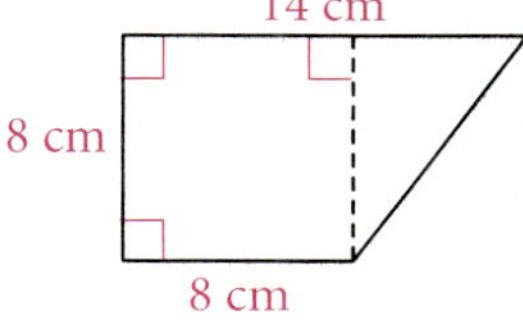

13. An isosceles right triangle has area of 98 cm^2. Find the length of each leg.

14. Algebra In a triangle, a base and a corresponding height are in the ratio 3 : 2. The area is 108 in.^{2}. Find the base and the corresponding height.

15. Technology Ki used geometry software to create the figure at the right. She constructed $\overleftrightarrow{AB}$ and a point C not on $\overleftrightarrow{AB}$. Then she constructed line k parallel to $\overleftrightarrow{AB}$ through point C. Next, Ki constructed point D on line k as well as $\overline{AD}$ and $\overline{BD}$. She dragged point D along line k to manipulate $\triangle ABD$. How does the area of $\triangle ABD$ change? Explain.

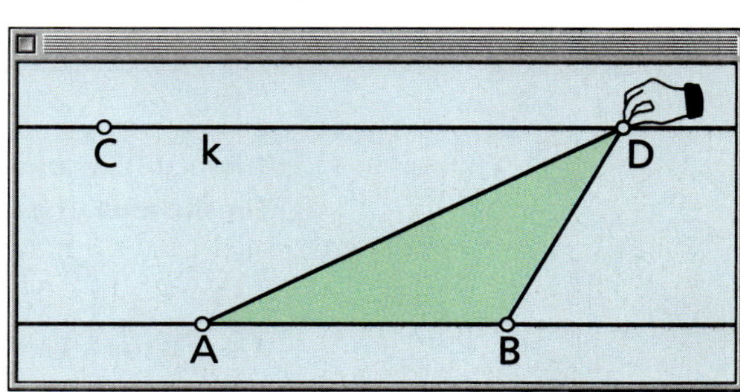

Homework Video Tutor

Visit: PHSchool.com
Web Code: bce-1001

16. Open-Ended Using graph paper, draw an acute triangle, an obtuse triangle, and a right triangle, each with area 12 units^2.

Find the area of each figure.

17. $\square ABJF$
18. $\triangle BDJ$
19. $\triangle DKJ$
20. $\square BDKJ$
21. $\square ADKF$
22. $\triangle BCJ$
23. $ADJF$

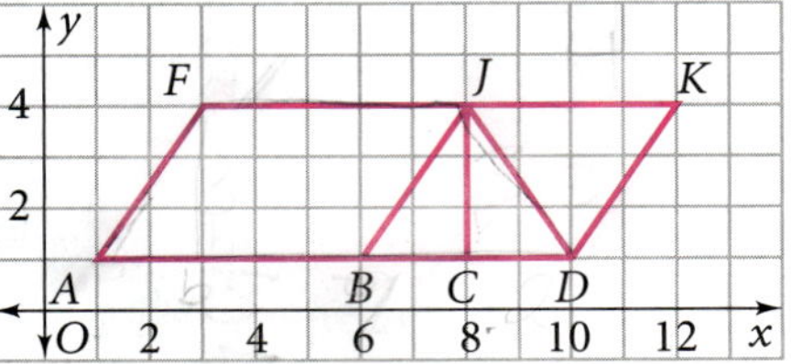

In Exercises 24–27, (a) graph the lines and (b) find the area of the triangle enclosed by the lines.

24. $y = x, x = 0, y = 7$

25. $y = x + 2, y = 2, x = 6$

26. $y = -\frac{1}{2}x + 3, y = 0, x = -2$

27. $y = \frac{3}{4}x - 2, y = -2, x = 4$

Exercise 28

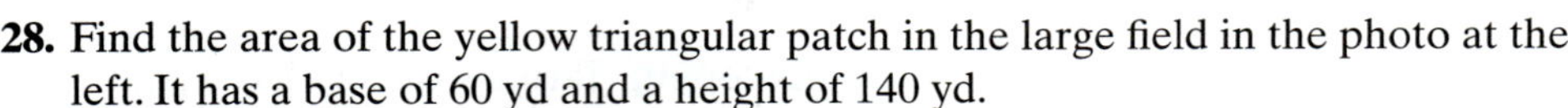

28. Find the area of the yellow triangular patch in the large field in the photo at the left. It has a base of 60 yd and a height of 140 yd.

29. **Probability** Ann drew these three figures on a grid. A fly lands at random at a point on the grid.

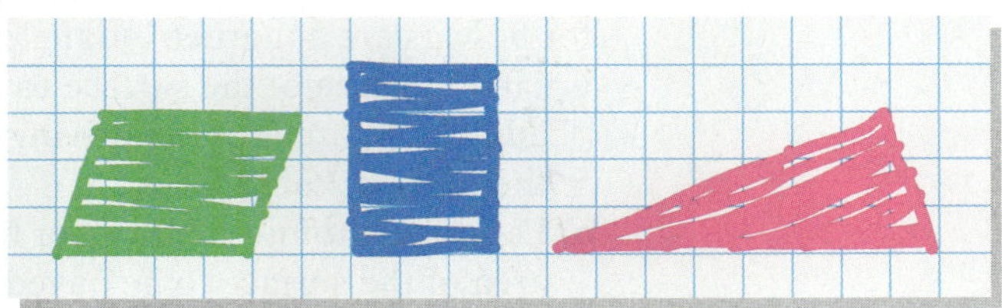

a. **Writing** Is the fly more likely to land on one of the figures or on the blank grid? Explain.

b. Suppose you know the fly lands on one of the figures. Is the fly more likely to land on one figure than on another? Explain.

Coordinate Geometry **Find the area of a polygon with the given vertices.**

30. $A(3, 9), B(8, 9), C(2, -3), D(-3, -3)$

31. $E(1, 1), F(4, 5), G(11, 5), H(8, 1)$

32. $D(0, 0), E(2, 4), F(6, 4), G(6, 0)$

33. $K(-7, -2), L(-7, 6), M(1, 6), N(7, -2)$

Find the area of each figure.

34.

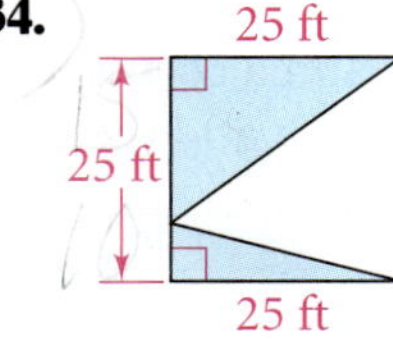

35.

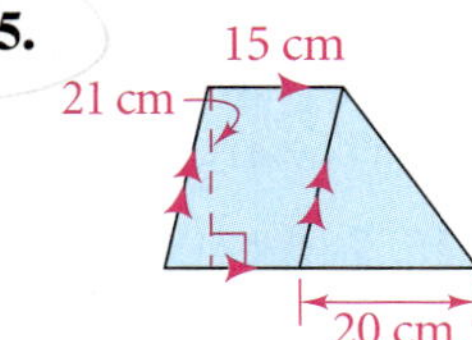

36.

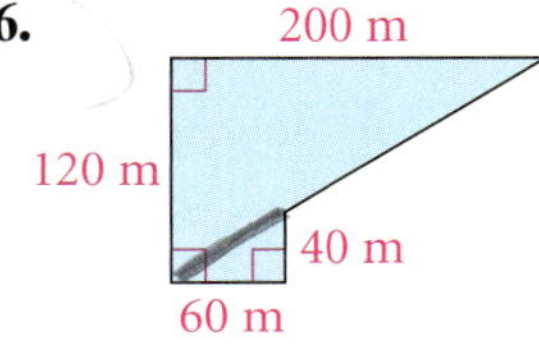

Challenge

History **The ancient Greek mathematician Heron is most famous for this formula for the area of a triangle in terms of the lengths of its sides *a*, *b*, and *c*.**

$$A = \sqrt{s(s - a)(s - b)(s - c)}, \text{ where } s = \frac{1}{2}(a + b + c)$$

Use Heron's Formula and a calculator to find the area of each triangle. Round your answer to the nearest whole number.

37. $a = 8$ in., $b = 9$ in., $c = 10$ in.

38. $a = 15$ m, $b = 17$ m, $c = 21$ m

39. a. Use Heron's Formula to find the area of this triangle.

b. Verify your answer to part (a) by using the formula $A = \frac{1}{2}bh$.

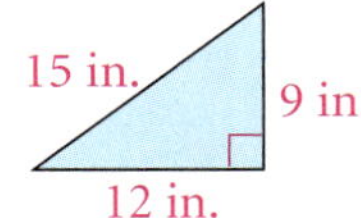

Multiple Choice Practice

For California Standards and CAHSEE Tutorials, visit PHSchool.com. Web Code: bcq-9045

GEOM 10.0

40. $\overline{NP}$ is the altitude to $\overline{ML}$ in the figure below. What is the area of $\triangle MNL$ in square centimeters?

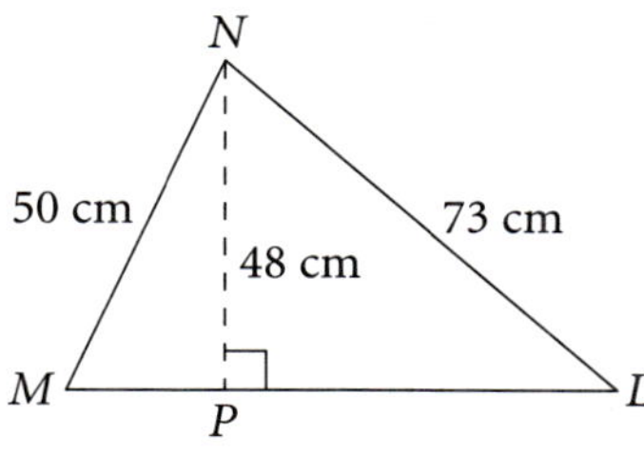

Ⓐ 3650 Ⓑ 1752 Ⓒ 1656 Ⓓ 1320

CAHSEE 7AF 1.1

41. Felicia borrowed money from her mother to buy a used car. Felicia agreed to pay her mother back by making equal-sized monthly payments. After 13 payments, her balance was \$6325. After 15 payments, it was \$5775. Which function can Felicia use to describe the relationship between b, the remaining balance, and p, the number of payments she has made?

Ⓐ $5775 + 275p = 6325$

Ⓑ $275p - 9900 = b$

Ⓒ $9900p + 275 = b$

Ⓓ $9900 - 275p = b$

Mixed Review

Lesson 9-7

List the symmetries in each tessellation.

42.

43.

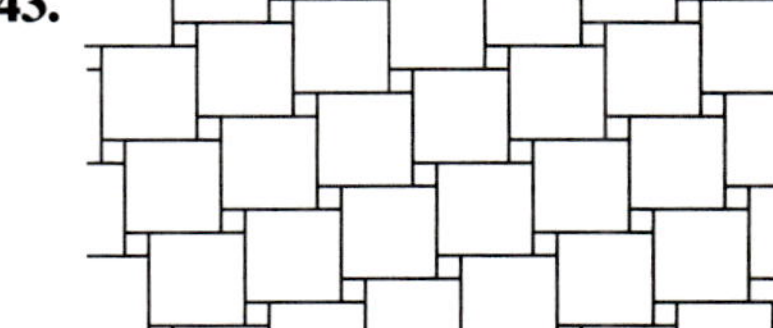

Lesson 4-5

The base of the isosceles triangle is a side of a regular pentagon *PENTA*. Find the measure of each angle.

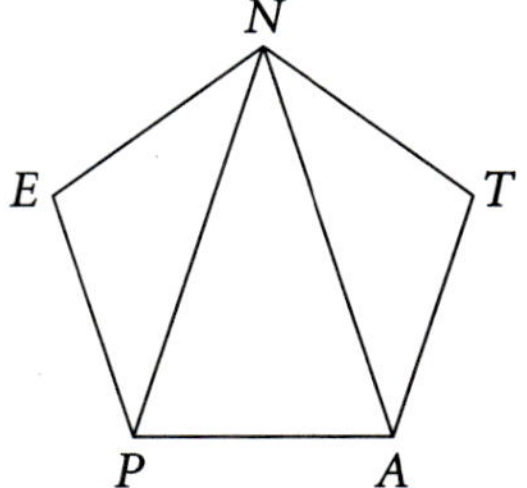

44. $\angle APE$ **45.** $\angle APN$

46. $\angle PAN$ **47.** $\angle PNA$

48. $\angle EPN$ **49.** $\angle ANT$

Lesson 3-8

Use a compass and straightedge for the following constructions.

50. Draw a segment and label it $\overline{AB}$. Construct $\overleftrightarrow{AD}$ so that $\overleftrightarrow{AD} \perp \overline{AB}$ at point A.

51. Draw a segment. Label it $\overline{EF}$. Construct a line $\overleftrightarrow{GH}$ so that $\overleftrightarrow{GH} \parallel \overline{EF}$.

52. Draw a segment and label it $\overline{KL}$. Draw a point X not on $\overleftrightarrow{KL}$. Construct a perpendicular from point X to $\overline{KL}$ (or to $\overleftrightarrow{KL}$).

Areas of Trapezoids, Rhombuses, and Kites

California Content Standards

GEOM 8.0 Students know, derive, and solve problems involving the area of common geometric figures. ***Develop***

GEOM 10.0 Students compute areas of polygons, including rhombi and trapezoids. ***Develop***

What You'll Learn

- To find the area of a trapezoid
- To find the area of a rhombus or a kite

. . . And Why

To use a map and the trapezoid area formula to approximate the area of Arkansas, as in Example 1

Check Skills You'll Need

GO for Help Lesson 10-1

Write the formula for the area of each type of figure.

1. a rectangle

2. a triangle

Find the area of each trapezoid by using the formulas for area of a rectangle and area of a triangle.

3.
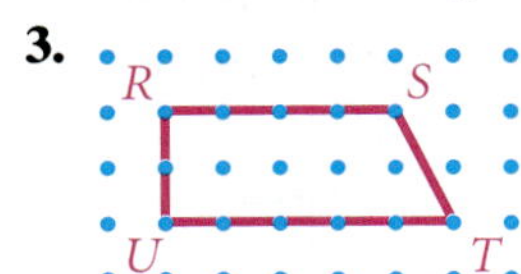

4.
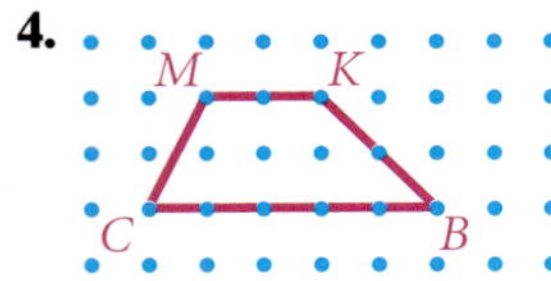

5.
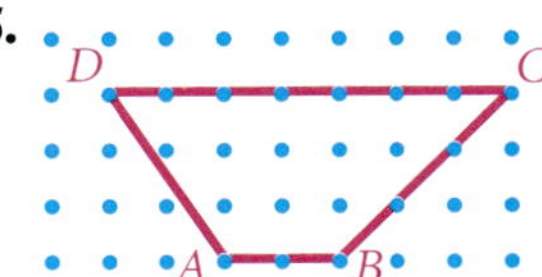

New Vocabulary • height of a trapezoid

1 Area of a Trapezoid

Vocabulary Tip

The term base can refer to either a line segment or its length.

In Lesson 6-5, you learned that the bases of a trapezoid are the parallel sides and the legs are the nonparallel sides. The **height of a trapezoid** is the perpendicular distance h between the bases.

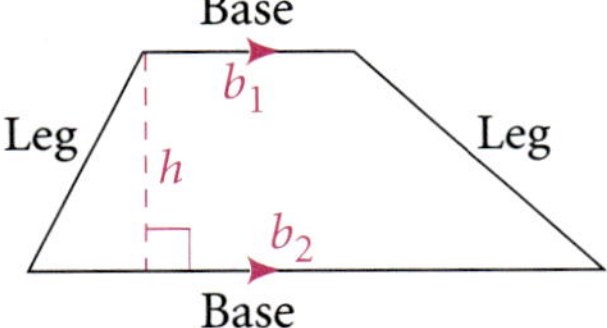

Key Concepts

Theorem 10-4 Area of a Trapezoid

The area of a trapezoid is half the product of the height and the sum of the bases.

$$A = \frac{1}{2}h(b_1 + b_2)$$

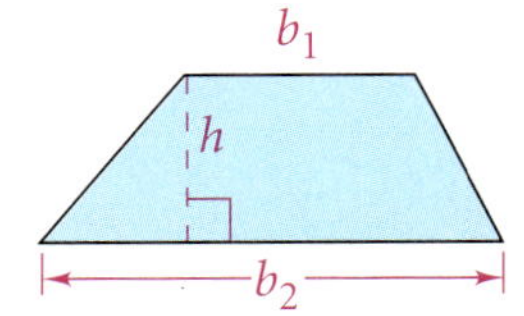

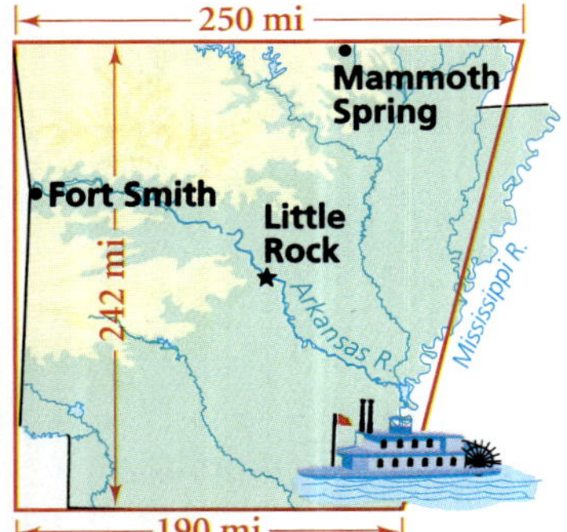

1 EXAMPLE **Real-World Connection**

Geography Approximate the area of Arkansas by finding the area of the trapezoid shown.

$A = \frac{1}{2}h(b_1 + b_2)$ **Use the formula for area of a trapezoid.**

$= \frac{1}{2}(242)(190 + 250)$ **Substitute 242 for h, 190 for b_1, and 250 for b_2.**

$= 53{,}240$ **Simplify.**

The area of Arkansas is about 53,240 mi^2.

1 Find the area of a trapezoid with height 7 cm and bases 12 cm and 15 cm.

Properties of special right triangles can help you find the area of a trapezoid.

2 EXAMPLE Finding Area Using a Right Triangle

Multiple Choice What is the area of trapezoid $PQRS$?

Ⓐ $6\ m^2$ Ⓑ $10\sqrt{3}\ m^2$ Ⓒ $12\sqrt{3}\ m^2$ Ⓓ $35\ m^2$

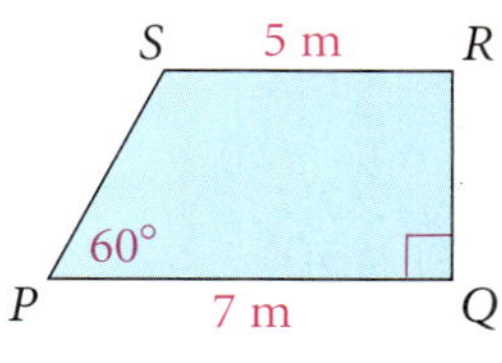

Test-Taking Tip

You can sometimes find the area of a figure by adding the areas of two simpler figures, such as the rectangle and triangle in Example 1.

You can draw an altitude that divides the trapezoid into a rectangle and a 30°-60°-90° triangle. Since the opposite sides of a rectangle are congruent, the longer base of the trapezoid is divided into segments of lengths 2 m and 5 m.

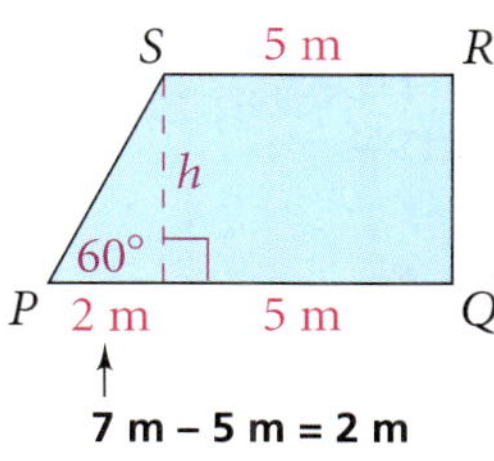

Find h.

$h = 2\sqrt{3}$ **longer leg = shorter leg · $\sqrt{3}$**

$A = \frac{1}{2}h(b_1 + b_2)$ **Use the trapezoid area formula.**

$= \frac{1}{2}(2\sqrt{3})(7 + 5)$ **Substitute.**

$= 12\sqrt{3}$ **Simplify.**

The area of trapezoid $PQRS$ is $12\sqrt{3}\ m^2$. The answer is C.

CA Standards Check **2** In Example 2, suppose h is made smaller so that $m\angle P = 45$ while bases and angles R and Q are unchanged. Find the area of trapezoid $PQRS$.

2 Finding Areas of Rhombuses and Kites

Rhombuses and kites have perpendicular diagonals. This property allows you to find areas using the following theorem.

Key Concepts

Theorem 10-5 **Area of a Rhombus or a Kite**

The area of a rhombus or a kite is half the product of the lengths of its diagonals.

$$A = \frac{1}{2}d_1d_2$$

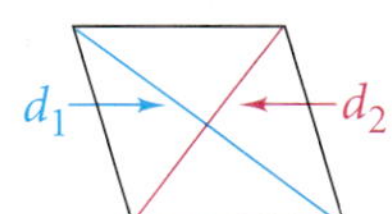

3 EXAMPLE Finding the Area of a Kite

Find the area of kite $KLMN$.

For the two diagonals, $KM = 2 + 5 = 7$ m and $LN = 3 + 3 = 6$ m.

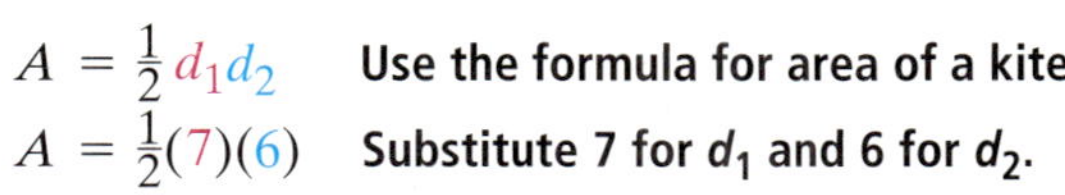

$A = \frac{1}{2}d_1d_2$ **Use the formula for area of a kite.**

$A = \frac{1}{2}(7)(6)$ **Substitute 7 for d_1 and 6 for d_2.**

$A = 21$ **Simplify.**

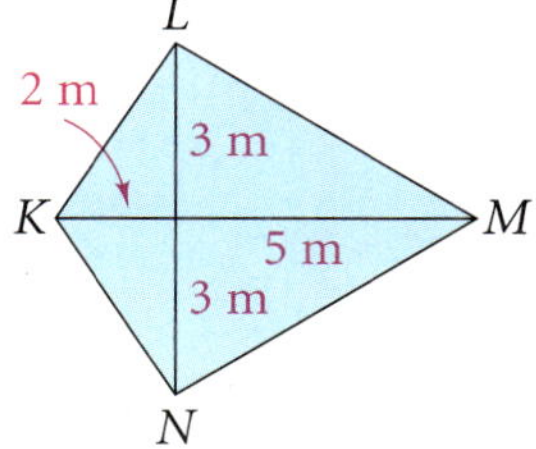

The area of kite $KLMN$ is $21\ m^2$.

CA Standards Check **3** Find the area of a kite with diagonals that are 12 in. and 9 in. long.

The fact that the diagonals of a rhombus bisect each other can help you find the area.

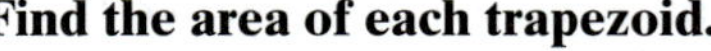

4 EXAMPLE Finding the Area of a Rhombus

Find the area of rhombus $ABCD$.

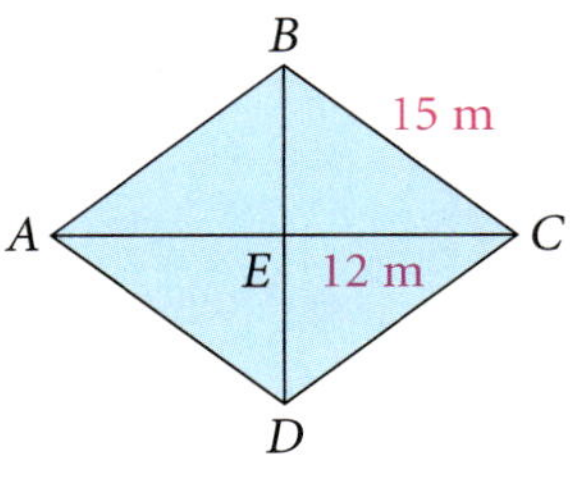

$\triangle BEC$ is a right triangle. Using a Pythagorean triple, $BE = 9$. Since the diagonals of a rhombus bisect each other, $AC = 24$ and $BD = 18$.

$A = \frac{1}{2}d_1d_2$ **Use the formula for area of a rhombus.**

$A = \frac{1}{2}(24)(18)$ **Substitute 24 for d_1 and 18 for d_2.**

$A = 216$ **Simplify.**

The area is 216 m^2.

CA Standards Check **4 Critical Thinking** In Example 4, explain how you can use a Pythagorean triple to conclude that $BE = 9$.

EXERCISES Standards Practice

For more exercises, see *Extra Skill, Word Problem, and Proof Practice.*

GEOM 8.0, 10.0

A Practice by Example

Example 1 (page 540)

Find the area of each trapezoid.

1.

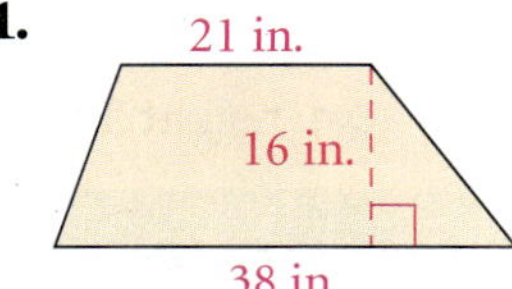

2.

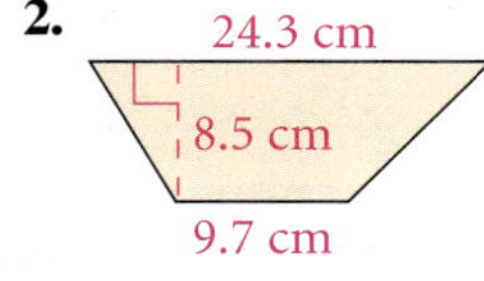

3.

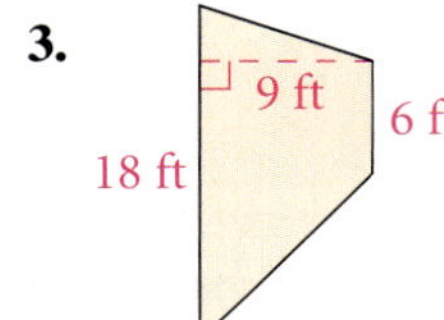

Video Tutor Help

Visit: PHSchool.com
Web Code: bce-0775

4. Geography Approximate the area of Nevada by finding the area of the trapezoid shown.

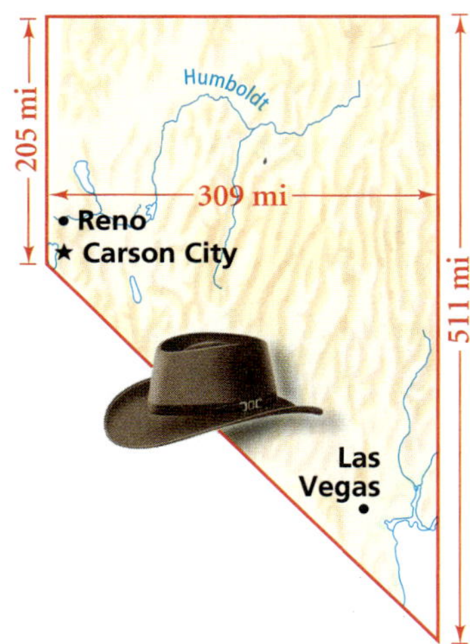

5. Find the area of a trapezoid with bases 12 cm and 18 cm and height 10 cm.

6. Find the area of a trapezoid with bases 2 ft and 3 ft and height $\frac{1}{3}$ ft.

7. Geography The border of Tennessee resembles a trapezoid with bases 342 mi and 438 mi, and height 111 mi. Approximate the area of Tennessee by finding the area of this trapezoid.

Example 2 (page 541)

Find the area of each trapezoid. If your answer is not an integer, leave it in simplest radical form.

8.

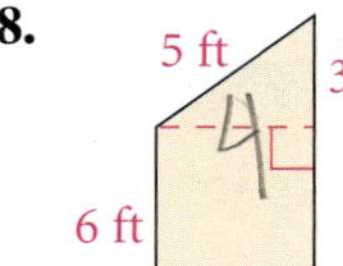

9.

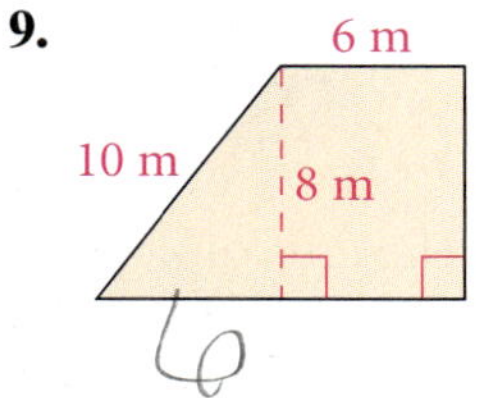

10.

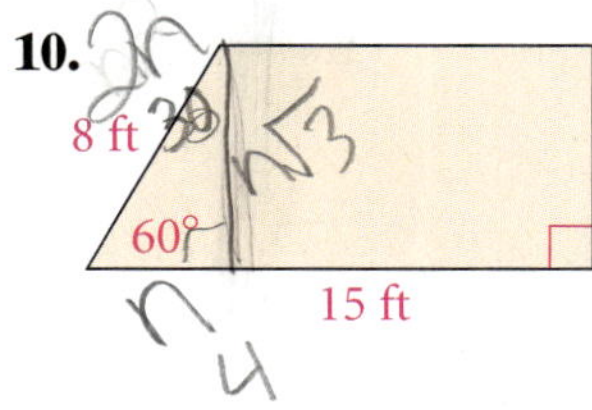

Example 3
(page 541)

Find the area of each kite.

11. 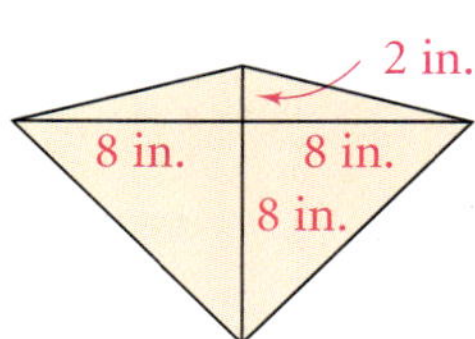

12. 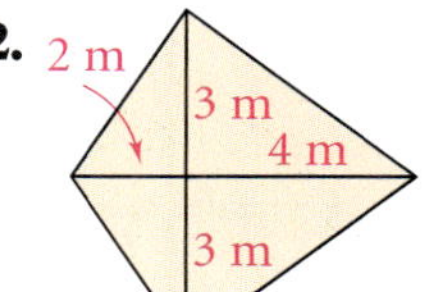

13. 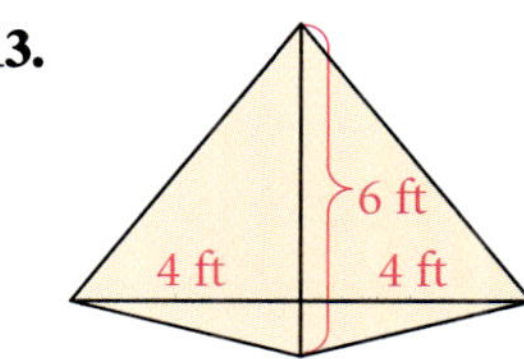

Example 4
(page 542)

Find the area of each rhombus.

14. 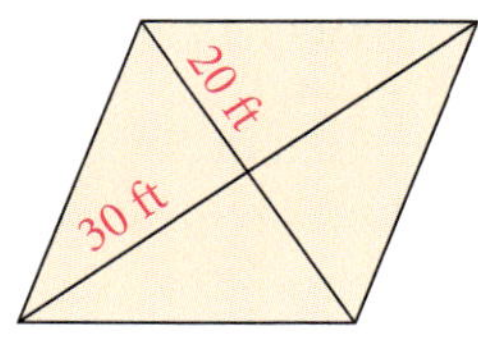

15. 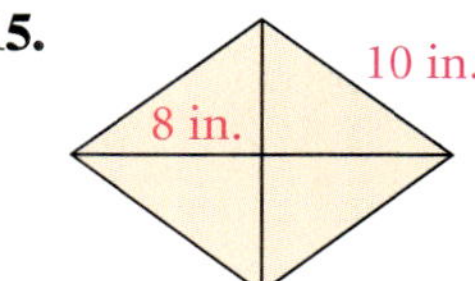

16. 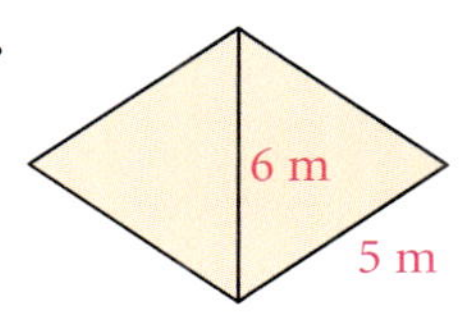

Apply Your Skills

17. The end of the rain gutter has the shape of a trapezoid with the measurements shown. Find the area of this end.

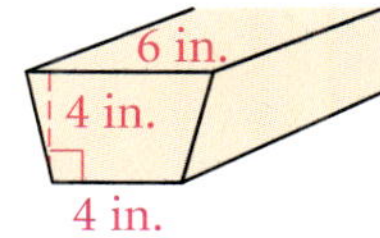

18. A trapezoid has two right angles, 12-m and 18-m bases, and 8-m height.
a. Sketch the trapezoid. **b.** Find the perimeter. **c.** Find the area.

19. **Open-Ended** Draw a kite. Measure the lengths of its diagonals. Find its area.

Gold Bars **Find the area of each trapezoidal face of the gold bars.**

20. End face: bases 4 cm and 2 cm, height 3 cm.

21. Side face: bases 8 cm and 5 cm, height 3 cm.

Real-World Connection

On each gold bar the four trapezoidal faces tip inwards. This simplifies the molding process.

Find the area of each trapezoid to the nearest tenth.

22. 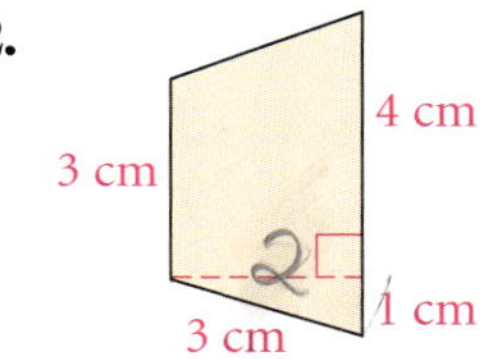

23. 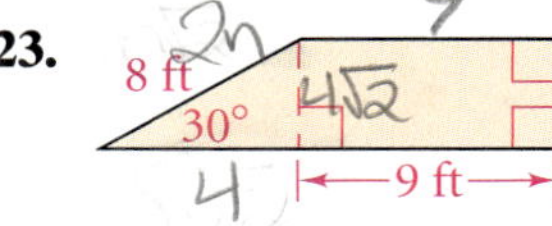

24. 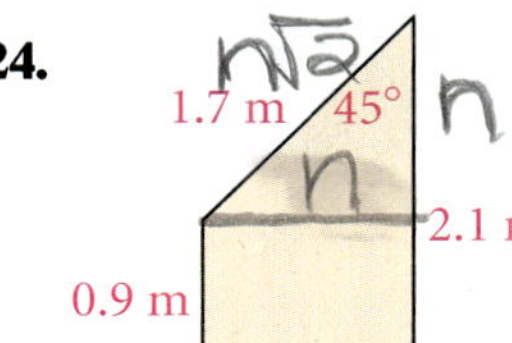

Coordinate Geometry **In Exercises 25–27, find the area of quadrilateral *QRST*.**

25. 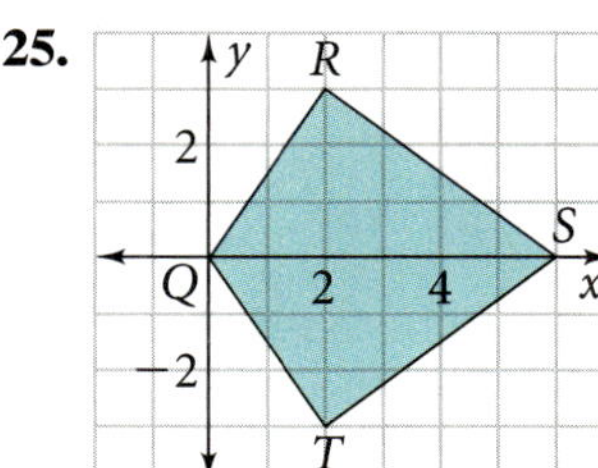

26. 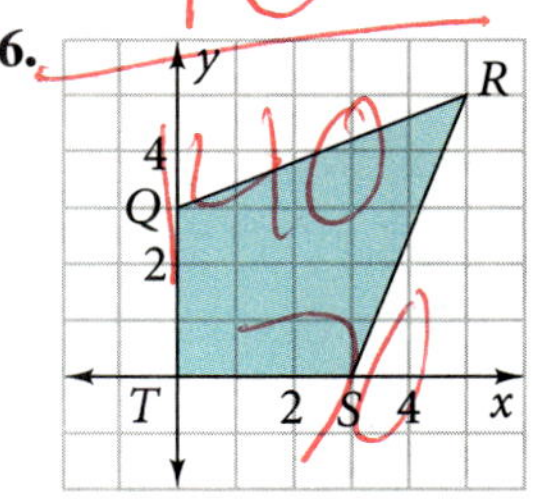

27. 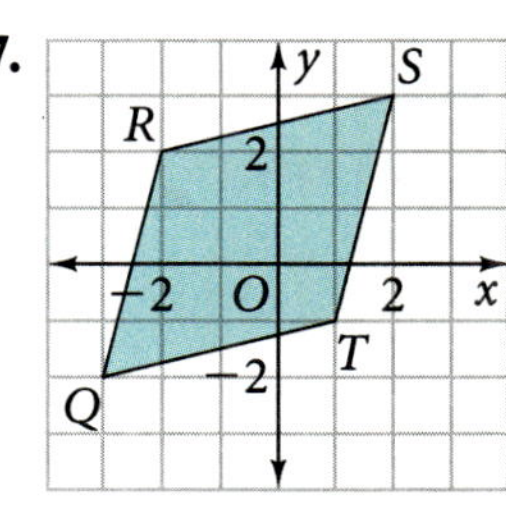

28. **Multiple Choice** What is the area of the kite at the right?

Ⓐ $90\ m^2$ Ⓒ $135\ m^2$
Ⓑ $108\ m^2$ Ⓓ $216\ m^2$

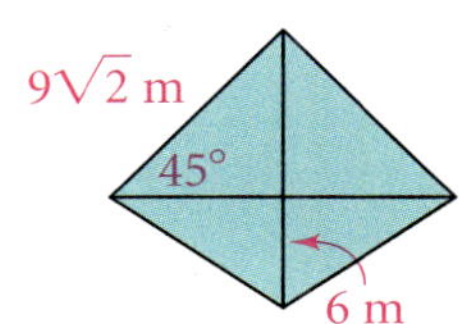

29. **a. Coordinate Geometry** Graph the lines $x = 0$, $x = 6$, $y = 0$, and $y = x + 4$.
b. What type of quadrilateral do the lines form?
c. Find the area of the quadrilateral.

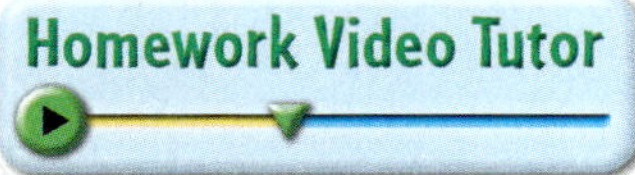

Visit: PHSchool.com
Web Code: bce-1002

Find the area of each rhombus. Leave your answer in simplest radical form.

30.

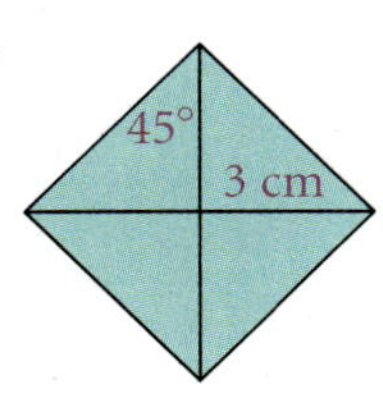

31.

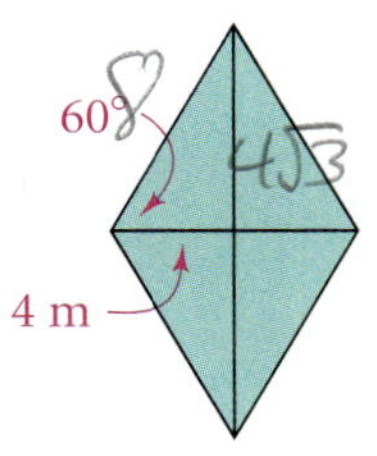

32.

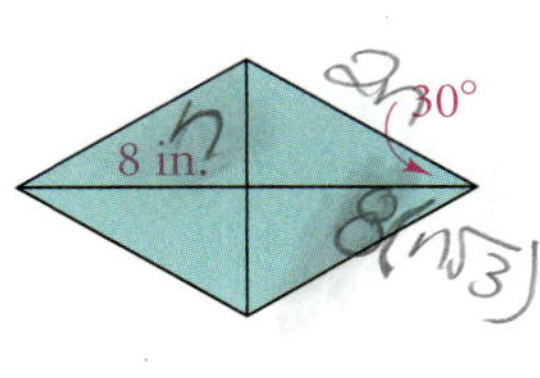

33. Draw a trapezoid. Label its bases and height b_1, b_2, and h, respectively. Then draw a diagonal of the trapezoid.

a. Write equations for the area of each of the two triangles formed.

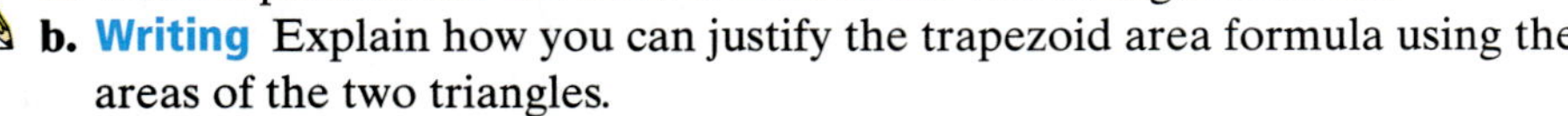

b. Writing Explain how you can justify the trapezoid area formula using the areas of the two triangles.

34. Visualization The kite has diagonals d_1 and d_2 congruent to the sides of the rectangle. Explain why the area of the kite is $\frac{1}{2}d_1d_2$.

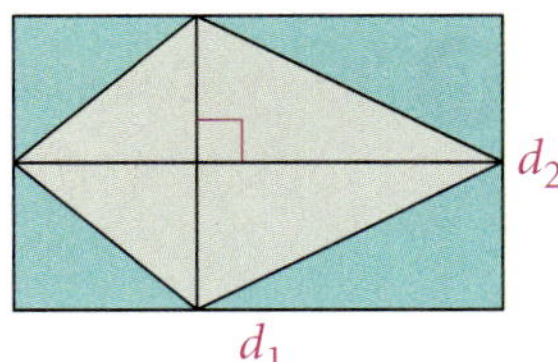

Challenge

35. Algebra One base of a trapezoid is twice the other. The height is the average of the two bases. The area is 324 cm^2. Find the height and the bases. (*Hint:* Let the smaller base be $2x$.)

36. Gravity Sports Ty wants to paint one end of his homemade skateboarding ramp. The ramp is 4 m wide. Its surface is modeled by the equation $y = 0.25x^2$. Use the trapezoids and triangles shown to estimate the area to be painted.

$y = 0.25x^2$

Real-World Connection

A skateboarder's speed is affected by the curve of the ramp.

37. In trapezoid $ABCD$, $\overline{AB} \parallel \overline{DC}$. Find the area of $ABCD$.

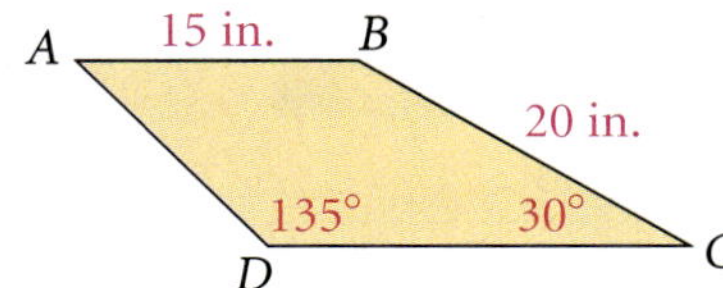

Multiple Choice Practice

For California Standards and CAHSEE Tutorials, visit PHSchool.com. Web Code: bcq-9045

GEOM 8.0

38. Compute the area of the rhombus below.

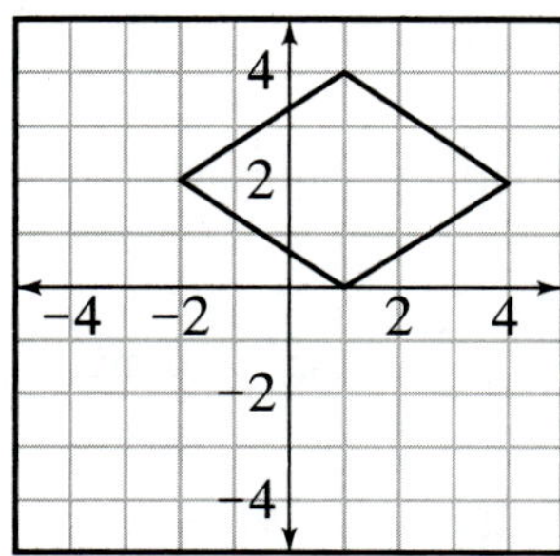

Ⓐ 9 units^2 Ⓑ 10 units^2 Ⓒ 12 units^2 Ⓓ 13 units^2

CAHSEE 7NS 1.7

39. An artist purchases red paint for \$24.50, blue paint for \$30, and yellow paint for \$31.10. The artist must also pay a 5% sales tax. Is \$90 enough to cover the total cost? Choose the best answer.

A. No, the artist needs \$0.60 more.

B. Yes, the artist has \$0.12 extra.

C. Yes, the artist has the exact amount needed.

D. Yes, the artist has \$4.40 extra.

Mixed Review

Lesson 10-1

40. Find the area of an isosceles right triangle that has one leg of length 12 cm.

41. An isosceles right triangle has area of 112.5 ft^2. Find the length of each leg.

Lesson 5-3

Fill in the blank with *always*, *sometimes*, or *never* to form a true statement.

42. The incenter of a triangle _?_ lies inside the triangle.

43. The orthocenter of a triangle _?_ lies outside the triangle.

44. The centroid of a triangle _?_ lies on the triangle.

Lesson 3-5

45. Find the measure of an interior angle of a regular 9-gon.

A Point in Time

1500 1600 1700 1800 1900 2000

Presidents are known more often for their foreign policy than for their mathematical creativity. James Garfield, the 20th President of the United States, is an exception. In 1876, Garfield demonstrated this proof of the Pythagorean Theorem.

In the diagram, $\triangle NRM$ and $\triangle RPQ$ are congruent right triangles with sides of lengths a, b, and c. The legs of isosceles right triangle NRP have length c. The three triangles form trapezoid $MNPQ$. The sum of the areas of the three triangles equals the area of trapezoid $MNPQ$.

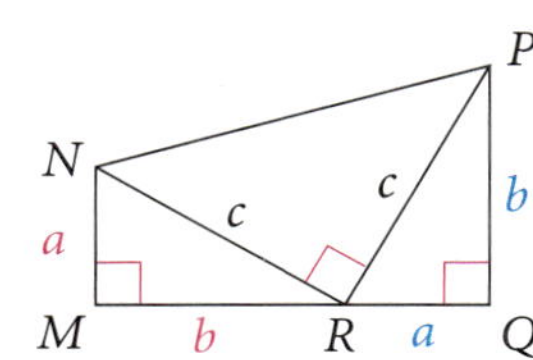

Areas of Triangles = Area of Trapezoid

$$\frac{1}{2}ab + \frac{1}{2}ab + \frac{1}{2}c^2 = \frac{1}{2}(a+b)(a+b)$$

$$ab + \frac{1}{2}c^2 = \frac{1}{2}a^2 + ab + \frac{1}{2}b^2$$

$$\frac{1}{2}c^2 = \frac{1}{2}a^2 + \frac{1}{2}b^2$$

$$c^2 = a^2 + b^2$$

For: Information about Pythagorean Theorem proofs
Web Code: bce-2032

10-3 Areas of Regular Polygons

California Content Standards

GEOM 8.0 Students solve problems involving the area of geometric figures. ***Master***
GEOM 10.0 Students compute areas of polygons, including equilateral triangles. ***Master***
GEOM 20.0 Students know and are able to use angle and side relationships in problems with special right triangles. ***Master***

What You'll Learn

- To find the area of a regular polygon

. . . And Why

To find the area of pieces of honeycomb material used to build boats, as in Example 3

Check Skills You'll Need

GO for Help Lesson 8-2

Find the area of each regular polygon. If your answer involves a radical, leave it in simplest radical form.

1.

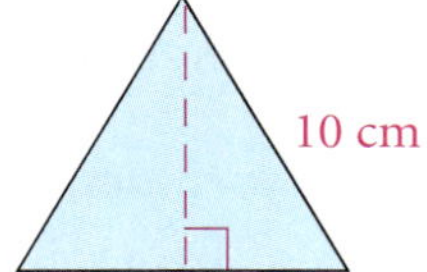

2. (square with diagonal 10 ft)

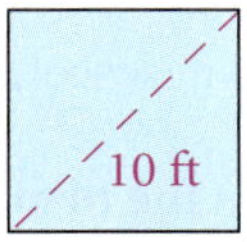

3.

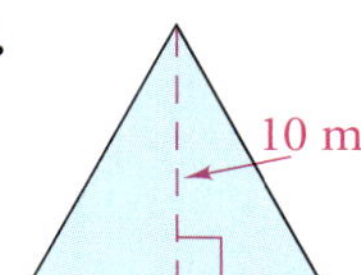

Find the perimeter of the regular polygon.

4. a hexagon with sides of 4 in.

5. an octagon with sides of $2\sqrt{3}$ cm

New Vocabulary • radius of a regular polygon • apothem

1 Areas of Regular Polygons

Vocabulary Tip

The terms radius and apothem (AP uh them) can each refer to either a segment or its length.

You can circumscribe a circle about any regular polygon. The center of a regular polygon is the center of the circumscribed circle. The **radius** is the distance from the center to a vertex. The **apothem** is the perpendicular distance from the center to a side.

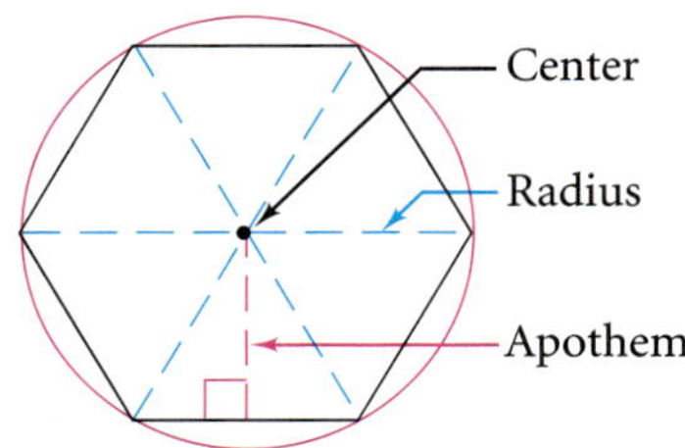

1 EXAMPLE Finding Angle Measures

The figure at the right is a regular pentagon with radii and an apothem drawn. Find the measure of each numbered angle.

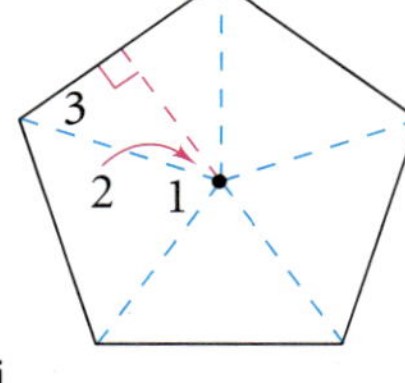

$m\angle 1 = \frac{360}{5} = 72$ **Divide 360 by the number of sides.**

$m\angle 2 = \frac{1}{2}m\angle 1$ **The apothem bisects the vertex angle of the isosceles triangle formed by the radii.**

$= \frac{1}{2}(72) = 36$

$90 + 36 + m\angle 3 = 180$ **The sum of the measures of the angles of a triangle is 180.**

$m\angle 3 = 54$

$m\angle 1 = 72, m\angle 2 = 36$, and $m\angle 3 = 54$

CA Standards Check **1** At the right, a portion of a regular octagon has radii and an apothem drawn. Find the measure of each numbered angle.

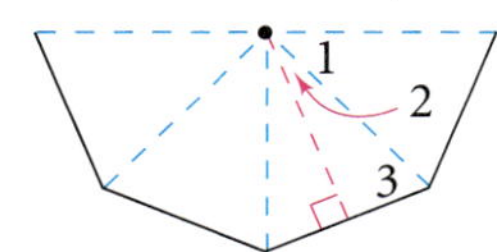

Suppose you have a regular n-gon with side s. The radii divide the figure into n congruent isosceles triangles. Each isosceles triangle has area equal to $\frac{1}{2}as$.

Since there are n congruent triangles, the area of the n-gon is $A = n \cdot \frac{1}{2}as$. The perimeter p of the n-gon is ns. Substituting p for ns results in a formula for the area in terms of a and p: $A = \frac{1}{2}ap$.

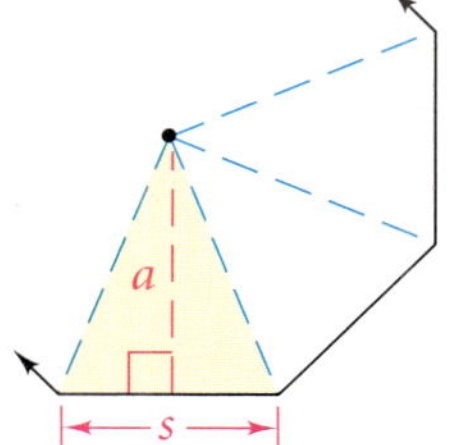

Key Concepts

Theorem 10-6 — Area of a Regular Polygon

The area of a regular polygon is half the product of the apothem and the perimeter.

$$A = \frac{1}{2}ap$$

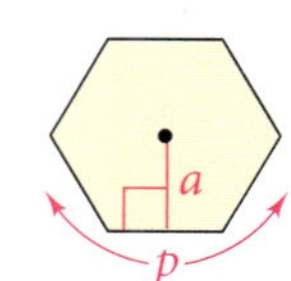

2 EXAMPLE Finding the Area of a Regular Polygon

Find the area of a regular decagon with a 12.3-in. apothem and 8-in. sides.

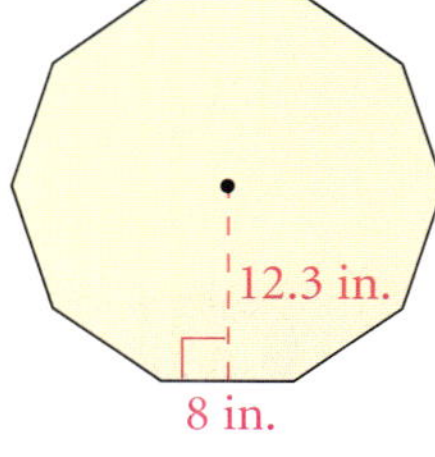

$p = ns$ — **Find the perimeter.**

$= 10(8) = 80$ in. — **A decagon has 10 sides, so $n = 10$.**

$A = \frac{1}{2}ap$ — **Use the formula for the area of a regular polygon.**

$= \frac{1}{2}(12.3)(80) = 492$

The regular decagon has area 492 in.^{2}.

CA Standards Check **2** Find the area of a regular pentagon with 11.6-cm sides and an 8-cm apothem.

3 EXAMPLE Real-World Connection

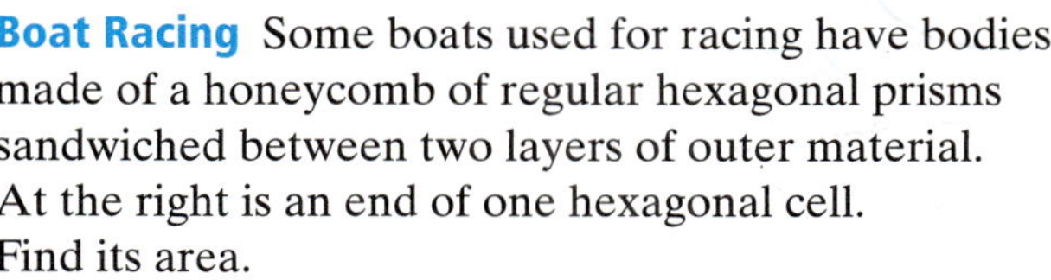

Boat Racing Some boats used for racing have bodies made of a honeycomb of regular hexagonal prisms sandwiched between two layers of outer material. At the right is an end of one hexagonal cell. Find its area.

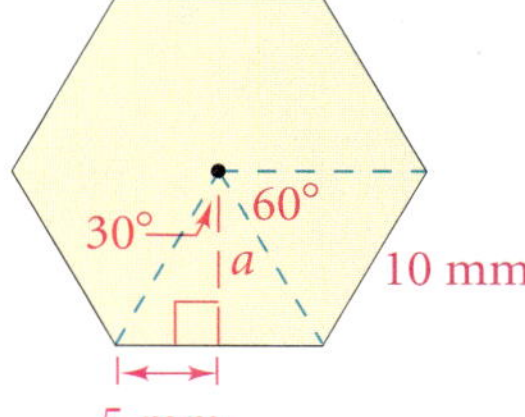

The radii form six 60° angles at the center. You can use a 30°-60°-90° triangle to find the apothem a.

$a = 5\sqrt{3}$ — **longer leg $= \sqrt{3} \cdot$ shorter leg**

$p = ns$ — **Find the perimeter of the hexagon.**

$= 6(10) = 60$ — **Substitute 6 for n and 10 for s.**

$A = \frac{1}{2}ap$ — **Find the area.**

$= \frac{1}{2}(5\sqrt{3})(60)$ — **Substitute $5\sqrt{3}$ for a and 60 for p.**

≈ 259.80762 — **Use a calculator.**

The area is about 260 mm^2.

CA Standards Check **3** The side of a regular hexagon is 16 ft. Find the area of the hexagon.

EXERCISES
For more exercises, see *Extra Skill, Word Problem, and Proof Practice.*

Standards Practice
GEOM 8.0, 10.0, 20.0

A Practice by Example

Example 1 (page 546)

Each regular polygon has radii and apothem as shown. Find the measure of each numbered angle.

1. 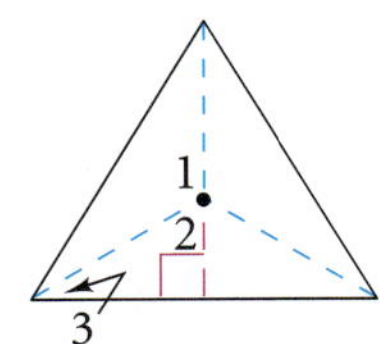

2. 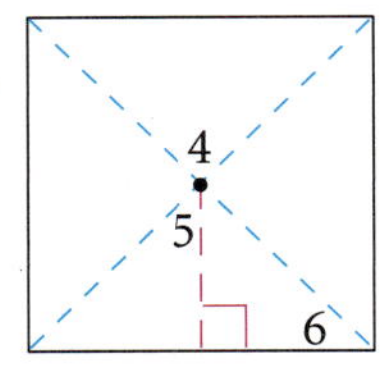

3. 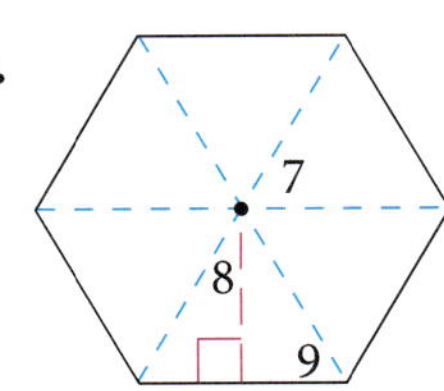

Example 2 (page 547)

Find the area of each regular polygon with the given apothem a and side length s.

4. pentagon, $a = 24.3$ cm, $s = 35.3$ cm

5. 7-gon, $a = 29.1$ ft, $s = 28$ ft

6. octagon, $a = 60.4$ in., $s = 50$ in.

7. nonagon, $a = 27.5$ in., $s = 20$ in.

8. decagon, $a = 19$ m, $s = 12.3$ m

9. dodecagon, $a = 26.1$ cm, $s = 14$ cm

Example 3 (page 547)

Find the area of each regular polygon. Round your answer to the nearest tenth.

10. 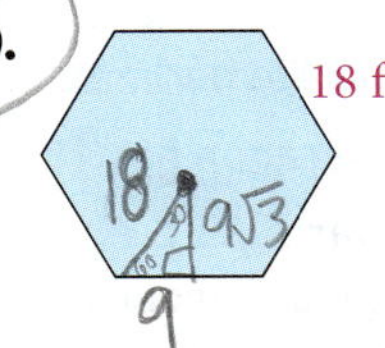

11. 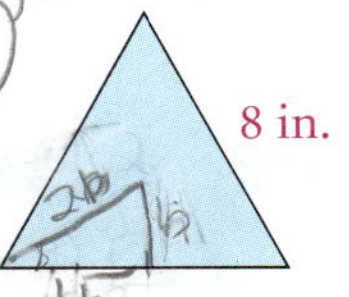

12. 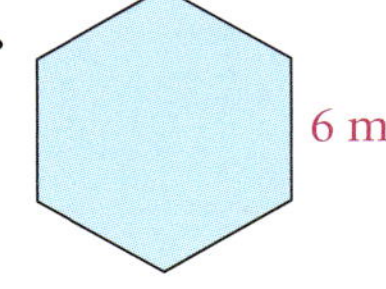

13. Art The smaller triangles in the Minneapolis sculpture at the left are equilateral. Each has a 12.7-in. radius. What is the area of each to the nearest square inch?

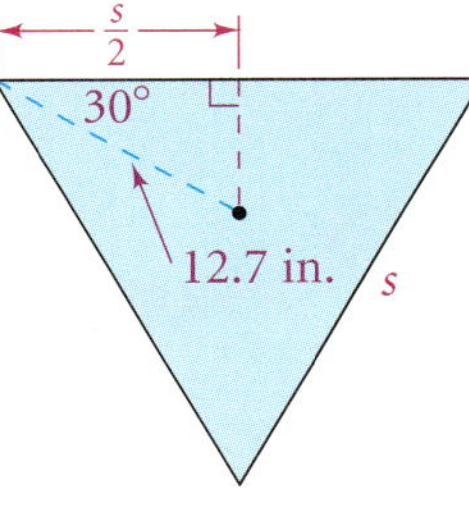

Exercise 13

Find the area of each regular polygon with the given radius or apothem. If your answer is not an integer, leave it in simplest radical form.

14. 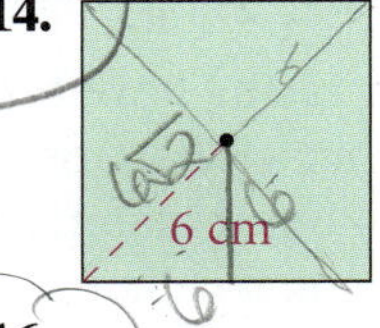

15.

16. 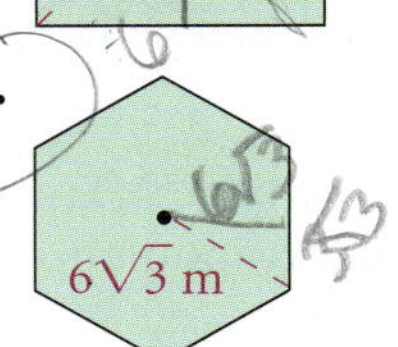

17. 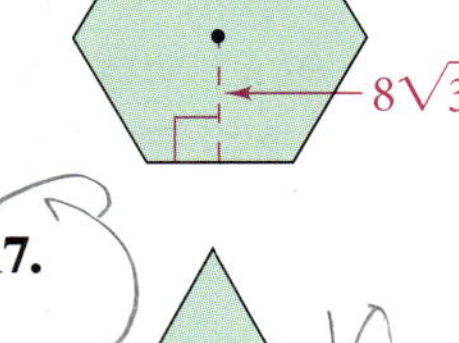

18. 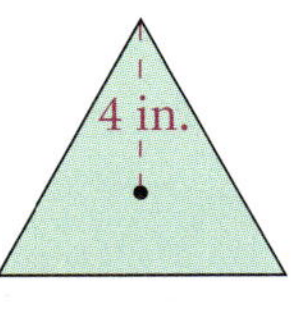

B Apply Your Skills

Find the measures of the angles formed by (a) two consecutive radii and (b) a radius and a side of the given regular polygon.

19. pentagon

20. octagon

21. nonagon

22. dodecagon

For a guide to solving Exercise 23, see p. 552.

23. Satellites One of the smallest space satellites ever developed has the shape of a pyramid. Each of the four faces of the pyramid is an equilateral triangle with sides about 13 cm long. What is the area of one equilateral triangular face of the satellite? Round your answer to the nearest whole number.

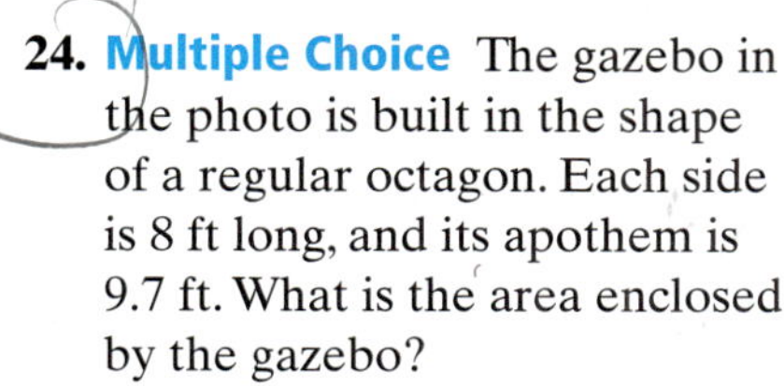

24. **Multiple Choice** The gazebo in the photo is built in the shape of a regular octagon. Each side is 8 ft long, and its apothem is 9.7 ft. What is the area enclosed by the gazebo?

Ⓐ 38.8 ft^2 Ⓒ 232.8 ft^2

Ⓑ 77.6 ft^2 Ⓓ 310.4 ft^2

25. The area of a regular polygon is 36 $in.^2$. Find the length of a side if the polygon has the given number of sides. Round your answer to the nearest tenth.

a. 3 **b.** 4 **c.** 6

d. **Estimation** Suppose the polygon is a pentagon. What would you expect the length of its side to be? Explain.

26. A portion of a regular decagon has radii and an apothem drawn. Find the measure of each numbered angle.

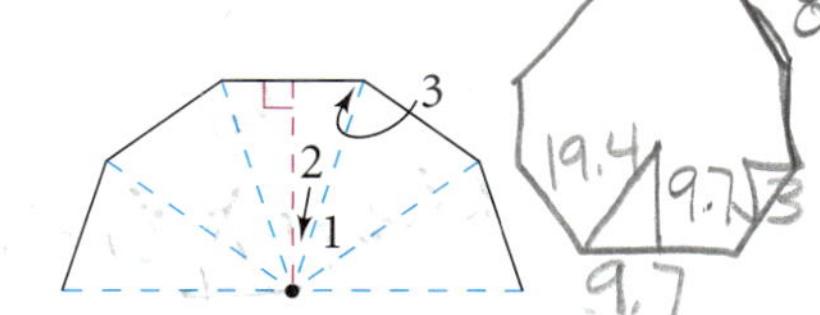

27. **Writing** Explain why the radius of a regular polygon is greater than the apothem.

28. **Constructions** Use a compass to construct a circle.

a. Construct two perpendicular diameters of the circle.

b. Construct diameters that bisect each of the four right angles.

c. Connect the consecutive points where the diameters intersect the circle. What regular polygon have you constructed?

d. **Critical Thinking** How can a circle help you construct a regular hexagon?

29. A regular hexagon has perimeter 120 m. Find its area.

30. **Open-Ended** Create a design using equilateral triangles and regular hexagons that have sides of the same length. Find the area of the completed design.

Find the area of each regular polygon. Show your answers in simplest radical form and rounded to the nearest tenth.

31.

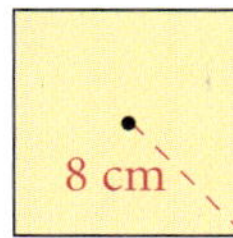

32.

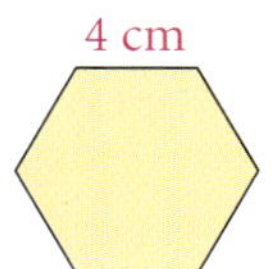

33.

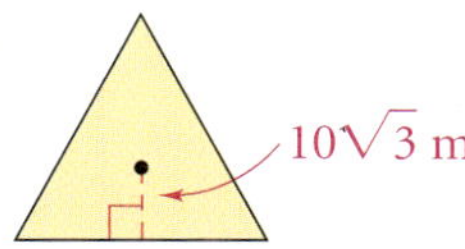

34. To find the area of an equilateral triangle, you can use the formula $A = \frac{1}{2}bh$ or $A = \frac{1}{2}ap$. A third way to find the area of an equilateral triangle is to use the formula $A = \frac{1}{4}s^2\sqrt{3}$. Verify the formula $A = \frac{1}{4}s^2\sqrt{3}$ in two ways as follows:

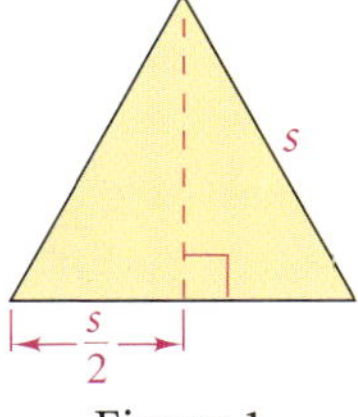

Figure 1

Figure 2

a. Find the area of Figure 1 using the formula $A = \frac{1}{2}bh$.

b. Find the area of Figure 2 using the formula $A = \frac{1}{2}ap$.

Homework Video Tutor

Visit: PHSchool.com
Web Code: bce-1003

Proof **35.** For Example 1 on page 546, write a proof that the apothem bisects the vertex angle of the isosceles triangle formed by the radii.

C **Challenge** **Proof** **36.** Prove that the bisectors of the angles of a regular polygon (given congruent sides and angles) are concurrent and that they are, in fact, radii of the polygon. (*Hint:* For regular n-gon $ABCDE\ldots$, let P be the intersection of the bisectors of $\angle ABC$ and $\angle BCD$. Show that $\overrightarrow{DP}$ must be the bisector of $\angle CDE$.)

Real-World Connection

Horizontal cross sections of the Wenfeng Pagoda in Yangzhou, China, are regular octagons.

37. **Coordinate Geometry** A regular octagon with center at the origin and radius 4 is graphed in the coordinate plane.

a. Since V_2 lies on the line $y = x$, its x- and y-coordinates are equal. Use the Distance Formula to find the coordinates of V_2 to the nearest tenth.

b. Use the coordinates of V_2 and the formula $A = \frac{1}{2}bh$ to find the area of $\triangle V_1OV_2$ to the nearest tenth.

c. Use your answer to part (b) to find the area of the octagon to the nearest whole number.

38. In $\triangle ABC$, $\angle C$ is acute.

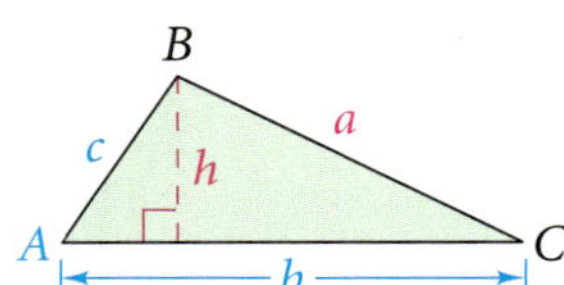

a. Show that the area of $\triangle ABC = \frac{1}{2}ab \sin C$.

b. Complete: The area of a triangle is half the product of __?__ and the sine of the __?__ angle.

c. Show that the area of a regular n-gon with radius r is $\frac{nr^2}{2} \sin\left(\frac{360}{n}\right)$.

Multiple Choice Practice

For California Standards and CAHSEE Tutorials, visit PHSchool.com. Web Code: bcq-9045

GEOM 8.0 **39.** Which of the following figures has an area of 15 square meters?

Ⓐ
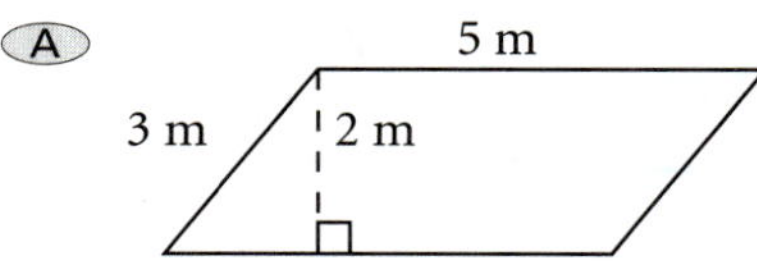

Ⓒ
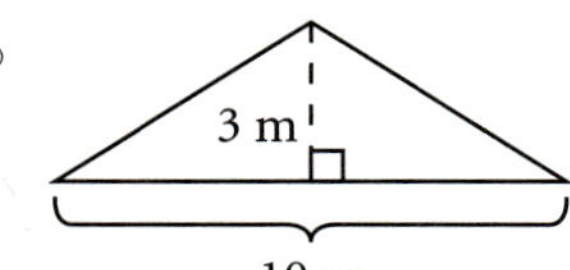

Ⓑ
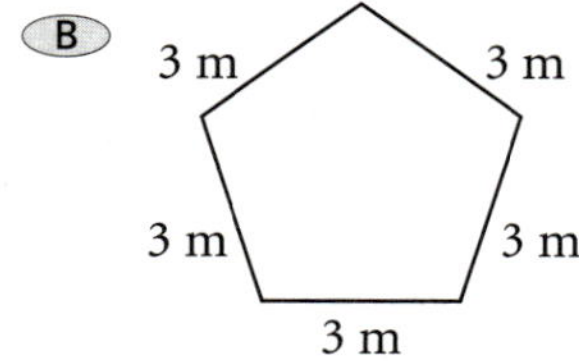

Ⓓ
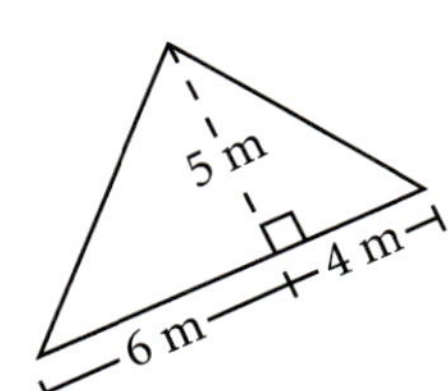

GEOM 10.0 **40.** What is the exact area of an equilateral triangle with 4-cm legs?

Ⓐ $2\sqrt{3}$ cm^2 Ⓒ $8\sqrt{3}$ cm^2

Ⓑ $4\sqrt{3}$ cm^2 Ⓓ 12 cm^2

CAHSEE 7AF 4.1 **41.** Solve for n: $3n - 8 > 10$.

Ⓐ $n > \frac{2}{3}$ Ⓒ $n \leq 15$

Ⓑ $n > 6$ Ⓓ $n > 18$

Mixed Review

Lesson 10-2

42. Find the area of a kite with diagonals 8 m and 11.5 m.

43. The area of a kite is 150 in.^{2}. The length of one diagonal is 10 in. Find the length of the other diagonal.

44. The area of a trapezoid is 42 m^2. The trapezoid has a height of 7 m and one base of 4 m. Find the length of the other base.

Lesson 4-4

Name the pairs of triangles you would have to prove congruent so that the indicated congruences are true by CPCTC.

Given: $\angle DAB \cong \angle CBA$,
$\overline{AD} \cong \overline{BC}$,
$\overline{DF}$ bisects $\angle ADB$,
$\overline{CG}$ bisects $\angle BCA$.

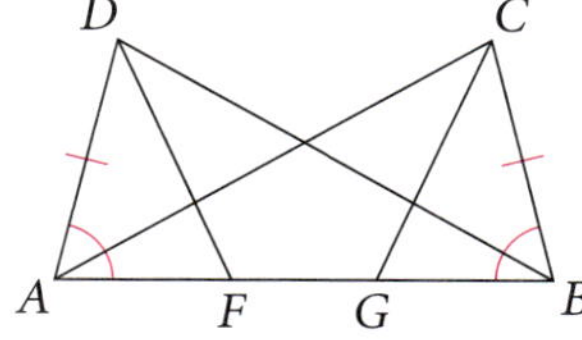

45. $\overline{AC} \cong \overline{BD}$ **46.** $\overline{AG} \cong \overline{BF}$ **47.** $\angle DFA \cong \angle CGB$

Lesson 1-9

48. a. Biology The size of a jaguar's territory depends on how much food is available. Where there is a lot of food, such as in a forest, jaguars have circular territories about 3 mi in diameter. Use 3.14 for π to estimate the area of such a region to the nearest tenth.

b. Where food is less available, a jaguar may need up to 200 mi^2. Estimate the radius of this circular territory.

Checkpoint Quiz 1 — Lessons 10-1 through 10-3

Find the area of each figure.

1.
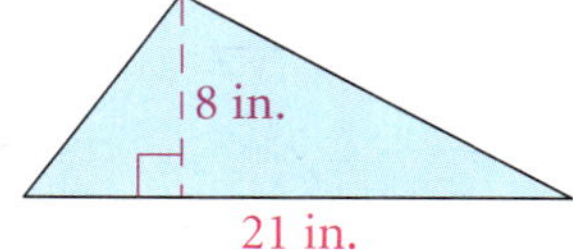

2.
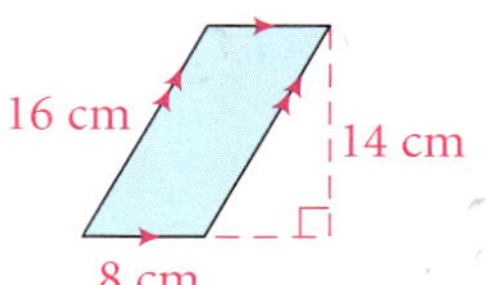

3.
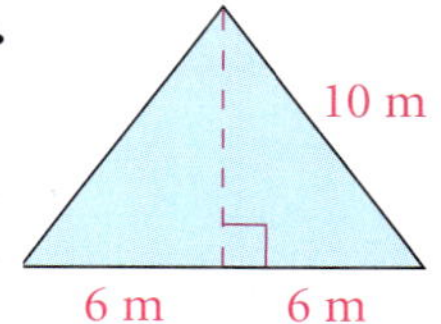

Find the area of each trapezoid, rhombus, or regular polygon. You may leave answers in simplest radical form.

4.
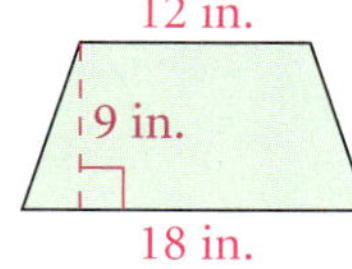

5.
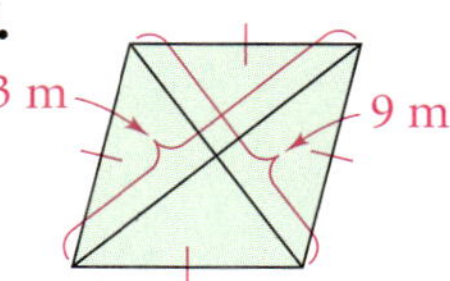

6.
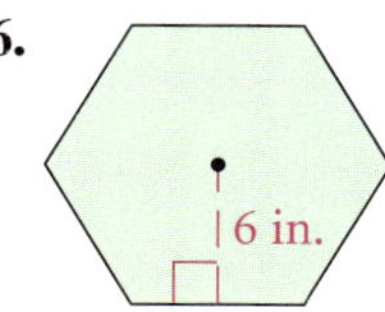

7.
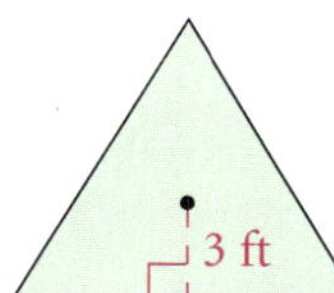

8.
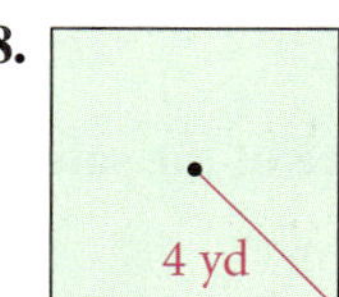

9.
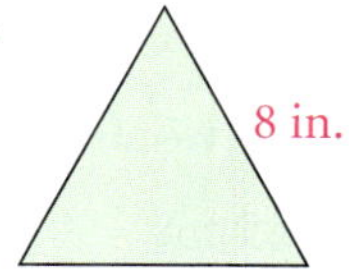

10. A regular hexagon has a radius of $3\sqrt{3}$ m. Find the area. Show your answer in simplest radical form and rounded to the nearest tenth.

GPS Guided Problem Solving

For Use With Page 548, Exercise 23

Understanding Word Problems **Read the problem below. Let Jake's thinking guide you through the solution. Check your understanding with the exercises at the bottom of the page.**

Satellites One of the smallest space satellites ever developed has the shape of a pyramid. Each of the four faces of the pyramid is an equilateral triangle with sides about 13 cm long. What is the area of one equilateral triangular face of the satellite? Round your answer to the nearest whole number.

What Jake Thinks	What Jake Writes
I'll make a sketch. First, the pyramid. Then a face. Each equilateral-triangle face has 13-cm sides.	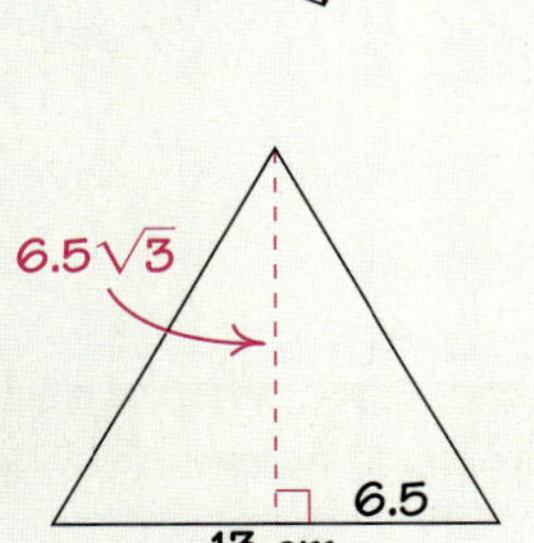
I have to find the area. I'll use red for an altitude. That gives me half of an equilateral triangle, or a 30°-60°-90° triangle. The short leg of the triangle is 6.5 cm. The altitude is $\sqrt{3}$ times the short leg.	
The area is $\frac{1}{2}$ base times height.	$A = \frac{1}{2}bh$
I'll use a calculator.	$A = \frac{1}{2} \cdot 13 \cdot 6.5\sqrt{3}$
I have to round to the nearest whole number.	$A \approx 73.18 \text{ cm}^2$ $A \approx 73 \text{ cm}^2$

EXERCISES

What would be the area of a face of the satellite for each side length given?

1. 6.5 cm **2.** 26 cm **3.** 1.3 m **4.** 2.6 m

5. If you change the length of each side of the satellite by a factor k to $13k$ cm, how does the area of a face change?

Perimeters and Areas of Similar Figures

California Content Standards

GEOM 11.0 Students determine how changes in dimensions affect the perimeter and area of common geometric figures. ***Introduce***

What You'll Learn

- To find the perimeters and areas of similar figures

. . . And Why

To find the expected yield of a garden, as in Example 3

Check Skills You'll Need

GO for Help Lesson 1-9

Find the perimeter and area of each figure.

1. 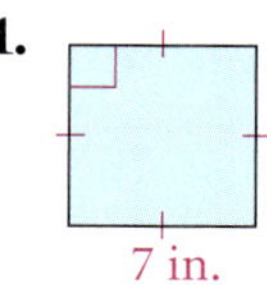

2. 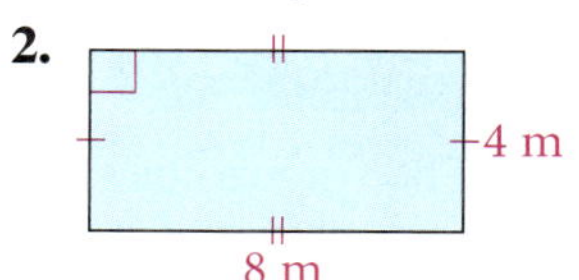

3. 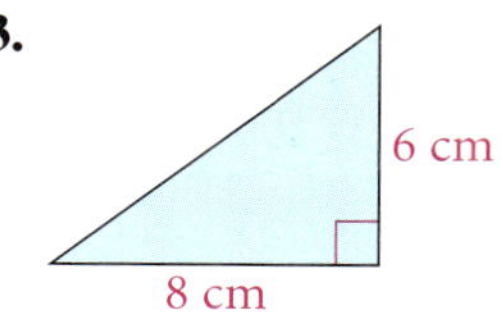

Find the perimeter and area of each rectangle with the given base and height.

4. $b = 1$ cm, $h = 3$ cm **5.** $b = 2$ cm, $h = 6$ cm **6.** $b = 3$ cm, $h = 9$ cm

1 Finding Perimeters and Areas of Similar Figures

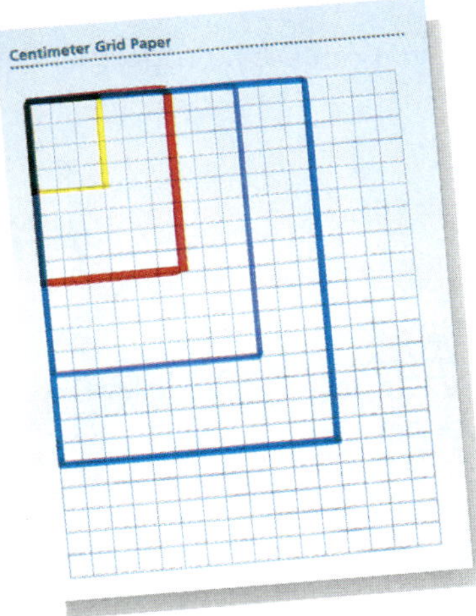

Hands-On Activity: Perimeters and Areas of Similar Rectangles

- On a piece of grid paper, draw a 3-unit by 4-unit rectangle.
- Draw three different rectangles, each similar to the original rectangle. Label them I, II, and III.

1. Use your drawings to complete a chart like this.

Rectangle	Perimeter	Area
Original		
I		
II		
III		

2. Use the information from the first chart to complete a chart like this.

Rectangle	Similarity Ratio	Ratio of Perimeters	Ratio of Areas
I to Original			
II to Original			
III to Original			

3. How do the ratios of perimeters and the ratios of areas compare with the similarity ratios?

To compare areas of similar figures, you can square the similarity ratio.

Key Concepts

Theorem 10-7 **Perimeters and Areas of Similar Figures**

If the similarity ratio of two similar figures is $\frac{a}{b}$, then
(1) the ratio of their perimeters is $\frac{a}{b}$ and
(2) the ratio of their areas is $\frac{a^2}{b^2}$.

1 EXAMPLE Finding Ratios in Similar Figures

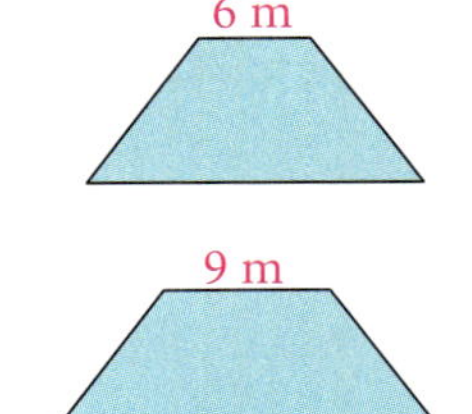

The trapezoids at the right are similar. The ratio of the lengths of corresponding sides is $\frac{6}{9}$, or $\frac{2}{3}$.

a. Find the ratio (smaller to larger) of the perimeters.

The ratio of the perimeters is the same as the ratio of corresponding sides, which is $\frac{2}{3}$.

b. Find the ratio (smaller to larger) of the areas.

The ratio of the areas is the square of the ratio of corresponding sides, which is $\frac{2^2}{3^2}$, or $\frac{4}{9}$.

CA Standards Check 1 Two similar polygons have corresponding sides in the ratio 5 : 7.
a. Find the ratio of their perimeters.
b. Find the ratio of their areas.

When you know the area of one of two similar polygons, you can use a proportion to find the area of the other polygon.

2 EXAMPLE Finding Areas Using Similar Figures

Multiple Choice The area of the smaller regular pentagon is about 27.5 cm^2. What is the best approximation for the area of the larger regular pentagon?

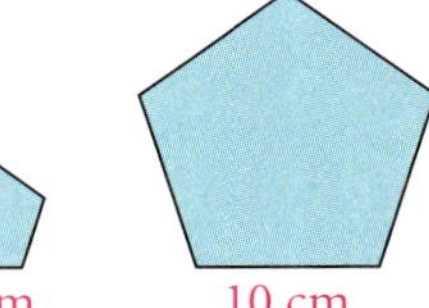

Ⓐ 11 cm^2 Ⓑ 69 cm^2 Ⓒ 172 cm^2 Ⓓ 275 cm^2

Test-Taking Tip

Look to eliminate choices. Here you can eliminate choice A immediately.
The area of the larger pentagon must be greater than 27.5 cm^2.

Regular pentagons are similar because all angles measure 108 and all sides in each are congruent. Here the ratio of corresponding-side lengths is $\frac{4}{10}$, or $\frac{2}{5}$. The ratio of the areas is $\frac{2^2}{5^2}$, or $\frac{4}{25}$.

$\frac{4}{25} = \frac{27.5}{A}$ **Write a proportion.**

$4A = 687.5$ **Cross-Product Property**

$A = \frac{687.5}{4} = 171.875$ **Solve for *A*.**

The area of the larger pentagon is about 172 cm^2. The answer is C.

CA Standards Check 2 The corresponding sides of two similar parallelograms are in the ratio $\frac{3}{4}$. The area of the larger parallelogram is 96 in.^{2}. Find the area of the smaller parallelogram.

Many cities make city land available to the community for gardening.

3 EXAMPLE Real-World Connection

Community Service During the summer, a group of high school students used a plot of city land and harvested 13 bushels of vegetables that they gave to a food pantry. Their project was so successful that next summer the city will let them use a larger, similar plot of land.

In the new plot, each dimension is 2.5 times the corresponding dimension of the original plot. How many bushels can they expect to harvest next year?

The ratio of the dimensions is $2.5 : 1$. So, the ratio of the areas is $(2.5)^2 : 1^2$, or $6.25 : 1$. With 6.25 times as much land next year, the students can expect to harvest 6.25(13), or about 81 bushels.

CA Standards Check **3** The similarity ratio of the dimensions of two similar pieces of window glass is $3 : 5$. The smaller piece costs \$2.50. What should be the cost of the larger piece?

When you know the ratio of the areas of two similar figures, you can work backward to find the ratio of their perimeters.

4 EXAMPLE Finding Similarity and Perimeter Ratios

For: Perimeter and Area Activity
Use: Interactive Textbook, 10-4

The areas of two similar triangles are 50 cm^2 and 98 cm^2. What is the similarity ratio? What is the ratio of their perimeters?

Find the similarity ratio $a : b$.

$\frac{a^2}{b^2} = \frac{50}{98}$ **The ratio of the areas is $a^2 : b^2$.**

$\frac{a^2}{b^2} = \frac{25}{49}$ **Simplify.**

$\frac{a}{b} = \frac{5}{7}$ **Take square roots.**

The ratio of the perimeters equals the similarity ratio $5 : 7$.

CA Standards Check **4** The areas of two similar rectangles are 1875 ft^2 and 135 ft^2. Find the ratio of their perimeters.

EXERCISES Standards Practice

For more exercises, see *Extra Skill, Word Problem, and Proof Practice.*

GEOM 11.0

A Practice by Example

Example 1 (page 554)

The figures in each pair are similar. Compare the first figure to the second. Give the ratio of the perimeters and the ratio of the areas.

1.

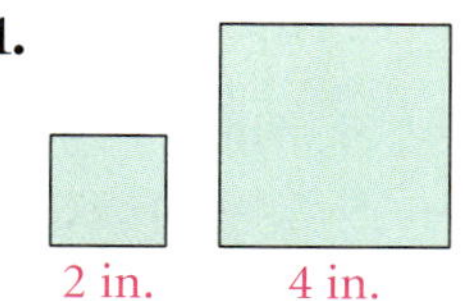

2.

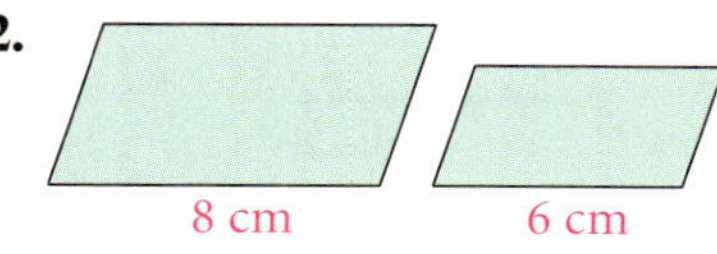

3.

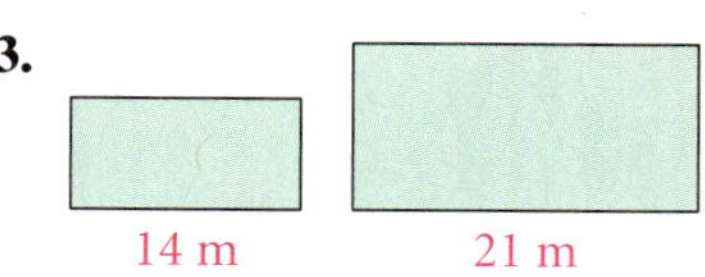

4.

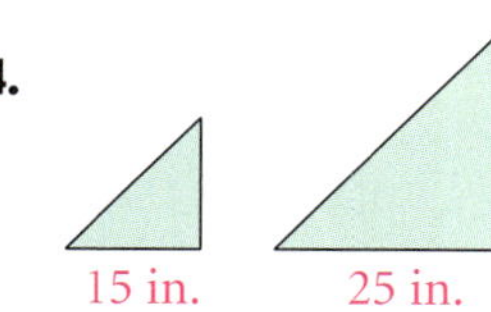

Example 2 (page 554)

The figures in each pair are similar. The area of one figure is given. Find the area of the other figure to the nearest whole number.

5.

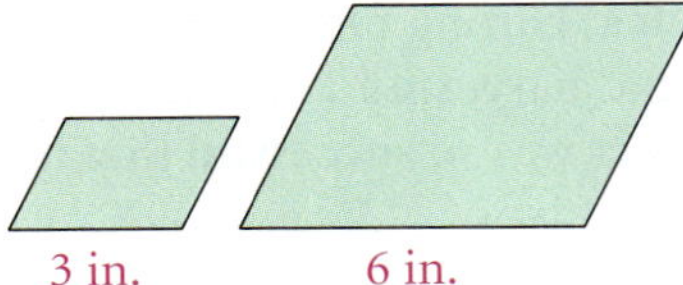

Area of smaller parallelogram = 6 in.^{2}

6.

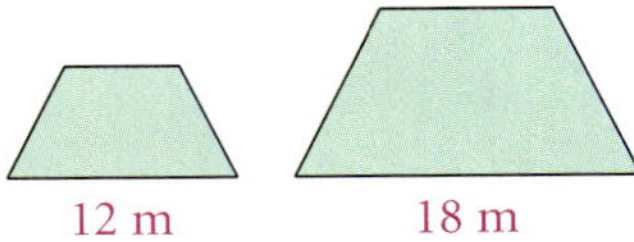

Area of larger trapezoid = 121 m^2

7.

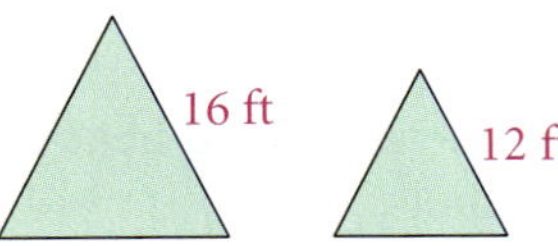

Area of larger triangle = 105 ft^2

8.

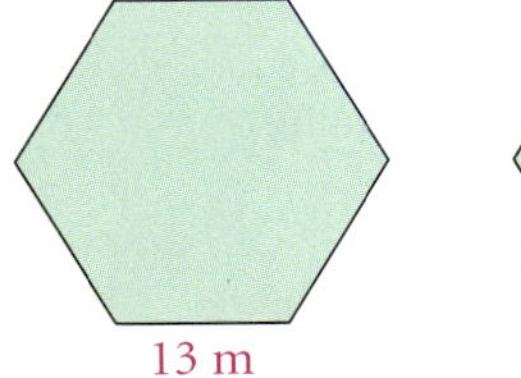

Area of smaller hexagon = 65 m^2

Example 3 (page 555)

9. **Remodeling** It costs a family \$216 to have a 9 ft-by-12 ft wooden floor refinished. At that rate, how much would it cost them to have a 12 ft-by-16 ft wooden floor refinished?

10. **Decorating** An embroidered placemat costs \$2.95. An embroidered tablecloth is similar to the placemat, but four times as long and four times as wide. How much would you expect to pay for the tablecloth?

Example 4 (page 555)

Find the similarity ratio and the ratio of perimeters for each pair of similar figures.

11. two regular octagons with areas 4 ft^2 and 16 ft^2

12. two triangles with areas 75 m^2 and 12 m^2

13. two trapezoids with areas 49 cm^2 and 9 cm^2

14. two parallelograms with areas 18 in.^{2} and 32 in.^{2}

15. two equilateral triangles with areas $16\sqrt{3}$ ft^2 and $\sqrt{3}$ ft^2

16. two circles with areas 2π cm^2 and 200π cm^2

The similarity ratio of two similar polygons is given. Find the ratio of their perimeters and the ratio of their areas.

17. 3 : 1 **18.** 2 : 5 **19.** $\frac{2}{3}$ **20.** $\frac{7}{4}$ **21.** 6 : 1

22. **Multiple Choice** The area of a regular decagon is 50 cm^2. What is the area of a regular decagon with sides four times the sides of the smaller decagon?

Ⓐ 200 cm^2 Ⓑ 500 cm^2 Ⓒ 800 cm^2 Ⓓ 2000 cm^2

23. **Error Analysis** A reporter used the graphic below to show that the number of houses with more than two televisions had doubled in the past few years. Explain why this graphic is misleading.

Visit: PHSchool.com
Web Code: bce-1004

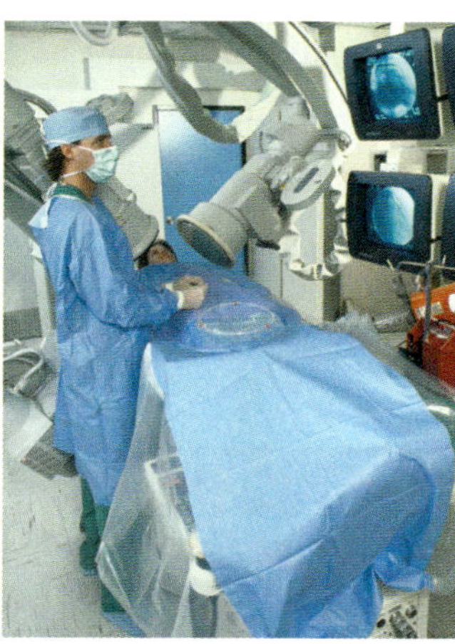

Real-World Connection

Doctors use enlarged images to aid in certain medical procedures.

Problem Solving Hint

For Exercise 34, recall the length of a diagonal of a square with 2-in. sides.

24. **Medicine** For some medical imaging, the scale of the image is 3 : 1. That means that if an image is 3 cm long, the corresponding length on the person's body is 1 cm. Find the actual area of a lesion if its image has area 2.7 cm^2.

25. The longer sides of a parallelogram are 5 m. The longer sides of a similar parallelogram are 15 m. The area of the smaller parallelogram is 28 m^2. What is the area of the larger parallelogram?

Algebra **Find the values of x and y when the smaller triangle shown here has the given area.**

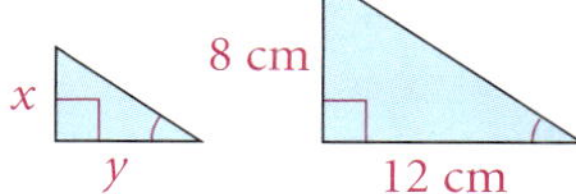

26. 3 cm^2 27. 6 cm^2 28. 12 cm^2

29. 16 cm^2 30. 24 cm^2 31. 48 cm^2

32. Two similar rectangles have areas 27 in.^{2} and 48 in.^{2}. The length of one side of the larger rectangle is 16 in. What are the dimensions of both rectangles?

33. In $\triangle RST$, $RS = 20$ m, $ST = 25$ m, and $RT = 40$ m.
   a. **Open-Ended** Choose a convenient scale. Then use a ruler and compass to draw $\triangle R'S'T' \sim \triangle RST$.
   b. **Constructions** Construct an altitude of $\triangle R'S'T'$ and measure its length. Find the area of $\triangle R'S'T'$.
   c. **Estimation** Estimate the area of $\triangle RST$.

34. **Drawing** Draw a square with an area of 8 in.^{2}. Draw a second square with an area that is four times as large. What is the ratio of their perimeters?

Compare the blue figure to the red figure. Find the ratios of (a) their perimeters and (b) their areas.

35.

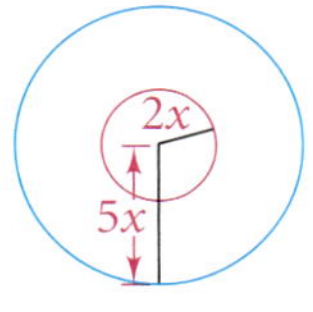

36.

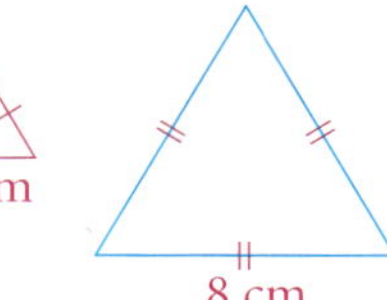

37.

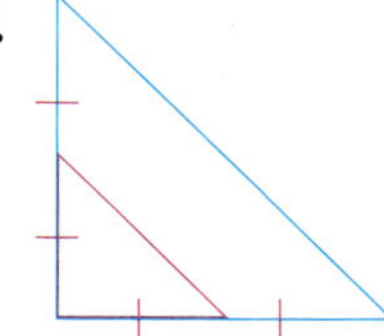

38. **Writing** The enrollment at an elementary school is going to increase from 200 students to 395 students. A parents' group is planning to increase the 100 ft-by-200 ft playground area to a larger area that is 200 ft by 400 ft. What would you tell the parents' group when they ask your opinion about whether the new playground will be large enough?

39. a. **Surveying** A surveyor measured one side and two angles of a field as shown in the diagram. Use a ruler and a protractor to draw a similar triangle.
   b. Measure the sides and altitude of your triangle and find its perimeter and area.
   c. **Estimation** Estimate the perimeter and area of the field.

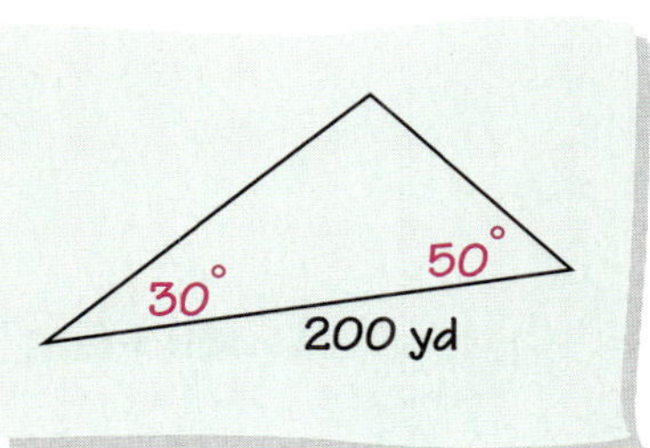

40. a. Find the area of a regular hexagon with sides 2 cm long. Leave your answer in simplest radical form.
   b. Use your answer to part (a) and Theorem 10-7 to find the areas of the regular polygons shown at the right.

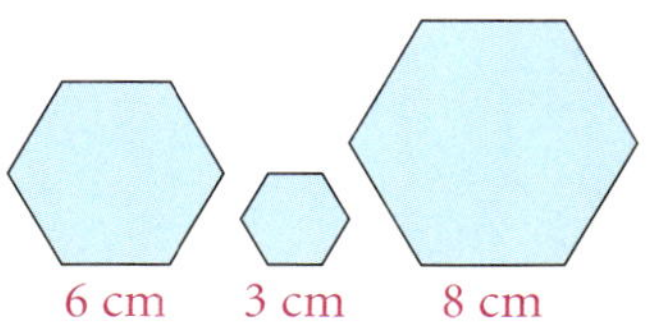

 Challenge

Complete each statement with *sometimes, always,* or *never.* Justify your answers.

41. Two similar rectangles with the same perimeter are _?_ congruent.

42. Two rectangles with the same area are _?_ similar.

43. Two rectangles with the same area and different perimeters are _?_ similar.

44. Similar figures _?_ have the same area.

Multiple Choice Practice

For California Standards and CAHSEE Tutorials, visit PHSchool.com. Web Code: bcq-9045

GEOM 11.0

45. The trapezoids below are similar. The area of the larger is 15 square feet. Find the area of the smaller trapezoid to the nearest tenth of a foot.

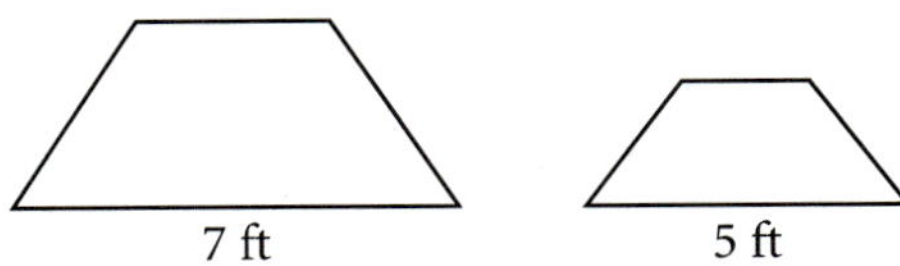

Ⓐ 29.4　Ⓑ 21　Ⓒ 10.7　Ⓓ 7.7

GEOM 12.0

46. Find the $m\angle x$ in the figure at the right.

Ⓐ 56°　Ⓒ 108°
Ⓑ 74°　Ⓓ 124°

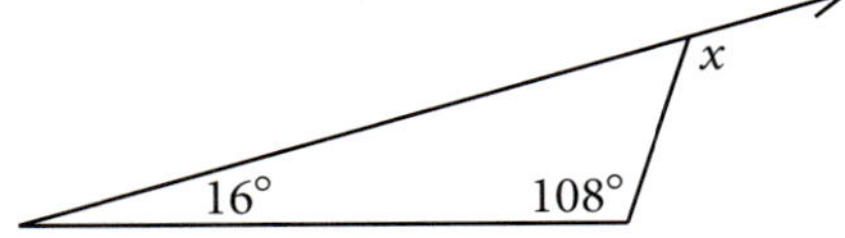

Mixed Review

Lesson 10-3

Find the area of each regular polygon.

47. a square with a 10-cm diagonal

48. a pentagon with apothem 13.8 and side length 20

49. an octagon with apothem 12 and side length 10

50. a 12-sided polygon with apothem 3.7 and side length 2.0

Lesson 7-5

51. Solve for x and y in the diagram at the right.

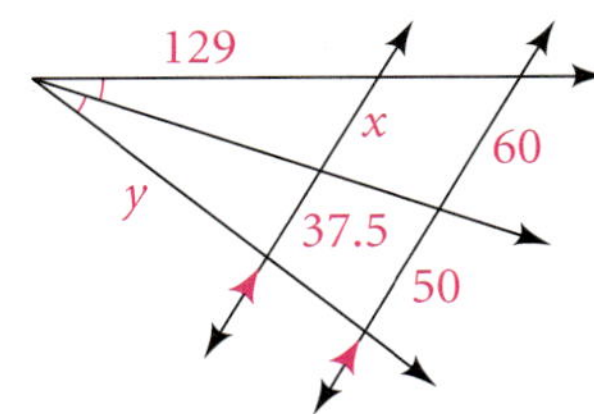

52. An angle bisector divides the opposite side of a triangle into segments 4 cm and 6 cm long. A second side of the triangle is 8 cm long. Find all possible lengths for the third side of the triangle.

Lesson 3-6

x^2 Algebra **Write the equation in slope-intercept form, and graph the line.**

53. $6x - 2y = 8$　**54.** $x + y = -2$　**55.** $3x = 4y$　**56.** $2x + \frac{1}{2}y = \frac{5}{2}$

x^2 Algebra **Write an equation of the line described.**

57. has slope -3 and contains point $(1, -2)$

58. contains points $(3, 7)$ and $(0, -1)$

10-5 Trigonometry and Area

California Content Standards

GEOM 8.0 Students know, derive, and solve problems involving the perimeter and area of common geometric figures. ***Develop***

GEOM 19.0 Students use trigonometric functions to solve for an unknown length of a side of a right triangle, given an angle and a length of a side. ***Master***

What You'll Learn

- To find the area of a regular polygon using trigonometry
- To find the area of a triangle using trigonometry

. . . And Why

To find the area of a courtyard, as in Example 2

Check Skills You'll Need

Find the area of each regular polygon.

1. 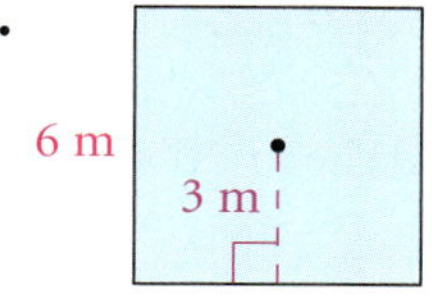

2. 36 in., 42 in.

3. 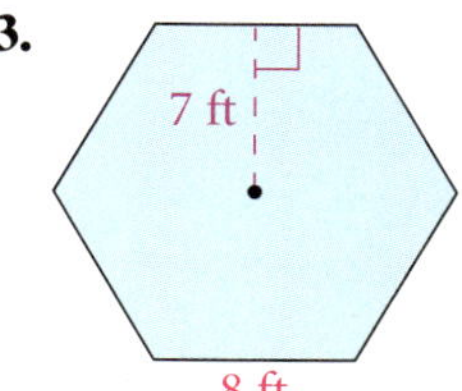

1 Finding the Area of a Regular Polygon

In Lesson 10-3, you learned to find the area of a regular polygon by using the formula $A = \frac{1}{2}ap$, where a is the apothem and p is the perimeter. By using this formula and trigonometric ratios, you can solve other types of problems.

Vocabulary Tip

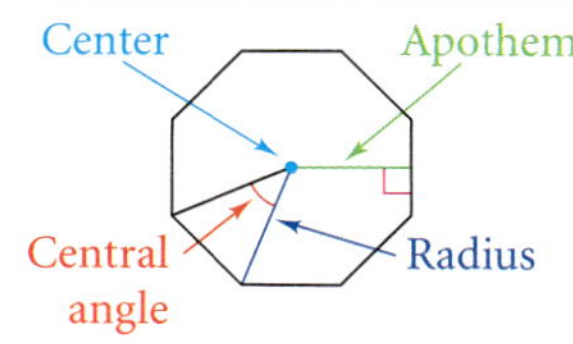

1 EXAMPLE Finding Area

Find the area of a regular pentagon with 8-cm sides.

To use the formula $A = \frac{1}{2}ap$, you need the apothem and perimeter. The perimeter is $5 \cdot 8$, or 40 cm.

To find the apothem, use trigonometry.

The measure of the central angle $\angle XCZ$ is $\frac{360}{5}$, or 72.

$m\angle XCY = \frac{1}{2}m\angle XCZ = 36.$

$XY = \frac{1}{2}XZ$, so $XY = 4$.

8 cm, C, a, X, Y, Z

C, 36°, a, X, 4, Y

$\tan 36° = \frac{4}{a}$ **Use the tangent ratio.**

$a = \frac{4}{\tan 36°}$ **Solve for a.**

Now substitute into the area formula.

$A = \frac{1}{2}ap$

$= \frac{1}{2} \cdot \frac{4}{\tan 36°} \cdot 40$ **Substitute for a and p.**

$= \frac{80}{\tan 36°}$ **Simplify.**

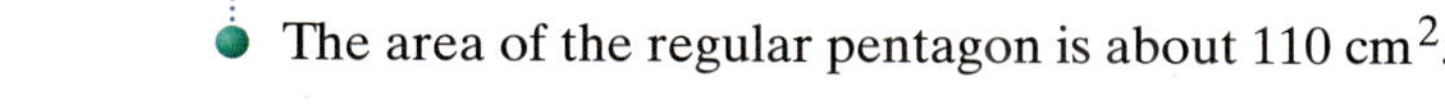

Use a calculator.

The area of the regular pentagon is about 110 cm^2.

CA Standards Check

1. Find the area of a regular octagon with a perimeter of 80 in. Give the area to the nearest tenth.

Sometimes you can use trigonometry to find both apothem and perimeter.

2 EXAMPLE **Real-World Connection**

Architecture The Castel del Monte, built on a hill in southern Italy circa 1240, makes extraordinary use of regular octagons. One regular octagon, the inner courtyard, has radius 16 m. Find the area of the courtyard.

The measure of a central angle of the octagon is $\frac{360}{8}$, or 45. So $m\angle C = \frac{1}{2}(45) = 22.5$.

Use the cosine ratio to find the apothem.

$$\cos 22.5^\circ = \frac{a}{16}$$
$$a = 16(\cos 22.5^\circ)$$

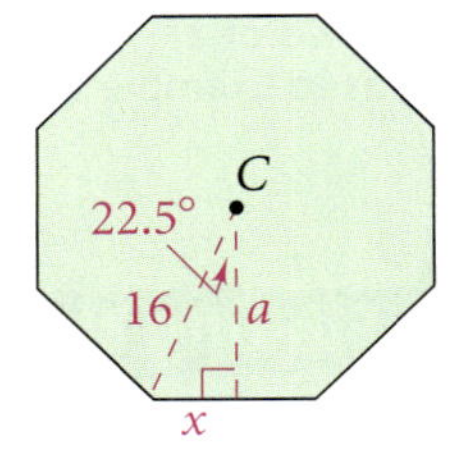

Use the sine ratio to find the perimeter.

$$\sin 22.5^\circ = \frac{x}{16}$$
$$x = 16(\sin 22.5^\circ)$$

$p = 8 \cdot$ length of a side

$= 8 \cdot 2x$ **The length of each side is 2*x*.**

$= 8 \cdot 2 \cdot 16(\sin 22.5^\circ)$ **Substitute for *x*.**

$= 256(\sin 22.5^\circ)$ **Simplify.**

Substitute into the area formula, $A = \frac{1}{2}ap$.

$A = \frac{1}{2} \cdot 16(\cos 22.5^\circ) \cdot 256(\sin 22.5^\circ)$ **Substitute for *a* and *p*.**

≈ 724.07734 **Use a calculator.**

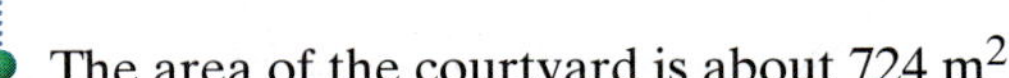

The area of the courtyard is about 724 m^2.

2 **Critical Thinking** If the radius of the main structure is twice the radius of the inner courtyard, how does the area it covers compare to the area of the courtyard?

2 Finding the Area of a Triangle

Suppose you want to find the area of $\triangle ABC$, but you know only $m\angle A$ and the lengths b and c. To use the formula Area $= \frac{1}{2}bh$, you need to know the height. You can find the height by using the sine ratio.

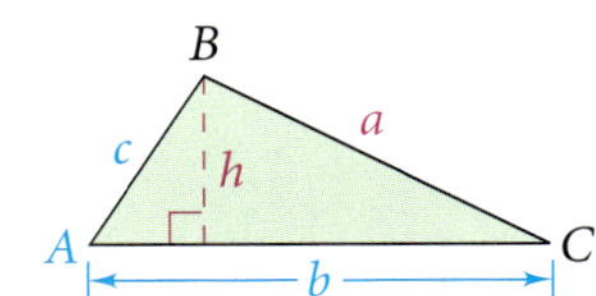

$\sin A = \frac{h}{c}$ **Use the sine ratio.**

$h = c(\sin A)$ **Solve for *h*.**

Now substitute for h in the formula Area $= \frac{1}{2}bh$.

$$\text{Area} = \frac{1}{2}bc(\sin A)$$

Your work at the bottom of page 560 completes a proof of the following theorem for the case in which $\angle A$ is acute.

Key Concepts

Theorem 10-8 **Area of a Triangle Given SAS**

The area of a triangle is one half the product of the lengths of two sides and the sine of the included angle.

$$\text{Area of } \triangle ABC = \tfrac{1}{2}bc(\sin A)$$

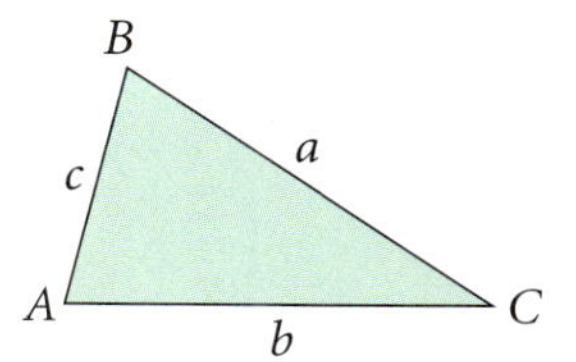

3 EXAMPLE **Real-World Connection**

Surveying The surveyed lengths of two adjacent sides of a triangular plot of land are 412 ft and 386 ft. The angle between the sides is 71°. Find the area of the plot.

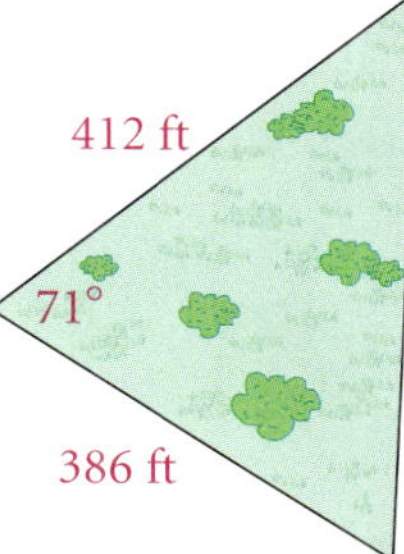

Area $= \frac{1}{2} \cdot$ side length $\cdot$ side length $\cdot$ sine of included angle

$= \frac{1}{2} \cdot 412 \cdot 386 \cdot \sin 71°$ **Substitute.**

$\approx$ 75183.855 **Use a calculator.**

The area of the plot is approximately 75,200 ft^2.

CA Standards Check **3** Two sides of a triangular building plot are 120 ft and 85 ft long. They include an angle of 85°. Find the area of the building plot to the nearest square foot.

EXERCISES
Standards Practice

For more exercises, see *Extra Skill, Word Problem, and Proof Practice.*

GEOM 8.0, 19.0

Practice by Example

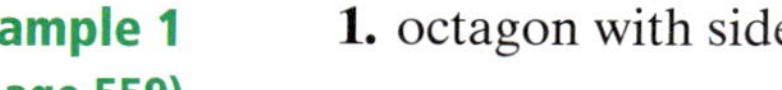

Example 1 (page 559)

Go for Help

Find the area of each regular polygon. Give answers to the nearest tenth.

1. octagon with side length 6 cm

2. pentagon with side length 7 in.

3. hexagon with perimeter 60 m

4. 15-gon with perimeter 180 yd

5. *PQRST* is a regular pentagon with center *O* and radius 10 in.

a. Find $m\angle POQ$.
b. Find $m\angle POX$.
c. Find OX.
d. Find PQ.
e. Find the perimeter.
f. Find the area.

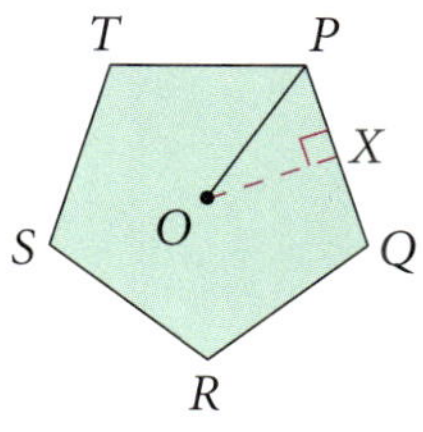

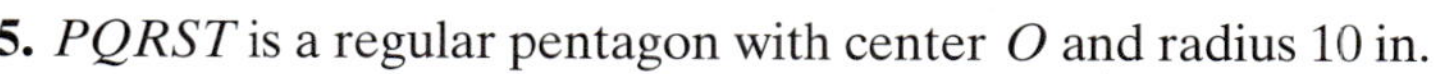

Example 2 (page 560)

Find the area of each regular polygon. Give answers to the nearest tenth.

6. hexagon with radius 10 ft

7. decagon with radius 4 in.

8. octagon with radius 20 cm

9. square with radius 2 ft

10. **Architecture** Each of the eight small towers around Castel del Monte (page 560) is a regular octagon. The radius is 7.3 m. Find the area each tower covers to the nearest square meter.

Example 3
(page 561)

Find the area of each triangle. Give answers to the nearest tenth.

11.
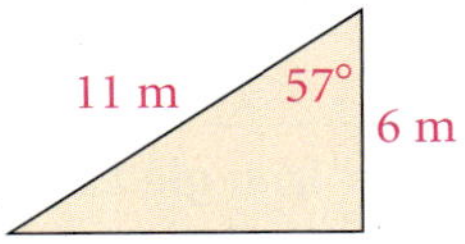

12.
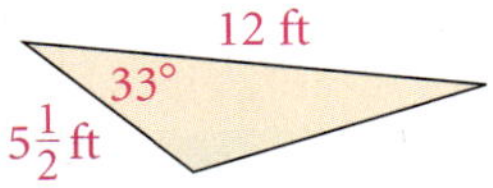

13.
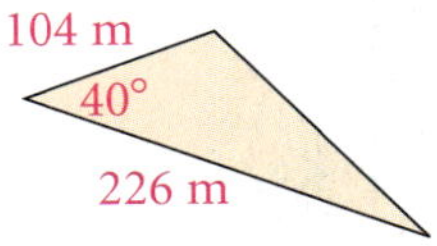

14.
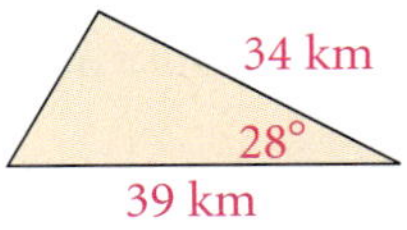

15.
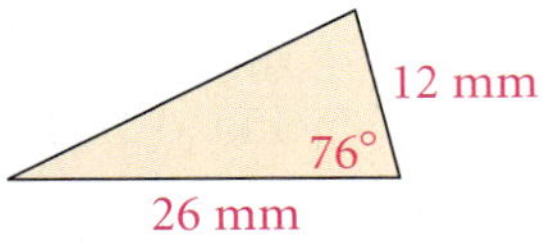

16.
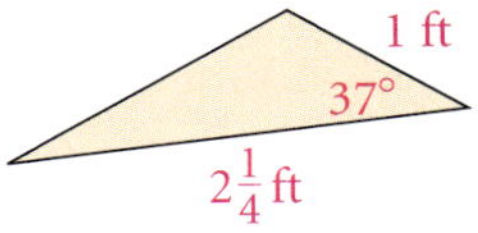

17. **Surveying** A surveyor marks off a triangular parcel of land. One side of the triangle extends 80 yd. A second side of 150 yd forms an angle of 67° with the first side. Determine the area of the parcel of land to the nearest square yard.

B Apply Your Skills

18. **Industrial Design** Refer to the diagram of the regular hexagonal nut. Round each answer to the nearest unit.

a. Find the area of the circular hole in the hexagonal nut.

b. Find the area of the hexagonal face.

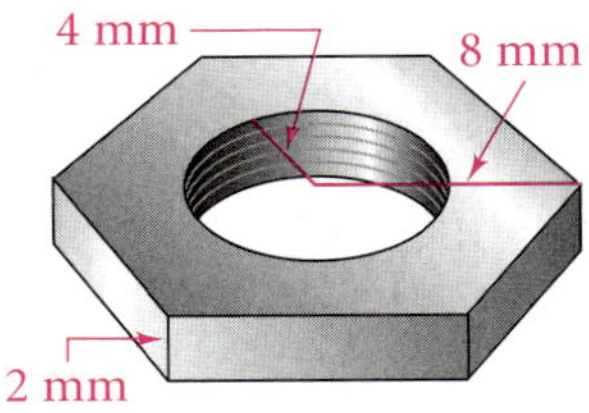

19. **Writing** Describe two ways to find the area of an equilateral triangle that has a 1-in. radius.

Real-World Connection

The length of each side of the Pentagon courtyard is 356 ft. A football field is 360 ft. long.

20. **Architecture** The Pentagon, in Arlington, Virginia, is one of the world's largest office buildings. It is a regular pentagon, and the length of each of its sides is 921 ft. Find the area of this pentagon to the nearest thousand square feet.

Find the perimeter and area of each regular polygon to the nearest tenth.

21.
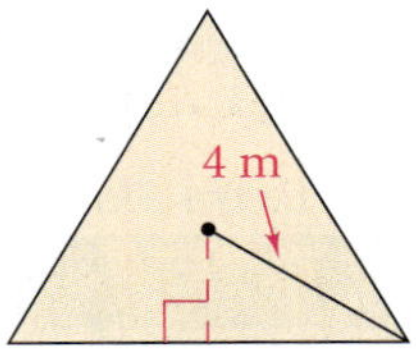

22.
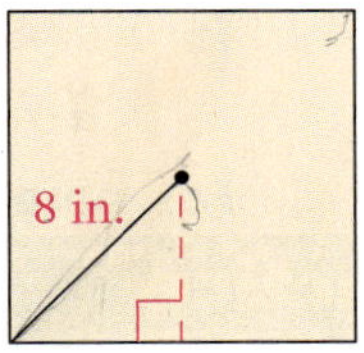

23.
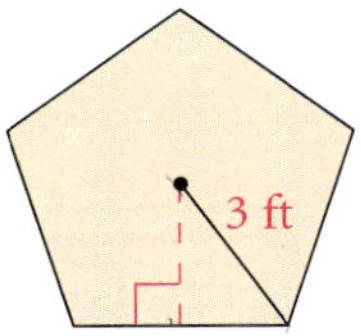

24.
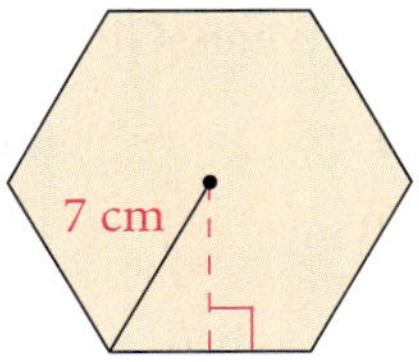

25.
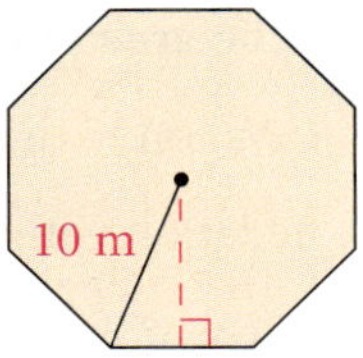

26.
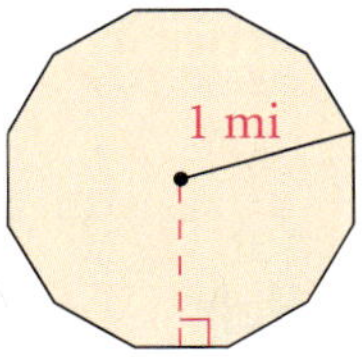

27. **Multiple Choice** Replacement glass for more energy efficient windows costs $\$5/\text{ft}^2$. Approximately how much will you pay for replacement glass for a regular hexagonal window with a radius of 2 ft?

Ⓐ \$10.39 Ⓑ \$27.78 Ⓒ \$45.98 Ⓓ \$51.96

Regular polygons A and B are similar. Compare their areas.

28. The radius of square A is twice the radius of square B.

29. The apothem of pentagon A equals the radius of pentagon B.

30. The length of a side of hexagon A equals the radius of hexagon B.

31. The radius of octagon A equals the apothem of octagon B.

32. The perimeter of decagon A equals the length of a side of decagon B.

Homework Video Tutor

Visit: PHSchool.com
Web Code: bce-1005

33. **Road Signs** The length of a side of the standard stop sign shown at left is 1 ft $\frac{1}{4}$ in. Find the area of the stop sign to the nearest tenth of a square foot.

34. Suppose a circle is inscribed in a regular n-gon with center C and apothem 1, as shown in the diagram. In parts (a)–(d), explain why each statement is true.

a. $m\angle C = \frac{1}{2}\left(\frac{360}{n}\right) = \frac{180}{n}$

b. $s = \tan C$

c. The perimeter of the n-gon is $2n(\tan C)$.

d. The area of the n-gon is $n\left(\tan \frac{180}{n}\right)$.

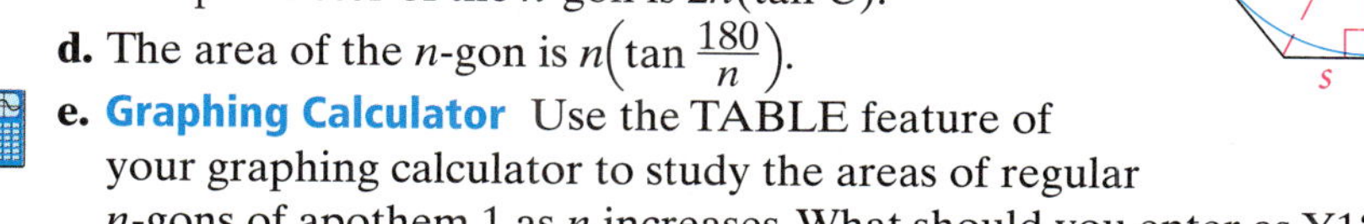

e. **Graphing Calculator** Use the TABLE feature of your graphing calculator to study the areas of regular n-gons of apothem 1 as n increases. What should you enter as Y1?

f. Use the **TBLSET** feature so that X starts at 3 and changes by 1. Access the **TABLE**. Tell what happens in the X and Y1 columns as you scroll down.

g. For what value of X does Y1 take on a new value for the last time? Explain why this is so. Also, in terms of the circle and the circumscribing regular n-gons, interpret what you observe.

35. Suppose a circle is circumscribed about a regular n-gon with center C and radius 1. Proceed with steps similar to those in Exercise 34 to study the areas of regular n-gons inscribed in a circle as n increases.

36. **Surveying** A surveyor wants to mark off a triangular parcel with an area of 1 acre (1 acre = 43,560 ft^2). One side of the triangle extends 300 ft along a straight road. A second side extends at an angle of 65° from one end of the first side. Draw a triangle to represent the piece of land. Determine the length of the second boundary line to the nearest foot.

37. Segments are drawn between the midpoints of consecutive sides of a regular pentagon to form another regular pentagon. Find, to the nearest hundredth, the ratio of the area of the smaller pentagon to the area of the larger pentagon.

Multiple Choice Practice

For California Standards and CAHSEE Tutorials, visit PHSchool.com. Web Code: bcq-9045

CAHSEE 7AF 1.5

38. A store sells fruit baskets containing yellow and red apples. Yellow apples are \$1.80 per pound and red apples are \$2.40 per pound. The apples in one basket weigh 6 pounds and cost \$12.00. The graphs of the equations $x + y = 6$ and $1.8x + 2.4y = 12$ are shown at the right. What information does the location of their intersection provide?

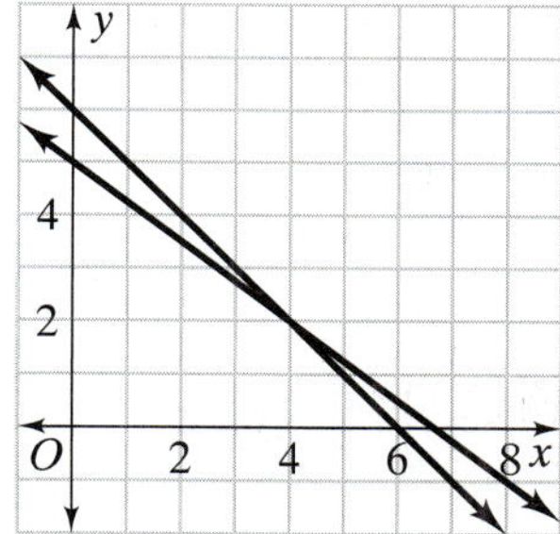

Ⓐ A two-pound basket of apples costs \$4.00.

Ⓑ A four-pound basket of apples costs \$2.00.

Ⓒ The basket contains 2 pounds of yellow apples and 4 pounds of red apples.

Ⓓ The basket contains 4 pounds of yellow apples and 2 pounds of red apples.

GEOM 8.0

39. Determine the perimeter of the regular pentagon at the right. Round your answer to the nearest tenth of an inch.

Ⓐ 4.7 Ⓑ 20 Ⓒ 23.5 Ⓓ 38

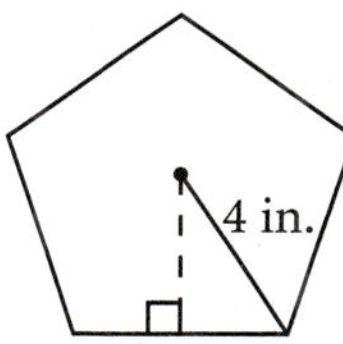

Online Lesson Quiz Visit: PHSchool.com, Web Code: bca-1005

Mixed Review

Lesson 10-4

40. Two regular octagons are pictured.

a. What is the similarity ratio of the smaller octagon to the larger octagon?

b. The area of the larger octagon is 391.1 in.^{2}. What is the area of the smaller octagon to the nearest tenth of a square inch?

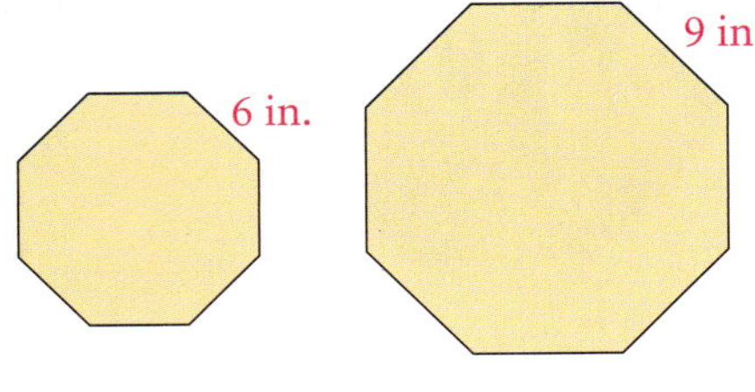

Lesson 8-6

41. Find the sum $\vec{\mathbf{a}} + \vec{\mathbf{c}}$. Give your answer as an ordered pair.

42. Describe a vector $\vec{\mathbf{d}}$ such that $\vec{\mathbf{a}} + \vec{\mathbf{d}} = \vec{\mathbf{c}}$.

43. Describe a vector $\vec{\mathbf{e}}$ such that $\vec{\mathbf{c}} + \vec{\mathbf{e}} = \vec{\mathbf{a}}$.

44. Which two vectors have $\vec{\mathbf{c}}$ as their sum?

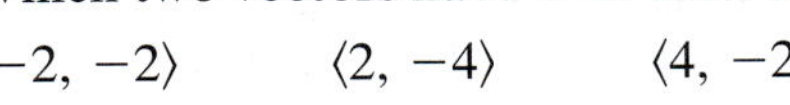

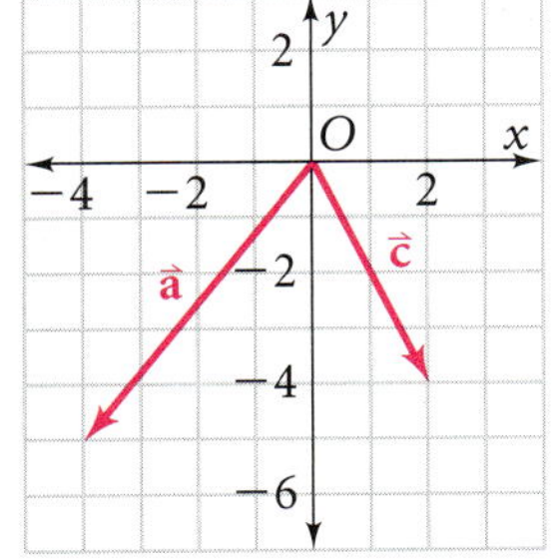

Lesson 6-6

Coordinate Geometry **Find the coordinates of the midpoint of $\overline{WZ}$. Then find WZ.**

45. rectangle

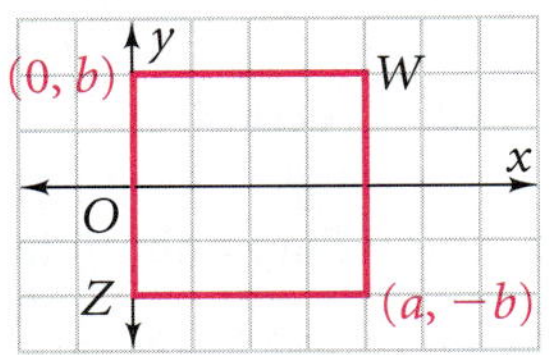

46. kite

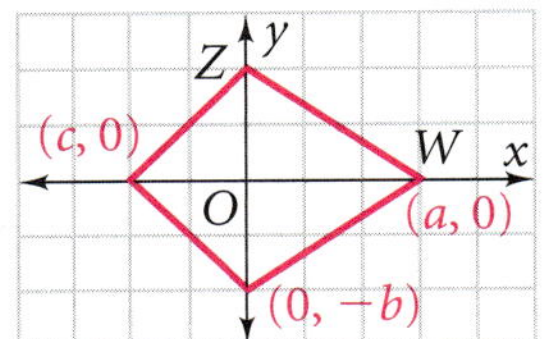

Geometry at Work

Surveyor

Surveyors calculate the locations, shapes, and areas of plots of land. A survey begins with a benchmark, a reference point whose latitude, longitude, and elevation are known. The surveyor uses a device called a transit to measure the boundary angles for the plot of land and the distances of key points from the benchmark. Using trigonometry, the surveyor finds the latitude, longitude, and elevation of each key point in the survey. These points are located on a map and an accurate sketch of the plot is drawn. Finally, the area is calculated.

One method involves dividing the plot into triangles and measuring the lengths of two sides and the included angle of each. The formula $A = \frac{1}{2}ab(\sin C)$ gives the area of each triangle. The area of the entire plot is the sum of the areas of the triangles.

For: Information about surveying
Web Code: bcb-2031

Extension Laws of Sines and Cosines

For Use With Lesson 10-5

You have seen *sine* and *cosine* defined in terms of acute angles in right triangles. In math courses after Algebra 2, you will learn how to extend the trigonometric relationships to angle measures in general—obtuse angle measures, right angle measures, even angle measures that are 180° or more, or 0° or less!

One result of this will be two special laws that can be applied to any triangle.

Law of Sines

$$\frac{\sin A}{a} = \frac{\sin B}{b} = \frac{\sin C}{c}$$

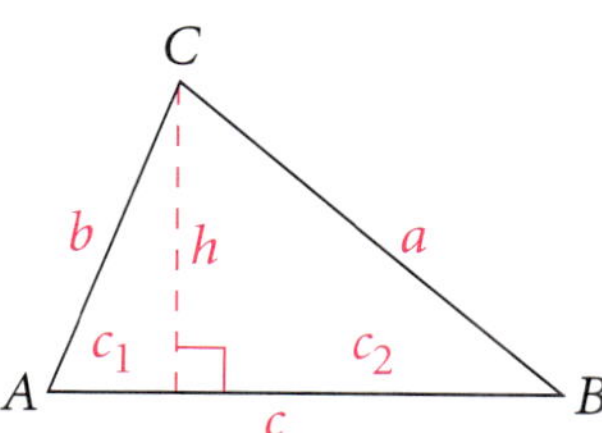

Law of Cosines

$$a^2 = b^2 + c^2 - 2bc\cos A$$
$$b^2 = a^2 + c^2 - 2ac\cos B$$
$$c^2 = a^2 + b^2 - 2ab\cos C$$

EXERCISES

1. Here is the first step of a proof that $\frac{\sin A}{a} = \frac{\sin B}{b}$:

1 $\frac{h}{ab} = \frac{h}{ab}$

Complete the proof.

2. Here are the first three steps of a proof that $a^2 = b^2 + c^2 - 2bc\cos A$:

1 $b^2 = {c_1}^2 + h^2$

2 $c^2 = (c_1 + c_2)^2 = {c_1}^2 + 2c_1c_2 + {c_2}^2$

3 $-2bc\cos A = -2b(c_1 + c_2)\frac{c_1}{b} = -2{c_1}^2 - 2c_1c_2$

Complete the proof.

The Law of Sines lets you find missing measures in a triangle when you know the measures of two angles and a side, or two sides and a nonincluded angle. Use it to find the missing measures in each triangle.

3.

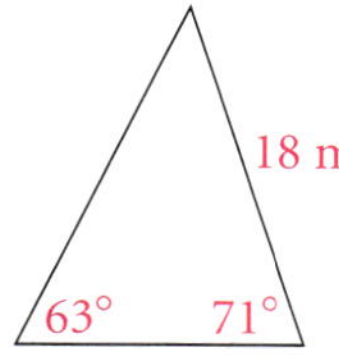

4.

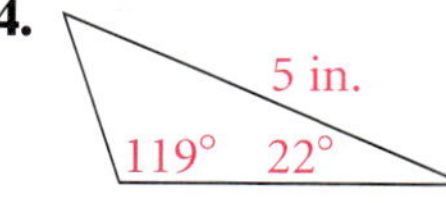

5.

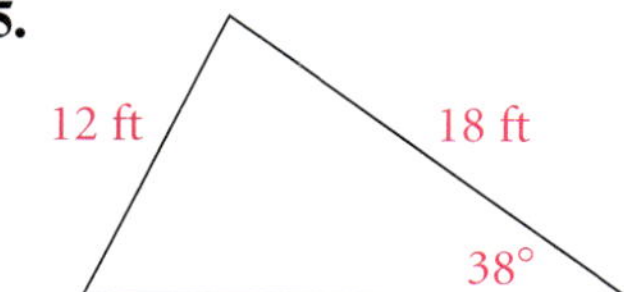

The Law of Cosines lets you find missing measures in a triangle when you know the measures of two sides and the included angle, or three sides. Use it to find the missing measures in each triangle.

6.

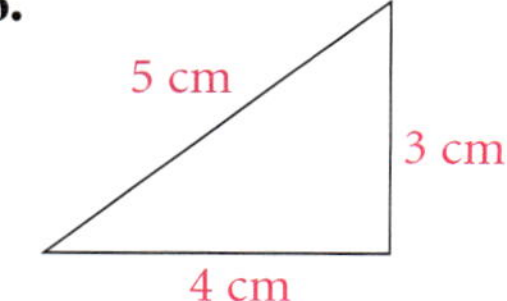

7.

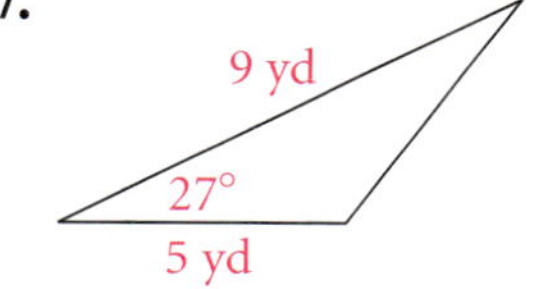

8.

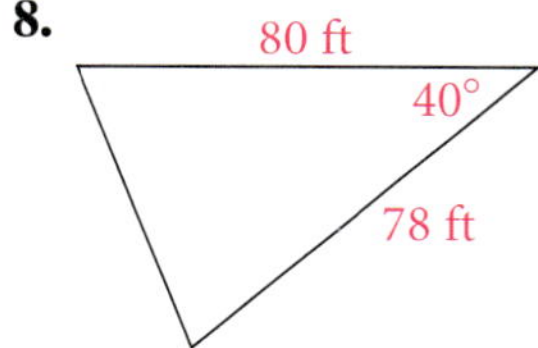

10-6 Circles and Arcs

California Content Standards

GEOM 7.0 Students prove and use theorems involving the properties of circles. ***Introduce***
GEOM 8.0 Students know, derive, and solve problems involving circumference. ***Introduce***

What You'll Learn

- To find the measures of central angles and arcs
- To find circumference and arc length

. . . And Why

To use the turning radius of a car to compare the distances that its tires travel, as in Example 4

Check Skills You'll Need  **for Help** Lesson 1-9 and Skills Handbook, p. 761

Find the diameter or radius of each circle.

1. $r = 7$ cm, $d = \blacksquare$ **2.** $r = 1.6$ m, $d = \blacksquare$

3. $d = 10$ ft, $r = \blacksquare$ **4.** $d = 5$ in., $r = \blacksquare$

Round to the nearest whole number.

5. 9% of 360 **6.** 38% of 360 **7.** 50% of 360 **8.** 21% of 360

New Vocabulary • circle • center • radius • congruent circles • diameter • central angle • semicircle • minor arc • major arc • adjacent arcs • circumference • pi • concentric circles • arc length • congruent arcs

1 Central Angles and Arcs

Vocabulary Tip

Diameter comes from the classical Greek words *dia*, meaning through, and *meter*, meaning measure.

In a plane, a **circle** is the set of all points equidistant from a given point called the **center.** You name a circle by its center. Circle P ($\odot P$) is shown at the right.

A **radius** is a segment that has one endpoint at the center and the other endpoint on the circle. $\overline{PC}$ is a radius. $\overline{PA}$ and $\overline{PB}$ are also radii. **Congruent circles** have congruent radii.

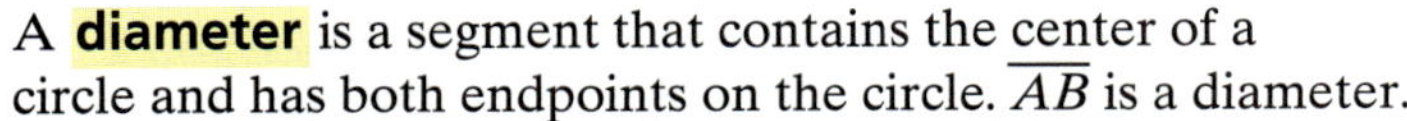

A **diameter** is a segment that contains the center of a circle and has both endpoints on the circle. $\overline{AB}$ is a diameter.

A **central angle** is an angle whose vertex is the center of the circle. $\angle CPA$ is a central angle.

1 EXAMPLE **Real-World Connection**

To learn how people really spend their time, a research firm studied the hour-by-hour activities of 3600 people. The participants were between 18 and 90 years old. Each participant was sent a 24-hour recording sheet every March for three years from 2000 to 2002.

Some information from the study is shown in this circle graph. What is the measure, in degrees, of the central angle used for the Entertainment part?

There are 360 degrees in a circle. To find the measure of a central angle in the circle graph, find the corresponding percent of 360. Entertainment is 18%, and 18% of 360 = $0.18 \cdot 360$, or 64.8.

You can use the same method to find the measures of the other central angles.

Sleep: 31% of 360 = 111.6 | Other: 15% of 360 = 54

Food: 9% of 360 = 32.4 | Must Do: 7% of 360 = 25.2

Work: 20% of 360 = 72

1 **a.** **Critical Thinking** Each section of the circle graph represents a measurable quantity. What is that quantity?

b. Each section of the circle graph represents an average. Explain.

Test-Taking Tip

You can find the measure *a* of a central angle by using a proportion. For Entertainment (18%) in Example 1: $\frac{18}{100} = \frac{a}{360}$.

An arc is a part of a circle. One type of arc, a **semicircle,** is half of a circle. A **minor arc** is smaller than a semicircle. A **major arc** is greater than a semicircle.

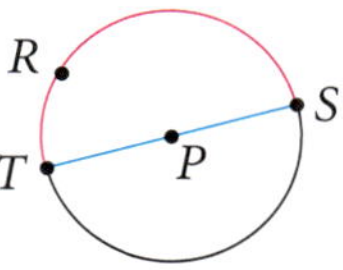

$\widehat{TRS}$ is a semicircle.
$m\widehat{TRS} = 180$

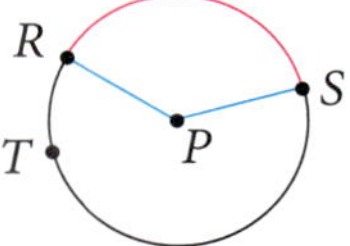

$\widehat{RS}$ is a minor arc.
$m\widehat{RS} = m\angle RPS$

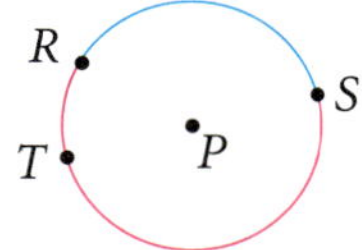

$\widehat{RTS}$ is a major arc.
$m\widehat{RTS} = 360 - m\widehat{RS}$

The measure of a semicircle is 180. The measure of a minor arc is the measure of its corresponding central angle. The measure of a major arc is 360 minus the measure of its related minor arc.

The water line separates a circle into a major arc and a minor arc.

2 EXAMPLE Identifying Arcs

Identify the following in $\odot O$.

A C O D E

a. the minor arcs

$\widehat{AD}$, $\widehat{CE}$, $\widehat{AC}$, and $\widehat{DE}$ are minor arcs.

b. the semicircles

$\widehat{ACE}$, $\widehat{CED}$, $\widehat{EDA}$, and $\widehat{DAC}$ are semicircles.

c. the major arcs that contain point *A*

$\widehat{ACD}$, $\widehat{CEA}$, $\widehat{EDC}$, and $\widehat{DAE}$ are major arcs that contain point *A*.

2 Identify the four major arcs of $\odot O$ that contain point *E*.

Adjacent arcs are arcs of the same circle that have exactly one point in common. You can add the measures of adjacent arcs just as you can add the measures of adjacent angles.

Key Concepts

Postulate 10-1 **Arc Addition Postulate**

The measure of the arc formed by two adjacent arcs is the sum of the measures of the two arcs.

$$m\widehat{ABC} = m\widehat{AB} + m\widehat{BC}$$

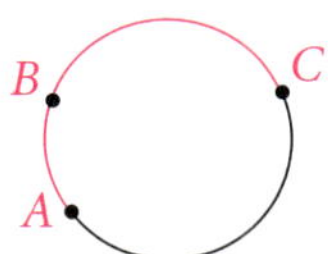

3 EXAMPLE Finding the Measures of Arcs

Find the measure of each arc.

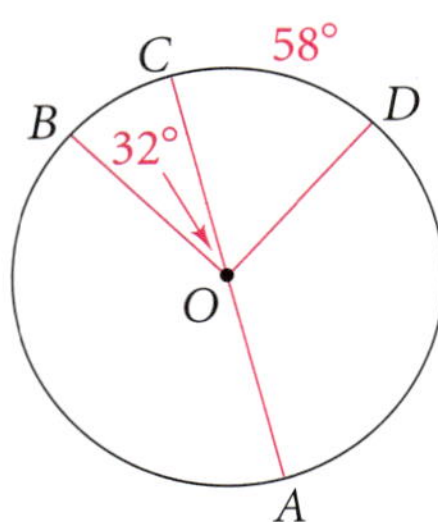

a. $\overset{\frown}{BC}$ $\quad m\overset{\frown}{BC} = m\angle BOC = 32$

b. $\overset{\frown}{BD}$ $\quad m\overset{\frown}{BD} = m\overset{\frown}{BC} + m\overset{\frown}{CD}$

$\quad m\overset{\frown}{BD} = 32 + 58 = 90$

c. $\overset{\frown}{ABC}$ $\quad \overset{\frown}{ABC}$ is a semicircle.

$\quad m\overset{\frown}{ABC} = 180$

d. $\overset{\frown}{AB}$ $\quad m\overset{\frown}{AB} = 180 - 32 = 148$

CA Standards Check 3 Find $m\angle COD$, $m\overset{\frown}{CDA}$, $m\overset{\frown}{AD}$ and $m\overset{\frown}{BAD}$.

2 Circumference and Arc Length

The **circumference** of a circle is the distance around the circle. The number **pi** (π) is the ratio of the circumference of a circle to its diameter.

Key Concepts

Theorem 10-9 **Circumference of a Circle**

The circumference of a circle is π times the diameter.

$$C = \pi d \text{ or } C = 2\pi r$$

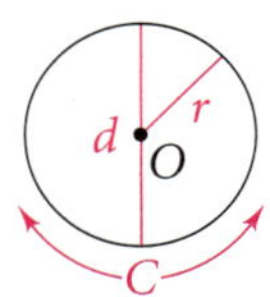

Since the number π is irrational, you cannot write it as a terminating or repeating decimal. To approximate π, you can use 3.14, $\frac{22}{7}$, or the π key on your calculator.

Circles that lie in the same plane and have the same center are **concentric circles.**

4 EXAMPLE Real-World Connection

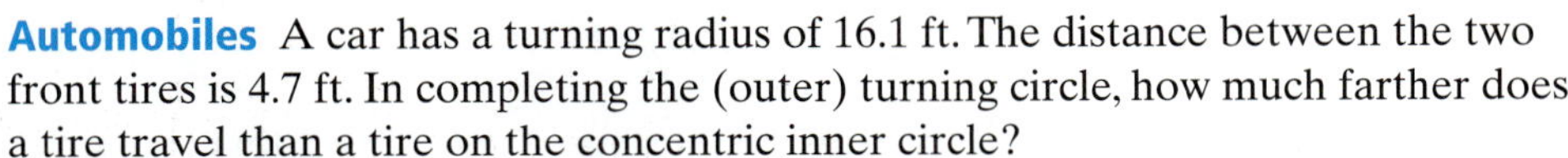

Automobiles A car has a turning radius of 16.1 ft. The distance between the two front tires is 4.7 ft. In completing the (outer) turning circle, how much farther does a tire travel than a tire on the concentric inner circle?

To find the radius of the inner circle, subtract 4.7 ft from the turning radius.

circumference of outer circle $= C = 2\pi r = 2\pi(16.1) = 32.2\pi$

radius of the inner circle $= 16.1 - 4.7 = 11.4$

circumference of inner circle $= C = 2\pi r = 2\pi(11.4) = 22.8\pi$

The difference in the two distances is $32.2\pi - 22.8\pi$, or 9.4π.

$9.4\pi \approx 29.530971$ **Use a calculator.**

A tire on the turning circle travels about 29.5 ft farther than a tire on the inner circle.

CA Standards Check 4 The diameter of a bicycle wheel is 22 in. To the nearest whole number, how many revolutions does the wheel make when the bicycle travels 100 ft?

The measure of an arc is in degrees while the **arc length** is a fraction of a circle's circumference. An arc of 60° represents $\frac{60}{360}$ or $\frac{1}{6}$ of the circle. Its arc length is $\frac{1}{6}$ the circumference of the circle. This observation suggests the following theorem.

Key Concepts

Theorem 10-10 **Arc Length**

The length of an arc of a circle is the product of the ratio $\frac{\text{measure of the arc}}{360}$ and the circumference of the circle.

$$\text{length of } \widehat{AB} = \frac{m\widehat{AB}}{360} \cdot 2\pi r$$

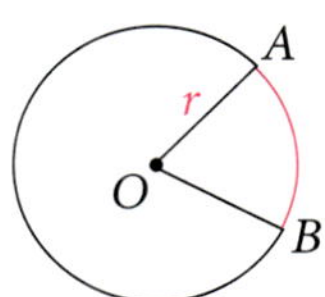

5 EXAMPLE Finding Arc Length

Find the length of each arc shown in red. Leave your answer in terms of π.

a.

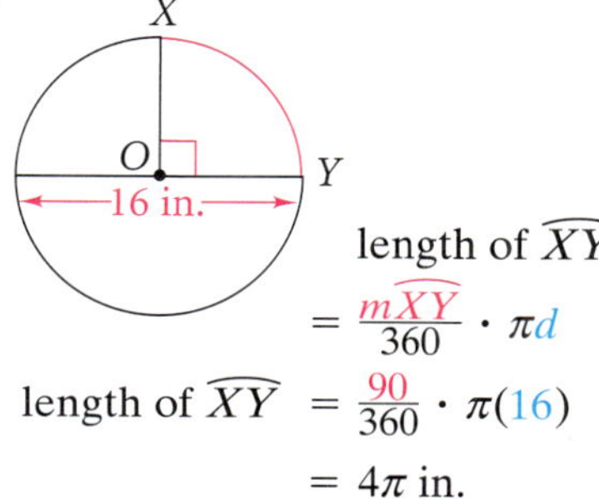

length of $\widehat{XY}$

$= \frac{m\widehat{XY}}{360} \cdot \pi d$

length of $\widehat{XY} = \frac{90}{360} \cdot \pi(16)$

$= 4\pi$ in.

b.

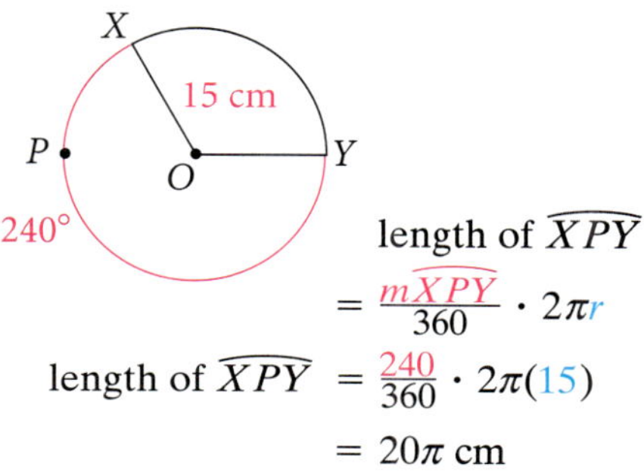

length of $\widehat{XPY}$

$= \frac{m\widehat{XPY}}{360} \cdot 2\pi r$

length of $\widehat{XPY} = \frac{240}{360} \cdot 2\pi(15)$

$= 20\pi$ cm

CA Standards Check **5** Find the length of a semicircle with radius 1.3 m. Leave your answer in terms of π.

60° 60°

It is possible for two arcs of different circles to have the same measure but different lengths, as shown at the left. It is also possible for two arcs of different circles to have the same length but different measures. **Congruent arcs** are arcs that have the same measure *and* are in the same circle or in congruent circles.

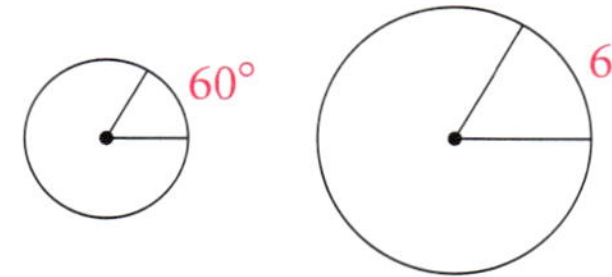

EXERCISES

For more exercises, see *Extra Skill, Word Problem, and Proof Practice.*

Standards Practice GEOM 7.0, 8.0

A Practice by Example

Example 1 (page 566)

GO for Help

Trash **The graph shows types of trash in a typical American city. Find the measure of each central angle to the nearest whole number.**

1. Glass
2. Metals
3. Plastics
4. Wood
5. Food Waste
6. Yard Waste
7. Other
8. Paper and Paperboard

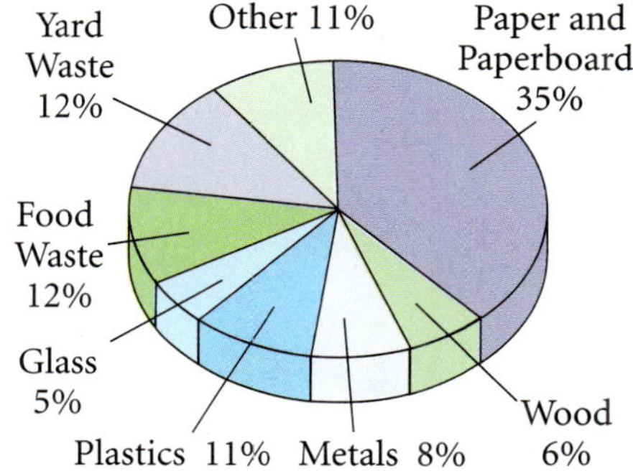

SOURCE: Environmental Protection Agency, 2003. Go to **www.PHSchool.com** for a data update. Web Code: bcg-9041

Example 2 (page 567)

Identify the following in $\odot O$.

9. a minor arc

10. a major arc

11. a semicircle

12. a pair of adjacent arcs

13. an acute central angle

14. a pair of congruent angles

Example 3 (page 568)

Find the measure of each arc in $\odot P$.

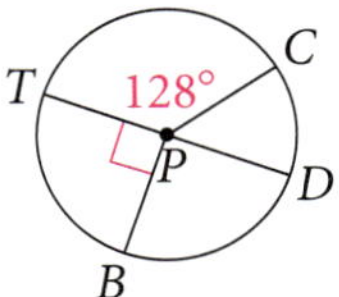

15. $\overset{\frown}{TC}$ **16.** $\overset{\frown}{TBD}$ **17.** $\overset{\frown}{BTC}$ **18.** $\overset{\frown}{TCB}$

19. $\overset{\frown}{CD}$ **20.** $\overset{\frown}{CBD}$ **21.** $\overset{\frown}{TCD}$ **22.** $\overset{\frown}{DB}$

23. $\overset{\frown}{TDC}$ **24.** $\overset{\frown}{TB}$ **25.** $\overset{\frown}{BC}$ **26.** $\overset{\frown}{BCD}$

Example 4 (page 568)

Find the circumference of each circle. Leave your answer in terms of π.

27.

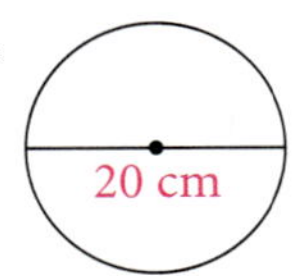

28.

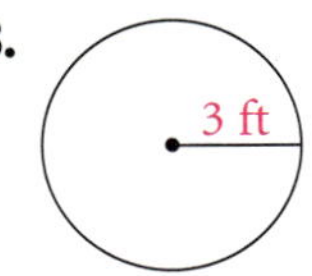

29.

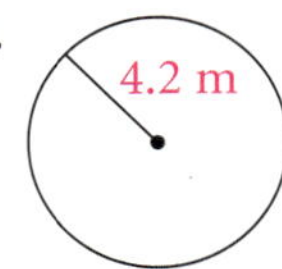

30.

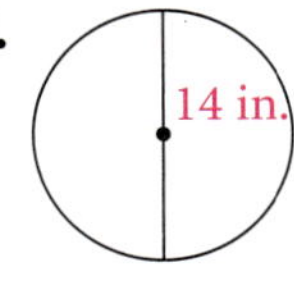

31.

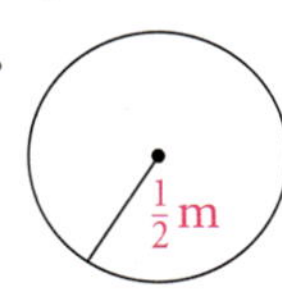

32.

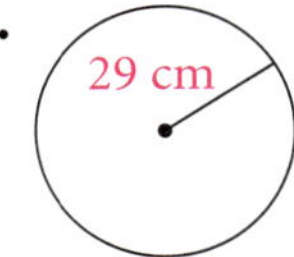

33. The wheel of an adult's bicycle has diameter 26 in. The wheel of a child's bicycle has diameter 18 in. To the nearest inch, how much farther does the larger bicycle wheel travel in one revolution than the smaller bicycle wheel?

Example 5 (page 569)

Find the length of each arc shown in red. Leave your answer in terms of π.

34.

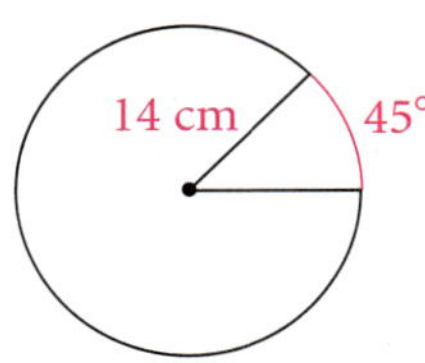

35.

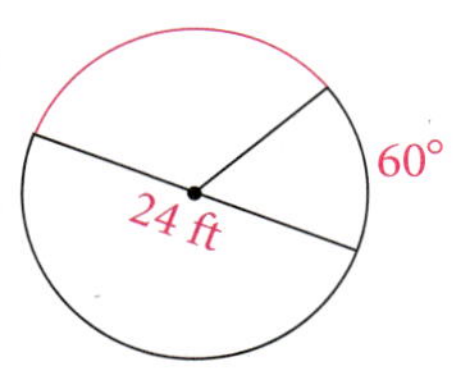

36.

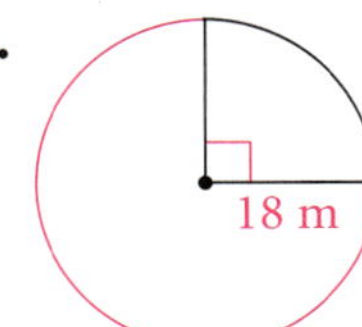

37.

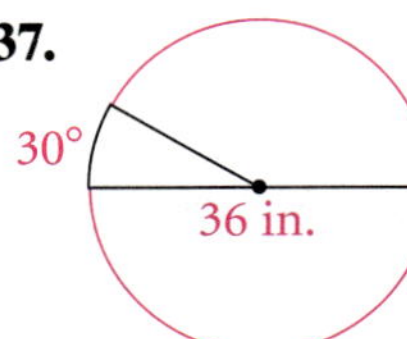

38.

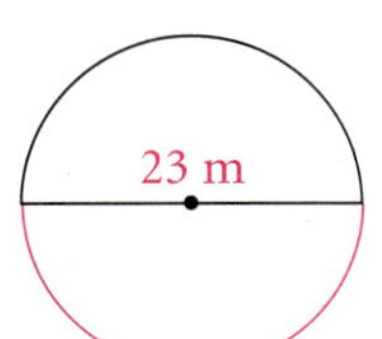

39.

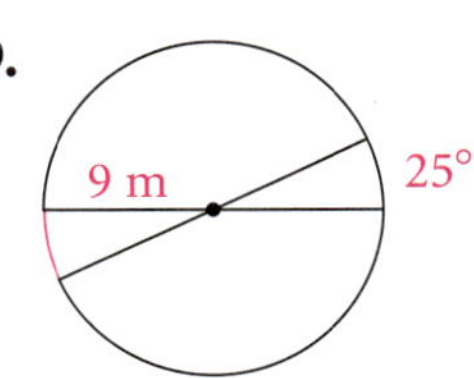

Apply Your Skills

40. Use a compass to draw $\odot A$ and $\odot B$ with different radii. Then use a protractor to draw $\overset{\frown}{XY}$ on $\odot A$ and $\overset{\frown}{ZW}$ on $\odot B$ so that $m\overset{\frown}{XY} = m\overset{\frown}{ZW}$. Is $\overset{\frown}{XY} \cong \overset{\frown}{ZW}$?

41. Surveys Use the data in the table to construct a circle graph.

Interest in Languages by Students at McClellan High School

Language	Percent
German	24%
Japanese	13%
Chinese	12%
French	25%
Spanish	26%

Visit: PHSchool.com
Web Code: bce-1006

Exercise 51

Find each indicated measure for $\odot O$.

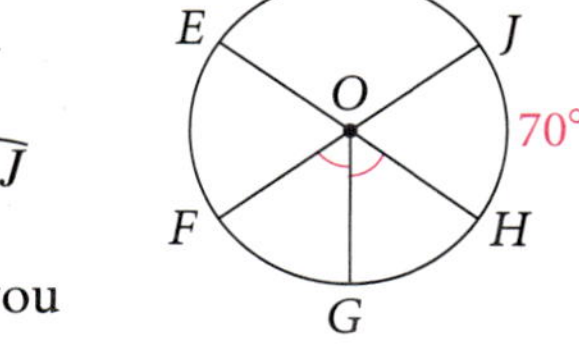

42. $m\angle EOF$ **43.** $m\widehat{EJH}$ **44.** $m\widehat{FH}$

45. $m\angle FOG$ **46.** $m\widehat{JEG}$ **47.** $m\widehat{HFJ}$

48. Open-Ended Make a circle graph showing how you spend a 24-hour weekday.

Time Hands of a clock suggest an angle whose measure is continually changing.

49. Through how many degrees does a minute hand move in each time interval?
a. 1 minute **b.** 5 minutes **c.** 20 minutes

50. Through how many degrees does an hour hand move in each time interval?
a. 1 minute **b.** 5 minutes **c.** 20 minutes

51. What is the measure of the angle formed by the hands of a clock at 7:20?

x^2 **Algebra Find the value of each variable.**

52.

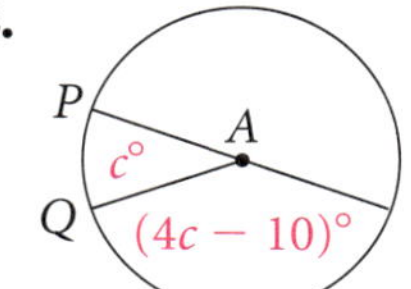

53.

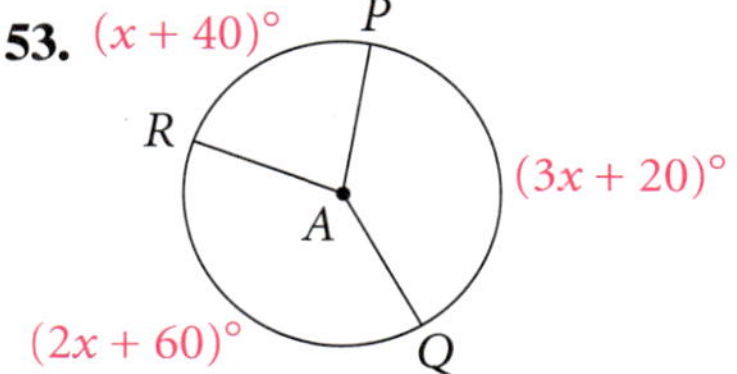

The circumference of a circle is 100π in. Find each of the following.

54. the diameter **55.** the radius

56. the length of an arc of 120°

57. Multiple Choice Five streets come together at a traffic circle. The diameter of the circle is 200 ft. If traffic travels counterclockwise, what is the approximate distance from East St. to Neponset St.?

Ⓐ 227 ft Ⓒ 454 ft
Ⓑ 244 ft Ⓓ 488 ft

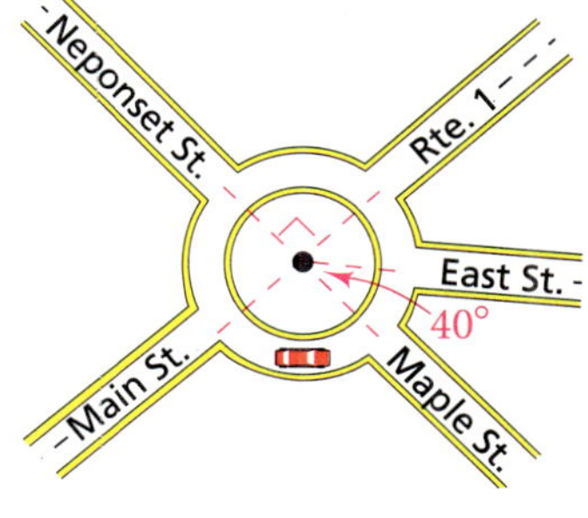

Problem Solving Hint

For Exercise 58, draw $\odot A$ and $\odot B$ concentric. Draw 60° and 45° angles that share a side. To have equal arc lengths, which circle must be larger?

58. A 60° arc of $\odot A$ has the same length as a 45° arc of $\odot B$. Find the ratio of the radius of $\odot A$ to the radius of $\odot B$.

59. Metalworking Nina designed an arch made of wrought iron for the top of a mall entrance. The 11 segments between the two concentric semicircles are each 3 ft long. Find the total length of wrought iron used to make this structure. Round your answer to the nearest foot.

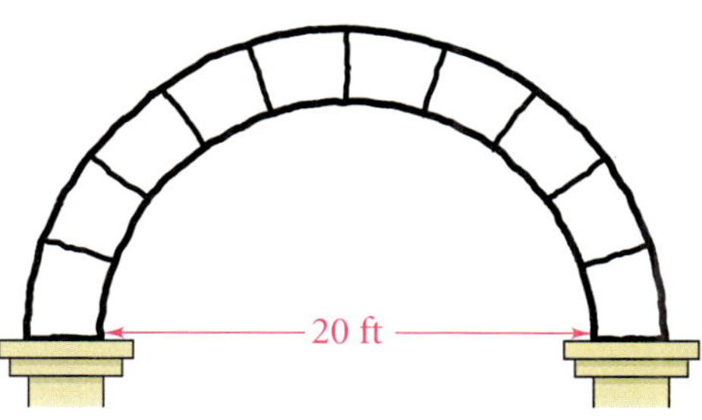

60. History In Exercise 31 on page 133, you learned that in 220 B.C., Eratosthenes estimated the circumference of Earth. He did so by finding that on a great circle of Earth, an arc of approximately 500 mi has a central angle of 7.2°.
a. Use Eratosthenes's measurements to estimate the circumference of Earth.
b. Compare your answer in part (a) to the actual circumference of Earth (at the equator) of 24,902 mi.

The Distance and Midpoint Formulas are on pages 53 and 55.

Coordinate Geometry A diameter of a circle has endpoints $A(1, 3)$ and $B(4, 7)$. Find each of the following.

61. the coordinates of the center

62. the circumference

Find the length of each arc shown in red. Leave your answer in terms of π.

63.

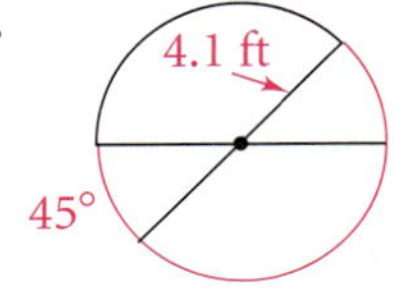

64.

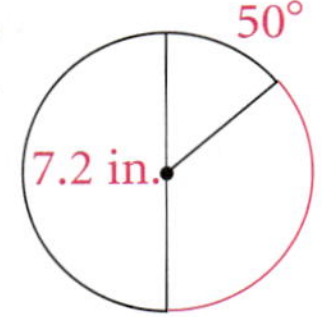

65.

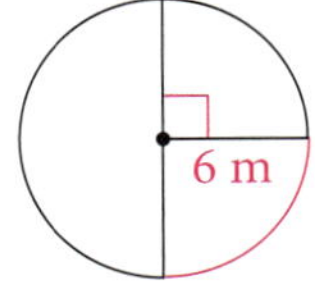

Use what you learn from Calvin's father to answer Exercises 66 and 67.

Calvin and Hobbes by Bill Watterson

BUT THE POINT ON THE RECORD'S EDGE HAS TO MAKE A BIGGER CIRCLE IN THE SAME TIME, SO IT GOES FASTER. SEE, TWO POINTS ON ONE DISK MOVE AT TWO SPEEDS, EVEN THOUGH THEY BOTH MAKE THE SAME REVOLUTIONS PER MINUTE!

66. In one revolution, how much farther does a point 10 cm from the center of the record travel than a point 3 cm from the center? Round your answer to the nearest tenth.

67. Writing Kendra and her mother plan to ride the carousel. Two horses on the carousel are side by side. For a more exciting ride, should Kendra sit on the inside or the outside? Explain your reasoning.

68. In $\odot O$, the length of $\widehat{AB}$ is 6π cm and $m\widehat{AB}$ is 120. What is the diameter of $\odot O$?

69. Coordinate Geometry Find the length of a semicircle with endpoints $(3, 7)$ and $(3, -1)$. Round your answer to the nearest tenth.

70. The two circles shown below are concentric.

a. Name two arcs that have the same measure.

b. Find the value of x.

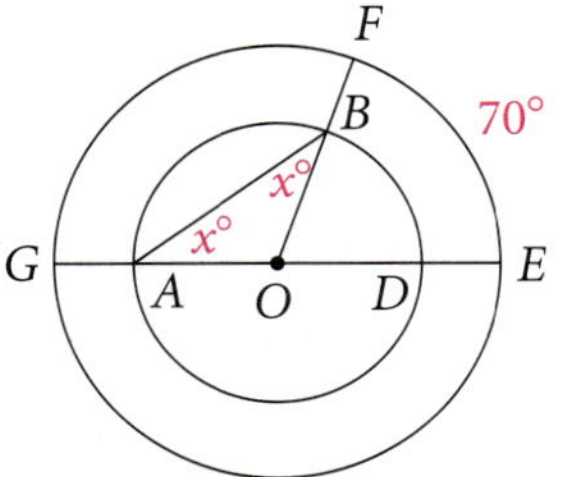

71. Find the perimeter of the shaded portion of the figure below. Leave your answer in terms of π. Explain your reasoning and state what assumptions you make.

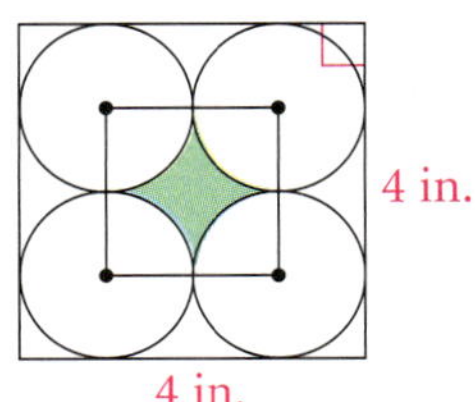

Real-World Connection

The track is longer for runners on the outside, so the start of the race is staggered.

72. Sports An athletic field is a rectangle, 100 yd by 40 yd, with a semicircle at each of the short sides. A running track 10 yd wide surrounds the field. If the track is divided into eight lanes of equal width, find the distance around the track along the inside edge of each lane.

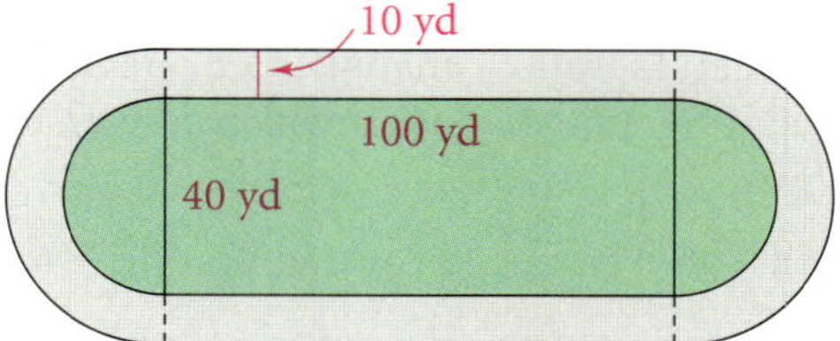

Multiple Choice Practice

For California Standards and CAHSEE Tutorials, visit PHSchool.com. Web Code: bcq-9045

GEOM 8.0

73. What is the circumference of a circle with a diameter of 8 meters?

(A) 16π m (B) 8π m (C) 4π m (D) 2π m

GEOM 11.0

74. Each side of a square with an area of 144 square centimeters is multiplied by 5. What is the perimeter of the new square?

(A) 120 cm (B) 240 cm (C) 360 cm (D) 720 cm

GEOM 8.0

75. Earth's equator is a circle around Earth halfway between the North Pole and the South Pole. An astronomer mathematically expressed the length of the equator. Which measure of Earth did the astronomer report?

(A) Radius (C) Surface area
(B) Volume (D) Circumference

Mixed Review

Lesson 10-5

Part of a regular 12-gon is shown at the right.

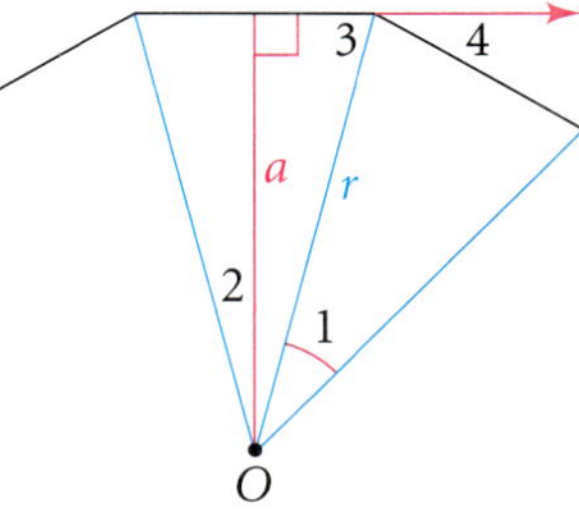

76. Find the measure of each numbered angle.

77. The radius is 19.3 mm. Find the apothem.

78. Find the area of the 12-gon to the nearest square millimeter.

Lesson 6-3

Can you conclude that the figure is a parallelogram? Explain.

79. 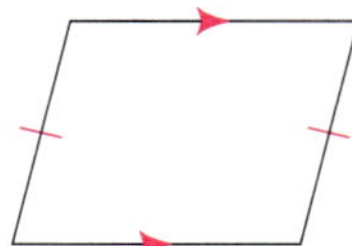

80. 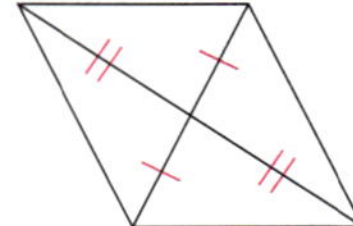

81. 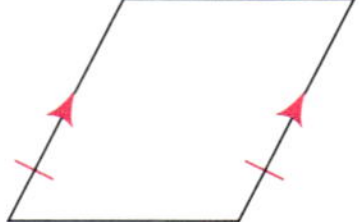

Lesson 3-7

Indicate whether each statement is *always, sometimes,* or *never* true.

82. Two nonvertical parallel lines have the same slope.

83. Two perpendicular lines have slopes that are reciprocals.

Online Lesson Quiz Visit: PHSchool.com, Web Code: bca-1006

Algebra 1 Review

Dimensional Analysis

For Use With Lesson 10-7

You can use conversion factors to change from one unit of measure to another. The process of analyzing units to decide which conversion factors to use is called *dimensional analysis.*

Since 60 min = 1 h, $\frac{60 \text{ min}}{1 \text{ h}}$ equals 1. You can use $\frac{60 \text{ min}}{1 \text{ h}}$ to convert hours to minutes.

$7 \cancel{\text{h}} \cdot \frac{60 \text{ min}}{1 \cancel{\text{h}}} = 420 \text{ min}$ **The hour units cancel, and the result is minutes.**

Sometimes you need to use a conversion factor more than once.

EXAMPLE

The area of the top of a circular table is 8 ft^2. Convert the area to square inches.

You need to convert feet to inches.

feet to inches ↓ feet to inches ↓

$8 \text{ ft} \cdot \text{ft} \cdot \frac{12 \text{ in.}}{1 \text{ ft}} \cdot \frac{12 \text{ in.}}{1 \text{ ft}} = 8 \cancel{\text{ft}} \cdot \cancel{\text{ft}} \cdot \frac{12 \text{ in.}}{1 \cancel{\text{ft}}} \cdot \frac{12 \text{ in.}}{1 \cancel{\text{ft}}}$ **The feet units cancel. The result is square inches.**

$= 8 \cdot 12 \cdot 12 \text{ in.}^2$ **Simplify.**

$= 1152 \text{ in.}^2$

The area of the table top is 1152 in.^{2}.

EXERCISES

Choose the correct conversion factor for changing the units.

1. centimeters to meters A. $\frac{100 \text{ cm}}{1 \text{ m}}$ B. $\frac{1 \text{ m}}{100 \text{ cm}}$

2. yards to feet A. $\frac{3 \text{ ft}}{1 \text{ yd}}$ B. $\frac{1 \text{ yd}}{3 \text{ ft}}$

3. inches to yards A. $\frac{36 \text{ in.}}{1 \text{ yd}}$ B. $\frac{1 \text{ yd}}{36 \text{ in.}}$

4. square feet to square inches A. $\frac{144 \text{ in.}^2}{1 \text{ ft}^2}$ B. $\frac{1 \text{ ft}^2}{144 \text{ in.}^2}$

5. cubic meters to cubic kilometers A. $\frac{1{,}000{,}000{,}000 \text{ m}^3}{1 \text{ km}^3}$ B. $\frac{1 \text{ km}^3}{1{,}000{,}000{,}000 \text{ m}^3}$

Write each quantity in the given unit.

6. 4 m = ■ cm

7. 360 in. = ■ yd

8. 9 mm = ■ cm

9. 17 yd = ■ ft

10. 2.5 ft = ■ in.

11. 35 m = ■ km

12. 2 yd = ■ in.

13. 23 cm = ■ mm

14. 2 ft^2 = ■ in.^{2}

15. 500 mm^2 = ■ cm^2

16. 840 in.^{2} = ■ ft^2

17. 3 km^2 = ■ cm^2

18. 7 m^2 = ■ km^2

19. 360 in.^{2} = ■ yd^2

20. 900 cm^3 = ■ m^3

21. 4 yd^3 = ■ ft^3

22. 75 m^2 = ■ km^2

23. 250 in.^{2} = ■ yd^2

24. 3 km^3 = ■ m^3

25. 2 km^2 = ■ mm^2

26. 60 in.^{2} = ■ yd^2

10-7 Areas of Circles and Sectors

California Content Standards

GEOM 8.0 Students know, derive, and solve problems involving the circumference and area of common geometric figures. ***Develop***

What You'll Learn

- To find the areas of circles, sectors, and segments of circles

. . . And Why

To compare the area of different-size pizzas, as in Example 1

Check Skills You'll Need

GO for Help Lesson 10-6

1. What is the radius of a circle with diameter 9 cm?
2. What is the diameter of a circle with radius 8 ft?
3. Find the circumference of a circle with diameter 12 in.
4. Find the circumference of a circle with radius 3 m.

New Vocabulary • sector of a circle • segment of a circle

1 Finding Areas of Circles and Parts of Circles

Hands-On Activity: Exploring the Area of a Circle

- Use a compass to draw a large circle. Fold the circle horizontally and vertically. Cut the circle into four wedges on the fold lines.

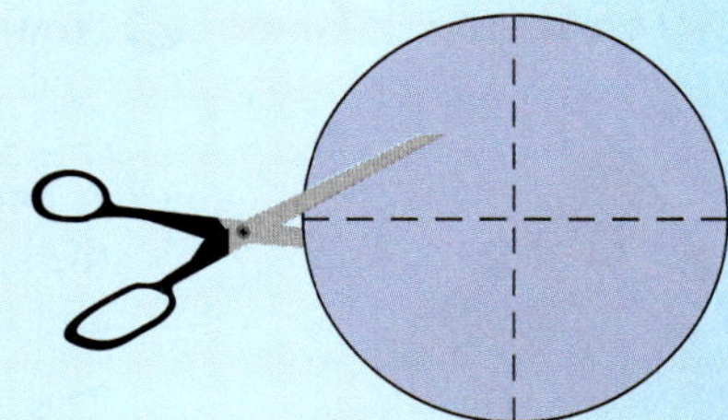

- Fold each wedge into quarters. Cut each wedge on the fold lines. You will have 16 wedges.

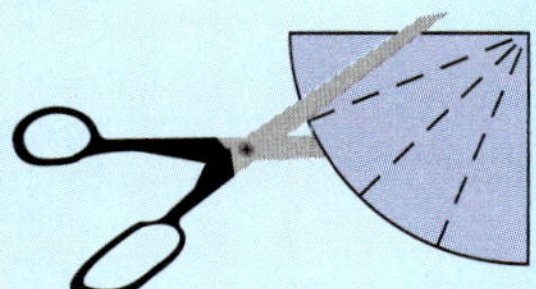

- Tape the wedges to a piece of paper to form the figure shown here.

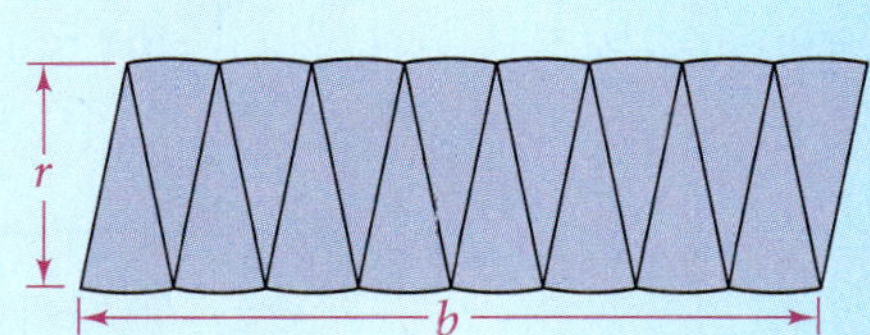

1. How does the area of the figure compare with area of the circle?
2. The base of the figure is formed by arcs of the circle. Explain how the length b relates to the circumference C of the circle.
3. Explain how the length b relates to the radius r of the circle.
4. If you increase the number of wedges, the figure you create becomes more and more like a rectangle with base b and height r. Write an expression for the area of the rectangle in terms of r.

In the diagrams on the preceding page,

area of a circle = area of a "parallelogram" $\approx b \cdot r \approx \frac{1}{2}C \cdot r = \pi r^2$,

and the approximations improve as the circle is cut into more pieces.

Theorem 10-11 Area of a Circle

The area of a circle is the product of π and the square of the radius.

$$A = \pi r^2$$

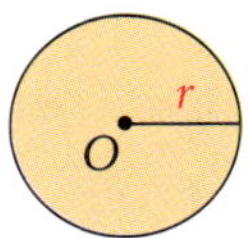

1 EXAMPLE Real-World Connection

Food How much more pizza is in a 12-in.-diameter pizza than in a 10-in. pizza?

radius of small pizza $= \frac{10}{2} = 5$ — **Find the radii.**

radius of medium pizza $= \frac{12}{2} = 6$

area of small pizza $= \pi(5)^2 = 25\pi$ — **Use the formula for area of a circle.**

area of medium pizza $= \pi(6)^2 = 36\pi$

difference in area $= 36\pi - 25\pi = 11\pi$

≈ 34.557519 — **Use a calculator.**

There is about 35 in.^{2} more pizza in the medium pizza.

1 How much more pizza is in a 14-in.-diameter pizza than in a 12-in. pizza?

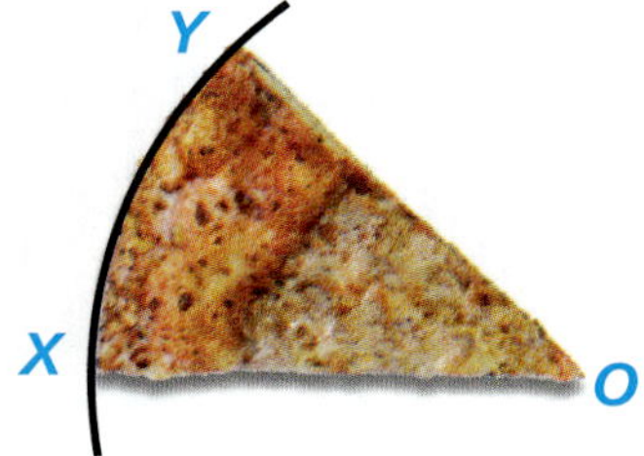

A **sector of a circle** is a region bounded by an arc of the circle and the two radii to the arc's endpoints. You name a sector using one arc endpoint, the center of the circle, and the other arc endpoint. The slice of pizza at the left is sector XOY of a circle O.

The area of a sector is a fractional part of the area of a circle. The ratio of a sector's area to a circle's area is $\frac{\text{measure of the arc}}{360}$.

Key Concepts

Theorem 10-12 Area of a Sector of a Circle

The area of a sector of a circle is the product of the ratio $\frac{\text{measure of the arc}}{360}$ and the area of the circle.

$$\text{Area of sector } AOB = \frac{m\overset{\frown}{AB}}{360} \cdot \pi r^2$$

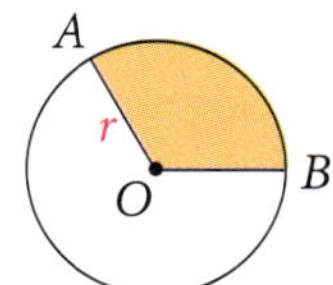

2 EXAMPLE Finding the Area of a Sector of a Circle

Find the area of sector ZOM. Leave your answer in terms of π.

area of sector $ZOM = \frac{m\overset{\frown}{ZM}}{360} \cdot \pi r^2$

$= \frac{72}{360} \cdot \pi(20)^2$

$= 80\pi$

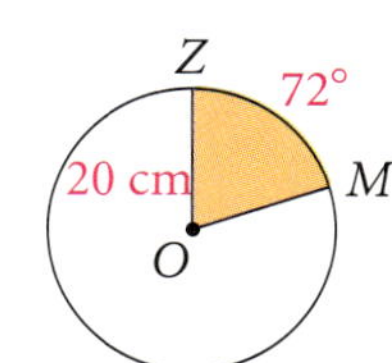

The area of sector ZOM is 80π cm^2.

2 **Critical Thinking** A circle has a diameter of 20 cm. What is the area of a sector bounded by a 208° major arc? Round your answer to the nearest tenth.

This pizza is cut into two segments.

A part of a circle bounded by an arc and the segment joining its endpoints is a **segment of a circle.** To find the area of a segment for a minor arc, draw radii to form a sector. The area of the segment equals the area of the sector minus the area of the triangle formed.

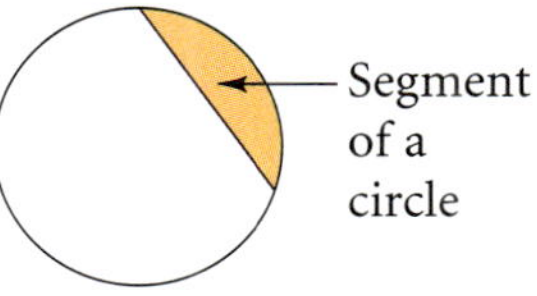

Area of sector − Area of triangle = Area of segment

3 EXAMPLE Finding the Area of a Segment of a Circle

Find the area of the shaded segment. Round your answer to the nearest tenth.

A, B, O, 10 in.

area of sector $AOB = \frac{m\overset{\frown}{AB}}{360} \cdot \pi r^2$ **Use the formula for area of a sector.**

$= \frac{90}{360} \cdot \pi(10)^2$ **Substitute.**

$= \frac{1}{4} \cdot 100\pi$

$= 25\pi$

area of $\triangle AOB = \frac{1}{2}bh$ **Use the formula for area of a triangle.**

$= \frac{1}{2}(10)(10)$ **Substitute.**

$= 50$

area of segment $= 25\pi - 50$

≈ 28.539816 **Use a calculator.**

The area of the segment is about 28.5 in.^{2}

CA Standards Check **3** A circle has a radius of 12 cm. Find the area of the smaller segment of the circle determined by a 60° arc. Round your answer to the nearest tenth.

EXERCISES Standards Practice

For more exercises, see *Extra Skill, Word Problem, and Proof Practice.* GEOM 8.0

A Practice by Example

Example 1 (page 576)

GO for Help

Find the area of each circle. Leave your answer in terms of π.

1. 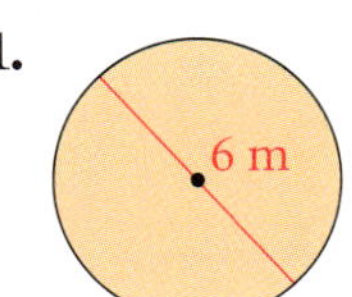

2. 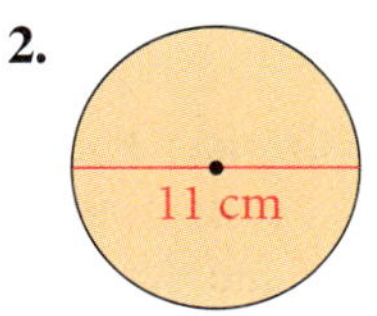

3. 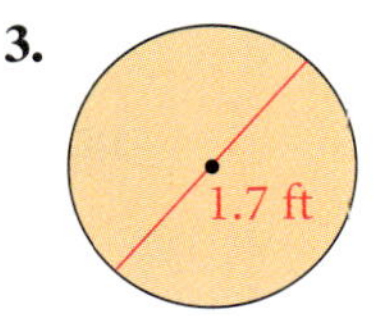

4. 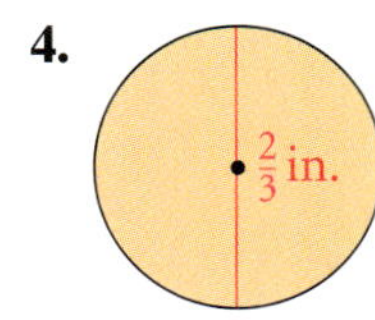

5. **Agriculture** Some farmers use a circular irrigation method. An irrigation arm acts as the radius of an irrigation circle. How much more land is covered with an irrigation arm of 300 ft than by an irrigation arm of 250 ft?

6. What is the difference in the areas of a circular table with diameter 6 ft and a circular table with diameter 8 ft?

Example 2 (page 576)

Find the area of each shaded sector of a circle. Leave your answer in terms of π.

7.

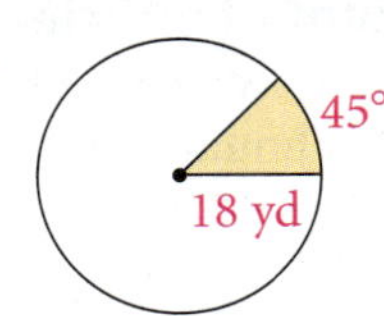

8.

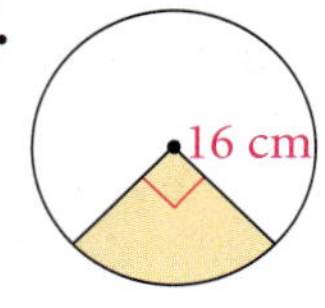

9.

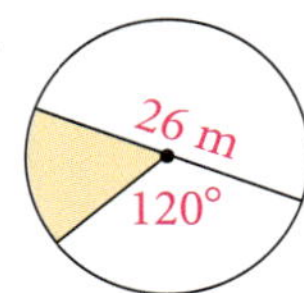

10.

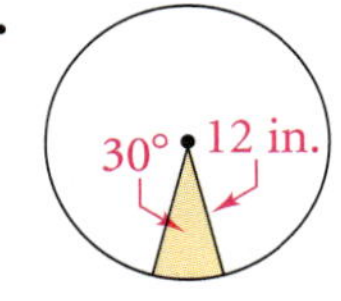

11.

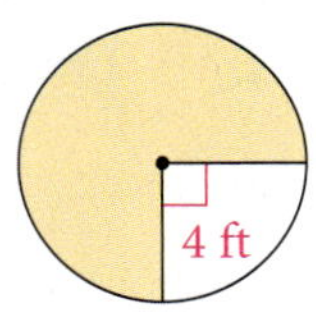

12.

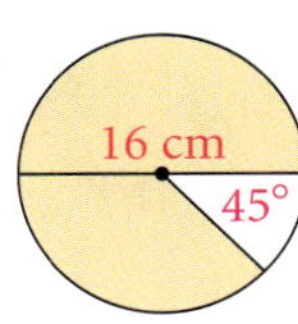

Find the area of sector TOP in $\odot O$ using the given information. Leave your answer in terms of π.

13. $r = 5$ m, $m\widehat{TP} = 90°$

14. $r = 6$ ft, $m\widehat{TP} = 15°$

15. $d = 16$ in., $m\widehat{PT} = 135°$

16. $d = 15$ cm, $m\widehat{PT} = 180°$

Example 3 (page 577)

Find the area of each shaded segment. Round your answer to the nearest tenth.

17.

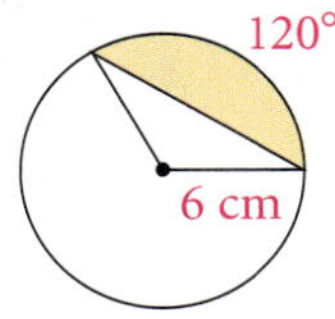

18.

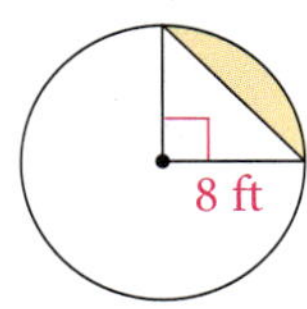

19.

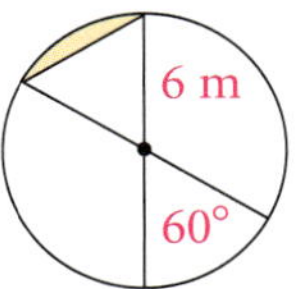

A circle has the given radius. Find the area of the smaller segment of the circle determined by an arc with the given measure. Round to the nearest tenth.

20. radius 15 m, arc 60°

21. radius 14 cm, arc 120°

B Apply Your Skills

Find the area of the shaded region. Leave your answer in terms of π and in simplest radical form.

22.

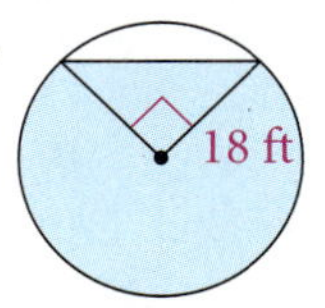

23.

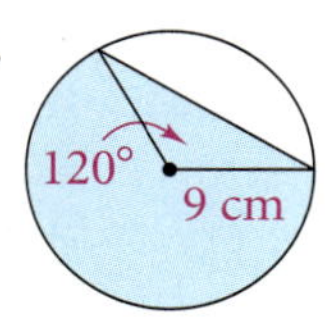

24.

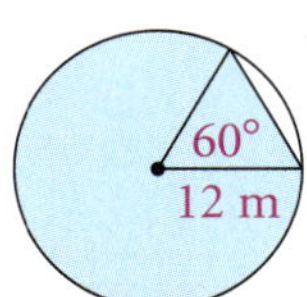

25.

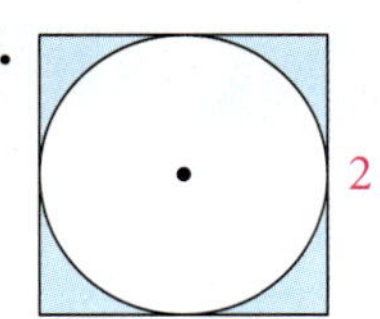

26.

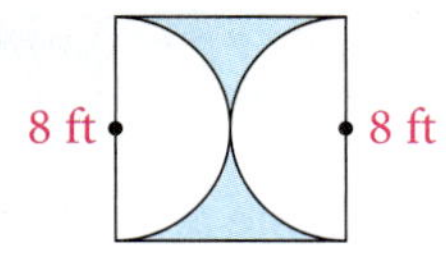

27.

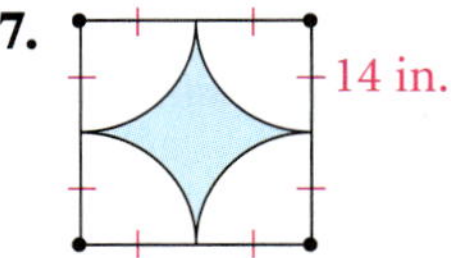

28. **Multiple Choice** The diver at the right is collecting samples from the ocean floor. The line to the diver is 100 ft long, and the diver is working at a depth of 80 ft. What is the approximate area of the circle that the diver can cover?

Ⓐ 11,300 ft^2 Ⓒ 31,400 ft^2

Ⓑ 25,400 ft^2 Ⓓ 51,400 ft^2

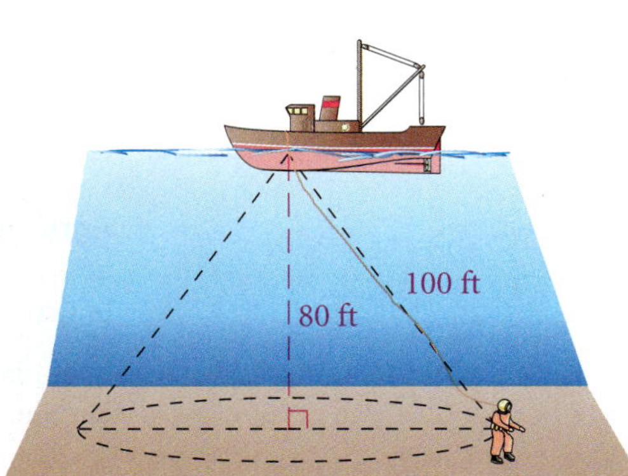

Homework Video Tutor

Visit: PHSchool.com
Web Code: bce-1007

29. **Writing** The American Institute of Baking suggests a technique for cutting and serving a tiered cake. The tiers of a cake have the same height and have radii 8 in. and 13 in. The top tier and the cake directly under it are each cut into 8 wedges as shown. The outer ring of the 13-inch tier is cut into 12 pieces. Which would be larger, a piece from the top or a piece from the outer ring? Explain.

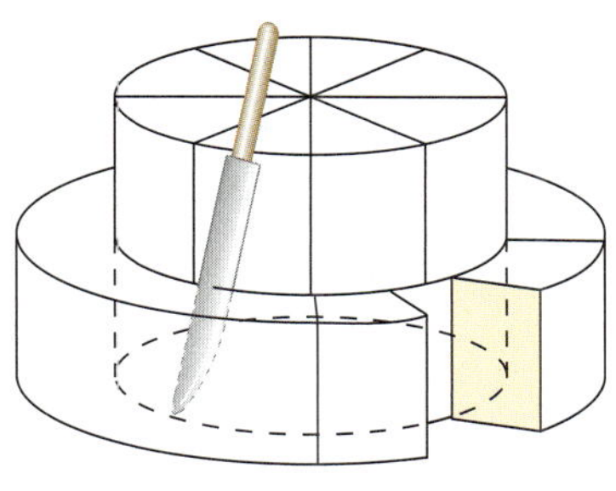

30. How many circles with radius 4 in. will have the same total area as a circle with radius 12 in.?

31. **Games** A dart board has diameter 20 in. and is divided into 20 congruent sectors. Find the area of one sector. Round your answer to the nearest tenth.

32. In a circle, a 90° sector has area 36π in.^{2}. What is the circle's radius?

33. **Open-Ended** Draw a circle and a sector so that the area of the sector is 16π cm^2. Give the radius of the circle and the measure of the arc of the sector.

34. A method for finding the area of a segment determined by a minor arc is described on page 577.
    a. Describe two ways to find the area of a segment determined by a major arc.
    b. If $m\widehat{AB} = 90$ in a circle of radius 10, find the areas of the two segments determined by $\widehat{AB}$.

Find the area of the shaded region. Leave your answer in terms of π.

35.

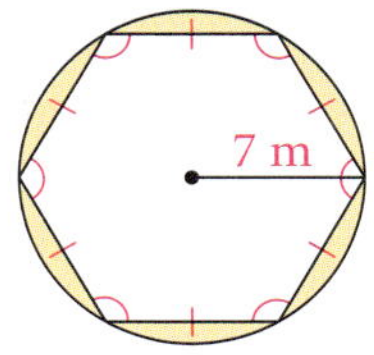

36.

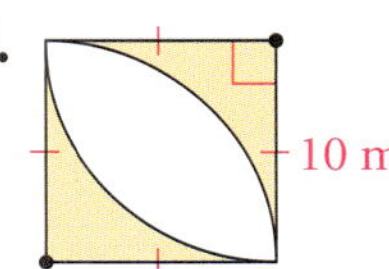

37.

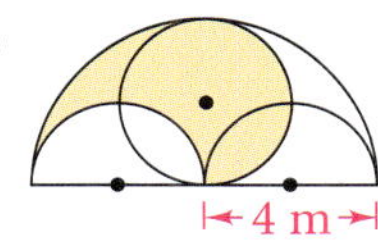

38. Circle O at the right is inscribed in square $ABCD$ and circumscribed about square $PQRS$. Which is smaller, the blue region or the yellow region? Explain.

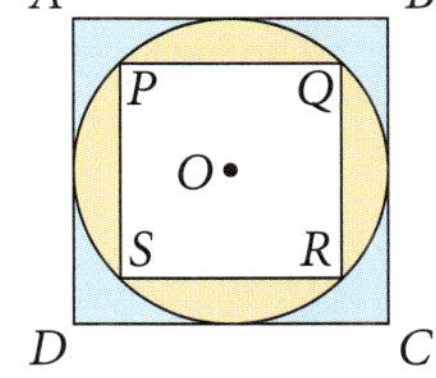

Problem Solving Hint

For Exercise 39, where must T and U lie in a diagram of $\odot T$ and $\odot U$?

39. Circles T and U each have a radius 10 and $TU = 10$. Find the area of the region that is contained inside both circles.

40. **Recreation** An 8 ft-by-10 ft floating dock is anchored in the middle of a pond. The bow of a canoe is tied to a corner of the dock with a 10-ft rope as shown in the picture below.
    a. Sketch a diagram of the region in which the bow of the canoe can travel.
    b. Write a plan for finding the area.
    c. Find the area. Round your answer to the nearest square foot.

Multiple Choice Practice

For California Standards and CAHSEE Tutorials, visit PHSchool.com. Web Code: bcq-9045

GEOM 8.0

41. Determine the area of the shaded region in the circle at the right. Leave your answer in terms of π.

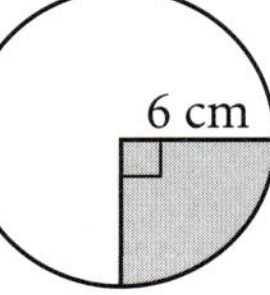

Ⓐ 6π cm^2 Ⓒ 12π cm^2
Ⓑ 9π cm^2 Ⓓ 36π cm^2

GEOM 11.0

42. The scale of two similar triangles is 1 : 3. If the perimeter of the smaller triangle is 24 centimeters, what is the perimeter of the larger triangle in centimeters?

Ⓐ 8 Ⓑ 26 Ⓒ 72 Ⓓ 216

Mixed Review

Lesson 10-6

Find the length of $\widehat{AB}$ in each circle. Leave your answers in terms of π.

43.

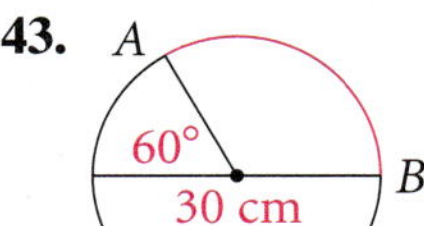

44.

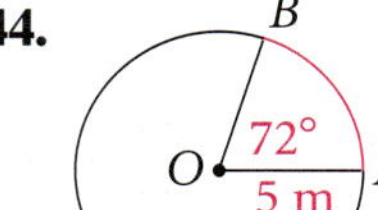

45.

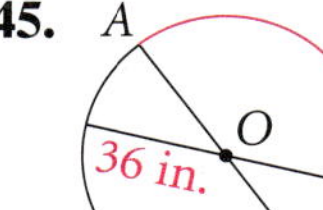

Lesson 6-5

46. Three sides of a trapezoid are congruent. The fourth side is 4 in. longer than each of the other three. The perimeter is 49 in. Find the length of each side.

Lesson 4-7

47. Write a proof. (*Hint:* First prove overlapping triangles are congruent.)

Given: $\overline{AC} \cong \overline{BC}$, $\angle A \cong \angle B$
Prove: $\triangle BDF \cong \triangle AEF$

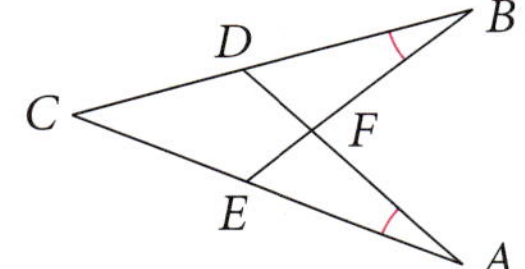

Checkpoint Quiz 2 — Lessons 10-4 through 10-7

The similarity ratio of $\triangle ABC$ to $\triangle DEF$ is 3 : 5. Fill in the missing information.

1. The perimeter of $\triangle ABC$ is 36 in. The perimeter of $\triangle DEF$ is __?__.

2. The area of $\triangle ABC$ is __?__. The area of $\triangle DEF$ is 125 in.^{2}.

3. The areas of two similar triangles are 1.44 and 1.00. Find their similarity ratio.

4. Find the area of $\triangle ABC$ if $a = 4$ ft, $b = 7$ ft and $m\angle C = 20$.

5. A regular 12-gon has perimeter 24 cm. Find its area to the nearest tenth.

Find the area of each shaded region. Leave answers in terms of π.

6.

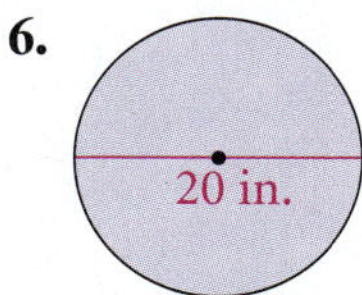

7.

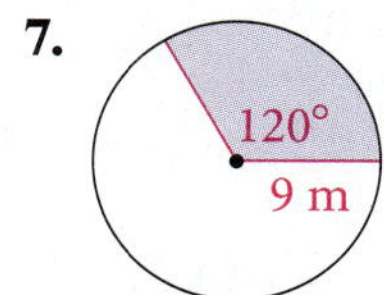

8.

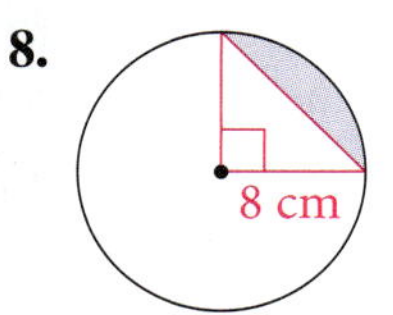

9.

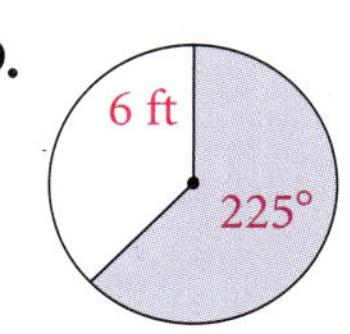

10. In a circle of radius 18 mm, $m\widehat{AB} = 45$. Find the length of $\widehat{AB}$ in terms of π.

Technology

Exploring Area and Circumferences

FOR USE WITH LESSON 10-7

Prepares for GEOM 21.0 Students prove and solve problems regarding relationships among inscribed polygons of circles.

A polygon that is *inscribed* in a circle has all its vertices on the circle. Work in pairs or small groups. Investigate the ratios of the perimeters and areas of inscribed regular polygons to the circumference and area of the circle in which they are inscribed.

Begin by making a table like this.

Regular Polygon			Circle		Ratios	
Sides	Perimeter	Area	Circumference	Area	$\frac{\text{Perimeter}}{\text{Circumference}}$	$\frac{\text{Polygon Area}}{\text{Circle Area}}$
3						

Construct

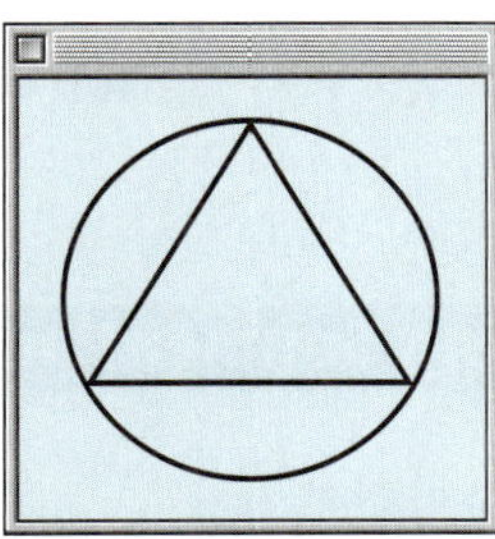

Use geometry software to construct a circle. Find its circumference and area and record them in your table. Inscribe an equilateral triangle in the circle. Your software may be able to do this for you automatically, or you can construct three points on the circle and move them so they are approximately evenly spaced on the circle. Then draw a triangle.

Investigate

Use your geometry software to measure the perimeter and area of the triangle and to calculate the ratios $\frac{\text{triangle perimeter}}{\text{circle circumference}}$ and $\frac{\text{triangle area}}{\text{circle area}}$. Record the results.

Manipulate the circle to change its size. Do the ratios you calculate stay the same or change?

Now inscribe a square in a circle and fill in your table for a polygon of four sides. Do the same for a regular pentagon.

EXERCISES

1. **Make a Conjecture** What will happen to the ratios $\frac{\text{perimeter}}{\text{circumference}}$ and $\frac{\text{polygon area}}{\text{circle area}}$ as you increase the number of sides of the polygon?

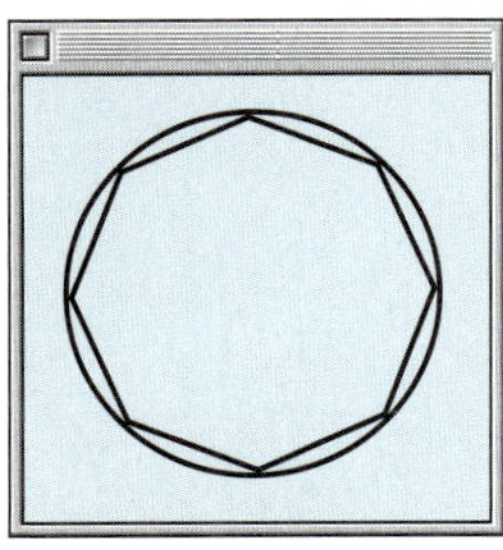

regular octagon

Extend

2. Extend your table to include polygons of 12 sides.
   a. Does your conjecture still hold?
   b. Compare the two columns of ratios in your table. How do they differ?
3. Estimate the perimeter and area of a polygon of 100 sides that is inscribed in a circle with a radius of 10 cm.

10-8 Geometric Probability

California Content Standards

GEOM 8.0 Students know, derive, and solve problems involving the area of common geometric figures. ***Master***
GEOM 10.0 Students compute areas of polygons. ***Master***

What You'll Learn

- To use segment and area models to find the probabilities of events

. . . And Why

To find the probability of winning a carnival game, as in Example 4

Check Skills You'll Need

GO for Help Skills Handbook pages 756 and 762

Find and simplify each ratio.

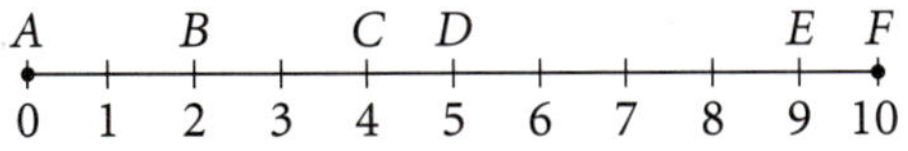

1. $\frac{BD}{AE}$ 2. $\frac{CE}{AF}$ 3. $\frac{AB}{BC}$

(Number line: A at 0, B at 2, C at 4, D at 5, E at 9, F at 10; ticks 0–10.)

4. Two circles have radii 1 m and 2 m, respectively. What is the simplest form of the fraction with numerator equal to the area of the smaller circle and denominator equal to the area of the larger circle?

You roll a number cube. Find the probability of rolling each of the following.

5. 4 6. an odd number 7. 2 or 5 8. a prime number

New Vocabulary • geometric probability

1 Using Segment and Area Models

Vocabulary Tip

P(event) is read "the probability of an event."

You may recall that the probability of an event is the ratio of the number of favorable outcomes to the number of possible outcomes.

$$P(\text{event}) = \frac{\text{favorable outcomes}}{\text{possible outcomes}}$$

Sometimes you can use a **geometric probability** model in which you let points represent outcomes. You find probabilities by comparing measurements of sets of points. For example, if points of segments represent outcomes, then

$$P(\text{event}) = \frac{\text{length of favorable segment}}{\text{length of entire segment}}.$$

1 EXAMPLE Finding Probability Using Segments

A gnat lands at a random point on the ruler's edge. Find the probability that the point is between 3 and 7.

$$P(\text{landing between 3 and 7}) = \frac{\text{length of favorable segment}}{\text{length of entire segment}} = \frac{4}{12}, \text{ or } \frac{1}{3}$$

CA Standards Check

1. A point on $\overline{AB}$ is selected at random. What is the probability that it is a point on $\overline{CD}$?

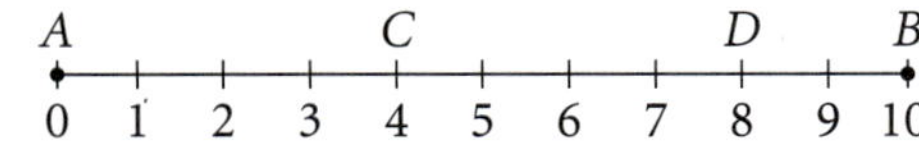

Real-World Connection

If the bus runs on schedule, Elena's average wait (Example 2) will be 12.5 min.

You can use a segment model to find the probability of how long you will wait for a bus.

2 EXAMPLE Real-World Connection

Commuting Elena's bus runs every 25 minutes. If she arrives at her bus stop at a random time, what is the probability that she will have to wait at least 10 minutes for the bus?

Assume that a stop takes very little time, and let $\overline{AB}$ represent the 25 minutes between buses.

A ———————— B
0 5 10 15 20 25

If Elena arrives at any time between A and C, she has to wait at least 10 minutes until B.

A ———— C ——— B
0 5 10 15 20 25

$$P(\text{waiting at least 10 min}) = \frac{\text{length of } \overline{AC}}{\text{length of } \overline{AB}} = \frac{15}{25}, \text{ or } \frac{3}{5}$$

The probability that Elena will have to wait at least 10 minutes for the bus is $\frac{3}{5}$ or 60%.

CA Standards Check 2 What is the probability that Elena will have to wait no more than 10 minutes for the bus?

If the points of a region represent equally-likely outcomes, then you can find probabilities by comparing areas.

$$P(\text{event}) = \frac{\text{area of favorable region}}{\text{area of entire region}}$$

3 EXAMPLE Finding Probability Using Area

Online active math

For: Probability Activity
Use: Interactive Textbook, 10-8

Target Game Assume that a dart you throw will land on the 1-ft square dartboard and is equally likely to land at any point on the board. Find the probability of hitting each of the blue, yellow, and red regions. The radii of the concentric circles are 1, 2, and 3 inches, respectively.

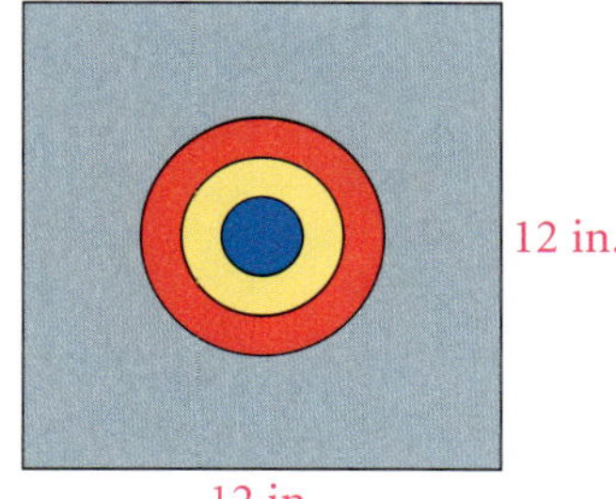

$$P(\text{blue}) = \frac{\text{area of blue region}}{\text{area of square}} = \frac{\pi(1)^2}{12^2} = \frac{\pi}{144} \approx 0.022, \text{ or } 2.2\%$$ **Use a calculator.**

$$P(\text{yellow}) = \frac{\text{area of yellow region}}{\text{area of square}} = \frac{\pi(2)^2 - \pi(1)^2}{12^2} = \frac{3\pi}{144} \approx 0.065, \text{ or } 6.5\%$$

$$P(\text{red}) = \frac{\text{area of red region}}{\text{area of square}} = \frac{\pi(3^2) - \pi(2)^2}{12^2} = \frac{5\pi}{144} \approx 0.109, \text{ or } 10.9\%$$

The probabilities of hitting the blue, yellow, and red regions are about 2.2%, 6.5%, and 10.9%, respectively.

CA Standards Check 3 If you change the blue circle as indicated, how does the probability of hitting the blue circle change? Explain.

a. Double the radius. **b.** Triple the radius.

As Example 3 suggests, you can apply geometric probability to some games. This can help you decide how easy or difficult it may be to win such games.

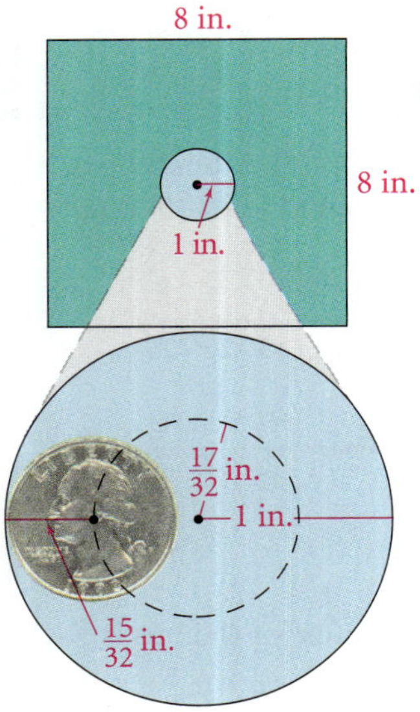

4 EXAMPLE Real-World Connection

Coin Toss To win a prize in a carnival game, you must toss a quarter so that it lands entirely within the circle as shown at the left. Find the probability of this happening on one toss. Assume that the center of a tossed quarter is equally likely to land at any point within the 8-in. square.

The radius of the circle is 1 in. The radius of a quarter is $\frac{15}{32}$ in. The favorable points are those that are less than $\frac{17}{32}$ in. from the center of the circle. They are the points within the dashed circle.

$$P(\text{quarter landing in circle}) = \frac{\text{area of dashed circle}}{\text{area of square}}$$

$$= \frac{\pi\left(\frac{17}{32}\right)^2}{8^2} \approx 0.014, \text{ or } 1.4\%$$

The probability of a quarter landing in the circle is about 1.4%.

4 Critical Thinking Suppose you toss 100 quarters. Would you expect to win a prize? Explain.

EXERCISES

Standards Practice GEOM 8.0, 10.0

For more exercises, see *Extra Skill, Word Problem, and Proof Practice.*

Practice by Example

Example 1 (page 582)

Find the probability that a point chosen at random from $\overline{AK}$ is on the given segment.

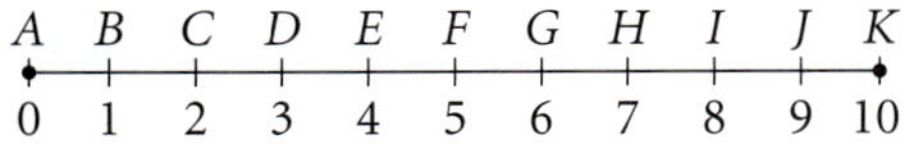

(Number line: A 0, B 1, C 2, D 3, E 4, F 5, G 6, H 7, I 8, J 9, K 10)

1. $\overline{CH}$
2. $\overline{FG}$
3. $\overline{DJ}$
4. $\overline{EI}$
5. $\overline{AK}$

6. Points M and N are on $\overline{ZB}$ with $ZM = 5$, $NB = 9$, and $ZB = 20$. A point is chosen at random from $\overline{ZB}$. What is the probability that the point is on $\overline{MN}$?

Example 2 (page 583)

7. **Transportation** A rapid transit line runs trains every 10 minutes. Draw a geometric model and find the probability that randomly arriving passengers will not have to wait more than 4 minutes.

Traffic Patterns **Main Street intersects each street below. The traffic lights on Main follow the cycles shown. As you travel along Main and approach the intersection, what is the probability that the first color you see is green?**

8. Durham Avenue: green 30 s, yellow 5 s, red 25 s
9. Martin Luther King Boulevard: green 20 s, yellow 5 s, red 50 s
10. Yonge Street: green 40 s, yellow 5 s, red 25 s
11. International Drive: green 25 s, yellow 5 s, red 45 s
12. Tamiami Trail: green 35 s, yellow 8 s, red 32 s
13. Flutie Pass: green 50 s, yellow 4 s, red 26 s

14. During May, a certain drawbridge over the Intracoastal Waterway is raised every half hour to allow boats to pass. It remains open for 5 min. What is the probability that a motorist arriving at the bridge in May will find it raised?

Examples 3, 4 (pages 583 and 584)

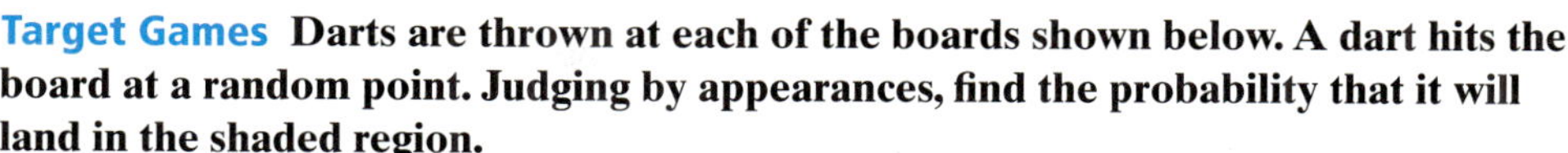

Target Games **Darts are thrown at each of the boards shown below. A dart hits the board at a random point. Judging by appearances, find the probability that it will land in the shaded region.**

15.

16.

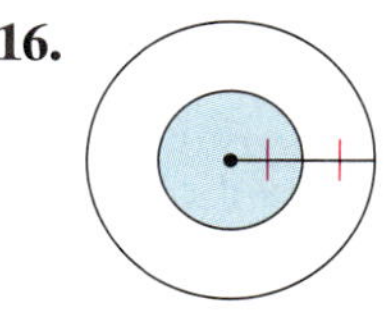

17.

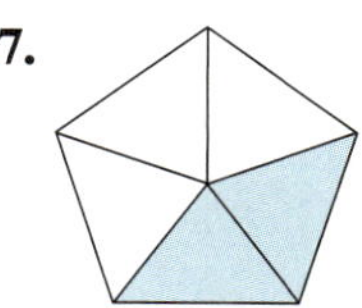

18.

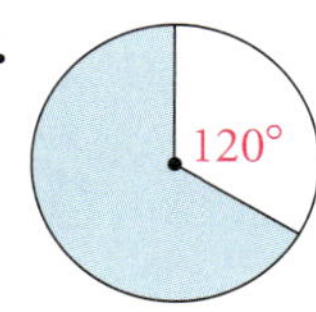

19.

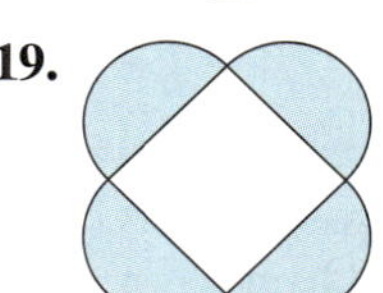

20.

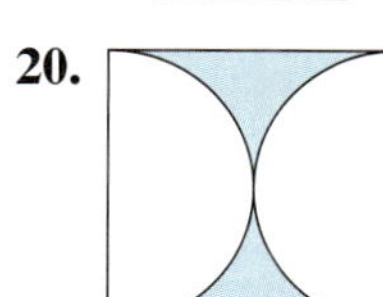

Real-World Connection

The probability of hitting the center circle is small, so it is worth the maximum number of points.

21. Archery An archery target with a radius of 61 cm has 5 scoring zones formed by concentric circles. The colors of the zones are yellow, red, blue, black, and white. The radius of the yellow circle is 12.2 cm. The width of each ring is also 12.2 cm. If an arrow hits the target at a random point, what is the probability that it hits the center yellow zone?

B Apply Your Skills

22. $\overline{BZ}$ contains $\overline{MN}$ and $BZ = 20$. A point is chosen at random from $\overline{BZ}$. The probability that the point is also on $\overline{MN}$ is 0.3, or 30%. Find MN.

Target Games **A dart hits each square dartboard at a random point. Find the probability that the dart lands inside a circle. Leave your answer in terms of π.**

23.

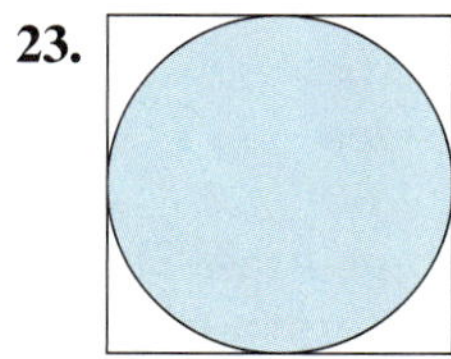

24.

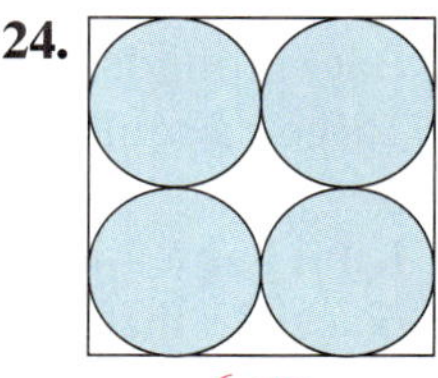

25.

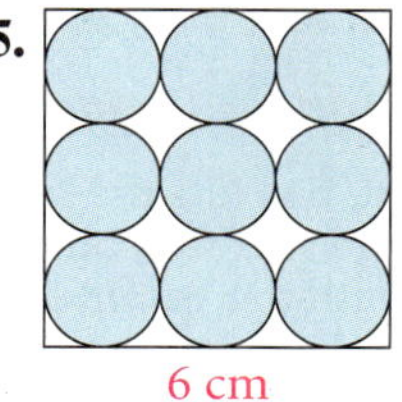

26. A dartboard is a square of radius 10 in. You throw a dart and hit the target. Find the probability that the dart lies within $\sqrt{10}$ in. of the center of the square.

27. Critical Thinking Use the information given in Example 4.

a. For each 1000 quarters tossed, about how many prizes would be won?

b. Suppose the game prize costs the carnival \$10. About how much profit would the carnival expect for every 1000 quarters tossed?

28. Commuting Suppose a bus arrives at a bus stop every 25 min and waits 5 min before leaving. Sketch a geometric model. Use it to find the probability that a person has to wait more than 10 min for a bus to leave.

29. Traffic Patterns The traffic lights at Fourth and Commercial Streets repeat themselves in 60-second cycles. Ms. Li regularly has students drive on Fourth Street through the Commercial Street intersection. By experience, she knows that they will face a red light 60% of the time. Use this information to estimate how long the Fourth Street light is red during each 1-min cycle.

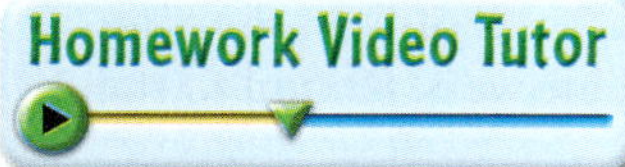

Visit: PHSchool.com
Web Code: bce-1008

For Exercises 30 and 31, sketch a geometric model and solve.

30. **Astronomy** Meteoroids (mostly dust-particle size) are continually bombarding Earth. The surface area of Earth is about 65.7 million square miles. The area of the United States is about 3.7 million square miles. What is the probability that a meteoroid landing on Earth will land in the United States?

31. **Tape Recording** Amy made a tape recording of a chorus rehearsal. The recording began 21 min into the 60-min tape and lasted 8 min. Later she accidentally erased a 15-min segment somewhere on the tape.

a. In your model show the possible starting times of the erasure. Explain how you know that the erasure did not start after the 45-min mark.

b. In your model show the starting times of the erasures that would erase the entire rehearsal. Find the probability that the entire rehearsal was erased.

Problem Solving Hint

$0 \le P(\text{event}) \le 1$

$P(\text{event}) = 0$ means the event will not occur.

$P(\text{event}) = 1$ means the event will occur.

x^2 **Algebra** **Find the probability that coordinate x of a point chosen at random from $\overline{AK}$ satisfies the inequality.**

A	B	C	D	E	F	G	H	I	J	K
0	1	2	3	4	5	6	7	8	9	10

32. $2 \le x \le 8$ **33.** $x \ge 7$ **34.** $2x \le 9$ **35.** $\frac{1}{2}x - 5 \ge 0$

36. $2 \le 4x \le 3$ **37.** $0 \le \frac{1}{3}x + 1 \le 5$ **38.** $|x - 6| \le 1.5$ **39.** $\sqrt{2} \le \pi x \le \sqrt{10}$

Real-World Connection

When the ball hits the red target, the seat falls out from underneath the volunteer.

Dunk Tank **At a fund-raiser, a volunteer sits on a platform above a tank of water. She gets dunked when you throw a ball and hit the red target. The radius of the ball is 3.6 cm. What is the probability that a ball heading randomly for the given background shape would hit the given target shape?**

40 cm

10 cm

40. Background (at right): a circle 40 cm across
Target: a circle with 10-cm radius

41. Background: a square with 40-cm sides
Target: a circle with 10-cm radius

42. Background: a circle 40-cm across
Target: a square with 20-cm sides

43. Background: a square with 40-cm sides
Target: a square with 20-cm sides

44. Kimi has a 4-in. straw and a 6-in. straw. She wants to cut the 6-in. straw into two pieces so that the three pieces form a triangle.

a. If she cuts the straw to get two 3-in. pieces, can she form a triangle?

b. If the two pieces are 1 in. and 5 in., can she form a triangle?

c. If Kimi cuts the straw at a random point, what is the probability that she can form a triangle?

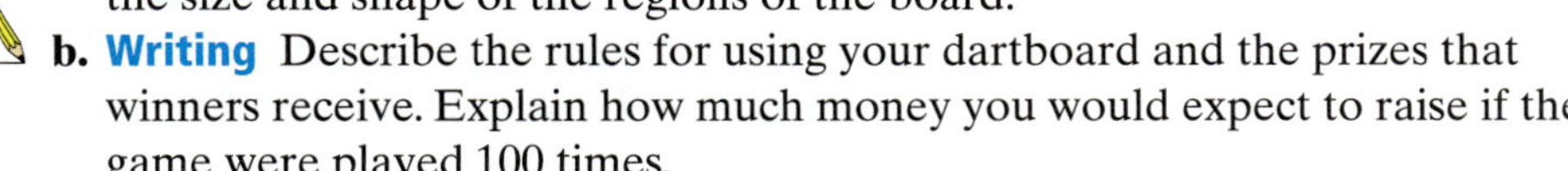

45. a. **Open-Ended** Design a dartboard game to be used at a charity fair. Specify the size and shape of the regions of the board.

b. **Writing** Describe the rules for using your dartboard and the prizes that winners receive. Explain how much money you would expect to raise if the game were played 100 times.

C Challenge

46. **Graphing Calculator** A circular dartboard has radius 1 m and a yellow circle in the center. Assume you hit the target at a random point. For what radius of the yellow center region would $P(\text{hitting yellow})$ equal each of the following? Use the table feature of a calculator to generate all six answers. Round to the nearest centimeter.

a. 0.2 **b.** 0.4 **c.** 0.5
d. 0.6 **e.** 0.8 **f.** 1.0

47. Target Game A target has a central circle and three concentric rings. The diameters of the circles are 2 cm, 6 cm, 10 cm, and 14 cm. Find the probability of landing in the gray region. Compare it with the probability of landing in *either* the blue or red region.

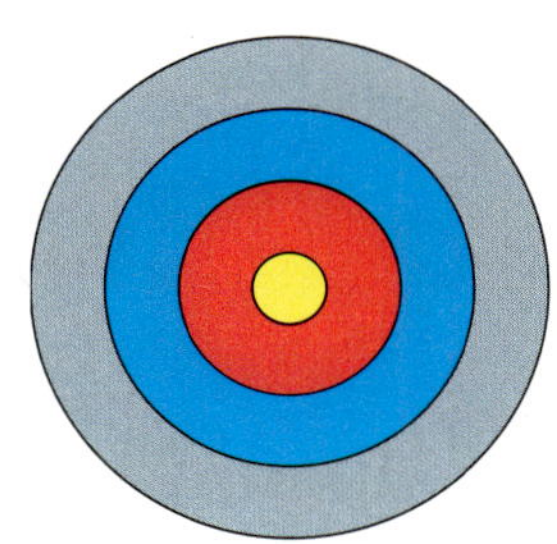

Multiple Choice Practice

For California Standards and CAHSEE Tutorials, visit PHSchool.com. Web Code: bcq-9045

CAHSEE 7SDAP 1.1

48. The graph at the right shows how a certain community spent its citizens' tax money. Which of the following statements is best supported by the graph?

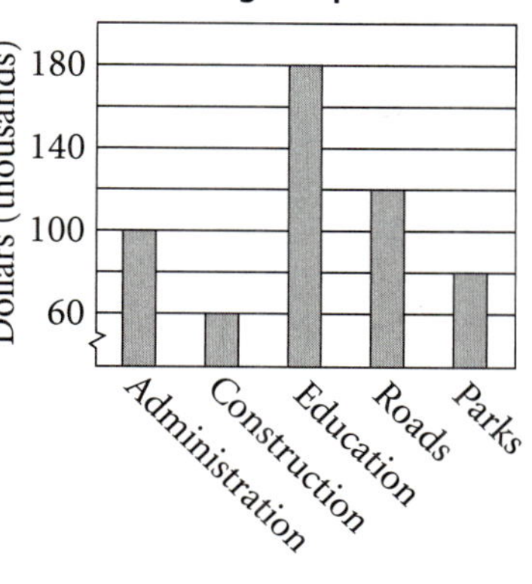

Ⓐ More money should be spent on construction.

Ⓑ More money was spent on construction than on parks.

Ⓒ About twice as much money was spent on roads as was spent on construction.

Ⓓ About twice as much money was spent on administration as was spent on construction.

GEOM 8.0

49. In the figure at the right, a square is circumscribed about a circle. What is the ratio of the area of the circle to the area of the square?

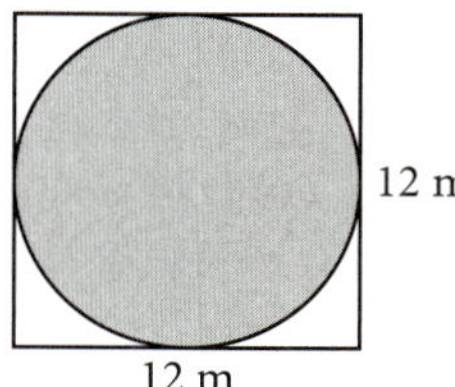

Ⓐ 36π Ⓒ $\frac{\pi}{6}$

Ⓑ $\frac{\pi}{4}$ Ⓓ $\frac{\pi}{24}$

Mixed Review

GO for Help

Lesson 10-7

50. A circle has circumference 20π ft. What is its area?

51. A circle has radius 12 cm. What is the area of a sector of the circle with a 30° central angle?

52. What is the area of a semicircle with diameter 20 ft?

Lesson 6-2

Algebra **Find the values of the variables in each parallelogram.**

53.

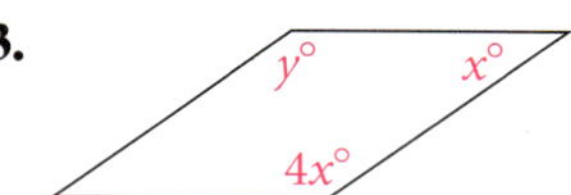

54.

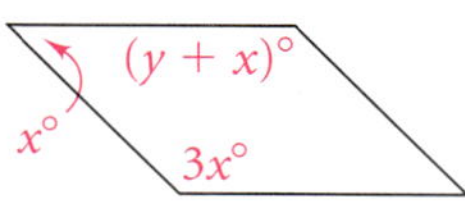

Lesson 5-1

55. The coordinates of the vertices of a triangle are $A(1, -4)$, $B(5, 6)$, and $C(-3, 2)$.

a. Find the coordinates of D, the midpoint of $\overline{AB}$, and E, the midpoint of $\overline{BC}$.

b. Find the slope of $\overline{DE}$ and the slope of $\overline{AC}$.

c. Verify that $\overline{DE} \parallel \overline{AC}$.

d. Find DE and AC.

e. Verify that $DE = \frac{1}{2}AC$.

Test-Taking Strategies

Choosing "Cannot Be Determined"

A question can have "Cannot be determined" as the fourth answer choice. You choose "Cannot be determined" when you are sure of the answer and it is not among the first three listed. Since the correct answer may not be shown, you should solve a "Cannot be determined" problem without regard for the choices.

EXAMPLE

The perimeter of kite $DEFG$ is 30 units. Kite $KLMN$ is similar to $DEFG$.

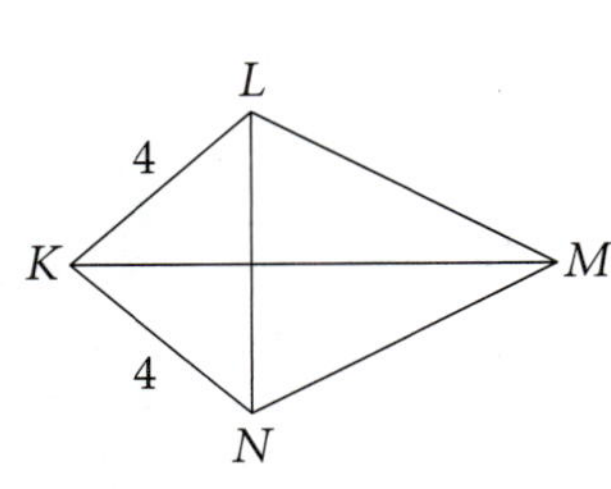

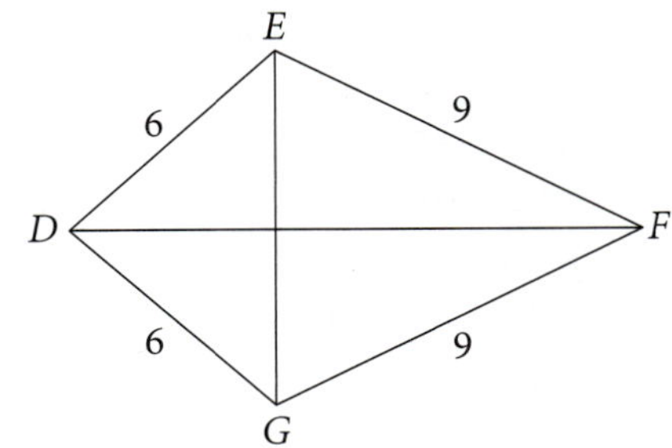

What is the perimeter of kite $KLMN$?

Ⓐ 6 units
Ⓑ 10 units
Ⓒ 28 units
Ⓓ Cannot be determined

The ratio of corresponding sides is $\frac{4}{6}$ or $\frac{2}{3}$. The ratio of perimeters is equal to the ratio of the sides, so the perimeter of $KLMN$ is $\frac{2}{3}(30)$, or 20 units. Since 20 is not one of the choices, the correct choice is D.

EXERCISES

1. The points T, U, and V, shown at the right, form three vertices of an isosceles trapezoid $STUV$. What could be the coordinates of S?

Ⓐ $(0, 3)$
Ⓑ $(0, 4)$
Ⓒ $(-1, 4)$
Ⓓ Cannot be determined

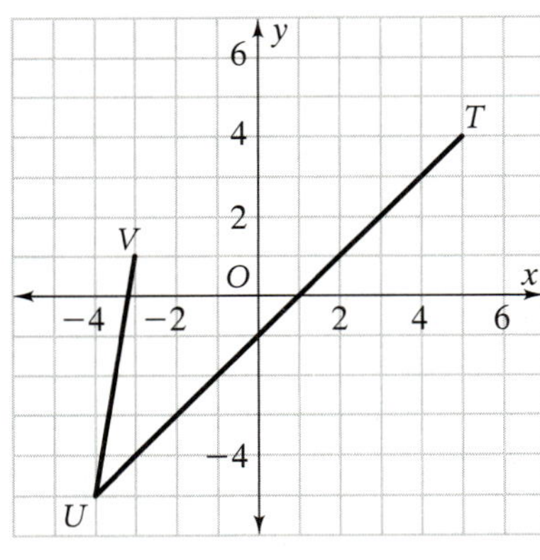

2. Which of the following shows all the x-intercepts of the graph of $f(x) = 24x^2 + 5x - 14$?

Ⓐ $2, -7$
Ⓑ $\frac{2}{3}, -\frac{7}{8}$
Ⓒ $-7, \frac{2}{3}$
Ⓓ Cannot be determined

3. Tim is designing a doghouse. The front will use a square and an isosceles triangle as shown in the diagram at the right. To the nearest tenth of a foot, how tall will the doghouse be?

Ⓐ 4.5 ft
Ⓑ 6.0 ft
Ⓒ 7.2 ft
Ⓓ Cannot be determined

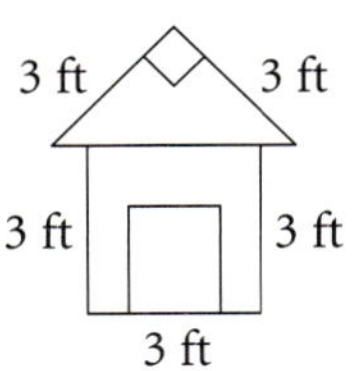

Chapter 10 Review

Vocabulary Review

English and Spanish Audio Online

For: Vocabulary quiz
Web Code: bcj-1051

Choose the correct term to complete each sentence.

1. You can use any side as the (*altitude, base*) of a triangle.
2. A (*sector, segment*) of a circle is a region bounded by two radii and the intercepted arc.
3. A segment that contains the center of a circle and has both endpoints on the circle is the (*diameter, circumference*) of a circle.
4. In a regular polygon, the perpendicular distance from the center to a side is the (*apothem, radius*) of the parallelogram.
5. Two arcs of a circle with exactly one point in common are (*congruent arcs, adjacent arcs*).

Skills and Concepts

Lessons 10-1, 10-2, and 10-3

- To find the area of a parallelogram (p. 534)
- To find the area of a triangle (p. 535)
- To find the area of a trapezoid (p. 540)
- To find the area of a rhombus or a kite (p. 541)
- To find the area of a regular polygon (p. 546)

GEOM 8.0, 10.0, 20.0

You can find the area of a rectangle, a parallelogram, or a triangle if you know the **base,** *b,* and **height,** *h.* The area of a rectangle or a parallelogram is $A = bh$.

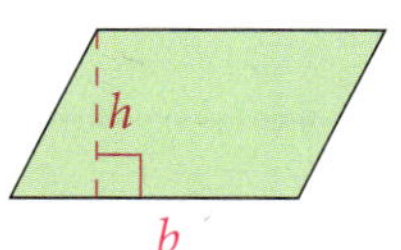

The area of a triangle is $A = \frac{1}{2}bh$.

The **height of a trapezoid,** *h,* is the perpendicular distance between the bases, b_1 and b_2. The area of a trapezoid is $A = \frac{1}{2}h(b_1 + b_2)$.

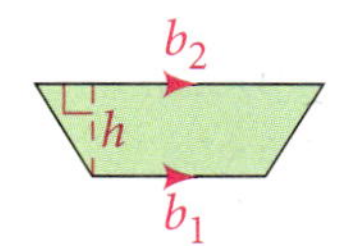

The **center of a regular polygon,** *C,* is the center of its circumscribed circle. The **radius,** *r,* is the distance from the center to a vertex. The **apothem,** *a,* is the perpendicular distance from the center to a side. The area of a regular polygon with apothem *a* and perimeter *p* is $A = \frac{1}{2}ap$.

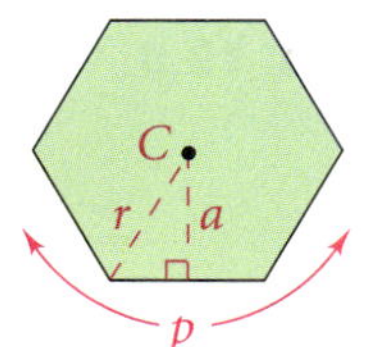

The area of a rhombus or kite is $A = \frac{1}{2}d_1d_2$.

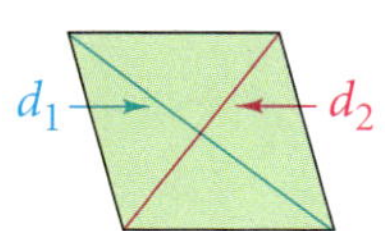

Find the area of each figure. If your answer is not an integer, leave it in simplest radical form.

6. 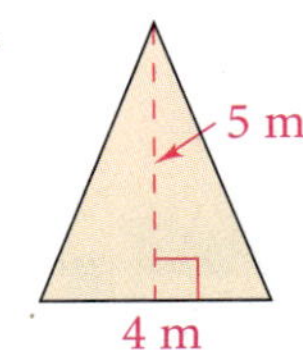

7. 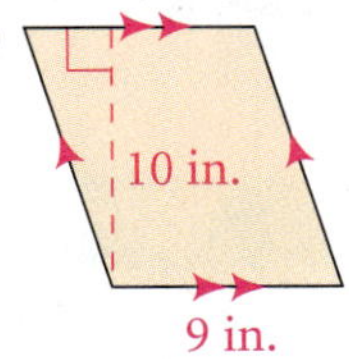

8. 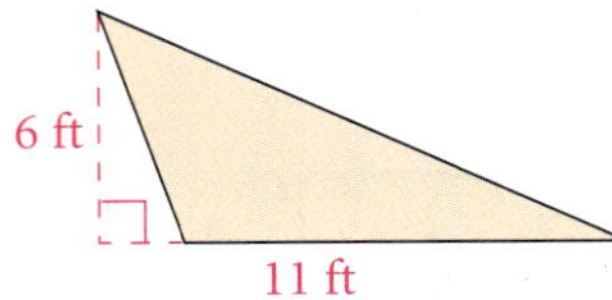

9. 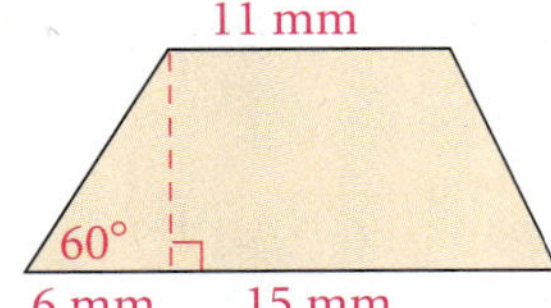

10. 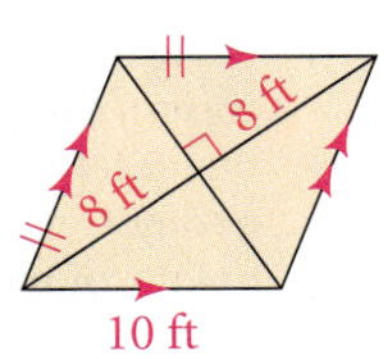

11. 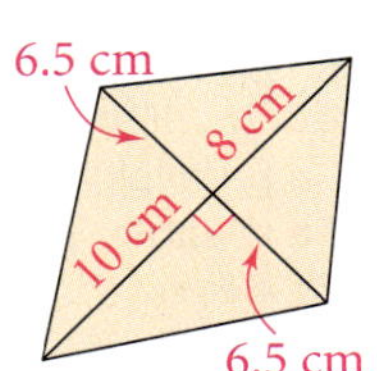

Sketch each regular polygon with the given radius. Then find its area. Round your answers to the nearest tenth.

12. triangle; radius 4 in. **13.** square; radius 8 mm **14.** hexagon; radius 7 cm

Lesson 10-4

- To find the perimeters and areas of similar figures (p. 553)

 **GEOM 11.0**

If the similarity ratio of two similar figures is $\frac{a}{b}$, then the ratio of their perimeters is $\frac{a}{b}$, and the ratio of their areas is $\frac{a^2}{b^2}$.

For each pair of similar figures, find the ratio of the area of the first figure to the area of the second.

15. 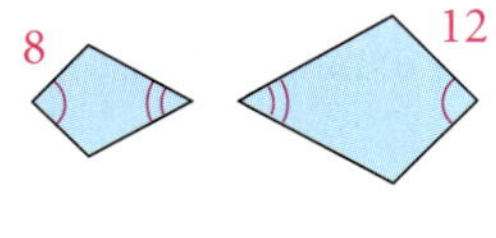

16. 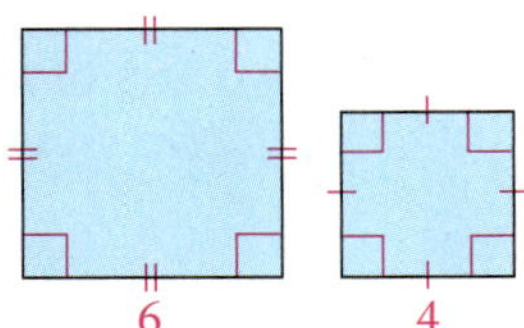

17. 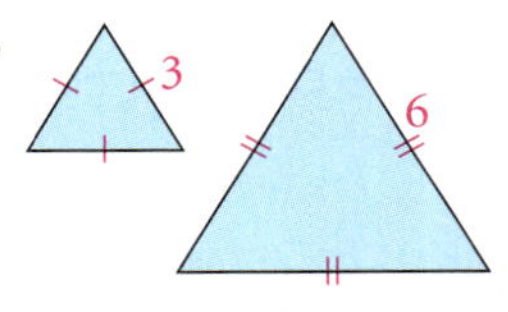

18. If the ratio of areas of two similar hexagons is 8 : 25, what is the ratio of their perimeters?

Lesson 10-5

- To find the area of a regular polygon using trigonometry (p. 559)
- To find the area of a triangle using trigonometry (p. 560)

 **GEOM 8.0, 19.0**

You can use trigonometry to find the areas of regular polygons.

You can also use trigonometry to find the area of a triangle when you know the lengths of two sides and the measure of the included angle.

Area of triangle $= \frac{1}{2} \cdot$ side length $\cdot$ side length $\cdot$ sine of included angle

Find the area of each polygon. Round your answers to the nearest tenth.

19. regular decagon with radius 5 ft

20. regular pentagon with apothem 8 cm

21. regular hexagon with apothem 6 in.

22. regular quadrilateral with radius 2 m

23. 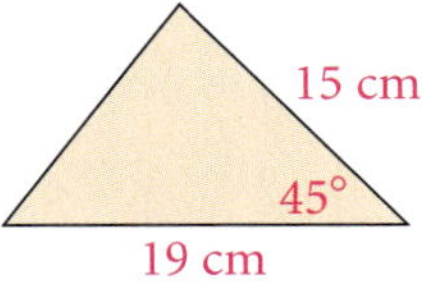

24. 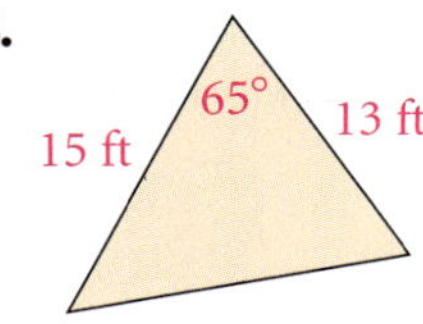

25. 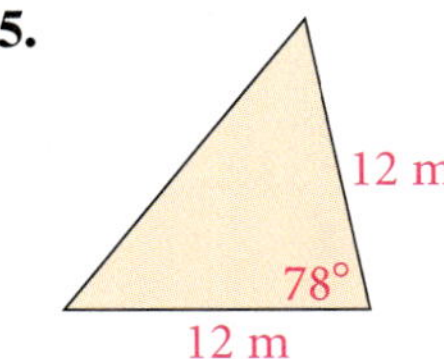

Lessons 10-6 and 10-7

- To find the measures of central angles and arcs (p. 566)
- To find circumference and arc length (p. 568)
- To find the areas of circles, sectors, and segments of circles (p. 575)

 **GEOM 7.0, 8.0**

A **circle** is the set of all points in a plane equidistant from one point called the **center.** The measure of a **minor arc** is the measure of its corresponding central angle. The measure of a **major arc** is 360 minus the measure of its related minor arc. **Adjacent arcs** have exactly one point in common.

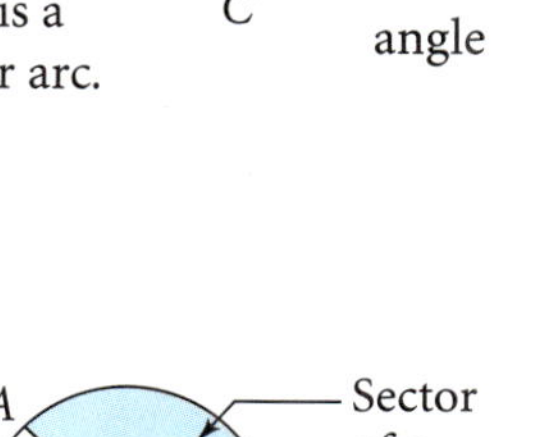

The **circumference** of a circle is $C = \pi d$ or $C = 2\pi r$. The area of a circle is $A = \pi r^2$.

Arc length is a fraction of a circle's circumference.
The length of $\widehat{AB} = \frac{m\widehat{AB}}{360} 2\pi r$.

A **sector of a circle** is a region bounded by two radii and their intercepted arc.
The area of sector $APB = \frac{m\widehat{AB}}{360} \pi r^2$.

A
B
P
C
Sector of a circle
Segment of a circle

A **segment of a circle** is the part of a circle bounded by an arc and the segment joining its endpoints. The area of a segment of a circle is the difference between the areas of the related sector and the related triangle.

Find each measure.

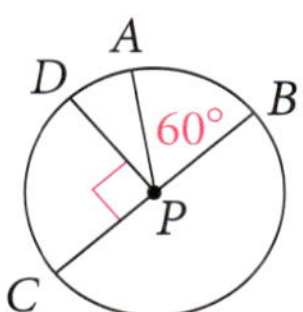

26. $m\angle APD$ **27.** $m\widehat{AC}$

28. $m\widehat{ABD}$ **29.** $m\angle CPA$

Find the length of each arc shown in red. Leave your answer in terms of π.

30.

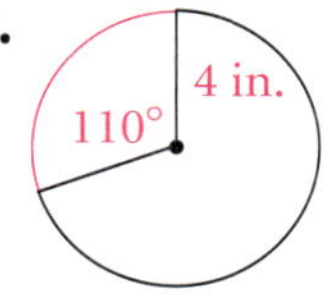

31.

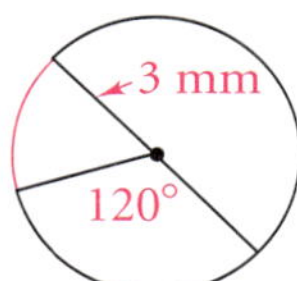

Find the area of each shaded region. Round your answer to the nearest tenth.

32.

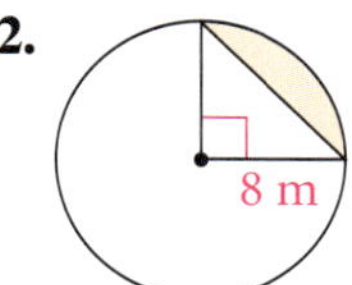

33.

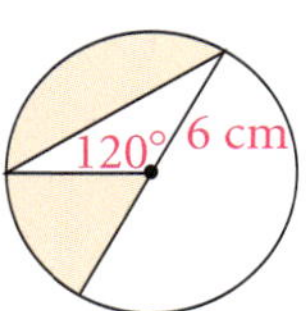

Lesson 10-8

- To use segment and area models to find the probabilities of events (p. 582)

 **GEOM 8.0, 10.0**

Geometric probability uses geometric figures to represent occurrences of events. You can use a segment model or an area model. Compare the part that represents favorable outcomes to the whole, which represents all outcomes.

A dart hits each dartboard at a random point. Find the probability that it lands in the shaded area.

34.

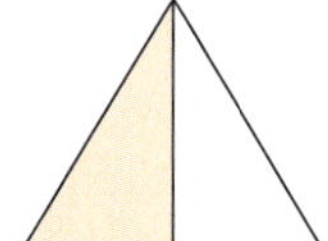

35.

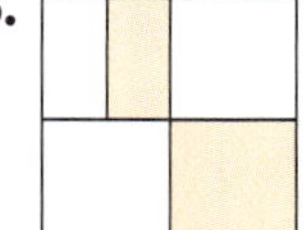

36.

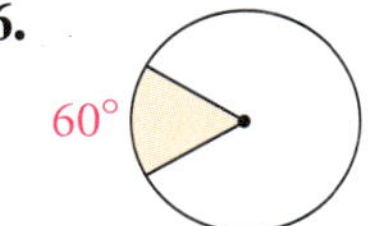

Chapter 10 Test

For: Chapter Test
Web Code: bca-1052

Find the area of each figure. If your answer is not an integer, round to the nearest tenth.

1.

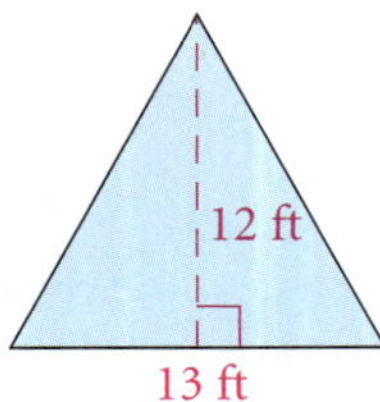

2.

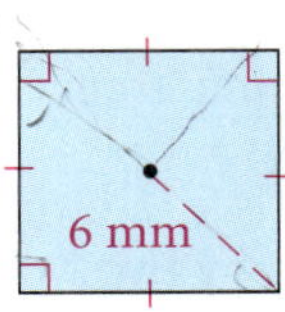

3.

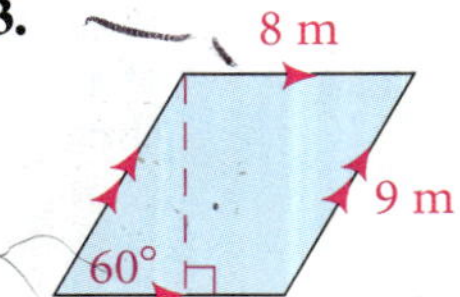

4.

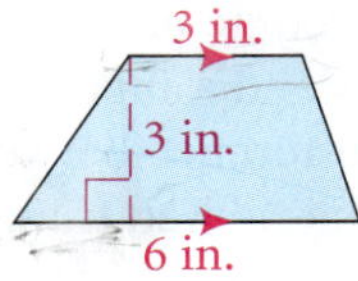

Find the area of each regular polygon. Round to the nearest tenth.

5.

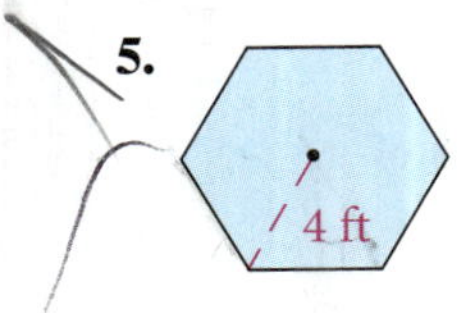

6.

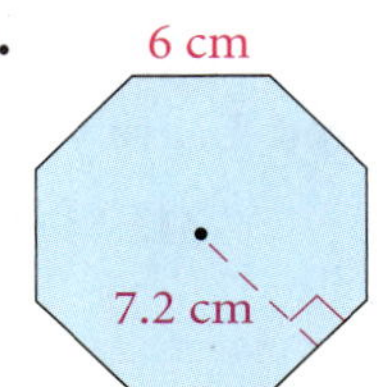

For each pair of similar figures, find the ratio of the area of the first figure to the area of the second.

7.

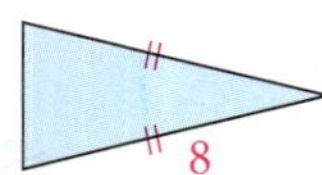

8.

Find the area of each polygon to the nearest tenth.

9.

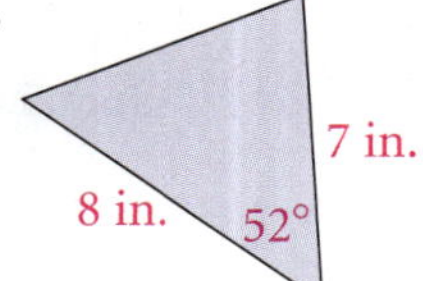

10.

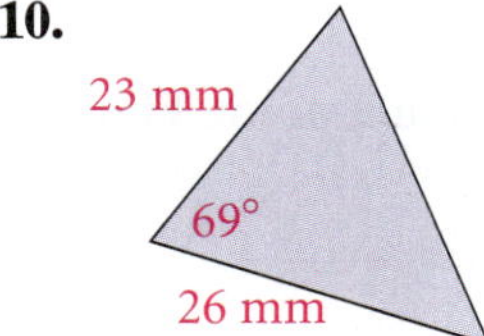

11. a regular hexagon with apothem 5 ft

12. a regular pentagon with radius 3 cm

Find each measure for $\odot P$.

13. $m\angle BPC$

14. $m\widehat{AB}$

15. $m\widehat{ADC}$

16. $m\widehat{ADB}$

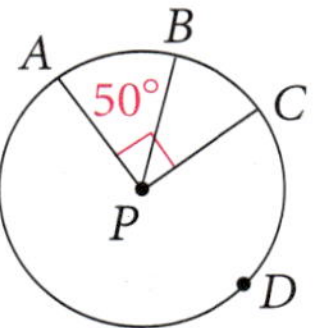

Find the length of each arc shown in red. Leave your answer in terms of π.

17.

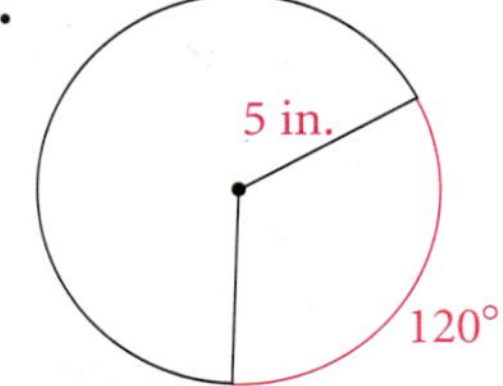

18.

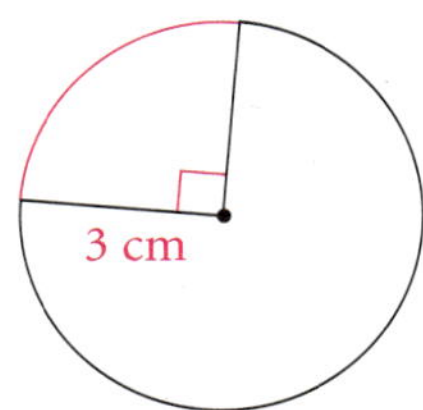

Find the area of each shaded region to the nearest hundredth.

19.

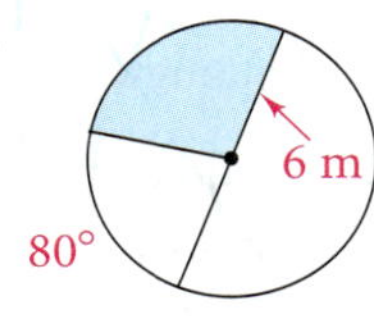

20.

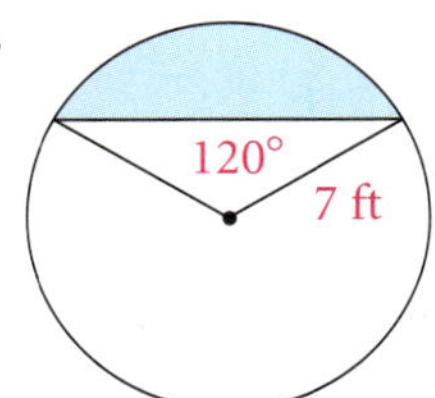

21. Open-Ended Use a compass to draw a circle. Shade a sector of the circle and find its area.

Find the area of each shaded region. Leave your answers in terms of π.

22.

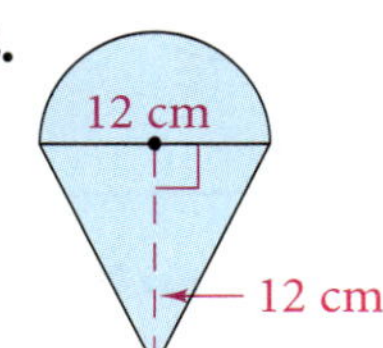

23.

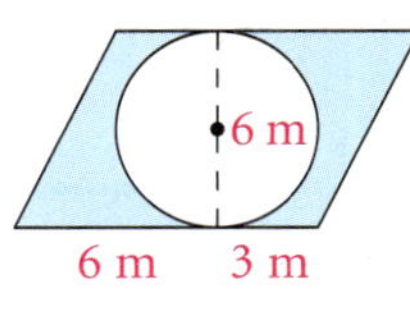

24. Probability Every 20 minutes from 4:00 P.M. to 7:00 P.M., a commuter train crosses Main Street. For three minutes a gate stops cars from passing as the train goes by. What is the probability that a motorist approaching the train crossing during this time interval will have to stop for the train?

Standards Mastery

Cumulative Practice

Some questions require you to relate changes in lengths to changes in areas of similar figures. Read the question at the right. Then follow the tips to answer it.

Tip 1
When you see length and area of two figures in the same problem, ask whether the figures are similar. (All squares are similar.)

The ratio of the areas of two squares is 4 : 9. The length of a side of the smaller square is 6 inches. What is the length of a side of the larger square?

- (A) 4 in.
- (B) 9 in.
- (C) 13.5 in.
- (D) 36 in.

Tip 2
For similar figures, the ratio of areas equals the square of the ratio of sides.

Think It Through

The ratio of the areas of the squares is 4 : 9, so the ratio of the side lengths is $\sqrt{4} : \sqrt{9}$, or 2 : 3.

Use a proportion to find the side length S of the larger square.

$$\frac{2}{3} = \frac{6}{S}$$

$S = 9$, so the correct answer is B.

Vocabulary Review

As you solve problems, you must understand the meanings of mathematical terms. Fill in the blank with the correct term to complete each sentence.

I. The distance around a polygon is the _?_ of the polygon.

II. A _?_ is a two-dimensional representation of a three-dimensional figure.

III. The _?_ of a point are its x- and y-values.

IV. In the expression πr^2, the number π is a constant and the letter r is a _?_.

V. Rewriting the expression $x^2 + 5x + 6$ in the form $(x + 2)(x + 3)$ is an example of _?_.

Multiple Choice

Read each question. Then write the letter of the correct answer on your paper.

1. The blueprint of a building plan uses a scale in which $\frac{1}{2}$ inch equals 1 foot. If the length of one side of the building is $17\frac{1}{2}$ inches on the blueprint, how many feet long is that side of the building? **(Lesson 7-2)**

   (A) 30 ft (B) $32\frac{1}{2}$ ft (C) 34 ft (D) 35 ft

2. The vertices of $\triangle ABC$ have coordinates $A(-2, 3)$, $B(3, 4)$, and $C(1, -2)$. The vertices of $\triangle PQR$ have coordinates $P(-2, -3)$, $Q(3, -4)$, and $R(1, 2)$. Which transformation can be used to justify that $\triangle ABC$ is congruent to $\triangle PQR$? **(Lesson 9-2)**

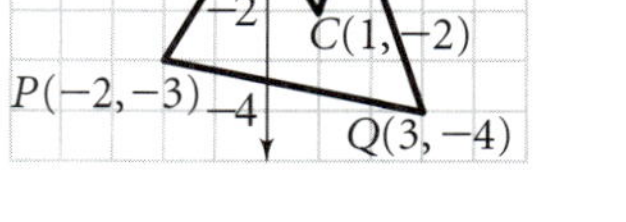

   (A) A reflection across the y-axis
   (B) A reflection across the x-axis
   (C) A rotation 180° clockwise around the origin
   (D) A translation 5 units left and 6 units down

3. Find the exact area of an equilateral triangle with sides of length 10 meters. **(Lesson 10-3)**

   (A) $25\sqrt{3}$ m^2 (B) 25 m^2 (C) $10\sqrt{3}$ m^2 (D) 10 m^2

4. If a circle has a radius of 2 feet, what is the circumference in feet? **(Lesson 10-6)**

   (A) 8π (B) 6π (C) 4π (D) 2π

5. Claire designed a pendant like the one shown at the right. It is a regular octagon set in a circle. Opposite vertices are connected by line segments. What is the measure of angle P in degrees? **(Lesson 10-3)**

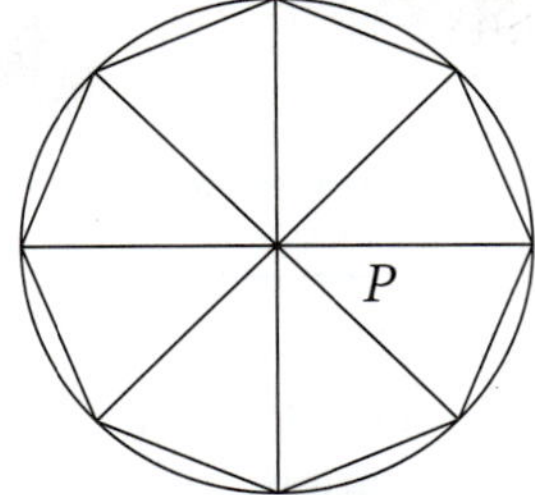

Ⓐ 22.5° Ⓒ 60°
Ⓑ 45° Ⓓ 72°

6. Mark wants to send a basketball in the mail. He uses a measuring tape to find that the circumference of the ball is about 80 cm. From this measurement, he estimates that the ball will fit into a packing box measuring 26 cm on each side. Which formula could Mark use to confirm his estimate?

(Lesson 10-6)

Ⓐ $C = \pi d$ Ⓒ $C = 2\pi d$
Ⓑ $A = \pi r^2$ Ⓓ $V = \frac{4}{3}\pi r^3$

7. What is the area of the unshaded part of the rectangle below? **(Lesson 10-8)**

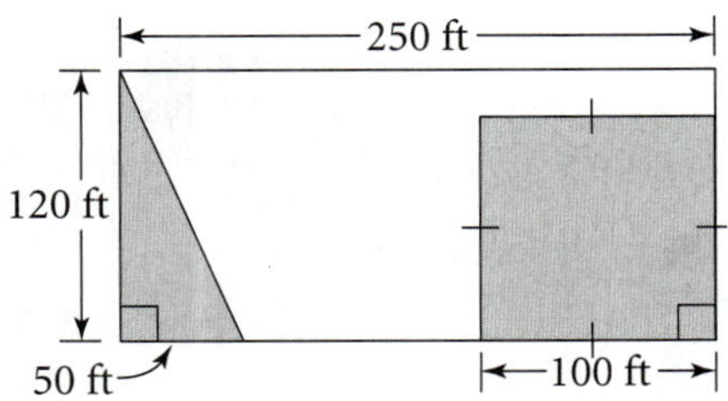

Ⓐ 30,000 ft^2 Ⓒ 19,000 ft^2
Ⓑ 13,000 ft^2 Ⓓ 17,000 ft^2

8. Every triangle in the figure at the right is an equilateral triangle. What is the total area of the shaded triangles? **(Lesson 10-3)**

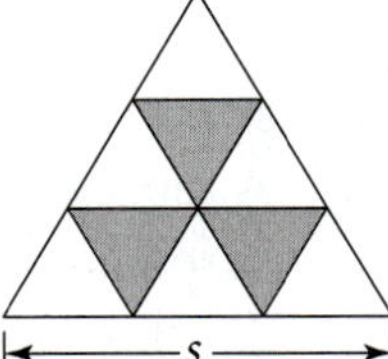

Ⓐ $\frac{s^2\sqrt{3}}{36}$

Ⓑ $\frac{s^2\sqrt{3}}{12}$

Ⓒ $\frac{s^2\sqrt{3}}{9}$

Ⓓ $\frac{s^2\sqrt{3}}{6}$

9. Which of the following could be the side lengths of a right triangle? **(Lesson 8-1)**

Ⓐ 4.1, 6.2, 7.3 Ⓒ 3.2, 5.4, 6.2
Ⓑ 40, 60, 72 Ⓓ 33, 56, 65

10. Which of the following diagrams is a correct model for the Pythagorean Theorem? **(Lesson 8-1)**

Ⓐ

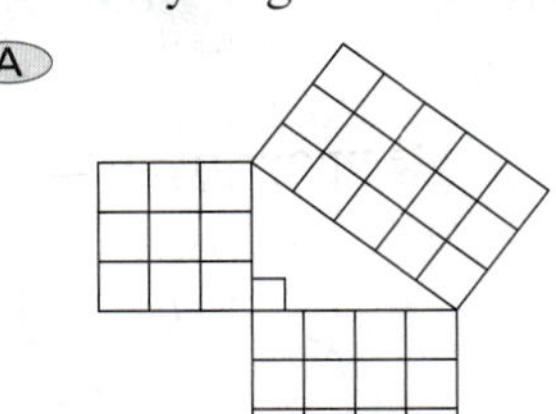

Ⓑ

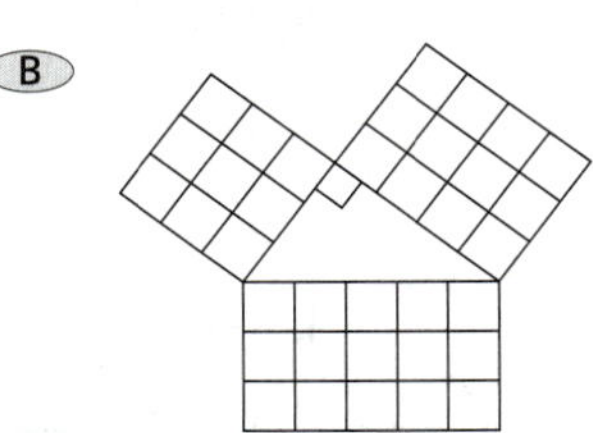

Ⓒ

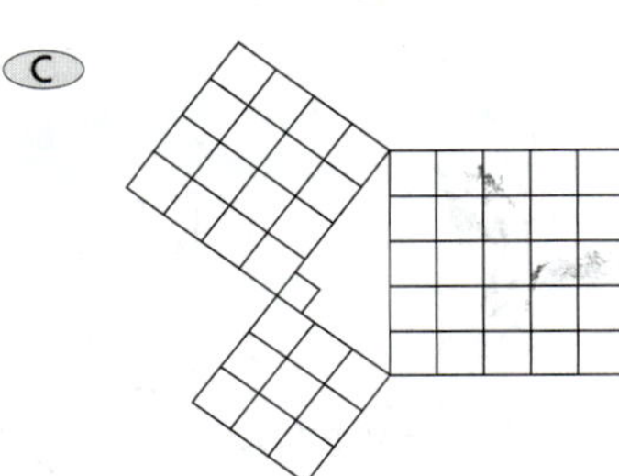

Ⓓ

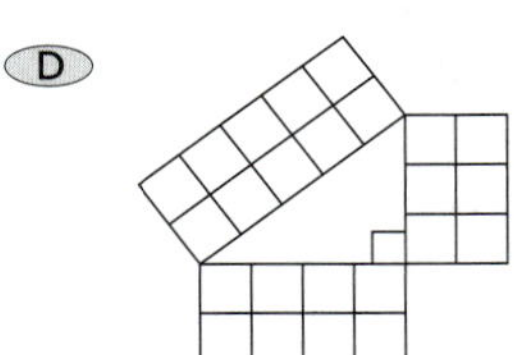

11. A triangle has a perimeter of 81 inches. If the length of each side is divided by 3, what will the perimeter of the new triangle be? **(Lesson 10-4)**

Ⓐ 3 in. Ⓑ 9 in. Ⓒ 18 in. Ⓓ 27 in.

12. A square has an area of 225 square centimeters. If the length of each side is doubled, what will the area of the new square be in square centimeters? **(Lesson 10-4)**

Ⓐ 450 cm^2 Ⓒ 1200 cm^2
Ⓑ 900 cm^2 Ⓓ 1800 cm^2

13. A rectangle has a perimeter of 12 meters. If the length of each side is multiplied by 5, what will be the perimeter in meters of the new rectangle? (Lesson 10-4)

Ⓐ 300 m Ⓑ 240 m Ⓒ 60 m Ⓓ 48 m

14. Find the image of the triangle under the translation $(x, y) \rightarrow (x + 4, y - 1)$. (Lesson 9-1)

Ⓐ
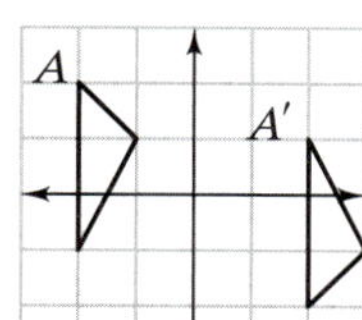

Ⓒ
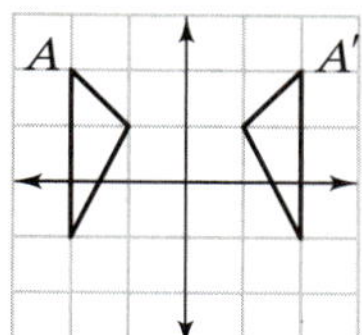

Ⓑ
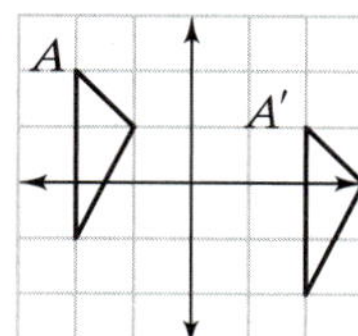

Ⓓ
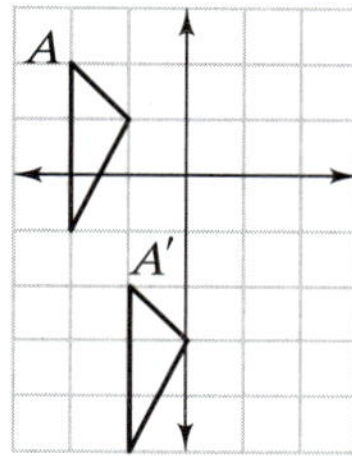

15. The vertices of triangle ART are $A(1, -2)$, $R(5, -1)$, and $T(4, -4)$. If the triangle is translated 2 units to the left and 5 units up, what are the coordinates of the image triangle $A'R'T'$? (Lesson 9-1)

Ⓐ $A'(5, -1), R'(4, -4), T'(1, -2)$

Ⓑ $A'(6, -4), R'(10, -3), T'(9, -6)$

Ⓒ $A'(3, 3), R'(7, 4), T'(6, 1)$

Ⓓ $A'(-1, 3), R'(3, 4), T'(2, 1)$

16. The shaded part of the circle below represents the portion of a garden where a landscaper has planted rose bushes.

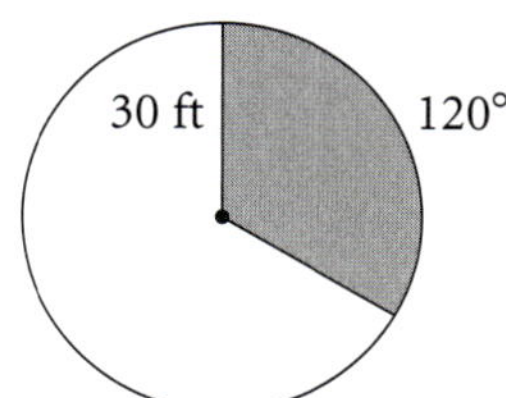

What is the approximate area of the part of the garden used for rose bushes? (Lesson 10-7)

Ⓐ 300 ft^2 Ⓑ 942 ft^2 Ⓒ 3769 ft^2 Ⓓ 8482 ft^2

17. In the figure below, what is the length of $\overline{AD}$? (Lesson 7-4)

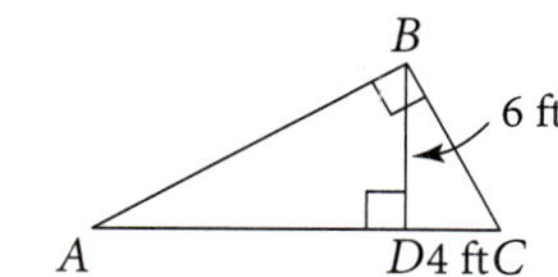

Ⓐ $2\sqrt{13}$ ft Ⓑ 9 ft Ⓒ 10 ft Ⓓ $3\sqrt{13}$ ft

18. If triangle CAT is rotated 90° around vertex C, what are the coordinates of A'? (Lesson 9-3)

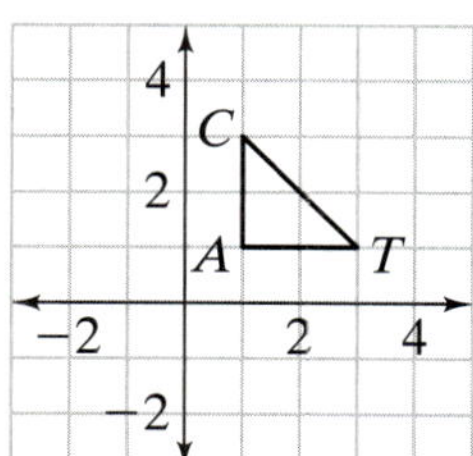

Ⓐ (1, 7) Ⓑ (3, 3) Ⓒ (−1, 1) Ⓓ (0, 0)

Standards Reference Guide

California Content Standard	Item Number(s)
GEOM 8.0	1, 4, 6, 16
GEOM 22.0	2, 14, 15, 18
GEOM 10.0	3, 7, 8
GEOM 12.0	5, 17
GEOM 15.0	9
GEOM 14.0	10
GEOM 11.0	11, 12, 13

For additional review and practice, see the *California Standards Review and Practice Workbook* or go online to use the

Visit: PHSchool.com, **Web Code:** bcq-9045

Surface Area and Volume

What You've Learned

California Content Standards

8.0 Students know, derive, and solve problems involving perimeter, circumference, and area of common geometric figures.

10.0 Students compute areas of polygons, including rectangles, scalene triangles, equilateral triangles, rhombi, parallelograms, and trapezoids.

12.0 Students find and use measures of sides and of angles of triangles and polygons to classify figures and solve problems.

The Pythagorean Theorem (Lesson 8-1)

Algebra Solve for a, b, or c in right $\triangle ABC$ where a and b are the lengths of the legs and c is the length of the hypotenuse.

1. $a = 8; b = 15; c = \underline{\ ?\ }$ **2.** $a = \underline{\ ?\ }; b = 4; c = 12$ **3.** $a = 2\sqrt{3}; b = 2\sqrt{6}; c = \underline{\ ?\ }$

Special Right Triangles (Lesson 8-2)

4. Find the length of the shorter leg of a 30°-60°-90° right triangle with hypotenuse $8\sqrt{5}$.

5. Find the length of the diagonal of a square whose perimeter is 24.

6. Find the height of an equilateral triangle with sides of length 8.

Area (Lessons 10-1, 10-2, 10-3)

Find the area of each figure. Leave your answers in simplest radical form.

7.
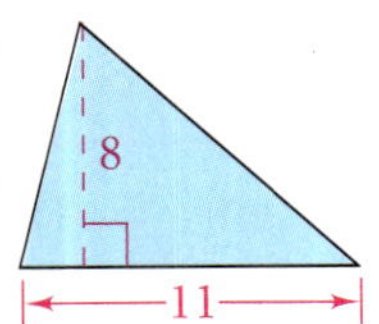

8.
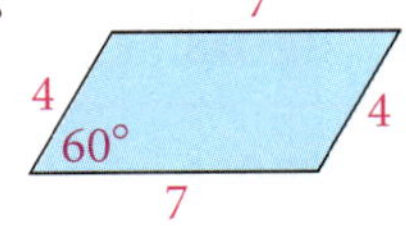

9.
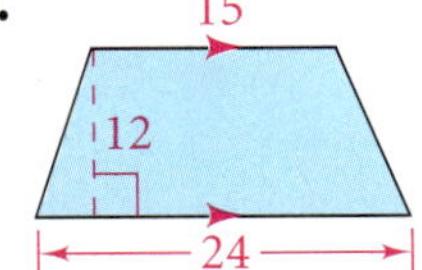

10.
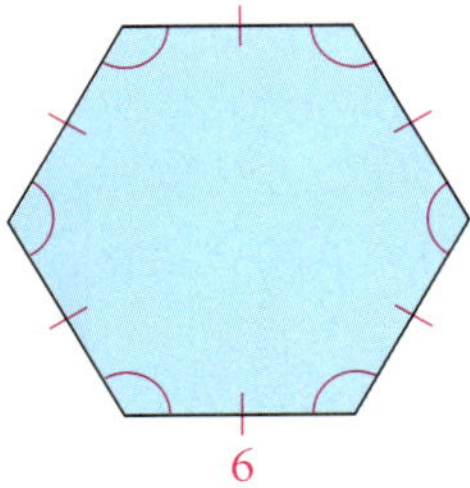

Perimeters and Areas of Similar Figures (Lesson 10-4)

11. Two similar triangles have corresponding sides in a ratio of 3 : 5. Find the perimeter of the smaller triangle if the larger triangle has perimeter 40.

12. Two regular hexagons have areas of 8 and 25. Find the ratio of corresponding sides.

What You'll Learn Next

▲ The wrinkled skin and large ears of elephants help keep them cool by maximizing the ratio of the surface area of their skin to their volume.

California Content Standards

8.0 Students know, derive, and solve problems involving volume, lateral area, and surface area of common geometric figures.

9.0 Students compute the volumes and surface areas of prisms, pyramids, cylinders, cones, and spheres; and students commit to memory the formulas for prisms, pyramids, and cylinders.

11.0 Students determine how changes in dimensions affect the area and volume of common geometric figures and solids.

CAHSEE 7NS 2.1, 7NS 2.4, 6SDAP 1.1, 6SDAP 2.5, 6SDAP 3.3, 7AF 1.2, 7AF 2.2, 7AF 3.1, 7MG 2.3, 7MG 3.2, 7MG 3.3, 7MR 2.3, Alg1 2.0, Alg1 3.0, Alg1 7.0

New Vocabulary

English and Spanish Audio Online

- **altitude** (pp. 608, 610, 617, 619)
- **base(s)** (pp. 608, 610, 617, 619)
- **cone** (p. 619)
- **cross section** (p. 600)
- **cylinder** (p. 610)
- **edge** (p. 598)
- **face** (p. 598)
- **height** (pp. 608, 610, 617, 619)
- **polyhedron** (p. 598)
- **prism** (p. 608)
- **pyramid** (p. 617)
- **similar solids** (p. 646)
- **sphere** (p. 638)
- **surface area** (pp. 608, 610, 618, 619)

Space Figures and Cross Sections

What You'll Learn

- To recognize polyhedra and their parts
- To visualize cross sections of space figures

. . . And Why

To learn about medical techniques, as in Exercise 44.

Check Skills You'll Need

GO for Help Lesson 1-3

For each exercise, make a copy of the cube at the right. Shade the plane that contains the indicated points.

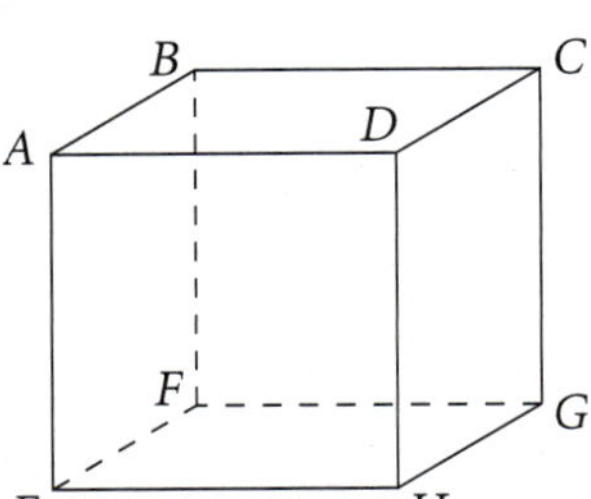

1. A, B, and C
2. A, B, and G
3. A, C, and G
4. A, D, and G
5. F, D, and G
6. B, D, and G
7. the midpoints of $\overline{AD}$, $\overline{CD}$, $\overline{EH}$, and $\overline{GH}$

New Vocabulary • polyhedron • face • edge • vertex • cross section

1 Identifying Parts of a Polyhedron

Vocabulary Tip

Polyhedron comes from the Greek *poly* for "many" and *hedron* for "side." A cube is a polyhedron with six sides, or faces, each of which is a square.

A **polyhedron** is a three-dimensional figure whose surfaces are polygons. Each polygon is a **face** of the polyhedron. An **edge** is a segment that is formed by the intersection of two faces. A **vertex** is a point where three or more edges intersect.

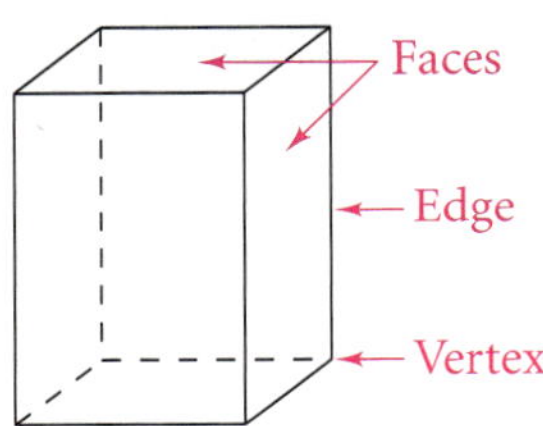

1 EXAMPLE Identifying Vertices, Edges, and Faces

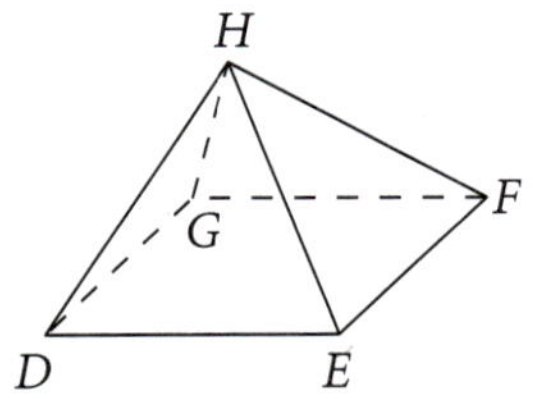

a. How many vertices are there in the polyhedron at the right? List them.

There are five vertices: D, E, F, G, and H.

b. How many edges are there? List them.

There are eight edges: $\overline{DE}$, $\overline{EF}$, $\overline{FG}$, $\overline{GD}$, $\overline{DH}$, $\overline{EH}$, $\overline{FH}$, and $\overline{GH}$.

c. How many faces are there? List them.

There are five faces: $\triangle DEH$, $\triangle EFH$, $\triangle FGH$, $\triangle GDH$, and the quadrilateral $DEFG$.

1 List the vertices, edges, and faces of the polyhedron.

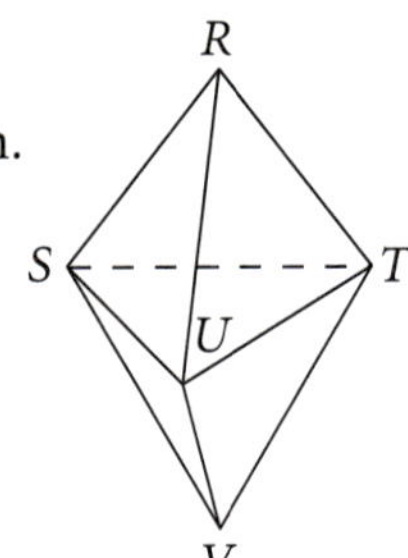

Leonhard Euler, a Swiss mathematician, discovered a relationship among the numbers of faces, vertices, and edges of any polyhedron. The result is known as Euler's Formula.

Key Concepts

Formula — **Euler's Formula**

The numbers of faces (F), vertices (V), and edges (E) of a polyhedron are related by the formula $F + V = E + 2$.

2 EXAMPLE Using Euler's Formula

Real-World Connection

Euler's Formula applies to the polyhedron suggested by the panels on a volleyball.

Count faces and edges. Then use Euler's Formula to find the number of vertices in the polyhedron at the right.

The polyhedron has 2 hexagons and 6 rectangles for a total of 8 faces.

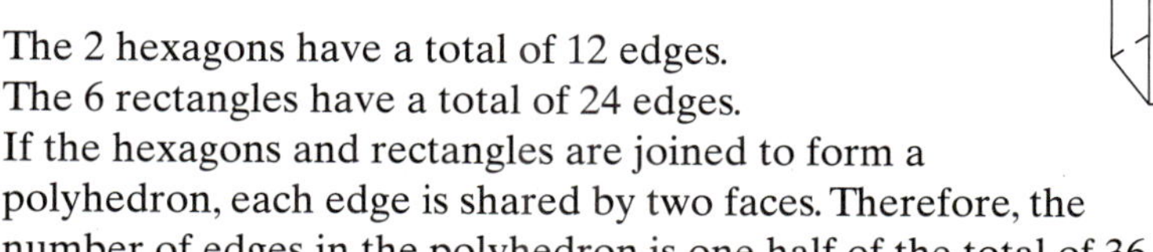

The 2 hexagons have a total of 12 edges.
The 6 rectangles have a total of 24 edges.
If the hexagons and rectangles are joined to form a polyhedron, each edge is shared by two faces. Therefore, the number of edges in the polyhedron is one half of the total of 36, or 18.

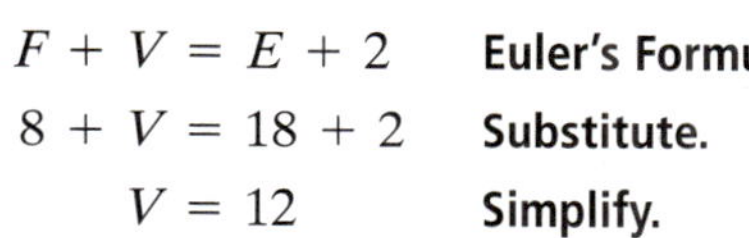

$F + V = E + 2$	**Euler's Formula**
$8 + V = 18 + 2$	**Substitute.**
$V = 12$	**Simplify.**

Count the number of vertices in the figure to verify the result.

CA Standards Check

2 Use Euler's Formula to find the number of edges on a polyhedron with eight triangular faces.

In two dimensions, Euler's Formula reduces to

$$F + V = E + 1$$

where F is the number of regions formed by V vertices linked by E segments.

3 EXAMPLE Verifying Euler's Formula

Verify Euler's Formula for a two-dimensional net of the solid in Example 2.

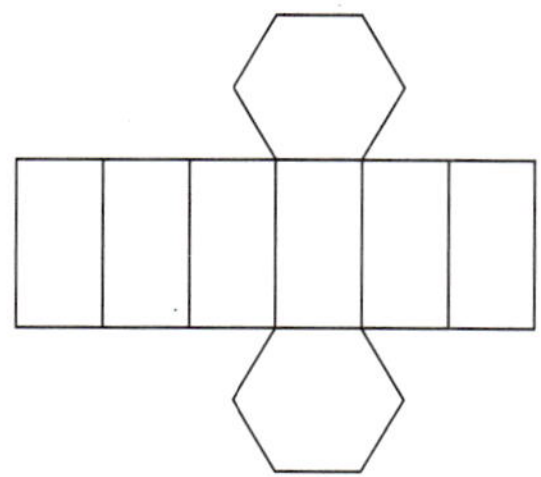

Draw a net:

Count the regions: $F = 8$
Count the vertices: $V = 22$
Count the segments: $E = 29$

$$8 + 22 = 29 + 1$$

CA Standards Check

3 The figure at the right is a trapezoidal prism.

a. Verify Euler's formula $F + V = E + 2$ for the prism.
b. Draw a net for the prism.
c. Verify Euler's formula $F + V = E + 1$ for your two-dimensional net.

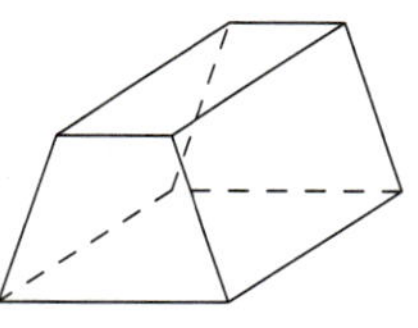

2 Describing Cross Sections

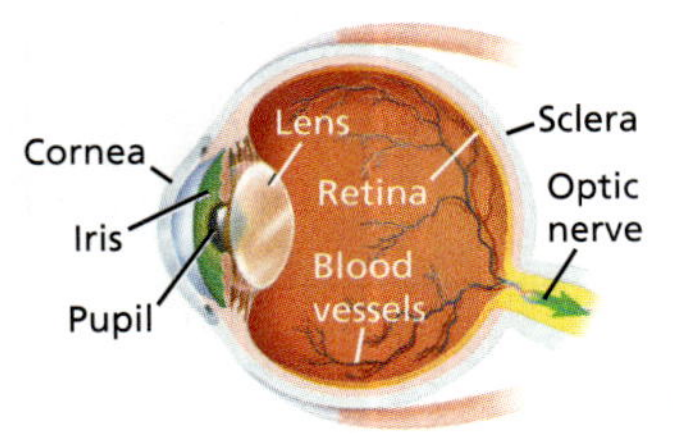

Cross sections are used to study the anatomy of the eye.

A **cross section** is the intersection of a solid and a plane. You can think of a cross section as a very thin slice of the solid.

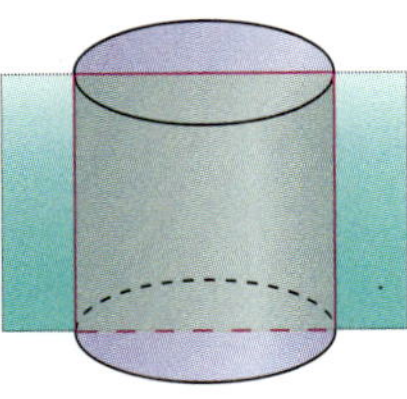

4 EXAMPLE Describing a Cross Section

Describe each cross section.

a.

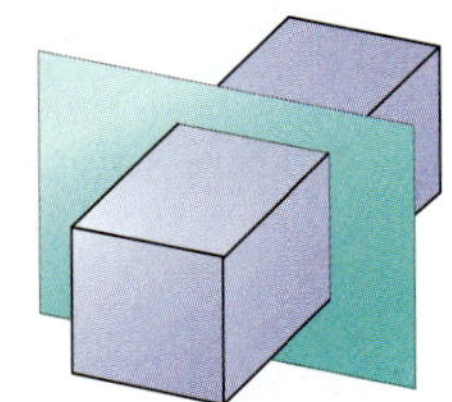

The cross section is a square.

b.

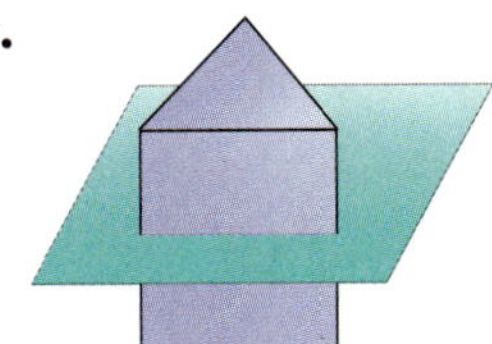

The cross section is a triangle.

4 For the funnel shown, sketch each of the following.

a. a horizontal cross section

b. a vertical cross section that contains the axis of symmetry

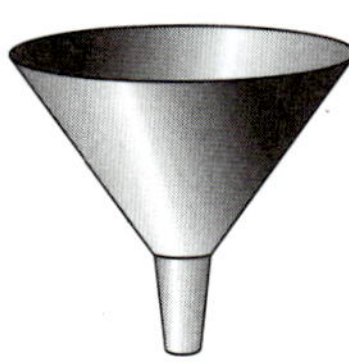

To draw a cross section, you can sometimes use the idea from Postulate 1-3 that the intersection of two planes is exactly one line.

5 EXAMPLE Drawing a Cross Section

Visualization Draw and describe a cross section formed by a vertical plane intersecting the front and right faces of the cube.

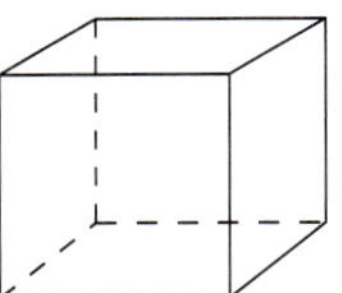

A vertical plane cuts the vertical faces of the cube in parallel segments.

Draw the parallel segments.

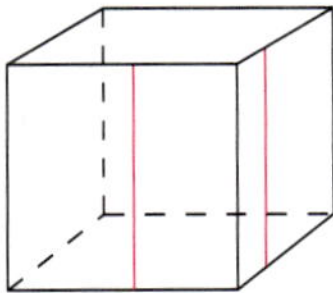

Join their endpoints. Shade the cross section.

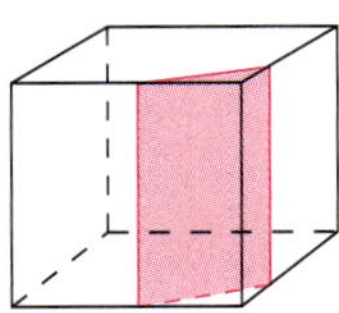

The cross section is a rectangle.

CA Standards Check

5 Draw and describe the cross section formed by a horizontal plane intersecting the left and right faces of the cube.

EXERCISES

Standards Practice

For more exercises, see *Extra Skill, Word Problem, and Proof Practice.*

Practice by Example

Example 1 (page 598)

For each polyhedron, how many vertices, edges, and faces are there? List them.

1.

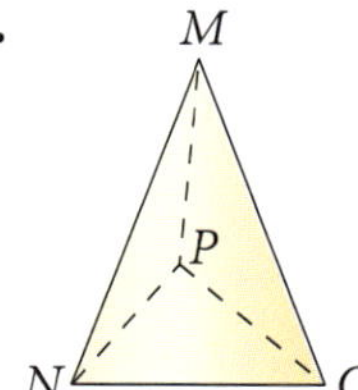

2.

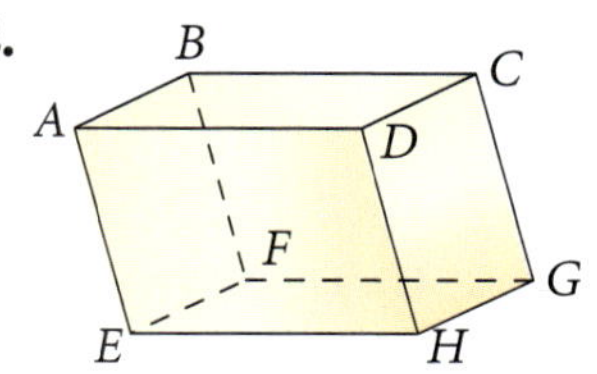

3.

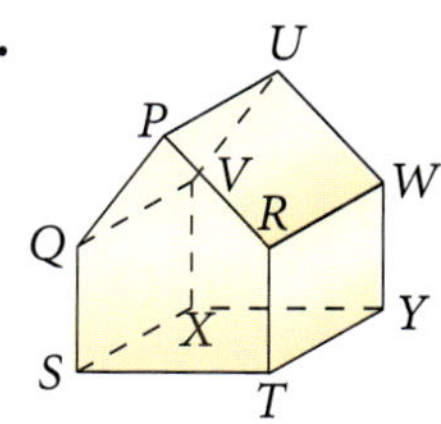

Example 2 (page 599)

Use Euler's Formula to find the missing number.

4. Faces: ■
Edges: 15
Vertices: 9

5. Faces: 8
Edges: ■
Vertices: 6

6. Faces: 20
Edges: 30
Vertices: ■

Use Euler's Formula to find the number of vertices in each polyhedron described below.

7. 6 square faces

8. 5 faces: 1 rectangle and 4 triangles

9. 9 faces: 1 octagon and 8 triangles

Example 3 (page 599)

Verify Euler's Formula for each polyhedron. Then draw a net for the figure and verify Euler's Formula for the two-dimensional figure.

10.

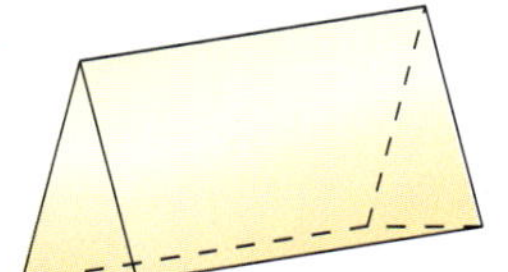

11.

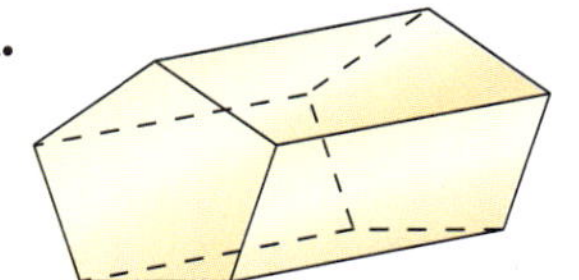

12.

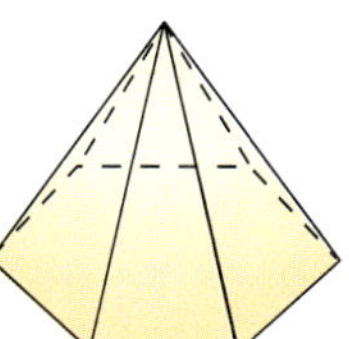

Example 4 (page 600)

Describe each cross section.

13.

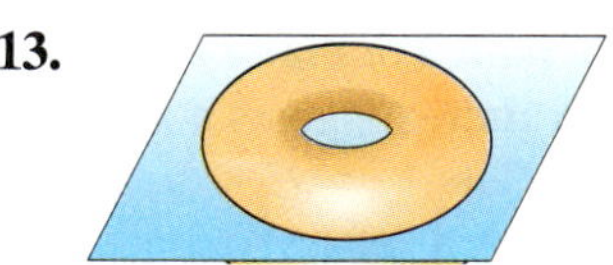

14.

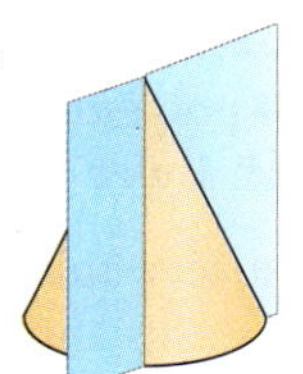

15.

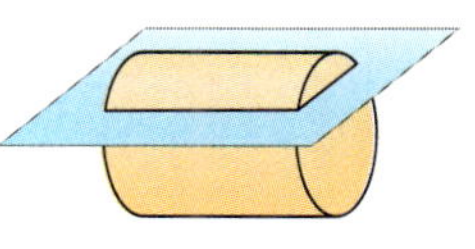

16. For the nut shown, sketch each of following.

a. a horizontal cross section

b. a vertical cross section that contains the vertical line of symmetry

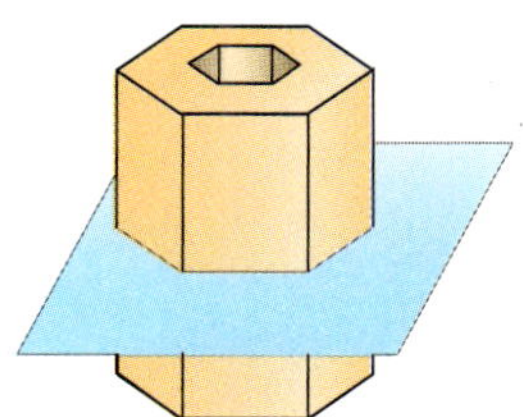

Example 5 (page 600)

Visualization **Draw and describe a cross section formed by a vertical plane intersecting the cube as follows.**

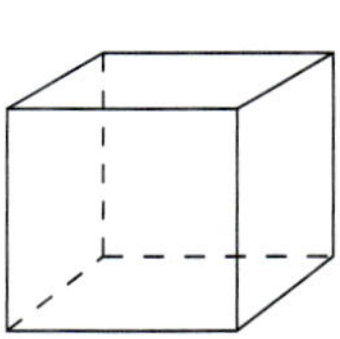

17. The vertical plane intersects the front and left faces of the cube.

18. The vertical plane intersects opposite faces of the cube.

19. The vertical plane contains opposite edges of the cube.

Apply Your Skills

20. a. Open-Ended Sketch a polyhedron whose faces are all rectangles. Label the lengths of its edges.

b. Use graph paper to draw two different nets for the polyhedron.

Visualization Draw and describe a cross section formed by a plane intersecting the cube as follows.

21. The plane is tilted and intersects the left and right faces of the cube.

22. The plane contains opposite horizontal edges of the cube.

23. The plane cuts off a corner of the cube.

Describe the cross section shown.

24.

25.

26.

Visualization A plane region that revolves completely about a line sweeps out a *solid of revolution.* Use the sample to help you describe the solid of revolution you get by revolving each region about line ℓ.

Sample: Revolve the rectangular region about the line ℓ and you get a cylinder as a solid of revolution.

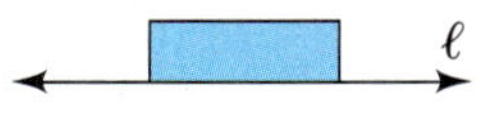

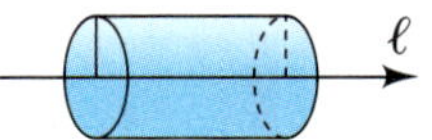

27.

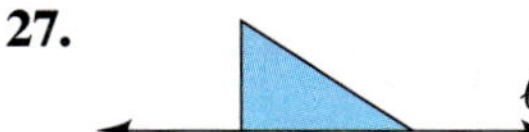

28.

29.

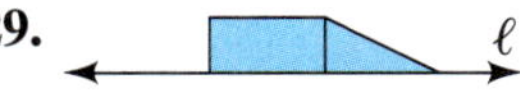

Sports Equipment Some balls are made from panels that suggest polygons. The ball then suggests a polyhedron to which Euler's Formula, $F + V = E + 2$, applies.

30. A soccer ball suggests a polyhedron with 20 regular hexagons and 12 regular pentagons. How many vertices does this polyhedron have?

31. Show how Euler's Formula applies to the polyhedron suggested by the volleyball pictured on page 599. (*Hint:* It has 6 sets of 3 panels.)

Euler's Formula $F + V = E + 1$ applies to any two-dimensional network where F is the number of regions formed by V vertices linked by E edges (or paths). Verify Euler's Formula for each network shown.

32.

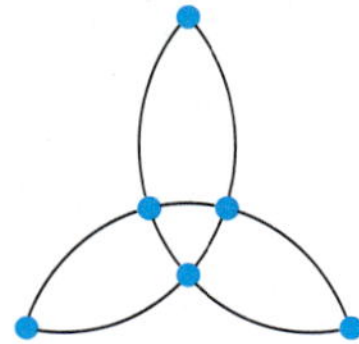

33.

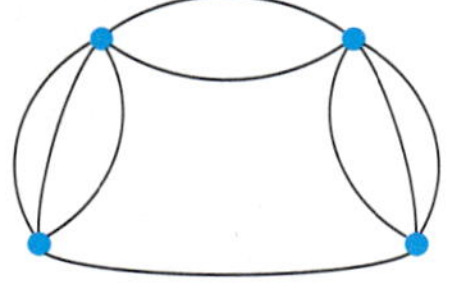

34.

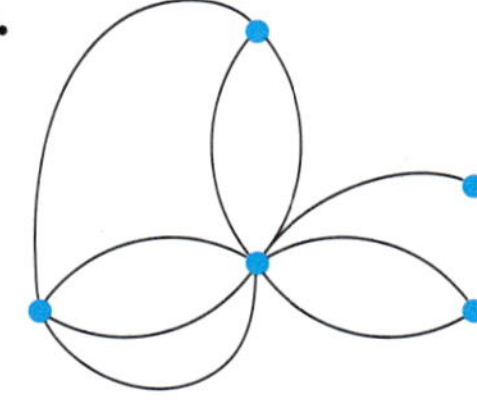

35. Draw a network of your own. Verify Euler's Formula for it.

Homework Video Tutor

Visit: PHSchool.com
Web Code: bce-1101

Real-World Connection

A fluorite crystal forms as a regular octahedron.

36. There are five regular polyhedrons. They are called *regular* because all their faces are congruent regular polygons, and the same number of faces meet at each vertex. They are also called Platonic Solids after the Greek philosopher Plato (427–347 B.C.).

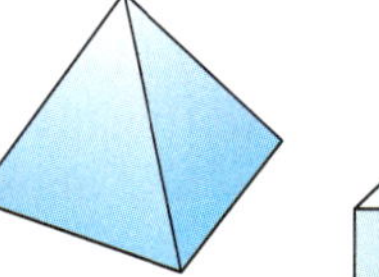

Tetrahedron

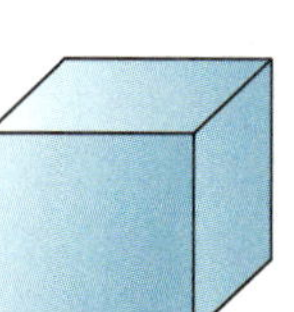

Hexahedron

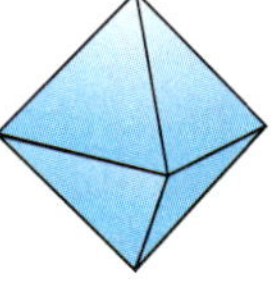

Octahedron

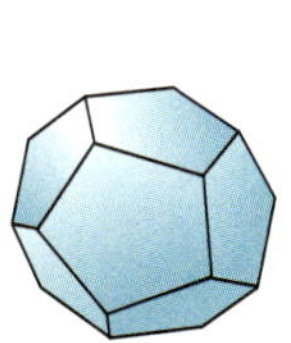

Dodecahedron

Icosahedron

a. Match each net below with a Platonic Solid.

A. 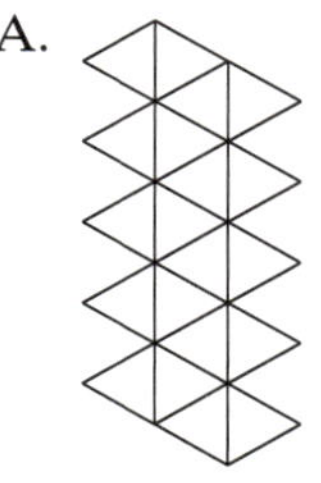
B. 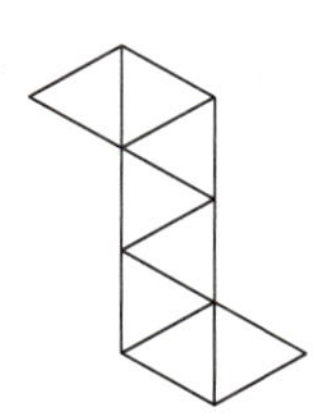
C. 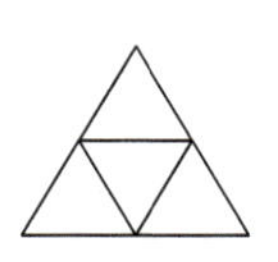
D. 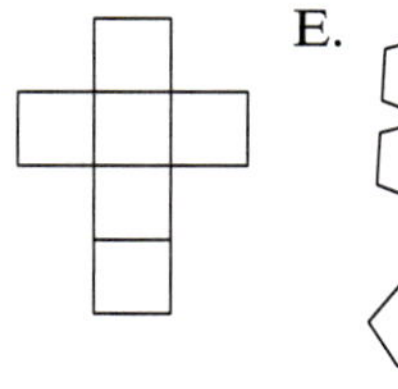
E. 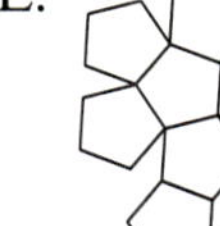

b. The first two Platonic solids have more familiar names. What are they?

c. Verify that Euler's Formula is true for the first three Platonic solids.

37. Multiple Choice A cube has a total surface area of 216 in.^{2}. How long is an edge of the cube?

(A) 6 in. (B) 15 in. (C) 36 in. (D) 54 in.

Draw each object. Then draw a horizontal and a vertical cross section.

38. a golf tee **39.** a football **40.** a baseball bat

41. a banana **42.** a pear **43.** a bagel

44. Writing Cross sections are used in medical training and research. Research and write a paragraph on how magnetic resonance imaging (MRI) is used to study cross sections of the brain.

Challenge

45. Draw a solid that has the following cross sections.

horizontal

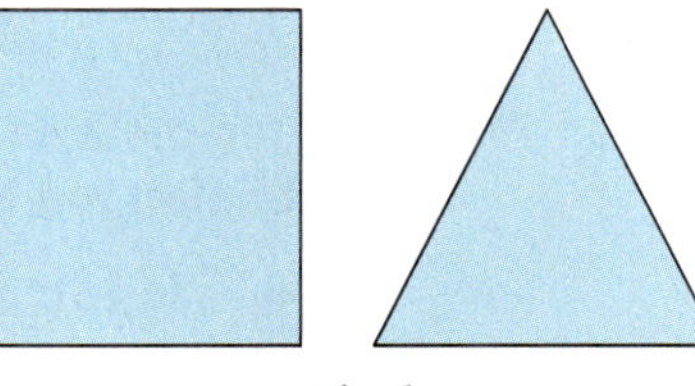

vertical

Visualization Draw a plane intersecting a cube to get the cross section indicated.

46. scalene triangle **47.** isosceles triangle **48.** equilateral triangle

49. trapezoid **50.** isosceles trapezoid **51.** parallelogram

52. rhombus **53.** pentagon **54.** hexagon

Multiple Choice Practice

For California Standards and CAHSEE Tutorials, visit PHSchool.com. Web Code: bcq-9045

GEOM 15.0

55. A 15-foot cable secures an antenna to a roof. One end of the cable is attached to the roof 11 feet from the antenna's base. The other end is secured to the antenna at the highest possible point. About how many feet above the roof is the cable secured?

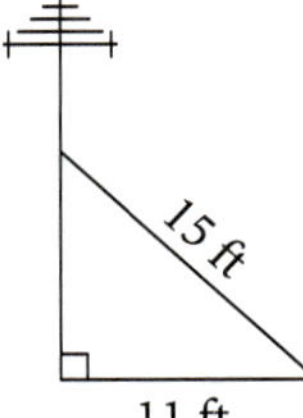

- (A) 4 feet
- (B) 10 feet
- (C) 12 feet
- (D) 19 feet

GEOM 22.0

56. Triangle ABC is translated so that A is mapped to A'. Which set of coordinates best identifies the locations of B' and C'?

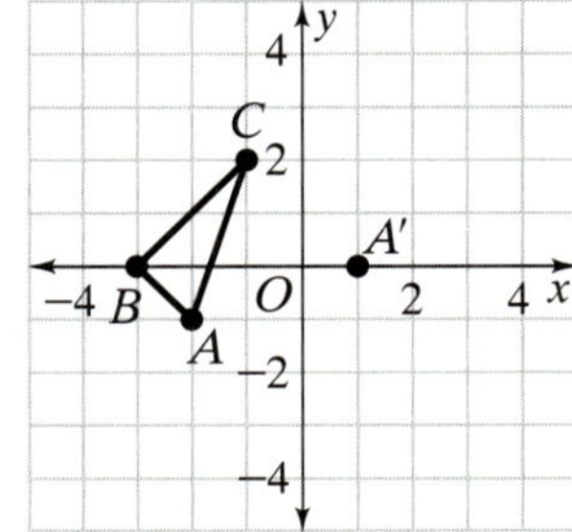

- (A) $(2, -1)$ and $(2, 3)$
- (B) $(2, -1)$ and $(3, 2)$
- (C) $(0, 1)$ and $(2, 3)$
- (D) $(0, 1)$ and $(3, 2)$

CAHSEE Alg1 2.0

57. The perimeter, P, of a square may be found by using the formula $P = 4\sqrt{A}$, where A is the area of the square. What is the perimeter of a square with an area of 49 square inches?

- (A) 196 inches
- (B) 28 inches
- (C) 12.25 inches
- (D) 7 inches

Mixed Review

Lesson 10-8

58. Probability A shuttle bus to an airport terminal leaves every 20 min from a remote parking lot. Draw a geometric model and find the probability that a traveler who arrives at a random time will have to wait at least 8 min for the bus to leave the parking lot.

59. Games A dartboard is a circle with a 12-in. radius. You throw a dart that hits the dartboard. What is the probability that the dart lands within 6 in. of the center of the dartboard?

Lesson 10-3

Find the area of each equilateral triangle with the given measure. Leave answers in simplest radical form.

60. side 2 ft

61. apothem 8 cm

62. radius 100 in.

Lesson 10-3

Find the value of x to the nearest tenth.

63.

65°
x
10

64.

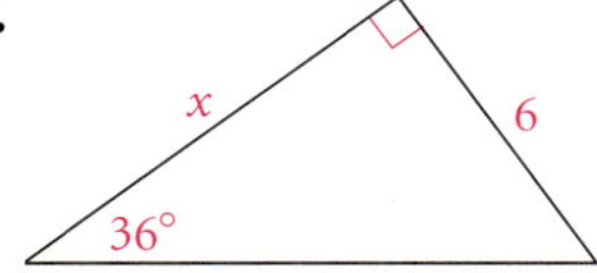

65. The lengths of the diagonals of a rhombus are 4 cm and 6 cm. Find the measures of the angles of the rhombus to the nearest degree.

Perspective Drawing

For Use With Lesson 11-1

You can create a three-dimensional space figure with a two-dimensional *perspective drawing*. Suppose two lines are parallel in three dimensions but recede from the viewer. You draw them—and create perspective—so that they meet at a *vanishing point* on a *horizon line*.

1 EXAMPLE

Draw a cube in one-point perspective.

Step 1: Draw a square. Then draw a horizon line and a vanishing point on the line.

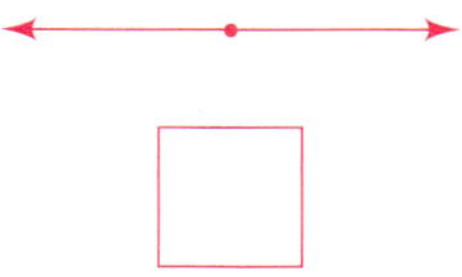

Step 2: Lightly draw segments from the vertices of the square to the vanishing point.

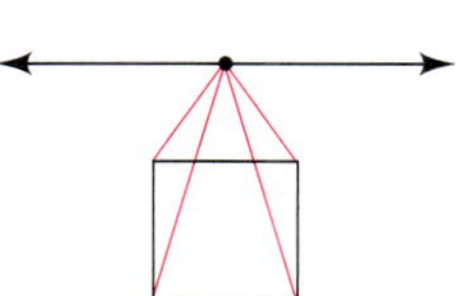

Step 3: Draw a square for the back of the cube. Each vertex should lie on a segment you drew in Step 2.

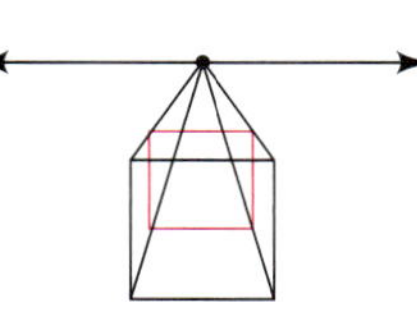

Step 4: Complete the figure by using dashes for hidden edges of the cube. Erase unneeded lines.

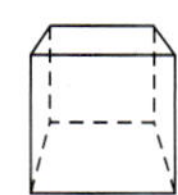

Two-point perspective involves the use of two vanishing points.

2 EXAMPLE

Draw a box in two-point perspective.

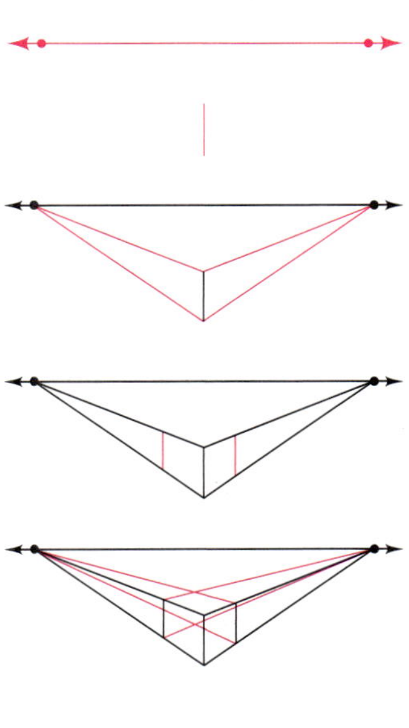

Step 1: Draw a vertical segment. Then draw a horizon line and two vanishing points on the line.

Step 2: Lightly draw segments from the endpoints of the vertical segment to each vanishing point.

Step 3: Draw two vertical segments between the segments of Step 2.

Step 4: Draw segments from the endpoints of the segments you drew in Step 3 to the vanishing points.

Step 5: Complete the figure by using dashes for hidden edges of the figure. Erase unneeded lines.

In one-point perspective, the front of the cube is parallel to the drawing surface. A two-point perspective drawing generally looks like a corner view. For either type of drawing, you should be able to envision each vanishing point.

EXERCISES

Is each object drawn in one- or two-point perspective?

1. 

2. 

3. 

4. 

5. 

6. 

Draw each object in one-point perspective and then in two-point perspective.

7. a shoe box

8. a building in your town that sits on a street corner

Draw each container using one-point perspective. Show a base at the front.

9. triangular carton

10. hexagonal box

Copy each figure and locate the vanishing point(s).

11. 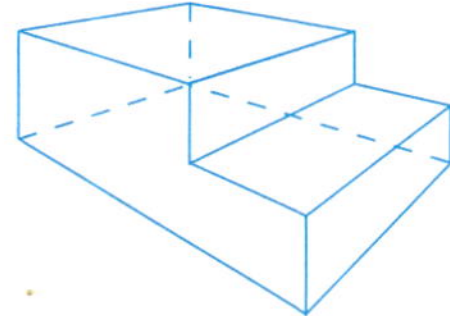

12. 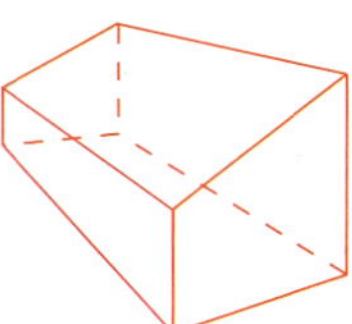

13. 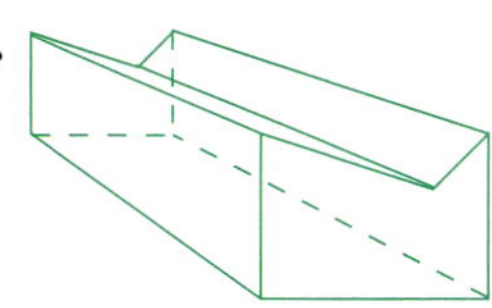

Optical Illusions **What is the optical illusion? Explain the role of perspective in each illusion.**

14. 

15. 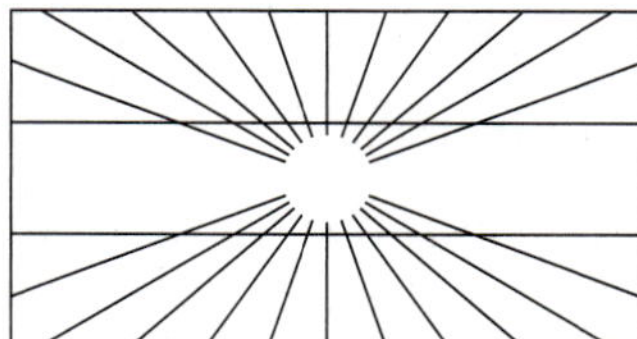

16. **Open Ended** You can draw block letters in either one-point perspective or two-point perspective. Write your initials in block letters using one-point perspective and two-point perspective.

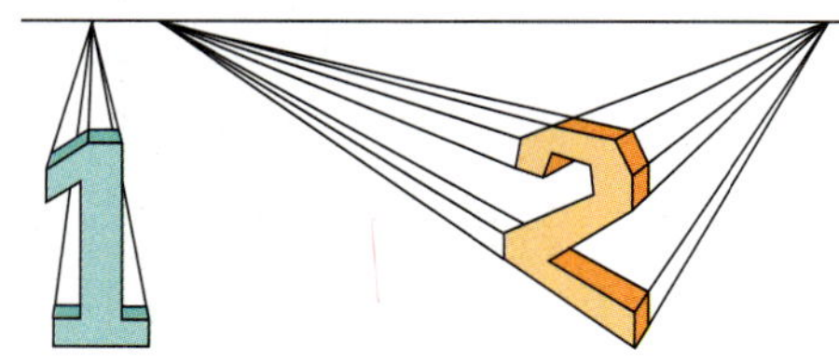

Algebra 1 Review

Literal Equations

FOR USE WITH LESSON 11-2

Prepares for GEOM 8.0 Students solve problems involving perimeter, circumference, area, and volume of common geometric figures.

A *literal equation* is an equation involving two or more variables. A formula is a special type of literal equation. You can transform a formula by solving for one variable in terms of the others.

1 EXAMPLE

Algebra The formula for the volume of a cylinder is $V = \pi r^2 h$. Find a formula for the height in terms of the radius and volume.

$V = \pi r^2 h$

$\frac{V}{\pi r^2} = \frac{\pi r^2 h}{\pi r^2}$ Divide each side by πr^2, $r \neq 0$.

$\frac{V}{\pi r^2} = h$ Simplify.

The formula for the height is $h = \frac{V}{\pi r^2}$.

Solving literal equations also allows you to build other formulas.

2 EXAMPLE

Algebra Find a formula for the area of a square in terms of its perimeter.

$P = 4s$ Use the formula for perimeter.

$\frac{P}{4} = s$ Solve for s in terms of P.

$A = s^2$ Use the formula for area.

$A = \left(\frac{P}{4}\right)^2$ Substitute.

$A = \frac{P^2}{16}$ Simplify.

The formula for the area is $A = \frac{P^2}{16}$.

EXERCISES

Algebra Solve each equation for the variable in red.

1. $C = 2\pi r$
2. $A = \frac{1}{2}bh$
3. $A = \pi r^2$

Algebra Solve for the variable in red. Then solve for the variable in blue.

4. $P = 2w + 2\ell$
5. $\tan A = \frac{y}{x}$
6. $A = \frac{1}{2}(b_1 + b_2)h$

Find a formula as stated.

7. the circumference of a circle in terms of its area
8. the area of an isosceles right triangle in terms of the hypotenuse
9. the apothem of a regular hexagon in terms of the area of the hexagon

10. Solve $A = \frac{1}{2}ab \sin C$ for $m\angle C$.

Surface Areas of Prisms and Cylinders

California Content Standards

GEOM 8.0 Students know, derive, and solve problems involving the lateral area and surface area of common geometric figures. ***Introduce***
GEOM 9.0 Students compute surface areas of prisms and cylinders; students commit to memory the formulas for prisms and cylinders. ***Introduce***

What You'll Learn

- To find the surface area of a prism
- To find the surface area of a cylinder

. . . And Why

To find the area covered by a drum on a roller used in road construction, as in Example 4

Check Skills You'll Need

Find the area of each net.

1.

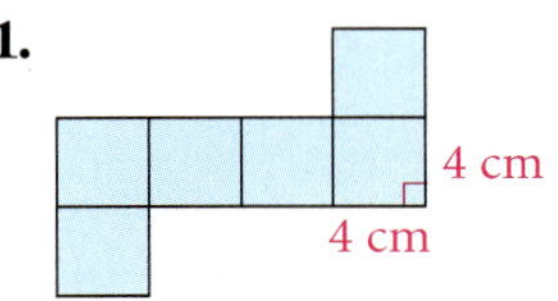

2.

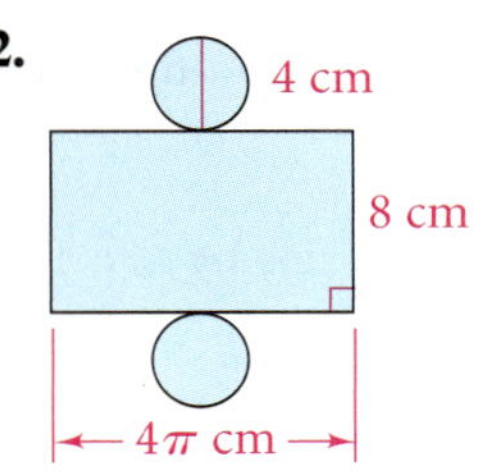

3.

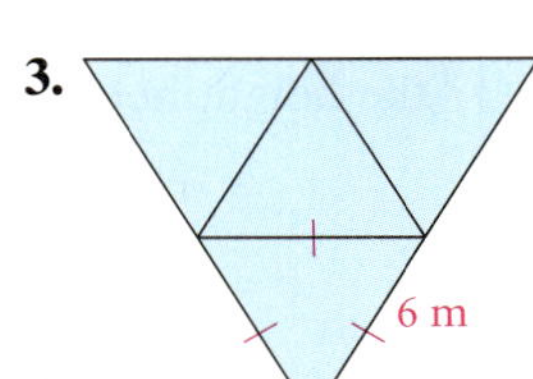

New Vocabulary • prism • bases, lateral faces, altitude, height, lateral area, surface area (of a prism) • right prism • oblique prism • cylinder • bases, altitude, height, lateral area, surface area (of a cylinder) • right cylinder • oblique cylinder

1 Finding Surface Area of a Prism

Real-World Connection

A triangular prism separates white light into its constituent colors.

A **prism** is a polyhedron with exactly two congruent, parallel faces, called **bases.** Other faces are **lateral faces.** You name a prism by the shape of its bases.

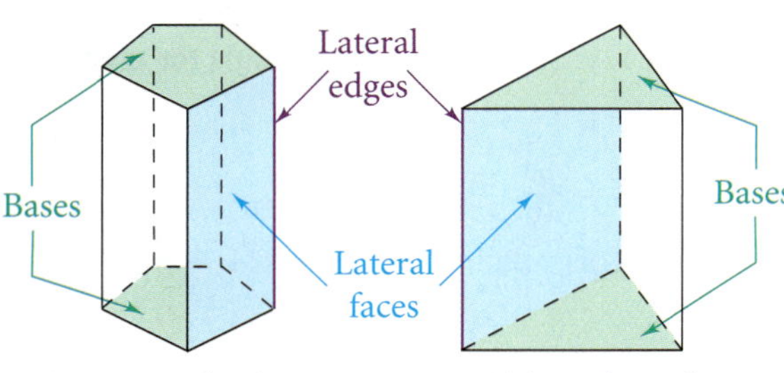

An **altitude** of a prism is a perpendicular segment that joins the planes of the bases. The **height** h of the prism is the length of an altitude.

A prism may either be right or oblique.

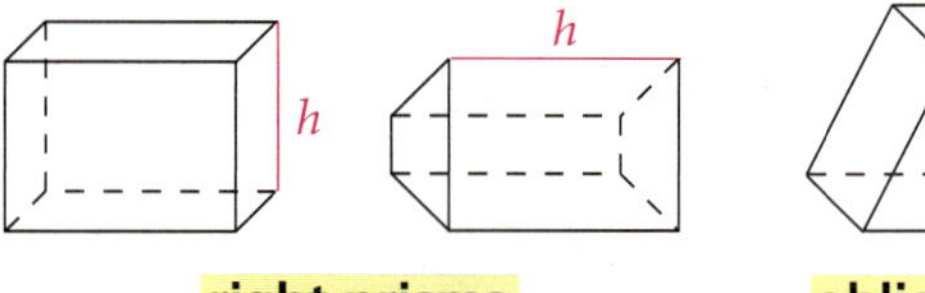

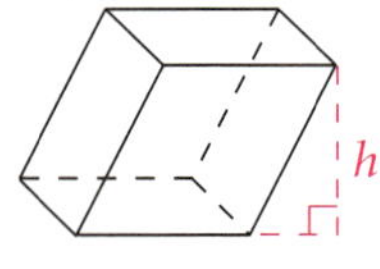

In a right prism the lateral faces are rectangles and a lateral edge is an altitude. In this book you may assume that a prism is a right prism unless stated or pictured otherwise.

The **lateral area** of a prism is the sum of the areas of the lateral faces. The **surface area** is the sum of the lateral area and the area of the two bases.

1 EXAMPLE Finding Surface Area of a Prism

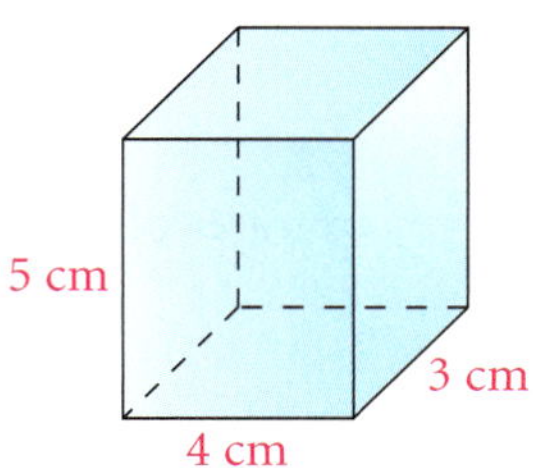

Use a net to find the surface area of the prism at the left.

Surface Area = Lateral Area + area of bases

$= \text{sum of areas of lateral faces} + \text{area of bases}$

$= (5 \cdot 4 + 5 \cdot 3 + 5 \cdot 4 + 5 \cdot 3) + 2(3)(4)$

$= 70 + 24$

$= 94$

The surface area of the prism is 94 cm^2.

1 Use a net to find the surface area of the triangular prism.

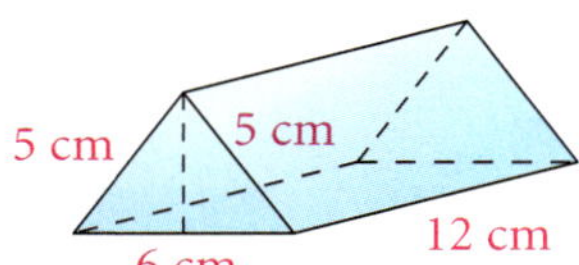

You can find formulas for lateral and surface areas by looking at a net for a prism.

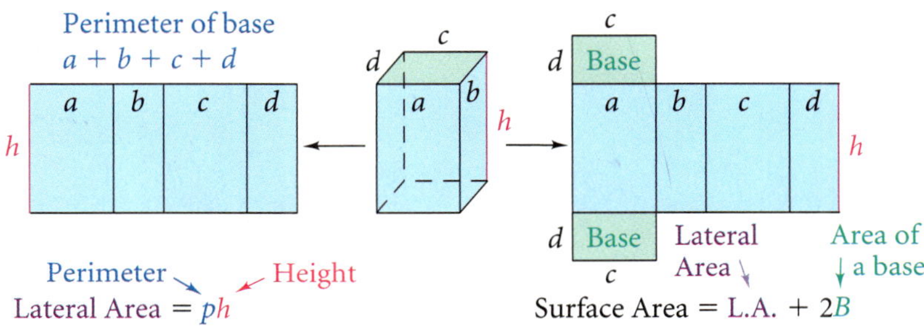

You can use the formulas with any right prism.

2 EXAMPLE Using Formulas to Find Surface Area

Test-Taking Tip

A question could ask for either surface area or lateral area of a solid. Read the question carefully.

Multiple Choice What is the surface area of the prism?

Ⓐ 72 cm^2 Ⓑ 78 cm^2 Ⓒ 84 cm^2 Ⓓ 96 cm^2

By the Pythagorean Theorem, the hypotenuse of the triangular base is 5 cm.

$\text{L.A.} = ph$ **Use the formula for lateral area.**

$= 12 \cdot 6$ **$p = 3 + 4 + 5 = 12$ cm**

$= 72$

The lateral area of the prism is 72 cm^2.

Now use the formula for surface area.

$\text{S.A.} = \text{L.A.} + 2B$

$= 72 + 2(6) = 84$ **$B = \frac{1}{2}(3 \cdot 4) = 6\ \text{cm}^2$**

The surface area of the prism is 84 cm^2. Choice C is the answer.

2 Use formulas to find the lateral area and surface area of the prism.

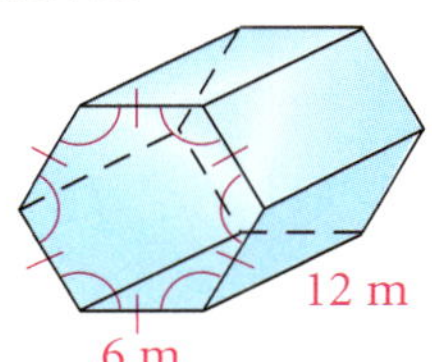

Theorem 11-1 Lateral and Surface Areas of a Prism

The lateral area of a right prism is the product of the perimeter of the base and the height.

$$\text{L.A.} = ph$$

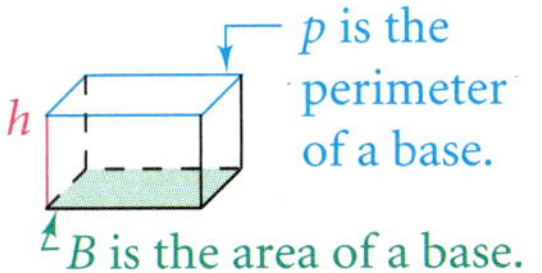

The surface area of a right prism is the sum of the lateral area and the areas of the two bases.

$$\text{S.A.} = \text{L.A.} + 2B$$

2 Finding Surface Area of a Cylinder

Like a prism, a **cylinder** has two congruent parallel **bases.** However, the bases of a cylinder are circles. An **altitude** of a cylinder is a perpendicular segment that joins the planes of the bases. The **height** h of a cylinder is the length of an altitude.

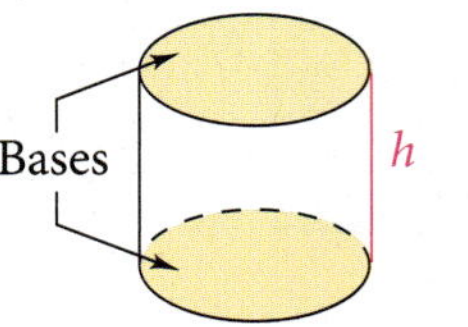

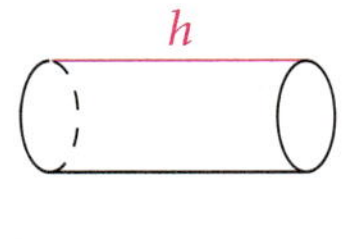

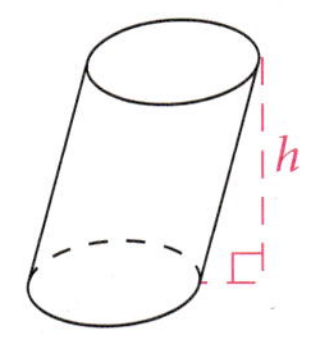

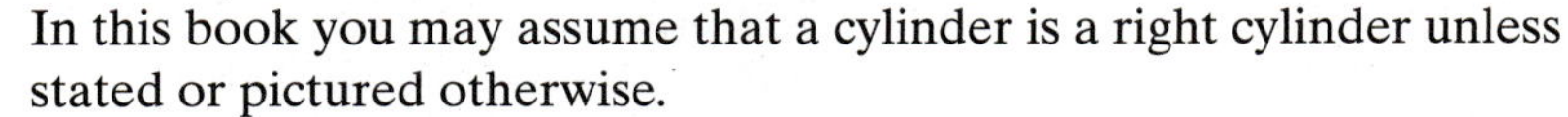

In this book you may assume that a cylinder is a right cylinder unless stated or pictured otherwise.

To find the area of the curved surface of a cylinder, visualize "unrolling" it. The area of the resulting rectangle is the **lateral area** of the cylinder. The **surface area** of a cylinder is the sum of the lateral area and the areas of the two circular bases. You can find formulas for these areas by looking at a net for a cylinder.

Real-World Connection

A full turn of the roller inks a rectangle with area equal to the roller's lateral area.

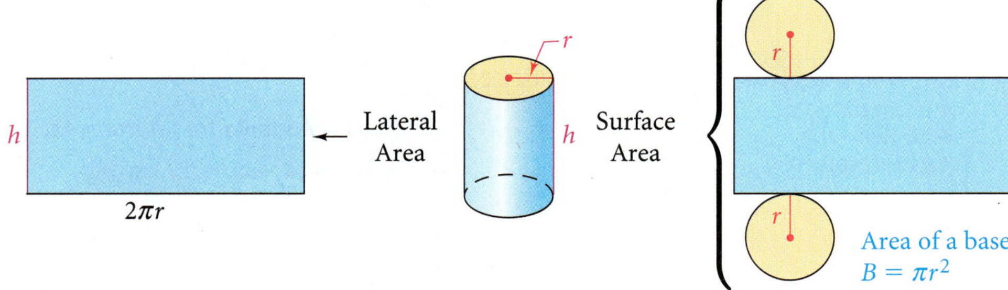

Key Concepts

Theorem 11-2 Lateral and Surface Areas of a Cylinder

The lateral area of a right cylinder is the product of the circumference of the base and the height of the cylinder.

$$\text{L.A.} = 2\pi rh, \text{ or L.A.} = \pi dh$$

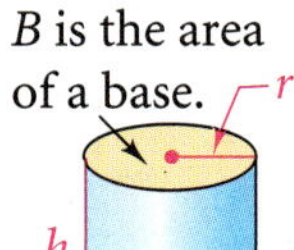

The surface area of a right cylinder is the sum of the lateral area and the areas of the two bases.

$$\text{S.A.} = \text{L.A.} + 2B, \text{ or S.A.} = 2\pi rh + 2\pi r^2$$

3 EXAMPLE Finding Surface Area of a Cylinder

The radius of the base of a cylinder is 4 in. and its height is 6 in. Find the surface area of the cylinder in terms of π.

$\text{S.A.} = \text{L.A.} + 2B$	**Use the formula for surface area of a cylinder.**
$= 2\pi rh + 2(\pi r^2)$	**Substitute the formulas for lateral area and area of a circle.**
$= 2\pi(4)(6) + 2\pi(4^2)$	**Substitute 4 for *r* and 6 for *h*.**
$= 48\pi + 32\pi$	**Simplify.**
$= 80\pi$	

The surface area of the cylinder is 80π in.^{2}.

CA Standards Check

3 Find the surface area of a cylinder with height 10 cm and radius 10 cm in terms of π.

4 EXAMPLE Real-World Connection

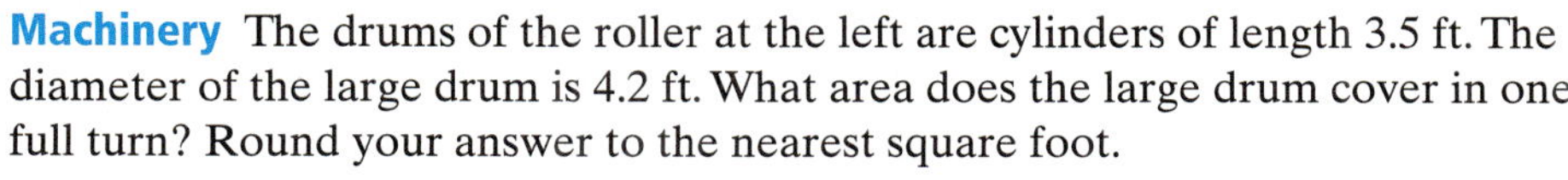

Machinery The drums of the roller at the left are cylinders of length 3.5 ft. The diameter of the large drum is 4.2 ft. What area does the large drum cover in one full turn? Round your answer to the nearest square foot.

The area covered is the lateral area of a cylinder that has a diameter of 4.2 ft and a height of 3.5 ft.

$\text{L.A.} = \pi dh$	**Use the formula for lateral area of a cylinder.**
$= \pi(4.2)(3.5)$	**Substitute.**
$= 46.181412$	**Use a calculator.**

In one full turn, the large drum covers about 46 ft^2.

CA Standards Check

4 The small drum has diameter 3 ft.

a. To the nearest square foot, what area does the small drum cover in one turn?

b. **Critical Thinking** What area does the small drum cover in one turn of the large drum?

EXERCISES

For more exercises, see *Extra Skill, Word Problem, and Proof Practice.*

Standards Practice

GEOM 8.0, 9.0

A Practice by Example

Example 1 (page 609)

GO for Help

Use a net to find the surface area of each prism.

1.

2.

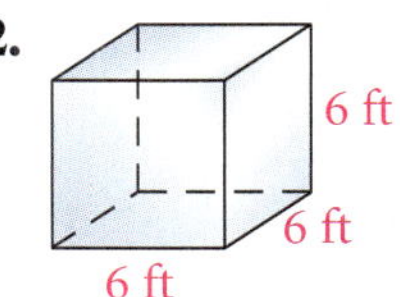

3.

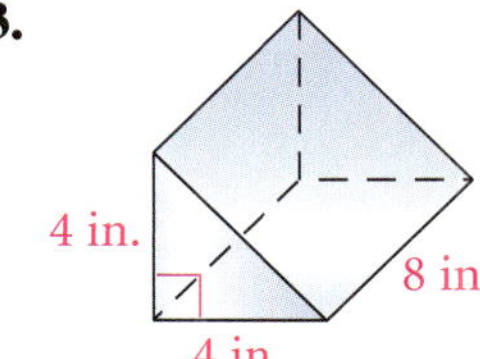

4. a. Classify the prism.

b. Find the lateral area of the prism.

c. The bases are regular hexagons. Find the sum of their areas.

d. Find the surface area of the prism.

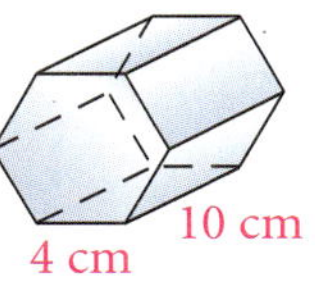

Example 2 (page 609)

Use formulas to find the lateral area and surface area of each prism. Show your answer to the nearest whole number.

5. 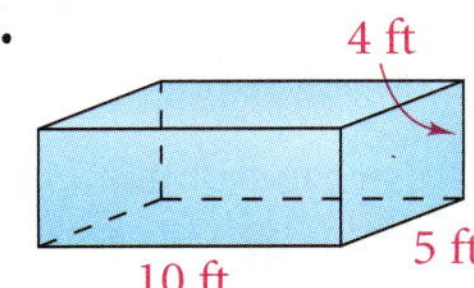

6. 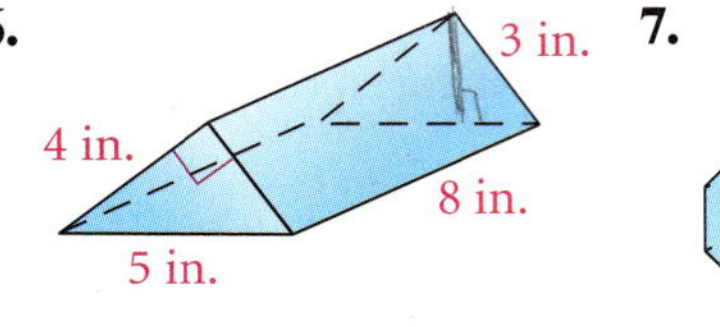

7. 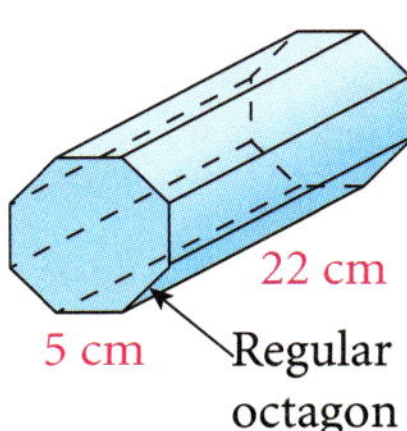

Example 3 (page 611)

Find the surface area of each cylinder in terms of π.

8. 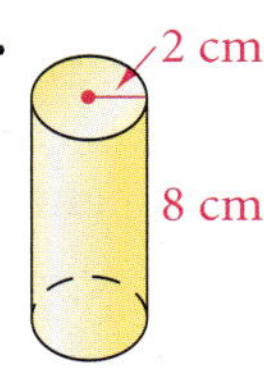

9. 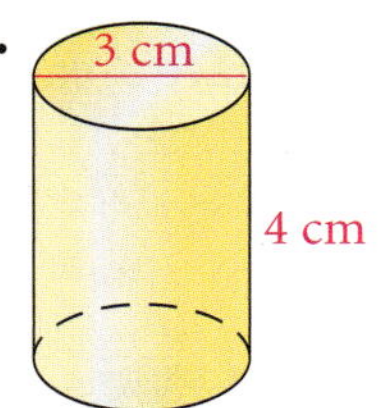

10. 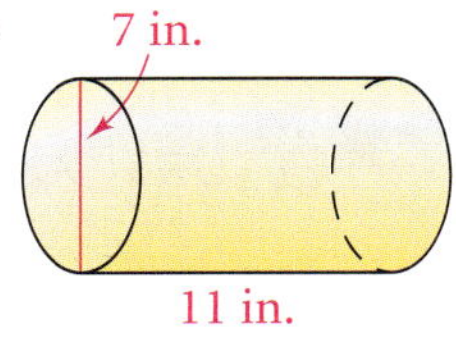

11. A standard drinking straw is 19.5 cm long and has a diameter of 0.6 cm. How many square centimeters of plastic are used in one straw? Round your answer to the nearest tenth.

Example 4 (page 611)

12. **Packaging** A cylindrical carton of oatmeal with radius 3.5 in. is 9 in. tall. If all surfaces except the top are made of cardboard, how much cardboard is used to make the oatmeal carton? Round your answer to the nearest square inch.

Find the surface area of each cylinder to the nearest whole number.

13. 

14. 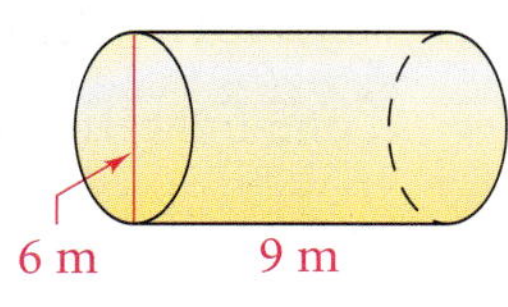

15. 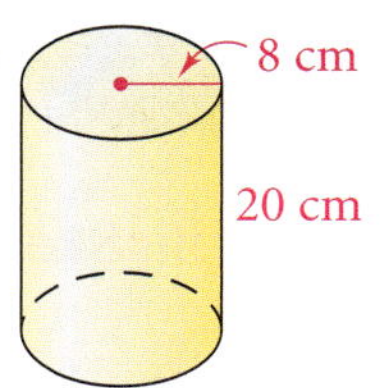

B Apply Your Skills

16. A triangular prism has base edges 4 cm, 5 cm, and 6 cm long. Its lateral area is 300 cm^2. What is the height of the prism?

17. **Estimation** Estimate the surface area of a cube with edges 4.95 cm long.

18. **Writing** Explain how a cylinder and a prism are alike and how they are different.

19. A hexagonal pencil is a hexagonal prism. A base edge of the pencil has length 4 mm. The pencil (without eraser) has height 170 mm. How much surface area of a hexagonal pencil gets painted?

20. **Open-Ended** Draw a net for a rectangular prism with a surface area of 220 cm^2.

21. Consider a box with dimensions 3, 4, and 5.
    a. Find its surface area.
    b. Double each dimension and then find the new surface area.
    c. Find the ratio of the new surface area to the original surface area.
    d. Repeat parts (a)–(c) for a box with dimensions 6, 9, and 11.
    e. **Make a Conjecture** How does doubling the dimensions of a rectangular prism affect the surface area?

Visit: PHSchool.com
Web Code: bce-1102

22. **Multiple Choice** A cylindrical can of cocoa has the dimensions shown at the right. What is the approximate lateral area available for the label?

Ⓐ 110 in.^{2} Ⓑ 148 in.^{2} Ⓒ 179 in.^{2} Ⓓ 219 in.^{2}

23. **Pest Control** A flour moth trap has the shape of a triangular prism that is open on both ends. An environmentally safe chemical draws the moth inside the prism, which is lined with an adhesive. Find the surface area of the trap.

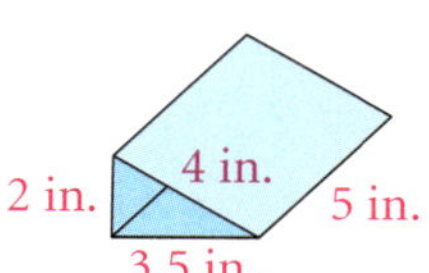

Exercise 24

24. **Packaging** A typical box for a videocassette tape is open on one side as pictured at the left. How many square inches of cardboard are in a typical box for a videocassette tape?

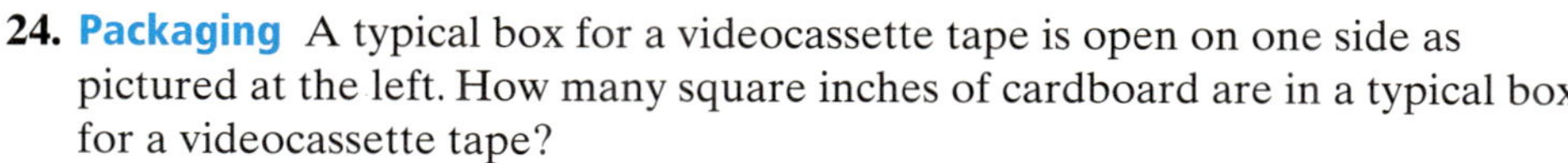

25. Suppose that a cylinder has a radius of r units, and that the height of the cylinder is also r units. The lateral area of the cylinder is 98π square units.

a. **Algebra** Find the value of r.

b. Find the surface area of the cylinder.

26. a. **Geometry in 3 Dimensions** Find the three coordinates of each vertex A, B, C, and D of the rectangular prism.

b. Find AB.

c. Find BC.

d. Find CD.

e. Find the surface area of the prism.

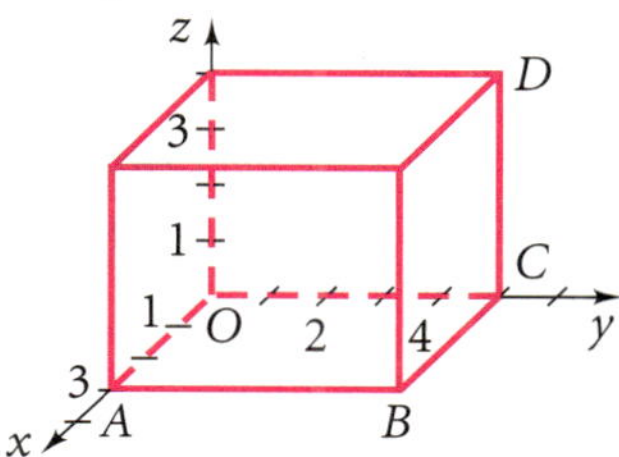

Visualization **The plane region is revolved completely about the given line to sweep out a solid of revolution. Describe the solid and find its surface area in terms of π.**

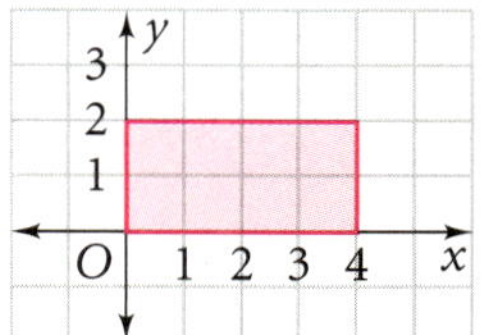

27. the y-axis

28. the x-axis

29. the line $y = 2$

30. the line $x = 4$

31. a. **Critical Thinking** Suppose you double the radius of a right cylinder. How does that affect the lateral area?

b. How does that affect the surface area?

c. Use the formula for surface area of a right cylinder to explain why the surface area in part (b) was not doubled.

For a guide to solving Exercise 32, see p. 615.

32. a. **Packaging** Some cylinders have wrappers with a spiral seam. Peeled off, the wrapper has the shape of a parallelogram. The wrapper for a biscuit container has base 7.5 in. and height 6 in. Find the radius and height of the container.

b. Find the surface area of the container.

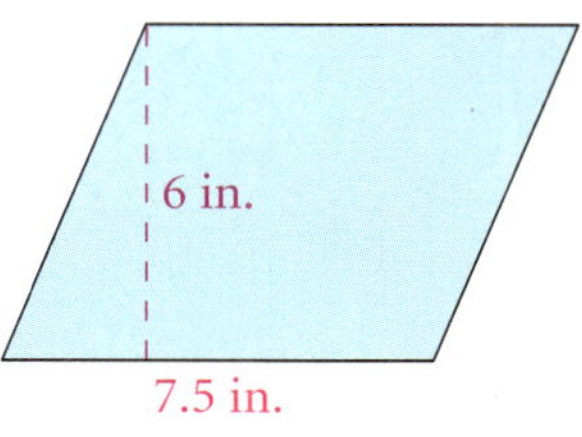

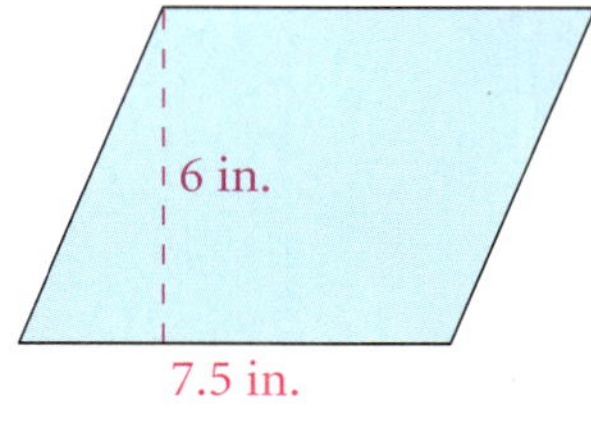

Challenge

Judging by appearances, what is the surface area of each solid?

33.

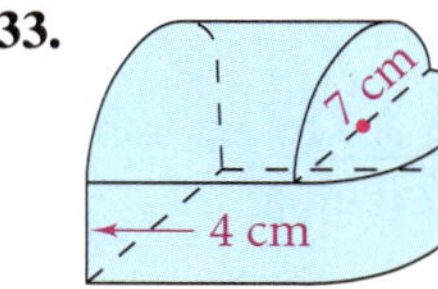

34.

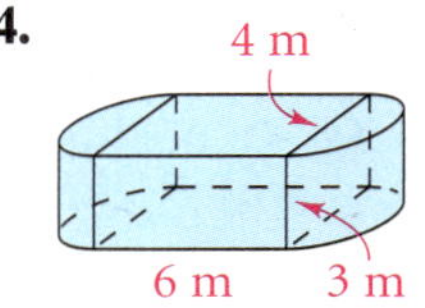

35.

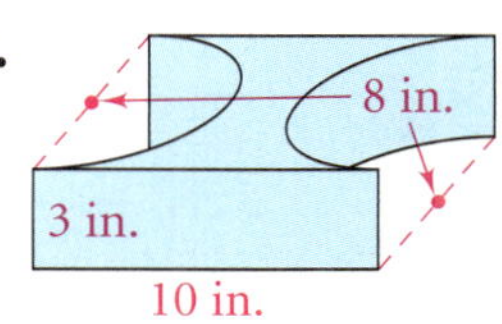

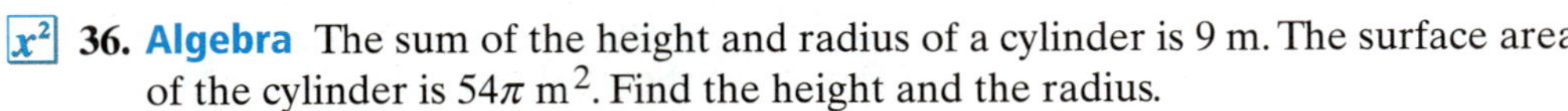

36. **Algebra** The sum of the height and radius of a cylinder is 9 m. The surface area of the cylinder is 54π m^2. Find the height and the radius.

37. Each edge of the large cube at the right is 12 inches long. The cube is painted on the outside, and then cut into 27 smaller cubes. Answer these questions about the 27 cubes.
   a. How many are painted on 4, 3, 2, 1, and 0 faces?
   b. What is the total surface area that is unpainted?

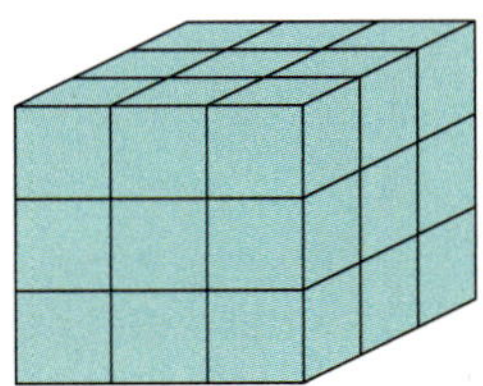

Multiple Choice Practice

For California Standards and CAHSEE Tutorials, visit PHSchool.com. Web Code: bcq-9045

GEOM 9.0

38. Find the surface area of the prism at the right.
   (A) 54 ft^2
   (B) 45 ft^2
   (C) $39 + 45\sqrt{2}$ ft^2
   (D) $39 + 15\sqrt{2}$ ft^2

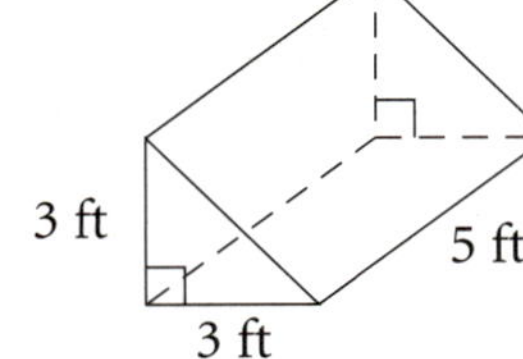

CAHSEE 7NS 2.1

39. Which expression is equivalent to $\dfrac{12x^{-2}z^5y^0}{32x^3z}$?
   (A) $\dfrac{8z^6}{3x^5}$
   (B) $\dfrac{3z^4y}{8x^5}$
   (C) $\dfrac{4z^3y}{11x}$
   (D) $\dfrac{3z^4}{8x^5}$

CAHSEE 7AF 1.2

40. Simplify the expression $3(2x - 1)(x + 1) - 2(3x^2 - 1)$.
   (A) $x - 1$
   (B) $3x - 1$
   (C) $3x - 5$
   (D) $x^2 + 3x - 1$

Mixed Review

Lesson 11-1

Sketch each space figure and then draw a net for it. Label the net with its dimensions.

41. a rectangular box with height 5 cm and a base 3 cm by 4 cm

42. a cube with 2-in. sides

Lesson 10-7

Find the area of each part of the circle to the nearest tenth.

43. sector QOP

44. the segment of the circle bounded by $\overline{QP}$ and $\widehat{QP}$

P Q 120° O 6 cm

Lesson 6-7

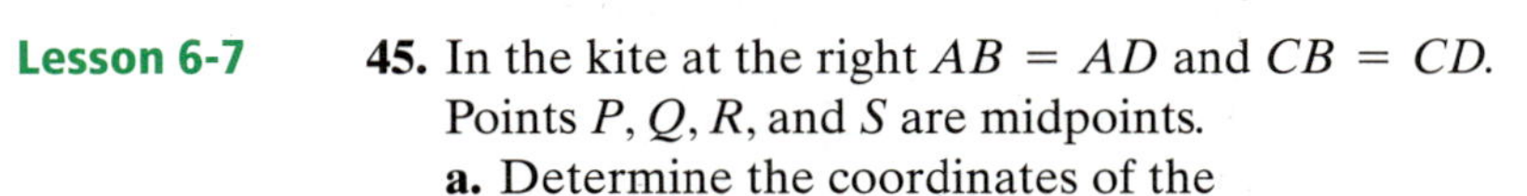

45. In the kite at the right $AB = AD$ and $CB = CD$. Points P, Q, R, and S are midpoints.
   a. Determine the coordinates of the midpoints.
   b. $RQ = ■$; $SP = ■$; $PQ = ■$; $SR = ■$
   c. Use your answers to part (b) to explain why $PQRS$ must be a parallelogram.

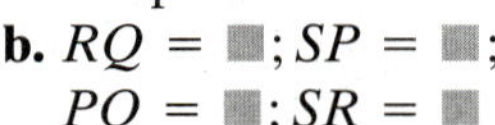

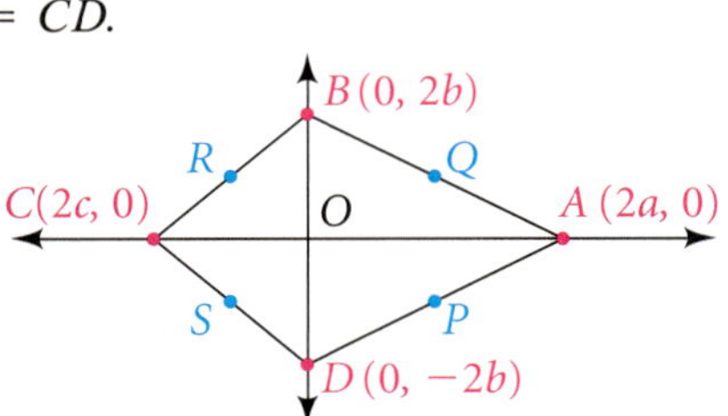

Guided Problem Solving

FOR USE WITH PAGE 613, EXERCISE 32

Understanding Word Problems **Read the problem below. Then let Ray's thinking guide you through the solution. Check your understanding with the exercises at the bottom of the page.**

a. **Packaging** Some cylinders have wrappers with a spiral seam. Peeled off, the wrapper has the shape of a parallelogram. The wrapper for a biscuit container has base 7.5 in. and height 6 in. Find the radius and height of the container.

b. Find the surface area of the container.

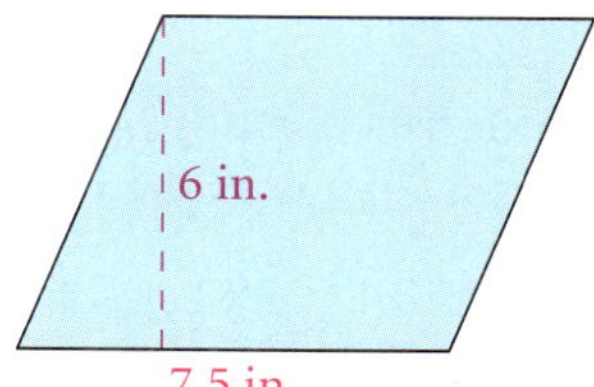

What Ray Thinks

A cutout model I made suggests that the diameter of the cylinder is about 2.3 in. When I roll the parallelogram into a cylinder, the height stays the same. The circumference, $2\pi r$, is the base of the parallelogram, 7.5.

I can solve for r. I'll use a calculator.

So, $r \approx 1.2$ in. and $h = 6$ in. The 1.2 in. radius agrees with what I found with my cutout model!

Surface area = lateral area + base areas.
I can replace $2\pi r$ with 7.5 from above.

What Ray Writes

a. $2\pi r = 7.5$

$r = \frac{7.5}{2\pi}$

≈ 1.19

$r \approx 1.2$ in. and $h = 6$ in.

b. S.A. = L.A. + 2B

$= 2\pi rh + 2\pi r^2$

$\approx (7.5) \cdot 6 + 2\pi(1.2)^2$

≈ 54.0

The surface area is about 54 in.^{2}.

EXERCISES

For Exercises 1–3, refer to the exercise above.

1. For the given base and height, how does the slant of the parallelogram affect the surface area of the biscuit cylinder?

2. The slanted edge has length 6.4 in. If it were used for the base, what would be the height, lateral area, and surface area of the cylinder formed?

3. How big should the small angle of the parallelogram be so that the two endpoints of the seam will be in line vertically when the can is wrapped?

Activity Lab: Exploring Surface Area

Technology

FOR USE WITH LESSON 11-2

> **Prepares for GEOM 11.0** Students determine how changes in dimensions affect the perimeter, area, and volume of common geometric figures and solids.

At room temperature, 1 L, 1000 mL, and 1000 cm^3 all represent the same amount of water. Thus, one type of model for a liter is any square prism that holds 1000 cm^3. The best model, perhaps, is a 10-cm cube as shown here. But there are many others.

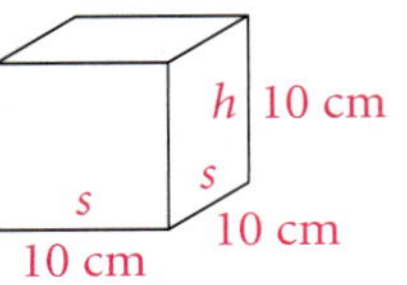

You can use graphing calculator lists to study how height (h) and surface area (S.A.) of a 1-L square prism change as the length (s) of each side of a base changes.

The volume of a prism equals the area of a base times its height ($V = Bh$ or $V = s^2h$). You can solve for h in each equation to find $h = \frac{V}{B} = \frac{V}{s^2}$. The surface area equals two times the area of a base plus four times the area of a face, or

$$\begin{aligned} \text{S.A.} = 2B + 4sh &= 2s^2 + 4sh \\ &= 2s^2 + 4s\frac{V}{s^2} \quad \textbf{Substitute.} \\ &= 2s^2 + \frac{4V}{s} \quad \textbf{Simplify.} \end{aligned}$$

Use the commands shown on the screens below to generate lists L_1, L_2, and L_3, for s, h, and S.A., respectively. The fourth screen shows what the lists should look like.

```
seq(N, N, 1, 50, .2)
→L1
{1 1.2 1.4 1.6 ...
■
```

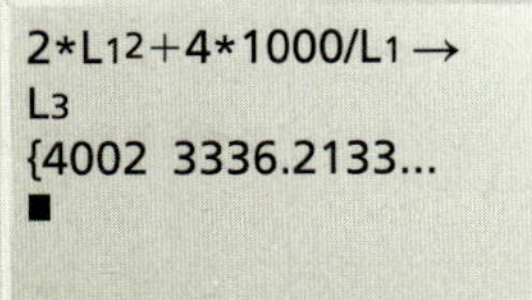

L1	L2	L3
1	1000	4002
1.2	694.44	3336.2
1.4	510.2	2861.1
1.6	390.63	2505.1
1.8	308.64	2228.7
2	250	2008
2.2	206.61	1827.9

L1(1)=1

EXERCISES

Generate the lists (shown above) on your graphing calculator. Scroll down to study them.

1. How small can the surface area be? How large can it be?

2. a. Which dimensions give a very large surface area?
b. Which dimensions give the smallest surface area?
c. How do s and h compare in the prism with the smallest surface area?
d. What is the shape of the prism that has the smallest surface area?

Extend

3. If a square prism must have a volume of 100 cm^3, what dimensions would give the smallest surface area?

4. A cereal manufacturer is designing a cereal box that has a capacity of 3000 cm^3. Surface area should be large to provide space for advertising. What else should be considered for the box design? Use a graphing calculator as needed to support your conclusions.

11-3 Surface Areas of Pyramids and Cones

California Content Standards

GEOM 8.0 Students know, derive, and solve problems involving the lateral area and surface area of common geometric figures. ***Develop***
GEOM 9.0 Students compute surface areas of pyramids and cones; students commit to memory the formula for pyramids. ***Develop***

What You'll Learn

- To find the surface area of a pyramid
- To find the surface area of a cone

. . . And Why

To find the lateral area of the Great Pyramid of Egypt, as in Example 2

Check Skills You'll Need

GO for Help Lesson 8-1

Find the length of the hypotenuse in simplest radical form.

1.
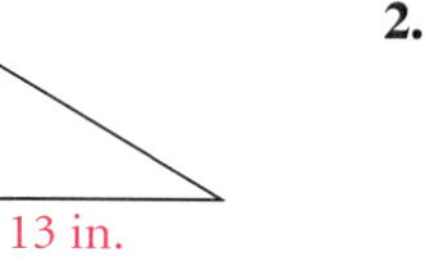

2.

3.
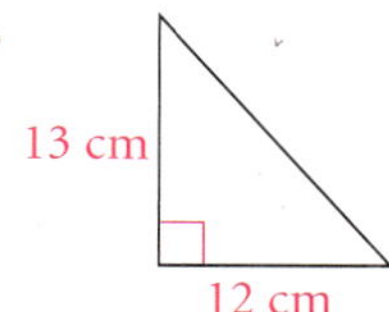

New Vocabulary

• pyramid • base, lateral faces, vertex, altitude, height, slant height, lateral area, surface area (of a pyramid) • regular pyramid • cone • base, altitude, vertex, height, slant height, lateral area, surface area (of a cone) • right cone

1 Finding Surface Area of a Pyramid

A **pyramid** is a polyhedron in which one face (the **base**) can be any polygon and the other faces (the **lateral faces**) are triangles that meet at a common vertex (called the **vertex** of the pyramid).

Vocabulary Tip

If the base is a hexagon, the pyramid is a hexagonal pyramid.

You can name a pyramid by the shape of its base. The **altitude** of a pyramid is the perpendicular segment from the vertex to the plane of the base. The length of the altitude is the **height** h of the pyramid.

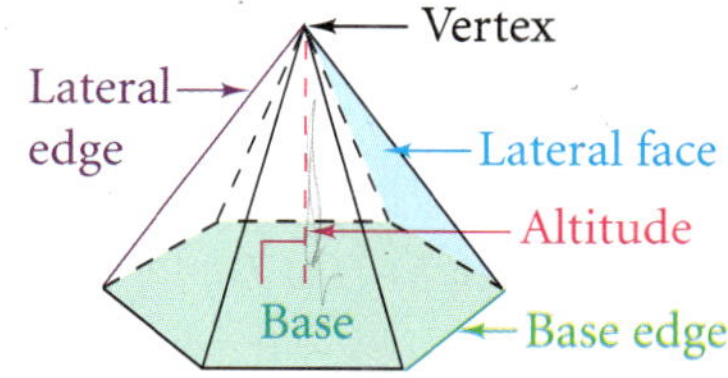

A **regular pyramid** is a pyramid whose base is a regular polygon and whose lateral faces are congruent isosceles triangles. The **slant height** ℓ is the length of the altitude of a lateral face of the pyramid.

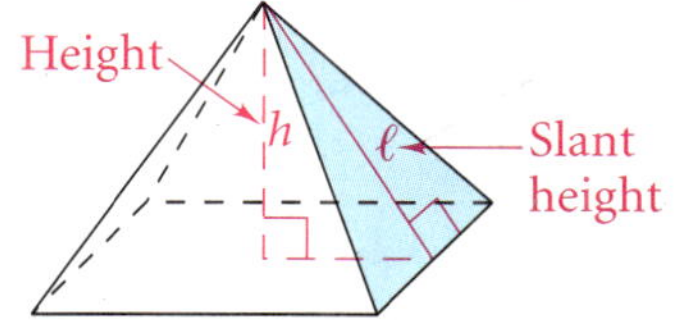

In this book, you can assume that a pyramid is regular unless stated otherwise.

The **lateral area** of a pyramid is the sum of the areas of the congruent lateral faces. You can find a formula for the lateral area of a pyramid by looking at its net.

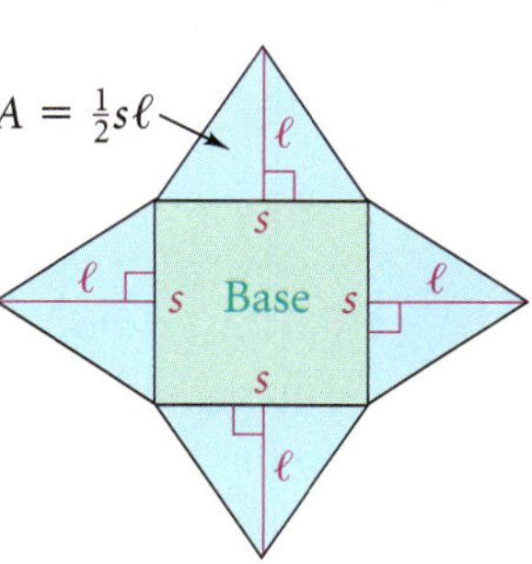

L.A. $= 4\left(\frac{1}{2}s\ell\right)$ — The area of each lateral face is $\frac{1}{2}s\ell$.

$= \frac{1}{2}(4s)\ell$ — Commutative and Associative Properties of Multiplication

$= \frac{1}{2}p\ell$ — The perimeter p of the base is $4s$.

To find the **surface area** of a pyramid, add the area of its base to its lateral area.

Key Concepts

Theorem 11-3 **Lateral and Surface Areas of a Regular Pyramid**

The lateral area of a regular pyramid is half the product of the perimeter of the base and the slant height.

$$\text{L.A.} = \frac{1}{2}p\ell$$

The surface area of a regular pyramid is the sum of the lateral area and the area of the base.

$$\text{S.A.} = \text{L.A.} + B$$

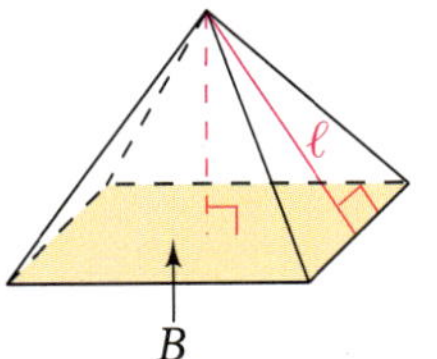

1 EXAMPLE Finding Surface Area of a Pyramid

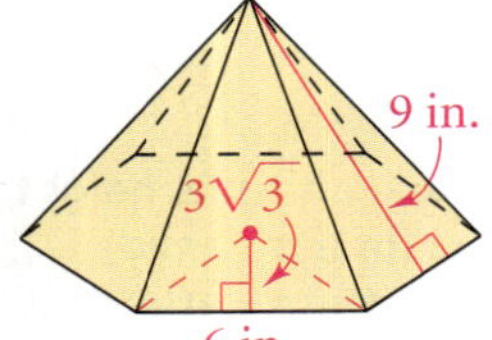

Find the surface area of the hexagonal pyramid at the left.

$\text{S.A.} = \text{L.A.} + B$	**Use the formula for surface area.**
$= \frac{1}{2}p\ell + \frac{1}{2}ap$	**Substitute the formulas for L.A. and *B*.**
$= \frac{1}{2}(36)(9) + \frac{1}{2}(3\sqrt{3})(36)$	**Substitute.**
≈ 255.53074	**Use a calculator.**

The surface area of the pyramid is about 256 in.^{2}.

CA Standards Check **1** Find the surface area of a square pyramid with base edges 5 m and slant height 3 m.

Sometimes the slant height of a pyramid is not given. You must calculate it before you can find the lateral or surface area.

2 EXAMPLE Real-World Connection

Social Studies The Great Pyramid at Giza, Egypt, pictured at the left, was built about 2580 B.C. as a final resting place for Pharaoh Khufu. At the time it was built, its height was about 481 ft. Each edge of the square base was about 756 ft long. What was the lateral area of the Great Pyramid?

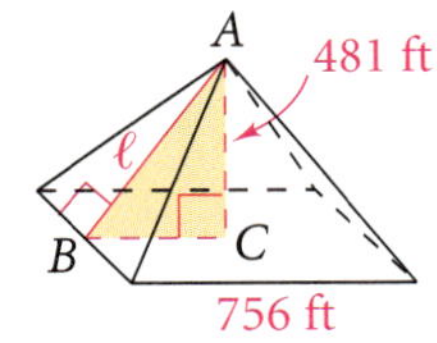

The legs of right $\triangle ABC$ are the height of the pyramid and the apothem of the base. The height of the pyramid was 481 ft. The apothem of the base was $\frac{756}{2}$, or 378 ft. You can use the Pythagorean Theorem to find the slant height ℓ.

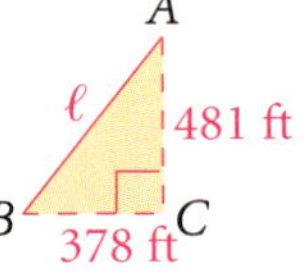

$\text{L.A.} = \frac{1}{2}p\ell$	**Use the formula for lateral area.**
$= \frac{1}{2}(4s)\sqrt{a^2 + b^2}$	**Substitute the formulas for *p* and ℓ.**
$= \frac{1}{2}(4 \cdot 756)\sqrt{378^2 + 481^2}$	**Substitute.**
≈ 924974.57	**Use a calculator.**

The lateral area of the Great Pyramid was about 925,000 ft^2.

CA Standards Check **2** Find the surface area of the Great Pyramid to the nearest square foot.

Finding Surface Area of a Cone

A **cone** is "pointed" like a pyramid, but its **base** is a circle. In a **right cone,** the **altitude** is a perpendicular segment from the **vertex** to the center of the base. The **height** h is the length of the altitude. The **slant height** ℓ is the distance from the vertex to a point on the edge of the base.

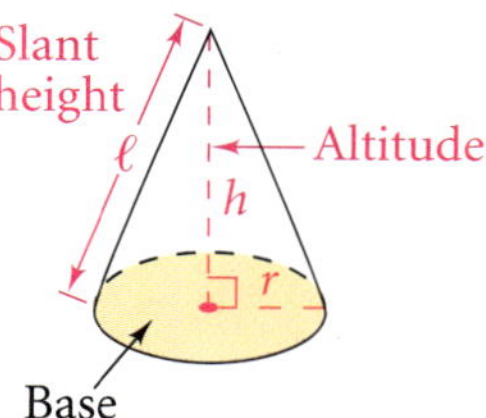

As with a pyramid, the **lateral area** is $\frac{1}{2}$ the perimeter (circumference) of the base times the slant height. The formulas for the lateral area and **surface area** of a cone are similar to those for a pyramid.

Key Concepts

Theorem 11-4 **Lateral and Surface Areas of a Cone**

The lateral area of a right cone is half the product of the circumference of the base and the slant height.

$$\text{L.A.} = \frac{1}{2} \cdot 2\pi r \cdot \ell, \text{ or L.A.} = \pi r \ell$$

The surface area of a right cone is the sum of the lateral area and the area of the base.

$$\text{S.A.} = \text{L.A.} + B$$

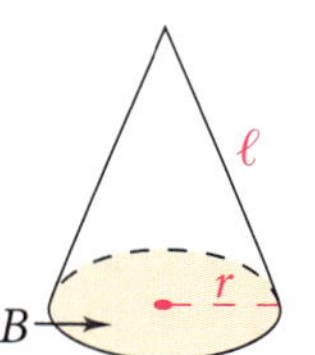

In this book, you can assume that a cone is a right cone unless stated or pictured otherwise.

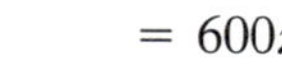

3 EXAMPLE Finding Surface Area of a Cone

Find the surface area of the cone in terms of π.

$\text{S.A.} = \text{L.A.} + B$	**Use the surface area formula.**
$= \pi r \ell + \pi r^2$	**Substitute the formulas for L.A. and *B*.**
$= \pi(15)(25) + \pi(15)^2$	**Substitute.**
$= 375\pi + 225\pi$	**Simplify.**
$= 600\pi$	

25 cm

15 cm

The surface area of the cone is 600π cm^2.

CA Standards Check **3** The radius of the base of a cone is 16 m. Its slant height is 28 m. Find the surface area in terms of π.

By cutting a cone and laying it out flat, you can see how the formula for lateral area of a cone $\left(\text{L.A.} = \frac{1}{2} \cdot C_{\text{base}} \cdot \ell\right)$ resembles that for the area of a triangle $\left(A = \frac{1}{2}bh\right)$.

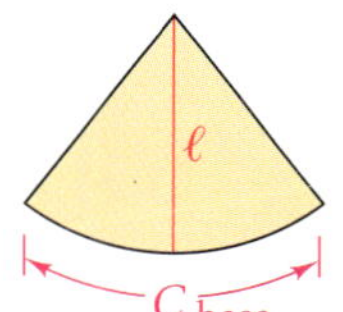

Chemists use funnels to safely and efficiently transfer fluids.

4 EXAMPLE — Real-World Connection

Chemistry The funnel is in the shape of a cone. How much filter paper do you need to line the funnel?

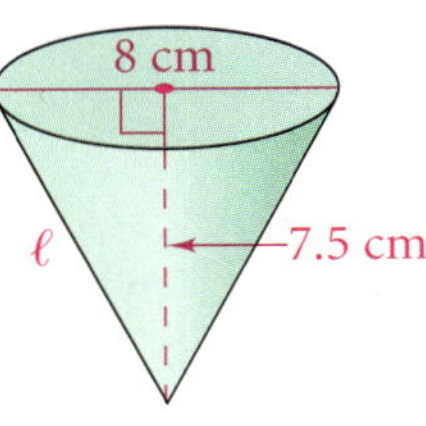

The area covered is the lateral area of a cone that has a diameter of 8 cm and a height of 7.5 cm.

Use the formula for the lateral area of a cone.

$\text{L.A.} = \pi r \ell$

$= \pi r\left(\sqrt{a^2 + b^2}\right)$ **To find the slant height use the Pythagorean Theorem.**

$= \pi(4)\left(\sqrt{4^2 + 7.5^2}\right)$ **Substitute. If $d = 8$, then $r = 4$.**

$= 106.81415$ **Use a calculator.**

You need about 107 cm^2 of filter paper to line the funnel.

CA Standards Check 4 Find the lateral area of a cone with radius 15 in. and height 20 in.

EXERCISES

Standards Practice

For more exercises, see *Extra Skill, Word Problem, and Proof Practice.*

GEOM 8.0, 9.0

A Practice by Example

Example 1 (page 618)

Find the surface area of each pyramid to the nearest whole number.

1.

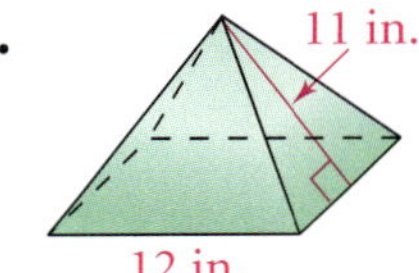

2.

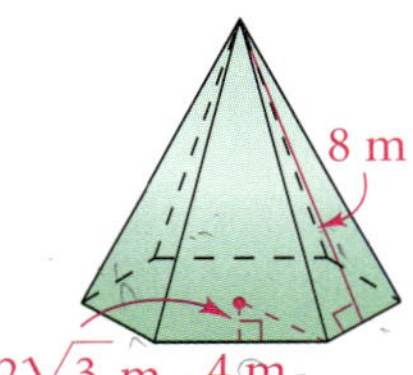

3.

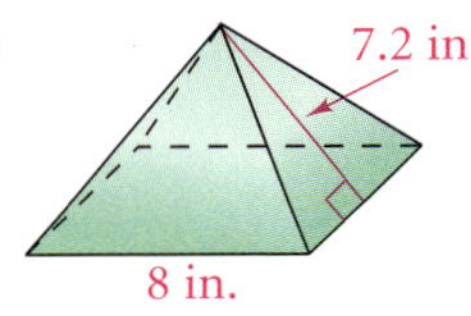

Example 2 (page 618)

Find the lateral area of each pyramid to the nearest whole number.

4.

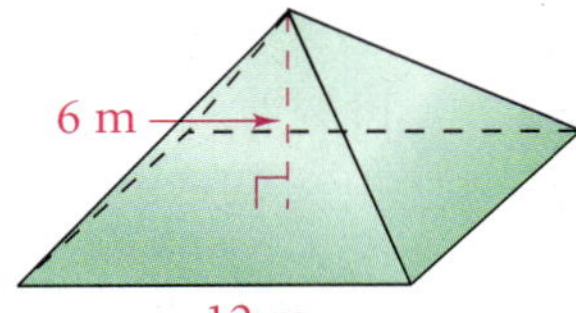

5.

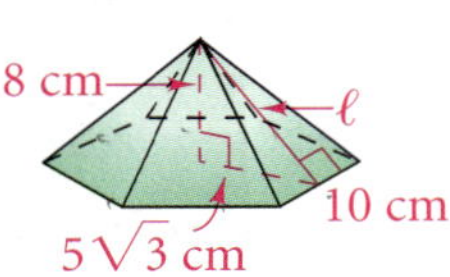

6.

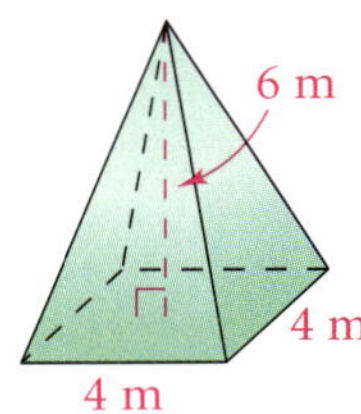

7. **Social Studies** The original height of the pyramid built for Khafre, next to the Great Pyramid, was about 471 ft. Each side of its square base was about 708 ft. What is the lateral area to the nearest square foot of a pyramid with those dimensions?

Example 3 (page 619)

Find the surface area of each cone in terms of π.

8.

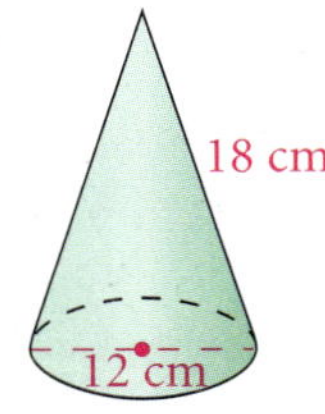

9.

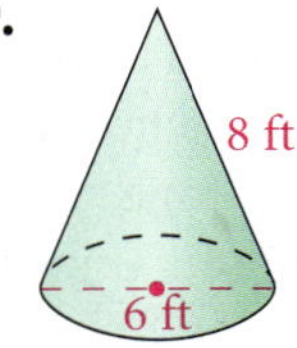

10.

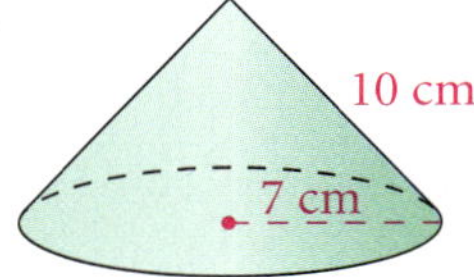

Example 4 (page 620)

Find the lateral area of each cone to the nearest whole number.

11.

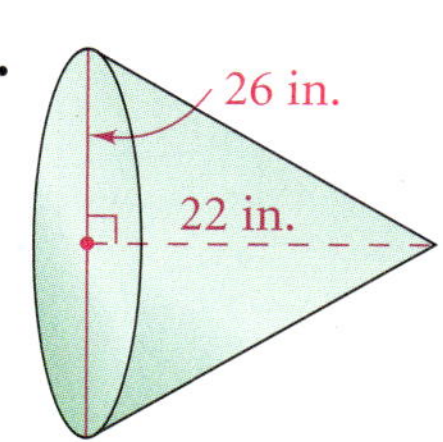

12.

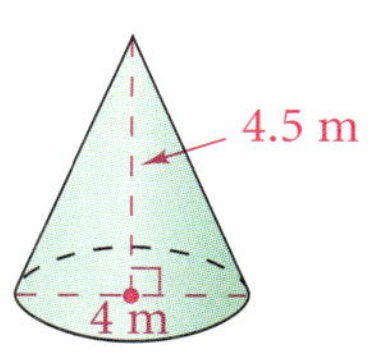

13.

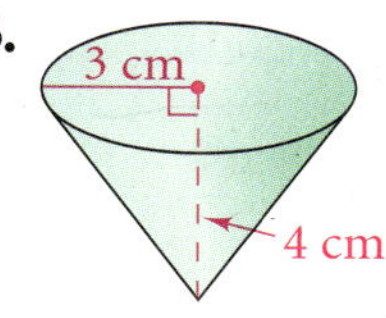

 **Apply Your Skills**

14. Writing Explain why the altitude $\overline{PT}$ in the pyramid at the right must be shorter than each edge $\overline{PA}$, $\overline{PB}$, $\overline{PC}$, and $\overline{PD}$.

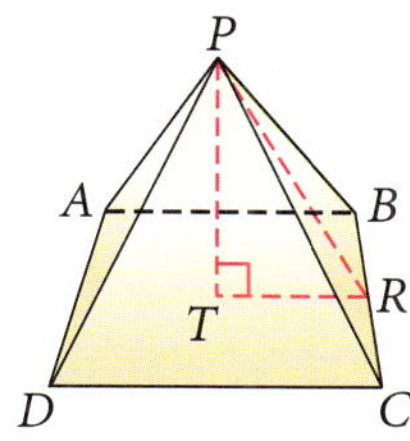

> **Problem Solving Hint**
> In Exercise 14, explain why $\overline{PT}$ is shorter than $\overline{PR}$, and then why $\overline{PR}$ is shorter than $\overline{PC}$.

15. Reasoning Suppose you could climb to the top of the Great Pyramid in Egypt. Which route would be shorter, a route along a lateral edge or along the altitude of a side? Which of these routes is steeper? Explain your answers.

16. A square pyramid has base edges 10 in. long and height 4 in. Sketch the pyramid and find its surface area. Round your answer to the nearest tenth.

 **17. Algebra** The lateral area of a pyramid with a square base is 240 ft^2. Its base edges are 12 ft long. Find the height of the pyramid.

Find the surface area to the nearest whole number.

18.

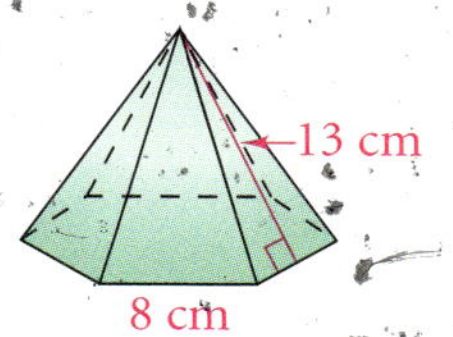

19.

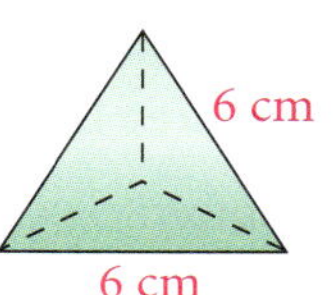

20.

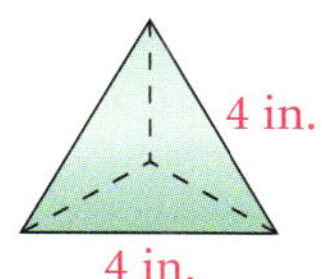

Exercise 23

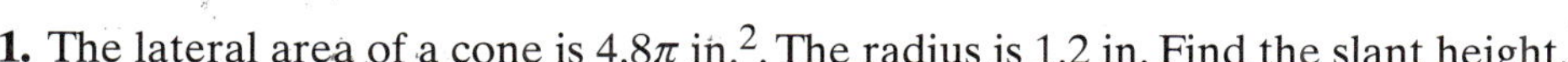

21. The lateral area of a cone is 4.8π in.^{2}. The radius is 1.2 in. Find the slant height.

22. Open-Ended Draw a square pyramid with a lateral area of 48 cm^2. Label its dimensions. Then find its surface area.

23. Architecture The roof of a tower in a castle is shaped like a cone. The height of the roof is 30 ft and the radius of the base is 15 ft. What is the area of the roof? Round your answer to the nearest tenth.

24. The hourglass shown at the right is made by connecting two glass cones inside a glass cylinder. Which has more glass, the two cones or the cylinder? Explain.

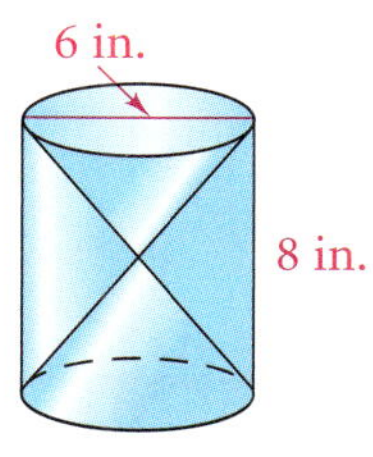

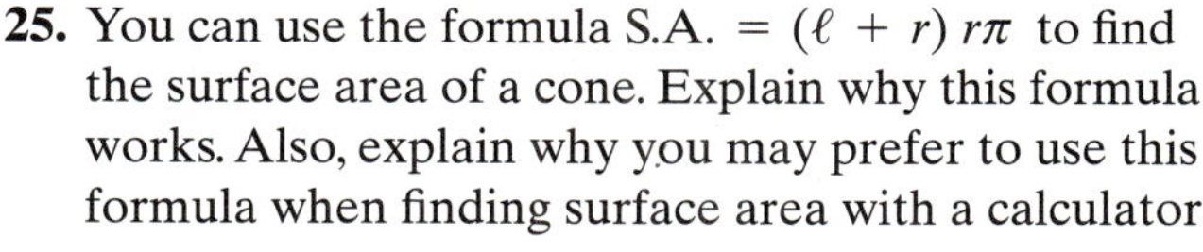

25. You can use the formula S.A. $= (\ell + r)r\pi$ to find the surface area of a cone. Explain why this formula works. Also, explain why you may prefer to use this formula when finding surface area with a calculator.

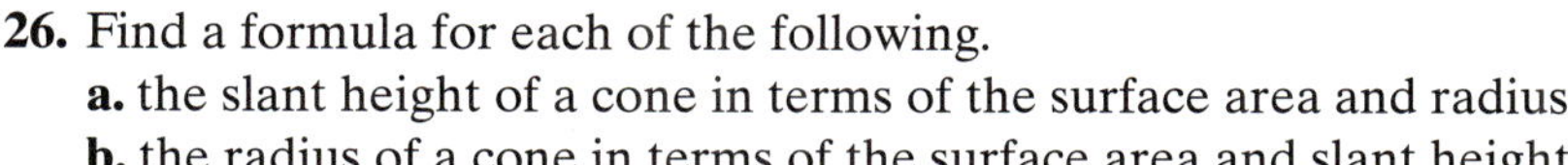

26. Find a formula for each of the following.
a. the slant height of a cone in terms of the surface area and radius
b. the radius of a cone in terms of the surface area and slant height

27. Multiple Choice The roof of a tower is a square pyramid with base edges 10 ft long. The height of the pyramid is 6 ft. What is the approximate area of the roofing material needed to cover the roof?

Ⓐ 156 ft^2 Ⓑ 233 ft^2 Ⓒ 256 ft^2 Ⓓ 333 ft^2

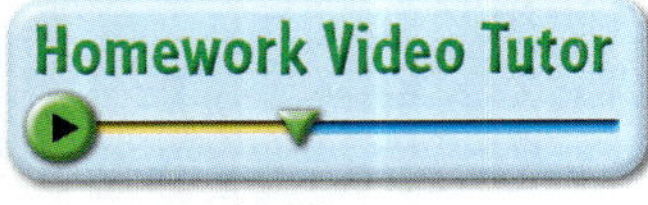

Visit: PHSchool.com
Web Code: bce-1103

Find the surface area to the nearest whole number.

28. 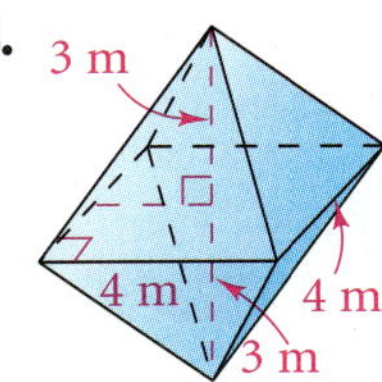

29. 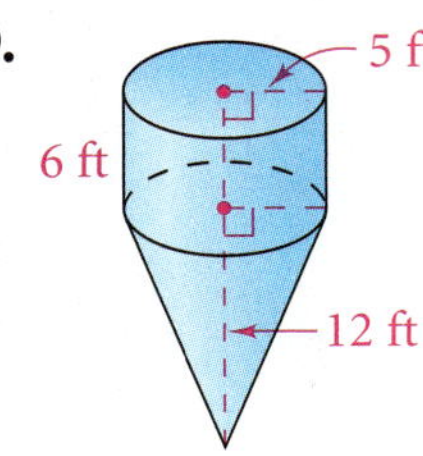

30. 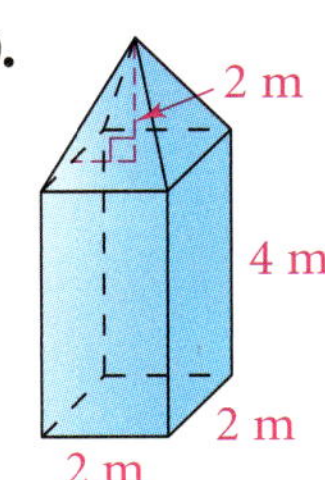

The length of a side of the base (s), slant height, height, lateral area, and surface area are measurements of a square pyramid. Given two of the measurements, find the other three to the nearest tenth.

31. $s = 3$ in., S.A. $= 39$ in.^{2}

32. $h = 8$ m, $\ell = 10$ m

33. $\ell = 5$ ft, L.A. $= 20$ ft^2

34. L.A. $= 118$ cm^2, S.A. $= 182$ cm^2

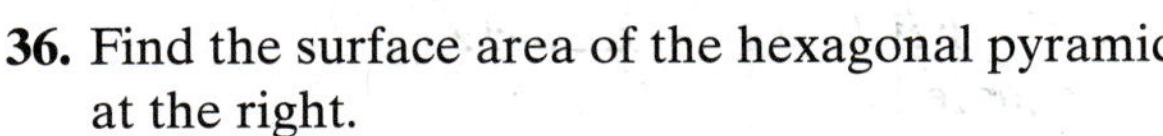

35. A cone with radius 9 cm has the same surface area as a cylinder with radius 6 cm and height 18 cm. What is the height of the cone to the nearest tenth?

36. Find the surface area of the hexagonal pyramid at the right.

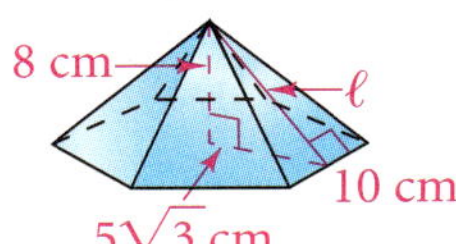

Visualization **The plane region is revolved completely about the given line to sweep out a solid of revolution. Describe the solid. Then find its surface area in terms of π.**

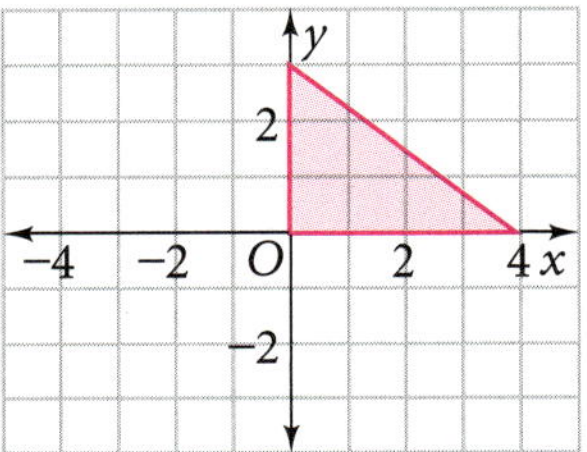

37. about the y-axis

38. about the x-axis

Challenge

39. about the line $x = 4$

40. about the line $y = 3$

The given figure fits inside a 10-cm cube. The figure's base is in one face of the cube and is as large as possible. The figure's vertex is in the opposite face of the cube. Draw a sketch and find the lateral and surface areas of the figure.

41. a square pyramid

42. a cone

43. A sector has been cut out of the disk. The radii of the part that remains are taped together, without overlapping, to form the cone. The cone has a lateral area of 64π cm^2. Find the measure of the central angle of the cut-out sector.

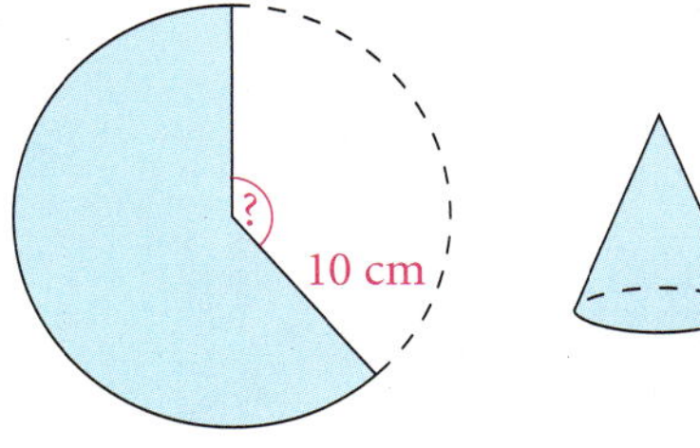

Multiple Choice Practice

For California Standards and CAHSEE Tutorials, visit PHSchool.com. Web Code: bcq-9045

GEOM 8.0

44. Find the lateral area of the cone at the right. Leave your answer in terms of π.

(A) 14π

(B) 10π

(C) 8π

(D) 7π

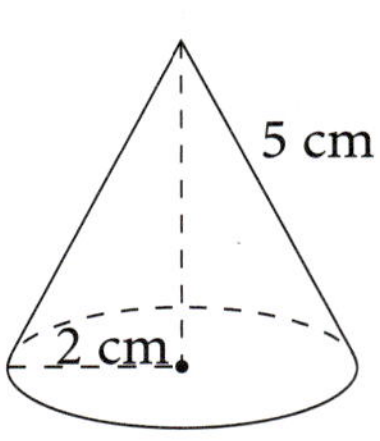

GEOM 9.0

45. The diagram at the right is a net for a square pyramid. What is the total surface area of the pyramid to the nearest square meter?

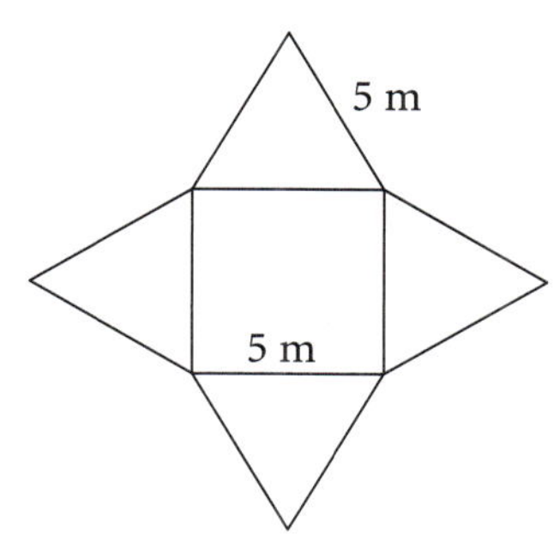

Ⓐ 75 m^2

Ⓑ 68 m^2

Ⓒ 60 m^2

Ⓓ 43 m^2

CAHSEE 7NS 2.4

46. The square root of 89 is between

Ⓐ 8 and 9. Ⓒ 10 and 11.

Ⓑ 9 and 10. Ⓓ 88 and 90.

Mixed Review

Lesson 11-2

47. How much cardboard do you need to make a closed box that is 4 ft by 5 ft by 2 ft?

48. How much posterboard do you need to make a cylinder, open at each end, with height 9 in. and diameter $4\frac{1}{2}$ in.? Round your answer to the nearest square inch.

Lesson 10-2

49. The area of a rhombus is 714 cm^2. One diagonal is 42 cm long. Find the length of the other diagonal.

50. A kite with area 195 in.^2 has a 15-in. diagonal. How long is the other diagonal?

Lesson 8-5

51. A TV camera views a tall building 400 m away with a 35° angle of elevation to the top. How tall is the building if the camera lens is 160 cm off the ground?

Checkpoint Quiz 1 — Lessons 11-1 through 11-3

Draw a net for each figure. Label the net with its dimensions.

1.

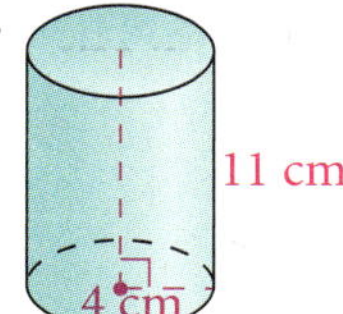

2.

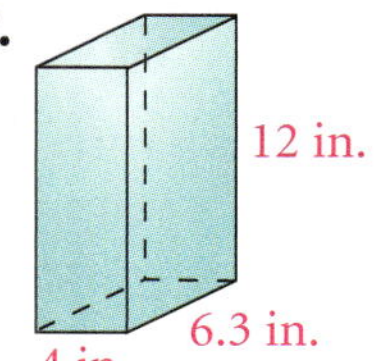

3.

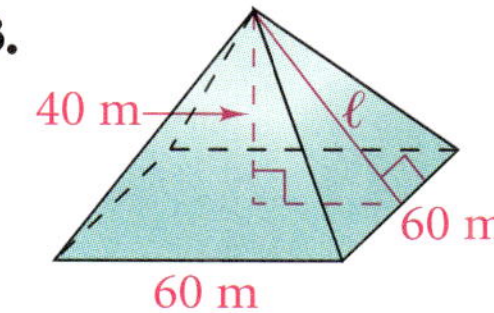

4. Find the surface area of the cylinder in Exercise 1.

5. Find the surface area of the prism in Exercise 2.

6. Find the surface area of the pyramid in Exercise 3.

7. A cone has a radius of 8 in. and a slant height of 13 in. Find the lateral area of the cone. Leave your answer in terms of π.

8. **Open-Ended** Draw a net for a regular hexagonal prism.

Draw a cube. Shade the cube to show each cross section.

9. a rectangle

10. a trapezoid

Volumes of Prisms and Cylinders

California Content Standards

GEOM 8.0 Students know, derive, and solve problems involving volume of common geometric figures. ***Introduce***
GEOM 9.0 Students compute the volumes of prisms and cylinders; and students commit to memory the formulas for prisms and cylinders. ***Introduce***

What You'll Learn

- To find the volume of a prism
- To find the volume of a cylinder

. . . And Why

To estimate the volume of a backpack, as in Example 4

Check Skills You'll Need

Lessons 1-9 and 10-1

Find the area of each figure. For answers that are not whole numbers, round to the nearest tenth.

1. a square with side length 7 cm
2. a circle with diameter 15 in.
3. a circle with radius 10 mm
4. a rectangle with length 3 ft and width 1 ft
5. a rectangle with base 14 in. and height 11 in.
6. a triangle with base 11 cm and height 5 cm
7. an equilateral triangle that is 8 in. on each side

New Vocabulary • volume • composite space figure

1 Finding Volume of a Prism

Hands-On Activity: Finding Volume

Explore the volume of a prism with unit cubes.

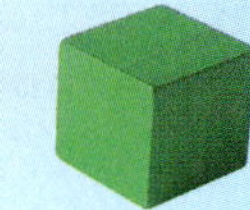

- Make a one-layer rectangular prism that is 4 cubes long and 2 cubes wide. The prism will be 4 units by 2 units by 1 unit.

1. How many cubes are in the prism?
2. Add a second layer to your prism to make a prism 4 units by 2 units by 2 units. How many cubes are in this prism?

3. Add a third layer to your prism to make a prism 4 units by 2 units by 3 units. How many cubes are in this prism?
4. How many cubes would be in the prism if you added two additional layers of cubes for a total of 5 layers?
5. How many cubes would be in the prism if there were 10 layers?

Volume is the space that a figure occupies. It is measured in cubic units such as cubic inches (in.^{3}), cubic feet (ft^3), or cubic centimeters (cm^3). The volume of a cube is the cube of the length of its edge, or $V = e^3$.

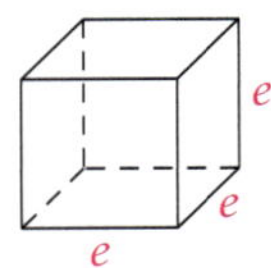

Both stacks of paper below contain the same number of sheets.

The first stack forms a right prism. The second forms an oblique prism. The stacks have the same height. The area of every cross section parallel to a base is the area of one sheet of paper. The stacks have the same volume. These stacks illustrate the following principle.

 **Key Concepts**

Theorem 11-5 **Cavalieri's Principle**

If two space figures have the same height and the same cross-sectional area at every level, then they have the same volume.

The area of each shaded cross section below is 6 cm^2. Since the prisms have the same height, their volumes must be the same by Cavalieri's Principle.

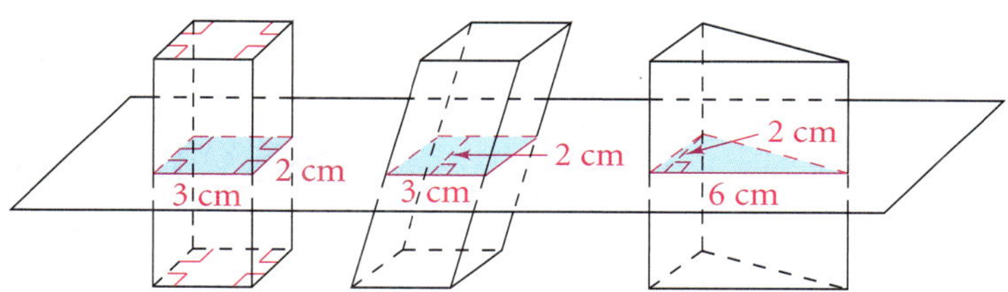

You can find the volume of a right prism by multiplying the area of the base by the height. Cavalieri's Principle lets you extend this idea to any prism.

 **Key Concepts**

Theorem 11-6 **Volume of a Prism**

The volume of a prism is the product of the area of a base and the height of the prism.

$V = Bh$

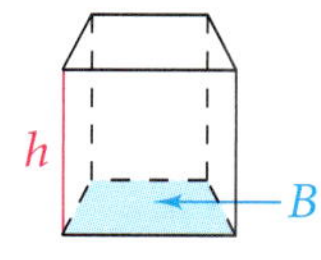

Online active math

For: Prism, Cylinder Activity
Use: Interactive Textbook, 11-4

1 EXAMPLE Finding Volume of a Rectangular Prism

Find the volume of the prism at the right.

$V = Bh$	**Use the formula for volume.**
$= 480 \cdot 10$	$B = 24 \cdot 20 = 480$ **cm^2**
$= 4800$	**Simplify.**

10 cm
20 cm
24 cm

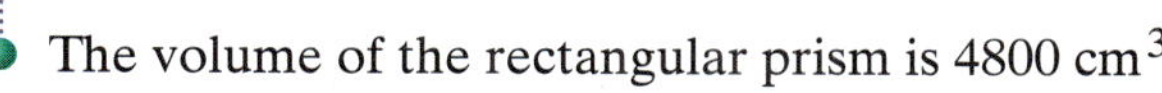

The volume of the rectangular prism is 4800 cm^3.

CA Standards Check

1 **Critical Thinking** Suppose the prism in Example 1 is turned so that the base is 20 cm by 10 cm and the height is 24 cm. Explain why the volume does not change.

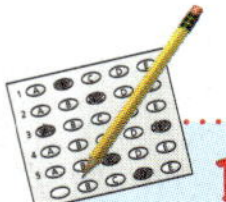

Test-Taking Tip

A volume question often requires you to find a base of a solid. A base does not have to be at the bottom (or top) of the solid.

2 EXAMPLE Finding Volume of a Triangular Prism

Multiple Choice Find the approximate volume of the triangular prism at the right.

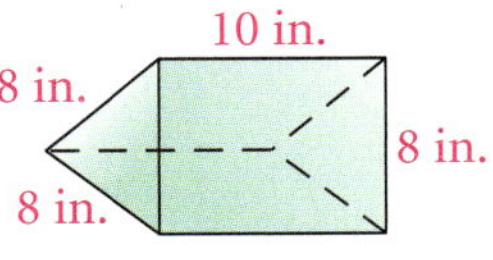

Ⓐ 188 in.^{3} Ⓑ 277 in.^{3} Ⓒ 295 in.^{3} Ⓓ 554 in.^{3}

Each base of the triangular prism is an equilateral triangle. An altitude of the triangle divides it into two 30°-60°-90° triangles. The area of the base is $\frac{1}{2} \cdot 8 \cdot 4\sqrt{3}$, or $16\sqrt{3}$ in.^{2}.

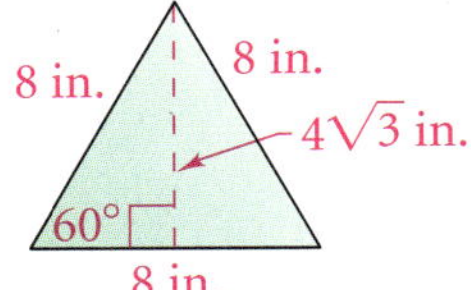

$V = Bh$ **Use the formula for the volume of a prism.**

$= 16\sqrt{3} \cdot 10$ **Substitute.**

$= 160\sqrt{3}$ **Simplify.**

$= 277.12813$ **Use a calculator.**

The volume of the triangular prism is about 277 in.^{3}. The answer is B.

CA Standards Check **2** Find the volume of the triangular prism at the right.

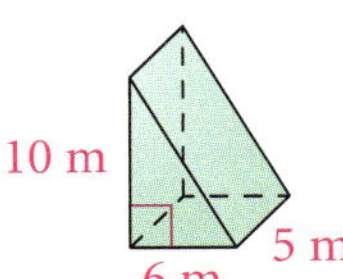

2 Finding Volume of a Cylinder

To find the volume of a cylinder, you use the same formula $V = Bh$ that you use to find the volume of a prism. Now, however, B is the area of the circle, so you use the formula $B = \pi r^2$ to find its value.

Key Concepts

Theorem 11-7 **Volume of a Cylinder**

The volume of a cylinder is the product of the area of the base and the height of the cylinder.

$$V = Bh, \text{or } V = \pi r^2 h$$

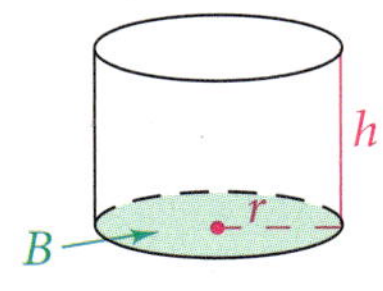

Video Tutor Help
Visit: PHSchool.com
Web Code: bce-0775

3 EXAMPLE Finding Volume of a Cylinder

Find the volume of the cylinder at the right. Leave your answer in terms of π.

$V = \pi r^2 h$ **Use the formula for the volume of a cylinder.**

$= \pi(3)^2(8)$ **Substitute.**

$= \pi(72)$ **Simplify.**

The volume of the cylinder is 72π cm^3.

CA Standards Check **3** The cylinder at the right is oblique.

a. Find its volume in terms of π.

b. Find its volume to the nearest tenth of a cubic meter.

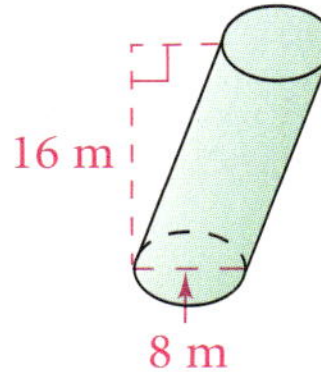

A **composite space figure** is a three-dimensional figure that is the combination of two or more simpler figures. A space probe, for example, might begin as a composite figure—a cylindrical rocket engine in combination with a nose cone.

You can find the volume of a composite space figure by adding the volumes of the figures that are combined.

4 EXAMPLE Finding Volume of a Composite Figure

17 in.
4 in.
12 in.

Estimation Use a composite space figure to estimate the volume of the backpack shown at the left.

Step 1: You can use a prism and half of a cylinder to approximate the shape, and therefore the volume, of the backpack.

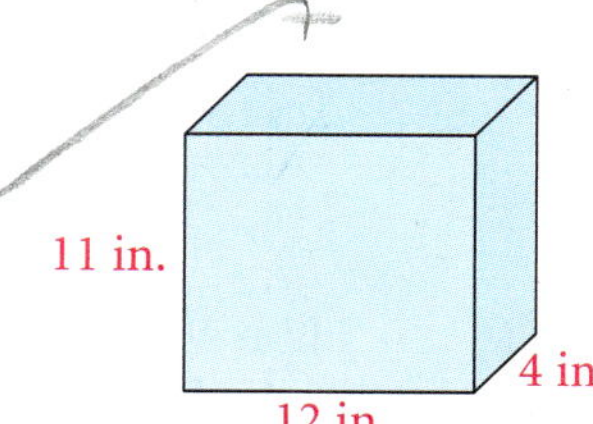

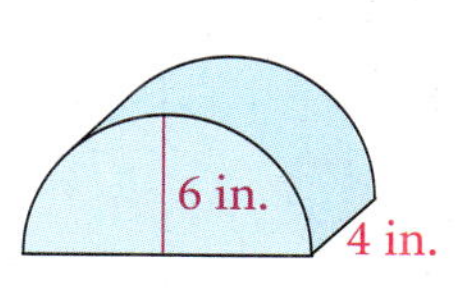

Step 2: Volume of the prism $= Bh = (12 \cdot 4)11 = 528$

Step 3: Volume of the half cylinder $= \frac{1}{2}(\pi r^2 h) = \frac{1}{2}\pi(6)^2(4)$

$= \frac{1}{2}\pi(36)(4) \approx 226$

Step 4: Sum of the two volumes $= 528 + 226 = 754$

The approximate volume of the backpack is 754 in.^{3}.

4 Find the volume of the composite space figure.

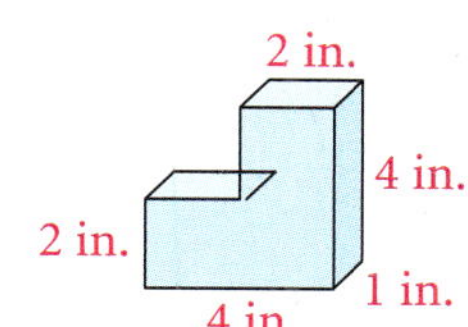

EXERCISES Standards Practice

For more exercises, see *Extra Skill, Word Problem, and Proof Practice.*

GEOM 8.0, 9.0

A Practice by Example

Example 1 (page 625)

In Exercises 1–8, find the volume of each prism.

1.

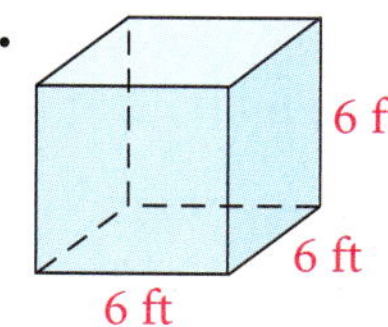

2.

5 in.
2 in.
8 in.

3.

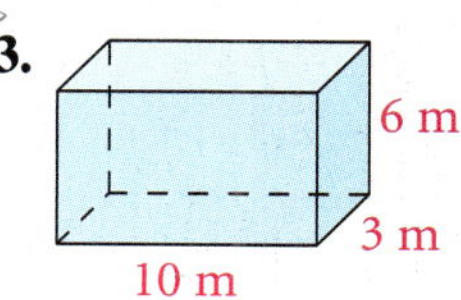

4. The base is a square, 2 cm on a side. The height is 3.5 cm.

Example 2 (page 626)

5.

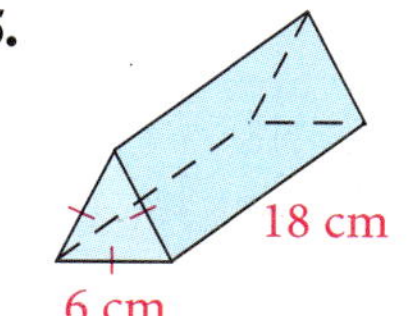

6.

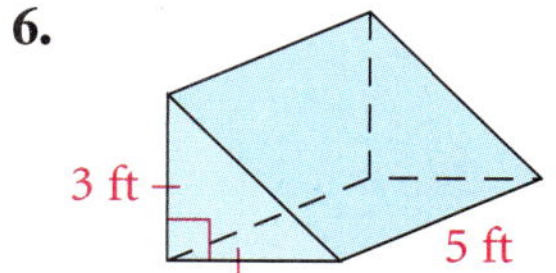

7.

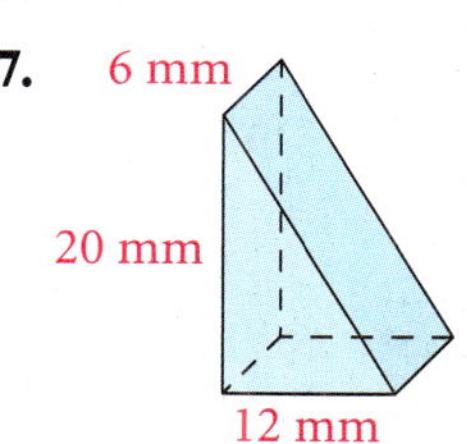

8. The base is a 45°-45°-90° triangle with a leg of 5 in. The height is 1.8 in.

Example 3 (page 626)

Find the volume of each cylinder in terms of π and to the nearest tenth.

9. 

10. 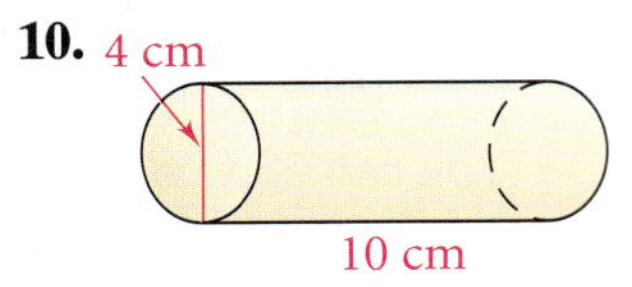

11. 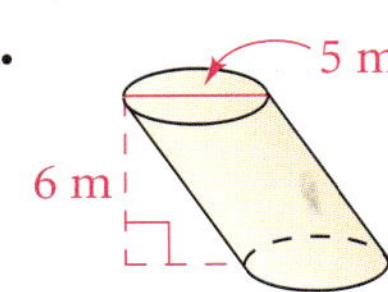

Example 4 (page 627)

Find the volume of each composite space figure to the nearest whole number.

12. 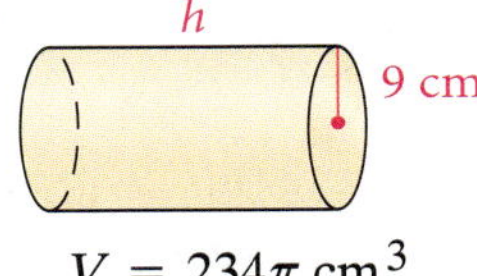

13. 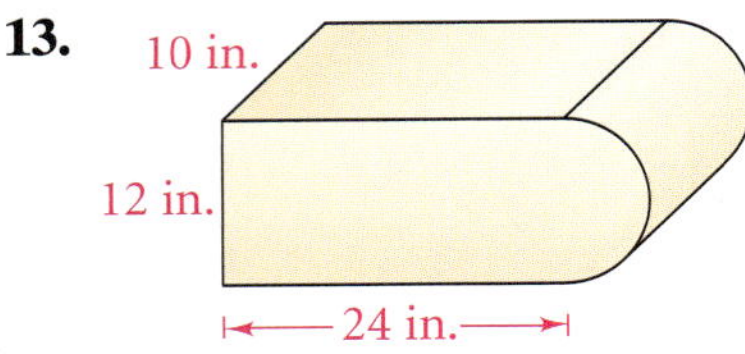

B Apply Your Skills

14. **a.** What is the volume of a waterbed mattress that is 7 ft by 4 ft by 1 ft?
    b. To the nearest pound, what is the weight of the water in a full mattress? (Water weighs 62.4 lb/ft^3.)

15. Find the volume of the lunch box shown at the right to the nearest cubic inch.

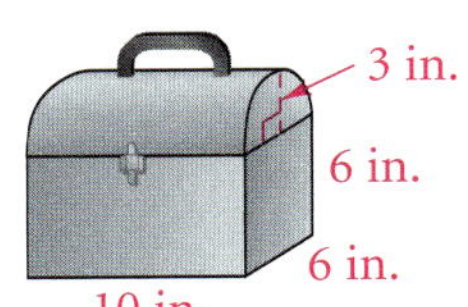

16. **Open-Ended** Give the dimensions of two rectangular prisms that have volumes of 80 cm^3 each but also have different surface areas.

Find the height of each figure with the given volume.

17. 

$V = 234\pi \text{ cm}^3$

18. 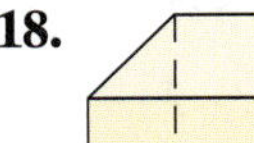
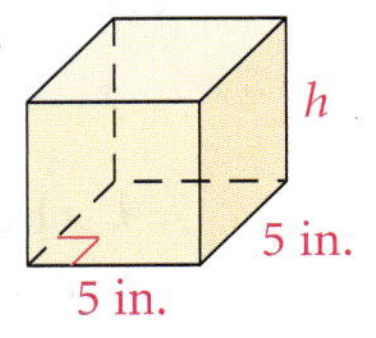

$V = 125 \text{ in.}^3$

19. 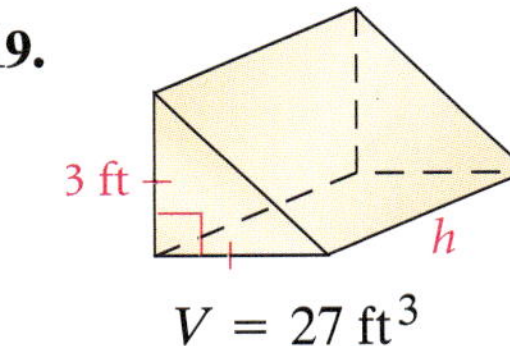

$V = 27 \text{ ft}^3$

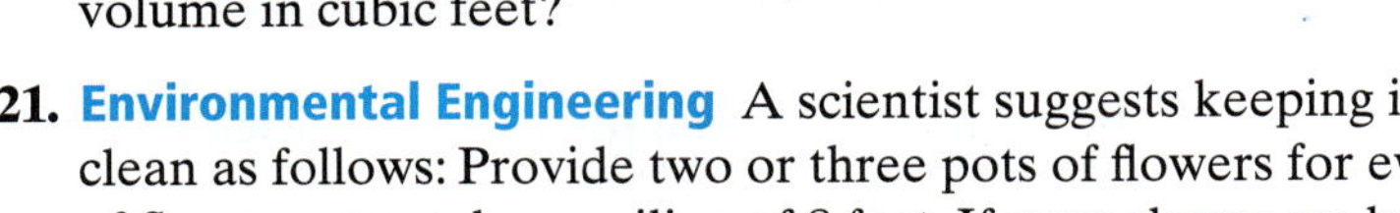

20. **Ecology** The isolation cube at the left measures 27 in. on each side. What is its volume in cubic feet?

21. **Environmental Engineering** A scientist suggests keeping indoor air relatively clean as follows: Provide two or three pots of flowers for every 100 square feet of floor space under a ceiling of 8 feet. If your classroom has an 8-ft ceiling and measures 35 ft by 40 ft, how many pots of flowers should it have?

22. Find the volume of the oblique prism pictured at the right.

23. **Tank Capacity** The main tank at an aquarium is a cylinder with diameter 203 ft and height 25 ft.
    a. Find the volume of the tank to the nearest cubic foot.
    b. Convert your answer to part (a) to cubic inches.
    c. If 1 gallon $\approx$ 231 in.^{3}, about how many gallons does the tank hold?

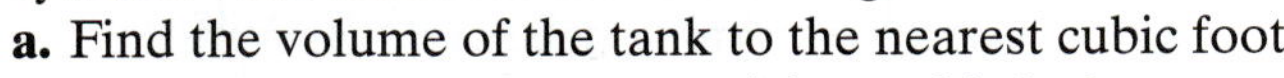

24. **Writing** The figures at the right can be covered by equal numbers of straws that are the same length. Describe how Cavalieri's Principle could be adapted to compare the areas of these figures.

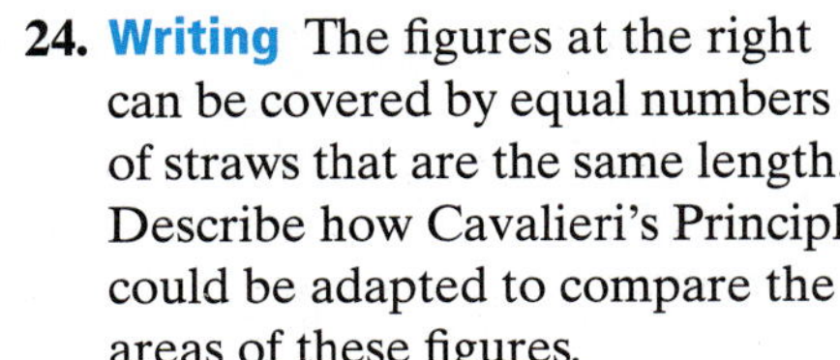
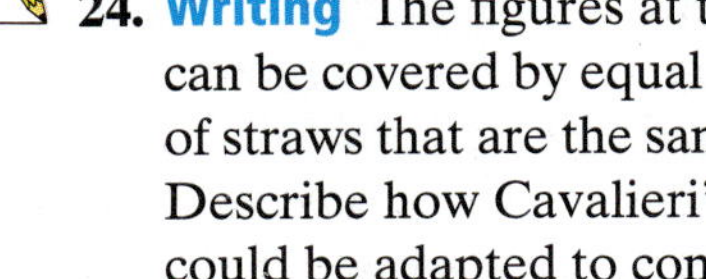
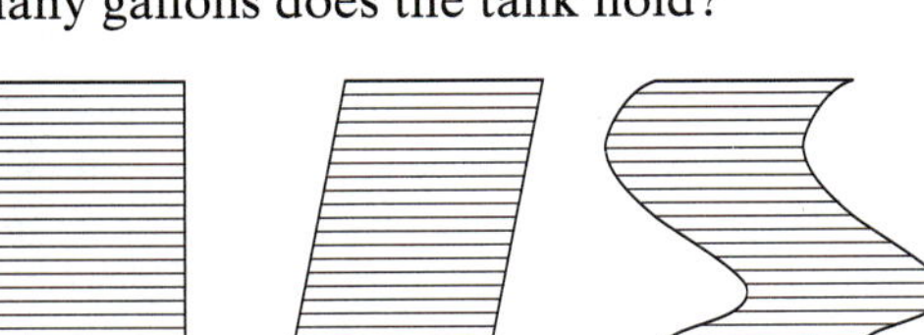

Real-World Connection

Scientists use isolation cubes to avoid contamination of lab specimens and to control the environment inside the cube.

Visit: PHSchool.com
Web Code: bce-1104

25. **Coordinate Geometry** Find the volume of the rectangular prism at the right.

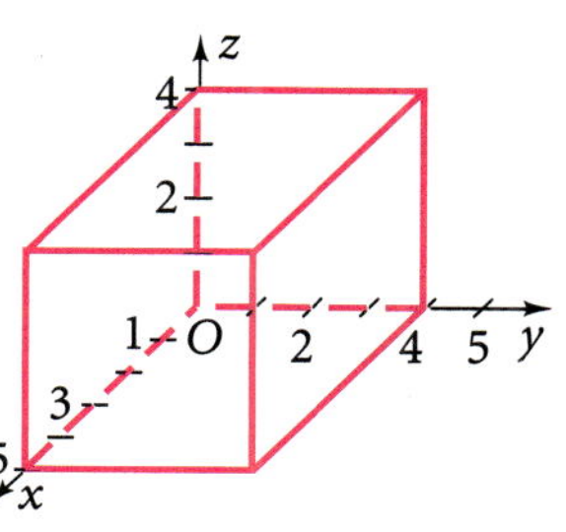

26. The volume of a cylinder is 600π cm^3. The radius of a base of the cylinder is 5 cm. What is the height of the cylinder?

27. The volume of a cylinder is 135π cm^3. The height of the cylinder is 15 cm. What is the radius of a base of the cylinder?

28. **Multiple Choice** A cylindrical water tank has a diameter of 9 inches and a height of 12 inches. The water surface is 2.5 inches from the top. About how much water is in the tank?

Ⓐ 604 in.^{3} Ⓒ 668 in.^{3}
Ⓑ 636 in.^{3} Ⓓ 763 in.^{3}

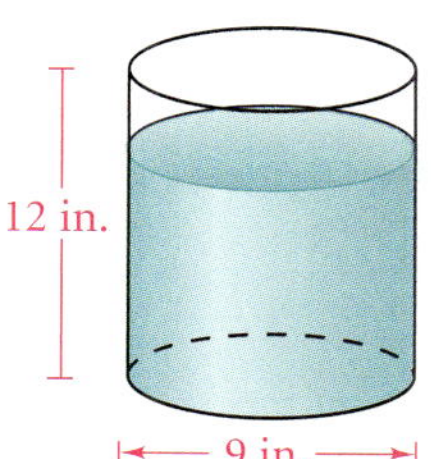

29. **Landscaping** To landscape her 70 ft-by-60 ft rectangular backyard, Joy is planning first to put down a 4-in. layer of topsoil. She can buy bags of topsoil at \$2.50 per 3-ft^3 bag, with free delivery. Or, she can buy bulk topsoil for \$22.00/yd^3, plus a \$20 delivery fee. Which option is less expensive? Explain.

Visualization **The plane region is revolved completely about the given line to sweep out a solid of revolution. Describe the solid and find its volume in terms of π.**

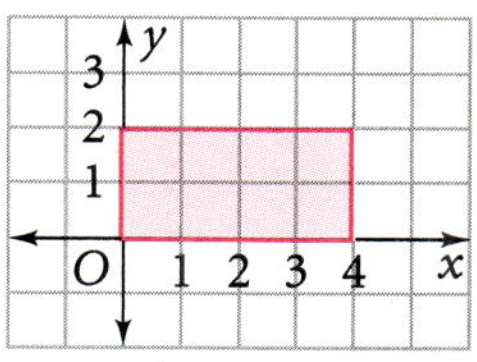

30. the x-axis
31. the y-axis

32. the line $y = 2$
33. the line $x = 5$

A cylinder has been cut out of each solid. Find the volume of the remaining solid. Round your answer to the nearest tenth.

34.

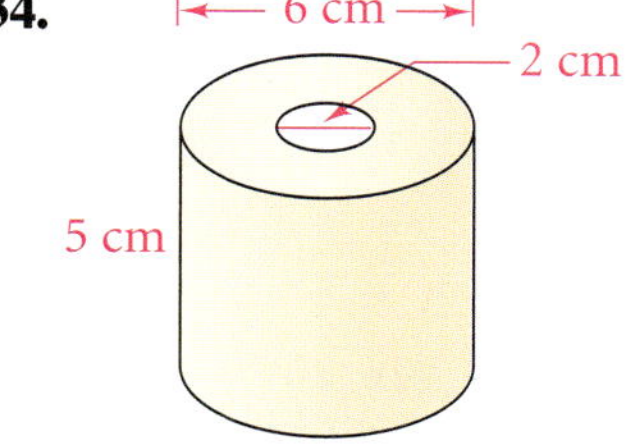

35.

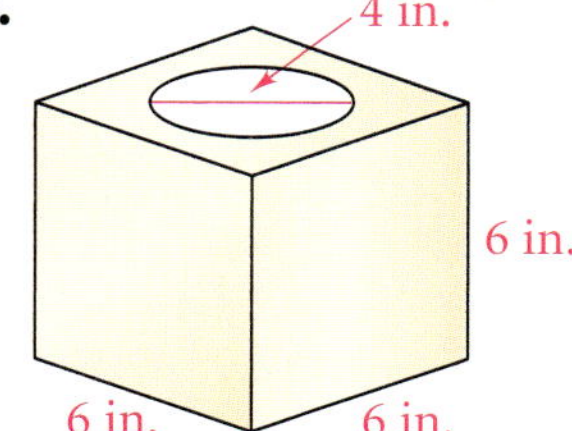

36. A closed box is 9 in. by 14 in. by 6 in. on the inside and 11 in. by 16 in. by 7 in. on the outside. Find each measurement.
   a. the outside surface area
   b. the inside surface area
   c. the inside volume
   d. the volume of the material needed to make the box

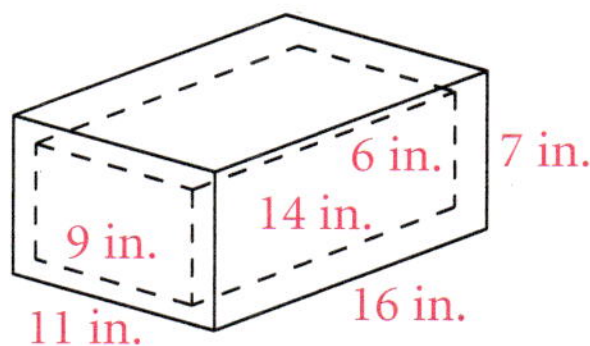

37. Any rectangular sheet of paper can be rolled into a right cylinder in two ways.
   a. Use ordinary sheets of paper to model the two cylinders. Compute the volume of each cylinder. How do they compare?
   b. Of all sheets of paper with perimeter 39 in., which size can be rolled into a right cylinder with greatest volume? (*Hint:* See Activity Lab, page 616.)

38. The outside diameter of a pipe is 5 cm. The inside diameter is 4 cm. The pipe is 4 m long. What is the volume of the material used for this length of pipe? Round your answer to the nearest cubic centimeter.

39. A cube has a volume of $2M$ cubic units and a total surface area of $3M$ square units. Find the length of an edge of the cube.

40. The radius of cylinder B is twice the radius of cylinder A. The height of cylinder B is half the height of cylinder A. Compare their volumes.

Multiple Choice Practice

For California Standards and CAHSEE Tutorials, visit PHSchool.com. Web Code: bcq-9045

GEOM 9.0

41. Find the volume of the cylinder to the nearest tenth of a cubic meter.

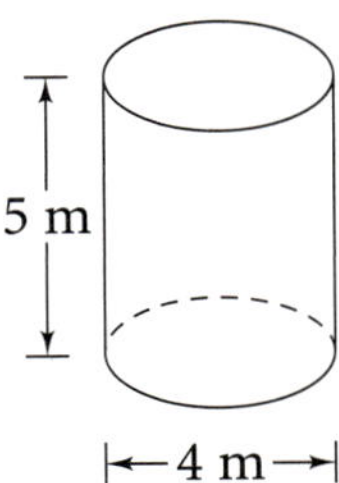

(A) 12.6 m^3
(B) 20 m^3
(C) 62.8 m^3
(D) 251.3 m^3

GEOM 11.0

42. Amparo multiplies the length and the width of a rectangle by the same value to make a new rectangle. The new rectangle has a perimeter 4 times that of the original rectangle. The area of the original rectangle is R. How can Amparo find the area of the new rectangle?

(A) She can multiply R by 4.
(B) She can multiply R by 16.
(C) She can multiply R by 64.
(D) She needs more information to find the area.

CAHSEE Alg1 7.0

43. What is the equation of the line that goes through points (0, 1) and (1, −2)?

(A) $y = 3x + 1$ (B) $y = -3x + 1$ (C) $y = \frac{1}{3}x + 1$ (D) $y = -\frac{1}{3}x + 1$

Mixed Review

Lesson 11-3

Find the lateral area of each figure to the nearest tenth.

44. a right circular cone with height 12 mm and radius 5 mm

45. a regular hexagonal pyramid with base edges 9.2 ft long and slant height 17 ft

Lesson 10-6

x^2 Algebra **Find the value of each variable and the measure of each labeled angle.**

46. $a°$, $(2a - 30)°$

47. $(6b + 10)°$, $(4b + 5)°$, $(4b - 5)°$

48. $(c - 36)°$, $(3c - 6)°$, $(3c - 2)°$, $(c + 4)°$

Lesson 7-3

49. You want to find the height of a tree near your school. Your shadow is three-fourths of your height. The tree's shadow is 57 feet. How tall is the tree?

Volumes of Pyramids and Cones

California Content Standards

GEOM 8.0 Students know, derive, and solve problems involving the volume of common geometric figures. ***Develop***

GEOM 9.0 Students compute the volumes of pyramids and cones; students commit to memory the formula for pyramids. ***Develop***

What You'll Learn

- To find the volume of a pyramid
- To find the volume of a cone

. . . And Why

To find the volume of a structure in the shape of a pyramid, as in Example 1

Check Skills You'll Need

Use the Pythagorean Theorem to find the value of the variable.

1. 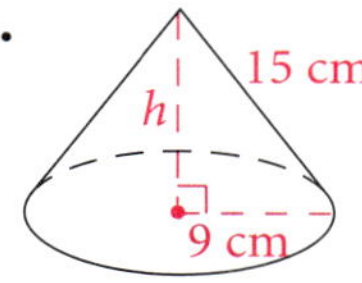

2. 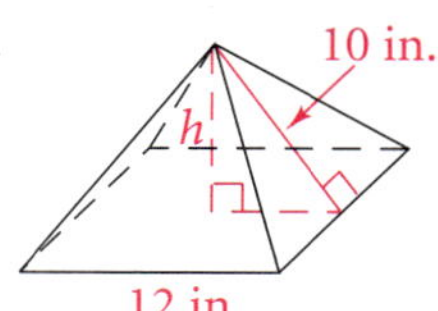

3. 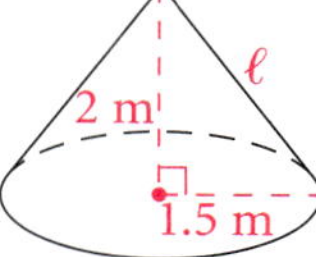

Finding Volume of a Pyramid

Hands-On Activity: Finding Volume

You know how to find the volume of a prism. Use the following to explore finding the volume of a pyramid.

- Draw the nets shown at the right on cardboard.
- Cut out the nets and tape them together to make a cube and a regular square pyramid. Each model will have one open face.

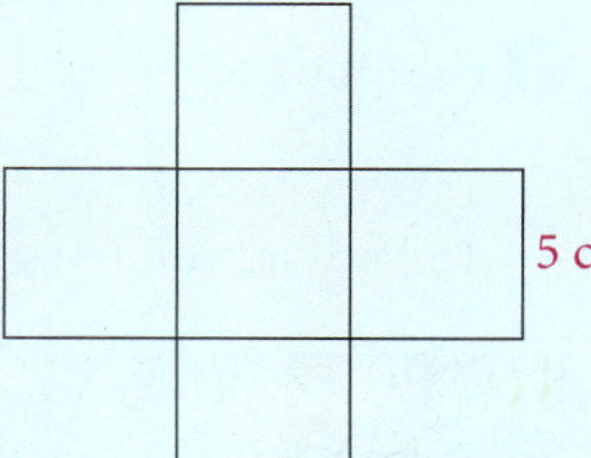

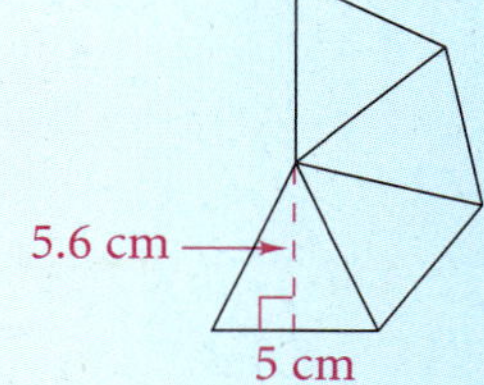

1. How do the areas of the bases of the cube and the pyramid compare?

2. How do the heights of the cube and pyramid compare?

3. Fill the pyramid with rice or other material. Then pour the rice from the pyramid into the cube. How many pyramids full of rice does the cube hold?

4. The volume of the pyramid is what fractional part of the volume of the cube?

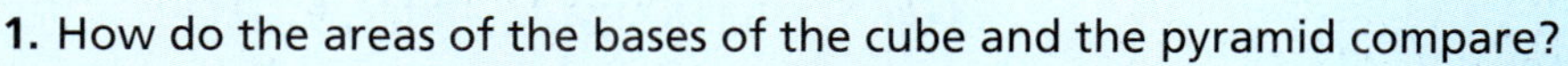

The volume of a pyramid is a particular fraction of the volume of a prism that has the same base and height as the pyramid. The fraction is shown and this fact is stated as Theorem 11-8 at the top of the next page.

Key Concepts

Theorem 11-8 **Volume of a Pyramid**

The volume of a pyramid is one third the product of the area of the base and the height of the pyramid.

$$V = \frac{1}{3}Bh$$

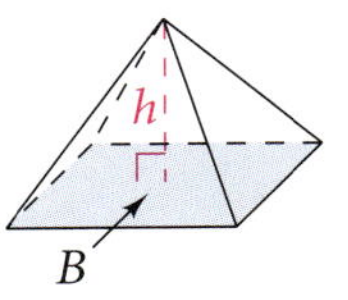

GO for Help

You can find Cavalieri's Principle on page 625.

Because of Cavalieri's Principle, the volume formula is true for all pyramids, including oblique pyramids. The height h of an oblique pyramid is the length of the perpendicular segment from the vertex to the plane of the base.

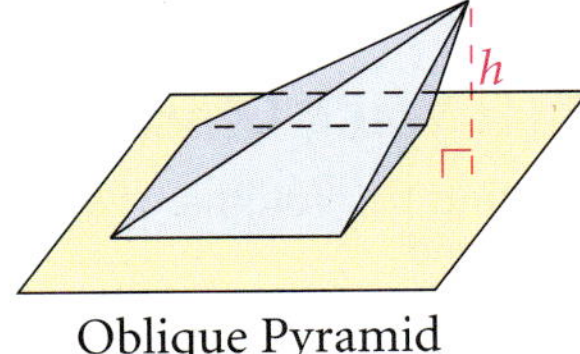

Oblique Pyramid

1 EXAMPLE Real-World Connection

Architecture The Pyramid is an arena in Memphis, Tennessee. The area of the base of The Pyramid is about 300,000 ft^2. Its height is 321 ft. What is the volume of The Pyramid?

$V = \frac{1}{3}Bh$	**Use the formula for volume of a pyramid.**
$= \frac{1}{3}(300{,}000)(321)$	**Substitute.**
$= 32{,}100{,}000$	**Simplify.**

The volume is about 32,100,000 ft^3.

CA Standards Check 1 Find the volume of a square pyramid with base edges 12 in. and height 8 in.

To find the volume of a pyramid you may first need to find its height.

2 EXAMPLE Finding Volume of a Pyramid

Find the volume in cubic feet of a square pyramid with base edges 40 ft and slant height 25 ft.

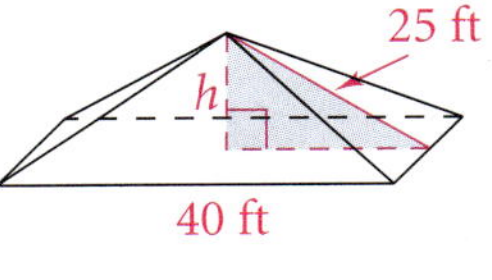

h
25 ft
20 ft

Step 1: Find the height of the pyramid.

$25^2 = h^2 + 20^2$	**Use the Pythagorean Theorem.**
$625 = h^2 + 400$	**Simplify.**
$h^2 = 225$	**Solve for h^2.**
$h = 15$	**Take square roots.**

Step 2: Find the volume of the pyramid.

$V = \frac{1}{3}Bh$	**Use the formula for volume of a pyramid.**
$= \frac{1}{3}(40 \cdot 40)15$	**Substitute.**
$= 8000$	**Simplify.**

The volume of the pyramid is 8000 ft^3.

CA Standards Check 2 Find the volume of a square pyramid with base edges 24 m and slant height 13 m.

2 Finding Volume of a Cone

You have seen that the volume of a pyramid is one third the volume of a prism with the same base and height. Similarly, the volume of a cone is one third the volume of a cylinder with the same base and height.

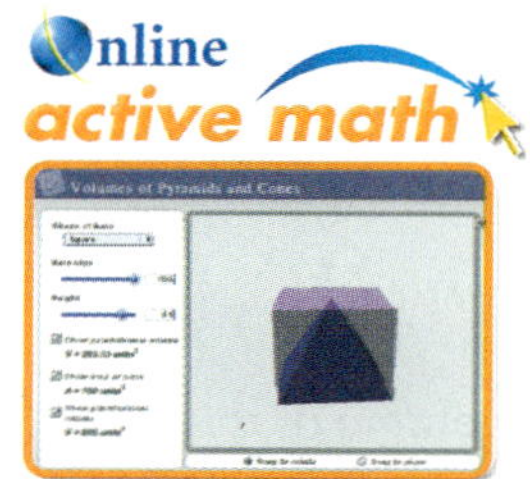

For: Pyramid, Cone Activity
Use: Interactive Textbook, 11-5

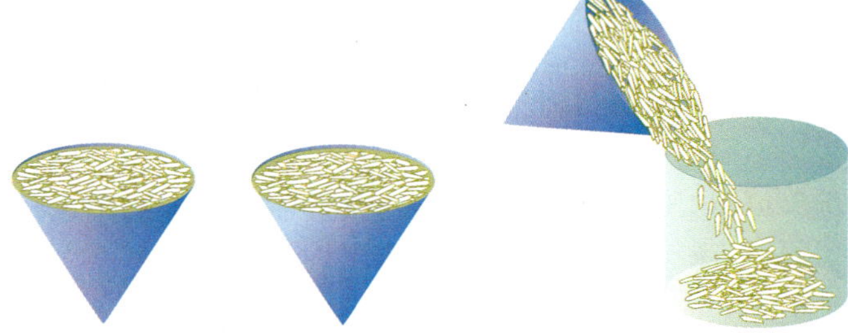

The cones and the cylinder have the same base and height. It takes three cones full of rice to fill the cylinder.

Key Concepts

Theorem 11-9 Volume of a Cone

The volume of a cone is one third the product of the area of the base and the height of the cone.

$$V = \frac{1}{3}Bh, \text{ or } V = \frac{1}{3}\pi r^2 h$$

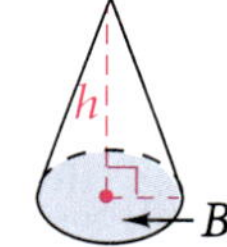

This volume formula applies to all cones, including oblique cones.

3 EXAMPLE Finding Volume of an Oblique Cone

Find the volume of an oblique cone with diameter 30 ft and height 25 ft. Give your answer in terms of π and also rounded to the nearest cubic foot.

$V = \frac{1}{3}\pi r^2 h$ **Use the formula for volume of a cone.**

$= \frac{1}{3}\pi(15)^2 25$ **Substitute 15 for *r* and 25 for *h*.**

$= 1875\pi$ **Simplify.**

$\approx$ 5890.4862 **Use a calculator.**

The volume of the cone is 1875π ft^3, or about 5890 ft^3.

Test-Taking Tip

The formula $V = \frac{1}{3}Bh$ can be used to find the volume of a pyramid or a cone. The only difference is how you calculate B, the area of the base.

CA Standards Check

3 Find the volume of each cone in terms of π and also rounded as indicated.

a. to the nearest cubic meter

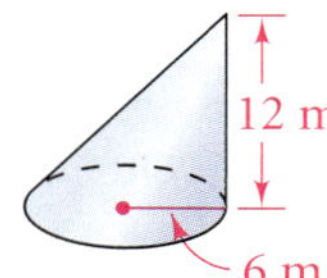

b. to the nearest cubic millimeter

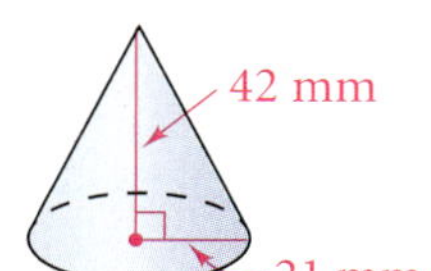

A cone-shaped structure can be particularly strong, as downward forces at the vertex are distributed to all points in its circular base.

Dwelling The covering on a teepee rests on poles that come together like concurrent lines. The resulting structure approximates a cone. If the teepee pictured is 12 ft. high with a base diameter 14 ft, estimate its volume.

$V = \frac{1}{3}\pi r^2 h$ **Use the formula for the volume of a cone.**

$V = \frac{1}{3}\pi(7)^2(12)$ **Substitute; $r = \frac{1}{2}(14) = 7$.**

$= 615.75216$ **Use a calculator.**

The volume of the teepee is approximately 616 ft^3.

4 A small child's teepee is 6 ft tall and 7 ft in diameter. Find the volume of the teepee to the nearest cubic foot.

EXERCISES

For more exercises, see *Extra Skill, Word Problem, and Proof Practice.*

Standards Practice

GEOM 8.0, 9.0

A Practice by Example

Example 1 (page 632)

1. The entrance to the Louvre Museum in Paris, France is a pyramid. The height of the pyramid is about 70 ft and the area of its base is about 10,000 ft^2. What is the volume of the pyramid?

Find the volume of a square pyramid with the following dimensions.

2. base edges 10 cm, height 6 cm
3. base edges 18 in., height 12 in.
4. base edges 5 m, height 6 m

Find the volume of each square pyramid.

5. 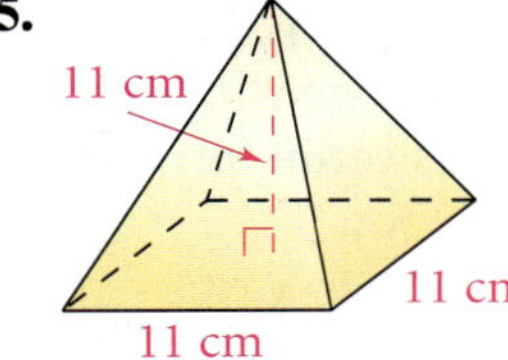

6. 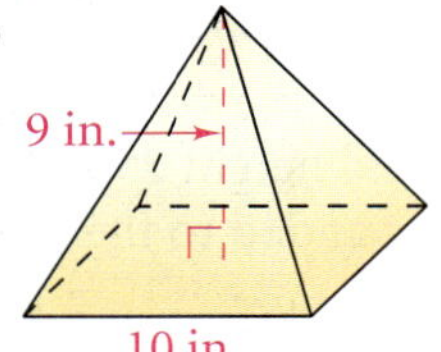

7. 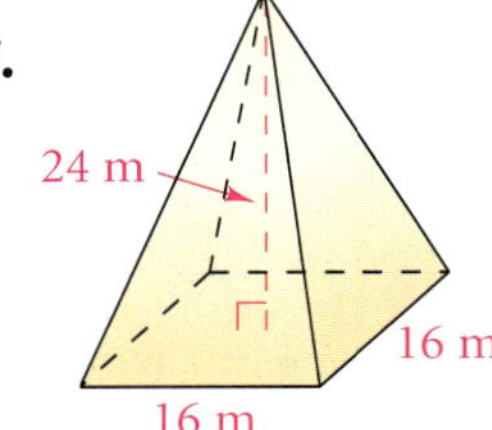

Example 2 (page 632)

8. 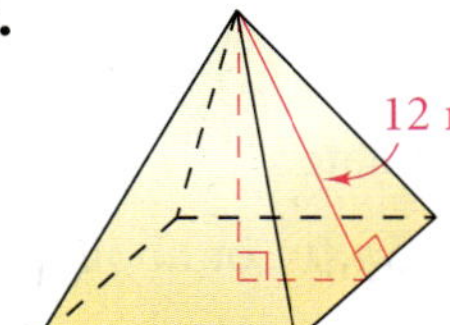

9. 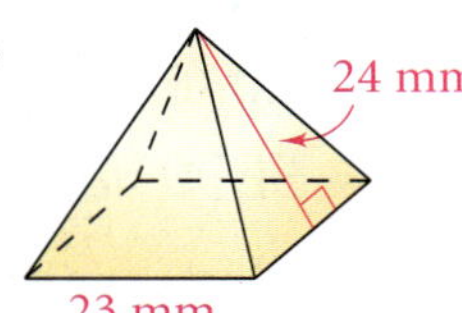

10. 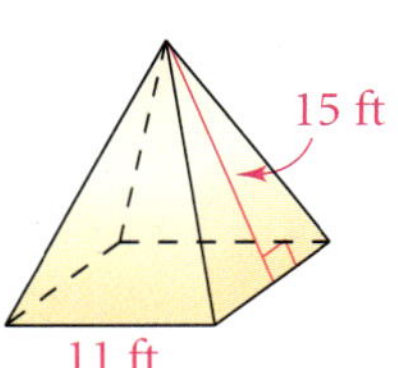

Example 3 (page 633)

Find the volume of each cone in terms of π and also rounded as indicated.

11. nearest cubic foot
12. nearest cubic inch
13. nearest cubic inch

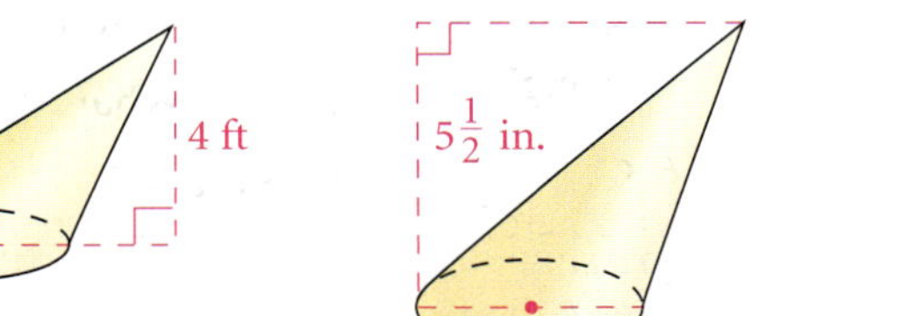

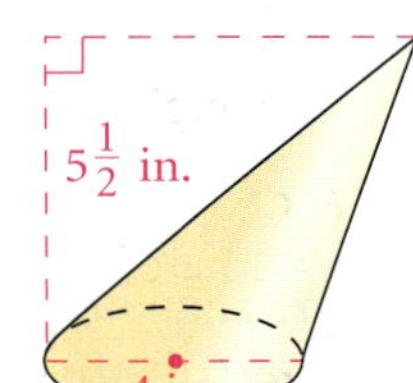

Example 4 (page 634)

14. **Chemistry** In a chemistry lab you use a filter paper cone to filter a liquid. The diameter of the cone is 6.5 cm and its height is 6 cm. How much liquid will the cone hold when it is full?

15. **Chemistry** This funnel has a filter that was being used to remove impurities from a solution but became clogged and stopped draining. The remaining solution is represented by the shaded region. How many cubic centimeters of the solution remain in the funnel?

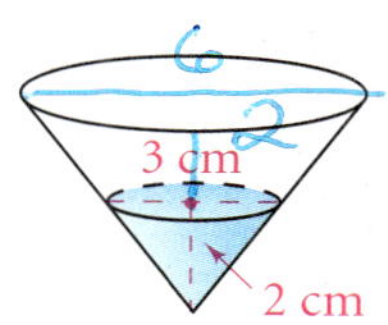

Find the volume to the nearest whole number.

16.
7.5 in.
7 in.
Square base

17.
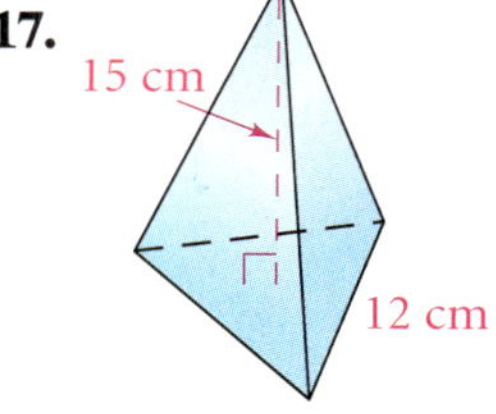

Equilateral base

18.
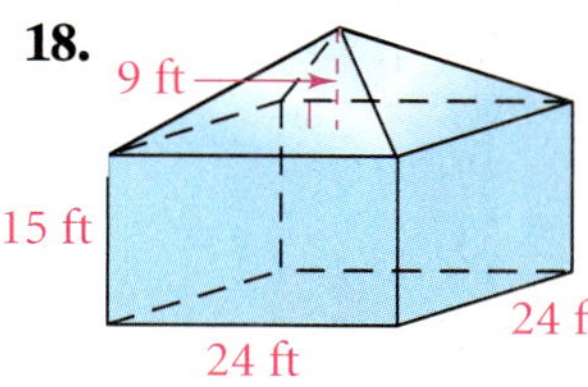

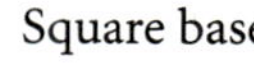
Square base

19. **Writing** The two cylinders pictured at the right are congruent. How does the volume of the larger cone compare to the total volume of the two smaller cones? Explain.

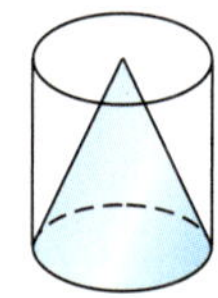
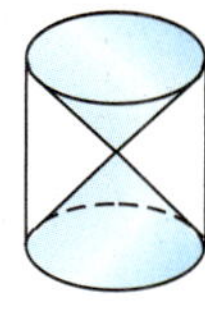

Exercise 20

20. **Architecture** The Transamerica Pyramid in San Francisco (see photo at left) is 853 ft tall with a square base that is 149 ft on each side.
   a. What is its volume to the nearest thousand cubic feet?
   b. How tall would a prism-shaped building with the same square base as the Pyramid have to be to have the same volume as the Pyramid?

x^2 **Algebra** **Find the value of the variable in each figure. Leave answers in simplest radical form. The diagrams are not to scale.**

21.
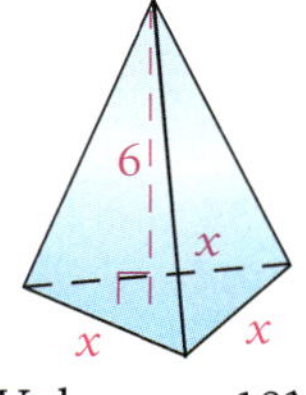

Volume = $18\sqrt{3}$

22.
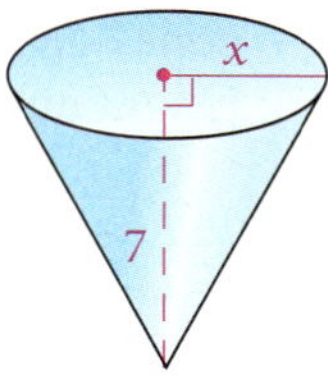

Volume = 21π

23.
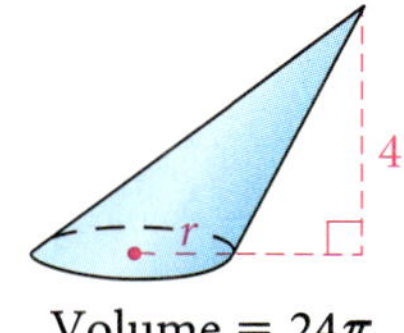

Volume = 24π

24. **Hardware** Builders use a plumb bob to find a vertical line. The plumb bob shown combines a regular hexagonal prism with a pyramid. Find its volume to the nearest cubic centimeter.

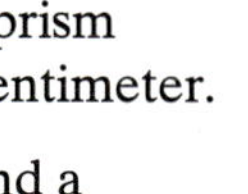

25. **Multiple Choice** A cone has a volume of 600π in.^{3} and a height of 50 in. What is the radius of the cone?
   (A) 3.5 in. (B) 6.0 in. (C) 10.6 in. (D) 36.0 in.

26. A cone with radius 1 fits snugly inside a square pyramid which fits snugly inside a cube. What are the volumes of the three figures?

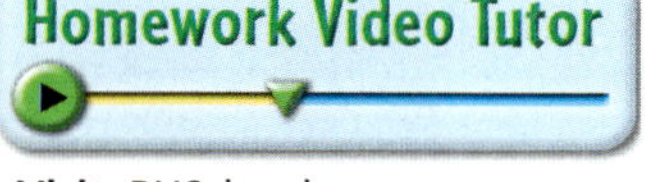

Visit: PHSchool.com
Web Code: bce-1105

27. A cone with radius 3 ft and height 10 ft has a volume of 30π ft^3. What is the volume of the cone formed when the following happens to the original cone?
   a. The radius is doubled.
   b. The height is doubled.
   c. The radius and the height are both doubled.

28. List the volumes of the cone, prism, and pyramid in order from least to greatest.

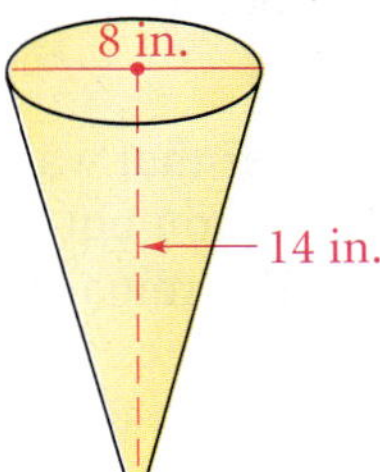

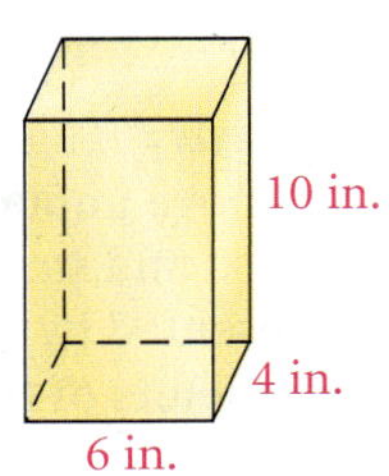

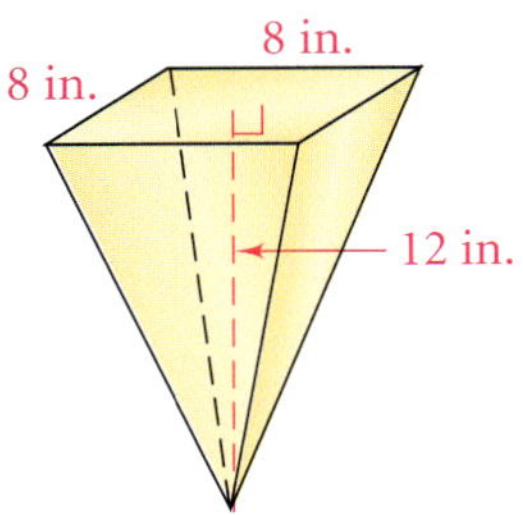

Visualization **The plane region is revolved completely about the given line to sweep out a solid of revolution. Describe the solid. Then find its volume in terms of π.**

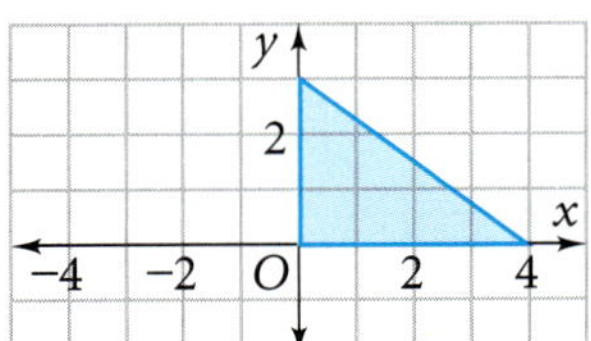

29. the y-axis

30. the x-axis

Challenge

31. the line $x = 4$

32. the line $y = -1$

33. A *frustum* of a cone is the part that remains when the vertex is cut off by a plane parallel to the base.

a. Explain how to use the formula for the volume of a cone to find the volume of a frustum of a cone.

b. **Containers** A 9-in. tall popcorn container is the frustum of a cone. Its small radius is 4.5 in. and its large radius is 6 in. What is its volume?

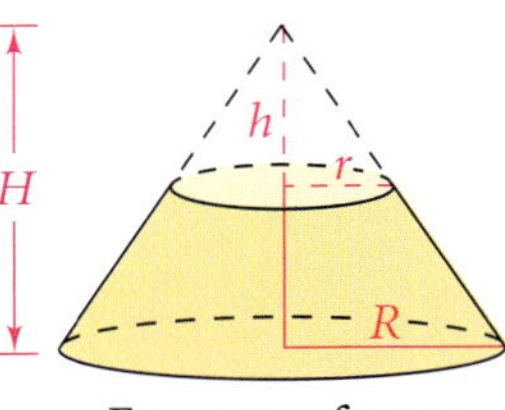

Frustum of cone

34. A disk has radius 10 m. A 90° sector is cut away, and a cone is formed.

a. What is the circumference of the base of the cone?

b. What is the area of the base of the cone?

c. What is the volume of the cone? (*Hint:* Use the slant height and the radius of the base to find the height.)

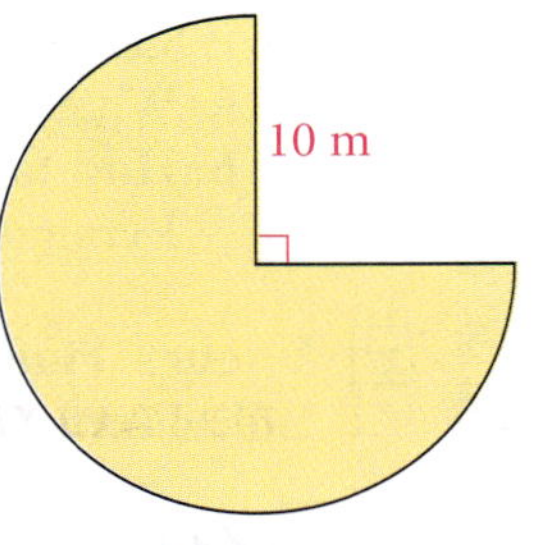

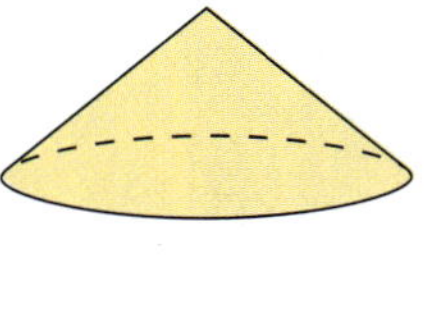

Graphing Calculator **In Exercises 35 and 36, the volume of the solid is 1000 cm^3. Use the Activity Lab on page 616 to help you complete each exercise.**

35. For a square pyramid, find the length of a side of the base for which the lateral area is as small as possible.

36. For a cone, find the radius for which the lateral area is as small as possible.

Multiple Choice Practice

For California Standards and CAHSEE Tutorials, visit PHSchool.com. Web Code: bcq-9045

GEOM 9.0

37. Find the volume of the cone to the nearest tenth of a cubic inch.

(A) 16 in.^{3}

(B) 33.5 in.^{3}

(C) 100.5 in.^{3}

(D) 134 in.^{3}

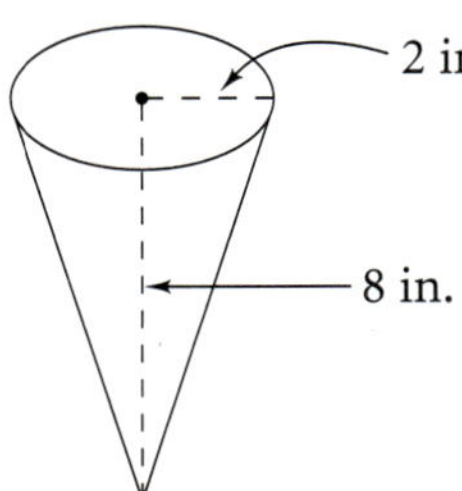

GEOM 17.0

38. If point A has coordinates $(3, 6)$ and point B has coordinates $(-7, 10)$, what are the coordinates of the midpoint of $\overline{AB}$?

(A) $(-2, 8)$ (B) $(5, 8)$ (C) $(-4, 16)$ (D) $(-5, 2)$

CAHSEE 6SDAP 3.3

39. Mandy tossed two number cubes 50 times. She kept track of whether both were even, both were odd, or one was even and the other odd. Based on the data in the Venn Diagram, what is the experimental probability of rolling one even number and one odd number?

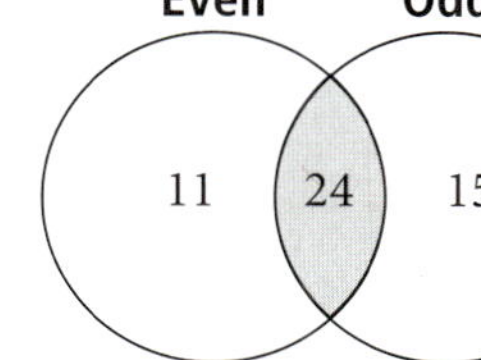

(A) 24% (C) 50%

(B) 48% (D) 52%

Lesson 11-4

40. **Sports** A cylindrical hockey puck is 1 in. high and 3 in. in diameter. What is its volume in cubic inches? Round your answer to the nearest tenth.

GO for Help

41. A triangular prism has height 30 cm. Its base is a right triangle with legs 10 cm and 24 cm. Find the volume of the prism.

Lesson 10-5

42. Find the area of a regular pentagon with a radius 5 in. Give your answer to the nearest tenth of a square inch.

Lesson 8-2

Find the area of each equilateral triangle to the nearest tenth of a square unit.

43. The triangle has 12 cm sides. **44.** The triangle has 10-in. altitudes.

45. Find the area of a 30°-60°-90° triangle with shorter leg of length 4 cm.

Geometry at Work

Package Designer

Each year, more than one trillion dollars in manufactured goods are packaged in containers. To create each new box, bag, or carton, package designers must balance such factors as safety, environmental impact, and attractiveness against cost of production.

Consider the three boxes of dishwasher detergent. All three boxes have standard volumes of 108 in.^{3}. The boxes have different shapes, however, and different surface areas. The box on the left has the greatest surface area and therefore costs the most to produce. Despite the higher cost, the box on the left has become standard. In this case, the least expensive package on the right is too difficult for a consumer to pick up and pour.

For: Information about package design
Web Code: bcb-2031

Surface Areas and Volumes of Spheres

California Content Standards

GEOM 8.0 Students know, derive, and solve problems involving the volume and surface area of common geometric figures. ***Develop, Master***

GEOM 9.0 Students compute the volumes and surface areas of spheres. ***Develop, Master***

What You'll Learn

- To find the surface area and volume of a sphere

. . . And Why

To approximate the surface area of Earth, as in Example 2

Check Skills You'll Need

GO for Help Lesson 1-9

Find the area and circumference of a circle with the given radius. Round your answers to the nearest tenth.

1. 6 in.	**2.** 5 cm	**3.** 2.5 ft
4. 1.2 m	**5.** 15 yd	**6.** 12 mm

New Vocabulary • sphere • center, radius, diameter, circumference (of a sphere) • great circle • hemisphere

1 Finding Surface Area and Volume of a Sphere

Real-World Connection

The diameter of the Hayden Sphere in New York City is 87 ft.

A **sphere** is the set of all points in space equidistant from a given point called the **center.** A **radius** is a segment that has one endpoint at the center and the other endpoint on the sphere. A **diameter** is a segment passing through the center with endpoints on the sphere.

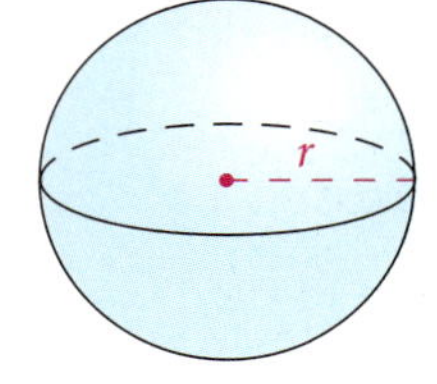

The length r is the *radius* of the sphere. $2r$ is the *diameter*.

When a plane and a sphere intersect in more than one point, the intersection is a circle. If the center of the circle is also the center of the sphere, the circle is called a **great circle** of the sphere. The circumference of a great circle is the **circumference** of the sphere. A great circle divides a sphere into two **hemispheres.**

A baseball is a model of a sphere. To approximate its surface area, you can take apart its covering. Each of the two sections suggests a pair of circles with radius r approximately the radius of the ball. The area of the four circles, $4\pi r^2$, suggests the surface area of the ball.

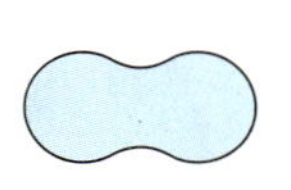

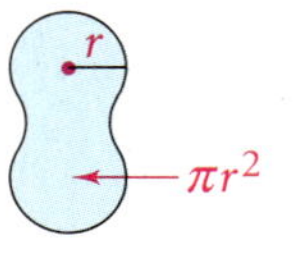

Key Concepts

Theorem 11-10 **Surface Area of a Sphere**

The surface area of a sphere is four times the product of π and the square of the radius of the sphere.

$$\text{S.A.} = 4\pi r^2$$

1 EXAMPLE Finding Surface Area

Find the surface area of this sphere. Leave your answer in terms of π.

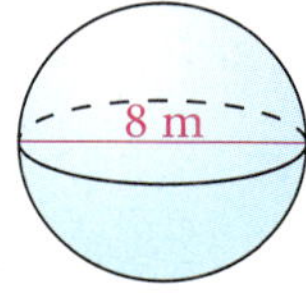

$\text{S.A.} = 4\pi r^2$ **Use the formula for surface area.**

$= 4\pi 4^2$ **Substitute $r = \frac{8}{2} = 4$.**

$= 64\pi$ **Simplify.**

The surface area is 64π m^2.

CA Standards Check **1** Find the surface area of a sphere with $d = 14$ in. Give your answer two ways, in terms of π and rounded to the nearest square inch.

You can use spheres to approximate the surface areas of real-world objects.

2 EXAMPLE Real-World Connection

Geography Earth's equator is about 24,902 mi long. Approximate the surface area of Earth by finding the surface area of a sphere with circumference 24,902 mi.

Step 1 Find the radius.

$C = 2\pi r$ **Use the formula for circumference.**

$24{,}902 = 2\pi r$ **Substitute.**

$\frac{24{,}902}{2\pi} = r$ **Solve for r.**

$r = 3963.2764$ **Use a calculator.**

Go Online PHSchool.com

For: Graphing calculator procedures

Web Code: bcb-2113

Step 2 Use the radius to find the surface area.

$\text{S.A.} = 4\pi r^2$ **Use the formula for surface area.**

$= 4\pi$ ANS x^2 ENTER **Use a calculator.**

$= 197387020$

The surface area of Earth is about 197,400,000 mi^2.

CA Standards Check **2** Find the surface area of a melon with circumference 18 in. Round your answer to the nearest ten square inches.

The following model suggests a formula for the volume of a sphere.

Fill a sphere with a large number n of small pyramids. The vertex of each pyramid is the center of the sphere. The height of each pyramid is approximately the radius r of the sphere. The sum of the areas of the bases of the n pyramids approximates the surface area of the sphere. The sum of the volumes of the n pyramids should approximate the volume of the sphere.

Volume of each pyramid $= \frac{1}{3}Bh$

Sum of the volumes of n pyramids $\approx n \cdot \frac{1}{3}Br$ **Substitute r for h.**

$= \frac{1}{3} \cdot (nB) \cdot r$

$\approx \frac{1}{3} \cdot (4\pi r^2) \cdot r$ **Replace nB with the surface area of a sphere.**

$= \frac{4}{3}\pi r^3$

It is reasonable to conjecture that the volume of a sphere is $\frac{4}{3}\pi r^3$.

Theorem 11-11 Volume of a Sphere

The volume of a sphere is four thirds the product of π and the cube of the radius of the sphere.

$$V = \frac{4}{3}\pi r^3$$

3 EXAMPLE Finding Volume

Find the volume of the sphere. Leave your answer in terms of π.

$V = \frac{4}{3}\pi r^3$ **Use the formula for volume.**

$= \frac{4}{3}\pi 6^3$ **Substitute.**

$= 288\pi$

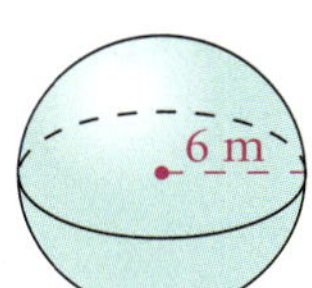

The volume of the sphere is 288π m^3.

3 Find the volume to the nearest cubic inch of a sphere with diameter 60 in.

If you know the volume of a sphere, you can find its surface area.

4 EXAMPLE Using Volume to Find Surface Area

The volume of a sphere is 5000 m^3. What is the surface area of the sphere?

Step 1 Find the radius r.

$V = \frac{4}{3}\pi r^3$ **Use the formula for volume of a sphere.**

$5000 = \frac{4}{3}\pi r^3$ **Substitute.**

$5000\left(\frac{3}{4\pi}\right) = r^3$ **Solve for r^3.**

$\sqrt[3]{5000\left(\frac{3}{4\pi}\right)} = r$ **Take cube roots.**

$r = 10.607844$ **Use a calculator.**

Problem Solving Hint

The cube root of x, $\sqrt[3]{x}$, is the number whose third power is x.

Step 2 Find the surface area of the sphere.

S.A. $= 4\pi r^2$ **Use the formula for the surface area of a sphere.**

$= 4\pi$ ANS x^2 ENTER **Use a calculator.**

$= 1414.0479$

The surface area of the sphere is about 1414 m^2.

4 The volume of a sphere is 4200 ft^3. Find the surface area to the nearest tenth.

EXERCISES

For more exercises, see *Extra Skill, Word Problem, and Proof Practice.*

Standards Practice GEOM 8.0, 9.0

A Practice by Example

Example 1 (page 639)

Find the surface area of the sphere with the given diameter or radius. Leave your answer in terms of π.

1. $d = 30$ m **2.** $r = 10$ in. **3.** $d = 32$ mm **4.** $r = 100$ yd

Find the surface area of each ball. Leave each answer in terms of π.

5.

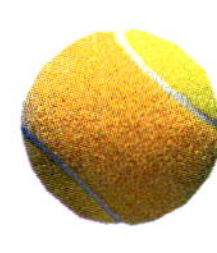

$d = 68$ mm

6.

$d = 24$ cm

7.

$d = 2\frac{3}{4}$ in.

Example 2 (page 639)

Use the given circumference to find the surface area of each spherical object. Round your answer to the nearest whole number.

8. a grapefruit with $C = 14$ cm

9. a bowling ball with $C = 27$ in.

10. a pincushion with $C = 8$ cm

11. a head of lettuce with $C = 22$ in.

Example 3 (page 640)

Find the volume of each sphere. Give each answer in terms of π and rounded to the nearest cubic unit.

12.

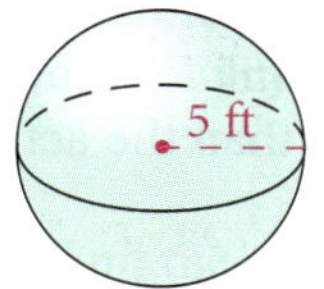

13.

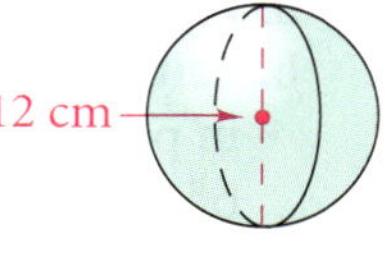

14.

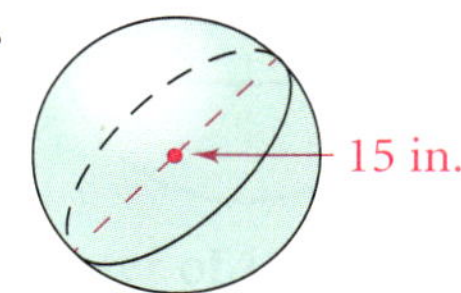

15.

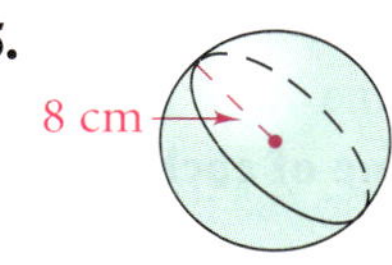

16.

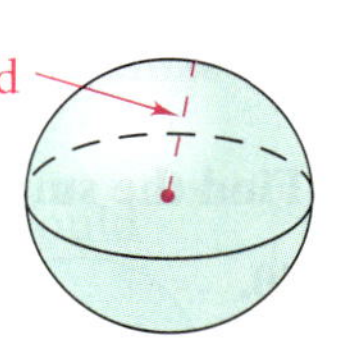

17.

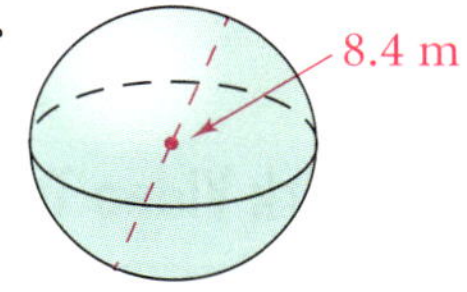

Example 4 (page 640)

A sphere has the volume given. Find its surface area to the nearest whole number.

18. $V = 900 \text{ in.}^3$

19. $V = 3000 \text{ m}^3$

20. $V = 140 \text{ cm}^3$

B Apply Your Skills

21. Mental Math Use $\pi \approx 3$ to estimate the surface area and volume of a sphere with radius 3 cm.

22. Visualization The region enclosed by the semicircle at the right is revolved completely about the x-axis.

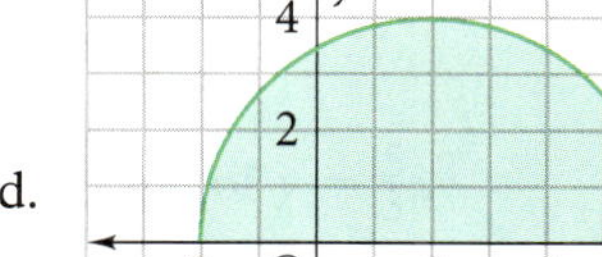

a. Describe the solid of revolution that is formed.

b. Find its volume in terms of π.

c. Find its surface area in terms of π.

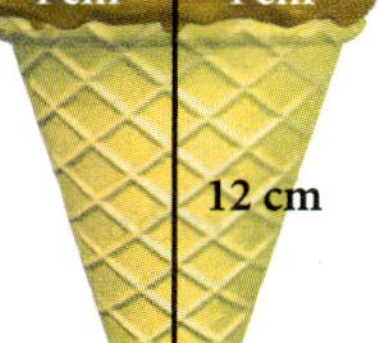

Exercise 23

23. Food A sphere of frozen yogurt was pressed into the cone as shown at the left. If the yogurt melts into the cone, would the cone overflow? Explain.

24. Multiple Choice The sphere at the right fits snugly inside a cube with 6-in. edges. What is the approximate volume of the space between the sphere and cube?

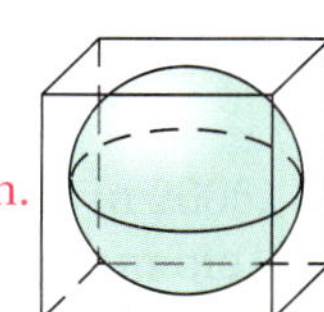

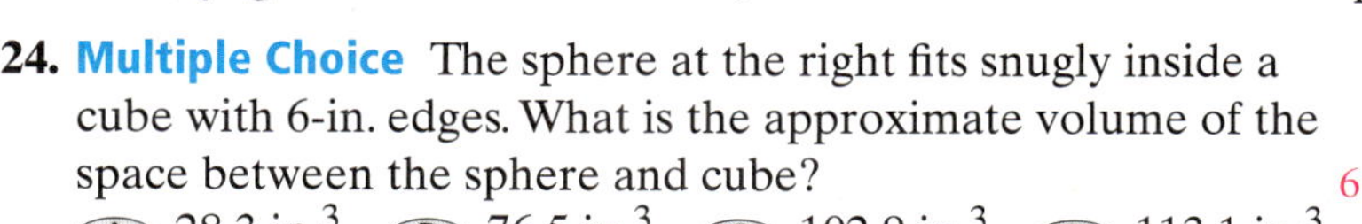

Ⓐ 28.3 in.^3 Ⓑ 76.5 in.^3 Ⓒ 102.9 in.^3 Ⓓ 113.1 in.^3

Geometry in 3 Dimensions **A sphere has center (0, 0, 0) and radius 5.**

25. Name the coordinates of six points on the sphere.

26. Tell whether each of the following points is inside, outside, or on the sphere. $A(0, -3, 4)$, $B(1, -1, -1)$, $C(4, -6, -10)$

Mixed Review

Lesson 11-5 **Find the volume of each figure to the nearest cubic unit.**

58. 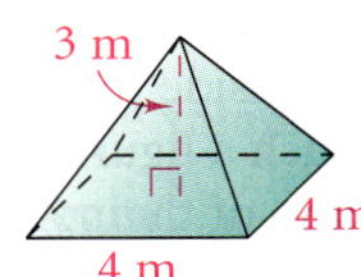

59. 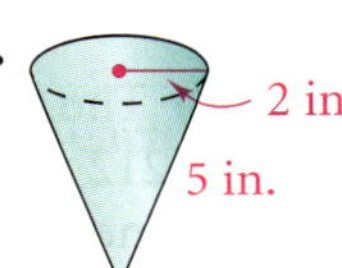

60. 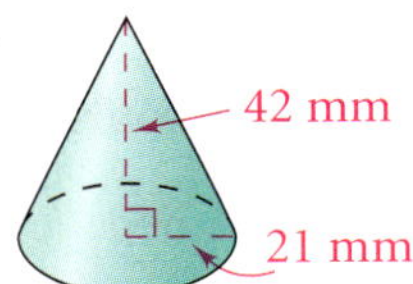

Lesson 10-4 **The similarity ratio of a pair of similar isosceles trapezoids is 2 : 3. A diagonal of the smaller figure has length 7 cm.**

61. Find the length of a diagonal in the larger trapezoid.

62. The perimeter of the larger trapezoid is 40.5 cm. What is the perimeter of the smaller trapezoid?

63. The area of the smaller trapezoid is 30 cm^2. What is the area of the larger trapezoid?

Lesson 8-4

64. A leg of a right triangle measures 4 cm and the hypotenuse measures 7 cm. Find the measure of each acute angle of the triangle to the nearest degree.

65. The length of each side of a rhombus is 16. The longer diagonal has length 26. Find the measures of the angles of the rhombus to the nearest degree.

Checkpoint Quiz 2 — Lessons 11-4 through 11-6

Find the surface area and volume of each figure to the nearest tenth.

1. 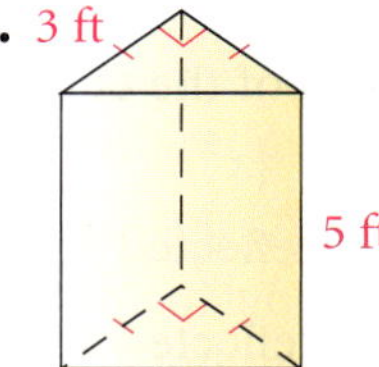

2. 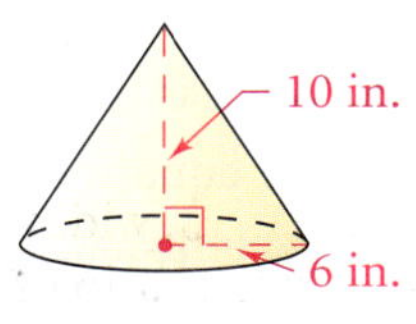

3. 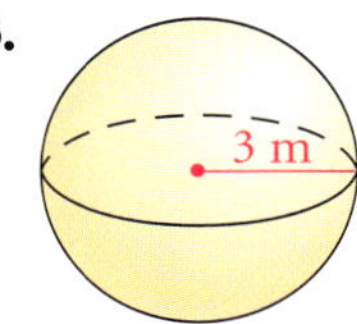

4. 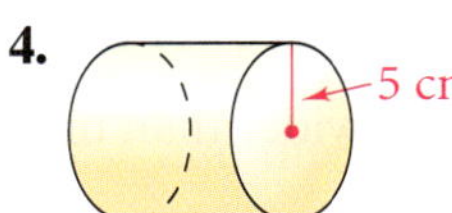

5. 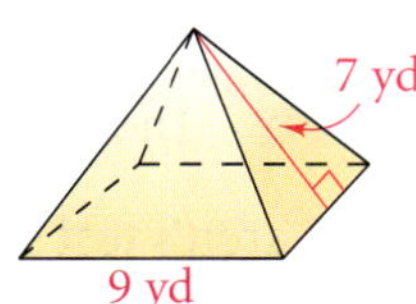

6. 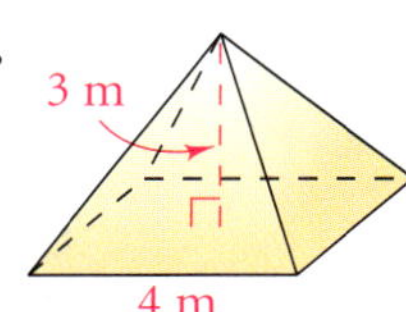

7. 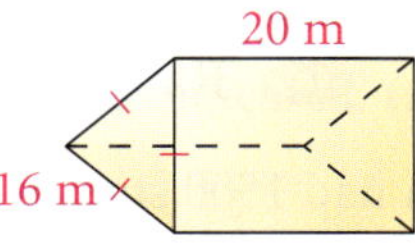

8. 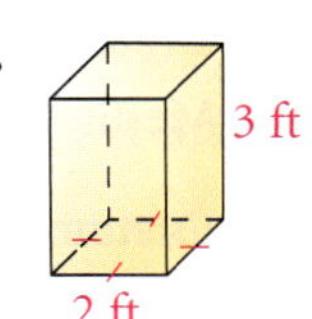

9. 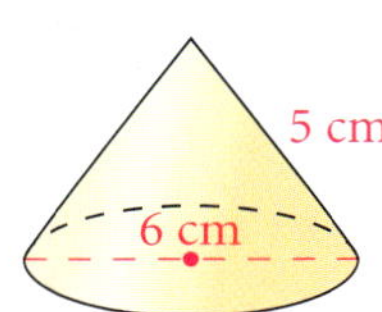

10. Critical Thinking Tennis balls fit snugly inside a cylinder as shown. Which is greater, the volume of the three tennis balls or the volume of the space around the balls? Explain.

Exploring Similar Solids

Technology

FOR USE WITH LESSON 11-7

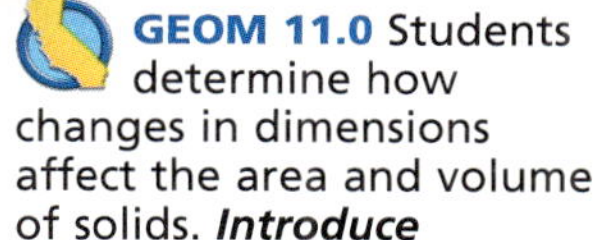

GEOM 11.0 Students determine how changes in dimensions affect the area and volume of solids. *Introduce*

To explore the surface areas and volumes of similar rectangular prisms, you can set up a spreadsheet like the one below. You choose the numbers for the length, width, height, and similarity ratio. The computer will use formulas to calculate all the other numbers.

	A	B	C	D	E	F	G	H	I
1					Surface		Similarity		
2		Length	Width	Height	Area	Volume	Ratio (II : I)	Ratio of	Ratio of
3	Rectangular Prism I	6	4	23	508	552	2	Surface	Volumes
4								Areas (II : I)	(II : I)
5	Similar Prism II	12	8	46	2032	4416		4	8

In cell E3 enter the formula =2*(B3*C3+B3*D3+C3*D3).
This will calculate the sum of the areas of the six faces of Prism I.
In cell F3 enter the formula =B3*C3*D3. This will calculate the volume of Prism I.

In cells B5, C5, and D5 enter the formulas =G3*B3, =G3*C3, and =G3*D3, respectively. These will calculate the dimensions of similar Prism II. Copy the formulas from E3 and F3 into E5 and F5 to calculate the surface area and volume of Prism II.

In cell H5 enter the formula =E5/E3 and in cell I5 enter the formula =F5/F3. These will calculate the ratios of the surface areas and volumes.

Investigate

In row 3, enter numbers for the length, width, height, and similarity ratio. Change those numbers to investigate how the ratio of the surface areas and the ratio of the volumes are each related to the similarity ratio.

EXERCISES

Make a Conjecture **State a relationship that seems to be true about the similarity ratio and each given ratio.**

1. the ratio of volumes

2. the ratio of surface areas

Extend

Set up spreadsheets that allow you to investigate the following ratios. State a conclusion from each investigation.

3. the volumes of similar cylinders

4. the lateral areas of similar cylinders

5. the surface areas of similar cylinders

6. the volumes of similar square pyramids

7. the lateral areas of similar square pyramids

8. the surface areas of similar square pyramids

Areas and Volumes of Similar Solids

California Content Standards

GEOM 11.0 Students determine how changes in dimensions affect the area and volume of solids. ***Develop, Master***

What You'll Learn

- To find relationships between the ratios of the areas and volumes of similar solids

. . . And Why

To use similarity ratios to find the weight of an object, as in Example 4

Check Skills You'll Need

Are the figures similar? Explain. Include the similarity ratio as appropriate.

1. two squares, one with 3-in. sides and the other with 1-in. sides
2. two isosceles right triangles, one with a 3-cm hypotenuse and the other with a 1-cm leg

Find the volume of each space figure.

3. a cube with a 3-in. edge
4. a 3 m-by-5 m-by-9 m rectangular prism
5. a cylinder with radius 4 cm and height 8 cm

New Vocabulary • similar solids • similarity ratio

1 Finding Relationships in Area and Volume

Similar solids have the same shape, and all their corresponding dimensions are proportional. The ratio of corresponding linear dimensions of two similar solids is the **similarity ratio.** Any two cubes are similar, as are any two spheres.

Real-World Connection

These Russian nesting dolls suggest similar solids.

1 EXAMPLE Identifying Similar Solids

Are the two rectangular prisms similar? If so, give the similarity ratio.

a.

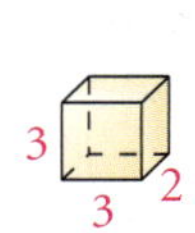

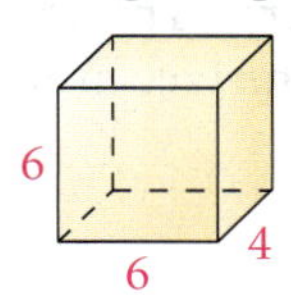

$\frac{3}{6} = \frac{3}{6} = \frac{2}{4}$

The rectangular prisms are similar because the ratios of the corresponding linear dimensions are equal.

The similarity ratio is $\frac{1}{2}$.

b.

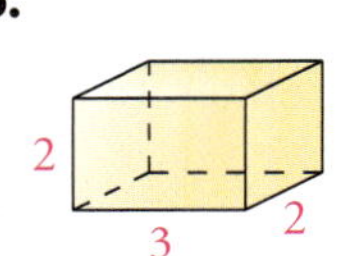

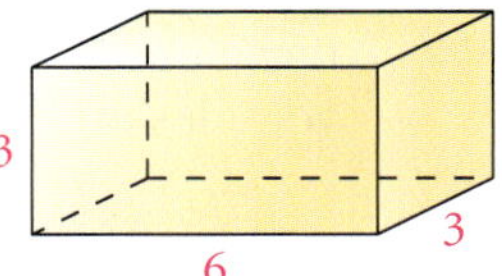

$\frac{2}{3} \neq \frac{3}{6}$

The rectangular prisms are not similar because the ratios of corresponding linear dimensions are not equal.

1 Are the two cylinders similar? If so, give the similarity ratio.

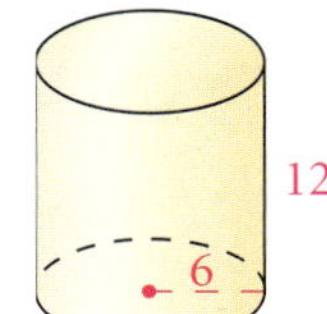

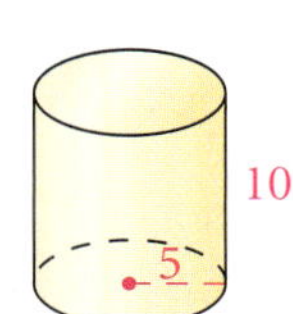

The two similar prisms shown here suggest two important relationships for similar solids.

The ratio of the side lengths is 1 : 2.
The ratio of the surface areas is 22 : 88, or 1 : 4.
The ratio of the volumes is 6 : 48, or 1 : 8.

The ratio of the surface areas equals the square of the similarity ratio. The ratio of the volumes equals the cube of the similarity ratio. These two facts apply to all similar solids.

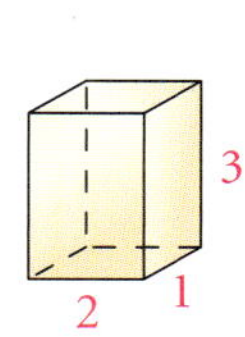

S.A. = 22 m^2
V = 6 m^3

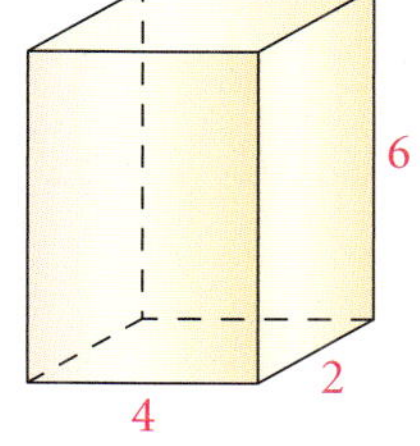

S.A. = 88 m^2
V = 48 m^3

Key Concepts

Theorem 11-12 **Areas and Volumes of Similar Solids**

If the similarity ratio of two similar solids is $a : b$, then
(1) the ratio of their corresponding areas is $a^2 : b^2$, and
(2) the ratio of their volumes is $a^3 : b^3$.

2 EXAMPLE Finding the Similarity Ratio

Find the similarity ratio of two cubes with volumes of 729 cm^3 and 1331 cm^3.

$\frac{a^3}{b^3} = \frac{729}{1331}$ **The ratio of the volumes is $a^3 : b^3$.**

$\frac{a}{b} = \frac{9}{11}$ **Take cube roots.**

The similarity ratio is 9 : 11.

CA Standards Check

2 Find the similarity ratio of two similar prisms with surface areas 144 m^2 and 324 m^2.

3 EXAMPLE Using a Similarity Ratio

Paint Cans The lateral areas of two similar paint cans are 1019 cm^2 and 425 cm^2. The volume of the small can is 1157 cm^3. Find the volume of the large can.

First find the similarity ratio $a : b$.

$\frac{a^2}{b^2} = \frac{1019}{425}$ **The ratio of the surface areas is $a^2 : b^2$.**

$\frac{a}{b} = \frac{\sqrt{1019}}{\sqrt{425}}$ **Take square roots.**

Vocabulary Tip

The subscripts "large" and "small" let you use the letter V for the volumes of both the large can (V_{large}) and small can (V_{small}).

Use the similarity ratio to find the volume.

$\frac{V_{\text{large}}}{V_{\text{small}}} = \frac{\sqrt{1019}^3}{\sqrt{425}^3}$ **The ratio of the volumes is $a^3 : b^3$.**

$\frac{V_{\text{large}}}{1157} = \frac{\sqrt{1019}^3}{\sqrt{425}^3}$ **Substitute 1157 for V_{small}.**

$V_{\text{large}} = 1157 \cdot \frac{\sqrt{1019}^3}{\sqrt{425}^3}$ **Solve for V_{large}.**

$V_{\text{large}} \approx 4295$ **Use a calculator.**

The volume of the large paint can is about 4295 cm^3.

CA Standards Check

3 The volumes of two similar solids are 128 m^3 and 250 m^3. The surface area of the larger solid is 250 m^2. What is the surface area of the smaller solid?

The weights of solid objects made of the same material are proportional to their volumes.

4 EXAMPLE Real-World Connection

Paperweights A marble paperweight shaped like a pyramid weighs 0.15 lb. How much does a similarly shaped marble paperweight weigh if each dimension is three times as large?

The similarity ratio is 1 : 3. The ratio of the volumes, and hence the ratio of the weights, is $1^3 : 3^3$, or 1 : 27.

$\frac{1}{27} = \frac{0.15}{x}$ **Let x = the weight of the larger paperweight.**

$x = 27(0.15)$ **Use the Cross Product Property.**

$x = 4.05$

The larger paperweight weighs about 4 lb.

4 Find the weight of a marble bead that is similar to the paperweight in Example 4 but has dimensions half as large.

For more exercises, see *Extra Skill, Word Problem, and Proof Practice.*

GEOM 11.0

Example 1 (page 646)

Are the two figures similar? If so, give the similarity ratio.

1.

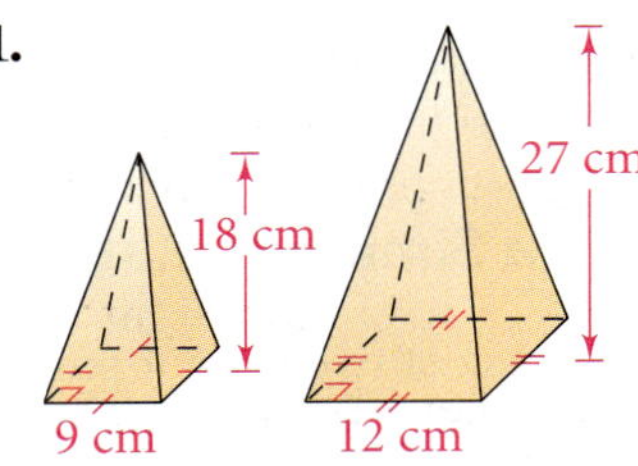

2.

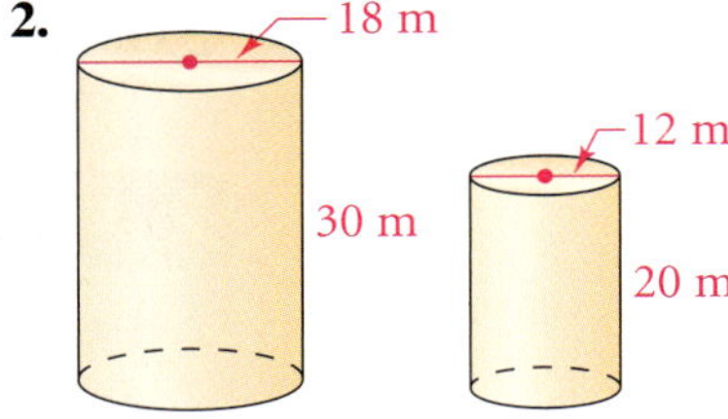

3.

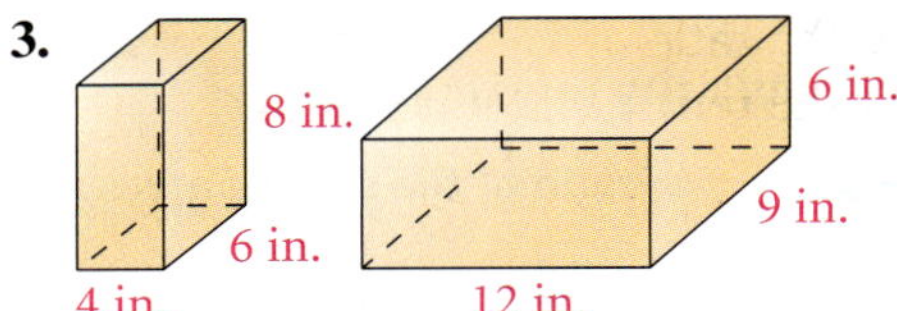

4.

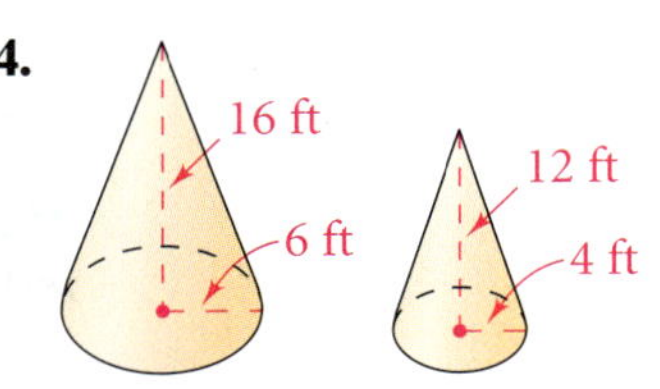

5. two cubes, one with 3-cm edges, the other with 4.5-cm edges

6. a cylinder and a square prism each with 3-in. radii and 1-in. heights

Example 2 (page 647)

Each pair of figures is similar. Use the given information to find the similarity ratio of the smaller figure to the larger figure.

7.

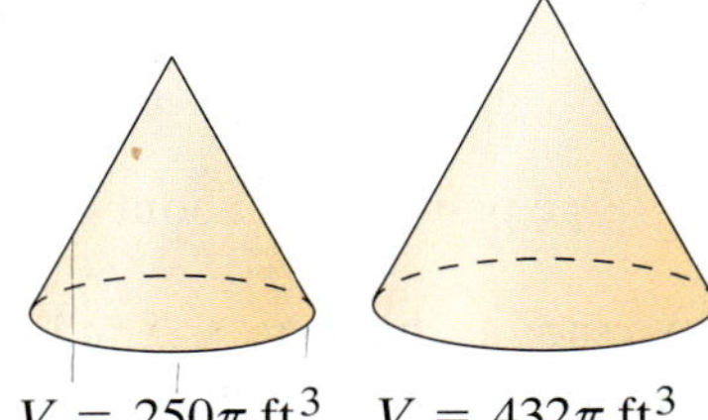

$V = 250\pi$ ft^3 $V = 432\pi$ ft^3

8.

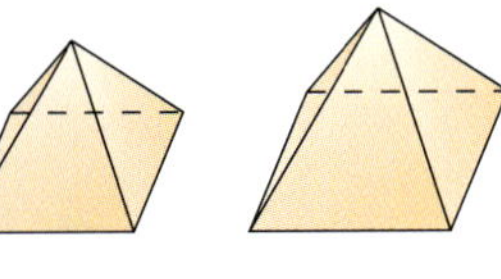

$V = 216$ in.^{3} $V = 343$ in.^{3}

9.

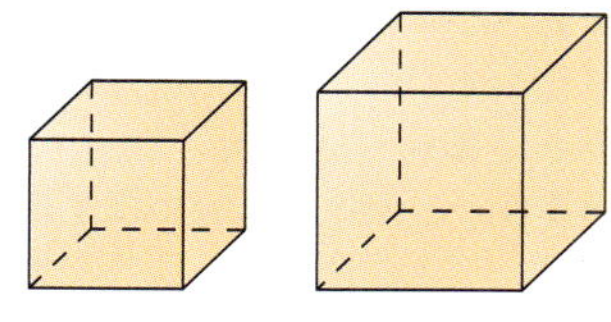

S.A. = 18 m^2 S.A. = 32 m^2

10.

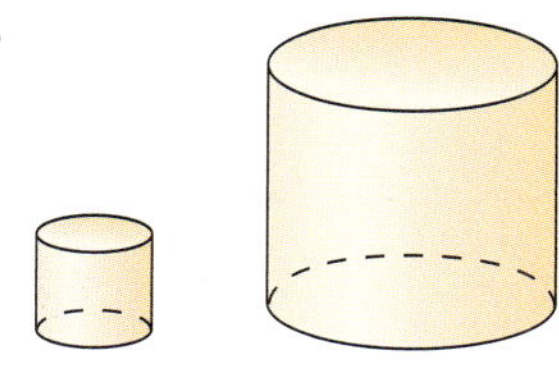

S.A. = 20π yd^2 S.A. = 125π yd^2

Example 3 (page 647)

The surface areas of two similar figures are given. The volume of the larger figure is given. Find the volume of the smaller figure.

11. S.A. = 248 in.^{2}
S.A. = 558 in.^{2}
V = 810 in.^{3}

12. S.A. = 192 m^2
S.A. = 1728 m^2
V = 4860 m^3

13. S.A. = 52 ft^2
S.A. = 208 ft^2
V = 192 ft^3

The volumes of two similar figures are given. The surface area of the smaller figure is given. Find the surface area of the larger figure.

14. V = 27 in.^{3}
V = 125 in.^{3}
S.A. = 63 in.^{2}

15. V = 27 m^3
V = 64 m^3
S.A. = 63 m^2

16. V = 2 yd^3
V = 250 yd^3
S.A. = 13 yd^2

Example 4 (page 648)

17. **Packaging** There are 750 toothpicks in a regular-sized box. If a jumbo box is made by doubling all the dimensions of the regular-sized box, how many toothpicks will the jumbo box hold?

18. **Packaging** A cylinder 4 in. in diameter and 6 in. high holds 1 lb of oatmeal. To the nearest ounce, how much oatmeal will a similar 10-in.-high cylinder hold? (*Hint:* 1 lb = 16 oz)

19. A regular pentagonal solid prism has 9-cm base edges. A larger, similar solid prism of the same material has 36-cm base edges. How does each indicated measurement for the larger prism compare to the same measurement for the smaller prism?
a. the volume **b.** the weight

B Apply Your Skills

20. Two similar prisms have heights 4 cm and 10 cm.
a. What is their similarity ratio?
b. What is the ratio of their surface areas?
c. What is the ratio of their volumes?

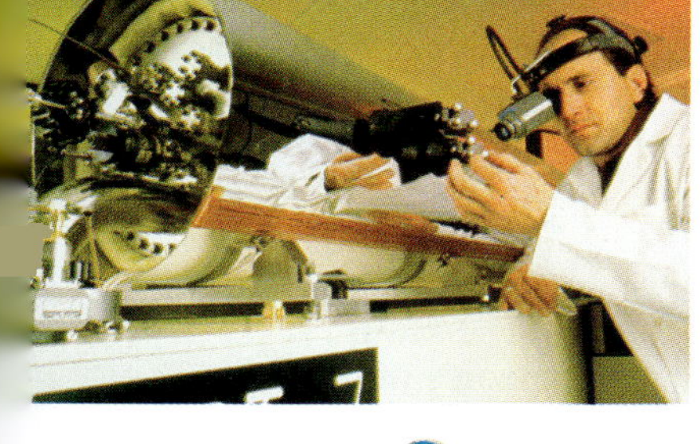

Real-World Connection
The time scale of an atomic clock is based on vibrations of atoms and molecules.

21. **Atomic Clock** A company announced that it had developed the technology to reduce the size of its atomic clock, which is used in electronic devices that transmit data. The company claims that the smaller clock will be similar to the existing clock made of the same material. It will be $\frac{1}{10}$ the size of its existing atomic clocks and $\frac{1}{100}$ the weight. Do these ratios make sense? Explain.

22. Is there a value of x for which the rectangular prisms below are similar? Explain.

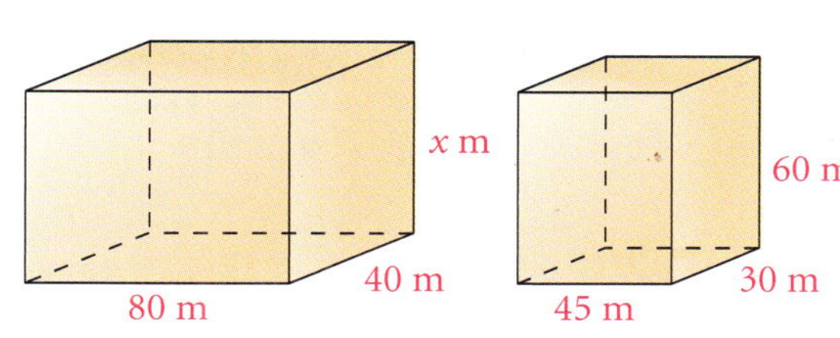

23. The volume of a spherical balloon with radius 3.1 cm is about 125 cm^3. Estimate the volume of a similar balloon with radius 6.2 cm.

Visit: PHSchool.com
Web Code: bce-1107

24. **Critical Thinking** A carpenter is making a blanket chest based on an antique chest. Both chests have the shape of a rectangular prism. The length, width, and height of the new chest will all be 4 in. greater than the respective dimensions of the antique. Will the chests be similar? Explain.

25. **Writing** Explain why all spheres are similar.

26. Two similar pyramids have lateral area 20 ft^2 and 45 ft^2. The volume of the smaller pyramid is 8 ft^3. Find the volume of the larger pyramid.

27. The volumes of two spheres are 729 $in.^3$ and 27 $in.^3$.
   a. Find the ratio of their radii.
   b. Find the ratio of their surface areas.

28. The volumes of two similar pyramids are 1331 cm^3 and 2744 cm^3.
   a. Find the ratio of their heights.
   b. Find the ratio of their surface areas.

29. A clown's face on a balloon is 4 in. high when the balloon holds 108 $in.^3$ of air. How much air must the balloon hold for the face to be 8 in. high?

Copy and complete the table for the similar solids.

	Similarity Ratio	Ratio of Surface Areas	Ratio of Volumes
30.	1 : 2	■ : ■	■ : ■
31.	3 : 5	■ : ■	■ : ■
32.	■ : ■	49 : 81	■ : ■
33.	■ : ■	■ : ■	125 : 512

34. **Literature** In *Gulliver's Travels* by Jonathan Swift, Gulliver first traveled to Lilliput. The Lilliputian average height was one twelfth of Gulliver's height.
   a. How many Lilliputian coats could be made from the material in Gulliver's coat? (*Hint:* Use the ratio of surface areas.)
   b. How many Lilliputian meals would be needed to make a meal for Gulliver? (*Hint:* Use the ratio of volumes.)

35. **Indirect Reasoning** Some stories say that Paul Bunyan was ten times as tall as the average human. Assume that Paul Bunyan's bone structure was proportional to that of ordinary people.
   a. Strength of bones is proportional to the area of their cross section. How many times as strong as the average person's bones would Paul Bunyan's bones be?
   b. Weights of objects made of like material are proportional to their volumes. How many times the average person's weight would Paul Bunyan's weight be?
   c. Human leg bones can support about 6 times the average person's weight. Use your answers to parts (a) and (b) to explain why Paul Bunyan could not exist with a bone structure that was proportional to that of ordinary people.

36. Square pyramids A and B are similar. In pyramid A, each base edge is 12 cm. In pyramid B, each base edge is 3 cm and the volume is 6 cm^3.
   a. Find the volume of pyramid A.
   b. Find the ratio of the surface area of A to the surface area of B.
   c. Find the surface area of each pyramid.

37. The cone is cut by a plane parallel to its base. The small cone on top is similar to the large cone. The ratio of the slant heights of the cones is 1 : 2. Find the ratio indicated.

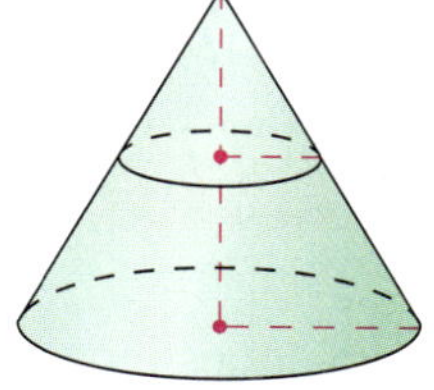

a. the surface area of the large cone to that of the small cone; the volume of the large cone to that of the small cone

b. the surface area of the frustum to that of the large cone; to that of the small cone

c. the volume of the frustum to that of the large cone; to that of the small cone

A frustum of a cone is defined on page 636.

Multiple Choice Practice

For California Standards and CAHSEE Tutorials, visit PHSchool.com. Web Code: bcq-9045

GEOM 11.0

38. One cylinder has height 4 cm. A larger cylinder has height 10 cm. The diameter of the larger cylinder is 2.5 times the diameter of the smaller cylinder. How many times the volume of the smaller cylinder is the volume of the larger cylinder?

(A) 2.5 (B) 4 (C) 6.25 (D) 15.625

GEOM 8.0

39. The rectangular prism has dimensions 5 cm by 7.5 cm by 20 cm. What is the surface area in square centimeters?

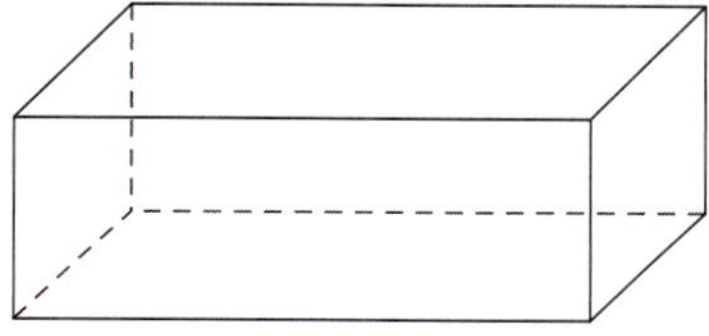

(A) 750 (C) 287.5

(B) 575 (D) 65

CAHSEE 6SDAP 3.3

40. Steven put 4 black marbles, 7 blue marbles, and 10 white marbles into a bag. Violet selects a marble at random, does not replace it, and selects again. What is the probability that both marbles are blue?

(A) 0.7 (B) 0.6 (C) 0.1 (D) 0.095

Mixed Review

Lesson 11-6

41. **Sports Equipment** The circumference of a regulation basketball is between 75 cm and 78 cm. What are the smallest and the largest surface areas that a basketball can have? Give your answers to the nearest whole unit.

Find the volume and surface area of each sphere to the nearest tenth.

42. diameter = 6 in. **43.** circumference = 2.5π m **44.** radius = 6 in.

Lesson 7-4

45. The altitude to the hypotenuse of a right triangle ABC divides the hypotenuse into 12-mm and 16-mm segments. Find the length of each of the following.

a. the altitude to the hypotenuse

b. the shorter leg of $\triangle ABC$

c. the longer leg of $\triangle ABC$

Lesson 7-1 x^2 Algebra **Solve each proportion.**

46. $\frac{25}{16} = \frac{x}{16}$ **47.** $\frac{21}{x} = \frac{8}{5}$ **48.** $\frac{3}{8} = \frac{n}{n + 4}$

GO Online Lesson Quiz Visit: PHSchool.com, Web Code: bca-1107

Test-Taking Strategies

Measuring to Solve

Some questions ask you to measure with a ruler before solving. You will be asked to measure in either centimeters or inches.

EXAMPLE

The net of a square pyramid is shown at the right. Use a ruler to measure the dimensions to the nearest tenth of a centimeter. What is the surface area of the pyramid to the nearest tenth of a square centimeter?

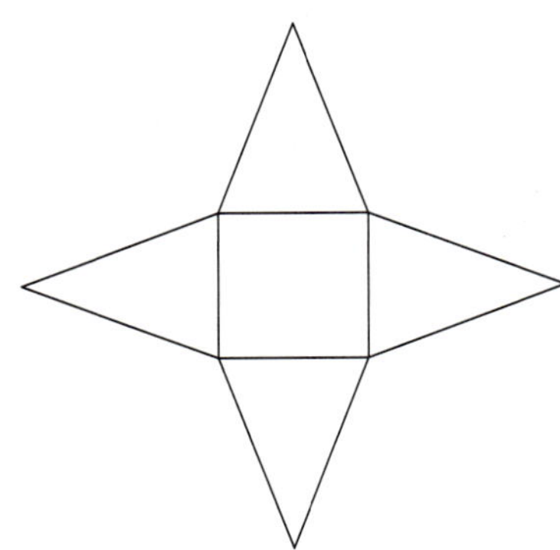

Ⓐ 1.7 cm^2 Ⓑ 4.4 cm^2 Ⓒ 5.2 cm^2 Ⓓ 6.1 cm^2

Measuring with a centimeter ruler:
Each side of the square base is 1.3 centimeters.
The height of each triangular face is 1.7 centimeters.

Surface Area = Lateral Area + Base Area

$$\begin{aligned} S.A. &= L.A. + B \\ &= \tfrac{1}{2}(4)(1.3)(1.7) + 1.3^2 \\ &\approx 6.1\ \text{cm}^2, \text{ so D is the correct answer.} \end{aligned}$$

EXERCISES

1. The net of a triangular prism is shown at the right. Use a ruler to measure the dimensions to the nearest $\frac{1}{8}$ inch. Which value is closest to the volume of the prism?

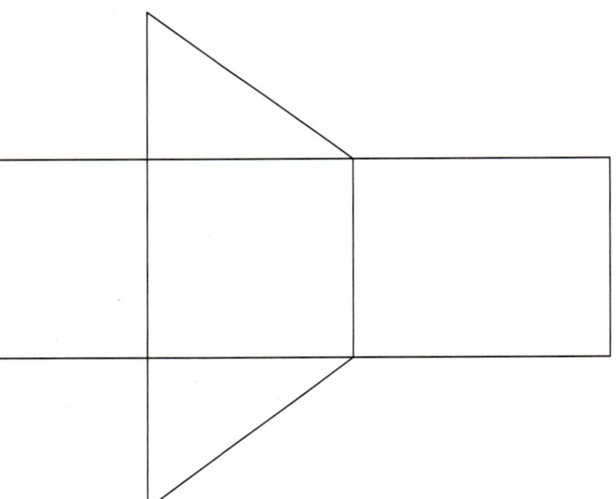

Ⓐ $\frac{3}{8}$ in.^{3} Ⓒ 1 in.^{3}
Ⓑ $\frac{1}{2}$ in.^{3} Ⓓ $1\frac{3}{8}$ in.^{3}

2. The net of a cube is shown below.

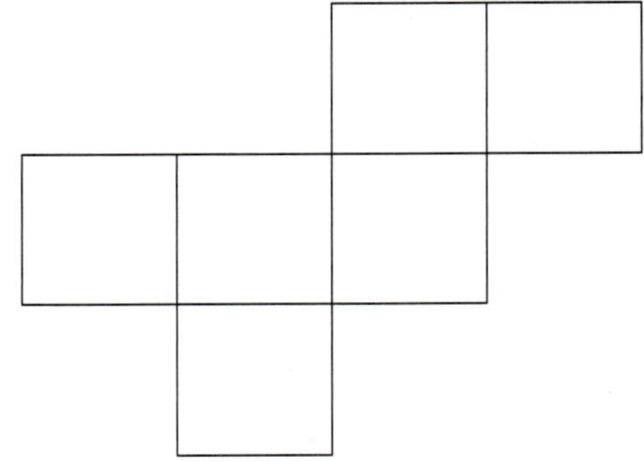

Which value is closest to the surface area of the cube?

Ⓐ 1.5 cm^2 Ⓒ 13.5 cm^2
Ⓑ 2.3 cm^2 Ⓓ 24 cm^2

Chapter 11 Review

Vocabulary Review

English and Spanish Audio Online

altitude (pp. 608, 610, 617, 619)
base(s) (pp. 608, 610, 617, 619)
center of a sphere (p. 638)
circumference of a sphere (p. 638)
composite space figure (p. 627)
cone (p. 619)
cross section (p. 600)
cylinder (p. 610)
diameter of a sphere (p. 638)
edge (p. 598)
face (p. 598)
great circle (p. 638)
height (pp. 608, 610, 617, 619)
hemisphere (p. 638)
lateral area (pp. 608, 610, 617, 619)
lateral faces (pp. 608, 617)
oblique cylinder (p. 610)
oblique prism (p. 608)
polyhedron (p. 598)
prism (p. 608)
pyramid (p. 617)
radius of a sphere (p. 638)
regular pyramid (p. 617)
right cone (p. 619)
right cylinder (p. 610)
right prism (p. 608)
similar solids (p. 646)
similarity ratio (p. 646)
slant height (p. 617, 619)
sphere (p. 638)
surface area (pp. 608, 610, 618, 619)
vertex (pp. 598, 617, 619)
volume (p. 624)

Choose the correct term to complete each sentence.

1. A set of points in space equidistant from a given point is called a (*circle, sphere*).
2. The (*height, altitude*) of a cylinder is a segment that joins the bases and is perpendicular to each base.
3. A (*pyramid, prism*) is a polyhedron in which one face can be any polygon and the lateral faces are triangles that meet at a common vertex.
4. If you slice a prism with a plane, the intersection of the prism and the plane is a (*lateral area, cross section*) of the prism.
5. In a(n) (*right, oblique*) prism, the lateral faces are rectangles and a lateral edge is an altitude.

For: Vocabulary quiz
Web Code: bcj-1151

Skills and Concepts

Lesson 11-1

- To recognize polyhedra and their parts (p. 598)
- To visualize cross sections of space figures (p. 600)

A **polyhedron** is a three-dimensional figure whose surfaces are polygons. The polygons are **faces** of the polyhedron. An **edge** is a segment that is the intersection of two faces. A **vertex** is a point where three or more edges intersect. A **net** is a two-dimensional pattern that folds to form a three-dimensional figure.

Draw a net for each three-dimensional figure.

6. 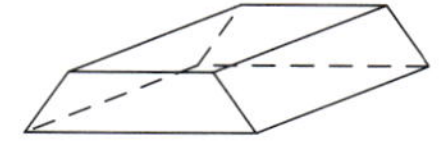

7. 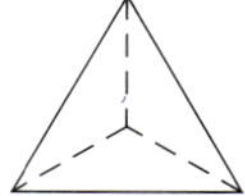

8. 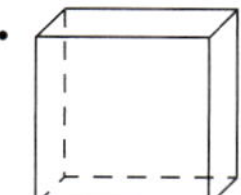

The number of faces (F), vertices (V), and edges (E) of a polyhedron are related by Euler's Formula $F + V = E + 2$.

Use Euler's Formula to find the missing number.

9. $F = 5, V = 5, E = \blacksquare$

10. $F = 6, V = \blacksquare, E = 12$

A **cross section** is the intersection of a solid and a plane.

11. Sketch a cube with an equilateral triangle cross section.

Lessons 11-2 and 11-4

- To find the surface area of a prism (p. 608)
- To find the surface area of a cylinder (p. 610)
- To find the volume of a prism (p. 624)
- To find the volume of a cylinder (p. 626)

GEOM 8.0, 9.0

The **lateral area** of a **right prism** is the product of the perimeter of the base and the height. The **lateral area** of a **right cylinder** is the product of the circumference of the base and the height of the cylinder. The **surface area** of each solid is the sum of the lateral area and the areas of the bases.

The **volume** of a space figure is the space that the figure occupies. Volume is measured in cubic units. The **volume** of a **prism** and the **volume** of a **cylinder** are the product of the area of a base and the height of the solid.

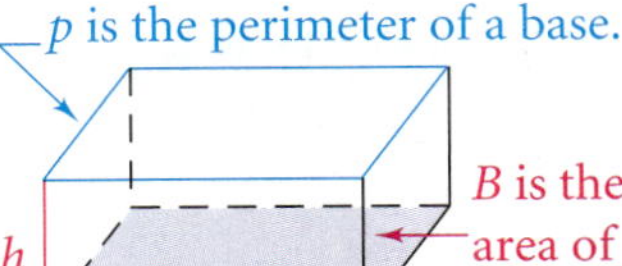

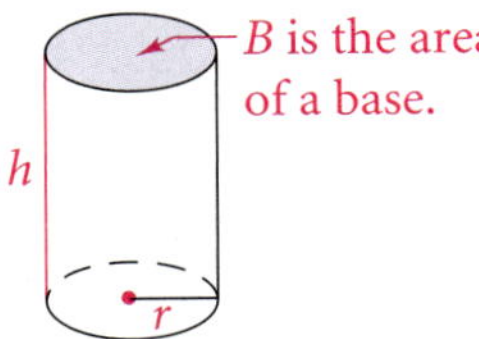

L.A. (prism) $= ph$

S.A. $=$ L.A. $+ 2B$

$V = Bh$

L.A. (cylinder) $= 2\pi rh$ or πdh

S.A. $=$ L.A. $+ 2B$

$V = Bh$

Find the surface area and volume of each figure. Leave your answers in terms of π.

12.

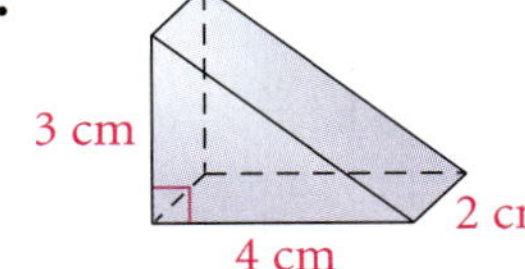

13.

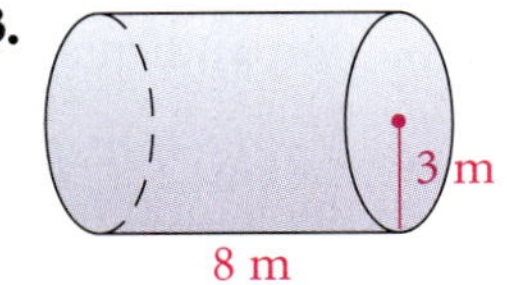

14.

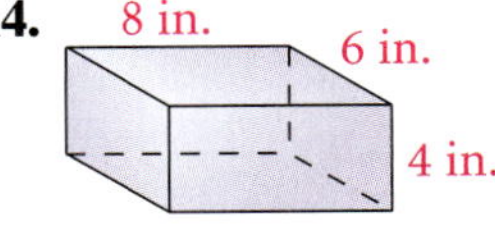

Lessons 11-3 and 11-5

- To find the surface area of a pyramid (p. 617)
- To find the surface area of a cone (p. 619)
- To find the volume of a pyramid (p. 631)
- To find the volume of a cone (p. 633)

GEOM 8.0, 9.0

The **lateral area** of a **regular pyramid** is half the product of the perimeter of the base and the slant height.

The **surface area** of a pyramid is the sum of the lateral area and the area of the base.

The **volume** of a pyramid is one third the product of the area of the base and the height of the solid.

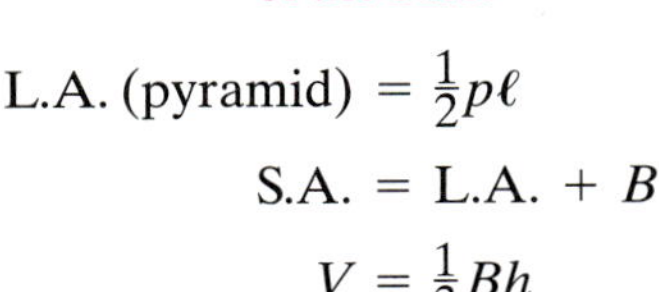

L.A. (pyramid) $= \frac{1}{2}p\ell$

S.A. $=$ L.A. $+ B$

$V = \frac{1}{3}Bh$

The **lateral area** of a **right cone** is half the product of the circumference of the base and the slant height.

The **surface area** of a cone is the sum of the lateral area and the area of the base.

The **volume** of a cone is one third the product of the area of the base and the height of the solid.

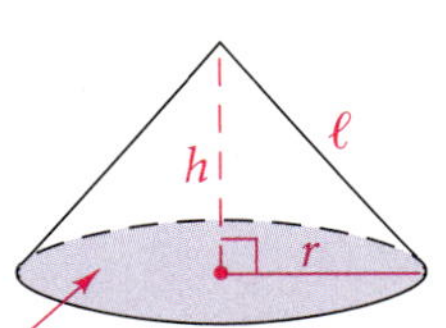

L.A. (cone) $= \pi r\ell$

S.A. $=$ L.A. $+ B$

$V = \frac{1}{3}Bh$

Find the surface area and volume of each figure. Round to the nearest tenth.

15.
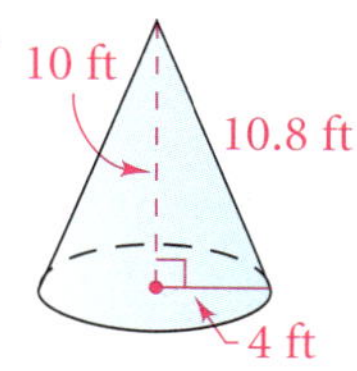

16.
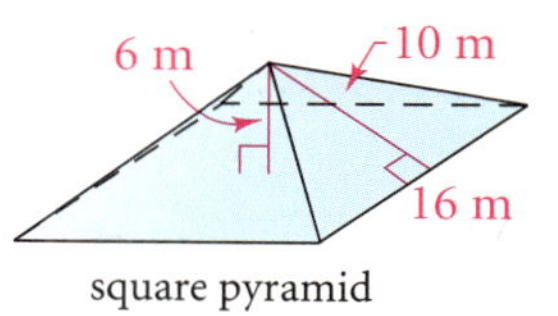

17.
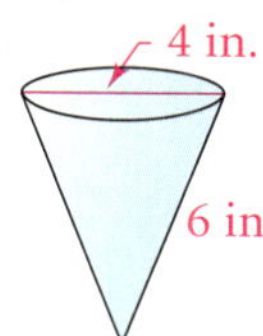

18.
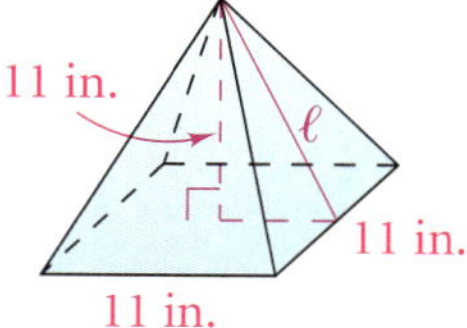

19.
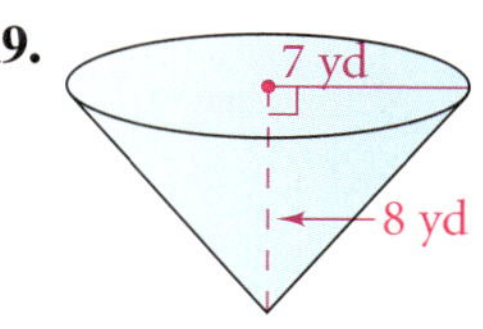

20.
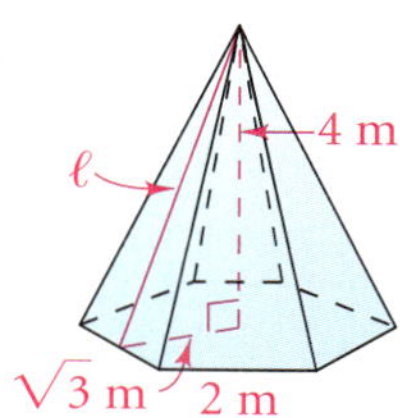

Find each formula as stated.

21. the base area of a prism in terms of the surface area and lateral area

22. the radius of a cylinder in terms of the surface area and height (*Hint:* Use the quadratic formula.)

Lesson 11-6

- To find the surface area and volume of a sphere (p. 638)

 **GEOM 8.0, 9.0**

The **surface area of a sphere** is four times the product of π and the square of the radius of the sphere. The **volume of a sphere** is $\frac{4}{3}$ the product of π and the cube of the radius of the sphere.

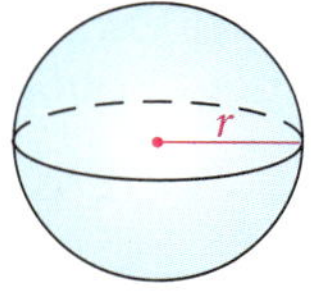

Find the surface area and volume of a sphere with the given radius or diameter. Round answers to the nearest tenth.

23. $r = 5$ in. **24.** $d = 7$ cm **25.** $d = 4$ ft **26.** $r = 0.8$ ft

27. Sports Equipment The circumference of a lacrosse ball is 8 in. Find its volume to the nearest tenth of a cubic inch.

Lesson 11-7

- To find relationships between the ratios of the areas and volumes of similar solids (p. 646)

 **GEOM 11.0**

Similar solids have the same shape and all their corresponding dimensions are proportional.

If the **similarity ratio** of two similar solids is $a : b$, then the ratio of their corresponding surface areas is $a^2 : b^2$, and the ratio of their volumes is $a^3 : b^3$.

28. Open-Ended Sketch two similar solids whose surface areas are in the ratio $16 : 25$. Include dimensions.

For each pair of similar solids, find the ratio of the volume of the first figure to the volume of the second.

29.
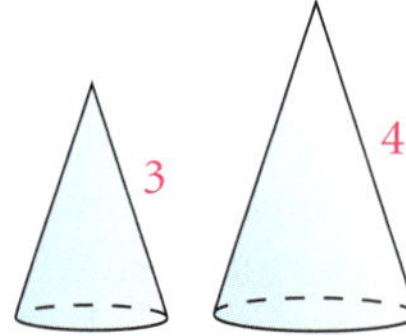

30.
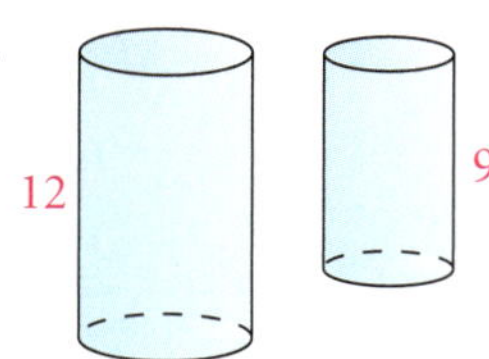

Chapter 11 Test

For: Chapter Test
Web Code: bca-1152

Draw a net for each figure. Label the net with appropriate dimensions.

1. 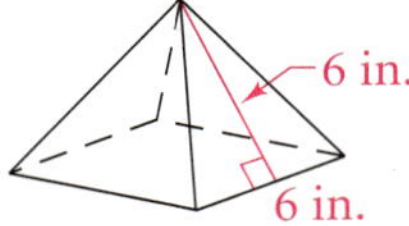

2. 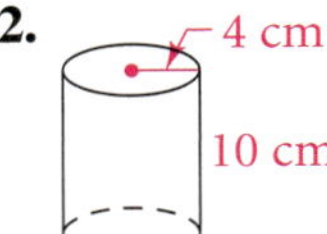

Use the polyhedron at the right for Exercises 3 and 4.

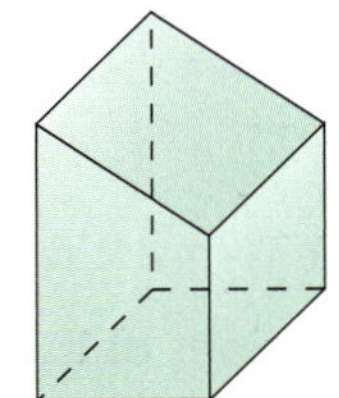

3. Verify Euler's Formula for the polyhedron.
4. Draw a net for the polyhedron. Verify $F + V = E + 1$ for the net.

5. Find the number of edges in a pyramid with seven faces.

Describe the cross section formed in each diagram.

6. 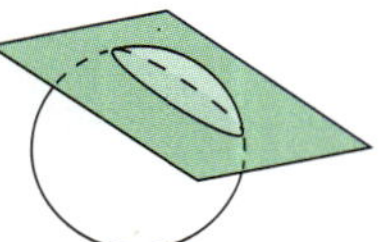

7. 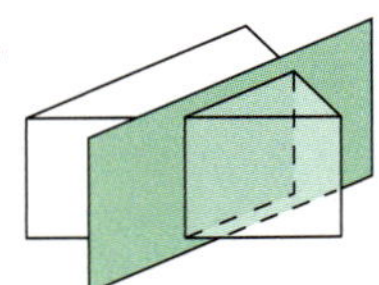

Aviation The "black box" data recorders on commercial airliners are rectangular prisms.

8. The base of a recorder is 15 in. by 8 in. Its height ranges from 15 in. to 22 in. What are the largest and smallest possible volumes for the recorder?

9. New flight data recorders are smaller and record more data. A new recorder might be 8 in. by 8 in. by 13 in. What is its volume?

Find the volume and surface area of each figure to the nearest tenth.

10. 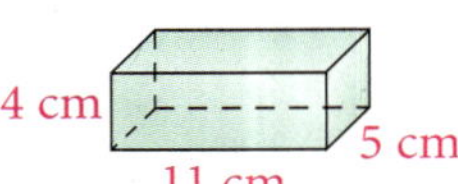

11. 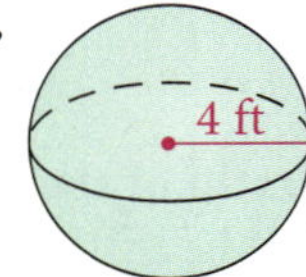

12. 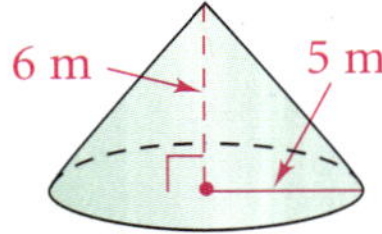

13. 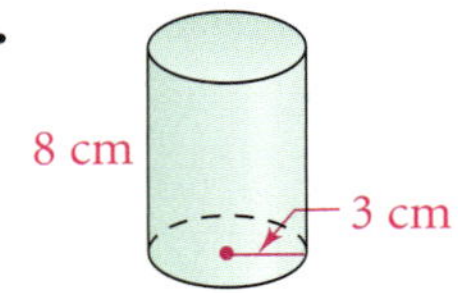

14. 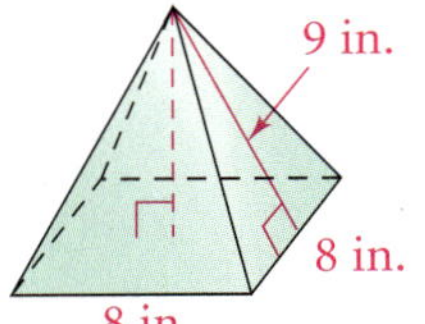

15. 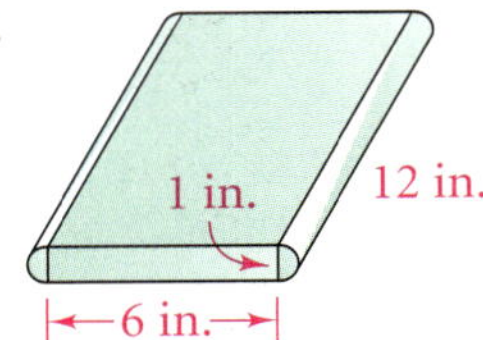

16. **Open-Ended** Draw two different space figures that have a volume of 100 in.^{3}. Label the dimensions of each figure.

17. **Visualization** The triangle is revolved completely about the y-axis.

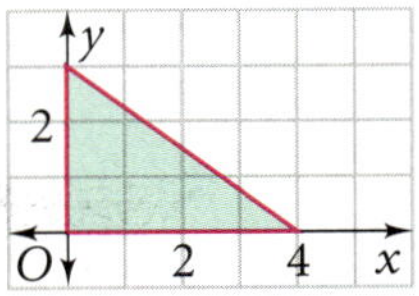

    a. Describe the solid of revolution that is formed.

    b. Find its lateral area and volume in terms of π.

18. **Painting** The floor of a bedroom is 12 ft by 15 ft and the walls are 7 ft high. One gallon of paint covers about 450 ft^2. How many gallons of paint do you need to paint the walls of the bedroom?

19. List these space figures in order from the one with least volume to the one with greatest volume.

    A. cube with an edge of 5 cm

    B. cylinder with radius 4 cm and height 4 cm

    C. square pyramid with base sides of 6 cm and height 6 cm

    D. cone with radius 4 cm and height 9 cm

    E. rectangular prism with a 5 cm-by-5 cm base and height 6 cm

20. **Writing** Describe a real-world situation in which you would need to know the volume of an object. Then describe another situation in which you would need to know the lateral area of an object.

21. The two solids are similar. Find the ratio of the volume of the first figure to the volume of the second.

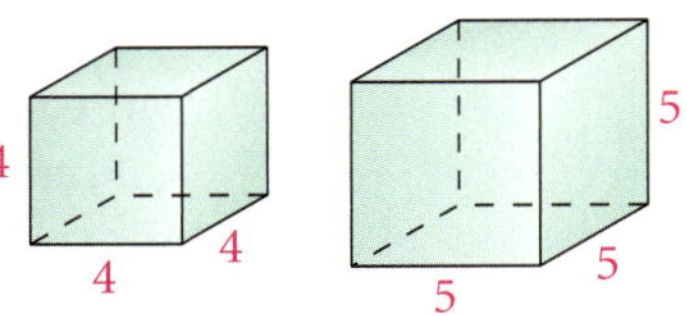

22. The volumes of two spheres are 327π mm^3 and 8829π mm^3. What is the ratio of their surface areas?

Standards Mastery

CAHSEE Practice

Some questions ask you to solve percent problems. Read the question at the right. Then follow the tips to answer it.

Miranda and five friends celebrate her birthday at a local restaurant. The total cost of the six dinners is \$112.85. The restaurant adds a 15% service charge to the cost of the dinners, and the sales tax on the dinners is 5%. About how much change will Miranda receive if she pays with three \$50 bills?

(A) Less than \$10

(B) More than \$10 but less than \$20

(C) More than \$20 but less than \$30

(D) More than \$30

Tip 1
Both percents apply to the same amount—the cost of the dinner. You can use 15 + 5 = 20%.

Tip 2
Round the cost of the dinner to \$110 for easy calculating and to \$113 for a final estimate.

Think It Through

The dinner costs about \$110.
20% of \$110 is \$22.

$$113 + 22 = 135$$

$$150 - 135 = 15$$

Since \$15 is more than \$10 but less than \$20, the correct answer is B.

Vocabulary Review

As you solve problems, you must understand the meanings of mathematical terms. Match each term with its mathematical meaning.

A. reduction

B. cubic units

C. line plot

D. similar

E. box-and-whisker plot

I. describes figures that have congruent corresponding angles and proportional corresponding sides

II. a dilation with a scale factor between 0 and 1

III. a graph along a number line that shows frequencies

IV. used to measure volume

V. a statistical graph showing median and quartiles

Multiple Choice

Read each question. Then write the letter of the correct answer on your paper.

1. Which dimensions represent a reduction of a rectangle that is 8 units by 10 units?

(A) 25 units by 35 units
(B) 14 units by 16 units
(C) 6 units by 7.5 units
(D) 5 units by 7 units

2. If $x = -5$, then $-x =$

(A) 5.
(B) -5.
(C) $\frac{1}{5}$.
(D) -1.

3. What is the reciprocal of $-\frac{4x^2}{3y}$?

(A) $\frac{4x^2}{3y}$
(B) $\frac{3y}{4x^2}$
(C) $-\frac{3y}{4x^2}$
(D) $4x^2 3y$

4. The volume of a cube is 729 cubic centimeters. A smaller cube has sides that are $\frac{1}{3}$ as long as the sides of the larger cube. What is the volume of the smaller cube?

(A) 243 cm^3
(B) 81 cm^3
(C) 27 cm^3
(D) 9 cm^3

CAHSEE Practice (continued)

5. If x is an integer, what is the solution to $|x - 5| = 9$?

(A) $\{-14, 14\}$
(B) $\{14\}$
(C) $\{-4, 14\}$
(D) $\{4, 14\}$

6. If z is an integer, what is the solution to $|z + 1| < 3$?

(A) $\{-4, 2\}$
(B) $\{0, 1, 2\}$
(C) $\{-4, -3, -2, -1, 0, 1, 2\}$
(D) $\{-3, -2, -1, 0, 1\}$

7. If k is an integer, which of the following is the solution set for $4|k| = 28$?

(A) $\{7\}$
(B) $\{-7, 7\}$
(C) $\{112\}$
(D) $\{-7, 0, 7\}$

8. If y is an integer, what is the solution set of $|y - 5| < -7$?

(A) $\{-2, 12\}$
(B) $\{-4, -3, -2, 0, 1, 2, 3, 4, 5, 6\}$
(C) $\varnothing$
(D) $\{0\}$

9. Which of the following could be the side lengths of a right triangle?

(A) 4.1, 6.2, 7.3
(B) 40, 60, 72
(C) 3.2, 5.4, 6.2
(D) 33, 56, 65

10. The stairway pictured below is made of solid concrete. The height and width of each step is 8 inches. The length is 24 inches. What is the volume of the concrete used to create this stairway?

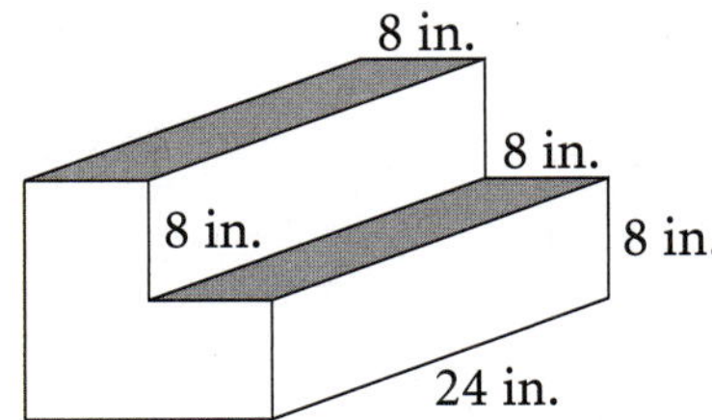

(A) 4608 in.^{3}
(B) 4096 in.^{3}
(C) 3072 in.^{3}
(D) 384 in.^{3}

11. Ms. Kostant made a box-and-whisker plot to show the results of a math test she gave her class. According to this figure,

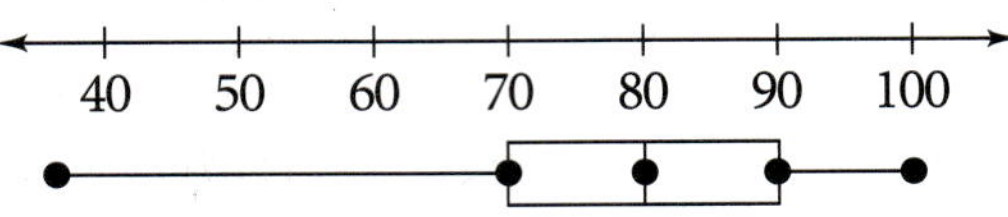

(A) only one student failed the test.
(B) no student scored a 50.
(C) more than half of the students earned a 70 or higher.
(D) only five students took the test.

12. The line plot shows the number of different states in which each of Natalie's classmates has lived. Which one of the following statements is true based on the information given in the plot?

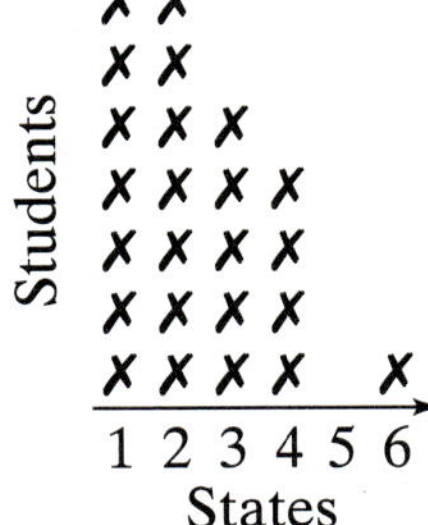

(A) Three students have lived in five different states.
(B) All of the students have lived in more than one state.
(C) Natalie's classmates have liked living in different states.
(D) Only one student has lived in six states.

13. The circle graph at the right shows how a park district budgets its money. Which one of the following statements is ***not*** true based on this graph?

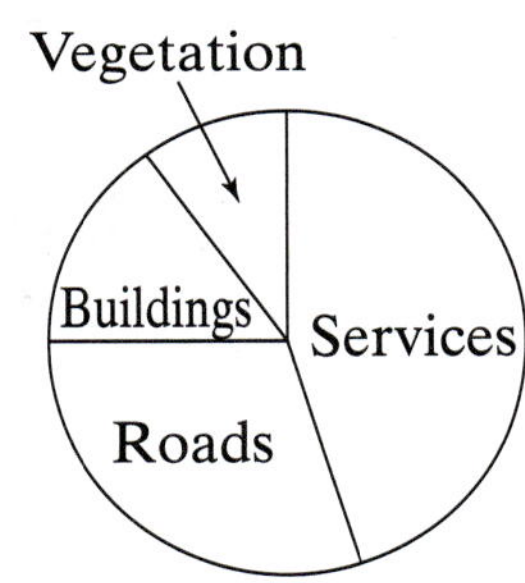

(A) The majority of the money is spent on Services and Roads.
(B) More money will be spent on Buildings this year than in any previous year.
(C) Of the four categories, the least amount of money is spent on Vegetation.
(D) A little more than $\frac{1}{4}$ of the money is spent on Roads.

14. Simplify the expression $\sqrt{25x^9}$.

(A) $5x^9$
(B) $5x^3$
(C) $5x$
(D) $5x^4\sqrt{x}$

Standards Mastery

15. Which of the following could be the graph of $y = 2x^3$?

Ⓐ 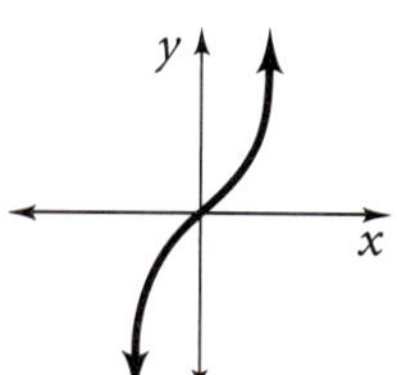

Ⓒ 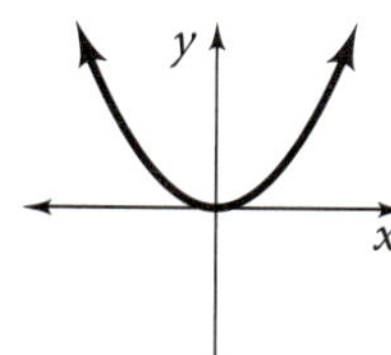

Ⓑ 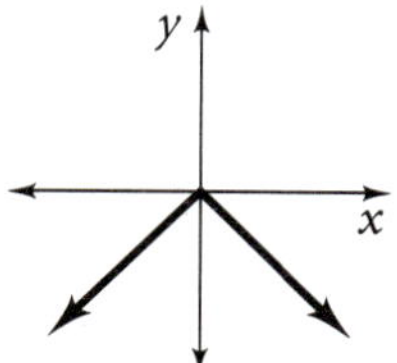

Ⓓ 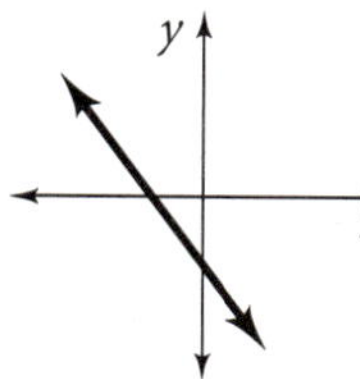

16. Simplify the expression $(4x^3y^5z)(3xy^4z)$.

Ⓐ $12x^3y^9z$
Ⓑ $12x^3y^{20}z$
Ⓒ $12x^2y$
Ⓓ $12x^4y^9z^2$

17. Which of the following triangles $C'A'T'$ is the image of triangle CAT that results from the translation $(x, y) \rightarrow (x + 4, y - 3)$?

Ⓐ 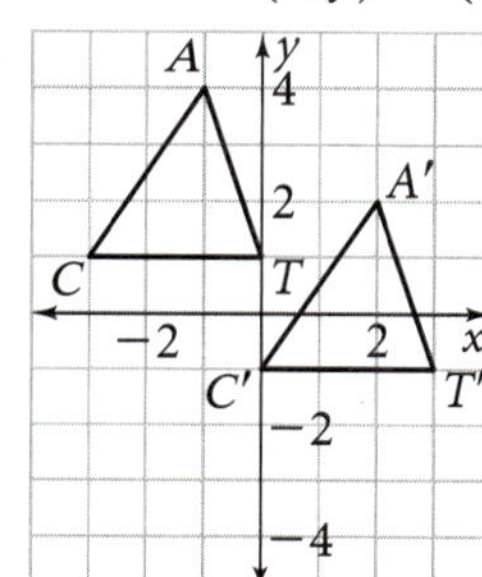

Ⓒ 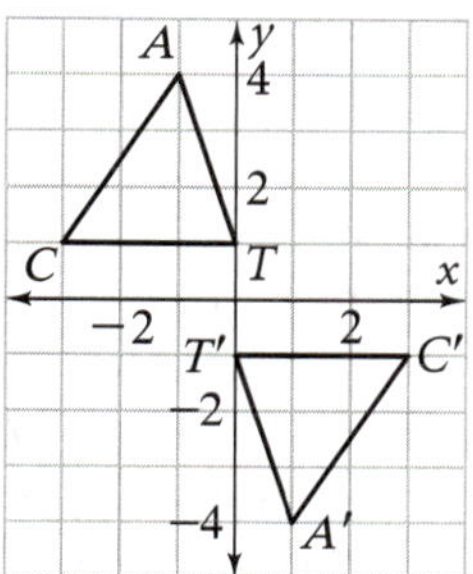

Ⓑ 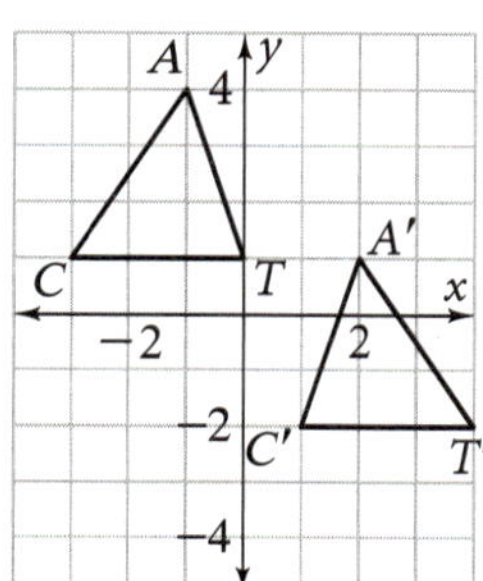

Ⓓ 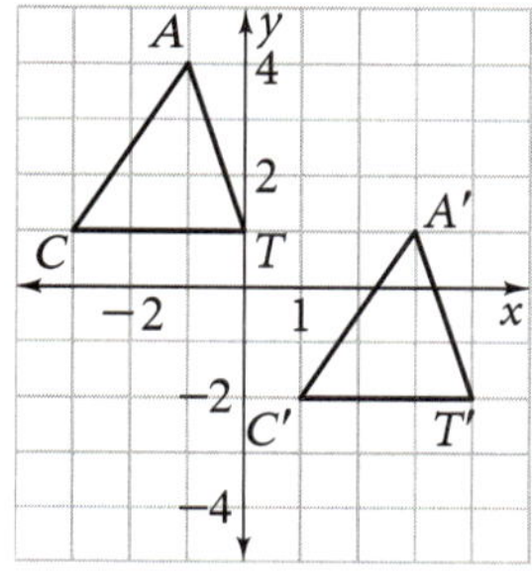

18. Which expression is equivalent to $\frac{63a^6bc}{9a^2b}$?

Ⓐ $7a^4bc$
Ⓑ $\frac{1}{7}a^4c$
Ⓒ $7a^4c$
Ⓓ $7a^4b^2c$

CAHSEE Reference Guide

California Content Standard	Item Number(s)
7MG 2.3	1, 4, 10
Alg1 2.0	2, 3
Alg1 3.0	5, 6, 7, 8
7MG 3.3	9
6SDAP 2.5	11, 12, 13
7AF 3.1	15
7AF 2.2	14, 16, 18
7MG 3.2	17

For additional review and practice, see the *California Standards Review and Practice Workbook* or go online to use the

Visit: PHSchool.com, **Web Code:** bcq-9045

Circles

What You've Learned

California Content Standards

7.0 Students prove and use theorems involving the properties of circles.

8.0 Students know, derive, and solve problems involving circumference and area of common geometric figures.

12.0 Students find and use measures of sides and of interior and exterior angles of triangles and polygons to classify figures and solve problems.

for Help to the Lesson in green.

Solving Equations (Skills Handbook Page 758)

Algebra Solve for x.

1. $\frac{1}{2}(x + 42) = 62$ **2.** $(5 + 3)8 = (4 + x)6$ **3.** $(9 + x)2 = (12 + 4)3$

Distance Formula (Lesson 1-8)

Find the distance between each pair of points.

4. $(13, 7), (6, 31)$ **5.** $(-4, 4), (2, -4)$ **6.** $(-3, -1), (0, 3)$ **7.** $(2\sqrt{3}, 5), (-\sqrt{3}, 2)$

Isosceles and Equilateral Triangles (Lesson 4-5)

Algebra Find the value of x.

8.
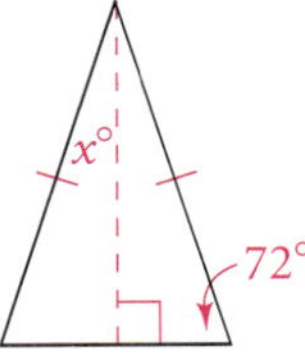

9.
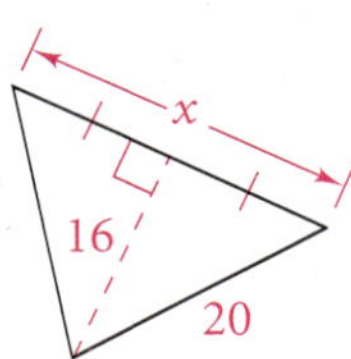

10.
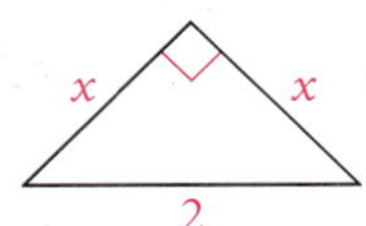

11.
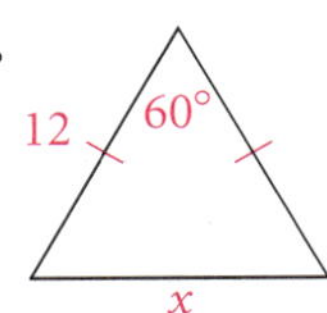

The Pythagorean Theorem (Lesson 8-1)

Algebra Find the value of x. Leave your answer in simplest radical form.

12.
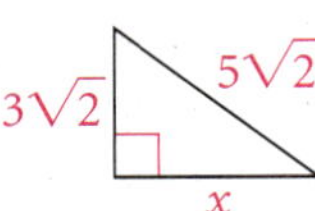

13.
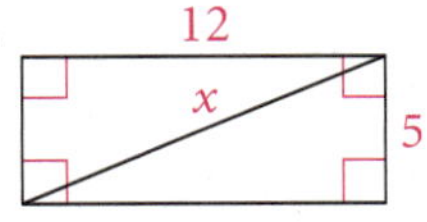

14.
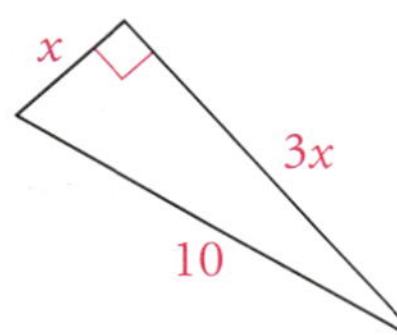

15.
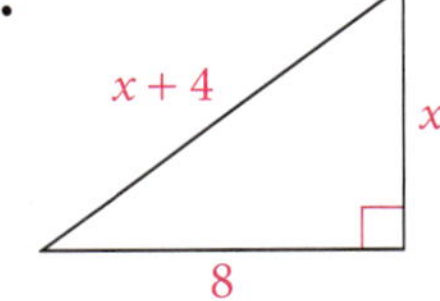

▲ Radio telescopes use radio frequency band to gather data about the universe.

What You'll Learn Next

California Content Standards

2.0 Students write geometric proofs.

7.0 Students prove and use theorems involving the properties of circles.

17.0 Students prove theorems by using coordinate geometry, including equations of circles.

21.0 Students prove and solve problems regarding relationships among chords, secants, tangents, inscribed angles, and inscribed and circumscribed polygons of circles.

CAHSEE 6SDAP 3.3, 7AF 1.2, 7AF 4.2, 7MG 2.4, Alg1 6.0

New Vocabulary

English and Spanish Audio Online

- **chord** (p. 670)
- **circumscribed about** (p. 664)
- **inscribed angle** (p. 678)
- **inscribed in** (p. 664)
- **intercepted arc** (p. 678)
- **locus** (p. 701)
- **point of tangency** (p. 662)
- **secant** (p. 687)
- **standard form of an equation of a circle** (p. 695)
- **tangent to a circle** (p. 662)

Tangent Lines

California Content Standards

GEOM 7.0 Students prove and use theorems involving the properties of circles. ***Develop***
GEOM 21.0 Students prove and solve problems regarding relationships among tangents and circumscribed polygons of circles. ***Introduce***

What You'll Learn

- To use the relationship between a radius and a tangent
- To use the relationship between two tangents from one point

. . . And Why

To find the distance between the centers of two dirt bike gears, as in Example 2

Check Skills You'll Need

GO for Help Skills Handbook page 754 and Lesson 8-1

Find each product.

1. $(p + 3)^2$ **2.** $(w + 10)^2$ **3.** $(m - 2)^2$

 Algebra **Find the value of x. Leave your answer in simplest radical form.**

4.

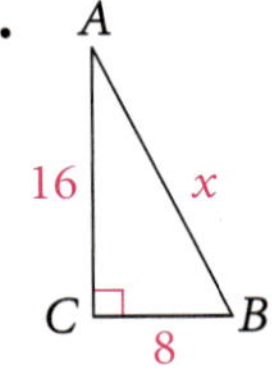

5.

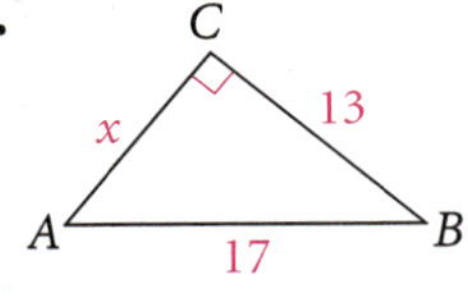

6.

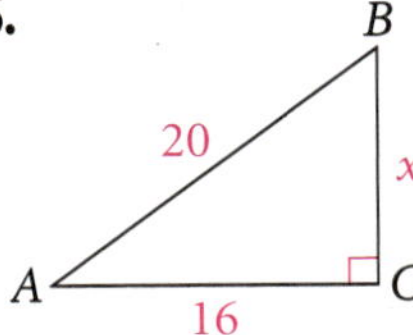

- tangent to a circle
- point of tangency
- inscribed in
- circumscribed about

1 Using the Radius-Tangent Relationship

In Chapter 8, you studied the tangent ratio in right triangles. The tangents you will study here relate to circles.

The word tangent may refer to a line, ray, or segment.

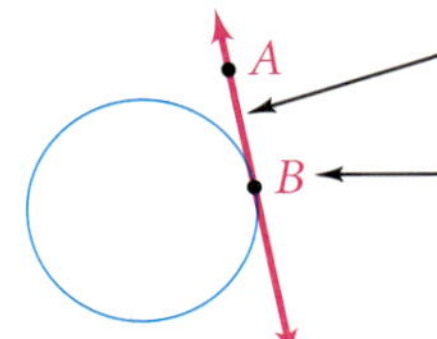

A **tangent to a circle** is a line in the plane of the circle that intersects the circle in exactly one point.

The point where a circle and a tangent intersect is the **point of tangency.**

$\overrightarrow{BA}$ is a tangent ray and $\overline{BA}$ is a tangent segment.

Theorem 12-1 relates a tangent and a radius in a given circle. You will write an indirect proof for Theorem 12-1 in Exercise 29.

Theorem 12-1

If a line is tangent to a circle, then the line is perpendicular to the radius drawn to the point of tangency.

$$\overleftrightarrow{AB} \perp \overline{OP}$$

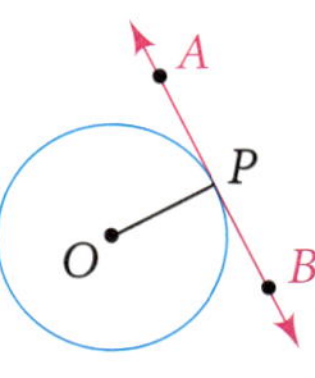

You can use Theorem 12-1 to solve problems involving tangents to circles.

1 EXAMPLE Finding Angle Measures

Test-Taking Tip

Remember that you can find the sum of the angles of a polygon with n sides using the formula $(n - 2)180$.

Multiple Choice $\overline{ML}$ and $\overline{MN}$ are tangent to $\odot O$. Find the value of x.

Ⓐ 58 Ⓑ 63 Ⓒ 90 Ⓓ 117

Since $\overline{ML}$ and $\overline{MN}$ are tangent to $\odot O$, $\angle L$ and $\angle N$ are right angles. $LMNO$ is a quadrilateral whose angle measures have a sum of 360.

$$m\angle L + m\angle M + m\angle N + m\angle O = 360$$
$$90 + x + 90 + 117 = 360 \quad \textbf{Substitute.}$$
$$297 + x = 360 \quad \textbf{Simplify.}$$
$$x = 63 \quad \textbf{Solve.}$$

The correct answer is B.

1 $\overline{ED}$ is tangent to $\odot O$. Find the value of x.

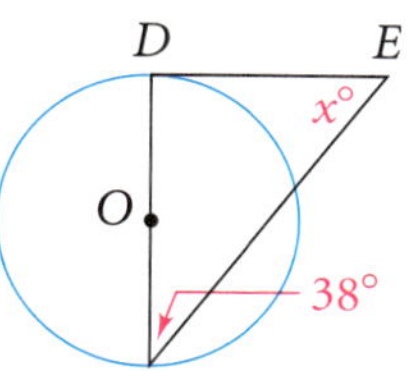

2 EXAMPLE Real-World Connection

Dirt Bikes A dirt bike chain fits tightly around two gears. The chain and gears form a figure like the one at the right. Find the distance between the centers of the gears.

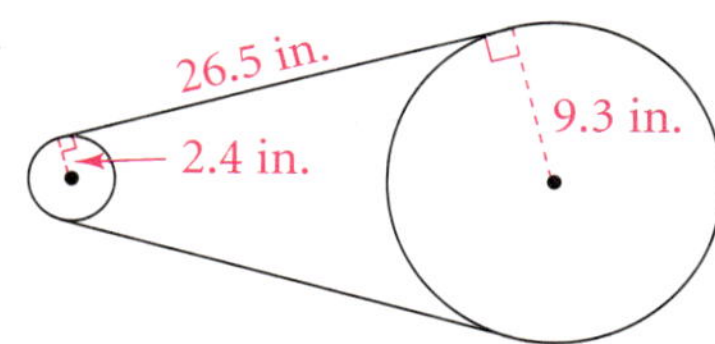

Real-World Connection

This motorcycle and many other two-wheeled vehicles have chain-drive systems like the one shown in Example 2.

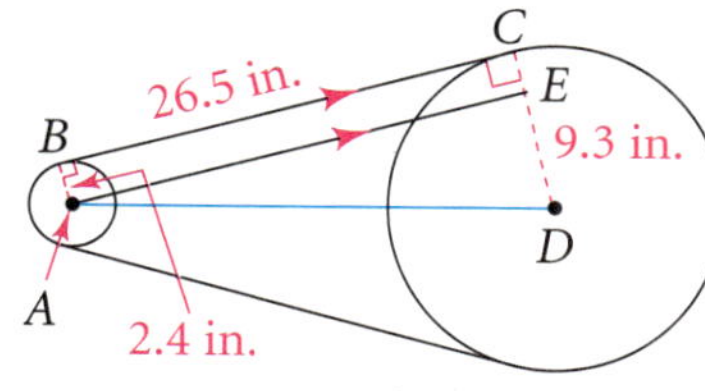

Label the diagram.

Draw $\overline{AE}$ parallel to $\overline{BC}$.

$ABCE$ is a rectangle. $\triangle AED$ is a right triangle with $AE = 26.5$ in. and $ED = 9.3 - 2.4 = 6.9$ in.

$$AD^2 = AE^2 + ED^2 \quad \textbf{Pythagorean Theorem}$$
$$AD^2 = 26.5^2 + 6.9^2 \quad \textbf{Substitute.}$$
$$AD^2 = 749.86 \quad \textbf{Simplify.}$$
$$AD \approx 27.383572 \quad \textbf{Use a calculator to find the square root.}$$

The distance between the centers is about 27.4 in.

CA Standards Check

2 A belt fits tightly around two circular pulleys, as shown at the right. Find the distance between the centers of the pulleys.

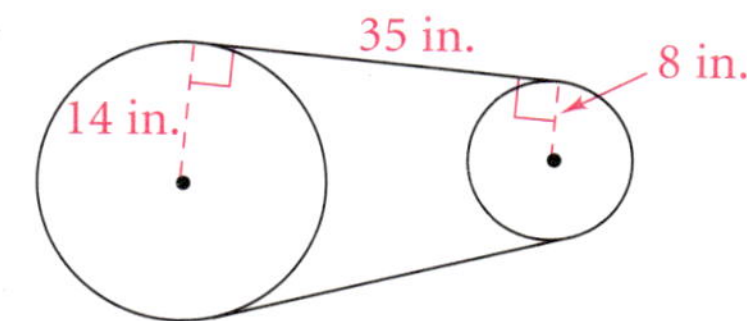

Theorem 12-2 (next page) is the converse of Theorem 12-1. You can use it to prove that a line or segment is tangent to a circle. You can also use it to construct a tangent to a circle (see Exercise 24). You will prove this theorem in Exercise 33.

Key Concepts

Theorem 12-2

If a line in the plane of a circle is perpendicular to a radius at its endpoint on the circle, then the line is tangent to the circle.

$\overleftrightarrow{AB}$ is tangent to $\odot O$.

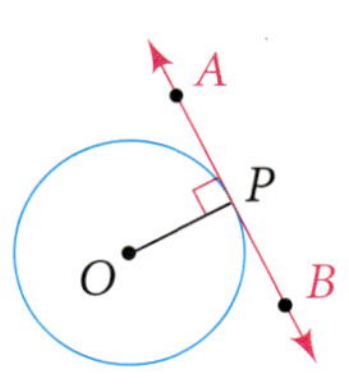

3 EXAMPLE Finding a Tangent

Is $\overline{ML}$ tangent to $\odot N$ at L? Explain.

$NL^2 + LM^2 \stackrel{?}{=} NM^2$ **Is △*MLN* a right triangle?**

$7^2 + 24^2 \stackrel{?}{=} 25^2$ **Substitute.**

$625 = 625$ ✓ **Simplify.**

By the Converse of the Pythagorean Theorem, $\triangle MLN$ is a right triangle with right $\angle L$. Therefore $\overline{ML} \perp \overline{NL}$, and $\overline{ML}$ is tangent to $\odot N$ at L by Theorem 12-2.

CA Standards Check

3 If $NL = 4$, $LM = 7$, and $NM = 8$, is $\overline{ML}$ tangent to $\odot N$ at L? Explain.

2 Using Multiple Tangents

If a circle is circumscribed about a triangle (Chapter 5), the triangle is inscribed in the circle. Similarly, when a circle is **inscribed in** a triangle, as in the diagram, the triangle is **circumscribed about** the circle. Each side of the triangle is tangent to the circle. The tangent segments from each vertex are congruent. You will prove this theorem in Exercise 30.

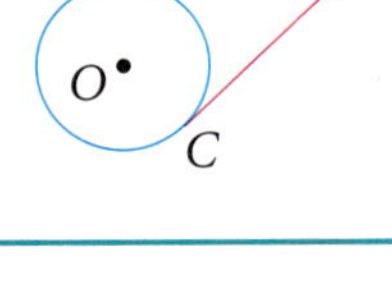

Key Concepts

Theorem 12-3

The two segments tangent to a circle from a point outside the circle are congruent.

$\overline{AB} \cong \overline{CB}$

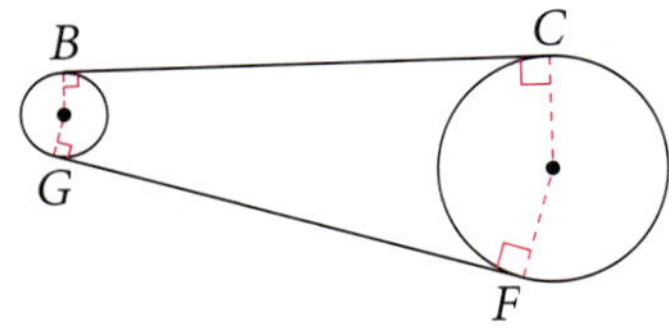

4 EXAMPLE Using Theorem 12-3

The diagram represents a chain drive system on a bicycle. Give a convincing argument that $BC = GF$.

Extend $\overline{BC}$ and $\overline{GF}$ to intersect in point H. By Theorem 12-3, $HC = HF$, or $HB + BC = HG + GF$. By Theorem 12-3 again, $HB = HG$, so by the Subtraction Property of Equality, $BC = GF$.

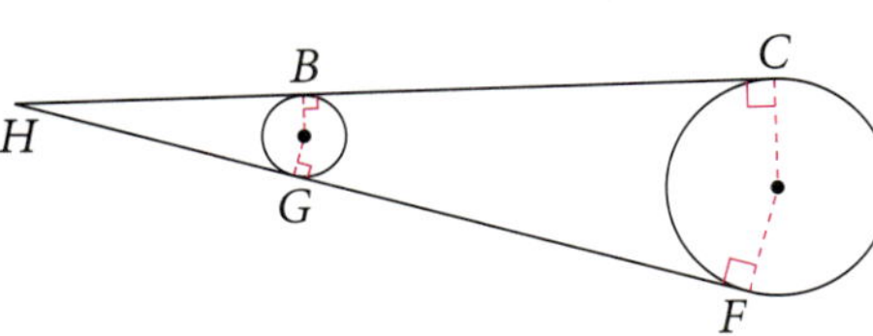

CA Standards Check

4 **Critical Thinking** Give a convincing argument that $BC = GF$ above if you know that $\overleftrightarrow{BC}$ and $\overleftrightarrow{GF}$ never intersect.

5 EXAMPLE Circles Inscribed in Polygons

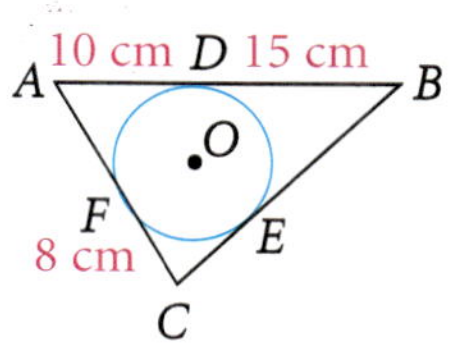

$\odot O$ is inscribed in $\triangle ABC$. Find the perimeter of $\triangle ABC$.

$AD = AF = 10$ cm	**The two segments tangent to a circle from a point outside the circle are congruent.**
$BD = BE = 15$ cm	
$CF = CE = 8$ cm	

$p = AB + BC + CA$	**Definition of perimeter p**
$= AD + DB + BE + EC + CF + FA$	**Segment Addition Postulate**
$= 10 + 15 + 15 + 8 + 8 + 10$	**Substitute.**
$= 66$	

The perimeter is 66 cm.

5 $\odot O$ is inscribed in $\triangle PQR$. $\triangle PQR$ has a perimeter of 88 cm. Find QY.

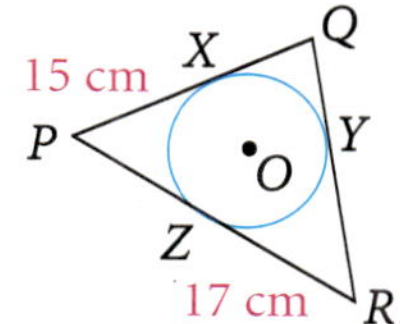

For more exercises, see *Extra Skill, Word Problem, and Proof Practice.*

GEOM 7.0, 21.0

Practice by Example

Algebra **Assume that lines that appear to be tangent are tangent. *O* is the center of each circle. Find the value of *x*.**

Example 1 (page 663)

1.

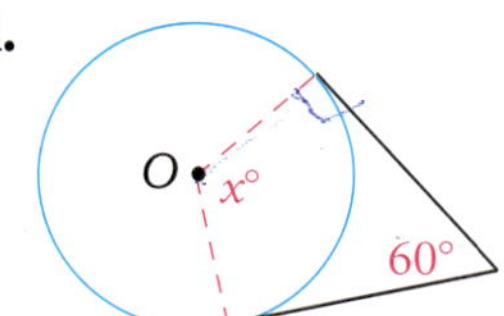

2.

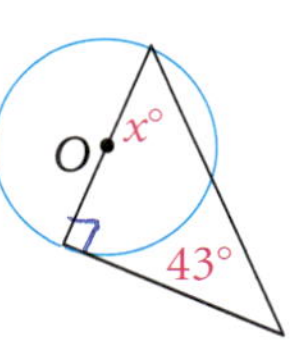

3.

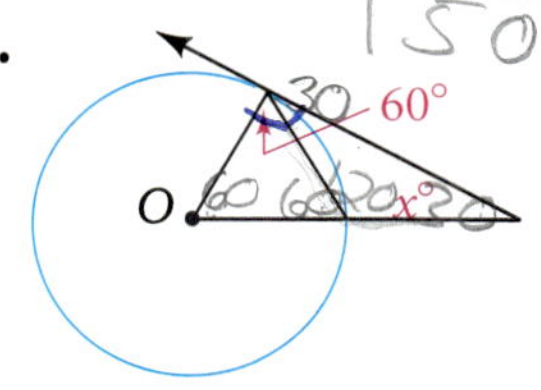

Examples 2, 4 (pages 663, 664)

A belt fits snugly around the two circular pulleys shown.

4. Find the distance between the centers of the pulleys. Round to the nearest hundredth.

5. Give a convincing argument why the belt lengths RS and QP are equal.

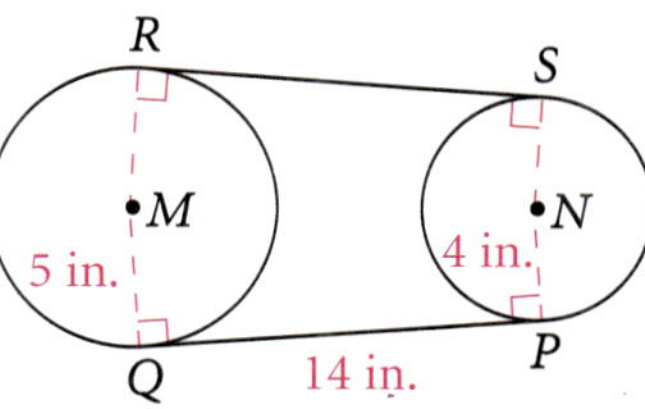

Exercises 4–7

Vocabulary Tip

$\overline{RS}$ and $\overline{QP}$ are common tangents.

For the pulley system shown, use the lengths given below. Find the missing length to the nearest tenth.

6. $MQ = 10$ cm, $NP = 4$ cm, $QP = 14$ cm, $MN = \blacksquare$ cm

7. $MQ = 5$ in., $NP = 4$ in., $QP = 20$ in., $MN = \blacksquare$ in.

Example 3 (page 664)

Determine whether a tangent line is shown in each diagram. Explain.

8. (diagram: 5, 15, 16)

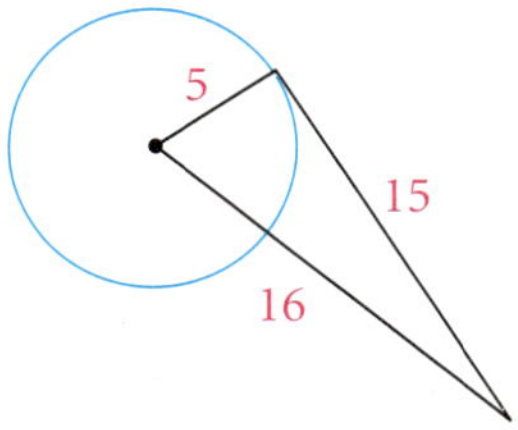

9.

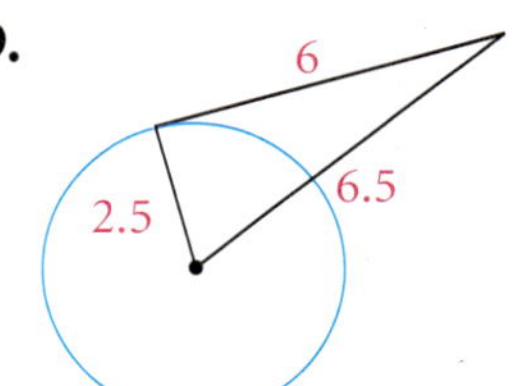

10.

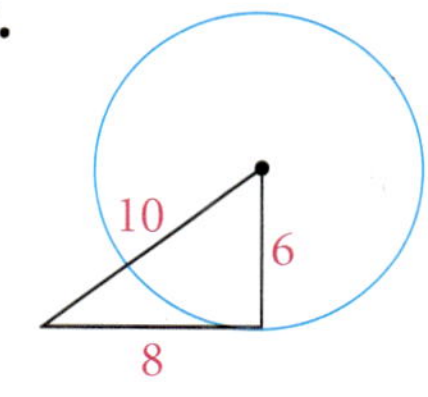

Example 5 (page 665)

Each polygon circumscribes a circle. Find the perimeter of the polygon.

11.

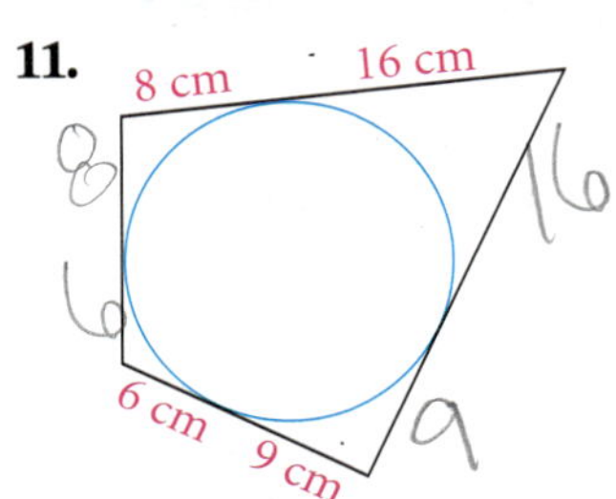

12.

B Apply Your Skills

x^2 **Algebra** **Assume that lines that appear to be tangent are tangent. O is the center of each circle. Find the value of x to the nearest tenth.**

13.

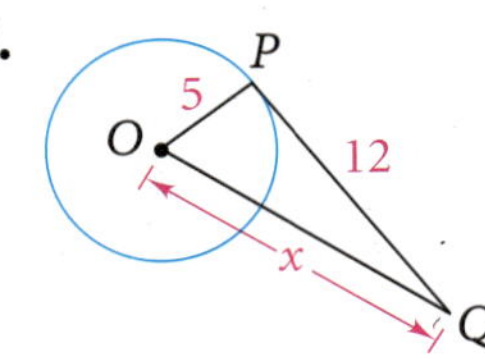

14.

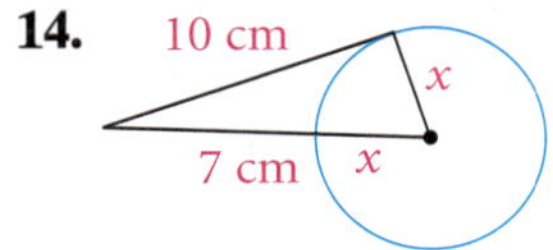

15.

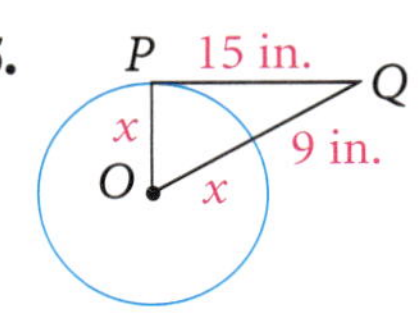

Real-World Connection This "diamond ring" effect in a solar eclipse may be seen by a person on Earth at the end of a common external tangent of the sun and moon. (See diagram at right.)

16. Solar Eclipse Common tangents to two circles may be *internal* or *external.* If you draw a segment joining the centers of the circles, a common internal tangent will intersect the segment. A common external tangent will not.

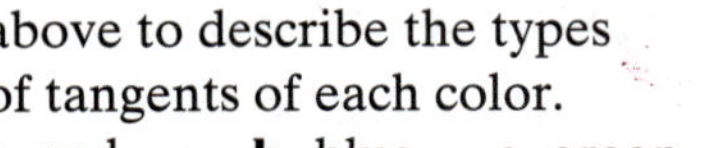

For this cross-sectional diagram of the sun, moon, and Earth during a solar eclipse, use the terms above to describe the types of tangents of each color.

a. red **b.** blue **c.** green

d. Which tangents show the extent on Earth's surface of total eclipse? Of partial eclipse?

e. Reasoning In general, does every pair of circles have common tangents of both types? Explain.

Earth **The circle at the right represents Earth. The radius of Earth is about 6400 km. Find the distance d that a person can see on a clear day from each of the following heights h above Earth. Round your answer to the nearest tenth of a kilometer.**

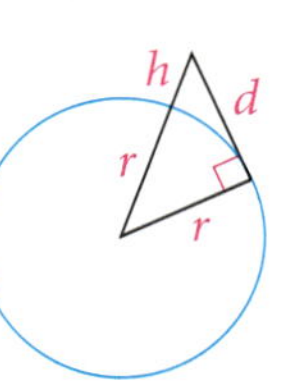

17. 100 m **18.** 500 m **19.** 1 km

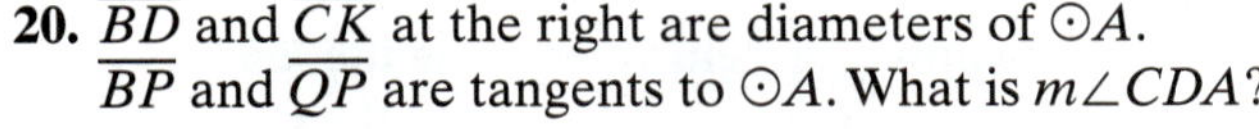

20. $\overline{BD}$ and $\overline{CK}$ at the right are diameters of $\odot A$. $\overline{BP}$ and $\overline{QP}$ are tangents to $\odot A$. What is $m\angle CDA$?

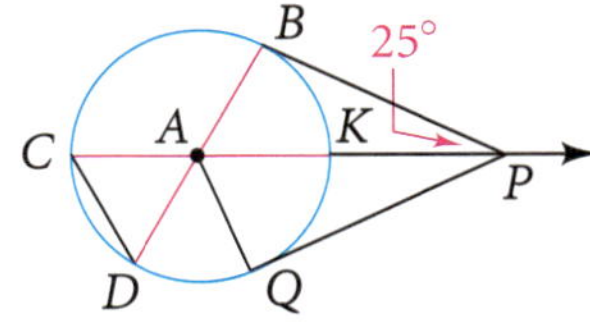

21. History Leonardo da Vinci wrote, "When each of two squares touch the same circle at four points, one is double the other."

a. Sketch a figure that illustrates this statement.

b. Writing Explain why the statement is true.

22. Multiple Choice A regular hexagon is circumscribed about the ring surrounding the clock face. The diameter of the ring is 10 in. What is the perimeter of the hexagon?

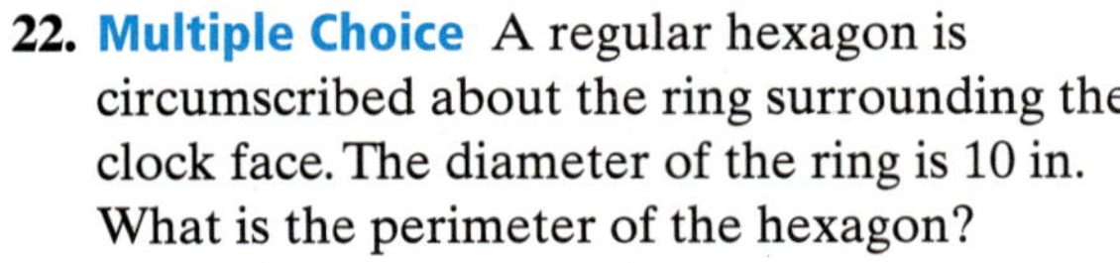

Ⓐ 30.0 in. Ⓒ 43.3 in.

Ⓑ 34.6 in. Ⓓ 51.7 in.

Visit: PHSchool.com
Web Code: bce-1201

23. **Critical Thinking** A nickel, a dime, and a quarter are touching as shown. Tangents are drawn from point A to both sides of each coin. What can you conclude about the four tangent segments? Explain.

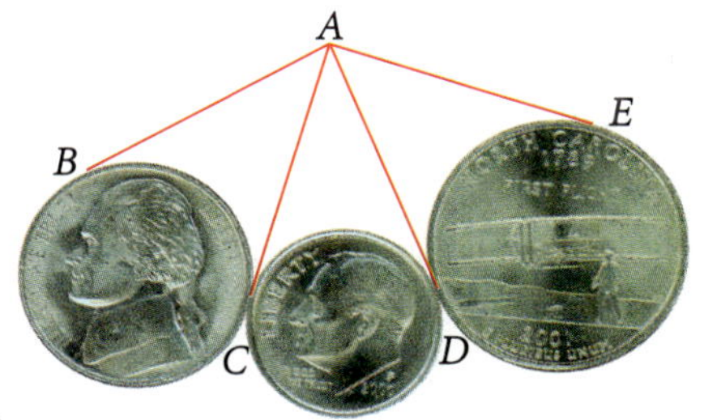

24. **Constructions** Draw a circle. Label the center T. Locate a point on the circle and label it R. Construct a tangent to $\odot T$ at R.

$\overline{AC}$ is tangent to $\odot O$ at A, and $m\angle 1 = 70$.

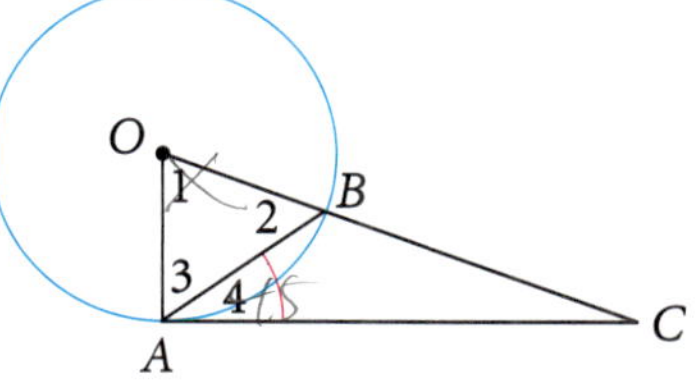

25. Find $m\angle 4$.

26. Let $m\angle 1 = x$. Find $m\angle 4$ in terms of x. What is the relationship between $\angle 1$ and $\angle 4$?

27. **Coordinate Geometry** Graph the equation $x^2 + y^2 = 9$. Then draw a segment from (0, 5) tangent to the circle. Find the length of the segment.

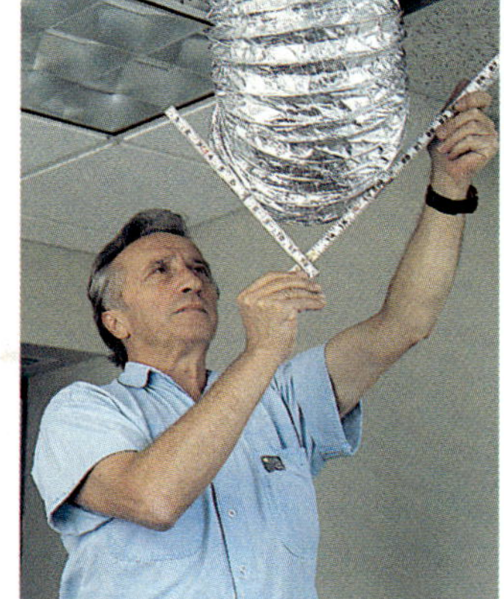

Real-World Connection

The radius of this air-conditioning duct can be determined by positioning two rulers as tangents.

28. **Maintenance** Mr. Gonzales is replacing a cylindrical air-conditioning duct. He estimates the radius of the duct by folding a ruler to form two 6-in. tangents to the duct. The tangents form an angle. Mr. Gonzales measures the angle bisector from the vertex to the duct. It is about $2\frac{3}{4}$ in. long. What is the radius of the duct?

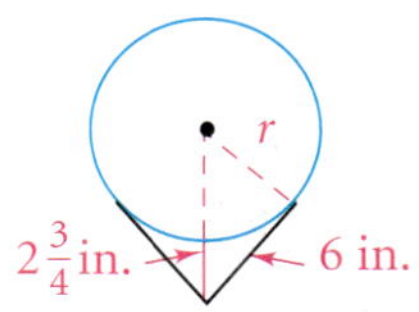

29. Complete the following indirect proof of Theorem 12-1.

Given: Line n is tangent to $\odot O$ at P.

Prove: line $n \perp \overline{OP}$

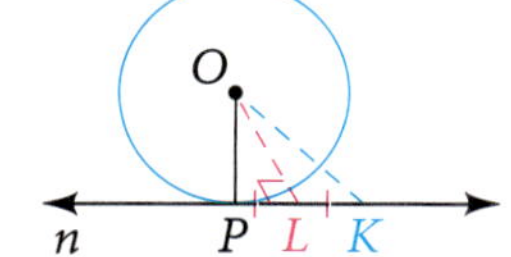

Step 1 Assume that line n is not perpendicular to $\overline{OP}$.

Step 2 If line n is not perpendicular to $\overline{OP}$, some other segment $\overline{OL}$ must be **a.** __?__ to line n. By the Ruler Postulate, there is a point K on $\overrightarrow{PL}$ such that $PK = 2PL$, so $PL =$ **b.** __?__. $\triangle OPL \cong \triangle OKL$ by **c.** __?__, so $OP = OK$ because **d.** __?__. Since P and K are the same distance from O, both K and P are on $\odot O$. This contradicts the given fact that line n is **e.** __?__ to $\odot O$ at P.

Step 3 The assumption that line n is not perpendicular to $\overline{OP}$ is **f.** __?__, so line $n \perp \overline{OP}$.

Proof 30. Prove Theorem 12-3.

Given: $\overline{BA}$ and $\overline{BC}$ are tangent to $\odot O$ at A and C, respectively.

Prove: $\overline{BA} \cong \overline{BC}$

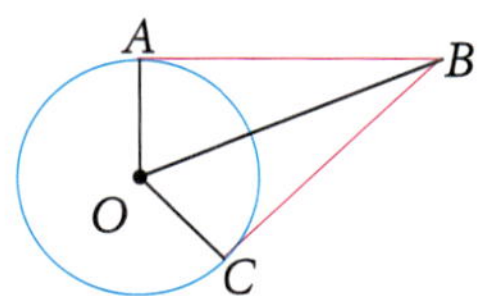

31. **Given:** $\overline{BC}$ is tangent to $\odot A$ at D.
$\overline{DB} \cong \overline{DC}$

Prove: $\overline{AB} \cong \overline{AC}$

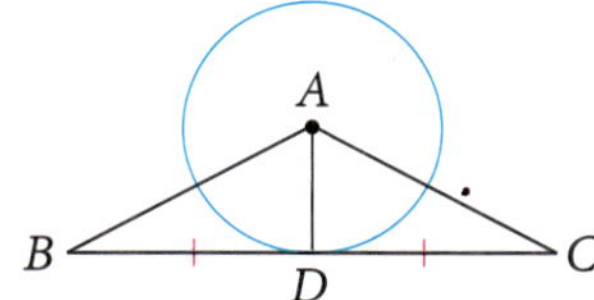

32. **Given:** $\odot A$ and $\odot B$ with common tangents $\overline{DF}$ and $\overline{CE}$

Prove: $\triangle GDC \sim \triangle GFE$

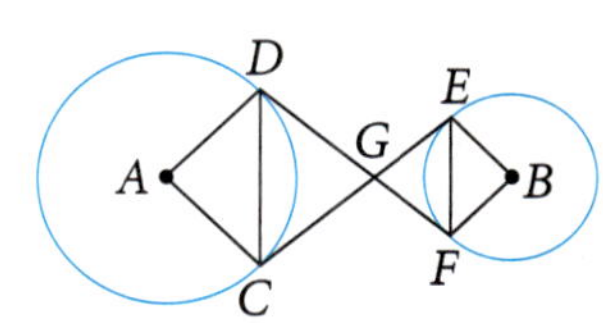

 Challenge Proof 33. Write an indirect proof of Theorem 12-2.

Given: $\overleftrightarrow{AB} \perp \overline{OP}$ at P.

Prove: $\overleftrightarrow{AB}$ is tangent to $\odot O$.

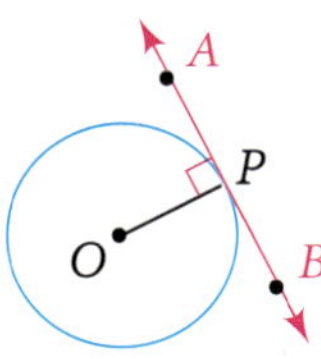

34. Two circles that have one point in common are *tangent circles*. Given any triangle, explain how to draw three circles that are centered at each vertex of the triangle and are tangent to each other.

Multiple Choice Practice

For California Standards and CAHSEE Tutorials, visit PHSchool.com. Web Code: bcq-9045

GEOM 21.0

35. The triangle circumscribes the circle. Find the perimeter of the triangle.

Ⓐ 24 cm
Ⓑ 32 cm
Ⓒ 34 cm
Ⓓ 60 cm

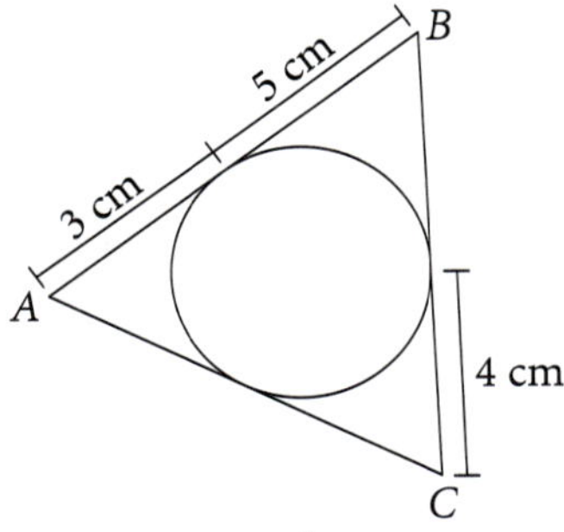

GEOM 9.0

36. Tran made a model of a square pyramid for a history presentation. If the height of the model pyramid at the right is 15 inches, what is the volume of the figure?

Ⓐ 1500 cubic inches
Ⓑ 750 cubic inches
Ⓒ 500 cubic inches
Ⓓ 250 cubic inches

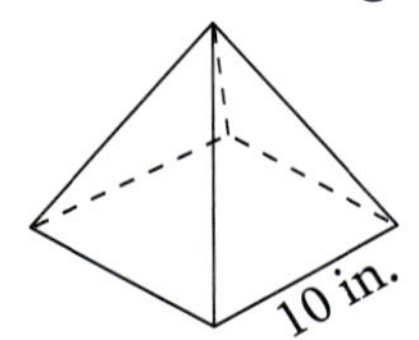

Mixed Review

Lesson 11-8

Two cubes have heights 6 in. and 8 in. Find each ratio.

37. similarity ratio **38.** ratio of surface areas **39.** ratio of volumes

Lesson 8-3 x^2 Algebra **Find the value of x. Round answers to the nearest tenth.**

40.

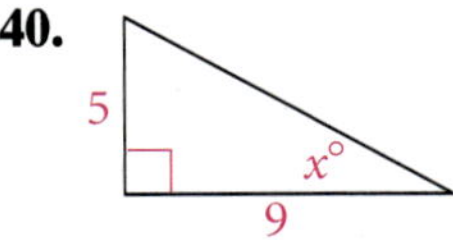

41.

42.

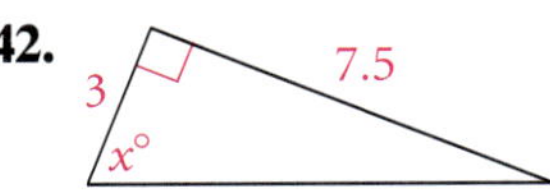

Lesson 7-2

The polygons are similar. (a) State the similarity ratio and (b) find the values of the variables.

43.

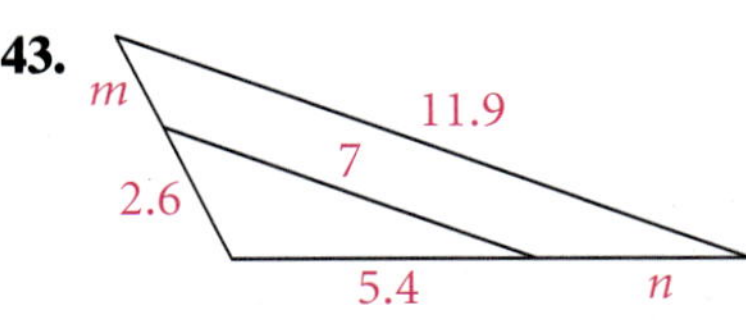

44.

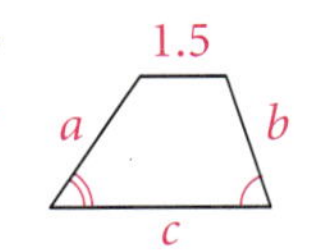

Paper Folding With Circles

Hands-On

For Use With Lesson 12-2

Prepares for GEOM 21.0 Students prove and solve problems regarding relationships among chords.

A *chord* is a segment whose endpoints are on a circle. In these activities, you will explore some properties of chords.

1 Activity

- Use a compass. Draw a circle on tracing paper.
- Use a straightedge. Draw two radii.
- Set your compass to a distance shorter than the radii. Place the compass point at the center of the circle. Mark two congruent segments, one on each radius.
- At the point you marked on each radius, fold a line perpendicular to the radius.

1. How do you measure the distance between a point and a line?

2. Each perpendicular contains a chord of the circle. Measure the length of each chord and compare. What do you notice?

3. Make a conjecture about the lengths of chords that are equidistant from the center of a circle.

2 Activity

- Use a compass. Draw a circle on tracing paper.
- Use a straightedge. Draw two chords of different lengths. Do not draw either one as a diameter.
- Fold the perpendicular bisector for each chord.

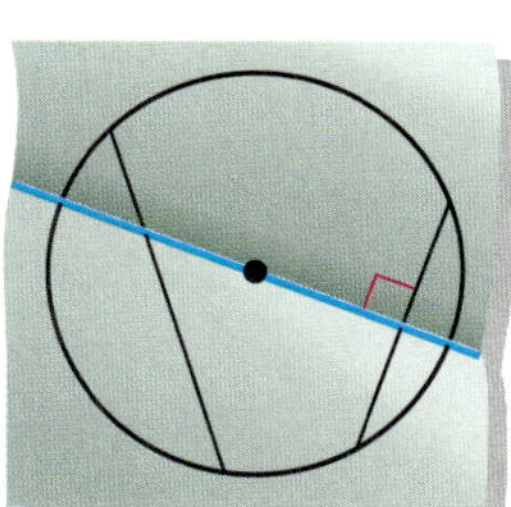

4. Where do the perpendicular bisectors appear to intersect?

5. Draw a third chord and fold its perpendicular bisector. Where does it appear to intersect the other two?

6. Make a conjecture about the perpendicular bisector of a chord of a circle.

EXERCISES

7. Write a proof of your conjecture from Exercise 3 or give a counterexample to prove it false.

8. What theorem provides a one-sentence proof of your conjecture from Exercise 6?

9. Suppose two chords have different lengths. Make a conjecture that compares their distances from the center of the circle.

10. You are building a circular table for a patio. For the umbrella, you have to drill a hole through the center of the tabletop. Explain how you can use a carpenter's square like the one shown to find the center.

Chords and Arcs

California Content Standards

GEOM 2.0 Students write geometric proofs. ***Master***
GEOM 7.0 Students prove and use theorems involving the properties of circles. ***Develop***
GEOM 21.0 Students prove and solve problems regarding relationships among chords. ***Introduce***

What You'll Learn

- To use congruent chords, arcs, and central angles
- To recognize properties of lines through the center of a circle

. . . And Why

To see how an archaeologist finds the center and radius of the rim of a jar, as in Exercise 20

Check Skills You'll Need

GO for Help Lesson 8-2

Find the value of each variable. Leave your answer in simplest radical form.

1. 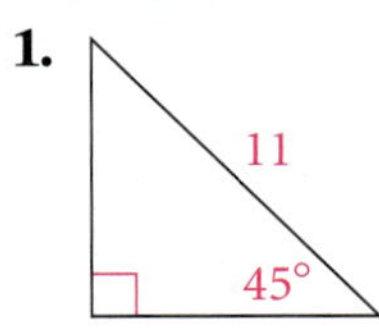

2. 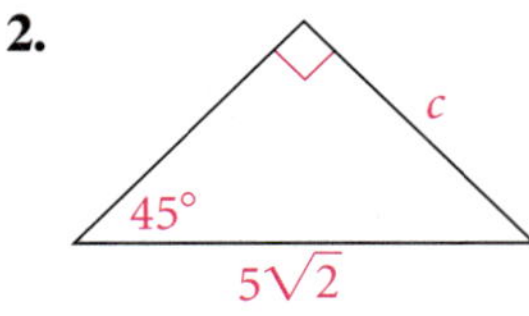

3. 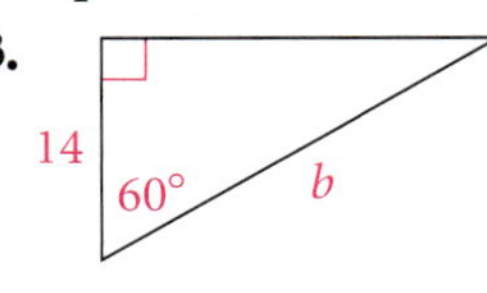

1 Using Congruent Chords, Arcs, and Central Angles

A segment whose endpoints are on a circle is called a **chord.** The diagram shows the related chord and arc, $\overline{PQ}$ and $\overparen{PQ}$.

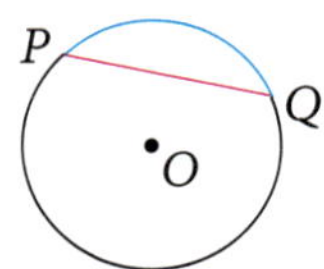

The following theorem is about related central angles, chords, and arcs. It says, for example, that if two central angles in a circle are congruent, then so are the two chords and two arcs that the angles intercept.

Key Concepts

Theorem 12-4

Within a circle or in congruent circles

(1) Congruent central angles have congruent chords.

(2) Congruent chords have congruent arcs.

(3) Congruent arcs have congruent central angles.

Online active math

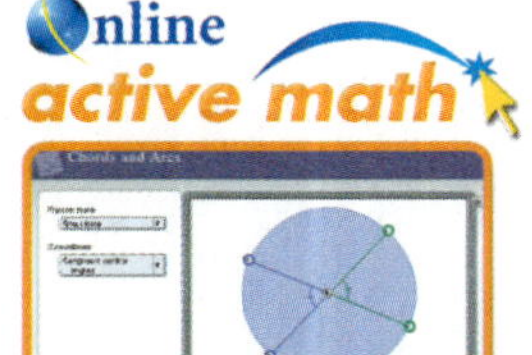

For: Chords and Arcs Activity
Use: Interactive Textbook, 12-2

You will prove Theorem 12-4 in Exercises 23, 24, and 35.

1 EXAMPLE Using Theorem 12-4

In the diagram, $\odot O \cong \odot P$. Given that $\overparen{BC} \cong \overparen{DF}$, what can you conclude?

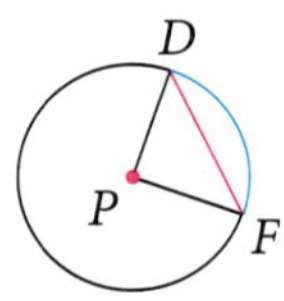

By Theorem 12-4, $\angle O \cong \angle P$ and $\overline{BC} \cong \overline{DF}$.

CA Standards Check

1 If you are instead given that $\overline{BC} \cong \overline{DF}$, what can you conclude?

Theorem 12-5 shows a relationship between two chords and their distances from the center of a circle. You will prove part (2) in Exercise 38.

Theorem 12-5

Within a circle or in congruent circles

(1) Chords equidistant from the center are congruent.

(2) Congruent chords are equidistant from the center.

Proof

Proof of Theorem 12-5, Part (1)

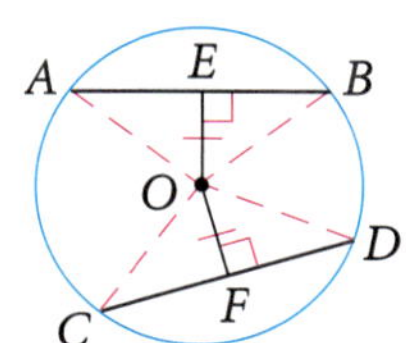

Given: $\odot O, \overline{OE} \cong \overline{OF}$,
$\overline{OE} \perp \overline{AB}, \overline{OF} \perp \overline{CD}$

Prove: $\overline{AB} \cong \overline{CD}$

Statements	Reasons
1. $\overline{OA} \cong \overline{OB} \cong \overline{OC} \cong \overline{OD}$	1. Radii of a circle are congruent.
2. $\overline{OE} \cong \overline{OF}, \overline{OE} \perp \overline{AB}, \overline{OF} \perp \overline{CD}$	2. Given
3. $\angle AEO$ and $\angle CFO$ are right angles.	3. Def. of perpendicular segments
4. $\triangle AEO \cong \triangle CFO$	4. HL Theorem
5. $\angle A \cong \angle C$	5. CPCTC
6. $\angle B \cong \angle A, \angle C \cong \angle D$	6. Isosceles Triangle Theorem
7. $\angle B \cong \angle D$	7. Transitive Property of Congruence
8. $\angle AOB \cong \angle COD$	8. If two ∠s of a △ are ≅ to two ∠s of another △, then the third ∠s are ≅.
9. $\overline{AB} \cong \overline{CD}$	9. ≅ central angles have ≅ chords.

Real-World Connection

Steel beams model congruent chords equidistant from the center to give the illusion of a circle.

You can use Theorem 12-5 to find missing lengths in circles.

2 EXAMPLE Using Theorem 12-5

Multiple Choice What is the value of a in the circle at the right?

(A) 9 (B) 12.5 (C) 18 (D) 25

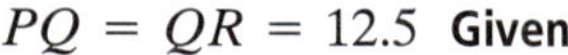

$PQ = QR = 12.5$ **Given**

$PQ + QR = PR$ **Segment Addition Postulate**

$25 = PR$ **Substitute.**

$a = PR$ **Chords equidistant from the center of a circle are congruent.**

$a = 25$ **Substitute.**

The correct answer is D.

Problem Solving Hint

In a circle, the length of the perpendicular segment from the center to a chord is the distance from the center to the chord.

2 Find the value of x in the circle at the right.

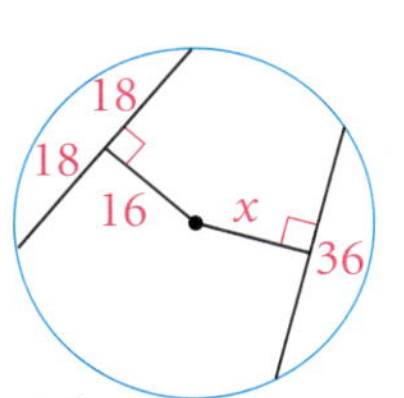

Lines Through the Center of a Circle

The Converse of the Perpendicular Bisector Theorem from Lesson 5-2 has special applications to a circle and its diameters, chords, and arcs.

Key Concepts

Theorem 12-6

In a circle, a diameter that is perpendicular to a chord bisects the chord and its arcs.

Theorem 12-7

In a circle, a diameter that bisects a chord (that is not a diameter) is perpendicular to the chord.

Theorem 12-8

In a circle, the perpendicular bisector of a chord contains the center of the circle.

Proof

Proof of Theorem 12-7

Given: $\odot T$ with diameter $\overline{QR}$ bisecting $\overline{SU}$ at V.

Prove: $\overline{QR} \perp \overline{SU}$

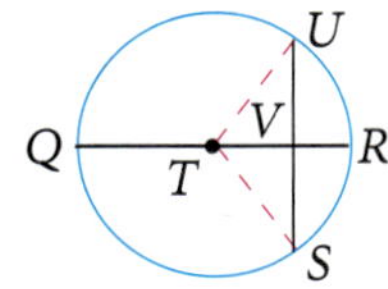

Proof: $TS = TU$ because the radii of a circle are congruent. $VS = VU$ by the definition of bisect. Thus, T and V are equidistant from S and U. By the Converse of the Perpendicular Bisector Theorem, T and V are on the perpendicular bisector of $\overline{SU}$. Since two points determine one line, $\overline{TV}$ is the perpendicular bisector of $\overline{SU}$. Another name for $\overline{TV}$ is $\overline{QR}$. Thus, $\overline{QR} \perp \overline{SU}$.

You will prove Theorems 12-6 and 12-8 in Exercises 25 and 36, respectively.

3 EXAMPLE Using Diameters and Chords

Algebra Find each missing length to the nearest tenth.

a.

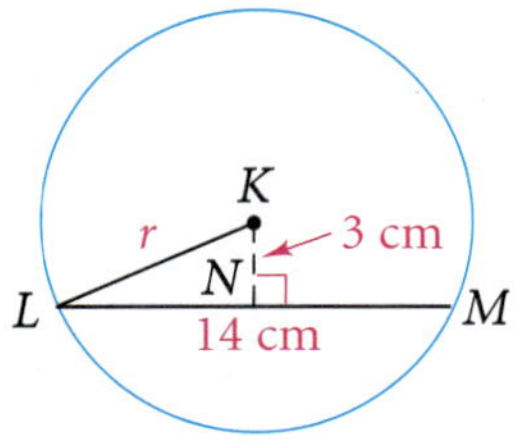

Real-World Connection

The center of the tire is located on the perpendicular bisector of the flat part.

$LN = \frac{1}{2}(14) = 7$	A diameter $\perp$ to a chord bisects the chord.
$r^2 = 3^2 + 7^2$	Use the Pythagorean Theorem.
$r \approx 7.6$	Find the square root of each side.

b.

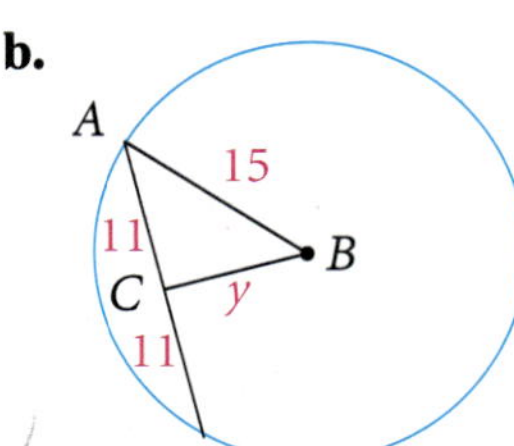

$\overline{BC} \perp \overline{AC}$	A diameter that bisects a chord that is not a diameter is $\perp$ to the chord.
$y^2 + 11^2 = 15^2$	Use the Pythagorean Theorem.
$y^2 = 104$	Solve for y^2.
$y \approx 10.2$	Find the square root of each side.

3 Use the circle at the right.

a. Find the length of the chord.

b. Find the distance from the midpoint of the chord to the midpoint of its minor arc.

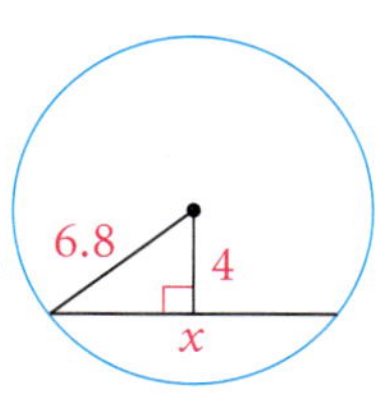

EXERCISES Standards Practice

For more exercises, see *Extra Skill, Word Problem, and Proof Practice.*

GEOM 2.0, 7.0, 21.0

A Practice by Example

Example 1 (page 670)

In Exercises 1 and 2, the circles are congruent. What can you conclude?

1.

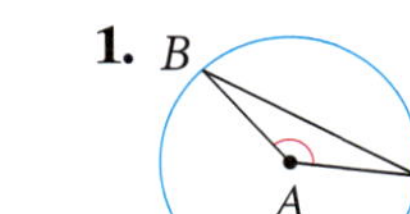

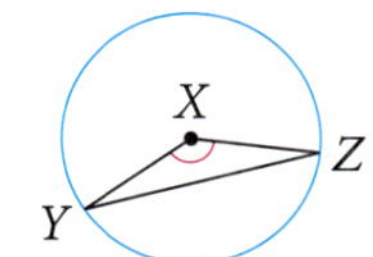

2.

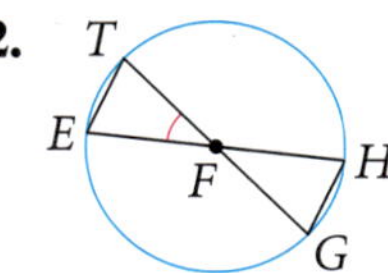

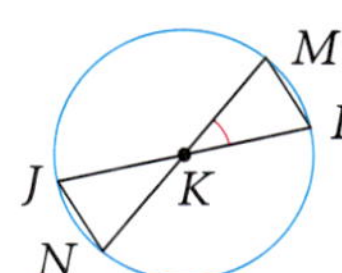

Example 2 (page 671)

Find the value of *x*.

3.

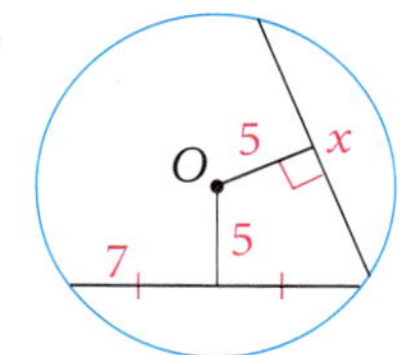

4.

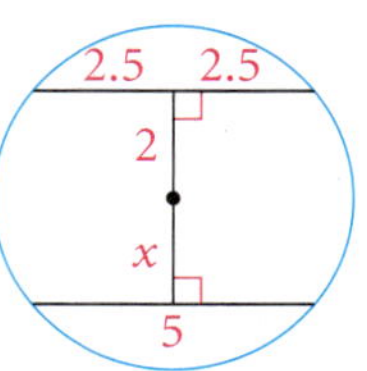

5.

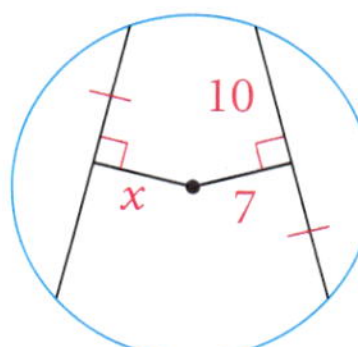

6.

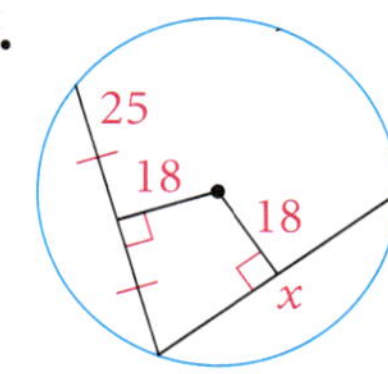

7.

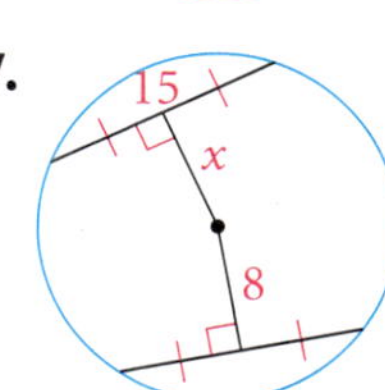

8.

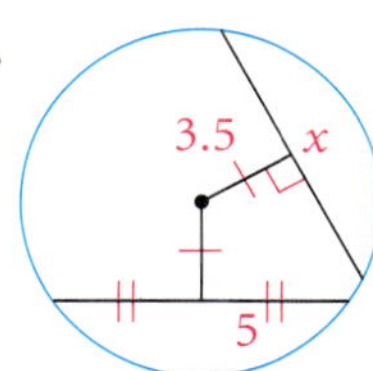

Example 3 (page 672)

Use the diagram at the right to complete Exercises 9 and 10.

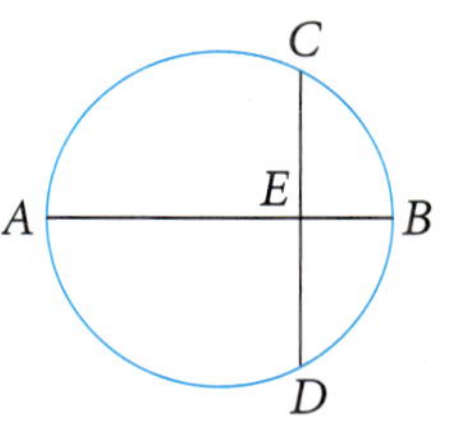

9. Given that $\overline{AB}$ is a diameter of the circle and $\overline{AB} \perp \overline{CD}$, then **a.** _?_ ≅ **b.** _?_ and **c.** _?_ ≅ **d.** _?_.

10. Given that $\overline{AB}$ is the perpendicular bisector of $\overline{CD}$, then $\overline{AB}$ contains _?_.

Algebra **Find the value of *x* to the nearest tenth.**

11.

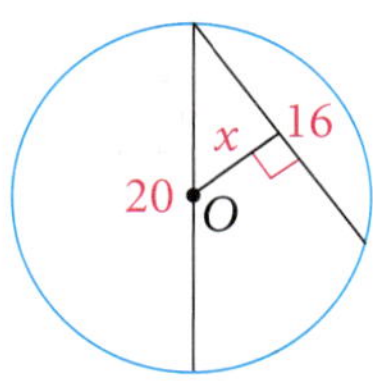

12.

13.

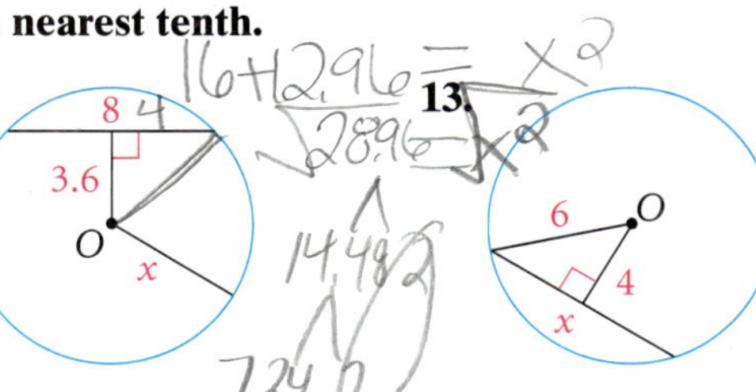

14.

15.

16.

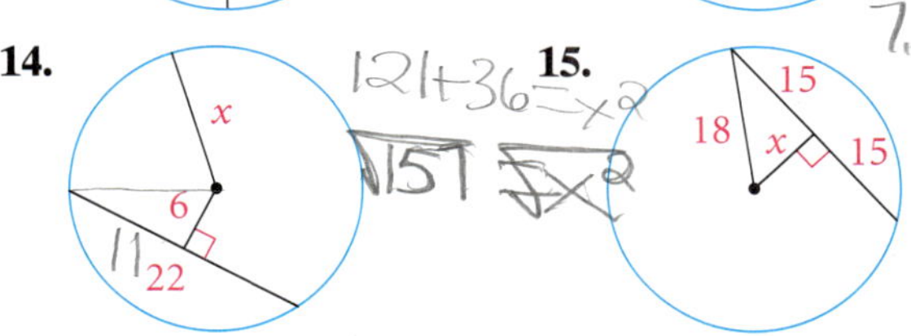

B Apply Your Skills

Find $m\widehat{AB}$. (*Hint:* You will need to use trigonometry in Exercise 19.)

17. 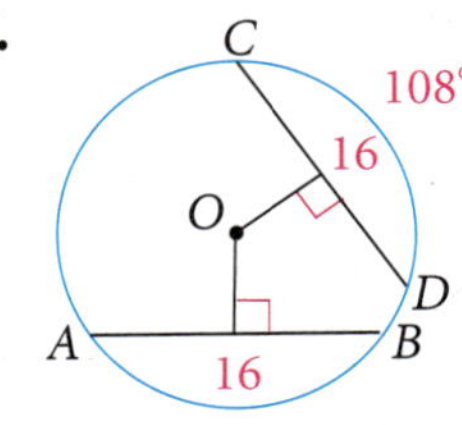

18. 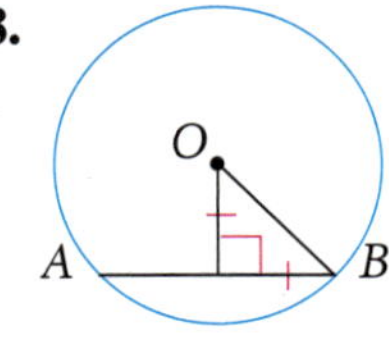

19. 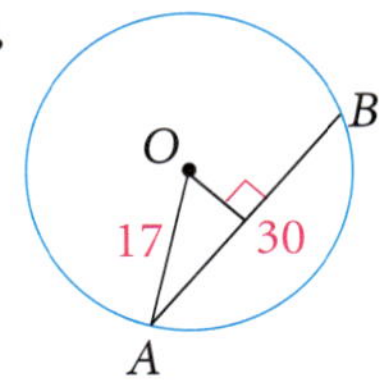

Real-World Connection

Field archaeologists analyze artifacts to provide glimpses of life in the past.

20. Archaeology An archaeologist found several jar fragments including a large piece of the circular rim. How can she find the center and radius of the rim to help her reconstruct the jar?

21. Geometry in 3 Dimensions In the figure at the right, sphere O with radius 13 cm is intersected by a plane 5 cm from center O. Find the radius of cross section $\odot A$.

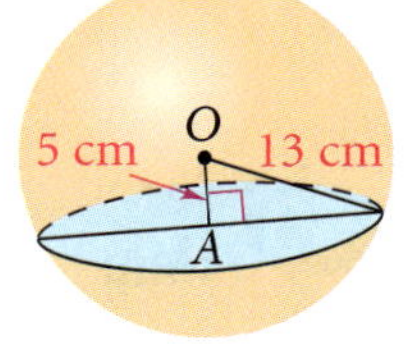

22. Geometry in 3 Dimensions A plane intersects a sphere that has radius 10 in. forming cross section $\odot B$ with radius 8 in. How far is the plane from the center of the sphere?

23. Complete this proof of Theorem 12-4, Part (1).

Given: $\odot P$ with $\angle KPM \cong \angle LPN$

Prove: $\overline{KM} \cong \overline{LN}$

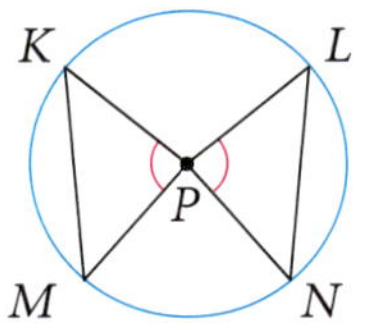

Proof: $\overline{KP} \cong$ **a.** _?_ $\cong$ **b.** _?_ $\cong \overline{NP}$ because **c.** _?_. $\triangle KPM \cong$ **d.** _?_ by **e.** _?_. $\overline{KM} \cong \overline{LN}$ by **f.** _?_.

24. Complete this proof of Theorem 12-4, Part (2).

Given: $\odot E$ with congruent chords $\overline{AB}$ and $\overline{CD}$

Prove: $\widehat{AB} \cong \widehat{CD}$

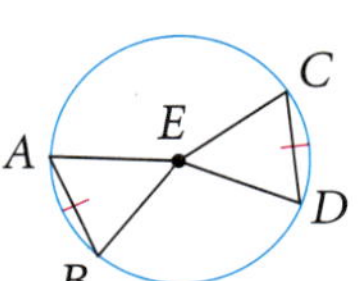

Problem Solving Hint

Recall that in a circle congruent central angles intercept congruent arcs.

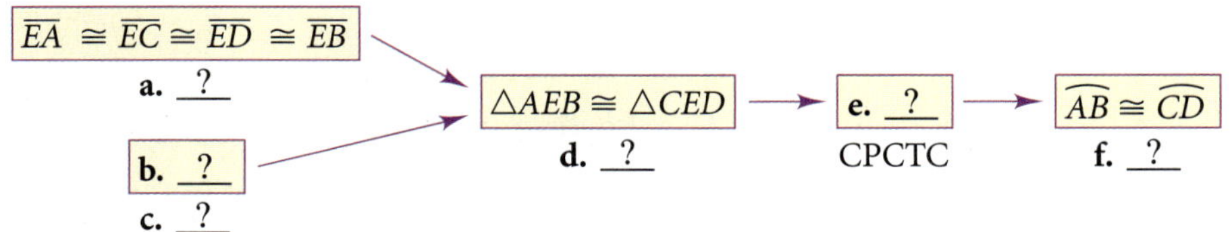

Proof **25.** Prove Theorem 12-6.

Given: $\odot O$ with diameter $\overline{ED} \perp \overline{AB}$ at C

Prove: $\overline{AC} \cong \overline{BC}$ and $\widehat{AD} \cong \widehat{BD}$

(*Hint*: Begin by drawing $\overline{OA}$ and $\overline{OB}$.)

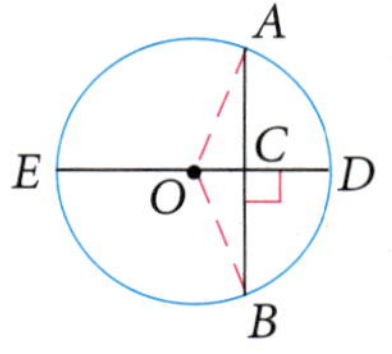

GO for Help

For a guide to solving Exercise 26, see p. 677.

26. Two concentric circles have radii of 4 cm and 8 cm. A segment tangent to the smaller circle is a chord of the larger circle. What is the length of the segment?

27. Error Analysis Scott looks at this figure and concludes that $\overline{ST} \cong \overline{PR}$. What is wrong with Scott's conclusion?

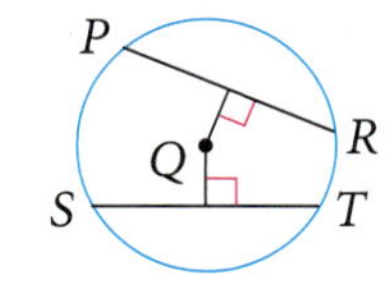

28. Open-Ended Use a circular object such as a can or a saucer to draw a circle. Construct the center of the circle.

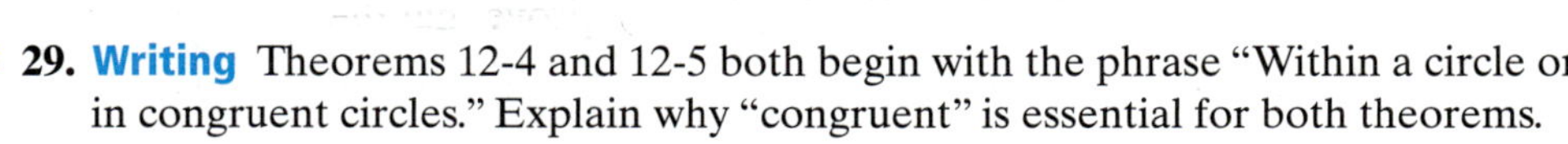

29. Writing Theorems 12-4 and 12-5 both begin with the phrase "Within a circle or in congruent circles." Explain why "congruent" is essential for both theorems.

Visit: PHSchool.com
Web Code: bce-1202

$\odot A$ and $\odot B$ are congruent. $\overline{CD}$ is a chord of both circles.

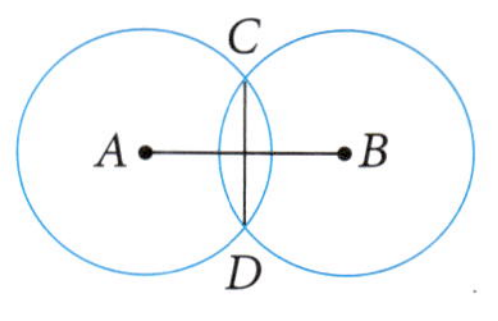

30. $AB = 8$ in., $CD = 6$ in. How long is a radius?

31. $AB = 24$ cm, radius $= 13$ cm. How long is $\overline{CD}$?

32. radius $= 13$ ft, $CD = 24$ ft. How long is $\overline{AB}$?

33. **Multiple Choice** In the diagram at the right, the endpoints of the chord are the points where the line $x = 2$ intersects the circle $x^2 + y^2 = 25$. What is the length of the chord?

Ⓐ 4.6

Ⓑ 8.0

Ⓒ 9.2

Ⓓ 10.0

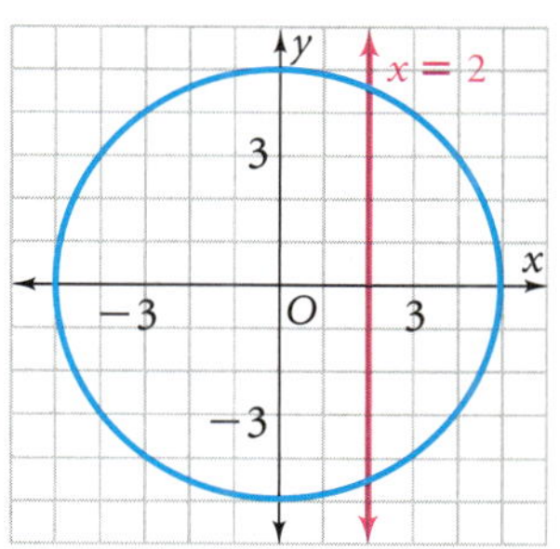

34. **Critical Thinking** The diameter of a circle is 20 cm. Two chords parallel to the diameter are 6 cm and 16 cm long. What are the possible distances between the chords to the nearest tenth of a centimeter?

Proof **35.** Prove Theorem 12-4, Part (3).

Given: $\odot P$ with $\overline{QS} \cong \overline{RT}$

Prove: $\angle QPS \cong \angle RPT$

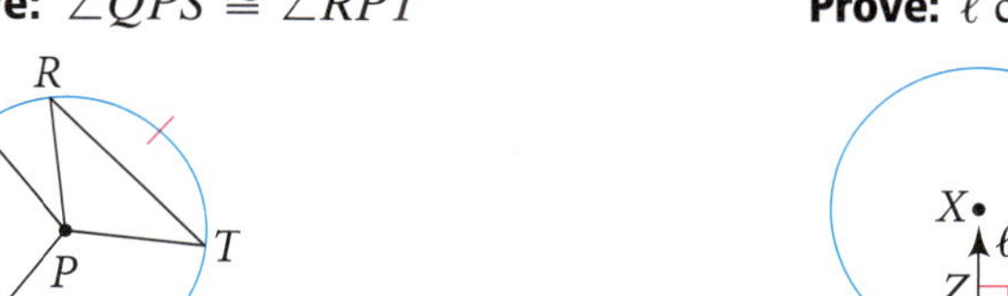

36. Prove Theorem 12-8.

Given: ℓ is the $\perp$ bisector of $\overline{WY}$.

Prove: ℓ contains the center of $\odot X$.

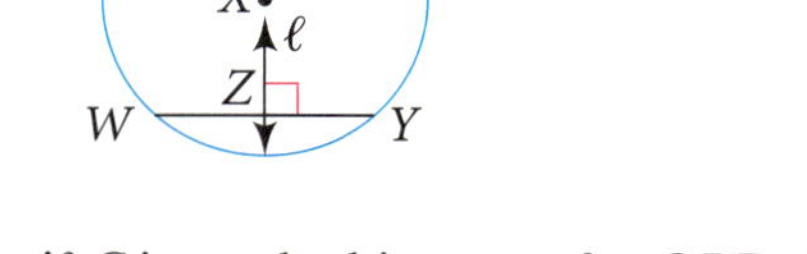

Proof **37.** $\overline{PQ}$ and $\overline{PR}$ are chords of $\odot C$. Prove that if C is on the bisector of $\angle QPR$, then $PQ = PR$.

Proof **38.** Prove Theorem 12-5, Part (2).

Given: $\odot O$ with $\overline{AB} \cong \overline{CD}$

Prove: $\overline{OE} \cong \overline{OF}$

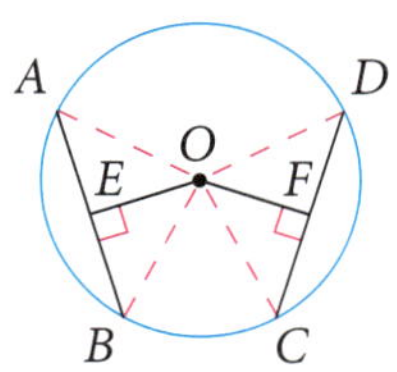

39. Given: $\odot A$ with $\overline{CE} \perp \overline{BD}$

Prove: $\overparen{BC} \cong \overparen{DC}$

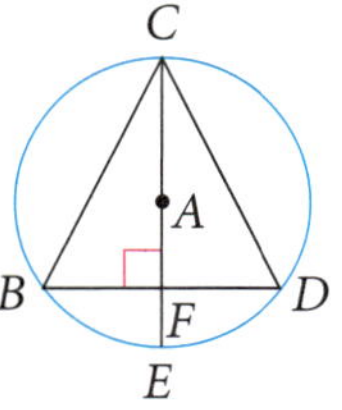

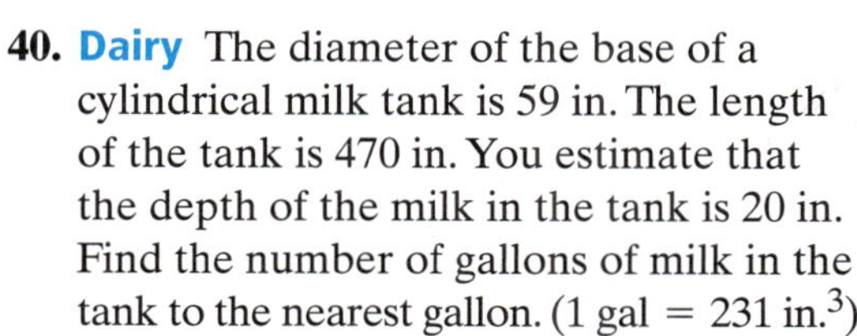

Real-World Connection

The cylinders used on milk tank trucks lie on the lateral surface and have a vertical base at each end.

40. **Dairy** The diameter of the base of a cylindrical milk tank is 59 in. The length of the tank is 470 in. You estimate that the depth of the milk in the tank is 20 in. Find the number of gallons of milk in the tank to the nearest gallon. (1 gal = 231 in.^{3})

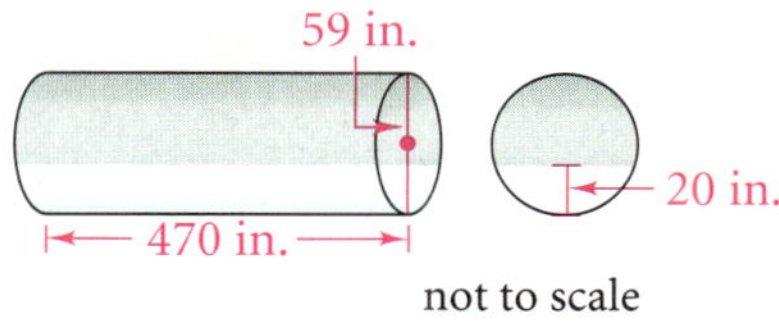

Proof **41.** If two circles are concentric and a chord of the larger circle is tangent to the smaller circle, prove that the point of tangency is the midpoint of the chord.

GO Online Lesson Quiz Visit: PHSchool.com, Web Code: bca-1202

Multiple Choice Practice

For California Standards and CAHSEE Tutorials, visit PHSchool.com. Web Code: bcq-9045

GEOM 21.0

42. Find the value of x in the figure at the right.

Ⓐ 1 Ⓑ 2 Ⓒ 3 Ⓓ 4

GEOM 11.0

43. Myra multiplies the length of each side of a triangle by $\frac{1}{5}$. By what factor can she multiply the perimeter of the original triangle to find the perimeter of the new triangle?

Ⓐ 0.008 Ⓑ 0.04 Ⓒ 0.2 Ⓓ 5

GEOM 11.0

44. A large box of laundry detergent has the shape of a rectangular prism. A similar box has length, width, and height that are each one half the length, width, and height, respectively, of the large box. How many times the volume of the small box is the volume of the large box?

Ⓐ 4 Ⓑ 8 Ⓒ 64 Ⓓ 512

CAHSEE Alg1 6.0

45. What is the x-intercept of the graph of the line below?

Ⓐ −3
Ⓑ 0.5
Ⓒ 1.5
Ⓓ 3

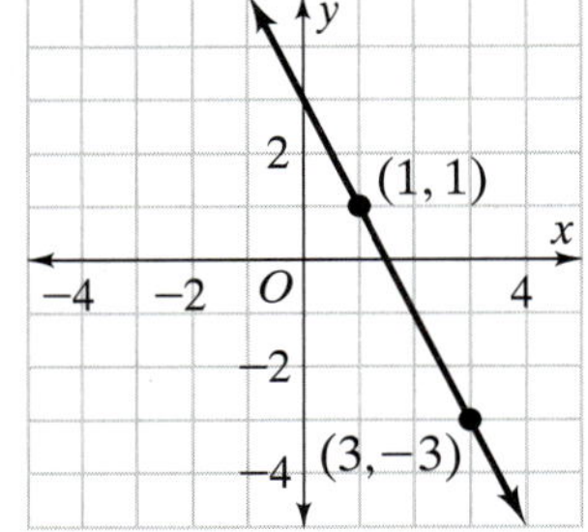

CAHSEE 7AF 4.2

46. A parking meter allots 8 min for one nickel and 16 min for one dime. At that rate, how many minutes would the meter allot for 8 quarters?

Ⓐ 40 Ⓑ 160 Ⓒ 200 Ⓓ 320

Mixed Review

Lesson 12-1

Assume that lines that appear to be tangent are tangent. *O* is the center of each circle. Find the value of *x* to the nearest tenth.

47.

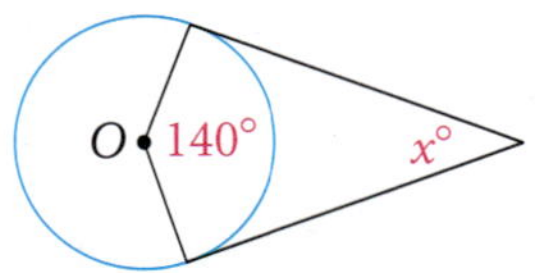

48.

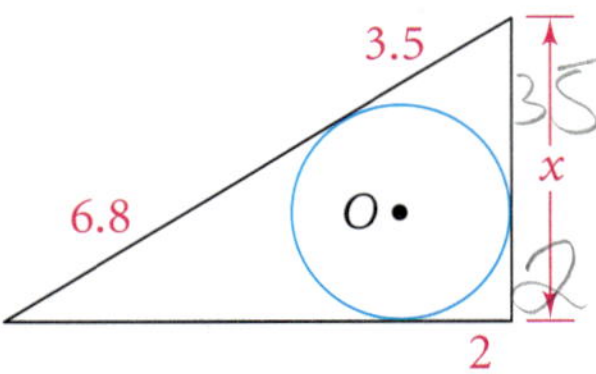

Lesson 8-5

49. From the top of a building you look down at an object on the ground. If your eyes are 50 feet above the ground and the angle of depression is 50°, how far is the object on the ground from the base of the building?

Lesson 7-5

50. The legs of a right triangle are 10 in. and 24 in. long. Find the lengths, to the nearest tenth, of the segments into which the bisector of the right angle divides the hypotenuse.

Guided Problem Solving

FOR USE WITH PAGE 674, EXERCISE 26

Understanding Math Problems **Read the problem below. Then let Jamal's thinking guide you through the solution. Check your understanding with the exercises at the bottom of the page.**

Two concentric circles have radii of 4 cm and 8 cm. A segment tangent to the smaller circle is a chord of the larger circle. What is the length of the segment?

What Jamal Thinks	What Jamal Writes
I need a picture to understand what this is about. I'll make a sketch. Concentric circles have the same center, like circles in a bull's eye. The larger circle has radius 8 cm. The smaller circle has radius 4 cm.	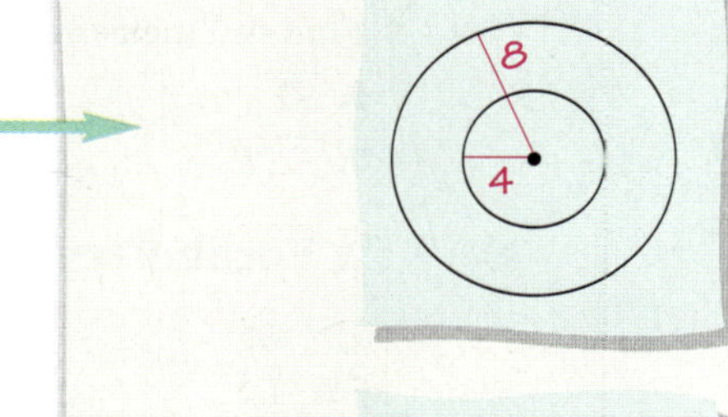
I need to draw a segment tangent to the smaller circle with both ends on the larger circle.	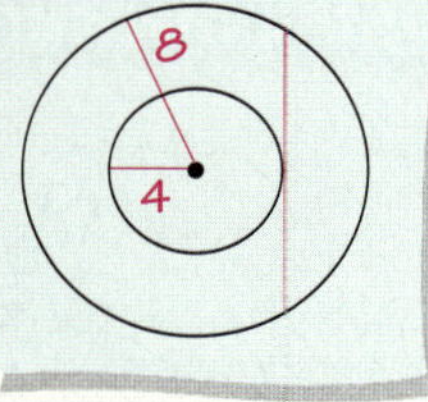
Now I need to find a relationship between these segments. I'll redo my sketch; move things around. Is a right triangle possible for my sketch so I can use the Pythagorean Theorem? Yes! Theorem 12-1 says a tangent to a circle and the radius to the point of tangency are perpendicular.	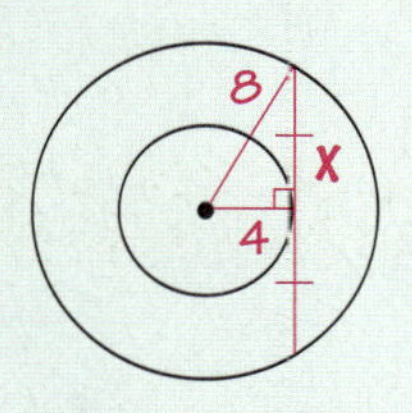
I'll use the Pythagorean Theorem . . . . . . and a calculator to find $\sqrt{48}$.	$4^2 + x^2 = 8^2$ $x^2 = 64 - 16$ $x = \sqrt{48} \approx 6.9$
To find the length of the chord, I multiply 6.9 by 2 because Theorem 12-6 says that a diameter perpendicular to a chord bisects the chord.	The chord is about 13.8 cm long.

EXERCISES

1. Two parallel chords of a circle are each 24 cm long. The distance between them is 10 cm. Find the circumference of the circle.

2. A chord 1 in. from the center of a circle is 6 in. long. How long is a chord that is 3 in. from the center of the circle?

12-3 Inscribed Angles

California Content Standards

2.0 Students write geometric proofs. ***Master***
7.0 Students prove and use theorems involving the properties of circles. ***Develop***
21.0 Students prove and solve problems regarding relationships among inscribed angles. ***Develop***

What You'll Learn

- To find the measure of an inscribed angle
- To find the measure of an angle formed by a tangent and a chord

. . . And Why

To use theorems and corollaries to find missing lengths in circles, as in Example 3

Check Skills You'll Need

GO for Help Lesson 10-6

Identify the following in $\odot P$ at the right.

1. a semicircle
2. a minor arc
3. a major arc
4. a central angle

Find the measure of each arc in $\odot P$.

5. $\overset{\frown}{ST}$
6. $\overset{\frown}{STQ}$
7. $\overset{\frown}{RST}$
8. $\overset{\frown}{TQ}$

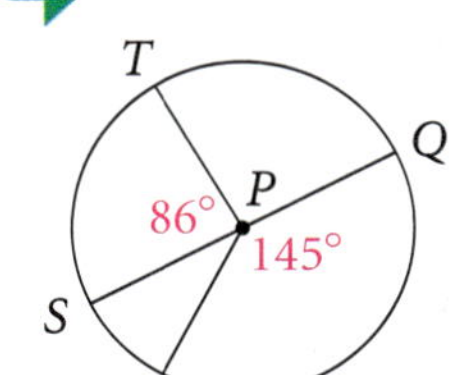

New Vocabulary • inscribed angle • intercepted arc

1 Finding the Measure of an Inscribed Angle

Hands-On Activity: Exploring Inscribed Angles

- Draw two large circles with a compass. Label the centers X and Y.
- On the circles, use a straightedge and copy the diagrams shown.

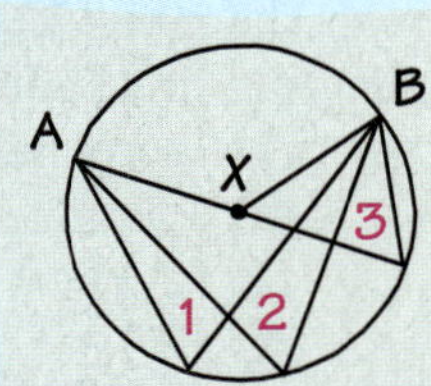

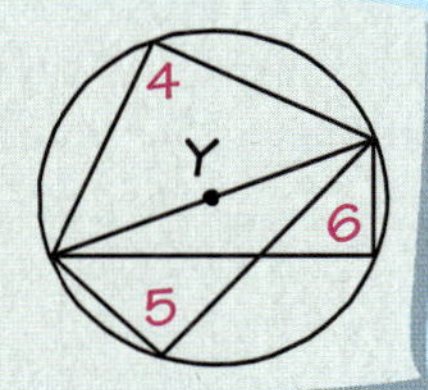

1. a. **Patterns** In $\odot X$, use a protractor to measure $\angle AXB$ and each numbered angle. Determine $m\overset{\frown}{AB}$. Record your results and look for patterns. Compare your results with others.
   b. Write a conjecture about the relationship between $m\angle 1$ and $m\overset{\frown}{AB}$.
   c. Write a conjecture about the measures of $\angle 1$, $\angle 2$, and $\angle 3$.
2. a. **Patterns** Use a protractor to measure the numbered angles in $\odot Y$. Record your results and look for patterns. Compare your results.
   b. Write a conjecture about an angle whose vertex is on a circle and whose sides intersect the endpoints of a diameter of the circle.

At the right, the vertex of $\angle C$ is on $\odot O$, and the sides of $\angle C$ are chords of the circle. $\angle C$ is an **inscribed angle.** $\overset{\frown}{AB}$ is the **intercepted arc** of $\angle C$.

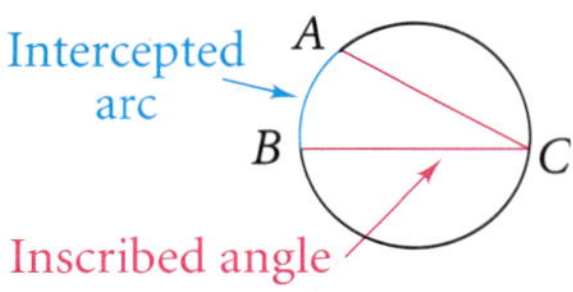

Theorem 12-9 describes the relationship between an inscribed angle and its intercepted arc.

Key Concepts

Theorem 12-9 **Inscribed Angle Theorem**

The measure of an inscribed angle is half the measure of its intercepted arc.

$$m\angle B = \frac{1}{2}m\overset{\frown}{AC}$$

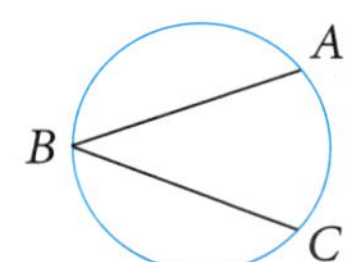

Vocabulary Tip

When different conditions are possible, each possibility can be called a <u>case</u>. You can prove a theorem by proving it for all possible cases.

To prove Theorem 12-9, there are three cases to consider.

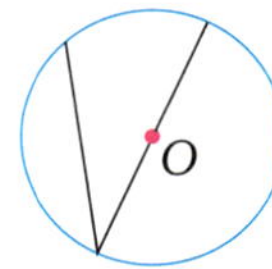

I: The center is on a side of the angle.

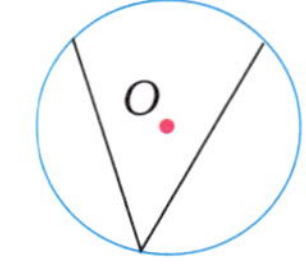

II: The center is inside the angle.

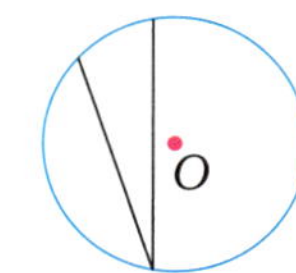

III: The center is outside the angle.

A proof of Case I is below. You will prove Cases II and III in Exercises 36 and 37.

Proof

Proof of Theorem 12-9, Case I

Given: $\odot O$ with inscribed $\angle B$ and diameter $\overline{BC}$

Prove: $m\angle B = \frac{1}{2}m\overset{\frown}{AC}$

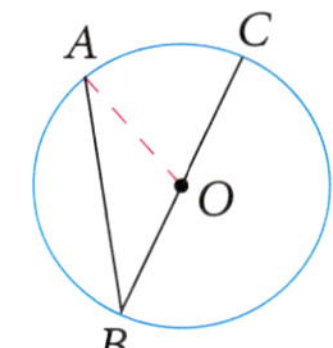

Draw radius $\overline{OA}$ to form isosceles $\triangle AOB$ with $OA = OB$ and, hence, $m\angle A = m\angle B$.

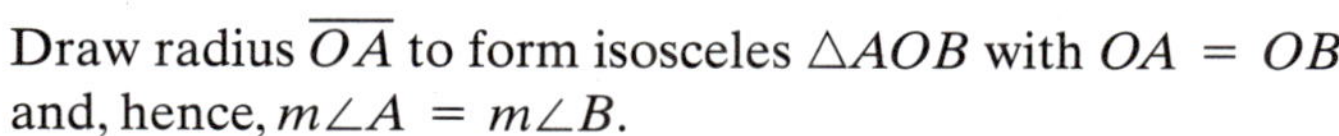

$m\overset{\frown}{AC}$	$= m\angle AOC$	**Definition of $m\overset{\frown}{AC}$**
	$= m\angle A + m\angle B$	**Triangle Exterior Angle Theorem**
	$= 2m\angle B$	**Substitute and simplify.**
$\frac{1}{2}m\overset{\frown}{AC}$	$= m\angle B$	**Solve for $m\angle B$.**

You can use the Inscribed Angle Theorem to find missing measures in circles.

1 EXAMPLE Using the Inscribed Angle Theorem

Find the values of a and b.

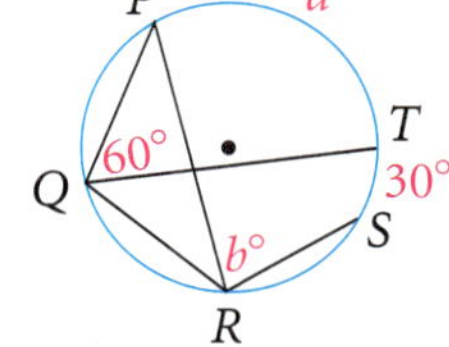

$m\angle PQT$	$= \frac{1}{2}m\overset{\frown}{PT}$	**Inscribed Angle Theorem**
60	$= \frac{1}{2}a$	**Substitute.**
120	$= a$	**Solve for a.**

$m\angle PRS$	$= \frac{1}{2}m\overset{\frown}{PS}$	**Inscribed Angle Theorem**
b	$= \frac{1}{2}\left(m\overset{\frown}{PT} + m\overset{\frown}{TS}\right)$	**Arc Addition Postulate**
	$= \frac{1}{2}(120 + 30)$	**Substitute.**
b	$= 75$	**Simplify.**

CA Standards Check

1 Find $m\angle PQR$ if $m\overset{\frown}{RS} = 60$.

You will use three corollaries to the Inscribed Angle Theorem to find measures of angles in circles. You will justify these corollaries in Exercises 38, 39, and 40.

Key Concepts

Corollaries | **Corollaries to the Inscribed Angle Theorem**

1. Two inscribed angles that intercept the same arc are congruent.
2. An angle inscribed in a semicircle is a right angle.
3. The opposite angles of a quadrilateral inscribed in a circle are supplementary.

2 EXAMPLE Using Corollaries to Find Angle Measures

For: Inscribed Angle Activity
Use: Interactive Textbook, 12-3

Find the measure of the numbered angle.

a.

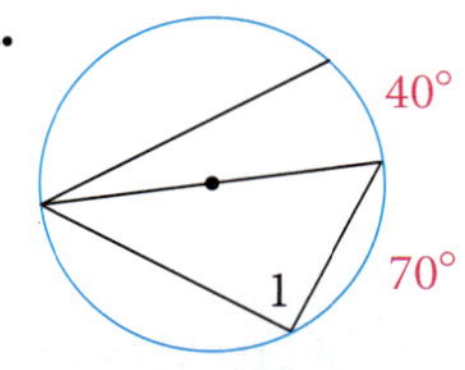

$\angle 1$ is inscribed in a semicircle.
By Corollary 2, $\angle 1$ is a right angle.
$m\angle 1 = 90$

b.

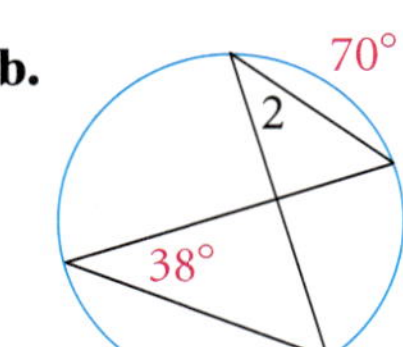

$\angle 2$ and the 38° angle intercept the same arc. By Corollary 1, the angles are congruent, so $m\angle 2 = 38$.

CA Standards Check **2** For the diagram at the right, find the measure of each numbered angle.

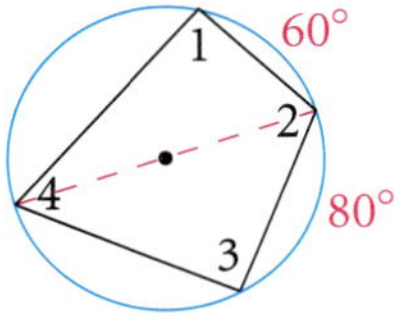

2 The Angle Formed by a Tangent and a Chord

In the diagram, B and C are fixed points, and point A moves along the circle. From the Inscribed Angle Theorem, you know that as A moves, $m\angle A$ remains the same and is $\frac{1}{2}m\widehat{BC}$. As the last diagram suggests, this is also true when A and C coincide.

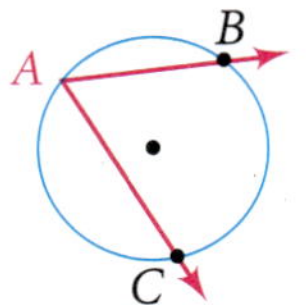

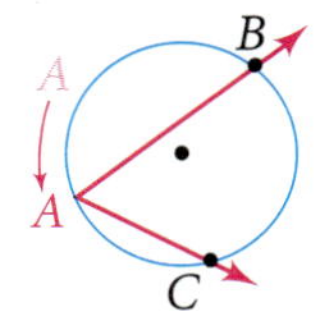

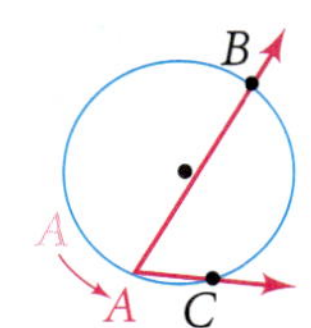

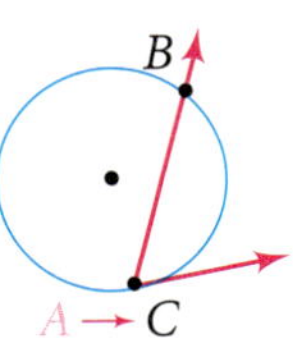

Key Concepts

Theorem 12-10

The measure of an angle formed by a tangent and a chord is half the measure of the intercepted arc.

$$m\angle C = \frac{1}{2}m\widehat{BDC}$$

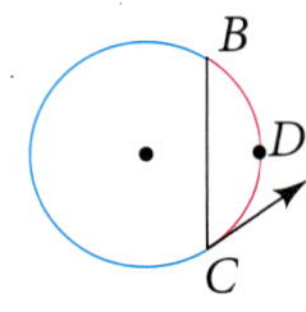

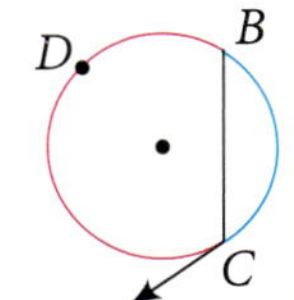

You will prove Theorem 12-10 in Exercise 41.

3 EXAMPLE Using Theorem 12-10

In the diagram at the right, $\overrightarrow{KJ}$ is tangent to the circle at J. Find the values of x and y.

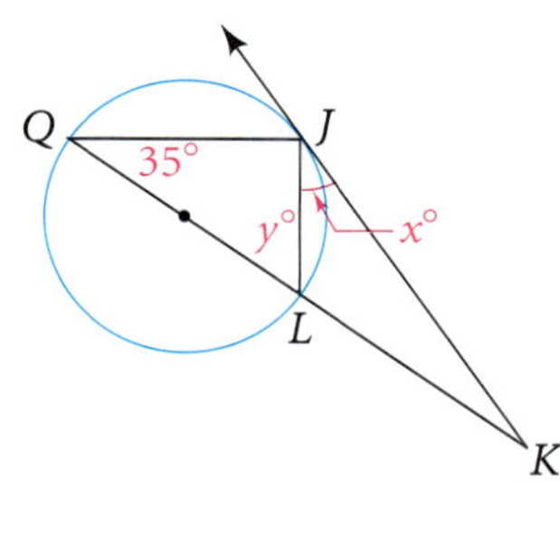

$x = \frac{1}{2}m\widehat{JL}$	Theorem 12-10
$= m\angle Q$	Inscribed Angle Theorem
$= 35$	Substitution
$y = \frac{1}{2}m\widehat{QJ}$	Theorem 12-10
$= \frac{1}{2}(m\widehat{QL} - m\widehat{JL})$	Arc Addition Postulate
$= \frac{1}{2}(180 - 70)$	Substitute.
$= 55$	Simplify.

CA Standards Check 3 Describe two ways to find $m\angle QJK$ using Theorem 12-10.

EXERCISES Standards Practice

For more exercises, see *Extra Skill, Word Problem, and Proof Practice.*

GEOM 2.0, 7.0, 21.0

A Practice by Example

Example 1 (page 679)

GO for Help

Identify the inscribed angle and its intercepted arc.

1. 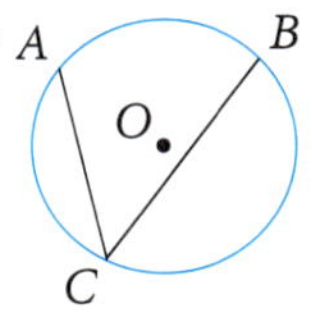

2. 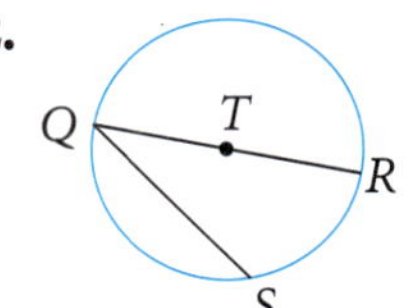

3. 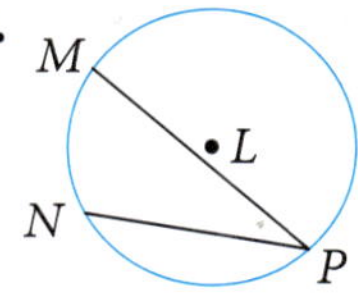

4. a. Name the four inscribed angles and their intercepted arcs.
   b. Which angles appear to intercept major arcs? What kind of angles do these appear to be?

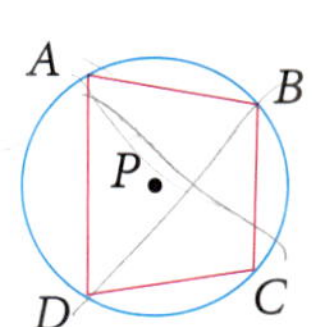

Find the value of each variable.

5. 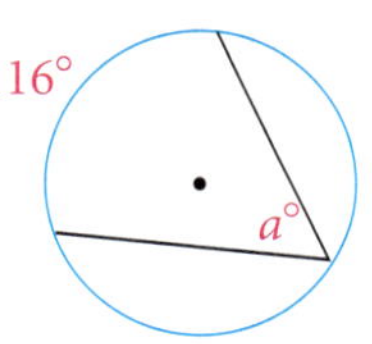

6. 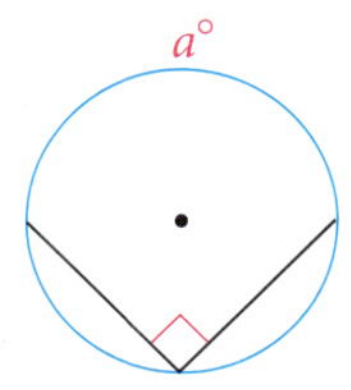

7. 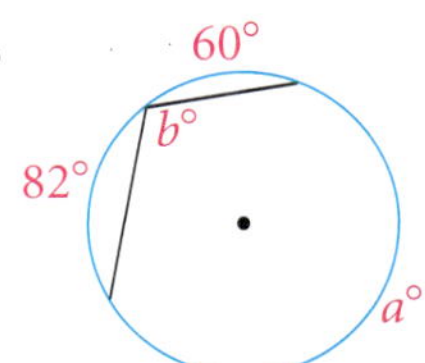

8. 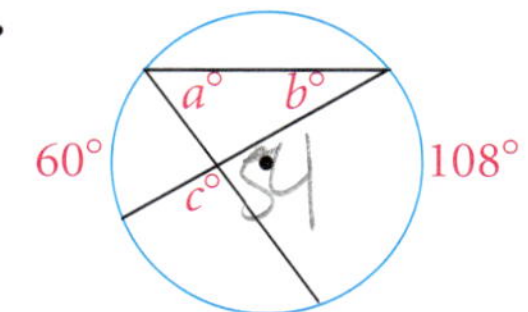

9. 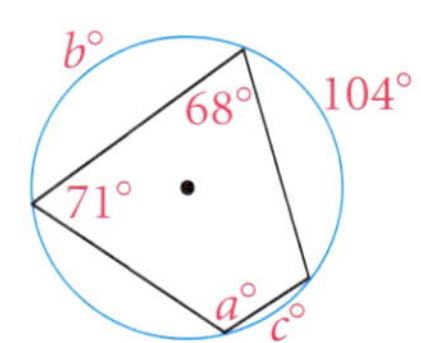

10. 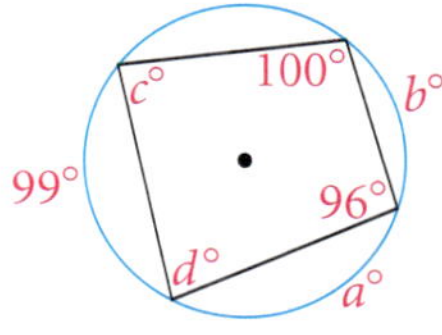

Example 2 (page 680)

11. 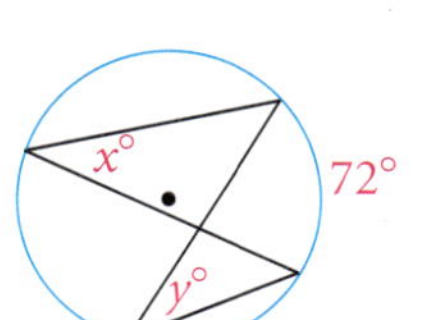

12. 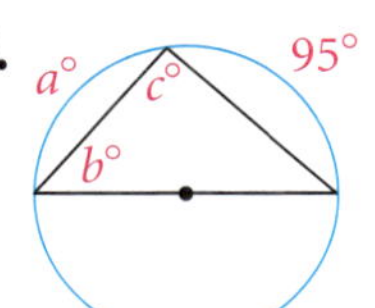

13. 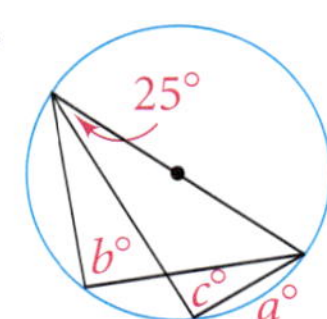

14. 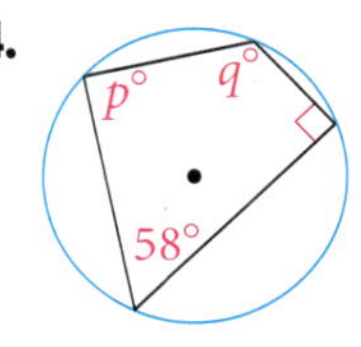

Example 3
(page 681)

Find the value of each variable. You may assume that rays that appear to be tangent are tangent.

15.

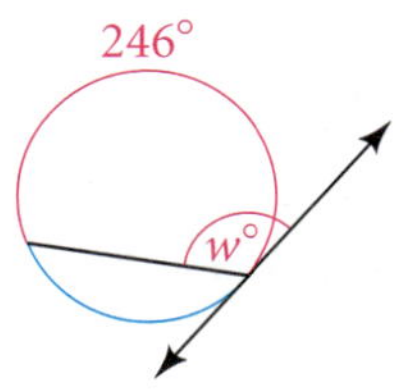

16.

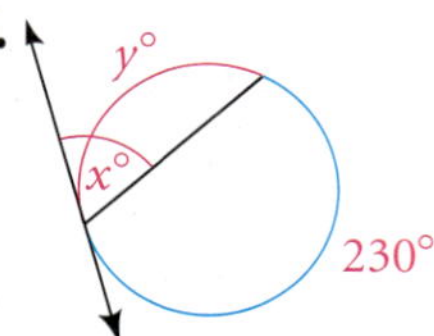

17.

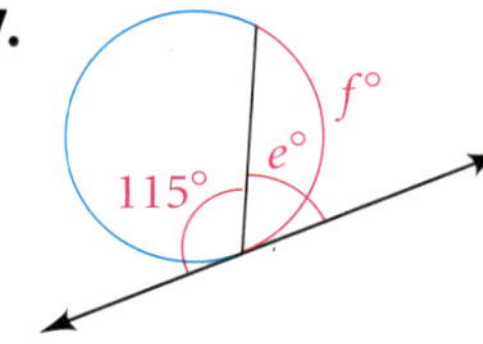

18.

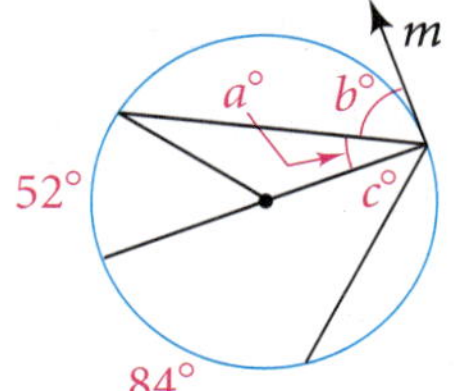

19.

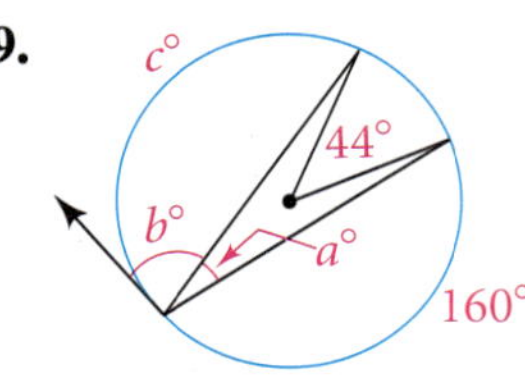

20.

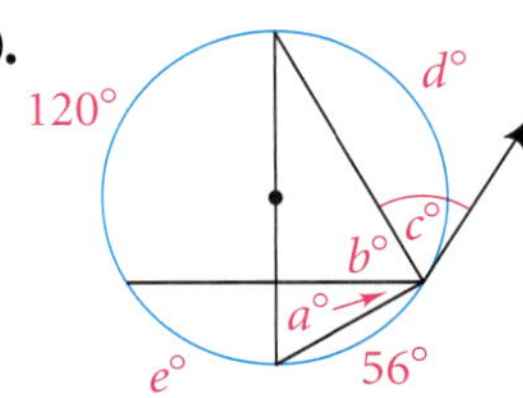

B Apply Your Skills

Find each indicated measure for ⊙*O*.

21. **a.** $m\widehat{BC}$
b. $m\angle B$
c. $m\angle C$
d. $m\widehat{AB}$

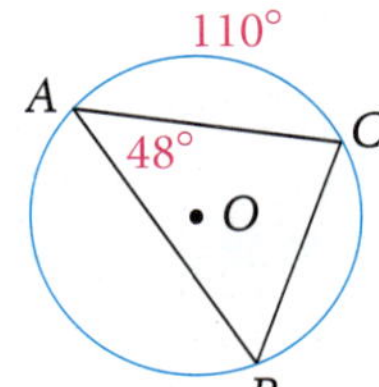

22. **a.** $m\angle A$
b. $m\widehat{CE}$
c. $m\angle C$
d. $m\angle D$
e. $m\angle ABE$

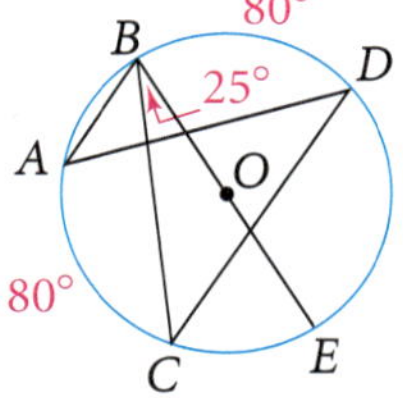

23. **Multiple Choice** What is the measure of $\widehat{EA}$ in the circle at the right?
Ⓐ 21 Ⓑ 30 Ⓒ 36 Ⓓ 51

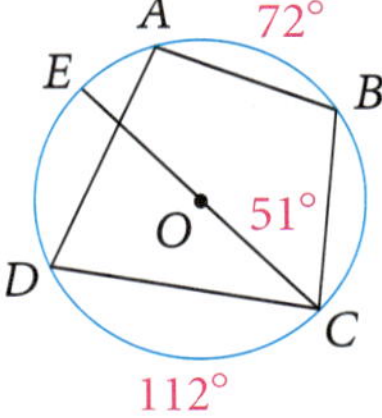

24. Find each indicated measure for the circle in Exercise 23.
a. $m\widehat{BC}$ **b.** $m\angle A$ **c.** $m\angle B$ **d.** $m\angle BCD$

25. **Writing** Copy the diagram at the right on your paper. Draw chord $\overline{RQ}$. Explain why $m\widehat{PR} = m\widehat{QS}$.

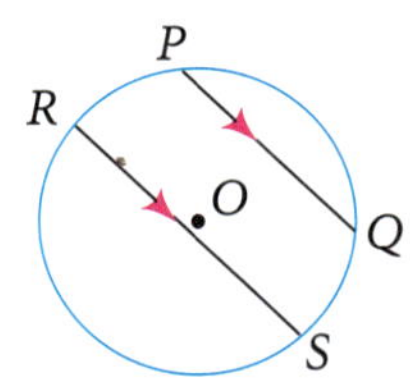

26. **a.** **Open-Ended** Sketch a trapezoid inscribed in a circle. Repeat several times using different circles.
b. **Make a Conjecture** What kind of trapezoid can be inscribed in a circle? Justify your response.

Vocabulary Tip

A regular star (Exercise 27) has congruent sides. Also, the angles "pointing out" are congruent, as are the angles "pointing in."

27. **Landscape Architecture** Some circular English gardens, like the one shown here, have paths in the shape of an inscribed regular star.
a. Find the measure of an inscribed angle formed by the star in the garden shown here.
b. What is the measure of an inscribed angle in a garden with a five-pointed star?

28. **Critical Thinking** A parallelogram inscribed in a circle must be what kind of parallelogram? Explain.

Graphing Calculator **The diameter of a circle is 10 cm. Find the dimensions of the largest figure of each type that can be inscribed in the circle. (*Hint:* Use techniques demonstrated in the Activity Lab on page 616.)**

29. a rectangle **30.** a triangle **31.** a right triangle

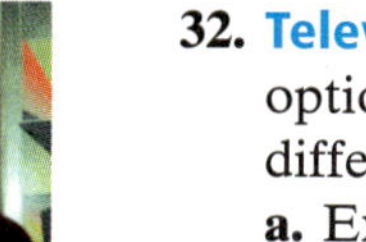

Real-World Connection

A big red "tally light" on each camera tells this anchorman which camera is "hot" (on the air).

32. Television The director of a telecast wants the option of showing the same scene from three different views.

a. Explain why cameras in the positions shown in the diagram will transmit the same scene.

b. **Critical Thinking** Will the scenes look the same to the director when she views them on the control room monitors? Explain.

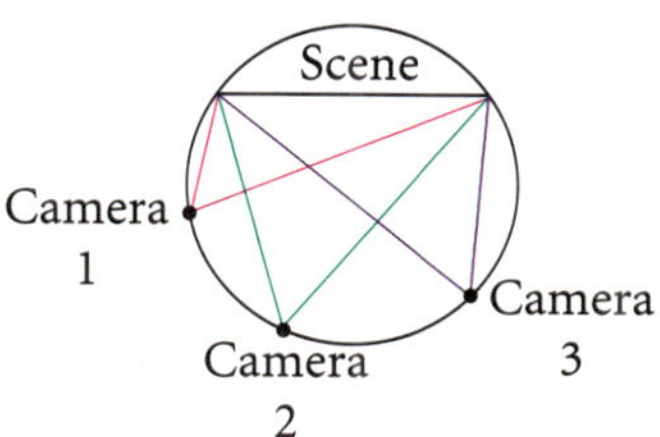

33. Constructions The diagrams below show the construction of a tangent to a circle from a point outside the circle. Explain why $\overleftrightarrow{BC}$ must be tangent to $\odot A$. (*Hint:* Copy the third diagram and draw $\overline{AC}$.)

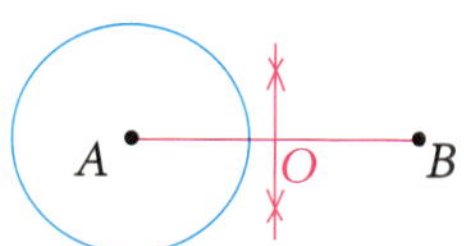

Given: $\odot A$ and point B. Construct the midpoint of $\overline{AB}$. Label the point O.

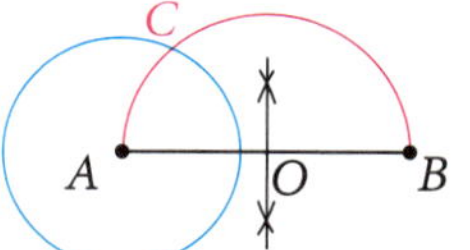

Construct a semicircle with radius OA and center O. Label its intersection with $\odot A$ as C.

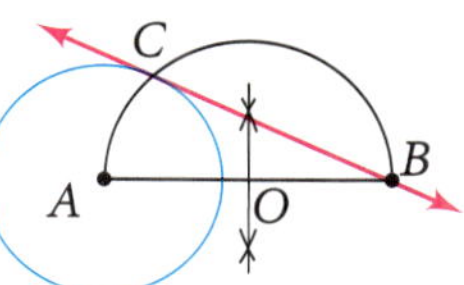

Draw $\overleftrightarrow{BC}$.

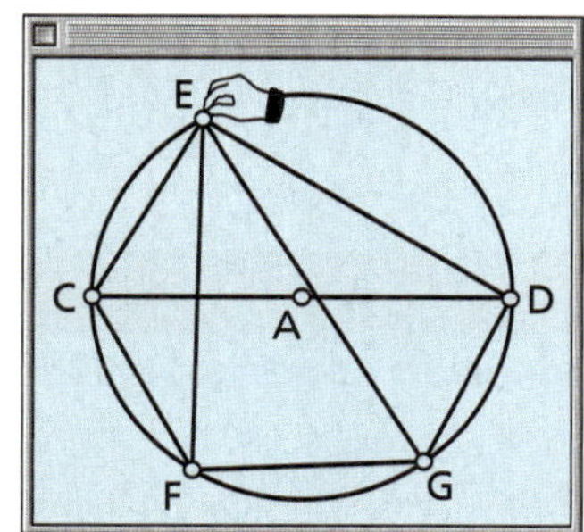

34. Technology Construct $\odot A$ and the chords shown with geometry software.

a. As you move E on $\overset{\frown}{CED}$ between C and D, which inscribed angles remain congruent?

b. Which inscribed angle remains a right angle?

c. Which inscribed angles remain supplementary in quadrilateral $EFGD$?

35. Constructions Use Corollary 2 of Theorem 12-9 to construct a right triangle given one leg and the hypotenuse.

Proof **Write a proof for each of Exercises 36–41.**

36. Inscribed Angle Theorem, Case II

Given: $\odot O$ with inscribed $\angle ABC$

Prove: $m\angle ABC = \frac{1}{2}m\overset{\frown}{AC}$

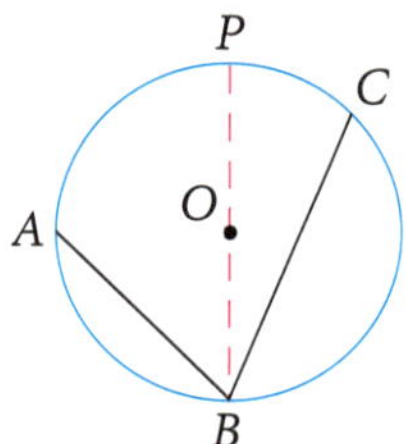

Hint: Use the Inscribed Angle Theorem, Case I.

37. Inscribed Angle Theorem, Case III

Given: $\odot S$ with inscribed $\angle PQR$

Prove: $m\angle PQR = \frac{1}{2}m\overset{\frown}{PR}$

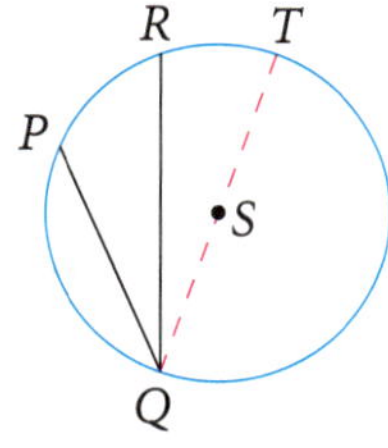

Hint: Use the Inscribed Angle Theorem, Case I.

38. Inscribed Angle Theorem, Cor. 1

Given: $\odot O$; $\angle A$ intercepts $\overset{\frown}{BC}$, and $\angle D$ intercepts $\overset{\frown}{BC}$.

Prove: $\angle A \cong \angle D$

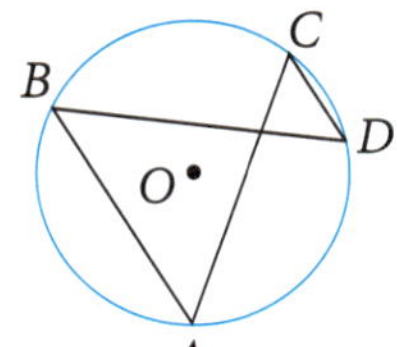

39. Inscribed Angle Theorem, Cor. 2

Given: $\odot O$ with $\angle CAB$ inscribed in a semicircle

Prove: $\angle CAB$ is a right angle.

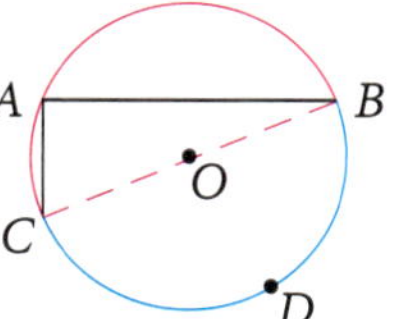

Visit: PHSchool.com
Web Code: bce-1203

Problem Solving Hint

In Exercise 41, let $\overline{GH}$ first be a diameter.

40. Inscribed Angle Theorem, Cor. 3

Given: quadrilateral $ABCD$ inscribed in $\odot O$

Prove: $\angle A$ and $\angle C$ are supplementary. $\angle B$ and $\angle D$ are supplementary.

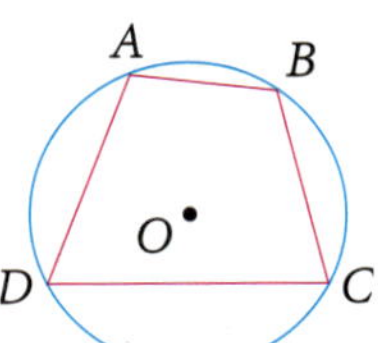

41. Theorem 12-10

Given: $\overline{GH}$ and tangent ℓ intersecting at H on $\odot E$

Prove: $m\angle GHI = \frac{1}{2}m\overset{\frown}{GFH}$

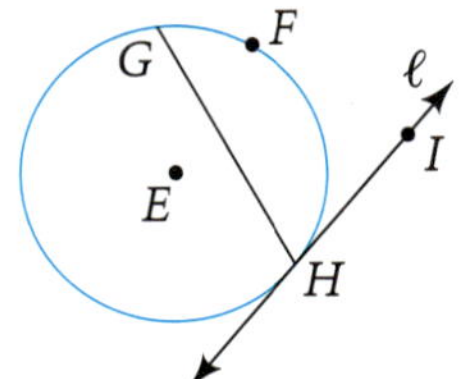

Challenge

Critical Thinking **Is the statement true or false? If true, give a convincing argument. If false, give a counterexample.**

42. If two angles inscribed in a circle are congruent, then they intercept the same arc.

43. If an inscribed angle is a right angle, then it is inscribed in a semicircle.

44. A circle can always be circumscribed about a quadrilateral whose opposite angles are supplementary.

45. **Constructions** Draw two segments. Label their lengths x and y. Construct the geometric mean of x and y. (*Hint:* Recall a theorem about a geometric mean.)

Multiple Choice Practice

For California Standards and CAHSEE Tutorials, visit PHSchool.com. Web Code: bcq-9045

GEOM 21.0

46. Find the value of the variables a and b in the figure at the right.

Ⓐ $a = 90°$ and $b = 130°$

Ⓑ $a = 50°$ and $b = 130°$

Ⓒ $a = 90°$ and $b = 65°$

Ⓓ $a = 65°$ and $b = 65°$

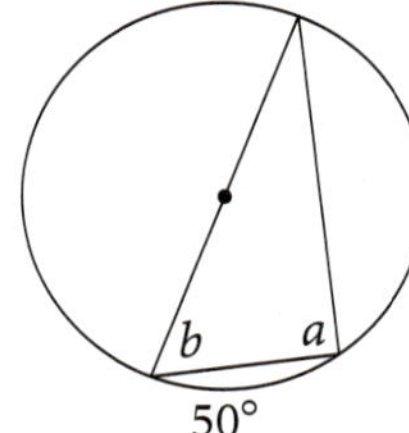

GEOM 9.0

47. The base and side views of a solid figure are shown at the right. What is the volume of the figure to the nearest cubic inch?

Ⓐ 603 in.^{3}　　Ⓒ 151 in.^{3}

Ⓑ 226 in.^{3}　　Ⓓ 113 in.^{3}

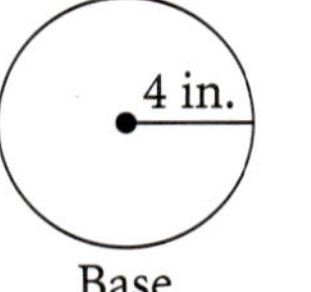

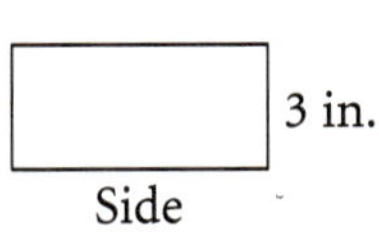

GEOM 9.0

48. An exercise ball is about 4 feet wide. Which is the best estimate for its volume?

Ⓐ 8 cubic feet　　Ⓒ 34 cubic feet

Ⓑ 16 cubic feet　　Ⓓ 64 cubic feet

CAHSEE 7MG 2.4

49. Rick measures the capacity of one type of juice cup and finds that the cup can hold 8 ounces of juice. Rick needs about 250 of these cups for juice at the school dance. How can Rick find the number of gallons of juice he should buy?

Ⓐ Divide 250 by 8. Then multiply the answer by 64.

Ⓑ Divide 250 by 64. Then multiply the answer by 8.

Ⓒ Multiply 250 by 8. Then divide the answer by 128.

Ⓓ Multiply 250 by 128. Then divide the answer by 8.

Mixed Review

Lesson 12-2 x^2 **Algebra** **Find the value of x to the nearest tenth.**

50. 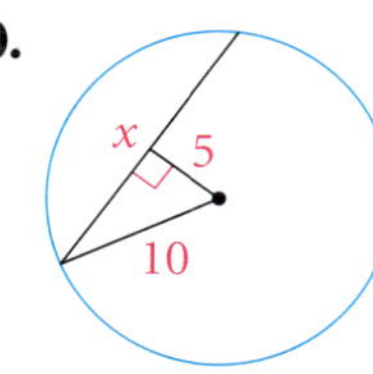

51. 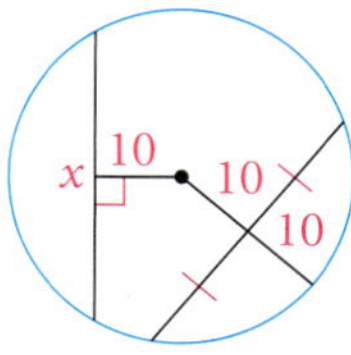

52. 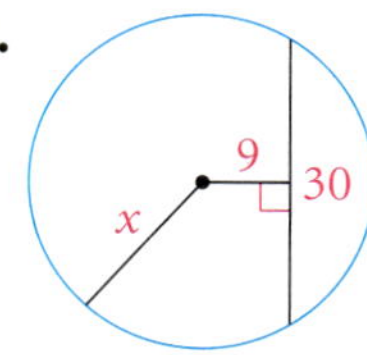

Lesson 10-5 **Find the area of each triangle. Give answers to the nearest tenth.**

53. 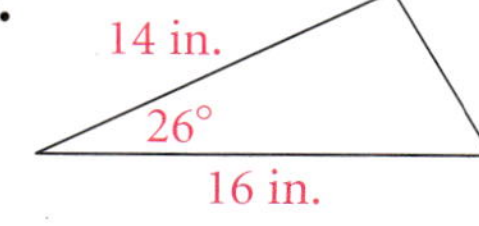

54. 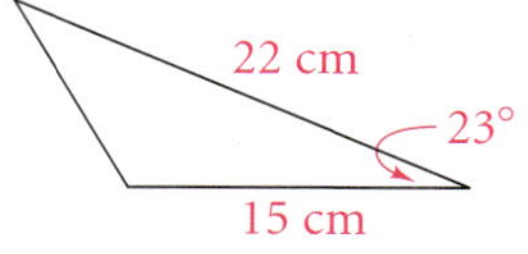

55. 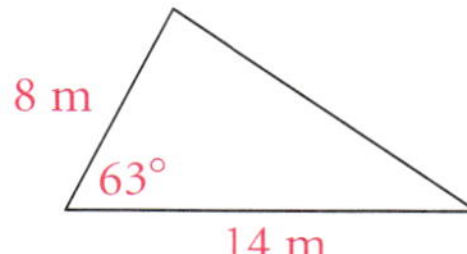

Lesson 7-3 **Indirect Measurement** To find the width of a river, you have made the measurements shown in the sketch.

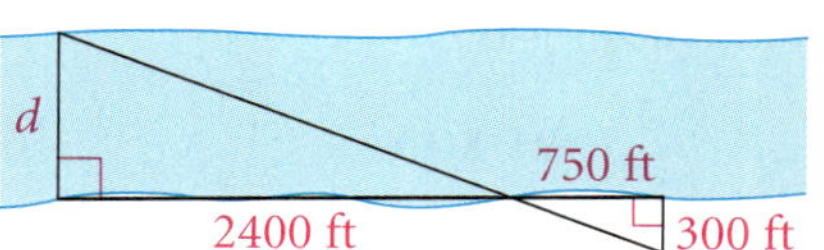

56. Explain why the triangles are similar.

57. a. Find the width of the river in feet.
b. Find the width of the river in miles.

Checkpoint Quiz 1 — Lessons 12-1 through 12-3

Each polygon below circumscribes the circle. Find the perimeter of the polygon.

1. 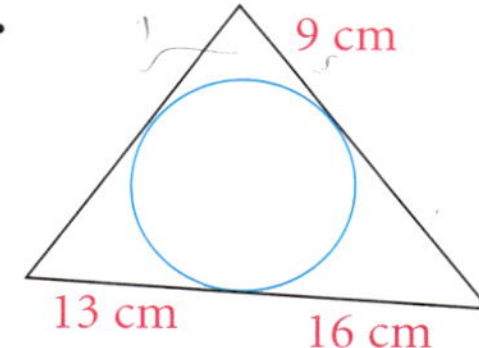

2. 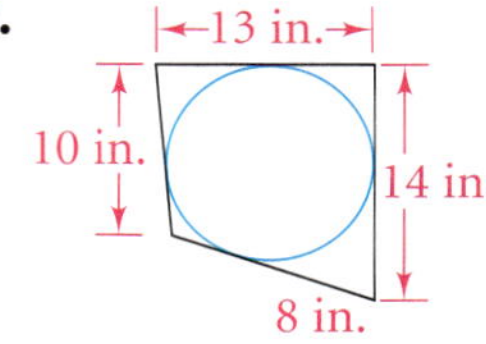

3. 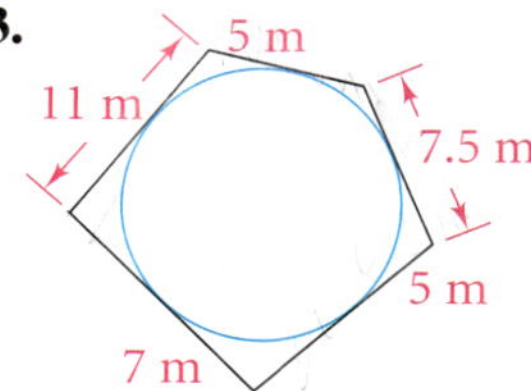

x^2 **Algebra** **Find the value of x.**

4. 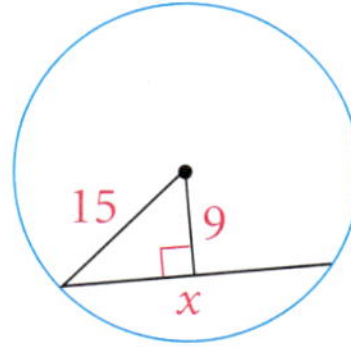

5. 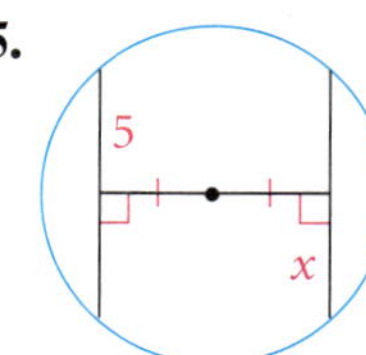

6. 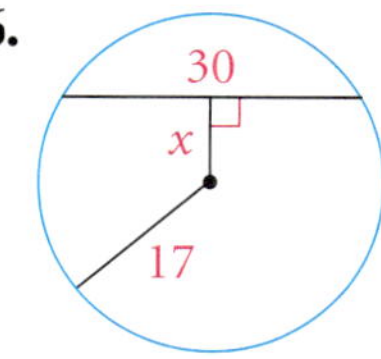

x^2 **Algebra** **Find the value of each variable. Lines that appear to be tangent are tangent.**

7. 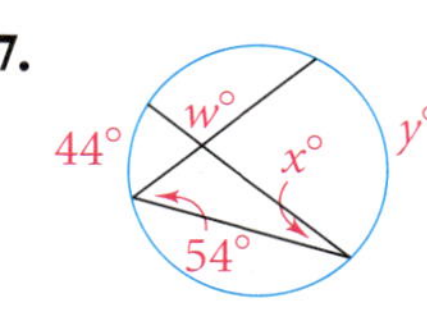

8. 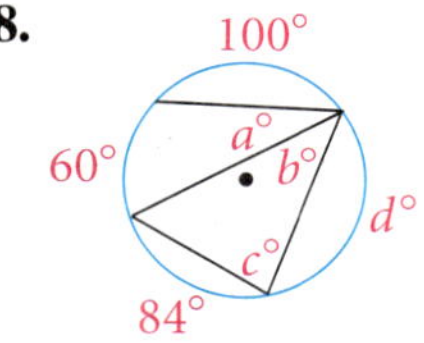

9. 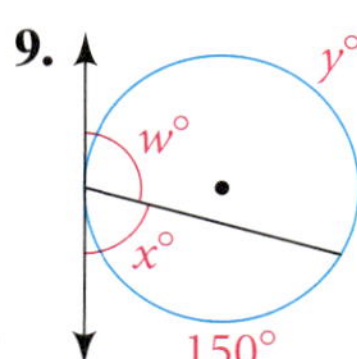

10. 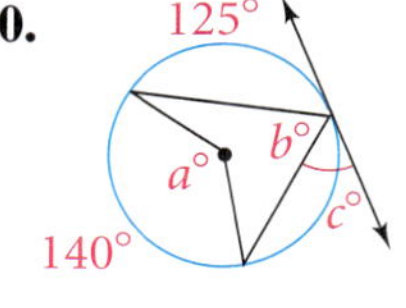

Exploring Chords and Secants

Technology

FOR USE WITH LESSON 12-4

Prepares for GEOM 21.0 Students prove and solve problems regarding relationships among chords and secants.

Construct

Construct a circle with center A and two chords $\overline{BC}$ and $\overline{DE}$ that intersect each other at F.

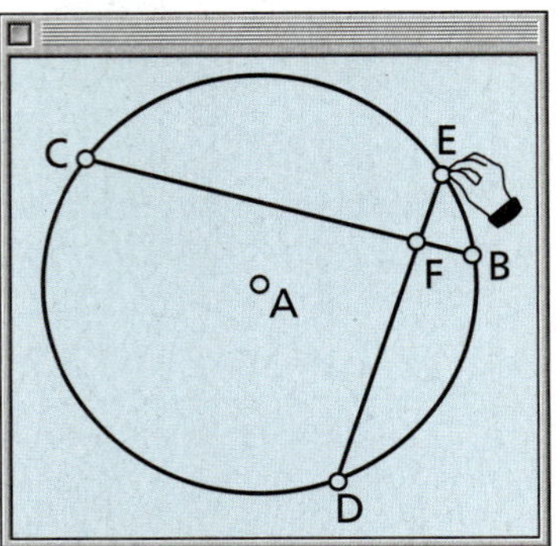

Investigate

Measure $\overline{BF}$, $\overline{FC}$, $\overline{EF}$, and $\overline{FD}$. Use the calculator program of your software to find the products $BF \cdot FC$ and $EF \cdot FD$. Manipulate your construction and observe the products. What do you discover? Use your discovery to answer Exercise 1.

Construct

A *secant* is a line that intersects a circle in two points. Construct another circle and two secants that intersect in a point outside the circle. Label your construction as shown in the diagram.

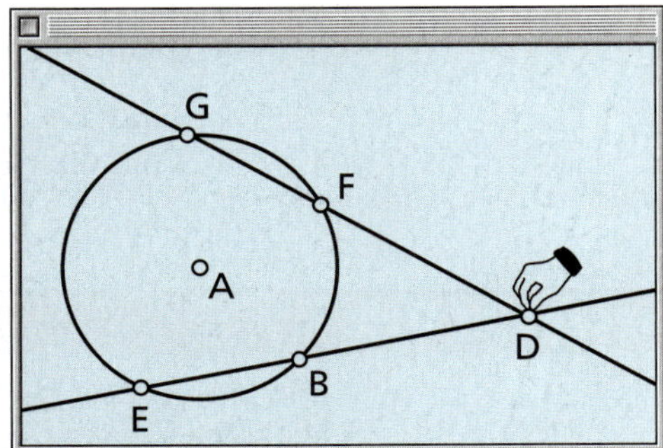

Investigate

Measure $\overline{DG}$, $\overline{DF}$, $\overline{DE}$, and $\overline{DB}$. Calculate the products $DG \cdot DF$ and $DE \cdot DB$. Manipulate your construction and observe the products. What do you discover? Use your discovery to answer Exercise 2.

EXERCISES

Use results from your two investigations above. Make conjectures about the products of the lengths of the segments to a circle from the point of intersection of the figures given below.

1. two chords inside a circle

2. two secants outside a circle

Extend

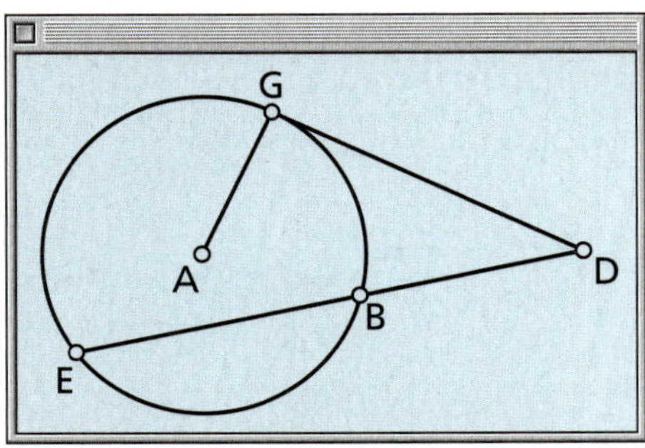

3. Construct circle A with radius $\overline{AG}$ as shown. Construct a segment $\overline{DG}$ perpendicular to radius $\overline{AG}$. Then construct a secant $\overline{DE}$ that does not cross $\overline{AG}$. Hide $\overline{AG}$. Measure $\overline{DG}$, $\overline{DE}$, and $\overline{DB}$. Calculate DG^2 and the product $DE \cdot DB$. What is true about the products you calculated? Can you explain how this special case is related to the case of two secants?

4. Combine conjectures from Exercises 1–3 into a conjecture about any two intersecting lines, each of which also intersects a circle.

Angle Measures and Segment Lengths

California Content Standards

2.0 Students write geometric proofs. ***Master***
7.0 Students prove and use theorems involving the properties of circles. ***Develop***
21.0 Students prove and solve problems regarding relationships among chords, secants, tangents, inscribed angles, and inscribed and circumscribed polygons of circles. ***Master***

What You'll Learn

- To find the measures of angles formed by chords, secants, and tangents
- To find the lengths of segments associated with circles

. . . And Why

To find the measure of an arc of a circular basin, as in Example 2

In the diagram at the right, $\overline{FE}$ and $\overline{FD}$ are tangents to $\odot C$. Find each arc measure, angle measure, or length.

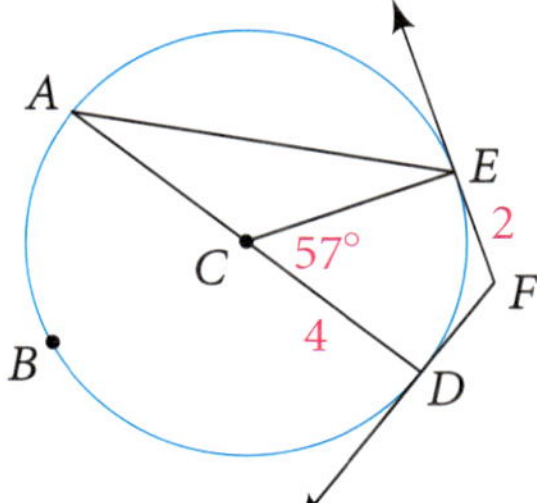

1. $m\widehat{DE}$	**2.** $m\widehat{AED}$	**3.** $m\widehat{EBD}$
4. $m\angle EAD$	**5.** $m\angle AEC$	**6.** CE
7. DF	**8.** CF	**9.** $m\angle EFD$

• secant

1 Finding Angle Measures

Vocabulary Tip

The word secant may refer to a line, ray, or segment.

A **secant** is a line that intersects a circle at two points. $\overleftrightarrow{AB}$ is a secant ray, and $\overline{AB}$ is a secant segment.

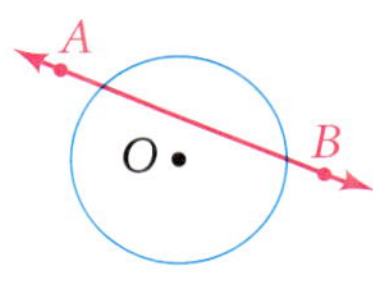

Angles formed by secants, tangents, and chords intercept arcs on circles. The measures of the intercepted arcs can help you find the measures of the angles.

Key Concepts

Theorem 12-11

The measure of an angle formed by two lines that

(1) intersect inside a circle is half the sum of the measures of the intercepted arcs.

$$m\angle 1 = \frac{1}{2}(x + y)$$

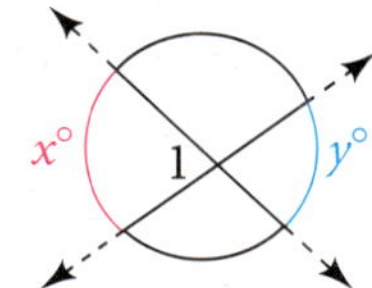

(2) intersect outside a circle is half the difference of the measures of the intercepted arcs.

$$m\angle 1 = \frac{1}{2}(x - y)$$

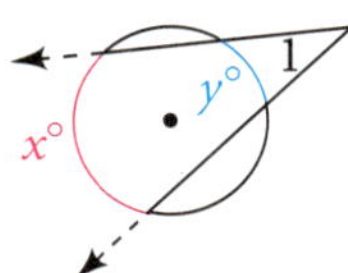

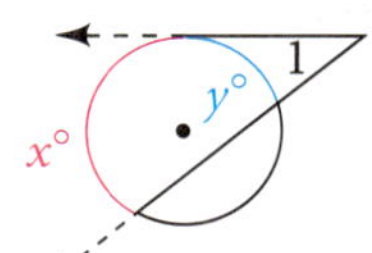

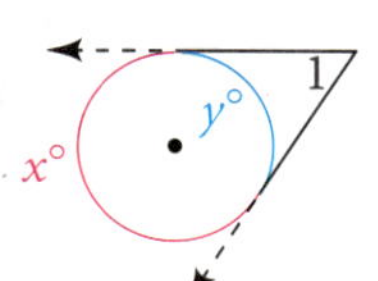

Part (1) is proved on the next page. The three cases of Part (2) are proved in Exercises 29 and 30.

Proof

Proof of Theorem 12-11, Part (1)

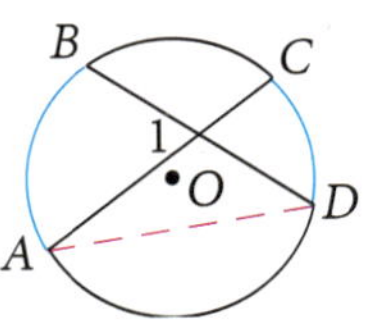

Given: $\odot O$ with intersecting chords $\overline{AC}$ and $\overline{BD}$

Prove: $m\angle 1 = \frac{1}{2}(m\widehat{AB} + m\widehat{CD})$

Begin by drawing $\overline{AD}$ as shown in the diagram.

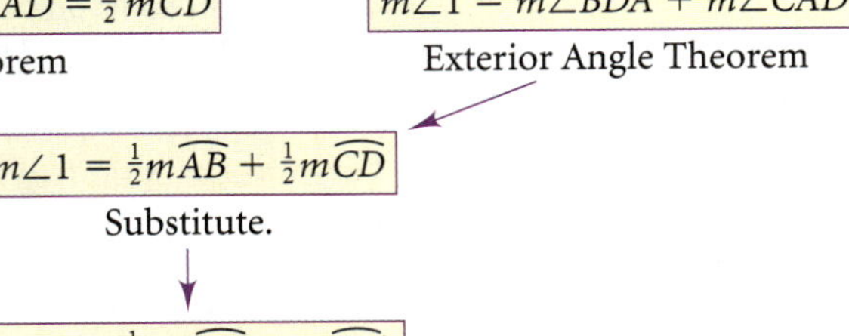

$m\angle 1 = \frac{1}{2}m\widehat{AB} + \frac{1}{2}m\widehat{CD}$

Substitute.

$m\angle 1 = \frac{1}{2}(m\widehat{AB} + m\widehat{CD})$

Distributive Property

You can use Theorem 12-11 to find the measures of angles and intercepted arcs.

1 EXAMPLE Finding Angle Measures

Algebra Find the value of each variable.

a.

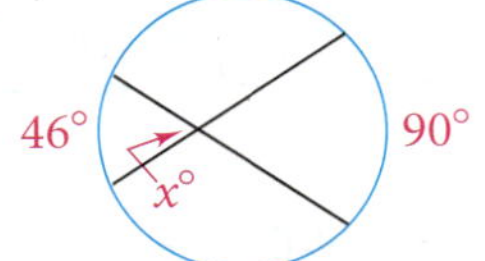

b. 95°

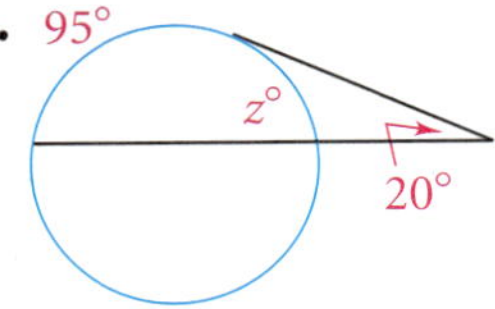

$x = \frac{1}{2}(46 + 90)$ **Theorem 12-11 (1)**

$x = 68$ **Simplify.**

$20 = \frac{1}{2}(95 - z)$ **Theorem 12-11 (2)**

$40 = 95 - z$ **Solve for *z*.**

$z = 55$

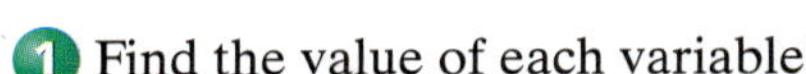

a.

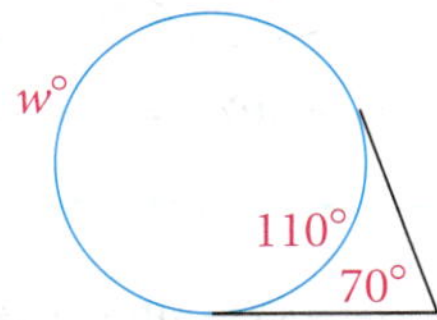

b.

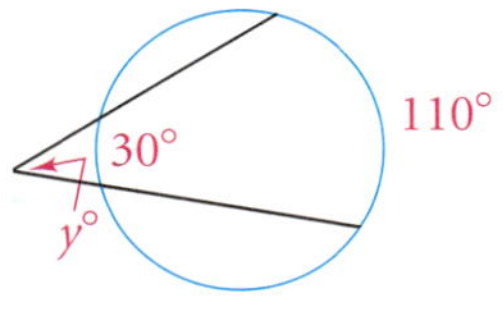

2 EXAMPLE Real-World Connection

Real-World Connection

Line-of-sight tangents to this fountain basin form a larger angle than do those to the distant basin.

You focus your camera on a fountain. Your camera is at the vertex of the angle formed by tangents to the fountain. You estimate that this angle is 40°.

What is the measure, in degrees, of the arc of the circular basin of the fountain that will be in the photograph?

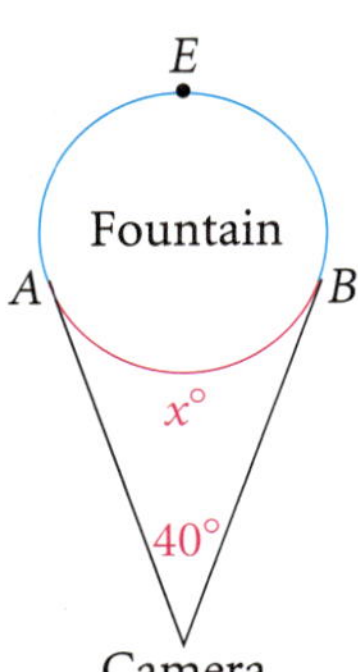

Let $m\widehat{AB} = x$.

Then $m\widehat{AEB} = 360 - x$.

$$40 = \frac{1}{2}(m\widehat{AEB} - m\widehat{AB})$$ **Theorem 12-11 (2)**

$$40 = \frac{1}{2}[(360 - x) - x]$$ **Substitute.**

$$40 = \frac{1}{2}(360 - 2x)$$ **Simplify.**

$$40 = 180 - x$$ **Distribute.**

$$x = 140$$ **Solve for x.**

A 140° arc will be in the photograph.

CA Standards Check **2** **Critical Thinking** To photograph a 160° arc of the basin, should you move towards or away from the fountain? What angle should the tangents form?

2 Finding Segment Lengths

From a given point P, you can draw two segments to a circle along infinitely many lines. For example, $\overline{PA_1}$ and $\overline{PB_1}$ lie along one such line. Theorem 12-12 states the surprising result that, no matter which line you use, the product $PA \cdot PB$ remains constant.

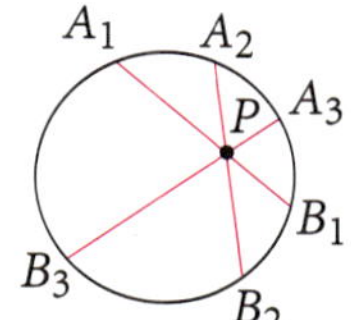

$PA_i \cdot PB_i$ is constant.

Key Concepts

Theorem 12-12

For a given point and circle, the product of the lengths of the two segments from the point to the circle is constant along any line through the point and circle.

I.

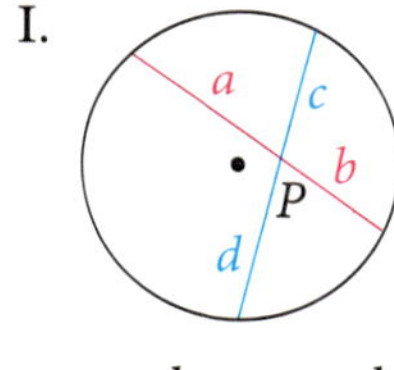

$a \cdot b = c \cdot d$

II.

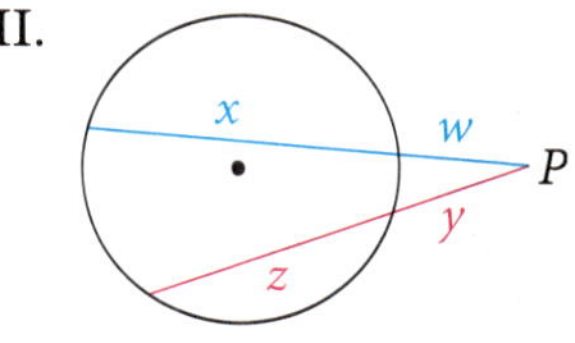

$(w + x)w = (y + z)y$

III.

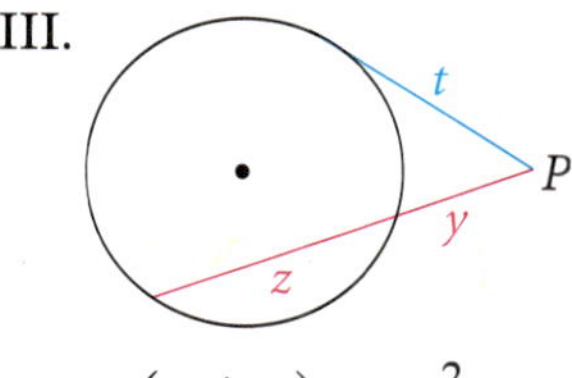

$(y + z)y = t^2$

Vocabulary Tip

The two segments to a circle along a secant are called secant segments.

Note in Case III that the tangent segment is used twice.

Here is a proof for Case I. You will prove II and III in Exercises 31 and 32.

Proof

Proof of Theorem 12-12 (I)

Given: a circle with chords $\overline{AB}$ and $\overline{CD}$ intersecting at P

Prove: $a \cdot b = c \cdot d$

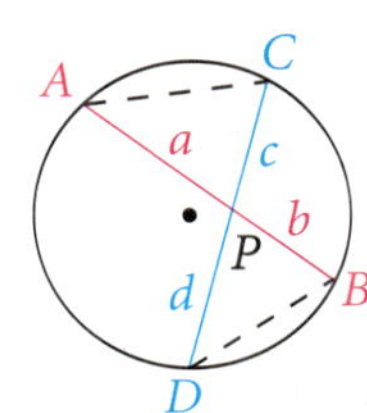

Draw $\overline{AC}$ and $\overline{BD}$. $\angle A \cong \angle D$ and $\angle C \cong \angle B$ because they are inscribed angles and each pair intercept the same arc. Thus, $\triangle APC \sim \triangle DPB$ by the Angle-Angle Similarity Postulate. The lengths of corresponding sides of similar triangles are proportional, so $\frac{a}{d} = \frac{c}{b}$. Therefore, $a \cdot b = c \cdot d$.

You can use Theorem 12-12 to find lengths of segments in circles.

3 EXAMPLE Finding Segment Lengths

Algebra Find the value of the variable. If the answer is not a whole number, round to the nearest tenth.

a.

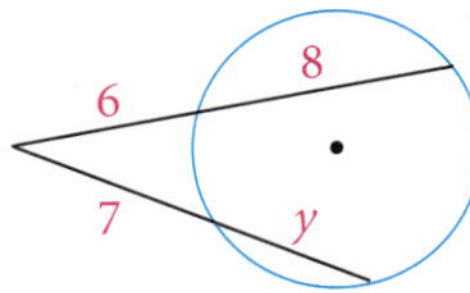

$$(6 + 8)6 = (7 + y)7 \quad \textbf{Thm. 12-12 (II)}$$
$$84 = 49 + 7y \quad \textbf{Solve for } y.$$
$$35 = 7y$$
$$5 = y$$

b.

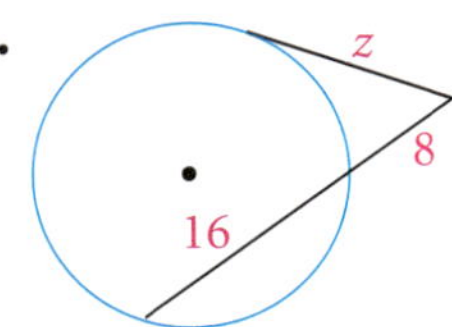

$$(8 + 16)8 = z^2 \quad \textbf{Thm. 12-12 (III)}$$
$$192 = z^2 \quad \textbf{Solve for } z.$$
$$13.9 \approx z$$

CA Standards Check 3 Find the value of the variable to the nearest tenth.

a.

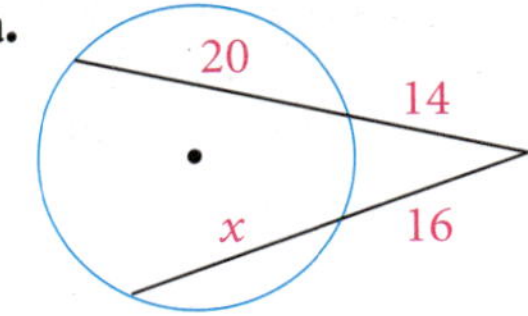

b.

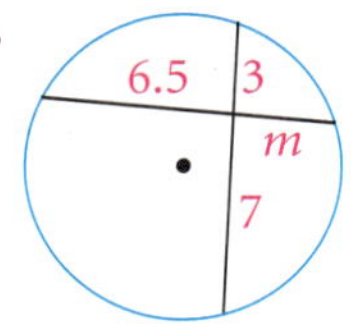

4 EXAMPLE Real-World Connection

Bridge Design The arch of the Taiko Bashi is an arc of a circle. A 14-ft chord is 4.8 ft from the edge of the circle. Find the radius of the circle.

Draw a diagram that shows a 14-ft chord 4.8 ft below the top of a circle. Let x represent the length of the part of the diameter from the chord to the bottom of the circle. Use x and Theorem 12-12 to find the radius.

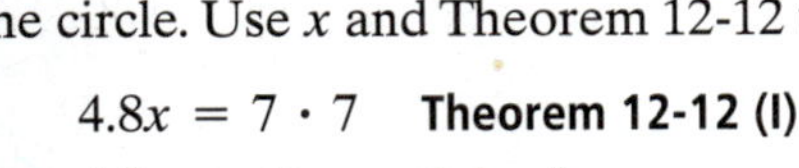

$$4.8x = 7 \cdot 7 \quad \textbf{Theorem 12-12 (I)}$$
$$4.8x = 49 \quad \textbf{Solve for } x.$$
$$x \approx 10.2$$
$$\text{diameter} \approx 10.2 + 4.8 = 15 \text{ ft} \quad \textbf{Add the segment lengths.}$$
$$\text{radius} \approx 7.5 \text{ ft}$$

The radius is about 7.5 ft.

Real-World Connection

The Taiko Bashi is a pedestrian bridge in the Japanese Tea Garden in San Francisco's Golden Gate Park.

CA Standards Check 4 The basis of a design of a rotor for a Wankel engine is an equilateral triangle. Each side of the triangle is a chord to an arc of a circle. The opposite vertex of the triangle is the center of the arc. In the diagram at the right, each side of the equilateral triangle is 8 in. long.

a. Use what you know about equilateral triangles and find the value of x.

b. **Critical Thinking** Copy the diagram and complete the circle with the given center. Then use Theorem 12-12 to find the value of x. Show that your answers to parts (a) and (b) are equal.

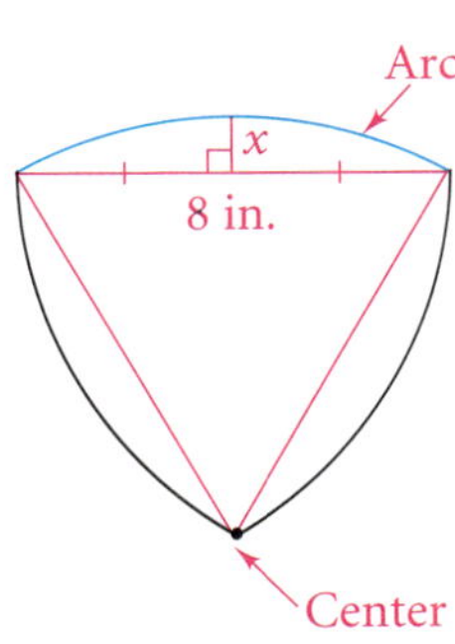

For more exercises, see *Extra Skill, Word Problem, and Proof Practice.*

GEOM 2.0, 7.0, 21.0

Practice by Example

Algebra Find the value of each variable.

Example 1 (page 688)

1.
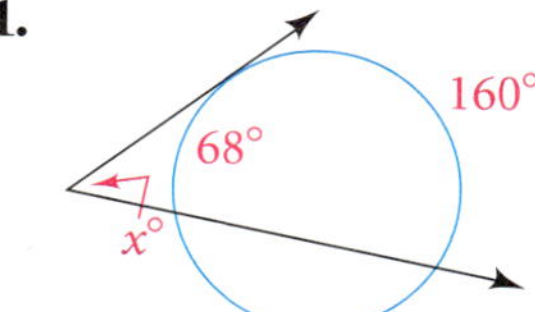

2.
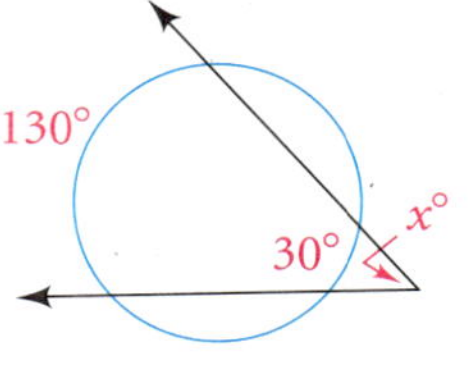

3.
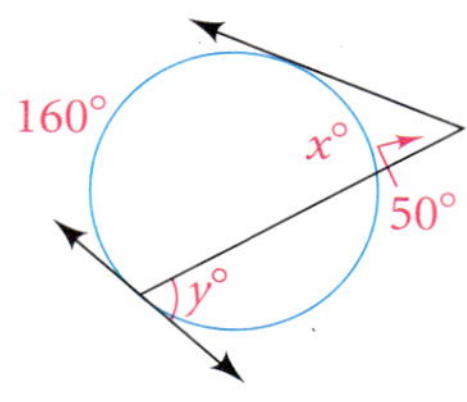

4.
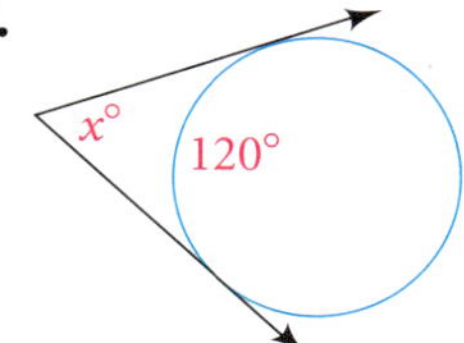

5.
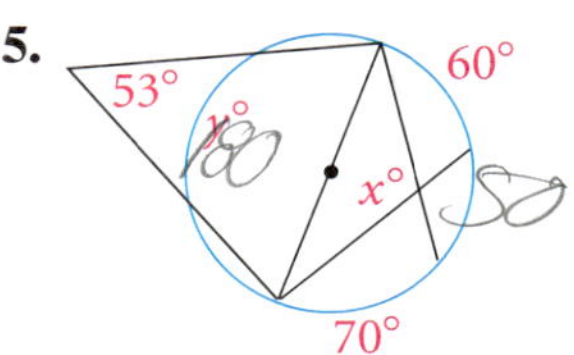

6.
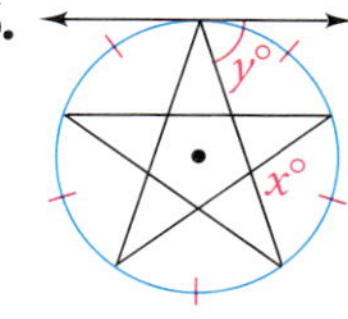

Example 2 (pages 688–689)

7. **Astroscience** A departing space probe sends back a picture of Earth as it crosses the plane of Earth's equator. The angle formed by the two tangents to the equator is 20°. What arc of the equator is visible to the space probe?

Equator
20°
Space probe
Earth

8. At the left, the cross section of the ball is a circle. About how many degrees is the arc of the circle that is below the points of contact with the hands?

Example 3 (page 690)

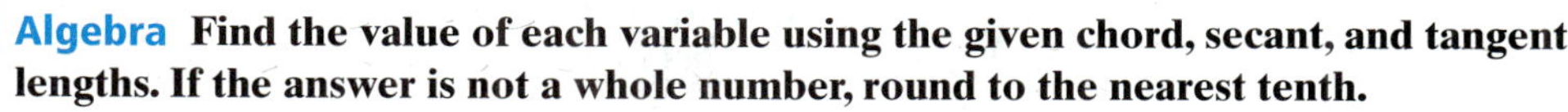

Algebra Find the value of each variable using the given chord, secant, and tangent lengths. If the answer is not a whole number, round to the nearest tenth.

9.
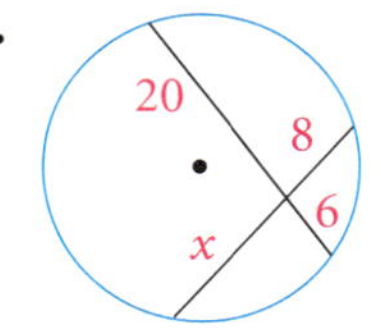

10.
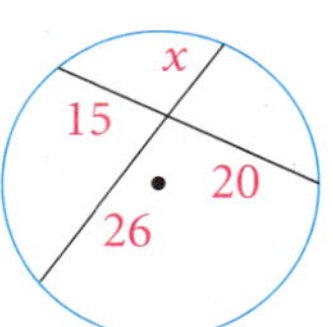

11.
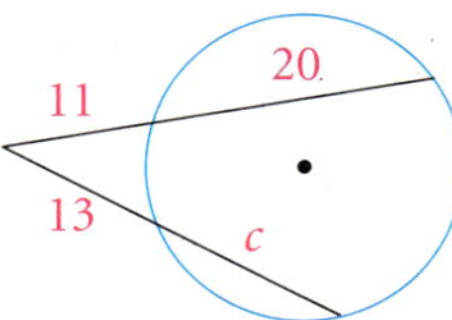

12.
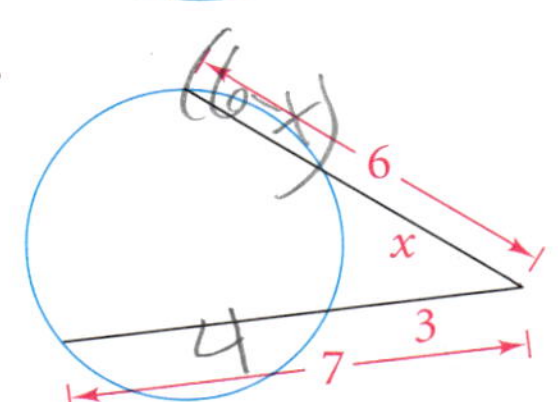

13.
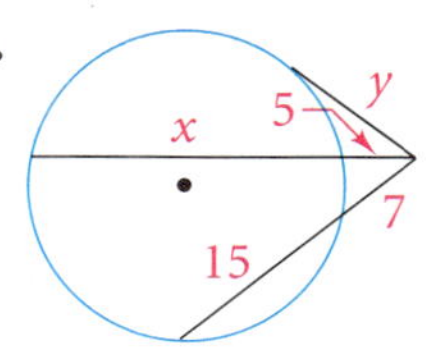

14.
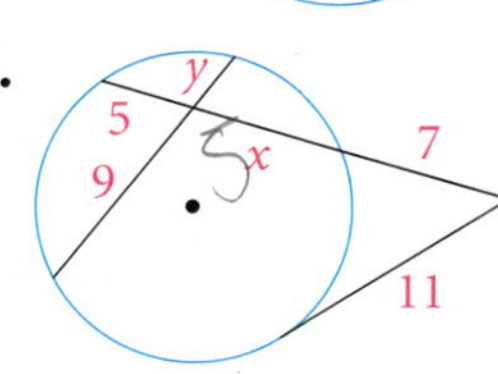

Example 4 (page 690)

Geology This natural arch, in Arches National Park, Utah, is an arc of a circle.

15. Find the diameter of the circle.

16. The chord length shown is rounded. It could range from 165 ft to 175 ft. Find the corresponding range for the diameter.

B Apply Your Skills

Algebra $\overline{CA}$ **and** $\overline{CB}$ **are tangents to** $\odot O$**. Write an expression for each arc or angle in terms of the given variable.**

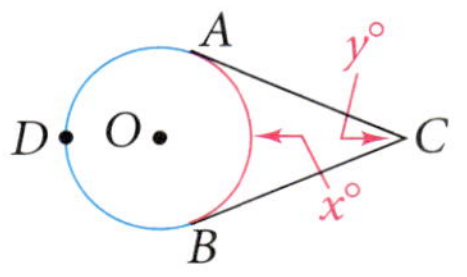

17. $m\widehat{ADB}$ using x 18. $m\angle C$ using x 19. $m\widehat{AB}$ using y

Online Lesson Quiz Visit: PHSchool.com, Web Code: bca-1204

Visit: PHSchool.com
Web Code: bce-1204

Find the diameter of $\odot O$. If your answer is not a whole number, round it to the nearest tenth.

20.

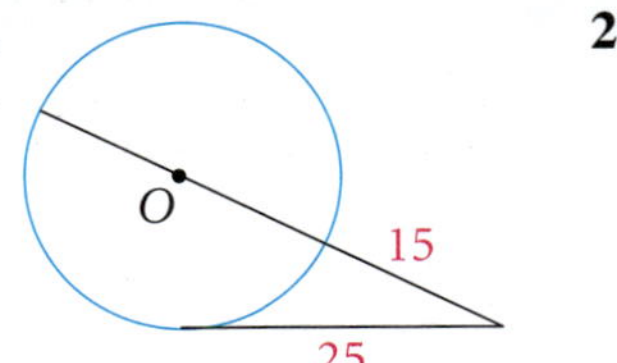

21.

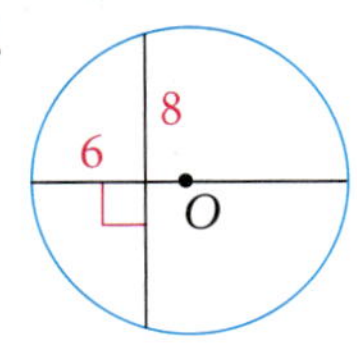

22.

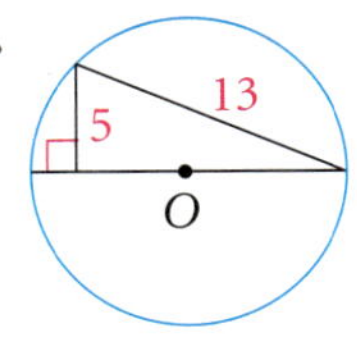

Algebra Find the values of x and y using the given chord, secant, and tangent lengths. If your answer is not a whole number, round it to the nearest tenth.

23.

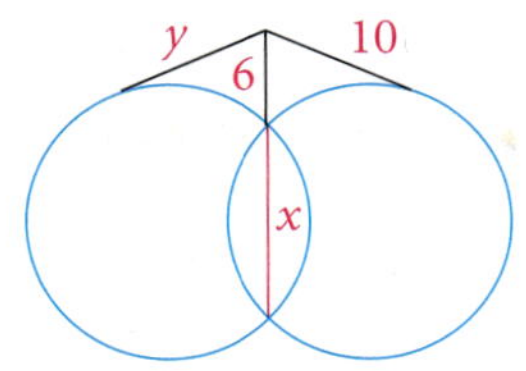

24.

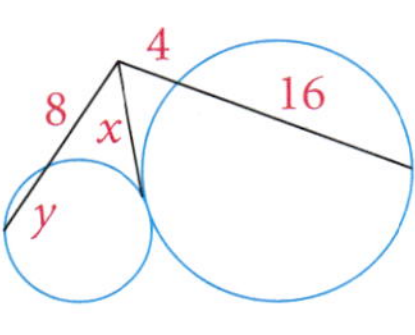

25.

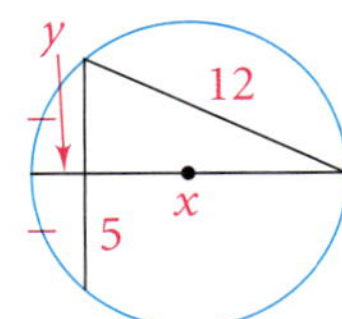

26. Error Analysis To find the value of x, a student wrote the equation $(7.5)6 = x^2$. What error did the student make?

27. A circle is inscribed in a quadrilateral whose four angles have measures 85, 76, 94, and 105. Find the measures of the four arcs between consecutive points of tangency.

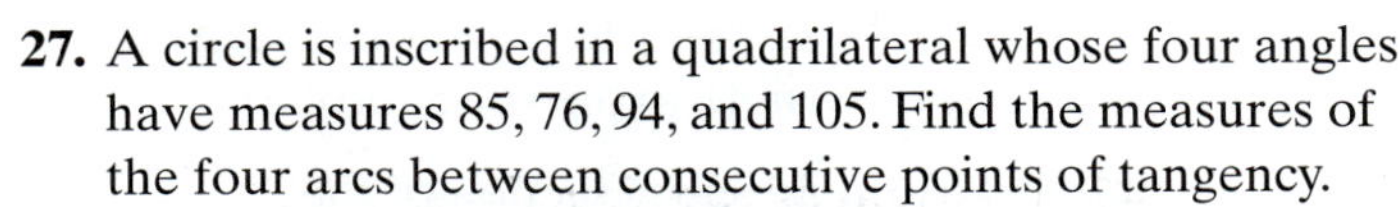

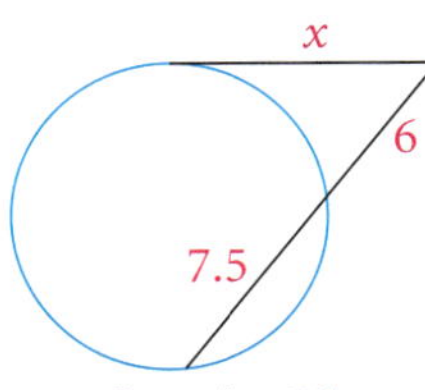

Exercise 26

28. Navigation The map at the left shows that the waters within $\widehat{AXB}$, a 300° arc, are unsafe. Here are what the letters represent.

A: a lighthouse B: a lighthouse X: locations of a ship on $\odot O$
Y: locations of a ship inside $\odot O$ Z: locations of a ship outside $\odot O$

a. Critical Thinking What measures are possible for $\angle X$? For $\angle Y$? For $\angle Z$?

b. Writing Using the angles a ship makes with the lighthouses (like angles X, Y, and Z), explain how a navigator can be sure the ship is in safe waters.

Proof **29.** Prove Theorem 12-11, Part (2), as it applies to two secants that intersect outside a circle.

Given: $\odot O$ with secants $\overline{CA}$ and $\overline{CE}$ intersecting at C

Prove: $m\angle ACE = \frac{1}{2}(m\widehat{AE} - m\widehat{BD})$

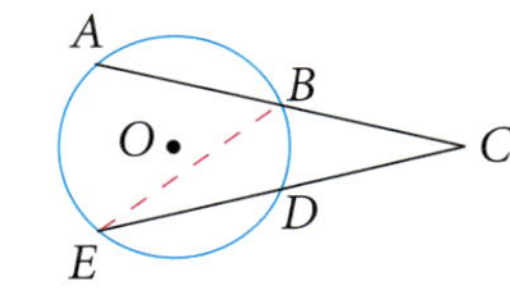

Proof **30.** Prove the other two cases of Theorem 12-11, Part (2). (See Exercise 29.)

Proof **For Exercises 31 and 32, write proofs that use similar triangles.**

31. Prove Theorem 12-12 (II). **32.** Prove Theorem 12-12 (III).

33. Explain why Theorem 12-12 is true when the given point is on the circle.

Challenge Proof **In Exercises 34–37, prove each statement or theorem.**

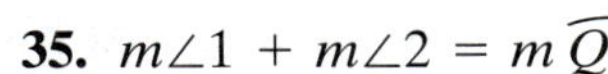

34. $m\angle 1 + m\widehat{PQ} = 180$ **35.** $m\angle 1 + m\angle 2 = m\widehat{QR}$

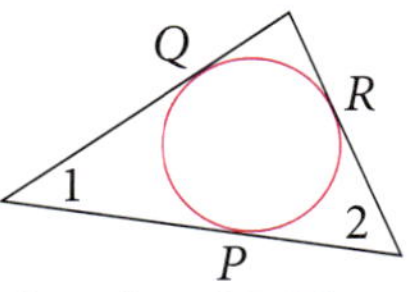

Exercises 34, 35

36. the Pythagorean Theorem (Use the diagram at the right and the theorems of this lesson.)

37. The tangents to a circle at the vertices of an inscribed equilateral triangle form an equilateral triangle.

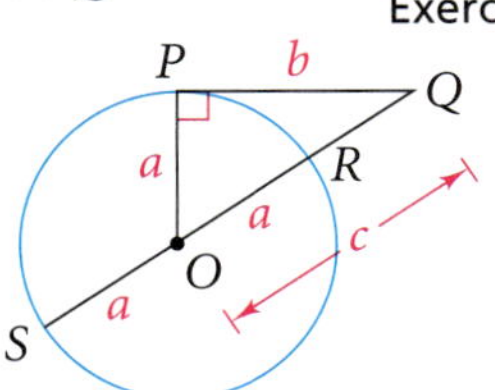

Multiple Choice Practice

For California Standards and CAHSEE Tutorials, visit PHSchool.com. Web Code: bcq-9045

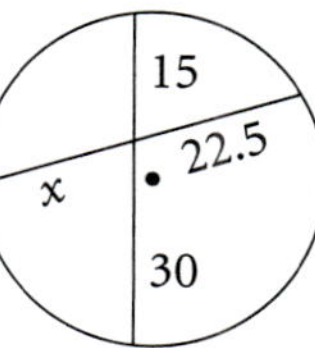

GEOM 7.0

38. Find the value of x in the circle at the right.

(A) 25 (B) 20 (C) 11.25 (D) 7.5

CAHSEE 7AF 1.2

39. Which expression is equivalent to $2x(4x - 1) - (3x^2 + 2x + 1)$?

(A) $5x^2 - 1$

(B) $5x^2 + 2x$

(C) $5x^2 - 4x - 1$

(D) $5x^2 - 4x - 2$

Mixed Review

GO for Help

Lesson 12-3

Find the value of each variable.

40.

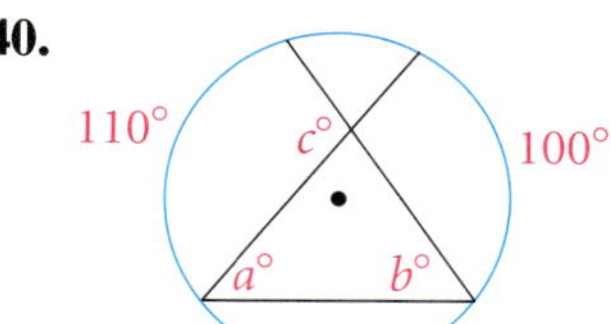

41.

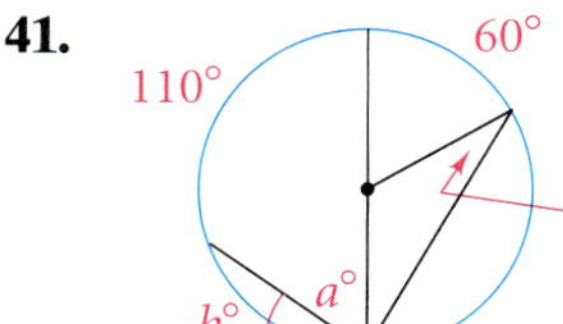

Lesson 10-4

42. The areas of two similar parallelograms are 20 cm^2 and 3.2 cm^2. Find the similarity ratio of the larger parallelogram to the smaller parallelogram.

Lesson 8-4

Find the value of x to the nearest degree.

43.

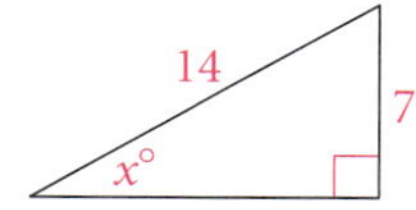

44.

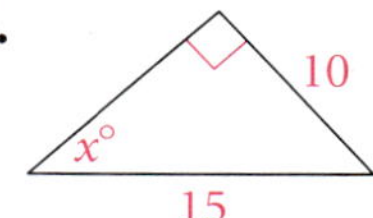

45.

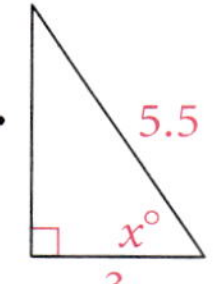

Geometry at Work

Aerospace Engineer

Aerospace engineers design and build all types of spacecraft, from the low-orbit space shuttle to interplanetary probes. Much of today's work involves communications satellites that relay television, telephone, computer, and other signals all over the world. The portion of Earth's surface that can communicate with a satellite increases as the height of the orbit increases.

Earth has a radius of about 3960 miles. The figure at the right shows a satellite 12,000 miles above Earth. $\widehat{AB}$ is the arc of Earth that is in the range of the satellite. You can find $m\widehat{AB}$ by finding $m\angle AEB$, which is twice $m\angle AES$.

$$m\widehat{AB} = m\angle AEB = 2m\angle AES = 2 \cdot \cos^{-1}\left(\frac{3960}{3960 + 12{,}000}\right) \approx 151.3$$

The measure of the arc of Earth in the range of the satellite is about 151.3.

For: Information about aerospace careers
Web Code: bcb-2031

Extension Tangent Lines, Tangent Ratios

For Use With Lesson 12-4

GEOM 18.0 Students know the definitions of the basic trigonometric functions defined by the angles of a right triangle. They also know and are able to use elementary relationships between them. For example, tan(x) = sin(x)/cos(x), (sin(x))^2 + (cos(x))^2 = 1. *Develop*

You learned about tangent ratios in Chapter 8 and tangent lines in this chapter. Are the two related? Yes, indeed! In fact, there are six trigonometric ratios that you can study in advanced mathematics.

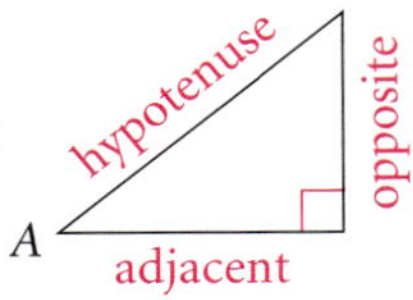

$$\text{sine } \angle A = \frac{\text{opposite}}{\text{hypotenuse}} \qquad \text{cosine } \angle A = \frac{\text{adjacent}}{\text{hypotenuse}} \qquad \text{tangent } \angle A = \frac{\text{opposite}}{\text{adjacent}}$$

$$\text{cosecant } \angle A = \frac{\text{hypotenuse}}{\text{opposite}} \qquad \text{secant } \angle A = \frac{\text{hypotenuse}}{\text{adjacent}} \qquad \text{cotangent } \angle A = \frac{\text{adjacent}}{\text{opposite}}$$

Each of these is related to the geometry of a unit circle, the circle shown in both figures below. Sine A, tangent A, and secant A are the segment lengths highlighted in Figure 1. Cosine A, cotangent A, and cosecant A are highlighted in Figure 2.

Figure 1:

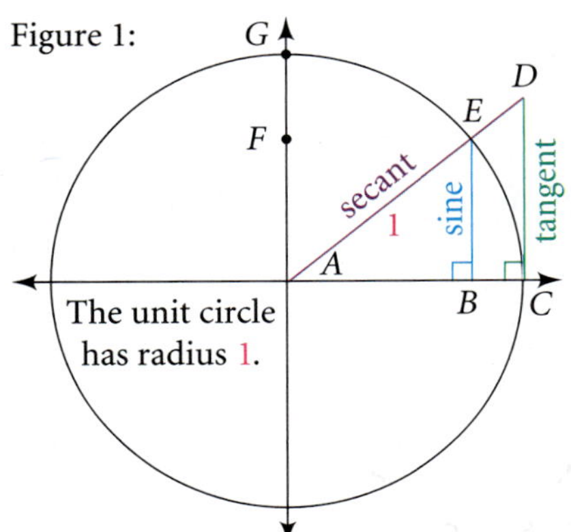

Figure 2:

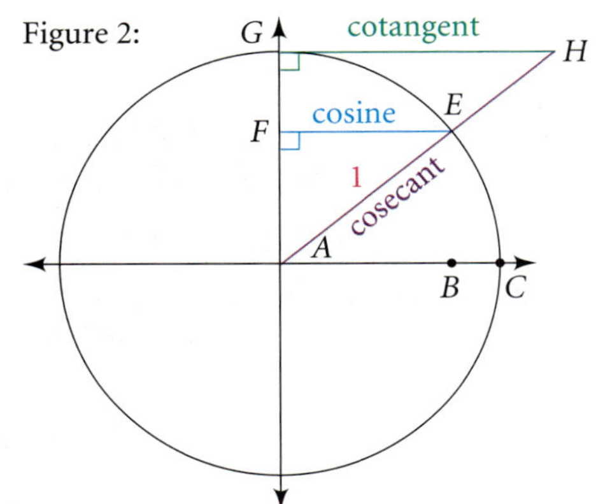

EXERCISES

1. For the four triangles shown above, complete these similarity statements: $\triangle ABE \sim \triangle$ _?_ $\sim \triangle$ _?_ $\sim \triangle$ _?_.
2. Explain why the sine A, tangent A, and secant A ratios have the same values as the segment lengths highlighted in Figure 1.
3. Explain why the cosine A, cotangent A, and cosecant A ratios have the same values as the segment lengths highlighted in Figure 2.

Describe the connection between each of the following.

4. the tangent A ratio and a tangent segment
5. the secant A ratio and a secant segment
6. $\angle EAB$ in Figure 1 and $\angle EAF$ in Figure 2; what are the first two letters in a word that is commonly used to describe this connection?

Show that each equation is true.

7. $(\text{tangent } A)^2 = (\text{secant } A)^2 - 1$
8. $\text{tangent } A = \frac{\text{sine } A}{\text{cosine } A}$
9. $\text{cotangent } A = \frac{1}{\text{tangent } A}$
10. $(\text{sine } A)^2 + (\text{cosine } A)^2 = 1$
11. $\text{secant } A = \frac{1}{\text{cosine } A}$
12. $\text{cosecant } A = \frac{1}{\text{sine } A}$
13. For acute $\angle A$, $0 < \text{sine } A < 1$, $0 < \text{cosine } A < 1$, and $\text{tangent } A > 0$. What values are possible for cosecant A, secant A, and cotangent A? Explain.

12-5 Circles in the Coordinate Plane

California Content Standards

GEOM 7.0 Students prove and use theorems involving the properties of circles. ***Develop***
GEOM 17.0 Students prove theorems by using coordinate geometry, including equations of circles. ***Develop***

What You'll Learn

- To write an equation of a circle
- To find the center and radius of a circle

. . . And Why

To describe the position and range of three cellular telephone towers, as in Example 4

Check Skills You'll Need

GO for Help Lesson 1-8

Find the length of each segment to the nearest tenth.

1. 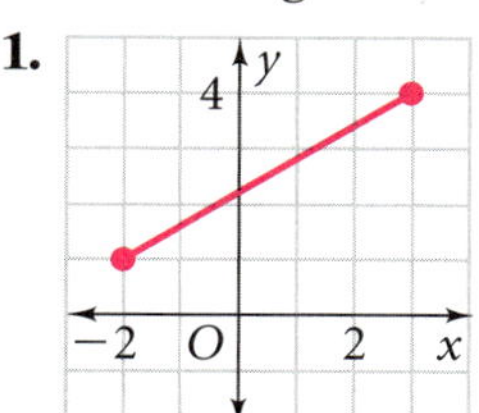

2. 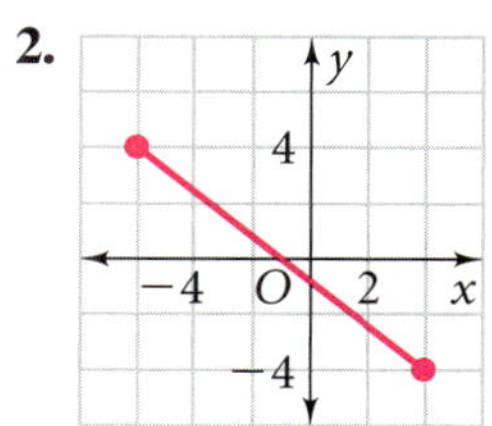

3. 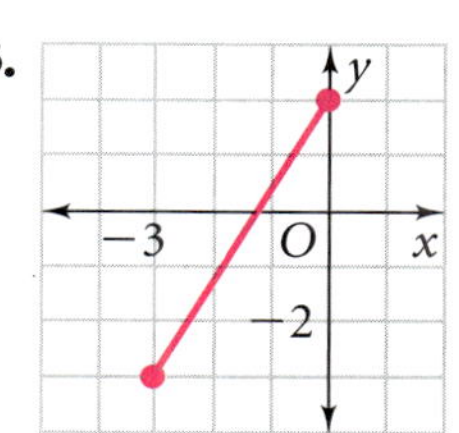

New Vocabulary • standard form of an equation of a circle

1 Writing an Equation of a Circle

You can use the Distance Formula to find an equation of a circle with center (h, k) and radius r. Let (x, y) be any point on the circle. Then the radius r is the distance from (h, k) to (x, y).

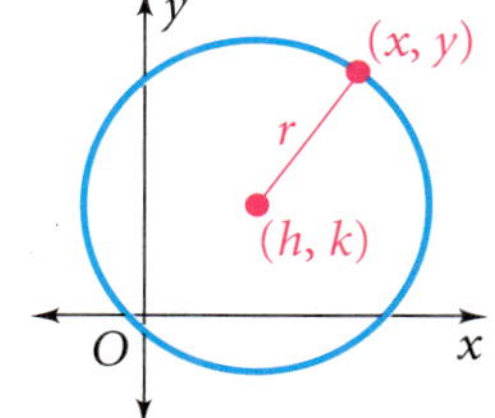

$$r = \sqrt{(x - h)^2 + (y - k)^2} \quad \textbf{Distance Formula}$$

$$r^2 = (x - h)^2 + (y - k)^2 \quad \textbf{Square both sides.}$$

This essentially proves the following theorem.

Key Concepts

Theorem 12-13

An equation of a circle with center (h, k) and radius r is $(x - h)^2 + (y - k)^2 = r^2$.

The equation $(x - h)^2 + (y - k)^2 = r^2$ is in **standard form.** You may also call it the *standard equation* of a circle.

Online active math

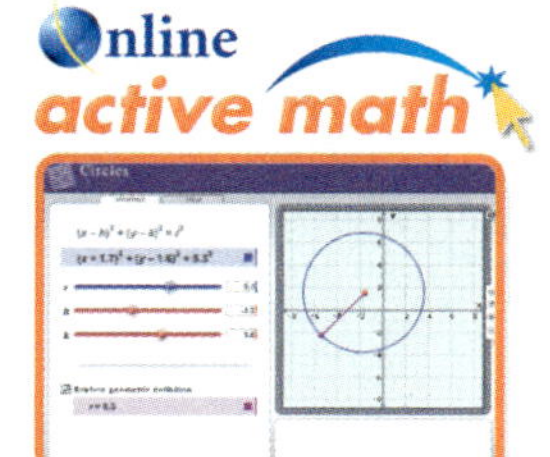

For: Circle Activity
Use: Interactive Textbook, 12-5

1 EXAMPLE Writing the Equation of a Circle

Write the standard equation of the circle with center $(5, -2)$ and radius 7.

$(x - h)^2 + (y - k)^2 = r^2$ **Use standard form.**

$(x - 5)^2 + [y - (-2)]^2 = 7^2$ **Substitute (5, −2) for (h, k), and 7 for r.**

$(x - 5)^2 + (y + 2)^2 = 49$ **Simplify.**

CA Standards Check

1. Write the standard equation of each circle.

a. center $(3, 5)$; radius 6

b. center $(-2, -1)$; radius $\sqrt{2}$

If you know the center of a circle and a point on the circle, you can write the standard equation of the circle.

2 EXAMPLE Using the Center and a Point on a Circle

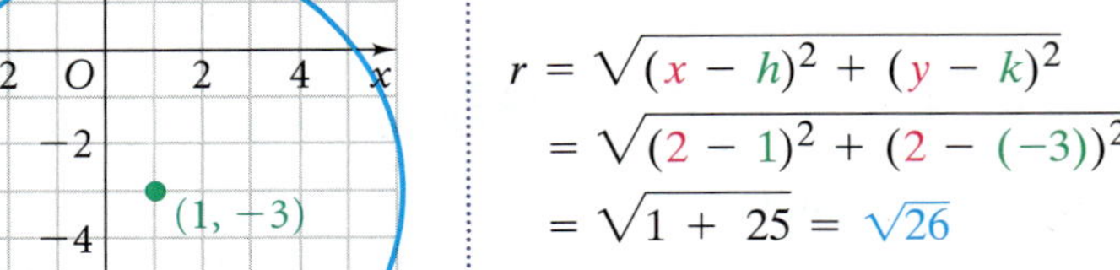

Write the standard equation of the circle with center $(1, -3)$ that passes through the point $(2, 2)$.

$r = \sqrt{(x - h)^2 + (y - k)^2}$ **Use the Distance Formula to find *r*.**

$= \sqrt{(2 - 1)^2 + (2 - (-3))^2}$ **Substitute (1, −3) for (*h*, *k*), and (2, 2) for (*x*, *y*).**

$= \sqrt{1 + 25} = \sqrt{26}$ **Simplify.**

$(x - h)^2 + (y - k)^2 = r^2$ **Use standard form.**

$(x - 1)^2 + [y - (-3)^2] = (\sqrt{26})^2$ **Substitute (1, −3) for (*h*, *k*), and $\sqrt{26}$ for *r*.**

$(x - 1)^2 + (y + 3)^2 = 26$ **Simplify.**

CA Standards Check 2 Write the standard equation of the circle with center $(2, 3)$ that passes through the point $(-1, 1)$.

2 Finding the Center and Radius of a Circle

If you know the standard equation of a circle, you can describe the circle by naming its center and radius. Then you can use this information to graph the circle.

3 EXAMPLE Graphing a Circle Given its Equation

Find the center and radius of the circle with equation $(x - 7)^2 + (y + 2)^2 = 64$. Then graph the circle.

$(x - 7)^2 + (y + 2)^2 = 64$

$(x - 7)^2 + (y - (-2))^2 = 8^2$ **Use standard form.**

$\uparrow h \qquad \uparrow k \qquad \uparrow r$

The center is $(7, -2)$ and the radius is 8.

To graph the circle, place the compass point at the center $(7, -2)$ and draw a circle with radius 8.

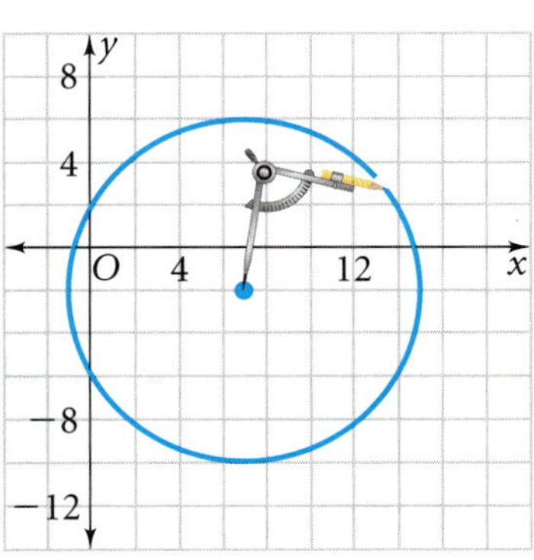

CA Standards Check 3 Find the center and radius of the circle with equation $(x - 2)^2 + (y - 3)^2 = 100$. Then graph the circle.

You can use equations of circles to model real-world situations.

4 EXAMPLE Real-World Connection

Communications When you make a call on a cellular phone, a tower receives the call. In the diagram, the centers of circles O and A are locations of cellular telephone towers.

a. The equation $(x - 16)^2 + (y - 10)^2 = 100$ models the position and range of Tower A. Describe the position and range of Tower A.

$(x - 16)^2 + (y - 10)^2 = 10^2$ is in standard form. It shows that tower A is located at $(16, 10)$ and has a range of 10 units.

b. A new tower is to be built at B with range indicated in the graph. Write an equation that describes the position and range of this tower.

$\odot B$ has center $(4, 20)$ and radius 10. Substitute these into the standard equation.

$(x - h)^2 + (y - k)^2 = r^2$

$(x - 4)^2 + (y - 20)^2 = 10^2$ **Substitute.**

$(x - 4)^2 + (y - 20)^2 = 100$ **This is an equation for Tower *B*.**

Real-World Connection

To blend into the landscape, cellular phone towers can be disguised as trees.

CA Standards Check 4 Write an equation that describes the position and range of Tower O.

EXERCISES

For more exercises, see *Extra Skill, Word Problem, and Proof Practice.*

Standards Practice GEOM 7.0, 17.0

A Practice by Example

GO for Help

Example 1 (page 695)

Write the standard equation of each circle.

1. center $(2, -8)$; $r = 9$ **2.** center $(0, 3)$; $r = 7$ **3.** center $(0.2, 1.1)$; $r = 0.4$

4. center $(5, -1)$; $r = 12$ **5.** center $(-6, 3)$; $r = 8$ **6.** center $(-9, -4)$; $r = \sqrt{5}$

7. center $(0, 0)$; $r = 4$ **8.** center $(-4, 0)$; $r = 3$ **9.** center $(-1, -1)$; $r = 1$

Example 2 (page 696)

Write the standard equation of the circle with the given center that passes through the given point.

10. center $(-2, 6)$; point $(-2, 10)$ **11.** center $(1, 2)$; point $(0, 6)$

12. center $(7, -2)$; point $(1, -6)$ **13.** center $(-10, -5)$; point $(-5, 5)$

14. center $(6, 5)$; point $(0, 0)$ **15.** center $(-1, -4)$; point $(-4, 0)$

Example 3 (page 696)

Find the center and radius of the circle with the given equation. Then graph the circle.

16. $(x + 7)^2 + (y - 5)^2 = 16$ **17.** $(x - 3)^2 + (y + 8)^2 = 100$

18. $(x + 4)^2 + (y - 1)^2 = 25$ **19.** $x^2 + y^2 = 36$

20. $(x - 0.3)^2 + y^2 = 0.04$ **21.** $(x + 5)^2 + (y + 2)^2 = 48$

Example 4 (page 697)

Use the diagram at the right. Write an equation that describes the position and radius of each circle.

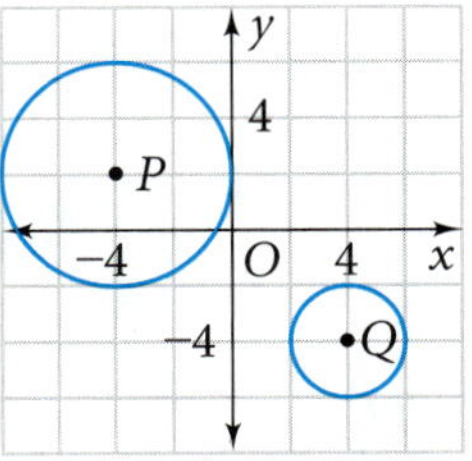

22. $\odot P$ **23.** $\odot Q$

24. Communications The plotted location of a cellular phone tower on a coordinate grid is $(-3, 2)$ and the range is 5 units. Write an equation that describes the position and range of the tower.

Each equation models the position and range of a tornado alert siren. Describe the position and range of each.

25. $(x - 5)^2 + (y - 7)^2 = 81$ **26.** $(x + 4)^2 + (y - 9)^2 = 144$

Apply Your Skills

Write the standard equation of each circle.

27.

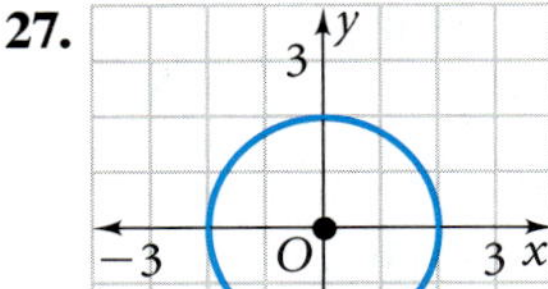

28.

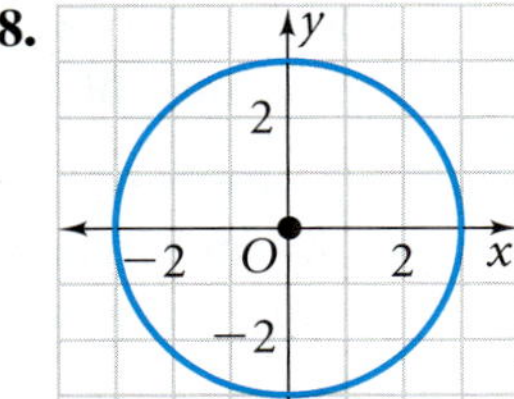

29.

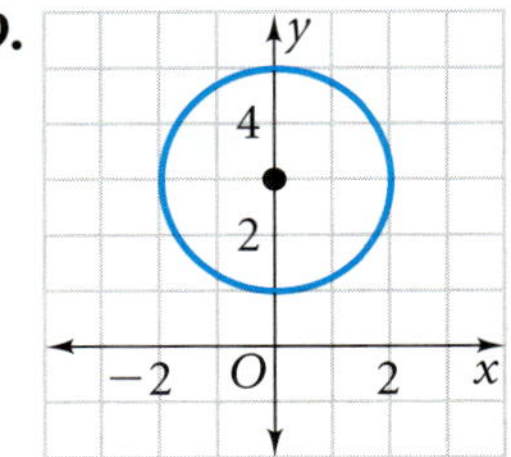

30.

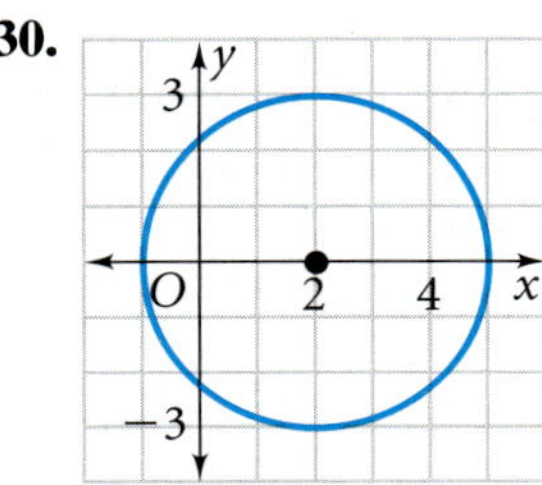

31.

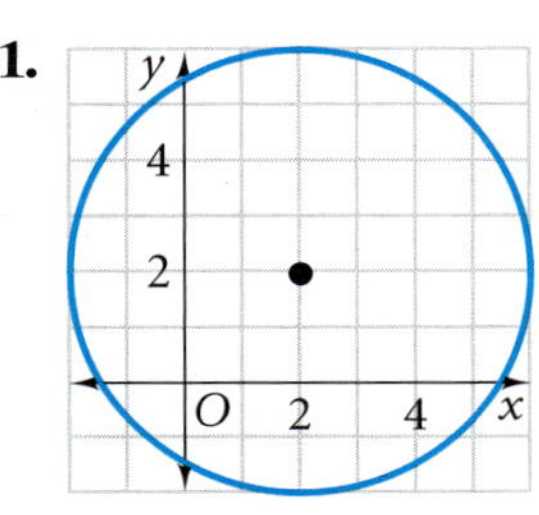

32.

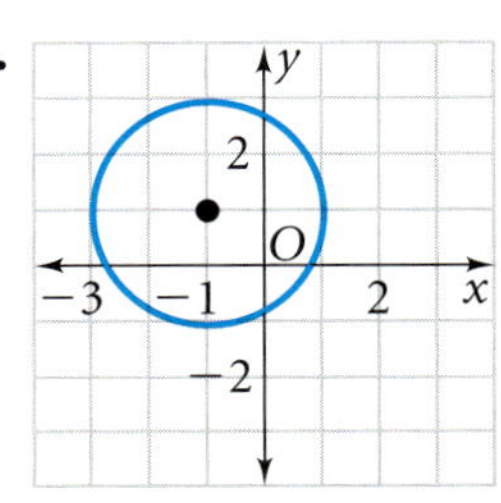

GO for Help

In Exercises 33–38, use the Midpoint Formula (p. 55) to find centers.

Write an equation of a circle with diameter $\overline{AB}$.

33. $A(0, 0), B(8, 6)$ **34.** $A(3, 0), B(7, 6)$ **35.** $A(1, 1), B(5, 5)$

36. $A(-1, 0), B(-5, -3)$ **37.** $A(-3, 1), B(0, 9)$ **38.** $A(-2, 3), B(6, -7)$

39. The *unit circle* has center $(0, 0)$ and radius 1. Write an equation for this circle.

40. Critical Thinking Describe the graph of $x^2 + y^2 = r^2$ when $r = 0$.

41. Open-Ended On graph paper, make a design that includes at least three circles. Write the standard equations of your circles.

Determine whether each equation is an equation of a circle. If not, explain.

42. $(x - 1)^2 + (y + 2)^2 = 9$ **43.** $x + y = 9$ **44.** $x + (y - 3)^2 = 9$

45. Find the circumference and area of the circle whose equation is $(x - 9)^2 + (y - 3)^2 = 64$. Leave your answers in terms of π.

46. Write an equation of a circle with area 36π and center $(4, 7)$.

47. What are the x- and y-intercepts of the line tangent to the circle $(x - 2)^2 + (y - 2)^2 = 5^2$ at the point $(5, 6)$?

48. For $(x - h)^2 + (y - k)^2 = r^2$, show that $y = \sqrt{r^2 - (x - h)^2} + k$ or $y = -\sqrt{r^2 - (x - h)^2} + k$.

Homework Video Tutor

Visit: PHSchool.com
Web Code: bce-1205

For: Graphing calculator procedures
Web Code: bcb-2113

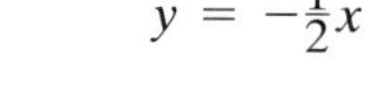

Graphing Calculator **Use a graphing calculator to graph each circle. (*Hint:* See Exercise 48.) View the plotting in both sequential mode and simultaneous mode.**

49. $(x - 3)^2 + (y - 2)^2 = 9$

50. $(x + 5)^2 + (y - 8)^2 = 1$

51. circle with center $(0, 0)$ and radius 7

52. circle with center $(-6, -3)$ and radius 2

Find all points of intersection of each pair of graphs. Make a sketch.

53. $x^2 + y^2 = 13$
$y = -x + 5$

54. $x^2 + y^2 = 17$
$y = -\frac{1}{4}x$

55. $x^2 + y^2 = 8$
$y = 2$

56. $x^2 + y^2 = 20$
$y = -\frac{1}{2}x + 5$

57. $(x + 1)^2 + (y - 1)^2 = 18$
$y = x + 8$

58. $(x - 2)^2 + (y - 2)^2 = 10$
$y = -\frac{1}{3}x + 6$

Graphing Calculator **Use a graphing calculator to convince yourself that the given line is not tangent to the circle $x^2 + y^2 = 25$. Explain what you did.**

59. $y = -5x + 26$

60. $3x + 5y = 29$

61. Writing Explain why it is not possible to conclude that a line and a circle are tangent by viewing their graphs.

62. Lines $y = \frac{2}{3}x + 3$ and $y = 5$ cut the ring formed by circles $(x - 3)^2 + (y - 5)^2 = 64$ and $(x - 3)^2 + (y - 5)^2 = 25$ into four parts. Find the area of each part.

63. Nautical Distance The radius of Earth's equator is about 3960 miles.

a. Write the equation of the equator with the center of Earth as the origin.

b. Find the length of a 1° arc on the equator to the nearest tenth of a mile.

c. A 1° arc along the equator is 60 nautical miles long. How many miles are in a nautical mile? Round to the nearest tenth.

d. History Columbus planned his trip to the East by going west. He thought each 1° arc was 45 miles long. He estimated that the trip would take 21 days. Use your answer to part (b) to find a better estimate.

64. Geometry in 3 Dimensions The equation of a sphere is similar to the equation of a circle. The equation of a sphere with center (h, j, k) and radius r is

$$(x - h)^2 + (y - j)^2 + (z - k)^2 = r^2.$$

a. $M(-1, 3, 2)$ is the center of a sphere passing through $T(0, 5, 1)$. What is the radius of the sphere?

b. Write an equation of the sphere.

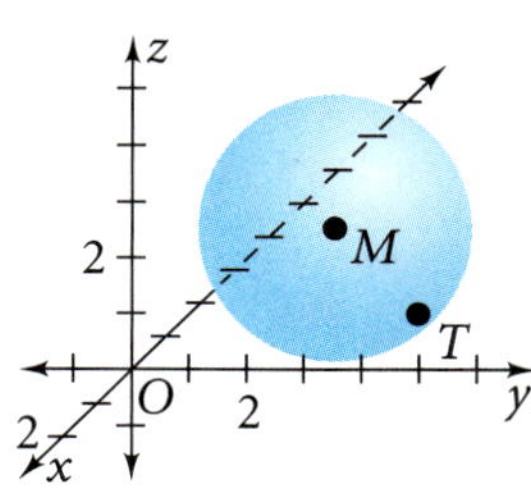

Multiple Choice Practice

For California Standards and CAHSEE Tutorials, visit PHSchool.com. Web Code: bcq-9045

GEOM 17.0

65. Write the standard equation of the circle.

(A) $x^2 + (y - 2)^2 = 9$

(B) $(x - 2)^2 + (y - 1)^2 = 3$

(C) $(x - 2)^2 + y^2 = 9$

(D) $(x - 2)^2 + y^2 = 3$

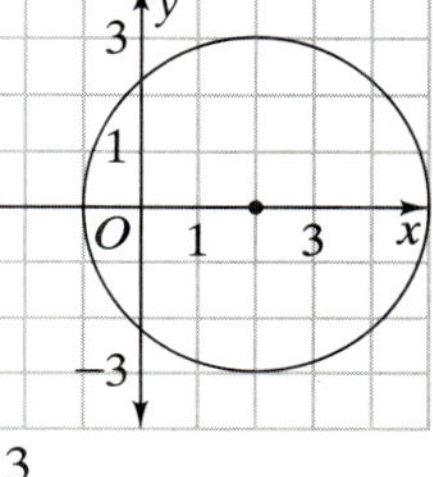

GEOM 17.0

66. Write an equation of a circle with diameter $\overline{AB}$, where $A(3, 2)$ and $B(3, -4)$.

(A) $(x - 3)^2 + (y - 1)^2 = 9$

(B) $(x + 1)^2 + (y - 3)^2 = 3$

(C) $(x - 3)^2 + (y + 1)^2 = 3$

(D) $(x - 3)^2 + (y + 1)^2 = 9$

CAHSEE 6SDAP 3.3

67. Luis placed 7 black chips, 5 red chips, and 4 white chips into a bag. Then he randomly removed 3 chips one at a time without replacing them. What is the probability that he pulled out a black chip, then a red chip, and then a white chip?

Ⓐ $\frac{1}{24}$ Ⓑ $\frac{4}{35}$ Ⓒ $\frac{3}{16}$ Ⓓ $\frac{13}{24}$

Mixed Review

Lesson 12-4

Find the value of each variable. Assume that lines that appear tangent are tangent.

68.

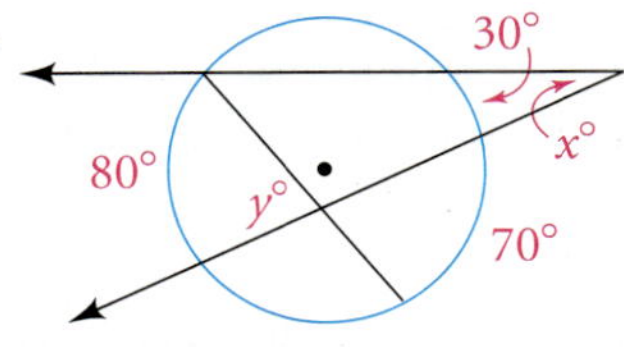

69.

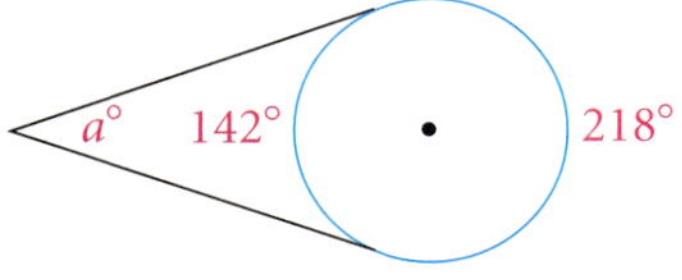

Lesson 8-6

For the given vectors $\vec{a}$ and $\vec{c}$, write the sum $\vec{a} + \vec{c}$ as an ordered pair.

70. $\vec{a} = \langle -2, 5\rangle$ and $\vec{c} = \langle 8, 7\rangle$

71. $\vec{a} = \langle -3, -4\rangle$ and $\vec{c} = \langle -2, 6\rangle$

72. $\vec{a} = \langle 3, 1\rangle$ and $\vec{c} = \langle 1, 3\rangle$

73. $\vec{a} = \langle 9, -6\rangle$ and $\vec{c} = \langle 2, -1\rangle$

Lesson 7-4

Find the geometric mean of each pair of numbers in simplest radical form.

74. 3 and 12

75. 9 and 27

76. 4 and 18

77. $\sqrt{3}$ and $\sqrt{27}$

78. $\sqrt{3}$ and $\sqrt{12}$

79. $\frac{3}{8}$ and $\frac{3}{2}$

Checkpoint Quiz 2

Lessons 12-4 through 12-5

Find the value of each variable. Assume that lines that appear tangent are tangent.

1.

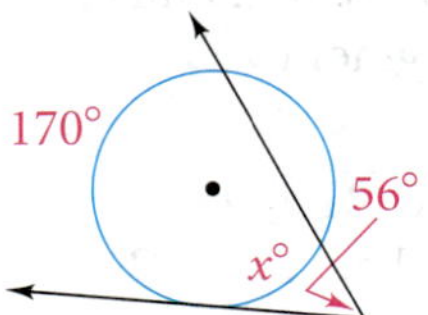

2.

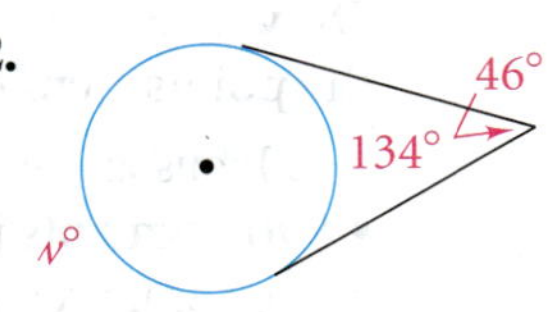

3.

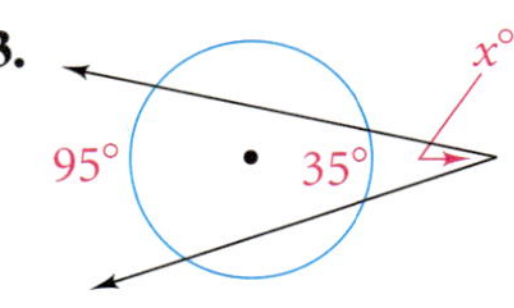

4.

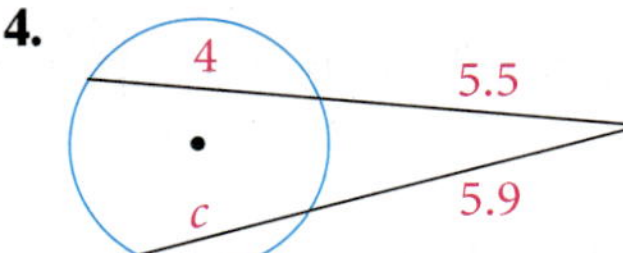

5.

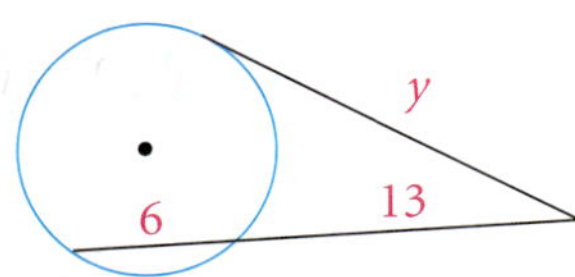

6. In the circle at the right, what is $m\widehat{BF}$?

7. Writing Explain the difference between a chord and a secant. Include a diagram.

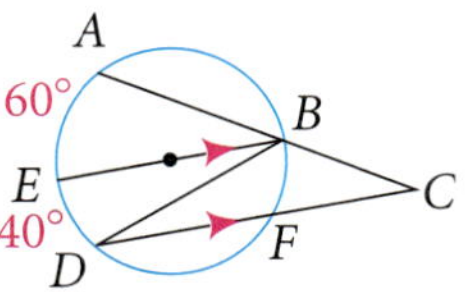

The endpoints of a diameter are given. Write an equation of the circle.

8. (3, 1) and (0, 0)

9. (−2, 5) and (9, −3)

10. (−4, −8) and (1, 0)

12-6 Locus: A Set of Points

What You'll Learn

- To draw and describe a locus

. . . And Why

To interpret a locus description of a geometric figure, as in Example 3

Check Skills You'll Need — GO for Help Lessons 1-7 and 3-8

Sketch each of the following.

1. the perpendicular bisector of $\overline{CD}$

2. $\angle EFG$ bisected by $\overrightarrow{FH}$

3. line k parallel to line m and perpendicular to line w, all in plane N

New Vocabulary • locus

1 Drawing and Describing a Locus

A **locus** is a set of points, all of which meet a stated condition. To sketch a locus, draw points of the locus until you see a pattern.

Real-World Connection

The locus of footprints of children pushing the merry-go-round is a circle.

1 EXAMPLE Describing a Locus in a Plane

a. Draw and describe the locus: In a plane, the points 1 cm from a given point C.

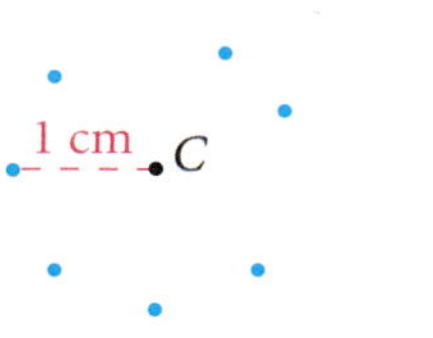

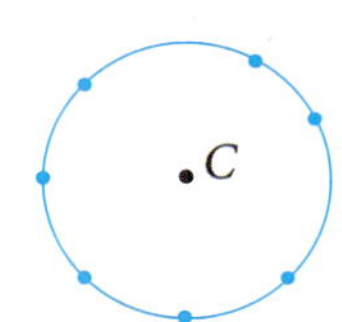

Draw a point C.
Sketch several points 1 cm from C.
Keep doing so until you see a pattern.
Draw the figure the pattern suggests.

The locus is a circle with center C and radius 1 cm.

b. Draw and describe the locus: In a plane, the points 1 cm from a segment $\overline{AB}$.

The locus is
- two segments parallel to $\overline{AB}$ and
- two semicircles centered at A and B

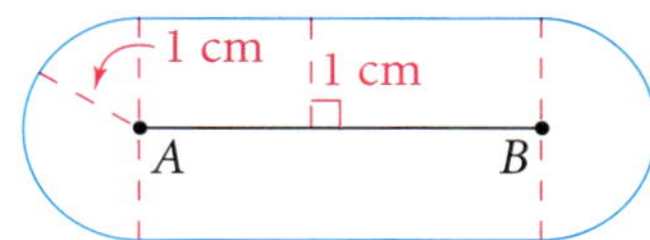

CA Standards Check **1** Draw and describe the locus: In a plane, the points 2 cm from a line $\overleftrightarrow{XY}$.

You can use locus descriptions for geometric terms.

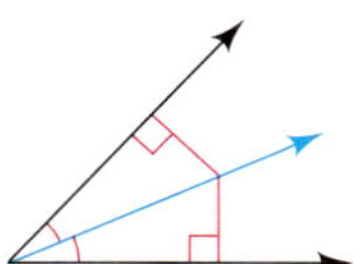

An angle bisector: The points in the interior of the angle that are equidistant from the sides of the angle.

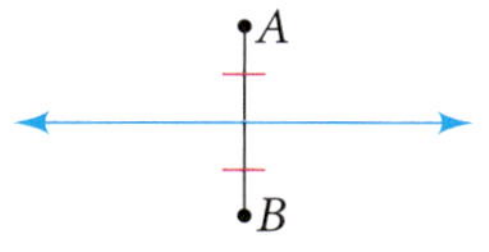

A perpendicular bisector of a segment: In a plane, the points that are equidistant from the segment endpoints.

Vocabulary Tip

The word locus comes from the Latin word for "location." Its plural is *loci* (LOH sy).

Sometimes a locus is described by two conditions. You can draw the locus by first drawing the points that satisfy each condition. Then find their intersection.

2 EXAMPLE Drawing a Locus for Two Conditions

Draw the locus: In a plane, the points equidistant from two lines k and m and 5 cm from the point where k and m intersect.

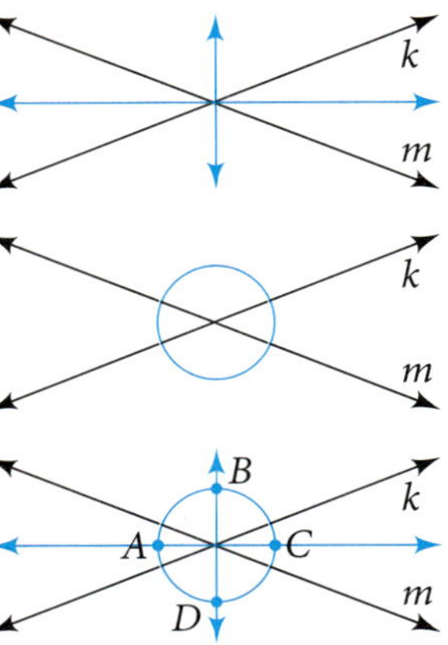

The points in a plane equidistant from lines k and m are two lines that bisect the vertical angles formed by k and m.

The points in a plane 5 cm from the point where k and m intersect is a circle.

The locus that satisfies both conditions is the set of points A, B, C, and D.

2 Draw the locus: In a plane, the points equidistant from two points X and Y and 2 cm from the midpoint of $\overline{XY}$.

A locus in a plane and a locus in space can be quite different.

Real-World Connection

A soap bubble is a sphere, a locus of points in space that are a given distance from a given point.

3 EXAMPLE Describing a Locus in Space

a. Draw and describe the locus: In space, the points that are c units from a point D.

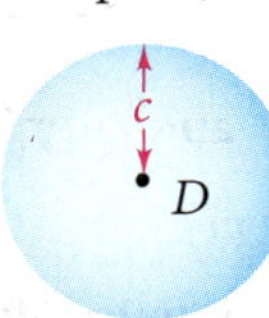

The locus is a sphere with center at point D and radius c.

b. Draw and describe the locus: In space, the points that are 3 cm from a line ℓ.

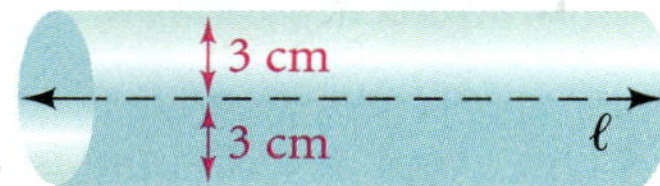

The locus is an endless cylinder with radius 3 cm and center-line ℓ.

3 Draw and describe each locus.

a. In a plane, the points that are equidistant from two parallel lines.

b. In space, the points that are equidistant from two parallel planes.

You can also think of a locus as a path. For example, the locus of the tip of a hand of a clock each day is the circle traced by the tip as it travels around the clock face. The locus of a point on the handle of a sliding-glass door when you enter a room is the line segment along which the point travels as the door slides back and forth.

EXERCISES

For more exercises, see *Extra Skill, Word Problem, and Proof Practice.*

Standards Practice

A Practice by Example

Draw and describe each locus in a plane.

Example 1 (page 701)

GO for Help

1. points 4 cm from a point X
2. points 2 in. from a segment $\overline{UV}$
3. points 3 mm from a line $\overleftrightarrow{LM}$
4. points 1 in. from a circle with radius 3 in.
5. points equidistant from the endpoints of $\overline{PQ}$
6. points in the interior of $\angle ABC$ and equidistant from the sides of $\angle ABC$
7. points equidistant from two perpendicular lines
8. midpoints of radii of a circle with radius 2 cm

Example 2 (page 702)

In a plane, draw the locus whose points satisfy the given conditions.

9. equidistant from points M and N and on a circle with center M and radius $= \frac{1}{2}MN$
10. 3 cm from $\overline{GH}$ and 5 cm from G, where $GH = 4.5$ cm
11. equidistant from the sides of $\angle PQR$ and on a circle with center P and radius PQ
12. equidistant from both points A and B and points C and D
13. equidistant from the sides of $\angle JKL$ and on $\odot C$

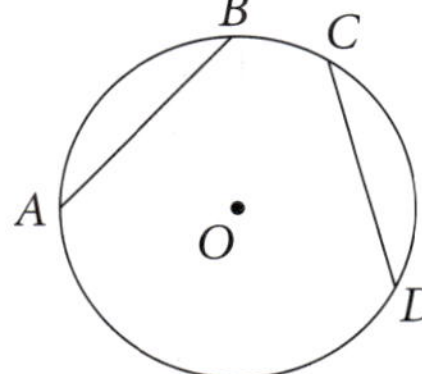

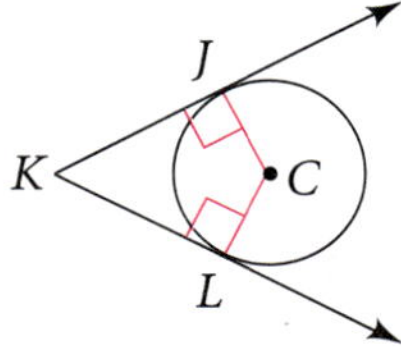

Example 3 (page 702)

Draw and describe each locus in space.

14. points 3 cm from a point F
15. points 4 cm from a line $\overleftrightarrow{DE}$
16. points 1 in. from plane M
17. points 5 mm from $\overrightarrow{PQ}$

B Apply Your Skills

Describe the locus that each blue figure represents.

18.

19. 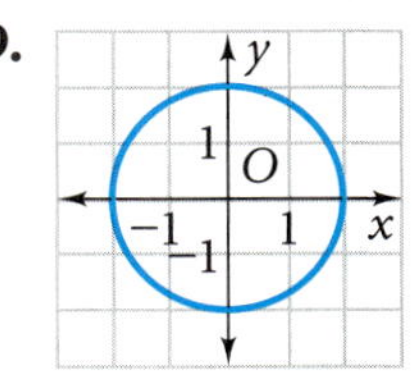

20. 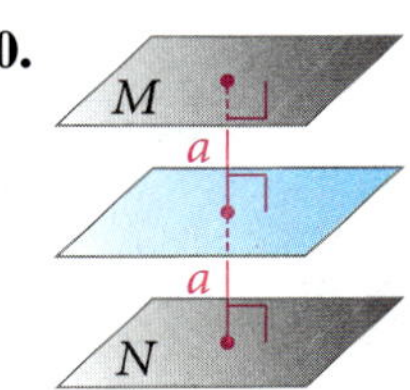

21. **Open-Ended** Give two examples of loci from everyday life, one in a plane and one in space.
22. **Reasoning** Rosie says that it is impossible to find a point equidistant from three collinear points. Is she correct? Explain.

Real-World Connection

A fingertip of the skater traces a locus as she twirls.

Coordinate Geometry **Write an equation for the locus: In the plane, the points equidistant from the two given points.**

23. $A(0, 2)$ and $B(2, 0)$
24. $P(1, 3)$ and $Q(5, 1)$
25. $T(2, -3)$ and $V(6, 1)$

Visit: PHSchool.com
Web Code: bce-1206

26. **Coordinate Geometry** Complete the following locus description of the points highlighted in blue at the right: in the coordinate plane, the points 2 units from the __?__ and 1 unit from the __?__-axis.

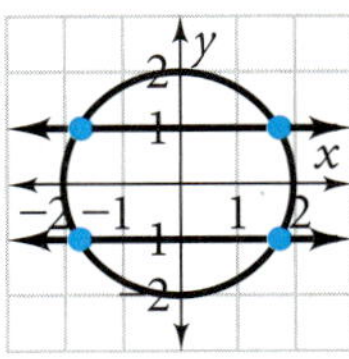

Make a drawing of each locus.

27. the path of a doorknob as a door opens

28. the path of a knot in the middle of a jump rope as it is being used

29. the path of the tip of your nose as you turn your head

30. the path of a fast-pitched softball

31. Jack and Julie Wilson take new jobs in Shrevetown and need a place to live. Jack says, "Let's try to move somewhere equidistant from both of our offices." Julie says, "Let's try to stay within three miles of downtown." Where on the map should the Wilsons look for a home?

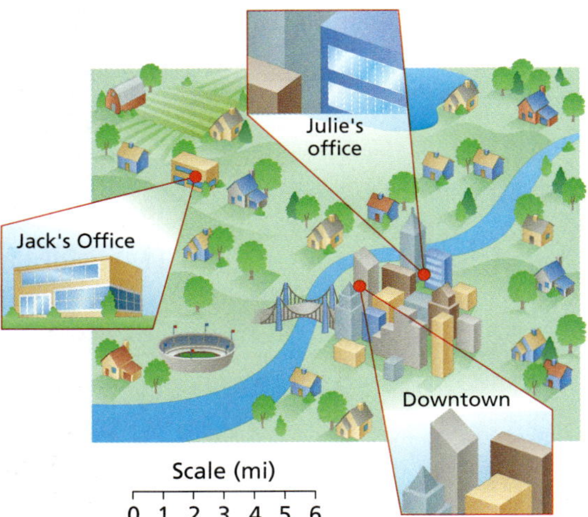

32. **Critical Thinking** Points A and B are 5 cm apart. Do the following loci in a plane have any points in common?
    the points 3 cm from A
    the points 4 cm from B
    Illustrate with a sketch.

Coordinate Geometry **Draw each locus on the coordinate plane.**

33. all points 3 units from the origin

34. all points 2 units from $(-1, 3)$

35. all points 4 units from the y-axis

36. all points 5 units from $x = 2$

37. all points equidistant from $y = 3$ and $y = -1$

38. all points equidistant from $x = 4$ and $x = 5$

39. all points equidistant from the x- and y-axes

40. all points equidistant from $x = 3$ and $y = 2$

Spinning cup   Axis

Real-World Connection

An anemometer measures wind speed.

Meteorology **In an anemometer, there are three cups mounted on an axis. Imagine a point on the edge of one of the cups.**

41. Describe the locus that this point traces as the cup spins in the wind.

42. Suppose the distance of the point from the axis of the anemometer is 2 in. Write an equation for the locus of part (a). Use the axis as the origin.

43. Robert draws a segment to use as the base of an isosceles triangle.
    a. Draw a segment to represent Robert's base. Locate three points that could be the vertex of the isosceles triangle.
    b. Describe the locus of possible vertices for Robert's isosceles triangle.
    c. **Writing** Explain why points in the locus you described are the only possibilities for the vertex of Robert's triangle.

44. Describe the locus: The points in space equidistant from the points of a circle.

Playground Equipment **Think about the path of a child on each piece of playground equipment. Draw the path from (a) a top view, (b) a front view, and (c) a side view.**

45. a swing

46. a straight slide

47. a corkscrew slide

48. a merry-go-round

49. a firefighters' pole

50. In the diagram, three students are seated at uniform distances around a circular table. Copy the diagram. Shade the points on the table that are closer to Moesha than to Jan or Leandra.

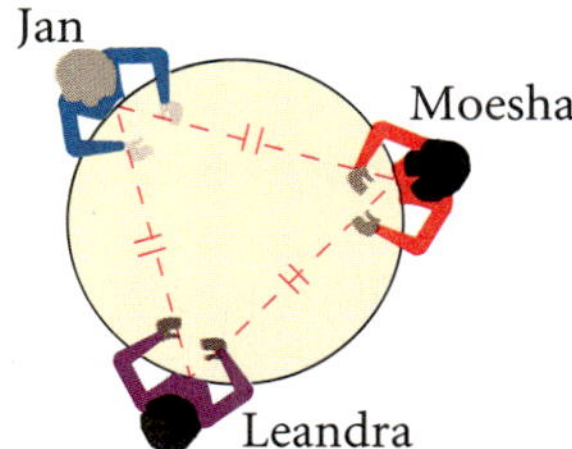

Multiple Choice Practice

For California Standards and CAHSEE Tutorials, visit PHSchool.com. Web Code: bcq-9045

GEOM 11.0

51. Two adjacent cubes form a climbing structure at a playground. Each side of the smaller cube is one-third the length of a side of the larger cube. How many times the volume of the smaller cube is the volume of the larger cube?

Ⓐ 3 Ⓑ 9 Ⓒ 18 Ⓓ 27

GEOM 9.0

52. What is the volume of the file box at the right?

Ⓐ 3 cubic feet

Ⓑ $\frac{1}{2}$ cubic foot

Ⓒ $\frac{1}{4}$ cubic foot

Ⓓ $\frac{1}{48}$ cubic foot

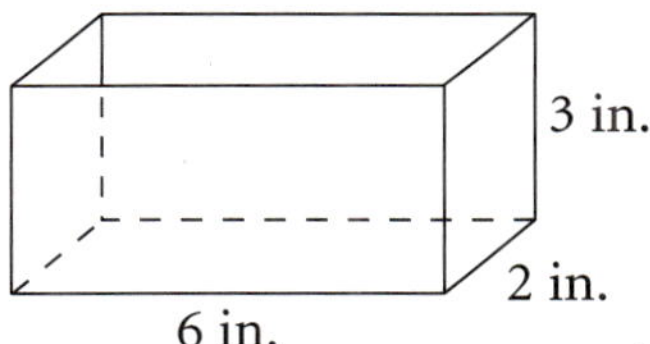

GEOM 15.0

53. Which of the following segment lengths could be used with the two segments at the right in order to form a right triangle?

2.5 in.

6.5 in.

Ⓐ 4 in. Ⓒ 7 in.

Ⓑ 6 in. Ⓓ 9 in.

Mixed Review

Lesson 12-5

Write an equation of the circle with center *C* and radius *r*.

54. $C(6, -10), r = 5$

55. $C(1, 7), r = 6$

56. $C(-8, 1), r = \sqrt{13}$

Lesson 11-2

Find the surface area of each figure to the nearest tenth.

57.

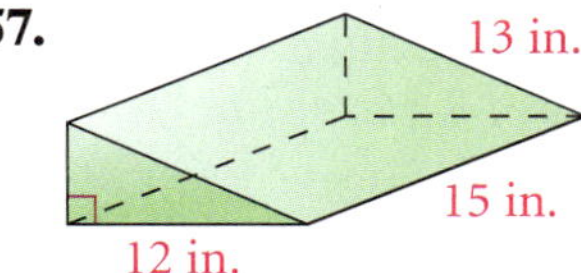

58.

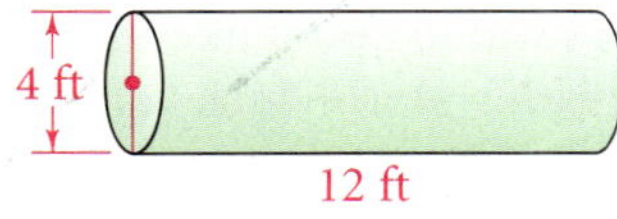

Lesson 10-7

In ⊙*O*, find the area of sector *AOB*. Leave your answer in terms of π.

59. $OA = 4, m\widehat{AB} = 90$

60. $OA = 8, m\widehat{AB} = 72$

61. $OA = 10, m\widehat{AB} = 36$

Test-Taking Strategies

Using Estimation

Estimation may help you find answers, check an answer, or eliminate one or more answer choices. Here are some decimal approximations that can be helpful. The symbol $\approx$ means "is approximately equal to."

$\pi \approx 3$ $\quad\quad$ $\frac{1}{\pi} \approx \frac{1}{3} \approx 0.3$ $\quad\quad$ $\sqrt{2} \approx 1.4$ $\quad\quad$ $\sqrt{3} \approx 1.7$

EXAMPLE

Two circles have the same center O. The radius of the larger circle is twice the radius of the smaller circle, and chord $\overline{DB}$ is tangent to the smaller circle at C. Which of the following is the greatest?

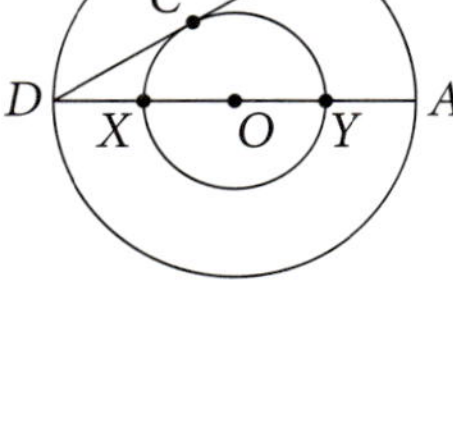

A. DB

B. OA

C. length of $\widehat{XCY}$

D. length of $\widehat{AB}$

Let $OX = OY = 1$. Then $OA = 2$.

The length of $\widehat{XCY}$ is one-half the circumference of the small circle.
The length of $\widehat{XCY} = \frac{1}{2}(2\pi r) = \pi \approx 3$. Thus, choice C is greater than choice B.

Draw $\overline{OC}$. $\triangle ODC$ has a right angle at C with $OD = 2$ and $OC = 1$.
By the Pythagorean Theorem, $DC = \sqrt{3}$. Therefore $DB = 2\sqrt{3} \approx 2(1.7) = 3.4$.
So, choice A is greater than choice C.

$\triangle OCD$ is a 30°-60°-90° triangle, so $m\widehat{AB} = 60$.
Thus, the length of $\widehat{AB}$ is $\frac{1}{6}$ the circumference of the larger circle.
The length of $\widehat{AB} = \frac{1}{6}(2\pi r) = \frac{1}{6}(4\pi) = \frac{2}{3}\pi \approx 2$, which is less than DB (choice A).

Therefore the greatest quantity is DB. The correct answer is A.

EXERCISES

1. Jake is building a square frame with a cross brace like the one shown at the right. Which of the following lengths is closest to the length of the cross brace?

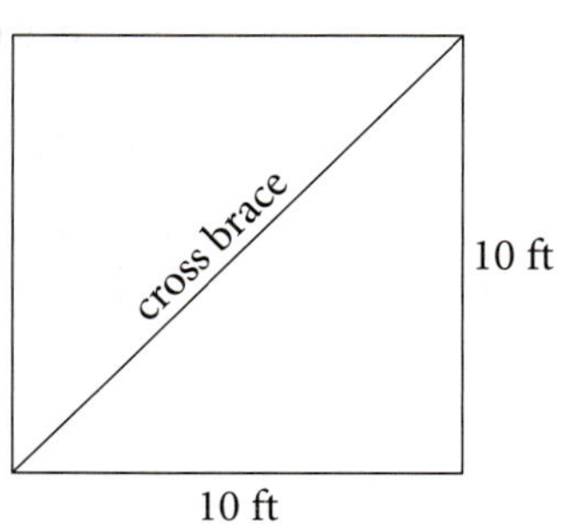

A. 10 ft

B. 12 ft

C. 14 ft

D. Not here

2. Brianna found that 5 of the 60 students in her ninth-grade class like asparagus. Which of the following is closest to the percent of students who do not like asparagus?

A. 10%

B. 30%

C. 60%

D. 90%

3. A wading pool like the one shown at the right is to be filled with water. Which value is closest to the amount of water needed to fill the pool?

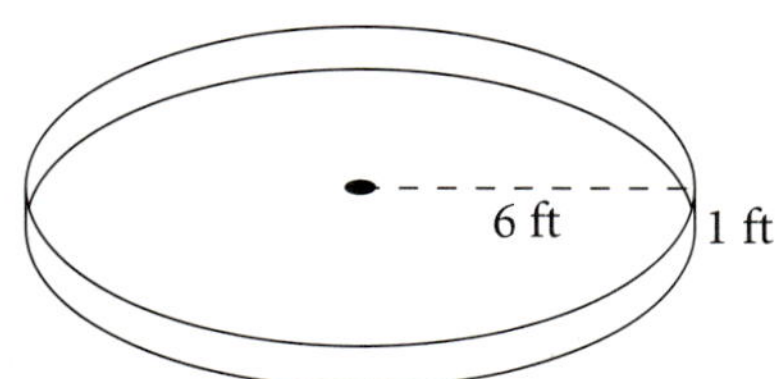

A. 6 ft^3

B. 36 ft^3

C. 100 ft^3

D. 216 ft^3

Chapter 12 Review

Vocabulary Review

English and Spanish Audio Online

chord (p. 670)
circumscribed about (p. 664)
inscribed angle (p. 678)
inscribed in (p. 664)
intercepted arc (p. 678)
locus (p. 701)
point of tangency (p. 662)
secant (p. 687)
standard form of an equation of a circle (p. 695)
tangent to a circle (p. 662)

Use the figure to choose the correct term to complete each sentence.

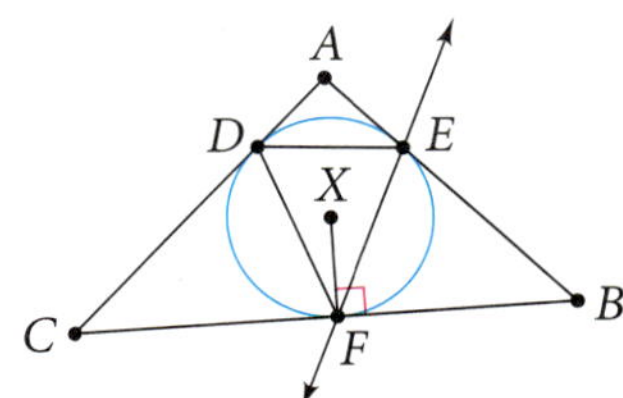

1. $\overline{CB}$ is *(a secant of, tangent to)* $\odot X$.
2. $\overline{DF}$ is a *(chord, locus)* of $\odot X$.
3. $\triangle DEF$ is *(inscribed in, circumscribed about)* $\odot X$.
4. $\angle DEF$ is an *(intercepted arc, inscribed angle)* of $\odot X$.
5. The set of "all points equidistant from the endpoints of $\overline{CB}$" is a *(locus, tangent)*.

For: Vocabulary quiz
Web Code: bcj-1251

Skills and Concepts

Lesson 12-1

- To use the relationship between a radius and a tangent (p. 662)
- To use the relationship between two tangents from one point (p. 664)

GEOM 7.0, 21.0

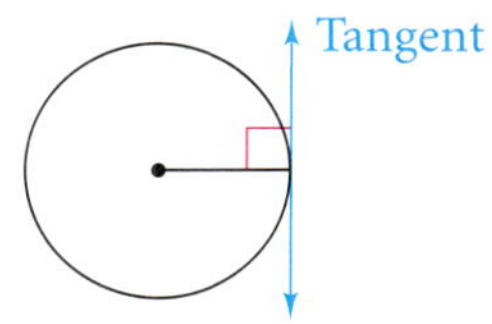

A **tangent to a circle** is a line, ray, or segment in the plane of the circle that intersects the circle in exactly one point, the **point of tangency**. Two segments tangent to a circle from a point outside the circle are congruent. If a line is tangent to a circle, then the line is perpendicular to the radius drawn to the point of tangency. The converse is also true.

A triangle is **inscribed in** a circle if all of the vertices lie on the circle. When a triangle is **circumscribed about** a circle, each side is tangent to the circle.

Each polygon circumscribes a circle. Find the perimeter of the polygon.

6.
7. 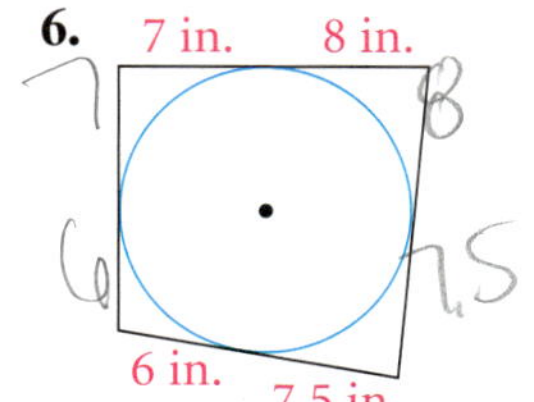

8. 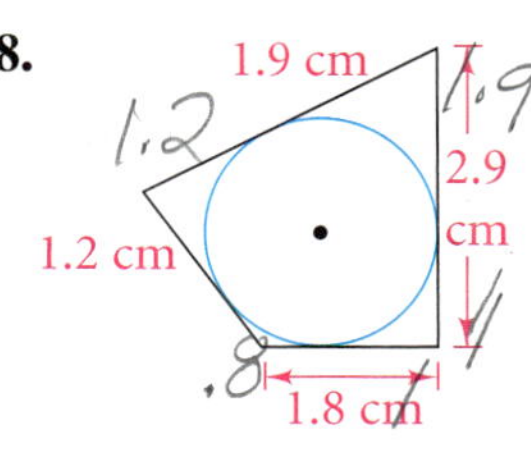

7 in. 8 in. 6 in. 7.5 in.

Lesson 12-2

- To use congruent chords, arcs, and central angles (p. 670)
- To recognize properties of lines through the center of a circle (p. 672)

GEOM 2.0, 7.0, 21.0

Segments with endpoints on a circle are called **chords**. Within a circle or in congruent circles,

- congruent central angles have congruent chords.
- congruent chords have congruent arcs.
- congruent arcs have congruent central angles.

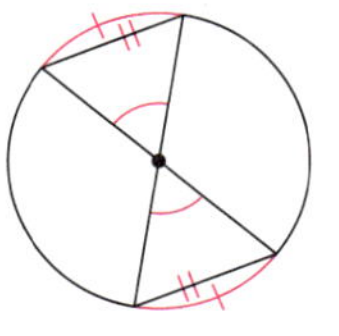

- chords equidistant from the center are congruent.
- congruent chords are equidistant from the center.

A diameter that is perpendicular to a chord bisects the chord and its arcs. A diameter that bisects a chord that is not a diameter is perpendicular to the chord. The perpendicular bisector of a chord contains the center of the circle.

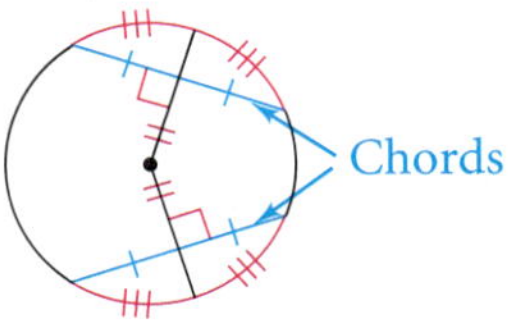

Find the value of x to the nearest tenth.

9. 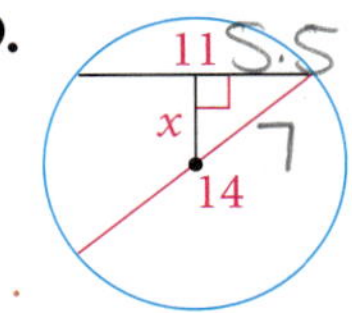

10. 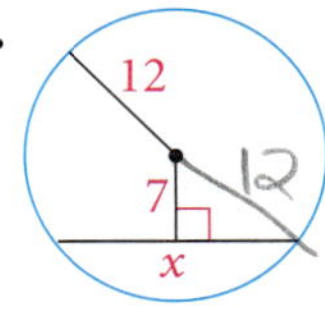

11. 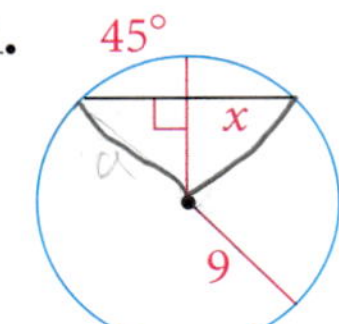

12. 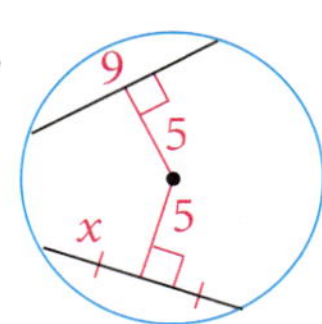

Lesson 12-3

- To find the measure of an inscribed angle (p. 678)
- To find the measure of an angle formed by a tangent and a chord (p. 680)

 **GEOM 2.0, 7.0, 21.0**

An angle is an **inscribed angle** if the vertex is on a circle and sides of the angle are chords of the circle. Its **intercepted arc** is the arc whose endpoints are on the sides of the angle and whose remaining points lie in the interior of the angle.

The measure of an inscribed angle is half the measure of its intercepted arc. The measure of an angle formed by a tangent and a chord that intersect on a circle is half the measure of the intercepted arc.

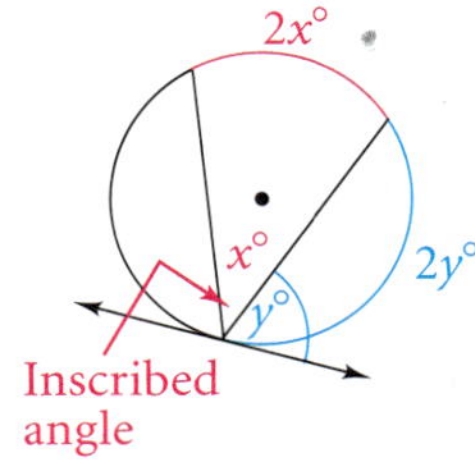

Two inscribed angles that intercept the same arc are congruent. An angle inscribed in a semicircle is a right angle. The opposite angles of a quadrilateral inscribed in a circle are supplementary.

Assume that lines that appear tangent are tangent. Find the value of each variable.

13. 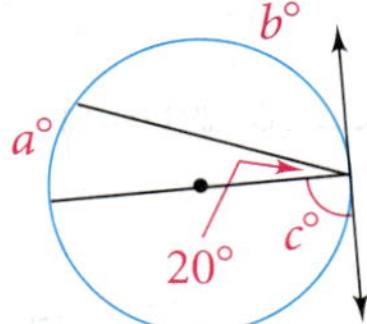

14. 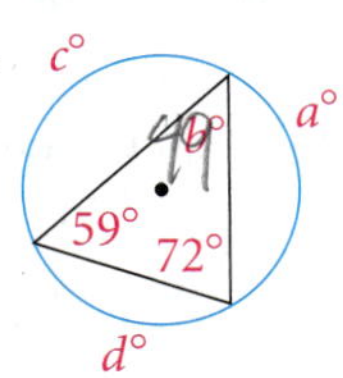

15. 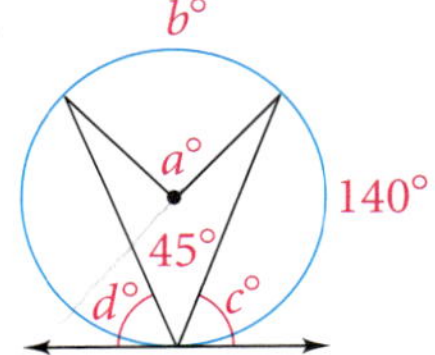

Lesson 12-4

- To find the measures of angles formed by chords, secants, and tangents (p. 687)
- To find the lengths of segments associated with circles (p. 689)

 **GEOM 2.0, 7.0, 21.0**

A **secant** is a line, ray, or segment that intersects a circle at two points.

The measure of an angle formed by two chords that intersect in a circle is half the sum of the measures of the intercepted arcs.

The measure of an angle formed by two secants, two tangents, or a secant and a tangent drawn from a point outside the circle is half the difference of the measures of the intercepted arcs.

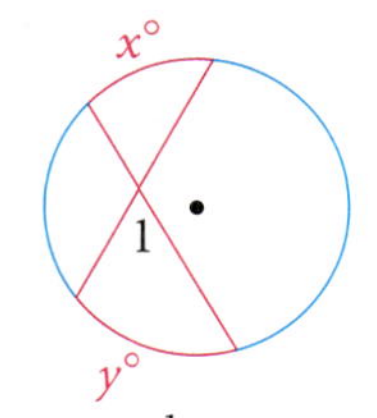

$m\angle 1 = \frac{1}{2}(x + y)$

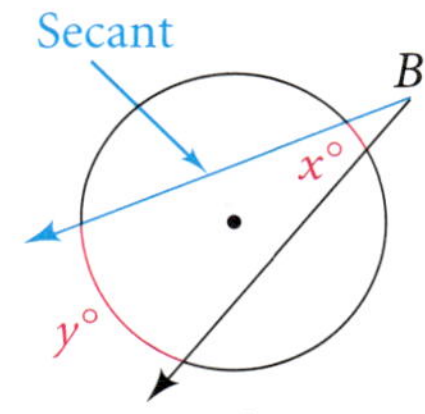

$m\angle B = \frac{1}{2}(y - x)$

For a given point and circle, the product of the lengths of the two segments from the point to the circle is constant along any line through the point and circle.

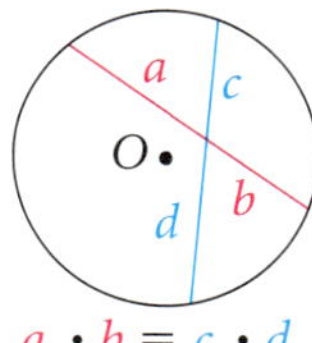

$a \cdot b = c \cdot d$

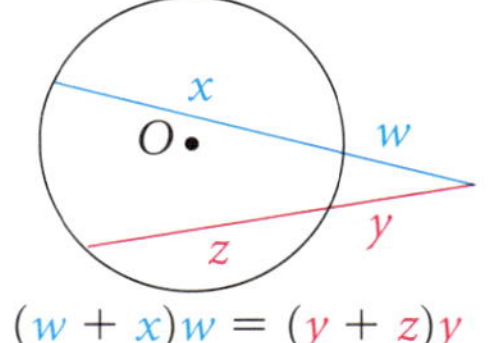

$(w + x)w = (y + z)y$

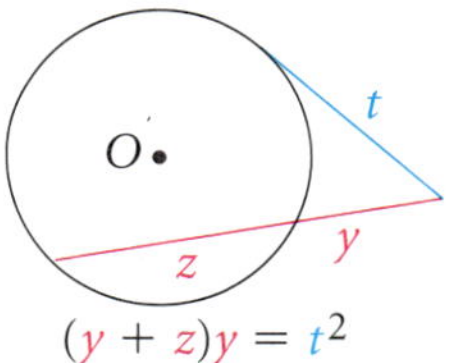

$(y + z)y = t^2$

Assume that lines that appear tangent are tangent. Find the value of each variable.

16.

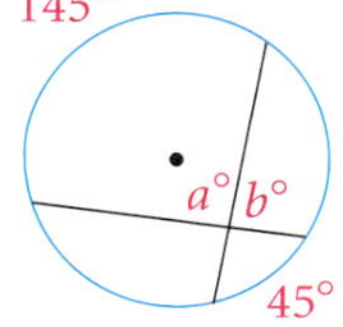

17.

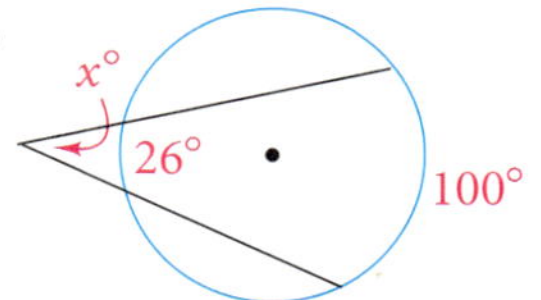

18.

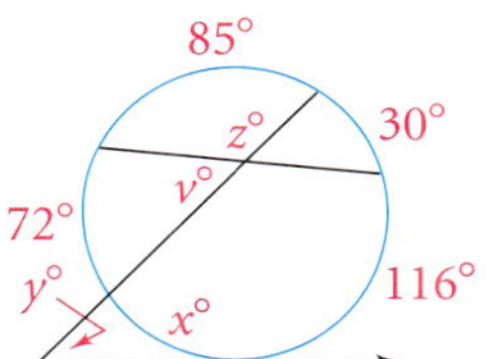

x^2 **Algebra** **Find the value of each variable using the given chords, secants, and tangents. If your answer is not an integer, round to the nearest tenth.**

19.

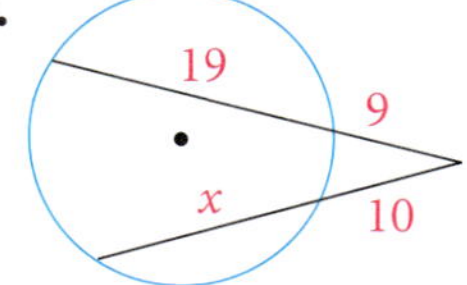

20.

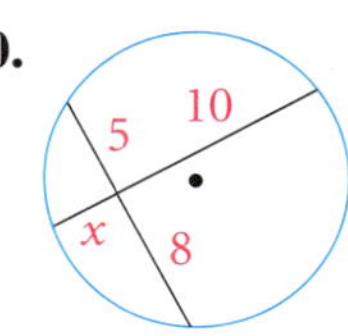

21.

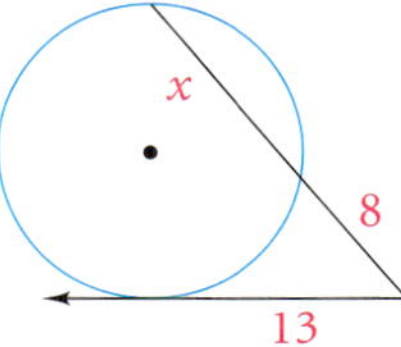

Lessons 12-5 and 12-6

- To write an equation of a circle (p. 695)
- To find the center and radius of a circle (p. 696)
- To draw and describe a locus (p. 701)

The equation $(x - h)^2 + (y - k)^2 = r^2$ is in **standard form.** You may also call it the *standard equation* of a circle.

If you know the center and a point on a circle, you can write the standard equation of the circle. Use the Distance Formula to find the radius. Then substitute the coordinates of the center for (h, k) and the radius for r in the equation of a circle. If you know the equation of a circle, you can identify the center and radius.

A set of points that meet a stated condition is a **locus**. Sometimes you can describe a figure as a locus.

Write the standard equation of the circle with center C and radius r.

22. $C(2, 5); r = 3.5$

23. $C(-3, 1); r = \sqrt{5}$

24. $C(9, -4); r = 4$

Write the standard equation of the circle with center C passing through point P.

25. $C(0, 1); P(4, 9)$

26. $C(-2, 3); P(4, -4)$

27. $C(10, 7); P(-8, -5)$

Describe the circle with the given equation.

28. $x^2 + (y - 8)^2 = 49$

29. $(x - 5)^2 + (y + 9)^2 = 40$

30. $(x + 1)^2 + y^2 = 9$

Sketch and label each locus.

31. all points in a plane 2 cm from a circle with radius 1 cm

32. all points in a plane equidistant from two points

33. all points in space a distance a from $\overline{DS}$

For: Chapter Test
Web Code: bca-1252

Algebra **Assume that lines that appear tangent are tangent. Find the value of x.**

1. 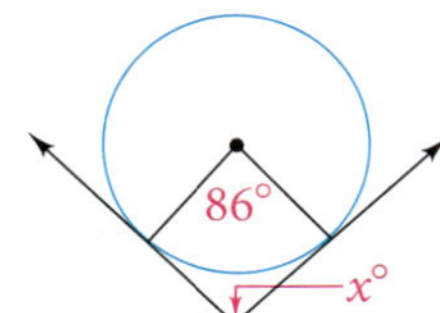

2. 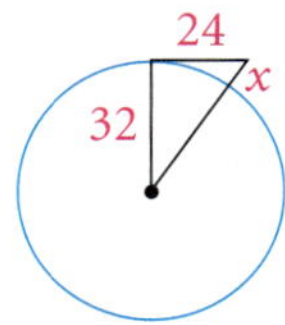

3. a. **Open-Ended** Draw a circle with two congruent chords that form an inscribed angle.
   b. **Constructions** Construct the bisector of the inscribed angle. What do you notice?

Algebra **Find the value of x. If your answer is not an integer, round to the nearest tenth.**

4. 

5.

6. **Coordinate Geometry** An equation of a circle is $(x - 4)^2 + (y - 3)^2 = 25$. Find the center and radius of the circle. Then graph the circle.

Find $m\widehat{AB}$.

7. 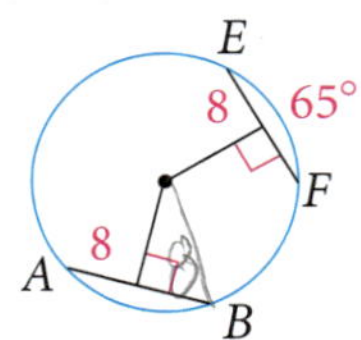

8. 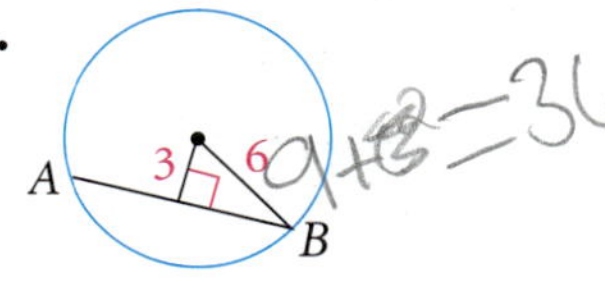

9. Find the value of z.

10. **Writing** What is special about a rhombus inscribed in a circle? Justify your answer.

11. A chord of a circle has length 4.2 cm and is 8 cm from the center of the circle. What is the radius of the circle to the nearest hundredth?

Find the center and radius of each circle.

12. $(x + 3)^2 + (y - 2)^2 = 9$

13. $x^2 + (y - 9)^2 = 225$

Algebra **For Exercises 14–19, lines that appear tangent are tangent. Find the value of each variable. If your answer is not an integer, round to the nearest tenth.**

14. 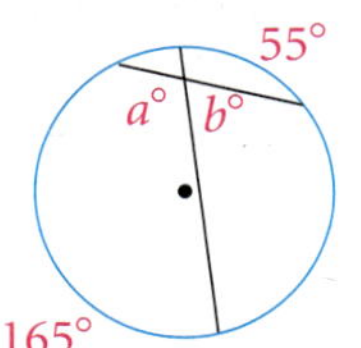

15. 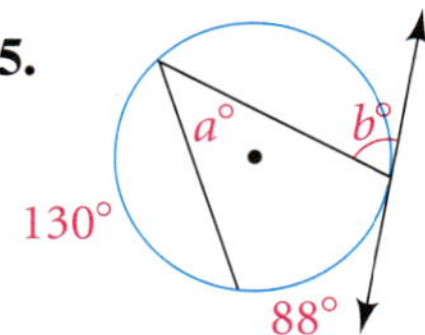

16. 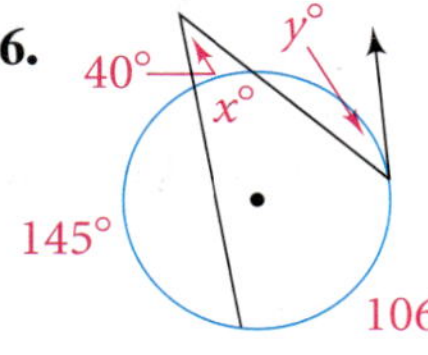

17. 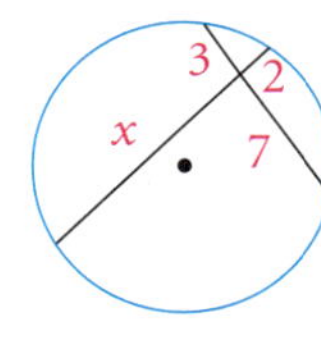

18. 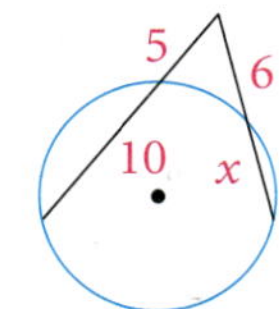

19. 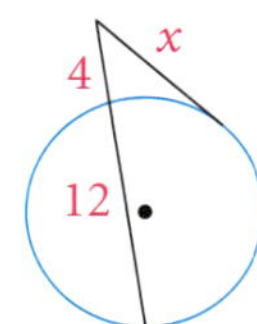

Write the standard equation of each circle.

20. 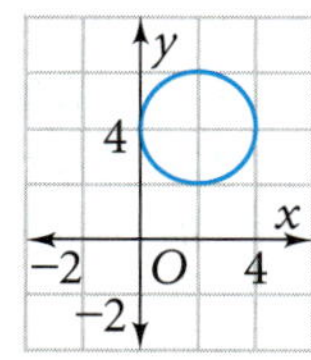

21. 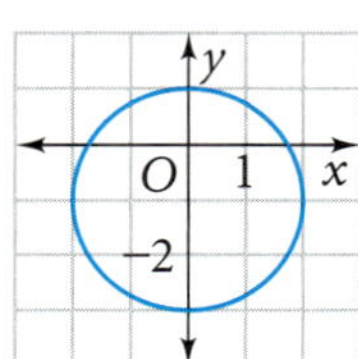

22. Write an equation of the circle with center $(3, 0)$ that passes through point $(-2, -4)$.

23. Find the circumference and area of the circle whose equation is $(x - 2)^2 + (y - 7)^2 = 81$. Round to the nearest tenth.

24. Write an equation for the locus: Points in the coordinate plane that are 4 units from $(-5, 2)$.

Coordinate Geometry **Sketch each locus on a coordinate plane.**

25. all points 6 units from the origin

26. all points 3 units from the line $y = -2$

27. all points equidistant from points $(2, 4)$ and $(0, 0)$

28. all points equidistant from the axes

Standards Mastery

Cumulative Practice

Multiple Choice

Read each question. Then write the letter of the correct answer on your paper.

1. All four angles of a quadrilateral have the same measure. Which statement is true? **(Lesson 6-1)**

(A) All four sides of the quadrilateral have the same length.

(B) All four angles of the quadrilateral are acute.

(C) Opposite sides of the quadrilateral are parallel.

(D) The quadrilateral must be a square.

2. A triangular park is bordered by three streets as shown in the map below.

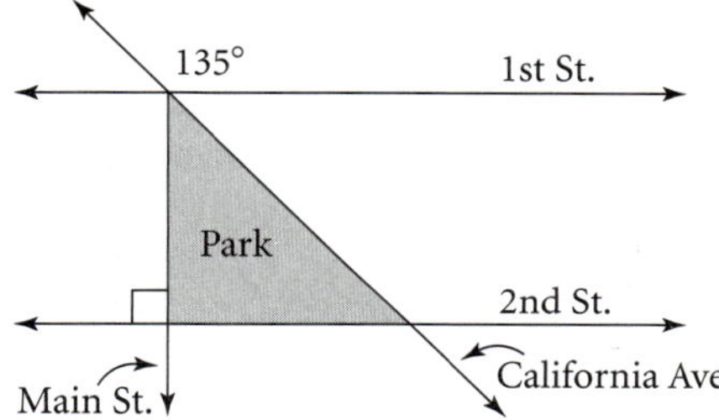

If 1st Street and 2nd Street are parallel, what are the measures of the three angles of the park? **(Lesson 3-1)**

(A) 90°, 45°, 45°

(B) 90°, 35°, 55°

(C) 90°, 25°, 65°

(D) 135°, 25°, 10°

3. What is the volume of the triangular prism shown at the right?  **(Lesson 11-4)**

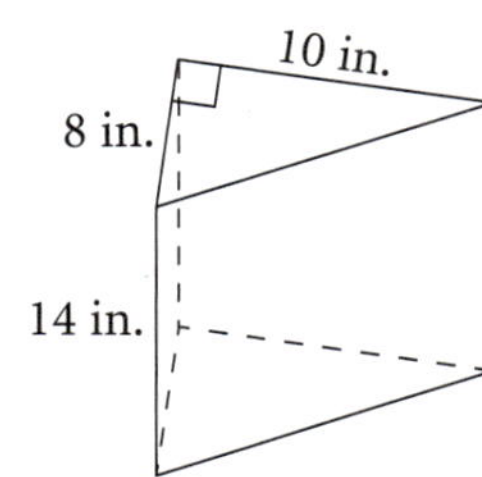

(A) 520 in.^{3}

(B) 560 in.^{3}

(C) 600 in.^{3}

(D) 1,120 in.^{3}

4. The endpoints of $\overline{PK}$ are $P(12, -3)$ and $K(4, 4)$. The segment is translated so that the image of P is $(8, 4)$. What is the image of K? **(Lesson 9-1)**

(A) $(4, -3)$

(B) $(0, 11)$

(C) $(0, -3)$

(D) $(4, 0)$

5. The diameter of a circle is 24 cm. What is the length of a 60° arc? **(Lesson 10-6)**

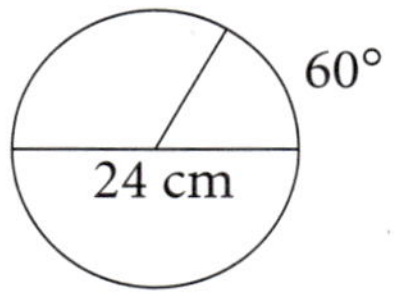

60°

24 cm

(A) 3π cm

(B) 4π cm

(C) 6π cm

(D) 8π cm

6. In the figure below, $\overline{PQ}$ is parallel to $\overline{RS}$, and segments $\overline{PS}$ and $\overline{QR}$ intersect at point T. What is the length of $\overline{PS}$? **(Lesson 7-3)**

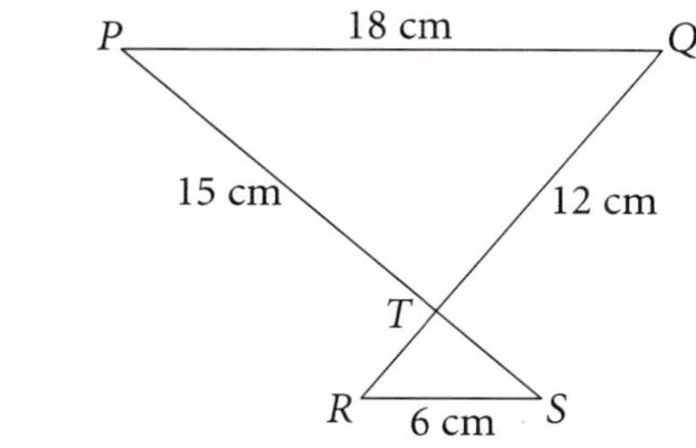

(A) 4 cm

(B) 5 cm

(C) 19 cm

(D) 20 cm

7. A wooden pole was broken during a windstorm. Before the pole broke, the height of the pole above the ground was 18 feet. When it broke, the top of the pole hit the ground 12 feet from the base. **(Lesson 8-1)**

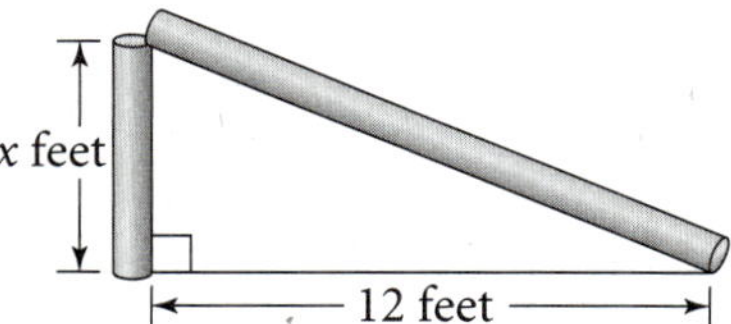

How tall is the part of the pole that remains standing?

(A) 5 ft

(B) 10 ft

(C) 13 ft

(D) 15 ft

8. In the map below, 1st Ave., 2nd Ave., and 3rd Ave. are parallel. **(Lesson 3-1)**

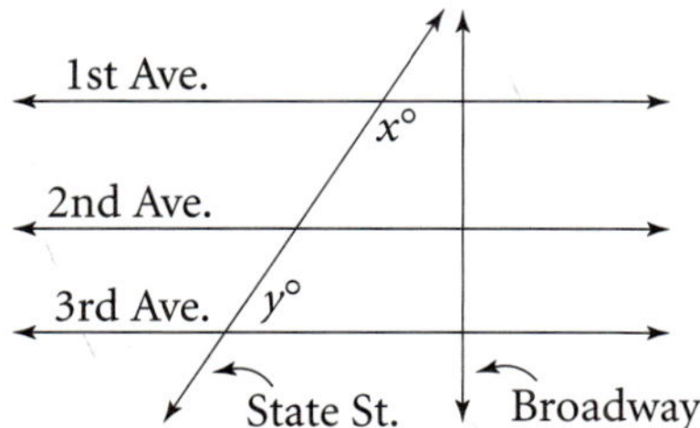

Which expression has the same value as y?

(A) $360 - x$

(B) $180 - x$

(C) $90 - x$

(D) $90 + x$

Cumulative Practice (continued)

9. The diameter of the circle below is 12 units. What is the minor arc length from Point C to Point D? **(Lesson 10-6)**

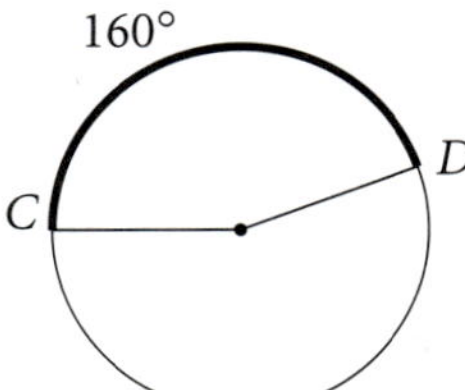

Ⓐ $\frac{16}{3}\pi$ units

Ⓑ 8π units

Ⓒ 16π units

Ⓓ $\frac{160}{3}\pi$ units

10. A party store sells cone-shaped cardboard party hats like the one shown below. Approximately how much cardboard is needed to make one of these party hats? **(Lesson 11-3)**

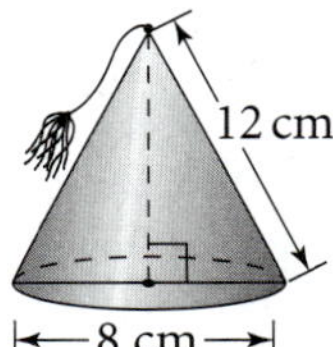

Ⓐ 151 cm^2

Ⓑ 301 cm^2

Ⓒ 603 cm^2

Ⓓ 804 cm^2

11. The surface area of Cube X is 216 square units. The edge length of Cube Y is $\frac{1}{12}$ the edge length of Cube X. By what factor can you multiply the surface area of Cube X to find the surface area of Cube Y? **(Lesson 11-7)**

Ⓐ $\sqrt{\frac{1}{12}}$ Ⓑ $\frac{1}{12}$ Ⓒ $\left(\frac{1}{12}\right)^2$ Ⓓ $\left(\frac{1}{12}\right)^3$

12. The lengths of two sides of a triangle are 4 centimeters and 8 centimeters. Which could be the length of the third side? **(Lesson 5-5)**

Ⓐ 4 cm

Ⓑ 8 cm

Ⓒ 12 cm

Ⓓ 16 cm

13. A frozen dinner is divided into three sections on a circular plate with a 12-inch diameter. What is the approximate arc length in inches of the section containing green beans? **(Lesson 10-6)**

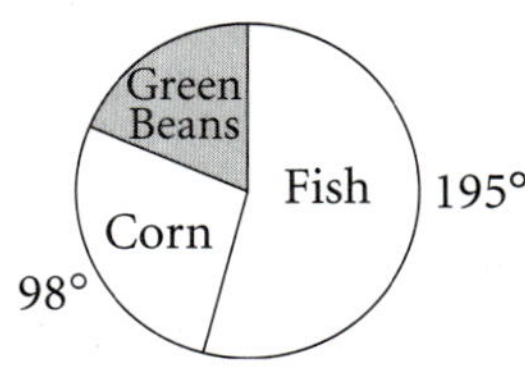

Ⓐ 7

Ⓑ 10

Ⓒ 16

Ⓓ 20

14. The net of a cereal box is shown at the right. What is the surface area? **(Lesson 11-2)**

3 in.
12 in.
10 in.

Ⓐ 180 $in.^2$

Ⓑ 186 $in.^2$

Ⓒ 360 $in.^2$

Ⓓ 372 $in.^2$

15. A cylindrical column in a building has a volume of 114 cubic feet. In another part of the building, another cylindrical column has the same base area, but three times the height. What is the volume of the taller column? **(Lesson 11-4)**

Ⓐ 38 cubic feet

Ⓑ 342 cubic feet

Ⓒ 1026 cubic feet

Ⓓ 3078 cubic feet

16. A square pyramid has a volume of 108 cubic meters. A second pyramid, which is similar in shape, has sides and height that are $\frac{1}{3}$ those of the original pyramid. What is the volume of the second pyramid? **(Lesson 11-7)**

Ⓐ 3 cubic meters

Ⓑ 4 cubic meters

Ⓒ 12 cubic meters

Ⓓ 36 cubic meters

17. In the figure below, what is the length of $\overline{CD}$? **(Lesson 8-2)**

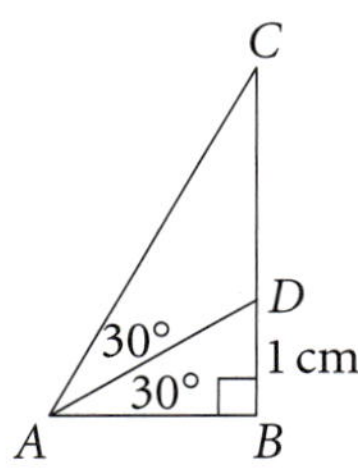

Ⓐ 2 cm

Ⓑ 3 cm

Ⓒ $\sqrt{3}$ cm

Ⓓ $2\sqrt{3}$ cm

18. The trapezoid at the right is divided into three triangles. Which expression could describe the area of the trapezoid? **(Lesson 10-1)**

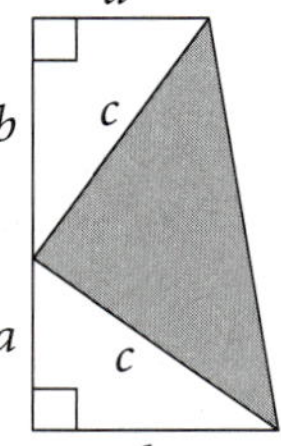

Ⓐ $\frac{1}{2}c^2 + 2ab$

Ⓑ $\frac{1}{2}c^2 + ab$

Ⓒ $a^2 + \frac{1}{2}ab + b^2$

Ⓓ $a^2 + 2ab + b^2$

19. In the figure at the right, $\overline{AB}$ is tangent to circle O, $AB = 15$ cm, and $BC = 9$ cm. What is the radius of circle O? **(Lesson 12-1)**

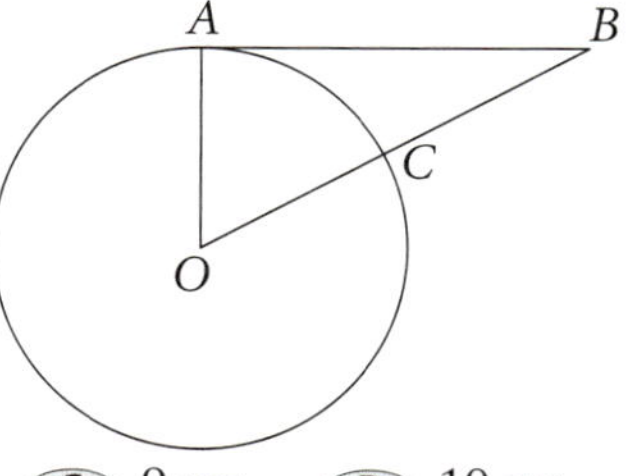

Ⓐ 7 cm Ⓑ 8 cm Ⓒ 9 cm Ⓓ 10 cm

20. Which equation can be solved to find the height h of the triangle at the right? **(Lesson 8-1)**

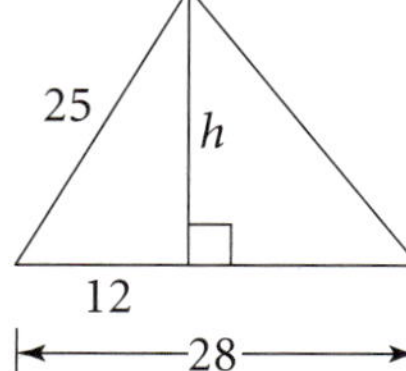

Ⓐ $h^2 = 28^2 - 25^2$

Ⓑ $h^2 = 25^2 + 12^2$

Ⓒ $25^2 = 12^2 + h^2$

Ⓓ $25^2 = 12^2 - h^2$

21. A rancher wants to enclose a pen for new calves using 60 feet of fencing. Which shape uses all of the fencing and encloses the greatest area? **(Lesson 10-1)**

Ⓐ A rectangle that is 12 feet by 18 feet

Ⓑ A square 15 feet on each side

Ⓒ A rectangle that is 20 feet by 30 feet

Ⓓ A circle with an 8-foot radius

22. The diagram shows a cylindrical tin of bath salts that Mitsuko bought for her mother's birthday present.

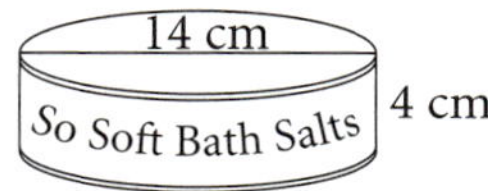

Find the approximate surface area of the tin in square centimeters. **(Lesson 11-2)**

Ⓐ 176 Ⓑ 242 Ⓒ 484 Ⓓ 1583

23. What is the area of the shaded sector? **(Lesson 10-7)**

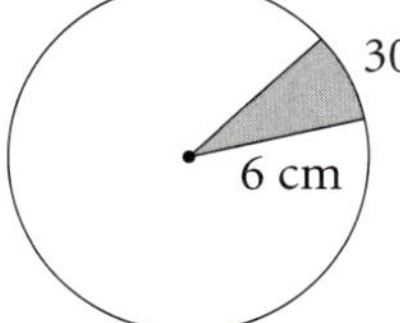

Ⓐ 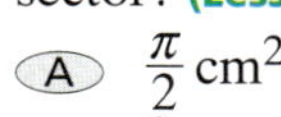 $\frac{\pi}{2}$ cm^2 Ⓒ π cm^2

Ⓑ 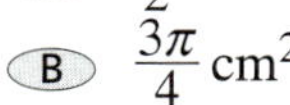 $\frac{3\pi}{4}$ cm^2 Ⓓ 3π cm^2

24. Gerald built a jewelry box for his sister. He wants to line the whole interior of the box with felt. If the dimensions of the interior are 3.5 inches by 2 inches by 4 inches, how much felt must Gerald purchase from the craft store? **(Lesson 11-2)**

Ⓐ 28 in.^{2} Ⓑ 29 in.^{2} Ⓒ 58 in.^{2} Ⓓ 70 in.^{2}

25. In the figure at the right, what is the length of $\overline{PS}$? **(Lesson 7-4)**

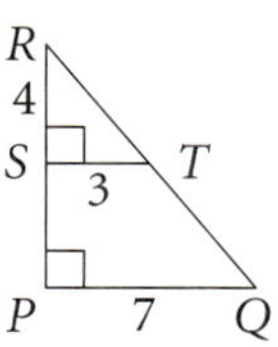

Ⓐ 3 Ⓒ 4

Ⓑ $3\sqrt{2}$ Ⓓ $5\frac{1}{3}$

26. What is the area of ΔRST in square feet? **(Lesson 8-2)**

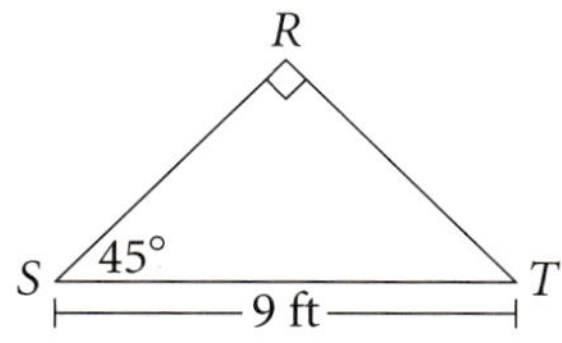

Ⓐ $9\sqrt{2}$ Ⓑ 20.25 Ⓒ 40.5 Ⓓ $81\sqrt{2}$

27. The lengths of the bases of an isosceles trapezoid are shown below. If the perimeter of this trapezoid is 44 units, what is its area? **(Lesson 10-2)**

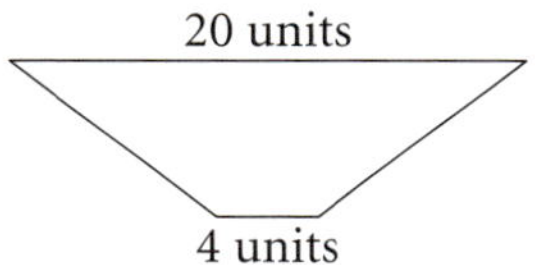

Ⓐ 60 square units Ⓒ 120 square units

Ⓑ 72 square units Ⓓ 144 square units

28. If quadrilateral $PQRS$ is reflected across the line $x = 1$ to become quadrilateral $P'Q'R'S'$, what will be the coordinates of P'? **(Lesson 9-2)**

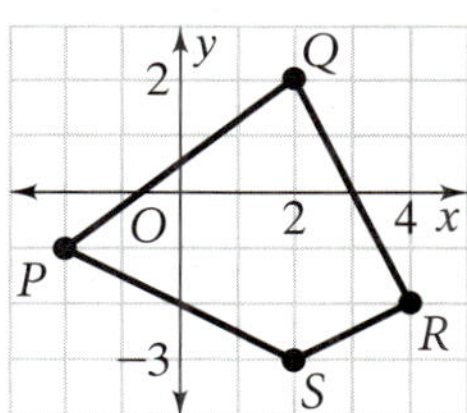

Ⓐ $(2, -1)$ Ⓒ $(-2, 3)$

Ⓑ $(4, 2)$ Ⓓ $(4, -1)$

29. Figure $QRST$ is shown in the coordinate plane. Which transformation creates an image with a vertex at the point $(-2, 1)$? **(Lesson 9-3)**

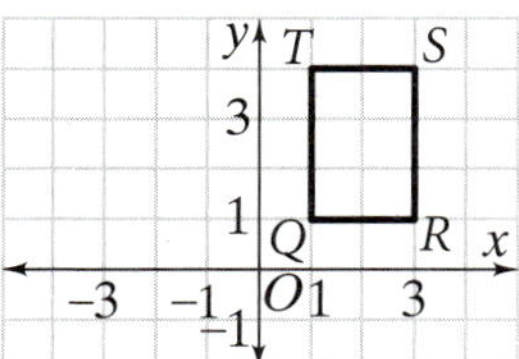

Ⓐ Rotate figure $QRST$ by 90° around R.

Ⓑ Reflect figure $QRST$ across the line $y = 1$.

Ⓒ Reflect figure $QRST$ across the line $x = 1$.

Ⓓ Rotate figure $QRST$ by 90° around Q.

Cumulative Practice (continued)

30. A manufacturer is comparing two packages for a new product. Package I is a rectangular prism that is 9 inches by 5 inches by 8 inches. Package II is a triangular prism with height 15 inches. Its bases are right triangles with 6-inch and 8-inch legs. Which statement best describes the relationship between the two prisms? **(Lesson 11-4)**

- (A) The rectangular prism has the greater surface area.
- (B) The surface areas are equal.
- (C) The rectangular prism has the greater volume.
- (D) The volumes are equal.

31. Which of the following could be the dimensions of the triangle at the right? **(Lesson 8-1)**

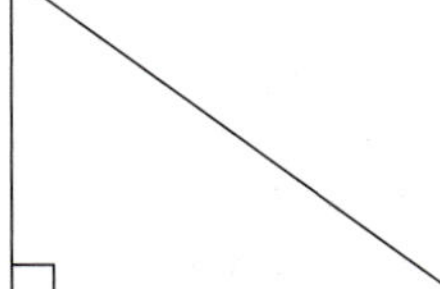

- (A) 3.2, 4.2, 5.2
- (B) 4, 5, 6
- (C) 4.5, 6, 7.5
- (D) 4, 4.8, 6

32. One side of a rectangular tablecloth measures $(n - 1)$ feet. The other side measures $(2n - 4)$ feet. The area of the tablecloth is 40 square feet. Doreen will add a $\frac{1}{2}$-inch wide lace trim to the edge. Approximately how many feet of lace should she order? **(Lesson 1-9)**

- (A) 12
- (B) 13
- (C) 24
- (D) 26

33. Rectangle $RSTU$ has diagonals that intersect at point Z. The coordinates of point R are $(-5, 3)$ and the coordinates of point T are $(8, -7)$. What are the coordinates of point Z? **(Lesson 6-7)**

- (A) $\left(\frac{3}{2}, -2\right)$
- (B) $\left(\frac{13}{2}, 4\right)$
- (C) $\left(-\frac{13}{2}, -4\right)$
- (D) $\left(-\frac{3}{2}, 2\right)$

34. The area of a rectangle is $108a^7b^6$ square units. If the length of the rectangle is $18a^4b^4$ units, which expression represents the width in units? **(Lesson 1-9)**

- (A) $1{,}944a^{11}b^{10}$
- (B) $126a^3b^2$
- (C) $90a^{11}b^{10}$
- (D) $6a^3b^2$

35. If the radius of a sphere with a volume of 2 cubic meters is multiplied by 4, what will be the volume of the new sphere in cubic meters? **(Lesson 11-7)**

- (A) 8
- (B) 32
- (C) 64
- (D) 128

36. $\triangle DEF$ is shown on the coordinate plane at the right. Find the coordinates of its image after $\triangle DEF$ is reflected across the x-axis. **(Lesson 9-2)**

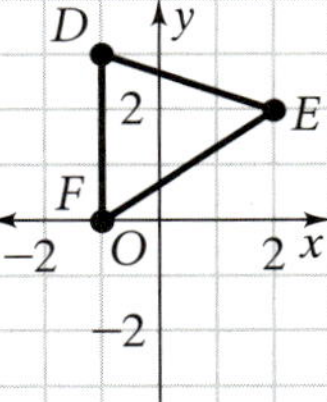

- (A) $(-1, -3), (2, -2), (-1, 0)$
- (B) $(3, -1), (-2, 2), (1, 0)$
- (C) $(1, -3), (-2, -2), (-1, 0)$
- (D) $(-1, 3), (2, 2), (1, 0)$

37. The formula for the volume of a cone is $V = \frac{1}{3}\pi r^2 h$. Which statement is true? **(Lesson 11-5)**

- (A) The volume depends on the mean of π, the radius, and the height.
- (B) The volume depends only on the square of the radius.
- (C) The radius depends on the product of the height and π.
- (D) The volume depends on both the height and the radius.

38. Which of the following facts would be sufficient to prove $\triangle ACB \cong \triangle DBC$? **(Lesson 4-3)**

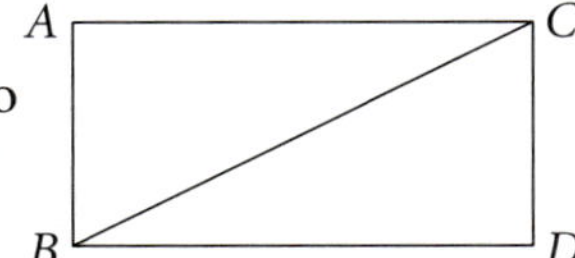

- (A) $\angle ABC$ and $\angle DCB$ are acute
- (B) $\angle A$ is a right angle
- (C) $\overline{BC} \cong \overline{BC}$
- (D) $\overline{AB} \parallel \overline{CD}$ and $\overline{AC} \parallel \overline{BD}$

39. The amount of water that a cylinder will hold best represents the __?__ of the cylinder. **(Lesson 11-4)**

- (A) perimeter
- (B) circumference
- (C) surface area
- (D) volume

40. Bob walked diagonally across a rectangular field that measured 240 feet by 320 feet. Which expression could be used to determine how far Bob walked? **(Lesson 8-1)**

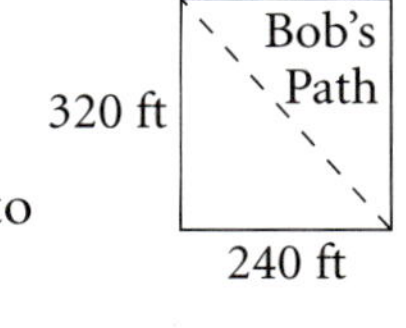

- (A) $2(240 + 320)$
- (B) $\sqrt{240} + \sqrt{320}$
- (C) $\frac{240 + 320}{2}$
- (D) $\sqrt{240^2 + 320^2}$

Standards Mastery

41. Anna has cereal in a cylindrical container that has a diameter of 6 inches and is 12 inches high. She wants to store the cereal in a rectangular prism container. Which prism dimensions give a volume that is closest to the volume of the cylindrical container? **(Lesson 11-4)**

- (A) 6 in. by 6 in. by 10 in.
- (B) 5 in. by 7 in. by 10 in.
- (C) 4 in. by 8 in. by 10 in.
- (D) 3 in. by 9 in. by 10 in.

42. Priscilla stands at point A directly below a kite that Kirk is flying from point B. Kirk has let out 130 feet of the string. Priscilla stands 40 feet from Kirk. To the nearest foot, how many feet from the ground is the kite? **(Lesson 8-1)**

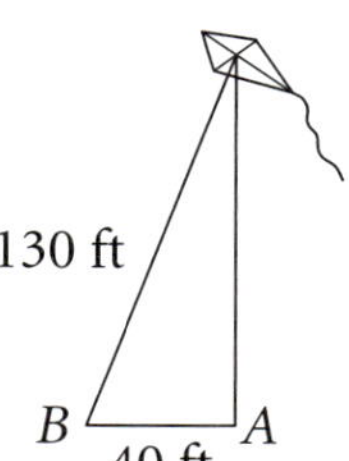

- (A) 156
- (B) 136
- (C) 124
- (D) 90

43. Solve for x in the figure below. Round your answer to the nearest tenth of a foot. **(Lesson 8-4)**

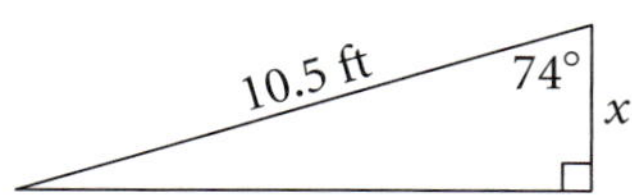

- (A) 2.9
- (B) 10.1
- (C) 38.1
- (D) 72.8

44. In $\triangle ABC$ and $\triangle DEF$, $\overline{AB} \cong \overline{DE}$, $\overline{AC} \cong \overline{EF}$ and $m\angle A = m\angle E$. What can you conclude? **(Lesson 4-2)**

- (A) $\triangle ABC \cong \triangle DEF$ by SAS
- (B) $\triangle ABC \cong \triangle DEF$ by AAS
- (C) $\triangle ABC \cong \triangle EDF$ by SAS
- (D) $\triangle ABC \cong \triangle EDF$ by ASA

45. *If each pair of opposite sides of a quadrilateral are equal in length and parallel, then the quadrilateral is a square.*
Which of the following serves as a *counterexample* to the assertion above? **(Lesson 1-1)**

- (A) kite
- (B) rhombus
- (C) square
- (D) trapezoid

Standards Reference Guide

California Content Standard	Item Number(s)
GEOM 7.0	1, 2, 8
GEOM 9.0	3, 10, 14, 15, 22, 24, 37, 39, 41
GEOM 22.0	4, 28, 29, 36
GEOM 21.0	5, 19
GEOM 5.0	6, 25
GEOM 15.0	7, 20, 31, 40, 42
GEOM 8.0	9, 13, 18, 21, 23, 27, 30, 32, 34
GEOM 11.0	11, 16, 35
GEOM 6.0	12
GEOM 20.0	17, 26
GEOM 17.0	33
GEOM 4.0	38
GEOM 19.0	43
GEOM 2.0	44
GEOM 3.0	45

For additional review and practice, see the *California Standards Review and Practice Workbook* or go online to use the

Visit: PHSchool.com, **Web Code:** bcq-9045

Chapter 1 Extra Practice: Skills, Word Problems, and Proof

Lesson 1-1

Find the next two terms in each sequence.

1. 12, 17, 22, 27, 32, . . .

2. 1, 1.1, 1.11, 1.111, 1.1111, . . .

3. 5000, 1000, 200, 40, . . .

4. 1, 12, 123, 1234, . . .

5. 3, 0.3, 0.03, 0.003, . . .

6. 1, 4, 9, 16, 25, . . .

Draw the next figure in each sequence.

7.

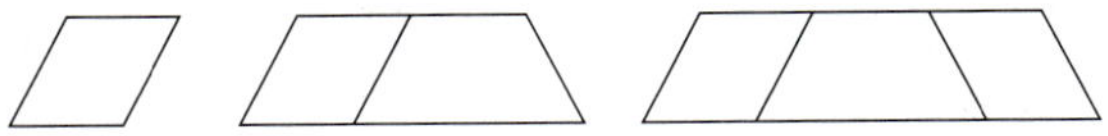

8.

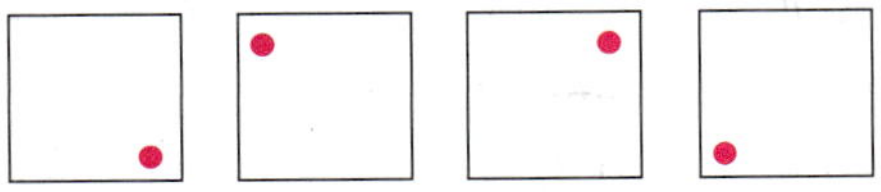

Lesson 1-2

Name the space figure that can be formed by folding each net.

9.

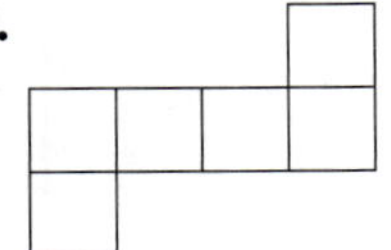

10.

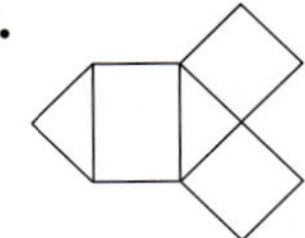

11.

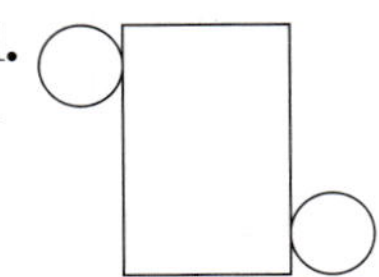

12.

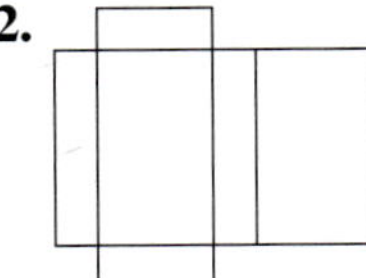

Make (a) an isometric drawing and (b) an orthographic drawing for each foundation drawing.

13.

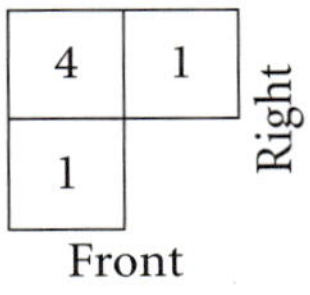

4	1
1	

Right

Front

14.

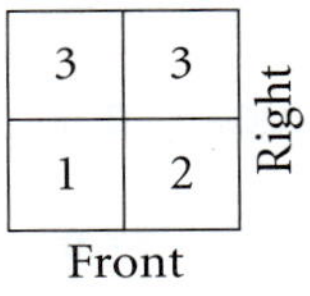

3	3
1	2

Right

Front

15.

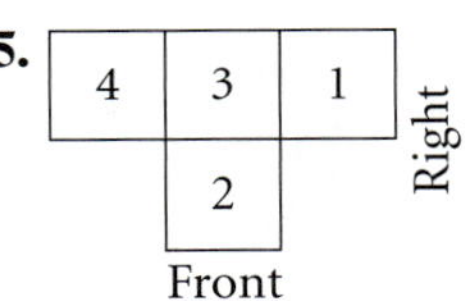

16.

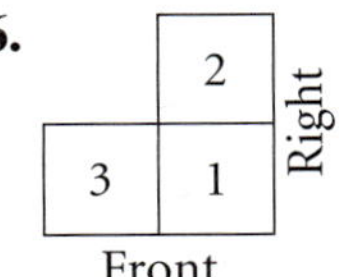

17. You can cut four of the lettered squares from the figure at the right and fold the remaining net to make a box that is open at one end. Write the letters of the squares you could remove to do this. List all the possibilities.

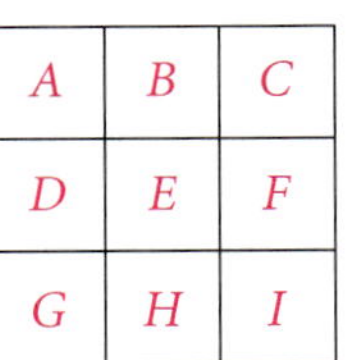

Lessons 1-3 and 1-4

Write *true* or *false*.

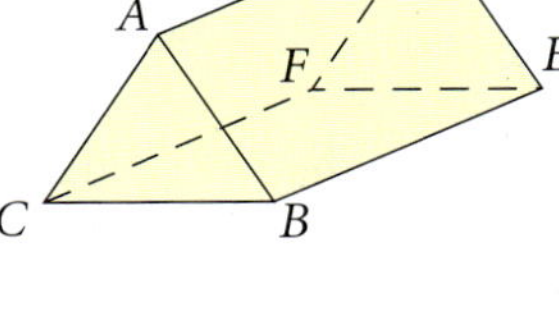

18. A, D, F are coplanar.

19. $\overleftrightarrow{AC}$ and $\overleftrightarrow{FE}$ are coplanar.

20. A, B, E are coplanar.

21. D, A, B, E are coplanar.

22. $\overleftrightarrow{FC} \parallel \overleftrightarrow{EF}$

23. plane $ABC \parallel$ plane FDE

24. $\overleftrightarrow{BC}$ and $\overleftrightarrow{DF}$ are skew lines.

25. $\overleftrightarrow{AD}$ and $\overleftrightarrow{EB}$ are skew lines.

26. $\overleftrightarrow{DE} \parallel \overleftrightarrow{CF}$

27. D, E, and B are collinear.

28. How many sets of four collinear points are there in a 4-by-4 geoboard as pictured at the right?

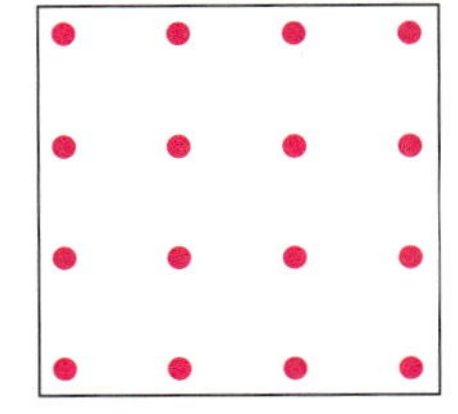

29. $\overline{AB}$ and $\overline{CD}$ do not intersect but $\overrightarrow{DC}$ intersects $\overline{AB}$ in one point. Make a sketch that shows this.

Lessons 1-5 and 1-6

Use the figure at the right for Exercises 30–35.

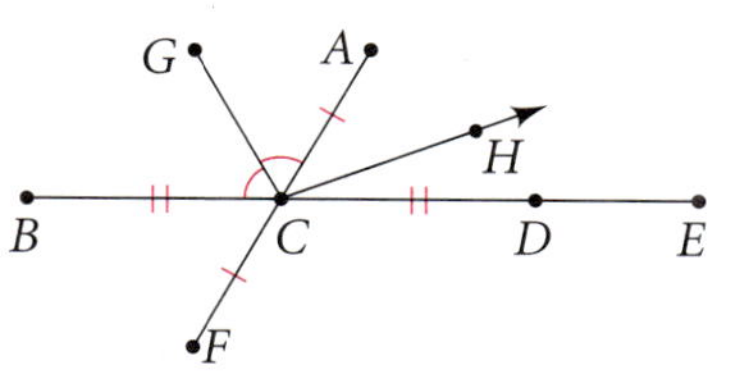

30. If $BC = 12$ and $CE = 15$, then $BE = $ ■.

31. ■ is the angle bisector of ■.

32. **Algebra** $BC = 3x + 2$ and $CD = 5x - 10$. Solve for x.

33. **Algebra** If $AC = 5x - 16$ and $CF = 2x - 4$, then $AF = $ ■.

34. $m\angle BCG = 60$, $m\angle GCA = $ ■, and $m\angle BCA = $ ■.

35. $m\angle ACD = 60$ and $m\angle DCH = 20$. Find $m\angle HCA$.

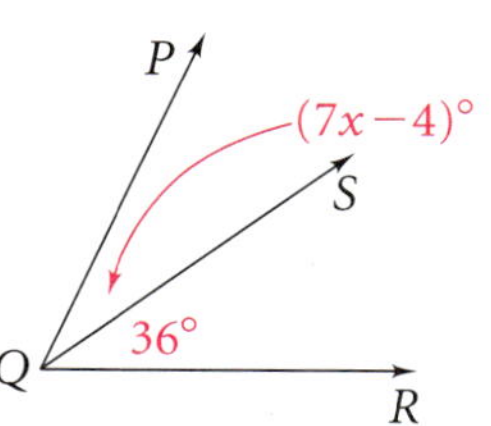

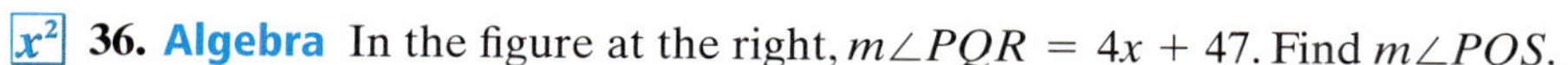

36. **Algebra** In the figure at the right, $m\angle PQR = 4x + 47$. Find $m\angle PQS$.

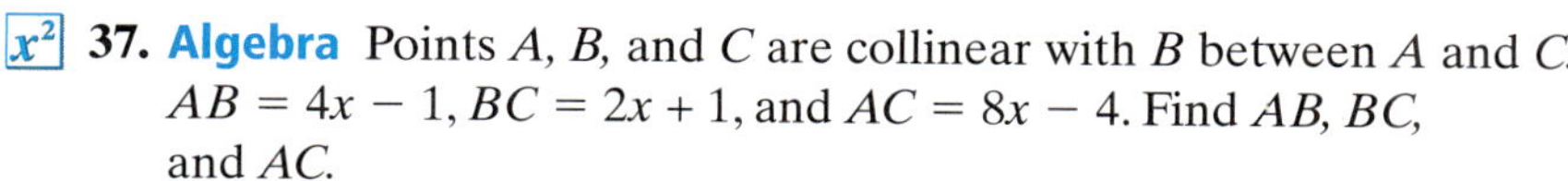

37. **Algebra** Points A, B, and C are collinear with B between A and C. $AB = 4x - 1$, $BC = 2x + 1$, and $AC = 8x - 4$. Find AB, BC, and AC.

Lesson 1-7

Make a diagram larger than the given one. Then do the construction.

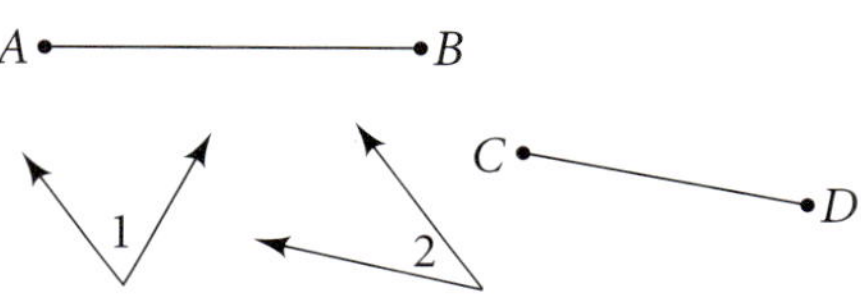

38. Construct $\angle A$ so that $m\angle A = m\angle 1 + m\angle 2$.

39. Construct the perpendicular bisector of $\overline{AB}$.

40. Construct the angle bisector of $\angle 1$.

41. Construct $\overline{FG}$ so that $FG = AB + CD$.

Lesson 1-8

(a) Find the distance between the points to the nearest tenth.
(b) Find the coordinates of the midpoint of the segments with the given endpoints.

42. $A(2, 1)$, $B(3, 0)$

43. $R(5, 2)$, $S(-2, 4)$

44. $Q(-7, -4)$, $T(6, 10)$

45. $C(-8, -1)$, $D(-5, -11)$

46. $J(0, -5)$, $N(3, 4)$

47. $Y(-2, 8)$, $Z(3, -5)$

48. A map of a city and suburbs shows an airport located at $A(25, 11)$. An ambulance is on a straight expressway headed from the airport to Grant Hospital at $G(1, 1)$. The ambulance gets a flat tire at the midpoint M of $\overline{AG}$. As a result, the ambulance crew calls for helicopter assistance.
a. What are the coordinates of point M?
b. How far does the helicopter have to fly to get from M to G? Assume all coordinates are in miles.

Lesson 1-9

Find the perimeter (or circumference) and area of each figure.

49. 14 in. 7 in.

50.

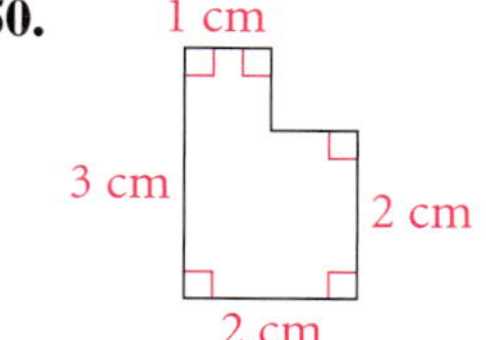

51.

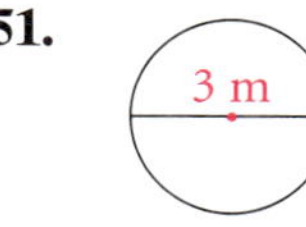

52.

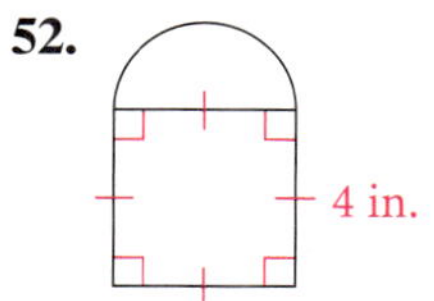

Extra Practice: Skills, Word Problems, and Proof

Lessons 2-1 and 2-2

For Exercises 1–3, identify the hypothesis and conclusion of each conditional.

1. If you can predict the future, then you can control the future.
2. If Dan is nearsighted, then Dan needs glasses.
3. If lines k and m are skew, then lines k and m are not perpendicular.
4. Write the converse of each statement in Exercises 1–3.

For each of the statements, write the conditional form and then the converse of the conditional. If the converse is true, combine the statements as a biconditional.

5. The number one is the smallest positive square.
6. Rectangles have four sides.
7. A square with area 100 m^2 has sides that measure 10 m.
8. Two numbers that add up to be less than 12 have a product less than 37.
9. Three points on the same line are collinear.

Lesson 2-2

Is each statement a good definition? If not, find a counterexample.

10. A real number is an even number if its last digit is 0, 2, 4, 6, or 8.
11. A circle with center O and radius r is defined by the set of points in a plane a distance r from the point O.
12. A plane is defined by two lines.
13. Segments with the same length are congruent.

For Exercises 14 and 15, write the two statements that form each biconditional. Tell whether each statement is *true* or *false*.

14. Lines m and n are skew if and only if lines m and n do not intersect.
15. A person can be president of the United States if and only if the person is a citizen of the United States.

Lesson 2-3

Using the statements below, apply the Law of Detachment or the Law of Syllogism to draw a conclusion.

16. If Jorge can't raise money, he can't buy a new car. Jorge can't raise money.
17. If Shauna is early for her meeting, she will gain a promotion. If Shauna wakes up early, she will be early for her meeting. Shauna wakes up early.
18. If Linda's band wins the contest, they will win $500. If Linda practices, her band will win the contest. Linda practices.
19. If Brendan learns the audition song, he will be selected for the chorus. If Brendan stays after school to practice, he will learn the audition song. Brendan stays after school to practice.

For Exercises 20–23, apply the Law of Detachment, the Law of Syllogism, or both to draw a conclusion. Tell which law(s) you used.

20. If you enjoy all foods, then you like cheese sandwiches. If you like cheese sandwiches, then you eat bread.

21. If you go to a monster movie, then you will have a nightmare. You go to a monster movie.

22. If Catherine is exceeding the speed limit, then she will get a speeding ticket. Catherine is driving at 80 mi/h. If Catherine is driving at 80 mi/h, then she is exceeding the speed limit.

23. If Carlos has more than $250, then he can afford the video game he wants. If Carlos worked more than 20 hours last week, then he has more than $250. If Carlos works 15 hours this week, then he worked more than 20 hours last week.

Lesson 2-4

x^2 **24. Algebra** You are given that $2c^2 = 2bc + \frac{ac}{2}$ with $c \neq 0$. Show that $4b = 4c - a$ by filling in the blanks.

a. $2c^2 = 2bc + \frac{ac}{2}$	**a.** Given
b. $4c^2 = 4bc + ac$	**b.** ___?___ and ___?___
c. $4c = 4b + a$	**c.** ___?___ and Distributive Property
d. ___?___	**d.** Subtraction Property
e. $4b = 4c - a$	**e.** ___?___

x^2 **25. Algebra** Solve for x. Show your work. Justify each step.

Given: $\overrightarrow{PF}$ bisects $\angle 1$.

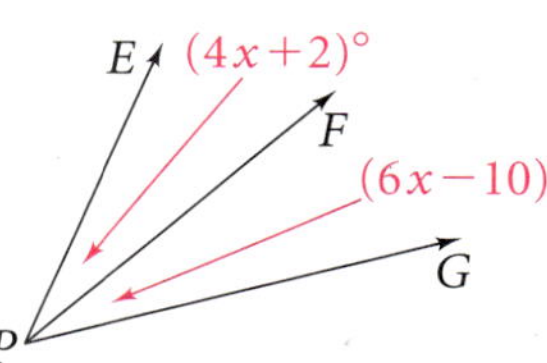

Lesson 2-5

x^2 **Algebra Find the value of x.**

26.

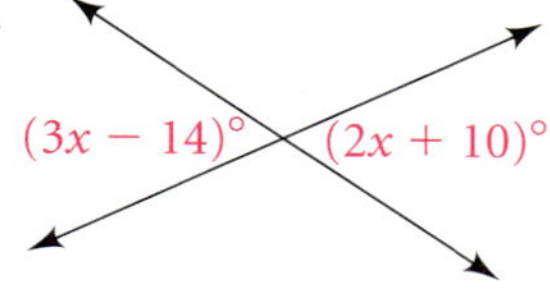

27.

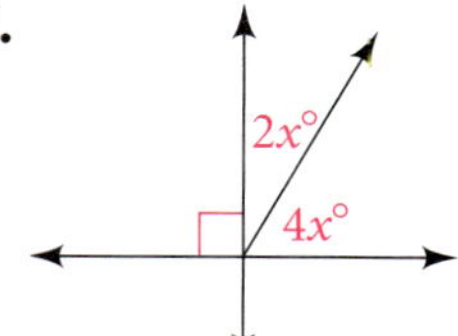

28.

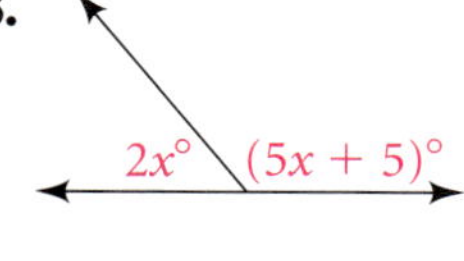

29. Given: $\angle 1$ and $\angle 2$ are complementary.
$\angle 3$ and $\angle 4$ are complementary.

Prove: $\angle 5 \cong \angle 6$

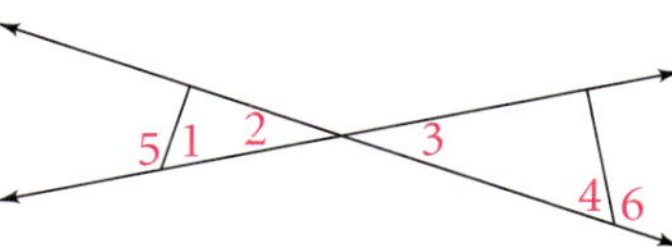

30. Prove or disprove the following statement.

If $\angle APB$ and $\angle CPD$ are vertical angles, $\angle APB$ and $\angle APE$ are complementary, and $\angle CPD$ and $\angle CPF$ are complementary, then $\angle APE$ and $\angle CPF$ are vertical angles.

Chapter 3 Extra Practice: Skills, Word Problems, and Proof

Lesson 3-1

Find $m\angle 1$ and then $m\angle 2$. State the theorems or postulates that justify your answers.

1. 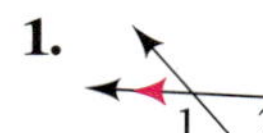
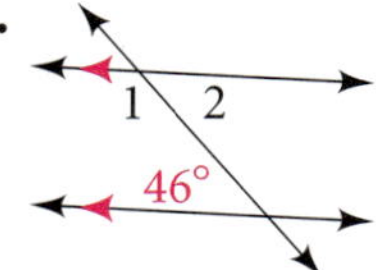

2. 
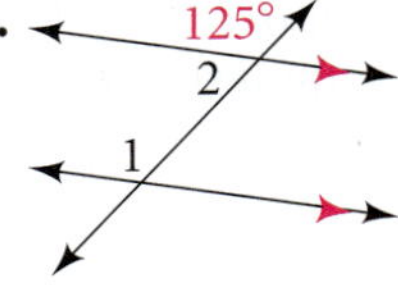

3. 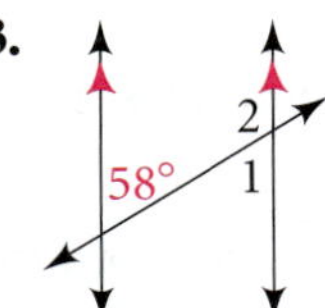

4. 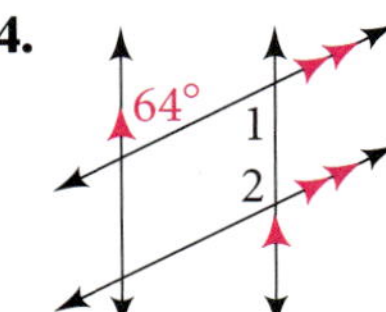

5. Complete the proof.

Given: $\ell \parallel m$, $a \parallel b$

Prove: $\angle 1 \cong \angle 5$

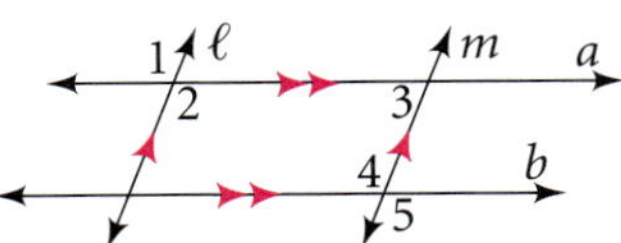

Statements	Reasons
1. $\ell \parallel m$, $a \parallel b$	**1.** Given
2. $\angle 1 \cong \angle 2$	**a.** __?__
3. $\angle 2$ and $\angle 3$ are supplementary.	**b.** __?__
4. $\angle 3$ and $\angle 4$ are supplementary.	**c.** __?__
5. $\angle 2 \cong \angle 4$	**d.** __?__
6. $\angle 1 \cong \angle 4$	**e.** __?__
7. $\angle 4 \cong \angle 5$	**f.** __?__
8. $\angle 1 \cong \angle 5$	**g.** __?__

Lessons 3-2 and 3-3

Refer to the diagram at the right. Use the given information to determine which lines, if any, must be parallel. If any lines are parallel, use a theorem or postulate to tell why.

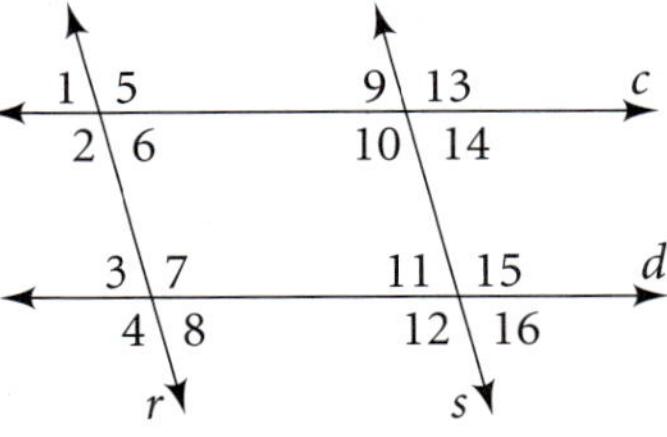

6. $\angle 9 \cong \angle 14$

7. $\angle 1 \cong \angle 9$

8. $\angle 2$ is supplementary to $\angle 3$.

9. $\angle 7 \cong \angle 10$

10. $m\angle 6 = 60, m\angle 13 = 120$

11. $\angle 4 \cong \angle 13$

12. $\angle 3$ is supplementary to $\angle 10$.

13. $\angle 10 \cong \angle 15$

14. Given: $\ell \parallel m$, $a \parallel b$, $a \perp \ell$

Prove: $b \perp m$

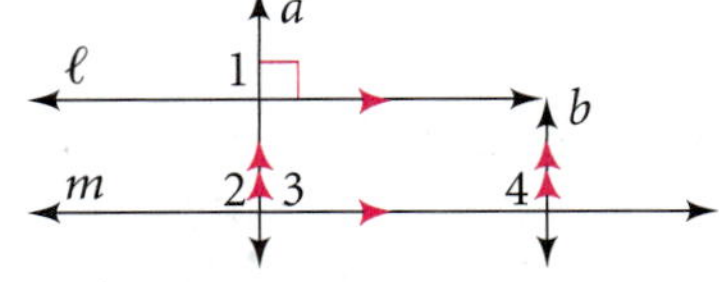

Lesson 3-4

Use a protractor and a centimeter ruler to measure the angles and the sides of each triangle. Classify each triangle by its angles and sides.

15. 

16. 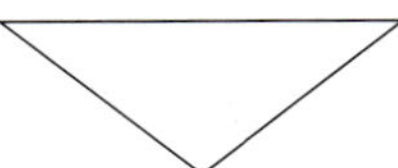

17. 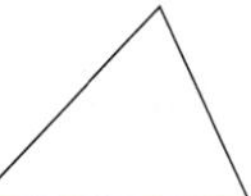

18. 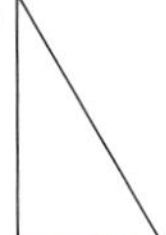

19. Use the figure at the right.
What is the relationship between $\overleftrightarrow{BC}$ and $\overleftrightarrow{DF}$? Justify your answer.

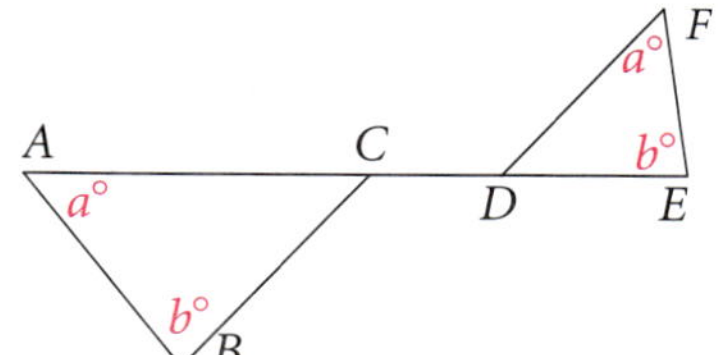

Lessons 3-4 and 3-5

Algebra Find the value of each variable.

20. 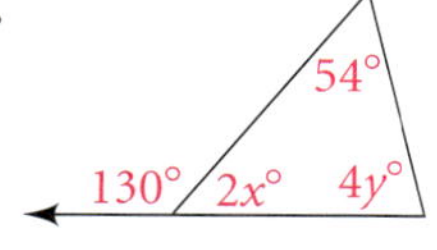

21. 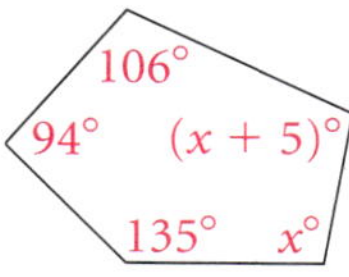

22. 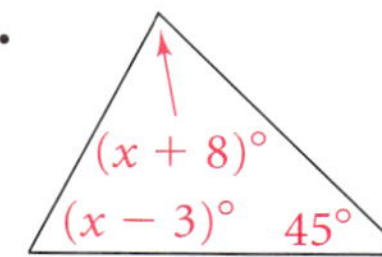

23. 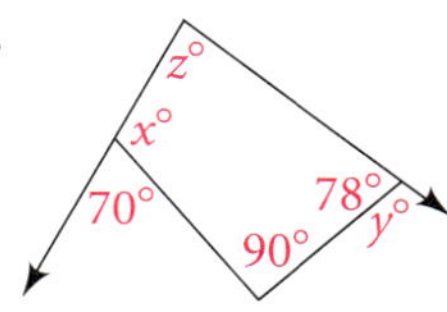

Algebra Find the missing angle measures.

24. 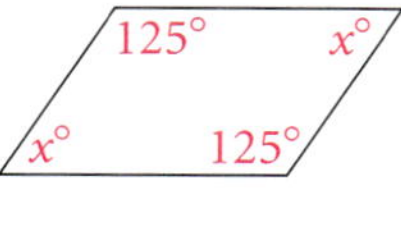

25. 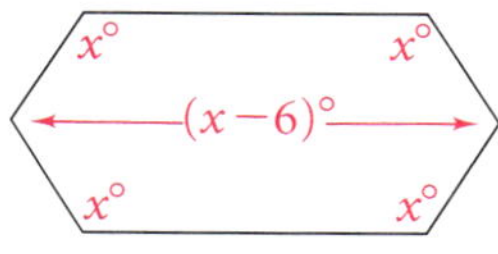

Lessons 3-6

Write an equation in point-slope form of the line that contains the given points.

26. $A(4, 2), B(6, -3)$
27. $C(-1, -1), D(1, 1)$
28. $F(3, -5), G(-5, 3)$
29. $K(5, 0), L(-5, 2)$

Write an equation in slope-intercept form of the line through the given points.

30. $H(2, 7), J(-3, 1)$
31. $M(-2, 4), N(5, -8)$
32. $P(0, 2), Q(6, 8)$
33. $K(5, 0), L(-5, 2)$

Lessons 3-6 and 3-7

Graph each pair of lines and state whether they are parallel, perpendicular, or neither. Explain.

34. $y = 4x - 8$
$y = 4x - 2$

35. $13y - x = 7$
$7 - \frac{y}{2} = x$

36. $y = \frac{-4}{3}x + 2$
$\frac{4}{3}y = x - 1$

37. $\frac{3}{5}y = -x + \frac{3}{2}$
$3x - \frac{15}{3}y = 0$

Without graphing, tell whether the lines are parallel, perpendicular, or neither. Explain.

38. $2x + 3y = 5$
$5x - 10y = 30$

39. $y = -2x + 7$
$x - 2y = 8$

40. $5x - 3y = 0$
$y = \frac{5}{3}x + 2$

41. $y = 3x + 8$
$x + 3y = 8$

42. On a city map, Washington Street is straight and passes through points at (7, 13) and (1, 5). Wellington Street is straight and passes through points at (3, 24) and (9, 32). Do Washington Street and Wellington Street intersect? How do you know?

Lesson 3-8

Use the segments for each construction.

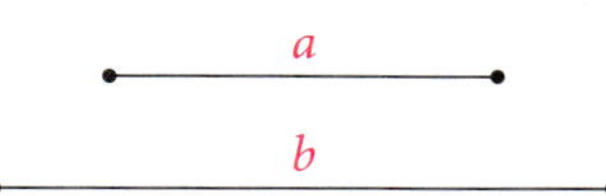

43. Construct a square with side length $2a$.

44. Construct a quadrilateral with one pair of parallel sides each of length $2b$.

45. Construct a rectangle with sides b and a.

Chapter 4 Extra Practice: Skills, Word Problems, and Proof

Lesson 4-1

$\triangle SAT \cong \triangle GRE$. Complete each congruence statement.

1. $\angle S \cong$ __?__ **2.** $\overline{GR} \cong$ __?__ **3.** $\angle E \cong$ __?__

4. $\overline{AT} \cong$ __?__ **5.** $\triangle ERG \cong$ __?__ **6.** $\overline{EG} \cong$ __?__

7. $\triangle REG \cong$ __?__ **8.** $\angle R \cong$ __?__

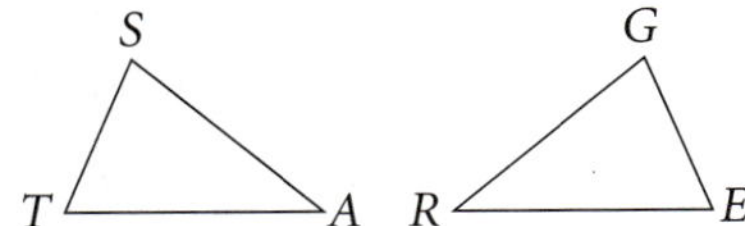

State whether the figures are congruent. Justify each answer.

9. $\triangle ABF$; $\triangle EDC$

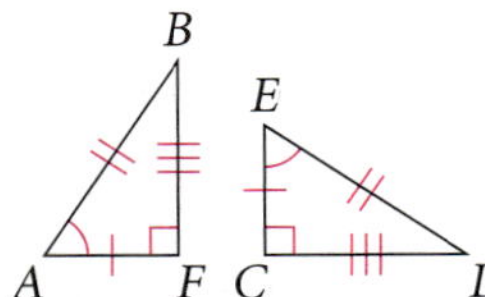

10. $\triangle TUV$; $\triangle UVW$

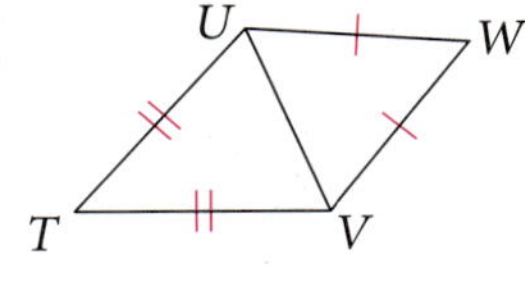

11. $\square XYZV$; $\square UTZV$

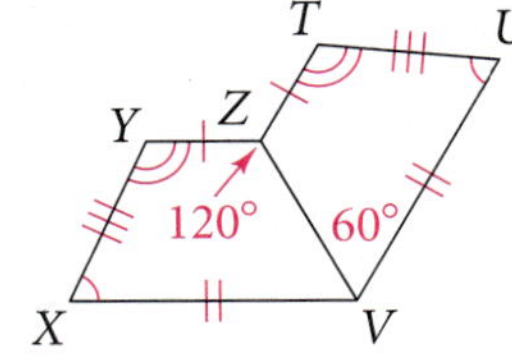

12. $\triangle ABD$; $\triangle EDB$

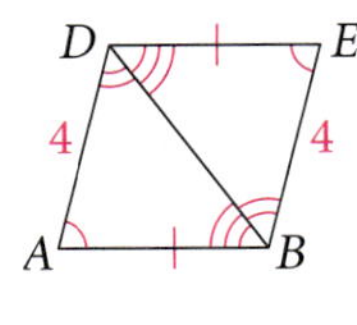

Lessons 4-2 and 4-3

Where possible, explain how you would use SSS, SAS, ASA, or AAS to prove the triangles congruent. If not possible, write *not possible.*

13.

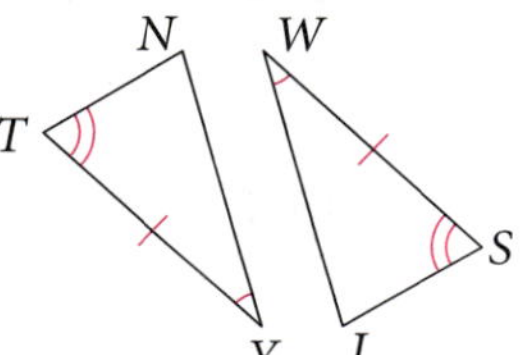

14.

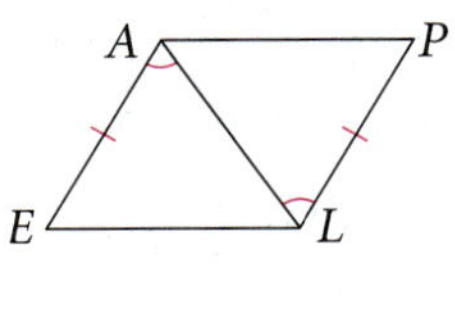

15.

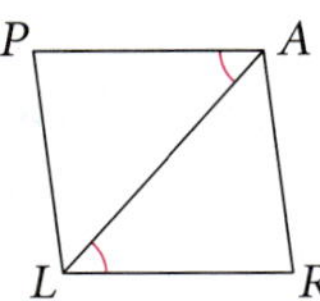

16.

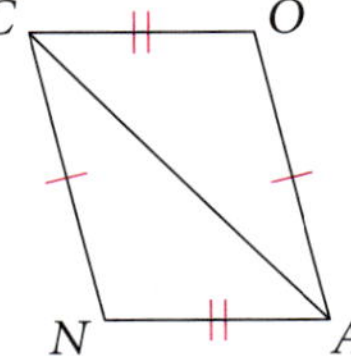

17. Given: $\overline{PX} \cong \overline{PY}$, $\overline{ZP}$ bisects $\overline{XY}$.

Prove: $\triangle PXZ \cong \triangle PYZ$

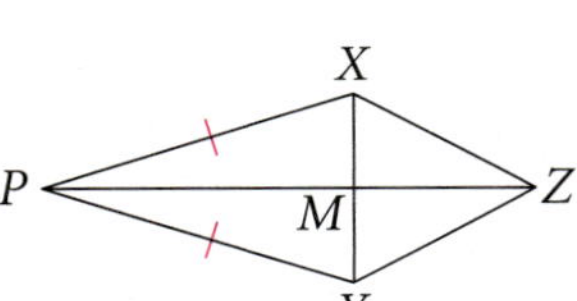

18. Given: $\angle 1 \cong \angle 2$, $\angle 3 \cong \angle 4$, $\overline{PD} \cong \overline{PC}$, P is the midpoint of $\overline{AB}$.

Prove: $\triangle ADP \cong \triangle BCP$

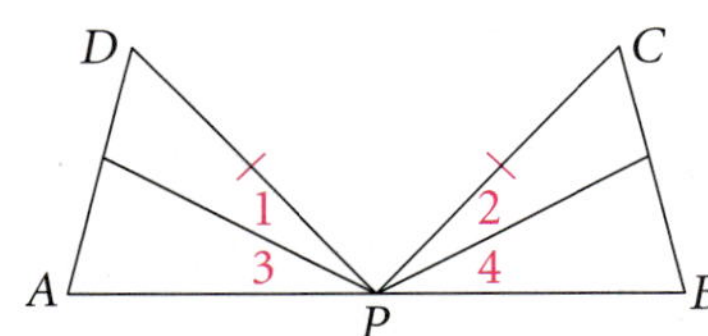

19. Given: $\angle 1 \cong \angle 2$, $\angle 3 \cong \angle 4$, $\overline{AP} \cong \overline{DP}$

Prove: $\triangle ABP \cong \triangle DCP$

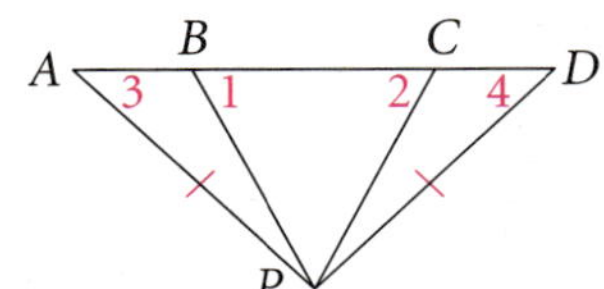

20. Given: $\overline{MP} \parallel \overline{NS}$, $\overline{RS} \parallel \overline{PQ}$, $\overline{MR} \cong \overline{NQ}$

Prove: $\triangle MQP \cong \triangle NRS$

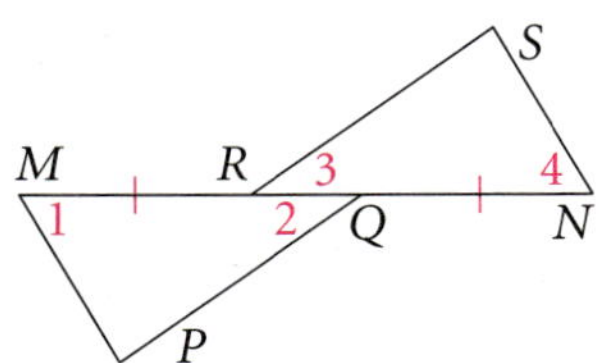

Lesson 4-4

Explain how you would use SSS, SAS, ASA, or HL with CPCTC to prove each statement.

21. $\angle MLN \cong \angle ONL$

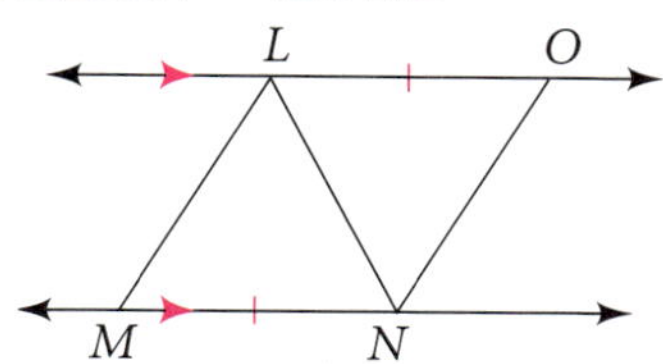

22. $\overline{TO} \cong \overline{ES}$

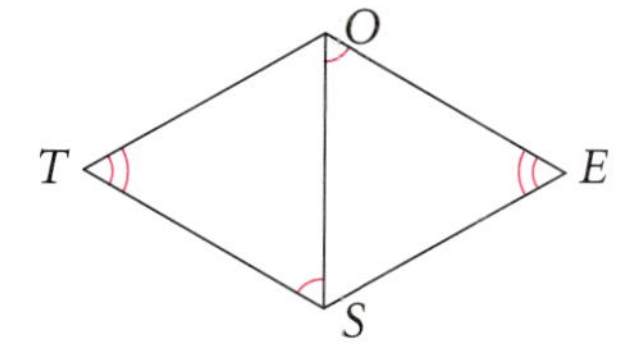

23. $\overline{MB} \cong \overline{RI}$

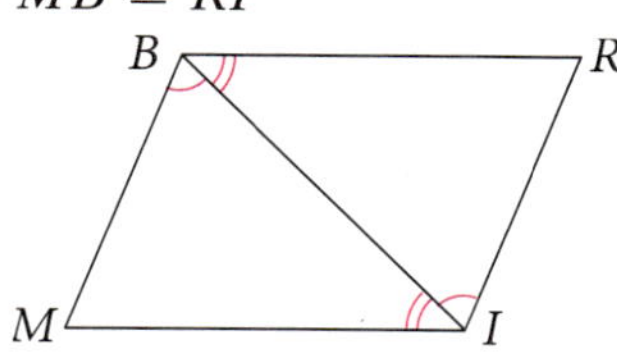

24. Given: $\angle 1 \cong \angle 2$, $\angle 3 \cong \angle 4$, M is the midpoint of $\overline{PR}$
Prove: $\triangle PMQ \cong \triangle RMQ$

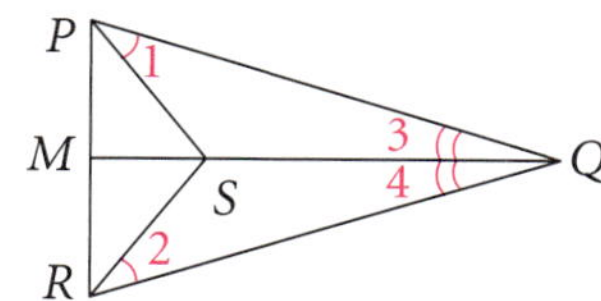

25. Given: $PO = QO$, $\angle 1 \cong \angle 2$,
Prove: $\angle A \cong \angle B$

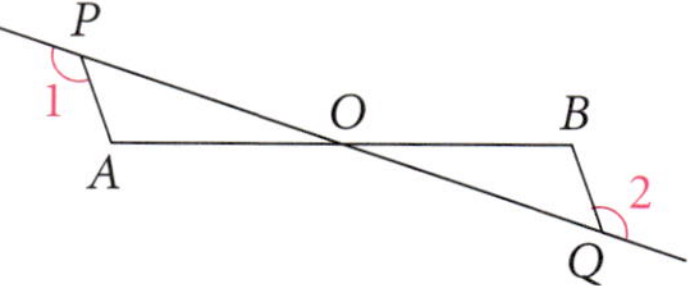

Lesson 4-5

Algebra **Find the value of each variable.**

26.

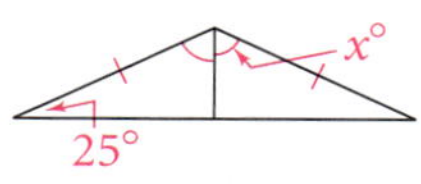

27.

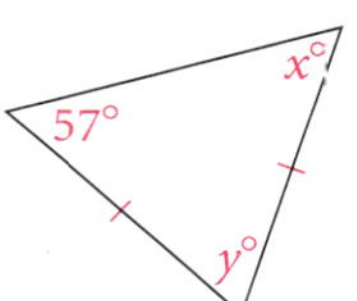

28.

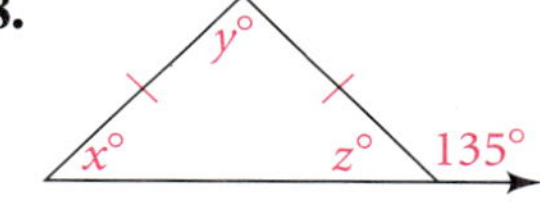

29. Given: $\angle 5 \cong \angle 6$, $\overline{PX} \cong \overline{PY}$
Prove: $\triangle PAB$ is isosceles.

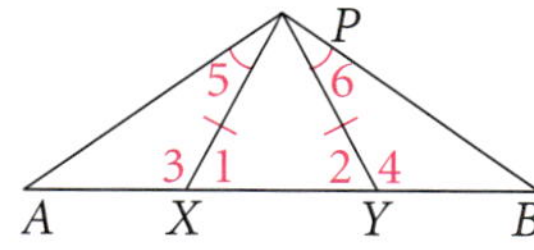

30. Given: $\overline{AP} \cong \overline{BP}$, $\overline{PC} \cong \overline{PD}$
Prove: $\triangle QCD$ is isosceles.

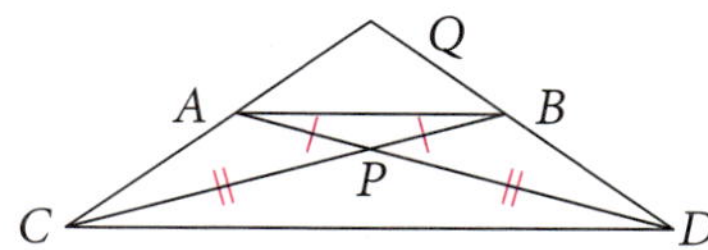

Lessons 4-6 and 4-7

Name a pair of overlapping congruent triangles in each diagram. State whether the triangles are congruent by SSS, SAS, ASA, AAS, or HL.

31.

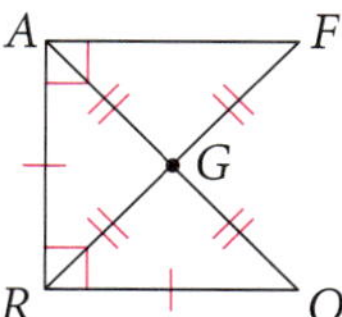

32.

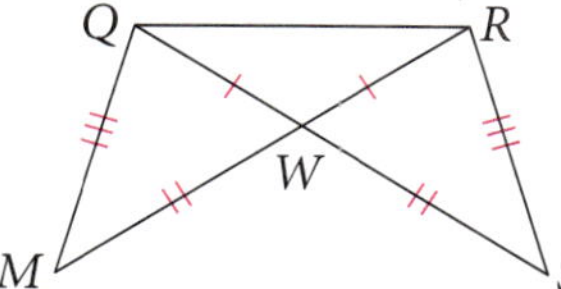

33.

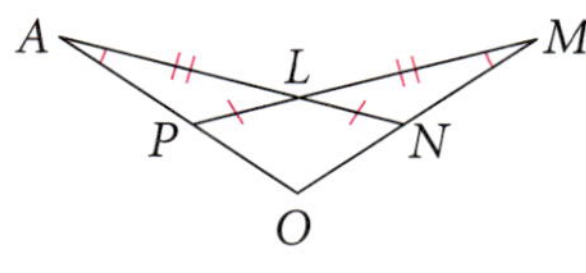

34. Given: M is the midpoint of $\overline{AB}$, $\overline{MC} \perp \overline{AC}$, $\overline{MD} \perp \overline{BD}$, $\angle 1 \cong \angle 2$
Prove: $\triangle ACM \cong \triangle BDM$

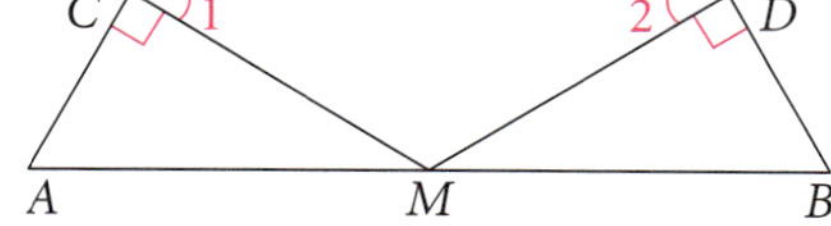

35. Given: $\triangle APQ \cong \triangle BQP$
$\overline{AP} \perp \overline{PQ}$, $\overline{BQ} \perp \overline{PQ}$
Prove: X is the midpoint of $\overline{AQ}$.

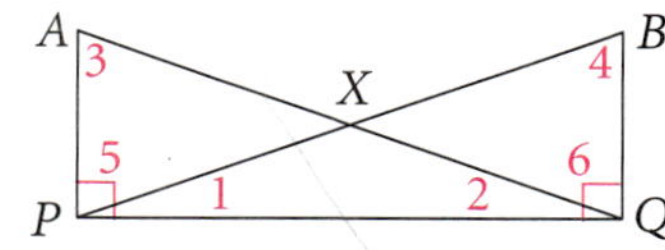

Chapter 5 Extra Practice: Skills, Word Problems, and Proof

Lesson 5-1

Algebra **Find the value of x.**

1. 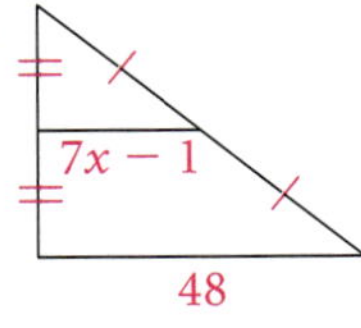

2. 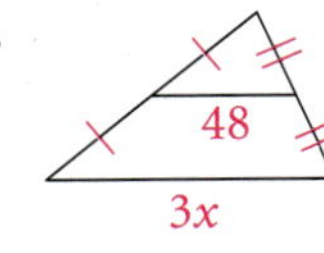

3. 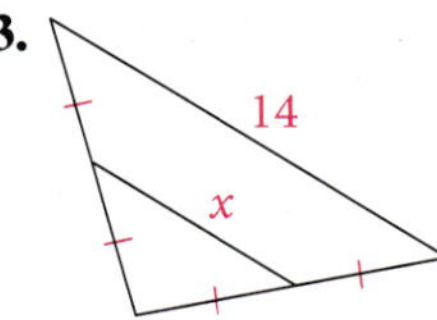

4. 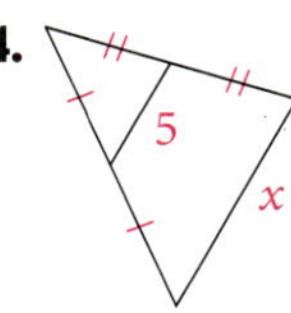

5. A sinkhole caused the sudden collapse of a large section of highway. Highway safety investigators paced out the triangle shown in the figure to help them estimate the distance across the sinkhole. What is the distance across the sinkhole?

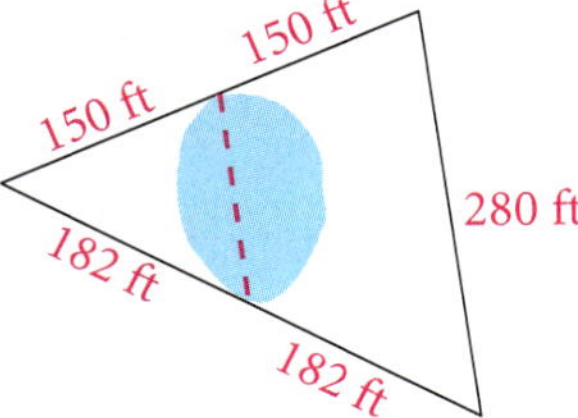

Lessons 5-1 and 5-2

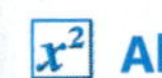 **Algebra** **Use the figure at the right.**

6. Find the value of x.

7. Find the length of $\overline{AD}$.

8. Find the value of y.

9. Find the length of $\overline{EG}$.

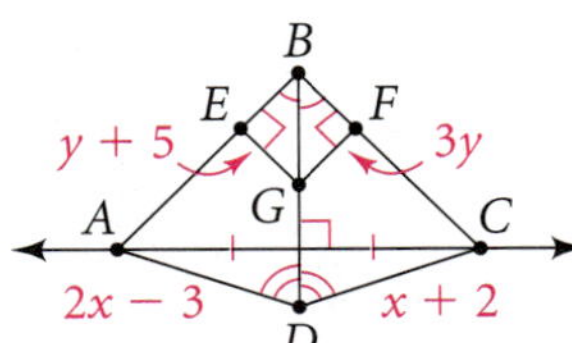

10. Given: $\overline{AP} \cong \overline{AQ}$, $\overline{BP} \cong \overline{BQ}$, $\overline{CP} \cong \overline{CQ}$

Prove: A, B, and C are collinear.

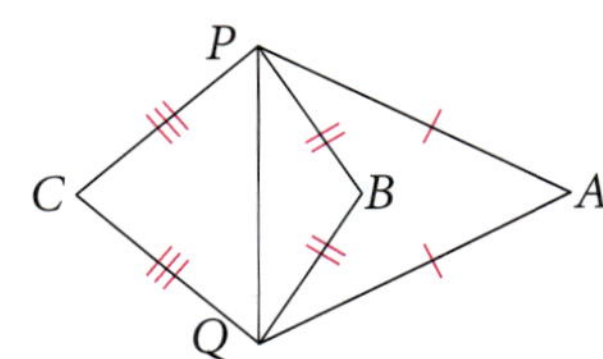

11. Given: $\overrightarrow{CX}$ bisects $\angle BCN$.
$\overrightarrow{BX}$ bisects $\angle CBM$.

Prove: X is on the bisector of $\angle A$

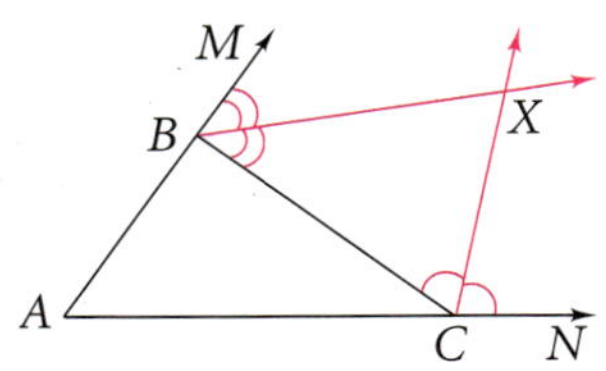

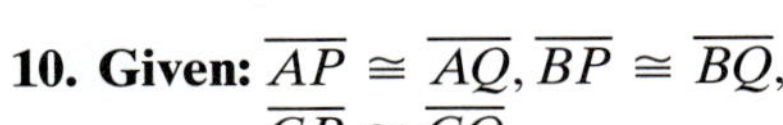

12. Find an equation in slope-intercept form for the perpendicular bisector of the segment with endpoints $H(-3, 2)$ and $K(7, -5)$.

Lesson 5-3

Find the center of the circle that you can circumscribe about $\triangle ABC$.

13. $A(2, 8)$
$B(0, 8)$
$C(2, 2)$

14. $A(-3, 6)$
$B(-3, -2)$
$C(7, 6)$

15. $A(4, 3)$
$B(-4, -3)$
$C(4, -3)$

16. $A(-10, -2)$
$B(-2, -2)$
$C(-2, -10)$

Is $\overline{AB}$ an angle bisector, altitude, median, or perpendicular bisector?

17. 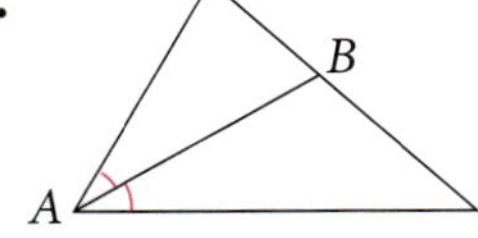

18. 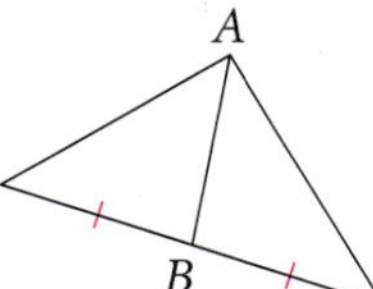

19. 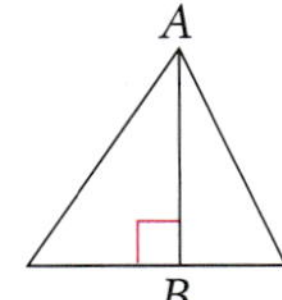

20. 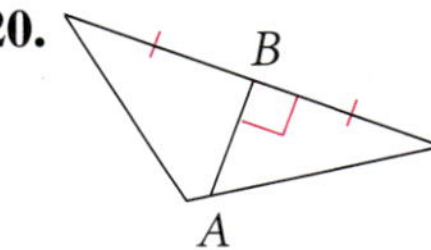

21. Find the center of the circle that you can circumscribe about the triangle with vertices $A(1, 3)$, $B(5, 8)$, and $C(6, 3)$.

22. Tell which line contains each point for $\triangle ABC$.
   a. the circumcenter
   b. the orthocenter
   c. the centroid
   d. the incenter

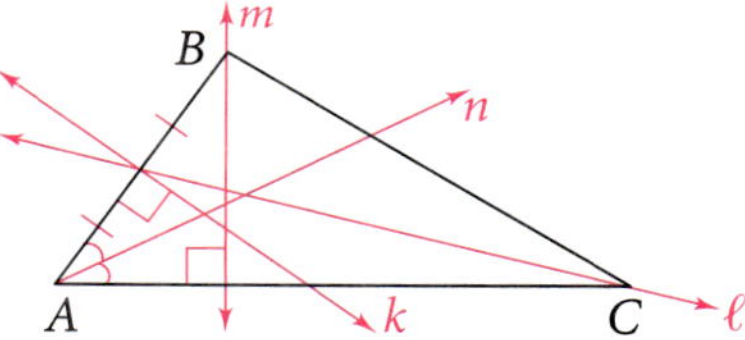

23. Draw an acute triangle and construct its inscribed circle.

Lesson 5-4

Write (a) the inverse and (b) the contrapositive of each statement.

24. If two angles are vertical, then they are congruent.

25. If figures are similar, then their side lengths are proportional.

26. If a car is blue, then it has no doors.

27. If a triangle is scalene, then it is not equiangular.

28. Suppose you know that $\angle A$ is an obtuse angle in $\triangle ABC$. You want to prove that $\angle B$ is an acute angle. What assumption would you make to give an indirect proof?

Write the first step of an indirect proof of each statement.

29. $\triangle ABC$ is a right triangle.

30. Points J, K, and L are collinear.

31. Lines ℓ and m are not parallel.

32. $\square XYZV$ is a square.

Lesson 5-5

List the sides of each triangle in order from shortest to longest.

33. 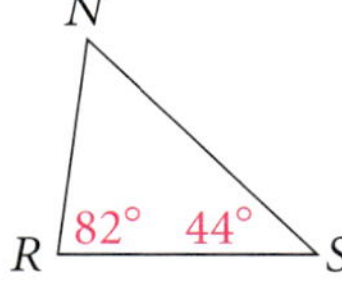

34. 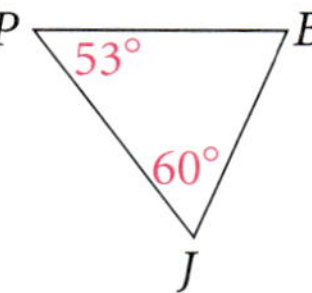

35. 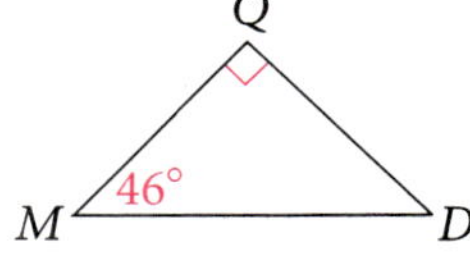

36. 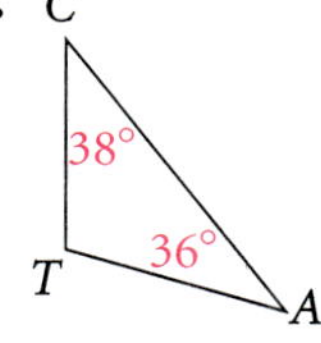

Can a triangle have sides with the given lengths? Explain.

37. 2 in., 3 in., 5 in.

38. 9 cm, 11 cm, 15 cm

39. 8 ft, 9 ft, 18 ft

40. In $\triangle PQR$, $m\angle P = 55$, $m\angle Q = 82$, and $m\angle R = 43$. List the sides of the triangle in order from shortest to longest.

41. In $\triangle MNS$, $MN = 7$, $NS = 5$, and $MS = 9$. List the angles of the triangle in order from smallest to largest.

42. Two sides of a triangle have side lengths 8 units and 17 units. Describe the lengths x that are possible for the third side.

Chapter 6

Extra Practice: Skills, Word Problems, and Proof

Lesson 6-1

Graph the given points. Use slope and the Distance Formula to determine the most precise name for quadrilateral $ABCD$.

1. $A(3, 5), B(6, 5), C(2, 1), D(1, 3)$

2. $A(-1, 1), B(3, -1), C(-1, -3), D(-5, -1)$

3. $A(2, 1), B(5, -1), C(4, -4), D(1, -2)$

4. $A(-4, 5), B(-1, 3), C(-3, 0), D(-6, 2)$

Lesson 6-2

Algebra **Find the values of the variables in each parallelogram.**

5.

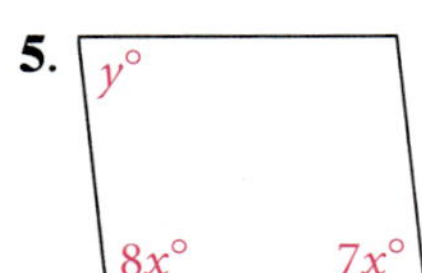

6.

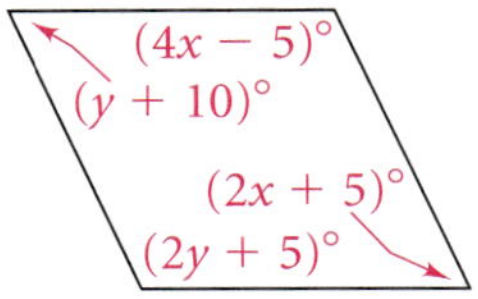

$(4x - 5)°$
$(y + 10)°$
$(2x + 5)°$
$(2y + 5)°$

7.

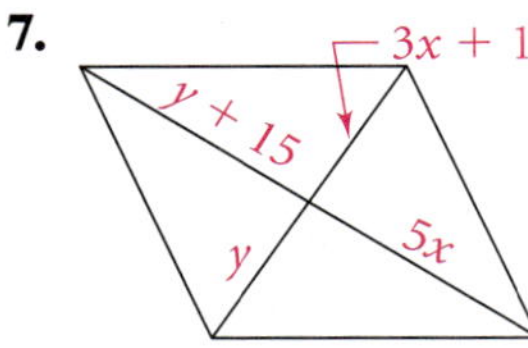

8.

$2y - 9$
$3x + 1$
$5x - 1$
$y - 2$

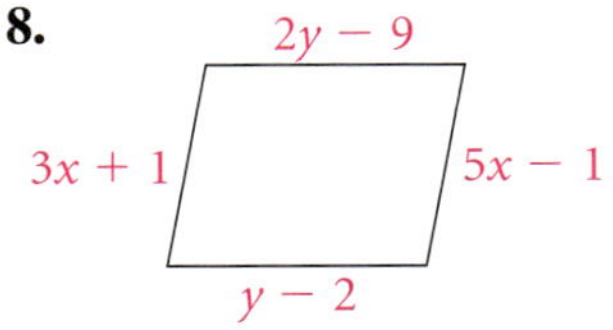

9. Given: $PQRS$ and $QDCA$ are parallelograms.

Prove: $AP = BS$

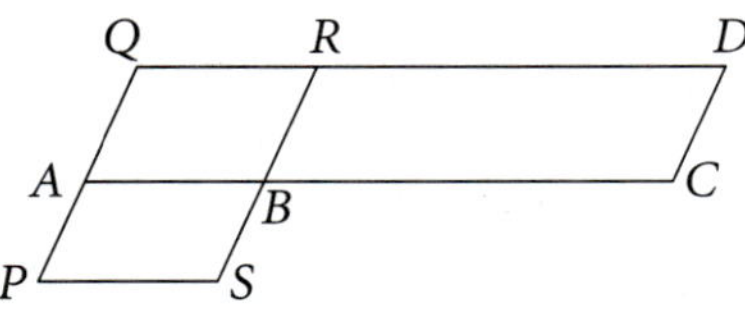

10. Given: $\square ABCD$
M is the midpoint of $\overline{CD}$.

Prove: $\overline{PM} \parallel \overline{AD}$

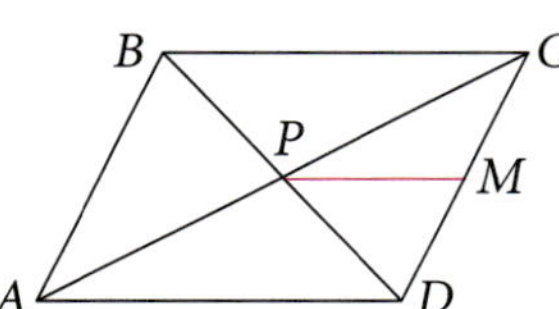

Lesson 6-3

Based on the markings, decide whether each figure must be a parallelogram.

11.

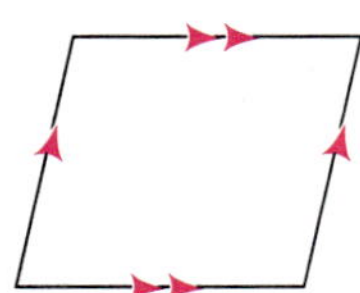

12.

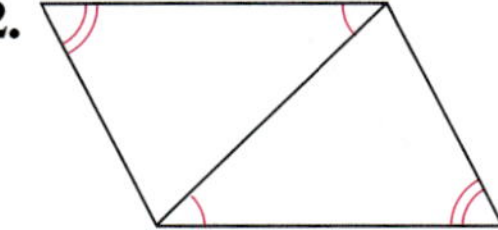

13.

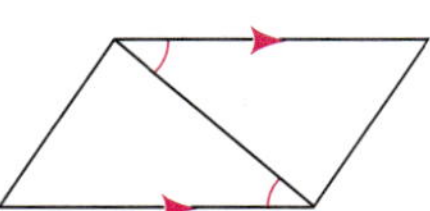

14.

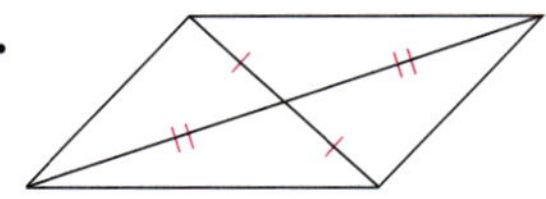

15. Describe how you can use what you know about parallelograms to construct a point halfway between a given pair of parallel lines.

16. Given: $\square ABCD$
$\overline{BX} \perp \overline{AC}$,
$\overline{DY} \perp \overline{AC}$

Prove: $BXDY$ is a parallelogram.

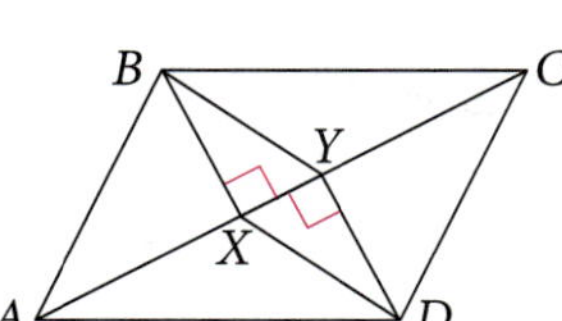

Lesson 6-4

For each parallelogram, determine the most precise name and find the measures of the numbered angles.

17. 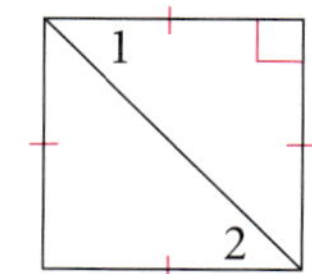

18. 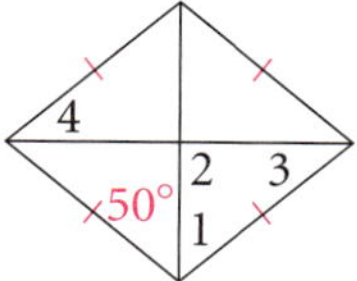

19. 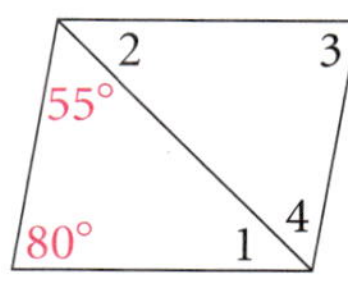

20. 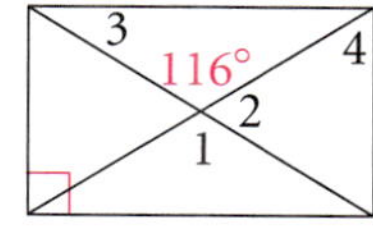

21. Use the information in the figure. Explain how you know that $ABCD$ is a rectangle.

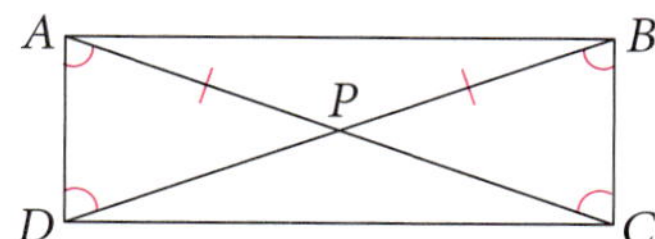

22. ▱$ABCD$ is a rhombus. What is the relationship between $\angle 1$ and $\angle 2$? Explain.

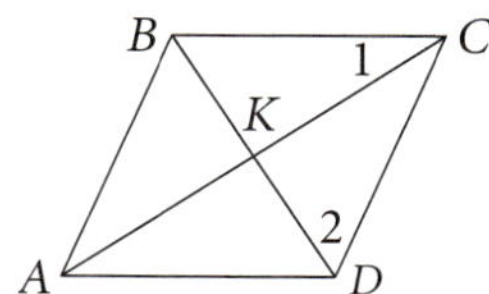

Lesson 6-5

Find $m\angle 1$ and $m\angle 2$.

23. 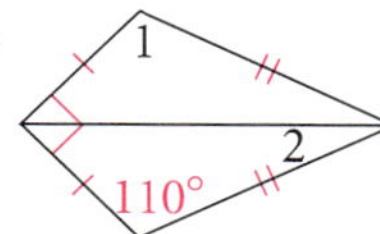

24. 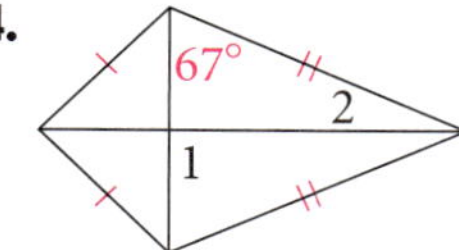

25. 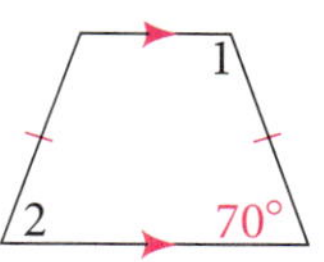

26. 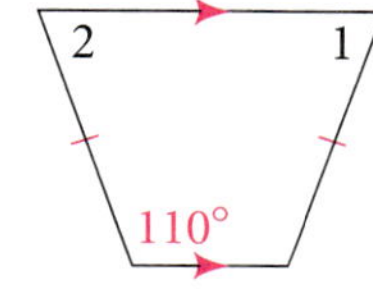

27. Suppose you manipulate the figure so that $\triangle PAB$, $\triangle PBC$, and $\triangle PCD$ are congruent isosceles triangles with their vertex angles at point P. What kind of figure is $ABCD$? Be sure to consider all the possibilities.

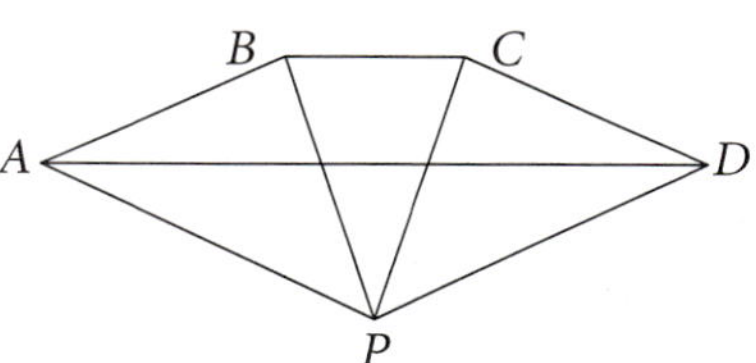

Lesson 6-6

Give coordinates for points D and S without using any new variables.

28. rectangle

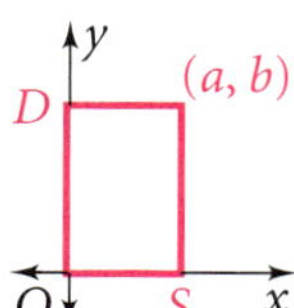

29. parallelogram

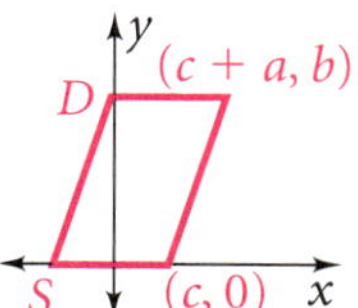

30. rhombus

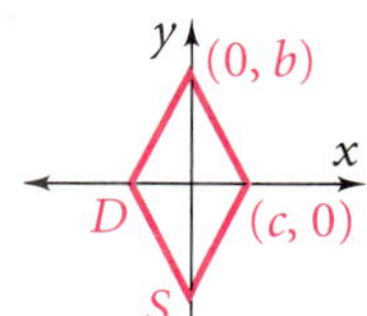

31. square

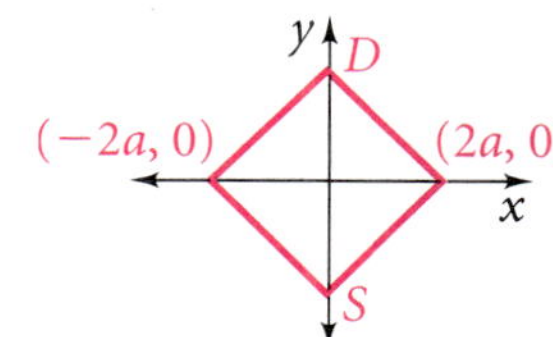

Lesson 6-7

32. For the figure in Exercise 31, use coordinate geometry to prove that the midpoints of the sides of a square determine a square.

33. In the figure, $\triangle PQR$ is an isosceles triangle. Points M and N are the midpoints of $\overline{PQ}$ and $\overline{PR}$, respectively. Give a coordinate proof that the medians of isosceles triangle PQR intersect at $H\left(0, \frac{2b}{3}\right)$.

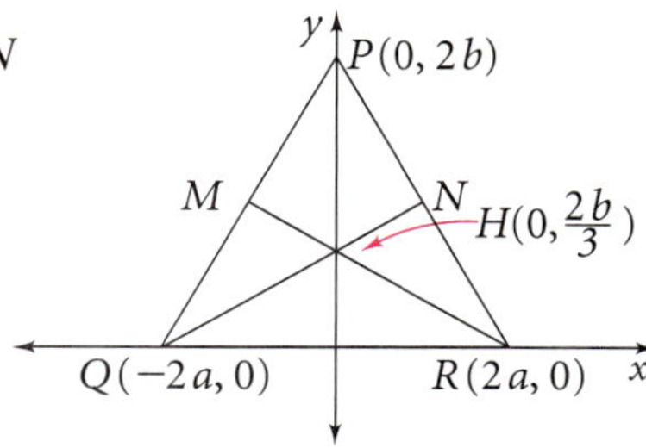

Chapter 7

Extra Practice: Skills, Word Problems, and Proof

Lesson 7-1

Algebra Solve each proportion.

1. $\frac{2}{3} = \frac{x}{15}$
2. $\frac{4}{9} = \frac{16}{x}$
3. $\frac{x}{4} = \frac{6}{12}$
4. $\frac{2}{x} = \frac{3}{9}$
5. $\frac{3}{4} = \frac{x}{6}$
6. $\frac{3}{7} = \frac{9}{x}$

Lesson 7-2

Algebra The polygons are similar. Find the values of the variables.

7.

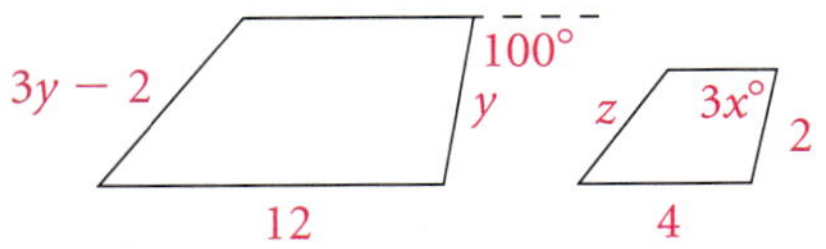

8.

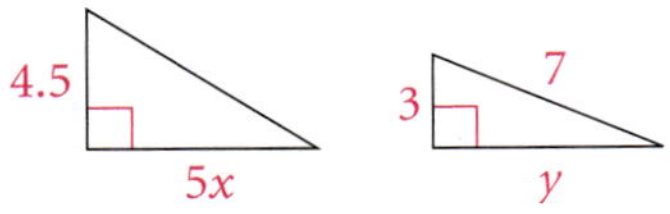

9.

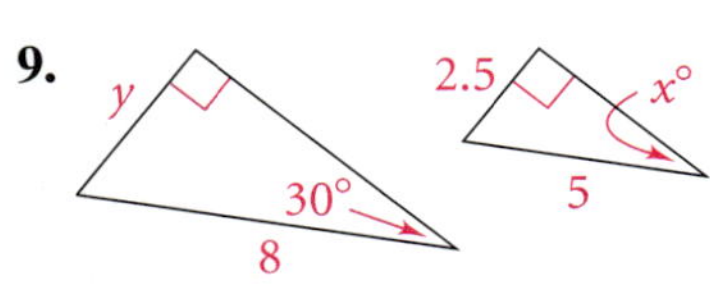

10.

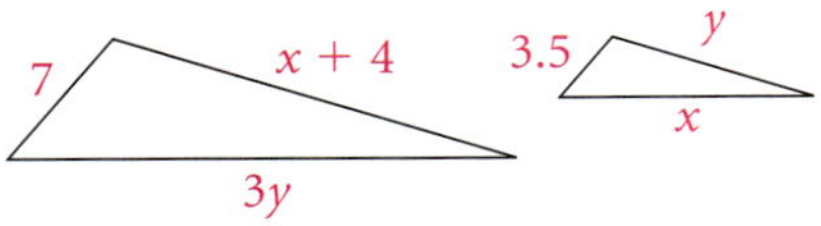

11. Are all equilateral quadrilaterals similar? Make a sketch to support your answer.

Lesson 7-3

Can you prove that the triangles are similar? If so, write a similarity statement and tell whether you would use AA~, SAS~, or SSS~.

12.

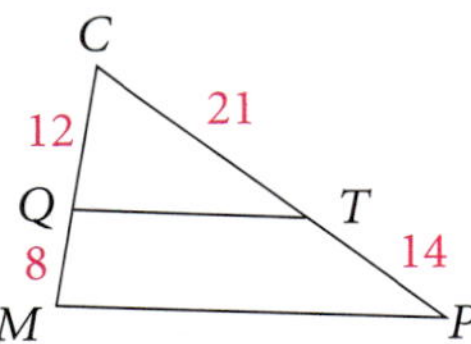

13.

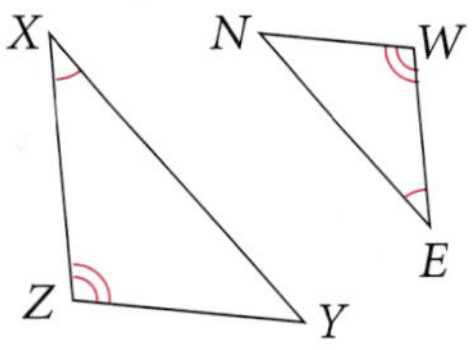

14.

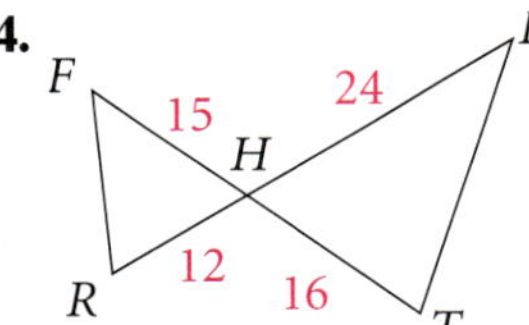

15.

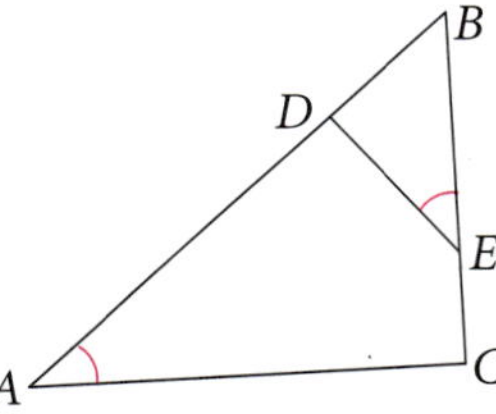

16.

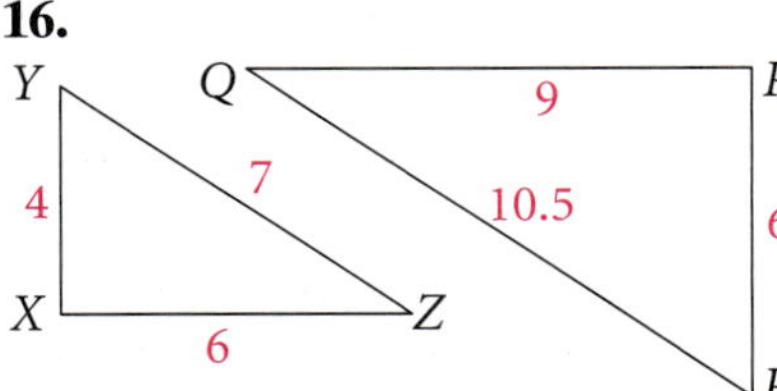

17.

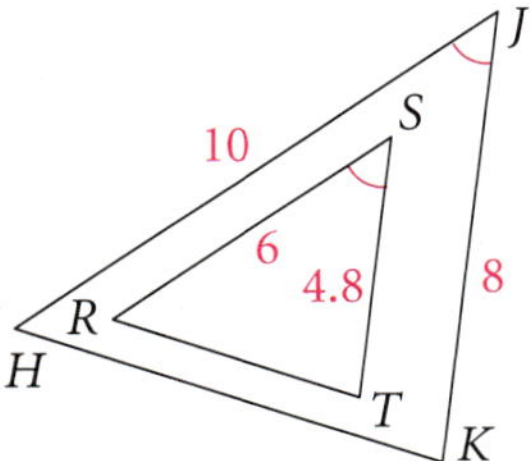

18. Refer to the figure at the right. Explain how you know that $\overline{AB} \parallel \overline{ED}$.

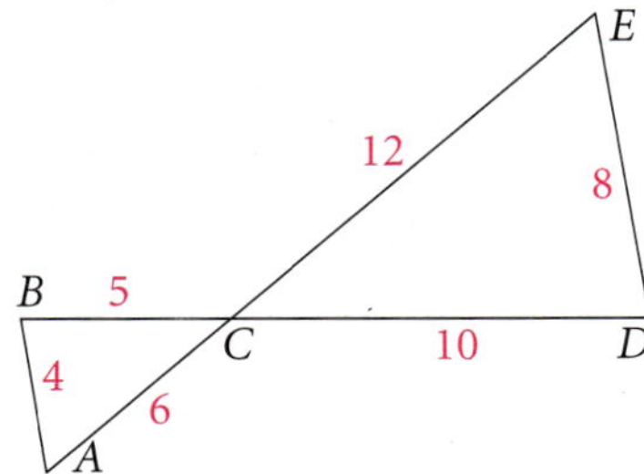

Skills, Word Problems, and Proof

Lesson 7-4

Algebra Find the value of each variable. If an answer is not a whole number, leave it in simplest radical form.

19.

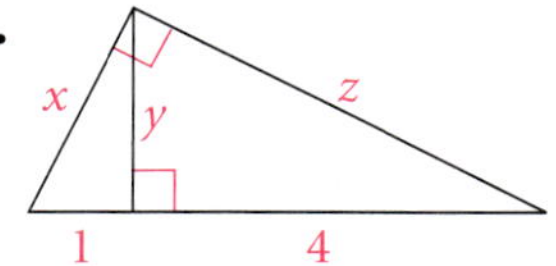

20.

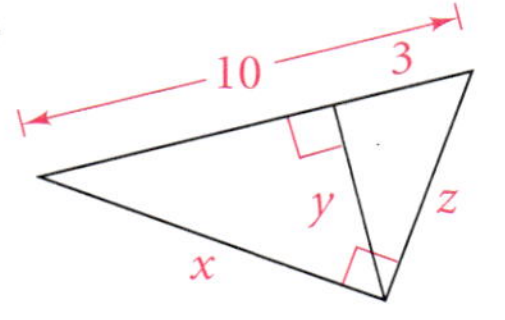

21.

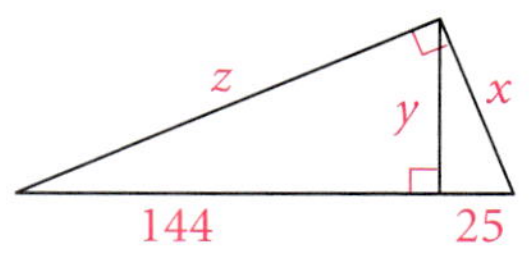

22.

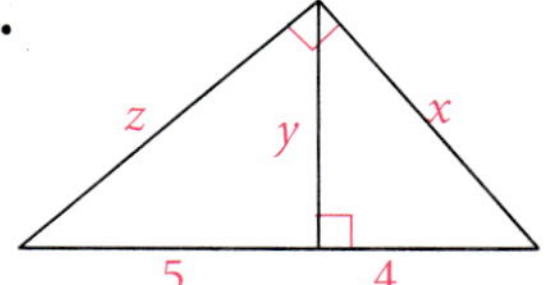

23. Give a coordinate proof of the converse of Corollary 1 to Theorem 7-3. That is, prove that if $\overline{CD}$ is the altitude from C to side $\overline{AB}$ of $\triangle ABC$, and if CD is the geometric mean of AD and DB, then $\triangle ABC$ is a right triangle with its right angle at C.

24. An artist is going to cut four similar right triangles from a rectangular piece of paper like the one shown below. What is the distance from B and D to the diagonal $\overline{AC}$?

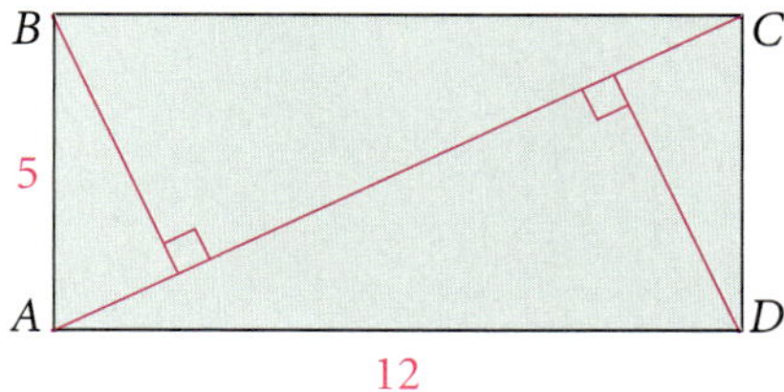

Lesson 7-5

Algebra Find the value of x.

25.

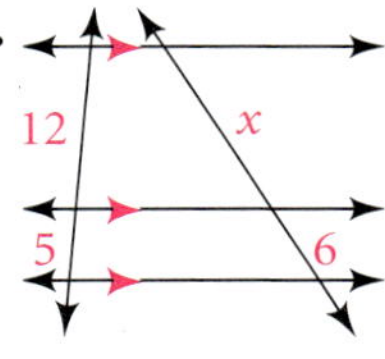

26.

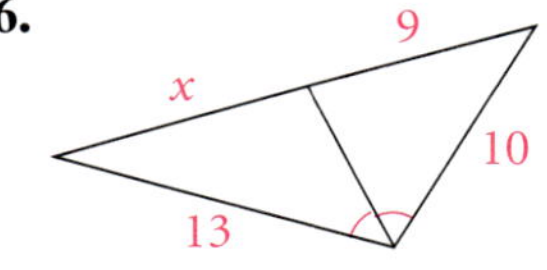

27.

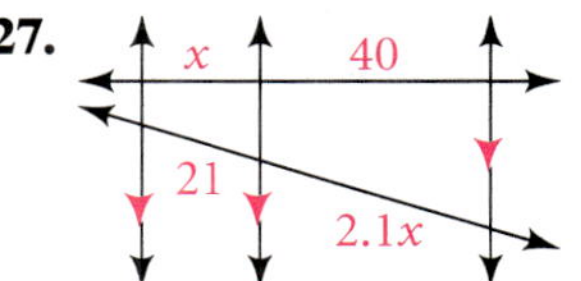

28.

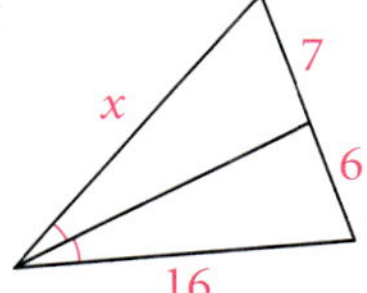

29. Suppose you are given a segment $\overline{AB}$ of length 1 unit and a segment $\overline{CD}$ of length x units. Show how you can apply the Side-Splitter Theorem to construct a segment of length $\frac{1}{x}$.

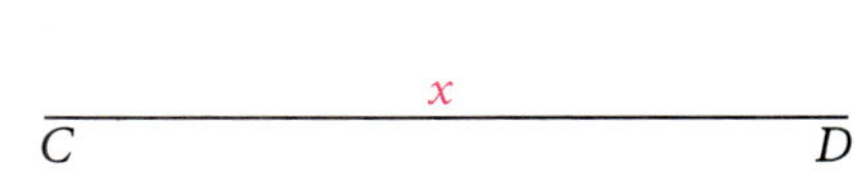

30. The figure below shows the locations of a high school, a computer store, a library, and a convention center. The street along which the computer store and library are located bisects the obtuse angle formed by two of the other streets. Use the information in the figure to find the distance from the library to the convention center.

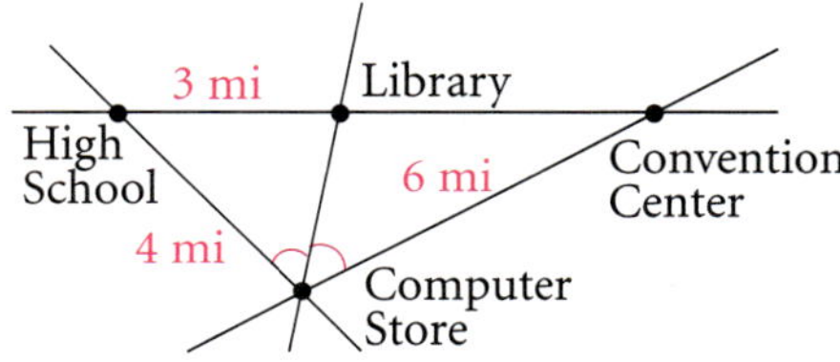

Chapter 8 Extra Practice: Skills, Word Problems, and Proof

Lessons 8-1 and 8-2

Find the value of x. If your answer is not a whole number, leave it in simplest radical form.

1. 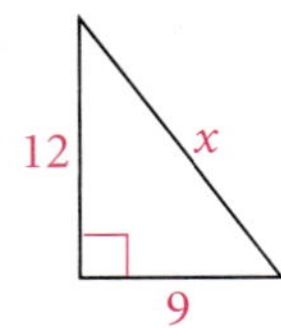

2. 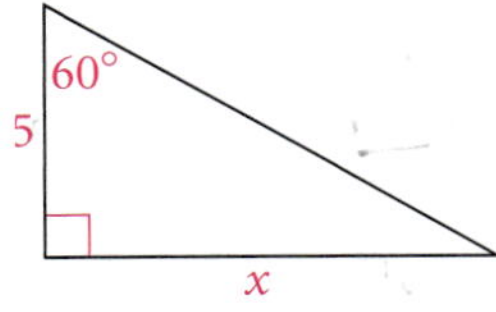

3. 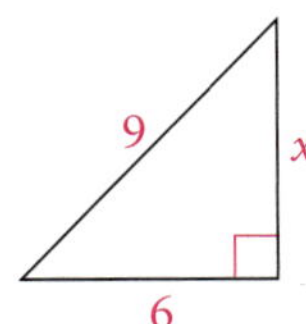

4. 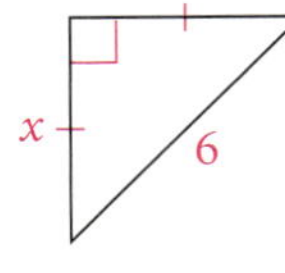

5. A rectangular lot is 165 feet long and 90 feet wide. How many feet of fencing are needed to make a diagonal fence for the lot? Round to the nearest foot.

Find the missing side lengths. Give answers in radical form if necessary.

6. 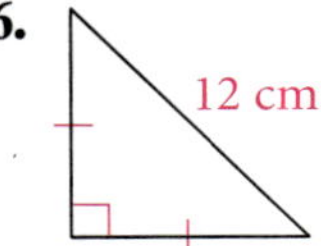

7. 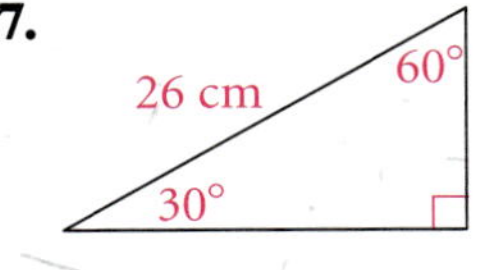

8. 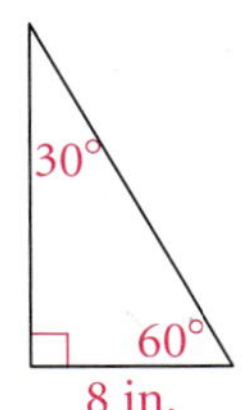

9. 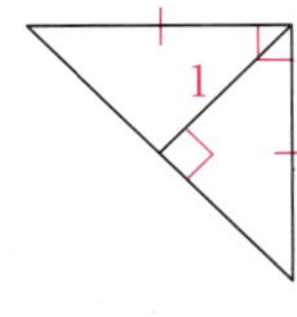

Lessons 8-3 and 8-4

Find the value of x. Round lengths of segments to the nearest tenth and angle measures to the nearest degree.

10. 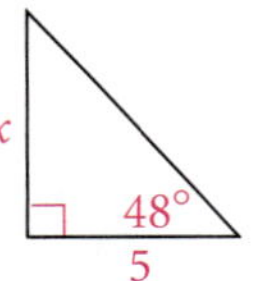

11. 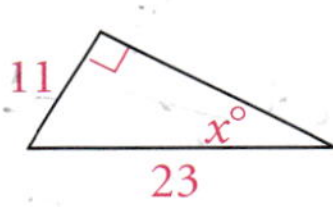

12. 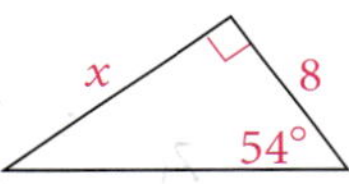

13. 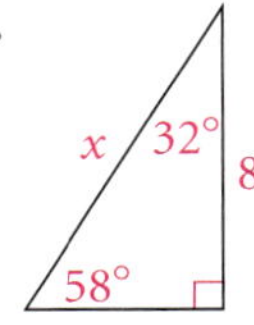

14. 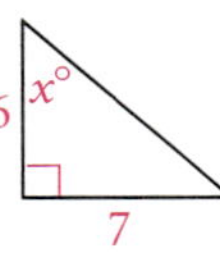

15. 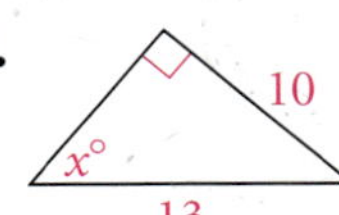

16. 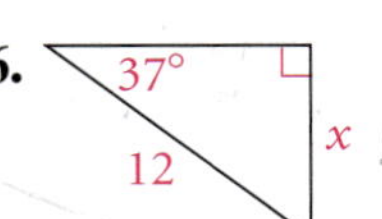

17. 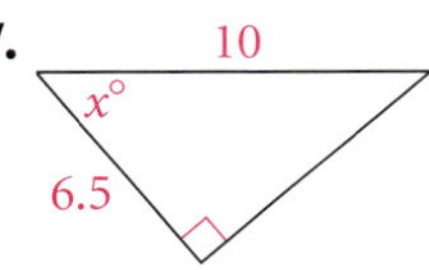

18. An architect includes wheelchair ramps in her plans for the entrance to a new museum. She wants the angle that the ramp makes with level ground to measure 4°. Will the dimensions shown in the figure work? If not, what change should she make?

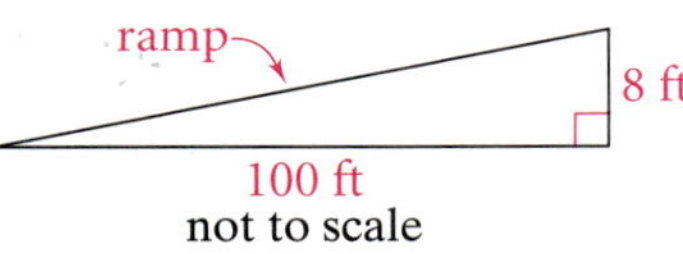

19. A 12-ft ladder is propped against a vertical wall. The top end is 11 ft above the ground. What is the measure of the angle formed by the ladder with the ground?

20. How long is the guy wire shown in the figure if it is attached to the top of a 50-ft antenna and makes a 70° angle with the ground? Round to the nearest tenth.

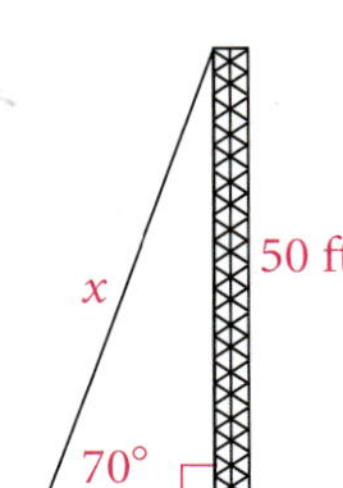

21. A 15-ft ladder is propped against a vertical wall and makes a 72° angle with the ground. How far is the foot of the ladder from the base of the wall? Round to the nearest tenth.

Lesson 8-5

Solve each problem. Round your answers to the nearest foot.

22. A couple is taking a balloon ride. After 25 minutes aloft, they measure the angle of depression from the balloon to its launch place as 16°. They are 180 ft above ground. Find the distance from the balloon to its launch place.

23. A surveyor is 300 ft from the base of an apartment building. The angle of elevation to the top of the building is 24°, and her angle-measuring device is 5 ft above the ground. Find the height of the building.

24. Oriana is flying a kite. She lets out 105 ft of string and anchors it to the ground. She determines that the angle of elevation of the kite is 48°. Find the height the kite is from the ground.

25. Two office buildings are 100 ft apart. From the edge of the shorter building, the angle of elevation to the top of the taller building is 28°, and the angle of depression to the bottom is 42°. How tall is each building? Round to the nearest foot.

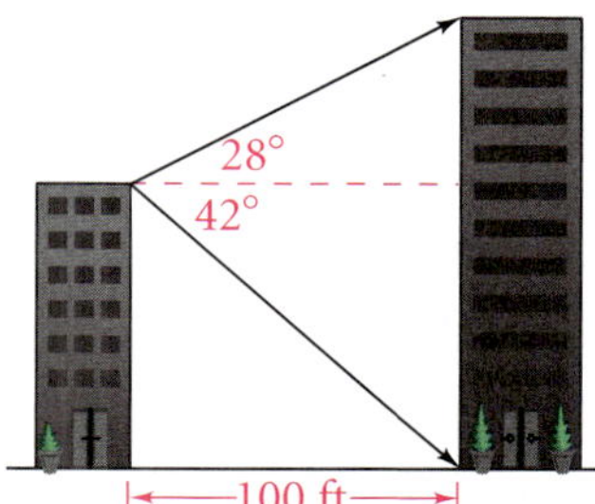

26. A plane flying at 10,000 ft spots a hot air balloon in the distance. The balloon is 9000 ft above ground. The angle of depression from the plane to the balloon is 30°. Find the distance from the plane to the balloon.

Lesson 8-6

(a) Describe each vector as an ordered pair. Give the coordinates to the nearest unit. (b) Write the resultant of each pair of vectors as an ordered pair.

27. 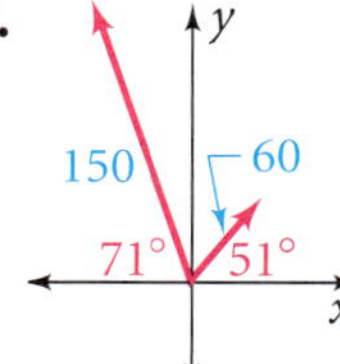

28. 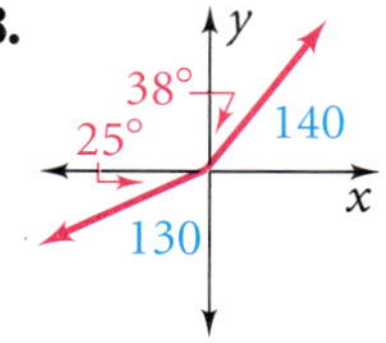

29. 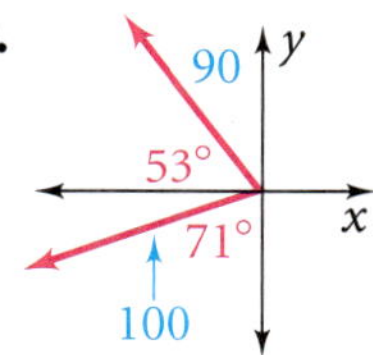

30. 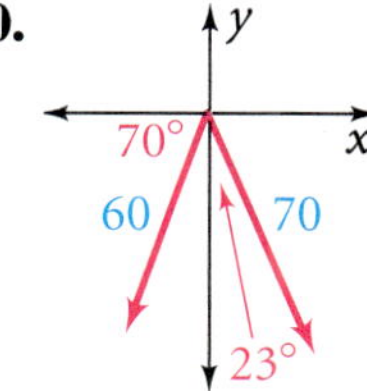

Write the sum of the two vectors as an ordered pair.

31. $\langle 5, 9\rangle$ and $\langle -3, 2\rangle$

32. $\langle -1, 0\rangle$ and $\langle 4, -6\rangle$

33. $\langle 2, 4\rangle$ and $\langle 0, 9\rangle$

34. $\langle 4, -2\rangle$ and $\langle -4, 2\rangle$

35. A helicopter lands 55 km west and 14 km north of the airport from which it departed. It followed a straight flight path. Find the magnitude and direction of the resultant vector $\langle -55, 14\rangle$.

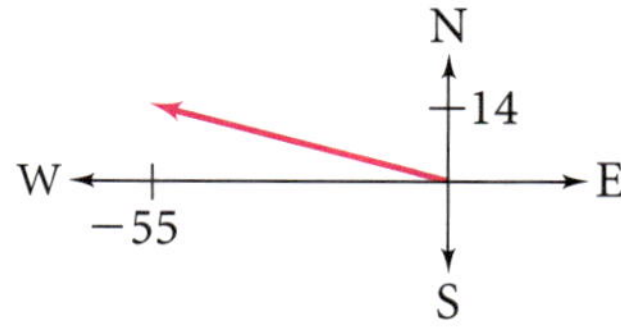

Chapter 9 Extra Practice: Skills, Word Problems, and Proof

Lesson 9-1

In Exercises 1–6, refer to the figure at the right.

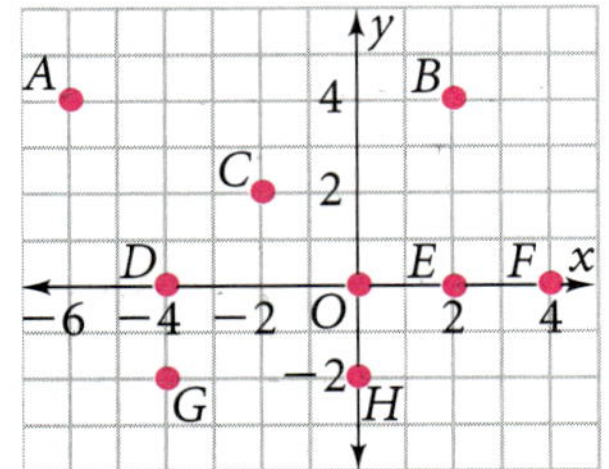

1. What is the image of C under $(x, y) \to (x + 4, y - 2)$?
2. What rule describes the translation $F \to B$?
3. What is the image of H under $(x, y) \to (x - 2, y + 4)$?
4. What rule describes the translation $D \to H$?
5. What is the image of C under $(x, y) \to (x - 2, y - 4)$?
6. What rule describes the translation $B \to A$?

Find the image of each figure under the given translation.

7. $\triangle ABC$ with vertices $A(-3, 4)$, $B(-1, -2)$, $C(1, 5)$; translation: $(x, y) \to (x-2, y + 5)$
8. $\triangle EFG$ with vertices $E(0, 3)$, $F(6, -1)$, $G(4, 2)$; translation: $(x, y) \to (x + 1, y - 3)$
9. $\triangle PQR$ with vertices $P(-9, -4)$, $Q(-5, 1)$, $R(2, 8)$; translation: $(x, y) \to (x - 6, y - 7)$

10. Write two translation rules of the form $(x, y) \to (x + a, y + b)$ that map the line $y = x - 1$ to the line $y = x + 3$.

Lesson 9-2

Given points $S(6, 1)$, $U(2, 5)$, and $B(-1, 2)$, draw $\triangle SUB$ and its reflection image across each line.

11. $y = 5$
12. $x = 7$
13. $y = -1$
14. the x-axis
15. $y = x$
16. $x = -1$
17. $y = 3$
18. the y-axis

19. What are the two shortest words in the English language that you can write with capital letters so that each word looks like its own reflection across a line?

20. The segments $\overline{AB}$ and $\overline{A'B'}$ are two different segments in the same plane. There is a translation such that $\overline{A'B'}$ is the translation image of $\overline{AB}$. There is also a line k in the plane such that $\overline{A'B'}$ is the reflection image of $\overline{AB}$ across line k. If $\overline{AB}$ and $\overline{A'B'}$ are opposite sides of a quadrilateral, what kind of quadrilateral is it?

Lesson 9-3

Copy each figure and point P. Draw the image of each figure for the given rotation about P. Label the vertices of the image.

21. 90°

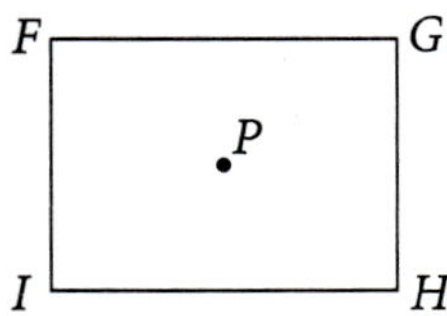

22. 60°

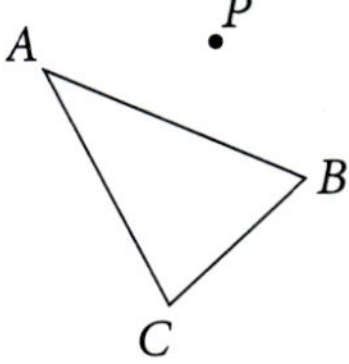

23. 45°

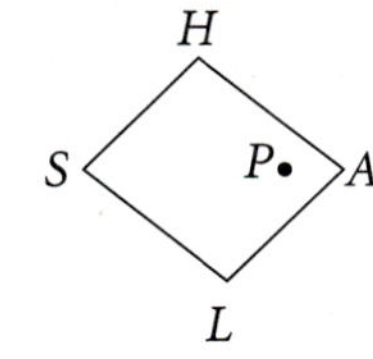

24. 180°

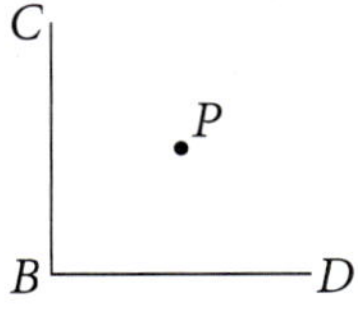

25. The right triangle ABC shown here has side lengths 3, 4, and 5. Point P is the incenter of the triangle. Copy the triangle and draw the image of the triangle for a 60° counterclockwise rotation about P.

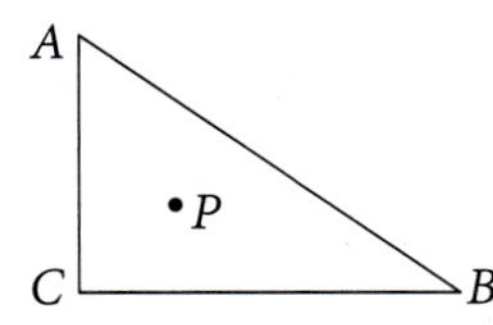

26. What is the smallest angle of rotation you can use to have the rotation image of the figure below exactly overlap the original figure?

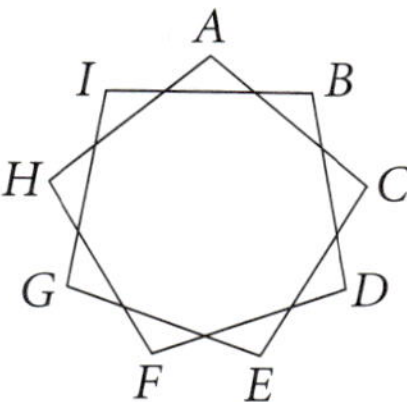

Lesson 9-4

State what kind of symmetry each figure has.

27.

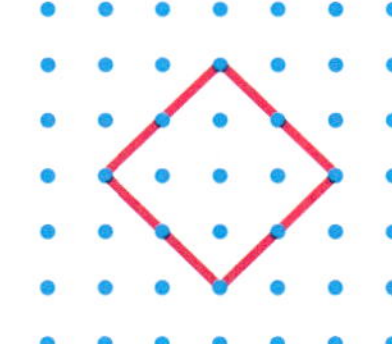

28.

29.

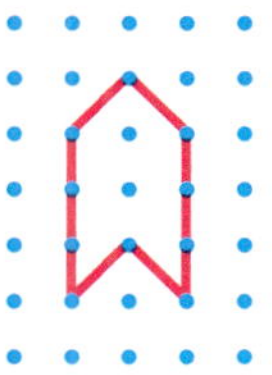

30.

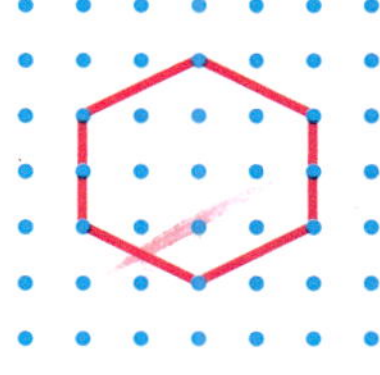

31. Armando is going to draw a triangle that he will put on his backpack.
   a. If the triangle has a line of symmetry, what kind of triangle must it be?
   b. If the triangle has two lines of symmetry, what kind of triangle must it be?

Lessons 9-5 and 9-6

The blue figure is the image of the gray figure. State whether the mapping is a reflection, rotation, translation, glide reflection, or dilation.

32.

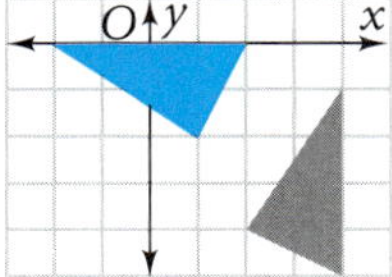

33.

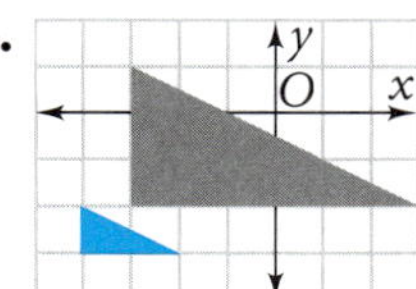

34.

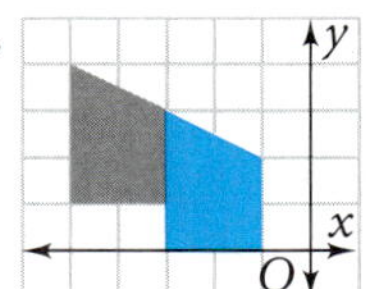

35.

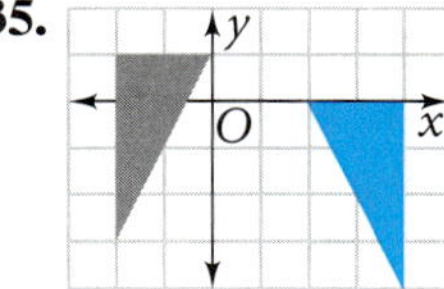

36. The vertices of trapezoid $ABCD$ are $A(-1, -1)$, $B(-1, 1)$, $C(2, 2)$, and $D(2, -1)$. Draw the trapezoid and its dilation image for a dilation with center $(0, 0)$ and scale factor 3.

37. Suppose you know the coordinates of the vertices of a polygon. Describe how you can use what you know about translations and dilations with respect to the origin to find the coordinates of the vertices of the image polygon if the center for the dilation is $(2, 5)$ and the scale factor is 3.

38. Find the image of the polygon for a reflection across line ℓ followed by a reflection across line m. Then use a separate diagram to repeat the process, but reflect across line m first and then across line ℓ. Each time, draw the intermediate image with dashed segments.

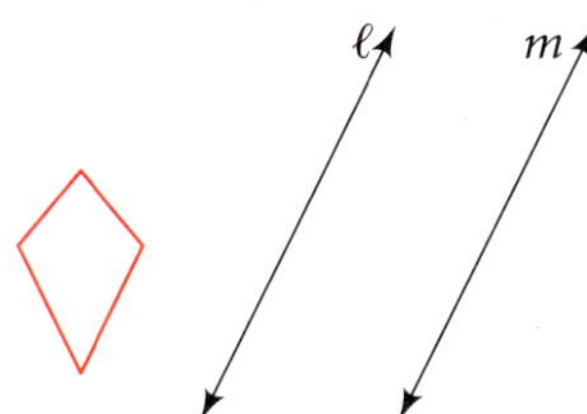

Lesson 9-7

39. Which of the four figures in Exercises 27–30 will tessellate a plane?

40. Use a square and an equilateral triangle to make a tessellation. The square and equilateral triangle should have congruent sides.

Chapter 10 Extra Practice: Skills, Word Problems, and Proof

Lesson 10-1

If possible, find the perimeter and area of each figure. If not possible, state why.

1. 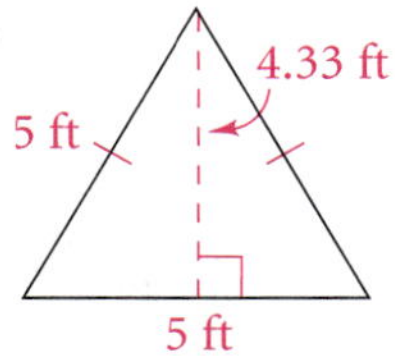

2. 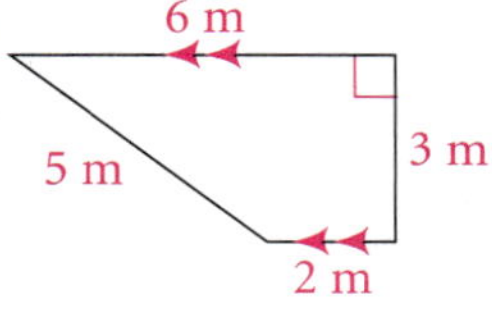

3. 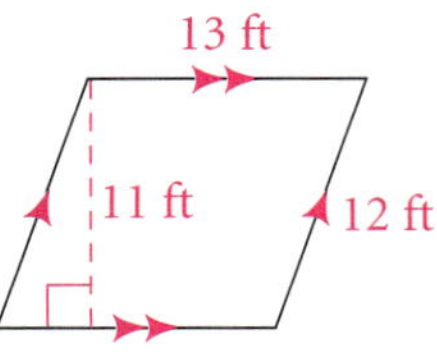

4. 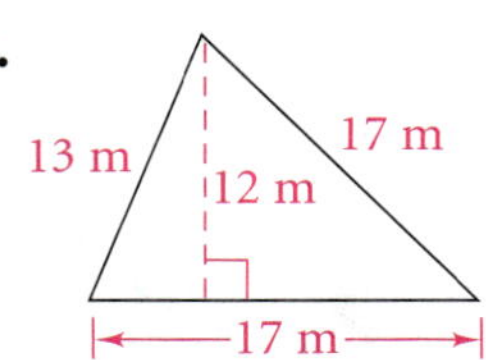

5. 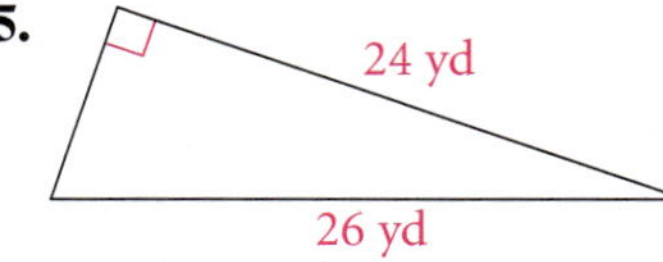

6. 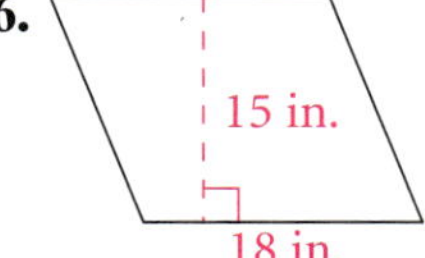

Lessons 10-2 and 10-3

Find the area of each trapezoid or regular polygon. Leave your answer in simplest radical form.

7. 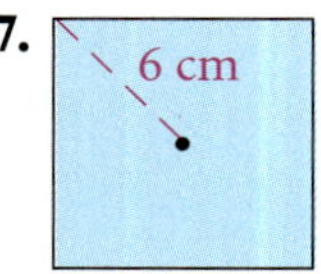

8. 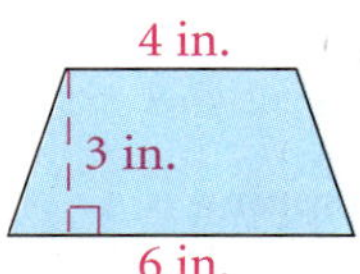

9. 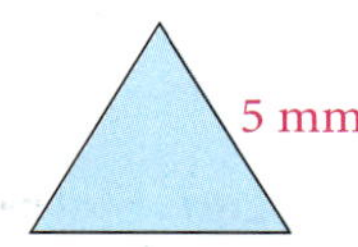

10. 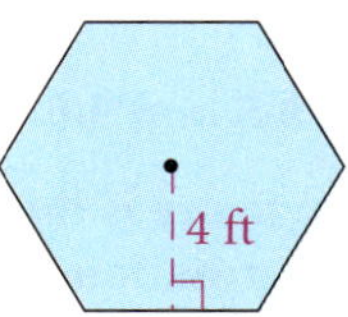

11. The patio section of a restaurant is a trapezoid with the dimensions shown in the figure. What is the area of the patio section?

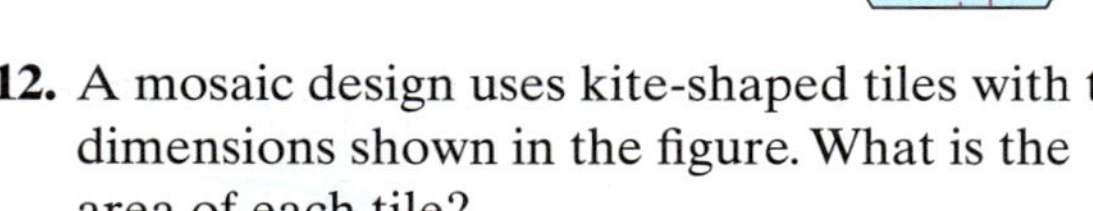

12. A mosaic design uses kite-shaped tiles with the dimensions shown in the figure. What is the area of each tile?

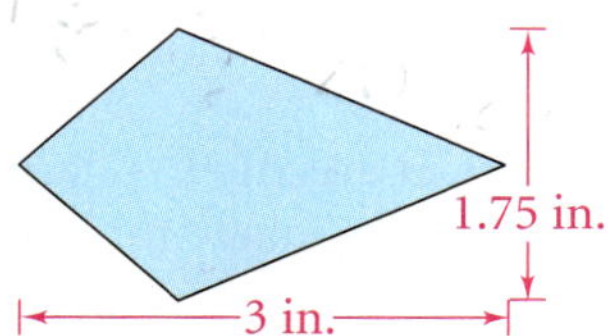

60 ft

50 ft

80 ft

13. The tiles for a bathroom floor are regular hexagons that are $\frac{5}{8}$ in. on each side. Find the area of an individual tile. Express the answer in radical form.

14. The floor of a gazebo is a regular hexagon with sides that are 9 ft long. What is the area of the floor? Round to the nearest square foot.

Lesson 10-4

Find the ratio of the perimeters and the ratio of the areas of the blue figure to the red figure.

15. 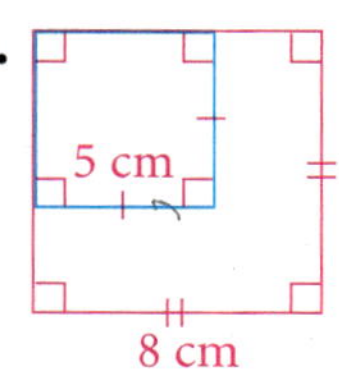

16. 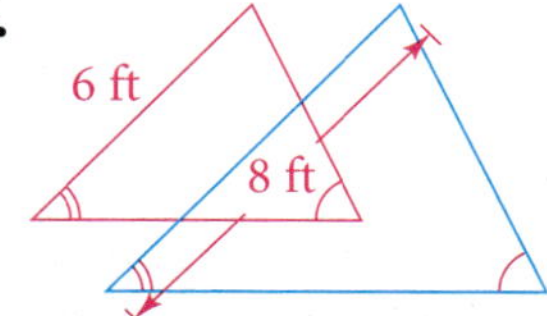

17. 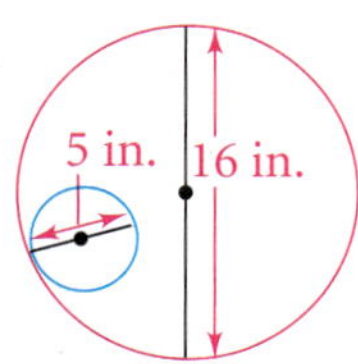

18. A triangular banner has an area of 315 in.^{2}. A similar banner has sides $1\frac{1}{3}$ times as long as those of the smaller banner. What is the area of the larger banner?

19. You want to enlarge the picture on the front of a postcard by 10%. If the perimeter of the postcard is 44 cm, what will be the perimeter of the enlargement?

Lesson 10-5

Find the area of each polygon. Round your answers to the nearest tenth.

20.

9 ft
7 ft
73°

21. 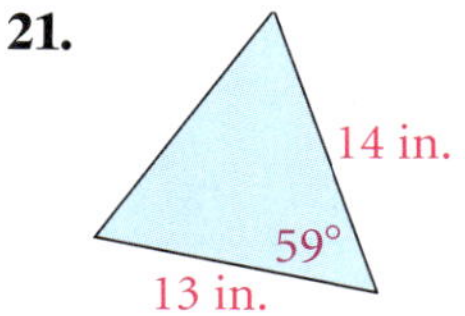

22. 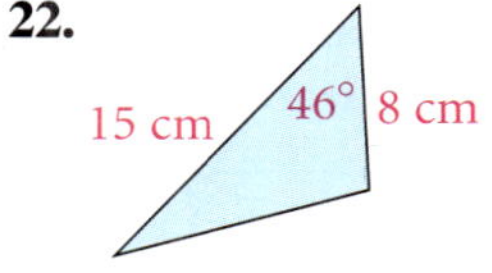

23. 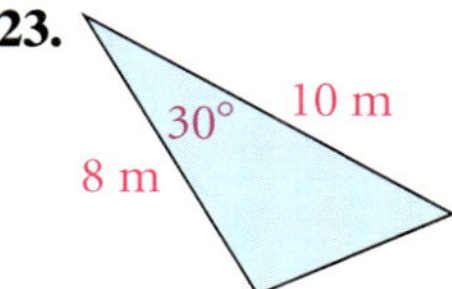

24. a regular hexagon with an apothem of 3 ft

25. a regular octagon with radius 5 ft

Lesson 10-6

(a) Find the circumference of each circle. (b) Find the length of the arc shown in red. Leave your answers in terms of π.

26. 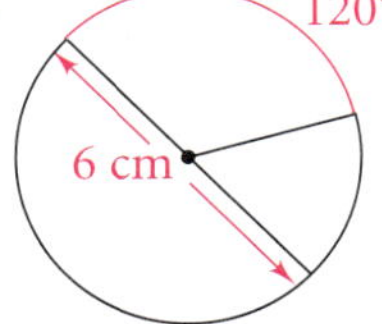

27. 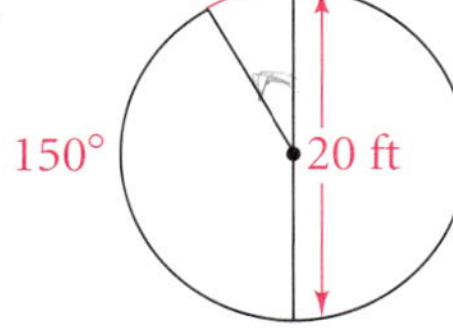

28. 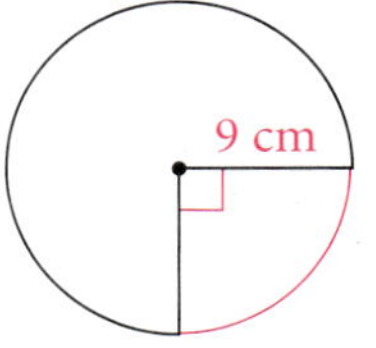

29. 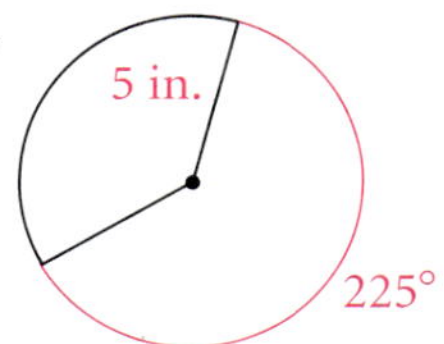

30. A bicycle wheel has a radius of 0.33 m. How many revolutions does the wheel make when the bicycle is ridden 1 km? Round to the nearest whole number.

Lesson 10-7

Find the area of each shaded sector or segment. Leave your answers in terms of π.

31. 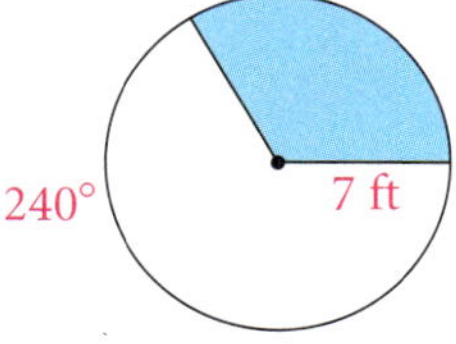

32. 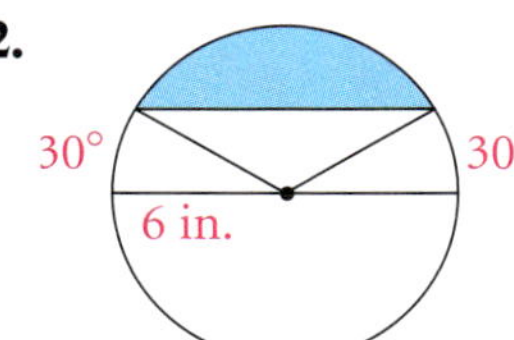

33. 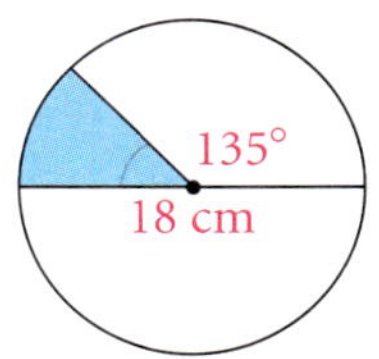

34. 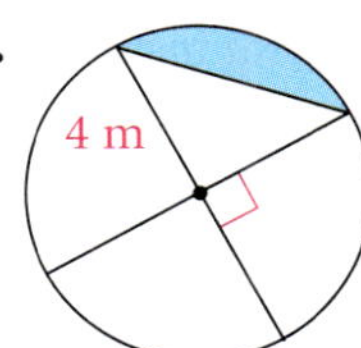

35. A 14-in. diameter pizza is cut into 6 equal slices. About how many square inches of pizza are in each slice? Round to the nearest square inch.

Lesson 10-8

Darts are thrown at random at each of the boards shown. If a dart hits the board, find the probability that it will land in the shaded area.

36. 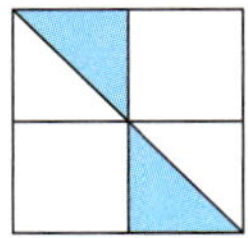

37. 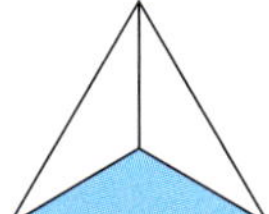

38. 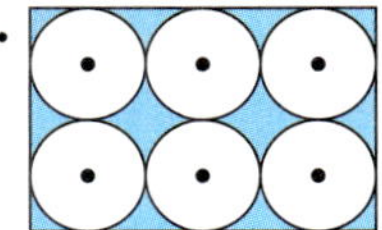

39. 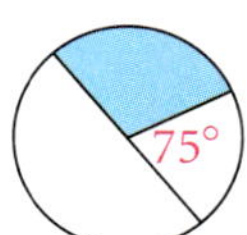

40. A square garden that is 80 ft on each side is surrounded by a cobblestone street that is 8 ft wide. If a child's balloon lands at random in the region formed by the garden and street, what is the probability that it lands on the street?

41. A dart hits the circular board shown in the figure at a random point. What is the probability that it does not hit the shaded square? Express your answer in terms of π.

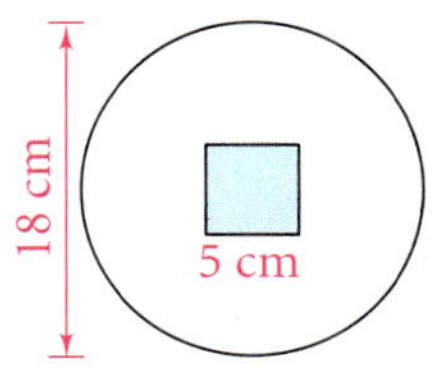

Chapter 11 Extra Practice: Skills, Word Problems, and Proof

Lesson 11-1

The diagrams in Exercises 1–4 each show a cube after part of it has been cut away. Identify the shape of the cross section formed by the cut. Also, verify Euler's Formula, $F + V = E + 2$, for the polyhedron that remains.

1. 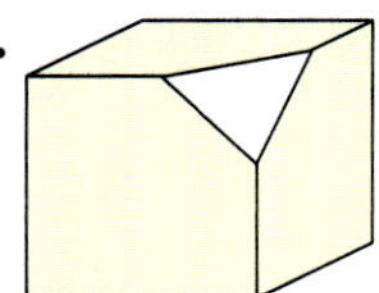

2. 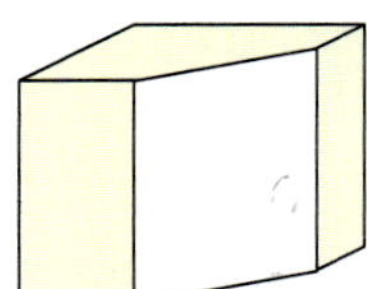

3. 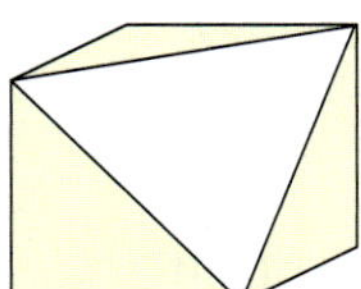

4. 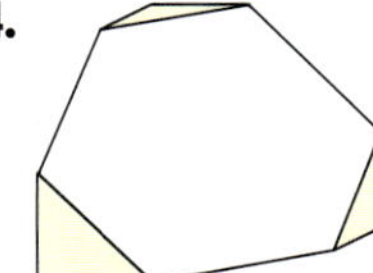

5. The bases of the prism shown at the right are equilateral triangles. Make a sketch that shows how you can have a plane intersect the prism to give a cross section that is an isosceles trapezoid.

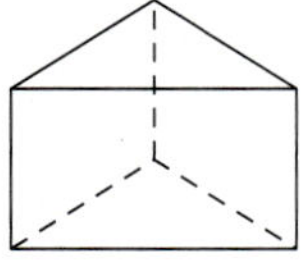

Lessons 11-2 and 11-3

Find the (a) lateral area and (b) surface area of each figure. Leave your answers in terms of π or in simplest radical form.

6. 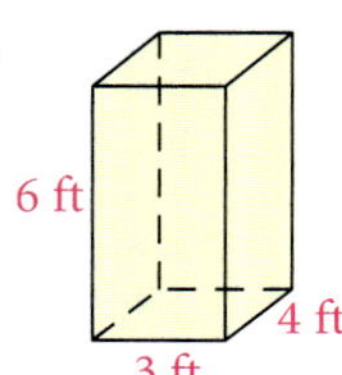

7. 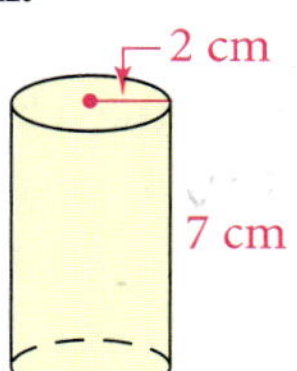

8. 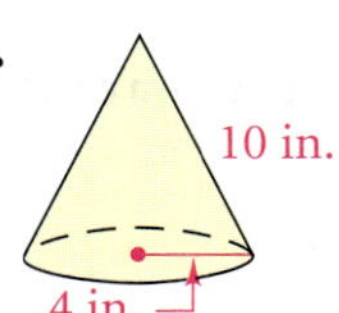

9. 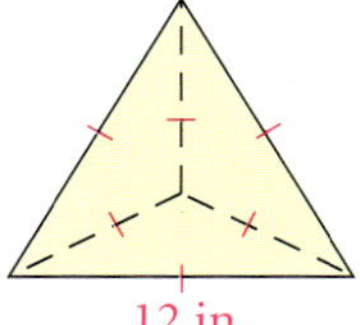

10. An optical instrument contains a triangular glass prism with the dimensions shown at the right. Find the lateral area and surface area of the prism. Round to the nearest tenth.

11. A company packages salt in a cylindrical box that has a diameter of 8 cm and a height of 13.5 cm. Find the lateral area and surface area of the box. Round to the nearest tenth.

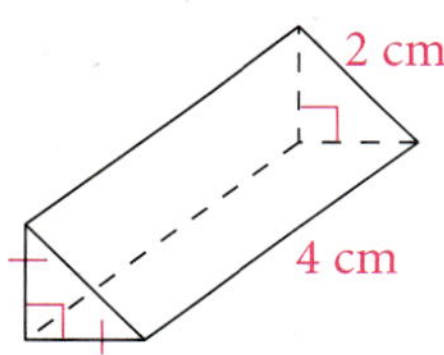

Find the (a) lateral area and (b) surface area of each pyramid or cone. Assume that the base of each pyramid is a regular polygon. Round your answers to the nearest tenth.

12. 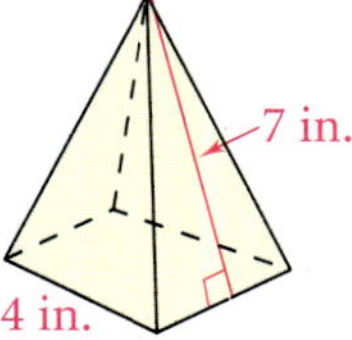

13. 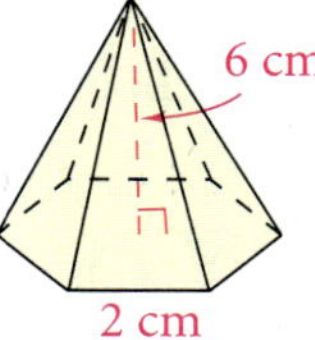

14. 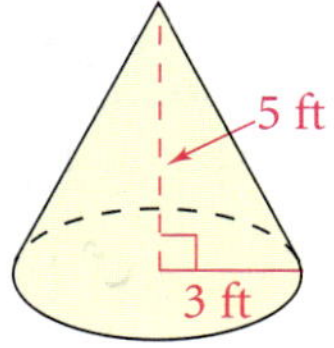

15. 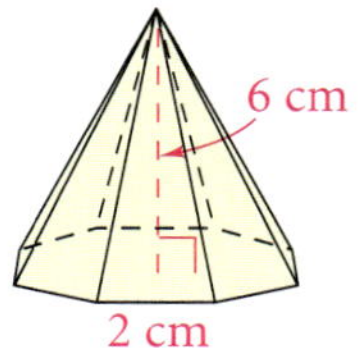

Lessons 11-4 and 11-5

Find the volume of each figure. Round your answers to the nearest tenth.

16. 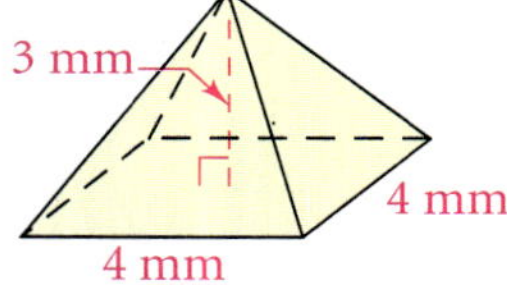

17. 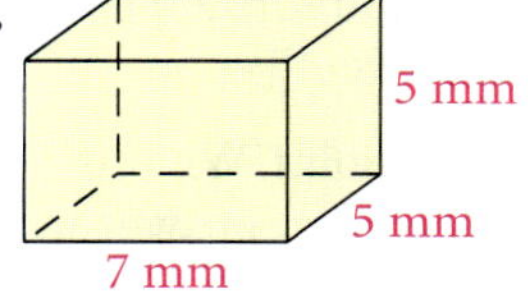

18. 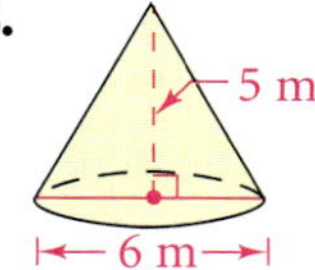

19. 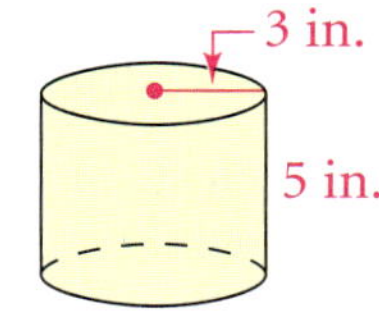

20. 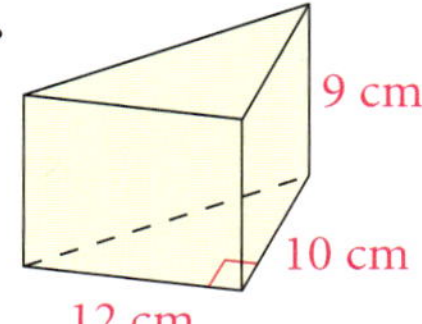

21. 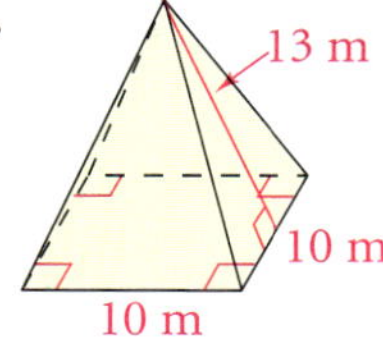

22. 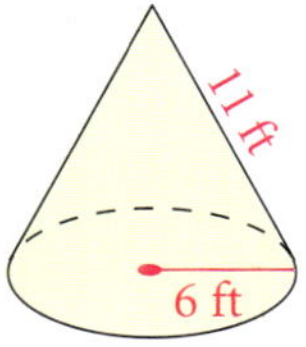

23. 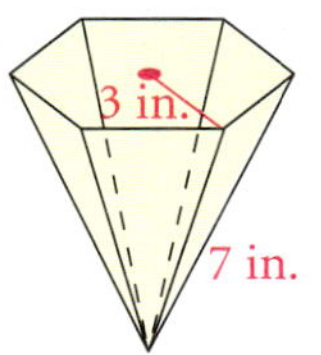

24. A greenhouse has the dimensions shown in the figure. What is the volume of the greenhouse? Round to the nearest cubic foot.

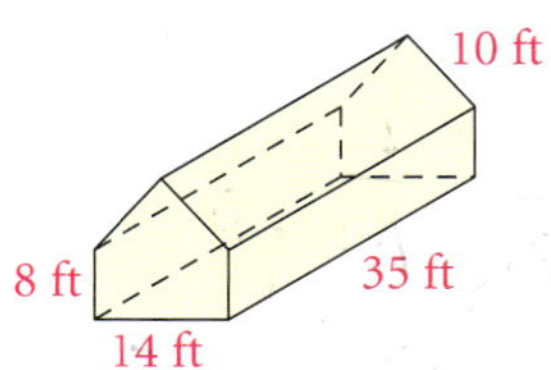

25. Find the volume of a can of chicken broth that has a diameter of 7.5 cm and a height of 11 cm. Round to the nearest tenth.

26. A paper drinking cup is a cone that has a diameter of $2\frac{1}{2}$ in. and a height of $3\frac{1}{2}$ in. How many cubic inches of water does the cup hold when it is full to the brim? Round to the nearest tenth.

Lesson 11-6

Find the volume and surface area of a sphere with the given radius or diameter. Give each answer in terms of π and rounded to the nearest whole number.

27. $r = 5$ cm **28.** $r = 3$ ft **29.** $d = 8$ in.

30. $d = 2$ ft **31.** $r = 0.5$ in. **32.** $d = 9$ m

The surface area of each sphere is given. Find the volume of each sphere in terms of π.

33. 64π m^2 **34.** 16π in.^{2} **35.** 49π ft^2

36. A spherical beach ball has a diameter of 1.75 ft when it is full of air. What is the surface area of the beach ball, and how many cubic feet of air does it contain? Round to the nearest hundredth.

Lesson 11-7

Copy and complete the table for three similar solids.

	Similarity Ratio	Ratio of Surface Areas	Ratio of Volumes
37.	2 : 3	■ : ■	■ : ■
38.	■ : ■	25 : 64	■ : ■
39.	■ : ■	■ : ■	27 : 64

40. How do the surface area and volume of a cylinder change if the radius and height are multiplied by $\frac{5}{4}$?

41. For two similar solids, how are the ratios of their volumes and surface areas related?

Chapter 12 Extra Practice: Skills, Word Problems, and Proof

Lesson 12-1

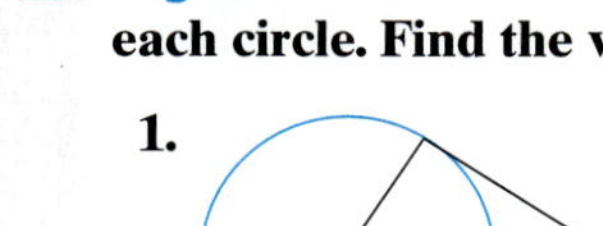

Algebra **Assume that lines that appear to be tangent are tangent. P is the center of each circle. Find the value of x.**

1. 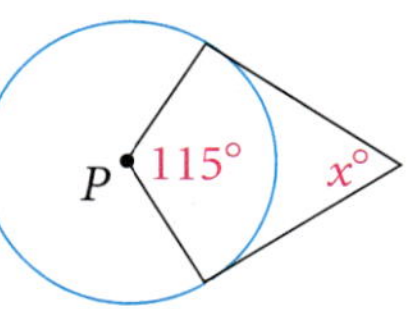

2. 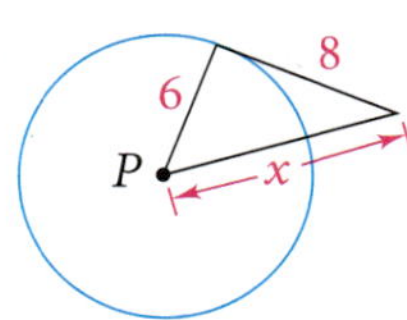

3. 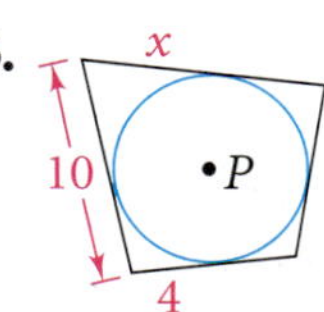

4. 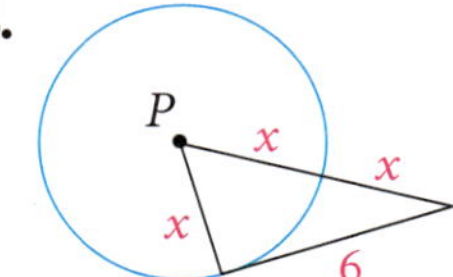

5. Given: Quadrilateral $ABCD$ is circumscribed about $\odot O$.

Prove: $AB + DC = BC + AD$

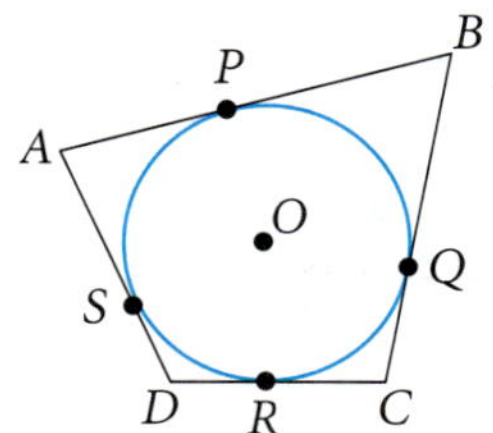

Lessons 12-2 and 12-3

Algebra **Find the value of each variable. If your answer is not a whole number, round it to the nearest tenth.**

6. 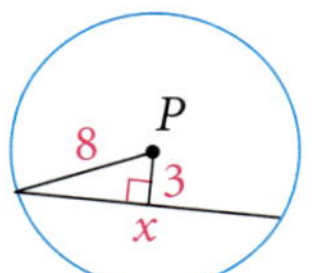

7. 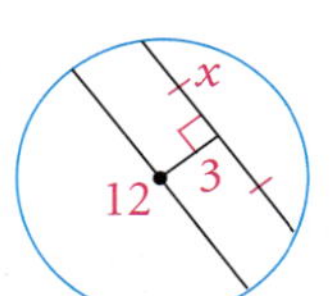

8. 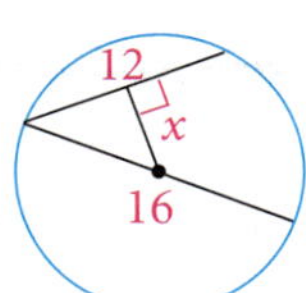

9. 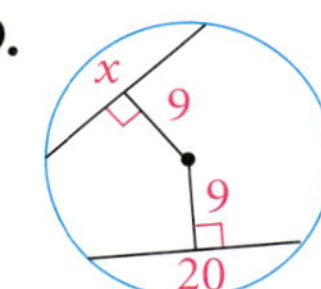

10.

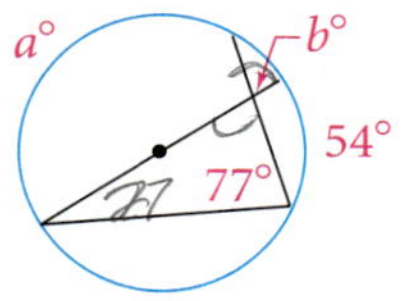

11. 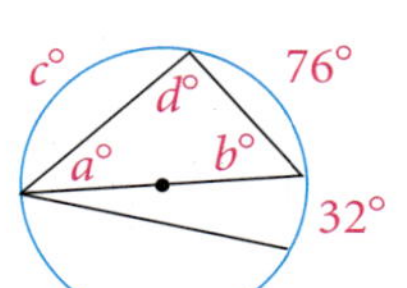

12. 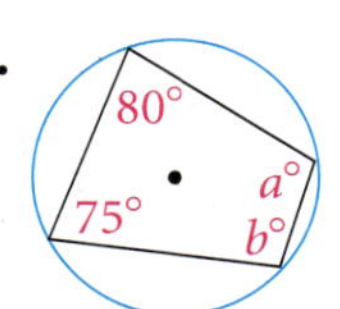

13. 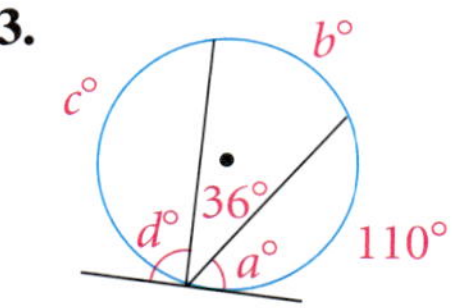

14. A polygon is inscribed in a circle. Are the perpendicular bisectors of the sides of the polygon concurrent? Explain.

15. A circle has a diameter of 4 units. A chord parallel to a diameter is 1.5 units from the center of the circle. The endpoints of the diameter and the chord are the vertices of an isosceles trapezoid. What is the distance from the center of the circle to each leg of the trapezoid? Round to the nearest hundredth.

16. Given: $\angle A$ and $\angle D$ are inscribed angles in $\odot O$ that intercept $\overset{\frown}{BC}$. $\overline{BD}$ and $\overline{AC}$ intersect at P.

Prove: $\triangle APB \sim \triangle DPC$

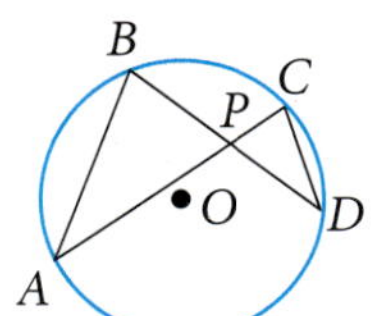

Lesson 12-4

Algebra **Assume that lines that appear to be tangent are tangent. Find the value of each variable. If your answer is not a whole number, round it to the nearest tenth.**

17.

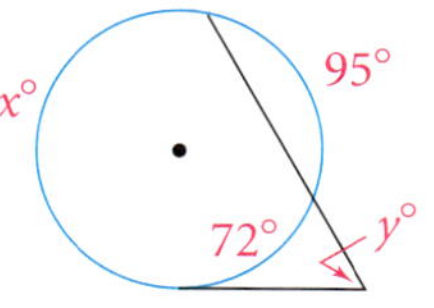

18.

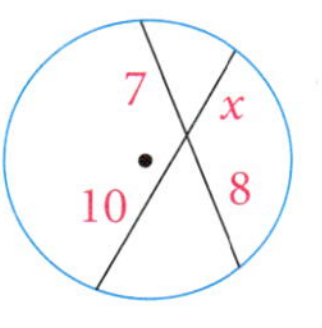

19.

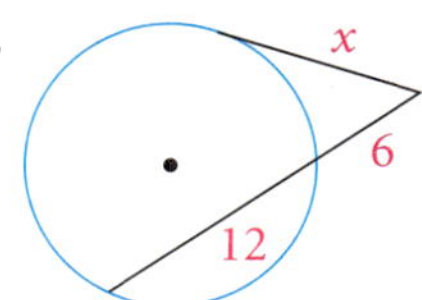

20.

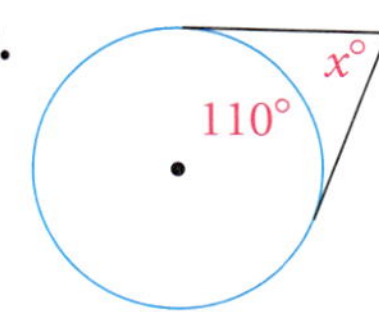

21.

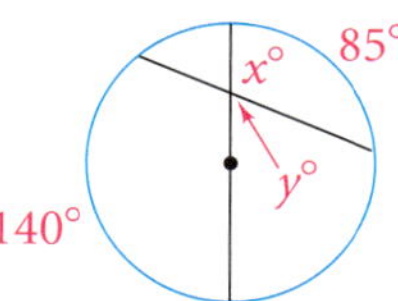

22.

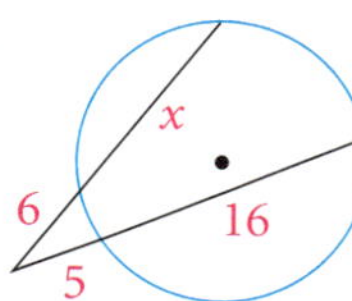

23.

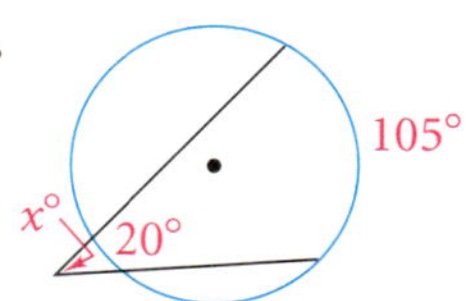

24.

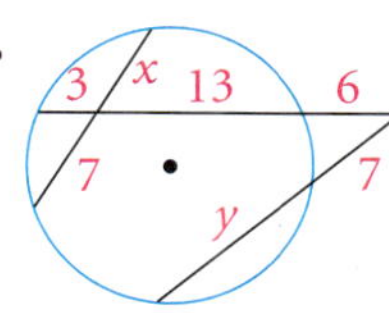

25. The outer rim of a circular garden will be planted with three colors of tulips. The landscaper has stretched two strings from a point P to help workers see how much of the circular rim should be planted with each color. Use the information in the figure at the right to find x and y.

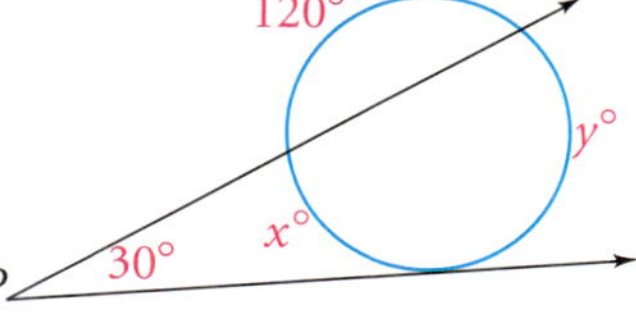

26. Planks are placed across the circular pool shown in the figure at the right. What is the length of the longer plank?

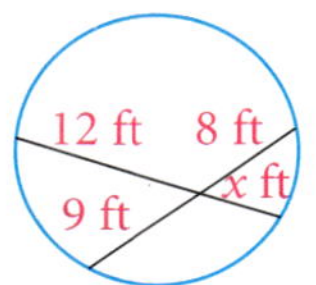

Lesson 12-5

Write the standard equation for each circle with center P.

27. $P = (0, 0); r = 4$

28. $P = (0, 5); r = 3$

29. $P = (9, -3); r = 7$

30. $P = (-4, 0)$; through $(2, 1)$

31. $P = (-6, -2)$; through $(-8, 1)$

32. $P = (-1, -3); r = 3$

33. When a coordinate grid is imposed over a map, the location of a radio station is given by (113, 215). A town located at (149, 138) is at the outermost edge of the circular region where clear reception is assured.

a. Write an equation that describes the boundary of the clear reception region.

b. If the radio station boosts power to increase the size of the clear-reception region by a factor of 4, what will be the equation for the new boundary for clear reception?

Lesson 12-6

Draw and describe each locus.

34. all points in a plane 3 cm from a circle with $r = 2$ cm

35. all points in a plane 2 cm from $\overrightarrow{AB}$

36. all points in space 1.5 in. from a point Q

37. A dog is on a 20-ft leash. The leash is attached to a pipe at the midpoint of the back wall of a 30 ft-by-30 ft house, as shown in the diagram. Sketch and use shading to indicate the region in which the dog can play while attached to the leash. Include measurements to describe the region.

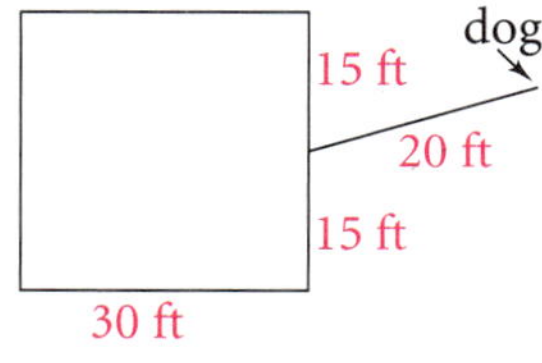

Skills Handbook

Problem Solving Strategies

You may find one or more of these strategies helpful in solving a word problem.

Strategy	When to Use it
Draw a Diagram	You need help in visualizing the problem.
Try, Check, Revise	Solving the problem directly is too complicated.
Make a Table	The problem has data that need organizing.
Look for a Pattern	The problem describes a relationship.
Solve a Simpler Problem	The problem is complex or has numbers that are too unmanageable to use at first.
Use Logical Reasoning	You need to reach a conclusion from some given information.
Work Backward	You undo various operations to arrive at the answer.

Problem Solving: Draw a Diagram

EXAMPLE

Antoine is 1.91 m tall. He measures his shadow and finds that it is 2.34 m long. He then measures the length of the shadow of a flagpole and finds that it is 13.2 m long. How tall is the flagpole?

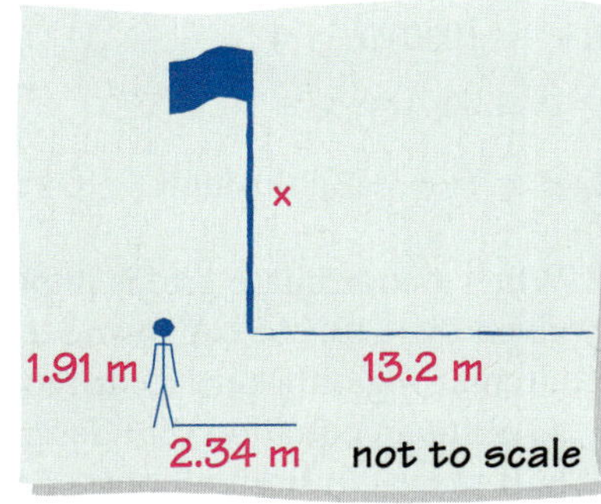

Start by drawing a diagram showing the given information. The diagram shows that you can solve the problem by using a proportion.

$\frac{1.91}{2.34} = \frac{x}{13.2}$ **Write a proportion.**

$x \approx 10.77$ **Solve for x.**

The flagpole is about 10.8 m tall.

EXERCISES

1. Five people meet and shake hands with one another. How many handshakes are there in all?
2. Three tennis balls fit snugly in an ordinary, cylindrical tennis ball container. Which is greater, the circumference of a ball or the height of the container?
3. Three lines that all intersect a circle can determine at most 7 regions within the circle, as shown in the diagram. What is the greatest number of regions that can be determined by 5 lines?

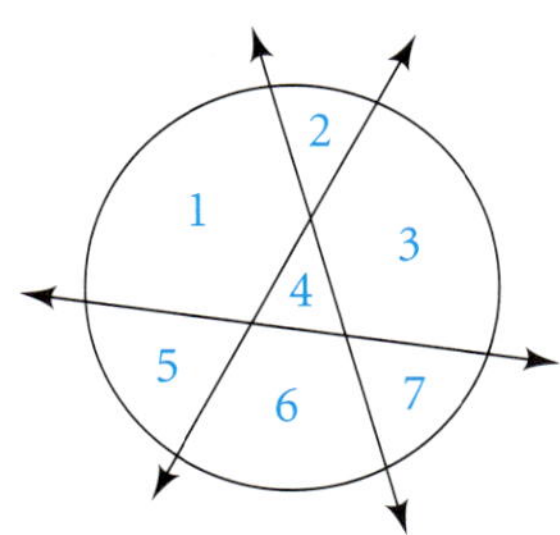

4. A triangle has vertices (1, 3), (2, 3), and (7, 5). Find its area.

Problem Solving: Try, Check, Revise

Have you ever weighed yourself on a balance scale at a doctor's office? You start by guessing your weight, and then you see if the scale balances. If it doesn't, you slide the weights back and forth until the scale does balance. This is an example of the *Try, Check, Revise* strategy, a strategy helpful for solving many types of problems.

EXAMPLE

You have 100 ft of fencing and want to build a fence in the shape of a rectangle to enclose the largest possible area. What should be the dimensions of the rectangle?

Try 1: 10 ft wide by 40 ft long
$10 \cdot 40 = 400 \text{ ft}^2$
Make an initial try with a perimeter of 100 ft. Find the area.

Try 2: 20 ft wide by 30 ft long
$20 \cdot 30 = 600 \text{ ft}^2$
Try again and find the area. The area is larger than the initial try.

Try 3: 35 ft wide by 15 ft long
$35 \cdot 15 = 525 \text{ ft}^2$
Continue trying and testing areas. This area is smaller than the last try.

Try 4: 22 ft wide by 28 ft long
$22 \cdot 28 = 616 \text{ ft}^2$
Notice that the areas are larger when the width and length are closer together.

Try 5: 25 ft wide by 25 ft long
$25 \cdot 25 = 625 \text{ ft}^2$
Choose dimensions that are as close together as possible.

The dimensions of your rectangle should be 25 ft by 25 ft.

EXERCISES

1. The product of three consecutive even integers is 480. Find the integers.

2. The combined ages of a father and his twin daughters are 54 years. The father was 24 years old when the twins were born. How old is each of the three people?

3. What numbers can x represent in the rectangle?

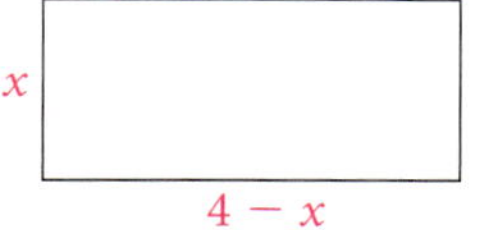

4. Alexandra has a collection of dimes and quarters. The number of dimes equals the number of quarters. She has a total of $2.80. How many of each coin does Alexandra have?

Use the Try, Check, Revise strategy to find the value of each variable.

5. $2a + 5 = 1$

6. $10 - 3c = -2$

7. $\frac{w}{-3} + 12 = -6$

8. $5y - 32 = 28$

9. $12b + 11 = 14$

10. $0.5x - 15 = -7$

11. Ruisa bought 7 rolls of film to take 192 pictures on a field trip. Some rolls had 36 exposures and the rest had 24 exposures. How many of each type did Ruisa buy?

12. The sum of five consecutive integers is 5. Find the integers.

13. Paul buys a coupon for $20 from a local theater that allows him to see movies for half price over the course of one year. The cost of seeing a movie is normally $7.50. What is the least number of movies Paul would have to see to pay less than the normal price per movie?

Problem Solving: Make a Table and Look for a Pattern

There are two important ways that making a table can help you solve a problem. First, a table is a handy method of organizing information. Second, once the information is in a table, it is easier for you to find patterns.

EXAMPLE

The squares below are made of toothpicks. How many toothpicks are in the square with 7 toothpicks on a side?

Use a table to organize the information.

Notice the pattern in the increases in the numbers of toothpicks in the squares. For each increase of 1 toothpick on a side, the increase increases by 4. The number of toothpicks in the 5th square is 40 + 20, or 60. The number in the 6th square is 60 + 24, or 84, and the number in the 7th square is 84 + 28, or 112.

Toothpicks on a side	1	2	3	4
Toothpicks in the square	4	12	24	40

+8 +12 +16

EXERCISES

1. The triangles are made of toothpicks. How many toothpicks are in Figure 10?

Figure 1

Figure 2

Figure 3

2. In each figure, the vertices of the smallest square are midpoints of the sides of the next larger square. Find the area of the ninth shaded square.

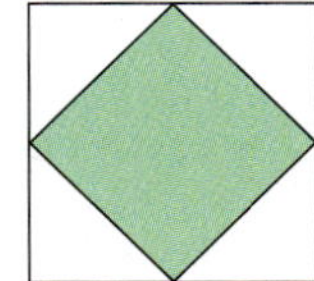

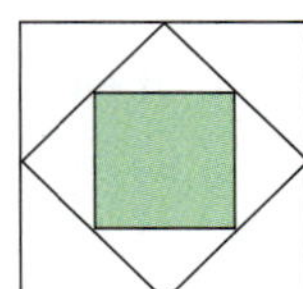

3. In each figure, the midpoints of the sides of the unshaded triangles are used as vertices of the shaded triangles. Find the total number of shaded triangles in Figure 8.

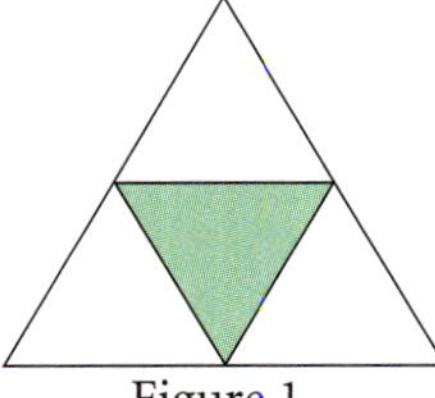

Figure 1

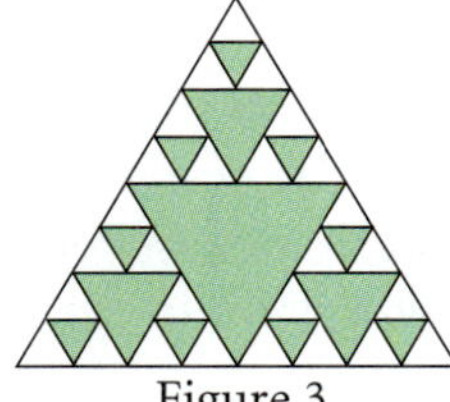

Figure 2

Figure 3

Problem Solving: Solve a Simpler Problem

Looking at a simpler version of a problem can be helpful in suggesting a problem solving approach.

EXAMPLE

A fence along the highway is 570 m long. There is a fence post every 10 m. How many fence posts are there?

You may be tempted to divide 570 by 10, getting 57 as an answer, but looking at a simpler problem suggests that this answer isn't right. Suppose you have just 10 or 20 m of fencing.

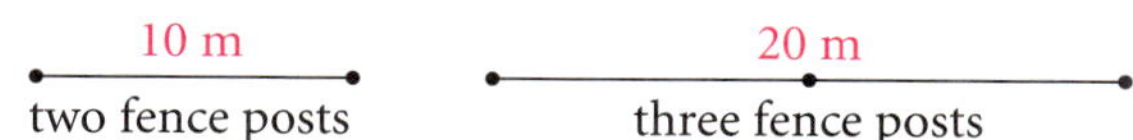

These easier problems suggest that there is always *one more* fence post than one tenth the length. So for a 570-m fence, there are $\frac{570}{10} + 1$, or 58 fence posts.

EXERCISES

1. A farmer wishes to fence in a square lot 70 yards by 70 yards. He will install a fence post every 10 yards. How many fence posts will he need?

2. A snail is trying to escape from a well 10 ft deep. The snail can climb 2 ft each day, but each night it slides back 1 ft. How many days will the snail take to climb out of the well?

3. Janette is planning to walk from her house to her friend Barbara's house. How many different paths can she take to get there? Assume that she walks only east and south (along the grid lines).

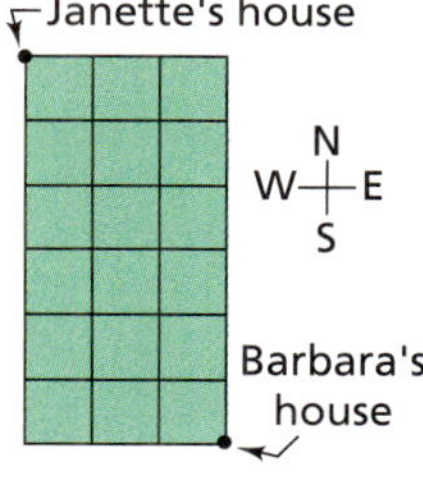

4. A square table can seat four people. For a banquet, a long rectangular table is formed by placing 14 such tables edge to edge in a straight line. How many people can sit at the long table?

5. Find the sum of the whole numbers from 1 to 999.

6. How many trapezoids are in the figure below? (*Hint:* Solve several simpler problems, and then look for a pattern.)

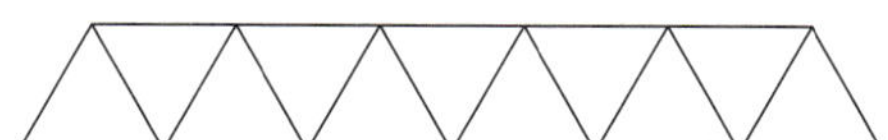

7. At a business luncheon, 424 handshakes took place. No two people shook hands with each other more than once. What is the least number of people in attendance at the luncheon?

8. On the occasion of his 50th birthday, the President was honored with a 21-gun salute. The sound of each gunshot lasted 1 second, and 4 seconds elapsed between shots. How long did the salute last?

9. In a tennis tournament, each athlete plays one match against each of the other athletes. There are 14 athletes scheduled to play in the tournament. How many matches will be played?

Problem Solving: Use Logical Reasoning

Some problems can be solved without the use of numbers. They can be solved by the use of logical reasoning, given some information.

Anna, Bill, Carla, and Doug are siblings. Each lives in a different state beginning with W. Use these clues to determine where each sibling lives:

(1) Neither sister lives in a state containing two words.

(2) Bill lives to the west of his sisters.

(3) Anna doesn't cross the Mississippi River when she visits Doug.

Make a table to organize what you know. Use an initial for each name.

State	A	B	C	D
West Virginia	✗	✗	✗	
Wisconsin		✗		
Wyoming		✗		
Washington	✗	✓	✗	✗

From clue 1, you know that neither Anna nor Carla lives in West Virginia.

Using clues 1 and 2, you know that Bill must live in Washington if he lives to the west of his sisters.

Use logical reasoning to complete the table.

State	A	B	C	D
West Virginia	✗	✗	✗	✓
Wisconsin	✓	✗	✗	✗
Wyoming	✗	✗	✓	✗
Washington	✗	✓	✗	✗

Doug lives in West Virginia because no other sibling does.

From clue 3, you know that Anna must live in Wisconsin.

Carla, therefore, lives in Wyoming.

EXERCISES

1. Harold has a dog, a parrot, a goldfish, and a hamster. Their names are J. T., Izzy, Arf, and Blinky. Izzy has neither feathers nor fins. Arf can't bark. J. T. weighs less than the four-legged pets. Neither the goldfish nor the dog has the longest name. Arf and Blinky don't get along well with the parrot. What is each pet's name?

2. At the state basketball championship tournament, 31 basketball games are played to determine the winner of the tournament. After each game, the loser is eliminated from the tournament. How many teams are in the tournament?

3. The sophomore class has 124 students. Of these students, 47 are involved in musical activities: 25 in band and 36 in choir. How many students are involved in both band and choir?

4. Tina's height is between Kimiko's and Ignacio's. Ignacio's height is between Jerome's and Kimiko's. Tina is taller than Jerome. List the people in order from shortest to tallest.

Problem Solving: Work Backward

In some situations it is easier to start with the end result and work backward to find the solution. You work backward in order to solve linear equations. The equation $2x + 3 = 11$ means "double x and add 3 to get 11." To find x, you "undo" those steps in reverse order.

$$2x + 3 = 11$$

$$2x = 8 \quad \textbf{Subtract 3 from each side.}$$

$$x = 4 \quad \textbf{Divide each side by 2.}$$

Another time it is convenient to work backward is when you want to "reverse" a set of directions.

EXAMPLE

Algebra Sandy spent $\frac{1}{10}$ of the money in her purse for lunch. She then spent \$23.50 for a gift for her brother, then half of what she had left on a new CD. If Sandy has \$13 left in her purse, how much money did she have in it before lunch?

Start with the \$13 that Sandy has left in her purse.
She spent half of what she had before the \$13 on a new CD, so she must have had twice \$13, or \$26, before she bought the CD.

She spent \$23.50 on a gift for her brother, so add \$23.50 to \$26. Before buying the gift for her brother, she had \$49.50.

She spent $\frac{1}{10}$ of the money for lunch and was left with \$49.50. That means $\frac{9}{10}$ of what she had is \$49.50. Set up an equation.

$$\frac{9}{10}x = 49.50$$

$$x = 55$$

Sandy had \$55 in her purse before lunch.

EXERCISES

1. To go from Bedford to Worcester, take Route 4 south, then Route 128 south, and then Route 90 west. How do you get from Worcester to Bedford?
2. Algae are growing on a pond's surface. The area covered doubles each day. It takes 24 days to cover the pond completely. After how many days will the pond be half covered with algae?
3. Don sold $\frac{1}{5}$ as many raffle tickets as Carlita. Carlita sold 3 times as many as Ranesha. Ranesha sold 7 fewer than Russell. If Russell sold 12 tickets, how many did Don sell?
4. At 6% interest compounded annually, the balance in a bank account will double about every 12 years. If such an account has a balance of \$16,000 now, how much was deposited when the account was opened 36 years ago?
5. Solve the puzzle that Yuan gave to Inez: I am thinking of a number. If I triple the number and then halve the result, I get 12. What number am I thinking of?
6. Carlos paid a \$14.60 taxi fare from a hotel to the airport, including a \$2.00 tip. Green Cab Co. charges \$1.20 per passenger plus \$0.20 for each $\frac{1}{5}$ mile. How many miles is the hotel from the airport?

Using a Ruler and Protractor

Knowing how to use a ruler and protractor is crucial for success in geometry.

EXAMPLE

Draw a triangle that has sides of length 5.2 cm and 3.0 cm and a 68° angle between these two sides.

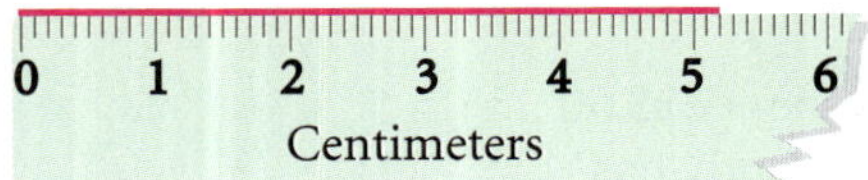

Step 1 Use a ruler to draw a segment 5.2 cm long.

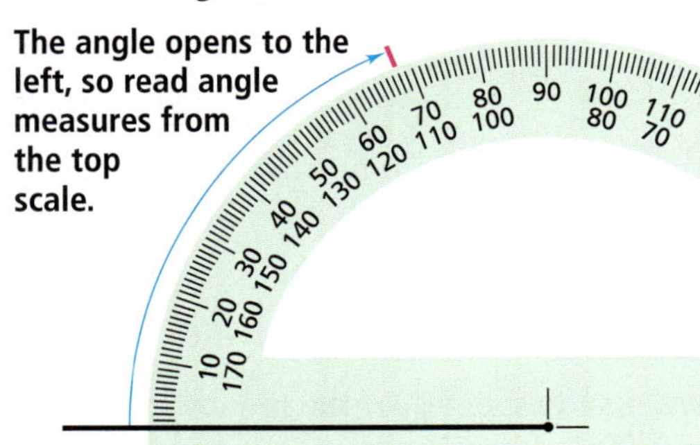

Step 2 Place the crosshairs of a protractor at one endpoint of the segment. Make a small mark at the 68° position along the protractor.

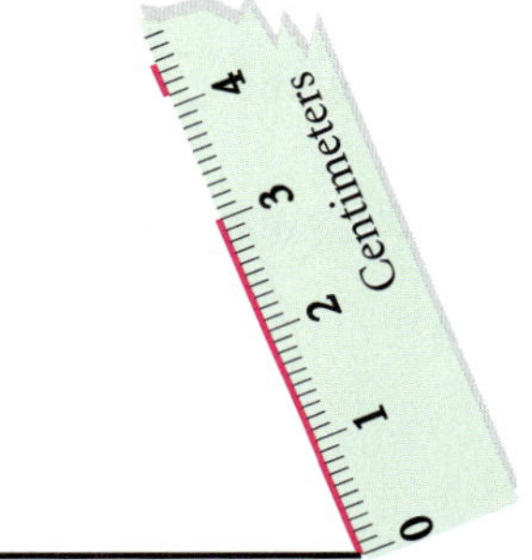

Step 3 Align the ruler along the small mark and the endpoint you used in Step 2. Place the zero point of the ruler at the endpoint. Draw a segment 3.0 cm long.

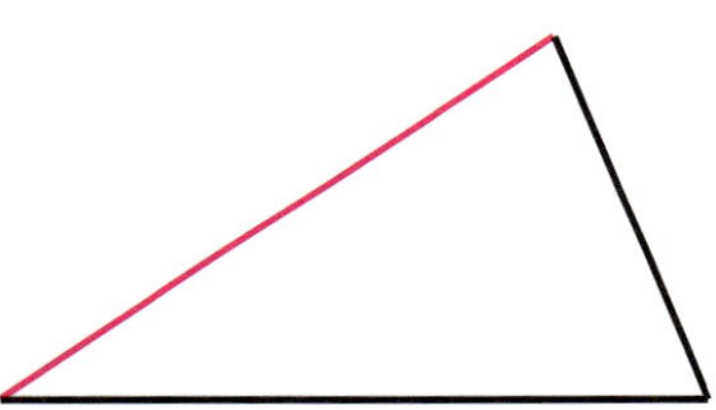

Step 4 Complete the triangle by connecting the endpoints of the first and second segments.

EXERCISES

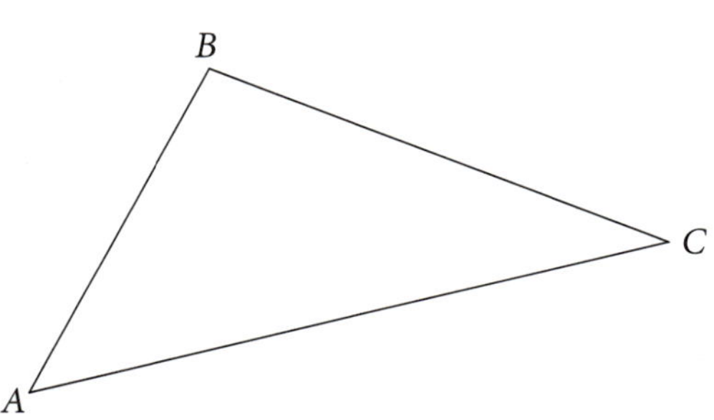

1. Measure sides $\overline{AB}$ and $\overline{BC}$ to the nearest millimeter.
2. Measure each angle of $\triangle ABC$ to the nearest degree.
3. Draw a triangle that has sides of length 4.8 cm and 3.7 cm and a 34° angle between these two sides.
4. Draw a triangle that has 43° and 102° angles and a side of length 5.4 cm between these two angles.
5. Draw a rhombus that has sides of length $2\frac{1}{4}$ in., and 68° and 112° angles.
6. Draw an isosceles trapezoid that has one pair of 48° base angles and a base of length 2 in. between these two base angles.
7. Draw an isosceles triangle that has two congruent sides $3\frac{1}{2}$ in. long and a 134° vertex angle.

Measurement Conversions

To convert from one unit of measure to another, you multiply by a conversion factor in the form of a fraction. The numerator and denominator are in different units, but they represent the same amount. So, you can think of this as multiplying by 1.

An example of a conversion factor is $\frac{1 \text{ ft}}{12 \text{ in.}}$. You can create other conversion factors using the table on page 766.

1 EXAMPLE

Complete each statement.

a. 88 in. = ■ ft

$88 \text{ in.} \cdot \frac{1 \text{ ft}}{12 \text{ in.}} = \frac{88}{12} \text{ ft} = 7\frac{1}{3} \text{ ft}$

b. 5.3 m = ■ cm

$5.3 \text{ m} \cdot \frac{100 \text{ cm}}{1 \text{ m}} = 5.3(100) \text{ cm} = 530 \text{ cm}$

c. 3700 mm = ■ cm

$3700 \text{ mm} \cdot \frac{1 \text{ cm}}{10 \text{ mm}} = 370 \text{ cm}$

d. 90 in. = ■ yd

$90 \text{ in.} \cdot \frac{1 \text{ ft}}{12 \text{ in.}} \cdot \frac{1 \text{ yd}}{3 \text{ ft}} = \frac{90}{36} \text{ yd} = 2\frac{1}{2} \text{ yd}$

Area is always in square units, and volume is always in cubic units.

3 ft

1 yd = 3 ft

3 ft

3 ft

$1 \text{ yd}^2 = 9 \text{ ft}^2$

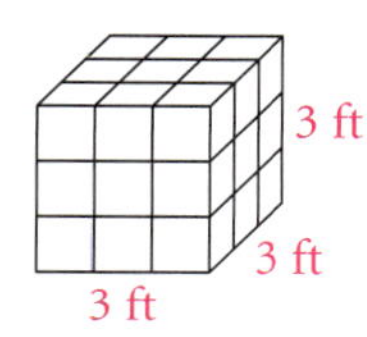

$1 \text{ yd}^3 = 27 \text{ ft}^3$

2 EXAMPLE

Complete each statement.

a. 300 in.^2 = ■ ft^2

$1 \text{ ft} = 12 \text{ in., so } 1 \text{ ft}^2 = (12 \text{ in.})^2 = 144 \text{ in.}^2$

$300 \text{ in.}^2 \cdot \frac{1 \text{ ft}^2}{144 \text{ in.}^2} = 2\frac{1}{12} \text{ ft}^2$

b. $200{,}000 \text{ cm}^3$ = ■ m^3

$1 \text{ m} = 100 \text{ cm, so } 1 \text{ m}^3 = (100 \text{ cm})^3 = 1{,}000{,}000 \text{ cm}^3$

$200{,}000 \text{ cm}^3 \cdot \frac{1 \text{ m}^3}{1{,}000{,}000 \text{ cm}^3} = 0.2 \text{ m}^3$

EXERCISES

Complete each statement.

1. 40 cm = ■ m
2. 1.5 kg = ■ g
3. 60 cm = ■ mm
4. 200 in. = ■ ft
5. 28 yd = ■ in.
6. 1.5 mi = ■ ft
7. 42 fl oz = ■ qt
8. 430 mg = ■ g
9. 34 L = ■ mL
10. 1.2 m = ■ cm
11. 43 mm = ■ cm
12. 3600 s = ■ min
13. 15 g = ■ mg
14. 12 qt = ■ c
15. 0.03 kg = ■ mg
16. 14 gal = ■ qt
17. 4500 lb = ■ t
18. 234 min = ■ h
19. 12 mL = ■ L
20. 2 pt = ■ fl oz
21. 20 m/s = ■ km/h
22. 3 ft^2 = ■ in.^2
23. 108 m^2 = ■ cm^2
24. 2100 mm^2 = ■ cm^2
25. 1.4 yd^2 = ■ ft^2
26. 0.45 km^2 = ■ m^2
27. 1300 ft^2 = ■ yd^2
28. 1030 in.^2 = ■ ft^2
29. $20{,}000{,}000 \text{ ft}^2$ = ■ mi^2
30. 1000 cm^3 = ■ m^3

Measurement, Rounding Error, and Reasonableness

There is no such thing as an *exact* measurement. Measurements are always approximate. No matter how precise it is, a measurement actually represents a range of values.

1 EXAMPLE

Chris's height, to the nearest inch, is 5 ft 8 in. Find the range of values this measurement represents.

The height is given to the nearest inch, so the error is $\frac{1}{2}$ in. Chris's height, then, is between 5 ft $7\frac{1}{2}$ in. and 5 ft $8\frac{1}{2}$ in., or 5 ft 8 in. $\pm \frac{1}{2}$ in. Within this range are all measures which, when rounded to the nearest inch, equal 5 ft 8 in.

As you calculate with measurements, errors can accumulate.

2 EXAMPLE

Jean drives 18 km to work each day. The distance is given to the nearest kilometer.

a. Find the range of values this measurement represents.

The driving distance is between 17.5 and 18.5 km, or 18 $\pm$ 0.5 km.

b. Find the error in the round-trip distance.

Double the lower limit, 17.5, and the upper limit, 18.5. Thus, the round trip can be anywhere between 35 and 37 km, or 36 $\pm$ 1 km. The error for the round trip is double the error of a single leg of the trip.

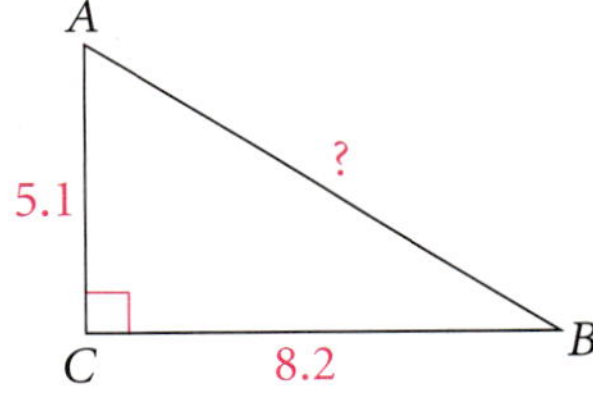

So that your answers will be reasonable, keep precision and error in mind as you calculate. For example, in finding AB, the length of the hypotenuse of $\triangle ABC$, it would be inappropriate to give the answer as 9.6566 if the sides are given to the nearest tenth. Round your answer to 9.7.

EXERCISES

Each measurement is followed by its unit of greatest precision. Find the range of values that each measurement represents.

1. 24 ft (ft)

2. 124 cm (cm)

3. 340 mL (mL)

4. $5\frac{1}{2}$ mi. $\left(\frac{1}{2}\text{ mi}\right)$

5. 73.2 mm (0.1 mm)

6. 34 yd^2 (yd^2)

7. 5.4 mi (0.1 mi)

8. 6 ft 5 in. (0.5 in.)

9. $15\frac{1}{2}$ yd $\left(\frac{1}{2}\text{ yd}\right)$

10. The lengths of the sides of $TJCM$ are given to the nearest tenth of a centimeter. Find the range of values for the figure's perimeter.

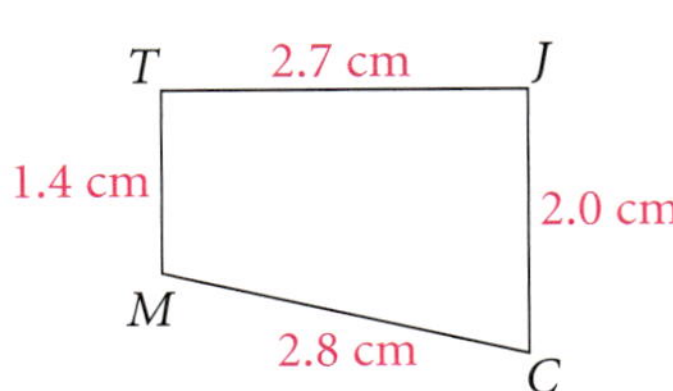

11. To the nearest degree, two angles of a triangle are 49° and 73°. What is the range of values for the measure of the third angle?

12. The lengths of the legs of a right triangle are measured as 131 m and 162 m. You use a calculator to find the length of the hypotenuse. The calculator display reads 208.33867. What should your answer be?

The Effect of Measurement Errors on Calculations

Measurements are always approximate, and calculations with these measurements produce error. Percent error is a measure of accuracy of a measurement or calculation. It is the ratio of the greatest possible error to the measurement.

$$\text{percent error} = \frac{\text{greatest possible error}}{\text{measurement}}$$

EXAMPLE

The dimensions of a box are measured as 18 in., 12 in., and 9 in. Find the percent error in calculating its volume.

The measurements are to the nearest inch, so the greatest possible length error is one half of one inch, or 0.5 in.

as measured

$$\begin{aligned} V &= \ell \cdot w \cdot h \\ &= 18 \cdot 12 \cdot 9 \\ &= 1944, \text{ or } 1944 \text{ in.}^3 \end{aligned}$$

maximum value

$$\begin{aligned} V &= \ell \cdot w \cdot h \\ &= 18.5 \cdot 12.5 \cdot 9.5 \\ &\approx 2196.9, \text{ or } 2196.9 \text{ in.}^3 \end{aligned}$$

minimum value

$$\begin{aligned} V &= \ell \cdot w \cdot h \\ &= 17.5 \cdot 11.5 \cdot 8.5 \\ &\approx 1710.6, \text{ or } 1710.6 \text{ in.}^3 \end{aligned}$$

Possible Error:

maximum − measured

$2196.9 - 1944 = 252.9$

measured − minimum

$1944 - 1710.6 = 233.4$

$$\begin{aligned} \text{percent error} &= \frac{\text{greatest possible error}}{\text{measurement}} \\ &= \frac{252.9}{1944} \\ &\approx 0.1300926 \\ &\approx 13\% \end{aligned}$$

The percent error is about 13%.

EXERCISES

Find the percent error in calculating the volume of each box given its dimensions. Round to the nearest percent.

1. 10 cm by 5 cm by 20 cm

2. 12 in. by 6 in. by 2 in.

3. 1.2 mm by 5.7 mm by 2.0 mm

4. 7.5 m by 6.4 m by 2.7 m

5. 22.5 cm by 16.4 cm by 26.4 cm

6. 1.24 cm by 4.45 cm by 5.58 cm

7. $8\frac{1}{4}$ in. by $17\frac{1}{2}$ in. by 5 in.

8. $7\frac{3}{4}$ in. by $22\frac{1}{8}$ in. by $6\frac{1}{4}$ in.

Find the percent error in calculating the perimeter of each figure.

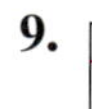

9.

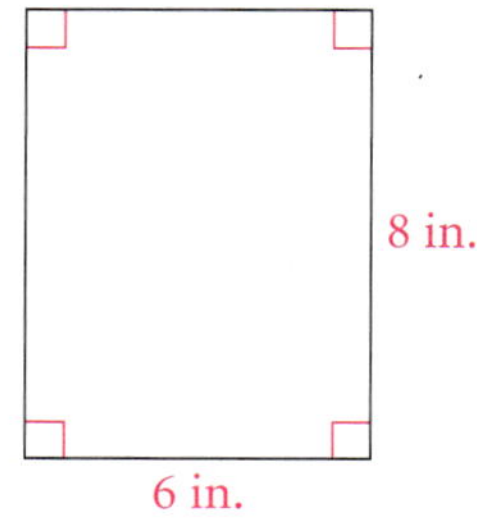

10.

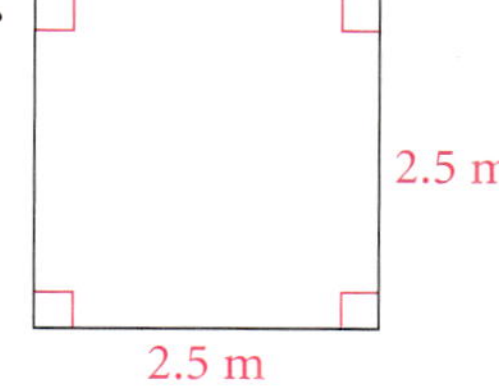

11.

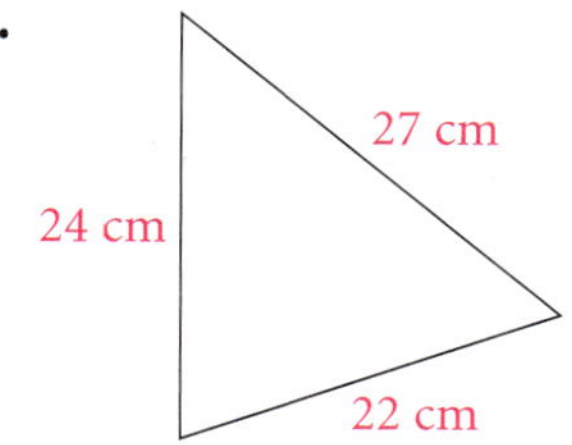

Mean, Median, and Mode

Measures of central tendency, such as mean, median, and mode, are numbers that describe a set of data.

The mean, sometimes called the average, is the sum of the data items divided by the number of data items.

The median is the middle number when data items are placed in order and there are an odd number of data items. For an even number of data items, the median is the mean of the middle two numbers.

The mode is the data item that appears most frequently.
A set of data may have more than one mode or no modes.

EXAMPLE

Eighteen students were asked to measure the angle formed by the three objects in the diagram. Their answers, in order from least to greatest, are as follows:

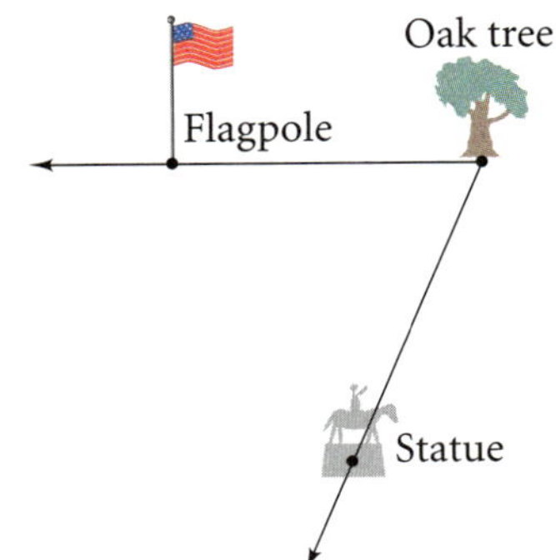

65, 66, 66, 66, 66, 66, 66, 67, 67, 67, 67, 67, 68, 68, 69, 70, 74, 113

Find the mean, median, and mode of the data.

Mean: $\frac{\text{sum of the 18 measures}}{18} = \frac{1258}{18} = 69\frac{8}{9}$

Median:
This data list is already ordered. The two middle numbers—the ninth and the tenth numbers on the list—are both 67. So the median is 67.

Mode:
There are more 66's than any other number, so the mode is 66.

EXERCISES

Find the mean, median, and mode of each set of data.

1. Numbers of students per school in Newtown: 234, 341, 253, 313, 273, 301, 760
2. Lunch expenses: \$4.50, \$3.26, \$5.02, \$3.58, \$1.25, \$3.05, \$4.24, \$3.56, \$3.31
3. Salaries at D. B. Widget & Co.: \$15,000; 18,000; \$18,000; \$21,700; \$26,500; \$27,000; \$29,300; \$31,100; \$43,000; \$47,800; \$69,000; \$140,000
4. Population of towns in Brower County: 567, 632, 781, 902, 1034, 1100, 1598, 2164, 2193, 3062, 3074, 3108, 3721, 3800, 4104
5. In Exercise 3, which measure or measures of central tendency do you think best represent the data? Explain.
6. Find the mean, median, and mode of the exam scores at the right.
7. In the example, the student who reported the angle measure as 113 most likely made an error. If this measure is dropped from the list, what are the mean, median, and mode of the remaining 17 scores?
8. In the example, if the measurement 65 were instead 51, would the mean decrease? Would the median? Would the mode?
9. In the example, if the two students who measured the angle at 68 both reduced their measurements to 67, would the mode be affected? How?

Final Exam Scores
34, 47, 53, 56, 57, 62, 62, 64, 67, 70, 74, 74, 74, 78, 82, 85, 85, 85, 85, 86, 88, 92, 93, 93, 94, 95, 97

Bar Graphs and Line Graphs

Data displayed in a table can be very useful, but a table is not always as easy to interpret as a graph. Bar graphs and line graphs can show the same data, but sometimes one type of graph has advantages over the other.

Make a bar graph and a line graph showing the data in the table at the right.

Revenue of HJL Co.

Year	Revenue
1998	$39,780
1999	$40,019
2000	$51,772
2001	$63,444
2002	$79,855

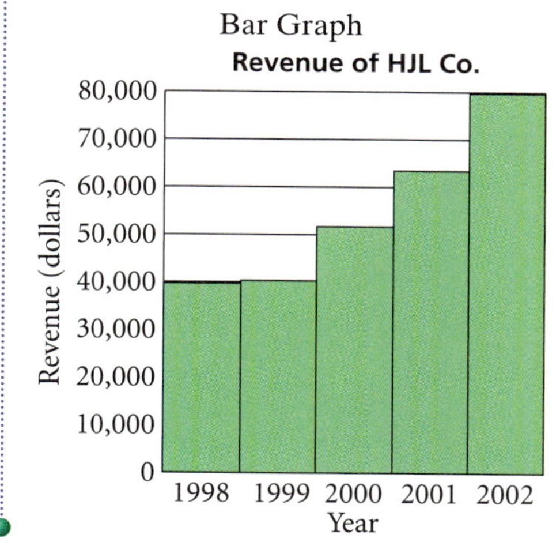

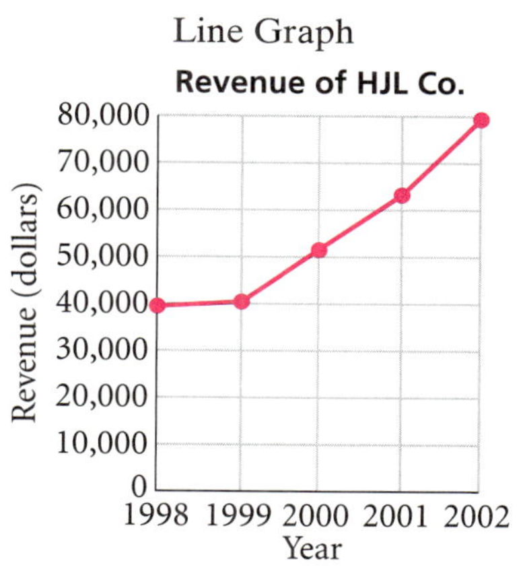

Bar graphs are useful when you wish to compare amounts. In the example above, the tallest bar is clearly twice the height of the shortest bar. At a glance, it is evident that in four years the revenue approximately doubled.

Line graphs allow you to see how a set of data changes over time. In the example, the slope of the line shows that revenue has increased at a steady rate since 1999.

Did revenue increase from 1998 to 1999? It is difficult to tell by looking at either graph; for that information, you should look back at the table.

EXERCISES

1. Create a bar graph and a line graph to display the data in the table below.

Sales of Rock Music (in millions of dollars)

Year	1995	1996	1997	1998	1999	2000
Sales	$4127	$4086	$3977	$3527	$3675	$3552

SOURCE: Recording Industry Association of America.
Go to **www.PHSchool.com** for a data update.
Web Code: bcg-2041

For Exercises 2–6, refer to the line graph at the right.

Temperatures in Grand Island, Nebraska, on February 2

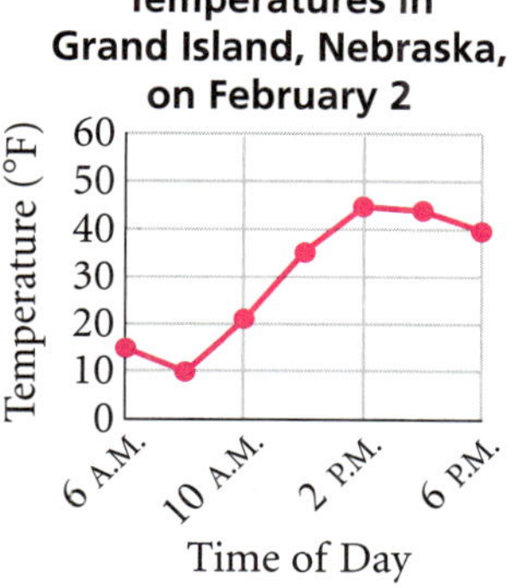

2. What was the lowest temperature recorded between 6 A.M. and 6 P.M.?

3. During which time periods did the temperature appear to increase?

4. Estimate the temperature at 11 A.M. and at 5 P.M.

5. Can you tell from the graph what the actual maximum and minimum temperatures were between 6 A.M. and 6 P.M.? Explain.

6. The same data could be presented in a bar graph. Which presentation is better for these data, a line graph or a bar graph? Explain why.

Box-and-Whisker Plots

A *box-and-whisker plot* is a way to display data on a number line. It provides a picture of how tightly the data cluster around the median and how wide a range the data have. The diagram below shows the various points associated with a box-and-whisker plot.

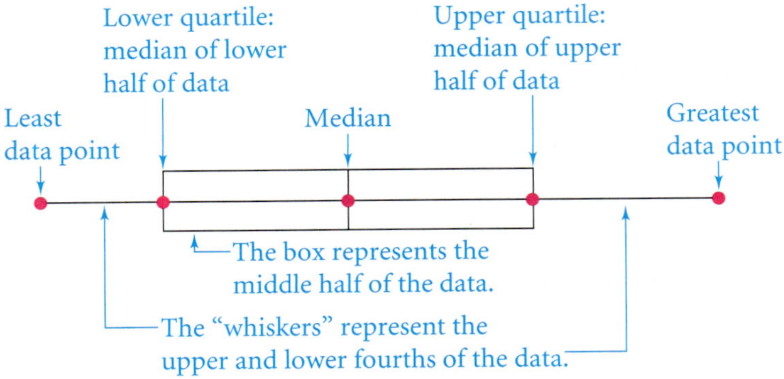

EXAMPLE

The heights, in inches, of 23 geometry students are as follows.

58, 61, 63, 63, 63, 64, 64, 65, 65, 65, 67, 68, 68, 68, 69, 70, 70, 70, 72, 72, 72, 74, 75

Draw a box-and-whisker plot.

The heights range from 58 in. to 75 in. Show 58 and 75 as endpoints on a line segment. The median is 68, so locate 68 in relation to 58 and 75. The lower quartile (the median of the lower eleven heights) is 64. The upper quartile (the median of the upper eleven heights) is 70. Locate 64 and 70 and draw a box enclosing them. Draw a vertical segment inside the box through the median.

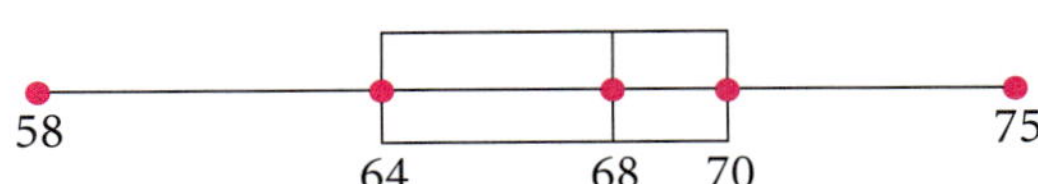

EXERCISES

1. All of the physical education students at Martin Luther King, Jr., High School were timed sprinting the 100-meter dash. The box-and-whisker plot below summarizes the data. Use it to find the following.
   a. median **b.** lower quartile **c.** upper quartile

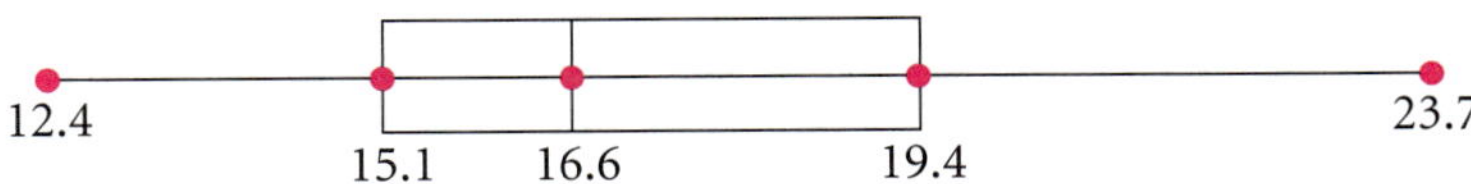

2. Make a box-and-whisker plot for the following data set, which lists the weights, in pounds, of the students trying out for the wrestling team at Benjamin Banneker High School.

   104, 121, 122, 130, 130, 131, 140, 144, 147, 147, 148, 155, 160, 163, 171

3. Make a box-and-whisker plot for the following set of data, which lists the numbers of pages in a set of books. (*Hint:* Order the data from smallest to largest.)

   205, 198, 312, 254, 185, 268, 297, 242, 356, 262

Squaring Numbers and Finding Square Roots

The square of a number is found by multiplying the number by itself. An exponent of 2 is used to indicate that a number is being squared.

1 EXAMPLE

Simplify.

a. 5^2

$5^2 = 5 \cdot 5$

$= 25$

b. $(-3.5)^2$

$(-3.5)^2 = (-3.5) \cdot (-3.5)$

$= 12.25$

c. $(\frac{2}{7})^2$

$(\frac{2}{7})^2 = \frac{2}{7} \cdot \frac{2}{7}$

$= \frac{4}{49}$

The square root of a number is itself a number that, when squared, results in the original number. A radical symbol ($\sqrt{\ }$) is used to represent the positive square root of a number.

2 EXAMPLE

Simplify. Round to the nearest tenth if necessary.

a. $\sqrt{36}$

$\sqrt{36} = 6$ since $6^2 = 36$

b. $\sqrt{174}$

$\sqrt{174} \approx 13.2$ since $13.2^2 \approx 174$

You can solve equations that include squared numbers.

3 EXAMPLE

Algebra Solve.

a. $x^2 = 144$

$x = 12$ or -12

b. $a^2 + 3^2 = 5^2$

$a^2 + 9 = 25$

$a^2 = 16$

$a = 4$ or -4

EXERCISES

Simplify.

1. 11^2 **2.** 16^2 **3.** $(-14)^2$ **4.** $(-21)^2$

5. 5.1^2 **6.** $(\frac{3}{7})^2$ **7.** $(\frac{8}{5})^2$ **8.** -6^2

Simplify. Round to the nearest tenth if necessary.

9. $\sqrt{100}$ **10.** $\sqrt{169}$ **11.** $\sqrt{74}$ **12.** $\sqrt{50}$

13. $\sqrt{400}$ **14.** $\sqrt{289}$ **15.** $\sqrt{\frac{4}{9}}$ **16.** $\sqrt{\frac{49}{81}}$

Algebra **Solve. Round to the nearest tenth if necessary.**

17. $x^2 = 49$ **18.** $a^2 = 9$ **19.** $y^2 + 7 = 8$ **20.** $5 + x^2 = 11$

21. $8^2 + b^2 = 10^2$ **22.** $5^2 + 4^2 = c^2$ **23.** $p^2 + 12^2 = 13^2$ **24.** $20^2 = 15^2 + a^2$

Evaluating and Simplifying Expressions

You evaluate an expression with variables by substituting a number for each variable. Then simplify the expression using the order of operations. Be especially careful with exponents and negative signs. For example, the expression $-x^2$ always yields a negative or zero value, and $(-x)^2$ is always positive or zero.

Order of Operations

1. Perform any operation(s) inside grouping symbols.
2. Simplify any term with exponents.
3. Multiply and divide in order from left to right.
4. Add and subtract in order from left to right.

1 EXAMPLE

Algebra Evaluate each expression for $r = 4$.

a. $-r^2$

$-r^2 = -(4^2) = -16$

b. $-3r^2$

$-3r^2 = -3(4^2) = -3(16) = -48$

c. $(-3r)^2$

$(-3r)^2 = (-3 \cdot 4)^2 = (-12)^2 = 144$

To simplify an expression, you eliminate any parentheses and combine like terms.

2 EXAMPLE

Algebra Simplify each expression.

a. $5r - 2r + 1$

Combine like terms.
$5r - 2r + 1 = 3r + 1$

b. $\pi(3r - 1)$

Use the distributive property.
$\pi(3r - 1) = 3\pi r - \pi$

c. $(r + \pi)(r - \pi)$

Multiply polynomials.
$(r + \pi)(r - \pi) = r^2 - \pi^2$

EXERCISES

Algebra Evaluate each expression for $x = 5$ and $y = -3$.

1. $-2x^2$ **2.** $-y + x$ **3.** $-xy$ **4.** $(x + 5y) \div x$

5. $x + 5y \div x$ **6.** $(-2y)^2$ **7.** $(2y)^2$ **8.** $(x - y)^2$

9. $\frac{x + 1}{y}$ **10.** $y - (x - y)$ **11.** $-y^x$ **12.** $\frac{2(1 - x)}{y - x}$

13. $x \cdot y - x$ **14.** $x - y \cdot x$ **15.** $\frac{y^3 - x}{x - y}$ **16.** $-y(x - 3)^2$

17. Which expression gives the area of the shaded figure at the right?

A. $\pi(r - s)^2$ **B.** $\pi(r^2 - s^2)$

C. $\pi(s^2 - r^2)$ **D.** $\pi r^2 - 2\pi s$

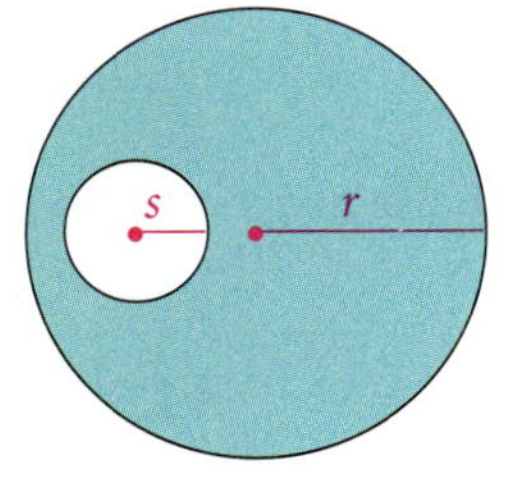

Algebra Simplify.

18. $6x - 4x + 8 - 5$ **19.** $2(\ell + w)$ **20.** $-(4x + 7)$

21. $-4x(x - 2)$ **22.** $3x - (5 + 2x)$ **23.** $2t^2 + 4t - 5t^2$ **24.** $(r - 1)^2$

25. $(1 - r)^2$ **26.** $(y + 1)(y - 3)$ **27.** $4h + 3h - 4 + 3$ **28.** $\pi r - (1 + \pi r)$

29. $(x + 4)(2x - 1)$ **30.** $2\pi h(1 - r)^2$ **31.** $3y^2 - (y^2 + 3y)$ **32.** $-(x + 4)^2$

Simplifying Radicals

A radical expression is in its simplest form when all three of the following statements are true.

1. The expression under the radical sign contains no perfect square factors (other than 1).
2. The expression under the radical sign does not contain a fraction.
3. The denominator does not contain a radical expression.

1 EXAMPLE

Simplify.

a. $\sqrt{\frac{4}{9}}$

$\sqrt{\frac{4}{9}} = \frac{\sqrt{4}}{\sqrt{9}} = \frac{2}{3}$

b. $\sqrt{12}$

$\sqrt{12} = \sqrt{4} \cdot \sqrt{3} = 2\sqrt{3}$

2 EXAMPLE

Find the length of the diagonal of rectangle *HJKL*.

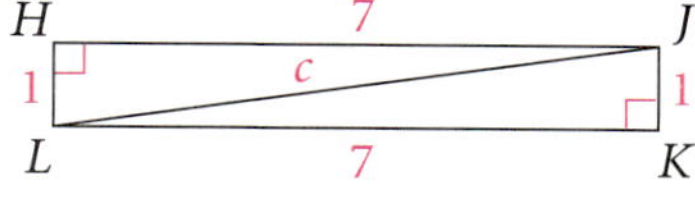

$c^2 = 7^2 + 1^2$	**Use the Pythagorean Theorem.**
$c^2 = 50$	**Simplify the right side.**
$c = \sqrt{50}$	**Find the square root of each side.**
$= \sqrt{25 \cdot 2}$	**Find a perfect square factor of 50.**
$= 5\sqrt{2}$	**Simplify the radical.**

3 EXAMPLE

Simplify $\frac{1}{\sqrt{3}}$.

$\frac{1}{\sqrt{3}} \cdot \frac{\sqrt{3}}{\sqrt{3}} = \frac{\sqrt{3}}{3}$ **Multiply by $\frac{\sqrt{3}}{\sqrt{3}}$, or 1, to eliminate the radical in the denominator.**

EXERCISES

Simplify each radical expression.

1. $\sqrt{27}$ **2.** $\sqrt{24}$ **3.** $\sqrt{150}$ **4.** $\sqrt{\frac{1}{9}}$ **5.** $\sqrt{\frac{72}{9}}$

6. $\frac{\sqrt{228}}{\sqrt{16}}$ **7.** $\sqrt{\frac{2}{5}}$ **8.** $\sqrt{\frac{27}{75}}$ **9.** $\frac{3}{\sqrt{8}}$ **10.** $\frac{6\sqrt{18}}{\sqrt{48}}$

x^2 **Algebra** **Find the value of *x*. Leave your answer in simplest radical form.**

11.

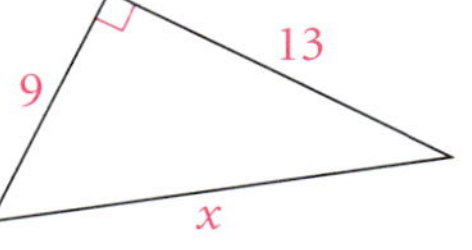

12.

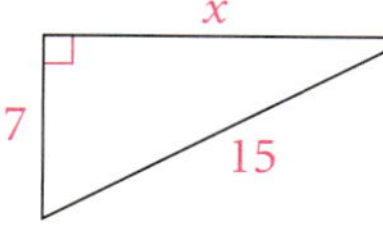

13.

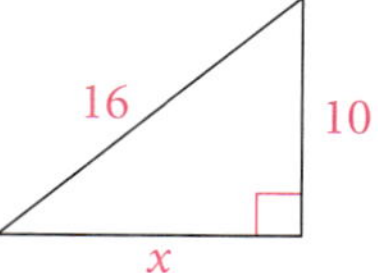

14.

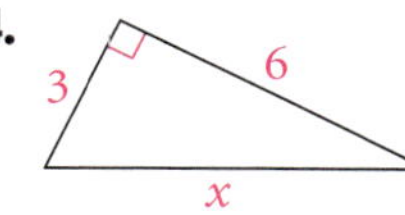

15.

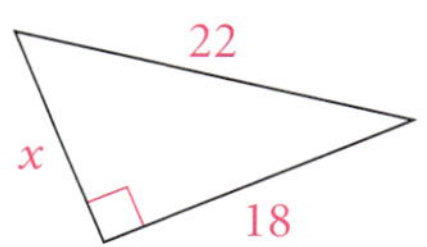

16.

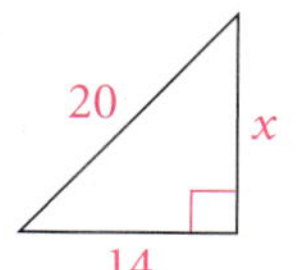

17.

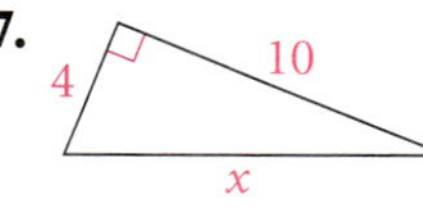

18.

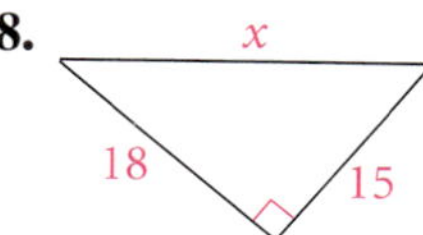

Simplifying Ratios

The ratio of the length of the shorter leg to the length of the longer leg for this right triangle is 4 to 6. This ratio can be written in several ways.

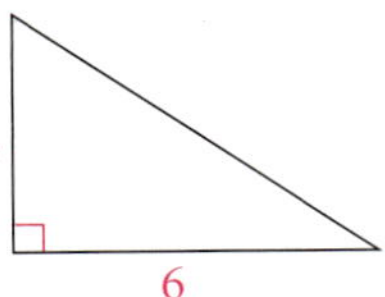

4 to 6 $\quad \frac{4}{6} \quad$ 4 : 6

Algebra Simplify each ratio.

a. 4 to 6

$$4 \text{ to } 6 = \frac{4}{6}$$
$$= \frac{2 \cdot 2}{2 \cdot 3}$$
$$= \frac{2}{3}$$

b. $3ab : 27ab$

$$3ab : 27ab = \frac{3ab}{27ab}$$
$$= \frac{3ab}{9 \cdot 3ab}$$
$$= \frac{1}{9}$$

← **Find and remove the common factor.** →

c. $\frac{4a + 4b}{a + b}$

$$\frac{4a + 4b}{a + b} = \frac{4(a + b)}{a + b}$$
$$= 4$$

Factor the numerator. The denominator cannot be factored. Remove the common factor ($a + b$).

EXERCISES

x^2 **Algebra Simplify each ratio.**

1. 25 to 15
2. $6 : 9$
3. $\frac{36}{54}$
4. 0.8 to 2.4
5. $\frac{7}{14x}$
6. $\frac{12c}{14c}$
7. $22x^2$ to $35x$
8. $0.5ab : 8ab$
9. $\frac{4xy}{0.25x}$
10. $1\frac{1}{2}x$ to $5x$
11. $\frac{x^2 + x}{2x}$
12. $\frac{1}{4}r^2$ to $6r$
13. $0.72t : 7.2t^2$
14. $(2x - 6) : (6x - 4)$
15. $12xy : 8xy$
16. $(9x - 9y)$ to $(x - y)$
17. $\frac{\pi r}{r^2 + \pi r}$
18. $\frac{8ab}{32xy}$

Express each ratio in simplest form.

19. shorter leg : longer leg
20. hypotenuse to shorter leg
21. $\frac{\text{shorter leg}}{\text{hypotenuse}}$
22. hypotenuse : longer leg
23. longer leg to shorter leg
24. $\frac{\text{longer leg}}{\text{hypotenuse}}$

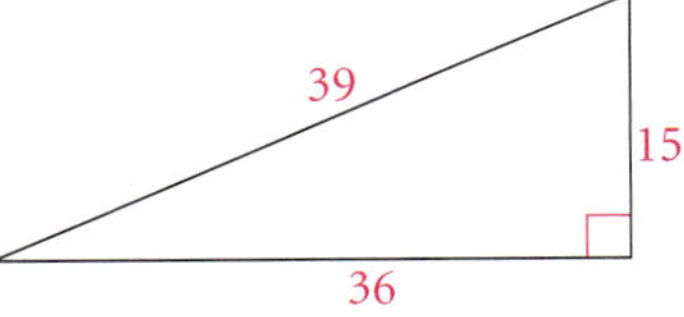

x^2 **Algebra Write an expression in simplest form for $\frac{\text{area of shaded figure}}{\text{area of blue figure}}$.**

25.

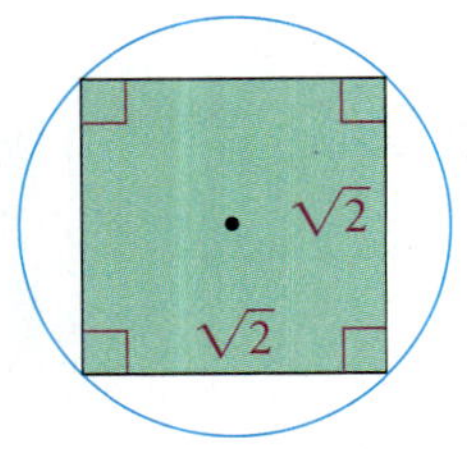

26.

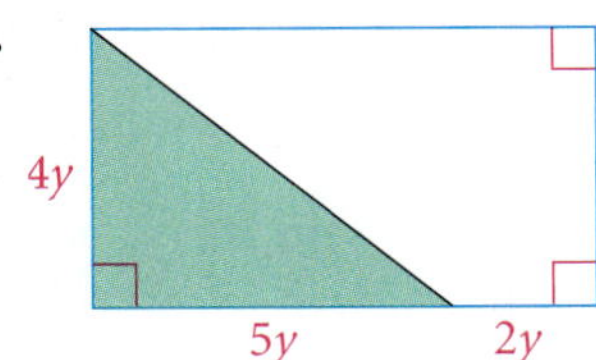

27.

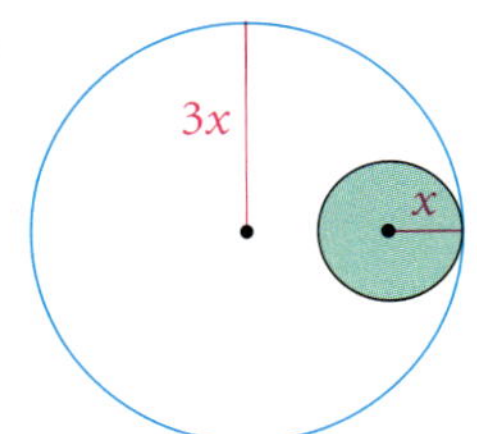

Absolute Value

Absolute value is used to represent the distance of a number from 0 on a number line. Since distance is always referred to as a nonnegative number, the absolute value of an expression is nonnegative.

On the number line at the right, both 4 and -4 are four units from zero. Therefore, $|4|$ and $|-4|$ are both equal to four.

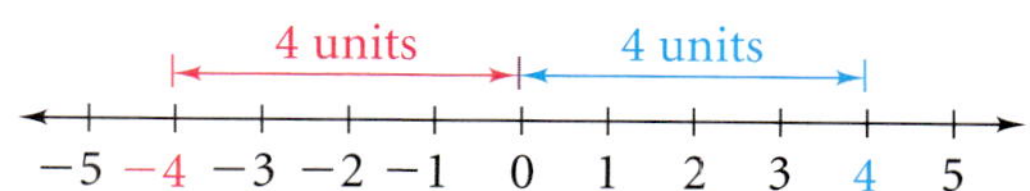

1 EXAMPLE

Simplify each expression.

a. $|-7|$

$|-7| = 7$

b. $|15|$

$|15| = 15$

c. $|4| + |-19|$

$|4| + |-19| = 4 + 19 = 23$

When working with more complicated expressions, always remember to simplify within absolute value symbols first.

2 EXAMPLE

Simplify each expression.

a. $|4 - 8|$

$|4 - 8| = |-4|$

$= 4$

b. $-3|-7 - 4|$

$-3|-7 - 4| = -3|-11|$

$= -3 \cdot 11$

$= -33$

To solve an equation involving absolute value, remember that absolute value symbols cause both negative and positive values to become positive.

3 EXAMPLE

Algebra Solve.

a. $|x| = 7$

$x = 7$ or -7

b. $|x| - 3 = 22$

$|x| - 3 = 22$

$|x| = 25$

$x = -25$ or 25

EXERCISES

Simplify each expression.

1. $|-8|$ **2.** $|11|$ **3.** $|16|$ **4.** $|-23|$

5. $|-7| + |15|$ **6.** $|-12| - |-12|$ **7.** $|5| - |10|$ **8.** $|4| + |2|$

9. $10 - |-20|$ **10.** $|-9 - 11|$ **11.** $2|-21 + 16|$ **12.** $-8|-9 + 4|$

Algebra Solve.

13. $|x| = 16$ **14.** $2 = |x|$ **15.** $|x| + 7 = 27$ **16.** $|x| - 9 = 15$

Skills Handbook

Solving and Writing Linear Equations

To solve a linear equation, use the properties of equality and properties of real numbers to find the value of the variable that satisfies the equation.

1 EXAMPLE

Algebra Solve each equation.

a. $5x - 3 = 2$

$5x - 3 = 2$

$5x = 5$ **Add 3 to each side.**

$x = 1$ **Divide each side by 5.**

b. $1 - 2(x + 1) = x$

$1 - 2(x + 1) = x$

$1 - 2x - 2 = x$ **Use the Distributive Property.**

$-1 - 2x = x$ **Simplify the left side.**

$-1 = 3x$ **Add $2x$ to each side.**

$-\frac{1}{3} = x$ **Divide each side by 3.**

You will sometimes need to translate word problems into equations. Look for words that suggest a relationship or some type of mathematical operation.

2 EXAMPLE

Algebra A student has grades of 80, 65, 78, and 92 on four tests. What is the minimum grade she must earn on her next test to ensure an average of 80?

Relate average of 80, 65, 78, 92, and next test, is 80 **Pull out the key words and numbers.**

Define Let x = the grade on the next test. **Let a variable represent what you are looking for.**

Write $\frac{80 + 65 + 78 + 92 + x}{5} = 80$ **Write an equation.**

$\frac{315 + x}{5} = 80$ **Combine like terms.**

$315 + x = 400$ **Multiply each side by 5.**

$x = 85$ **Subtract 315 from each side.**

The student must earn 85 on the next test for an average of 80.

EXERCISES

Algebra Solve each equation.

1. $3n + 2 = 17$

2. $5a - 2 = -12$

3. $2x + 4 = 10$

4. $3(n - 4) = 15$

5. $4 - 2y = 8$

6. $-6z + 1 = 13$

7. $6 - (3t + 4) = -17$

8. $7 = -2(4n - 4.5)$

9. $(w + 5) - (2w + 5) = 5$

10. $\frac{5}{7}p - 10 = 30$

11. $\frac{m}{-2} - 3 = 1$

12. $5k + 2(k + 1) = 23$

Algebra Write an equation and solve the problem.

13. Twice a number subtracted from 35 is 9. What is the number?

14. A new tenant pays the landlord the amount of the first month's rent, the same amount for the last month's rent, and half a month's rent for a security deposit. The total is \$2437.50. How much is the monthly rent?

15. The Johnsons pay \$9.95 a month plus \$0.035 per minute for local phone service. Last month, they paid \$12.75. How many minutes of local calls did they make?

Solving Literal Equations

An equation with two or more variables is called a literal equation. It is often necessary to solve a literal equation for a particular variable.

1 EXAMPLE

Algebra The formula $P = 2(\ell + w)$ gives the perimeter P of a rectangle with length ℓ and width w. Solve the equation for ℓ.

$$P = 2(\ell + w)$$

$$P = 2\ell + 2w \quad \text{Use the Distributive Property.}$$

$$P - 2w = 2\ell \quad \text{Subtract } 2w \text{ from each side.}$$

$$\frac{P - 2w}{2} = \ell \quad \text{Divide each side by 2.}$$

2 EXAMPLE

Algebra The formula $A = \frac{1}{2}(b_1 + b_2)h$ gives the area A of a trapezoid with bases b_1 and b_2 and height h. Solve for h.

$$A = \frac{1}{2}(b_1 + b_2)h$$

$$2A = h(b_1 + b_2) \quad \text{Multiply each side by 2.}$$

$$\frac{2A}{b_1 + b_2} = h \quad \text{Divide each side by } (b_1 + b_2).$$

3 EXAMPLE

Algebra The formula for converting from degrees Celsius C to degrees Fahrenheit F is $F = \frac{9}{5}C + 32$. Solve for C.

$$F = \frac{9}{5}C + 32$$

$$F - 32 = \frac{9}{5}C \quad \text{Subtract 32 from each side.}$$

$$\frac{5}{9}(F - 32) = C \quad \text{Multiply each side by } \frac{5}{9}.$$

EXERCISES

Algebra **Solve each equation for the variable in red.**

1. Perimeter of rectangle: $P = 2w + 2\ell$
2. Volume of prism: $V = \ell wh$
3. Surface area of sphere: $S = 4\pi r^2$
4. Lateral area of cylinder: $A = 2\pi rh$
5. Area of kite or rhombus: $A = \frac{1}{2}d_1d_2$
6. Area of circle: $A = \pi r^2$
7. Area of regular polygon: $A = \frac{1}{2}ap$
8. Volume of cylinder: $V = \pi r^2 h$
9. Area of triangle: $A = \frac{1}{2}bh$
10. Tangent of $\angle A$: $\tan A = \frac{y}{x}$
11. Euler's Formula: $F + V = E + 2$
12. Circumference of circle: $C = 2\pi r$
13. Cosine of $\angle A$: $\cos A = \frac{b}{c}$
14. Volume of cone: $V = \frac{1}{3}\pi r^2 h$
15. Surface area of right cone: $S = \pi r^2 + \pi r\ell$
16. Area of trapezoid: $A = \frac{1}{2}(b_1 + b_2)h$
17. Volume of pyramid: $V = \frac{1}{3}Bh$
18. Pythagorean Theorem: $a^2 + b^2 = c^2$
19. Surface area of regular pyramid: $S = B + \frac{1}{2}p\ell$
20. Surface area of right cylinder: $S = 2\pi r^2 + 2\pi rh$

Systems of Linear Equations

Normally, there are many ordered pairs that satisfy a given equation. For example, (3, 4), (4, 5), (5, 6), and infinitely many other pairs all satisfy the equation $y = x + 1$. In solving a system of two linear equations, however, you need to find ordered pairs that satisfy both equations at once. Ordinarily, there is just one such ordered pair; it is the point where the graphs of the two lines intersect.

One method you can always use to solve a system of linear equations is the substitution method.

EXAMPLE

Algebra Solve the system.
$$2x - y = -10$$
$$-3x - 2y = 1$$

Solve one of the equations for a variable. Looking at the two equations, it seems easiest to solve the first equation for y.

$2x - y = -10$

$-y = -2x - 10$ **Subtract $2x$ from each side.**

$y = 2x + 10$ **Multiply each side by −1.**

Now substitute $2x + 10$ for y in the other equation.

$-3x - 2y = 1$ **Write the other equation.**

$-3x - 2(2x + 10) = 1$ **Substitute $(2x + 10)$ for y.**

$-3x - 4x - 20 = 1$ **Use the Distributive Property.**

$-7x = 21$ **Simplify and add 20 to each side.**

$x = -3$ **Divide each side by −7.**

So $x = -3$. To find y, substitute -3 for x in either equation.

$2x - y = -10$ **Write one of the equations.**

$2(-3) - y = -10$ **Substitute −3 for x.**

$-6 - y = -10$ **Simplify.**

$-y = -4$ **Add 6 to each side.**

$y = 4$ **Multiply each side by −1.**

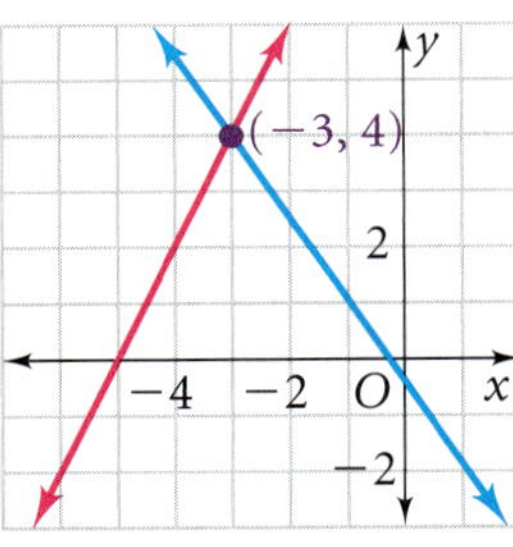

So the solution is $x = -3$ and $y = 4$, or $(-3, 4)$. If you graph $2x - y = -10$ and $-3x - 2y = 1$, you'll find that the lines intersect at $(-3, 4)$.

EXERCISES

Algebra Solve each system.

1. $x + y = 3$
   $x - y = 5$
2. $y - x = 4$
   $x + 3 = y$
3. $y = 1$
   $5x - 2y = 18$
4. $3x - y = 5$
   $x = -2$
5. $4y - x = -3$
   $2x - 6 = 8y$
6. $8x - 1 = 4y$
   $3x = y + 1$
7. $2x + 2y = -4$
   $-x + 3y = 6$
8. $12y - 3x = 11$
   $x - 2y = -2$
9. $5x + 7y = 1$
   $4x - 2y = 16$
10. Give an example of a system of linear equations with no solution. What do you know about the slopes of the lines of such a system?

Percents

A percent is a ratio in which a number is compared to 100. For example, the expression *60 percent* means "60 out of 100." The symbol % stands for "percent."

A percent can be written in decimal form by first writing it in ratio form, and then writing the ratio as a decimal. For example, 25% is equal to the ratio $\frac{25}{100}$ or $\frac{1}{4}$. As a decimal, $\frac{1}{4}$ is equal to 0.25. Note that 25% can also be written directly as a decimal by moving the decimal point two places to the left.

1 EXAMPLE

Convert each percent to a decimal.

a. 42%

42% = 0.42

b. 157%

157% = 1.57

c. 12.4%

12.4% = 0.124

d. 4%

4% = 0.04

To calculate a percent of a number, write the percent as a decimal and multiply.

2 EXAMPLE

Simplify. Where necessary, round to the nearest tenth.

a. 30% of 242

30% of 242 = $0.3 \cdot 242$

= 72.6

b. 7% of 38

7% of 38 = $0.07 \cdot 38$

= $2.66 \approx 2.7$

For a percent problem, it is a good idea to check that your answer is reasonable by estimating it.

3 EXAMPLE

Estimate 23% of 96.

23% ≈ 25% and 96 ≈ 100. 25% $\left(\text{or } \frac{1}{4}\right)$ of 100 = 25. A reasonable estimate is 25.

EXERCISES

Convert each percent to a decimal.

1. 50% **2.** 75% **3.** 27% **4.** 6%

5. 32.5% **6.** 84.6% **7.** 9% **8.** 2.5%

Simplify. Where necessary, round to the nearest tenth.

9. 21% of 40 **10.** 45% of 200 **11.** 6% of 120 **12.** 2% of 54

13. 80.4% of 52 **14.** 23.8% of 176 **15.** 7.5% of 32 **16.** 9.25% of 89

Estimate.

17. 12% of 70 **18.** 48% of 87 **19.** 73% of 64 **20.** 77% of 42

Probability

Probability is a measure of the likelihood of an event occurring. All probabilities range from 0 to 1 where 0 is the probability of an event that cannot happen and 1 is the probability of an event that is certain to happen. An event with probability 0.5 or 50% has an equal chance of happening or not happening.

The formula $P(E) = \frac{\text{number of favorable outcomes}}{\text{number of possible outcomes}}$ is used to calculate the probability of event E.

1 EXAMPLE

The numbers 2 through 21 are written on pieces of paper and placed in a hat. One piece of paper is drawn at random. Determine the probability of selecting a perfect square.

The total number of outcomes, 2, 3, 4, . . . , 21, for this event is 20.

There are 3 favorable outcomes: 4, 9, 16.

$P(\text{selecting a perfect square}) = \frac{3}{20}$

2 EXAMPLE

Determine the probability of getting exactly two heads when two coins are tossed.

The total number of outcomes, (H, H), (H, T), (T, H), (T, T), for this event is 4.

There is 1 favorable outcome, (H, H).

$P(\text{two heads}) = \frac{1}{4}$

EXERCISES

A jar contains 3 white balls, 7 red balls, and 4 green balls. A ball is selected at random from the jar. Determine the probability of selecting a ball with the given color.

1. red **2.** white **3.** green **4.** green or white

5. A red ball is removed from the jar. Determine the probability that the next ball selected will be green.

6. Two green balls are removed from the jar. Determine the probability that the next ball selected will be green.

You roll a 12-sided polyhedron with the numbers 1–12 on its congruent faces. Determine the probability of each outcome.

7. rolling a 2 **8.** rolling a 4 or a 5 **9.** rolling an even number

10. rolling an odd number **11.** rolling a prime number **12.** rolling a factor of 8

A coin is flipped three times. Determine the probability of each outcome.

13. exactly two tails **14.** two heads and one tail **15.** no more than two tails

16. no more than one head **17.** at least one tail **18.** all tails or all heads

Tables

Table 1 Math Symbols

Symbol	Meaning	Page
$\ldots$	and so on	p. 4
$=$	is equal to, equality	p. 4
$\cdot, \times$	times (multiplication)	p. 4
n^2	square of n	p. 4
$+$	plus (addition)	p. 5
$-a$	opposite of a	p. 6
$^\circ$	degree(s)	p. 7
$-$	minus (subtraction)	p. 16
$\overleftrightarrow{AB}$	line through points A and B	p. 17
$\overline{AB}$	segment with endpoints A and B	p. 23
$\overrightarrow{AB}$	ray with endpoint A and through point B	p. 23
$\parallel$	is parallel to	p. 24
$(\ )$	parentheses for grouping	p. 30
AB	length of $\overline{AB}$	p. 31
$\lvert a \rvert$	absolute value of a	p. 31
$\cong$	is congruent to	p. 31
$\angle A$	angle with vertex A	p. 36
$\angle ABC$	angle with sides $\overrightarrow{BA}$ and $\overrightarrow{BC}$	p. 36
$m\angle A$	measure of angle A	p. 36
⌝	right angle symbol	p. 37
$\perp$	is perpendicular to	p. 45
d	distance	p. 53
(a, b)	ordered pair with x-coordinate a and y-coordinate b	p. 53
$\sqrt{x}$	nonnegative square root of x	p. 53
A	area	p. 62
s	length of a side	p. 62
b	base length	p. 62
h	height	p. 62
$\approx$	is approximately equal to	p. 62
d	diameter	p. 62
r	radius	p. 62
P	perimeter	p. 62
π	pi, ratio of the circumference of a circle to its diameter	p. 62
C	circumference	p. 62
$p \to q$	If p, then q	p. 82
$>$	is greater than	p. 85
$<$	is less than	p. 85
$p \leftrightarrow q$	p if and only if q	p. 88
$\neq$	is not equal to	p. 103
$\measuredangle$s	angles	p. 135
$\triangle ABC$	triangle with vertices A, B, and C	p. 147
n-gon	polygon with n sides	p. 158
m	slope of a linear function	p. 166
b	y-intercept of a linear function	p. 166
$[\ ]$	brackets for grouping	p. 168
$\ncong$	is not congruent to	p. 218
$\triangle$s	triangles	p. 237
$\sim$	not	p. 281
$\geq$	is greater than or equal to	p. 288
$\leq$	is less than or equal to	p. 288
$\ngtr$	is not greater than	p. 291
$\nless$	is not less than	p. 291
$\square ABCD$	parallelogram with vertices A, B, C, and D	p. 312
$\square$s	parallelograms	p. 312
$a\!:\!b, \frac{a}{b}$	ratio of a to b	p. 366
$\pm$	plus or minus	p. 372
$\sim$	is similar to	p. 373
$\stackrel{?}{=}$	Is this statement true?	p. 419
$\tan A$	tangent of $\angle A$	p. 432
$\sin A$	sine of $\angle A$	p. 439
$\cos A$	cosine of $\angle A$	p. 439
$\overrightarrow{AB}$	vector with initial point A and terminal point B	p. 452
$\langle x, y \rangle$	ordered pair notation for a vector	p. 452
$\vec{\mathbf{v}}$	vector $\mathbf{v}$	p. 454
A'	image of A, A prime	p. 471
$\to$	maps to	p. 471
$\begin{bmatrix} 1 & 2 \\ 3 & 4 \end{bmatrix}$	matrix	p. 504
b_1, b_2	bases of a trapezoid	p. 540
d_1, d_2	lengths of diagonals	p. 541
a	apothem	p. 547
$\odot A$	circle with center A	p. 566
$\%$	percent	p. 566
$\widehat{AB}$	arc with endpoints A and B	p. 567
$\widehat{ABC}$	arc with endpoints A and C and containing B	p. 567
$m\widehat{AB}$	measure of $\widehat{AB}$	p. 567
P(event)	probability of the event	p. 582
h	length of an altitude	p. 608
B	area of a base	p. 609
L.A.	lateral area	p. 609
S.A.	surface area	p. 609
ℓ	slant height	p. 617
V	volume	p. 624

Table 2 Formulas

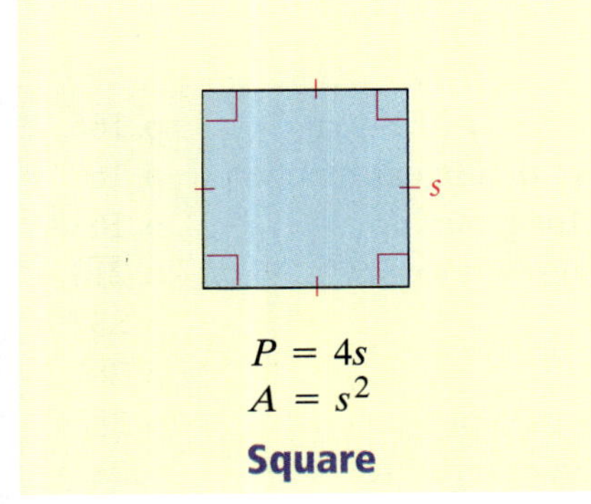

$P = 4s$
$A = s^2$

Square

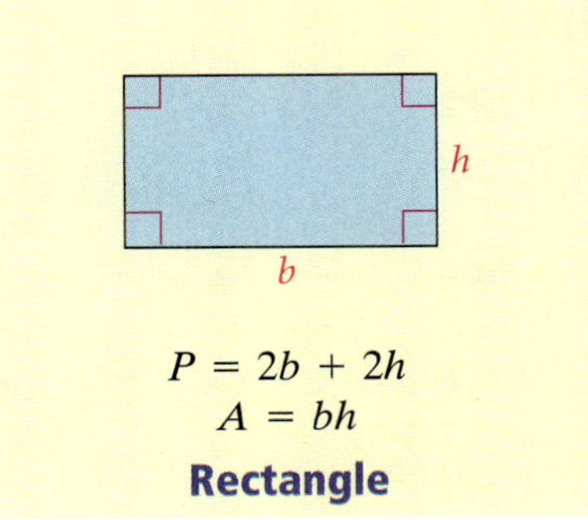

$P = 2b + 2h$
$A = bh$

Rectangle

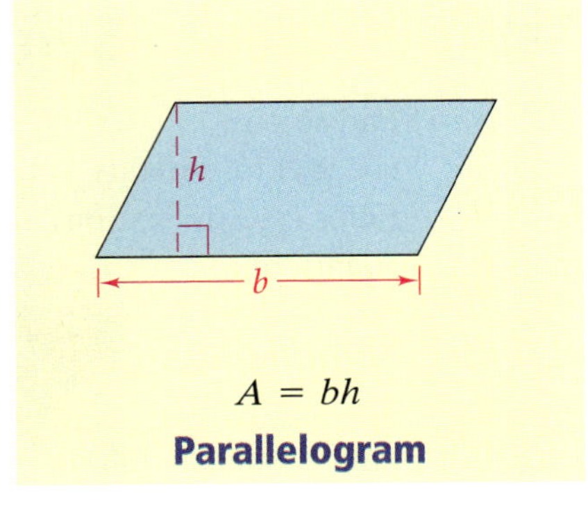

$A = bh$

Parallelogram

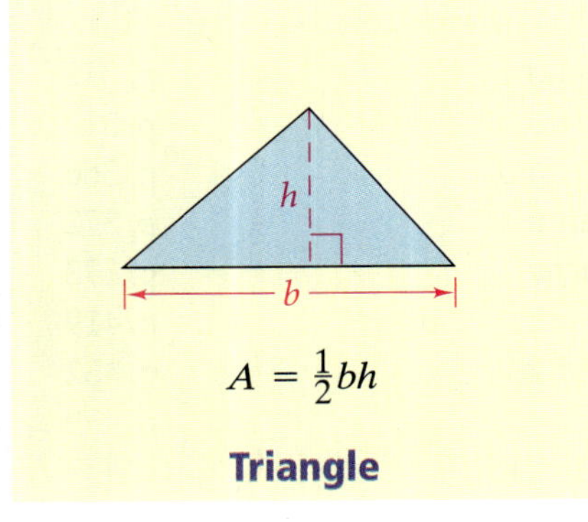

$A = \frac{1}{2}bh$

Triangle

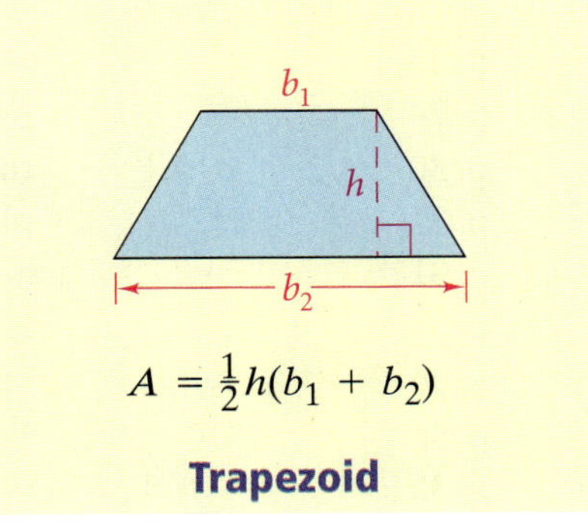

$A = \frac{1}{2}h(b_1 + b_2)$

Trapezoid

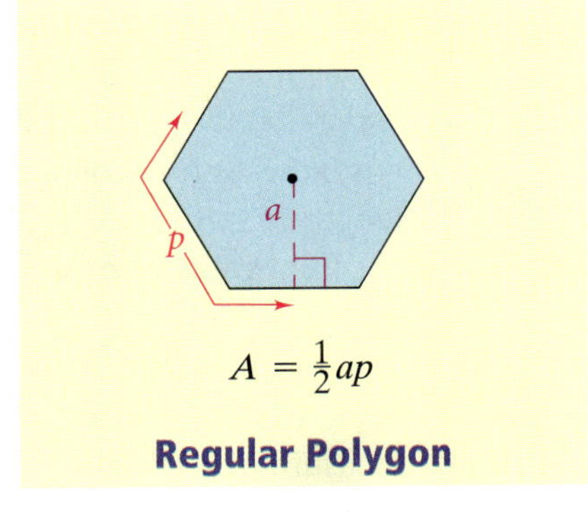

$A = \frac{1}{2}ap$

Regular Polygon

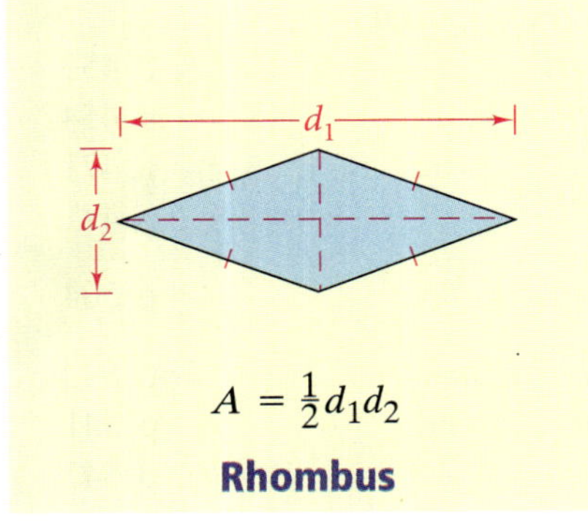

$A = \frac{1}{2}d_1d_2$

Rhombus

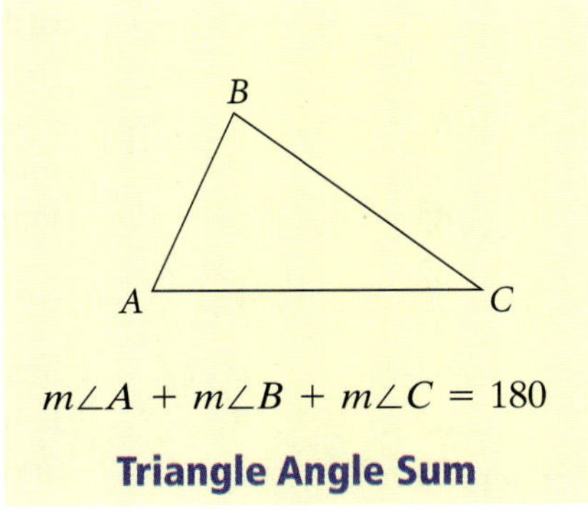

$m\angle A + m\angle B + m\angle C = 180$

Triangle Angle Sum

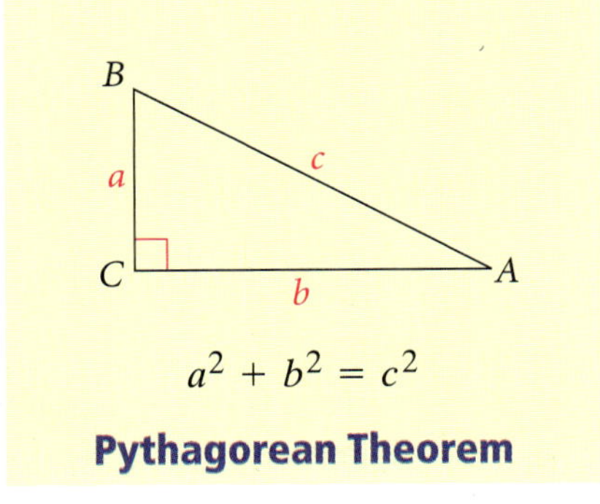

$a^2 + b^2 = c^2$

Pythagorean Theorem

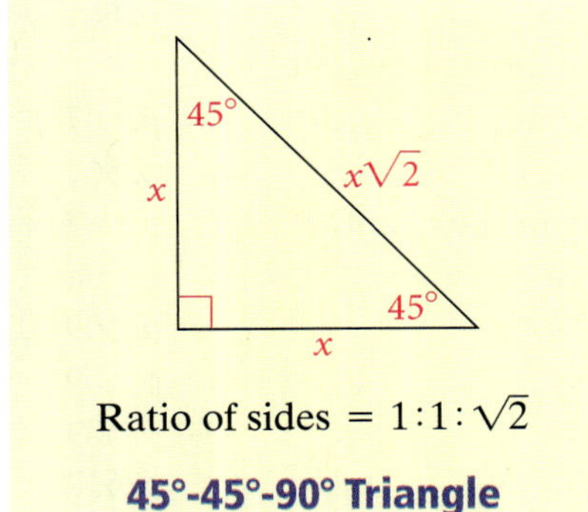

Ratio of sides $= 1:1:\sqrt{2}$

45°-45°-90° Triangle

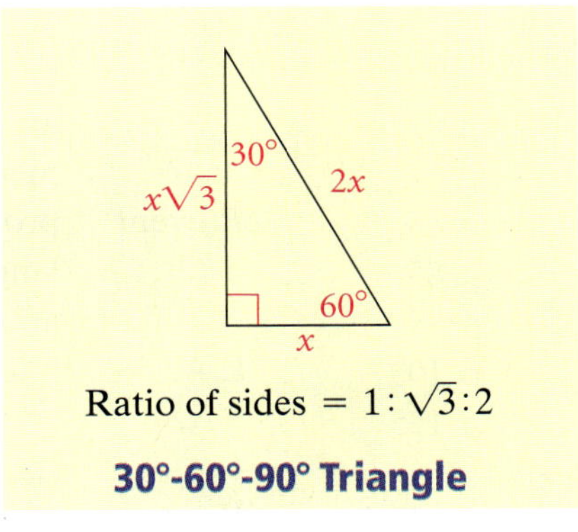

Ratio of sides $= 1:\sqrt{3}:2$

30°-60°-90° Triangle

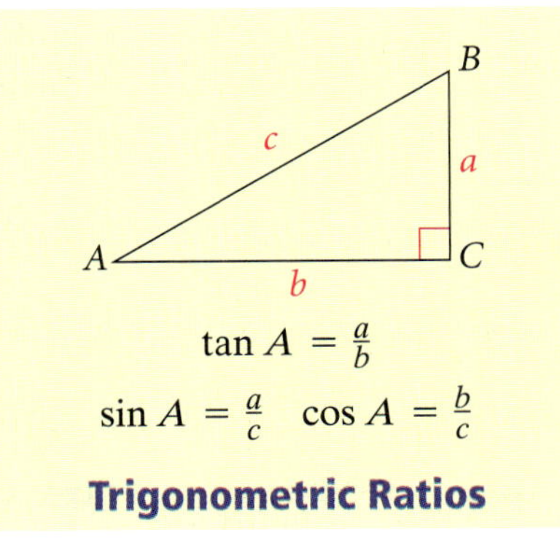

$\tan A = \frac{a}{b}$

$\sin A = \frac{a}{c} \quad \cos A = \frac{b}{c}$

Trigonometric Ratios

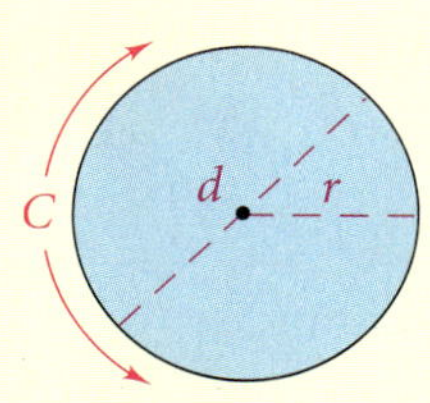

$C = \pi d$ or $C = 2\pi r$
$A = \pi r^2$

Circle

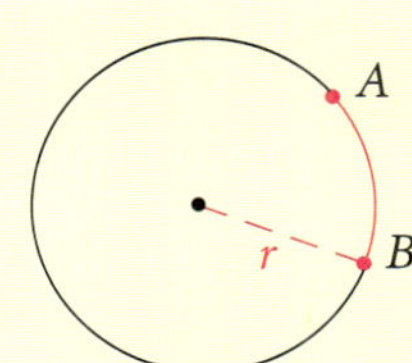

$$\text{Length of } \widehat{AB} = \frac{m\widehat{AB}}{360} \cdot 2\pi r$$

Arc

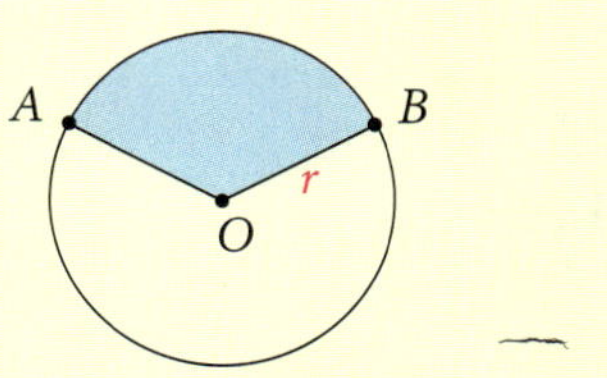

$$\text{Area of sector } AOB = \frac{m\widehat{AB}}{360} \cdot \pi r^2$$

Sector of a Circle

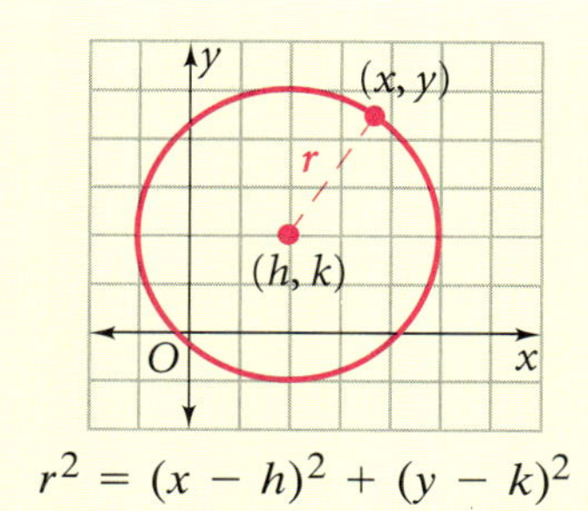

$$r^2 = (x - h)^2 + (y - k)^2$$

Equation of Circle

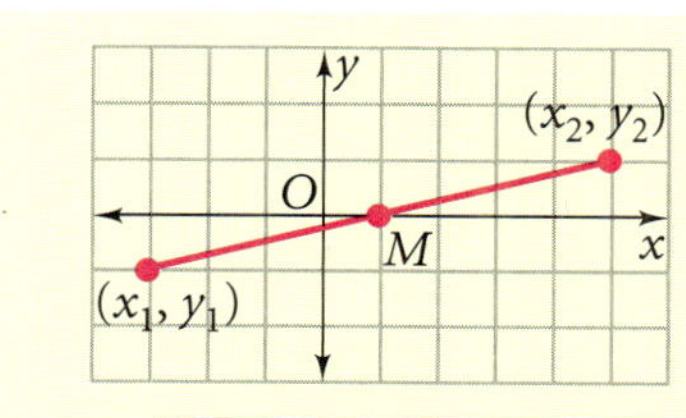

$$d = \sqrt{(x_2 - x_1)^2 + (y_2 - y_1)^2}$$

$$M = \left(\frac{x_1 + x_2}{2}, \frac{y_1 + y_2}{2}\right)$$

Distance and Midpoint

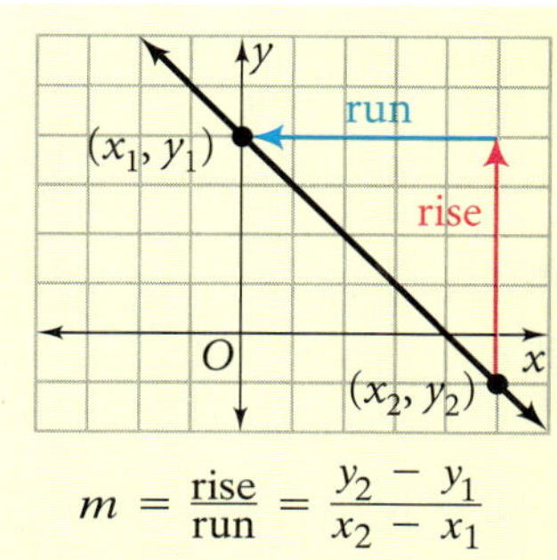

$$m = \frac{\text{rise}}{\text{run}} = \frac{y_2 - y_1}{x_2 - x_1}$$

Slope

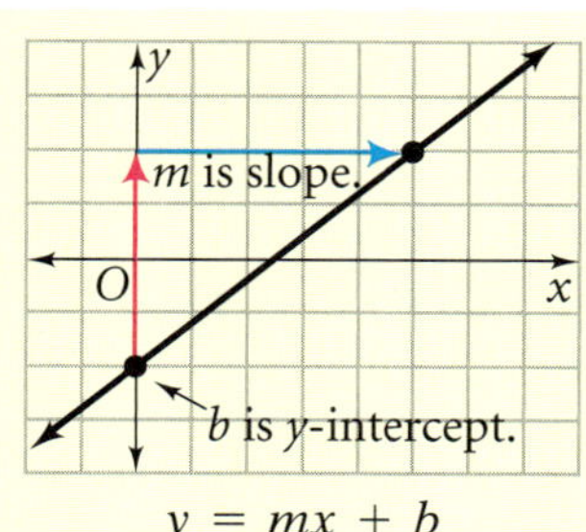

$y = mx + b$

Slope-intercept Form of a Linear Equation

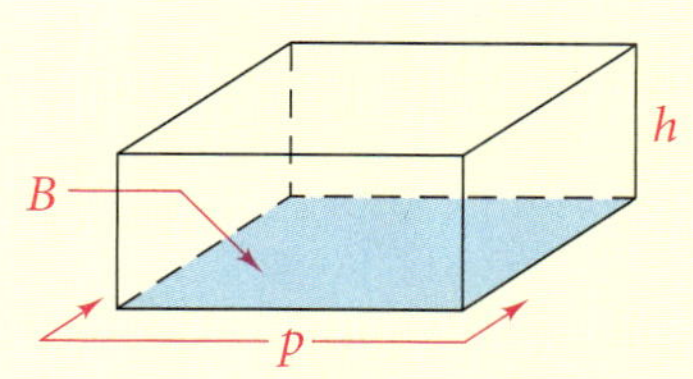

L.A. $= ph$
S.A. $=$ L.A. $+ 2B$
$V = Bh$

Right Prism

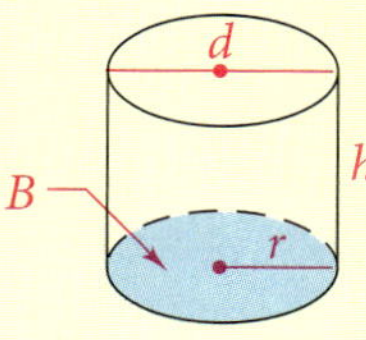

L.A. $= 2\pi rh$ or L.A. $= \pi dh$
S.A. $=$ L.A. $+ 2B$
$V = Bh$ or $V = \pi r^2 h$

Right Cylinder

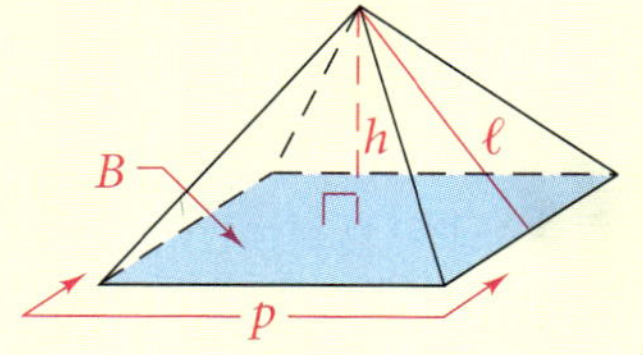

L.A. $= \frac{1}{2}p\ell$
S.A. $=$ L.A. $+ B$
$V = \frac{1}{3}Bh$

Regular Pyramid

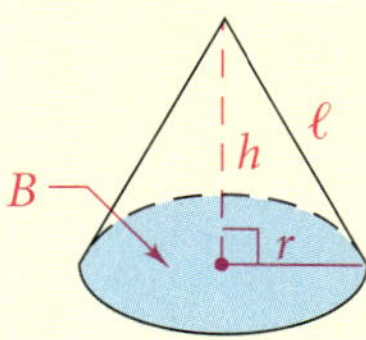

L.A. $= \pi r\ell$
S.A. $=$ L.A. $+ B$
$V = \frac{1}{3}Bh$ or $V = \frac{1}{3}\pi r^2 h$

Right Cone

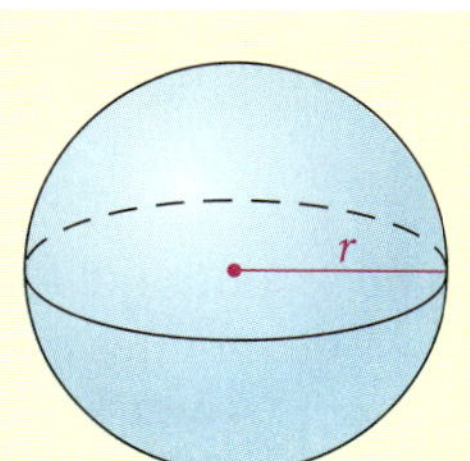

S.A. $= 4\pi r^2$
$V = \frac{4}{3}\pi r^3$

Sphere

Table 3 Measures

United States Customary	Metric
Length	
12 inches (in.) = 1 foot (ft)	10 millimeters (mm) = 1 centimeter (cm)
36 in. = 1 yard (yd)	100 cm = 1 meter (m)
3 ft = 1 yard	1000 mm = 1 meter
5280 ft = 1 mile (mi)	1000 m = 1 kilometer (km)
1760 yd = 1 mile	
Area	
144 square inches (in.^2) = 1 square foot (ft^2)	100 square millimeters (mm^2) = 1 square centimeter (cm^2)
9 ft^2 = 1 square yard (yd^2)	10,000 cm^2 = 1 square meter (m^2)
43,560 ft^2 = 1 acre	10,000 m^2 = 1 hectare (ha)
4840 yd^2 = 1 acre	
Volume	
1728 cubic inches (in.^3) = 1 cubic foot (ft^3)	1000 cubic millimeters (mm^3) = 1 cubic centimeter (cm^3)
27 ft^3 = 1 cubic yard (yd^3)	1,000,000 cm^3 = 1 cubic meter (m^3)
Liquid Capacity	
8 fluid ounces (fl oz) = 1 cup (c)	1000 milliliters (mL) = 1 liter (L)
2 c = 1 pint (pt)	1000 L = 1 kiloliter (kL)
2 pt = 1 quart (qt)	
4 qt = 1 gallon (gal)	
Weight or Mass	
16 ounces (oz) = 1 pound (lb)	1000 milligrams (mg) = 1 gram (g)
2000 pounds = 1 ton (t)	1000 g = 1 kilogram (kg)
	1000 kg = 1 metric ton
Temperature	
32°F = freezing point of water	0°C = freezing point of water
98.6°F = normal body temperature	37°C = normal body temperature
212°F = boiling point of water	100°C = boiling point of water

Time	
60 seconds (s) = 1 minute (min)	365 days = 1 year (yr)
60 minutes = 1 hour (h)	52 weeks (approx.) = 1 year
24 hours = 1 day (d)	12 months = 1 year
7 days = 1 week (wk)	10 years = 1 decade
4 weeks (approx.) = 1 month (mo)	100 years = 1 century

Table 4 Properties of Real Numbers

Unless otherwise stated, a, b, c, and d are real numbers.

Identity Properties

Addition $a + 0 = a$ and $0 + a = a$

Multiplication $a \cdot 1 = a$ and $1 \cdot a = a$

Commutative Properties

Addition $a + b = b + a$

Multiplication $a \cdot b = b \cdot a$

Associative Properties

Addition $(a + b) + c = a + (b + c)$

Multiplication $(a \cdot b) \cdot c = a \cdot (b \cdot c)$

Inverse Properties

Addition

The sum of a number and its *opposite,* or *additive inverse,* is zero.

$a + (-a) = 0$ and $-a + a = 0$

Multiplication

The reciprocal, or multiplicative inverse, of a rational number $\frac{a}{b}$ is $\frac{b}{a}$ $(a, b \neq 0)$.

$a \cdot \frac{1}{a} = 1$ and $\frac{1}{a} \cdot a = 1$ $(a \neq 0)$

Distributive Properties

$a(b + c) = ab + ac$ $(b + c)a = ba + ca$

$a(b - c) = ab - ac$ $(b - c)a = ba - ca$

Properties of Equality

Addition If $a = b$, then $a + c = b + c$.

Subtraction If $a = b$, then $a - c = b - c$.

Multiplication If $a = b$, then $a \cdot c = b \cdot c$.

Division If $a = b$ and $c \neq 0$, then $\frac{a}{c} = \frac{b}{c}$.

Substitution If $a = b$, then b can replace a in any expression.

Reflexive $a = a$

Symmetric If $a = b$, then $b = a$.

Transitive If $a = b$ and $b = c$, then $a = c$.

Properties of Proportions

$\frac{a}{b} = \frac{c}{d}$ $(a, b, c, d \neq 0)$ is equivalent to

(1) $ad = bc$ (2) $\frac{b}{a} = \frac{d}{c}$

(3) $\frac{a}{c} = \frac{b}{d}$ (4) $\frac{a + b}{b} = \frac{c + d}{d}$

Zero-Product Property

If $ab = 0$, then $a = 0$ or $b = 0$.

Properties of Inequality

Addition If $a > b$ and $c \geq d$, then $a + c > b + d$.

Multiplication If $a > b$ and $c > 0$, then $ac > bc$. If $a > b$ and $c < 0$, then $ac < bc$.

Transitive If $a > b$ and $b > c$, then $a > c$.

Comparison If $a = b + c$ and $c > 0$, then $a > b$.

Properties of Exponents

For any nonzero numbers a and b, any positive number c, and any integers m and n,

Zero Exponent $a^0 = 1$

Negative Exponent $a^{-n} = \frac{1}{a^n}$

Product of Powers $a^m \cdot a^n = a^{m+n}$

Quotient of Powers $\frac{a^m}{a^n} = a^{m-n}$

Power to a Power $(c^m)^n = c^{mn}$

Product to a Power $(ab)^n = a^n b^n$

Quotient to a Power $\left(\frac{a}{b}\right)^n = \frac{a^n}{b^n}$

Properties of Square Roots

For any nonnegative numbers a and b, and any positive number c,

Product of Square Roots $\sqrt{a} \cdot \sqrt{b} = \sqrt{ab}$

Quotient of Square Roots $\frac{\sqrt{a}}{\sqrt{c}} = \sqrt{\frac{a}{c}}$

Tables

Table 5 Squares and Square Roots

Number n	Square n^2	Positive Square Root $\sqrt{n}$
1	1	1.000
2	4	1.414
3	9	1.732
4	16	2.000
5	25	2.236
6	36	2.449
7	49	2.646
8	64	2.828
9	81	3.000
10	100	3.162
11	121	3.317
12	144	3.464
13	169	3.606
14	196	3.742
15	225	3.873
16	256	4.000
17	289	4.123
18	324	4.243
19	361	4.359
20	400	4.472
21	441	4.583
22	484	4.690
23	529	4.796
24	576	4.899
25	625	5.000
26	676	5.099
27	729	5.196
28	784	5.292
29	841	5.385
30	900	5.477
31	961	5.568
32	1024	5.657
33	1089	5.745
34	1156	5.831
35	1225	5.916
36	1296	6.000
37	1369	6.083
38	1444	6.164
39	1521	6.245
40	1600	6.325
41	1681	6.403
42	1764	6.481
43	1849	6.557
44	1936	6.633
45	2025	6.708
46	2116	6.782
47	2209	6.856
48	2304	6.928
49	2401	7.000
50	2500	7.071
51	2601	7.141
52	2704	7.211
53	2809	7.280
54	2916	7.348
55	3025	7.416
56	3136	7.483
57	3249	7.550
58	3364	7.616
59	3481	7.681
60	3600	7.746
61	3721	7.810
62	3844	7.874
63	3969	7.937
64	4096	8.000
65	4225	8.062
66	4356	8.124
67	4489	8.185
68	4624	8.246
69	4761	8.307
70	4900	8.367
71	5041	8.426
72	5184	8.485
73	5329	8.544
74	5476	8.602
75	5625	8.660
76	5776	8.718
77	5929	8.775
78	6084	8.832
79	6241	8.888
80	6400	8.944
81	6561	9.000
82	6724	9.055
83	6889	9.110
84	7056	9.165
85	7225	9.220
86	7396	9.274
87	7569	9.327
88	7744	9.381
89	7921	9.434
90	8100	9.487
91	8281	9.539
92	8464	9.592
93	8649	9.644
94	8836	9.695
95	9025	9.747
96	9216	9.798
97	9409	9.849
98	9604	9.899
99	9801	9.950
100	10,000	10.000
101	10,201	10.050
102	10,404	10.100
103	10,609	10.149
104	10,816	10.198
105	11,025	10.247
106	11,236	10.296
107	11,449	10.344
108	11,664	10.392
109	11,881	10.440
110	12,100	10.488
111	12,321	10.536
112	12,544	10.583
113	12,769	10.630
114	12,996	10.677
115	13,225	10.724
116	13,456	10.770
117	13,689	10.817
118	13,924	10.863
119	14,161	10.909
120	14,400	10.954
121	14,641	11.000
122	14,884	11.045
123	15,129	11.091
124	15,376	11.136
125	15,625	11.180
126	15,876	11.225
127	16,129	11.269
128	16,384	11.314
129	16,641	11.358
130	16,900	11.402
131	17,161	11.446
132	17,424	11.489
133	17,689	11.533
134	17,956	11.576
135	18,225	11.619
136	18,496	11.662
137	18,769	11.705
138	19,044	11.747
139	19,321	11.790
140	19,600	11.832
141	19,881	11.874
142	20,164	11.916
143	20,449	11.958
144	20,736	12.000
145	21,025	12.042
146	21,316	12.083
147	21,609	12.124
148	21,904	12.166
149	22,201	12.207
150	22,500	12.247

Table 6 Trigonometric Ratios

Angle	Sine	Cosine	Tangent
1°	0.0175	0.9998	0.0175
2°	0.0349	0.9994	0.0349
3°	0.0523	0.9986	0.0524
4°	0.0698	0.9976	0.0699
5°	0.0872	0.9962	0.0875
6°	0.1045	0.9945	0.1051
7°	0.1219	0.9925	0.1228
8°	0.1392	0.9903	0.1405
9°	0.1564	0.9877	0.1584
10°	0.1736	0.9848	0.1763
11°	0.1908	0.9816	0.1944
12°	0.2079	0.9781	0.2126
13°	0.2250	0.9744	0.2309
14°	0.2419	0.9703	0.2493
15°	0.2588	0.9659	0.2679
16°	0.2756	0.9613	0.2867
17°	0.2924	0.9563	0.3057
18°	0.3090	0.9511	0.3249
19°	0.3256	0.9455	0.3443
20°	0.3420	0.9397	0.3640
21°	0.3584	0.9336	0.3839
22°	0.3746	0.9272	0.4040
23°	0.3907	0.9205	0.4245
24°	0.4067	0.9135	0.4452
25°	0.4226	0.9063	0.4663
26°	0.4384	0.8988	0.4877
27°	0.4540	0.8910	0.5095
28°	0.4695	0.8829	0.5317
29°	0.4848	0.8746	0.5543
30°	0.5000	0.8660	0.5774
31°	0.5150	0.8572	0.6009
32°	0.5299	0.8480	0.6249
33°	0.5446	0.8387	0.6494
34°	0.5592	0.8290	0.6745
35°	0.5736	0.8192	0.7002
36°	0.5878	0.8090	0.7265
37°	0.6018	0.7986	0.7536
38°	0.6157	0.7880	0.7813
39°	0.6293	0.7771	0.8098
40°	0.6428	0.7660	0.8391
41°	0.6561	0.7547	0.8693
42°	0.6691	0.7431	0.9004
43°	0.6820	0.7314	0.9325
44°	0.6947	0.7193	0.9657
45°	0.7071	0.7071	1.0000

Angle	Sine	Cosine	Tangent
46°	0.7193	0.6947	1.0355
47°	0.7314	0.6820	1.0724
48°	0.7431	0.6691	1.1106
49°	0.7547	0.6561	1.1504
50°	0.7660	0.6428	1.1918
51°	0.7771	0.6293	1.2349
52°	0.7880	0.6157	1.2799
53°	0.7986	0.6018	1.3270
54°	0.8090	0.5878	1.3764
55°	0.8192	0.5736	1.4281
56°	0.8290	0.5592	1.4826
57°	0.8387	0.5446	1.5399
58°	0.8480	0.5299	1.6003
59°	0.8572	0.5150	1.6643
60°	0.8660	0.5000	1.7321
61°	0.8746	0.4848	1.8040
62°	0.8829	0.4695	1.8807
63°	0.8910	0.4540	1.9626
64°	0.8988	0.4384	2.0503
65°	0.9063	0.4226	2.1445
66°	0.9135	0.4067	2.2460
67°	0.9205	0.3907	2.3559
68°	0.9272	0.3746	2.4751
69°	0.9336	0.3584	2.6051
70°	0.9397	0.3420	2.7475
71°	0.9455	0.3256	2.9042
72°	0.9511	0.3090	3.0777
73°	0.9563	0.2924	3.2709
74°	0.9613	0.2756	3.4874
75°	0.9659	0.2588	3.7321
76°	0.9703	0.2419	4.0108
77°	0.9744	0.2250	4.3315
78°	0.9781	0.2079	4.7046
79°	0.9816	0.1908	5.1446
80°	0.9848	0.1736	5.6713
81°	0.9877	0.1564	6.3138
82°	0.9903	0.1392	7.1154
83°	0.9925	0.1219	8.1443
84°	0.9945	0.1045	9.5144
85°	0.9962	0.0872	11.4301
86°	0.9976	0.0698	14.3007
87°	0.9986	0.0523	19.0811
88°	0.9994	0.0349	28.6363
89°	0.9998	0.0175	57.2900
90°	1.0000	0.0000	

Postulates, Theorems, and Constructions

Chapter 1: Tools of Geometry

Postulate 1-1
Through any two points there is exactly one line. (p. 18)

Postulate 1-2
If two lines intersect, then they intersect in exactly one point. (p. 18)

Postulate 1-3
If two planes intersect, then they intersect in exactly one line. (p. 18)

Postulate 1-4
Through any three noncollinear points there is exactly one plane. (p. 19)

Postulate 1-5
Ruler Postulate
The points of a line can be put into one-to-one correspondence with the real numbers so that the distance between any two points is the absolute value of the difference of the corresponding numbers. (p. 31)

Postulate 1-6
Segment Addition Postulate
If three points A, B, and C are collinear and B is between A and C, then $AB + BC = AC$. (p. 32)

Postulate 1-7
Protractor Postulate
Let $\overrightarrow{OA}$ and $\overrightarrow{OB}$ be opposite rays in a plane. $\overrightarrow{OA}$, $\overrightarrow{OB}$, and all the rays with endpoint O that can be drawn on one side of $\overleftrightarrow{AB}$ can be paired with the real numbers from 0 to 180 so that

a. $\overrightarrow{OA}$ is paired with 0 and $\overrightarrow{OB}$ is paired with 180.

b. If $\overrightarrow{OC}$ is paired with x and $\overrightarrow{OD}$ is paired with y, then $m\angle COD = |x - y|$. (p. 37)

Postulate 1-8
Angle Addition Postulate
If point B is in the interior of $\angle AOC$, then $m\angle AOB + m\angle BOC = m\angle AOC$.
If $\angle AOC$ is a straight angle, then $m\angle AOB + m\angle BOC = 180$. (p. 38)

The Distance Formula
The distance d between two points $A(x_1, y_1)$ and $B(x_2, y_2)$ is $d = \sqrt{(x_2 - x_1)^2 + (y_2 - y_1)^2}$. (p. 53)
- Proof on p. 421, Exercise 34

The Midpoint Formula
The coordinates of the midpoint M of $\overline{AB}$ with endpoints $A(x_1, y_1)$ and $B(x_2, y_2)$ are the following.
$M = \left(\frac{x_1 + x_2}{2}, \frac{y_1 + y_2}{2}\right)$ (p. 55)

The Distance Formula (Three Dimensions)
In a three-dimensional coordinate system, the distance between two points (x_1, y_1, z_1) and (x_2, y_2, z_2) can be found using this extension of the Distance Formula.
$d=\sqrt{(x_2 - x_1)^2 + (y_2 - y_1)^2 + (z_2 - z_1)^2}$ (p. 58)

Postulate 1-9
If two figures are congruent, then their areas are equal. (p. 64)

Postulate 1-10
The area of a region is the sum of the areas of its nonoverlapping parts. (p. 64)

Chapter 2: Reasoning and Proof

Law of Detachment
If a conditional is true and its hypothesis is true, then its conclusion is true. In symbolic form: If $p \rightarrow q$ is a true statement and p is true, then q is true. (p. 95)

Law of Syllogism
If $p \rightarrow q$ and $q \rightarrow r$ are true statements, then $p \rightarrow r$ is a true statement. (p. 95)

Properties of Congruence
Reflexive Property
$\overline{AB} \cong \overline{AB}$ and $\angle A \cong \angle A$
Symmetric Property
If $\overline{AB} \cong \overline{CD}$, then $\overline{CD} \cong \overline{AB}$.
If $\angle A \cong \angle B$, then $\angle B \cong \angle A$.
Transitive Property
If $\overline{AB} \cong \overline{CD}$ and $\overline{CD} \cong \overline{EF}$, then $\overline{AB} \cong \overline{EF}$.
If $\angle A \cong \angle B$ and $\angle B \cong \angle C$, then $\angle A \cong \angle C$. (p. 105)

Theorem 2-1
Vertical Angles Theorem
Vertical angles are congruent. (p. 110)
- Proof on p. 111

Theorem 2-2
Congruent Supplements Theorem
If two angles are supplements of the same angle (or of congruent angles), then the two angles are congruent. (p. 111)
- Proofs on p. 112, Example 2; p. 114, Exercise 27

Theorem 2-3
Congruent Complements Theorem
If two angles are complements of the same angle (or of congruent angles), then the two angles are congruent. (p. 112)
- Proofs on p. 113, Exercise 7; p. 114, Exercise 28

Theorem 2-4
All right angles are congruent. (p. 112)
- Proof on p. 113, Exercise 14

Theorem 2-5
If two angles are congruent and supplementary, then each is a right angle. (p. 112)
- Proof on p. 114, Exercise 21

Chapter 3: Parallel and Perpendicular Lines

Postulate 3-1
Corresponding Angles Postulate
If a transversal intersects two parallel lines, then corresponding angles are congruent. (p. 128)

Theorem 3-1
Alternate Interior Angles Theorem
If a transversal intersects two parallel lines, then alternate interior angles are congruent. (p. 128)
- Proof on p. 129

Theorem 3-2
Same-Side Interior Angles Theorem
If a transversal intersects two parallel lines, then same-side interior angles are supplementary. (p. 128)
- Proof on p. 132, Exercise 29

Theorem 3-3
Alternate Exterior Angles Theorem
If a transversal intersects two parallel lines, then alternate exterior angles are congruent. (p. 130)
- Proof on p. 129, Example 3

Theorem 3-4
Same-Side Exterior Angles Theorem
If a transversal intersects two parallel lines, then same-side exterior angles are supplementary. (p. 130)
- Proof on p. 129, Quick Check 3

Postulate 3-2
Converse of the Corresponding Angles Postulate
If two lines and a transversal form corresponding angles that are congruent, then the two lines are parallel. (p. 134)

Theorem 3-5
Converse of the Alternate Interior Angles Theorem
If two lines and a transversal form alternate interior angles that are congruent, then the two lines are parallel. (p. 135)
- Proof on p. 135

Theorem 3-6
Converse of the Same-Side Interior Angles Theorem
If two lines and a transversal form same-side interior angles that are supplementary, then the two lines are parallel. (p. 135)
- Proofs on p. 138, Exercise 22 and p. 139, Exercise 40

Theorem 3-7
Converse of the Alternate Exterior Angles Theorem
If two lines and a transversal form alternate exterior angles that are congruent, then the lines are parallel. (p. 136)
- Proof on p. 136

Theorem 3-8
Converse of the Same-Side Exterior Angles Theorem
If two lines and a transversal form same-side exterior angles that are supplementary, then the lines are parallel. (p. 136)
- Proof on p. 138, Exercise 27

Theorem 3-9
If two lines are parallel to the same line, then they are parallel to each other. (p. 141)
- Proofs on p. 143, Exercise 3; p. 179, Exercise 37

Theorem 3-10
In a plane, if two lines are perpendicular to the same line, then they are parallel to each other. (p. 141)
- Proofs on p. 141; p. 143, Exercise 12; p. 179, Exercise 38

Theorem 3-11
In a plane, if a line is perpendicular to one of two parallel lines, then it is also perpendicular to the other. (p. 142)
- Proof on p. 143, Exercise 11

Theorem 3-12

Triangle Angle-Sum Theorem

The sum of the measures of the angles of a triangle is 180. (p. 147)

- Proof on p. 147

Theorem 3-13

Triangle Exterior Angle Theorem

The measure of each exterior angle of a triangle equals the sum of the measures of its two remote interior angles. (p. 149)

- Proof on p. 152, Exercise 35

  Corollary

  The measure of an exterior angle of a triangle is greater than the measure of each of its remote interior angles. (p. 290)

  - Proof on p. 290

Parallel Postulate

Through a point not on a line, there is one and only one line parallel to the given line. (p. 154)

Spherical Geometry Parallel Postulate

Through a point not on a line, there is no line parallel to the given line. (p. 154)

Theorem 3-14

Polygon Angle-Sum Theorem

The sum of the measures of the angles of an n-gon is $(n - 2)180$. (p. 159)

- Proof on p. 163, Exercise 54

Theorem 3-15

Polygon Exterior Angle-Sum Theorem

The sum of the measures of the exterior angles of a polygon, one at each vertex, is 360. (p. 160)

- Proofs on p. 156 (using a computer) and p. 162, Exercise 46

Slopes of Parallel Lines

If two nonvertical lines are parallel, their slopes are equal. If the slopes of two distinct nonvertical lines are equal, the lines are parallel. Any two vertical lines are parallel. (p. 174)

- Proofs on pp. 387–388, Exercises 42, 43

Slopes of Perpendicular Lines

If two nonvertical lines are perpendicular, the product of their slopes is -1. If the slopes of two lines have a product of -1, the lines are perpendicular. Any horizontal line and vertical line are perpendicular. (p. 175)

- Proofs on p. 346, Exercise 29 and p. 353, Exercise 41

Chapter 4: Congruent Triangles

Theorem 4-1

If the two angles of one triangle are congruent to two angles of another triangle, then the third angles are congruent. (p. 199)

- Proof on p. 202, Exercise 45

Postulate 4-1

Side-Side-Side (SSS) Postulate

If the three sides of one triangle are congruent to the three sides of another triangle, then the two triangles are congruent. (p. 205)

Postulate 4-2

Side-Angle-Side (SAS) Postulate

If two sides and the included angle of one triangle are congruent to two sides and the included angle of another triangle, then the two triangles are congruent. (p. 206)

Postulate 4-3

Angle-Side-Angle (ASA) Postulate

If two angles and the included side of one triangle are congruent to two angles and the included side of another triangle, then the two triangles are congruent. (p. 213)

Theorem 4-2

Angle-Angle-Side (AAS) Theorem

If two angles and a nonincluded side of one triangle are congruent to two angles and the corresponding nonincluded side of another triangle, then the triangles are congruent. (p. 214)

- Proof on p. 214

Theorem 4-3

Isosceles Triangle Theorem

If two sides of a triangle are congruent, then the angles opposite those sides are congruent. (p. 228)

- Proofs on p. 229; p. 231, Exercise 15

  Corollary

  If a triangle is equilateral, then the triangle is equiangular. (p. 230)

  - Proof on p. 231, Exercise 18

Theorem 4-4

Converse of the Isosceles Triangle Theorem

If two angles of a triangle are congruent, then the sides opposite the angles are congruent. (p. 228)

- Proof on p. 231, Exercise 16

  Corollary

  If a triangle is equiangular, then the triangle is equilateral. (p. 230)

  - Proof on p. 231, Exercise 18

Theorem 4-5
The bisector of the vertex angle of an isosceles triangle is the perpendicular bisector of the base. (p. 228)
- Proof on p. 232, Exercise 29

Theorem 4-6
Hypotenuse-Leg (HL) Theorem
If the hypotenuse and a leg of one right triangle are congruent to the hypotenuse and a leg of another right triangle, then the triangles are congruent. (p. 235)
- Proof on p. 235

Chapter 5: Relationships Within Triangles

Theorem 5-1
Triangle Midsegment Theorem
If a segment joins the midpoints of two sides of a triangle, then the segment is parallel to the third side, and is half its length. (p. 260)
- Proof on p. 260

Theorem 5-2
Perpendicular Bisector Theorem
If a point is on the perpendicular bisector of a segment, then it is equidistant from the endpoints of the segment. (p. 265)
- Proof on p. 269, Exercise 40

Theorem 5-3
Converse of the Perpendicular Bisector Theorem
If a point is equidistant from the endpoints of a segment, then it is on the perpendicular bisector of the segment. (p. 265)
- Proof on p. 269, Exercise 41

Theorem 5-4
Angle Bisector Theorem
If a point is on the bisector of an angle, then the point is equidistant from the sides of the angle. (p. 266)
- Proof on p. 269, Exercise 43

Theorem 5-5
Converse of the Angle Bisector Theorem
If a point in the interior of an angle is equidistant from the sides of the angle, then the point is on the angle bisector. (p. 266)
- Proof on p. 269, Exercise 44

Theorem 5-6
The perpendicular bisectors of the sides of a triangle are concurrent at a point equidistant from the vertices. (p. 273)
- Proof on p. 277, Exercise 29

Theorem 5-7
The bisectors of the angles of a triangle are concurrent at a point equidistant from the sides. (p. 273)
- Proof on p. 277, Exercise 30

Theorem 5-8
The medians of a triangle are concurrent at a point that is two thirds the distance from each vertex to the midpoint of the opposite side. (p. 274)
- Proof on p. 352, Exercise 35

Theorem 5-9
The lines that contain the altitudes of a triangle are concurrent. (p. 275)
- Proof on p. 352, Exercise 36

Comparison Property of Inequality
If $a = b + c$ and $c > 0$, then $a > b$. (p. 289)
- Proof on p. 289

Theorem 5-10
If two sides of a triangle are not congruent, then the larger angle lies opposite the longer side. (p. 290)
- Proof on p. 294, Exercise 33

Theorem 5-11
If two angles of a triangle are not congruent, then the longer side lies opposite the larger angle. (p. 291)
- Proof on p. 291

Theorem 5-12
Triangle Inequality Theorem
The sum of the lengths of any two sides of a triangle is greater than the length of the third side. (p. 292)
- Proof on p. 294, Exercise 41

Chapter 6: Quadrilaterals

Theorem 6-1
Opposite sides of a parallelogram are congruent. (p. 312)
- Proofs on p. 312; p. 317, Exercise 35

Theorem 6-2
Opposite angles of a parallelogram are congruent. (p. 313)
- Proof on p. 317, Exercise 36

Theorem 6-3
The diagonals of a parallelogram bisect each other. (p. 314)
- Proofs on p. 314; p. 350, Exercise 2

Theorem 6-4
If three (or more) parallel lines cut off congruent segments on one transversal, then they cut off congruent segments on every transversal. (p. 315)
- Proof on p. 318, Exercise 52

Theorem 6-5
If both pairs of opposite sides of a quadrilateral are congruent, then the quadrilateral is a parallelogram. (p. 321)
- Proof on p. 321

Theorem 6-6
If both pairs of opposite angles of a quadrilateral are congruent, then the quadrilateral is a parallelogram. (p. 321)
- Proof on p. 325, Exercise 12

Theorem 6-7
If the diagonals of a quadrilateral bisect each other, then the quadrilateral is a parallelogram. (p. 322)
- Proof on p. 322

Theorem 6-8
If one pair of opposite sides of a quadrilateral are both congruent and parallel, then the quadrilateral is a parallelogram. (p. 322)
- Proof on p. 325, Exercise 18

Theorem 6-9
Each diagonal of a rhombus bisects two angles of the rhombus. (p. 329)
- Proof on p. 329

Theorem 6-10
The diagonals of a rhombus are perpendicular. (p. 330)
- Proof on p. 333, Exercise 22

Theorem 6-11
The diagonals of a rectangle are congruent. (p. 330)
- Proof on p. 330

Theorem 6-12
If one diagonal of a parallelogram bisects two angles of the parallelogram, then the parallelogram is a rhombus. (p. 331)
- Proof on p. 334, Exercise 39

Theorem 6-13
If the diagonals of a parallelogram are perpendicular, then the parallelogram is a rhombus. (p. 331)
- Proof on p. 333, Exercise 23

Theorem 6-14
If the diagonals of a parallelogram are congruent, then the parallelogram is a rectangle. (p. 331)
- Proof on p. 334, Exercise 40

Theorem 6-15
The base angles of an isosceles trapezoid are congruent. (p. 336)
- Proof on p. 340, Exercise 38

Theorem 6-16
The diagonals of an isosceles trapezoid are congruent. (p. 337)
- Proofs on p. 337; p. 350, Exercise 3

Theorem 6-17
The diagonals of a kite are perpendicular. (p. 338)
- Proof on p. 338

Theorem 6-18
Trapezoid Midsegment Theorem

(1) The midsegment of a trapezoid is parallel to the bases.

(2) The length of a midsegment of a trapezoid is half the sum of the lengths of the bases. (p. 348)

- Proof on p. 349, Quick Check 1

Chapter 7: Similarity

Postulate 7-1
Angle-Angle Similarity (AA ~) Postulate

If two angles of one triangle are congruent to two angles of another triangle, then the triangles are similar. (p. 382)

Theorem 7-1
Side-Angle-Side Similarity (SAS ~) Theorem

If an angle of one triangle is congruent to an angle of a second triangle, and the sides including the two angles are proportional, then the triangles are similar. (p. 383)
- Proof on p. 383

Theorem 7-2
Side-Side-Side Similarity (SSS ~) Theorem

If the corresponding sides of two triangles are proportional, then the triangles are similar. (p. 383)
- Proof on p. 383

Theorem 7-3

The altitude to the hypotenuse of a right triangle divides the triangle into two triangles that are similar to the original triangle and to each other. (p. 392)

- Proof on p. 392

Corollary 1

The length of the altitude to the hypotenuse of a right triangle is the geometric mean of the lengths of the segments of the hypotenuse. (p. 392)

- Proof on p. 392

Corollary 2

The altitude to the hypotenuse of a right triangle separates the hypotenuse so that the length of each leg of the triangle is the geometric mean of the length of the adjacent hypotenuse segment and the length of the hypotenuse. (p. 393)

- Proof on p. 393

Theorem 7-4

Side-Splitter Theorem

If a line is parallel to one side of a triangle and intersects the other two sides, then it divides those sides proportionally. (p. 398)

- Proof on p. 398

Converse

If a line divides two sides of a triangle proportionally, then it is parallel to the third side.

- Proof on p. 402, Exercise 34

Corollary

If three parallel lines intersect two transversals, then the segments intercepted on the transversals are proportional. (p. 399)

- Proof on p. 402, Exercise 35

Theorem 7-5

Triangle-Angle-Bisector Theorem

If a ray bisects an angle of a triangle, then it divides the opposite side into two segments that are proportional to the other two sides of the triangle. (p. 400)

- Proof on p. 400

Chapter 8: Right Triangles and Trigonometry

Theorem 8-1

Pythagorean Theorem

In a right triangle, the sum of the squares of the lengths of the legs is equal to the square of the length of the hypotenuse.

$a^2 + b^2 = c^2$ (p. 417)

- Proofs on p. 395, Exercise 53; p. 416; p. 422, Exercise 48; p. 545; p. 692, Exercise 36

Theorem 8-2

Converse of the Pythagorean Theorem

If the square of the length of one side of a triangle is equal to the sum of the squares of the lengths of the other two sides, then the triangle is a right triangle. (p. 419)

- Proof on p. 423, Exercise 58

Theorem 8-3

If the square of the length of the longest side of a triangle is greater than the sum of the squares of the lengths of the other two sides, the triangle is obtuse. (p. 419)

Theorem 8-4

If the square of the length of the longest side of a triangle is less than the sum of the squares of the lengths of the other two sides, the triangle is acute. (p. 419)

Theorem 8-5

45°-45°-90° Triangle Theorem

In a 45°-45°-90° triangle, both legs are congruent and the length of the hypotenuse is $\sqrt{2}$ times the length of a leg.

hypotenuse $= \sqrt{2} \cdot$ leg (p. 425)

- Proof on p. 425

Theorem 8-6

30°-60°-90° Triangle Theorem

In a 30°-60°-90° triangle, the length of the hypotenuse is twice the length of the shorter leg. The length of the longer leg is $\sqrt{3}$ times the length of the shorter leg.

hypotenuse $= 2 \cdot$ shorter leg

longer leg $= \sqrt{3} \cdot$ shorter leg (p. 426)

- Proof on p. 427

Chapter 9: Transformations

Theorem 9-1

A translation or rotation is a composition of two reflections. (p. 506)

Theorem 9-2

A composition of reflections across two parallel lines is a translation. (p. 507)

Theorem 9-3

A composition of reflections across two intersecting lines is a rotation. (p. 507)

Theorem 9-4

Fundamental Theorem of Isometries

In a plane, one of two congruent figures can be mapped onto the other by a composition of at most three reflections. (p. 508)

Theorem 9-5

Isometry Classification Theorem

There are only four isometries. They are reflection, translation, rotation, and glide reflection. (p. 509)

Theorem 9-6

Every triangle tessellates. (p. 516)

Theorem 9-7

Every quadrilateral tessellates. (p. 516)

Chapter 10: Area

Theorem 10-1

Area of a Rectangle

The area of a rectangle is the product of its base and height.

$A = bh$ (p. 534)

Theorem 10-2

Area of a Parallelogram

The area of a parallelogram is the product of a base and the corresponding height.

$A = bh$ (p. 534)

Theorem 10-3

Area of a Triangle

The area of a triangle is half the product of a base and the corresponding height.

$A = \frac{1}{2}bh$ (p. 535)

Theorem 10-4

Area of a Trapezoid

The area of a trapezoid is half the product of the height and the sum of the bases.

$A = \frac{1}{2}h(b_1 + b_2)$ (p. 540)

Theorem 10-5

Area of a Rhombus or a Kite

The area of a rhombus or a kite is half the product of the lengths of its diagonals.

$A = \frac{1}{2}d_1d_2$ (p. 541)

Theorem 10-6

Area of a Regular Polygon

The area of a regular polygon is half the product of the apothem and the perimeter.

$A = \frac{1}{2}ap$ (p. 547)

Theorem 10-7

Perimeters and Areas of Similar Figures

If the similarity ratio of two similar figures is $\frac{a}{b}$, then

(1) the ratio of their perimeters is $\frac{a}{b}$ and

(2) the ratio of their areas is $\frac{a^2}{b^2}$. (p. 554)

Theorem 10-8

Area of a Triangle Given SAS

The area of a triangle is one half the product of the lengths of two sides and the sine of the included angle.

Area of $\triangle ABC = \frac{1}{2}bc(\sin A)$ (p. 561)

- Proof on p. 560

Postulate 10-1

Arc Addition Postulate

The measure of the arc formed by two adjacent arcs is the sum of the measures of the two arcs. (p. 567)

Theorem 10-9

Circumference of a Circle

The circumference of a circle is π times the diameter.

$C = \pi d$ or $C = 2\pi r$ (p. 568)

Theorem 10-10

Arc Length

The length of an arc of a circle is the product of the ratio $\frac{\text{measure of the arc}}{360}$ and the circumference of the circle.

length of $\widehat{AB} = \frac{m\widehat{AB}}{360} \cdot 2\pi r$ (p. 569)

Theorem 10-11

Area of a Circle

The area of a circle is the product of π and the square of the radius. $A = \pi r^2$ (p. 576)

Theorem 10-12

Area of a Sector of a Circle

The area of a sector of a circle is the product of the ratio $\frac{\text{measure of the arc}}{360}$ and the area of the circle.

Area of sector $AOB = \frac{m\widehat{AB}}{360} \cdot \pi r^2$ (p. 576)

Chapter 11: Surface Area and Volume

Theorem 11-1

Lateral and Surface Areas of a Prism

The lateral area of a right prism is the product of the perimeter of the base and the height.

L.A. $= ph$

The surface area of a right prism is the sum of the lateral area and the areas of the two bases.

S.A. $=$ L.A. $+ 2B$ (p. 610)

Theorem 11-2

Lateral and Surface Areas of a Cylinder

The lateral area of a right cylinder is the product of the circumference of the base and the height of the cylinder.

L.A. $= 2\pi rh$, or L.A. $= \pi dh$

The surface area of a right cylinder is the sum of the lateral area and the areas of the two bases.

S.A. $=$ L.A. $+ 2B$, or S.A. $= 2\pi rh + 2\pi r^2$ (p. 610)

Theorem 11-3

Lateral and Surface Areas of a Regular Pyramid

The lateral area of a regular pyramid is half the product of the perimeter of the base and the slant height.

$\text{L.A.} = \frac{1}{2}p\ell$

The surface area of a regular pyramid is the sum of the lateral area and the area of the base.

$\text{S.A.} = \text{L.A.} + B$ (p. 618)

Theorem 11-4

Lateral and Surface Areas of a Cone

The lateral area of a right cone is half the product of the circumference of the base and the slant height.

$\text{L.A.} = \frac{1}{2} \cdot 2\pi r\ell$, or $\text{L.A.} = \pi r\ell$

The surface area of a right cone is the sum of the lateral area and the area of the base.

$\text{S.A.} = \text{L.A.} + B$ (p. 619)

Theorem 11-5

Cavalieri's Principle

If two space figures have the same height and the same cross-sectional area at every level, then they have the same volume. (p. 625)

Theorem 11-6

Volume of a Prism

The volume of a prism is the product of the area of a base and the height of the prism.

$V = Bh$ (p. 625)

Theorem 11-7

Volume of a Cylinder

The volume of a cylinder is the product of the area of the base and the height of the cylinder.

$V = Bh$, or $V = \pi r^2 h$ (p. 626)

Theorem 11-8

Volume of a Pyramid

The volume of a pyramid is one third the product of the area of the base and the height of the pyramid.

$V = \frac{1}{3}Bh$ (p. 632)

Theorem 11-9

Volume of a Cone

The volume of a cone is one third the product of the area of the base and the height of the cone.

$V = \frac{1}{3}Bh$, or $V = \frac{1}{3}\pi r^2 h$ (p. 633)

Theorem 11-10

Surface Area of a Sphere

The surface area of a sphere is four times the product of π and the square of the radius of the sphere.

$\text{S.A.} = 4\pi r^2$ (p. 638)

Theorem 11-11

Volume of a Sphere

The volume of a sphere is four thirds the product of π and the cube of the radius of the sphere.

$V = \frac{4}{3}\pi r^3$ (p. 640)

Theorem 11-12

Areas and Volumes of Similar Solids

If the similarity ratio of two similar solids is $a : b$, then (1) the ratio of their corresponding areas is $a^2 : b^2$, and (2) the ratio of their volumes is $a^3 : b^3$. (p. 647)

Chapter 12: Circles

Theorem 12-1

If a line is tangent to a circle, then the line is perpendicular to the radius drawn to the point of tangency. (p. 662)

- Proof on p. 667, Exercise 36

Theorem 12-2

If a line in the plane of a circle is perpendicular to a radius at its endpoint on the circle, then the line is tangent to the circle. (p. 663)

- Proof on p. 667, Exercise 37

Theorem 12-3

The two segments tangent to a circle from a point outside the circle are congruent. (p. 664)

- Proof on p. 667, Exercise 38

Theorem 12-4

Within a circle or in congruent circles

(1) Congruent central angles have congruent chords.

(2) Congruent chords have congruent arcs.

(3) Congruent arcs have congruent central angles. (p. 670)

- Proofs on p. 674, Exercises 23, 24; p. 675, Exercise 35

Theorem 12-5

Within a circle or in congruent circles

(1) Chords equidistant from the center are congruent.

(2) Congruent chords are equidistant from the center. (p. 671)

- Proofs on p. 671; p. 675, Exercise 37

Theorem 12-6

In a circle, a diameter that is perpendicular to a chord bisects the chord and its arcs. (p. 672)

- Proof on p. 674, Exercise 25

Theorem 12-7

In a circle, a diameter that bisects a chord (that is not a diameter) is perpendicular to the chord. (p. 672)

- Proof on p. 672

Theorem 12-8

In a circle, the perpendicular bisector of a chord contains the center of the circle. (p. 672)

- Proof on p. 678, Exercise 36

Theorem 12-9

Inscribed Angle Theorem

The measure of an inscribed angle is half the measure of its intercepted arc. (p. 679)

- Proofs on p. 679; p. 683, Exercises 40, 41

Corollary 1

Two inscribed angles that intercept the same arc are congruent. (p. 680)

- Proof on p. 684, Exercise 42

Corollary 2

An angle inscribed in a semicircle is a right angle. (p. 680)

- Proof on p. 684, Exercise 43

Corollary 3

The opposite angles of a quadrilateral inscribed in a circle are supplementary. (p. 680)

- Proof on p. 684, Exercise 44

Theorem 12-10

The measure of an angle formed by a tangent and a chord is half the measure of the intercepted arc. (p. 680)

- Proof on p. 684, Exercise 45

Theorem 12-11

The measure of an angle formed by two lines that
(1) intersect inside a circle is half the sum of the measures of the intercepted arcs.
(2) intersect outside a circle is half the difference of the measures of the intercepted arcs. (p. 687)

- Proofs on p. 688; p. 692, Exercises 29, 30

Theorem 12-12

For a given point and circle, the product of the lengths of the two segments from the point to the circle is constant along any line through the point and circle. (p. 689)

- Proofs on p. 689; p. 692, Exercises 31–33

Theorem 12-13

An equation of a circle with center (h, k) and radius r is $(x - h)^2 + (y - k)^2 = r^2$. (p. 695)

Constructions

Construction 1

Congruent Segments

Construct a segment congruent to a given segment. (p. 44)

Construction 2

Congruent Angles

Construct an angle congruent to a given angle. (p. 45)

Construction 3

Perpendicular Bisector

Construct the perpendicular bisector of a segment. (p. 46)

Construction 4

Angle Bisector

Construct the bisector of an angle. (p. 47)

Construction 5

Parallel Through a Point Not on a Line

Construct a line parallel to a given line and through a given point that is not on the line. (p. 181)

Construction 6

Perpendicular Through a Point on a Line

Construct the perpendicular to a given line at a given point on the line. (p. 182)

Construction 7

Perpendicular Through a Point Not on a Line

Construct the perpendicular to a given line through a given point not on the line. (p. 183)

English/Spanish Illustrated Glossary

EXAMPLES

Acute angle (p. 37) An acute angle is an angle whose measure is between 0 and 90.

Ángulo agudo (p. 37) Un ángulo agudo es un ángulo que mide entre 0 y 90 grados.

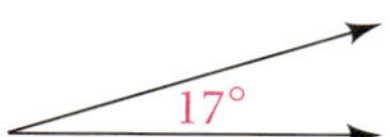

Acute triangle (p. 148) An acute triangle has three acute angles.

Triángulo acutángulo (p. 148) Un triángulo acutángulo tiene los tres ángulos agudos.

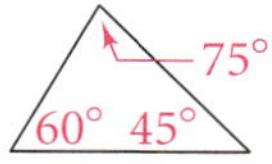

Adjacent angles (p. 38) Adjacent angles are two coplanar angles that have a common side and a common vertex but no common interior points.

Ángulos adyacentes (p. 38) Los ángulos adyacentes son dos ángulos coplanarios que tienen un lado común y el mismo vértice, pero no tienen puntos interiores comunes.

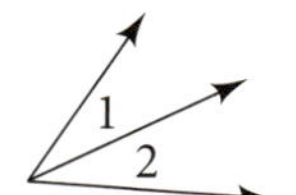

$\angle 1$ and $\angle 2$ are adjacent.

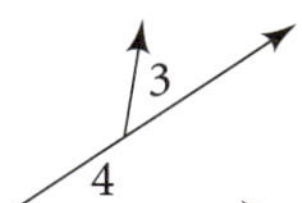

$\angle 3$ and $\angle 4$ are *not* adjacent.

Adjacent arcs (p. 567) Adjacent arcs are on the same circle and have exactly one point in common.

Arcos adyacentes (p. 567) Los arcos adyacentes están en el mismo círculo y tienen exactamente un punto en común.

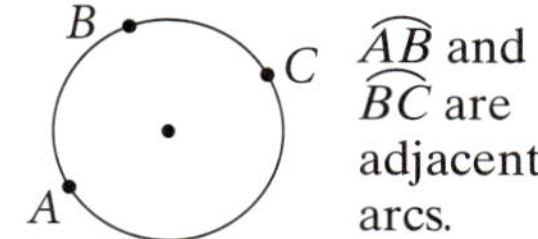

$\widehat{AB}$ and $\widehat{BC}$ are adjacent arcs.

Alternate interior (exterior) angles (pp. 127, 129) Alternate interior (exterior) angles are nonadjacent interior (exterior) angles that lie on opposite sides of the transversal.

Ángulos alternos internos (externos) (pp. 127, 129) Los ángulos alternos internos (externos) son ángulos internos (externos) no adyacentes situados en lados opuestos de la transversal.

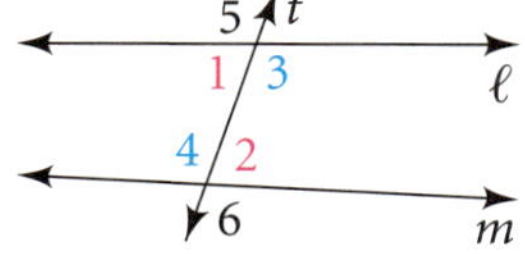

$\angle 1$ and $\angle 2$ are alternate interior angles, as are $\angle 3$ and $\angle 4$. $\angle 5$ and $\angle 6$ are alternate exterior angles.

Altitude *See* **cone; cylinder; parallelogram; prism; pyramid; trapezoid; triangle.**

Altura *Ver* **cone; cylinder; parallelogram; prism; pyramid; trapezoid; triangle.**

Altitude of a triangle (p. 275) An altitude of a triangle is a perpendicular segment from a vertex to the line containing the side opposite that vertex.

Altura de un triángulo (p. 275) Una altura de un triángulo es el segmento perpendicular que va desde un vértice hasta la recta que contiene el lado opuesto a ese vértice.

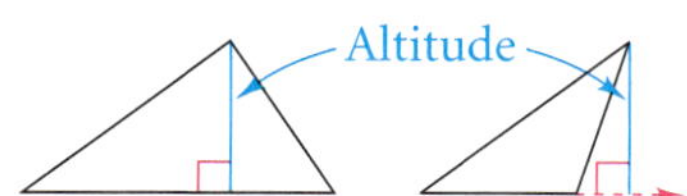

English/Spanish Glossary

EXAMPLES

Angle (p. 36) An angle is formed by two rays with the same endpoint. The rays are the *sides* of the angle and the common endpoint is the *vertex* of the angle.

Ángulo (p. 36) Un ángulo está formado por dos rayos que convergen en un mismo extremo. Los rayos son los *lados* del ángulo y los extremos en común son el vértice.

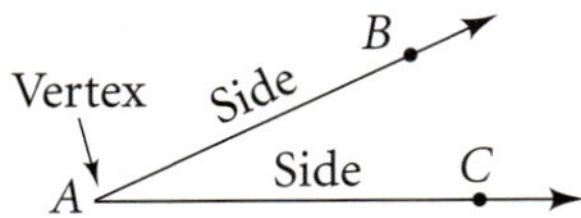

This angle could be named $\angle A$, $\angle BAC$, or $\angle CAB$.

Angle bisector (p. 46) An angle bisector is a ray that divides an angle into two congruent angles.

Bisectriz de un ángulo (p. 46) La bisectriz de un ángulo es un rayo que divide al ángulo en dos ángulos congruentes.

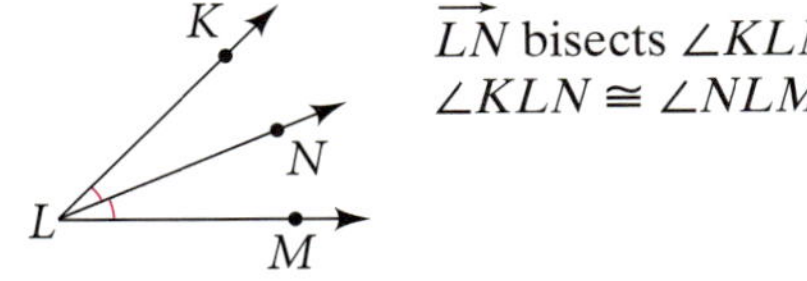

$\overrightarrow{LN}$ bisects $\angle KLM$.
$\angle KLN \cong \angle NLM$.

Angle of elevation or depression (p. 445) An angle of elevation (depression) is the angle formed by a horizontal line and the line of sight to an object above (below) the horizontal line.

Ángulo de elevación o depresión (p. 445) Un ángulo de elevación (depresión) es el ángulo formado por una línea horizontal y la recta que va de esa línea a un objeto situado arriba (debajo) de ella.

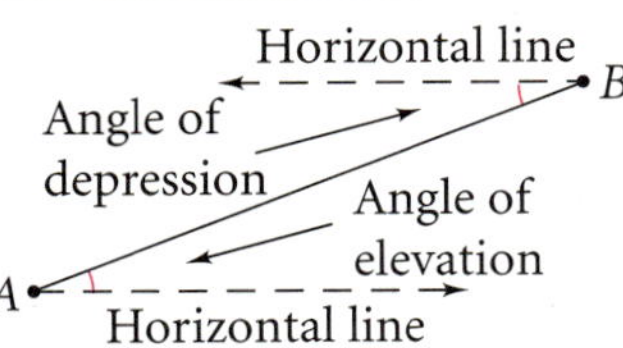

Apothem (p. 546) The apothem of a regular polygon is the distance from the center to a side.

Apotema (p. 546) La apotema de un polígono regular es la distancia desde el centro hasta un lado.

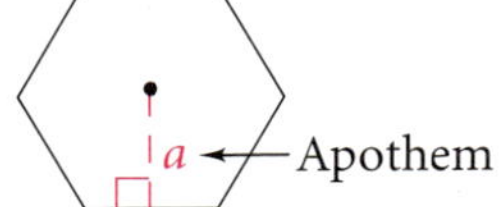

Arc *See* **major arc; minor arc.** *See also* **arc length; measure of an arc.**

Arco *Ver* **major arc; minor arc.** *Ver también* **arc length; measure of an arc.**

Arc length (p. 569) The length of an arc of a circle is the product of the ratio $\frac{\text{measure of the arc}}{360}$ and the circumference of the circle.

Longitud de un arco (p. 569) La longitud del arco de un círculo es el producto del cociente $\frac{\text{medida del arco}}{360}$ por la circunferencia del círculo.

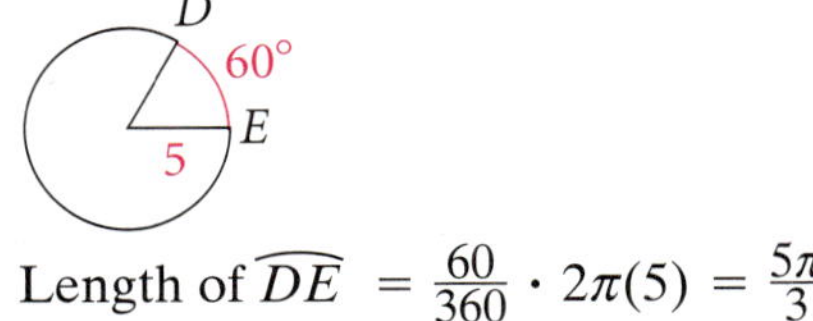

Length of $\widehat{DE} = \frac{60}{360} \cdot 2\pi(5) = \frac{5\pi}{3}$

Area (pp. 534, 535, 540, 541, 547, 576) The area of a plane figure is the number of square units enclosed by the figure. A list of area formulas is on pp. 764–765.

Área (pp. 534, 535, 540, 541, 547, 576) El área de una figura plana es la cantidad de unidades cuadradas que contiene la figura. Una lista de fórmulas para calcular áreas está en las págs. 764–765.

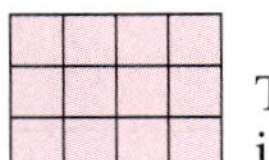

The area of the rectangle is 12 square units, or 12 units^2.

EXAMPLES

Axes (p. 53) *See* **coordinate plane.**

Ejes (p. 53) *Ver* **coordinate plane.**

Axiom (p. 18) *See* **postulate.**

Axioma (p. 18) *Ver* **postulate.**

B

Base(s) *See* **cone; cylinder; isosceles triangle; parallelogram; prism; pyramid; trapezoid; triangle.**

Base(s) *Ver* **cone; cylinder; isosceles triangle; parallelogram; prism; pyramid; trapezoid; triangle.**

Base angles *See* **trapezoid; isosceles triangle.**

Ángulos de base *Ver* **trapezoid; isosceles triangle.**

Biconditional (p. 87) A biconditional statement is the combination of a conditional statement and its converse. A biconditional contains the words "if and only if."

Bicondicional (p. 87) Un enunciado bicondicional es la combinación de un enunciado condicional y su recíproco. El enunciado bicondicional incluye las palabras "si y solo si".

This biconditional statement is true: Two angles are congruent *if and only if* they have the same measure.

Bisector *See* **segment bisector; angle bisector.**

Bisectriz *Ver* **segment bisector; angle bisector.**

C

Center *See* **circle; dilation; regular polygon; sphere.**

Centro *Ver* **circle; dilation; regular polygon; sphere.**

Central angle of a circle (p. 566) A central angle of a circle is an angle whose vertex is the center of the circle.

Ángulo central de un círculo (p. 566) Un ángulo central de un círculo es un ángulo cuyo vértice es el centro del círculo.

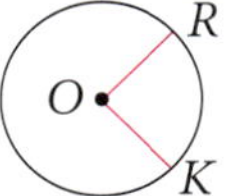

$\angle ROK$ is a central angle of $\odot O$.

English/Spanish Glossary

EXAMPLES

Central angle of a regular polygon (p. 559) A central angle of a regular polygon is an angle formed by two consecutive radii.

Ángulo central de un polígono regular (p. 559) Un ángulo central de un polígono regular es el ángulo formado por dos radios consecutivos.

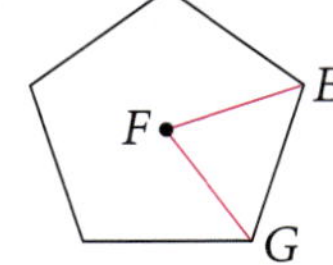

$\angle EFG$ is a central angle of the regular pentagon.

Centroid (p. 274) The centroid of a triangle is the point of intersection of the medians of that triangle.

EXAMPLE P is the centroid of $\triangle ABC$.

Centroide (p. 274) El centroide de un triángulo es el punto de intersección de las medianas del triángulo.

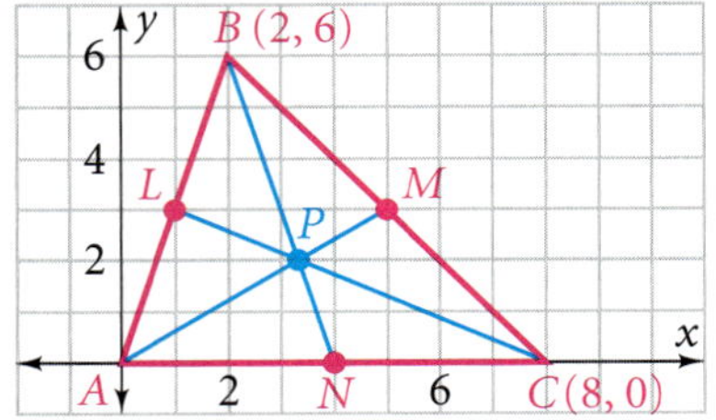

Chord (p. 680) A chord of a circle is a segment whose endpoints are on the circle.

Cuerda (p. 680) Una cuerda de un círculo es un segmento cuyos extremos son dos puntos del círculo.

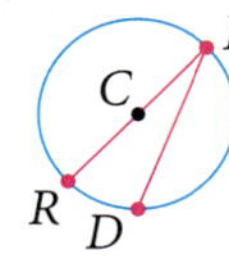

$\overline{HD}$ and $\overline{HR}$ are chords of $\odot C$.

Circle (pp. 566, 695) A circle is the set of all points in a plane that are a given distance, the *radius*, from a given point, the *center*. The standard form for an equation of a circle with center (h, k) and radius r is $(x - h)^2 + (y - k)^2 = r^2$.

Círculo (pp. 566, 695) Un círculo es el conjunto de todos los puntos de un plano situados a una distancia dada, el *radio*, de un punto dado, el *centro*. La fórmula normal de la ecuación de un círculo con centro (h, k) y radio r es $(x - h)^2 + (y - k)^2 = r^2$.

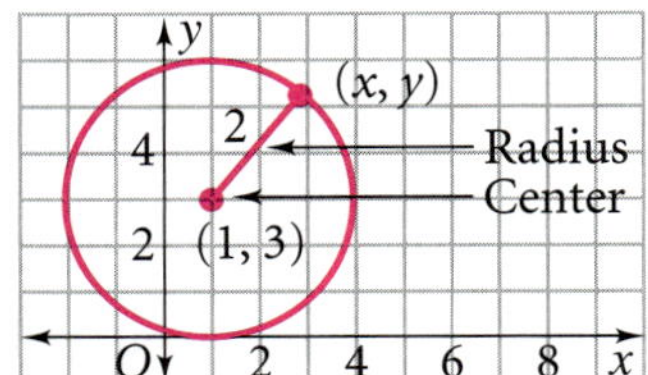

The equation of the circle whose center is $(1, 3)$ and whose radius is 2 is $(x - 1)^2 + (y - 3)^2 = 4$.

Circumcenter (p. 273) A circumcenter is the point of concurrency of the perpendicular bisectors of the sides of a triangle.

Circuncentro (p. 273) El circuncentro es el punto de intersección de las mediatrices perpendiculares a los lados de un triángulo.

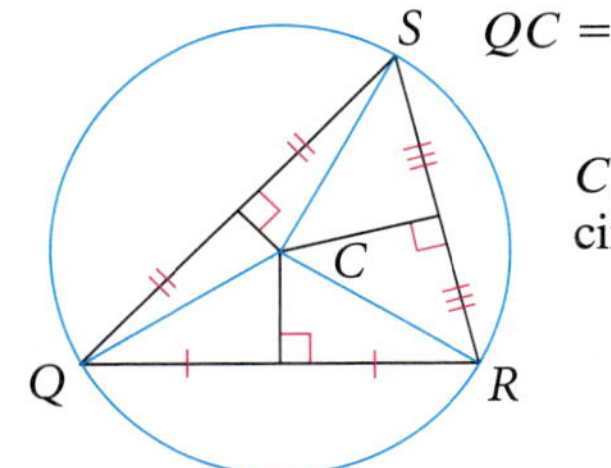

$QC = SC = RC$

C is the circumcenter.

Circumference (p. 568) The circumference of a circle is the distance around the circle. Given the radius r of a circle, you can find its circumference C by using the formula $C = 2\pi r$.

Circunferencia (p. 568) La circunferencia de un círculo es la distancia alrededor del círculo. Dado el radio r de un círculo, se puede hallar la circunferencia C usando la fórmula $C = 2\pi r$.

$C = 2\pi r$
$= 2\pi(4)$
$= 8\pi$

4

Circumference is the distance around the circle.

EXAMPLES

Circumference of a sphere (p. 638) *See* **sphere.**

Circunferencia de una esfera (p. 638) *Ver* **sphere.**

Circumscribed about (pp. 273, 664) A circle is circumscribed about a polygon if the vertices of the polygon are on the circle. A polygon is circumscribed about a circle if all the sides of the polygon are tangent to the circle.

Circunscrito en (pp. 273, 664) Un círculo está circunscrito en un polígono si los vértices del polígono están en el círculo. Un polígono está circunscrito en un círculo si todos los lados del polígono son tangentes al círculo.

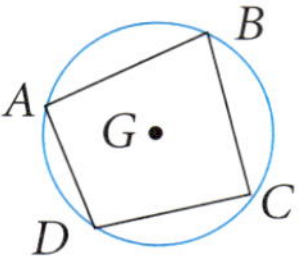

$\odot G$ is circumscribed about $ABCD$.

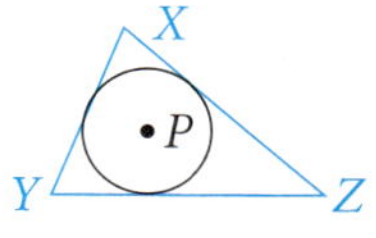

$\triangle XYZ$ is circumscribed about $\odot P$.

Collinear points (p. 17) Collinear points lie on the same line.

Puntos colineales (p. 17) Los puntos colineales son los que están sobre la misma recta.

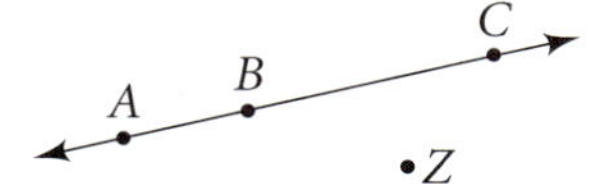

Points A, B, and C are collinear, but points A, B, and Z are noncollinear.

Compass (p. 44) A compass is a geometric tool used to draw circles and parts of circles, called arcs.

Compás (p. 44) El compás es un instrumento usado para dibujar círculos y partes de círculos, llamados arcos.

Complementary angles (p. 38) Two angles are complementary angles if the sum of their measures is 90.

Ángulos complementarios (p. 38) Dos ángulos son complementarios si la suma de sus medidas es igual a 90 grados.

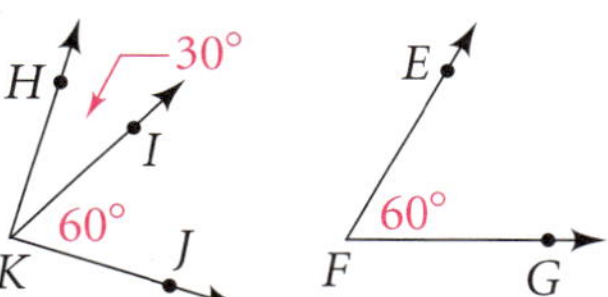

$\angle HKI$ and $\angle IKJ$ are complementary angles, as are $\angle HKI$ and $\angle EFG$.

Composite space figures (p. 627) A composite space figure is the combination of two or more figures into one object.

Figuras geométricas compuestas (p. 627) Una figura geométrica compuesta es la combinación de dos o más figuras en un mismo objeto.

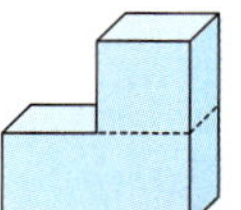

Composition of transformations (p. 472) A composition of two transformations is a transformation in which a second transformation is performed on the image of a first transformation.

Composición de transformaciones (p. 472) Una composición de dos transformaciones es una transformación en la cual una segunda transformación se realiza a partir de la imagen de la primera.

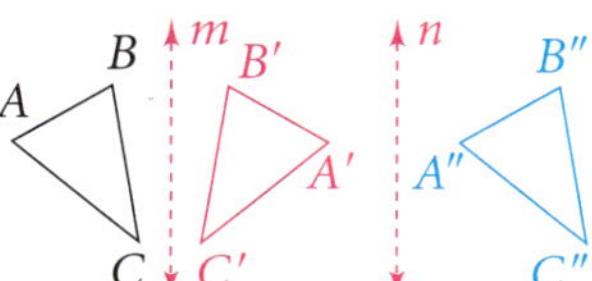

If you reflect $\triangle ABC$ across line m to get $\triangle A'B'C'$ and then reflect $\triangle A'B'C'$ across line n to get $\triangle A''B''C''$, you perform a composition of transformations.

English/Spanish Glossary

EXAMPLES

Concave polygon (p. 158) *See* **polygon.**

Polígono cóncavo (p. 158) *Ver* **polygon.**

Concentric circles (p. 568) Concentric circles lie in the same plane and have the same center.

Círculos concéntricos (p. 568) Los círculos concéntricos están en el mismo plano y tienen el mismo centro.

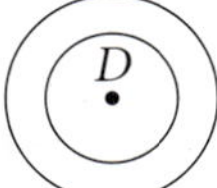

The two circles both have center D and are therefore concentric.

Conclusion (p. 80) The conclusion is the part of an *if-then* statement (conditional) that follows *then*.

Conclusión (p. 80) La conclusión es lo que sigue a la palabra *entonces* en un enunciado (condicional), *si..., entonces....*

In the statement, "If it rains, then I will go outside," the *conclusion* is "I will go outside."

Concurrent lines (p. 273) Concurrent lines are three or more lines that meet in one point. The point at which they meet is the *point of concurrency*.

Rectas concurrentes (p. 273) Las rectas concurrentes son tres o más rectas que se unen en un punto. El punto en que se unen es el *punto de concurrencia.*

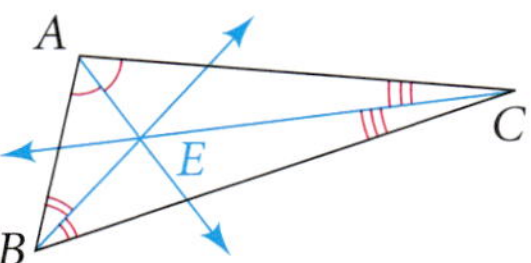

Point E is the point of concurrency of the bisectors of the angles of $\triangle ABC$. The bisectors are concurrent.

Conditional (p. 80) A conditional is an *if-then* statement.

Condicional (p. 80) Un enunciado condicional es del tipo *si..., entonces...*

If you act politely, *then* you will earn respect.

Cone (p. 619) A cone is a three-dimensional figure that has a circular *base,* a *vertex* not in the plane of the circle, and a curved lateral surface, as shown in the diagram. The *altitude* of a cone is the perpendicular segment from the vertex to the plane of the base. The *height* is the length of the altitude. In a *right cone,* the altitude contains the center of the base. The *slant height* of a right cone is the distance from the vertex to the edge of the base.

Cono (p. 619) Un cono es una figura tridimensional que tiene una *base* circular, un *vértice* que no está en el plano del círculo y una superficie lateral curvada (indicada en el diagrama). La *altura* de un cono es el segmento perpendicular desde el vértice hasta el plano de la base. La *altura*, por extensión, es la longitud de la altura. Un *cono recto* es un cono cuya altura contiene el centro de la base. La *longitud de la generatriz* de un cono recto es la distancia desde el vértice hasta el borde de la base.

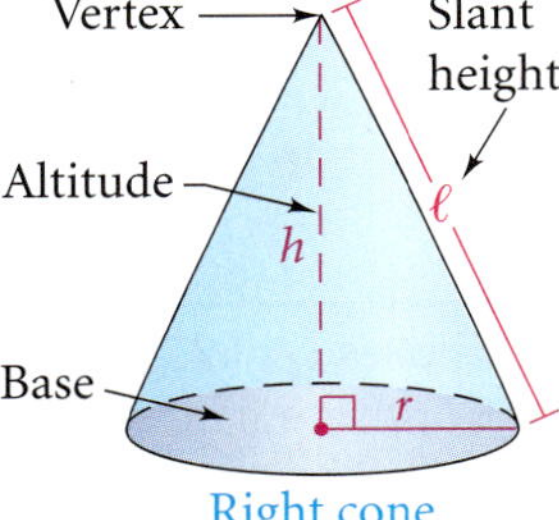

Right cone

Congruence transformation (p. 470) *See* **isometry.**

Transformación de congruencia (p. 470) *Ver* **isometry.**

EXAMPLES

Congruent angles (p. 37) Congruent angles are angles that have the same measure.

Ángulos congruentes (p. 37) Los ángulos congruentes son ángulos que tienen la misma medida.

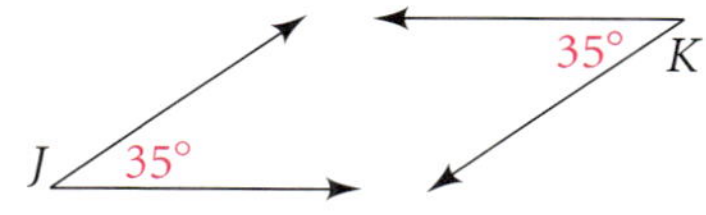

$m\angle J = m\angle K$, so $\angle J \cong \angle K$.

Congruent arcs (p. 569) Congruent arcs are arcs that have the same measure and are in the same circle or congruent circles.

Arcos congruentes (p. 569) Arcos congruentes son arcos que tienen la misma medida y están en el mismo círculo o en círculos congruentes.

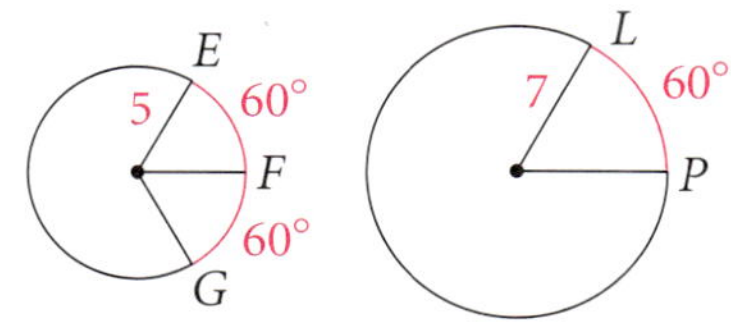

$\widehat{EF} \cong \widehat{FG}$ $\widehat{EF} \ncong \widehat{LP}$

Congruent circles (p. 566) Congruent circles are circles whose radii are congruent.

Círculos congruentes (p. 566) Los círculos congruentes son círculos cuyos radios son congruentes.

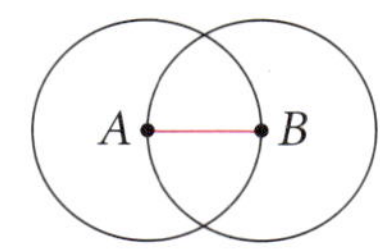

$\odot A$ and $\odot B$ have the same radius, so $\odot A \cong \odot B$.

Congruent polygons (p. 198) Congruent polygons are polygons that have corresponding sides congruent and corresponding angles congruent.

Polígonos congruentes (p. 198) Los polígonos congruentes son polígonos cuyos lados correspondientes son congruentes y cuyos ángulos correspondientes son congruentes.

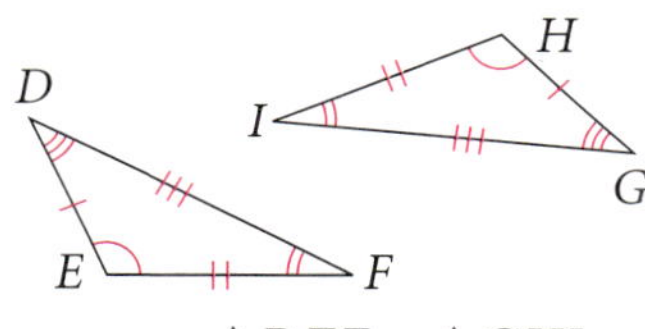

$\triangle DEF \cong \triangle GHI$

Congruent segments (p. 31) Congruent segments are segments that have the same length.

Segmentos congruentes (p. 31) Los segmentos congruentes son segmentos que tienen la misma longitud.

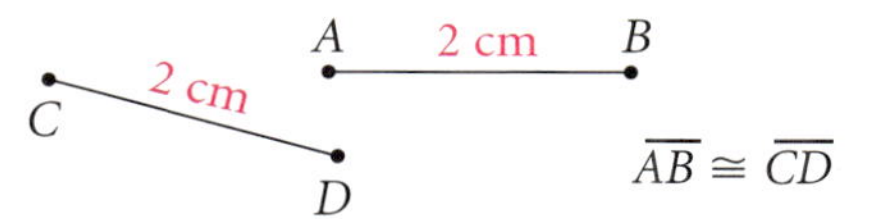

$\overline{AB} \cong \overline{CD}$

Conjecture (p. 5) A conjecture is a conclusion reached by using inductive reasoning.

Conjetura (p. 5) Una conjetura es una conclusión obtenida usando el razonamiento inductivo.

As you walk down the street, you see many people holding unopened umbrellas. You conjecture that the forecast must call for rain.

Consecutive angles (p. 312) Consecutive angles of a polygon share a common side.

Ángulos consecutivos (p. 312) Los ángulos consecutivos de un polígono tienen un lado común.

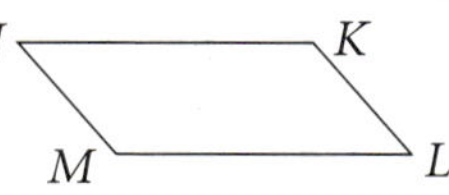

In $\square JKLM$, $\angle J$ and $\angle M$ are consecutive angles, as are $\angle J$ and $\angle K$. $\angle J$ and $\angle L$ are *not* consecutive.

EXAMPLES

Construction (p. 44) A construction is a geometric figure made with only a straightedge and compass.

Construcción (p. 44) Una construcción es una figura geométrica trazada solamente con una regla sin graduación y un compás.

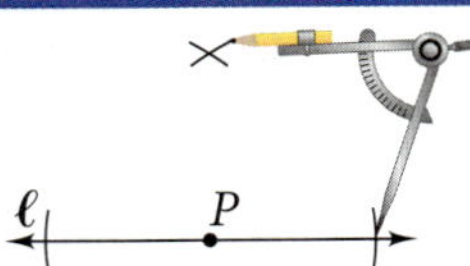

The diagram shows the construction (in progress) of a line perpendicular to a line ℓ through a point P on ℓ.

Contrapositive (p. 280) The contrapositive of the conditional "if p, then q" is the conditional "if not q, then not p." A conditional and its contrapositive always have the same truth value.

Contrapositivo (p. 280) El contrapositivo del condicional "si p, entonces q" es el condicional "si no q, entonces no p." Un condicional y su contrapositivo siempre tienen el mismo valor verdadero.

Conditional: If a figure is a triangle, then it is a polygon.
Contrapositive: If a figure is not a polygon, then it is not a triangle.

Converse (p. 81) The converse of the conditional "if p, then q" is the conditional "if q, then p."

Recíproco (p. 81) El recíproco del condicional "si p, entonces q" es el condicional "si q, entonces p".

Conditional: If you live in Cheyenne, then you live in Wyoming.
Converse: If you live in Wyoming, then you live in Cheyenne.

Convex polygon (p. 158) *See* **polygon.**

Polígono convexo (p. 158) *Ver* **polygon.**

Coordinate(s) of a point (pp. 31, 53) The coordinate of a point is its distance and direction from the origin of a number line. The coordinates of a point on a coordinate plane are in the form (x, y), where x is the x-coordinate and y is the y-coordinate.

Coordenada(s) de un punto (pp. 31, 53) La coordenada de un punto es su distancia y dirección desde el origen en una recta numérica. Las coordenadas de un punto en un plano de coordenadas se expresan como (x, y), donde x es la coordenada x, e y es la coordenada y.

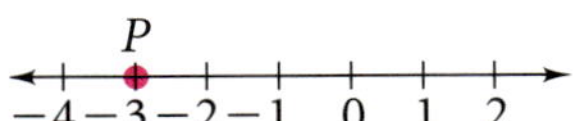

The coordinate of P is -3.

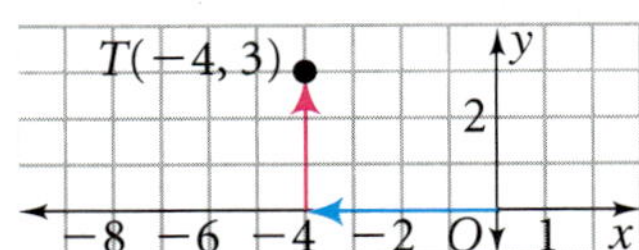

The coordinates of T are $(-4, 3)$.

Coordinate plane (p. 53) The coordinate plane is formed by two number lines, called the *axes*, intersecting at right angles. The x-axis is the horizontal axis, and the y-axis is the vertical axis. The two axes meet at the *origin*, $O(0, 0)$. The axes divide the plane into four *quadrants*.

Plano de coordenadas (p. 53) El plano de coordenadas se forma con dos rectas numéricas, llamadas *ejes*, que se cortan en ángulos rectos. El eje x es el eje horizontal y el eje y es el eje vertical. Los dos ejes se unen en el *origen*, $O(0, 0)$. Los ejes dividen el plano de coordenadas en cuatro *cuadrantes*.

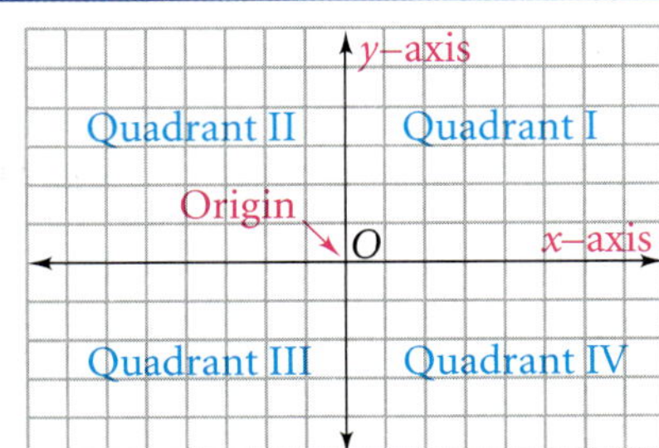

EXAMPLES

Coordinate proof (p. 260) *See* **proof.**

Prueba de coordenadas (p. 260) *Ver* **proof.**

Coplanar figures (p. 17) Coplanar figures are figures in the same plane.

Figuras coplanarias (p. 17) Las figuras coplanarias son las figuras que están localizadas en el mismo plano.

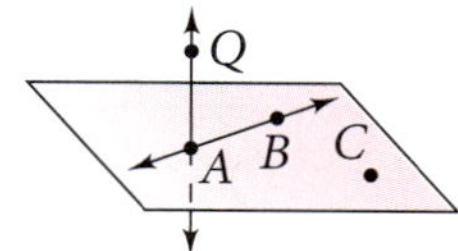

Point C and $\overleftrightarrow{AB}$ are coplanar but points A, B, C, and Q are noncoplanar.

Corollary (p. 229) A corollary is a statement that follows directly from a theorem.

Corolario (p. 229) Un corolario es un enunciado que procede directamente de un teorema.

Theorem: If two sides of a triangle are congruent, then the angles opposite those sides are congruent.
Corollary: If a triangle is equilateral, then it is equiangular.

Corresponding angles (p. 127) Corresponding angles lie on the same side of the transversal t and in corresponding positions relative to ℓ and m.

Ángulos correspondientes (p. 127) Los ángulos correspondientes están en el mismo lado de la transversal t y en las correspondientes posiciones relativas a ℓ y m.

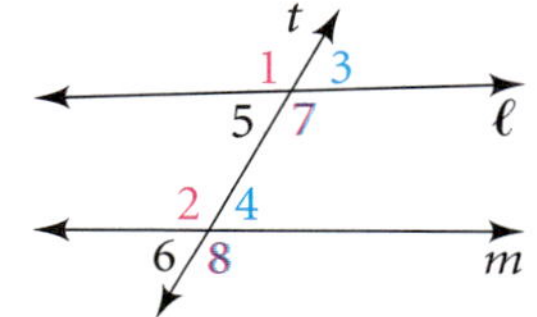

$\angle 1$ and $\angle 2$ are corresponding angles, as are $\angle 3$ and $\angle 4$, $\angle 5$ and $\angle 6$, and $\angle 7$ and $\angle 8$.

Cosine ratio (p. 439) *See* **trigonometric ratios.**

Razón coseno (p. 439) *Ver* **trigonometric ratios.**

Counterexample (pp. 5, 81) A counterexample to a statement is a particular example or instance of the statement that makes the statement false.

Contraejemplo (pp. 5, 81) Un contraejemplo de un enunciado es un ejemplo particular o caso que demuestra que el enunciado no es verdadero.

Statement: If the name of a state begins with W, then that state does not border an ocean.
Counterexample: Washington

CPCTC (p. 221) CPCTC is an abbreviation for "corresponding parts of congruent triangles are congruent."

EXAMPLE By the SAS Congruence Postulate, $\triangle KLM \cong \triangle QPR$. By CPCTC, you also know that $\angle L \cong \angle P$, $\angle M \cong \angle R$, and $\overline{LM} \cong \overline{PR}$.

PCTCC (p. 221) PCTCC es la abreviatura para "partes correspondientes de triángulos congruentes son congruentes".

English/Spanish Glossary

EXAMPLES

Cross-Product Property (p. 367) The product of the extremes of a proportion is equal to the product of the means.

Propiedad del producto cruzado (p. 367) El producto de los extremos de una proporción es igual al producto de los medios.

If $\frac{x}{3} = \frac{12}{21}$, then $21x = 3 \cdot 12$.

Cross section (p. 600) A cross section is the intersection of a solid and a plane.

Sección de corte (p. 600) Una sección de corte es la intersección de un plano y un cuerpo geométrico.

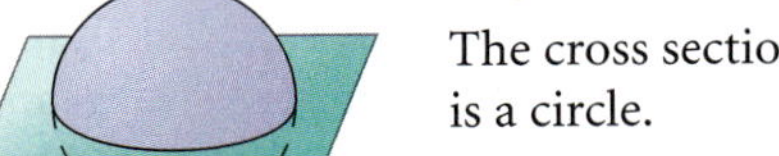

The cross section is a circle.

Cube (p. 598) A cube is a polyhedron with six faces, each of which is a square.

Cubo (p. 598) Un cubo es un poliedro de seis caras, cada una de las caras es un cuadrado.

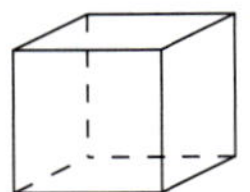

Cylinder (p. 610) A cylinder is a three-dimensional figure with two congruent circular *bases* that lie in parallel planes. An *altitude* of a cylinder is a perpendicular segment that joins the planes of the bases. Its length is the *height* of the cylinder. In a *right cylinder,* the segment joining the centers of the bases is an altitude. In an *oblique cylinder,* the segment joining the centers of the bases is not perpendicular to the planes containing the bases.

Cilindro (p. 610) Un cilindro es una figura tridimensional con dos *bases* congruentes circulares en planos paralelos. Una *altura* de un cilindro es un segmento perpendicular que une los planos de las bases. Su longitud es, por extensión, la *altura* del cilindro. En un *cilindro recto,* el segmento que une los centros de las bases es una altura. En un *cilindro oblicuo,* el segmento que une los centros de las bases no es perpendicular a los planos que contienen las bases.

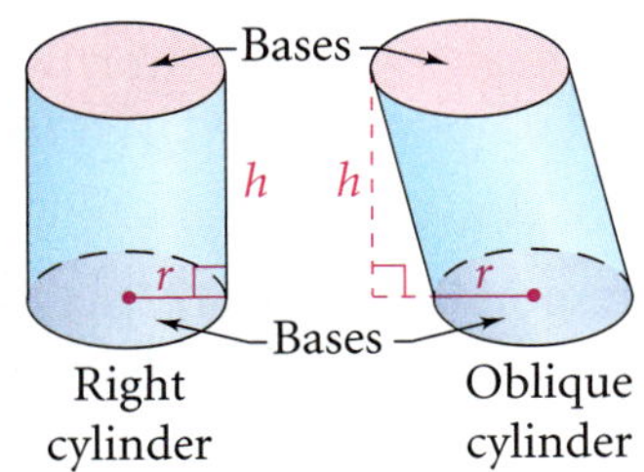

Decagon (p. 158) A decagon is a polygon with ten sides.

Decágono (p. 158) Un decágono es un polígono de diez lados.

Deductive reasoning (p. 94) Deductive reasoning is a process of reasoning logically from given facts to a conclusion.

Razonamiento deductivo (p. 94) El razonamiento deductivo es un proceso de razonamiento lógico que parte de hechos dados hasta llegar a una conclusión.

Based on the fact that the sum of any two even numbers is even, you can deduce that the product of any whole number and any even number is even.

EXAMPLES

Diagonal (p. 158) *See* **polygon.**

Diagonal (p. 158) *Ver* **polygon.**

Diameter of a circle (p. 566) A diameter of a circle is a segment that contains the center of the circle and whose endpoints are on the circle. The term *diameter* can also mean the length of this segment.

Diámetro de un círculo (p. 566) Un diámetro de un círculo es un segmento que contiene el centro del círculo y cuyos extremos están en el círculo. El término *diámetro* también puede referirse a la longitud de este segmento.

$\overline{DM}$ is a diameter of $\odot C$.

Diameter of a sphere (p. 638) The diameter of a sphere is a segment passing through the center, with endpoints on the sphere.

Diámetro de una esfera (p. 638) El diámetro de una esfera es un segmento que contiene el centro de la esfera y cuyos extremos están en la esfera.

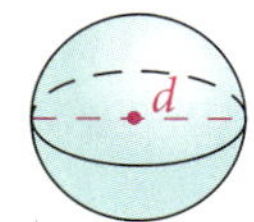

Dilation (p. 498) A dilation, or *similarity transformation,* is a transformation that has *center C* and *scale factor n*, where $n > 0$, and maps a point R to R' in such a way that R' is on $\overrightarrow{CR}$ and $CR' = n \cdot CR$. The center of a dilation is its own image. If $n > 1$, the dilation is an *enlargement,* and if $0 < n < 1$, the dilation is a *reduction.*

Dilatación (p. 498) Una dilatación, o *transformación de semejanza,* tiene *centro C* y *factor de escala n* para $n > 0$, y asocia un punto R a R' de tal modo que R' está en $\overrightarrow{CR}$ y $CR' = n \cdot CR$. El centro de una dilatación es su propia imagen. Si $n > 1$, la dilatación es un *aumento,* y si $0 < n < 1$, la dilatación es una *reducción.*

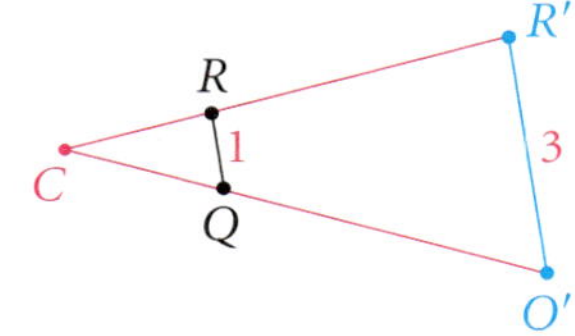

$\overline{R'Q'}$ is the image of $\overline{RQ}$ under a dilation with center C and scale factor 3.

Direction of a vector (p. 452) *See* **vector.**

Dirección de un vector (p. 452) *Ver* **vector.**

Distance from a point to a line (p. 266) The distance from a point to a line is the length of the perpendicular segment from the point to the line.

Distancia desde un punto hasta una recta (p. 266) La distancia desde un punto hasta una recta es la longitud del segmento perpendicular que va desde el punto hasta la recta.

The distance from point P to a line ℓ is PT.

Dodecagon (p. 158) A dodecagon is a polygon with twelve sides.

Dodecágono (p. 158) Un dodecágono es un polígono de doce lados.

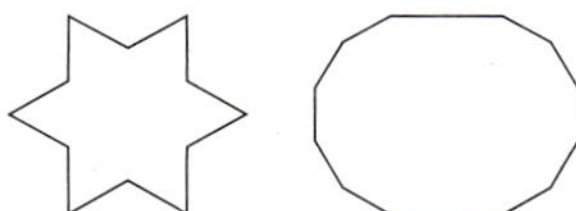

EXAMPLES

Edge (p. 598) *See* **polyhedron.**

Arista (p. 598) *Ver* **polyhedron.**

Endpoint (p. 23) *See* **ray; segment.**

Extremo (p. 23) *Ver* **ray; segment.**

Enlargement (p. 498) *See* **dilation.**

Aumento (p. 498) *Ver* **dilation.**

Equiangular triangle or polygon (pp. 148, 160) An equiangular triangle (polygon) is a triangle (polygon) whose angles are all congruent.

Triángulo o polígono equiángulo (pp. 148, 160) Un triángulo (polígono) equiángulo es un triángulo (polígono) cuyos ángulos son todos congruentes.

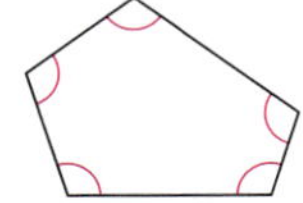

Each angle of the pentagon is a 108° angle.

Equilateral triangle or polygon (pp. 148, 160) An equilateral triangle (polygon) is a triangle (polygon) whose sides are all congruent.

Triángulo o polígono equilátero (pp. 148, 160) Un triángulo (polígono) equilátero es un triángulo (polígono) cuyos lados son todos congruentes.

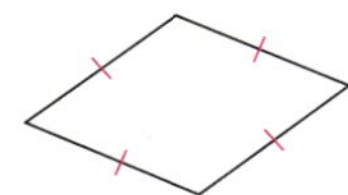

Each side of the quadrilateral is 1.2 cm long.

Equivalent statements (p. 281) Equivalent statements are statements with the same truth value.

Enunciados equivalentes (p. 281) Los enunciados equivalentes son enunciados con el mismo valor verdadero.

The following statements are equivalent:
If a figure is a square, then it is a rectangle.
If a figure is not a rectangle, then it is not a square.

Euclidean geometry (p. 154) Euclidean geometry is a geometry of the plane in which Euclid's Parallel Postulate is accepted as true.

Geometría euclidiana (p. 154) La geometría euclidiana es una geometría del plano en donde el postulado paralelo de Euclides es verdadero.

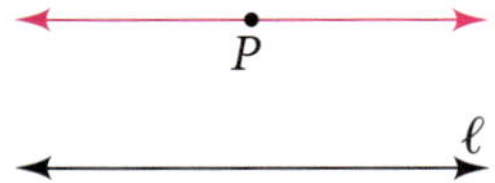

In Euclidean geometry, there is exactly one line parallel to line ℓ through point P.

Extended proportion (p. 367) *See* **proportion.**

Proporción extendida (p. 367) *Ver* **proportion.**

EXAMPLES

Exterior angle of a polygon (p. 149) An exterior angle of a polygon is an angle formed by a side and an extension of an adjacent side.

Ángulo exterior de un polígono (p. 149) El ángulo exterior de un polígono es un ángulo formado por un lado y una extensión de un lado adyacente.

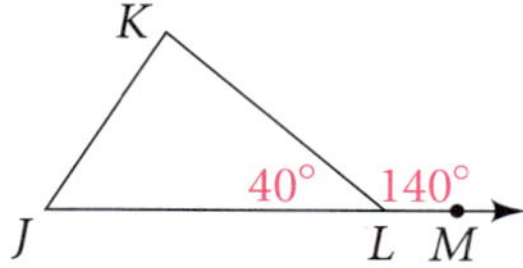

$\angle KLM$ is an exterior angle of $\triangle JKL$.

F

Face (p. 598) *See* **polyhedron.**

Cara (p. 598) *Ver* **polyhedron.**

Flip (p. 478) *See* **reflection.**

Flow proof (p. 135) *See* **proof.**

Prueba de flujo (p. 135) *Ver* **proof.**

Foundation drawing (p. 11) A foundation drawing shows the base of a structure and the height of each part.

Dibujo de fundación (p. 11) Un dibujo de fundación muestra la base de una estructura y la altura de cada parte.

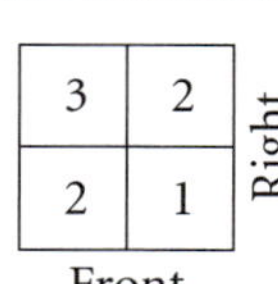

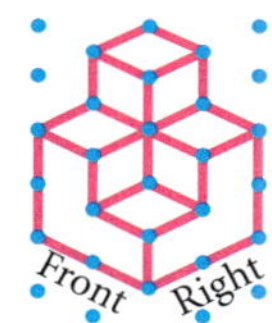

The first drawing is a foundation drawing, and the second is an isometric drawing based on the foundation drawing.

G

Geometric mean (p. 392) The geometric mean is the number x such that $\frac{a}{x} = \frac{x}{b}$, where a, b and x are positive numbers.

Media geométrica (p. 392) La media geométrica es el número x tanto que $\frac{a}{x} = \frac{x}{b}$, donde a, b y x son números positivos.

The geometric mean of 6 and 24 is 12.

$\frac{6}{x} = \frac{x}{24} \rightarrow x^2 = 144 \rightarrow x = 12$

Geometric probability (p. 582) Geometric probability is a probability that uses a geometric model in which points represent outcomes.

Probabilidad geométrica (p. 582) La probabilidad geométrica es una probabilidad que utiliza un modelo geométrico donde se usan puntos para representar resultados.

English/Spanish Glossary

EXAMPLES

Glide reflection (p. 508) A glide reflection is the composition of a translation followed by a reflection across a line parallel to the translation vector.

Reflexión deslizada (p. 508) Una reflexión deslizada es la composición de una traslación seguida de una reflexión en una recta paralela al vector de traslación.

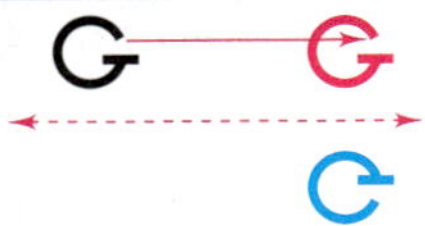

The blue G in the diagram is a glide reflection image of the black G.

Glide reflectional symmetry (p. 516) Glide reflectional symmetry is the type of symmetry for which there is a glide reflection that maps a figure onto itself.

Simetría por reflexión deslizada (p. 516) La simetría por reflexión deslizada es un tipo de simetría en la que una reflexión deslizada vuelve a trazar una figura sobre sí misma.

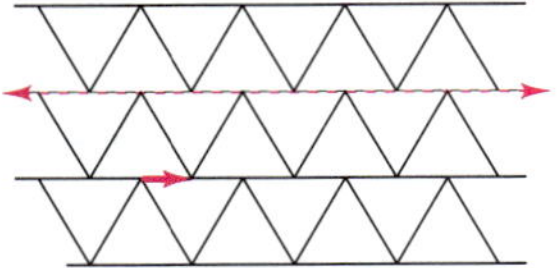

The tessellation shown can be mapped onto itself by a glide reflection for the given glide vector and reflection line.

Golden rectangle, Golden ratio (p. 375) A *golden rectangle* is a rectangle that can be divided into a square and a rectangle that is similar to the original rectangle. The *golden ratio* is the ratio of the length of a golden rectangle to its width. The value of the golden ratio is $\frac{1 + \sqrt{5}}{2}$, or about 1.62.

Rectángulo áureo, razón áurea (p. 375) Un *rectángulo áureo* es un rectángulo que se puede dividir en un cuadrado y un rectángulo semejante al rectángulo original. La *razón áurea* es la razón de la longitud de un rectángulo áureo en relación a su ancho. El valor de la razón áurea es $\frac{1 + \sqrt{5}}{2}$ o aproximadamente 1.62.

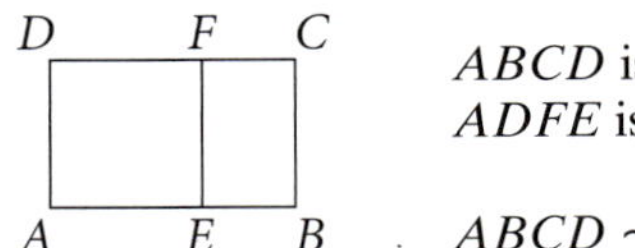

$ABCD$ is a rectangle.
$ADFE$ is a square.

$ABCD \sim BCFE$

Great circle (p. 638) A great circle is the intersection of a sphere and the plane containing the center of the sphere. A great circle divides a sphere into two *hemispheres.*

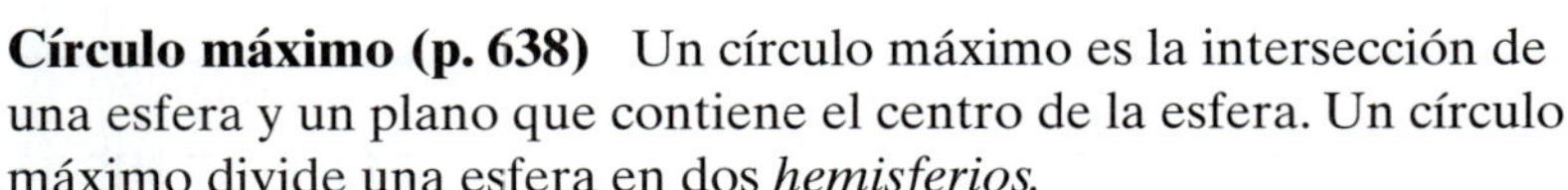

Círculo máximo (p. 638) Un círculo máximo es la intersección de una esfera y un plano que contiene el centro de la esfera. Un círculo máximo divide una esfera en dos *hemisferios.*

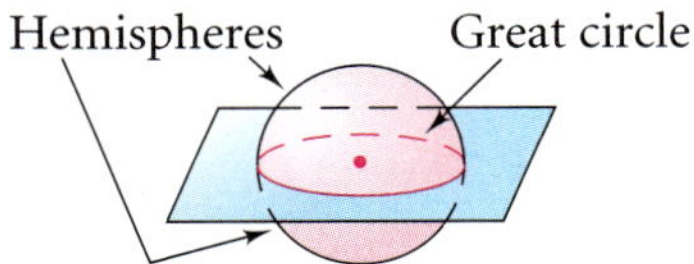

H

Height *See* **cone; cylinder; parallelogram; prism; pyramid; trapezoid; triangle.**

Altura *Ver* **cone; cylinder; parallelogram; prism; pyramid; trapezoid; triangle.**

EXAMPLES

Hemisphere (p. 638) *See* **great circle.**

Hemisferio (p. 638) *Ver* **great circle.**

Heron's Formula (p. 538) Heron's Formula is a formula for finding the area of a triangle given the lengths of its sides.

Fórmula de Herón (p. 538) La fórmula de Herón se usa para hallar el área de un triángulo, dadas las longitudes de sus lados.

$A = \sqrt{s(s-a)(s-b)(s-c)}$, where s is half the perimeter (semi-perimeter) of the triangle and a, b, and c are the lengths of its sides.

Hexagon (p. 158) A hexagon is a polygon with six sides.

Hexágono (p. 158) Un hexágono es un polígono de seis lados.

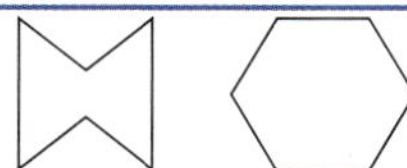

Hypotenuse (p. 235) *See* **right triangle.**

Hipotenusa (p. 235) *Ver* **right triangle.**

Hypothesis (p. 80) The hypothesis is the part that follows *if* in an *if-then* statement (conditional).

Hipótesis (p. 80) La hipótesis es lo que sigue a la palabra *si* en un enunciado (condicional), *si..., entonces....*

In the statement "If she leaves, then I will go with her," the hypothesis is "she leaves."

I

Identity (p. 440) An identity is an equation that is true for all allowed values of the variable.

Identidad (p. 440) Una identidad es una ecuación que es verdadera para todos los valores posibles de las variables.

$\sin x^\circ = \cos(90 - x)^\circ$

Incenter of a triangle (p. 273) The incenter of a triangle is the point of concurrency of the angle bisectors of the triangle.

Incentro de un triángulo (p. 273) El incentro de un triángulo es el punto donde concurren las tres bisectrices de los ángulos del triángulo.

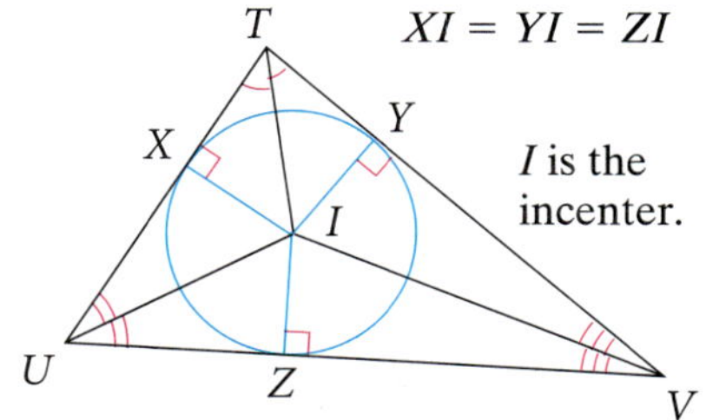

$XI = YI = ZI$

I is the incenter.

Image (p. 470) *See* transformation.

Imagen (p. 470) *Ver* transformation.

English/Spanish Glossary

EXAMPLES

Indirect measurement (p. 384) Indirect measurement is a way of measuring things that are difficult to measure directly.

EXAMPLE By measuring the distances shown in the diagram and using proportions of similar figures, you can find the height of the taller tower. $\frac{196}{540} = \frac{x}{1300} \rightarrow x \approx 472$ ft

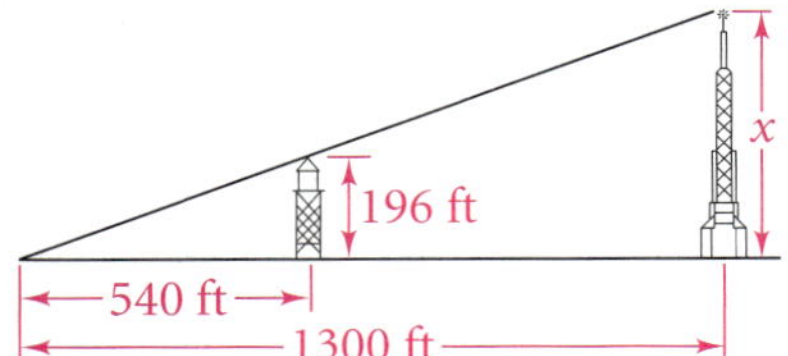

Medición indirecta (p. 384) La medición indirecta es un modo de medir cosas difíciles de medir directamente.

Indirect proof (p. 281) *See* **indirect reasoning; proof.**

Prueba indirecta (p. 281) *Ver* **indirect reasoning; proof.**

Indirect reasoning (p. 281) Indirect reasoning is a type of reasoning in which all possibilities are considered and then the unwanted ones are proved false. The remaining possibilities must be true.

Razonamiento indirecto (p. 281) El razonamiento indirecto es un tipo de razonamiento en el que todas las posibilidades se consideran, y luego las no deseadas resultan falsas. Las posibilidades que quedan deben ser verdaderas.

Eduardo spent more than $60 on two books at a store. Prove that at least one book costs more than $30.
Proof: Suppose neither costs more than $30. Then he spent no more than $60 at the store. Since this contradicts the given information, at least one book costs more than $30.

Inductive reasoning (p. 4) Inductive reasoning is a type of reasoning that reaches conclusions based on a pattern of specific examples or past events.

Razonamiento inductivo (p. 4) El razonamiento inductivo es un tipo de razonamiento en el cual se llega a conclusiones con base en un patrón de ejemplos específicos o sucesos pasados.

You see four people walk into a building. Each person emerges with a small bag containing hot food. You use inductive reasoning to conclude that this building contains a restaurant.

Initial point of a vector (p. 452) *See* **vector.**

Punto inicial de un vector (p. 452) *Ver* **vector.**

Inscribed in (pp. 273, 664) A circle is inscribed in a polygon if the sides of the polygon are tangent to the circle. A polygon is inscribed in a circle if the vertices of the polygon are on the circle.

Inscrito en (pp. 273, 664) Un círculo está inscrito en un polígono si los lados del polígono son tangentes al círculo. Un polígono está inscrito en un círculo si los vértices del polígono están en el círculo.

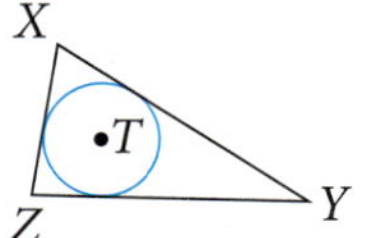

$\odot T$ is inscribed in $\triangle XYZ$.

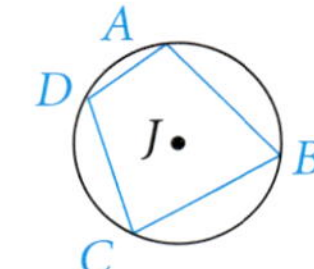

$ABCD$ is inscribed in $\odot J$.

Inscribed angle (p. 678) An angle is inscribed in a circle if the vertex of the angle is on the circle and the sides of the angle are chords of the circle.

Ángulo inscrito (p. 678) Un ángulo está inscrito en un círculo si el vértice del ángulo está en el círculo y los lados del ángulo son cuerdas del círculo.

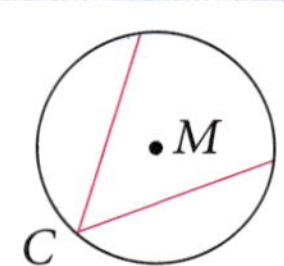

$\angle C$ is inscribed in $\odot M$.

EXAMPLES

Intercepted arc (p. 678) An intercepted arc is an arc of a circle having endpoints on the sides of an inscribed angle, and its other points in the interior of the angle.

Arco interceptor (p. 678) Un arco interceptor es un arco de un círculo cuyos extremos están en los lados de un ángulo inscrito y los puntos restantes están en el interior del ángulo.

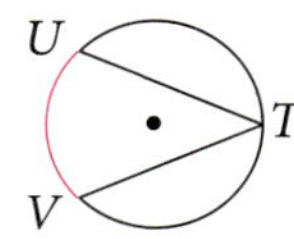

$\widehat{UV}$ is the intercepted arc of inscribed angle $\angle T$.

Inverse (p. 280) The inverse of the conditional "if p, then q" is the conditional "if not p, then not q."

Inverso (p. 280) El inverso del condicional "si p, entonces q," es el condicional "si no p, entonces no q".

Conditional: If a figure is a square, then it is a parallelogram.
Inverse: If a figure is not a square, then it is not a parallelogram.

Isometric drawing (p. 10) An isometric drawing of a three-dimensional object shows a corner view of a figure. It is not drawn in perspective and distances are not distorted.

Dibujo isométrico (p. 10) Un dibujo isométrico de un objeto tridimensional muestra una vista desde una esquina de la figura. No se muestra en perspectiva y las distancias no aparacen distorcionadas.

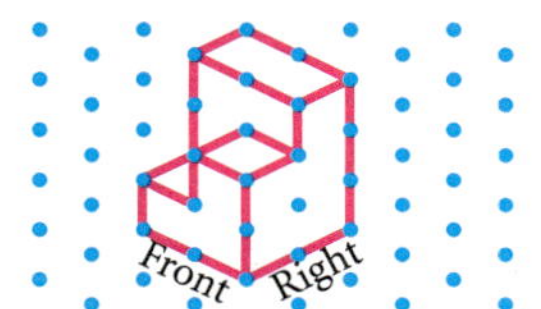

Isometry (p. 470) An isometry, also known as a *congruence transformation,* is a transformation in which an original figure and its image are congruent.

Isometría (p. 470) Una isometría, conocida también como una *transformación de congruencia,* es una transformación en donde una figura original y su imagen son congruentes.

The four isometries are reflections, rotations, translations, and glide reflections.

Isosceles trapezoid (p. 306) An isosceles trapezoid is a trapezoid whose nonparallel opposite sides are congruent.

Trapecio isósceles (p. 306) Un trapecio isósceles es un trapecio cuyos lados opuestos no paralelos son congruentes.

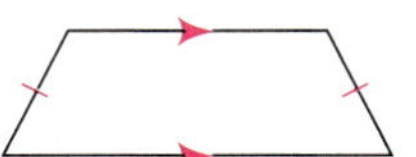

Isosceles triangle (pp. 148, 228) An isosceles triangle is a triangle that has at least two congruent sides. If there are two congruent sides, they are called *legs.* The *vertex angle* is between them. The third side is called the *base* and the other two angles are called the *base angles.*

Triángulo isósceles (pp. 148, 228) Un triángulo isósceles es un triángulo que tiene por lo menos dos lados congruentes. Si tiene dos lados congruentes, éstos se llaman *catetos*. Entre ellos se encuentra el *ángulo del vértice*. El tercer lado se llama *base* y los otros dos ángulos se llaman *ángulos de base.*

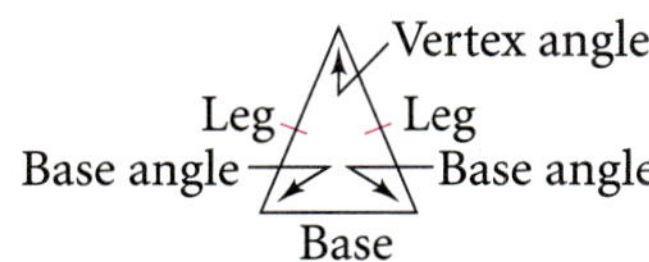

English/Spanish Glossary

EXAMPLES

Kite (p. 306) A kite is a quadrilateral with two pairs of congruent adjacent sides and no opposite sides congruent.

Cometa (p. 306) Una cometa es un cuadrilátero con dos pares de lados adyacentes congruentes, pero sin lados opuestos congruentes.

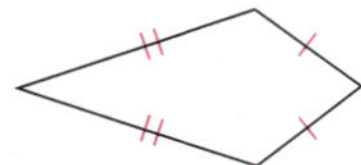

Lateral area (pp. 608, 610, 617, 619) The lateral area of a prism or pyramid is the sum of the areas of the lateral faces. The lateral area of a cylinder or cone is the area of the curved surface. A list of lateral area formulas is on p. 765.

Área lateral (pp. 608, 610, 617, 619) El área lateral de un prisma o pirámide es la suma de las áreas de sus caras laterales. El área lateral de un cilindro o de un cono es el área de la superficie curvada. Una lista de las fórmulas de áreas laterales está en la p. 765.

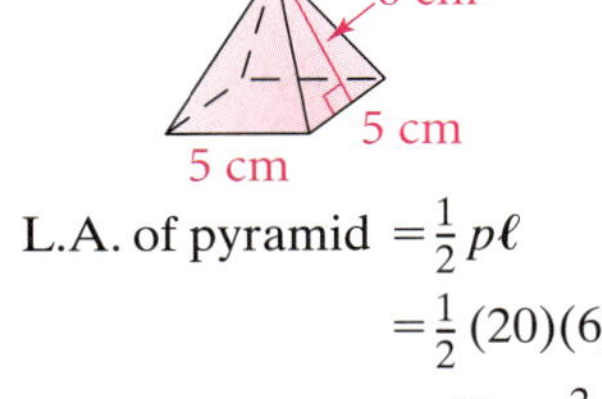

L.A. of pyramid $= \frac{1}{2}p\ell$
$= \frac{1}{2}(20)(6)$
$= 60 \text{ cm}^2$

Lateral face *See* **prism; pyramid.**

Cara lateral *Ver* **prism; pyramid.**

Leg *See* **isosceles triangle; right triangle; trapezoid.**

Cateto *Ver* **isosceles triangle; right triangle; trapezoid.**

Line (pp. 17, 154) In Euclidean geometry, a line is undefined. You can think of a line as a series of points that extend in two directions without end. In spherical geometry, you can think of a line as a great circle of a sphere.

Recta (pp. 17, 154) En la geometría euclidiana, una recta es indefinida. Puedes imaginarte a una recta como una serie de puntos que se extienden en dos direcciones sin fin. En la geometría esférica, puedes imaginarte a una recta como un círculo máximo de una esfera.

Line symmetry (p. 492) *See* **reflectional symmetry.**

Simetría axial (p. 492) *Ver* **reflectional symmetry.**

Locus (p. 701) A locus is a set of points, all of which meet a stated condition.

Lugar geométrico (p. 701) Un lugar geométrico es un conjunto de puntos que cumplen una condición dada.

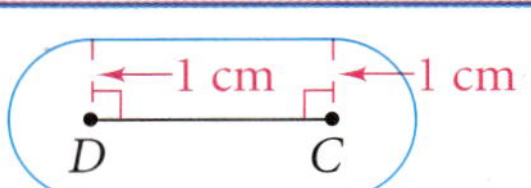

The points in blue are the locus of points in a plane 1 cm from $\overline{DC}$.

EXAMPLES

Magnitude of a vector (p. 452) *See* **vector.**

Magnitud de un vector (p. 452) *Ver* **vector.**

Major arc (p. 567) A major arc of a circle is an arc that is larger than a semicircle.

Arco mayor (p. 567) Un arco mayor de un círculo es cualquier arco más grande que un semicírculo.

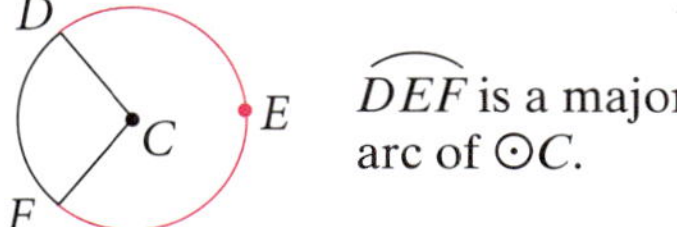

$\widehat{DEF}$ is a major arc of $\odot C$.

Map (p. 471) *See* **transformation.**

Trazar (p. 471) *Ver* **transformation.**

Matrix (p. 504) A matrix is a rectangular array of numbers. Each number in a matrix is called an *entry.*

Matriz (p. 504) Una matriz es un conjunto de números dispuestos en forma de rectángulo. Cada número de una matriz se llama *elemento* de la matriz.

The matrix $\begin{bmatrix} 1 & -2 \\ 0 & 13 \end{bmatrix}$ has dimensions 2×2. The number 1 is the entry in the first row and first column.

Measure of an angle (p. 36) The measure of an angle is a number of degrees greater than 0 and less than or equal to 180. An angle can be measured with a protractor.

Medida de un ángulo (p. 36) La medida de un ángulo es un número de grados mayor de 0 y menor o igual a 180. Un ángulo se puede medir con un transportador.

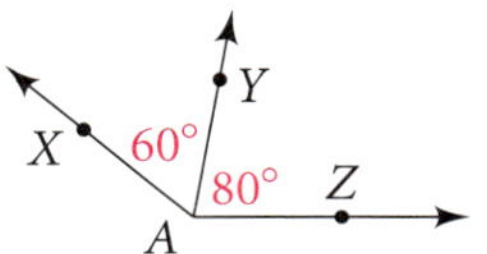

$m\angle ZAY = 80$, $m\angle YAX = 60$, and $m\angle ZAX = 140$.

Measure of an arc (p. 567) The measure of a minor arc is the measure of its central angle. The measure of a major arc is 360 minus the measure of its related minor arc.

Medida de un arco (p. 567) La medida de un arco menor es la medida de su ángulo central. La medida de un arco mayor es 360 menos la medida en grados de su arco menor correspondiente.

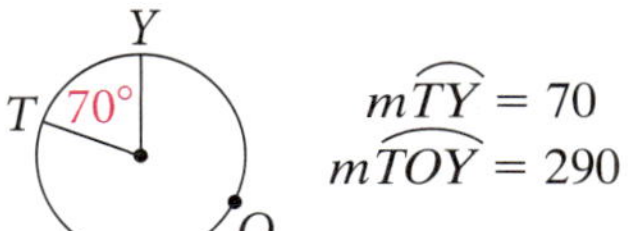

$m\widehat{TY} = 70$
$m\widehat{TOY} = 290$

Median of a triangle (p. 274) A median of a triangle is a segment that has as its endpoints a vertex of the triangle and the midpoint of the opposite side.

Mediana de un triángulo (p. 274) Una mediana de un triángulo es un segmento que tiene en sus extremos el vértice del triángulo y el punto medio del lado opuesto.

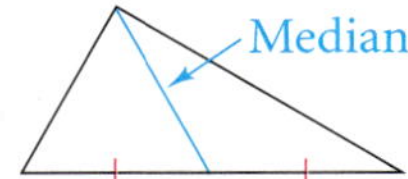

Midpoint of a segment (p. 32) A midpoint of a segment is the point that divides the segment into two congruent segments.

Punto medio de un segmento (p. 32) El punto medio de un segmento es el punto que divide el segmento en dos segmentos congruentes.

Midpoint of $\overline{AB}$

A M B

EXAMPLES

Midsegment of a trapezoid (p. 348) The midsegment of a trapezoid is the segment that joins the midpoints of the nonparallel opposite sides of a trapezoid.

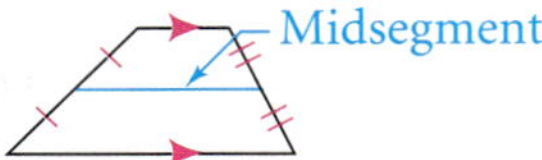

Segmento medio de un trapecio (p. 348) El segmento medio de trapecio es el segmento que une los puntos medios de los lados no paralelos de un trapecio.

Midsegment of a triangle (p. 259) A midsegment of a triangle is the segment that joins the midpoints of two sides of the triangle.

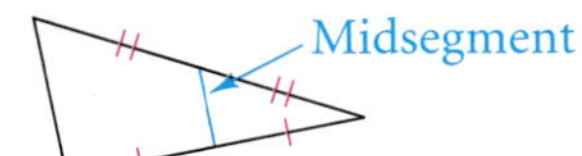

Segmento medio de un triángulo (p. 259) Un segmento medio de un triángulo es el segmento que une los puntos medios de dos lados del triángulo.

Minor arc (p. 567) A minor arc is an arc that is smaller than a semicircle.

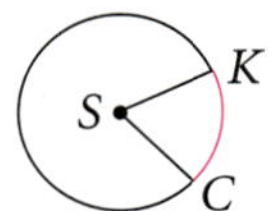

$\widehat{KC}$ is a minor arc of $\odot S$.

Arco menor (p. 567) Un arco menor de un círculo es un arco más corto que un semicírculo.

N

Negation (p. 280) The negation of a statement has the opposite meaning of the original statement.

Statement: The angle is obtuse.
Negation: The angle is not obtuse.

Negación (p. 280) La negación de un enunciado tiene el sentido opuesto del enunciado original.

Net (p. 12) A net is a two-dimensional pattern that you can fold to form a three-dimensional figure.

Plantilla (p. 12) Una plantilla es una figura bidimensional que se puede doblar para formar una figura tridimensional.

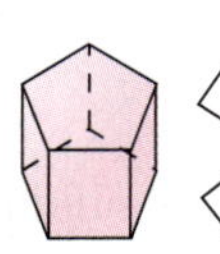
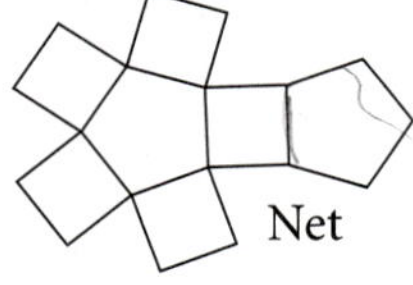

The net shown can be folded into a prism with pentagonal bases.

n-gon (p. 158) An n-gon is a polygon with n sides.

n-ágono (p. 158) Un n-ágono es un polígono de n lados.

Nonagon (p. 158) A nonagon is a polygon with nine sides.

Nonágono (p. 158) Un nonágono es un polígono de nueve lados.

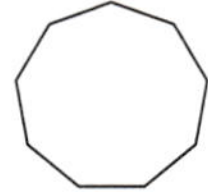

EXAMPLES

Oblique cylinder or prism *See* **cylinder; prism.**

Cilindro oblicuo o prisma *Ver* **cylinder; prism.**

Obtuse angle (p. 37) An obtuse angle is an angle whose measure is between 90 and 180.

Ángulo obtuso (p. 37) Un ángulo obtuso es un ángulo que mide entre 90 y 180 grados.

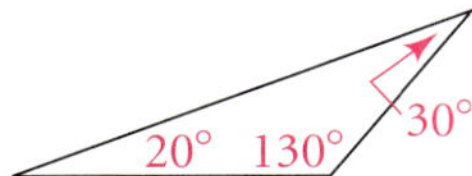

Obtuse triangle (p. 148) An obtuse triangle has one obtuse angle.

Triángulo obtusángulo (p. 148) Un triángulo obtusángulo tiene un ángulo obtuso.

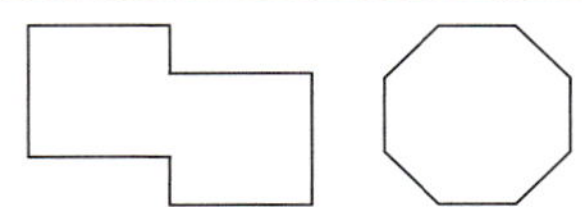

Octagon (p. 158) An octagon is a polygon with eight sides.

Octágono (p. 158) Un octágono es un polígono de ocho lados.

Opposite rays (p. 23) Opposite rays are collinear rays with the same endpoint. They form a line.

Rayos opuestos (p. 23) Los rayos opuestos son rayos colineales con el mismo extremo. Forman una recta.

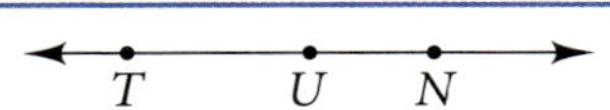

$\overrightarrow{UT}$ and $\overrightarrow{UN}$ are opposite rays.

Orientation (p. 478) Two congruent figures have *opposite* orientation if a reflection is needed to map one onto the other. If a reflection is not needed to map one figure onto the other, the figures have the same orientation.

Orientación (p. 478) Dos figuras congruentes tienen orientación *opuesta* si una reflexión es necesaria para trazar una sobre la otra. Si una reflexión no es necesaria para trazar una figura sobre la otra, las figuras tiene la misma orientación.

R Я The two R's have opposite orientation.

Origin (p. 53) *See* **coordinate plane.**

Origen (p. 53) *Ver* **coordinate plane.**

Orthocenter (p. 275) The orthocenter of a triangle is the point of intersection of the lines containing the altitudes of the triangle.

Ortocentro (p. 275) El ortocentro de un triángulo es el punto donde concurren las tres alturas del triángulo.

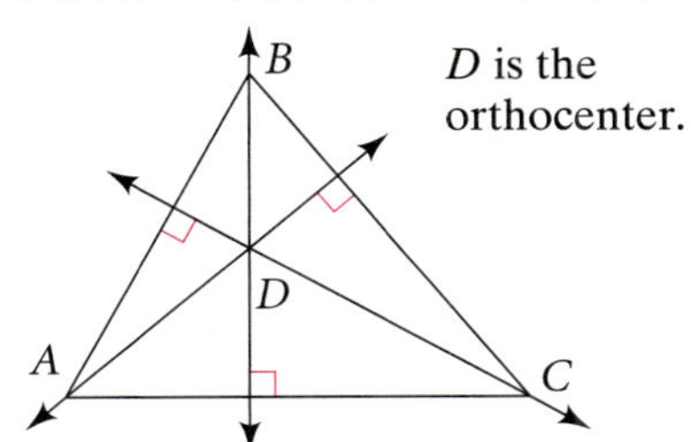

D is the orthocenter.

English/Spanish Glossary

EXAMPLES

Orthographic drawing (p. 11) An orthographic drawing is the top view, front view, and right-side view of a three-dimensional figure.

EXAMPLE The diagram shows an isometric drawing (upper right) and the three views that make up an orthographic drawing.

Dibujo ortográfico (p. 11) Un dibujo ortográfico es la vista desde arriba, la vista de frente y la vista del lado derecho de una figura tridimensional.

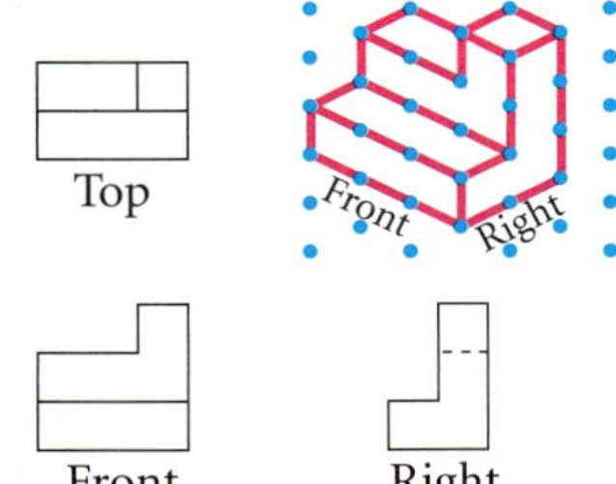

Paragraph proof (p. 111) *See* **proof.**

Prueba de párrafo (p. 111) *Ver* **proof.**

Parallel lines (p. 24) Two lines are parallel if they lie in the same plane and do not intersect. The symbol || means "is parallel to."

Rectas paralelas (p. 24) Dos rectas son paralelas si están en el mismo plano y no se cortan. El símbolo || significa "es paralelo a".

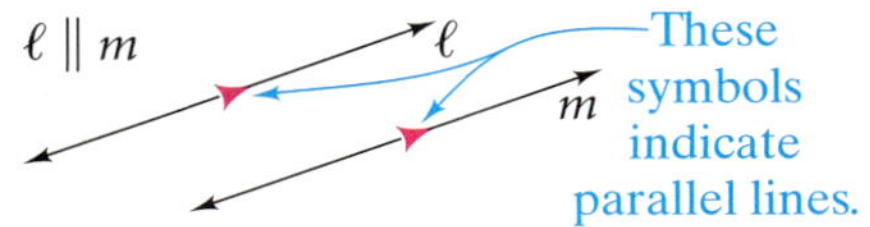

Parallel planes (p. 24) Parallel planes are planes that do not intersect.

Planos paralelos (p. 24) Planos paralelos son planos que no se cortan.

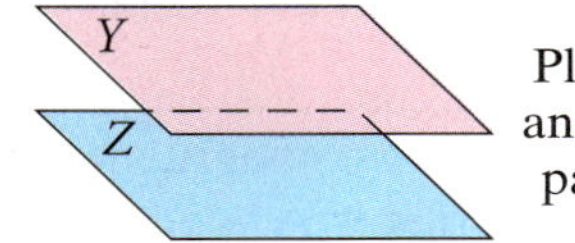

Planes Y and Z are parallel.

Parallelogram (p. 306) A parallelogram is a quadrilateral with two pairs of parallel sides. You can choose any side to be the *base.* An *altitude* is any segment perpendicular to the line containing the base drawn from the side opposite the base. The *height* is the length of an altitude.

Paralelogramo (p. 306) Un paralelogramo es un cuadrilátero con dos pares de lados paralelos. Se puede escoger cualquier lado como la *base.* Una *altura* es un segmento perpendicular a la recta que contiene la base, trazada desde el lado opuesto a la base. La *altura*, por extensión, es la longitud de una altura.

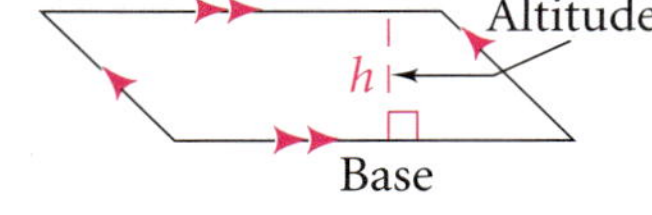

Pentagon (p. 158) A pentagon is a polygon with five sides.

Pentágono (p. 158) Un pentágono es un polígono de cinco lados.

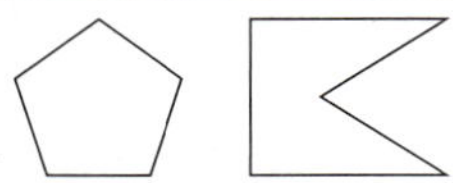

Perimeter of a polygon (p. 61) The perimeter of a polygon is the sum of the lengths of its sides.

Perímetro de un polígono (p. 61) El perímetro de un polígono es la suma de las longitudes de sus lados.

4 in.

4 in. 3 in.

5 in.

$$P = 4 + 4 + 5 + 3$$
$$= 16 \text{ in.}$$

EXAMPLES

Perpendicular bisector (p. 45) The perpendicular bisector of a segment is a line, segment, or ray that is perpendicular to the segment at its midpoint.

Mediatriz (p. 45) La mediatriz de un segmento es una recta, segmento o rayo que es perpendicular al segmento en su punto medio.

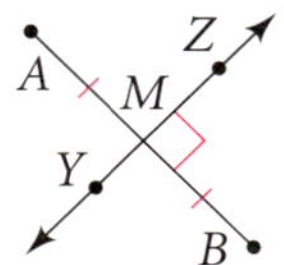

$\overleftrightarrow{YZ}$ is the perpendicular bisector of $\overline{AB}$. It is perpendicular to $\overline{AB}$ and intersects $\overline{AB}$ at midpoint M.

Perpendicular lines (p. 45) Perpendicular lines are lines that intersect and form right angles. The symbol $\perp$ means "is perpendicular to."

Rectas perpendiculares (p. 45) Las rectas perpendiculares son rectas que se cortan y forman ángulos rectos. El símbolo $\perp$ significa "es perpendicular a".

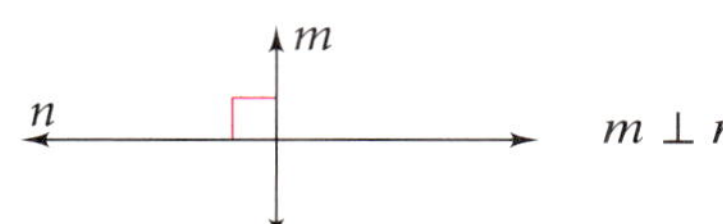

$m \perp n$

Perspective drawing (p. 606) Perspective drawing is a way of drawing objects on a flat surface so that they look the same way as they appear to the eye. In *one-point perspective,* there is one *vanishing point.* In *two-point perspective,* there are two vanishing points.

Dibujar en perspectiva (p. 606) Dibujar en perspectiva es una manera de dibujar objetos en una superficie plana de modo que se vean como los percibe el ojo humano. En la *perspectiva de un punto* hay un *punto de fuga.* En la *perspectiva de dos puntos* hay dos puntos de fuga.

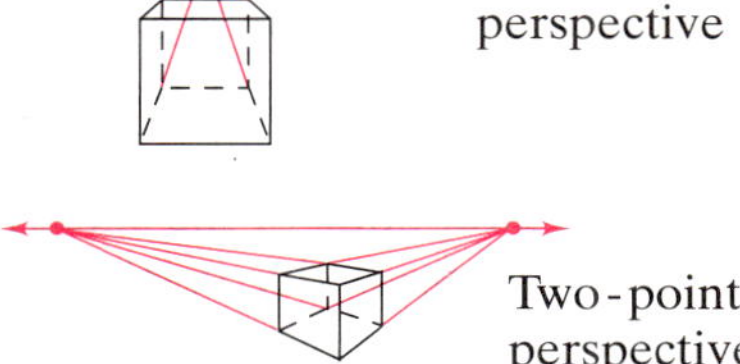

One-point perspective

Two-point perspective

Pi (p. 568) Pi (π) is the ratio of the circumference of any circle to its diameter. The number π is irrational and is approximately 3.14159.

Pi (p. 568) Pi (π) es la razón de la circunferencia de cualquier círculo a su diámetro. El número π es irracional y se aproxima a $\pi \approx 3.14159$.

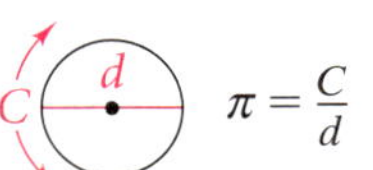

$\pi = \frac{C}{d}$

Plane (p. 17) In Euclidean geometry, a plane is undefined. You can think of a plane as a flat surface that has no thickness. A plane contains many lines and extends without end in the directions of its lines.

Plano (p. 17) En la geometría euclidiana, un plano es indefinido. Puedes imaginarte a un plano como una superficie plana que no tiene grosor. Un plano tiene muchas rectas y se extiende sin fin en la misma dirección que todas las rectas.

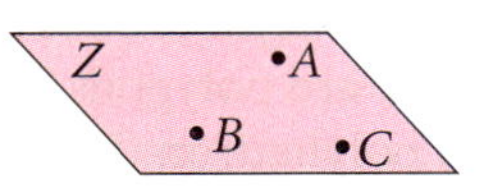

Plane ABC or plane Z

Point (p. 17) In Euclidean geometry, a point is undefined. You can think of a point as a location. A point has no size.

Punto (p. 17) En la geometría euclidiana, un punto es indefinido. Puedes imaginarte a un punto como un lugar. Un punto no tiene dimensión.

• P

Point of concurrency (p. 273) *See* **concurrent lines.**

Punto de concurrencia (p. 273) *Ver* **concurrent lines.**

EXAMPLES

Pyramid (p. 617) A pyramid is a polyhedron in which one face, the *base,* is a polygon and the other faces, the *lateral faces,* are triangles with a common vertex, called the *vertex* of the pyramid. An *altitude* of a pyramid is the perpendicular segment from the *vertex* to the plane of the base. Its length is the *height* of the pyramid. A *regular pyramid* is a pyramid whose base is a regular polygon and whose lateral faces are congruent isosceles triangles. The *slant height* of a regular pyramid is the length of an altitude of a lateral face.

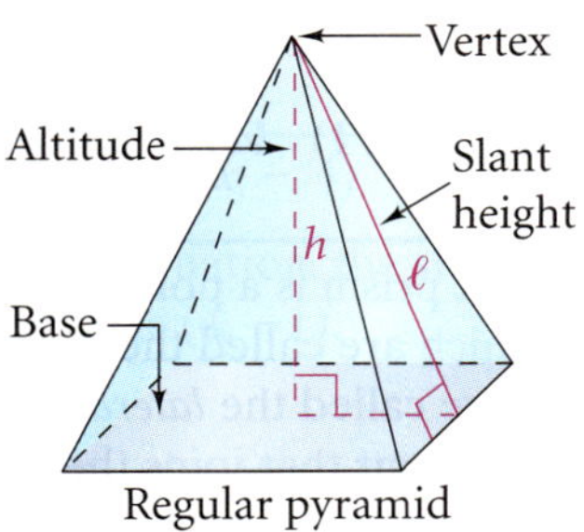

Pirámide (p. 617) Una pirámide es un poliedro en donde una cara, la *base,* es un polígono y las otras caras, las *caras laterales,* son triángulos con un vértice común, llamado el *vértice* de la pirámide. Una *altura* de una pirámide es el segmento perpendicular que va del *vértice* hasta el plano de la base. Su longitud es, por extensión, la *altura* de la pirámide. Una *pirámide regular* es una pirámide cuya base es un polígono regular y cuyas caras laterales son triángulos isósceles congruentes. La *apotema* de una pirámide regular es la longitud de la altura de la cara lateral.

Pythagorean triple (p. 417) A Pythagorean triple is a set of three nonzero whole numbers a, b, and c, that satisfy the equation $a^2 + b^2 = c^2$.

The numbers 5, 12, and 13 form a Pythagorean triple because $5^2 + 12^2 = 13^2 = 169$.

Tripleta de Pitágoras (p. 417) Una tripleta de Pitágoras es un conjunto de tres números enteros positivos a, b, and c que satisfacen la ecuación $a^2 + b^2 = c^2$.

Quadrant (p. 53) *See* **coordinate plane.**

Cuadrante (p. 53) *Ver* **coordinate plane.**

Quadrilateral (p. 158) A quadrilateral is a polygon with four sides.

Cuadrilátero (p. 158) Un cuadrilátero es un polígono de cuatro lados.

Radius of a circle (p. 566) A radius of a circle is any segment with one endpoint on the circle and the other endpoint at the center of the circle. *Radius* can also mean the length of this segment.

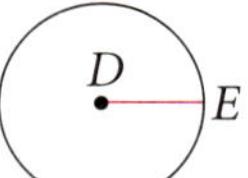

$\overline{DE}$ is a radius of $\odot D$.

Radio de un círculo (p. 566) Un radio de un círculo es cualquier segmento con un extremo en el círculo y el otro extremo en el centro del círculo. *Radio* también se refiere a la longitud de este segmento.

EXAMPLES

Radius of a regular polygon (p. 546) The radius of a regular polygon is the distance from the center to a vertex.

Radio de un polígono regular (p. 546) El radio de un polígono regular es la distancia desde el centro hasta un vértice.

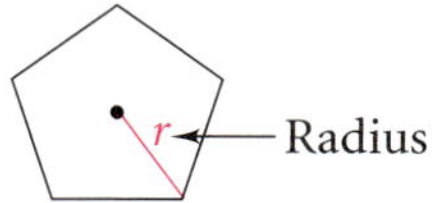

Radius of a sphere (p. 638) The radius of a sphere is a segment that has one endpoint at the center and the other endpoint on the sphere.

Radio de una esfera (p. 638) El radio de una esfera es un segmento con un extremo en el centro y otro en la esfera.

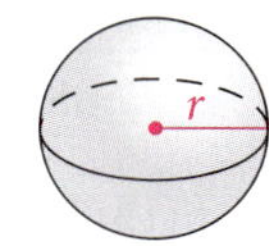

Ratio (p. 366) A ratio is the comparison of two quantities by division.

Razón (p. 366) Una razón es la comparación de dos cantidades por medio de una división.

5 to 7
5 : 7
$\frac{5}{7}$

Ray (p. 23) A ray is the part of a line consisting of one *endpoint* and all the points of the line on one side of the endpoint.

Rayo (p. 23) Un rayo es una parte de una recta que contiene un *extremo* y todos los puntos de la recta a un lado del extremo.

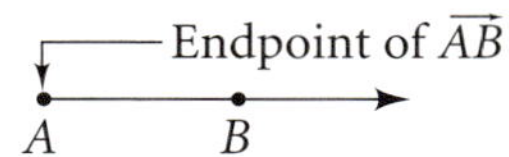

Rectangle (p. 306) A rectangle is a parallelogram with four right angles.

Rectángulo (p. 306) Un rectángulo es un paralelogramo con cuatro ángulos rectos.

Reduction (p. 498) *See* **dilation.**

Reducción (p. 498) *Ver* **dilation.**

Reflection (p. 478) A reflection (*flip*) across line r is a transformation such that if a point A is on line r, then the image of A is itself, and if a point B is not on line r, then its image B' is the point such that r is the perpendicular bisector of $\overline{BB'}$.

Reflexión (p. 478) Una reflexión en la recta r es una transformación tal que si un punto A está en la recta r, entonces la imagen de A es ella misma, y si un punto B no está en la recta r, entonces su imagen B' es el punto tal que r es la mediatriz de $\overline{BB'}$.

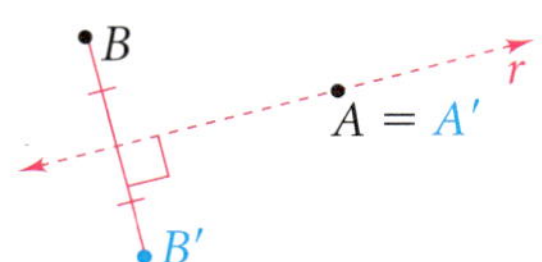

Reflectional symmetry (p. 492) Reflectional symmetry, or *line symmetry,* is the type of symmetry for which there is a reflection that maps a figure onto itself. The reflection line is the line of symmetry.

Simetría por reflexión (p. 492) La simetría por reflexión, o *simetría axial*, es un tipo de simetría en la que la reflexión vuelve a trazar la figura sobre sí misma. La recta de reflexión es el eje de simetría.

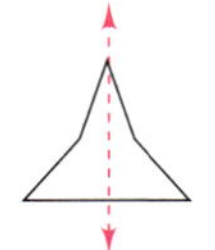

A reflection across the given line maps the figure onto itself.

English/Spanish Glossary

EXAMPLES

Regular polygon (p. 160) A regular polygon is a polygon that is both equilateral and equiangular. Its *center* is the center of the circumscribed circle.

Polígono regular (p. 160) Un polígono regular es un polígono que es equilátero y equiángulo. Su *centro* es el centro del círculo circunscrito.

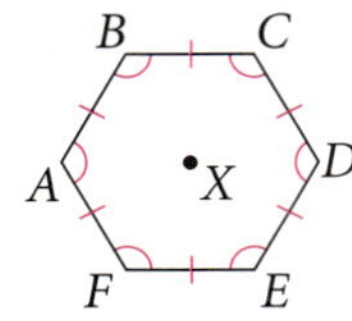

$ABCDEF$ is a regular hexagon. Point X is its center.

Regular pyramid (p. 617) *See* **pyramid.**

Pirámide regular (p. 617) *Ver* **pyramid.**

Remote interior angles (p. 149) Remote interior angles are the two nonadjacent interior angles corresponding to each exterior angle of a triangle.

Ángulos interiores remotos (p. 149) Los ángulos interiores remotos son los dos ángulos interiores no adyacentes que corresponden a cada ángulo exterior de un triángulo.

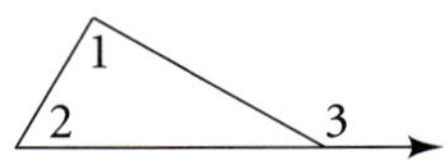

$\angle 1$ and $\angle 2$ are remote interior angles of $\angle 3$.

Resultant vector (p. 454) The sum of two vectors is a resultant.

Vector resultante (p. 454) La suma de dos vectores es el vector resultante.

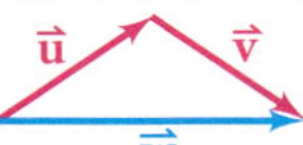

$\vec{w}$ is the resultant of $\vec{u} + \vec{v}$.

Rhombus (p. 306) A rhombus is a parallelogram with four congruent sides.

Rombo (p. 306) Un rombo es un paralelogramo de cuatro lados congruentes.

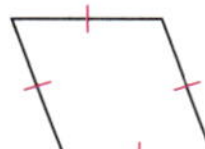

Right angle (p. 37) A right angle is an angle whose measure is 90.

Ángulo recto (p. 37) Un ángulo recto es un ángulo que mide 90.

90°

This symbol indicates a right angle.

Right cone (p. 619) *See* **cone.**

Cono recto (p. 619) *Ver* **cone.**

Right cylinder (p. 610) *See* **cylinder.**

Cilindro recto (p. 610) *Ver* **cylinder.**

Right prism (p. 608) *See* **prism.**

Prisma rectangular (p. 608) *Ver* **prism.**

EXAMPLES

Right triangle (pp. 148, 235) A right triangle contains one right angle. The side opposite the right angle is the *hypotenuse* and the other two sides are the *legs*.

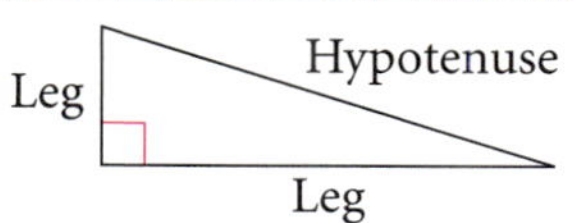

Triángulo rectángulo (pp. 148, 235) Un triángulo rectángulo contiene un ángulo recto. El lado opuesto del ángulo recto es la *hipotenusa* y los otros dos lados son los *catetos*.

Rigid motion (p. 470) A rigid motion is a motion that preserves distance and angles.

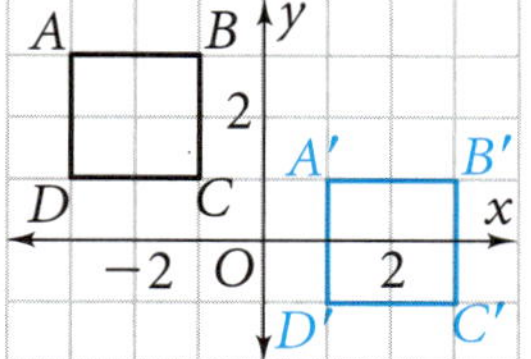

The translation of square $ABCD$ to $A'B'C'D'$ is a rigid motion.

Movimiento rígido (p. 470) Un movimiento rígido es un movimiento que mantiene la distancia y los ángulos.

Rotation (p. 484) A rotation (*turn*) of $x°$ about a point R is a transformation such that for any point V, its image is the point V', where $RV = RV'$ and $m\angle VRV' = x$. The image of R is itself.

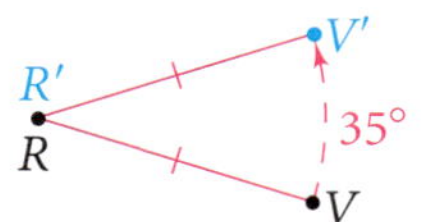

Rotación (p. 484) Una rotación de $x°$ alrededor de un punto R es una transformación de modo que para cualquier punto V, su imagen es el punto V', donde $RV = RV'$ y $m\angle VRV' = x$. La imagen de R es R misma.

Rotational symmetry (p. 493) Rotational symmetry is the type of symmetry for which there is a rotation of 180° or less that maps a figure onto itself.

The figure has 120° rotational symmetry.

Simetría rotacional (p. 493) La simetría rotacional es un tipo de simetría en la que una rotación de 180° o menos vuelve a trazar una figura sobre sí misma.

S

Same-side interior (exterior) angles (pp. 127, 129) Same side interior (exterior) angles lie on the same side of the transversal t and between (outside of) ℓ and m.

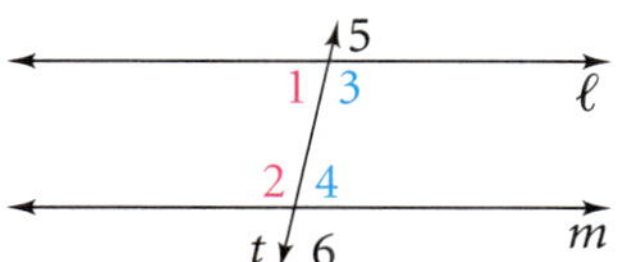

$\angle 1$ and $\angle 2$ are same-side interior angles, as are $\angle 3$ and $\angle 4$. $\angle 5$ and $\angle 6$ are same-side exterior angles.

Ángulos internos (externos) del mismo lado (pp. 127, 129) Los ángulos internos (externos) del mismo lado están situados en el mismo lado de la transversal t y dentro de (fuera de) ℓ y m.

Scalar multiplication (p. 504) Scalar multiplication is the multiplication of each entry in a matrix by the same number, the *scalar*.

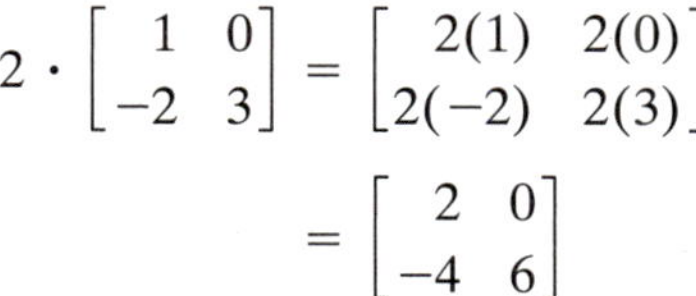

$$2 \cdot \begin{bmatrix} 1 & 0 \\ -2 & 3 \end{bmatrix} = \begin{bmatrix} 2(1) & 2(0) \\ 2(-2) & 2(3) \end{bmatrix} = \begin{bmatrix} 2 & 0 \\ -4 & 6 \end{bmatrix}$$

Multiplicación escalar (p. 504) La multiplicación escalar es la multiplicación de cada elemento de una matriz por el mismo número, el *escalar*.

English/Spanish Glossary

EXAMPLES

Scale (p. 368) A scale is the ratio of any length in a scale drawing to the corresponding actual length. The lengths may be in different units.

Escala (p. 368) Una escala es la razón de cualquier longitud en un dibujo a escala en relación a la longitud verdadera correspondiente. Las longitudes pueden expresarse en distintas unidades.

1 cm to 1 ft
1 cm = 1 ft
1 cm : 1 ft

Scale drawing (p. 368) A scale drawing is a drawing in which all lengths are proportional to corresponding actual lengths.

Dibujo a escala (p. 368) Un dibujo a escala es un dibujo en el que todas las longitudes son proporcionales a las longitudes verdaderas correspondientes.

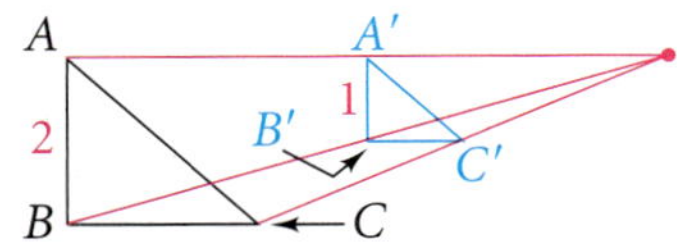

Scale factor (p. 498) The scale factor of a dilation is the number that describes the size change from an original figure to its image. *See also* **dilation.**

Factor de escala (p. 498) El factor de escala de una dilatación es el número que describe el cambio de tamaño de una figura original a su imagen. *Ver también* **dilation.**

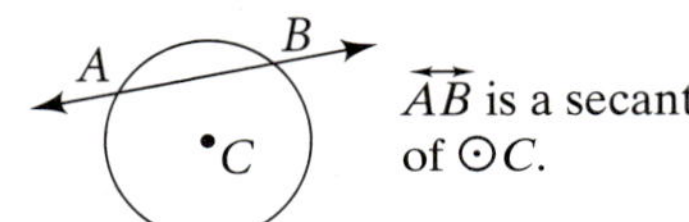

The scale factor of the dilation that maps $\triangle ABC$ to $\triangle A'B'C'$ is $\frac{1}{2}$.

Scalene triangle (p. 148) A scalene triangle has no sides congruent.

Triángulo escaleno (p. 148) Un triángulo escaleno no tiene lados congruentes.

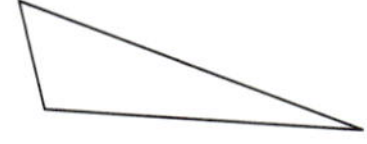

Secant (p. 687) A secant is a line, ray, or segment that intersects a circle at two points.

Secante (p. 687) Una secante es una recta, rayo o segmento que corta un círculo en dos puntos.

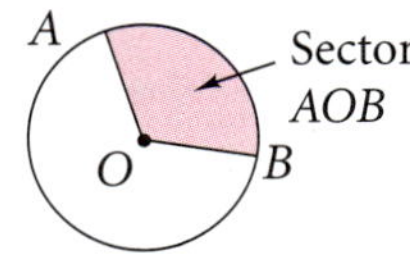

$\overleftrightarrow{AB}$ is a secant of $\odot C$.

Sector of a circle (p. 576) A sector of a circle is the region bounded by two radii and their intercepted arc.

Sector de un círculo (p. 576) Un sector de un círculo es la región limitada por dos radios y el arco abarcado por ellos.

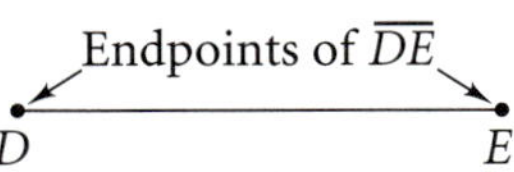

Segment (p. 23) A segment is the part of a line consisting of two points, called *endpoints,* and all points between them.

Segmento (p. 23) Un segmento es una parte de una recta que consiste en dos puntos, llamados *extremos,* y todos los puntos entre los extremos.

Endpoints of $\overline{DE}$

D E

Segment bisector (p. 32) A segment bisector is a line, segment, ray, or plane that intersects a segment at its midpoint.

Bisectriz de un segmento (p. 32) La bisectriz de un segmento es una recta, segmento, rayo o plano que corta un segmento en su punto medio.

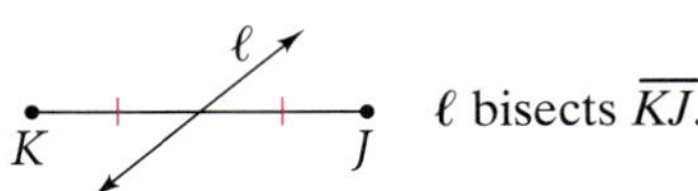

ℓ bisects $\overline{KJ}$.

EXAMPLES

Segment of a circle (p. 577) A segment of a circle is the part of a circle bounded by an arc and the segment joining its endpoints.

Segmento de un círculo (p. 577) Un segmento de un círculo es la parte de un círculo bordeada por un arco y el segmento que une sus extremos.

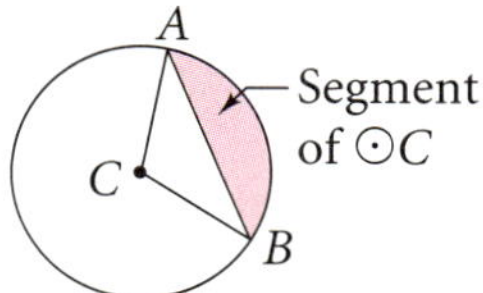

Semicircle (p. 567) A semicircle is half a circle.

Semicírculo (p. 567) Un semicírculo es la mitad de un círculo.

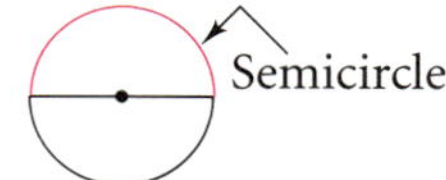

Side *See* **angle; polygon.**

Lado *Ver* **angle; polygon.**

Similar polygons (p. 373) Similar polygons are polygons having corresponding angles congruent and corresponding sides proportional. You denote similarity by $\sim$.

Polígonos semejantes (p. 373) Los polígonos semejantes son polígonos cuyos ángulos correspondientes son congruentes y los lados correspondientes son proporcionales. El símbolo $\sim$ significa "es semejante a".

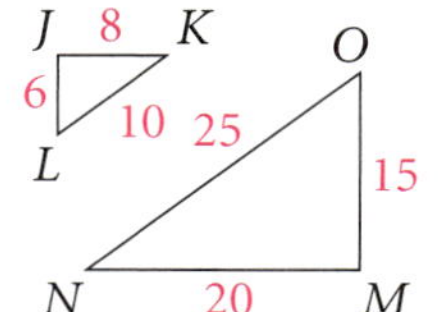

$\triangle JKL \sim \triangle MNO$

Similarity ratio $= \frac{2}{5}$

Similar solids (p. 646) Similar solids have the same shape and have all their corresponding dimensions proportional.

Cuerpos geométricos semejantes (p. 646) Los cuerpos geométricos semejantes tienen la misma forma y todas sus dimensiones correspondientes son proporcionales.

Similarity ratio (pp. 373, 646) The similarity ratio is the ratio of the lengths of corresponding sides of similar figures.

Razón de semejanza (pp. 373, 646) La razón de semejanza es la razón de la longitud de los lados correspondientes de figuras semejantes.

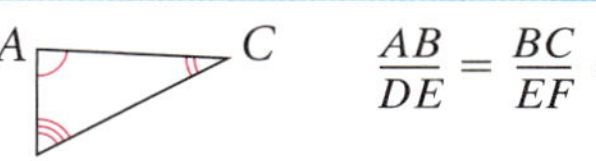

$\frac{AB}{DE} = \frac{BC}{EF} = \frac{CA}{FD}$

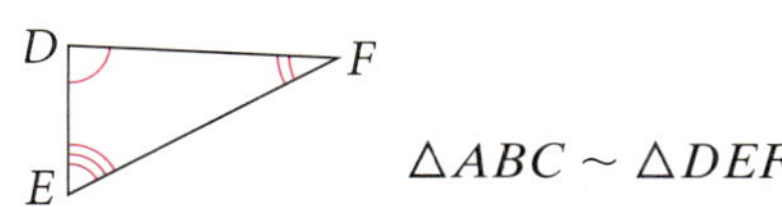

$\triangle ABC \sim \triangle DEF$

Similarity transformation (p. 498) *See* **dilation.**

Transformación de semejanza (p. 498) *Ver* **dilation.**

Sine ratio (p. 439) *See* **trigonometric ratios.**

Razón seno (p. 439) *Ver* **trigonometric ratios.**

English/Spanish Glossary

EXAMPLES

Skew lines (p. 24) Skew lines are lines that do not lie in the same plane.

Rectas cruzadas (p. 24) Las rectas cruzadas son rectas que no están en el mismo plano.

$\overleftrightarrow{AB}$ and $\overleftrightarrow{EF}$ are skew.

Slant height *See* **cone; pyramid.**

Generatriz (cono) o apotema (pirámide) *Ver* **cone; pyramid.**

Slide (p. 471) *See* **translation.**

Traslación (p. 471) *Ver* **translation.**

Slope-intercept form (p. 166) The slope-intercept form of a linear equation is $y = mx + b$, where m is the slope of the line and b is the y-intercept.

Forma pendiente-intercepto (p. 166) La forma pendiente-intercepto es la ecuación lineal $y = mx + b$, en la que m es la pendiente de la recta y b es el punto de intersección de esa recta con el eje y.

$y = \frac{1}{2}x - 3$

In this equation, the slope is $\frac{1}{2}$ and the y-intercept is -3.

Slope of a line (p. 165) The slope of a line is the ratio of its vertical change in the coordinate plane to the corresponding horizontal change. If (x_1, y_1) and (x_2, y_2) are points on a nonvertical line, then the slope is $\frac{y_2 - y_1}{x_2 - x_1}$. The slope of a horizontal line is 0 and the slope of a vertical line is undefined.

Pendiente de una recta (p. 165) La pendiente de una recta es la razón del cambio vertical en el plano de coordenadas en relación al cambio horizontal correspondiente. Si (x_1, y_1) y (x_2, y_2) son puntos en una recta no vertical, entonces la pendiente es $\frac{y_2 - y_1}{x_2 - x_1}$. La pendiente de una recta horizontal es 0, y la pendiente de una recta vertical es indefinida.

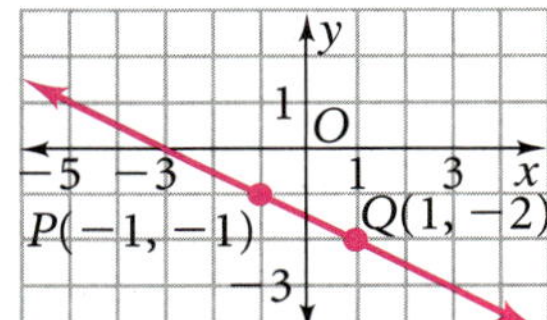

The line containing $P(-1, -1)$ and $Q(1, -2)$ has slope

$\frac{-2 - (-1)}{1 - (-1)} = \frac{-1}{2} = -\frac{1}{2}$.

Space (p. 17) Space is the set of all points.

Espacio (p. 17) El espacio es el conjunto de todos los puntos.

Sphere (p. 638) A sphere is the set of all points in space that are a given distance r, the *radius,* from a given point C, the *center.* A *great circle* is the intersection of a sphere with a plane containing the center of the sphere. The *circumference* of a sphere is the circumference of any great circle of the sphere.

Esfera (p. 638) Una esfera es el conjunto de los puntos del espacio que están a una distancia dada r, el *radio,* de un punto dado C, el *centro.* Un *círculo máximo* es la intersección de una esfera y un plano que contiene el centro de la esfera. La *circunferencia* de una esfera es la circunferencia de cualquier círculo máximo de la esfera.

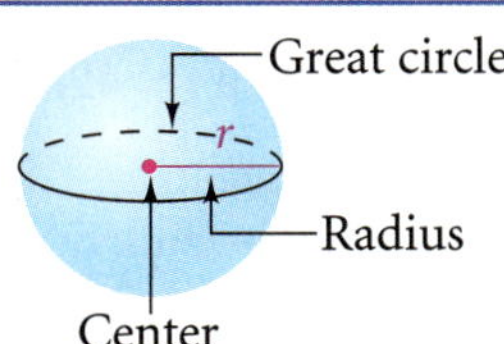

EXAMPLES

Spherical geometry (p. 154) In spherical geometry, a plane is considered to be the surface of a sphere and a line is considered to be a great circle of the sphere. In spherical geometry, through a point not on a given line there is no line parallel to the given line.

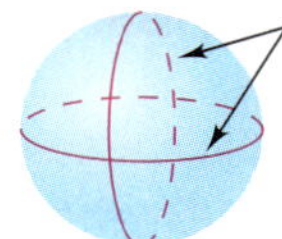

In spherical geometry, lines are represented by great circles of a sphere.

Geometría esférica (p. 154) En la geometría esférica, un plano es la superficie de una esfera y una recta es un círculo máximo de la esfera. En la geometría esférica, a través de un punto que no está en una recta dada, no hay recta paralela a la recta dada.

Square (p. 306) A square is a parallelogram with four congruent sides and four right angles.

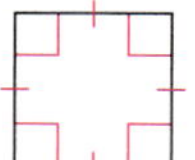

Cuadrado (p. 306) Un cuadrado es un paralelogramo con cuatro lados congruentes y cuatro ángulos rectos.

Standard form of an equation of a circle (p. 695) The standard form of an equation of a circle is $(x - h)^2 + (y - k)^2 = r^2$, where (h, k) is the center of the circle.

In $(x + 5)^2 + (y + 2)^2 = 48$, $(-5, -2)$ is the center of the circle.

Forma normal de la ecuación de un círculo (p. 695) La forma normal de la ecuación de un círculo es $(x - h)^2 + (y - k)^2 = r^2$, donde (h, k) son las coordenadas del centro del círculo.

Standard form of a linear equation (p. 167) The standard form of a linear equation is $Ax + By = C$, where A, B, and C are integers and A and B are not both zero.

$6x - y = 3$

Forma normal de una ecuación lineal (p. 167) La forma normal de una ecuación lineal es $Ax + By = C$, donde A, B y C son números reales, y donde A y B no son iguales a cero.

Straight angle (p. 37) A straight angle is an angle whose measure is 180.

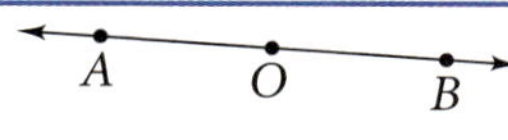

$m\angle AOB = 180$

Ángulo llano (p. 37) Un ángulo llano es un ángulo que mide 180.

Straightedge (p. 44) A straightedge is a ruler with no markings on it.

Regla sin graduación (p. 44) Una regla sin graduación no tiene marcas.

Supplementary angles (p. 38) Two angles are supplementary if the sum of their measures is 180.

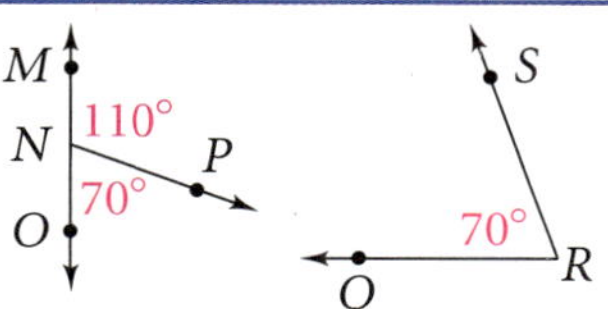

$\angle MNP$ and $\angle ONP$ are supplementary, as are $\angle MNP$ and $\angle QRS$.

Ángulos suplementarios (p. 38) Dos ángulos son suplementarios cuando sus medidas suman 180.

EXAMPLES

Surface area (pp. 608, 610, 618, 619, 638) The surface area of a prism, cylinder, pyramid, or cone is the sum of the lateral area and the areas of the bases. The surface area of a sphere is four times the area of a great circle. A list of surface area formulas is on p. 765.

Área (pp. 608, 610, 618, 619, 638) El área de un prisma, pirámide, cilindro o cono es la suma del área lateral y las áreas de las bases. El área de una esfera es igual a cuatro veces el área de un círculo máximo. Una lista de fórmulas de áreas está en la p. 765.

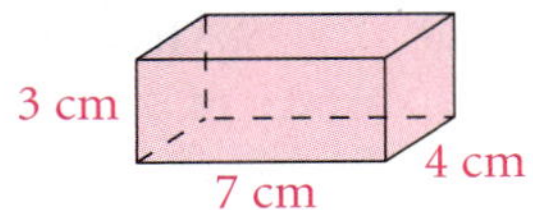

S.A. of prism $= \text{L.A.} + 2B$
$= 66 + 2(28)$
$= 122 \text{ cm}^2$

Symmetry (pp. 492, 493, 516) A figure has symmetry if there is an isometry that maps the figure onto itself. *See* **glide reflectional symmetry; point symmetry; reflectional symmetry; rotational symmetry; translational symmetry.**

Simetría (pp. 492, 493, 516) Una figura tiene simetría si hay una isometría que traza la figura sobre sí misma. *Ver* **glide reflectional symmetry; point symmetry; reflectional symmetry; rotational symmetry; translational symmetry.**

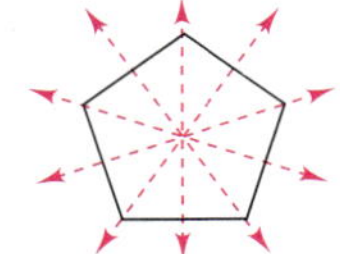

A regular pentagon has reflectional symmetry and 72° rotational symmetry.

Tangent ratio (p. 432) *See* **trigonometric ratios.**

Razón tangente (p. 432) *Ver* **trigonometric ratios.**

Tangent to a circle (p. 598) A tangent to a circle is a line, segment, or ray in the plane of the circle that intersects the circle in exactly one point. That point is the *point of tangency.*

Tangente de un círculo (p. 598) Una tangente de un círculo es una recta, segmento o rayo en el plano del círculo que corta el círculo en exactamente un punto. Ese punto es el *punto de tangencia.*

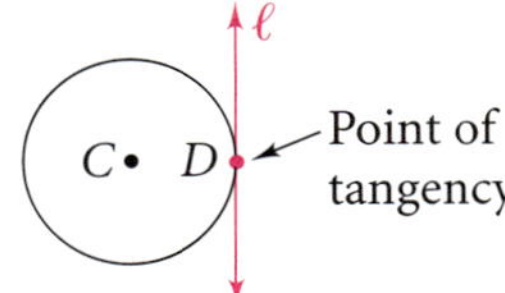

Line ℓ is tangent to $\odot C$. Point D is the point of tangency.

Terminal point of a vector (p. 452) *See* **vector.**

Punto terminal de un vector (p. 452) *Ver* **vector.**

Tessellation (p. 515) A tessellation, or *tiling,* is a repeating pattern of figures that completely covers a plane without gaps or overlap. A *pure tessellation* is a tessellation that consists of congruent copies of one figure.

Teselado (p. 515) Un teselado o *reticulado* es un patrón repetitivo de figuras que cubre completamente una superficie plana sin dejar espacios vacíos ni traslaparse. Un *teselado puro* consiste en copias congruentes de una figura.

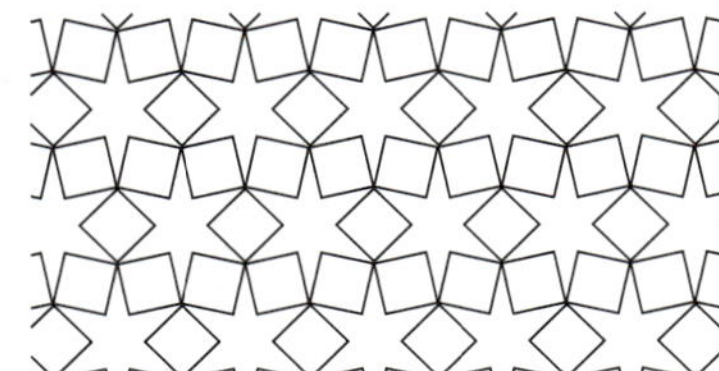

Theorem (p. 110) A theorem is a conjecture that is proven.

Teorema (p. 110) Un teorema es una conjetura que se demuestra.

The theorem "Vertical angles are congruent" can be proven by using postulates, definitions, properties, and previously stated theorems.

Tiling (p. 515) *See* **tessellation.**

Reticulado (p. 515) *Ver* **tessellation.**

Transformation (p. 470) A transformation is a change in the position, size, or shape of a geometric figure. The given figure is called the *preimage* and the resulting figure is called the *image.* A transformation *maps* a figure onto its image. *Prime notation* is sometimes used to identify image points. In the diagram, X' (read "X prime") is the image of X.

Transformación (p. 470) Una transformación es un cambio en la posición, tamaño o forma de una figura. La figura dada se llama la *preimagen* y la figura resultante se llama la *imagen.* Una transformación *traza* la figura sobre su propia imagen. La *notación prima* a veces se utiliza para identificar los puntos de la imagen. En el diagrama de la derecha, X' (leído X prima) es la imagen de X.

Preimage Image

X Y Z Y′ X′ Z′

$\triangle XYZ \rightarrow \triangle X'Y'Z'$

Translation (p. 471) A translation (*slide*) is a transformation that moves points the same distance and in the same direction. A translation in the coordinate plane is described by a vector.

Traslación (p. 471) Una traslación es la transformación que mueve puntos a la misma distancia y en la misma dirección. Un vector puede describir la traslación en un plano de coordenadas.

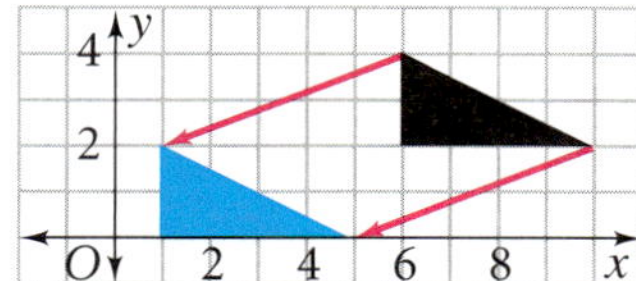

The blue triangle is the image of the black triangle under the translation $\langle -5, -2 \rangle$.

Translational symmetry (p. 516) Translational symmetry is the type of symmetry for which there is a translation that maps a figure onto itself.

Simetría traslacional (p. 516) La simetría traslacional es un tipo de simetría en la que la traslación vuelve a trazar la figura sobre sí misma.

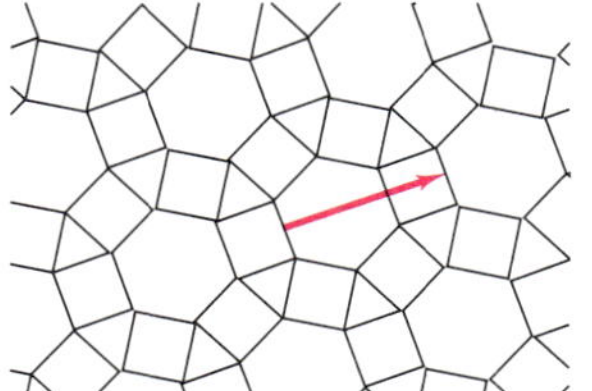

The tessellation shown can be mapped onto itself by the given translation.

Transversal (p. 127) A transversal is a line that intersects two coplanar lines in two points.

Transversal (p. 127) Una transversal es una recta que corta dos rectas coplanarias en dos puntos.

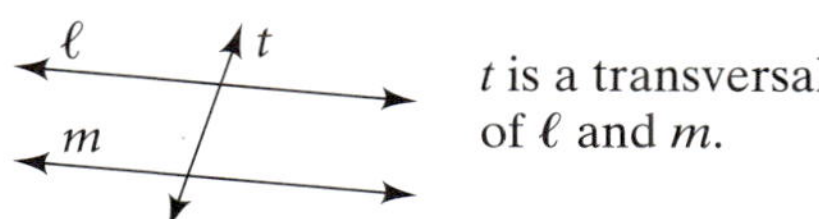

t is a transversal of ℓ and m.

EXAMPLES

Trapezoid (pp. 306, 336) A trapezoid is a quadrilateral with exactly one pair of parallel sides, the *bases.* The nonparallel sides are called the *legs* of the trapezoid. Each pair of angles adjacent to a base are *base angles* of the trapezoid. An *altitude* of a trapezoid is a perpendicular segment from one base to the line containing the other base. Its length is called the *height* of the trapezoid.

Trapecio (pp. 306, 336) Un trapecio es un cuadrilátero con exactamente un par de lados paralelos, las *bases.* Los lados no paralelos se llaman los *catetos* del trapecio. Cada par de ángulos adyacentes a la base son los *ángulos de base* del trapecio. Una *altura* del trapecio es un segmento perpendicular que va de una base a la recta que contiene la otra base. Su longitud se llama, por extensión, la *altura* del trapecio.

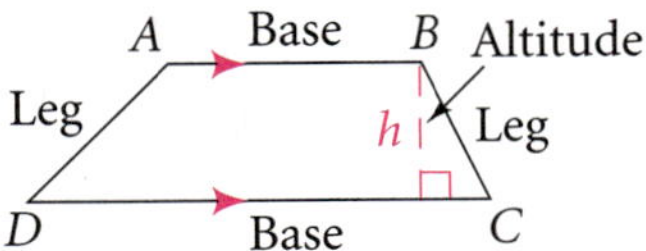

In trapezoid $ABCD$, $\angle ADC$ and $\angle BCD$ are one pair of base angles, and $\angle DAB$ and $\angle ABC$ are the other.

Triangle (pp. 158, 275, 536) A triangle is a polygon with three sides. You can choose any side to be a *base.* The *height* is the length of the altitude drawn to the line containing that base.

Triángulo (pp. 158, 275, 536) Un triángulo es un polígono con tres lados. Se puede escoger cualquier lado como *base.* La *altura,* entonces, es la longitud de la altura trazada hasta la recta que contiene la base.

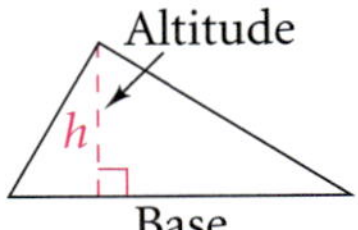

Trigonometric ratios (pp. 432, 439) In right triangle $\triangle ABC$ with acute angle $\angle A$

$$\text{sine } \angle A = \sin A = \frac{\text{leg opposite } \angle A}{\text{hypotenuse}}$$

$$\text{cosine } \angle A = \cos A = \frac{\text{leg adjacent } \angle A}{\text{hypotenuse}}$$

$$\text{tangent } \angle A = \tan A = \frac{\text{leg opposite } \angle A}{\text{leg adjacent } \angle A}$$

Razones trigonométricas (pp. 432, 439) En un triángulo rectángulo $\triangle ABC$ con ángulo agudo $\angle A$

$$\text{seno } \angle A = \text{sen } A = \frac{\text{cateto opuesto a } \angle A}{\text{hipotenusa}}$$

$$\text{coseno } \angle A = \cos A = \frac{\text{cateto adyecente a } \angle A}{\text{hipotenusa}}$$

$$\text{tangente } \angle A = \tan A = \frac{\text{cateto opuesto a } \angle A}{\text{cateto adyecente a } \angle A}$$

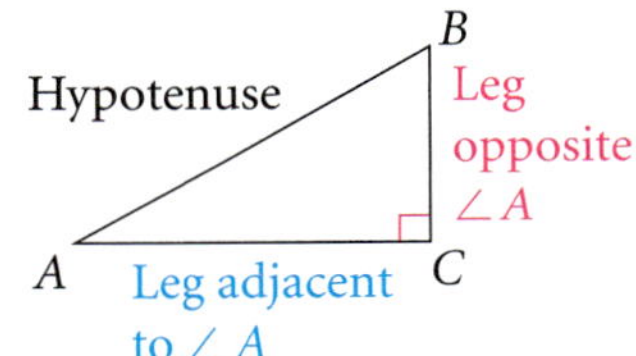

Truth value (p. 81) The truth value of a statement is “true” or “false” according to whether the statement is true or false, respectively.

Valor verdadero (p. 81) El valor verdadero de un enunciado es “verdadero” o “falso” según el enunciado sea verdadero o falso, respectivamente.

The truth value of the statement “If a figure is a triangle, then it has four sides” is *false*.

Turn (p. 484) *See* **rotation.**

Rotación (p. 484) *Ver* **rotation.**

EXAMPLES

Two-column proof (p. 129) *See* **proof.**

Prueba de dos columnas (p. 129) *Ver* **proof.**

Vector (p. 452) A vector is any quantity that has *magnitude* (size) and *direction.* You can represent a vector as an arrow that starts at one point, the *initial point,* and points to a second point, the *terminal point.* A vector can be described by *ordered pair notation* $\langle x, y\rangle$, where x represents horizontal change from the initial point to the terminal point and y represents vertical change from the initial point to the terminal point.

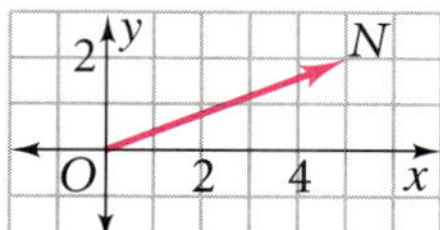

Vector ON has initial point O and terminal point N. The ordered pair notation for the vector is $\langle 5, 2\rangle$.

Vector (p. 452) Un vector es cualquier cantidad que tiene *magnitud* (tamaño) y *dirección.* Se puede representar un vector como una flecha que empieza en un punto, el *punto inicial,* y se dirige a un segundo punto, el *punto terminal.* Un vector se puede describir mediante la *notación de pares ordenados* $\langle x, y\rangle$, donde x representa el cambio horizontal desde el punto inicial hasta el punto final, e y representa el cambio vertical desde el punto inicial hasta el punto final.

Vertex *See* **angle; cone; polygon; polyhedron; pyramid.** The plural form of *vertex* is *vertices.*

Vértice *Ver* **angle; cone; polygon; polyhedron; pyramid.**

Vertex angle (p. 228) *See* **isosceles triangle.**

Ángulo del vértice (p. 228) *Ver* **isosceles triangle.**

Vertical angles (p. 38) Vertical angles are two angles whose sides form two pairs of opposite rays.

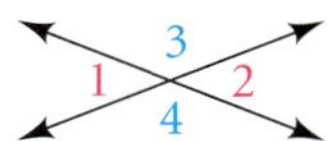

$\angle 1$ and $\angle 2$ are vertical angles, as are $\angle 3$ and $\angle 4$.

Ángulos opuestos por el vértice (p. 38) Dos ángulos son ángulos opuestos por el vértice si sus lados son rayos opuestos.

Volume (p. 624) Volume is a measure of the space a figure occupies. A list of volume formulas is on p. 765.

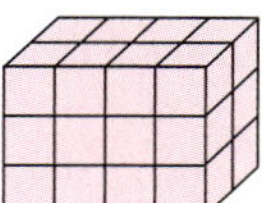

The volume of this prism is 24 cubic units, or 24 units^3.

Volumen (p. 624) El volumen es una medida del espacio que ocupa una figura. Una lista de las fórmulas de volumen está en la p. 765.

Answers to California Check System

Chapter 1

Check Your Readiness — p. 2

1. 9 **2.** 16 **3.** 121 **4.** 37 **5.** 78.5 **6.** 13 **7.** 1 **8.** $-\frac{3}{5}$ **9.** 5 **10.** 8 **11.** 4 **12.** 3 **13.** 3 **14.** 6 **15.** 1

Lesson 1-1 — pp. 4–6

Check Skills You'll Need **1.** 2, 4, 6, 8, 10, . . . **2.** 1, 3, 5, 7, 9, . . . **3.** $1^2 = 1, 2^2 = 4, 3^2 = 9, 4^2 = 16, 5^2 = 25, 6^2 = 36, 7^2 = 49, 8^2 = 64, 9^2 = 81, 10^2 = 100$ **4.** It is odd.

CA Standards Check **1a.** 29, 37 **b.** Thursday, Friday **c.**

2. The sum of the first 35 odd numbers is 35^2, or 1225.

3.

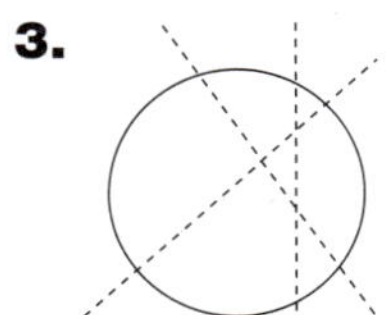

4a. 39 skateboards **b.** Not confident; December is too far away.

Lesson 1-2 — pp. 10–12

Check Skills You'll Need
1.

2.

CA Standards Check
1. **2.**

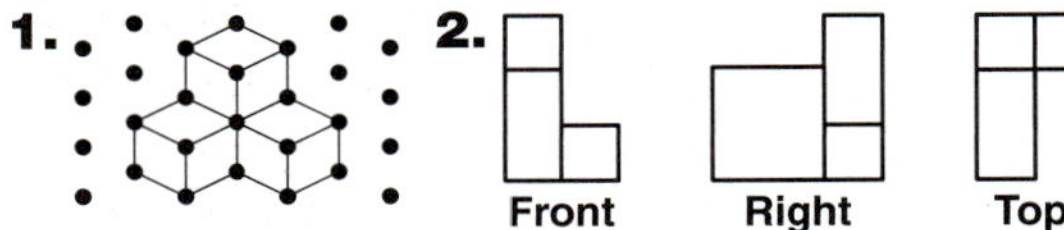

3a. 9 cubes **b.** Answers may vary. Sample: The foundation drawing; you can just add the five numbers. **4.** E, C **5.**

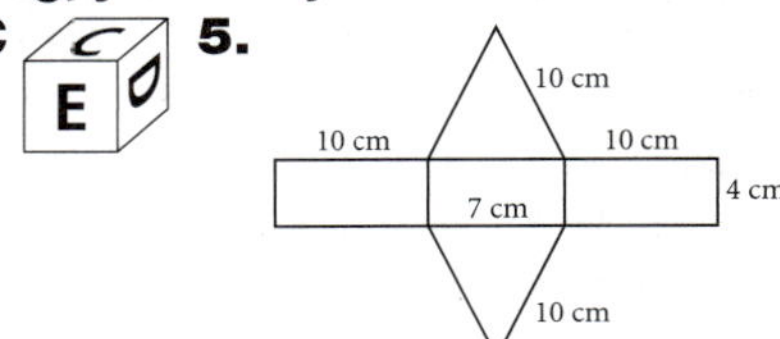

Lesson 1-3 — pp. 16–19

Check Skills You'll Need **1.** (1, 6) **2.** (3, 2) **3.** (5, 10)

CA Standards Check **1a.** no **b.** Answers may vary. Sample: $\overleftrightarrow{EF}$, $\overleftrightarrow{FC}$, $\overleftrightarrow{CE}$ **c.** Arrowheads are used to show that the line extends in opposite directions without end. **2.** Answers may vary. Sample: *HEF*, *HEFG*, *FGH* **3.** *ABF* and *CBF* **4a.** *D* **b.** *B*

Lesson 1-4 — pp. 23–25, 28

Check Skills You'll Need **1.** no **2.** yes **3.** no **4–9.** Answers may vary. Samples are given. **4.** *NMR* **5.** *PQL* **6.** *NKL* **7.** *PQR* **8.** *PKN* **9.** *LQR*

CA Standards Check **1.** No, they do not have the same endpoint. **2a.** $\overline{HI}$, $\overline{DN}$ **b.** $\overline{AB}$, $\overline{CD}$, $\overline{CH}$ **c.** $\overline{DN}$, $\overline{HI}$; $\overline{DN}$, $\overline{HC}$ **3a.** *PSWT* ∥ *RQVU*, *PRUT* ∥ *SQVW*, *PSQR* ∥ *TWVU* **b.** Answers may vary. Sample: $\overleftrightarrow{PS}$

Checkpoint Quiz 1 **1.** 29, 31.5 **2.** 3.45678, 3.456789 **3.** For 1: Add 2.5. For 2: Extend the decimal to one more place with a digit that is 1 more than the one to its left. **4.** yes, plane *AEF* **5.** yes, plane *DCEF* **6.** No; *H*, *G*, and *F* are in the front plane, *B* is not. **7.** No; *A*, *E*, and *B* are in the top plane, *C* is not. **8.** $\overline{CD}$, $\overline{AB}$, $\overline{EF}$ **9.** Answers may vary. Sample: $\overleftrightarrow{AE}$ and $\overleftrightarrow{BC}$ **10.**

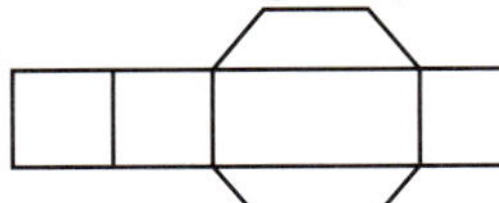

Lesson 1-5 — pp. 31–33

Check Skills You'll Need **1.** 6 **2.** 3.5 **3.** 3 **4.** 6 **5.** 2 **6.** 9 **7.** 4 **8.** 9 **9.** $\frac{1}{3}$

CA Standards Check **1a.** $CD = DE$; yes **b.** yes; $|-5 - (-8)| = |3| = 3$ **2.** 15; $EF = 40$, $FG = 60$ **3.** 13.5

Lesson 1-6 — pp. 36–39

Check Skills You'll Need **1.** 80 **2.** 130 **3.** 45 **4.** 135 **5.** 110 **6.** 45

CA Standards Check **1a.** ∠2, ∠*DEC* **b.** No; 3 ∠s have *E* for a vertex, so you need more info. in the name to distinguish them from one another. **2a.** 30; acute **b.** 90; right **c.** 140; obtuse **3.** 35 **4a.** Answers may vary. Sample: ∠*AFB* and ∠*BFC*; ∠*BFD* and ∠*DFE* **b.** 153 **5a.** Yes; the congruent segments are marked. **b.** No; there are no markings. **c.** No; there are no markings. **d.** No; there are no markings.
e. Yes; the congruent segments are marked.

Lesson 1-7 — pp. 44–47

Check Skills You'll Need **1.** C D

2. G H **3.** A B

4. **5.** **6.**

7. 10 **8.** **9.**

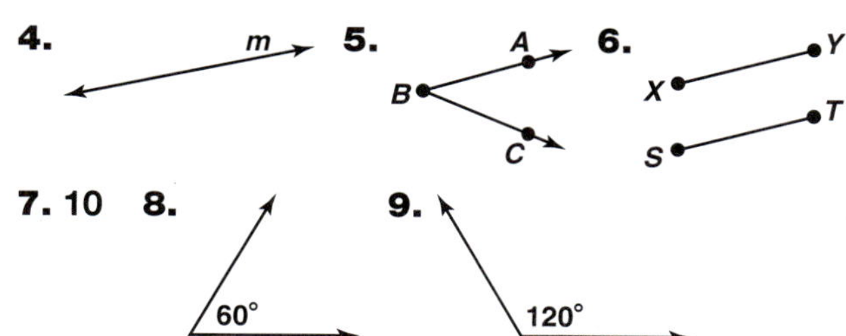

CA Standards Check **1.**

2.

B F

3.

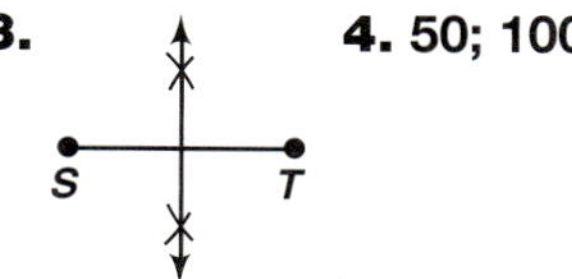

4. 50; 100

5a.

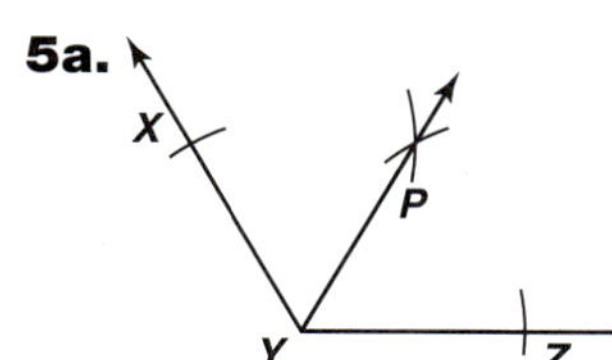

b. Measure $\angle XYP$ and $\angle PYZ$ to see that they are $\cong$.

Lesson 1-8 pp. 53–55

Check Skills You'll Need **1.** 5.0 **2.** 4.1 **3.** 11.1 **4.** 100 **5.** 100 **6.** 58 **7.** 196 **8.** 10 **9.** -1

CA Standards Check **1a.** 8.6 **b.** Yes; the differences are opposites, and the square of a number and the square of its opposite are the same. **2a.** about 8.9 mi **b.** about 3.2 mi **3.** (4, 4) **4.** (6, -9)

Checkpoint Quiz 2 **1.** 17 **2.** 110 **3.** 140 **4a.** 90 **b.** 60 **5.** $\angle APT \cong \angle RPT$ **6.** 45

7.

8.

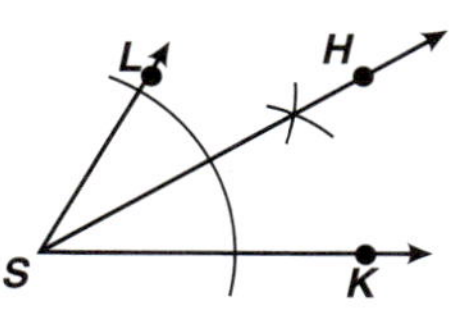

9. 12.2 units **10.** (1, 1.5)

Lesson 1-9 pp. 61–64

Check Skills You'll Need **1.** 4 **2.** 15 **3.** 8 **4.** 6.7 **5.** 3.2 **6.** 7.8 **7.** 4.5 **8.** 13.0 **9.** 7.8

CA Standards Check **1a.** 26 in. **b.** 30 in. **2a.** 36π m **b.** 56.5 m **3.**

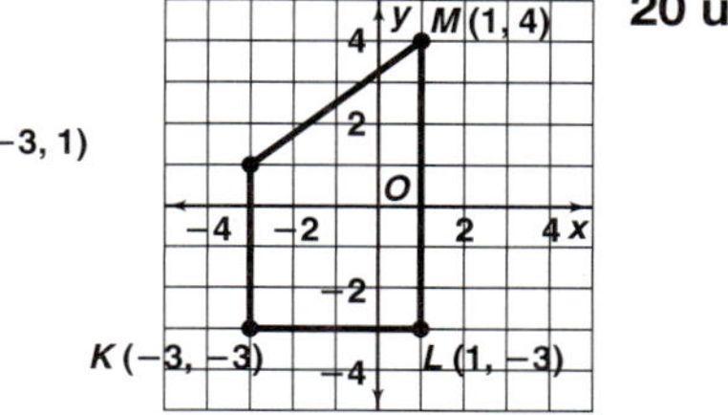

20 units

4. $9\frac{1}{3}$ yd^2; $9\frac{1}{3}$ is one-ninth of 84.

5a. $\frac{25}{4}\pi$ ft^2 **b.** 19.6 ft^2

6.

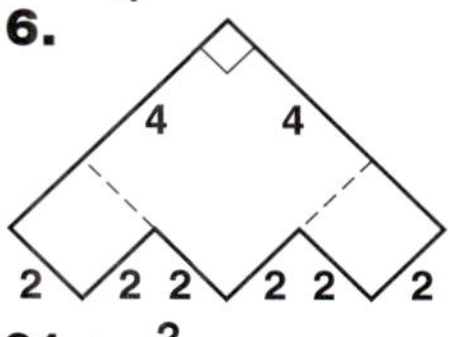

24 cm^2

Chapter 2

Check Your Readiness p. 78

1. 50 **2.** -3 **3.** $25\frac{1}{2}$ **4.** 10.5 **5.** 15 **6.** 11 **7.** 20 **8.** -5 **9.** -4 **10.** 5 **11.** 6 **12.** 7 **13.** 18 **14.** $\angle ACD$, $\angle DCA$ **15.** C **16.** 3 **17.** $\angle ADB$ or $\angle BDA$ **18.** $\overline{CD}$ **19.** 45 **20.** 48, 42

Lesson 2-1 pp. 80–82

Check Skills You'll Need **1.** -1 **2.** $-2, 2$ **3.** 2 **4.** -4 **5.** 0 **6.** -1

CA Standards Check **1.** Hypothesis: $T - 38 = 3$, Conclusion: $T = 41$ **2a.** If an integer ends with 0, then it is divisible by 5. **b.** If a figure is a square, then it has 4 congruent sides. **3.** The conditional is false; New Mexico is a counterexample.

4.

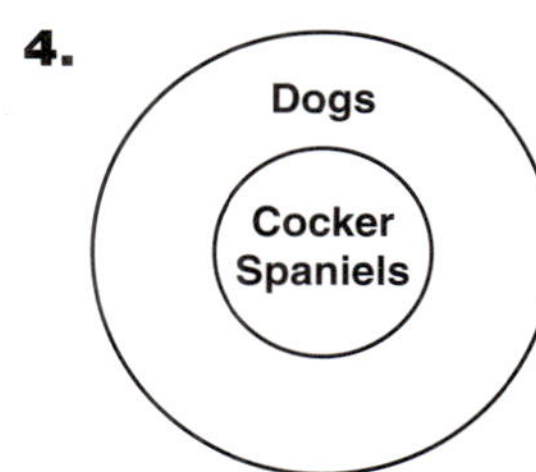

5. If two lines are skew, then they are not parallel and do not intersect. **6a.** If two lines are parallel, then they do not intersect. The conditional is false and the converse is true. **b.** If $|x| = 2$, then $x = 2$. The conditional is true and the converse is false. **7.** Answers may vary. Sample: The statement "I breathe when I sleep" can be rewritten as "If I sleep, then I breathe." The statement "I sleep when I breathe" can be rewritten as "If I breathe, then I sleep." The two statements are converses, and do not have the same meaning.

Lesson 2-2 pp. 87–89

Check Skills You'll Need **1.** Hypothesis: $x > 10$, Conclusion: $x > 5$ **2.** Hypothesis: You live in Milwaukee. Conclusion: You live in Wisconsin. **3.** If a figure is a square, then it has four sides. **4.** If something is a butterfly, then it has wings. **5.** If we go on a picnic, then the sun shines. **6.** If two lines do not intersect, then they are skew. **7.** If $x^3 = -27$, then $x = -3$.

CA Standards Check **1.** If three points lie on the same line, then they are collinear. The converse is also true. Three points are collinear if and only if they lie on the same line. **2.** If a number is prime, then it has only two distinct factors, 1 and itself. If a number has only two distinct factors, 1 and itself, then it is prime. **3.** Conditional: If an angle is a right angle, then its measure is 90. Converse: If an angle has measure 90, then it is a right angle. The two statements are true. An angle is a right angle if and only if its measure is 90. **4.** It is not a good definition because a rectangle has four right angles and is not necessarily a square.

Lesson 2-3 pp. 94–96, 100

Check Skills You'll Need **1.** If your grades suffer, then you don't sleep enough. **2.** If you must start early, then you want to arrive on time. **3.** If a year is a leap year, then it has 366 days. **4.** If students do not complete their homework, then they will have lower grades. **5.** If two lines are perpendicular, then they meet to form right angles. **6.** If a person is 16 years old, then that person is a teenager.

CA Standards Check **1.** No, there could be other things wrong with the car, such as a faulty starter. **2.** Answers may vary. Sample: Vladimir Nuñez should not pitch a complete game on Tuesday. **3.** Not possible; you do not know that the hypothesis is true. **4a.** If a number ends in 0, then it is divisible by 5. **b.** Not possible; the conclusion of one statement is not the hypothesis of the other statement. **5.** The Volga River is less than 2300 miles long. The Volga River is not one of the world's ten longest rivers.

Checkpoint Quiz 1 **1.** Hypothesis: $x > 5$, Conclusion: $x^2 > 25$ **2.** If something is a rose, then it is a beautiful flower. **3.** If an integer is divisible by 2, then the integer ends with 0. **4.** Answers may vary. Sample: 42 is divisible by 2, but it does not end with 0. **5.** If an angle is an acute angle, then its measure is between 0 and 90. If an angle's measure is between 0 and 90, then it is an acute angle. **6.** Points are collinear if and only if they lie on the same line. **7.** Answers may vary. Sample: A graphing calculator has a keyboard and a memory. **8.** Theresa has passing grades. **9.** If a student studies geometry, then the student's mind is expanded. **10.** not possible

Lesson 2-4 pp. 103–105

Check Skills You'll Need **1.** $\angle AOB$, $\angle BOA$ **2.** O **3.** $\overrightarrow{OB}$ **4.** 45 **5.** $\overrightarrow{OA}$ and $\overrightarrow{OC}$

CA Standards Check **1.** Subst. Prop.; Subtr. Prop. of =; Div. Prop. of = **2.** $AB = 12$; $BC = 9$; $AB + BC = 12 + 9 = 21$ **3a.** Reflexive Prop. of ≅ **b.** Transitive or Subst. Prop. of ≅

Lesson 2-5 pp. 110–112

Check Skills You'll Need **1.** 50 **2.** 90 **3.** 35 **4.** right ∠s **5.** vertex

CA Standards Check **1a.** 140; 140 **b.** 40; 40 **c.** 140 + 40 = 180 **2.** Subtraction Property of Equality

Chapter 3

Check Your Readiness p. 124

1. 72 **2.** 1260 **3.** 2700 **4.** 4 **5.** 15 **6.** −24 **7.** $2x + x + x = 180$; 45, 45, 90

8. $\frac{1}{2}x + x + x = 180$; 36, 72, 72

9.

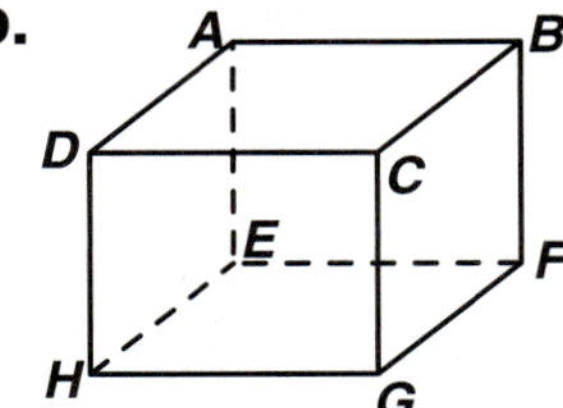

$\overleftrightarrow{AB}$ and $\overleftrightarrow{GH}$; $\overleftrightarrow{AE}$ and $\overleftrightarrow{EF}$

10–12. Check students' work.

Lesson 3-1 pp. 127–130

Check Skills You'll Need **1.** 30 **2.** 30 **3.** 60 **4.** 9 **5.** $m\angle 1 + 2(90 - m\angle 1) = 146$; $m\angle 1 = 34$ **6.** 72 and 108

CA Standards Check **1.** ∠5 and ∠4; ∠6 and ∠2; ∠3 and ∠8 **2.** same-side int. ∠s **3.** 1. $a \parallel b$ (Given) 2. $m\angle 3 + m\angle 2 = 180$ (∠ Add. Post.) 3. $m\angle 1 = m\angle 2$ (Corr. ∠s Post.) 4. $m\angle 1 + m\angle 3 = 180$ (Subst.) 5. ∠1 and ∠3 are suppl. (Def. of Suppl. ∠s) **4a.** 130; corr. ∠s are ≅. **b.** 130; vert. ∠s are ≅. **c.** 50; alt. int. ∠s are ≅. **d.** 50; alt. int. ∠s are ≅. **e.** 130; same-side int. ∠s are suppl. **f.** 50; corr. ∠s are ≅ or vert. ∠s are ≅. **5.** $x = 45$, $y = 115$; 90, 90, 115, 65

Lesson 3-2 pp. 134–137

Check Skills You'll Need **1.** 11 **2.** 4 **3.** 26 **4.** 6 **5.** If a △ has a 90° ∠, then it is a right △; true. **6.** If two ∠s are ≅, then they are vert. ∠s; false. **7.** If two ∠s are suppl., then they are same-side int. ∠s; false.

CA Standards Check **1.** $\overleftrightarrow{EC} \parallel \overleftrightarrow{DK}$; Conv. of Corr. ∠s Post. **2.** 18; $7 \cdot 18 - 8 = 118$, and $62 + 118 = 180$. **3.** $m\angle 3 = 120$; the lines will be ∥ by the Conv. of Same-Side Ext. ∠s Thm.

Lesson 3-3 pp. 141–142

Check Skills You'll Need **1.** sometimes **2.** always **3.** sometimes **4.** never

CA Standards Check **1.** Yes; 30° + 60° = 90° **2.** Yes; $b \parallel d$ by Thm. 3-10.

Lesson 3-4 pp. 147–150

Check Skills You'll Need **1.** right **2.** acute **3.** acute **4.** 60 **5.** 20 **6.** 32 **7.** 58

CA Standards Check **1.** 55

2a. 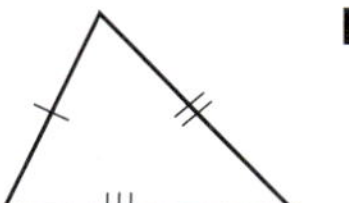

b. **c.** Not possible; an equilateral △ has all acute ∠s.

3. 135, 135, 90 **4a.** 130 **b.** Answers may vary. Sample: Find the measure of the third ∠ of the triangle. Subtract this from 180.

Checkpoint Quiz 1 **1.** Corr. ∠s Post. **2.** Conv. of Corr. ∠s Post. **3.** Same-Side Int. ∠s Thm. **4.** Conv. of the Alt. Int. ∠s Thm. **5.** Vert. ∠s Thm. **6.** Alt. Int. ∠s Thm. **7.** Conv. of Alt. Ext. ∠s Thm. **8.** Corr. ∠s Post. **9.** If a line is ⊥ to 1 of 2 parallel lines, it is ⊥ to both. **10.** 38, 55, 87; acute, 55, 26, 99; obtuse

Lesson 3-5 pp. 157–160

Check Skills You'll Need **1.** $m\angle DAB = 77$; $m\angle B = 65$; $m\angle BCD = 131$; $m\angle D = 87$ **2.** $m\angle D = m\angle B = 60$; $m\angle DAB = m\angle DCB = 120$ **3.** $m\angle A = 70$; $m\angle ABC = 85$; $m\angle C = 125$; $m\angle ADC = 80$

CA Standards Check **1.** *ABE*; sides: $\overline{AB}, \overline{BE}, \overline{EA}$; ∠s: $\angle A, \angle ABE, \angle BEA$; *BCDE*; sides: $\overline{BC}, \overline{CD}, \overline{DE}, \overline{EB}$; ∠s: $\angle EBC, \angle C, \angle D, \angle DEB$; *ABCDE*; sides: $\overline{AB}, \overline{BC}, \overline{CD}, \overline{DE}, \overline{EA}$; ∠s: $\angle A, \angle ABC, \angle C, \angle D, \angle AED$ **2a.** hexagon; convex **b.** octagon; concave **c.** 24-gon; concave **3a.** 1980 **b.** You can solve the equation $(n - 2)180 = 720$. **4.** 108 **5a.** 60 **b.** 30; no, it is not formed by extending one side of the polygon.

Lesson 3-6 pp. 166–168

Check Skills You'll Need **1.** $-\frac{2}{3}$ **2.** $\frac{5}{3}$ **3.** 0 **4.** undefined or no slope **5.** $\frac{2}{3}$ **6.** $-\frac{3}{2}$ **7.** $\frac{4}{5}$ **8.** -1

CA Standards Check

1. 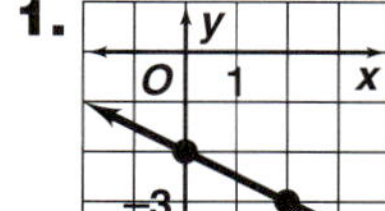

2. 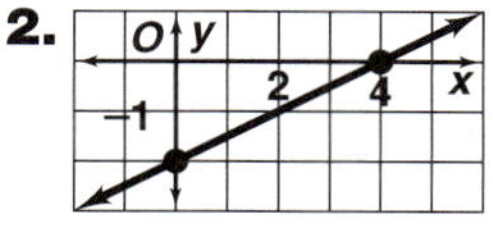

3. 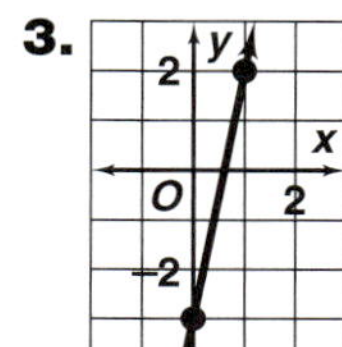

4. $y + 4 = -1(x - 2)$
5. $y - 0 = -\frac{3}{2}(x - 5)$ or $y + 3 = -\frac{3}{2}(x - 7)$
6. $y = -1$; $x = 5$

Lesson 3-7 pp. 174–177, 180

Check Skills You'll Need **1.** $\frac{1}{2}$ **2.** $\frac{5}{2}$ **3.** -5 **4.** 2 **5.** -1 **6.** $\frac{2}{3}$ **7.** 1 **8.** 0 **9.** $\frac{2}{3}$

CA Standards Check **1.** No; the slope of $\ell_3 = -\frac{1}{7}$, and the slope of $\ell_4 = -\frac{1}{6}$. **2a.** Yes; both slopes $= -\frac{1}{2}$, and the *y*-intercepts are different. **b.** No; the lines are the same. **3.** $y = -x + 3$ **4.** No; the slope of $\ell_3 = \frac{4}{9}$, and the slope of $\ell_4 = -\frac{7}{3}$, and $\frac{4}{9} \cdot -\frac{7}{3} \neq -1$. **5.** $y + 4 = -5(x - 15)$ **6.** $y - 8 = \frac{3}{2}(x - 2)$ or $y = \frac{3}{2}x + 5$

Checkpoint Quiz 2 **1.** octagon; $n = 125$ **2.** hexagon; $x = 122$; $w = 90$ **3.** quad.; $a = 105$; $m = 116$

4. 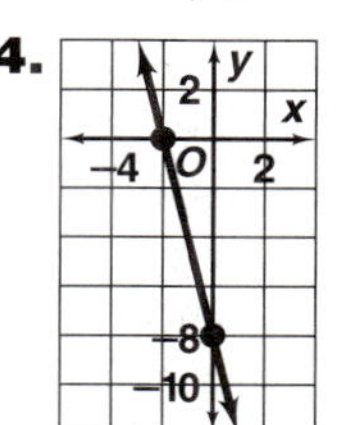

5. 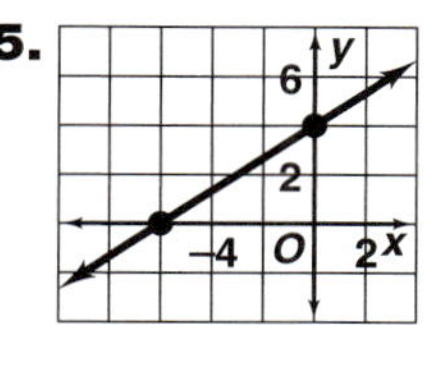

6. 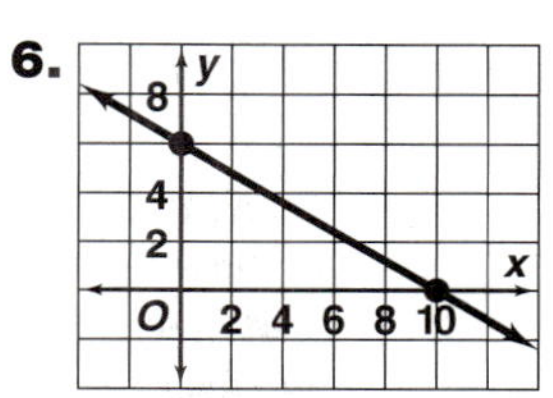

7. slope of $\overleftrightarrow{RS} = -\frac{2}{5}$; slope of $\overleftrightarrow{TV} = \frac{5}{3}$; neither
8. slope of $\overleftrightarrow{RS} = 1$; slope of $\overleftrightarrow{TV} = -1$; ⊥
9. slope of $\overleftrightarrow{RS} = -\frac{5}{4}$; slope of $\overleftrightarrow{TV} = \frac{4}{5}$; ⊥
10. slope of $\overleftrightarrow{RS} = \frac{2}{9}$; slope of $\overleftrightarrow{TV} = \frac{2}{9}$; ∥

Lesson 3-8 pp. 181–183

Check Skills You'll Need **1.** $\overline{AB} \cong \overline{CD}$

2. 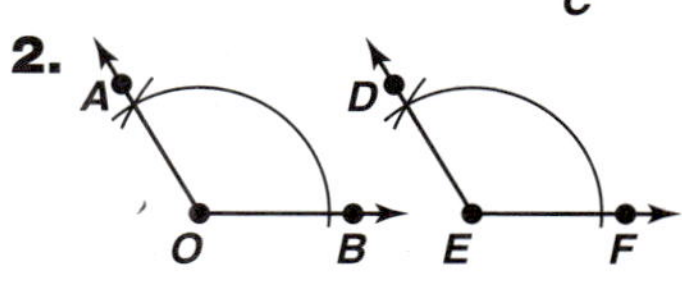

$\angle AOB \cong \angle DEF$

3. 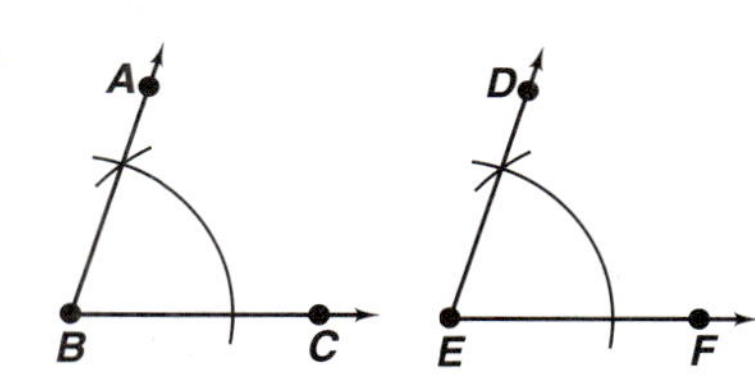

$\angle ABC \cong \angle DEF$

4. **5.** 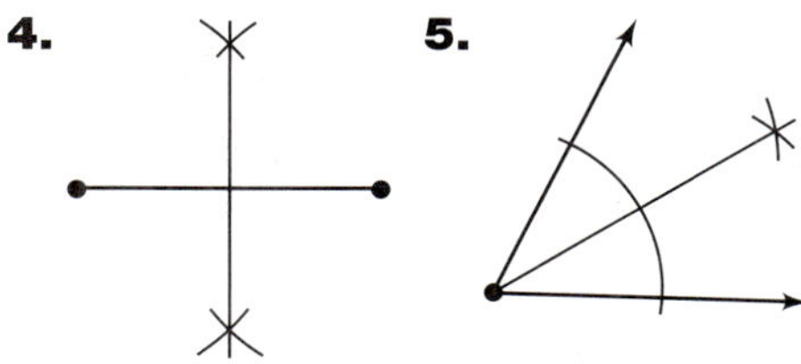

6. 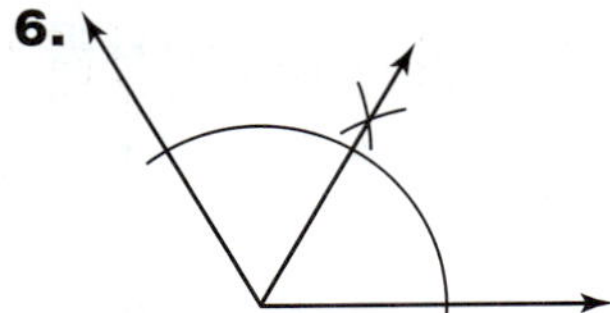

CA Standards Check **1.** If corr. $\angle$s are $\cong$, the lines are $\parallel$ by the Conv. of Corr. $\angle$s Post.

2. **3.** **4.**

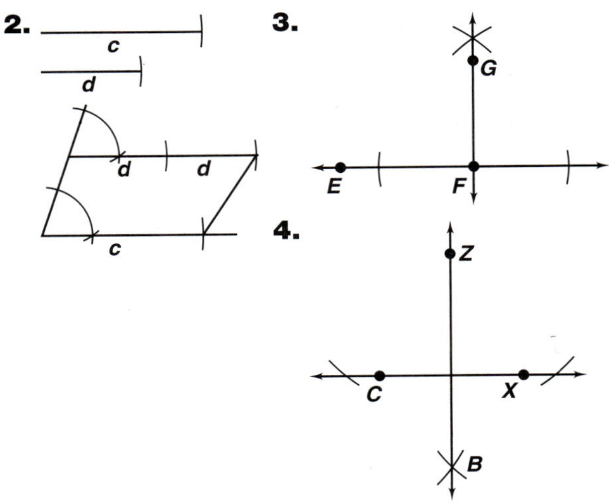

Chapter 4

Check Your Readiness p. 196

1. $AB = 4$, $BC = 3$, $AC = 5$ **2.** $AB = 8$, $BC = \sqrt{265}$, $AC = \sqrt{137}$ **3.** $AB = \sqrt{58}$, $BC = 4\sqrt{2}$, $AC = \sqrt{58}$ **4.** $\angle A \cong \angle C$ **5.** $m\angle A = 90 = m\angle B$ or $\angle A$ and $\angle B$ are rt. $\angle$s. **6.** $\angle B$ is a rt. $\angle$. **7.** $\angle AFB \cong \angle DFC$ **8.** $\angle A \cong \angle D$ and $\angle B \cong \angle C$ **9.** $\angle ACD \cong \angle CAB$ and $\angle DAC \cong \angle BCA$ **10.** $x = 12$

Lesson 4-1 pp. 198–200

Check Skills You'll Need **1.** 19 **2.** 13 **3.** 108 **4.** 10 **5.** 50

CA Standards Check **1.** $\angle WSY \cong \angle MVK$; $\angle SWY \cong \angle VMK$; $\angle WYS \cong \angle MKV$; $\overline{WY} \cong \overline{MK}$; $\overline{WS} \cong \overline{MV}$; $\overline{YS} \cong \overline{KV}$ **2.** $m\angle K = 35$; corr. $\angle$s are $\cong$. **3.** No; corr. sides are not necessarily $\cong$. **4a.** $\angle A \cong \angle D$; $\angle E \cong \angle C$ (Given) **b.** $\angle ABE \cong \angle DBC$ (Vert. $\angle$s are $\cong$.) **c.** $\overline{AE} \cong \overline{CD}$; $\overline{AB} \cong \overline{BD}$; $\overline{EB} \cong \overline{BC}$ (Given) **d.** $\triangle ABE \cong \triangle DBC$ (Def. of $\cong$ $\triangle$s)

Lesson 4-2 pp. 205–207

Check Skills You'll Need **1.** $\overline{AB} \cong \overline{DE}$; $\angle C \cong \angle F$ **2.** $\angle Q \cong \angle S$; $\angle QPR \cong \angle SRP$; $\overline{PR} \cong \overline{PR}$ **3.** $\angle M \cong \angle S$; $\angle MON \cong \angle SVT$; $\overline{TO} \cong \overline{NV}$; $\overline{MO} \parallel \overline{VS}$

CA Standards Check **1.** You are given that $\overline{HF} \cong \overline{HJ}$ and $\overline{FG} \cong \overline{JK}$. H is the midpoint of $\overline{GK}$, so $\overline{GH} \cong \overline{KH}$ by def. of midpoint. $\triangle FGH \cong \triangle JKH$ by SSS. **2.** $\angle DCA \cong \angle BAC$ **3.** No; you don't know that $\angle E \cong \angle DBC$ or that $\overline{AB} \cong \overline{DC}$.

Lesson 4-3 pp. 213–215, 219

Check Skills You'll Need **1.** $\overline{JH}$ **2.** $\overline{HK}$ **3.** $\angle L$ **4.** $\angle N$ **5.** Reflexive Prop. of $\cong$ **6.** If 2 $\angle$s of a $\triangle$ are $\cong$ to 2 $\angle$s of another $\triangle$, the third $\angle$s are $\cong$.

CA Standards Check **1.** No; the $\cong$ side is not the included side. **2.** It is given that $\overline{NM} \cong \overline{NP}$ and $\angle M \cong \angle P$. $\angle MNL \cong \angle PNO$ because vert. $\angle$s are $\cong$. $\triangle NML \cong \triangle NPO$ by ASA.

3. 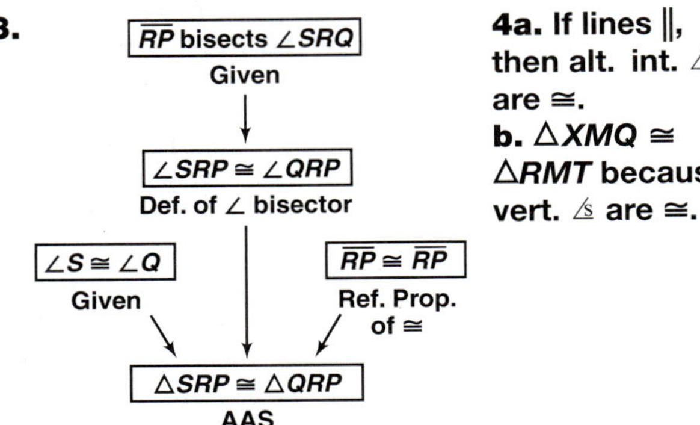

4a. If lines $\parallel$, then alt. int. $\angle$s are $\cong$. **b.** $\triangle XMQ \cong \triangle RMT$ because vert. $\angle$s are $\cong$.

Checkpoint Quiz 1 **1.** $\overline{RS} \cong \overline{JK}$; $\overline{ST} \cong \overline{KL}$; $\overline{RT} \cong \overline{JL}$; $\angle R \cong \angle J$; $\angle S \cong \angle K$; $\angle T \cong \angle L$ **2.** ASA **3.** SSS **4.** SAS **5.** not possible **6.** AAS **7.** not possible **8.** If $\parallel$ lines, then alt. int. $\angle$s are $\cong$. **9.** Vert. $\angle$s are $\cong$. **10.** ASA or AAS

Lesson 4-4 pp. 221–222

Check Skills You'll Need **1.** $\angle J \cong \angle H$; $\angle R \cong \angle V$; $\angle C \cong \angle G$ **2.** $\overline{JR} \cong \overline{HV}$; $\overline{RC} \cong \overline{VG}$; $\overline{JC} \cong \overline{HG}$ **3.** $\angle T \cong \angle L$; $\angle I \cong \angle O$; $\angle C \cong \angle K$ **4.** $\overline{TI} \cong \overline{LO}$; $\overline{IC} \cong \overline{OK}$; $\overline{TC} \cong \overline{LK}$

CA Standards Check **1.** It is given that $\angle Q \cong \angle R$ and $\angle QPS \cong \angle RSP$. $\overline{PS} \cong \overline{SP}$ by the Reflexive Prop. of $\cong$. $\triangle QPS \cong \triangle RSP$ by ASA. $\overline{SQ} \cong \overline{PR}$ by CPCTC. **2.** 50 ft

Lesson 4-5 pp. 228–230

Check Skills You'll Need **1.** $\angle C$ **2.** $\angle A$ **3.** $\overline{BC}$ **4.** $\overline{BA}$ **5.** 105

CA Standards Check **1.** No, point U could be anywhere between R and T. **2.** 6 **3.** 150

Lesson 4-6 pp. 235–237, 240

Check Skills You'll Need **1.** yes **2.** yes **3.** no **4.** yes **5.** yes **6.** no **7.** yes; SAS **8.** yes; SAS

CA Standards Check **1.** $\triangle LMN \cong \triangle OQP$ **2.** In the two rt. $\triangle$s, the hypotenuses are given to be $\cong$. Legs

$\overline{CB}$ and $\overline{EB}$ are $\cong$ by def. of $\perp$ bis. $\triangle CBD \cong \triangle EBA$ by HL. **3.** In the two rt. $\triangle$s, hypotenuses $\overline{JZ} \cong \overline{JZ}$ are $\cong$ by the Reflex. Prop. of $\cong$. Two legs are given to be $\cong$. $\triangle JWZ \cong \triangle ZKJ$ by HL.

Checkpoint Quiz 2 **1.** $\overline{PR} \cong \overline{SQ}$; $\angle P \cong \angle S$; $\angle PRQ \cong \angle SQR$ **2a.** Isosc. $\triangle$ **b.** $\cong$ **c.** Converse of the Isosc. $\triangle$ Thm. **3.** $\triangle AED$; $\angle EAB \cong \angle EDC$ (Given), $\triangle EBC$; $\angle EBC \cong \angle ECB$ (Suppl. of $\cong$ $\angle$s are $\cong$.) **4.** HL **5.** $\triangle GTW \cong \triangle SWT$ by SAS since $\overline{WT} \cong \overline{WT}$, $\angle WTG \cong \angle TWS$, and $\overline{GT} \cong \overline{SW}$. So $\overline{GW} \cong \overline{ST}$ by CPCTC.

Lesson 4-7 pp. 241–243

Check Skills You'll Need **1.** 15; 31 **2a.** yes; SAS **b.** yes; AAS **c.** yes; Trans. Prop. of $\cong$

CA Standards Check **1a.** $\overline{CD}$ **b.** Answers may vary. Sample: $\triangle ABD$ and $\triangle CBD$; $\overline{BD}$ **2.** 1. $\triangle ACD \cong \triangle BDC$ (Given) 2. $\angle ADC \cong \angle BCD$ (CPCTC) 3. $\overline{CE} \cong \overline{DE}$ (If base $\angle$s are $\cong$, the opp. sides are $\cong$.) **3.** 1. $\overline{PS} \cong \overline{RS}$; $\angle PSQ \cong \angle RSQ$ (Given) 2. $\overline{QS} \cong \overline{QS}$ (Reflexive Prop. of $\cong$) 3. $\triangle PSQ \cong \triangle RSQ$ (SAS)
4. $\overline{PQ} \cong \overline{RQ}$ (CPCTC) 5. $\angle PQT \cong \angle RQT$ (CPCTC) 6. $\overline{QT} \cong \overline{QT}$ (Reflexive Prop. of $\cong$) 7. $\triangle PQT \cong \triangle RQT$ (SAS) **4.** 1. $\angle CAD \cong \angle EAD$; $\angle C \cong \angle E$ (Given) 2. $\overline{AD} \cong \overline{AD}$ (Reflexive Prop. of $\cong$) 3. $\triangle ACD \cong \triangle AED$ (AAS) 4. $\overline{CD} \cong \overline{ED}$ (CPCTC) 5. $\angle BDC \cong \angle FDE$ (Vert. $\angle$s are $\cong$.) 6. $\triangle BDC \cong \triangle FDE$ (ASA) 7. $\overline{BD} \cong \overline{FD}$ (CPCTC)

Chapter 5

Check Your Readiness p. 256

1. $x \le 4$ **2.** $x > \frac{15}{2}$ **3.** $x \le 1$

4. **5.**

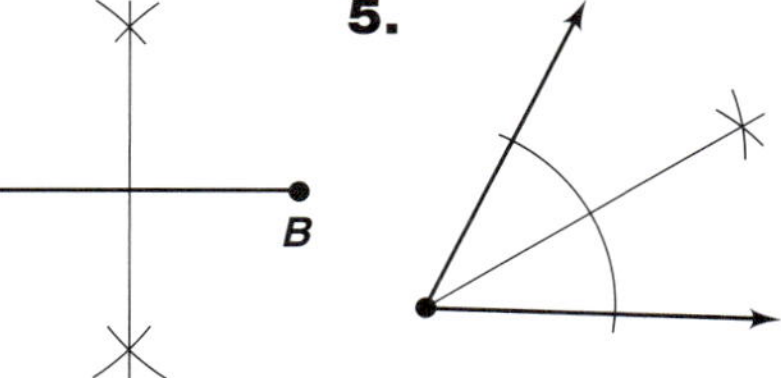

6. 5 **7.** 13 **8.** $4\sqrt{5}$ **9.** (5, 7) **10.** $\left(-3, -\frac{7}{2}\right)$ **11.** $\left(-\frac{9}{2}, \frac{5}{2}\right)$ **12.** −9 **13.** $-\frac{8}{3}$ **14.** 0

Lesson 5-1 pp. 259–261

Check Skills You'll Need **1.** (1, 2) **2.** $\left(\frac{3}{2}, \frac{11}{2}\right)$ **3.** $\left(-\frac{1}{2}, 8\right)$ **4.** (1, 1) **5.** $-\frac{2}{5}$ **6.** $\frac{1}{3}$ **7.** $\frac{4}{7}$ **8.** $\frac{3}{2}$

CA Standards Check **1.** $EB = 9$; $BC = 10$; $AC = 20$ **2.** 65; the lines are $\parallel$ so $\angle VUZ$ and $\angle YXZ$ are corr. and $\cong$. **3a.** 1320 ft **b.** $\frac{1}{4}$ mi

Lesson 5-2 pp. 265–267

Check Skills You'll Need

1.

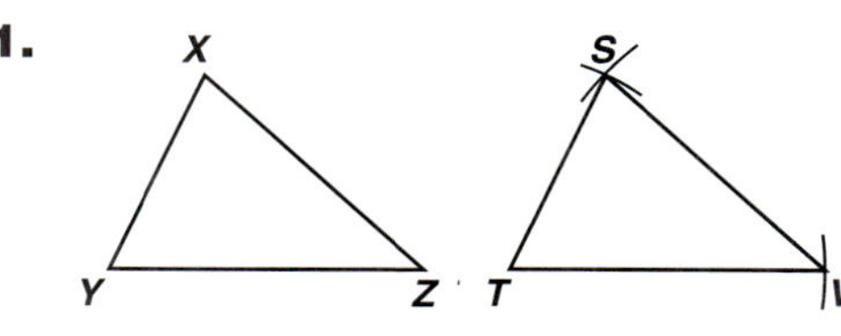

2.

3. **4.**

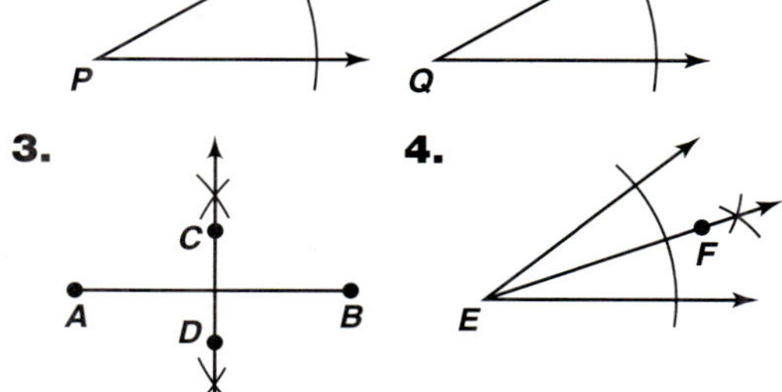

5. 6 **6.** 68

CA Standards Check **1.** $CA = 5$; $DB = 6$; $\overleftrightarrow{CD}$ is the $\perp$ bis. of $\overline{AB}$; therefore $CA = CB$ and $DA = DB$. **2a.** 10; 10 **b.** $\overrightarrow{EK}$ is on the $\angle$ bis. of $\angle DEH$. **c.** 20 **d.** 80

Lesson 5-3 pp. 272–275, 279

Check Skills You'll Need

1–2.

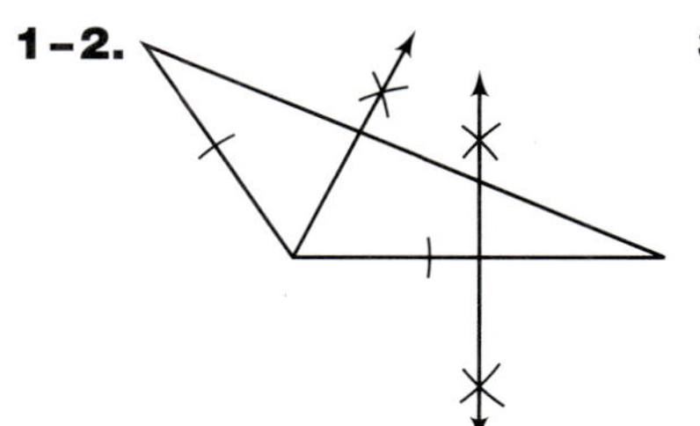

3.

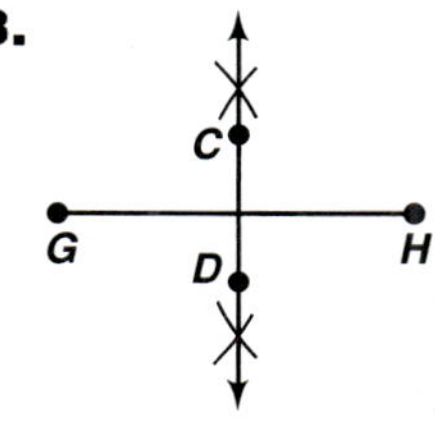

4.

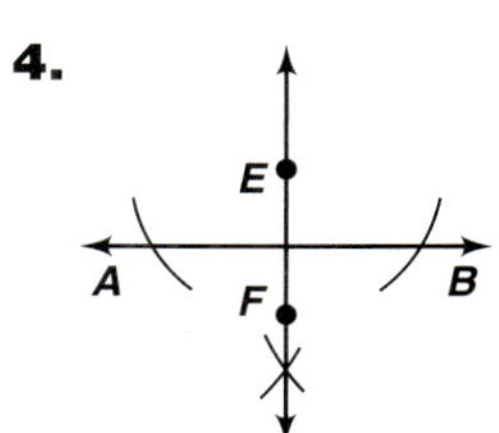

CA Standards Check **1a.** (−4, 3) **b.** Thm. 5-6: All of the $\perp$ bis. of the sides of a $\triangle$ are concurrent. **2a.** Draw segments connecting the towns. Build the library at the inters. pt. of the $\perp$ bisectors of the segments. **b.** The $\perp$ bisectors of the sides of a $\triangle$ are concurrent at a point equidistant from the vertices. **3.** 12 **4.** Median; $\overline{UW}$ is a segment drawn from vertex U to the midpt. of the opp. side.

Checkpoint Quiz 1 **1.** 6 **2.** 3 **3a.** 104 **b.** 228 **4.** right $\angle$; suppl. to $\angle ADB$ **5.** $\triangle ABD \cong \triangle CBD$; HL **6.** $\overline{AD} \cong \overline{DC}$; CPCTC **7.** $\overrightarrow{XY}$ bisects $\angle ZXW$; Y is equidist. from $\overrightarrow{XZ}$ and $\overrightarrow{XW}$. **8.** 21; $\triangle XYZ \cong \triangle XYW$ by HL, so $XZ = 21$ by CPCTC. **9.** Answers may vary. Sample: Bisect a side of a $\triangle$. Connect the opp. vertex with the midpt. **10.** Use the procedure for constructing a $\perp$ to a line from a point not on the line.

Lesson 5-4 pp. 280–283

Check Skills You'll Need **1.** If we go skiing, then it snows tomorrow. **2.** If 2 lines do not intersect, then they are parallel. **3.** If $x^2 = 1$, then $x = -1$. **4.** If a point is on the bisector of an angle, then it is equidistant from the sides of the angle. If a point is equidistant from the sides of an angle, then it is on the bisector of the angle. **5.** If a point is on the $\perp$ bis. of a segment, then it is equidistant from the endpoints of the segment. If a point is equidistant from the endpoints of a segment, then it is on the $\perp$ bis. of the segment. **6.** If you will pass a geometry course, then you are successful with your homework. If you are successful with your homework, then you will pass a geometry course.

CA Standards Check **1a.** The measure of $\angle XYZ$ is not more than 70. **b.** Today is Tuesday. **2a.** If you stand for something, you won't fall for anything. **b.** If you won't fall for anything, then you stand for something. **3a.** Assume that the shoes cost more than \$20. **b.** Assume that $m\angle A \leq m\angle B$. **4.** I and II **5.** $\angle X$ could be a right $\angle$.

Lesson 5-5 pp. 289–292

Check Skills You'll Need

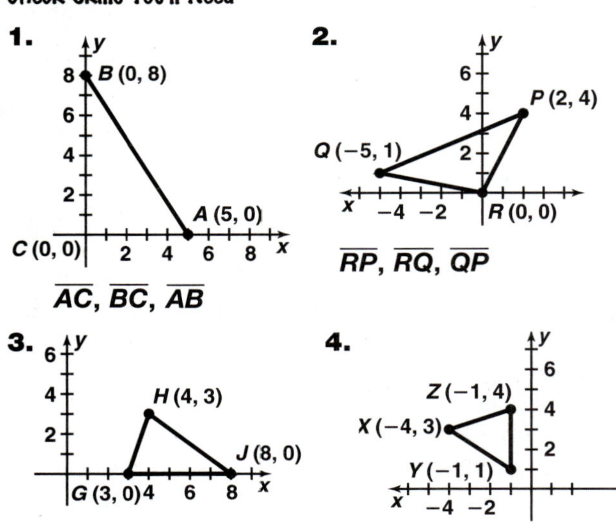

5. Assume that $m\angle A \leq m\angle B$. **6.** $AB < AC$

CA Standards Check **1.** $m\angle OTY > m\angle 2$ by the Comparison Prop. of Ineq. Since it was proven that $m\angle 2 > m\angle 3$, then by the Trans. Prop. $m\angle OTY > m\angle 3$. **2.** $\angle A$, $\angle C$, $\angle B$ **3.** $\overline{YZ}$, $\overline{XY}$, $\overline{XZ}$; $m\angle Y = 80$. **4a.** No; $2 + 7 \not> 9$. **b.** Yes; $4 + 6 > 9$; $6 + 9 > 4$; $4 + 9 > 6$. **5.** $9 < x < 15$

Chapter 6

Check Your Readiness p. 304

1. 30 **2.** 42 **3.** 22 **4.** yes **5.** no **6.** yes **7.** parallel **8.** perpendicular **9.** neither **10.** ASA **11.** SAS **12.** AAS

Lesson 6-1 pp. 306–308

Check Skills You'll Need **1.** 10.8 **2.** 7.8 **3.** 12.7 **4.** $\frac{3}{4}$ **5.** $-\frac{8}{3}$ **6.** 1

CA Standards Check **1a.** quad., $\square$, rhombus **b.** Rhombus;
it is a $\square$ and quad. with 4 sides that are $\cong$.
2. square **3.** $a = 2, b = 4$; $LN = ST = NT = SL = 14$

Lesson 6-2 pp. 312–315

Check Skills You'll Need **1.** ASA **2a.** $\angle HGE$ **b.** $\angle GHE$ **c.** $\angle HEG$ **d.** $\overline{GH}$ **e.** $\overline{HE}$ **f.** $\overline{EG}$ **3.** They are $\parallel$.

CA Standards Check **1.** Yes; by the Converse of the Same-Side Int. $\angle$s Thm., both pairs of opp. sides are $\parallel$. **2.** 11; $m\angle E = 70$, $m\angle G = 70$, $m\angle F = 110$, $m\angle H = 110$ **3.** $a = 16$, $b = 14$ **4.** 7.5

Lesson 6-3 pp. 321–324, 327

Check Skills You'll Need **1.** $\left(\frac{5}{2}, \frac{3}{2}\right)$, $\left(\frac{5}{2}, \frac{3}{2}\right)$; they bisect each other. **2.** Slope of $\overline{BC} = \frac{1}{3}$, slope of $\overline{AD} = \frac{1}{3}$. The slopes are =. **3.** Yes; they are vertical lines. **4.** parallelogram

CA Standards Check **1.** 70, 2 **2a.** Yes; $\overline{PQ}$ and $\overline{SR}$ are congruent and parallel, so $PQRS$ is a parallelogram. **b.** No; it is possible that the diagonals do not bisect each other, so $DEFG$ would not be a prallelogram. **3.** Once in place, both rulers show the direction and remain $\parallel$. Keep the second ruler in place and move the first ruler to get the compass reading.

Checkpoint Quiz 1 **1.** $m\angle 1 = 59$, $m\angle 2 = 121$, $m\angle 3 = 59$ **2.** $m\angle 1 = 43$, $m\angle 2 = 62$, $m\angle 3 = 62$ **3.** $m\angle 1 = 106$, $m\angle 2 = 74$, $m\angle 3 = 26$ **4.** trapezoid, isos. trapezoid **5.** rectangle, $\square$ **6.** rectangle, $\square$ **7.** $x = 45$, $y = 60$ **8.** $x = 1$, $y = 2$ **9.** 20.6 **10.** kite

Lesson 6-4 pp. 329–332

Check Skills You'll Need **1.** 4.5 **2.** 7 **3.** 109 **4.** 4.75 **5.** 3.5 **6.** 9.5 **7.** 71 **8.** 71

9.

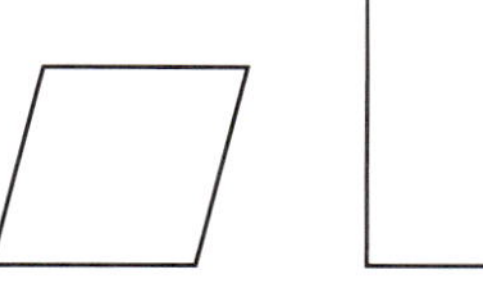

The rhombus is not a square because it has no right ⦜. The rectangle is not a square because all 4 sides aren't ≅.

CA Standards Check **1.** $\angle 1 = 90$, $\angle 2 = 50$, $\angle 3 = 50$, $\angle 4 = 40$ **2.** $8\frac{1}{2}$ **3.** Sample: No; there is not enough information to conclude that the parallelogram is a rhombus. It cannot be a rectangle because it has no right ⦜. **4.** Yes; if the ropes are ⊥ to each other, then the endpoints of the ropes determine a square.

Lesson 6-5 pp. 336–338

Check Skills You'll Need **1.** $a = 5.6$, $b = 6.8$; 4.5, 4.2, 4.5, 4.2 **2.** 3; 4.8, 16.4, 18, 18 **3.** $m = 5$, $n = 15$; 15, 15, 21, 21

CA Standards Check **1.** 110, 110, 70 **2.** 85, 95 **3.** 90, 46, 44

Lesson 6-6 pp. 343–344, 347

Check Skills You'll Need

1.

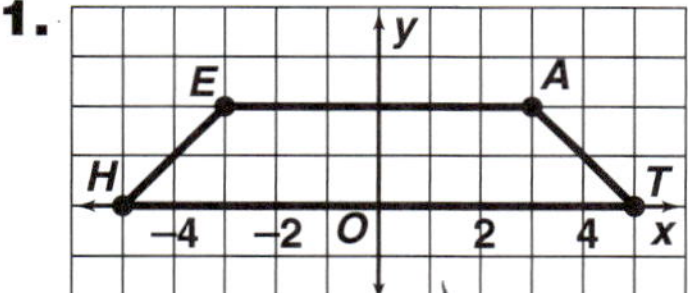

isosc. trapezoid

2.

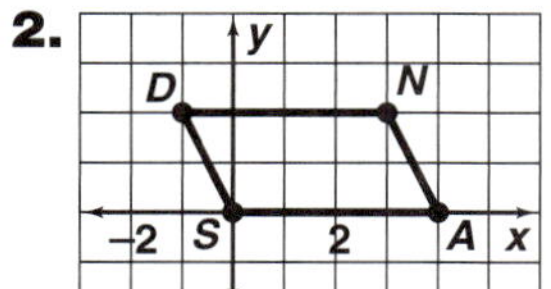

parallelogram

3.

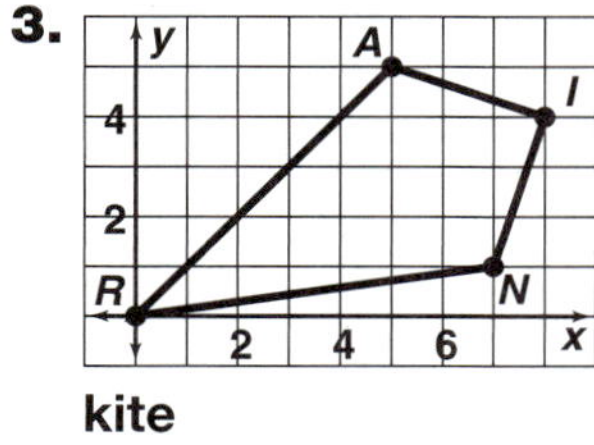

kite

4.

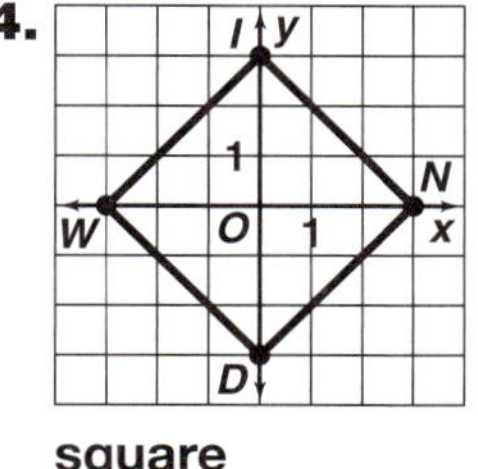

square

CA Standards Check **1.** Both $= \frac{b}{a - e}$, so $\overline{WV} \parallel \overline{TU}$. **2.** $Q\ (s + b, c)$.

Checkpoint Quiz 2 **1.** $x = 51$, $y = 51$ **2.** $x = 58$, $y = 32$ **3.** $x = 2$, $y = 4$ **4.** 3 **5.** $x = \frac{5}{3}$, $y = \frac{9}{2}$, $b = 90$ **6–8.** Counterexamples may vary.

6. false;

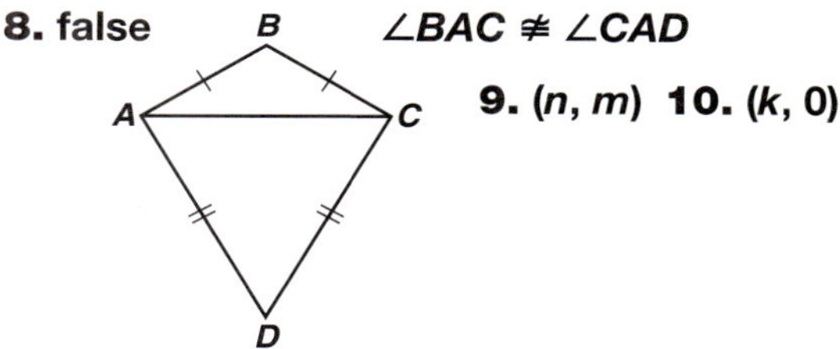

7. False; a kite can have ≅ ⊥ diags.

8. false $\angle BAC \not\cong \angle CAD$

9. (n, m) **10.** $(k, 0)$

Lesson 6-7 pp. 348–349

Check Skills You'll Need **1.** The quad. is a rectangle. **2.** (a, c) **3.** $(-a, 0)$

CA Standards Check **1a.** $M(b, c)$, $N(a + d, c)$; by starting with multiples of 2 you eliminate fractions when using the midpoint formula. **b.** 0, 0, 0; they are =. **c.** $MN = d + a - b$, $TP = 2a$, $RA = 2d - 2b$; so $RA + TP = 2d + 2a - 2b$ which is twice MN. So the midsegment is half the sum of the lengths of the bases. **d.** The base along the x-axis allows us to calculate length by subtracting x-values.
2. Using multiples of 2 in the coordinates for M, N, P, and O eliminates the use of fractions when finding midpoints since finding midpoints requires division by 2.

Chapter 7

Check Your Readiness p. 364

1. $\frac{2}{3}$ **2.** 5 **3.** $\frac{4}{3a^3}$ **4.** $\frac{3x - 12}{x^2 - x}$ **5.** 108 **6.** 135 **7.** 144 **8.** $166\frac{2}{3}$ **9.** $\overline{DL}$ **10.** $\angle A$ **11.** $\angle DLH$ **12.** $\triangle APC$ **13.** SAS **14.** AAS **15.** SSS **16.** 6; 6 **17.** 4.7; 9.4

Lesson 7-1 pp. 366–368

Check Skills You'll Need **1–5.** Answers may vary. Samples are given. **1.** $\frac{1}{2}$ **2.** $\frac{2}{3}$ **3.** $\frac{3}{4}$ **4.** 1 **5.** $\frac{2}{3}$ **6.** 4 **7.** $\frac{1}{4}$ **8.** $\frac{4}{3}$ **9.** Each side of the smaller △ is $\frac{1}{2}$ the length of a side of the larger △.

CA Standards Check **1.** 1 : 3 **2.** Answers may vary. Sample: $\frac{4}{m} = \frac{11}{n}$, $\frac{m + 4}{4} = \frac{n + 11}{11}$ **3a.** 0.75 **b.** 3 **4.** 10 ft

Lesson 7-2 pp. 373–375, 379

Check Skills You'll Need **1.** $\overline{AB} \cong \overline{HI}$; $\overline{BC} \cong \overline{IJ}$; $\overline{AC} \cong \overline{HJ}$ **2.** 6 **3.** 6 **4.** 3 **5.** 5

CA Standards Check **1.** $m\angle F = 127$; BC **2.** Yes; corr. ∠s are ≅ and corr. sides are prop. **3.** 3.8 **4.** 7.2 in. by 12 in. **5.** about 32.4 cm

Checkpoint Quiz 1 **1.** 1 : 8 **2.** $\frac{b}{10}$ **3.** no; $\frac{4}{6} \neq \frac{8}{10}$ **4.** 2 **5.** $\frac{34}{7}$ **6.** 12.5 ft **7.** $\angle BDF$ **8.** BF **9.** 3 ft by 2 ft **10.** $3\frac{1}{3}$

Lesson 7-3 pp. 382–385

Check Skills You'll Need **1.** SSS **2.** SAS **3.** ASA

CA Standards Check **1.** No; none of the side lengths are known. **2.** $\frac{AC}{EG} = \frac{CB}{GF} = \frac{AB}{EF} = \frac{3}{4}$, so the △s are ~ by SSS ~ Thm.; $\triangle ABC \sim \triangle EFG$. **3.** 9 **4.** 13.5 ft

Lesson 7-4 pp. 391–393

Check Skills You'll Need **1.** 6 **2.** $\frac{14}{3}$ **3.** $\frac{24}{5}$ **4.** 39 **5.** 2 **6.** $\frac{8}{3}$ **7.** $\frac{40}{9}$ **8.** 18 **9.** Sample:

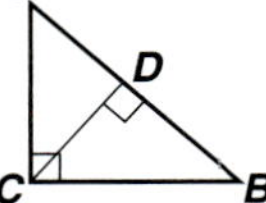

$\triangle ADC$ and $\triangle BCD$

CA Standards Check **1.** $10\sqrt{3}$ **2.** $x = 8$; $y = 4\sqrt{3}$ **3.** 240 m

Lesson 7-5 pp. 398–400, 404

Check Skills You'll Need **1.** 28 cm **2.** $3\frac{3}{7}$ mm **3.** 9.8 in. **4.** 11.25 ft

CA Standards Check **1.** 1.5 **2.** $x = \frac{225}{13}$; $y = 28.6$ **3.** 5.76

Checkpoint Quiz 2 **1.** $\triangle ABC \sim \triangle XYZ$; AA ~ Post. **2.** $\triangle WST \sim \triangle HJG$; SAS ~ Thm. **3.** $x = \frac{3\sqrt{13}}{2}$; $w = 4.5$ **4.** 12 **5.** 15 **6.** 7.5 **7.** $4\sqrt{5}$ **8.** 3.6 **9.** 17.5 **10.** 77

Chapter 8

Check Your Readiness p. 414

1. 4.648 **2.** 40.970 **3.** 6149.090 **4.** −5 **5.** AA~ Post. **6.** SSS~ Thm. **7.** SAS~ Thm. **8.** 12 **9.** 8 **10.** $2\sqrt{13}$ **11.** 9

Lesson 8-1 pp. 417–419

Check Skills You'll Need **1.** $3^2 + 4^2 = 5^2$ **2.** $5^2 + 12^2 = 13^2$ **3.** $6^2 + 8^2 = 10^2$ **4.** $4^2 + 4^2 = (4\sqrt{2})^2$

CA Standards Check **1.** $5\sqrt{21}$; no **2.** $6\sqrt{3}$ **3.** You want to know the nearest whole number value, which may not be apparent in a radical expression. **4.** no **5.** acute

Lesson 8-2 pp. 425–427, 430

Check Skills You'll Need **1.** 45, 45, 90 **2.** 30, 60, 90 **3.** 45, 45, 90

CA Standards Check **1.** $5\sqrt{6}$ **2.** $5\sqrt{2}$ **3.** 141 ft **4.** $x = 4$, $y = 4\sqrt{3}$ **5.** about 0.9 m

Checkpoint Quiz 1 **1.** 12 **2.** $x = 10$; $y = 10\sqrt{2}$ **3.** $x = 12\sqrt{3}$, $y = 24$ **4.** 9 **5.** $x = 4\sqrt{3}$; $y = 6$ **6.** $x = 5\sqrt{3}$; $y = 10\sqrt{3}$ **7.** acute **8.** right **9.** obtuse **10.** 28.3 cm

Lesson 8-3 pp. 432–434

Check Skills You'll Need **1.** 0.71, 0.71, 1.00 **2.** 0.71; 0.71; 1.00 **3.** 0.87; 0.5; 1.73 **4.** $\frac{12}{7}$ **5.** $\frac{54}{11}$ **6.** $\frac{15}{2}$ **7.** $\frac{60}{7}$

CA Standards Check **1a.** $\frac{3}{7}$; $\frac{7}{3}$ **b.** They are reciprocals. **2a.** 13.8 **b.** 1.9 **c.** 3.8 **3.** 68

Lesson 8-4 pp. 439–440

Check Skills You'll Need **1.** 9; 12 **2.** 7; $2\sqrt{78}$ **3.** 10; $3\sqrt{29}$

CA Standards Check **1a.** $\sin X = \frac{64}{80}$; $\cos X = \frac{48}{80}$; $\sin Y = \frac{48}{80}$; $\cos Y = \frac{64}{80}$ **b.** $\sin X = \cos Y$ when $\angle X$ and $\angle Y$ are complementary. **2a.** about 0.72 AU **b.** 66,960,000 mi; 35,340,000 mi **3a.** 41 **b.** 68

Lesson 8-5 pp. 445–446, 450

Check Skills You'll Need **1.** ∠7 **2.** ∠11 **3.** ∠6 **4.** 90 **5.** 180 **6.** ∠8

CA Standards Check **1a.** ∠ of elevation from hiker to peak **b.** ∠ of depression from hiker to hut **2.** about 1179 ft **3.** about 6.2 km

Checkpoint Quiz 1 **1.** $\tan A = \frac{5}{4}$; $\sin A = \frac{25}{32}$; $\cos A = \frac{5}{8}$; $\tan B = \frac{4}{5}$; $\sin B = \frac{5}{8}$; $\cos B = \frac{25}{32}$ **2.** $\tan A = \frac{5}{12}$; $\sin A = \frac{5}{13}$; $\cos A = \frac{12}{13}$; $\tan B = \frac{12}{5}$; $\sin B = \frac{12}{13}$; $\cos B = \frac{5}{13}$ **3.** $\tan A = \frac{57}{40}$; $\sin A = \frac{57}{70}$; $\cos A = \frac{4}{7}$; $\tan B = \frac{40}{57}$; $\sin B = \frac{4}{7}$; $\cos B = \frac{57}{70}$ **4.** 15.0 **5.** 61 **6.** 20.8 **7.** about 13.1 ft **8.** about 393 m

9. Answers may vary. Sample: Identify the unknown you want to find in a right triangle. Then find two pieces of known information that will let you write a trigonometric-ratio equation you can solve for the unknown. **10.** about 22.7 m

Lesson 8-6 pp. 452–455

Check Skills You'll Need **1.** $4\sqrt{41}$ **2.** $\sqrt{13}$ **3.** $10\sqrt{65}$

CA Standards Check **1.** $\langle -21.6, 46.2 \rangle$

2a.

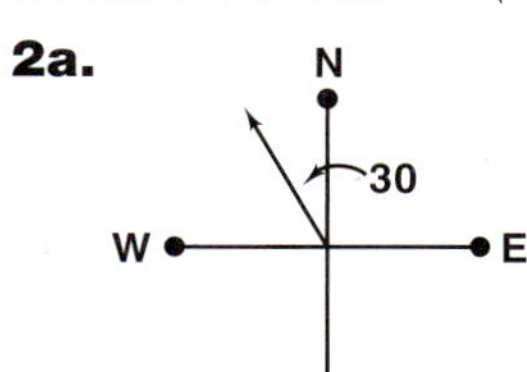

b. 60° north of west **3.** about 257 mi at 17° north of east **4.** $\langle -2, 1 \rangle$ **5.** about 16° north of west

Chapter 9

Check Your Readiness p. 468

1. 108 **2.** 135 **3.** 144 **4.** 160 **5.** $\triangle RTS$ **6.** $\triangle LJK$ **7.** $\triangle ADC$ **8.** $\triangle LHC$ **9.** always **10.** never **11.** sometimes **12.** always **13.** 50 ft **14.** 0.25 in. **15.** $\langle 1, 5 \rangle$ **16.** $\langle -6, 2 \rangle$ **17.** $\langle -6, 3 \rangle$

Lesson 9-1 pp. 470–472

Check Skills You'll Need **1.** $\overline{EF}$ **2.** $\overline{AC}$ **3.** $\overline{BC}$ **4.** $\angle G$ **5.** $\angle A$ **6.** $\angle F$ **7.** $\overline{GL}$; $\angle GRL$

CA Standards Check **1a.** Yes; the figures are ≅ by a flip.
b. Yes; the figures are ≅ by a flip and a slide.
2a. $\angle U$; P **b.** $\overline{NI}$ and $\overline{SU}$; $\overline{ID}$ and $\overline{UP}$; $\overline{ND}$ and $\overline{SP}$
3. $X'(2, 2)$, $Y'(-1, 4)$, $Z'(-5, 4)$
4. $(x, y) \rightarrow (x + 7, y - 1)$ **5.** 1 block east and 3 blocks north of her hotel

Lesson 9-2 pp. 478–479

Check Skills You'll Need **1.** $y = -2$ **2.** $x = -1$ **3.** $y = -x + 1$

CA Standards Check **1.** (0, −1)

2.

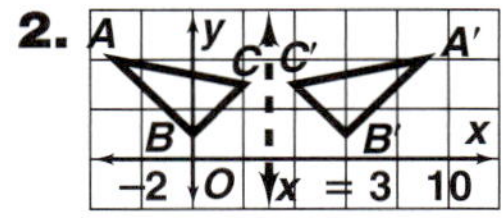

3. Yes, the intersection of $\overline{DW'}$ and ℓ is the same point P.

Lesson 9-3 pp. 483–485, 488

Check Skills You'll Need **1-7.** Check students' work.

CA Standards Check

1.

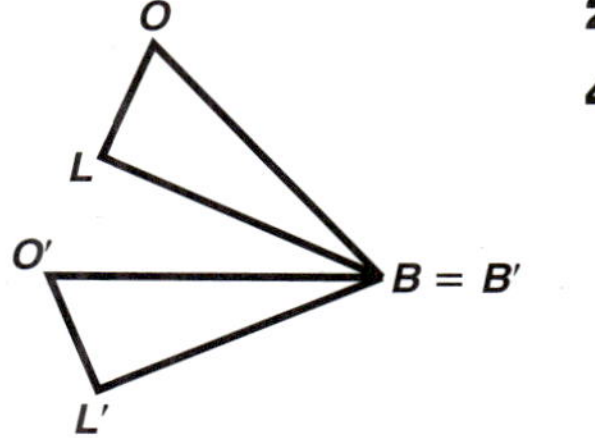

2. E **3.** 135°

4.

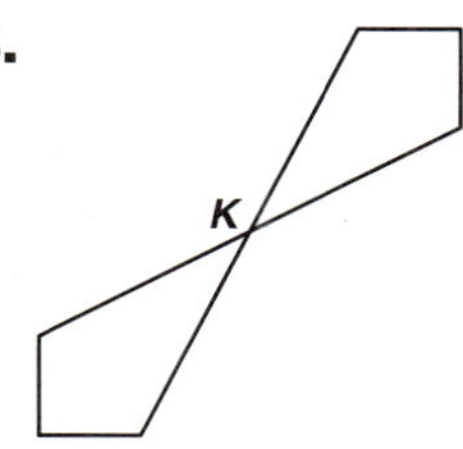

Checkpoint Quiz 1 **1.** No; the figures are not ≅. **2.** Yes; the figures are ≅ and the transf. is a translation. **3.** Yes; the figures are ≅ and the transf. is a translation, rotation or reflection. **4.** a translation of 3 units left and 5 units up. **5.** $(x, y) \rightarrow (x - 5, y + 10)$ **6.** a translation 4 units right

7.

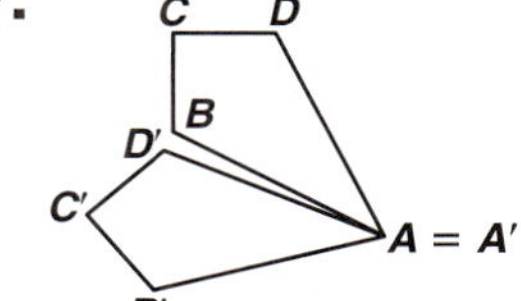

8.

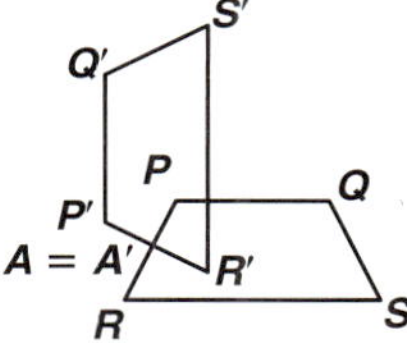

9.

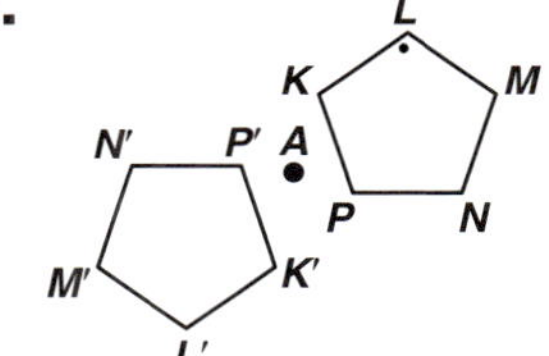

10. $W'(-6, 6)$, $X'(0, -2)$, $Y'(-2, 2)$

Lesson 9-4 pp. 492–493

Check Skills You'll Need **1.** G **2.** D **3.** $\overline{AH}$ **4.** $\overline{CB}$

CA Standards Check

1.

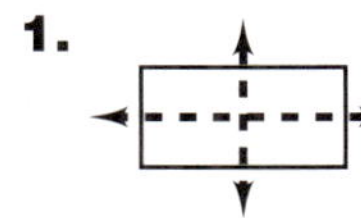

2a. yes; 180° **b.** yes **3.** rotational and reflectional symmetry

Lesson 9-5 pp. 498–499

Check Skills You'll Need **1.** 3 in. by 4 in. **2.** 2 in. by $2\frac{1}{2}$ in. **3.** $1\frac{1}{2}$ in. by $2\frac{1}{4}$ in. **4.** $1\frac{1}{4}$ in. by $1\frac{3}{4}$ in.

CA Standards Check **1.** The dilation is a reduction with center (0, 0) and scale factor $\frac{1}{2}$. **2.** 8 cm

3. $P'(1, 0)$, $Z'(-\frac{1}{2}, \frac{1}{4})$, $G'(\frac{1}{2}, -1)$

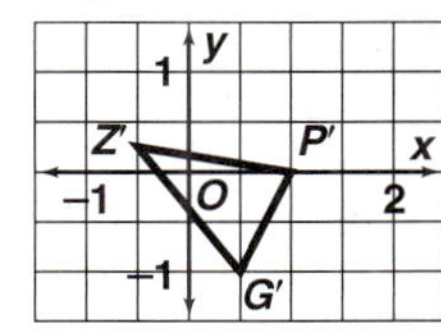

California Check System Answers

Lesson 9-6 pp. 506–509

Check Skills You'll Need

1. 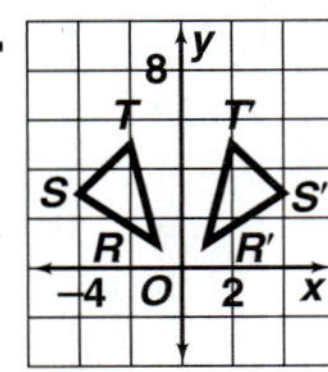

2. 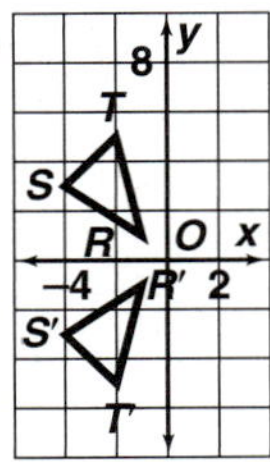

3. 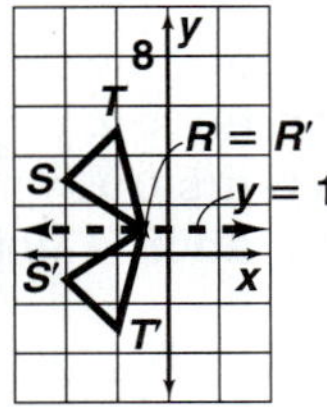

4. 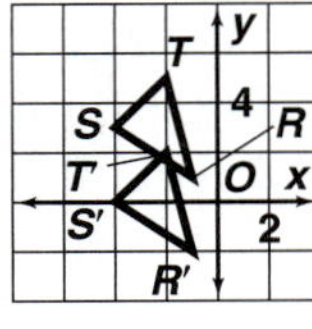

5. 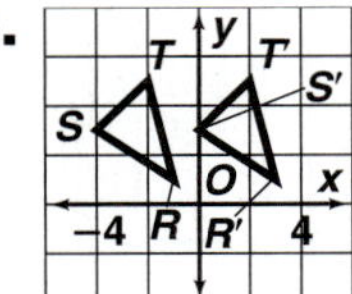

6. 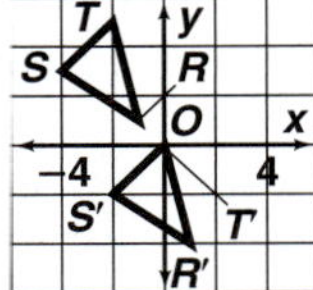

7. 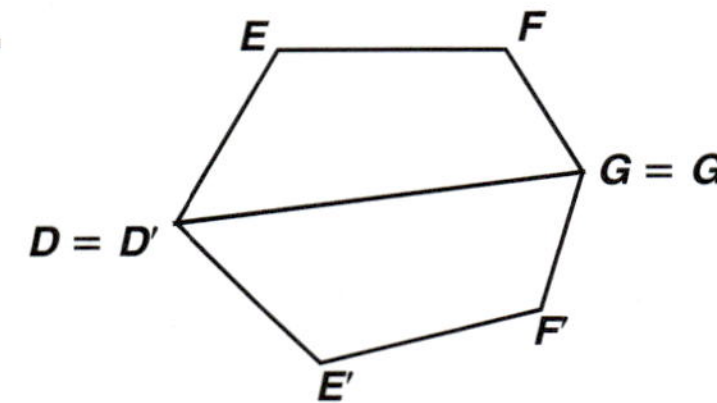

CA Standards Check **1.** Neither; the figures do not have the same orientation.

2. 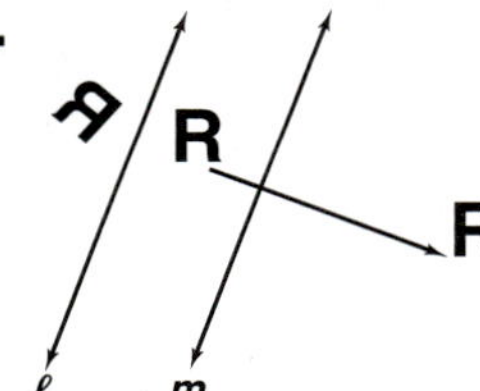

R is translated the distance and direction shown by the arrow. The length of the arrow is twice the distance between ℓ and *m*.

3. Answers may vary. Sample:

4a. 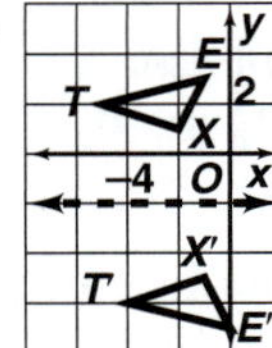

b. Yes; if you reflected it then moved it right, the result would be the same.

5. rotation

Lesson 9-7 pp. 515–517, 521

Check Skills You'll Need **1.** pentagon **2.** octagon **3.** dodecagon

CA Standards Check **1a.** rotation; one fish **b.** translation, glide reflection; horse and rider **2.** The interior ∠s of an equilateral △ measure 60. 60 divides 360, so it will tessellate. **3.** line symmetry, rotational symmetry, glide reflectional symmetry, translational symmetry

Checkpoint Quiz 2 **1.** Rotation; the image appears rotated ≈ 90°. **2.** reflection; reverse orientation **3.** enlargement; center (0, 0); scale factor 2 **4.** reduction; center (0, 0); scale factor 5 **5.** line, point **6.** point **7.** line, rotational: 120

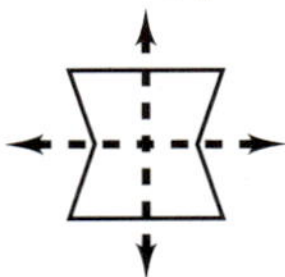

8. point **9.** rotational, point, reflectional, glide reflectional, and translational **10.** rotational, point, reflectional, glide reflectional, and translational

Chapter 10

Check Your Readiness p. 530

1. 9 **2.** 64 **3.** 144 **4.** 225 **5.** 4 **6.** 8 **7.** 10 **8.** 13 **9.** ±6 **10.** ±10.2 **11.** ±6.9 **12.** ±8.1 **13.** $2\sqrt{2}$ **14.** $3\sqrt{3}$ **15.** $4\sqrt{3}$ **16.** $36\sqrt{2}$ **17.** $\frac{3}{5}$ **18.** $\frac{2}{5}$ **19.** 0 **20.** 1 **21.** rhombus **22.** parallelogram **23.** rhombus

Lesson 10-1 pp. 534–536

Check Skills You'll Need **1.** 25 cm^2 **2.** 28 $in.^2$ **3.** 11.5 m^2 **4.** $\frac{3}{2}$ ft^2 **5.** 6 $units^2$ **6.** 2 $units^2$ **7.** 8 $units^2$

CA Standards Check **1.** 108 m^2 **2.** 7.5 cm **3.** 30 cm^2 **4.** The force is doubled.

Lesson 10-2 pp. 540–542

Check Skills You'll Need **1.** $A = bh$ or $A = \ell w$ **2.** $A = \frac{1}{2}bh$ **3.** 9 $units^2$ **4.** 7 $units^2$ **5.** 13.5 $units^2$

CA Standards Check **1.** 94.5 cm^2 **2.** 12 m^2 **3.** 54 $in.^2$ **4.** $9^2 + 12^2 = 15^2$

Lesson 10-3 pp. 546–547, 551

Check Skills You'll Need **1.** $25\sqrt{3}$ cm^2 **2.** 50 ft^2 **3.** $\frac{100\sqrt{3}}{3}$ m^2 **4.** 24 in. **5.** $16\sqrt{3}$ cm

CA Standards Check **1.** $m\angle 1 = 45$; $m\angle 2 = 22.5$; $m\angle 3 = 67.5$ **2.** 232 cm^2 **3.** $384\sqrt{3}$ ft^2

Checkpoint Quiz 1 **1.** 84 in.^{2} **2.** 112 cm^2 **3.** 48 m^2 **4.** 135 in.^{2} **5.** 58.5 m^2 **6.** $72\sqrt{3}$ ft^2 **7.** $27\sqrt{3}$ ft^2 **8.** 32 yd^2 **9.** $16\sqrt{3}$ in.^{2} **10.** $\frac{81\sqrt{3}}{2}$ m^2; 70.1 m^2

Lesson 10-4 pp. 553–555

Check Skills You'll Need **1.** 28 in.; 49 in.^{2} **2.** 24 m; 32 m^2 **3.** 24 cm; 24 cm^2 **4.** 8 cm; 3 cm^2 **5.** 16 cm; 12 cm^2 **6.** 24 cm; 27 cm^2

CA Standards Check **1a.** 5 : 7 **b.** 25 : 49 **2.** 54 in.^{2} **3.** \$6.94 **4.** $5\sqrt{5}$: 3

Lesson 10-5 pp. 559–561

Check Skills You'll Need **1.** 36 m^2 **2.** 4536 in.^{2} **3.** 168 ft^2

CA Standards Check **1.** 482.8 in.^{2} **2.** It is 4 times as large. **3.** 5081 ft^2

Lesson 10-6 pp. 566–569

Check Skills You'll Need **1.** 14 cm **2.** 3.2 m **3.** 5 ft **4.** 2.5 in. **5.** 32 **6.** 137 **7.** 180 **8.** 76

CA Standards Check **1a.** number of hours spent doing an activity **b.** Each section represents the average of the 3600 participants' answers. **2.** $\overset{\frown}{CEA}$, $\overset{\frown}{DAE}$, $\overset{\frown}{ACD}$, $\overset{\frown}{EDC}$ **3.** 58; 180; 122; 270 **4.** 17 revolutions **5.** 1.3π m

Lesson 10-7 pp. 575–577, 580

Check Skills You'll Need **1.** 4.5 cm **2.** 16 ft **3.** 12π or about 37.7 in. **4.** 6π or about 18.8 m

CA Standards Check **1.** about 41 in.^{2} **2.** 181.5 cm^2 **3.** 13.0 cm^2

Checkpoint Quiz 2 **1.** 60 in. **2.** 45 in.^{2} **3.** 6 : 5 **4.** about 4.8 ft^2 **5.** 44.8 cm^2 **6.** 100π in.^{2} **7.** 27π m^2 **8.** $(16\pi - 32)$ cm^2 **9.** 22.5π ft^2 **10.** $\frac{9\pi}{2}$ mm

Lesson 10-8 pp. 582–584

Check Skills You'll Need **1.** $\frac{1}{3}$ **2.** $\frac{1}{2}$ **3.** 1 **4.** $\frac{1}{4}$ **5.** $\frac{1}{6}$ **6.** $\frac{1}{2}$ **7.** $\frac{1}{3}$ **8.** $\frac{1}{2}$

CA Standards Check **1.** $\frac{2}{5}$ **2.** $\frac{2}{5}$ **3a.** It becomes about 8.7%, or about 4 times greater. **b.** It becomes 19.6%, or about 9 times greater. **4.** Yes; theoretically you should win 1.4 times out of 100.

Chapter 11

Check Your Readiness p. 596

1. 17 **2.** $8\sqrt{2}$ **3.** 6 **4.** $4\sqrt{5}$ **5.** $6\sqrt{2}$ **6.** $4\sqrt{3}$ **7.** 44 units^2 **8.** $14\sqrt{3}$ units^2 **9.** 234 units^2 **10.** $54\sqrt{3}$ units^2 **11.** 24 **12.** $2\sqrt{2}$: 5

Lesson 11-1 pp. 598–600

Check Skills You'll Need

1. 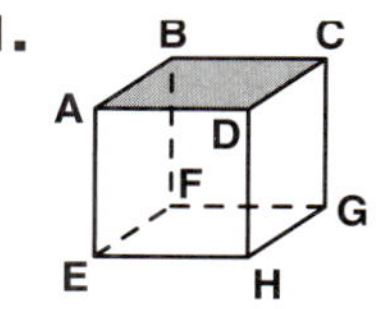

2. 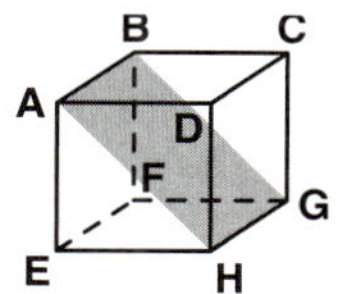

3. 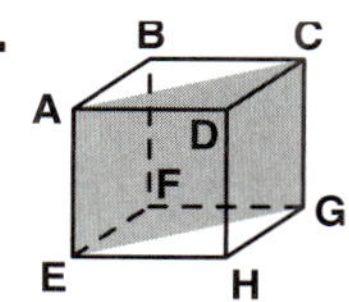

4. 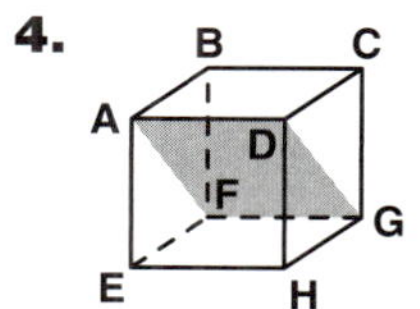

5. 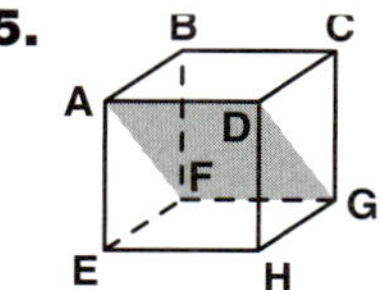

6. 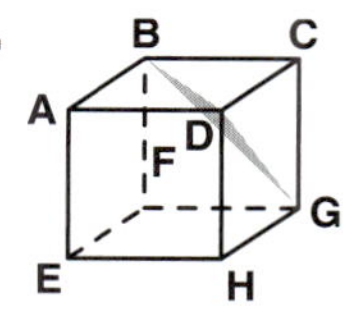

7. 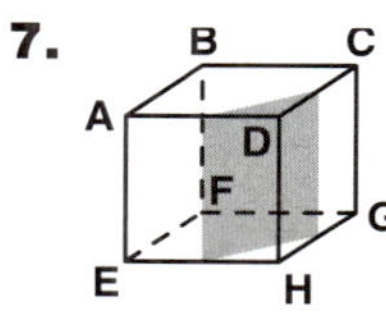

CA Standards Check
1. R, S, T, U, V; $\overline{RS}$, $\overline{RU}$, $\overline{RT}$, $\overline{VS}$, $\overline{VU}$, $\overline{VT}$, $\overline{SU}$, $\overline{UT}$, $\overline{TS}$; $\triangle RSU$, $\triangle RUT$, $\triangle RTS$, $\triangle VSU$, $\triangle VUT$ $\triangle VTS$
2. 12 edges
3a. 6 + 8 = 12 + 2
b.

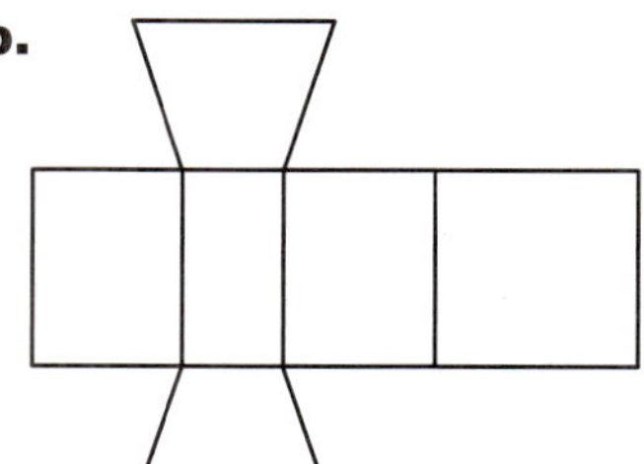

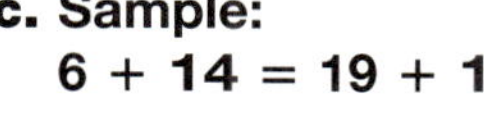

c. Sample: 6 + 14 = 19 + 1

4a. **b.**

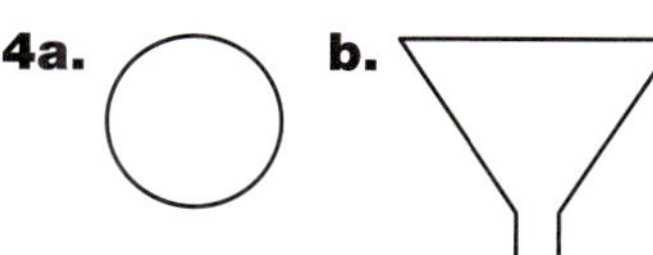

5. square

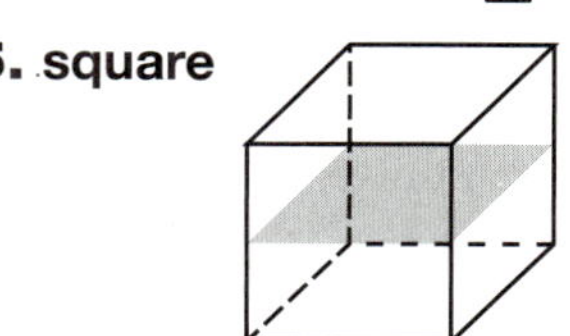

Lesson 11-2 pp. 608–611

Check Skills You'll Need **1.** 96 cm^2 **2.** 40π cm^2 **3.** $36\sqrt{3}$ m^2

CA Standards Check

1. 216 cm^2

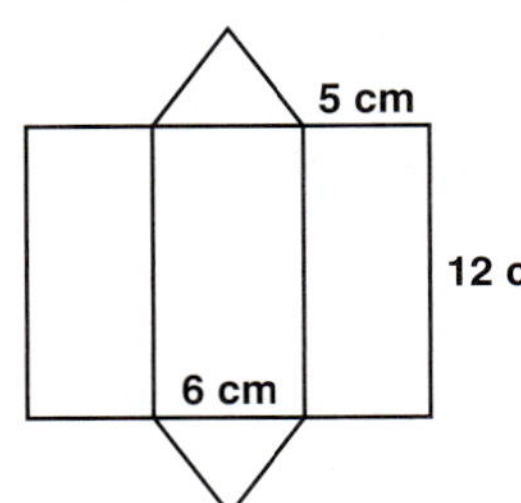

2. 432 m^2; about 619 m^2
3. 400π cm^2
4a. 33 ft^2
b. same as large drum (about 46 ft^2)

Lesson 11-3 pp. 617–620, 623

Check Skills You'll Need **1.** $\sqrt{233}$ in. **2.** $\sqrt{130}$ m **3.** $\sqrt{313}$ cm

CA Standards Check **1.** 55 m^2 **2.** 1,496,511 ft^2 **3.** 704π m^2 **4.** 1178 $in.^2$

Checkpoint Quiz 1

1. **2.**

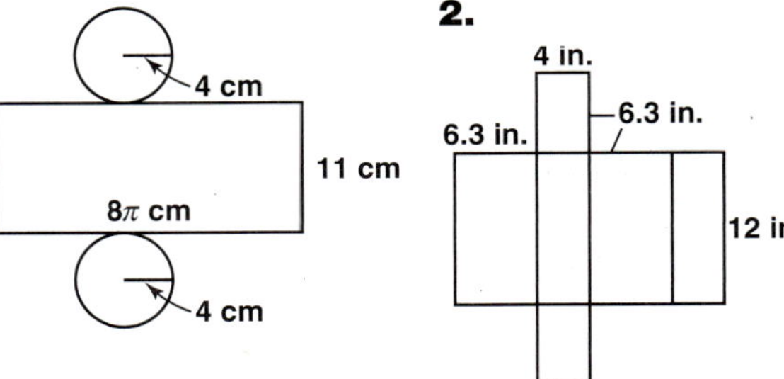

3.

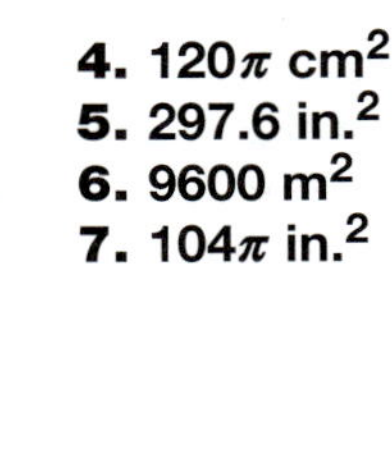

4. 120π cm^2
5. 297.6 $in.^2$
6. 9600 m^2
7. 104π $in.^2$

8.

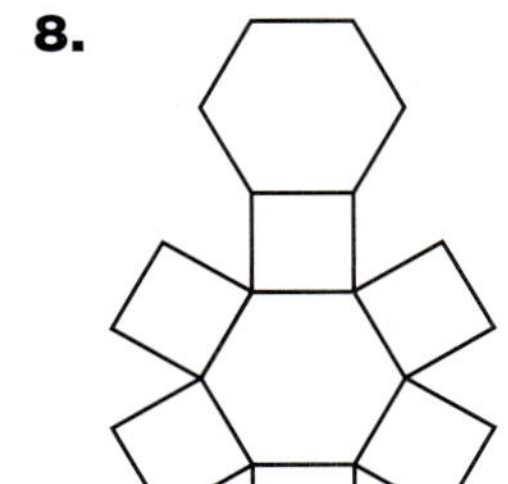

9–10. Answers may vary. Samples:
9.
10.

Lesson 11-4 pp. 624–627

Check Skills You'll Need **1.** 49 cm^2 **2.** 176.7 $in.^2$ **3.** 314.2 mm^2 **4.** 3 ft^2 **5.** 154 $in.^2$ **6.** 27.5 cm^2 **7.** 27.7 $in.^2$

CA Standards Check **1.** Answers may vary. Sample: Multiplication is commutative. **2.** 150 m^3 **3a.** 256π m^3 **b.** 804.2 m^3 **4.** 12 $in.^3$

Lesson 11-5 pp. 631–634

Check Skills You'll Need **1.** 12 cm **2.** 8 in. **3.** 2.5 m

CA Standards Check **1.** 384 $in.^3$ **2.** 960 m^3 **3a.** 144π m^3; 452 m^3 **b.** 6174π mm^3; 19,396 mm^3 **4.** 77 ft^3

Lesson 11-6 pp. 638–640, 644

Check Skills You'll Need **1.** 113.1 $in.^2$; 37.7 in. **2.** 78.5 cm^2; 31.4 cm **3.** 19.6 ft^2; 15.7 ft **4.** 4.5 m^2; 7.5 m **5.** 706.9 yd^2; 94.2 yd **6.** 452.4 mm^2; 75.4 mm

CA Standards Check **1.** 196π $in.^2$; 616 $in.^2$ **2.** 100 $in.^2$ **3.** 113,097 $in.^3$ **4.** 1258.9 ft^2

Checkpoint Quiz 2 **1.** 60.2 ft^2; 22.5 ft^3 **2.** 332.9 $in.^2$; 377.0 $in.^3$ **3.** 113.1 m^2; 113.1 m^3 **4.** 439.8 cm^2; 706.9 cm^3 **5.** 207 yd^2; 144.8 yd^3 **6.** 44.8 m^2; 16 m^3 **7.** 181.7 m^2; 2217.0 m^3 **8.** 32 ft^2; 12 ft^3 **9.** 75.4 cm^2; 37.7 cm^3 **10.** The balls; the volume of the space is $2\pi r^3$ and the volume of the balls is $4\pi r^3$.

Lesson 11-7 pp. 646–648

Check Skills You'll Need **1.** Yes; all corr. $\angle$s are $\cong$ and corr. sides are prop.; 3 : 1. **2.** Yes; all corr. $\angle$s are $\cong$ and corr. sides are prop.; $3:\sqrt{2}$. **3.** 27 $in.^3$ **4.** 135 m^3 **5.** 402.1 cm^3

CA Standards Check **1.** yes; 6 : 5 **2.** 2 : 3 **3.** 160 m^2 **4.** 0.01875 lb

Chapter 12

Check Your Readiness p. 660

1. 82 **2.** $6\frac{2}{3}$ **3.** 15 **4.** 25 **5.** 10 **6.** 5 **7.** 6 **8.** 18 **9.** 24 **10.** $\sqrt{2}$ **11.** 12 **12.** $4\sqrt{2}$ **13.** 13 **14.** $\sqrt{10}$ **15.** 6

Lesson 12-1 pp. 662–665

Check Skills You'll Need **1.** $p^2 + 6p + 9$ **2.** $w^2 + 20w + 100$ **3.** $m^2 - 4m + 4$ **4.** $8\sqrt{5}$ **5.** $2\sqrt{30}$ **6.** 12

CA Standards Check **1.** 52 **2.** about 35.5 in. **3.** No; $4^2 + 7^2 \neq 8^2$ **4.** If $\overleftrightarrow{BC}$ and $\overleftrightarrow{GF}$ never intersect, then *BCFG* is a rectangle. **5.** 12 cm

Lesson 12-2 pp. 670–673

Check Skills You'll Need **1.** $\frac{11\sqrt{2}}{2}$ **2.** 5 **3.** 28

CA Standards Check **1.** $\angle O \cong \angle P$; $\widehat{BC} \cong \widehat{DF}$ **2.** 16 **3a.** about 11 **b.** 2.8

Lesson 12-3 pp. 678–681, 685

Check Skills You'll Need **1–4.** Answers may vary. Samples are given. **1.** $\widehat{STQ}$ **2.** $\widehat{SR}$ **3.** $\widehat{RTQ}$ **4.** $\angle TPQ$ **5.** 86 **6.** 180 **7.** 121 **8.** 94

CA Standards Check **1.** 105 **2.** $m\angle 1 = 90, m\angle 2 = 110, m\angle 3 = 90, m\angle 4 = 70$ **3.** $m\angle QJK = m\angle LJK + m\angle QJL = 35 + 90 = 125$; $m\angle QJK + \frac{1}{2}m\widehat{QLJ} = \frac{1}{2}(70 + 180) = 125$

Checkpoint Quiz 1 **1.** 76 cm **2.** 48 in. **3.** 51 m **4.** 24 **5.** 5 **6.** 8 **7.** $w = 104; x = 22; y = 108$ **8.** $a = 30; b = 42; c = 80; d = 116$ **9.** $w = 105; x = 75; y = 210$ **10.** $a = 140; b = 70; c = 47.5$

Lesson 12-4 pp. 687–690

Check Skills You'll Need **1.** 57 **2.** 180 **3.** 303 **4.** 28.5 **5.** 28.5 **6.** 4 **7.** 2 **8.** about 4.5 **9.** 123

CA Standards Check **1a.** 250 **b.** 40 **2.** away; 20° **3a.** 13.8 **b.** 3.2 **4a.** $(8 - 4\sqrt{3})$ in. **b.** $\frac{16}{8 + 4\sqrt{3}}$ in.

Lesson 12-5 pp. 695–697, 700

Check Skills You'll Need **1.** 5.8 **2.** 12.8 **3.** 5.8

CA Standards Check **1a.** $(x - 3)^2 + (y - 5)^2 = 36$ **b.** $(x + 2)^2 + (y + 1)^2 = 2$ **2.** $(x - 2)^2 + (y - 3)^2 = 13$ **3.** center: (2, 3); radius: 10

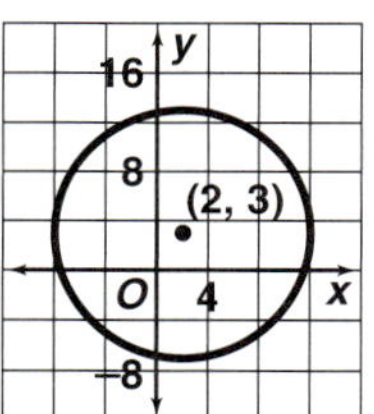

4. $(x - 0)^2 + (y - 0)^2 = x^2 + y^2 = 144$

Checkpoint Quiz 2 **1.** 58 **2.** 226 **3.** 30 **4.** about 3.0 **5.** about 15.7 **6.** 40 **7.** A chord is a segment whose endpoints are on the circle. A secant is a line, ray, or segment that intersects a circle at two points.

$\overline{AB}$ is a chord
$\overleftrightarrow{CD}$ is a secant

8. $(x - 1.5)^2 + (y - 0.5)^2 = 2.5$
9. $(x - 3.5)^2 + (y - 1)^2 = 46.25$
10. $(x + 1.5)^2 + (y + 4)^2 = 22.25$

Lesson 12-6 pp. 701–702

Check Skills You'll Need

1.

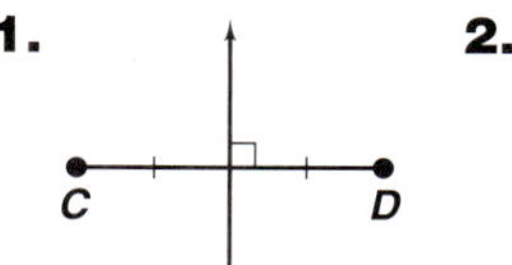

2.

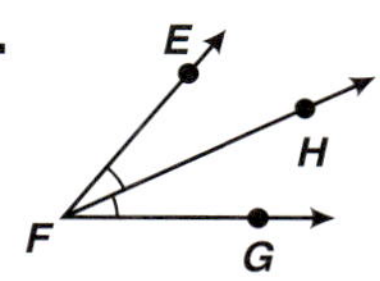

3.

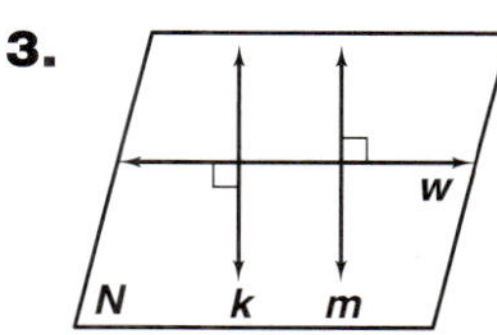

CA Standards Check **1.** Two lines ∥ to $\overleftrightarrow{XY}$, each 2 cm from $\overleftrightarrow{XY}$.

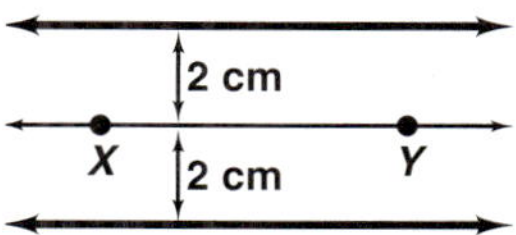

2.

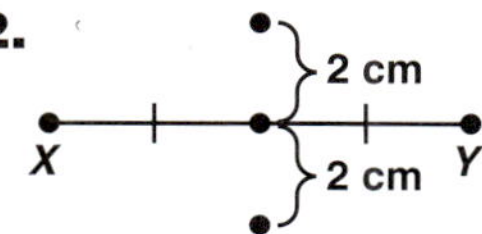

3a. Consider two ∥ lines in a plane. The locus is the line ∥ to and equidist. from the ∥ lines.

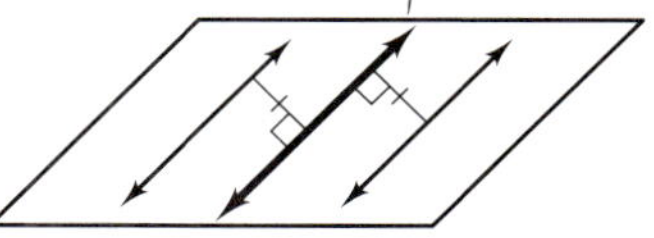

b.

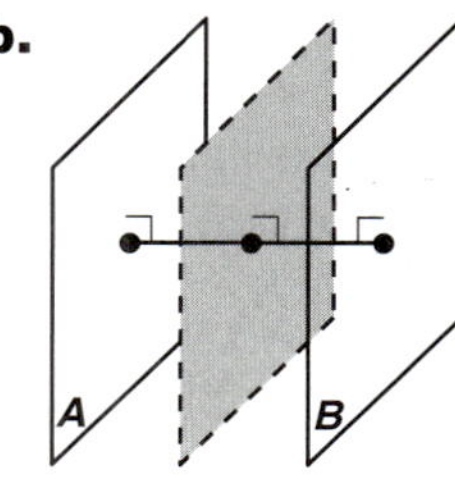

Consider a segment ⊥ to both planes, with one endpoint on plane *A* and the other on plane *B*. The locus is the plane ⊥ to this segment at the segment's midpoint.

Selected Answers

Chapter 1

Lesson 1-1 pp. 6–9

EXERCISES 1. 80, 160 **3.** −3, 4 **5.** 3, 0 **7.** N, T **9.** 720, 5040 **11.** $\frac{1}{36}$, $\frac{1}{49}$ **13.** James, John **15.** Andrew, Ulysses

17. **19.** The sum of the first 6 pos. even numbers is 6 · 7, or 42.

21. The sum of the first 100 pos. even numbers is 100 · 101, or 10,100. **23.** 555,555,555

25–27. Answers may vary. Samples are given.

25. $8 + (-5) = 3$ and $3 \ngtr 8$ **27.** $-6 - (-4) \nless -6$ and $-6 - (-4) \nless -4$ **29.** 75°F **31.** 31, 43 **33.** 0.0001, 0.00001 **35.** 63, 127 **37.** J, S **39.** B, C

41. You would get points on a third line between and parallel to the first two lines.

43. **45.**

47. Answers may vary. Samples are given. **a.** Women may soon outrun men in running competitions. **b.** The conclusion was based on continuing the trend shown in past records. **c.** The conclusions are based on fairly recent records for women, and those rates of improvement may not continue. The conclusion about the marathon is most suspect because records date only from 1955. **49.** Answers may vary. Sample: 1, 3, 9, 27, 81, . . . 1, 3, 5, 7, 9, . . .

51.a. **b.** *H* and *I* **c.** a circle

53.a. Leap years are years that are divisible by 4. **b.** 2020, 2100, and 2400 **c.** Leap years are years divisible by 4, except the final year of a century which must be divisible by 400. So, 2100 will not be a leap year, but 2400 will be.

55.a. 1, 3, 6, 10, 15, 21 **b.** They are the same. **c.** The diagram shows the product of n and $n + 1$ divided by 2 when $n = 3$. The result is 6.

d.

57. D **59.** B **61.** $\angle D$: 59°; $\angle E$: 80°; $\angle F$: 41°

Lesson 1-2 pp. 13–15

1. **3.**

5.a. **b.**

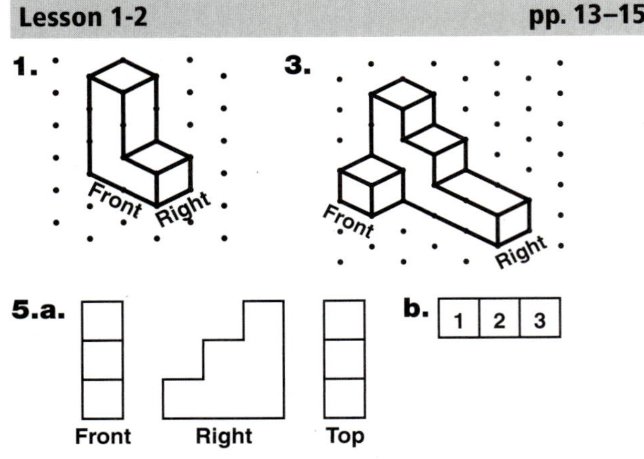

7. 6 **9.** 8 **11.** C **13.** B

15. Answers may vary. Sample:

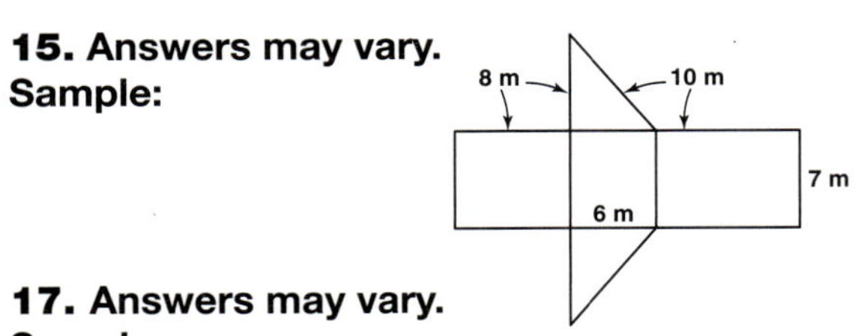

17. Answers may vary. Sample:

a. **b.**

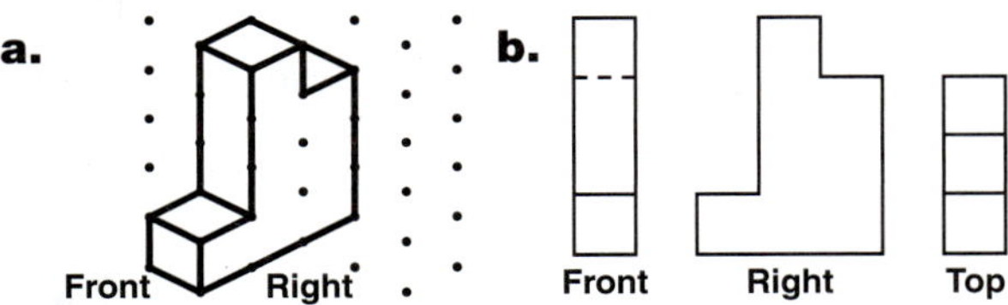

c. 3 4 1 Right Front

19.a.

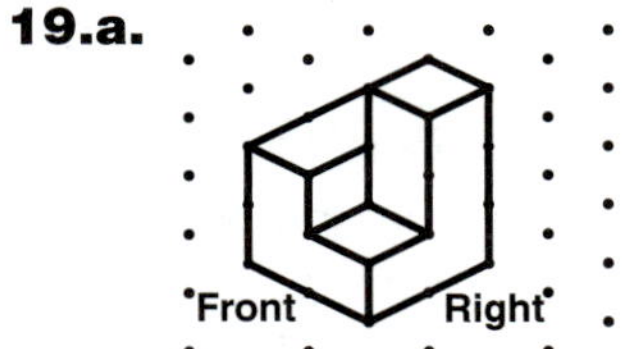

b.

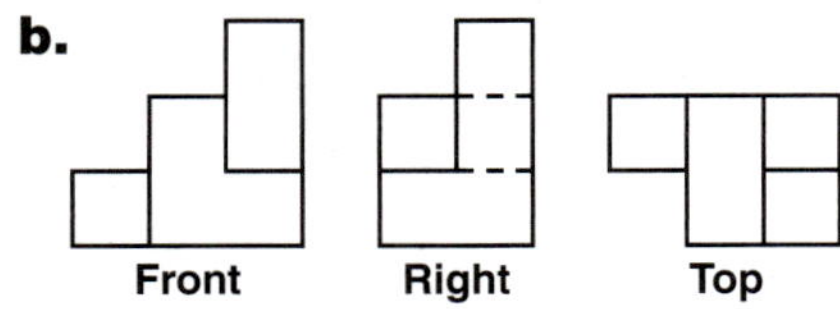

21. orthographic top view **23.** blue **25.** orange

27.a. Answers may vary. Sample:

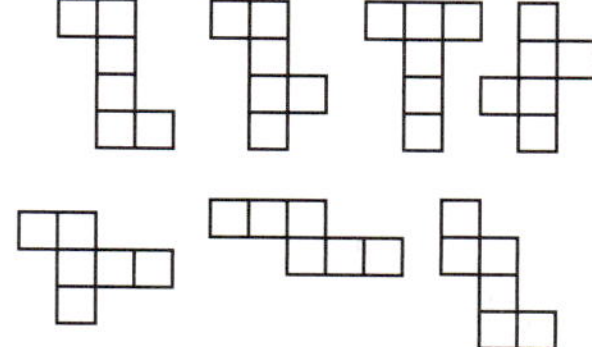

b. Answers may vary. Sample: the net shown in Exercise 23; it is easy to cut and fold.

29.

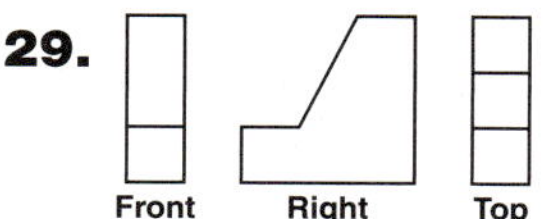

31. Answers may vary. Sample:

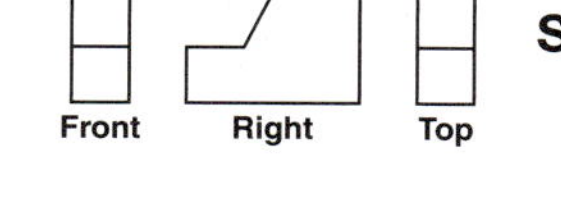

33.a.

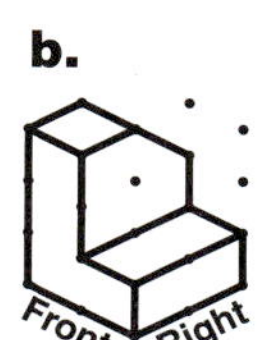

b. Front Right

c.

d. Front Right

Top Front Right

35. B **37.**

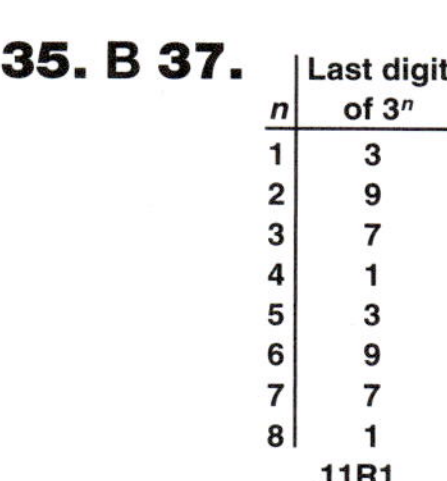

n	Last digit of 3^n
1	3
2	9
3	7
4	1
5	3
6	9
7	7
8	1

$4\overline{)45}$ 11R1

The pattern 3, 9, 7, 1 repeats 11 times for $n = 1$ to 44. For $n = 45$, the last digit is 3.

39. 42, 56 **41.** 25, −5 **43.** 44

Lesson 1-3 pp. 19–22

EXERCISES **1.** no **3.** yes; line n **5.** yes; line n **7.** no **9.** Answers may vary. Sample: $\overleftrightarrow{AE}$, $\overleftrightarrow{EC}$, $\overleftrightarrow{GA}$ **11.** *ABCD* **13.** *ABHF* **15.** *EFAD* **17.** $\overleftrightarrow{RS}$ **19.** $\overleftrightarrow{UV}$ **21.** planes *QUX* and *QUV* **23.** planes *UXT* and *WXT* **25.**

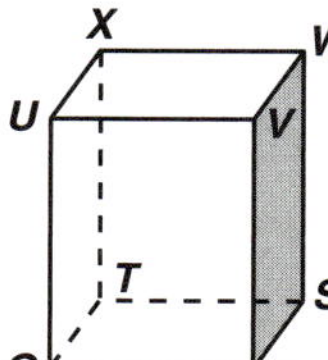

27.

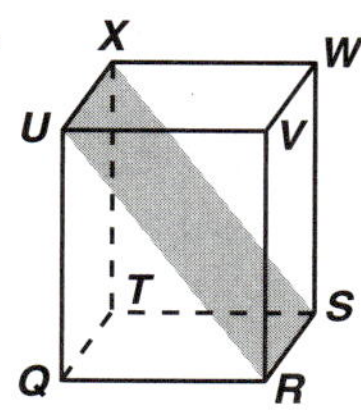

29. X W U V T S Q R

31. *X* **33.** *Q* **35.** no **37.** no **39.** coplanar **41.** coplanar **43.** noncoplanar **45.** Answers may vary. Sample:

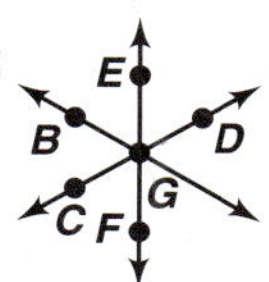

47. not possible

49. not possible

51.

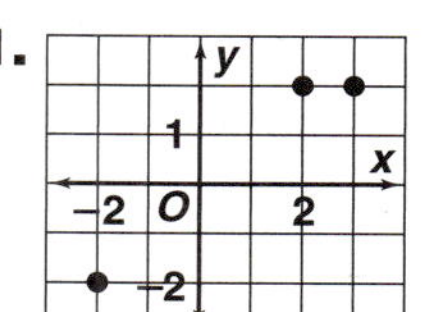

no **53.**

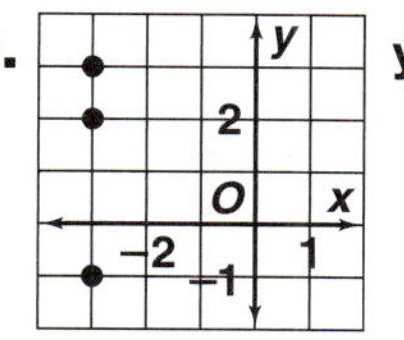

yes

55. always **57.** always **59.** sometimes **61.a.** 1 **b.** 1 **c.** 1 **d.** 1 **e.** A line and a point not on the line are always coplanar.

63. Answers may vary. Sample: Post. 1-3: If two planes intersect, then they intersect in exactly one line.

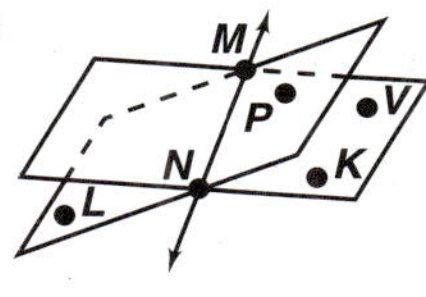

65. Post. 1-1: Through any two points there is exactly one line.

67. The end of one leg might not be coplanar with the ends of the other three legs. (Post. 1-4)

69.

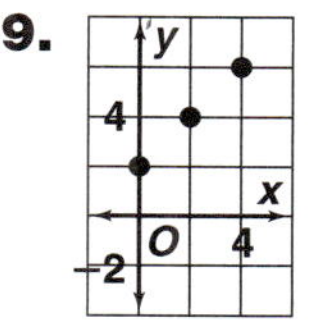

yes **71.**

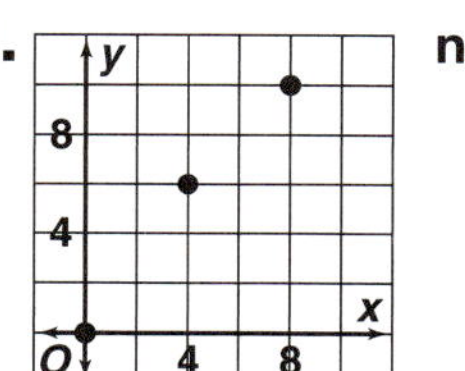

no

73. no

75.

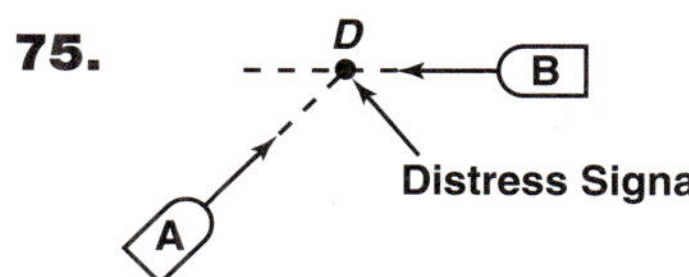

By Post. 1-1, points *D* and *B* determine a line and points *A* and *D* determine a line. The distress signal is on both lines and, by Post. 1-2, there can be only one location for the distress signal.

77. 1 **79.** 1 **81.** C

83. **85.**

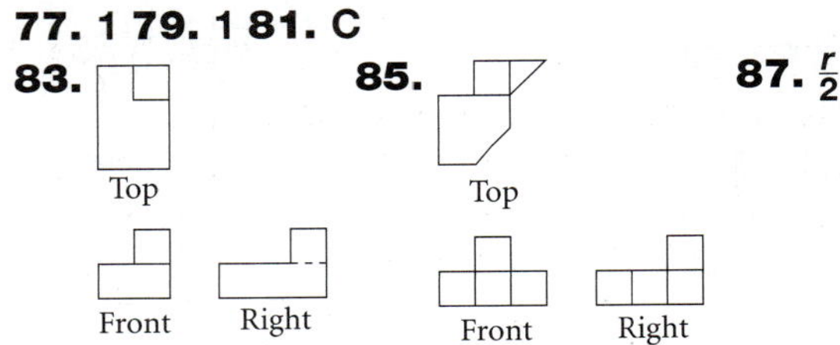

87. $\frac{r}{2}$

Lesson 1-4 pp. 25–28

EXERCISES **1.** $\overline{RS}, \overline{RT}, \overline{RW}, \overline{ST}, \overline{SW}, \overline{TW}$

3.a. $\overrightarrow{TS}$ or $\overrightarrow{TR}, \overrightarrow{TW}$ **b.** $\overrightarrow{SR}, \overrightarrow{ST}$ **5.** $\overline{BC}$ **7.** $\overline{DE}, \overline{EF}, \overline{BE}$ **9.** $\overline{BC}, \overline{EF}$ **11.** Sample: *DEF*, $\overleftrightarrow{BC}$ **13.** Answers may vary. Sample: $\overleftrightarrow{CD}, \overleftrightarrow{AB}$ **15.** $\overleftrightarrow{AF}$ **17.** False; they are skew. **19.** False; they intersect above $\overline{CG}$. **21.** False; they intersect above pt. *A*. **23.** False; they are $\parallel$. **25.** always **27.** always **29.** never **31.** always **33.** sometimes **35.** Answers may vary. Sample: (0, 0); Check students' graphs. **37.** Two lines can be parallel, skew, or intersecting in one point. Sample: train tracks—parallel; vapor trail of a northbound jet and an eastbound jet at different altitudes—skew; streets that cross—intersecting **39.a.** The lines of intersection are parallel. **b.** Examples may vary. Sample: The floor and ceiling are parallel. A wall intersects both. The lines of intersection are parallel. **41.** No; two different planes cannot intersect in more than one line. **43.** Answers may vary. Sample: $\overleftrightarrow{VR}, \overleftrightarrow{QR}, \overleftrightarrow{SR}$ **45.** Yes; no; yes; explanations may vary. **47.** B **49–55.** Answers may vary. Samples are given. **49.** $\overleftrightarrow{EF}$ **51.** C **53.** *ABH* **55.** $\overleftrightarrow{FG}$

57. T R **59.** N V

61. −22, −29 **63.** P, S

Algebra 1 Review p. 30

1. 6 **3.** $-\frac{7}{3}$ or $-2\frac{1}{3}$ **5.** 4 **7.** −6 **9.** 15 **11.** 4 **13.** 17 **15.** −5 **17.** −20 **19.** 11

Lesson 1-5 pp. 33–35

EXERCISES **1.** 9; 9; yes **3.** 11; 13; no **5.** $XY = ZW = 4$; yes **7.** $YZ = 4, XW = 12$; no **9.** 25 **11.a.** 7 **b.** $RS = 60, ST = 36, RT = 96$ **13.** 33 **15.** 130 **17.** 6 **19.** 1 **21.** −3.5, 3.5 **23.a.** 54 mi **b.** Artois **25–27.** Check students' work. **29.** true; $AB = 2, CD = 2$ **31.** false; $AC = 9, BD = 9, AD = 11$, and $9 + 9 \neq 11$ **33.** 2, 12 **35.** $ED = 10, DB = 10, EB = 20$ **37.a.** $5x$ **b.** 9, 15 **39.** B **41.** A **43.** D **45.** never **47.** never **49.** always **51.** never **53.** 3125; 15,625

Lesson 1-6 pp. 40–42

1. $\angle XYZ, \angle ZYX, \angle Y$ **3.** $\angle ABC, \angle CBA$

5 and **7.** Drawings may vary. Samples are given.

5. **7.**

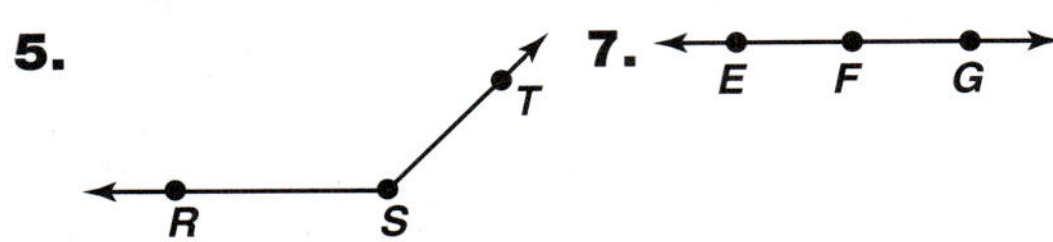

9. 60; acute **11.** 90; right **13.** 34 **15.** $\angle AOB$ or $\angle DOC$ **17.** $\angle EOC$ **19.** $\angle AOB$ and $\angle DOC$ or $\angle BOC$ and $\angle AOD$ **21.** 30 **23.** 30 **25.** No; there are no markings. **27.** No; there are no markings. **29.** Yes; there are markings. **31.** Yes; you can conclude that $\angle$s are vert. from the diagram. **33.** 115 **35.** 180 **37.** 30 **39.** 40 **41.** 45, 75 and 165, or 135, 105, and 15 **43.** 8; $m\angle AOB = 30, m\angle BOC = 50, m\angle COD = 30$ **45.** 7; $m\angle AOB = 31, m\angle BOC = 49, m\angle AOD = 111$ **47.a.** 19.5 **b.** 43; 137 **c.** Answers may vary. Sample: The sum of the $\angle$ measures should be 180. **49.** Angle Add. Post. **51.** B **53.** 47 **55.** Answers may vary. Sample: An orthographic drawing shows top, front, and side views of a figure. An isometric drawing uses isometric dot paper and shows three sides of a figure from a corner view. **57.** $10 = 2 \times 5$, so 10 has 5 as a factor and 10 is even.

Lesson 1-7 pp. 47–50

EXERCISES **1.**

3. **5.**

7.

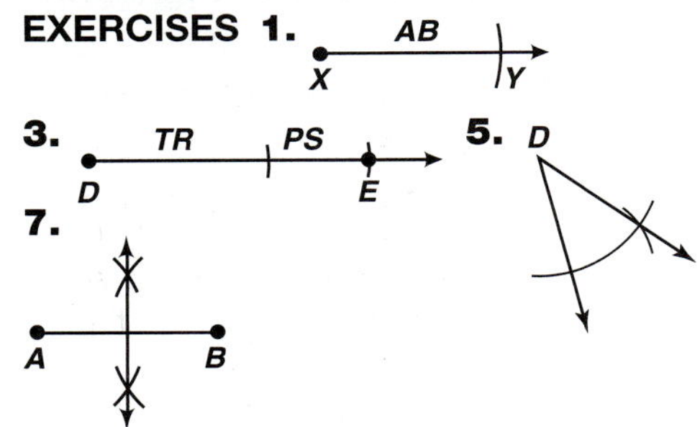

9.a. 11; 30 **b.** 30 **c.** 60 **11.** 15; 48

13. **15.**

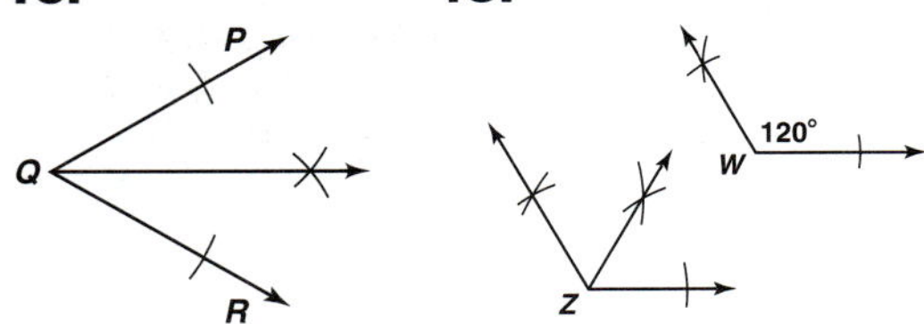

17. Find a segment on $\overleftrightarrow{SQ}$ so that you can construct $\overleftrightarrow{SP}$ as its $\perp$ bisector. Then bisect $\angle PSQ$.

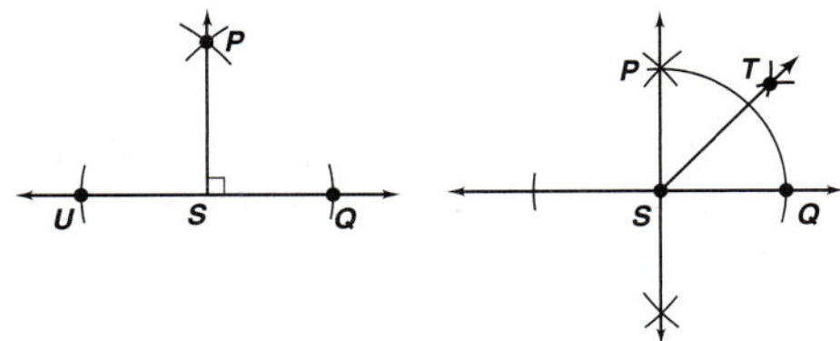

19.a–b.

Angle of incidence
Angle of reflection
Mirror

21. Explanations may vary. Samples are given. **a.** One midpt.; a midpt. divides a segment into two ≅ segments. If there were more than one midpt. the segments wouldn't be ≅. **b.** Infinitely many; there's only 1 midpt. but there exist infinitely many lines through the midpoint. A segment has exactly one ⊥ bisecting line because there can be only one line ⊥ to a segment at its midpt. **c.** There are an infinite number of lines in space that are ⊥ to a segment at its midpt. The lines are coplanar.

23.

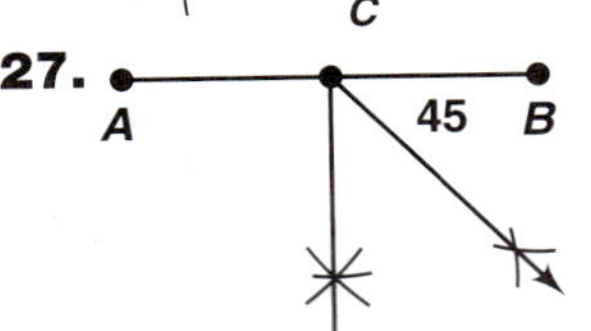

25. They are both correct. If you mult. each side of Lani's eq. by 2, the result is Denyse's eq.

27. A 45 B

29.

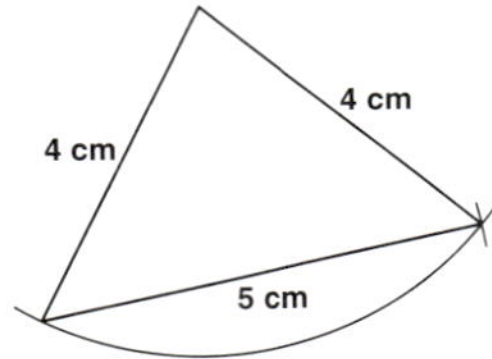

31. Impossible; the short segments are not long enough to form a △.

33.a. X Y **b.** They are all 60°. **c.** Answers may vary. Sample: Mark a pt., *A*. Swing a long arc from A. From a pt. *P* on the arc, swing another arc the same size that intersects the arc at a second pt., *Q*. Draw ∠*PAQ*. To construct a 30° ∠, bisect the 60° ∠.

35.a–b. **c.** Point *O* is the center of the circle.

37. A **39.** A

41. 72°

43. 20 and 180 **45.** 10 **47.** 3 **49.** No; they do not have the same endpoint.

Lesson 1-8 pp. 56–59

EXERCISES **1.** 6 **3.** 8 **5.** 23.3 **7.** 25 **9.** 12.0 **11.** about 4.5 mi **13.** 6.4 **15.** 15.8 **17.** B, C, D, E, F **19.** (3, 1) **21.** (6, 1) **23.** $\left(3\frac{7}{8}, -3\right)$ **25.** (5, −1) **27.** (12, −24) **29.** (5.5, −13.5) **31.** (4, −11) **33.** 5.8; (1.5, 0.5) **35.** 5.4; (−2.5, 3) **37.** 2.8; (−4, −4) **39.** 5.4; (3, 0.5) **41.** IV **43.** B **45.** 10.8 units; (3, −4) **47.** 165 units; The dist. *TV* is less than the dist. *TU*, so the airplane should fly from *T* to *V* to *U* for the shortest route. **49.** 934 mi **51.** 2693 mi **53** and **55.** Answers may vary. Samples are given. **53.** (3, 6), (0, 4.5) **55.** (1, 0), (−1, 4) **57.** exactly one pt., *H* (−5, 2)

59.a–f. Answers may vary. Samples are given. **a.** *BC* = *AD* **b.** If two opp. sides of a quad. are both ∥ and ≅, then the other two opp. sides are ≅. **c.** The midpts. are the same. **d.** If one pair of opp. sides of a quad. are both ∥ and ≅, then its diagonals bisect each other. **e.** *EF* = *AB* **f.** If a pair of opp. sides of a quad. are both ∥ and ≅, then the segment joining the midpts. of the other two sides has the same length as each of the first pair of sides.

61. z R y x

63. 11.7 units **65.** A **67.** B

69. A B P Q

71.

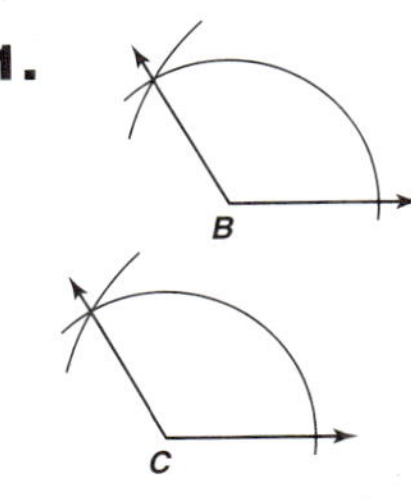

73. ∠*TAP*, ∠*PAT* **75.** 10 **77.** 48

Lesson 1-9 pp. 65–68

EXERCISES **1.** 22 in. **3.** 56 in. **5.** 120 m **7.** 38 ft **9.** 10π ft **11.** $\frac{1}{2}\pi$m **13.** 22.9 m **15.** 351.9 cm

17.

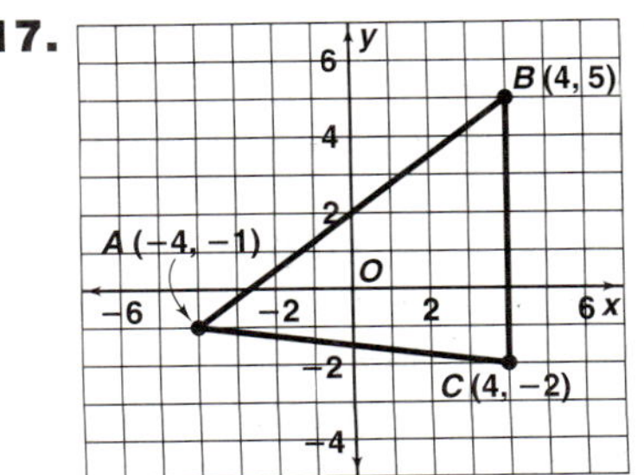

≈ 25.1 units

Selected Answers

59. 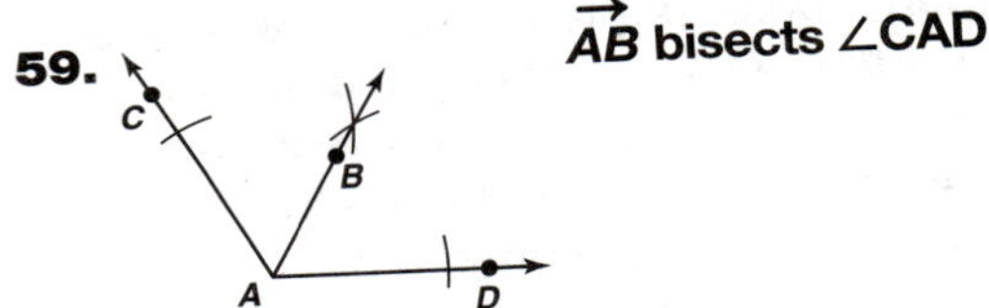

$\overrightarrow{AB}$ bisects $\angle CAD$

61. $\overleftrightarrow{AB}$ and $\overleftrightarrow{CG}$. **63.** *ABC*, *EFG* **65.** *AEF*, *BFG* **67.** $\overleftrightarrow{BC}$

Lesson 2-3 pp. 96–99

EXERCISES **1.** Felicia will pass the music theory course. **3.** Line ℓ and line *m* do not intersect. **5.** Figure *ABCD* has two pairs of parallel sides. **7.** Points *X*, *Y*, and *Z* are collinear. **9.** Nadine Muzerall attended the University of Minnesota. **11.** If two planes are not parallel, then they intersect in a line. **13.** If you are studying botany, then you are studying a science. **15.** Answers may vary. Sample: If you live in Lubbock, then you live in the 28th state to enter the Union. Levon lives in the 28th state to enter the Union. **17.** Must be true; by (E) and (A), it is breakfast time. Then by (C), Curtis drinks water. **19.** Is not true; by (E) and (A), it is breakfast time. By (C), Curtis drinks water and nothing else. **21.** Is not true; by (A) and (E), it is breakfast time. By (D), Julio drinks juice and nothing else. **23.** If you are in Key West, Florida, then the temperature is always above 32°F; not possible. **25.** If a figure is a square, then it is a rectangle; *ABCD* is a rectangle. **27.** No; red cars can never park. **29.** yes **31.** yes **33.a.**

Gills
Fish
Turtles

b. Turtles are not in the circle of animals with gills, so a turtle is not a fish. **35.** D **37.** B **39.** good definition **41.** Answers may vary. Sample: $\frac{1}{2}$. **43.** never

Lesson 2-4 pp. 105–108

EXERCISES **1.a.** $\angle$ Add. Post. **b.** Subst. Prop. **c.** Simplify. **d.** Subtr. Prop. of = **e.** Div. Prop. of = **3.a.** Mult. Prop. of = **b.** Distr. Prop. **c.** Add. Prop. of = **5.** Reflexive Prop. of ≅ **7.** Div. Prop. of = **9.** Mult. Prop. of = **11.** Subtr. Prop. of = **13.** Sub. Prop. **15.** Transitive Prop. of = **17.** 5*x* **19.** $\angle K$ **21.** 3 **23.** $\angle XYZ \cong \angle WYT$ **25.** Answers may vary. Sample: $\overline{LR}$ and $\overline{RL}$ are different ways to name the same segment and $\angle CBA$ and $\angle ABC$ are different ways to name the same $\angle$. **27.a.** Given **b.** Def. of midpoint **c.** Subst. Prop. of = **d.** Subtr. Prop. of = **e.** Div. Prop. of = **29.a.1.** $m\angle GFE + m\angle EFI = m\angle GFI$ ($\angle$ Add. Post.) **2.** $9x - 2 + 4x = 128$ (Subst. Prop.) **3.** $13x - 2 = 128$ (Simplify.) **4.** $13x = 130$ (Add. Prop. of =) **5.** $x = 10$ (Div. Prop. of =) **b.** 40 **31.** In the fifth step, each side is divided by $(b - a)$. But $b - a = 0$ and division by 0 is not defined. **33.** transitive **35.** symmetric **37.** transitive **39.** B **41.** B **43.** If a person has a job, then that person can save money each week **45.** 125 **47.** 90 **49.** 29, 31.5 **51.** −162, 486

Lesson 2-5 pp. 112–115

1. 20 **3.** 30 **5.** 75, 105 **7.a.** 90 **b.** 90 **c.** Subst. **d.** $m\angle 3$ **9.** Answers may vary. Sample: scissors. **11.** The two acute $\angle$s have measure 72. The two obtuse $\angle$s have measure 108. **13.** $x = 14$, $y = 15$; 50, 50, 130 **15.** C **17.** $\angle EIG \cong \angle FIH$ since all rt. $\angle$s are ≅; $\angle EIF \cong \angle HIG$ since they are compl. of the same $\angle$. **19.** Answers may vary. Sample: (−5, −1) **21.a.** V **b.** 180 **c.** Division **d.** right **23.** $m\angle A = 60$, $m\angle B = 30$ **25.** $m\angle A = 120$, $m\angle B = 60$ **27.** By the def. of suppl. $\angle$s, $m\angle 1 + m\angle 2 = 180$ and $m\angle 3 + m\angle 4 = 180$. By the Subst. Prop. $m\angle 1 + m\angle 2 = m\angle 3 + m\angle 4$. It is given that $\angle 2 \cong \angle 4$, so $m\angle 2 = m\angle 4$. Then by the Subtr. Prop. of =, $m\angle 1 = m\angle 3$, or $\angle 1 \cong \angle 3$. **29.a–b.** It is the bisector of both angles. **c.** Sample: perpendicular; bisectors of two adjacent supplementary angles form two adjacent angles whose measures add to $\frac{1}{2}(180)$, or 90. **31.** $x = 35$, $y = 70$; 70, 110, 70 **33.** B **35.** 12 **37.** $3 + NP = 15$ **39.** $\angle 1$ and $\angle 2$ are congruent **41.** If you live south of the equator, then you live in Australia.

Chapter Review pp. 117–119

1. Reflexive **2.** hypothesis **3.** Transitive **4.** biconditional **5.** converse **6.** Symmetric **7.** conclusion **8.** truth value **9.** deductive reasoning **10.** theorem **11.a.** If you are younger than 20, then you are a teenager. **b.** conditional: true, converse: false **12.a.** If an angle has measure greater than 90 and less than 180, then it is obtuse. **b.** conditional: true, converse: true **c.** An angle is obtuse if and only if it has measure greater than 90 and less than 180. **13.a.** If a figure has four sides, then it is a square. **b.** conditional: true, converse: false **14.** If something is a flower, then it is beautiful. **15.** Rico's definition is not reversible. A magazine is a counterexample. You read a magazine, but it is not a book. **16.** A phrase is an oxymoron if and only if it contains contradictory terms. **17.** If two angles are complementary, then the sum of their measures is 90. If the sum of the measures of two angles is 90, then the angles are complementary.

18. Lucy will become a better player. **19.** Lines ℓ and m intersect to form right angles. **20.** The sum of the measures of $\angle 1$ and $\angle 2$ is 180. **21.** If Kate studies, then she will graduate. **22.** If a, then c. **23.** If the weather is wet, then Nathan can stop at the ice cream shop. **24.a.** Segment Add. Post. **b.** Subst. Prop. **c.** Simplify. **d.** Subtr. Prop. of Equality **e.** Div. Prop. of Equality **25.** 8 **26.** BY **27.** $m\angle Y$ **28.** $RS = XY$ **29.** y **30.** 10 **31.** $p - 2q$ **32.** $\overline{NM}$ **33.** 18 **34.a.** 74 **b.** 74 **c.** 106 **35.** $\angle 2$, $\angle 3$, transitive

Chapter 3

Lesson 3-1 pp. 131–133

EXERCISES **1.** $\overleftrightarrow{PQ}$ and $\overleftrightarrow{SR}$ with transversal $\overleftrightarrow{SQ}$; alt. int. $\angle$s **3.** $\overleftrightarrow{PS}$ and $\overleftrightarrow{QR}$ with transversal $\overleftrightarrow{PQ}$; same-side int. $\angle$s **5.** $\angle 1$ and $\angle 2$: corr. $\angle$s, $\angle 3$ and $\angle 4$: alt. int. $\angle$s, $\angle 5$ and $\angle 6$: corr. $\angle$s **7.** $\angle 1$ and $\angle 2$: corr. $\angle$s, $\angle 3$ and $\angle 4$: same-side int. $\angle$s, $\angle 5$ and $\angle 6$: alt. int. $\angle$s **9.** 2. Same-Side Int. Angles Thm., 4. Same-Side Int. Angles Thm., 5. Congruent Supplements Thm. **11.** $m\angle 1 = 75$ because corr. $\angle$s of $\|$ lines are $\cong$; $m\angle 2 = 105$ because same-side int. $\angle$s of $\|$ lines are suppl. **13.** $m\angle 1 = 100$ because same-side int. $\angle$s of $\|$ lines are suppl.; $m\angle 2 = 70$ because alt. int. $\angle$s of $\|$ lines have = measure. **15.** 25; 65, 65 **17.** $m\angle 1 = m\angle 3 = m\angle 6 = m\angle 8 = m\angle 9 = m\angle 11 = m\angle 13 = m\angle 15 = 52$; $m\angle 2 = m\angle 4 = m\angle 5 = m\angle 7 = m\angle 10 = m\angle 12 = m\angle 14 = 128$ **19.** two **21.** two **23.** 32 **25.** $x = 135$, $y = 45$ **27.** Answers may vary. Sample: E illustrates corr. $\angle$s ($\angle 1$ and $\angle 3$, $\angle 2$ and $\angle 4$) and same-side int. $\angle$s ($\angle 1$ and $\angle 2$, $\angle 3$ and $\angle 4$); I illustrates alt. int. $\angle$s ($\angle 1$ and $\angle 4$, $\angle 2$ and $\angle 3$) and same-side int. $\angle$s ($\angle 1$ and $\angle 3$, $\angle 2$ and $\angle 4$).

29. Since $a \| b$, $\angle 1 \cong \angle 3$ because they are corr. $\angle$s. Also $\angle 3$ and $\angle 2$ are supplementary by the $\angle$Add. Post. So by Subst., $\angle 1$ and $\angle 2$ are supplementary. **31.a.** alt. int. $\angle$s **b.** He knew that alt. int. $\angle$s of $\|$ lines are $\cong$. **33.** Never; the two planes do not intersect. **35.** Sometimes; they may be skew. **37.** B **39.** 121 **41.** 29.5 **43.** $(-0.5, 3.5)$ **45.** Add 4; 20, 24 **47.** Subtract 7; -5, -12.

Lesson 3-2 pp. 137–140

EXERCISES **1.** $\overleftrightarrow{BE} \| \overleftrightarrow{CG}$; Conv. of Corr. $\angle$s Post. **3.** $\overline{JO} \| \overline{LM}$; if two lines and a transversal form same-side int. $\angle$s that are suppl., then the lines are $\|$. **5.** 30 **7.** 59 **9.** The corr. $\angle$s are $\cong$, so the lines are $\|$ by the Conv. of Corr. $\angle$s Post. **11.** $a \| b$; if two lines and a transversal form same-side int. $\angle$s that are suppl., then the lines are $\|$. **13.** none **15.** none **17.** $\ell \| m$; Conv. of Corr. $\angle$s Post. **19.** $a \| b$; Conv. of Corr. $\angle$s Post. **21.** $\ell \| m$; Conv. of Alt. Int. $\angle$s Thm. **23.** The corr. $\angle$s he draws are $\cong$. **25.** 20 **27.** $m\angle 1 + m\angle 3 = 180$ (Given). $m\angle 1 + m\angle 2 = 180$ ($\angle$ Add. Post.). $m\angle 1 + m\angle 3 = m\angle 1 + m\angle 2$ (Substitution). $m\angle 3 = m\angle 2$ (Subt. Prop. of =). $\ell \| n$ (Conv. of Corres. $\angle$s Post.) **29.** 5; $m\angle 1 = m\angle 2 = 50$ **31.** 1.25; $m\angle 1 = m\angle 2 = 10$ **33.** Answers may vary. Sample: $\angle 3 \cong \angle 9$; $j \| k$ by Conv. of the Alt. Int. $\angle$s Thm. **35.** Answers may vary. Sample: $\angle 3 \cong \angle 11$; $\ell \| m$ by Conv. of the Alt. Int. $\angle$s Thm. and $j \| k$ by Conv. of Corr. $\angle$s Post. **37.** Vert. $\angle$s Thm. and Conv. of Corr. $\angle$s Post.
39. 1. $j \| k$ 1. Given 2. $m\angle 9 + m\angle 4 = 180$ 2. Same-Side Int. $\angle$s Thm. 3. $m\angle 8 + m\angle 9 = 180$ 3. Given 4. $m\angle 9 + m\angle 4 = m\angle 8 + m\angle 9$ 4. Trans. Prop. of = 5. $m\angle 4 = m\angle 8$ 5. Subt. Prop. of = 6. $\ell \| n$ 6. Conv. of Corres. $\angle$s Post.
41. $\overline{PL} \| \overline{NA}$ and $\overline{PN} \| \overline{LA}$ by Conv. of Same-Side Int. $\angle$s Thm. **43.** none
45.a. Answers may vary. Sample:

b. Given: $a \| b$ with transversal e, c bisects $\angle AOB$, d bisects $\angle AXZ$.
c. Prove: $c \| d$
d. To prove that $c \| d$, show that $\angle 1 \cong \angle 3$. $\angle 1 \cong \angle 3$ if $\angle AOB \cong \angle OXZ$. $\angle AOB \cong \angle OXZ$ by the Corr. $\angle$s Post.
e. **1.** $a \| b$ (Given)
2. $\angle AOB \cong \angle AXZ$ (Corr. $\angle$s Post.)
3. $m\angle AOB = m\angle AXZ$ (Def. of $\cong$ $\angle$s)
4. $m\angle AOB = m\angle 1 + m\angle 2$; $m\angle AXZ = m\angle 3 + m\angle 4$ ($\angle$ Add. Post.)
5. c bisects $\angle AOB$; d bisects $\angle AXZ$. (Given)
6. $m\angle 1 = m\angle 2$; $m\angle 3 = m\angle 4$ (Def. of $\angle$ bisector)
7. $m\angle 1 + m\angle 2 = m\angle 3 + m\angle 4$ (Trans. Prop. of $\cong$)
8. $m\angle 1 + m\angle 1 = m\angle 3 + m\angle 3$ (Subst.)
9. $2m\angle 1 = 2m\angle 3$ (Add. Prop.)
10. $m\angle 1 = m\angle 3$ (Div. Prop.)
11. $c \| d$ (Conv. of Corr. $\angle$s Post.)
47. A **49.** $m\angle 1 = 70$ since it is a supp. of the $110°\angle$. $m\angle 2 = 110$ since same-side int. $\angle$s are supp. **51.** If you are west of the Mississippi River, then you are in Nebraska. Original statement is true, converse is false. **53.** If same-side int. $\angle$s are supp., then a line intersects a pair of $\|$ lines. Both are true. **55.** If there are clouds in the sky, then it is raining. Original statement is true, converse is false. **57.** 28.3 cm^2 **59.** 78.5 $in.^2$ **61.** 4.5 m^2 **63.** 1.1 m^2

Lesson 3-3 pp. 143–144

EXERCISES **1.a.** $\angle 1 \cong \angle 2 \cong \angle 3$ **b.** Slat D is perp. to slats B and C. Explanations may vary. Sample: Slat D is perp. to slat A. Slats A, B, and C are parallel, so by Theorem 3-11, slat D is also perp. to B and C. **3.a.** corr. ∠s **b, c.** $\angle 1$, $\angle 3$ (any order) **d.** Converse of Corr. **5-11.** Answers may vary. Samples are given. **5.** All of the rungs are perpendicular to one side. The side is perp. to the top rung, and because all of the rungs are parallel to each other, the side is perp. to all of the rungs. **7.** The rungs are parallel to each other because they are all perpendicular to one side. The sides are parallel because they are both perpendicular to one rung. **9.** All of the rungs are parallel. All of the rungs are parallel to one rung, so they are all parallel to each other. **11.** In the diagram, $a \perp b$ means the marked ∠ is a rt. ∠. $b \parallel c$ means that the corres. ∠ formed by a and c is a rt. ∠, so $a \perp c$. **13.** Answers may vary. Sample: In the diagram, $\overline{AB} \perp \overline{BH}$ and $\overline{AB} \perp \overline{BD}$, but $\overline{BH} \nparallel \overline{BD}$. They intersect. **15.** $a \perp d$ **17.** $a \perp d$ **19.** $a \parallel d$ **21.** $a \perp d$ **23.** Reflexive: $a \perp a$; false; ⊥ lines are two lines that intersect to form right ∠s. Symmetric: If $a \perp b$, then $b \perp a$; true; b and a intersect to form right ∠s. Transitive: If $a \perp b$, and $b \perp c$, then $a \perp c$; false; in a plane, two lines ⊥ to the same line are ∥. **25.** D **27.** 46 **29.** If two lines in a plane are parallel, they do not meet. Two lines in a plane do not meet if and only if the lines are parallel.

Lesson 3-4 pp. 150–153

EXERCISES **1.** 30 **3.** 90 **5.** $x = 80$; $y = 80$ **7.** right, scalene **9.** obtuse, isosceles **11.** Not possible; a right △ will always have one longest side opp. the right ∠. **13.** **15.**

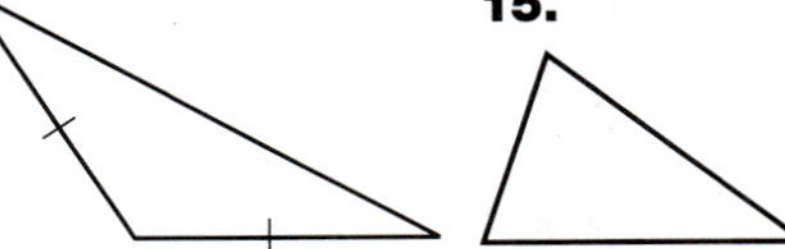

17.a. 2 **b.** 6 **19.** 115.5 **21.** $x = 147$, $y = 33$ **23.** $x = 7$; 55, 35, 90; right **25.** $x = 38$, $y = 36$, $z = 90$; $\triangle ABD$: 36, 90, 54; right; $\triangle BCD$: 90, 52, 38; right; $\triangle ABC$: 74; 52, 54; acute. **27.** 60; $180 \div 3 = 60$ **29.** eight

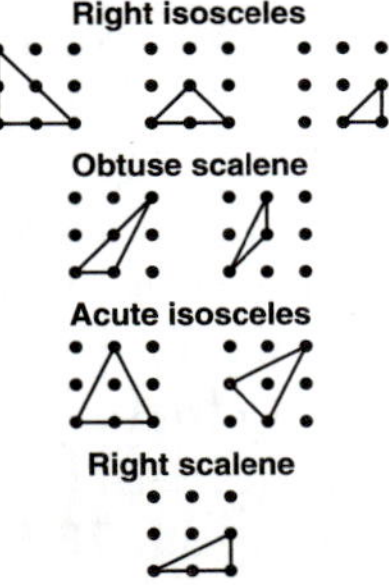

31. 30 and 60 **33.** Check students' work. Answers may vary. Sample: The two ext. ∠s formed at vertex A are vert. ∠s and thus have the same measure. **35.** $m\angle 1 + m\angle 4 = 180$ by the ∠ Add. Postulate. $m\angle 2 + m\angle 3 + m\angle 4 = 180$ by the △ ∠-Sum Theorem. $m\angle 1 + m\angle 4 = m\angle 2 + m\angle 3 + m\angle 4$ by the Trans. Property of Equality. $m\angle 1 = m\angle 2 + m\angle 3$ by the Subtr. Property of Equality. **37.** Check students' work. **39.a–b.** There are no such triangles. **c.** isosceles triangle. **41.** Answers may vary. Sample: The measure of the ext. ∠ is = to the sum of the measures of the two remote int. ∠s. Since these ∠s are ≅, the ∠s formed by the bisector of the ext. ∠ are ≅ to each of them. Therefore, the bisector is ∥ to the included side of the remote int. ∠s by the Conv. of the Alt Int. ∠s Thm. **43.** B **45.** $a \parallel b$; two lines ∥ to the same line are ∥ to each other.
47.

Lesson 3-5 pp. 161–164

EXERCISES **1.** yes **3.** No; it is not a plane figure. **5.** *MWBFX*; sides: $\overline{MW}$, $\overline{WB}$, $\overline{BF}$, $\overline{FX}$, $\overline{XM}$; ∠s: $\angle M$, $\angle W$, $\angle B$, $\angle F$, $\angle X$ **7.** *HEPTAGN*; sides: $\overline{HE}$, $\overline{EP}$, $\overline{PT}$, $\overline{TA}$, $\overline{AG}$, $\overline{GN}$, $\overline{NH}$; ∠s: $\angle H$, $\angle E$, $\angle P$, $\angle T$, $\angle A$, $\angle G$, $\angle N$ **9.** decagon; concave **11.** 1080 **13.** 1440 **15.** 180,000 **17.** 103 **19.** 37 **21.** 113, 119 **23.** 150; 30 **25.** 176.4; 3.6
27. 90 **29.**

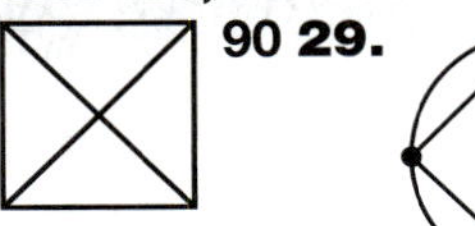

31.

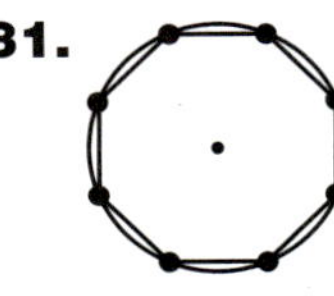

33. 8 **35.** 18 **37.** octagon; $m\angle 1 = 135$; $m\angle 2 = 45$ **39.** 20-80-80; 50-50-80 **41.** 144; 10 **43.** 150; 12 **45.** $\frac{4}{5}$ **47.** $y = 103$; $z = 70$; quad. **49.** $x = 36$, $2x = 72$, $3x = 108$, $4x = 144$; quad. **51 and 53.** Answers may vary. Samples are given.

51. **53.**

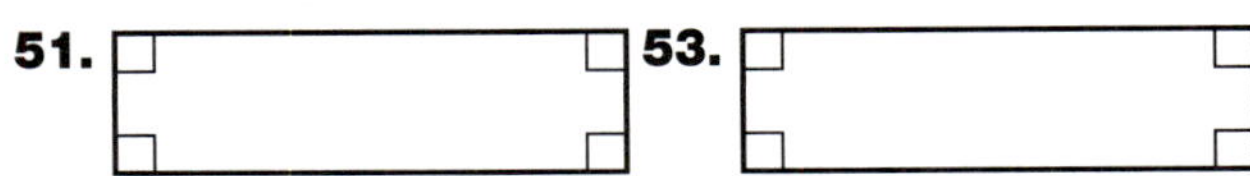

55. Answers may vary. Sample: The figure is a convex equilateral quad. The sum of its ∠s is $2 \cdot 180$ or 360. **57.a.** (20, 162), (40, 171), (60, 174), (80, 175.5), (100, 176.4), (120, 177), (140, 177.4), (160, 177.75), (180, 178), (200, 178.2)
b.

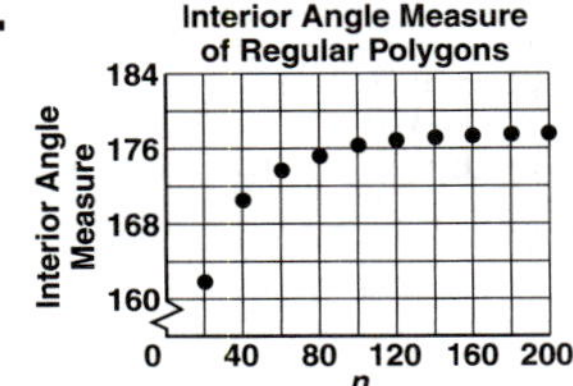

c. It is very close to 180. **d.** No, two sides cannot be collinear.

59. 36 **61 and 63.** Answers may vary. Samples are given. **61.** Not possible; opp. sides would overlap. **63.** Not possible; opp. and adj. sides would overlap. **65.** A **67.** 120, 25 **69.** 104, 76, 35, 69 **71.** Subst. Prop. **73.** Symm. Prop. of ≅ **75.** Trans. Prop. **77.** ∠*BRT*, ∠*BRK* **79.** Answers may vary. Sample: ∠*BRM* **81.** ∠*TRK*

Algebra 1 Review p. 165

1. $\frac{8}{3}$; it is steeper than line ℓ, but has the same tilt.
3. $-\frac{3}{2}$; it is steeper than line *r*, but has the same tilt.
5. −1; it is the same steepness and tilt as line *r*.
7. −5; it is steeper than line *r*, but has the same tilt.
9. Undefined; the line is vertical.

11. 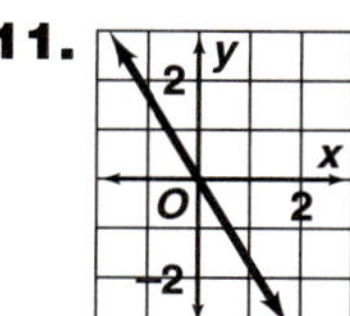 The line is steeper than line *r*, but has the same tilt.

13. The line is horizontal. 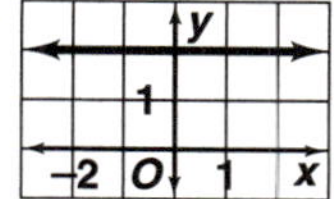

15. Sample: For a vertical line, the horizontal change between any two points is 0, and $\frac{\text{nonzero vertical change}}{0}$ is undefined.

Lesson 3-6 pp. 169–171

EXERCISES

1. 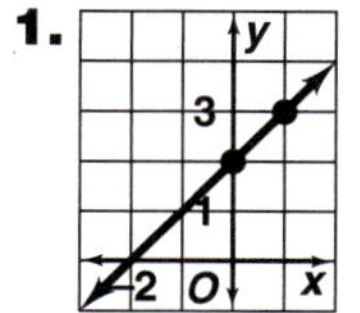 **3.** 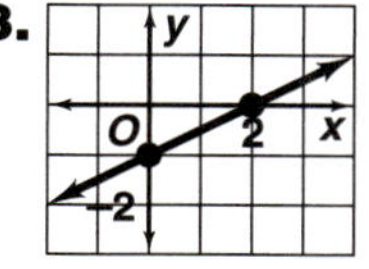

5. 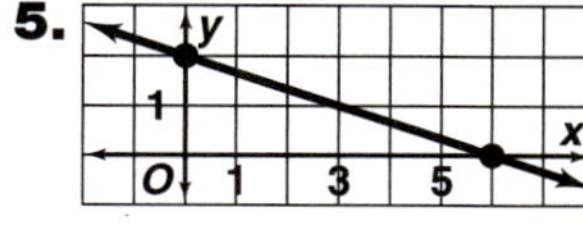 **7.** 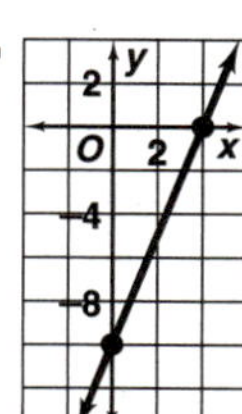

9. 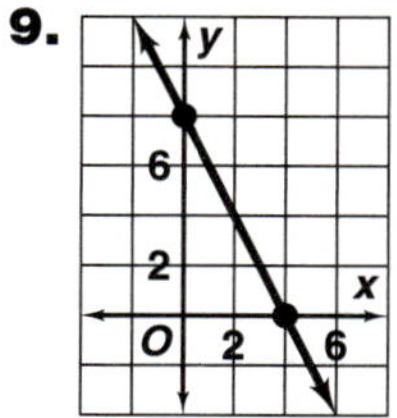 **11.** 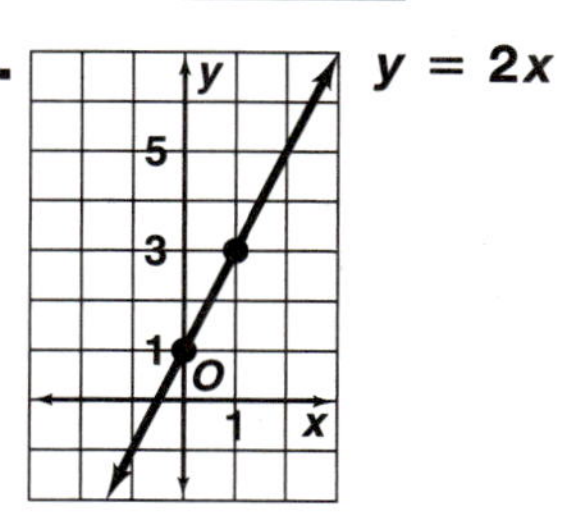 $y = 2x + 1$

13. $y = -2x + 4$ **15.** $y = -\frac{1}{3}x + 1$

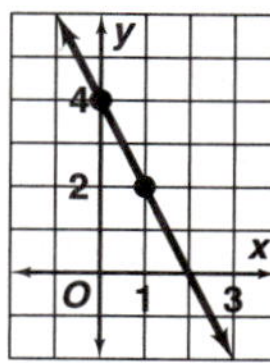 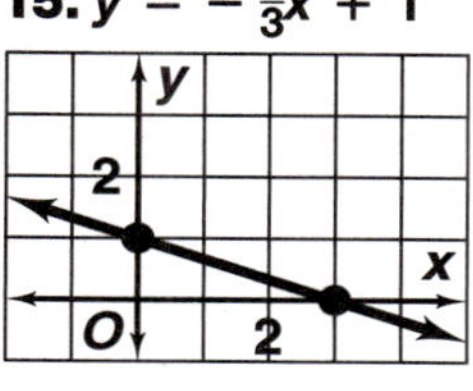

17. $y - 3 = 2(x - 2)$ **19.** $y - 5 = -1(x + 3)$ **21.** $y - 1 = \frac{1}{2}(x - 6)$ **23–27.** Equations may vary from the pt. chosen. Samples are given.

23. $y - 5 = \frac{3}{5}(x - 0)$ **25.** $y - 6 = 1(x - 2)$

27. $y - 0 = \frac{1}{2}(x + 1)$ **29.a.** $y = 7$ **b.** $x = 4$

31.a. $y = -1$ **b.** $x = 0$

33. 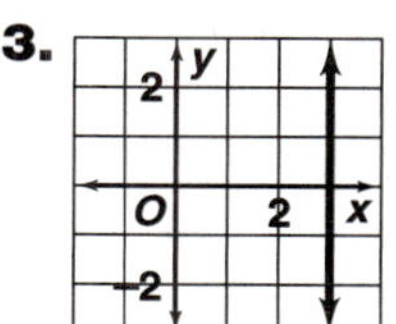 **35.** 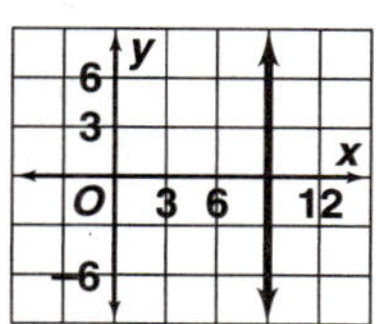

37. 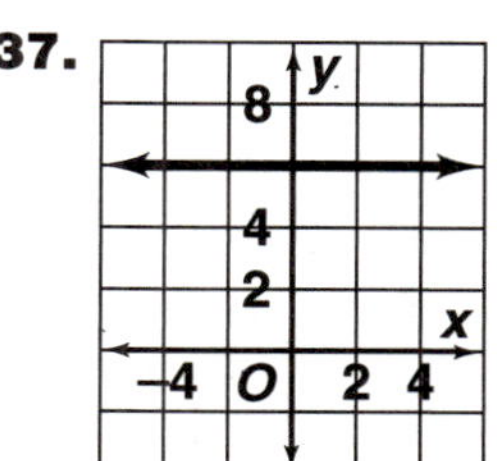

39. No; a line with no slope is a vertical line. 0 slope is a horizontal line. **41.a.** Undefined; it is a vertical line. **b.** $x = 0$ **43.** Answers may vary. Sample: The eq. is in slope-int. form; use slope-int. form, because the eq. is already in that form.

45. 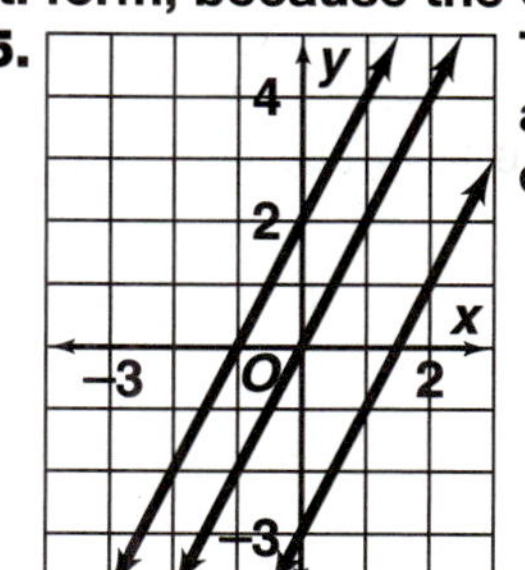 The slopes are the same, and the *y*-intercepts are different.

47. Check students' work.

49. 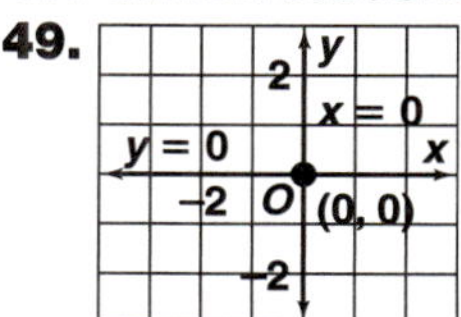

51. 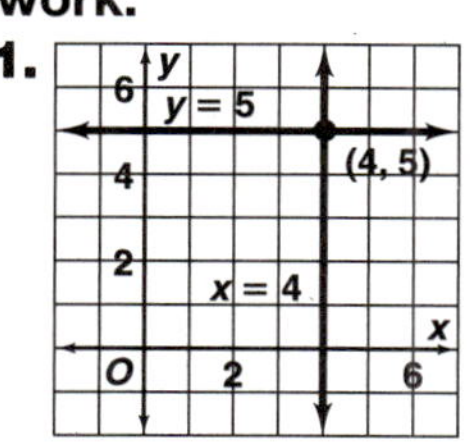

53. The *y*-intercepts are the same, and the lines have the same steepness. One line rises from left to right while the other falls from left to right.

Selected Answers

55. (2, 0), (0, 4); $m = \frac{0-4}{2-0} = \frac{-4}{2} = -2$
$y - 0 = -2(x - 2)$, $2x + y = 4$ or $y = -2x + 4$
57. Yes; the slope of $\overline{AB}$ = the slope of $\overline{BC}$.
59. Yes; the slope of $\overline{GH}$ = the slope of $\overline{HI}$.
61. $y - 2 = 3(x + 2)$; $3x - y = -8$
63. $y - 6 = \frac{2}{3}(x - 2)$; $2x - 3y = -14$
65. A **67.** 540 **69.** 2160 **71.** No; Parallel lines never intersect, but they are not skew **73.** $a = 5$; $m\angle MPR = 30$ **75.** $a = 2$; $m\angle QPR = 8$

Lesson 3-7 pp. 177–180

EXERCISES 1. Yes; both slopes = $-\frac{1}{2}$. **3.** No; the slope of $\ell_1 = \frac{3}{2}$, and the slope of $\ell_2 = 2$. **5.** Yes; both slopes = 0. **7.** Yes; the lines both have a slope of $\frac{3}{4}$ but different y-intercepts. **9.** No; one slope = 7 and the other slope = −7. **11.** Yes; the lines both have a slope of $-\frac{2}{5}$ but different y-intercepts. **13.** $y - 0 = \frac{1}{3}(x - 6)$ or $y = \frac{1}{3}(x - 6)$ **15.** $y + 2 = -\frac{3}{2}(x - 6)$ **17.** Yes; the slope of $\ell_1 = -\frac{3}{2}$, and the slope of $\ell_2 = \frac{2}{3}$; $-\frac{3}{2} \cdot \frac{2}{3} = -1$. **19.** Yes; the slope of $\ell_1 = -1$, and the slope of $\ell_2 = 1$; $-1 \cdot 1 = -1$. **21 and 23.** Answers may vary. Samples are given. **21.** $y = -2(x - 4)$ **23.** $y = \frac{4}{5}x$ **25.** Yes; $1 \cdot (-1) = -1$. **27.** No; $\frac{2}{7} \cdot (-\frac{7}{4}) \neq -1$. **29.** slope of $\overline{AB}$ = slope of $\overline{CD} = \frac{2}{3}$; $\overline{AB} \parallel \overline{CD}$ slope of $\overline{BC}$ = slope of $\overline{AD} = -3$; $\overline{BC} \parallel \overline{AD}$ **31.** slope of $\overline{AB} = \frac{1}{2}$; slope of $\overline{CD} = \frac{1}{4}$; $\overline{AB} \nparallel \overline{CD}$; slope of $\overline{BC} = -1$; slope of $\overline{AD} = -\frac{1}{2}$; $\overline{BC} \nparallel \overline{AD}$ **33.** Answers may vary. Sample: $y = \frac{4}{5}x + 5$, $y = -\frac{5}{4}x + 5$ **35.** $\overline{RS}$ and $\overline{VU}$ are horizontal with slope = 0; $\overline{RS} \parallel \overline{VU}$; slope of $\overline{RW}$ = slope of $\overline{UT}$ = 1; $\overline{RW} \parallel \overline{UT}$; slope of $\overline{WV}$ = slope of $\overline{ST} = -1$; $\overline{WV} \parallel \overline{ST}$ **37.** The lines will have the same slope. **39.a.** $y + 20 = \frac{3}{4}(x - 35)$ **b.** because you are given a point and can quickly find the slope **41.** $\perp$ **43.** $\perp$
45. $\overline{AC}: d = \sqrt{(7-9)^2 + (11-1)^2} = \sqrt{104}$
$\overline{BD}: d = \sqrt{(13-3)^2 + (7-5)^2} = \sqrt{104}$
$\overline{AC} \cong \overline{BD}$
47.a–b. Answers may vary. Sample:

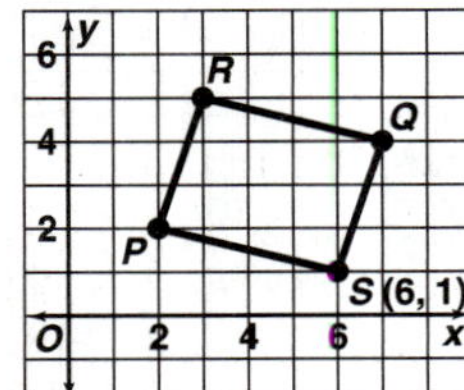

c. The other possible locations for S are (−2, 3) and (8, 7). **49.** C **51.** D **53.** $y - 2 = \frac{5}{3}(x + 4)$ **55.** Refl. Prop. of ≅ **57.** Dist. Prop. **59.** If you are in geometry class, then you are at school.

Lesson 3-8 pp. 184–186

EXERCISES 1. $\ell \parallel \overleftrightarrow{AB}$

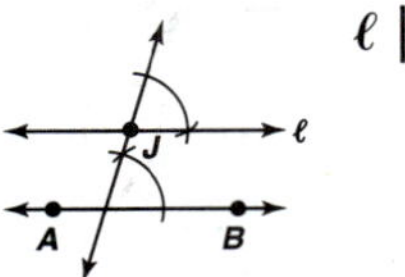

3. $\ell \parallel \overline{AB}$

5-7. Constructions may vary. Samples using the following segments are shown:

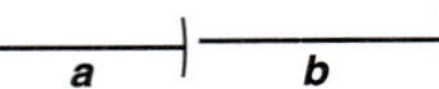

5.

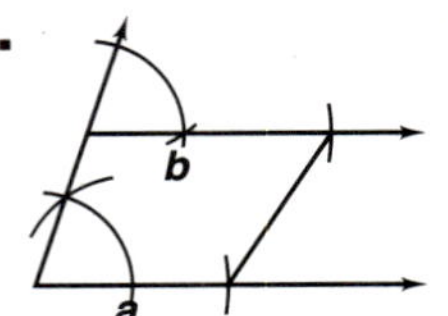

7.

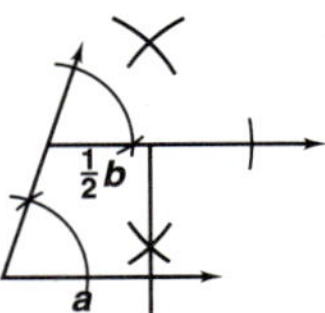

9. **11.**

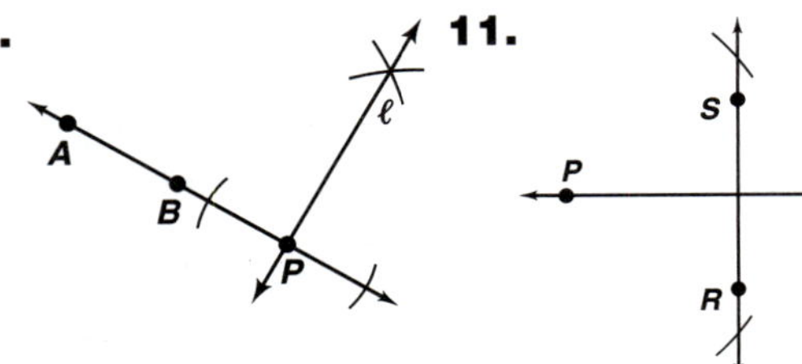

$\ell \perp \overleftrightarrow{AB}$ $\ell \perp \overleftrightarrow{RS}$

13.

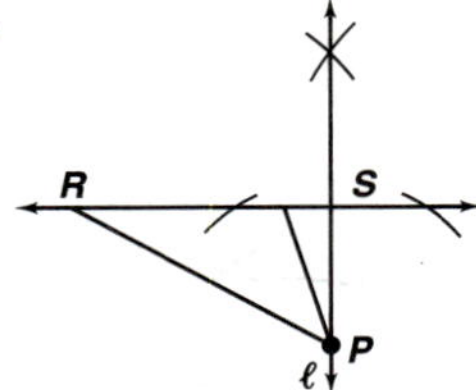

$\ell \perp \overleftrightarrow{RS}$

15. Construct a ≅ alt. int. ∠; then draw the ∥ line.
17–23. Constructions may vary. Samples are given.
17. **19.**

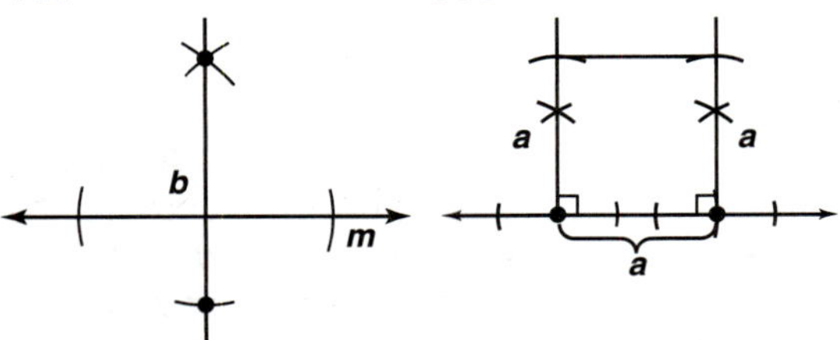

21.a.

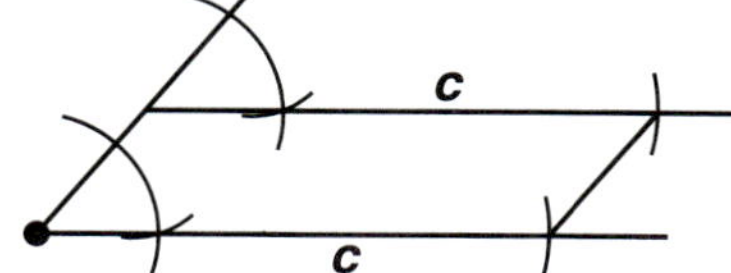

b. The sides are ∥ and ≅.
c. Check students' work.

23.

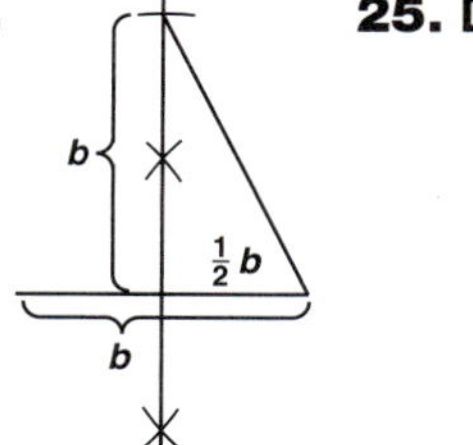

25. D

27.

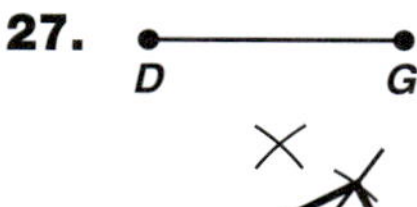

Answers may vary. Sample:

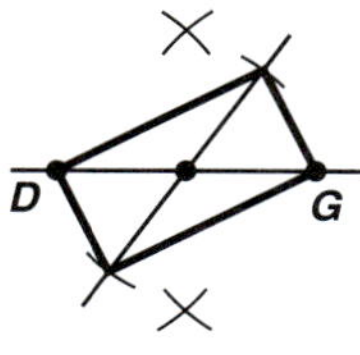

The quad is a rectangle.

29.

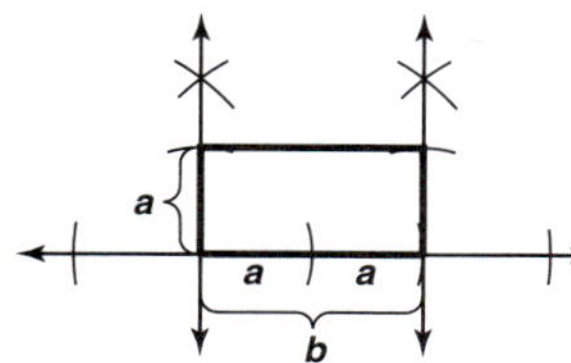

31.

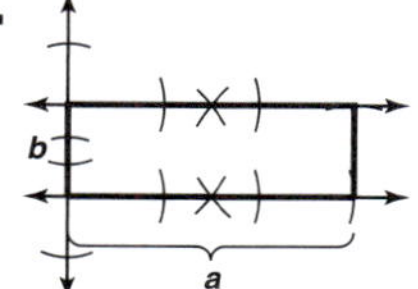

33.

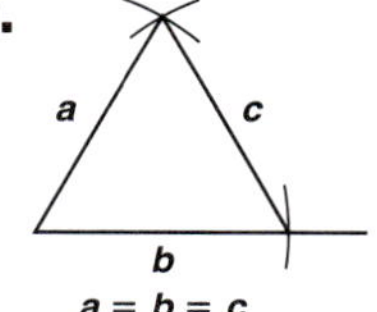

35. Not possible; if $a = 2b = 2c$, then $2a = 2b + 2c$ or $a = b + c$. The smaller sides would meet at the midpoint of the longer side, forming a segment. **37.** A **39.** D **41.** No; the slopes are different **43.** Yes; the slopes are both $-\frac{1}{3}$, and the y-intercepts are different. **45.** 8.9 **47.** $\overleftrightarrow{DF}$

Chapter Review pp. 189–191

1. acute **2.** obtuse **3.** corr. ∠s **4.** exterior **5.** convex **6.** equiangular **7.** regular **8.** point-slope **9.** slope-int. **10.** alt. int. ∠s **11.** $m\angle 2 = 121$, $m\angle 3 = 59$, $m\angle 4 = 59$ **12.** $m\angle 1 = 120$; corr. ∠s are ≅. $m\angle 2 = 120$; vert. ∠s are ≅ or alt. ext. ∠s are ≅. **13.** $m\angle 1 = 75$; same side int. ∠s are suppl. $m\angle 2 = 105$; alt. int. ∠s are ≅ or two ∠s that form a straight ∠ are suppl. **14.** $m\angle 1 = 55$; same side int. ∠s are suppl. $m\angle 2 = 90$; alt. int. ∠s are ≅. **15.** 20 **16.** 20 **17.** 24

18. **19.**

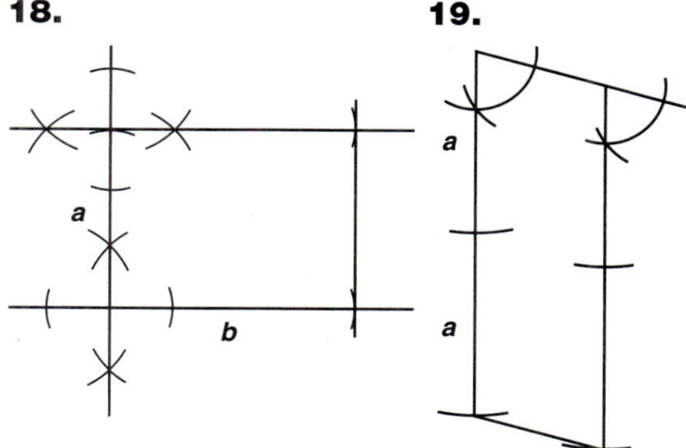

20. If 2 lines are ∥ to the same line, then the lines are ∥ (Thm. 3–10). **21.** If a line is ⊥ to one of two ∥ lines, then it is also ⊥ to the other (Thm. 3–11). **22.** 61; scalene, acute **23.** $x = 60$; $y = 60$; equilateral, acute **24.** $x = 45$; $y = 45$; isosc., right **25.** 55; acute **26.** 30; right **27.** 3; acute **28.** 8; obtuse **29.** One ∠ is 90; the remaining 2 ∠s are compl. **30.** 120; 60 **31.** 135; 45 **32.** 144; 36 **33.** 165; 15 **34.** 360

35. $m = 2$; y-int. $= -1$

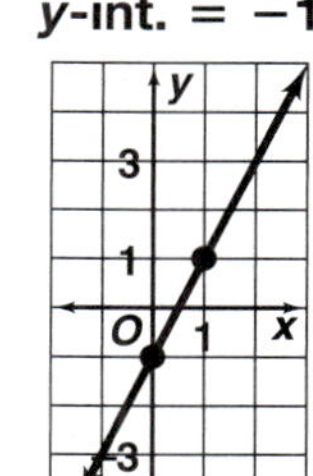

36. $m = -2$; point $= (-5, 3)$

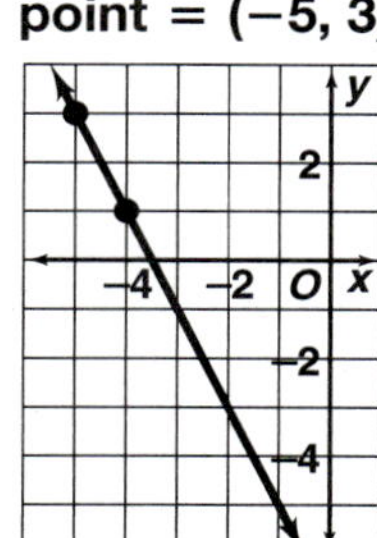

37.

38.

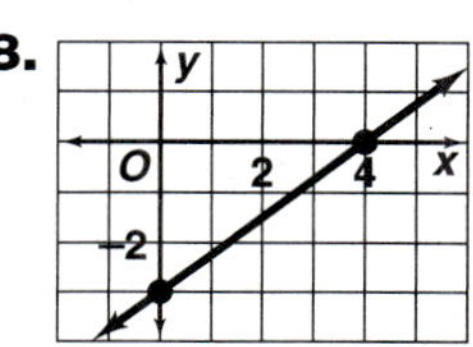

39. $x = 6$ **40.** neither **41.** ∥ **42.** ⊥ **43.** ∥ **44.** 0; the difference of y-coordinates is always zero.

Chapter 4

Lesson 4-1 pp. 200–203

EXERCISES **1.** $\angle CAB \cong \angle DAB$; $\angle C \cong \angle D$; $\angle ABC \cong \angle ABD$; $\overline{AC} \cong \overline{AD}$; $\overline{AB} \cong \overline{AB}$; $\overline{CB} \cong \overline{DB}$ **3.** $\overline{BK}$ **5.** $\overline{ML}$ **7.** ∠C **9.** △KJB **11.** △JBK **13.** E, K, G, N **15.** $\angle P \cong \angle S$; $\angle O \cong \angle I$; $\angle L \cong \angle D$; $\angle Y \cong \angle E$ **17.** 54 in. **19.** 77 **21.** 34 in. **23.** 103 **25.** No; the corr. sides are not ≅. **27.** Yes; all corr. sides and ∠s are ≅. **29.** B **31.** 5 **33.** $m\angle B = m\angle E = 21$ **35.** $AC = DF = 19$ **37.** Answers may vary. Sample: She could arrange them in a neat pile and pull out the ones of like sizes. **39.** △BCE ≅ △ADE **41.** △JLM ≅ △NRZ; △JLM ≅ △ZRN

43. Answers may vary. Sample: $\triangle TKR \cong \triangle MJL$: $\overline{TK} \cong \overline{MJ}$; $\overline{TR} \cong \overline{ML}$; $\overline{KR} \cong \overline{JL}$; $\angle TKR \cong \angle MJL$; $\angle TRK \cong \angle MLJ$; $\angle KTR \cong \angle JML$
45. $\angle A \cong \angle D$, $\angle B \cong \angle E$ (Given), $m\angle A + m\angle B + m\angle C = 180$, $m\angle D + m\angle E + m\angle F = 180$ ($\triangle$-$\angle$ Sum Thm.), $m\angle A + m\angle B + m\angle C = m\angle D + m\angle E + m\angle F$ (Subst. Prop.), $m\angle D + m\angle E + m\angle C = m\angle D + m\angle E + m\angle F$ (Subst. Prop.), $m\angle C = m\angle F$ (Subtr.) **47.** 2; either (3, 1) or (3, −7)
49. C **51.** **53.** $RS = PQ$ **55.** 12

Lesson 4-2 pp. 208–211

EXERCISES **1.a.** Given **b.** Reflexive **c.** $\triangle JKM$ **d.** $\triangle LMK$ **3.** It is given that $\overline{WZ} \cong \overline{ZS} \cong \overline{SD} \cong \overline{DW}$. $\overline{ZD} \cong \overline{ZD}$ by the Reflexive Prop. of Congruence. Therefore, $\triangle WZD \cong \triangle SDZ$ by SSS. **5.** Yes; the legs have equal lengths and are joined at midpts. So $\overline{AE} \cong \overline{CE}$ and $\overline{BE} \cong \overline{DE}$; $\angle AEB \cong \angle CED$ by vert. $\angle$s are $\cong$; $\triangle AEB \cong \triangle CED$ by SAS. **7.** Yes; since $\overline{AC} \cong \overline{AC}$ by the Refl. Prop., the $\triangle$s are $\cong$ by SAS. **9.** $\angle T \cong \angle V$ or $\overline{RS} \cong \overline{WU}$ **11.** $\angle W$ **13.** $\overline{WU}$ **15.** $\overline{XZ}$, $\overline{YZ}$ **17.** Yes; $\triangle PVQ \cong \triangle STR$ by SSS. **19.** A **21.** $\triangle KLJ \cong \triangle MON$; SSS **23.** $\triangle JEF \cong \triangle SVF$ or $\triangle JEF \cong \triangle SFV$; SSS **25.** $\triangle PQR \cong \triangle NMO$; SAS **27.** $\overline{AE}$ and $\overline{BD}$ bisect each other, so $\overline{AC} \cong \overline{CE}$ and $\overline{BC} \cong \overline{CD}$. $\angle ACB \cong \angle DCE$ because vert. $\angle$s are $\cong$. $\triangle ACB \cong \triangle ECD$ by SAS. **29.** No; you would need $\angle H \cong \angle K$ or $\overline{GI} \cong \overline{JL}$.

31.

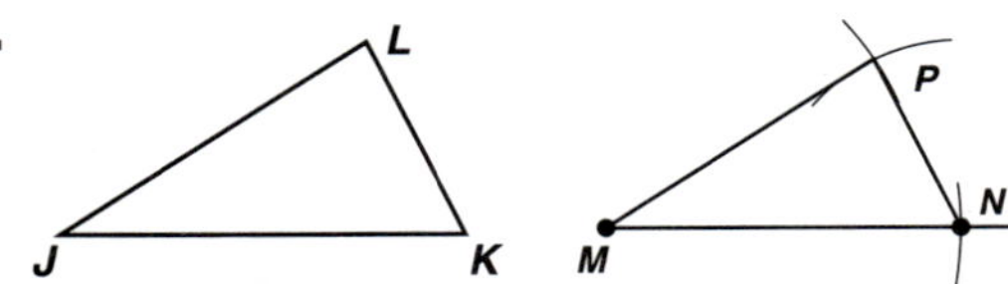

33.a–b. Answers may vary. Sample: **a.** wallpaper designs; ironwork on a bridge; highway warning signs **b.** $\cong$ $\triangle$s produce a well-balanced, symmetric appearance. In construction, $\cong$ $\triangle$s enhance designs. Highway warning signs are more easily identified if they are $\cong$. **35.** $\overline{IP} \cong \overline{PO}$; $\triangle ISP \cong \triangle OSP$ by SSS. **37.** Yes; $\triangle ABC \cong \triangle CDA$ by SAS; $\angle DAC \cong \angle ACB$ because if $\parallel$ lines, then alt. int. $\angle$s are $\cong$. **39.** $\overline{AM} \cong \overline{MB}$ because M is the midpt. of $\overline{AB}$. $\angle B \cong \angle AMC$ because all right $\angle$s are $\cong$. $\overline{CM} \cong \overline{DB}$ is given. $\triangle AMC \cong \triangle MBD$ by SAS.
41. $\triangle MNO \cong \triangle OLM$ by SAS. Therefore $\angle NMO \cong \angle LOM$ by def. of $\cong$ $\triangle$s, so $\overline{MN} \parallel \overline{LO}$ by the Conv. of the Alt. Int. $\angle$s Thm. **43.a.** No; the angles are not necessarily $\cong$. **b.** No; sample explanation: the $\angle$s can be changed without changing the side lengths. **c.** Answers may vary. Sample: a diagonal **45.** D **47.** $\angle E$ **49.** $\overline{FG}$ **51.** The product of the slopes of two lines is −1 if and only if the lines are $\perp$. **53.** If $2x = 6$, then $x = 3$. The statement and the converse are both true.

Lesson 4-3 pp. 215–219

EXERCISES **1.** $\triangle PQR \cong \triangle VXW$ **3.** $\overline{RS}$ **5.a.** Reflexive, **b.** ASA **7.** $\overline{QR} \cong \overline{TS}$: given, $\overline{QR} \parallel \overline{ST}$: given, $\angle TQR \cong \angle QTS$: Alt. Int. $\angle$s Thm., $\angle QTR \cong \angle TQS$: Alt. Int. $\angle$s Thm., $\triangle QRT \cong \triangle TSQ$: AAS
9. It is given that $\angle UWT$ and $\angle UWV$ are right $\angle$s and that $\angle T \cong \angle V$. $\angle UWT \cong \angle UWV$ since all right $\angle$s are congruent. $\overline{UW} \cong \overline{UW}$ by the Reflexive Property of Congruence, so $\triangle UWT \cong \triangle UWV$ by AAS. **11.1.** $\angle V \cong \angle Y$ (given) **2.** $\overline{WZ}$ bisects $\angle VWY$ (given) **3.** $\overline{WZ} \cong \overline{WZ}$ (Refl. Prop. $\cong$) **4.** $\triangle VWZ \cong \triangle YWZ$ (AAS) **13.** $\triangle PMO \cong \triangle NMO$; ASA **15.** $\triangle ZVY \cong \triangle WVY$; AAS **17.** Yes; if 2 $\angle$s of a $\triangle$ are $\cong$ to 2 $\angle$s of another $\triangle$, then the 3rd $\angle$s are $\cong$. So, an AAS proof can be rewritten as an ASA proof. **19.** No; you also need one pair of corres. sides $\cong$. **21.** $\triangle FGJ \cong \triangle HJG$ by AAS, since $\angle FGJ \cong \angle HJG$ because when lines are $\parallel$, then alt. int. $\angle$s are $\cong$, and $\overline{GJ} \cong \overline{GJ}$ by the Reflexive Prop. of $\cong$. **23.** $\triangle BDH \cong \triangle FDH$ by ASA since $\angle BDH \cong \angle FDH$ by def. of $\angle$bis. and $\overline{DH} \cong \overline{DH}$ by the Reflexive Prop. of $\cong$. **25.** $\overline{AB} \parallel \overline{DC}$, $\overline{AD} \parallel \overline{BC}$ (Given), $\angle DAC \cong \angle BCA$ (Alt. Int. $\angle$s Thm.), $\angle ACD \cong \angle CAB$ (Alt. Int. $\angle$s Thm.), $\overline{AC} \cong \overline{AC}$ (Reflexive Prop.), so $\triangle ABC \cong \triangle CDA$ by ASA. **27.a.** Check students' work. **b.** Answers may vary; most likely ASA. **29.** $\triangle AEB \cong \triangle CED$, $\triangle BEC \cong \triangle DEA$, $\triangle ABC \cong \triangle CDA$, $\triangle ABD \cong \triangle DCA$, $\triangle BAD \cong \triangle DCB$, $\triangle ABD \cong \triangle DCB$, $\triangle CBA \cong \triangle DAB$, $\triangle BCD \cong \triangle ADC$ **31.** $\frac{13}{20}$ **33.** C **35.** D **37.** $\triangle ONL \cong \angle MLN$; SAS **39.** $\overline{AC}$ and $\overline{CB}$ **41.** 56 **43.** 60% more paper

Lesson 4-4 pp. 222–225

EXERCISES **1.** SAS; $\triangle KLJ \cong \triangle OMN$; $\angle K \cong \angle O$; $\angle J \cong \angle N$; $\overline{JK} \cong \overline{NO}$ **3.** $\triangle MOE \cong \triangle REO$ by SSS because $\overline{OE} \cong \overline{OE}$ by Reflexive Prop. of $\cong$; $\angle M \cong \angle R$ by CPCTC. **5.** The $\triangle$s are $\cong$ by SAS so the distance across the sinkhole is 26.5 yd by CPCTC. **7.** $\overline{YT} \cong \overline{YP}$, $\angle C \cong \angle R$, $\angle T \cong \angle P$ (Given), $\angle CYT \cong \angle RYP$ (If 2 $\angle$s of a $\triangle$ are $\cong$ to 2 $\angle$s of another, the 3rd $\angle$s are $\cong$.), $\triangle CYT \cong \triangle RYP$ (ASA), $\overline{CT} \cong \overline{RP}$ (CPCTC) **9.** $\overline{KL} \cong \overline{KL}$ by Reflexive Prop. of $\cong$; $\overline{PL} \cong \overline{LQ}$ by Def. of $\perp$ bis.; $\angle KLP \cong \angle KLQ$ by Def. of $\perp$; the $\triangle$s are $\cong$ by SAS. **11.** $\angle QPS \cong \angle RSP$, $\angle Q \cong \angle R$ (Given), $\angle QSP \cong \angle RPS$ (If 2 $\angle$s of a $\triangle$ are $\cong$ to 2 $\angle$s of another, the

3rd $\triangle$s are $\cong$.), $\overline{PS} \cong \overline{PS}$ (Reflexive Prop.), $\triangle QPS \cong \triangle RSP$ (ASA), $\overline{PQ} \cong \overline{SR}$ (CPCTC) **13.a.** $\overline{AP} \cong \overline{PB}$; $\overline{AC} \cong \overline{BC}$ **b.** The diagram is constructed in such a way that the $\triangle$s are $\cong$ by SSS. $\angle CPA \cong \angle CPB$ by CPCTC. Since these $\angle$s are $\cong$ and suppl., they are right $\angle$s. Thus, $\overline{CP}$ is $\perp$ to ℓ. **15.** $\overline{BA} \cong \overline{BC}$ is given; $\overline{BD} \cong \overline{BD}$ by the Reflexive Prop. of $\cong$ and since $\overline{BD}$ bisects $\angle ABC$, $\angle ABD \cong \angle CBD$ by def. of an $\angle$ bisector; thus, $\triangle ABD \cong \triangle CBD$ by SAS; $\overline{AD} \cong \overline{DC}$ by CPCTC so $\overline{BD}$ bisects $\overline{AC}$ by def. of a bis.; $\angle ADB \cong \angle CDB$ by CPCTC and $\angle ADB$ and $\angle CDB$ are suppl.; thus, $\angle ADB$ and $\angle CDB$ are right $\angle$s and $\overline{BD} \perp \overline{AC}$ by def. of $\perp$. **17.** $\triangle ABX \cong \triangle ACX$ by SSS, so $\angle BAX \cong \angle CAX$ by CPCTC. Thus $\overrightarrow{AX}$ bisects $\angle BAC$ by the def. of $\angle$ bis. **19.** Prove $\triangle KJM \cong \triangle QPM$ by ASA since $\angle P \cong \angle J$ and $\angle K \cong \angle Q$ by alt. int. $\angle$s are $\cong$. **21.** Since $\triangle PGM \cong \triangle GPR$ (or $\triangle PMR \cong \triangle GRM$), then $\overline{PR} \cong \overline{MG}$ and $\overline{MP} \cong \overline{GR}$ by CPCTC. **23.** A **25.** ASA **27.** 95; 85 **29.** not possible

Lesson 4-5 pp. 230–233

EXERCISES **1.** $\overline{VX}$; Conv. of the Isosc. $\triangle$ Thm. **3.** $\overline{VY}$; $VT = VX$ (Ex. 1) and $UT = YX$ (Ex. 2), so $VU = VY$ by the Subtr. Prop. of $=$. **5.** $x = 80$; $y = 40$ **7.** $x = 38$; $y = 4$ **9.** 24, 48, 72, 96, 120 **11.** $2\frac{1}{2}$ **13.** 35 **15.a.** $\overline{KM}$ **b.** $\overline{KM}$ **c.** By construction **d.** Def. of segment bisector **e.** Reflexive Prop. of $\cong$ **f.** SSS **g.** CPCTC

17.a.

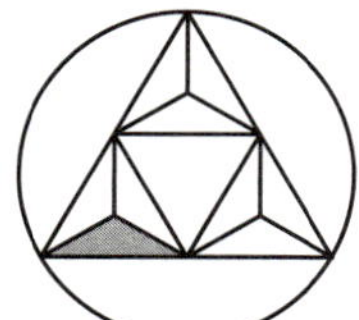

30, 30, 120

b. 5; 30, 60, 90, 120, 150

19. C

21. $x = 36$; $y = 36$ **23.** Two sides of a $\triangle$ are $\cong$ if and only if the $\angle$s opp. those sides are $\cong$. **25.a.** isosc. $\triangle$s **b.** 900 ft; 1100 ft **c.** The tower is the $\perp$ bis. of the base of each $\triangle$. **27.** 45; they are $=$ and have sum 90. **29.** $\overline{AC} \cong \overline{CB}$ and $\angle ACD \cong \angle DCB$ are given. $\overline{CD} \cong \overline{CD}$ by the Refl. Prop. of $\cong$, so $\triangle ACD \cong \triangle BCD$ by SAS. So $\overline{AD} \cong \overline{DB}$ by CPCTC, and $\overline{CD}$ bisects $\overline{AB}$. Also $\angle ADC \cong \angle BDC$ by CPCTC, $m\angle ADC + m\angle BDC = 180$ by $\angle$ Add. Post., so $m\angle ADC = m\angle BDC = 90$ by the Subst. Prop. So $\overline{CD}$ is the $\perp$ bis. of $\overline{AB}$. **31.** $m = 60$; $n = 30$ **33.** (0, 0), (4, 4), (−4, 0), (0, −4), (8, 4), (4, 8) **35.** (5, 3); (2, 6); (2, 9); (8, 3); (−1, 6); (5, 0) **37.** The $\perp$ bis. of the base of an isosc. $\triangle$ is the bis. of the vertex $\angle$; given isosc. $\triangle ABC$ with $\perp$ bis. $\overline{CD}$, $\angle ADC \cong \angle BDC$ and $\overline{AD} \cong \overline{DB}$ by def. of $\perp$ bis. Since $\overline{CD} \cong \overline{CD}$ by Refl. Prop., $\triangle ACD \cong \triangle BCD$ by SAS. So $\angle ACD \cong \angle BCD$ by CPCTC, and $\overline{CD}$ bisects $\angle ACB$. **39.** $0 <$ measure of base $\angle < 45$ **41.** B **43.** A **45.** AAS **47.** 24 sides

Algebra Review p. 234

1. (−3, −7) **3.** no sol. **5.** inf. many sol. **7.** inf. many sol. **9.** $\left(\frac{3}{2}, -\frac{1}{2}\right)$

Lesson 4-6 pp. 237–240

EXERCISES **1.** $\triangle ABC \cong \triangle DEF$ by HL. Both $\triangle$s are rt. $\triangle$s, $\overline{AC} \cong \overline{DF}$, and $\overline{CB} \cong \overline{FE}$. **3.** $\angle T$ and $\angle Q$ are rt. $\angle$s. **5.a.** $\cong$ suppl. $\angle$s are rt. $\angle$s **b.** Def. of rt. $\triangle$ **c.** Given **d.** Reflexive Prop. of $\cong$ **e.** HL **7.a.** Given **b.** Def. of $\perp$ **c.** $\triangle MLJ$ and $\triangle KJL$ are rt. $\triangle$s. **d.** Given **e.** $\overline{LJ} \cong \overline{LJ}$ **f.** HL **9.** HL; each rt. $\triangle$ has a $\cong$ hyp. and side. **11.** $x = -1$; $y = 3$ **13.** It is given that $\overline{RS} \cong \overline{TU}$, $\overline{RS} \perp \overline{ST}$, $\overline{TU} \perp \overline{UV}$, and that T is the midpoint of $\overline{RV}$. $\triangle RST$ and $\triangle TUV$ are both right triangles by the definition of a right triangle. $\overline{RT} \cong \overline{TV}$ by the definition of midpoint. Therefore, $\triangle RST \cong \triangle TUV$ by HL. **15.** $\overline{PS} \cong \overline{PT}$ so $\angle S \cong \angle T$ by the Isosc. $\triangle$ Thm. $\angle PRS \cong \angle PRT$. $\triangle PRS \cong \triangle PRT$ by AAS.

17.

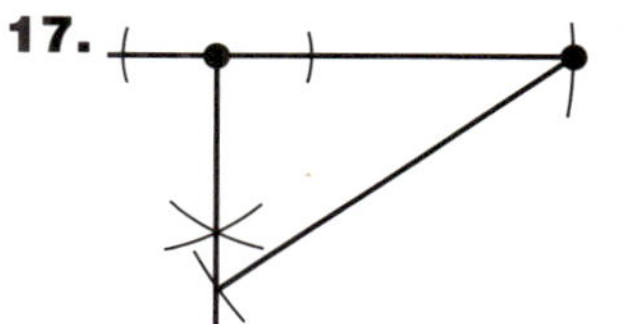

19.

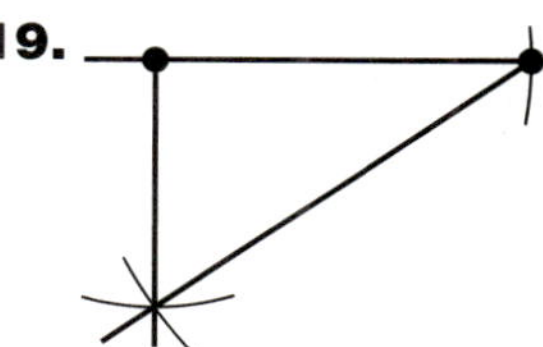

21.1. $\overline{LO}$ bisects $\angle MLN$, $\overline{OM} \perp \overline{LM}$, $\overline{ON} \perp \overline{LN}$, (Given) **2.** $\angle M$ and $\angle N$ are rt. $\angle$s (Def. of $\perp$) **3.** $\angle MLO \cong \angle NLO$ (Def. of $\angle$ bis.) **4.** $\angle M \cong \angle N$ (All rt. $\angle$s are $\cong$.) **5.** $\overline{LO} \cong \overline{LO}$ (Reflexive Prop. of $\cong$) **6.** $\triangle LMO \cong \triangle LNO$ (AAS)

23.a.

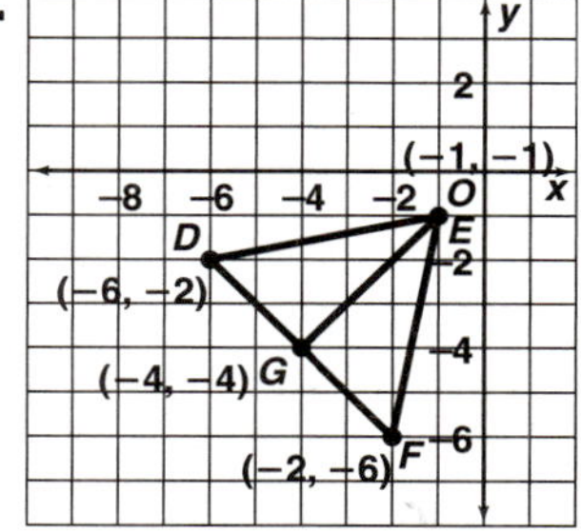

b. slope of $\overline{DG} = -1$; slope of $\overline{GF} = -1$; slope of $\overline{GE} = 1$ **c.** $\angle EGD$ and $\angle EGF$ are rt. $\angle$s. **d.** $DE = \sqrt{26}$; $FE = \sqrt{26}$ **e.** $\triangle EGD \cong \triangle EGF$ by HL. Both $\triangle$s are rt. $\triangle$s, $\overline{DE} \cong \overline{FE}$, and $\overline{EG} \cong \overline{EG}$. **25.** Since $\overline{BE} \perp \overline{EA}$ and $\overline{BE} \perp \overline{EC}$, $\triangle AEB$ and $\triangle CEB$ are both rt. $\triangle$s. $\overline{AB} \cong \overline{BC}$ because $\triangle ABC$ is equilateral, and $\overline{BE} \cong \overline{BE}$. $\triangle AEB \cong \triangle CEB$ by HL. **27.** A **29.** D **31.** equilateral **33.** If $\parallel$ lines then alt. int. $\angle$s $\cong$ **35.** Vert. $\angle$s are $\cong$ **37.** If two lines are $\parallel$, then corr. $\angle$s are $\cong$.

Lesson 4-7 pp. 243–246

EXERCISES **1.** $\angle M$ **3.** $\overline{XY}$ **5.**

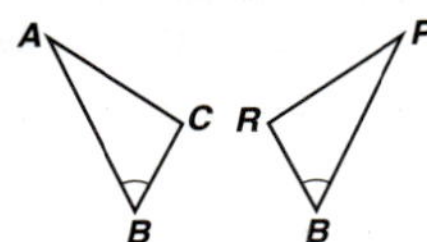

7.a. Given **b.** Reflexive Prop. of $\cong$ **c.** Given **d.** AAS **e.** CPCTC **9.** Plan: Two sides and two angles are $\cong$. The other included sides are the same segment. Use SAS. Proof: It is given that $\overline{QD} \cong \overline{UA}$ and $\angle QDA \cong \angle UAD$. $\overline{DA} \cong \overline{DA}$ by the Reflex. Prop. of $\cong$. $\triangle QDA \cong \triangle UAD$ by SAS. **11.** $\triangle ADC \cong \triangle EDG$ by ASA if $\angle A \cong \angle E$. $\angle A$ and $\angle E$ are corr. parts in $\triangle ADB$ and $\triangle EDF$, which are $\cong$ by SAS. **13–15.** Answers may vary. Sample answers are given. **13.**

15.a. 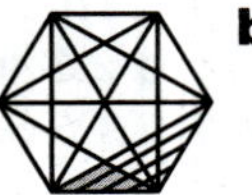 **b.**

17. 1. $\overline{AC} \cong \overline{EC}$; $\overline{CB} \cong \overline{CD}$ (Given) **2.** $\angle C \cong \angle C$ (Reflexive Prop. of $\cong$) **3.** $\triangle ACD \cong \triangle ECB$ (SAS) **4.** $\angle A \cong \angle E$ (CPCTC) **19.** $m\angle 1 = 56$; $m\angle 2 = 56$; $m\angle 3 = 34$; $m\angle 4 = 90$; $m\angle 5 = 22$; $m\angle 6 = 34$; $m\angle 7 = 34$; $m\angle 8 = 68$; $m\angle 9 = 112$ **21.** $\overline{TD} \cong \overline{RO}$ if $\triangle TDI \cong \triangle ROE$ by AAS. $\angle TID \cong \angle REO$ if $\triangle TEI \cong \triangle RIE$. $\triangle TEI \cong \triangle RIE$ by SSS. **23.a.** $\overline{AD} \cong \overline{BC}$; $\overline{AB} \cong \overline{DC}$; $\overline{AE} \cong \overline{EC}$; $\overline{DE} \cong \overline{EB}$ **b.** Use $\overline{DB} \cong \overline{DB}$ (refl.) and alt. int. $\angle$s to show $\triangle ADB \cong \triangle CBD$ (ASA). $\overline{AB} \cong \overline{DC}$ and $\overline{AD} \cong \overline{BC}$ (CPCTC). $\triangle AEB \perp \triangle CED$ (ASA) and $\triangle AED \cong \triangle CEB$ (ASA). Then $\overline{AE} \cong \overline{EC}$ and $\overline{DE} \cong \overline{EB}$ (CPCTC). **25.** $\triangle WYX \cong \triangle ZXY$ by HL; $\overline{WY} \perp \overline{YX}$, $\overline{ZX} \perp \overline{YX}$, $\overline{WX} \cong \overline{ZY}$ (Given) $\angle WYX$ and $\angle ZXY$ are rt. $\angle$s (Def. of $\perp$) $\overline{XY} \cong \overline{XY}$ (Reflexive Prop. of $\cong$) $\triangle WYX \cong \triangle ZXY$ (HL) **27.** D **29.** D **31.**

33. $y + 6 = \frac{1}{2}(x - 2)$
35. $y - 6 = -2(x + 3)$
37. $y - 4 = 2(x - 1)$
39. $y + 3 = -\frac{5}{6}(x + 4)$

Chapter Review pp. 249–251

1. legs **2.** vertex angle **3.** CPCTC **4.** hypotenuse **5.** base angles **6.** corollary **7.** legs **8.** Congruent polygons **9.** base **10.** $\overline{ML}$ **11.** $\angle U$ **12.** $\overline{ST}$ **13.** *ONMLK* **14.** 80 **15.** 3 **16.** 5 **17.** 35 **18.** 100 **19.** SSS **20.** not possible **21.** SAS **22.** not possible **23.** AAS **24.** ASA **25.** $\triangle AWC \cong \triangle RCW$; AAS **26.** $\triangle JKL \cong \triangle UVT$; SAS **27.** $\triangle RGB \cong \triangle DCS$; ASA **28.** $\triangle VTY \cong \triangle WYX$ by AAS so $\overline{TV} \cong \overline{YW}$ by CPCTC. **29.** $\triangle BCE \cong \triangle DCE$ by ASA so $\overline{BE} \cong \overline{DE}$ by CPCTC. **30.** $\triangle KNM \cong \triangle MLK$ by SAS so $\overline{KN} \cong \overline{ML}$ by CPCTC. **31.** $x = 4, y = 65$ **32.** $x = 55, y = 62.5$ **33.** $x = 65, y = 90$

34. Since $\overline{PS} \perp \overline{SQ}$ and $\overline{RQ} \perp \overline{QS}$, $\triangle PSQ$ and $\triangle RQS$ are rt. $\triangle$s. $\overline{PQ} \cong \overline{RS}$ and $\overline{QS} \cong \overline{SQ}$ so $\triangle PSQ \cong \triangle RQS$ by HL. **35.** Since $\overline{LN} \perp \overline{KM}$, $m\angle LNK = m\angle LNM = 90$. $\overline{KL} \cong \overline{ML}$ and $\overline{LN} \cong \overline{LN}$ so $\triangle KLN \cong \triangle MLN$ by HL. **36.** $\triangle AEC \cong \triangle ABD$ by SAS. **37.** $\triangle FIH \cong \triangle GHI$ by SAS. **38.** $\triangle PTS \cong \triangle RTA$ by ASA. **39.** $\triangle CFE \cong \triangle DEF$ by ASA.

Chapter 5

Lesson 5-1 pp. 262–264

EXERCISES **1.** 9 **3.** 14 **5.** 11 **7.** 40 **9.** 160 **11.** $\overline{UW} \parallel \overline{TX}$; $\overline{UY} \parallel \overline{VX}$; $\overline{YW} \parallel \overline{TV}$ **13.a.** $\overline{ST} \parallel \overline{PR}$; $\overline{SU} \parallel \overline{QR}$; $\overline{UT} \parallel \overline{PQ}$ **b.** $m\angle QPR = 40$ **15.** $\overline{FG}$ **17.** $\overline{EG}$ **19.** $\overline{CB}$ **21.a.** 114 ft 9 in. **b.** Answers may vary. Sample: The highlighted segment is the midsegment of the triangular face of the building. **23.** 45 **25.** 55 **27.** $18\frac{1}{2}$ **29.** C **31.** 50 **33.** $x = 6$; $y = 6\frac{1}{2}$ **35.** $x = 3$; $DF = 24$ **37.** Answers may vary. Sample: Draw $\overline{CA}$ and extend $\overrightarrow{CA}$ to P so that $CA = AP$. Find B, the midpt. of $\overline{PD}$. Then, by the $\triangle$ Midsegment Thm., $\overline{AB} \parallel \overline{CD}$ and $AB = \frac{1}{2}CD$. **39.** $\triangle UTS$; Proofs may vary. Sample: $\overline{VS} \cong \overline{SY}$, $\overline{YT} \cong \overline{TZ}$, and $\overline{VU} \cong \overline{UZ}$ because S, T, and U are midpts. of the respective sides; $ST = \frac{1}{2}VZ$ so $\overline{ST} \cong \overline{VU} \cong \overline{UZ}$; $SU = \frac{1}{2}YZ$ so $\overline{SU} \cong \overline{YT} \cong \overline{TZ}$; and $TU = \frac{1}{2}VY$ so $\overline{TU} \cong \overline{SY} \cong \overline{SV}$; therefore $\triangle YST \cong \triangle TUZ \cong \triangle SVU \cong \triangle UTS$ by SSS. **41.** $\triangle SXT \cong \triangle TYS$; SAS **43.** $\triangle KLQ \cong \triangle PNR$; HL

45. The coefficient of x is 3, so the slope is 3. The y-intercept is -2. Graph $(0, -2)$ and plot another point 3 units up and 1 unit right at $(1, 1)$. Draw a line through the two points. $y = 3x - 2$

47. $46\frac{2}{3}$ **49.** 40

Lesson 5-2 pp. 267–270

EXERCISES **1.** $\overline{AC}$ is the $\perp$ bis. of $\overline{BD}$. **3.** 18 **5.** The set of points equidistant from H and S is the $\perp$ bis. of $\overline{HS}$. **7.** $y = 3$; $ST = 15$; $TU = 15$ **9.** $\overrightarrow{HL}$ is the $\angle$ *bis.* of $\angle KHF$ because a point on $\overrightarrow{HL}$ is equidistant from $\overrightarrow{HK}$ and $\overrightarrow{HF}$. **11.** 54; 54 **13.** 10 **15.** Isosceles; it has 2 $\cong$ sides. **17.** A point is on the $\perp$ bis. of a segment if and only if it is equidistant from the endpts. of the segment. **19.** 4 **21.** 16 **23.** 10 **25.** 14 **27.** Answers may vary. Sample: The student needs to know that $\overline{QS}$ bisects $\overline{PR}$. **29.** Yes AX bis. $\angle TXR$. **31.** the pitcher's plate

33.a.

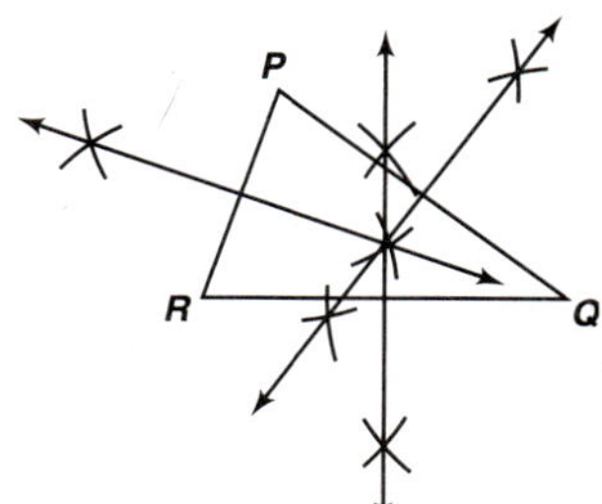

b. The $\perp$ bisectors intersect at the same point.
c. Check students' work.
35. $C(3, 2)$, $D(3, 0)$; $AC = BC = 3$, $AD = BD = \sqrt{13}$
37. $C(0, 0)$, $D(1, 1)$; $AC = BC = 3$, $AD = BD = \sqrt{5}$
39. $C(\frac{5}{2}, \frac{5}{2})$, $D(5, 3)$; $AC = BC = \frac{\sqrt{26}}{2}$, $AD = BD = \sqrt{13}$
41. $\triangle ABP$ and $\triangle ABQ$ are right triangles with a common leg and congruent hypotenuses. Thus, $\triangle BAP \cong \triangle BAQ$ by the HL Theorem. $\overline{PB} \cong \overline{BQ}$ using CPCTC, so $\overline{AB}$ bisects $\overline{PQ}$ by the definition of bisector. Hence, $\overline{AB}$ is the perpendicular bisector of $\overline{PQ}$. **43.** $\overline{BP} \perp \overrightarrow{AB}$ and $\overline{PC} \perp \overrightarrow{AC}$, thus $\angle ABP$ and $\angle ACP$ are rt. $\angle$s. Since $\overrightarrow{AP}$ bisects $\angle BAC$, $\angle BAP \cong \angle CAP$. $\overline{AP} \cong \overline{AP}$ by the Reflexive Prop. of $\cong$. Thus $\triangle ABP \cong \triangle ACP$ by AAS and $\overline{PB} \cong \overline{PC}$ by CPCTC. Therefore, $PB = PC$. **45.** D **47.** $y = -(x - 2)$ **49.** Line ℓ through the midpoints of 2 sides of $\triangle ABC$ is equidistant from A, B, and C. This is because $\triangle 1 \cong \triangle 2$ and $\triangle 3 \cong \triangle 4$ by ASA.

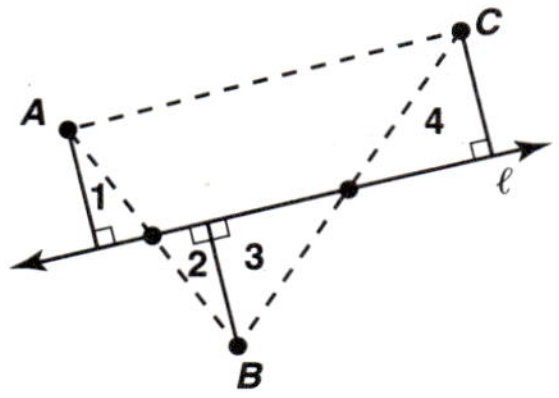

51. C **53.** 8 **55.** 6 **57.** Div. Prop. of = **59.** Distr. Prop. **61.** Trans Prop of $\cong$ **63.** $C(0, \frac{7}{2})$; $AB = \sqrt{97}$, $AC = BC = \frac{\sqrt{17}}{2}$

Lesson 5-3 pp. 275–279

EXERCISES **1.** $(-2, -3)$ **3.** $(1\frac{1}{2}, 1)$ **5.** $(-3, 1\frac{1}{2})$ **7.** $(3\frac{1}{2}, 3)$ **9.** Z **11.** $TY = 18$; $TW = 27$ **13.** $VY = 6$; $YX = 3$ **15.** Neither; it's not a segment drawn from a vertex. **17.**

19. $\overline{BE}$ **21.** $\overrightarrow{CA}$ **23.** 1:2 or 2:1 **25.** Check students' work. **27.** D **29.a.** $\overline{AB}$ **b.** $\overline{BC}$ **c.** X
d. $\perp$ bis. **31.** A right triangle; check students' explanations. **33.** I-D; II-B; III-C; IV-A
35. Answers may vary. Sample: Let $\triangle ABC$ be isosc. with base $\angle$s B and C. If $\overline{AD}$ bisects $\angle A$, then it is $\perp$ to $\overline{BC}$, and therefore the altitude from $\angle A$. So, $\overleftrightarrow{AD}$ contains the circumcenter, incenter, centroid, and orthocenter.
37. B **39.** D **41.** Yes, point B is equidistant from the sides. **43.** right **45.** $\overleftrightarrow{AB}$ **47.** ABC and ADE **49.** $\overleftrightarrow{BC}$

Lesson 5-4 pp. 283–286

EXERCISES **1.** Two angles are not congruent. **3.** The angle is obtuse. **5.** The figure is not a triangle. **7.a.** If you don't eat all of your vegetables, then you won't grow. **b.** If you won't grow, then you don't eat all of your vegetables. **9.a.** If a figure isn't a rectangle, then it doesn't have four sides. **b.** If a figure doesn't have four sides, then it isn't a rectangle. **11.** Assume that $\angle J$ is a right angle. **13.** Assume that none of the angles is obtuse. **15.** Assume that $m\angle 2 \leq 90$. **17.** I and II **19.** II and III **21.a.** right angle **b.** right angles **c.** 90 **d.** 180 **e.** 90 **f.** 90 **g.** 0 **h.** more than one right angle **i.** at most one right angle **23.** Assume one base $\angle$ is a right $\angle$. Then the other base $\angle$ is also a right $\angle$ since the base $\angle$s of an isosceles $\triangle$ are congruent. But a $\triangle$ can have at most one right $\angle$. So neither base $\angle$ is a right $\angle$. **25.a.** If four points aren't collinear, then they aren't coplanar; false **b.** If four points aren't coplanar, then they aren't collinear; true
27–29. Answers may vary. Samples are given.
27. If today is Sunday, then tomorrow is Monday.
29. If two sides of a triangle are congruent, then the triangle is isosceles. **31.** Assume that the temperature outside is more than 32°F. Then ice would not be forming on the sidewalk. This contradicts the fact that ice is forming. Thus the statement that the temperature must be 32°F or less is true. **33.** Assume $\overleftrightarrow{XY}$ and $\overleftrightarrow{XZ}$ are two different lines $\perp$ to $\overleftrightarrow{AX}$, with Y and Z on the same side of $\overleftrightarrow{AX}$. If B is on $\overleftrightarrow{AX}$ opp. pt. A from X, then $m\angle AXY + m\angle YXZ + m\angle ZXB = 180$. But $m\angle AXY = m\angle ZXB = 90$, so $m\angle YXZ = 0$. Thus X, Y, and Z are collinear. **35.** If the angle measures 120, then it is obtuse. If the angle isn't obtuse, then it doesn't measure 120. **37.** Angie assumed that the inverse of the statement was true, but a conditional and its inverse may not have the same truth value. **39.** The culprit entered the

room through a hole in the roof; the other possibilities were eliminated. **41.** Assume $\overline{XB} \perp \overline{AC}$. Then $\angle AXB$ and $\angle CXB$ are right ∠s. Since $m\angle ABX = m\angle CBX = 36$, then $\angle A \cong \angle C$ because if two ∠s of a △ are ≅, the third ∠s are ≅. Then $AB = BC$ since sides opp. ≅ ∠s are ≅ and $\triangle ABC$ is an isosceles △. But this contradicts the given statement that $\triangle ABC$ is scalene. Thus, $\overline{XB}$ is not ⊥ to $\overline{AC}$. **43.** A **45.** B **47.** $\overline{XY}$ is an altitude since it goes through a vertex and is ⊥ to the opp. side. **49.** corr. ∠s **51.** alt. int. ∠s **53.** 35

Algebra 1 Review — p. 288

1. $x \leq -1$ **3.** $x > -\frac{21}{2}$ **5.** $a \leq 18$ **7.** $z > \frac{1}{2}$ **9.** $n \leq -\frac{7}{2}$ **11.** $x > -5$ **13.** $n \geq -119$ **15.** $x > -8$ **17.** $b \geq 8$ **19.** $r \leq -58$ **21.** $x < -5$ **23.** $x \geq -9.7$ **25.** $a \geq -1$ **27.** $x \geq -362$ **29.** $x > 5$ **31.** $x > -7$ **33.** $c < -2$

Lesson 5-5 — pp. 292–294

EXERCISES **1.** $\angle 3 \cong \angle 2$ because they are vertical ∠s and $m\angle 1 > m\angle 3$ by Corollary to the Ext. ∠ Thm. So, $m\angle 1 > m\angle 2$ by subst. **3.** $m\angle 1 > m\angle 4$ by Corollary to the Ext. ∠ Thm. and $\angle 4 \cong \angle 2$ because if ∥ lines, then alt. int. ∠s are ≅. **5.** $\angle D, \angle C, \angle E$ **7.** $\angle A, \angle B, \angle C$ **9.** $\angle Z, \angle X, \angle Y$ **11.** $\overline{FH}, \overline{GF}, \overline{GH}$ **13.** $\overline{AC}, \overline{AB}, \overline{CB}$ **15.** $\overline{ZY}, \overline{XZ}, \overline{XY}$ **17.** Yes; $11 + 12 > 15$; $12 + 15 > 11$; $11 + 15 > 12$. **19.** Yes; $1 + 15 > 15$; $15 + 15 > 1$. **21.** No; $4 + 5 \not> 9$ **23.** $11 < s < 21$ **25.** $5 < s < 41$ **27.** $15 < s < 55$ **29.** Let the distance between the peaks be d and the distances from the hiker to each of the peaks be a and b. Then $d + a > b$ and $d + b > a$. Thus, $d > b - a$ and $d > a - b$.
31. Answers may vary. Sample: The shortcut across the grass is shorter than the sum of the two paths. **33.a.** $m\angle OTY$ **b.** $m\angle 3$ **c.** Base ∠s of an isosc. △ are ≅. **d.** ∠ Add. Post. **e.** Comparison Prop. of Ineq. **f.** Subst. (step 2) **g.** An ext. ∠ of a △ is greater than either remote int. ∠. **h.** Trans. Prop. of Ineq. **35.** $\overline{RS}$ **37.** $\overline{XY}$ **39.** (2, 4), (2, 5), (2, 6), (3, 3), (3, 4), (3, 6), (3, 7), (4, 3), (4, 4), (4, 5), (4, 6), (4, 7), (4, 8) **41.**

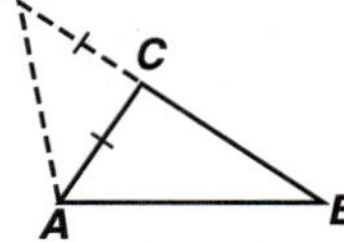

$CD = AC$ was given so $\triangle ACD$ is isosc. by def. of isosc. △. This means $m\angle D = m\angle CAD$. Then $m\angle DAB > m\angle CAD$ by the Comparison Prop. of Ineq. So by subst., $m\angle DAB > m\angle D$ and by Thm. 5-11 $DB > AB$. Since $DC + CB = DB$, by subst. $DC + CB > AB$. Using subst. again, $AC + CB > AB$.

43. D **45.** C **47.** $m\angle A > m\angle B$ **49.** The angle is not a right angle. **51.** 90 **53.** 145 **55.** 8.0 ft^2 **57.** 0.8 m^2

Chapter Review — pp. 297–299

1. median of a △ **2.** distance from the point to the line **3.** ⊥ Bis. Thm. **4.** altitude **5.** contrapositive **6.** indirect proof **7.** △ Ineq. Thm. **8.** incenter **9.** ∠ Bis. Thm. **10.** point of concurrency **11.** 15 **12.** 11 **13.** 40 **14.** 7 **15.** 14 **16.** 80 **17.** (−1, 0) **18.** (0, −1) **19.** (2, −3) **20.** ∠ bisector; it bisects an ∠. **21.** altitude; it is ⊥ to a side. **22.** median; it goes through a midpoint. **23.** Inverse: If it is not snowing, then it is not cold outside. Contrapositive: If it is not cold outside, then it is not snowing. **24.** Inverse: If an angle is not obtuse, then its measure is not greater than 90 and less than 180. Contrapositive: If an angle's measure is not greater than 90 and less than 180, then it is not obtuse. **25.** Inverse: If a figure is not a square, then its sides are not congruent. Contrapositive: If a figure's sides are not congruent, then it is not a square. **26.** Inverse: If you are not in Australia, then you are not south of the equator. Contrapositive: If you are not south of the equator, then you are not in Australia. **27.** Assume that both numbers are odd. The product of 2 odd numbers is always odd, which contradicts that the product is even. Therefore, at least one number must be even. **28.** Assume a right ∠ can be formed by non-perp. lines. Then by the def. of ⊥, the lines are ⊥. Therefore, the assumption is false. **29.** Assume that an △ has 2 obtuse ∠s. Then these ∠s by def. are greater than 90, which makes their sum greater than 180. But the sum of the measures of the ∠s of a △ = 180, so the assumption must be false. **30.** Assume an ∠ is obtuse, and therefore has measure greater than 90. Since the △ is equilateral, it is equiangular, and each ∠ measures 60. **31.** $\angle T, \angle R, \angle S$; $\overline{RS}, \overline{TS}, \overline{TR}$ **32.** $\angle G, \angle O, \angle F$; $\overline{OF}, \overline{FG}, \overline{OG}$ **33.** No; $5 + 8 \not> 15$. **34.** Yes; each pair > 3rd. **35.** Yes; each pair > 3rd. **36.** Yes; each pair > 3rd. **37.** No; $1 + 1 \not> 3$. **38.** Yes; each pair > 3rd. **39.** $3 < x < 11$ **40.** $7 < x < 23$ **41.** $6 < x < 10$ **42.** $1 < x < 25$

Chapter 6

Lesson 6-1 — pp. 308–311

EXERCISES **1.** ▱, rectangle, rhombus, square **3.** trapezoid **5.** kite **7.** rhombus **9.** rhombus **11.** kite **13.** rhombus

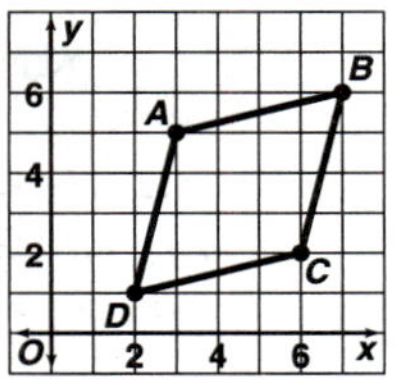

15. trapezoid

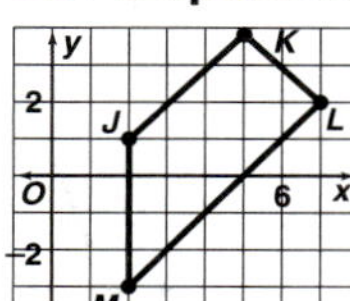

17. quadrilateral

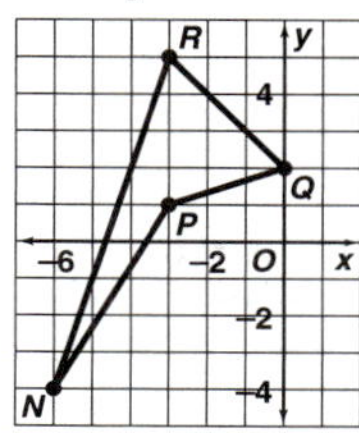

19. $x = 11, y = 29$; 13, 13, 23, 23 **21.** $x = 2, y = 6$; 2, 7, 7, 2 **23.** $x = 3, y = 5$; 15, 15, 15, 15 **25.** 40, 40, 140, 140; 11, 11, 15, 32 **27.** rectangle, square, trapezoid **29–33.** Answers may vary. Samples are given.

29.

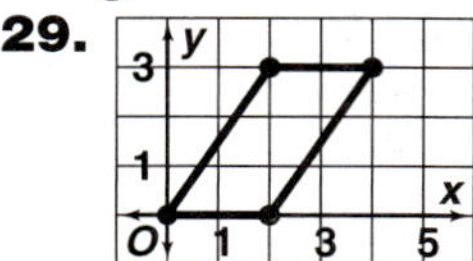

31. Impossible; a trapezoid with one rt. ∠ must have another, since two sides are ∥.

33.

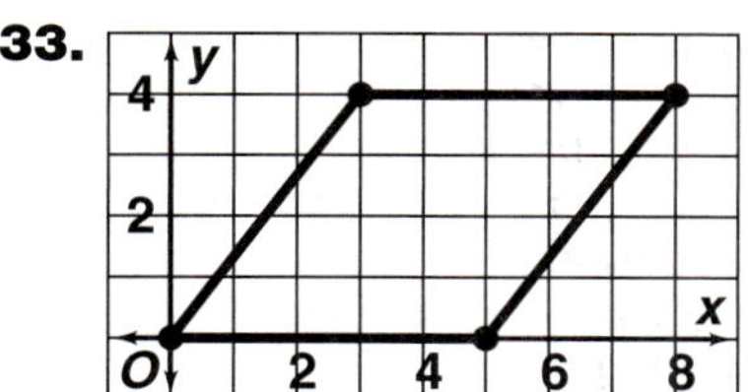

35.

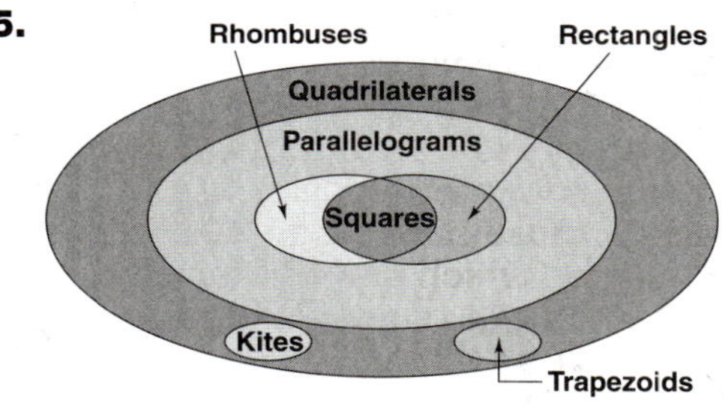

37. False; a trapezoid only has one pair of ∥ sides. **39.** True; all squares are □s. **41.** False; only rhombuses with rt. ∠s are squares. **43.** Check students' work. **45–47.** Check students' sketches. **45.** some isos. trapezoids, some trapezoids **47.** rectangle, square **49.** A trapezoid has only one pair of ∥ sides. **51–53.** Check students' sketches. **51.** rhombus, ▱ **53.** rhombus, ▱, kite **53–55.** Check students' work. **57–59.** Explanations may vary. Samples are given. **57.** ▱, kite, rhombus, trapezoid, isos. trapezoid **59.** ▱, rectangle, square, rhombus, kite, trapezoid **61.** D **63.** Yes; the sum of the lengths of any 2 sides is greater than the third side. **65.** No; $3 + 5 \not> 8$. **67.** 13 mm **69.** 82 **71.** 58

Lesson 6-2 pp. 315–318

EXERCISES **1.** 127 **3.** 76 **5.** 100 **7.** 3; 10, 20, 20 **9.** 20 **11.** 17 **13.** 6; $m\angle H = m\angle J = 30$, $m\angle I = m\angle K = 150$ **15.** $x = 5, y = 7$ **17.** $x = 6$, $y = 9$ **19.** 12; 24 **21.** 3 **23.** 6 **25.** 9 **27.** 2.25 **29.** 4.5 **31.** $BC = AD = 14.5$ in.; $AB = CD = 9.$ **33.** A **35.a.** $\overline{DC}$ **b.** $\overline{AD}$ **c.** ≅ **d.** Reflexive **e.** ASA **f.** CPCTC **37.** 38, 32, 110 **39.** 95, 37, 37 **41.** 18, 162 **43.** Answers may vary. Sample: In □ *LENS* and *NGTH*, $\overline{GT} \parallel \overline{EH}$ and $\overline{EH} \parallel \overline{LS}$ by the def. of a ▱. Therefore $\overline{LS} \parallel \overline{GT}$ because if 2 lines are ∥ to the same line then they are ∥ to each other. **45.** $x = 12, y = 4$ **47.** $x = 9, y = 6$ **49.** In ▱ *RSTW* and ▱ *XYTZ*, $\overline{XY} \parallel \overline{TW}$ and $\overline{RS} \parallel \overline{TW}$ by the def. of a ▱. Then $\overline{XY} \parallel \overline{RS}$ because if 2 lines are ∥ to the same line, then they are ∥ to each other. **51.a.** Answers may vary. Check students' work. **b.** No; the corr. sides can be ≅ but the ∠s may not be. **53.a.** Given: 2 sides and the included ∠ of ▱*ABCD* are ≅ to the corr. parts of ▱*WXYZ*. Let $\angle A \cong \angle W$, $\overline{AB} \cong \overline{WX}$ and $\overline{AD} \cong \overline{WZ}$. Since opp. ∠s of a ▱ are ≅, $\angle A \cong \angle C$ and $\angle W \cong \angle Y$. Thus $\angle C \cong \angle Y$ by the Trans. Prop. of ≅. Similarly, opp. sides of a ▱ are ≅, thus $\overline{AB} \cong \overline{CD}$ and $\overline{WX} \cong \overline{ZY}$. Using the Trans. Prop. of ≅, $\overline{CD} \cong \overline{ZY}$. The same can be done to prove $\overline{BC} \cong \overline{XY}$. Since consec. ∠s of a ▱ are suppl., ∠*A* is suppl. to ∠*D*, and ∠*W* is suppl. to ∠*Z*. Suppls. of ≅ ∠s are ≅, thus $\angle D \cong \angle Z$. The same can be done to prove $\angle B \cong \angle X$. Therefore, since all corr. ∠s and sides are ≅, ▱*ABCD* ≅ ▱*WXYZ*. **b.** No; opp. ∠s and sides are not necessarily ≅ in a trapezoid. **55.** C **57.** rhombus **59.** $\overline{AC} \perp \overline{DB}$ **61.** 131 **63.** 131

Lesson 6-3 pp. 324–326

EXERCISES **1.** 5 **3.** $x = 1.6, y = 1$ **5.** 5 **7.** Yes; both pairs of opp. sides are ≅. **9.** Yes; both pairs

EXERCISES **1.** 5 **3.** $x = 1.6, y = 1$ **5.** 5 **7.** Yes; both pairs of opp. sides are ≅. **9.** Yes; both pairs of opp. ∠s are ≅. **11.** A quad. is a ▱ if and only if opp. sides are ≅ (6-1 and 6-5); opp. ∠s are ≅ (6-2 and 6-6); diags. bis. each other (6-3 and 6-7). **13.** Draw diagonals $\overline{TX}$ and $\overline{WY}$ intersecting at *R*. **a.** $\overline{TW} \cong \overline{YX}$ (Given) **b.** $\angle TWR \cong \angle XYR$ (Alt. Int. ∠s ≅) **c.** $\angle WTR \cong \angle YXR$ (Alt. Int. ∠s ≅) **d.** $\triangle TWR \cong \triangle YXR$ (ASA) **e.** $\overline{WR} \cong \overline{YR}$ (CPCTC) **f.** $\overline{TR} \cong \overline{XR}$ (CPCTC) **g.** The diagonals bisect each other. (def. of bis.) **h.** *TWXY* is a ▱ (Thm. 6-7). **15.** $x = 3$, $y = 11$ **17.** $k = 9, m = 23.4$ **19.** Answers may vary. Sample:

A B C D E F G H

21. $\triangle TRS \cong \triangle RTW$ (given), so $\overline{ST} \cong \overline{RW}$ and $\overline{SR} \cong \overline{TW}$. $RSTW$ is a ▱ because opp. sides are ≅ (Thm. 6-5) **23.** (6, 6) **25.** You can show a quad. is a ▱ if both pairs of opp. sides are ∥ or ≅, if both pairs of opp. ∠s are ≅, if diagonals bisect each other, if all consecutive ∠s are suppl., or if one pair of opp. sides is both ∥ and ≅. **27.** Answers may vary. Sample: **1.** $\overline{AB} \cong \overline{CD}$, $\overline{AC} \cong \overline{BD}$ (Given) **2.** $ACDB$ is a ▱. (If opp. sides of a quad. are ≅, then it is a ▱.) **3.** M is the midpoint of $\overline{BC}$. (The diag. of a ▱ bisect each other.) **4.** $\overline{AM}$ is a median. (Def. of a median) **29.** D **31.** $a = 8$, $h = 30$, $k = 120$ **33.** $e = 12$, $f = 11$, $c = 204$ **35.** If a quad is a ▱ then the diagonals bisect each other; if the diagonals of a quad. bisect each other, then it is a ▱. **37.** If the prod. of the slopes of two nonvertical lines is -1, then they are ⊥; if two nonvertical lines are ⊥, then the prod. of their slopes is -1.

Lesson 6-4 pp. 332–335

EXERCISES **1.** 38, 38, 38, 38 **3.** 118, 31, 31 **5.** 32, 90, 58, 32 **7.** 55, 35, 55, 90 **9.** 90, 55, 90 **11.** 3; $LN = MP = 7$ **13.** 9; $LN = MP = 67$ **15.** $\frac{5}{2}$; $LN = MP = 12\frac{1}{2}$ **17.** rhombus; the diags. are ⊥. **19.** The pairs of opp. sides of the frame remain ≅, so the frame remains a ▱. **21.** Square; a square is both a rectangle and a rhombus, so its diag. have the properties of both. **23.** Answers may vary. Sample: The diagonals of a ▱ bisect each other so $\overline{AE} \cong \overline{CE}$. Both $\angle AED$ and $\angle CED$ are right ∠s because $\overline{AC} \perp \overline{BD}$, and since $\overline{DE} \cong \overline{DE}$ by the Reflexive Prop., $\triangle AED \cong \triangle CED$ by SAS. By CPCTC $\overline{AD} \cong \overline{CD}$, and since opp. sides of a ▱ are ≅, $\overline{AB} \cong \overline{BC} \cong \overline{CD} \cong \overline{AD}$. So $ABCD$ is a rhombus because it has 4 ≅ sides. **25–33.** Symbols may vary. Samples are given: parallelogram: ▱, rhombus: [R], rectangle: ▭, square: [S] **25.** [R], [S] **27.** ▱, [R], ▭, [S] **29.** ▭, [S] **31.** ▱, [R], ▭, [S] **33.** [R], [S]

35. Diag. are ≅, diag. are ⊥.

37.  Diag. are ≅, diag. are ⊥.

39.1. $ABCD$ is a parallelogram. (Given) $\overline{AC}$ bisects $\angle BAD$ and $\angle BCD$. (Given) **2.** $\angle 1 \cong \angle 2$, $\angle 3 \cong \angle 4$ (Def. of bisect) **3.** $\overline{AC} \cong \overline{AC}$ (Refl. Prop. of ≅) **4.** $\triangle ABC \cong \triangle ADC$ (ASA) **5.** $\overline{AB} \cong \overline{AD}$ (CPCTC) **6.** $\overline{AB} \cong \overline{DC}$, $\overline{AD} \cong \overline{BC}$ (Opp. sides of a ▱ are ≅.) **7.** $\overline{AB} \cong \overline{BC} \cong \overline{CD} \cong \overline{AD}$ (Trans. Prop. of ≅) **8.** $ABCD$ is a rhombus. (Def. of rhomb.) **41.** Yes; since all right ∠s are ≅, the opp. ∠s are ≅ and it is a ▱. Since it has all right ∠s, it is a rectangle. **43.** Yes; a quad. with 4 ≅ sides is a ▱ and a ▱ with 4 ≅ sides and 4 right ∠s is a square. **45.** $x = 5$, $y = 32$, $z = 7.5$ **47–49.** Drawings may vary. Samples are given. **47.** Square, rectangle, isosceles trapezoid, kite

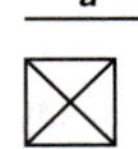

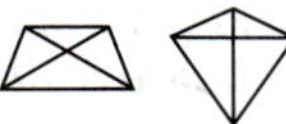

49. For $a < b$: trapezoid, isosc. trapezoid ($a > \frac{1}{2}b$), ▱, rhombus, kite

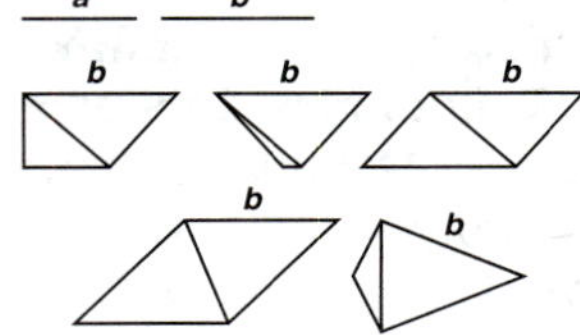

For $a > b$: trapezoid, isosc. trapezoid, ▱, rhombus ($a < 2b$), kite, rectangle, square (if $a = \sqrt{2}b$).

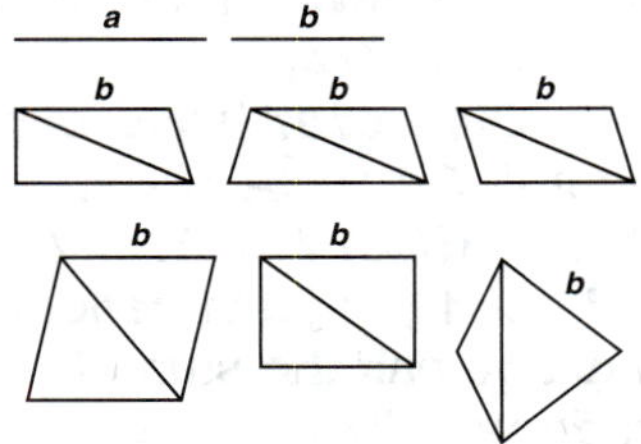

51. 2, 2 **53.** 1, 1 **55–59.** Answers may vary. Samples are given. **55.** Construct a rt. ∠, and draw diag. 1 from its vertex. Construct the ⊥ from the opp. end of diag. 1 to a side of the rt. ∠. Repeat to other side. **57.** Draw diagonal 1, and construct its midpt. Draw a line ⊥ to diag. 1 at its midpt. Construct segments of length diag. 1 in opp. directions from midpt. Then, bisect these segments. Connect these midpts. with the endpts. of diag. 1. **59.** Draw an acute ∠. Use the compass to mark the length of diag. 1 on one side of the angle. The other side will be a base for the trap. Construct a line ∥ to the base through the non-vertex endpt. of diag. 1. Set the compass to the length of diag. 2 and place the point on the non-vertex endpt. of the base. Draw an arc that intersects the line ∥ to the base. Draw diag. 2 through these two points. Finish by drawing the non-∥ sides of the trap. **61.** Yes; ≅ diag. in a ▱ mean it can be a rectangle with 2 opp. sides 2 cm long. **63.** Given ▱$ABCD$ with diag. $\overline{AC}$. Let $\overline{AC}$ bisect $\angle BAD$. Because $\triangle ABC = \triangle DAC$, $AB = DA$ by CPCTC. But since opp. sides of a ▱ are ≅, $AB = CD$ and $BC = DA$. So $AB = BC = CD = DA$, and ▱$ABCD$ is a rhombus. The new statement is true. **65.** B **67.** No **69.** 6 **71.** 5 **73.** $\overline{RP}$ **75.** 89

Lesson 6-5 pp. 338–341

EXERCISES **1.** 77, 103, 103 **3.** 49, 131, 131 **5.** 115, 115, 65 **7.a.** isosc. trapezoids **b.** 69, 69, 111, 111 **9.** 90, 45, 45 **11.** 90, 26, 90 **13.** 90, 55, 90, 55, 35 **15.** 90, 90, 90, 90, 46, 34, 56, 44, 56, 44 **17.** Answers may vary. Sample:

19. Explanations may vary. Sample: If both ∠s are bisected, then this combined with $\overline{KM} \cong \overline{KM}$ by the Reflexive Prop. means $\triangle KLM \cong \triangle KNM$ by SAS. By CPCTC, $\angle L \cong \angle N$. $\angle L$ and $\angle N$ are opp. ∠s, but if *KLMN* is isos., both pairs of base ∠s are also ≅. By the Trans. Prop., all 4 angles are ≅, so *KLMN* must be a rect. or a square. This contradicts what is given, so $\overline{KM}$ cannot bisect $\angle LMN$ and $\angle LKN$. **21.** 15 **23.** 3 **25.** 1 **27.** $x = 35, y = 30$ **29.** Isosc. trapezoid; all the large rt. △s appear to be ≅. **31.** Yes, the ≅ ∠s can be obtuse. **33.** Yes; if 2 ≅ ∠s are rt. ∠s, they are suppl. The other 2 ∠s are also suppl. **35.** Yes; the ≅ ∠s may be 45° each. **37.** Rhombuses and squares would be kites since opp. sides can be ≅ also. **39.** Answers may vary. Sample: Draw $\overline{TA}$ and $\overline{RP}$. **1.** isosc. trapezoid *TRAP* (Given) **2.** $\overline{TA} \cong \overline{PR}$ (Diag. of an isosc. trap. are ≅.) **3.** $\overline{TR} \cong \overline{PA}$ (Given) **4.** $\overline{RA} \cong \overline{RA}$ (Refl. Prop. of ≅) **5.** $\triangle TRA \cong \triangle PAR$ (SSS) **6.** $\angle RTA \cong \angle APR$ (CPCTC) **41.** Check students justifications. A sample is given. **41.** It is one half the sum of the lengths of the bases; draw a diag. of the trap. to form 2 △s. The bases *B* and *b* of the trap. are each a base of a △. Then the segment joining the midpts. of the non-∥ sides is the sum of the midsegments of the △s. This sum is $\frac{1}{2}B + \frac{1}{2}b = \frac{1}{2}(B + b)$. **43.** *D* is any point on $\overleftrightarrow{BN}$ such that $ND \neq BN$ and *D* is below *N*. **45.** B **47.** B **49.** 27 **51.a.** 4 **b.** 5 **c.** 5 **53.** SAS

Lesson 6-6 pp. 344–346

EXERCISES **1.a.** $(2a, 0)$ **b.** $(0, 2b)$ **c.** (a, b) **d.** $\sqrt{b^2 + a^2}$ **e.** $\sqrt{b^2 + a^2}$ **f.** $\sqrt{b^2 + a^2}$ **g.** $MA = MB = MC$ **3.** $W(a, a)$; $Z(a, 0)$ **5.** $W(0, b)$; $Z(a, 0)$ **7.** $W(-b, c)$; $Z(0, c)$ **9.a.** Diag. of a rhombus are ⊥. **b.** Diag. of a ▱ that is not a rhombus are not ⊥. **11.** *B, D, H, F* **13.** *A, C, G, E* **15.** *A, D, G, F* **17.** $W(2a, 2a)$; $Z(2a, 0)$ **19.** $W(0, b)$; $Z(2a, 0)$ **21.** $W(-2b, 2c)$; $Z(0, 2c)$ **23.** $(c–a, b)$ **25.** $(–b, 0)$

27.a. y (b, 2c) x O (2b, 0) **b.**

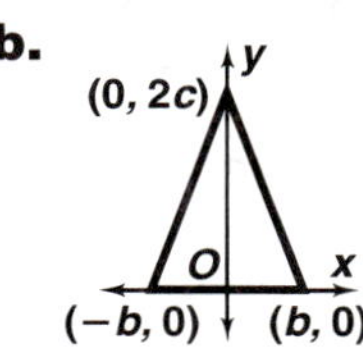

c. $\sqrt{b^2 + 4c^2}$ **d.** $\sqrt{b^2 + 4c^2}$ **e.** The lengths are =. **29.** Step 1: (0, 0) Step 2: (*a*, 0) Step 3: Since $m\angle 1 + m\angle 2 + 90 = 180$, ∠1 and ∠2 must be compl. ∠3 and ∠2 are the acute ∠s of a rt. △. Step 4: $(-b, 0)$ Step 5: $(-b, a)$ Step 6: Using the formula for slope, the slope for $\ell_1 = \frac{b}{a}$ and the slope for $\ell_2 = -\frac{a}{b}$. Mult. the slopes, $\frac{b}{a} \cdot -\frac{a}{b} = -1$. **31.** B **33.** (3,2) **35.** $\angle ACD \cong \angle ACB$ and $\angle D \cong \angle B$ (Given), $\overline{AC} \cong \overline{AC}$ (Reflex. Prop. of ≅), therefore, $\triangle ADC \cong \angle ABC$ (AAS).

Lesson 6-7 pp. 349–353

EXERCISES **1.a.** $W\left(\frac{a}{2}, \frac{b}{2}\right)$; $Z\left(\frac{c+e}{2}, \frac{d}{2}\right)$ **b.** $W(a, b)$; $Z(c + e, d)$ **c.** $W(2a, 2b)$; $Z(2c + 2e, 2d)$; **d.** *c*; it uses multiples of 2 to name the coordinates of *W* and *Z*. **3.a.** *y*-axis **b.** Distance **5.a.** isos. **b.** *x*-axis **c.** *y*-axis **d.** Midpts. **e.** ≅ sides **f.** slopes **g.** the Distance Formula **7.a.** $\sqrt{a^2 + b^2}$ **b.** $2\sqrt{a^2 + b^2}$ **9.a.** (a, b) **b.** (a, b) **c.** the same point **11.** The vertices of *KLMN* are $L(b, a + c)$, $M(b, c)$, $N(-b, c)$, and $K(-b, a + c)$. The slopes of $\overline{KL}$ and $\overline{MN}$ are zero, so these segments are horizontal. The endpoints of $\overline{KN}$ have equal *x*-coordinates, and so do the endpoints of $\overline{LM}$. So these segments are vertical. Hence opposite sides of *KLMN* are parallel and consecutive sides are ⊥. It follows that *KLMN* is a rectangle. **13–23.** Answers may vary. Samples are given. **13.** yes; same slope **15.** no; may not have intersection pt. **17.** no; may need ∠ measures **19.** yes; Dist. Formula **21.** no; may need ∠ measures **23.** yes; Dist. Formula, $AB = BC = CD = AD$ **25.** 1, 4, 7 **27.** −0.8, 0.4, 1.6, 2.8, 4, 5.2, 6.4, 7.6, 8.8 **29.** $-2 + \frac{12}{n}, -2 + 2\left(\frac{12}{n}\right), -2 + 3\left(\frac{12}{n}\right), \ldots, -2 + (n - 1)\left(\frac{12}{n}\right)$ **31.** $\left(-1, 6\frac{2}{3}\right), \left(1, 8\frac{1}{3}\right), (3, 10), \left(5, 11\frac{2}{3}\right), \left(7, 13\frac{1}{3}\right)$ **33.** $(-2.76, 5.2), (-2.52, 5.4), (-2.28, 5.6), \ldots, (8.52, 14.6), (8.76, 14.8)$ **35.a.** $L(b, d)$, $M(b + c, d)$, $N(c, 0)$

b. $\overleftrightarrow{AM}$: $y = \frac{d}{b + c}x$; $\overleftrightarrow{BN}$: $y = \frac{2d}{2b - c}(x - c)$; $\overleftrightarrow{CL}$: $y = \frac{d}{b - 2c}(x - 2c)$; **c.** $P\left(\frac{2(b + c)}{3}, \frac{2d}{3}\right)$

d. Pt. *P* satisfies the eqs. for $\overleftrightarrow{AM}$ and $\overleftrightarrow{CL}$.

e. $AM = \sqrt{(b + c)^2 + d^2}$;

$$AP = \sqrt{\left(\frac{2(b + c)}{3}\right)^2 + \left(\frac{2d}{3}\right)^2} =$$

$$\sqrt{\left(\frac{2}{3}\right)^2\left((b + c)^2 + d^2\right)} = \frac{2}{3}\sqrt{(b + c)^2 + d^2}$$

$= \frac{2}{3}AM$. The other 2 distances are found similarly.

37. Assume $b > a$. $a + \frac{b - a}{n}, a + 2\left(\frac{b - a}{n}\right), \ldots, a + (n - 1)\left(\frac{b - a}{n}\right)$

Selected Answers

39.a. The $\triangle$s with bases d and b, and heights c and a, respectively, have the same area. They share the small right $\triangle$ with base d and height c, and the remaining areas are $\triangle$s with base c and height $(b - d)$. So $\frac{1}{2}ad = \frac{1}{2}bc$. Mult. both sides by 2 gives $ad = bc$. **b.** The diagram shows that $\frac{a}{b} = \frac{c}{d}$, since both represent the slope of the top segment of the $\triangle$. So by (a), $ad = bc$. **41.a.** Horiz. lines have slope 0, and vert. lines have undef. slope. Neither could be mult. to get -1. **b.** Assume the lines do not intersect. Then they have the same slope, say m. Then $m \cdot m = m^2 = -1$, which is impossible. So the lines must intersect. **c.** Let the eq. for ℓ_1 be $y = \frac{b}{a}x$, and for ℓ_2 be $y = -\frac{a}{b}x$, and the origin be the int. point.

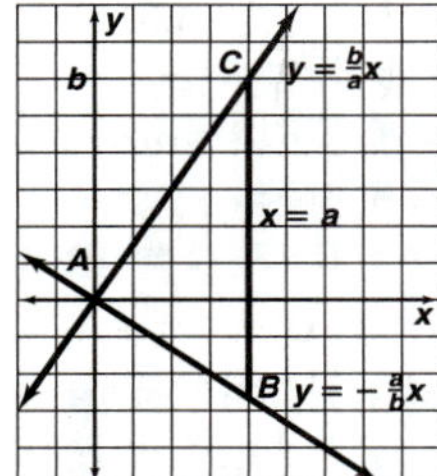

Define: $C(a, b)$, $A(0, 0)$, and $B(a, -\frac{a^2}{b})$.
Using the Distance Formula,
$AC = \sqrt{a^2 + b^2}$, $BA = \sqrt{a^2 + \frac{a^4}{b^2}}$, and
$CB = b + \frac{a^2}{b}$. Then $AC^2 + BA^2 = CB^2$, and $m\angle A = 90$ by the Conv. of the Pythagorean Thm. So $\ell_1 \perp \ell_2$. **43.** C **45.a.** If the sum of the $\angle$s of a polygon is 360°, then the polygon is a quad. **b.** If a polygon is a quad., then the sum of its $\angle$s is 360° **47.a.** If $a \neq 5$, then $a^2 \neq 25$. **b.** If $a^2 \neq 25$, then $a \neq 5$. **49.a.** If $c \leq 0$, then c is not pos. **b.** If c is not pos., then $c \leq 0$. **51.** $\overline{HE} \cong \overline{FG}$, $\overline{EF} \cong \overline{GH}$, and $\overline{HF} \cong \overline{HF}$ by the Reflexive Prop. of $\cong$, so $\triangle HEF \cong \triangle FGH$ by SSS. Then by CPCTC $\angle 1 \cong \angle 2$.

Algebra 1 Review pp. 354–355

1.

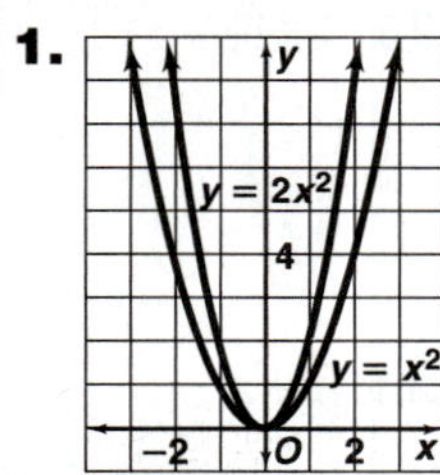

3.

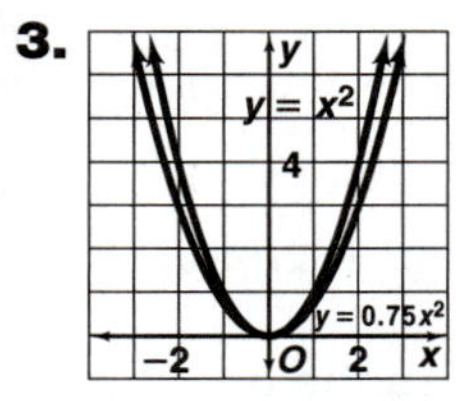

5.

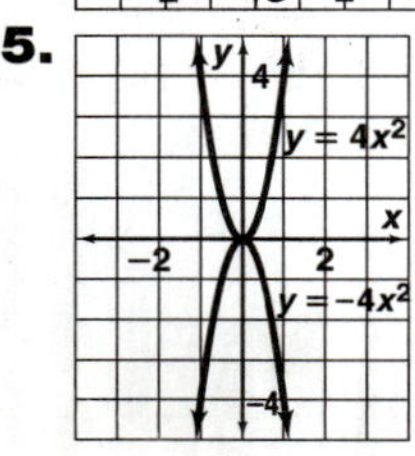

7.

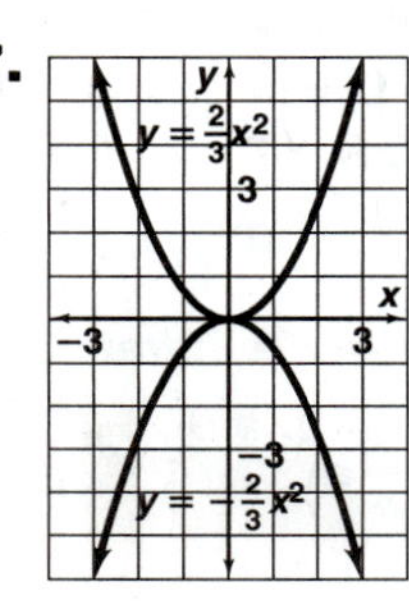

9. $y = x^2 + 1$ is a translation of $y = x^2$ by 1 unit up.

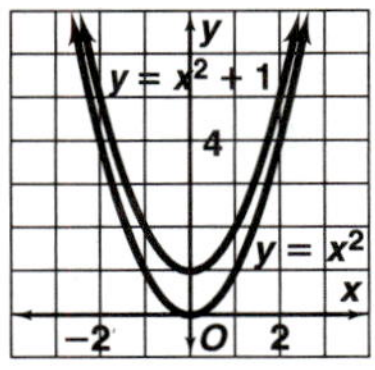

11. $y = x^2 + \frac{5}{2}$ is a translation of $y = x^2$ by $\frac{5}{2}$ units up.

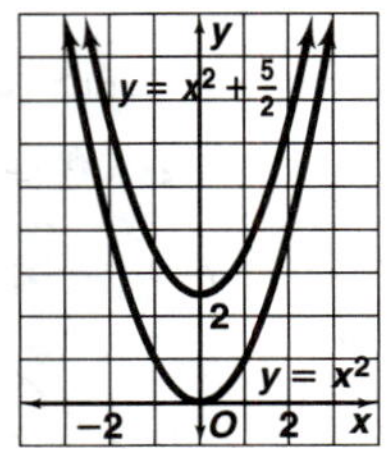

13. 0,4 **15.** $-5, 5$ **17.** $-3, -1$ **19.** $-6, \frac{1}{2}$

Chapter Review pp. 357–359

1. F **2.** H **3.** G **4.** B **5.** I **6.** J **7.** A **8.** C **9.** E **10.** D

11.

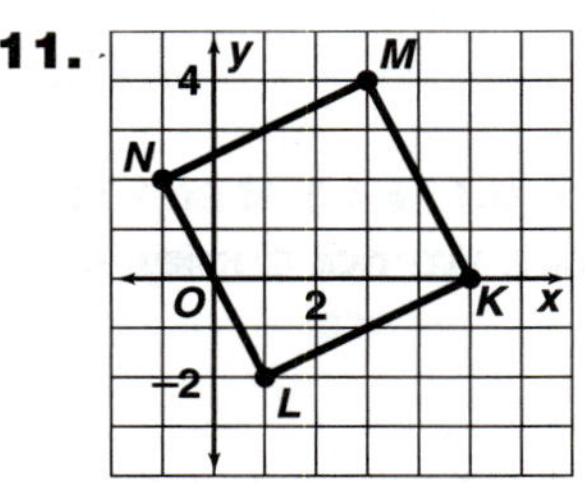

square

12.

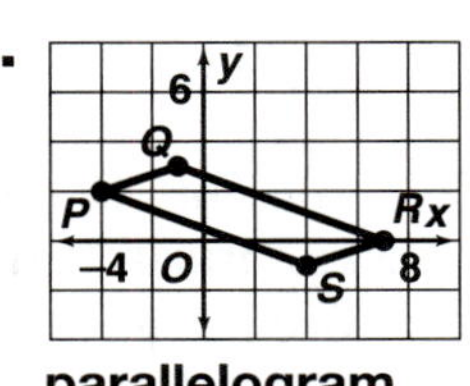

parallelogram

13. $x = 8$; 9, 14, 9, 7 **14.** $m = 4$, $t = 5$; 7, 14, 14, 7 **15.** 101, 79, 101 **16.** 38, 43, 99 **17.** 37, 26, 26 **18.** yes **19.** yes **20.** no **21.** yes **22.** $x = 29$, $y = 28$ **23.** $x = 4$, $y = 5$ **24.** 124, 28, 62 **25.** 60, 90, 30 **26.** 90, 25 **27.** 26 in. **28.** 20 cm **29.** 19 ft **30.** (a, b) **31.** $(0, c)$ **32.** $(a - b, c)$ **33.a.** -1 **b.** 1 **c.** The prod. of the slopes is -1. **34.a.** a **b.** $(0, b)$ **c.** $\sqrt{a^2 + b^2}$ **d.** $\sqrt{a^2 + b^2}$ **e.** BD

Chapter 7

Lesson 7-1 pp. 368–371

EXERCISES **1.** 1 : 1000 **3.** $3b$ **5.** $\frac{b}{4}$ **7.** $\frac{3}{a}$ **9.** $\frac{7}{4}$

11. $\frac{b + 4}{4}$ **13.** $1\frac{2}{3}$ **15.** 6.875 **17.** 7.2 **19.** 14

21. 125 mi **23.** about 135 mi **25.** 18 in. by 22.5 in. **27.** 5 : 4 **29.** A **31.** $\frac{9}{4}$ **33.** $\frac{b}{2}$ **35.** 6 **37.** 16.5 **39.** 4

41. 4, -3 **43.** 16 cm **45.** 9; 18 **47.** 8; 21

49–51. Answers may vary. Samples are given.

49.

Scale 1 cm = 10 ft

51.

Scale 1 cm = 32 ft

53. $\frac{b}{d}$ or $\frac{a}{c}$ **55.** $\frac{c + 2d}{d}$

57. $\frac{a}{b} = \frac{c}{d}$ (Given); $ad = bc$ (Cross-Prod. Prop.); $\frac{ad}{cd} = \frac{bc}{cd}$ (Div. Prop. of =); $\frac{a}{c} = \frac{b}{d}$ (Simplify) **59.** $x = 5; y = 24$ **61.** $x = 3; y = 21$ **63.** C **65.** $(a + c, b)$ **67.** trapezoid

69. rhombus, parallelogram

71. II and III **73.a.** If two lines are not ∥, then they are not coplanar. **b.** If two lines are not coplanar, then they are not ∥.

Algebra 1 Review pp. 372

1. −7, 2 **3.** −0.5, −3 **5.** 0.40, −2.07 **7.** 9, −3 **9.** 0.75, −0.5

Lesson 7-2 pp. 375–379

EXERCISES 1. ∠*JHY* **3.** ∠*JXY* **5.** *JT* **7.** no; $\frac{20}{30} \neq \frac{36}{52}$ **9.** yes; *KLMJ* ~ *PQNO*; $\frac{3}{5}$ **11.** No; corr. ∠s are not ≅. **13.** $x = 4; y = 3$ **15.** $x = 16; y = 4.5; z = 7.5$ **17.** 6.6 in. by 11 in. **19.** 70 mm **21.** 2 : 3 **23.** 50 **25.** 70 **27.** 7.5 m **29.** Yes; corresponding ∠s are ≅ and corresponding sides are proportional with a ratio of 1 : 1. **31.** C **33.** 2.6 cm **35.** 3 : 1 **37.** 1 : 2 **39.** 2 : 3 **41.** sides of 2 cm; ∠s of 60° and 120° **43.** sides of 0.8 cm; ∠s of 60° and 120° **45.** sides of 3 cm; ∠s of 60° and 120° **47.** 6.2 in. **49.** Yes; explanations may vary. Sample: The ratios of radii, diameters, and circumferences of 2 circles are =. **51.a.(1)** Corr. sides of ~ polygons are proport. **(2)** Subst. **(3)** Cross-Product Prop. **(4)** Subtr. Prop. **b.** The value of a length cannot be negative. **c.** 1.6180 **53.** D **55.** D **57.** 7*y* **59.** $\frac{y + 9}{9}$ **61.** No; only one pair of ∥ sides **63.** △*CEA*, △*FED*, △*BCD* **65.** 8

Extension pp. 380–381

1. **3.** $\frac{64}{27}; \frac{256}{81}$ **5.** **9.**

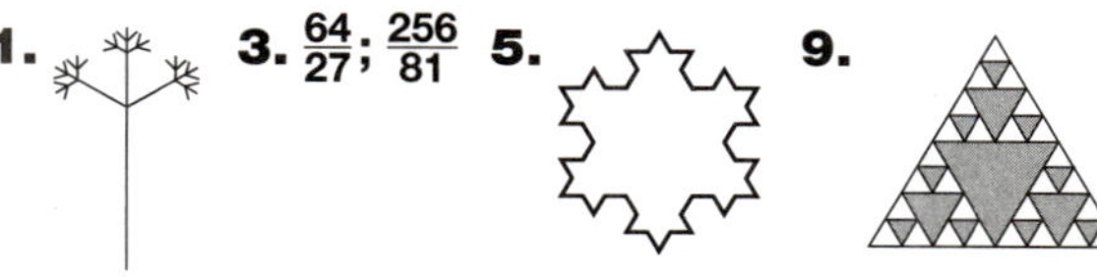

Lesson 7-3 pp. 385–388

EXERCISES 1. Yes; △*ABC* ~ △*FED*; SSS ~ Thm. **3.** Ex. 1: $\frac{2}{3}$ (for △*ABC* to △*FED*); Ex. 2: Not possible; the △s aren't necessarily similar. **5.** No; $\frac{6}{3} \neq \frac{10}{4}$. **7.** Yes; △*APJ* ~ △*ABC*; SSS ~ Thm. or SAS ~ Thm. **9.** No; $\frac{32}{22} \neq \frac{45}{30}$. **11.** AA ~ Post.; 2.5 **13.** AA ~ Post.; 12 **15.** AA ~ Post.; 15 **17.** AA ~ Post.; 220 yd **19.** AA ~ Post.; 90 ft **21.** 151 m **23.a.** No; the corr. ∠s may not be ≅. **b.** Yes; every isosc. rt. △ is a 45°-45°-90° △. Therefore, by AA ~ Thm. they are all ~. **25.** Yes; △*AWV* ~ △*AST*; SAS ~ Thm. **27.** No; there is only one pair of ≅ ∠s. **29.** Check students' work. **31.** 2 : 1 **33.** 4 : 3 **35.** 3:2 **37.** 2:1 **39.** 6:1 **41.1.** $RT \cdot TQ = MT \cdot TS$ (Given) **2.** $\frac{MT}{TQ} = \frac{RT}{TS}$ (Prop. of Proportions) **3.** ∠*RTM* ≅ ∠*STQ* (Vert. ∠s are ≅.) **4.** △*RTM* ~ △*STQ* (SAS ~ Thm.) **43.1.** $\frac{BC}{AC} = \frac{EF}{DF}$, $\overline{EF} \perp \overline{AF}$, $\overline{BC} \perp \overline{AF}$ (Given) **2.** ∠*ACB* and ∠*DFE* are rt. ∠s. (Def. of ⊥) **3.** ∠*ACB* ≅ ∠*DFE* (All rt. ∠s are ≅.) **4.** △*ABC* ~ △*DEF* (SAS ~) **5.** ∠*BAC* ≅ ∠*EDF* (Def. of similar) **6.** $\ell_1 \parallel \ell_2$ (If corr. ∠s are ≅, then ∥ lines.) **45.** B **47.** B **49.** ∠*P* **51.** *ZY* **53.** *YZ* **55.** *x*-values may vary. Sample:

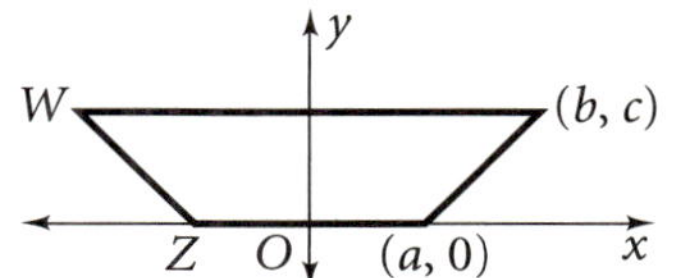

Isosceles trapezoid *W*(−*b*, *c*); *Z*(−*a*, 0)

Algebra 1 Review pp. 390

1. $5\sqrt{2}$ **3.** 8 **5.** $4\sqrt{3}$ **7.** $6\sqrt{2}$ **9.** 6 **11.** $7\sqrt{2}$ **13.** $2\sqrt{6}$ **15.** 5 **17.** $2\sqrt{17}$ **19.** 2 **21.** $\frac{\sqrt{5}}{2}$ **23.** $\sqrt{2}$ **25.** 1

Lesson 7-4 pp. 394–396

EXERCISES 1. 6 **3.** $4\sqrt{3}$ **5.** $14\sqrt{2}$ **7.** $6\sqrt{6}$ **9.** *s* **11.** *c* **13.** *h* **15.** 9 **17.** 10 **19.** 12 **21.a.** 18 mi **b.** 24 mi **23.a.** 4 cm **b.**

4 cm
2 cm 8 cm

Selected Answers

c. Answers may vary. Sample: Draw a 10-cm segment. 2 cm from one endpoint, construct a $\perp$ of length 4 cm. Connect to form a $\triangle$. **25.** (10, 6), (−2, 6) **27.** 14 **29.** $\sqrt{14}$ **31.** 2.5 **33.** 121 **35.** $x = 12\sqrt{5}; y = 12;\ z = 6\sqrt{5}$ **37.** $12\sqrt{2}$ **39.** $\ell_1 = \sqrt{2}, \ell_2 = \sqrt{2}, a = 1, h_2 = 1$ **41.** $\ell_1 = \ell_2 = 6\sqrt{2}, h = 12, h_2 = 6$ **43.** $\ell_1 = 5, a = \frac{60}{13}, h_1 = \frac{25}{13}, h_2 = \frac{144}{13}$ **45.** $\ell_1 = 8\sqrt{5}, \ell_2 = 4\sqrt{5}, h_1 = 4, h_2 = 20$ **47.** *C* is equidistant from *A* and *B* so *C* is on the $\perp$ bisector of $\overline{AB}$ ($\perp$ Bis. Thm.) which thus must be $\overline{CM}$, the altitude to the hypotenuse. Since *M* is the midpt. of $\overline{AB}$, $AM = \frac{1}{2}AB$. Also, by Corollary 2 to Thm. 7-3, *x* is the geometric mean of *AM* and *AB*, so $\frac{\frac{1}{2}AB}{x} = \frac{AM}{x} = \frac{x}{AB}$. By the Cross-Product Property, $\frac{1}{2}AB^2 = x^2$, so $AB = x\sqrt{2}$. **49.** 3 **51.** 4.5 **53.a.** By Corollary 2 to Thm. 7-3 $\frac{c}{a} = \frac{a}{r}$ and $\frac{c}{b} = \frac{b}{q}$. Combined with $c = q + r$, the resulting system can be reduced to $c^2 = a^2 + b^2$ **b.** The square of the hypotenuse equals the sum of the squares of the legs of the triangle. **55.** B **57.** AA ~ Post. **59.** not similar **61.** $x = 6; y = 9$ **63.** 8.75

Lesson 7-5	pp. 400–404

EXERCISES **1.** 7.5 **3.** 5.2 **5.** *c* **7.** *d* **9.** $3\frac{1}{3}$ **11.** 6 **13.** 35 **15.** $\frac{40}{7}$ **17.** *KS* **19.** *JP* **21.** *KM* **23.** *JP* **25.** 559 ft **27.** 2.4 cm and 2.6 cm; 3.3 cm and 8.7 cm; 3.8 cm and 9.2 cm **29.** $x = 18$ m; $y = 12$ m **31.** 20 **33.** $\frac{2}{7}$, 3 **35.a.** $\frac{AB}{BC}$ **b.** $\frac{WX}{XY}$ **c.** $\frac{AB}{BC} = \frac{WX}{XY}$ **37.** No; $\frac{28}{12} \neq \frac{24}{10}$ **39.** Measure $\overline{AC}$, $\overline{CE}$, and $\overline{BD}$. Use the Side-Splitter Thm. Write the proport. $\frac{AC}{CE} = \frac{AB}{BD}$ and solve for *AB*. **41.** 6 **43.** 19.5 **45.a.** A midsegment of a $\square$ connects the midpts. of 2 opp. sides.

b.

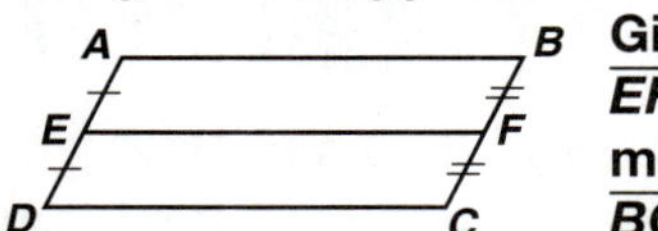

Given: $\square ABCD$ with $\overline{EF}$ connecting the midpts. of $\overline{AD}$ and $\overline{BC}$ Prove: $\overline{AB} \parallel \overline{EF}$; $\overline{EF} \parallel \overline{CD}$ **1.** $\square ABCD$ (Given) **2.** $\overline{AE} \parallel \overline{BF}$ and $\overline{ED} \parallel \overline{FC}$ (Def. of $\square$) **3.** $\overline{AD} \cong \overline{BC}$ (Opp. sides of $\square$ are $\cong$.) **4.** *E* and *F* are midpts. of $\overline{AD}$ and $\overline{BC}$. (Given) **5.** $AE = ED = \frac{1}{2}AD; BF = FC = \frac{1}{2}BC$ (Def. of midpt.) **6.** $AE = BF, ED = FC$ (Subst.) **7.** *ABFE* and *EFCD* are $\square$s. (If one pair of opp. sides of a quad. is $\cong$ and $\parallel$, it is a $\square$.) **8.** $\overline{AB} \parallel \overline{EF}$ and $\overline{EF} \parallel \overline{CD}$ (Opp. sides of a $\square$ are $\parallel$.)

c.

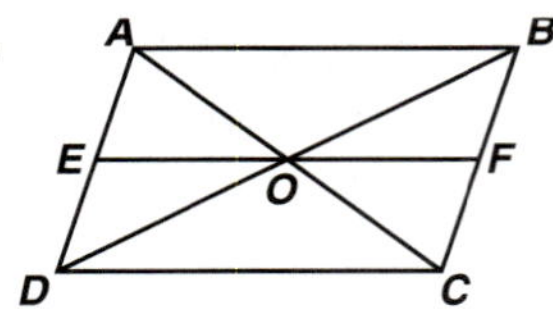

Given: $\square ABCD$ with midsegment $\overline{EF}$ Prove: $\overline{EF}$ bisects $\overline{AC}$ and $\overline{BD}$. Since $\overline{AB} \parallel \overline{EF} \parallel \overline{DC}$ by part (b), and $\overline{EF}$ bisects $\overline{AD}$, by the Side-Splitter Thm., $\overline{EF}$ bisects $\overline{AC}$ and $\overline{BD}$. **47.a.** 14 **b.** 11 **49.** D **51.** *m* **53.** *c* **55.** $RT = SV = 38$ **57.** $RT = SV = 48$ **59.** (3, −3) **61.** (1.5, 2.5) **63.** incenter

Chapter Review	pp. 407–409

1. similar **2.** Cross-Product Property **3.** golden rectangle **4.** similarity ratio **5.** proportion **6.** indirect measurement **7.** golden ratio **8.** 1 : 48 **9.** 1 : 24 **10.** True; use the Cross-Product Prop. **11.** True; the cross product is equivalent to the original proportion. **12.** False; the cross product is *not* equivalent to the given equation. **13.** True; the cross product is equivalent to the original proportion. **14.** $\angle M \cong \angle R, \angle N \cong \angle S, \angle P \cong \angle T; \frac{MN}{RS} = \frac{MP}{RT} = \frac{NP}{ST}$ **15.** 39 in. or 15 in. **16.** 2 : 3 **17.** 2.5 : 1 or 5 : 2 **18.** 9 **19.** $x = 12; y = 15$ **20.** $\triangle XYZ \sim \triangle JKL$; SAS ~ Thm. **21.** No; corr. sides are not in prop. **22.** AA ~ Post. **23.** 13.5 ft **24.** 10 **25.** 30 **26.** $2\sqrt{15}$ **27.** $\frac{2\sqrt{7}}{3}$ **28.** 1.2 **29.** $\sqrt{12}$ **30.** $x = 15; y = 12; z = 20$ **31.** $x = 2\sqrt{21};\ y = 4\sqrt{3}; z = 4\sqrt{7}$ **32.** $x = 2\sqrt{3}; y = 6; z = 4\sqrt{3}$ **33.** 7.5 **34.** 5.5 **35.** 37.5 **36.** 11.25 **37.** 17.5 **38.** 12

Chapter 8

Lesson 8-1	pp. 420–423

EXERCISES **1.** 10 **3.** 34 **5.** 65 **7.** No; $4^2 + 5^2 \neq 6^2$. **9.** Yes; $15^2 + 20^2 = 25^2$. **11.** $\sqrt{33}$ **13.** $2\sqrt{89}$ **15.** $5\sqrt{2}$ **17.** 17.0 m **19.** No; $8^2 + 24^2 \neq 25^2$ **21.** acute **23.** acute **25.** right **27.** 10 **29.** $2\sqrt{2}$ **31.** B **33.** Yes; $7^2 + 24^2 = 25^2$, so $\angle RST$ is a rt. $\angle$. **35.** Answers may vary. Sample: Using 2 segments of length 1, construct the hyp. of the right $\triangle$ formed by these segments. Using the hyp. found as one leg and a segment of length 1 as the other leg, construct the hyp. of the $\triangle$ formed by those legs. Continue this process until constructing a hypotenuse of length $\sqrt{n}$. **37.** 50 **39.** 35 **41–47.** Answers may vary. Samples are given. **41.** 4; 5 **43.** 11; 12 **45.** 14; 16 **47.** 39; 42 **49.** 2830 km **51.** 12.5 cm **53.a.** Answers may vary. Sample: $n = 6$; 12, 35, 37 **b.** $12^2 + 35^2 = 37^2$ **c.** $(2n)^2 + (n^2 - 1)^2 = 4n^2 + n^4 - 2n^2 + 1 = n^4 + 2n^2 + 1 = (n^2 + 1)^2$ **55.** $\sqrt{14}$

57. $\sqrt{17}$ **59.** B **61.** 15 **63.** $x = 21, y = 12$ **65.** 3; 20 **67.** 3 **69.** 7

Lesson 8-2 pp. 428–429

EXERCISES **1.** $x = 8; y = 8\sqrt{2}$ **3.** $y = 60\sqrt{2}$ **5.** $4\sqrt{2}$ **7.** 14.1 cm **9.** $x = 20; y = 20\sqrt{3}$ **11.** $x = 5; y = 5\sqrt{3}$ **13.** $x = 4; y = 2$ **15.** 50 ft; **17.** $a = 7; b = 14; c = 7; d = 7\sqrt{3}$ **19.** $a = 10\sqrt{3}; b = 5\sqrt{3}; c = 15; d = 5$ **21.** $a = 3; b = 7$ **23.** Rika; Sandra marked the shorter leg as opposite the 60° angle. **25.a.** 28 ft **b.** about 0.28 min **27.a.** $\sqrt{3}$ units **b.** $2\sqrt{3}$ units **c.** $s\sqrt{3}$ units **29.** A **31.** $\sqrt{11}$ in. **33.** no

35. no; an isosceles trapezoid **37.** no **39.** yes; ASA Post.

Lesson 8-3 pp. 434–437

EXERCISES **1.** $\frac{1}{2}$; 2 **3.** 1; 1 **5.** 12.3 **7.** 2.5 **9.** 21.4 **11.** 32 **13.** 48 **15.** 63 **17.** 74.1 **19.** 114.5 **21.** 44 and 136 **23.** D **25.** $\frac{\sqrt{2}}{\sqrt{2}} = 1$, so we have to show $\tan^{-1} 1 = 45°$. This is equivalent to showing $1 = \tan 45°$. Consider a 45-45-90 △. Let the lengths of the shorter sides be a. Thus, $\tan 45° = \frac{a}{a} = 1$. **27.** $w = 5; x \approx 4.7$ **29.** $w \approx 59$; $x \approx 36$ **31.** about 51° **33.** about 296 ft **35.** 71.6 **37.** 45.0 **39.** 22.4 **41.** 6.0 **43.** 1.6 **45.a.** 57.290 **b.** 572.96 **c.** Answers may vary. Sample: $\tan X° \approx 572{,}958$ for $X° = 89.9999$ **d.** Answers may vary. Sample: In a rt. △, as an acute ∠ approaches 90°, the opp. side gets longer. **47.** 42 **49.** 6 **51.** x **53.** 26.6 **55.** 78.7 **57.** 36.9 **59.** B **61.** $5\sqrt{2}$ units **63.** acute **65.** $R(a, b)$; $S(a, -b)$; $T(c, -b)$; $V(c, b)$

Lesson 8-4 pp. 441–443

EXERCISES **1.** $\frac{7}{25}$; $\frac{24}{25}$ **3.** $\frac{1}{2}$; $\frac{\sqrt{3}}{2}$ **5.** 8.3 **7.** 17.0 **9.** 106.5 **11.** 21 **13.** 46 **15.** 24 **17.** about 17 ft 8 in. **19.** $\cos X \cdot \tan X = \frac{\text{adj.}}{\text{hyp.}} \cdot \frac{\text{opp.}}{\text{adj.}} = \frac{\text{opp.}}{\text{hyp.}} = \sin X$ **21.** No; the △s are ~ and the sine ratio for 35° is constant. **23.** $w \approx 37$; $x \approx 7.5$ **25.a.** They are equal; yes; the sine and cosine of complementary △s are equal. **b.** ∠B; ∠A **c.** Answers may vary. Sample: cosine of ∠A = sine of the compl. of ∠A. **27.** Yes; use any trig. function and the known measures to find one other side. Use the Pythagorean Thm. to find the 3rd side. Subtract the acute ∠ measure from 90 to get the other ∠ measure. **29.** Answers may vary. Samples are given. **a.** Since $\sin A = \frac{\text{opp.}}{\text{hyp.}}$, if $\sin A \geq 1$, then opp. ≥ hyp., which is impossible. **b.** Since $\cos A = \frac{\text{adj.}}{\text{hyp.}}$, if $\cos A \geq 1$, then adj. ≥ hyp., which is impossible. **31.** $(\sin A)^2 + (\cos A)^2 = \left(\frac{a}{c}\right)^2 + \left(\frac{b}{c}\right)^2 = \frac{a^2}{c^2} + \frac{b^2}{c^2} = \frac{a^2 + b^2}{c^2} = \frac{c^2}{c^2} = 1$

33. $\frac{1}{(\cos A)^2} - (\tan A)^2 = \left(1 \div \frac{b^2}{c^2}\right) - \frac{a^2}{b^2} = \frac{c^2}{b^2} - \frac{a^2}{b^2} = \frac{c^2 - a^2}{b^2} = \frac{b^2}{b^2} = 1$ **35.** $(\tan A)^2 - (\sin A)^2 = \left(\frac{a}{b}\right)^2 - \left(\frac{a}{c}\right)^2 = \frac{a^2}{b^2} - \frac{a^2}{c^2} = \frac{a^2c^2}{b^2c^2} - \frac{a^2b^2}{b^2c^2} = \frac{a^2c^2 - a^2b^2}{b^2c^2} = \frac{a^2(c^2 - b^2)}{b^2c^2} = \frac{a^2 \cdot a^2}{b^2c^2} = \left(\frac{a}{b}\right)^2\left(\frac{a}{c}\right)^2 = (\tan A)^2(\sin A)^2$

37. B **39.** 6.9 **41.** 18 **43.** $\frac{3}{4}$ **45.** 4

Lesson 8-5 pp. 447–449

EXERCISES **1.** ∠ of elevation from sub to boat **3.** ∠ of elevation from boat to lighthouse **5.** ∠ of elevation from Jim to top of waterfall **7.** ∠ of depression from top of waterfall to Jim **9.** 34.2 ft **11.** about 986 m **13.** 0.6 km **15.** 64° **17.** about 193 m **19.** 72, 72 **21.** 27, 27 **23.** B **25.** 5 **27.** 0.5; about 84.9 **29.** about 28 ft **31.** A **33.** C **35.** 85.2 m **37.** 45 **39.** $y = 3, x = 2$; 16, 10, 10, 16

Lesson 8-6 pp. 455–459

EXERCISES **1.** ⟨602.2, 668.8⟩ **3.** ⟨37.5, −65.0⟩ **5.** 20° west of south

7.

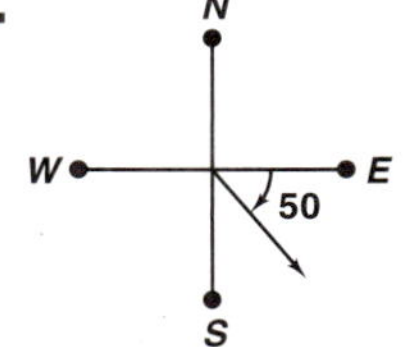

9.

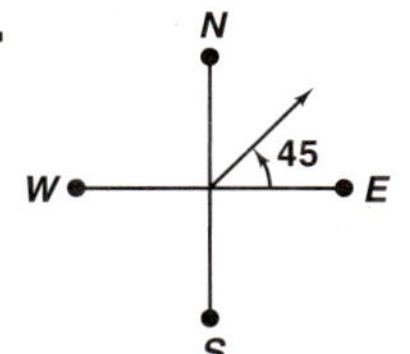

11.

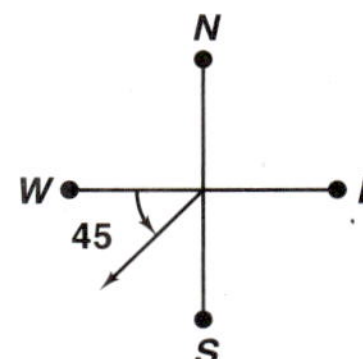

13. about 97 mi at about 41° south of west **15.** about 54 mi/h; about 22° north of east

17.a. ⟨−9, −9⟩ **b.**

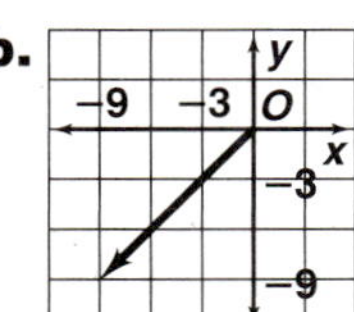

19.a. ⟨−1, 0⟩ **b.**

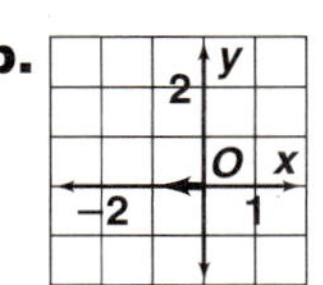

21.a. ⟨−8, 6⟩ **b.**

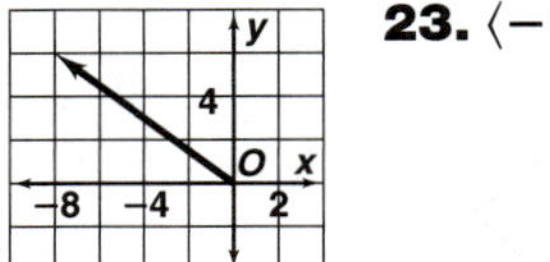

23. ⟨−1, 3⟩

25. ⟨−2, 3⟩ **27.** about 13.2° north of west **29.** Yes; both vectors have the same direction, but could have diff. mag.

31.a.

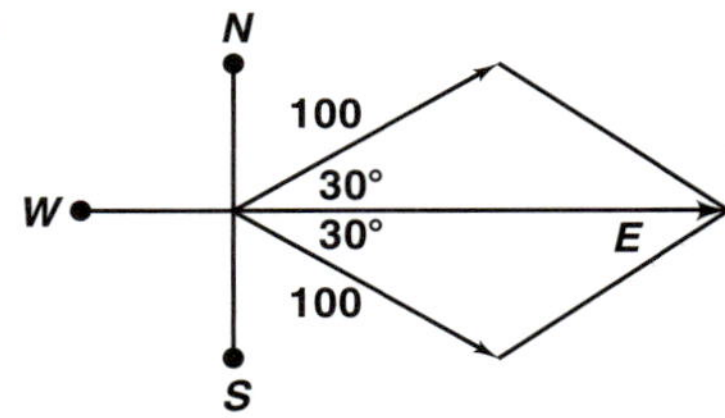

b. about 173 due east **33.** Vectors are ∥ if they have the same or opp. directions. **35.a.** ⟨0, 0⟩ **b.** $\vec{a}$ and $\vec{c}$ have = mag. and opp. direction.

37. $\begin{bmatrix} -1 \\ -2 \end{bmatrix}$ **39.** $\begin{bmatrix} -1 \\ 0 \end{bmatrix}$ **41.a.**

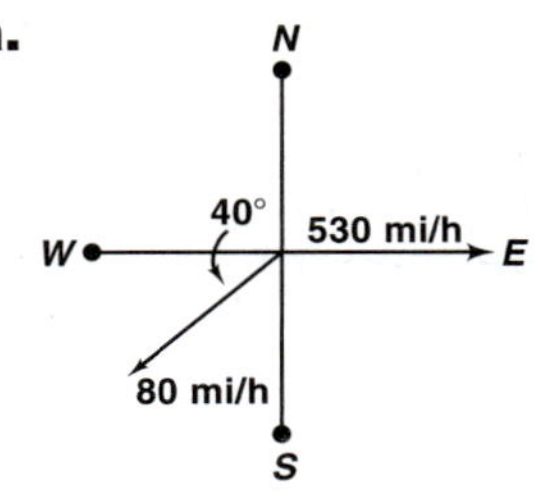

b. ⟨530, 0⟩; ⟨−61.3, −51.4⟩ **c.** ⟨468.7, −51.4⟩ **d.** 471.5 mi/h at 6.3° south of east

43. ⟨0, −4⟩

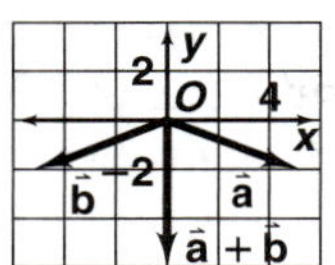

45. The vectors have the same mag.; the vectors have opp. directions. **47.a.** about 15° south of west **b.** about 6.7 h **49.** about 2229 ft; about 10° **51.** Answers may vary. Sample: zero vector = ⟨0, 0⟩; it has mag. 0 and no direction. **53.** B **55.** 4492 ft **57.** $6\sqrt{2}$ **59.** parallelogram

Chapter Review pp. 461–463

1. vector **2.** angle of elevation **3.** cosine **4.** sine **5.** identity **6.** resultant vector **7.** Pythagorean triple **8.** 16 **9.** $2\sqrt{113}$ **10.** 17 **11.** $x = 9\sqrt{3}$; $y = 18$; **12.** $12\sqrt{2}$ **13.** $x = \frac{20\sqrt{3}}{3}$; $y = \frac{40\sqrt{3}}{3}$; **14.** 42 **15.** 2 **16.** 2 **17.** 67 **18.** 23 **19.** 42 **20.** 12 **21.** 8 **22.** $\frac{\sqrt{3}}{2}, \frac{1}{2}, \sqrt{3}$ **23.** $\frac{\sqrt{19}}{10}, \frac{9}{10}, \frac{\sqrt{19}}{9}$ **24.** $\frac{4}{5}, \frac{3}{5}, \frac{4}{3}$ **25.** 51 **26.** 16.5 **27.** 33 **28.** 1410 ft **29.** 280 ft **30.** about 38.2 yd **31.** ⟨125.8, 81.7⟩ **32.** ⟨37.5, −92.7⟩ **33.** ⟨−21.8, 33.5⟩ **34.** about 167.7 mi; about 26.6° east of south **35.** about 206.2 km; about 14.0° west of south **36.** about 503.1 mi/h; about 26.6° north of west **37.** ⟨6, 8⟩, ⟨9, 12⟩, ⟨30, 40⟩; ⟨x, y⟩ and ⟨nx, ny⟩ have the same direction for $n > 0$.

38. **39.**

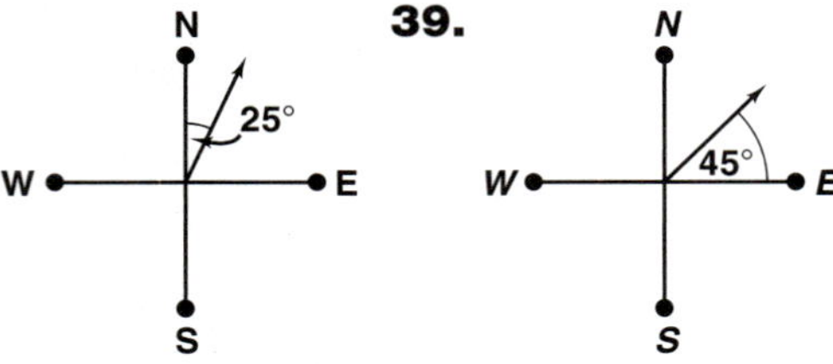

40.

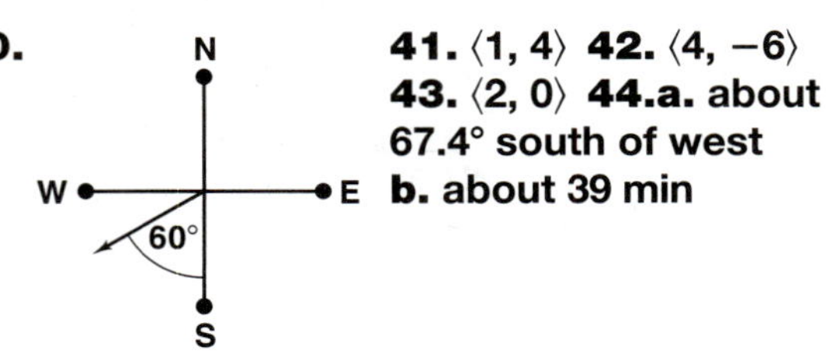

41. ⟨1, 4⟩ **42.** ⟨4, −6⟩ **43.** ⟨2, 0⟩ **44.a.** about 67.4° south of west **b.** about 39 min

Chapter 9

Lesson 9-1 pp. 473–476

EXERCISES 1. Yes; the trans. is a slide. **3.** No; the figures are not ≅. **5.a.** Answers may vary. Sample: $\angle R \to \angle R'$ **b.** $\overline{RI}$ and $\overline{R'I'}$; $\overline{IT}$ and $\overline{I'T'}$; $\overline{RT}$ and $\overline{R'T'}$ **7.** (−6, 5), (2, 4), (−3, 1) **9.** (−7, 5), (−7, 8), (−4, 8), (−1, 2) **11.** $(x, y) \to (x + 1, y - 3)$ **13.** $(x, y) \to (x - 5, y - 2)$

15.a.

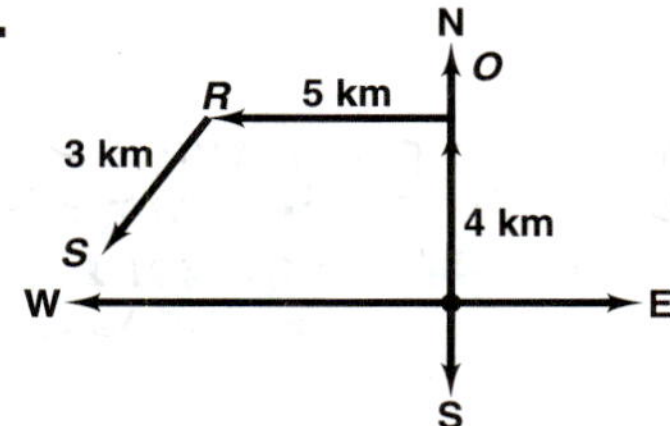

b. about 7.1 km west, 1.9 km north **17.** $(x, y) \to (x + 2, y + 2)$ **19.** D

21.

23.a. at least 5 ft east, 10 ft north **b.** Sample: $(x, y) \to (x + 5, y + 10)$ **25.** U' (1, 16), G' (2, 12) **27.** No; $\triangle HYP \to \triangle Y'H'P'$ is the translation. **29.** $(x, y) \to (x + 13, y - 2.5)$

31.

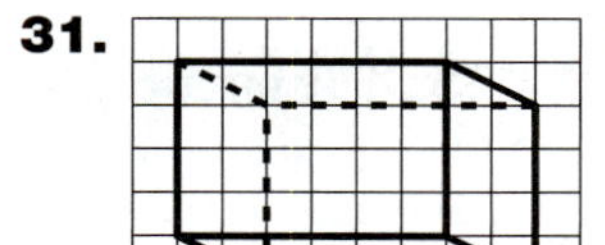

33.

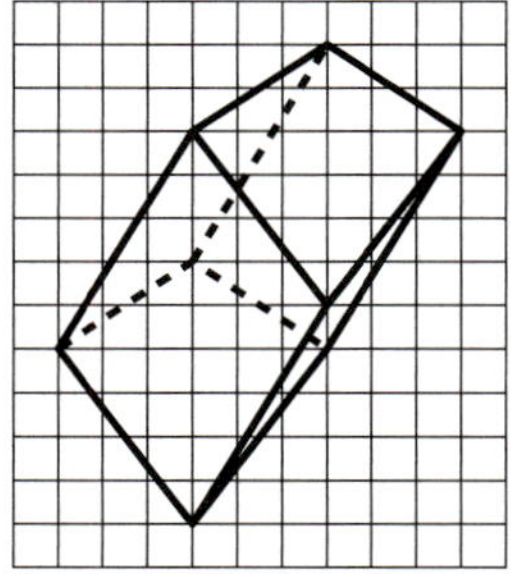

35. $B'(3, 6, 4)$, $C'(-1, 6, 4)$, $D'(-1, 2, 4)$, $E'(-1, 2, 0)$, $F'(3, 2, 0)$, $G'(3, 6, 0)$, $H'(-1, 6, 0)$ **37.** A **39.** about 431.7 mi at 76.6° south of east **41.** 4.5 **43.** $\overline{AB} = \overline{CB}$, so $\triangle ABC$ is isos., $\angle A \cong \angle C$ (base ∠s of an isos. △ are ≅), so $\triangle ABD \cong \triangle CBD$ by AAS. $\angle ABD \cong \angle CBD$ by CPCTC.

Lesson 9-2 pp. 480–482

EXERCISES **1.** $(-1, 2)$ **3.** $(-3, 2)$ **5.** $(-5, -3)$

7.

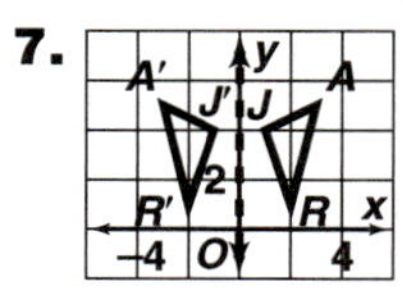

9.

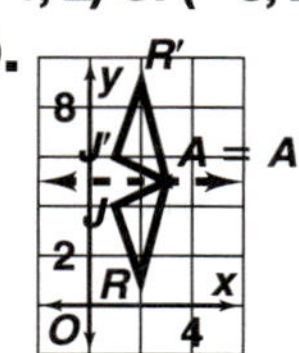

11.

13.

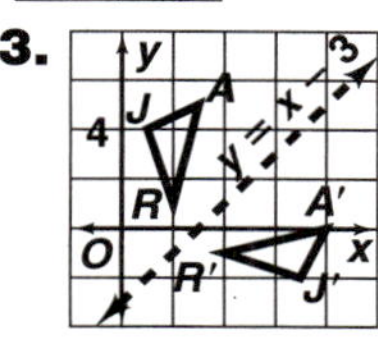

15. Reflect point *D* over the mirrored wall. Connect this point and *C*. The intersection of the segment and the wall is the point to focus the camera.

17.

19. A **21.**

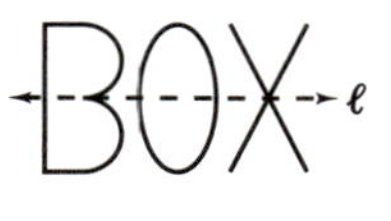

23. Answers may vary. Sample: scissors, a baseball glove, a guitar **25.** (x, y) has image $(-x, y)$.
27.a. Leonardo da Vinci was left-handed.
b. Answers may vary. Sample: His writing hand would not cover what was written so far.
29. (4, 0) **31.** (–4, 6) **33.** (0, –4) **35.** (0, 2*a*)
37. $\overline{AB} \cong \overline{A'B'}$; $\overline{BC} \cong \overline{B'C'}$; $\overline{AC} \cong \overline{A'C'}$; $A \to A'$; $B \to B'$; $C \to C'$; $\angle A \cong \angle A'$; $\angle B \cong \angle B'$; $\angle C \cong \angle C'$
39–45. Answers may vary. Samples are given.
39. yes; follow Exercise 38 steps, first using one side of the triangle and again using a second side. **41.** yes; reflect an isosc. △ across its base. **43.** yes; reflect an isosc. rt. △ across its hyp.
45. for $b \neq d$, $y = \left(\frac{a-c}{d-b}\right)x - \frac{a^2 + b^2 - c^2 - d^2}{2(d-b)}$; for $b = d$, $x = \frac{a+c}{2}$ **47.** A **49.** C **51.** $(x, y) \to (x + 4, y - 2)$ **53.a.** 277.5 km **b.** about 172 **55.** $y = 5x + 4$

Lesson 9-3 pp. 485–488

EXERCISES

1.

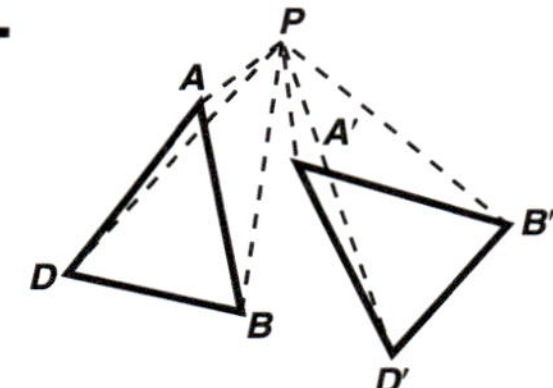

3.

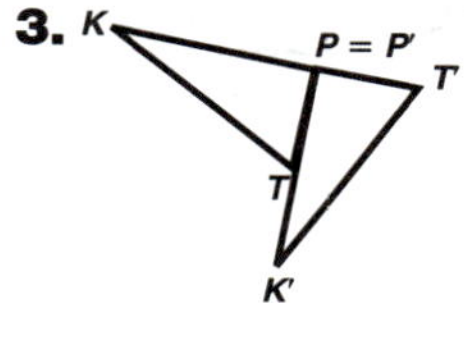

5. **7.**

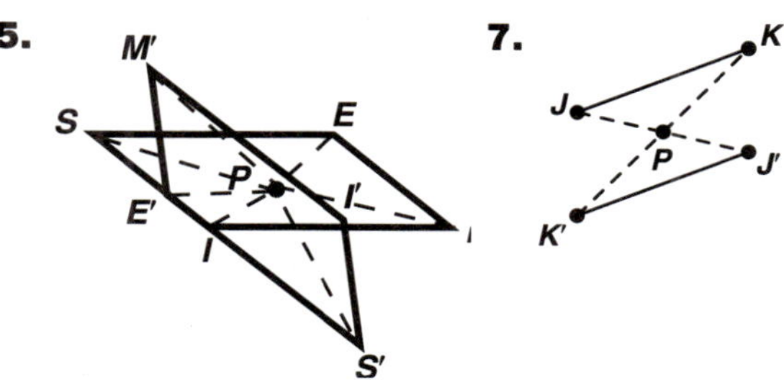

9.

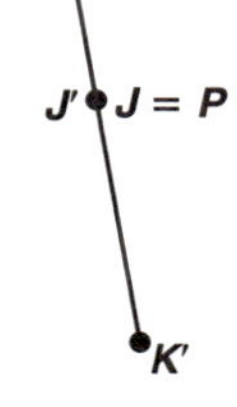

11. *M* **13.** $\overline{BC}$ **15.** $\overline{LM}$ **17.** *K*
19. 108°; 252°

21.

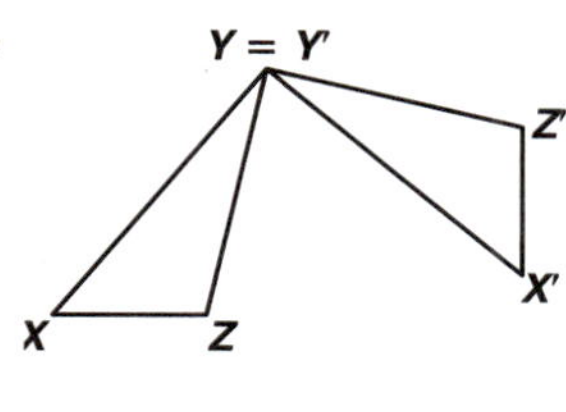

23. **25.**

27. 180° rotation about its center **29.** 110

31.a–c.

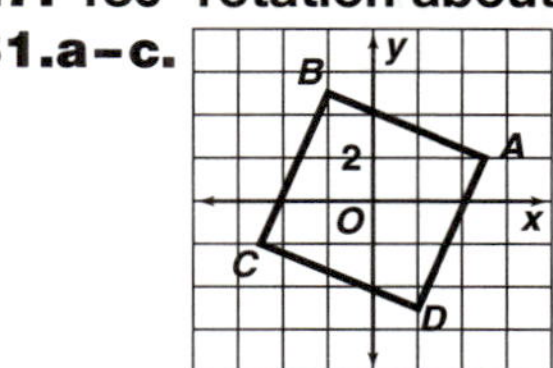

d. Square; all sides are ≅ and all ∠s are 90° **33.** Answers may vary. Sample: a 90° and a 270° rotation **35.** B **37.** B

39.

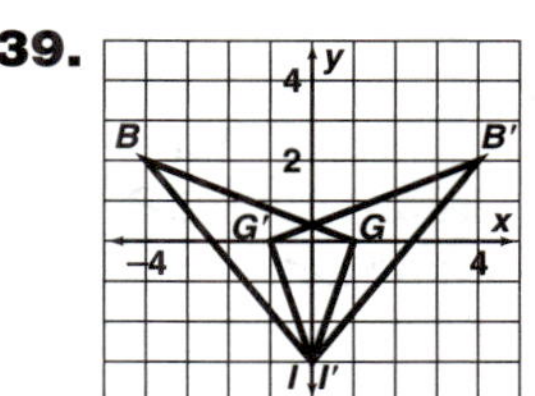

41.

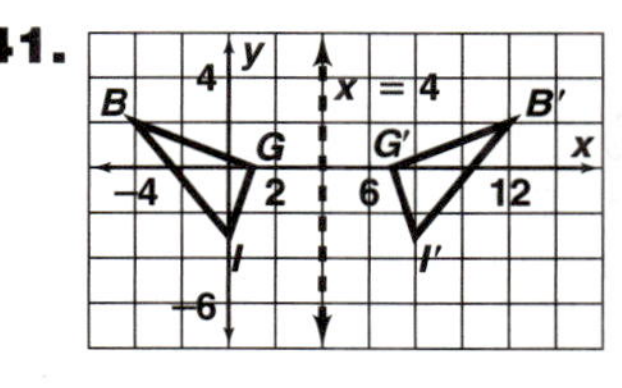

43. 86.6 ft **45.** Parallel; both lines are vertical.

Lesson 9-4 pp. 494–496

EXERCISES

1. line; rotational 180° **3.** rotational: 90°

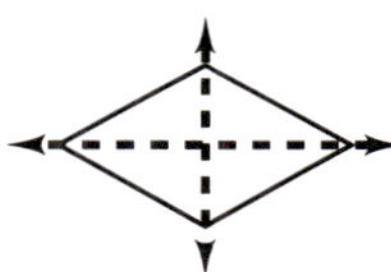

5. rotational: 180° **7.** no symmetry **9.** rotational: 60°

11. line, rotational: 180°

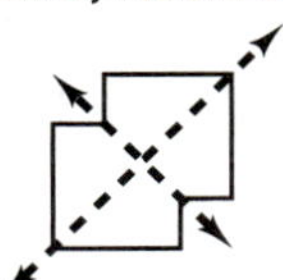

13.

15.

17. rotational and reflectional **19.** Answer may vary. Samples: CODE, HOOD, DOCK

21.a.

Language	Horiz. line	Vert. line	Point
English	B, C, D, E, H, I, K, O, X	A, H, I, M, O, T, U, V, W, X, Y	H, I, N, O S, X, Z
Greek	B, E, H, Θ I, K, Ξ, O, Σ, Φ, X	A, Δ, H, Θ, I, Λ, M, Ξ, O, Π, T, Υ, Φ, X, Ψ, Ω	Z, H, Θ, I, N, Ξ, O, Φ, X

b. Answers may vary. Sample: Greek; Greek alphabet has more letters with at least one kind of symmetry and more letters with multiple symmetries. **23.** Sketches may vary. rotational: 90°; reflectional **25.** reflectional; rotational **27.** point **29.** reflectional; rotational **31.** reflectional, rotational **33.** Yes; the bisector divides the ∠ into 2 ≅ ∠s with one side of the ∠ being the reflection of the other. **35.** Not necessarily; the bisector divides the segment into 2 ≅ parts but one part cannot be the reflection of the other unless the bisector is the ⊥ bisector. **37.** (−3, 4) **39.** (−3, −4) **41.** point symmetry about any pt. on the line; reflectional in any member of the family of lines $y = -x + b$

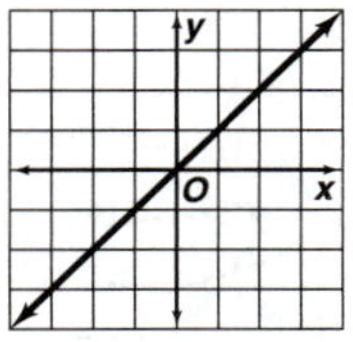

43. reflectional in x-axis

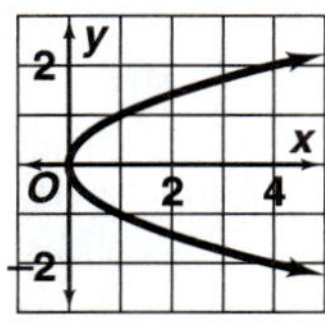

45. reflectional in $x = -2$

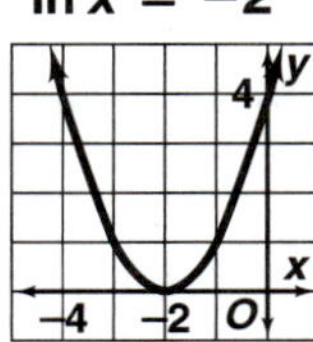

47. reflectional in y-axis

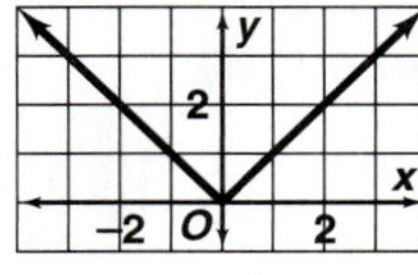

49. Answers may vary. Samples are given.

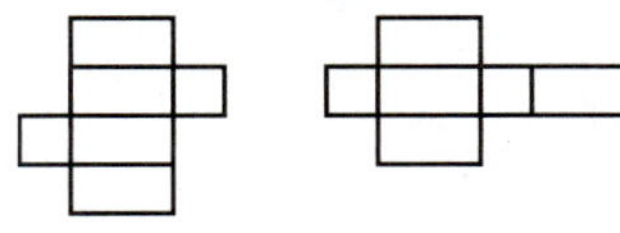

51. A **53.** D **55.** Sample: yes; rotate equilateral △*ABC* 180° about its center. **57.** Sample: yes; reflect *ABCD* across a line halfway between $\overline{BC}$ and $\overline{YZ}$.

Lesson 9-5 pp. 500–503

EXERCISES **1.** enlargement; center *A*, scale factor $\frac{3}{2}$ **3.** enlargement; center *R*, scale factor $\frac{3}{2}$ **5.** reduction; center *L*, scale factor $\frac{1}{3}$ **7.** reduction; center (0, 0), scale factor $\frac{1}{2}$ **9.** enlargement; center (0, 0), scale factor $\frac{3}{2}$ **11.** 512 in. **13.** 1.25 ft **15.** $P'(6, -3)$, $Q'(6, 12)$, $R'(12, -3)$

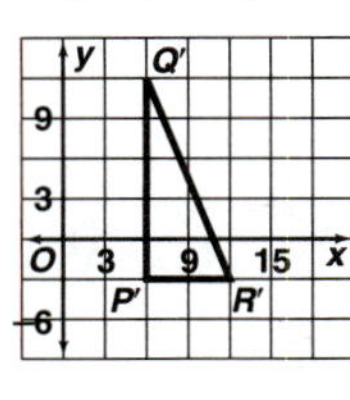

17. $P'\left(-\frac{9}{4}, 0\right)$, $Q'\left(0, \frac{9}{4}\right)$, $R'\left(\frac{3}{4}, -\frac{9}{4}\right)$

19. $L'(-15, 0)$ **21.** $T'(0, 18)$ **23.** $N'(-0.4, -0.7)$ **25.** $B'\left(\frac{1}{8}, -\frac{1}{15}\right)$ **27.** $Q'(-9, 12)$, $W'(9, 15)$, $T'(9, 3)$, $R'(-6, -3)$ **29.** $Q'\left(-\frac{3}{2}, 2\right)$, $W'\left(\frac{3}{2}, \frac{5}{2}\right)$, $T'\left(\frac{3}{2}, \frac{1}{2}\right)$, $R'\left(-1, -\frac{1}{2}\right)$ **31.** $Q'(-1.8, 2.4)$, $W'(1.8, 3)$, $T'(1.8, 0.6)$, $R'(-1.2, -0.6)$ **33.** $Q'(-30, 40)$, $W'(30, 50)$, $T'(30, 10)$, $R'(-20, -10)$ **35.** The image has side lengths 10 in. and ∠measures 60.

37.

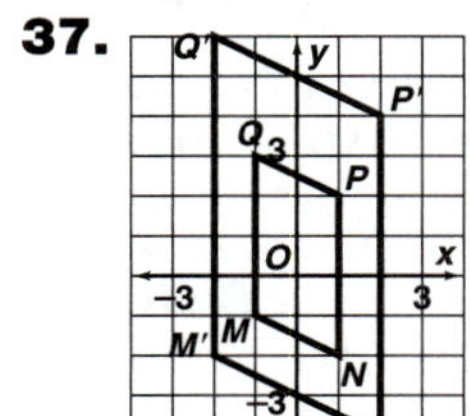

39.

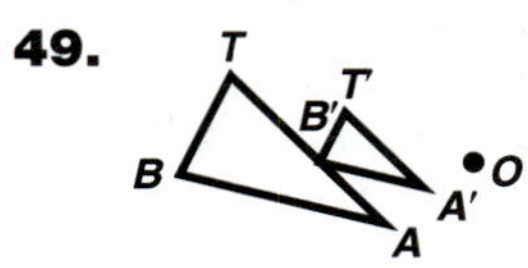

41. Check students' work. **43.** $I'J' = 10$; $H'J' = 12$ **45.** $HI = 32$; $I'J' = 7.5$ **47.** $x = 3$; $y = 60$

49.

51.

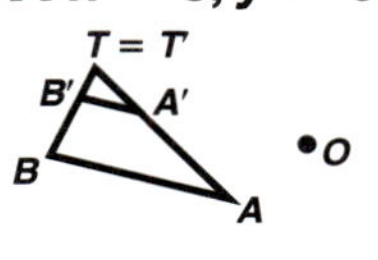

53.

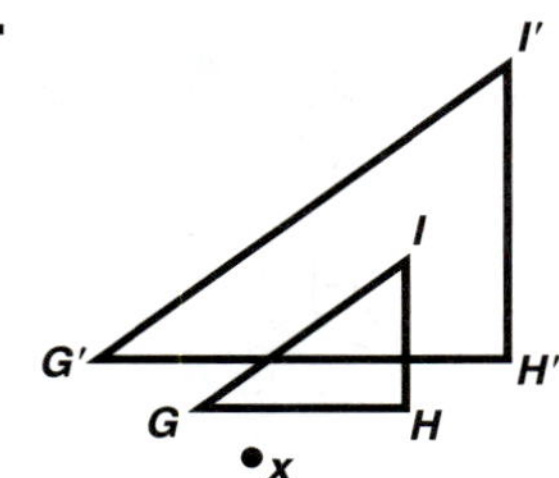

55. 60 cm **57.** False; a dilation doesn't map a segment to a ≅ segment unless the scale factor is 1. **59.** False; a dilation with a scale factor greater than 1 is an enlargement. **61.** False; if the center of dilation is on the preimage, it is also on the image. **63.** Connect corresponding points *A* and *A′* and *B* and *B′*. Extend $\overline{AA'}$ and $\overline{BB'}$ until they intersect at the center of dilation. The scale factor is the length of $\overline{A'B'}$ divided by the length of $\overline{AB}$. **65.a.** *P′*(−1, −2), *Q′*(−3, −4), *R′*(−4, −1) **b.** Each point of the △ is reflected in the origin, which is the point of reflection. Two figures are symmetrical with respect to a pt. *P* if *P* is the midpoint of each segment that connects two corr. points of the figures. **67.** D **69.** A **71.** (2,7) **73.** (7, −2) **75.** $(x, y) \rightarrow (x - 2, y - 1)$ **77.** $(x, y) \rightarrow (x + 8, y + 10)$ **79.** 777.9 m

Extension p. 504–505

1. *A′*(4, −2), *B′*(8, 2), *C′*(9, −1) **3.** *J′*(−5, −2), *K′*(−1, 0), *L′*(0, −4), *M′*(−6, −4) **5.** ⟨−10, 7⟩ **7.** ⟨1, 16⟩ **9.** *H′*(2, −1), *J′*(8, −1), *K′*(3, −5)

11.

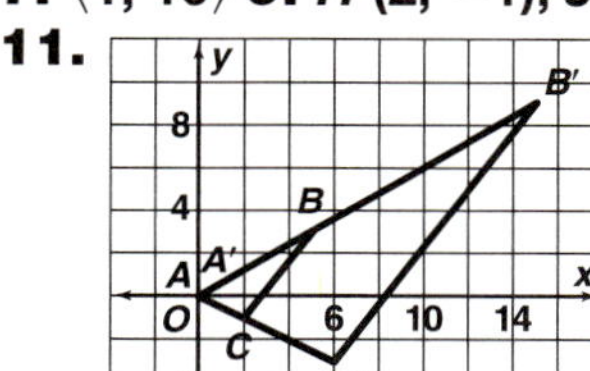

13.

Lesson 9-6 pp. 509–511

EXERCISES 1. rotation **3.** Neither; the figures do not have the same orientation.

5.

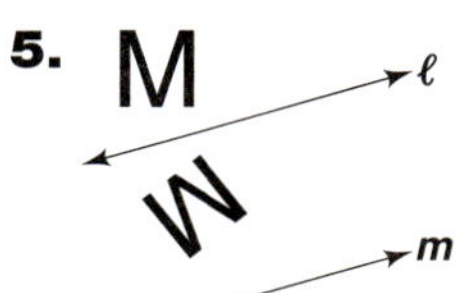

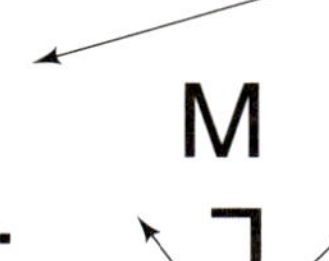

M is translated across line *m* twice the distance between ℓ and *m*.

7.

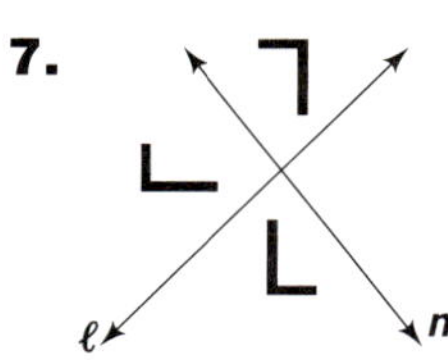

L is rotated clockwise about 180°.

9.

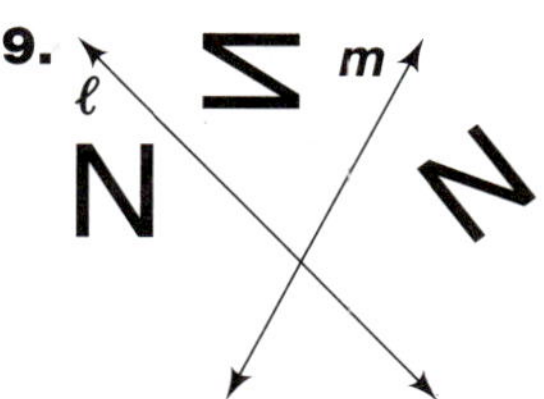

N is rotated clockwise about 160°.

11.

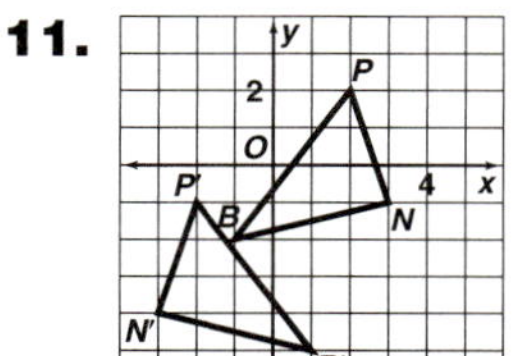

13.

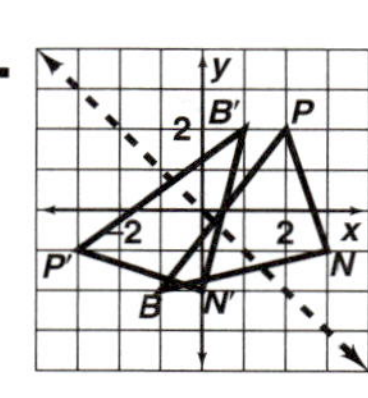

15.

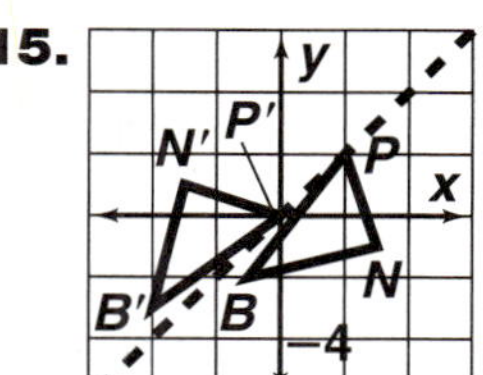

17. opp.; glide reflection **19.** same; rotation **21.** same; translation **23.** opp.; glide reflection **25.** rotation; 180° about the point $\left(\frac{1}{2}, 0\right)$ **27.** Odd isometries can be expressed as the composition of an odd number of reflections. Even isometries are the composition of an even number of reflections. **29.** Yes; a rotation of *x*° followed by a rotation of *y*° is equivalent to a rotation of $(x + y)°$. **31.** 60° **33.** $51\frac{3}{7}°$ **35.** rotation; center *C*, ∠of rotation 180° **37.** translation; $(x, y) \rightarrow (x - 9, y)$ **39.** reflection; $x = 4$ **41.** rotation; center (3, 0), ∠ of rotation 180° **43.** translation; $(x, y) \rightarrow (x - 11, y - 4)$ **45.** Sample: Translate the red R so that one point moves to its corresponding point on the blue R. Then reflect across a line passing through that point. **47.** Answers may vary. A sample is given: $\overline{XX'} \parallel \overline{YY'}$ and $\overline{XX'} \cong \overline{YY'}$, so *XX′Y′Y* is a ▱. Therefore, $\overline{XY} \cong \overline{X'Y'}$. **49.** Answers may vary. Sample: Since a reflection moves a pt. in the direction ⊥ to the translation, the order does not matter. **51.** (6, 5) **53.** (2, 6) **55.** B **57.** D **59.** *A′*(0, 12), *B′*(0, 0), *C′*(9, 0) **61.** *A′*(4, 6), *B′*(−8, −4), *C′*(10, −6) **63.** $8\sqrt{3}$ **65.** 10 **67.** I and III

Lesson 9-7 pp. 518–520

EXERCISES 1–3. Answers may vary. Samples are given. **1.** yes; translation; two ⊥ rectangles **3.** yes; translation; four rectangles in a square shape **5.** yes **7.** no **9.** no **11.** rotational, reflectional, glide reflectional, and translational **13.** rotational, reflectional, glide reflectional, and translational

15.

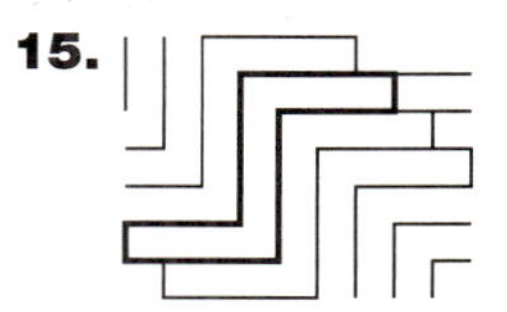

17.

19–21. Answers may vary. Samples are given.

19.

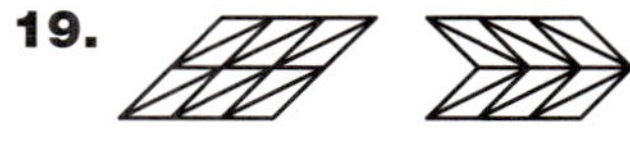

21.

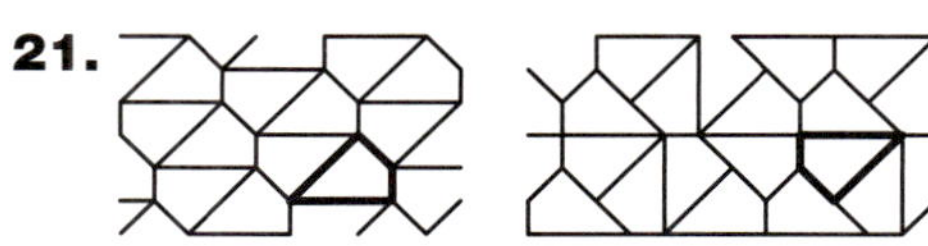

23. no **25.** yes;

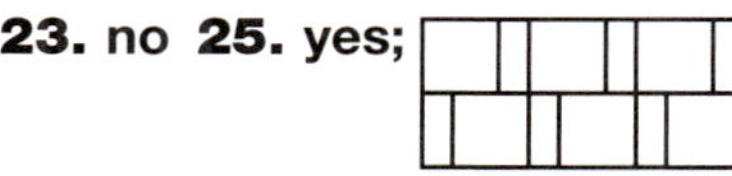

Selected Answers

27. reflectional, glide reflectional, rotational, translational

29–39.

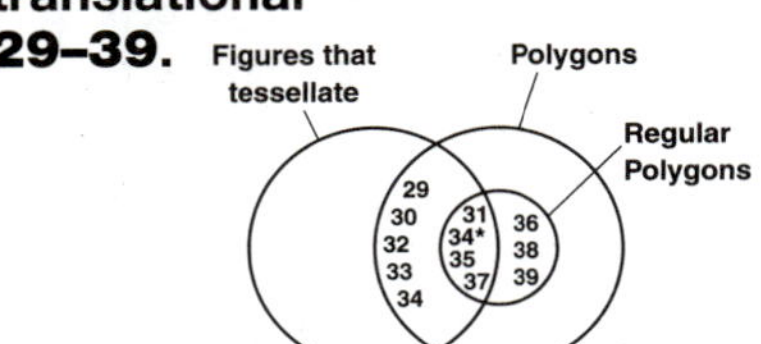

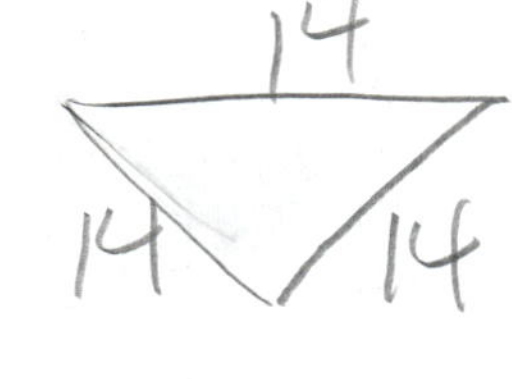

41. Answers may vary. Sample: Draw $\triangle ABC$. Locate M, the mdpt. of $\overline{AB}$, and N, the mdpt. of $\overline{BC}$. Draw the images of $\triangle ABC$ under 180° rotations about M and N. Draw the image of $\triangle ABC$ under the translation that maps A to C. 2nd way: Draw $\triangle ABC$. Draw the reflection image of pt. C over $\overline{AB}$, C'. Now use the steps from Ex. 38 for quad. $ACBC'$. **43.** C **45.** $A'(-3, 3)$, $B'(-4, 2)$, $C'(-4, 4)$ **47.** $10\sqrt{2}$ **49.** $10 < s < 42$ **51.** $0 < s < 18$ **53.** $3\frac{1}{2} < s < 12\frac{1}{2}$

Chapter Review pp. 523–525

1. F **2.** D **3.** E or B **4.** G **5.** B **6.** C **7.** A **8.** $A'(7, 12)$, $B'(8, 6)$, $C'(3, 5)$ **9.** $R'(-4, 3)$, $S'(-6, 6)$, $T'(-10, 8)$ **10.** $(x, y) \rightarrow (x - 2, y + 1)$ **11.** $(x, y) \rightarrow (x + 11, y - 4)$

12.

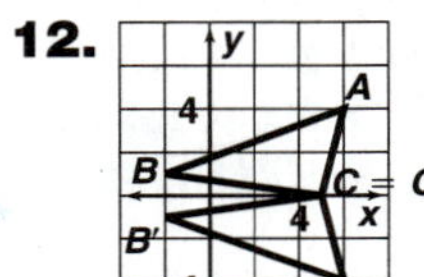

13.

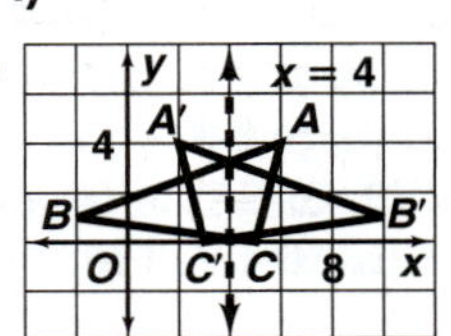

14.

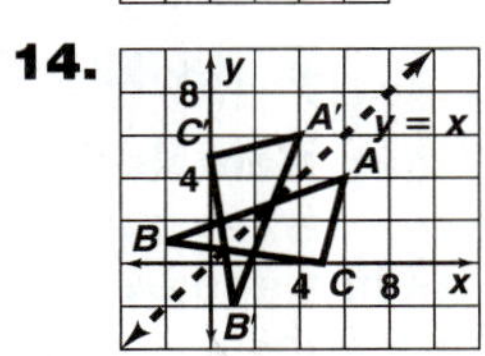

15.

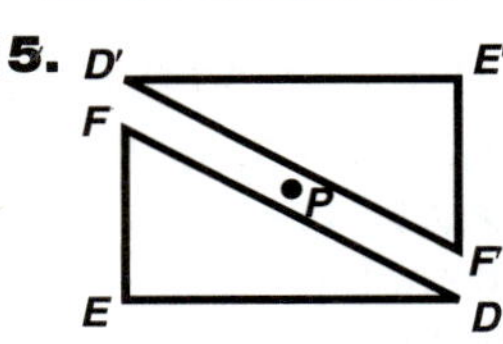

16.

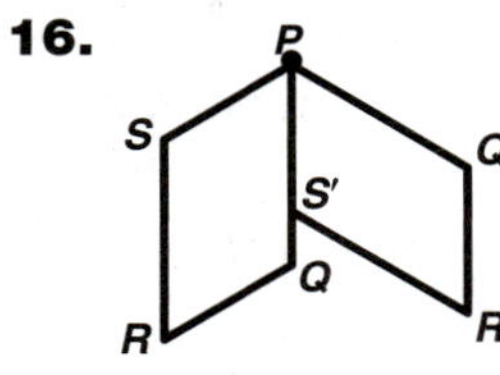

17.

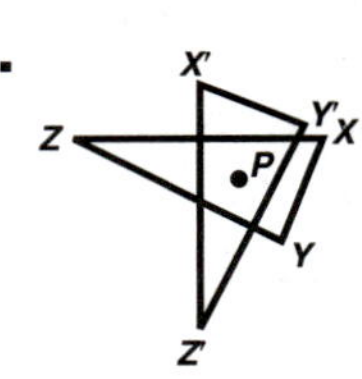

18. $(-2, 5)$ **19.** $(-3, 0)$ **20.** $(-1, -4)$ **21.** $(0, 7)$ **22.** $(8, -2)$ **23.** reflectional **24.** rotational; 72° **25.** $A'(0, 12)$ **26.** $B'(-1, 3)$ **27.** $C'(15, -20)$ **28.** $M'(-15, 20)$, $A'(-30, -5)$, $T'(0, 0)$, $H'(15, 10)$ **29.** $F'(-2, 0)$, $U'\left(\frac{5}{2}, 0\right)$, $N'\left(-1, -\frac{5}{2}\right)$ **30.** same; rotation **31.** opposite; reflection **32.** same; translation **33.** opposite; glide reflection **34.** $T'(-4, -9)$, $A'(0, -5)$, $M'(-1, -10)$

35.a.

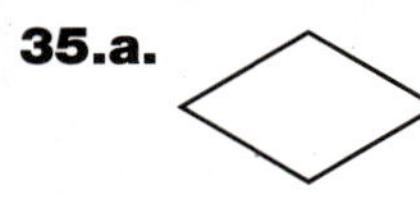

b. rotational, point, reflectional, translational, glide reflectional

36.a.

b. rotational, point, reflectional, translational, glide reflectional

Chapter 10

Lesson 10-1 pp. 536–538

EXERCISES **1.** 240 cm^2 **3.** 20.3 m^2 **5.** 0.24 **7.** 14 m^2 **9.** 3 ft^2 **11.** 4 in. **13.** 14 cm **15.** The area does not change; the height and base AB do not change. **17.** 15 $units^2$ **19.** 6 $units^2$ **21.** 27 $units^2$ **23.** 21 $units^2$

25.a.

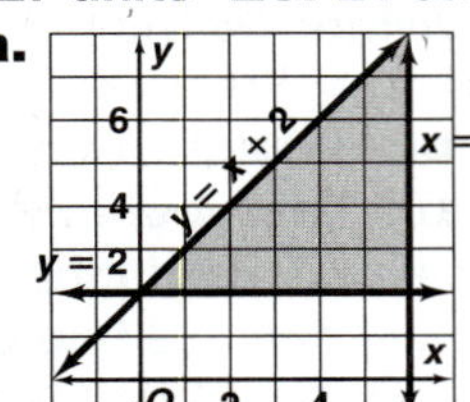

b. 18 $units^2$

27.a.

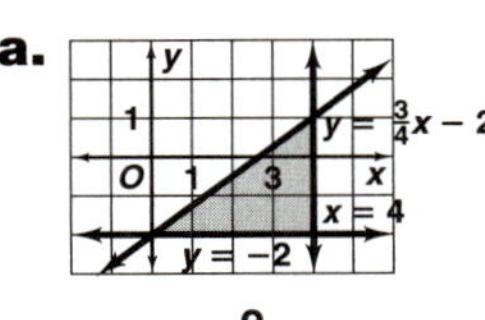

b. 6 $units^2$

29.a. Blank grid; area is 84 $units^2$ while figures are 36 $units^2$. **b.** No; the figures have the same area. **31.** 28 $units^2$ **33.** 88 $units^2$ **35.** 525 cm^2 **37.** 34 $in.^2$ **39.a.** 54 $in.^2$ **b.** 54 $in.^2$ **41.** D **43.** rotational, point, translational **45.** 72 **47.** 36 **49.** 36 **51.**

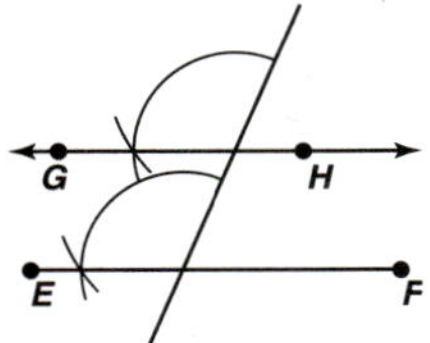

Lesson 10-2 pp. 542–544

EXERCISES **1.** 472 $in.^2$ **3.** 108 ft^2 **5.** 150 cm^2 **7.** about 43,290 mi^2 **9.** 72 m^2 **11.** 80 $in.^2$ **13.** 24 ft^2 **15.** 96 $in.^2$ **17.** 20 $in.^2$ **19.** Check students' work. **21.** 19.5 cm^2 **23.** 49.9 ft^2 **25.** 18 $units^2$ **27.** 15 $units^2$

29.a.

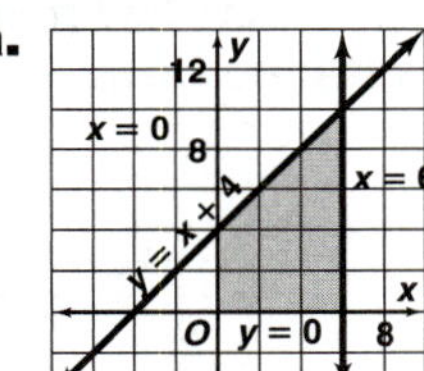

b. trapezoid **c.** 42 $units^2$

31. $32\sqrt{3}$ m^2 **33.a.** $A = \frac{1}{2}b_1h$; $A = \frac{1}{2}b_2h$; **b.** The area of the trapezoid is the sum of the areas of the triangles, so $A = \frac{1}{2}b_1h + \frac{1}{2}b_2h = \frac{1}{2}h(b_1 + b_2)$. **35.** $b_1 = 12$ cm, $b_2 = 24$ cm, $h = 18$ cm **37.** $100 + 50\sqrt{3}$ or about 186.6 $in.^2$ **39.** B **41.** 15 ft **43.** sometimes **45.** 140

Lesson 10-3 pp. 548–550

EXERCISES **1.** $m\angle 1 = 120$; $m\angle 2 = 60$; $m\angle 3 = 30$ **3.** $m\angle 7 = 60$; $m\angle 8 = 30$; $m\angle 9 = 60$ **5.** 2851.8 ft^2 **7.** 2475 $in.^2$ **9.** 2192.4 cm^2 **11.** 27.7 $in.^2$ **13.** 210 $in.^2$ **15.** $384\sqrt{3}$ $in.^2$ **17.** $75\sqrt{3}$ m^2 **19.a.** 72 **b.** 54 **21.a.** 40 **b.** 70 **23.** 73 cm^2 **25.a.** 9.1 in. **b.** 6 in. **c.** 3.7 in.

d. Answers may vary. Sample: About 4.6 in.; the length of a side of a pentagon should be between 3.7 in. and 6 in. **27.** The apothem is one leg of a rt. △, and the radius is the hypotenuse. **29.** $600\sqrt{3}$ m^2 **31.** 128 cm^2 **33.** $900\sqrt{3}$ m^2; 1558.8 m^2 **35.** The apothem is ⊥ to a side of the pentagon. Two right ⚠ are formed with the radii of the pentagon. So the ⚠ are ≅ by HL. Therefore, the ⚠ formed by the apothem and radii are ≅ by CPCTC, and the apothem bisects the vertex ∠. **37.a.** (2.8, 2.8) **b.** 5.6 units^2 **c.** 45 units^2 **39.** C **41.** B **43.** 30 in. **45.** △*DAB* and △*CBA* **47.** △*DFA* and △*CGB*

Lesson 10-4 pp. 555–558

EXERCISES **1.** 1:2; 1:4 **3.** 2:3; 4:9 **5.** 24 in.^{2} **7.** 59 ft^2 **9.** $384 **11.** 1:2; 1:2 **13.** 7:3; 7:3 **15.** 4:1; 4:1 **17.** 3:1; 9:1 **19.** 2:3; 4:9 **21.** 6:1; 36:1 **23.** While the ratio of lengths is 2:1, the ratio of the areas is 4 : 1. **25.** 252 m^2 **27.** $x = 2\sqrt{2}$ cm, $y = 3\sqrt{2}$ cm **29.** $x = \frac{8\sqrt{3}}{3}$ cm, $y = 4\sqrt{3}$ cm **31.** $x = 8$ cm, $y = 12$ cm **33.a.** Check students' work. **b.** Check students' work. **c.** Estimates may vary. Sample: 205 m^2 **35.a.** $\frac{5}{2}$ **b.** $\frac{25}{4}$ **37.a.** $\frac{2}{1}$ **b.** $\frac{4}{1}$ **39.a.** Answers may vary. Sample:

39 mm 25 mm 19 mm 50 mm

b. 114 mm; 475 mm^2; **c.** 456 yd; 7600 yd^2

41. Always; ~ rectangles with = perimeters have a similarity ratio of 1, so they are ≅. **43.** Never; if they were ≅, both measures would be the same. If they were ~, but not ≅, their areas would not be =. **45.** D **47.** 50 cm^2 **49.** 480 units^2 **51.** $x = 45$; $y = 107.5$ **53.** $y = 3x - 4$; **55.** $y = \frac{3}{4}x$;

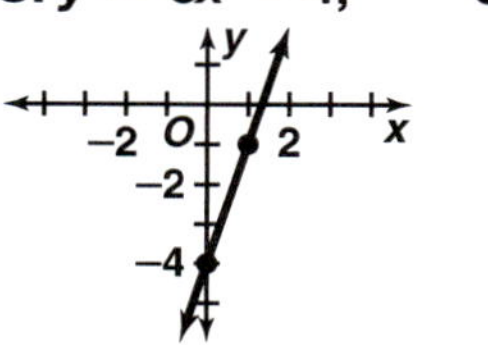

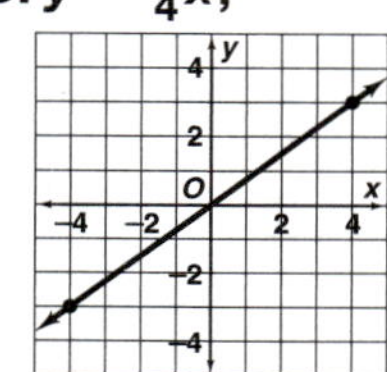

57. Eq. forms may vary. Sample is given. $y = -3x + 1$

Lesson 10-5 pp. 561–563

EXERCISES **1.** 173.8 cm^2 **3.** 259.8 m^2 **5.a.** 72 **b.** 36 **c.** about 8.1 in. **d.** about 11.8 in. **e.** about 58.8 in. **f.** about 238 in.^{2} **7.** 47.0 in.^{2} **9.** 8 ft^2 **11.** 27.7 m^2 **13.** 7554.0 m^2 **15.** 151.4 mm^2 **17.** 5523 yd^2 **19.** Answers may vary. Sample: **1.** Find the apothem and the side ⊥ to the apothem using a 30-60-90 △ with hyp. 1. Then use the formula $A = \frac{1}{2}ap$. **2.** After finding the apothem and the ⊥ side, the height of the equil. △ is the apothem + 1. Then use the formula $A = \frac{1}{2}bh$. **21.** 20.8 m; 20.8 m^2 **23.** 17.6 ft; 21.4 ft^2 **25.** 61.2 m; 282.8 m^2 **27.** D **29.** (area of pent. A) ≈ 1.53 · (area of pent. B) **31.** (area of oct. B) ≈ 1.17 · (area of oct. A) **33.** 5.0 ft^2 **35.** Using steps similar to Ex. 34, $A = n\left(\cos\frac{180}{n}\right)\left(\sin\frac{180}{n}\right)$, which also appr. π as *n* increases. **37.** (0.65) **39.** C **41.** (−2, −9) **43.** (−6, −1) **45.** $(\frac{a}{2}, 0)$; $\sqrt{a^2 + 4b^2}$

Lesson 10-6 pp. 569–573

EXERCISES **1.** 18 **3.** 40 **5.** 43 **7.** 40 **9–13.** Answers may vary. Samples are given. **9.** $\overset{\frown}{ED}$ **11.** $\overset{\frown}{BFE}$ **13.** ∠*FOE* **15.** 128 **17.** 218 **19.** 52 **21.** 180 **23.** 232 **25.** 142 **27.** 20π cm **29.** 8.4π m **31.** π m **33.** 25 in. **35.** 8π ft **37.** 33π in. **39.** $\frac{5\pi}{4}$ m **41.**

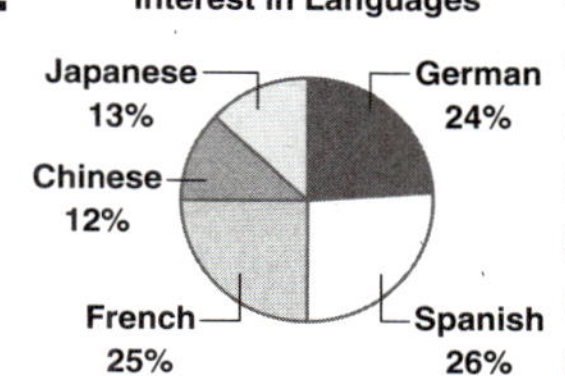

43. 180 **45.** 55 **47.** 290 **49.a.** 6 **b.** 30 **c.** 120 **51.** 100 **53.** 40 **55.** 50 in. **57.** B **59.** 105 ft **61.** (2.5, 5) **63.** 5.125π ft **65.** 3π m **67.** Outside; a point on the outside travels farther in the same time, so it goes faster. **69.** 12.6 units **71.** 2π in.; assumptions may vary. **73.** B **75.** D **77.** 18.6 mm **79.** No; it could be an isosc. trap. **81.** Yes; if one pair of sides is both ≅ and ∥, it is a ▱. **83.** never

Algebra 1 Review pp. 574

1. B **3.** B **5.** B **7.** 10 **9.** 51 **11.** 0.035 **13.** 230 **15.** 5 **17.** 30,000,000,000 **19.** $\frac{5}{18}$ **21.** 108 **23.** $\frac{125}{648}$ **25.** 2,000,000,000,000

Lesson 10-7 pp. 577–579

EXERCISES **1.** 9π m^2 **3.** 0.7225π ft^2 **5.** about 86,394 ft^2 **7.** 40.5π yd^2 **9.** $\frac{169\pi}{6}$ m^2 **11.** 12π ft^2 **13.** $\frac{25\pi}{4}$ m^2 **15.** 24π in.^{2} **17.** 22.1 cm^2 **19.** 3.3 m^2 **21.** 120.4 cm^2 **23.** $(54\pi + 20.25\sqrt{3})$ cm^2 **25.** $(4 - \pi)$ ft^2 **27.** $(784 - 196\pi)$ in.^{2} **29.** Lower outside; the lower inside and top pieces have base areas 8π in.^{2}, but the lower outside pieces have base areas 8.75π in.^{2}. **31.** 15.7 in.^{2} **33.** Answers may vary. Sample: 8 cm radius; 90° arc

8 cm

35. $(49\pi - 73.5\sqrt{3})$ m^2 **37.** 4π m^2 **39.** $\left(\frac{200\pi}{3} - 50\sqrt{3}\right)$ units^2 **41.** B **43.** 10π cm **45.** 28π in. **47.** Show that △*BCE* ≅ △*ACD* by ASA. Then use $\overline{CE} \cong \overline{CD}$ and segment subtraction to show that $\overline{BD} \cong \overline{AE}$. Now with vertical ⚠ ∠*BFD* ≅ ∠*AFE*, △*BDF* ≅ △*AEF* by ASA.

Lesson 10-8 pp. 584–587

EXERCISES **1.** $\frac{1}{2}$ **3.** $\frac{3}{5}$ **5.** 1 **7.** $\frac{2}{5}$ or 40%

0 1 2 3 4 5 6 7 8 9 10 **9.** $\frac{4}{15}$ or about 27%

11. $\frac{1}{3}$ or about 33% **13.** $\frac{5}{8}$ or 62.5% **15.** $\frac{1}{4}$ or 25% **17.** $\frac{2}{5}$ or 40% **19.** $\frac{\pi}{2+\pi}$ or about 61% **21.** 4% **23.** $\frac{\pi}{4}$ **25.** $\frac{\pi}{4}$ **27.a.** 14 prizes **b.** $110 **29.** 36 s

31.a. 0 10 20 30 40 50 60

If it starts after 45 min, you cannot erase 15 min of a 60 min tape.

b. 14 21 0 10 20 30 40 50 60 $\frac{7}{45}$ or about 16%

33. $\frac{3}{10}$ **35.** 0 **37.** 1 **39.** $\frac{\sqrt{10}-\sqrt{2}}{10\pi} \approx 0.06$ **41.** about 36% **43.** about 46% **45.a–b.** Check students' work. **47.** $\frac{24}{49} \approx 49\%$; the probability is the same. **49.** B **51.** 12π cm^2 **53.** $x = 36$; $y = 144$ **55.a.** $D(3, 1)$; $E(1, 4)$ **b.** slope $\overline{DE} = -\frac{3}{2}$; slope $\overline{AC} = -\frac{3}{2}$ **c.** $\overline{DE}$ and $\overline{AC}$ have the same slope **d.** $DE = \sqrt{13}$; $AC = 2\sqrt{13}$ **e.** $\sqrt{13} = \frac{1}{2} \cdot 2\sqrt{13}$

Chapter Review pp. 589–591

1. base **2.** sector **3.** diameter **4.** apothem **5.** adjacent arcs **6.** 10 m^2 **7.** 90 in.^{2} **8.** 33 ft^2 **9.** $96\sqrt{3}$ mm^2 **10.** 96 ft^2 **11.** 117 cm^2

12. 20.8 in.^{2} **13.** 128 mm^2

4 in.

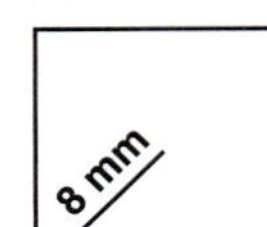

14. 127.3 cm^2 **15.** 4 : 9 **16.** 9 : 4

17. 1 : 4 **18.** $2\sqrt{2} : 5$ **19.** 73.5 ft^2 **20.** 232.5 cm^2 **21.** 124.7 in.^{2} **22.** 8 m^2 **23.** 100.8 cm^2 **24.** 88.4 ft^2 **25.** 70.4 m^2 **26.** 30 **27.** 120 **28.** 330 **29.** 120 **30.** $\frac{22}{9}\pi$ in.

7cm

31. π mm **32.** 18.3 m^2 **33.** 41.0 cm^2 **34.** $\frac{1}{2}$ or 50% **35.** $\frac{3}{8}$ or 37.5% **36.** $\frac{1}{6}$ or about 16.7%

Chapter 11

Lesson 11-1 pp. 601–603

EXER[illegible]SES **1.** 4, 6, 4; M, N, O, P; $\overline{MN}$, $\overline{MP}$, $\overline{MO}$, [illegible] $\overline{OP}$; $\triangle MNP$, $\triangle MOP$, $\triangle MNO$, $\triangle PNO$ [illegible]7; $P, Q, R, S, T, U, V, W, X, Y$; $\overline{UV}$, $\overline{VX}$, $\overline{XY}$, [illegible]$\overline{P}$, $\overline{VQ}$, $\overline{XS}$, $\overline{YT}$, $\overline{WR}$, $\overline{PQ}$, $\overline{QS}$, $\overline{ST}$, $\overline{TR}$, $\overline{RP}$; [illegible]als $UPQV$, $VQSX$, $XSTY$, $YTRW$, $UWRP$ and pentagons $UVXYW$ and $PQSTR$ **5.** 12 **7.** 8 **9.** 9 **11.** 7 + 10 = 15 + 2;

13. two concentric circles **15.** rectangle

17. rectangle

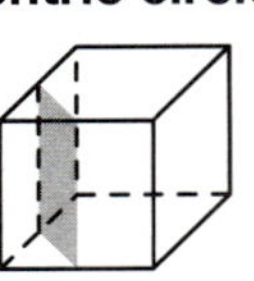

19. rectangle

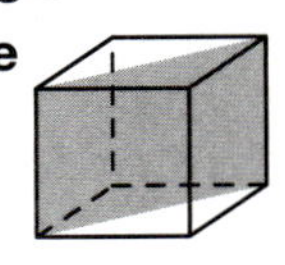

21. rectangle

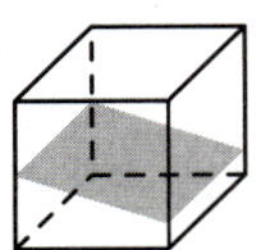

23. triangle

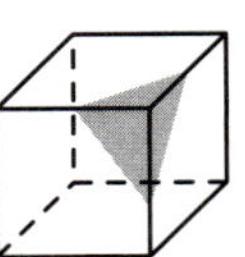

25. circle **27.** cone **29.** cylinder attached to a cone **31.** 18 + 32 = 48 + 2 **33.** 6 + 4 = 9 + 1 **35.** Check students' work. **37.** A **39–43.** Check students' work.

45.

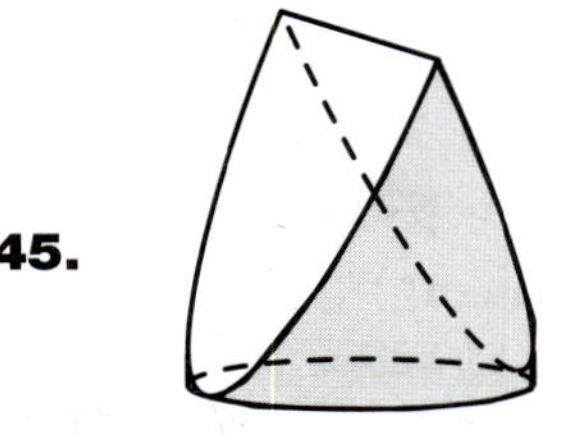

47.

49.

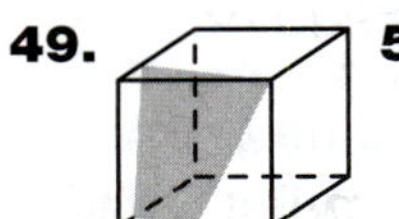

51.

53.

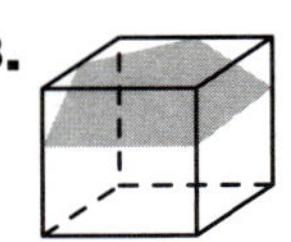

55. B **57.** B **59.** 25% **61.** $192\sqrt{3}$ cm^2 **63.** 4.7 **65.** 67 and 113

Extension p. 606

1. two-point **3.** one-point **5.** one-point

7.

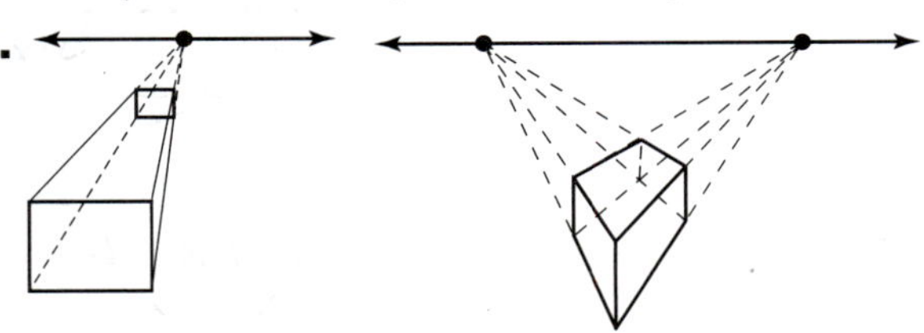

9.

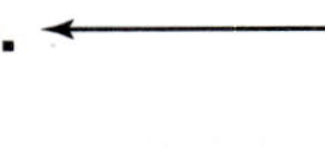

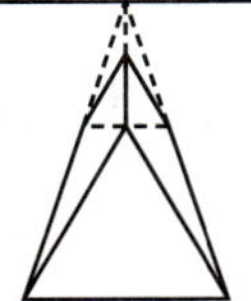

15. Answer may vary. Sample: The horizontal lines appear to be curved; the slanted lines that would meet at the vanishing pt. create a cylinder effect.

Algebra 1 Review — p. 607

1. $r = \frac{C}{2\pi}$ **3.** $r = \sqrt{\frac{A}{\pi}}$ **5.** $y = x \tan A$; $x = \frac{y}{\tan A}$ **7.** $C = 2\sqrt{\pi A}$ **9.** $a = \frac{\sqrt{6A\sqrt{3}}}{6}$ or $\frac{\sqrt{6A} \cdot \sqrt[4]{3}}{6}$

Lesson 11-2 — pp. 611–614

EXERCISES **1.** 1726 cm^2

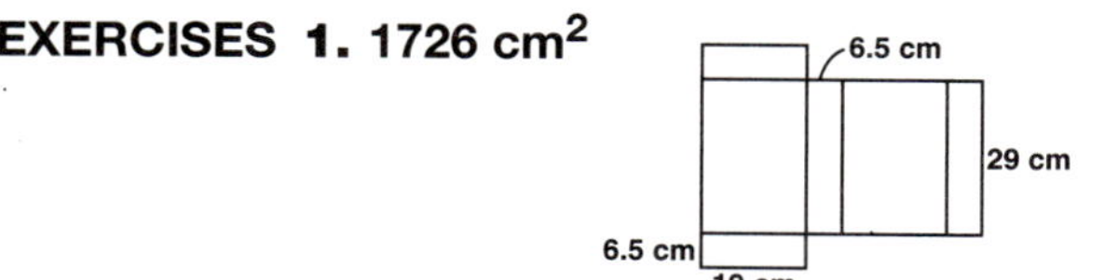

3. $(80 + 32\sqrt{2})$ in.^{2} or about 125.3 in.^{2}

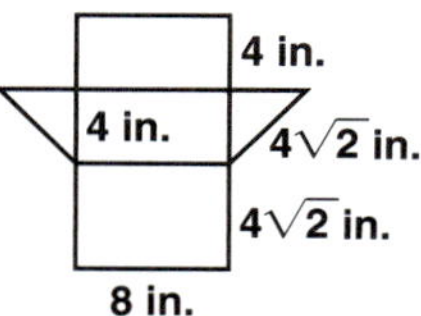

5. 120 ft^2; 220 ft^2 **7.** 880 cm^2; 1121 cm^2 **9.** 16.5π cm^2 **11.** 36.8 cm^2 **13.** 107 in.^{2} **15.** 1407 cm^2 **17.** 150 cm^2 **19.** 4080 mm^2 **21.a.** 94 units^2 **b.** 376 units^2 **c.** 4 : 1 **d.** 438 units^2; 1752 units^2; 4 : 1 e. The surface area becomes 4 times as large. **23.** 47.5 in.^{2} **25.a.** 7 units **b.** 196π units^2 **27.** cylinder of radius 4 and height 2; 48π units^2 **29.** cylinder of radius 2 and height 4; 24π units^2 **31.a.** Lateral area is doubled. **b.** Surface area is more than doubled. **c.** S.A. = $2\pi r^2 + 2\pi rh$; if r doubles: S.A. = $2(4\pi r^2 + 2\pi rh)$. Since r is squared, surface area is more than doubled. **33.** $(148 + 66.5\pi)$ cm^2 **35.** $(220 - 8\pi)$ in.^{2} **37.a.** 0, 8, 12, 6, 1 **b.** 1728 in.^{2} **39.** D **41.** Answers may vary. Sample:

3 cm
4 cm
4 cm
5 cm
3 cm

43. 37.7 cm^2 **45. a.** $R(c, b)$, $S(c, -b)$, $P(a, -b)$, $Q(a, b)$ **b.** $a - c$; $a - c$; $2b$; $2b$ **c.** Opposite sides are ≅.

Lesson 11-3 — pp. 620–622

EXERCISES **1.** 408 in.^{2} **3.** 179 in.^{2} **5.** 354 cm^2 **7.** 834,308 ft^2 **9.** 33π ft^2 **11.** 1044 in^2 **13.** 47 cm^2 **15.** Altitude; altitude; the altitude is shorter because it is one leg of a right △ with the lateral edge as the hyp., and is steeper because it rises the same vert. distance over less horiz. distance. **17.** 8 ft **19.** 62 cm^2 **21.** 4 in. **23.** 1580.6 ft^2 **25.** $(\ell + r)r\pi = \pi r\ell + \pi r^2$, which is the lateral area plus the circular base area. This formula may require fewer keystrokes and doesn't use exponents. **27.** A **29.** 471 ft^2 **31.** L.A. = 30 in.^{2}, $h \approx 4.8$ in.; $\ell = 5$ in. **33.** $s = 2$ ft, $h \approx 4.9$ ft, S.A. = 24 ft^2 **35.** 21.2 cm **37.** cone with $r = 4$ and $h = 3$; 36π **39.** cylinder with cone-shaped hole; 60π **41.** $100\sqrt{5}$ cm^2; $100\sqrt{5} + 100$ cm^2 **43.** 129.6 **45.** B **47.** 76 ft^2 **49.** 34 cm **51.** about 281.7 m

Lesson 11-4 — pp. 627–630

EXERCISES **1.** 216 ft^3 **3.** 180 m^3 **5.** about 280.6 cm^3 **7.** 720 mm^3 **9.** 288π in.^{3}, 904.8 in.^{3} **11.** 37.5π m^3, 117.8 m^3 **13.** 3445 in.^{3} **15.** 501 in.^{3} **17.** $\frac{26}{9}$ cm **19.** 6 ft **21.** 28–42 pots **23.a.** 809,137 ft^3 **b.** 1,398,188,736 in.^{3} **c.** 6,052,765 gal **25.** 80 units^3 **27.** 3 cm **29.** Bulk; cost of bags ≈ \$1167, cost of bulk is ≈ \$1161. **31.** cylinder with $r = 4$ and $h = 2$; 32π units^3 **33.** cylinder with $r = 5$, $h = 2$, and a hole of radius 1; 48π units^3 **35.** 140.6 in.^{3} **37.a.** circumference $8\frac{1}{2}$ in. and height 11 in.: $V \approx 63.2$ in.^{3}; circumference 11 in. and height $8\frac{1}{2}$ in.: $V \approx 81.8$ in.^{3}; one is about 0.8 times the volume of the other. **b.** about 6.5 in. by 13.0 in. **39.** 4 units **41.** C **43.** B **45.** 469.2 ft^2 **47.** 25; 160, 105, 95 **49.** 76 ft

Lesson 11-5 — pp. 634–636

EXERCISES **1.** about 233,333 ft^3 **3.** 1296 in.^{3} **5.** about 443.7 cm^3 **7.** 2048 m^3 **9.** about 3714.5 mm^3 **11.** $\frac{16}{3}\pi$ ft^3; 17 ft^3 **13.** 36.75π in.^{3}; 115 in.^{3} **15.** about 4.7 cm^3 **17.** 312 cm^3 **19.** They are equal; both volumes are $\frac{1}{3}\pi r^2 h$. **21.** 6 **23.** $3\sqrt{2}$ **25.** B **27.a.** 120π ft^3 **b.** 60π ft^3 **c.** 240π ft^3 **29.** cone with $r = 4$ and $h = 3$; 16π **31.** cylinder with $r = 4$, $h = 3$, with a cone of $r = 4$, $h = 3$ removed from it; 32π **33.a.** The frustum has vol. $V = \frac{1}{3}\pi R^2 H - \frac{1}{3}\pi r^2 h = \frac{1}{3}\pi(R^2 H - r^2 h)$. Now if $h_1 = H - h$ is the frustum's height, $V = \frac{1}{3}\pi(R^2(h_1 + h) - r^2 h) = \frac{1}{3}\pi(R^2 h_1 + h(R^2 - r^2))$. By similar △s, $\frac{h}{r} = \frac{h_1 + h}{R}$, or $h = \frac{rh_1}{R - r}$. Simplifying, $V = \frac{1}{3}\pi h_1(r^2 + rR + R^2)$. **b.** about 784.6 in.^{3} **35.** about 16.2 cm **37.** B **39.** B **41.** 3600 cm^3 **43.** 62.4 cm^2 **45.** 13.9 cm^2

Lesson 11-6 — pp. 640–643

EXERCISES **1.** 900π m^2 **3.** 1024π mm^2 **5.** 4624π mm^2 **7.** $\frac{121}{16}\pi$ in.^{2} **9.** 232 in.^{2} **11.** 154 in.^{2}

13. 288π cm^3; 905 cm^3 **15.** $\frac{2048}{3}\pi$ cm^3; 2145 cm^3 **17.** 98.784π m^3; 310 m^3 **19.** 1006 m^2 **21.** S.A. $\approx$ 108 cm^2; $V \approx$ 108 cm^3 **23.** Yes; the volume of the frozen yogurt is $\frac{256}{3}\pi$ cm^3, and the volume of the cone is 64π cm^3. **25.** Answers may vary. Sample: (5, 0, 0), (0, 5, 0), (0, 0, 5), (−5, 0, 0), (0, −5, 0), (0, 0, −5) **27.** 1.7 lb **29.** $\frac{4}{3}\pi$ m^3 **31.** $\frac{9}{2}\pi$ ft^3 **33.** $\frac{125}{6}\pi$ yd^3 **35.** $\frac{343}{6}\pi$ m^3 **37.a.** $457\frac{1}{3}\pi$ in.^{3} **b.** $228\frac{2}{3}\pi$ in.^{3} **c.** 11 in. **39.** Answers may vary. Sample: sphere radius 3 in.; cylinder radius 3 in., height 4 in. **41.** 26π cm^2; $\frac{62}{3}\pi$ cm^3 **43.** Answers may vary. Sample: You could lift the small ball because it weighs about 75 lb. The big ball would be much harder to lift since it weighs about 253 lb. **45.** $r \approx 7.1$ cm, $h \approx 14.1$ cm **47.** 707 cm^2 **49.** 3 m **51.** 2 : 3 **53.** C **55.** D **57.** B **59.** 19 in.^{3} **61.** 10.5 cm **63.** 67.5 cm^2 **65.** 109, 109, 71, 71

Lesson 11-7 pp. 648–651

EXERCISES **1.** no **3.** yes; 2 : 3 **5.** yes; 2 : 3 **7.** 5 : 6 **9.** 3 : 4 **11.** 240 in.^{3} **13.** 24 ft^3 **15.** 112 m^2 **17.** 6000 toothpicks **19.a.** It is 64 times the smaller prism. **b.** It is 64 times the smaller prism. **21.** No; explanations may vary. Sample: If "size" refers to the vol., then the new clock should be at $\frac{1}{10}$ the weight. **23.** about 1000 cm^3 **25.** Answers may vary. Sample: A sphere has only one measure, r, so there's only one possible ratio. **27.a.** 3 : 1 **b.** 9 : 1 **29.** 864 in.^{3} **31.** 9 : 25; 27 : 125 **33.** 5 : 8; 25 : 64 **35.a.** 100 times **b.** 1000 times **c.** His weight is 1000 times the weight of an average person, but his bones can only support 600 times the weight. **37.a.** 4 : 1; 8 : 1 **b.** Let r = radius, ℓ = slant height of small cone. $3\ell + 5r : 4\ell + 4r$; $3\ell + 5r : \ell + r$ **c.** 7 : 8; 7 : 1 **39.** B **41.** 1790 cm^2 and 1937 cm^2 **43.** 8.2 m^3; 19.6 m^2 **45.a.** $8\sqrt{3}$ mm **b.** $4\sqrt{21}$ mm **c.** $8\sqrt{7}$ mm **47.** 13.125

Chapter Review pp. 653–655

1. sphere **2.** altitude **3.** pyramid **4.** cross section **5.** right

6.

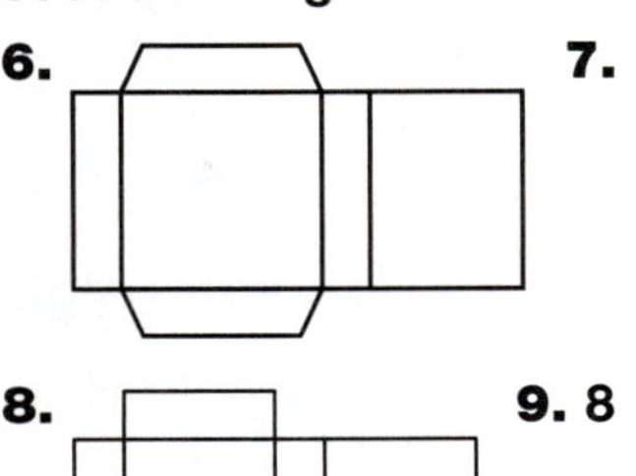

7.

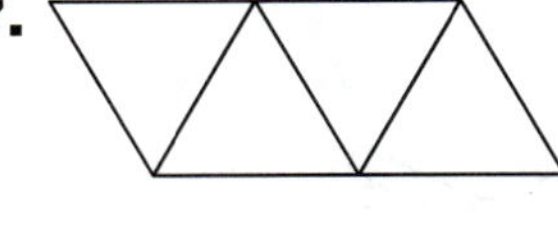

8.

9. 8 **10.** 8

11.

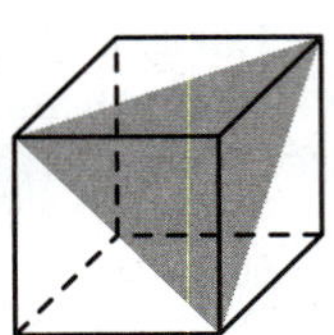

12. 36 cm^2; 12 cm^3 **13.** 66π m^2; 72π m^3 **14.** 208 in.^{2}; 192 in.^{3} **15.** 186.0 ft^2; 167.6 ft^3 **16.** 576 m^2; 512 m^3 **17.** 50.3 in.^{2}; 23.7 in.^{3} **18.** 391.6 in.^{2}; 443.7 in.^{3} **19.** 282.8 yd^2; 410.5 yd^3 **20.** 36.5 in.^{2}; 13.9 in.^{3} **21.** $B = \frac{S.A. - L.A.}{2}$. **22.** $r = -\frac{1}{2}h + \sqrt{\frac{1}{4}h^2 + \frac{S.A.}{2\pi}}$ **23.** 314.2 in.^{2}; 523.6 in.^{3} **24.** 153.9 cm^2; 179.6 cm^3 **25.** 50.3 ft^2; 33.5 ft^3 **26.** 8.0 ft^2; 2.1 ft^3 **27.** 8.6 in.^{3} **28.** Answers may vary. Sample:

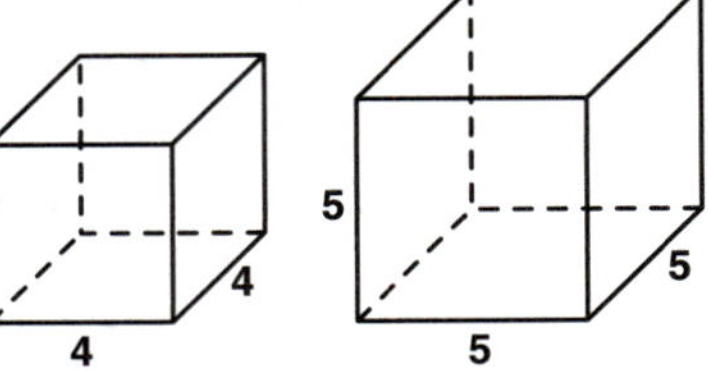

29. 27 : 64 **30.** 64 : 27

Chapter 12

Lesson 12-1 pp. 665–668

EXERCISES **1.** 120 **3.** 30 **5.** Extend $\overline{RS}$ and $\overline{QP}$ until they meet at a point, H. By Thm. 11-3, $RH = QH$, or $SH + RS = QP + PH$. By 11-3 again, $SH = PH$. Thus, $RS = QP$. **7.** 20.0 in. **9.** Yes; $2.5^2 + 6^2 = 6.5^2$ **11.** 78 cm **13.** 13 **15.** 8 in. **17.** 35.8 km **19.** 113.1 km

21.a.

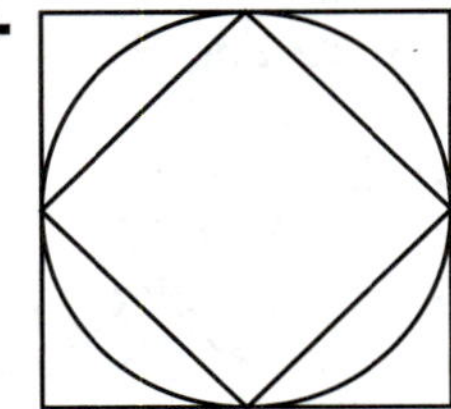

b. Answers may vary. Sample: If you draw diagonals in the small square, 8 ≅ △s are formed in the large square with 4 in the small square.

23. All four are ≅; the two tangents to each coin from A are ≅, so by the Trans. Prop., all are ≅. **25.** 35 **27.**

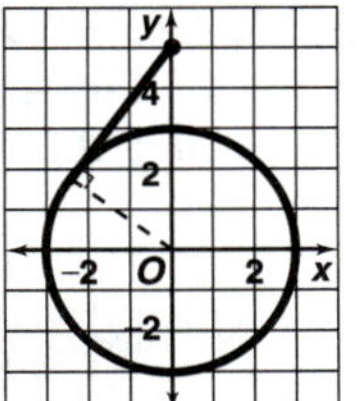

4 units

29.a. ⊥; **b.** LK; **c.** SAS; **d.** CPCTC; **e.** tangent; **f.** false **31.1.** $\overline{BC}$ is tangent to ⊙A at D. (Given) **2.** $\overline{DB} \cong \overline{DC}$ (Given) **3.** $\overline{AD} \perp \overline{BC}$ (If a line is tan. to a circle, it is ⊥ to the radius.) **4.** $\triangle ADB$ and $\triangle ADC$ are rt. △s (Def. of rt. △) **5.** $\overline{AD} \cong \overline{AD}$ (Refl. Prop. of ≅) **6.** $\triangle ADB \cong ADC$ (SAS) **7.** $\overline{AB} \cong \overline{AC}$ (CPCTC) **33.** Assume $\overleftrightarrow{AB}$ is not tangent to ⊙O.

Then either $\overleftrightarrow{AB}$ does not intersect $\odot O$ or $\overleftrightarrow{AB}$ intersects $\odot O$ at two pts. If $\overleftrightarrow{AB}$ does not intersect $\odot O$, then P is not on $\odot O$, which contradicts $\overline{OP}$ being a radius. If $\overleftrightarrow{AB}$ intersects $\odot O$ at two pts., P and Q, then $\overline{OP} \cong \overline{OQ}$ ($\cong$ radii), $\triangle OPQ$ is isosc., and $\angle OPQ \cong \angle OQP$. But $\angle OPQ$ is a rt. $\angle$, since $\overline{AB} \perp \overline{OP}$, and $\triangle OPQ$ has two rt. $\angle$s. This is a contradiction also, so $\overleftrightarrow{AB}$ is tangent to $\odot O$. **35.** A **37.** 3:4 **39.** 27:64 **41.** 28.1 **43.a.** 10 : 17 **b.** $m = 1.82$; $n = 3.78$

Lesson 12-2 — pp. 673–676

EXERCISES **1.** $\overset{\frown}{BC} \cong \overset{\frown}{YZ}$; $\overline{BC} \cong \overline{YZ}$ **3.** 14 **5.** 7 **7.** 8 **9.** Answers may vary. Samples are given. **a.** $\overline{CE}$ **b.** $\overline{DE}$ **c.** $\angle CEB$ **d.** $\angle DEA$ **11.** 6 **13.** 8.9 **15.** 9.9 **17.** 108 **19.** about 123.9 **21.** 12 cm **23.a.** $\overline{PL}$ **b.** $\overline{PM}$ **c.** All radii of a circle are $\cong$. **d.** $\triangle LPN$ **e.** SAS **f.** CPCTC **25.** $\triangle OAC \cong \triangle OBC$ by HL. $\overline{AC} \cong \overline{BC}$ since CPCTC. Also, $\angle AOC \cong \angle BOC$ (CPCTC), so $\overset{\frown}{AD} \cong \overset{\frown}{BD}$ since $\cong$ central $\angle$s intercept $\cong$ arcs. **27.** He doesn't know that the chords are equidistant from the center. **29.** If the chords, arcs, or central $\angle$s are in different circles and the circles have unequal radii, the theorems do not apply. **31.** 10 cm **33.** C **35.1.** $\odot P$ with $\overset{\frown}{QS} \cong \overset{\frown}{RT}$ (Given) **2.** $m\overset{\frown}{QS} = m\angle QPS$ and $m\overset{\frown}{RT} = m\angle RPT$ (Arc measure = central $\angle$ measure.) **3.** $m\overset{\frown}{QS} = m\overset{\frown}{RT}$ (Def. of $\cong$) **4.** $\angle QPS \cong \angle RPT$ (Subst.) **37.** C on the bisector of $\angle QPR$ means that C is equidistant from $\overline{PQ}$ and $\overline{PR}$, or $\overline{PQ}$ and $\overline{PR}$ are equidistant from C. Therefore, $PQ = PR$ since chords equidistant from the center of a circle are $\cong$. **39.1.** $\odot A$ with $\overline{CE} \perp \overline{BD}$ (Given) **2.** $\overline{CF} \cong \overline{CF}$ (Refl. Prop. of $\cong$) **3.** $\overline{BF} \cong \overline{FD}$ (A diameter $\perp$ to a chord bisects the chord.) **4.** $\angle CFB$ and $\angle CFD$ are rt. $\angle$s (Def. of $\perp$). **5.** $\triangle CFB \cong \triangle CFD$ (SAS) **6.** $\overline{BC} \cong \overline{CD}$ (CPCTC) **7.** $\overset{\frown}{BC} \cong \overset{\frown}{DC}$ ($\cong$ chords have $\cong$ arcs.) **41.** Let O be the center of the circles, and P be the pt. of tangency of the larger circle's chord to the smaller circle. Then $\overline{OP}$ is $\perp$ to the chord, and therefore bisects it. So P is the mdpt. of the chord. **43.** C **45.** C **47.** 40 **49.** about 42 ft

Lesson 12-3 — pp. 681–685

EXERCISES **1.** $\angle ACB$; $\overset{\frown}{AB}$ **3.** $\angle MPN$; $\overset{\frown}{MN}$ **5.** 58 **7.** $a = 218$; $b = 109$ **9.** $a = 112$; $b = 120$; $c = 38$ **11.** $x = 36$; $y = 36$ **13.** $a = 50$; $b = 90$; $c = 90$ **15.** $w = 123$ **17.** $e = 65$; $f = 130$ **19.** $a = 22$; $b = 78$; $c = 156$ **21.a.** 96 **b.** 55 **c.** 77 **d.** 154 **23.** B **25.** $\angle PQR \cong \angle QRS$ since they are alt. int. $\angle$s. Since $\cong$ inscribed $\angle$s intercept $\cong$ arcs, $m\overset{\frown}{PR} = m\overset{\frown}{QS}$. **27.a.** $77\frac{1}{7}$ **b.** 36 **29.** about 7.1 cm by 7.1 cm **31.** about 7.1 cm legs, and a 10 cm base **33.** $\angle ACB$ is a rt. $\angle$ because it is inscribed in semicircle $\overset{\frown}{ACB}$, and if a line is $\perp$ to a radius at its endpoint, it is tangent to the circle.

35.

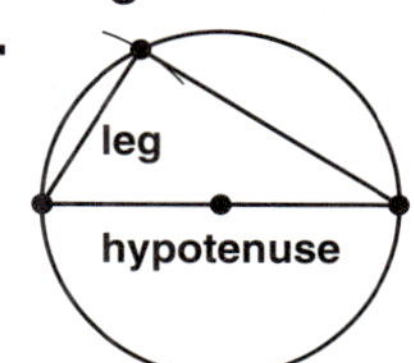

37.1. $\odot S$ with inscribed $\angle PQR$ (Given) **2.** $m\angle PQT = \frac{1}{2}m\overset{\frown}{PT}$ (Inscribed $\angle$ Thm., Case I) **3.** $m\angle RQT = \frac{1}{2}m\overset{\frown}{RT}$ (Inscr. $\angle$ Thm., Case I) **4.** $m\overset{\frown}{PR} = m\overset{\frown}{PT} - m\overset{\frown}{RT}$ (Arc Add. Post.) **5.** $m\angle PQR = m\angle PQT - m\angle RQT$ ($\angle$ Add. Post.) **6.** $m\angle PQR = \frac{1}{2}m\overset{\frown}{PT} - \frac{1}{2}m\overset{\frown}{RT}$ (Subst.) **7.** $m\angle PQR = \frac{1}{2}m\overset{\frown}{PR}$ (Subst.) **39.1.** $\odot O$ with inscribed $\angle CAB$ in a semicircle (Given) **2.** $m\angle CAB = \frac{1}{2}m\overset{\frown}{BDC}$ (Inscr. $\angle$ Thm.) **3.** $m\overset{\frown}{BDC} = 180$ (Meas. of semicircle = 180.) **4.** $m\angle CAB = 90$ (Subst.) **5.** $\angle CAB$ is a rt. $\angle$. (Def. of rt. $\angle$) **41.1.** $\overline{GH}$ and tangent ℓ intersecting at H on $\odot E$ (Given) **2.** Construct diameter $\overline{HD}$ intersecting circle E at D. (Constr.) **3.** $\angle DHI$ is a rt. $\angle$. (Tangent and radius are $\perp$.) **4.** $\overset{\frown}{DGH}$ is a semicircle of measure 180. (Def. of semicircle) **5.** $m\angle DHG + m\angle GHI = m\angle DHI$ ($\angle$ Add. Post.) **6.** $m\overset{\frown}{DG} + m\overset{\frown}{GFH} = m\overset{\frown}{DGH}$ (Arc Add. Post.) **7.** $90 = m\angle DHG + m\angle GHI$ (Subst.) **8.** $180 = m\overset{\frown}{DG} + m\overset{\frown}{GFH}$ (Subst.) **9.** $90 = \frac{1}{2}(m\overset{\frown}{DG} + m\overset{\frown}{GFH})$ (Div. Prop.) **10.** $m\angle DHG + m\angle GHI = \frac{1}{2}m\overset{\frown}{DG} + \frac{1}{2}m\overset{\frown}{GFH}$ (Subst. and Distr. Prop.) **11.** $m\angle DHG = \frac{1}{2}m\overset{\frown}{DG}$ (Inscr. $\angle$ Thm.) **12.** $m\angle GHI = \frac{1}{2}m\overset{\frown}{GFH}$ (Subtr. Prop.) **43.** True; the angle's intercepted arc has measure 180, so the angle is inscribed in a semicircle. **45.** In the construction below, $RS = x$, $ST = y$, O is the mdpt. of $\overline{RT}$, and $\overline{QS} \perp \overline{RT}$.

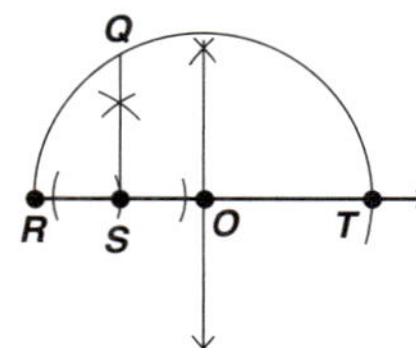

47. C **49.** C **51.** 34.6 **53.** 49.1 in.^{2} **55.** 49.9 m^2 **57.** about 0.18 mi

Lesson 12-4 — pp. 691–693

EXERCISES **1.** 46 **3.** $x = 60$; $y = 70$ **5.** $x = 115$; $y = 74$ **7.** 160° **9.** 15 **11.** 13.2 **13.** $x = 25.8$; $y = 12.4$ **15.** about 270.8 ft **17.** $360 - x$ **19.** $180 - y$ **21.** 16.7 **23.** $x = 10.7$; $y = 10$ **25.** $x = 10.9$; $y = 2.3$ **27.** 95, 104, 86, 75 **29.** Answers may vary. Sample: Since they are inscribed ∡: $m\angle BED = \frac{1}{2}m\widehat{BD}$ and $m\angle ABE = \frac{1}{2}m\widehat{AE}$. Apply the Ext. ∠ Thm. to $\triangle BCE$ to prove that $m\angle C = \frac{1}{2}(m\widehat{AE} - m\widehat{BD})$. **31.1.** Construct

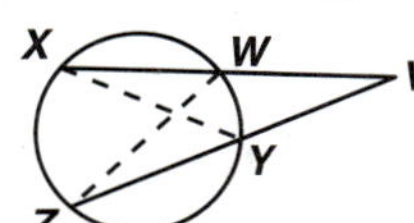

$\overline{XY}$ and $\overline{ZW}$. **2.** $\angle XVY \cong \angle ZVW$ (Reflexive Prop. of ≅) **3.** $\angle VXY \cong \angle WZV$ (2 inscribed ∡ that intercept the same arc are ≅.) **4.** $\triangle XVY \sim \triangle ZVW$ (AA~) **5.** $\frac{XV}{ZV} = \frac{YV}{WV}$ (In similar figures, corr. sides are proport.) **6.** $XV \cdot WV = YV \cdot ZV$ (Prop. of Proport.) **33.** Answers may vary. Sample: If the given pt. is on the circle, then a line through the given pt. can only intersect the circle tangentially or one other place. It follows that one segment has length zero, so the product of the segments is always zero.
35.1. $m\angle 1 = \frac{1}{2}m\widehat{QRP} - \frac{1}{2}m\widehat{PQ}$ ($\frac{1}{2}$ diff. of intercepted arcs) **2.** $m\angle 2 = \frac{1}{2}m\widehat{RQP} - \frac{1}{2}m\widehat{RP}$ ($\frac{1}{2}$ diff. of intercepted arcs) **3.** $m\angle 1 + m\angle 2 = \frac{1}{2}m\widehat{QRP} + \frac{1}{2}m\widehat{RQP} - \frac{1}{2}m\widehat{PQ} - \frac{1}{2}m\widehat{RP}$ (Subst.) **4.** $m\angle 1 + m\angle 2 = \frac{1}{2}m\widehat{QR} + \frac{1}{2}m\widehat{RP} + \frac{1}{2}m\widehat{QR} + \frac{1}{2}m\widehat{PQ} - \frac{1}{2}m\widehat{PQ} - \frac{1}{2}m\widehat{RP}$ (Arc Add. Post. and Distr. Prop.) **5.** $m\angle 1 + m\angle 2 = m\widehat{QR}$ (Distr. Prop.)

37.

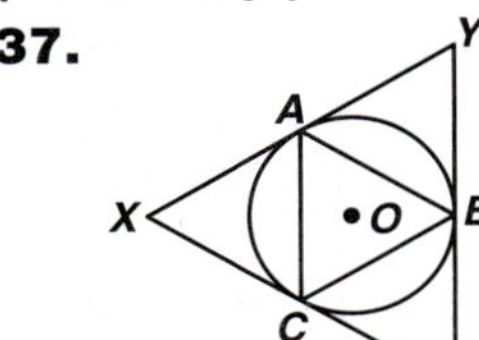

$m\widehat{AB} = m\widehat{BC} = m\widehat{AC} = 120$, since chords $\overline{AB}$, $\overline{BC}$, and $\overline{CA}$ are all ≅. So the measures of $\angle X$, $\angle Y$, and $\angle Z$ are $\frac{1}{2}(240 - 120) = 60$, and $\triangle XYZ$ is equilateral.

39. C **41.** $a = 55$; $b = 35$; $c = 30$ **43.** 30 **45.** 57

Extension p. 694

1. *ACD*; *HGA*; *EFA* **3.** The radius of the circle is 1, which is the denominator in each of the ratios. **5.** On the unit circle, secant $A = \frac{\text{hyp.}}{1} = \text{hyp.} =$ length of $\overline{DA}$, the secant segment.

7. $(\text{tangent } A)^2 = \left(\frac{DC}{CA}\right)^2 = \left(\frac{EB}{BA}\right)^2 = \frac{(EB)^2}{(BA)^2} = \frac{(EA)^2 - (BA)^2}{(BA)^2} = \frac{(EA)^2}{(BA)^2} - 1 = (\text{secant } A)^2 - 1$

9. $\text{cotangent } A = \frac{HG}{GA} = \frac{CA}{CD} = \frac{1}{\frac{CD}{CA}} = \frac{1}{\text{tangent } A}$

11. $\text{secant } A = \frac{EA}{AB} = \frac{EA}{EF} = \frac{1}{\frac{EF}{EA}} = \frac{1}{\text{cosine } A}$

13. Consider the possible lengths of each segment in the diagrams: cosecant *A* and secant *A* can be any value > 1; cotangent *A* can be any value > 0.

Lesson 12-5 pp. 697–700

EXERCISES **1.** $(x - 2)^2 + (y + 8)^2 = 81$ **3.** $(x - 0.2)^2 + (y - 1.1)^2 = 0.16$ **5.** $(x + 6)^2 + (y - 3)^2 = 64$ **7.** $x^2 + y^2 = 16$ **9.** $(x + 1)^2 + (y + 1)^2 = 1$ **11.** $(x - 1)^2 + (y - 2)^2 = 17$ **13.** $(x + 10)^2 + (y + 5)^2 = 125$ **15.** $(x + 1)^2 + (y + 4)^2 = 25$

17. center: (3, −8); radius: 10

19. center: (0, 0); radius: 6

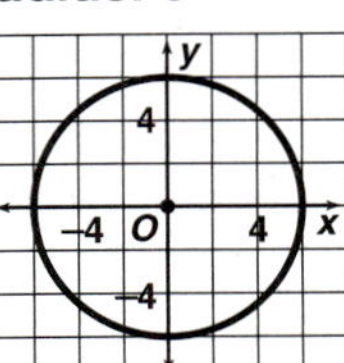

21. center: (−5, −2); radius: $4\sqrt{3}$

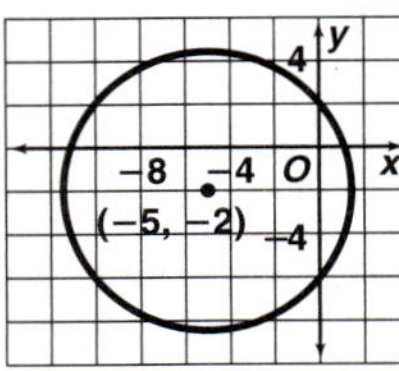

23. $(x - 4)^2 + (y + 4)^2 = 4$ **25.** position: (5, 7); range: 9 units **27.** $x^2 + y^2 = 4$ **29.** $x^2 + (y - 3)^2 = 4$ **31.** $(x - 2)^2 + (y - 2)^2 = 16$ **33.** $(x - 4)^2 + (y - 3)^2 = 25$ **35.** $(x - 3)^2 + (y - 3)^2 = 8$ **37.** $(x + 1.5)^2 + (y - 5)^2 = 18.25$ **39.** $x^2 + y^2 = 1$ **41.** Check students' work. **43.** No; the *x* and *y* terms are not squared. **45.** circumference: 16π; area: 64π **47.** *x*-int. = 13, *y*-int. = $\frac{39}{4}$

49.

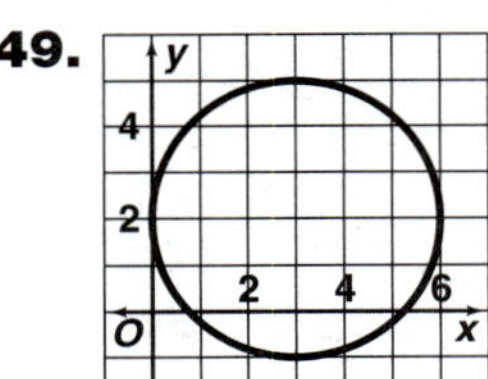

51.

53. (3, 2); (2, 3)

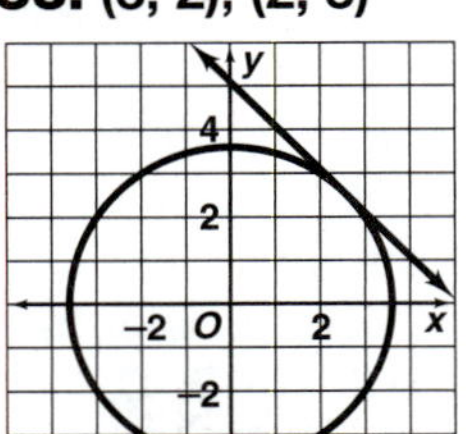

55. (2, 2); (−2, 2)

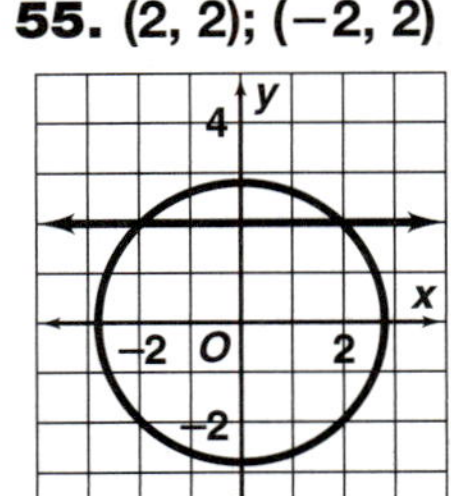

57. (−4, 4)

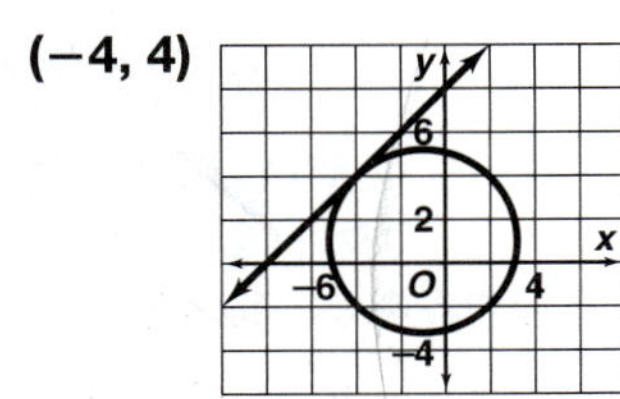

59–60. Explanations may vary. Sample: Solve the circle and line eqs. for *y*, enter in the calc., and use the zooming feature. **61.** Answers may vary. Sample: Lines can appear tangent on a graph, but may not be. **63.a.** $x^2 + y^2 = 15{,}681{,}600$

b. 69.1 mi **c.** 1.2 mi **d.** about 32 days **65.** C **67.** A **69.** 38 **71.** (−5, 2) **73.** (11, −7) **75.** $9\sqrt{3}$ **77.** 3 **79.** $\frac{3}{4}$

Lesson 12-6 pp. 703–705

EXERCISES **1.** a circle of radius 4 cm with center *X*

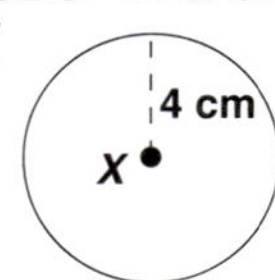

3. two distinct lines ∥ to $\overleftrightarrow{LM}$ and 3 mm from $\overleftrightarrow{LM}$

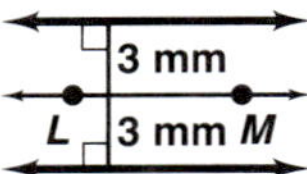

5. a line ⊥ to $\overline{PQ}$ and through the midpoint of $\overline{PQ}$

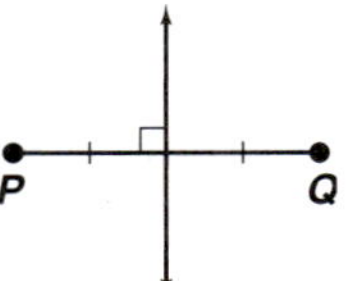

7. two ⊥ lines, meeting at the intersection of the orig. lines, making 45° ∡s with orig. lines

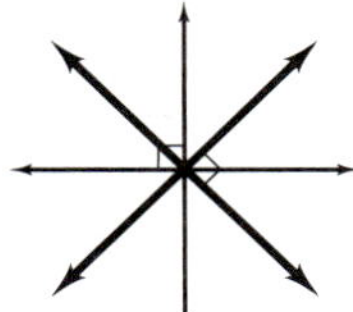

9. the single pt. *L*

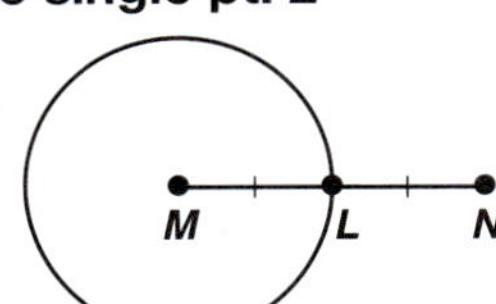

11. the single pt. *N*

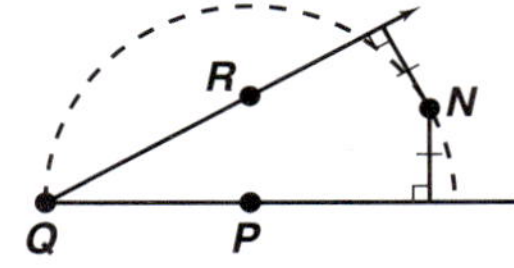

13. the pts. *B* and *D*

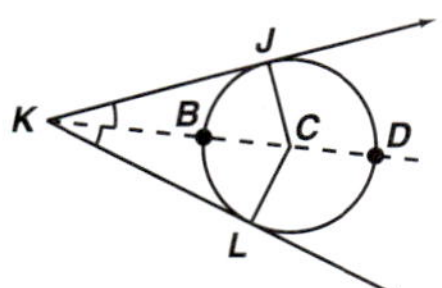

15. an endless cylinder with radius 4 cm and center-line $\overleftrightarrow{DE}$

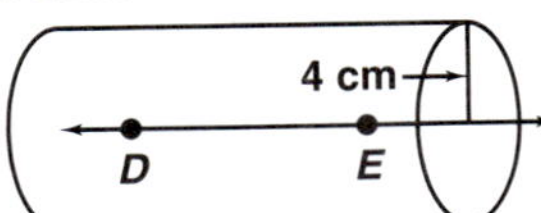

17. an endless cylinder with radius 5 mm and center-ray $\overrightarrow{PQ}$ and a hemisphere of radius 5 mm centered at *P*

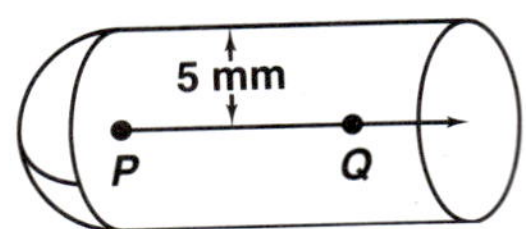

19. the set of all points 2 units from the origin **21.** Check students' work. **23.** $y = x$ **25.** $y = -x + 3$

27. **29.**

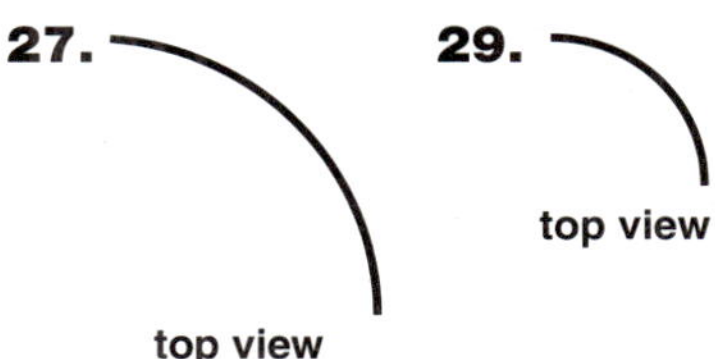

31. Draw a circle with radius 3 mi and center at downtown area. Connect Jack's and Julie's offices with a segment and construct the ⊥ bis. Locations will lie on the ⊥ bis. and on or inside the circle.

33. **35.**

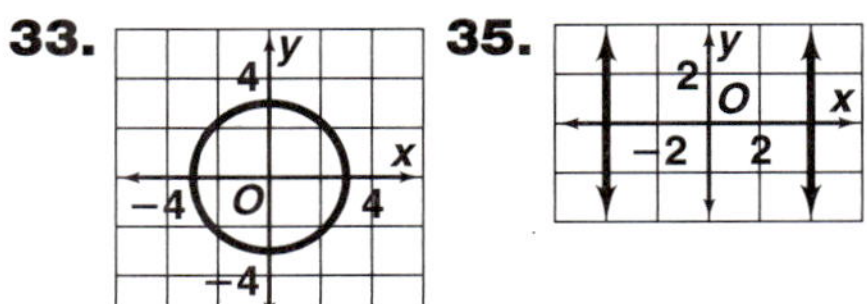

37. **39.**

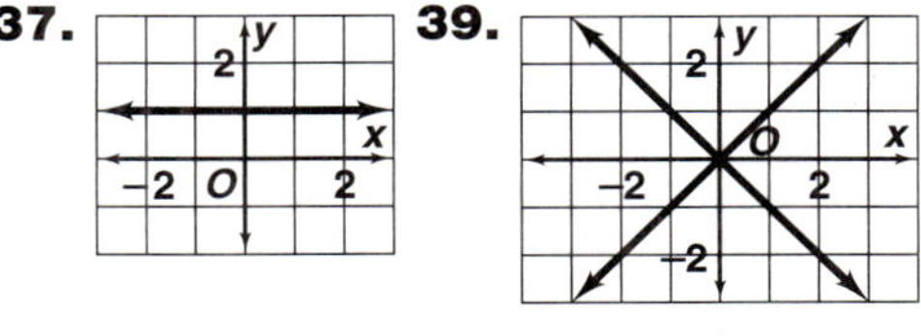

41. a circle **43.a.**

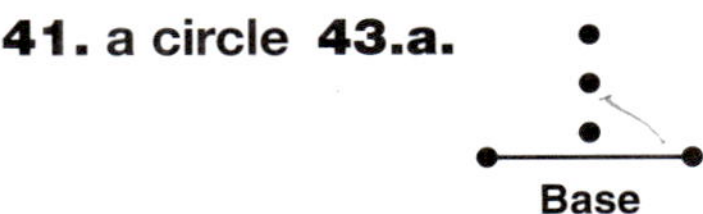

b. ⊥ bis. of the base, except the pt. on the base **c.** The vertex must be equidist. from the endpoints of the base. These points lie only on the ⊥ bis.

45.

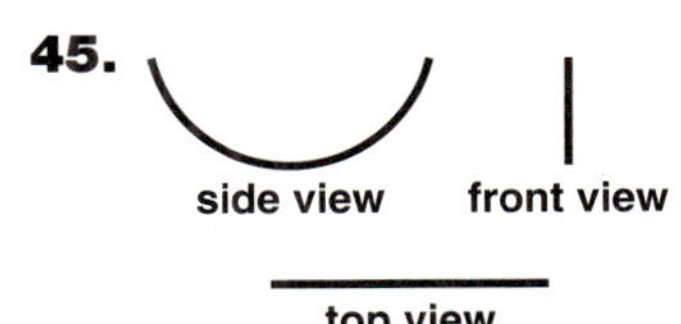

47. **49.**

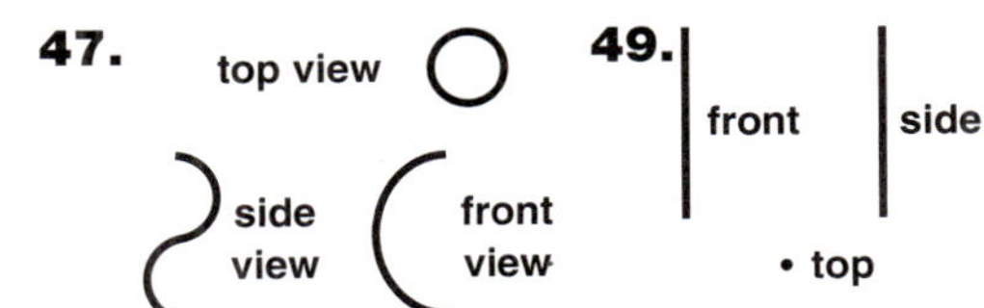

51. D **53.** B **55.** $(x - 1)^2 + (y - 7)^2 = 36$ **57.** 510 in.^{2} **59.** 4π **61.** 10π

Chapter Review pp. 707–709

1. tangent to **2.** chord **3.** inscribed in **4.** inscribed angle **5.** locus **6.** 57 in. **7.** 72 mm **8.** 9.8 cm **9.** 4.3 **10.** 19.5 **11.** 6.4 **12.** 4.5 **13.** $a = 40$; $b = 140$; $c = 90$ **14.** $a = 118$; $b = 49$; $= 144$; $d = 98$ **15.** $a = 90$; $b = 90$; $c = 70$; $d = 65$

and a rhombus **13.** Yes; $\triangle XZY \sim \triangle EWN$ by AA~. **15.** Yes; $\triangle ABC \sim \triangle EBD$ by AA~. **17.** Yes; $\triangle HJK \sim \triangle RST$ by SAS~. **19.** $x = \sqrt{5}$; $y = 2$; $z = 2\sqrt{5}$ **21.** $x = 65$; $y = 60$; $z = 156$ **23.** Place $\triangle ABC$ in the coordinate plane with $A(-a, 0)$, $B(b, 0)$, $C(0, \sqrt{ab})$ (given), and $D(0, 0)$. Slope of $\overleftrightarrow{AC} = \frac{\sqrt{ab}}{a}$. Slope of $\overleftrightarrow{BC} = \frac{\sqrt{ab}}{-b}$. The product $\frac{\sqrt{ab}}{a} \cdot \frac{\sqrt{ab}}{-b} = \frac{ab}{-ab}$ -1, so $\overleftrightarrow{AC} \perp \overleftrightarrow{BC}$. **25.** $\frac{72}{5}$ **27.** 20

29. Construct $\overline{AC}$ with $AB = x$ and $BC = 1$. On another line from A, construct $\overline{AD}$ of length 1. Construct a line through $C \parallel$ to $\overline{BD}$ and intersecting $\overline{AD}$ in E. By the Side-Splitter Thm., $\frac{x}{1} = \frac{1}{DE}$, so $DE = \frac{1}{x}$.

Chapter 8

1. 15 **3.** $3\sqrt{5}$ **5.** 188 ft **7.** 13 cm, $13\sqrt{3}$ cm **9.** $\sqrt{2}$ units, $\sqrt{2}$ units, 2 units **11.** 29 **13.** 9.4 **15.** 50 **17.** 49 **19.** about 66.4° **21.** 4.6 ft **23.** 139 ft **25.** 90 ft; 143 ft **27.a.** $\langle -49, 142 \rangle$, $\langle 38, 47 \rangle$ **b.** $\langle -11, 189 \rangle$ **29.a.** $\langle -54, 72 \rangle$, $\langle -95, -33 \rangle$ **b.** $\langle -149, 39 \rangle$ **31.** $\langle 2, 11 \rangle$ **33.** $\langle 2, 13 \rangle$ **35.** about 56.8 km; about 14.3° north of west

Chapter 9

1. *E* **3.** *C* **5.** *G* **7.** $A'(-5, 9)$, $B'(-3, 3)$, $C'(-1, 10)$ **9.** $P'(-15, -11)$, $Q'(-11, -6)$, $R'(-4, 1)$

11.

13.

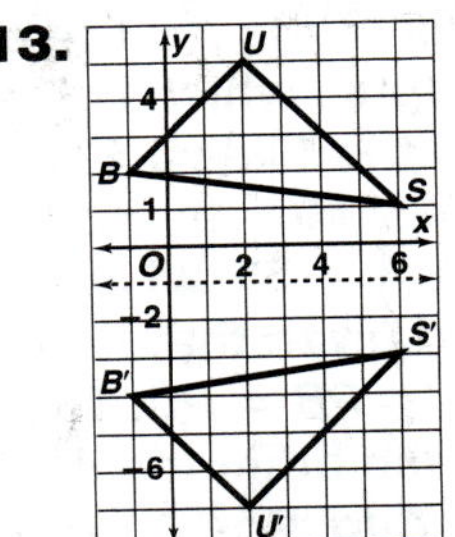

15.

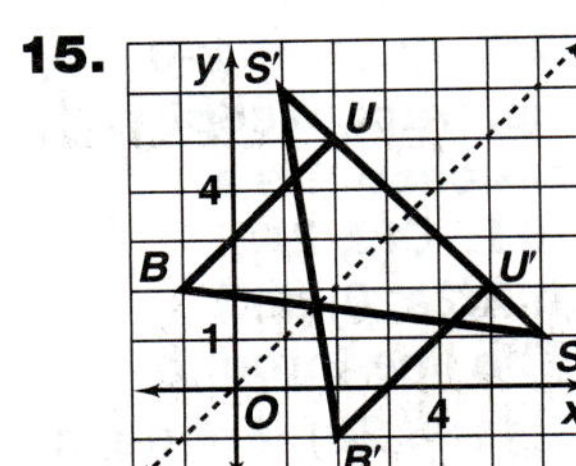

17.

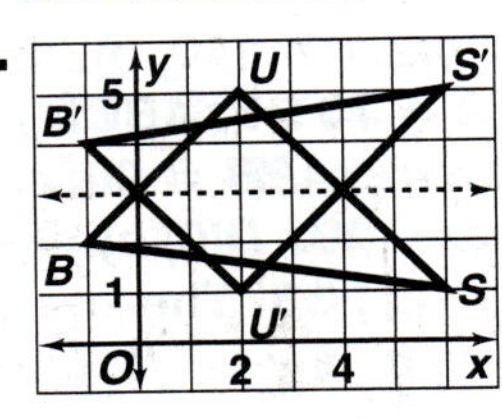

19. A and I

21.

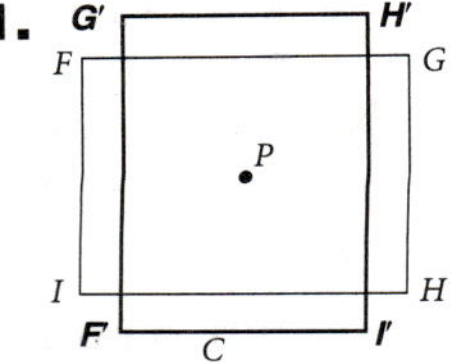

23.

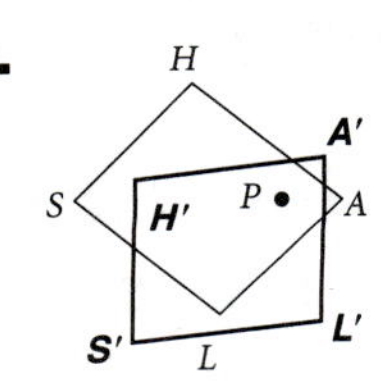

25.

27. line, rotation, point **29.** line **31.a.** isosceles **b.** equilateral **33.** dilation **35.** glide reflection

37. Translate the polygon using $(x, y) \rightarrow (x - 2, y - 5)$. Then dilate with center (0, 0) and scale factor 3. Then translate using $(x, y) \rightarrow (x + 2, y + 5)$. **39.** Reflect over m, then over ℓ.

Chapter 10

1. 15 ft; 10.825 ft^2 **3.** 50 ft; 143 ft^2 **5.** 60 yd; 120 yd^2 **7.** 72 cm^2 **9.** $\frac{25}{4}\sqrt{3}$ mm^2 **11.** 3500 ft^2 **13.** $\frac{73\sqrt{3}}{128}$ in.^{2} **15.** 5 : 8; 25 : 64 **17.** 5 : 16; 25 : 256 **19.** 48.4 cm **21.** 78.0 in.^{2} **23.** 20 m^2 **25.** 70.7 ft^2 **27.a.** 20π ft **b.** $\frac{5}{3}\pi$ ft **29.a.** 10π in. **b.** $\frac{25}{4}\pi$ in. **31.** $\frac{49}{3}\pi$ ft^2 **33.** $\frac{81}{8}\pi$ cm^2 **35.** 26 in.^{2} **37.** $\frac{1}{3}$ **39.** $\frac{7}{24}$ **41.** $1 - \frac{25}{81\pi}$

Chapter 11

1. equilateral $\triangle$; $7 + 10 = 15 + 2$ **3.** equilateral $\triangle$; $7 + 7 = 12 + 2$

5.

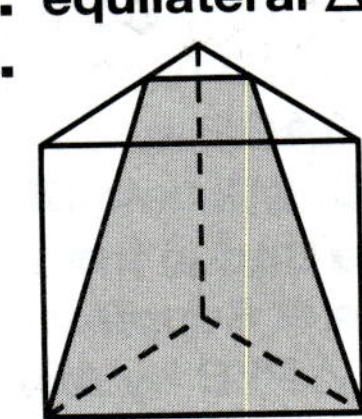

7. 28π cm^2; 36π cm^2 **9.** $108\sqrt{3}$ in.^{2}; $144\sqrt{3}$ in.^{2} **11.** 339.3 cm^2; 439.8 cm^2 **13.** 37.5 cm^2; 47.9 cm^2 **15.** 51.7 cm^2; 71.0 cm^2 **17.** 175 mm^3 **19.** 45π in.^{3} **21.** 400 in.^{3} **23.** 49.3 in.^{3} **25.** 486.0 cm^3 **27.** $\frac{500\pi}{3}$ cm^3, 524 cm^3; 100π cm^2, 314 cm^2 **29.** $\frac{256\pi}{3}$ in.^{3}, 268 in.^{3}; 64π in.^{2}, 201 in.^{2} **31.** $\frac{\pi}{6}$ in.^{3}, 1 in.^{3}; π in.^{2}, 3 in.^{2} **33.** $\frac{256\pi}{3}$ m^3 **35.** $\frac{343\pi}{6}$ ft^3 **37.** 4:9; 8:27 **39.** 3:4; 9:16 **41.** $\left(\frac{V_1}{V_2}\right)^2 = \left(\frac{A_1}{A_2}\right)^3$

Chapter 12

1. 65 **3.** 6 **5.** Tangents to a $\odot$ from a point outside the $\odot$ are $\cong$, so $AS = AP$, $BP = BQ$, $CQ = CR$, and $DR = DS$. By the Segment Add. Post. and various Props. or =, AB + DC = AP + BP + DR + CR = AS + BQ + DS + CQ = BQ + CQ + AS + DS = BC + AD **7.** 5.2 **9.** 20 **11.** $a = 38$; $b = 52$; $c = 104$; $d = 90$ **13.** $a = 55$; $b = 72$; $c = 178$; $d = 89$ **15.** 1.82 units **17.** $x = 193$; $y = 60.5$ **19.** $x \approx 10.4$ **21.** $x = 112.5$; $y = 67.5$ **23.** $x = 42.5$

25. 90; 150 **27.** $x^2 + y^2 = 16$ **29.** $(x - 9)^2 + (y + 3)^2 = 49$ **31.** $(x + 6)^2 + (y + 2)^2 = 13$ **33.a.** $(x - 113)^2 + (y - 215)^2 = 85^2$ **b.** $(x - 113)^2 + (y - 215)^2 = 170^2$

35. two rays ∥ to and 2 cm from $\overrightarrow{AB}$, and the semicircle of radius 2 with center A, opp. pt. B.

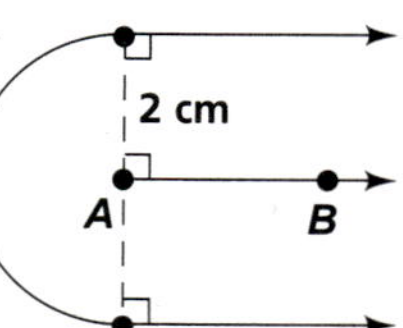

37.

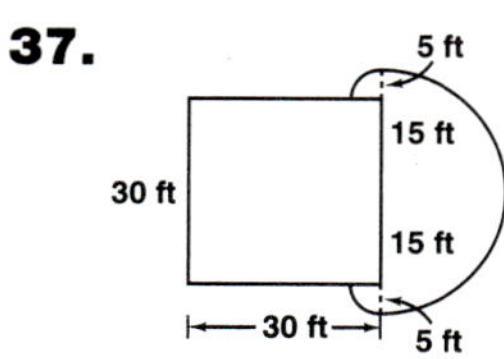

Skills Handbook

p. 740 **1.** 10 handshakes **3.** 16 regions

p. 741 **1.** 6, 8, 10 **3.** any pos. number less than 4 **5.** –2 **7.** 54 **9.** $\frac{1}{4}$ **11.** 2 rolls of 36 exposures and 5 rolls of 24 exposures **13.** 6 movies

p. 742 **1.** 165 toothpicks **3.** 3280 triangles

p. 743 **1.** 28 posts **3.** 84 paths **5.** 499,500 **7.** 30 people **9.** 91 matches

p. 744 **1.** J.T. is the parrot; Izzy is the dog; Arf is the goldfish; Blinky is the hamster. **3.** 14 students

p. 745 **1.** Route 90 east, Route 128 north, Route 4 north **3.** 3 tickets **5.** 8

p. 746 Answers may vary slightly. Sample is given. **1.** 35 mm; 46 mm **3.**

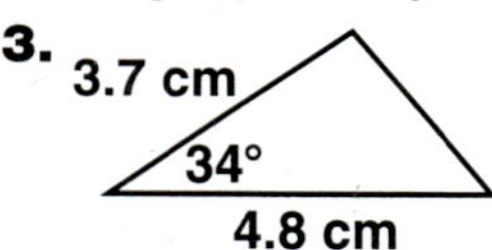

5.

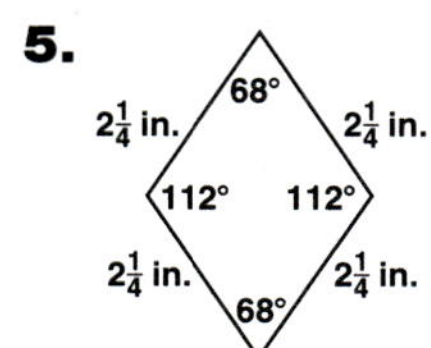

7.

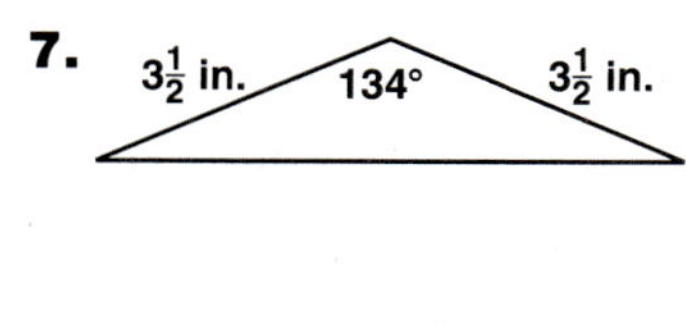

p. 747 **1.** 0.4 **3.** 600 **5.** 1008 **7.** $1\frac{5}{16}$ **9.** 34,000 **11.** 4.3 **13.** 15,000 **15.** 30,000, **17.** $2\frac{1}{4}$ **19.** 0.012 **21.** 72 **23.** 1,080,000 **25.** 12.6 **27.** $144\frac{4}{9}$ **29.** $\frac{3125}{4356}$

p. 748 **1.** $23\frac{1}{2}$ ft to $24\frac{1}{2}$ ft **3.** $339\frac{1}{2}$ mL to $340\frac{1}{2}$ mL **5.** 73.15 mm to 73.25 mm **7.** 5.35 mi to 5.45 mi **9.** $15\frac{1}{4}$ yd to $15\frac{3}{4}$ yd **11.** 57 to 59

p. 749 **1.** 18% **3.** 8% **5.** 1% **7.** 5% **9.** ≈7% **11.** ≈2%

p. 750 **1.** 353.6; 301; no mode **3.** $40,533; $28,150; $18,000 **5.** Median; the mode does not represent most of the data; the mean is significantly affected by the single highest salary. **7.** $67\frac{6}{17}$, 67, 66 **9.** Yes; the mode would be 67.

p. 751 **1.**

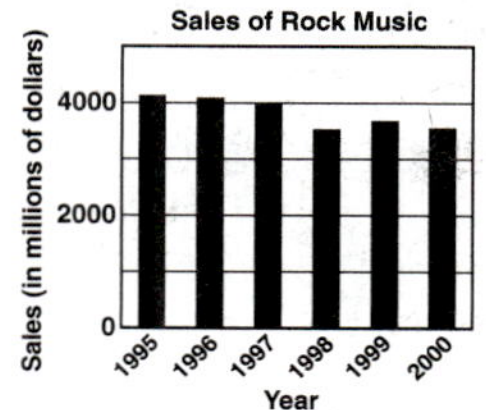

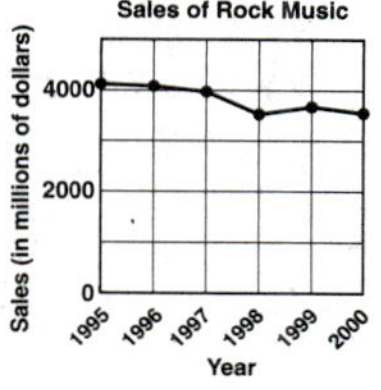

3. 8 A.M.–2 P.M. **5.** No; you cannot tell how the temperature changed between the measurements.

p. 752 **1.a.** 16.6 **b.** 15.1 **c.** 19.4

3. Pages in Books

185, 205, 258, 297, 356

p. 753 **1.** 121 **3.** 196 **5.** 26.01 **7.** $\frac{64}{25}$ **9.** 10 **11.** 8.6 **13.** 20 **15.** $\frac{2}{3}$ **17.** ±7 **19.** ±1 **21.** ±6 **23.** ±5

p. 754 **1.** −50 **3.** 15 **5.** 2 **7.** 36 **9.** −2 **11.** 243 **13.** −20 **15.** −4 **17.** B **19.** $2\ell + 2w$ **21.** $-4x^2 + 8x$ **23.** $-3t^2 + 4t$ **25.** $r^2 - 2r + 1$ **27.** $7h - 1$ **29.** $2x^2 + 7x - 4$ **31.** $2y^2 - 3y$

p. 755 **1.** $3\sqrt{3}$ **3.** $5\sqrt{6}$ **5.** $2\sqrt{2}$ **7.** $\frac{\sqrt{10}}{5}$ **9.** $\frac{3\sqrt{2}}{4}$ **11.** $5\sqrt{10}$ **13.** $2\sqrt{39}$ **15.** $4\sqrt{10}$ **17.** $2\sqrt{29}$

p. 756 **1.** $\frac{5}{3}$ **3.** $\frac{2}{3}$ **5.** $\frac{1}{2x}$ **7.** $\frac{22x}{35}$ **9.** $16y$ **11.** $\frac{x+1}{2}$ **13.** $\frac{1}{10t}$ **15.** $\frac{3}{2}$ **17.** $\frac{\pi}{r+\pi}$ **19.** $\frac{5}{12}$ **21.** $\frac{5}{13}$ **23.** $\frac{12}{5}$ **25.** $\frac{2}{\pi}$ **27.** $\frac{1}{9}$

p. 757 **1.** 8 **3.** 16 **5.** 22 **7.** −5 **9.** −10 **11.** 10 **13.** −16 or 16 **15.** −20 or 20

p. 758 **1.** 5 **3.** 3 **5.** −2 **7.** $\frac{19}{3}$ **9.** −5 **11.** −8 **13.** $35 - 2x = 9$; 13 **15.** $9.95 + 0.035m$ = $12.75; 80 min

p. 759 **1.** $w = \frac{P - 2\ell}{2}$ **3.** $r = \frac{1}{2}\sqrt{\frac{S}{\pi}} = \frac{\sqrt{\pi S}}{2\pi}$ **5.** $d_2 = \frac{2A}{d_1}$ **7.** $a = \frac{2A}{p}$ **9.** $h = \frac{2A}{b}$ **11.** $V = E + 2 - F$ **13.** $b = c \cos A$ **15.** $\ell = \frac{S - \pi r^2}{\pi r} = \frac{S}{\pi r} - r$ **17.** $B = \frac{3V}{h}$ **19.** $\ell = \frac{2S - 2B}{p}$

p. 760 **1.** (4, −1) **3.** (4, 1) **5.** inf. many sol.: all pts on the line $y = \frac{1}{4}x - \frac{3}{4}$ **7.** (−3, 1) **9.** (3, −2)

p. 761 **1.** 0.5 **3.** 0.27 **5.** 0.325 **7.** 0.09 **9.** 8.4 **11.** 7.2 **13.** 41.8 **15.** 2.4 **17–19.** Answers may vary. Samples are given. **17.** 7 **19.** 45

p. 762 **1.** $\frac{1}{2}$ **3.** $\frac{2}{7}$ **5.** $\frac{4}{13}$ **7.** $\frac{1}{12}$ **9.** $\frac{1}{2}$ **11.** $\frac{5}{12}$ **13.** $\frac{3}{8}$ **15.** $\frac{7}{8}$ **17.** $\frac{7}{8}$

Index

A

B

Index

C

D

H

I

J

K

L

M

N

O

Index

Q

R

Index

S

T

U

V

W

X

Y

Acknowledgments

Staff Credits

The people who made up the California Mathematics team—representing design services, editorial, editorial services, education technology, image services, marketing, market research, production services, publishing processes, and strategic markets—are listed below. Bold type denotes the core team members.

Dan Anderson, **Scott Andrews,** Carolyn Artin, Judith D. Buice, Kerry Cashman, Sarah Castrignano, Allison Cook, Carl Cottrell, Bob Craton, Patrick Culleton, Sheila DeFazio, Kathleen J. Dempsey, Frederick Fellows, **Suzanne Finn,** Patricia Fromkin, David J. George, **Patricia K. Gilbert,** Sandy Graff, **Ellen Welch Granter,** Richard Heater, Jayne Holman, Jennifer King, Betsy Krieble, Christopher Langley, Lisa LaVallee, Christine Lee, **Elizabeth Lehnertz,** Catherine Maglio, Cheryl Mahan, **Ann Mahoney,** Constance McCarty, **Carolyn McGuire,** Anne McLaughlin, Richard McMahon, Eve Melnechuk, Terri Mitchell, Michael Oster, Jeffrey Paulhus, Marcy Rose, Rashid Ross, Siri Schwartzman, Vicky Shen, **Dennis Slattery,** Nancy Smith, Mark Tricca, Paula Vergith, Teresa Whitney, Merce Wilczek, Joe Will, Kristin Winters, Heather Wright, Helen Young

Additional Credits: J. J. Andrews, Sarah J. Aubry, Deborah Belanger, Beth Blumberg, Casey Clark, Patty Fagan, Tom Greene, Karmyn Guthrie, Gillian Kahn, Jonathan Kier, Mary Landry, Mary Beth McDaniel, Anakin Steuart Michele, Hope Morley, Carol Roy, Jewel Simmons, Ted Smykal, Richard Sullivan, Dan Tanguay, Steve Thomas, Michael Torocsik, Alwyn Velasquez, Allison Wyss

Cover Design

Nancy Smith

Cover Photos

Building, Chris Daniels/Corbis; Trees, Panoramic Images/Getty Images.

Technical Illustration

Network Graphics

Illustration

Andrea G. Maginnis: 41
John Edwards: 24, 132, 133t, 200, 207, 290
JB Woolsey: 21, 88, 162, 201, 230, 332, 385, 386, 445, 542
Dennis Harms: 8, 57, 222, 223, 239, 274, 556, 566
Function through Form: 52, 266, 472, 611, 641
Leo Abbett: 92, 606 all
Jim Delapine: 579, 638, 639
Seymour Levy: 643
Kenneth Batelman: 395, 433, 650, 701, 704
Christine Graham: 487, 682
Peter Bollinger: 469
Brucie Rosch: 69, 81, 269, 472, 479
Gary Torrisi: 447 all, 448, 449
Lois Leonard Stock: 517
Ortelius Design, Inc.: 540
Gary Phillips: 37, 41

Photography

Page CA7, Richard Cummins/Corbis; **CA8,** Johnny Johnson/DRK Photo; **CA9,** Fotopic/Omni-Photo; **CA10,** Peter Menzel/Stock Boston; **CA11,** Russ Lappa; **CA12,** Timothy Hursley/SuperStock, Inc.; **CA13,** Kunio Owaki/Corbis; **CA14,** David Young-Wolff/PhotoEdit; **CA15,** Scott T. Smith/Corbis; **CA16,** Roy Ooms/Masterfile; **CA17,** David Young-Wolff/PhotoEdit; **CA18,** Felicia Martinez/PhotoEdit; **CA19,** Reza Estakhrian/Getty Images, Inc.; **CA32,** Phil Schermeister/Corbis; **CA35,** Neil Setchfield/Alamy; **CA42–43,** Joseph Sohm/Visions of America/Corbis.

Chapter 1: Pages 2, 3, Richard Cummins/SuperStock; **4,** Julie Houck/Stock Boston; **6,** Brett Froomer/Getty Images, Inc.; **7,** Dennis O'Clair/Getty Images, Inc.; **9,** PhotoDisc, Inc./Getty Images, Inc.; **11,** Barry Durand/Odyssey/Chicago; **14,** Reprinted by Permission of Tribune Media Services.; **19,** Russ Lappa; **20,** Johnny Johnson/DRK Photo; **23,** Daryl Benson/Masterfile; **24,** Stacy Pick/Stock Boston; **26,** Tim O'Hara/Index Stock Imagery, Inc.; **33,** AP Photo/Cliff Schiappa; **36,** Bill Nation/Corbis, **36** inset, Chad Slattery/Getty Images, Inc.; **37,** Corbis; **40,**William Sallaz/Duomo; **41,** Cameramann/The Image Works; **45,** Jon Feingersh/Corbis; **46,** Bob Daemmrich/The Image Works; **48 t,** ©1997 Richard Megna/Fundamental Photographs, NYC; **48 b,** Peanuts reprinted by permission of United Feature Syndicate, Inc.; **50,** Dan McCoy/Corbis; **55,** Michael Newman/PhotoEdit; **57,** Laura Dwight/ PhotoEdit; **58,** Richard Eller/Aerial Images Photography; **63,** Gail Mooney/Corbis; **66,** Hans Georg Roth/Corbis; **67,** Michael Rosenfeld/Getty Images, Inc.

Chapter 2: Pages 78, 79, Stewart Cohen/Getty Images, Inc.; **82,** Jon Chomitz; **83 t,** Frank and Ernest reprinted with permission of Newspaper Enterprise Association, Inc.; **83 bl,** AP Photo/Richard Drew; **83 br,** Tony Vaccaro/Archive Photos/Getty Images, Inc.; **85 t,** Hulton-Deutsch Collection/Corbis; **85 b,** Mug Shots/Corbis; **89,** Richard Haynes; **91 l all,** Russ Lappa; **91 r,** National Association for the Deaf; **92 l,** Andy Sacks/Getty Images, Inc.; **92 r,** C. Squared Studios/PhotoDisc, Inc./PictureQuest; **94,** Tony Freeman/PhotoEdit; **96,** InterNetwork Media/PhotoDisc, Inc./PictureQuest; **97,** Damian Strohmey/Sports Illustrated; **98 t,** Fotopic/Omni-Photo Communications, Inc.; **98 m,** Close to Home by John McPherson. Reprinted with permission of Universal Press Syndicate. All rights reserved.; **98 b,** Gregory Scott/Index Stock Imagery, Inc.; **100 l,** Courtesy of WGBH; **100 r,** King Collection/Retna, Ltd.; **106 l,** PhotoDisc, Inc./Getty Images, Inc.; **106 r,** Tony Freeman/PhotoEdit; **107 t,** The Granger Collection, NY; **107 m,** John Lopinot/Black Star Publishing/PictureQuest; **107 b,** Neal Preston-Andy Kent/Still Bill Productions; **113,** Corbis Digital Images/PictureQuest; **114,** Paul A. Souder/Corbis

Chapter 3: Pages 124, 125, Imtek Imagineering/Masterfile; **128** composite, H.P. Merten/Corbis, Joe Towers/Corbis; **131,** Peter Menzel/Stock Boston; **134,** Gary Kufner/Corbis; **138 both,** Jon Chomitz; **139,** Richard Pasley/Stock Boston; **142,** Jon Chomitz; **143,** John M. Roberts/Corbis; **149,** Stewart Cohen/Getty Images, Inc.; **151,** John Coletti/The Picture Cube/Index Stock Imagery, Inc.; **152 both,** Jerry Jacka Photography; **157,** Jim Cummins/Getty Images, Inc.; **158,** Arthur Thevenart/Corbis; **160,** Ken O'Donoghue; **161 l,** Russ Lappa; **161 m,** Nawrocki Stock Photo/Picture Perfect; **161 r,** Russ Lappa; **162,** Courtesy of the Theatre in the Round Players, Inc., Minneapolis, MN; **163,** Michael Newman/PhotoEdit; **169,** NASA; **170,** Michael Newman/PhotoEdit; **174,** Bob Daemmrich/Stock Boston; **177,** PhotoDisc, Inc./Getty Images, Inc.; **179,** Getty Images, Inc.; **181,** Michael Newman/PhotoEdit; **183,** Sol Lewitt, Installation view of photo of students with Sol Lewitt, from Sol Lewitt: Twenty-Five Years of Wall Drawings, 1968-1993, Addison Gallery of American Art, Phillips Academy, Andover, Massachusetts; **185 all,** Jon Chomitz

Chapter 4: Pages 196, 197, Chateau de Versailles, France/Peter Willi/SuperStock, Inc.; **199,** NASA; **200,** John Elk III/Stock Boston; **202,** Richard Haynes; **203,** Hans Gregory Roth/Corbis; **206,** Tom Alexander/ Getty Images, Inc.; **208 l,** Tony Freeman Photographs; **208 r,** Russ Lappa; **209 both,** Prentice Hall; **210,** Chuck Pefley/Getty Images, Inc.; **214,** Mike Greenlar Photography; **217,** Richard Haynes; **218,** Photo Researchers, Inc.; **221,** Russ Lappa; **223,** Leif Skoogfors/Woodfin Camp & Associates; **229,** Pat O'Hara/Corbis; **231,** Fernando Serna/ Department of the Air Force; **232,** Richard Hutchins/PhotoEdit; **233,** Corbis; **236,** Tony Freeman Photographs; **238,** Jeff Greenberg/Omni-Photo Communications, Inc.; **241,** Jim Rudnick/Corbis; **242,** Miriam Nathan-Roberts; **243 t,** Russ Lappa; **243 b,**Walter Hodges/Getty Images, Inc.; **245,** Corbis

Chapter 5: Pages 256, 257, Jeff Greenberg/PhotoEdit; **261,** Clyde Lockwood/Animals Animals/Earth Scenes; **263,** Timothy Hursley/SuperStock, Inc.; **268,** Jim Cummins/Corbis; **281,** AP Photo/Dean Times Herald; **282,** Michael Newman/PhotoEdit; **284,** AP/Wide World Photos; **285,** Corbis; **286,** Tom Pantages; **289,** Russ Lappa; **290,** David Young-Wolff/PhotoEdit; **293,** Dave Schiefelbein/Getty Images, Inc.; **294,** Fred Wood/Summer Productions

Chapter 6: Pages 304, 305, Jupiter Images; **306,** The Image Works; **309,** Charles Demuth, My Egypt, 1927 Oil on composition board 35 3/4 x 30 in. Whitney Museum of American Art, New York; Purchase, with funds from Gertrude Vanderbilt Whitney 31.172; **310,** Russ Lappa; **313,** Richard Haynes; **314,** Tony Freeman/ PhotoEdit; **316,** Richard Haynes; **320,** Ken O'Donoghue; **321,** Tony Freeman Photographs; **322 t,** Photo of Russell W.M. and Ruth Kraus residence, S.340 in The Architecture of Frank Lloyd Wright, A Complete Catalog, 3rd Edition, ©2002/William Allin Storrer, PhD; **322 b,** Plan of Russell W.M. and Ruth Kraus residence, S.340 in The Frank Lloyd Wright Companion, ©1973/William Allin Storrer, PhD; **324 l,** Dave Bartruff/Stock Boston; **324 r,** Mark Thayer; **325,** Russ Lappa; **330,**Walter Bibikow/Index Stock Imagery, Inc.; **332,** Ron Sherman/ Stone/Getty Images, Inc.; **333,** Tony Freeman Photographs; **333 inset,** Russ Lappa; **336,** archivebroehandesign.com, New York; **337,** The Purcell Team/Corbis; **339,** Mark E. Gibson/Visuals Unlimited; **340,** Wallace Garrison/Index Stock Imagery, Inc.; **343,** Richard Haynes; **345,** ©National Park Service, photo by Bill Hudson, Biscayne National Park; **347 both,** Geoffrey Clifford/Woodfin Camp & Associates; **349,** SuperStock, Inc.; **352,** Kunio Owaki/Corbis

Chapter 7: Pages 364, 365, Steve Satushek/Getty Images; **366 l,** David Young-Wolff/PhotoEdit; **366 r,** David Young-Wolff/ PhotoEdit; **368,** Prentice Hall; **369,** ©AAA reprinted with permission.; **370,** Michael Dwyer/Stock Boston; **371,** NASA; **374,** Miro Vintoniv/Stock Boston; **375,** Telegraph Colour Library/Getty Images, Inc.; **377 t,** Peanuts reprinted by permission of United Feature Syndicate, Inc.; **377 m,** Ralf-Finn Hestoft/Index Stock Imagery, Inc.; **377 b,** Russ Lappa; **382,** Lee Snider/Photo Images; **385,** Francis Lepine/Animals Animals/Earth Scenes; **387,** AP/Wide World Photos; **393,** Paul Chesley/Getty Images, Inc.; **395,** Kevin Miller/Getty Images, Inc.; **399,** Corbis; **401,** Vincent Hobbs/SuperStock, Inc.; **402,** Mark Richards/PhotoEdit; **403,** David Woodfall/Getty Images, Inc.

Chapter 8: Pages 414, 415, Peter Hendrie/Getty Images, Inc.; **419,** Tom Tracy/Getty Images, Inc.; **422,** The Observatories of the Carnegie Institution of Washington; **426,** Rhoda Sidney/PhotoEdit; **427,** Stephen Saks/Index Stock Imagery, Inc./PictureQuest; **428,** Russ Lappa; **429,** Macduff Everton/Corbis; **433,** Scott T. Smith/Corbis; **436,** Courtesy of Katoomba Scenic Railway; **439,** Richard Hamilton/Corbis; **441,** David Young-Wolff/PhotoEdit; **443 both,** Tom Pantages; **446,** Ron Dahlquist/Getty Images, Inc.; **448,** Dwayne Newton/PhotoEdit; **450,** Giulio Andreini/Getty Images, Inc.; **452,** Lowell Georgia/Corbis; **453,** Getty Images, Inc.; **454,** Bob Daemmrich/Stock Boston; **455,**William J. Weber/Visuals Unlimited; **457,** NOAA; **458,** The Far Side ® by Gary Larson © 1993 Farworks, Inc. All rights reserved. Used with permission.

Chapter 9: Pages 468, 469 cave,Tom Till/Getty Images, Inc.; **470 tl,** Tom Rosenthal/SuperStock, Inc.; **470 bl,** Tom Rosenthal/ SuperStock, Inc.; **470 r all,** Russ Lappa; **471,** Adam Smith Productions/Corbis; **475,** Roy Ooms/Masterfile; **478,** Gabe Palmer/Corbis; **481,** Seth Joel/Corbis; **485,** Jerry Jacka Photography; **486 both,** Jerry Jacka Photography; **492,** Brian Parker/Tom Stack & Associates, Inc.; **493 both,** Prentice Hall; **494 l,** Andrew J. Martinez © 1993/Photo Researchers, Inc.; **494 r,** Guido Alberto Rossi/Image Bank/Getty Images, Inc.; **495 l,** Tom Salyer/Stock Connection/PictureQuest; **495 r,** Getty Images, Inc.; **498,** Frank T. Awbrey/Visuals Unlimited; **502,** David Young-Wolff/ PhotoEdit; **507,** Michael Newman/PhotoEdit; **508,** PhotoDisc, Inc./Getty Images, Inc.; **511 tl,** Viviane Moos/Corbis; **511 tr,** Alfred Pasieka/Science Photo Library/Photo Researchers, Inc.; **511 bl,** Paul Jablonka/International Stock Photo/ImageState; **511 br,** Adam Peirport/Corbis; **515 t,** ©1996 M.C. Escher Heirs/Cordon Art, Baarn, Holland. All rights reserved.; **515 bl,** ©2001 Cordon Art, Baarn, Holland. All rights reserved.; **515 br,** Symmetry Drawing E67 by M.C. Escher. ©2002 Cordon Art, Baarn, Holland. All rights reserved.; **516,** Syracuse Newspapers/The Image Works; **518 tl,** Ira Kirschenbaum/Stock Boston; **518 tr,** Suzanne Murphy-Larronde; **518 bl,** Margaret Courtney-Clarke/Corbis; **518 br,** M. Angelo/Corbis; **519,** From The Grammar of Ornament by Owen Jones, 1856 Edition/1998 Octavo Corporation; **521,** Robert Frerck/Odyssey/Chicago; **524 l,** Patti Murray/Animals Animals/Earth Scenes; **524 r,** Russ Lappa; **526 l,** Don & Pat Valenti/DRK Photo; **526 r,** Jeff Foott/DRK Photo

Chapter 10: Pages 530, 531 AP Photo/NASA; **536,** Mick Roessler/SuperStock, Inc.; **538,** Larry Lefever/Grant Heilman Photography, Inc.; **543,** Simon Battensby/Getty Images, Inc.; **544,** Tony Donaldson/SportsChrome-USA; **545,** The Granger Collection, NY; **547,** Photo courtesy of NidaCore; **548,** Richard Cummins/Corbis; **549,** Howard G. Ross; **550,** F. Dewey Webster/Sovfoto/Eastfoto/PictureQuest; **553,** Russ Lappa; **555,** David Young-Wolff/PhotoEdit; **557,** Bob Daemmrich/Stock Boston/PictureQuest; **560,** George Gerster/Photo Researchers, Inc.; **562,** Corbis; **563,** PhotoDisc, Inc./Getty Images, Inc.; **564,** Ralph Cowan/Getty Images, Inc.; **567,** PhotoDisc, Inc./Getty Images, Inc.; **571,** Bill Horsman/Stock Boston; **572,** Calvin and Hobbs ©Watterson. Reprinted with permission of Universal Press Syndicate. All rights reserved.; **573,** Bob Daemmrich/Stock Boston; **576,** Russ Lappa; **577,** Russ Lappa; **579,** Bill Bachmann/Photo Network/PictureQuest; **583,** David Young-Wolff/PhotoEdit; **585,** Frank Fournier/Corbis; **586,** AP/Wide World Photos

Chapter 11: 596, 597, Dorling Kindersley; **599,** Michael Kevin Daly/Corbis; **599 inset,** Prentice Hall; **602 t all,** Russ Lappa; **602 b,** PhotoDisc, Inc./Getty Images, Inc.; **603,** ©Breck P. Kent; **608,** Uniphoto, Inc./Pictor International/ImageState; **610,** Russ Lappa; **611,** Tony Freeman/PhotoEdit; **612,** Russ Lappa; **618,** Will & Deni McIntyre/Getty Images, Inc.; **620,** Roger Allyn Lee/SuperStock, Inc.; **621,** Michael Busselle/Corbis; **624 both,** Russ Lappa; **628,** James King-Holmes/Science Photo Library/Photo Researchers, Inc.; **631,** Ken O'Donoghue; **632,**

Pictor International/ImageState; **634,** Carolyn Ross/Index Stock Imagery, Inc.; **635,** John Elk/Stock, Boston/PictureQuest; **637,** Alan Klehr/Getty Images, Inc.; **638,** Rafael Macia/Photo Researchers, Inc.; **641 l,** Richard Hutchings/Photo Researchers, Inc.; **641 m,** Mark C. Burnett/Photo Researchers, Inc.; **641 r,** Tony Freeman/PhotoEdit; **642,** David Young-Wolff/PhotoEdit; **642,** Gene Moore/Phototake/PictureQuest; **644,** Ken O'Donoghue; **646,** Felicia Martinez/PhotoEdit; **647,** Russ Lappa; **649,** U.S. Department of Commerce

Chapter 12: Page 660, 661, Ray Ooms/Masterfile; **663,** Patrick Ward/Corbis; **666 t,** Carl F. Romney; **666 b,** Russ Lappa; **667 both,** Russ Lappa; **671,** Steve Vidler/SuperStock, Inc.; **672,** Tony Freeman Photographs; **674,** Robert Frerck/Odyssey/Chicago; **675,** Grant Heilman Photography, Inc.; **683,** AP/Wide World Photos; **688,** Shaun Egan/Getty Images, Inc.; **690,** Carol Simowitz; **691,** Getty Images, Inc.; **693,** Rosenfeld Images Ltd/Science Photo Library/Photo Researchers, Inc.; **697,** Tony Freeman Photographs; **699,** AFP/Corbis; **702,** Reza Estakhrian/Stone/Getty Images, Inc.; **703,** Steven E. Sutton/Duomo; **704,** Christian Grzimek/Okapia/Photo Researchers, Inc.